$149.95 per copy (in United States).
Price subject to change without prior notice.

3016

RSMeans

Metric Construction Cost Data

64th Annual Edition

2006

RSMeans
Construction Publishers & Consultants
63 Smiths Lane
Kingston, MA 02364-0800
(781) 422-5000

Copyright© 2005 by Reed Construction Data, Inc.
All rights reserved.

Printed in the United States of America.
ISSN 1069-0549
ISBN 0-87629-798-X

Senior Editors
John H. Chiang, PE
Eugene R. Spencer

Contributing Editors
Christopher Babbitt
Ted Baker
Barbara Balboni
Robert A. Bastoni
John Kane
Robert J. Kuchta
Robert C. McNichols
Robert W. Mewis, CCC
Melville J. Mossman, PE
John J. Moylan
Jeannene D. Murphy
Stephen C. Plotner
Marshall J. Stetson
Phillip R. Waier, PE

Senior Engineering Operations Manager
John H. Ferguson, PE

Senior Vice President & General Manager
John Ware

Vice President of Sales
John M. Shea

Production Manager
Michael Kokernak

Technical Support
Wayne D. Anderson
Thomas J. Dion
Jonathan Forgit
Mary Lou Geary
Gary L. Hoitt
Genevieve Medeiros
Paula Reale-Camelio
Kathryn S. Rodriguez
Sheryl A. Rose
Laurie Thom

Book & Cover Design
Norman R. Forgit

Editorial Advisory Board

 This book is recyclable.

 This book is printed on recycled stock.

 Reed Construction Data®

First Printing

Foreword

RSMeans is a product line of Reed Construction Data, Inc., a leading provider of construction information, products, and services in North America and globally. Reed Construction Data's project information products include more than 100 regional editions, national construction data, sales leads, and local plan rooms in major business centers. Reed Construction Data's PlansDirect provides surveys, plans, and specifications. The First Source suite of products consists of *First Source for Products*, SPEC-DATA™, MANU-SPEC™, CADBlocks, Manufacturer Catalogs, and First Source Exchange (www.firstsourceexchange.com) for the selection of nationally available building products. Reed Construction Data also publishes ProFile, a database of more than 20,000 U.S. architectural firms. RSMeans provides construction cost data, training, and consulting services in print, CD-ROM, and online. Reed Construction Data, headquartered in Atlanta, is owned by Reed Business Information (www.reedconstructiondata.com), a leading provider of critical information and marketing solutions to business professionals in the media, manufacturing, electronics, construction, and retail industries. Its market-leading properties include more than 135 business-to-business publications and over 125 Webzines and Web portals as well as online services, custom publishing, directories, research, and direct-marketing lists. Reed Business Information is a member of the Reed Elsevier plc group (NYSE: RUK and ENL)—a world-leading publisher and information provider operating in the science and medical, legal, education, and business-to-business industry sectors.

Our Mission

Since 1942, RSMeans has been actively engaged in construction cost publishing and consulting throughout North America.

Today, over 60 years after RSMeans began, our primary objective remains the same: to provide you, the construction and facilities professional, with the most current and comprehensive construction cost data possible.

Whether you are a contractor, an owner, an architect, an engineer, a facilities manager, or anyone else who needs a fast and reliable construction cost estimate, you'll find this publication to be a highly useful and necessary tool.

Today, with the constant flow of new construction methods and materials, it's difficult to find the time to look at and evaluate all the different construction cost possibilities. In addition, because labor and material costs keep changing, last year's cost information is not a reliable basis for today's estimate or budget.

That's why so many construction professionals turn to RSMeans. We keep track of the costs for you, along with a wide range of other key information, from city cost indexes . . . to productivity rates . . . to crew composition . . . to contractor's overhead and profit rates.

RSMeans performs these functions by collecting data from all facets of the industry and organizing it in a format that is instantly accessible to you. From the preliminary budget to the detailed unit price estimate, you'll find the data in this book useful for all phases of construction cost determination.

The Staff, the Organization, and Our Services

When you purchase one of RSMeans' publications, you are, in effect, hiring the services of a full-time staff of construction and engineering professionals.

Our thoroughly experienced and highly qualified staff works daily at collecting, analyzing, and disseminating comprehensive cost information for your needs. These staff members have years of practical construction experience and engineering training prior to joining the firm. As a result, you can count on them not only for the cost figures, but also for additional background reference information that will help you create a realistic estimate.

The RSMeans organization is always prepared to help you solve construction problems through its five major divisions: Construction and Cost Data Publishing, Electronic Products and Services, Consulting Services, Insurance Services, and Educational Services.

Besides a full array of construction cost estimating books, RSMeans also publishes a number of other reference works for the construction industry. Subjects include construction estimating and project and business management; special topics such as HVAC, roofing, plumbing, and hazardous waste remediation; and a library of facility management references.

In addition, you can access all of our construction cost data through your computer with *Means CostWorks 2006* CD-ROM, an electronic tool that offers over 50,000 lines of RSMeans detailed construction cost data, along with assembly and whole building cost data. You can also access RSMeans cost information from our Web site at www.rsmeans.com.

What's more, you can increase your knowledge and improve your construction estimating and management performance with an RSMeans Construction Seminar or In-House Training Program. These two-day seminar programs offer unparalleled opportunities for everyone in your organization to get updated on a wide variety of construction-related issues.

RSMeans also is a worldwide provider of construction cost management and analysis services for commercial and government owners and of claims and valuation services for insurers.

In short, RSMeans can provide you with the tools and expertise for constructing accurate and dependable construction estimates and budgets in a variety of ways.

Robert Snow Means Established a Tradition of Quality That Continues Today

Robert Snow Means spent years building RSMeans, making certain he always delivered a quality product.

Today, at RSMeans, we do more than talk about the quality of our data and the usefulness of our books. We stand behind all of our data, from historical cost indexes to construction materials and techniques to current costs.

If you have any questions about our products or services, please call us toll-free at 1-800-334-3509. Our customer service representatives will be happy to assist you. You can also visit our Web site at www.rsmeans.com.

Table of Contents

UNIT PRICES

GENERAL REQUIREMENTS	1
SITE CONSTRUCTION	2
CONCRETE	3
MASONRY	4
METALS	5
WOOD & PLASTICS	6
THERMAL & MOISTURE PROTECTION	7
DOORS & WINDOWS	8
FINISHES	9
SPECIALTIES	10
EQUIPMENT	11
FURNISHINGS	12
SPECIAL CONSTRUCTION	13
CONVEYING SYSTEMS	14
MECHANICAL	15
ELECTRICAL	16

REFERENCE INFORMATION

REFERENCE TABLES

CREWS/EQUIPMENT

COST INDEXES

SQUARE METER COSTS

Related RSMeans Products and Services

John H. Chiang, P.E., and Eugene R. Spencer, Senior Editors of this cost data book, suggest the following RSMeans products and services as **companion** information resources to *Metric Construction Cost Data*:

Construction Cost Data Books

Assemblies Cost Data 2006
Building Construction Cost Data 2006
Square Foot Costs 2006

Reference Books

Unit Price Estimating Methods, 3rd Edition
ADA Compliance Pricing Guide, 2nd Edition
Building Security: Strategies & Costs
Designing & Building with the IBC
Estimating Handbook, 2nd Edition
Green Building: Project Planning & Estimating
How to Estimate with Means Data and CostWorks
Plan Reading & Material Takeoff
Project Scheduling and Management for Construction

Seminars and In-House Training

Unit Price Estimating
Square Foot & Assemblies Estimating
Means CostWorks Training
Means Data for Job Order Contracting (JOC)
Plan Reading & Material Takeoff
Scheduling & Project Management

RSMeans on the Internet

Visit RSMeans at **http://www.rsmeans.com.** The site contains useful *interactive* cost and reference material. Request or download *FREE* estimating software demos. Visit our bookstore for convenient ordering and to learn more about new publications and companion products.

RSMeans Data on CD-ROM

Get the information found in RSMeans traditional cost books on RSMeans' new, easy-to-use CD-ROM, *CostWorks 2006. Means CostWorks* users can now enhance their Costlist with the *Means CostWorks Estimator.* For more information see the special full-color brochure inserted in this book.

RSMeans Business Solutions

RSMeans Business Solutions provides construction cost-related services including: Customized Databases, Benchmark Studies, Estimating Audits, Feasibility Studies, Litigation Support, and Predictive Cost Models.

New! RSMeans for Job Order Contracting (JOC)

New methods and "best practices" for estimating and project management to reduce administrative costs for a $147 billion+ renovation market in the U.S. Finish renovation projects faster in schools and universities, on government building projects, and in medical facilities.

* RSMeans Job Order Contracting (JOC) Cost Data
* JOCWorks Software (Basic, Advanced, PRO)
* Consultation in Contracting Methods
* RSMeans Data™ Licensing Program

Construction Costs for Software Applications

Over 25 unit price and assemblies cost databases are available through a number of leading estimating and facilities management software providers (listed below). For more information see the "yellow" pages at the back of this publication.

RSMeansData™ is also available to federal, state, and local government agencies as multi-year, multi-seat licenses.

- 3D International
- 4Clicks-Solutions, LLC
- Aepco, Inc.
- Applied Flow Technology
- ArenaSoft Estimating
- Best Software–Timberline
- BSD – Building Systems Design, Inc.
- CMS – Construction Management Software
- Corecon Technologies, Inc.
- CorVet Systems
- Estimating Systems, Inc.
- Maximus Asset Solutions
- MC^2 – Management Computer Controls
- Shaw Beneco Enterprises, Inc.
- US Cost, Inc.
- VFA–Vanderweil Facility Advisers
- WinEstimator, Inc.

How the Book Is Built: An Overview

A Powerful Construction Tool

You have in your hands one of the most powerful construction tools available today. A successful project is built on the foundation of an accurate and dependable estimate. This book will enable you to construct just such an estimate.

For the casual user the book is designed to be:

- quickly and easily understood so you can get right to your estimate.
- filled with valuable information so you can understand the necessary factors that go into the cost estimate.

For the regular user, the book is designed to be:

- a handy desk reference that can be quickly referred to for key costs.
- a comprehensive, fully reliable source of current construction costs and productivity rates so you'll be prepared to estimate any project.
- a source book for preliminary project cost, product selections, and alternate materials and methods.

To meet all of these requirements we have organized the book into the following clearly defined sections.

How To Use the Book: The Details

This section contains an in-depth explanation of how the book is arranged . . . and how you can use it to determine a reliable construction cost estimate. It includes information about how we develop our cost figures and how to completely prepare your estimate.

Unit Price Section

All cost data has been divided into the 16 divisions according to the MasterFormat system of classification and numbering as developed by the Construction Specifications Institute (CSI) and Construction Specifications Canada (CSC). For a listing of these divisions and an outline of their subdivisions, see the Unit Price Section Table of Contents.

Estimating tips are included at the beginning of each division.

Assemblies Section

The cost data in this section has been organized in an "Assemblies" format. These assemblies are the functional elements of a building and are arranged according to the 7 divisions of the UNIFORMAT II classification system. For a complete explanation of a typical "Assemblies" page, see "How To Use the Assemblies Cost Tables."

Reference Section

This section includes information on Equipment Rental Costs, Crew Listings, Historical Cost Indexes, City Cost Indexes, Location Factors, Reference Tables, Change Orders, Square Meter Costs, Metric Units Used in the Construction Trades, and a listing of Abbreviations.

Equipment Rental Costs: This section contains the average costs to rent and operate hundreds of pieces of construction equipment.

Crew Listings: This section lists all the crews referenced in the book. For the purposes of this book, a crew is composed of more than one trade classification and/or the addition of power equipment to any trade classification. Power equipment is included in the cost of the crew. Costs are shown both with the bare labor rates and with the installing contractor's overhead and profit added. For each, the total crew cost per eight-hour day and the composite cost per labor-hour are listed.

Historical Cost Indexes: These indexes provide you with data to adjust construction costs over time.

City Cost Indexes: All costs in this book are U.S. national averages. Costs vary because of the regional economy. You can adjust costs by CSI Division to over 719 locations throughout the U.S. and Canada by using the data in this section.

Location Factors: You can adjust total project costs to over 900 locations throughout the U.S. and Canada by using the data in this section.

Reference Tables: At the beginning of selected major classifications in the Unit Price section are "reference numbers" shown in a bold box. These numbers refer you to related information in the Reference Section. In this section, you'll find reference tables, explanations, and estimating information that support how we develop the unit price data, technical data, and estimating procedures.

Change Orders: This section includes information on the factors that influence the pricing of change orders.

Square Meter Costs: Formerly Division 17, this section contains costs for 59 different building types that allow you to make a rough estimate for the overall cost of a project or its major components.

Metric Units: The units of measure used in the construction trades and their symbols are listed.

Abbreviations: A listing of abbreviations used throughout this book, along with the terms they represent, is included in this section.

Index

A comprehensive listing of all terms and subjects in this book will help you quickly find what you need when you are not sure where it falls in MasterFormat.

The Scope of This Book

This book is designed to be as comprehensive and as easy to use as possible. To that end we have made certain assumptions and limited its scope in three key ways:

1. We have established material prices based on a national average.
2. We have computed labor costs based on a 30-city national average of union wage rates.
3. We have targeted the data for projects of a certain size range.

For a more detailed explanation of how the cost data is developed, see "How To Use the Book: The Details."

Project Size

This book is aimed primarily at commercial and industrial projects costing $1,000,000 and up, or large multi-family housing projects. Costs are primarily for new construction or major renovation of buildings rather than repairs or minor alterations.

With reasonable exercise of judgment the figures can be used for any building work. *However, for civil engineering structures such as bridges, dams, highways, or the like, please refer to RSMeans Heavy Construction Cost Data.*

v

How to Use the Book: The Details

What's Behind the Numbers? The Development of Cost Data

The staff at RSMeans continuously monitors developments in the construction industry in order to ensure reliable, thorough, and up-to-date cost information.

While *overall* construction costs may vary relative to general economic conditions, price fluctuations within the industry are dependent upon many factors. Individual price variations may, in fact, be opposite to overall economic trends. Therefore, costs are continually monitored and complete updates are published yearly. Also, new items are frequently added in response to changes in materials and methods.

Costs—$ (U.S.)

All costs represent U.S. national averages and are given in U.S. dollars. The RSMeans City Cost Indexes can be used to adjust costs to a particular location. The City Cost Indexes for Canada can be used to adjust U.S. national averages to local costs in Canadian dollars. No exchange rate is necessary.

Material Costs

The RSMeans staff contacts manufacturers, dealers, distributors, and contractors all across the U.S. and Canada to determine national average material costs. If you have access to current material costs for your specific location, you may wish to make adjustments to reflect differences from the national average. Included within material costs are fasteners for a normal installation. RSMeans engineers use manufacturers' recommendations, written specifications and/or standard construction practice for size and spacing of fasteners. Adjustments to material costs may be required for your specific application or location. Material costs do not include sales tax.

Labor Costs

Labor costs are based on the average of wage rates from 30 major U.S. cities. Rates are determined from labor union agreements or prevailing wages for construction trades for the current year. Rates, along with overhead and profit markups, are listed on the inside back cover of this book.

- If wage rates in your area vary from those used in this book, or if rate increases are expected within a given year, labor costs should be adjusted accordingly.

Labor costs reflect productivity based on actual working conditions. These figures include time spent during a normal workday on tasks other than actual installation, such as material receiving and handling, mobilization at site, site movement, breaks, and cleanup.

Productivity data is developed over an extended period so as not to be influenced by abnormal variations and reflects a typical average.

Equipment Costs

Equipment costs include not only rental, but also operating costs for equipment under normal use. The operating costs include parts and labor for routine servicing such as repair and replacement of pumps, filters, and worn lines. Normal operating expendables, such as fuel, lubricants, tires, and electricity (where applicable), are also included. Extraordinary operating expendables with highly variable wear patterns, such as diamond bits and blades, are excluded. These costs are included under materials. Equipment rental rates are obtained from industry sources throughout North America—contractors, suppliers, dealers, manufacturers, and distributors.

Crew Equipment Cost/Day—The power equipment required for each crew is included in the crew cost. The daily cost for crew equipment is based on dividing the weekly bare rental rate by 5 (number of working days per week) and then adding the hourly operating cost times 8 (hours per day). This "Crew Equipment Cost/Day" is listed in the Reference Section.

Mobilization/Demobilization—The cost to move construction equipment from an equipment yard or rental company to the job site and back again is not included in equipment costs. Mobilization (to the site) and demobilization (from the site) costs can be found in Section 02305-250. If a piece of equipment is already at the job site, it is not appropriate to utilize mob/demob costs again in an estimate.

General Conditions

Cost data in this book is presented in two ways: Bare Costs and Total Cost including O&P (Overhead and Profit). General Conditions, when applicable, should also be added to the Total Cost including O&P. The costs for General Conditions are listed in Division 1 of the Unit Price Section and the Reference Section of this book. General Conditions for the *Installing Contractor* may range from 0% to 10% of the Total Cost including O&P. For the *General* or *Prime Contractor*, costs for General Conditions may range from 5% to 15% of the Total Cost including O&P, with a figure of 10% as the most typical allowance.

Overhead and Profit

Total Cost including O&P for the *Installing Contractor* is shown in the last column on both the Unit Price and the Assemblies pages of this book. This figure is the sum of the bare material cost plus 10% for profit, the base labor cost plus total overhead and profit, and the bare equipment cost plus 10% for profit. Details for the calculation of Overhead and Profit on labor are shown on the inside back cover and in the Reference Section of this book. (See the "How To Use the Unit Price Pages" for an example of this calculation.)

Factors Affecting Costs

Costs can vary depending upon a number of variables. Here's how we have handled the main factors affecting costs.

Quality—The prices for materials and the workmanship upon which productivity is based represent sound construction work. They are also in line with U.S. government specifications.

Overtime—We have made no allowance for overtime. If you anticipate premium time or work beyond normal working hours, be sure to make an appropriate adjustment to your labor costs.

Productivity—The productivity, daily output, and labor-hour figures for each line item are based on working an eight-hour day in daylight hours in moderate temperatures. For work that extends beyond normal work hours or is performed under adverse conditions, productivity may decrease. (See the section in "How To Use the Unit Price Pages" for more on productivity.)

Size of Project—The size, scope of work, and type of construction project will have a significant impact on cost. Economies of scale can reduce costs for large projects. Unit costs can often run higher for small projects. Costs in this book are intended for the size and type of project as previously described in "How the Book Is Built: An Overview." Costs for projects of a significantly different size or type should be adjusted accordingly.

Location—Material prices in this book are for metropolitan areas. However, in dense urban areas, traffic and site storage limitations may increase costs. Beyond a 32-km radius of large cities, extra trucking or transportation charges may also increase the material costs slightly. On the other hand, lower wage rates may be in effect. Be sure to consider both of these factors when preparing an estimate, particularly if the job site is located in a central city or remote rural location.

In addition, highly specialized subcontract items may require travel and per-diem expenses for mechanics.

Other Factors —
- season of year
- contractor management
- weather conditions
- local union restrictions
- building code requirements
- availability of:
  - adequate energy
  - skilled labor
  - building materials
- owner's special requirements/restrictions
- safety requirements
- environmental considerations

Unpredictable Factors—General business conditions influence "in-place" costs of all items. Substitute materials and construction methods may have to be employed. These may affect the installed cost and/or life cycle costs. Such factors may be difficult to evaluate and cannot necessarily be predicted on the basis of the job's location in a particular section of the country. Thus, where these factors apply, you may find significant but unavoidable cost variations for which you will have to apply a measure of judgment to your estimate.

Rounding of Costs

In general, all unit prices in excess of $5.00 have been rounded to make them easier to use and still maintain adequate precision of the results. The rounding rules we have chosen are in the following table.

Prices from . . .	Rounded to the nearest . . .
$.01 to $5.00	$.01
$5.01 to $20.00	$.05
$20.01 to $100.00	$.50
$100.01 to $300.00	$1.00
$300.01 to $1,000.00	$5.00
$1,000.01 to $10,000.00	$25.00
$10,000.01 to $50,000.00	$100.00
$50,000.01 and above	$500.00

Final Checklist

Estimating can be a straightforward process provided you remember the basics. Here's a checklist of some of the steps you should remember to complete before finalizing your estimate.

Did you remember to . . .
- factor in the City Cost Index for your locale?
- take into consideration which items have been marked up and by how much?
- mark up the entire estimate sufficiently for your purposes?
- read the background information on techniques and technical matters that could impact your project time span and cost?
- include all components of your project in the final estimate?
- double check your figures for accuracy?
- call RSMeans if you have any questions about your estimate or the data you've found in our publications?

Remember, RSMeans stands behind its publications. If you have any questions about your estimate . . . about the costs you've used from our books . . . or even about the technical aspects of the job that may affect your estimate, feel free to call your RSMeans representative at 1-800-334-3509.

About the Metric Edition

Metric is Here!

With the advent of NAFTA (North American Free Trade Agreement) and the strengthening of the E.C. (European Community), metric is here!

Two co-signers of NAFTA, Canada and Mexico, use metric standards of measure. If the United States, the third co-signer of NAFTA, is to have free trade with Canada and Mexico, and if architects, engineers, specifiers, material suppliers, and contractors are to communicate and trade in the construction industry, the United States must convert to the metric system of measure.

Much of the preliminary work for metrication has been done. The U.S. building codes, BOCA and UBC, now have dual units (inch-pound and SI) and the standards of NFPA (National Fire Protection Association) and ASTM (American Society for Testing and Materials) are also written in dual units.

The GSA (General Services Administration) and several other U.S. Government agencies have many multi-million dollar projects in progress using metric plans and specifications. This book was prepared by the editors at RSMeans to fulfill the needs of all construction professionals requiring cost information related to the construction industry. As more items are manufactured and become available in hard metric sizes, the editors will be adding these to the cost data file on a continuing basis.

What Is SI Metric?

SI (System International) Metric is the name given to the modern measurement system that has been adopted as a worldwide standard for the preferred system of weights and measures.

In the SI Metric System, only one unit is used to describe a measured quantity. It is infinitely simple in that there are no conversions to remember (inches, feet, yards, fathoms, rods, chains, furlongs, miles, etc.). All quantities are derived using decimal arithmetic from the base units. The following is a list of SI Metric base units used in construction.

Base Units:

There are seven metric base units of measurement, six of which are used in design and construction as follows.

Quantity or Measurement	SI Unit	Symbol
length	meter	m
mass (weight)	kilogram	kg
time	second	s
electric current	ampere	A
temperature	kelvin	K
luminous intensity	candela	cd

All other SI units are derived from these base units.

Note: In using symbols for SI Units, the use of the proper upper case or lower case letters is most important.

SI Decimal Prefixes:

Factor		Prefix	Symbol
$1\,000\,000\,000 = 10^9$		giga	G
$1\,000\,000 = 10^6$		mega	M
$1\,000 = 10^3$		kilo	k
$100 = 10^2$		hecto	h
$10 = 10^1$		deka	da
$0.1 = 10^{-1}$		deci	d
$0.01 = 10^{-2}$		centi	c
$0.001 = 10^{-3}$		milli	m

The most commonly used SI prefixes in construction are kilo (k) and milli (m).

A complete list of all units of measure and conversions used in this book is provided in the Reference Section.

Formats and Conventions:

Commas are not used when writing SI Metric numbers. Use spaces for writing numbers over four digits (27 548).

Use a 0 before a decimal when writing a number less than one (0.275).

Leave a space between a number and a symbol (81 kg, not 81kg).

Do not use the plural of symbols (54 m, not 54 ms).

A center dot is used to indicate the product of two units (N · m) rather than x or * .

Soft Conversion:

Over 95% of the products used in building construction in the United States today will undergo no physical change in material composition or standards of size and configuration during the SI Metric transition.

What will occur is that the dimensions of the product shown on drawings, in specifications, and on product literature will be expressed in metric units. This is called *Soft Conversion*.

Most building product dimensions and installation productivities in *RSMeans Metric Construction Cost Data* book have been developed from soft conversion.

Where building products are readily available from domestic sources in standard metric dimensions, unit material costs and installation costs have been expressed in *Hard Metric* units. Anchor bolts are readily available in *Hard Metric* sizes (Line 03150-082-1000 [M10, 10 mm dia x 125 mm long]).

Transition Guidelines:

For the convenience of the user during this transition period, the following conventions have been used.

* Gage (or gauge) remains the same (16 ga, 18 ga) for all products such as sheet metal and wire.

* Soft metric conversion of pipe sizes is followed by the inch dimension [3″].

* The M after A615 indicates a metric specification. A615M reinforcing bars come in Grade 400, indicating 400 MPa yield strength.

* Structural steel members have been expressed in their nominal SI size, followed by their in-lb size (W 360 x 39 [W 14 x 26]).

* Concrete strengths have been expressed in nominal SI units (2500 psi = 20 MPa, 3500 psi = 25 MPa).

* Electrical conduit sizes have been expressed in nominal SI (mm) sizes, with the inch size following [3″].

Unit Price Section

Table of Contents

How to Use the Unit Price Pages

The following is a detailed explanation of a sample entry in the Unit Price Section. Next to each bold number below is the described item with the appropriate component of the sample entry following in parentheses. Some prices are listed as bare costs; others as costs that include overhead and profit of the installing contractor. In most cases, if the work is to be subcontracted, the general contractor will need to add an additional markup (RSMeans suggests using 10%) to the figures in the column "Total Incl. O&P."

1 Division Number/Title (03400/Precast Concrete)

Use the Unit Price Section Table of Contents to locate specific items. The sections are classified according to the CSI MasterFormat (1995 Edition).

2 Line Numbers (03410 620 0100)

Each unit price line item has been assigned a unique 12-digit code based on the CSI MasterFormat classification.

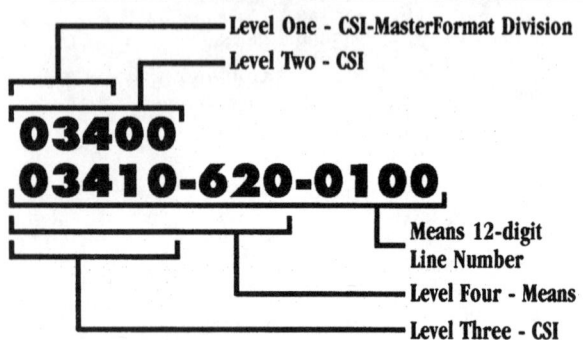

Level One - CSI-MasterFormat Division
Level Two - CSI

03400
03410-620-0100

Means 12-digit Line Number
Level Four - Means
Level Three - CSI

3 Description (SLABS, etc.)

Each line item is described in detail. Sub-items and additional sizes are indented beneath the appropriate line items. The first line or two after the main item (in boldface) may contain descriptive information that pertains to all line items beneath this boldface listing.

4 Reference Number Information

R034105 -30

You'll see reference numbers shown in bold rectangles at the beginning of some sections. These refer to related items in the Reference Section, visually identified by a vertical gray bar on the page edges.

The relation may be: (1) an estimating procedure that should be read before estimating, (2) an alternate pricing method, or (3) technical information.

The "R" designates the Reference Section. The numbers refer to the MasterFormat classification system.

It is strongly recommended that you review all reference numbers that appear within the section in which you are working.

03400 | Precast Concrete

		03410	Plant-Precast Structural Concrete	CREW	DAILY OUTPUT	LABOR-HOURS	UNIT	MAT.	LABOR	EQUIP.	TOTAL	TOTAL INCL O&P	
210	0010		**PRECAST COLUMNS**										210
	0020		Rectangular to 3.7 m high, small columns	C-11	36	.969	m	140	77	42	259	335	
	0050		Large columns		29	1.61		244	96	52.50	393		
	0300		7.3 m high, small columns		58.52	1.230		140	48	26.50	214.50	267	
400	0010		**PRECAST JOISTS**										400
	0015		195 kg/m² L.L., 150 mm deep for 3.7 m span	C-12	183	.262		23.50	9.15	3.48	36.13	43.50	
	0050		200 mm deep for 4.9 m spans		175	.274		39	9.60	3.64	52.24	61.50	
620	0010		**PRECAST SLAB PLANKS**										620
	0020		Prestressed roof/floor, grouted, solid, 100 mm thick	C-11	223	.323	m²	48	12.65	6.90	67.55	82	
	0050		150 mm thick		260	.277		59	10.85	5.90	75.75	90.50	
	0100		Hollow, 200 mm thick		297	.242		62	9.50	5.20	76.70	90	
	0150		250 mm thick		334	.216		65.50	8.45	4.61	78.56	92	
	0200		300 mm thick		372	.194		74.50	7.55	4.14	86.19	100	

Crew (C-11)

The "Crew" column designates the typical trade or crew used to install the item. If an installation can be accomplished by one trade and requires no power equipment, that trade and the number of workers are listed (for example, "2 Struc. Steel Workers"). If an installation requires a composite crew, a crew code designation is listed (for example, "C-11"). You'll find full details on all composite crews in the Crew Listings.

* For a complete list of all trades utilized in this book and their abbreviations, see the inside back cover.

Crews

Crew No.	Bare Costs		Incl. Subs O & P		Cost Per Labor-Hour	
Crew C-11	Hr.	Daily	Hr.	Daily	Bare Costs	Incl. O&P
1 Struc. Steel Foreman	$41.95	$335.60	$75.50	$604.00	$39.13	$68.15
6 Struc. Steel Workers	39.95	1917.60	71.90	3451.20		
1 Equip. Oper. (crane)	38.10	304.80	57.50	460.00		
1 Equip. Oper. Oiler	32.45	259.60	48.95	391.60		
1 Truck Crane, 136 metric ton		1539.00		1692.90	21.38	23.51
72 L.H., Daily Totals		$4356.60		$6599.70	$60.51	$91.66

Productivity: Daily Output (297)/ Labor-Hours (0.242)

The "Daily Output" represents the typical number of units the designated crew will install in a normal 8-hour day. To find out the number of days the given crew would require to complete the installation, divide your quantity by the daily output. For example:

Quantity	÷	Daily Output	=	Duration
1500 m²	÷	297 m²/ Crew Day	=	5.05 Crew Days

The "Labor-Hours" figure represents the number of labor-hours required to install one unit of work. To find out the number of labor-hours required for your particular task, multiply the quantity of the item times the number of labor-hours shown. For example:

Quantity	x	Productivity Rate	=	Duration
1500 m²	x	0.242 Labor-Hours/ m²	=	363 Labor-Hours

Unit (m²)

The abbreviated designation indicates the unit of measure upon which the price, production, and crew are based (m² = Square Meter). For a complete listing of abbreviations refer to the Abbreviations Listing in the Reference Section of this book.

Bare Costs:

Mat. (Bare Material Cost) (62)

The unit material cost is the "bare" material cost with no overhead and profit included. *Costs shown reflect national average material prices for January of the current year and include delivery to the job site. No sales taxes are included.*

Labor (9.50)

The unit labor cost is derived by multiplying bare labor-hour costs for Crew C-11 by labor-hour units. The bare labor-hour cost is found in the Crew Section under C-11. (If a trade is listed, the hourly labor cost—the wage rate—is found on the inside back cover.)

Labor-Hour Cost Crew C-11	x	Labor-Hour Units	=	Labor
$39.13	x	0.242	=	$9.50

Equip. (Equipment) (5.20)

Equipment costs for each crew are listed in the description of each crew. Tools or equipment whose value justifies purchase or ownership by a contractor are considered overhead as shown on the inside back cover. The unit equipment cost is derived by multiplying the bare equipment hourly cost by the labor-hour units.

Equipment Cost Crew C-1	x	Labor-Hour Units	=	Equip.
$21.38	x	0.242	=	$5.20

Total (76.70)

The total of the bare costs is the arithmetic total of the three previous columns: mat., labor, and equip.

Material	+	Labor	+	Equip.	=	Total
$62.00	+	$9.50	+	$5.20	=	$76.70

Total Costs Including O&P

This figure is the sum of the bare material cost plus 10% for profit; the bare labor cost plus total overhead and profit (per the inside back cover or, if a crew is listed, from the crew listings); and the bare equipment cost plus 10% for profit.

Material is Bare Material Cost + 10% = $62.00 + $6.20	=	$68.20
Labor for Crew C-11 = Labor-Hour Cost ($68.15) x Labor-Hour Units (0.242)	=	$16.49
Equip. is Bare Equip. Cost + 10% = $5.20 + $0.52	=	$ 5.72
Total (Rounded)	=	$90.00

Division 1
General Requirements

Estimating Tips

The General Requirements of any contract are very important to both the bidder and the owner. These lay the ground rules under which the contract will be executed and have a significant influence on the cost of operations. Therefore, it is extremely important to thoroughly read and understand the General Requirements both before preparing an estimate and when the estimate is complete, to ascertain that nothing in the contract is overlooked. Caution should be exercised when applying items listed in Division 1 to an estimate. Many of the items are included in the unit prices listed in the other divisions such as mark-ups on labor and company overhead.

01200 Price & Payment Procedures

- When estimating historic preservation projects (depending on the condition of the existing structure and the owner's requirements), a 15-20% contingency or allowance is recommended, regardless of the stage of the drawings.

01300 Administrative Requirements

- Before determining a final cost estimate, it is a good practice to review all the items listed in subdivision 01300 to make final adjustments for items that may need customizing to specific job conditions.
- Historic preservation projects may require specialty labor and methods, as well as extra time to protect existing materials that must be preserved and/or restored. Some additional expenses may be incurred in architectural fees for facility surveys and other special inspections and analyses.
- Requirements for initial and periodic submittals can represent a significant cost to the General Requirements of a job. Thoroughly check the submittal specifications when estimating a project to determine any costs that should be included.

01400 Quality Requirements

- All projects will require some degree of Quality Control. This cost is not included in the unit cost of construction listed in each division. Depending upon the terms of the contract, the various costs of inspection and testing can be the responsibility of either the owner or the contractor. Be sure to include the required costs in your estimate.

01500 Temporary Facilities & Controls

- Barricades, access roads, safety nets, scaffolding, security and many more requirements for the execution of a safe project are elements of direct cost. These costs can easily be overlooked when preparing an estimate. When looking through the major classifications of this subdivision, determine which items apply to each division in your estimate.

01740 Execution Requirements

- When preparing an estimate, thoroughly read the specifications to determine the requirements for Contract Closeout. Final cleaning, record documentation, operation and maintenance data, warranties and bonds, and spare parts and maintenance materials can all be elements of cost for the completion of a contract. Do not overlook these in your estimate.

01830 Operations & Maintenance

- If maintenance and repair are included in your contract, they require special attention. Estimating the cost to remove and replace any unit usually requires a site visit to determine the accessibility and the specific difficulty at that location. Obstructions, dust control, safety, and often, overtime hours must be considered when preparing your estimate.

Reference Numbers

Reference numbers are shown in bold squares at the beginning of some major classifications. These numbers refer to related items in the Reference Section. The reference information may be an estimating procedure, an alternate pricing method, or technical information.

Note: Not all subdivisions listed here necessarily appear in this publication.

01103	Models & Renderings	CREW	DAILY OUTPUT	LABOR-HOURS	UNIT	2006 BARE COSTS				TOTAL INCL O&P	
						MAT.	LABOR	EQUIP.	TOTAL		
200	0010	**MODELS**										**200**
	0020	Cardboard & paper, 1 building, minimum				Ea.	640			640	705
	0050	Maximum					1,475			1,475	1,600
	0100	2 buildings, minimum					860			860	945
	0150	Maximum				↓	1,950			1,950	2,125
	0200	Plexiglass and metal, basic layout				m² Flr.	.65			.65	.75
	0210	Including equipment and personnel				"	3.12			3.12	3.44
	0300	Site plan layout, minimum				Ea.	1,225			1,225	1,350
	0350	Maximum				"	2,050			2,050	2,250
500	0010	**RENDERINGS** Color, matted, 508 mm x 762 mm, eye level,										**500**
	0020	1 building, minimum				Ea.	1,850			1,850	2,050
	0050	Average					2,650			2,650	2,925
	0100	Maximum					4,250			4,250	4,675
	1000	5 buildings, minimum					3,725			3,725	4,075
	1100	Maximum					7,425			7,425	8,175
	2000	Aerial perspective, color, 1 building, minimum					2,650			2,650	2,925
	2100	Maximum					7,425			7,425	8,175
	3000	5 buildings, minimum					5,300			5,300	5,825
	3100	Maximum				↓	10,600			10,600	11,700

01107	Professional Consultant	CREW	DAILY OUTPUT	LABOR-HOURS	UNIT	MAT.	LABOR	EQUIP.	TOTAL	TOTAL INCL O&P	
100	0011	**ARCHITECTURAL FEES**	R011110 -10									**100**
	0020	For new construction									
	0060	Minimum				Project					4.90%
	0090	Maximum									16%
	0100	For alteration work, to $500,000, add to fee									50%
	0150	Over $500,000, add to fee	↓			↓					25%
200	0010	**CONSTRUCTION MANAGEMENT FEES**										**200**
	0020	$1,000,000 job, minimum				Project					4.50%
	0050	Maximum									7.50%
	0300	$5,000,000 job, minimum									2.50%
	0350	Maximum				↓					4%
300	0010	**ENGINEERING FEES**	R011110 -30									**300**
	0020	Educational planning consultant, minimum				Project					.50%
	0100	Maximum				"					2.50%
	0200	Electrical, minimum				Contrct					4.10%
	0300	Maximum									10.10%
	0400	Elevator & conveying systems, minimum									2.50%
	0500	Maximum									5%
	0600	Food service & kitchen equipment, minimum									8%
	0700	Maximum									12%
	0800	Landscaping & site development, minimum									2.50%
	0900	Maximum									6%
	1000	Mechanical (plumbing & HVAC), minimum									4.10%
	1100	Maximum				↓					10.10%
	1200	Structural, minimum				Project					1%
	1300	Maximum	↓			"					2.50%
700	0010	**TOPOGRAPHICAL SURVEYS**										**700**
	0020	Conventional, topographical, minimum	A-7	1.34	17.971	Hectare	41.50	700	43.50	785	1,175
	0100	Maximum	A-8	.24	131		125	4,975	240	5,340	8,025
	0300	Lot location and lines, minimum, for large quantities	A-7	.81	29.652		65	1,150	72	1,287	1,925
	0320	Average	"	.51	47.442		117	1,850	115	2,082	3,100
	0400	Maximum, for small quantities	A-8	.40	79.071	↓	187	2,975	144	3,306	4,950

01100 | Summary

01107 | Professional Consultant

			CREW	DAILY OUTPUT	LABOR-HOURS	UNIT	MAT.	LABOR	EQUIP.	TOTAL	TOTAL INCL O&P	
							2006 BARE COSTS					
700	0600	Monuments, 914 mm long	A-7	10	2.400	Ea.	20.50	93	5.85	119.35	173	700
	0800	Property lines, perimeter, cleared land	"	305	.079	m	.10	3.06	.19	3.35	5.05	
	0900	Wooded land	A-8	267	.120	"	.16	4.52	.22	4.90	7.40	
	1100	Crew for layout of building, trenching or pipe laying, 2 person crew	A-6	1	16	Day		565	58	623	935	
	1200	3 person crew	A-7	1	24			930	58.50	988.50	1,525	
	1400	Crew for roadway layout, 4 person crew	A-8	1	32	▼		1,200	58	1,258	1,925	
	1500	Aerial surveying, including ground control, minimum fee, 4 hectares				Total					5,700	
	1510	40 hectares									9,500	
	1550	From existing photography, deduct				▼					1,370	
	1600	610 mm contours, 4 hectares				Hectare					1,140	
	1650	8 hectares									780	
	1800	20 hectares									235	
	1850	40 hectares									210	
	2000	405 hectares									44.10	
	2050	4050 hectares				▼					28.40	
	2150	For 305 mm contours and										
	2160	dense urban areas, add to above				Hectare					40%	
	3000	Inertial guidance system for										
	3010	locating coordinates, rent/day				Ea.					4,000	

01200 | Price & Payment Procedures

01250 | Contract Modification Procedures

			CREW	DAILY OUTPUT	LABOR-HOURS	UNIT	MAT.	LABOR	EQUIP.	TOTAL	TOTAL INCL O&P	
							2006 BARE COSTS					
200	0010	CONTINGENCIES										200
	0020	For estimate at conceptual stage				Project					20%	
	0050	Schematic stage									15%	
	0100	Preliminary working drawing stage (Design Dev.)									10%	
	0150	Final working drawing stage				▼					3%	
300	0010	CREWS For building construction, see How To Use This Book										300
500	0010	JOB CONDITIONS Modifications to total										500
	0020	project cost summaries										
	0100	Economic conditions, favorable, deduct				Project					2%	
	0200	Unfavorable, add									5%	
	0300	Hoisting conditions, favorable, deduct									2%	
	0400	Unfavorable, add									5%	
	0500	General Contractor management, experienced, deduct									2%	
	0600	Inexperienced, add									10%	
	0700	Labor availability, surplus, deduct									1%	
	0800	Shortage, add									10%	
	0900	Material storage area, available, deduct									1%	
	1000	Not available, add									2%	
	1100	Subcontractor availability, surplus, deduct									5%	
	1200	Shortage, add									12%	
	1300	Work space, available, deduct									2%	
	1400	Not available, add				▼					5%	
600	0010	OVERTIME For early completion of projects or where	R012909 -90									600
	0020	labor shortages exist, add to usual labor, up to				Costs		100%				

01200 | Price & Payment Procedures

01255	Cost Indexes		CREW	DAILY OUTPUT	LABOR-HOURS	UNIT	2006 BARE COSTS				TOTAL INCL O&P	
							MAT.	LABOR	EQUIP.	TOTAL		
200	0010	**CONSTRUCTION COST INDEX** (Reference) Over 930 zip code locations in										**200**
	0020	the U.S. and Canada, total bldg. cost, min. (Clarksdale, MS)				%					67%	
	0050	Average				↓					100%	
	0100	Maximum (New York, NY)				↓					131.90%	
400	0010	**HISTORICAL COST INDEXES** (Reference) Back to 1956										**400**
500	0010	**LABOR INDEX** (Reference) For over 930 zip code locations in										**500**
	0020	the U.S. and Canada, minimum (Clarksdale, MS)				%		30.80%				
	0050	Average				↓		100%				
	0100	Maximum (New York, NY)				↓		164.40%				
600	0010	**MATERIAL INDEX** (Reference) For over 930 zip code locations in										**600**
	0020	the U.S. and Canada, minimum (Elizabethtown, KY)				%	91.10%					
	0040	Average				↓	100%					
	0060	Maximum (Ketchikan, AK)				↓	141.90%					

01290	Payment Procedures											
800	0010	**TAXES**	R012909 -80									**800**
	0020	Sales tax, state, average				%	4.83%					
	0050	Maximum	R012909 -85				7%					
	0200	Social Security, on first $90,000 of wages						7.65%				
	0300	Unemployment, combined Federal and State, minimum						.80%				
	0350	Average						6.20%				
	0400	Maximum	↓			↓		11.76%				

01300 | Administrative Requirements

01310	Project Management/Coordination		CREW	DAILY OUTPUT	LABOR-HOURS	UNIT	2006 BARE COSTS				TOTAL INCL O&P	
							MAT.	LABOR	EQUIP.	TOTAL		
150	0010	**PERMITS**										**150**
	0020	Rule of thumb, most cities, minimum				Job					.50%	
	0100	Maximum				"					2%	
200	0010	**PERFORMANCE BOND**	R013113 -80									**200**
	0020	For buildings, minimum				Job					.60%	
	0100	Maximum									2.50%	
	0200	Maximum									1.50%	
	0300	Highways & Bridges, resurfacing, minimum									.40%	
	0350	Maximum	↓			↓					.94%	
300	0010	**CONSTRUCTION TIME** Requirements	R012157 -20									**300**
350	0010	**INSURANCE**	R013113 -40									**350**
	0020	Builders risk, standard, minimum				Job					.24%	
	0050	Maximum	R024119 -10								.64%	
	0200	All-risk type, minimum									.25%	
	0250	Maximum				↓					.62%	
	0400	Contractor's equipment floater, minimum				Value					.50%	
	0450	Maximum				"					1.50%	
	0600	Public liability, average				Job					2.02%	
	0800	Workers' compensation & employer's liability, average										
	0850	by trade, carpentry, general	↓			Payroll		18.39%				

		01310	Project Management/Coordination	CREW	DAILY OUTPUT	LABOR-HOURS	UNIT	2006 BARE COSTS				TOTAL INCL O&P	
								MAT.	LABOR	EQUIP.	TOTAL		
350	0900		Clerical				Payroll		.65%				350
	0950	R013113 -40	Concrete						15.82%				
	1000		Electrical						6.84%				
	1050	R024119 -10	Excavation						10.63%				
	1100		Glazing						13.87%				
	1150		Insulation						15.76%				
	1200		Lathing						11.44%				
	1250		Masonry						14.92%				
	1300		Painting & decorating						13.29%				
	1350		Pile driving						22.07%				
	1400		Plastering						14%				
	1450		Plumbing						8.22%				
	1500		Roofing						32.40%				
	1550		Sheet metal work (HVAC)						11.76%				
	1600		Steel erection, structural						39.65%				
	1650		Tile work, interior ceramic						9.52%				
	1700		Waterproofing, brush or hand caulking						7.04%				
	1800		Wrecking						40.06%				
	2000		Range of 35 trades in 50 states, excl. wrecking, min.						2.40%				
	2100		Average						16.30%				
	2200		Maximum						137.60%				
400	0010		**MAIN OFFICE EXPENSE** Average for General Contractors										400
	0020	R013113 -50	As a percentage of their annual volume										
	0030		Annual volume to $50,000, minimum				% Vol.				20%		
	0040		Maximum								30%		
	0060		To $100,000, minimum								17%		
	0070		Maximum								22%		
	0080		To $250,000, minimum								16%		
	0090		Maximum								19%		
	0110		To $500,000, minimum								14%		
	0120		Maximum								16%		
	0125		Annual volume under 1 million dollars								13.60%		
	0130		To $1,000,000, minimum								8%		
	0140		Maximum								10%		
	0145		Up to 2.5 million dollars								8%		
	0150		Up to 4.0 million dollars								6.80%		
	0200		Up to 7.0 million dollars								5.60%		
	0250		Up to 10 million dollars								5.10%		
	0300		Over 10 million dollars								3.90%		
500	0010		**MARK-UP** For General Contractors for change										500
	0100	R012909 -80	of scope of job as bid										
	0200		Extra work, by subcontractors, add				%					10%	
	0250		By General Contractor, add									15%	
	0400		Omitted work, by subcontractors, deduct all but									5%	
	0450		By General Contractor, deduct all but									7.50%	
	0600		Overtime work, by subcontractors, add									15%	
	0650		By General Contractor, add									10%	
	1000		Installing contractors, on his own labor, minimum						46.90%				
	1100		Maximum						87.30%				
	1150		Overhead markup, see division 01310-620										
600	0010		**OVERHEAD**										600
	0020	R013113 -50	As percent of direct costs, minimum				%				5%		
	0050		Average								13%		
	0100		Maximum								30%		

1 GENERAL REQUIREMENTS

01310 | Project Management/Coordination

			CREW	DAILY OUTPUT	LABOR-HOURS	UNIT	2006 BARE COSTS				TOTAL INCL O&P	
							MAT.	LABOR	EQUIP.	TOTAL		
620	0010	**OVERHEAD & PROFIT** Allowance to add to items in this										620
	0020	book that do not include Subs' O&P, average				%				25%		
	0100	Allowance to add to items in this book that										
	0110	do include Subs' O&P, minimum				%					5%	
	0150	Average									10%	
	0200	Maximum									15%	
	0300	Typical, by size of project, under $100,000									30%	
	0350	$500,000 project									25%	
	0400	$2,000,000 project									20%	
	0450	Over $10,000,000 project				↓					15%	
700	0010	**FIELD PERSONNEL**										700
	0020	Clerk average				Week		335		335	515	
	0100	Field engineer, minimum						800		800	1,225	
	0120	Average						1,040		1,040	1,600	
	0140	Maximum						1,200		1,200	1,850	
	0160	General purpose laborer, average						1,100		1,100	1,675	
	0180	Project manager, minimum						1,500		1,500	2,300	
	0200	Average						1,700		1,700	2,600	
	0220	Maximum						1,950		1,950	3,000	
	0240	Superintendent, minimum						1,450		1,450	2,225	
	0260	Average						1,575		1,575	2,425	
	0280	Maximum						1,800		1,800	2,750	
	0290	Timekeeper, average				↓		930		930	1,425	

01320 | Const. Progress Documentation

			CREW	DAILY OUTPUT	LABOR-HOURS	UNIT	MAT.	LABOR	EQUIP.	TOTAL	TOTAL INCL O&P	
200	0010	**SCHEDULING**										200
	0020	Critical path, as % of architectural fee, minimum				%					.50%	
	0100	Maximum				"					1%	
	0300	Computer update, micro, no plots, minimum				Ea.					450	
	0400	Including plots, maximum				"					1,550	
	0600	Rule of thumb, CPM scheduling, small job ($10 Million)				Job					.05%	
	0650	Large job ($50 Million +)									.03%	
	0700	Including cost control, small job									.08%	
	0750	Large job				↓					.04%	

01321 | Construction Photos

			CREW	DAILY OUTPUT	LABOR-HOURS	UNIT	MAT.	LABOR	EQUIP.	TOTAL	TOTAL INCL O&P	
500	0010	**PHOTOGRAPHS**										500
	0020	203 mm x 254 mm, 4 shots, 2 prints ea., std. mounting				Set	300			300	330	
	0100	Hinged linen mounts					325			325	360	
	0200	203 mm x 254 mm, 4 shots, 2 prints each, in color					350			350	385	
	0300	For I.D. slugs, add to all above					4.21			4.21	4.63	
	0500	Aerial photos, 1st fly-over, 6 shots, 1 print ea., 203 mm x 254 mm					710			710	780	
	0550	279 mm x 356 mm prints					790			790	870	
	0600	406 mm x 508 mm prints					1,025			1,025	1,125	
	0700	For full color prints, add					40%				40%	
	0750	Add for traffic control area				↓	294			294	325	
	0900	For over 48 km from airport, add per				km	3.30			3.30	3.63	
	1000	Vertical photography, 4 to 6 shots with										
	1010	different scales, 1 print each				Set	1,100			1,100	1,200	
	1500	Time lapse equipment, camera and projector, buy					3,775			3,775	4,175	
	1550	Rent/month				↓	565			565	620	
	1700	Cameraman and film, including processing, B.&W.				Day	1,375			1,375	1,525	
	1720	Color				"	1,375			1,375	1,525	

							2006 BARE COSTS				TOTAL	
01450		**Quality Control**		CREW	DAILY OUTPUT	LABOR-HOURS	UNIT	MAT.	LABOR	EQUIP.	TOTAL	INCL O&P	
500	0010	**TESTING** and Inspectional Services	R312323 -30										**500**
	0015	For concrete building costing $1,000,000, minimum					Project					5,200	
	0020	Maximum										41,800	
	0050	Steel building, minimum										5,200	
	0070	Maximum										16,300	
	0100	For building costing, $10,000,000, minimum										33,100	
	0150	Maximum					▼					53,000	
	0200	Asphalt testing, compressive strength Marshall stability, set of 3					Ea.					165	
	0220	Density, set of 3										95	
	0250	Extraction, individual tests on sample										150	
	0300	Penetration										45	
	0350	Mix design, 5 specimens										200	
	0360	Additional specimen										40	
	0400	Specific gravity										45	
	0420	Swell test										70	
	0450	Water effect and cohesion, set of 6										200	
	0470	Water effect and plastic flow										70	
	0600	Concrete testing, aggregates, abrasion, ASTM C131										150	
	0650	Absorption, ASTM C 127										46	
	0800	Petrographic analysis, ASTM C 295										850	
	0900	Specific gravity, ASTM C 127										55	
	1000	Sieve analysis, washed, ASTM C 136										65	
	1050	Unwashed										65	
	1200	Sulfate soundness										125	
	1300	Weight/cubic mm										40	
	1500	Cement, physical tests, ASTM C 150										350	
	1600	Chemical tests, ASTM C 150										270	
	1800	Compressive test, cylinder, delivered to lab, ASTM C 39										13	
	1900	Picked up by lab, minimum										15	
	1950	Average										20	
	2000	Maximum										30	
	2200	Compressive strength, cores (not incl. drilling), ASTM C 42					▼					40	
	2250	Core drilling, 102 mm diameter (plus technician) [4"]					mm					1	
	2260	Technician for core drilling					Hr.					50	
	2300	Patching core holes					Ea.					24	
	2400	Drying shrinkage at 28 days										260	
	2500	Flexural test beams, ASTM C 78										65	
	2600	Mix design, one batch mix										285	
	2650	Added trial batches										132	
	2800	Modulus of elasticity, ASTM C 469										180	
	2900	Tensile test, cylinders, ASTM C 496										50	
	3000	Water-Cement ratio curve, 3 batches										155	
	3100	4 batches										205	
	3300	Masonry testing, absorption, per 5 brick, ASTM C 67										50	
	3350	Chemical resistance, per 2 brick										55	
	3400	Compressive strength, per 5 brick, ASTM C 67										75	
	3420	Efflorescence, per 5 brick, ASTM C 67										75	
	3440	Imperviousness, per 5 brick										96	
	3470	Modulus of rupture, per 5 brick										95	
	3500	Moisture, block only										35	
	3550	Mortar, compressive strength, set of 3										25	
	4100	Reinforcing steel, bend test										61	
	4200	Tensile test, up to #25 bar										40	
	4220	#30 to #35 bar										45	
	4240	#45 bar and larger										70	
	4400	Soil testing, Atterberg limits, liquid and plastic limits					▼					65	

01400 | Quality Requirements

		01450	Quality Control	CREW	DAILY OUTPUT	LABOR-HOURS	UNIT	MAT.	LABOR	EQUIP.	TOTAL	TOTAL INCL O&P	
									2006 BARE COSTS				
500	4510	Hydrometer analysis	R312323 -30				Ea.					120	500
	4530	Specific gravity, ASTM D 354										48	
	4600	Sieve analysis, washed, ASTM D 422										60	
	4700	Unwashed, ASTM D 422										65	
	4710	Consolidation test (ASTM D2435), minimum										275	
	4715	Maximum										475	
	4720	Density and classification of undisturbed sample										80	
	4735	Soil density, nuclear method, ASTM D2922										38.67	
	4740	Sand cone method ASTM D1556										30.17	
	4750	Moisture content, ASTM D 2216										10	
	4780	Permeability test, double ring infiltrometer										550	
	4800	Permeability, var. or constant head, undist., ASTM D 2434										250	
	4850	Recompacted										275	
	4900	Proctor compaction, 102 mm standard mold, ASTM D 698										135	
	4950	152 mm modified mold										75	
	5100	Shear tests, triaxial, minimum										450	
	5150	Maximum										600	
	5300	Direct shear, minimum, ASTM D 3080										350	
	5350	Maximum										450	
	5550	Technician for inspection, per day, earthwork										220	
	5570	Concrete										270	
	5650	Bolting										280	
	5750	Roofing										255	
	5790	Welding					▼					270	
	5820	Non-destructive testing, dye penetrant					Day					330	
	5840	Magnetic particle										330	
	5860	Radiography										495	
	5880	Ultrasonic					▼					340	
	5900	Vibration monitoring, seismograph and technician					▼					495	
	5910	Seismograph, rental only					Week					275	
	6000	Welding certification, minimum					Ea.					100	
	6100	Maximum					"					275	
	7000	Underground storage tank											
	7500	Volumetric tightness test, <=405 m³					Ea.					400	
	7510	<=1010 m³					"					660	
	7600	Vadose zone (soil gas) sampling, 10-40 samples, min.					Day					1,500	
	7610	Maximum					"					2,500	
	7700	Ground water monitoring incl. drilling 3 wells, min.					Total					5,000	
	7710	Maximum					"					7,000	
	8000	X-ray concrete slabs					Ea.					200	

01500 | Temporary Facilities & Controls

		01510	Temporary Utilities	CREW	DAILY OUTPUT	LABOR-HOURS	UNIT	MAT.	LABOR	EQUIP.	TOTAL	TOTAL INCL O&P	
									2006 BARE COSTS				
050	0010	**TEMPORARY POWER EQUIP (PRO-RATED PER JOB)**											050
	0020	Service, overhead feed, 3 use											
	0030	100 A	1 Elec	1.25	6.400	Ea.	580	269		849	1,050	
	0040	200 A		1	8		745	335		1,080	1,325	
	0050	400 A		.75	10.667		1,450	450		1,900	2,250	
	0060	600 A	▼	.50	16	▼	1,800	670		2,470	2,975	

01510	Temporary Utilities	CREW	DAILY OUTPUT	LABOR-HOURS	UNIT	2006 BARE COSTS				TOTAL INCL O&P
						MAT.	LABOR	EQUIP.	TOTAL	
0100	Underground feed, 3 use									
0110	100 A	1 Elec	2	4	Ea.	555	168		723	865
0120	200 A		1.15	6.957		760	292		1,052	1,275
0130	400 A		1	8		1,425	335		1,760	2,075
0140	600 A		.75	10.667		1,650	450		2,100	2,475
0150	800 A		.50	16		2,750	670		3,420	4,025
0160	1000 A		.35	22.857		2,850	960		3,810	4,575
0170	1200 A		.25	32		3,125	1,350		4,475	5,425
0180	2000 A	▼	.20	40	▼	4,000	1,675		5,675	6,900
0200	Transformers, 3 use									
0210	30 kVA	1 Elec	1	8	Ea.	845	335		1,180	1,425
0220	45 kVA		.75	10.667		970	450		1,420	1,750
0230	75 kVA		.50	16		1,475	670		2,145	2,625
0240	112.5 kVA	▼	.40	20	▼	2,000	840		2,840	3,450
0250	Feeder, PVC, CU wire									
0260	60 A w/trench	1 Elec	29.26	.273	m	13	11.50		24.50	31.50
0270	100 A w/trench		25.91	.309		17.45	12.95		30.40	38.50
0280	200 A w/trench		17.98	.445		36.50	18.70		55.20	68.50
0290	400 A w/trench	▼	12.80	.625	▼	77	26.50		103.50	124
0300	Feeder, PVC, aluminum wire									
0310	60 A w/trench	1 Elec	29.26	.273	m	12.15	11.50		23.65	30.50
0320	100 A w/trench		25.91	.309		14.05	12.95		27	35
0330	200 A w/trench		17.98	.445		29.50	18.70		48.20	60.50
0340	400 A w/trench	▼	12.80	.625	▼	48.50	26.50		75	92.50
0350	Feeder, EMT, CU wire									
0360	60 A	1 Elec	27.43	.292	m	12.25	12.25		24.50	31.50
0370	100 A		24.38	.328		23	13.80		36.80	45.50
0380	200 A		18.29	.437		44	18.35		62.35	76
0390	400 A	▼	10.67	.750	▼	57.50	31.50		89	111
0400	Feeder, EMT, Al wire									
0410	60 A	1 Elec	27.43	.292	m	9.85	12.25		22.10	29
0420	100 A		24.38	.328		19.90	13.80		33.70	42.50
0430	200 A		18.29	.437		37	18.35		55.35	68.50
0440	400 A	▼	10.67	.750	▼	50.50	31.50		82	103
0500	Equipment, 3 use									
0510	Spider box 50 A	1 Elec	8	1	Ea.	168	42		210	248
0520	Lighting cord 30.5 m		8	1		36.50	42		78.50	103
0530	Light stanchion	▼	8	1	▼	74	42		116	144
0540	Temporary cords, 30.5 m, 3 use									
0550	Feeder cord, 50 A	1 Elec	16	.500	Ea.	156	21		177	204
0560	Feeder cord, 100 A		12	.667		450	28		478	535
0570	Tap cord, 50 A		12	.667		430	28		458	515
0580	Tap cord, 100 A	▼	6	1.333	▼	615	56		671	765
0590	Temporary cords, 15.2 m, 3 use									
0600	Feeder cord, 50 A	1 Elec	16	.500	Ea.	91	21		112	132
0610	Feeder cord, 100 A		12	.667		224	28		252	289
0620	Tap cord, 50 A		12	.667		195	28		223	256
0630	Tap cord, 100 A	▼	6	1.333	▼	320	56		376	440
0700	Connections									
0710	Compressor or pump									
0720	30 A	1 Elec	7	1.143	Ea.	18.30	48		66.30	91.50
0730	60 A		5.30	1.509		42	63.50		105.50	141
0740	100 A	▼	4	2	▼	54	84		138	185
0750	Tower crane									
0760	60 A	1 Elec	4.50	1.778	Ea.	42	74.50		116.50	157
0770	100 A	"	3	2.667	"	54	112		166	227

GENERAL REQUIREMENTS 1

13

01500 | Temporary Facilities & Controls

GENERAL REQUIREMENTS | **1**

01510 | Temporary Utilities

			CREW	DAILY OUTPUT	LABOR-HOURS	UNIT	2006 BARE COSTS MAT.	LABOR	EQUIP.	TOTAL	TOTAL INCL O&P	
050	0780	Manlift										**050**
	0790	Single	1 Elec	3	2.667	Ea.	43	112		155	215	
	0800	Double	"	2	4	"	55.50	168		223.50	310	
	0810	Welder										
	0820	50 A w/disconnect	1 Elec	5	1.600	Ea.	270	67		337	395	
	0830	100 A w/disconnect		3.80	2.105		415	88.50		503.50	585	
	0840	200 A w/disconnect		2.50	3.200		660	134		794	925	
	0850	400 A w/disconnect		1	8		2,100	335		2,435	2,800	
800	0010	**TEMPORARY UTILITIES**										**800**
	0100	Heat, incl. fuel and operation, per week, 12 hrs./day	1 Skwk	929	.009	m² Flr.	.78	.31		1.09	1.34	
	0200	24 hrs./day	"	557	.014		1.16	.52		1.68	2.10	
	0350	Lighting, incl. service lamps, wiring & outlets, minimum	1 Elec	316	.025		.26	1.06		1.32	1.87	
	0360	Maximum	"	158	.051		.56	2.13		2.69	3.79	
	0400	Power for temp lighting only, per month, min/month 24 MJ								.08	.13	
	0450	Maximum/month 85 MJ								.29	.32	
	0600	Power for job duration incl. elevator, etc., minimum								5.05	5.55	
	0650	Maximum								11.85	13	
	0700	Temporary construction water bill per mo. average				Month	60			60	66	
	1000	Toilet, portable, see Equipment Rental in Reference section										

01520 | Construction Facilities

			CREW	DAILY OUTPUT	LABOR-HOURS	UNIT	MAT.	LABOR	EQUIP.	TOTAL	TOTAL INCL O&P	
500	0010	**OFFICE**										**500**
	0020	Trailer, furnished, no hookups, 6000 mm x 2400 mm, buy	2 Skwk	1	16	Ea.	7,550	585		8,135	9,200	
	0250	Rent/month					167			167	183	
	0300	9600 mm x 2400 mm, buy	2 Skwk	.70	22.857		10,400	835		11,235	12,700	
	0350	Rent/month					183			183	201	
	0400	15 000 mm x 3000 mm, buy	2 Skwk	.60	26.667		18,100	975		19,075	21,500	
	0450	Rent/month					293			293	320	
	0500	15 000 mm x 3600 mm, buy	2 Skwk	.50	32		23,300	1,175		24,475	27,400	
	0550	Rent/month					340			340	375	
	0700	For air conditioning, rent/month, add					41			41	45	
	0800	For delivery, add/km				km	2.17			2.17	2.39	
	1000	Portable buildings, prefab, on skids, economy, 2.5 m x 2.5 m	2 Carp	24.62	.650	m²	885	23		908	1,000	
	1100	Deluxe, 2.5 m x 3.6 m	"	13.94	1.148	"	970	41		1,011	1,150	
	1200	Storage boxes, 6000 mm x 2400 mm, buy	2 Skwk	1.80	8.889	Ea.	3,250	325		3,575	4,075	
	1250	Rent/month					75			75	82.50	
	1300	12000 mm x 2400 mm, buy	2 Skwk	1.40	11.429		4,275	415		4,690	5,350	
	1350	Rent/month					109			109	120	
	5000	Air supported structures, see Division 13011-200										
550	0010	**FIELD OFFICE EXPENSE**										**550**
	0100	Office equipment rental, average				Month	145			145	160	
	0120	Office supplies, average				"	90			90	99	
	0125	Office trailer rental, see Division 01520-500										
	0140	Telephone bill; avg. bill/month incl. long dist.				Month	210			210	231	
	0160	Field office lights & HVAC				"	110			110	121	
900	0010	**WEATHER STATION**										**900**
	0020	Remote recording, minimum				Ea.	5,500			5,500	6,050	
	0100	Maximum				"	27,000			27,000	29,700	

01530 | Temporary Construction

			CREW	DAILY OUTPUT	LABOR-HOURS	UNIT	MAT.	LABOR	EQUIP.	TOTAL	TOTAL INCL O&P	
700	0010	**PROTECTION**										**700**
	0020	Stair tread, 50 mm x 300 mm planks, 1 use	1 Carp	75	.107	Tread	4.80	3.79		8.59	11.20	
	0100	Exterior plywood, 13 mm thick, 1 use		65	.123		1.84	4.38		6.22	8.85	
	0200	19 mm thick, 1 use		60	.133		2.46	4.74		7.20	10.10	

14

01530	Temporary Construction	CREW	DAILY OUTPUT	LABOR-HOURS	UNIT	2006 BARE COSTS				TOTAL INCL O&P	
						MAT.	LABOR	EQUIP.	TOTAL		
900	**0010**	**WINTER PROTECTION** Reinforced plastic on wood										**900**
0100	framing to close openings	2 Clab	69.68	.230	m²	4.20	6.30		10.50	14.45
0200	Tarpaulins hung over scaffolding, 8 uses, not incl. scaffolding		139	.115		2.05	3.15		5.20	7.15
0250	Tarpaulin polyester reinf. w/integral fastening system 0.279 mm thk		149	.107		8.30	2.94		11.24	13.75
0300	Prefab. fiberglass panels, steel frame, 8 uses		111	.144		8.05	3.95		12	15.10

01540	Construction Aids	CREW	DAILY OUTPUT	LABOR-HOURS	UNIT	MAT.	LABOR	EQUIP.	TOTAL	TOTAL INCL O&P	
500	**0010**	**PERSONNEL PROTECTIVE EQUIPMENT**										**500**
0015	Hazardous waste protection									
0020	Respirator mask only, full face, silicone				Ea.	197			197	217
0030	Half face, silicone					26.50			26.50	29.50
0040	Respirator cartridges, 2 req'd/mask, dust or asbestos					4.80			4.80	5.30
0050	Chemical vapor					4.61			4.61	5.05
0060	Combination vapor and dust					10			10	11
0100	Emergency escape breathing apparatus, 5 min.					430			430	475
0110	10 min.					490			490	535
0150	Self contained breathing apparatus with full face piece, 30 min.					1,725			1,725	1,900
0160	60 min.					3,125			3,125	3,450
0200	Encapsulating suits, limited use, level A					895			895	980
0210	Level B					230			230	253
0300	Over boots, latex				Pr.	4.20			4.20	4.62
0310	PVC					12.50			12.50	13.75
0320	Neoprene					51.50			51.50	57
0400	Gloves, nitrile/PVC					5.40			5.40	5.95
0410	Neoprene coated					24			24	26.50
550	**0010**	**PUMP STAGING**, Aluminum	R015423 -20									**550**
0200	6.12 m long pole section, buy				Ea.	360			360	400
0300	4.6 m long pole section, buy					278			278	305
0400	3.7 m long pole section, buy					191			191	210
0500	1.8 m long pole section, buy					100			100	110
0600	1.8 m long splice joint section, buy					73.50			73.50	81
0700	Pump jack					119			119	131
0900	Foldable brace					51.50			51.50	56.50
1000	Workbench/back safety rail support					63			63	69
1100	Scaffolding planks/workbench, 357 mm wide x 7.34 m long					580			580	640
1200	Plank end safety rail					196			196	215
1250	Safety net, 6.7 m long					285			285	315
1300	System in place, 15 m working height, per use based on 50 uses	2 Carp	788	.020	m²	.57	.72		1.29	1.75
1400	100 uses		788	.020		.29	.72		1.01	1.43
1500	150 uses		788	.020		.19	.72		.91	1.33
700	**0010**	**SAFETY NETS**										**700**
0020	No supports, stock sizes, nylon, 100 mm mesh				m²	11.10			11.10	12.15
0100	Polypropylene, 150 mm mesh					17.10			17.10	18.85
0200	Small mesh debris nets, 6 mm & 20 mm mesh, stock sizes					7.95			7.95	8.70
0220	Combined 100 mm mesh and 6 mm mesh, stock sizes					22			22	24.50
0300	Monthly rental, 100 mm mesh, stock sizes, 1st month					4.31			4.31	4.74
0320	2nd month rental					2.15			2.15	2.37
0340	Maximum rental/year					9.70			9.70	10.65
750	**0010**	**SCAFFOLDING**	R015423 -10									**750**
0015	Steel tube, reg., rent/mo., no plank, incl. erect or dismantle									
0090	Building ext, wall face, 1 to 5 stories, 1900 x 1350 mm frames	3 Carp	223	.108	m²	2.75	3.83		6.58	8.95
0200	6 to 12 stories	4 Carp	197	.162		2.75	5.75		8.50	12
0310	13 to 20 stories	5 Carp	186	.215		2.75	7.65		10.40	14.90
0460	Building interior, wall face area, up to 5 m high	3 Carp	232	.103		2.75	3.68		6.43	8.75

01540	Construction Aids		CREW	DAILY OUTPUT	LABOR-HOURS	UNIT	2006 BARE COSTS				TOTAL INCL O&P	
							MAT.	LABOR	EQUIP.	TOTAL		
0560	5 m to 12 m high	R015423 -10	3 Carp	214	.112	m²	2.75	3.99		6.74	9.20	750
0800	Building interior floor area, up to 9 m high		↓	883	.027	m³	.95	.97		1.92	2.55	
0900	Over 9 m high		4 Carp	779	.041	"	.95	1.46		2.41	3.32	
0910	Steel tubular, heavy duty shoring, buy											
0920	Frames 1500 mm high, 600 mm wide					Ea.	86.50			86.50	95.50	
0925	1500 mm high, 1200 mm wide						98			98	108	
0930	1800 mm high, 600 mm wide						99.50			99.50	109	
0935	1800 mm high, 1200 mm wide					↓	117			117	128	
0940	Accessories											
0945	Cross braces					Ea.	18.50			18.50	20.50	
0950	U-head, 200 x 200 mm						20			20	22	
0955	J-head, 100 x 200 mm						14.80			14.80	16.25	
0960	Base plate, 200 x 200 mm						16.40			16.40	18.05	
0965	Leveling jack					↓	35			35	39	
1000	Steel tubular, regular, buy											
1100	Frames 900 mm high 1500 mm wide					Ea.	67			67	73.50	
1150	1500 mm high 1500 mm wide						77.50			77.50	85	
1200	1900 mm high 1500 mm wide						97			97	107	
1350	2250 mm high 1800 mm wide						167			167	184	
1500	Accessories cross braces						17.35			17.35	19.05	
1550	Guardrail post						17.35			17.35	19.05	
1600	Guardrail 2100 mm section						8.40			8.40	9.20	
1650	Screw jacks & plates						27.50			27.50	30.50	
1700	Sidearm brackets						32.50			32.50	35.50	
1750	200 mm casters						38			38	42	
1800	Plank 50 mm x 250 mm x 4800 mm						49			49	54	
1900	Stairway section						283			283	310	
1910	Stairway starter bar						33.50			33.50	37	
1920	Stairway inside handrail						61			61	67.50	
1930	Stairway outside handrail						84.50			84.50	93	
1940	Walk-thru frame guardrail					↓	42.50			42.50	46.50	
2000	Steel tubular, regular, rent/mo.											
2100	Frames 900 mm high 1500 mm wide					Ea.	3.94			3.94	4.33	
2150	1500 mm high 1500 mm wide						3.94			3.94	4.33	
2200	1900 mm high 1500 mm wide						3.94			3.94	4.33	
2250	2250 mm high 1800 mm wide						7.35			7.35	8.10	
2500	Accessories, cross braces						.63			.63	.69	
2550	Guardrail post						1.05			1.05	1.16	
2600	Guardrail 2100 mm section						.79			.79	.87	
2650	Screw jacks & plates						1.58			1.58	1.74	
2700	Sidearm brackets						1.58			1.58	1.74	
2750	200 mm casters						6.30			6.30	6.95	
2800	Outrigger for rolling tower						3.15			3.15	3.47	
2850	Plank 50 mm x 250 mm x 4800 mm						5.25			5.25	5.80	
2900	Stairway section						10.50			10.50	11.55	
2910	Stairway starter bar						.11			.11	.12	
2920	Stairway inside handrail						5.25			5.25	5.80	
2930	Stairway outside handrail						5.25			5.25	5.80	
2940	Walk-thru frame guardrail					↓	2.10			2.10	2.31	
3000	Steel tubular, heavy duty shoring, rent/month											
3250	1500 mm high 600/1200 mm wide					Ea.	5.25			5.25	5.80	
3300	1800 mm high 600/1200 mm wide						5.25			5.25	5.80	
3500	Accessories, cross braces						1.05			1.05	1.16	
3600	U - head, 200 x 200 mm						1.05			1.05	1.16	
3650	J - head, 100 x 200 mm						1.05			1.05	1.16	
3700	Base plate, 200 x 200 mm					↓	1.05			1.05	1.16	

01540 | Construction Aids

		CREW	DAILY OUTPUT	LABOR-HOURS	UNIT	MAT.	LABOR	EQUIP.	TOTAL	TOTAL INCL O&P	
750							2006 BARE COSTS				**750**
3750	Leveling jack				Ea.	2.10			2.10	2.31	
4000	Scaffolding, reg., rent/mo, no plank, w/ erect or dismantle R015423-10										
4100	Building exterior 2 stories	3 Carp	232	.103	m²	2.75	3.68		6.43	8.75	
4150	4 stories	"	214	.112		2.75	3.99		6.74	9.20	
4200	6 stories	4 Carp	223	.144		2.75	5.10		7.85	10.95	
4250	8 stories		204	.157		2.75	5.60		8.35	11.70	
4300	10 stories		186	.172		2.75	6.10		8.85	12.50	
4350	12 stories		167	.192		2.75	6.80		9.55	13.60	
4500	1st tier, 900x1500 mm frames, 1 use/mo, erect or dismantle	1 Carp	22.11	.362		4.23	12.85		17.08	24.50	
4550	2 uses/month		22.11	.362		2.11	12.85		14.96	22.50	
4600	4 uses/month		22.11	.362		1.06	12.85		13.91	21	
4650	8 uses/month		22.11	.362		.53	12.85		13.38	20.50	
5000	1500 mm H x 1500 mm W frames, 1 use/month		36.88	.217		3.08	7.70		10.78	15.40	
5050	2 uses/month		36.88	.217		1.54	7.70		9.24	13.70	
5100	4 uses/month		36.88	.217		.77	7.70		8.47	12.85	
5150	8 uses/month		36.88	.217		.39	7.70		8.09	12.40	
5500	1900 mm H x 1500 mm W frames, 1 use/month		46.64	.172		2.75	6.10		8.85	12.50	
5550	2 uses/month		46.64	.172		1.37	6.10		7.47	11	
5600	4 uses/month		46.64	.172		.69	6.10		6.79	10.25	
5650	8 uses/month		46.64	.172		.34	6.10		6.44	9.90	
5700	Planks, 50x250x4800 mm, labor , erect or remove to 15 m ht	3 Carp	144	.167	Ea.		5.95		5.95	9.25	
5800	Over 15 m high	4 Carp	160	.200	"		7.10		7.10	11.05	
6000	Heavy duty shoring for elevated slab forms to 2.5 m H, floor area										
6010	incl. erection or dismantle labor										
6100	1 use/month	4 Carp	334	.096	m²	3.34	3.41		6.75	9	
6150	2 uses/month		334	.096		1.67	3.41		5.08	7.15	
6200	3 uses/month		334	.096		1.11	3.41		4.52	6.50	
6500	To 4.5 m high										
6600	1 use/month	4 Carp	167	.192	m²	4.88	6.80		11.68	15.95	
6650	2 uses/month		167	.192		2.44	6.80		9.24	13.30	
6700	3 uses/month		167	.192		1.62	6.80		8.42	12.40	
755	**SCAFFOLDING SPECIALTIES**										**755**
0010											
1200	Sidewalk bridge, heavy duty steel posts & beams, including										
1210	parapet protection & waterproofing										
1220	2.5 m to 3.0 m wide, 2 posts	3 Carp	4.57	5.249	m	110	187		297	410	
1230	3 posts	"	3.05	7.874	"	169	280		449	620	
1500	Sidewalk bridge using tubular steel										
1510	scaffold frames, including planking	3 Carp	13.72	1.750	m	16.25	62		78.25	115	
1600	For 2 uses/month, deduct from all above					50%					
1700	For 1 use every 2 months, add to all above					100%					
1900	Catwalks, 500 mm wide, no guardrails, 2.1 m span, buy				Ea.	144			144	159	
2000	3 m span, buy				"	185			185	203	
2800	Hand winch-operated masons scaffolding, no plank										
2810	plank moving not required										
2900	30 m long, 3.2 m high, buy				Ea.	30,300			30,300	33,300	
3000	Rent/month					1,200			1,200	1,325	
3100	8.7 m high, buy					37,200			37,200	40,900	
3200	Rent/month					1,475			1,475	1,625	
3400	60 m long, 8.7 m high, buy					72,000			72,000	79,000	
3500	Rent/month					2,875			2,875	3,150	
3600	20 m high, buy					98,500			98,500	108,500	
3700	Rent/month					3,950			3,950	4,325	
3720	Putlog, standard, 2.5 m span, with hangers, buy					70.50			70.50	77.50	
3730	Rent/month					10.50			10.50	11.55	
3750	3.5 m span, buy					106			106	117	

		01540	Construction Aids	CREW	DAILY OUTPUT	LABOR-HOURS	UNIT	2006 BARE COSTS MAT.	LABOR	EQUIP.	TOTAL	TOTAL INCL O&P	
755	3755		Rent per month				Ea.	15.75			15.75	17.35	**755**
	3760		Trussed type, 4.8 m span, buy					243			243	267	
	3770		Rent/month					21			21	23	
	3790		6.6 m span, buy					291			291	320	
	3795		Rent/month					31.50			31.50	34.50	
	3800		Rolling ladders with handrails, 762 mm wide, buy, 2 step					197			197	216	
	4000		7 step					660			660	725	
	4050		10 step					870			870	955	
	4100		Rolling towers, buy, 1.5 mm wide, 2.15 m long, 3.0 m high					1,325			1,325	1,450	
	4200		For 1.5 m high added sections, to buy, add					219			219	241	
	4300		Complete incl. wheels, railings, outriggers,										
	4350		6.3 m high, to buy				Ea.	2,225			2,225	2,450	
	4400		Rent/month				"	167			167	184	
760	0010		**STAGING AIDS** and fall protection equipment										**760**
	0100		Sidewall staging bracket, tubular, buy				Ea.	34			34	37	
	0110		Cost each per day, based on 250 days' use				Day	.14			.14	.15	
	0200		Guard post, buy				Ea.	16.50			16.50	18.15	
	0210		Cost each per day, based on 250 days' use				Day	.07			.07	.07	
	0300		End guard chains, buy per pair				Pair	27.50			27.50	30.50	
	0310		Cost per set per day, based on 250 days' use				Day	.14			.14	.15	
	1000		Roof shingling bracket, steel, buy				Ea.	7.15			7.15	7.85	
	1010		Cost each per day, based on 250 days' use				Day	.03			.03	.03	
	1100		Wood bracket, buy				Ea.	14.20			14.20	15.60	
	1110		Cost each per day, based on 250 days' use				Day	.06			.06	.06	
	2000		Ladder jack, aluminum, buy per pair				Pair	89.50			89.50	98.50	
	2010		Cost per pair per day, based on 250 days' use				Day	.36			.36	.39	
	2100		Steel siderail jack, buy per pair				Pair	69.50			69.50	76	
	2110		Cost per pair per day, based on 250 days' use				Day	.28			.28	.30	
	3000		Laminated wood plank, 50 mm x 250 mm x 4800 mm, buy				Ea.	47			47	51.50	
	3010		Cost each per day, based on 250 days' use				Day	.19			.19	.21	
	3100		Aluminum scaffolding plank, 500 mm wide x 7200 mm long, buy				Ea.	725			725	795	
	3110		Cost each per day, based on 250 days' use				Day	2.89			2.89	3.18	
	4000		Nylon full-body harness, lanyard and rope grab				Ea.	218			218	239	
	4010		Cost each per day, based on 250 days' use				Day	.87			.87	.96	
	4100		Rope for safety line, 16 mm x 30 000 mm, buy				Ea.	45			45	49.50	
	4110		Cost each per day, based on 250 days' use				Day	.18			.18	.20	
	4200		Permanent U-Bolt roof anchor, buy				Ea.	32.50			32.50	36	
	4300		Temporary (one use) roof ridge anchor, buy				"	26.50			26.50	29	
	5000		Installation (setup and removal) of staging aids										
	5010		Sidewall staging bracket	2 Carp	64	.250	Ea.		8.90		8.90	13.85	
	5020		Guard post with 2 wood rails	"	64	.250			8.90		8.90	13.85	
	5030		End guard chains, set	1 Carp	64	.125			4.44		4.44	6.90	
	5100		Roof shingling bracket		96	.083			2.96		2.96	4.61	
	5200		Ladder jack		64	.125			4.44		4.44	6.90	
	5300		Wood plank, 50 mm x 250 mm x 4800 mm	2 Carp	80	.200			7.10		7.10	11.05	
	5310		Aluminum scaffold plank, 500 mm x 7200 mm	"	40	.400			14.20		14.20	22	
	5410		Safety rope	1 Carp	40	.200			7.10		7.10	11.05	
	5420		Permanent U-Bolt roof anchor (install only)	2 Carp	40	.400			14.20		14.20	22	
	5430		Temporary roof ridge anchor (install only)	1 Carp	64	.125			4.44		4.44	6.90	
780	0010		**SWING STAGING**, 225 kg cap., 600 mm W to 7.3 m L, hand-operated										**780**
	0020		steel cable type, with 18 m cables, buy				Ea.	4,675			4,675	5,150	
	0030		Rent per month				"	470			470	515	
	0600		Lightweight (not for masons) 7.3 m long for 45 m height										
	0610		manual type, buy				Ea.	4,950			4,950	5,450	
	0620		Rent/month					495			495	545	

GENERAL REQUIREMENTS

1

		01540	Construction Aids	CREW	DAILY OUTPUT	LABOR-HOURS	UNIT	2006 BARE COSTS				TOTAL INCL O&P	
								MAT.	LABOR	EQUIP.	TOTAL		
780	0700		Powered, electric or air, to 45 m high, buy				Ea.	17,500			17,500	19,300	780
	0710		Rent/month					1,225			1,225	1,350	
	0780		To 90 m high, buy					20,600			20,600	22,600	
	0800		Rent/month					1,450			1,450	1,575	
	1000		Bosun's chair or work basket 900x1050 mm, to 90 m H, elec, buy					7,725			7,725	8,500	
	1010		Rent/month					540			540	595	
	2200		Move swing staging (setup and remove)	E-4	2	16	Move		645	44.50	689.50	1,225	
790	0010		**SURVEYOR STAKES**										790
	0020		Hardwood, 25 mm x 25 mm x 1.2 m long				h	48			48	53	
	0100		50 mm x 50 mm x 460 mm long					63			63	69.50	
	0150		50 mm x 50 mm x 610 mm long					75			75	82.50	
	0200		50 mm x 50 mm x 760 mm long					74			74	81.50	
800	0010		**TARPAULINS**										800
	0020		Cotton duck, 285 gram to 445 gram/m², minimum				m²	5.40			5.40	5.90	
	0050		Maximum					6.25			6.25	6.90	
	0100		Polyvinyl-coated nylon, 400 gram to 500 gram, minimum					5.15			5.15	5.70	
	0150		Maximum					7.30			7.30	8.05	
	0200		Reinforced polyethylene 0.075 mm thick, white					1.18			1.18	1.29	
	0300		0.100 mm thick, white, clear or black					1.51			1.51	1.61	
	0400		0.140 mm thick, clear					2.15			2.15	2.37	
	0500		White, fire retardant					1.94			1.94	2.15	
	0600		0.190 mm, oil resistant, fire retardant					2.05			2.05	2.26	
	0700		0.215 mm, black					2.58			2.58	2.80	
	0710		Woven polyethylene, 0.150 mm thick					5.15			5.15	5.70	
	0730		Polyester reinforced w/integral fastening system 0.280 mm thick					11.50			11.50	12.70	
	0740		Mylar polyester, non-reinforced, 0.180 mm thick					12.60			12.60	13.90	
820	0010		**SMALL TOOLS** R013113 -50										820
	0020		As % of contractor's work, minimum				Total					.50%	
	0100		Maximum				"					2%	

		01550	**Vehicular Access & Parking**										
700	0010		**ROADS AND SIDEWALKS** Temporary										700
	0050		Roads, gravel fill, no surfacing, 100 mm gravel depth	B-14	598	.080	m²	3.95	2.33	.38	6.66	8.35	
	0100		200 mm gravel depth	"	514	.093		7.90	2.71	.44	11.05	13.40	
	1000		Ramp, 19 mm plywood on 50 mm x 150 mm joists, 400 mm O.C.	2 Carp	27.87	.574		15.20	20.50		35.70	48.50	
	1100		On 50 mm x 250 mm joists, 400 mm O.C.	"	25.55	.626		22.50	22.50		45	59.50	
	2200		Sidewalks, 50 mm x 300 mm planks, 2 uses	1 Carp	32.52	.246		8.60	8.75		17.35	23	
	2300		Exterior plywood, 2 uses, 13 mm thick		69.68	.115		3.34	4.08		7.42	10	
	2400		16 mm thick		60.39	.132		3.55	4.71		8.26	11.35	
	2500		19 mm thick		55.74	.144		4.41	5.10		9.51	12.80	

		01560	**Barriers & Enclosures**										
100	0010		**BARRICADES**										100
	0020		1500 mm high, 3 rail @ 50 mm x 200 mm, fixed	2 Carp	6.10	2.625	m	16.65	93.50		110.15	163	
	0150		Movable	"	9.14	1.750	"	14.35	62		76.35	113	
	0300		Stock units, 2 m high, 2.5 m wide, plain, buy				Ea.	435			435	480	
	0350		With reflective tape, buy				"	525			525	580	
	0400		Break-a-way 75 mm PVC pipe barricade										
	0410		with 3 ea. 300 mm x 1200 mm reflectorized panels, buy				Ea.	305			305	335	
	0500		Plywood with steel legs, 800 mm wide					72			72	79	
	0600		Telescoping Christmas tree, 2.7 m high, 5 flags, buy					122			122	134	
	0800		Traffic cones, PVC, 450 mm high					6.25			6.25	6.85	

GENERAL REQUIREMENTS

1

01500 | Temporary Facilities & Controls

01560 | Barriers & Enclosures

			CREW	DAILY OUTPUT	LABOR-HOURS	UNIT	2006 BARE COSTS				TOTAL INCL O&P	
							MAT.	LABOR	EQUIP.	TOTAL		
100	0850	700 mm high				Ea.	19			19	21	100
	1000	Guardrail, wood, 0.9 m H, 25 x 150 mm, on 50 x 100 mm post	2 Carp	60.96	.262	m	3.41	9.35		12.76	18.30	
	1100	50 mm x 150 mm, on 100 mm x 100 mm posts	"	50.29	.318		6.25	11.30		17.55	24.50	
	1200	Portable metal with base pads, buy					52.50			52.50	57.50	
	1250	Typical installation, assume 10 reuses	2 Carp	183	.087		5.40	3.11		8.51	10.80	
	1300	Barricade tape, polyethelyne, 7 mil, 75 mm wide x 152 m long roll				Ea.	25			25	27.50	
	5000	Barricades, see also Equipment Rental in Reference Section										
250	0010	TEMPORARY FENCING										250
	0020	Chain link, 11 ga., 1500 mm high	2 Clab	122	.131	m	15.70	3.59		19.29	23	
	0100	1800 mm high		91.44	.175		15.15	4.79		19.94	24	
	0200	Rented chain link, 1.8 m high, to 300 m (up to 12 mo.)		122	.131		9.95	3.59		13.54	16.55	
	0250	Over 300 m (up to 12 mo.)		91.44	.175		7.20	4.79		11.99	15.35	
	0350	Plywood, painted, 50 x 100 mm frame, 1.2 m ht.	A-4	41.15	.583		16.95	20		36.95	49.50	
	0400	100 mm x 100 mm frame, 2.5 m high	"	33.53	.716		32.50	24.50		57	73.50	
	0500	Wire mesh on 100 mm x 100 mm posts, 1.2 m high	2 Carp	30.48	.525		26.50	18.65		45.15	58.50	
	0550	2.5 m high	"	24.38	.656		41.50	23.50		65	82	
400	0010	TEMPORARY CONSTRUCTION See also Division 01530										400
800	0010	WATCHMAN										800
	0020	Service, monthly basis, uniformed person, minimum				Hr.					20	
	0100	Maximum									50	
	0200	Person and command dog, minimum									28	
	0300	Maximum									60	
	0500	Sentry dog, leased, with job patrol (yard dog), 1 dog				Week					250	
	0600	2 dogs				"					320	
	0800	Purchase, trained sentry dog, minimum				Ea.					1,500	
	0900	Maximum				"					3,000	

01580 | Project Signs

			CREW	DAILY OUTPUT	LABOR-HOURS	UNIT	MAT.	LABOR	EQUIP.	TOTAL	INCL O&P	
700	0010	SIGNS										700
	0020	Hi Intensity reflectorized, no posts, buy				m²	178			178	196	

01700 | Execution Requirements

01740 | Cleaning

			CREW	DAILY OUTPUT	LABOR-HOURS	UNIT	2006 BARE COSTS				TOTAL INCL O&P	
							MAT.	LABOR	EQUIP.	TOTAL		
500	0010	CLEANING UP										500
	0020	After job completion, allow, minimum				Job					.30%	
	0040	Maximum				"					1%	
	0050	Cleanup of floor area, continuous, per day, during const.	A-5	2230	.008	m²	.02	.22	.01	.25	.38	
	0100	Final by GC at end of job	"	1068	.017	"	.03	.46	.03	.52	.78	
	0200	Rubbish removal, see Division 02220-350										

				DAILY	LABOR-		2006 BARE COSTS				TOTAL	
	01810	**Commissioning**	CREW	OUTPUT	HOURS	UNIT	MAT.	LABOR	EQUIP.	TOTAL	INCL O&P	
100	0010	**COMMISSIONING** Including documentation of design intent										**100**
	0100	performance verification, O&M, training, min				Project					.50%	
	0150	Maximum				"					.75%	

For information about Means Estimating Seminars, see yellow pages 12 and 13 in back of book

Division Notes

	CREW	DAILY OUTPUT	LABOR-HOURS	UNIT	2006 BARE COSTS				TOTAL INCL O&P
					MAT.	LABOR	EQUIP.	TOTAL	

Division 2
Site Construction

Estimating Tips

02200 Site Preparation

- If possible visit the site and take an inventory of the type, quantity and size of the trees. Certain trees may have a landscape resale value or firewood value. Stump disposal can be very expensive, particularly if they cannot be buried at the site. Consider using a bulldozer in lieu of hand cutting trees.

- Estimators should visit the site to determine the need for haul road, access, storage of materials, and security considerations. When estimating for access roads on unstable soil, consider using a geotextile stabilization fabric. It can greatly reduce the quantity of crushed stone or gravel. Sites of limited size and access can cause cost overruns due to lost productivity. Theft and damage is another consideration if the location is isolated. A temporary fence or security guards may be required. Investigate the site thoroughly.

02210 Subsurface Investigation

In preparing estimates on structures involving earthwork or foundations, all information concerning soil characteristics should be obtained. Look particularly for hazardous waste, evidence of prior dumping of debris, and previous stream beds.

02220 Selective Demolition

The costs shown for selective demolition do not include rubbish handling or disposal. These items should be estimated separately using Means Data or other sources.

- Historic preservation often requires that the contractor remove materials from the existing structure, rehab them, and replace them. The estimator must be aware of any related measures and precautions that must be taken when doing selective demolition and cutting and patching. Requirements may include special handling and storage, as well as security.

- In addition to Section 02220, you can find selective demolition items in each division. Example: Roofing demolition is in division 7.

02300 Earthwork

- Estimating the actual cost of performing earthwork requires careful consideration of the variables involved. This includes items such as type of soil, whether water will be encountered, dewatering, whether banks need bracing, disposal of excavated earth, length of haul to fill or spoil sites, etc. If the project has large quantities of cut or fill, consider raising or lowering the site to reduce costs, while paying close attention to the effect on site drainage and utilities.

- If the project has large quantities of fill, creating a borrow pit on the site can significantly lower the costs.

- It is very important to consider what time of year the project is scheduled for completion. Bad weather can create large cost overruns from dewatering, site repair and lost productivity from cold weather.

02500 Utility Services
02600 Drainage & Containment

- Never assume that the water, sewer and drainage lines will go in at the early stages of the project. Consider the site access needs before dividing the site in half with open trenches, loose pipe, and machinery obstructions. Always inspect the site to establish that the site drawings are complete. Check off all existing utilities on your drawing as you locate them. If you find any discrepancies, mark up the site plan for further research. Differing site conditions can be very costly if discovered later in the project.

- See also Section 02955 for restoration of pipe where removal/replacement may be undesirable. Use of new types of piping materials can reduce the overall project cost. Owners/design engineers should consider the installing construction as a valuable source of current information on piping products that could lend to significant utility cost savings.

02700 Bases, Ballasts, Pavements/Appurtenances

- When estimating paving, keep in mind the project schedule. If an asphaltic paving project is in a colder climate and runs through to the spring, consider placing the base course in the autumn and then topping it in the spring just prior to completion. This could save considerable costs in spring repair. Keep in mind that prices for asphalt and concrete are generally higher in the cold seasons.

- See also Sections 02960/02965.

02900 Planting

- The timing of planting and guarantee specifications often dictate the costs for establishing tree and shrub growth and a stand of grass or ground cover. Establish the work performance schedule to coincide with the local planting season. Maintenance and growth guarantees can add from 20% to 100% to the total landscaping cost. The cost to replace trees and shrubs can be as high as 5% of the total cost depending on the planting zone, soil conditions and time of year.

02960 & 02965 Flexible Pavement Surfacing Recovery

- Recycling of asphalt pavement is becoming very popular and is an alternative to removal and replacement of asphalt pavement. It can be a very good value engineering proposal if removed pavement can be recycled either at the site or another site that is reasonably close to the project site.

Reference Numbers

Reference numbers are shown in bold squares at the beginning of some major classifications. These numbers refer to related items in the Reference Section. The reference information may be an estimating procedure, an alternate pricing method, or technical information.

Note: Not all subdivisions listed here necessarily appear in this publication.

		02055	Soils	CREW	DAILY OUTPUT	LABOR-HOURS	UNIT	2006 BARE COSTS MAT.	LABOR	EQUIP.	TOTAL	TOTAL INCL O&P	
150	0010	**BORROW**											150
	0020	Spread, with 150 kW dozer, no compaction, 3.2 km haul	R312316 -40										
	0200	Common borrow		B-15	459	.061	m³	8	1.87	4.08	13.95	16.15	
	0700	Screened loam			459	.061		28.50	1.87	4.08	34.45	38.50	
	0800	Topsoil, weed free			459	.061		27.50	1.87	4.08	33.45	37.50	
	0900	For 8 km haul, add		B-34B	153	.052			1.48	3.12	4.60	5.70	

		02060	Aggregate										
150	0010	**BORROW**											150
	0020	Spread, with 150 kW dozer, no compaction, 3.2 km RT haul	R312316 -40										
	0100	Bank run gravel		B-15	459	.061	m³	23.50	1.87	4.08	29.45	33.50	
	0300	Crushed stone, 38 mm			459	.061		36	1.87	4.08	41.95	47.50	
	0320	19 mm			459	.061		36	1.87	4.08	41.95	47.50	
	0340	13 mm			459	.061		35.50	1.87	4.08	41.45	46.50	
	0360	10 mm			459	.061		35	1.87	4.08	40.95	46	
	0400	Sand, washed, concrete			459	.061		35	1.87	4.08	40.95	46	
	0500	Dead or bank sand			459	.061		5.50	1.87	4.08	11.45	13.40	
	0600	Select structural fill			459	.061		10.50	1.87	4.08	16.45	18.90	
	0900	For 8 km haul, add		B-34B	153	.052			1.48	3.12	4.60	5.70	

		02065	Cement & Concrete										
300	0010	**PLANT MIXED BITUMINOUS CONCRETE**											300
	0020	Asphaltic concrete, plant mix (2323 kg/m³)					Met. Ton	41.50			41.50	46	
	0040	Asphaltic concrete less than 272 metric tons add trucking											
	0050	See Div. 02315-490 for hauling costs											
	0200	All weather patching mix, hot						41.50			41.50	46	
	0250	Cold patch						45			45	49.50	
	0300	Berm mix						41.50			41.50	46	
	0400	Base mix						41.50			41.50	46	
	0500	Binder mix						41.50			41.50	46	
	0600	Sand or sheet mix						45			45	49.50	
400	0010	**RECYCLED PLANT MIXED BITUMINOUS CONCRETE**											400
	0200	Reclaimed pavement in stockpile					Met. Ton	15.80			15.80	17.35	
	0400	Recycled pavement, at plant, ratio old: new, 70:30						23.50			23.50	26	
	0600	Ratio old: new, 30:70						34			34	37.50	

		02080	Utility Materials										
100	0010	**FIRE HYDRANTS**											100
	0020	Mechanical joints unless otherwise noted											
	1000	Fire hydrants, two way; excavation and backfill not incl.											
	1100	114 mm valve size, depth 600 mm		B-21	10	2.800	Ea.	1,100	90	15.25	1,205.25	1,375	
	1120	750 mm			10	2.800		1,100	90	15.25	1,205.25	1,375	
	1140	900 mm			10	2.800		1,125	90	15.25	1,230.25	1,375	
	1160	1050 mm			9	3.111		1,250	100	16.95	1,366.95	1,550	
	1200	1375 mm			9	3.111		1,250	100	16.95	1,366.95	1,575	
	1220	1525 mm			8	3.500		1,275	112	19.10	1,406.10	1,600	
	1240	1675 mm			8	3.500		1,275	112	19.10	1,406.10	1,600	
	1260	1825 mm			7	4		1,225	128	22	1,375	1,575	
	1280	1975 mm			7	4		1,275	128	22	1,425	1,625	
	1300	2125 mm			6	4.667		1,300	150	25.50	1,475.50	1,700	
	1340	2450 mm			6	4.667		1,400	150	25.50	1,575.50	1,775	
	1420	3050 mm			5	5.600		1,475	180	30.50	1,685.50	1,925	
	2000	133 mm valve size, depth 600 mm			10	2.800		880	90	15.25	985.25	1,125	
	2080	1200 mm			9	3.111		995	100	16.95	1,111.95	1,275	
	2160	1825 mm			7	4		1,100	128	22	1,250	1,450	

SITE CONSTRUCTION

2

02080	Utility Materials	CREW	DAILY OUTPUT	LABOR-HOURS	UNIT	2006 BARE COSTS				TOTAL INCL O&P
						MAT.	LABOR	EQUIP.	TOTAL	
100 2240	2450 mm	B-21	6	4.667	Ea.	1,225	150	25.50	1,400.50	1,600	**100**
2320	3050 mm	↓	5	5.600		1,350	180	30.50	1,560.50	1,775
2350	For threeway valves, add					7%				
2400	Lower barrel extensions with stems, 300 mm	B-20	14	1.714		350	53.50		403.50	470
2440	600 mm		13	1.846		430	57.50		487.50	560
2480	900 mm		12	2		505	62		567	650
2520	1200 mm	↓	10	2.400	↓	580	74.50		654.50	755
5000	Indicator post									
5020	Adjustable, valve size 100 mm to 350 mm, 1200 mm bury	B-21	10	2.800	Ea.	385	90	15.25	490.25	580
5060	2450 mm bury		7	4		495	128	22	645	765
5080	3050 mm bury		6	4.667		505	150	25.50	680.50	815
5100	3650 mm bury		5	5.600		760	180	30.50	970.50	1,150
5120	4250 mm bury		4	7		830	225	38	1,093	1,300
5500	Non-adjustable, valve size 100 mm to 350 mm, 900 mm bury		10	2.800		385	90	15.25	490.25	580
5520	1050 mm bury		10	2.800		385	90	15.25	490.25	580
5540	1200 mm bury	↓	9	3.111	↓	385	100	16.95	501.95	600
400 0010	**UTILITY BOXES** Precast concrete, 150 mm thick										**400**
0040	1200 mm x 1800 mm x 1800 mm high, I.D.	B-13	2	28	Ea.	1,250	840	325	2,415	3,025
0050	1500 mm x 3000 mm x 1800 mm high, I.D.		2	28		1,550	840	325	2,715	3,375
0100	1800 mm x 3000 mm x 1800 mm high, I.D.		2	28		1,600	840	325	2,765	3,425
0150	1500 mm x 3700 mm x 1800 mm high, I.D.		2	28		1,700	840	325	2,865	3,525
0200	1800 mm x 3700 mm x 1800 mm high, I.D.		1.80	31.111		1,925	930	360	3,215	3,925
0250	1800 mm x 4000 mm x 1800 mm high, I.D.		1.50	37.333		2,525	1,125	435	4,085	4,975
0300	2400 mm x 4300 mm x 2100 mm high, I.D.	↓	1	56	↓	2,725	1,675	650	5,050	6,300
0350	Hand hole, precast concrete, 38 mm thick									
0400	300 mm x 600 mm x 530 mm, I.D., light duty	B-1	4	6	Ea.	265	168		433	555
0450	1400 mm x 1000 mm x 600 mm, O.D., heavy duty	B-6	3	8		815	240	75.50	1,130.50	1,350
0460	Meter pit, 1200 mm x 1200 mm, 1200 mm deep		2	12		695	360	113	1,168	1,450
0470	1800 mm deep		1.60	15		990	450	141	1,581	1,925
0480	2400 mm deep		1.40	17.143		1,300	515	162	1,977	2,425
0490	3000 mm deep		1.20	20		1,650	600	189	2,439	2,950
0500	4600 mm deep		1	24		2,425	720	226	3,371	4,025
0510	1800 mm x 1800 mm, 1200 mm deep		1.40	17.143		1,025	515	162	1,702	2,100
0520	1800 mm deep		1.20	20		1,550	600	189	2,339	2,825
0530	2400 mm deep		1	24		2,050	720	226	2,996	3,600
0540	3000 mm deep		.80	30		2,575	900	283	3,758	4,500
0550	4600 mm deep	↓	.60	40	↓	3,900	1,200	375	5,475	6,575
500 0010	**VALVES** Water distribution, see also div. 15110 [R221113 -50]										**500**
3000	Butterfly valves with boxes, cast iron, mech. jt.									
3100	100 mm diameter [4"]	B-20	6	4	Ea.	218	124		342	435
3180	200 mm diameter [8"]	B-21	4	7		305	225	38	568	725
3340	300 mm diameter [12"]		3	9.333		585	300	51	936	1,150
3400	350 mm diameter [14"]		2	14		1,000	450	76.50	1,526.50	1,875
3460	450 mm diameter [18"]		1.50	18.667		1,675	600	102	2,377	2,900
3480	500 mm diameter [20"]		1	28		2,100	900	153	3,153	3,900
3500	600 mm diameter [24"]	↓	.50	56	↓	2,950	1,800	305	5,055	6,325
3600	With lever operator									
3610	100 mm diameter [4"]	B-20	6	4	Ea.	118	124		242	325
3616	200 mm diameter [8"]	B-21	4	7		240	225	38	503	655
3620	300 mm diameter [12"]		3	9.333		715	300	51	1,066	1,300
3624	400 mm diameter [16"]		2	14		1,550	450	76.50	2,076.50	2,475
3630	600 mm diameter [24"]	↓	.50	56	↓	2,550	1,800	305	4,655	5,900
3700	Check valves, flanged									
3710	100 mm diameter [4"]	B-20	6	4	Ea.	435	124		559	670
3714	150 mm diameter [6"]	"	5	4.800	↓	735	149		884	1,050

For expanded coverage of these items see *Means Site Work & Landscape Cost Data 2006*

SITE CONSTRUCTION **2**

2 SITE CONSTRUCTION

02080	Utility Materials		CREW	DAILY OUTPUT	LABOR-HOURS	UNIT	2006 BARE COSTS				TOTAL INCL O&P	
							MAT.	LABOR	EQUIP.	TOTAL		
500	3716	200 mm diameter [8"]		B-21	4	7	Ea.	1,500	225	38	1,763	2,050	500
	3720	300 mm diameter [12"] R221113-50			3	9.333		3,650	300	51	4,001	4,550	
	3724	400 mm diameter [16"]			2	14		8,050	450	76.50	8,576.50	9,625	
	3726	450 mm diameter [18"]			1.50	18.667		11,900	600	102	12,602	14,100	
	3730	600 mm diameter [24"]		↓	.50	56	↓	19,900	1,800	305	22,005	25,000	
	3800	Gate valves, C.I., 1723 kPa, mechanical joint, w/boxes											
	3810	100 mm diameter [4"]		B-21A	8	5	Ea.	169	172	67	408	520	
	3814	150 mm diameter [6"]			6.80	5.882		219	202	79	500	640	
	3816	200 mm diameter [8"]			5.60	7.143		345	245	96	686	860	
	3820	300 mm diameter [12"]		↓	4.08	9.804		740	335	131	1,206	1,475	
	3824	400 mm diameter [16"]		B-21	1	28		915	900	153	1,968	2,575	
	3828	500 mm diameter [20"]			.80	35		1,425	1,125	191	2,741	3,525	
	3830	600 mm diameter [24"]			.50	56		1,750	1,800	305	3,855	5,025	
	3831	750 mm diameter [30"]			.35	80		2,125	2,575	435	5,135	6,775	
	3832	900 mm diameter [36"]		↓	.30	93.333		2,525	3,000	510	6,035	8,000	
	3880	Sleeve, for tapping mains, 200 mm x 100 mm, add						530			530	580	
	3884	250 mm x 150 mm, add						1,025			1,025	1,125	
	3888	300 mm x 150 mm, add						1,025			1,025	1,125	
	3892	300 mm x 200 mm, add					↓	1,100			1,100	1,200	
	3900	Globe valves, flanged, iron body, class 125											
	3910	100 mm diameter [4"]		B-20	10	2.400	Ea.	1,125	74.50		1,199.50	1,375	
	3914	150 mm diameter [6"]		"	9	2.667		1,850	83		1,933	2,175	
	3916	200 mm diameter [8"]		B-21	6	4.667		3,450	150	25.50	3,625.50	4,050	
	3920	300 mm diameter [12"]			4	7		6,425	225	38	6,688	7,475	
	3924	400 mm diameter [16"]			2	14		8,025	450	76.50	8,551.50	9,600	
	3928	500 mm diameter [20"]			.60	46.667		8,700	1,500	254	10,454	12,200	
	3930	600 mm diameter [24"]		↓	.50	56	↓	10,300	1,800	305	12,405	14,500	
600	0010	**UTILITY ACCESSORIES**											600
	0400	Underground tape, detectable, reinforced, alum. foil core, 50 mm		1 Clab	4572	.002	m	.05	.05		.10	.12	
	0500	150 mm		"	4267	.002	"	.12	.05		.17	.21	

02110	Hazmat Removal & Handling		CREW	DAILY OUTPUT	LABOR-HOURS	UNIT	2006 BARE COSTS				TOTAL INCL O&P	
							MAT.	LABOR	EQUIP.	TOTAL		
300	0010	**HAZARDOUS WASTE CLEANUP/PICKUP/DISPOSAL**											300
	0100	For contractor equipment, i.e., dozer,											
	0110	Front end loader, dump truck, etc., see Reference Section											
	1000	Solid pickup											
	1100	200 L drums					Ea.					220	
	1120	Bulk material, minimum					Met. Ton					182	
	1130	Maximum					"					605	
	1200	Transportation to disposal site											
	1220	Truckload = 80 drums or 19 m³ or 16 metric ton											
	1260	Minimum					km					1.57	
	1270	Maximum					"					2.75	
	3000	Liquid pickup, vacuum truck, stainless steel tank											
	3100	Minimum charge, 4 hours											
	3110	1 compartment, 8.3 kL					Hr.					110	

Important: See the Reference Section for supporting data - Crews, Rental Equipment, City Cost Indexes and Reference Data

02100 | Site Remediation

02110 | Hazmat Removal & Handling

		CREW	DAILY OUTPUT	LABOR-HOURS	UNIT	MAT.	LABOR	EQUIP.	TOTAL	TOTAL INCL O&P	
300	3120	2 compartment, 18.9 kL				Hr.					110	300
	3400	Transportation in 26.1 kL bulk truck				km					2.94	
	3410	In teflon lined truck				"					3.41	
	5000	Heavy sludge or dry vacuumable material				Hr.					110	
	6000	Dumpsite disposal charge, minimum				Met. Ton					121	
	6020	Maximum				"					485	

02115 | Underground Storage Tank Removal

		CREW	DAILY OUTPUT	LABOR-HOURS	UNIT	MAT.	LABOR	EQUIP.	TOTAL	TOTAL INCL O&P	
200	0010	**REMOVAL OF UNDERGROUND STORAGE TANKS** R026510 -20										200
	0011	Petroleum storage tanks, non-leaking										
	0100	Excavate & load onto trailer										
	0110	11 350 L to 19 000 L tank	B-14	4	12	Ea.		350	56.50	406.50	600	
	0120	22 700 L to 30 300 L tank	B-3A	3	13.333			390	241	631	870	
	0130	34 000 L to 45 400 L tank	"	2	20			585	360	945	1,300	
	0190	Known leaking tank add				%				100%	100%	
	0200	Remove sludge, water and remaining product from tank bottom										
	0201	of tank with vacuum truck										
	0300	11 350 L to 19 000 L tank	A-13	5	1.600	Ea.		56.50	100	156.50	195	
	0310	22 700 L to 30 300 L tank		4	2			70.50	125	195.50	244	
	0320	34 000 L to 45 400 L tank		3	2.667			94	167	261	325	
	0390	Dispose of sludge off-site, average				liter					1.15	
	0400	Insert inert solid CO_2 "dry ice" into tank										
	0401	For cleaning/transporting tanks (0.68 kg/380 L)	1 Clab	227	.035	kg	3.02	.97		3.99	4.83	
	1020	Haul tank to certified salvage dump, 160 km round trip										
	1023	11 350 L to 19 000 L tank				Ea.				525	690	
	1026	22 700 L to 30 300 L tank								625	825	
	1029	34 000 L to 45 400 L tank								850	1,100	
	1100	Disposal of contaminated soil to landfill										
	1110	Minimum				m³					143	
	1111	Maximum				"					429	
	1120	Disposal of contaminated soil to										
	1121	bituminous concrete batch plant										
	1130	Minimum				m³					71.50	
	1131	Maximum				"					144	
	2010	Decontamination of soil on site incl poly tarp on top/bottom										
	2011	Soil containment berm, and chemical treatment										
	2020	Minimum	B-11C	76.46	.209	m³	7.45	6.70	2.96	17.11	21.50	
	2021	Maximum	"	76.46	.209		9.70	6.70	2.96	19.36	24	
	2050	Disposal of decontaminated soil, minimum									86	
	2055	Maximum									182	

02200 | Site Preparation

02210 | Subsurface Investigation

		CREW	DAILY OUTPUT	LABOR-HOURS	UNIT	MAT.	LABOR	EQUIP.	TOTAL	TOTAL INCL O&P	
120	0010	**BORING AND EXPLORATORY DRILLING**										120
	0020	Borings, initial field stake out & determination of elevations	A-6	1	16	Day		565	58	623	935	
	0100	Drawings showing boring details				Total		185		185	270	
	0200	Report and recommendations from P.E.						415		415	595	
	0300	Mobilization and demobilization, minimum	B-55	4	6			165	202	367	475	
	0350	For over 160 km, per added km		724	.033	km		.91	1.12	2.03	2.64	

SITE CONSTRUCTION 2

2 SITE CONSTRUCTION

02210 | Subsurface Investigation

		CREW	DAILY OUTPUT	LABOR-HOURS	UNIT	2006 BARE COSTS				TOTAL INCL O&P
						MAT.	LABOR	EQUIP.	TOTAL	
120 0600	Auger holes in earth, no samples, 63 mm diameter	B-55	23.96	1.002	m		27.50	33.50	61	79.50	**120**
0650	100 mm diameter		20.57	1.167			32	39.50	71.50	92.50
0800	Cased borings in earth, with samples, 63 mm diameter		16.92	1.419		50.50	39	48	137.50	169
0850	100 mm diameter		9.94	2.415		79	66.50	81.50	227	280
1000	Drilling in rock, "BX" core, no sampling	B-56	10.64	1.504			47	100	147	182
1050	With casing & sampling		9.66	1.656		50.50	52	111	213.50	257
1200	"NX" core, no sampling		7.90	2.025			63.50	135	198.50	246
1250	With casing and sampling		7.62	2.100		61.50	65.50	140	267	325
1400	Drill rig and crew with truck mounted auger	B-55	1	24	Day		660	810	1,470	1,925
1450	With crawler type drill	B-56	1	16	"		500	1,075	1,575	1,950
1500	For inner city borings add, minimum									10%
1510	Maximum									20%
200 0010	**CORE DRILLING**										**200**
0020	Reinf. conc. slab, up to 150 mm thick, incl. bit, layout & set up									
0100	25 mm diameter core [1"]	B-89A	28	.571	Ea.	2.59	18.25	4.13	24.97	36
0150	Each added 25 mm thick, add		300	.053		.46	1.70	.39	2.55	3.58
0300	80 mm diameter core [3"]		23	.696		5.75	22	5	32.75	46.50
0350	Each added 25 mm thick, add		186	.086		1.04	2.75	.62	4.41	6.10
0500	100 mm diameter core [4"]		19	.842		5.75	27	6.10	38.85	55
0550	Each added 25 mm thick, add		170	.094		1.31	3.01	.68	5	6.85
0700	150 mm diameter core [6"]		14	1.143		9.50	36.50	8.25	54.25	76.50
0750	Each added 25 mm thick, add		140	.114		1.61	3.65	.83	6.09	8.40
0900	200 mm diameter core [8"]		11	1.455		12.95	46.50	10.50	69.95	98.50
0950	Each added 25 mm thick, add		95	.168		2.18	5.40	1.22	8.80	12.15
1100	250 mm diameter core [10"]		10	1.600		17.30	51	11.55	79.85	111
1150	Each added 25 mm thick, add		80	.200		2.86	6.40	1.44	10.70	14.70
1300	300 mm diameter core [12"]		9	1.778		21	57	12.85	90.85	126
1350	Each added 25 mm thick, add		68	.235		3.43	7.50	1.70	12.63	17.35
1500	350 mm diameter core [14"]		7	2.286		25	73	16.50	114.50	160
1550	Each added 25 mm thick, add		55	.291		4.36	9.30	2.10	15.76	21.50
1700	450 mm diameter core [18"]		4	4		32.50	128	29	189.50	267
1750	Each added 25 mm thick, add		28	.571		5.70	18.25	4.13	28.08	39.50
1760	For horizontal holes, add to above								30%	30%
1770	Prestressed hollow core plank, 150 mm thick									
1780	25 mm diameter core [1"]	B-89A	52	.308	Ea.	1.72	9.85	2.22	13.79	19.65
1790	Each added 25 mm thick, add		350	.046		.30	1.46	.33	2.09	2.96
1800	75 mm diameter core [3"]		50	.320		3.79	10.20	2.31	16.30	22.50
1810	Each added 25 mm thick, add		240	.067		.63	2.13	.48	3.24	4.54
1820	100 mm diameter core [4"]		48	.333		5.05	10.65	2.41	18.11	25
1830	Each added mm thick, add		216	.074		.87	2.37	.53	3.77	5.25
1840	150 mm diameter core [6"]		44	.364		6.25	11.60	2.63	20.48	28
1850	Each added 25 mm thick, add		175	.091		1.04	2.92	.66	4.62	6.40
1860	200 mm diameter core [8"]		32	.500		8.35	16	3.61	27.96	38
1870	Each added 25 mm thick, add		118	.136		1.45	4.33	.98	6.76	9.45
1880	250 mm diameter core [10"]		28	.571		11.30	18.25	4.13	33.68	45.50
1890	Each added 25 mm thick, add		99	.162		1.56	5.15	1.17	7.88	11.05
1900	300 mm diameter core [12"]		22	.727		13.75	23	5.25	42	57
1910	Each added 25 mm thick, add		85	.188		2.29	6	1.36	9.65	13.35
1950	Minimum charge for above, 75 mm diameter core [3"]		7	2.286	Total		73	16.50	89.50	132
2000	100 mm diameter core [4"]		6.80	2.353			75	17	92	136
2050	150 mm diameter core [6"]		6	2.667			85	19.25	104.25	154
2100	200 mm diameter core [8"]		5.50	2.909			93	21	114	168
2150	250 mm diameter core [10"]		4.75	3.368			108	24.50	132.50	195
2200	300 mm diameter core [12"]		3.90	4.103			131	29.50	160.50	237
2250	350 mm diameter core [14"]		3.38	4.734			151	34	185	273
2300	450 mm diameter core [18"]		3.15	5.079			162	36.50	198.50	294

02200 | Site Preparation

02210 | Subsurface Investigation

		CREW	DAILY OUTPUT	LABOR-HOURS	UNIT	2006 BARE COSTS				TOTAL INCL O&P	
						MAT.	LABOR	EQUIP.	TOTAL		
200	3010	Bits for core drill, diamond, premium, 25 mm dia. [1"]				Ea.	116			116	127	200
	3020	80 mm diameter [3"]					286			286	315	
	3040	100 mm diameter [4"]					320			320	350	
	3050	150 mm diameter [6"]					510			510	560	
	3080	200 mm diameter [8"]					695			695	765	
	3120	300 mm diameter [12"]					1,100			1,100	1,200	
	3180	450 mm diameter [18"]					2,275			2,275	2,500	
	3240	600 mm diameter [24"]					3,025			3,025	3,325	
700	0010	**TEST PITS**										700
	0020	Hand digging, light soil	1 Clab	3.44	2.325	m³		63.50		63.50	99	
	0100	Heavy soil	"	1.91	4.185			115		115	179	
	0120	Loader-backhoe, light soil	B-11M	21.41	.747			24	12.95	36.95	51	
	0130	Heavy soil	"	15.29	1.046			33.50	18.15	51.65	71.50	
	1000	Subsurface exploration, mobilization				km				3.41	3.93	
	1010	Difficult access for rig, add				Hr.				110	126	
	1020	Auger borings, drill rig, incl. samples				M				41.25	47.85	
	1030	Hand auger								61.35	70.55	
	1050	Drill and sample every 1.5 m, split spoon								53.90	62.15	
	1060	Extra samples				Ea.				22	25.50	

02220 | Site Demolition

		CREW	DAILY OUTPUT	LABOR-HOURS	UNIT	2006 BARE COSTS				TOTAL INCL O&P	
110	0010	**BUILDING DEMOLITION** Large urban projects, incl. 32 km haul R024119-10										110
	0011	No foundation or dump fees, m³ is vol of building standing										
	0012	Steel	B-8	608	.105	m³		3.25	3.94	7.19	9.30	
	0050	Concrete		433	.148			4.56	5.55	10.11	13.05	
	0080	Masonry		569	.112			3.47	4.21	7.68	9.95	
	0100	Mixture of types, average		569	.112			3.47	4.21	7.68	9.95	
	0500	Small bldgs, or single bldgs, no salvage included, steel	B-3	419	.115			3.39	4.17	7.56	9.80	
	0600	Concrete		320	.150			4.44	5.45	9.89	12.85	
	0650	Masonry		419	.115			3.39	4.17	7.56	9.80	
	0700	Wood		419	.115			3.39	4.17	7.56	9.80	
	1000	Single family, one story house, wood, minimum				Ea.				2,525	2,975	
	1020	Maximum								4,400	5,275	
	1200	Two family, two story house, wood, minimum								3,300	3,950	
	1220	Maximum								6,375	7,700	
	1300	Three family, three story house, wood, minimum								4,400	5,275	
	1320	Maximum								7,700	9,250	
	5000	For buildings with no interior walls, deduct								50%		
120	0010	**EXPLOSIVE/IMPLOSIVE DEMOLITION** Large projects, R024119-10										120
	0020	no disposal fee based on building volume, steel building	B-5B	478	.100	m³		3.26	3.96	7.22	9.35	
	0100	Concrete building		478	.100			3.26	3.96	7.22	9.35	
	0200	Masonry building		478	.100			3.26	3.96	7.22	9.35	
	0400	Disposal of material, minimum	B-3	340	.141			4.18	5.15	9.33	12.10	
	0500	Maximum	"	279	.172			5.10	6.25	11.35	14.75	
130	0010	**BLDG. FOOTINGS AND FOUNDATIONS DEMOLITION** R024119-10										130
	0200	Floors, concrete slab on grade,										
	0240	100 mm thick, plain concrete	B-9C	46.45	.861	m²		24	3.17	27.17	41	
	0280	Reinforced, wire mesh		43.66	.916			25.50	3.37	28.87	43	
	0300	Rods		37.16	1.076			30	3.96	33.96	51	
	0400	150 mm thick, plain concrete		34.84	1.148			32	4.23	36.23	54	
	0420	Reinforced, wire mesh		31.59	1.266			35	4.66	39.66	60	
	0440	Rods		27.87	1.435			40	5.30	45.30	68	
	1000	Footings, concrete, 300 mm thick, 600 mm wide	B-5	91.44	.612	m		18.60	10.30	28.90	40	
	1080	450 mm thick, 600 mm wide		76.20	.735			22.50	12.35	34.85	48	

		02220	Site Demolition	CREW	DAILY OUTPUT	LABOR-HOURS	UNIT	2006 BARE COSTS				TOTAL INCL O&P	
								MAT.	LABOR	EQUIP.	TOTAL		
130	1120		900 mm wide	B-5	60.96	.919	m		28	15.45	43.45	60	130
	1140		600 mm thick, 900 mm wide		53.34	1.050			32	17.65	49.65	68.50	
	1200		Average reinforcing, add								10%	10%	
	1220		Heavy reinforcing, add								20%	20%	
	2000		Walls, block, 100 mm thick	1 Clab	16.72	.478	m²		13.10		13.10	20.50	
	2040		150 mm thick		15.79	.507			13.90		13.90	21.50	
	2080		200 mm thick		13.94	.574			15.75		15.75	24.50	
	2100		300 mm thick		13.94	.574			15.75		15.75	24.50	
	2200		For horizontal reinforcing, add								10%	10%	
	2220		For vertical reinforcing, add								20%	20%	
	2400		Concrete, plain concrete, 150 mm thick	B-9	14.86	2.691			75	9.90	84.90	127	
	2420		200 mm thick		13.01	3.076			85.50	11.30	96.80	145	
	2440		250 mm thick		11.15	3.588			100	13.20	113.20	170	
	2500		300 mm thick		9.29	4.306			120	15.85	135.85	203	
	2600		For average reinforcing, add								10%	10%	
	2620		For heavy reinforcing, add								20%	20%	
	4000		For congested sites or small quantities, add up to								200%	200%	
	4200		Add for disposal, on site	B-11A	177	.090	m³		2.90	5.20	8.10	10.15	
	4250		To 8 km	B-30	168	.143	"		4.45	10	14.45	17.70	
220	0010		**FENCING DEMOLITION**										220
	1600		Fencing, barbed wire, 3 strand	2 Clab	131	.122	m		3.35		3.35	5.20	
	1650		5 strand	"	85.34	.187			5.15		5.15	8	
	1700		Chain link, posts & fabric, remove only, 2400 mm to 3000 mm high	B-6	136	.176			5.30	1.66	6.96	10	
	1750		Remove and reset	"	21.34	1.125			34	10.60	44.60	63.50	
	1775		Fencing, wood, all types, 1200 mm to 1800 mm high	2 Clab	132	.121			3.32		3.32	5.15	
	1790		Remove and store	B-80	71.63	.447			13.35	7.45	20.80	28.50	
230	0010		**HYDRODEMOLITION**										230
	0015		Concrete pavement, 28 MPa, 50 mm depth	B-5	46.45	1.206	m²		36.50	20.50	57	79	
	0120		100 mm depth		41.81	1.340			40.50	22.50	63	87.50	
	0130		150 mm depth		37.16	1.507			45.50	25.50	71	98.50	
	0410		41 MPa, 50 mm depth		38.09	1.470			44.50	24.50	69	95.50	
	0420		100 mm depth		32.52	1.722			52.50	29	81.50	113	
	0430		150 mm depth		27.87	2.009			61	34	95	131	
	0510		55 MPa, 50 mm depth		30.66	1.827			55.50	30.50	86	120	
	0520		100 mm depth		26.01	2.153			65.50	36	101.50	141	
	0530		150 mm depth		22.30	2.512			76	42	118	164	
240	0010		**MINOR SITE DEMOLITION**										240
	0015		No hauling, abandon catch basin or manhole	B-6	7	3.429	Ea.		103	32.50	135.50	194	
	0020		Remove existing catch basin or manhole, masonry		4	6			180	56.50	236.50	340	
	0030		Catch basin or manhole frames and covers, stored		13	1.846			55.50	17.40	72.90	104	
	0040		Remove and reset		7	3.429			103	32.50	135.50	194	
	0100		Roadside delineators, remove only	B-80	175	.183			5.45	3.04	8.49	11.75	
	0110		Remove and reset	"	100	.320			9.55	5.30	14.85	20.50	
	0800		Guiderail, remove only	B-80A	30.48	.787	m		21.50	5.80	27.30	40	
	0850		Remove and reset	"	12.19	1.969	"		54	14.45	68.45	100	
	0860		Guide posts, remove only	B-80B	120	.267	Ea.		7.85	1.96	9.81	14.20	
	0870		Remove and reset	B-55	50	.480			13.20	16.15	29.35	38.50	
	0900		Hydrants, fire, remove only	B-21A	5	8			275	107	382	540	
	0950		Remove and reset	"	2	20			685	268	953	1,350	
	1000		Masonry walls, block or tile, solid, remove	B-5	50.94	1.099	m³		33.50	18.50	52	72	
	1100		Cavity wall		62.26	.899			27.50	15.10	42.60	58.50	
	1200		Brick, solid		25.47	2.199			66.50	37	103.50	144	
	1300		With block back up		31.98	1.751			53	29.50	82.50	115	
	1400		Stone, with mortar		25.47	2.199			66.50	37	103.50	144	

Reference boxes: R024119-10 (appears at rows 130/1120, 220/0010, 230/0010, 240/0010)

Important: See the Reference Section for supporting data - Crews, Rental Equipment, City Cost Indexes and Reference Data

02220 | Site Demolition

			CREW	DAILY OUTPUT	LABOR-HOURS	UNIT	2006 BARE COSTS MAT.	LABOR	EQUIP.	TOTAL	TOTAL INCL O&P	
240	1500	Dry set	B-5	42.45	1.319	m³		40	22	62	86	240
	1600	Median barrier, precast concrete, remove and store	B-3	131	.366	m		10.85	13.35	24.20	31.50	
	1610	Remove and reset	"	119	.403			11.95	14.70	26.65	34.50	
	2900	Pipe removal, sewer/water, no excavation, 300 mm diameter [12"]	B-6	53.34	.450			13.50	4.24	17.74	25.50	
	2930	375-450 mm diameter		45.72	.525			15.75	4.95	20.70	29.50	
	2960	525-600 mm diameter		36.58	.656			19.70	6.20	25.90	37.50	
	3000	675-900 mm diameter		27.43	.875			26.50	8.25	34.75	49.50	
	3200	Steel, welded connections, 100 mm diameter		48.77	.492			14.75	4.64	19.39	27.50	
	3300	250 mm diameter	▼	24.38	.984	▼		29.50	9.30	38.80	55.50	
	3500	Railroad track removal, ties and track	B-13	101	.554			16.60	6.45	23.05	32.50	
	3600	Ballast	B-14	382	.126	m³		3.65	.59	4.24	6.30	
	3700	Remove and re-install, ties & track using new bolts & spikes		15.24	3.150	m		91.50	14.85	106.35	157	
	3800	Turnouts using new bolts and spikes	▼	1	48	Ea.		1,400	226	1,626	2,400	
	3850	Runways, remove rubber skid marks, 4-6 passes	B-59A	3252	.007	m²	.30	.20	.10	.60	.76	
	3860	6-10 passes	"	3252	.007		.45	.20	.10	.75	.92	
	4000	Sidewalk removal, bituminous, 63 mm thick	B-6	272	.088			2.65	.83	3.48	4.99	
	4050	Brick, set in mortar		155	.155			4.65	1.46	6.11	8.75	
	4100	Concrete, plain, 100 mm		134	.179			5.35	1.69	7.04	10.10	
	4200	Mesh reinforced	▼	125	.192			5.75	1.81	7.56	10.85	
	5000	Slab on grade removal, plain	B-5	34.41	1.628	m³		49.50	27.50	77	106	
	5100	Mesh reinforced		25.23	2.219			67.50	37.50	105	145	
	5200	Rod reinforced	▼	19.12	2.930			89	49.50	138.50	191	
	5500	For congested sites or small quantities, add up to								200%	200%	
	5550	For disposal on site, add	B-11A	177	.090			2.90	5.20	8.10	10.15	
	5600	To 8 km, add	B-34D	58.11	.138	▼		3.90	7.20	11.10	13.85	
250	0010	**DEMOLISH, REMOVE PAVEMENT AND CURB**										250
	5010	Pavement removal, bituminous roads, 75 mm thick	B-38	577	.069	m²		2.16	1.45	3.61	4.91	
	5050	100 mm to 150 mm thick		351	.114			3.56	2.38	5.94	8.05	
	5100	Bituminous driveways		535	.075			2.33	1.56	3.89	5.30	
	5200	Concrete to 150 mm thick, hydraulic hammer mesh reinforced		213	.188			5.85	3.92	9.77	13.30	
	5300	Rod reinforced		167	.240			7.50	5	12.50	17	
	5400	Concrete, 175 mm to 600 mm thick, plain		25.23	1.585	m³		49.50	33	82.50	113	
	5500	Reinforced	▼	18.35	2.180	"		68	45.50	113.50	154	
	5600	With hand held air equipment, bituminous, to 150 mm thick	B-39	177	.271	m²		7.85	.83	8.68	13.10	
	5700	Concrete to 150 mm thick, no reinforcing		149	.322			9.35	.99	10.34	15.55	
	5800	Mesh reinforced		130	.369			10.70	1.13	11.83	17.85	
	5900	Rod reinforced	▼	71.07	.675	▼		19.60	2.07	21.67	33	
	6000	Curbs, concrete, plain	B-6	110	.218	m		6.55	2.06	8.61	12.30	
	6100	Reinforced		83.82	.286			8.60	2.70	11.30	16.15	
	6200	Granite		110	.218			6.55	2.06	8.61	12.30	
	6300	Bituminous		161	.149			4.47	1.41	5.88	8.45	
	6500	Site demo, berms under 100 mm in height, bituminous		161	.149			4.47	1.41	5.88	8.45	
	6600	100 mm or over in height	▼	91.44	.262	▼		7.85	2.48	10.33	14.80	
310	0010	**SELECTIVE DEMOLITION, CUTOUT**										310
	0020	Concrete, elev. slab, light reinforcement, under 0.17 m³	B-9C	1.84	21.745	m³		605	80	685	1,025	
	0050	Light reinforcing, over 0.17 m³	"	2.12	18.846	"		525	69.50	594.50	890	
	0200	Slab on grade to 150 mm thick, not reinforced, under 0.74 m²	B-9	7.90	5.066	m²		141	18.65	159.65	240	
	0250	0.74 - 1.49 m²	"	16.26	2.460	"		68.50	9.05	77.55	116	
	0255	For over 1.49 m² see 02220-130-0400										
	0600	Walls, not reinforced, under 0.17 m³	B-9	1.70	23.557	m³		655	86.50	741.50	1,125	
	0650	0.17 - 0.34 m³	"	2.26	17.668	"		490	65	555	835	
	0655	For over 0.34 m³ see 02220-130-2500										
	1000	Concrete, elevated slab, bar reinforced, under 0.17 m³	B-9C	1.27	31.410	m³		875	116	991	1,475	
	1050	Bar reinforced, over 0.17 m³	"	1.42	28.269	"		785	104	889	1,350	
	1200	Slab on grade to 150 mm thick, bar reinforced, under 0.74 m²	B-9	6.97	5.741	m²		160	21	181	272	

Note reference boxes:
- R024119 -10 (near line 1500)
- R024119 -10 (near line 0010, 250)
- R024119 -10 (near line 0010, 310)

02220 | Site Demolition

		CREW	DAILY OUTPUT	LABOR-HOURS	UNIT	MAT.	LABOR	EQUIP.	TOTAL	TOTAL INCL O&P	
310											**310**
1250	0.74 - 1.49 m²	B-9	13.94	2.870	m²	80	10.55		90.55	136	
1255	For over 1.49 m² see 02220-130-0440 [R024119-10]										
1400	Walls, bar reinforced, under 0.17 m³	B-9C	1.42	28.269	m³	785	104		889	1,350	
1450	0.17 - 0.34 m³	"	1.98	20.192	"	560	74.50		634.50	955	
1455	For over 0.34 m³ see 02220-130-2500 & 2600										
2000	Brick, to 0.37 m² opening, not including toothing										
2040	100 mm thick	B-9C	30	1.333	Ea.	37	4.91		41.91	63	
2060	200 mm thick		18	2.222		62	8.20		70.20	105	
2080	300 mm thick		10	4		111	14.70		125.70	189	
2400	Concrete block, to 0.37 m² opening, 50 mm thick		35	1.143		32	4.21		36.21	54	
2420	100 mm thick		30	1.333		37	4.91		41.91	63	
2440	200 mm thick		27	1.481		41	5.45		46.45	70	
2460	300 mm thick	▼	24	1.667		46.50	6.15		52.65	79	
2600	Gypsum block, to 0.37 m² opening, 50 mm thick	B-9	80	.500		13.90	1.84		15.74	23.50	
2620	100 mm thick		70	.571		15.90	2.10		18	27	
2640	200 mm thick		55	.727		20	2.68		22.68	34.50	
2800	Terra cotta, to 0.37 m² opening, 100 mm thick		70	.571		15.90	2.10		18	27	
2840	200 mm thick		65	.615		17.10	2.26		19.36	29	
2880	300 mm thick	▼	50	.800	▼	22	2.94		24.94	37.50	
3000	Toothing masonry cutouts, brick, soft old mortar	1 Brhe	12.19	.656	m		18.20		18.20	27.50	
3100	Hard mortar		9.14	.875			24.50		24.50	37	
3200	Block, soft old mortar		21.34	.375			10.40		10.40	15.85	
3400	Hard mortar	▼	15.24	.525	▼		14.55		14.55	22	
6000	Walls, interior, not including re-framing,										
6010	openings to 0.47 m²										
6100	Drywall to 16 mm thick	1 Clab	24	.333	Ea.		9.15		9.15	14.20	
6200	Paneling to 19 mm thick		20	.400			10.95		10.95	17.05	
6300	Plaster, on gypsum lath		20	.400			10.95		10.95	17.05	
6340	On wire lath	▼	14	.571	▼		15.65		15.65	24.50	
7000	Wood frame, not including re-framing, openings to 0.47 m²										
7200	Floors, sheathing and flooring to 50 mm thick	1 Clab	5	1.600	Ea.		44		44	68	
7310	Roofs, sheathing to 25 mm thick, not including roofing		6	1.333			36.50		36.50	57	
7410	Walls, sheathing to 25 mm thick, not including siding	▼	7	1.143	▼		31.50		31.50	48.50	
320 0010	**SELECTIVE DEMOLITION, DISPOSAL ONLY** [R024119-10]										**320**
0015	Urban building w/salvage value allowed										
0020	Including loading and 8 km haul to dump										
0200	Steel frame	B-3	329	.146	m³		4.32	5.30	9.62	12.50	
0300	Concrete frame		279	.172			5.10	6.25	11.35	14.75	
0400	Masonry construction		340	.141			4.18	5.15	9.33	12.10	
0500	Wood frame	▼	189	.254	▼		7.50	9.25	16.75	21.50	
330 0010	**SELECTIVE DEMOLITION, DUMP CHARGES** [R024119-10]										**330**
0020	Dump charges, typical urban city, tipping fees only										
0100	Building construction materials				Met. Ton					77	
0200	Trees, brush, lumber									55	
0300	Rubbish only									65	
0500	Reclamation station, usual charge	▼			▼					94	
340 0010	**SELECTIVE DEMOLITION, GUTTING** [R024119-10]										**340**
0020	Building interior, including disposal, dumpster fees not incl.										
0500	Residential building										
0560	Minimum	B-16	37.16	.861	m² Flr.		24	12.80	36.80	51.50	
0580	Maximum	"	33.44	.957	"		27	14.25	41.25	57	
0900	Commercial building										
1000	Minimum	B-16	32.52	.984	m² Flr.		27.50	14.65	42.15	59	
1020	Maximum	"	23.23	1.378	"		39	20.50	59.50	82.50	

Important: See the Reference Section for supporting data - Crews, Rental Equipment, City Cost Indexes and Reference Data

02200 | Site Preparation

02220 | Site Demolition

		Description	CREW	DAILY OUTPUT	LABOR-HOURS	UNIT	2006 BARE COSTS				TOTAL INCL O&P
							MAT.	LABOR	EQUIP.	TOTAL	
350	0010	**SELECTIVE DEMOLITION, RUBBISH HANDLING** R024119-10									350
	0020	The following are to be added to the demolition prices									
	0400	Chute, circular, prefabricated steel, 450 mm diameter	B-1	12.19	1.969	m	94	55.50		149.50	189
	0440	750 mm diameter	"	9.14	2.625	"	126	73.50		199.50	253
	0725	Dumpster, weekly rental, 1 dump/wk, 15.3 m³ cap. (7.25 Met. Ton)				Week	420			420	462
	0800	23 m³ capacity (9.07 Met. Tons)					605			605	665
	0840	30 m³ capacity (11.8 Met. Tons)				↓	775			775	825
	0900	Alternate pricing for dumpsters									
	0910	Delivery, average for all sizes				Ea.	50			50	55
	0920	Haul, average for all sizes					150			150	165
	0930	Rent per day, average for all sizes					5			5	5.50
	0940	Rent per month, average for all sizes					45			45	49.50
	0950	Disposal fee per ton, average for all sizes				Met. Ton	44			44	48.40
	1000	Dust partition, 6 mil polyethylene, 25 mm x 75 mm frame	2 Carp	186	.086	m²	1.94	3.06		5	6.90
	1080	50 mm x 100 mm frame	"	186	.086	"	3.44	3.06		6.50	8.55
	2000	Load, haul, and dump, 15 m haul	2 Clab	18.35	.872	m³		24		24	37
	2040	30 m haul		12.62	1.268			35		35	54
	2080	Over 30 m haul, add per 30 m		27.14	.589			16.15		16.15	25
	2120	In elevators, per 10 floors, add	↓	107	.150			4.10		4.10	6.40
	3000	Loading & trucking, including 3.2 km haul, chute loaded	B-16	34.41	.930			26	13.85	39.85	56
	3040	Hand loaded, 15 000 mm haul	"	36.70	.872			24.50	13	37.50	52.50
	3080	Machine loaded	B-17	91.75	.349			10.30	6.05	16.35	22.50
	5000	Haul, per km, up to 6 m³ truck	B-34B	891	.009			.25	.54	.79	.98
	5100	Over 6 m³ truck	"	1185	.007	↓		.19	.40	.59	.73
360	0010	**SELECTIVE DEMOLITION, SAW CUTTING** R024119-10									360
	0015	Asphalt, up to 75 mm deep	B-89	320	.050	m	.92	1.57	.93	3.42	4.42
	0020	Each additional 25 mm of depth		549	.029		.20	.92	.54	1.66	2.22
	0400	Concrete slabs, mesh reinforcing, up to 80 mm deep		299	.054		1.25	1.68	.99	3.92	5
	0420	Each additional mm of depth	↓	488	.033		.43	1.03	.61	2.07	2.69
	0800	Concrete walls, hydraulic saw, plain, per 25 mm of depth	B-89B	76.20	.210		1.12	6.60	6.10	13.82	17.90
	0820	Rod reinforcing, per 25 mm of depth		45.72	.350		1.54	11	10.15	22.69	29.50
	1200	Masonry walls, hydraulic saw, brick, per 25 mm of depth		91.44	.175		1.12	5.50	5.10	11.72	15.15
	1220	Block walls, solid, per 25 mm of depth	↓	76.20	.210		1.18	6.60	6.10	13.88	18
	2000	Brick or masonry w/hand held saw, per inch of depth	A-1	38.10	.210	↓	.95	5.75	1.29	7.99	11.40
	3020	Blades for saw, diamond, 300 mm diameter [12"]				Ea.	620			620	680
	3040	450 mm diameter [18"]					1,000			1,000	1,100
	3080	600 mm diameter [24"]					1,525			1,525	1,675
	3120	760 mm diameter [30"]					1,625			1,625	1,800
	3160	900 mm diameter [36"]					2,300			2,300	2,525
	3200	1070 mm diameter [42"]				↓	3,325			3,325	3,650
	5000	Wood sheathing to 25 mm thick, on walls	1 Carp	60.96	.131	m		4.67		4.67	7.25
	5020	On roof	"	76.20	.105	"		3.73		3.73	5.80
	9950	See also Div. 02210-200 core drilling	↓								
370	0010	**SELECTIVE DEMOLITION, TORCH CUTTING** R024119-10									370
	0020	Steel, 25 mm thick plate	1 Clab	110	.073	m	.59	1.99		2.58	3.76
	0040	25 mm diameter bar [1"]	"	210	.038	Ea.		1.04		1.04	1.63
	1000	Oxygen lance cutting, reinforced concrete walls									
	1040	300 mm to 400 mm thick walls	1 Clab	3.05	2.625	m		72		72	112
	1080	600 mm thick walls	"	1.83	4.374	"		120		120	187
372	0010	**SELECTIVE DEMOLITION, UTILITY MATERIALS**									372
	0020	See other utility items in 02220-240-0010									
	0100	Fire Hydrant extensions	B-20	14	1.714	Ea.		53.50		53.50	83
	0200	Precast Utility boxes up to 2.4 M x 4.2 M x 2.1 M	B-13	2	28			840	325	1,165	1,650
	0300	Handholes and meter pits	B-6	2	12			360	113	473	680
	0400	Utility valves 100 mm- 300 mm	B-20	4	6	↓		187		187	291

		02220	Site Demolition	CREW	DAILY OUTPUT	LABOR-HOURS	UNIT	2006 BARE COSTS				TOTAL INCL O&P	
								MAT.	LABOR	EQUIP.	TOTAL		
372	0500		350-600 mm	B-21	2	14	Ea.		450	76.50	526.50	780	372
374	0010	**SELECTIVE DEMOLITION, RIP-RAP & ROCK LINING**											374
	0100		Slope protection, broken stone	B-13	47.41	1.181	m³		35.50	13.75	49.25	69.50	
	0200		0.29 to 0.19 m³ pieces		50.17	1.116	m²		33.50	12.95	46.45	66	
	0300		450 mm depth		50.17	1.116	"		33.50	12.95	46.45	66	
	0400		Dumped stone		84.37	.664	Met. Ton		19.85	7.70	27.55	39	
	0500		Gabions, 150-300 mm deep		50.17	1.116	m²		33.50	12.95	46.45	66	
	0600		450 - 900 mm deep	↓	25.08	2.233	"		67	26	93	132	
376	0010	**SELECTIVE DEMOLITION, SHORE PROTECT/MOORING STRUCTURES**											376
	0100		Breakwaters, bulkheads, concrete, maximun	B-9	3.66	10.936	m		305	40.50	345.50	520	
	0200		Breakwaters, bulkheads, concrete, 3.6 M, minimum		3.05	13.123			365	48.50	413.50	625	
	0300		Maximum	↓	2.74	14.582			405	53.50	458.50	690	
	0400		Steel, shore driven	B-40B	16.46	2.916			88.50	68.50	157	212	
	0500		Barge driven	B-76A	9.14	6.999	↓		207	169	376	505	
	0600		Jetties, docks, floating	B-21B	55.74	.718	m²		21.50	11.40	32.90	45.50	
	0700		Pier supported, 75-100 mm decking		27.87	1.435			43	23	66	91.50	
	0800		Floating, prefab, small boat, minimum		55.74	.718			21.50	11.40	32.90	45.50	
	0900		Maximum	↓	27.87	1.435	↓		43	23	66	91.50	
	1000		Floating, prefab, per slip, minimum		3	13.333	Ea.		400	212	612	850	
	1010		Maximum	↓	3	13.333	"		400	212	612	850	
378	0010	**SELECTIVE DEMOLITION, PILES**											378
	0100		Cast in place piles, 200-250 mm, corrugated	B-19	183	.350	m		12.40	8.60	21	29.50	
	0200		1300-350 mm corrugated		152	.421			14.95	10.35	25.30	35.50	
	0300		400 mm corrugated		122	.525			18.65	12.90	31.55	43.50	
	0400		300 mm fluted		183	.350			12.40	8.60	21	29.50	
	0500		350-450 mm fluted		152	.421			14.95	10.35	25.30	35.50	
	0600		300 mm end bearing		183	.350			12.40	8.60	21	29.50	
	0700		350-450 mm end bearing		152	.421			14.95	10.35	25.30	35.50	
	0800		Precast prestressed piles, 300-350 mm diameter		213	.300			10.65	7.40	18.05	25	
	0900		400-600 mm diameter		183	.350			12.40	8.60	21	29.50	
	1000		900-1650 mm diameter		91.44	.700			25	17.20	42.20	58.50	
	1100		250-350 mm thick		183	.350			12.40	8.60	21	29.50	
	1200		400-600 mm thick		152	.421			14.95	10.35	25.30	35.50	
	1300		Pressure grouted pile, 125 mm		45.72	1.400			49.50	34.50	84	117	
	1400		Steel piles, 200-300 mm tip		183	.350			12.40	8.60	21	29.50	
	1500		H sections HP8 to HP12	↓	183	.350			12.40	8.60	21	29.50	
	1600		H section HP14	B-19A	183	.350			12.40	11	23.40	32	
	1700		Steel pipe piles, 200-300 mm	B-19	183	.350			12.40	8.60	21	29.50	
	1800		350-450 mm plain		152	.421			14.95	10.35	25.30	35.50	
	1900		350-450 mm concrete filled		122	.525			18.65	12.90	31.55	43.50	
	2000		Timber piles to 350 mm diameter	↓	183	.350	↓		12.40	8.60	21	29.50	
380	0010	**SELECTIVE DEMOLITION, UTILITY VALVES & ACCESSORIES**											380
	0015		Excludes excavation										
	0100		Utility valves 100-300 mm diameter	B-20	4	6	Ea.		187		187	291	
	0200		350-600 mm diameter	B-21	2	14			450	76.50	526.50	780	
	0300		Crosses 100-300 mm	B-20	8	3			93.50		93.50	145	
	0400		350-600 mm	B-21	4	7			225	38	263	390	
	0500		Utility cut-in valves 100-300 mm diameter	B-20	20	1.200			37.50		37.50	58	
	0600		Curb boxes	"	20	1.200	↓		37.50		37.50	58	
381	0010	**SELECTIVE DEMOLITION, WATER & SEWER PIPING AND FITTINGS**											381
	0015		Excludes excavation										
	0020		See other utility items in 02220-240										
	0090		Concrete pipe 100-250 mm diameter	B-6	76.20	.315	m		9.45	2.97	12.42	17.80	

02220 | Site Demolition

		CREW	DAILY OUTPUT	LABOR-HOURS	UNIT	2006 BARE COSTS				TOTAL INCL O&P
						MAT.	LABOR	EQUIP.	TOTAL	
381 0100	1050-1200 mm diameter	B-13B	29.26	1.914	m		57.50	32.50	90	125
0200	1500-2100 mm diameter	"	24.38	2.297			69	39	108	149
0300	2400 mm diameter	B-13C	24.38	2.297			69	65	134	178
0400	2700-3600 mm diameter	"	19.51	2.871			86	81.50	167.50	222
0450	Concrete fittings 300 mm diameter	B-6	24	1	Ea.		30	9.45	39.45	56.50
0480	Concrete end pieces 300 mm diameter		60.96	.394	m		11.80	3.71	15.51	22
0485	375 mm diameter		45.72	.525			15.75	4.95	20.70	29.50
0490	450 mm diameter		45.72	.525			15.75	4.95	20.70	29.50
0500	600-900 mm diameter		30.48	.787			23.50	7.45	30.95	44.50
0600	Concrete fittings 600-900 mm diameter		12	2	Ea.		60	18.85	78.85	113
0700	1200-2100 mm diameter	B-13B	12	4.667			140	79.50	219.50	300
0800	2400 mm diameter	"	8	7			210	119	329	455
0900	2700-3600 mm diameter	B-13C	4	14			420	395	815	1,075
1000	Ductile iron pipe 100 mm diameter	B-21B	60.96	.656	m		19.65	10.45	30.10	42
1100	150-300 mm diameter		53.34	.750			22.50	11.95	34.45	47.50
1200	350-600 mm diameter		36.58	1.094			32.50	17.40	49.90	69.50
1300	Ductile iron fittings 100-300 mm diameter		24	1.667	Ea.		50	26.50	76.50	106
1400	350-400 mm diameter		18	2.222			66.50	35.50	102	142
1500	450-600 mm diameter		12	3.333			100	53	153	213
1600	Plastic pipe 18-100 mm diameter	B-20	213	.113	m		3.50		3.50	5.45
1700	150-200 mm diameter		152	.158			4.91		4.91	7.65
1800	250-450 mm diameter		91.44	.262			8.15		8.15	12.70
1900	500-900 mm diameter		60.96	.394			12.25		12.25	19.05
1910	1050-1200 mm diameter		54.86	.437			13.60		13.60	21
1920	1350-1500 mm diameter		48.77	.492			15.30		15.30	24
2000	Plastic fittings 100-200 mm diameter	B-6	75	.320	Ea.		9.60	3.02	12.62	18.05
2100	250-350 mm diameter		50	.480			14.40	4.53	18.93	27
2200	400-600 mm diameter		20	1.200			36	11.30	47.30	68
2210	750-900 mm diameter		15	1.600			48	15.10	63.10	90.50
2220	1050-1200 mm diameter		12	2			60	18.85	78.85	113
2300	Copper pipe 18-50 mm diameter	Q-1	152	.105	m		4.05		4.05	6.10
2400	63-75 mm diameter		91.44	.175			6.70		6.70	10.10
2500	100-150 mm diameter		60.96	.262			10.10		10.10	15.20
2600	Copper fittings 18-50 mm diameter		15	1.067	Ea.		41		41	61.50
2700	Cast iron pipe 100 mm diameter		60.96	.262	m		10.10		10.10	15.20
2800	125-150 mm diameter	Q-2	60.96	.394			15.70		15.70	23.50
2900	200-300 mm diameter	Q-3	60.96	.525			21.50		21.50	32
3000	Cast iron fittings 100 mm diameter	Q-1	30	.533	Ea.		20.50		20.50	31
3100	125-150 mm diameter	Q-2	30	.800			32		32	48
3200	200-375 mm diameter	Q-3	30	1.067			43.50		43.50	65.50
3300	Vent cast iron pipe 100-200 mm diameter	Q-1	60.96	.262	m		10.10		10.10	15.20
3400	250-375 mm diameter	Q-3	60.96	.525	"		21.50		21.50	32
3500	Vent cast iron fittings 100-200 mm diameter	Q-2	30	.800	Ea.		32		32	48
3600	250-375 mm diameter	"	20	1.200	"		48		48	72
382 0010	**SELECTIVE DEMOLITION, METAL DRAINAGE PIPING**									
0015	Excludes excavation									
0020	See other utility items in 02220-240, 02220-380									
0025	02220-381									
0100	CMP pipe, aluminium, 150-250 mm diameter	B-21	244	.115	m		3.68	.63	4.31	6.40
0110	300 mm diameter		183	.153			4.91	.83	5.74	8.50
0120	450 mm diameter		183	.153			4.91	.83	5.74	8.50
0140	Steel, 150-250 mm diameter		244	.115			3.68	.63	4.31	6.40
0150	300 mm diameter		183	.153			4.91	.83	5.74	8.50
0160	450 mm diameter		122	.230			7.35	1.25	8.60	12.80
0170	600 mm diameter	B-13	91.44	.612			18.35	7.10	25.45	36
0180	750-900 mm diameter		76.20	.735			22	8.55	30.55	43.50

02220 | Site Demolition

		CREW	DAILY OUTPUT	LABOR-HOURS	UNIT	MAT.	LABOR	EQUIP.	TOTAL	TOTAL INCL O&P	
382							**2006 BARE COSTS**					**382**
0190	1200-1500 mm diameter	B-13	60.96	.919	m		27.50	10.65	38.15	54.50	
0200	steel, 1800 mm diameter	B-13B	30.48	1.837	↓		55	31	86	119	
0210	CMP end sections, steel, 250-450 mm diameter	B-21	40	.700	Ea.		22.50	3.82	26.32	39	
0220	600-900 mm diameter	B-13	30	1.867			56	21.50	77.50	110	
0230	1200 mm diameter		20	2.800			84	32.50	116.50	165	
0240	1500 mm diameter	↓	10	5.600			168	65	233	330	
0250	1800 mm diameter	B-13B	10	5.600			168	95	263	365	
0260	CMP fittings, 200-300 mm diameter	B-21	60	.467			15	2.54	17.54	26	
0270	450 mm diameter	"	40	.700			22.50	3.82	26.32	39	
0280	600-1200 mm diameter	B-13	30	1.867			56	21.50	77.50	110	
0290	1500 mm diameter	"	20	2.800			84	32.50	116.50	165	
0300	1800 mm diameter	B-13B	10	5.600	↓		168	95	263	365	
0310	Oval arch 425 x 325, 525 x 375, 375-450 mm equivalent	B-21	122	.230	m		7.35	1.25	8.60	12.80	
0320	700 mm x 500 mm, 600 mm equivalent	B-13	91.44	.612			18.35	7.10	25.45	36	
0330	875 x 325 , 1050 x 725 , 750-900 mm equivalent		76.20	.735			22	8.55	30.55	43.50	
0340	1225 x 825, 1425 x 950, 1050-1200 mm equivalent	↓	60.96	.919	↓		27.50	10.65	38.15	54.50	
0350	Oval arch 425 mm x 325 mm end piece, 375 mm equivalent	B-21	40	.700	Ea.		22.50	3.82	26.32	39	
0360	1050 mm x 725 mm end piece, 900 mm equivalent	B-13	30	1.867	"		56	21.50	77.50	110	
383 0010	**SELECTIVE DEMOLITION, MANHOLES & CATCH BASINS**										**383**
0015	Excludes excavation										
0020	See other utility items in 02220-240										
0100	Manholes, precast or brick over 2.4 m deep	B-6	2.44	9.843	m		295	93	388	555	
0200	Cast in place 1.2-2.4 m deep	B-9	11.80	3.390	m² Face		94.50	12.50	107	161	
0300	Over 2.4 m deep	"	9.29	4.306	"		120	15.85	135.85	203	
0400	Top, precast, 200 mm thick, 1200-1800 mm diameter	B-6	8	3	Ea.		90	28.50	118.50	169	
0500	Steps	1 Clab	60	.133	"		3.65		3.65	5.70	
384 0010	**SELECTIVE DEMOLITION, BOX CULVERT**										**384**
0015	Excludes excavation										
0100	Box culvert 2.4 m x 1.8 m x 0 .9 m to 2.4 m x 2.4 m x 2.4 m	B-69	91.44	.525	m		15.95	11.60	27.55	37.50	
0200	2.4 m x 3 m x 0.9 m to 2.4 m x 3.6 m x 2.4 m	"	60.96	.787	"		24	17.35	41.35	56	
386 0010	**SELECTIVE DEMOLITION, WATER WELLS**										**386**
0100	Well, 12 M deep with casing & gravel pack, 600 mm-900 mm dia	B-23	.25	160	Ea.		4,450	14,200	18,650	22,500	
0200	Riser pipe, 31 mm, for observation well	"	91.44	.437	m		12.15	39	51.15	62	
0300	Pump, 0.375 kW to 3.75 kW up to 30 m depth	Q-1	3	5.333	Ea.		205		205	310	
0400	Up to 45 m well 18.75 kW pump	Q-22	1.50	10.667	↓		410	425	835	1,075	
0500	Up to 150 m well 22.5 kW pump	"	1	16	↓		615	635	1,250	1,625	
0600	Well screen 50 mm to 200 mm	B-23	91.44	.437	m		12.15	39	51.15	62	
0700	250 mm to 400 mm		60.96	.656			18.25	58.50	76.75	92.50	
0800	450 mm to 650 mm		45.72	.875			24.50	78	102.50	124	
0900	Slotted PVC for wells 31 mm-200mm		183	.219			6.10	19.45	25.55	31	
1000	Well screen and casing 150 mm to 400 mm		91.44	.437			12.15	39	51.15	62	
1100	450 mm to 650 mm		45.72	.875			24.50	78	102.50	124	
1200	750 mm to 900 mm	↓	30.48	1.312	↓		36.50	117	153.50	185	
388 0010	**SELECTIVE DEMOLITION, SEPTIC TANKS & RELATED COMPONENTS**										**388**
0020	Excludes excavation										
0100	Septic tanks, precast, 4-4.5 kL	B-21	8	3.500	Ea.		112	19.10	131.10	195	
0200	5.5 kL		7	4			128	22	150	223	
0300	7.7-9.5 kL		5	5.600			180	30.50	210.50	310	
0400	15 kL	↓	4	7			225	38	263	390	
0500	Precast, 19 kL, 4 piece	B-13	3	18.667			560	217	777	1,100	
0600	57 kL	B-13B	1.70	32.941			985	560	1,545	2,150	
0700	95 kL		1.10	50.909			1,525	865	2,390	3,300	
0800	150 kL		.80	70			2,100	1,200	3,300	4,525	
0900	Precast, 190 kL, 5 piece	B-13C	.60	93.333			2,800	2,650	5,450	7,225	
1000	Cast-in-place, 285 kL	B-9	.06	666	↓		18,500	2,450	20,950	31,600	

Important: See the Reference Section for supporting data - Crews, Rental Equipment, City Cost Indexes and Reference Data

02220	Site Demolition	CREW	DAILY OUTPUT	LABOR-HOURS	UNIT	2006 BARE COSTS				TOTAL INCL O&P	
						MAT.	LABOR	EQUIP.	TOTAL		
388	1100	380 kL	B-9	.05	800	Ea.	22,200	2,950		25,150	37,900	**388**
	1200	HDPE, 4 kL	B-21	9	3.111			100	16.95	116.95	174	
	1300	5.5 kL		8	3.500			112	19.10	131.10	195	
	1400	Galley, 1200 mm x 1200 mm x 1200 mm	↓	16	1.750			56	9.55	65.55	97.50	
	1500	Distribution boxes, concrete, 7 outlets	2 Clab	16	1			27.50		27.50	42.50	
	1600	9 outlets	"	8	2			55		55	85.50	
	1700	Leaching chambers 3900 mm x 1075 mm x 400 mm, standard	B-13	16	3.500			105	40.50	145.50	206	
	1800	2.4 m x 1.075 m x 0.45 m, heavy duty		14	4			120	46.50	166.50	235	
	1900	3.9 m x 1.125 m x 0.45 m		12	4.667			140	54	194	275	
	2100	6 m x 1.2 m x 0.45 m		5	11.200			335	130	465	660	
	2200	Leaching pit 1.95 m x 1.8 m	B-21	5	5.600			180	30.50	210.50	310	
	2300	1.95 m x 2.4 m		4	7			225	38	263	390	
	2400	2.4 m x 1.8 m H20		4	7			225	38	263	390	
	2500	2.4 m x 2.4 m H20		3	9.333			300	51	351	520	
	2600	Velocity reducing pit, precast 1.8 m x 0.9 m deep	↓	4.70	5.957	↓		191	32.50	223.50	330	
390	0010	**SELECTIVE DEMOLITION, STEEL PIPE WITH INSULATION**										**390**
	0020	Excludes excavation										
	0100	Steel pipe, with insulation, 18 mm-100 mm	B-1A	122	.197	m		5.50	1.33	6.83	10.05	
	0200	125 mm-250 mm	B-1B	110	.291			8.90	7.25	16.15	21.50	
	0300	300 mm-400 mm		73.15	.437			13.40	10.90	24.30	32.50	
	0400	450 mm-600 mm		48.77	.656			20	16.40	36.40	49	
	0450	650 mm-900 mm	↓	30.48	1.050	↓		32	26	58	78.50	
	0500	Steel gland seal, with insulation, 18 mm-100 mm	B-1A	100	.240	Ea.		6.75	1.62	8.37	12.30	
	0600	125 mm-250 mm	B-1B	75	.427			13.05	10.65	23.70	31.50	
	0700	300 mm-400 mm		60	.533			16.30	13.30	29.60	39.50	
	0800	450 mm-600 mm		50	.640			19.55	15.95	35.50	47.50	
	0850	650 mm-900 mm	↓	40	.800			24.50	19.95	44.45	59.50	
	0900	Demo steel fittings with insulation 18 mm-100 mm	B-1A	60	.400			11.25	2.70	13.95	20.50	
	1000	125 mm-250 mm	B-1B	40	.800			24.50	19.95	44.45	59.50	
	1100	300 mm-400 mm		30	1.067			32.50	26.50	59	80	
	1200	450 mm-600 mm		20	1.600			49	40	89	120	
	1300	650 mm-900 mm		15	2.133			65	53.50	118.50	160	
	1400	Steel pipe anchors, 125 mm-250 mm		40	.800			24.50	19.95	44.45	59.50	
	1500	300 mm-400 mm		30	1.067			32.50	26.50	59	80	
	1600	450 mm-600 mm		20	1.600			49	40	89	120	
	1700	650 mm-900 mm	↓	15	2.133	↓		65	53.50	118.50	160	
392	0010	**SELECTIVE DEMOLITION, GASOLINE CONTAINMENT PIPING**										**392**
	0020	Excludes excavation										
	0030	Excludes environmental site remediation										
	0100	Gasoline plastic primary containment piping 50 mm to 100 mm	Q-6	244	.098	m		3.95		3.95	5.95	
	0200	Fittings 50 mm to 100 mm		40	.600	Ea.		24		24	36.50	
	0300	Gasoline plastic secondary containment piping 75 mm to 150 mm		244	.098	m		3.95		3.95	5.95	
	0400	Fittings 75 mm to 150 mm	↓	40	.600	Ea.		24		24	36.50	
394	0010	**SELECTIVE DEMOLITION, NATURAL GAS, PE PIPE**										**394**
	0020	Excludes excavation										
	0100	Natural gas coils, PE, 31 mm to 75 mm	Q-6	244	.098	m		3.95		3.95	5.95	
	0200	Joints, 12 m, PE, 75-100 mm		244	.098			3.95		3.95	5.95	
	0300	150-200 mm	↓	183	.131	↓		5.25		5.25	7.95	
396	0010	**SELECTIVE DEMOLITION, NATURAL GAS, STEEL PIPE**										**396**
	0020	Excludes excavation										
	0100	Natural gas steel pipe 25-100 mm	B-1A	244	.098	m		2.76	.66	3.42	5.05	
	0200	125-250 mm	B-1B	110	.291	↓		8.90	7.25	16.15	21.50	

SITE CONSTRUCTION 2

For expanded coverage of these items see *Means Site Work & Landscape Cost Data 2006*

		02220	Site Demolition	CREW	DAILY OUTPUT	LABOR-HOURS	UNIT	2006 BARE COSTS				TOTAL INCL O&P	
								MAT.	LABOR	EQUIP.	TOTAL		
396	0300		300-400 mm	B-1B	73.15	.437	m		13.40	10.90	24.30	32.50	396
	0400		450-600 mm	↓	48.77	.656	↓		20	16.40	36.40	49	
	0500		Natural gas steel fittings 25-100 mm	B-1A	160	.150	Ea.		4.21	1.01	5.22	7.65	
	0600		125-250 mm	B-1B	160	.200			6.10	4.99	11.09	14.95	
	0700		300-400 mm		108	.296			9.05	7.40	16.45	22	
	0800		450-600 mm	↓	70	.457	↓		14	11.40	25.40	34	
398	0010	**SELECTIVE DEMO, NATURAL GAS, VALVES, FITTINGS, REGULATORS**											398
	0100		Gas stops 31 mm - 50 mm	1 Plum	22	.364	Ea.		15.55		15.55	23.50	
	0200		Gas regulator 38 mm - 50 mm	"	22	.364			15.55		15.55	23.50	
	0300		75 mm - 100 mm	Q-1	22	.727			28		28	42	
	0400		Gas plug valve, 18 mm - 50 mm	1 Plum	22	.364			15.55		15.55	23.50	
	0500		63 mm - 75 mm	Q-1	10	1.600	↓		61.50		61.50	92.50	
401	0010	**SELECTIVE DEMOLITION, RADIO TOWERS**											401
	0100		Radio tower, guyed, 15 m	2 Skwk	16	1	Ea.		36.50		36.50	57	
	0200		57 m, 18.2 kg section	K-2	.70	34.286			1,250	252	1,502	2,450	
	0300		60 m 31.8 kg section		.70	34.286			1,250	252	1,502	2,450	
	0400		90 m 31.8 kg section		.40	60			2,200	440	2,640	4,275	
	0500		81 m 40.9 kg section		.40	60			2,200	440	2,640	4,275	
	0600		120 m		.30	80			2,925	585	3,510	5,700	
	0700		Self supported, 18 m		.90	26.667			975	196	1,171	1,900	
	0800		36 m		.80	30			1,100	220	1,320	2,150	
	0900		57 m	↓	.40	60	↓		2,200	440	2,640	4,275	
402	0010	**SELECTIVE DEMOLITION, ELECTRIC DUCTS & FITTINGS**											402
	0020		Excludes excavation										
	0100		Plastic conduit 13-50 mm	1 Elec	183	.044	m		1.84		1.84	2.74	
	0200		75-150 mm	2 Elec	122	.131	"		5.50		5.50	8.20	
	0300		Fittings 13-50 mm	1 Elec	50	.160	Ea.		6.70		6.70	10	
	0400		75-150 mm	"	40	.200	"		8.40		8.40	12.50	
404	0010	**SELECTIVE DEMOLITION, ELECTRIC DUCT BANKS**											404
	0020		Excludes excavation										
	0100		Hand holes sized to 1.2 m x 1.2 m x 1.2 m	R-3	7	2.857	Ea.		118	22	140	201	
	0200		Manholes sized to 1.8 m x 3 m x 2.1 m	B-13	6	9.333	"		279	108	387	550	
	0300		Conduit 1 @ 50 mm diameter, EB plastic, no concrete	2 Elec	305	.052	m		2.20		2.20	3.28	
	0400		2 @ 50 mm diameter		152	.105			4.42		4.42	6.60	
	0500		4 @ 50 mm diameter		76.20	.210			8.80		8.80	13.15	
	0600		1 @ 75 mm diameter		244	.066			2.75		2.75	4.10	
	0700		2 @ 75 mm diameter		122	.131			5.50		5.50	8.20	
	0800		4 @ 75 mm diameter		60.96	.262			11		11	16.45	
	0900		1 @ 100 mm diameter		244	.066			2.75		2.75	4.10	
	1000		2 @ 100 mm diameter		122	.131			5.50		5.50	8.20	
	1100		4 @ 100 mm diameter		60.96	.262			11		11	16.45	
	1200		6 @ 100 mm diameter		30.48	.525			22		22	33	
	1300		1 @ 125 mm diameter		152	.105			4.42		4.42	6.60	
	1400		2 @ 125 mm diameter		76.20	.210			8.80		8.80	13.15	
	1500		4 @ 125 mm diameter		48.77	.328			13.80		13.80	20.50	
	1600		6 @ 125 mm diameter		30.48	.525			22		22	33	
	1700		1 @ 150 mm diameter		152	.105			4.42		4.42	6.60	
	1800		2 @ 150 mm diameter		76.20	.210			8.80		8.80	13.15	
	1900		4 @ 150 mm diameter		48.77	.328			13.80		13.80	20.50	
	2000		6 @ 150 mm diameter	↓	30.48	.525			22		22	33	
	2100		Conduit 1 EB plastic, with concrete, 0.087 m³/m	B-9	45.72	.875			24.50	3.22	27.72	41.50	
	2200		2 EB plastic 0.144 m³/m		30.48	1.312			36.50	4.83	41.33	62.50	
	2300		2 x 2 EB plastic 0.237 m³/m		24.38	1.640			45.50	6.05	51.55	77.50	
	2400		2 x 3 EB plastic 0.336 m³/m	↓	18.29	2.187			61	8.05	69.05	103	
	2500		Conduit 2 @ 50 mm diameter, steel, no concrete	2 Elec	122	.131			5.50		5.50	8.20	
	2600		4 @ 50 mm diameter	↓	60.96	.262	↓		11		11	16.45	

Important: See the Reference Section for supporting data - Crews, Rental Equipment, City Cost Indexes and Reference Data

SITE CONSTRUCTION 2

02220 | Site Demolition

		CREW	DAILY OUTPUT	LABOR-HOURS	UNIT	MAT.	LABOR	EQUIP.	TOTAL	TOTAL INCL O&P	
							2006 BARE COSTS				
404	2700	2 @ 75 mm diameter	2 Elec	60.96	.262	m		11		11	16.45	**404**
	2800	4 @ 75 mm diameter		30.48	.525			22		22	33	
	2900	2 @ 100 mm diameter		60.96	.262			11		11	16.45	
	3000	4 @ 100 mm diameter		30.48	.525			22		22	33	
	3100	6 @ 100 mm diameter		15.24	1.050			44		44	65.50	
	3200	2 @ 125 mm diameter		36.58	.437			18.35		18.35	27.50	
	3300	4 @ 125 mm diameter		18.29	.875			37		37	55	
	3400	6 @ 125 mm diameter		12.19	1.312			55		55	82	
	3500	2 @ 150 mm diameter		36.58	.437			18.35		18.35	27.50	
	3600	4 @ 150 mm diameter		18.29	.875			37		37	55	
	3700	6 @ 150 mm diameter	▼	12.19	1.312			55		55	82	
	3800	Conduit 2 steel, with concrete, 0.144 m³/m	B-9	24.38	1.640			45.50	6.05	51.55	77.50	
	3900	2 x 2 EB steel 0.237 m³/m		18.29	2.187	↓		61	8.05	69.05	103	
	4000	2 x 3 steel 0.336 m³/m	▼	15.24	2.625	▼		73	9.65	82.65	125	
	4100	Conduit fittings, PVC type EB, 50-75 mm	1 Elec	30	.267	Ea.		11.20		11.20	16.70	
	4200	100-150 mm	"	20	.400	"		16.80		16.80	25	
406	0010	**SELECTIVE DEMOLITION, UTILITY POLES & CROSS ARMS**										**406**
	0100	Utility poles, wood, 6-9 m high	R-3	6	3.333	Ea.		138	25.50	163.50	234	
	0200	10.5-13.5 M high	"	5	4			166	30.50	196.50	282	
	0300	Cross arms, wood, 1.2-1.8 m long	1 Elec	5	1.600	▼		67		67	100	
408	0010	**SELECTIVE REMOVAL, PAVEMENT LINES & MARKINGS**										**408**
	0015	Does not include traffic control costs										
	0020	See other items in 02760-300										
	0100	Remove painted traffic lines and markings permanent	B-78A	15240	.001	m		.02	.05	.07	.09	
	0200	Temporary traffic line tape	2 Clab	457	.035			.96		.96	1.49	
	0300	Thermoplastic traffic lines and markings	B-79A	15240	.001			.03	.07	.10	.12	
	0400	Painted pavement markings	B-78B	46.45	.388	m²		10.95	5.30	16.25	23	
410	0010	**SELECTIVE DEMOLITION, WALKS, STEPS AND PAVERS**										**410**
	0020	See other utility items in 02220-240										
	0100	Splash blocks	1 Clab	27.87	.287	m²		7.85		7.85	12.25	
	0200	Tree grates	"	50	.160	Ea.		4.38		4.38	6.80	
	0300	Walks, limestone pavers	2 Clab	13.94	1.148	m²		31.50		31.50	49	
	0400	Redwood sections		55.74	.287			7.85		7.85	12.25	
	0500	Redwood planks		44.59	.359			9.85		9.85	15.30	
	0600	Shale paver		27.87	.574			15.75		15.75	24.50	
	0700	Tile thinset paver	▼	62.71	.255	▼		7		7	10.90	
	0800	Wood round	B-1	350	.069	Ea.		1.92		1.92	3	
	0900	Asphalt block	2 Clab	41.81	.383	m²		10.50		10.50	16.30	
	1000	Bluestone		41.81	.383			10.50		10.50	16.30	
	1100	Slate, 25 mm or thinner		62.71	.255			7		7	10.90	
	1200	Granite blocks		27.87	.574			15.75		15.75	24.50	
	1300	Precast patio blocks		41.81	.383			10.50		10.50	16.30	
	1400	Planter blocks		55.74	.287			7.85		7.85	12.25	
	1500	Brick paving, dry set		27.87	.574			15.75		15.75	24.50	
	1600	Mortar set		16.72	.957			26		26	41	
	1700	Dry set on edge		22.30	.718	▼		19.65		19.65	30.50	
	1800	Steps, brick		60.96	.262	m		7.20		7.20	11.20	
	1900	Railroad tie		45.72	.350			9.60		9.60	14.95	
	2000	Bluestone		54.86	.292			8		8	12.45	
	2100	Wood/steel edging for steps		305	.052			1.44		1.44	2.24	
	2200	Timber or railroad tie edging for steps	▼	122	.131	▼		3.59		3.59	5.60	
414	0010	**SELECTIVE DEMOLITION, ATHLETIC SURFACES**										**414**
	0100	Synthetic grass	2 Clab	186	.086	m²		2.36		2.36	3:67	
	0200	Surface coat latex rubber	"	186	.086	"		2.36		2.36	3.67	
	0300	Tennis court posts	B-11C	16	1	Ea.		32	14.15	46.15	64.50	

2 SITE CONSTRUCTION

		CREW	DAILY OUTPUT	LABOR-HOURS	UNIT	MAT.	LABOR	EQUIP.	TOTAL	TOTAL INCL O&P
02220	**Site Demolition**					2006 BARE COSTS				
418	**0010 SELECTIVE DEMOLITION, LAWN SPRINKLER SYSTEMS**									**418**
0100	Golf course sprinkler system, 9 hole	4 Skwk	.10	320	Ea.		11,700		11,700	18,200
0200	Sprinkler system, 7.2 m diam. @ 4.5 m O.C., per head	B-20	110	.218	Head		6.80		6.80	10.55
0300	18 m diam. @ 7.2 m O.C., per head	"	52	.462	"		14.35		14.35	22.50
0400	Sprinkler heads, plastic	2 Skwk	150	.107	Ea.		3.89		3.89	6.05
0500	Impact circle pattern, 8.4-22.8 m diam.		75	.213			7.80		7.80	12.10
0600	Pop-up, 12.6-22.8 m diam.		50	.320			11.70		11.70	18.20
0700	311.7-29.7 M diameter		50	.320			11.70		11.70	18.20
0800	Sprinkler valves		40	.400			14.60		14.60	22.50
0900	Valve boxes		40	.400			14.60		14.60	22.50
1000	Controls		2	8			292		292	455
1100	Backflow preventer		4	4			146		146	227
1200	Vacuum breaker	↓	4	4	↓		146		146	227
420	**0010 SELECTIVE DEMOLITION, CHAIN LINK FENCES & GATES**									**420**
0020	See other fence items in 02220-220									
0100	Chain link, gates, 900-1200 mm width	B-6	30	.800	Ea.		24	7.55	31.55	45.50
0200	3000-3600 mm width		16	1.500			45	14.15	59.15	84.50
0300	4200 mm width		15	1.600			48	15.10	63.10	90.50
0400	6000 mm width		10	2.400	↓		72	22.50	94.50	136
0500	5400 mm width overhead & cantilever		24.38	.984	m		29.50	9.30	38.80	55.50
0510	Sliding		24.38	.984			29.50	9.30	38.80	55.50
0520	Cantilever to 12 m wide	↓	24.38	.984			29.50	9.30	38.80	55.50
0530	Motor operators	2 Skwk	1	16	Ea.		585		585	910
0540	Transmitter systems	"	15	1.067	"		39		39	60.50
0600	Chain link, fence, 1500 mm high	B-6	271	.089	m		2.66	.84	3.50	5
0650	1200-1500 mm high		305	.079			2.36	.74	3.10	4.45
0675	3600 mm high		122	.197			5.90	1.86	7.76	11.10
0700	Snow fence, 1200 mm high	↓	305	.079	↓		2.36	.74	3.10	4.45
0800	Chain link, fence, braces	B-1	2000	.012	Ea.		.34		.34	.52
0900	Privacy slats	"	2000	.012			.34		.34	.52
1000	Fence posts, steel, in concrete	B-6	80	.300			9	2.83	11.83	16.95
1100	Fence fabric & accessories, fabric to 2.4 m high		244	.098	m		2.95	.93	3.88	5.55
1200	Barbed wire		1524	.016	"		.47	.15	.62	.89
1300	Extension arms & eye tops		300	.080	Ea.		2.40	.75	3.15	4.52
1400	Fence rails		610	.039	m		1.18	.37	1.55	2.22
1500	Reinforcing wire	↓	1524	.016	"		.47	.15	.62	.89
421	**0010 SELECTIVE DEMOLITION, VINYL FENCES & GATES**									**421**
0100	Vinyl fence up to 1.8 m high	B-6	305	.079	m		2.36	.74	3.10	4.45
0200	Gates, up to 1.8 m high	"	40	.600	Ea.		18	5.65	23.65	33.50
422	**0010 SELECTIVE DEMOLITION, MISC METAL FENCES & GATES**									**422**
0020	See other fence items in 02220-220									
0100	Misc steel mesh fences, 1.2 - 1.8 m high	B-6	183	.131	m		3.93	1.24	5.17	7.40
0200	Kennels, 1.8-3.6 m long	2 Clab	8	2	Ea.		55		55	85.50
0300	Tops, 1.8-3.6 m long	"	30	.533	"		14.60		14.60	23
0400	Security fences 3.6-4.8 m high	B-6	30.48	.787	m		23.50	7.45	30.95	44.50
0500	Metal tubular picket fences 1.2-1.8 m high		152	.158	"		4.74	1.49	6.23	8.95
0600	Gates 900-1200 mm wide	↓	20	1.200	Ea.		36	11.30	47.30	68
424	**0010 SELECTIVE DEMOLITION, WOOD FENCES & GATES**									**424**
0020	See other fence items in 02220-220									
0100	Wood fence gates 900-1200 mm wide	2 Clab	20	.800	Ea.		22		22	34
0200	Wood fence, open rail, to 1.2 m high		780	.021	m		.56		.56	.87
0300	Wood fences 2.4 m high		112	.143	"		3.91		3.91	6.10
0400	Post, in concrete	↓	50	.320	Ea.		8.75		8.75	13.65

Important: See the Reference Section for supporting data - Crews, Rental Equipment, City Cost Indexes and Reference Data

02220	Site Demolition	CREW	DAILY OUTPUT	LABOR-HOURS	UNIT	2006 BARE COSTS				TOTAL INCL O&P	
						MAT.	LABOR	EQUIP.	TOTAL		
426	**0010**	**SELECTIVE DEMOLITION, RETAINING WALLS**									**426**
	0020	See other retaining wall items in 02220-240									
	0100	Concrete retaining wall, 1.8 m high, no reinforcing	B-9	3.87	10.333	m		287	38	325	485
	0200	2.4 m high		3.05	13.123			365	48.50	413.50	625
	0300	3 m high		2.38	16.825			470	62	532	800
	0400	With reinforcing, 1.8 m high		3.51	11.412			315	42	357	540
	0500	2.4 m high		2.74	14.582			405	53.50	458.50	690
	0600	3 m high		2.13	18.748			520	69	589	885
	0700	6 m high		1.22	32.808			910	121	1,031	1,550
	0800	Concrete cribbing, 3.6 m high, open/closed face	↓	11.71	3.417	m²		95	12.60	107.60	162
	0900	Interlocking segmental retaining wall	B-62	74.32	.323			9.70	2.07	11.77	17.20
	1000	Wall caps	"	55.74	.431			12.90	2.76	15.66	23
	1100	Metal bin retaining wall, 3 m wide, 1.2-3.6 m high	B-13	111	.505			15.10	5.85	20.95	30
	1200	3 m wide, 4.8-8.4 m high		92.90	.603	↓		18.05	7	25.05	35.50
	1300	Stone gabions, 1800 mm x 900 mm x 300 mm, stone filled		170	.329	Ea.		9.85	3.83	13.68	19.40
	1400	1800 mm x 900 mm x 450 mm		75	.747			22.50	8.70	31.20	44
	1500	1800 mm x 900 mm x 900 mm		25	2.240			67	26	93	132
	1600	2700 mm x 900 mm x 300 mm		75	.747			22.50	8.70	31.20	44
	1700	2700 mm x 900 mm x 450 mm		33	1.697			51	19.70	70.70	100
	1800	2700 mm x 900 mm x 900 mm		12	4.667			140	54	194	275
	1900	3600 mm x 900 mm x 300 mm		42	1.333			40	15.50	55.50	78.50
	2000	3600 mm x 900 mm x 450 mm		20	2.800			84	32.50	116.50	165
	2100	3600 mm x 900 mm x 900 mm	↓	6	9.333	↓		279	108	387	550
428	**0010**	**SELECTIVE DEMOLITION, HIGHWAY GUARD RAILS & BARRIERS**									**428**
	0100	Guard rail, corrugated steel	B-6	183	.131	m		3.93	1.24	5.17	7.40
	0200	End sections		40	.600	Ea.		18	5.65	23.65	33.50
	0300	Wrap around		40	.600	"		18	5.65	23.65	33.50
	0400	Timber 100 mm x 200 mm		183	.131	m		3.93	1.24	5.17	7.40
	0500	Three 19 mm cables		183	.131	"		3.93	1.24	5.17	7.40
	0600	Wood posts	↓	240	.100	Ea.		3	.94	3.94	5.65
	0700	Guide rail, 150 mm x 150 mm box beam	B-80B	36.58	.875	m		25.50	6.45	31.95	46.50
	0800	Median barrier, 150 mm x 200 mm box beam		73.15	.437			12.85	3.22	16.07	23.50
	0850	Precast concrete 1050 mm high x 600 mm wide		91.44	.350	↓		10.25	2.57	12.82	18.70
	0900	Impact barrier, UTMCD, barrel type	B-16	60	.533	Ea.		15	7.95	22.95	32.50
	1000	Resilient guide fence and light shield 1.8 m high	"	36.58	.875	m		24.50	13.05	37.55	52.50
	1100	Concrete posts, 1925 mm triangular	B-6	200	.120	Ea.		3.60	1.13	4.73	6.80
	1200	Speed bumps 263 x 56 x 1200 mm		91.44	.262	m		7.85	2.48	10.33	14.80
	1300	Pavement marking channelizing		200	.120	Ea.		3.60	1.13	4.73	6.80
	1400	Barrier and curb delineators		300	.080			2.40	.75	3.15	4.52
	1500	Rumble strips 600 mm x 88 mm x 13 mm	↓	150	.160	↓		4.80	1.51	6.31	9.05
430	**0010**	**SELECTIVE DEMOLITION, PARKING APPURTENANCES**									**430**
	0100	Bumper rails, garage, 150 mm wide	B-6	91.44	.262	m		7.85	2.48	10.33	14.80
	0200	300 mm channel rail		91.44	.262			7.85	2.48	10.33	14.80
	0300	Parking bumper, timber	↓	305	.079	↓		2.36	.74	3.10	4.45
	0400	Folding, with locks	B-1	100	.240	Ea.		6.75		6.75	10.50
	0500	Flexible fixed garage stanchion	B-6	150	.160			4.80	1.51	6.31	9.05
	0600	Wheel stops, precast concrete		120	.200			6	1.89	7.89	11.30
	0700	Thermoplastic		120	.200			6	1.89	7.89	11.30
	0800	Ppe bollards, 150 mm - 300 mm diameter	↓	80	.300	↓		9	2.83	11.83	16.95
432	**0010**	**SELECTIVE DEMOLITION, PREFAB PEDESTRIAN BRIDGES**									**432**
	0100	Bridges, pedestrian, precast, 18 m-45 m long	B-21C	23.23	2.411	m²		72	68.50	140.50	187
	0200	Steel, 15-48 m long, 2.4-3 m wide	"	46.45	1.206			36	34	70	93
	0300	Laminated wood, 24-39 m long	C-12	27.87	1.722	↓		60	23	83	118

SITE CONSTRUCTION 2

For expanded coverage of these items see *Means Site Work & Landscape Cost Data 2006*

		Site Clearing	CREW	DAILY OUTPUT	LABOR-HOURS	UNIT	2006 BARE COSTS				TOTAL INCL O&P	
		02230					MAT.	LABOR	EQUIP.	TOTAL		
100	0010	**CLEAR AND GRUB**										100
	0020	Cut & chip light trees to 150 mm diam. [6"]	B-7	.40	118	Hectare		3,475	2,550	6,025	8,175	
	0150	Grub stumps and remove	B-30	.81	29.652			925	2,075	3,000	3,675	
	0200	Cut and chip medium, trees to 300 mm diam. [12"]	B-7	.28	169			4,950	3,650	8,600	11,700	
	0250	Grub stumps and remove	B-30	.40	59.303			1,850	4,150	6,000	7,350	
	0300	Cut & chip heavy, trees to 600 mm diam. [24"]	B-7	.12	395			11,600	8,525	20,125	27,300	
	0350	Grub stumps and remove	B-30	.20	118			3,700	8,275	11,975	14,700	
	0400	If burning is allowed, reduce cut & chip				↓					40%	
	3000	Chipping stumps, to 450 mm deep, 300 mm diam. [18"-12"]	B-86	20	.400	Ea.		14.70	3.18	17.88	25.50	
	3040	450 mm diameter [18"]		16	.500			18.35	3.97	22.32	32	
	3080	600 mm diameter [24"]		14	.571			21	4.54	25.54	36.50	
	3100	760 mm diameter [30"]		12	.667			24.50	5.30	29.80	43	
	3120	900 mm diameter [36"]		10	.800			29.50	6.35	35.85	51.50	
	3160	1200 mm diameter [48"]	↓	8	1	↓		36.50	7.95	44.45	64.50	
	5000	Tree thinning, feller buncher, conifer										
	5080	Up to 200 mm diameter [8"]	B-93	240	.033	Ea.		1.22	2.05	3.27	4.10	
	5120	300 mm diameter [12"]		160	.050			1.84	3.07	4.91	6.15	
	5240	Hardwood, up to 100 mm diameter [4"]		240	.033			1.22	2.05	3.27	4.10	
	5280	200 mm diameter [8"]		180	.044			1.63	2.73	4.36	5.45	
	5320	300 mm diameter [12"]	↓	120	.067	↓		2.45	4.10	6.55	8.20	
	7000	Tree removal, congested area, aerial lift truck										
	7040	200 mm diameter [8"]	B-85	7	5.714	Ea.		168	99.50	267.50	370	
	7080	300 mm diameter [12"]		6	6.667			196	116	312	430	
	7120	450 mm diameter [18"]		5	8			236	139	375	520	
	7160	600 mm diameter [24"]		4	10			295	174	469	645	
	7240	900 mm diameter [36"]		3	13.333			395	232	627	860	
	7280	1200 mm diameter [48"]	↓	2	20			590	350	940	1,300	
	9000	Site clearing with 250 kW dozer, trees to 150 mm diameter [6"]	B-10M	280	.043			1.44	4.28	5.72	6.90	
	9010	To 300 mm diameter [12"]		150	.080			2.69	8	10.69	12.90	
	9020	To 600 mm diameter [24"]		100	.120			4.03	11.95	15.98	19.30	
	9030	To 900 mm diameter [36"]		50	.240			8.05	24	32.05	39	
	9100	Grub stumps, trees to 150 mm diameter [6"]		400	.030			1.01	2.99	4	4.82	
	9103	To 900 mm diameter [36"]	↓	195	.062	↓		2.07	6.15	8.22	9.90	
200	0010	**SELECTIVE CLEARING**										200
	0020	Clearing brush with brush saw	A-1C	.10	79.067	Hectare		2,175	202	2,377	3,600	
	0100	By hand	1 Clab	.05	164			4,525		4,525	7,025	
	0300	With dozer, ball and chain, light clearing	B-11A	.81	19.768			635	1,150	1,785	2,225	
	0400	Medium clearing		.61	26.357			845	1,525	2,370	2,975	
	0500	With dozer and brush rake, light		4.05	3.954			127	228	355	445	
	0550	Medium brush to 100 mm diameter [4"]		3.24	4.942			158	284	442	555	
	0600	Heavy brush to 100 mm diameter [4"]	↓	2.59	6.177	↓		198	355	553	695	
	1000	Brush mowing, tractor w/rotary mower, no removal										
	1020	Light density	B-84	.81	9.884	Hectare		365	285	650	865	
	1040	Medium density		.61	13.178			485	380	865	1,150	
	1080	Heavy density	↓	.40	19.768	↓		725	570	1,295	1,725	
300	0010	**SELECTIVE TREE REMOVAL** with tractor, large tract, firm										300
	0020	level terrain, no boulders, less than 300 mm diam. trees [12"]										
	0300	224 kW dozer, up to 1000 trees/ha, 0 to 25% hardwoods	B-10M	.30	39.535	Hectare		1,325	3,950	5,275	6,375	
	0340	25% to 50% hardwoods		.24	49.419			1,650	4,925	6,575	7,950	
	0370	75% to 100% hardwoods		.18	65.891			2,225	6,575	8,800	10,600	
	0400	1250 trees/ha, 0% to 25% hardwoods		.24	49.419			1,650	4,925	6,575	7,950	
	0440	25% to 50% hardwoods		.19	61.773			2,075	6,150	8,225	9,925	
	0470	75% to 100% hardwoods		.15	82.367			2,775	8,225	11,000	13,300	
	0500	More than 1500 trees/ha, 0 to 25% hardwoods		.21	57.023			1,925	5,700	7,625	9,175	
	0540	25% to 50% hardwoods	↓	.17	70.601			2,375	7,050	9,425	11,400	

2 SITE CONSTRUCTION

Important: See the Reference Section for supporting data - Crews, Rental Equipment, City Cost Indexes and Reference Data

SITE CONSTRUCTION 2

		02230	Site Clearing	CREW	DAILY OUTPUT	LABOR-HOURS	UNIT	MAT.	LABOR	EQUIP.	TOTAL	TOTAL INCL O&P	
300	0570		75% to 100% hardwoods	B-10M	.13	95.648	Hectare		3,225	9,550	12,775	15,400	300
	0900	Large tract clearing per tree											
	1500	224 kW dozer, to 300 mm diameter, softwood [12″]		B-10M	320	.038	Ea.		1.26	3.74	5	6.05	
	1550	Hardwood			100	.120			4.03	11.95	15.98	19.30	
	1600	300 mm to 600 mm diameter, softwood [12″-24″]			200	.060			2.02	6	8.02	9.65	
	1650	Hardwood			80	.150			5.05	14.95	20	24	
	1700	600 mm to 900 mm diameter, softwood [24″-36″]			100	.120			4.03	11.95	15.98	19.30	
	1750	Hardwood			50	.240			8.05	24	32.05	39	
	1800	900 mm to 1200 mm diameter, softwood [36″-48″]			70	.171			5.75	17.10	22.85	27.50	
	1850	Hardwood			35	.343			11.50	34	45.50	55	
	2000	Stump removal on site by hydraulic backhoe, 1.15 m³											
	2040	100 mm to 150 mm diameter [4″ to 6″]		B-17	60	.533	Ea.		15.80	9.20	25	34	
	2050	200 mm to 300 mm diameter [8″ - 12″]		B-30	33	.727			22.50	51	73.50	90.50	
	2100	350 mm to 600 mm diameter [14″ - 24″]			25	.960			30	67	97	120	
	2150	650 mm to 900 mm diameter [26″ - 36″]			16	1.500			46.50	105	151.50	186	
	3000	Remove selective trees, on site using chain saws and chipper,											
	3050	not incl. stumps, up to 150 mm diameter [6″]		B-7	18	2.667	Ea.		78	57.50	135.50	184	
	3100	200 mm to 300 mm diameter [8″ - 12″]			12	4			117	86	203	276	
	3150	350 mm to 600 mm diameter [14″ - 24″]			10	4.800			141	103	244	330	
	3200	650 mm to 900 mm diameter [26″ - 36″]			8	6			176	129	305	415	
	3300	Machine load, 3 km haul to dump, 300 mm diam. tree, add [12″]									160	240	
500	0010	**STRIPPING & STOCKPILING OF SOIL**											500
	0020	149 kW dozer, ideal conditions		B-10B	1759	.007	m³		.23	.52	.75	.93	
	0100	Adverse conditions		″	879	.014			.46	1.05	1.51	1.85	
	0200	224 kW dozer, ideal conditions		B-10M	2294	.005			.18	.52	.70	.84	
	0300	Adverse conditions		″	1262	.010			.32	.95	1.27	1.53	
	0400	298 kW dozer, ideal conditions		B-10X	2982	.004			.14	.51	.65	.78	
	0500	Adverse conditions		″	1529	.008			.26	1	1.26	1.51	
	0600	Clay, dry and soft, 149 kW dozer, ideal conditions		B-10B	1223	.010			.33	.75	1.08	1.33	
	0700	Adverse conditions		″	612	.020			.66	1.50	2.16	2.66	
	1000	Medium hard, 224 kW dozer, ideal conditions		B-10M	1529	.008			.26	.78	1.04	1.26	
	1100	Adverse conditions		″	841	.014			.48	1.42	1.90	2.30	
	1200	Very hard, 298 kW dozer, ideal conditions		B-10X	1988	.006			.20	.77	.97	1.16	
	1300	Adverse conditions		″	1025	.012			.39	1.50	1.89	2.25	
	1400	Loam or topsoil, remove and stockpile on site											
	1420	150 mm deep, 60 m haul		B-10B	661	.018	m³		.61	1.39	2	2.46	
	1430	90 m haul			398	.030			1.01	2.31	3.32	4.09	
	1440	150 m haul			172	.070			2.34	5.35	7.69	9.45	
	1450	Alternate method: 150 mm deep, 60 m haul			4256	.003	m²		.09	.22	.31	.38	
	1460	150 m haul			1108	.011	″		.36	.83	1.19	1.46	

		02240	Dewatering	CREW	DAILY OUTPUT	LABOR-HOURS	UNIT	MAT.	LABOR	EQUIP.	TOTAL	TOTAL INCL O&P	
330	0010	**CUT DRAINAGE DITCH**											330
	0020	Cut drainage ditch, common earth, 10 m w x 30 cm deep		B-11L	1829	.009	m		.28	.25	.53	.71	
	0200	Clay and till			1280	.013			.40	.36	.76	1	
	0250	Clean wet drainage ditch, 10 m wide			3048	.005			.17	.15	.32	.43	
500	0010	**DEWATERING**											500
	0020	Excavate drainage trench, 600 mm wide, 600 mm deep		B-11C	68.81	.233	m³		7.45	3.29	10.74	15	
	0100	600 mm wide, 900 mm deep, with backhoe loader		″	103	.155			4.98	2.20	7.18	10	
	0200	Excavate sump pits by hand, light soil		1 Clab	5.43	1.474			40.50		40.50	63	
	0300	Heavy soil		″	2.68	2.989			82		82	128	
	0500	Pumping 8 hr., attended 2 hrs./day, including 6 m											
	0550	of suction hose & 30 m discharge hose											
	0600	50 mm diaphragm pump used for 8 hours		B-10H	4	3	Day		101	14.50	115.50	169	

02240 | Dewatering

	Line	Description	CREW	DAILY OUTPUT	LABOR-HOURS	UNIT	MAT.	LABOR	EQUIP.	TOTAL	TOTAL INCL O&P	
500	0620	Add per additional pump				Day			37.50	37.50	43	500
	0650	100 mm diaphragm pump used for 8 hours	B-10I	4	3			101	21	122	176	
	0670	Add per additional pump							80	80	91	
	0800	8 hrs. attended, 50 mm diaphragm pump	B-10H	1	12			405	58	463	680	
	0820	Add per additional pump							37.50	37.50	43	
	0900	75 mm centrifugal pump	B-10J	1	12			405	64.50	469.50	685	
	0920	Add per additional pump							53	53	58.50	
	1000	100 mm diaphragm pump	B-10I	1	12			405	83	488	705	
	1020	Add per additional pump							91	91	105	
	1100	150 mm centrifugal pump	B-10K	1	12			405	267	672	910	
	1120	Add per additional pump							118	118	134	
	1300	CMP, incl. excavation 1 m deep, 300 mm diameter [12"]	B-6	35.05	.685	m	33	20.50	6.45	59.95	74.50	
	1400	450 mm diameter [18"]		30.48	.787	"	41	23.50	7.45	71.95	89.50	
	1600	Sump hole construction, incl. excavation and gravel, pit		35.38	.678	m³	26.50	20.50	6.40	53.40	68	
	1700	With 300 mm gravel collar, 300 mm pipe, corr., 1.5 mm T		21.34	1.125	m	53.50	34	10.60	98.10	123	
	1800	380 mm pipe, corrugated, 1.5 mm T		16.76	1.432		68.50	43	13.50	125	156	
	1900	450 mm pipe, corrugated, 1.5 mm T		15.24	1.575		80	47	14.85	141.85	177	
	2000	600 mm pipe, corrugated, 1.9 mm T		12.19	1.969		96	59	18.55	173.55	218	
	2200	Wood lining, up to 1.2 m x 1.2 m, add		27.87	.861	m²CA	149	26	8.10	183.10	212	
	9950	See Div. 02240-900 for wellpoints										
	9960	See Div. 02240-700 for deep well systems										
	9970	See Div. 15440 for pumps										
700	0010	WELLS For dewatering 3 m to 6 m deep, 600 mm diameter										700
	0020	with steel casing, minimum	B-6	50.29	.477	m	8.35	14.30	4.50	27.15	36	
	0050	Average		29.87	.803		16.55	24	7.60	48.15	63.50	
	0100	Maximum		14.94	1.607		44	48	15.15	107.15	139	
	0300	For pumps for dewatering, see Reference Section										
	0500	For domestic water wells, see Reference Section										
900	0010	WELLPOINTS For wellpoint equip. rental, see Reference Section [R312319-90]										900
	0100	Installation and removal of single stage system										
	0110	Labor only, 2.46 labor-hours/m, minimum	1 Clab	3.26	2.453	m Hdr		67		67	105	
	0200	6.56 labor-hours/m, maximum	"	1.22	6.562	"		180		180	280	
	0400	Pump operation, 4 @ 6 hr. shifts										
	0410	Per 24 hour day	4 Eqlt	1.27	25.197	Day		885		885	1,350	
	0500	Per 168 hour week, 160 hr. straight, 8 hr. double time		.18	177	Week		6,250		6,250	9,450	
	0550	Per 4.3 week month		.04	800	Month		28,200		28,200	42,500	
	0600	Complete installation, operation, equipment rental, fuel &										
	0610	removal of system with 50 mm wellpoints 1.5 m O.C.										
	0700	30 m long header, 150 mm diameter, first month [6"]	4 Eqlt	.98	32.504	m Hdr	440	1,150		1,590	2,200	
	0800	Thereafter, per month		1.26	25.421		355	895		1,250	1,750	
	1000	60 m long header, 200 mm diameter, first month [8"]		1.83	17.498		390	615		1,005	1,350	
	1100	Thereafter, per month		2.56	12.513		199	440		639	885	
	1300	150 m long header, 200 mm diameter, first month [8"]		3.24	9.876		154	350		504	695	
	1400	Thereafter, per month		6.37	5.021		110	177		287	390	
	1600	300 m long header, 250 mm diameter, first month [10"]		3.54	9.035		132	320		452	625	
	1700	Thereafter, per month		12.74	2.511		66	88.50		154.50	206	
	1900	Note: above figures include pumping 168 hrs./week										
	1910	and include the pump operator and one stand-by pump.										

02250 | Shoring & Underpinning

	Line	Description	CREW	DAILY OUTPUT	LABOR-HOURS	UNIT	MAT.	LABOR	EQUIP.	TOTAL	TOTAL INCL O&P	
100	0010	GROUTING, PRESSURE										100
	0020	Grouting, pressure, cement & sand, 1:1 mix, minimum	B-61	124	.323	Bag	9.50	9.45	2.10	21.05	27.50	
	0100	Maximum		51	.784	"	9.50	23	5.10	37.60	51.50	
	0200	Cement and sand, 1:1 mix, minimum		7.08	5.654	m³	670	166	37	873	1,050	
	0300	Maximum		2.83	14.134		1,000	415	92	1,507	1,850	
	0400	Epoxy cement grout, minimum		3.88	10.317		5,000	305	67	5,372	6,050	

02200 | Site Preparation

02250 | Shoring & Underpinning

			CREW	DAILY OUTPUT	LABOR-HOURS	UNIT	MAT.	LABOR	EQUIP.	TOTAL	TOTAL INCL O&P	
100	0500	Maximum	B-61	1.61	24.797	m³	5,000	730	161	5,891	6,800	100
	0600	Structural epoxy grout				liter	13.10			13.10	14.40	
	0700	Alternate pricing method: (Add for materials)										
	0710	5 person crew and equipment	B-61	1	40	Day		1,175	260	1,435	2,100	
400	0010	**SHEET PILING**										400
	0020	Sheet piling steel, not incl. wales, 107 kg/m², 4.5 m excav.,	B-40	9.81	6.526	Met. Ton	1,025	232	281	1,538	1,800	
	0100	Pull & salvage		5.44	11.758		455	420	505	1,380	1,725	
	0300	6 m deep excavation, 132 kg/m², left in place	R314116-40	11.75	5.448		1,025	194	235	1,454	1,700	
	0400	Pull & salvage		5.94	10.770		455	385	465	1,305	1,625	
	0600	7.5 m deep excavation, 185 kg/m², left in place	R314116-45	17.24	3.713		1,025	132	160	1,317	1,500	
	0700	Pull & salvage		9.53	6.719		455	239	289	983	1,200	
	0900	12 m deep excavation, 185 kg/m², left in place		19.23	3.328		1,025	118	143	1,286	1,475	
	1000	Pull & salvage		11.11	5.759	m²	455	205	248	908	1,100	
	1200	4.5 m deep excavation, 100 kg/m², left in place		91.32	.701		117	25	30	172	202	
	1300	Pull & salvage		50.63	1.264		50	45	54.50	149.50	187	
	1500	6 m deep excavation, 130 kg/m², left in place		89.18	.718		147	25.50	31	203.50	236	
	1600	Pull & salvage		45.06	1.420		65	50.50	61	176.50	220	
	1800	7.6 m deep excavation, 185 kg/m², left in place		92.90	.689		216	24.50	29.50	270	310	
	1900	Pull & salvage		51.37	1.246	Met. Ton	89	44.50	53.50	187	228	
	2100	Rent steel sheet piling and wales, first month					245			245	269	
	2200	Per added month					24.50			24.50	27	
	2300	Rental piling left in place, add to rental					815			815	900	
	2500	Wales, connections & struts, 2/3 salvage					250			250	275	
	2700	High strength piling, 345 MPa, add					55.50			55.50	61	
	2800	380 MPa, add					59			59	65	
	3000	Tie rod, not upset, 40 mm to 100 mm diameter with turnbuckle					1,825			1,825	2,000	
	3100	No turnbuckle					1,400			1,400	1,550	
	3300	Upset, 45 mm to 100 mm diameter with turnbuckle					2,050			2,050	2,250	
	3400	No turnbuckle					1,775			1,775	1,950	
	3600	Lightweight, 450 mm to 700 mm wide, 7 ga., 45 kg/m², and										
	3610	9 ga., 42 kg/m², minimum				kg	1.50			1.50	1.65	
	3700	Average					1.63			1.63	1.79	
	3750	Maximum					1.87			1.87	2.07	
	3900	Wood, solid sheeting, incl. wales, braces and spacers,										
	3910	pull & salvage, 2.4 m deep excavation	B-31	30.66	1.305	m²	19.15	38.50	4.45	62.10	86	
	4000	3.0 m deep, 4.65 m²/hr in & 13.94 m²/hr out		27.87	1.435		19.70	42	4.89	66.59	93	
	4100	3.6 m deep, 4.18 m²/hr in & 12.54 m²/hr out		25.08	1.595		20	47	5.45	72.45	102	
	4200	4.2 m deep, 3.90 m²/hr in & 11.71 m²/hr out		23.23	1.722		21	50.50	5.85	77.35	108	
	4300	4.8 m deep, 3.72 m²/hr in & 11.15 m²/hr out		22.30	1.794		21.50	53	6.10	80.60	112	
	4400	5.4 m deep, 3.53 m²/hr in & 10.59 m²/hr out		21.37	1.872		22	55	6.40	83.40	118	
	4500	6.0 m deep, 3.25 m²/hr in & 9.76 m²/hr out		19.51	2.050		23	60.50	7	90.50	127	
	4520	Left in place, 2.4 m deep, 5.11 m²/hr		40.88	.979		34.50	29	3.34	66.84	86.50	
	4540	3.0 m deep, 4.65 m²/hr		37.16	1.076		36.50	31.50	3.67	71.67	93.50	
	4560	3.7 m deep, 4.18 m²/hr		33.44	1.196		38.50	35	4.08	77.58	101	
	4565	4.3 m deep, 3.9 m²/hr.		31.12	1.285		40.50	38	4.38	82.88	108	
	4570	4.9 m deep, 3.72 m²/hr		29.73	1.346		43	39.50	4.59	87.09	114	
	4580	5.5 m deep, 3.53 m²/hr		28.33	1.412		46	41.50	4.81	92.31	120	
	4590	6.0 m deep, 3.25 m²/hr		26.01	1.538		49	45.50	5.25	99.75	130	
	4700	Alternate pricing, left in place, 2.4 m deep		4.15	9.631	m³	305	283	33	621	810	
	4800	Pull and salvage, 2.4 m deep		3.11	12.842	"	271	380	44	695	935	
	5000	For treated lumber add cost of treatment to lumber										
	5010	See Division 06070-400										
500	0010	**SHORING**										500
	0020	Shoring, existing building, with timber, no salvage allowance	B-51	5.19	9.246	m³	340	257	25	622	805	
	1000	On cribbing with 32 metric ton screw jacks, per box and jack	"	3.60	13.333	Jack	48	370	35.50	453.50	670	
	1100	Masonry openings in walls, see Div. 02220-310										

SITE CONSTRUCTION 2

		02250	Shoring & Underpinning	CREW	DAILY OUTPUT	LABOR-HOURS	UNIT	2006 BARE COSTS				TOTAL INCL O&P	
								MAT.	LABOR	EQUIP.	TOTAL		
600	0010		**SLABJACKING**										600
	0100		100 mm thick slab	D-4	139	.230	m²	2.80	7.30	1.04	11.14	15.25	
	0150		150 mm thick slab		111	.288		3.88	9.15	1.31	14.34	19.60	
	0200		200 mm thick slab		92.90	.344		4.63	10.95	1.56	17.14	23.50	
	0250		250 mm thick slab		83.61	.383		4.95	12.15	1.73	18.83	26	
	0300		300 mm thick slab		78.97	.405		5.40	12.90	1.84	20.14	27.50	
800	0010		**UNDERPINNING FOUNDATIONS** Including excavation,										800
	0020		forming, reinforcing, concrete and equipment										
	0100		1.5 to 5 m below grade, 76 to 380 m³	B-52	1.76	31.844	m³	315	1,025	226	1,566	2,200	
	0200		Over 380 m³		1.91	29.296		283	955	208	1,446	2,025	
	0400		5 m to 7.6 m below grade, 76 to 380 m³		1.53	36.620		345	1,200	260	1,805	2,525	
	0500		Over 380 m³		1.61	34.877		325	1,125	247	1,697	2,375	
	0700		8 m to 12 m below grade, 76 to 380 m³		1.22	45.776		375	1,500	325	2,200	3,075	
	0800		Over 380 m³		1.38	40.689		345	1,325	288	1,958	2,750	
	0900		For under 40 m³, add					10%	40%				
	1000		For 40 m³ to 76 m³, add					5%	20%				
900	0010		**VIBROFLOTATION** R314513-90										900
	0900		Vibroflotation compacted sand cylinder, minimum	B-60	229	.245	m		7.85	5.30	13.15	17.80	
	0950		Maximum		99.06	.565			18.20	12.25	30.45	41.50	
	1100		Vibro replacement compacted stone cylinder, minimum		152	.368			11.85	8	19.85	27	
	1150		Maximum		76.20	.735			23.50	15.90	39.40	53.50	
	1300		Mobilization and demobilization, minimum		.47	119	Total		3,825	2,575	6,400	8,675	
	1400		Maximum		.14	400	"		12,900	8,650	21,550	29,100	

		02260	Excavation Support/Protection										
200	0010		**COFFERDAMS**, incl. mobilization and temporary sheeting										200
	0020		Shore driven	B-40	89.18	.718	m²	144	25.50	31	200.50	233	
	0060		Barge driven	"	51.10	1.253	"	144	44.50	54	242.50	289	
	0080		Soldier beams & lagging H piles with 75 mm wood sheeting										
	0090		horizontal between piles, including removal of wales & braces										
	0100		No hydrostatic head, 5 m deep, 1 line of braces, minimum	B-50	50.63	2.212	m²	86	74.50	34.50	195	252	
	0200		Maximum		45.99	2.436		95.50	82	38	215.50	278	
	0400		5 m to 7 m deep with 2 lines of braces, 250 mm H, minimum		33.44	3.349		101	113	52.50	266.50	350	
	0500		Maximum		30.66	3.653		115	123	57	295	385	
	0700		7 m to 11 m deep with 3 lines of braces, 300 mm H, minimum		30.19	3.710		133	125	58	316	410	
	0800		Maximum		27.41	4.087		143	138	64	345	450	
	1000		11 m to 14 m deep with 4 lines of braces, 350 mm H, minimum		26.94	4.157		148	140	65	353	460	
	1100		Maximum		24.62	4.549		157	154	71	382	495	
	1300		No hydrostatic head, left in place, 5 m dp., 1 line of braces, min.		58.99	1.899		115	64	29.50	208.50	261	
	1400		Maximum		53.42	2.097		123	71	33	227	284	
	1600		5 m to 7 m deep with 2 lines of braces, minimum		42.27	2.650		172	89.50	41.50	303	380	
	1700		Maximum		38.55	2.905		191	98	45.50	334.50	420	
	1900		7 m to 11 m deep with 3 lines of braces, minimum		39.02	2.870		205	97	45	347	430	
	2000		Maximum		35.30	3.173		227	107	49.50	383.50	475	
	2200		11 m to 14 m deep with 4 lines of braces, minimum		35.77	3.131		246	106	49	401	495	
	2300		Maximum		32.52	3.445		287	116	54	457	560	
	2350		Lagging only, 75 mm thk. wood between piles 2.4 m OC, min.	B-46	37.16	1.292		19.15	40.50	.90	60.55	87	
	2370		Maximum		23.23	2.067		28.50	65	1.45	94.95	137	
	2400		Open sheeting no bracing, for trenches to 3 m deep, min.		161	.298		8.60	9.35	.21	18.16	24.50	
	2450		Maximum		140	.343		9.60	10.75	.24	20.59	28	
	2500		Tie-back method, add to open sheeting, add, minimum								20%	20%	
	2550		Maximum								60%	60%	
	2700		Tie-backs only, based on tie-backs total length, minimum	B-46	26.46	1.814	m	35	57	1.27	93.27	131	
	2750		Maximum		11.73	4.090	"	61.50	128	2.86	192.36	277	
	3500		Tie-backs only, typical average, 8 m long		2	24	Ea.	470	755	16.80	1,241.80	1,725	

Important: See the Reference Section for supporting data - Crews, Rental Equipment, City Cost Indexes and Reference Data

			DAILY	LABOR-		2006 BARE COSTS				TOTAL	
02260	**Excavation Support/Protection**	CREW	OUTPUT	HOURS	UNIT	MAT.	LABOR	EQUIP.	TOTAL	INCL O&P	
200	3600	11 m long	B-46	1.58	30.380	Ea.	625	955	21.50	1,601.50	2,250	**200**
	4500	Trench box, 2.1 m deep, 4.8 m x 2.4 m, see reference section				Day				175	193	
	4600	6 m x 3 m, see 01590-400-7070, see reference section				"				295	320	
	5200	Wood sheeting, in trench, jacks at 1.2 m O.C., 2.4 m deep	B-1	74.32	.323	m²	6.65	9.05		15.70	21.50	
	5250	3.7 m deep		65.03	.369		7.85	10.35		18.20	25	
	5300	4.6 m deep	↓	55.74	.431	↓	10.75	12.10		22.85	30.50	
	6000	See also Div. 02250-400										
720	0010	**ROCK BOLTING**										**720**
	2020	Hollow core, prestressable anchor, 25 mm diameter, 1.5 m long	2 Skwk	32	.500	Ea.	81	18.25		99.25	118	
	2025	3 m long		24	.667		162	24.50		186.50	216	
	2060	50 mm diameter, 1.5 m long		32	.500		300	18.25		318.25	365	
	2065	3 m long		24	.667		605	24.50		629.50	705	
	2100	Super high-tensile, 20 mm diameter, 1.5 m long		32	.500		18.35	18.25		36.60	48.50	
	2105	3 m long		24	.667		36.50	24.50		61	78.50	
	2160	50 mm diameter, 1.5 m long		32	.500		164	18.25		182.25	209	
	2165	3 m long	↓	24	.667		325	24.50		349.50	400	
	4400	Drill hole for rock bolt, 45 mm diam., 1.5 m long (for 20 mm bolt)	B-56	17	.941			29.50	63	92.50	114	
	4405	3 m long		9	1.778			55.50	119	174.50	215	
	4420	50 mm diameter, 1.5 m long (for 25 mm bolt)		13	1.231			38.50	82	120.50	150	
	4425	3 m long		7	2.286			71.50	153	224.50	277	
	4460	90 mm diameter, 1.5 m long (for 50 mm bolt)		10	1.600			50	107	157	194	
	4465	3 m long	↓	5	3.200	↓		100	214	314	390	
730	0010	**SOIL NAILING**										**730**
	0020	Soil nailing does not include guniting of surfaces										
	0030	See 03370-300 for guniting of surfaces										
	0035	Layout and vertical and horizontal control per day	A-6	1	16	Day		565	58	623	935	
	0038	Layout and vertical and horizontal control average holes per day	"	60	.267	Ea.		9.45	.97	10.42	15.55	
	0050	For A625m Grade 1000 soil nail add $ 4.52/ M to base material cost										
	0060	Material delivery add $ 1.14 to $ 1.26 per truck km for shipping										
	0090	Average Soil nailing, grade 500Mpa, 15 min setup per hole & 24 m/hr.										
	0100	Soil nailing, drill hole, install # 25 M nail, grout 6 m depth	B-47G	16	2	Ea.	93.50	59.50	74.50	227.50	277	
	0110	7.5 m depth		14.20	2.254		117	67.50	84	268.50	325	
	0120	9 m depth		12.80	2.500		191	74.50	93	358.50	430	
	0130	10.5 m depth		11.60	2.759		223	82.50	103	408.50	485	
	0140	12 m depth		10.70	2.991		254	89.50	111	454.50	540	
	0150	13.5 m depth		9.90	3.232		286	96.50	120	502.50	595	
	0160	15 m depth		9.10	3.516		355	105	131	591	695	
	0170	16.5 m depth		8.50	3.765		385	112	140	637	750	
	0180	18 m depth		8	4		415	119	149	683	810	
	0190	19.5 m depth		7.50	4.267		450	127	159	736	865	
	0200	21 m depth		7.10	4.507		480	135	168	783	925	
	0210	22.5 m depth	↓	6.70	4.776	↓	510	143	178	831	980	
	0290	Average Soil nailing, grade 500Mpa, 15 min setup per hole & 27 m/hr										
	0300	Soil nailing, drill hole, install # 25 M nail, grout 6 m depth	B-47G	16.60	1.928	Ea.	93.50	57.50	72	223	271	
	0310	7.5 m depth		15	2.133		117	63.50	79.50	260	315	
	0320	9 m depth		13.70	2.336		191	69.50	87	347.50	415	
	0330	10.5 m depth		12.30	2.602		223	77.50	97	397.50	470	
	0340	12 m depth		11.40	2.807		254	84	105	443	525	
	0350	13.5 m depth		10.70	2.991		286	89.50	111	486.50	575	
	0360	15 m depth		9.80	3.265		355	97.50	122	574.50	675	
	0370	16.5 m depth		9.20	3.478		385	104	130	619	730	
	0380	18 m depth		8.70	3.678		415	110	137	662	780	
	0390	19.5 m depth		8.10	3.951		450	118	147	715	840	
	0400	21 m depth	↓	7.70	4.156	↓	480	124	155	759	890	

SITE CONSTRUCTION **2**

	02260	Excavation Support/Protection	CREW	DAILY OUTPUT	LABOR-HOURS	UNIT	2006 BARE COSTS				TOTAL INCL O&P	
							MAT.	LABOR	EQUIP.	TOTAL		
730	0410	22.5 m depth	B-47G	7.40	4.324	Ea.	510	129	161	800	940	730
	0490	Average Soil nailing, grade 500Mpa, 15 min setup per hole & 30 m/hr.										
	0500	Soil nailing, drill hole, install # 25 M nail, grout 6 m depth	B-47G	17.80	1.798	Ea.	93.50	53.50	67	214	260	
	0510	7.5 m depth		16	2		117	59.50	74.50	251	300	
	0520	9 m depth		14.60	2.192		191	65.50	81.50	338	400	
	0530	10.5 m depth		13.30	2.406		223	72	89.50	384.50	455	
	0540	12 m depth		12.30	2.602		254	77.50	97	428.50	505	
	0550	13.5 m depth		11.40	2.807		286	84	105	475	560	
	0560	15 m depth		10.70	2.991		355	89.50	111	555.50	650	
	0570	16.5 m depth		10	3.200		385	95.50	119	599.50	705	
	0580	18 m depth		9.40	3.404		415	102	127	644	755	
	0590	19.5 m depth		8.90	3.596		450	107	134	691	810	
	0600	21 m depth		8.40	3.810		480	114	142	736	860	
	0610	22.5 m depth	▼	8	4	▼	510	119	149	778	915	
	0690	Average Soil nailing, grade 500Mpa, 15 min setup per hole & 33 m/hr.										
	0700	Soil nailing, drill hole, install # 25 M nail, grout 6 m depth	B-47G	18.50	1.730	Ea.	93.50	51.50	64.50	209.50	254	
	0710	7.5 m depth		16.60	1.928		117	57.50	72	246.50	296	
	0720	9 m depth		15.50	2.065		191	61.50	77	329.50	390	
	0730	10.5 m depth		14.10	2.270		223	67.50	84.50	375	445	
	0740	12 m depth		13	2.462		254	73.50	91.50	419	495	
	0750	13.5 m depth		12	2.667		286	79.50	99.50	465	545	
	0760	15 m depth		11.40	2.807		355	84	105	544	635	
	0770	16.5 m depth		10.70	2.991		385	89.50	111	585.50	685	
	0780	18 m depth		10	3.200		415	95.50	119	629.50	740	
	0790	19.5 m depth		9.40	3.404		450	102	127	679	790	
	0800	21 m depth		9.10	3.516		480	105	131	716	835	
	0810	22.5 m depth	▼	8.60	3.721	▼	510	111	139	760	890	
	0890	Average Soil nailing, grade 500 Mpa, 15 min setup per hole & 36 m/hr.										
	0900	Soil nailing, drill hole, install # 25 M nail, grout 6 m depth	B-47G	19.20	1.667	Ea.	93.50	50	62	205.50	249	
	0910	7.5 m depth		17.10	1.871		117	56	70	243	291	
	0920	9 m depth		16	2		191	59.50	74.50	325	385	
	0930	10.5 m depth		14.60	2.192		223	65.50	81.50	370	435	
	0940	12 m depth		13.70	2.336		254	69.50	87	410.50	485	
	0950	13.5 m depth		12.60	2.540		286	76	94.50	456.50	535	
	0960	15 m depth		12	2.667		355	79.50	99.50	534	620	
	0970	16.5 m depth		11.20	2.857		385	85.50	106	576.50	675	
	0980	18 m depth		10.70	2.991		415	89.50	111	615.50	720	
	0990	19.5 m depth		10	3.200		450	95.50	119	664.50	775	
	1000	21 m depth		9.60	3.333		480	99.50	124	703.50	820	
	1010	22.5 m depth	▼	9.10	3.516	▼	510	105	131	746	870	
	1190	Difficult Soil nailing, grade 500Mpa, 20 min setup per hole & 24 m/hr.										
	1200	Soil nailing, drill hole, install # 25 M nail, grout 6 m depth	B-47G	13.70	2.336	Ea.	93.50	69.50	87	250	305	
	1210	7.5 m depth		12.30	2.602		117	77.50	97	291.50	355	
	1220	9 m depth		11.20	2.857		191	85.50	106	382.50	460	
	1230	10.5 m depth		10.40	3.077		223	92	115	430	515	
	1240	12 m depth		9.60	3.333		254	99.50	124	477.50	570	
	1250	13.5 m depth		8.90	3.596		286	107	134	527	630	
	1260	15 m depth		8.30	3.855		355	115	144	614	725	
	1270	16.5 m depth		7.90	4.051		385	121	151	657	780	
	1280	18 m depth		7.40	4.324		415	129	161	705	835	
	1290	19.5 m depth		7	4.571		450	136	170	756	895	
	1300	21 m depth		6.60	4.848		480	145	181	806	950	
	1310	22.5 m depth	▼	6.30	5.079	▼	510	152	189	851	1,000	
	1390	Difficult soil nailing, grade 500Mpa, 20 min setup per hole & 27 m/hr.										
	1400	Soil nailing, drill hole, install # 25 M nail, grout 6 m depth	B-47G	14.10	2.270	Ea.	93.50	67.50	84.50	245.50	300	
	1410	7.5 m depth	▼	13	2.462	▼	117	73.50	91.50	282	340	

Important: See the Reference Section for supporting data - Crews, Rental Equipment, City Cost Indexes and Reference Data

		02260	Excavation Support/Protection	CREW	DAILY OUTPUT	LABOR-HOURS	UNIT	2006 BARE COSTS				TOTAL INCL O&P	
								MAT.	LABOR	EQUIP.	TOTAL		
730	1420		9 m depth	B-47G	12	2.667	Ea.	191	79.50	99.50	370	440	730
	1430		10.5 m depth		10.90	2.936		223	87.50	109	419.50	500	
	1440		12 m depth		10.20	3.137		254	93.50	117	464.50	555	
	1450		13.5 m depth		9.60	3.333		286	99.50	124	509.50	605	
	1460		15 m depth		8.90	3.596		355	107	134	596	705	
	1470		16.5 m depth		8.40	3.810		385	114	142	641	755	
	1480		18 m depth		8	4		415	119	149	683	810	
	1490		19.5 m depth		7.50	4.267		450	127	159	736	865	
	1500		21 m depth		7.20	4.444		480	133	166	779	915	
	1510		22.5 m depth	▼	6.90	4.638	▼	510	138	173	821	970	
	1590		Difficult soil nailing, grade 500Mpa, 20 min setup per hole & 30 m/hr.										
	1600		Soil nailing, drill hole, install # 25 M nail, grout 6 m depth	B-47G	15	2.133	Ea.	93.50	63.50	79.50	236.50	289	
	1610		7.5 m depth		13.70	2.336		117	69.50	87	273.50	330	
	1620		9 m depth		12.60	2.540		191	76	94.50	361.50	430	
	1630		10.5 m depth		11.70	2.735		223	81.50	102	406.50	485	
	1640		12 m depth		10.90	2.936		254	87.50	109	450.50	535	
	1650		13.5 m depth		10.20	3.137		286	93.50	117	496.50	590	
	1660		15 m depth		9.60	3.333		355	99.50	124	578.50	680	
	1670		16.5 m depth		9.10	3.516		385	105	131	621	730	
	1680		18 m depth		8.60	3.721		415	111	139	665	785	
	1690		19.5 m depth		8.10	3.951		450	118	147	715	840	
	1700		21 m depth		7.70	4.156		480	124	155	759	890	
	1710		22.5 m depth	▼	7.40	4.324	▼	510	129	161	800	940	
	1790		Difficult soil nailing, grade 500Mpa, 20 min setup per hole & 33 m/hr.										
	1800		Soil nailing, drill hole, install # 25 M nail, grout 6 m depth	B-47G	15.50	2.065	Ea.	93.50	61.50	77	232	283	
	1810		7.5 m depth		14.10	2.270		117	67.50	84.50	269	325	
	1820		9 m depth		13.30	2.406		191	72	89.50	352.50	420	
	1830		10.5 m depth		12.30	2.602		223	77.50	97	397.50	470	
	1840		12 m depth		11.40	2.807		254	84	105	443	525	
	1850		13.5 m depth		10.70	2.991		286	89.50	111	486.50	575	
	1860		15 m depth		10.20	3.137		355	93.50	117	565.50	665	
	1870		16.5 m depth		9.60	3.333		385	99.50	124	608.50	715	
	1880		18 m depth		9.10	3.516		415	105	131	651	765	
	1890		19.5 m depth		8.60	3.721		450	111	139	700	820	
	1900		21 m depth		8.30	3.855		480	115	144	739	865	
	1910		22.5 m depth	▼	7.90	4.051	▼	510	121	151	782	920	
	1990		Difficult soil nailing, grade 500Mpa, 20 min setup per hole & 36 m/hr.										
	2000		Soil nailing, drill hole, install # 25 M nail, grout 6 m depth	B-47G	16	2	Ea.	93.50	59.50	74.50	227.50	277	
	2010		7.5 m depth		14.60	2.192		117	65.50	81.50	264	320	
	2020		9 m depth		13.70	2.336		191	69.50	87	347.50	415	
	2030		10.5 m depth		12.60	2.540		223	76	94.50	393.50	465	
	2040		12 m depth		12	2.667		254	79.50	99.50	433	510	
	2050		13.5 m depth		11.20	2.857		286	85.50	106	477.50	565	
	2060		15 m depth		10.70	2.991		355	89.50	111	555.50	650	
	2070		16.5 m depth		10	3.200		385	95.50	119	599.50	705	
	2080		18 m depth		9.60	3.333		415	99.50	124	638.50	750	
	2090		19.5 m depth		9.10	3.516		450	105	131	686	800	
	2100		21 m depth		8.70	3.678		480	110	137	727	850	
	2110		22.5 m depth	▼	8.30	3.855	▼	510	115	144	769	900	
	2990		Very difficult soil nailing, grade 500Mpa, 25 min setup & 24 m/hr.										
	3000		Soil nailing, drill ,install # 25 M nail, grout 6 m depth	B-47G	12	2.667	Ea.	93.50	79.50	99.50	272.50	335	
	3010		7.5 m depth		10.90	2.936		117	87.50	109	313.50	385	
	3020		9 m depth		10	3.200		191	95.50	119	405.50	490	
	3030		10.5 m depth		9.40	3.404		223	102	127	452	540	
	3040		12 m depth		8.70	3.678		254	110	137	501	600	
	3050		13.5 m depth	▼	8.10	3.951	▼	286	118	147	551	660	

SITE CONSTRUCTION 2

For expanded coverage of these items see Means Site Work & Landscape Cost Data 2006

02260	Excavation Support/Protection	CREW	DAILY OUTPUT	LABOR-HOURS	UNIT	2006 BARE COSTS				TOTAL INCL O&P
						MAT.	LABOR	EQUIP.	TOTAL	
3060	15 m depth	B-47G	7.60	4.211	Ea.	355	126	157	638	755
3070	16.5 m depth		7.30	4.384		385	131	163	679	805
3080	18 m depth		6.90	4.638		415	138	173	726	865
3090	19.5 m depth		6.50	4.923		450	147	183	780	925
3100	21 m depth		6.20	5.161		480	154	192	826	980
3110	22.5 m depth		5.90	5.424		510	162	202	874	1,025
3190	Very difficult soil nailing, grade 500Mpa, 25 min setup & 27 m/hr.									
3200	Soil nailing, drill ,install # 25 M nail, grout 6 m depth	B-47G	12.30	2.602	Ea.	93.50	77.50	97	268	330
3210	7.5 m depth		11.40	2.807		117	84	105	306	370
3220	9 m depth		10.70	2.991		191	89.50	111	391.50	470
3230	10.5 m depth		9.80	3.265		223	97.50	122	442.50	530
3240	12 m depth		9.20	3.478		254	104	130	488	585
3250	13.5 m depth		8.70	3.678		286	110	137	533	635
3260	15 m depth		8.10	3.951		355	118	147	620	735
3270	16.5 m depth		7.70	4.156		385	124	155	664	785
3280	18 m depth		7.40	4.324		415	129	161	705	835
3290	19.5 m depth		7	4.571		450	136	170	756	895
3300	21 m depth		6.70	4.776		480	143	178	801	945
3310	22.5 m depth		6.40	5		510	149	186	845	1,000
3390	Very difficult soil nailing, grade 500Mpa, 25 min setup & 30 m/hr.									
3400	Soil nailing, drill ,install # 25 M nail, grout 6 m depth	B-47G	13	2.462	Ea.	93.50	73.50	91.50	258.50	315
3410	7.5 m depth		12	2.667		117	79.50	99.50	296	360
3420	9 m depth		11.20	2.857		191	85.50	106	382.50	460
3430	10.5 m depth		10.40	3.077		223	92	115	430	515
3440	12 m depth		9.80	3.265		254	97.50	122	473.50	565
3450	13.5 m depth		9.20	3.478		286	104	130	520	620
3460	15 m depth		8.70	3.678		355	110	137	602	710
3470	16.5 m depth		8.30	3.855		385	115	144	644	760
3480	18 m depth		7.90	4.051		415	121	151	687	815
3490	19.5 m depth		7.50	4.267		450	127	159	736	865
3500	21 m depth		7.20	4.444		480	133	166	779	915
3510	22.5 m depth		6.90	4.638		510	138	173	821	970
3590	Very difficult soil nailing, grade 500Mpa, 25 min setup & 33 m/hr.									
3600	Soil nailing, drill ,install # 25 M nail, grout 6 m depth	B-47G	13.30	2.406	Ea.	93.50	72	89.50	255	315
3610	7.5 m depth		13.70	2.336		117	69.50	87	273.50	330
3620	9 m depth		11.70	2.735		191	81.50	102	374.50	450
3630	10.5 m depth		10.90	2.936		223	87.50	109	419.50	500
3640	12 m depth		10.20	3.137		254	93.50	117	464.50	555
3650	13.5 m depth		9.60	3.333		286	99.50	124	509.50	605
3660	15 m depth		9.20	3.478		355	104	130	589	695
3670	16.5 m depth		8.70	3.678		385	110	137	632	745
3680	18 m depth		8.30	3.855		415	115	144	674	795
3690	19.5 m depth		7.90	4.051		450	121	151	722	850
3700	21 m depth		7.60	4.211		480	126	157	763	895
3710	22.5 m depth		7.30	4.384		510	131	163	804	945
3790	Very difficult soil nailing, grade 500Mpa, 25 min setup & 36 m/hr.									
3800	Soil nailing, drill ,install # 25 M nail, grout 6 m depth	B-47G	13.70	2.336	Ea.	93.50	69.50	87	250	305
3810	7.5 m depth		12.60	2.540		117	76	94.50	287.50	350
3820	9 m depth		12	2.667		191	79.50	99.50	370	440
3830	10.5 m depth		11.20	2.857		223	85.50	106	414.50	495
3840	12 m depth		10.70	2.991		254	89.50	111	454.50	540
3850	13.5 m depth		10	3.200		286	95.50	119	500.50	595
3860	15 m depth		9.60	3.333		355	99.50	124	578.50	680
3870	16.5m depth		9.10	3.516		385	105	131	621	730
3880	18 m depth		8.70	3.678		415	110	137	662	780
3890	19.5 m depth		8.30	3.855		450	115	144	709	830

730

Important: See the Reference Section for supporting data - Crews, Rental Equipment, City Cost Indexes and Reference Data

	02260	Excavation Support/Protection	CREW	DAILY OUTPUT	LABOR-HOURS	UNIT	2006 BARE COSTS				TOTAL INCL O&P	
							MAT.	LABOR	EQUIP.	TOTAL		
730	3900	21 m depth	B-47G	8	4	Ea.	480	119	149	748	880	**730**
	3910	22.5 m depth	↓	7.60	4.211	↓	510	126	157	793	930	
	3990	Severe soil nailing, grade 500Mpa, 30 min setup per hole & 24 m/hr.										
	4000	Soil nailing , drill hole, install # 25 M nail, grout 6 m depth	B-47G	10.70	2.991	Ea.	93.50	89.50	111	294	365	
	4010	7.5 m depth		9.80	3.265		117	97.50	122	336.50	410	
	4020	9 m depth		9.10	3.516		191	105	131	427	515	
	4030	10.5 m depth		8.60	3.721		223	111	139	473	570	
	4040	12 m depth		8	4		254	119	149	522	630	
	4050	13.5 m depth		7.50	4.267		286	127	159	572	685	
	4060	15 m depth		7.10	4.507		355	135	168	658	785	
	4070	16.5 m depth		6.80	4.706		385	140	175	700	835	
	4080	18 m depth		6.40	5		415	149	186	750	895	
	4090	19.5 m depth		6.10	5.246		450	157	196	803	950	
	4100	21 m depth		5.80	5.517		480	165	206	851	1,000	
	4110	22.5 m depth	↓	5.60	5.714	↓	510	171	213	894	1,050	
	4190	Severe soil nailing, grade 500Mpa, 30 min setup per hole & 27 m/hr.										
	4200	Soil nailing , drill hole, install # 25 M nail, grout 6 m depth	B-47G	10.90	2.936	Ea.	93.50	87.50	109	290	360	
	4210	7.5 m depth		10.20	3.137		117	93.50	117	327.50	400	
	4220	9 m depth		9.60	3.333		191	99.50	124	414.50	500	
	4230	10.5 m depth		8.90	3.596		223	107	134	464	560	
	4240	12 m depth		8.40	3.810		254	114	142	510	610	
	4250	13.5 m depth		8	4		286	119	149	554	665	
	4260	15 m depth		7.50	4.267		355	127	159	641	760	
	4270	16.5 m depth		7.20	4.444		385	133	166	684	810	
	4280	18 m depth		6.90	4.638		415	138	173	726	865	
	4290	19.5 m depth		6.50	4.923		450	147	183	780	925	
	4300	21 m depth		6.20	5.161		480	154	192	826	980	
	4310	22.5 m depth	↓	6	5.333	↓	510	159	199	868	1,025	
	4390	Severe soil nailing, grade 500Mpa, 30 min setup per hole & 30 m/hr.										
	4400	Soil nailing , drill hole, install # 25 M nail, grout 6 m depth	B-47G	11.40	2.807	Ea.	93.50	84	105	282.50	345	
	4410	7.5 m depth		10.70	2.991		117	89.50	111	317.50	390	
	4420	9 m depth		10	3.200		191	95.50	119	405.50	490	
	4430	10.5 m depth		9.40	3.404		223	102	127	452	540	
	4440	12 m depth		8.90	3.596		254	107	134	495	595	
	4450	13.5 m depth		8.40	3.810		286	114	142	542	645	
	4460	15 m depth		8	4		355	119	149	623	740	
	4470	16.5 m depth		7.60	4.211		385	126	157	668	790	
	4480	18 m depth		7.30	4.384		415	131	163	709	840	
	4490	19.5 m depth		7	4.571		450	136	170	756	895	
	4500	21 m depth		6.70	4.776		480	143	178	801	945	
	4510	22.5 m depth	↓	6.40	5	↓	510	149	186	845	1,000	
	4590	Severe soil nailing, grade 500Mpa, 30 min setup per hole & 33 m/hr.										
	4600	Soil nailing , drill hole, install # 25 M nail, grout 6 m depth	B-47G	11.70	2.735	Ea.	93.50	81.50	102	277	340	
	4610	7.5 m depth		10.90	2.936		117	87.50	109	313.50	385	
	4620	9 m depth		10.40	3.077		191	92	115	398	480	
	4630	10.5 m depth		9.80	3.265		223	97.50	122	442.50	530	
	4640	12 m depth		9.20	3.478		254	104	130	488	585	
	4650	13.5 m depth		8.70	3.678		286	110	137	533	635	
	4660	15 m depth		8.40	3.810		355	114	142	611	720	
	4670	16.5 m depth		8	4		385	119	149	653	775	
	4680	18 m depth		7.60	4.211		415	126	157	698	825	
	4690	19.5 m depth		7.30	4.384		450	131	163	744	875	
	4700	21 m depth		7.10	4.507		480	135	168	783	925	
	4710	grout 22.5 m depth	↓	6.80	4.706	↓	510	140	175	825	975	
	4790	Severe soil nailing, grade 500Mpa, 30 min setup per hole & 36 m/hr.										
	4800	Soil nailing, drill hole, install # 25 M nail, grout 6 m depth	B-47G	12	2.667	Ea.	93.50	79.50	99.50	272.50	335	

SITE CONSTRUCTION 2

	02260	Excavation Support/Protection	CREW	DAILY OUTPUT	LABOR-HOURS	UNIT	2006 BARE COSTS				TOTAL INCL O&P	
							MAT.	LABOR	EQUIP.	TOTAL		
730	4810	7.5 m depth	B-47G	11.20	2.857	Ea.	117	85.50	106	308.50	375	730
	4820	grout 9 m depth		10.70	2.991		191	89.50	111	391.50	470	
	4830	10.5 m depth		10	3.200		223	95.50	119	437.50	525	
	4840	12 m depth		9.60	3.333		254	99.50	124	477.50	570	
	4850	13.5 m depth		9.10	3.516		286	105	131	522	620	
	4860	15 m depth		8.70	3.678		355	110	137	602	710	
	4870	16.5 m depth		8.30	3.855		385	115	144	644	760	
	4880	18 m depth		8	4		415	119	149	683	810	
	4890	19.5 m depth		7.60	4.211		450	126	157	733	860	
	4900	21 m depth		7.40	4.324		480	129	161	770	905	
	4910	22.5 m depth		7.10	4.507	m²	510	135	168	813	960	

	02305	Equipment	CREW	DAILY OUTPUT	LABOR-HOURS	UNIT	2006 BARE COSTS				TOTAL INCL O&P	
							MAT.	LABOR	EQUIP.	TOTAL		
250	0010	**MOBILIZATION OR DEMOB.** (One or the other, unless noted) R015433-10										250
	0015	Up to 40 km haul dist (80 km RT for mob/demob crew)										
	0020	Dozer, loader, backhoe, excav., grader, paver, roller, 52 to 112 kW	B-34N	4	2	Ea.		56.50	112	168.50	210	
	0100	Above 112 kW	B-34K	3	2.667			75.50	175	250.50	305	
	0300	Scraper, towed type (incl. tractor), 4.58 m³ capacity		3	2.667			75.50	175	250.50	305	
	0400	7.63 m³		2.50	3.200			90.50	210	300.50	370	
	0600	Self-propelled scraper, 11.46 m³		2.50	3.200			90.50	210	300.50	370	
	0700	18.34 m³		2	4			113	262	375	460	
	0900	Shovel or dragline, 0.57 m³		3.60	2.222			63	146	209	256	
	1000	1.15 m³		3	2.667			75.50	175	250.50	305	
	1100	Small equipment, placed in rear of, or towed by pickup truck	A-3A	8	1			27.50	10.65	38.15	53.50	
	1150	Equip up to 52 kW, on flatbed trailer behind pickup truck	A-3D	4	2			55	43.50	98.50	132	
	2000	Crane, truck-mtd, up to 68 metric ton , driver only	1 Eqhv	3.60	2.222			84.50		84.50	128	
	2100	Crane, tuck-mounted, over 68 metric ton	A-3E	2.50	6.400			213	34	247	365	
	2200	Crawler-mounted, up to 68 metric ton	A-3F	2	8			266	278	544	710	
	2300	Over 68 metric ton	A-3G	1.50	10.667			355	405	760	985	
	2500	For each additional 8 km haul distance, add						10%	10%			
	3000	For large pieces of equipment, allow for assembly/knockdown										
	3001	For mob/demob of vibrofloatation equip, see section 02250-900										
	3100	For mob/demob of micro-tunneling equip, see section 02441-400										
	3200	For mob/demob of pile driving equip, see section 02455-650										
	3300	For mob/demob of caisson drilling equip, see section 02465-950										

	02310	Grading	CREW	DAILY OUTPUT	LABOR-HOURS	UNIT	MAT.	LABOR	EQUIP.	TOTAL	INCL O&P	
100	0010	**FINISH GRADING**										100
	0012	Finish grading area to be paved with grader, small area	B-11L	334	.048	m²		1.54	1.37	2.91	3.86	
	0100	Large area		1672	.010			.31	.27	.58	.77	
	0200	Grade subgrade for base course, roadways		2926	.005			.18	.16	.34	.44	
	1020	For large parking lots	B-32C	4181	.011			.37	.38	.75	.98	
	1050	For small irregular areas	"	1672	.029			.93	.94	1.87	2.46	
	1100	Fine grade for slab on grade, machine	B-11L	870	.018			.59	.53	1.12	1.48	
	1150	Hand grading	B-18	585	.041			1.15	.06	1.21	1.86	
	1200	Fine grade granular base for sidewalks and bikeways	B-62	1003	.024			.72	.15	.87	1.27	
	2550	Hand grade select gravel	2 Clab	557	.029			.79		.79	1.23	

02310	Grading	CREW	DAILY OUTPUT	LABOR-HOURS	UNIT	2006 BARE COSTS				TOTAL INCL O&P	
						MAT.	LABOR	EQUIP.	TOTAL		
100 3000	Hand grade select gravel, including compaction, 100 mm deep	B-18	464	.052	m²		1.45	.07	1.52	2.34	**100**
3100	150 mm deep		334	.072			2.02	.10	2.12	3.25	
3120	200 mm deep		251	.096			2.68	.14	2.82	4.33	
3300	Finishing grading slopes, gentle	B-11L	7441	.002			.07	.06	.13	.18	
3310	Steep slopes	"	5936	.003			.09	.08	.17	.21	

02315	Excavation and Fill										
110 0010	**BACKFILL, GENERAL**	R312323 -30									**110**
0015	By hand, no compaction, light soil	1 Clab	10.70	.747	Lm³		20.50		20.50	32	
0100	Heavy soil		8.41	.951	"		26		26	40.50	
0300	Compaction in 150 mm layers, hand tamp, add to above		15.75	.508	Em³		13.90		13.90	21.50	
0400	Roller compaction operator walking, add	B-10A	76.46	.157			5.25	1.64	6.89	9.85	
0500	Air tamp, add	B-9D	145	.276			7.65	1.28	8.93	13.35	
0600	Vibrating plate, add	A-1D	45.88	.174			4.78	.58	5.36	8.10	
0800	Compaction in 300 mm layers, hand tamp, add to above	1 Clab	26	.308			8.45		8.45	13.10	
0900	Roller compaction operator walking, add	B-10A	115	.104			3.51	1.09	4.60	6.55	
1000	Air tamp, add	B-9	218	.183			5.10	.68	5.78	8.70	
1100	Vibrating plate, add	A-1E	68.81	.116			3.19	.50	3.69	5.50	
1300	Dozer backfilling, bulk, up to 90 m haul, no compaction	B-10B	918	.013	Lm³		.44	1	1.44	1.77	
1400	Air tamped, add	B-11B	61.17	.262	Em³		8.20	3.21	11.41	16.05	
1600	Compacting backfill, 150 mm to 300 mm lifts, vibrating roller	B-10C	612	.020			.66	2.25	2.91	3.48	
1700	Sheepsfoot roller	B-10D	573	.021			.70	2.44	3.14	3.75	
1900	Dozer backfilling, trench, up to 90 m haul, no compaction	B-10B	688	.017	Lm³		.59	1.34	1.93	2.36	
2000	Air tamped, add	B-11B	61.17	.262	Em³		8.20	3.21	11.41	16.05	
2200	Compacting backfill, 150 mm to 300 mm lifts, vibrating roller	B-10C	535	.022			.75	2.58	3.33	3.98	
2300	Sheepsfoot roller	B-10D	497	.024			.81	2.81	3.62	4.33	
2350	Spreading in 203 mm layers, small dozer	B-10B	810	.015	Lm³		.50	1.14	1.64	2.01	
3000	For flowable fill, see div. 03310-220										
120 0010	**BACKFILL, STRUCTURAL** Dozer or F.E. loader										**120**
0020	From existing stockpile, no compaction										
2000	60 kW, 15 m haul, sand & gravel	B-10L	841	.014	Lm³		.48	.38	.86	1.15	
2020	Common earth		745	.016			.54	.43	.97	1.29	
2040	Clay		650	.018			.62	.49	1.11	1.48	
2200	45 m haul, sand & gravel		421	.029			.96	.76	1.72	2.29	
2220	Common earth		375	.032			1.08	.85	1.93	2.58	
2240	Clay		325	.037			1.24	.98	2.22	2.97	
2400	90 m haul, sand & gravel		283	.042			1.42	1.13	2.55	3.41	
2420	Common earth		252	.048			1.60	1.27	2.87	3.83	
2440	Clay		222	.054			1.82	1.44	3.26	4.34	
3000	78 kW, 15 m haul, sand & gravel	B-10W	1032	.012			.39	.45	.84	1.08	
3020	Common earth		937	.013			.43	.49	.92	1.20	
3040	Clay		841	.014			.48	.55	1.03	1.34	
3200	45 m haul, sand & gravel		512	.023			.79	.90	1.69	2.19	
3220	Common earth		466	.026			.87	.99	1.86	2.41	
3240	Clay		421	.029			.96	1.10	2.06	2.67	
3300	90 m haul, sand & gravel		356	.034			1.13	1.30	2.43	3.15	
3320	Common earth		317	.038			1.27	1.46	2.73	3.55	
3340	Clay		283	.042			1.42	1.64	3.06	3.97	
4000	149 kW, 15 m haul, sand & gravel	B-10B	1912	.006			.21	.48	.69	.85	
4020	Common earth		1682	.007			.24	.55	.79	.96	
4040	Clay		1491	.008			.27	.62	.89	1.09	
4200	45 m haul, sand & gravel		937	.013			.43	.98	1.41	1.74	
4220	Common earth		841	.014			.48	1.09	1.57	1.93	
4240	Clay		745	.016			.54	1.24	1.78	2.18	

SITE CONSTRUCTION 2

		02315	Excavation and Fill	CREW	DAILY OUTPUT	LABOR-HOURS	UNIT	2006 BARE COSTS				TOTAL INCL O&P	
								MAT.	LABOR	EQUIP.	TOTAL		
120	4400		90 m haul, sand & gravel	B-10B	616	.019	Lm³		.65	1.49	2.14	2.64	120
	4420		Common earth		562	.021			.72	1.64	2.36	2.89	
	4440		Clay	↓	505	.024			.80	1.82	2.62	3.23	
	5000	224 kW, 15 m haul, sand & gravel		B-10M	2424	.005			.17	.49	.66	.79	
	5020		Common earth		2217	.005			.18	.54	.72	.87	
	5040		Clay		2064	.006			.20	.58	.78	.94	
	5200	45 m haul, sand & gravel			1682	.007			.24	.71	.95	1.14	
	5220		Common earth		1491	.008			.27	.80	1.07	1.29	
	5240		Clay		1300	.009			.31	.92	1.23	1.48	
	5400	90 m haul, sand & gravel			1147	.010			.35	1.04	1.39	1.69	
	5420		Common earth		1032	.012			.39	1.16	1.55	1.87	
	5440		Clay	↓	937	.013	↓		.43	1.28	1.71	2.07	
	6000	For compaction, see div. 02315-310											
	6010	For trench backfill, see div. 02315-610 & 02315-620											
	6100	For compaction, see div. 02315-320											
210	0010	**BORROW, LOADING AND/OR SPREADING**											210
	4000	Common earth, shovel, 0.76 m³ bucket		B-12N	642	.025	Bm³	10.80	.82	1.46	13.08	14.75	
	4010		1.15 m³ bucket	B-12O	868	.018		10.80	.60	1.16	12.56	14.10	
	4020		2.29 m³ bucket	B-12T	1376	.012	↓	10.80	.38	.96	12.14	13.55	
	4030	Front end loader, wheel mounted											
	4050		0.57 m³ bucket	B-10R	421	.029	Bm³	10.80	.96	.46	12.22	13.85	
	4060		1.15 m³ bucket	B-10S	742	.016		10.80	.54	.33	11.67	13.10	
	4070		2.29 m³ bucket	B-10T	1204	.010		10.80	.34	.25	11.39	12.70	
	4080		3.82 m³ bucket	B-10U	1988	.006		10.80	.20	.35	11.35	12.60	
	5000	Select granular fill, shovel, 0.76 m³ bucket		B-12N	707	.023		10.50	.74	1.33	12.57	14.15	
	5010		1.15 m³ bucket	B-12O	956	.017		10.50	.55	1.05	12.10	13.55	
	5020		2.29 m³ bucket	B-12T	1514	.011	↓	10.50	.35	.87	11.72	13.05	
	5030	Front end loader, wheel mounted											
	5050		0.57 m³ bucket	B-10R	612	.020	Bm³	10.50	.66	.32	11.48	12.90	
	5060		1.15 m³ bucket	B-10S	814	.015		10.50	.50	.30	11.30	12.65	
	5070		2.29 m³ bucket	B-10T	1327	.009		10.50	.30	.23	11.03	12.25	
	5080		3.82 m³ bucket	B-10U	2179	.006		10.50	.19	.32	11.01	12.20	
	6000	Clay, till, or blasted rock, shovel, 0.76 m³ bucket		B-12N	547	.029		8	.96	1.72	10.68	12.15	
	6010		1.15 m³ bucket	B-12O	738	.022		8	.71	1.36	10.07	11.40	
	6020		2.29 m³ bucket	B-12T	1170	.014	↓	8	.45	1.13	9.58	10.75	
	6030	Front end loader, wheel mounted											
	6035		0.57 m³ bucket	B-10R	356	.034	Bm³	8	1.13	.55	9.68	11.10	
	6040		1.15 m³ bucket	B-10S	631	.019		8	.64	.38	9.02	10.20	
	6045		2.29 m³ bucket	B-10T	1025	.012		8	.39	.30	8.69	9.75	
	6050		3.82 m³ bucket	B-10U	1682	.007	↓	8	.24	.42	8.66	9.60	
	6060	Front end loader, track mounted											
	6065		1.15 m³ bucket	B-10N	547	.022	Bm³	8	.74	.58	9.32	10.55	
	6070		2.29 m³ bucket	B-10P	910	.013		8	.44	.87	9.31	10.45	
	6075		3.82 m³ bucket	B-10Q	1403	.009		8	.29	.79	9.08	10.10	
	7000	Topsoil or loam from stockpile, shovel, 0.76 m³ bucket		B-12N	642	.025		27.50	.82	1.46	29.78	33	
	7010		1.15 m³ bucket	B-12O	868	.018		27.50	.60	1.16	29.26	32	
	7020		2.29 m³ bucket	B-12T	1376	.012	↓	27.50	.38	.96	28.84	31.50	
	7030	Front end loader, wheel mounted											
	7050		0.57 m³ bucket	B-10R	421	.029	Bm³	27.50	.96	.46	28.92	32	
	7060		1.15 m³ bucket	B-10S	742	.016		27.50	.54	.33	28.37	31	
	7070		2.29 m³ bucket	B-10T	1204	.010		27.50	.34	.25	28.09	31	
	7080		3.82 m³ bucket	B-10U	1988	.006	↓	27.50	.20	.35	28.05	30.50	
	8900	For larger hauling units, deduct from above									30%		
	9000	Hauling only, excavated or borrow material, see Div. 02315-490											
	9200	For flowable fill, see section 03310-220											

		CREW	DAILY OUTPUT	LABOR-HOURS	UNIT	2006 BARE COSTS				TOTAL INCL O&P	
	02315	Excavation and Fill				MAT.	LABOR	EQUIP.	TOTAL		
310	**0010**	**COMPACTION, GENERAL** R312323 -30									**310**
	5000	Riding, vibrating roller, 150 mm lifts, 2 passes	B-10Y	2294	.005	Em³		.18	.18	.36	.47
	5020	3 passes		1759	.007			.23	.24	.47	.61
	5040	4 passes		1453	.008			.28	.29	.57	.74
	5060	300 mm lifts, 2 passes		3976	.003			.10	.11	.21	.27
	5080	3 passes		2676	.004			.15	.16	.31	.40
	5100	4 passes		1988	.006			.20	.21	.41	.54
	5600	Sheepsfoot or wobbly wheel roller, 150 mm lifts, 2 passes	B-10G	1835	.007			.22	.46	.68	.84
	5620	3 passes		1327	.009			.30	.64	.94	1.16
	5640	4 passes		994	.012			.41	.85	1.26	1.56
	5680	300 mm lifts, 2 passes		3976	.003			.10	.21	.31	.38
	5700	3 passes		2676	.004			.15	.32	.47	.58
	5720	4 passes		1988	.006			.20	.43	.63	.78
	6000	Towed sheepsfoot or wobbly wheel roller, 150 mm lifts, 2 passes	B-10D	7646	.002			.05	.18	.23	.28
	6020	3 passes		1529	.008			.26	.92	1.18	1.41
	6030	4 passes		1147	.010			.35	1.22	1.57	1.88
	6050	300 mm lifts, 2 passes		4588	.003			.09	.31	.40	.47
	6060	3 passes		3058	.004			.13	.46	.59	.70
	6070	4 passes		2294	.005			.18	.61	.79	.94
	6200	Vibrating roller, 150 mm lifts, 2 passes	B-10C	1988	.006			.20	.69	.89	1.07
	6210	3 passes		1327	.009			.30	1.04	1.34	1.60
	6220	4 passes		994	.012			.41	1.39	1.80	2.15
	6250	300 mm lifts, 2 passes		3976	.003			.10	.35	.45	.53
	6260	3 passes		2649	.005			.15	.52	.67	.80
	6270	4 passes		1988	.006			.20	.69	.89	1.07
	7000	Walk behind, vibrating plate 450 mm wide, 150 mm lifts, 2 passes	A-1D	153	.052			1.43	.17	1.60	2.42
	7020	3 passes		141	.057			1.55	.19	1.74	2.63
	7040	4 passes		107	.075			2.05	.25	2.30	3.46
	7200	300 mm lifts, 2 passes	A-1E	428	.019			.51	.08	.59	.89
	7220	3 passes		287	.028			.76	.12	.88	1.32
	7240	4 passes		214	.037			1.02	.16	1.18	1.77
	7500	Vibrating roller 600 mm wide, 150 mm lifts, 2 passes	B-10A	321	.037			1.26	.39	1.65	2.34
	7520	3 passes		214	.056			1.88	.59	2.47	3.52
	7540	4 passes		161	.075			2.50	.78	3.28	4.67
	7600	300 mm lifts, 2 passes		642	.019			.63	.20	.83	1.18
	7620	3 passes		428	.028			.94	.29	1.23	1.75
	7640	4 passes		321	.037			1.26	.39	1.65	2.34
	8000	Rammer tamper, 150 mm to 280 mm, 100 mm lifts, 2 passes	A-1F	99.40	.080			2.21	.37	2.58	3.84
	8050	3 passes		74.17	.108			2.96	.50	3.46	5.15
	8100	4 passes		49.70	.161			4.41	.75	5.16	7.65
	8200	200 mm lifts, 2 passes		199	.040			1.10	.19	1.29	1.91
	8250	3 passes		149	.054			1.47	.25	1.72	2.56
	8300	4 passes		99.40	.080			2.21	.37	2.58	3.84
	8400	300 mm to 450 mm, 100 mm lifts, 2 passes	A-1G	298	.027			.74	.12	.86	1.29
	8450	3 passes		222	.036			.99	.17	1.16	1.72
	8500	4 passes		149	.054			1.47	.25	1.72	2.56
	8600	200 mm lifts, 2 passes		596	.013			.37	.06	.43	.64
	8650	3 passes		447	.018			.49	.08	.57	.85
	8700	4 passes		298	.027			.74	.12	.86	1.29
	9000	Water, 11 355 L truck, 4.8 km haul	B-45	1444	.011		.26	.36	.37	.99	1.25
	9010	9.7 km haul		1104	.014		.26	.47	.48	1.21	1.54
	9020	19.3 km haul		765	.021		.26	.68	.70	1.64	2.09
	9030	22 710 L wagon, 4.8 km haul	B-59	1529	.005		.26	.15	.21	.62	.75
	9040	9.7 km haul	"	1223	.007		.26	.19	.27	.72	.86
320	**0010**	**COMPACTION, STRUCTURAL** R312323 -30									**320**
	0020	Steel wheel tandem roller, 4.5 t	B-10E	8	1.500	Hr.		50.50	14.30	64.80	92

02315 | Excavation and Fill

			CREW	DAILY OUTPUT	LABOR-HOURS	UNIT	2006 BARE COSTS				TOTAL INCL O&P	
							MAT.	LABOR	EQUIP.	TOTAL		
320	0100	9.1 metric ton	B-10F	8	1.500	Hr.		50.50	24.50	75	104	320
	0300	Sheepsfoot or wobbly whl. roller, 200 mm lifts, common fill	B-10G	994	.012	Em³		.41	.85	1.26	1.56	
	0400	Select fill	"	1147	.010			.35	.74	1.09	1.35	
	0600	Vibratory plate, 200 mm lifts, common fill	A-1D	153	.052			1.43	.17	1.60	2.42	
	0700	Select fill	"	165	.048			1.33	.16	1.49	2.25	
412	0010	**DRILLING ONLY ROCK**										412
	0015	Drilling only rock, 50 mm hole for rock bolts	B-47	96.32	.249	m		7.65	11.30	18.95	24	
	0800	65 mm hole for pre-splitting		183	.131			4.02	5.95	9.97	12.75	
	4600	Quarry operations, 65 mm to 90 mm diameter [2-1/2", 3-1/2"]		218	.110			3.38	4.98	8.36	10.70	
	4610	150 mm diameter drill holes [6"]	B-47A	411	.058			1.95	2	3.95	5.15	
416	0010	**DRILLING AND BLASTING ROCK**										416
	0020	Rock, open face, under 1000 m³	B-47	172	.140	Bm³	2.67	4.28	6.30	13.25	16.50	
	0100	Over 1000 m³		229	.105		2.67	3.21	4.74	10.62	13.05	
	0200	Areas where blasting mats are required, under 1000 m³		134	.179		2.67	5.50	8.10	16.27	20.50	
	0250	Over 1000 m³		191	.126		2.67	3.85	5.70	12.22	15.15	
	0300	Bulk drilling and blasting, can vary greatly, average				m³					7.20	
	0500	Pits, average				"					28.50	
	1300	Deep hole method, up to 1000 m³	B-47	38.23	.628	Bm³	2.67	19.25	28.50	50.42	64	
	1400	Over 1000 m³		50.46	.476		2.67	14.60	21.50	38.77	49	
	1900	Restricted areas, up to 1000 m³		9.94	2.415		2.67	74	109	185.67	237	
	2000	Over 1000 m³		15.29	1.569		2.67	48	71	121.67	155	
	2200	Trenches, up to 1000 m³		16.82	1.427		7.75	44	64.50	116.25	147	
	2300	Over 1000 m³		19.88	1.207		7.75	37	54.50	99.25	126	
	2500	Pier holes, up to 1000 m³		16.82	1.427		2.67	44	64.50	111.17	141	
	2600	Over 1000 m³		23.70	1.013		2.67	31	46	79.67	101	
	2800	Boulders under 0.38 m³, loaded on truck, no hauling	B-100	61.17	.196			6.60	9.10	15.70	20	
	2900	Boulders, drilled, blasted	B-47	76.46	.314		2.67	9.65	14.20	26.52	33.50	
	3100	Jackhammer operators with foreman, compressor, air tools	B-9	1	40	Day		1,100	147	1,247	1,875	
	3300	Track drill, compressor, operator and foreman	B-47	1	24	"		735	1,075	1,810	2,325	
	3500	Blasting caps				Ea.	3.70			3.70	4.07	
	3700	Explosives					.31			.31	.34	
	3900	Blasting mats, rent, for first day					102			102	112	
	4000	Per added day					34			34	37.50	
	4200	Preblast survey for 6 room house, individual lot, minimum	A-6	2.40	6.667			236	24.50	260.50	385	
	4300	Maximum	"	1.35	11.852			420	43	463	695	
	4500	City block within zone of influence, minimum	A-8	2341	.014	m²		.52	.02	.54	.82	
	4600	Maximum	"	1403	.023	"		.86	.04	.90	1.37	
	5000	Excavate and load boulders, less than 0.38 m³	B-10T	61.17	.196	Bm³		6.60	4.99	11.59	15.55	
	5020	0.38 m³ to 0.76 m³	B-10U	76.46	.157			5.25	9.15	14.40	18.15	
	5200	Excavate and load blasted rock, 2.29 m³ power shovel	B-12T	1170	.014			.45	1.13	1.58	1.93	
	5400	Haul boulders, 23 metric ton off-highway dump, 1.6 km round trip	B-34E	252	.032			.90	3.62	4.52	5.35	
	5420	3.2 km round trip		210	.038			1.08	4.34	5.42	6.45	
	5440	4.8 km round trip		172	.047			1.32	5.30	6.62	7.85	
	5460	6.4 km round trip		153	.052			1.48	5.95	7.43	8.80	
	5600	Bury boulders on site, less than 0.38 m³, 224 kW dozer										
	5620	45 m haul	B-10M	237	.051	Bm³		1.70	5.05	6.75	8.15	
	5640	90 m haul		161	.075			2.50	7.45	9.95	12	
	5800	0.5 to 0.76 m³, 224 kW dozer, 45 m haul		229	.052			1.76	5.25	7.01	8.45	
	5820	90 m haul		153	.078			2.64	7.80	10.44	12.60	
418	0010	**RIPPING**										418
	0020	Ripping, trap rock, soft, 224 kW dozer, ideal conditions	B-11S	535	.022	Bm³		.75	2.37	3.12	3.76	
	1500	Adverse conditions		505	.024			.80	2.52	3.32	3.99	
	1600	Medium hard, 224 kW dozer, ideal conditons		459	.026			.88	2.77	3.65	4.38	

Reference boxes: R312323 -30, R312316 -40, R312316 -45

Important: See the Reference Section for supporting data - Crews, Rental Equipment, City Cost Indexes and Reference Data

02315 | Excavation and Fill

			CREW	DAILY OUTPUT	LABOR-HOURS	UNIT	MAT.	LABOR	EQUIP.	TOTAL	TOTAL INCL O&P	
								2006 BARE COSTS				
418	1700	Adverse conditions	B-11S	413	.029	Bm³		.98	3.08	4.06	4.87	**418**
	2000	Very hard, 306 kW dozer, ideal conditions	B-11T	268	.045			1.50	6.05	7.55	8.95	
	2100	Adverse conditions	"	237	.051			1.70	6.85	8.55	10.10	
	2200	Shale, soft, 224 kW dozer, ideal conditons	B-11S	1147	.010			.35	1.11	1.46	1.76	
	2300	Adverse conditions	"	1032	.012			.39	1.23	1.62	1.94	
	2310	Grader rear ripper, 134 kW ideal conditions	B-11J	566	.028			.91	.94	1.85	2.42	
	2320	Adverse conditions	"	482	.033			1.06	1.10	2.16	2.84	
	2400	Medium hard, 224 kW dozer, ideal conditons	B-11S	918	.013			.44	1.38	1.82	2.19	
	2500	Adverse conditions	"	826	.015			.49	1.54	2.03	2.43	
	2510	Grader rear ripper, 134 kW ideal conditions	B-11J	478	.033			1.07	1.11	2.18	2.86	
	2520	Adverse conditions	"	405	.040			1.27	1.31	2.58	3.38	
	2600	Very hard, 306 kW dozer, ideal conditons	B-11T	612	.020			.66	2.64	3.30	3.91	
	2700	Adverse conditions	"	551	.022			.73	2.94	3.67	4.34	
	2800	Till, boulder clay/hardpan, soft, 224 kW dozer, ideal conditions	B-11S	5352	.002			.08	.24	.32	.37	
	2810	Adverse conditions	"	4817	.002			.08	.26	.34	.42	
	2815	Grader rear ripper, 134 kW ideal conditions	B-11J	1147	.014			.45	.46	.91	1.19	
	2816	Adverse conditions	"	975	.016			.53	.54	1.07	1.40	
	2820	Medium hard, 224 kW dozer, ideal conditions	B-11S	4588	.003			.09	.28	.37	.44	
	2830	Adverse conditions	"	4129	.003			.10	.31	.41	.49	
	2835	Grader rear ripper, 134 kW ideal conditions	B-11J	566	.028			.91	.94	1.85	2.42	
	2836	Adverse conditions	"	482	.033			1.06	1.10	2.16	2.84	
	2840	Very hard, 306 kW dozer, ideal conditions	B-11T	3823	.003			.11	.42	.53	.63	
	2850	Adverse conditions	"	3441	.003			.12	.47	.59	.70	
	3000	Dozing ripped material, 149 kW, 30 m haul	B-10B	535	.022			.75	1.72	2.47	3.04	
	3050	90 m haul	"	191	.063			2.11	4.82	6.93	8.50	
	3200	224 kW, 30 m haul	B-10M	879	.014			.46	1.36	1.82	2.20	
	3250	90 m haul	"	306	.039			1.32	3.91	5.23	6.30	
	3400	306 kW, 30 m haul	B-10X	1285	.009			.31	1.20	1.51	1.80	
	3450	90 m haul	"	459	.026			.88	3.35	4.23	5	
424	0010	**EXCAVATING, BULK BANK MEASURE** Common earth piled										**424**
	0020	For loading onto trucks, add								15%	15%	
	0050	For mobilization and demobilization, see Division 02305-250										
	0100	For hauling, see Division 02315-490										
	0200	Backhoe, hydraulic, crawler mtd., 0.76 m³ cap. = 57 m³/hr	B-12A	459	.035	Bm³		1.14	1.22	2.36	3.09	
	0250	1.15 m³ cap. = 75 m³/hr	B-12B	612	.026			.86	1.18	2.04	2.61	
	0260	1.53 m³ cap. = 100 m³/hr	B-12C	795	.020			.66	1.16	1.82	2.28	
	0300	2.29 m³ cap. = 120 m³/hr	B-12D	979	.016			.54	2.07	2.61	3.09	
	0310	Wheel mounted, 0.38 m³ cap. = 23 m³/hr	B-12E	184	.087			2.85	1.82	4.67	6.35	
	0360	0.57 m³ cap. = 34 m³/hr	B-12F	275	.058			1.91	1.74	3.65	4.82	
	0500	Clamshell, 0.38 m³ cap. = 15 m³/hr	B-12G	122	.131			4.30	4.66	8.96	11.70	
	0550	0.76 m³ cap. = 27 m³/hr	B-12H	214	.075			2.45	4.32	6.77	8.50	
	0950	Dragline, 0.38 m³ cap. = 23 m³/hr	B-12I	184	.087			2.85	3.71	6.56	8.45	
	1000	0.57 m³ cap. = 27 m³/hr	"	214	.075			2.45	3.19	5.64	7.25	
	1050	1.15 m³ cap. = 50 m³/hr	B-12P	398	.040			1.32	2.43	3.75	4.68	
	1100	2.29 m³ cap. = 85 m³/hr	B-12V	688	.023			.76	1.85	2.61	3.20	
	1200	Front end loader, track mtd., 1.15 m³ cap. = 55 m³/hr	B-10N	428	.028			.94	.74	1.68	2.24	
	1250	1.91 m³ cap. = 75 m³/hr	B-10O	581	.021			.69	.96	1.65	2.11	
	1300	2.29 m³ cap. = 100 m³/hr	B-10P	795	.015			.51	1	1.51	1.87	
	1350	3.82 m³ cap. = 120 m³/hr	B-10Q	979	.012			.41	1.13	1.54	1.88	
	1500	Wheel mounted, 0.57 m³ cap. = 35 m³/hr	B-10R	275	.044			1.47	.71	2.18	3.01	
	1550	1.15 m³ cap. = 60 m³/hr	B-10S	489	.025			.82	.49	1.31	1.80	
	1600	1.72 m³ cap. = 75 m³/hr	B-10T	612	.020			.66	.50	1.16	1.55	
	1601	2.29 m³ cap. = 105 m³/hr	"	856	.014			.47	.36	.83	1.11	
	1650	3.82 m³ cap. = 141 m³/hr	B-10U	1132	.011			.36	.62	.98	1.22	
	1800	Hydraulic excavator, truck mtd, 0.38 m³ = 23 m³/hr	B-12J	184	.087			2.85	4.57	7.42	9.40	

R312316-45
R312316-40
R312316-45

SITE CONSTRUCTION 2

02315 | Excavation and Fill

				DAILY	LABOR-		2006 BARE COSTS				TOTAL	
			CREW	OUTPUT	HOURS	UNIT	MAT.	LABOR	EQUIP.	TOTAL	INCL O&P	
424	1850	1219 mm bucket, 0.76 m³ = 35 m³/hr	R312316 -40	B-12K	275	.058	Bm³		1.91	3.54	5.45	6.80	**424**
	3700	Shovel, 0.38 m³ capacity = 40 m³/hr		B-12L	336	.048			1.56	1.74	3.30	4.29	
	3750	0.57 m³ capacity = 65 m³/hr	R312316 -45	B-12M	520	.031			1.01	1.38	2.39	3.06	
	3800	0.76 m³ capacity = 90 m³/hr		B-12N	734	.022			.71	1.28	1.99	2.50	
	3850	1.15 m³ capacity = 120 m³/hr		B-120	979	.016			.54	1.02	1.56	1.95	
	3900	2.29 m³ cap. = 190 m³/hr		B-12T	1529	.010			.34	.86	1.20	1.47	
	4000	For soft soil or sand, deduct					m³				15%	15%	
	4100	For heavy soil or stiff clay, add									60%	60%	
	4200	For wet excavation with clamshell or dragline, add									100%	100%	
	4250	All other equipment, add									50%	50%	
	4400	Clamshell in sheeting or cofferdam, minimum		B-12H	122	.131	Bm³		4.30	7.55	11.85	14.90	
	4450	Maximum		"	45.88	.349	"		11.40	20	31.40	39.50	
	8000	For hauling excavated material, see Div. 02315-490											
432	0010	**EXCAVATING, BULK, DOZER** Open site	R312316 -40										**432**
	2000	60 kW, 15 m haul, sand & gravel		B-10L	352	.034	Bm³		1.15	.91	2.06	2.74	
	2020	Common earth			306	.039			1.32	1.04	2.36	3.16	
	2040	Clay			191	.063			2.11	1.67	3.78	5.05	
	2200	45 m haul, sand & gravel			176	.068			2.29	1.81	4.10	5.50	
	2220	Common earth			153	.078			2.64	2.08	4.72	6.30	
	2240	Clay			95.58	.126			4.22	3.34	7.56	10.05	
	2400	90 m haul, sand & gravel			91.75	.131			4.39	3.48	7.87	10.50	
	2420	Common earth			76.46	.157			5.25	4.17	9.42	12.65	
	2440	Clay			49.70	.241			8.10	6.40	14.50	19.40	
	3000	78 kW, 15 m haul, sand & gravel		B-10W	535	.022			.75	.87	1.62	2.10	
	3020	Common earth			466	.026			.87	.99	1.86	2.41	
	3040	Clay			294	.041			1.37	1.57	2.94	3.82	
	3200	45 m haul, sand & gravel			237	.051			1.70	1.95	3.65	4.74	
	3220	Common earth			206	.058			1.96	2.25	4.21	5.45	
	3240	Clay			130	.092			3.10	3.56	6.66	8.65	
	3300	90 m haul, sand & gravel			107	.112			3.77	4.33	8.10	10.50	
	3320	Common earth			91.75	.131			4.39	5.05	9.44	12.25	
	3340	Clay			76.46	.157			5.25	6.05	11.30	14.70	
	4000	149 kW, 15 m haul, sand & gravel		B-10B	1070	.011			.38	.86	1.24	1.52	
	4020	Common earth			940	.013			.43	.98	1.41	1.73	
	4040	Clay			589	.020			.68	1.56	2.24	2.76	
	4200	45 m haul, sand & gravel			455	.026			.89	2.02	2.91	3.58	
	4220	Common earth			395	.030			1.02	2.33	3.35	4.11	
	4240	Clay			248	.048			1.63	3.71	5.34	6.55	
	4400	90 m haul, sand & gravel			237	.051			1.70	3.88	5.58	6.85	
	4420	Common earth			206	.058			1.96	4.47	6.43	7.90	
	4440	Clay			130	.092			3.10	7.10	10.20	12.50	
	5000	224 kW, 15 m haul, sand & gravel		B-10M	1453	.008			.28	.82	1.10	1.33	
	5020	Common earth			1262	.010			.32	.95	1.27	1.53	
	5040	Clay			784	.015			.51	1.53	2.04	2.46	
	5200	45 m haul, sand & gravel			703	.017			.57	1.70	2.27	2.74	
	5220	Common earth			612	.020			.66	1.96	2.62	3.15	
	5240	Clay			382	.031			1.06	3.13	4.19	5.05	
	5400	90 m haul, sand & gravel			359	.033			1.12	3.33	4.45	5.40	
	5420	Common earth			313	.038			1.29	3.82	5.11	6.15	
	5440	Clay			191	.063			2.11	6.25	8.36	10.10	
	5500	343 kW, 15 m haul, sand & gravel		B-10X	1476	.008			.27	1.04	1.31	1.56	
	5510	Common earth			1285	.009			.31	1.20	1.51	1.80	
	5520	Clay			803	.015			.50	1.91	2.41	2.86	
	5530	45 m haul, sand & gravel			986	.012			.41	1.56	1.97	2.33	
	5540	Common earth			856	.014			.47	1.79	2.26	2.69	

		02315 | Excavation and Fill	CREW	DAILY OUTPUT	LABOR-HOURS	UNIT	2006 BARE COSTS				TOTAL INCL O&P	
							MAT.	LABOR	EQUIP.	TOTAL		
432	5550	Clay	B-10X	535	.022	Bm³	.75	2.87	3.62		4.31	**432**
	5560	90 m haul, sand & gravel	R312316 -40	505	.024		.80	3.04	3.84		4.57	
	5570	Common earth		440	.027		.92	3.49	4.41		5.25	
	5580	Clay		268	.045		1.50	5.75	7.25		8.60	
	6000	522 kW, 15 m haul, sand & gravel	B-10V	2676	.004		.15	1.18	1.33		1.52	
	6010	Common earth		2321	.005		.17	1.36	1.53		1.75	
	6020	Clay		1472	.008		.27	2.14	2.41		2.77	
	6030	45 m haul, sand & gravel		1548	.008		.26	2.03	2.29		2.64	
	6040	Common earth		1338	.009		.30	2.35	2.65		3.05	
	6050	Clay		841	.014		.48	3.74	4.22		4.85	
	6060	90 m haul, sand & gravel		788	.015		.51	4	4.51		5.15	
	6070	Common earth		688	.017		.59	4.58	5.17		5.95	
	6080	Clay		421	.029		.96	7.50	8.46		9.65	
	6090	For dozer with ripper, see Div. 02315-418										
442	0010	**EXCAVATION, BULK, DRAG LINE**										**442**
	0011	Excavate and load on truck, bank measure										
	0012	Bucket drag line, 0.57 m³, sand/gravel	B-12I	336	.048	Bm³	1.56	2.03	3.59		4.61	
	0100	Light clay		237	.068		2.21	2.88	5.09		6.55	
	0110	Heavy clay		191	.084		2.74	3.57	6.31		8.15	
	0120	Unclassified soil		214	.075		2.45	3.19	5.64		7.25	
	0200	1.15 m³ bucket, sand/gravel	B-12P	440	.036		1.19	2.19	3.38		4.23	
	0210	Light clay		336	.048		1.56	2.87	4.43		5.55	
	0220	Heavy clay		269	.059		1.95	3.59	5.54		6.95	
	0230	Unclassified soil		229	.070		2.29	4.22	6.51		8.15	
	0300	2.29 m³, sand/gravel	B-12V	551	.029		.95	2.31	3.26		3.99	
	0310	Light clay		535	.030		.98	2.38	3.36		4.12	
	0320	Heavy clay		459	.035		1.14	2.78	3.92		4.80	
	0330	Unclassified soil		421	.038		1.24	3.03	4.27		5.25	
452	0010	**EXCAVATION, BULK, SCRAPERS**										**452**
	0100	Elev. scraper 8.4 m³, sand & gravel 450 m haul, 1/4 dozer	B-33F	528	.027	Bm³	.90	2.13	3.03		3.71	
	0150	900 m haul	R312316 -40	466	.030		1.02	2.41	3.43		4.20	
	0200	1500 m haul	R312316 -45	386	.036		1.23	2.91	4.14		5.10	
	0300	Common earth, 450 m haul		459	.031		1.04	2.44	3.48		4.27	
	0350	900 m haul		405	.035		1.18	2.77	3.95		4.84	
	0400	1500 m haul		336	.042		1.42	3.34	4.76		5.85	
	0500	Clay, 450 m haul		287	.049		1.66	3.91	5.57		6.80	
	0550	900 m haul		252	.056		1.89	4.45	6.34		7.80	
	0600	1500 m haul		210	.067		2.27	5.35	7.62		9.35	
	1000	Self propelled scraper, 10.7 m³ 1/4 push dozer, sand										
	1050	Sand and gravel, 450 m haul	B-33D	703	.020	Bm³	.68	2.67	3.35		3.97	
	1100	900 m haul		616	.023		.77	3.05	3.82		4.53	
	1200	1500 m haul		493	.028		.97	3.81	4.78		5.65	
	1300	Common earth, 450 m haul		612	.023		.78	3.07	3.85		4.55	
	1350	900 m haul		535	.026		.89	3.51	4.40		5.20	
	1400	1500 m haul		428	.033		1.11	4.38	5.49		6.50	
	1500	Clay, 450 m haul		382	.037		1.25	4.91	6.16		7.30	
	1550	900 m haul		336	.042		1.42	5.60	7.02		8.30	
	1600	1500 m haul		268	.052		1.78	7	8.78		10.40	
	2000	16 m³, 1/4 push dozer, sand & gravel, 450 m haul	B-33E	902	.016		.53	2.94	3.47		4.03	
	2100	900 m haul		696	.020		.68	3.81	4.49		5.25	
	2200	1500 m haul		573	.024		.83	4.62	5.45		6.35	
	2300	Common earth, 450 m haul		788	.018		.60	3.36	3.96		4.62	
	2350	900 m haul		604	.023		.79	4.39	5.18		6.05	
	2400	1500 m haul		497	.028		.96	5.35	6.31		7.30	
	2500	Clay, 450 m haul		493	.028		.97	5.35	6.32		7.35	
	2550	900 m haul		378	.037		1.26	7	8.26		9.60	

For expanded coverage of these items see *Means Site Work & Landscape Cost Data 2006*

SITE CONSTRUCTION **2**

	02315	Excavation and Fill	CREW	DAILY OUTPUT	LABOR-HOURS	UNIT	MAT.	LABOR	EQUIP.	TOTAL	TOTAL INCL O&P	
452	2600	1500 m haul	B-33E	310	.045	Bm³		1.54	8.55	10.09	11.75	**452**
	2700	Towed, 9 m³, 1/4 push dozer, sand & gravel, 450 m haul [R312316-40]	B-33B	428	.033			1.11	3.92	5.03	6	
	2720	900 m haul [R312316-45]		344	.041			1.39	4.87	6.26	7.45	
	2730	1500 m haul		279	.050			1.71	6	7.71	9.20	
	2750	Common earth, 450 m haul		321	.044			1.48	5.20	6.68	8	
	2770	900 m haul		306	.046			1.56	5.50	7.06	8.40	
	2780	1500 m haul		237	.059			2.01	7.05	9.06	10.85	
	2800	Clay, 450 m haul		241	.058			1.98	6.95	8.93	10.65	
	2820	900 m haul		229	.061			2.08	7.30	9.38	11.20	
	2840	1500 m haul		172	.081			2.77	9.75	12.52	14.90	
	2900	11.46 m³, 1/4 push dozer, sand & gravel, 450 m haul	B-33C	612	.023			.78	2.74	3.52	4.19	
	2920	900 m haul		489	.029			.97	3.43	4.40	5.25	
	2940	1500 m haul		398	.035			1.20	4.21	5.41	6.45	
	2960	Common earth, 450 m haul		459	.031			1.04	3.65	4.69	5.60	
	2980	900 m haul		428	.033			1.11	3.92	5.03	6	
	3000	1500 m haul		336	.042			1.42	4.99	6.41	7.65	
	3020	Clay, 450 m haul		344	.041			1.39	4.87	6.26	7.45	
	3040	900 m haul		321	.044			1.48	5.20	6.68	8	
	3060	1500 m haul		245	.057			1.95	6.85	8.80	10.50	
462	0010	**EXCAVATION, STRUCTURAL**										**462**
	0015	Hand, pits to 1.8 m deep, sandy soil [R312316-40]	1 Clab	6.12	1.308	Bm³		36		36	56	
	0100	Heavy soil or clay	"	3.06	2.616			71.50		71.50	112	
	0200	Pits to 600 mm deep, normal soil [R312316-45]	B-2	18.35	2.180			60.50		60.50	94.50	
	0300	Pits 2 m to 4 m deep, sandy soil	1 Clab	3.82	2.093			57.50		57.50	89.50	
	0500	Heavy soil or clay		2.29	3.488			95.50		95.50	149	
	0700	Pits 4 m to 6 m deep, sandy soil		3.06	2.616			71.50		71.50	112	
	0900	Heavy soil or clay		1.53	5.231			143		143	223	
	1000	Hand trimming, bottom of excavation	B-2	223	.179	m²		4.99		4.99	7.75	
	1010	Slopes and sides		223	.179	"		4.99		4.99	7.75	
	1030	Around obstructions		6.12	6.539	Bm³		182		182	283	
	1100	Hand loading trucks from stock pile, sandy soil	1 Clab	9.18	.872			24		24	37	
	1300	Heavy soil or clay	"	6.12	1.308			36		36	56	
	1500	For wet or muck hand excavation, add to above				%				50%	50%	
	1550	Excavation rock by hand/air tool	B-9	2.60	15.387	Bm³		430	56.50	486.50	730	
	6000	Machine excavation, for spread and mat footings, elevator pits,										
	6001	and small building foundations										
	6030	Common earth, hydraulic backhoe, 0.38 m³ bucket	B-12E	42.05	.380	Bm³		12.45	7.95	20.40	28	
	6035	0.57 m³ bucket	B-12F	68.81	.233			7.60	6.95	14.55	19.30	
	6040	0.76 m³ bucket	B-12A	82.58	.194			6.35	6.80	13.15	17.15	
	6050	1.15 m³ bucket	B-12B	110	.145			4.76	6.55	11.31	14.55	
	6060	1.53 m³ bucket	B-12C	153	.105			3.43	6	9.43	11.85	
	6070	Sand and gravel, 0.57 m³ bucket	B-12F	76.46	.209			6.85	6.25	13.10	17.35	
	6080	0.76 m³ bucket	B-12A	91.75	.174			5.70	6.10	11.80	15.45	
	6090	1.15 m³ bucket	B-12B	122	.131			4.30	5.95	10.25	13.05	
	6100	1.53 m³ bucket	B-12C	168	.095			3.12	5.45	8.57	10.75	
	6110	Clay, till, or blasted rock, 0.57 m³ bucket	B-12F	61.17	.262			8.55	7.80	16.35	21.50	
	6120	0.76 m³ bucket	B-12A	72.64	.220			7.20	7.70	14.90	19.55	
	6130	1.15 m³ bucket	B-12B	99.40	.161			5.25	7.25	12.50	16.05	
	6140	1.53 m³ bucket	B-12C	134	.119			3.91	6.85	10.76	13.55	
	9010	For mobilization or demobilization, see Div. 02305-250										
	9020	For dewatering, see Div. 02240-500										
	9022	For larger structures, see Bulk Excavation, Div. 02315-424										
	9024	For loading onto trucks, add								15%		
	9026	For hauling, see Div. 02315-490										
	9030	For sheeting or soldier bms/lagging, see 02250 & 02260										

		02315	Excavation and Fill	CREW	DAILY OUTPUT	LABOR-HOURS	UNIT	2006 BARE COSTS MAT.	LABOR	EQUIP.	TOTAL	TOTAL INCL O&P	
462	9040		For trench excavation of strip ftgs, see 02315-610 R312316 -40										462
490	0010		**HAULING**, excavated or borrow, loose cubic yards										490
	0012		no loading included, highway haulers										
	0020		4.58 m³ dump truck, 0.5 km round trip, 5.0 loads/hr.	B-34A	149	.054	Lm³		1.52	2.19	3.71	4.73	
	0030		1 km round trip, 4.1 loads/hr.		122	.066			1.86	2.68	4.54	5.80	
	0040		1.5 km round trip, 3.3 loads/hr.		99.40	.080			2.28	3.29	5.57	7.10	
	0100		3 km round trip, 2.6 loads/hr.		76.46	.105			2.97	4.27	7.24	9.25	
	0150		5 km round trip, 2.1 loads/hr.		61.17	.131			3.71	5.35	9.06	11.50	
	0200		6 km round trip, 1.8 loads/hr.		53.52	.149			4.24	6.10	10.34	13.15	
	0310		9.2 m³ dump truck, 0.4 km round trip 3.7 loads/hr.	B-34B	220	.036			1.03	2.17	3.20	3.95	
	0320		1 km round trip, 3.2 loads/hr.		191	.042			1.19	2.50	3.69	4.55	
	0330		1.5 km round trip 2.7, loads/hr.		161	.050			1.41	2.96	4.37	5.40	
	0400		3 km round trip, 2.2 loads/hr.		138	.058			1.64	3.45	5.09	6.30	
	0450		4.8 km round trip, 1.9 loads/hr.		130	.062			1.74	3.67	5.41	6.70	
	0500		6 km round trip, 1.6 loads/hr.		95.58	.084			2.37	4.99	7.36	9.10	
	0540		8 km round trip, 1 load/hr.		59.64	.134			3.80	8	11.80	14.60	
	0550		16 km round trip, 0.60 load/hr.		44.35	.180			5.10	10.75	15.85	19.60	
	0560		32 km round trip, 0.4 load/hr.		29.82	.268			7.60	16	23.60	29	
	0600		12.6 m³ dump trailer, 1.5 km round trip, 2.6 loads/hr.	B-34C	214	.037			1.06	1.89	2.95	3.70	
	0700		3 km round trip, 2.1 loads/hr.		172	.047			1.32	2.35	3.67	4.60	
	1000		5 km round trip, 1.8 loads/hr.		148	.054			1.53	2.73	4.26	5.35	
	1100		6 km round trip, 1.6 loads/hr.		132	.061			1.72	3.07	4.79	6	
	1110		8 km round trip, 1 load/hr.		82.58	.097			2.75	4.90	7.65	9.60	
	1120		16 km round trip, 0.60 load/hr.		61.17	.131			3.71	6.60	10.31	12.95	
	1130		32 km round trip, 0.4 load/hr.		41.29	.194			5.50	9.80	15.30	19.20	
	1150		15.28 m³ dump trailer, 1.5 km round trip, 2.5 loads/hr.	B-34D	248	.032			.91	1.68	2.59	3.25	
	1200		3 km round trip, 2 loads/hr.		199	.040			1.14	2.10	3.24	4.05	
	1220		5 km round trip, 1.7 loads/hr.		169	.047			1.34	2.47	3.81	4.77	
	1240		6 km round trip, 1.5 loads/hr.		149	.054			1.52	2.80	4.32	5.40	
	1245		8 km round trip, 1.1 load/hr.		109	.073			2.08	3.83	5.91	7.40	
	1250		16 km round trip, 0.75 load/hr.		84.11	.095			2.70	4.96	7.66	9.55	
	1255		32 km round trip, 0.5 load/hr.		59.64	.134			3.80	7	10.80	13.50	
	1300		Hauling in medium traffic, add				m³				20%	20%	
	1400		Heavy traffic, add				"				30%	30%	
	1600		Grading at dump, or embankment if required, by dozer	B-10B	765	.016	Lm³		.53	1.20	1.73	2.12	
	1800		Spotter at fill or cut, if required	1 Clab	8	1	Hr.	27.50			27.50	42.50	
	2000		Off highway haulers										
	2010		16.8 m³ rear/bottom dump, 300 m rnd trip, 4.5 loads/hr.	B-34F	493	.016	Lm³		.46	1.89	2.35	2.78	
	2020		1 km round trip, 4.2 loads/hr.		459	.017			.49	2.03	2.52	2.99	
	2030		1.5 km round trip, 3.9 loads/hr.		424	.019			.54	2.20	2.74	3.24	
	2040		3 km round trip, 3.3 loads/hr.		359	.022			.63	2.60	3.23	3.82	
	2050		26 m³ rear or bottom dump, 300 m round trip, 4 loads/hr.	B-34G	677	.012			.34	1.75	2.09	2.43	
	2060		1 km round trip, 3.8 loads/hr.		642	.012			.35	1.84	2.19	2.56	
	2070		1.5 km round trip, 3.5 loads/hr.		593	.013			.38	1.99	2.37	2.77	
	2080		3 km round trip, 3.0 loads/hr.		508	.016			.45	2.33	2.78	3.24	
	2090		32.1 m³ rear or bottom bump, 300 m round trip, 3.8 loads/hr.	B-34H	795	.010			.29	1.60	1.89	2.21	
	2100		1 km round trip, 3.6 loads/hr.		749	.011			.30	1.70	2	2.33	
	2110		1.5 km round trip, 3.3 loads/hr.		688	.012			.33	1.86	2.19	2.54	
	2120		3 km round trip, 2.8 loads/hr.		585	.014			.39	2.18	2.57	2.99	
	2130		45.8 m³ rear or bottom dump, 300 m round trip, 3.6 loads/hr.	B-34J	1070	.007			.21	1.54	1.75	2.02	
	2140		1 km round trip, 3.4 loads/hr.		1013	.008			.22	1.63	1.85	2.13	
	2150		1.5 km round trip, 3.1 loads/hr.		918	.009			.25	1.80	2.05	2.35	
	2160		3 km round trip, 2.6 loads/hr.		776	.010			.29	2.13	2.42	2.79	
	3000		Rough terrain or steep grades, add to above								100%		
	4500		Dust control, light	B-59	1	8	Day	227	325		552	705	

SITE CONSTRUCTION **2**

02315 | Excavation and Fill

		Description	CREW	DAILY OUTPUT	LABOR-HOURS	UNIT	2006 BARE COSTS				TOTAL INCL O&P	
							MAT.	LABOR	EQUIP.	TOTAL		
490	4501	Heavy	B-59	.50	16	Day		455	650	1,105	1,400	**490**
	4600	Haul road maintenance	B-86A	1	8	↓		294	455	749	950	
	4700	Highway hauling beyond 32 km, per loaded Kilometer, minimum				km				.74	.81	
	4750	Maximum				"				1.49	1.64	
510	0010	**FILL BY BORROW**, load, 1.6 km haul, compact and shape										**510**
	0020	for embankments	B-15	918	.031	Lm³	8	.93	2.04	10.97	12.45	
	0100	Select fill for shoulders & embankments	"	918	.031	"	10.50	.93	2.04	13.47	15.20	
	0201	For hauling over 1.6 km, add to above/m³, see Div 02315-490				km				.44	.58	
520	0010	**FILL**, spread dumped material, no compaction										**520**
	0020	By dozer, no compaction	B-10B	765	.016	Lm³		.53	1.20	1.73	2.12	
	0100	By hand	1 Clab	9.18	.872	"		24		24	37	
	0150	Spread fill, from stockpile with 1.91 m³ F.E. loader										
	0170	97 kW, 90 m haul	B-10P	459	.026	Lm³		.88	1.73	2.61	3.24	
	0190	With dozer 224 kW, 90 m haul	B-10M	459	.026	"		.88	2.61	3.49	4.21	
	0400	For compaction of embankment, see div. 02315-310										
	0500	Gravel fill, compacted, under floor slabs, 100 mm deep	B-37	929	.052	m²	1.72	1.50	.12	3.34	4.40	
	0600	150 mm deep		799	.060		2.58	1.74	.14	4.46	5.65	
	0700	230 mm deep		669	.072		4.31	2.08	.17	6.56	8.15	
	0800	300 mm deep		557	.086	↓	6.05	2.50	.21	8.76	10.75	
	1000	Alternate pricing method, 100 mm deep		91.75	.523	Em³	15.75	15.20	1.25	32.20	42	
	1100	150 mm deep		122	.393		15.75	11.40	.94	28.09	36	
	1200	230 mm deep		153	.314		15.75	9.10	.75	25.60	32.50	
	1300	300 mm deep	↓	168	.286	↓	15.75	8.30	.68	24.73	31	
	1500	For fill under exterior paving, see Div. 02720-200										
	1600	For flowable fill, see Division 03310-220										
610	0010	**EXCAVATING, TRENCH** or continuous footing, common earth [R312316 -40]										**610**
	0020	No sheeting or dewatering included										
	0050	0.3 m to 1.2 m deep, 0.29 m³ tractor loader/backhoe	B-11C	115	.139	Bm³		4.46	1.97	6.43	8.95	
	0060	0.38 m³ tractor loader/backhoe	B-11M	153	.105			3.35	1.81	5.16	7.15	
	0062	0.57 m³ hydraulic backhoe	B-12F	206	.078			2.54	2.32	4.86	6.45	
	0090	1.2 m to 2 m deep, 0.38 m³ tractor loader/backhoe	B-11M	153	.105			3.35	1.81	5.16	7.15	
	0100	0.48 m³ hydraulic backhoe	B-12Q	191	.084			2.74	2.28	5.02	6.70	
	0110	0.57 m³ hydraulic backhoe	B-12F	229	.070			2.29	2.09	4.38	5.80	
	0120	0.76 m³ hydraulic backhoe	B-12A	306	.052			1.71	1.83	3.54	4.63	
	0130	1.15 m³ hydraulic backhoe	B-12B	413	.039			1.27	1.75	3.02	3.87	
	0300	0.38 m³ hydraulic excavator, truck mounted	B-12J	153	.105			3.43	5.50	8.93	11.30	
	0500	2 m to 3 m deep, 0.57 m³ hydraulic backhoe	B-12F	172	.093			3.05	2.78	5.83	7.70	
	0510	0.76 m³ hydraulic backhoe	B-12A	306	.052			1.71	1.83	3.54	4.63	
	0600	0.76 m³ hydraulic excavator, truck mounted	B-12K	306	.052			1.71	3.18	4.89	6.10	
	0610	1.15 m³ hydraulic backhoe	B-12B	459	.035			1.14	1.58	2.72	3.48	
	0620	1.91 m³ hydraulic backhoe	B-12S	765	.021			.69	1.59	2.28	2.80	
	0900	3 m to 4 m deep, 0.57 m³ hydraulic backhoe	B-12F	153	.105			3.43	3.12	6.55	8.70	
	0910	0.76 m³ hydraulic backhoe	B-12A	275	.058			1.91	2.04	3.95	5.15	
	1000	1.15 m³ hydraulic backhoe	B-12B	413	.039			1.27	1.75	3.02	3.87	
	1020	1.91 m³ hydraulic backhoe	B-12S	765	.021			.69	1.59	2.28	2.80	
	1030	2.29 m³ hydraulic backhoe	B-12D	1070	.015			.49	1.89	2.38	2.83	
	1300	4 m to 6 m deep, 0.76 m³ hydraulic backhoe	B-12A	245	.065			2.14	2.29	4.43	5.80	
	1310	1.15 m³ hydraulic backhoe	B-12B	367	.044			1.43	1.97	3.40	4.35	
	1320	1.91 m³ hydraulic backhoe	B-12S	650	.025			.81	1.87	2.68	3.29	
	1330	2.29 m³ hydraulic backhoe	B-12D	765	.021			.69	2.65	3.34	3.96	
	1400	By hand with pick and shovel to 2 m deep, light soil	1 Clab	6.12	1.308			36		36	56	
	1500	Heavy soil	"	3.06	2.616	↓		71.50		71.50	112	
	1700	For tamping backfilled trenches, air tamp, add	A-1G	76.46	.105	Em³		2.87	.48	3.35	4.99	
	1900	Vibrating plate, add	B-18	138	.174	"		4.88	.25	5.13	7.90	
	2100	Trim sides and bottom for concrete pours, common earth	↓	139	.173	m²		4.85	.25	5.10	7.80	

Important: See the Reference Section for supporting data - Crews, Rental Equipment, City Cost Indexes and Reference Data

02315 | Excavation and Fill

			CREW	DAILY OUTPUT	LABOR-HOURS	UNIT	2006 BARE COSTS MAT.	LABOR	EQUIP.	TOTAL	TOTAL INCL O&P	
610	2300	Hardpan	B-18	55.74	.431	m²		12.10	.62	12.72	19.50	**610**
	2400	Pier and spread footing excavation, add to above	R312316 -40			m³				30%	30%	
	3000	Backfill trench, F.E. loader, wheel mtd., 0.76 m³ bucket										
	3020	Minimal haul	B-10R	306	.039	Lm³		1.32	.64	1.96	2.71	
	3040	30 m haul		153	.078			2.64	1.27	3.91	5.40	
	3060	60 m haul	↓	76.46	.157			5.25	2.55	7.80	10.85	
	3080	1.72 m³ bucket, minimum haul	B-10T	459	.026			.88	.67	1.55	2.07	
	3090	30 m haul		229	.052			1.76	1.33	3.09	4.15	
	3100	60 m haul	↓	115	.104	↓		3.51	2.66	6.17	8.25	
	4000	For backfill with dozer, see Div. 02315-120										
	4010	For compaction of backfill, see div. 02315-310										
620	0010	**EXCAVATING, UTILITY TRENCH** Common earth										**620**
	0050	Trenching with chain trencher, 9 kW, operator walking										
	0100	100 mm wide trench, 300 mm deep	B-53	244	.033	m		1.15	.20	1.35	1.95	
	0150	450 mm deep		229	.035			1.23	.21	1.44	2.08	
	0200	600 mm deep		213	.038			1.32	.22	1.54	2.24	
	0300	150 mm wide trench, 300 mm deep		198	.040			1.42	.24	1.66	2.41	
	0350	450 mm deep		183	.044			1.54	.26	1.80	2.61	
	0400	600 mm deep		168	.048			1.68	.28	1.96	2.84	
	0450	900 mm deep		137	.058			2.06	.35	2.41	3.48	
	0600	200 mm wide trench, 300 mm deep		145	.055			1.94	.33	2.27	3.29	
	0650	450 mm deep		122	.066			2.31	.39	2.70	3.91	
	0700	600 mm deep		107	.075			2.63	.44	3.07	4.46	
	0750	900 mm deep	↓	91.44	.087	↓		3.08	.52	3.60	5.20	
	0830	Fly wheel trencher, 1200 mm wide trench, 1.8 m deep, light soil	B-54A	1523	.006	Bm³		.22	.54	.76	.93	
	0840	Medium soil		1219	.008			.27	.68	.95	1.16	
	0850	Heavy soil	↓	990	.009			.34	.83	1.17	1.43	
	0860	600 mm wide trench, 2.7 m deep, light soil	B-54B	3808	.003			.09	.42	.51	.60	
	0870	Medium soil		3058	.003			.12	.52	.64	.74	
	0880	Heavy soil	↓	2475	.004	↓		.14	.64	.78	.92	
	1000	Backfill by hand including compaction, add										
	1050	100 mm wide trench, 300 mm deep	A-1G	244	.033	m		.90	.15	1.05	1.57	
	1100	450 mm deep		162	.049			1.35	.23	1.58	2.36	
	1150	600 mm deep		122	.066			1.80	.30	2.10	3.13	
	1300	150 mm wide trench, 300 mm deep		165	.048			1.33	.22	1.55	2.32	
	1350	450 mm deep		123	.065			1.78	.30	2.08	3.10	
	1400	600 mm deep		82.30	.097			2.66	.45	3.11	4.64	
	1450	900 mm deep		54.86	.146			4	.68	4.68	6.95	
	1600	200 mm wide trench, 300 mm deep		122	.066			1.80	.30	2.10	3.13	
	1650	450 mm deep		80.77	.099			2.71	.46	3.17	4.72	
	1700	600 mm deep		60.96	.131			3.60	.61	4.21	6.25	
	1750	900 mm deep	↓	41.15	.194	↓		5.35	.90	6.25	9.30	
	2000	Chain trencher, 30 kW operator riding										
	2050	150 mm wide trench and backfill, 300 mm deep	B-54	366	.022	m		.77	.59	1.36	1.81	
	2100	450 mm deep		305	.026			.92	.71	1.63	2.17	
	2150	600 mm deep		297	.027			.95	.73	1.68	2.23	
	2200	900 mm deep		274	.029			1.03	.79	1.82	2.42	
	2250	1200 mm deep		229	.035			1.23	.95	2.18	2.89	
	2300	1500 mm deep		198	.040			1.42	1.09	2.51	3.35	
	2400	200 mm wide trench and backfill, 300 mm deep		305	.026			.92	.71	1.63	2.17	
	2450	450 mm deep		290	.028			.97	.75	1.72	2.29	
	2500	600 mm deep		274	.029			1.03	.79	1.82	2.42	
	2550	900 mm deep		244	.033			1.15	.89	2.04	2.72	
	2600	1200 mm deep		198	.040			1.42	1.09	2.51	3.35	
	2700	300 mm wide trench and backfill, 300 mm deep	↓	297	.027	↓		.95	.73	1.68	2.23	

SITE CONSTRUCTION 2

02315 | Excavation and Fill

			CREW	DAILY OUTPUT	LABOR-HOURS	UNIT	MAT.	2006 BARE COSTS LABOR	EQUIP.	TOTAL	TOTAL INCL O&P	
620	2750	450 mm deep	B-54	262	.031	m		1.07	.83	1.90	2.53	**620**
	2800	600 mm deep		244	.033			1.15	.89	2.04	2.72	
	2850	900 mm deep		221	.036			1.27	.98	2.25	3	
	3000	400 mm wide trench and backfill, 300 mm deep		255	.031			1.10	.85	1.95	2.61	
	3050	450 mm deep		229	.035			1.23	.95	2.18	2.89	
	3100	600 mm deep		213	.038			1.32	1.02	2.34	3.11	
	3200	Compaction with vibratory plate, add								50%	50%	
	9100	For clay or till, add up to								150%	150%	
630	0010	**EXCAVATING, UTILITY TRENCH, PLOW**										**630**
	0100	Single cable, plowed into fine material	B-11Q	1158	.010	m		.35	.49	.84	1.07	
	0200	Two cable		975	.012			.41	.59	1	1.28	
	0300	Single cable, plowed into coarse material		610	.020			.66	.94	1.60	2.04	
640	0010	**UTILITY BEDDING** For pipe and conduit, not incl. compaction										**640**
	0050	Crushed or screened bank run gravel	B-6	115	.209	Lm³	27	6.25	1.97	35.22	41.50	
	0100	Crushed stone 20 mm to 15 mm		115	.209		36	6.25	1.97	44.22	52	
	0200	Sand, dead or bank		115	.209		5.50	6.25	1.97	13.72	17.85	
	0500	Compacting bedding in trench	A-1D	68.81	.116	Em³		3.19	.38	3.57	5.40	
	0600	If material source exceeds 3.2 km, add for extra mileage.										
	0610	See 02315-490 for hauling kilometer add.										

02325 | Dredging

			CREW	DAILY OUTPUT	LABOR-HOURS	UNIT	MAT.	LABOR	EQUIP.	TOTAL	TOTAL INCL O&P	
250	0010	**DREDGING**										**250**
	0020	Dredging mobilization and demobilization., add to below, minimum	B-8	.53	120	Total		3,725	4,525	8,250	10,700	
	0100	Maximum	"	.10	640	"		19,700	24,000	43,700	56,500	
	0300	Barge mounted clamshell excavation into scows,										
	0310	Dumped 30 km at sea, minimum	B-57	237	.203	Bm³		6.40	5.60	12	15.95	
	0400	Maximum	"	163	.294	"		9.30	8.15	17.45	23	
	0500	Barge mounted dragline or clamshell, hopper dumped,										
	0510	pumped 300 m to shore dump, minimum	B-57	260	.185	Bm³		5.85	5.10	10.95	14.55	
	0525	All pumping uses 7571 Liters of water per cubic meter										
	0600	Maximum	B-57	186	.258	Bm³		8.15	7.15	15.30	20.50	
	1000	Hydraulic method, pumped 300 m to shore dump, minimum		352	.136			4.32	3.76	8.08	10.75	
	1100	Maximum		237	.203			6.40	5.60	12	15.95	
	1400	Into scows dumped 30 km, minimum		325	.148			4.68	4.08	8.76	11.65	
	1500	Maximum		186	.258			8.15	7.15	15.30	20.50	
	1600	For inland rivers and canals in South, deduct				m³				30%	30%	

02340 | Soil Stabilization

			CREW	DAILY OUTPUT	LABOR-HOURS	UNIT	MAT.	LABOR	EQUIP.	TOTAL	TOTAL INCL O&P	
100	0010	**ASPHALT SOIL STABILIZATION** Including scarifying and compaction										**100**
	0020	Asphalt, 40 mm deep, 2.3 L/m²	B-75	3344	.017	m²	1.24	.56	1.05	2.85	3.36	
	0040	3.4 L/m²		3344	.017		1.85	.56	1.05	3.46	4.05	
	0100	75 mm deep, 4.5 L/m²		2926	.019		2.48	.63	1.20	4.31	5	
	0140	6.8 L/m²		2926	.019		3.72	.63	1.20	5.55	6.40	
	0200	150 mm deep, 9.0 L/m²		2508	.022		4.95	.74	1.40	7.09	8.10	
	0240	14 L/m²		2508	.022		7.45	.74	1.40	9.59	10.80	
	0300	200 mm deep, 12 L/m²		2341	.024		6.60	.79	1.50	8.89	10.10	
	0340	18 L/m²		2341	.024		9.90	.79	1.50	12.19	13.75	
	0540	27 L/m²		2174	.026		14.85	.85	1.61	17.31	19.40	
200	0010	**CEMENT SOIL STABILIZATION** Including scarifying and compaction										**200**
	1020	Cement, 4% mix, by volume, 150 mm deep	B-74	920	.070	m²	1.22	2.26	4.14	7.62	9.35	
	1030	200 mm deep		878	.073		1.59	2.37	4.34	8.30	10.15	
	1060	300 mm deep		803	.080		2.38	2.59	4.74	9.71	11.75	
	1100	6% mix, 150 mm deep		920	.070		1.75	2.26	4.14	8.15	9.90	
	1120	200 mm deep		878	.073		2.27	2.37	4.34	8.98	10.90	

Important: See the Reference Section for supporting data - Crews, Rental Equipment, City Cost Indexes and Reference Data

SITE CONSTRUCTION 2

02300 | Earthwork

02340 | Soil Stabilization

			CREW	DAILY OUTPUT	LABOR-HOURS	UNIT	MAT.	LABOR	EQUIP.	TOTAL	TOTAL INCL O&P	
200	1160	300 mm deep	B-74	803	.080	m²	3.43	2.59	4.74	10.76	12.95	**200**
	1200	9% mix, 150 mm deep		920	.070		2.64	2.26	4.14	9.04	10.90	
	1220	200 mm deep		878	.073		3.43	2.37	4.34	10.14	12.15	
	1260	300 mm deep		803	.080		5.20	2.59	4.74	12.53	14.85	
	1300	12% mix, 150 mm deep		920	.070		3.43	2.26	4.14	9.83	11.80	
	1320	200 mm deep		878	.073		4.60	2.37	4.34	11.31	13.45	
	1360	300 mm deep	↓	803	.080	↓	6.90	2.59	4.74	14.23	16.70	
300	0010	**GEOTEXTILE SOIL STABILIZATION**										**300**
	1500	Geotextile fabric, woven, 90 kg tensil strength	2 Clab	2090	.008	m²	2.48	.21		2.69	3.06	
	1510	Heavy Duty, 270 kg tensile strength		2007	.008		2.21	.22		2.43	2.78	
	1550	Non-woven, 55 kg tensile strength	↓	2090	.008	↓	1.26	.21		1.47	1.72	
500	0010	**LIME SOIL STABILIZATION** Including scarifying and compaction										**500**
	2020	Hydrated lime, for base, 2% mix by weight, 150 mm deep	B-74	1505	.043	m²	3.25	1.38	2.53	7.16	8.45	
	2030	200 mm deep		1421	.045		4.33	1.47	2.68	8.48	9.95	
	2060	300 mm deep		1296	.049		3.27	1.61	2.94	7.82	9.25	
	2100	4% mix, 150 mm deep		1505	.043		3.31	1.38	2.53	7.22	8.55	
	2120	200 mm deep		1421	.045		4.49	1.47	2.68	8.64	10.10	
	2160	300 mm deep		1296	.049		6.70	1.61	2.94	11.25	13.10	
	2200	6% mix, 150 mm deep		1505	.043		5	1.38	2.53	8.91	10.45	
	2220	200 mm deep		1421	.045		6.70	1.47	2.68	10.85	12.60	
	2260	300 mm deep	↓	1296	.049		9.95	1.61	2.94	14.50	16.65	
700	0010	**CALCIUM CHLORIDE**										**700**
	0020	Calcium chloride, delivered 45.36 kg bags, truckload lots				Met. Ton	545			545	600	
	0030	Solution, 0.479 kg flake/L, tank truck delivery				liter	.28			.28	.31	

02360 | Soil Treatment

			CREW	DAILY OUTPUT	LABOR-HOURS	UNIT	MAT.	LABOR	EQUIP.	TOTAL	TOTAL INCL O&P	
200	0010	**TERMITE PRETREATMENT**										**200**
	0020	Slab and walls, residential	1 Skwk	111	.072	m² Flr.	3.12	2.63		5.75	7.55	
	0100	Commercial, minimum		232	.034		3.34	1.26		4.60	5.60	
	0200	Maximum		153	.052	↓	5.05	1.91		6.96	8.45	
	0400	Insecticides for termite control, minimum		53.75	.149	liter	3.13	5.45		8.58	11.90	
	0500	Maximum	↓	41.64	.192	"	5.35	7		12.35	16.80	
	3000	Soil poisoning (sterilization)	1 Clab	418	.019	m²	3.01	.52		3.53	4.16	
	3100	Herbicide application from truck	B-59	15886	.001	"		.01	.02	.03	.04	

02370 | Erosion & Sedimentation Control

			CREW	DAILY OUTPUT	LABOR-HOURS	UNIT	MAT.	LABOR	EQUIP.	TOTAL	TOTAL INCL O&P	
450	0010	**RIP-RAP & ROCK LINING**, Random, broken stone										**450**
	0100	Machine placed for slope protection	B-12G	47.41	.338	Lm³	33	11.05	12	56.05	66.50	
	0110	0.29 to 0.19 m³ pieces, grouted	B-13	66.89	.837	m²	59	25	9.75	93.75	114	
	0200	450 mm minimum thickness, not grouted	"	44.31	1.264	"	18.95	38	14.70	71.65	95.50	
	0300	Dumped, 23 kg average	B-11A	726	.022	Met. Ton	20	.71	1.27	21.98	24.50	
	0350	45 kg average		635	.025		29	.81	1.45	31.26	34.50	
	0370	135 kg average	↓	544	.029	↓	33.50	.94	1.69	36.13	40.50	
	0400	Gabions, galv. steel mesh mats or boxes, stone filled, 150 mm deep	B-13	167	.335	m²	20.50	10.05	3.90	34.45	42	
	0500	225 mm deep		136	.412		31.50	12.35	4.78	48.63	59.50	
	0600	300 mm deep		128	.438		33.50	13.10	5.10	51.70	62	
	0700	450 mm deep		85.28	.657		42.50	19.65	7.65	69.80	86	
	0800	900 mm deep	↓	50.17	1.116	↓	72.50	33.50	12.95	118.95	146	
700	0010	**SYNTHETIC EROSION CONTROL**										**700**
	0020	Jute mesh, 85 m² per roll, 1200 mm wide, stapled	B-80A	2007	.012	m²	.90	.33	.09	1.32	1.60	
	0060	Nylon, 3 dimensional geomatrix, 9 mil thick		585	.041		5.25	1.12	.30	6.67	7.85	
	0062	0.305 mm thick		431	.056		7.10	1.53	.41	9.04	10.60	
	0064	0.457 mm thick	↓	385	.062		7.45	1.71	.46	9.62	11.35	
	0070	Paper biodegradable mesh	B-1	2090	.011	↓	.10	.32		.42	.61	

02300 | Earthwork

02370 | Erosion & Sedimentation Control

		CREW	DAILY OUTPUT	LABOR-HOURS	UNIT	2006 BARE COSTS MAT.	LABOR	EQUIP.	TOTAL	TOTAL INCL O&P	
700	0080	Paper mulch	B-64	16722	.001	m²	.06	.03	.01	.10	.13	700
	0100	Plastic netting, stapled, 50 mm x 25 mm mesh, 0.50 mm	B-1	2090	.011		.81	.32		1.13	1.40	
	0120	Revegetation mat, webbed	2 Clab	836	.019		6.50	.52		7.02	7.95	
	0200	Polypropylene mesh, stapled, 220 gram/m²	B-1	2090	.011		1.67	.32		1.99	2.34	
	0300	Tobacco netting, or jute mesh #2, stapled	"	2090	.011		.10	.32		.42	.61	
	0400	Soil sealant, liquid sprayed from truck	B-81	4181	.006	↓	.55	.18	.10	.83	.99	
	1000	Silt fence, polypropylene, 9 m high, ideal conditions	2 Clab	488	.033	m	1.12	.90		2.02	2.61	
	1100	Adverse conditions	"	290	.055	"	1.12	1.51		2.63	3.56	
	1200	Place and remove hay bales	A-2	2.72	8.818	Met. Ton	62	242	47.50	351.50	495	
	1250	Hay bales, staked	"	762	.032	m	7.40	.87	.17	8.44	9.65	

02390 | Shore Protect/Mooring Structures

		CREW	DAILY OUTPUT	LABOR-HOURS	UNIT	2006 BARE COSTS MAT.	LABOR	EQUIP.	TOTAL	TOTAL INCL O&P	
110	0010	**BREAKWATERS, BULKHEADS** Reinforced concrete,										110
	0015	include footing and tie-backs										
	0020	Up to 1.8 m high, minimum	C-17C	8.53	9.725	m	143	360	46.50	549.50	770	
	0060	Maximum		7.39	11.229		228	415	53.50	696.50	955	
	0100	3.7 m high, minimum		6.10	13.615		370	505	65	940	1,250	
	0160	Maximum	↓	5.64	14.719	↓	430	545	70.50	1,045.50	1,400	
	0180	Precast bulkhead, complete, including										
	0190	vertical and battered piles, face panels, and cap										
	0195	Using 4.9 m vertical piles				M				770	910	
	0196	Using 6 m vertical piles				"				805	965	
	0200	Steel sheeting, 1200 x 1200 x 200 mm concrete deadmen,										
	0210	3 m O.C., 3.7 m high, shore driven	B-40	8.23	7.777	m	204	276	335	815	1,025	
	0260	Barge driven	B-76	4.57	15.748	"	305	560	510	1,375	1,775	
	6000	Crushed stone placed behind bulkhead by clam bucket	B-12H	91.75	.174	Lm³	25	5.70	10.05	40.75	47.50	
120	0010	**BREAKWATERS, BULKHEADS, RESIDENTIAL CANAL**										120
	0020	Aluminum panel sheeting, incl. concrete cap and anchor										
	0030	Coarse compact sand, 1.2 m high, 600 mm embedment	B-40	60.96	1.050	m	151	37.50	45	233.50	276	
	0040	1 m embedment		42.67	1.500		178	53.50	64.50	296	350	
	0060	1.8 m embedment		27.43	2.333		227	83	101	411	495	
	0120	1.8 m high, 760 mm embedment		51.82	1.235		199	44	53	296	350	
	0140	1.2 m embedment		38.10	1.680		235	59.50	72.50	367	435	
	0160	1.7 m embedment		28.96	2.210		310	78.50	95	483.50	570	
	0220	2.4 m high, 1 m embedment		42.67	1.500		257	53.50	64.50	375	440	
	0240	1.5 m embedment		30.48	2.100		257	74.50	90.50	422	500	
	0420	Medium compact sand, 900 mm high, 600 mm embedment		71.63	.894		226	31.50	38.50	296	340	
	0440	1.2 m embedment		45.72	1.400		277	49.50	60.50	387	450	
	0460	1.7 m embedment		35.05	1.826		335	65	78.50	478.50	560	
	0520	1.5 m high, 1 m embedment		50.29	1.273		320	45	55	420	490	
	0540	1.5 m embedment		36.58	1.750		370	62	75.50	507.50	590	
	0560	2 m embedment		32	2		485	71	86	642	740	
	0620	2.1 m high, 1.4 m embedment		41.15	1.555		485	55.50	67	607.50	690	
	0640	1.8 m embedment		33.53	1.909		535	68	82	685	785	
	0720	Loose silty sand, 900 mm high, 900 mm embedment		62.48	1.024		330	36.50	44	410.50	465	
	0740	1.4 m embedment		47.24	1.355		395	48	58.50	501.50	575	
	0760	1.8 m embedment		38.10	1.680		460	59.50	72.50	592	685	
	0820	1.4 m high, 1.4 m embedment		47.24	1.355		445	48	58.50	551.50	630	
	0840	1.8 m embedment		38.10	1.680		515	59.50	72.50	647	740	
	0860	2.1 m embedment		35.05	1.826		625	65	78.50	768.50	875	
	0920	1.8 m high, 1.7 m embedment		39.62	1.615		585	57.50	69.50	712	810	
	0940	2.1 m embedment	↓	35.05	1.826	↓	645	65	78.50	788.50	895	
310	0010	**JETTIES, DOCK ACCESSORIES**										310
	0100	Cleats, aluminum, "S" type, 300 mm long	1 Clab	8	1	Ea.	15.70	27.50		43.20	60	
	0140	250 mm long		6.70	1.194		38	32.50		70.50	93	
	0180	380 mm long	↓	6	1.333	↓	50	36.50		86.50	112	

SITE CONSTRUCTION

2

Important: See the Reference Section for supporting data - Crews, Rental Equipment, City Cost Indexes and Reference Data

02390 | Shore Protect/Mooring Structures

		CREW	DAILY OUTPUT	LABOR-HOURS	UNIT	2006 BARE COSTS MAT.	LABOR	EQUIP.	TOTAL	TOTAL INCL O&P
310	0400	Dock wheel for corners and piles, vinyl, 300 mm diameter [12"]	1 Clab	4	2	Ea.	68	55		123	161	**310**
	1000	Electrical receptacle with circuit breaker,									
	1020	Pile mounted, double 30 A, 125 volt	1 Elec	2	4	Unit	445	168		613	735
	1060	Double 50 A, 125/240 volt		1.60	5		600	210		810	970
	1120	Free standing, add		4	2		154	84		238	294
	1140	Double free standing, add		2.70	2.963		615	124		739	860
	1160	Light, 2 louvered, with photo electric switch, add		8	1		102	42		144	175
	1180	Telephone jack on stanchion		8	1		74	42		116	144
	1300	Fender, Vinyl, 100 mm high	1 Clab	48.77	.164	m	9.95	4.49		14.44	17.95
	1380	Corner piece		80	.100	Ea.	13.65	2.74		16.39	19.25
	1400	Hose holder, cast aluminum		16	.500	"	29	13.70		42.70	53.50
	1500	Ladder, aluminum, heavy duty									
	1520	Crown top, 5 to 7 step, minimum	1 Clab	5.30	1.509	Ea.	191	41.50		232.50	275
	1560	Maximum		2	4		241	110		351	435
	1580	Bracket for portable clamp mounting		8	1		99.50	27.50		127	152
	1800	Line holder, treated wood, small		16	.500		11.80	13.70		25.50	34.50
	1840	Large		13.30	.602		20	16.50		36.50	47.50
	2000	Mooring whip, fiberglass bolted to dock,									
	2020	550 kg boat	1 Clab	8.80	.909	Pr.	287	25		312	355
	2040	4550 kg boat		6.70	1.194		580	32.50		612.50	690
	2080	27 200 kg boat		4	2		680	55		735	835
	2400	Shock absorbing tubing, vertical bumpers									
	2420	75 mm diam., vinly, white [3"]	1 Clab	24.38	.328	m	47	9		56	65.50
	2440	Polybutyl, clear		24.38	.328	"	52	9		61	71.50
	2480	Mounts, polybutyl		20	.400	Ea.	16.75	10.95		27.70	35.50
	2490	Deluxe		20	.400	"	37.50	10.95		48.45	58
320	0010	**JETTIES, FLOATING DOCK ACCESSORIES**										**320**
	0200	Dock connectors, stressed cables with rubber spacers									
	0220	635 mm long, 900 mm wide dock	1 Clab	2	4	Joint	575	110		685	800
	0240	1500 mm wide dock		2	4		670	110		780	905
	0400	965 mm long, 1200 mm wide dock		1.75	4.571		630	125		755	890
	0440	1800 mm wide dock		1.45	5.517		715	151		866	1,025
	1000	Gangway, aluminum, one end rolling, no hand rails									
	1020	900 mm wide, minimum	1 Clab	20.42	.392	m	525	10.75		535.75	595
	1040	Maximum		9.75	.820		605	22.50		627.50	700
	1100	1200 mm wide, minimum		12.19	.656		560	18		578	645
	1140	Maximum		7.32	1.094		640	30		670	750
	1180	For handrails, add					122			122	134
	2000	Pile guides, beads on stainless cable	1 Clab	4	2	Ea.	375	55		430	500
	2020	Rod type, 200 mm diameter piles, minimum [8"]		4	2		37.50	55		92.50	127
	2040	Maximum		2	4		55	110		165	232
	2100	250 mm to 350 mm diameter piles minimum [10"-14"]		3.20	2.500		90.50	68.50		159	207
	2140	Maximum		1.75	4.571		158	125		283	370
	2200	Roller type, 4 rollers, minimum		4	2		340	55		395	460
	2240	Maximum		1.75	4.571		485	125		610	725
330	0010	**JETTIES, DOCKS, FIXED**										**330**
	0020	Pile supported, treated wood,									
	0030	1500 mm x 6000 mm platform, minimum	F-3	10.68	3.744	m²	130	135	59.50	324.50	420
	0060	Maximum		6.97	5.741		275	207	91.50	573.50	720
	0100	1800 mm x 6000 mm platform, minimum		12.08	3.312		115	119	52.50	286.50	370
	0160	Maximum		7.90	5.066		271	183	80.50	534.50	670
	0200	2400 mm x 6000 mm platform, minimum		13.94	2.870		108	104	45.50	257.50	330
	0260	Maximum		9.29	4.306		249	155	68.50	472.50	590
	0420	1500 mm x 9000 mm platform, minimum		14.40	2.778		121	100	44	265	335
	0460	Maximum		9.29	4.306		281	155	68.50	504.50	625

SITE CONSTRUCTION 2

02390 | Shore Protect/Mooring Structures

		CREW	DAILY OUTPUT	LABOR-HOURS	UNIT	MAT.	LABOR	EQUIP.	TOTAL	TOTAL INCL O&P	
330	**0500** For Greenhart lumber, add								40%		**330**
	0550 Diagonal planking, add								25%		
	1000 Pipe supported dock, 25 mm aluminum pipe										
	1020 Alum. planks, galv. stl. framing, 600 mm 200 mm wide, minimum	F-3	29.73	1.346	m²	240	48.50	21.50	310	365	
	1060 Maximum		11.15	3.588		300	129	57	486	595	
	1100 1500 mm wide, minimum		23.23	1.722		135	62	27.50	224.50	274	
	1160 Maximum		14.86	2.691		148	97	43	288	360	
	1200 Wood deck and galv. steel framing										
	1220 900 mm wide, minimum	F-3	29.73	1.346	m²	240	48.50	21.50	310	365	
	1260 Maximum		11.15	3.588		289	129	57	475	580	
	1300 1800 mm wide, minimum		24.15	1.656		221	59.50	26.50	307	365	
	1360 Maximum		9.29	4.306		272	155	68.50	495.50	615	
	1400 3000 mm wide, minimum		18.58	2.153		240	77.50	34.50	352	420	
	1460 Maximum		7.43	5.382		296	194	85.50	575.50	720	
	1800 38 mm galv. steel pipe, treated wood dock and framing										
	1820 1200 mm wide, minimum	F-3	25.55	1.566	m²	160	56.50	25	241.50	291	
	1860 Maximum		15.79	2.533		262	91.50	40.50	394	475	
	1900 1500 mm wide, minimum		23.23	1.722		139	62	27.50	228.50	279	
	1960 Maximum		14.86	2.691		227	97	43	367	445	
	2000 1800 mm wide, minimum		20.90	1.914		132	69	30.50	231.50	287	
	2060 Maximum		13.94	2.870		188	104	45.50	337.50	420	
340	**0010 JETTIES, DOCKS** Floating, recreational, prefabricated galvanized steel with										**340**
	0020 polyethylene encased polystyrene, no pilings included	F-3	30.66	1.305	m²	320	47	21	388	445	
	0200 Pile supported, shore constructed, bare, 75 mm decking		12.08	3.312		219	119	52.50	390.50	485	
	0250 100 mm decking		11.15	3.588		211	129	57	397	495	
	0400 Floating, small boat, prefab, no shore facilities, minimum		23.23	1.722		108	62	27.50	197.50	245	
	0500 Maximum		13.94	2.870		365	104	45.50	514.50	610	
	0700 Per slip, minimum (16.72 m² each)		1.59	25.157	Ea.	2,150	905	400	3,455	4,200	
	0800 Maximum		1.40	28.571	"	6,375	1,025	455	7,855	9,100	
350	**0010 JETTIES, DOCKS,** floating including anchors										**350**
	1000 Polystyrene flotation, minimum	F-3	18.58	2.153	m²	146	77.50	34.50	258	320	
	1040 Maximum		12.54	3.189	"	189	115	51	355	440	
	1100 Alternate method of figuring, minimum		1.13	35.398	Slip	3,375	1,275	565	5,215	6,325	
	1140 Maximum		.70	57.143	"	4,850	2,050	910	7,810	9,500	
	1200 Galv. steel frame and wood deck, 900 mm wide, minimum		29.73	1.346	m²	124	48.50	21.50	194	235	
	1240 Maximum		18.58	2.153		199	77.50	34.50	311	375	
	1300 1200 mm wide, minimum		29.73	1.346		117	48.50	21.50	187	227	
	1340 Maximum		18.58	2.153		187	77.50	34.50	299	365	
	1500 2400 mm wide, minimum		23.23	1.722		98.50	62	27.50	188	234	
	1540 Maximum		14.86	2.691		149	97	43	289	360	
	1700 Treated wood frames and deck, 900 mm wide, minimum		23.23	1.722		128	62	27.50	217.50	267	
	1740 Maximum		11.61	3.445		440	124	55	619	740	
	2000 Polyethylene drums, treated wood frame and deck										
	2100 1800 mm wide, minimum	F-3	23.23	1.722	m²	146	62	27.50	235.50	286	
	2140 Maximum		11.61	3.445		194	124	55	373	465	
	2200 2400 mm wide, minimum		21.65	1.848		131	66.50	29.50	227	280	
	2240 Maximum		11.15	3.588		179	129	57	365	460	
	2300 3000 mm wide, minimum		18.58	2.153		113	77.50	34.50	225	283	
	2340 Maximum		10.22	3.914		157	141	62.50	360.50	460	
360	**0010 JETTIES, PIERS,** Municipal w/75mm x 300mm frmng. & 75mm decking,										**360**
	0020 wood piles and cross bracing, alternate bents battered	B-76	5.57	12.917	m²	1,025	460	415	1,900	2,325	
	0200 Treated piles, not including mobilization										
	0210 15 m long, 9 kg creosote, shore driven	B-19	165	.388	m	33	13.80	9.50	56.30	69	
	0220 Barge driven	B-76	97.54	.738		33	26	24	83	105	
	0230 1.134 kg CCA, shore driven	B-19	165	.388		41	13.80	9.50	64.30	77.50	

Important: See the Reference Section for supporting data - Crews, Rental Equipment, City Cost Indexes and Reference Data

02300 | Earthwork

02390 | Shore Protect/Mooring Structures

			CREW	DAILY OUTPUT	LABOR-HOURS	UNIT	2006 BARE COSTS MAT.	LABOR	EQUIP.	TOTAL	TOTAL INCL O&P	
360	0240	Barge driven	B-76	97.54	.738	m	41	26	24	91	113	**360**
	0250	9 m long, 9 kg creosote, shore driven	B-19	165	.388		31	13.80	9.50	54.30	66.50	
	0260	Barge driven	B-76	97.54	.738		31	26	24	81	102	
	0270	1.134 kg CCA, shore driven	B-19	165	.388		23	13.80	9.50	46.30	58	
	0280	Barge driven	B-76	97.54	.738	↓	23	26	24	73	93.50	
	0300	Mobilization, barge, by tug boat	B-83	40.23	.398	km		12.75	11.75	24.50	32.50	
	0350	Standby time for shore pile driving crew				Hr.				430	535	
	0360	Standby time for barge driving rig				"				590	695	

02400 | Tunneling, Boring & Jacking

02410 | Tunnel Excavation

			CREW	DAILY OUTPUT	LABOR-HOURS	UNIT	2006 BARE COSTS MAT.	LABOR	EQUIP.	TOTAL	TOTAL INCL O&P	
410	0010	ROCK EXCAVATION, TUNNEL BORING GENERAL	R317100 -10									**410**
	0020	Bored Tunnels, incl. mucking 6 m diameter				M				1,275	1,450	
	0100	Rock excavation, minimum								1,275	1,450	
	0110	Average								2,700	3,100	
	0120	Maximum								7,200	8,300	
	0200	Mixed rock and earth, minimum								2,175	2,400	
	0210	Average								5,425	6,225	
	0220	Maximum				↓				7,950	8,750	
500	0010	SOFT GROUND SHIELD DRIVEN BORING										**500**
	0300	Earth excavation, minimum				M				1,075	1,300	
	0310	Average								2,525	2,775	
	0320	Maximum				↓				6,500	7,150	

02425 | Tunnel Linings

100	0010	CAST IN PLACE CONCRETE TUNNEL LININGS										**100**
	0500	Tunnel liner invert, incl. reinf., bored tunnels, 6 m diam., rock				M				1,275	1,400	
	0510	Earth								815	900	
	0520	Arch, incl. reinf., for bored tunnels, rock								1,175	1,300	
	0530	Earth				↓				1,150	1,275	

02430 | Tunnel Grouting

100	0010	TUNNEL LINER GROUTING										**100**
	0800	Contact grouting incl. drilling and connecting, minimum				m³				243	268	
	0820	Maximum				"				1,550	1,700	

02441 | Microtunneling

400	0010	MICROTUNNELING Not including excavation, backfill, shoring,										**400**
	0020	or dewatering, average 15 m/day, slurry method										
	0100	600 mm to 1200 mm outside diameter, minimum				m					2,325	
	0110	Adverse conditions, add				%					50%	
	1000	Rent microtunneling machine, average monthly lease				Month					94,000	
	1010	Operating technician				Day					700	
	1100	Mobilization and demobilization, minimum				Job					47,000	
	1110	Maximum				"					450,000	

SITE CONSTRUCTION 2

For expanded coverage of these items see _Means Site Work & Landscape Cost Data 2006_

69

02400 | Tunneling, Boring & Jacking

02442 | Cut and Cover Tunnels

			CREW	DAILY OUTPUT	LABOR-HOURS	UNIT	2006 BARE COSTS MAT.	LABOR	EQUIP.	TOTAL	TOTAL INCL O&P	
150	0010	**CUT AND COVER TUNNELS**										150
	2000	Cut and cover, excavation, not incl. hauling or backfill	B-57	61.17	.785	m³		25	21.50	46.50	62	
	2100	Decking on steel frames, concrete				m²				226	249	
	2110	Steel plates								296	325	
	2120	Wood				↓				160	177	
	2500	Shotcrete applied in tunnels				m³				288	330	

02444 | Shaft Construction

			CREW	DAILY OUTPUT	LABOR-HOURS	UNIT	2006 BARE COSTS MAT.	LABOR	EQUIP.	TOTAL	TOTAL INCL O&P	
150	0010	**SHAFT CONSTRUCTION**										150
	0400	Shaft excavation, rock, minimum				m³				72.50	79	
	0410	Average								100	111	
	0420	Maximum								201	223	
	0430	Earth, minimum								43	47.50	
	0440	Average								54.50	71.50	
	0450	Maximum				↓				179	201	
	3000	Ventilation for tunnel construction										
	3100	Duct, 1200 mm diameter, 20 ga., spun on site [48"]				M				48.50	54	
	3200	Fan, 1200 mm diameter, 93 kW, incl. starter [48"]				Ea.				19,000	22,000	

02445 | Boring or Jacking Conduits

			CREW	DAILY OUTPUT	LABOR-HOURS	UNIT	2006 BARE COSTS MAT.	LABOR	EQUIP.	TOTAL	TOTAL INCL O&P	
300	0010	**HORIZONTAL BORING** Casing only, 30 m minimum,										300
	0020	not incl. jacking pits or dewatering										
	0100	Roadwork, 13 mm thick wall, 600 mm diameter casing [1/2",24"]	B-42	6.10	10.499	m	238	325	186	749	985	
	0200	900 mm diameter [36"]		4.88	13.123		375	410	232	1,017	1,325	
	0300	1200 mm diameter [48"]		4.57	13.998		555	435	248	1,238	1,575	
	0500	Railroad work, 600 mm diameter [24"]		4.57	13.998		238	435	248	921	1,225	
	0600	900 mm diameter [36"]		4.27	14.998		375	470	265	1,110	1,450	
	0700	1200 mm diameter [48"]	↓	3.66	17.498		555	545	310	1,410	1,825	
	0900	For ledge, add								510	615	
	1000	Small diameter boring, 75 mm, sandy soil [3"]	B-82	274	.058		65.50	1.83	.24	67.57	75	
	1040	Rocky soil	"	152	.105	↓	65.50	3.29	.43	69.22	77.50	
	1100	Prepare jacking pits, incl. mobilization & demobilization, minimum				Ea.				2,850	3,375	
	1101	Maximum				"				16,100	19,300	

02450 | Foundation & Load Bearing Elements

02455 | Driven Piles

			CREW	DAILY OUTPUT	LABOR-HOURS	UNIT	2006 BARE COSTS MAT.	LABOR	EQUIP.	TOTAL	TOTAL INCL O&P	
100	0010	**CAST-IN-PLACE CONCRETE PILES**, 200 piles, 18 m long	R316000									100
	0020	unless specified otherwise, not incl. pile caps or mobilization	-20									
	0050	Cast-in-place augered piles, no casing or reinforcing										
	0060	200 mm diameter [8"]	B-43	165	.291	m	10.70	8.85	20.50	40.05	48	
	0065	250 mm diameter [10"]		146	.329		17	10	23	50	59.50	
	0070	300 mm diameter [12"]		128	.375		24	11.40	26.50	61.90	73	
	0075	350 mm diameter [14"]		110	.436		32.50	13.25	30.50	76.25	90	
	0080	400 mm diameter [16"]		91.44	.525		43.50	15.95	37	96.45	113	
	0085	450 mm diameter [18"]	↓	73.15	.656	↓	54	19.90	46	119.90	141	
	0100	Cast-in-place, thin wall shell pile, straight sided,										
	0110	not incl. reinforcing, 200 mm diam., 16 ga., 8.5 kg/m [8"]	B-19	213	.300	m	20	10.65	7.40	38.05	47	
	0200	250 mm diameter, 16 ga. corrugated, 11 kg/m [10"]	↓	198	.323	↓	26.50	11.50	7.95	45.95	56	

Important: See the Reference Section for supporting data - Crews, Rental Equipment, City Cost Indexes and Reference Data

02455	Driven Piles	CREW	DAILY OUTPUT	LABOR-HOURS	UNIT	2006 BARE COSTS				TOTAL INCL O&P
						MAT.	LABOR	EQUIP.	TOTAL	
100 0300	300 mm diameter, 16 ga. corrugated, 13 kg/m [12"] R316000-20	B-19	183	.350	m	34	12.40	8.60	55	67 **100**
0400	350 mm diameter, 16 ga. corrugated, 15 kg/m [14"]		168	.381		40	13.55	9.35	62.90	76
0500	400 mm diameter, 16 ga. corrugated, 17 kg/m [16"]		152	.421		49	14.95	10.35	74.30	89.50
0800	Cast-in-place friction pile, 15 m long, fluted,									
0810	tapered steel, 28 MPa concrete, no reinforcing									
0900	300 mm diameter, 7 ga. [12"]	B-19	183	.350	m	60	12.40	8.60	81	95.50
1000	350 mm diameter, 7 ga. [14"]		171	.374		65.50	13.30	9.20	88	103
1100	400 mm diameter, 7 ga. [16"]		158	.405		77	14.40	9.95	101.35	118
1200	450 mm diameter, 7 ga. [18"]		146	.438		90.50	15.55	10.75	116.80	136
1300	End bearing, fluted, constant diameter,									
1320	28 MPa concrete, no reinforcing									
1340	300 mm diameter, 7 ga. [12"]	B-19	183	.350	m	63	12.40	8.60	84	98.50
1360	350 mm diameter, 7 ga. [14"]		171	.374		78.50	13.30	9.20	101	118
1380	400 mm diameter, 7 ga. [16"]		158	.405		91	14.40	9.95	115.35	134
1400	450 mm diameter, 7 ga. [18"]		146	.438		101	15.55	10.75	127.30	148
1500	For reinforcing steel, add				kg	1.90			1.90	2.07
1700	For ball or pedestal end, add	B-19	8.41	7.609	m³	129	270	187	586	780
1900	For lengths above 18 m, concrete, add	"	8.41	7.609	"	134	270	187	591	785
2000	For steel thin shell, pipe only				kg	1.52			1.52	1.68
450 0010	**PRESTRESSED CONCRETE PILES**, 200 piles R316000-20									**450**
0020	unless specified otherwise, not incl. pile caps or mobilization									
2200	Precast, prestressed, 15 m long, 300 mm dia., 59 mm wall [12"]	B-19	219	.292	m	41	10.40	7.15	58.55	70
2300	350 mm diameter, 65 mm wall [14"]		207	.309		54	11	7.60	72.60	85.50
2500	400 mm diameter, 75 mm wall [16"]		195	.328		75	11.65	8.05	94.70	110
2600	450 mm diameter, 75 mm wall [18"]	B-19A	183	.350		94.50	12.40	11	117.90	136
2800	500 mm diameter, 90 mm wall [20"]		171	.374		109	13.30	11.75	134.05	154
2900	600 mm diameter, 90 mm wall [24"]		158	.405		133	14.40	12.70	160.10	184
2920	900 mm diameter, 115 mm wall [36"]	B-19	122	.525		188	18.65	12.90	219.55	251
2940	1370 mm diameter, 130 mm wall [54"]		104	.615		252	22	15.10	289.10	330
2960	1680 mm diameter, 150 mm wall [66"]		67.06	.954		380	34	23.50	437.50	495
3100	Precast, prestressed, 12 m long, 250 mm thick, square [10"]		213	.300		30	10.65	7.40	48.05	58
3200	300 mm thick, square [12"]		207	.309		37.50	11	7.60	56.10	67.50
3275	Shipping for 18 m long concrete piles, add to material cost					1.64			1.64	1.80
3400	350 mm thick, square [14"]	B-19	183	.350		44.50	12.40	8.60	65.50	78.50
3500	Octagonal		195	.328		61.50	11.65	8.05	81.20	95
3700	400 mm thick, square [16"]		171	.374		70	13.30	9.20	92.50	108
3800	Octagonal		183	.350		73.50	12.40	8.60	94.50	110
4000	450 mm thick, square [18"]	B-19A	158	.405		84.50	14.40	12.70	111.60	130
4100	Octagonal	B-19	171	.374		87	13.30	9.20	109.50	127
4300	500 mm thick, square [20"]	B-19A	146	.438		104	15.55	13.75	133.30	154
4400	Octagonal	B-19	158	.405		96.50	14.40	9.95	120.85	140
4600	600 mm thick, square [24"]	B-19A	134	.478		149	16.95	15	180.95	208
4700	Octagonal	B-19	146	.438		139	15.55	10.75	165.30	190
4730	Precast, prestressed, 18 m long, 250 mm thick, square [10"]		213	.300		31.50	10.65	7.40	49.55	59.50
4740	300 mm thick, square [12"] (18 m long)		207	.309		39	11	7.60	57.60	69
4750	Mobilization for 3000 m pile job, add		1006	.064			2.26	1.56	3.82	5.30
4800	7600 m pile job, add		2591	.025			.88	.61	1.49	2.07
4850	Mobilization by water for barge driving rig, add								100%	
5000	Pressure grouted pin pile, 125 mm diam., cased, to 45 met. ton,									
5040	End bearing, less than 6 m	B-48	27.43	2.041	m	93	63.50	133	289.50	345
5080	More than 12 m		41.15	1.361		54.50	42.50	88.50	185.50	223
5120	Friction, loose sand and gravel		32.61	1.717		93	53.50	112	258.50	305
5160	Dense sand and gravel		41.15	1.361		54.50	42.50	88.50	185.50	223
5200	Uncased, up to 9.1 metric ton capacity, 6 m		41.15	1.361		55	42.50	88.50	186	223

SITE CONSTRUCTION 2

02455	Driven Piles		CREW	DAILY OUTPUT	LABOR-HOURS	UNIT	MAT.	LABOR	EQUIP.	TOTAL	TOTAL INCL O&P	
600												**600**
0010	**STEEL PILES**, Not including mobilization or demobilization	R316000 -20										
0100	Step tapered, round, concrete filled											
0110	200 mm tip, 55 metric ton capacity, 9 m depth		B-19	232	.276	m	25	9.80	6.75	41.55	50.50	
0120	18 m depth			226	.283		28	10.05	6.95	45	54.50	
0130	24 m depth			213	.300		29	10.65	7.40	47.05	57	
0150	250 mm tip, 80 metric ton capacity, 9 m depth			213	.300		30.50	10.65	7.40	48.55	58.50	
0160	18 m depth			210	.305		31.50	10.85	7.50	49.85	60	
0170	24 m depth			204	.314		34	11.15	7.70	52.85	63.50	
0190	300 mm tip, 100 metric ton capacity, 9 m depth			201	.318		42	11.30	7.80	61.10	72.50	
0200	18 m depth, 300 mm diameter			192	.333		42	11.85	8.20	62.05	74	
0210	24 m depth			180	.356		37.50	12.65	8.75	58.90	70.50	
0250	"H" Sections, 15 m long, 200 mm x 200 mm, 55 kg/m		B-19	195	.328	m	42	11.65	8.05	61.70	73.50	
0400	250 mm x 250 mm, 65 kg/m			186	.344		49	12.20	8.45	69.65	83	
0500	85 kg/m			186	.344		66.50	12.20	8.45	87.15	102	
0700	300 mm x 300 mm, 80 kg/m			180	.356		62.50	12.65	8.75	83.90	98.50	
0800	110 kg/m		B-19A	180	.356		87.50	12.65	11.15	111.30	129	
1000	350 mm x 350 mm, 110 kg/m			165	.388		87	13.80	12.20	113	131	
1100	130 kg/m			165	.388		106	13.80	12.20	132	151	
1300	350 mm x 350 mm, 150 kg/m			155	.413		121	14.65	12.95	148.60	171	
1400	175 kg/m			155	.413		139	14.65	12.95	166.60	191	
1600	Splice on std. points, not in leads, 200 mm or 250 mm		1 Sswl	5	1.600	Ea.	84	64		148	207	
1700	300 mm or 350 mm			4	2		122	80		202	278	
1900	Heavy duty points, not in leads, 250 mm wide			4	2		130	80		210	286	
2100	350 mm wide			3.50	2.286		168	91.50		259.50	350	
2600	Pipe piles, 15 m long, 200 mm diam., 45 kg/m, no concrete [8"]		B-19	152	.421	m	45.50	14.95	10.35	70.80	85.50	
2700	Concrete filled			140	.457		48	16.25	11.20	75.45	91.50	
2900	250 mm diameter, 50 kg/m, no concrete [10"]			152	.421		56.50	14.95	10.35	81.80	97.50	
3000	Concrete filled			137	.467		62.50	16.60	11.45	90.55	108	
3200	300 mm diameter, 65 kg/m, no concrete [12"]			145	.441		70	15.70	10.85	96.55	114	
3300	Concrete filled			126	.508		72.50	18.05	12.45	103	122	
3500	350 mm diameter, 70 kg/m, no concrete [14"]			131	.489		75	17.35	12	104.35	123	
3600	Concrete filled			108	.593		80.50	21	14.55	116.05	138	
3800	400 mm diameter, 75 kg/m, no concrete [16"]			117	.547		83.50	19.45	13.45	116.40	137	
3900	Concrete filled			102	.627		92.50	22.50	15.40	130.40	154	
4100	450 mm diameter, 90 kg/m, no concrete [18"]			108	.593		109	21	14.55	144.55	170	
4200	Concrete filled			94.49	.677		109	24	16.65	149.65	176	
4400	Splice, pipe pile, not in leads 200 mm diameter [8"]		1 Sswl	4.67	1.713	Ea.	62	68.50		130.50	192	
4500	350 mm diameter [14"]			3.79	2.111		81.50	84.50		166	242	
4600	400 mm diameter [16"]			3.03	2.640		101	105		206	300	
4650	450 mm diameter [18"]			4.50	1.778		133	71		204	274	
4800	Points, standard, 200 mm diameter [8"]			4.61	1.735		69.50	69.50		139	202	
4900	350 mm diameter [14"]			4.05	1.975		96.50	79		175.50	248	
5000	400 mm diameter [16"]			3.37	2.374		118	95		213	300	
5050	450 mm diameter [18"]			5	1.600		184	64		248	315	
5200	Points, heavy duty, 250 mm diameter [10"]			2.89	2.768		48	111		159	252	
5300	350 mm or 400 mm diameter [14",16"]			2.02	3.960		77	158		235	370	
5500	For reinforcing steel, add			522	.015	kg	1.26	.61		1.87	2.49	
5700	For thick wall sections, add					"	1.43			1.43	1.59	
6020	Steel pipe pile end plates, 200 mm diameter [8"]		1 Sswl	14	.571	Ea.	53	23		76	99	
6050	250 mm diameter [10"]			14	.571		55.50	23		78.50	102	
6100	300 mm diameter [12"]			12	.667		59	26.50		85.50	113	
6150	350 mm diameter [14"]			10	.800		61	32		93	125	
6200	400 mm diameter [16"]			9	.889		65	35.50		100.50	136	
6250	450 mm diameter [18"]			8	1		73	40		113	153	

Important: See the Reference Section for supporting data - Crews, Rental Equipment, City Cost Indexes and Reference Data

02455	Driven Piles	CREW	DAILY OUTPUT	LABOR-HOURS	UNIT	2006 BARE COSTS MAT.	LABOR	EQUIP.	TOTAL	TOTAL INCL O&P	
600	6300	Steel pipe pile shoes, 200 mm diameter [8"] R316000-20	1 Sswl	12	.667	Ea.	113	26.50		139.50	172	600
	6350	250 mm diameter [10"]		12	.667		117	26.50		143.50	176	
	6400	300 mm diameter [12"]		10	.800		125	32		157	195	
	6450	350 mm diameter [14"]		9	.889		144	35.50		179.50	222	
	6500	400 mm diameter [16"]		8	1		160	40		200	248	
	6550	450 mm diameter [18"]		6	1.333		179	53.50		232.50	293	
650	0010	**TIMBER PILES**, Untreated, friction or end bearing, not including R316000-20										650
	0050	mobilization or demobilization										
	0100	Up to 9 m long, 300 mm butts, 200 mm points	B-19	191	.335	m	21.50	11.90	8.25	41.65	51.50	
	0200	9 m to 12 m long, 300 mm butts, 200 mm points		213	.300		21.50	10.65	7.40	39.55	48.50	
	0300	12 m to 15 m long, 300 mm butts, 180 mm points		219	.292		21.50	10.40	7.15	39.05	48	
	0400	15 m to 18 m long, 330 mm butts, 180 mm points		244	.262		22	9.30	6.45	37.75	46	
	0500	18 m to 21 m long, 330 mm butts, 180 mm points		256	.250		24.50	8.90	6.15	39.55	48	
	0600	21 m to 24 m long, 330 mm butts, 150 mm points		256	.250		27	8.90	6.15	42.05	51	
	0700	21 m to 24 m long, 400 mm butts, 150 mm points		800	.080	Ea.	11.90	2.84	1.96	16.70	19.80	
	0800	Treated piles, 5.44 kg per m³,										
	0810	friction or end bearing, ASTM class B										
	1000	Up to 9 m long, 300 mm butts, 200 mm points	B-19	191	.335	m	32	11.90	8.25	52.15	63	
	1100	9 m to 12 m long, 300 mm butts, 200 mm points		213	.300		31	10.65	7.40	49.05	59	
	1200	12 m to 15 m long, 300 mm butts, 175 mm points		219	.292		33	10.40	7.15	50.55	61	
	1300	15 m to 18 m long, 325 mm butts, 175 mm points		244	.262		36.50	9.30	6.45	52.25	62.50	
	1400	18 m to 21 m long, 325 mm butts, 150 mm points	B-19A	256	.250		50	8.90	7.85	66.75	78	
	1500	21 m to 24 m long, 325 mm butts, 150 mm points	"	256	.250		63	8.90	7.85	79.75	92.50	
	1600	Treated piles, C.C.A., 40 kg/m³										
	1610	200 mm butts, 3 m long	B-19	122	.525	m	23	18.65	12.90	54.55	69	
	1620	3 m to 5 m long		152	.421		23	14.95	10.35	48.30	61	
	1630	5 m to 6 m long		175	.366		23	13	9	45	56	
	1640	250 mm butts, 3 m to 5 m long		152	.421		26.50	14.95	10.35	51.80	64.50	
	1650	5 m to 6 m long		175	.366		26.50	13	9	48.50	59.50	
	1660	6 m to 12 m long		213	.300		26.50	10.65	7.40	44.55	54	
	1670	300 mm butts, 3 m to 6 m long		175	.366		28.50	13	9	50.50	62	
	1680	6 m to 11 m long		198	.323		28.50	11.50	7.95	47.95	58.50	
	1690	11 m to 12 m long		213	.300		28.50	10.65	7.40	46.55	56.50	
	1695	350 mm butts. to 12 m long		213	.300		41	10.65	7.40	59.05	70	
	1700	Boot for pile tip, minimum	1 Pile	27	.296	Ea.	19.85	10.30		30.15	39	
	1800	Maximum		21	.381		59.50	13.20		72.70	87	
	2000	Point for pile tip, minimum		20	.400		19.80	13.90		33.70	45	
	2100	Maximum		15	.533		71.50	18.50		90	109	
	2300	Splice for piles over 15 m long, minimum	B-46	35	1.371		49.50	43	.96	93.46	125	
	2400	Maximum		20	2.400		59.50	75.50	1.68	136.68	188	
	2600	Concrete encasement with wire mesh and tube		101	.475	m	30.50	14.90	.33	45.73	58	
	2700	Mobilization for 3000 m pile job, add	B-19	1006	.064			2.26	1.56	3.82	5.30	
	2800	7600 m pile job, add	"	2591	.025			.88	.61	1.49	2.07	
	2900	Mobilization by water for barge driving rig, add								100%		
800	0010	**PILING SPECIAL COSTS**										800
	0011	Pile caps, see Division 03310-240										
	0500	Cutoffs, concrete piles, plain	1 Pile	5.50	1.455	Ea.		50.50		50.50	83	
	0600	With steel thin shell, add		38	.211			7.30		7.30	12	
	0700	Steel pile or "H" piles		19	.421			14.60		14.60	24	
	0800	Wood piles		38	.211			7.30		7.30	12	
	0900	Pre-augering up to 9 m deep, average soil, 600 mm diameter [24"]	B-43	54.86	.875	m		26.50	61.50	88	109	
	0920	900 mm diameter [36"]		35.05	1.369			41.50	96.50	138	170	
	0960	1200 mm diameter [48"]		21.34	2.250			68.50	158	226.50	279	
	0980	1500 mm diameter [60"]		15.24	3.150			95.50	222	317.50	390	
	1000	Testing, any type piles, test load is twice the design load										
	1050	45 metric ton design load, 90 metric ton test				Ea.				15,000	16,050	

SITE CONSTRUCTION 2

SITE CONSTRUCTION **2**

02455 | Driven Piles

			CREW	DAILY OUTPUT	LABOR-HOURS	UNIT	2006 BARE COSTS MAT.	LABOR	EQUIP.	TOTAL	TOTAL INCL O&P	
800	1100	91 metric ton design load, 181 metric ton test				Ea.				19,250	20,300	800
	1150	135 metric ton design load, 270 metric ton test								24,100	25,700	
	1200	181 metric ton design load, 363 metric ton test								26,200	28,800	
	1250	360 metric ton design load, 720 metric ton test				↓				30,300	33,700	
	1500	Wet conditions, soft damp ground										
	1600	Requiring mats for crane, add								40%	40%	
	1700	Barge mounted driving rig, add								30%	30%	

			CREW	DAILY OUTPUT	LABOR-HOURS	UNIT	MAT.	LABOR	EQUIP.	TOTAL	TOTAL INCL O&P	
900	0010	**MOBILIZATION**										900
	0020	Set up & remove, air compressor, 0.283 m³/s	A-5	3.30	5.455	Ea.	150	9.75		159.75	243	
	0100	0.566 m³/s	"	2.20	8.182		224	14.65		238.65	365	
	0200	Crane, with pile leads and pile hammer, 70 metric ton	B-19	.60	106			3,800	2,625	6,425	8,900	
	0300	140 metric ton	"	.36	177			6,325	4,375	10,700	14,900	
	0500	Drill rig, for caissons, to 900 mm, minimum	B-43	2	24			730	1,700	2,430	2,975	
	0520	Maximum		.50	96			2,925	6,750	9,675	11,900	
	0600	Up to 2100 mm	↓	1	48			1,450	3,375	4,825	5,975	
	0800	Auxiliary boiler, for steam small	A-5	1.66	10.843			297	19.40	316.40	480	
	0900	Large	"	.83	21.687			595	39	634	970	
	1100	Rule of thumb: complete pile driving set up, small	B-19	.45	142			5,050	3,500	8,550	11,900	
	1200	Large	"	.27	237	↓		8,425	5,825	14,250	19,800	
	1300	Mobilization by water for barge driving rig										
	1310	Minimum				Ea.				6,250	6,900	
	1320	Maximum				"				40,700	44,900	

02465 | Bored Piles

			CREW	DAILY OUTPUT	LABOR-HOURS	UNIT	MAT.	LABOR	EQUIP.	TOTAL	TOTAL INCL O&P	
800	0010	**DRILLED CAISSONS** Incl. excav., concrete, 30 kg/m³ reinf.	R316326 -60									800
	0020	per C.Y., not incl. mobiliz., boulder removal, disposal										
	0100	Open style, machine drilled, to 15 m deep, in stable grd., no										
	0110	casings or grd. water, 450 mm diam., 0.16 m³/m	B-43	60.96	.787	m	23	24	55.50	102.50	124	
	0200	600 mm diameter, 0.29 m³/m		57.91	.829		41	25	58.50	124.50	148	
	0300	750 mm diameter, 0.46 m³/m		45.72	1.050		64.50	32	74	170.50	202	
	0400	900 mm diameter, 0.66 m³/m		38.10	1.260		93	38.50	88.50	220	259	
	0500	1200 mm diameter, 1.17 m³/m		30.48	1.575		165	48	111	324	380	
	0600	1500 mm diameter, 1.82 m³/m		27.43	1.750		259	53	123	435	500	
	0700	1800 mm diameter, 2.63 m³/m		24.38	1.969		375	60	139	574	655	
	0800	2100 mm diameter, 3.59 m³/m	↓	22.86	2.100	↓	510	64	148	722	820	
	1000	For bell excavation and concrete, add										
	1020	1200 mm bell diameter, 600 mm shaft, 0.34 m³	B-43	20	2.400	Ea.	38.50	73	169	280.50	340	
	1040	1800 mm bell diameter, 750 mm shaft, 1.20 m³		5.70	8.421		137	256	595	988	1,200	
	1060	2400 mm bell diameter, 900 mm shaft, 2.84 m³		2.40	20		325	605	1,400	2,330	2,850	
	1080	2700 mm bell diameter, 1200 mm shaft, 3.42 m³		2	24		390	730	1,700	2,820	3,400	
	1100	3000 mm bell diameter, 1500 mm shaft, 4.00 m³		1.70	28.235		455	855	2,000	3,310	4,000	
	1120	3600 mm bell diameter, 1800 mm shaft, 6.68 m³		1	48		760	1,450	3,375	5,585	6,800	
	1140	4200 mm bell diameter, 2100 mm shaft, 10.39 m³	↓	.70	68.571	↓	1,175	2,075	4,825	8,075	9,800	
	1200	Open style, machine drilled, to 15 m deep, in wet ground, pulled										
	1300	casing and pumping, 450 mm diameter, 0.16 m³/m	B-48	48.77	1.148	m	23	35.50	74.50	133	162	
	1400	600 mm diameter, 0.29 m³/m		38.10	1.470		41	45.50	95.50	182	221	
	1500	750 mm diameter, 0.46 m³/m		25.91	2.161		64.50	67	140	271.50	330	
	1600	900 mm diameter, 0.66 m³/m	↓	18.29	3.062		93	95	199	387	465	
	1700	1200 mm diameter, 1.17 m³/m	B-49	16.76	5.249		165	171	256	592	725	
	1800	1500 mm diameter, 1.82 m³/m		10.67	8.249		259	268	400	927	1,150	
	1900	1800 mm diameter, 2.63 m³/m		9.14	9.624		375	315	470	1,160	1,400	
	2000	2100 mm diameter, 3.59 m³/m	↓	7.62	11.549	↓	510	375	565	1,450	1,750	
	2100	For bell excavation and concrete, add										
	2120	1200 mm bell diameter, 600 mm shaft, 0.34 m³	↓ B-48	19.80	2.828	Ea.	38.50	88	184	310.50	380	

Important: See the Reference Section for supporting data - Crews, Rental Equipment, City Cost Indexes and Reference Data

	02465 | Bored Piles		CREW	DAILY OUTPUT	LABOR-HOURS	UNIT	2006 BARE COSTS MAT.	LABOR	EQUIP.	TOTAL	TOTAL INCL O&P	
800	2140	1800 mm bell diam, 750 mm shaft, 1.20 m³ (R316326-60)	B-48	5.70	9.825	Ea.	137	305	640	1,082	1,325	800
	2160	2400 mm bell diam, 900 mm shaft, 2.84 m³		2.40	23.333		325	725	1,525	2,575	3,125	
	2180	2700 mm bell diam, 1200 mm shaft, 3.42 m³	B-49	3.30	26.667		390	865	1,300	2,555	3,200	
	2200	3000 mm bell diam, 1500 mm shaft, 4.00 m³		2.80	31.429		455	1,025	1,525	3,005	3,750	
	2220	3600 mm bell diam, 1800 mm shaft, 6.68 m³		1.60	55		760	1,775	2,675	5,210	6,550	
	2240	4200 mm bell diam, 2100 mm shaft, 10.39 m³		1	88		1,175	2,850	4,300	8,325	10,500	
	2300	Open style, machine drilled, to 15 m deep, in soft rocks and										
	2400	medium hard shales, 450 mm diameter, 0.16 m³/m	B-49	15.24	5.774	m	23	188	281	492	625	
	2500	600 mm diameter, 0.29 m³/m		9.14	9.624		41	315	470	826	1,050	
	2600	750 mm diameter, 0.46 m³/m		6.10	14.436		64.50	470	705	1,239.50	1,575	
	2700	900 mm diameter, 0.66 m³/m		4.57	19.248		93	625	940	1,658	2,100	
	2800	1200 mm diameter, 1.17 m³/m		3.05	28.871		165	940	1,400	2,505	3,175	
	2900	1500 mm diameter, 1.82 m³/m		2.13	41.245		259	1,350	2,000	3,609	4,550	
	3000	1800 mm diameter, 2.63 m³/m		1.83	48.119		375	1,575	2,350	4,300	5,400	
	3100	2100 mm diameter, 3.59 m³/m		1.52	57.743		510	1,875	2,825	5,210	6,550	
	3200	For bell excavation and concrete, add										
	3220	1200 mm bell diameter, 600 mm shaft, 0.34 m³	B-49	10.90	8.073	Ea.	38.50	262	395	695.50	885	
	3240	1800 mm bell diameter, 750 mm shaft, 1.20 m³		3.10	28.387		137	920	1,375	2,432	3,100	
	3260	2400 mm bell diameter, 900 mm shaft, 2.84 m³		1.30	67.692		325	2,200	3,300	5,825	7,375	
	3280	2700 mm bell diameter, 1200 mm shaft, 3.42 m³		1.10	80		390	2,600	3,900	6,890	8,750	
	3300	3000 mm bell diameter, 1500 mm shaft, 4.00 m³		.90	97.778		455	3,175	4,775	8,405	10,700	
	3320	3600 mm bell diameter, 1800 mm shaft, 6.68 m³		.60	146		760	4,775	7,150	12,685	16,100	
	3340	4200 mm bell diameter, 2100 mm shaft, 10.39 m³		.40	220		1,175	7,150	10,700	19,025	24,200	
	3600	For rock excavation, sockets, add, minimum		3.40	25.913	m³		840	1,275	2,115	2,700	
	3650	Average		2.69	32.732			1,075	1,600	2,675	3,400	
	3700	Maximum		1.36	64.782			2,100	3,150	5,250	6,725	
	3900	For 15 m to 30 m deep, add				m				7%	7%	
	4000	For 30 m to 45 m deep, add								25%	25%	
	4100	For 45 m to 60 m deep, add								30%	30%	
	4200	For casings left in place, add				kg	1.59			1.59	1.74	
	4300	For other than 30 kg/m³, add or deduct				"	1.70			1.70	1.87	
	4400	For steel "I" beam cores, add	B-49	7.53	11.687	Met. Ton	1,675	380	570	2,625	3,075	
	4500	Load and haul excess excavation, 3.2 km	B-34B	136	.059	Lm³		1.67	3.50	5.17	6.40	
	4600	For mobilization, 80 km radius, rig to 900 mm	B-43	2	24	Ea.		730	1,700	2,430	2,975	
	4650	Rig to 2100 mm	B-48	1.75	32			995	2,075	3,070	3,800	
	4700	For low headroom, add								50%		
	4750	For difficult access, add								25%		
	5000	Bottom inspection	1 Skwk	1.20	6.667			243		243	380	
950	0010	**DRILLED CONCRETE PIERS AND SHAFTS** or Displacement										950
	0100	Caissons, incl. mobilization and demobilization, up to 80 km										
	0200	Uncased shafts, 30 to 70 met. ton cap., 430 mm diam., 3 m depth	B-44	26.82	2.386	m	57	83.50	37	177.50	237	
	0300	8 m depth		50.29	1.273		41	44.50	19.60	105.10	138	
	0400	70 metric ton capacity, 560 mm diameter, 3 m depth		24.38	2.625		71.50	91.50	40.50	203.50	269	
	0500	6 m depth		39.62	1.615		57	56.50	25	138.50	181	
	0700	Cased shafts, 9 to 27 metric ton cap., 270 mm diam., 6 m depth		53.34	1.200		41	42	18.50	101.50	133	
	0800	9 m depth		73.15	.875		38	30.50	13.50	82	106	
	0850	30 to 50 metric ton cap., 300 mm diameter, 6 m depth		48.77	1.312		57	46	20	123	158	
	0900	12 m depth		70.10	.913		44	32	14.05	90.05	115	
	1000	75 to 90 metric ton cap., 400 mm diameter, 6 m depth		48.77	1.312		81.50	46	20	147.50	185	
	1100	12 m depth		70.10	.913		76	32	14.05	122.05	150	
	1200	100 to 130 metric ton cap., 450 mm dia., 6 m depth		48.77	1.312		88	46	20	154	192	
	1300	12 m depth		70.10	.913		81.50	32	14.05	127.55	156	
	1400	130 to 160 met. ton cap., 480 mm diam., 6 m depth		39.62	1.615		95	56.50	25	176.50	223	
	1500	12 m depth		64.01	1		88	35	15.40	138.40	169	
	1700	Over 9 m long, mm cost tends to be lower										
	1900	Maximum depth is about 27 m										

SITE CONSTRUCTION 2

SITE CONSTRUCTION · **2**

02510	Water Distribution	CREW	DAILY OUTPUT	LABOR-HOURS	UNIT	2006 BARE COSTS				TOTAL INCL O&P
						MAT.	LABOR	EQUIP.	TOTAL	
600	**0010**	**VALVES**									**600**
9000	Valves, gate valve, N.R.S. post type, 100 mm diameter [4"]	B-21	32	.875	Ea.	365	28	4.77	397.77	455
9040	200 mm diameter [8"]		16	1.750		705	56	9.55	770.55	875
9080	300 mm diameter [12"]		13	2.154		1,425	69	11.75	1,505.75	1,675
9120	O.S.&Y., 100 mm diameter [4"]		32	.875		291	28	4.77	323.77	370
9160	200 mm diameter [8"]		16	1.750		695	56	9.55	760.55	865
9200	300 mm diameter [12"]		13	2.154		1,475	69	11.75	1,555.75	1,750
9220	350 mm diameter [14"]		11	2.545		4,150	81.50	13.85	4,245.35	4,725
9400	Check valves, rubber disc, 65 mm diameter [2-1/2"]	B-20	44	.545		350	16.95		366.95	410
9440	100 mm diameter [4"]	B-21	32	.875		430	28	4.77	462.77	525
9500	200 mm diameter [8"]		16	1.750		1,050	56	9.55	1,115.55	1,250
9540	300 mm diameter [12"]		13	2.154		2,675	69	11.75	2,755.75	3,075
9700	Detector check valves, reducing, 100 mm diameter [4"]		32	.875		705	28	4.77	737.77	825
9740	200 mm diameter [8"]		16	1.750		1,775	56	9.55	1,840.55	2,050
9800	Galvanized, 100 mm diameter [4"]		32	.875		920	28	4.77	952.77	1,075
9840	200 mm diameter [8"]		16	1.750		2,150	56	9.55	2,215.55	2,450
710	**0010**	**TAPPING, CROSSES AND SLEEVES**									**710**
4000	Drill and tap pressurized main (labor only)									
4100	150 mm main, 25 mm to 50 mm service	Q-1	3	5.333	Ea.		205		205	310
4150	200 mm main, 25 mm to 50 mm service	"	2.75	5.818	"		224		224	335
4500	Tap and insert gate valve									
4600	200 mm main, 100 mm branch	B-21	3.20	8.750	Ea.		281	47.50	328.50	490
4650	150 mm branch		2.70	10.370			335	56.50	391.50	575
4651	Piping, drill, tap & insert gate valve, 200 mm main, 150 mm branch		2.70	10.370			335	56.50	391.50	575
4700	250 mm Main, 100 mm branch		2.70	10.370			335	56.50	391.50	575
4750	150 mm branch		2.35	11.915			380	65	445	660
4800	300 mm main, 150 mm branch		2.35	11.915			380	65	445	660
4850	200 mm branch		2.35	11.915			380	65	445	660
7000	Tapping crosses, sleeves, valves; with rubber gaskets									
7020	Crosses, 100 mm x 100 mm	B-21	37	.757	Ea.	460	24.50	4.12	488.62	550
7060	200 mm x 150 mm		21	1.333		675	43	7.25	725.25	815
7080	200 mm x 200 mm		21	1.333		730	43	7.25	780.25	880
7100	250 mm x 150 mm		21	1.333		1,325	43	7.25	1,375.25	1,550
7160	300 mm x 300 mm		18	1.556		1,675	50	8.50	1,733.50	1,925
7180	350 mm x 150 mm		16	1.750		3,325	56	9.55	3,390.55	3,775
7240	400 mm x 250 mm		14	2		3,625	64	10.90	3,699.90	4,075
7280	450 mm x 150 mm		10	2.800		5,350	90	15.25	5,455.25	6,025
7320	450 mm x 450 mm		10	2.800		4,925	90	15.25	5,030.25	5,550
7340	500 mm x 150 mm		8	3.500		4,325	112	19.10	4,456.10	4,950
7360	500 mm x 300 mm		8	3.500		4,625	112	19.10	4,756.10	5,300
7420	600 mm x 300 mm		6	4.667		5,650	150	25.50	5,825.50	6,475
7440	600 mm x 450 mm		6	4.667		8,450	150	25.50	8,625.50	9,525
7600	Cut-in sleeves with rubber gaskets, 100 mm		18	1.556		153	50	8.50	211.50	256
7640	200 mm		10	2.800		265	90	15.25	370.25	445
7680	300 mm		9	3.111		435	100	16.95	551.95	655
7800	Cut-in valves with rubber gaskets, 100 mm		18	1.556		390	50	8.50	448.50	510
7840	200 mm		10	2.800		810	90	15.25	915.25	1,050
7880	300 mm		9	3.111		1,225	100	16.95	1,341.95	1,525
7900	Tapping Valve 100 mm, MJ, ductile iron		18	1.556		410	50	8.50	468.50	540
7920	Tapping Valve 150 mm, MJ, ductile iron		12	2.333		490	75	12.70	577.70	670
8000	Sleeves with rubber gaskets, 100 mm x 100 mm		37	.757		815	24.50	4.12	843.62	935
8030	200 mm x 100 mm		21	1.333		505	43	7.25	555.25	630
8040	200 mm x 150 mm		21	1.333		540	43	7.25	590.25	670
8060	200 mm x 200 mm		21	1.333		620	43	7.25	670.25	760
8070	250 mm x 100 mm		21	1.333		505	43	7.25	555.25	630
8080	250 mm x 150 mm		21	1.333		565	43	7.25	615.25	700

Important: See the Reference Section for supporting data - Crews, Rental Equipment, City Cost Indexes and Reference Data

		02510	Water Distribution	CREW	DAILY OUTPUT	LABOR-HOURS	UNIT	2006 BARE COSTS				TOTAL INCL O&P	
								MAT.	LABOR	EQUIP.	TOTAL		
710	8090		250 mm x 200 mm	B-21	21	1.333	Ea.	905	43	7.25	955.25	1,075	710
	8140		300 mm x 300 mm		18	1.556		1,075	50	8.50	1,133.50	1,250	
	8160		350 mm x 150 mm		16	1.750		2,600	56	9.55	2,665.55	2,975	
	8220		400 mm x 250 mm		14	2		2,850	64	10.90	2,924.90	3,225	
	8260		450 mm x 150 mm		10	2.800		4,175	90	15.25	4,280.25	4,750	
	8300		450 mm x 450 mm		10	2.800		4,325	90	15.25	4,430.25	4,925	
	8320		500 mm x 150 mm		8	3.500		3,375	112	19.10	3,506.10	3,925	
	8340		500 mm x 300 mm		8	3.500		3,625	112	19.10	3,756.10	4,200	
	8400		600 mm x 300 mm		6	4.667		4,450	150	25.50	4,625.50	5,125	
	8420		600 mm x 450 mm		6	4.667		6,625	150	25.50	6,800.50	7,525	
	8800		Curb box, 1.8 m long	B-20	20	1.200		140	37.50		177.50	212	
	8820		2.4 m long	"	18	1.333		144	41.50		185.50	224	
720	0010		**WATER SUPPLY, CONCRETE PIPE**										720
	0020		Not including excavation or backfill, without gaskets										
	3000		Conc cylinder pipe(CCP),1034 kPa,12.2m, 300mm dia, [12″]	B-13	58.52	.957	m	181	28.50	11.10	220.60	255	
	3010		600 mm diameter [24″]	"	39.01	1.435		225	43	16.70	284.70	330	
	3040		900 mm diameter [36″]	B-13B	29.26	1.914		278	57.50	32.50	368	430	
	3050		1200 mm diameter, [48″]		19.51	2.871		380	86	49	515	605	
	3060		Prestressed(PCCP),1034 kPa,7.5m, 1500 mm dia, [60″]		18.29	3.062		625	91.50	52	768.50	890	
	3070		1800 mm diameter, [72″]		18.29	3.062		785	91.50	52	928.50	1,050	
	3080		2100 mm diameter, [84″]		12.19	4.593		1,025	138	78	1,241	1,425	
	3090		2400 mm diameter, [96″]	B-13C	12.19	4.593		1,500	138	130	1,768	2,025	
	3100		2700 mm diameter, [108″]		9.75	5.741		2,075	172	163	2,410	2,725	
	3102		3000 mm diameter, [120″]		4.88	11.483		3,050	345	325	3,720	4,275	
	3104		3600 mm diameter, [144″]		4.88	11.483		3,775	345	325	4,445	5,050	
	3110		Conc cylinder pipe(CCP),1034 kPa, elbow, 90°, 300 mm dia, [12″]	B-13	32	1.750	Ea.	192	52.50	20.50	265	315	
	3140		600 mm diameter, [24″]	"	6	9.333		345	279	108	732	930	
	3150		900 mm diameter, [36″]	B-13B	15	3.733		610	112	63.50	785.50	915	
	3160		1200 mm diameter, [48″]		12	4.667		675	140	79.50	894.50	1,050	
	3170		Prestressed(PCCP),1034 kPa, elbow, 90°, 1500 mm dia, [60″]		10	5.600		1,050	168	95	1,313	1,525	
	3180		1800 mm diameter, [72″]		4	14		10,700	420	238	11,358	12,700	
	3190		2100 mm diameter, [84″]		3.20	17.500		15,200	525	297	16,022	17,800	
	3200		2400 mm diameter, [96″]		2	28		29,600	840	475	30,915	34,400	
	3210		2700 mm diameter, [108″]	B-13C	1.20	46.667		35,000	1,400	1,325	37,725	42,100	
	3220		3000 mm diameter, [120″]		.40	140		36,300	4,200	3,975	44,475	51,000	
	3225		3600 mm diameter, [144″]		.30	184		49,400	5,525	5,225	60,150	69,000	
	3230		Conc cylinder pipe(CCP), 1034 kPa, elbow, 45°, 300 mm dia, [12″]	B-13	24	2.333		161	70	27	258	315	
	3250		600 mm diameter, [24″]	"	6	9.333		345	279	108	732	930	
	3260		900 mm diameter, [36″]	B-13B	4	14		1,850	420	238	2,508	2,950	
	3270		1200 mm diameter, [48″]		3	18.667		3,350	560	315	4,225	4,900	
	3280		Prestressed(PCCP), 1034 kPa, elbow, 45°, 1500 mm dia, [60″]		2	28		4,675	840	475	5,990	6,975	
	3290		1800 mm diameter, [72″]		1.60	35		8,300	1,050	595	9,945	11,400	
	3300		2100 mm diameter, [84″]		1.30	42.945		10,400	1,275	730	12,405	14,200	
	3310		2400 mm diameter, [96″]		1	56		12,400	1,675	950	15,025	17,200	
	3320		2700 mm diameter, [108″]	B-13C	.66	84.337		15,600	2,525	2,400	20,525	23,700	
	3330		3000 mm diameter, [120″]		.40	140		17,600	4,200	3,975	25,775	30,100	
	3340		3600 mm diameter, [144″]		.30	184		17,700	5,525	5,225	28,450	33,800	
730	0010		**WATER SUPPLY, DUCTILE IRON PIPE** cement lined	R331113 -80									730
	0020		Not including excavation or backfill										
	2000		Pipe, class 50 water piping, 5.4 m lengths										
	2020		Mechanical joint, 100 mm diameter [4″]	B-21A	60.96	.656	m	44.50	22.50	8.80	75.80	92.50	
	2040		150 mm diameter [6″]		48.77	.820		51.50	28	11	90.50	112	
	2060		200 mm diameter [8″]		40.64	.984		57	34	13.20	104.20	129	
	2080		250 mm diameter [10″]		34.84	1.148		77	39.50	15.40	131.90	161	
	2100		300 mm diameter [12″]		32.08	1.247		94.50	43	16.70	154.20	187	

SITE CONSTRUCTION **2**

For expanded coverage of these items see *Means Site Work & Landscape Cost Data 2006*

02510 | Water Distribution

		CREW	DAILY OUTPUT	LABOR-HOURS	UNIT	2006 BARE COSTS				TOTAL INCL O&P
						MAT.	LABOR	EQUIP.	TOTAL	
2120	350 mm diameter [14"] R331113-80	B-21A	30.48	1.312	m	121	45	17.60	183.60	221
2140	400 mm diameter [16"]		22.17	1.804		132	62	24	218	266
2160	450 mm diameter [18"]		21.02	1.903		165	65.50	25.50	256	310
2170	500 mm diameter [20"]		17.42	2.297		194	79	31	304	365
2180	600 mm diameter [24"]		14.34	2.789		248	96	37.50	381.50	460
3000	Tyton, push-on joint, 100 mm diameter [4"]		122	.328		25	11.25	4.40	40.65	49.50
3020	150 mm diameter [6"]		102	.392		29	13.45	5.25	47.70	58.50
3040	200 mm diameter [8"]		60.96	.656		39.50	22.50	8.80	70.80	87.50
3060	250 mm diameter [10"]		55.42	.722		60.50	25	9.70	95.20	115
3080	300 mm diameter [12"]		48.77	.820		64	28	11	103	125
3100	350 mm diameter [14"]		40.64	.984		70	34	13.20	117.20	144
3120	400 mm diameter [16"]		34.84	1.148		100	39.50	15.40	154.90	187
3140	450 mm diameter [18"]		30.48	1.312		111	45	17.60	173.60	210
3160	500 mm diameter [20"]		27.09	1.476		122	50.50	19.80	192.30	233
3180	600 mm diameter [24"]	▼	23.45	1.706	▼	162	58.50	23	243.50	294
8000	Fittings, mechanical joint									
8006	90° bend, 100 mm diameter [4"]	B-20A	16	2	Ea.	174	67		241	293
8020	150 mm diameter [6"]		12.80	2.500		210	83.50		293.50	360
8040	200 mm diameter [8"]	▼	10.67	2.999		340	100		440	525
8060	250 mm diameter [10"]	B-21A	11.43	3.500		355	120	47	522	630
8080	300 mm diameter [12"]		10.53	3.799		610	130	51	791	930
8100	350 mm diameter [14"]		10	4		965	137	53.50	1,155.50	1,325
8120	400 mm diameter [16"]		7.27	5.502		1,150	189	74	1,413	1,650
8140	450 mm diameter [18"]		6.90	5.797		1,150	199	77.50	1,426.50	1,675
8160	500 mm diameter [20"]		5.71	7.005		2,475	241	94	2,810	3,200
8180	600 mm diameter [24"]	▼	4.70	8.511		4,625	292	114	5,031	5,650
8200	Wye or tee, 100 mm diameter [4"]	B-20A	10.67	2.999		268	100		368	450
8220	150 mm diameter [6"]		8.53	3.751		360	125		485	585
8240	200 mm diameter [8"]	▼	7.11	4.501		510	150		660	795
8260	250 mm diameter [10"]	B-21A	7.62	5.249		880	180	70.50	1,130.50	1,325
8280	300 mm diameter [12"]		7.02	5.698		1,125	196	76.50	1,397.50	1,600
8300	350 mm diameter [14"]		6.67	5.997		1,250	206	80.50	1,536.50	1,775
8320	400 mm diameter [16"]		4.85	8.247		1,300	283	111	1,694	1,975
8340	450 mm diameter [18"]		4.60	8.696		3,375	299	117	3,791	4,300
8360	500 mm diameter [20"]		3.81	10.499		5,300	360	141	5,801	6,525
8380	600 mm diameter [24"]	▼	3.14	12.739		6,275	440	171	6,886	7,750
8398	45° bends, 100 mm diameter	B-20A	16	2		157	67		224	275
8400	150 mm diameter [6"]	"	12.80	2.500		213	83.50		296.50	365
8410	300 mm diameter [12"]	B-21A	10.53	3.799		585	130	51	766	900
8420	400 mm diameter [16"]		7.27	5.502		585	189	74	848	1,025
8430	500 mm diameter [20"]		5.71	7.005		2,375	241	94	2,710	3,100
8440	600 mm diameter [24"]	▼	4.70	8.511		3,350	292	114	3,756	4,250
8450	Decreaser, 150 mm x 100 mm diameter [6"-4"]	B-20A	14.22	2.250		191	75		266	325
8460	200 mm x 150 mm diameter [8"-6"]	"	11.64	2.749		279	92		371	445
8470	250 mm x 150 mm diameter [10"-6"]	B-21A	13.33	3.001		340	103	40	483	575
8480	300 mm x 150 mm diameter [12"-6"]		12.70	3.150		450	108	42	600	705
8490	400 mm x 150 mm diameter [16"-6"]		10	4		850	137	53.50	1,040.50	1,200
8500	500 mm x 150 mm diameter [20"-6"]	▼	8.42	4.751	▼	1,675	163	63.50	1,901.50	2,175
8550	Butterfly valves with boxes, cast iron									
8560	100 mm diameter [4"]	B-20	6	4	Ea.	760	124		884	1,025
8570	150 mm diameter [6"]	"	5	4.800		460	149		609	735
8580	200 mm diameter [8"]	B-21	4	7		760	225	38	1,023	1,225
8590	250 mm diameter [10"]		3.50	8		1,050	257	43.50	1,350.50	1,600
8600	300 mm diameter [12"]		3	9.333		1,350	300	51	1,701	2,000
8610	350 mm diameter [14"]		2	14		1,875	450	76.50	2,401.50	2,825
8620	400 mm diameter [16"]	▼	2	14	▼	2,400	450	76.50	2,926.50	3,425

Important: See the Reference Section for supporting data - Crews, Rental Equipment, City Cost Indexes and Reference Data

02510 | Water Distribution

		CREW	DAILY OUTPUT	LABOR-HOURS	UNIT	2006 BARE COSTS MAT.	LABOR	EQUIP.	TOTAL	TOTAL INCL O&P	
730	9600	Steel sleeve and tap, 100 mm diameter [4"] R331113-80	B-20	3	8	Ea.	430	249		679	855	**730**
	9620	150 mm diameter [6"]		2	12		505	375		880	1,125	
	9630	200 mm diameter [8"]		2	12		680	375		1,055	1,325	
740	0010	**WATER SUPPLY, POLYETHYLENE PIPE, C901**										**740**
	0020	Not including excavation or backfill										
	1000	Piping, 1103 kPa, 20 mm diameter [3/4"]	B-20	160	.150	m	1.02	4.67		5.69	8.35	
	1120	25 mm diameter [1"]		148	.162		2.33	5.05		7.38	10.40	
	1140	40 mm diameter [1-1/2"]		137	.175		3.25	5.45		8.70	12.10	
	1160	50 mm diameter [2"]		111	.216		5.10	6.70		11.80	16.05	
	2000	Fittings, insert type, nylon, 160 & 250 psi, cold water										
	2220	Clamp ring, stainless steel, 20 mm diameter [3/4"]	B-20	345	.070	Ea.	1.02	2.16		3.18	4.49	
	2240	25 mm diameter [1"]		321	.075		1.07	2.33		3.40	4.80	
	2260	40 mm diameter [1-1/2"]		285	.084		1.31	2.62		3.93	5.50	
	2280	50 mm diameter [2"]		255	.094		1.49	2.93		4.42	6.20	
	2300	Coupling, 20 mm diameter [3/4"]		66	.364		.71	11.30		12.01	18.40	
	2320	25 mm diameter [1"]		57	.421		.94	13.10		14.04	21.50	
	2340	40 mm diameter [1-1/2"]		51	.471		2.19	14.65		16.84	25.50	
	2360	50 mm diameter [2"]		48	.500		2.63	15.55		18.18	27	
	2400	Elbow, 90°, 20 mm diameter [3/4"]		66	.364		1.26	11.30		12.56	19	
	2420	25 mm diameter [1"]		57	.421		1.34	13.10		14.44	22	
	2440	40 mm diameter [1-1/2"]		51	.471		3.34	14.65		17.99	26.50	
	2460	50 mm diameter [2"]		48	.500		5.60	15.55		21.15	30	
	2500	Tee, 20 mm diameter [3/4"]		42	.571		1.34	17.75		19.09	29	
	2520	25 mm diameter [1"]		39	.615		2.21	19.15		21.36	32.50	
	2540	40 mm diameter [1-1/2"]		33	.727		4.97	22.50		27.47	40.50	
	2560	50 mm diameter [2"]		30	.800		6.70	25		31.70	46	
750	0010	**WATER SUPPLY, POLYVINYL CHLORIDE PIPE**										**750**
	0020	Not including excavation or backfill, unless specified										
	2100	AWWA Class 160, S.D.R. 26, 40 mm diameter [1-1/2"]	B-20	229	.105	m	1.51	3.26		4.77	6.70	
	2120	50 mm diameter [2"]		209	.115		3.64	3.57		7.21	9.55	
	2140	65 mm diameter [2-1/2"]		152	.158		5.40	4.91		10.31	13.60	
	2160	80 mm diameter [3"]		131	.183		7.75	5.70		13.45	17.40	
	2180	100 mm diameter [4"]		114	.211		12.65	6.55		19.20	24	
	2200	150 mm diameter [6"]		96.32	.249		27.50	7.75		35.25	42	
	2210	200mm diameter [8"]		79.25	.303		46	9.40		55.40	65.50	
	3010	AWWA C905, PR 100, DR 41										
	3030	350 mm diameter [14"]	B-20A	64.92	.493	m	42	16.45		58.45	71.50	
	3040	400mm diameter [16"]		60.96	.525		56	17.55		73.55	88.50	
	3050	450mm diameter [18"]		48.77	.656		70	22		92	111	
	3060	500mm diameter [20"]		40.54	.789		87.50	26.50		114	137	
	3070	600mm diameter [24"]		32.61	.981		123	33		156	185	
	3080	750mm diameter [30"]		24.38	1.312		190	44		234	276	
	3090	900mm diameter [36"]		24.38	1.312		268	44		312	360	
	3100	1050 mm diameter		18.29	1.750		365	58.50		423.50	490	
	3200	1200 mm diameter		18.29	1.750		475	58.50		533.50	615	
	4520	Class 150, SDR 18, AWWA C900, 100 mm [4"]		116	.276		8.45	9.20		17.65	23.50	
	4530	150 mm [6"]		96.32	.332		16.65	11.10		27.75	35.50	
	4540	200 mm [8"]		80.47	.398		29	13.30		42.30	52	
	4550	250 mm [10"]		67.06	.477		43.50	15.95		59.45	72.50	
	4560	300 mm [12"]		56.69	.564		61.50	18.85		80.35	97	
	8000	Fittings with rubber gasket										
	8003	Class 150, D.R. 18										
	8006	90° Bent, 100mm diameter [4"]	B-20	100	.240	Ea.	51	7.45		58.45	68	
	8020	150mm diameter [6"]		90	.267		91	8.30		99.30	113	

SITE CONSTRUCTION 2

SITE CONSTRUCTION **2**

02510	Water Distribution	CREW	DAILY OUTPUT	LABOR-HOURS	UNIT	2006 BARE COSTS				TOTAL INCL O&P
						MAT.	LABOR	EQUIP.	TOTAL	
750 8040	200mm diameter [8"]	B-20	80	.300	Ea.	176	9.35		185.35	209	**750**
8060	250mm diameter [10"]		50	.480		269	14.95		283.95	320
8080	300mm diameter [12"]		30	.800		410	25		435	490
8100	Tee, 100mm diameter [4"]		90	.267		70.50	8.30		78.80	90.50
8120	150mm diameter [6"]		80	.300		158	9.35		167.35	189
8140	200mm diameter [8"]		70	.343		257	10.65		267.65	300
8160	250mm diameter [10"]		40	.600		440	18.65		458.65	515
8180	300mm diameter [12"]		20	1.200		615	37.50		652.50	740
8200	45° Bend, 100mm diameter [4"]		100	.240		50.50	7.45		57.95	67
8220	150mm diameter [6"]		90	.267		89	8.30		97.30	110
8240	200mm diameter [8"]		50	.480		270	14.95		284.95	320
8260	250mm diameter [10"]		50	.480		410	14.95		424.95	480
8280	300mm diameter [12"]		30	.800		545	25		570	640
8300	Reducing tee 150mm x 100mm [6"x4"]		100	.240		153	7.45		160.45	181
8320	200mm x 150mm [8" x 6"]		90	.267		249	8.30		257.30	287
8330	250mm x 150mm [10"x6"]		90	.267		288	8.30		296.30	330
8340	250mm x 200mm [10"x8"]		90	.267		310	8.30		318.30	355
8350	300mm x 150mm [12"x6"]		90	.267		385	8.30		393.30	440
8360	300mm x 200mm [12"x8"]		90	.267		430	8.30		438.30	485
8400	Tapped service tee (threaded) 150mm x 20mm [6" x 6"x 3/4"]		100	.240		54.50	7.45		61.95	71.50
8420	150mm x 150mm x 20 mm [6" x 6" x 3/4"]		90	.267		54.50	8.30		62.80	73
8430	150mm x 150mm x 25mm [6" x 6" x 1]		90	.267		54.50	8.30		62.80	73
8440	150mm x 150mm x 40mm [6" x 6" x 1 1/2"]		90	.267		54.50	8.30		62.80	73
8450	150mm x 150mm x 50mm [6" x 6" x 2"]		90	.267		54.50	8.30		62.80	73
8460	200mm x 200mm x 20mm [8" x 8" x 3/4"]		90	.267		230	8.30		238.30	266
8470	200mm x 200mm x 25mm [8" x 8" x 1"]		90	.267		230	8.30		238.30	266
8480	200mm x 200mm x 40mm [8" x 8" x 1 1/2"]		90	.267		230	8.30		238.30	266
8490	200mm x 200mm x 50mm [8" x 8" x 2"]		90	.267		230	8.30		238.30	266
8500	Repair coupling 100mm [4"]		100	.240		32.50	7.45		39.95	47
8520	150mm diameter [6"]		90	.267		50	8.30		58.30	68
8540	200mm diameter [8"]		50	.480		93.50	14.95		108.45	126
8560	250mm diameter [10"]		50	.480		143	14.95		157.95	180
8580	300 mm diameter [12"]		50	.480		208	14.95		222.95	252
8600	Plug end 100mm [4"]		100	.240		27.50	7.45		34.95	42
8620	150mm diameter [6"]		90	.267		49.50	8.30		57.80	67.50
8640	200mm diameter [8"]		50	.480		70.50	14.95		85.45	101
8660	250mm diameter [10"]		50	.480		91.50	14.95		106.45	124
8680	300mm diameter [12"]		50	.480		118	14.95		132.95	152
760 0010	**WATER SUPPLY, HDPE**, butt fusion joints, SDR 21, 12 m lengths										**760**
0100	100 mm diameter	B-22A	122	.311	m	6.55	9.80	4.98	21.33	28
0200	150 mm diameter		116	.328		9.40	10.30	5.25	24.95	32
0300	200 mm diameter		97.54	.390		15.90	12.25	6.25	34.40	43.50
0400	250 mm diameter [10"]		91.44	.416		24.50	13.05	6.65	44.20	54.50
0500	300 mm diameter [12"]		79.25	.480		34.50	15.05	7.65	57.20	70
0600	350 mm diameter [14"]		67.06	.567		42	17.80	9.05	68.85	83.50
0700	400 mm diameter [16"]		54.86	.693		54.50	22	11.05	87.55	106
0800	450 mm diameter [18"]		42.67	.891		69	28	14.25	111.25	135
0900	600 mm diameter [24"]		30.48	1.247		123	39	19.90	181.90	218
1000	Fittings									
1100	Elbows, 90 degrees									
1200	100 mm diameter [4"]	B-22B	32	.500	Ea.	27.50	16	4.64	48.14	60
1300	150 mm diameter [6"]		28	.571		72.50	18.25	5.30	96.05	114
1400	200 mm diameter [8"]		24	.667		182	21.50	6.20	209.70	240
1500	250 mm diameter [10"]		18	.889		315	28.50	8.25	351.75	400
1600	300 mm diameter [12"]		12	1.333		530	42.50	12.35	584.85	665
1700	350 mm diameter [14"]		9	1.778		530	57	16.50	603.50	685

Important: See the Reference Section for supporting data - Crews, Rental Equipment, City Cost Indexes and Reference Data

02510 | Water Distribution

		CREW	DAILY OUTPUT	LABOR-HOURS	UNIT	2006 BARE COSTS				TOTAL INCL O&P
						MAT.	LABOR	EQUIP.	TOTAL	
760	**1800**	400 mm diameter [16"]	B-22B	6	2.667	Ea.	635	85	25	745	860	**760**
	1900	450 mm diameter [18"]		4	4		860	128	37	1,025	1,175
	2000	600 mm diameter [24"]		3	5.333		1,750	170	49.50	1,969.50	2,250
	2100	Tees									
	2200	100 mm diameter [4"]	B-22B	30	.533	Ea.	38	17.05	4.95	60	74
	2300	150 mm diameter [6"]		26	.615		101	19.65	5.70	126.35	148
	2400	200 mm diameter [8"]		22	.727		255	23	6.75	284.75	325
	2500	250 mm diameter [10"]		15	1.067		335	34	9.90	378.90	435
	2600	300 mm diameter [12"]		10	1.600		460	51	14.85	525.85	600
	2700	350 mm diameter [14"]		8	2		545	64	18.55	627.55	715
	2800	400 mm diameter [16"]		6	2.667		640	85	25	750	865
	2900	450 mm diameter [18"]		4	4		860	128	37	1,025	1,175
	3000	600 mm diameter [24"]		2	8		1,775	256	74	2,105	2,450
770	**0010**	**WATER SUPPLY, BLACK STEEL PIPE**										**770**
	0011	Not including excavation or backfill									
	1000	Pipe, black steel, plain end, welded, 8 mm wall thk, 200 mm diam.	B-35A	32.92	1.701	m	29.50	57	40	126.50	164
	1010	250 mm diam., [10"]		31.70	1.767		36.50	59	41.50	137	177
	1020	300 mm diam., [12"]		59.44	.942		44	31.50	22	97.50	122
	1030	450 mm diam., [18"]		53.34	1.050		67.50	35	24.50	127	155
	1040	9 mm wall thickness, 300 mm diam., [12"]		59.44	.942		56.50	31.50	22	110	136
	1050	450 mm diam., [18"]		18.04	3.104		99	104	73	276	350
	1060	900 mm diam., [36"]		8.83	6.344		198	212	149	559	705
	1070	10 mm wall thickness, 450 mm diam., [18"]		13.17	4.253		99	142	100	341	435
	1080	600 mm diam., [24"]		10.97	5.104		133	171	120	424	540
	1090	750 mm diam., [30"]		9.27	6.044		175	202	142	519	660
	1100	15 mm wall thickness, 900 mm, [36"] diam.		7.95	7.045		282	235	165	682	850
	1110	1200 mm, [48"] diameter		6.61	8.475		420	283	199	902	1,125
	1135	11 mm wall thickness, 1200 mm diameter		6.34	8.833		365	295	207	867	1,100
	1140	18 mm wall thickness, 1200 mm, [48"] diameter		6.61	8.475		520	283	199	1,002	1,225
780	**0010**	**WATER SUPPLY, COPPER PIPE** R221113 -50										**780**
	0020	Not including excavation or backfill									
	2000	Tubing, type K, 6 m joints, 20 mm diameter [3/4"]	Q-1	122	.131	m	10.55	5.05		15.60	19.20
	2200	25 mm diameter [1"]		97.54	.164		15	6.30		21.30	26
	3000	40 mm diameter [1-1/2"]		80.77	.198		23	7.60		30.60	37
	3020	50 mm diameter [2"]		70.10	.228		36	8.75		44.75	52.50
	3040	65 mm diameter [2-1/2"]		44.50	.360		53.50	13.80		67.30	79.50
	3060	80 mm diameter [3"]		40.84	.392		74.50	15.05		89.55	105
	4012	100 mm diameter [4"]		28.96	.553		124	21		145	168
	4016	150 mm diameter [6"]	Q-2	28.96	.829		375	33		408	460
	5000	Tubing, type L									
	5108	50 mm diameter [2"]	Q-1	32	.500	m	29.50	19.20		48.70	61.50
	6010	80 mm diameter [3"]		42.67	.375		60.50	14.40		74.90	88
	6012	100 mm diameter [4"]		28.96	.553		101	21		122	143
	6016	150 mm diameter [6"]	Q-2	30.48	.787		345	31.50		376.50	425
	7020	Fittings, brass, corporation stops, 20 mm diameter [3/4"]	1 Plum	19	.421	Ea.	16.65	18		34.65	45.50
	7040	25 mm diameter [1"]		16	.500		29	21.50		50.50	64
	7060	40 mm diameter [1-1/2"]		13	.615		86.50	26.50		113	135
	7080	50 mm diameter [2"]		11	.727		144	31		175	205
	7100	Curb stops, 20 mm diameter [3/4"]		19	.421		41	18		59	72
	7120	25 mm diameter [1"]		16	.500		56.50	21.50		78	94.50
	7140	40 mm diameter [1-1/2"]		13	.615		126	26.50		152.50	178
	7160	50 mm diameter [2"]		11	.727		206	31		237	273
	7180	Curb box, cast iron, 15 mm to 25 mm curb stops		12	.667		59	28.50		87.50	108
	7200	32 mm to 50 mm curb stops		8	1		108	42.50		150.50	184
	7220	Saddles, 20 mm diameter, add [3/4"]					27.50			27.50	30.50

SITE CONSTRUCTION 2

For expanded coverage of these items see *Means Site Work & Landscape Cost Data 2006*

02510 | Water Distribution

				DAILY OUTPUT	LABOR-HOURS	UNIT	2006 BARE COSTS				TOTAL INCL O&P	
			CREW				MAT.	LABOR	EQUIP.	TOTAL		
780	7240	50 mm diameter, add [2"]	R221113-50			Ea.	32.50			32.50	36	780
	7250	For copper fittings, see Div. 15107-460										

02520 | Wells

			CREW	DAILY OUTPUT	LABOR-HOURS	UNIT	MAT.	LABOR	EQUIP.	TOTAL	INCL O&P	
510	0010	**WELLS & ACCESSORIES**, domestic										510
	0100	Drilled, 100 mm to 150 mm diameter	B-23	36.58	1.094	m		30.50	97	127.50	155	
	0200	200 mm diameter [8"]	"	29.02	1.379	"		38.50	123	161.50	195	
	0400	Gravel pack well, 12 m deep, incl. gravel & casing, complete										
	0500	600 mm diameter casing x 450 mm diameter screen [24",18"]	B-23	.13	307	Total	24,800	8,550	27,300	60,650	70,500	
	0600	900 mm diameter casing x 450 mm diameter screen [36",18"]	"	.12	333	"	26,600	9,275	29,600	65,475	76,500	
	0800	Observation wells, 32 mm riser pipe	↓	49.68	.805	m	44.50	22.50	71.50	138.50	163	
	0900	For flush Buffalo roadway box, add	1 Skwk	16.60	.482	Ea.	37	17.60		54.60	68.50	
	1200	Test well, 65 mm diameter, up to 15 m deep (1 to 3 L/s) [2-1/2"]	B-23	1.51	26.490	"	555	735	2,350	3,640	4,375	
	1300	Over 15 m deep, add	"	37.12	1.077	m	49	30	96	175	205	
	1500	Pumps, installed in wells to 30 m deep, 100 mm submersible										
	1510	373 W	Q-1	3.22	4.969	Ea.	345	191		536	665	
	1520	560 W		2.66	6.015		400	231		631	790	
	1600	746 W	↓	2.29	6.987		425	269		694	870	
	1700	1.1 kW	Q-22	1.60	10		1,100	385	400	1,885	2,225	
	1800	1.5 kW		1.33	12.030		1,275	460	480	2,215	2,625	
	1900	2.2 kW		1.14	14.035		1,475	540	560	2,575	3,050	
	2000	3.7 kW		1.14	14.035		2,000	540	560	3,100	3,625	
	2050	Remove and install motor only, 3 kW		1.14	14.035		755	540	560	1,855	2,250	
	3000	Pump, 150 mm submersible, 8 - 45 m deep, 19 kW, 16 - 19 L/s		.89	17.978		4,600	690	715	6,005	6,875	
	3100	8 m - 150 m deep, 22 kW, 6 - 19 L/s	↓	.73	21.918	↓	5,225	840	870	6,935	7,975	
	8000	Steel well casing	B-23A	1370	.018	kg	1.39	.55	2.52	4.46	5.15	
	8110	Well screen assembly, stainless steel, 50 mm diameter [2"]		83.21	.288	m	171	9	41.50	221.50	247	
	8120	80 mm diameter [3"]		77.11	.311		236	9.70	45	290.70	325	
	8130	100 mm diameter [4"]		60.96	.394		271	12.25	56.50	339.75	380	
	8140	125 mm diameter [5"]		51.21	.469		315	14.60	67.50	397.10	445	
	8150	150 mm diameter [6"]		38.40	.625		380	19.50	90	489.50	545	
	8160	200 mm diameter [8"]		30.02	.799		495	25	115	635	705	
	8170	250 mm diameter [10"]		22.25	1.079		620	33.50	155	808.50	905	
	8180	300 mm diameter [12"]		19.05	1.260		730	39.50	181	950.50	1,075	
	8190	350 mm diameter [14"]		16.55	1.450		830	45	209	1,084	1,200	
	8200	400 mm diameter [16"]		14.72	1.630		910	51	235	1,196	1,325	
	8210	450 mm diameter [18"]		11.95	2.009		1,125	62.50	289	1,476.50	1,675	
	8220	500 mm diameter [20"]		9.51	2.524		1,300	78.50	365	1,743.50	1,950	
	8230	600 mm diameter [24"]		7.25	3.308		1,600	103	475	2,178	2,425	
	8240	650 mm diameter [26"]		6.40	3.750		1,800	117	540	2,457	2,750	
	8300	Slotted PVC, 32 mm diameter [1-1/4"]		159	.151		6	4.70	21.50	32.20	38	
	8310	40 mm diameter [1-1/2"]		149	.161		8.75	5	23	36.75	43	
	8320	50 mm diameter [2"]		83.21	.288		12.05	9	41.50	62.55	72.50	
	8330	80 mm diameter [3"]		77.11	.311		12.85	9.70	45	67.55	78.50	
	8340	100 mm diameter [4"]		60.96	.394		14.85	12.25	56.50	83.60	97.50	
	8350	125 mm diameter [5"]		51.21	.469		15.55	14.60	67.50	97.65	114	
	8360	150 mm diameter [6"]		38.40	.625		21.50	19.50	90	131	153	
	8370	200 mm diameter [8"]	↓	30.02	.799		32.50	25	115	172.50	202	
	8400	Artificial gravel pack, 50 mm screen, 150 mm casing [6"]	B-23B	53.04	.453		9.80	14.10	69.50	93.40	109	
	8405	200 mm casing [8"]		33.83	.709		13.40	22	109	144.40	169	
	8410	250 mm casing [10"]		22.71	1.057		16.85	33	163	212.85	248	
	8415	300 mm casing [12"]		18.29	1.312		22.50	41	202	265.50	310	
	8420	350 mm casing [14"]		15.30	1.569		29.50	49	242	320.50	375	
	8425	400 mm casing [16"]		12.41	1.935		40.50	60.50	298	399	470	
	8430	450 mm casing [18"]		10.97	2.187		47	68	335	450	525	
	8435	500 mm casing [20"]	↓	8.99	2.669	↓	54.50	83	410	547.50	640	

2 SITE CONSTRUCTION

02520	Wells	CREW	DAILY OUTPUT	LABOR-HOURS	UNIT	2006 BARE COSTS				TOTAL INCL O&P	
						MAT.	LABOR	EQUIP.	TOTAL		
510 8440	600 mm casing [24"]	B-23B	7.83	3.064	m	60	95.50	470	625.50	735	**510**
8445	650 mm casing [26"]		7.50	3.201		67.50	100	495	662.50	765	
8450	750 mm casing [30"]		6.10	3.937		77	123	605	805	940	
8455	900 mm casing [36"]		5	4.801		83	150	740	973	1,125	
8500	Develop well		8	3	Hr.	224	93.50	460	777.50	900	
8550	Pump test well		8	3		59.50	93.50	460	613	720	
8560	Standby well	B-23A	8	3		58	93.50	430	581.50	685	
8570	Standby, drill rig		8	3			93.50	430	523.50	620	
8580	Surface seal well, concrete filled		1	24	Ea.	565	750	3,450	4,765	5,575	
8590	Well test pump, install & remove	B-23	1	40			1,100	3,550	4,650	5,625	
8600	Well sterilization, chlorine	2 Clab	1	16		415	440		855	1,150	
9950	See Div. 02240-900 for wellpoints										
9960	See Div. 02240-700 for drainage wells										
520 0010	**WATER SUPPLY WELLS, PUMPS** with pressure control										**520**
1000	Deep well, jet, 159 L galvanized tank										
1040	559 W	1 Plum	.80	10	Ea.	635	425		1,060	1,350	
3000	Shallow well, jet, 114 L galvanized tank										
3040	373 W	1 Plum	2	4	Ea.	400	171		571	695	

02530	Sanitary Sewerage	CREW	DAILY OUTPUT	LABOR-HOURS	UNIT	MAT.	LABOR	EQUIP.	TOTAL	TOTAL INCL O&P	
300 0010	**PACKAGED LIFT STATIONS**										**300**
2500	Sewage lift station, 757 000 L/day	E-8	.20	520	Ea.	129,000	20,400	8,925	158,325	186,500	
2510	1 892 500 L/day		.15	684		144,500	26,800	11,700	183,000	218,500	
2520	3 028 000 L/day		.13	812		173,500	31,800	13,900	219,200	261,500	
720 0010	**SEWAGE COLLECTION, VENT CAST IRON PIPE**										**720**
0020	Not including excavation or backfill										
2022	Sewage vent cast iron, B & S, 100 mm diameter [4"]	Q-1	20.12	.795	m	27.50	30.50		58	76	
2024	125 mm diameter [5"]	Q-2	26.82	.895		39	35.50		74.50	96	
2026	150 mm diameter [6"]	"	25.60	.937		47	37.50		84.50	108	
2028	200 mm diameter [8"]	Q-3	21.34	1.500		75	61		136	174	
2030	250 mm diameter [10"]		20.12	1.591		124	64.50		188.50	234	
2032	300 mm diameter [12"]		17.37	1.842		178	75		253	310	
2034	375 mm diameter [15"]		14.94	2.143		256	87		343	415	
8001	Fittings, bends and elbows										
8110	100 mm diameter [4"]	Q-1	13	1.231	Ea.	27	47.50		74.50	101	
8112	125 mm diameter [5"]	Q-2	18	1.333		39	53		92	123	
8114	150 mm diameter [6"]	"	17	1.412		46.50	56.50		103	136	
8116	200 mm diameter [8"]	Q-3	11	2.909		133	118		251	325	
8118	250 mm diameter [10"]		10	3.200		196	130		326	410	
8120	300 mm diameter [12"]		9	3.556		263	145		408	505	
8122	375 mm diameter [15"]		7	4.571		765	186		951	1,125	
8500	Wyes and tees										
8510	100 mm diameter [4"]	Q-1	8	2	Ea.	44	77		121	165	
8512	125 mm diameter [5"]	Q-2	12	2		72.50	79.50		152	200	
8514	150 mm diameter [6"]	"	11	2.182		93	87		180	233	
8516	200 mm diameter [8"]	Q-3	7	4.571		223	186		409	525	
8518	250 mm diameter [10"]		6	5.333		355	217		572	720	
8520	300 mm diameter [12"]		4	8		585	325		910	1,125	
8522	375 mm diameter [15"]		3	10.667		1,175	435		1,610	1,950	
730 0010	**SEWAGE COLLECTION, CONCRETE PIPE**										**730**
0020	See 02630-530 for sewage/drainage collection, concrete pipe										
770 0010	**SEWAGE COLLECTION, PLASTIC PIPE**										**770**
0020	Not including excavation & backfill										

SITE CONSTRUCTION 2

		02530 | **Sanitary Sewerage**	CREW	DAILY OUTPUT	LABOR-HOURS	UNIT	2006 BARE COSTS				TOTAL INCL O&P	
							MAT.	LABOR	EQUIP.	TOTAL		
770	1100	Piping, DWV Sch 40 ABS, 100 mm diameter [4"]	B-20	114	.211	m	5.80	6.55		12.35	16.55	**770**
	1110	150 mm diameter [6"]		107	.224	"	49.50	7		56.50	65.50	
	1120	Fitting, 1/4 bend, 100 mm [4"]		19	1.263	Ea.	22.50	39.50		62	85.50	
	1130	150 mm [6"]		15	1.600		15.10	50		65.10	94	
	1140	Tee, 100 mm [4"]		12	2	▼	22.50	62		84.50	122	
	3000	Piping HDPE Corr. Type S w/ watertight gaskets, 100 mm diameter [4"]		130	.185	m	2.95	5.75		8.70	12.20	
	3020	150 mm diameter [6"]		122	.197		6.75	6.10		12.85	16.95	
	3040	200 mm diameter [8"]		116	.207		12.95	6.45		19.40	24.50	
	3060	250 mm diameter [10"]		113	.212		17.90	6.60		24.50	30	
	3080	300 mm diameter [12"]		104	.231		19.95	7.20		27.15	33	
	3100	375 mm diameter [15"]	▼	91.44	.262		27	8.15		35.15	42.50	
	3120	450 mm diameter [18"]	B-21	83.82	.334		38.50	10.70	1.82	51.02	61	
	3140	600 mm diameter [24"]		76.20	.367		60	11.80	2	73.80	86.50	
	3160	750 mm diameter [30"]		60.96	.459		95	14.75	2.50	112.25	131	
	3180	900 mm diameter [36"]		54.86	.510		120	16.40	2.78	139.18	161	
	3200	1050 mm diameter [42"]		53.34	.525		168	16.85	2.86	187.71	214	
	3220	1200 mm diameter [48"]		51.82	.540		218	17.35	2.95	238.30	270	
	3240	1350 mm diameter [54"]		48.77	.574		335	18.45	3.13	356.58	400	
	3260	1500 mm diameter [60"]	▼	45.72	.612	▼	395	19.65	3.34	417.99	465	
	3300	Watertight elbows, 300 mm diameter	B-20	11	2.182	Ea.	67.50	68		135.50	180	
	3320	375 mm diameter	"	9	2.667		104	83		187	243	
	3340	450 mm diameter	B-21	9	3.111		171	100	16.95	287.95	360	
	3360	600 mm diameter		9	3.111		365	100	16.95	481.95	575	
	3380	750 mm diameter		8	3.500		580	112	19.10	711.10	835	
	3400	900 mm diameter		8	3.500		745	112	19.10	876.10	1,025	
	3420	1050 mm diameter		6	4.667		945	150	25.50	1,120.50	1,300	
	3440	1200 mm diameter	▼	6	4.667		1,450	150	25.50	1,625.50	1,825	
	3460	Watertight tee, 300 mm diameter	B-20	7	3.429		151	107		258	330	
	3480	375 mm diameter	"	6	4		180	124		304	390	
	3500	450 mm diameter	B-21	6	4.667		264	150	25.50	439.50	550	
	3520	600 mm diameter		5	5.600		345	180	30.50	555.50	690	
	3540	750 mm diameter		5	5.600		655	180	30.50	865.50	1,025	
	3560	900 mm diameter		4	7		850	225	38	1,113	1,325	
	3580	1050 mm diameter		4	7		1,475	225	38	1,738	2,000	
	3600	1200 mm diameter	▼	4	7	▼	2,425	225	38	2,688	3,050	
780	0010	**SEWAGE COLLECTION, POLYVINYL CHLORIDE PIPE**										**780**
	0020	Not including excavation or backfill										
	2000	3 m lengths, S.D.R. 35, B&S, 100 mm diameter [4"]	B-20	114	.211	m	8.30	6.55		14.85	19.30	
	2040	150 mm diameter [6"]		107	.224		17.45	7		24.45	30	
	2080	200 mm diameter [8"]	▼	102	.235		26.50	7.30		33.80	40.50	
	2120	250 mm diameter [10"]	B-21	101	.277		38	8.90	1.51	48.41	57.50	
	2160	300 mm diameter [12"]		97.54	.287		38	9.20	1.56	48.76	57.50	
	2200	375 mm diameter [15"]	▼	57.91	.483	▼	55.50	15.50	2.64	73.64	88	
	3040	Fittings, bends or elbows, 100 mm diameter [4"]	B-20	19	1.263	Ea.	5.35	39.50		44.85	67	
	3080	150 mm diameter [6"]		15	1.600		16.95	50		66.95	96	
	3120	Tees, 100 mm diameter [4"]	▼	12	2		5.55	62		67.55	103	
	3160	150 mm diameter [6"]	2 Skwk	10	1.600		23	58.50		81.50	117	
	3200	Wyes, 100 mm diameter [4"]	"	12	1.333		7.35	48.50		55.85	83.50	
	3240	150 mm diameter [6"]	B-20	10	2.400	▼	31	74.50		105.50	151	
	4000	Piping, DWV PVC, no excav/backfill, 3 m long, Sch. 40, 100 mm dia.		114	.211	m	6.75	6.55		13.30	17.60	
	4010	150 mm diameter		107	.224		14.55	7		21.55	27	
	4020	200 mm diameter	▼	102	.235	▼	43	7.30		50.30	59	

2

SITE CONSTRUCTION

02540	Septic Tank Systems	CREW	DAILY OUTPUT	LABOR-HOURS	UNIT	2006 BARE COSTS				TOTAL INCL O&P
						MAT.	LABOR	EQUIP.	TOTAL	
400	**0010** **SEPTIC TANKS**									**400**
0015	Septic tanks, not incl excav. or piping, precast, 4 kL	B-21	8	3.500	Ea.	585	112	19.10	716.10	840
0020	4.5 kL		8	3.500		820	112	19.10	951.10	1,100
0060	5.5 kL		7	4		965	128	22	1,115	1,275
0100	7.5 kL		5	5.600		1,175	180	30.50	1,385.50	1,575
0140	9.5 kL		5	5.600		1,550	180	30.50	1,760.50	2,000
0180	15 kL	▼	4	7		4,525	225	38	4,788	5,375
0200	19 kL	B-13	3.50	16		5,900	480	186	6,566	7,450
0220	19 kL, 4 piece	"	3	18.667		7,225	560	217	8,002	9,050
0300	57 kL, 4 piece	B-13B	1.70	32.941		16,400	985	560	17,945	20,100
0400	95 kL, 4 piece		1.10	50.909		34,600	1,525	865	36,990	41,400
0500	150 kL, 4 piece	▼	.80	70		40,800	2,100	1,200	44,100	49,300
0520	190 kL, 5 piece	B-13C	.60	93.333		46,900	2,800	2,650	52,350	58,500
0540	Cast-in-place, 285 kL	C-14C	.25	448		57,000	15,200	85	72,285	87,000
0560	380 kL	"	.15	746		70,500	25,300	142	95,942	117,500
0600	High density polyethylene, 4 kL	B-21	6	4.667		910	150	25.50	1,085.50	1,250
0700	5.5 kL		4	7		1,175	225	38	1,438	1,675
0900	Galley, 1200 mm x 1200 mm x 1200 mm	▼	16	1.750		216	56	9.55	281.55	335
1000	Distribution boxes, concrete, 7 outlets	2 Clab	16	1		109	27.50		136.50	163
1100	9 outlets	"	8	2		295	55		350	410
1150	Leaching field chambers, 3900 x 1025 x 400 mm, std.	B-13	16	3.500		590	105	40.50	735.50	855
1200	Heavy duty, 2400 mm x 1200 mm x 460 mm		14	4		410	120	46.50	576.50	685
1300	3900 mm x 1125 mm x 450 mm		12	4.667		1,450	140	54	1,644	1,850
1350	6000 mm x 1200 mm x 460 mm	▼	5	11.200		955	335	130	1,420	1,700
1400	Leaching pit, precast concrete, 900 mm diameter, 900 mm deep	B-21	8	3.500		385	112	19.10	516.10	620
1500	1800 mm diameter, 900 mm section		4.70	5.957		680	191	32.50	903.50	1,075
1600	Leaching pit, 1950 mm diameter, 1800 mm deep		5	5.600		655	180	30.50	865.50	1,025
1620	2400 mm deep		4	7		770	225	38	1,033	1,250
1700	2400 mm diameter, H-20 load, 1800 mm deep		4	7		1,050	225	38	1,313	1,575
1720	2400 mm deep		3	9.333		1,300	300	51	1,651	1,975
2000	Velocity reducing pit, precast conc., 1800 mm diam, 900 mm dp.	▼	4.70	5.957	▼	325	191	32.50	548.50	690
2200	Excavation for septic tank, 0.573 m³ backhoe	B-12F	111	.144	m³		4.72	4.30	9.02	11.95
2400	1200 mm trench for disposal field, 0.573 m³ backhoe	"	102	.157	m		5.15	4.68	9.83	13
2600	Gravel fill, run of bank	B-6	115	.209	m³	21.50	6.25	1.97	29.72	35.50
2800	Crushed stone, 20 mm	"	115	.209	"	30	6.25	1.97	38.22	45

02550	Piped Energy Distribution	CREW	DAILY OUTPUT	LABOR-HOURS	UNIT	MAT.	LABOR	EQUIP.	TOTAL	INCL O&P
100	**0010** **CHILLED/HVAC HOT WATER DISTRIBUTION**									**100**
1005	Pipe, black steel w/50 mm polyurethane insul, 6 m lengths									
1010	Align & tackweld on sleepers (NIC), 32 mm (1-1/4")	B-35	263	.183	m	44	6.30	2.08	52.38	60
1020	40 mm, [1-1/2"]		251	.191		48.50	6.60	2.18	57.28	66
1030	50 mm, [2"]		207	.232		51.50	8	2.64	62.14	71.50
1040	65 mm, [2-1/2"]		171	.281		55.50	9.65	3.20	68.35	79.50
1050	80 mm, [3"]		161	.298		56.50	10.25	3.40	70.15	82
1060	100 mm, [4"]		117	.410		62	14.15	4.67	80.82	94.50
1070	125 mm, [5"]		110	.436		91.50	15	4.97	111.47	129
1080	150 mm, [6"]		90.22	.532		100	18.30	6.05	124.35	145
1090	200 mm, [8"]		80.47	.597		134	20.50	6.80	161.30	186
1100	300 mm, [12"]		65.84	.729		231	25	8.30	264.30	300
1110	On trench bottom, 450 mm, [18"]	▼	53.68	.894		360	31	10.20	401.20	455
1120	600 mm, [24"]	B-35A	44.20	1.267		535	42.50	30	607.50	690
1130	750 mm, [30"]		36.88	1.518		770	51	35.50	856.50	960
1140	900 mm, [36"]	▼	30.48	1.837	▼	1,025	61.50	43	1,129.50	1,275
1150	Fittings, elbows, on sleepers, 40 mm [1-1/2"]	Q-17	21	.762	Ea.	345	29.50	3.29	377.79	430
1160	80 mm, [3"]	▼	9.36	1.709	▼	450	66	7.40	523.40	605

SITE CONSTRUCTION **2**

			DAILY OUTPUT	LABOR-HOURS	UNIT	2006 BARE COSTS				TOTAL INCL O&P
02550	**Piped Energy Distribution**	CREW				MAT.	LABOR	EQUIP.	TOTAL	
1170	100 mm, [4"]	Q-17	8	2	Ea.	555	77.50	8.65	641.15	735
1180	150 mm, [6"]		6	2.667		720	103	11.50	834.50	960
1190	200 mm, [8"]		4.64	3.448		870	134	14.90	1,018.90	1,175
1200	Fittings, tees, 40 mm [1-1/2"]		17	.941		690	36.50	4.07	730.57	820
1210	80 mm, [3"]		8.50	1.882		855	73	8.15	936.15	1,050
1220	100 mm, [4"]	↓	6	2.667		925	103	11.50	1,039.50	1,200
1230	150 mm, [6"]	Q-17A	6.72	3.571		1,150	138	105	1,393	1,600
1240	200 mm, [8"]	"	6.40	3.750		1,350	144	110	1,604	1,825
1250	Fittings, reducer, 80 mm [3"]	Q-17	16	1		182	39	4.32	225.32	264
1260	100 mm, [4"]		12	1.333		217	51.50	5.75	274.25	325
1270	150 mm, [6"]		12	1.333		310	51.50	5.75	367.25	425
1280	200 mm, [8"]	↓	10	1.600		400	62	6.90	468.90	540
1290	Fittings, anchor, 100 mm [4"]	Q-17A	12	2		530	77	59	666	760
1300	150 mm, [6"]		10.50	2.286		610	88	67	765	875
1310	200 mm, [8"]	↓	10	2.400		680	92.50	70.50	843	965
1320	Fittings, cap, 40 mm [1-1/2"]	Q-17	42	.381		59.50	14.75	1.65	75.90	89.50
1330	80 mm, [3"]		14.64	1.093		75.50	42.50	4.72	122.72	152
1340	100 mm, [4"]		16	1		86	39	4.32	129.32	158
1350	150 mm, [6"]		16	1		148	39	4.32	191.32	226
1360	200 mm, [8"]		13.50	1.185		189	46	5.10	240.10	283
1365	300 mm [12"]	↓	11	1.455		288	56.50	6.30	350.80	405
1370	Elbow, on trench bottom, 300 mm	Q-17A	12	2		1,400	77	59	1,536	1,700
1380	450 mm, [18"]		8	3		2,475	116	88	2,679	3,000
1390	600 mm, [24"]		6	4		3,525	154	118	3,797	4,225
1400	750 mm, [30"]		5.36	4.478		4,825	173	132	5,130	5,700
1410	900 mm, [36"]		4	6		6,550	231	176	6,957	7,750
1420	Fittings, tee, 300 mm		6.72	3.571		1,825	138	105	2,068	2,325
1430	450 mm		6	4		3,050	154	118	3,322	3,700
1440	600 mm		4.64	5.172		4,750	199	152	5,101	5,700
1450	750 mm		4.16	5.769		7,275	222	170	7,667	8,550
1460	900 mm		3.36	7.143		10,700	275	210	11,185	12,400
1470	Fittings, reducer, 300 mm		10.64	2.256		580	87	66.50	733.50	845
1480	450 mm		9.68	2.479		925	95.50	73	1,093.50	1,250
1490	600 mm		8	3		1,425	116	88	1,629	1,850
1500	750 mm		7.04	3.409		2,200	131	100	2,431	2,725
1510	900 mm		5.36	4.478		3,150	173	132	3,455	3,850
1520	Fittings, anchor, 300 mm		11	2.182		735	84	64	883	1,000
1530	450 mm		6.72	3.571		785	138	105	1,028	1,200
1540	600 mm		6.32	3.797		1,100	146	112	1,358	1,550
1550	750 mm		4.64	5.172		1,325	199	152	1,676	1,925
1560	900 mm	↓	4	6	↓	1,625	231	176	2,032	2,325
1565	Weld in place and install shrink collar									
1570	On sleepers, 40 mm	Q-17A	18.50	1.297	Ea.	15.95	50	38	103.95	135
1580	80 mm, [3"]		6.72	3.571		27	138	105	270	355
1590	100 mm, [4"]		5.36	4.478		27	173	132	332	435
1600	150 mm, [6"]		4	6		32	231	176	439	580
1610	200 mm, [8"]		3.36	7.143		49.50	275	210	534.50	700
1620	300 mm, [12"]		2.64	9.091		64	350	267	681	895
1630	On trench bottom, 450 mm, [18"]		2	12		96	460	355	911	1,200
1640	600 mm, [24"]		1.36	17.647		128	680	520	1,328	1,725
1650	750 mm, [30"]		1.04	23.077		160	890	680	1,730	2,275
1660	900 mm, [36"]	↓	1	24	↓	192	925	705	1,822	2,375
0010	**PIPE CONDUIT, PREFABRICATED / PREINSULATED**									
0020	Does not include trenching, fittings or crane.									
0300	For cathodic protection, add 12 to 14%									
0310	of total built-up price (casing plus service pipe)									

Important: See the Reference Section for supporting data - Crews, Rental Equipment, City Cost Indexes and Reference Data

02550	Piped Energy Distribution	CREW	DAILY OUTPUT	LABOR-HOURS	UNIT	2006 BARE COSTS				TOTAL INCL O&P
						MAT.	LABOR	EQUIP.	TOTAL	
200 0580	Polyurethane insulated system, 394° K. max. temp.									**200**
0620	Black steel service pipe, standard wt., 13 mm insulation [1/2"]									
0660	19 mm diam. pipe size [3/4"]	Q-17	16.46	.972	m	86.50	37.50	4.20	128.20	157
0670	25 mm diam. pipe size [1"]		15.24	1.050		95	40.50	4.54	140.04	171
0680	32 mm diam. pipe size [1-1/4"]		14.33	1.117		106	43.50	4.82	154.32	187
0690	38 mm diam. pipe size [1-1/2"]		13.72	1.167		115	45	5.05	165.05	201
0700	51 mm diam. pipe size [2"]		12.80	1.250		120	48.50	5.40	173.90	211
0710	64 mm diam. pipe size [2-1/2"]		10.36	1.544		121	60	6.65	187.65	231
0720	76 mm diam. pipe size [3"]		8.53	1.875		141	72.50	8.10	221.60	273
0730	102 mm diam. pipe size [4"]		6.71	2.386		177	92.50	10.30	279.80	345
0740	127 mm diam. pipe size [5"]	▼	5.49	2.916		226	113	12.60	351.60	435
0750	152 mm diam. pipe size [6"]	Q-18	7.01	3.423		263	138	9.85	410.85	505
0760	203 mm diam. pipe size [8"]		5.79	4.144		385	167	11.95	563.95	690
0770	254 mm diam. pipe size [10"]		4.88	4.921		510	198	14.15	722.15	875
0780	305 mm diam. pipe size [12"]		3.96	6.057		630	243	17.45	890.45	1,075
0790	356 mm diam. pipe size [14"]		3.35	7.158		705	288	20.50	1,013.50	1,225
0800	406 mm diam. pipe size [16"]		3.05	7.874		810	315	22.50	1,147.50	1,400
0810	457 mm diam. pipe size [18"]		2.44	9.843		930	395	28.50	1,353.50	1,650
0820	508 mm diam. pipe size [20"]		2.13	11.249		1,025	450	32.50	1,507.50	1,850
0830	610 mm diam. pipe size [24"]	▼	1.83	13.123		1,275	525	38	1,838	2,225
0900	For 25 mm thick insulation, add [1"]					10%				
0940	For 38 mm thick insulation, add [1-1/2"]					13%				
0980	For 51 mm thick insulation, add [2"]				▼	20%				
1500	Gland seal for system, 19 mm diam. pipe size [3/4"]	Q-17	32	.500	Ea.	430	19.40	2.16	451.56	500
1510	25 mm diam. pipe size [1"]		32	.500		430	19.40	2.16	451.56	500
1540	32 mm diam. pipe size [1-1/4"]		30	.533		460	20.50	2.30	482.80	540
1550	38 mm diam. pipe size [1-1/2"]		30	.533		460	20.50	2.30	482.80	540
1560	51 mm diam. pipe size [2"]		28	.571		545	22	2.47	569.47	635
1570	64 mm diam. pipe size [2-1/2"]		26	.615		585	24	2.66	611.66	685
1580	76 mm diam. pipe size [3"]		24	.667		630	26	2.88	658.88	735
1590	102 mm diam. pipe size [4"]		22	.727		745	28	3.14	776.14	865
1600	127 mm diam. pipe size [5"]	▼	19	.842		915	32.50	3.64	951.14	1,050
1610	152 mm diam. pipe size [6"]	Q-18	26	.923		975	37	2.66	1,014.66	1,125
1620	203 mm diam. pipe size [8"]		25	.960		1,125	38.50	2.76	1,166.26	1,300
1630	254 mm diam. pipe size [10"]		23	1.043		1,375	42	3.01	1,420.01	1,600
1640	305 mm diam. pipe size [12"]		21	1.143		1,525	46	3.29	1,574.29	1,750
1650	356 mm diam. pipe size [14"]		19	1.263		1,700	51	3.64	1,754.64	1,950
1660	406 mm diam. pipe size [16"]		18	1.333		2,000	53.50	3.84	2,057.34	2,275
1670	457 mm diam. pipe size [18"]		16	1.500		2,125	60.50	4.32	2,189.82	2,450
1680	508 mm diam. pipe size [20"]		14	1.714		2,425	69	4.94	2,498.94	2,775
1690	610 mm diam. pipe size [24"]	▼	12	2	▼	2,700	80.50	5.75	2,786.25	3,100
2000	Elbow, 45° for system									
2020	19 mm diam. pipe size [3/4"]	Q-17	14	1.143	Ea.	270	44.50	4.94	319.44	370
2040	25 mm diam. pipe size [1"]		13	1.231		278	47.50	5.30	330.80	385
2050	32 mm diam. pipe size [1-1/4"]		11	1.455		315	56.50	6.30	377.80	435
2060	38 mm diam. pipe size [1-1/2"]		9	1.778		330	69	7.70	406.70	475
2070	51 mm diam. pipe size [2"]		6	2.667		345	103	11.50	459.50	550
2080	64 mm diam. pipe size [2-1/2"]		4	4		375	155	17.30	547.30	660
2090	76 mm diam. pipe size [3"]		3.50	4.571		430	177	19.75	626.75	765
2100	102 mm diam. pipe size [4"]		3	5.333		500	207	23	730	885
2110	127 mm diam. pipe size [5"]	▼	2.80	5.714		645	221	24.50	890.50	1,075
2120	152 mm diam. pipe size [6"]	Q-18	4	6		730	241	17.30	988.30	1,200
2130	203 mm diam. pipe size [8"]		3	8		1,050	320	23	1,393	1,675
2140	254 mm diam. pipe size [10"]		2.40	10		1,350	400	29	1,779	2,100
2150	305 mm diam. pipe size [12"]		2	12		1,775	480	34.50	2,289.50	2,725
2160	356 mm diam. pipe size [14"]	▼	1.80	13.333		2,200	535	38.50	2,773.50	3,275

SITE CONSTRUCTION **2**

02550	Piped Energy Distribution	CREW	DAILY OUTPUT	LABOR-HOURS	UNIT	2006 BARE COSTS				TOTAL INCL O&P	
						MAT.	LABOR	EQUIP.	TOTAL		
200	2170	406 mm diam. pipe size [16"]	Q-18	1.60	15	Ea.	2,625	605	43	3,273	3,850	**200**
	2180	457 mm diam. pipe size [18"]		1.30	18.462		3,300	740	53	4,093	4,800	
	2190	508 mm diam. pipe size [20"]		1	24		4,150	965	69	5,184	6,100	
	2200	610 mm diam. pipe size [24"]		.70	34.286		5,225	1,375	98.50	6,698.50	7,925	
	2260	For elbow, 90°, add					25%					
	2300	For tee, straight, add					85%	30%				
	2340	For tee, reducing, add					170%	30%				
	2380	For weldolet, straight, add					50%					
	2800	Calcium silicate insulated system, high temp. (922° K)										
	2840	Steel casing with protective exterior coating										
	2850	168 mm diameter [6-5/8"]	Q-18	15.85	1.514	m	162	61	4.36	227.36	274	
	2860	219 mm diameter [8-5/8"]		15.24	1.575		177	63.50	4.54	245.04	295	
	2870	273 mm diameter [10-3/4"]		14.33	1.675		207	67.50	4.82	279.32	335	
	2880	324 mm diameter [12-3/4"]		13.41	1.790		226	72	5.15	303.15	360	
	2890	356 mm diameter [14"]		12.50	1.920		254	77	5.55	336.55	400	
	2900	406 mm diameter [16"]		11.89	2.019		273	81	5.80	359.80	430	
	2910	457 mm diameter [18"]		10.97	2.187		300	88	6.30	394.30	470	
	2920	508 mm diameter [20"]		10.36	2.316		340	93	6.65	439.65	515	
	2930	559 mm diameter [22"]		9.75	2.461		470	99	7.10	576.10	670	
	2940	610 mm diameter [24"]		8.84	2.715		535	109	7.80	651.80	765	
	2950	660 mm diameter [26"]		7.92	3.028		610	122	8.70	740.70	870	
	2960	711 mm diameter [28"]		7.01	3.423		760	138	9.85	907.85	1,050	
	2970	762 mm diameter [30"]		6.40	3.750		810	151	10.80	971.80	1,125	
	2980	813 mm diameter [32"]		5.79	4.144		910	167	11.95	1,088.95	1,275	
	2990	864 mm diameter [34"]		5.49	4.374		920	176	12.60	1,108.60	1,300	
	3000	914 mm diameter [36"]		4.88	4.921		990	198	14.15	1,202.15	1,400	
	3040	For multi-pipe casings, add					10%					
	3060	For oversize casings, add					2%					
	3400	Steel casing gland seal, single pipe										
	3420	168 mm diameter [6-5/8"]	Q-18	25	.960	Ea.	690	38.50	2.76	731.26	815	
	3440	219 mm diameter [8-5/8"]		23	1.043		805	42	3.01	850.01	950	
	3450	273 mm diameter [10-3/4"]		21	1.143		905	46	3.29	954.29	1,075	
	3460	324 mm diameter [12-3/4"]		19	1.263		1,075	51	3.64	1,129.64	1,250	
	3470	356 mm diameter [14"]		17	1.412		1,175	56.50	4.07	1,235.57	1,400	
	3480	406 mm diameter [16"]		16	1.500		1,375	60.50	4.32	1,439.82	1,625	
	3490	457 mm diameter [18"]		15	1.600		1,525	64.50	4.61	1,594.11	1,775	
	3500	508 mm diameter [20"]		13	1.846		1,700	74	5.30	1,779.30	1,975	
	3510	559 mm diameter [22"]		12	2		1,900	80.50	5.75	1,986.25	2,225	
	3520	610 mm diameter [24"]		11	2.182		2,125	87.50	6.30	2,218.80	2,500	
	3530	660 mm diameter [26"]		10	2.400		2,425	96.50	6.90	2,528.40	2,825	
	3540	711 mm diameter [28"]		9.50	2.526		2,750	102	7.30	2,859.30	3,175	
	3550	762 mm diameter [30"]		9	2.667		2,800	107	7.70	2,914.70	3,275	
	3560	813 mm diameter [32"]		8.50	2.824		3,150	113	8.15	3,271.15	3,650	
	3570	864 mm diameter [34"]		8	3		3,450	121	8.65	3,579.65	3,975	
	3580	914 mm diameter [36"]		7	3.429		3,650	138	9.85	3,797.85	4,250	
	3620	For multi-pipe casings, add					5%					
	4000	Steel casing anchors, single pipe										
	4020	168 mm diameter [6-5/8"]	Q-18	8	3	Ea.	615	121	8.65	744.65	870	
	4040	219 mm diameter [8-5/8"]		7.50	3.200		645	129	9.20	783.20	915	
	4050	273 mm diameter [10-3/4"]		7	3.429		845	138	9.85	992.85	1,150	
	4060	324 mm diameter [12-3/4"]		6.50	3.692		905	148	10.65	1,063.65	1,225	
	4070	356 mm diameter [14"]		6	4		1,050	161	11.50	1,222.50	1,425	
	4080	406 mm diameter [16"]		5.50	4.364		1,225	175	12.55	1,412.55	1,625	
	4090	457 mm diameter [18"]		5	4.800		1,400	193	13.80	1,606.80	1,825	
	4100	508 mm diameter [20"]		4.50	5.333		1,525	214	15.35	1,754.35	2,025	
	4110	559 mm diameter [22"]		4	6		1,700	241	17.30	1,958.30	2,250	

02550	Piped Energy Distribution	CREW	DAILY OUTPUT	LABOR-HOURS	UNIT	2006 BARE COSTS				TOTAL INCL O&P
						MAT.	LABOR	EQUIP.	TOTAL	
4120	610 mm diameter [24"]	Q-18	3.50	6.857	Ea.	1,875	276	19.75	2,170.75	2,475
4130	660 mm diameter [26"]		3	8		2,100	320	23	2,443	2,825
4140	711 mm diameter [28"]		2.50	9.600		2,300	385	27.50	2,712.50	3,150
4150	762 mm diameter [30"]		2	12		2,450	480	34.50	2,964.50	3,475
4160	813 mm diameter [32"]		1.50	16		2,950	645	46	3,641	4,250
4170	864 mm diameter [34"]		1	24		3,300	965	69	4,334	5,150
4180	914 mm diameter [36"]	▼	1	24	▼	3,575	965	69	4,609	5,475
4220	For multi-pipe, add					5%	20%			
4800	Steel casing elbow									
4820	168 mm diameter [6-5/8"]	Q-18	15	1.600	Ea.	845	64.50	4.61	914.11	1,025
4830	219 mm diameter [8-5/8"]		15	1.600		905	64.50	4.61	974.11	1,100
4850	273 mm diameter [10-3/4"]		14	1.714		1,100	69	4.94	1,173.94	1,300
4860	324 mm diameter [12-3/4"]		13	1.846		1,300	74	5.30	1,379.30	1,550
4870	356 mm diameter [14"]		12	2		1,375	80.50	5.75	1,461.25	1,650
4880	406 mm diameter [16"]		11	2.182		1,500	87.50	6.30	1,593.80	1,800
4890	457 mm diameter [18"]		10	2.400		1,725	96.50	6.90	1,828.40	2,050
4900	508 mm diameter [20"]		9	2.667		1,825	107	7.70	1,939.70	2,200
4910	559 mm diameter [22"]		8	3		1,975	121	8.65	2,104.65	2,375
4920	610 mm diameter [24"]		7	3.429		2,175	138	9.85	2,322.85	2,625
4930	660 mm diameter [26"]		6	4		2,375	161	11.50	2,547.50	2,875
4940	711 mm diameter [28"]		5	4.800		2,575	193	13.80	2,781.80	3,125
4950	762 mm diameter [30"]		4	6		2,575	241	17.30	2,833.30	3,225
4960	813 mm diameter [32"]		3	8		2,950	320	23	3,293	3,725
4970	864 mm diameter [34"]		2	12		3,225	480	34.50	3,739.50	4,325
4980	914 mm diameter [36"]	▼	2	12	▼	3,450	480	34.50	3,964.50	4,550
5500	Black steel service pipe, std. wt., 25 mm thick insulation [1"]									
5510	19 mm diameter pipe size [3/4"]	Q-17	16.46	.972	m	70.50	37.50	4.20	112.20	139
5540	25 mm diameter pipe size [1"]		15.24	1.050		73.50	40.50	4.54	118.54	146
5550	32 mm diameter pipe size [1-1/4"]		14.33	1.117		82.50	43.50	4.82	130.82	161
5560	38 mm diameter pipe size [1-1/2"]		13.72	1.167		89.50	45	5.05	139.55	172
5570	51 mm diameter pipe size [2"]		12.80	1.250		99.50	48.50	5.40	153.40	189
5580	64 mm diameter pipe size [2-1/2"]		10.36	1.544		105	60	6.65	171.65	212
5590	76 mm diameter pipe size [3"]		8.53	1.875		119	72.50	8.10	199.60	249
5600	102 mm diameter pipe size [4"]		6.71	2.386		153	92.50	10.30	255.80	320
5610	127 mm diameter pipe size [5"]	▼	5.49	2.916		207	113	12.60	332.60	410
5620	152 mm diameter pipe size [6"]	Q-18	7.01	3.423	▼	226	138	9.85	373.85	465
6000	Black steel service pipe, std. wt., 38 mm thick insul. [1-1/2"]									
6010	19 mm diameter pipe size [3/4"]	Q-17	16.46	.972	m	71.50	37.50	4.20	113.20	140
6040	25 mm diameter pipe size [1"]		15.24	1.050		80	40.50	4.54	125.04	154
6050	32 mm diameter pipe size [1-1/4"]		14.33	1.117		89.50	43.50	4.82	137.82	169
6060	38 mm diameter pipe size [1-1/2"]		13.72	1.167		98	45	5.05	148.05	182
6070	51 mm diameter pipe size [2"]		12.80	1.250		106	48.50	5.40	159.90	196
6080	64 mm diameter pipe size [2-1/2"]		10.36	1.544		113	60	6.65	179.65	222
6090	76 mm diameter pipe size [3"]		8.53	1.875		127	72.50	8.10	207.60	258
6100	102 mm diameter pipe size [4"]		6.71	2.386		162	92.50	10.30	264.80	330
6110	127 mm diameter pipe size [5"]	▼	5.49	2.916		207	113	12.60	332.60	410
6120	152 mm diameter pipe size [6"]	Q-18	7.01	3.423		235	138	9.85	382.85	475
6130	203 mm diameter pipe size [8"]		5.79	4.144		340	167	11.95	518.95	635
6140	254 mm diameter pipe size [10"]		4.88	4.921		440	198	14.15	652.15	800
6150	305 mm diameter pipe size [12"]	▼	3.96	6.057		525	243	17.45	785.45	965
6190	For 51 mm thick insulation, add [2"]					15%				
6220	For 64 mm thick insulation, add [2-1/2"]					25%				
6260	For 76 mm thick insulation, add [3"]				▼	30%				
6800	Black steel service pipe, ex. hvy. wt., 25 mm thick insul. [1"]									
6820	19 mm diameter pipe size [3/4"]	Q-17	15.24	1.050	m	74	40.50	4.54	119.04	147
6840	25 mm diameter pipe size [1"]	▼	14.33	1.117	▼	79	43.50	4.82	127.32	157

02550 | Piped Energy Distribution

		CREW	DAILY OUTPUT	LABOR-HOURS	UNIT	2006 BARE COSTS				TOTAL INCL O&P
						MAT.	LABOR	EQUIP.	TOTAL	
200	6850	32 mm diameter pipe size [1-1/4"]	Q-17	13.41	1.193	m	91	46	5.15	142.15	175
	6860	38 mm diameter pipe size [1-1/2"]		12.80	1.250		97	48.50	5.40	150.90	186
	6870	51 mm diameter pipe size [2"]		12.19	1.312		103	51	5.65	159.65	196
	6880	64 mm diameter pipe size [2-1/2"]		9.45	1.693		127	65.50	7.30	199.80	247
	6890	76 mm diameter pipe size [3"]		8.23	1.944		143	75.50	8.40	226.90	279
	6900	102 mm diameter pipe size [4"]		6.40	2.500		188	97	10.80	295.80	365
	6910	127 mm diameter pipe size [5"]		5.18	3.088		263	120	13.35	396.35	485
	6920	152 mm diameter pipe size [6"]	Q-18	6.71	3.579		292	144	10.30	446.30	545
	7400	Black steel service pipe, ex. hvy. wt., 38 mm thick insul. [1-1/2"]									
	7420	19 mm diameter pipe size [3/4"]	Q-17	15.24	1.050	m	74	40.50	4.54	119.04	147
	7440	25 mm diameter pipe size [1"]		14.33	1.117		84.50	43.50	4.82	132.82	163
	7450	32 mm diameter pipe size [1-1/4"]		13.41	1.193		98.50	46	5.15	149.65	183
	7460	38 mm diameter pipe size [1-1/2"]		12.80	1.250		108	48.50	5.40	161.90	197
	7470	51 mm diameter pipe size [2"]		12.19	1.312		109	51	5.65	165.65	202
	7480	64 mm diameter pipe size [2-1/2"]		9.45	1.693		129	65.50	7.30	201.80	249
	7490	76 mm diameter pipe size [3"]		8.23	1.944		152	75.50	8.40	235.90	289
	7500	102 mm diameter pipe size [4"]		6.40	2.500		198	97	10.80	305.80	375
	7510	127 mm diameter pipe size [5"]		5.18	3.088		273	120	13.35	406.35	495
	7520	152 mm diameter pipe size [6"]	Q-18	6.71	3.579		300	144	10.30	454.30	555
	7530	203 mm diameter pipe size [8"]		5.49	4.374		450	176	12.60	638.60	775
	7540	254 mm diameter pipe size [10"]		4.57	5.249		535	211	15.10	761.10	920
	7550	305 mm diameter pipe size [12"]		3.96	6.057		660	243	17.45	920.45	1,100
	7590	For 51 mm thick insulation, add [2"]					13%				
	7640	For 64 mm thick insulation, add [2-1/2"]					18%				
	7680	For 76 mm thick insulation, add [3"]					24%				
464	0010	**PIPING, GAS SERVICE & DISTRIBUTION, POLYETHYLENE**									
	0020	not including excavation or backfill									
	1000	414 kPa coils, 15 mm diam, comp cplg@30 m, SDR 9.3 [1/2"]	B-20A	185	.173	m	1.67	5.80		7.47	10.70
	1040	32 mm diameter, SDR 11 [1-1/4"]		166	.193		3.12	6.45		9.57	13.25
	1100	50 mm diameter, SDR 11 [2"]		149	.215		3.84	7.20		11.04	15.20
	1160	80 mm diameter, SDR 11 [3"]		124	.258		7.95	8.60		16.55	22
	1500	414 kPa 12 m joints with coupling, 80 mm diameter, SDR 11 [3"]	B-21A	124	.323		7.95	11.10	4.33	23.38	30.50
	1540	100 mm diameter, SDR 11 [4"]		107	.374		18.30	12.85	5	36.15	45
	1600	150 mm diameter, SDR 11 [6"]		99.97	.400		57	13.75	5.35	76.10	89.50
	1640	200 mm diameter, SDR 11 [8"]		82.91	.482		77.50	16.55	6.45	100.50	118
466	0010	**PIPING, GAS SERVICE & DISTRIBUTION, STEEL**									
	0020	not including excavation or backfill, tar coated and wrapped									
	4000	Schedule 40, plain end									
	4040	25 mm diameter [1"]	Q-4	91.44	.350	m	9.35	14.25	.76	24.36	32.50
	4080	50 mm diameter [2"]		85.34	.375		14.65	15.25	.81	30.71	40
	4120	80 mm diameter [3"]		79.25	.404		24.50	16.45	.87	41.82	52
	4160	100 mm diameter [4"]	B-35	77.72	.618		31.50	21.50	7.05	60.05	75.50
	4200	125 mm diameter [5"]		67.06	.716		46	24.50	8.15	78.65	97
	4240	150 mm diameter [6"]		54.86	.875		56	30	9.95	95.95	118
	4280	200 mm diameter [8"]		42.67	1.125		89	38.50	12.80	140.30	171
	4320	250 mm diameter [10"]		30.48	1.575		305	54	17.95	376.95	440
	4360	300 mm diameter [12"]		24.38	1.969		385	68	22.50	475.50	555
	4400	350 mm diameter [14"]		22.86	2.100		410	72.50	24	506.50	590
	4440	400 mm diameter [16"]		21.34	2.250		450	77.50	25.50	553	640
	4480	450 mm diameter [18"]		19.81	2.423		575	83.50	27.50	686	795
	4520	500 mm diameter [20"]		18.29	2.625		895	90.50	30	1,015.50	1,150
	4560	600 mm diameter [24"]		15.24	3.150		1,025	108	36	1,169	1,325
	5000	Threaded and coupled									
	5002	100 mm diameter [4"]	B-20	43.89	.547	m	89.50	17		106.50	125
	5004	125 mm diameter [5"]	"	42.67	.562		119	17.50		136.50	158

		02550 Piped Energy Distribution	CREW	DAILY OUTPUT	LABOR-HOURS	UNIT	2006 BARE COSTS				TOTAL INCL O&P	
							MAT.	LABOR	EQUIP.	TOTAL		
466	5006	150 mm diameter [6"]	B-21	38.40	.729	m	189	23.50	3.97	216.47	248	466
	5008	200 mm diameter [8"]		32.92	.851		287	27.50	4.64	319.14	365	
	5012	300 mm diameter [12"]	↓	21.95	1.276	↓	415	41	6.95	462.95	525	
	6000	Schedule 80, plain end										
	6002	100 mm diameter [4"]	B-35	43.89	1.094	m	138	37.50	12.45	187.95	223	
	6006	150 mm diameter [6"]		38.40	1.250		230	43	14.25	287.25	335	
	6008	200 mm diameter [8"]		32.92	1.458		305	50	16.60	371.60	435	
	6012	300 mm diameter [12"]	↓	21.95	2.187	↓	615	75.50	25	715.50	820	
	8008	Elbow, weld joint, standard weight										
	8020	100 mm diameter [4"]	Q-16	6.80	3.529	Ea.	62.50	141	10.15	213.65	292	
	8026	200 mm diameter [8"]		3.40	7.059		259	281	20.50	560.50	735	
	8030	300 mm diameter [12"]		2.30	10.435		580	415	30	1,025	1,300	
	8034	400 mm diameter [16"]		1.50	16		1,100	640	46	1,786	2,200	
	8038	500 mm diameter [20"]		1.20	20		2,075	795	57.50	2,927.50	3,575	
	8040	600 mm diameter [24"]	↓	1.02	23.529	↓	2,750	940	68	3,758	4,525	
	8100	Extra heavy										
	8102	100 mm diameter [4"]	Q-16	5.30	4.528	Ea.	125	180	13.05	318.05	425	
	8108	200 mm diameter [8"]		2.60	9.231		390	370	26.50	786.50	1,025	
	8112	300 mm diameter [12"]		1.80	13.333		775	530	38.50	1,343.50	1,700	
	8116	400 mm diameter [16"]		1.20	20		1,475	795	57.50	2,327.50	2,900	
	8120	500 mm diameter [20"]		.94	25.532		2,775	1,025	73.50	3,873.50	4,650	
	8122	600 mm diameter [24"]	↓	.80	30		3,675	1,200	86.50	4,961.50	5,950	
	8200	Malleable, standard weight										
	8202	100 mm diameter [4"]	B-20	12	2	Ea.	228	62		290	350	
	8208	200 mm diameter [8"]		6	4		625	124		749	885	
	8212	300 mm diameter [12"]	↓	4	6	↓	690	187		877	1,050	
	8300	Extra heavy										
	8302	100 mm diameter [4"]	B-20	12	2	Ea.	455	62		517	600	
	8308	200 mm diameter [8"]	B-21	6	4.667		785	150	25.50	960.50	1,125	
	8312	300 mm diameter [12"]	"	4	7	↓	920	225	38	1,183	1,400	
	8500	Tee weld, standard weight										
	8510	100 mm diameter [4"]	Q-16	4.50	5.333	Ea.	101	213	15.35	329.35	450	
	8514	150 mm diameter [6"]		3	8		173	320	23	516	695	
	8516	200 mm diameter [8"]		2.30	10.435		305	415	30	750	995	
	8520	300 mm diameter [12"]		1.50	16		840	640	46	1,526	1,925	
	8524	400 mm diameter [16"]		1	24		1,650	955	69	2,674	3,350	
	8528	500 mm diameter [20"]		.80	30		4,050	1,200	86.50	5,336.50	6,350	
	8530	600 mm diameter [24"]	↓	.70	34.286	↓	5,225	1,375	98.50	6,698.50	7,900	
	8810	Malleable, standard weight										
	8812	100 mm diameter [4"]	B-20	8	3	Ea.	390	93.50		483.50	575	
	8818	200 mm diameter [8"]	B-21	4	7	"	670	225	38	933	1,125	
	8900	Extra heavy										
	8902	100 mm diameter [4"]	B-20	8	3	Ea.	585	93.50		678.50	790	
	8908	200 mm diameter [8"]	B-21	4	7		835	225	38	1,098	1,300	
	8912	300 mm diameter [12"]	"	2.70	10.370	↓	1,275	335	56.50	1,666.50	1,975	
468	0010	**PIPING, VALVES & METERS, GAS DISTRIBUTION**										468
	0020	not including excavation or backfill										
	0100	Gas stops, with or without checks										
	0140	32 mm size	1 Plum	12	.667	Ea.	49	28.50		77.50	97	
	0180	40 mm size		10	.800		66	34		100	125	
	0200	50 mm size	↓	8	1	↓	99.50	42.50		142	174	
	0600	Pressure regulator valves, iron and bronze										
	0680	50 mm diameter [2"]	1 Plum	11	.727	Ea.	224	31		255	293	
	0700	80 mm diameter [3"]	Q-1	13	1.231		455	47.50		502.50	570	
	0740	100 mm diameter [4"]	"	8	2	↓	1,300	77		1,377	1,550	

SITE CONSTRUCTION   **2**

For expanded coverage of these items see Means Site Work & Landscape Cost Data 2006

02550	Piped Energy Distribution	CREW	DAILY OUTPUT	LABOR-HOURS	UNIT	2006 BARE COSTS				TOTAL INCL O&P	
						MAT.	LABOR	EQUIP.	TOTAL		
468	**2000**	Lubricated semi-steel plug valve									**468**
2040	20 mm diameter [3/4"]	1 Plum	16	.500	Ea.	84.50	21.50		106	125	
2080	25 mm diameter [1"]		14	.571		82.50	24.50		107	128	
2100	32 mm diameter [1-1/4"]		12	.667		85	28.50		113.50	137	
2140	40 mm diameter [1-1/2"]		11	.727		85	31		116	140	
2180	50 mm diameter [2"]		8	1		79.50	42.50		122	152	
2300	65 mm diameter [2-1/2"]	Q-1	5	3.200		127	123		250	325	
2340	80 mm diameter [3"]	"	4.50	3.556		196	137		333	420	
550	**0010**	**GAS STATION PRODUCT LINE**									**550**
0020	Primary containment pipe, fiberglass-reinforced										
0030	Plastic pipe 4.5 m & 9 m lengths										
0040	50 mm diameter [2"]	Q-6	130	.185	m	11.45	7.40		18.85	24	
0050	80 mm diameter [3"]		122	.197		15	7.90		22.90	28.50	
0060	100 mm diameter [4"]		114	.211		19.30	8.45		27.75	34.50	
0100	Fittings										
0110	Elbows, 90° & 45°, bell-ends, 50 mm	Q-6	24	1	Ea.	35.50	40		75.50	99.50	
0120	80 mm diameter [3"]		22	1.091		37	44		81	107	
0130	100 mm diameter [4"]		20	1.200		49	48		97	127	
0200	Tees, bell ends, 50 mm		21	1.143		43	46		89	116	
0210	80 mm diameter [3"]		18	1.333		43	53.50		96.50	128	
0220	100 mm diameter [4"]		15	1.600		59	64.50		123.50	162	
0230	Flanges bell ends, 50 mm		24	1		14.15	40		54.15	76	
0240	80 mm diameter [3"]		22	1.091		17.85	44		61.85	85.50	
0250	100 mm diameter [4"]		20	1.200		24.50	48		72.50	99	
0260	Sleeve couplings, 50 mm		21	1.143		9	46		55	79	
0270	80 mm diameter [3"]		18	1.333		12.80	53.50		66.30	94.50	
0280	100 mm diameter [4"]		15	1.600		17.65	64.50		82.15	116	
0290	Threaded adapters 50 mm		21	1.143		11.85	46		57.85	82	
0300	80 mm diameter [3"]		18	1.333		21	53.50		74.50	104	
0310	100 mm diameter [4"]		15	1.600		28	64.50		92.50	128	
0320	Reducers, 50 mm		27	.889		15.75	35.50		51.25	71.50	
0330	80 mm diameter [3"]		22	1.091		18.25	44		62.25	86	
0340	100 mm diameter [4"]		20	1.200		23.50	48		71.50	98.50	
1010	Gas station product line for secondary containment										
1100	Fiberglass reinforced plastic pipe 7.6 m lengths										
1120	Pipe, plain end, 80 mm [3"]	Q-6	114	.211	m	19.70	8.45		28.15	34.50	
1130	100 mm diameter [4"]		107	.224		32.50	9		41.50	49	
1140	125 mm diameter [5"]		99.06	.242		42	9.75		51.75	60.50	
1150	150 mm diameter [6"]		91.44	.262		43	10.55		53.55	63.50	
1200	Fittings										
1230	Elbows, 90° & 45°, 80 mm diam. [3"]	Q-6	18	1.333	Ea.	43	53.50		96.50	128	
1240	100 mm diameter [4"]		16	1.500		75.50	60.50		136	174	
1250	125 mm diameter [5"]		14	1.714		175	69		244	296	
1260	150 mm diameter [6"]		12	2		177	80.50		257.50	315	
1270	Tees, 80 mm diam. [3"]		15	1.600		64	64.50		128.50	168	
1280	100 mm diameter [4"]		12	2		94	80.50		174.50	224	
1290	125 mm diameter [5"]		9	2.667		190	107		297	370	
1300	150 mm diameter [6"]		6	4		199	161		360	460	
1310	Couplings, 80 mm diam. [3"]		18	1.333		30.50	53.50		84	114	
1320	100 mm diameter [4"]		16	1.500		78.50	60.50		139	177	
1330	125 mm diameter [5"]		14	1.714		163	69		232	283	
1340	150 mm diameter [6"]		12	2		169	80.50		249.50	305	
1350	Cross-over nipples, 80 mm diam. [3"]		18	1.333		6.90	53.50		60.40	88	
1360	100 mm diameter [4"]		16	1.500		8.10	60.50		68.60	99.50	
1370	125 mm diameter [5"]		14	1.714		12.05	69		81.05	117	
1380	150 mm diameter [6"]		12	2		12.65	80.50		93.15	135	

2 SITE CONSTRUCTION

02550 | Piped Energy Distribution

			CREW	DAILY OUTPUT	LABOR-HOURS	UNIT	MAT.	LABOR	EQUIP.	TOTAL	TOTAL INCL O&P	
								2006 BARE COSTS				
550	1400	Telescoping, reducers, concentric 100mm x 80 mm [4",3"]	Q-6	18	1.333	Ea.	23	53.50		76.50	106	550
	1410	125 mm x 100 mm		17	1.412		60	56.50		116.50	152	
	1420	150 mm x 125 mm	↓	16	1.500	↓	144	60.50		204.50	250	

02580 | Elec/Communication Structures

			CREW	DAILY OUTPUT	LABOR-HOURS	UNIT	MAT.	LABOR	EQUIP.	TOTAL	TOTAL INCL O&P	
100	0010	**RADIO TOWERS**										100
	0020	Guyed, 15 m H, 18 kg sec., 113 km/hr basic wind speed	2 Sswk	1	16	Ea.	2,400	640		3,040	3,800	
	0100	Wind load 145 km/hr basic wind speed	"	1	16		2,400	640		3,040	3,800	
	0300	58 m H, 18 kg section, wind load 113 km/hr basic wind speed	K-2	.33	72.727		6,500	2,650	535	9,685	12,300	
	0400	61 m H, 32 kg section, wind load 145 km/hr basic wind speed		.33	72.727		13,100	2,650	535	16,285	19,600	
	0600	91 m H, 32 kg section, wind load 117 km/hr basic wind speed		.20	120		18,500	4,375	880	23,755	28,900	
	0700	82 m H, 41 kg section, wind load 145 km/hr basic wind speed		.20	120		21,400	4,375	880	26,655	32,000	
	0800	122 m H, 45 kg section, wind load 117 km/hr basic wind speed		.14	171		31,200	6,250	1,250	38,700	46,500	
	0900	Self-supporting, 18 m H, wind load 113 km/hr basic wind speed		.80	30		5,000	1,100	220	6,320	7,650	
	0910	18 m H, wind load 145 km/hr basic wind speed		.45	53.333		9,025	1,950	390	11,365	13,700	
	1000	37 m H, wind load 113 km/hr basic wind speed		.40	60		12,300	2,200	440	14,940	17,800	
	1200	58 m H, wind load 145 km/hr basic wind speed	↓	.20	120		29,800	4,375	880	35,055	41,300	
	2000	For states west of Rocky Mountains, add for shipping				↓	10%					
410	0010	**UNDERGROUND DUCTS AND MANHOLES**, In slab or duct bank										410
	0011	Not including excavation, backfill and cast in place concrete										
	1000	Direct burial										
	1010	PVC, schedule 40, w/coupling, 15 mm diameter	1 Elec	104	.077	m	1.25	3.23		4.48	6.20	
	1020	20 mm diameter		88.39	.091		1.64	3.80		5.44	7.45	
	1030	25 mm diameter		79.25	.101		2.43	4.24		6.67	8.95	
	1040	40 mm diameter		64.01	.125		3.94	5.25		9.19	12.15	
	1050	50 mm diameter	↓	54.86	.146		5.20	6.10		11.30	14.85	
	1060	80 mm diameter	2 Elec	73.15	.219		9.90	9.20		19.10	24.50	
	1070	100 mm diameter		48.77	.328		13.85	13.80		27.65	35.50	
	1080	125 mm diameter		36.58	.437		20	18.35		38.35	49.50	
	1090	150 mm diameter	↓	27.43	.583	↓	26.50	24.50		51	65.50	
	1110	Elbows, 15 mm diameter	1 Elec	48	.167	Ea.	1.58	7		8.58	12.20	
	1120	20 mm diameter		38	.211		1.59	8.85		10.44	14.95	
	1130	25 mm diameter		32	.250		2.69	10.50		13.19	18.60	
	1140	40 mm diameter		21	.381		5.20	16		21.20	29.50	
	1150	50 mm diameter		16	.500		7.55	21		28.55	40	
	1160	80 mm diameter		12	.667		23	28		51	67	
	1170	100 mm diameter		9	.889		40	37.50		77.50	99.50	
	1180	125 mm diameter		8	1		70	42		112	140	
	1190	150 mm diameter		5	1.600		119	67		186	231	
	1210	Adapters, 15 mm diameter		52	.154		.51	6.45		6.96	10.20	
	1220	20 mm diameter		43	.186		.94	7.80		8.74	12.70	
	1230	25 mm diameter		39	.205		1.19	8.60		9.79	14.15	
	1240	40 mm diameter		35	.229		1.85	9.60		11.45	16.35	
	1250	50 mm diameter		26	.308		2.66	12.90		15.56	22	
	1260	80 mm diameter		20	.400		6.75	16.80		23.55	32.50	
	1270	100 mm diameter		14	.571		11.80	24		35.80	49	
	1280	125 mm diameter		12	.667		23	28		51	67	
	1290	150 mm diameter		9	.889		28	37.50		65.50	86.50	
	1340	Bell end & cap, 40 mm diameter		35	.229		9	9.60		18.60	24	
	1350	Bell end & plug, 50 mm diameter		26	.308		9.50	12.90		22.40	29.50	
	1360	80 mm diameter		20	.400		11.90	16.80		28.70	38	
	1370	100 mm diameter		14	.571		13.90	24		37.90	51.50	
	1380	125 mm diameter		12	.667		21	28		49	64.50	
	1390	150 mm diameter	↓	9	.889	↓	24	37.50		61.50	81.50	

SITE CONSTRUCTION 2

For expanded coverage of these items see *Means Site Work & Landscape Cost Data 2006*

02580	Elec/Communication Structures	CREW	DAILY OUTPUT	LABOR-HOURS	UNIT	2006 BARE COSTS				TOTAL INCL O&P	
						MAT.	LABOR	EQUIP.	TOTAL		
410	1450	Base spacer, 50 mm diameter	1 Elec	56	.143	Ea.	1.61	6		7.61	10.70	**410**
	1460	80 mm diameter		46	.174		1.78	7.30		9.08	12.85
	1470	100 mm diameter		41	.195		1.88	8.20		10.08	14.25
	1480	125 mm diameter		37	.216		2.31	9.10		11.41	16.10
	1490	150 mm diameter		34	.235		3.24	9.90		13.14	18.30
	1550	Intermediate spacer, 50 mm diameter		60	.133		1.40	5.60		7	9.90
	1560	80 mm diameter		46	.174		1.78	7.30		9.08	12.85
	1570	100 mm diameter		41	.195		1.88	8.20		10.08	14.25
	1580	125 mm diameter		37	.216		2.31	9.10		11.41	16.10
	1590	150 mm diameter	▼	34	.235	▼	3.24	9.90		13.14	18.30
420	0010	**ELECTRIC & TELEPHONE UNDERGROUND**, Not including excavation,										**420**
	0200	backfill and cast in place concrete									
	0400	Hand holes, precast concrete with concrete cover									
	0600	610 mm x 610 mm x 915 mm deep	R-3	2.40	8.333	Ea.	268	345	63.50	676.50	880
	0800	915 mm x 915 mm x 915 mm deep		1.90	10.526		350	435	80.50	865.50	1,125
	1000	1220 mm x 1220 mm x 1220 mm deep	▼	1.40	14.286	▼	695	590	109	1,394	1,775
	1200	Manholes, precast, with iron racks, pulling irons, C.I. frame									
	1400	and cover, 1220 mm x 1830 mm x 2135 mm deep	B-13	2	28	Ea.	1,325	840	325	2,490	3,125
	1600	1830 mm x 2450 mm x 2135 mm deep		1.90	29.474		1,625	880	340	2,845	3,525
	1800	1830 mm x 3050 mm x 2135 mm deep	▼	1.80	31.111	▼	1,825	930	360	3,115	3,850
	4200	Underground duct, banks ready for concrete fill, min. of 190 mm									
	4400	between conduits, ctr. to ctr.(for wire & cable see Div. 16120)									
	4580	PVC, type EB, 1 @ 50 mm diameter [2"]	2 Elec	146	.110	m	2.46	4.60		7.06	9.55
	4600	2 @ 50 mm diameter [2"]		73.15	.219		4.92	9.20		14.12	19.10
	4800	4 @ 50 mm diameter [2"]		36.58	.437		9.85	18.35		28.20	38.50
	4900	1 @ 80 mm diameter [3"]		122	.131		3.38	5.50		8.88	11.90
	5000	2 @ 80 mm diameter [3"]		60.96	.262		6.75	11		17.75	24
	5200	4 @ 80 mm diameter [3"]		30.48	.525		13.50	22		35.50	48
	5300	1 @ 100 mm diameter [4"]		97.54	.164		5.20	6.90		12.10	15.95
	5400	2 @ 100 mm diameter [4"]		48.77	.328		10.35	13.80		24.15	32
	5600	4 @ 100 mm diameter [4"]		24.38	.656		20.50	27.50		48	64
	5800	6 @ 100 mm diameter [4"]		16.46	.972		31	41		72	95
	5810	1 @ 125 mm diameter [5"]		79.25	.202		7.70	8.50		16.20	21
	5820	2 @ 125 mm diameter [5"]		39.62	.404		15.40	16.95		32.35	42.50
	5840	4 @ 125 mm diameter [5"]		21.34	.750		31	31.50		62.50	81
	5860	6 @ 125 mm diameter [5"]		15.24	1.050		46.50	44		90.50	117
	5870	1 @ 150 mm diameter [6"]		60.96	.262		11	11		22	28.50
	5880	2 @ 150 mm diameter [6"]		30.48	.525		22	22		44	57.50
	5900	4 @ 150 mm diameter [6"]		15.24	1.050		44	44		88	114
	5920	6 @ 150 mm diameter [6"]		9.14	1.750		66	73.50		139.50	183
	6200	Rigid galvanized steel, 2 @ 50 mm diameter [2"]		54.86	.292		48.50	12.25		60.75	72
	6400	4 @ 50 mm diameter [2"]		27.43	.583		97	24.50		121.50	144
	6800	2 @ 80 mm diameter [3"]		30.48	.525		114	22		136	159
	7000	4 @ 80 mm diameter [3"]		15.24	1.050		229	44		273	315
	7200	2 @ 100 mm diameter [4"]		21.34	.750		158	31.50		189.50	221
	7400	4 @ 100 mm diameter [4"]		10.36	1.544		315	65		380	445
	7600	6 @ 100 mm diameter [4"]		6.71	2.386		475	100		575	675
	7620	2 @ 125 mm diameter [5"]		18.29	.875		345	37		382	430
	7640	4 @ 125 mm diameter [5"]		9.14	1.750		685	73.50		758.50	865
	7660	6 @ 125 mm diameter [5"]		5.49	2.916		1,025	122		1,147	1,300
	7680	2 @ 150 mm diameter [6"]		12.19	1.312		500	55		555	630
	7700	4 @ 150 mm diameter [6"]		6.10	2.625		995	110		1,105	1,275
	7720	6 @ 150 mm diameter [6"]	▼	4.27	3.750	▼	1,500	157		1,657	1,875
	8000	Fittings, PVC type EB, elbow, 50 mm diameter [2"]	1 Elec	16	.500	Ea.	11	21		32	43.50
	8200	80 mm diameter [3"]		14	.571		11.60	24		35.60	49
	8400	100 mm diameter [4"]	▼	12	.667	▼	16.45	28		44.45	59.50

Important: See the Reference Section for supporting data - Crews, Rental Equipment, City Cost Indexes and Reference Data

SITE CONSTRUCTION 2

02580	Elec/Communication Structures	CREW	DAILY OUTPUT	LABOR-HOURS	UNIT	2006 BARE COSTS MAT.	LABOR	EQUIP.	TOTAL	TOTAL INCL O&P	
420											**420**
8420	125 mm diameter [5"]	1 Elec	10	.800	Ea.	39	33.50		72.50	93	
8440	150 mm diameter [6"]	↓	9	.889		74	37.50		111.50	137	
8500	Coupling, 50 mm diameter [2"]					.79			.79	.87	
8600	80 mm diameter [3"]					2.74			2.74	3.01	
8700	100 mm diameter [4"]					4.30			4.30	4.73	
8720	125 mm diameter [5"]					7.75			7.75	8.55	
8740	150 mm diameter [6"]					22.50			22.50	25	
8800	Adapter, 50 mm diameter [2"]	1 Elec	26	.308		1.40	12.90		14.30	21	
9000	80 mm diameter [3"]		20	.400		3.90	16.80		20.70	29.50	
9200	100 mm diameter [4"]		16	.500		5.20	21		26.20	37.50	
9220	125 mm diameter [5"]		13	.615		13	26		39	53	
9240	150 mm diameter [6"]		10	.800		17.15	33.50		50.65	69	
9400	End bell, 50 mm diameter [2"]		16	.500		6.80	21		27.80	39	
9600	80 mm diameter [3"]		14	.571		8.15	24		32.15	45	
9800	100 mm diameter [4"]		12	.667		9.60	28		37.60	52	
9810	125 mm diameter [5"]		10	.800		14.50	33.50		48	66	
9820	150 mm diameter [6"]		8	1		27.50	42		69.50	93	
9830	5° angle coupling, 50 mm diameter [2"]		26	.308		11.45	12.90		24.35	32	
9840	80 mm diameter [3"]		20	.400		14.50	16.80		31.30	41	
9850	100 mm diameter [4"]		16	.500		17.15	21		38.15	50.50	
9860	125 mm diameter [5"]		13	.615		18.80	26		44.80	59	
9870	150 mm diameter [6"]		10	.800		19.25	33.50		52.75	71	
9880	Expansion joint, 50 mm diameter [2"]		16	.500		33.50	21		54.50	68.50	
9890	80 mm diameter [3"]		18	.444		59	18.65		77.65	93	
9900	100 mm diameter [4"]		12	.667		85.50	28		113.50	136	
9910	125 mm diameter [5"]		10	.800		133	33.50		166.50	196	
9920	150 mm diameter [6"]	↓	8	1		179	42		221	260	
9930	Heat bender, 50 mm diameter [2"]					440			440	485	
9940	150 mm diameter [6"]					1,225			1,225	1,350	
9950	Cement, 0.95 L				↓	14.70			14.70	16.15	
500	**UTILITY POLES**										**500**
6200	Poles, wood, preservative treatment, 6.1 m high, see also Div. 16520	R-3	3.10	6.452	Ea.	251	267	49	567	730	
6400	7.6 m high		2.90	6.897		265	286	52.50	603.50	775	
6600	9.1 m high		2.60	7.692		292	320	58.50	670.50	860	
6800	10.7 m high		2.40	8.333		380	345	63.50	788.50	1,000	
7000	12.2 m high		2.30	8.696		465	360	66.50	891.50	1,125	
7200	13.7 m high	↓	1.70	11.765	↓	570	485	90	1,145	1,450	
7400	Cross arms with hardware & insulators										
7600	1.2 m long	1 Elec	2.50	3.200	Ea.	119	134		253	330	
7800	1.5 m long		2.40	3.333		138	140		278	360	
8000	1.8 m long	↓	2.20	3.636	↓	159	153		312	405	

02600 | Drainage & Containment

02620	Subdrainage	CREW	DAILY OUTPUT	LABOR-HOURS	UNIT	2006 BARE COSTS MAT.	LABOR	EQUIP.	TOTAL	TOTAL INCL O&P	
300	**0010** **GEOTEXTILES FOR SUBSURFACE DRAINAGE**										**300**
0100	Fabric, laid in trench, polypropylene, ideal conditions	2 Clab	2007	.008	m²	1.65	.22		1.87	2.16	
0110	Adverse conditions		1338	.012		1.65	.33		1.98	2.33	
0170	Fabric ply bonded to 3 dimen. nylon mat, 10 mm thk, ideal conditions	↓	186	.086	↓	10.35	2.36		12.71	15.05	

SITE CONSTRUCTION 2

02620	Subdrainage	CREW	DAILY OUTPUT	LABOR-HOURS	UNIT	2006 BARE COSTS MAT.	LABOR	EQUIP.	TOTAL	TOTAL INCL O&P	
300	0180	Adverse conditions	2 Clab	111	.144	m²	12.90	3.95		16.85	20.50	300
	0185	Soil drainage mat on vertical wall, 11 mm thick		222	.072		20	1.97		21.97	25	
	0188	6 mm thick		251	.064		12.25	1.75		14	16.15	
	0190	20 mm thick, ideal conditions		223	.072		16.25	1.97		18.22	21	
	0200	Adverse conditions		149	.107		21.50	2.94		24.44	28	
	0300	Drainage material, 20 mm gravel fill in trench	B-6	199	.121	m³	71	3.62	1.14	75.76	85.50	
	0400	Pea stone	"	199	.121	"	25	3.62	1.14	29.76	34.50	
610	0010	**PIPING, SUBDRAINAGE, CONCRETE**										610
	0021	Not including excavation and backfill										
	3000	Porous wall concrete underdrain, std. strength, 100 mm diam. [4"]	B-20	102	.235	m	6.75	7.30		14.05	18.80	
	3020	150 mm diameter [6"]	"	96.01	.250		8.75	7.75		16.50	21.50	
	3040	200 mm diameter [8"]	B-21	94.49	.296		10.80	9.50	1.62	21.92	28.50	
	3060	300 mm diameter [12"]		86.87	.322		23	10.35	1.76	35.11	43	
	3080	375 mm diameter [15"]		70.10	.399		26.50	12.80	2.18	41.48	51.50	
	3100	450 mm diameter [18"]		50.29	.557		34.50	17.85	3.03	55.38	69	
	4000	Extra strength, 150 mm diameter [6"]	B-20	96.01	.250		8.85	7.75		16.60	22	
	4020	200 mm diameter [8"]	B-21	94.49	.296		13.30	9.50	1.62	24.42	31	
	4040	250 mm diameter [10"]		86.87	.322		26.50	10.35	1.76	38.61	47	
	4060	300 mm diameter [12"]		70.10	.399		29	12.80	2.18	43.98	54	
	4080	375 mm diameter [15"]		60.96	.459		32	14.75	2.50	49.25	61	
	4100	450 mm diameter [18"]		50.29	.557		46.50	17.85	3.03	67.38	82	
620	0010	**PIPING, SUBDRAINAGE, CORRUGATED METAL**										620
	0021	Not including excavation and backfill										
	2010	Aluminum, perforated										
	2020	150 mm diameter, 1.2 mm T [6"]	B-14	116	.414	m	9.95	12	1.95	23.90	31.50	
	2200	200 mm diameter, 1.5 mm T [8"]		113	.425		14.25	12.35	2	28.60	37	
	2220	250 mm diameter, 1.5 mm T [10"]		110	.436		17.80	12.65	2.06	32.51	41.50	
	2240	300 mm diameter, 1.5 mm T [12"]		86.87	.553		19.95	16.05	2.60	38.60	50	
	2260	450 mm diameter, 1.5 mm T [18"]		62.48	.768		30	22.50	3.62	56.12	71.50	
	3000	Uncoated galvanized, perforated										
	3020	150 mm diameter, 1.2 mm [6"]	B-20	116	.207	m	15.80	6.45		22.25	27.50	
	3200	200 mm diameter, 1.5 mm [8"]	"	113	.212		21.50	6.60		28.10	34.50	
	3220	250 mm diameter, 1.5 mm [10"]	B-21	110	.255		32.50	8.15	1.39	42.04	50	
	3240	300 mm diameter, 1.5 mm [12"]		86.87	.322		34	10.35	1.76	46.11	55.50	
	3260	450 mm diameter, 1.5 mm [18"]		62.48	.448		52	14.40	2.44	68.84	82.50	
	4000	Steel, perforated, asphalt coated										
	4020	150 mm diameter 1.2 mm [6"]	B-20	116	.207	m	12.60	6.45		19.05	24	
	4030	200 mm diameter 1.2 mm [8"]	"	113	.212		19.75	6.60		26.35	32	
	4040	250 mm diameter 1.5 mm [10"]	B-21	110	.255		22.50	8.15	1.39	32.04	39	
	4050	300 mm diameter 1.5 mm [12"]		86.87	.322		26	10.35	1.76	38.11	46.50	
	4060	450 mm diameter 1.5 mm [18"]		62.48	.448		35.50	14.40	2.44	52.34	64	
630	0010	**PIPING, SUBDRAINAGE, PLASTIC**										630
	0020	Not including excavation and backfill										
	1110	Piping, subdrainage, corr polthn, coupler, 250 mm				Ea.	3.51			3.51	3.86	
	2100	Perforated PVC, 100 mm diameter	B-14	95.71	.502	m	2.72	14.55	2.36	19.63	28	
	2110	150 mm diameter		91.44	.525		5.05	15.25	2.47	22.77	32	
	2120	200 mm diameter		88.39	.543		5.70	15.75	2.56	24.01	33.50	
	2130	250 mm diameter		85.34	.562		8.85	16.35	2.65	27.85	38	
	2140	300 mm diameter		82.30	.583		12.25	16.95	2.75	31.95	42.50	
660	0010	**PIPING, SUBDRAINAGE, CORR. PLASTIC TUBING, PERF. OR PLAIN**										660
	0020	In rolls, not including excavation and backfill										
	0030	80 mm diameter [3"]	2 Clab	366	.044	m	1.64	1.20		2.84	3.66	
	0040	100 mm diameter [4"]		366	.044		2.07	1.20		3.27	4.12	
	0041	With silt sock		366	.044		2.39	1.20		3.59	4.48	
	0060	150 mm diameter [6"]		274	.058		6.55	1.60		8.15	9.70	

Important: See the Reference Section for supporting data - Crews, Rental Equipment, City Cost Indexes and Reference Data

		02620	Subdrainage	CREW	DAILY OUTPUT	LABOR-HOURS	UNIT	2006 BARE COSTS MAT.	LABOR	EQUIP.	TOTAL	TOTAL INCL O&P	
660	0080		200 mm diameter [8"]	2 Clab	213	.075	m	9.25	2.06		11.31	13.35	660
	0200		Fittings										
	0230		Elbows, 80 mm diameter [3"]	1 Clab	32	.250	Ea.	4.89	6.85		11.74	16.05	
	0240		100 mm diameter [4"]		32	.250		5.40	6.85		12.25	16.55	
	0250		125 mm diameter [5"]		32	.250		6.35	6.85		13.20	17.65	
	0260		150 mm diameter [6"]		32	.250		8.55	6.85		15.40	20	
	0280		200 mm diameter [8"]		32	.250		10.50	6.85		17.35	22	
	0330		Tees, 80 mm diameter [3"]		27	.296		4.77	8.10		12.87	17.90	
	0340		100 mm diameter [4"]		27	.296		5	8.10		13.10	18.15	
	0350		125 mm diameter [5"]		27	.296		6.20	8.10		14.30	19.45	
	0360		150 mm diameter [6"]		27	.296		8.35	8.10		16.45	22	
	0370		150 mm x 150 mm x 100 mm		27	.296		8.60	8.10		16.70	22	
	0380		200 mm diameter [8"]		27	.296		10	8.10		18.10	23.50	
	0390		200 mm x 200 mm x 150 mm		27	.296		17.90	8.10		26	32.50	
	0430		End cap, 80 mm diameter [3"]		32	.250		1.85	6.85		8.70	12.70	
	0440		100 mm diameter [4"]		32	.250		2.04	6.85		8.89	12.90	
	0460		150 mm diameter [6"]		32	.250		3.05	6.85		9.90	14	
	0480		200 mm diameter [8"]		32	.250		6.40	6.85		13.25	17.65	
	0530		Coupler, 80 mm diameter [3"]		32	.250		1.52	6.85		8.37	12.30	
	0540		100 mm diameter [4"]		32	.250		1.60	6.85		8.45	12.40	
	0550		125 mm diameter [5"]		32	.250		1.98	6.85		8.83	12.80	
	0560		150 mm diameter [6"]		32	.250		2.66	6.85		9.51	13.60	
	0580		200 mm diameter [8"]	▼	32	.250		4.94	6.85		11.79	16.10	
	0590		Heavy duty highway type, add					10%					
	0660		Reducer, 150 mm to 100 mm	1 Clab	32	.250		3.42	6.85		10.27	14.40	
	0680		200 mm to 150 mm		32	.250		5.45	6.85		12.30	16.65	
	0730		"Y" fitting, 80 mm diameter [3"]		27	.296		6	8.10		14.10	19.25	
	0740		100 mm diameter [4"]		27	.296		6.90	8.10		15	20.50	
	0750		125 mm diameter [5"]		27	.296		8.70	8.10		16.80	22	
	0760		150 mm diameter [6"]		27	.296		10.50	8.10		18.60	24	
	0780		200 mm diameter [8"]	▼	27	.296	▼	12.90	8.10		21	27	
	0860		Silt sock only for above tubing, 150 mm dia. [6"]				m	2.23			2.23	2.46	
	0880		200 mm diameter [8"]				"	3.67			3.67	4.04	
670	0010	**PIPING, SUBDRAINAGE, POLYVINYL CHLORIDE**											670
	0020		Perforated, price as solid pipe, Division 02530-780										

		02630	Storm Drainage										
110	0010	**CATCH BASIN GRATES AND FRAMES** not including footing, excavation											110
	1580		Curb inlet frame, grate, and curb box										
	1582		Large 600 mm x 900 mm heavy duty	B-24	2	12	Ea.	475	390		865	1,125	
	1590		Small 250 mm x 525 mm medium duty	"	2	12		335	390		725	965	
	1600		Frames & covers, C.I., 600 mm square, 227 kg	B-6	7.80	3.077		268	92.50	29	389.50	470	
	1700		650 mm D shape, 272 kg		7	3.429		460	103	32.50	595.50	700	
	1800		Light traffic, 450 mm diameter, 45 kg [18"]		10	2.400		150	72	22.50	244.50	300	
	1900		600 mm diameter, 136 kg [24"]		8.70	2.759		232	83	26	341	410	
	2000		900 mm diameter, 408 kg [36"]		5.80	4.138		465	124	39	628	745	
	2100		Heavy traffic, 600 mm diameter, 181 kg [24"]		7.80	3.077		224	92.50	29	345.50	420	
	2200		900 mm diameter, 522 kg [36"]		3	8		740	240	75.50	1,055.50	1,275	
	2300		Mass. State standard, 650 mm diameter, 215 kg [26"]		7	3.429		560	103	32.50	695.50	810	
	2400		750 mm diameter, 281 kg [30"]		7	3.429		355	103	32.50	490.50	585	
	2500		Watertight, 600 mm diameter, 159 kg [24"]		7.80	3.077		380	92.50	29	501.50	595	
	2600		650 mm diameter, 227 kg [26"]		7	3.429		360	103	32.50	495.50	590	
	2700		800 mm diameter, 261 kg [32"]	▼	6	4	▼	805	120	37.50	962.50	1,100	
	2800		3 piece cover & frame, 250 mm deep,										
	2900		544 kg, for heavy equipment	B-6	3	8	Ea.	1,150	240	75.50	1,465.50	1,700	

SITE CONSTRUCTION **2**

02630	Storm Drainage	CREW	DAILY OUTPUT	LABOR-HOURS	UNIT	MAT.	LABOR	EQUIP.	TOTAL	TOTAL INCL O&P
110										**110**
3000	Raised for paving 32 mm to 50 mm high,									
3100	4 piece expansion ring									
3200	500 mm to 650 mm diameter [20″,26″]	1 Clab	3	2.667	Ea.	126	73		199	252
3300	750 mm to 900 mm diameter [30″,36″]	″	3	2.667	″	175	73		248	305
3320	Frames and covers, existing, raised for paving, 50 mm, including									
3340	row of brick, concrete collar, up to 300 mm wide frame	B-6	18	1.333	Ea.	39.50	40	12.55	92.05	118
3360	500 mm to 650 mm wide frame		11	2.182		62	65.50	20.50	148	192
3380	750 mm to 900 mm wide frame	▼	9	2.667		76.50	80	25	181.50	235
3400	Inverts, single channel brick	D-1	3	5.333		86	171		257	355
3500	Concrete		5	3.200		67.50	103		170.50	231
3600	Triple channel, brick		2	8		131	257		388	535
3700	Concrete	▼	3	5.333	▼	115	171		286	390
400	**0010**	**STORM DRAINAGE MANHOLES, FRAMES & COVERS** not including								**400**
	0020	footing, excavation, backfill (See line items for frame & cover)								
	0050	Brick, 1200 mm inside diameter, 1200 mm deep	D-1	1	16	Ea.	360	515		875	1,175
	0100	1800 mm deep		.70	22.857		505	735		1,240	1,675
	0150	2400 mm deep		.50	32	▼	645	1,025		1,670	2,275
	0200	For depths over 2400 mm, add		1.22	13.123	m	600	420		1,020	1,300
	0400	Conc. blocks (radial), 1200 mm I.D., 1200 mm deep		1.50	10.667	Ea.	300	345		645	850
	0500	1800 mm deep		1	16		400	515		915	1,225
	0600	2400 mm deep		.70	22.857		500	735		1,235	1,675
	0700	For depths over 2400 mm, add	▼	1.68	9.544	m	168	305		473	650
	0800	Conc.C.I.P., 1200 mm x 1200 mm, 200 mm thk., 1200 mm deep	C-14H	2	24	Ea.	470	840	10.80	1,320.80	1,825
	0900	1800 mm deep		1.50	32		685	1,125	14.40	1,824.40	2,525
	1000	2400 mm deep		1	48	▼	975	1,675	21.50	2,671.50	3,725
	1100	For depths over 2400 mm, add	▼	2.44	19.685	m	370	690	8.85	1,068.85	1,500
	1110	Precast, 1200 mm I.D., 1200 mm deep	B-22	4.10	7.317	Ea.	710	238	56	1,004	1,200
	1120	1800 mm deep		3	10		915	325	76.50	1,316.50	1,575
	1130	2400 mm deep		2	15	▼	1,075	490	114	1,679	2,050
	1140	For depths over 2400 mm, add	▼	4.88	6.152	m	490	200	47	737	900
	1150	1500 mm I.D., 1200 mm deep	B-6	3	8	Ea.	755	240	75.50	1,070.50	1,275
	1160	1800 mm deep		2	12		1,025	360	113	1,498	1,800
	1170	2400 mm deep		1.50	16	▼	1,275	480	151	1,906	2,300
	1180	For depths over 2400 mm, add		3.66	6.562	m	550	197	62	809	980
	1190	1800 mm I.D., 1200 mm deep		2	12	Ea.	1,225	360	113	1,698	2,025
	1200	1800 mm deep		1.50	16		1,600	480	151	2,231	2,675
	1210	2400 mm deep		1	24	▼	1,975	720	226	2,921	3,525
	1220	For depths over 2400 mm, add	▼	2.44	9.843	m	850	295	93	1,238	1,475
	1250	Slab tops, precast, 200 mm thick									
	1300	1200 mm diameter manhole	B-6	8	3	Ea.	173	90	28.50	291.50	360
	1400	1500 mm diameter manhole		7.50	3.200		340	96	30	466	555
	1500	1800 mm diameter manhole	▼	7	3.429		425	103	32.50	560.50	665
	3800	Steps, heavyweight cast iron, 175 mm x 225 mm	1 Bric	40	.200		13.95	7.30		21.25	26.50
	3900	200 mm x 230 mm		40	.200		21	7.30		28.30	34
	3928	300 mm x 270 mm		40	.200		19.80	7.30		27.10	33
	4000	Standard sizes, galvanized steel		40	.200		16.85	7.30		24.15	29.50
	4100	Aluminum	▼	40	.200	▼	23	7.30		30.30	36
510	**0010**	**PIPING, STORM DRAINAGE, CORRUGATED METAL**								**510**
	0020	Not including excavation or backfill								
	2000	Corrugated metal pipe, galvanized and coated								
	2020	Bituminous coated with paved invert, 6 m lengths								
	2040	200 mm diameter, 1.5 mm [8″]	B-14	101	.475	m	30.50	13.80	2.24	46.54	57.50
	2060	250 mm diameter, 1.5 mm [10″]		79.25	.606		36.50	17.60	2.85	56.95	70.50
	2080	300 mm diameter, 1.5 mm [12″]		64.01	.750		44	22	3.53	69.53	86
	2100	375 mm diameter, 1.5 mm [15″]	▼	60.96	.787		53.50	23	3.71	80.21	98

02630 | Storm Drainage

		CREW	DAILY OUTPUT	LABOR-HOURS	UNIT	2006 BARE COSTS MAT.	LABOR	EQUIP.	TOTAL	TOTAL INCL O&P	
510	2120	450 mm diameter, 1.5 mm [18"]	B-14	57.91	.829	m	68.50	24	3.90	96.40	116	510
	2140	600 mm diameter, 1.9 mm [24"]	↓	48.77	.984		83.50	28.50	4.64	116.64	141	
	2160	750 mm diameter, 1.9 mm [30"]	B-13	36.58	1.531		111	46	17.80	174.80	212	
	2180	900 mm diameter, 2.7 mm [36"]		36.58	1.531		162	46	17.80	225.80	268	
	2200	1200 mm diameter, 2.7 mm [48"]	↓	30.48	1.837		247	55	21.50	323.50	380	
	2220	1500 mm diameter, 3.4 mm [60"]	B-13B	22.86	2.450		320	73.50	41.50	435	510	
	2240	1800 mm diameter, 4.2 mm[72"]	"	13.72	4.083	↓	475	122	69.50	666.50	790	
	2250	End sections, 200 mm diameter, 1.5 mm [8"]	B-14	20	2.400	Ea.	54.50	69.50	11.30	135.30	180	
	2255	250 mm diameter, 1.5 mm [10"]		20	2.400		62	69.50	11.30	142.80	189	
	2260	300 mm diameter, 1.5 mm [12"]		18	2.667		76.50	77.50	12.55	166.55	218	
	2265	375 mm diameter, 1.5 mm [15"]		18	2.667		94.50	77.50	12.55	184.55	238	
	2270	450 mm diameter, 1.5 mm [18"]	↓	16	3		120	87	14.15	221.15	282	
	2275	600 mm diameter, 1.5 mm [24"]	B-13	16	3.500		171	105	40.50	316.50	395	
	2280	750 mm diameter, 1.5 mm [30"]		14	4		325	120	46.50	491.50	590	
	2285	900 mm diameter, 1.9 mm [36"]		14	4		425	120	46.50	591.50	705	
	2290	1200 mm diameter, 1.9 mm [48"]		10	5.600		885	168	65	1,118	1,300	
	2292	1500 mm diameter, 1.9 mm [60"]	↓	6	9.333		1,200	279	108	1,587	1,875	
	2294	1800 mm diameter, 1.9 mm [72"]	B-13B	5	11.200		2,125	335	190	2,650	3,050	
	2300	Bends or elbows, 200 mm diameter [8"]	B-14	28	1.714		87.50	50	8.05	145.55	182	
	2320	250 mm diameter [10"]		25	1.920		108	55.50	9.05	172.55	215	
	2340	300 mm diameter [12"]		30	1.600		125	46.50	7.55	179.05	218	
	2342	450 mm diameter, 1.5 mm [18"]		20	2.400		181	69.50	11.30	261.80	320	
	2344	600 mm diameter, 1.9 mm [24"]		16	3		252	87	14.15	353.15	430	
	2346	750 mm diameter, 1.9 mm [30"]	↓	15	3.200		315	93	15.05	423.05	505	
	2348	900 mm diameter, 1.9 mm [36"]	B-13	15	3.733		425	112	43.50	580.50	685	
	2350	1200 mm diameter, 12.7 mm[48"]	"	12	4.667		605	140	54	799	940	
	2352	1500 mm diameter, 3.4 mm [60"]	B-13B	10	5.600		810	168	95	1,073	1,250	
	2354	1800 mm diameter, 3.4 mm [72"]	"	6	9.333		1,000	279	159	1,438	1,700	
	2360	Wyes or tees, 200 mm diameter [8"]	B-14	25	1.920		122	55.50	9.05	186.55	230	
	2380	250 mm diameter [10"]		21	2.286		152	66.50	10.75	229.25	282	
	2400	300 mm diameter [12"]		22.48	2.135		127	62	10.05	199.05	246	
	2410	450 mm diameter, 1.5 mm [18"]		16	3		241	87	14.15	342.15	415	
	2412	600 mm diameter, 1.9 mm [24"]	↓	16	3		365	87	14.15	466.15	550	
	2414	750 mm diameter, 1.9 mm [30"]	B-13	12	4.667		470	140	54	664	795	
	2416	900 mm diameter, 1.9 mm [36"]		11	5.091		580	152	59	791	940	
	2418	1200 mm diameter, 2.7 mm [48"]	↓	10	5.600		880	168	65	1,113	1,300	
	2420	1500 mm diameter, 3.4 mm [60"]	B-13B	8	7		1,275	210	119	1,604	1,875	
	2422	1800 mm diameter, 3.4 mm [72"]	"	5	11.200	↓	1,550	335	190	2,075	2,425	
	2500	Galvanized, uncoated, 6 m lengths										
	2520	200 mm diameter, 1.5 mm [8"]	B-14	108	.444	m	25	12.90	2.09	39.99	50	
	2540	250 mm diameter, 1.5 mm [10"]		85.34	.562		27.50	16.35	2.65	46.50	58.50	
	2560	300 mm diameter, 1.5 mm [12"]		67.06	.716		31.50	21	3.37	55.87	70	
	2580	375 mm diameter, 1.5 mm [15"]		67.06	.716		40	21	3.37	64.37	79	
	2600	450 mm diameter, 1.5 mm [18"]		62.48	.768		52.50	22.50	3.62	78.62	96.50	
	2620	600 mm diameter, 1.9 mm [24"]	↓	53.34	.900		76.50	26	4.24	106.74	129	
	2640	750 mm diameter, 1.9 mm [30"]	B-13	39.62	1.413		97	42.50	16.40	155.90	190	
	2660	900 mm diameter, 2.7 mm [36"]		39.62	1.413		159	42.50	16.40	217.90	258	
	2680	1200 mm diameter, 2.7 mm [48"]	↓	33.53	1.670		213	50	19.40	282.40	335	
	2690	1500 mm diameter, 3.4 mm [60"]	B-13B	23.77	2.355		335	70.50	40	445.50	525	
	2695	1800 mm diameter, 3.4 mm [72"]	"	18.29	3.062	↓	425	91.50	52	568.50	665	
	2711	Bends or elbows, 300 mm diameter, 1.5 mm [12"]	B-14	30	1.600	Ea.	108	46.50	7.55	162.05	199	
	2712	375 mm diameter, 1.5 mm [15"]		25.04	1.917		134	55.50	9.05	198.55	243	
	2714	450 mm diameter, 1.5 mm [18"]		20	2.400		150	69.50	11.30	230.80	285	
	2716	600 mm diameter, 1.9 mm [24"]		16	3		219	87	14.15	320.15	390	
	2718	750 mm diameter, 1.9 mm [30"]	↓	15	3.200		274	93	15.05	382.05	460	
	2720	900 mm drainage, 1.9 mm [36"]	B-13	15	3.733	↓	375	112	43.50	530.50	635	

SITE CONSTRUCTION 2

02630 | Storm Drainage

		CREW	DAILY OUTPUT	LABOR-HOURS	UNIT	MAT.	LABOR	EQUIP.	TOTAL	TOTAL INCL O&P	
510							2006 BARE COSTS				**510**
2722	1200 mm diameter, 2.7 mm [48"]	B-13	12	4.667	Ea.	515	140	54	709	840	
2724	1500 mm diameter, 3.4 mm [60"]		10	5.600		790	168	65	1,023	1,200	
2726	1800 mm diameter, 3.4 mm [72"]	↓	6	9.333		980	279	108	1,367	1,625	
2728	Wyes or tees, 300 mm diameter, 1.5 mm [12"]	B-14	22.48	2.135		144	62	10.05	216.05	265	
2730	450 mm diameter, 1.5 mm [18"]		15	3.200		209	93	15.05	317.05	390	
2732	600 mm diameter, 1.9 mm [24"]		15	3.200		325	93	15.05	433.05	515	
2734	750 mm diameter, 1.9 mm [30"]	↓	14	3.429		425	99.50	16.15	540.65	635	
2736	900 mm diameter, 1.9 mm [36"]	B-13	14	4		550	120	46.50	716.50	840	
2738	1200 mm diameter, 2.7 mm [48"]		12	4.667		810	140	54	1,004	1,175	
2740	1500 mm diameter, 3.4 mm [60"]		10	5.600		1,175	168	65	1,408	1,625	
2742	1800 mm diameter, 3.4 mm [72"]	↓	6	9.333		1,400	279	108	1,787	2,100	
2780	End sections, 200 mm diameter [8"]	B-14	24	2		74	58	9.40	141.40	182	
2785	250 mm diameter [10"]		22	2.182		76	63.50	10.30	149.80	193	
2790	300 mm diameter [12"]		35	1.371		86	40	6.45	132.45	163	
2800	450 mm diameter [18"]	↓	30	1.600		102	46.50	7.55	156.05	192	
2810	600 mm diameter [24"]	B-13	25	2.240		149	67	26	242	296	
2820	750 mm diameter [30"]		25	2.240		282	67	26	375	440	
2825	900 mm diameter [36"]		20	2.800		375	84	32.50	491.50	575	
2830	1200 mm diameter [48"]	↓	10	5.600		775	168	65	1,008	1,175	
2835	1500 mm diameter [60"]	B-13B	5	11.200		1,000	335	190	1,525	1,825	
2840	1800 mm diameter [72"]	"	4	14		1,825	420	238	2,483	2,900	
2850	Couplings, 300 mm diameter [12"]					16.60			16.60	18.30	
2855	450 mm diameter [18"]					24.50			24.50	26.50	
2860	600 mm diameter [24"]					33			33	36	
2865	750 mm diameter [30"]					40			40	44.50	
2870	900 mm diameter [36"]					56.50			56.50	62	
2875	1200 mm diameter [48"]					76.50			76.50	84	
2880	1500 mm diameter [60"]					120			120	132	
2885	1800 mm diameter [72"]	↓			↓	166			166	183	
520	0010 **PIPING, DRAINAGE & SEWAGE, CORRUGATED HDPE TYPE S**										**520**
	0020 Not including excavation & backfill, bell & spigot										
	1000 With gaskets, 100 mm diameter	B-20	130	.185	m	2.56	5.75		8.31	11.75	
	1010 150 mm diameter		122	.197		5.85	6.10		11.95	16	
	1020 200 mm diameter		116	.207		11.25	6.45		17.70	22.50	
	1030 250 mm diameter		113	.212		15.55	6.60		22.15	27.50	
	1040 300 mm diameter		104	.231		17.35	7.20		24.55	30.50	
	1050 375 mm diameter	↓	91.44	.262		23.50	8.15		31.65	38.50	
	1060 450 mm diameter	B-21	83.82	.334		33.50	10.70	1.82	46.02	55.50	
	1070 600 mm diameter		76.20	.367		52	11.80	2	65.80	78	
	1080 750 mm diameter		60.96	.459		82.50	14.75	2.50	99.75	117	
	1090 900 mm diameter		54.86	.510		105	16.40	2.78	124.18	144	
	1100 1050 mm diameter		53.34	.525		146	16.85	2.86	165.71	190	
	1110 1200 mm diameter		51.82	.540		190	17.35	2.95	210.30	239	
	1120 1350 mm diameter		48.77	.574		293	18.45	3.13	314.58	350	
	1130 1500 mm diameter	↓	45.72	.612	↓	340	19.65	3.34	362.99	410	
	1135 Add 15% to material pipe cost for water tight connection bell & spigot										
	1140 HDPE type s, elbows, 300 mm diam	B-20	11	2.182	Ea.	58.50	68		126.50	171	
	1150 HDPE type s, elbows, 375 mm diam	"	9	2.667		90.50	83		173.50	229	
	1160 HDPE type s, elbows, 450 mm diam	B-21	9	3.111		149	100	16.95	265.95	340	
	1170 HDPE type s, elbows, 600 mm diam		9	3.111		315	100	16.95	431.95	525	
	1180 HDPE type s, elbows, 750 mm diam		8	3.500		505	112	19.10	636.10	750	
	1190 HDPE type s, elbows, 900 mm diam	↓	8	3.500		650	112	19.10	781.10	910	
	1240 HDPE type s, Tee, 300 mm diam	B-20	7	3.429		132	107		239	310	
	1260 HDPE type s, Tee, 375 mm diam	"	6	4		156	124		280	365	
	1280 HDPE type s, Tee, 450 mm diam	B-21	6	4.667	↓	229	150	25.50	404.50	510	

SITE CONSTRUCTION — 2

		02630 | Storm Drainage	CREW	DAILY OUTPUT	LABOR-HOURS	UNIT	2006 BARE COSTS				TOTAL INCL O&P	
							MAT.	LABOR	EQUIP.	TOTAL		
520	1300	HDPE type s, Tee, 600 mm diam	B-21	5	5.600	Ea.	300	180	30.50	510.50	640	**520**
	1320	HDPE type s, Tee, 750 mm diam	↓	5	5.600	↓	565	180	30.50	775.50	935	
	1400	Add to basic installation cost for each split coupling joint										
	1402	HDPE type s, split coupling, 300 mm diam	B-20	17	1.412	Ea.	4.27	44		48.27	73	
	1420	HDPE type s, split coupling, 375 mm diam	"	15	1.600	"	5.65	50		55.65	84	
530	0010	**SEWAGE/DRAINAGE COLLECTION, CONCRETE PIPE**										**530**
	0020	Not including excavation or backfill										
	0050	Box culvert, cast in place, 1.8 m x 1.8 m	C-15	4.88	14.764	m	355	490		845	1,150	
	0060	2.4 m x 2.4 m		4.27	16.873		520	560		1,080	1,450	
	0070	3.7 m x 3.7 m	↓	3.05	23.622		1,025	785		1,810	2,350	
	0100	Box culvert, precast, base price, 2.4 m long, 1.8 m x 0.9 m	B-69	42.67	1.125		555	34	25	614	690	
	0150	1.8 m x 2.1 m		38.10	1.260		835	38.50	28	901.50	1,000	
	0200	2.4 m x 0.9 m		40.54	1.184		760	36	26	822	920	
	0250	2.4 m x 2.4 m		30.48	1.575		1,025	48	34.50	1,107.50	1,225	
	0300	3.0 m x 0.9 m		33.53	1.432		1,125	43.50	31.50	1,200	1,350	
	0350	3.0 m x 2.4 m		24.38	1.969		1,275	60	43.50	1,378.50	1,550	
	0400	3.7 m x 0.9 m		30.48	1.575		1,100	48	34.50	1,182.50	1,300	
	0450	3.7 m x 2.4 m	↓	20.42	2.350	↓	1,875	71.50	52	1,998.50	2,250	
	0500	Set up charge at plant, add to base price				Job	3,200			3,200	3,525	
	0510	Inserts and keyway, add				Ea.	281			281	310	
	0520	Sloped or skewed end, add				"	445			445	490	
	1000	Non-reinforced pipe, extra strength, B&S or T&G joints										
	1010	150 mm diameter [6"]	B-14	80.78	.594	m	14.65	17.25	2.80	34.70	45.50	
	1020	200 mm diameter [8"]		68.28	.703		16.10	20.50	3.31	39.91	53	
	1030	250 mm diameter [10"]		65.84	.729		17.85	21	3.43	42.28	56.50	
	1040	300 mm diameter [12"]		60.96	.787		22	23	3.71	48.71	63.50	
	1050	375 mm diameter [15"]		54.86	.875		25.50	25.50	4.12	55.12	72	
	1060	450 mm diameter [18"]		43.89	1.094		31.50	32	5.15	68.65	89	
	1070	530 mm diameter [21"]		34.14	1.406		39	41	6.60	86.60	113	
	1080	600 mm diameter [24"]	↓	30.48	1.575	↓	47.50	45.50	7.40	100.40	131	
	1560	Reinforced culvert, class 2, no gaskets										
	1590	675 mm diameter [27"]	B-21	26.82	1.044	m	95	33.50	5.70	134.20	162	
	1592	750 mm diameter [30"]	B-13	24.38	2.297		103	69	26.50	198.50	249	
	1594	900 mm diameter [36"]	"	21.95	2.552	↓	120	76.50	29.50	226	284	
	2000	Reinforced culvert, class 3, no gaskets										
	2010	300 mm diameter [12"]	B-14	45.72	1.050	m	40	30.50	4.94	75.44	96	
	2020	375 mm diameter [15"]		45.72	1.050		51	30.50	4.94	86.44	108	
	2030	450 mm diameter [18"]		40.23	1.193		53.50	34.50	5.60	93.60	119	
	2035	525 mm diameter [21"]		36.58	1.312		69.50	38	6.20	113.70	142	
	2040	600 mm diameter [24"]	↓	30.48	1.575		79	45.50	7.40	131.90	165	
	2045	675 mm diameter [27"]	B-13	28.04	1.997		99.50	60	23	182.50	227	
	2050	750 mm diameter [30"]		26.82	2.088		108	62.50	24.50	195	242	
	2060	900 mm diameter [36"]	↓	21.95	2.552		154	76.50	29.50	260	320	
	2070	1050 mm diameter [42"]	B-13B	21.95	2.552		210	76.50	43.50	330	395	
	2080	1200 mm diameter [48"]		19.51	2.871		261	86	49	396	475	
	2090	1500 mm diameter [60"]		14.63	3.828		415	115	65	595	710	
	2100	1800 mm diameter [72"]		12.19	4.593		590	138	78	806	950	
	2120	2100 mm diameter [84"]		9.75	5.741		995	172	97.50	1,264.50	1,475	
	2140	2400 mm diameter [96"]	↓	7.32	7.655		1,200	229	130	1,559	1,800	
	2200	With gaskets, class 3, 300 mm diameter [12"]	B-21	51.21	.547		43.50	17.55	2.98	64.03	78.50	
	2220	375 mm diameter [15"]		48.77	.574		52	18.45	3.13	73.58	89.50	
	2230	450 mm diameter [18"]		46.33	.604		65	19.40	3.29	87.69	105	
	2240	600 mm diameter [24"]	↓	41.45	.675		98	21.50	3.68	123.18	146	
	2260	750 mm diameter [30"]	B-13	26.82	2.088		130	62.50	24.50	217	266	
	2270	900 mm diameter [36"]	"	21.95	2.552	↓	196	76.50	29.50	302	365	

SITE CONSTRUCTION **2**

For expanded coverage of these items see *Means Site Work & Landscape Cost Data 2006*

02630 | Storm Drainage

			CREW	DAILY OUTPUT	LABOR-HOURS	UNIT	MAT.	LABOR	EQUIP.	TOTAL	TOTAL INCL O&P	
								2006 BARE COSTS				
530	2290	1200 mm diameter [48"]	B-13B	19.51	2.871	m	320	86	49	455	535	**530**
	2310	1800 mm diameter [72"]	"	12.19	4.593		825	138	78	1,041	1,200	
	2330	Flared ends, 1830 mm long, 300 mm diameter [12"]	B-21	57.91	.483		120	15.50	2.64	138.14	159	
	2340	375 mm diameter [15"]		47.24	.593		137	19	3.23	159.23	183	
	2400	1880 mm long, 450 mm diameter [18"]		37.19	.753		142	24	4.10	170.10	199	
	2420	600 mm diameter [24"]		26.82	1.044		164	33.50	5.70	203.20	238	
	2440	900 mm diameter [36"]	B-13	18.29	3.062		299	91.50	35.50	426	510	
	2500	Class 4										
	2510	300 mm diameter [12"]	B-21	51.21	.547	m	47	17.55	2.98	67.53	82	
	2512	375 mm diameter [15"]		48.77	.574		55	18.45	3.13	76.58	92.50	
	2514	450 mm diameter [18"]		46.33	.604		65.50	19.40	3.29	88.19	106	
	2516	525 mm diameter [21"]		43.89	.638		81	20.50	3.48	104.98	124	
	2518	600 mm diameter [24"]		41.45	.675		95.50	21.50	3.68	120.68	143	
	2520	675 mm diameter [27"]		36.58	.766		119	24.50	4.17	147.67	174	
	2522	750 mm diameter [30"]	B-13	26.82	2.088		131	62.50	24.50	218	267	
	2524	900 mm diameter [36"]	"	21.95	2.552		194	76.50	29.50	300	365	
	2600	Class 5										
	2610	300 mm diameter [12"]	B-21	51.21	.547	m	47.50	17.55	2.98	68.03	83	
	2612	375 mm diameter [15"]		48.77	.574		56	18.45	3.13	77.58	93.50	
	2614	450 mm diameter [18"]		46.33	.604		67	19.40	3.29	89.69	107	
	2616	525 mm diameter [21"]		43.89	.638		83.50	20.50	3.48	107.48	127	
	2618	600 mm diameter [24"]		41.45	.675		97.50	21.50	3.68	122.68	145	
	2620	675 mm diameter [27"]		36.58	.766		125	24.50	4.17	153.67	181	
	2622	750 mm diameter [30"]	B-13	26.82	2.088		138	62.50	24.50	225	275	
	2624	900 mm diameter [36"]	"	21.95	2.552		194	76.50	29.50	300	365	
	2800	Add for rubber joints,					12%					
	3080	Radius pipe, add, 300 mm to 1500 mm diameter [12",60"]					50%					
	3090	Over 1500 mm diameter, add [60"]					20%					
	3500	Reinforced elliptical, 2400 mm lengths, C507 class 3										
	3520	350 mm x 575 mm inside, [14",23"]	B-21	24.99	1.120	m	94	36	6.10	136.10	165	
	3530	600 mm x 950 mm inside, [24",38"]	B-13	17.68	3.168		166	95	37	298	370	
	3540	725 mm x 1125 mm inside, [29",45"]		15.85	3.533		214	106	41	361	445	
	3550	950 mm x 1500 mm inside, [38",60"]		11.58	4.835		279	145	56	480	590	
	3560	1200 mm x 1900 mm inside, [48",76"]		7.92	7.066		425	212	82	719	885	
	3570	1450 mm x 2300 mm inside, [58",91"]		6.71	8.351		605	250	97	952	1,150	
	3780	Concrete slotted pipe, class 4 mortar joint										
	3800	300 mm diameter [12"]	B-21	51.21	.547	m	46.50	17.55	2.98	67.03	82	
	3840	450 mm diameter [18"]	"	46.33	.604	"	72	19.40	3.29	94.69	113	
	3900	Class 4 O-ring										
	3940	300 mm diameter [12"]	B-21	51.21	.547	m	49	17.55	2.98	69.53	84	
	3960	450 mm diameter [18"]	"	46.33	.604	"	65.50	19.40	3.29	88.19	106	
	6200	Gasket, conc. pipe joint, 300 mm, [12"]				Ea.	3.41			3.41	3.75	
	6220	600 mm, [24"]					6.25			6.25	6.85	
	6240	900 mm, [36"]					9.05			9.05	10	
	6260	1200 mm, [48"]					14.35			14.35	15.75	
	6270	1500 mm, [60"]					20.50			20.50	22.50	
	6280	1800 mm, [72"]					22.50			22.50	25	

02640 | Culverts/Manuf Const

			CREW	DAILY OUTPUT	LABOR-HOURS	UNIT	MAT.	LABOR	EQUIP.	TOTAL	TOTAL INCL O&P	
810	0010	**OVAL ARCH CULVERTS**										**810**
	3000	Corrugated galvanized or aluminum, coated & paved										
	3020	430 mm x 330 mm, 1.5 mm, 375 mm equivalent [15"]	B-14	60.96	.787	m	95.50	23	3.71	122.21	145	
	3040	525 mm x 375 mm, 1.5 mm, 450 mm equivalent [18"]		45.72	1.050		123	30.50	4.94	158.44	187	
	3060	700 mm x 500 mm, 1.9 mm, 600 mm equivalent [24"]		38.10	1.260		176	36.50	5.95	218.45	257	
	3080	900 mm x 600 mm, 1.9 mm, 750 mm equivalent [30"]		30.48	1.575		215	45.50	7.40	267.90	315	

Important: See the Reference Section for supporting data - Crews, Rental Equipment, City Cost Indexes and Reference Data

2
SITE CONSTRUCTION

02600 | Drainage & Containment

02640 | Culverts/Manuf Const

			CREW	DAILY OUTPUT	LABOR-HOURS	UNIT	2006 BARE COSTS				TOTAL INCL O&P	
							MAT.	LABOR	EQUIP.	TOTAL		
810	3100	1050 mm x 725 mm, 2.7 mm, 900 mm equivalent [36"]	B-13	30.48	1.837	m	320	55	21.50	396.50	460	810
	3120	1250 mm x 840 mm, 2.7 mm, 1050 mm equivalent [42"]		27.43	2.041		385	61	23.50	469.50	545	
	3140	1425 mm x 950 mm, 2.7 mm, 1200 mm equivalent [48"]	↓	22.86	2.450	↓	430	73.50	28.50	532	615	
	3160	Steel, plain oval arch culverts, plain										
	3180	430 mm x 330 mm, 1.5 mm, 375 mm equivalent [15"]	B-14	68.58	.700	m	51.50	20.50	3.30	75.30	92	
	3200	525 mm x 375 mm, 1.5 mm, 450 mm equivalent [18"]		53.34	.900		61	26	4.24	91.24	112	
	3220	710 mm x 510 mm, 1.9 mm, 600 mm equivalent [24"]	↓	45.72	1.050		98	30.50	4.94	133.44	160	
	3240	875 mm x 600 mm, 1.9 mm, 750 mm equivalent [30"]	B-13	32.92	1.701		123	51	19.75	193.75	235	
	3260	1050 mm x 725 mm, 12.7 mm 900 mm equivalent [36"]		32.92	1.701		203	51	19.75	273.75	325	
	3280	1250 mm x 840 mm, 2.7 mm, 1050 mm equivalent [42"]		28.04	1.997		240	60	23	323	385	
	3300	1425 mm x 950 mm, 2.7 mm, 1200 mm equivalent [48"]	↓	22.86	2.450	↓	212	73.50	28.50	314	380	
	3320	End sections, 430 mm x 330 mm		22	2.545	Ea.	86	76	29.50	191.50	245	
	3340	1050 mm x 725 mm	↓	17	3.294	"	355	98.50	38.50	492	585	
	3360	Multi-plate arch, steel	B-20	767	.031	kg	2.18	.97		3.15	3.92	

02660 | Ponds & Reservoirs

			CREW	DAILY OUTPUT	LABOR-HOURS	UNIT	MAT.	LABOR	EQUIP.	TOTAL	TOTAL INCL O&P	
600	0010	**SLURRY TRENCH** Excavated slurry trench in wet soils										600
	0020	backfilled with 21 MPa concrete, no reinforcing steel										
	0050	Minimum	C-7	9.42	7.640	m³	227	229	107	563	720	
	0100	Maximum		5.66	12.721	"	380	380	179	939	1,200	
	0200	Alternate pricing method, minimum		13.94	5.167	m²	139	155	72.50	366.50	470	
	0300	Maximum	↓	11.15	6.459		208	194	90.50	492.50	625	
	0500	Reinforced slurry trench, minimum	B-48	16.44	3.406		104	106	221	431	520	
	0600	Maximum	"	6.41	8.736	↓	345	271	570	1,186	1,425	
	0800	Haul for disposal, 3.2 km haul, excavated material, add	B-34B	75.70	.106	m³		3	6.30	9.30	11.55	
	0900	Haul bentonite castings for disposal, add	"	30.58	.262	"		7.40	15.60	23	28.50	
610	0010	**POND AND RESERVOIR LINERS, HDPE,** 9290 m² or more										610
	1100	0.762 mm thick	3 Skwk	172	.140	m²	3.34	5.10		8.44	11.60	
	1200	2 mm thick		149	.161		6.25	5.90		12.15	16.05	
	1300	3 mm thick	↓	134	.179	↓	14	6.55		20.55	25.50	

02700 | Bases, Ballasts, Pavements & Appurtenances

02710 | Bound Base Courses

			CREW	DAILY OUTPUT	LABOR-HOURS	UNIT	2006 BARE COSTS				TOTAL INCL O&P	
							MAT.	LABOR	EQUIP.	TOTAL		
100	0010	**AGGREGATE-BITUMINOUS BASE COURSE** for roadways										100
	0020	and large paved areas										
	0500	Bituminous concrete, 100 mm thick	B-25	3800	.023	m²	10.20	.70	.56	11.46	12.90	
	0550	150 mm thick		3094	.028		14.95	.86	.68	16.49	18.50	
	0560	200 mm thick		2508	.035		19.90	1.06	.84	21.80	24.50	
	0570	250 mm thick	↓	2128	.041	↓	24.50	1.25	.99	26.74	30	
	8900	For small and irregular areas, add						50%	50%			
200	0010	**ASPHALT-TREATED PERMEABLE BASE COURSE** for roadways										200
	0020	and large paved areas										
	0700	Liquid application to gravel base, asphalt emulsion	B-45	22712	.001	liter	.79	.02	.02	.83	.93	
	0800	Prime and seal, cut back asphalt		22712	.001	"	.94	.02	.02	.98	1.09	
	1000	Macadam penetration crushed stone, 9 L/m², 100 mm thick		5017	.003	m²	7.20	.10	.11	7.41	8.20	
	1100	150 mm thick, 14 L/m²		3344	.005		10.75	.16	.16	11.07	12.25	
	1200	200 mm thick, 18 L/m²	↓	2508	.006	↓	14.35	.21	.21	14.77	16.35	
	8900	For small and irregular areas, add						50%	50%			

SITE CONSTRUCTION **2**

		02720	Unbound Base Courses & Ballasts	CREW	DAILY OUTPUT	LABOR-HOURS	UNIT	2006 BARE COSTS				TOTAL INCL O&P	
								MAT.	LABOR	EQUIP.	TOTAL		
200	0010		**AGGREGATE BASE COURSE** For roadways and large areas										200
	0050		20 mm stone compacted, 75 mm deep	B-36C	4348	.009	m²	3.22	.31	.61	4.14	4.68	
	0100		150 mm deep		4181	.010		6.45	.32	.63	7.40	8.30	
	0200		225 mm deep		3846	.010		9.65	.35	.69	10.69	11.90	
	0300		300 mm deep	▼	3512	.011		12.90	.38	.75	14.03	15.55	
	0301		Crushed 40 mm stone base, compacted to 100 mm deep	B-36B	5017	.013		5.10	.41	.60	6.11	6.90	
	0302		150 mm deep		4515	.014		7.65	.46	.67	8.78	9.85	
	0303		200 mm deep		3762	.017		10.20	.55	.80	11.55	12.90	
	0304		300 mm deep	▼	3177	.020	▼	15.25	.65	.95	16.85	18.85	
	0350		Bank run gravel, spread and compacted										
	0370		150 mm deep	B-32	5017	.006	m²	4.23	.22	.31	4.76	5.35	
	0390		230 mm deep		4097	.008		6.35	.27	.38	7	7.80	
	0400		300 mm deep	▼	3512	.009	▼	8.45	.31	.45	9.21	10.25	
	0600		Cold laid asphalt pavement, see div. 02740-400										
	1500		Alternate method to figure base course										
	1510		Crushed stone, 20 mm, compacted, 75 mm deep	B-36C	333	.120	Em³	36	4.03	7.95	47.98	55	
	1511		150 mm deep	B-36B	638	.100		36	3.25	4.73	43.98	50	
	1512		230 mm deep		879	.073		36	2.36	3.44	41.80	47.50	
	1513		300 mm deep		1070	.060		36	1.94	2.82	40.76	46	
	1520		Crushed stone, 40 mm , compacted, 100 mm deep		508	.126		36	4.08	5.95	46.03	53	
	1521		150 mm deep		688	.093		36	3.02	4.39	43.41	49.50	
	1522		200 mm deep		765	.084		36	2.71	3.95	42.66	48.50	
	1523		300 mm deep	▼	967	.066		36	2.15	3.12	41.27	46.50	
	1530		Gravel, bank run, compacted, 150 mm deep	B-36C	638	.063		23.50	2.10	4.15	29.75	34	
	1531		230 mm deep		879	.046		23.50	1.53	3.01	28.04	31.50	
	1532		300 mm deep	▼	1070	.037	▼	23.50	1.25	2.47	27.22	30.50	
	2000		Alternate method to figure base course										
	2005		Bituminous concrete, 100 mm thick	B-25	907	.097	Met. Ton	41.50	2.92	2.33	46.75	53	
	2006		150 mm thick		1107	.079		41.50	2.39	1.91	45.80	52	
	2007		200 mm thick		1198	.073		41.50	2.21	1.77	45.48	51.50	
	2008		250 mm thick	▼	1270	.069		41.50	2.09	1.67	45.26	51	
	2010		Crushed stone, 20 mm maximum size, 75 mm deep	B-36	490	.082		18.40	2.57	2.35	23.32	26.50	
	2011		150 mm deep		1474	.027		18.40	.86	.78	20.04	22	
	2012		230 mm deep		1619	.025		18.40	.78	.71	19.89	22	
	2013		300 mm deep		1769	.023		18.40	.71	.65	19.76	22	
	2020		Crushed stone, 40 mm maximum size, 100 mm deep		653	.061		18.40	1.93	1.76	22.09	25	
	2021		150 mm deep		739	.054		18.40	1.71	1.56	21.67	24.50	
	2022		200 mm deep		758	.053		18.40	1.66	1.52	21.58	24	
	2023		300 mm deep	▼	885	.045		18.40	1.42	1.30	21.12	23.50	
	2030		Bank run gravel, 150 mm deep	B-32A	794	.030		13.45	1.02	1.23	15.70	17.70	
	2031		230 mm deep		880	.027		13.45	.92	1.11	15.48	17.40	
	2032		300 mm deep	▼	962	.025	▼	13.45	.84	1.02	15.31	17.20	
	6000		Stabilization fabric, polypropylene, 200 gram/m²	B-6	8361	.003	m²	1.12	.09	.03	1.24	1.39	
	6900		For small and irregular areas, add						50%	50%			
	7000		Prepare and roll sub-base, small areas to 2100 m²	B-32A	1254	.019	m²		.64	.78	1.42	1.84	
	8000		Large areas over 2100 m²	B-32	3094	.010	"		.36	.51	.87	1.10	

		02740	Flexible Pavement										
310	0010		**ASPHALTIC CONCRETE PAVEMENT, HIGHWAYS**										310
	0020		and large paved areas										
	0080		Binder course, 40 mm thick	B-25	6459	.014	m²	3.46	.41	.33	4.20	4.79	
	0120		50 mm thick		5305	.017		4.60	.50	.40	5.50	6.25	
	0160		75 mm thick		4101	.021		6.85	.65	.52	8.02	9.05	
	0200		100 mm thick	▼	3461	.025		9.15	.77	.61	10.53	11.90	
	0300		Wearing course, 25 mm thick	B-25B	8842	.011		2.44	.33	.26	3.03	3.49	
	0340		40 mm thick	▼	6459	.015	▼	3.72	.46	.36	4.54	5.20	

02740 | Flexible Pavement

			CREW	DAILY OUTPUT	LABOR-HOURS	UNIT	2006 BARE COSTS MAT.	LABOR	EQUIP.	TOTAL	TOTAL INCL O&P	
310	0380	50 mm thick	B-25B	5305	.018	m²	5	.56	.44	6	6.85	**310**
	0420	65 mm thick		4582	.021		6.15	.64	.51	7.30	8.35	
	0460	75 mm thick	↓	4097	.023		7.35	.72	.56	8.63	9.80	
	0500	Open graded friction course	B-25C	4181	.011	↓	2.21	.35	.44	3	3.47	
	0800	Alternate method of figuring paving costs										
	0810	Binder course, 40 mm thick	B-25	572	.154	Met. Ton	41.50	4.63	3.70	49.83	57	
	0811	50 mm thick		626	.141		41.50	4.23	3.38	49.11	56	
	0812	75 mm thick		726	.121		41.50	3.65	2.92	48.07	55	
	0813	100 mm thick	↓	771	.114		41.50	3.44	2.75	47.69	54.50	
	0850	Wearing course, 25 mm thick	B-25B	522	.184		45	5.65	4.43	55.08	63	
	0851	40 mm thick		572	.168		45	5.15	4.05	54.20	62	
	0852	50 mm thick		626	.153		45	4.70	3.70	53.40	61	
	0853	65 mm thick		676	.142		45	4.36	3.42	52.78	60	
	0854	75 mm thick	↓	726	.132	↓	45	4.06	3.19	52.25	59.50	
	1000	Pavement replacement over trench, 50 mm thick	B-37	75.25	.638	m²	5.10	18.50	1.52	25.12	36	
	1050	100 mm thick		58.53	.820		10.05	24	1.95	36	50	
	1080	150 mm thick	↓	45.99	1.044	↓	16	30.50	2.48	48.98	67.50	
	1200	For paving projects 272 metric tons or less add for trucking										
	1300	See 02315-490-0010 for hauling costs										
315	0011	**ASPHALTIC CONCRETE PAVEMENT, LOTS & DRIVEWAYS**										**315**
	0020	150 mm stone base, 50 mm binder course, 25 mm topping	B-25C	836	.057	m²	15.30	1.77	2.22	19.29	22	
	0300	Binder course, 40 mm thick		3252	.015		3.66	.46	.57	4.69	5.40	
	0400	50 mm thick		2323	.021		4.84	.64	.80	6.28	7.10	
	0500	75 mm thick		1394	.034		7.45	1.06	1.33	9.84	11.30	
	0600	100 mm thick		1003	.048		9.70	1.48	1.85	13.03	14.95	
	0800	Sand finish course, 20 mm thick		3809	.013		1.94	.39	.49	2.82	3.29	
	0900	25 mm thick	↓	3159	.015		2.37	.47	.59	3.43	3.95	
	1000	Fill pot holes, hot mix, 50 mm thick	B-16	390	.082		5.05	2.31	1.22	8.58	10.40	
	1100	100 mm thick		325	.098		7.30	2.77	1.47	11.54	13.95	
	1120	150 mm thick	↓	288	.111		9.90	3.13	1.65	14.68	17.50	
	1140	Cold patch, 50 mm thick	B-51	279	.172		5.05	4.78	.46	10.29	13.50	
	1160	100 mm thick		251	.191		9.70	5.30	.51	15.51	19.45	
	1180	150 mm thick	↓	177	.271	↓	15.05	7.55	.73	23.33	29	
400	0010	**COLD MIXED BITUMINOUS PAVEMENT** 1.9 L asphalt/m², 25 mm depth										**400**
	0020	Well graded granular aggregate										
	0100	Blade mixed in windrows, spread & compacted 100 mm course	B-90A	1338	.042	m²	6	1.38	1.02	8.40	9.85	
	0200	Traveling plant mixed in windrows, compacted 100 mm course	B-90B	2508	.019		6	.62	.86	7.48	8.50	
	0300	Rotary plant mixed in place, compacted 100 mm course	"	2926	.016		6	.53	.74	7.27	8.20	
	0400	Central stationary plant, mixed, compacted 100 mm course	B-36	6020	.007	↓	12.05	.21	.19	12.45	13.80	

02750 | Rigid Pavement

			CREW	DAILY OUTPUT	LABOR-HOURS	UNIT	MAT.	LABOR	EQUIP.	TOTAL	TOTAL INCL O&P	
300	0010	**PLAIN CEMENT CONCRETE PAVEMENT**										**300**
	0015	Including joints, finishing and curing										
	0020	Fixed form, 3.7 m pass, unreinforced, 150 mm thick	B-26	2508	.035	m²	25.50	1.09	1.07	27.66	31	
	0030	180 mm thick		2383	.037		32	1.15	1.13	34.28	38.50	
	0100	200 mm thick		2299	.038		35.50	1.19	1.17	37.86	42	
	0110	200 mm thick, small area		1150	.077		35.50	2.37	2.34	40.21	45.50	
	0200	230 mm thick		2090	.042		41	1.31	1.29	43.60	48.50	
	0300	250 mm thick		1756	.050		44.50	1.55	1.53	47.58	53	
	0310	250 mm thick, small area		878	.100		44.50	3.11	3.07	50.68	57	
	0400	300 mm thick		1505	.058		48	1.81	1.79	51.60	57.50	
	0500	375 mm thick	↓	1254	.070	↓	53.50	2.18	2.15	57.83	64	
	0510	For small irregular areas, add						100%				
	0600	For continuous welded steel reinforcement over 3050 mm wide, add				m²				4.60	5.35	
	0610	For under 3000 mm pass, add								8.60	10.15	

02700 | Bases, Ballasts, Pavements & Appurtenances

02750 | Rigid Pavement

			CREW	DAILY OUTPUT	LABOR-HOURS	UNIT	2006 BARE COSTS				TOTAL INCL O&P	
							MAT.	LABOR	EQUIP.	TOTAL		
300	0650	For 7.3 m pass, deduct					20%	20%	20%	20%		300
	0655	For slipforming, deduct						10%	10%	10%		
	0700	Finishing, broom finish small areas	2 Cefi	100	.160	m²		5.50		5.50	8.10	
	0710	Transverse joint support dowels	C-1	350	.091	Ea.	3.24	3.06		6.30	8.35	
	0720	Transverse contraction joints, saw cut & grind	A-1B	36.58	.219	m		6	2.93	8.93	12.55	
	0730	Transverse expansion joints, incl. premolded bit. jt. filler	C-1	45.72	.700		5.15	23.50		28.65	42	
	0740	Transverse construction joint using bulkhead	"	22.25	1.438	▼	7.50	48		55.50	83	
	0750	Longitudinal joint tie bars, grouted	B-23	70	.571	Ea.	3.53	15.90	51	70.43	84.50	
	1000	Curing, with sprayed membrane by hand	2 Clab	1254	.013	m²	.60	.35		.95	1.20	
	1650	For integral coloring, see div. 03310-220										
	3200	Conc grooving, continuous for roadways	B-71	585	.096	m²		3.03	9.20	12.23	14.80	

02760 | Paving Specialties

			CREW	DAILY OUTPUT	LABOR-HOURS	UNIT	MAT.	LABOR	EQUIP.	TOTAL	TOTAL INCL O&P	
300	0010	**PAINTED TRAFFIC LINES AND MARKINGS**										300
	0020	Acrylic waterborne, white or yellow, 100 mm wide	B-78	6096	.008	m	.56	.22	.06	.84	1.03	
	0200	150 mm wide		3353	.014		.52	.40	.12	1.04	1.34	
	0500	200 mm wide		3048	.016		.72	.44	.13	1.29	1.61	
	0600	300 mm wide		1219	.039	▼	1.35	1.09	.32	2.76	3.53	
	0620	Arrows or gore lines		214	.224	m²	6.90	6.25	1.82	14.97	19.20	
	0640	Temporary paint, white or yellow	▼	4572	.011	m	.62	.29	.09	1	1.23	
	0660	Removal	1 Clab	91.44	.087			2.40		2.40	3.73	
	0680	Temporary tape	2 Clab	457	.035		6.05	.96		7.01	8.20	
	0710	Thermoplastic, white or yellow, 100 mm wide	B-79	4572	.009		2.43	.24	.19	2.86	3.25	
	0730	150 mm wide		4267	.009		3.51	.26	.21	3.98	4.50	
	0740	200 mm wide		3658	.011		4.72	.30	.24	5.26	5.95	
	0750	300 mm wide		1829	.022	▼	7.05	.61	.49	8.15	9.20	
	0760	Arrows		61.31	.652	m²	21.50	18.15	14.50	54.15	67.50	
	0770	Gore lines		232	.172		14.10	4.80	3.83	22.73	27	
	0780	Letters	▼	61.31	.652	▼	17.65	18.15	14.50	50.30	63.50	
	0782	Thermoplastic material				Met. Ton	1,150			1,150	1,275	
	0784	Glass beads, add				km	30.50			30.50	33.50	
500	0010	**PAVEMENT MARKINGS**										500
	0790	Layout of pavement marking	A-2	7620	.003	m		.09	.02	.11	.15	
	0800	Parking stall, paint, white	B-78	440	.109	Stall	3.43	3.03	.88	7.34	9.45	
	1000	Street letters and numbers	"	149	.322	m²	7.10	8.95	2.61	18.66	24.50	
	1100	Pavement marking letter, 150 mm	2 Pord	400	.040	Ea.	1.32	1.27		2.59	3.36	
	1110	300 mm letter		272	.059		1.93	1.86		3.79	4.93	
	1120	600 mm letter		160	.100		3.22	3.17		6.39	8.30	
	1130	900 mm letter		84	.190		6.50	6.05		12.55	16.25	
	1140	1050 mm letter		84	.190		15.75	6.05		21.80	26.50	
	1150	1800 mm letter		40	.400		44.50	12.70		57.20	67.50	
	1200	Handicap symbol	▼	40	.400	▼	27.50	12.70		40.20	49	

02770 | Curbs and Gutters

			CREW	DAILY OUTPUT	LABOR-HOURS	UNIT	MAT.	LABOR	EQUIP.	TOTAL	TOTAL INCL O&P	
100	0010	**BITUMINOUS CONCRETE CURBS**										100
	0012	Curbs, asphaltic, mach formed, 200 mm W, 150 mm H, 11 m/met ton	B-27	305	.105	m	2.69	2.93	.76	6.38	8.35	
	0100	200 mm wide, 200 mm high, 8 m/metric ton		274	.117		3.08	3.26	.85	7.19	9.40	
	0150	Asphaltic berm, 300 mm W, 1900 mm H, before pavement	▼	213	.150		3.51	4.19	1.09	8.79	11.60	
	0200	300 mm W, 38 mm to 100 mm H, laid with pavement	B-2	320	.125	▼	2.13	3.48		5.61	7.75	
300	0010	**CEMENT CONCRETE CURBS**										300
	0300	Concrete, wood forms, 150 mm x 450 mm, straight	C-2A	152	.316	m	9.50	10.85		20.35	27	
	0400	150 mm x 450 mm, radius		60.96	.787		9.85	27		36.85	52.50	
	0410	Steel forms, 150 mm x 450 mm, straight	▼	213	.225	▼	12.45	7.75		20.20	25.50	

Important: See the Reference Section for supporting data - Crews, Rental Equipment, City Cost Indexes and Reference Data

2 SITE CONSTRUCTION

SITE CONSTRUCTION | **2**

02770 | Curbs and Gutters

			CREW	DAILY OUTPUT	LABOR-HOURS	UNIT	2006 BARE COSTS				TOTAL INCL O&P	
							MAT.	LABOR	EQUIP.	TOTAL		
300	0411	150 mm x 40 mm, radius	C-2A	122	.393	m	14.25	13.50		27.75	36.50	300
	0415	Machine formed, 150 mm x 450 mm, straight	B-69A	610	.079		12.45	2.40	.97	15.82	18.45	
	0416	150 mm x 450 mm, radius	"	274	.175		13	5.35	2.17	20.52	25	
	0421	Curb and gutter, straight										
	0422	with 150 mm high curb and 150 mm thick gutter, wood forms										
	0430	600 mm wide, 0.14 m³/m	C-2A	114	.421	m	47	14.45		61.45	74	
	0435	760 mm wide, 0.17 m³/m		104	.462		51	15.85		66.85	80.50	
	0440	Steel forms, 600 mm wide, straight		213	.225		20.50	7.75		28.25	34.50	
	0441	Radius		91.44	.525		20.50	18		38.50	50.50	
	0442	760 mm wide, straight		213	.225		24.50	7.75		32.25	39	
	0443	Radius		91.44	.525		24.50	18		42.50	55	
	0445	Machine formed, 600 mm wide, straight	B-69A	610	.079		20.50	2.40	.97	23.87	27	
	0446	Radius		274	.175		20.50	5.35	2.17	28.02	33	
	0447	760 mm wide, straight		610	.079		24.50	2.40	.97	27.87	31.50	
	0448	Radius		274	.175		24.50	5.35	2.17	32.02	37.50	
	0451	Median mall, 600 mm x 230 mm high, straight		671	.072		21.50	2.18	.88	24.56	28	
	0452	Radius	B-69B	274	.175		21.50	5.35	2.69	29.54	34.50	
	0453	1200 mm x 230 mm high, straight		610	.079		42.50	2.40	1.21	46.11	52	
	0454	Radius		244	.197		42.50	6	3.03	51.53	59.50	
	0550	Precast, 150 mm x 450 mm, straight	B-29	213	.263		26	7.85	3.95	37.80	45	
	0600	150 mm x 450 mm, radius	"	99.06	.565		30	16.95	8.50	55.45	68	
500	0010	**STONE CURBS**										500
	1000	Granite, split face, straight, 125 mm x 400 mm	D-13	152	.316	m	33.50	10.75	3.53	47.78	57.50	
	1100	150 mm x 450 mm	"	137	.350		44	11.95	3.92	59.87	71	
	1300	Radius curbing, 150 mm x 450 mm, over 3 m radius	B-29	79.25	.707		54	21	10.60	85.60	103	
	1400	Corners, 600 mm radius		80	.700	Ea.	55	21	10.50	86.50	105	
	1600	Edging, 115 mm x 300 mm, straight		91.44	.612	m	16.75	18.35	9.20	44.30	56.50	
	1800	Curb inlets, (guttermouth) straight		41	1.366	Ea.	122	41	20.50	183.50	221	
	2000	Indian granite (belgian block)										
	2100	Jumbo, 265 mm x 190 mm x 100 mm, grey	D-1	45.72	.350	m	6.10	11.25		17.35	24	
	2150	Pink		45.72	.350		7.90	11.25		19.15	26	
	2200	Regular, 225 mm x 115 mm x 115 mm, grey		48.77	.328		5.50	10.55		16.05	22	
	2250	Pink		48.77	.328		7.60	10.55		18.15	24.50	
	2300	Cubes, 100 mm x 100 mm x 100 mm, grey		53.34	.300		5.25	9.65		14.90	20.50	
	2350	Pink		53.34	.300		5.50	9.65		15.15	21	
	2400	150 mm x 150 mm x 150 mm, pink		47.24	.339		13.70	10.90		24.60	31.50	
	2500	Alternate pricing method for indian granite										
	2550	Jumbo, 265 mm x 190 mm x 100 mm, (13.6 kg), grey				Met. Ton	117			117	128	
	2600	Pink					155			155	170	
	2650	Regular, 225 mm x 115 mm x 115 mm (9 kg), grey					131			131	144	
	2700	Pink					179			179	196	
	2750	Cubes, 100 mm x 100 mm x 100 mm, (6.8 kg), grey					214			214	236	
	2800	Pink					238			238	262	
	2850	150 mm x 150 mm x 150 mm, (11.3kg), pink					179			179	196	
	2900	For pallets, add					18.15			18.15	19.95	

02775 | Sidewalks

			CREW	DAILY OUTPUT	LABOR-HOURS	UNIT	MAT.	LABOR	EQUIP.	TOTAL	TOTAL INCL O&P	
275	0010	**SIDEWALKS, DRIVEWAYS, & PATIOS** No base										275
	0020	Asphaltic concrete, 50 mm thick	B-37	602	.080	m²	4.80	2.31	.19	7.30	9.05	
	0100	65 mm thick	"	552	.087		6.10	2.52	.21	8.83	10.85	
	0110	Bedding for brick or stone, mortar, 25 mm thick	D-1	27.87	.574		7	18.45		25.45	36	
	0120	50 mm thick	"	18.58	.861		17.55	27.50		45.05	61.50	
	0130	Sand, 50 mm thick	B-18	743	.032		1.72	.91	.05	2.68	3.40	
	0140	100 mm thick	"	372	.065		3.44	1.81	.09	5.34	6.80	
	0300	Concrete, 20 MPa, C.I.P., 150 x 150 - #10/10 mesh,										

2 SITE CONSTRUCTION

02775 | Sidewalks

		CREW	DAILY OUTPUT	LABOR-HOURS	UNIT	MAT.	LABOR	EQUIP.	TOTAL	TOTAL INCL O&P	
275											275
0310	broomed finish, no base, 100 mm thick	B-24	55.74	.431	m²	15.60	13.95		29.55	38.50	
0350	125 mm thick		50.63	.474		21	15.40		36.40	46.50	
0400	150 mm thick	↓	47.38	.507		24.50	16.45		40.95	51.50	
0450	For bank run gravel base, 100 mm thick, add	B-18	232	.103		4.52	2.90	.15	7.57	9.75	
0520	200 mm thick, add	"	149	.161		9.15	4.52	.23	13.90	17.40	
0550	Exposed aggregate finish, add to above, minimum	B-24	174	.138		1.18	4.48		5.66	8.15	
0600	Maximum		42.27	.568		3.77	18.40		22.17	32	
0700	Patterned surface, add to above min.	↓	111	.216	↓		7		7	10.70	
0800	For integral colors, see Div. 03310-220										
0950	Concrete tree grate, 1500 mm square	B-6	25	.960	Ea.	345	29	9.05	383.05	435	
0960	Cast iron tree grate with frame, 2 piece, round, 1500 mm diameter		25	.960		950	29	9.05	988.05	1,100	
0980	Square, 1500 mm side	↓	25	.960	↓	980	29	9.05	1,018.05	1,125	
1000	Crushed stone, 25 mm thick, white marble	2 Clab	158	.101	m²	2.15	2.77		4.92	6.70	
1050	Bluestone		158	.101		2.26	2.77		5.03	6.90	
1070	Granite chips	↓	158	.101		2.26	2.77		5.03	6.80	
1200	For 50 mm asphaltic conc base and tack coat, add to above	B-37	669	.072		6.65	2.08	.17	8.90	10.70	
1660	Limestone pavers, 75 mm thick	D-1	6.69	2.392		86.50	77		163.50	212	
1670	100 mm thick		6.50	2.460		115	79		194	246	
1680	125 mm thick	↓	6.32	2.533		143	81.50		224.50	282	
1700	Redwood, prefabricated, 1200 mm x 1200 mm sections	2 Carp	29.36	.545		87	19.40		106.40	126	
1750	Redwood planks, 25 mm thick, on sleepers	"	22.30	.718		61	25.50		86.50	107	
1830	38 mm thick	B-28	15.51	1.547		42.50	51		93.50	126	
1840	50 mm thick		15.51	1.547		54.50	51		105.50	139	
1850	75 mm thick		13.94	1.722		80.50	56.50		137	177	
1860	100 mm thick		13.94	1.722		105	56.50		161.50	204	
1870	125 mm thick	↓	13.94	1.722	↓	134	56.50		190.50	235	
2100	River or beach stone, stock	B-1	16.33	1.470	Met. Ton	29.50	41.50		71	96.50	
2150	Quarried	"	16.33	1.470	"	51.50	41.50		93	121	
2160	Load, dump, and spread stone with skid steer, 30 m haul	B-62	18.35	1.308	m³		39	8.40	47.40	70	
2165	60 m haul		13.76	1.744			52.50	11.20	63.70	93	
2168	91 m haul	↓	9.18	2.616	↓		78.50	16.80	95.30	139	
2170	Shale paver, 55 mm thick	D-1	18.58	.861	m²	29.50	27.50		57	74.50	
2200	Course washed sand bed, 25 mm thick	B-62	1129	.021		1.45	.64	.14	2.23	2.72	
2250	Stone dust, 100 mm thick	"	752	.032		3.52	.96	.20	4.68	5.55	
2300	Tile thinset pavers, 10 mm thick	D-1	27.87	.574		32.50	18.45		50.95	64	
2350	20 mm thick	"	26.01	.615	↓	52	19.80		71.80	87.50	
2400	Wood rounds, cypress	B-1	175	.137	Ea.	9.10	3.85		12.95	16	
8000	For temporary barricades, see div. 01560-100										

02780 | Unit Pavers

		CREW	DAILY OUTPUT	LABOR-HOURS	UNIT	MAT.	LABOR	EQUIP.	TOTAL	TOTAL INCL O&P	
100	0010 **ASPHALT BLOCKS**										100
0020	Rectangular, 150 mm x 300 mm x 32 mm, w/bed & neoprene adhesive	D-1	12.54	1.276	m²	50	41		91	118	
0100	75 mm thick		12.08	1.325		70	42.50		112.50	142	
0300	Hexagonal tile, 200 mm wide, 32 mm thick		12.54	1.276		50	41		91	118	
0400	50 mm thick		12.08	1.325		70	42.50		112.50	142	
0500	Square, 200 mm x 200 mm, 32 mm thick		12.54	1.276		50	41		91	118	
0600	50 mm thick	↓	12.08	1.325		70	42.50		112.50	142	
0900	For exposed aggregate (ground finish) add					3.44			3.44	3.77	
0910	For colors, add				↓	2.58			2.58	2.80	
200	0010 **BRICK PAVING**										200
0012	100 x 200 x 38 mm, w/o joints (48.5 brick/m²)	D-1	10.22	1.566	m²	27.50	50.50		78	107	
0100	Grouted, 10 mm joint (42 brick/m²)		8.36	1.914		32.50	61.50		94	129	
0200	100 mm x 200 mm x 57 mm, without joints (48.5 bricks/m²)		10.22	1.566		36	50.50		86.50	116	
0300	Grouted, 10 mm joint (42 brick/m²)	↓	8.36	1.914		33	61.50		94.50	130	
0500	Bedding, asphalt, 19 mm thick	B-25	477	.184	↓	4.09	5.55	4.44	14.08	17.95	

Important: See the Reference Section for supporting data - Crews, Rental Equipment, City Cost Indexes and Reference Data

SITE CONSTRUCTION 2

		02780	Unit Pavers	CREW	DAILY OUTPUT	LABOR-HOURS	UNIT	2006 BARE COSTS MAT.	LABOR	EQUIP.	TOTAL	TOTAL INCL O&P	
200	0540		Course washed sand bed, 25 mm thick	B-18	465	.052	m²	1.94	1.45	.07	3.46	4.49	200
	0580		Mortar, 25 mm thick	D-1	27.87	.574		5.80	18.45		24.25	34.50	
	0620		50 mm thick		18.58	.861		11.75	27.50		39.25	55	
	1500		Brick on 25 mm thick sand bed laid flat, 48.5/m²		9.29	1.722		27	55.50		82.50	115	
	2000		Brick pavers, laid on edge, 77.5/m²		6.50	2.460		26.50	79		105.50	150	
	2500		For 100 mm thick concrete bed and joints, add		55.28	.289		11.10	9.30		20.40	26.50	
	2800		For steam cleaning, add	A-1H	88.26	.091		.54	2.48	.65	3.67	5.25	
400	0010	**PRECAST CONCRETE PAVING SLABS**											400
	0710		Precast conc. patio blk., 60 mm th., colors, 200 mm x 400 mm	D-1	24.62	.650	m²	13.35	21		34.35	46.50	
	0715		300 mm x 300 mm		27.87	.574		17.35	18.45		35.80	47	
	0720		400 mm x 400 mm		31.12	.514		19.05	16.55		35.60	46	
	0730		600 mm x 600 mm		47.38	.338		24	10.85		34.85	42.50	
	0740		Green, 200 mm x 400 mm		24.62	.650		17.20	21		38.20	51	
	0750		Exposed local aggregate, natural	2 Bric	23.23	.689		57.50	25		82.50	102	
	0800		Colors		23.23	.689		62.50	25		87.50	107	
	0850		Exposed granite or limestone aggregate		23.23	.689		69	25		94	115	
	0900		Exposed white tumblestone aggregate		23.23	.689		41	25		66	83.50	
410	0010	**PLANTER BLOCKS** Precast concrete, interlocking											410
	0020		"V" blocks for retaining soil	D-1	19.04	.840	m²	49.50	27		76.50	95.50	
600	0010	**STONE PAVERS**											600
	1100		Flagging, bluestone, irregular, 25 mm thick,	D-1	7.52	2.126	m²	51	68.50		119.50	160	
	1110		38 mm thick		8.36	1.914		60	61.50		121.50	160	
	1120		Pavers, 13 mm thick		10.22	1.566		84.50	50.50		135	170	
	1130		19 mm thick		8.83	1.813		108	58.50		166.50	208	
	1140		25 mm thick		7.52	2.126		115	68.50		183.50	231	
	1150		Snapped random rectangular, 25 mm thick		8.55	1.872		77	60		137	176	
	1200		38 mm thick		7.90	2.026		92.50	65		157.50	201	
	1250		50 mm thick		7.71	2.075		108	66.50		174.50	221	
	1300		Slate, natural cleft, irregular, 19 mm thick		8.55	1.872		56.50	60		116.50	154	
	1310		25 mm thick		7.90	2.026		66	65		131	172	
	1350		Random rectangular, gauged, 13 mm thick		9.75	1.640		123	52.50		175.50	216	
	1400		Random rectangular, butt joint, gauged, 6 mm thick		13.94	1.148		132	37		169	201	
	1450		For sand rubbed finish, add					61.50			61.50	67.50	
	1500		For interior setting, add								25%	25%	
	1550		Granite blocks, 90 mm x 90 mm x 90 mm	D-1	8.55	1.872	m²	68.50	60		128.50	167	
	1560		100 mm x 100 mm x 100 mm		8.83	1.813		72.50	58.50		131	168	
	1600		100 to 300 mm L, 75 to 125 mm W, 75 to 125 mm T		9.10	1.757		57.50	56.50		114	149	
	1650		150 to 380 mm L, 75 to 150 mm W, 75 to 125 mm T		9.75	1.640		30.50	52.50		83	114	
700	0010	**STEPS** Incl. excav., borrow & concrete base, where applicable											700
	0100		Bricks	B-24	10.67	2.250	m Riser	34	73		107	150	
	0200		Railroad ties	2 Clab	7.62	2.100		9.95	57.50		67.45	100	
	0300		Bluestone treads, 300 mm x 50 mm or 300 mm x 40 mm	B-24	9.14	2.625		75.50	85		160.50	213	
	0500		Concrete, cast in place, see Division 03310-240										
	0600		Precast concrete, see Division 03480-800										
	4025		Steel edge strips, incl. stakes, 8 mm x 125 mm	B-1	119	.202	m	11.90	5.65		17.55	22	
	4050		Edging, landscape timber or railroad ties, 152 mm x 203 mm	2 Carp	51.82	.309	"	8.85	11		19.85	27	

02785 | Flexible Pavement Coating

				CREW	DAILY OUTPUT	LABOR-HOURS	UNIT	MAT.	LABOR	EQUIP.	TOTAL	TOTAL INCL O&P	
250	0010	**FOG SEAL**											250
	0012		Sealcoating, 2 coat coal tar pitch emulsion over 8400 m²	B-45	4181	.004	m²	.63	.12	.13	.88	1.02	
	0030		840 to 8400 m²	"	2508	.006		.63	.21	.21	1.05	1.23	
	0100		Under 840 m²	B-1	878	.027		.63	.77		1.40	1.88	
	0300		Petroleum resistant, over 8400 m²	B-45	4181	.004		.79	.12	.13	1.04	1.20	
	0320		840 to 8400 m²	"	2508	.006		.79	.21	.21	1.21	1.41	

For expanded coverage of these items see Means Site Work & Landscape Cost Data 2006

SITE CONSTRUCTION · **2**

02785	Flexible Pavement Coating	CREW	DAILY OUTPUT	LABOR-HOURS	UNIT	2006 BARE COSTS MAT.	LABOR	EQUIP.	TOTAL	TOTAL INCL O&P
250 0400	Under 840 m²	B-1	878	.027	m²	.79	.77		1.56	2.06
0600	Non-skid pavement renewal, over 8400 m²	B-45	4181	.004		.92	.12	.13	1.17	1.35
0620	840 to 8400 m²	"	2508	.006		.92	.21	.21	1.34	1.56
0700	Under 840 m²	B-1	878	.027		.92	.77		1.69	2.21
0800	Prepare and clean surface for above	A-2	7144	.003			.09	.02	.11	.16
1000	Hand seal asphalt curbing	B-1	1347	.018	m	1.28	.50		1.78	2.19
1900	Asphalt surface treatment, single course, small area									
1901	1.4 L/m² asphalt material, 11 kg/m² aggregate	B-91	4181	.015	m²	1.09	.50	.35	1.94	2.35
1910	Roadway or large area		8361	.008		1	.25	.18	1.43	1.67
1950	Asphalt surface treatment, dbl. course for small area		2508	.026		2.01	.83	.59	3.43	4.12
1960	Roadway or large area		5017	.013		1.81	.41	.29	2.51	2.94
1980	Asphalt surface treatment, single course, for shoulders		6271	.010		1.16	.33	.24	1.73	2.05
300 0010	**LATEX MODIFIED EMULSION**									
3600	Waterproofing, membrane, tar and fabric, small area	B-63	195	.205	m²	7.45	5.95	.79	14.19	18.20
3640	Large area		1200	.033		6.85	.97	.13	7.95	9.15
3680	Preformed rubberized asphalt, small area		83.61	.478		10.05	13.85	1.84	25.74	34.50
3720	Large area		307	.130		9.15	3.77	.50	13.42	16.45
3780	Rubberized asphalt (latex) seal	B-45	4181	.004		1.45	.12	.13	1.70	1.92
400 0010	**SAND SEAL**									
2080	Sand sealing, sharp sand, asphalt emulsion, small area	B-91	8361	.008	m²	.79	.25	.18	1.22	1.44
2120	Roadway or large area	"	15050	.004	"	.68	.14	.10	.92	1.06
3000	Sealing random cracks, min 13 mm wide, to 38 mm, 300 m	B-77	853	.047	m	1.71	1.31	.45	3.47	4.38
3040	3000 m		1219	.033	"	1.31	.91	.31	2.53	3.20
3080	Alternate method, 300 m		757	.053	liter	2.07	1.47	.50	4.04	5.10
3120	3000 m		1230	.033	"	1.67	.91	.31	2.89	3.57
3200	Multi-cracks (flooding), 1 coat, small area	B-92	385	.083	m²	2.82	2.32	.85	5.99	7.65
3240	Large area		2383	.013		2.51	.37	.14	3.02	3.49
3280	2 coat, small area		192	.167		9.30	4.65	1.70	15.65	19.30
3320	Large area		1191	.027		8.45	.75	.27	9.47	10.70
3360	Alternate method, small area		435	.074	liter	2.14	2.05	.75	4.94	6.40
3400	Large area		2707	.012	"	2.01	.33	.12	2.46	2.85
500 0010	**SLURRY SEAL**									
0100	Slurry seal, type I, 3.6 kg agg./m², 1 coat, small or irregular area	B-90	2341	.027	m²	1.08	.82	.81	2.71	3.33
0150	Roadway or large area		8361	.008		1.08	.23	.23	1.54	1.78
0200	Type II, 5.4 kg aggregate/m², 2 coats, small or irregular area		1672	.038		2.15	1.14	1.14	4.43	5.35
0250	Roadway or large area		6689	.010		2.15	.29	.29	2.73	3.12
0300	Type III, 9.1 kg aggregate/m², 2 coats, small or irregular area		1505	.043		2.54	1.27	1.27	5.08	6.15
0350	Roadway or large area		5017	.013		2.54	.38	.38	3.30	3.80
0400	Slurry seal, thermoplastic coal-tar, type I, small or irregular area		2007	.032		2.24	.95	.95	4.14	4.97
0450	Roadway or large area		6689	.010		2.24	.29	.29	2.82	3.21
0500	Type II, small or irregular area		2007	.032		2.86	.95	.95	4.76	5.65
0550	Roadway or large area		6522	.010		2.86	.29	.29	3.44	3.92
0600	Average mobilization cost				Ea.				3,750	3,750
600 0010	**SURFACE TREATMENT**									
3000	Pavement overlay, polypropylene									
3040	170 gram per m², ideal conditions	B-63	8361	.005	m²	1.50	.14	.02	1.66	1.88
3080	Adverse conditions		836	.048		2.01	1.39	.18	3.58	4.54
3120	113 gram per m², ideal conditions		8361	.005		1.02	.14	.02	1.18	1.35
3160	Adverse conditions		836	.048		1.32	1.39	.18	2.89	3.79
3200	Tack coat, emulsion, 0.19 L per m², 840 m²	B-45	2090	.008		.29	.25	.26	.80	.97
3240	8400 m²		8361	.002		.23	.06	.06	.35	.41
3270	0.37 L per m², 840 m²		2090	.008		.54	.25	.26	1.05	1.26
3275	0.37 L per m², 8400 m²		8361	.002		.54	.06	.06	.66	.76
3280	0.56 L per m², 840 m²		2090	.008		.79	.25	.26	1.30	1.53
3320	8400 m²		8361	.002		.63	.06	.06	.75	.85

02790		Athletic/Recreational Surfaces	CREW	DAILY OUTPUT	LABOR-HOURS	UNIT	2006 BARE COSTS				TOTAL INCL O&P	
							MAT.	LABOR	EQUIP.	TOTAL		
400	0010	**SYNTHETIC GRASS SURFACING**										400
	0015	Not including asphalt base or drainage,										
	0020	but including cushion pad, over 4600 m²										
	0200	13 mm pile and 8 mm cushion pad, standard	C-17	297	.269	m²	78	9.95		87.95	101	
	0300	Deluxe		238	.336		92.50	12.40		104.90	121	
	0500	13 mm pile and 16 mm cushion pad, standard		264	.303		114	11.20		125.20	142	
	0600	Deluxe	↓	216	.370	↓	125	13.65		138.65	159	
	0800	For asphaltic concrete base, 64 mm thick,										
	0900	with 150 mm crushed stone sub-base, add	B-25	1115	.079	m²	13.55	2.38	1.90	17.83	20.50	
500	0010	**SYNTHETIC RUNNING TRACK SURFACING**										500
	0020	Running track, asphalt, base, 80 mm thick	B-37	251	.191	m²	16.55	5.55	.46	22.56	27.50	
	0100	Surface, latex rubber system, 10 mm thick, black	B-20	105	.229		7.45	7.10		14.55	19.25	
	0150	Colors		105	.229		13.25	7.10		20.35	25.50	
	0300	Urethane rubber system, 10 mm thick, black		100	.240		19.85	7.45		27.30	33.50	
	0400	Color coating	↓	96.15	.250	↓	23.50	7.75		31.25	38	
600	0010	**TENNIS COURT SURFACING**										600
	0020	Tennis court, asphalt, incl base, 65 mm thick, one court	B-37	376	.128	m²	19.35	3.71	.30	23.36	27.50	
	0200	Two courts		564	.085		10.35	2.47	.20	13.02	15.45	
	0300	Clay courts		301	.159		37.50	4.63	.38	42.51	48.50	
	0400	Pulverized natural greenstone with 100 mm base, fast dry		209	.230		34.50	6.65	.55	41.70	48.50	
	0800	Rubber-acrylic base resilient pavement	↓	502	.096		38	2.78	.23	41.01	46.50	
	1000	Colored sealer, acrylic emulsion, 3 coats	2 Clab	669	.024		4.99	.66		5.65	6.50	
	1100	3 coat, 2 colors	"	752	.021		6.90	.58		7.48	8.50	
	1200	For preparing old courts, add	1 Clab	690	.012	↓		.32		.32	.49	
	1400	Posts for nets, 90 mm diameter with eye bolts [3-1/2"]	B-1	3.40	7.059	Pr.	175	198		373	505	
	1500	With pulley & reel		3.40	7.059	"	249	198		447	585	
	1700	Net, 12.6 m long, nylon thread with binder		50	.480	Ea.	231	13.45		244.45	275	
	1800	All metal	↓	6.50	3.692	"	415	104		519	615	
	2000	Paint markings on asphalt, 2 coats	1 Pord	1.78	4.494	Court	69.50	142		211.50	292	
	2200	Complete court with fence, etc., asphaltic conc., minimum	B-37	.20	240		14,500	6,975	570	22,045	27,300	
	2300	Maximum		.16	300		18,000	8,700	715	27,415	34,100	
	2800	Clay courts, minimum		.20	240		17,000	6,975	570	24,545	30,100	
	2900	Maximum	↓	.16	300	↓	20,400	8,700	715	29,815	36,800	

02800 | Site Improvements and Amenities

02810		Irrigation System	CREW	DAILY OUTPUT	LABOR-HOURS	UNIT	2006 BARE COSTS				TOTAL INCL O&P	
							MAT.	LABOR	EQUIP.	TOTAL		
300	0010	**SPRINKLER IRRIGATION SYSTEM** For lawns										300
	0100	Golf course with fully automatic system	C-17	.05	1600	9 holes	86,000	59,000		145,000	186,500	
	0200	7.3 m diam. head at 4.6 m O.C incl. piping, auto oper., minimum	B-20	70	.343	Head	18.95	10.65		29.60	37.50	
	0300	Maximum		40	.600		43.50	18.65		62.15	77	
	0500	18 m diam. head at 12 m O.C. incl. piping, auto oper., minimum		28	.857		57.50	26.50		84	105	
	0600	Maximum	↓	23	1.043	↓	161	32.50		193.50	228	
	0800	Residential system, custom, 25 mm supply [1"]		186	.129	m²	3.01	4.01		7.02	9.60	
	0900	38 mm supply [1-1/2"]	↓	167	.144	"	3.55	4.47		8.02	10.85	
	1020	Pop up spray head w/risers, hi-pop, full circle pattern, 100 mm	2 Skwk	76	.211	Ea.	3.42	7.70		11.12	15.70	
	1030	1/2 circle pattern, 100 mm		76	.211		3.50	7.70		11.20	15.80	
	1040	Full circle pattern, 150 mm		76	.211		8.75	7.70		16.45	21.50	
	1050	1/2 circle pattern, 150 mm	↓	76	.211		8.75	7.70		16.45	21.50	

SITE CONSTRUCTION **2**

02810	Irrigation System	CREW	DAILY OUTPUT	LABOR-HOURS	UNIT	2006 BARE COSTS				TOTAL INCL O&P	
						MAT.	LABOR	EQUIP.	TOTAL		
1060	Full circle pattern, 300 mm	2 Skwk	76	.211	Ea.	12.50	7.70		20.20	25.50	300
1070	1/2 circle pattern, 300 mm		76	.211		12.55	7.70		20.25	26	
1080	Pop up bubbler head w/risers, 100 mm hi-pop bubbler head		76	.211		3.57	7.70		11.27	15.90	
1090	150 mm		76	.211		8.75	7.70		16.45	21.50	
1100	300 mm		76	.211		12.50	7.70		20.20	25.50	
1110	Impact full/part circle sprinklers, spaced 9-16 m @ 172-414 kPa		37	.432		11.70	15.80		27.50	37.50	
1120	Spaced 11-15 m @ 172-345 kPa		37	.432		25.50	15.80		41.30	52.50	
1130	Spaced 13-19 m @ 207-414 kPa		37	.432		46.50	15.80		62.30	75.50	
1140	Spaced 16-24 m @ 276-552 kPa	▼	37	.432	▼	107	15.80		122.80	142	
1145	Impact rotor pop-up full/part commercial circle sprinklers										
1150	Spaced 13-20 m @ 241-552 kPa	2 Skwk	25	.640	Ea.	23.50	23.50		47	62.50	
1160	Spaced 15-23 m @ 310-586 kPa	"	25	.640	"	9.75	23.50		33.25	47	
1165	Impact rotor pop-up part circle comm.16-23 m @ 379-690 kPa										
1170	Plastic case, metal cover	2 Skwk	25	.640	Ea.	163	23.50		186.50	217	
1180	Rubber cover		25	.640		123	23.50		146.50	172	
1190	Iron case, metal cover		22	.727		149	26.50		175.50	206	
1200	Rubber cover		22	.727		162	26.50		188.50	220	
1250	Plastic case, 2 nozzle, metal cover		25	.640		119	23.50		142.50	168	
1260	Rubber cover		25	.640		126	23.50		149.50	176	
1270	Iron case, 2 nozzle, metal cover		22	.727		173	26.50		199.50	233	
1280	Rubber cover	▼	22	.727	▼	177	26.50		203.50	237	
1282	Impact rotor pop-up full circle comm., 12-30 m, 207-690 kPa										
1284	Plastic case, metal cover	2 Skwk	25	.640	Ea.	122	23.50		145.50	171	
1286	Rubber cover		25	.640		135	23.50		158.50	186	
1288	Iron case, metal cover		22	.727		175	26.50		201.50	235	
1290	Rubber cover		22	.727		184	26.50		210.50	245	
1292	Plastic case, 2 nozzle, metal cover		22	.727		131	26.50		157.50	186	
1294	Rubber cover		22	.727		136	26.50		162.50	192	
1296	Iron case, 2 nozzle, metal cover		20	.800		169	29		198	232	
1298	Rubber cover		20	.800		181	29		210	245	
1305	Electric remote control valve, plastic, 20 mm		18	.889		18.40	32.50		50.90	71	
1310	25 mm		18	.889		32.50	32.50		65	86	
1320	40 mm		18	.889		62.50	32.50		95	119	
1330	50 mm	▼	18	.889	▼	92.50	32.50		125	153	
1335	Quick coupling valves, brass, locking cover										
1340	Inlet coupling valve, 20 mm	2 Skwk	18.75	.853	Ea.	45	31		76	98	
1350	25 mm		18.75	.853		55	31		86	109	
1360	Controller valve boxes, 150mm round boxes		18.75	.853		4.02	31		35.02	53	
1370	250 mm round boxes		14.25	1.123		8.80	41		49.80	73.50	
1380	300 mm square box	▼	9.75	1.641	▼	17.70	60		77.70	113	
1388	Electromech. controls, 14 day, 3-60 min., auto start to 23/day										
1390	4 station	2 Skwk	1.04	15.385	Ea.	191	560		751	1,075	
1400	7 station		.64	25		247	915		1,162	1,700	
1410	12 station		.40	40		305	1,450		1,755	2,600	
1420	Dual programs, 18 station		.24	66.667		1,525	2,425		3,950	5,450	
1430	23 station	▼	.16	100	▼	1,975	3,650		5,625	7,850	
1435	Backflow preventer, bronze, 0-1207 kPa, w/valves, test cocks										
1440	20 mm	2 Skwk	2	8	Ea.	125	292		417	590	
1450	25 mm		2	8		130	292		422	600	
1460	40 mm		2	8		258	292		550	740	
1470	50 mm	▼	2	8	▼	269	292		561	750	
1475	Pressure vacuum breaker, brass, 103-1034 kPa										
1480	20 mm	2 Skwk	2	8	Ea.	75	292		367	540	
1490	25 mm		2	8		98.50	292		390.50	565	
1500	40 mm		2	8		194	292		486	670	
1510	50 mm	▼	2	8	▼	180	292		472	655	

SITE CONSTRUCTION 2

SITE CONSTRUCTION 2

	02820	Fences & Gates	CREW	DAILY OUTPUT	LABOR-HOURS	UNIT	MAT.	LABOR	EQUIP.	TOTAL	TOTAL INCL O&P	
							2006 BARE COSTS					
130	0010	**FENCE, CHAIN LINK INDUSTRIAL**, Schedule 40										**130**
	0020	3 strands barb wire, 50mm post @ 3m O.C., set in conc, 1800 mm H										
	0200	9 ga. wire, galv. steel	B-80C	73.15	.328	m	42	9	1.83	52.83	62	
	0300	Aluminized steel		73.15	.328		53.50	9	1.83	64.33	75	
	0500	6 ga. wire, galv. steel		73.15	.328		65.50	9	1.83	76.33	88	
	0600	Aluminized steel		73.15	.328		74.50	9	1.83	85.33	98	
	0800	6 ga. wire, 1.8 m high but omit barbed wire, galv. steel		76.20	.315		63.50	8.65	1.76	73.91	85.50	
	0900	Aluminized steel		76.20	.315		89	8.65	1.76	99.41	113	
	0920	2.4 m H, 6 ga. wire, 65 mm line post, galv. steel		54.86	.437		101	12	2.45	115.45	132	
	0940	Aluminized steel		54.86	.437	↓	124	12	2.45	138.45	157	
	1100	Add for corner posts, 75 mm diam., galv. steel [3"]		40	.600	Ea.	88.50	16.50	3.35	108.35	127	
	1200	Aluminized steel		40	.600		106	16.50	3.35	125.85	146	
	1300	Add for braces, galv. steel		80	.300		24	8.25	1.68	33.93	41	
	1350	Aluminized steel		80	.300		32	8.25	1.68	41.93	50	
	1400	Gate for 1.8 m fence, 40 mm frame, 900 mm wide, galv.		10	2.400		141	66	13.40	220.40	273	
	1500	Aluminized steel	↓	10	2.400	↓	174	66	13.40	253.40	310	
	2000	1.5 m high fence, 9 ga., no barbed wire, 50 mm line post,										
	2010	3 m O.C., 40 mm top rail										
	2100	Galvanized steel	B-80C	91.44	.262	m	36.50	7.20	1.47	45.17	53	
	2200	Aluminized steel		91.44	.262	"	43	7.20	1.47	51.67	60	
	2400	Gate, 1200 mm wide, 1.5 m high, 50 mm frame, galv. steel		10	2.400	Ea.	161	66	13.40	240.40	294	
	2500	Aluminized steel		10	2.400	"	181	66	13.40	260.40	315	
	3100	Overhead slide gate, chain link, 1.8 m high, to 5.5 m wide	↓	11.58	2.072	m	465	57	11.60	533.60	610	
	3105	2.4 m high	B-80	9.14	3.500		465	105	58	628	740	
	3108	3 m high		7.32	4.374		470	131	73	674	800	
	3110	Cantilever type		14.63	2.187		201	65.50	36.50	303	360	
	3120	2.4 m high		7.32	4.374		291	131	73	495	600	
	3130	3 m high	↓	5.49	5.833	↓	345	174	97	616	755	
	5000	Double swing gates, incl. posts & hardware										
	5010	1.5 m high, 3.7 m opening	B-80C	3.40	7.059	Opng.	430	194	39.50	663.50	820	
	5020	6 m opening		2.80	8.571		590	235	48	873	1,075	
	5060	1.8 m high, 3.7 m opening		3.20	7.500		725	206	42	973	1,175	
	5070	6 m opening	↓	2.60	9.231		1,000	254	51.50	1,305.50	1,550	
	5080	2.4 m high, 3.7 m opening	B-80	2.13	15.002		1,125	450	250	1,825	2,225	
	5090	6 m opening		1.45	22.069		1,475	660	365	2,500	3,050	
	5100	3 m high, 3.7 m opening		1.31	24.427		1,300	730	405	2,435	3,000	
	5110	6 m opening		1.03	31.068		1,925	930	515	3,370	4,125	
	5120	3.7 m high, 3.7 m opening		1.05	30.476		1,875	910	505	3,290	4,025	
	5130	6 m opening	↓	.85	37.647	↓	2,425	1,125	625	4,175	5,075	
	5190	For aluminized steel add					20%					
	7001	Snow fence on steel posts 3 m O.C., 1200 mm high	B-1	152	.158	m	7.80	4.43		12.23	15.50	
	7055	Braces, galv. steel	B-80A	293	.082		5.85	2.24	.60	8.69	10.60	
	7056	Aluminized steel	"	293	.082	↓	7.05	2.24	.60	9.89	11.90	
140	0010	**FENCE, CHAIN LINK RESIDENTIAL**, sch. 20, 3 mm wire, 40 mm post										**140**
	0020	3 m O.C., 35 mm top rail, 50 mm corner, galv. 1 m high	B-80C	152	.158	m	12.65	4.34	.88	17.87	21.50	
	0050	1.2 m high		122	.197		14.35	5.40	1.10	20.85	25.50	
	0100	1.8 m high		60.96	.394	↓	17.45	10.80	2.20	30.45	38.50	
	0150	Add for gate 1 m wide, 35 mm frame, 1 m high		12	2	Ea.	57.50	55	11.20	123.70	161	
	0170	1.2 m high		10	2.400		66.50	66	13.40	145.90	190	
	0190	1.8 m high		10	2.400		91	66	13.40	170.40	217	
	0200	Add for gate 1.2 m wide, 35 mm frame, 1 m high		9	2.667		65	73.50	14.90	153.40	201	
	0220	1.2 m high		9	2.667		73.50	73.50	14.90	161.90	210	
	0240	1.8 m high		8	3		86	82.50	16.75	185.25	240	
	0350	Aluminized steel, 11 ga. wire, 1 m high		152	.158	m	16.75	4.34	.88	21.97	26	
	0380	1.2 m high	↓	122	.197	↓	21.50	5.40	1.10	28	33	

02820 | Fences & Gates

		CREW	DAILY OUTPUT	LABOR-HOURS	UNIT	2006 BARE COSTS				TOTAL INCL O&P
						MAT.	LABOR	EQUIP.	TOTAL	
140 0400	1.8 m high	B-80C	60.96	.394	m	30	10.80	2.20	43	52 **140**
0450	Add for gate 1 m wide, 35 mm frame, 1 m high		12	2	Ea.	70.50	55	11.20	136.70	175
0470	1.2 m high		10	2.400		116	66	13.40	195.40	244
0490	1.8 m high		10	2.400		146	66	13.40	225.40	277
0500	Add for gate 1.2 m wide, 35 mm frame, 1 m high		10	2.400		80	66	13.40	159.40	205
0520	1.2 m high		9	2.667		88.50	73.50	14.90	176.90	227
0540	1.8 m high		8	3		102	82.50	16.75	201.25	257
0620	Vinyl covered, 9 ga. wire, 1 m high		152	.158	m	13.70	4.34	.88	18.92	22.50
0640	1.2 m high		122	.197		16	5.40	1.10	22.50	27
0660	1.8 m high		60.96	.394		20	10.80	2.20	33	41.50
0720	Add for gate 1 m wide, 35 mm frame, 1 m high		12	2	Ea.	80.50	55	11.20	146.70	186
0740	1.2 m high		10	2.400		91	66	13.40	170.40	217
0760	1.8 m high		10	2.400		141	66	13.40	220.40	272
0780	Add for gate 1.2 m wide, 35 mm frame, 1 m high		10	2.400		92.50	66	13.40	171.90	219
0800	1.2 m high		9	2.667		101	73.50	14.90	189.40	240
0820	1.8 m high		8	3		114	82.50	16.75	213.25	270
150 0010	**FENCE, CHAIN LINK, GATES & POSTS** (1/3 post length in ground)									**150**
6580	Line posts, galvanized, 65 mm OD, set in conc., 1200 mm	B-80	80	.400	Ea.	22	11.95	6.65	40.60	49.50
6585	1500 mm		76	.421		26.50	12.60	7	46.10	56
6590	1800 mm		74	.432		27	12.95	7.20	47.15	58
6595	2100 mm		72	.444		29.50	13.30	7.40	50.20	61
6600	2400 mm		69	.464		32.50	13.85	7.70	54.05	65.50
6610	H-beam, 47 mm, 1200 mm		83	.386		25	11.55	6.40	42.95	52.50
6615	1500 mm		81	.395		28.50	11.80	6.55	46.85	57
6620	1800 mm		78	.410		33.50	12.25	6.85	52.60	63.50
6625	2100 mm		75	.427		36.50	12.75	7.10	56.35	68
6630	2400 mm		73	.438		40	13.10	7.30	60.40	72
6635	Vinyl coated, 65 mm OD, set in conc., 1200 mm		79	.405		43	12.10	6.75	61.85	73.50
6640	1500 mm		77	.416		49.50	12.45	6.90	68.85	81
6645	1800 mm		74	.432		62	12.95	7.20	82.15	96.50
6650	2100 mm		72	.444		66.50	13.30	7.40	87.20	102
6655	2400 mm		69	.464		72.50	13.85	7.70	94.05	110
6660	End gate post, stl, 80 mm OD, set in conc., 1200 mm		68	.471		50.50	14.05	7.85	72.40	85.50
6665	1500 mm		65	.492		64	14.70	8.20	86.90	102
6670	1800 mm		63	.508		65	15.20	8.45	88.65	104
6675	2100 mm		61	.525		87	15.70	8.75	111.45	129
6685	Vinyl, 1200 mm		68	.471		84.50	14.05	7.85	106.40	123
6690	1500 mm		65	.492		93	14.70	8.20	115.90	134
6695	1800 mm		63	.508		115	15.20	8.45	138.65	160
6705	2400 mm		59	.542		137	16.20	9.05	162.25	186
6710	Corner post, galv. stl, 100 mm OD, set in conc., 1200 mm		65	.492		85	14.70	8.20	107.90	125
6715	1800 mm		63	.508		120	15.20	8.45	143.65	165
6720	2100 mm		61	.525		135	15.70	8.75	159.45	182
6725	2400 mm		65	.492		155	14.70	8.20	177.90	203
6730	Vinyl, 1500 mm		65	.492		148	14.70	8.20	170.90	195
6735	1800 mm		63	.508		197	15.20	8.45	220.65	250
6740	2100 mm		61	.525		180	15.70	8.75	204.45	231
6745	2400 mm		59	.542		235	16.20	9.05	260.25	293
7031	For corner, end, & pull post bracing, add					20%	15%			
7795	Cantilever, manual, exp. roller, (pr), 12.2 m w x 2400 mm high	B-22	1	30	Ea.	3,575	975	229	4,779	5,675
7800	9.1 m wide x 2400 mm high		1	30		3,000	975	229	4,204	5,050
7805	7.3 m wide x 2400 mm high		1	30		2,225	975	229	3,429	4,200
7810	Motor operators for gates, (no elec wiring), 900 mm w swing	2 Skwk	.50	32		1,225	1,175		2,400	3,150
7815	Up to 6.1 m wide swing		.50	32		4,050	1,175		5,225	6,275
7820	Up to 13.7 m sliding		.50	32		4,125	1,175		5,300	6,350
7825	Overhead gate, 1800 mm to 5400 mm, slide/cantilever		13.72	1.167	m	470	42.50		512.50	585

Important: See the Reference Section for supporting data - Crews, Rental Equipment, City Cost Indexes and Reference Data

2 SITE CONSTRUCTION

02820 | Fences & Gates

			CREW	DAILY OUTPUT	LABOR-HOURS	UNIT	2006 BARE COSTS				TOTAL INCL O&P	
							MAT.	LABOR	EQUIP.	TOTAL		
150	7830	Gate operators, digital receiver	2 Skwk	7	2.286	Ea.	435	83.50		518.50	605	**150**
	7835	Two button transmitter		24	.667		55.50	24.50		80	99	
	7840	3 button station		14	1.143		68.50	41.50		110	141	
	7845	Master slave system	↓	4	4		244	146		390	495	
	7900	Auger fence post hole, 900 mm deep, medium soil, by hand	1 Clab	30	.267			7.30		7.30	11.35	
	7925	By machine	B-80	175	.183			5.45	3.04	8.49	11.75	
	7950	Rock, with jackhammer	B-9C	32	1.250			35	4.60	39.60	59	
	7975	With rock drill	B-47C	65	.246	↓		7.70	17.05	24.75	30.50	
170	0010	**FENCE, FABRIC & ACCESSORIES**										**170**
	1000	Fabric, 9 ga., galv., 34.02 gram coat, 50 mm chain link, 1200 mm	B-80A	92.66	.259	m	11.30	7.10	1.90	20.30	25.50	
	1150	1500 mm		86.87	.276		13.70	7.55	2.03	23.28	29	
	1200	1800 mm		81.08	.296		21.50	8.10	2.17	31.77	39	
	1250	2100 mm		75.29	.319		32	8.75	2.34	43.09	51	
	1300	2400 mm		69.49	.345		39.50	9.45	2.53	51.48	61	
	1400	3.3 mm, fused, 1200 mm		92.66	.259		18.55	7.10	1.90	27.55	33.50	
	1450	1500 mm		86.87	.276		23	7.55	2.03	32.58	39.50	
	1500	1800 mm		81.08	.296		28	8.10	2.17	38.27	45.50	
	1550	2100 mm		75.29	.319		32.50	8.75	2.34	43.59	52	
	1600	2400 mm		69.49	.345		46.50	9.45	2.53	58.48	68.50	
	1650	Barbed wire, galv., cost per strand		695	.035		.36	.95	.25	1.56	2.14	
	1700	Vinyl coated		695	.035	↓	.85	.95	.25	2.05	2.70	
	1750	Barbed wire extension arms 3 strands		143	.168	Ea.	7.85	4.60	1.23	13.68	17.15	
	1800	6 strands, 59 mm		119	.202		9.95	5.55	1.48	16.98	21	
	1850	Eye tops, 59 mm		143	.168	↓	3.26	4.60	1.23	9.09	12.10	
	1900	Top rail, incl. tie wires, 44 mm, galv.		278	.086	m	10.15	2.37	.63	13.15	15.60	
	1950	Vinyl coated		278	.086		16	2.37	.63	19	22	
	2100	Rail, middle/bottom, w/tie wire, 44 mm, galv.		278	.086		8.60	2.37	.63	11.60	13.85	
	2150	Vinyl coated		278	.086		11.60	2.37	.63	14.60	17.15	
	2200	Reinforcing wire, coiled spring, 4.6 mm, galv.		695	.035		.52	.95	.25	1.72	2.34	
	2250	3.3 mm, vinyl coated	↓	696	.034	↓	.79	.94	.25	1.98	2.60	
300	0010	**FENCE, VINYL**, white, steel reinforced, stainless steel fasteners										**300**
	0020	Picket, 100 mm x 100 mm posts @ 1800 mm OC, 900 mm high	B-1	42.67	.562	m	56.50	15.80		72.30	86.50	
	0030	1200 mm high		39.62	.606		65	17		82	98	
	0040	1500 mm high		36.58	.656		74.50	18.40		92.90	111	
	0100	Board (semi-privacy), 125 mm posts @ 2286 mm OC, 1500 mm high		39.62	.606		77	17		94	112	
	0120	1800 mm high		38.10	.630		88	17.70		105.70	124	
	0200	Basketweave, 125 mm posts @ 2286 mm OC, 1500 mm high		48.77	.492		68.50	13.80		82.30	97	
	0220	1800 mm high		45.72	.525		81	14.75		95.75	112	
	0300	Privacy, 125 mm posts @ 2286 mm OC, 1500 mm high		39.62	.606		83.50	17		100.50	118	
	0320	1800 mm high		45.72	.525	↓	95	14.75		109.75	127	
	0350	Gate, 1500 mm high		9	2.667	Ea.	360	75		435	510	
	0360	1800 mm high		9	2.667		365	75		440	520	
	0400	For posts set in concrete, add	↓	25	.960	↓	7.30	27		34.30	50	
410	0010	**FENCES, MISC. METAL**										**410**
	0012	Chicken wire, posts 1.2 m OC, 25 mm mesh, 1.2 m high	B-80C	125	.192	m	5.75	5.25	1.07	12.07	15.70	
	0100	50 mm mesh, 1.8 m high		107	.224		5.20	6.15	1.25	12.60	16.70	
	0200	Galv. steel, 12 ga., 50 mm x 100 mm mesh, 1 m high		91.44	.262		7.85	7.20	1.47	16.52	21.50	
	0300	1.5 m high		91.44	.262		10.45	7.20	1.47	19.12	24.50	
	0400	14 ga., 25 mm x 50 mm mesh, 1 m high		91.44	.262		8.35	7.20	1.47	17.02	22	
	0500	1.5 m high	↓	91.44	.262	↓	11.50	7.20	1.47	20.17	25.50	
	1000	Kennel fencing, 38 mm mesh, 1.8 m x 1 m, 1.8 m high	2 Clab	4	4	Ea.	400	110		510	610	
	1050	3.7 m long		4	4		480	110		590	695	
	1200	Top covers, 38 mm mesh, 1.8 m long		15	1.067		81	29		110	135	
	1250	3.7 m long	↓	12	1.333	↓	130	36.50		166.50	200	
	1300	For kennel doors, see Division 08310-400										

SITE CONSTRUCTION 2

02820	Fences & Gates	CREW	DAILY OUTPUT	LABOR-HOURS	UNIT	2006 BARE COSTS				TOTAL INCL O&P
						MAT.	LABOR	EQUIP.	TOTAL	
410										**410**
4500	Security fence, prison grade, set in concrete, 3.7 m high	B-80	7.62	4.199	m	115	126	70	311	395
4600	4.9 m high	"	6.10	5.249		138	157	87.50	382.50	490
5300	Tubular picket, steel, 1830 mm sections, 40 mm posts, 1220 mm H	B-80C	91.44	.262		81	7.20	1.47	89.67	102
5400	50 mm posts, 1530 mm high		73.15	.328		113	9	1.83	123.83	140
5600	50 mm posts, 1830 mm high		60.96	.394		127	10.80	2.20	140	159
5700	Staggered picket 40 mm posts, 1220 mm high		91.44	.262		73	7.20	1.47	81.67	93.50
5800	50 mm posts, 1530 mm high		73.15	.328		120	9	1.83	130.83	149
5900	50 mm posts, 1830 mm high		60.96	.394		125	10.80	2.20	138	156
6200	Gates, 1220 mm high, 900 mm wide	B-1	10	2.400	Ea.	215	67.50		282.50	340
6300	1530 mm high, 900 mm wide		10	2.400		279	67.50		346.50	410
6400	1830 mm high, 900 mm wide		10	2.400		287	67.50		354.50	420
6500	1220 mm wide		10	2.400		335	67.50		402.50	475
420	0010	**WIRE FENCING, GENERAL**								**420**
0015	Barbed wire, galvanized, domestic steel, hi-tensile 15-1/2 ga.				km	95			95	105
0020	Standard, 12-3/4 ga.					128			128	140
0210	Barbless wire, 2-strand galvanized, 12-1/2 ga.					127			127	140
0500	Helical razor ribbon, stainless steel, 450 mm dia x 450 mm spacing				m	3.70			3.70	4.07
0600	Hardware cloth, galv., 6 mm mesh, 23 ga., 600 mm wide				m²	5.45			5.45	6
0700	900 mm wide					5.30			5.30	5.85
0900	13 mm mesh, 19 ga., 600 mm wide					4.81			4.81	5.30
1000	1200 mm wide					4.72			4.72	5.20
1200	Chain link fabric, steel, 50 mm mesh, 6 ga, galvanized					18			18	19.80
1300	9 ga, galvanized					8.90			8.90	9.75
1350	Vinyl coated					7.45			7.45	8.20
1360	Aluminized					11.55			11.55	12.70
1400	56 mm mesh, 11.5 ga, galvanized					6			6	6.60
1600	44 mm mesh (tennis courts), 11.5 ga (core), vinyl coated					8.50			8.50	9.35
1700	9 ga, galvanized					7.65			7.65	8.45
2100	Welded wire fabric, galvanized, 25 mm x 50 mm, 14 ga.					5.10			5.10	5.60
2200	50 mm x 100 mm, 12-1/2 ga.					1.74			1.74	1.91
510	0010	**FENCE, WOOD** Basket weave, 9 x 100 mm boards, 50 x 100 mm									**510**
0020	stringers on spreaders, 100 mm x 100 mm posts									
0050	No. 1 cedar, 1800 mm high	B-80C	48.77	.492	m	27.50	13.50	2.75	43.75	54.50
0070	Treated pine, 1800 mm high	"	45.72	.525	"	33.50	14.40	2.93	50.83	62.50
0200	Board, 25x100mm boards, 50x100mm rails, 100x100mm post									
0220	Preservative treated, 2 rail, 900 mm high	B-80C	44.20	.543	m	20.50	14.90	3.04	38.44	49
0240	1200 mm high		41.15	.583		22.50	16	3.26	41.76	53.50
0260	3 rail, 1500 mm high		39.62	.606		25.50	16.65	3.39	45.54	57
0300	1800 mm high		38.10	.630		29	17.30	3.52	49.82	63
0320	No. 2 grade western cedar, 2 rail, 900 mm high		44.20	.543		22.50	14.90	3.04	40.44	51
0340	1200 mm high		41.15	.583		26.50	16	3.26	45.76	58
0360	3 rail, 1500 mm high		39.62	.606		30.50	16.65	3.39	50.54	62.50
0400	1800 mm high		38.10	.630		33.50	17.30	3.52	54.32	68
0420	No. 1 grade cedar, 2 rail, 900 mm high		44.20	.543		33.50	14.90	3.04	51.44	63.50
0440	1200 mm high		41.15	.583		38.50	16	3.26	57.76	70.50
0460	3 rail, 1500 mm high		39.62	.606		44.50	16.65	3.39	64.54	78
0500	1800 mm high		38.10	.630		49.50	17.30	3.52	70.32	85.50
0860	Open rail fence, split rails, 2 rail, 900 mm high, no. 1 cedar		48.77	.492		18.65	13.50	2.75	34.90	44.50
0870	No. 2 cedar		48.77	.492		14.50	13.50	2.75	30.75	40
0880	3 rail, 1200 mm high, no. 1 cedar		45.72	.525		25	14.40	2.93	42.33	53
0890	No. 2 cedar		45.72	.525		16.55	14.40	2.93	33.88	44
0920	Rustic rails, 2 rail 900 mm high, no. 1 cedar		48.77	.492		11.60	13.50	2.75	27.85	37
0930	No. 2 cedar		48.77	.492		11.15	13.50	2.75	27.40	36.50
0940	3 rail, 1200 mm high		45.72	.525		15.60	14.40	2.93	32.93	43
0950	No. 2 cedar		45.72	.525		11.80	14.40	2.93	29.13	38.50
1240	Stockade fence, no. 1 cedar, 83 mm rails, 1.8 m high		48.77	.492		34	13.50	2.75	50.25	61

2 SITE CONSTRUCTION

02820 | Fences & Gates

			CREW	DAILY OUTPUT	LABOR-HOURS	UNIT	2006 BARE COSTS MAT.	LABOR	EQUIP.	TOTAL	TOTAL INCL O&P	
510	1260	2.4 m high	B-80C	47.24	.508	m	43.50	13.95	2.84	60.29	72.50	510
	1270	Gate, 1 m wide		9	2.667	Ea.	175	73.50	14.90	263.40	320	
	1300	No. 2 cedar, treated wood rails, 1.8 m high		48.77	.492	m	34	13.50	2.75	50.25	61	
	1320	Gate, 1 m wide		8	3	Ea.	60.50	82.50	16.75	159.75	212	
	1360	Treated pine, treated rails, 1.8 m high		48.77	.492	m	33	13.50	2.75	49.25	60.50	
	1400	2.4 m high		45.72	.525	"	49.50	14.40	2.93	66.83	80	
	1420	Gate, 1 m wide	↓	9	2.667	Ea.	66.50	73.50	14.90	154.90	203	
520	0010	**FENCE, WOOD RAIL**										520
	0012	Picket, No. 2 cedar, Gothic, 2 rail, 900 mm high	B-1	48.77	.492	m	17.75	13.80		31.55	41	
	0050	Gate, 1070 mm wide	B-80C	9	2.667	Ea.	46.50	73.50	14.90	134.90	180	
	0400	3 rail, 1220 mm high		45.72	.525	m	20.50	14.40	2.93	37.83	48	
	0500	Gate, 1070 mm wide		9	2.667	Ea.	56	73.50	14.90	144.40	191	
	0600	Open rail, rustic, No. 1 cedar, 2 rail, 900 mm high		48.77	.492	m	15.90	13.50	2.75	32.15	41.50	
	0650	Gate, 900 mm wide		9	2.667	Ea.	54	73.50	14.90	142.40	189	
	0700	3 rail, 1220 mm high		45.72	.525	m	18.20	14.40	2.93	35.53	45.50	
	0900	Gate, 900 mm wide		9	2.667	Ea.	69.50	73.50	14.90	157.90	206	
	1200	Stockade, No. 2 cedar, treated wood rails, 1830 mm high		48.77	.492	m	22	13.50	2.75	38.25	48.50	
	1250	Gate, 900 mm wide		9	2.667	Ea.	55.50	73.50	14.90	143.90	190	
	1300	No. 1 cedar, 85 mm cedar rails, 1830 mm high		48.77	.492	m	55.50	13.50	2.75	71.75	85	
	1500	Gate, 900 mm wide		9	2.667	Ea.	141	73.50	14.90	229.40	284	
	1520	Open rail, split, No. 1 cedar, 2 rail, 900 mm high		48.77	.492	m	16.25	13.50	2.75	32.50	42	
	1540	3 rail, 1220 mm high		45.72	.525		21	14.40	2.93	38.33	48.50	
	3300	Board, shadow box, 25 mm x 150 mm, treated pine, 1830 mm high		48.77	.492		32.50	13.50	2.75	48.75	60	
	3400	No. 1 cedar, 1830 mm high		45.72	.525		64.50	14.40	2.93	81.83	96.50	
	3900	Basket weave, No. 1 cedar, 1800 mm high	↓	48.77	.492	↓	64	13.50	2.75	80.25	94.50	
	3950	Gate, 1070 mm wide	B-1	8	3	Ea.	133	84		217	277	
	4000	Treated pine, 1800 mm high		45.72	.525	m	38	14.75		52.75	64.50	
	4200	Gate, 1070 mm wide		9	2.667	Ea.	62	75		137	186	
	5000	Fence rail, redwood, 51 mm x 102 mm, merch grade 2438 mm		732	.033	m	3.48	.92		4.40	5.25	
	5050	Select grade, 2438 mm		732	.033	"	10.75	.92		11.67	13.30	
	6000	Fence post, redwood, earthpacked & treated, 102mmx102mmx1.8m		96	.250	Ea.	10.50	7		17.50	22.50	
	6010	102 mm x 102 mm x 2438 mm		96	.250		12.45	7		19.45	24.50	
	6020	Set in concrete, 100 mm x 100 mm x 1.8 m		50	.480		13.50	13.45		26.95	36	
	6030	102 mm x 102 mm x 2438 mm		50	.480		16.15	13.45		29.60	39	
	6040	Wood post, 1219 mm high, set in concrete, incl. concrete		50	.480		7.90	13.45		21.35	29.50	
	6050	Earth packed		96	.250		5.40	7		12.40	16.85	
	6060	1829 mm high, set in concrete, incl. concrete		50	.480		10.10	13.45		23.55	32	
	6070	Earth packed	↓	96	.250	↓	7.35	7		14.35	19	

02830 | Retaining Walls

			CREW	DAILY OUTPUT	LABOR-HOURS	UNIT	MAT.	LABOR	EQUIP.	TOTAL	TOTAL INCL O&P	
100	0010	**CAST-IN-PLACE RETAINING WALLS**										100
	1800	Concrete gravity wall with vertical face including excavation & backfill										
	1850	No reinforcing										
	1900	1.8 m high, level embankment	C-17C	10.97	7.564	m	209	279	36	524	705	
	2000	33° slope embankment		9.75	8.510		190	315	40.50	545.50	745	
	2200	2.4 m high, no surcharge		8.23	10.086		259	375	48	682	920	
	2300	33° slope embankment		7.32	11.346		315	420	54	789	1,050	
	2500	3.0 m high, level embankment		5.79	14.332		370	530	68.50	968.50	1,300	
	2600	33° slope embankment	↓	5.49	15.128	↓	515	560	72.50	1,147.50	1,525	
	2800	Reinforced concrete cantilever, incl. excavation, backfill & reinf.										
	2900	1.8 m high, 33° slope embankment	C-17C	10.67	7.780	m	190	287	37	514	695	
	3000	2.4 m high, 33° slope embankment		8.84	9.390		220	345	45	610	830	
	3100	3.0 m high, 33° slope embankment		6.10	13.615		285	505	65	855	1,175	
	3200	6.1 m high, 750 kg/m surcharge	↓	2.29	36.308	↓	855	1,350	174	2,379	3,200	

SITE CONSTRUCTION 2

02830 | Retaining Walls

			CREW	DAILY OUTPUT	LABOR-HOURS	UNIT	2006 BARE COSTS MAT.	LABOR	EQUIP.	TOTAL	TOTAL INCL O&P	
100	3500	Concrete cribbing, incl. excavation and backfill										**100**
	3700	3.7 m high, open face	B-13	19.51	2.870	m²	310	86	33.50	429.50	515	
	3900	Closed face	"	19.51	2.870	"	293	86	33.50	412.50	490	
	4100	Concrete filled slurry trench, see Div. 02660-600										
200	0010	**INTERLOCKING SEGMENTAL RETAINING WALLS**										**200**
	7100	Segmental Retaining Wall system, incl pins, and void fill										
	7120	base not included										
	7140	Large unit, 200 x 460 x 508 mm, 3 plane split	B-62	27.87	.861	m²	99.50	26	5.55	131.05	156	
	7150	Straight split		27.87	.861		99.50	26	5.55	131.05	156	
	7160	Med., lightweight, 200 x 450 x 300 mm deep, 3 plane split		37.16	.646		87.50	19.40	4.15	111.05	131	
	7170	Straight split		37.16	.646		87.50	19.40	4.15	111.05	131	
	7180	Small unit, 100 x 450 x 250 mm deep, 3 plane split		37.16	.646		110	19.40	4.15	133.55	156	
	7190	Straight split		37.16	.646		110	19.40	4.15	133.55	156	
	7200	Cap unit, 3 plane split		27.87	.861		123	26	5.55	154.55	182	
	7210	Cap unit, st split		27.87	.861		123	26	5.55	154.55	182	
	7260	For reinforcing, add								23.65	29.75	
	8000	For higher walls, add components as necessary										
600	0010	**METAL BIN RETAINING WALLS**, Aluminized steel bin, excavation										**600**
	0020	and backfill not included, 3.0 m wide										
	0100	1.2 m high, 1.7 m deep	B-13	60.39	.927	m²	193	28	10.80	231.80	267	
	0200	2.4 m high, 1.6 m deep		57.13	.980		222	29.50	11.40	262.90	300	
	0300	3.0 m high, 2.3 m deep		53.88	1.039		234	31	12.10	277.10	320	
	0400	3.7 m high, 2.3 m deep		49.24	1.137		252	34	13.20	299.20	345	
	0500	4.8 m high, 2.3 m deep		47.84	1.170		266	35	13.60	314.60	360	
	0600	4.9 m high, 3.0 m deep		46.45	1.206		281	36	14	331	380	
	0700	6 m high, 3.0 m deep		43.66	1.283		315	38.50	14.90	368.40	420	
	0800	6.1 m high, 3.7 m deep		42.73	1.310		335	39	15.25	389.25	445	
	0900	7.2 m high, 3.6 m deep		42.27	1.325		360	39.50	15.40	414.90	475	
	1000	7.3 m high, 4.4 m deep		41.81	1.340		375	40	15.55	430.55	490	
	1100	8.4 m high, 4.3 m deep		40.88	1.370		390	41	15.90	446.90	505	
	1300	For plain galvanized bin type walls, deduct					10%					
700	0010	**STONE GABION RETAINING WALLS**										**700**
	4300	Stone filled gabions, not incl. excavation,										
	4310	Stone, delivered, 900 mm wide										
	4340	Galvanized, 1800 mm long, 300 mm high	B-13	113	.496	Ea.	63.50	14.85	5.75	84.10	99.50	
	4350	Galvanized, 1.8 m high, 33° slope embankment		14.94	3.750	m	62	112	43.50	217.50	290	
	4400	450 mm high		50	1.120	Ea.	80.50	33.50	13	127	154	
	4490	900 mm high		13	4.308	"	133	129	50	312	400	
	4500	Highway surcharge		8.23	6.805	m	132	204	79	415	545	
	4590	2700 mm long, 300 mm high		50	1.120	Ea.	90.50	33.50	13	137	165	
	4600	2.7 m high, up to 33° slope embankment		7.32	7.655	m	140	229	89	458	605	
	4650	450 mm high		22	2.545	Ea.	119	76	29.50	224.50	281	
	4690	900 mm high		6	9.333	"	203	279	108	590	770	
	4700	Highway surcharge		4.88	11.483	m	230	345	133	708	930	
	4890	3600 mm long, 300 mm high		28	2	Ea.	121	60	23	204	251	
	4900	3.7 m high, up to 33° slope embankment		4.27	13.123	m	217	395	152	764	1,000	
	4950	450 mm high		13	4.308	Ea.	155	129	50	334	425	
	4990	900 mm high		3	18.667	"	267	560	217	1,044	1,400	
	5000	Highway surcharge		3.35	16.702	m	330	500	194	1,024	1,350	
	5200	PVC coated, 1800 mm long, 300 mm high		113	.496	Ea.	72.50	14.85	5.75	93.10	109	
	5250	450 mm high		50	1.120		88.50	33.50	13	135	163	
	5300	900 mm high		13	4.308		148	129	50	327	415	
	5500	2700 mm long, 300 mm high		50	1.120		95.50	33.50	13	142	171	
	5550	450 mm high		22	2.545		129	76	29.50	234.50	292	
	5600	900 mm high		6	9.333		219	279	108	606	790	

02830 | Retaining Walls

		CREW	DAILY OUTPUT	LABOR-HOURS	UNIT	2006 BARE COSTS				TOTAL INCL O&P
						MAT.	LABOR	EQUIP.	TOTAL	
700	5800	3600 mm long, 300 mm high	B-13	28	2	Ea.	133	60	23	216	264
	5850	450 mm high		13	4.308		170	129	50	349	440
	5900	900 mm high	▼	3	18.667	▼	287	560	217	1,064	1,425
	5950	For PVC coating, add				m	20%				
	6000	Galvanized, 1800 mm long, 300 mm high	B-13	57.35	.977	m³	82.50	29	11.35	122.85	148
	6010	450 mm high		38.23	1.465		106	44	17	167	202
	6020	900 mm high		19.12	2.930		87.50	87.50	34	209	269
	6030	2700 mm long, 300 mm high		38.23	1.465		118	44	17	179	216
	6040	450 mm high		25.46	2.199		104	66	25.50	195.50	243
	6050	900 mm high		12.77	4.386		88.50	131	51	270.50	355
	6060	3600 mm long, 300 mm high		28.67	1.953		119	58.50	22.50	200	246
	6070	450 mm high		19.12	2.930		102	87.50	34	223.50	285
	6080	900 mm high		9.56	5.859		87.50	175	68	330.50	440
	6100	PVC coated, 1800 mm long, 300 mm high		57.35	.977		142	29	11.35	182.35	214
	6110	450 mm high		38.23	1.465		116	44	17	177	213
	6120	900 mm high		19.12	2.930		96.50	87.50	34	218	279
	6130	2700 mm long, 300 mm high		38.23	1.465		125	44	17	186	223
	6140	450 mm high		25.46	2.199		113	66	25.50	204.50	253
	6150	900 mm high		12.75	4.394		95.50	132	51	278.50	365
	6160	3600 mm long, 300 mm high		28.67	1.953		130	58.50	22.50	211	258
	6170	450 mm high		19.12	2.930		101	87.50	34	222.50	284
	6180	900 mm high	▼	9.56	5.859	▼	93.50	175	68	336.50	450
800	0010	**STONE RETAINING WALLS**									
	0015	Including excavation, concrete footing and									
	0020	stone 900 mm below grade. Price is exposed face area.									
	0200	Decorative random stone, to 2 m high, 450 mm thick, dry set	D-1	3.25	4.921	m²	385	158		543	665
	0300	Mortar set		3.72	4.306		385	138		523	635
	0500	Cut stone, to 2 m high, 450 mm thick, dry set		3.25	4.921		385	158		543	665
	0600	Mortar set		3.72	4.306		385	138		523	635
	0800	Random stone, 2 m to 3 m H, 600 mm T, dry set		4.18	3.827		385	123		508	610
	0900	Mortar set		4.65	3.445		385	111		496	595
	1100	Cut stone, 2 m to 3 m high, 600 mm thick, dry set		4.18	3.827		385	123		508	610
	1200	Mortar set		4.65	3.445	▼	385	111		496	595
	5100	Setting stone, dry		2.83	5.654	m³		182		182	277
	5600	With mortar	▼	3.40	4.711	"		151		151	231

02840 | Walk/Road/Parking Appurtenances

		CREW	DAILY OUTPUT	LABOR-HOURS	UNIT	MAT.	LABOR	EQUIP.	TOTAL	TOTAL INCL O&P
200	0010	**GUIDE/GUARD RAIL**									
	0012	Corrugated stl, galv. steel posts, 2 m O.C.	B-80	259	.124	m	54	3.69	2.06	59.75	67.50
	0100	Double face		174	.184	"	78.50	5.50	3.06	87.06	98
	0200	End sections, galvanized, flared		50	.640	Ea.	68	19.15	10.65	97.80	116
	0300	Wrap around end		50	.640		102	19.15	10.65	131.80	154
	0350	Anchorage units	▼	15	2.133		820	64	35.50	919.50	1,025
	0365	End section, flared	B-80A	78	.308		29.50	8.45	2.26	40.21	48
	0370	End section, wrap-around, corrugated steel	"	78	.308	▼	36.50	8.45	2.26	47.21	56
	0400	Timber guide rail, 100 x 200 w/ 150 x 200 mm treated posts	B-80	293	.109	m	71	3.27	1.82	76.09	85.50
	0600	Cable guide rail, 3 at 19 mm cables, steel posts, single face		274	.117		24	3.49	1.94	29.43	34
	0650	Double face		194	.165		56	4.93	2.74	63.67	72.50
	0700	Wood posts		290	.110		33	3.30	1.84	38.14	43.50
	0750	Double face		198	.162	▼	58	4.83	2.69	65.52	74.50
	0760	Breakaway wood posts		195	.164	Ea.	284	4.91	2.73	291.64	320
	0800	Anchorage units, breakaway		15	2.133	"	745	64	35.50	844.50	955
	0900	Guide rail, steel box beam, 150 mm x 150 mm		36.58	.875	m	81.50	26	14.55	122.05	146
	0950	End assembly	B-80A	48	.500	Ea.	246	13.70	3.67	263.37	297
	1100	Median barrier, steel box beam, 150 mm x 200 mm	B-80	65.53	.488	m	101	14.60	8.15	123.75	142

SITE CONSTRUCTION 2

02840 | Walk/Road/Parking Appurtenances

		CREW	DAILY OUTPUT	LABOR-HOURS	UNIT	2006 BARE COSTS				TOTAL INCL O&P
						MAT.	LABOR	EQUIP.	TOTAL	
200	1120	Shop curved	B-80A	28.04	.856	m	120	23.50	6.30	149.80	175	200
	1140	End assembly		48	.500	Ea.	330	13.70	3.67	347.37	390
	1150	Corrugated beam	↓	122	.197	m	64.50	5.40	1.44	71.34	81
	1400	Resilient guide fence and light shield, 1.8 m high	B-2	39.62	1.009	"	78.50	28		106.50	130
	1500	Concrete posts, individual, 1925 mm, triangular	B-80	110	.291	Ea.	51	8.70	4.84	64.54	74.50
	1550	Square		110	.291		54.50	8.70	4.84	68.04	78.50
	1600	Wood guide posts	↓	150	.213	↓	35	6.40	3.55	44.95	52
	2000	Median, precast concrete, 1050 mm H, 600 mm W, single face	B-29	116	.483	m	130	14.45	7.25	151.70	174
	2200	Double face	"	104	.538		149	16.10	8.10	173.20	197
	2300	Cast-in-place, steel forms	C-2	51.82	.926		284	32		316	360
	2320	Slipformed	C-7	107	.673	↓	118	20	9.45	147.45	171
	2400	Speed bumps, thermoplastic, 270 mm x 57 mm x 1200 mm long	B-2	120	.333	Ea.	97	9.25		106.25	120
	3000	Energy absorbing terminal, 10 bay, 900 mm wide	B-80B	.10	320		33,200	9,400	2,350	44,950	53,500
	3010	7 bay, 750 mm wide		.20	160		24,400	4,700	1,175	30,275	35,400
	3020	5 bay, 600 mm wide	↓	.20	160		18,900	4,700	1,175	24,775	29,400
	3030	Impact barrier, UTMCD, barrel type	B-16	30	1.067		330	30	15.90	375.90	430
	3100	Wide hazard protection, foam sandwich, 7 bay, 2250 mm wide	B-80B	.14	228		24,400	6,700	1,675	32,775	39,100
	3110	1500 mm wide		.15	213		21,500	6,250	1,575	29,325	35,000
	3120	900 mm wide	↓	.18	177	↓	22,200	5,225	1,300	28,725	33,900
	5000	For bridge railing, see div. 02850-205									
400	0010	**FENDERS**										400
	0015	Bumper rails for garages, 12 ga. rail, 150 mm wide, with steel									
	0020	posts 3.8 m O.C., minimum	E-4	57.91	.553	m	43	22.50	1.54	67.04	89
	0030	Average	↓	50.29	.636	↓	54	25.50	1.78	81.28	108
	0100	Maximum		42.67	.750		64.50	30.50	2.09	97.09	128
	0300	300 mm channel rail, minimum		48.77	.656		54	26.50	1.83	82.33	110
	0400	Maximum	↓	36.58	.875	↓	81	35.50	2.44	118.94	155
800	0010	**PARKING BUMPERS**										800
	0020	Parking barriers, timber w/saddles, treated type									
	0100	100 mm x 100 mm for cars	B-2	158	.253	m	8.35	7.05		15.40	20
	0200	150 mm x 150 mm for trucks		158	.253	"	17.40	7.05		24.45	30
	0600	Flexible fixed stanchion, 600 mm high, 75 mm diameter [3"]		100	.400	Ea.	25.50	11.10		36.60	45.50
	1000	Wheel stops, precast incl. dowels, 150 mm x 250 mm x 1800 mm		120	.333		40	9.25		49.25	58.50
	1100	200 mm x 325 mm x 1800 mm		120	.333		46.50	9.25		55.75	65.50
	1200	Thermoplastic, 150 mm x 250 mm x 1800 mm	↓	120	.333		74.50	9.25		83.75	96.50
	1300	Pipe bollards, conc. fill/paint, 2.4m L x 1.2m D hole, 150 mm diam.	B-6	20	1.200		300	36	11.30	347.30	400
	1400	200 mm diam.		15	1.600		455	48	15.10	518.10	590
	1500	300 mm diam.		12	2		595	60	18.85	673.85	765
	1592	Bollards, steel, 1 m H, retractable, incl hydraulic controls, min		4	6		27,800	180	56.50	28,036.50	30,900
	1594	Max	↓	2	12		31,000	360	113	31,473	34,800
	2030	Folding with individual padlocks	B-2	50	.800	↓	730	22		752	835
	9000	Parking lot control, see Div. 11156-600									
910	0010	**TRAFFIC BARRIERS, TRAFFIC CONTROL DEVICES**										910
	0020	Crash barriers									
	0100	Traffic channelizing pavement markers, layout only	A-7	2000	.012	Ea.		.47	.03	.50	.75
	0110	330 x 190 x 60 mm high, non-plowable install	2 Clab	96	.167		19.85	4.57		24.42	29
	0200	200 x 200 x 80 mm high, non-plowable, install		96	.167		18.05	4.57		22.62	27
	0230	100 x 100 x 20 mm high, non-plowable, install	↓	120	.133		2.41	3.65		6.06	8.35
	0240	235 x 145 x 6 mm high, plowable, concrete pavement	A-2A	70	.343		14.45	9.40	3.37	27.22	34
	0250	235 x 145 x 6 mm high, plowable, asphalt pavement	"	120	.200		2.76	5.50	1.97	10.23	13.70
	0300	Barrier and curb delineators,reflectorized, 50 x 100 mm	2 Clab	150	.107		1.57	2.92		4.49	6.30
	0310	75 x 150 mm	"	150	.107	↓	3.19	2.92		6.11	8.05
	0500	Rumble strip, polycarbonate									
	0510	600 x 90 x 13 mm high	2 Clab	50	.320	Ea.	6	8.75		14.75	20.50

02840	Walk/Road/Parking Appurtenances	CREW	DAILY OUTPUT	LABOR-HOURS	UNIT	2006 BARE COSTS				TOTAL INCL O&P
						MAT.	LABOR	EQUIP.	TOTAL	
920	**0010**	**TRAFFIC BARRIERS, HIGHWAY SOUND BARRIERS**										920
	0020	Highway sound barriers, not including footing									
	0100	Precast concrete, concrete columns @ 9 m OC, 200 mm T, 2.4 m H	C-12	122	.393	m	291	13.75	5.20	309.95	345
	0110	3.6 m H		80.77	.594		435	21	7.90	463.90	520
	0120	4.8 m H		60.96	.787		585	27.50	10.45	622.95	695
	0130	6 m H	↓	48.77	.984		730	34.50	13.05	777.55	870
	0400	Lt. wt. composite panel, cementitious face, St posts @ 3m OC,2.4 m H	B-80B	57.91	.553		335	16.20	4.06	355.26	400
	0410	3.6 m H		38.10	.840		505	24.50	6.15	535.65	600
	0420	4.8 m H		28.96	1.105		675	32.50	8.10	715.60	800
	0430	6 m H	↓	22.86	1.400	↓	845	41	10.30	896.30	1,000

02850 | Prefabricated Bridges

			CREW	DAILY OUTPUT	LABOR-HOURS	UNIT	MAT.	LABOR	EQUIP.	TOTAL	INCL O&P
205	**0010**	**BRIDGES, HIGHWAY**										205
	0020	Structural steel, rolled beams	E-5	7.71	10.375	Met. Ton	2,100	410	197	2,707	3,225
	0500	Built up, plate girders	E-6	9.53	13.437	"	2,500	530	172	3,202	3,875
	1000	Concrete in place, no reinforcing, abutment footings	C-17B	22.94	3.575	m³	178	132	11.80	321.80	415
	1050	Abutment		17.59	4.663		178	172	15.40	365.40	480
	1100	Walls, stems and wing walls		15.29	5.362		240	198	17.70	455.70	595
	1150	Parapets		9.18	8.937		240	330	29.50	599.50	810
	1170	Pier footings	↓	30.58	2.681	↓	178	99	8.85	285.85	360
	1180	Piers and columns, see div. 03310-240									
	1190	Pier caps including shoring	C-17B	7.65	10.725	m³	293	395	35.50	723.50	980
	1210	Beams, including shoring	C-14	11.47	12.556		360	435	92.50	887.50	1,175
	1220	Haunches		7.65	18.833	↓	420	650	138	1,208	1,625
	1230	Bridge sidewalks	↓	69.68	2.067	m²	64.50	71.50	15.20	151.20	200
	1250	Decks, including finish and cure, 200 mm thick									
	1260	Shored forms	C-14	279	.516	m²	52	17.85	3.79	73.64	89.50
	1270	Beam supported forms		325	.443		38	15.30	3.26	56.56	69.50
	1280	Stay in place metal slab forms, 20 gauge	↓	465	.310	↓	52	10.70	2.28	64.98	77
	1500	Precast, prestressed concrete									
	1510	Box girders, 11 m to 15 m span				Ea.				11,200	12,900
	1520	15 m to 20 m span								15,000	17,200
	1530	20 m to 26 m span				↓				19,300	22,100
	1540	Deck beams, 300 mm deep				m²				460	530
	1541	375 mm deep								520	600
	1542	450 mm deep								580	670
	1543	500 mm deep				↓				630	725
	1550	Double T, 12 m to 14 m span				Ea.				8,125	9,350
	1555	14 m to 17 m span								8,550	9,850
	1560	17 m to 20 m span				↓				8,875	9,050
	1580	Price per m²				m²				198	230
	1600	I beams, 18 m to 24 m span				Ea.				8,775	10,100
	1610	24 m to 30 m span								10,400	11,900
	1620	30 m to 37 m span								13,900	16,000
	1630	37 m to 44 m span				↓				16,600	19,100
	2000	Reinforcing, in place	4 Rodm	2.72	11.758	Met. Ton	1,675	465		2,140	2,625
	2050	Galvanized coated		2.72	11.758		2,100	465		2,565	3,075
	2100	Epoxy coated	↓	2.72	11.758	↓	3,100	465		3,565	4,200
	2110	See also Div. 03200									
	3000	Exp. dams, st., double upset 100 mm x 200 mm angles welded to									
	3010	double 200 mm x 200 mm angles, 44 mm compression seal	C-22	9.14	4.593	m	1,375	182	8.90	1,565.90	1,825
	3040	Double 200 mm x 200 mm angles only, 44 mm comp. seal		10.67	3.937		1,025	156	7.65	1,188.65	1,400
	3050	Galvanized		10.67	3.937		1,275	156	7.65	1,438.65	1,675
	3060	Double 200 mm x 150 mm angles only, 44 mm comp. seal		10.67	3.937		790	156	7.65	953.65	1,125
	3100	Double 250 mm channels, 44 mm comp. seal	↓	10.67	3.937		720	156	7.65	883.65	1,050
	3420	For 76 mm compression seal, add				↓	149			149	164

For expanded coverage of these items see *Means Site Work & Landscape Cost Data 2006*

SITE CONSTRUCTION **2**

02850 | Prefabricated Bridges

		CREW	DAILY OUTPUT	LABOR-HOURS	UNIT	2006 BARE COSTS MAT.	LABOR	EQUIP.	TOTAL	TOTAL INCL O&P	
205	3440	For double slotted extrusions with seal strip, add				m	690			690	755	205
	3490	For galvanizing, add				kg	1.70			1.70	1.87	
	4000	Approach railings, steel, galv. pipe, 2 line	C-22	42.67	.984	m	320	39	1.91	360.91	420	
	4200	Bridge railings, steel, galv. pipe, 3 line w/screen		25.91	1.621		1,000	64.50	3.15	1,067.65	1,200	
	4220	4 line w/screen		22.86	1.837		870	73	3.56	946.56	1,075	
	4300	Aluminum, pipe, 3 line w/screen		28.96	1.450		350	57.50	2.81	410.31	485	
	8000	For structural excavation, see div. 02315-462										
	8010	For dewatering, see div. 02240										
210	0010	**BRIDGES, PEDESTRIAN,** spans over streams, roadways, etc.										210
	0020	including erection, not including foundations										
	0050	Precast concrete, complete in place, 2400 mm wide, 18 m span	E-2	19.97	2.804	m²	560	109	71.50	740.50	880	
	0100	30 m span		17.19	3.258		610	127	83	820	980	
	0150	37 m span		14.86	3.767		665	147	96	908	1,100	
	0200	46 m span		13.47	4.157		690	162	106	958	1,150	
	0300	Steel, trussed or arch spans, compl. in place, 2400 mm wide, 12 m		29.73	1.884		725	73.50	48	846.50	980	
	0400	15 m span		36.70	1.526		650	59.50	39	748.50	860	
	0500	18 m span		43.20	1.296		650	50.50	33	733.50	840	
	0600	24 m span		52.95	1.058		775	41	27	843	955	
	0700	30 m span		43.20	1.296		1,100	50.50	33	1,183.50	1,325	
	0800	37 m span		33.91	1.652		1,375	64	42	1,481	1,675	
	0900	46 m span		28.80	1.945		1,475	75.50	49.50	1,600	1,775	
	1000	49 m span		23.69	2.364		1,475	92	60	1,627	1,825	
	1100	3 m wide, 24 m span		59.46	.942		805	36.50	24	865.50	975	
	1200	37 m span		38.55	1.453		1,050	56.50	37	1,143.50	1,300	
	1300	46 m span		41.34	1.355		1,175	52.50	34.50	1,262	1,425	
	1400	61 m span		19.04	2.940		1,250	114	75	1,439	1,650	
	1600	Wood, laminated type, complete in place, 24 m span	C-12	18.86	2.545		480	89	34	603	700	
	1700	40 m span	"	14.21	3.377		500	118	45	663	785	

02870 | Site Furnishings

		CREW	DAILY OUTPUT	LABOR-HOURS	UNIT	MAT.	LABOR	EQUIP.	TOTAL	TOTAL INCL O&P	
310	0010	**BENCHES**										310
	0012	Seating, benches, park, prec. conc, w/backs, wood rails, 1200 mm	2 Clab	5	3.200	Ea.	335	87.50		422.50	505	
	0100	2400 mm long		4	4		700	110		810	940	
	0300	Fiberglass, without back, one piece, 1200 mm long		10	1.600		440	44		484	555	
	0400	2400 mm long		7	2.286		905	62.50		967.50	1,100	
	0500	Steel pedestals w/backs, 50 mm x 75 mm wood rails, 1200 mm long		10	1.600		835	44		879	985	
	0510	2400 mm long		7	2.286		985	62.50		1,047.50	1,175	
	0520	75 mm x 200 mm wood plank, 1200 mm long		10	1.600		840	44		884	995	
	0530	2400 mm long		7	2.286		875	62.50		937.50	1,075	
	0540	Backless, 100 mm x 100 mm wood plank, 1200 mm square		10	1.600		815	44		859	965	
	0550	2400 mm long		7	2.286		775	62.50		837.50	950	
	0600	Aluminum pedestals, with backs, aluminum slats, 2400 mm long		8	2		405	55		460	530	
	0610	4600 mm long		5	3.200		400	87.50		487.50	575	
	0620	Portable, aluminum slats, 2400 mm long		8	2		355	55		410	475	
	0630	4600 mm long		5	3.200		520	87.50		607.50	705	
	0800	Cast iron pedestals, back & arms, wood slats, 1200 mm long		8	2		310	55		365	425	
	0820	2400 mm long		5	3.200		895	87.50		982.50	1,125	
	0840	Backless, wood slats, 1200 mm long		8	2		535	55		590	675	
	0860	2400 mm long		5	3.200		530	87.50		617.50	720	
	1700	Steel frame, fir seat, 3000 mm long		10	1.600		190	44		234	277	
510	0010	**TRASH RECEPTACLE**										510
	0020	Trash receptacle, fiberglass, 600 mm square, 450 mm high	2 Clab	30	.533	Ea.	266	14.60		280.60	315	
	0100	600 mm square, 760 mm high		30	.533		370	14.60		384.60	435	
	0300	Circular, 600 mm diameter, 460 mm high		30	.533		239	14.60		253.60	286	
	0400	600 mm diameter, 760 mm high		30	.533		320	14.60		334.60	375	

02800 | Site Improvements and Amenities

02870 | Site Furnishings

			CREW	DAILY OUTPUT	LABOR-HOURS	UNIT	MAT.	LABOR	EQUIP.	TOTAL	TOTAL INCL O&P	
520	0010	**TRASH CLOSURE** Steel with pullover cover										**520**
	0020	690 mm wide, 1400 mm high, 1880 mm long	2 Clab	5	3.200	Ea.	950	87.50		1,037.50	1,175	
	0100	330 mm long		4	4		1,250	110		1,360	1,550	
	0300	Wood, 3050 mm wide, 1830 mm high, 3050 mm long	↓	1.20	13.333	↓	990	365		1,355	1,675	

02880 | Playfield Equipment

			CREW	DAILY OUTPUT	LABOR-HOURS	UNIT	MAT.	LABOR	EQUIP.	TOTAL	TOTAL INCL O&P	
100	0010	**ATHLETIC OR RECREATIONAL SCREENING**										**100**
	0015	Backstops, baseball, prefab, 9000 mm W, 3600 mm H & 1 overhang	B-1	1	24	Ea.	2,675	675		3,350	4,000	
	0100	12 200 mm wide, 3660 mm high & 2 overhangs	"	.75	32		3,350	900		4,250	5,100	
	0300	Basketball, steel, single goal	B-13	3.04	18.421		1,050	550	214	1,814	2,225	
	0400	Double goal	"	1.92	29.167	↓	640	875	340	1,855	2,425	
	0600	Tennis, wire mesh with pair of ends	B-1	2.48	9.677	Set	1,650	272		1,922	2,225	
	0700	Enclosed court	"	1.30	18.462	Ea.	4,675	520		5,195	5,950	
	0900	Handball or squash court, outdoor, wood	2 Carp	.50	32		3,850	1,150		5,000	6,000	
	1000	Masonry handball/squash court	D-1	.30	53.333	↓	22,400	1,725		24,125	27,300	
110	0010	**TENNIS COURT FENCES**										**110**
	0860	Tennis courts, 3 mm wire, 65 mm post set										
	0870	set in concrete, 3 m O.C., 44 mm top rail										
	0900	3000 mm high	B-80	57.91	.553	m	41	16.50	9.20	66.70	81	
	0920	3600 mm high		51.82	.618	"	48.50	18.45	10.30	77.25	93	
	1000	Add for gate 1.2 m wide, 40 mm frame 2.1 m high		10	3.200	Ea.	118	95.50	53.50	267	335	
	1040	Aluminized steel, 11 ga. wire 3 m high		57.91	.553	m	50	16.50	9.20	75.70	90.50	
	1100	3.7 m high		51.82	.618	"	61.50	18.45	10.30	90.25	107	
	1140	Add for gate 1.2 m wide, 40 mm frame, 2.1 m high		10	3.200	Ea.	138	95.50	53.50	287	360	
	1250	Vinyl covered, 3.3 mm wire, 3000 mm high		57.91	.553	m	48.50	16.50	9.20	74.20	88.50	
	1300	3600 mm high	↓	51.82	.618	"	80	18.45	10.30	108.75	128	
	1310	Fence, CL, tennis ct, gate,sgl galv,1200 mm x 2100 mm x 900 mm	B-80A	8.72	2.752	Ea.	256	75.50	20	351.50	420	
	1400	Add for gate 1.2 m wide, 40 mm frame, 2.1 m high	B-80	10	3.200	"	138	95.50	53.50	287	355	
210	0010	**PLAYGROUND EQUIPMENT** See also individual items										**210**
	0200	Bike rack, 3000 mm long, permanent	B-1	12	2	Ea.	505	56		561	650	
	0400	Horizontal monkey ladder, 4200 mm long, 1800 mm high		4	6		660	168		828	985	
	0590	Parallel bars, 3 m long		4	6		284	168		452	575	
	0600	Posts, tether ball set, 59 mm O.D.		12	2	↓	188	56		244	295	
	0800	Poles, multiple purpose, 3.2 m long		12	2	Pr.	141	56		197	243	
	1000	Ground socket for movable posts, 59 mm post		10	2.400		99.50	67.50		167	214	
	1100	90 mm post		10	2.400	↓	139	67.50		206.50	258	
	1300	See-saw, spring, steel, 2 units		6	4	Ea.	980	112		1,092	1,250	
	1400	4 units		4	6		1,250	168		1,418	1,625	
	1500	6 units		3	8		2,500	225		2,725	3,100	
	1700	Shelter, fiberglass golf tee, 3 person		4.60	5.217		2,750	146		2,896	3,250	
	1900	Slides, stainless steel bed, 3600 mm long, 1800 mm high		3	8		2,450	225		2,675	3,050	
	2000	6100 mm long, 3050 mm high		2	12		3,575	335		3,910	4,475	
	2200	Swings, plain seats, 2400 mm high, 4 seats		2	12		1,100	335		1,435	1,750	
	2300	8 seats		1.30	18.462		2,025	520		2,545	3,025	
	2500	3600 mm high, 4 seats		2	12		1,500	335		1,835	2,150	
	2600	8 seats		1.30	18.462		1,950	520		2,470	2,950	
	2800	Whirlers, 2400 mm diameter		3	8		2,700	225		2,925	3,300	
	2900	3000 mm diameter	↓	3	8	↓	3,275	225		3,500	3,950	
220	0010	**MODULAR PLAYGROUND** Basic components										**220**
	0100	Deck, square, steel, 1.2 m x 1.2 m	B-1	1	24	Ea.	770	675		1,445	1,900	
	0110	Recycled polyurethane		1	24		695	675		1,370	1,825	
	0120	Triangular, steel, 1.2 m side		1	24		565	675		1,240	1,675	
	0130	Post, steel, 125 mm square		5.49	4.374	m	84	123		207	284	
	0140	Aluminum, 59 mm square	↓	6.10	3.937	↓	75.50	111		186.50	255	

SITE CONSTRUCTION 2

For expanded coverage of these items see *Means Site Work & Landscape Cost Data 2006*

SITE CONSTRUCTION

2

02880 | Playfield Equipment

		CREW	DAILY OUTPUT	LABOR-HOURS	UNIT	2006 BARE COSTS MAT.	LABOR	EQUIP.	TOTAL	TOTAL INCL O&P	
220	0150	125 mm square	B-1	5.49	4.374	m	102	123		225	305	220
	0160	Roof, square poly		18	1.333	Ea.	965	37.50		1,002.50	1,100	
	0170	Wheelchair transfer module, for 1 m high deck		3	8	"	2,525	225		2,750	3,125	
	0180	Guardrail, pipe, 1 m high		18.29	1.312	m	945	37		982	1,100	
	0190	Steps, deck-to-deck, 3 - 200 mm steps		8	3	Ea.	1,375	84		1,459	1,625	
	0200	Activity panel, crawl through panel		2	12		395	335		730	955	
	0210	Alphabet/spelling panel		2	12		745	335		1,080	1,350	
	0360	with guardrails		3	8		2,125	225		2,350	2,675	
	0370	Crawl tunnel, straight, 1400 mm long		4	6		1,250	168		1,418	1,625	
	0380	90°, 1.2 m long		4	6		1,525	168		1,693	1,925	
	1200	Slide, tunnel, for 1.4 m high deck		8	3		1,725	84		1,809	2,025	
	1210	Straight, poly		8	3		245	84		329	400	
	1220	Stainless steel, 1350 mm high deck		6	4		253	112		365	455	
	1230	Curved, poly, 1 m high deck		6	4		1,275	112		1,387	1,575	
	1240	Spiral slide, 1.4 m - 1.8 m high		5	4.800		4,450	135		4,585	5,100	
	1300	Ladder, vertical, for 600 mm - 1800 mm high deck		5	4.800		565	135		700	830	
	1310	Horizontal, 2400 mm long		5	4.800		785	135		920	1,075	
	1320	Corkscrew climber, 1800 mm high		3	8		605	225		830	1,025	
	1330	Fire pole for 1800 mm high deck		6	4		288	112		400	490	
	1340	Bridge, ring climber, 2.4 m long		4	6		1,150	168		1,318	1,500	
	1350	Suspension	▼	1.22	19.685	m	1,075	555		1,630	2,025	
330	0010	**GOAL POSTS**										330
	0020	Goal posts, steel, football, double post	B-1	1.50	16	Pr.	1,650	450		2,100	2,525	
	0100	Deluxe, single post		1.50	16		2,675	450		3,125	3,625	
	0300	Football, convertible to soccer		1.50	16		2,650	450		3,100	3,625	
	0500	Soccer, regulation	▼	2	12	▼	2,100	335		2,435	2,825	
350	0010	**PLATFORM PADDLE TENNIS COURT** Complete with lighting, etc.										350
	0100	Aluminum slat deck with aluminum frame	B-1	.08	300	Court	43,600	8,425		52,025	61,000	
	0500	Aluminum slat deck and wood frame	C-1	.12	266		43,600	8,925		52,525	62,000	
	0800	Aluminum deck heater, add	B-1	1.18	20.339		4,250	570		4,820	5,575	
	0900	Douglas fir planking and wood frame 50 mm x 150 mm x 9150 mm	C-1	.12	266		42,800	8,925		51,725	61,000	
	1000	Plywood deck with steel frame		.12	266		42,800	8,925		51,725	61,000	
	1100	Steel slat deck with wood frame	▼	.12	266	▼	30,100	8,925		39,025	47,000	

02890 | Traffic Signs & Signals

		CREW	DAILY OUTPUT	LABOR-HOURS	UNIT	MAT.	LABOR	EQUIP.	TOTAL	TOTAL INCL O&P	
100	0010	**SIGNS**										100
	0012	SIGNS Stock, 600 mm x 600 mm, no posts, 2 mm alum. reflectorized	B-80	70	.457	Ea.	55	13.65	7.60	76.25	90	
	0100	High intensity		70	.457		55	13.65	7.60	76.25	90	
	0300	750 mm x 750 mm, reflectorized		70	.457		112	13.65	7.60	133.25	152	
	0400	High intensity		70	.457		112	13.65	7.60	133.25	152	
	0600	Guide and directional signs, 300 mm x 450 mm, reflectorized		70	.457		35.50	13.65	7.60	56.75	69	
	0700	High intensity		70	.457		34	13.65	7.60	55.25	67	
	0900	450 mm x 600 mm, stock signs, reflectorized		70	.457		40	13.65	7.60	61.25	73.50	
	1000	High intensity		70	.457		40	13.65	7.60	61.25	73.50	
	1200	600 mm x 600 mm, stock signs, reflectorized		70	.457		49.50	13.65	7.60	70.75	84	
	1300	High intensity		70	.457		49.50	13.65	7.60	70.75	84	
	1500	Add to above for steel posts, galvanized, 3 m upright, bolted		200	.160		17.90	4.78	2.66	25.34	30	
	1600	3.7 m upright, bolted		140	.229	▼	24	6.85	3.80	34.65	40.50	
	1800	Highway road signs, aluminum, over 2 m², reflectorized		32.52	.984	m²	245	29.50	16.40	290.90	330	
	2000	High intensity		32.52	.984		245	29.50	16.40	290.90	330	
	2200	Highway, suspended over road, 7.4 m² min., reflectorized		15.33	2.088		245	62.50	34.50	342	405	
	2300	High intensity		15.33	2.088	▼	245	62.50	34.50	342	405	
	2350	Roadway delineators and reference markers	▼	500	.064	Ea.	23	1.91	1.07	25.98	29	

02890	Traffic Signs & Signals	CREW	DAILY OUTPUT	LABOR-HOURS	UNIT	2006 BARE COSTS				TOTAL INCL O&P	
						MAT.	LABOR	EQUIP.	TOTAL		
100	2360	Delineator post only, 1.8 m	B-80	500	.064	Ea.	7.25	1.91	1.07	10.23	12.05	**100**
	2400	Highway sign bridge structure, 14 m to 24 m								24,600	26,800	
	2410	Cantilever structure, add				▼					20%	
	5000	Removal of signs, including supports										
	5020	To 1 m²	B-80B	16	2	Ea.		58.50	14.70	73.20	107	
	5030	1 m² to 2 m²	"	5	6.400			188	47	235	340	
	5040	2 m² to 4 m²	B-14	1.80	26.667	▼		775	126	901	1,350	
	5050	4 m² to 9 m²	B-13	1.30	43.077	▼		1,300	500	1,800	2,525	
	5200	Remove and relocate signs, including supports										
	5210	Remove and relocate signs, to 1 m²	B-80B	5	6.400	Ea.	267	188	47	502	635	
	5220	1 m² to 2 m²	"	1.70	18.824		595	550	138	1,283	1,650	
	5230	2 m² to 4 m²	B-14	.56	85.714		625	2,500	405	3,530	4,975	
	5240	4 m² to 9 m²	B-13	.32	175		1,025	5,250	2,025	8,300	11,400	
	5300	Remove traffic posts to 3.6 M high	B-6	100	.240	▼		7.20	2.26	9.46	13.55	
	8000	For temporary barricades and lights, see div. 01560-100										
300	0010	**TRAFFIC SIGNALS** Mid block pedestrian crosswalk,										**300**
	0020	with pushbutton and mast arms	R-11	.30	186	Total	57,000	7,350	2,575	66,925	76,500	
	0600	Traffic signals, school flashing system, solar powered, remote controlled	"	1	56	Signal	14,000	2,200	770	16,970	19,500	
	1000	Intersection traffic signals, LED, mast, programmable, no lane control										
	1010	Includes all labor, material and equip. for complete installation				Ea.	157,500			157,500	173,500	
	1100	Intersection traffic signals, LED, mast, programmable, R/L lane control										
	1110	Includes all labor, material and equip. for complete installation				Ea.	210,000			210,000	231,000	
	1200	Add protective/permissive left turns to existing traffic light										
	1210	Includes all labor, material and equip. for complete installation				Ea.	63,000			63,000	69,500	
	1300	Replace existing light heads with LED Heads wire hung										
	1310	Includes all labor, material and equip. for complete installation				Ea.	42,000			42,000	46,200	
	1400	Replace existing light heads with LED Heads mast arm hung										
	1410	Includes all labor, material and equip. for complete installation				Ea.	84,000			84,000	92,500	

02900 | Planting

02905	Plants, Planting, Transplanting	CREW	DAILY OUTPUT	LABOR-HOURS	UNIT	2006 BARE COSTS				TOTAL INCL O&P	
						MAT.	LABOR	EQUIP.	TOTAL		
725	0010	**PLANTING**										**725**
	0012	Moving shrubs on site, 300 mm ball	B-62	28	.857	Ea.		25.50	5.50	31	45.50	
	0100	600 mm ball	"	22	1.091	"		32.50	7	39.50	58	
	0300	Moving trees on site, 900 mm ball	B-6	3.75	6.400	Ea.		192	60.50	252.50	360	
	0400	1500 mm ball	"	1	24	"		720	226	946	1,350	
925	0010	**TREE REMOVAL**										**925**
	0100	Dig & lace, shrubs, broadleaf evergreen, 450 mm - 600 mm	B-1	55	.436	Ea.		12.25		12.25	19.05	
	0200	600 mm - 900 mm	"	35	.686			19.25		19.25	30	
	0300	900 mm - 1200 mm	B-6	30	.800			24	7.55	31.55	45.50	
	0400	1200 mm - 1500 mm	"	20	1.200			36	11.30	47.30	68	
	1000	Deciduous, 300 mm - 380 mm	B-1	110	.218			6.10		6.10	9.55	
	1100	450 mm - 600 mm		65	.369			10.35		10.35	16.15	
	1200	600 mm - 900 mm	▼	55	.436	▼		12.25		12.25	19.05	

SITE CONSTRUCTION 2

02905 | Plants, Planting, Transplanting

		CREW	DAILY OUTPUT	LABOR-HOURS	UNIT	MAT.	LABOR	EQUIP.	TOTAL	TOTAL INCL O&P	
925	1300	900 mm - 1200 mm	B-6	50	.480	Ea.		14.40	4.53	18.93	27	925
	2000	Evergreeen, 450 mm - 600 mm	B-1	55	.436			12.25		12.25	19.05	
	2100	600 mm to 760 mm		50	.480			13.45		13.45	21	
	2200	760 mm to 900 mm		35	.686			19.25		19.25	30	
	2300	900 mm to 1070 mm		20	1.200			33.50		33.50	52.50	
	3000	Trees, deciduous, small, 600 mm - 900 mm		55	.436			12.25		12.25	19.05	
	3100	900 mm - 1200 mm	B-6	50	.480			14.40	4.53	18.93	27	
	3200	1200 mm - 1500 mm		35	.686			20.50	6.45	26.95	38.50	
	3300	1500 mm - 1800 mm		30	.800			24	7.55	31.55	45.50	
	4000	Shade, 1500 mm - 1800 mm		50	.480			14.40	4.53	18.93	27	
	4100	1800 mm - 2400 mm		35	.686			20.50	6.45	26.95	38.50	
	4200	2400 mm - 3000 mm		25	.960			29	9.05	38.05	54.50	
	4300	50 mm caliper		12	2			60	18.85	78.85	113	
	5000	Evergreen, 1200 mm - 1500 mm		35	.686			20.50	6.45	26.95	38.50	
	5100	1500 mm - 1800 mm		25	.960			29	9.05	38.05	54.50	
	5200	1800 mm - 2100 mm		19	1.263			38	11.90	49.90	71.50	
	5300	2100 mm - 2400 mm		15	1.600			48	15.10	63.10	90.50	
	5400	2400 mm - 3000 mm		11	2.182			65.50	20.50	86	124	

02910 | Plant Preparation

		CREW	DAILY OUTPUT	LABOR-HOURS	UNIT	MAT.	LABOR	EQUIP.	TOTAL	TOTAL INCL O&P	
500	0010	**MULCHING**										500
	0100	Aged barks, 75 mm deep, hand spread	1 Clab	83.61	.096	m²	2.48	2.62		5.10	6.80	
	0150	Skid steer loader	B-63	1254	.032		2.48	.92	.12	3.52	4.30	
	0200	Hay, 25 mm deep, hand spread	1 Clab	397	.020		.68	.55		1.23	1.61	
	0250	Power mulcher, small	B-64	16723	.001		.68	.03	.01	.72	.81	
	0350	Large	B-65	49239	.001		.68	.01	.01	.70	.77	
	0400	Humus peat, 25 mm deep, hand spread	1 Clab	585	.014		2.60	.37		2.97	3.44	
	0450	Push spreader	"	2090	.004		2.60	.10		2.70	3.02	
	0550	Tractor spreader	B-66	65032	.001		2.60			2.60	2.87	
	0600	Oat straw, 25 mm deep, hand spread	1 Clab	397	.020		.39	.55		.94	1.29	
	0650	Power mulcher, small	B-64	16723	.001		.40	.03	.01	.44	.50	
	0700	Large	B-65	49239	.001		.40	.01	.01	.42	.46	
	0750	Add for asphaltic emulsion	B-45	6700	.002	liter	.47	.08	.08	.63	.73	
	0800	Peat moss, 25 mm deep, hand spread	1 Clab	752	.011	m²	2.06	.29		2.35	2.71	
	0850	Push spreader	"	2090	.004		2.06	.10		2.16	2.42	
	0950	Tractor spreader	B-66	65032	.001		2.06			2.06	2.28	
	1000	Polyethylene film, 0.150 mm	2 Clab	1672	.010		.20	.26		.46	.64	
	1010	0.100 mm		1923	.008		.17	.23		.40	.53	
	1020	0.038 mm		2090	.008		.12	.21		.33	.46	
	1050	Filter fabric weed barrier		1672	.010		.92	.26		1.18	1.43	
	1100	Redwood nuggets, 75 mm deep, hand spread	1 Clab	125	.064		5.50	1.75		7.25	8.80	
	1150	Skid steer loader	B-63	1254	.032		5.55	.92	.12	6.59	7.65	
	1200	Stone mulch, hand spread, ceramic chips, economy	1 Clab	105	.076		7.45	2.09		9.54	11.45	
	1250	Deluxe	"	79.43	.101		11.50	2.76		14.26	16.95	
	1300	Granite chips	B-1	7.65	3.139	m³	41	88		129	182	
	1400	Marble chips		7.65	3.139		153	88		241	305	
	1600	Pea gravel		21.41	1.121		76	31.50		107.50	133	
	1700	Quartz		7.65	3.139		197	88		285	355	
	1800	Tar paper, 6.8 kg felt	1 Clab	669	.012	m²	.37	.33		.70	.92	
	1900	Wood chips, 50 mm deep, hand spread	"	184	.043		2.12	1.19		3.31	4.18	
	1950	Skid steer loader	B-63	1886	.021		2.12	.61	.08	2.81	3.38	
710	0010	**LAWN BED PREPARATION**										710
	0100	Rake topsoil, site material, harley rock rake, ideal	B-6	3066	.008	m²		.23	.07	.30	.44	
	0200	Adverse	"	650	.037			1.11	.35	1.46	2.08	
	0300	Screened loam, york rake and finish, ideal	B-62	2230	.011			.32	.07	.39	.58	

SITE CONSTRUCTION

SITE CONSTRUCTION **2**

	02910	Plant Preparation	CREW	DAILY OUTPUT	LABOR-HOURS	UNIT	2006 BARE COSTS MAT.	LABOR	EQUIP.	TOTAL	TOTAL INCL O&P	
710	0400	Adverse	B-62	1858	.013	m²		.39	.08	.47	.69	710
	1000	Remove topsoil & pile on site, 56 kW dozer, 150 mm D, 15 m haul	B-10L	2787	.004			.14	.11	.25	.35	
	1050	90 m haul		567	.021			.71	.56	1.27	1.70	
	1100	300 mm deep, 15 m haul		1440	.008			.28	.22	.50	.67	
	1150	90 m haul		288	.042			1.40	1.11	2.51	3.35	
	1200	149 kW dozer, 150 mm deep, 15 m haul	B-10B	11613	.001			.03	.08	.11	.14	
	1250	90 m haul		2852	.004			.14	.32	.46	.58	
	1300	300 mm deep, 15 m haul		5760	.002			.07	.16	.23	.29	
	1350	90 m haul		1431	.008			.28	.64	.92	1.14	
	1400	Alternate method, 56 kW dozer, 15 m haul	B-10L	658	.018	m³		.61	.48	1.09	1.46	
	1450	90 m haul	"	87.16	.138			4.63	3.66	8.29	11.05	
	1500	149 kW dozer, 15 m haul	B-10B	2034	.006			.20	.45	.65	.80	
	1600	90 m haul	"	436	.028			.92	2.11	3.03	3.73	
	1800	Rolling topsoil, hand push roller	1 Clab	297	.027	m²		.74		.74	1.15	
	1850	Tractor drawn roller	B-66	991	.008			.28	.18	.46	.63	
	1900	Remove rocks & debris from grade, by hand	B-62	7432	.003			.10	.02	.12	.17	
	1920	With rock picker	B-10S	13006	.001			.03	.02	.05	.07	
	2000	Root raking and loading, residential, no boulders	B-6	4952	.005			.15	.05	.20	.27	
	2100	With boulders		2973	.008			.24	.08	.32	.45	
	2200	Municipal, no boulders		18581	.001			.04	.01	.05	.07	
	2300	With boulders		11148	.002			.06	.02	.08	.12	
	2400	Large commercial, no boulders	B-10B	37161	.001			.01	.02	.03	.05	
	2500	With boulders	"	22297	.001			.02	.04	.06	.08	
	2600	Rough grade & scarify subsoil to receive topsoil, common earth										
	2610	149 kW dozer with scarifier	B-11A	7432	.002	m²		.07	.12	.19	.25	
	2620	134 kW grader with scarifier	B-11L	10219	.002			.05	.04	.09	.13	
	2700	Clay and till, 149 kW dozer with scarifier	B-11A	4645	.003			.11	.20	.31	.39	
	2710	134 kW grader with scarifier	B-11L	3716	.004			.14	.12	.26	.35	
	3000	Scarify subsoil, residential, skid steer loader w/scarifiers, 37 kW	B-66	2973	.003			.09	.06	.15	.21	
	3050	Municipal, skid steer loader w/scarifiers, 37 kW	"	11148	.001			.03	.02	.05	.06	
	3100	Large commercial, 56 kW, dozer w/scarifier	B-10L	22297	.001			.02	.01	.03	.05	
	3200	Grader with scarifier, 101 kW	B-11L	26013	.001			.02	.02	.04	.05	
	3500	Screen topsoil from stockpile, vibrating screen, wet material (organic)	B-10P	153	.078	m³		2.64	5.20	7.84	9.70	
	3550	Dry material	"	229	.052			1.76	3.47	5.23	6.50	
	3600	Mixing with conditioners, manure and peat	B-10R	421	.029			.96	.46	1.42	1.97	
	3650	Mobilization add for 2 days or less operation	B-34K	3	2.667	Job		75.50	175	250.50	305	
	3800	Spread conditioned topsoil, 150 mm deep, by hand	B-1	301	.080	m²	5.55	2.24		7.79	9.60	
	3850	223 kW dozer	B-10M	2508	.005		5.45	.16	.48	6.09	6.70	
	3900	100 mm deep, by hand	B-1	393	.061		5	1.71		6.71	8.15	
	3920	223 kW dozer	B-10M	3159	.004		4.07	.13	.38	4.58	5.10	
	3940	134 kW grader	B-11L	3437	.005		4.07	.15	.13	4.35	4.86	
	4000	Spread soil cond., alum. sulfate, 0.54 kg/m², hand push spreader	1 Clab	14632	.001		18.30	.02		18.32	20	
	4050	Tractor spreader	B-66	65032	.001		18.35			18.35	20	
	4100	Fertilizer, 0.11 kg/m², push spreader	1 Clab	14632	.001		.10	.02		.12	.13	
	4150	Tractor spreader	B-66	65032	.001		.10			.10	.12	
	4200	Ground limestone, 0.54 kg/m², push spreader	1 Clab	14632	.001		.11	.02		.13	.14	
	4250	Tractor spreader	B-66	65032	.001		.11			.11	.13	
	4300	Lusoil, 1.62 kg/m², push spreader	1 Clab	14632	.001		.56	.02		.58	.64	
	4350	Tractor spreader	B-66	65032	.001		.56			.56	.63	
	4400	Manure, 9.72/m², push spreader	1 Clab	2090	.004		3.34	.10		3.44	3.83	
	4450	Tractor spreader	B-66	26013	.001		3.35	.01	.01	3.37	3.71	
	4500	Perlite, 25 mm deep, push spreader	1 Clab	14632	.001		10.30	.02		10.32	11.35	
	4550	Tractor spreader	B-66	65032	.001		10.35			10.35	11.40	
	4600	Vermiculite, push spreader	1 Clab	14632	.001		3.15	.02		3.17	3.48	
	4650	Tractor spreader	B-66	65032	.001		3.16			3.16	3.48	
	5000	Spread topsoil, skid steer loader and hand dress	B-62	206	.117	m³	27.50	3.50	.75	31.75	36	

02910 | Plant Preparation

		Description	CREW	DAILY OUTPUT	LABOR-HOURS	UNIT	MAT.	LABOR	EQUIP.	TOTAL	TOTAL INCL O&P	
710	5100	Articulated loader and hand dress	B-100	245	.049	m³	27.50	1.65	2.27	31.42	35	710
	5200	Articulated loader and 56 kW dozer	B-10M	382	.031		27.50	1.06	3.13	31.69	35	
	5300	Road grader and hand dress	B-11L	765	.021		27.50	.67	.60	28.77	31.50	
	6000	Tilling topsoil, 15 kW tractor, disk harrow, 50 mm deep	B-66	41806	.001	m²		.01		.01	.01	
	6050	100 mm deep		33445	.001			.01	.01	.02	.02	
	6100	150 mm deep		25084	.001			.01	.01	.02	.03	
	6150	660 mm rototiller, 50 mm deep	A-1J	1045	.008			.21	.10	.31	.44	
	6200	100 mm deep		836	.010			.26	.12	.38	.55	
	6250	150 mm deep		627	.013			.35	.17	.52	.72	
720	0010	**PLANT BED PREPARATION, SHRUB & TREE**										720
	0100	Backfill planting pit, by hand, on site topsoil	2 Clab	13.76	1.163	m³		32		32	49.50	
	0200	Prepared planting mix	"	18.35	.872			24		24	37	
	0300	Skid steer loader, on site topsoil	B-62	260	.092			2.77	.59	3.36	4.91	
	0400	Prepared planting mix	"	313	.077			2.30	.49	2.79	4.08	
	1000	Excavate planting pit, by hand, sandy soil	2 Clab	12.23	1.308			36		36	56	
	1100	Heavy soil or clay	"	6.12	2.616			71.50		71.50	112	
	1200	0.382 m³ backhoe, sandy soil	B-11C	115	.139			4.46	1.97	6.43	8.95	
	1300	Heavy soil or clay	"	87.93	.182			5.85	2.57	8.42	11.75	
	2000	Mix planting soil, incl. loam, manure, peat, by hand	2 Clab	45.88	.349		48	9.55		57.55	67.50	
	2100	Skid steer loader	B-62	115	.209		48	6.25	1.34	55.59	63.50	
	3000	Pile sod, skid steer loader	"	2341	.010	m²		.31	.07	.38	.54	
	3100	By hand	2 Clab	334	.048			1.31		1.31	2.04	
	4000	Remove sod, front-end loader	B-10S	1672	.007			.24	.14	.38	.53	
	4100	Sod cutter	B-12K	2676	.006			.20	.36	.56	.70	
	4200	By hand	2 Clab	201	.080			2.18		2.18	3.39	
810	0010	**LOAM & TOPSOIL**										810
	0300	Fine grade, base course for paving, see div. 02720-200										
	0400	Spread from pile to rough finish grade, F.E. loader, 1.15 m³	B-10S	153	.078	Em³		2.64	1.58	4.22	5.75	
	0500	Up to 60 m radius, by hand	1 Clab	10.70	.747			20.50		20.50	32	
	0600	Top dress by hand, 0.76 m³ for 56 m²	"	8.79	.910		28.50	25		53.50	70	
	0700	Furnish and place, truck dumped, screened, 100 mm deep	B-10S	1087	.011	m²	3.24	.37	.22	3.83	4.36	
	0800	150 mm deep	"	686	.017	"	4.14	.59	.35	5.08	5.85	

02915 | Shrub and Tree Transplanting

		Description	CREW	DAILY OUTPUT	LABOR-HOURS	UNIT	MAT.	LABOR	EQUIP.	TOTAL	TOTAL INCL O&P	
200	0010	**GROUND COVER**										200
	0012	Plants, pachysandra, in prepared beds	B-1	15	1.600	h	26	45		71	98.50	
	0200	Vinca minor, 1 yr, bare root		12	2	"	27	56		83	117	
	0600	Stone chips, in 23 kg bags, Georgia marble		520	.046	Bag	2.44	1.30		3.74	4.70	
	0700	Onyx gemstone		260	.092		17.70	2.59		20.29	23.50	
	0800	Quartz		260	.092		6.60	2.59		9.19	11.30	
	0900	Pea gravel, truckload lots		25.40	.945	Met. Ton	27.50	26.50		54	71.50	
400	0010	**PLANTING** Trees, shrubs and ground cover										400
	0100	Light soil										
	0110	Bare root seedlings, 75 mm to 125 mm	1 Clab	960	.008	Ea.		.23		.23	.36	
	0120	150 mm to 250 mm		520	.015			.42		.42	.66	
	0130	275 mm to 400 mm		370	.022			.59		.59	.92	
	0140	425 mm to 600 mm		210	.038			1.04		1.04	1.63	
	0200	Potted, 57 mm diameter [2-1/4"]		840	.010			.26		.26	.41	
	0210	75 mm diameter [3"]		700	.011			.31		.31	.49	
	0220	100 mm diameter [4"]		620	.013			.35		.35	.55	
	0300	Container, 3.8 L	2 Clab	84	.190			5.20		5.20	8.10	
	0310	7.6 L		52	.308			8.45		8.45	13.10	
	0320	11 L		40	.400			10.95		10.95	17.05	

2 SITE CONSTRUCTION

Important: See the Reference Section for supporting data - Crews, Rental Equipment, City Cost Indexes and Reference Data

02915 | Shrub and Tree Transplanting

		CREW	DAILY OUTPUT	LABOR-HOURS	UNIT	2006 BARE COSTS				TOTAL INCL O&P
						MAT.	LABOR	EQUIP.	TOTAL	
400	0330	19 L	2 Clab	29	.552	Ea.		15.10		15.10	23.50	400
	0400	Bagged and burlapped, 300 mm diameter ball, by hand [12"]	↓	19	.842			23		23	36
	0410	Backhoe/loader, 36 kW	B-6	40	.600			18	5.65	23.65	33.50
	0415	380 mm diameter, by hand [15"]	2 Clab	16	1			27.50		27.50	42.50
	0416	Backhoe/loader, 36 kW	B-6	30	.800			24	7.55	31.55	45.50
	0420	450 mm diameter by hand [18"]	2 Clab	12	1.333			36.50		36.50	57
	0430	Backhoe/loader, 36 kW	B-6	27	.889			26.50	8.40	34.90	50
	0440	600 mm diameter by hand [24"]	2 Clab	9	1.778			48.50		48.50	76
	0450	Backhoe/loader 36 kW	B-6	21	1.143			34.50	10.80	45.30	64.50
	0470	900 mm diameter, backhoe/loader, 36 kW [36"]	"	17	1.412	↓		42.50	13.30	55.80	79.50
	0550	Medium soil									
	0560	Bare root seedlings, 75 mm to 125 mm	1 Clab	672	.012	Ea.		.33		.33	.51
	0561	150 mm to 250 mm		364	.022			.60		.60	.94
	0562	275 mm to 400 mm		260	.031			.84		.84	1.31
	0563	425 mm to 600 mm		145	.055			1.51		1.51	2.35
	0570	Potted, 57 mm diameter [2-1/4"]		590	.014			.37		.37	.58
	0572	75 mm diameter [3"]		490	.016			.45		.45	.70
	0574	100 mm diameter [4"]	↓	435	.018			.50		.50	.78
	0590	Container, 3.8 L	2 Clab	59	.271			7.45		7.45	11.55
	0592	7.6 L		36	.444			12.20		12.20	18.95
	0594	11 L		28	.571			15.65		15.65	24.50
	0595	19 L		20	.800			22		22	34
	0600	Bagged and burlapped, 300 mm diameter ball, by hand [12"]	↓	13	1.231			33.50		33.50	52.50
	0605	Backhoe/loader, 36 kW	B-6	28	.857			25.50	8.10	33.60	48.50
	0607	380 mm diameter, by hand [15"]	2 Clab	11.20	1.429			39		39	61
	0608	Backhoe/loader, 36 kW	B-6	21	1.143			34.50	10.80	45.30	64.50
	0610	450 mm diameter, by hand [18"]	2 Clab	8.50	1.882			51.50		51.50	80.50
	0615	Backhoe/loader, 36 kW	B-6	19	1.263			38	11.90	49.90	71.50
	0620	600 mm diameter, by hand [24"]	2 Clab	6.30	2.540			69.50		69.50	108
	0625	Backhoe/loader, 36 kW	B-6	14.70	1.633			49	15.40	64.40	92.50
	0630	900 mm diameter, backhoe/loader, 36 kW [36"]	"	12	2	↓		60	18.85	78.85	113
	0700	Heavy or stoney soil									
	0710	Bare root seedlings, 75 mm to 125 mm	1 Clab	470	.017	Ea.		.47		.47	.73
	0711	150 mm to 250 mm		255	.031			.86		.86	1.34
	0712	275 mm to 400 mm		182	.044			1.20		1.20	1.87
	0713	425 mm to 600 mm		101	.079			2.17		2.17	3.38
	0720	Potted, 57 mm diameter [2-1/4"]		360	.022			.61		.61	.95
	0722	75 mm diameter [3"]		343	.023			.64		.64	.99
	0724	100 mm diameter [4"]	↓	305	.026			.72		.72	1.12
	0730	Container, 3.8 L	2 Clab	41.30	.387			10.60		10.60	16.50
	0732	7.6 L		25.20	.635			17.40		17.40	27
	0734	11 L		19.60	.816			22.50		22.50	35
	0735	19 L		14	1.143			31.50		31.50	48.50
	0750	Bagged and burlapped, 300 mm diameter ball, by hand [12"]	↓	9.10	1.758			48		48	75
	0751	Backhoe/loader	B-6	19.60	1.224			36.50	11.55	48.05	69
	0752	380 mm diameter, by hand [15"]	2 Clab	7.80	2.051			56		56	87.50
	0753	Backhoe/loader, 36 kW	B-6	14.70	1.633			49	15.40	64.40	92.50
	0754	450 mm diameter, by hand [18"]	2 Clab	5.60	2.857			78.50		78.50	122
	0755	Backhoe/loader, 36 kW	B-6	13.30	1.805			54	17	71	102
	0756	600 mm diameter, by hand [24"]	2 Clab	4.40	3.636			99.50		99.50	155
	0757	Backhoe/loader, 36 kW	B-6	10.30	2.330			70	22	92	131
	0758	900 mm diameter backhoe/loader, 36 kW [36"]	"	8.40	2.857			85.50	27	112.50	162
	2000	Stake out tree and shrub locations	2 Clab	220	.073	↓		1.99		1.99	3.10

02920 | Lawns & Grasses

			CREW	DAILY OUTPUT	LABOR-HOURS	UNIT	2006 BARE COSTS				TOTAL INCL O&P
							MAT.	LABOR	EQUIP.	TOTAL	
310	**0010**	**SEEDING, GENERAL**									**310**
	0020	Mechanical seeding,240kg/hectacre	B-66	.61	13.178	Hectare	1,300	465	295	2,060	2,450
	0100	0.024 kg/m²	"	2090	.004	m²	.18	.13	.09	.40	.49
	0300	Fine grading and seeding incl. lime, fertilizer & seed,									
	0310	with equipment	B-14	836	.057	m²	.19	1.67	.27	2.13	3.10
	0600	Limestone hand push spreader, 0.245 kg/m²	1 Clab	16723	.001		.04	.01		.05	.06
	0800	Grass seed hand push spreader, 0.022 kg/m²	"	16723	.001		.18	.01		.19	.22
	1000	Hydro or air seeding for large areas, incl. seed and fertilizer	B-81	7441	.003		.18	.10	.05	.33	.41
	1100	With wood fiber mulch added	"	7441	.003		.33	.10	.05	.48	.58
	1300	Seed only, over 45 kg, field seed, minimum				kg	2.49			2.49	2.73
	1400	Maximum					8.30			8.30	9.15
	1500	Lawn seed, minimum					3.86			3.86	4.25
	1600	Maximum					10.10			10.10	11.10
	1800	Aerial operations, seeding only, field seed	B-58	20.24	1.186	Hectare	905	35.50	122	1,062.50	1,200
	1900	Lawn seed		20.24	1.186		1,400	35.50	122	1,557.50	1,750
	2100	Seed and liquid fertilizer, field seed		20.24	1.186		1,100	35.50	122	1,257.50	1,425
	2200	Lawn seed		20.24	1.186		1,600	35.50	122	1,757.50	1,975
320	**0010**	**SEEDING, ATHLETIC FIELDS**									**320**
	0020	Seeding, athletic fields, athletic field mix, 3.91 kg/100 m²,push spreader	1 Clab	743	.011	m²	.24	.30		.54	.72
	0100	Tractor spreader	B-66	4831	.002		.24	.06	.04	.34	.39
	0200	Hydro or air seeding, with mulch & fertil.	B-81	7432	.003		.26	.10	.05	.41	.50
	0400	Birdsfoot trefoil, 0.22 kg/100m², push spreader	1 Clab	743	.011		.08	.30		.38	.55
	0500	Tractor spreader	B-66	4831	.002		.08	.06	.04	.18	.22
	0600	Hydro or air seeding,with mulch & fertil.	B-81	7432	.003		.15	.10	.05	.30	.37
	0800	Bluegrass, 1.95 kg/100m², common, push spreader	1 Clab	743	.011		.16	.30		.46	.64
	0900	Tractor spreader	B-66	4831	.002		.16	.06	.04	.26	.31
	1000	Hydro or air seeding, with mulch & fertil.	B-81	7432	.003		.27	.10	.05	.42	.51
	1100	Baron, push spreader	1 Clab	743	.011		.22	.30		.52	.70
	1200	Tractor spreader	B-66	4831	.002		.22	.06	.04	.32	.37
	1300	Hydro or air seeding, with mulch & fertil.	B-81	7432	.003		.30	.10	.05	.45	.54
	1500	Clover, 0.33 kg/100m², white, push spreader	1 Clab	743	.011		.01	.30		.31	.48
	1600	Tractor spreader	B-66	4831	.002		.01	.06	.04	.11	.15
	1700	Hydro or air seeding, with mulch and fertil.	B-81	7432	.003		.08	.10	.05	.23	.30
	1800	Ladino, push spreader	1 Clab	743	.011		.06	.30		.36	.52
	1900	Tractor spreader	B-66	4831	.002		.06	.06	.04	.16	.19
	2000	Hydro or air seeding, with mulch and fertil.	B-81	7432	.003		.25	.10	.05	.40	.48
	2200	Fescue 2.69 kg/100m², tall, push spreader	1 Clab	743	.011		.11	.30		.41	.58
	2300	Tractor spreader	B-66	4831	.002		.11	.06	.04	.21	.25
	2400	Hydro or air seeding, with mulch and fertilizer	B-81	7432	.003		.37	.10	.05	.52	.61
	2500	Chewing, push spreader	1 Clab	743	.011		.11	.30		.41	.58
	2600	Tractor spreader	B-66	4831	.002		.11	.06	.04	.21	.25
	2700	Hydro or air seeding, with mulch and fertil.	B-81	7432	.003		.37	.10	.05	.52	.61
	2900	Crown vetch, 1.95 kg/100m², push spreader	1 Clab	743	.011		.41	.30		.71	.91
	3000	Tractor spreader	B-66	4831	.002		.41	.06	.04	.51	.58
	3100	Hydro or air seeding, with mulch and fertilizer	B-81	7432	.003		.56	.10	.05	.71	.83
	3300	Rye, 4.88 kg/100m², annual, push spreader	1 Clab	743	.011		.07	.30		.37	.54
	3400	Tractor spreader	B-66	4831	.002		.07	.06	.04	.17	.21
	3500	Hydro or air seeding, with mulch and fertilizer	B-81	7432	.003		.16	.10	.05	.31	.39
	3600	Fine textured, push spreader	1 Clab	743	.011		.07	.30		.37	.54
	3700	Tractor spreader	B-66	4831	.002		.07	.06	.04	.17	.21
	3800	Hydro or air seeding, with mulch and fertilizer	B-81	7432	.003		.16	.10	.05	.31	.39
	4000	Shade mix, 2.93 kg/100m², push spreader	1 Clab	743	.011		.11	.30		.41	.58
	4100	Tractor spreader	B-66	4831	.002		.11	.06	.04	.21	.25
	4200	Hydro or air seeding, with mulch and fertilizer	B-81	7432	.003		.23	.10	.05	.38	.47
	4400	Slope mix, 2.93 kg/100m², push spreader	1 Clab	743	.011		.11	.30		.41	.58

R329219-50

Important: See the Reference Section for supporting data - Crews, Rental Equipment, City Cost Indexes and Reference Data

SITE CONSTRUCTION 2

02920	Lawns & Grasses	CREW	DAILY OUTPUT	LABOR-HOURS	UNIT	2006 BARE COSTS MAT.	LABOR	EQUIP.	TOTAL	TOTAL INCL O&P	
320	4500	Tractor spreader	B-66	4831	.002	m²	.11	.06	.04	.21	.25	320
	4600	Hydro or air seeding, with mulch and fertilizer	B-81	7432	.003		.27	.10	.05	.42	.50	
	4800	Turf mix, 1.95 kg/100m², push spreader	1 Clab	743	.011		.07	.30		.37	.54	
	4900	Tractor spreader	B-66	4831	.002		.07	.06	.04	.17	.21	
	5000	Hydro or air seeding, with mulch and fertilizer	B-81	7432	.003		.18	.10	.05	.33	.41	
	5200	Utility mix, 3.42 kg/100m², push spreader	1 Clab	743	.011		.12	.30		.42	.60	
	5300	Tractor spreader	B-66	4831	.002		.12	.06	.04	.22	.27	
	5400	Hydro or air seeding, with mulch and fertilizer	B-81	7432	.003		.47	.10	.05	.62	.72	
	5600	Wildflower, 0.05 kg/100m², push spreader	1 Clab	743	.011		.04	.30		.34	.51	
	5700	Tractor spreader	B-66	4831	.002		.04	.06	.04	.14	.18	
	5800	Hydro or air seeding, with mulch and fertilizer	B-81	7432	.003	↓	.24	.10	.05	.39	.47	
	7000	Apply fertilizer, 145 kg/hectare	B-66	3.63	2.205	Met. Ton	390	77.50	49.50	517	600	
	7025	Limestone, mechanical spread	1 Clab	.71	11.296	Hectare	8.55	310		318.55	490	
	7100	Apply mulch, see div. 02910-500										
400	0010	**SODDING**										400
	0020	Sodding, 25 mm deep, bluegrass sod, on level ground, over 750 m²	B-63	2044	.020	m²	2.40	.57	.08	3.05	3.60	
	0200	370 m²		1579	.025		2.69	.73	.10	3.52	4.19	
	0300	100 m²		1254	.032		2.93	.92	.12	3.97	4.80	
	0500	Sloped ground, over 750 m²		557	.072		2.40	2.08	.28	4.76	6.15	
	0600	370 m²		465	.086		2.69	2.49	.33	5.51	7.15	
	0700	100 m²		372	.108		2.93	3.11	.41	6.45	8.50	
	1000	Bent grass sod, level ground, over 550 m²		1858	.022		5.40	.62	.08	6.10	7	
	1100	280 m²		1672	.024		5.95	.69	.09	6.73	7.70	
	1200	Sodding, 92.9 m² or less		1301	.031		6.85	.89	.12	7.86	9	
	1500	sloped ground, over 550 m²		1394	.029		5.40	.83	.11	6.34	7.35	
	1600	280 m²		1254	.032		5.95	.92	.12	6.99	8.10	
	1700	100 m²	↓	1115	.036	↓	6.85	1.04	.14	8.03	9.25	
500	0010	**STOLENS, SPRIGGING**										500
	0100	150 mm OC, by hand	1 Clab	372	.022	m²	.15	.59		.74	1.09	
	0110	Walk behind sprig planter	"	7432	.001		.15	.03		.18	.22	
	0120	Towed sprig planter	B-66	32516	.001		.15	.01	.01	.17	.19	
	0130	225 mm OC, by hand	1 Clab	483	.017		.11	.45		.56	.83	
	0140	Walk behind sprig planter	"	8547	.001		.11	.03		.14	.16	
	0150	Towed sprig planter	B-66	39019	.001		.11	.01		.12	.14	
	0160	300 mm OC, by hand	1 Clab	557	.014		.07	.39		.46	.69	
	0170	Walk behind sprig planter	"	10219	.001		.07	.02		.09	.11	
	0180	Towed sprig planter	B-66	46452	.001		.07	.01		.08	.09	
	0200	Broadcast, by hand, 0.07 m³ per m²	1 Clab	1394	.006		.06	.16		.22	.31	
	0210	0.14 m³ per m²		929	.009		.12	.24		.36	.51	
	0220	0.21 m³ per m²	↓	604	.013		.19	.36		.55	.77	
	0300	Hydro planter, 0.21 m³ per m²	B-64	9290	.002		.19	.05	.03	.27	.30	
	0320	Manure spreader planting 0.21 m³ per m²	B-66	18581	.001	↓	.19	.02	.01	.22	.23	

02930	Exterior Plants	CREW	DAILY OUTPUT	LABOR-HOURS	UNIT	MAT.	LABOR	EQUIP.	TOTAL	TOTAL INCL O&P	
050	0010	**TRAVEL** To all nursery items, for 16 to 32 km, add				All					5%	050
	0100	32 to 80 km, add				"					10%	
310	0010	**SHRUBS AND TREES** Evergreen, in prepared beds, B & B										310
	0100	Arborvitae pyramidal, 1200 mm to 1500 mm	B-17	30	1.067	Ea.	44	31.50	18.45	93.95	117	
	0150	Globe, 300 mm to 375 mm	B-1	96	.250		11.05	7		18.05	23	
	0300	Cedar, blue, 2400 mm to 2540 mm	B-17	18	1.778		183	52.50	30.50	266	315	
	0500	Hemlock, canadian, 760 mm to 900 mm	B-1	36	.667		23	18.70		41.70	54.50	
	0550	Holly, Savannah, 2.4 m - 3 m H		9.68	2.479		530	69.50		599.50	690	
	0600	Juniper, andorra, 460 mm to 600 mm		80	.300		15.90	8.40		24.30	30.50	
	0620	Wiltoni, 360 mm to 460 mm	↓	80	.300	↓	15.25	8.40		23.65	30	

02930 | Exterior Plants

		CREW	DAILY OUTPUT	LABOR-HOURS	UNIT	2006 BARE COSTS				TOTAL INCL O&P	
						MAT.	LABOR	EQUIP.	TOTAL		
310	0640	Skyrocket, 1370 mm to 1500 mm	B-17	55	.582	Ea.	48.50	17.20	10.05	75.75	91	**310**
	0660	Blue pfitzer, 600 mm to 760 mm	B-1	44	.545		25	15.30		40.30	51.50	
	0680	Ketleerie, 760 mm to 900 mm		50	.480		32	13.45		45.45	56	
	0700	Pine, black, 750 mm to 900 mm		50	.480		39	13.45		52.45	64	
	0720	Mugo, 760 mm to 600 mm		60	.400		34	11.25		45.25	55	
	0740	White, 1200 mm to 1500 mm	B-17	75	.427		50.50	12.65	7.35	70.50	83	
	0800	Spruce, blue, 460 mm to 600 mm	B-1	60	.400		38	11.25		49.25	59.50	
	0840	Norway, 1200 mm to 1500 mm	B-17	75	.427		85.50	12.65	7.35	105.50	122	
	0900	Yew, denisforma, 300 mm to 375 mm	B-1	60	.400		23	11.25		34.25	42.50	
	1000	Capitata, 460 mm to 600 mm		30	.800		19.65	22.50		42.15	56.50	
	1100	Hicksi, 600 mm to 760 mm		30	.800		29	22.50		51.50	67	
320	0010	**SHRUBS** Broadleaf evergreen, planted in prepared beds										**320**
	0100	Andromeda, 380 mm to 460 mm, container	B-1	96	.250	Ea.	22.50	7		29.50	36	
	0200	Azalea, 380 mm to 460 mm, container		96	.250		25	7		32	38.50	
	0300	Barberry, 230 mm to 300 mm, container		130	.185		10.05	5.20		15.25	19.10	
	0400	Boxwood, 380 mm to 460 mm, B & B		96	.250		27	7		34	40.50	
	0500	Euonymus, emerald gaiety, 300 mm to 380 mm, container		115	.209		16.20	5.85		22.05	27	
	0600	Holly, 380 mm to 460 mm, B & B		96	.250		15.80	7		22.80	28.50	
	0900	Mount laurel, 460 to 600 mm, B & B		80	.300		50	8.40		58.40	68	
	1000	Paxistema, 225 mm to 300 mm high		130	.185		16.20	5.20		21.40	26	
	1100	Rhododendron, 460 mm to 600 mm, container		48	.500		28	14.05		42.05	52.50	
	1200	Rosemary, 3.785 L container		600	.040		59	1.12		60.12	67	
	2000	Deciduous, amelanchier, 600 mm to 900 mm, B & B		57	.421		80.50	11.80		92.30	107	
	2100	Azalea, 375 mm to 450 mm, B & B		96	.250		21.50	7		28.50	34.50	
	2300	Bayberry, 600 mm to 900 mm, B & B		57	.421		24.50	11.80		36.30	45.50	
	2600	Cotoneaster, 380 mm to 460 mm, B & B		80	.300		14.70	8.40		23.10	29.50	
	2800	Dogwood, 900 mm to 1200 mm, B & B	B-17	40	.800		23	23.50	13.80	60.30	77	
	2900	Euonymus, alatus compacta, 380 mm to 460 mm, container	B-1	80	.300		19.30	8.40		27.70	34	
	3200	Forsythia, 600 mm to 900 mm, container	"	60	.400		17	11.25		28.25	36	
	3300	Hibiscus, 900 mm to 1200 mm, B & B	B-17	75	.427		13.25	12.65	7.35	33.25	42	
	3400	Honeysuckle, 900 mm to 1200 mm, B & B	B-1	60	.400		18.95	11.25		30.20	38.50	
	3500	Hydrangea, 600 mm to 900 mm, B & B	"	57	.421		22	11.80		33.80	42.50	
	3600	Lilac, 900 mm to 1200 mm, B & B	B-17	40	.800		23	23.50	13.80	60.30	76.50	
	3900	Privet, bare root, 460 mm to 600 mm	B-1	80	.300		11.75	8.40		20.15	26	
	4100	Quince, 600 mm to 900 mm, B & B	"	57	.421		19.50	11.80		31.30	40	
	4200	Russian olive, 900 mm to 1200 mm, B & B	B-17	75	.427		21	12.65	7.35	41	51	
	4400	Spirea, 900 mm to 1200 mm, B & B	B-1	70	.343		24	9.60		33.60	41.50	
	4500	Viburnum, 900 to 1200 mm, B & B	B-17	40	.800		25	23.50	13.80	62.30	79	
410	0010	**TREES** Deciduous, in prep. beds, balled & burlapped (B&B)										**410**
	0100	Ash, 50 mm caliper	B-17	8	4	Ea.	113	118	69	300	385	
	0200	Beech, 1500 mm to 1800 mm		50	.640		221	18.95	11.05	251	285	
	0300	Birch, 1860 mm to 2400 mm, 3 stems		20	1.600		122	47.50	27.50	197	237	
	0500	Crabapple, 1800 mm to 2400 mm		20	1.600		159	47.50	27.50	234	278	
	0600	Dogwood, 1200 mm to 1500 mm		40	.800		68	23.50	13.80	105.30	126	
	0700	Eastern redbud 1200 mm to 1500 mm		40	.800		133	23.50	13.80	170.30	198	
	0800	Elm, 2400 mm to 3000 mm		20	1.600		112	47.50	27.50	187	226	
	0900	Ginkgo, 1800 mm to 2100 mm		24	1.333		164	39.50	23	226.50	267	
	1000	Hawthorn, 2400 mm to 3000 mm, 25 mm caliper		20	1.600		125	47.50	27.50	200	241	
	1100	Honeylocust, 3000 mm to 3600 mm, 38 mm caliper		10	3.200		149	94.50	55.50	299	370	
	1300	Larch, 2400 mm		32	1		96.50	29.50	17.30	143.30	171	
	1400	Linden, 2400 mm to 3000 mm, 25 mm caliper		20	1.600		104	47.50	27.50	179	217	
	1500	Magnolia, 1200 mm to 1500 mm		20	1.600		71.50	47.50	27.50	146.50	182	
	1600	Maple, red, 2400 mm to 3000 mm, 38 mm caliper		10	3.200		159	94.50	55.50	309	380	
	1700	Mountain ash, 2400 mm to 3000 mm, 25 mm caliper		16	2		164	59	34.50	257.50	310	

Important: See the Reference Section for supporting data - Crews, Rental Equipment, City Cost Indexes and Reference Data

02930	Exterior Plants	CREW	DAILY OUTPUT	LABOR-HOURS	UNIT	2006 BARE COSTS				TOTAL INCL O&P	
						MAT.	LABOR	EQUIP.	TOTAL		
410	1800	Oak, 60 mm to 75 mm caliper	B-17	6	5.333	Ea.	248	158	92	498	615	**410**
	2100	Planetree, 2750 mm to 3350 mm, 32 mm caliper		10	3.200		102	94.50	55.50	252	320	
	2200	Plum, 1800 mm to 2400 mm, 25 mm caliper		20	1.600		90.50	47.50	27.50	165.50	203	
	2300	Poplar, 2750 mm to 3250 mm, 32 mm caliper		10	3.200		47	94.50	55.50	197	258	
	2500	Sumac, 600 mm to 900 mm		75	.427		22.50	12.65	7.35	42.50	52.50	
	2700	Tulip, 1500 mm to 2400 mm		40	.800		49.50	23.50	13.80	86.80	106	
	2800	Willow, 1800 mm to 2400 mm, 25 mm caliper	↓	20	1.600	↓	60.50	47.50	27.50	135.50	170	
430	0010	**TREES, DECIDUOUS** zones 2-6										**430**
	0100	Acer campestre, (Hedge Maple), zone 4, B&B										
	0110	1200 mm to 1500 mm				Ea.	51.50			51.50	56.50	
	0120	1500 mm to 1800 mm					62			62	68	
	0130	40 mm to 50 mm caliper					139			139	153	
	0140	50 mm to 65 mm caliper					156			156	171	
	0150	65 mm to 80 mm caliper				↓	184			184	202	
	0200	Acer ginnala, (Amur Maple), zone 2, container or B&B										
	0210	2400 mm to 3000 mm				Ea.	143			143	157	
	0220	3 m to 3.5 m					156			156	171	
	0230	3.5 m to 4 m				↓	189			189	208	
	0600	Acer platanoides, (Norway Maple), zone 4, B&B										
	0610	2400 mm to 3000 mm				Ea.	167			167	184	
	0620	40 mm to 50 mm caliper					124			124	136	
	0630	50 mm to 65 mm caliper					149			149	164	
	0640	65 mm to 80 mm caliper					183			183	202	
	0650	80 mm to 90 mm caliper					239			239	263	
	0660	Bare root, 2400 mm to 3000 mm					208			208	229	
	0670	3 m to 3.5 m					244			244	269	
	0680	3.5 m to 4 m				↓	174			174	191	
	0700	Acer platanoides columnare, (Column maple), zone 4, B&B										
	0710	50 mm to 65 mm caliper				Ea.	184			184	203	
	0720	65 mm to 80 mm caliper					229			229	252	
	0730	80 mm to 90 mm caliper					229			229	252	
	0740	90 mm mm to 100 mm caliper					283			283	310	
	0750	100 mm to 115 mm caliper					410			410	450	
	0760	115 mm to 125 mm caliper					575			575	630	
	0770	125 mm to 140 mm caliper				↓	770			770	845	
	0800	Acer rubrum, (Red Maple), zone 4, B&B										
	0810	40 mm to 50 mm caliper				Ea.	121			121	133	
	0820	50 mm to 65 mm caliper					159			159	175	
	0830	65 mm to 80 mm caliper					191			191	210	
	0840	Bare Root, 2400 mm to 3000 mm					153			153	169	
	0850	3 m to 3.5 m				↓	153			153	169	
	0900	Acer saccharum, (Sugar Maple), zone 3, B&B										
	0910	40 mm to 50 mm caliper				Ea.	127			127	139	
	0920	50 mm to 65 mm caliper					168			168	185	
	0930	65 mm to 80 mm caliper					176			176	194	
	0940	80 mm to 90 mm caliper					226			226	249	
	0950	90 mm mm to 100 mm caliper					279			279	305	
	0960	Bare root, 2400 mm to 3000 mm					140			140	154	
	0970	3 m to 3.5 m					210			210	231	
	0980	3.5 m to 4 m				↓	283			283	310	

02935	Plant Maintenance										
100	0010	**FERTILIZE**										**100**
	0100	Dry granular, 2 kg/100m², hand spread	1 Clab	2230	.004	m²	.02	.10		.12	.17	

02935 | Plant Maintenance

		CREW	DAILY OUTPUT	LABOR-HOURS	UNIT	2006 BARE COSTS				TOTAL INCL O&P
						MAT.	LABOR	EQUIP.	TOTAL	
100	0110	Push rotary	1 Clab	13006	.001	m²	.02	.02		.04	.05
	0112	Push rotary, per 100 m²	↓	130	.062	Ea.	2.06	1.69		3.75	4.89
	0120	Tractor towed spreader, 2.4 m	B-66	46452	.001	m²	.02	.01		.03	.03
	0130	3.7 m spreader		74322	.001		.02			.02	.03
	0140	Truck whirlwind spreader	↓	111484	.001		.02			.02	.02
	0180	Water soluble, hydro spread, 7.3 kg/1000m²	B-64	55742	.001	↓	.02	.01		.03	.04
300	0010	**MOWING**									
	1650	Mowing brush, tractor with rotary mower									
	1660	Light density	B-84	2044	.004	m²		.14	.11	.25	.34
	1670	Medium density		1208	.007			.24	.19	.43	.58
	1680	Heavy density	↓	836	.010			.35	.28	.63	.83
	4000	Lawn mowing, improved areas, 400 mm hand push	1 Clab	4459	.002			.05		.05	.08
	4050	Power mower, 450 to 550 mm		6039	.001			.04		.04	.06
	4100	550 to 750 mm		10219	.001			.02		.02	.03
	4150	750 to 800 mm	↓	13006	.001			.02		.02	.03
	4160	Riding mower, 900 to 1120 mm	B-66	27871	.001			.01	.01	.02	.03
	4170	1200 to 1470 mm	"	44593	.001	↓		.01		.01	.01
	4175	Mowing with tractor & attachments									
	4200	Cutter or sickle-bar, 1500 mm, rough terrain	B-66	19510	.001	m²		.01	.01	.02	.03
	4210	Cutter or sickle-bar, 1500 mm, smooth terrain	"	31587	.001	"		.01	.01	.02	.02
	4220	Drainage channel, 1500 mm sickle bar	↓	8.05	.994	km		35	22	57	77.50
	4250	Lawnmower, rotary type, sharpen (all sizes)	1 Clab	10	.800	Ea.		22		22	34
	4260	Repair or replace part		7	1.143	"		31.50		31.50	48.50
	5000	Edge trimming with weed whacker	↓	1756	.005	m		.12		.12	.19
410	0010	**TREE PRUNING**									
	0020	38 mm caliper	1 Clab	84	.095	Ea.		2.61		2.61	4.06
	0030	50 mm caliper		70	.114			3.13		3.13	4.87
	0040	64 mm caliper		50	.160			4.38		4.38	6.80
	0050	75 mm caliper	↓	30	.267			7.30		7.30	11.35
	0060	100 mm caliper, by hand	2 Clab	21	.762			21		21	32.50
	0070	Aerial lift equipment	B-85	38	1.053			31	18.30	49.30	67.50
	0100	150 mm caliper, by hand	2 Clab	12	1.333			36.50		36.50	57
	0110	Aerial lift equipment	B-85	20	2			59	35	94	129
	0200	230 mm caliper, by hand	2 Clab	7.50	2.133			58.50		58.50	91
	0210	Aerial lift equipment	B-85	12.50	3.200			94	55.50	149.50	207
	0300	300 mm caliper, by hand	2 Clab	6.50	2.462			67.50		67.50	105
	0310	Aerial lift equipment	B-85	10.80	3.704			109	64.50	173.50	239
	0400	450 mm caliper by hand	2 Clab	5.60	2.857			78.50		78.50	122
	0410	Aerial lift equipment	B-85	9.30	4.301			127	75	202	278
	0500	600 mm caliper, by hand	2 Clab	4.60	3.478			95.50		95.50	148
	0510	Aerial lift equipment	B-85	7.70	5.195			153	90.50	243.50	335
	0600	760 mm caliper, by hand	2 Clab	3.70	4.324			118		118	184
	0610	Aerial lift equipment	B-85	6.20	6.452			190	112	302	415
	0700	900 mm caliper, by hand	2 Clab	2.70	5.926			162		162	253
	0710	Aerial lift equipment	B-85	4.50	8.889			262	155	417	575
	0800	1200 mm caliper, by hand	2 Clab	1.70	9.412			258		258	400
	0810	Aerial lift equipment	B-85	2.80	14.286	↓		420	249	669	925
420	0010	**SHRUB PRUNING**									
	6700	Prune, shrub bed	1 Clab	650	.012	m²		.34		.34	.53
	6710	Shrub under 900 mm height		190	.042	Ea.		1.15		1.15	1.80
	6720	1200 mm height		90	.089			2.44		2.44	3.79
	6730	Over 1800 mm		50	.160			4.38		4.38	6.80
	7350	Prune trees from ground		20	.400			10.95		10.95	17.05
	7360	High work	↓	8	1	↓		27.50		27.50	42.50

02935 | Plant Maintenance

			CREW	DAILY OUTPUT	LABOR-HOURS	UNIT	2006 BARE COSTS MAT.	LABOR	EQUIP.	TOTAL	TOTAL INCL O&P	
500	0010	**WATERING**										**500**
	4900	Water lawn or planting bed with hose, 25 mm of water	1 Clab	1486	.005	m²		.15		.15	.23	
	4910	15 m soaker hoses, in place		7618	.001			.03		.03	.04	
	4920	18 m soaker hoses, in place		8268	.001	↓		.03		.03	.04	
	7500	Water trees or shrubs, under 25 mm caliper		32	.250	Ea.		6.85		6.85	10.65	
	7550	25 to 75 mm caliper		17	.471			12.90		12.90	20	
	7600	75 to 100 mm caliper		12	.667			18.25		18.25	28.50	
	7650	Over 100 mm caliper	↓	10	.800	↓		22		22	34	
	9000	For sprinkler irrigation systems, see Div. 02810										
610	0010	**WEEDING**										**610**
	0100	Weed planting bed	1 Clab	669	.012	m²		.33		.33	.51	

02945 | Planting Accessories

			CREW	DAILY OUTPUT	LABOR-HOURS	UNIT	2006 BARE COSTS MAT.	LABOR	EQUIP.	TOTAL	TOTAL INCL O&P	
120	0010	**EDGING**										**120**
	0050	Aluminum alloy, including stakes, 3 mm x 102 mm, mill finish	B-1	119	.202	m	7.95	5.65		13.60	17.55	
	0051	Black paint		119	.202		9.25	5.65		14.90	18.95	
	0052	Black anodized	↓	119	.202		10.70	5.65		16.35	20.50	
	0100	Brick, set horizontally, 5 bricks per m	D-1	113	.142		3.38	4.55		7.93	10.65	
	0150	Set vertically, 10 per m	"	41.15	.389		9.05	12.50		21.55	29	
	0200	Corrugated aluminum, roll, 100 mm wide	1 Carp	198	.040		1.21	1.44		2.65	3.59	
	0250	150 mm wide	"	168	.048		1.51	1.69		3.20	4.31	
	0600	Railroad ties, 152 mm x 203 mm	2 Carp	51.82	.309		8.85	11		19.85	27	
	0650	178 mm x 229 mm		41.45	.386		9.80	13.70		23.50	32.50	
	0750	51 mm x 102 mm	↓	101	.158		7.20	5.65		12.85	16.70	
	0800	Steel edge strips, incl. stakes, 8 mm x 125 mm	B-1	119	.202		11.90	5.65		17.55	22	
	0850	5 mm x 100 mm	"	119	.202	↓	9.40	5.65		15.05	19.15	
300	0010	**PLANTERS**										**300**
	0012	Conc., sandblasted, precast, 1200 mm diam., 600 mm high	2 Clab	15	1.067	Ea.	585	29		614	685	
	0100	Fluted, precast, 2100 mm diameter, 900 mm high	"	10	1.600	"	985	44		1,029	1,150	
	0280	Jersey barriers, conc, 1067mm H, for bldg perimeter security, excl frt	B-29	107	.523	m	130	15.65	7.85	153.50	176	
	0282	Jersey barr., plstc, wtr-filled, 1067mm H x 2438mm L x 610mm W bse	2 Clab	40	.400	Ea.	480	10.95		490.95	545	
	0300	Fiberglass, circular, 900 mm diameter, 600 mm high		15	1.067		415	29		444	505	
	0320	900 mm diameter, 675 mm high		12	1.333		330	36.50		366.50	420	
	0330	825 mm high		15	1.067		435	29		464	520	
	0335	600 mm diameter, 900 mm high		15	1.067		385	29		414	465	
	0340	1500 mm diameter, 975 mm high		8	2		990	55		1,045	1,175	
	0400	1500 mm diameter, 600 mm high		10	1.600		705	44		749	845	
	0600	Square, 600 mm side, 900 mm high		15	1.067		405	29		434	490	
	0610	600 mm side, 675 mm high		12	1.333		265	36.50		301.50	350	
	0620	600 mm side, 400 mm high		20	.800		293	22		315	355	
	0700	1200 mm side, 900 mm high		15	1.067		905	29		934	1,050	
	0900	Planter/bench, 1830 mm square, 900 mm high		5	3.200		1,125	87.50		1,212.50	1,350	
	1000	2440 mm square, 690 mm high		5	3.200		1,775	87.50		1,862.50	2,075	
	1200	Wood, square, 1200 mm side, 600 mm high		15	1.067		905	29		934	1,050	
	1300	Circular, 1200 mm diameter, 760 mm high		10	1.600		745	44		789	890	
	1500	1830 mm diameter, 760 mm high		10	1.600		1,325	44		1,369	1,525	
	1600	Planter/bench, 1800 mm	↓	5	3.200	↓	2,925	87.50		3,012.50	3,325	
510	0010	**TREE GUYING**										**510**
	0015	Tree guying Including stakes, guy wire and wrap										
	0100	Less than 75 mm caliper, 2 stakes	2 Clab	35	.457	Ea.	15.50	12.55		28.05	36.50	
	0200	75 mm to 100 mm caliper, 3 stakes	"	21	.762	"	18.30	21		39.30	52.50	

SITE CONSTRUCTION 2

02900 | Planting

02945	Planting Accessories	CREW	DAILY OUTPUT	LABOR-HOURS	UNIT	2006 BARE COSTS				TOTAL INCL O&P
						MAT.	LABOR	EQUIP.	TOTAL	
510 1000	Including arrowhead anchor, cable, turnbuckles and wrap									**510**
1100	Less than 75 mm caliper, 75 mm anchors	2 Clab	20	.800	Ea.	48.50	22		70.50	87.50
1200	75 mm to 150 mm caliper, 100 mm anchors		15	1.067		69.50	29		98.50	122
1300	150 mm caliper, 150 mm anchors		12	1.333		86	36.50		122.50	152
1400	200 mm caliper, 200 mm anchors	↓	9	1.778	↓	98.50	48.50		147	185

02950 | Site Restoration & Rehabilitation

02955	Restoration of Underground Piping	CREW	DAILY OUTPUT	LABOR-HOURS	UNIT	2006 BARE COSTS				TOTAL INCL O&P
						MAT.	LABOR	EQUIP.	TOTAL	
210 0010	**PIPE INTERNAL CLEANING & INSPECTION**									**210**
0100	Cleaning, pressure pipe systems									
0120	Pig method, lengths 300 m to 3000 m									
0140	100 mm diameter thru 600 mm diameter, minimum [4"-24"]				M				8.75	10.10
0160	Maximum				"				17.50	21
6000	Sewage/sanitary systems									
6100	Power rodder with header & cutters									
6110	Mobilization charge, minimum				Total				320	375
6120	Mobilization charge, maximum				"				750	850
6140	100 mm diameter [4"]				M				3.50	4.05
6150	150 mm diameter [6"]								4.40	5.10
6160	200 mm diameter [8"]								5.25	6.15
6170	250 mm diameter [10"]								6.15	7
6180	300 mm diameter [12"]								6.70	7.70
6190	350 mm diameter [14"]								7	8.10
6200	400 mm diameter [16"]								7.90	9.15
6210	450 mm diameter [18"]								8.75	10
6220	500 mm diameter [20"]								9.15	10.55
6230	600 mm diameter [24"]								10.15	11.75
6240	750 mm diameter [30"]								12.30	14
6250	900 mm diameter [36"]								14	16.15
6260	1200 mm diameter [48"]								15.10	17.40
6270	1500 mm diameter [60"]								19.65	22.50
6280	1800 mm diameter [72"]				↓				22	25.10
9000	Inspection, television camera with film									
9060	150 meters				Total				1,075	1,230
220 0010	**PIPEBURSTING**, 91.5 m runs, replace with HDPE pipe									**220**
0020	Not including excavation, backfill, shoring, or dewatering									
0100	150 mm to 375 mm diameter, minimum				M					262
0200	Maximum									525
0300	450 mm to 900 mm diameter, minimum									615
0400	Maximum				↓					1,050
0500	Mobilize and demobilize, minimum				Job					2,675
0600	Maximum				"					26,800
230 0010	**LINING PIPE** with cement, incl. bypass and cleaning									**230**
0020	Less than 3050 m, urban, 150 mm to 250 mm	C-17E	39.62	2.019	m	23.50	74.50	2.04	100.04	144
0050	250 mm to 300 mm		38.10	2.100		29	77.50	2.12	108.62	155
0070	300 mm to 400 mm		35.05	2.282		30	84	2.31	116.31	167
0100	400 mm to 500 mm		28.96	2.763		35	102	2.79	139.79	201
0200	600 mm to 900 mm	↓	27.43	2.916	↓	38	108	2.95	148.95	212

Important: See the Reference Section for supporting data - Crews, Rental Equipment, City Cost Indexes and Reference Data

02955	Restoration of Underground Piping	CREW	DAILY OUTPUT	LABOR-HOURS	UNIT	2006 BARE COSTS				TOTAL INCL O&P	
						MAT.	LABOR	EQUIP.	TOTAL		
230	0300	1200 mm to 1800 mm	C-17E	24.38	3.281	m	60	121	3.31	184.31	258	**230**
	0500	Rural, 150 mm to 250 mm		54.86	1.458		23.50	54	1.47	78.97	111	
	0550	250 mm to 300 mm		53.34	1.500		29	55.50	1.51	86.01	120	
	0570	300 mm to 400 mm		48.77	1.640		30.50	60.50	1.66	92.66	129	
	0600	400 mm to 500 mm		41.15	1.944		32	71.50	1.96	105.46	150	
	0700	600 mm to 900 mm		38.10	2.100		38.50	77.50	2.12	118.12	165	
	0800	1200 mm to 1800 mm		30.48	2.625		60	97	2.65	159.65	220	
	1000	Greater than 3050 m, urban, 150 mm to 250 mm		48.77	1.640		23.50	60.50	1.66	85.66	122	
	1050	250 mm to 300 mm		47.24	1.693		28.50	62.50	1.71	92.71	130	
	1070	300 mm to 400 mm		42.67	1.875		30	69	1.89	100.89	143	
	1100	400 mm to 500 mm		36.58	2.187		32	80.50	2.21	114.71	164	
	1200	600 mm to 900 mm		35.05	2.282		38.50	84	2.31	124.81	176	
	1300	1200 mm to 1800 mm		28.96	2.763		60	102	2.79	164.79	228	
	1500	Rural, 150 mm to 250 mm		65.53	1.221		23.50	45	1.23	69.73	97.50	
	1550	250 mm to 300 mm		64.01	1.250		29	46	1.26	76.26	105	
	1570	300 mm to 400 mm		56.39	1.419		30	52.50	1.43	83.93	116	
	1600	400 mm to 500 mm		45.72	1.750		32	64.50	1.77	98.27	137	
	1700	600 mm to 900 mm		42.67	1.875		38.50	69	1.89	109.39	152	
	1800	1200 mm to 1800 mm	▼	36.58	2.187	▼	60.50	80.50	2.21	143.21	195	
	2000	Cured in place pipe, non-pressure, flexible felt resin, 122 m runs										
	2100	150 mm diameter				M					105	
	2200	200 mm diameter									140	
	2300	250 mm diameter									193	
	2400	300 mm diameter									246	
	2500	380 mm diameter									335	
	2600	450 mm diameter									400	
	2700	530 mm diameter									475	
	2800	600 mm diameter									525	
	2900	750 mm diameter									670	
	3000	900 mm diameter									875	
	3100	1200 mm diameter									1,175	
	3200	1500 mm diameter									1,475	
	3300	1800 mm diameter				▼					1,750	
240	0010	**HDPE PIPE LINING** , excludes cleaning and video inspection										**240**
	0020	Pipe relined with one pipe size smaller than orig. (100 mm for 150 mm)										
	0100	150 mm diameter, original size	B-6B	183	.262	m	6.55	7.35	4.27	18.17	23.50	
	0150	200 mm diameter, original size		183	.262		9.40	7.35	4.27	21.02	26.50	
	0200	250 mm diameter, original size		183	.262		15.90	7.35	4.27	27.52	33.50	
	0250	300 mm diameter, original size		122	.393		24.50	11.05	6.40	41.95	51.50	
	0300	350 mm diameter, original size	▼	122	.393	▼	34.50	11.05	6.40	51.95	62.50	
310	0010	**CORROSION RESISTANCE**										**310**
	0012	Wrap & coat, add to pipe, 100 mm dia. [4"]				m	5.20			5.20	5.70	
	0020	125 mm diameter [5"]					5.40			5.40	5.95	
	0040	150 mm diameter [6"]					5.40			5.40	5.95	
	0060	200 mm diameter [8"]					8.40			8.40	9.25	
	0080	250 mm diameter [10"]					12.55			12.55	13.80	
	0100	300 mm diameter [12"]					12.55			12.55	13.80	
	0120	350 mm diameter [14"]					18.35			18.35	20	
	0140	400 mm diameter [16"]					18.35			18.35	20	
	0160	450 mm diameter [18"]					18.55			18.55	20.50	
	0180	500 mm diameter [20"]					19.40			19.40	21.50	
	0200	600 mm diameter [24"]				▼	24			24	26	

SITE CONSTRUCTION 2

02955	Restoration of Underground Piping	CREW	DAILY OUTPUT	LABOR-HOURS	UNIT	2006 BARE COSTS				TOTAL INCL O&P	
						MAT.	LABOR	EQUIP.	TOTAL		
310											**310**
0220	Small diameter pipe, 25 mm diameter, add [1"]				m	3.90			3.90	4.30	
0240	50 mm diameter [2"]					4.30			4.30	4.72	
0260	65 mm diameter [2-1/2"]					4.72			4.72	5.20	
0280	80 mm diameter [3"]					4.72			4.72	5.20	
0300	Fittings, field covered, add				m²	81.50			81.50	89.50	
0500	Coating, bituminous, per diameter mm, 1 coat, add				m	.95			.95	1.05	
0540	3 coat					1.48			1.48	1.64	
0560	Coal tar epoxy, per diameter mm, 1 coat, add					.56			.56	.62	
0600	3 coat					1.21			1.21	1.35	
1000	Polyethylene H.D. extruded, 0.64 mm thk., 13 mm diam. add					.85			.85	.95	
1020	20 mm diameter [3/4"]					.89			.89	.98	
1040	25 mm diameter [1"]					.98			.98	1.08	
1060	32 mm diameter [1-1/4"]					1.21			1.21	1.35	
1080	40 mm diameter [1-1/2"]					1.35			1.35	1.48	
1100	0.75 mm thk., 50 mm diameter [2"]					1.44			1.44	1.57	
1120	65 mm diameter [2-1/2"]					1.71			1.71	1.87	
1140	0.89 mm thk., 80 mm diameter [3"]					2.13			2.13	2.36	
1160	90 mm diameter [3-1/2"]					2.39			2.39	2.62	
1180	100 mm diameter [4"]					2.69			2.69	2.95	
1200	1 mm thk, 125 mm diameter [5"]					3.71			3.71	4.07	
1220	150 mm diameter [6"]					3.87			3.87	4.27	
1240	200 mm diameter [8"]					5.05			5.05	5.55	
1260	250 mm diameter [10"]					6.15			6.15	6.80	
1280	300 mm diameter [12"]					7.40			7.40	8.15	
1300	2 mm thk., 350 mm diameter [14"]					10.15			10.15	11.20	
1320	400 mm diameter [16"]					11.90			11.90	13.10	
1340	450 mm diameter [18"]					14.05			14.05	15.50	
1360	500 mm diameter [20"]					15.80			15.80	17.40	
1380	Fittings, field wrapped, add				m²	53.50			53.50	59	
400	0010	**PIPE REPAIR**									**400**
	0020	Not including excavation or backfill									
	0100	Clamp, stainless steel, lightweight, for steel pipe									
	0110	80 mm long, 13 mm diameter pipe [3"-1/2"]	1 Plum	34	.235	Ea.	11.30	10.05		21.35	27.50
	0120	20 mm diameter pipe [3/4"]		32	.250		11.65	10.70		22.35	29
	0130	25 mm diameter pipe [1"]		30	.267		12.50	11.40		23.90	31
	0140	32 mm diameter pipe [1-1/4"]		28	.286		13.30	12.20		25.50	33
	0150	40 mm diameter pipe [1-1/2"]		26	.308		13.70	13.15		26.85	35
	0160	50 mm diameter pipe [2"]		24	.333		15	14.25		29.25	38
	0170	65 mm diameter pipe [2-1/2"]		23	.348		16.30	14.85		31.15	40.50
	0180	80 mm diameter pipe [3"]		22	.364		18	15.55		33.55	43.50
	0190	90 mm diameter pipe [3-1/2"]		21	.381		19.30	16.25		35.55	45.50
	0200	100 mm diameter pipe [4"]	B-20	56	.429		20	13.35		33.35	43
	0210	125 mm diameter pipe [5"]		53	.453		23	14.10		37.10	47.50
	0220	150 mm diameter pipe [6"]		48	.500		25.50	15.55		41.05	52.50
	0230	200 mm diameter pipe [8"]		30	.800		30	25		55	71.50
	0240	250 mm diameter pipe [10"]		28	.857		77	26.50		103.50	127
	0250	300 mm diameter pipe [12"]		24	1		82	31		113	139
	0260	350 mm diameter pipe [14"]		22	1.091		85.50	34		119.50	147
	0270	400 mm diameter pipe [16"]		20	1.200		92.50	37.50		130	160
	0280	450 mm diameter pipe [18"]		18	1.333		97.50	41.50		139	172
	0290	500 mm diameter pipe [20"]		16	1.500		108	46.50		154.50	192
	0300	600 mm diameter pipe [24"]		14	1.714		120	53.50		173.50	215
	0360	For 150 mm long, add					100%	40%			
	0370	For 230 mm long, add					200%	100%			
	0380	For 300 mm long, add					300%	150%			

2

SITE CONSTRUCTION

02955	Restoration of Underground Piping	CREW	DAILY OUTPUT	LABOR-HOURS	UNIT	MAT.	LABOR	EQUIP.	TOTAL	TOTAL INCL O&P
						2006 BARE COSTS				
0390	For 450 mm long, add				Ea.	500%	200%			
0400	Pipe Freezing for live repairs of systems, each side									
0410	Note: Pipe Freezing can also used to install a valve into a live system									
0420	9 mm	2 Skwk	8	2	Ea.	520	73		593	685
0425	9 mm, second location same kit		8	2		19.75	73		92.75	136
0430	19 mm		8	2		445	73		518	605
0435	19 mm, second location same kit		8	2		19.75	73		92.75	136
0440	37 mm		6	2.667		445	97.50		542.50	640
0445	37 mm, second location same kit		6	2.667		19.75	97.50		117.25	173
0450	50 mm		6	2.667		660	97.50		757.50	875
0455	50 mm, second location same kit		6	2.667		19.75	97.50		117.25	173
0460	62 mm-75 mm		6	2.667		925	97.50		1,022.50	1,175
0465	62-75 mm, second loc same kit		6	2.667		39.50	97.50		137	195
0470	100 mm		4	4		1,500	146		1,646	1,875
0475	100 mm, second location same kit		4	4		62.50	146		208.50	296
0480	125-150 mm		4	4		3,625	146		3,771	4,200
0485	125-150 mm, 2nd location same kit		4	4		187	146		333	435
0490	extra 9 kg CO_2 cyl. (9-50 mm - 1 ea, 75 mm -2 ea)					197			197	216
0500	extra 22.5 kg CO_2 cyl (100 mm-2 ea, 125-150 mm-6 ea)					440			440	485
1000	Clamp, stainless steel, with threaded service tap									
1040	Full seal for iron, steel, PVC pipe									
1100	150 mm long, 50 mm diameter pipe [2"]	1 Plum	17	.471	Ea.	146	20		166	190
1110	65 mm diameter pipe [2-1/2"]		16	.500		154	21.50		175.50	202
1120	80 mm diameter pipe [3"]		15.60	.513		164	22		186	214
1130	90 mm diameter pipe [3-1/2"]		15	.533		171	23		194	223
1140	100 mm diameter pipe [4"]	B-20	40	.600		188	18.65		206.65	236
1150	150 mm diameter pipe [6"]		34	.706		218	22		240	273
1160	200 mm diameter pipe [8"]		21	1.143		271	35.50		306.50	355
1170	250 mm diameter pipe [10"]		20	1.200		300	37.50		337.50	390
1180	300 mm diameter pipe [12"]		17	1.412		345	44		389	445
1205	For 230 mm long, add					20%	45%			
1210	For 300 mm long, add					40%	80%			
1220	For 450 mm long, add					70%	110%			
1600	Clamp, stainless steel, single section									
1640	Full seal for iron, steel, PVC pipe									
1700	150 mm long, 50 mm diameter pipe [2"]	1 Plum	17	.471	Ea.	75.50	20		95.50	113
1710	65 mm diameter pipe [2-1/2"]		16	.500		82	21.50		103.50	123
1720	80 mm diameter pipe [3"]		15.60	.513		85.50	22		107.50	127
1730	90 mm diameter pipe [3-1/2"]		15	.533		94	23		117	139
1740	100 mm diameter pipe [4"]	B-20	40	.600		103	18.65		121.65	142
1750	150 mm diameter pipe [6"]		34	.706		128	22		150	175
1760	200 mm diameter pipe [8"]		21	1.143		171	35.50		206.50	244
1770	250 mm diameter pipe [10"]		20	1.200		197	37.50		234.50	275
1780	300 mm diameter pipe [12"]		17	1.412		214	44		258	305
1800	For 230 mm long, add					40%	45%			
1810	For 300 mm long, add					60%	80%			
1820	For 450 mm long, add					120%	110%			
2000	Clamp, stainless steel, two section									
2040	Full seal, for iron, steel, PVC pipe									
2100	150 mm long, 100 mm diameter pipe [4"]	B-20	24	1	Ea.	146	31		177	209
2110	150 mm diameter pipe [6"]		20	1.200		171	37.50		208.50	246
2120	200 mm diameter pipe [8"]		13	1.846		188	57.50		245.50	297
2130	250 mm diameter pipe [10"]		12	2		274	62		336	395
2140	300 mm diameter pipe [12"]		10	2.400		300	74.50		374.50	445
2200	230 mm long, 100 mm diameter pipe [4"]		16	1.500		188	46.50		234.50	280
2210	150 mm diameter pipe [6"]		13	1.846		214	57.50		271.50	325

400

SITE CONSTRUCTION 2

2 SITE CONSTRUCTION

02955 | Restoration of Underground Piping

		CREW	DAILY OUTPUT	LABOR-HOURS	UNIT	2006 BARE COSTS MAT.	LABOR	EQUIP.	TOTAL	TOTAL INCL O&P	
400	2220	200 mm diameter pipe [8"]	B-20	9	2.667	Ea.	223	83		306	375	**400**
	2230	250 mm diameter pipe [10"]		8	3		310	93.50		403.50	485	
	2240	300 mm diameter pipe [12"]		7	3.429		345	107		452	540	
	2250	350 mm diameter pipe [14"]		6.40	3.750		515	117		632	745	
	2260	400 mm diameter pipe [16"]		6	4		540	124		664	790	
	2270	450 mm diameter pipe [18"]		5	4.800		555	149		704	840	
	2280	500 mm diameter pipe [20"]		4.60	5.217		600	162		762	915	
	2290	600 mm diameter pipe [24"]		4	6		1,025	187		1,212	1,425	
	2320	For 300 mm long, add to 230 mm					15%	25%				
	2330	For 450 mm long, add to 230 mm					70%	55%				
	8000	For internal cleaning and inspection, see Div. 02955-210										
	8100	For pipe testing, see Div. 15955-700										

02960 | Flex. Pavement Surfacing Recovery

		CREW	DAILY OUTPUT	LABOR-HOURS	UNIT	MAT.	LABOR	EQUIP.	TOTAL	TOTAL INCL O&P	
100	0010	**PAVEMENT MILLING AND PAVEMENT COLD PLANING**										**100**
	5200	Cold planing & cleaning, 25 - 75 mm asph. pavmt., over 21 000 m²	B-71	5017	.011	m²		.35	1.07	1.42	1.72	
	5280	4200 to 8400 m²	"	3344	.017	"		.53	1.61	2.14	2.58	
	5300	Asphalt pavement removal from conc. base, no haul										
	5320	Rip, load & sweep 25 mm to 75 mm	B-70	6689	.008	m²		.27	.19	.46	.62	
	5330	75 mm to 150 mm deep	"	4181	.013			.42	.30	.72	.98	
	5340	Profile grooving, asphalt pavement load & sweep, 25 mm deep	B-71	10451	.005	m²		.17	.52	.69	.83	
	5350	75 mm deep		7525	.007			.24	.72	.96	1.15	
	5360	150 mm deep		4181	.013			.42	1.29	1.71	2.07	
	5400	Mixing material in windrow, 134 kW grader	B-11L	7187	.002	m³		.07	.06	.13	.18	
	5450	For cold laid asphalt pavement, see Div. 02740-400										

02965 | Flex./Bit. Pavement Recycling

		CREW	DAILY OUTPUT	LABOR-HOURS	UNIT	MAT.	LABOR	EQUIP.	TOTAL	TOTAL INCL O&P	
200	0010	**COLD IN-PLACE RECYCLED BITUMINOUS PAVEMENT COURSES**										**200**
	5000	Reclamation, pulverizing and blending with existing base										
	5040	Aggregate base, 100 mm thick pavement, over 12 500 m²	B-73	2007	.032	m²		1.07	2.02	3.09	3.85	
	5080	4200 to 12 500 m²		1839	.035			1.16	2.21	3.37	4.21	
	5120	200 mm thick pavement, over 12 500 m²		1839	.035			1.16	2.21	3.37	4.21	
	5160	4200 to 12 500 m²		1672	.038			1.28	2.43	3.71	4.62	
	5180	Add for mobilization and demobilization for crew B-73				Ea.				1,700	2,150	
400	0010	**HOT IN-PLACE RECYCLED BITUMINOUS PAVEMENT COURSES**										**400**
	5500	Recycle asphalt pavement at site										
	5520	Remove, rejuvenate and spread 100 mm deep	B-72	2090	.031	m²	3.05	.99	4.26	8.30	9.55	
	5521	150 mm deep	"	1672	.038	"	4.49	1.24	5.30	11.03	12.65	

02985 | Site Maintenance

		CREW	DAILY OUTPUT	LABOR-HOURS	UNIT	MAT.	LABOR	EQUIP.	TOTAL	TOTAL INCL O&P	
700	0010	**SITE MAINTENANCE**										**700**
	1550	General site work maintenance										
	1560	Clearing brush with brush saw & rake	1 Clab	472	.017	m²		.46		.46	.72	
	1570	By hand	"	234	.034			.94		.94	1.46	
	1580	With dozer, ball and chain, light clearing	B-11A	3073	.005			.17	.30	.47	.59	
	1590	Medium clearing	"	2600	.006			.20	.35	.55	.69	
	3000	Lawn maintenance										
	3010	Aerate lawn, 450 mm cultivating width, walk behind	A-1K	8826	.001	m²		.02		.02	.04	
	3040	1200 mm cultivating width	B-66	69677	.001	"					.01	
	3100	Edge lawn, by hand at walks	1 Clab	488	.016	m		.45		.45	.70	
	3150	At planting beds		213	.038			1.03		1.03	1.60	
	3200	Using gas powered edger at walks		2682	.003			.08		.08	.13	
	3250	At planting beds		732	.011			.30		.30	.47	
	3260	Vacuum, 760 mm gas, outdoors with hose	A-1L	29.26	.273	km		7.50	.77	8.27	12.50	

SITE CONSTRUCTION 2

02985 | Site Maintenance

		CREW	DAILY OUTPUT	LABOR-HOURS	UNIT	2006 BARE COSTS				TOTAL INCL O&P
						MAT.	LABOR	EQUIP.	TOTAL	
700 3400	Weed lawn, by hand	1 Clab	279	.029	m²		.79		.79	1.22 **700**
4500	Rake leaves or lawn, by hand, (cavex)	"	697	.011			.31		.31	.49
4510	Power rake	A-1L	4181	.002	↓		.05	.01	.06	.09
4700	Seeding lawn, see Division 02920-320									
4750	Sodding, see Division 02920-400									
5900	Road & walk maintenance									
5915	De-icing roads and walks									
5920	Calcium Chloride in truckload lots see Division 02340-700									
6000	Ice melting comp., 90% Calc. Chlor., effec. to -34°C									
6010	22 to 36 kg poly bags, med. applic. 0.1 kg/m², by hand	1 Clab	5574	.001	m²	.17	.04		.21	.25
6050	With hand operated rotary spreader		10219	.001		.17	.02		.19	.22
6100	Rock salt, med. applic. on road & walkway, by hand		5574	.001		.04	.04		.08	.11
6110	With hand operated rotary spreader	↓	10219	.001	↓	.04	.02		.06	.08
6600	Shrub maintenance									
6630	Mulch-see Div. 02910-500									
6640	Shrub bed fertilize dry granular 0.015 kg/m²	1 Clab	7897	.001	m²	.01	.03		.04	.05
6800	Weed, by handhoe		743	.011			.30		.30	.46
6810	Spray out		2973	.003			.07		.07	.11
6820	Spray after mulch	↓	4459	.002	↓		.05		.05	.08
6840	Shrub purchase and installation see Div. 02915, 02930									
7100	Tree maintenance									
7140	Clear and grub trees, see Div. 02230-100									
7160	Cutting and piling trees, see Div. 02230-300									
7200	Fertilize, tablets, slow release, 30 gram/tree	1 Clab	100	.080	Ea.	.34	2.19		2.53	3.78
7280	Guying, including stakes, guy wire & wrap, see Div. 02945-510									
7300	Planting, trees, Deciduous, in prep. beds, see Div. 02915-400									
7400	Removal, trees see Div. 02905-925									
7420	Pest control, spray	1 Clab	24	.333	Ea.	15.45	9.15		24.60	31
7430	Systemic	"	48	.167	"	14.35	4.57		18.92	23

02990 | Structure Moving

		CREW	DAILY OUTPUT	LABOR-HOURS	UNIT	MAT.	LABOR	EQUIP.	TOTAL	TOTAL INCL O&P
300 0010	**MOVING BUILDINGS** One day move, up to 7.2 m wide									**300**
0020	Reset on new foundation, patch & hook-up, average move				Total					9,300
0040	Wood or steel frame bldg., based on ground floor area	B-4	17.19	2.793	m²		78	19.10	97.10	142
0060	Masonry bldg., based on ground floor area	"	12.73	3.771	↓		105	26	131	192
0200	For 7.3 m to 13 m wide, add									15%
0220	For each additional day on road, add	B-4	1	48	Day		1,350	330	1,680	2,425
0240	Construct new basement, move building, 1 day									
0300	move, patch & hook-up, based on ground floor area	B-3	14.40	3.333	m²	69.50	98.50	121	289	360

For information about Means Estimating Seminars, see yellow pages 12 and 13 in back of book

For expanded coverage of these items see Means Site Work & Landscape Cost Data 2006

	CREW	DAILY OUTPUT	LABOR-HOURS	UNIT	2006 BARE COSTS				TOTAL INCL O&P
					MAT.	LABOR	EQUIP.	TOTAL	

Division 3
Concrete

Estimating Tips

General

- Carefully check all the plans and specifications. Concrete often appears on drawings other than structural drawings, including mechanical and electrical drawings for equipment pads. The cost of cutting and patching is often difficult to estimate. See Subdivisions 02220 and 03055 for demolition costs.
- Always obtain concrete prices from suppliers near the job site. A volume discount can often be negotiated depending upon competition in the area. Remember to add for waste, particularly for slabs and footings on grade.

03100 Concrete Forms & Accessories

- A primary cost for concrete construction is forming. Most jobs today are constructed with prefabricated forms. The selection of the forms best suited for the job and the total square meters of forms required for efficient concrete forming and placing are key elements in estimating concrete construction. Enough forms must be available for erection to make efficient use of the concrete placing equipment and crew.

- Concrete accessories for forming and placing depend upon the systems used. Study the plans and specifications to assure that all special accessory requirements have been included in the cost estimate such as anchor bolts, inserts and hangers.

03200 Concrete Reinforcement

- Ascertain that the reinforcing steel supplier has included all accessories, cutting, bending and an allowance for lapping, splicing and waste. A good rule of thumb is 10% for lapping, splicing and waste. Also, 10% waste should be allowed for welded wire fabric.

03300 Cast-in-Place Concrete

- When estimating structural concrete, pay particular attention to requirements for concrete additives, curing methods and surface treatments. Special consideration for climate, hot or cold, must be included in your estimate. Be sure to include requirements for concrete placing equipment and concrete finishing.

03400 Precast Concrete
03500 Cementitious Decks & Toppings

- The cost of hauling precast concrete structural members is often an important factor. For this reason, it is important to get a quote from the nearest supplier. It may become economically feasible to set up precasting beds on the site if the hauling costs are prohibitive.

Reference Numbers

Reference numbers are shown in bold squares at the beginning of some major classifications. These numbers refer to related items in the Reference Section. The reference information may be an estimating procedure, an alternate pricing method, or technical information.

Note: Not all subdivisions listed here necessarily appear in this publication.

03055 | Selective Demolition

		CREW	DAILY OUTPUT	LABOR-HOURS	UNIT	MAT.	LABOR	EQUIP.	TOTAL	TOTAL INCL O&P	
110	**0010 SELECTIVE CONCRETE DEMOLITION** R024119-10										110
	0012 Excludes saw cutting, torch cutting, loading or hauling										
	0050 Break up into small pieces, minimum reinforcing	B-9	18.35	2.180	m³		60.50	8	68.50	103	
	0060 Average reinforcing		12.23	3.270			91	12.05	103.05	155	
	0070 Maximum reinforcing	↓	6.12	6.539	↓		182	24	206	310	
	0150 Remove whole pieces, up to 2 tonnes per pc	E-18	36	1.111	Ea.		44	26.50	70.50	106	
	0160 2 - 5 tonnes per piece		30	1.333			53	31.50	84.50	127	
	0170 5 - 10 tonnes per piece		24	1.667			66	39.50	105.50	160	
	0180 10 - 15 tonnes per piece	↓	18	2.222			88	52.50	140.50	212	
	0250 Precast unit embedded in masonry, up to 0.03 m³	D-1	16	1			32		32	49	
	0260 0.03 - 0.06 m³		12	1.333			43		43	65.50	
	0270 0.06 - 0.14 m³		10	1.600			51.50		51.50	78.50	
	0280 0.14 - 0.28 m³	↓	8	2	↓		64.50		64.50	98	

03060 | Basic Concrete Materials

		CREW	DAILY OUTPUT	LABOR-HOURS	UNIT	MAT.	LABOR	EQUIP.	TOTAL	TOTAL INCL O&P	
100	**0010 CONCRETE ADMIXTURES & SURFACE TREATMENTS**										100
	0040 Abrasives, aluminum oxide, over 18 metric ton				kg	2.60			2.60	2.87	
	0070 Under 1 metric ton					2.82			2.82	3.11	
	0100 Silicon carbide, black, over 18 metric ton					3.62			3.62	3.97	
	0120 Under 1 metric ton				↓	3.90			3.90	4.30	
	0200 Air entraining agent, 20 to 40 gram/bag, 200 L lots				liter	2.37			2.37	2.61	
	0220 20 L lots					2.17			2.17	2.39	
	0300 Bonding agent, acrylic latex, 6.1 m²/L					4.70			4.70	5.15	
	0320 Epoxy resin, 2.0 m²/L				↓	10.25			10.25	11.30	
	0400 Calcium chloride, 22 kg bags				Met. Ton	455			455	505	
	0420 Less than truckload lots				Bag	29.50			29.50	32.50	
	0500 Carbon black, liquid, 0.91 to 3.63 kg/bag of cement				kg	9.40			9.40	10.35	
	0600 Colors, integral, 0.91 to 4.54 kg/bag of cement, minimum					5.10			5.10	5.60	
	0610 Average					6.90			6.90	7.55	
	0620 Maximum				↓	8.70			8.70	9.60	
	0700 Curing compound, solvent based, 10 m²/L, 200 L lots				liter	3.26			3.26	3.59	
	0720 20 L lots					3.75			3.75	4.13	
	0800 Water based, 23 m²/L, 200 L lots					3			3	3.30	
	0820 20 L lots					3.43			3.43	3.78	
	0900 Dustproofing compound, (4.9 m²/L), 200 L lots					1.01			1.01	1.11	
	0920 20 L lots				↓	1.51			1.51	1.66	
	1000 Epoxy dustproof coating, colors, (28 to 37 m²/coat),										
	1010 or transparent, (37 to 57 m²/coat)				liter	7.75			7.75	8.50	
	1100 Hardeners, metallic, 25 kg bags, natural (grey)				kg	1.54			1.54	1.70	
	1200 Colors					2.14			2.14	2.36	
	1300 Non-metallic, 25 kg bags, natural grey					.66			.66	.73	
	1320 Colors				↓	1.21			1.21	1.34	
	1550 Release agent, for tilt slabs				liter	2.44			2.44	2.68	
	1570 For forms, average					1.50			1.50	1.65	
	1600 Sealer, hardener and dustproofer, epoxy-based, 4.91m²/L, minimum					7.75			7.75	8.50	
	1620 Maximum					8.50			8.50	9.35	
	1630 Solvent-based, 6.14m²/L, minimum					3.26			3.26	3.59	
	1640 Maximum					3.75			3.75	4.13	
	1650 Water based, 8.60 m²/L, minimum					3			3	3.30	
	1660 Maximum					3.43			3.43	3.78	
	1700 Colors (7.4 to 9.8 m²/L)					12.90			12.90	14.20	
	1800 Set accelerator for below freezing, 5 to 7.4 L/m³					2.64			2.64	2.91	
	1900 Set retarder, 60 to 120 mL/bag of cement				↓	3.52			3.52	3.87	
	2000 Waterproofing, integral 0.454 kg/bag of cement				kg	2.16			2.16	2.38	
	2100 Powdered metallic, 2 kg/m², minimum				↓	2.69			2.69	2.95	

		03060	Basic Concrete Materials		CREW	DAILY OUTPUT	LABOR-HOURS	UNIT	2006 BARE COSTS				TOTAL INCL O&P	
									MAT.	LABOR	EQUIP.	TOTAL		
100	2120		Maximum					kg	5.40			5.40	5.90	100
110	0010	**AGGREGATE**		R033105 -20										110
	0020	Expanded shale, C.L. lots, 833 kg/m³, minimum						Met. Ton	39.50			39.50	43.50	
	0050	Maximum		R033105 -30				"	53			53	58	
	0100	Lightweight vermiculite or perlite, 0.11 m³ bag, C.L. lots						Bag	6.80			6.80	7.50	
	0150	L.C.L. lots		R033105 -40				"	7.50			7.50	8.25	
	0250	Sand & stone, loaded at pit, crushed bank gravel						Met. Ton	21			21	23	
	0350	Sand, washed, for concrete		R033105 -50					28			28	31	
	0400	For plaster or brick							14.10			14.10	15.50	
	0450	Stone, 20 mm to 40 mm							20.50			20.50	22.50	
	0500	10 mm roofing stone & 13 mm pea stone							17.55			17.55	19.30	
	0550	For trucking 16 km (8 km R/T), add to the above			B-34B	106	.075			2.14	4.50	6.64	8.20	
	0600	48 km (24 km R/T), add to the above			"	65.32	.122	↓		3.47	7.30	10.77	13.35	
	0850	Sand & stone, loaded at pit, crushed bank gravel						m³	27.50			27.50	30.50	
	0950	Sand, washed, for concrete							16.95			16.95	18.60	
	1000	For plaster or brick							48.50			48.50	53.50	
	1050	Stone, 20 mm to 40 mm							30			30	33	
	1100	10 mm roofing stone & 13 mm pea stone							43.50			43.50	48	
	1150	For trucking 16 km (8 km R/T), add to the above			B-34B	59.64	.134			3.80	8	11.80	14.60	
	1200	48 km (24 km R/T), add to the above			"	36.70	.218	↓		6.20	13	19.20	24	
	1310	Quartz chips, 22.68 kg bags						Met. Ton	375			375	410	
	1330	Silica chips, 22.68 kg bags							206			206	226	
	1410	White marble, 10 mm to 13 mm, 22.68 kg bags							310			310	340	
	1430	20 mm		↓				↓	78			78	86	
200	0010	**CEMENT** Material only		R033105 -20										200
	0240	Portland, Type I/II, TL lots, 68 kg bags						Bag	8.25			8.25	9.10	
	0250	LTL/LCL lots		R033105 -30				"	10.45			10.45	11.50	
	0300	Trucked in bulk, per 45 kg						Met. Ton	97.50			97.50	107	
	0400	Type III, high early strength, TL lots, 68 kg bags		R033105 -40				Bag	9.50			9.50	10.45	
	0420	L.T.L. or L.C.L. lots							10.30			10.30	11.35	
	0500	White, type III, high early strength, T.L. or C.L. lots, bags		R033105 -50					21.50			21.50	24	
	0520	L.T.L. or L.C.L. lots							23			23	25	
	0600	White, type I, T.L. or C.L. lots, bags							20.50			20.50	22.50	
	0620	L.T.L. or L.C.L. lots		↓				↓	21.50			21.50	23.50	
210	0010	**CRIBBING** See under Retaining Walls, division 02830-100												210
220	0010	**CUTTING** Concrete see division 02220-360												220
250	0010	**DAMPPROOFING** See division 07100												250
300	0010	**EQUIPMENT** For placing concrete see Reference Section		R015433 -10										300
400	0010	**LIFT SLAB** See division 03310-240		R033105 -85										400
700	0010	**SAWING CONCRETE** See division 02220-360												700
850	0010	**WATERPROOFING AND DAMPPROOFING** See division 07100												850
	0050	Integral waterproofing, add to cost of regular concrete						m³	7.70			7.70	8.45	
870	0010	**WINTER PROTECTION**												870
	0012	For heated ready mix, add, minimum						m³	6.20			6.20	6.80	
	0050	Maximum						"	7.75			7.75	8.50	
	0100	Temporary heat to protect concrete, 24 hours, minimum			2 Clab	4645	.003	m²	1.17	.09		1.26	1.43	
	0150	Maximum			"	2323	.007	"	1.52	.19		1.71	1.96	
	0200	Temporary shelter for slab on grade, wood frame/polyethylene sheeting												

CONCRETE 3

03060 | Basic Concrete Materials

			CREW	DAILY OUTPUT	LABOR-HOURS	UNIT	MAT.	LABOR	EQUIP.	TOTAL	TOTAL INCL O&P	
							2006 BARE COSTS					
870	0201	Build or remove, minimum	2 Carp	929	.017	m²	3.07	.61		3.68	4.32	870
	0210	Maximum	"	279	.057	"	3.69	2.04		5.73	7.20	
	0300	See Also Division 03390-200										

03100 | Concrete Forms & Accessories

03110 | Structural C.I.P. Forms

			CREW	DAILY OUTPUT	LABOR-HOURS	UNIT	MAT.	LABOR	EQUIP.	TOTAL	TOTAL INCL O&P	
							2006 BARE COSTS					
300	0010	**EXPANSION JOINT** See division 03150-250										300
405	0010	**FORMS IN PLACE, BEAMS AND GIRDERS**										405
	0020	See also Elevated Slabs, Division 03110-420										
	0500	Exterior spandrel, job-built plywood, 300 mm wide, 1 use	C-2	20.90	2.296	m²CA	40.50	79.50		120	168	
	0550	2 use		25.55	1.879		21.50	65		86.50	125	
	0600	3 use		27.41	1.751		16.15	60.50		76.65	112	
	0650	4 use		28.80	1.667		13.15	57.50		70.65	104	
	1000	450 mm wide, 1 use		23.23	2.067		34.50	71.50		106	149	
	1050	2 use		25.55	1.879		19.05	65		84.05	122	
	1100	3 use		28.33	1.694		13.90	58.50		72.40	106	
	1150	4 use		29.26	1.640		11.30	56.50		67.80	101	
	1500	600 mm wide, 1 use		24.62	1.950		32	67.50		99.50	140	
	1550	2 use		26.94	1.782		17.85	61.50		79.35	116	
	1600	3 use		29.26	1.640		12.70	56.50		69.20	102	
	1650	4 use		30.19	1.590		10.35	55		65.35	97	
	2000	Interior beam, job-built plywood, 300 mm wide, 1 use		27.87	1.722		44	59.50		103.50	141	
	2050	2 use		31.59	1.520		22	52.50		74.50	106	
	2100	3 use		33.82	1.419		17.65	49		66.65	96	
	2150	4 use		35.02	1.371		14.30	47.50		61.80	89.50	
	2500	600 mm wide, 1 use		29.73	1.615		32.50	56		88.50	123	
	2550	2 use		33.91	1.416		18.20	49		67.20	96	
	2600	3 use		35.77	1.342		12.90	46.50		59.40	86	
	2650	4 use		36.70	1.308		10.45	45		55.45	82	
	3000	Encasing steel beam, hung, job-built plywood, 1 use		30.19	1.590		31.50	55		86.50	121	
	3050	2 use		36.23	1.325		17.45	46		63.45	90	
	3100	3 use		38.55	1.245		12.70	43		55.70	81	
	3150	4 use		39.95	1.202		10.35	41.50		51.85	76	
	3500	Bottoms only, to 750 mm wide, job-built plywood, 1 use		21.37	2.246		48	77.50		125.50	174	
	3550	2 use		24.62	1.950		27	67.50		94.50	135	
	3600	3 use		26.01	1.845		19.25	63.50		82.75	120	
	3650	4 use		26.94	1.782		15.70	61.50		77.20	113	
	4000	Sides only, vertical, 900 mm high, job-built plywood, 1 use		31.12	1.542		50.50	53.50		104	139	
	4050	2 use		37.62	1.276		28	44		72	99	
	4100	3 use		39.95	1.202		20	41.50		61.50	86.50	
	4150	4 use		41.34	1.161		16.35	40		56.35	80.50	
	4500	Sloped sides, 900 mm high, 1 use		28.33	1.694		49.50	58.50		108	146	
	4550	2 use		34.37	1.396		27.50	48		75.50	106	
	4600	3 use		37.62	1.276		19.80	44		63.80	90	
	4650	4 use		39.48	1.216		16.05	42		58.05	83	
	5000	Upstanding beams, 900 mm high, 1 use		20.90	2.296		60.50	79.50		140	190	
	5050	2 use		23.69	2.026		34	70		104	146	

03110	Structural C.I.P. Forms		CREW	DAILY OUTPUT	LABOR-HOURS	UNIT	2006 BARE COSTS				TOTAL INCL O&P	
							MAT.	LABOR	EQUIP.	TOTAL		
405	5100	3 use		C-2	25.55	1.879	m²CA	24.50	65		89.50	128	**405**
	5150	4 use		↓	26.01	1.845	↓	19.80	63.50		83.30	121	
410	0010	**FORMS IN PLACE, COLUMNS**	R031113 -40										**410**
	0500	Round fiberglass, 4 use/mo., rent, 300 mm diameter		C-1	48.77	.656	m	21	22		43	57	
	0550	400 mm diameter	R031113 -60		45.72	.700		25	23.50		48.50	64	
	0600	450 mm diameter			42.67	.750		28	25		53	69.50	
	0650	600 mm diameter			41.15	.778		34.50	26		60.50	78.50	
	0700	700 mm diameter			39.62	.808		38.50	27		65.50	84.50	
	0800	760 mm diameter			38.10	.840		40.50	28		68.50	88.50	
	0850	900 mm diameter			36.58	.875		54	29.50		83.50	105	
	1500	Round fiber tube, 1 use, 200 mm diameter			47.24	.677		4.95	22.50		27.45	41	
	1550	250 mm diameter			47.24	.677		6.35	22.50		28.85	42.50	
	1600	300 mm diameter			45.72	.700		7.60	23.50		31.10	45	
	1650	350 mm diameter			44.20	.724		9.95	24.50		34.45	49	
	1700	400 mm diameter			42.67	.750		11.60	25		36.60	52	
	1750	500 mm diameter			41.15	.778		18.45	26		44.45	61	
	1800	600 mm diameter			39.62	.808		24	27		51	68.50	
	1850	750 mm diameter			38.10	.840		34	28		62	81.50	
	1900	900 mm diameter			35.05	.913		43	30.50		73.50	95	
	1950	1050 mm diameter			30.48	1.050		104	35		139	169	
	2000	1200 mm diameter		↓	25.91	1.235	↓	132	41.50		173.50	210	
	2200	For seamless type, add						15%					
	3000	Round, steel, 4 use/mo., rent, regular duty, 300 mm diameter [12"]		C-1	44.20	.724	m	34.50	24.50		59	75.50	
	3050	400 mm diameter [16"]			38.10	.840		38.50	28		66.50	86.50	
	3100	Heavy duty, 500 mm diameter [20"]			32	1		42.50	33.50		76	98.50	
	3150	600 mm diameter [24"]			25.91	1.235		46.50	41.50		88	116	
	3200	760 mm diameter [30"]			21.34	1.500		53.50	50.50		104	137	
	3250	900 mm diameter [36"]			18.29	1.750		57	58.50		115.50	154	
	3300	1200 mm diameter [48"]			15.24	2.100		78.50	70.50		149	197	
	3350	1500 mm diameter [60"]			13.72	2.333	↓	104	78		182	237	
	4000	Column capitals, steel, 4 use/mo., 600 mm col, 1200 mm cap			12	2.667	Ea.	198	89.50		287.50	355	
	4050	1500 mm cap diameter			11	2.909		226	97.50		323.50	400	
	4100	1800 mm cap diameter			10	3.200		310	107		417	505	
	4150	2100 mm cap diameter		↓	9	3.556	↓	380	119		499	600	
	4500	For second and succeeding months, deduct						50%					
	5000	Job-built plywood, 200 mm x 200 mm columns, 1 use		C-1	15.33	2.088	m²CA	23.50	70		93.50	135	
	5050	2 use			18.12	1.766		13.45	59		72.45	107	
	5100	3 use			19.51	1.640		9.35	55		64.35	96	
	5150	4 use			19.97	1.602		7.75	53.50		61.25	92	
	5500	300 mm x 300 mm columns, 1 use			16.72	1.914		23.50	64		87.50	126	
	5550	2 use			19.51	1.640		12.90	55		67.90	99.50	
	5600	3 use			20.44	1.566		9.35	52.50		61.85	92	
	5650	4 use			20.90	1.531		7.65	51.50		59.15	88.50	
	6000	400 mm x 400 mm columns, 1 use			17.19	1.862		24	62.50		86.50	124	
	6050	2 use			19.97	1.602		12.60	53.50		66.10	97.50	
	6100	3 use			21.37	1.498		9.60	50		59.60	88.50	
	6150	4 use			21.83	1.466		7.85	49		56.85	85	
	6500	600 mm x 600 mm columns, 1 use			17.65	1.813		27.50	61		88.50	125	
	6550	2 use			20.07	1.595		15.05	53.50		68.55	99.50	
	6600	3 use			21.37	1.498		11	50		61	90	
	6650	4 use			22.11	1.447		8.95	48.50		57.45	85.50	
	7000	900 mm x 900 mm columns, 1 use			18.58	1.722		24.50	57.50		82	117	
	7050	2 use			21.37	1.498		13.65	50		63.65	93	
	7100	3 use			22.76	1.406		9.70	47		56.70	84.50	
	7150	4 use			23.23	1.378		7.95	46		53.95	80.50	
	7500	Steel framed plywood, 4 use/mo., rent, 200 mm x 200 mm		↓	31.59	1.013	↓	35	34		69	91.50	

For expanded coverage of these items see *Means Concrete & Masonry Cost Data 2006*

03110 | Structural C.I.P. Forms

		CREW	DAILY OUTPUT	LABOR-HOURS	UNIT	MAT.	LABOR	EQUIP.	TOTAL	TOTAL INCL O&P	
410	7550	250 mm x 250 mm R031113-40	C-1	32.52	.984	m²CA	26.50	33		59.50	80.50	**410**
	7600	300 mm x 300 mm		34.37	.931		33.50	31		64.50	85.50	
	7650	400 mm x 400 mm R031113-60		37.16	.861		36.50	29		65.50	85.50	
	7700	500 mm x 500 mm		39.02	.820		17.65	27.50		45.15	62.50	
	7750	600 mm x 600 mm		40.88	.783		16.60	26		42.60	59	
	7755	750 mm x 750 mm		40.88	.783		14.85	26		40.85	57.50	
415	0010	**FORMS IN PLACE, CULVERT** R031113-40										**415**
	0015	1.5 m to 2.4 m square or rect, 1 use	C-1	15.79	2.026	m²CA	56.50	68		124.50	168	
	0050	2 use R031113-60		16.72	1.914		19.50	64		83.50	122	
	0100	3 use		17.65	1.813		15.40	61		76.40	111	
	0150	4 use		18.58	1.722		13.35	57.50		70.85	105	
420	0010	**FORMS IN PLACE, ELEVATED SLABS** R031113-40										**420**
	0050	See also corrugated form deck, Division 05310-300										
	1000	Flat plate, job-built plywood, to 4.6 m high, 1 use R031113-60	C-2	43.66	1.099	m²	46.50	38		84.50	110	
	1050	2 use		48.31	.994		25.50	34.50		60	81.50	
	1100	3 use		50.63	.948		18.50	32.50		51	71.50	
	1150	4 use		52.02	.923		15.05	32		47.05	66	
	1500	4.6 m to 6.1 m high ceilings, 4 use		45.99	1.044		17.85	36		53.85	75.50	
	1600	6.2 m to 11 m high ceilings, 4 use		41.81	1.148		25	39.50		64.50	89	
	2000	Flat slab, drop panels, job-built plywood, to 4.6 m high, 1 use		41.71	1.151		51.50	39.50		91	119	
	2050	2 use		47.29	1.015		28.50	35		63.50	85.50	
	2100	3 use		49.42	.971		20.50	33.50		54	74.50	
	2150	4 use		50.54	.950		16.80	33		49.80	69.50	
	2250	4.6 m to 6.1 m high ceilings, 4 use		44.59	1.076		45	37		82	108	
	2350	6.1 m to 11 m high ceilings, 4 use		40.41	1.188		52.50	41		93.50	122	
	3000	Floor slab hung from steel beams, 1 use		45.06	1.065		25	37		62	84.50	
	3050	2 use		49.70	.966		17.75	33.50		51.25	71.50	
	3100	3 use		51.10	.939		15.40	32.50		47.90	67.50	
	3150	4 use		52.49	.914		14.20	31.50		45.70	64.50	
	3500	Floor slab, with 500 mm metal pans, 1 use		38.55	1.245		65.50	43		108.50	139	
	3550	2 use		41.34	1.161		42.50	40		82.50	109	
	3600	3 use		44.13	1.088		34.50	37.50		72	96.50	
	3650	4 use		46.45	1.033		31	35.50		66.50	89.50	
	3700	Floor slab with 750 mm pans, 1 use		38.83	1.236		65	42.50		107.50	138	
	3720	2 use		42.27	1.136		41.50	39		80.50	107	
	3740	3 use		43.66	1.099		34	38		72	96	
	3760	4 use		44.59	1.076		30	37		67	91	
	3801	Rent 760 mm metal pans, 1 use					52.50			52.50	58	
	3821	2 use					52.50			52.50	58	
	3841	3 use					52.50			52.50	58	
	3861	4 use					52.50			52.50	58	
	4000	Floor slab with 475 mm metal domes, 1 use	C-2	37.62	1.276		69.50	44		113.50	145	
	4050	2 use		40.41	1.188		46	41		87	115	
	4100	3 use		43.20	1.111		38.50	38.50		77	102	
	4150	4 use		45.99	1.044		34.50	36		70.50	94	
	4500	With 760 mm fiberglass domes, 1 use		37.62	1.276		67.50	44		111.50	143	
	4520	2 use		41.81	1.148		44.50	39.50		84	111	
	4530	3 use		42.73	1.123		36.50	39		75.50	101	
	4550	4 use		43.66	1.099		32.50	38		70.50	95	
	4701	Rent 760 mm fiberglass domes, 1 use					35.50			35.50	39	
	4721	2 use					35.50			35.50	39	
	4741	3 use					35.50			35.50	39	
	4761	4 use					35.50			35.50	39	
	5000	Box out for slab openings, over 400 mm deep, 1 use	C-2	17.65	2.719	m²CA	34.50	94		128.50	184	
	5050	2 use		22.30	2.153	"	19.05	74.50		93.55	137	

03110 | Structural C.I.P. Forms

			CREW	DAILY OUTPUT	LABOR-HOURS	UNIT	2006 BARE COSTS				TOTAL INCL O&P	
							MAT.	LABOR	EQUIP.	TOTAL		
420	5500	Shallow slab box outs, to 1 m² R031113-40	C-2	42	1.143	Ea.	10.70	39.50		50.20	73.50	**420**
	5550	Over 1 m² (use perimeter)		183	.262	m	4.69	9.05		13.74	19.25	
	6000	Bulkhead forms for slab, with keyway, 1 use, 2 piece R031113-60		152	.316		5.20	10.90		16.10	22.50	
	6100	3 piece (see also edge forms)		140	.343		5.70	11.85		17.55	24.50	
	6200	Bulkhead form, 114 mm high, exp metal, keyway & stakes	C-1	366	.087		2.76	2.93		5.69	7.60	
	6210	140 mm high		335	.096		2.95	3.20		6.15	8.25	
	6215	190 mm high		293	.109		3.54	3.66		7.20	9.60	
	6220	240 mm high		256	.125		6.15	4.19		10.34	13.30	
	6500	Curb forms, wood, 150 mm to 300 mm H, on elevated slabs, 1 use		16.72	1.914	m²CA	16.60	64		80.60	118	
	6550	2 use		19.04	1.680		9.15	56.50		65.65	97.50	
	6600	3 use		20.44	1.566		6.65	52.50		59.15	89	
	6650	4 use		20.90	1.531		5.40	51.50		56.90	86	
	7000	Edge forms to 150 mm high, on elevated slab, 4 use		152	.211	m	.62	7.05		7.67	11.70	
	7070	175 mm to 300 mm high, 1 use		15.05	2.126	m²CA	12.60	71.50		84.10	125	
	7080	2 use		18.39	1.740		6.90	58.50		65.40	98.50	
	7090	3 use		20.62	1.552		5.05	52		57.05	86.50	
	7101	4 use		32.52	.984		4.09	33		37.09	56	
	7500	Depressed area forms to 300 mm high, 4 use		91.44	.350	m	2.36	11.75		14.11	21	
	7550	300 mm to 600 mm high, 4 use		53.34	.600		3.18	20		23.18	35	
	8000	Perimeter deck and rail for elevated slabs, straight		27.43	1.167		48	39		87	114	
	8050	Curved		19.81	1.615		65.50	54		119.50	157	
	8500	Void forms, round fiber, 75 mm diameter		137	.234		2.10	7.85		9.95	14.50	
	8550	100 mm diameter		130	.246		2.46	8.25		10.71	15.55	
	8600	150 mm diameter		122	.262		3.71	8.80		12.51	17.75	
	8650	200 mm diameter		114	.281		5.95	9.40		15.35	21	
	8700	250 mm diameter		107	.299		7.70	10		17.70	24	
	8750	300 mm diameter		91.44	.350		9.95	11.75		21.70	29	
425	0010	**FORMS IN PLACE, EQUIPMENT FOUNDATIONS** job built R031113-40										**425**
	0020	1 use	C-2	14.86	3.229	m²CA	27.50	112		139.50	204	
	0050	2 use R031113-60		17.65	2.719		15.05	94		109.05	163	
	0100	3 use		18.58	2.583		11	89		100	151	
	0150	4 use		19.04	2.520		8.95	87		95.95	145	
430	0010	**FORMS IN PLACE, FOOTINGS** R031113-40										**430**
	0020	Continuous wall, plywood, 1 use	C-1	34.84	.919	m²CA	28	31		59	79	
	0050	2 use R031113-60		40.88	.783		15.50	26		41.50	58	
	0100	3 use		43.66	.733		11.30	24.50		35.80	50.50	
	0150	4 use		45.06	.710		9.15	24		33.15	47	
	0500	Dowel supports for footings or beams, 1 use		152	.211	m	2.76	7.05		9.81	14	
	1000	Integral starter wall, to 100 mm high, 1 use		122	.262		2.92	8.80		11.72	16.90	
	1500	Keyway, 4 use, tapered wood, 50 mm x 100 mm	1 Carp	162	.049		.66	1.76		2.42	3.45	
	1550	50 mm x 150 mm		152	.053		.98	1.87		2.85	3.99	
	2000	Tapered plastic, 50 mm x 75 mm		162	.049		2.10	1.76		3.86	5.05	
	2050	50 mm x 100 mm		152	.053		2.69	1.87		4.56	5.85	
	2250	For keyway hung from supports, add		45.72	.175		2.76	6.20		8.96	12.70	
	3000	Pile cap, square or rectangular, job-built plywood, 1 use	C-1	26.94	1.188	m²CA	26.50	40		66.50	91	
	3050	2 use		32.14	.996		14.55	33.50		48.05	68	
	3100	3 use		34.47	.928		10.55	31		41.55	60	
	3150	4 use		35.58	.899		8.60	30		38.60	56.50	
	4000	Triangular or hexagonal, 1 use		20.90	1.531		31	51.50		82.50	114	
	4050	2 use		26.01	1.230		17.10	41		58.10	83	
	4100	3 use		28.33	1.129		12.50	38		50.50	72.50	
	4150	4 use		29.26	1.094		10.10	36.50		46.60	68	
	5000	Spread footings, job-built lumber, 1 use		28.33	1.129		19.70	38		57.70	80.50	
	5050	2 use		34.47	.928		10.85	31		41.85	60.50	

CONCRETE 3

03110 | Structural C.I.P. Forms

		03110 Structural C.I.P. Forms	Ref.	Crew	Daily Output	Labor-Hours	Unit	2006 Bare Costs Mat.	Labor	Equip.	Total	Total Incl O&P	
430	5100	3 use	R031113-40	C-1	37.25	.859	m²CA	7.85	29		36.85	53.50	430
	5150	4 use			38.46	.832		6.35	28		34.35	50.50	
	6000	Support for dowels, plinths or templates, 600 mm x 600 mm ftg			25	1.280	Ea.	5.45	43		48.45	73	
	6050	1200 mm x 1200 mm footing			22	1.455		10.95	48.50		59.45	88	
	6100	2400 mm x 2400 mm footing	R031113-60		20	1.600		22	53.50		75.50	108	
	6150	3650 mm x 3650 mm footing			17	1.882		35	63		98	137	
	7000	Plinths, job-built plywood, 1 use			23.23	1.378	m²CA	39	46		85	115	
	7100	4 use			25.08	1.276	"	12.80	43		55.80	80.50	
435	0010	**FORMS IN PLACE, GRADE BEAM**	R031113-40										435
	0020	Job-built plywood, 1 use		C-2	49.24	.975	m²CA	18.75	33.50		52.25	73	
	0050	2 use	R031113-60		53.88	.891		10.35	31		41.35	59.50	
	0100	3 use			55.74	.861		7.55	29.50		37.05	55	
	0150	4 use			56.20	.854		6.15	29.50		35.65	52.50	
440	0010	**FORMS IN PLACE, MAT FOUNDATION**	R031113-40										440
	0020	Job-built plywood, 1 use		C-2	26.94	1.782	m²CA	27.50	61.50		89	126	
	0050	2 use	R031113-60		28.80	1.667		10.85	57.50		68.35	101	
	0100	3 use			30.66	1.566		7.65	54		61.65	92.50	
	0120	4 use			32.52	1.476		6.35	51		57.35	86.50	
445	0010	**FORMS IN PLACE, SLAB ON GRADE**	R031113-40										445
	1000	Bulkhead forms w/keyway, wood, 150 mm high, 1 use		C-1	155	.206	m	3.18	6.90		10.08	14.25	
	1050	2 uses	R031113-60		122	.262		1.74	8.80		10.54	15.65	
	1100	4 uses			107	.299		1.05	10		11.05	16.75	
	1400	Bulkhead form, 114 mm high, exp metal, incl keyway & stakes			366	.087		2.76	2.93		5.69	7.60	
	1410	140 mm high			335	.096		2.95	3.20		6.15	8.25	
	1420	190 mm high	R031113-80		293	.109		3.54	3.66		7.20	9.60	
	1430	240 mm high			256	.125		6.15	4.19		10.34	13.30	
	2000	Curb forms, wood, 150 mm to 300 mm high, on grade, 1 use			19.97	1.602	m²CA	24	53.50		77.50	110	
	2050	2 use			23.23	1.378		13.25	46		59.25	86.50	
	2100	3 use			24.62	1.300		9.60	43.50		53.10	78.50	
	2150	4 use			25.55	1.253		7.75	42		49.75	74	
	3000	Edge forms, wood, 4 use, on grade, to 150 mm high			183	.175	m	.95	5.85		6.80	10.15	
	3050	175 mm to 300 mm high			40.41	.792	m²CA	8.20	26.50		34.70	50.50	
	3060	Over 300 mm			32.52	.984	"	9.05	33		42.05	61.50	
	3500	For depressed slabs, 4 use, to 300 mm high			91.44	.350	m	2.26	11.75		14.01	20.50	
	3550	To 600 mm high			53.34	.600		2.99	20		22.99	35	
	4000	For slab blockouts, to 300 mm high, 1 use			60.96	.525		2.43	17.60		20.03	30	
	4050	To 600 mm high, 1 use			36.58	.875		3.08	29.50		32.58	49	
	4100	Plastic (extruded), to 150 mm high, multiple use, on grade			244	.131		1.21	4.39		5.60	8.20	
	5000	Screed, 24 ga. metal key joint, see Division 03150-250											
	5020	Wood, incl. wood stakes, 25 mm x 75 mm		C-1	274	.117	m	1.87	3.91		5.78	8.15	
	5050	50 mm x 100 mm			274	.117	"	2.17	3.91		6.08	8.45	
	6000	Trench forms in floor, wood, 1 use			14.86	2.153	m²CA	27	72		99	142	
	6050	2 use			16.26	1.968		13.65	66		79.65	118	
	6100	3 use			16.72	1.914		10	64		74	111	
	6150	4 use			17.19	1.862		8.05	62.50		70.55	106	
	8760	Void form, corrugated fiberboard, 150 mm x 300 mm, 3 m long			22.30	1.435	m²	8.30	48		56.30	84	
450	0010	**FORMS IN PLACE, STAIRS**	R031113-40										450
	0015	(Slant length x width), 1 use		C-2	15.33	3.131	m²	63.50	108		171.50	238	
	0050	2 use	R031113-60		15.79	3.039		37.50	105		142.50	204	
	0100	3 use			16.72	2.870		28.50	99		127.50	186	
	0150	4 use			17.65	2.719		24.50	94		118.50	173	
	1000	Alternate pricing method (3.28 m/m²), 1 use			30.48	1.575	m Rsr	19.35	54.50		73.85	106	
	1050	2 use			32	1.500		11.40	52		63.40	93	
	1100	3 use			33.53	1.432		8.75	49.50		58.25	86.50	

		03110	Structural C.I.P. Forms	CREW	DAILY OUTPUT	LABOR-HOURS	UNIT	2006 BARE COSTS				TOTAL INCL O&P	
								MAT.	LABOR	EQUIP.	TOTAL		
450	1150		4 use	C-2	35.05	1.369	m Rsr	7.45	47.50		54.95	81.50	450
	2000		Stairs, cast on sloping ground (length x width), 1 use		20.44	2.349	m²	23.50	81		104.50	152	
	2100		4 use		22.30	2.153	"	7.55	74.50		82.05	124	
455	0010	**FORMS IN PLACE, WALLS**											455
	0100		Box out for wall openings, to 400 mm thick, to 1 m²	C-2	24	2	Ea.	23.50	69		92.50	134	
	0150		Over 1 m² (use perimeter)	"	85.34	.562	m	6.65	19.40		26.05	37.50	
	0250		Brick shelf, 100 mm w, add to wall form, wall area abv shelf										
	0260		1 use	C-2	22.30	2.153	m²CA	23.50	74.50		98	142	
	0300		2 use		25.55	1.879		13	65		78	115	
	0350		4 use		27.87	1.722		9.45	59.50		68.95	103	
	0500		Bulkhead, with keyway, 1 use, 2 piece		80.77	.594	m	10.40	20.50		30.90	43.50	
	0550		3 piece		53.34	.900		13.10	31		44.10	63	
	0600		Bulkhead forms with keyway, 1 piece expanded metal, 200 mm wall	C-1	305	.105		3.54	3.52		7.06	9.35	
	0610		250 mm wall		244	.131		6.15	4.39		10.54	13.65	
	0620		300 mm wall		160	.200		7.50	6.70		14.20	18.70	
	0700		Buttress, to 2.4 m high, 1 use	C-2	32.52	1.476	m²CA	43	51		94	127	
	0750		2 use		39.95	1.202		23.50	41.50		65	90.50	
	0800		3 use		42.73	1.123		17.35	39		56.35	79.50	
	0850		4 use		44.59	1.076		14.20	37		51.20	73.50	
	1000		Corbel or haunch, to 300 mm wide, add to wall forms, 1 use		45.72	1.050	m	7	36.50		43.50	64	
	1050		2 use		51.82	.926		3.84	32		35.84	54	
	1100		3 use		53.34	.900		2.79	31		33.79	51.50	
	1150		4 use		54.86	.875		2.26	30		32.26	49.50	
	2000		Wall, below grade, job-built plywood, to 2.4 m high, 1 use		27.87	1.722	m²CA	26.50	59.50		86	122	
	2050		2 use		33.91	1.416		16.90	49		65.90	94.50	
	2100		3 use		39.48	1.216		12.25	42		54.25	79	
	2150		4 use		40.41	1.188		10	41		51	75	
	2400		Over 2400 mm to 4800 mm high, 1 use		26.01	1.845		53.50	63.50		117	158	
	2420		2 use		32.05	1.498		21.50	51.50		73	104	
	2430		3 use		34.84	1.378		18.10	47.50		65.60	94	
	2440		4 use		36.70	1.308		16.25	45		61.25	88.50	
	2445		Exterior wall, 2.4 to 4.8 m high, 1 use		26.01	1.845		23.50	63.50		87	125	
	2450		2 use		32.05	1.498		12.80	51.50		64.30	94.50	
	2500		3 use		34.84	1.378		9.15	47.50		56.65	84	
	2550		4 use		36.70	1.308		7.55	45		52.55	79	
	2700		Over 4.8 m high, 1 use		21.83	2.199		26	76		102	147	
	2750		2 use		26.94	1.782		14.30	61.50		75.80	112	
	2800		3 use		29.26	1.640		10.45	56.50		66.95	99.50	
	2850		4 use		30.66	1.566		8.50	54		62.50	93.50	
	3000		For architectural finish, add		169	.284		58.50	9.80		68.30	80	
	4000		Radial, smooth curved, job-built plywood, 1 use		22.76	2.109		24.50	73		97.50	140	
	4050		2 use		27.87	1.722		13.45	59.50		72.95	107	
	4100		3 use		30.19	1.590		9.80	55		64.80	96.50	
	4150		4 use		31.12	1.542		7.95	53.50		61.45	91.50	
	4200		Below grade, job-built plywood, 1 use		20.90	2.296		33.50	79.50		113	160	
	4210		2 use		20.90	2.296		18.60	79.50		98.10	144	
	4220		3 use		20.90	2.296		15.05	79.50		94.55	140	
	4230		4 use		20.90	2.296		11	79.50		90.50	135	
	4300		Curved, 600 mm chords, job-built plywood, 1 use		26.94	1.782		20.50	61.50		82	119	
	4350		2 use		32.98	1.455		11.40	50.50		61.90	90.50	
	4400		3 use		35.77	1.342		8.30	46.50		54.80	81	
	4450		4 use		37.16	1.292		6.65	44.50		51.15	77	
	4500		Over 2.4 m high, 1 use		26.94	1.782		8.85	61.50		70.35	106	
	4525		2 use		32.98	1.455		4.84	50.50		55.34	83.50	
	4550		3 use		35.77	1.342		3.55	46.50		50.05	76	

R031113-40
R031113-60
R031113-10
R031113-40
R031113-60

CONCRETE 3

			DAILY	LABOR-		2006 BARE COSTS				TOTAL	
03110	**Structural C.I.P. Forms**	CREW	OUTPUT	HOURS	UNIT	MAT.	LABOR	EQUIP.	TOTAL	INCL O&P	
455	4575	4 use	C-2	37.16	1.292	m²CA	2.91	44.50		47.41	72.50	455
	4600	Retaining wall, battered, job-built plywood, to 2.4 m H, 1 use		27.87	1.722		19.50	59.50		79	114	
	4650	2 use		32.98	1.455		10.75	50.50		61.25	90	
	4700	3 use	R031113-10	34.84	1.378		7.75	47.50		55.25	82.50	
	4750	4 use		36.23	1.325		5.80	46		51.80	77.50	
	4900	Over 2.4 m to 4.8 m high, 1 use	R031113-40	22.30	2.153		21.50	74.50		96	140	
	4950	2 use		27.41	1.751		11.75	60.50		72.25	107	
	5000	3 use	R031113-60	28.33	1.694		8.50	58.50		67	100	
	5050	4 use		29.73	1.615		6.90	56		62.90	94.50	
	5100	Retaining wall form, plywood, smooth curve, 1 use		18.58	2.583		31.50	89		120.50	174	
	5120	2 use		21.83	2.199		17.35	76		93.35	137	
	5130	3 use		23.23	2.067		12.60	71.50		84.10	125	
	5140	4 use	▼	24.15	1.987		10.35	68.50		78.85	118	
	5500	For gang wall forming, 18 m² sections, deduct					10%	10%				
	5550	36 m² sections, deduct				▼	20%	20%				
	5750	Liners for forms (add to wall forms), A.B.S. plastic										
	5800	Aged wood, 100 mm wide, 1 use	1 Carp	23.23	.344	m²CA	69	12.25		81.25	95	
	5820	2 use		37.16	.215		38	7.65		45.65	54	
	5840	4 use		69.68	.115		22.50	4.08		26.58	31.50	
	5900	Fractured rope rib, 1 use		23.23	.344		41.50	12.25		53.75	64.50	
	6000	4 use		69.68	.115		13.45	4.08		17.53	21	
	6100	Ribbed look, 13 mm & 19 mm deep, 1 use		27.87	.287		58.50	10.20		68.70	80.50	
	6200	4 use		74.32	.108		19.05	3.83		22.88	27	
	6300	Rustic brick pattern, 1 use		23.23	.344		41.50	12.25		53.75	64.50	
	6400	4 use		69.68	.115		13.45	4.08		17.53	21	
	6500	Striated, random, 10 mm x 10 mm deep, 1 use		27.87	.287		41.50	10.20		51.70	61.50	
	6600	4 use	▼	74.32	.108	▼	13.45	3.83		17.28	20.50	
	6800	Rustication strips, A.B.S. plastic, 2 piece snap-on										
	6850	25 mm deep x 35 mm wide, 1 use	C-2	122	.393	m	14	13.60		27.60	36.50	
	6900	2 use		183	.262		7.70	9.05		16.75	22.50	
	6950	4 use		244	.197		4.53	6.80		11.33	15.55	
	7050	Wood, beveled edge, 19 mm deep, 1 use		183	.262		.33	9.05		9.38	14.45	
	7100	25 mm deep, 1 use		137	.350	▼	.49	12.10		12.59	19.40	
	7200	For solid board finish, uniform, 1 use, add to wall forms		27.87	1.722	m²CA	9.70	59.50		69.20	103	
	7300	Non-uniform finish	▼	23.23	2.067		9.05	71.50		80.55	121	
	7500	Lintel or sill forms, 1 use	1 Carp	2.79	2.870		31.50	102		133.50	194	
	7520	2 use		3.16	2.533		17.35	90		107.35	159	
	7540	3 use		3.34	2.392		12.60	85		97.60	146	
	7560	4 use	▼	3.44	2.327		10.25	82.50		92.75	140	
	7800	Modular prefabricated plywood, to 2.4 m high, 1 use	C-2	110	.436		20.50	15.05		35.55	46	
	7820	2 use		111	.432		11.30	14.95		26.25	35.50	
	7840	3 use		115	.417		8.20	14.40		22.60	31.50	
	7860	4 use		117	.410		6.80	14.15		20.95	29.50	
	8000	To 4.8 m high, 1 use		66.42	.723		27	25		52	68.50	
	8020	2 use		68.75	.698		14.85	24		38.85	54	
	8040	3 use		71.53	.671		10.75	23		33.75	48	
	8060	4 use		73.39	.654		8.95	22.50		31.45	45	
	8100	Over 4.8 m high, 1 use		66.42	.723		32.50	25		57.50	74.50	
	8120	2 use		68.75	.698		17.75	24		41.75	57	
	8140	3 use		71.53	.671		12.90	23		35.90	50	
	8160	4 use		73.39	.654		10.75	22.50		33.25	47	
	8600	Pilasters, 1 use		25.08	1.914		31	66		97	137	
	8620	2 use		30.66	1.566		17	54		71	103	
	8640	3 use		34.37	1.396		12.40	48		60.40	88.50	
	8660	4 use	▼	35.77	1.342	▼	10	46.50		56.50	83	
	9010	Steel framed plywood, based on 100 uses of purchased										

03110 | Structural C.I.P. Forms

			CREW	DAILY OUTPUT	LABOR-HOURS	UNIT	MAT.	LABOR	EQUIP.	TOTAL	TOTAL INCL O&P	
455	9020	forms, and 4 uses of bracing lumber	R031113 -10					4.31	29.50		33.81	51.50	455
	9060	To 2.4 m high		C-2	55.74	.861	m²CA	4.31	29.50		33.81	51.50	
	9260	Over 2.4 m to 4.8 m high	R031113 -40		41.81	1.148		4.31	39.50		43.81	66.50	
	9460	Over 4.8 m to 6.0 m high			37.16	1.292		4.31	44.50		48.81	74.50	
	9475	For elevated walls, add	R031113 -60						10%				
	9480	For battered walls, 1 side battered, add						10%	10%				
	9485	For battered walls, 2 sides battered, add						15%	15%				
460	0010	**FORMS IN PLACE, INSULATING CONCRETE**											460
	0020	Forms left in place, M2 is for both sides											
	1000	Panel system, flat cavity, minimum		2 Carp	89.18	.179	m²	19.80	6.40		26.20	31.50	
	1010	Maximum			89.18	.179		29.50	6.40		35.90	42.50	
	1020	Grid cavity, minimum			89.18	.179		21.50	6.40		27.90	33.50	
	1030	Maximum			89.18	.179		32.50	6.40		38.90	45.50	
	1040	Post and beam cavity, minimum			89.18	.179		23.50	6.40		29.90	36	
	1050	Maximum			89.18	.179		35.50	6.40		41.90	49	
	1060	Plank system, flat cavity, minimum			178	.090		26	3.20		29.20	33.50	
	1070	Maximum			178	.090		39	3.20		42.20	47.50	
	1120	Block system, flat cavity, minimum			44.59	.359		26	12.75		38.75	48.50	
	1130	Maximum			44.59	.359		39	12.75		51.75	63	
	1140	Grid cavity, minimum			44.59	.359		26	12.75		38.75	48.50	
	1150	Maximum			44.59	.359		39	12.75		51.75	63	
	1160	Post and beam cavity, minimum			44.59	.359		26	12.75		38.75	48.50	
	1170	Maximum			44.59	.359		39	12.75		51.75	63	
500	0010	**GAS STATION FORMS** Curb fascia, with template,		1 Carp	15.24	.525	m	32.50	18.65		51.15	65	500
	0050	12 ga. steel, left in place, 230 mm high											
	1000	Sign or light bases, 457 mm diameter, 230 mm high [18", 9"]			9	.889	Ea.	62.50	31.50		94	118	
	1050	762 mm diameter, 330 mm high [30", 13"]			8	1		99.50	35.50		135	166	
	2000	Island forms, 3050 mm long, 230 mm high, 900 mm-150 mm wide		C-1	10	3.200		278	107		385	470	
	2050	1200 mm wide			9	3.556		287	119		406	500	
	2500	6100 mm long, 230 mm high, 1200 mm wide			6	5.333		460	179		639	790	
	2550	1500 mm wide			5	6.400		480	214		694	865	
750	0010	**REGLET** See Division 07710-750											750
800	0010	**SCAFFOLDING** See Division 01540-750											800
820	0010	**CONCRETE SLIP FORMING**	R031113 -30										820
	0020	Silos, minimum		C-17E	361	.222	m²CA	34	8.20	.22	42.42	50.50	
	0050	Maximum			102	.784		48.50	29	.79	78.29	99.50	
	1000	Buildings, minimum			340	.235		22.50	8.70	.24	31.44	39	
	1050	Maximum			81.29	.984		42	36.50	.99	79.49	104	

03150 | Concrete Accessories

			CREW	DAILY OUTPUT	LABOR-HOURS	UNIT	MAT.	LABOR	EQUIP.	TOTAL	TOTAL INCL O&P	
080	0010	**ACCESSORIES, ANCHOR BOLTS**										080
	0015	J-type, plain, incl. nut and washer										
	0020	13 mm diameter, 150 mm long [1/2"]	1 Carp	90	.089	Ea.	1.03	3.16		4.19	6.05	
	0050	250 mm long		85	.094		1.17	3.35		4.52	6.50	
	0100	300 mm long		85	.094		1.29	3.35		4.64	6.60	
	0200	16 mm diameter, 300 mm long [5/8"]		80	.100		1.31	3.56		4.87	7	
	0250	450 mm long		70	.114		1.54	4.06		5.60	8.05	
	0300	600 mm long		60	.133		1.77	4.74		6.51	9.35	
	0350	19 mm diameter, 203 mm long [3/4"]		80	.100		1.54	3.56		5.10	7.25	
	0400	300 mm long		70	.114		1.92	4.06		5.98	8.45	
	0450	450 mm long		60	.133		2.50	4.74		7.24	10.15	
	0500	600 mm long		50	.160		3.26	5.70		8.96	12.45	

	03150	Concrete Accessories	CREW	DAILY OUTPUT	LABOR-HOURS	UNIT	2006 BARE COSTS				TOTAL INCL O&P	
							MAT.	LABOR	EQUIP.	TOTAL		
080	0600	22 mm diameter, 300 mm long [7/8"]	1 Carp	60	.133	Ea.	2.52	4.74		7.26	10.20	**080**
	0650	450 mm long		50	.160		3.38	5.70		9.08	12.55	
	0700	600 mm long		40	.200		3.49	7.10		10.59	14.90	
	0800	25 mm diameter, 300 mm long [1"]		55	.145		3.65	5.15		8.80	12.05	
	0850	450 mm long		45	.178		4.40	6.30		10.70	14.70	
	0900	600 mm long		35	.229		5.35	8.15		13.50	18.55	
	0950	900 mm long		25	.320		7.35	11.40		18.75	26	
	1200	38 mm diameter, 450 mm long [1-1/2"]		22	.364		13	12.95		25.95	34.50	
	1250	600 mm long		18	.444		15.50	15.80		31.30	41.50	
	1300	900 mm long	▼	12	.667	▼	19.40	23.50		42.90	58.50	
	1350	For larger sizes see Division 05090-080										
	2000	Galvanized, incl. nut and washer										
	2100	13 mm diameter, 150 mm long [1/2"]	1 Carp	90	.089	Ea.	1.81	3.16		4.97	6.90	
	2150	250 mm long		85	.094		2.05	3.35		5.40	7.45	
	2200	300 mm long		85	.094		2.26	3.35		5.61	7.70	
	2500	16 mm diameter, 300 mm long [5/8"]		80	.100		2.28	3.56		5.84	8.05	
	2550	450 mm long		70	.114		2.69	4.06		6.75	9.30	
	2600	600 mm long		60	.133		3.09	4.74		7.83	10.80	
	2700	19 mm diameter, 203 mm long [3/4"]		80	.100		2.69	3.56		6.25	8.50	
	2750	300 mm long		70	.114		3.36	4.06		7.42	10.05	
	2800	450 mm long		60	.133		4.37	4.74		9.11	12.20	
	2850	600 mm long		50	.160		5.70	5.70		11.40	15.15	
	3000	22 mm diameter, 300 mm long [7/8"]		60	.133		4.42	4.74		9.16	12.25	
	3050	450 mm long		50	.160		5.90	5.70		11.60	15.35	
	3100	600 mm long		40	.200		6.10	7.10		13.20	17.75	
	3200	25 mm diameter, 300 mm long [1"]		55	.145		6.40	5.15		11.55	15.10	
	3250	450 mm long		45	.178		7.70	6.30		14	18.35	
	3300	600 mm long		35	.229		9.40	8.15		17.55	23	
	3350	900 mm long		25	.320		12.85	11.40		24.25	32	
	3500	38 mm diameter, 450 mm long [1-1/2"]		22	.364		23	12.95		35.95	45	
	3550	600 mm long		18	.444		27	15.80		42.80	54.50	
	3600	900 mm long	▼	12	.667	▼	39.50	23.50		63	80.50	
	3700	For larger sizes, galvanized, see Division 05090-080										
	8000	Sleeves, see Division 03150-620										
	8800	Templates, steel, 200 mm bolt spacing	2 Carp	16	1	Ea.	7.75	35.50		43.25	64	
	8850	300 mm bolt spacing		15	1.067		8.05	38		46.05	68	
	8900	400 mm bolt spacing		14	1.143		9.70	40.50		50.20	74	
	8950	600 mm bolt spacing		12	1.333		12.90	47.50		60.40	88	
	9100	Wood, 200 mm bolt spacing		16	1		.79	35.50		36.29	56.50	
	9150	300 mm bolt spacing		15	1.067		1.04	38		39.04	60	
	9200	400 mm bolt spacing		14	1.143		1.30	40.50		41.80	65	
	9250	600 mm bolt spacing	▼	16	1	▼	1.82	35.50		37.32	57.50	
082	0010	**ANCHOR BOLTS** w/ hex nuts and washers, zinc plated										**082**
	0800	M8, 8 mm diameter x 100 mm long	1 Carp	100	.080	Ea.	.30	2.84		3.14	4.76	
	0850	160 mm long		100	.080		.42	2.84		3.26	4.89	
	0870	200 mm long		100	.080		.54	2.84		3.38	5	
	1000	M10, 10 mm diameter x 125 mm long		100	.080		.53	2.84		3.37	5	
	1050	200 mm long		100	.080		.72	2.84		3.56	5.20	
	1070	250 mm long		100	.080		.90	2.84		3.74	5.40	
	1200	M12, 12 mm diameter x 160 mm long		100	.080		1.29	2.84		4.13	5.85	
	1250	200 mm long		100	.080		1.45	2.84		4.29	6.05	
	1270	320 mm long		85	.094		1.61	3.35		4.96	6.95	
	1600	M16, 16 mm diameter x 160 mm long		85	.094		1.36	3.35		4.71	6.70	
	1650	200 mm long		85	.094		1.52	3.35		4.87	6.85	
	1670	320 mm long		80	.100		1.67	3.56		5.23	7.40	
	2000	M20, 20 mm diameter x 200 mm long	▼	80	.100	▼	1.92	3.56		5.48	7.65	

3
CONCRETE

03150 | Concrete Accessories

			DAILY OUTPUT	LABOR-HOURS	UNIT	2006 BARE COSTS				TOTAL INCL O&P	
		CREW				MAT.	LABOR	EQUIP.	TOTAL		
082	2050	320 mm long	1 Carp	70	.114	Ea.	2.40	4.06		6.46	9	**082**
	2070	400 mm long		60	.133		3	4.74		7.74	10.70	
	2400	M24, 24 mm diameter x 200 mm long		80	.100		3.36	3.56		6.92	9.25	
	2450	320 mm long		70	.114		4.56	4.06		8.62	11.35	
	2470	400 mm long	↓	60	.133	↓	5.35	4.74		10.09	13.30	
085	0013	**ANCHOR BOLTS** See Divisions 04080-070 and 05090-080										**085**
160	0010	**ACCESSORIES, CHAMFER STRIPS**										**160**
	2000	Polyvinyl chloride, 13 mm wide with leg	1 Carp	163	.049	m	.75	1.74		2.49	3.54	
	2200	19 mm wide with leg		160	.050		1.05	1.78		2.83	3.92	
	2400	25 mm radius with leg		157	.051		2.20	1.81		4.01	5.25	
	2800	38 mm radius with leg		152	.053		7.20	1.87		9.07	10.85	
	5000	Wood, 13 mm wide		163	.049		.33	1.74		2.07	3.08	
	5200	19 mm wide		160	.050		.33	1.78		2.11	3.13	
	5400	25 mm wide	↓	157	.051	↓	.49	1.81		2.30	3.38	
170	0010	**ACCESSORIES, COLUMN FORM**										**170**
	1000	Column clamps, adjustable to 610 mm x 610 mm, buy				Set	89.50			89.50	98.50	
	1100	Rent/month					7.60			7.60	8.35	
	1300	For sizes to 762 mm x 762 mm, buy					125			125	137	
	1400	Rent/month					10.30			10.30	11.30	
	1600	For sizes to 914 mm x 914 mm, buy					123			123	136	
	1700	Rent/month					13.20			13.20	14.50	
	2000	Bull winch (band iron) 914 mm x 914 mm, buy					48.50			48.50	53.50	
	2100	Rent/month					4.44			4.44	4.88	
	2300	1219 mm x 1219 mm, buy					50			50	55	
	2400	Rent/month					5.60			5.60	6.15	
	3000	Chain & wedge type 914 mm x 914 mm, buy					68.50			68.50	75	
	3100	Rent/month					7.65			7.65	8.40	
	3300	1524 mm x 1524 mm, buy					92			92	101	
	3400	Rent/month					10.70			10.70	11.80	
	4000	Friction collars 762 mm dia., buy					695			695	765	
	4100	Rent/month					68.50			68.50	75	
	4300	1219 mm dia., buy					790			790	870	
	4400	Rent/month				↓	86.50			86.50	95.50	
200	0010	**ACCESSORIES, DOVETAIL ANCHOR SYSTEM**										**200**
	0500	Anchor slot, galv., filled, 24 ga.	1 Carp	130	.062	m	1.97	2.19		4.16	5.60	
	0600	20 ga.		122	.066		2.46	2.33		4.79	6.35	
	0625	12 ga.		122	.066		3.77	2.33		6.10	7.80	
	0800	450 gram copper, foam filled		114	.070		7.60	2.49		10.09	12.25	
	0900	26 ga. stainless steel, foam filled	↓	114	.070	↓	3.15	2.49		5.64	7.35	
	1200	Brick anchor, corr., galv., 90 mm long, 16 ga.	1 Bric	10.50	.762	h	27	28		55	72.50	
	1300	12 ga.		10.50	.762		38	28		66	84	
	1500	Flat, galv., 90 mm long, 16 ga.		10.50	.762		27.50	28		55.50	73	
	1600	12 ga.		10.50	.762		50	28		78	97.50	
	2000	Cavity wall, corr., galv., 127 mm long, 16 ga.		10.50	.762		35	28		63	81	
	2100	12 ga.		10.50	.762		49.50	28		77.50	97	
	3000	Furring anchors, corr., galv., 38 mm long, 16 ga.		10.50	.762		13.90	28		41.90	58	
	3100	12 ga.		10.50	.762		23.50	28		51.50	68	
	6000	Stone anchors, 90 mm long, galv., 3 mm x 25 mm wide		10.50	.762		112	28		140	166	
	6100	6 mm x 25 mm wide	↓	10.50	.762	↓	174	28		202	234	
250	0010	**EXPANSION JOINTS**										**250**
	0020	Keyed, cold, 24 ga, incl. stakes, 90 mm high	1 Carp	60.96	.131	m	2.26	4.67		6.93	9.75	
	0050	115 mm high		60.96	.131		2.76	4.67		7.43	10.25	
	0100	140 mm high	↓	59.44	.135	↓	2.95	4.79		7.74	10.70	

3 CONCRETE

3 CONCRETE

03150		Concrete Accessories	CREW	DAILY OUTPUT	LABOR-HOURS	UNIT	MAT.	LABOR	EQUIP.	TOTAL	TOTAL INCL O&P	
250	0150	190 mm high	1 Carp	57.91	.138	m	3.54	4.91		8.45	11.55	**250**
	0200	241 mm high	↓	56.39	.142		7.50	5.05		12.55	16.10	
	0300	Poured asphalt, plain, 13 mm x 25 mm	1 Clab	137	.058		1.28	1.60		2.88	3.90	
	0350	25 mm x 50 mm		122	.066		5.10	1.80		6.90	8.40	
	0500	Neoprene, liquid, cold applied, 13 mm x 25 mm		137	.058		6.05	1.60		7.65	9.20	
	0550	25 mm x 50 mm		122	.066		20.50	1.80		22.30	25.50	
	0700	Polyurethane, poured, 2 part, 13 mm x 25 mm		122	.066		4.17	1.80		5.97	7.40	
	0750	25 mm x 50 mm		107	.075		16.75	2.05		18.80	21.50	
	0900	Rubberized asphalt, hot or cold applied, 13 mm x 25 mm		137	.058		1.21	1.60		2.81	3.84	
	0950	25 mm x 50 mm		122	.066		4.46	1.80		6.26	7.70	
	1100	Hot applied, fuel resistant, 13 mm x 25 mm		137	.058		3.41	1.60		5.01	6.25	
	1150	25 mm x 50 mm	↓	122	.066		6.55	1.80		8.35	10	
	2000	Premolded, bituminous fiber, 13 mm x 150 mm	1 Carp	114	.070		1.21	2.49		3.70	5.25	
	2050	25 mm x 300 mm		91.44	.087		4.95	3.11		8.06	10.30	
	2250	Cork with resin binder, 13 mm x 150 mm		114	.070		5.50	2.49		7.99	9.95	
	2300	25 mm x 300 mm		91.44	.087		20	3.11		23.11	27	
	2500	Neoprene sponge, closed cell, 13 mm x 150 mm		114	.070		4.66	2.49		7.15	9	
	2550	25 mm x 300 mm		91.44	.087		21.50	3.11		24.61	28.50	
	2750	Polyethylene foam, 13 mm x 150 mm		114	.070		1.54	2.49		4.03	5.60	
	2800	25 mm x 300 mm		91.44	.087		5.55	3.11		8.66	10.95	
	3000	Polyethylene backer rod, 10 mm diameter		140	.057		.10	2.03		2.13	3.26	
	3050	19 mm diameter		140	.057		.23	2.03		2.26	3.39	
	3100	25 mm diameter		140	.057		.36	2.03		2.39	3.55	
	3500	Polyurethane foam, with polybutylene, 13 mm x 13 mm		145	.055		3.02	1.96		4.98	6.35	
	3550	25 mm x 25 mm		137	.058		5.95	2.08		8.03	9.80	
	3750	Polyurethane foam, regular, closed cell, 13 mm x 150 mm		114	.070		2.03	2.49		4.52	6.10	
	3800	25 mm x 300 mm		91.44	.087		4.07	3.11		7.18	9.30	
	4000	Polyvinyl chloride foam, closed cell, 13 mm x 150 mm		114	.070		7.95	2.49		10.44	12.65	
	4050	25 mm x 300 mm		91.44	.087		20	3.11		23.11	27.50	
	4250	Rubber, gray sponge, 13 mm x 150 mm		114	.070		8.70	2.49		11.19	13.50	
	4300	25 mm x 300 mm	↓	91.44	.087	↓	33.50	3.11		36.61	42	
	4500	Lead wool for joints, 0.9 metric ton lots				kg	4.19			4.19	4.61	
	5000	For installation in walls, add						75%				
	5250	For installation in boxouts, add						25%				
350	0010	**ACCESSORIES, HANGERS**										**350**
	0020	Slab and beam form										
	0500	Banding iron, 19 mm x 22 ga, 9.4 m/kg or										
	0550	13 mm x 14 ga, 4.7 m/kg				kg	2.60			2.60	2.87	
	1000	Fascia ties, coil type add to frame ties below				h	85.50			85.50	94	
	1500	Frame ties to 206 mm					213			213	235	
	1550	206 mm to 257 mm					227			227	250	
	5000	Snap tie hanger, to 762 mm overall length, 1814 kg					505			505	555	
	5050	762 mm to 914 mm overall length					550			550	605	
	5100	1067 mm to 1219 mm overall length				↓	630			630	690	
	5500	Steel beam hanger										
	5600	Flange to 206 mm				h	410			410	455	
	5650	206 mm to 257 mm					340			340	375	
	6000	Tie hangers to 610 mm overall length, 2722 kg					470			470	515	
	6100	762 mm to 914 mm overall length					555			555	610	
	6500	Tie back hanger, up to 308 mm flange				↓	440			440	485	
	8500	Wire, black annealed, 9 ga				Met. Ton	2,675			2,675	2,925	
	8600	16 ga				"	2,800			2,800	3,075	
400	0010	**ACCESSORIES, INSERTS**										**400**
	1000	All size nut insert, 16 mm & 19 mm, incl. nut	1 Carp	84	.095	Ea.	3.90	3.39		7.29	9.55	
	2000	Continuous slotted, 41 mm x 35 mm										
	2100	12 ga., 76 mm long	1 Carp	65	.123	Ea.	3.54	4.38		7.92	10.70	

03150	Concrete Accessories	CREW	DAILY OUTPUT	LABOR-HOURS	UNIT	2006 BARE COSTS				TOTAL INCL O&P
						MAT.	LABOR	EQUIP.	TOTAL	
400										**400**
2150	152 mm long	1 Carp	65	.123	Ea.	4.58	4.38		8.96	11.85
2200	8 ga., 305 mm long		65	.123		11	4.38		15.38	18.90
2250	610 mm long		65	.123		18	4.38		22.38	26.50
2300	914 mm long		60	.133		25	4.74		29.74	35
2350	1524 mm long	▼	55	.145	▼	38.50	5.15		43.65	50.50
7000	Threaded cast									
7100	6 mm diameter bolt [1/4"]	1 Carp	84	.095	Ea.	6.75	3.39		10.14	12.65
7350	22 mm diameter bolt [7/8"]	"	84	.095	"	9.45	3.39		12.84	15.65
9000	Wedge									
9050	For 16 mm diameter bolt [5/8"]	1 Carp	60	.133	Ea.	4.27	4.74		9.01	12.10
9100	For 19 mm diameter bolt [3/4"]	"	60	.133	"	9.15	4.74		13.89	17.45
9800	Cut washers, "Black"									
9850	16 mm bolt				Ea.	.26			.26	.29
9950	For galvanized inserts, add					30%				
600	0010	**SHORES** Erect and strip, by hand, horizontal members									**600**
0500	Aluminum joists and stringers	2 Carp	60	.267	Ea.		9.50		9.50	14.75
0600	Steel, adjustable beams		45	.356			12.65		12.65	19.70
0700	Wood joists		50	.320			11.40		11.40	17.70
0800	Wood stringers		30	.533			18.95		18.95	29.50
1000	Vertical members to 3 m high		55	.291			10.35		10.35	16.10
1050	To 4 m high		50	.320			11.40		11.40	17.70
1100	To 5 m high	▼	45	.356	▼		12.65		12.65	19.70
1500	Reshoring	▼	130	.123	m²	4.52	4.38		8.90	11.75
1600	Flying truss system	C-17D	892	.094	m²CA		3.48	.59	4.07	6.05
1760	Horizontal, aluminum joists, 159 mm high x 1.5 to 6.3 m span, buy				m	110			110	121
1770	Beams, 184 mm high x 1.2 to 9 m span				"	153			153	169
1810	Horizontal, steel beam, adjustable, 1.2 m to 2.1 m span				Ea.	535			535	590
1830	1.8 m to 3.0 m span					690			690	760
1920	2.7 m to 4.6 m span					875			875	965
1940	3.7 m to 6.1 m span				▼	895			895	985
1970	Steel stringer, W200x15, 1.2 to 4.8 m span, buy				m	63			63	69.50
3000	Rent for job duration, aluminum joist @ 150 mm OC, per month				m² Flr.	9.05			9.05	10
3050	Steel W200x15					5.15			5.15	5.70
3060	Steel adjustable				▼	18.60			18.60	20.50
3500	#1 post shore, steel, 1700 mm to 2900 mm high, 4500 kg cap., buy				Ea.	355			355	390
3550	#2 post shore, 2200 mm to 3900 mm high, 3500 kg capacity					380			380	420
3600	#3 post shore, 2700 mm to 4900 mm high, 1720kg capacity				▼	415			415	455
5010	Frame shoring systems, steel, 5400 kg per leg, buy									
5040	Frame, 600 mm wide x 1800 mm high				Ea.	330			330	365
5250	X-brace					52.50			52.50	57.50
5550	Base plate					41.50			41.50	45.50
5600	Screw jack					132			132	145
5650	U-head, 200 x 200 mm				▼	21			21	23
620	0010	**ACCESSORIES, SLEEVES AND CHASES**									**620**
0100	Plastic, 1 use, 225 mm long, 50 mm diameter	1 Carp	100	.080	Ea.	.53	2.84		3.37	5
0150	100 mm diameter		90	.089		1.56	3.16		4.72	6.65
0200	150 mm diameter		75	.107		2.75	3.79		6.54	8.95
0250	300 mm diameter		60	.133		18.05	4.74		22.79	27.50
5000	Sheet metal, 50 mm diameter		100	.080		.97	2.84		3.81	5.50
5100	100 mm diameter		90	.089		1.21	3.16		4.37	6.25
5150	150 mm diameter		75	.107		1.76	3.79		5.55	7.85
5200	300 mm diameter		60	.133		3.52	4.74		8.26	11.25
6000	Steel pipe, 50 mm diameter		100	.080		6.25	2.84		9.09	11.35
6100	100 mm diameter		90	.089		18.15	3.16		21.31	25
6150	150 mm diameter	▼	75	.107	▼	38.50	3.79		42.29	48.50

For expanded coverage of these items see *Means Concrete & Masonry Cost Data 2006*

		03150	Concrete Accessories	CREW	DAILY OUTPUT	LABOR-HOURS	UNIT	2006 BARE COSTS MAT.	LABOR	EQUIP.	TOTAL	TOTAL INCL O&P	
620	6200		300 mm diameter	1 Carp	60	.133	Ea.	84.50	4.74		89.24	100	**620**
640	0010	**ACCESSORIES, SNAP TIES, FLAT WASHER**											**640**
	0100		1400 kg, to 200 mm				h	113			113	124	
	0150		225 mm & 250 mm					129			129	142	
	0200		275 mm & 300 mm					135			135	149	
	0250		400 mm					143			143	158	
	0300		450 mm					141			141	155	
	0500		With plastic cone, to 200 mm					107			107	117	
	0550		225 mm & 250 mm					118			118	130	
	0600		275 mm & 300 mm					127			127	139	
	0650		400 mm					134			134	147	
	0700		450 mm					139			139	153	
	1000		2300 kg, to 200 mm					142			142	156	
	1100		225 mm & 250 mm					155			155	171	
	1150		275 mm & 300 mm					166			166	182	
	1200		400 mm					182			182	200	
	1250		450 mm					178			178	196	
	1500		With plastic cone, to 200 mm					183			183	201	
	1550		225 mm & 250 mm					199			199	219	
	1600		275 mm & 300 mm					217			217	238	
	1650		400 mm					240			240	264	
	1700		450 mm					248			248	273	
660	0010	**STAIR TREAD INSERTS**											**660**
	0015		Cast iron, abrasive, 75 mm wide	1 Carp	27.43	.292	m	26.50	10.35		36.85	45	
	0020		100 mm wide		24.38	.328		35	11.65		46.65	56.50	
	0040		150 mm wide		22.86	.350		53	12.45		65.45	77.50	
	0050		230 mm wide		21.34	.375		79	13.35		92.35	108	
	0100		300 mm wide		19.81	.404		106	14.35		120.35	139	
	0300		Cast aluminum, compared to cast iron, deduct					10%					
	0500		Extruded aluminum safety tread, 75 mm wide	1 Carp	22.86	.350		30	12.45		42.45	52.50	
	0550		100 mm wide		22.86	.350		40	12.45		52.45	63.50	
	0600		150 mm wide		22.86	.350		60	12.45		72.45	85.50	
	0650		230 mm wide to resurface stairs		21.34	.375		90	13.35		103.35	120	
	1700		Cement fill for pan-type metal treads, plain	1 Cefi	10.68	.749	m²	17.35	26		43.35	57	
	1750		Non-slip	"	9.29	.861	"	19.05	29.50		48.55	64.50	
850	0010	**ACCESSORIES, WALL AND FOUNDATION**											**850**
	2000		Footings, form braces, solid steel, adjustable				Ea.	15.60			15.60	17.15	
	2050		Spreaders for footer, adjustable				"	5.35			5.35	5.85	
	3000		Form oil, coverage varies greatly, minimum				liter	1.32			1.32	1.45	
	3050		Maximum				"	2			2	2.20	
	3500		Form patches, 44 mm diameter [1-3/4″]				h	75			75	82.50	
	3550		70 mm diameter [2-3/4″]				"	95.50			95.50	105	
	4000		Nail stakes, 19 mm diameter, 450 mm long [3/4″]				Ea.	2.35			2.35	2.59	
	4050		600 mm long					2.70			2.70	2.97	
	4200		760 mm long					3.20			3.20	3.52	
	4250		900 mm long					3.70			3.70	4.07	
860	0010	**WATERSTOP**											**860**
	0020		PVC, ribbed 5 mm thick, 100 mm wide	1 Carp	47.24	.169	m	2.59	6		8.59	12.20	
	0050		150 mm wide		44.20	.181		4.27	6.45		10.72	14.70	
	0500		Ribbed, PVC, with center bulb, 150 mm wide, 5 mm thick		41.15	.194		3.90	6.90		10.80	15.05	
	0550		10 mm thick		39.62	.202		6.10	7.20		13.30	17.95	
	0800		Dumbbell type, PVC, 150 mm wide, 5 mm thick		45.72	.175		3.77	6.20		9.97	13.85	

03150	Concrete Accessories	CREW	DAILY OUTPUT	LABOR-HOURS	UNIT	2006 BARE COSTS				TOTAL INCL O&P
						MAT.	LABOR	EQUIP.	TOTAL	
860 0850	10 mm thick	1 Carp	44.20	.181	m	7.45	6.45		13.90	18.20 **860**
1000	225 mm wide, 10 mm thick, PVC, plain		39.62	.202		10.70	7.20		17.90	23
1050	Center bulb		39.62	.202		22	7.20		29.20	35
1250	Split PVC, 10 mm thick, 150 mm wide		44.20	.181		4.30	6.45		10.75	14.70
1300	225 mm wide		39.62	.202		8.30	7.20		15.50	20.50
2000	Rubber, flat dumbbell, 10 mm thick, 150 mm wide		44.20	.181		20.50	6.45		26.95	32.50
2050	225 mm wide		41.15	.194		30.50	6.90		37.40	45
2500	Flat dumbbell split, 10 mm thick, 150 mm wide		44.20	.181		4.30	6.45		10.75	14.70
2550	225 mm wide		41.15	.194		8.30	6.90		15.20	19.85
3000	Center bulb, 6 mm thick, 150 mm wide		44.20	.181		17.85	6.45		24.30	29.50
3050	225 mm wide		41.15	.194		35	6.90		41.90	49.50
3500	Center bulb split, 10 mm thick, 150 mm wide		44.20	.181		21.50	6.45		27.95	34
3550	225 mm wide		41.15	.194		37.50	6.90		44.40	52.50
5000	Waterstop fittings, rubber, flat									
5010	Dumbbell or center bulb, 10 mm thick,									
5200	Field union, 150 mm wide	1 Carp	50	.160	Ea.	21	5.70		26.70	32
5250	225 mm wide		50	.160		25	5.70		30.70	36
5500	Flat cross, 150 mm wide		30	.267		43	9.50		52.50	62.50
5550	225 mm wide		30	.267		61.50	9.50		71	82.50
6000	Flat tee, 150 mm wide		30	.267		40	9.50		49.50	59
6050	225 mm wide		30	.267		52.50	9.50		62	72.50
6500	Flat ell, 150 mm wide		40	.200		36	7.10		43.10	50.50
6550	225 mm wide		40	.200		48.50	7.10		55.60	64
7000	Vertical tee, 150 mm wide		25	.320		35	11.40		46.40	56
7050	225 mm wide		25	.320		44.50	11.40		55.90	66.50
7500	Vertical ell, 150 mm wide		35	.229		31	8.15		39.15	46.50
7550	225 mm wide		35	.229		42.50	8.15		50.65	59.50

03210	Reinforcing Steel	CREW	DAILY OUTPUT	LABOR-HOURS	UNIT	2006 BARE COSTS				TOTAL INCL O&P
						MAT.	LABOR	EQUIP.	TOTAL	
100 0010	**ACCESSORIES**	R032110-80								**100**
0020	See also Form Accessories, Division 03150									
0100	Beam bolsters, (BB) standard, lower, up to 38 mm high, plain				m	2.06			2.06	2.26
0102	Galvanized					3.16			3.16	3.48
0104	Stainless					5.95			5.95	6.55
0106	Plastic					3.58			3.58	3.94
0108	Epoxy					4.90			4.90	5.40
0110	64 mm to 75 mm high, plain					2.69			2.69	2.96
0120	Galvanized					3.58			3.58	3.94
0140	Stainless					8.15			8.15	9
0160	Plastic					3.74			3.74	4.12
0162	Epoxy					5.20			5.20	5.75
0200	Upper, standard (BBU) to 38 mm high, plain					5.90			5.90	6.50
0210	64 mm to 75 mm high					10.70			10.70	11.75
0300	Beam bolster with plate (BBP) to 38 mm high, plain					6.80			6.80	7.50
0310	64 mm to 75 mm high					13.20			13.20	14.50
0500	Slab bolsters, continuous, plain (SB) 19 mm to 25 mm H, plain					1.69			1.69	1.86
0502	Galvanized					1.95			1.95	2.15

CONCRETE 3

For expanded coverage of these items see *Means Concrete & Masonry Cost Data 2006*

	03210	Reinforcing Steel		CREW	DAILY OUTPUT	LABOR-HOURS	UNIT	MAT.	LABOR	EQUIP.	TOTAL	TOTAL INCL O&P	
100	0504	Stainless					m	2.74			2.74	3.01	100
	0506	Plastic	R032110 -80					2.32			2.32	2.55	
	0510	25 mm to 50 mm high, plain						2.11			2.11	2.32	
	0515	Galvanized						2.48			2.48	2.72	
	0520	Stainless						4.69			4.69	5.15	
	0525	Plastic						2.79			2.79	3.07	
	0530	For bolsters with wire runners (SBR), add						2.37			2.37	2.61	
	0540	For bolsters with plates (SBP), add						5.75			5.75	6.30	
	0700	Clip or bar ties, 16 ga., plain, 75 mm long					h	14.30			14.30	15.75	
	0710	100 mm long						15.10			15.10	16.60	
	0720	150 mm long						17.70			17.70	19.45	
	0730	200 mm long						18.50			18.50	20.50	
	0900	Flange clips, expand. flanges, 10 ga., 300 mm O.C., cont.,											
	0910	galvanized, over 152 m, 100 mm to 200 mm					m	1.74			1.74	1.91	
	0920	230 mm to 300 mm						2.37			2.37	2.61	
	0930	430 mm to 600 mm						2.42			2.42	2.67	
	1200	High chairs, individual, no plates (1 HC), to 75 mm high, plain					h	77.50			77.50	85	
	1202	Galvanized						108			108	119	
	1204	Stainless						221			221	243	
	1206	Plastic						103			103	113	
	1210	125 mm high, plain						119			119	131	
	1212	Galvanized						155			155	171	
	1214	Stainless						335			335	370	
	1216	Plastic						144			144	159	
	1220	200 mm high, plain						257			257	283	
	1222	Galvanized						345			345	380	
	1224	Stainless						565			565	625	
	1226	Plastic						325			325	355	
	1230	300 mm high, plain						535			535	585	
	1232	Galvanized						680			680	750	
	1234	Stainless						1,025			1,025	1,125	
	1236	Plastic						600			600	660	
	1240	380 mm high, plain						1,000			1,000	1,100	
	1242	Galvanized						1,175			1,175	1,275	
	1244	Stainless						1,725			1,725	1,900	
	1246	Plastic						1,075			1,075	1,200	
	1250	For each added 25 mm up to 600 mm high, plain, add						74			74	81.50	
	1252	Galvanized, add						86			86	94.50	
	1254	Stainless, add						96			96	106	
	1256	Plastic, add						86			86	94.50	
	1400	Individual high chairs, with plates, (HCP), to 125 mm high, add						340			340	375	
	1410	Over 125 mm high, add						390			390	430	
	1500	Bar chair (BC) for up to 44 mm high, plain						44			44	48	
	1520	Galvanized						50.50			50.50	55.50	
	1530	Stainless						135			135	148	
	1540	Plastic						89.50			89.50	98	
	1550	Joist chair (JC), joists up to 150 mm, plain						53			53	58.50	
	1580	Galvanized						67.50			67.50	74	
	1600	Stainless						96			96	106	
	1620	Plastic						91.50			91.50	101	
	1630	Epoxy						222			222	245	
	1700	Continuous high chairs, (CHC) to 100 mm high, plain					m	3.37			3.37	3.71	
	1705	Galvanized						4.14			4.14	4.56	
	1710	Stainless						8.85			8.85	9.75	
	1715	Plastic						4.16			4.16	4.58	
	1718	Epoxy						6.30			6.30	6.95	

Important: See the Reference Section for supporting data - Crews, Rental Equipment, City Cost Indexes and Reference Data

03210	Reinforcing Steel		CREW	DAILY OUTPUT	LABOR-HOURS	UNIT	2006 BARE COSTS				TOTAL INCL O&P	
							MAT.	LABOR	EQUIP.	TOTAL		
100	1720	150 mm high, plain	R032110 -80				m	4.85			4.85	5.35	100
	1725	Galvanized						6.95			6.95	7.65	
	1730	Stainless						10.30			10.30	11.30	
	1735	Plastic						6.65			6.65	7.30	
	1738	Epoxy						8.45			8.45	9.30	
	1740	200 mm high, plain						7.05			7.05	7.75	
	1745	Galvanized						9.80			9.80	10.75	
	1750	Stainless						12.20			12.20	13.45	
	1755	Plastic						9.20			9.20	10.15	
	1758	Epoxy						12.75			12.75	14.05	
	1760	300 mm high, plain						17.35			17.35	19.05	
	1765	Galvanized						21			21	23	
	1770	Stainless						29			29	32	
	1775	Plastic						19.75			19.75	21.50	
	1778	Epoxy						30.50			30.50	33.50	
	1780	380 mm high, plain						19.45			19.45	21.50	
	1785	Galvanized						23.50			23.50	26	
	1790	Stainless						23.50			23.50	25.50	
	1795	Plastic						22			22	24.50	
	1798	Epoxy						40			40	44	
	1800	For each added 25 mm up to 600 mm high, plain, add						1.66			1.66	1.82	
	1820	Galvanized, add						2.04			2.04	2.25	
	1840	Stainless, add						1.82			1.82	2.01	
	1860	Plastic, add						1.82			1.82	2.01	
	1900	For continuous bottom plate, (CHCP), add						7.95			7.95	8.75	
	1940	For upper continuous high chairs, (CHCU), add						7.95			7.95	8.75	
	1960	For galvanized wire runners, add						7.45			7.45	8.20	
	2100	Paper tubing, 1200 mm lengths, for #10M bar						2.99			2.99	3.29	
	2120	For #20M bar						3.72			3.72	4.09	
	2200	Screed base, adjustable, (SCBA), 64 mm high, plain						h	235			235	259	
	2210	Galvanized						244			244	268	
	2220	140 mm high, plain						284			284	310	
	2250	Galvanized						300			300	330	
	2300	19 mm diameter, 64 mm high, plain [3/4″,2-1/2″]						292			292	320	
	2310	Galvanized						310			310	340	
	2320	140 mm high, plain						360			360	395	
	2350	Galvanized						385			385	425	
	2400	Screed holder, (SCCH) for 25 mm I.D. pipe, 150 mm long						244			244	268	
	2420	300 mm long						405			405	445	
	2500	For 38 mm I.D. pipe, 150 mm long						440			440	480	
	2520	300 mm long						720			720	795	
	2700	Screw anchor for bolts, plain, 13 mm diameter [1/2″]						160			160	176	
	2720	25 mm diameter [1″]						480			480	525	
	2740	38 mm diameter [1-1/2″]						790			790	870	
	2800	Screw eye bolts, 13 mm x 125 mm long						182			182	200	
	2820	25 mm x 230 mm long						720			720	795	
	2840	38 mm x 350 mm long						1,750			1,750	1,925	
	2900	Screw anchor bolts, 13 mm x up to 175 mm long						730			730	805	
	2920	25 mm x up to 300 mm long						2,375			2,375	2,600	
	3000	Slab lifting inserts, single, galv., 100 mm high						490			490	540	
	3010	150 mm high						600			600	660	
	3030	175 mm high						685			685	755	
	3100	25 mm diameter, 125 mm high [1″,5″]						775			775	850	
	3120	175 mm high						815			815	900	
	3200	Double lifting inserts, 125 mm high						1,525			1,525	1,675	
	3220	175 mm high						1,625			1,625	1,775	

For expanded coverage of these items see *Means Concrete & Masonry Cost Data 2006*

			CREW	DAILY OUTPUT	LABOR-HOURS	UNIT	2006 BARE COSTS				TOTAL INCL O&P	
	03210	**Reinforcing Steel**					MAT.	LABOR	EQUIP.	TOTAL		
100	3330	32 mm diameter, 125 mm high [1-1/4",5"]	R032110 -80			h	1,675			1,675	1,850	**100**
	3500	Sleeper clips for wood sleepers, 20 ga., galv., 50 mm wide				k	585			585	645	
	3520	100 mm wide					720			720	795	
	3600	Spacers, plastic for 25 mm bar clearance, average					89.50			89.50	98.50	
	3620	For 50 mm bar clearance, average				▼	108			108	119	
	3800	Subgrade chairs, 13 mm diameter, 90 mm high				h	480			480	530	
	3850	300 mm high					1,350			1,350	1,475	
	3900	19 mm diameter, 90 mm high [3/4",3-1/2"]					620			620	680	
	3950	300 mm high					1,450			1,450	1,600	
	4200	Subgrade stakes, 19 mm diameter, 300 mm long					500			500	550	
	4250	600 mm long					670			670	735	
	4300	25 mm diameter, 400 mm long					750			750	820	
	4350	600 mm long				▼	1,100			1,100	1,200	
	4500	Tie wire, 16 ga. annealed steel, under 225 kg				Met. Ton	3,175			3,175	3,500	
	4520	900 to 1800 kg				"	3,000			3,000	3,300	
	4550	Tie wire holder, plastic case				Ea.	56.50			56.50	62.50	
	4600	Aluminum case				"	68.50			68.50	75.50	
200	0010	**COATED REINFORCING** Add to fabricated & delivered price										**200**
	0100	Epoxy coated, A775				Met. Ton	405			405	445	
	0150	Galvanized, #10M [#3]					805			805	885	
	0200	#10M [#4]					805			805	885	
	0250	#15M					790			790	865	
	0300	#20M or over					790			790	865	
	1000	For over 18 metric ton, #20M or larger, minimum					730			730	805	
	1500	Maximum				▼	875			875	965	
500	0010	**REINFORCING STEEL, A615**	R032110 -10									**500**
	0150	Reinforcing, A615 grade 40, mill base				Met. Ton	700			700	770	
	0200	Detailed, cut, bent, and delivered, average	R032110 -20				940			940	1,025	
	0650	Reinforcing steel, A615 grade 60, mill base					700			700	770	
	0700	Detailed, cut, bent, and delivered, average	R032110 -25			▼	940			940	1,025	
	1000	Reinforcing steel, extras, add to mill base										
	1005	Mill extra, add for delivery to shop, average	R032110 -40			Met. Ton	31			31	34	
	1010	Shop extra, add for for handling & storage, average					25.50			25.50	28	
	1020	Shop extra, add for bending, limited percent of bars	R032110 -50				25			25	27.50	
	1030	Average percent of bars					49.50			49.50	54.50	
	1050	Large percent of bars					99			99	109	
	1200	Shop extra, add for detailing, under 45 tonnes	R032110 -70				45.50			45.50	50	
	1250	45 to 136 metric tons					34			34	37.50	
	1300	150 to 454 metric tons	R032110 -80				32.50			32.50	35.50	
	1350	Over 454 metric tons					31			31	34	
	1700	Shop extra, add for listing, average					5.50			5.50	6.05	
	2000	Mill extra, add for quantity, under 18 metric tons					5.50			5.50	6.05	
	2100	Shop extra, add for quantity, under 18 metric tons					46.50			46.50	51	
	2200	18 to 44 metric tons					34.50			34.50	38	
	2250	45 to 90 metric tons					23			23	25.50	
	2300	90 to 272 metric tons					13.90			13.90	15.30	
	2500	Shop extra, add for size, #10M					154			154	170	
	2600	#15M					38.50			38.50	42.50	
	2650	#20M					34.50			34.50	38	
	2700	#25M to #35M					46.50			46.50	51	
	2750	#45M					58			58	63.50	
	2800	#55M					65.50			65.50	72	
	2900	Shop extra, add for delivery to job, average				▼	25			25	27.50	
600	0010	**REINFORCING IN PLACE** A615M Grade 400, incl. access. lbr	R032110 -10									**600**
	0100	Beams & Girders, #10M to #25M		4 Rodm	1.45	22.046	Met. Ton	940	870		1,810	2,450	

03210 | Reinforcing Steel

		CREW	DAILY OUTPUT	LABOR-HOURS	UNIT	MAT.	LABOR	EQUIP.	TOTAL	TOTAL INCL O&P	
600	0150	#25M to #55M R032110-20	4 Rodm	2.45	13.064	Met. Ton	940	515		1,455	1,875	**600**
	0200	Columns, #10M to #25M		1.36	23.516		940	930		1,870	2,550	
	0250	#25M to #55M R032110-25		2.09	15.336		940	605		1,545	2,025	
	0300	Spirals, hot rolled, 200 mm to 375 mm diam.		2	16.033		1,250	635		1,885	2,425	
	0320	375 mm to 600 mm diameter R032110-40		2	16.033		1,200	635		1,835	2,350	
	0330	600 mm to 900 mm diameter		2.09	15.336		1,125	605		1,730	2,250	
	0340	900 mm to 1200 mm diameter R032110-50		2.18	14.697		1,075	580		1,655	2,125	
	0360	1200 mm to 1600 mm diameter		2.27	14.109		1,200	555		1,755	2,225	
	0380	1600 mm to 2100 mm diameter R032110-70		2.36	13.567		1,250	535		1,785	2,250	
	0390	2100 mm to 2400 mm diameter		2.45	13.064		1,300	515		1,815	2,275	
	0400	Elevated slabs, #10M to #25M R032110-80		2.63	12.163		1,000	480		1,480	1,900	
	0500	Footings, #10M to #25M		1.91	16.797		895	665		1,560	2,075	
	0550	#25M to #55M		3.27	9.798		850	385		1,235	1,575	
	0600	Slab on grade, #10M to #25M		2.09	15.336		895	605		1,500	1,975	
	0700	Walls, #10M to #25M		2.72	11.758		895	465		1,360	1,750	
	0750	#25M to #55M		3.63	8.818		895	350		1,245	1,550	
	0770	Walls, #10M to #25M		2.72	11.758		895	465		1,360	1,750	
	0900	Use the following for a rough estimate guide										
	1000	Typical in place, average, under 9 metric ton job, #10M to #25M	4 Rodm	1.63	19.596	Met. Ton	975	775		1,750	2,350	
	1010	#25M to #55M		2.45	13.064		995	515		1,510	1,950	
	1050	9 - 45 metric ton job, #10M to #25M		1.91	16.797		955	665		1,620	2,150	
	1060	#25M to #55M		2.72	11.758		975	465		1,440	1,850	
	1100	45 - 90 metric ton job, #10M to #25M		2	16.033		935	635		1,570	2,075	
	1110	#25M to #55M		2.81	11.379		955	450		1,405	1,800	
	1150	Over 90 metric ton job, #10M to #25M		2.09	15.336		920	605		1,525	2,025	
	1160	#25M - #55M		2.90	11.023		940	435		1,375	1,750	
	1200	High strength steel, Grade 600, #45M bars only, add					68.50			68.50	75.50	
	2000	Unloading & sorting, add to above	C-5	90.72	.617			24	7.15	31.15	46.50	
	2200	Crane cost for handling, add to above, minimum		122	.459			17.70	5.35	23.05	34.50	
	2210	Average		83.46	.671			26	7.80	33.80	50	
	2220	Maximum		31.75	1.764			68	20.50	88.50	133	
	2300	For epoxy coated rebar, add					45%					
	2350	For stainless steel rebar, add					300%					
	2400	Dowels, 600 mm long, deformed, #10M	2 Rodm	520	.031	Ea.	.35	1.22		1.57	2.39	
	2410	#13M		480	.033		.63	1.32		1.95	2.86	
	2420	#15M		435	.037		.98	1.45		2.43	3.47	
	2430	#20M		360	.044		1.41	1.76		3.17	4.44	
	2450	Longer and heavier dowels, add		329	.049	kg	1.04	1.92		2.96	4.31	
	2500	Smooth dowels, 300 mm long, 10 mm diam.		140	.114	Ea.	.72	4.51		5.23	8.25	
	2520	16 mm diameter		125	.128		1.25	5.05		6.30	9.75	
	2530	19 mm diameter		110	.145		1.55	5.75		7.30	11.15	
	2600	Dowel sleeves for CIP concrete, 2-part system										
	2610	Sleeve base, plastic, for #15M bar, fasten to edge form	1 Rodm	200	.040	Ea.	.36	1.58		1.94	3	
	2615	Sleeve, plastic, #15M bar x 225 mm L, snap onto base		400	.020		1.03	.79		1.82	2.43	
	2620	Sleeve base, for #20M bar		175	.046		.36	1.81		2.17	3.37	
	2625	Sleeve, for #20M bar		350	.023		1.06	.90		1.96	2.66	
	2630	Sleeve base, for #25M bar		150	.053		.44	2.11		2.55	3.95	
	2635	Sleeve, for #25M bar		300	.027		1.18	1.05		2.23	3.03	
	2700	Dowel caps, visual warning only, #10M - #25M	2 Rodm	800	.020		.27	.79		1.06	1.60	
	2720	#25M - #45M		750	.021		.35	.84		1.19	1.77	
	2750	Impalement protective, #10M - #25M		800	.020		1.59	.79		2.38	3.05	
	2760	#10M - #35M		775	.021		2.21	.82		3.03	3.77	
	2770	#35M - 50M		750	.021		2.25	.84		3.09	3.87	
	3000	For epoxy dowel anchoring, see Div 05090-300										
650	0010	**GLASS FIBER REINFORCED POLYMER** reinforcing bars										**650**
	0050	#2 bar, .065 kg/m				m	1.15			1.15	1.28	

CONCRETE 3

		03210	Reinforcing Steel	CREW	DAILY OUTPUT	LABOR-HOURS	UNIT	2006 BARE COSTS				TOTAL INCL O&P	
								MAT.	LABOR	EQUIP.	TOTAL		
650	0100		#3 bar, .138 kg/m				m	1.57			1.57	1.74	650
	0150		#4 bar, .24 kg/m					2.26			2.26	2.49	
	0200		#5 bar, .375 kg/m					2.95			2.95	3.25	
	0250		#6 bar, .56 kg/m					4.10			4.10	4.53	
	0300		#7 bar, .75 kg/m					5.25			5.25	5.75	
	0350		#8 bar, .93 kg/m					6.15			6.15	6.80	
	0400		#9 bar, 1.2 kg/m					7.55			7.55	8.30	
	0450		#10 bar, 1.62 kg/m					9.85			9.85	10.85	
700	0010	**SPLICING REINFORCING BARS**											700
	0020		Including holding bars in place while splicing	R032110 -70									
	0100		Butt weld columns #10M bars	C-5	190	.295	Ea.	4.09	11.35	3.42	18.86	26.50	
	0110		#20M bars		150	.373		5.75	14.40	4.34	24.49	34	
	0130		#35M bars		95	.589		14.10	22.50	6.85	43.45	59.50	
	0150		#45M bars		65	.862		22.50	33	10	65.50	89	
	0280		Column splices, bar to bar end bearing										
	0300		#20M or #25M bars	C-5	190	.295	Ea.	37	11.35	3.42	51.77	62.50	
	0310		#30M or #35M bars		170	.329		40.50	12.70	3.83	57.03	69	
	0320		#35M bars		160	.350		49	13.50	4.07	66.57	80.50	
	0330		#45M bars		150	.373		61.50	14.40	4.34	80.24	95.50	
	0340		#55M bars		140	.400		88.50	15.45	4.65	108.60	128	
	0500		Transition bar to bar end bearing, #45M to #55M bar					87			87	96	
	0520		#45M to #35M bar					67			67	74	
	0540		#35M to #30M bar					50.50			50.50	55.50	
	0550		#35M to #30M bar					42			42	46.50	
	0560		#30M to #25M bar					39			39	43	
	0580		#25M to #20M bar					38			38	42	
	0600		For bolted speed sleeve type, deduct						15%				
	0800		Mechanical butt splice, sleeve type w/ filler metal, compression										
	0810		only, all grades, columns only #35M bars	C-5	68	.824	Ea.	43	32	9.55	84.55	109	
	0900		#45M bars		62	.903		70	35	10.50	115.50	145	
	0920		#55M bars		62	.903		128	35	10.50	173.50	209	
	1000		125% yield point, grade 400, columns only, #20M bars		68	.824		34.50	32	9.55	76.05	99.50	
	1020		#20M or #25M bars		68	.824		37	32	9.55	78.55	102	
	1030		#30M bars		68	.824		38	32	9.55	79.55	104	
	1040		#35M bars		68	.824		40.50	32	9.55	82.05	106	
	1050		#35M bars		68	.824		49	32	9.55	90.55	116	
	1060		#45M bars		62	.903		61.50	35	10.50	107	135	
	1070		#55M bars		62	.903		88.50	35	10.50	134	165	
	1200		Full tension, grade 400 steel, columns,										
	1220		slabs or beams, #20M, #25M bars	C-5	68	.824	Ea.	11.65	32	9.55	53.20	74.50	
	1230		#30M bars		68	.824		12.70	32	9.55	54.25	75.50	
	1240		#35M bars		68	.824		14.10	32	9.55	55.65	77	
	1250		#35M bars		68	.824		15	32	9.55	56.55	78	
	1260		#45M bars		62	.903		22.50	35	10.50	68	92	
	1270		#55M bars		62	.903		34	35	10.50	79.50	105	
	1400		If equipment handling not required, deduct						50%				
	1600		Mechanical threaded type, bar threading not included,										
	1700		Straight bars, #32M & #35M	C-5	140	.400	Ea.	15	15.45	4.65	35.10	46.50	
	1750		#45M bars		130	.431		22.50	16.60	5	44.10	57	
	1800		#55M bars		75	.747		34	29	8.70	71.70	93.50	
	2100		#35M to #55M & #45M to #55M transition		75	.747		36.50	29	8.70	74.20	96	
	2400		Bent bars, #35M		105	.533		15	20.50	6.20	41.70	56.50	
	2500		#45M		90	.622		25.50	24	7.25	56.75	74.50	
	2600		#55M		70	.800		42	31	9.30	82.30	106	
	2800		#35M to #45M transition		75	.747		24.50	29	8.70	62.20	82.50	
	2900		#35M to #55M & #45M to #55M transition		70	.800		40	31	9.30	80.30	104	

Important: See the Reference Section for supporting data - Crews, Rental Equipment, City Cost Indexes and Reference Data

03220 | Welded Wire Fabric

		CREW	DAILY OUTPUT	LABOR-HOURS	UNIT	2006 BARE COSTS				TOTAL INCL O&P
						MAT.	LABOR	EQUIP.	TOTAL	
200	**0010 WELDED WIRE FABRIC** ASTM A185 R032205 -30									**200**
0050	Sheets									
0100	150x150 - W1.4 x W1.4 (10 x 10) 1.025 kg/m²	2 Rodm	325	.049	m²	1.29	1.94		3.23	4.62
0200	150x150 - W2.1 x W2.1 (8 x 8) 1.464 kg/m²		288	.056		1.54	2.19		3.73	5.30
0300	150x150 - W2.9 x W2.9 (6 x 6) 2.05 kg/m²		269	.059		1.84	2.35		4.19	5.90
0400	150x150 - W4 x W4 (4 x 4) 2.83 kg/m²		251	.064		2.52	2.52		5.04	6.90
0500	100x100 - W1.4 x W1.4 (10 x 10) 1.513 kg/m²		288	.056		1.90	2.19		4.09	5.70
0600	100x100 - W2.1 x W2.1 (8 x 8) 2.147 kg/m²		269	.059		2.27	2.35		4.62	6.35
0650	100x100 - W2.9 x W2.9 (6 x 6) 2.977 kg/m²		251	.064		2.96	2.52		5.48	7.40
0700	100x100 - W4 x W4 (4 x 4) 4.148 kg/m²		232	.069		4.18	2.72		6.90	9.10
0750	Rolls									
0800	50x50 - #14 galv. @ 9.5 kg, beam & column wrap	2 Rodm	60.39	.265	m²	3.30	10.45		13.75	21
0900	50x50 - #12 galv. for gunite reinforcing	"	60.39	.265	"	3.77	10.45		14.22	21.50

03230 | Stressing Tendons

		CREW	DAILY OUTPUT	LABOR-HOURS	UNIT	MAT.	LABOR	EQUIP.	TOTAL	TOTAL INCL O&P
600	**0010 PRESTRESSING STEEL**									**600**
0100	Grouted strand, post-tensioned in field, 15 m span, 45 metric ton	C-3	544	.118	kg	3.17	4.26	.20	7.63	10.55
0150	135 metric ton		1225	.052		1.57	1.89	.09	3.55	4.89
0300	30 m span, 45 metric ton R034136 -90		771	.083		3.17	3	.14	6.31	8.50
0350	135 metric ton		1452	.044		2.78	1.60	.07	4.45	5.70
0500	60 m span, 45 metric ton	-	1225	.052		3.15	1.89	.09	5.13	6.65
0550	135 metric ton		1588	.040		2.78	1.46	.07	4.31	5.45
0800	Grouted bars, 15 m span, 19 metric ton		1179	.054		1.54	1.96	.09	3.59	4.97
0850	65 metric ton		1452	.044		1.50	1.60	.07	3.17	4.28
1000	25 m span, 19 metric ton		1452	.044		1.54	1.60	.07	3.21	4.35
1050	65 metric ton		1905	.034		1.37	1.22	.06	2.65	3.52
1200	Ungrouted strand, 15 m span, 45 metric ton	C-4	578	.055		1.04	2.21	.05	3.30	4.86
1250	135 metric ton		669	.048		1.04	1.91	.05	3	4.35
1400	30 m span, 45 metric ton		680	.047		1.04	1.88	.05	2.97	4.30
1450	135 metric ton		748	.043		1.04	1.71	.04	2.79	4.02
1600	60 m span, 45 metric ton		680	.047		1.04	1.88	.05	2.97	4.30
1650	135 metric ton		771	.042		1.04	1.66	.04	2.74	3.92
1800	Ungrouted bars, 15 m span, 19 metric ton		635	.050		.95	2.02	.05	3.02	4.41
1850	65 metric ton		771	.042		.95	1.66	.04	2.65	3.81
2000	25 m span, 19 metric ton		816	.039		.95	1.57	.04	2.56	3.66
2050	65 metric ton		998	.032		.95	1.28	.03	2.26	3.18
2220	Ungrouted single strand, 30 m slab, 11 metric ton		544	.059		1.04	2.35	.06	3.45	5.10
2250	16 metric ton		669	.048		1.04	1.91	.05	3	4.35

03240 | Fibrous Reinforcing

		CREW	DAILY OUTPUT	LABOR-HOURS	UNIT	MAT.	LABOR	EQUIP.	TOTAL	TOTAL INCL O&P
300	**0010 FIBROUS REINFORCING**									**300**
0100	Synthetic fibers, add to concrete				kg	8.75			8.75	9.65
0110	0.89 kg/m³				m³	8			8	8.85
0150	Steel fibers, add to concrete				kg	1.01			1.01	1.12
0155	15 kg. per m³				m³	15.05			15.05	16.55
0160	30 kg/m³					30			30	33
0170	45 kg/m³					46.50			46.50	51
0180	59 kg/m³					60			60	66

CONCRETE 3

03310 | Structural Concrete

			CREW	DAILY OUTPUT	LABOR-HOURS	UNIT	2006 BARE COSTS MAT.	LABOR	EQUIP.	TOTAL	TOTAL INCL O&P	
200	0010	**CONCRETE, FIELD MIX**	R033105 -65									**200**
	0015	FOB forms 15.5 MPa				m³	102			102	112	
	0020	21 MPa	▼			"	106			106	116	
220	0010	**CONCRETE, READY MIX** Normal weight	R033105 -10									**220**
	0012	Includes local aggregate, sand, portland cement, and water										
	0015	Excludes all additives and treatments	R033105 -20									
	0020	13 MPa				m³	110			110	120	
	0100	17 MPa	R033105 -30				111			111	122	
	0150	21 MPa					114			114	125	
	0200	24 MPa	R033105 -40				116			116	128	
	0300	28 MPa					119			119	131	
	0350	31 MPa	R033105 -50				122			122	134	
	0400	34 MPa					126			126	138	
	0411	41 MPa					143			143	157	
	0412	55 MPa					234			234	257	
	0413	69 MPa					330			330	365	
	0414	83 MPa					400			400	440	
	0900	Roller compacted concrete, 38-50 mm agg., 120 kgm cement/M3					83.50			83.50	92	
	1000	For high early strength cement, add					10%					
	1010	For structural lightweight with regular sand, add					25%					
	1300	For winter concrete, add					6.20			6.20	6.80	
	1400	For hot weather concrete, add					8.50			8.50	9.35	
	1500	For Saturday delivery, add					9.15			9.15	10.05	
	2000	For all lightweight aggregate, add				▼	45%					
	3000	For integral colors, 17 MPa (5 bag mix)										
	3100	Red, yellow or brown, 0.82 kg/bag, add				m³	27.50			27.50	30	
	3200	4.26 kg/bag, add					143			143	157	
	3400	Black, 0.82 kg/bag, add					36.50			36.50	40.50	
	3500	3.40 kg/bag, add					153			153	168	
	3700	Green, 0.82 kg/bag, add					46.50			46.50	51	
	3800	3.40 kg/bag, add				▼	194			194	213	
	4000	Flowable fill: ash, cement, aggregate, water										
	4100	.27 - .55 MPa				m³	90			90	99	
	4150	Structural: ash, cement, aggregate, water & sand										
	4200	.34 MPa				m³	96			96	106	
	4250	.96 MPa					100			100	110	
	4300	3.5 MPa					104			104	114	
	4350	7 MPa					110			110	121	
	5000	Concrete ready mix, for hot water, add	▼			▼	6.20			6.20	6.80	
240	0010	**CONCRETE IN PLACE**	R033053 -10									**240**
	0020	Including forms (4 uses), concrete, placement, reinforcing										
	0050	steel and finishing unless otherwise indicated	R033053 -50									
	0300	Beams, 7.441 kg/mm, 3 m span		C-14A	11.94	16.746	m³	375	595	60.50	1,030.50	1,425	
	0350	7.6 m span	R033053 -60	"	14.18	14.101		390	500	51	941	1,275	
	0500	Chimney foundations, industrial, minimum		C-14C	24.64	4.546		169	154	.86	323.86	425	
	0510	Maximum	R033105 -80	"	18.13	6.178		199	209	1.17	409.17	545	
	0700	Columns, square, 300 mm x 300 mm, minimum reinforcing		C-14A	9.14	21.871		400	780	79	1,259	1,750	
	0720	Average reinforcing	R033105 -85		7.75	25.822		630	920	93	1,643	2,250	
	0740	Maximum reinforcing			6.90	28.967		950	1,025	105	2,080	2,800	
	0800	400 mm x 400 mm, minimum reinforcing			12.40	16.127		320	575	58	953	1,325	
	0820	Average reinforcing			9.61	20.809		540	740	75	1,355	1,825	
	0840	Maximum reinforcing			7.84	25.519		835	910	92	1,837	2,450	
	0900	600 mm x 600 mm, minimum reinforcing			18.09	11.056		270	395	40	705	955	
	0920	Average reinforcing			13.54	14.770		480	525	53.50	1,058.50	1,425	
	0940	Maximum reinforcing	▼	▼	10.82	18.486	▼	770	660	66.50	1,496.50	1,950	

Important: See the Reference Section for supporting data - Crews, Rental Equipment, City Cost Indexes and Reference Data

03300 | Cast-In-Place Concrete

	03310	Structural Concrete		CREW	DAILY OUTPUT	LABOR-HOURS	UNIT	MAT.	LABOR	EQUIP.	TOTAL	TOTAL INCL O&P	
								2006 BARE COSTS					
240	1000	900 mm x 900 mm, minimum reinforcing	R033053 -10	C-14A	25.76	7.764	m³	239	276	28	543	730	240
	1020	Average reinforcing			17.83	11.217		425	400	40.50	865.50	1,125	
	1040	Maximum reinforcing	R033053 -50		13.63	14.679		715	525	53	1,293	1,675	
	1100	Columns, round, tied, 300 mm diameter, min. reinforcing			16.03	12.474		355	445	45	845	1,125	
	1120	Average reinforcing	R033053 -60		11.68	17.130		585	610	62	1,257	1,675	
	1140	Maximum reinforcing			9.26	21.600		890	770	78	1,738	2,250	
	1200	400 mm diameter, minimum reinforcing [16"]	R033105 -80		24.08	8.307		310	296	30	636	840	
	1220	Average reinforcing			14.62	13.681		545	485	49.50	1,079.50	1,425	
	1240	Maximum reinforcing	R033105 -85		10.53	18.996		825	675	68.50	1,568.50	2,025	
	1300	500 mm diameter, minimum reinforcing [20"]			31.38	6.374		310	227	23	560	720	
	1320	Average reinforcing			18.39	10.876		525	385	39.50	949.50	1,225	
	1340	Maximum reinforcing			13.01	15.378		825	550	55.50	1,430.50	1,825	
	1400	600 mm diameter, minimum reinforcing [24"]			39.64	5.045		292	180	18.20	490.20	620	
	1420	Average reinforcing			20.69	9.666		525	345	35	905	1,150	
	1440	Maximum reinforcing			13.98	14.302		810	510	51.50	1,371.50	1,750	
	1500	900 mm diameter, minimum reinforcing [36"]			57.38	3.486		293	124	12.60	429.60	530	
	1520	Average reinforcing			28.66	6.977		500	248	25	773	970	
	1540	Maximum reinforcing		▼	17.46	11.452		785	410	41.50	1,236.50	1,550	
	1900	Elev. slabs, flat w/ drops, 600 kg/m² Sup. Load, 6.1 m span		C-14B	29.40	7.075		315	252	24.50	591.50	770	
	1950	9.2 m span			38.99	5.335		325	190	18.50	533.50	680	
	2100	Flat plate, 600 kg/m² Sup. Load, 4.6 m span			23.12	8.996		288	320	31	639	850	
	2150	7.6 m span			37.92	5.485		295	195	19.05	509.05	650	
	2300	Waffle, 760 mm domes, 600 kg/m² Sup. Load, 6.1 m span			28.34	7.338		435	261	25.50	721.50	915	
	2350	9.2 m span			33.70	6.173		385	220	21.50	626.50	795	
	2500	One way, 760 mm pans, 600 kg/m² Sup. Load, 4.6 m span			20.93	9.936		535	355	34.50	924.50	1,175	
	2550	7.6 m span			23.82	8.733		490	310	30.50	830.50	1,050	
	2700	One way beam & slab, 600 kg/m² Sup. Load, 4.6 m span			15.74	13.212		320	470	46	836	1,150	
	2750	7.6 m span			21.68	9.592		294	340	33.50	667.50	895	
	2900	Two way beam & slab, 600 kg/m² Sup. Load, 4.6 m span			18.38	11.316		305	400	39.50	744.50	1,000	
	2950	7.6 m span		▼	27.43	7.584	▼	256	270	26.50	552.50	730	
	3100	Elevated slabs including finish, not											
	3110	including forms or reinforcing											
	3150	Regular concrete, 100 mm slab		C-8	243	.230	m²	12.70	7.15	2.88	22.73	28	
	3200	150 mm slab			240	.233		18.75	7.25	2.92	28.92	34.50	
	3250	65 mm thick floor fill			249	.225		8.20	6.95	2.81	17.96	22.50	
	3300	Lightweight, 1750 kg/m³, 64 mm thick floor fill			240	.233		11.20	7.25	2.92	21.37	26.50	
	3400	Cellular concrete, 40 mm fill, under 465 m²			186	.301		7.55	9.35	3.77	20.67	26.50	
	3450	Over 930 m²			204	.275		7.10	8.50	3.43	19.03	24.50	
	3500	Add/floor for 3 to 6 stories high			2954	.019			.59	.24	.83	1.16	
	3520	For 7 to 20 stories high		▼	1969	.028	▼		.88	.36	1.24	1.73	
	3800	Footings, spread under 0.75 m³		C-14C	29.11	3.848	m³	228	130	.73	358.73	455	
	3850	Over 4 m³			61.96	1.808		315	61	.34	376.34	445	
	3900	Footings, strip, 450 mm x 230 mm, unreinforced			30.58	3.662		143	124	.70	267.70	350	
	3920	450 mm x 225 mm, reinforced			26.76	4.185		169	142	.80	311.80	410	
	3925	500 mm x 250 mm, unreinforced			34.41	3.255		138	110	.62	248.62	325	
	3930	500 mm x 250 mm, reinforced			30.58	3.662		160	124	.70	284.70	370	
	3935	600 mm x 300 mm, unreinforced			42.05	2.663		135	90	.51	225.51	291	
	3940	600 mm x 300 mm, reinforced			36.70	3.052		158	103	.58	261.58	335	
	3945	900 mm x 300 mm, unreinforced			53.52	2.093		130	71	.40	201.40	254	
	3950	900 mm x 300 mm, reinforced			45.88	2.441		150	82.50	.46	232.96	295	
	4000	Foundation mat, under 7.5 m³			29.57	3.788		234	128	.72	362.72	460	
	4050	Over 15 m³		▼	43.12	2.597		204	88	.49	292.49	365	
	4200	Grade walls, 200 mm thick, 2400 mm high		C-14D	35.04	5.708		205	201	20.50	426.50	565	
	4250	4275 mm high			20.84	9.596		266	340	34.50	640.50	860	
	4260	300 mm thick, 2400 mm high			49.18	4.067		183	144	14.70	341.70	440	
	4270	4275 mm high		▼	30.59	6.538	▼	208	231	23.50	462.50	615	

CONCRETE 3

For expanded coverage of these items see *Means Concrete & Masonry Cost Data 2006*

3 CONCRETE

03310 | Structural Concrete

				CREW	DAILY OUTPUT	LABOR-HOURS	UNIT	2006 BARE COSTS MAT.	LABOR	EQUIP.	TOTAL	TOTAL INCL O&P	
240	4300	380 mm thick, 2400 mm high	R033053 -10	C-14D	61.18	3.269	m³	172	115	11.80	298.80	380	**240**
	4350	3650 mm high			39.19	5.103		183	180	18.40	381.40	505	
	4500	5500 mm high	R033053 -50		37.35	5.355		203	189	19.35	411.35	540	
	4520	Handicap access ramp, railing both sides, 1 m wide		C-14H	4.44	10.801	m	670	380	4.86	1,054.86	1,325	
	4525	1.52 m wide	R033053 -60		3.72	12.887		695	450	5.80	1,150.80	1,475	
	4530	With 150 mm curb and rails both sides, 1 m wide			2.61	18.419		695	645	8.30	1,348.30	1,775	
	4535	1.52 m wide	R033105 -80		2.23	21.543		705	755	9.70	1,469.70	1,950	
	4650	Slab on grade, not including finish, 100 mm thick		C-14E	46.45	1.895	m³	139	66	.45	205.45	258	
	4700	150 mm thick	R033105 -85	"	70.34	1.251	"	133	43.50	.30	176.80	215	
	4751	Slab on grade, incl. troweled finish, not incl. forms											
	4760	or reinforcing, over 930 m², 100 mm thick		C-14F	318	.226	m²	12.50	7.30	.07	19.87	24.50	
	4820	150 mm thick			311	.232		18.20	7.50	.07	25.77	31.50	
	4840	200 mm thick			296	.243		25	7.85	.07	32.92	39.50	
	4900	300 mm thick			254	.283		37.50	9.15	.09	46.74	55	
	4950	380 mm thick			233	.309		47	10	.09	57.09	66.50	
	5000	Slab on grade, incl. textured finish, not incl. forms											
	5001	or reinforcing, 100 mm thick		C-14G	267	.210	m²	12.15	6.65	.08	18.88	23.50	
	5010	150 mm thick			241	.232		19.05	7.35	.09	26.49	32	
	5020	200 mm thick			216	.259		25	8.20	.10	33.30	40	
	5200	Lift slab in place above the foundation, incl. forms,											
	5210	reinforcing, concrete and columns, minimum		C-14B	196	1.061	m²	57.50	37.50	3.68	98.68	126	
	5250	Average			153	1.359		62.50	48.50	4.72	115.72	149	
	5300	Maximum			139	1.496		67.50	53	5.20	125.70	163	
	5500	Lightweight, ready mix, including screed finish only,											
	5510	not including forms or reinforcing											
	5550	1:4 for structural roof decks		C-14B	199	1.045	m³	159	37	3.63	199.63	237	
	5600	1:6 for ground slab with radiant heat		C-14F	70.34	1.024		161	33	.31	194.31	227	
	5650	1:3:2 with sand aggregate, roof deck		C-14B	199	1.045		157	37	3.63	197.63	235	
	5700	Ground slab		C-14F	81.81	.880		157	28.50	.26	185.76	216	
	5900	Pile caps, incl. forms and reinf., square or rect., under 4 m³		C-14C	41.40	2.706		198	91.50	.51	290.01	360	
	5950	Over 7.5 m³			57.35	1.953		182	66	.37	248.37	305	
	6000	Triangular or hexagonal, under 4 m³			40.52	2.764		139	93.50	.53	233.03	300	
	6050	Over 7.5 m³			64.99	1.723		157	58.50	.33	215.83	265	
	6200	Retaining walls, gravity, 1200 mm high see division 02830-100		C-14D	50.62	3.951		163	139	14.25	316.25	415	
	6250	3000 mm high			95.58	2.093		153	74	7.55	234.55	291	
	6300	Cantilever, level backfill loading, 2400 mm high			53.52	3.737		178	132	13.50	323.50	415	
	6350	4800 mm high			69.58	2.874		170	101	10.40	281.40	355	
	6800	Stairs, not including safety treads, free standing, 1.06 m wide		C-14H	25.30	1.897	m Nose	17.45	66.50	.85	84.80	124	
	6850	Cast on ground			38.10	1.260	"	13.20	44	.57	57.77	84	
	7000	Stair landings, free standing			18.58	2.583	m²	47	90.50	1.16	138.66	194	
	7050	Cast on ground			44.13	1.088	"	33	38	.49	71.49	96	
450	0010	**INSULATING CONCRETE** See division 03520-250											**450**
700	0010	**PLACING CONCRETE**	R033105 -70										**700**
	0020	Includes labor and equipment to place and vibrate											
	0050	Beams, elevated, small beams, pumped		C-20	45.88	1.395	m³		41.50	16.20	57.70	81.50	
	0100	With crane and bucket		C-7	34.41	2.093			63	29.50	92.50	129	
	0200	Large beams, pumped		C-20	68.81	.930			27.50	10.80	38.30	54.50	
	0250	With crane and bucket		C-7	49.70	1.449			43.50	20.50	64	89	
	0400	Columns, square or round, 300 mm thick, pumped		C-20	45.88	1.395			41.50	16.20	57.70	81.50	
	0450	With crane and bucket		C-7	30.58	2.354			70.50	33	103.50	145	
	0600	450 mm thick, pumped		C-20	68.81	.930			27.50	10.80	38.30	54.50	
	0650	With crane and bucket		C-7	42.05	1.712			51.50	24	75.50	106	
	0800	600 mm thick, pumped		C-20	70.34	.910			27	10.55	37.55	53	
	0850	With crane and bucket		C-7	53.52	1.345			40.50	18.90	59.40	83	

			CREW	DAILY OUTPUT	LABOR-HOURS	UNIT	MAT.	LABOR	EQUIP.	TOTAL	TOTAL INCL O&P	
700	**03310**	**Structural Concrete**						**2006 BARE COSTS**				**700**
700	1000	900 mm thick, pumped	R033105	C-20	107	.598	m³		17.75	6.95	24.70	35	
	1050	With crane and bucket	-70	C-7	76.46	.942			28	13.20	41.20	58	
	1400	Elevated slabs, less than 150 mm thick, pumped		C-20	107	.598			17.75	6.95	24.70	35	
	1450	With crane and bucket		C-7	72.64	.991			29.50	13.90	43.40	61	
	1500	150 mm to 250 mm thick, pumped		C-20	122	.525			15.60	6.10	21.70	30.50	
	1550	With crane and bucket		C-7	84.11	.856			25.50	12	37.50	52.50	
	1600	Slabs over 250 mm thick, pumped		C-20	138	.464			13.75	5.40	19.15	27	
	1650	With crane and bucket		C-7	99.40	.724			21.50	10.15	31.65	44.50	
	1900	Footings, continuous, shallow, direct chute		C-6	91.75	.523			15.10	.47	15.57	24	
	1950	Pumped		C-20	115	.557			16.50	6.45	22.95	32.50	
	2000	With crane and bucket		C-7	68.81	1.046			31.50	14.70	46.20	64	
	2100	Deep continuous footings, direct chute		C-6	107	.449			12.95	.40	13.35	20.50	
	2150	Pumped		C-20	122	.525			15.60	6.10	21.70	30.50	
	2200	With crane and bucket		C-7	84.11	.856			25.50	12	37.50	52.50	
	2400	Footings, spread, under 0.75 m³, direct chute		C-6	42.05	1.141			33	1.02	34.02	52	
	2450	Pumped		C-20	49.70	1.288			38	14.95	52.95	75.50	
	2500	With crane and bucket		C-7	34.41	2.093			63	29.50	92.50	129	
	2600	Over 4 m³, direct chute		C-6	91.75	.523			15.10	.47	15.57	24	
	2650	Pumped		C-20	115	.557			16.50	6.45	22.95	32.50	
	2700	With crane and bucket		C-7	76.46	.942			28	13.20	41.20	58	
	2900	Foundation mats, over 15 m³, direct chute		C-6	268	.179			5.20	.16	5.36	8.20	
	2950	Pumped		C-20	306	.209			6.20	2.43	8.63	12.20	
	3000	With crane and bucket		C-7	229	.314			9.45	4.41	13.86	19.30	
	3200	Grade beams, direct chute		C-6	115	.417			12.05	.37	12.42	19	
	3250	Pumped		C-20	138	.464			13.75	5.40	19.15	27	
	3300	With crane and bucket		C-7	91.75	.785			23.50	11	34.50	48	
	3500	High rise, for more than 5 stories, pumped, add/story		C-20	1606	.040			1.18	.46	1.64	2.33	
	3510	With crane and bucket, add/story		C-7	1606	.045			1.34	.63	1.97	2.75	
	3700	Pile caps, under 4 m³, direct chute		C-6	68.81	.698			20	.62	20.62	31.50	
	3750	Pumped		C-20	84.11	.761			22.50	8.85	31.35	44	
	3800	With crane and bucket		C-7	61.17	1.177			35.50	16.55	52.05	72	
	3850	Pile cap, 4 m³ to 8 m³, direct chute		C-6	134	.358			10.35	.32	10.67	16.30	
	3900	Pumped		C-20	153	.418			12.40	4.86	17.26	24.50	
	3950	With crane and bucket		C-7	115	.626			18.80	8.80	27.60	38.50	
	4000	Over 8 m³, direct chute		C-6	164	.293			8.45	.26	8.71	13.35	
	4050	Pumped		C-20	184	.348			10.35	4.04	14.39	20.50	
	4100	With crane and bucket		C-7	141	.511			15.30	7.15	22.45	31.50	
	4300	Slab on grade, 100 mm thick, direct chute		C-6	84.11	.571			16.50	.51	17.01	26	
	4350	Pumped		C-20	99.40	.644			19.10	7.50	26.60	38	
	4400	With crane and bucket		C-7	84.11	.856			25.50	12	37.50	52.50	
	4600	Over 150 mm thick, direct chute		C-6	126	.381			11	.34	11.34	17.30	
	4650	Pumped		C-20	141	.454			13.50	5.25	18.75	26.50	
	4700	With crane and bucket		C-7	111	.649			19.45	9.10	28.55	40	
	4900	Walls, 200 mm thick, direct chute		C-6	68.81	.698			20	.62	20.62	31.50	
	4950	Pumped		C-20	76.46	.837			25	9.75	34.75	48.50	
	5000	With crane and bucket		C-7	61.17	1.177			35.50	16.55	52.05	72	
	5050	300 mm thick, direct chute		C-6	76.46	.628			18.15	.56	18.71	28.50	
	5100	Pumped		C-20	84.11	.761			22.50	8.85	31.35	44	
	5200	With crane and bucket		C-7	68.81	1.046			31.50	14.70	46.20	64	
	5300	375 mm thick, direct chute		C-6	80.28	.598			17.30	.53	17.83	27	
	5350	Pumped		C-20	91.75	.698			20.50	8.10	28.60	41	
	5400	With crane and bucket		C-7	72.64	.991			29.50	13.90	43.40	61	
	5600	Wheeled concrete dumping, add to placing costs above											
	5610	Walking cart, 15 m haul, add		C-18	24.47	.368	m³		10.15	2.13	12.28	18.15	
	5620	46 m haul, add			18.35	.490			13.55	2.84	16.39	24	
	5700	76 m haul, add			13.76	.654			18.05	3.79	21.84	32	

For expanded coverage of these items see *Means Concrete & Masonry Cost Data 2006*

			DAILY OUTPUT	LABOR-HOURS	UNIT	2006 BARE COSTS				TOTAL INCL O&P	
		03310 | **Structural Concrete**				MAT.	LABOR	EQUIP.	TOTAL		
			CREW								
700	5800	Riding cart, 15 m haul, add	C-19	61.17	.147	m³		4.06	1.22	5.28	7.70	**700**
	5810	46 m haul, add		45.88	.196			5.40	1.62	7.02	10.25	
	5900	76 m haul, add		34.41	.262			7.20	2.16	9.36	13.65	

R033105-70

03350 | Concrete Finishing

			DAILY OUTPUT	LABOR-HOURS	UNIT	MAT.	LABOR	EQUIP.	TOTAL	TOTAL INCL O&P	
300	0010	**FINISHING FLOORS**									**300**
	0020	Monolithic, screed finish	1 Cefi	83.61	.096	m²		3.29		3.29	4.86	
	0100	Screed and bull float (darby) finish		67.35	.119			4.09		4.09	6.05	
	0150	Screed, float, and broom finish		58.53	.137			4.70		4.70	6.95	
	0200	Screed, float, and hand trowel		55.74	.144			4.94		4.94	7.30	
	0250	Machine trowel		51.10	.157			5.40		5.40	7.95	
	0400	Integral topping and finish, using 1:1:2 mix, 5 mm thick	C-10B	92.90	.431		.65	13	1.89	15.54	22.50	
	0450	13 mm thick		88.26	.453		1.83	13.70	1.99	17.52	25	
	0500	19 mm thick		78.97	.507		2.80	15.30	2.23	20.33	29	
	0600	25 mm thick		69.68	.574		3.77	17.35	2.53	23.65	33.50	
	0800	Granolithic topping, laid after, 1:1:1-1/2 mix, 13 mm thick		54.81	.730		2.05	22	3.21	27.26	39.50	
	0820	19 mm thick		53.88	.742		3.12	22.50	3.27	28.89	41	
	0850	25 mm thick		53.42	.749		4.20	22.50	3.29	29.99	43	
	0950	50 mm thick		46.45	.861		8.30	26	3.79	38.09	53	
	1200	Heavy duty, 1:1:2, 19 mm thick, preshrunk, gray, 2000 m²		29.73	1.346		2.80	40.50	5.90	49.20	71.50	
	1300	10 000 m²		35.30	1.133		2.80	34	4.99	41.79	60.50	
	1600	Exposed local aggregate finish, minimum	1 Cefi	58.06	.138		1.72	4.74		6.46	8.95	
	1650	Maximum		43.20	.185		2.58	6.35		8.93	12.20	
	1800	Floor abrasives, 1.22 kg/m², aluminum oxide		78.97	.101		3.44	3.49		6.93	8.90	
	1850	Silicon carbide		78.97	.101		4.74	3.49		8.23	10.40	
	2000	Floor hardeners, metallic, light service, 2.44 kg/m², add		78.97	.101		4.09	3.49		7.58	9.65	
	2050	Medium service, 3.66 kg/m²		69.68	.115		6.15	3.95		10.10	12.50	
	2100	Heavy service, 4.88 kg/m²		60.39	.132		8.05	4.56		12.61	15.65	
	2150	Extra heavy, 7.32 kg/m²		53.42	.150		12.15	5.15		17.30	21	
	2300	Non-metallic, light service, 2.44 kg/m²		78.97	.101		1.40	3.49		4.89	6.65	
	2350	Medium service, 3.66 kg/m²		69.68	.115		2.05	3.95		6	8.10	
	2400	Heavy service, 4.88 kg/m²		60.39	.132		2.69	4.56		7.25	9.70	
	2450	Extra heavy, 7.32 kg/m²		53.42	.150		4.09	5.15		9.24	12.10	
	2800	Trap rock wearing surface for monolithic floors										
	2810	9.77 kg/m²	C-10B	116	.345	m²	.22	10.40	1.52	12.14	17.70	
	3000	Floor coloring, dusted on, minimum (2.93 kg/m²), add to above	1 Cefi	121	.066		6.45	2.27		8.72	10.45	
	3050	Maximum (4.88 kg/m²), add to above	"	58.06	.138		10.75	4.74		15.49	18.85	
	3100	Colored powder only				kg	2.20			2.20	2.43	
	3600	13 mm topping using 2.93 kg/m² powdered color	C-10B	54.81	.730	m²	51	22	3.21	76.21	93.50	
	3650	13 mm topping using 4.88 kg/m² powdered color	"	54.81	.730		55.50	22	3.21	80.71	98	
	3800	Dustproofing, solvent-based, 1 coat	1 Cefi	177	.045		2.58	1.55		4.13	5.10	
	3850	2 coats		121	.066		9.25	2.27		11.52	13.60	
	4000	Epoxy-based, 1 coat		139	.058		1.61	1.98		3.59	4.64	
	4050	2 coats		139	.058		3.12	1.98		5.10	6.35	
	4400	Stair finish, float		25.55	.313			10.75		10.75	15.90	
	4500	Steel trowel finish		18.58	.431			14.80		14.80	22	
	4600	Silicon carbide finish, 1.22 kg/m²		13.94	.574		3.44	19.75		23.19	33	
325	0010	**CONTROL JOINTS, SAW CUT**										**325**
	0100	Sawcut in green concrete										
	0120	25 mm depth	C-27	610	.026	m	.26	.90	.18	1.34	1.82	
	0140	38 mm depth		549	.029		.43	1	.19	1.62	2.15	
	0160	50 mm depth		488	.033		.56	1.13	.22	1.91	2.49	
	0200	Clean out control joint of debris	C-28	1829	.004			.15	.01	.16	.23	
	0300	Joint sealant										
	0320	Backer rod, polyethylene, 6 mm diameter	1 Cefi	140	.057	m	.07	1.97		2.04	2.97	

R032110-80

Important: See the Reference Section for supporting data - Crews, Rental Equipment, City Cost Indexes and Reference Data

		03350	Concrete Finishing		CREW	DAILY OUTPUT	LABOR-HOURS	UNIT	2006 BARE COSTS				TOTAL INCL O&P	
									MAT.	LABOR	EQUIP.	TOTAL		
325	0340		Sealant, polyurethane											325
	0360		6 mm x 6 mm (24.8 m/L)	R032110 -80	1 Cefi	82.30	.097	m	.52	3.34		3.86	5.50	
	0380		6 mm x 13 mm (12.4 m/L)		"	77.72	.103	"	1.05	3.54		4.59	6.35	
350	0010	**FINISHING WALLS**												350
	0020		Break ties and patch voids		1 Cefi	50.17	.159	m²	.32	5.50		5.82	8.40	
	0050		Burlap rub with grout			41.81	.191		.32	6.60		6.92	10	
	0100		Carborundum rub, dry			25.08	.319			10.95		10.95	16.20	
	0150		Wet rub			16.26	.492			16.95		16.95	25	
	0300		Bush hammer, green concrete		B-39	92.90	.517			15	1.59	16.59	25	
	0350		Cured concrete		"	60.39	.795			23	2.44	25.44	38	
	0600		Float finish, 2 mm thick		1 Cefi	27.87	.287		2.58	9.85		12.43	17.35	
	0700		Sandblast, light penetration		C-10	102	.235		2.58	7.55		10.13	14.20	
	0750		Heavy penetration		"	34.84	.689		5.15	22		27.15	38.50	
600	0010	**SLAB TEXTURE STAMPING**												600
	0020		Approx. 0.28 m²- 0.47 m² each, buy, minimum					Ea.	99			99	109	
	0030		Average						139			139	153	
	0120		Maximum						1,825			1,825	2,000	
	0200		Commonly used chemicals for texture systems											
	0210		Hardener, colored powder					m²	5.50			5.50	6.05	
	0220		Release agent, colored powder						.86			.86	.97	
	0230		Curing & sealing compound, solvent based						.54			.54	.54	
	0300		Broadcasting hardener & release agent, stamping		2 Cefi	92.90	.172			5.90		5.90	8.75	

		03370	Specially Placed Concrete											
300	0010	**GUNITE (DRY-MIX)**												300
	0020		Applied in 25 mm layers, no mesh included		C-8	186	.301	m²	2.91	9.35	3.77	16.03	21.50	
	0100		Mesh for gunite 50 x 50, #12, to 75 mm thick		2 Rodm	74.32	.215		3.77	8.50		12.27	18.20	
	0150		Over 75 mm thick		"	46.45	.344		3.77	13.60		17.37	26.50	
	0300		Typical in place, including mesh, 50 mm thick, minimum		C-16	92.90	.775		9.45	25.50	7.55	42.50	58.50	
	0350		Maximum			46.45	1.550		9.45	51	15.10	75.55	107	
	0500		100 mm thick, minimum			69.68	1.033		15.20	34	10.05	59.25	81	
	0550		Maximum			32.52	2.214		15.20	73	21.50	109.70	153	
	0900		Prepare old walls, no scaffolding, minimum		C-10	92.90	.258			8.30		8.30	12.40	
	0950		Maximum		"	25.55	.939			30		30	45	
	1100		For high finish requirement or close tolerance, add, minimum							50%				
	1150		Maximum							110%				
700	0010	**ROLLER COMPACTED CONCRETE**												700
	0100		Mass placement, 300 mm lift, 300 mm layer		B-10C	979	.012	m³		.41	1.41	1.82	2.18	
	0200		600 mm lift, 150 mm layer		"	1223	.010			.33	1.13	1.46	1.74	
	0210		Vertical face, formed, 300 mm lift		B-11V	306	.078			2.15	.41	2.56	3.80	
	0220		150 mm lift		"	153	.157			4.30	.82	5.12	7.60	
	0300		Sloped face, nonformed, 300 mm lift		B-11L	294	.054			1.74	1.56	3.30	4.38	
	0360		150 mm lift		"	147	.109			3.49	3.11	6.60	8.75	
	0400		Surface preparation, vacuum truck		B-6A	2742	.007	m²		.23	.12	.35	.49	
	0450		Water clean		B-9A	2508	.010			.27	.13	.40	.55	
	0460		Water blast		B-9B	669	.036			.99	.57	1.56	2.16	
	0500		Joint bedding placement, 25 mm thick		B-11C	815	.020			.63	.28	.91	1.27	
	0510		Conveyance of materials, 14 m³ truck, 5 min. cycle		B-34F	1566	.005	m³		.14	.60	.74	.88	
	0520		10 min. cycle			783	.010			.29	1.19	1.48	1.75	
	0540		15 min. cycle			520	.015			.44	1.79	2.23	2.64	
	0550		With crane and bucket		C-23A	1223	.033			1.01	1.69	2.70	3.41	
	0560		With 3 m³. loader, 4 min. cycle		B-10U	367	.033			1.10	1.91	3.01	3.77	
	0570		8 min. cycle			184	.065			2.19	3.81	6	7.55	
	0580		12 min. cycle			122	.098			3.30	5.75	9.05	11.35	

CONCRETE 3

03370 | Specially Placed Concrete

		CREW	DAILY OUTPUT	LABOR-HOURS	UNIT	MAT.	LABOR	EQUIP.	TOTAL	TOTAL INCL O&P	
700	0590	With belt conveyor	C-7D	459	.122	m³		3.54	.33	3.87	5.85	**700**
	0600	With 17 m³. scraper, 5 min. cycle	B-33J	1101	.007			.27	1.43	1.70	1.98	
	0610	10 min. cycle		551	.015			.53	2.86	3.39	3.95	
	0620	15 min. cycle		367	.022			.80	4.30	5.10	5.95	
	0630	20 min. cycle	▼	275	.029	▼		1.07	5.75	6.82	7.90	
	0640	Water cure, small job, < 400 m	B-94C	8	1	Hr.		27.50	8.05	35.55	51.50	
	0650	Large job, over 400 m³	B-59A	8	3	"		83	40.50	123.50	174	
	0660	RCC paving, with ashpalt paver including material	B-25C	765	.063	m³	71	1.93	2.43	75.36	83.50	
	0670	200 mm thick layers		3512	.014	m²	16.95	.42	.53	17.90	19.90	
	0680	300 mm thick layers	▼	2341	.021	"	25	.63	.79	26.42	29.50	
	0700	Roller compacted concrete, 38-50 mm agg., 120 kgm cement/m³				m³	83.50			83.50	92	
800	0010	**SHOTCRETE (WET-MIX)**										**800**
	0020	Wet mix, placed @ 7.6 m³ per hour, 21 MPa	C-8C	61.17	.785	m³	113	24	3.83	140.83	165	
	0100	2nd pour	C-6	11.47	4.185	"	101	121	3.72	225.72	300	
	1010	Fiber reinforced, 25 mm thick	C-8E	108	.296	m²	6.90	9.35	7.65	23.90	30.50	
	1020	50 mm thick		55.74	.574		13.80	18.15	14.85	46.80	59	
	1030	75 mm thick		51.10	.626		20.50	19.80	16.20	56.50	70.50	
	1040	100 mm thick	▼	46.45	.689	▼	27.50	22	17.85	67.35	83	

03390 | Concrete Curing

		CREW	DAILY OUTPUT	LABOR-HOURS	UNIT	MAT.	LABOR	EQUIP.	TOTAL	TOTAL INCL O&P	
200	0010	**WATER CURING**										**200**
	0015	With burlap, 4 uses assumed, 213 gram	2 Clab	511	.031	m²	.76	.86		1.62	2.17	
	0100	280 gram		511	.031		1.36	.86		2.22	2.84	
	0200	Curing blanket, burlap/poly, 2-ply		650	.025		1.62	.67		2.29	2.83	
	0300	Sprayed membrane curing compound	▼	883	.018		.60	.50		1.10	1.43	
	0400	Curing blankets, 25 mm to 50 mm thick, buy, minimum					2.91			2.91	3.23	
	0450	Maximum					4.41			4.41	4.84	
	0500	Electrically heated pads, 110 V, 15 W/m², buy					58			58	63.50	
	0600	20 W/m², buy					77			77	84.50	
	0710	Electrically, heated pads, 15 W/m², 20 uses, minimum					2.05			2.05	2.26	
	0800	Maximum				▼	3.44			3.44	3.77	

03410 | Plant-Precast Structural Concrete

			CREW	DAILY OUTPUT	LABOR-HOURS	UNIT	MAT.	LABOR	EQUIP.	TOTAL	TOTAL INCL O&P	
100	0010	**PRECAST BEAMS**										**100**
	0011	L-shaped, 6.1 m span, 300 mm x 500 mm	R034105 -30	C-11	32	2.250	Ea.	1,050	88	48	1,186	1,350	
	1000	Inverted tee beams, add to above, small beams						15%					
	1050	Large beams						20%					
	1200	Rectangular, 6.1 m span, 300 mm x 500 mm		C-11	32	2.250		780	88	48	916	1,050	
	1250	450 mm x 900 mm			24	3		885	117	64	1,066	1,250	
	1300	600 mm x 1120 mm			22	3.273		1,100	128	70	1,298	1,525	
	1400	9 m span, 300 mm x 900 mm			24	3		1,200	117	64	1,381	1,575	
	1450	450 mm x 1120 mm			20	3.600		1,500	141	77	1,718	1,975	
	1500	600 mm x 1320 mm			16	4.500		1,925	176	96	2,197	2,525	
	1600	12 m span, 300 mm x 1320 mm			20	3.600		2,200	141	77	2,418	2,750	
	1650	450 mm x 1320 mm			16	4.500		2,625	176	96	2,897	3,300	

03410 | Plant-Precast Structural Concrete

		CREW	DAILY OUTPUT	LABOR-HOURS	UNIT	MAT.	LABOR	EQUIP.	TOTAL	TOTAL INCL O&P	
100	**1700**	600 mm x 1320 mm	R034105 -30	C-11	12	6	Ea.	2,875	235	128	3,238	3,725	**100**
	2000	"T" shaped, 6 m span, 300 mm x 500 mm			32	2.250		1,325	88	48	1,461	1,650	
	2050	450 mm x 900 mm			24	3		1,500	117	64	1,681	1,925	
	2100	600 mm x 1120 mm			22	3.273		1,875	128	70	2,073	2,375	
	2200	9 m span, 300 mm x 900 mm			24	3		2,025	117	64	2,206	2,500	
	2250	450 mm x 1120 mm			20	3.600		2,550	141	77	2,768	3,125	
	2300	600 mm x 1320 mm			16	4.500		3,275	176	96	3,547	4,000	
	2500	12 m span, 300 mm x 1320 mm			20	3.600		3,750	141	77	3,968	4,425	
	2550	450 mm x 1320 mm			16	4.500		4,475	176	96	4,747	5,325	
	2600	600 mm x 1320 mm			12	6		4,900	235	128	5,263	5,950	
210	**0010**	**PRECAST COLUMNS**	R034105 -30	C-11									**210**
	0020	Rectangular to 3.7 m high, small columns			36.58	1.969	m	140	77	42	259	335	
	0050	Large columns			29.26	2.461		244	96.50	52.50	393	495	
	0300	7.3 m high, small columns			58.52	1.230		140	48	26.50	214.50	267	
	0350	Large columns			43.89	1.640		244	64	35	343	420	
	0700	7.3 m high, 1 haunch, 300 mm x 300 mm			32	2.250	Ea.	1,025	88	48	1,161	1,325	
	0800	500 mm x 500 mm			28	2.571	"	1,800	101	55	1,956	2,200	
400	**0010**	**PRECAST JOISTS**	R034105 -30	C-12									**400**
	0015	195 kg/m² L.L., 150 mm deep for 3.7 m spans			183	.262	m	23.50	9.15	3.48	36.13	43.50	
	0050	200 mm deep for 4.9 m spans			175	.274		39	9.60	3.64	52.24	61.50	
	0100	250 mm deep for 6.1 m spans			168	.286		68	10	3.79	81.79	94	
	0150	300 mm deep for 7.3 m spans			160	.300		93	10.50	3.98	107.48	123	
620	**0010**	**PRECAST SLAB PLANKS**	R034105 -30	C-11									**620**
	0020	Prestressed roof/floor, grouted, solid, 100 mm thick			223	.323	m²	48	12.65	6.90	67.55	82	
	0050	150 mm thick			260	.277		59	10.85	5.90	75.75	90.50	
	0100	Hollow, 200 mm thick			297	.242		62	9.50	5.20	76.70	90	
	0150	250 mm thick			334	.216		65.50	8.45	4.61	78.56	92	
	0200	300 mm thick			372	.194		74.50	7.55	4.14	86.19	100	
650	**0010**	**PRESTRESSED CONCRETE** post-tensioned in place	R034105 -30										**650**
	0020	See also Division 03230-600											
	0100	Post-tensioned in place, small job	R034136 -90	C-17B	6.50	12.617	m³	770	465	41.50	1,276.50	1,625	
	0200	Large job		"	7.65	10.725	"	575	395	35.50	1,005.50	1,300	
660	**0010**	**PRESTRESSED ROOF AND FLOOR MEMBERS**	R034105 -30										**660**
750	**0010**	**TEES PRESTRESSED**	R034105 -30	C-11									**750**
	0020	Quad tee, short spans, roof			669	.108	m²	61	4.21	2.30	67.51	77	
	0050	Floor			669	.108		61	4.21	2.30	67.51	77	
	0200	Double tee, floor members, 18 m span			780	.092		79.50	3.61	1.97	85.08	96	
	0250	24 m span			743	.097		93	3.79	2.07	98.86	111	
	0300	Roof members, 9 m span			446	.161		67	6.30	3.45	76.75	89	
	0350	15 m span			595	.121		77	4.74	2.59	84.33	95.50	
	0400	Wall members, up to 17 m high			334	.216		102	8.45	4.61	115.06	132	
	0500	Single tee roof members, 12 m span			297	.242		80.50	9.50	5.20	95.20	111	
	0550	24 m span			476	.151		99	5.90	3.23	108.13	123	
	0600	30 m span			557	.129		149	5.05	2.76	156.81	176	
	0650	37 m span			557	.129		163	5.05	2.76	170.81	191	
	1000	Double tees, floor members											
	1100	Lightweight, 500 mm x 2400 mm wide, 14 m span		C-11	20	3.600	Ea.	2,425	141	77	2,643	2,975	
	1150	600 mm x 2400 mm wide, 15 m span			18	4		2,625	157	85.50	2,867.50	3,250	
	1200	800 mm x 3050 mm wide, 18 m span			16	4.500		4,425	176	96	4,697	5,275	
	1250	Standard weight, 300 mm x 2400 mm wide, 6 m span			22	3.273		905	128	70	1,103	1,300	
	1300	400 mm x 2400 mm wide, 7.6 m span			20	3.600		1,200	141	77	1,418	1,650	
	1350	450 mm x 2400 mm wide, 9 m span			20	3.600		1,525	141	77	1,743	2,000	
	1400	500 mm x 2400 mm wide, 14 m span			18	4		1,725	157	85.50	1,967.50	2,275	

For expanded coverage of these items see *Means Concrete & Masonry Cost Data 2006*

3 CONCRETE

03410 | Plant-Precast Structural Concrete

			CREW	DAILY OUTPUT	LABOR-HOURS	UNIT	MAT.	LABOR	EQUIP.	TOTAL	TOTAL INCL O&P	
750	1450	600 mm x 2400 mm wide, 15 m span	C-11	16	4.500	Ea.	2,225	176	96	2,497	2,850	750
	1500	800 mm x 3050 mm wide, 18 m span (R034105-30)		14	5.143		4,075	201	110	4,386	4,950	
	2000	Roof members										
	2050	Lightweight, 500 mm x 2400 mm wide, 12 m span	C-11	20	3.600	Ea.	2,025	141	77	2,243	2,550	
	2100	600 mm x 2400 mm wide, 15 m span		18	4		2,700	157	85.50	2,942.50	3,350	
	2150	800 mm x 3050 mm wide, 18 m span		16	4.500		4,475	176	96	4,747	5,325	
	2200	Standard weight, 300 mm x 2400 mm wide, 9 m span		22	3.273		1,350	128	70	1,548	1,800	
	2250	400 mm x 2400 mm wide, 9 m span		20	3.600		1,425	141	77	1,643	1,900	
	2300	450 mm x 2400 mm wide, 9 m span		20	3.600		1,575	141	77	1,793	2,075	
	2350	500 mm x 2400 mm wide, 12 m span		18	4		1,625	157	85.50	1,867.50	2,175	
	2400	600 mm x 2400 mm wide, 15 m span		16	4.500		2,175	176	96	2,447	2,775	
	2450	800 mm x 3050 mm wide, 18 m span		14	5.143		3,800	201	110	4,111	4,650	

03450 | Plant-Precast Architectural Concrete

			CREW	DAILY OUTPUT	LABOR-HOURS	UNIT	MAT.	LABOR	EQUIP.	TOTAL	TOTAL INCL O&P	
850	0011	**WALL PANELS**										850
	0050	Uninsulated 100 mm thick, smooth gray (R034513-10)										
	0150	Low rise, 1200 mm x 2400 mm x 100 mm thick	C-11	29.73	2.422	m²	135	95	52	282	370	
	0200	2400 mm x 2400 mm x 100 mm thick		53.51	1.346		134	52.50	29	215.50	270	
	0250	2400 mm x 4800 mm x 100 mm thick		95.13	.757		132	29.50	16.20	177.70	215	
	0400	2400 mm x 2400 mm, 100 mm thick, smooth gray		53.51	1.346		134	52.50	29	215.50	270	
	0500	Exposed aggregate		53.51	1.346		171	52.50	29	252.50	310	
	0600	High rise, 1200 mm x 2400 mm x 100 mm		26.76	2.691		135	105	57.50	297.50	395	
	0650	2400 mm x 2400 mm x 100 mm thick		47.56	1.514		134	59	32.50	225.50	286	
	0700	2400 mm x 4800 mm x 100 mm		71.35	1.009		132	39.50	21.50	193	239	
	0800	Insulated panel, 50 mm polystyrene, add					10			10	11	
	0850	50 mm urethane, add					7.95			7.95	8.70	
	1000	6100 mm x 3050 mm, 150 mm thick, smooth gray	C-11	130	.554		229	21.50	11.85	262.35	305	
	1100	Exposed aggregate	"	167	.431		256	16.85	9.20	282.05	320	
	1200	Finishes, white, add					24.50			24.50	27	
	1250	Exposed aggregate, add					12.15			12.15	13.35	
	1300	Granite faced, domestic, add					266			266	292	
	2200	Precast fiberglass reinforced cement wall panels										
	2210	2400 mm x 2400 mm, minimum	E-2	69.68	.804	m²	183	31.50	20.50	235	278	
	2220	Maximum	"	55.74	1.005	"	205	39	25.50	269.50	320	
855	0010	**PRECAST WINDOW SILLS**										855
	0600	Precast concrete, 100 mm tapers to 75 mm, 230 mm wide	D-1	21.34	.750	m	30.50	24		54.50	70	
	0650	280 mm wide		18.29	.875		41	28		69	88	
	0700	330 mm wide, 90 mm tapers to 65 mm, 300 mm wall		15.24	1.050		44.50	34		78.50	100	

03470 | Tilt-Up, Precast Concrete

			CREW	DAILY OUTPUT	LABOR-HOURS	UNIT	MAT.	LABOR	EQUIP.	TOTAL	TOTAL INCL O&P	
600	0010	**TILT-UP WALL PANELS**										600
	0015	Wall panel construction, walls only, 140 mm thick (R034713-20)	C-14	149	.966	m²	48.50	33.50	7.10	89.10	114	
	0100	190 mm thick		144	1		61.50	34.50	7.35	103.35	130	
	0500	Walls and columns, 140 mm Thk. walls, 300 mm x 300 mm col.		145	.993		75.50	34.50	7.30	117.30	145	
	0550	190 mm thick wall, 300 mm x 300 mm columns		127	1.134		93	39	8.35	140.35	173	
	0800	Columns only, site precast, minimum		60.96	2.362	m	39	81.50	17.35	137.85	190	
	0850	Maximum		32	4.499	"	49	156	33	238	335	

03480 | Precast Concrete Specialties

			CREW	DAILY OUTPUT	LABOR-HOURS	UNIT	MAT.	LABOR	EQUIP.	TOTAL	TOTAL INCL O&P	
200	0010	**CURBS** Roadway type, see Division 02770-300										200
400	0010	**LINTELS**										400
	0800	Precast concrete, 100 mm x 200 mm, to 1.5 m long	D-10	28	1.429	Ea.	26	48	19.15	93.15	123	
	0850	1.5 m - 3.6 m long		24	1.667		66.50	56	22.50	145	183	
	1000	150 mm wide, 200 mm high, to 1.5 m long		26	1.538		32.50	52	20.50	105	137	

03400 | Precast Concrete

03480 | Precast Concrete Specialties

			CREW	DAILY OUTPUT	LABOR-HOURS	UNIT	2006 BARE COSTS				TOTAL INCL O&P	
							MAT.	LABOR	EQUIP.	TOTAL		
400	1050	1.5 m - 3.6 m long	D-10	22	1.818	Ea.	90	61.50	24.50	176	219	**400**
	1200	200 mm wide, 200 mm high, to 1.5 m long		24	1.667		31.50	56	22.50	110	145	
	1250	1.5 m - 3.6 m long		20	2		80.50	67.50	27	175	221	
	1400	250 mm wide, 200 mm high U-Shape, to 4.3 m long		18	2.222		168	75	30	273	330	
	1450	300 mm wide, 200 mm high U-Shape, to 5.8 m long	↓	16	2.500	↓	219	84.50	33.50	337	405	
800	0010	**PRECAST STAIRS**										**800**
	0020	Precast concrete treads on steel stringers, 900 mm wide	C-12	75	.640	Riser	101	22.50	8.50	132	155	
	0300	Front entrance, 1500 mm wide with 1200 mm platform, 2 risers		16	3	Flight	315	105	40	460	555	
	0350	5 risers		12	4		515	140	53	708	840	
	0500	1800 mm wide, 2 risers		15	3.200		335	112	42.50	489.50	590	
	0550	5 risers		11	4.364		530	153	58	741	885	
	0700	2100 mm wide, 2 risers		14	3.429		420	120	45.50	585.50	695	
	0750	5 risers	↓	10	4.800		680	168	63.50	911.50	1,075	
	1200	Basement entrance stairs, steel bulkhead doors, minimum	B-51	22	2.182		700	60.50	5.85	766.35	870	
	1250	Maximum	"	11	4.364	↓	1,200	121	11.70	1,332.70	1,525	

03500 | Cementitious Decks & Underlayments

03510 | Cementitious Roof Deck

				CREW	DAILY OUTPUT	LABOR-HOURS	UNIT	2006 BARE COSTS				TOTAL INCL O&P	
								MAT.	LABOR	EQUIP.	TOTAL		
200	0010	**WOOD FIBER** Lightweight cement system	R051223 -50										**200**
	0050	Plank, beveled, 25 mm thick		2 Carp	92.90	.172	m²	19.90	6.10		26	31.50	
	0100	Plank, T & G, 38 mm thick			90.58	.177		21	6.30		27.30	33.50	
	0150	51 mm thick			88.26	.181		28	6.45		34.45	41	
	0200	64 mm thick			85.93	.186		31	6.60		37.60	44.50	
	0250	76 mm thick			83.61	.191		35	6.80		41.80	49.50	
	0300	89 mm thick			81.29	.197		51.50	7		58.50	67.50	
	0350	102 mm thick		↓	78.97	.203		57.50	7.20		64.70	74.50	
	1000	Bulb tee, sub-purlin and grout, 1.83 m span, add		E-1	465	.052		17.75	2.01	.19	19.95	23	
	1100	2.44 m span		"	390	.062	↓	20	2.40	.23	22.63	26.50	
250	0010	**PRECAST CONCRETE CHANNEL SLABS**											**250**
	0020	Lightweight channel slab, 70 mm or 90 mm thick, straight		C-12	146	.329	m²	50	11.50	4.36	65.86	77.50	
	0050	Chopped up			72.93	.658		52.50	23	8.75	84.25	103	
	0200	150 mm thick, span to 6 m			121	.397		55.50	13.85	5.25	74.60	89	
	0300	200 mm thick, span to 7 m		↓	102	.471	↓	64.50	16.45	6.25	87.20	103	
270	0010	**PRECAST LIGHTWEIGHT CONCRETE PLANK**											**270**
	0015	Lightweight, nailable, T&G, 50 mm thick		C-12	167	.287	m²	34.50	10.05	3.81	48.36	57.50	
	0050	70 mm thick			146	.329		35.50	11.50	4.36	51.36	61.50	
	0100	95 mm thick		↓	128	.375		50	13.10	4.97	68.07	81	
	0150	For premium ceiling finish, add					↓	10%					
	0200	For sloping roofs, slope over 4 in 12, add							25%				
	0250	Slope over 6 in 12, add							150%				
350	0010	**FORMBOARD** Including sub-purlins	R051223 -50										**350**
	0050	Non-asbestos fiber cement, 3 mm thick		C-13	274	.088	m²	23	3.37	.33	26.70	31.50	
	0070	6 mm thick			274	.088		26.50	3.37	.33	30.20	35.50	
	0100	Fiberglass, 25 mm thick, economy		↓	251	.096		29	3.68	.36	33.04	38.50	
	1000	Poured gypsum, 50 mm thick, add to formboard above		C-8	557	.101		17.20	3.12	1.26	21.58	25	
	1100	75 mm thick		"	446	.126	↓	26	3.89	1.57	31.46	36	

03500 | Cementitious Decks & Underlayments

03510 | Cementitious Roof Deck

		CREW	DAILY OUTPUT	LABOR-HOURS	UNIT	2006 BARE COSTS				TOTAL INCL O&P	
						MAT.	LABOR	EQUIP.	TOTAL		
900	0010	**WOOD PLANK** Roof decks, see Divisions 06150-600 & 06170-550										900

03520 | Lightweight Concrete Roof Insulation

		CREW	DAILY OUTPUT	LABOR-HOURS	UNIT	MAT.	LABOR	EQUIP.	TOTAL	INCL O&P	
250	0010	**INSULATING ROOF FILL,** lightweight cellular concrete R035216-10									250
	0020	Portland cement and foaming agent	C-8	38.23	1.465	m³	110	45.50	18.30	173.80	210
	0100	Poured vermiculite or perlite, field mix,									
	0110	1:6 field mix	C-8	38.23	1.465	m³	121	45.50	18.30	184.80	222
	0200	Ready mix, 1:6 mix, roof fill, 50 mm thick		929	.060	m²	5.50	1.87	.75	8.12	9.75
	0250	75 mm thick		715	.078		8.30	2.43	.98	11.71	13.95
	0400	Expanded volcanic glass rock, with binder, minimum	2 Carp	139	.115		3.12	4.09		7.21	9.80
	0450	Maximum	"	111	.144		9.45	5.10		14.55	18.45

03600 | Grouts

03610 | Construction Grout

		CREW	DAILY OUTPUT	LABOR-HOURS	UNIT	2006 BARE COSTS				TOTAL INCL O&P	
						MAT.	LABOR	EQUIP.	TOTAL		
400	0010	**GROUT**									400
	0020	Column & machine bases, non-shrink, metallic, 25 mm deep	1 Cefi	3.25	2.460	m²	62	84.50		146.50	193
	0050	50 mm deep		2.32	3.445		124	118		242	310
	0300	Non-shrink, non-metallic, 25 mm deep		3.25	2.460		46.50	84.50		131	176
	0350	50 mm deep		2.32	3.445		93	118		211	277

03900 | Concrete Restoration & Cleaning

03920 | Concrete Resurfacing

		CREW	DAILY OUTPUT	LABOR-HOURS	UNIT	2006 BARE COSTS				TOTAL INCL O&P	
						MAT.	LABOR	EQUIP.	TOTAL		
600	0010	**PATCHING CONCRETE**									600
	0100	Floors, 6 mm thick, small areas, regular grout	1 Cefi	15.79	.507	m²	10.35	17.45		27.80	37
	0150	Epoxy grout	"	9.29	.861	"	67	29.50		96.50	117
	0300	Slab on grade, cut outs, up to 1.4 m³	2 Cefi	1.42	11.307	m³	227	390		617	825
	2000	Walls, including chipping, cleaning and epoxy grout									
	2100	6 mm deep	1 Cefi	6.04	1.325	m²	81	45.50		126.50	156
	2150	13 mm deep		4.65	1.722		162	59.50		221.50	266
	2200	19 mm deep		3.72	2.153		243	74		317	375
	2510	Underlayment, P.C. based self-leveling 28.27 MPa, pumped, 6 mm	C-8	1858	.030		14.55	.93	.38	15.86	17.80
	2520	13 mm		1765	.032		29	.98	.40	30.38	34
	2530	19 mm		1672	.033		43.50	1.04	.42	44.96	50
	2540	25 mm		1579	.035		58	1.10	.44	59.54	65.50
	2550	38 mm		1394	.040		87	1.25	.50	88.75	98
	2560	Hand mix, 13 mm	C-18	372	.024		29	.67	.14	29.81	33
	2610	Topping, P.C. based self-level/dry 42 MPa, pumped, 6 mm	C-8	1858	.030		22.50	.93	.38	23.81	26.50
	2620	13 mm		1765	.032		45	.98	.40	46.38	51.50
	2630	19 mm		1672	.033		67	1.04	.42	68.46	76
	2660	25 mm		1579	.035		89.50	1.10	.44	91.04	101

Important: See the Reference Section for supporting data - Crews, Rental Equipment, City Cost Indexes and Reference Data

		03920	Concrete Resurfacing	CREW	DAILY OUTPUT	LABOR-HOURS	UNIT	2006 BARE COSTS				TOTAL INCL O&P	
								MAT.	LABOR	EQUIP.	TOTAL		
600	2670		38 mm	C-8	1394	.040	m²	134	1.25	.50	135.75	150	600
	2680		Hand mix, 13 mm	C-18	372	.024	↓	45	.67	.14	45.81	50.50	
		03930	**Concrete Rehabilitation**										
300	0010	**CRACK REPAIR**, incl. chipping, sand blasting and cleaning											300
	0100		Epoxy injection, 3 mm wide, 300 mm deep	B-9	24.38	1.640	m	14.30	45.50	6.05	65.85	93.50	
	0110		6mm wide, 300 mm deep		18.29	2.187		28.50	61	8.05	97.55	135	
	0200		Latex injection, 3 mm wide, 300 mm deep		30.48	1.312		5.75	36.50	4.83	47.08	68.50	
	0210		Latex injection, 6 mm wide, 300 mm deep	↓	22.86	1.750	↓	11.50	48.50	6.45	66.45	95.50	

For information about Means Estimating Seminars, see yellow pages 12 and 13 in back of book

CONCRETE 3

Division Notes

	CREW	DAILY OUTPUT	LABOR-HOURS	UNIT	2006 BARE COSTS				TOTAL INCL O&P
					MAT.	LABOR	EQUIP.	TOTAL	

Division 4
Masonry

Estimating Tips

04050 Basic Masonry Materials & Methods

- The terms *mortar* and *grout* are often used interchangeably, and incorrectly. Mortar is used to bed masonry units, seal the entry of air and moisture, provide architectural appearance, and allow for size variations in the units. Grout is used primarily in reinforced masonry construction and is used to bond the masonry to the reinforcing steel. Common mortar types are M(17.24 MPa), S(12.41 MPa), N(5.17 MPa), and O(2.41 MPa), and conform to ASTM C270. Grout is either fine or coarse, conforms to ASTM C476, and in-place strengths generally exceed 17.24 MPa. Mortar and grout are different components of masonry construction and are placed by entirely different methods. An estimator should be aware of their unique uses and costs.

- Waste, specifically the loss/droppings of mortar and the breakage of brick and block, is included in all masonry assemblies in this division. A factor of 25% is added for mortar and 3% for brick and concrete masonry units.

- Scaffolding or staging is not included in any of the Division 4 costs. Refer to section 01540 for scaffolding and staging costs.

04800 Masonry Assemblies

- The most common types of unit masonry are brick and concrete masonry. The major classifications of brick are building brick (ASTM C62), facing brick (ASTM C216), and glazed brick, fire brick and pavers. Many varieties of texture and appearance can exist within these classifications, and the estimator would be wise to check local custom and availability within the project area. On repair and remodeling jobs, matching the existing brick may be the most important criteria.

- Brick and concrete block are priced by the piece and then converted into a price per square meter of wall. Openings less than 0.2 square meters are generally ignored by the estimator because any savings in units used is offset by the cutting and trimming required.

- It is often difficult and expensive to find and purchase small lots of historic brick. Costs can vary widely. Many design issues affect costs, selection of mortar mix, and repairs or replacement of masonry materials. Cleaning techniques must be reflected in the estimate.

- All masonry walls, whether interior or exterior, require bracing. The cost of bracing walls during construction should be included by the estimator, and this bracing must remain in place until permanent bracing is complete. Permanent bracing of masonry walls is accomplished by masonry itself, in the form of pilasters or abutting wall corners, or by anchoring the walls to the structural frame. Accessories in the form of anchors, anchor slots and ties are used, but their supply and installation can be by different trades. For instance, anchor slots on spandrel beams and columns are supplied and welded in place by the steel fabricator, but the ties from the slots into the masonry are installed by the bricklayer. Regardless of the installation method the estimator must be certain that these accessories are accounted for in pricing.

Reference Numbers

Reference numbers are shown in bold squares at the beginning of some major classifications. These numbers refer to related items in the Reference Section. The reference information may be an estimating procedure, an alternate pricing method, or technical information.

Note: Not all subdivisions listed here necessarily appear in this publication.

04055	Selective Demolition		CREW	DAILY OUTPUT	LABOR-HOURS	UNIT	2006 BARE COSTS				TOTAL INCL O&P
							MAT.	LABOR	EQUIP.	TOTAL	
110	0010	**SELECTIVE DEMOLITION, MASONRY**										110
	0300	Concrete block walls, unreinforced, 50 mm thick	R024119 -10	2 Clab	111	.144	m²		3.95		3.95	6.15
	0310	100 mm thick			107	.150			4.10		4.10	6.40
	0320	150 mm thick			102	.157			4.30		4.30	6.70
	0330	200 mm thick			97.55	.164			4.49		4.49	7
	0340	250 mm thick			92.90	.172			4.72		4.72	7.35
	0360	300 mm thick			88.26	.181			4.97		4.97	7.75
	0380	Reinforced alternate courses, 50 mm thick			105	.152			4.18		4.18	6.50
	0390	100 mm thick			100	.160			4.38		4.38	6.80
	0400	150 mm thick			96.15	.166			4.56		4.56	7.10
	0410	200 mm thick			91.97	.174			4.77		4.77	7.40
	0420	250 mm thick			87.33	.183			5		5	7.80
	0430	300 mm thick			82.68	.194			5.30		5.30	8.25
	0440	Reinf alternate courses & vertically 1200 mm OC, 100 mm thick			83.61	.191			5.25		5.25	8.15
	0450	150 mm thick			78.97	.203			5.55		5.55	8.65
	0460	200 mm thick			74.32	.215			5.90		5.90	9.20
	0480	250 mm thick			69.68	.230			6.30		6.30	9.80
	0490	300 mm thick			65.03	.246			6.75		6.75	10.50
	1000	Chimney, 400 mm x 400 mm, soft old mortar		1 Clab	1.56	5.140	m³		141		141	219
	1020	Hard mortar			1.13	7.067			194		194	300
	1030	400 mm x 500 mm, soft old mortar			1.56	5.140			141		141	219
	1040	Hard mortar			1.13	7.067			194		194	300
	1050	400 mm x 600 mm, soft old mortar			1.56	5.140			141		141	219
	1060	Hard mortar			1.13	7.067			194		194	300
	1080	500 mm x 500 mm, soft old mortar			1.56	5.140			141		141	219
	1100	Hard mortar			1.13	7.067			194		194	300
	1110	500 mm x 600 mm, soft old mortar			1.56	5.140			141		141	219
	1120	Hard mortar			1.13	7.067			194		194	300
	1140	500 mm x 800 mm, soft old mortar			1.56	5.140			141		141	219
	1160	Hard mortar			1.13	7.067			194		194	300
	1200	1200 mm x 1200 mm, soft old mortar			1.56	5.140			141		141	219
	1220	Hard mortar			1.13	7.067			194		194	300
	1250	Metal, high temp steel jacket, 600 mm diameter		E-2	39.62	1.413	m		55	36	91	135
	1260	1524 mm diameter		"	18.29	3.062			119	78	197	291
	1280	Flue lining, up to 300 mm x 300 mm		1 Clab	60.96	.131			3.60		3.60	5.60
	1282	Up to 600 mm x 600 mm			45.72	.175			4.79		4.79	7.45
	2000	Columns, 200 mm x 200 mm, soft old mortar			14.63	.547			15		15	23.50
	2020	Hard mortar			12.19	.656			18		18	28
	2060	400 mm x 400 mm, soft old mortar			4.88	1.640			45		45	70
	2100	Hard mortar			4.27	1.875			51.50		51.50	80
	2140	600 mm x 600 mm, soft old mortar			2.44	3.281			90		90	140
	2160	Hard mortar			1.83	4.374			120		120	187
	2200	900 mm x 900 mm, soft old mortar			1.22	6.562			180		180	280
	2220	Hard mortar			.91	8.749			240		240	375
	2230	Alternate pricing method, soft old mortar			.85	9.423	m³		258		258	400
	2240	Hard mortar			.65	12.291	"		335		335	525
	3000	Copings, precast or masonry, to 200 mm wide										
	3020	Soft old mortar		1 Clab	54.86	.146	m		4		4	6.20
	3040	Hard mortar		"	48.77	.164	"		4.49		4.49	7
	3100	To 300 mm wide										
	3120	Soft old mortar		1 Clab	48.77	.164	m		4.49		4.49	7
	3140	Hard mortar		"	42.67	.187	"		5.15		5.15	8
	4000	Fireplace, brick, 760 mm x 600 mm opening										
	4020	Soft old mortar		1 Clab	2	4	Ea.		110		110	171
	4040	Hard mortar			1.25	6.400			175		175	273
	4100	Stone, soft old mortar			1.50	5.333			146		146	227

		04055	Selective Demolition		CREW	DAILY OUTPUT	LABOR-HOURS	UNIT	2006 BARE COSTS				TOTAL INCL O&P	
									MAT.	LABOR	EQUIP.	TOTAL		
110	4120		Hard mortar	R024119 -10	1 Clab	1	8	Ea.		219		219	340	110
	5000		Veneers, brick, soft old mortar			13.01	.615	m²		16.85		16.85	26	
	5020		Hard mortar			11.61	.689			18.90		18.90	29.50	
	5100		Granite and marble, 50 mm thick			16.72	.478			13.10		13.10	20.50	
	5120		100 mm thick			15.79	.507			13.90		13.90	21.50	
	5140		Stone, 100 mm thick			16.72	.478			13.10		13.10	20.50	
	5160		200 mm thick			16.26	.492	▼		13.50		13.50	21	
	5400		Alternate pricing method, stone, 100 mm thick			1.70	4.711	m³		129		129	201	
	5420		200 mm thick			2.41	3.326	"		91		91	142	

		04060	Masonry Mortar											
200	0010	**CEMENT**												200
	0020		Gypsum 36 kg bag, T.L. lots					Bag	16.70			16.70	18.35	
	0050		L.T.L. lots						22.50			22.50	25	
	0100		Masonry, 32 kg bag, T.L. lots						6.45			6.45	7.10	
	0150		L.T.L. lots						6.85			6.85	7.50	
	0200		White, 32 kg bag, T.L. lots						13.40			13.40	14.75	
	0250		L.T.L. lots					▼	14.30			14.30	15.75	
400	0010	**LIME**												400
	0020		Masons, hydrated, 23 kg bag, T.L. lots					Bag	6.55			6.55	7.20	
	0050		L.T.L. lots						7.40			7.40	8.15	
	0200		Finish, double hydrated, 23 kg bag, T.L. lots						8.40			8.40	9.25	
	0250		L.T.L. lots					▼	9.30			9.30	10.25	
500	0010	**MORTAR**		R040513 -10										500
	0020		With masonry cement											
	0100		Type M, 1:1:6 mix		1 Brhe	4.05	1.977	m³	121	55		176	217	
	0200		Type N, 1:3 mix			4.05	1.977		110	55		165	205	
	0300		Type O, 1:3 mix			4.05	1.977		185	55		240	288	
	0400		Type PM, 1:1:6 mix, 17 MPa			4.05	1.977		197	55		252	300	
	0500		Type S, 1/2:1:4 mix		▼	4.05	1.977	▼	208	55		263	315	
	2000		With portland cement and lime											
	2100		Type M, 1:1/4:3 mix		1 Brhe	4.05	1.977	m³	233	55		288	340	
	2200		Type N, 1:1:6 mix, 5 MPa			4.05	1.977		195	55		250	299	
	2300		Type O, 1:2:9 mix (Pointing Mortar)			4.05	1.977		208	55		263	315	
	2400		Type PL, 1:1/2:4 mix, 17 MPa			4.05	1.977		214	55		269	320	
	2500		Type K, 1:3:12 mix, 0.5 MPa			4.05	1.977		185	55		240	287	
	2600		Type S, 1:1/2:4 mix, 12 MPa		▼	4.05	1.977		245	55		300	355	
	2650		Pre-mixed, type S or N						157			157	173	
	2700		Mortar for glass block		1 Brhe	4.05	1.977		287	55		342	400	
	2800		Gypsum cement mortar					▼	289			289	315	
	2900		Mortar for Fire Brick, 36 kg bag, T.L. lots					Bag	21			21	23	
520	0010	**MORTAR** (See also Division 04060-500)												520
	0020		Masonry restoration mix					kg	3.53			3.53	3.88	
	0050		White					"	4.06			4.06	4.45	
540	0010	**MORTAR PIGMENTS**, 23 kg bags (2 bags/k bricks),		R040513 -10										540
	0020		range 1 to 5 kg/bag of cement, minimum					kg	13.60			13.60	15	
	0050		Average						17.95			17.95	19.80	
	0100		Maximum					▼	35.50			35.50	39	
750	0010	**SAND**, screened and washed at the pit												750
	0020		For mortar, per ton					Met. Ton	13.70			13.70	15.05	
	0050		With 16 km haul						19.05			19.05	21	
	0100		With 50 km haul					▼	21.50			21.50	23.50	

MASONRY 4

MASONRY

4

04060	Masonry Mortar	CREW	DAILY OUTPUT	LABOR-HOURS	UNIT	2006 BARE COSTS				TOTAL INCL O&P	
						MAT.	LABOR	EQUIP.	TOTAL		
750	0200	Screened and washed, at the pit				m³	22.50			22.50	25	**750**
	0250	With 16 km haul					31.50			31.50	34.50	
	0300	With 50 km haul					35			35	38.50	
770	0010	**SURFACE BONDING** CMU walls with fiberglass mortar,										**770**
	0020	gray or white colors, not incl. block work	1 Bric	50.17	.159	m²	1.18	5.85		7.03	10.15	
900	0010	**WATERPROOFING ADMIXTURE**										**900**
	0020	Admixture, 1 L to 2 bags of masonry cement				liter	7.60			7.60	8.35	

04070 | Masonry Grout

			CREW	DAILY OUTPUT	LABOR-HOURS	UNIT	MAT.	LABOR	EQUIP.	TOTAL	INCL O&P	
420	0010	**GROUTING** Bond bms & lintels, 200 mm dp,pumped,excl block R040513 -10										**420**
	0020	200 mm thick, 0.2 m³/mm	D-4	427	.075	m	2.69	2.38	.34	5.41	6.95	
	0050	250 mm thick, 0.25 m³/mm		366	.087		2.99	2.78	.40	6.17	7.95	
	0060	300 mm thick, 0.3 m³/mm		317	.101		3.58	3.21	.46	7.25	9.30	
	0200	Concrete block cores, solid, 100 mm thk., by hand, 0.02 m³/m²	D-8	102	.392	m²	2.58	12.95		15.53	22.50	
	0210	150 mm thick, pumped, 0.053 m³/m²	D-4	66.89	.478		6.90	15.20	2.17	24.27	33	
	0250	200 mm thick, pumped, 0.078 m³/m²		63.17	.507		10.10	16.10	2.29	28.49	38	
	0300	250 mm thick, pumped, 0.102 m³/m²		61.31	.522		13.35	16.60	2.36	32.31	42.50	
	0350	300 mm thick, pumped, 0.127 m³/m²		59.46	.538		16.60	17.10	2.44	36.14	47	
	0500	Cavity walls, 50 mm space, pumped, 0.05 m³/m²		158	.203		6.55	6.45	.92	13.92	18	
	0550	75 mm space, 0.075 m³/m²		111	.288		9.80	9.15	1.31	20.26	26	
	0600	100 mm space, 0.1 m³/m²		107	.299		13	9.50	1.35	23.85	30.50	
	0700	150 mm space, 0.151 m³/m²		74.32	.431		19.60	13.70	1.95	35.25	44.50	
	0800	Door frames, 900 mm x 2100 mm opening, 0.07 m³ per opening		60	.533	Opng.	9.10	16.95	2.42	28.47	38.50	
	0850	1.8 m x 2.1 m opening, 0.1 m³ per opening		45	.711	"	12.75	22.50	3.22	38.47	52	
	2000	Grout, C476, for bond beams, lintels and CMU cores		9.91	3.231	m³	129	103	14.65	246.65	315	

04080 | Anchorage & Reinforcement

			CREW	DAILY OUTPUT	LABOR-HOURS	UNIT	MAT.	LABOR	EQUIP.	TOTAL	INCL O&P	
070	0010	**ANCHOR BOLTS**										**070**
	0020	Hooked w/ nut and washer, 13 mm diam., 200 mm long	1 Bric	200	.040	Ea.	.68	1.46		2.14	2.98	
	0030	300 mm long		190	.042		1.29	1.54		2.83	3.76	
	0040	16 mm diameter, 200 mm long		180	.044		1.01	1.62		2.63	3.58	
	0050	300 mm long		170	.047		1.11	1.72		2.83	3.84	
	0060	19 mm diameter, 200 mm long		160	.050		1.54	1.83		3.37	4.47	
	0070	300 mm long		150	.053		1.92	1.95		3.87	5.10	
200	0010	**REINFORCING** R040519 -50										**200**
	0015	Steel bars A615, placed horiz., #10M bars	1 Bric	204	.039	kg	.95	1.43		2.38	3.22	
	0020	#15M & #20M bars		363	.022		.95	.81		1.76	2.27	
	0050	Placed vertical, #10M bars		159	.050		.95	1.84		2.79	3.84	
	0060	#15M & #20M bars		295	.027		.95	.99		1.94	2.55	
	0200	Joint reinforcing, regular truss, to 152 mm wide, mill std galvanized		914	.009	m	.39	.32		.71	.92	
	0250	305 mm wide		610	.013		.43	.48		.91	1.20	
	0400	Cavity truss with drip section, to 152 mm wide		914	.009		.38	.32		.70	.91	
	0450	305 mm wide		610	.013		.39	.48		.87	1.16	
	0500	Joint reinforcing, ladder type, mill std galvanized										
	0600	9 ga. sides, 9 ga. ties, 100 mm wall	1 Bric	914	.009	m	.31	.32		.63	.83	
	0650	150 mm wall		914	.009		.35	.32		.67	.87	
	0700	200 mm wall		762	.011		.36	.38		.74	.98	
	0750	250 mm wall		610	.013		.38	.48		.86	1.15	
	0800	300 mm wall		610	.013		.40	.48		.88	1.18	
	1000	Truss type										
	1100	9 ga. sides, 9 ga. ties, 100 mm wall	1 Bric	914	.009	m	.38	.32		.70	.90	
	1150	150 mm wall		914	.009		.39	.32		.71	.92	
	1200	200 mm wall		762	.011		.41	.38		.79	1.04	
	1250	250 mm wall		610	.013		.44	.48		.92	1.21	

04080	Anchorage & Reinforcement	CREW	DAILY OUTPUT	LABOR-HOURS	UNIT	2006 BARE COSTS				TOTAL INCL O&P	
						MAT.	LABOR	EQUIP.	TOTAL		
200 1300	300 mm wall	1 Bric	610	.013	m	.46	.48		.94	1.24	**200**
1500	5 mm sides, 9 ga. ties, 100 mm wall R040519-50		914	.009		.55	.32		.87	1.10	
1550	150 mm wall		914	.009		.56	.32		.88	1.10	
1600	200 mm wall		762	.011		.58	.38		.96	1.21	
1650	250 mm wall		610	.013		.60	.48		1.08	1.39	
1700	300 mm wall		610	.013		.63	.48		1.11	1.42	
2000	5 mm sides, 5 mm ties, 100 mm wall		914	.009		.65	.32		.97	1.20	
2050	150 mm wall		914	.009		.65	.32		.97	1.20	
2100	200 mm wall		762	.011		.67	.38		1.05	1.32	
2150	250 mm wall		610	.013		.71	.48		1.19	1.51	
2200	300 mm wall		610	.013		.74	.48		1.22	1.54	
2500	Cavity truss type, galvanized										
2600	9 ga. sides, 9 ga. ties, 100 mm wall	1 Bric	762	.011	m	.94	.38		1.32	1.61	
2650	150 mm wall		762	.011		.95	.38		1.33	1.62	
2700	200 mm wall		610	.013		.97	.48		1.45	1.79	
2750	250 mm wall		457	.018		.99	.64		1.63	2.06	
2800	300 mm wall		457	.018		1.03	.64		1.67	2.10	
3000	5 mm sides, 9 ga. ties, 100 mm wall		762	.011		1.15	.38		1.53	1.84	
3050	150 mm wall		762	.011		1.16	.38		1.54	1.85	
3100	200 mm wall		610	.013		1.18	.48		1.66	2.02	
3150	250 mm wall		457	.018		1.20	.64		1.84	2.29	
3200	300 mm wall		457	.018		1.24	.64		1.88	2.33	
3500	For hot dip galvanizing, add					80%					
650 0010	**WALL TIES**										**650**
0020	To brick veneer, galv., corrugated, 20 mm x 180 mm, 22 Ga.	1 Bric	10.50	.762	h	8.35	28		36.35	51.50	
0100	24 Ga.		10.50	.762		7.20	28		35.20	50.50	
0150	16 Ga.		10.50	.762		18.25	28		46.25	62.50	
0200	Buck anchors, galv., 16 ga.,50 mm bend, 200 mm x 50 mm		10.50	.762		121	28		149	176	
0250	200 mm x 75 mm		10.50	.762		107	28		135	161	
0300	Adjustable, rectangular, 105 mm wide										
0350	Anchor and tie, 5 mm wire, mill galv.										
0400	70 mm eye, 83 mm tie	1 Bric	1.05	7.619	k	430	278		708	895	
0500	121 mm tie		1.05	7.619		475	278		753	945	
0520	140 mm tie		1.05	7.619		500	278		778	975	
0550	121 mm eye, 83 mm tie		1.05	7.619		480	278		758	950	
0570	120 mm tie		1.05	7.619		540	278		818	1,025	
0580	140 mm tie		1.05	7.619		565	278		843	1,050	
0660	Cavity wall, Z-type, galvanized, 150 mm long, 3 mm diam.		10.50	.762	h	16	28		44	60	
0670	5 mm diameter		10.50	.762		25.50	28		53.50	70.50	
0680	6 mm diameter		10.50	.762		33	28		61	79	
0850	200 mm long, 5 mm diameter		10.50	.762		28.50	28		56.50	74	
0855	6 mm diameter		10.50	.762		78.50	28		106.50	129	
1000	Rectangular type, galvanized, 6 mm diameter, 50 mm x 150 mm		10.50	.762		35.50	28		63.50	81.50	
1050	100 mm x 150 mm		10.50	.762		39.50	28		67.50	86	
1100	5 mm diameter, 50 mm x 150 mm		10.50	.762		28.50	28		56.50	73.50	
1150	100 mm x 150 mm		10.50	.762		30.50	28		58.50	76	
1200	Mesh wall tie, 13 mm mesh, hot dip galvanized										
1400	16 gauge, 300 mm long, 75 mm wide	1 Bric	9	.889	h	73.50	32.50		106	131	
1420	150 mm wide		9	.889		112	32.50		144.50	173	
1440	300 mm wide		8.50	.941		180	34.50		214.50	251	
1500	Rigid partition anchors, plain, 200 mm long, 25 mm x 3 mm		10.50	.762		69	28		97	119	
1550	25 mm x 6 mm		10.50	.762		114	28		142	168	
1580	40 mm x 3 mm		10.50	.762		92	28		120	144	
1600	38 mm x 6 mm		10.50	.762		187	28		215	249	
1650	50 mm x 3 mm		10.50	.762		119	28		147	174	

MASONRY 4

For expanded coverage of these items see *Means Concrete & Masonry Cost Data 2006*

4 MASONRY

			DAILY OUTPUT	LABOR-HOURS	UNIT	2006 BARE COSTS				TOTAL INCL O&P	
	04080	**Anchorage & Reinforcement**				MAT.	LABOR	EQUIP.	TOTAL		
			CREW								
650	1700	50 mm x 6 mm	1 Bric	10.50	.762	h	219	28		247	284	650
	2000	Column flange ties, wire, galvanized										
	2300	5 mm diameter, up to 75 mm wide [3/16"]	1 Bric	10.50	.762	h	73.50	28		101.50	124	
	2350	To 125 mm wide		10.50	.762		80	28		108	131	
	2400	To 180 mm wide		10.50	.762		85.50	28		113.50	137	
	2600	To 230 mm wide		10.50	.762		92	28		120	144	
	2650	6 mm diameter, up to 75 mm wide		10.50	.762		99	28		127	152	
	2700	To 125 mm wide		10.50	.762		109	28		137	163	
	2800	To 180 mm wide		10.50	.762		118	28		146	172	
	2850	To 230 mm wide		10.50	.762		128	28		156	184	
	2900	For hot dip galvanized, add					35%					
	4000	Channel slots, 35 mm x 13 mm x 200 mm										
	4100	12 gauge, plain	1 Bric	10.50	.762	h	188	28		216	249	
	4150	16 gauge, galvanized	"	10.50	.762	"	104	28		132	157	
	4200	Channel slot anchors										
	4300	16 gauge, galvanized, 32 mm x 90 mm				h	41.50			41.50	45.50	
	4350	32 mm x 140 mm					48			48	53	
	4400	32 mm x 190 mm					54.50			54.50	60	
	4500	3 mm plain, 32 mm x 90 mm					67			67	73.50	
	4550	32 mm x 140 mm					72			72	79	
	4600	32 mm x 190 mm					78			78	86	
	4700	For corrugation, add					39			39	43	
	4750	For hot dip galvanized, add					35%					
	5000	Dowels										
	5100	Plain, 6 mm diameter, 75 mm long				h	28			28	31	
	5150	100 mm long					31			31	34	
	5200	150 mm long					38			38	42	
	5300	10 mm diameter, 75 mm long					37			37	40.50	
	5350	100 mm long					46			46	50.50	
	5400	150 mm long					53			53	58.50	
	5500	13 mm diameter, 75 mm long					54			54	59.50	
	5550	100 mm long					64			64	70.50	
	5600	150 mm long					85			85	93.50	
	5700	16 mm diameter, 75 mm long					74			74	81.50	
	5750	100 mm long					91			91	100	
	5800	150 mm long					125			125	138	
	6000	19 mm diameter, 75 mm long					92			92	101	
	6100	100 mm long					118			118	130	
	6150	150 mm long					167			167	184	
	6300	For hot dip galvanized, add					35%					
	04090	**Masonry Accessories**										
170	0010	**CONTROL JOINT**										170
	0020	Rubber, 100 mm and wider wall	1 Bric	122	.066	m	5.90	2.40		8.30	10.15	
	0050	PVC, 100 mm wall		122	.066		3.61	2.40		6.01	7.60	
	0100	Rubber, 150 mm wall		97.54	.082		7.60	3		10.60	12.90	
	0120	PVC, 150 mm wall		97.54	.082		4.86	3		7.86	9.90	
	0140	Rubber, 200 mm and wider wall		85.34	.094		9.80	3.43		13.23	16	
	0160	PVC, 200 mm wall		85.34	.094		5.50	3.43		8.93	11.25	
	0180	300 mm wall		73.15	.109		11.35	4		15.35	18.60	
860	0010	**VENT BOX**										860
	0020	Extruded aluminum, 100 mm deep, 60 mm x 206 mm	1 Bric	30	.267	Ea.	22.50	9.75		32.25	39.50	
	0050	127 mm x 206 mm		25	.320		28.50	11.70		40.20	49	
	0100	57 mm x 635 mm		25	.320		58	11.70		69.70	81.50	
	0150	127 mm x 419 mm		22	.364		52.50	13.30		65.80	78	
	0200	152 mm x 419 mm		22	.364		68	13.30		81.30	95	

04090	**Masonry Accessories**	CREW	DAILY OUTPUT	LABOR-HOURS	UNIT	2006 BARE COSTS				TOTAL INCL O&P	
						MAT.	LABOR	EQUIP.	TOTAL		
860											**860**
0250	197 mm x 419 mm	1 Bric	20	.400	Ea.	66.50	14.60		81.10	96	
0400	For baked enamel finish, add					35%					
0500	For cast aluminum, painted, add					60%					
1000	Stainless steel ventilators, 150 mm x 150 mm	1 Bric	25	.320		98.50	11.70		110.20	126	
1050	200 mm x 200 mm		24	.333		104	12.20		116.20	133	
1100	300 mm x 300 mm		23	.348		120	12.70		132.70	151	
1150	300 mm x 150 mm		24	.333		105	12.20		117.20	134	
1200	Fdn block vent, galv., 32 mm T, 200 mm H, 400 mm L, no damper		30	.267		19.25	9.75		29	36	
1250	For damper, add					6.40			6.40	7.05	
900	**WALL PLUGS** for nailing to brickwork										**900**
0010											
0020	26 ga., galvanized, plain	1 Bric	10.50	.762	h	27	28		55	72	
0050	Wood filled	"	10.50	.762	"	90	28		118	142	

04200 | Masonry Units

04210	**Clay Masonry Units**	CREW	DAILY OUTPUT	LABOR-HOURS	UNIT	2006 BARE COSTS				TOTAL INCL O&P	
						MAT.	LABOR	EQUIP.	TOTAL		
100	**COMMON BUILDING BRICK** C62, TL lots, material only [R042110-20]										**100**
0010											
0020	Standard, minimum				k	315			315	345	
0050	Average (select)				"	385			385	425	
300	**FACE BRICK** C216, TL lots, material only [R042110-20]										**300**
0010											
0300	Standard modular, 100 mm x 68 mm x 200 mm, minimum				k	400			400	440	
0350	Maximum					515			515	565	
0450	Economy, 100 mm x 100 mm x 200 mm, minimum					775			775	855	
0500	Maximum					1,000			1,000	1,100	
0510	Economy, 100 mm x 100 mm x 300 mm, minimum					1,075			1,075	1,200	
0520	Maximum					1,400			1,400	1,550	
0550	Jumbo, 150 mm x 100 mm x 300 mm, minimum					1,225			1,225	1,350	
0600	Maximum					1,550			1,550	1,700	
0610	Jumbo, 200 mm x 100 mm x 300 mm, minimum					1,225			1,225	1,350	
0620	Maximum					1,550			1,550	1,700	
0650	Norwegian, 100 mm x 80 mm x 300 mm, minimum					785			785	860	
0700	Maximum					1,025			1,025	1,125	
0710	Norwegian, 150 mm x 81 mm x 300 mm, minimum					1,175			1,175	1,300	
0720	Maximum					1,525			1,525	1,675	
0850	Standard, glazed, plain colors, 100 mm x 68 mm x 200 mm, min.					1,225			1,225	1,350	
0900	Maximum					1,600			1,600	1,750	
1000	Deep trim shades, 100 mm x 70 mm x 200 mm, minimum					1,250			1,250	1,375	
1050	Maximum					1,400			1,400	1,525	
1080	Jumbo utility, 100 mm x 100 mm x 300 mm					1,125			1,125	1,250	
1120	100 mm x 200 mm x 200 mm					1,175			1,175	1,275	
1140	100 mm x 200 mm x 400 mm					2,450			2,450	2,700	
1260	Engineer, 100 mm x 81 mm x 200 mm, minimum					500			500	550	
1270	Maximum					530			530	580	
1350	King, 100 mm x 70 mm x 250 mm, minimum					465			465	515	
1360	Maximum					500			500	550	
1400	Norman, 100 mm x 70 mm x 300 mm					465			465	515	
1450	Roman, 100 mm x 50 mm x 300 mm					465			465	515	

MASONRY 4

For expanded coverage of these items see *Means Concrete & Masonry Cost Data 2006*

04210	Clay Masonry Units		CREW	DAILY OUTPUT	LABOR-HOURS	UNIT	2006 BARE COSTS				TOTAL INCL O&P	
							MAT.	LABOR	EQUIP.	TOTAL		
300	1500	SCR, 150 mm x 68 mm x 300 mm	R042110				k	465			465	515	**300**
	1550	Double, 100 mm x 135 mm x 200 mm	-20					465			465	515	
	1600	Triple, 100 mm x 135 mm x 300 mm						465			465	515	
	1770	Standard modular, double glazed, 100 mm x 70 mm x 200 mm						1,050			1,050	1,150	
	1850	Jumbo, colored glazed ceramic, 150 mm x 100 mm x 300 mm						1,550			1,550	1,700	
	2050	Jumbo utility, glazed, 100 mm x 100 mm x 300 mm						1,175			1,175	1,275	
	2100	100 mm x 200 mm x 200 mm						1,600			1,600	1,775	
	2150	100 mm x 400 mm x 200 mm						2,775			2,775	3,050	
	2170	For less than truck load lots, add						10			10	11	
	2180	For buff or gray brick, add						15			15	16.50	
	3050	Used brick, minimum						375			375	410	
	3100	Maximum						490			490	540	
	3150	Add for brick to match existing work, minimum						5%					
	3200	Maximum						50%					

04220	Concrete Masonry Units		CREW	DAILY OUTPUT	LABOR-HOURS	UNIT	MAT.	LABOR	EQUIP.	TOTAL	TOTAL INCL O&P	
210	0010	**CONCRETE BLOCK** material only											**210**
	0020	50 mm x 200 mm x 400 mm solid, normal-weight, 13 MPa					Ea.	.67			.67	.74	
	0050	24 MPa						.71			.71	.78	
	0100	34 MPa						.84			.84	.92	
	0150	Lightweight, std.						.80			.80	.88	
	0300	75 mm x 200 mm x 400 mm solid, normal-weight, 13 MPa						.70			.70	.77	
	0350	24 MPa						1.27			1.27	1.40	
	0400	34 MPa						1.29			1.29	1.42	
	0450	Lightweight, std.						.83			.83	.91	
	0600	100 mm x 200 mm x 400 mm hollow, normal-weight, 13 MPa						.76			.76	.84	
	0650	24 MPa						.82			.82	.90	
	0700	34 MPa						.93			.93	1.02	
	0750	Lightweight, std.						.94			.94	1.03	
	1300	Solid, normal-weight, 13 MPa						1.01			1.01	1.11	
	1350	24 MPa						1.15			1.15	1.27	
	1400	34 MPa						1.05			1.05	1.16	
	1450	Lightweight, std.						1.43			1.43	1.57	
	1600	150 mm x 200 mm x 400 mm hollow, normal-weight, 13 MPa						1.16			1.16	1.28	
	1650	24 MPa						1.24			1.24	1.36	
	1700	34 MPa						1.22			1.22	1.34	
	1750	Lightweight, std.						1.29			1.29	1.42	
	2300	Solid, normal-weight, 13 MPa						1.22			1.22	1.34	
	2350	24 MPa						1.34			1.34	1.47	
	2400	34 MPa						1.70			1.70	1.87	
	2450	Lightweight, std.						1.60			1.60	1.76	
	2600	200 mm x 200 mm x 400 mm hollow, normal-weight, 13 MPa						1.21			1.21	1.33	
	2650	24 MPa						1.41			1.41	1.55	
	2700	34 MPa						1.45			1.45	1.60	
	2750	Lightweight, std.						1.57			1.57	1.73	
	3200	Solid, normal-weight, 13 MPa						1.76			1.76	1.94	
	3250	24 MPa						2.06			2.06	2.27	
	3300	34 MPa						2.34			2.34	2.57	
	3350	Lightweight, std.						2.14			2.14	2.35	
	3400	250 mm x 200 mm x 400 mm hollow, normal-weight, 13 MPa						1.21			1.21	1.33	
	3410	24 MPa						1.41			1.41	1.55	
	3420	34 MPa						1.45			1.45	1.60	
	3430	Lightweight, std.						1.57			1.57	1.73	
	3480	Solid, normal-weight, 13 MPa						1.76			1.76	1.94	
	3490	24 MPa						2.06			2.06	2.27	
	3500	34 MPa						2.34			2.34	2.57	

04200 | Masonry Units

04220	Concrete Masonry Units	CREW	DAILY OUTPUT	LABOR-HOURS	UNIT	2006 BARE COSTS				TOTAL INCL O&P	
						MAT.	LABOR	EQUIP.	TOTAL		
210	3510	Lightweight, std.				Ea.	2.14			2.14	2.35	210
	3600	300 mm x 200 mm x 400 mm hollow, normal-weight, 13 MPa					1.78			1.78	1.96	
	3650	24 MPa					1.83			1.83	2.01	
	3700	34 MPa					2.02			2.02	2.22	
	3750	Lightweight, std.					2.06			2.06	2.27	
	4300	Solid, normal-weight, 13 MPa					2.74			2.74	3.01	
	4350	24 MPa					2.48			2.48	2.73	
	4400	34 MPa					3.52			3.52	3.87	
	4500	Lightweight, std.				▼	3.74			3.74	4.11	

04500 | Refractories

04550	Flue Liners	CREW	DAILY OUTPUT	LABOR-HOURS	UNIT	2006 BARE COSTS				TOTAL INCL O&P	
						MAT.	LABOR	EQUIP.	TOTAL		
250	0010	**FLUE LINING**, including mortar joints										250
	0020	200 mm x 200 mm	D-1	38.10	.420	m	11.20	13.50		24.70	33	
	0100	200 mm x 300 mm		31.39	.510		17.40	16.40		33.80	44	
	0200	300 mm x 300 mm		28.35	.564		22	18.15		40.15	51.50	
	0300	300 mm x 450 mm		25.60	.625		39	20		59	73.50	
	0400	450 mm x 450 mm		22.86	.700		56	22.50		78.50	96	
	0500	500 mm x 500 mm		20.12	.795		102	25.50		127.50	151	
	0600	600 mm x 600 mm		17.07	.937		125	30		155	184	
	1000	Round, 450 mm diameter [18″]		20.12	.795		82.50	25.50		108	130	
	1100	600 mm diameter [24″]	▼	14.33	1.117	▼	161	36		197	232	

04580	Refractory Brick	CREW	DAILY OUTPUT	LABOR-HOURS	UNIT	2006 BARE COSTS				TOTAL INCL O&P	
						MAT.	LABOR	EQUIP.	TOTAL		
250	0010	**FIRE BRICK**										250
	0012	225 mm x 65 mm x 115 mm, low duty, 1400 K	D-1	.60	26.667	k	860	855		1,715	2,250	
	0050	High duty, 1900 K	″	.60	26.667	″	1,625	855		2,480	3,100	
260	0010	**FIRE CLAY**										260
	0020	Gray, high duty, 45 kg bag				Bag	46			46	50.50	
	0050	45 kg drum, premixed (400 brick/drum)				Drum	56.50			56.50	62.50	

04700 | Simulated Masonry

04710	Simulated Brick	CREW	DAILY OUTPUT	LABOR-HOURS	UNIT	2006 BARE COSTS				TOTAL INCL O&P	
						MAT.	LABOR	EQUIP.	TOTAL		
600	0010	**SIMULATED BRICK**										600
	0020	Aluminum, baked on colors	1 Carp	18.58	.431	m²	29.50	15.30		44.80	56.50	
	0050	Fiberglass panels		18.58	.431		32.50	15.30		47.80	59.50	
	0100	Urethane pieces cemented in mastic		13.94	.574		56	20.50		76.50	93.50	
	0150	Vinyl siding panels	▼	18.58	.431		22.50	15.30		37.80	48.50	
	0160	Cement base, brick, incl. mastic	D-1	9.29	1.722	▼	35	55.50		90.50	124	

MASONRY 4

For expanded coverage of these items see *Means Concrete & Masonry Cost Data 2006*

04700 | Simulated Masonry

04710 | Simulated Brick

			CREW	DAILY OUTPUT	LABOR-HOURS	UNIT	2006 BARE COSTS				TOTAL INCL O&P	
							MAT.	LABOR	EQUIP.	TOTAL		
600	0170	Corner	D-1	15.24	1.050	m	27	34		61	81	600
	0180	Stone face, incl. mastic		9.29	1.722	m²	85	55.50		140.50	178	
	0190	Corner	↓	15.24	1.050	m	28.50	34		62.50	83	

04730 | Simulated Stone

			CREW	DAILY OUTPUT	LABOR-HOURS	UNIT	MAT.	LABOR	EQUIP.	TOTAL	TOTAL INCL O&P	
600	0010	**SIMULATED STONE**										600
	0100	Insulated fiberglass panels, 16 mm ply backer	L-4	18.58	1.292	m²	106	43		149	183	

04800 | Masonry Assemblies

04810 | Unit Masonry Assemblies

			CREW	DAILY OUTPUT	LABOR-HOURS	UNIT	2006 BARE COSTS				TOTAL INCL O&P	
							MAT.	LABOR	EQUIP.	TOTAL		
040	0010	**ADOBE BRICK** Semi-stabilized, with cement mortar										040
	0060	Brick, 250 x 100 x 350 mm, 28/m²	D-8	52.02	.769	m²	24	25.50		49.50	65	
	0080	300 mm x 100 mm x 400 mm, 24/m²		53.88	.742		61.50	24.50		86	105	
	0100	250 x 100 x 400 mm, 25/m²		54.81	.730		40	24		64	80.50	
	0120	200 x 100 x 400 mm, 25/m²		52.02	.769		42.50	25.50		68	85.50	
	0140	100 x 100 x 400 mm, 25/m²		50.17	.797		32	26.50		58.50	75	
	0160	150 x 100 x 400 mm, 25/m²		50.17	.797		26.50	26.50		53	69	
	0180	100 x 100 x 400 mm, 32/m²		48.31	.828		26	27.50		53.50	70	
	0200	200 x 100 x 300 mm, 32/m²	↓	48.31	.828	↓	30.50	27.50		58	75	
050	0010	**AUTOCLAVED AERATED CONCRETE BLOCK** Scaffolding not incl										050
	0050	Solid, 100 x 300 x 600 mm	D-8	55.74	.718	m²	20	23.50		43.50	58	
	0060	150 x 300 x 600 mm		55.74	.718		27	23.50		50.50	65.50	
	0070	200 x 200 x 600 mm		53.42	.749		34	24.50		58.50	75	
	0080	250 x 300 x 600 mm		53.42	.749		44	24.50		68.50	86	
	0090	300 x 300 x 600 mm	↓	51.10	.783	↓	51.50	26		77.50	96	
100	0010	**BRICK VENEER** Scaffolding not included, truck load lots [R042110-20]										100
	0015	Material costs incl. 3% brick and 25% mortar waste										
	0020	Standard, sel. common, 100 x 68 x 200 mm (73/m²) [R042110-50]	D-8	1.50	26.667	k	470	880		1,350	1,875	
	0050	Red, 102 x 68 x 203 mm, running bond		1.50	26.667		485	880		1,365	1,875	
	0100	Full header every 6th course (85/m²)		1.45	27.586		485	910		1,395	1,900	
	0150	English, full header every 2nd course (109/m²)		1.40	28.571		485	945		1,430	1,950	
	0200	Flemish, alternate header every course (96/m²)		1.40	28.571		485	945		1,430	1,950	
	0250	Flemish, alt. header every 6th course (77/m²)		1.45	27.586		485	910		1,395	1,900	
	0300	Full headers throughout (145/m²)		1.40	28.571		485	945		1,430	1,950	
	0350	Rowlock course (145/m²)		1.35	29.630		485	980		1,465	2,025	
	0400	Rowlock stretcher (48/m²)		1.40	28.571		490	945		1,435	1,975	
	0450	Soldier course (73/m²)		1.40	28.571		485	945		1,430	1,950	
	0500	Sailor course (48/m²)		1.30	30.769		490	1,025		1,515	2,100	
	0601	Buff or gray face, running bond, (73/m²)		1.50	26.667		485	880		1,365	1,875	
	0700	Glazed face, 100 mm x 70 mm x 200 mm, running bond		1.40	28.571		1,400	945		2,345	2,975	
	0750	Full header every 6th course (85/m²)		1.35	29.630		1,350	980		2,330	2,975	
	1000	Jumbo, 150 mm x 100 mm x 300 mm, (32/m²)		1.30	30.769		1,475	1,025		2,500	3,175	
	1051	Norman, 102 mm x 68 mm x 305 mm, (48/m²)		1.45	27.586		945	910		1,855	2,400	
	1100	Norwegian, 102 mm x 81 mm x 305 mm (40/m²)		1.40	28.571		915	945		1,860	2,425	
	1150	Economy, 102 mm x 102 mm x 203 mm, (48/m²)		1.40	28.571		885	945		1,830	2,400	
	1201	Engineer, 102 mm x 81 mm x 203 mm, (61/m²)		1.45	27.586		595	910		1,505	2,025	
	1251	Roman, 102 mm x 51 mm x 305 mm, (65/m²)	↓	1.50	26.667	↓	870	880		1,750	2,300	

Important: See the Reference Section for supporting data - Crews, Rental Equipment, City Cost Indexes and Reference Data

04810	Unit Masonry Assemblies	CREW	DAILY OUTPUT	LABOR-HOURS	UNIT	MAT.	LABOR	EQUIP.	TOTAL	TOTAL INCL O&P	
100							2006 BARE COSTS				**100**
1300	S.C.R. 152 mm x 68 mm x 305 mm (48/m²)	D-8	1.40	28.571	k	1,050	945		1,995	2,575	
1350	Utility, 102 mm x 102 mm x 305 mm (32/m²)	↓	1.35	29.630	↓	1,275	980		2,255	2,900	
1360	For less than truck load lots, add				↓	10			10	11	
1400	For battered walls, add						30%				
1450	For corbels, add						75%				
1500	For curved walls, add						30%				
1550	For pits and trenches, deduct						20%				
1999	Alternate method of figuring by m²										
2000	Standard, sel. common, 100 mm x 70 mm x 200 mm, (73/m²)	D-8	21.37	1.872	m²	35.50	62		97.50	133	
2020	Std., red, 100 mm x 68 mm x 200 mm, running bond (73/m²)		20.44	1.957		35.50	64.50		100	138	
2050	Full header every 6th course (85/m²)		17.19	2.327		41	77		118	162	
2100	English, full header every 2nd course (109/m²)		13.01	3.076		52.50	102		154.50	213	
2150	Flemish, alternate header every course (97/m²)		13.94	2.870		47	95		142	196	
2200	Flemish, alt. header every 6th course (77/m²)		19.04	2.100		37	69.50		106.50	147	
2250	Full headers throughout (145/m²)		9.75	4.101		70	135		205	283	
2300	Rowlock course (145/m²)		9.29	4.306		70	142		212	294	
2350	Rowlock stretcher (48/m²)		28.80	1.389		23.50	46		69.50	96	
2400	Soldier course (73/m²)		18.58	2.153		35.50	71		106.50	147	
2450	Sailor course (48/m²)		26.94	1.485		23.50	49		72.50	101	
2600	Buff or gray face, running bond, (73/m²)		20.44	1.957		37.50	64.50		102	140	
2700	Glazed face, brick, running bond		19.51	2.050		97.50	67.50		165	210	
2750	Full header every 6th course (85/m²)		15.79	2.533		113	83.50		196.50	252	
3000	Jumbo, 152 mm x 102 mm x 305 mm running bond (32/m²)		40.41	.990		43.50	32.50		76	97.50	
3050	Norman, 100 mm x 68 mm x 300 mm running bond, (48/m²)		29.73	1.346		47.50	44.50		92	120	
3100	Norwegian, 102 mm x 81 mm x 305 mm (40/m²)		34.84	1.148		35.50	38		73.50	97	
3150	Economy, 102 mm x 102 mm x 203 mm (8/m²)		28.80	1.389		42.50	46		88.50	117	
3200	Engineer, 102 mm x 81 mm x 203 mm (61/m²)		24.15	1.656		36	54.50		90.50	123	
3250	Roman, 102 mm x 51 mm x 305 mm (65/m²)		23.23	1.722		55	57		112	147	
3300	SCR, 152 mm x 68 mm x 305 mm (48/m²)		28.80	1.389		50	46		96	125	
3350	Utility, 102 mm x 102 mm x 305 mm (32/m²)	↓	41.81	.957	↓	40.50	31.50		72	92.50	
3400	For cavity wall construction, add						15%				
3450	For stacked bond, add						10%				
3500	For interior veneer construction, add						15%				
3550	For curved walls, add						30%				
110	0010 **OVERSIZED BRICK**, scaffolding not included										**110**
0100	Veneer, 100 mm x 57 mm x 300 mm	D-8	35.95	1.113	m²	48	37		85	109	
0105	100 mm x 70 mm x 300 mm		38.27	1.045		43.50	34.50		78	101	
0110	100 mm x 100 mm x 400 mm		42.73	.936		42.50	31		73.50	94	
0120	100 mm x 200 mm x 400 mm		49.52	.808		100	26.50		126.50	151	
0125	Loadbearing, 150 mm x 100 mm x 400 mm, grouted and reinforced		35.95	1.113		74	37		111	137	
0130	200 mm x 100 mm x 400 mm, grouted and reinforced		30.38	1.317		77.50	43.50		121	151	
0135	150 mm x 200 mm x 400 mm, grouted and reinforced		40.88	.979		70.50	32.50		103	127	
0140	200 mm x 200 mm x 400 mm, grouted and reinforced		37.16	1.076		78.50	35.50		114	141	
0145	Curtainwall / reinforced veneer, 150 mm x 100 mm x 400 mm		35.95	1.113		112	37		149	180	
0150	200 mm x 100 mm x 400 mm		30.38	1.317		136	43.50		179.50	215	
0155	150 mm x 200 mm x 400 mm		40.88	.979		114	32.50		146.50	174	
0160	200 mm x 200 mm x 400 mm	↓	37.16	1.076	↓	137	35.50		172.50	204	
0200	For 1 to 3 slots in face, add				Ea.	25%					
0210	For 4 to 7 slots in face, add				m²	15%					
0220	For bond beams, add					20%					
0230	For bullnose shapes, add					20%					
0240	For open end knockout, add					10%					
0250	For white or gray color group, add					10%					
0260	For 135 degree corner, add				↓	250%					
160	0010 **CHIMNEY** See Div. 03310-240 for foundation, add to prices below										**160**
0100	Brick, 400 mm x 400 mm, 200 mm flue, scaff. not incl.	D-1	5.55	2.884	m	57.50	92.50		150	204	

For expanded coverage of these items see Means Concrete & Masonry Cost Data 2006

04810	Unit Masonry Assemblies		CREW	DAILY OUTPUT	LABOR-HOURS	UNIT	2006 BARE COSTS				TOTAL INCL O&P	
							MAT.	LABOR	EQUIP.	TOTAL		
160	0150	400 mm x 500 mm with one 200 mm x 300 mm flue		D-1	4.88	3.281	m	88	105		193	258	**160**
	0200	400 mm x 600 mm with two 200 mm x 200 mm flues			4.27	3.750		126	121		247	320	
	0250	500 mm x 500 mm with one 300 mm x 300 mm flue			4.18	3.832		103	123		226	300	
	0300	500 mm x 600 mm with two 200 mm x 300 mm flues			3.66	4.374		144	141		285	370	
	0350	500 mm x 800 mm with two 300 mm x 300 mm flues			3.05	5.249		180	169		349	455	
	1800	Metal, high temp. steel jacket, factory lining, 600 mm diam.		E-2	19.81	2.827		625	110	72	807	960	
	1900	1520 mm diameter		"	9.14	6.124		2,275	238	156	2,669	3,075	
	2100	Poured concrete, brick lining, 60 m high x 3.05 m diam.						20,000			20,000	22,000	
	2800	150 m high x 6.1 m diameter						35,000			35,000	38,600	
165	0010	**CHIMNEY BLOCK** Scaffolding not included											**165**
	0220	1 piece, with 203 mm x 203 mm flue, 406 mm x 406 mm		D-1	8.53	1.875	m	35.50	60.50		96	131	
	0230	2 piece, 406 mm x 406 mm			7.92	2.019		41	65		106	144	
	0240	2 piece, with 203 mm x 305 mm flue, 406 mm x 508 mm			7.32	2.187		73.50	70.50		144	188	
170	0010	**COLUMNS** Face brick, includes mortar, scaffolding not included	R042110-10										**170**
	0050	200 mm x 200mm, 9 brick per course		D-1	17.07	.937	m	13.80	30		43.80	61	
	0100	300 mm x 200 mm, 13.5 brick			11.28	1.419		20.50	45.50		66	92.50	
	0200	305 mm x 305 mm, 20 brick			7.62	2.100		30.50	67.50		98	137	
	0300	406 mm x 305 mm, 27 brick			5.79	2.763		41.50	89		130.50	181	
	0400	406 mm x 406 mm, 36 brick			4.27	3.750		55.50	121		176.50	245	
	0500	508 mm x 406 mm, 45 brick			3.35	4.772		69	153		222	310	
	0600	508 mm x 508 mm, 56 brick			2.74	5.833		86	188		274	380	
	0700	610 mm x 508 mm, 68 brick			2.13	7.499		104	241		345	480	
	0800	610 mm x 610 mm, 81 brick			1.83	8.749		124	281		405	565	
	1000	915 mm x 915 mm, 182 brick			.91	17.498		279	565		844	1,150	
172	0010	**CONCRETE BLOCK, BACK-UP,** C90, 2000 psi	R042210-20										**172**
	0020	Normal weight, 200 x 400 mm units, tooled jnt 1 side											
	0050	Non-reinforced, 13 MPa, 50 mm thick		D-8	44.13	.906	m²	10.85	30		40.85	57.50	
	0200	100 mm thick			42.73	.936		13	31		44	61.50	
	0300	150 mm thick			40.88	.979		18.95	32.50		51.45	70	
	0350	200 mm thick			37.16	1.076		21	35.50		56.50	77	
	0400	250 mm thick			30.66	1.305		26	43		69	94	
	0450	300 mm thick		D-9	28.80	1.667		30	53.50		83.50	115	
	1000	Reinforced, alternate courses, 100 mm thick		D-8	41.81	.957		13.90	31.50		45.40	63.50	
	1100	150 mm thick			39.95	1.001		19.90	33		52.90	72.50	
	1150	200 mm thick			36.70	1.090		21.50	36		57.50	79	
	1200	250 mm thick			29.73	1.346		27	44.50		71.50	97	
	1250	300 mm thick		D-9	27.87	1.722		31	55.50		86.50	119	
	2000	Lightweight, not reinforced, 100 mm thick		D-8	42.73	.936		15.20	31		46.20	63.50	
	2100	150 mm thick			41.34	.968		20.50	32		52.50	71	
	2150	200 mm thick			40.41	.990		25	32.50		57.50	77.50	
	2200	250 mm thick			38.09	1.050		32.50	34.50		67	89	
	2250	300 mm thick		D-9	36.23	1.325		33.50	42.50		76	102	
	3000	Lightweight, reinforced, alternate courses, 100 mm thick		D-8	41.81	.957		16.15	31.50		47.65	66	
	3100	150 mm thick			39.95	1.001		21.50	33		54.50	74	
	3150	200 mm thick			39.02	1.025		26	34		60	80	
	3200	250 mm thick			37.16	1.076		33.50	35.50		69	91	
	3250	300 mm thick		D-9	35.30	1.360		34.50	43.50		78	105	
175	0010	**CONCRETE BLOCK BOND BEAM** C90, 2000 psi											**175**
	0020	Not including grout or reinforcing											
	0130	200 mm high, 200 mm thick		D-8	172	.233	m	6.55	7.70		14.25	18.90	
	0150	300 mm thick		D-9	155	.310		9.10	9.95		19.05	25	
	0525	Lightweight, 150 mm thick		D-8	180	.222		6.05	7.35		13.40	17.85	
	0530	200 mm high, 200 mm thick		"	175	.229		6.15	7.55		13.70	18.25	

04810	Unit Masonry Assemblies	CREW	DAILY OUTPUT	LABOR-HOURS	UNIT	MAT.	LABOR	EQUIP.	TOTAL	TOTAL INCL O&P	
175	0550	300 mm thick	D-9	158	.304	m	10.15	9.75		19.90	26	**175**
	2000	Including grout and 2 #20M bars										
	2100	Regular block, 200 mm high, 200 mm thick	D-8	91.44	.437	m	12.35	14.45		26.80	35.50	
	2150	300 mm thick	D-9	76.20	.630		16.45	20.50		36.95	49	
	2500	Lightweight, 200 mm high, 200 mm thick	D-8	92.96	.430		13.20	14.20		27.40	36	
	2550	300 mm thick	D-9	77.72	.618	↓	17.50	19.85		37.35	49.50	
180	0010	**CONCRETE BLOCK COLUMN** or pilaster, scaffolding not included										**180**
	0160	1 piece unit, 400 mm x 400 mm	D-1	7.92	2.019	m	52.50	65		117.50	157	
	0170	2 piece units, 400 mm x 500 mm		7.32	2.187		70	70.50		140.50	184	
	0180	500 mm x 500 mm		6.71	2.386		102	76.50		178.50	229	
	0190	550 mm x 600 mm		5.49	2.916		143	94		237	300	
	0200	500 mm x 800 mm	↓	4.27	3.750	↓	155	121		276	355	
182	0010	**CONCRETE BLOCK, DECORATIVE** C90, 2000 psi										**182**
	0020	Embossed, simulated brick face										
	0100	203 mm x 406 mm units, 102 mm thick	D-8	37.16	1.076	m²	29	35.50		64.50	86	
	0200	203 mm thick		31.59	1.266		40.50	42		82.50	108	
	0250	305 mm thick	↓	27.87	1.435	↓	53	47.50		100.50	130	
	0400	Embossed both sides										
	0500	203 mm thick	D-8	27.87	1.435	m²	45	47.50		92.50	122	
	0550	305 mm thick	"	25.55	1.566	"	57	51.50		108.50	141	
	1000	Fluted high strength										
	1100	200 mm x 400 mm x 100 mm thick, flutes 1 side,	D-8	32.05	1.248	m²	34.50	41		75.50	101	
	1150	Flutes 2 sides		31.12	1.285		42	42.50		84.50	111	
	1200	203 mm thick	↓	27.87	1.435		54.50	47.50		102	132	
	1250	For special colors, add				↓	3.44			3.44	3.77	
	1400	Deep grooved, smooth face										
	1450	203 mm x 406 mm x 102 mm thick	D-8	32.05	1.248	m²	22.50	41		63.50	88	
	1500	203 mm thick	"	27.87	1.435	"	39	47.50		86.50	115	
	2000	Formblock, incl. inserts & reinforcing										
	2100	203 mm x 406 mm x 203 mm thick	D-8	32.05	1.248	m²	39.50	41		80.50	107	
	2150	305 mm thick	"	28.80	1.389	"	49.50	46		95.50	125	
	2500	Ground face										
	2600	200 mm x 400 mm x 100 mm thick	D-8	32.05	1.248	m²	38.50	41		79.50	106	
	2650	150 mm thick		30.19	1.325		48	44		92	119	
	2700	200 mm thick	↓	27.87	1.435		55.50	47.50		103	133	
	2750	300 mm thick	D-9	24.62	1.950	↓	70.50	62.50		133	173	
	2900	For special colors, add, minimum					15%					
	2950	For special colors, add, maximum					45%					
	4000	Slump block										
	4100	102 mm face height x 406 mm x 102 mm thick	D-1	15.33	1.044	m²	37	33.50		70.50	91.50	
	4150	152 mm thick		14.86	1.076		52	34.50		86.50	110	
	4200	203 mm thick		14.40	1.111		56	35.50		91.50	116	
	4250	254 mm thick		13.01	1.230		105	39.50		144.50	175	
	4300	305 mm thick		12.08	1.325		109	42.50		151.50	185	
	4400	152 mm face height x 406 mm x 152 mm thick		14.40	1.111		46.50	35.50		82	106	
	4450	203 mm thick		13.94	1.148		62.50	37		99.50	125	
	4500	254 mm thick		12.08	1.325		93	42.50		135.50	167	
	4550	305 mm thick	↓	11.15	1.435	↓	99.50	46		145.50	180	
	5000	Split rib profile units, 25 mm deep ribs, 8 ribs										
	5100	200 mm x 400 mm x 100 mm thick	D-8	32.05	1.248	m²	26.50	41		67.50	92.50	
	5150	150 mm thick		30.19	1.325		31	44		75	101	
	5200	200 mm thick	↓	27.87	1.435		35.50	47.50		83	111	
	5250	300 mm thick	D-9	25.55	1.879		42.50	60.50		103	139	
	5400	For special deeper colors, 102 mm thick, add					9.45			9.45	10.45	
	5450	305 mm thick, add					8.20			8.20	8.95	
	5600	For white, 102 mm thick, add				↓	9.45			9.45	10.45	

MASONRY 4

04810	Unit Masonry Assemblies	CREW	DAILY OUTPUT	LABOR-HOURS	UNIT	2006 BARE COSTS				TOTAL INCL O&P	
						MAT.	LABOR	EQUIP.	TOTAL		
182 5650	152 mm thick, add				m²	9.45			9.45	10.45	**182**
5700	203 mm thick, add					8.85			8.85	9.80	
5750	305 mm thick, add					8.20			8.20	8.95	
6000	Split face										
6100	200 mm x 400 mm x 100 mm thick	D-8	32.52	1.230	m²	25.50	40.50		66	90	
6150	150 mm thick		30.19	1.325		29.50	44		73.50	99	
6200	200 mm thick		27.87	1.435		34.50	47.50		82	110	
6250	300 mm thick	D-9	25.08	1.914		41	61.50		102.50	139	
6300	For scored, add					2.69			2.69	3.01	
6400	For special deeper colors, 102 mm thick, add					5.50			5.50	6.05	
6450	152 mm thick, add					5.50			5.50	6.05	
6500	203 mm thick, add					4.84			4.84	5.40	
6550	305 mm thick, add					4.09			4.09	4.52	
6650	For white, 102 mm thick, add					9.45			9.45	10.45	
6700	152 mm thick, add					9.45			9.45	10.45	
6750	203 mm thick, add					8.85			8.85	9.80	
6800	305 mm thick, add					8.20			8.20	8.95	
7000	Scored ground face, 2 to 5 scores										
7100	203 mm x 406 mm x 102 mm thick	D-8	31.59	1.266	m²	44	42		86	112	
7150	152 mm thick		28.80	1.389		51	46		97	127	
7200	203 mm thick		26.94	1.485		59.50	49		108.50	140	
7250	305 mm thick	D-9	24.62	1.950		70.50	62.50		133	173	
8000	Hexagonal face profile units, 200 mm x 400 mm units										
8100	100 mm thick, hollow	D-8	31.59	1.266	m²	27.50	42		69.50	94	
8200	Solid		31.59	1.266		37	42		79	105	
8300	150 mm thick, hollow		28.80	1.389		31	46		77	104	
8350	200 mm thick, hollow		26.94	1.485		42.50	49		91.50	121	
8500	For stacked bond, add						26%				
8550	For high rise construction, add per story	D-8	6299	.006	m²		.21		.21	.32	
8600	For scored block, add					10%					
8650	For honed or ground face, per face, add				Ea.	.32			.32	.35	
8700	For honed or ground end, per end, add				"	2.49			2.49	2.73	
8750	For bullnose block, add					10%					
8800	For special color, add					13%					
184 0010	**CONCRETE BLOCK, EXTERIOR** C90, 2000 psi										**184**
0020	Reinforced alt courses, tooled joints 2 sides										
0100	Normal weight, 200 mm x 400 mm x 150 mm thick	D-8	36.70	1.090	m²	21	36		57	78.50	
0200	200 mm thick		33.44	1.196		31	39.50		70.50	94.50	
0250	250 mm thick		26.94	1.485		38	49		87	116	
0300	300 mm thick	D-9	23.23	2.067		38.50	66.50		105	144	
0500	Lightweight, 200 mm x 400 mm x 150 mm thick	D-8	41.81	.957		24.50	31.50		56	75	
0600	200 mm thick		39.95	1.001		32	33		65	86	
0650	250 mm thick		36.70	1.090		37.50	36		73.50	96.50	
0700	300 mm thick	D-9	32.52	1.476		54	47.50		101.50	132	
186 0010	**CONCRETE BLOCK FOUNDATION WALL** C90/C145										**186**
0050	Normal-weight, cut joints, horiz joint reinf, no vert reinf										
0200	Hollow, 200 mm x 400 mm x 150 mm thick	D-8	42.27	.946	m²	23	31.50		54.50	73	
0250	200 mm thick		39.48	1.013		25	33.50		58.50	78.50	
0300	250 mm thick		32.52	1.230		30.50	40.50		71	95.50	
0350	300 mm thick	D-9	27.87	1.722		34	55.50		89.50	122	
0500	Solid, 200 mm x 400 mm block, 150 mm thick	D-8	40.88	.979		23.50	32.50		56	75	
0550	200 mm thick	"	38.55	1.038		31.50	34.50		66	87	
0600	300 mm thick	D-9	32.52	1.476		46	47.50		93.50	123	
1000	Reinforced, #13 vert @ 1.2 m										
1125	150 mm thick	D-8	41.34	.968	m²	31	32		63	82.50	
1150	200 masm thick		38.55	1.038		36	34.50		70.50	91.50	

04810	Unit Masonry Assemblies	CREW	DAILY OUTPUT	LABOR-HOURS	UNIT	2006 BARE COSTS				TOTAL INCL O&P	
						MAT.	LABOR	EQUIP.	TOTAL		
186											**186**
1200	250 mm thick	D-8	31.59	1.266	m²	44.50	42		86.50	113	
1250	300 mm thick	D-9	26.94	1.782		52	57.50		109.50	144	
1500	Solid, 200 mm x 400 mm block, 150 mm thick	D-8	39.95	1.001		23.50	33		56.50	76.50	
1600	200 mm thick	"	37.62	1.063		31.50	35		66.50	88.50	
1650	300 mm thick	D-9	31.59	1.520		46	49		95	125	
187	0010	**CONCRETE BLOCK, HIGH STRENGTH** Normal weight									**187**
	0050	Hollow, reinforced alternate courses, 200 mm x 400 mm units									
	0200	21 MPa, 100 mm thick	D-8	40.88	.979	m²	17.35	32.50		49.85	68
	0250	150 mm thick	↓	36.70	1.090		20.50	36		56.50	77.50
	0300	200 mm thick	↓	33.44	1.196	↓	30.50	39.50		70	93.50
	0350	300 mm thick	D-9	23.23	2.067		37.50	66.50		104	143
	0500	34 MPa, 100 mm thick	D-8	40.88	.979		16.45	32.50		48.95	67
	0550	150 mm thick	↓	36.70	1.090		25.50	36		61.50	83
	0600	200 mm thick	↓	33.44	1.196		34.50	39.50		74	98
	0650	300 mm thick	D-9	27.87	1.722	↓	51	55.50		106.50	141
	1000	For 75% solid block, add					30%				
	1050	For 100% solid block, add					50%				
188	0010	**CONCRETE BLOCK INSULATION INSERTS**									**188**
	0100	Inserts, styrofoam, plant installed, add to block prices									
	0200	200 mm x 400 mm units, 150 mm thick				m²	11			11	12.05
	0250	200 mm thick					11			11	12.05
	0300	250 mm thick					12.90			12.90	14.20
	0350	300 mm thick					13.55			13.55	14.95
	0500	200 mm x 200 mm units, 200 mm thick					9.05			9.05	9.90
	0550	300 mm thick				↓	11			11	12.05
189	0010	**CONCRETE BLOCK, INTERLOCKING**									**189**
	0100	Not including grout or reinforcing									
	0200	203 mm x 406 mm units, 13 790 kPa, 203 mm thick	D-1	22.76	.703	m²	23	22.50		45.50	60
	0300	305 mm thick		20.44	.783		34	25		59	76
	0350	406 mm thick	↓	17.19	.931		51.50	30		81.50	102
	0400	Including grout & reinforcing, 203 mm thick	D-4	22.76	1.406		66	44.50	6.35	116.85	148
	0450	305 mm thick		20.44	1.566		79	50	7.10	136.10	170
	0500	406 mm thick	↓	17.19	1.862	↓	98.50	59	8.45	165.95	207
190	0010	**CONCRETE BLOCK, LINTELS** C90, normal weight									**190**
	0100	Including grout and horizontal reinforcing									
	0200	203 mm x 203 mm x 203 mm, 1 #10 bar [#4]	D-4	91.44	.350	m	13.10	11.15	1.59	25.84	33
	0250	2 #10 bars [#4]		89.92	.356		13.70	11.30	1.61	26.61	34
	0400	203 mm x 406 mm x 203 mm, 1 #10 bar [#4]		83.82	.382		22.50	12.15	1.73	36.38	45
	0450	2 #10 bars [#4]		82.30	.389		23	12.35	1.76	37.11	46
	1000	305 mm x 203 mm x 203 mm, 1 #10 bar [#4]		83.82	.382		18.35	12.15	1.73	32.23	40.50
	1100	2 #10 bars [#4]		82.30	.389		19	12.35	1.76	33.11	41.50
	1150	2 #15 bars		82.30	.389		19.70	12.35	1.76	33.81	42
	1200	2 #20 bars		80.77	.396		20.50	12.60	1.79	34.89	43.50
	1500	305 mm x 406 mm x 203 mm, 1 #10 bar [#4]		76.20	.420		29	13.35	1.90	44.25	54.50
	1600	2 #10 bars [#3]		74.68	.429		29	13.65	1.94	44.59	54.50
	1650	2 #10 bars [#4]		74.68	.429		30	13.65	1.94	45.59	55
	1700	2 #15 bars	↓	73.15	.437	↓	30.50	13.90	1.98	46.38	56.50
210	0010	**CONCRETE BLOCK, PARTITIONS**, scaffolding not included	R042210 -20									**210**
	0100	Acoustical slotted block									
	0200	100 mm thick, type A-1	D-8	29.26	1.367	m²	29	45		74	101
	0210	200 mm thick		25.55	1.566		42.50	51.50		94	126
	0250	200 mm thick, type Q		25.55	1.566		58	51.50		109.50	143
	0260	100 mm thick, type RSC		29.26	1.367		42.50	45		87.50	116
	0270	150 mm thick		27.41	1.460		42.50	48		90.50	121
	0280	200 mm thick	↓	25.55	1.566	↓	42.50	51.50		94	126

MASONRY 4

For expanded coverage of these items see *Means Concrete & Masonry Cost Data 2006*

04810 | Unit Masonry Assemblies

			CREW	DAILY OUTPUT	LABOR-HOURS	UNIT	2006 BARE COSTS				TOTAL INCL O&P	
							MAT.	LABOR	EQUIP.	TOTAL		
210	0290	300 mm thick	R042210 -20	D-8	23.23	1.722	m²	42.50	57		99.50	134	**210**
	0300	200 mm thick, type RSR			25.55	1.566		42.50	51.50		94	126	
	0400	200 mm thick, type RSC/RF			25.55	1.566		53	51.50		104.50	137	
	0410	250 mm thick			24.15	1.656		57	54.50		111.50	147	
	0420	300 mm thick			23.23	1.722		62	57		119	155	
	0430	300 mm thick, type RSC/RF-4			23.23	1.722		73	57		130	167	
	0500	NRC .60 type R, 203 mm thick			24.62	1.625		48.50	53.50		102	135	
	0600	NRC .65 type RR, 203 mm thick			24.62	1.625		54.50	53.50		108	142	
	0700	NRC .65 type 4R-RF, 203 mm thick			24.62	1.625		62	53.50		115.50	150	
	0710	NRC .70 type R, 305 mm thick		▼	22.76	1.757	▼	74	58		132	170	
	1000	Lightweight block, tooled joints, 2 sides, hollow											
	1100	Not reinforced, 200 mm x 400 mm x 100 mm thick		D-8	40.88	.979	m²	14.40	32.50		46.90	65	
	1150	150 mm thick			38.09	1.050		19.80	34.50		54.30	74.50	
	1200	200 mm thick			35.77	1.118		24.50	37		61.50	83	
	1250	250 mm thick		▼	34.37	1.164		32	38.50		70.50	93.50	
	1300	300 mm thick		D-9	32.52	1.476		32.50	47.50		80	108	
	2000	Not reinforced, 203 mm x 610 mm x 102 mm thick, hollow			42.73	1.123		10.55	36		46.55	66.50	
	2100	152 mm thick			40.88	1.174		14.30	38		52.30	73	
	2150	203 mm thick			38.55	1.245		18	40		58	80.50	
	2200	254 mm thick			35.77	1.342		23.50	43		66.50	91	
	2250	305 mm thick		▼	33.91	1.416		24	45.50		69.50	96	
	2800	Solid, not reinforced, 200 mm x 400 mm x 50 mm thick		D-8	40.88	.979		11.65	32.50		44.15	62	
	2900	100 mm thick			39.02	1.025		20.50	34		54.50	74	
	2950	150 mm thick			36.23	1.104		23.50	36.50		60	81.50	
	3000	200 mm thick			33.91	1.180		31.50	39		70.50	94	
	3050	250 mm thick		▼	32.52	1.230		36.50	40.50		77	103	
	3100	300 mm thick		D-9	30.66	1.566	▼	53	50.50		103.50	135	
	4000	Regular block, tooled joints, 2 sides, hollow											
	4100	Not reinforced, 200 mm x 400 mm x 100 mm thick		D-8	39.95	1.001	m²	12.25	33		45.25	64	
	4150	150 mm thick			37.16	1.076		18.20	35.50		53.70	74	
	4200	200 mm thick			34.84	1.148		19.90	38		57.90	79.50	
	4250	250 mm thick		▼	33.44	1.196		25	39.50		64.50	88	
	4300	300 mm thick		D-9	31.59	1.520		29	49		78	107	
	4500	Reinforced alternate courses, 200 x 400 x 100 mm thick		D-8	39.48	1.013		13.15	33.50		46.65	65.50	
	4550	152 mm thick			36.70	1.090		19.15	36		55.15	76	
	4600	203 mm thick			34.37	1.164		21	38.50		59.50	81.50	
	4650	254 mm thick		▼	32.98	1.213		29	40		69	93	
	4700	305 mm thick		D-9	31.12	1.542		30.50	49.50		80	109	
	4900	Solid, not reinforced, 50 mm thick		D-8	40.41	.990		10.45	32.50		42.95	61.50	
	5000	75 mm thick			39.95	1.001		11.10	33		44.10	62.50	
	5050	100 mm thick			38.55	1.038		15.30	34.50		49.80	69	
	5100	150 mm thick			35.77	1.118		18.95	37		55.95	77	
	5150	200 mm thick		▼	33.44	1.196		26.50	39.50		66	89.50	
	5200	300 mm thick		D-9	30.19	1.590		41	51		92	123	
	5500	Solid, reinforced alternate courses, 100 mm thick		D-8	39.02	1.025		16.05	34		50.05	69	
	5550	152 mm thick			35.30	1.133		19.80	37.50		57.30	78.50	
	5600	203 mm thick		▼	32.98	1.213		27.50	40		67.50	91.50	
	5650	305 mm thick		D-9	29.73	1.615	▼	39	52		91	122	
225	0010	**CONCRETE BRICK** C55, grade N, type I											**225**
	0100	Regular, 100 mm x 57 mm x 200 mm		D-8	220	.182	Ea.	.35	6		6.35	9.55	
	0125	Rusticated, 100 mm x 57 mm x 200 mm			220	.182		.39	6		6.39	9.60	
	0150	Frog, 100 mm x 57 mm x 200 mm			220	.182		.38	6		6.38	9.55	
	0200	Double, 100 mm x 124 mm x 200 mm		▼	180	.222	▼	.61	7.35		7.96	11.85	
240	0010	**CONCRETE SCREEN BLOCK**											**240**
	0200	203 mm x 406 mm, 102 mm thick		D-8	30.66	1.305	m²	19.40	43		62.40	87	

04810	Unit Masonry Assemblies	CREW	DAILY OUTPUT	LABOR-HOURS	UNIT	2006 BARE COSTS				TOTAL INCL O&P	
						MAT.	LABOR	EQUIP.	TOTAL		
240											**240**
0300	203 mm thick	D-8	25.08	1.595	m²	29.50	52.50		82	113	
0350	305 mm x 305 mm, 102 mm thick		26.94	1.485		23	49		72	100	
0500	203 mm thick	▼	23.23	1.722	▼	30	57		87	120	
250	0010	**COPING** Stock units									**250**
0050	Precast conc, 255 mm wide, 100 mm tapers to 90 mm, 200 mm wall	D-1	22.86	.700	m	49.50	22.50		72	88.50	
0100	305 mm wide, 90 mm tapers to 75 mm, 255 mm wall		21.34	.750		34.50	24		58.50	74.50	
0110	355 mm wide, 100 mm tapers to 90 mm, 305 mm wall		19.81	.808		65	26		91	111	
0150	400 mm wide, 100 mm tapers to 90 mm, 355 mm wall		18.29	.875	▼	47.50	28		75.50	95.50	
0250	Precast concrete corners		40	.400	Ea.	24	12.85		36.85	46	
0300	Limestone for 305 mm wall, 100 mm thick		27.43	.583	m	44.50	18.75		63.25	77	
0350	150 mm thick		24.38	.656		51.50	21		72.50	89	
0500	Marble, to 100 mm thick, no wash, 230 mm wide		27.43	.583		63	18.75		81.75	98	
0550	305 mm wide		24.38	.656		95	21		116	137	
0700	Terra cotta, 230 mm wide		27.43	.583		15.60	18.75		34.35	45.50	
0750	305 mm wide		24.38	.656		25.50	21		46.50	60	
0800	Aluminum, for 305 mm wall	▼	24.38	.656	▼	37	21		58	72.50	
300	0010	**FACING PANELS** Stone aggregate mounted on plywood,									**300**
0020	See Division 07440-200										
325	0010	**GLASS BLOCK**									**325**
0100	Plain, 100 mm thick, under 90 m², 150 mm x 150 mm	D-8	10.68	3.744	m²	193	124		317	400	
0150	200 mm x 200 mm		14.86	2.691		129	89		218	277	
0160	end block		14.86	2.691		365	89		454	535	
0170	90 deg corner		14.86	2.691		350	89		439	520	
0180	45 deg corner		14.86	2.691		151	89		240	300	
0200	300 mm x 300 mm		16.26	2.460		154	81.50		235.50	293	
0210	100 mm x 200 mm		14.86	2.691		86	89		175	230	
0220	150 mm x 200 mm		14.86	2.691		101	89		190	246	
0300	90 to 465 m², 150 mm x 150 mm		12.54	3.189		189	105		294	370	
0350	200 mm x 200 mm		17.65	2.266		126	75		201	253	
0400	300 mm x 300 mm		19.97	2.003		150	66		216	267	
0410	100 mm x 200 mm		19.97	2.003		40	66		106	145	
0420	150 mm x 200 mm		19.97	2.003		44.50	66		110.50	150	
0500	Over 465 m², 150 mm x 150 mm		13.47	2.969		184	98		282	350	
0550	200 mm x 200 mm		19.97	2.003		122	66		188	236	
0600	300 mm x 300 mm		22.30	1.794		146	59.50		205.50	250	
0610	100 mm x 200 mm		22.30	1.794		40.50	59.50		100	135	
0620	150 mm x 200 mm	▼	22.30	1.794	▼	41.50	59.50		101	136	
0700	For solar reflective blocks, add					100%					
1000	Thinline, plain, 79 mm thick, under 90 m², 152 mm x 152 mm	D-8	10.68	3.744	m²	212	124		336	420	
1050	203 mm x 203 mm		14.86	2.691		106	89		195	252	
1200	Over 465 m², 152 mm x 152 mm		13.47	2.969		134	98		232	297	
1250	203 mm x 203 mm		19.97	2.003		77.50	66		143.50	186	
1400	For cleaning block after installation (both sides), add	▼	92.90	.431	▼	1.08	14.20		15.28	22.50	
4000	Accessories										
4100	Anchors, 20 ga. galv., 44 mm wide x 610 mm long				Ea.	2.57			2.57	2.83	
4200	Emulsion asphalt				liter	2.54			2.54	2.79	
4300	Expansion joint, fiberglass				m	2			2	2.20	
4400	Steel mesh, double galvanized				"	.82			.82	.92	
350	0010	**GLAZED CONCRETE BLOCK** C744									**350**
0100	Single face, 200 mm x 400 mm units, 50 mm thick	D-8	33.44	1.196	m²	69.50	39.50		109	137	
0200	100 mm thick		32.05	1.248		70.50	41		111.50	141	
0250	150 mm thick		30.66	1.305		76	43		119	149	
0300	200 mm thick		28.80	1.389		81.50	46		127.50	160	
0350	250 mm thick	▼	27.41	1.460	▼	91.50	48		139.50	174	

MASONRY 4

04810 | Unit Masonry Assemblies

			CREW	DAILY OUTPUT	LABOR-HOURS	UNIT	MAT.	LABOR	EQUIP.	TOTAL	TOTAL INCL O&P	
350	0400	300 mm thick	D-9	26.01	1.845	m²	95.50	59.50		155	196	350
	0700	Double face, 200 mm x 400 mm units, 100 mm thick	D-8	31.59	1.266		110	42		152	185	
	0750	150 mm thick		29.73	1.346		121	44.50		165.50	201	
	0800	200 mm thick		27.87	1.435		127	47.50		174.50	211	
	1000	Jambs, bullnose or square, sgl. face, 200 mm x 400 mm, 50 mm T		315	.127	Ea.	11.55	4.19		15.74	19.10	
	1050	100 mm thick		285	.140	"	12.75	4.64		17.39	21	
	1200	Caps, bullnose or square, 200 mm x 400 mm, 50 mm thick		128	.313	m	34	10.30		44.30	52.50	
	1250	100 mm thick		116	.345		46	11.40		57.40	68	
	1500	Cove base, 200 mm x 400 mm, 50 mm thick		96.01	.417		21	13.75		34.75	44.50	
	1550	100 mm thick		86.87	.460		20.50	15.20		35.70	45.50	
	1600	150 mm thick		80.77	.495		22.50	16.35		38.85	49.50	
	1650	200 mm thick		74.68	.536		24.50	17.70		42.20	54	
500	0010	**TERRA COTTA COPING**										500
	0020	Coping, split type, not glazed, 229 mm wide	D-1	27.43	.583	m	24.50	18.75		43.25	55.50	
	0100	330 mm wide		24.38	.656		37	21		58	73	
	0200	Split type, glazed, 229 mm wide		27.43	.583		41.50	18.75		60.25	74.50	
	0250	330 mm wide		24.38	.656		54.50	21		75.50	92	
	0500	Partition or back-up blocks, scored, in C.L. lots										
	0700	Non-load bearing, 305 mm x 305 mm, 76 mm thick	D-8	51.10	.783	m²	133	26		159	186	
	0750	102 mm thick		46.45	.861		52.50	28.50		81	101	
	0800	152 mm thick		41.81	.957		62	31.50		93.50	116	
	0850	203 mm thick		37.16	1.076		80	35.50		115.50	142	
	1000	Load bearing, 305 mm x 305 mm, 102 mm thick, in walls		46.45	.861		55	28.50		83.50	104	
	1050	In floors		69.68	.574		55	18.95		73.95	89.50	
	1200	152 mm thick, in walls		41.81	.957		62	31.50		93.50	116	
	1250	In floors		62.71	.638		62	21		83	100	
	1400	203 mm thick, in walls		37.16	1.076		80	35.50		115.50	142	
	1450	In floors		53.42	.749		80	24.50		104.50	126	
	1600	254 mm thick, in walls, special order		32.52	1.230		99.50	40.50		140	172	
	1650	In floors, special order		46.45	.861		99.50	28.50		128	154	
	1800	305 mm thick, in walls, special order		27.87	1.435		199	47.50		246.50	291	
	1850	In floors, speical order		41.81	.957		199	31.50		230.50	267	
	2000	For reinforcing with steel rods, add to above					15%	5%				
	2100	For smooth tile instead of scored, add					26			26	28.50	
	2200	For L.C.L. quantities, add					10%	10%				
510	0010	**TERRA COTTA TILE** On walls, dry set, 13 mm thick										510
	0100	m², hexagonal or lattice shapes, unglazed	1 Tilf	12.54	.638	m²	51.50	22		73.50	88.50	
	0300	Glazed, plain colors		12.08	.662		71	22.50		93.50	112	
	0400	Intense colors		11.61	.689		83	23.50		106.50	126	
540	0010	**WALLS** Cavity, brick and CMU incl joint reinf and z-ties										540
	0200	100 mm face brick, 100 mm block	D-8	15.33	2.610	m²	50	86		136	186	
	0400	150 mm block		13.47	2.969		55	98		153	210	
	0600	200 mm block		11.61	3.445		56	114		170	235	
650	0010	**WALLS** Building brick, including mortar	R042110 -20									650
	0140	100 mm thick, face, 100 x 70 x 200 mm	D-8	1.45	27.586	k	390	910		1,300	1,800	
	0150	100 mm thick, as back-up, 73 bricks/m²		1.60	25		390	825		1,215	1,675	
	0204	200 mm thick, 145 bricks/m²		1.80	22.222		405	735		1,140	1,575	
	0250	300 mm thick, 218 bricks/m²		1.90	21.053		410	695		1,105	1,500	
	0304	400 mm thick, 291 bricks/m²		2	20		415	660		1,075	1,450	
	0500	Reinforced, 100 mm wall, 100 mm x 68 mm x 200 mm		1.40	28.571		390	945		1,335	1,850	
	0550	200 mm thick, 13.50 bricks/m²		1.75	22.857		480	755		1,235	1,675	
	0600	300 mm thick, 218 bricks/m²		1.85	21.622		480	715		1,195	1,600	
	0650	400 mm thick, 290 bricks/m²		1.95	20.513		485	680		1,165	1,550	
	0660	100 mm thick, select common, face, 100 x 70 x 200 mm		1.45	27.586		465	910		1,375	1,875	
	0790	Alternate method of figuring by m²										

Important: See the Reference Section for supporting data - Crews, Rental Equipment, City Cost Indexes and Reference Data

			DAILY	LABOR-		2006 BARE COSTS				TOTAL	
04810	**Unit Masonry Assemblies**	CREW	OUTPUT	HOURS	UNIT	MAT.	LABOR	EQUIP.	TOTAL	INCL O&P	
650	0800	Common, 100 mm x 70 mm x 200 mm, 100 mm wall, as face brick	D-8	19.97	2.003	m²	35	66		101	140	**650**
	0850	100 mm thick, as back up, 73 bricks/m²		22.30	1.794		28.50	59.50		88	122	
	0900	200 mm thick wall, 150 brick/m²		12.54	3.189		59	105		164	225	
	1000	300 mm thick wall, 218 bricks/m²		8.83	4.532		89	150		239	325	
	1050	400 mm thick wall, 291 bricks/m²		6.97	5.741		121	190		311	420	
	1200	Reinforced, 100 mm x 70 mm x 200 mm, 100 mm wall		19.04	2.100		28.50	69.50		98	138	
	1250	200 mm thick wall, 13.50 brick/m²		12.08	3.312		59	109		168	232	
	1300	300 mm thick wall, 20.25 bricks/m²		8.36	4.784		89	158		247	340	
	1350	400 mm thick wall, 27.00 bricks/m²		6.50	6.151		121	203		324	445	
670	0010	**STEPS**, with select common bricks										**670**
	0012	Entry steps, brick	D-1	.30	53.333	k	385	1,725		2,110	3,025	
700	0010	**STRUCTURAL BRICK** C652, Grade SW, incl mortar, scaffolding not incl										**700**
	0100	Standard unit, 117 mm x 70 mm x 244 mm	D-8	22.76	1.757	m²	50.50	58		108.50	144	
	0120	Bond beam		20.90	1.914		95.50	63		158.50	201	
	0140	V cut bond beam		20.90	1.914		98.50	63		161.50	204	
	0160	Stretcher quoin, 143 mm x 70 mm x 244 mm		22.76	1.757		95	58		153	193	
	0180	Corner quoin		22.76	1.757		106	58		164	206	
	0200	Corner, 45 deg, 117 mm x 70 mm x 265 mm		21.83	1.832		110	60.50		170.50	213	
750	0010	**STRUCTURAL FACING TILE** Scaffolding not incl, standard colors										**750**
	0020	6T series, 133 x 300 mm, 25 pieces/m², glazed 1 side, 50 mm T	D-8	20.90	1.914	m²	64.50	63		127.50	167	
	0100	100 mm thick		20.44	1.957		76	64.50		140.50	182	
	0150	Glazed 2 sides		18.12	2.208		120	73		193	243	
	0250	150 mm thick		19.51	2.050		118	67.50		185.50	232	
	0300	Glazed 2 sides		17.19	2.327		138	77		215	269	
	0400	200 mm thick		16.72	2.392		147	79		226	282	
	0500	Special shapes, group 1		400	.100	Ea.	5.40	3.30		8.70	10.95	
	0550	Group 2		375	.107		8.20	3.52		11.72	14.35	
	0600	Group 3		350	.114		10.15	3.78		13.93	16.90	
	0650	Group 4		325	.123		17.50	4.07		21.57	25.50	
	0700	Group 5		300	.133		23	4.40		27.40	31.50	
	0750	Group 6		275	.145		27.50	4.80		32.30	38	
	1000	Fire rated, 100 mm thick, 1 hr. rating		19.51	2.050	m²	130	67.50		197.50	246	
	1300	Acoustic, 100 mm thick		19.51	2.050	"	289	67.50		356.50	425	
	1400	For designer colors, add					25%					
	2000	8W series, 200 mm x 400 mm, 12.11 pieces/m²										
	2050	50 mm thick, glazed 1 side	D-8	33.44	1.196	m²	70	39.50		109.50	137	
	2100	100 mm thick, glazed 1 side		32.05	1.248		76.50	41		117.50	147	
	2150	Glazed 2 sides		30.19	1.325		114	44		158	193	
	2200	150 mm thick, glazed 1 side		30.66	1.305		108	43		151	185	
	2250	200 mm thick, glazed 1 side		28.80	1.389		125	46		171	208	
	2500	Special shapes, group 1		300	.133	Ea.	8.70	4.40		13.10	16.25	
	2550	Group 2		280	.143		11.45	4.72		16.17	19.80	
	2600	Group 3		260	.154		11.95	5.10		17.05	21	
	2650	Group 4		250	.160		26.50	5.30		31.80	37.50	
	2700	Group 5		240	.167		43.50	5.50		49	56.50	
	2750	Group 6		230	.174		51	5.75		56.75	65	
	3000	100 mm thick, glazed 1 side		32.05	1.248	m²	89	41		130	161	
	3100	Acoustic, 102 mm thick		32.05	1.248	"	118	41		159	193	
	3120	4W series, 200 mm x 200 mm, 24.2 pieces per m²										
	3125	50 mm thick, glazed 1 side	D-8	33.44	1.196	m²	82	39.50		121.50	151	
	3130	100 mm thick, glazed 1 side		32.05	1.248		82	41		123	154	
	3135	Glazed 2 sides		30.19	1.325		118	44		162	196	
	3140	150 mm thick, glazed 1 side		30.66	1.305		110	43		153	187	
	3150	200 mm thick, glazed 1 side		28.80	1.389		159	46		205	244	

R042110 -20

MASONRY 4

For expanded coverage of these items see *Means Concrete & Masonry Cost Data 2006*

04800 | Masonry Assemblies

04810 | Unit Masonry Assemblies

			CREW	DAILY OUTPUT	LABOR-HOURS	UNIT	2006 BARE COSTS				TOTAL INCL O&P	
							MAT.	LABOR	EQUIP.	TOTAL		
750	3155	Special shapes, group I	D-8	300	.133	Ea.	5.40	4.40		9.80	12.65	750
	3160	Group II	↓	280	.143	"	6.55	4.72		11.27	14.40	
	3200	For designer colors, add					25%					
	3300	For epoxy mortar joints, add				m²	13.45			13.45	14.85	

04840 | Prefabricated Masonry Panels

			CREW	DAILY OUTPUT	LABOR-HOURS	UNIT	MAT.	LABOR	EQUIP.	TOTAL	INCL O&P	
900	0010	**WALL PANELS**										900
	0020	Prefabricated, 100 mm thick, minimum	C-11	72	1	m²	98	39	21.50	158.50	200	
	0100	Maximum	"	46.45	1.550	↓	126	60.50	33	219.50	282	
	0200	100 mm brick & 50 mm concrete back-up, add					50%					
	0300	100 mm brick & 25 mm urethane & 75 mm conc back-up, add				↓	70%					

04850 | Stone Assemblies

			CREW	DAILY OUTPUT	LABOR-HOURS	UNIT	MAT.	LABOR	EQUIP.	TOTAL	INCL O&P	
050	0011	**ASHLAR VENEER** 100 mm +/- thk, random or random rectangular										050
	0150	Sawn face, split joints, low priced stone	D-8	13.01	3.076	m²	117	102		219	283	
	0200	Medium priced stone		12.08	3.312		144	109		253	325	
	0300	High priced stone		11.15	3.588		203	119		322	405	
	0600	Seam face, split joints, medium price stone		11.61	3.445		151	114		265	340	
	0700	High price stone		11.15	3.588		215	119		334	415	
	1000	Split or rock face, split joints, medium price stone		11.61	3.445		151	114		265	340	
	1100	High price stone	↓	11.15	3.588	↓	215	119		334	415	
100	0010	**BLUESTONE** Cut to size										100
	0100	Paving, natural cleft, to 0.9 m x 1.2 m, 25 mm thick	D-8	13.94	2.870	m²	76	95		171	228	
	0150	38 mm thick		13.47	2.969		69	98		167	225	
	0200	Smooth finish, 25 mm thick		13.94	2.870		81.50	95		176.50	234	
	0250	38 mm thick		13.47	2.969		87	98		185	245	
	0300	Thermal finish, 25 mm thick		13.94	2.870		99.50	95		194.50	253	
	0350	38 mm thick	↓	13.47	2.969	↓	112	98		210	272	
	0500	Sills, natural cleft, 250 mm wide to 1829 mm long, 38 mm thick	D-11	21.34	1.125	m	40.50	38.50		79	103	
	0550	50 mm thick	"	19.20	1.250		46	43		89	116	
	1000	Stair treads, natural cleft, 300 mm wide, 2 m long, 38 mm T	D-10	35.05	1.141		59.50	38.50	15.30	113.30	141	
	1050	50 mm thick		32	1.250		62.50	42	16.75	121.25	151	
	1100	Smooth finish, 38 mm thick		35.05	1.141		72	38.50	15.30	125.80	155	
	1150	50 mm thick		32	1.250		72	42	16.75	130.75	162	
	1300	Thermal finish		35.05	1.141		90	38.50	15.30	143.80	174	
	1350	50 mm thick	↓	32	1.250	↓	102	42	16.75	160.75	194	
300	0010	**GRANITE** Cut to size										300
	0050	Veneer, polished face, 20 mm to 50 mm thick										
	0150	Low price, gray, light gray, etc.	D-10	12.08	3.312	m²	242	112	44.50	398.50	485	
	0220	High price, red, black, etc.	"	12.08	3.312	"	405	112	44.50	561.50	665	
	0300	40 mm to 65 mm thick, veneer										
	0350	Low price, gray, light gray, etc.	D-10	12.08	3.312	m²	266	112	44.50	422.50	510	
	0550	High price, red, black, etc.	"	12.08	3.312	"	460	112	44.50	616.50	725	
	0700	65 mm to 100 mm thick, veneer										
	0750	Low price, gray, light gray, etc.	D-10	10.22	3.914	m²	330	132	52.50	514.50	625	
	0950	High price, red, black, etc.	"	10.22	3.914		530	132	52.50	714.50	845	
	1000	For bush hammered finish, deduct					5%					
	1050	Coarse rubbed finish, deduct					10%					
	1100	Honed finish, deduct					5%					
	1150	Thermal finish, deduct				↓	18%					
	1800	Carving or bas-relief, from templates or plaster molds										
	1850	Minimum	D-10	2.26	17.668	m³	4,950	595	237	5,782	6,625	
	1900	Maximum	"	2.26	17.668	"	14,900	595	237	15,732	17,500	
	2000	Intricate or hand finished pieces										

Important: See the Reference Section for supporting data - Crews, Rental Equipment, City Cost Indexes and Reference Data

04850 | Stone Assemblies

		CREW	DAILY OUTPUT	LABOR-HOURS	UNIT	2006 BARE COSTS MAT.	LABOR	EQUIP.	TOTAL	TOTAL INCL O&P	
300	2010	Mouldings, radius cuts, bullnose edges, etc.										**300**
	2050	Add, minimum					30%					
	2100	Add, maximum					300%					
	2450	For radius under 1.5 m, add				m	100%					
	2500	Steps, copings, etc., finished on more than one surface										
	2550	Minimum	D-10	1.42	28.269	m³	2,975	955	380	4,310	5,150	
	2600	Maximum	"	1.42	28.269	"	4,750	955	380	6,085	7,100	
	2700	Pavers, Belgian block, 200 mm long, 100 mm wide, 100 mm deep	D-11	11.15	2.153	m²	229	74		303	365	
	2800	Pavers, 100 mm x 100 mm x 100 mm blocks, split face and joints										
	2850	Minimum	D-11	7.43	3.229	m²	133	111		244	315	
	2900	Maximum	"	7.43	3.229	"	266	111		377	460	
	3000	Pavers, 100 mm x 100 mm x 100 mm, thermal face, sawn joints										
	3050	Minimum	D-11	6.04	3.975	m²	242	136		378	475	
	3100	Maximum	"	6.04	3.975	"	320	136		456	555	
	3500	Curbing, city street type, See Division 02770-300										
	3700	Slope, 115 mm x 300 mm, split face,										
	3710	sawn top, 600 mm to 2 m lengths	D-10	91.44	.437	m	31.50	14.75	5.85	52.10	63.50	
	3800	Radius curbs, over 1.5 m radius, add					50%					
	3850	Under 1.5 m radius, add					100%					
	4000	Soffits, 50 mm thick, minimum	D-13	3.25	14.762	m²	375	505	165	1,045	1,350	
	4100	Maximum		3.25	14.762		785	505	165	1,455	1,800	
	4200	100 mm thick, minimum		3.25	14.762		545	505	165	1,215	1,550	
	4300	Maximum		3.25	14.762		1,025	505	165	1,695	2,075	
350	0011	**LIGHTWEIGHT NATURAL STONE** Lava type										**350**
	0100	Veneer, rubble face, sawed back, irregular shapes	D-10	12.08	3.312	m²	65	112	44.50	221.50	291	
	0200	Sawed face and back, irregular shapes	"	12.08	3.312	"	65	112	44.50	221.50	291	
400	0010	**LIMESTONE,** Cut to size										**400**
	0020	Veneer facing panels										
	0500	Texture finish, light stick, 114 mm thick, 1.5 m x 3.6 m	D-4	27.87	1.148	m²	500	36.50	5.20	541.70	610	
	0750	127 mm thick, 1.5 m x 4.3 m panels	D-10	25.55	1.566		540	53	21	614	700	
	1000	Medium ribbed, textured finish, 114 mm thick, 1.5 m x 3.7 m		25.55	1.566		269	53	21	343	400	
	1050	127 mm thick, 1.5 m x 4.3 m panels		25.55	1.566		405	53	21	479	550	
	1200	Deep ribbed, textured finish, 114 mm thick, 1.5 m x 3.1 m		25.55	1.566		425	53	21	499	575	
	1400	Sugar cube, textured finish, 114 mm thick, 1.5 m x 3.7 m		25.55	1.566		470	53	21	544	620	
	1450	127 mm thick, 1.5 m x 4.3 m panels		25.55	1.566		550	53	21	624	710	
	2000	Coping, smooth finish, top & 2 sides		.85	47.114	m³	4,425	1,600	630	6,655	7,975	
	2100	Sills, lintels, jambs, smooth finish, average		.57	70.671		4,425	2,375	950	7,750	9,525	
	2150	Detailed		.57	70.671		4,425	2,375	950	7,750	9,525	
	2300	Steps, extra hard, 356 mm wide, 152 mm rise		15.24	2.625	m	74	88.50	35	197.50	255	
	3000	Quoins, plain finish, 152 mm x 305 mm x 305 mm	D-12	25	1.280	Ea.	125	42		167	202	
	3050	152 mm x 406 mm x 610 mm	"	25	1.280	"	167	42		209	247	
500	0011	**MARBLE,** ashlar, split face, 100 mm + or - thick, random										**500**
	0040	lengths 300 mm to 1.2 m & heights 50 mm to 200 mm, average	D-8	16.26	2.460	m²	157	81.50		238.50	296	
	0100	Base, polished, 19 mm or 22 mm thick, polished, 150 mm high	D-10	19.81	2.019	m	43.50	68	27	138.50	181	
	0300	Carvings or bas relief, from templates, average		7.43	5.382	m²	1,275	182	72	1,529	1,750	
	0350	Maximum		7.43	5.382	"	2,975	182	72	3,229	3,625	
	0600	Columns, cornices, mouldings, etc.										
	0650	Hand or special machine cut, average	D-10	.99	40.384	m³	4,175	1,375	540	6,090	7,275	
	0700	Maximum	"	.99	40.384	"	8,950	1,375	540	10,865	12,500	
	1000	Facing, polished finish, cut to size, 20 mm to 25 mm thick										
	1050	Average	D-10	12.08	3.312	m²	209	112	44.50	365.50	450	
	1100	Maximum		12.08	3.312		485	112	44.50	641.50	755	
	1300	32 mm thick, average		11.61	3.445		305	116	46	467	565	

MASONRY 4

For expanded coverage of these items see *Means Concrete & Masonry Cost Data 2006*

4

MASONRY

04850 | Stone Assemblies

			CREW	DAILY OUTPUT	LABOR-HOURS	UNIT	2006 BARE COSTS				TOTAL INCL O&P	
							MAT.	LABOR	EQUIP.	TOTAL		
500	1350	Maximum	D-10	11.61	3.445	m²	605	116	46	767	895	**500**
	1500	50 mm thick, average		11.15	3.588		350	121	48	519	620	
	1550	Maximum	↓	11.15	3.588	↓	610	121	48	779	905	
	2200	Window sills, 150 mm x 20 mm thick	D-1	25.91	.618	m	24	19.85		43.85	56.50	
	2500	Flooring, polished tiles, 305 mm x 3050 mm x 10 mm thick										
	2510	Thin set, average	D-11	8.36	2.870	m²	105	98.50		203.50	266	
	2600	Maximum		8.36	2.870		1,625	98.50		1,723.50	1,925	
	2700	Mortar bed, average		6.04	3.975		107	136		243	325	
	2740	Maximum	↓	6.04	3.975		1,625	136		1,761	1,975	
	2780	Travertine, 10 mm thick, average	D-10	12.08	3.312		134	112	44.50	290.50	365	
	2790	Maximum	"	12.08	3.312		325	112	44.50	481.50	580	
	2800	Patio tile, non-slip, 13 mm thick, flame finish	D-11	6.97	3.445	↓	136	118		254	330	
	2900	Shower or toilet partitions, 22 mm thick partitions										
	3050	22 mm or 32 mm thick stiles, polished 2 sides, average	D-11	6.97	3.445	m²	405	118		523	625	
	3201	Soffits, add to above prices				"	20%	100%				
	3210	Stairs, risers, 22 mm thick x 152 mm high	D-10	35.05	1.141	m	42	38.50	15.30	95.80	121	
	3360	Treads, 305 mm wide x 32 mm thick	"	35.05	1.141	"	61	38.50	15.30	114.80	142	
	3500	Thresholds, 914 mm L, 22 mm t, 100 mm to 150 mm W, plain	D-12	24	1.333	Ea.	14.05	43.50		57.55	82	
	3550	Beveled		24	1.333	"	16.30	43.50		59.80	84.50	
	3700	Window stools, polished, 22 mm thick, 127 mm wide	↓	25.91	1.235	m	43	40.50		83.50	109	
600	0011	**ROUGH STONE WALL,** Dry										**600**
	0100	Random fieldstone, under 450 mm thick	D-12	1.70	18.846	m³	315	615		930	1,275	
	0150	Over 450 mm thick	"	1.78	17.948	"	380	585		965	1,300	
700	0011	**SANDSTONE OR BROWNSTONE**										**700**
	0100	Sawed face veneer, 65 mm thick, to 610 mm x 1220 mm panels	D-10	12.08	3.312	m²	175	112	44.50	331.50	410	
	0150	1219 mm thick, to 10678 mm x 2438 mm panels		9.29	4.306		175	145	57.50	377.50	480	
	0300	Split face, random sizes	↓	9.29	4.306	↓	126	145	57.50	328.50	425	
	0350	Cut stone trim (limestone)										
	0360	Ribbon stone, 100 mm thick, 1525 mm pieces	D-8	120	.333	Ea.	144	11		155	175	
	0370	Cove stone, 100 mm thick, 1525 mm pieces		105	.381		144	12.60		156.60	177	
	0380	Cornice stone, 255 mm to 300 mm wide		90	.444		178	14.70		192.70	219	
	0390	Band stone, 100 mm thick, 1525 mm pieces		145	.276		92	9.10		101.10	115	
	0410	Window and door trim, 75 mm to 100 mm wide		160	.250		78	8.25		86.25	98.50	
	0420	Key stone, 455 mm long	↓	60	.667	↓	82	22		104	124	
800	0010	**SLATE** Pennsylvania, blue gray to gray black; Vermont,										**800**
	0050	Unfading green, mottled green & purple, gray & purple										
	0100	Virginia, blue black										
	0200	Exterior paving, natural cleft, 25 mm thick										
	0250	150 mm x 150 mm Pennsylvania	D-12	9.29	3.445	m²	72.50	112		184.50	251	
	0300	Vermont		9.29	3.445		95	112		207	275	
	0350	Virginia		9.29	3.445		126	112		238	310	
	0500	600 mm x 600 mm, Pennsylvania		11.15	2.870		117	93.50		210.50	272	
	0550	Vermont		11.15	2.870		147	93.50		240.50	305	
	0600	Virginia		11.15	2.870		168	93.50		261.50	325	
	0700	450 mm x 760 mm Pennsylvania		11.15	2.870		126	93.50		219.50	282	
	0750	Vermont		11.15	2.870		147	93.50		240.50	305	
	0800	Virginia	↓	11.15	2.870	↓	150	93.50		243.50	310	
	1000	Interior flooring, natural cleft, 13 mm thick										
	1300	610 mm x 610 mm Pennsylvania	D-12	11.15	2.870	m²	75.50	93.50		169	226	
	1350	Vermont		11.15	2.870		129	93.50		222.50	285	
	1400	Virginia	↓	11.15	2.870	↓	121	93.50		214.50	276	
	2000	Facing panels, 32 mm thick, to 1.2 m x 1.2 m panels										
	2100	Natural cleft finish, Pennsylvania	D-10	16.72	2.392	m²	280	80.50	32	392.50	470	
	2110	Vermont	↓	16.72	2.392	↓	249	80.50	32	361.50	435	

04850 | Stone Assemblies

		CREW	DAILY OUTPUT	LABOR-HOURS	UNIT	2006 BARE COSTS				TOTAL INCL O&P
						MAT.	LABOR	EQUIP.	TOTAL	
800 2120	Virginia	D-10	16.72	2.392	m²	273	80.50	32	385.50	460	**800**
2150	Sand rubbed finish, surface, add					54			54	59
2200	Honed finish, add					27			27	29.50
2500	Ribbon, natural cleft finish, 25 mm thick, to 0.836 m²	D-10	7.43	5.382		108	182	72	362	475
2700	40 mm thick	↓	7.25	5.520	↓	140	186	74	400	520
2850	50 mm thick		7.06	5.665		168	191	76	435	560
3000	Roofing, see division 07310-800									
3500	Stair treads, sand finish, 25 mm thick x 300 mm wide									
3550	Under 900 mm	D-10	25.91	1.544	m	58	52	20.50	130.50	166
3600	1 m to 1.8 m	"	36.58	1.094	"	59	37	14.65	110.65	137
3700	Ribbon, sand finish, 25 mm thick x 305 mm wide									
3750	To 1.8 m	D-10	36.58	1.094	m	48.50	37	14.65	100.15	126
4000	Stools or sills, sand finish, 25 mm thick, 150 mm wide	D-12	48.77	.656		30.50	21.50		52	66
4200	255 mm wide		27.43	1.167		46	38		84	109
4400	50 mm thick, 150 mm wide		42.67	.750		50	24.50		74.50	92.50
4600	255 mm wide	↓	27.43	1.167		76.50	38		114.50	142
4800	For lengths over 915 mm, add				↓	25%				
900 0010	**WINDOW SILL**										**900**
0020	Bluestone, thermal top, 250 mm wide, 40 mm thick	D-1	7.90	2.026	m²	160	65		225	275
0050	50 mm thick		6.97	2.296	"	187	74		261	315
0100	Cut stone, 125 mm x 200 mm plain		14.63	1.094	m	33.50	35		68.50	90.50
0200	Face brick on edge, brick, 200 mm wide		24.38	.656		7.05	21		28.05	40
0400	Marble, 230 mm wide, 25 mm thick		25.91	.618		24.50	19.85		44.35	57
0900	Slate, colored, unfading, honed, 300 mm wide, 25 mm thick		25.91	.618		50	19.85		69.85	85
0950	50 mm thick	↓	21.34	.750	↓	69.50	24		93.50	113

04880 | Masonry Fireplaces

		CREW	DAILY OUTPUT	LABOR-HOURS	UNIT	MAT.	LABOR	EQUIP.	TOTAL	TOTAL INCL O&P
600 0010	**FIREPLACE** For prefabricated fireplace, see Div. 10305-100										**600**
0100	Brick fireplace, not incl. foundations or chimneys									
0110	762 mm x 737 mm opening, incl. chamber, plain brickwork	D-1	.40	40	Ea.	415	1,275		1,690	2,400
0200	Fireplace box only (110 brick)	"	2	8	"	136	257		393	540
0300	For elaborate brickwork and details, add					35%	35%			
0400	For hearth, brick & stone, add	D-1	2	8	Ea.	153	257		410	560
0410	For steel angle, damper, cleanouts, add		4	4		107	129		236	315
0600	Plain brickwork, incl. metal circulator		.50	32	↓	790	1,025		1,815	2,450
0800	Face brick only, standard size, 203 mm x 68 mm x 102 mm	↓	.30	53.333	k	415	1,725		2,140	3,050

04900 | Masonry Restoration and Cleaning

04910 | Unit Masonry Restoration

		CREW	DAILY OUTPUT	LABOR-HOURS	UNIT	2006 BARE COSTS				TOTAL INCL O&P
						MAT.	LABOR	EQUIP.	TOTAL	
200 0010	**CAULKING MASONRY** 13 mm x 13 mm joint										**200**
0050	Re-caulk only, oil base	1 Bric	68.58	.117	m	.79	4.26		5.05	7.35
0100	Acrylic latex		62.48	.128		.85	4.68		5.53	8.05
0200	Polyurethane		60.96	.131		1.12	4.80		5.92	8.55
0300	Silicone		59.44	.135		1.64	4.92		6.56	9.30
1000	Cut out and re-caulk, oil base		44.20	.181		.79	6.60		7.39	10.90
1050	Acrylic latex		39.62	.202		.85	7.40		8.25	12.20
1100	Polyurethane	↓	38.10	.210	↓	1.12	7.65		8.77	12.95

MASONRY 4

For expanded coverage of these items see *Means Concrete & Masonry Cost Data 2006*

4

MASONRY

04910 | Unit Masonry Restoration

			CREW	DAILY OUTPUT	LABOR-HOURS	UNIT	2006 BARE COSTS				TOTAL INCL O&P	
							MAT.	LABOR	EQUIP.	TOTAL		
200	1150	Silicone	1 Bric	36.58	.219	m	1.64	8		9.64	13.95	200
600	0010	**NEEDLE BEAM MASONRY** Incl. wood shoring 3 m x 3 m opening										600
	0400	Block, concrete, 200 mm thick	B-9	7.10	5.634	Ea.	36.50	157	20.50	214	310	
	0420	300 mm thick		6.70	5.970		42.50	166	22	230.50	330	
	0800	Brick, 100 mm thick with 200 mm backup block		5.70	7.018		42.50	195	26	263.50	380	
	1000	Brick, solid, 200 mm thick		6.20	6.452		36.50	179	23.50	239	345	
	1040	300 mm thick		4.90	8.163		42.50	227	30	299.50	435	
	1080	400 mm thick	↓	4.50	8.889		53.50	247	32.50	333	480	
	2000	Add for additional floors of shoring	B-1	6	4	↓	36.50	112		148.50	216	
720	0010	**POINTING MASONRY**										720
	0300	Cut and repoint brick, hard mortar, running bond	1 Bric	7.43	1.076	m²	4.63	39.50		44.13	65	
	0320	Common bond		7.15	1.118		4.63	41		45.63	67	
	0360	Flemish bond		6.50	1.230		4.84	45		49.84	74	
	0400	English bond		6.04	1.325		4.84	48.50		53.34	79	
	0600	Soft old mortar, running bond		9.29	.861		4.74	31.50		36.24	53.50	
	0620	Common bond		8.92	.897		4.63	33		37.63	55	
	0640	Flemish bond		8.36	.957		4.84	35		39.84	59	
	0680	English bond	↓	7.62	1.050		4.84	38.50		43.34	64	
	0700	Stonework, hard mortar		42.67	.187	m	1.87	6.85		8.72	12.50	
	0720	Soft old mortar		48.77	.164	"	1.87	6		7.87	11.20	
	1000	Repoint, mask and grout method, running bond		8.83	.906	m²	6.15	33		39.15	57.50	
	1020	Common bond		8.36	.957		6.15	35		41.15	60.50	
	1040	Flemish bond		7.99	1.001		6.15	36.50		42.65	62.50	
	1060	English bond		7.15	1.118		6.15	41		47.15	69	
	2000	Scrub coat, sand grout on walls, minimum		11.15	.718		30	26		56	72.50	
	2020	Maximum	↓	9.10	.879	↓	21.50	32		53.50	72.50	
750	0010	**SAWING**										750
	0050	Brick or block by hand, per mm depth	A-1	38.10	.210	m		5.75	1.29	7.04	10.35	
800	0010	**TOOTHING MASONRY**										800
	0500	Brickwork, soft old mortar	1 Clab	12.19	.656	m		18		18	28	
	0520	Hard mortar		9.14	.875			24		24	37.50	
	0700	Blockwork, soft old mortar		21.34	.375			10.25		10.25	16	
	0720	Hard mortar	↓	15.24	.525	↓		14.40		14.40	22.50	

04930 | Unit Masonry Cleaning

			CREW	DAILY OUTPUT	LABOR-HOURS	UNIT	MAT.	LABOR	EQUIP.	TOTAL	TOTAL INCL O&P	
220	0010	**MASONRY CLEANING**										220
	0200	Chemical cleaning, new construction, brush and wash, minimum	D-1	92.90	.172	m²	.43	5.55		5.98	8.90	
	0220	Average		74.32	.215		.65	6.90		7.55	11.20	
	0240	Maximum		55.74	.287		.86	9.25		10.11	14.90	
	0260	Light restoration, minimum		74.32	.215		.75	6.90		7.65	11.40	
	0270	Average		37.16	.431		1.18	13.85		15.03	22.50	
	0280	Maximum		30.66	.522		1.61	16.80		18.41	27	
	0300	Heavy restoration, minimum		55.74	.287		.86	9.25		10.11	15	
	0310	Average		37.16	.431		1.29	13.85		15.14	22.50	
	0320	Maximum	↓	23.23	.689		1.72	22		23.72	35.50	
	0400	High pressure water only, minimum	B-9	186	.215			6	.79	6.79	10.15	
	0420	Average		139	.288			8	1.06	9.06	13.60	
	0440	Maximum		92.90	.431			11.95	1.58	13.53	20.50	
	0800	High pressure water and chemical, minimum		167	.240		.97	6.65	.88	8.50	12.40	
	0820	Average		111	.360		1.40	10	1.33	12.73	18.55	
	0840	Maximum		74.32	.538		1.94	14.95	1.98	18.87	27.50	
	1200	Sandblast, wet system, minimum		163	.245		1.72	6.80	.90	9.42	13.55	
	1220	Average	↓	102	.392	↓	2.58	10.90	1.44	14.92	21.50	

		04930	Unit Masonry Cleaning	CREW	DAILY OUTPUT	LABOR-HOURS	UNIT	MAT.	LABOR	EQUIP.	TOTAL	TOTAL INCL O&P	
									2006 BARE COSTS				
220	1240		Maximum	B-9	65.03	.615	m²	3.44	17.10	2.26	22.80	33	220
	1400		Dry system, minimum		232	.172		1.72	4.79	.63	7.14	10.10	
	1420		Average		163	.245		2.58	6.80	.90	10.28	14.50	
	1440		Maximum		92.90	.431		3.44	11.95	1.58	16.97	24.50	
	1800		For walnut shells, add					4.41			4.41	4.84	
	1820		For corn chips, add					4.41			4.41	4.84	
	2000		Steam cleaning, minimum	B-9	279	.143			3.99	.53	4.52	6.80	
	2020		Average		232	.172			4.79	.63	5.42	8.15	
	2040		Maximum		139	.288			8	1.06	9.06	13.60	
	4000		Add for masking doors and windows									8.60	
	4200		Add for pedestrian protection				Job					10%	
	4300		Add for common face brick				m²					2.26	
	4400		Add for wire cut face brick				"	1.61			1.61	1.72	
750	0010	**STEAM CLEANING MASONRY**											750
	0020		Steam, minimum	B-9	279	.143	m²		3.99	.53	4.52	6.80	
	0100		Maximum		139	.288			8	1.06	9.06	13.60	
	0300		Common face brick		163	.245			6.80	.90	7.70	11.60	
	0400		Wire cut face brick		116	.345			9.60	1.27	10.87	16.30	
900	0010	**WASHING MASONRY**	R040130 -10										900
	0012		Acid, smooth brick	1 Bric	52.02	.154	m²	.32	5.60		5.92	8.85	
	0050		Rough brick		37.16	.215		.32	7.85		8.17	12.45	
	0060		Stone, acid wash		55.74	.144		.43	5.25		5.68	8.55	
	1000		Muriatic acid, price per liter in 5 liter lots				liter	1.38			1.38	1.52	

MASONRY 4

For information about Means Estimating Seminars, see yellow pages 12 and 13 in back of book

		CREW	DAILY OUTPUT	LABOR-HOURS	UNIT	2006 BARE COSTS				TOTAL INCL O&P
						MAT.	LABOR	EQUIP.	TOTAL	

Division 5
Metals

Estimating Tips

05050 Basic Metal
Materials & Methods

- Nuts, bolts, washers, connection angles and plates can add a significant amount to both the tonnage of a structural steel job as well as the estimated cost. As a rule of thumb, add 10% to the total weight to account for these accessories.

- Type 2 steel construction, commonly referred to as "simple construction," consists generally of field bolted connections with lateral bracing supplied by other elements of the building, such as masonry walls or x-bracing. The estimator should be aware, however, that shop connections may be accomplished by welding or bolting. The method may be particular to the fabrication shop and may have an impact on the estimated cost.

05200 Metal Joists

- In any given project the total weight of open web steel joists is determined by the loads to be supported and the design. However, economies can be realized in minimizing the amount of labor used to place the joists. This is done by maximizing the joist spacing and therefore minimizing the number of joists required to be installed on the job. Certain spacings and locations may be required by the design, but in other cases maximizing the spacing and keeping it as uniform as possible will keep the costs down.

05300 Metal Deck

- The takeoff and estimating of metal deck involves more than simply the area of the floor or roof and the type of deck specified or shown on the drawings. Many different sizes and types of openings may exist. Small openings for individual pipes or conduits may be drilled after the floor/roof is installed, but larger openings may require special deck lengths as well as reinforcing or structural support. The estimator should determine who will be supplying this reinforcing. Additionally, some deck terminations are part of the deck package, such as screed angles and pour stops, and others will be part of the steel contract, such as angles attached to structural members and cast-in-place angles and plates. The estimator must ensure that all pieces are accounted for in the complete estimate.

05500 Metal Fabrications

- The most economical steel stairs are those that use common materials, standard details and most importantly, a uniform and relatively simple method of field assembly. Commonly available A36M channels and plates are very good choices for the main stringers of the stairs, as are angles and tees for the carrier members. Risers and treads are usually made by specialty shops, and it is most economical to use a typical detail in as many places as possible. The stairs should be pre-assembled and shipped directly to the site. The field connections should be simple and straightforward to be accomplished efficiently and with a minimum of equipment and labor.

Reference Numbers

Reference numbers are shown in bold squares at the beginning of some major classifications. These numbers refer to related items in the Reference Section. The reference information may be an estimating procedure, an alternate pricing method, or technical information.

Note: Not all subdivisions listed here necessarily appear in this publication.

			CREW	DAILY OUTPUT	LABOR-HOURS	UNIT	2006 BARE COSTS				TOTAL INCL O&P	
	05060	**Selective Demolition**					MAT.	LABOR	EQUIP.	TOTAL		
110	0010	**SELECTIVE METALS DEMOLITION**										110
	0015	Excludes shores, bracing, cutting, loading, hauling, dumping	R024119 -10									
	0020	Remove nuts only up to 19 mm diameter	1 Sswk	480	.017	Ea.		.67		.67	1.20	
	0030	22 to 31 mm diameter		240	.033			1.33		1.33	2.40	
	0040	34 to 50 mm diameter		160	.050			2		2	3.60	
	0060	Unbolt and remove structural bolts up to 19 mm diameter		240	.033			1.33		1.33	2.40	
	0070	22 to 31 mm diameter		160	.050			2		2	3.60	
	0140	Light weight framing members, remove whole or cut up, up to 9 kg	▼	240	.033			1.33		1.33	2.40	
	0150	10 - 18 kg	2 Sswk	210	.076			3.04		3.04	5.50	
	0160	19 - 36 kg	3 Sswk	180	.133			5.35		5.35	9.60	
	0170	37 - 54 kg	4 Sswk	150	.213			8.50		8.50	15.35	
	0230	Structural members, remove whole or cut up, up to 227 kg	E-19	48	.500			19.50	19.85	39.35	55.50	
	0240	1/4 - 2 tonnes	E-18	36	1.111			44	26.50	70.50	106	
	0250	2 - 5 tonnes	E-24	30	1.067			42	21.50	63.50	96.50	
	0260	5 - 10 tonnes	E-20	24	2.667			104	47.50	151.50	232	
	0270	10 - 15 tonnes	E-2	18	3.111			121	79.50	200.50	296	
	0340	Fabricated item, remove whole or cut up, up to 9 kg	1 Sswk	96	.083			3.33		3.33	6	
	0350	10 - 18 kg	2 Sswk	84	.190			7.60		7.60	13.70	
	0360	19 - 36 kg	3 Sswk	72	.333			13.30		13.30	24	
	0370	37 - 54 kg	4 Sswk	60	.533			21.50		21.50	38.50	
	0380	55 - 227 kg	▼ E-19	48	.500	▼		19.50	19.85	39.35	55.50	

	05090	**Metal Fastenings**										
080	0010	**ANCHOR BOLTS**										080
	0020	See also Division 04080-070										
	0100	J-type, incl. hex nut & washer, 13 mm diameter x 150 mm long	2 Carp	70	.229	Ea.	1.03	8.15		9.18	13.80	
	0110	300 mm long		65	.246		1.29	8.75		10.04	15	
	0120	450 mm long		60	.267		1.68	9.50		11.18	16.60	
	0130	19 mm diameter x 200 mm long		50	.320		1.54	11.40		12.94	19.40	
	0140	300 mm long		45	.356		1.92	12.65		14.57	22	
	0150	450 mm long		40	.400		2.50	14.20		16.70	25	
	0160	25 mm diameter x 300 mm long		35	.457		3.65	16.25		19.90	29.50	
	0170	450 mm long		30	.533		4.40	18.95		23.35	34.50	
	0180	600 mm long		25	.640		5.35	23		28.35	41.50	
	0190	900 mm long		20	.800		7.35	28.50		35.85	52.50	
	0200	38 mm diameter x 450 mm long		22	.727		13	26		39	55	
	0210	600 mm long		16	1		15.50	35.50		51	72.50	
	0300	L-type, incl. hex nut & washer, 19 mm diameter x 300 mm long		45	.356		1.44	12.65		14.09	21.50	
	0310	450 mm long		40	.400		1.85	14.20		16.05	24	
	0320	600 mm long		35	.457		2.25	16.25		18.50	28	
	0330	750 mm long		30	.533		2.86	18.95		21.81	32.50	
	0340	900 mm long		25	.640		3.26	23		26.26	39	
	0350	25 mm diameter x 300 mm long		35	.457		2.50	16.25		18.75	28.50	
	0360	450 mm long		30	.533		3.12	18.95		22.07	33	
	0370	600 mm long		25	.640		3.85	23		26.85	39.50	
	0380	750 mm long		23	.696		4.54	24.50		29.04	43.50	
	0390	900 mm long		20	.800		5.20	28.50		33.70	50	
	0400	1050 mm long		18	.889		6.30	31.50		37.80	56	
	0410	1200 mm long		15	1.067		7.10	38		45.10	67	
	0420	31 mm diameter x 450 mm long		25	.640		5.60	23		28.60	41.50	
	0430	600 mm long		22	.727		6.65	26		32.65	48	
	0440	750 mm long		20	.800		7.75	28.50		36.25	53	
	0450	900 mm long		18	.889		8.80	31.50		40.30	58.50	
	0460	1050 mm long		16	1		9.90	35.50		45.40	66.50	
	0470	1200 mm long	▼	14	1.143	▼	11.35	40.50		51.85	76	

05090	Metal Fastenings	CREW	DAILY OUTPUT	LABOR-HOURS	UNIT	2006 BARE COSTS				TOTAL INCL O&P	
						MAT.	LABOR	EQUIP.	TOTAL		
080 0480	1350 mm long	2 Carp	12	1.333	Ea.	13.35	47.50		60.85	88.50	**080**
0490	1500 mm long		10	1.600		14.65	57		71.65	105	
0500	38 mm diameter x 450 mm long		22	.727		8.65	26		34.65	50	
0510	600 mm long		19	.842		10.10	30		40.10	57.50	
0520	750 mm long		17	.941		11.40	33.50		44.90	64.50	
0530	900 mm long		16	1		13.10	35.50		48.60	70	
0540	1050 mm long		15	1.067		14.90	38		52.90	75.50	
0550	1200 mm long		13	1.231		16.75	44		60.75	86.50	
0560	1350 mm long		11	1.455		20.50	51.50		72	103	
0570	1500 mm long		9	1.778		22.50	63		85.50	123	
0580	44 mm diameter x 450 mm long		20	.800		13.05	28.50		41.55	59	
0590	600 mm long		18	.889		15.30	31.50		46.80	66	
0600	750 mm long		17	.941		17.80	33.50		51.30	71.50	
0610	900 mm long		16	1		20.50	35.50		56	78	
0620	1050 mm long		14	1.143		23	40.50		63.50	88.50	
0630	1200 mm long		12	1.333		25	47.50		72.50	102	
0640	1350 mm long		10	1.600		31	57		88	123	
0650	1500 mm long		8	2		33.50	71		104.50	148	
0660	50 mm diameter x 600 mm long		17	.941		19.50	33.50		53	73.50	
0670	750 mm long		15	1.067		22	38		60	83	
0680	900 mm long		13	1.231		24	44		68	94.50	
0690	1050 mm long		11	1.455		27	51.50		78.50	110	
0700	1200 mm long		10	1.600		31	57		88	123	
0710	1350 mm long		9	1.778		37	63		100	139	
0720	1500 mm long		8	2		39.50	71		110.50	155	
0730	1650 mm long		7	2.286		42.50	81.50		124	174	
0740	1800 mm long	▼	6	2.667	▼	46.50	95		141.50	199	
0990	For galvanized, add					75%					
150 0010	**BOLTS & HEX NUTS** Steel, A307										**150**
0100	6 mm diameter, 13 mm long	1 Sswk	140	.057	Ea.	.07	2.28		2.35	4.19	
0200	25 mm long		140	.057		.08	2.28		2.36	4.20	
0300	50 mm long		130	.062		.11	2.46		2.57	4.54	
0400	75 mm long		130	.062		.16	2.46		2.62	4.60	
0500	100 mm long		120	.067		.18	2.66		2.84	4.98	
0600	9 mm diameter, 25 mm long		130	.062		.12	2.46		2.58	4.55	
0700	50 mm long		130	.062		.15	2.46		2.61	4.59	
0800	75 mm long		120	.067		.20	2.66		2.86	5	
0900	100 mm long		120	.067		.25	2.66		2.91	5.05	
1000	125 mm long		115	.070		.31	2.78		3.09	5.35	
1100	13 mm diameter, 38 mm long		120	.067		.22	2.66		2.88	5.05	
1200	50 mm long		120	.067		.25	2.66		2.91	5.05	
1300	100 mm long		115	.070		.38	2.78		3.16	5.40	
1400	150 mm long		110	.073		.52	2.91		3.43	5.80	
1500	200 mm long		105	.076		.68	3.04		3.72	6.25	
1600	16 mm diameter, 38 mm long		120	.067		.45	2.66		3.11	5.30	
1700	50 mm long		120	.067		.49	2.66		3.15	5.35	
1800	100 mm long		115	.070		.68	2.78		3.46	5.75	
1900	150 mm long		110	.073		.85	2.91		3.76	6.20	
2000	200 mm long		105	.076		1.23	3.04		4.27	6.85	
2100	250 mm long		100	.080		1.53	3.20		4.73	7.45	
2200	19 mm diameter, 50 mm long		120	.067		.70	2.66		3.36	5.55	
2300	100 mm long		110	.073		.97	2.91		3.88	6.30	
2400	150 mm long		105	.076		1.23	3.04		4.27	6.85	
2500	200 mm long		95	.084		1.83	3.36		5.19	8.05	
2600	250 mm long		85	.094		2.38	3.76		6.14	9.35	
2700	300 mm long	▼	80	.100	▼	2.78	4		6.78	10.25	

05090 | Metal Fastenings

		CREW	DAILY OUTPUT	LABOR-HOURS	UNIT	2006 BARE COSTS				TOTAL INCL O&P	
						MAT.	LABOR	EQUIP.	TOTAL		
150	2800	25 mm diameter, 75 mm long	1 Sswk	105	.076	Ea.	1.82	3.04		4.86	7.50	**150**
	2900	150 mm long		90	.089		2.80	3.55		6.35	9.50	
	3000	300 mm long		75	.107		5.25	4.26		9.51	13.45	
	3100	For galvanized, add					75%					
	3200	For stainless, add					350%					
300	0010	**CHEMICAL ANCHORS**, Includes layout & drilling										**300**
	1430	Chemical anchor, w/rod & epoxy cartridge, 19 mm dia x 238 mm long	B-89A	27	.593	Ea.	12.80	18.95	4.28	36.03	48.50	
	1435	25 mm diameter x 294 mm long		24	.667		25	21.50	4.81	51.31	66	
	1440	32 mm diameter x 350 mm long		21	.762		47	24.50	5.50	77	96	
	1445	44 mm dia x 375 mm long		20	.800		89.50	25.50	5.80	120.80	145	
	1450	450 mm long		17	.941		107	30	6.80	143.80	172	
	1455	50 mm diameter x 450 mm long		16	1		137	32	7.20	176.20	207	
	1460	600 mm long		15	1.067		179	34	7.70	220.70	258	
	1500	Chemical anchoring, epoxy cartridge, excludes layout, drilling, fastener										
	1530	For fastener 19 mm dia x 150 mm embedment	B-89A	27	.593	Ea.	5.05	18.95	4.28	28.28	40	
	1535	25 mm dia x 200 mm embedment		24	.667		7.55	21.50	4.81	33.86	46.50	
	1540	31 mm dia x 250 mm embedment		21	.762		15.10	24.50	5.50	45.10	60.50	
	1545	44 mm dia x 300 mm embedment		20	.800		25	25.50	5.80	56.30	74	
	1550	350 mm embedment		17	.941		30	30	6.80	66.80	88	
	1555	50 mm dia x 300 mm embedment		16	1		40.50	32	7.20	79.70	102	
	1560	450 mm embedment		15	1.067		50.50	34	7.70	92.20	117	
340	0010	**DRILLING**										**340**
	0050	Up to 100 mm D in conc/brick flr/wall, incl. bit & layout, no anchor										
	0100	Holes, 6 mm diameter	1 Carp	75	.107	Ea.	.09	3.79		3.88	6	
	0150	For each additional 25 mm of depth, add		430	.019		.02	.66		.68	1.06	
	0200	9 mm diameter		63	.127		.09	4.51		4.60	7.15	
	0250	For each additional 25 mm of depth, add		340	.024		.02	.84		.86	1.32	
	0300	13 mm diameter		50	.160		.09	5.70		5.79	8.95	
	0350	For each additional 25 mm of depth, add		250	.032		.02	1.14		1.16	1.79	
	0400	16 mm diameter		48	.167		.16	5.95		6.11	9.45	
	0450	For each additional 25 mm of depth, add		240	.033		.04	1.18		1.22	1.88	
	0500	19 mm diameter		45	.178		.19	6.30		6.49	10.05	
	0550	For each additional 25 mm of depth, add		220	.036		.05	1.29		1.34	2.06	
	0600	22 mm diameter		43	.186		.23	6.60		6.83	10.55	
	0650	For each additional 25 mm of depth, add		210	.038		.06	1.35		1.41	2.17	
	0700	25 mm diameter		40	.200		.27	7.10		7.37	11.35	
	0750	For each additional 25 mm of depth, add		190	.042		.07	1.50		1.57	2.40	
	0800	31 mm diameter		38	.211		.38	7.50		7.88	12.05	
	0850	For each additional 25 mm of depth, add		180	.044		.10	1.58		1.68	2.56	
	0900	38 mm diameter		35	.229		.58	8.15		8.73	13.30	
	0950	For each additional 25 mm of depth, add		165	.048		.14	1.72		1.86	2.84	
	1000	For ceiling installations, add						40%				
	1100	Drilling & layout for drywall/plaster walls, up to 25 mm deep, no anchor										
	1200	Holes, 6 mm diameter	1 Carp	150	.053	Ea.	.01	1.90		1.91	2.96	
	1300	9 mm diameter		140	.057		.01	2.03		2.04	3.17	
	1400	13 mm diameter		130	.062		.01	2.19		2.20	3.42	
	1500	19 mm diameter		120	.067		.02	2.37		2.39	3.72	
	1600	25 mm diameter		110	.073		.03	2.59		2.62	4.07	
	1700	31 mm diameter		100	.080		.05	2.84		2.89	4.48	
	1800	39 mm diameter		90	.089		.07	3.16		3.23	5	
	1900	For ceiling installations, add						40%				
	1910	Drilling & layout for steel, up to 6 mm deep, no anchor										
	1920	Holes, 6 mm diameter	1 Sswk	112	.071	Ea.	.12	2.85		2.97	5.30	
	1925	For each additional 6 mm depth, add		336	.024		.12	.95		1.07	1.85	
	1930	9 mm diameter		104	.077		.14	3.07		3.21	5.70	

5
METALS

05090 | Metal Fastenings

		CREW	DAILY OUTPUT	LABOR-HOURS	UNIT	2006 BARE COSTS				TOTAL INCL O&P
						MAT.	LABOR	EQUIP.	TOTAL	
340	1935	For each additional 6 mm depth, add	1 Sswk	312	.026	Ea.	.14	1.02		1.16	1.99
	1940	13 mm diameter		96	.083		.16	3.33		3.49	6.15
	1945	For each additional 6 mm depth, add		288	.028		.16	1.11		1.27	2.17
	1950	16 mm diameter		88	.091		.26	3.63		3.89	6.85
	1955	For each additional 6 mm depth, add		264	.030		.26	1.21		1.47	2.46
	1960	19 mm diameter		80	.100		.29	4		4.29	7.50
	1965	For each additional 6 mm depth, add		240	.033		.29	1.33		1.62	2.72
	1970	22 mm diameter		72	.111		.34	4.44		4.78	8.40
	1975	For each additional 6 mm depth, add		216	.037		.34	1.48		1.82	3.04
	1980	25 mm diameter		64	.125		.39	4.99		5.38	9.45
	1985	For each additional 6 mm depth, add	▼	192	.042		.39	1.66		2.05	3.43
	1990	For drilling up, add				▼		40%			
380	0010	**EXPANSION ANCHORS**									
	0100	Anchors for concrete, brick or stone, no layout and drilling									
	0200	Expansion shields, zinc, 6 mm diameter, 33 mm long, single	1 Carp	90	.089	Ea.	1.06	3.16		4.22	6.10
	0300	34 mm long, double		85	.094		1.16	3.35		4.51	6.50
	0400	9 mm diameter, 38 mm long, single		85	.094		1.74	3.35		5.09	7.10
	0500	50 mm long, double		80	.100		2.15	3.56		5.71	7.90
	0600	13 mm diameter, 52 mm long, single		80	.100		2.88	3.56		6.44	8.70
	0700	63 mm long, double		75	.107		2.78	3.79		6.57	8.95
	0800	16 mm diameter, 66 mm long, single		75	.107		4.12	3.79		7.91	10.45
	0900	69 mm long, double		70	.114		4.12	4.06		8.18	10.90
	1000	19 mm diameter, 69 mm long, single		70	.114		6.10	4.06		10.16	13.10
	1100	98 mm long, double		65	.123		8.15	4.38		12.53	15.80
	1500	Self drilling anchor, snap-off, for 6 mm diameter bolt		26	.308		.91	10.95		11.86	18.05
	1600	9 mm diameter bolt		23	.348		1.32	12.35		13.67	20.50
	1700	13 mm diameter bolt		20	.400		2.03	14.20		16.23	24
	1800	16 mm diameter bolt		18	.444		3.39	15.80		19.19	28
	1900	19 mm diameter bolt	▼	16	.500	▼	5.70	17.80		23.50	34
	2100	Hollow wall anchors for gypsum wall board, plaster or tile									
	2300	3 mm diameter, short	1 Carp	160	.050	Ea.	.29	1.78		2.07	3.09
	2400	Long		150	.053		.34	1.90		2.24	3.32
	2500	5 mm diameter, short		150	.053		.61	1.90		2.51	3.62
	2600	Long		140	.057		.65	2.03		2.68	3.88
	2700	6 mm diameter, short		140	.057		.74	2.03		2.77	3.97
	2800	Long		130	.062		.84	2.19		3.03	4.33
	3000	Toggle bolts, bright steel, 6 mm diameter, 50 mm long		85	.094		.27	3.35		3.62	5.50
	3100	100 mm long		80	.100		.41	3.56		3.97	6
	3200	5 mm diameter, 75 mm long		80	.100		.46	3.56		4.02	6.05
	3300	150 mm long		75	.107		.64	3.79		4.43	6.60
	3400	6 mm diameter, 75 mm long		75	.107		.51	3.79		4.30	6.45
	3500	150 mm long		70	.114		.73	4.06		4.79	7.15
	3600	9 mm diameter, 75 mm long		70	.114		.99	4.06		5.05	7.45
	3700	150 mm long		60	.133		1.73	4.74		6.47	9.30
	3800	13 mm diameter, 100 mm long		60	.133		2.57	4.74		7.31	10.25
	3900	150 mm long	▼	50	.160	▼	4.23	5.70		9.93	13.50
	4000	Nailing anchors									
	4100	Nylon nailing anchor, 6 mm diameter, 25 mm long	1 Carp	3.20	2.500	h	19.80	89		108.80	160
	4200	38 mm long		2.80	2.857		25.50	102		127.50	186
	4300	50 mm long		2.40	3.333		42.50	119		161.50	232
	4400	Metal nailing anchor, 6 mm diameter, 25 mm long		3.20	2.500		30.50	89		119.50	172
	4500	38 mm long		2.80	2.857		41.50	102		143.50	204
	4600	50 mm long	▼	2.40	3.333	▼	52.50	119		171.50	243
	5000	Screw anchors for concrete, masonry,									
	5100	stone & tile, no layout or drilling included									
	5200	Jute fiber, #6, #8, & #10, 25 mm long	1 Carp	240	.033	Ea.	.25	1.18		1.43	2.12

METALS 5

05090 | Metal Fastenings

		CREW	DAILY OUTPUT	LABOR-HOURS	UNIT	2006 BARE COSTS MAT.	LABOR	EQUIP.	TOTAL	TOTAL INCL O&P	
380											**380**
5300	#12, 38 mm long	1 Carp	200	.040	Ea.	.37	1.42		1.79	2.62	
5400	#14, 50 mm long		160	.050		.57	1.78		2.35	3.40	
5500	#16, 50 mm long		150	.053		.60	1.90		2.50	3.61	
5600	#20, 50 mm long		140	.057		.96	2.03		2.99	4.22	
5700	Lag screw shields, 6 mm diameter, short		90	.089		.46	3.16		3.62	5.45	
5800	Long		85	.094		.54	3.35		3.89	5.80	
5900	9 mm diameter, short		85	.094		.84	3.35		4.19	6.10	
6000	Long		80	.100		.99	3.56		4.55	6.65	
6100	13 mm diameter, short		80	.100		1.17	3.56		4.73	6.85	
6200	Long		75	.107		1.46	3.79		5.25	7.50	
6300	19 mm diameter, short		70	.114		3.28	4.06		7.34	9.95	
6400	Long		65	.123		3.97	4.38		8.35	11.15	
6600	Lead, #6 & #8, 19 mm long		260	.031		.16	1.09		1.25	1.88	
6700	#10 - #14, 38 mm long		200	.040		.24	1.42		1.66	2.47	
6800	#16 & #18, 38 mm long		160	.050		.33	1.78		2.11	3.13	
6900	Plastic, #6 & #8, 19 mm long		260	.031		.11	1.09		1.20	1.82	
7000	#8 & #10, 22 mm long		240	.033		.05	1.18		1.23	1.90	
7100	#10 & #12, 25 mm long		220	.036		.14	1.29		1.43	2.16	
7200	#14 & #16, 38 mm long	▼	160	.050	▼	.08	1.78		1.86	2.86	
8000	Wedge anchors, not including layout or drilling										
8050	Carbon steel, 6 mm diameter, 44 mm long	1 Carp	150	.053	Ea.	.46	1.90		2.36	3.46	
8100	81 mm long		140	.057		.61	2.03		2.64	3.83	
8150	9 mm diameter, 56 mm long		145	.055		.69	1.96		2.65	3.81	
8200	125 mm long		140	.057		1.21	2.03		3.24	4.49	
8250	13 mm diameter, 69 mm long		140	.057		1.06	2.03		3.09	4.32	
8300	175 mm long		125	.064		1.81	2.28		4.09	5.55	
8350	16 mm diameter, 88 mm long		130	.062		2.09	2.19		4.28	5.70	
8400	213 mm long		115	.070		4.45	2.47		6.92	8.75	
8450	19 mm diameter, 106 mm long		115	.070		2.52	2.47		4.99	6.65	
8500	250 mm long		95	.084		5.75	2.99		8.74	10.95	
8550	25 mm diameter, 150 mm long		100	.080		8.45	2.84		11.29	13.75	
8575	225 mm long		85	.094		11	3.35		14.35	17.30	
8600	300 mm long		75	.107		11.85	3.79		15.64	18.95	
8650	31 mm diameter, 225 mm long		70	.114		15.40	4.06		19.46	23.50	
8700	300 mm long	▼	60	.133	▼	19.70	4.74		24.44	29	
8750	For type 303 stainless steel, add					350%					
8800	For type 316 stainless steel, add					450%					
8950	Self-drilling concrete screw, hex washer head, 5 mm dia x 44 mm long	1 Carp	300	.027	Ea.	.35	.95		1.30	1.87	
8960	56 mm long		250	.032		.54	1.14		1.68	2.36	
8970	Phillips flat head, 5 mm dia x 44 mm long		300	.027		.36	.95		1.31	1.88	
8980	56 mm long	▼	250	.032	▼	.53	1.14		1.67	2.35	
420											**420**
0010	**HIGH STRENGTH BOLTS** R050523 -10										
0020	A325 Type 1, structural steel, bolt-nut-washer set										
0100	13 mm diameter x 38 mm long	1 Sswk	130	.062	Ea.	.36	2.46		2.82	4.81	
0120	50 mm long		125	.064		.38	2.56		2.94	5	
0150	75 mm long		120	.067		.49	2.66		3.15	5.35	
0170	16 mm diameter x 38 mm long		125	.064		.55	2.56		3.11	5.20	
0180	50 mm long		120	.067		.58	2.66		3.24	5.45	
0190	75 mm long		115	.070		.68	2.78		3.46	5.75	
0200	19 mm diameter x 50 mm long		120	.067		.81	2.66		3.47	5.70	
0220	75 mm long		115	.070		.93	2.78		3.71	6	
0250	100 mm long		110	.073		1.09	2.91		4	6.45	
0300	150 mm long		105	.076		1.36	3.04		4.40	7	
0350	200 mm long		95	.084		2.48	3.36		5.84	8.80	
0360	22 mm diameter x 50 mm long	▼	115	.070	▼	1.21	2.78		3.99	6.35	

Important: See the Reference Section for supporting data - Crews, Rental Equipment, City Cost Indexes and Reference Data

05090	Metal Fastenings		CREW	DAILY OUTPUT	LABOR-HOURS	UNIT	2006 BARE COSTS				TOTAL INCL O&P
							MAT.	LABOR	EQUIP.	TOTAL	
420	0365	75 mm long	R050523-10	1 Sswk	110	.073	Ea.	1.38	2.91		4.29	6.75	**420**
	0370	100 mm long			105	.076		1.61	3.04		4.65	7.25
	0380	150 mm long			100	.080		1.97	3.20		5.17	7.90
	0390	200 mm long			90	.089		2.98	3.55		6.53	9.70
	0400	25 mm diameter x 50 mm long			105	.076		1.69	3.04		4.73	7.35
	0420	75 mm long			100	.080		1.85	3.20		5.05	7.80
	0450	100 mm long			95	.084		2.04	3.36		5.40	8.30
	0500	150 mm long			90	.089		2.57	3.55		6.12	9.20
	0550	200 mm long			85	.094		4.13	3.76		7.89	11.30
	0600	31 mm diameter x 75 mm long			85	.094		3.81	3.76		7.57	10.95
	0650	100 mm long			80	.100		1.73	4		5.73	9.10
	0700	150 mm long			75	.107		4.96	4.26		9.22	13.10
	0750	200 mm long		▼	70	.114	▼	6.05	4.57		10.62	14.85
	1020	A490, bolt-nut-washer set										
	1170	16 mm diameter x 38 mm long		1 Sswk	125	.064	Ea.	.79	2.56		3.35	5.45
	1180	50 mm long			120	.067		.90	2.66		3.56	5.80
	1190	75 mm long			115	.070		1.05	2.78		3.83	6.15
	1200	19 mm diameter x 50 mm long			120	.067		1.05	2.66		3.71	5.95
	1220	75 mm long			115	.070		1.19	2.78		3.97	6.30
	1250	100 mm long			110	.073		1.34	2.91		4.25	6.75
	1300	150 mm long			105	.076		1.85	3.04		4.89	7.55
	1350	200 mm long			95	.084		2.97	3.36		6.33	9.30
	1360	22 mm diameter x 50 mm long			115	.070		1.42	2.78		4.20	6.55
	1365	75 mm long			110	.073		1.61	2.91		4.52	7
	1370	100 mm long			105	.076		1.91	3.04		4.95	7.60
	1380	150 mm long			100	.080		2.54	3.20		5.74	8.55
	1390	200 mm long			90	.089		3.53	3.55		7.08	10.30
	1400	25 mm diameter x 50 mm long			105	.076		1.93	3.04		4.97	7.65
	1420	75 mm long			100	.080		2.21	3.20		5.41	8.20
	1450	100 mm long			95	.084		2.46	3.36		5.82	8.75
	1500	150 mm long			90	.089		3.09	3.55		6.64	9.80
	1550	200 mm long			85	.094		4.55	3.76		8.31	11.75
	1600	31 mm diameter x 75 mm long			85	.094		4.44	3.76		8.20	11.65
	1650	100 mm long			80	.100		4.92	4		8.92	12.60
	1700	150 mm long			75	.107		6.30	4.26		10.56	14.60
	1750	200 mm long	▼	▼	70	.114	▼	7.85	4.57		12.42	16.85
460	0010	**LAG SCREWS**											**460**
	0020	Steel, 6 mm diameter, 50 mm long		1 Carp	200	.040	Ea.	.09	1.42		1.51	2.31
	0100	9 mm diameter, 75 mm long			150	.053		.26	1.90		2.16	3.24
	0200	13 mm diameter, 75 mm long			130	.062		.42	2.19		2.61	3.87
	0300	16 mm diameter, 75 mm long		▼	120	.067	▼	.82	2.37		3.19	4.59
500	0010	**MACHINE SCREWS**											**500**
	0020	Steel, round head, #8 x 25 mm long		1 Carp	4.80	1.667	h	2.51	59.50		62.01	95.50
	0110	#8 x 50 mm			2.40	3.333		5.45	119		124.45	191
	0200	#10 x 25 mm long			4	2		3.57	71		74.57	115
	0300	#10 x 50 mm long		▼	2	4	▼	6.70	142		148.70	228
540	0010	**MACHINERY ANCHORS**, heavy duty, incl. sleeve, lfloating base nut,											**540**
	0020	lower stud & coupling nut, fiber plug, connecting stud, washer & nut.										
	0030	For flush mounted embedment in poured concrete heavy equip. pads.										
	0200	13 mm diameter stud & bolt		E-16	40	.400	Ea.	58	16.40	2.23	76.63	96
	0300	16 mm diameter [5/8"]			35	.457		64	18.70	2.55	85.25	107
	0500	19 mm diameter			30	.533		74	22	2.98	98.98	124
	0600	22 mm diameter			25	.640		81	26	3.57	110.57	140
	0800	25 mm diameter		▼	20	.800	▼	85	33	4.46	122.46	157

METALS 5

05090	Metal Fastenings	CREW	DAILY OUTPUT	LABOR-HOURS	UNIT	2006 BARE COSTS				TOTAL INCL O&P	
						MAT.	LABOR	EQUIP.	TOTAL		
540	**0900**	31 mm diameter	E-16	15	1.067	Ea.	113	43.50	5.95	162.45	209	**540**
580	**0010**	**POWDER ACTUATED TOOLS & FASTENERS**										**580**
	0020	Stud driver, .22 caliber, buy, minimum				Ea.	340			340	370	
	0100	Maximum				"	545			545	600	
	0300	Powder charges for above, low velocity				h	17.40			17.40	19.15	
	0400	Standard velocity					25			25	27.50	
	0600	Drive pins & studs, 6 mm & 9 mm diam.,to 75 mm long, min.	1 Carp	4.80	1.667		13.05	59.50		72.55	107	
	0700	Maximum	"	4	2		51	71		122	167	
	0800	Pneumatic stud driver for 3 mm diameter studs [1/8"]				Ea.	2,350			2,350	2,600	
	0900	Drive pins for above, 13 mm to 19 mm long	1 Carp	1	8	k	540	284		824	1,050	
600	**0010**	**RIVETS**										**600**
	0100	Aluminum rivet & mandrel, 13 mm grip length x 3 mm diameter	1 Carp	4.80	1.667	h	5.25	59.50		64.75	98.50	
	0200	5 mm diameter		4	2		8.10	71		79.10	120	
	0300	Aluminum rivet, steel mandrel, 3 mm diameter		4.80	1.667		8.05	59.50		67.55	101	
	0400	5 mm diameter		4	2		7.35	71		78.35	119	
	0500	Copper rivet, steel mandrel, 3 mm diameter		4.80	1.667		7.30	59.50		66.80	101	
	0600	Monel rivet, steel mandrel, 3 mm diameter		4.80	1.667		26	59.50		85.50	121	
	0700	5 mm diameter		4	2		75	71		146	194	
	0800	Stainless rivet & mandrel, 3 mm diameter		4.80	1.667		13.05	59.50		72.55	107	
	0900	5 mm diameter		4	2		24.50	71		95.50	138	
	1000	Stainless rivet, steel mandrel, 3 mm diameter		4.80	1.667		10.20	59.50		69.70	104	
	1100	5 mm diameter		4	2		18.45	71		89.45	132	
	1200	Steel rivet and mandrel, 3 mm diameter		4.80	1.667		6.30	59.50		65.80	99.50	
	1300	5 mm diameter		4	2		9.50	71		80.50	121	
	1400	Hand riveting tool, minimum				Ea.	122			122	134	
	1500	Maximum					233			233	257	
	1600	Power riveting tool, minimum					880			880	970	
	1700	Maximum					2,225			2,225	2,450	
820	**0010**	**VIBRATION PADS**										**820**
	0300	Laminated synthetic rubber impregnated cotton duck, 13 mm thick	2 Sswk	2.23	7.176	m²	635	287		922	1,225	
	0400	25 mm thick		1.86	8.611		1,275	345		1,620	2,025	
	0600	Neoprene bearing pads, 13 mm thick		2.23	7.176		240	287		527	780	
	0700	25 mm thick		1.86	8.611		490	345		835	1,150	
	0900	Fabric reinforced neoprene, 35 MPa, 13 mm thick		2.23	7.176		108	287		395	635	
	1000	25 mm thick		1.86	8.611		216	345		561	855	
	1200	Felt surfaced vinyl pads, cork and sisal, 16 mm thick		2.23	7.176		277	287		564	820	
	1300	25 mm thick		1.86	8.611		505	345		850	1,175	
	1500	Teflon bonded to 10 ga. carbon steel, 0.8 mm layer		2.23	7.176		500	287		787	1,075	
	1600	2 mm layer		2.23	7.176		750	287		1,037	1,350	
	1800	Bonded to 10 ga. stainless steel, 0.8 mm layer		2.23	7.176		890	287		1,177	1,500	
	1900	2 mm layer		2.23	7.176		1,150	287		1,437	1,800	
	2100	Circular machine leveling pad & stud				Met. Ton	14.85			14.85	16.35	
840	**0010**	**WELD SHEAR CONNECTORS**										**840**
	0020	19 mm diam., 80 mm long	E-10	960	.017	Ea.	.41	.68	.28	1.37	1.98	
	0030	86 mm long		950	.017		.43	.69	.28	1.40	2.02	
	0200	98 mm long		945	.017		.46	.69	.28	1.43	2.07	
	0300	106 mm long		935	.017		.48	.70	.28	1.46	2.10	
	0500	124 mm long		930	.017		.54	.70	.29	1.53	2.17	
	0600	132 mm long		920	.017		.56	.71	.29	1.56	2.22	
	0800	137 mm long		910	.018		.57	.72	.29	1.58	2.24	
	0900	157 mm long		905	.018		.62	.72	.29	1.63	2.30	
	1000	183 mm long		895	.018		.77	.73	.30	1.80	2.50	
	1100	208 mm long		890	.018		.85	.74	.30	1.89	2.59	
	1500	22 mm diameter, 94 mm long		920	.017		.67	.71	.29	1.67	2.33	

		05090	Metal Fastenings	CREW	DAILY OUTPUT	LABOR-HOURS	UNIT	2006 BARE COSTS MAT.	LABOR	EQUIP.	TOTAL	TOTAL INCL O&P	
840	1600		106 mm long	E-10	910	.018	Ea.	.71	.72	.29	1.72	2.41	840
	1700		132 mm long		905	.018		.81	.72	.29	1.82	2.51	
	1800		157 mm long		895	.018		.91	.73	.30	1.94	2.65	
	1900		183 mm long		890	.018		1.01	.74	.30	2.05	2.77	
	2000		208 mm long	▼	880	.018	▼	1.10	.74	.30	2.14	2.88	
860	0010	**WELD STUDS**											860
	0020		6 mm diameter, 68 mm long	E-10	1120	.014	Ea.	.27	.59	.24	1.10	1.60	
	0100		105 mm long		1080	.015		.25	.61	.25	1.11	1.64	
	0200		10 mm diameter, 105 mm long		1080	.015		.29	.61	.25	1.15	1.68	
	0300		156 mm long		1040	.015		.38	.63	.26	1.27	1.82	
	0400		13 mm diameter, 54 mm long		1040	.015		.27	.63	.26	1.16	1.71	
	0500		79 mm long		1025	.016		.33	.64	.26	1.23	1.80	
	0600		105 mm long		1010	.016		.39	.65	.26	1.30	1.89	
	0700		135 mm long		990	.016		.48	.66	.27	1.41	2.01	
	0800		156 mm long		975	.016		.52	.67	.27	1.46	2.08	
	0900		206 mm long		960	.017		.73	.68	.28	1.69	2.34	
	1000		16 mm diameter, 68 mm long		1000	.016		.47	.66	.27	1.40	1.99	
	1010		106 mm long		990	.016		.58	.66	.27	1.51	2.12	
	1100		167 mm long		975	.016		.76	.67	.27	1.70	2.35	
	1200		208 mm long	▼	960	.017	▼	1.02	.68	.28	1.98	2.65	
880	0010	**WELD ROD**											880
	0020		Steel, type 6011, 3 mm dia, less than 225 kg				kg	4.19			4.19	4.61	
	0100		225 kg to 900 kg					3.77			3.77	4.14	
	0200		900 kg to 2200 kg					3.55			3.55	3.90	
	0300		4 mm dia, less than 225 kg					4.01			4.01	4.41	
	0310		225 kg to 900 kg					3.62			3.62	3.97	
	0320		900 kg to 2200 kg					3.40			3.40	3.75	
	0400		5 mm diameter, less than 225 kg					4.03			4.03	4.43	
	0500		225 kg to 900 kg					3.64			3.64	4.01	
	0600		900 kg to 2200 kg					3.42			3.42	3.77	
	0620		Steel, type 6010), 3 mm dia, less than 225 kg					4.61			4.61	5.05	
	0630		225 kg to 900 kg					4.14			4.14	4.56	
	0640		900 kg to 2200 kg					3.90			3.90	4.28	
	0650		Steel, type 7018 Low Hydrogen, 3 mm dia, less than 225 kg					3.95			3.95	4.34	
	0660		225 kg to 900 kg					3.55			3.55	3.90	
	0670		900 kg to 2200 kg					3.33			3.33	3.66	
	0700		Steel, type 7024 Jet Weld, 3 mm dia, less than 225 kg					4.17			4.17	4.59	
	0710		225 kg to 900 kg					3.75			3.75	4.12	
	0720		900 kg to 2200 kg					3.53			3.53	3.88	
	1550		Aluminum, type 4043 TIG, 3 mm dia, less than 4.5 kg					14.40			14.40	15.85	
	1560		4.5 kg to 27 kg					13			13	14.30	
	1570		Over 27 kg					12.20			12.20	13.45	
	1600		Aluminum, type 5356 TIG, 3 mm dia, less than 4.5 kg					16			16	17.60	
	1610		4.5 kg to 27 kg					14.40			14.40	15.85	
	1620		Over 27 kg					13.55			13.55	14.90	
	1900		Cast iron, type 8 Nickel, 3 mm dia, less than 225 kg					49.50			49.50	54.50	
	1910		225 kg to 450 kg					44.50			44.50	49	
	1920		Over 900 kg					42			42	46	
	2000		Stainless steel, type 316/316L, 3 mm dia, less than 225 kg					23			23	25	
	2100		225 kg to 450 kg					20.50			20.50	22.50	
	2220		Over 450 kg				▼	19.25			19.25	21	
900	0010	**WELDING STRUCTURAL**											900
	0020		Field welding, 3 mm E6011, cost/welder, no oper. engr	R050521 -20	E-14	8	1	Hr.	3.80	42	11.15	56.95	92

METALS 5

		05090	**Metal Fastenings**	CREW	DAILY OUTPUT	LABOR-HOURS	UNIT	2006 BARE COSTS MAT.	LABOR	EQUIP.	TOTAL	TOTAL INCL O&P	
900	0200	With 1/2 operating engineer	R050521-20	E-13	8	1.500	Hr.	3.80	59.50	11.15	74.45	118	900
	0300	With 1 operating engineer		E-12	8	2		3.80	77	11.15	91.95	145	
	0500	With no operating engineer, 0.9 kg per metric ton		E-14	7.26	1.102	Met. Ton	4.19	46	12.30	62.49	101	
	0600	3.6 kg E6011 per metric ton		"	1.81	4.409		16.75	185	49	250.75	405	
	0800	With one operating engr per welder, 0.9 kg E6011 per metric ton		E-12	7.26	2.205		4.19	85	12.30	101.49	160	
	0900	3.6 kg E6011 per metric ton		"	1.81	8.818		16.75	340	49	405.75	635	
	1200	Continuous fillet, stick welding, incl. equipment											
	1300	Single pass, 3 mm thick, 0.1 kg/m		E-14	45.72	.175	m	.62	7.35	1.95	9.92	16.05	
	1400	5 mm thick, 0.3 kg/m			22.86	.350		1.25	14.70	3.90	19.85	32	
	1500	6 mm thick, 0.4 kg/m			15.24	.525		1.87	22	5.85	29.72	48	
	1610	8 mm thick, 0.6 kg/m			11.58	.691		2.49	29	7.70	39.19	63	
	1800	3 passes, 10 mm thick, 0.7 kg/m			9.14	.875		3.12	36.50	9.75	49.37	80	
	2010	4 passes, 13 mm thick, 1.0 kg/m			6.71	1.193		4.36	50	13.30	67.66	109	
	2200	5 to 6 passes, 19 mm thick, 1.9 kg/m			3.66	2.187		8.10	92	24.50	124.60	201	
	2400	8 to 11 passes, 25 mm thick, 3.6 kg/m			1.83	4.374		14.95	184	49	247.95	400	
	2600	For all position welding, add, minimum							20%				
	2700	Maximum							300%				
	2900	For semi-automatic welding, deduct, minimum							5%				
	3000	Maximum							15%				
	4000	Cleaning and welding plates, bars, or rods											
	4010	to existing beams, columns, or trusses		E-14	3.66	2.187	m	3.12	92	24.50	119.62	195	
920	0010	**STEEL CUTTING**											920
	0020	Hand burning, incl. preparation, torch cutting & grinding, no staging											
	0100	Steel to 13 mm thick		E-25	97.54	.082	m		3.44	.83	4.27	7.10	
	0150	19 mm thick			79.25	.101			4.23	1.02	5.25	8.70	
	0200	25 mm thick			60.96	.131			5.50	1.33	6.83	11.35	

05100 | Structural Metal Framing

		05120	**Structural Steel**	CREW	DAILY OUTPUT	LABOR-HOURS	UNIT	2006 BARE COSTS MAT.	LABOR	EQUIP.	TOTAL	TOTAL INCL O&P	
140	0010	**SUBPURLINS**	R051223-50										140
	0020	Bulb tees, painted, 816 mm O.C., 195 kg/m² L.L.											
	0100	Type 178, max 2.67 m span, 3.20 kg/m, 50 mm H x 41 mm W		E-1	390	.062	m²	14.85	2.40	.23	17.48	20.50	
	0200	Type 218, max 3.10 m span, 4.75 kg/m, 53 mm H x 53 mm W		"	288	.083		17.20	3.25	.31	20.76	25	
	1420	For 616 mm spacing, add						33%	33%				
	1430	For 1216 mm spacing, deduct						50%	50%				
180	0010	**CANOPY FRAMING**											180
	0020	150 mm and 200 mm members		E-4	1361	.024	kg	2.51	.95	.07	3.53	4.54	
220	0010	**CEILING SUPPORTS**											220
	1000	Entrance door/folding partition supports		E-4	18.29	1.750	m	62.50	71	4.88	138.38	201	
	1100	Linear accelerator door supports			4.27	7.499		284	305	21	610	880	
	1200	Lintels or shelf angles, hung, exterior hot dipped galv.			81.38	.393		42.50	15.90	1.10	59.50	76.50	
	1250	Two coats primer paint instead of galv.			81.38	.393		37	15.90	1.10	54	70	
	1400	Monitor support, ceiling hung, expansion bolted			4	8	Ea.	300	325	22.50	647.50	935	
	1450	Hung from pre-set inserts			6	5.333		325	216	14.90	555.90	760	
	1600	Motor supports for overhead doors			4	8		153	325	22.50	500.50	775	
	1700	Partition support for heavy folding partitions, without pocket			7.32	4.374	m	142	177	12.20	331.20	490	
	1750	Supports at pocket only			3.66	8.749		284	355	24.50	663.50	970	

05120 | Structural Steel

			CREW	DAILY OUTPUT	LABOR-HOURS	UNIT	2006 BARE COSTS MAT.	LABOR	EQUIP.	TOTAL	TOTAL INCL O&P	
220	2000	Rolling grilles & fire door supports	E-4	10.36	3.088	m	122	125	8.60	255.60	370	220
	2100	Spider-leg light supports, expansion bolted to ceiling slab		8	4	Ea.	124	162	11.15	297.15	440	
	2150	Hung from pre-set inserts		12	2.667	"	133	108	7.45	248.45	350	
	2400	Toilet partition support		10.97	2.916	m	142	118	8.15	268.15	375	
	2500	X-ray travel gantry support	▼	3.66	8.749	"	485	355	24.50	864.50	1,200	
250	0010	**COLUMNS, LIGHTWEIGHT**										250
	1000	Lightweight units (lally), 90 mm diameter	E-2	238	.235	m	9.80	9.15	6	24.95	33	
	1050	100mm diameter	"	274	.204	"	14.45	7.95	5.20	27.60	35.50	
	5800	Adjustable jack post, 2.4m maximum height, 70mm diameter				Ea.	32			32	35	
	5850	100mm diameter				"	51			51	56	
260	0010	**COLUMNS, STRUCTURAL**										260
	0020	Shop fab'd for 91-tonne, 1-2 story project, bolted conn's.	R051223 -10									
	0800	Steel, concrete filled, extra strong pipe, 90 mm diameter	E-2	201	.279	m	103	10.85	7.10	120.95	140	
	0830	100 mm diameter		238	.235		115	9.15	6	130.15	148	
	0890	125 mm diameter		311	.180		137	7	4.59	148.59	167	
	0930	150 mm diameter		366	.153		181	5.95	3.90	190.85	214	
	0940	8″ diameter	▼	335	.167	▼	181	6.50	4.26	191.76	215	
	1100	For galvanizing, add				kg	.49			.49	.55	
	1300	For web ties, angles, etc., add per added kg	1 Sswk	429	.019		2.09	.75		2.84	3.65	
	1500	Steel pipe, extra strong, no concrete, 75 mm to 125 mm diameter	E-2	7258	.008		2.09	.30	.20	2.59	3.05	
	1600	150 mm to 300 mm diameter		6350	.009	▼	2.09	.34	.22	2.65	3.15	
	1700	Steel pipe, extra strong, no concrete, 75 mm diameter x 3.7 m		60	.933	Ea.	117	36.50	24	177.50	218	
	1750	100 mm diameter x 3.7 m		58	.966		171	37.50	24.50	233	280	
	1800	150 mm diameter x 3.7 m		54	1.037		325	40.50	26.50	392	460	
	1850	200 mm diameter x 4.3 m		50	1.120		575	43.50	28.50	647	740	
	1900	250 mm diameter x 4.9 m		48	1.167		830	45.50	29.50	905	1,025	
	1950	310 mm diameter x 5.5 m		45	1.244	▼	1,125	48.50	31.50	1,205	1,350	
	3300	Structural tubing, square, A500GrB, 100 mm to 150 mm, light section		5112	.011	kg	2.09	.43	.28	2.80	3.35	
	3600	Heavy section	▼	14515	.004	"	2.09	.15	.10	2.34	2.68	
	4000	Concrete filled, add				m	11.40			11.40	12.50	
	4500	Structural tubing, sq, 100 mm x 100 mm x 6.4 mm x 3.7 m	E-2	58	.966	Ea.	157	37.50	24.50	219	264	
	4550	150 mm x 150 mm x 6.4 mm x 3.7 m		54	1.037		257	40.50	26.50	324	380	
	4600	200 mm x 200 mm x 9.5 mm x 4.3 m		50	1.120		555	43.50	28.50	627	715	
	4650	250 mm x 250 mm x 12.7 mm x 4.9 m		48	1.167	▼	1,025	45.50	29.50	1,100	1,225	
	5100	Structural tubing, rect, 127 mm to 152 mm wide, light section		3629	.015	kg	2.09	.60	.39	3.08	3.78	
	5200	Heavy section		5443	.010		2.09	.40	.26	2.75	3.29	
	5300	178 mm to 254 mm wide, light section		6804	.008		2.09	.32	.21	2.62	3.09	
	5400	Heavy section		8165	.007	▼	2.09	.27	.17	2.53	2.96	
	5500	Structural tubing, rect, 125 mm x 75 mm x 6.4 mm x 3.7 m		58	.966	Ea.	152	37.50	24.50	214	259	
	5550	150 mm x 100 mm x 7.9 mm x 3.7 m		54	1.037		238	40.50	26.50	305	360	
	5600	200 mm x 100 mm x 9.5 mm x 3.7 m		54	1.037		345	40.50	26.50	412	480	
	5650	250 mm x 150 mm x 9.5 mm x 4.3 m		50	1.120		555	43.50	28.50	627	715	
	5700	310 mm x 200 mm x 12.7 mm x 4.9 m		48	1.167	▼	1,025	45.50	29.50	1,100	1,225	
	6800	W Shape, A992 steel, 2 tier, W200 x 36 [W8 x 24]		329	.170	m	82.50	6.60	4.34	93.44	107	
	6850	W200 x 46 [W8 x 31]		329	.170		106	6.60	4.34	116.94	133	
	6900	W200 x 71 [W8 x 48]		315	.178		165	6.90	4.53	176.43	198	
	6950	W200 x 100 [W8 x 67]		300	.187		230	7.25	4.76	242.01	271	
	7000	W250 x 67 [W10 x 45]		315	.178		154	6.90	4.53	165.43	187	
	7050	W250 x 101 [W10 x 68]		300	.187		233	7.25	4.76	245.01	274	
	7100	W250 x 167 [W10 x 112]		293	.191		385	7.45	4.87	397.32	440	
	7150	W310 x 74 [W 12 x 50]		315	.178		171	6.90	4.53	182.43	206	
	7200	W310 x 129 [W12 x 87]		300	.187		298	7.25	4.76	310.01	350	
	7250	W310 x 179 [W12 x 120]		293	.191		410	7.45	4.87	422.32	475	
	7300	W310 x 283 [W12 x 190]	▼	278	.201	▼	650	7.85	5.15	663	735	

METALS 5

215

05120 | Structural Steel

			CREW	DAILY OUTPUT	LABOR-HOURS	UNIT	2006 BARE COSTS				TOTAL INCL O&P	
							MAT.	LABOR	EQUIP.	TOTAL		
260	7350	W360 x 110 [W14 x 74]	R051223-10	E-2	300	.187	m	254	7.25	4.76	266.01	297	**260**
	7400	W360 x 179 [W14 x 120]			293	.191		410	7.45	4.87	422.32	475	
	7450	W360 x 262 [W14 x 176]			278	.201		605	7.85	5.15	618	685	
	8090	For projects 68 to 90 tonne, add					All	10%					
	8092	46 to 67 tonne, add						20%					
	8094	23 to 45 tonne, add						30%	10%				
	8096	9 to 22 tonne, add						50%	25%				
	8098	2 to 8 tonne, add						75%	50%				
	8099	Less than 2 tonne, add						100%	100%				
300	0010	**CURB EDGING**											**300**
	0020	Steel angle w/anchors, on forms, 25 mm x 25 mm,1.2 kg/m		E-4	107	.299	m	5.30	12.10	.83	18.23	28.50	
	0100	50 mm x 50 mm angles, 5.8 kg/m			101	.317		16.95	12.80	.88	30.63	42.50	
	0200	75 mm x 75 mm angles, 9.1 kg/m			91.44	.350		27	14.15	.98	42.13	56	
	0300	100 mm x 100 mm angles, 12.2 kg/m			83.82	.382		34.50	15.45	1.07	51.02	67	
	1000	150 mm x 100 mm angles, 18.3 kg/m			76.20	.420		50	17	1.17	68.17	87	
	1050	Steel channels with anchors, on forms, 75 mm channel, 7.4 kg/m			88.39	.362		21	14.65	1.01	36.66	50.50	
	1100	100 mm channel, 8 kg/m			82.30	.389		22.50	15.75	1.08	39.33	54	
	1200	150 mm channel, 12 kg/m			77.72	.412		34.50	16.65	1.15	52.30	69.50	
	1300	200 mm channel, 17 kg/m			68.58	.467		47	18.85	1.30	67.15	87	
	1400	250 mm channel, 23 kg/m			54.86	.583		61	23.50	1.63	86.13	112	
	1500	300 mm channel, 31 kg/m			42.67	.750		81.50	30.50	2.09	114.09	146	
	2000	For curved edging, add						35%	10%				
440	0010	**LIGHTWEIGHT FRAMING**	R051223-35										**440**
	0200	For load-bearing steel studs see Division 05410-400											
	0400	Angle framing, field fabricated, 100 mm and larger	R051223-45	E-3	200	.120	kg	1.21	4.87	.45	6.53	10.60	
	0450	Less than 100 mm angles			120	.200	"	1.26	8.10	.74	10.10	16.80	
	0460	13 mm x 13 mm x 3 mm			60.96	.394	m	.36	16	1.46	17.82	31	
	0462	19 mm x 19 mm x 3 mm			48.77	.492		1.05	20	1.83	22.88	39	
	0464	25 mm x 25 mm x 3 mm			41.15	.583		1.51	23.50	2.17	27.18	46.50	
	0466	31 mm x 31 mm x 5 mm			35.05	.685		2.76	28	2.55	33.31	56	
	0468	38 mm x 38 mm x 5 mm			30.48	.787		3.38	32	2.93	38.31	64.50	
	0470	50 mm x 50 mm x 6 mm			27.43	.875		5.95	35.50	3.25	44.70	74	
	0472	63 mm x 63 mm x 6 mm			21.95	1.094		7.70	44.50	4.07	56.27	93	
	0474	75 mm x 50 mm x 9 mm			19.81	1.211		11	49	4.51	64.51	106	
	0476	75 mm x 75 mm x 9 mm			17.37	1.381		13.45	56	5.15	74.60	121	
	0600	Channel framing, field fabricated, 200 mm and larger			227	.106	kg	1.26	4.29	.39	5.94	9.55	
	0650	Less than 200 mm channels			152	.158	"	1.26	6.40	.59	8.25	13.60	
	0660	C50 x 2.6			35.05	.685	m	3.31	28	2.55	33.86	56.50	
	0662	C75 x 6.1			24.38	.984		7.70	40	3.66	51.36	84.50	
	0664	C100 x 8			20.12	1.193		10.10	48.50	4.44	63.04	103	
	0666	C125 x 10			17.37	1.381		12.55	56	5.15	73.70	120	
	0668	C150 x 12			16.76	1.432		14.85	58	5.35	78.20	127	
	0670	C175 x 15			12.19	1.969		18.35	80	7.30	105.65	172	
	0672	C200 x 17			10.97	2.187		21.50	89	8.15	118.65	192	
	0710	Structural bar tee, field fabricated, 19 mm x 19 mm x 3 mm			48.77	.492		1.05	20	1.83	22.88	39	
	0712	25 mm x 25 mm x 3 mm			41.15	.583		1.51	23.50	2.17	27.18	46.50	
	0714	38 mm x 38 mm x 6 mm			34.75	.691		4.36	28	2.57	34.93	58	
	0716	50 mm x 50 mm x 6 mm			27.13	.885		5.95	36	3.29	45.24	74.50	
	0718	63 mm x 63 mm x 9 mm			21.95	1.094		11	44.50	4.07	59.57	96.50	
	0720	75 mm x 75 mm x 9 mm			17.37	1.381		13.45	56	5.15	74.60	121	
	0730	Structural zee, field fabricated, 31 mm x 44 mm x 44 mm			34.75	.691		1.41	28	2.57	31.98	55	
	0732	67 mm x 75 mm x 67 mm			34.75	.691		3.31	28	2.57	33.88	57	
	0734	77 mm x 100 mm x 77 mm			40.54	.592		5	24	2.20	31.20	51.50	
	0736	81 mm x 125 mm x 81 mm			40.54	.592		6.85	24	2.20	33.05	53.50	

METALS 5

05120 | Structural Steel

		Description	CREW	DAILY OUTPUT	LABOR-HOURS	UNIT	MAT.	LABOR	EQUIP.	TOTAL	TOTAL INCL O&P	
440	0738	88 mm x 150 mm x 88 mm R051223-35	E-3	48.77	.492	m	10.35	20	1.83	32.18	49.50	**440**
	0740	Junior beam, field fabricated, 75 mm		24.38	.984		10.65	40	3.66	54.31	87.50	
	0742	100 mm R051223-45		21.95	1.094		14.40	44.50	4.07	62.97	100	
	0744	125 mm		20.42	1.175		18.70	47.50	4.37	70.57	111	
	0746	150 mm		18.90	1.270		23.50	51.50	4.72	79.72	124	
	0748	175 mm		17.37	1.381		28.50	56	5.15	89.65	138	
	0750	200 mm	▼	16.15	1.486	▼	34.50	60.50	5.55	100.55	153	
	1000	Continuous slotted channel framing system, shop fab, min	2 Sswk	1089	.015	kg	6.50	.59		7.09	8.20	
	1200	Maximum	"	726	.022		7.35	.88		8.23	9.65	
	1250	Plate & bar stock for reinforcing beams and trusses					2.31			2.31	2.54	
	1300	Cross bracing, rods, shop fabricated, 19 mm diameter	E-3	318	.075		2.51	3.07	.28	5.86	8.55	
	1310	22 mm diameter		386	.062		2.51	2.53	.23	5.27	7.55	
	1320	25 mm diameter		454	.053		2.51	2.15	.20	4.86	6.85	
	1330	Angle, 125 mm x 125 mm x 9 mm		1270	.019		2.51	.77	.07	3.35	4.22	
	1350	Hanging lintels, shop fabricated, average	▼	386	.062		2.51	2.53	.23	5.27	7.55	
	1380	Roof frames, shop fabricated, 0.9 m square, 1.5 m span	E-2	1905	.029		2.51	1.14	.75	4.40	5.55	
	1400	Tie rod, not upset, 38 mm to 100 mm diam., with turnbuckle	2 Sswk	363	.044		2.73	1.76		4.49	6.15	
	1420	No turnbuckle		318	.050		2.62	2.01		4.63	6.50	
	1500	Upset, 44 mm to 100 mm diameter, with turnbuckle		363	.044		2.73	1.76		4.49	6.15	
	1520	No turnbuckle	▼	318	.050	▼	2.62	2.01		4.63	6.50	
480	0010	**LINTELS**										**480**
	0020	Plain steel angles, under 225 kg	1 Bric	249	.032	kg	1.61	1.17		2.78	3.55	
	0100	225 to 450 kg		290	.028		1.57	1.01		2.58	3.26	
	0200	450 to 900 kg		290	.028		1.52	1.01		2.53	3.22	
	0300	900 to 1800 kg	▼	290	.028		1.48	1.01		2.49	3.17	
	0500	For built-up angles and plates, add to above					.53			.53	.57	
	0700	For engineering, add to above					.22			.22	.22	
	0900	For galvanizing, add to above, under 225 kg					.66			.66	.73	
	0950	225 - 900 kg					.60			.60	.66	
	1000	Over 900 kg				▼	.49			.49	.55	
	2000	Steel angles, 89 mm x 76 mm, 6.4 mm thick, 0.8 m long	1 Bric	47	.170	Ea.	10.25	6.20		16.45	21	
	2100	1.4 m long		26	.308		18.45	11.25		29.70	37.50	
	2600	102 mm x 89 mm, 6.4 mm thick, 1.5 m long		21	.381		23.50	13.90		37.40	47	
	2700	2.7 m long	▼	12	.667	▼	42.50	24.50		67	83.50	
	3500	For precast concrete lintels, see Div. 03480-400										
520	0010	**PIPE SUPPORT FRAMING**										**520**
	0020	Under 15 kg/m	E-4	1769	.018	kg	2.80	.73	.05	3.58	4.47	
	0200	15.1 to 25 kg/m		1950	.016		2.76	.66	.05	3.47	4.28	
	0400	25.1 to 30 kg/m		2177	.015		2.73	.59	.04	3.36	4.12	
	0600	Over 30 kg/m	▼	2449	.013	▼	2.69	.53	.04	3.26	3.94	
560	0010	**PLATES**										**560**
	0020	For connections & stiffener plates, shop fabricated R051223-80										
	0050	3 mm thick (25 kg/m²)				m²	52			52	57.50	
	0100	6 mm thick (50 kg/m²)					104			104	115	
	0300	9 mm thick (75 kg/m²)					157			157	172	
	0400	13 mm thick (100 kg/m²)					209			209	229	
	0450	19 mm thick (149 kg/m²)					315			315	345	
	0500	25 mm thick (199 kg/m²)				▼	415			415	460	
	2010	1.2 - 1.5 m wide, 3 - 18 m long, 4.5 - 5 tonnes, mill prices										
	2100	6 mm				Met. Ton	935			935	1,025	
	2150	8 mm thick					935			935	1,025	
	2200	9 mm thick					885			885	975	
	2250	13 mm thick					885			885	975	
	2300	19 mm thick	▼				885			885	975	

						2006 BARE COSTS				TOTAL	
	05120	**Structural Steel**		CREW	DAILY OUTPUT	LABOR-HOURS	UNIT	MAT.	LABOR	EQUIP.	TOTAL	INCL O&P	
560	2350	25 mm thick	R051223-80				Met. Ton	885			885	975	**560**
	2400	50 mm thick						910			910	1,000	
	2450	100 mm thick						960			960	1,050	
	2500	150 mm thick						1,000			1,000	1,100	
	2550	200 mm thick						1,000			1,000	1,100	
600	0010	**STRESSED SKIN ROOF & CEILING SYSTEM**											**600**
	0020	Double panel flat roof, spans to 30 m		E-2	107	.523	m²	82	20.50	13.35	115.85	140	
	0100	Double panel convex roof, spans to 60 m			89.18	.628		133	24.50	16	173.50	206	
	0200	Double panel arched roof, spans to 90 m			70.60	.793		205	31	20	256	300	
640	0010	**STRUCTURAL STEEL MEMBERS**	R051223-10										**640**
	0020	Shop fab'd for 91-tonne, 1-2 story project, bolted conn's.											
	0100	W 150 x 13 [W 6 x 9]		E-2	183	.306	m	31	11.90	7.80	50.70	63	
	0120	x 24 [W 6 x 16]			183	.306		55	11.90	7.80	74.70	89.50	
	0140	x 30 [W 6 x 20]			183	.306		68.50	11.90	7.80	88.20	105	
	0300	W 200 x 15 [W 8 x 10]			183	.306		34.50	11.90	7.80	54.20	66.50	
	0320	x 22 [W 8 x 15]			183	.306		51.50	11.90	7.80	71.20	85.50	
	0350	x 31 [W 8 x 21]			183	.306		72	11.90	7.80	91.70	108	
	0360	x 36 [W 8 x 24]			168	.333		82.50	12.95	8.50	103.95	122	
	0370	x 42 [W 8 x 28]			168	.333		96	12.95	8.50	117.45	138	
	0500	x 46 [W 8 x 31]			168	.333		106	12.95	8.50	127.45	149	
	0520	x 52 [W 8 x 35]			168	.333		120	12.95	8.50	141.45	164	
	0540	x 71 [W 8 x 48]			168	.333		165	12.95	8.50	186.45	213	
	0600	W 250 x 18 [W 10 x 12]			183	.306		41	11.90	7.80	60.70	74	
	0620	x 22 [W 10 x 15]			183	.306		51.50	11.90	7.80	71.20	85.50	
	0700	x 33 [W 10 x 22]			183	.306		75.50	11.90	7.80	95.20	112	
	0720	x 39 [W 10 x 26]			183	.306		89	11.90	7.80	108.70	127	
	0740	x 49 [W 10 x 33]			168	.333		113	12.95	8.50	134.45	156	
	0900	x 73 [W 10 x 49]			168	.333		168	12.95	8.50	189.45	217	
	1100	W 310 x 21 [W 12 x 14]			268	.209		48	8.15	5.30	61.45	73	
	1300	x 33 [W 12 x 22]			268	.209		75.50	8.15	5.30	88.95	103	
	1500	x 39 [W 12 x 26]			268	.209		89	8.15	5.30	102.45	118	
	1520	x 52 [W 12 x 35]			247	.227		120	8.80	5.80	134.60	154	
	1560	x 74 [W 12 x 50]			229	.245		171	9.50	6.25	186.75	212	
	1580	x 86 [W 12 x 58]			229	.245		199	9.50	6.25	214.75	242	
	1700	x 107 [W 12 x 72]			195	.287		247	11.15	7.30	265.45	299	
	1740	x 129 [W 12 x 87]			195	.287		298	11.15	7.30	316.45	355	
	1900	W 360 x 39 [W 14 x 26]			302	.185		89	7.20	4.72	100.92	116	
	2100	x 44 [W 14 x 30]			274	.204		103	7.95	5.20	116.15	132	
	2300	x 51 [W 14 x 34]			247	.227		117	8.80	5.80	131.60	150	
	2320	x 64 [W 14 x 43]			247	.227		147	8.80	5.80	161.60	184	
	2340	x 79 [W 14 x 53]			244	.230		182	8.95	5.85	196.80	222	
	2360	x 110 [W 14 x 74]			232	.241		254	9.40	6.15	269.55	300	
	2380	x 134 [W 14 x 90]			226	.248		310	9.65	6.30	325.95	365	
	2500	x 179 [W 14 x 120]			219	.256		410	9.95	6.50	426.45	480	
	2700	W 410 x 39 [W 16 x 26]			305	.184		89	7.15	4.68	100.83	115	
	2900	x 46 [W 16 x 31]			274	.204		106	7.95	5.20	119.15	136	
	3100	x 60 [W 16 x 40]			244	.230		137	8.95	5.85	151.80	173	
	3120	x 75 [W 16 x 50]			244	.230		171	8.95	5.85	185.80	211	
	3140	x 100 [W 16 x 67]			232	.241		230	9.40	6.15	245.55	276	
	3300	W 460 x 52 [W 18 x 35]		E-5	293	.273		120	10.75	5.15	135.90	157	
	3500	x 60 [W 18 x 40]			293	.273		137	10.75	5.15	152.90	176	
	3520	x 68 [W 18 x 46]			293	.273		158	10.75	5.15	173.90	198	
	3700	x 74 [W 18 x 50]			278	.288		171	11.35	5.45	187.80	215	
	3900	x 82 [W 18 x 55]			278	.288		189	11.35	5.45	205.80	233	
	3920	x 97 [W 18 x 65]			274	.292		223	11.50	5.55	240.05	271	

	05120	Structural Steel		CREW	DAILY OUTPUT	LABOR-HOURS	UNIT	MAT.	2006 BARE COSTS LABOR	EQUIP.	TOTAL	TOTAL INCL O&P	
640	3940	x 113 [W 18 x 76]	R051223 -10	E-5	274	.292	m	261	11.50	5.55	278.05	315	640
	3960	x 128 [W 18 x 86]			274	.292		295	11.50	5.55	312.05	350	
	3980	x 158 [W 18 x 106]			274	.292		365	11.50	5.55	382.05	425	
	4100	W 530 x 66 [W 21 x 44]			324	.247		151	9.75	4.68	165.43	188	
	4300	x 74 [W 21 x 50]			324	.247		171	9.75	4.68	185.43	211	
	4500	x 92 [W 21 x 62]			316	.253		213	10	4.80	227.80	257	
	4700	x 101 [W 21 x 68]			316	.253		233	10	4.80	247.80	279	
	4720	x 123 [W 21 x 83]			305	.262		285	10.35	4.97	300.32	340	
	4740	x 138 [W 21 x 93]			305	.262		320	10.35	4.97	335.32	375	
	4760	x 150 [W 21 x 101]			305	.262		345	10.35	4.97	360.32	405	
	4780	x 182 [W 21 x 122]			305	.262		420	10.35	4.97	435.32	485	
	4900	W 610 x 82 [W 24 x 55]			338	.237		189	9.35	4.49	202.84	228	
	5100	x 92 [W 24 x 62]			338	.237		213	9.35	4.49	226.84	255	
	5300	x 101 [W 24 x 68]			338	.237		233	9.35	4.49	246.84	277	
	5500	x 113 [W 24 x 76]			338	.237		261	9.35	4.49	274.84	310	
	5700	x 125 [W 24 x 84]			329	.243		288	9.60	4.61	302.21	335	
	5720	x 140 [W 24 x 94]			329	.243		320	9.60	4.61	334.21	375	
	5740	x 155 [W 24 x 104]			320	.250		355	9.85	4.74	369.59	410	
	5760	x 174 [W 24 x 117]			320	.250		400	9.85	4.74	414.59	460	
	5780	x 217 [W 24 x 146]			320	.250		500	9.85	4.74	514.59	570	
	5800	W 690 x 125 [W 27 x 84]			363	.220		288	8.70	4.18	300.88	335	
	5900	x 140 [W 27 x 94]			363	.220		320	8.70	4.18	332.88	375	
	5920	x 170 [W 27 x 114]			351	.228		390	9	4.32	403.32	450	
	5940	x 217 [W 27 x 146]			351	.228		500	9	4.32	513.32	570	
	5960	x 240 [W 27 x 161]			351	.228		550	9	4.32	563.32	625	
	6100	W 760 x 147 [W 30 x 99]			366	.219		340	8.60	4.14	352.74	395	
	6300	x 161 [W 30 x 108]			366	.219		370	8.60	4.14	382.74	425	
	6500	x 173 [W 30 x 116]			354	.226		400	8.90	4.28	413.18	455	
	6520	x 196 [W 30 x 132]			354	.226		455	8.90	4.28	468.18	520	
	6540	x 220 [W 30 x 148]			354	.226		505	8.90	4.28	518.18	580	
	6560	x 257 [W 30 x 173]			341	.235		595	9.25	4.45	608.70	670	
	6580	x 284 [W 30 x 191]			341	.235		655	9.25	4.45	668.70	740	
	6700	W 840 x 176 [W 33 x 118]			358	.223		405	8.80	4.23	418.03	465	
	6900	x 193 [W 33 x 130]			346	.231		445	9.10	4.38	458.48	510	
	7100	x 210 [W 33 x 141]			346	.231		485	9.10	4.38	498.48	550	
	7120	x 251 [W 33 x 169]			335	.239		580	9.40	4.53	593.93	655	
	7140	x 299 [W 33 x 201]			335	.239		690	9.40	4.53	703.93	780	
	7300	W 920 x 201 [W 36 x 135]			357	.224		465	8.85	4.25	478.10	530	
	7500	x 223 [W 36 x 150]			357	.224		515	8.85	4.25	528.10	585	
	7600	x 253 [W 36 x 170]			351	.228		585	9	4.32	598.32	660	
	7700	x 289 [W 36 x 194]			343	.233		665	9.20	4.42	678.62	750	
	7900	x 342 [W 36 x 230]			343	.233		790	9.20	4.42	803.62	885	
	7920	x 387 [W 36 x 260]			315	.254		890	10	4.81	904.81	1,000	
	8100	x 446 [W 36 x 300]			315	.254		1,025	10	4.81	1,039.81	1,150	
	8490	For projects 68 to 90 tonne, add						10%					
	8492	46 to 67 tonne, add						20%					
	8494	23 to 45 tonne, add						30%	10%				
	8496	9 to 22 tonne, add						50%	25%				
	8498	2 to 8 tonne, add						75%	50%				
	8499	Less than 2 tonne, add						100%	100%				
680	0010	**STRUCTURAL STEEL PROJECTS**	R050516 -30										680
	0020	Shop fab'd for 91-tonne, 1-2 story project, bolted conn's.											
	0200	Apartments, nursing homes, etc., 1 to 2 stories	R050523 -10	E-5	9.34	8.562	Met. Ton	2,100	335	162	2,597	3,075	
	0300	3 to 6 stories		"	9.16	8.731		2,125	345	165	2,635	3,125	
	0400	7 to 15 stories	R051223 -10	E-6	12.88	9.936		2,175	390	127	2,692	3,225	
	0500	Over 15 stories		"	12.61	10.151		2,250	400	130	2,780	3,350	

		DAILY OUTPUT	LABOR-HOURS	UNIT	2006 BARE COSTS				TOTAL INCL O&P	
05120	**Structural Steel** CREW				MAT.	LABOR	EQUIP.	TOTAL		
680	0700 Offices, hospitals, etc., steel bearing, 1 to 2 stories	E-5	9.34	8.562	Met. Ton	2,100	335	162	2,597	3,075	680
	0800 3 to 6 stories	E-6	13.06	9.798		2,125	385	125	2,635	3,175	
	0900 7 to 15 stories		12.88	9.936		2,175	390	127	2,692	3,225	
	1000 Over 15 stories	↓	12.61	10.151		2,250	400	130	2,780	3,350	
	1100 For multi-story masonry wall bearing construction, add						30%				
	1300 Industrial bldgs., 1 story, beams & girders, steel bearing	E-5	11.70	6.836		2,100	269	130	2,499	2,925	
	1400 Masonry bearing	"	9.07	8.818	↓	2,100	350	167	2,617	3,100	
	1500 Industrial bldgs., 1 story, under 10 metric tons,										
	1510 steel from warehouse, trucked	E-2	6.80	8.230	Met. Ton	2,525	320	210	3,055	3,550	
	1600 1 story with roof trusses, steel bearing	E-5	9.62	8.319		2,475	330	158	2,963	3,475	
	1700 Masonry bearing	"	7.53	10.625		2,475	420	201	3,096	3,675	
	1900 Monumental structures, banks, stores, etc., minimum	E-6	11.79	10.853		2,100	430	139	2,669	3,200	
	2000 Maximum	"	8.16	15.677		3,475	620	201	4,296	5,125	
	2200 Churches, minimum	E-5	10.52	7.602		1,950	300	144	2,394	2,825	
	2300 Maximum	"	4.72	16.958		2,600	670	320	3,590	4,375	
	2800 Power stations, fossil fuels, minimum	E-6	9.98	12.827		2,100	505	164	2,769	3,375	
	2900 Maximum		5.17	24.753		3,150	975	315	4,440	5,500	
	2950 Nuclear fuels, non-safety steel, minimum		6.35	20.156		2,100	795	258	3,153	3,975	
	3000 Maximum		4.99	25.653		3,150	1,000	330	4,480	5,575	
	3040 Safety steel, minimum		2.27	56.437		3,050	2,225	720	5,995	8,075	
	3070 Maximum	↓	1.36	94.062		4,025	3,700	1,200	8,925	12,300	
	3100 Roof trusses, minimum	E-5	11.79	6.783		2,925	267	129	3,321	3,825	
	3200 Maximum		7.53	10.625		3,550	420	201	4,171	4,875	
	3210 Schools, minimum		13.15	6.082		2,100	240	115	2,455	2,850	
	3220 Maximum	↓	7.53	10.625		3,050	420	201	3,671	4,325	
	3400 Welded construction, simple commercial bldgs., 1 to 2 stories	E-7	6.89	11.603		2,125	455	233	2,813	3,400	
	3500 7 to 15 stories	E-9	7.53	16.999		2,475	670	249	3,394	4,175	
	3700 Welded rigid frame, 1 story, minimum	E-7	14.33	5.581		2,175	220	112	2,507	2,900	
	3800 Maximum	"	4.99	16.033	↓	2,825	630	320	3,775	4,550	
	3810 Fabrication shop costs (included in project material cost, above)										
	3820 Mini mill base price, A992				Met. Ton	580			580	635	
	3830 Mill extras and delivery to shop					204			204	224	
	3840 Shop extra, add for shop drawings and detailing					248			248	273	
	3850 Add for shop fabricating and handling					840			840	920	
	3860 Add for sandblasting and primer coat of paint					132			132	146	
	3870 Add for shop delivery to the job site					93.50			93.50	103	
	3880 Total material cost, shop fabricated, primed, delivered				↓	2,100			2,100	2,300	
	3900 High strength steel mill spec extras: A242, A441,										
	3950 A529, A572 (42 ksi) and A992: same as A36 steel										
	4000 Add to A992 price for A572 (345, 415, 450 MPa)				Met. Ton	33			33	36.50	
	4100 A588 Weathering				"	66			66	73	
	4200 Mill size extras for W-Shapes: 0 to 45 kg/m: no extra charge										
	4210 Member sizes 46 to 97 kg/m, add				Met. Ton	1.10			1.10	1.21	
	4220 Member sizes 98 to 149 kg/m, add					1.10			1.10	1.21	
	4230 Member sizes 150 to 577 kg/m, add				↓	33			33	36.50	
	4300 Column base plates, light, up to 68 kg	2 Sswk	907	.018	kg	2.31	.70		3.01	3.81	
	4400 Heavy, over 68 kg	E-2	3402	.016	"	2.40	.64	.42	3.46	4.21	
	4600 Castellated beams, light sections, to 75 kg/m, minimum		9.71	5.769	Met. Ton	2,200	224	147	2,571	2,975	
	4700 Maximum		6.35	8.818		2,400	345	225	2,970	3,475	
	4900 Heavy sections, over 75 kg/m, minimum		10.61	5.276		2,300	205	134	2,639	3,025	
	5000 Maximum	↓	7.08	7.914		2,525	310	202	3,037	3,525	
	5390 For projects 68 to 90 tonne, add					10%					
	5392 46 to 67 tonne, add					20%					
	5394 23 to 45 tonne, add					30%	10%				
	5396 9 to 22 tonne, add					50%	25%				
	5398 2 to 8 tonne, add					75%	50%				

Reference boxes: R051223-20, R051223-25, R051223-30

5 METALS

05100 | Structural Metal Framing

05120 | Structural Steel

		CREW	DAILY OUTPUT	LABOR-HOURS	UNIT	2006 BARE COSTS				TOTAL INCL O&P	
						MAT.	LABOR	EQUIP.	TOTAL		
680	5399	Less than 2 tonne, add				Met. Ton	100%	100%				680
	5500	Steel domes - see R133423-30	R050516-30									
	5700	Steel estimating weights per m² - see R051223-20	R050523-10									

05140 | Structural Aluminum

		CREW	DAILY OUTPUT	LABOR-HOURS	UNIT	MAT.	LABOR	EQUIP.	TOTAL	TOTAL INCL O&P	
080	0010	**ALUMINUM**										080
	0020	Struct. shapes, 25 mm to 250 mm, under 1 metric ton	E-2	476	.118	kg	5.05	4.58	3	12.63	16.80	
	0050	1 to 5 metric tons		603	.093		4.83	3.61	2.37	10.81	14.15	
	0100	Over 5 metric tons		603	.093		4.70	3.61	2.37	10.68	14	
	0300	Extrusions, over 5 metric tons, stock shapes		603	.093		5.05	3.61	2.37	11.03	14.45	
	0400	Custom shapes	↓	603	.093	↓	5.20	3.61	2.37	11.18	14.55	
	0600	For formed aluminum columns, see Div. 05580-200										

05150 | Wire Rope Assemblies

		CREW	DAILY OUTPUT	LABOR-HOURS	UNIT	MAT.	LABOR	EQUIP.	TOTAL	TOTAL INCL O&P	
800	0010	**STEEL WIRE ROPE**										800
	0020	6 x 19, , bright, fiber core, 1500 m rolls, 13 mm diameter				m	2.76			2.76	3.05	
	0050	Steel core					3.64			3.64	4	
	0100	Fiber core, 25 mm diameter					9.30			9.30	10.25	
	0150	Steel core					10.65			10.65	11.70	
	0300	6 x 19, galvanized, fiber core, 13 mm diameter					4.07			4.07	4.49	
	0350	Steel core					4.66			4.66	5.10	
	0400	Fiber core, 25 mm diameter					11.95			11.95	13.10	
	0450	Steel core					12.55			12.55	13.80	
	0500	6 x 7, bright, IPS, fiber core, <152 m w/acc., 6 mm diameter	E-17	1951	.008		2.07	.34		2.41	2.86	
	0510	13 mm diameter		640	.025		5	1.02	.01	6.03	7.35	
	0520	19 mm diameter		293	.055		9.10	2.24	.02	11.36	14.05	
	0550	6 x 19, bright, IPS, IWRC, <152 m w/acc., 6 mm diameter		1756	.009		3.05	.37		3.42	4.05	
	0560	13 mm diameter		527	.030		4.95	1.24	.01	6.20	7.70	
	0570	19 mm diameter		235	.068		8.60	2.79	.02	11.41	14.50	
	0580	25 mm diameter		128	.125		14.55	5.10	.04	19.69	25.50	
	0590	31 mm diameter		88.39	.181		24	7.40	.06	31.46	40	
	0600	38 mm diameter	↓	58.52	.273		29.50	11.20	.09	40.79	52.50	
	0610	44 mm diameter	E-18	73.15	.547		47.50	21.50	12.95	81.95	104	
	0620	50 mm diameter		48.77	.820		60.50	32.50	19.40	112.40	146	
	0630	56 mm diameter	↓	48.77	.820		81.50	32.50	19.40	133.40	168	
	0650	6 x 37, bright, IPS, IWRC, <152 m w/acc., 6 mm diameter	E-17	1951	.008		3.87	.34		4.21	4.87	
	0660	13 mm diameter		527	.030		6.55	1.24	.01	7.80	9.50	
	0670	19 mm diameter		235	.068		10.65	2.79	.02	13.46	16.75	
	0680	25 mm diameter		131	.122		16.85	5	.04	21.89	27.50	
	0690	31 mm diameter		88.39	.181		25.50	7.40	.06	32.96	41.50	
	0700	38 mm diameter	↓	57.91	.276		36.50	11.30	.09	47.89	60.50	
	0710	44 mm diameter	E-18	79.25	.505		58	20	11.95	89.95	112	
	0720	50 mm diameter		60.96	.656		75	26	15.55	116.55	145	
	0730	56 mm diameter	↓	48.77	.820		99	32.50	19.40	150.90	188	
	0800	6 x 19 & 6 x 37, swaged, 13 mm diameter	E-17	372	.043		11.80	1.76	.01	13.57	16.20	
	0810	14 mm diameter		341	.047		13.70	1.92	.02	15.64	18.60	
	0820	16 mm diameter		283	.057		16.25	2.32	.02	18.59	22	
	0830	19 mm diameter		195	.082		20.50	3.36	.03	23.89	29	
	0840	22 mm diameter		146	.110		26	4.49	.04	30.53	37	
	0850	25 mm diameter		107	.150		32	6.10	.05	38.15	46	
	0860	28 mm diameter		87.78	.182		39	7.45	.06	46.51	56.50	
	0870	31 mm diameter		70.10	.228		47.50	9.35	.08	56.93	69.50	
	0880	34 mm diameter	↓	58.52	.273		55	11.20	.09	66.29	80.50	
	0890	38 mm diameter	E-18	91.44	.437	↓	66.50	17.35	10.35	94.20	115	

METALS 5

05100 | Structural Metal Framing

05150 | Wire Rope Assemblies

		CREW	DAILY OUTPUT	LABOR-HOURS	UNIT	2006 BARE COSTS MAT.	LABOR	EQUIP.	TOTAL	TOTAL INCL O&P	
800	**1500**	Thimbles, heavy duty, 6 mm	E-17	160	.100	Ea.	.54	4.10	.03	4.67	8	**800**
	1510	13 mm		160	.100		2.39	4.10	.03	6.52	10	
	1520	19 mm		105	.152		5.45	6.25	.05	11.75	17.30	
	1530	25 mm		52	.308		10.85	12.60	.10	23.55	34.50	
	1540	31 mm		38	.421		16.75	17.25	.14	34.14	49.50	
	1550	38 mm		13	1.231		47	50.50	.42	97.92	142	
	1560	44 mm		8	2		97	82	.68	179.68	255	
	1570	50 mm		6	2.667		141	109	.91	250.91	355	
	1580	56 mm		4	4		191	164	1.36	356.36	505	
	1600	Clips, 6 mm diameter		160	.100		2.45	4.10	.03	6.58	10.10	
	1610	9 mm diameter		160	.100		2.68	4.10	.03	6.81	10.35	
	1620	13 mm diameter		160	.100		4.31	4.10	.03	8.44	12.15	
	1630	19 mm diameter		102	.157		7	6.40	.05	13.45	19.30	
	1640	25 mm diameter		64	.250		11.65	10.25	.09	21.99	31.50	
	1650	31 mm diameter		35	.457		19.10	18.70	.16	37.96	54.50	
	1670	38 mm diameter		26	.615		26	25	.21	51.21	74	
	1680	44 mm diameter		16	1		60	41	.34	101.34	140	
	1690	50 mm diameter		12	1.333		67	54.50	.45	121.95	172	
	1700	56 mm diameter		10	1.600		98.50	65.50	.54	164.54	227	
	2200	Jaw & jaw turnbuckles, 6 mm x 100 mm		160	.100		19.50	4.10	.03	23.63	29	
	2250	13 mm x 150 mm		96	.167		24.50	6.85	.06	31.41	39.50	
	2260	13 mm x 225 mm		77	.208		33	8.50	.07	41.57	51.50	
	2270	13 mm x 300 mm		66	.242		37	9.95	.08	47.03	58.50	
	2300	19 mm x 150 mm		38	.421		48	17.25	.14	65.39	84	
	2310	19 mm x 225 mm		30	.533		53.50	22	.18	75.68	98	
	2320	19 mm x 300 mm		28	.571		68.50	23.50	.19	92.19	118	
	2330	19 mm x 450 mm		23	.696		82	28.50	.24	110.74	142	
	2350	25 mm x 150 mm		17	.941		93.50	38.50	.32	132.32	173	
	2360	25 mm x 300 mm		13	1.231		103	50.50	.42	153.92	204	
	2370	25 mm x 450 mm		10	1.600		154	65.50	.54	220.04	288	
	2380	25 mm x 600 mm		9	1.778		169	73	.60	242.60	320	
	2400	31 mm x 300 mm		7	2.286		172	93.50	.78	266.28	360	
	2410	31 mm x 450 mm		6.50	2.462		213	101	.84	314.84	415	
	2420	31 mm x 600 mm		5.60	2.857		287	117	.97	404.97	525	
	2450	38 mm x 300 mm		5.20	3.077		289	126	1.05	416.05	550	
	2460	38 mm x 450 mm		4	4		310	164	1.36	475.36	635	
	2470	38 mm x 600 mm		3.20	5		415	205	1.70	621.70	825	
	2500	44 mm x 450 mm		3.20	5		625	205	1.70	831.70	1,050	
	2510	44 mm x 600 mm		2.80	5.714		715	234	1.94	950.94	1,200	
	2550	50 mm x 600 mm		1.60	10		965	410	3.40	1,378.40	1,800	

05200 | Metal Joists

05210 | Steel Joists

		CREW	DAILY OUTPUT	LABOR-HOURS	UNIT	2006 BARE COSTS MAT.	LABOR	EQUIP.	TOTAL	TOTAL INCL O&P	
600	**0010**	**OPEN WEB JOISTS**									**600**
	0020	K series, 36-tonne job lots, horiz. bridging, spans to 10 m, minimum	E-7	13.61	5.879	Met. Ton	1,300	232	118	1,650	1,950	
	0050	Average		10.89	7.349		1,450	290	147	1,887	2,275	
	0080	Maximum		8.16	9.798		1,750	385	197	2,332	2,825	
	0130	8K1, 7.6 kg/m		366	.219	m	11	8.60	4.39	23.99	32	
	0140	10K1, 7.5 kg/m		366	.219		10.85	8.60	4.39	23.84	32	

5 METALS

		05210	Steel Joists	CREW	DAILY OUTPUT	LABOR-HOURS	UNIT	2006 BARE COSTS				TOTAL INCL O&P	
								MAT.	LABOR	EQUIP.	TOTAL		
600	0160		12K3, 8.5 kg/m	E-7	457	.175	m	12.35	6.90	3.51	22.76	29.50	600
	0180		14K3, 8.9 kg/m		457	.175		13	6.90	3.51	23.41	30	
	0200		16K3, 9.4 kg/m		549	.146		13.65	5.75	2.92	22.32	28.50	
	0220		16K6, 12.1 kg/m		549	.146		17.50	5.75	2.92	26.17	32.50	
	0240		18K5, 11.5 kg/m		610	.131		16.65	5.15	2.63	24.43	30.50	
	0260		18K9, 15.2 kg/m		610	.131		22	5.15	2.63	29.78	36.50	
	0410		Span 10 m to 15 m, minimum		15.42	5.187	Met. Ton	1,275	204	104	1,583	1,875	
	0440		Average		15.42	5.187		1,425	204	104	1,733	2,050	
	0460		Maximum		9.07	8.818		1,525	350	177	2,052	2,475	
	0500		20K5, 12.2 kg/m		610	.131	m	17.50	5.15	2.63	25.28	31	
	0520		20K9, 16.1 kg/m		610	.131		23	5.15	2.63	30.78	37.50	
	0540		22K5, 13.1 kg/m		610	.131		18.75	5.15	2.63	26.53	32.50	
	0560		22K9, 16.8 kg/m		610	.131		24	5.15	2.63	31.78	38.50	
	0580		24K6, 14.5 kg/m		671	.119		20.50	4.70	2.39	27.59	34	
	0600		24K10, 19.5 kg/m		671	.119		28	4.70	2.39	35.09	41.50	
	0620		26K6, 15.8 kg/m		671	.119		22.50	4.70	2.39	29.59	36	
	0640		26K10, 20.6 kg/m		671	.119		29.50	4.70	2.39	36.59	43.50	
	0660		28K8, 18.9 kg/m		732	.109		27	4.31	2.19	33.50	40	
	0680		28K12, 25.5 kg/m		732	.109		36.50	4.31	2.19	43	50	
	0700		30K8, 19.7 kg/m		732	.109		28	4.31	2.19	34.50	41	
	0720		30K12, 26.2 kg/m		732	.109		37.50	4.31	2.19	44	51.50	
	1010		CS series, horizontal bridging										
	1020		Spans to 10 m, minimum	E-7	13.61	5.879	Met. Ton	1,350	232	118	1,700	2,000	
	1040		Average		10.89	7.349		1,500	290	147	1,937	2,325	
	1060		Maximum		8.16	9.798		1,750	385	197	2,332	2,850	
	1100		10CS2, 11.2 kg/m		366	.219	m	16.65	8.60	4.39	29.64	38	
	1120		12CS2, 11.9 kg/m		457	.175		17.80	6.90	3.51	28.21	35.50	
	1140		14CS2, 11.9 kg/m		457	.175		17.80	6.90	3.51	28.21	35.50	
	1160		16CS2, 12.7 kg/m		549	.146		18.90	5.75	2.92	27.57	34.50	
	1180		16CS4, 21.6 kg/m		549	.146		32	5.75	2.92	40.67	49	
	1200		18CS2, 13.4 kg/m		610	.131		20	5.15	2.63	27.78	34	
	1220		18CS4, 22.4 kg/m		610	.131		33.50	5.15	2.63	41.28	48.50	
	1240		20CS2, 14.2 kg/m		610	.131		21	5.15	2.63	28.78	35	
	1260		20CS4, 24.6 kg/m		610	.131		36.50	5.15	2.63	44.28	52.50	
	1280		22CS2, 14.9 kg/m		610	.131		22	5.15	2.63	29.78	36.50	
	1300		22CS4, 24.6 kg/m		610	.131		36.50	5.15	2.63	44.28	52.50	
	1320		24CS2, 14.9 kg/m		671	.119		22	4.70	2.39	29.09	35.50	
	1340		24CS4, 24.6 kg/m		671	.119		36.50	4.70	2.39	43.59	51.50	
	1360		26CS2, 14.9 kg/m		671	.119		22	4.70	2.39	29.09	35.50	
	1380		26CS4, 24.6 kg/m		671	.119		36.50	4.70	2.39	43.59	51.50	
	1400		28CS2, 15.6 kg/m		732	.109		23.50	4.31	2.19	30	35.50	
	1420		28CS4, 24.6 kg/m		732	.109		36.50	4.31	2.19	43	50.50	
	1440		30CS2, 16.4 kg/m		732	.109		24.50	4.31	2.19	31	37	
	1460		30CS4, 24.6 kg/m		732	.109		36.50	4.31	2.19	43	50.50	
	2000		LH series, bolted cross bridging										
	2020		Spans to 30 m, minimum	E-7	14.52	5.511	Met. Ton	1,475	217	111	1,803	2,125	
	2040		Average		11.79	6.783		1,600	267	136	2,003	2,400	
	2080		Maximum		9.98	8.017		1,900	315	161	2,376	2,825	
	2200		18LH04, 18 kg/m		427	.187	m	29	7.40	3.76	40.16	48.50	
	2220		18LH08, 28 kg/m		427	.187		45.50	7.40	3.76	56.66	67	
	2240		20LH04, 18 kg/m		427	.187		29	7.40	3.76	40.16	48.50	
	2260		20LH08, 28 kg/m		427	.187		45.50	7.40	3.76	56.66	67	
	2280		24LH05, 19 kg/m		427	.187		31	7.40	3.76	42.16	51.50	
	2300		24LH10, 34 kg/m		427	.187		55	7.40	3.76	66.16	77.50	
	2320		28LH06, 24 kg/m		549	.146		38.50	5.75	2.92	47.17	55.50	
	2340		28LH11, 37 kg/m		549	.146		60	5.75	2.92	68.67	79.50	

METALS 5

05200 | Metal Joists

05210 | Steel Joists

		CREW	DAILY OUTPUT	LABOR-HOURS	UNIT	MAT.	LABOR	EQUIP.	TOTAL	TOTAL INCL O&P	
2360	32LH08, 25 kg/m	E-7	549	.146	m	41	5.75	2.92	49.67	58.50	600
2380	32LH13, 45 kg/m		549	.146		72	5.75	2.92	80.67	92.50	
2400	36LH09, 31 kg/m		549	.146		50.50	5.75	2.92	59.17	69	
2420	36LH14, 54 kg/m		549	.146		86.50	5.75	2.92	95.17	108	
2440	40LH10, 31 kg/m		671	.119		50.50	4.70	2.39	57.59	66.50	
2460	44LH11, 54 kg/m		671	.119		86.50	4.70	2.39	93.59	106	
2480	44LH11, 33 kg/m		671	.119		53	4.70	2.39	60.09	69	
2500	44LH16, 63 kg/m		671	.119		101	4.70	2.39	108.09	122	
2520	48LH11, 33 kg/m		671	.119		53	4.70	2.39	60.09	69	
2540	48LH16, 63 kg/m		671	.119		101	4.70	2.39	108.09	122	
3010	DLH series, bolted cross bridging										
3020	Spans to 45 m (shipped in 2 pieces), minimum	E-7	14.52	5.511	Met. Ton	1,600	217	111	1,928	2,250	
3040	Average		11.79	6.783		1,725	267	136	2,128	2,500	
3100	Maximum		9.98	8.017		2,075	315	161	2,551	3,000	
3200	52DLH11, 39 kg/m		610	.131	m	64	5.15	2.63	71.78	82	
3220	52DLH16, 67 kg/m		610	.131		115	5.15	2.63	122.78	138	
3240	56DLH11, 39 kg/m		610	.131		66.50	5.15	2.63	74.28	85	
3260	56DLH16, 69 kg/m		610	.131		117	5.15	2.63	124.78	141	
3280	60DLH12, 43 kg/m		610	.131		74	5.15	2.63	81.78	93.50	
3300	60DLH17, 77 kg/m		610	.131		133	5.15	2.63	140.78	158	
3320	64DLH12, 46 kg/m		671	.119		79	4.70	2.39	86.09	98	
3340	64DLH17, 77 kg/m		671	.119		133	4.70	2.39	140.09	157	
3360	68DLH13, 55 kg/m		671	.119		94.50	4.70	2.39	101.59	115	
3380	68DLH18, 91 kg/m		671	.119		156	4.70	2.39	163.09	182	
3400	72DLH14, 61 kg/m		671	.119		105	4.70	2.39	112.09	126	
3420	72DLH19, 104 kg/m		671	.119		178	4.70	2.39	185.09	207	
4010	SLH series, bolted cross bridging										
4020	Spans to 60 m (shipped in 3 pieces), minimum	E-7	14.52	5.511	Met. Ton	1,575	217	111	1,903	2,225	
4040	Average		11.79	6.783		1,750	267	136	2,153	2,550	
4060	Maximum		9.98	8.017		2,100	315	161	2,576	3,025	
4200	80SLH15, 60 kg/m		457	.175	m	105	6.90	3.51	115.41	131	
4220	80SLH20, 112 kg/m		457	.175		197	6.90	3.51	207.41	232	
4240	88SLH16, 69 kg/m		457	.175		121	6.90	3.51	131.41	149	
4260	88SLH21, 133 kg/m		457	.175		233	6.90	3.51	243.41	273	
4280	96SLH17, 77 kg/m		457	.175		136	6.90	3.51	146.41	166	
4300	96SLH22, 152 kg/m		457	.175		267	6.90	3.51	277.41	310	
4320	104SLH18, 88 kg/m		549	.146		155	5.75	2.92	163.67	183	
4340	104SLH23, 162 kg/m		549	.146		286	5.75	2.92	294.67	330	
4360	112SLH19, 100 kg/m		549	.146		176	5.75	2.92	184.67	206	
4380	112SLH24, 195 kg/m		549	.146		345	5.75	2.92	353.67	395	
4400	120SLH20, 115 kg/m		549	.146		202	5.75	2.92	210.67	235	
4420	120SLH25, 226 kg/m		549	.146		400	5.75	2.92	408.67	455	
6000	For welded cross bridging, add						30%				
6100	For less than 36-tonne job lots										
6102	For 27 to 35 tonnes, add					10%					
6104	18 to 26 tonnes, add					20%					
6106	9 to 17 tonnes, add					30%					
6107	5 to 8 tonnes, add					50%	25%				
6108	1 to 4 tonnes, add					75%	50%				
6109	Less than 1 tonne, add					100%	100%				
6200	For shop prime paint other than mfrs. standard, add					20%					
6300	For bottom chord extensions, add/chord				Ea.	28.50			28.50	31.50	
7000	Joist girders, minimum	E-5	13.61	5.879	Met. Ton	1,325	232	111	1,668	1,975	
7020	Average		11.79	6.783		1,450	267	129	1,846	2,200	
7040	Maximum		9.98	8.017		1,525	315	152	1,992	2,400	
8000	Trusses, factory fabricated WT chords, average		9.98	8.017		4,750	315	152	5,217	5,975	

Important: See the Reference Section for supporting data - Crews, Rental Equipment, City Cost Indexes and Reference Data

			CREW	DAILY OUTPUT	LABOR-HOURS	UNIT	MAT.	LABOR	EQUIP.	TOTAL	TOTAL INCL O&P
05310	**Steel Deck**							**2006 BARE COSTS**			
300	0010	**METAL DECKING** Steel decking	R053100 -10									300
0200	Cellular units, galv, 51 mm deep, 20-20 gauge [2"]		E-4	136	.235	m²	63	9.50	.66	73.16	87.50
0250	18-20 gauge			132	.242		71.50	9.80	.68	81.98	97.50
0300	18-18 gauge			129	.248		73.50	10.05	.69	84.24	100
0320	16-18 gauge			126	.254		87.50	10.25	.71	98.46	115
0340	16-16 gauge			124	.258		97.50	10.45	.72	108.67	127
0400	75 mm deep, galvanized, 20-20 gauge			128	.250		69.50	10.10	.70	80.30	95
0500	18-20 gauge			125	.256		83.50	10.35	.71	94.56	111
0600	18-18 gauge			120	.267		83.50	10.80	.74	95.04	112
0700	16-18 gauge			114	.281		94	11.35	.78	106.13	124
0800	16-16 gauge			107	.299		103	12.10	.83	115.93	136
1000	113 mm deep, galvanized, 20-18 gauge			102	.314		96.50	12.70	.88	110.08	130
1100	18-18 gauge			96.62	.331		96	13.40	.92	110.32	131
1200	16-18 gauge			91.04	.351		108	14.20	.98	123.18	146
1300	16-16 gauge			86.86	.368		118	14.90	1.03	133.93	158
1500	For acoustical deck, add						15%				
1700	For cells used for ventilation, add						15%				
1900	For multi-story or congested site, add							50%			
2100	Open type, galv., 38 mm deep, 22 gauge, under 450 m²		E-4	418	.077	m²	15.80	3.10	.21	19.11	23
2200	450-4500 m²			455	.070		12.25	2.84	.20	15.29	18.75
2400	Over 4500 m²			474	.068		11.40	2.73	.19	14.32	17.60
2600	20 gauge, under 450 m²			359	.089		18.60	3.61	.25	22.46	27.50
2650	450-4500 m²			387	.083		14.85	3.34	.23	18.42	22.50
2700	Over 4500 m²			399	.080		13.35	3.24	.22	16.81	21
2900	18 gauge, under 450 m²			353	.091		24	3.67	.25	27.92	33.50
2950	450-4500 m²			381	.084		19.25	3.40	.23	22.88	27.50
3000	Over 4500 m²			399	.080		17.35	3.24	.22	20.81	25
3050	16 gauge, under 450 m²			344	.093		32.50	3.76	.26	36.52	42.50
3060	450-4500 m²			372	.086		26	3.48	.24	29.72	35
3100	Over 4500 m²			390	.082		23.50	3.32	.23	27.05	31.50
3150	For intermediate rib instead of wide rib, deduct						.32			.32	.32
3160	For narrow rib instead of wide rib, add						4.84			4.84	5.40
3200	75 mm deep, 22 gauge, under 450 m²		E-4	334	.096		21.50	3.88	.27	25.65	31.50
3250	450-4500 m²			353	.091		17.45	3.67	.25	21.37	26
3260	over 4500 m²			372	.086		15.70	3.48	.24	19.42	23.50
3300	20 gauge, under 450 m²			316	.101		25.50	4.10	.28	29.88	35.50
3350	450-4500 m²			334	.096		20	3.88	.27	24.15	30
3360	over 4500 m²			353	.091		18.30	3.67	.25	22.22	27
3400	18 gauge, under 450 m²			297	.108		33	4.36	.30	37.66	44
3450	450-4500 m²			316	.101		26.50	4.10	.28	30.88	36.50
3460	over 4500 m²			334	.096		23.50	3.88	.27	27.65	33.50
3500	16 gauge, under 450 m²			279	.115		43.50	4.64	.32	48.46	56
3550	450-4500 m²			297	.108		34.50	4.36	.30	39.16	46
3560	over 4500 m²			316	.101		31	4.10	.28	35.38	41.50
3700	113 mm deep, long span roof, over 450 m², 20 gauge			251	.127		40.50	5.15	.36	46.01	54.50
3800	18 gauge			229	.140		52.50	5.65	.39	58.54	68.50
3900	16 gauge			218	.147		39	5.95	.41	45.36	54
4100	150 mm deep, long span, 18 gauge			186	.172		75	6.95	.48	82.43	95.50
4200	16 gauge			179	.179		56	7.25	.50	63.75	75
4300	14 gauge			173	.185		72	7.50	.52	80.02	93.50
4500	188 mm deep, long span, 18 gauge			157	.204		82.50	8.25	.57	91.32	106
4600	16 gauge			148	.216		61.50	8.75	.60	70.85	84
4700	14 gauge			138	.232		79.50	9.40	.65	89.55	105
4800	For painted instead of galvanized, deduct						2%				
5000	For acoustical perforated, with fiberglass, add					m²	10.75			10.75	11.85
5200	Non-cellular composite deck, galv., 50 mm deep, 22 gauge		E-4	359	.089		15.05	3.61	.25	18.91	23.50

METALS 5

05310 | Steel Deck

			CREW	DAILY OUTPUT	LABOR-HOURS	UNIT	MAT.	LABOR	EQUIP.	TOTAL	TOTAL INCL O&P	
							2006 BARE COSTS					
300	5300	20 gauge R053100-10	E-4	334	.096	m²	16.70	3.88	.27	20.85	25.50	**300**
	5400	18 gauge		314	.102		21	4.12	.28	25.40	31	
	5500	16 gauge		297	.108		26.50	4.36	.30	31.16	37	
	5700	75 mm deep, galv., 22 gauge		297	.108		16.45	4.36	.30	21.11	26.50	
	5800	20 gauge		279	.115		18.40	4.64	.32	23.36	28.50	
	5900	18 gauge		265	.121		22.50	4.88	.34	27.72	34	
	6000	16 gauge		251	.127		30	5.15	.36	35.51	42.50	
	6100	Slab form, steel, 28 gauge, 14 mm deep, uncoated		372	.086		10.55	3.48	.24	14.27	18.15	
	6200	Galvanized		372	.086		9.35	3.48	.24	13.07	16.85	
	6220	24 gauge, 25 mm deep, uncoated		362	.088		11.50	3.58	.25	15.33	19.40	
	6240	Galvanized		362	.088		13.55	3.58	.25	17.38	21.50	
	6300	24 gauge, 33 mm deep, uncoated		353	.091		12.25	3.67	.25	16.17	20.50	
	6400	Galvanized		353	.091		14.40	3.67	.25	18.32	22.50	
	6500	22 gauge, 33 mm deep, uncoated		344	.093		15.40	3.76	.26	19.42	24	
	6600	Galvanized		344	.093		15.70	3.76	.26	19.72	24.50	
	6700	22 gauge, 50 mm deep uncoated		334	.096		20.50	3.88	.27	24.65	30	
	6800	Galvanized		334	.096		19.90	3.88	.27	24.05	29.50	
	7000	Sheet metal edge closure form, 310 mm wide with 2 bends, galv										
	7100	18 gauge	E-14	110	.073	m	9.95	3.05	.81	13.81	17.35	
	7200	16 gauge	"	110	.073	"	13.50	3.05	.81	17.36	21	
	8000	Metal deck and trench, 51 mm thick, 20 gauge, combination										
	8010	60% cellular, 40% non-cellular, inserts and trench	R-4	102	.392	m²	126	16	.87	142.87	168	

05400 | Cold-Formed Metal Framing

05410 | Load-Bearing Metal Studs

			CREW	DAILY OUTPUT	LABOR-HOURS	UNIT	MAT.	LABOR	EQUIP.	TOTAL	TOTAL INCL O&P	
							2006 BARE COSTS					
100	0010	**BRACING**, shear wall X-bracing, per 3 m x 3 m bay, one face										**100**
	0120	Metal strap, 20 ga x 100 mm wide	2 Carp	18	.889	Ea.	18	31.50		49.50	69	
	0130	150 mm wide		18	.889		28	31.50		59.50	79.50	
	0160	18 ga x 100 mm wide		16	1		26	35.50		61.50	84	
	0170	150 mm wide		16	1		38.50	35.50		74	97.50	
	0410	Continuous strap bracing, per horizontal row on both faces										
	0420	Metal strap, 20 ga x 50 mm wide, studs 300 mm O.C.	1 Carp	213	.038	m	1.52	1.34		2.86	3.75	
	0430	400 mm O.C.		244	.033		1.52	1.17		2.69	3.48	
	0440	600 mm O.C.		305	.026		1.52	.93		2.45	3.12	
	0450	18 ga x 50 mm wide, studs 300 mm O.C.		183	.044		2.09	1.55		3.64	4.72	
	0460	400 mm O.C.		213	.038		2.09	1.34		3.43	4.38	
	0470	600 mm O.C.		244	.033		2.09	1.17		3.26	4.11	
120	0010	**BRIDGING**, solid between studs w/ 31 mm leg track, per stud bay										**120**
	0200	Studs 100 mm O.C., 18 ga x 63 mm wide	1 Carp	125	.064	Ea.	.77	2.28		3.05	4.39	
	0210	91 mm wide		120	.067		.93	2.37		3.30	4.71	
	0220	100 mm wide		120	.067		.99	2.37		3.36	4.78	
	0230	150 mm wide		115	.070		1.29	2.47		3.76	5.25	
	0240	200 mm wide		110	.073		1.63	2.59		4.22	5.85	
	0300	16 ga x 63 mm wide		115	.070		.97	2.47		3.44	4.91	
	0310	91 mm wide		110	.073		1.18	2.59		3.77	5.35	
	0320	100 mm wide		110	.073		1.26	2.59		3.85	5.40	
	0330	150 mm wide		105	.076		1.61	2.71		4.32	6	
	0340	200 mm wide		100	.080		2.06	2.84		4.90	6.70	
	1200	Studs 400 mm O.C., 18 ga x 63 mm wide		125	.064		.99	2.28		3.27	4.63	

5 **METALS**

05410 | Load-Bearing Metal Studs

		CREW	DAILY OUTPUT	LABOR-HOURS	UNIT	2006 BARE COSTS				TOTAL INCL O&P
						MAT.	LABOR	EQUIP.	TOTAL	
120										**120**
1210	91 mm wide	1 Carp	120	.067	Ea.	1.19	2.37		3.56	5
1220	100 mm wide		120	.067		1.27	2.37		3.64	5.10
1230	150 mm wide		115	.070		1.65	2.47		4.12	5.65
1240	200 mm wide		110	.073		2.10	2.59		4.69	6.35
1300	16 ga x 63 mm wide		115	.070		1.24	2.47		3.71	5.20
1310	91 mm wide		110	.073		1.52	2.59		4.11	5.70
1320	100 mm wide		110	.073		1.62	2.59		4.21	5.80
1330	150 mm wide		105	.076		2.06	2.71		4.77	6.50
1340	200 mm wide		100	.080		2.64	2.84		5.48	7.35
2200	Studs 600 mm O.C., 18 ga x 63 mm wide		125	.064		1.43	2.28		3.71	5.10
2210	91 mm wide		120	.067		1.72	2.37		4.09	5.60
2220	100 mm wide		120	.067		1.84	2.37		4.21	5.70
2230	150 mm wide		115	.070		2.39	2.47		4.86	6.50
2240	200 mm wide		110	.073		3.03	2.59		5.62	7.35
2300	16 ga x 63 mm wide		115	.070		1.79	2.47		4.26	5.80
2310	91 mm wide		110	.073		2.20	2.59		4.79	6.45
2320	100 mm wide		110	.073		2.34	2.59		4.93	6.60
2330	150 mm wide		105	.076		2.98	2.71		5.69	7.50
2340	200 mm wide	▼	100	.080	▼	3.82	2.84		6.66	8.65
3000	Continuous bridging, per row									
3100	16 ga x 38 mm channel thru studs 300 mm O.C.	1 Carp	183	.044	m	1.42	1.55		2.97	3.98
3110	16 ga x 38 mm channel thru studs 400 mm O.C.		213	.038		1.42	1.34		2.76	3.64
3120	16 ga x 38 mm channel thru studs 600 mm O.C.		268	.030		1.42	1.06		2.48	3.21
4100	50 x 50 mm angle x 18 ga, studs 300 mm O.C.		213	.038		2.06	1.34		3.40	4.34
4110	400 mm O.C.		274	.029		2.06	1.04		3.10	3.88
4120	600 mm O.C.		366	.022		2.06	.78		2.84	3.47
4200	16 ga, studs 300 mm O.C.		152	.053		2.63	1.87		4.50	5.80
4210	400 mm O.C.		213	.038		2.63	1.34		3.97	4.98
4220	600 mm O.C.	▼	305	.026	▼	2.63	.93		3.56	4.35
300										**300**
0010	**FRAMING, BOXED HEADERS/BEAMS**									
0200	Double, 18 ga x 150 mm deep	2 Carp	67.06	.239	m	14.75	8.50		23.25	29.50
0210	200 mm deep		64.01	.250		16.35	8.90		25.25	32
0220	250 mm deep		60.96	.262		19.80	9.35		29.15	36.50
0230	300 mm deep		57.91	.276		22	9.80		31.80	39.50
0300	16 ga x 200 mm deep		54.86	.292		18.80	10.35		29.15	36.50
0310	250 mm deep		51.82	.309		22.50	11		33.50	42
0320	300 mm deep		48.77	.328		24.50	11.65		36.15	45
0400	14 ga x 250 mm deep		42.67	.375		26	13.35		39.35	50
0410	300 mm deep		39.62	.404		29	14.35		43.35	54
1210	Triple, 18 ga x 200 mm deep		51.82	.309		23.50	11		34.50	43
1220	250 mm deep		50.29	.318		28.50	11.30		39.80	48.50
1230	300 mm deep		48.77	.328		31.50	11.65		43.15	52.50
1300	16 ga x 200 mm deep		44.20	.362		27.50	12.85		40.35	50
1310	250 mm deep		42.67	.375		32.50	13.35		45.85	57
1320	300 mm deep		41.15	.389		35.50	13.80		49.30	60.50
1400	14 ga x 250 mm deep		35.05	.456		36	16.25		52.25	65
1410	300 mm deep	▼	33.53	.477	▼	39.50	16.95		56.45	70
400										**400**
0010	**FRAMING, STUD WALLS** w/ top & bottom track, no openings,									
0020	headers, beams, bridging or bracing									
4100	2400 mm high walls, 18 ga x 63 mm wide, studs 300 mm O.C.	2 Carp	16.46	.972	m	25	34.50		59.50	81.50
4110	400 mm O.C.		23.47	.682		19.80	24		43.80	59.50
4120	600 mm O.C.		32.61	.491		14.80	17.45		32.25	43.50
4130	91 mm wide, studs 300 mm O.C.		16.15	.990		29.50	35		64.50	87.50
4140	400 mm O.C.		23.16	.691		23.50	24.50		48	64
4150	600 mm O.C.	▼	32	.500	▼	17.60	17.75		35.35	47

METALS 5

05410 | Load-Bearing Metal Studs

		CREW	DAILY OUTPUT	LABOR-HOURS	UNIT	2006 BARE COSTS				TOTAL INCL O&P
						MAT.	LABOR	EQUIP.	TOTAL	
400	4160	100 mm wide, studs 300 mm O.C.	2 Carp	15.85	1.009	m	31	36		67	90
	4170	400 mm O.C.		22.56	.709		24.50	25		49.50	66.50
	4180	600 mm O.C.		31.39	.510		18.55	18.10		36.65	48.50
	4190	150 mm wide, studs 300 mm O.C.		15.54	1.029		39	36.50		75.50	100
	4200	400 mm O.C.		22.25	.719		31.50	25.50		57	74.50
	4210	600 mm O.C.		30.78	.520		23.50	18.50		42	55
	4220	200 mm wide, studs 300 mm O.C.		15.24	1.050		47.50	37.50		85	111
	4230	400 mm O.C.		21.95	.729		38.50	26		64.50	82.50
	4240	600 mm O.C.		30.48	.525		29	18.65		47.65	61
	4300	16 ga x 63 mm wide, studs 300 mm O.C.		14.33	1.117		29	39.50		68.50	94
	4310	400 mm O.C.		20.73	.772		23	27.50		50.50	67.50
	4320	600 mm O.C.		28.65	.558		16.90	19.85		36.75	49.50
	4330	91 mm wide, studs 300 mm O.C.		14.02	1.141		34.50	40.50		75	101
	4340	400 mm O.C.		20.12	.795		27.50	28.50		56	74
	4350	600 mm O.C.		28.04	.571		20	20.50		40.50	53.50
	4360	100 mm wide, studs 300 mm O.C.		13.72	1.167		36.50	41.50		78	105
	4370	400 mm O.C.		19.81	.808		29	28.50		57.50	76.50
	4380	600 mm O.C.		27.43	.583		21.50	20.50		42	56
	4390	150 mm wide, studs 300 mm O.C.		13.41	1.193		46	42.50		88.50	117
	4400	400 mm O.C.		19.51	.820		36.50	29		65.50	86
	4410	600 mm O.C.		26.82	.597		27	21		48	63
	4420	200 mm wide, studs 300 mm O.C.		13.11	1.221		56.50	43.50		100	130
	4430	400 mm O.C.		19.20	.833		45	29.50		74.50	95.50
	4440	600 mm O.C.		26.21	.610		33.50	21.50		55	71
	5100	3000 mm high walls, 18 ga x 63 mm wide, studs 300 mm O.C.		16.46	.972		30	34.50		64.50	86.50
	5110	400 mm O.C.		23.47	.682		23.50	24		47.50	63.50
	5120	600 mm O.C.		32.61	.491		17.30	17.45		34.75	46
	5130	91 mm wide, studs 300 mm O.C.		16.15	.990		35.50	35		70.50	94
	5140	400 mm O.C.		23.16	.691		28	24.50		52.50	68.50
	5150	600 mm O.C.		32	.500		20.50	17.75		38.25	50
	5160	100 mm wide, studs 300 mm O.C.		15.85	1.009		37	36		73	97
	5170	400 mm O.C.		22.56	.709		29.50	25		54.50	72
	5180	600 mm O.C.		31.39	.510		21.50	18.10		39.60	52
	5190	150 mm wide, studs 300 mm O.C.		15.54	1.029		47	36.50		83.50	109
	5200	400 mm O.C.		22.25	.719		37	25.50		62.50	81
	5210	600 mm O.C.		30.78	.520		27.50	18.50		46	59
	5220	200 mm wide, studs 300 mm O.C.		15.24	1.050		57	37.50		94.50	121
	5230	400 mm O.C.		21.95	.729		45.50	26		71.50	90.50
	5240	600 mm O.C.		30.48	.525		33.50	18.65		52.15	66
	5300	16 ga x 63 mm wide, studs 300 mm O.C.		14.33	1.117		35	39.50		74.50	101
	5310	400 mm O.C.		20.73	.772		27.50	27.50		55	72.50
	5320	600 mm O.C.		28.65	.558		19.90	19.85		39.75	53
	5330	91 mm wide, studs 300 mm O.C.		14.02	1.141		42	40.50		82.50	109
	5340	400 mm O.C.		20.12	.795		33	28.50		61.50	80
	5350	600 mm O.C.		28.04	.571		24	20.50		44.50	57.50
	5360	100 mm wide, studs 300 mm O.C.		13.72	1.167		44.50	41.50		86	113
	5370	400 mm O.C.		19.81	.808		34.50	28.50		63	82.50
	5380	600 mm O.C.		27.43	.583		25	20.50		45.50	60.50
	5390	150 mm wide, studs 300 mm O.C.		13.41	1.193		55.50	42.50		98	127
	5400	400 mm O.C.		19.51	.820		43.50	29		72.50	93.50
	5410	600 mm O.C.		26.82	.597		32	21		53	68
	5420	200 mm wide, studs 300 mm O.C.		13.11	1.221		68.50	43.50		112	143
	5430	400 mm O.C.		19.20	.833		54	29.50		83.50	105
	5440	600 mm O.C.		26.21	.610		39	21.50		60.50	77
	6190	3600 mm high walls, 18 ga x 150 mm wide, studs 300 mm O.C.		12.50	1.280		54.50	45.50		100	131
	6200	400 mm O.C.		17.68	.905		43	32		75	97

	05410	Load-Bearing Metal Studs	CREW	DAILY OUTPUT	LABOR-HOURS	UNIT	MAT.	LABOR	EQUIP.	TOTAL	TOTAL INCL O&P	
400	6210	600 mm O.C.	2 Carp	24.69	.648	m	31.50	23		54.50	70.50	**400**
	6220	200 mm wide, studs 300 mm O.C.		12.19	1.312		66.50	46.50		113	146	
	6230	400 mm O.C.		17.37	.921		52.50	32.50		85	109	
	6240	600 mm O.C.		24.38	.656		38.50	23.50		62	78.50	
	6390	16 ga x 150 mm wide, studs 300 mm O.C.		10.67	1.500		65	53.50		118.50	155	
	6400	400 mm O.C.		15.54	1.029		51	36.50		87.50	113	
	6410	600 mm O.C.		21.34	.750		36.50	26.50		63	82	
	6420	200 mm wide, studs 300 mm O.C.		10.36	1.544		80	55		135	174	
	6430	400 mm O.C.		15.24	1.050		62.50	37.50		100	127	
	6440	600 mm O.C.		21.03	.761		45	27		72	91.50	
	6530	14 ga x 91 mm wide, studs 300 mm O.C.		10.36	1.544		61.50	55		116.50	154	
	6540	400 mm O.C.		14.63	1.094		48	39		87	114	
	6550	600 mm O.C.		19.81	.808		34.50	28.50		63	82.50	
	6560	100 mm wide, studs 300 mm O.C.		10.06	1.591		65.50	56.50		122	160	
	6570	400 mm O.C.		14.33	1.117		51	39.50		90.50	118	
	6580	600 mm O.C.		19.51	.820		36.50	29		65.50	86	
	6730	12 ga x 91 mm wide, studs 300 mm O.C.		9.45	1.693		85.50	60		145.50	188	
	6740	400 mm O.C.		13.11	1.221		66	43.50		109.50	140	
	6750	600 mm O.C.		17.98	.890		46.50	31.50		78	101	
	6760	100 mm wide, studs 300 mm O.C.		9.14	1.750		91.50	62		153.50	198	
	6770	400 mm O.C.		12.80	1.250		71	44.50		115.50	147	
	6780	600 mm O.C.		17.68	.905		50	32		82	105	
	7390	4800 mm high walls, 16 ga x 150 mm wide, studs 300 mm O.C.		10.06	1.591		84	56.50		140.50	181	
	7400	400 mm O.C.		14.63	1.094		65	39		104	132	
	7410	600 mm O.C.		20.42	.783		46	28		74	94	
	7420	200 mm wide, studs 300 mm O.C.		9.75	1.640		103	58.50		161.50	204	
	7430	400 mm O.C.		14.33	1.117		80	39.50		119.50	150	
	7440	600 mm O.C.		20.12	.795		56.50	28.50		85	107	
	7560	14 ga x 100 mm wide, studs 300 mm O.C.		9.45	1.693		84.50	60		144.50	187	
	7570	400 mm O.C.		13.72	1.167		65.50	41.50		107	137	
	7580	600 mm O.C.		18.59	.861		46	30.50		76.50	98.50	
	7590	150 mm wide, studs 300 mm O.C.		9.14	1.750		106	62		168	214	
	7600	400 mm O.C.		13.41	1.193		82	42.50		124.50	157	
	7610	600 mm O.C.		18.29	.875		58	31		89	113	
	7760	12 ga x 100 mm wide, studs 300 mm O.C.		8.84	1.810		120	64.50		184.50	232	
	7770	400 mm O.C.		12.19	1.312		91.50	46.50		138	174	
	7780	600 mm O.C.		16.76	.954		64	34		98	123	
	7790	150 mm wide, studs 300 mm O.C.		8.53	1.875		151	66.50		217.50	270	
	7800	400 mm O.C.		11.89	1.346		116	48		164	202	
	7810	600 mm O.C.		16.46	.972		80.50	34.50		115	143	
	8590	6000 mm high walls, 14 ga x 150 mm wide, studs 300 mm O.C.		8.84	1.810		130	64.50		194.50	243	
	8600	400 mm O.C.		12.80	1.250		100	44.50		144.50	179	
	8610	600 mm O.C.		17.37	.921		70	32.50		102.50	128	
	8620	200 mm wide, studs 300 mm O.C.		8.53	1.875		159	66.50		225.50	279	
	8630	400 mm O.C.		12.50	1.280		123	45.50		168.50	206	
	8640	600 mm O.C.		17.07	.937		86	33.50		119.50	147	
	8790	12 ga x 150 mm wide, studs 300 mm O.C.		8.23	1.944		186	69		255	310	
	8800	400 mm O.C.		11.28	1.419		142	50.50		192.50	235	
	8810	600 mm O.C.		15.54	1.029		98	36.50		134.50	165	
	8820	200 mm wide, studs 300 mm O.C.		7.92	2.019		227	72		299	360	
	8830	400 mm O.C.		10.97	1.458		173	52		225	272	
	8840	600 mm O.C.		15.24	1.050		120	37.50		157.50	190	

	05420	Cold-Formed Metal Joists										
100	0010	**BRACING**, continuous, per row, top & bottom										**100**
	0120	Flat strap, 20 ga x 50 mm wide, joists at 300 mm O.C.	1 Carp	142	.056	m	1.58	2		3.58	4.86	

5

METALS

229

		05420	Cold-Formed Metal Joists	CREW	DAILY OUTPUT	LABOR-HOURS	UNIT	2006 BARE COSTS				TOTAL INCL O&P	
								MAT.	LABOR	EQUIP.	TOTAL		
100	0130		400 mm O.C.	1 Carp	162	.049	m	1.53	1.76		3.29	4.41	**100**
	0140		600 mm O.C.		203	.039		1.47	1.40		2.87	3.80	
	0150		18 ga x 50 mm wide, joists at 300 mm O.C.		122	.066		2.07	2.33		4.40	5.90	
	0160		400 mm O.C.		142	.056		2.04	2		4.04	5.35	
	0170		600 mm O.C.	▼	162	.049	▼	2	1.76		3.76	4.93	
120	0010		**BRIDGING**, solid between joists w/ 31 mm leg track, per joist bay										**120**
	0230		Joists 300 mm O.C., 18 ga track x 150 mm wide	1 Carp	80	.100	Ea.	1.29	3.56		4.85	6.95	
	0240		200 mm wide		75	.107		1.63	3.79		5.42	7.70	
	0250		250 mm wide		70	.114		2.02	4.06		6.08	8.55	
	0260		300 mm wide		65	.123		2.33	4.38		6.71	9.35	
	0330		16 ga track x 150 mm wide		70	.114		1.61	4.06		5.67	8.10	
	0340		200 mm wide		65	.123		2.06	4.38		6.44	9.05	
	0350		250 mm wide		60	.133		2.54	4.74		7.28	10.20	
	0360		300 mm wide		55	.145		2.91	5.15		8.06	11.25	
	0440		14 ga track x 200 mm wide		60	.133		2.59	4.74		7.33	10.25	
	0450		250 mm wide		55	.145		3.18	5.15		8.33	11.55	
	0460		300 mm wide		50	.160		3.68	5.70		9.38	12.90	
	0550		12 ga track x 250 mm wide		45	.178		4.67	6.30		10.97	15	
	0560		300 mm wide		40	.200		5.30	7.10		12.40	16.85	
	1230		400 mm O.C., 18 ga track x 150 mm wide		80	.100		1.65	3.56		5.21	7.35	
	1240		200 mm wide		75	.107		2.10	3.79		5.89	8.20	
	1250		250 mm wide		70	.114		2.59	4.06		6.65	9.20	
	1260		300 mm wide		65	.123		2.99	4.38		7.37	10.10	
	1330		16 ga track x 150 mm wide		70	.114		2.06	4.06		6.12	8.60	
	1340		200 mm wide		65	.123		2.64	4.38		7.02	9.70	
	1350		250 mm wide		60	.133		3.25	4.74		7.99	11	
	1360		300 mm wide		55	.145		3.73	5.15		8.88	12.15	
	1440		14 ga track x 200 mm wide		60	.133		3.32	4.74		8.06	11.05	
	1450		250 mm wide		55	.145		4.08	5.15		9.23	12.55	
	1460		300 mm wide		50	.160		4.72	5.70		10.42	14.05	
	1550		12 ga track x 250 mm wide		45	.178		6	6.30		12.30	16.45	
	1560		300 mm wide		40	.200		6.80	7.10		13.90	18.50	
	2230		600 mm O.C., 18 ga track x 150 mm wide		80	.100		2.39	3.56		5.95	8.20	
	2240		200 mm wide		75	.107		3.03	3.79		6.82	9.25	
	2250		250 mm wide		70	.114		3.75	4.06		7.81	10.45	
	2260		300 mm wide		65	.123		4.32	4.38		8.70	11.55	
	2330		16 ga track x 150 mm wide		70	.114		2.98	4.06		7.04	9.65	
	2340		200 mm wide		65	.123		3.82	4.38		8.20	11	
	2350		250 mm wide		60	.133		4.70	4.74		9.44	12.55	
	2360		300 mm wide		55	.145		5.40	5.15		10.55	14	
	2440		14 ga track x 200 mm wide		60	.133		4.80	4.74		9.54	12.70	
	2450		250 mm wide		55	.145		5.90	5.15		11.05	14.55	
	2460		300 mm wide		50	.160		6.85	5.70		12.55	16.35	
	2550		12 ga track x 250 mm wide		45	.178		8.65	6.30		14.95	19.40	
	2560		300 mm wide	▼	40	.200	▼	9.80	7.10		16.90	22	
200	0010		**FRAMING, BAND JOIST** (track) fastened to bearing wall										**200**
	0220		18 ga track x 150 mm deep	2 Carp	305	.052	m	3.44	1.87		5.31	6.70	
	0230		200 mm deep		280	.057		4.36	2.03		6.39	8	
	0240		250 mm deep		262	.061		5.40	2.17		7.57	9.35	
	0320		16 ga track x 150 mm deep		274	.058		4.30	2.08		6.38	7.95	
	0330		200 mm deep		256	.063		5.50	2.22		7.72	9.50	
	0340		250 mm deep		238	.067		6.80	2.39		9.19	11.20	
	0350		300 mm deep		226	.071		7.80	2.52		10.32	12.45	
	0430		14 ga track x 200 mm deep		229	.070		6.90	2.48		9.38	11.45	
	0440		250 mm deep	▼	219	.073	▼	8.50	2.60		11.10	13.40	

Important: See the Reference Section for supporting data - Crews, Rental Equipment, City Cost Indexes and Reference Data

			CREW	DAILY OUTPUT	LABOR-HOURS	UNIT	2006 BARE COSTS				TOTAL INCL O&P	
05420		**Cold-Formed Metal Joists**					MAT.	LABOR	EQUIP.	TOTAL		
200	0450	300 mm deep	2 Carp	213	.075	m	9.85	2.67		12.52	15	**200**
	0540	12 ga track x 250 mm deep		204	.078		12.50	2.79		15.29	18.10	
	0550	300 mm deep	↓	198	.081	↓	14.15	2.87		17.02	20	
300	0010	**FRAMING, BOXED HEADERS/BEAMS**										**300**
	0200	Double, 18 ga x 150 mm deep	2 Carp	67.06	.239	m	14.75	8.50		23.25	29.50	
	0210	200 mm deep		64.01	.250		16.35	8.90		25.25	32	
	0220	250 mm deep		60.96	.262		19.80	9.35		29.15	36.50	
	0230	300 mm deep		57.91	.276		22	9.80		31.80	39.50	
	0300	16 ga x 200 mm deep		54.86	.292		18.80	10.35		29.15	36.50	
	0310	250 mm deep		51.82	.309		22.50	11		33.50	42	
	0320	300 mm deep		48.77	.328		24.50	11.65		36.15	45	
	0400	14 ga x 250 mm deep		42.67	.375		26	13.35		39.35	50	
	0410	300 mm deep		39.62	.404		29	14.35		43.35	54	
	0500	12 ga x 250 mm deep		33.53	.477		34.50	16.95		51.45	64.50	
	0510	300 mm deep		30.48	.525		38.50	18.65		57.15	71.50	
	1210	Triple, 18 ga x 200 mm deep		51.82	.309		23.50	11		34.50	43	
	1220	250 mm deep		50.29	.318		28.50	11.30		39.80	48.50	
	1230	300 mm deep		48.77	.328		31.50	11.65		43.15	52.50	
	1300	16 ga x 200 mm deep		44.20	.362		27.50	12.85		40.35	50	
	1310	250 mm deep		42.67	.375		32.50	13.35		45.85	57	
	1320	300 mm deep		41.15	.389		35.50	13.80		49.30	60.50	
	1400	14 ga x 250 mm deep		35.05	.456		38	16.25		54.25	67.50	
	1410	300 mm deep		33.53	.477		42	16.95		58.95	72.50	
	1500	12 ga x 250 mm deep		27.43	.583		50.50	20.50		71	88.50	
	1510	300 mm deep	↓	25.91	.618	↓	56.50	22		78.50	96	
410	0010	**FRAMING, JOISTS,** no band joists (track), web stiffeners, headers,										**410**
	0020	beams, bridging or bracing										
	0030	Joists (50 mm flange) and fasteners, materials only										
	0220	18 ga x 150 mm deep				m	4.56			4.56	4.99	
	0230	200 mm deep					5.40			5.40	5.95	
	0240	250 mm deep					6.35			6.35	7	
	0320	16 ga x 150 mm deep					5.60			5.60	6.15	
	0330	200 mm deep					6.70			6.70	7.35	
	0340	250 mm deep					7.80			7.80	8.55	
	0350	300 mm deep					8.85			8.85	9.75	
	0430	14 ga x 200 mm deep					8.45			8.45	9.30	
	0440	250 mm deep					9.70			9.70	10.70	
	0450	300 mm deep					11.05			11.05	12.15	
	0540	12 ga x 250 mm deep					14.15			14.15	15.60	
	0550	300 mm deep				↓	16.10			16.10	17.75	
	1010	Installation of joists to band joists, beams & headers, labor only										
	1220	18 ga x 150 mm deep	2 Carp	110	.145	Ea.		5.15		5.15	8.05	
	1230	200 mm deep		90	.178			6.30		6.30	9.85	
	1240	250 mm deep		80	.200			7.10		7.10	11.05	
	1320	16 ga x 150 mm deep		95	.168			6		6	9.30	
	1330	200 mm deep		70	.229			8.15		8.15	12.65	
	1340	250 mm deep		60	.267			9.50		9.50	14.75	
	1350	300 mm deep		55	.291			10.35		10.35	16.10	
	1430	14 ga x 200 mm deep		65	.246			8.75		8.75	13.60	
	1440	250 mm deep		45	.356			12.65		12.65	19.70	
	1450	300 mm deep		35	.457			16.25		16.25	25.50	
	1540	12 ga x 250 mm deep		40	.400			14.20		14.20	22	
	1550	300 mm deep	↓	30	.533	↓		18.95		18.95	29.50	
500	0010	**FRAMING, WEB STIFFENERS** at joist bearing, fabricated from										**500**
	0020	stud piece (41 mm flange) to stiffen joist (50 mm flange)										

METALS **5**

05400 | Cold-Formed Metal Framing

05420	Cold-Formed Metal Joists	CREW	DAILY OUTPUT	LABOR-HOURS	UNIT	2006 BARE COSTS				TOTAL INCL O&P	
						MAT.	LABOR	EQUIP.	TOTAL		
500 2120	For 150 mm deep joist, with 18 ga x 63 mm stud	1 Carp	120	.067	Ea.	1.67	2.37		4.04	5.55	**500**
2130	91 mm stud		110	.073		1.85	2.59		4.44	6.05	
2140	100 mm stud		105	.076		1.79	2.71		4.50	6.20	
2150	150 mm stud		100	.080		1.95	2.84		4.79	6.55	
2160	200 mm stud		95	.084		2	2.99		4.99	6.85	
2220	200 mm deep joist, with 63 mm stud		120	.067		1.83	2.37		4.20	5.70	
2230	91 mm stud		110	.073		1.99	2.59		4.58	6.20	
2240	100 mm stud		105	.076		1.95	2.71		4.66	6.35	
2250	150 mm stud		100	.080		2.13	2.84		4.97	6.80	
2260	200 mm stud		95	.084		2.30	2.99		5.29	7.20	
2320	250 mm deep joist, with 63 mm stud		110	.073		2.59	2.59		5.18	6.85	
2330	91 mm stud		100	.080		2.84	2.84		5.68	7.55	
2340	100 mm stud		95	.084		2.81	2.99		5.80	7.75	
2350	150 mm stud		90	.089		3.04	3.16		6.20	8.25	
2360	200 mm stud		85	.094		3.09	3.35		6.44	8.60	
2420	300 mm deep joist, with 63 mm stud		110	.073		2.74	2.59		5.33	7.05	
2430	91 mm stud		100	.080		2.97	2.84		5.81	7.70	
2440	100 mm stud		95	.084		2.91	2.99		5.90	7.85	
2450	150 mm stud		90	.089		3.19	3.16		6.35	8.40	
2460	200 mm stud		85	.094		3.43	3.35		6.78	9	
3130	For 150 mm deep joist, with 16 ga x 91 mm stud		100	.080		1.93	2.84		4.77	6.55	
3140	100 mm stud		95	.084		1.91	2.99		4.90	6.75	
3150	150 mm stud		90	.089		2.10	3.16		5.26	7.25	
3160	200 mm stud		85	.094		2.21	3.35		5.56	7.65	
3230	200 mm deep joist, with 91 mm stud		100	.080		2.14	2.84		4.98	6.80	
3240	100 mm stud		95	.084		2.10	2.99		5.09	6.95	
3250	150 mm stud		90	.089		2.33	3.16		5.49	7.50	
3260	200 mm stud		85	.094		2.49	3.35		5.84	7.95	
3330	250 mm deep joist, with 91 mm stud		85	.094		2.92	3.35		6.27	8.40	
3340	100 mm stud		80	.100		2.98	3.56		6.54	8.85	
3350	150 mm stud		75	.107		3.25	3.79		7.04	9.45	
3360	200 mm stud		70	.114		3.38	4.06		7.44	10.05	
3430	300 mm deep joist, with 91 mm stud		85	.094		3.19	3.35		6.54	8.70	
3440	100 mm stud		80	.100		3.13	3.56		6.69	9	
3450	150 mm stud		75	.107		3.48	3.79		7.27	9.75	
3460	200 mm stud		70	.114		3.72	4.06		7.78	10.45	
4230	For 200 mm deep joist, with 14 ga x 91 mm stud		90	.089		2.77	3.16		5.93	7.95	
4240	100 mm stud		85	.094		2.84	3.35		6.19	8.30	
4250	150 mm stud		80	.100		3.07	3.56		6.63	8.90	
4260	200 mm stud		75	.107		3.29	3.79		7.08	9.50	
4330	250 mm deep joist, with 91 mm stud		75	.107		3.89	3.79		7.68	10.20	
4340	100 mm stud		70	.114		3.88	4.06		7.94	10.60	
4350	150 mm stud		65	.123		4.25	4.38		8.63	11.50	
4360	200 mm stud		60	.133		4.44	4.74		9.18	12.30	
4430	300 mm deep joist, with 91 mm stud		75	.107		4.14	3.79		7.93	10.45	
4440	100 mm stud		70	.114		4.23	4.06		8.29	11	
4450	150 mm stud		65	.123		4.58	4.38		8.96	11.85	
4460	200 mm stud		60	.133		4.91	4.74		9.65	12.80	
5330	For 250 mm deep joist, with 12 ga x 91 mm stud		65	.123		4.11	4.38		8.49	11.30	
5340	100 mm stud		60	.133		4.24	4.74		8.98	12.05	
5350	150 mm stud		55	.145		4.67	5.15		9.82	13.20	
5360	200 mm stud		50	.160		5.15	5.70		10.85	14.50	
5430	300 mm deep joist, with 91 mm stud		65	.123		4.55	4.38		8.93	11.80	
5440	100 mm stud		60	.133		4.47	4.74		9.21	12.30	
5450	150 mm stud		55	.145		5.10	5.15		10.25	13.65	
5460	200 mm stud		50	.160		5.85	5.70		11.55	15.30	

Important: See the Reference Section for supporting data - Crews, Rental Equipment, City Cost Indexes and Reference Data

05425	Cold-Formed Roof Framing	CREW	DAILY OUTPUT	LABOR-HOURS	UNIT	2006 BARE COSTS				TOTAL INCL O&P
						MAT.	LABOR	EQUIP.	TOTAL	
100	**0010 FRAMING, BRACING**									**100**
	0020 Continuous bracing, per row									
	0100 16 ga x 38 mm channel trafters/trusses @ 400 mm O.C.	1 Carp	137	.058	m	1.42	2.08		3.50	4.79
	0120 600 mm O.C.		183	.044		1.42	1.55		2.97	3.98
	0300 50 x 50 mm angle x 18 ga, rafters/trusses @ 400 mm O.C.		183	.044		2.06	1.55		3.61	4.68
	0320 600 mm O.C.		244	.033		2.06	1.17		3.23	4.07
	0400 16 ga, rafters/trusses @ 400 mm O.C.		137	.058		2.63	2.08		4.71	6.15
	0420 600 mm O.C.		198	.040		2.63	1.44		4.07	5.15
200	**0010 FRAMING, BRIDGING**									**200**
	0020 Solid, between rafters w/ 31 mm leg track, per rafter bay									
	1200 Rafters 400 mm O.C., 18 ga x 100 mm deep	1 Carp	60	.133	Ea.	1.27	4.74		6.01	8.80
	1210 150 mm deep		57	.140		1.65	4.99		6.64	9.55
	1220 200 mm deep		55	.145		2.10	5.15		7.25	10.35
	1230 250 mm deep		52	.154		2.59	5.45		8.04	11.35
	1240 300 mm deep		50	.160		2.99	5.70		8.69	12.15
	2200 600 mm O.C., 18 ga x 100 mm deep		60	.133		1.84	4.74		6.58	9.40
	2210 150 mm deep		57	.140		2.39	4.99		7.38	10.40
	2220 200 mm deep		55	.145		3.03	5.15		8.18	11.40
	2230 250 mm deep		52	.154		3.75	5.45		9.20	12.60
	2240 300 mm deep		50	.160		4.32	5.70		10.02	13.60
500	**0010 FRAMING, PARAPETS**									**500**
	0100 900 mm high inst. on 1st sty, 18 ga x 100 mm W stds, 300 mm O.C.	2 Carp	30.48	.525	m	15.45	18.65		34.10	46
	0110 400 mm O.C.		45.72	.350		13.10	12.45		25.55	34
	0120 600 mm O.C.		60.96	.262		10.85	9.35		20.20	26.50
	0200 150 mm wide studs, 300 mm O.C.		30.48	.525		19.65	18.65		38.30	50.50
	0210 400 mm O.C.		45.72	.350		16.75	12.45		29.20	38
	0220 600 mm O.C.		60.96	.262		13.85	9.35		23.20	30
	1100 Installed on 2nd story, 18 ga x 100 mm wide studs, 300 mm O.C.		28.96	.553		15.45	19.65		35.10	47.50
	1110 400 mm O.C.		44.20	.362		13.10	12.85		25.95	34.50
	1120 600 mm O.C.		57.91	.276		10.85	9.80		20.65	27
	1200 150 mm wide studs, 300 mm O.C.		28.96	.553		19.65	19.65		39.30	52
	1210 400 mm O.C.		44.20	.362		16.75	12.85		29.60	38.50
	1220 600 mm O.C.		57.91	.276		13.85	9.80		23.65	30.50
	2100 Installed on gable, 18 ga x 100 mm wide studs, 300 mm O.C.		25.91	.618		15.45	22		37.45	51
	2110 400 mm O.C.		39.62	.404		13.10	14.35		27.45	37
	2120 600 mm O.C.		51.82	.309		10.85	11		21.85	29
	2200 150 mm wide studs, 300 mm O.C.		25.91	.618		19.65	22		41.65	55.50
	2210 400 mm O.C.		39.62	.404		16.75	14.35		31.10	41
	2220 600 mm O.C.		51.82	.309		13.85	11		24.85	32.50
550	**0010 FRAMING, ROOF RAFTERS**									**550**
	0100 Boxed ridge beam, double, 18 ga x 150 mm deep	2 Carp	48.77	.328	m	14.75	11.65		26.40	34.50
	0110 200 mm deep		45.72	.350		16.35	12.45		28.80	37.50
	0120 250 mm deep		42.67	.375		19.80	13.35		33.15	43
	0130 300 mm deep		39.62	.404		22	14.35		36.35	46.50
	0200 16 ga x 150 mm deep		45.72	.350		16.70	12.45		29.15	37.50
	0210 200 mm deep		42.67	.375		18.80	13.35		32.15	41.50
	0220 250 mm deep		39.62	.404		22.50	14.35		36.85	47.50
	0230 300 mm deep		36.58	.437		24.50	15.55		40.05	51
	1100 Rafters, 50 mm flange, material only, 18 ga x 150 mm deep					4.56			4.56	4.99
	1110 200 mm deep					5.40			5.40	5.95
	1120 250 mm deep					6.35			6.35	7
	1130 300 mm deep					7.40			7.40	8.10
	1200 16 ga x 150 mm deep					5.60			5.60	6.15
	1210 200 mm deep					6.70			6.70	7.35
	1220 250 mm deep					7.80			7.80	8.55

METALS 5

05425 | Cold-Formed Roof Framing

		CREW	DAILY OUTPUT	LABOR-HOURS	UNIT	2006 BARE COSTS MAT.	LABOR	EQUIP.	TOTAL	TOTAL INCL O&P	
550	1230	300 mm deep				m	8.85			8.85	9.75	**550**
	2100	Installation only, ordinary rafter to 4:12 pitch, 18 ga x 150 mm D	2 Carp	35	.457	Ea.		16.25		16.25	25.50	
	2110	200 mm deep		30	.533			18.95		18.95	29.50	
	2120	250 mm deep		25	.640			23		23	35.50	
	2130	300 mm deep		20	.800			28.50		28.50	44.50	
	2200	16 ga x 150 mm deep		30	.533			18.95		18.95	29.50	
	2210	200 mm deep		25	.640			23		23	35.50	
	2220	250 mm deep		20	.800			28.50		28.50	44.50	
	2230	300 mm deep	▼	15	1.067	▼		38		38	59	
	8100	Add to labor, ordinary rafters on steep roofs						25%				
	8110	Dormers & complex roofs						50%				
	8200	Hip & valley rafters to 4:12 pitch						25%				
	8210	Steep roofs						50%				
	8220	Dormers & complex roofs						75%				
	8300	Hip & valley jack rafters to 4:12 pitch						50%				
	8310	Steep roofs						75%				
	8320	Dormers & complex roofs						100%				
600	0010	**FRAMING, ROOF TRUSSES**										**600**
	0020	Fabrication of trusses on ground, Fink (W) or King Post, to 4:12 pitch										
	0120	18 ga x 100 mm chords, 4.8 m span	2 Carp	12	1.333	Ea.	52.50	47.50		100	132	
	0130	6.0 m span		11	1.455		66	51.50		117.50	153	
	0140	7.2 m span		11	1.455		79	51.50		130.50	168	
	0150	8.4 m span		10	1.600		92	57		149	190	
	0160	9.6 m span		10	1.600		105	57		162	205	
	0250	150 mm chords, 8.4 m span		9	1.778		116	63		179	226	
	0260	9.6 m span		9	1.778		132	63		195	244	
	0270	10.8 m span		8	2		149	71		220	275	
	0280	12.0 m span		8	2		165	71		236	293	
	1120	5:12 to 8:12 pitch, 18 ga x 100 mm chords, 4.8 m span		10	1.600		60	57		117	155	
	1130	6.0 m span		9	1.778		75	63		138	181	
	1140	7.2 m span		9	1.778		90	63		153	198	
	1150	8.4 m span		8	2		105	71		176	227	
	1160	9.6 m span		8	2		120	71		191	243	
	1250	150 mm chords, 8.4 m span		7	2.286		132	81.50		213.50	272	
	1260	9.6 m span		7	2.286		151	81.50		232.50	293	
	1270	10.8 m span		6	2.667		170	95		265	335	
	1280	12.0 m span		6	2.667		189	95		284	355	
	2120	9:12 to 12:12 pitch, 18 ga x 100 mm chords, 4.8 m span		8	2		75	71		146	194	
	2130	6.0 m span		7	2.286		94	81.50		175.50	230	
	2140	7.2 m span		7	2.286		113	81.50		194.50	251	
	2150	8.4 m span		6	2.667		132	95		227	293	
	2160	9.6 m span		6	2.667		150	95		245	315	
	2250	150 mm chords, 8.4 m span		5	3.200		165	114		279	360	
	2260	9.6 m span		5	3.200		189	114		303	385	
	2270	10.8 m span		4	4		212	142		354	455	
	2280	12.0 m span	▼	4	4		236	142		378	480	
	5120	Erection only of roof trusses, to 4:12 pitch, 4.8 m span	F-6	48	.833			27.50	13.25	40.75	57	
	5130	6.0 m span		46	.870			28.50	13.85	42.35	59.50	
	5140	7.2 m span		44	.909			30	14.45	44.45	62	
	5150	8.4 m span		42	.952			31	15.15	46.15	65	
	5160	9.6 m span		40	1			33	15.90	48.90	68	
	5170	10.8 m span		38	1.053			34.50	16.75	51.25	72	
	5180	12.0 m span		36	1.111			36.50	17.70	54.20	76	
	5220	5:12 to 8:12 pitch, 4.8 m span		42	.952			31	15.15	46.15	65	
	5230	6.0 m span	▼	40	1	▼		33	15.90	48.90	68	

05400 | Cold-Formed Metal Framing

05425 | Cold-Formed Roof Framing

		CREW	DAILY OUTPUT	LABOR-HOURS	UNIT	MAT.	LABOR	EQUIP.	TOTAL	TOTAL INCL O&P	
600	5240	7.2 m span	F-6	38	1.053	Ea.		34.50	16.75	51.25	72	**600**
	5250	8.4 m span		36	1.111			36.50	17.70	54.20	76	
	5260	9.6 m span		34	1.176			38.50	18.75	57.25	80	
	5270	10.8 m span		32	1.250			41	19.90	60.90	85.50	
	5280	12.0 m span		30	1.333			43.50	21	64.50	91	
	5320	9:12 to 12:12 pitch, 4.8 m span		36	1.111			36.50	17.70	54.20	76	
	5330	6.0 m span		34	1.176			38.50	18.75	57.25	80	
	5340	7.2 m span		32	1.250			41	19.90	60.90	85.50	
	5350	8.4 m span		30	1.333			43.50	21	64.50	91	
	5360	9.6 m span		28	1.429			47	22.50	69.50	97.50	
	5370	10.8 m span		26	1.538			50.50	24.50	75	105	
	5380	12.0 m span		24	1.667			54.50	26.50	81	114	
650	0010	**FRAMING, SOFFITS & CANOPIES**										**650**
	0130	Cont. ledger track @ wall, studs @ 400 mm O.C., 18 ga x 100 mm W	2 Carp	163	.098	m	2.79	3.49		6.28	8.50	
	0140	150 mm wide		152	.105		3.61	3.74		7.35	9.80	
	0150	200 mm wide		142	.113		4.59	4.01		8.60	11.30	
	0160	250 mm wide		131	.122		5.70	4.34		10.04	13	
	0230	Studs @ 600 mm O.C., 18 ga x 100 mm wide		244	.066		2.66	2.33		4.99	6.55	
	0240	150 mm wide		229	.070		3.44	2.48		5.92	7.70	
	0250	200 mm wide		213	.075		4.36	2.67		7.03	9	
	0260	250 mm wide		198	.081		5.40	2.87		8.27	10.40	
	1000	Horizontal soffit and canopy members, material only										
	1030	41 mm flange studs, 18 ga x 100 mm deep				m	3.71			3.71	4.07	
	1040	150 mm deep					4.66			4.66	5.10	
	1050	200 mm deep					5.65			5.65	6.20	
	1140	50 mm flange studs, 18 ga x 150 mm deep					5.20			5.20	5.70	
	1150	200 mm deep					6.15			6.15	6.80	
	1160	250 mm deep					7.25			7.25	7.95	
	4030	Installation only, 18 ga, 41 mm flange x 100 mm deep	2 Carp	130	.123	Ea.		4.38		4.38	6.80	
	4040	150 mm deep		110	.145			5.15		5.15	8.05	
	4050	200 mm deep		90	.178			6.30		6.30	9.85	
	4140	50 mm flange, 18 ga x 150 mm deep		110	.145			5.15		5.15	8.05	
	4150	200 mm deep		90	.178			6.30		6.30	9.85	
	4160	250 mm deep		80	.200			7.10		7.10	11.05	
	6010	Clips to attach fascia to rafter tails, 50 x 50 mm x 18 ga angle	1 Carp	120	.067		.74	2.37		3.11	4.51	
	6020	16 ga angle	"	100	.080		.95	2.84		3.79	5.45	

05500 | Metal Fabrications

05514 | Ladders

		CREW	DAILY OUTPUT	LABOR-HOURS	UNIT	MAT.	LABOR	EQUIP.	TOTAL	TOTAL INCL O&P	
500	0010	**LADDER**, shop fabricated										**500**
	0020	Steel, 500 mm wide, bolted to concrete, with cage	E-4	15.24	2.100	m	223	85	5.85	313.85	405	
	0100	Without cage		25.91	1.235		104	50	3.45	157.45	208	
	0300	Aluminum, bolted to concrete, with cage		15.24	2.100		300	85	5.85	390.85	490	
	0400	Without cage		25.91	1.235		174	50	3.45	227.45	286	
	1350	Alternating tread stair, 56/68°, steel, standard paint color	2 Sswk	15.24	1.050		495	42		537	620	
	1360	Non-standard paint color		15.24	1.050		560	42		602	690	
	1370	Galvanized steel		15.24	1.050		560	42		602	690	
	1380	Stainless steel		15.24	1.050		830	42		872	990	
	1390	68°, aluminum		15.24	1.050		610	42		652	745	

METALS 5

5 METALS

05517	Metal Stairs	CREW	DAILY OUTPUT	LABOR-HOURS	UNIT	2006 BARE COSTS				TOTAL INCL O&P
						MAT.	LABOR	EQUIP.	TOTAL	
300 0010	**FIRE ESCAPE**, shop fabricated									**300**
0200	610 mm wide balcony, 25 mm x 6 mm bars 38 mm O.C.	1 Sswk	1.52	5.249	m	150	210		360	540
0400	1st story cantilevered stair, standard		.09	88.889	Ea.	1,900	3,550		5,450	8,500
0700	Platform & fixed stair, 910 mm x 1020 mm	↓	.17	47.059	Flight	845	1,875		2,720	4,300
0900	For 1.1 m wide escapes, add to above					100%	150%			
350 0010	**FIRE ESCAPE STAIRS**									**350**
0020	One story, disappearing, stainless steel	2 Sswk	6.10	2.625	m	600	105		705	850
0100	Portable ladder				Ea.	55.50			55.50	61.50
700 0010	**STAIR**, shop fabricated, steel stringers, safety nosing on treads									**700**
0020	Grating tread and pipe railing, 1050 mm wide	E-4	35	.914	Riser	218	37	2.55	257.55	310
0100	1200 mm wide		30	1.067		283	43	2.98	328.98	390
0200	Cement fill metal pan, picket rail, 1050 mm wide		35	.914		325	37	2.55	364.55	430
0300	1200 mm wide		30	1.067		370	43	2.98	415.98	485
0350	Wall rail, both sides, 1050 mm wide		53	.604		250	24.50	1.68	276.18	320
0400	Cast iron tread and pipe rail, 1050 mm wide		35	.914		350	37	2.55	389.55	455
0500	Checkered plate tread, industrial, 1050 mm wide		28	1.143		218	46	3.19	267.19	325
0550	Circular, for tanks, 900 mm wide	↓	33	.970		240	39	2.71	281.71	335
0600	For isolated stairs, add						100%			
0800	Custom steel stairs, 1050 mm wide, minimum	E-4	35	.914		325	37	2.55	364.55	430
0810	Average		30	1.067		435	43	2.98	480.98	560
0900	Maximum	↓	20	1.600		545	64.50	4.46	613.96	720
1100	For 1200 mm wide stairs, add					5%	5%			
1300	For 1500 mm wide stairs, add	↓			↓	10%	10%			
1500	Landing, steel pan, conventional	E-4	14.86	2.153	m²	470	87	6	563	680
1600	Pre-erected	"	23.69	1.351	"	820	54.50	3.77	878.27	1,000
1700	Pre-erected, steel pan tread, 1050 mm wide, 2 line pipe rail	E-2	87	.644	Riser	360	25	16.40	401.40	455
1810	Spiral aluminum, 1500 mm diameter, stock units	E-4	45	.711		395	29	1.98	425.98	490
1820	Custom units		45	.711		745	29	1.98	775.98	875
1900	Spiral, cast iron, 1200 mm diameter, ornamental, minimum		45	.711		350	29	1.98	380.98	445
1920	Maximum		25	1.280		480	52	3.57	535.57	625
2000	Spiral, steel, industrial checkered plate, 1200 mm diameter		45	.711		350	29	1.98	380.98	445
2200	Stock units, 1800 mm diameter	↓	40	.800	↓	425	32.50	2.23	459.73	530
3110	Spiral steel, stock units, primed, flat metal tread, 1050 mm diameter	2 Carp	1.60	10	Flight	1,050	355		1,405	1,725
3120	1200 mm diameter		1.45	11.034		1,225	390		1,615	1,950
3130	1350 diameter		1.35	11.852		1,350	420		1,770	2,125
3140	1500 mm diameter		1.25	12.800		1,450	455		1,905	2,300
3210	Galvanized, 1050 mm diameter		1.60	10		1,850	355		2,205	2,600
3220	1200 mm diameter		1.45	11.034		2,075	390		2,465	2,900
3230	1350 diameter		1.35	11.852		2,275	420		2,695	3,150
3240	1500 mm diameter		1.25	12.800		2,450	455		2,905	3,400
3310	Checkered plate tread, 1050 mm diameter		1.45	11.034		1,300	390		1,690	2,025
3320	1200 mm diameter		1.35	11.852		1,475	420		1,895	2,275
3330	1350 diameter		1.25	12.800		1,625	455		2,080	2,475
3340	1500 mm diameter		1.15	13.913		1,750	495		2,245	2,700
3410	Galvanized, 1050 mm diameter		1.45	11.034		2,125	390		2,515	2,925
3420	1200 mm diameter		1.35	11.852		2,375	420		2,795	3,275
3430	1350 diameter		1.25	12.800		2,575	455		3,030	3,550
3440	1500 mm diameter		1.15	13.913		2,775	495		3,270	3,850
3510	Red oak tread on flat metal, 1050 mm diameter		1.35	11.852		1,775	420		2,195	2,600
3520	1200 mm diameter		1.25	12.800		1,975	455		2,430	2,850
3530	1350 diameter		1.15	13.913		2,125	495		2,620	3,125
3540	1500 mm diameter	↓	1.05	15.238	↓	2,300	540		2,840	3,375
3900	Industrial ships ladder, 900 mm W, grating treads, 2 line pipe rail	E-4	30	1.067	Riser	142	43	2.98	187.98	237
4000	Aluminum	"	30	1.067	"	218	43	2.98	263.98	320

05520 | Handrails & Railings

		CREW	DAILY OUTPUT	LABOR-HOURS	UNIT	MAT.	LABOR	EQUIP.	TOTAL	TOTAL INCL O&P	
700	0010	**RAILING, PIPE**, shop fabricated										**700**
	0020	Aluminum, 2 rail, satin finish, 31 mm diameter	E-4	48.77	.656	m	65	26.50	1.83	93.33	121	
	0030	Clear anodized		48.77	.656		80	26.50	1.83	108.33	138	
	0040	Dark anodized		48.77	.656		90.50	26.50	1.83	118.83	150	
	0080	38 mm diameter, satin finish		48.77	.656		77.50	26.50	1.83	105.83	135	
	0090	Clear anodized		48.77	.656		86.50	26.50	1.83	114.83	145	
	0100	Dark anodized		48.77	.656		96	26.50	1.83	124.33	155	
	0140	Aluminum, 3 rail, 31 mm diam., satin finish		41.76	.766		99	31	2.14	132.14	167	
	0150	Clear anodized		41.76	.766		124	31	2.14	157.14	195	
	0160	Dark anodized		41.76	.766		137	31	2.14	170.14	209	
	0200	38 mm diameter, satin finish		41.76	.766		119	31	2.14	152.14	189	
	0210	Clear anodized		41.76	.766		135	31	2.14	168.14	206	
	0220	Dark anodized		41.76	.766		148	31	2.14	181.14	221	
	0500	Steel, 2 rail, on stairs, primed, 31 mm diameter		48.77	.656		53	26.50	1.83	81.33	109	
	0520	38 mm diameter		48.77	.656		58.50	26.50	1.83	86.83	114	
	0540	Galvanized, 31 mm diameter		48.77	.656		73.50	26.50	1.83	101.83	131	
	0560	38 mm diameter		48.77	.656		82.50	26.50	1.83	110.83	141	
	0580	Steel, 3 rail, primed, 31 mm diameter		41.76	.766		79.50	31	2.14	112.64	146	
	0600	38 mm diameter		41.76	.766		84	31	2.14	117.14	151	
	0620	Galvanized, 31 mm diameter		41.76	.766		111	31	2.14	144.14	180	
	0640	38 mm diameter		41.76	.766		131	31	2.14	164.14	202	
	0700	Stainless steel, 2 rail, 31 mm diam. #4 finish		41.76	.766		192	31	2.14	225.14	269	
	0720	High polish		41.76	.766		310	31	2.14	343.14	400	
	0740	Mirror polish		41.76	.766		385	31	2.14	418.14	485	
	0760	Stainless steel, 3 rail, 38 mm diam., #4 finish		36.58	.875		289	35.50	2.44	326.94	385	
	0770	High polish		36.58	.875		475	35.50	2.44	512.94	590	
	0780	Mirror finish		36.58	.875		580	35.50	2.44	617.94	705	
	0900	Wall rail, alum. pipe, 31 mm diam., satin finish		64.92	.493		37	19.95	1.38	58.33	78.50	
	0905	Clear anodized		64.92	.493		45	19.95	1.38	66.33	87	
	0910	Dark anodized		64.92	.493		54.50	19.95	1.38	75.83	97.50	
	0915	38 mm diameter, satin finish		64.92	.493		41	19.95	1.38	62.33	82.50	
	0920	Clear anodized		64.92	.493		51.50	19.95	1.38	72.83	94	
	0925	Dark anodized		64.92	.493		63.50	19.95	1.38	84.83	108	
	0930	Steel pipe, 31 mm diameter, primed		64.92	.493		32.50	19.95	1.38	53.83	73	
	0935	Galvanized		64.92	.493		46.50	19.95	1.38	67.83	89	
	0940	38 mm diameter		53.64	.597		33	24	1.66	58.66	82	
	0945	Galvanized		64.92	.493		47	19.95	1.38	68.33	89	
	0955	Stainless steel pipe, 38 mm diam., #4 finish		32.61	.981		153	39.50	2.74	195.24	243	
	0960	High polish		32.61	.981		310	39.50	2.74	352.24	420	
	0965	Mirror polish		32.61	.981		365	39.50	2.74	407.24	480	
780	0010	**RAILINGS, INDUSTRIAL** Welded, shop fabricated										**780**
	0020	2 rail, 1.1 m high, 38 mm pipe	E-4	77.72	.412	m	70.50	16.65	1.15	88.30	109	
	0100	50 mm angle rail	"	77.72	.412		64.50	16.65	1.15	82.30	102	
	0200	For 100 mm high kick plate, 10 gauge, add					14.70			14.70	16.15	
	0300	6 mm thick, add					18.90			18.90	21	
	0500	For curved rails, add					30%	30%				

05530 | Gratings

		CREW	DAILY OUTPUT	LABOR-HOURS	UNIT	MAT.	LABOR	EQUIP.	TOTAL	TOTAL INCL O&P	
300	0010	**FLOOR GRATING, ALUMINUM**, field fabricated from panels										**300**
	0110	Bearing bars @ 30 mm O.C., cross bars @ 100 mm O.C.,										
	0111	Up to 28 m², 25 mm x 3 mm bar	E-4	83.61	.383	m²	105	15.50	1.07	121.57	144	
	0112	Over 28 m²		78.97	.405		95	16.40	1.13	112.53	136	
	0113	31 mm x 3 mm bar, up to 28 m²		74.32	.431		119	17.40	1.20	137.60	164	
	0114	Over 28 m²		92.90	.344		108	13.95	.96	122.91	145	
	0122	31 mm x 5 mm bar, up to 28 m²		69.68	.459		153	18.60	1.28	172.88	203	
	0124	Over 28 m²		92.90	.344		139	13.95	.96	153.91	179	

METALS 5

05530 | Gratings

		CREW	DAILY OUTPUT	LABOR-HOURS	UNIT	2006 BARE COSTS				TOTAL INCL O&P
						MAT.	LABOR	EQUIP.	TOTAL	
300 0132	38 mm x 3 mm bar, up to 28 m²	E-4	65.03	.492	m²	95.50	19.90	1.37	116.77	143
0134	Over 28 m²		92.90	.344		86.50	13.95	.96	101.41	122
0136	44 mm x 5 mm bar, up to 28 m²		46.45	.689		205	28	1.92	234.92	277
0138	Over 28 m²		92.90	.344		186	13.95	.96	200.91	231
0146	56 mm x 5 mm bar, up to 28 m²		55.74	.574		241	23	1.60	265.60	310
0148	Over 28 m²		92.90	.344		219	13.95	.96	233.91	267
0162	Cross bars @ 50 mm O.C., 25 mm x 3 mm, up to 28 m²		55.74	.574		132	23	1.60	156.60	189
0164	Over 28 m²		92.90	.344		120	13.95	.96	134.91	158
0172	31 mm x 5 mm bar, up to 28 m²		55.74	.574		183	23	1.60	207.60	245
0174	Over 28 m²		92.90	.344		166	13.95	.96	180.91	209
0182	38 mm x 3 mm bar, up to 28 m²		55.74	.574		131	23	1.60	155.60	188
0184	Over 28 m²		92.90	.344		119	13.95	.96	133.91	157
0186	44 mm x 5 mm bar, up to 28 m²		55.74	.574		190	23	1.60	214.60	253
0188	Over 28 m²		92.90	.344		173	13.95	.96	187.91	216
0200	For straight cuts, add				m	8.05			8.05	8.85
0212	Close mesh, 25 mm x 3 mm, up to 28 m²	E-4	48.31	.662	m²	335	27	1.85	363.85	415
0214	Over 28 m²		85.47	.374		305	15.15	1.04	321.19	365
0222	31 mm x 5 mm, up to 28 m²		48.31	.662		415	27	1.85	443.85	505
0224	Over 28 m²		85.47	.374		380	15.15	1.04	396.19	445
0232	38 mm x 3 mm, up to 28 m²		48.31	.662		168	27	1.85	196.85	235
0234	Over 28 m²		85.47	.374		153	15.15	1.04	169.19	197
0300	For curved cuts, add				m	10.95			10.95	12.05
0400	For straight banding, add					9.65			9.65	10.60
0500	For curved banding, add					13.40			13.40	14.75
0600	For aluminum checkered plate nosings, add					13.20			13.20	14.55
0700	For straight toe plate, add					24			24	26.50
0800	For curved toe plate, add					30			30	33
1000	For cast aluminum abrasive nosings, add					16.10			16.10	17.70
1200	Expanded aluminum, 3.1 kg/m²	E-4	97.55	.328	m²	46	13.25	.92	60.17	75.50
1400	Extruded I bars are 10% less than 5 mm bars									
1600	Heavy duty, all extruded plank, 19 mm deep, 8.8 kg/m²	E-4	102	.314	m²	129	12.70	.88	142.58	166
1700	31 mm deep, 14.2 kg/m²		92.90	.344		137	13.95	.96	151.91	177
1800	44 mm deep, 20.5 kg/m²		85.93	.372		164	15.05	1.04	180.09	208
1900	56 mm deep, 24.4 kg/m²		81.29	.394		187	15.90	1.10	204	236
2100	For safety serrated surface, add					15%				
320 0010	**FLOOR GRATING PLANKS**, field fabricated from planks									
0020	Aluminum, 240 mm wide, 14 ga., 50 mm rib	E-4	290	.110	m	46	4.46	.31	50.77	59.50
0200	Galvanized steel, 240 mm wide, 14 ga., 64 mm rib		290	.110		30	4.46	.31	34.77	41.50
0300	100 mm rib		290	.110		36	4.46	.31	40.77	48
0500	12 gauge, 64 mm rib		290	.110		39.50	4.46	.31	44.27	52
0600	75 mm rib		290	.110		46.50	4.46	.31	51.27	59.50
0800	Stainless steel, type 304, 16 ga., 50 mm rib		290	.110		80	4.46	.31	84.77	96.50
0900	Type 316		290	.110		96.50	4.46	.31	101.27	114
340 0010	**FLOOR GRATING, STEEL**, field fabricated from panels									
0050	Labor for installing, from ground/floor	E-4	78.50	.408	m²		16.50	1.14	17.64	31
0100	Elevated		42.73	.749	"		30.50	2.09	32.59	57
0300	Platforms, to 3600 mm high, rectangular		1429	.022	kg	3.06	.91	.06	4.03	5.05
0400	Circular		1043	.031	"	3.40	1.24	.09	4.73	6.05
0410	Painted bearing bars @ 30 mm									
0412	Cross bars @ 100 mm O.C., 19 mm x 3 mm bar, up to 28 m²	E-2	46.45	1.206	m²	63	47	30.50	140.50	185
0414	Over 28 m²		69.68	.804		57.50	31.50	20.50	109.50	140
0422	31 mm x 5 mm, up to 28 m²		37.16	1.507		85	58.50	38.50	182	237
0424	Over 28 m²		55.74	1.005		77.50	39	25.50	142	181
0432	38 mm x 3 mm, up to 28 m²		37.16	1.507		76	58.50	38.50	173	227
0434	Over 28 m²		55.74	1.005		69	39	25.50	133.50	172

05530 | Gratings

		CREW	DAILY OUTPUT	LABOR-HOURS	UNIT	2006 BARE COSTS				TOTAL INCL O&P
						MAT.	LABOR	EQUIP.	TOTAL	
340	0436	44 mm x 5 mm, up to 28 m²	E-2	37.16	1.507	m²	111	58.50	38.50	208	265	340
	0438	Over 28 m²		55.74	1.005		101	39	25.50	165.50	207
	0452	56 mm x 5 mm, up to 28 m²		27.87	2.009		131	78	51	260	335
	0454	Over 28 m²		41.81	1.340		119	52	34	205	259
	0462	Cross bars @ 50 mm O.C., 19 mm x 3 mm, up to 28 m²		46.45	1.206		121	47	30.50	198.50	249
	0464	Over 28 m²		69.68	.804		101	31.50	20.50	153	188
	0472	31 mm x 5 mm, up to 28 m²		37.16	1.507		182	58.50	38.50	279	345
	0474	Over 28 m²		55.74	1.005		152	39	25.50	216.50	263
	0482	38 mm x 3 mm, up to 28 m²		37.16	1.507		141	58.50	38.50	238	298
	0484	Over 28 m²		55.74	1.005		117	39	25.50	181.50	225
	0486	44 mm x 5 mm, up to 28 m²		37.16	1.507		190	58.50	38.50	287	350
	0488	Over 28 m²		55.74	1.005		158	39	25.50	222.50	270
	0502	56 mm x 5 mm, up to 28 m²		27.87	2.009		226	78	51	355	440
	0504	Over 28 m²		41.81	1.340		189	52	34	275	335
	0601	Painted bearing bars @ 23 mm O.C., cross bars @ 100 mm O.C.,									
	0612	Up to 28 m², 19 mm x 3 mm bars	E-4	78.97	.405	m²	78.50	16.40	1.13	96.03	117
	0622	31 mm x 5 mm bars		55.74	.574		111	23	1.60	135.60	166
	0632	38 mm x 3 mm bars		51.10	.626		101	25.50	1.75	128.25	158
	0636	44 mm x 5 mm bars		41.81	.765		146	31	2.14	179.14	218
	0652	56 mm x 5 mm bars	E-2	27.87	2.009		191	78	51	320	400
	0662	Cross bars @ 50 mm O.C., up to 28 m², 19 mm x 3 mm		46.45	1.206		134	47	30.50	211.50	262
	0672	31 mm x 5 mm bars		37.16	1.507		144	58.50	38.50	241	300
	0682	38 mm x 3 mm bars		37.16	1.507		137	58.50	38.50	234	294
	0686	44 mm x 5 mm bars		27.87	2.009		187	78	51	316	395
	0690	For galvanized grating, add					25%				
	0800	For straight cuts, add				m	8.55			8.55	9.40
	0900	For curved cuts, add					11.75			11.75	12.95
	1000	For straight banding, add					9.80			9.80	10.75
	1100	For curved banding, add					15.10			15.10	16.65
	1200	For checkered plate nosings, add					17.25			17.25	19
	1300	For straight toe or kick plate, add					24.50			24.50	27
	1400	For curved toe or kick plate, add					29.50			29.50	32.50
	1500	For abrasive nosings, add					17.90			17.90	19.70
	1510	For stair treads, see Division 05550-700									
	1600	For safety serrated surface, minimum, add					15%				
	1700	Maximum, add					25%				
	2000	Stainless steel gratings, close spaced, 25 x 3 mm bars, up to 28 m²	E-4	41.81	.765	m²	355	31	2.14	388.14	455
	2100	Standard spacing, 19 x 3 mm bars		46.45	.689		281	28	1.92	310.92	360
	2200	31 x 5 mm bars		37.16	.861		445	35	2.40	482.40	555
	2400	Expanded steel grating, at ground, 14.5 kg/m²		83.61	.383		35	15.50	1.07	51.57	67.50
	2500	15.1 kg/m²		83.61	.383		37	15.50	1.07	53.57	70
	2600	19.3 kg/m²		78.97	.405		46	16.40	1.13	63.53	81
	2650	20.6 kg/m²		78.97	.405		49.50	16.40	1.13	67.03	85
	2700	24.1 kg/m²		74.32	.431		59	17.40	1.20	77.60	98
	2800	30.1 kg/m²		69.68	.459		74.50	18.60	1.28	94.38	117
	2900	33.8 kg/m²		65.03	.492		84	19.90	1.37	105.27	130
	3100	For flattened expanded steel grating, add					8%				
	3300	For elevated installation above 5 m, add						15%			
360	0010	**GRATING FRAME**, field fabricated										360
	0020	Aluminum, for gratings 25 mm to 40 mm deep	1 Sswk	21.34	.375	m	24.50	15		39.50	54
	0100	For each corner, add				Ea.	5.85			5.85	6.45

05540 | Floor Plates

		CREW	DAILY OUTPUT	LABOR-HOURS	UNIT	2006 BARE COSTS				TOTAL INCL O&P
200	0010	**CHECKERED PLATE**, field fabricated										200
	0020	6 mm & 10 mm, 175 to 475 m², bolted	E-4	1315	.024	kg	1.76	.98	.07	2.81	3.78

METALS 5

05540	Floor Plates	CREW	DAILY OUTPUT	LABOR-HOURS	UNIT	2006 BARE COSTS				TOTAL INCL O&P	
						MAT.	LABOR	EQUIP.	TOTAL		
200 0100	Welded	E-4	1996	.016	kg	1.65	.65	.04	2.34	3.05	**200**
0300	Pit or trench cover and frame, 6 mm plate, 600 mm to 900 mm wide	↓	9.29	3.445	m²	226	139	9.60	374.60	510	
0400	For galvanizing, add				kg	.95			.95	1.04	
0500	Platforms, 6 mm plate, no handrails included, rectangular	E-4	1905	.017	↓	2.80	.68	.05	3.53	4.36	
0600	Circular	"	1134	.028	↓	3.90	1.14	.08	5.12	6.45	
700 0010	**TRENCH COVER**, field fabricated										**700**
0020	Cast iron grating with bar stops and angle frame, to 450 mm wide	1 Sswk	6.10	1.312	m	183	52.50		235.50	297	
0100	Frame only (both sides of trench), 25 mm grating		13.72	.583		40	23.50		63.50	86	
0150	50 mm grating	↓	10.67	.750	↓	62.50	30		92.50	123	
0200	Aluminum, stock units, including frames and										
0210	9 mm plain cover plate, 100 mm opening	E-4	62.48	.512	m	115	20.50	1.43	136.93	166	
0300	150 mm opening		56.39	.568		142	23	1.58	166.58	200	
0400	250 mm opening		51.82	.618		196	25	1.72	222.72	263	
0500	400 mm opening	↓	47.24	.677		269	27.50	1.89	298.39	350	
0700	Add/mm for additional widths to 600 mm					10.75			10.75	11.80	
0900	For custom fabrication, add					50%					
1100	For 6 mm plain cover plate, deduct					12%					
1500	For cover recessed for tile, 6 mm thick, deduct					12%					
1600	9 mm thick, add					5%					
1800	For checkered plate cover, 6 mm thick, deduct					12%					
1900	9 mm thick, add					2%					
2100	For slotted or round holes in cover, 6 mm thick, add					3%					
2200	9 mm thick, add					4%					
2300	For abrasive cover, add	↓				12%					

05550	Stair Treads & Nosings	CREW	DAILY OUTPUT	LABOR-HOURS	UNIT	MAT.	LABOR	EQUIP.	TOTAL	TOTAL INCL O&P	
700 0010	**STAIR TREADS**										**700**
0020	Alum. grating, 900 mm L, 38 x 5 mm rect. bars, 150 mm W	1 Sswk	24	.333	Ea.	49.50	13.30		62.80	78.50	
0100	300 mm wide		22	.364		73	14.55		87.55	106	
0200	38 mm x 5 mm I-bars, 150 mm wide		24	.333		41.50	13.30		54.80	69.50	
0300	300 mm wide	↓	22	.364		67	14.55		81.55	99.50	
0400	For abrasive nosings, add					12.95			12.95	14.25	
0500	For narrow mesh, add				↓	60%					
0600	Stair treads, not incl. stringers. See also Div. 03150-660										
0700	Cast aluminum, abrasive, 900 mm L x 300 mm W, 8 mm thk	1 Sswk	15	.533	Ea.	110	21.50		131.50	160	
0800	9 mm thick		15	.533		117	21.50		138.50	167	
0900	13 mm thick		15	.533		128	21.50		149.50	180	
1000	Cast bronze, abrasive, 9 mm thick		8	1		475	40		515	595	
1100	13 mm thick		8	1		625	40		665	760	
1200	Cast iron, abrasive, 900 mm long x 300 mm wide, 9 mm thick		15	.533		96.50	21.50		118	145	
1300	13 mm thick	↓	15	.533	↓	111	21.50		132.50	162	
1400	Fiberglass reinforced plastic with safety nosing,										
1500	38 mm thick, 300 mm wide, 600 mm long	1 Sswk	22	.364	Ea.	104	14.55		118.55	141	
1600	750 mm long		22	.364		109	14.55		123.55	146	
1700	900 mm long		22	.364		119	14.55		133.55	157	
2000	Steel grating, painted, 900 mm long, 31 x 5 mm bars, 150 mm wide		20	.400		42	16		58	75	
2010	300 mm wide	↓	18	.444		48.50	17.75		66.25	85.50	
2100	Add for abrasive nosing, 900 mm long, painted					13.10			13.10	14.40	
2200	Galvanized					17.60			17.60	19.35	
2300	Painting 900 mm treads, in shop, nonstandard paint					2.31			2.31	2.54	
2400	Added coats, standard paint					.99			.99	1.09	
2500	Expanded steel, 63 mm deep, 225 mm x 900 mm long, 18 ga.	1 Sswk	20	.400		23	16		39	54.50	
2600	14 gauge	"	20	.400	↓	28.50	16		44.50	60.50	

Important: See the Reference Section for supporting data - Crews, Rental Equipment, City Cost Indexes and Reference Data

05560	Metal Castings	CREW	DAILY OUTPUT	LABOR-HOURS	UNIT	2006 BARE COSTS				TOTAL INCL O&P
						MAT.	LABOR	EQUIP.	TOTAL	
200	**0010** **CONSTRUCTION CASTINGS**									**200**
	0020 Manhole covers and frames see Division 02630-110									
	0100 Column bases, cast iron, 400 mm x 400 mm, approx. 30 kg	E-4	46	.696	Ea.	107	28	1.94	136.94	171
	0200 800 mm x 800 mm, approx. 115 kg	"	23	1.391		400	56.50	3.88	460.38	545
	0400 Cast aluminum for wood columns, 200 mm x 200 mm	1 Carp	32	.250		34.50	8.90		43.40	52
	0500 300 mm x 300 mm	"	32	.250		74.50	8.90		83.40	96
	0600 Miscellaneous C.I. castings, light sections, less than 68 kg	E-4	1452	.022	kg	3.66	.89	.06	4.61	5.70
	1100 Heavy sections, more than 68 kg		1905	.017		1.79	.68	.05	2.52	3.23
	1300 Special low volume items		1452	.022		6.35	.89	.06	7.30	8.65
	1500 For ductile iron, add					100%				

05580	Formed Metal Fabrications	CREW	DAILY OUTPUT	LABOR-HOURS	UNIT	MAT.	LABOR	EQUIP.	TOTAL	TOTAL INCL O&P
150	**0010** **ALLOY STEEL CHAIN**, Grade 80									**150**
	0015 Self-colored, cut lengths, w/accessories, 6 mm	E-17	122	.131	m	15.40	5.35	.04	20.79	26.50
	0020 9 mm		60.96	.262		19.85	10.75	.09	30.69	41.50
	0030 13 mm		36.58	.437		31	17.90	.15	49.05	66.50
	0040 16 mm		21.95	.729		49	30	.25	79.25	108
	0050 19 mm		14.63	1.094		72	45	.37	117.37	160
	0060 22 mm		12.19	1.312		106	53.50	.45	159.95	214
	0070 25 mm		10.67	1.500		174	61.50	.51	236.01	305
	0080 31 mm		7.32	2.187		325	89.50	.74	415.24	515
	0110 Clevis slip hook, 6 mm				Ea.	13.60			13.60	14.95
	0120 9 mm					20.50			20.50	22.50
	0130 13 mm					34.50			34.50	38
	0140 16 mm					51.50			51.50	57
	0150 19 mm					108			108	119
	0160 Eye/sling hook w/ hammerlock coupling, 22 mm					305			305	335
	0170 25 mm					430			430	470
	0180 31 mm					645			645	710
200	**0010** **ALUMINUM COLUMNS**									**200**
	0020 Aluminum, extruded, stock units, no cap or base, 150 mm diameter	E-4	73.15	.437	m	28.50	17.70	1.22	47.42	65
	0100 200 mm diameter		51.82	.618		38.50	25	1.72	65.22	89
	0200 250 mm diameter		45.72	.700		50.50	28.50	1.95	80.95	109
	0300 310 mm diameter		42.67	.750		88	30.50	2.09	120.59	154
	0400 380 mm diameter		36.58	.875		107	35.50	2.44	144.94	183
	0410 Caps and bases, plain, 150 mm diameter				Set	18.95			18.95	21
	0420 200 mm diameter					23.50			23.50	25.50
	0430 250 mm diameter					31			31	34
	0440 310 mm diameter					57.50			57.50	63
	0450 380 mm diameter					86			86	95
	0460 Caps, ornamental, minimum					218			218	239
	0470 Maximum					950			950	1,050
	0500 For square columns, add to column prices above				m	50%				
	0700 Residential, flat, 2.5 m high, plain	E-4	20	1.600	Ea.	65	64.50	4.46	133.96	192
	0720 Fancy		20	1.600		132	64.50	4.46	200.96	266
	0740 Corner type, plain		20	1.600		115	64.50	4.46	183.96	248
	0760 Fancy		20	1.600		232	64.50	4.46	300.96	375
600	**0010** **LAMP POSTS**									**600**
	0020 Aluminum, 2150 mm high, stock units, post only	1 Carp	16	.500	Ea.	32	17.80		49.80	63
	0100 Mild steel, plain	"	16	.500	"	28	17.80		45.80	58.50
900	**0010** **WINDOW GUARDS**, shop fabricated									**900**
	0015 Expanded metal, steel angle frame, permanent	E-4	32.52	.984	m²	212	40	2.75	254.75	310
	0025 Steel bars, 13 mm x 13 mm, spaced 130 mm O.C.	"	26.94	1.188	"	147	48	3.31	198.31	251
	0030 Hinge mounted, add				Opng.	39.50			39.50	43.50

METALS 5

05580 | Formed Metal Fabrications

		CREW	DAILY OUTPUT	LABOR-HOURS	UNIT	2006 BARE COSTS				TOTAL INCL O&P
						MAT.	LABOR	EQUIP.	TOTAL	
900	0040	Removable type, add				Opng.	25			25	27.50	**900**
	0050	For galvanized guards, add				m²	35%				
	0070	For pivoted or projected type, add					105%	40%			
	0100	Mild steel, stock units, economy	E-4	37.62	.851		57.50	34.50	2.37	94.37	128
	0200	Deluxe		37.62	.851	↓	118	34.50	2.37	154.87	195
	0400	Woven wire units,10 mm chan. frame,910 mm x 1520 mm opng.		40	.800	Opng.	144	32.50	2.23	178.73	219
	0500	1220 mm x 1830 mm opening	↓	38	.842		231	34	2.35	267.35	320
	0800	Basket guards for above, add					197			197	217
	1000	Swinging guards for above, add				↓	68			68	74.50

5 METALS

05655 | Railroad Trackwork

			CREW	DAILY OUTPUT	LABOR-HOURS	UNIT	2006 BARE COSTS				TOTAL INCL O&P
							MAT.	LABOR	EQUIP.	TOTAL	
700	0010	**RAILROAD SIDINGS**	R347216-10									**700**
	0020	Car bumpers, standard	B-14	2	24	Ea.	2,650	695	113	3,458	4,100
	0100	Heavy duty		2	24		5,000	695	113	5,808	6,700
	0200	Derails hand throw (sliding)	R347216-20	10	4.800	↓	835	139	22.50	996.50	1,150
	0300	Hand throw with standard timbers, open stand & target		8	6	↓	910	174	28.50	1,112.50	1,300
	0400	Resurface and realign existing track		60.96	.787	m		23	3.71	26.71	39.50
	0600	For crushed stone ballast, add	↓	152	.316	"	42	9.15	1.49	52.64	62
	0800	Siding, yard spur, level grade									
	0808	Wood ties and ballast, 36 kg new rail	B-14	17.37	2.763	m	268	80	13	361	430
	0809	36 kg relay rail		17.37	2.763		196	80	13	289	355
	0810	45 kg rail, new material on wood ties		17.37	2.763		305	80	13	398	475
	0812	41 kg new rail		17.37	2.763		286	80	13	379	455
	0813	41 kg relay rail		17.37	2.763		198	80	13	291	355
	0820	45 kg new rail		17.37	2.763		305	80	13	398	475
	0822	45 kg relay rail		17.37	2.763		206	80	13	299	365
	0830	50 kg new rail		17.37	2.763		325	80	13	418	500
	0832	50 kg relay rail		17.37	2.763		219	80	13	312	380
	1000	Steel ties in concrete w/45 kg new rail	↓	6.71	7.158	↓	420	208	33.50	661.50	815
	1002	Steel ties in concrete, incl. fasteners & plates									
	1003	36 kg new rail	B-14	6.71	7.158	m	390	208	33.50	631.50	785
	1005	36 kg relay rail		6.71	7.158		325	208	33.50	566.50	715
	1012	41 kg new rail		6.71	7.158		405	208	33.50	646.50	800
	1015	41 kg relay rail		6.71	7.158		335	208	33.50	576.50	720
	1020	45 kg new rail		6.71	7.158		420	208	33.50	661.50	815
	1025	45 kg relay rail		6.71	7.158		340	208	33.50	581.50	730
	1030	50 kg new rail		6.71	7.158		440	208	33.50	681.50	840
	1035	50 kg relay rail		6.71	7.158	↓	350	208	33.50	591.50	740
	1200	Switch timber, for a #8 switch, pressure treated		8.73	5.498	m³	950	160	26	1,136	1,325
	1300	Complete set of timbers, 8.73 m³ for #8 switch		1	48	Total	8,650	1,400	226	10,276	11,900
	1400	Ties, concrete, 2550 mm long, 750 mm O.C.		80	.600	Ea.	89.50	17.40	2.83	109.73	128
	1600	Wood, treated, 150 mm x 200 mm x 2550 mm, C.L. lots		90	.533		37.50	15.50	2.51	55.51	68
	1700	L.C.L. lots		90	.533		39.50	15.50	2.51	57.51	70
	1900	Heavy duty, 180 x 230 x 2600 mm, C.L. lots		70	.686		41	19.90	3.23	64.13	80
	2000	L.C.L. lots	↓	70	.686	↓	41	19.90	3.23	64.13	80
	2200	Turnouts, #8, incl. 45 kg rails, plates, bars, frog, switch point,									
	2300	Timbers and ballast 150 mm below bottom of tie	B-14	.50	96	Ea.	10,000	2,775	450	13,225	15,900

05650 | Railroad Track & Accessories

05655 | Railroad Trackwork

				CREW	DAILY OUTPUT	LABOR-HOURS	UNIT	2006 BARE COSTS MAT.	LABOR	EQUIP.	TOTAL	TOTAL INCL O&P	
700	2400	Wheel stops, fixed	R347216 -10	B-14	18	2.667	Pr.	595	77.50	12.55	685.05	790	700
	2450	Hinged		↓	14	3.429	"	800	99.50	16.15	915.65	1,050	
750	0010	**RAILROAD TRACK**	R347216 -10										750
	0020	Track bolts					Ea.	2.08			2.08	2.29	
	0100	Joint bars					Pr.	63			63	69.50	
	0200	Spikes					Ea.	.48			.48	.53	
	0300	Tie plates					"	10.50			10.50	11.55	
	1000	Rail, 45.36 kg prime grade					m	77.50			77.50	85.50	
	1500	Relay rail	↓				"	39			39	42.50	

05700 | Ornamental Metal

05720 | Ornamental Handrails & Railings

			CREW	DAILY OUTPUT	LABOR-HOURS	UNIT	2006 BARE COSTS MAT.	LABOR	EQUIP.	TOTAL	TOTAL INCL O&P	
700	0010	**RAILINGS, ORNAMENTAL,** shop fabricated										700
	0020	Aluminum, bronze or stainless, minimum	1 Sswk	7.32	1.094	m	90.50	43.50		134	178	
	0100	Maximum		2.74	2.916		855	117		972	1,150	
	0200	Aluminum ornamental rail, minimum		4.57	1.750		90.50	70		160.50	226	
	0300	Maximum		2.44	3.281		271	131		402	535	
	0400	Hand-forged wrought iron, minimum		3.66	2.187		282	87.50		369.50	465	
	0500	Maximum		2.44	3.281		845	131		976	1,175	
	0550	Steel, minimum		3.66	2.187		87.50	87.50		175	253	
	0560	Maximum		2.44	3.281		262	131		393	525	
	0600	Composite metal/wood/glass, minimum		1.83	4.374		540	175		715	910	
	0700	Maximum	↓	1.52	5.249	↓	1,075	210		1,285	1,575	

05730 | Ornamental Formed Metal

				CREW	DAILY OUTPUT	LABOR-HOURS	UNIT	2006 BARE COSTS MAT.	LABOR	EQUIP.	TOTAL	TOTAL INCL O&P	
200	0010	**COLUMNS, ORNAMENTAL,** shop fabricated	R051223 -10										200
	6400	Mild steel, flat, 230mm wide, stock units, painted, plain		E-4	48.77	.656	m	20	26.50	1.83	48.33	72	
	6450	Fancy			48.77	.656		40.50	26.50	1.83	68.83	94.50	
	6500	Corner columns, painted, plain			48.77	.656		35.50	26.50	1.83	63.83	89	
	6550	Fancy	↓		48.77	.656	↓	71.50	26.50	1.83	99.83	129	

05800 | Expansion Control

05810 | Exp. Joint Cover Assemblies

			CREW	DAILY OUTPUT	LABOR-HOURS	UNIT	2006 BARE COSTS MAT.	LABOR	EQUIP.	TOTAL	TOTAL INCL O&P	
350	0010	**EXPANSION JOINT ASSEMBLIES** Custom units										350
	0200	Floor cover assemblies, 25 mm space, aluminum	1 Sswk	11.58	.691	m	53	27.50		80.50	108	
	0300	Bronze		11.58	.691		108	27.50		135.50	169	
	0500	50 mm space, aluminum		11.58	.691		64.50	27.50		92	121	
	0600	Bronze		11.58	.691		116	27.50		143.50	178	
	0800	Wall and ceiling assemblies, 25 mm space, aluminum	↓	11.58	.691	↓	32	27.50		59.50	84.50	

METALS 5

05800 | Expansion Control

05810 | Exp. Joint Cover Assemblies

			CREW	DAILY OUTPUT	LABOR-HOURS	UNIT	MAT.	LABOR	EQUIP.	TOTAL	TOTAL INCL O&P	
350	0900	Bronze	1 Sswk	11.58	.691	m	101	27.50		128.50	161	350
	1100	50 mm space, aluminum		11.58	.691		54	27.50		81.50	109	
	1200	Bronze		11.58	.691		124	27.50		151.50	187	
	1400	Floor to wall assemblies, 25 mm space, aluminum		11.58	.691		48.50	27.50		76	103	
	1500	Bronze or stainless		11.58	.691		122	27.50		149.50	185	
	1700	Gym floor angle covers, aluminum, 75 mm x 75 mm angle		14.02	.571		42	23		65	87	
	1800	75 mm x 100 mm angle		14.02	.571		49.50	23		72.50	95.50	
	2000	Roof closures, aluminum, flat roof, low profile, 25 mm space		17.37	.460		97	18.40		115.40	140	
	2100	High profile		17.37	.460		119	18.40		137.40	164	
	2300	Roof to wall, low profile, 25 mm space		17.37	.460		53.50	18.40		71.90	92	
	2400	High profile	↓	17.37	.460	↓	69	18.40		87.40	109	

05900 | Metal Restoration & Cleaning

05910 | Metal Cleaning

			CREW	DAILY OUTPUT	LABOR-HOURS	UNIT	MAT.	LABOR	EQUIP.	TOTAL	TOTAL INCL O&P	
500	0010	**METAL CLEANING**										500
	6125	Steel surface treatments										
	6170	Wire brush, hand (SSPC-SP2)	1 Psst	22.30	.359	m²	.22	11.65		11.87	22	
	6180	Power tool (SSPC-SP3)	"	55.74	.144		.54	4.66		5.20	9.25	
	6215	Pressure washing, 260-555 m²/day	1 Pord	325	.025			.78		.78	1.18	
	6220	Steam cleaning, 260-370 m²/day		223	.036			1.14		1.14	1.71	
	6225	Water blasting	↓	297	.027			.85		.85	1.29	
	6230	Brush-off blast (SSPC-SP7)	E-11	557	.057		.86	1.83	.26	2.95	4.37	
	6235	Com'l blast (SSPC-SP6), loose scale, fine pwder rust, 10 kg/m² sand		372	.086		1.72	2.74	.40	4.86	7.05	
	6240	Tight mill scale, little/no rust, 15 kg/m² sand		297	.108		2.58	3.43	.50	6.51	9.30	
	6245	Exist coat blistered/pitted, 20 kg/m² sand		195	.164		3.44	5.25	.76	9.45	13.60	
	6250	Exist coat badly pitted/nodules, 33 kg/m² sand		121	.264		5.80	8.45	1.22	15.47	22	
	6255	Near white blast (SSPC-SP10), loose scale, fine rust, 27 kg/m² sand		158	.203		4.84	6.45	.93	12.22	17.40	
	6260	Tight mill scale, little/no rust, 34 kg/m² sand		130	.246		6.05	7.85	1.13	15.03	21	
	6265	Exist coat blistered/pitted, 44 kg/m² sand		102	.314		7.85	10	1.45	19.30	27	
	6270	Exist coat badly pitted/nodules, 55 kg/m² sand	↓	74.32	.431	↓	9.80	13.75	1.98	25.53	36.50	

05950 | Paints & Protective Coatings

				CREW	DAILY OUTPUT	LABOR-HOURS	UNIT	MAT.	LABOR	EQUIP.	TOTAL	TOTAL INCL O&P	
650	0010	**PAINTS & PROTECTIVE COATINGS**	R050516 -30										650
	5900	Galvanizing structural steel in shop, under 1 tonne					Met. Ton	595			595	655	
	5950	1 tonne to 18 tonnes						495			495	545	
	6000	Over 18 tonnes					↓	435			435	480	
	6100	Cold galvanizing, brush in field		1 Psst	102	.078	m²	.75	2.55		3.30	5.50	
	6510	Paints & protective coatings, sprayed in field											
	6520	Alkyds, primer		2 Psst	334	.048	m²	.65	1.55		2.20	3.65	
	6540	Gloss topcoats			297	.054		.54	1.75		2.29	3.80	
	6560	Silicone alkyd			297	.054		1.08	1.75		2.83	4.44	
	6610	Epoxy, primer			279	.057		1.83	1.86		3.69	5.50	
	6630	Intermediate or topcoat			260	.062		1.51	2		3.51	5.35	
	6650	Enamel coat			260	.062		1.83	2		3.83	5.75	
	6700	Epoxy ester, primer			260	.062		3.66	2		5.66	7.80	
	6720	Topcoats			260	.062		1.18	2		3.18	5.10	
	6810	Latex primer			334	.048		.54	1.55		2.09	3.55	
	6830	Topcoats	↓	↓	297	.054	↓	.65	1.75		2.40	4.01	

Important: See the Reference Section for supporting data - Crews, Rental Equipment, City Cost Indexes and Reference Data

	05950	Paints & Protective Coatings		CREW	DAILY OUTPUT	LABOR-HOURS	UNIT	2006 BARE COSTS				TOTAL INCL O&P	
								MAT.	LABOR	EQUIP.	TOTAL		
650	6910	Universal primers, one part, phenolic, modified alkyd	R050516 -30	2 Psst	186	.086	m²	.86	2.79		3.65	6.15	650
	6940	Two part, epoxy spray			186	.086		2.15	2.79		4.94	7.55	
	7000	Zinc rich primers, self cure, spray, inorganic			167	.096		5.50	3.11		8.61	11.85	
	7010	Epoxy, spray, organic			167	.096		1.51	3.11		4.62	7.40	
	7020	Above one story, spray painting simple structures, add							25%				
	7030	Intricate structures, add							50%				

For information about Means Estimating Seminars, see yellow pages 12 and 13 in back of book

METALS **5**

	CREW	DAILY OUTPUT	LABOR-HOURS	UNIT	2006 BARE COSTS				TOTAL INCL O&P
					MAT.	LABOR	EQUIP.	TOTAL	

Division 6
Wood & Plastics

Estimating Tips

06050 Basic Wood & Plastic Materials & Methods

- Common to any wood framed structure are the accessory connector items such as screws, nails, adhesives, hangers, connector plates, straps, angles and holdowns. For typical wood framed buildings, such as residential projects, the aggregate total for these items can be significant, especially in areas where seismic loading is a concern. For floor and wall framing, the material cost is based on 2.0 to 5.0 kg per cubic meter. Holdowns, hangers and other connectors should be taken off by the piece.

06100 Rough Carpentry

- Lumber is a traded commodity and therefore sensitive to supply and demand in the marketplace. Even in "budgetary" estimating of wood framed projects, it is advisable to call local suppliers for the latest market pricing.

- Common quantity units for wood framed projects are "cubic meters".
- Waste is an issue of concern at the quantity takeoff for any area of construction. Framing lumber is sold in even lengths, and depending on spans, wall heights and the grade of lumber, waste is inevitable. A rule of thumb for lumber waste is 5% to 10% depending on material quality and the complexity of the framing.
- Wood in various forms and shapes is used in many projects, even where the main structural framing is steel, concrete or masonry. Plywood as a back-up partition material and boards used as blocking and cant strips around roof edges are two common examples. The estimator should ensure that the costs of all wood materials are included in the final estimate.

06200 Finish Carpentry

- It is necessary to consider the grade of workmanship when estimating labor costs for erecting millwork and interior finish. In practice, there are three grades: premium, custom and economy. The Means daily output for base and case moldings is in the range of 60 to 75 meters per carpenter per day. This is appropriate for most average custom-grade projects. For premium projects an adjustment to productivity of 25% to 50% should be made depending on the complexity of the job.

Reference Numbers

Reference numbers are shown in bold squares at the beginning of some major classifications. These numbers refer to related items in the Reference Section. The reference information may be an estimating procedure, an alternate pricing method, or technical information.

Note: Not all subdivisions listed here necessarily appear in this publication.

06052	Selective Demolition		CREW	DAILY OUTPUT	LABOR-HOURS	UNIT	2006 BARE COSTS				TOTAL INCL O&P
							MAT.	LABOR	EQUIP.	TOTAL	
110	0010	**SELECTIVE DEMOLITION, WOOD FRAMING** R024119-10										110
	0100	Timber connector, nailed, small	1 Clab	96	.083	Ea.		2.28		2.28	3.55
	0110	Medium		60	.133			3.65		3.65	5.70
	0120	Large		48	.167			4.57		4.57	7.10
	0130	Bolted, small		48	.167			4.57		4.57	7.10
	0140	Medium		32	.250			6.85		6.85	10.65
	0150	Large		24	.333			9.15		9.15	14.20
	2958	Beams, 50 mm x 150 mm	2 Clab	335	.048	m		1.31		1.31	2.04
	2960	50 mm x 200 mm		251	.064			1.75		1.75	2.72
	2965	50 mm x 250 mm		203	.079			2.16		2.16	3.36
	2970	50 mm x 300 mm		168	.095			2.61		2.61	4.06
	2972	50 mm x 350 mm		143	.112			3.07		3.07	4.77
	2975	100 mm x 200 mm	B-1	126	.190			5.35		5.35	8.30
	2980	100 mm x 250 mm		101	.238			6.65		6.65	10.40
	2985	100 mm x 300 mm		83.82	.286			8.05		8.05	12.50
	3000	150 mm x 200 mm		83.82	.286			8.05		8.05	12.50
	3040	150 mm x 250 mm		67.06	.358			10.05		10.05	15.65
	3080	150 mm x 300 mm		56.39	.426			11.95		11.95	18.60
	3120	200 mm x 300 mm		42.67	.562			15.80		15.80	24.50
	3160	250 mm x 300 mm		33.53	.716			20		20	31.50
	3162	Alternate pricing method		2.60	9.246	m³		260		260	405
	3170	Blocking, in 400 mm OC wall framing, 50 mm x 100 mm	1 Clab	183	.044	m		1.20		1.20	1.86
	3172	50 mm x 150 mm		122	.066			1.80		1.80	2.80
	3174	In 600 mm OC wall framing, 50 mm x 100 mm		183	.044			1.20		1.20	1.86
	3176	50 mm x 150 mm		122	.066			1.80		1.80	2.80
	3178	Alt method, wood blocking removal from wood framimg		.94	8.476	m³		232		232	360
	3179	Wood blocking removal from steel framimg		.85	9.417	"		258		258	400
	3180	Bracing, let in, 25 mm x 75 mm, studs 400 mm OC		320	.025	m		.69		.69	1.07
	3181	Bracing, let in, 25 mm x 75 mm, studs 600 mm OC		329	.024			.67		.67	1.04
	3182	25 mm x 100 mm, studs 400 mm OC		320	.025			.69		.69	1.07
	3183	Studs 600 mm OC		329	.024			.67		.67	1.04
	3184	25 mm x 150 mm, studs 400 mm OC		320	.025			.69		.69	1.07
	3185	Studs 600 mm OC		329	.024			.67		.67	1.04
	3186	50 mm x 75 mm, studs 400 mm OC		244	.033			.90		.90	1.40
	3187	Studs 600 mm OC		253	.032			.87		.87	1.35
	3188	50 mm x 100 mm, studs 400 mm OC		244	.033			.90		.90	1.40
	3189	Studs 600 mm OC		253	.032			.87		.87	1.35
	3190	50 mm x 150 mm, studs 400 mm OC		244	.033			.90		.90	1.40
	3191	Studs 600 mm OC		253	.032			.87		.87	1.35
	3192	50 mm x 200 mm, studs 400 mm OC		244	.033			.90		.90	1.40
	3193	Studs 600 mm OC		253	.032			.87		.87	1.35
	3194	"T" shaped metal bracing, studs at 400 mm OC		323	.025			.68		.68	1.06
	3195	Studs at 600 mm OC		366	.022			.60		.60	.93
	3196	Metal straps, studs at 400 mm OC		366	.022			.60		.60	.93
	3197	Studs at 600 mm OC		378	.021			.58		.58	.90
	3200	Columns, round, 2.4 m to 4.2 m tall		40	.200	Ea.		5.50		5.50	8.55
	3202	Dimensional lumber sizes	2 Clab	2.60	6.164	m³		169		169	263
	3250	Blocking, between joists	1 Clab	320	.025	Ea.		.69		.69	1.07
	3252	Bridging, metal strap, between joists		320	.025	Pr.		.69		.69	1.07
	3254	Wood, between joists		320	.025	"		.69		.69	1.07
	3260	Door buck,studs,Hdr/access,2.4 m H 50 x 100 mm wall, 900 mm W		32	.250	Ea.		6.85		6.85	10.65
	3261	1200 mm wide		32	.250			6.85		6.85	10.65
	3262	1500 mm wide		32	.250			6.85		6.85	10.65
	3263	1800 mm wide		32	.250			6.85		6.85	10.65
	3264	2438 mm wide		30	.267			7.30		7.30	11.35
	3265	3050 mm wide		30	.267			7.30		7.30	11.35

	06052	Selective Demolition	CREW	DAILY OUTPUT	LABOR-HOURS	UNIT	2006 BARE COSTS				TOTAL INCL O&P	
							MAT.	LABOR	EQUIP.	TOTAL		
110	3266	3658 mm wide	1 Clab	30	.267	Ea.		7.30		7.30	11.35	110
	3267	50 mm x 150 mm wall, 900 mm wide	R024119 -10	32	.250			6.85		6.85	10.65	
	3268	1200 mm wide		32	.250			6.85		6.85	10.65	
	3269	1500 mm wide		32	.250			6.85		6.85	10.65	
	3270	1800 mm wide		32	.250			6.85		6.85	10.65	
	3271	2438 mm wide		30	.267			7.30		7.30	11.35	
	3272	3050 mm wide		30	.267			7.30		7.30	11.35	
	3273	3658 mm wide		30	.267			7.30		7.30	11.35	
	3274	Window buck,studs, Hdr/acc.,2.4 m H 50 x 100 mm wall, 600 mm W		24	.333			9.15		9.15	14.20	
	3275	900 mm wide		24	.333			9.15		9.15	14.20	
	3276	1200 mm wide		24	.333			9.15		9.15	14.20	
	3277	1500 mm wide		24	.333			9.15		9.15	14.20	
	3278	1800 mm wide		24	.333			9.15		9.15	14.20	
	3279	2133 mm wide		24	.333			9.15		9.15	14.20	
	3280	2438 mm wide		22	.364			9.95		9.95	15.50	
	3281	3050 mm wide		22	.364			9.95		9.95	15.50	
	3282	3658 mm wide		22	.364			9.95		9.95	15.50	
	3283	50 mm x 150 mm wall, 600 mm wide		24	.333			9.15		9.15	14.20	
	3284	900 mm wide		24	.333			9.15		9.15	14.20	
	3285	1200 mm wide		24	.333			9.15		9.15	14.20	
	3286	1500 mm wide		24	.333			9.15		9.15	14.20	
	3287	1800 mm wide		24	.333			9.15		9.15	14.20	
	3288	2133 mm wide		24	.333			9.15		9.15	14.20	
	3289	2438 mm wide		22	.364			9.95		9.95	15.50	
	3290	3050 mm wide		22	.364			9.95		9.95	15.50	
	3291	3658 mm wide		22	.364	▼		9.95		9.95	15.50	
	3400	Fascia boards, 25 mm x 150 mm		152	.053	m		1.44		1.44	2.24	
	3440	25 mm x 200 mm		137	.058			1.60		1.60	2.49	
	3480	25 mm x 250 mm		122	.066			1.80		1.80	2.80	
	3490	51 mm x 152 mm		137	.058			1.60		1.60	2.49	
	3500	51 mm x 203 mm		122	.066			1.80		1.80	2.80	
	3510	51 mm x 254 mm		107	.075	▼		2.05		2.05	3.19	
	3610	Furring, on wood walls or ceiling		372	.022	m²		.59		.59	.92	
	3620	On masonry or concrete walls or ceiling		111	.072	"		1.97		1.97	3.07	
	3800	Headers over openings, 2 @ 50 mm x 150 mm		33.53	.239	m		6.55		6.55	10.20	
	3840	2 @ 50 mm x 200 mm		30.48	.262			7.20		7.20	11.20	
	3880	2 @ 50 mm x 250 mm		27.43	.292	▼		8		8	12.45	
	3885	Alternate pricing method		.62	12.990	m³		355		355	555	
	3920	Joists, 25 mm x 100 mm		381	.021	m		.58		.58	.90	
	3930	25 mm x 150 mm		346	.023			.63		.63	.99	
	3940	25 mm x 200 mm		305	.026			.72		.72	1.12	
	3950	25 mm x 250mm		273	.029			.80		.80	1.25	
	3960	25 mm x 300 mm		233	.034			.94		.94	1.46	
	4200	50 mm x 100 mm	2 Clab	305	.052			1.44		1.44	2.24	
	4230	50 mm x 150 mm		296	.054			1.48		1.48	2.31	
	4240	50 mm x 200 mm		287	.056			1.53		1.53	2.38	
	4250	50 mm x 250 mm		277	.058			1.58		1.58	2.46	
	4280	50 mm x 300 mm		268	.060			1.64		1.64	2.55	
	4281	50 mm x 350 mm		259	.062			1.69		1.69	2.63	
	4282	Composite joists, 240 mm		293	.055			1.50		1.50	2.33	
	4283	300 mm		283	.057			1.55		1.55	2.41	
	4284	356 mm		273	.059			1.61		1.61	2.50	
	4285	406 mm		264	.061	▼		1.66		1.66	2.59	
	4290	Wood joists, alternate pricing method		3.54	4.520	m³		124		124	193	
	4300	Plates, single		305	.052	m		1.44		1.44	2.24	
	4320	Double		244	.066			1.80		1.80	2.80	

WOOD & PLASTICS 6

			DAILY	LABOR-			2006 BARE COSTS			TOTAL	
06052	**Selective Demolition**	CREW	OUTPUT	HOURS	UNIT	MAT.	LABOR	EQUIP.	TOTAL	INCL O&P	
110 4500	Open web joist, 305 mm deep	2 Clab	152	.105	m		2.88		2.88	4.49	110
4505	350 mm deep		145	.110			3.02		3.02	4.71	
4510	405 mm deep		137	.117			3.20		3.20	4.98	
4520	455 mm deep		130	.123			3.37		3.37	5.25	
4530	610 mm deep	↓	122	.131			3.59		3.59	5.60	
4550	25 mm x 50 mm	1 Clab	366	.022			.60		.60	.93	
4560	25 mmm x 75 mm		366	.022			.60		.60	.93	
4570	25 mm x 100 mm		366	.022			.60		.60	.93	
4580	50 mm x 50 mm		335	.024			.65		.65	1.02	
4600	50 mm x 150 mm		305	.026			.72		.72	1.12	
4601	50 mm x 200 mm or 50 mm x 250 mm		244	.033			.90		.90	1.40	
4602	100 mm x 150 mm		183	.044			1.20		1.20	1.86	
4604	100 mm x 200 mm	↓	137	.058			1.60		1.60	2.49	
5400	Posts, 100 mm x 100 mm	2 Clab	244	.066			1.80		1.80	2.80	
5405	100 mm x 150 mm		168	.095			2.61		2.61	4.06	
5410	50 mm x 102 mm		134	.119			3.27		3.27	5.10	
5425	100 mm x 254 mm		119	.134			3.68		3.68	5.75	
5430	100 mm x 305 mm		107	.150			4.10		4.10	6.40	
5440	150 mm x 150 mm		122	.131			3.59		3.59	5.60	
5445	150 mm x 204 mm		107	.150			4.10		4.10	6.40	
5450	150 mm x 254 mm		97.54	.164			4.49		4.49	7	
5455	150 mm x 305 mm		88.39	.181			4.96		4.96	7.70	
5480	200 mm x 200 mm		91.44	.175			4.79		4.79	7.45	
5500	250 mm x 250 mm		73.15	.219	↓		6		6	9.35	
5660	Tongue and groove floor planks		4.72	3.390	m³		93		93	145	
5750	Rafters, ordinary, 400 mm OC, 50 mm x 100 mm		81.75	.196	m²		5.35		5.35	8.35	
5755	50 mm x 150 mm		78.04	.205			5.60		5.60	8.75	
5760	50 mm x 200 mm		76.18	.210			5.75		5.75	8.95	
5770	50 mm x 250 mm		76.18	.210			5.75		5.75	8.95	
5780	50 mm x 300 mm		75.25	.213			5.85		5.85	9.05	
5785	600 mm OC, 50 mm x 100 mm		109	.147			4.02		4.02	6.25	
5786	50 mm x 150 mm		104	.154			4.22		4.22	6.55	
5787	50 mm x 200 mm		101	.158			4.34		4.34	6.75	
5788	50 mm x 250 mm		101	.158			4.34		4.34	6.75	
5789	50 mm x 300 mm		100	.160	↓		4.38		4.38	6.80	
5795	Rafters, ordinary, 50 mm x 100 mm (alternate method)		263	.061	m		1.67		1.67	2.59	
5800	50 mm x 150 mm (alternate method)		259	.062			1.69		1.69	2.63	
5840	50 mm x 200 mm (alternate method)		255	.063			1.72		1.72	2.68	
5855	50 mm x 254 mm (alternate method)		251	.064			1.75		1.75	2.72	
5865	50 mm x 305 mm (alternate method)	↓	247	.065			1.78		1.78	2.76	
5870	Sill plate, 50 mm x 100 mm	1 Clab	357	.022			.61		.61	.96	
5871	50 mm x 150 mm		238	.034			.92		.92	1.43	
5872	50 mm x 200 mm	↓	179	.045	↓		1.22		1.22	1.91	
5873	Alternate pricing method	↓	1.84	4.346	m³		119		119	185	
5885	Ridge board, 25 mm x 100 mm	2 Clab	274	.058	m		1.60		1.60	2.49	
5886	25 mm x 150 mm		267	.060			1.64		1.64	2.56	
5887	25 mm x 200 mm		259	.062			1.69		1.69	2.63	
5888	25 mm x 250 mm		251	.064			1.75		1.75	2.72	
5889	25 mm x 300 mm		244	.066			1.80		1.80	2.80	
5890	50 mm x 100 mm		274	.058			1.60		1.60	2.49	
5892	50 mm x 150 mm		267	.060			1.64		1.64	2.56	
5894	50 mm x 200 mm		259	.062			1.69		1.69	2.63	
5896	50 mm x 250 mm		251	.064			1.75		1.75	2.72	
5898	50 mm x 300 mm		244	.066			1.80		1.80	2.80	
6050	Rafter tie, 25 mm x 100 mm		381	.042			1.15		1.15	1.79	
6052	25 mm x 150 mm	↓ ↓	346	.046	↓		1.27		1.27	1.97	

R024119
-10

6

WOOD & PLASTICS

06052	Selective Demolition		CREW	DAILY OUTPUT	LABOR-HOURS	UNIT	2006 BARE COSTS				TOTAL INCL O&P	
							MAT.	LABOR	EQUIP.	TOTAL		
110	6054	50 mm x 100 mm	R024119 -10	2 Clab	305	.052	m		1.44		1.44	2.24	110
	6056	50 mm x 150 mm		↓	296	.054			1.48		1.48	2.31	
	6070	Sleepers, on concrete, 25 mm x 50 mm		1 Clab	1433	.006			.15		.15	.24	
	6075	25 mm x 75 mm			1219	.007			.18		.18	.28	
	6080	50 mm x 100 mm			914	.009			.24		.24	.37	
	6085	50 mm x 150 mm		↓	792	.010	↓		.28		.28	.43	
	6086	Sheathing from roof, 8 mm		2 Clab	149	.107	m²		2.94		2.94	4.58	
	6088	10 mm			142	.113			3.09		3.09	4.81	
	6090	13 mm			130	.123			3.37		3.37	5.25	
	6092	16 mm			121	.132			3.62		3.62	5.65	
	6094	19 mm			111	.144			3.95		3.95	6.15	
	6096	Board sheathing from roof			130	.123			3.37		3.37	5.25	
	6100	Sheathing, from walls, 6 mm			111	.144			3.95		3.95	6.15	
	6110	8 mm			109	.147			4.02		4.02	6.25	
	6120	10 mm			107	.150			4.10		4.10	6.40	
	6130	13 mm			105	.152			4.18		4.18	6.50	
	6140	16 mm			102	.157			4.30		4.30	6.70	
	6150	19 mm			99.87	.160			4.39		4.39	6.85	
	6152	Board sheathing from walls			139	.115			3.15		3.15	4.91	
	6158	Subfloor, with boards			97.55	.164			4.49		4.49	7	
	6160	Plywood, 13 mm thick			71.35	.224			6.15		6.15	9.55	
	6162	16 mm thick			70.60	.227			6.20		6.20	9.65	
	6164	19 mm thick			69.68	.230			6.30		6.30	9.80	
	6165	29 mm thick		↓	66.89	.239			6.55		6.55	10.20	
	6166	Underlayment, particle board, 10 mm thick		1 Clab	72.46	.110			3.03		3.03	4.71	
	6168	13 mm thick			71.35	.112			3.07		3.07	4.78	
	6170	16 mm thick			70.60	.113			3.10		3.10	4.83	
	6172	19 mm thick		↓	69.68	.115	↓		3.15		3.15	4.90	
	6200	Stairs and stringers, minimum		2 Clab	40	.400	Riser		10.95		10.95	17.05	
	6240	Maximum		"	26	.615	"		16.85		16.85	26.50	
	6300	Components, tread		1 Clab	110	.073	Ea.		1.99		1.99	3.10	
	6320	Riser			80	.100	"		2.74		2.74	4.27	
	6390	Stringer, 50 mm x 250 mm			79.25	.101	m		2.77		2.77	4.31	
	6400	50 mm x 300 mm			79.25	.101			2.77		2.77	4.31	
	6410	75 mm x 250 mm			76.20	.105			2.88		2.88	4.48	
	6420	75 mm x 300 mm		↓	76.20	.105			2.88		2.88	4.48	
	6590	Wood studs, 50 mm x 75 mm		2 Clab	938	.017			.47		.47	.73	
	6600	50 mm x 100 mm			610	.026			.72		.72	1.12	
	6640	50 mm x 150 mm		↓	488	.033	↓		.90		.90	1.40	
	6720	Wall framing, including studs plates and blocking, 50 mm x 100 mm		1 Clab	55.74	.144	m²		3.93		3.93	6.10	
	6740	50 mm x 152 mm			44.59	.179	"		4.92		4.92	7.65	
	6750	Headers, 50 mm x 100 mm			343	.023	m		.64		.64	.99	
	6755	50 mm x 150 mm			343	.023			.64		.64	.99	
	6760	50 mm x 200 mm			320	.025			.69		.69	1.07	
	6765	50 mm x 250 mm			320	.025			.69		.69	1.07	
	6770	50 mm x 300 mm			305	.026			.72		.72	1.12	
	6780	100 mm x 250 mm			160	.050			1.37		1.37	2.13	
	6785	100 mm x 300 mm			152	.053			1.44		1.44	2.24	
	6790	150 mm x 200 mm			171	.047			1.28		1.28	2	
	6795	150 mm x 250 mm			160	.050			1.37		1.37	2.13	
	6797	150 mm x 300 mm		↓	152	.053	↓		1.44		1.44	2.24	
	7000	Trusses											
	7050	3.7 m span		2 Clab	74	.216	Ea.		5.90		5.90	9.20	
	7150	7.3 m span		F-3	66	.606			22	9.65	31.65	44.50	
	7350	9.7 m span			56	.714			26	11.35	37.35	52.50	
	7450	11 m span		↓	52	.769	↓		27.50	12.25	39.75	56.50	

WOOD & PLASTICS 6

		06052	Selective Demolition		CREW	DAILY OUTPUT	LABOR-HOURS	UNIT	2006 BARE COSTS				TOTAL INCL O&P	
									MAT.	LABOR	EQUIP.	TOTAL		
110	8000		Soffit, T & G wood	R024119 -10	1 Clab	48.31	.166	m²		4.54		4.54	7.05	110
	8010		Hardboard, vinyl or aluminum		"	59.46	.135			3.69		3.69	5.75	
	8030		Plywood		2 Carp	29.26	.547			19.45		19.45	30.50	
	9500		See Div. 02220-350 for rubbish handling											
120	0010		**SELECTIVE DEMOLITION, MILLWORK AND TRIM**	R024119 -10										120
	1000		Cabinets, wood, base cabinets, per meter		2 Clab	24.38	.656	m		18		18	28	
	1020		Wall cabinets, per meter		"	24.38	.656	"		18		18	28	
	1060		Remove and reset, base cabinets		2 Carp	18	.889	Ea.		31.50		31.50	49	
	1070		Wall cabinets			20	.800			28.50		28.50	44.50	
	1072		Oven cabinet, 2134 mm high			11	1.455			51.50		51.50	80.50	
	1074		Cabinet door, up to 600 mm high		1 Clab	66	.121			3.32		3.32	5.15	
	1076		600 mm to 1200 mm high		"	46	.174			4.77		4.77	7.40	
	1100		Steel, painted, base cabinets		2 Clab	18.29	.875	m		24		24	37.50	
	1120		Wall cabinets		"	18.29	.875	"		24		24	37.50	
	1200		Casework, large area			29.73	.538	m²		14.75		14.75	23	
	1220		Selective			18.58	.861	"		23.50		23.50	36.50	
	1500		Counter top, minimum			60.96	.262	m		7.20		7.20	11.20	
	1510		Maximum			36.58	.437			12		12	18.65	
	1550		Remove and reset, minimum		2 Carp	15.24	1.050			37.50		37.50	58	
	1560		Maximum		"	12.19	1.312			46.50		46.50	72.50	
	2000		Paneling, 1219 mm x 2438 mm sheets		2 Clab	186	.086	m²		2.36		2.36	3.67	
	2100		Boards, 25 mm x 100 mm			65.03	.246			6.75		6.75	10.50	
	2120		25 mm x 150 mm			69.68	.230			6.30		6.30	9.80	
	2140		25 mm x 200 mm			74.32	.215			5.90		5.90	9.20	
	3000		Trim, baseboard, to 150 mm wide			366	.044	m		1.20		1.20	1.86	
	3040		Greater than 150 mm and up to 300 mm wide			305	.052			1.44		1.44	2.24	
	3080		Remove and reset, minimum		2 Carp	122	.131			4.66		4.66	7.25	
	3090		Maximum		"	91.44	.175			6.20		6.20	9.70	
	3100		Ceiling trim		2 Clab	305	.052			1.44		1.44	2.24	
	3120		Chair rail			366	.044			1.20		1.20	1.86	
	3140		Railings with balusters			73.15	.219			6		6	9.35	
	3160		Wainscoting			65.03	.246	m²		6.75		6.75	10.50	

06070 | Lumber Treatment

					CREW	DAILY OUTPUT	LABOR-HOURS	UNIT	MAT.	LABOR	EQUIP.	TOTAL	INCL O&P	
400	0011		**LUMBER TREATMENT**											400
	0400		Fire retardant, wet					m³	132			132	145	
	0500		KDAT						126			126	138	
	0700		Salt treated, water borne, 0.18 kg retention						60			60	66	
	0800		Oil borne, 3.63 kg retention						70.50			70.50	77.50	
	1000		Kiln dried lumber, 25 mm & 50 mm thick, softwoods						40			40	44.50	
	1100		Hardwoods						43			43	47	
	1500		For small size 25 mm stock, add						5.45			5.45	5.95	
	1700		For full size rough lumber, add						20%					
600	0010		**PLYWOOD TREATMENT**											600
	0020		Fire retardant, 6 mm thick					m²	2.56			2.56	2.82	
	0030		10 mm thick						2.81			2.81	3.10	
	0050		13 mm thick						3.02			3.02	3.32	
	0070		16 mm thick						3.20			3.20	3.52	
	0100		19 mm thick						3.52			3.52	3.87	
	0200		For KDAT, add						.77			.77	.84	
	0500		Salt treated water borne, 0.11 kg, wet, 6 mm thick						1.41			1.41	1.55	
	0530		10 mm thick						1.47			1.47	1.62	
	0550		13 mm thick						1.53			1.53	1.69	
	0570		16 mm thick						1.67			1.67	1.84	
	0600		19 mm thick						1.73			1.73	1.91	

6 WOOD & PLASTICS

06070	Lumber Treatment	CREW	DAILY OUTPUT	LABOR-HOURS	UNIT	MAT.	LABOR	EQUIP.	TOTAL	TOTAL INCL O&P	
						2006 BARE COSTS					
600	0800	For KDAT add				m²	.77			.77	.84	600
	0900	For 0.18 kg, per m³ retention, add					.64			.64	.71	
	1000	For certification stamp, add				↓	.38			.38	.41	

06090	Wood & Plastic Fastenings										
600	0010	**NAILS**, material only, based upon 23 kg box purchase										600
	0020	Copper nails, plain				kg	14.50			14.50	15.95	
	0400	Stainless steel, plain					13.10			13.10	14.40	
	0500	Box, 5g to 30g, bright					2.07			2.07	2.27	
	0520	Galvanized					3.04			3.04	3.35	
	0600	Common, 5g to 90g, plain					1.72			1.72	1.90	
	0700	Galvanized					2.25			2.25	2.47	
	0800	Aluminum					9.80			9.80	10.80	
	1000	Annular or spiral thread, 6g to 90g, plain					4.14			4.14	4.56	
	1200	Galvanized					4.30			4.30	4.74	
	1400	Drywall nails, plain					1.74			1.74	1.92	
	1600	Galvanized					3.40			3.40	3.73	
	1800	Finish nails, 6g to 15g, plain					2.31			2.31	2.56	
	2000	Galvanized					2.91			2.91	3.20	
	2100	Aluminum					9.05			9.05	9.95	
	2300	Flooring nails, hardened steel, 3g to 15g, plain					3.55			3.55	3.90	
	2400	Galvanized					4.96			4.96	5.45	
	2500	Gypsum lath nails, 29 mm, 13 ga. flathead, blued					3.40			3.40	3.73	
	2600	Masonry nails, hardened steel, 19 mm to 75 mm long, plain					3.26			3.26	3.59	
	2700	Galvanized					3.42			3.42	3.77	
	2900	Roofing nails, threaded, galvanized					2.91			2.91	3.20	
	3100	Aluminum					10.60			10.60	11.65	
	3300	Compressed lead head, threaded, galvanized					3.31			3.31	3.64	
	3600	Siding nails, plain shank, galvanized					3.20			3.20	3.53	
	3800	Aluminum					9.05			9.05	9.95	
	5000	Add to prices above for cement coating					.22			.22	.24	
	5200	Zinc or tin plating					.29			.29	.31	
	5500	Vinyl coated sinkers, 12g to 25g				↓	1.26			1.26	1.39	
650	0010	**NAILS**, pneumatic tools										650
	0020	Framing, per carton of 5000, 50 mm				Ea.	37			37	41	
	0100	60 mm					42.50			42.50	46.50	
	0200	Per carton of 4000, 75 mm					37.50			37.50	41.50	
	0300	82 mm					40			40	44	
	0400	Per carton of 5000, 60 mm, galv.					57.50			57.50	63.50	
	0500	Per carton of 4000, 75 mm, galv.					65			65	71.50	
	0600	82 mm, galv.					80.50			80.50	88.50	
	0700	Roofing, per carton of 7200, 25 mm					35			35	38.50	
	0800	32 mm					32.50			32.50	36	
	0900	40 mm					37.50			37.50	41.50	
	1000	45 mm				↓	45.50			45.50	50	
700	0010	**SHEET METAL SCREWS**										700
	0020	Steel, standard, #8 x 19 mm, plain				h	3.29			3.29	3.62	
	0100	Galvanized					3.44			3.44	3.78	
	0300	#10 x 25 mm, plain					4.49			4.49	4.94	
	0400	Galvanized					4.35			4.35	4.79	
	0600	With washers, #14 x 25 mm, plain					10.20			10.20	11.20	
	0700	Galvanized					10.50			10.50	11.55	
	0900	#14 x 50 mm, plain					18			18	19.80	
	1000	Galvanized					18.15			18.15	19.95	
	1500	Self-drilling, with washers, (pinch point) #8 x 19 mm, plain					7.10			7.10	7.80	

WOOD & PLASTICS   6

For expanded coverage of these items see *Means Interior Cost Data 2006*

	06090	Wood & Plastic Fastenings	CREW	DAILY OUTPUT	LABOR-HOURS	UNIT	2006 BARE COSTS				TOTAL INCL O&P	
							MAT.	LABOR	EQUIP.	TOTAL		
700	1600	Galvanized				h	7.20			7.20	7.90	**700**
	1800	#10 x 19 mm, plain					9.30			9.30	10.20	
	1900	Galvanized					9.40			9.40	10.35	
	3000	Stainless steel w/alum. or neoprene washers, #14 x 25 mm, plain					18.35			18.35	20	
	3100	#14 x 50 mm, plain					25			25	27.50	
750	0010	**WOOD SCREWS**										**750**
	0020	#8, 25 mm long, steel				h	3.29			3.29	3.62	
	0100	Brass					11.50			11.50	12.65	
	0200	#8, 50 mm long, steel					5.60			5.60	6.15	
	0300	Brass					19.50			19.50	21.50	
	0400	#10, 25 mm long, steel					5.90			5.90	6.50	
	0500	Brass					14			14	15.40	
	0600	#10, 50 mm long, steel					6.30			6.30	6.90	
	0700	Brass					23			23	25.50	
	0800	#10, 75 mm long, steel					10.50			10.50	11.55	
	1000	#12, 50 mm long, steel					7.50			7.50	8.25	
	1100	Brass					30			30	33	
	1500	#12, 75 mm long, steel					11.50			11.50	12.65	
	2000	#12, 100 mm long, steel					43			43	47.50	
800	0010	**TIMBER CONNECTORS** Add up cost of each part for total										**800**
	0020	cost of connection										
	0100	Connector plates, steel, with bolts, straight	2 Carp	75	.213	Ea.	22.50	7.60		30.10	37	
	0110	Tee		50	.320		33	11.40		44.40	54	
	0120	T- Strap, 14 gauge, 300 x 200 x 50 mm		50	.320		33	11.40		44.40	54	
	0150	Anchor plate, 7 gauge, 225 mm x 75 mm		75	.213		22.50	7.60		30.10	37	
	0200	Bolts, machine, sq. hd. with nut & washer, 13 mm dia.,100 mm long	1 Carp	140	.057		.38	2.03		2.41	3.58	
	0300	188 mm long		130	.062		.67	2.19		2.86	4.15	
	0500	19 mm diameter, 190 mm long [3/4"]		130	.062		1.83	2.19		4.02	5.40	
	0610	Mach. bolts, w/nut, washer, 19 mm dia, 381 mm L, HD's & beam hgrs.		95	.084		3.36	2.99		6.35	8.35	
	0800	Drilling bolt holes in timber, 13 mm diameter [1/2"]		11430	.001	mm		.02		.02	.04	
	0900	25 mm diameter [1"]		8890	.001	"		.03		.03	.05	
	1100	Framing anchors, 2 or 3 dimensional, 10 gauge, no nails incl.		175	.046	Ea.	.56	1.63		2.19	3.15	
	1150	Framing anchors, 18 gauge, 114 mm x 75 mm		175	.046		.56	1.63		2.19	3.15	
	1160	Framing anchors, 18 gauge, 114 mmx 75 mm		175	.046		.56	1.63		2.19	3.15	
	1170	Clip anchors plates, 18 gauge, 300 mm x 29 mm		175	.046		.56	1.63		2.19	3.15	
	1250	Holdowns, 3 gauge base, 10 gauge body		8	1		18.75	35.50		54.25	76	
	1260	Holdowns, 7 gauge 280 mm x 83 mm		8	1		18.75	35.50		54.25	76	
	1270	Holdowns, 7 gauge 365 mm x 79 mm		8	1		18.75	35.50		54.25	76	
	1275	Holdowns, 12 gauge 200 mm x 64 mm		8	1		18.75	35.50		54.25	76	
	1300	Joist and beam hangers, 18 ga. galv., for 50 mm x 100 mm joist		175	.046		.70	1.63		2.33	3.30	
	1400	50 mm x 150 mm to 50 mm x 250 mm joist		165	.048		.80	1.72		2.52	3.56	
	1600	16 ga. galv., 75 x 150 mm to 75 x 250 mm joist		160	.050		3.16	1.78		4.94	6.25	
	1700	75 mm x 250 mm to 75 mm x 350 mm joist		160	.050		3.67	1.78		5.45	6.80	
	1800	100 mm x 150 mm to 100 mm x 250 mm joist		155	.052		2.78	1.83		4.61	5.90	
	1900	100 mm x 250 mm to 100 mm x 350 mm joist		155	.052		3.77	1.83		5.60	7	
	2000	Two-50 mm x 150 mm to two 50 mm x 250 mm joists		150	.053		3.04	1.90		4.94	6.30	
	2100	Two-50 mm x 250 mm to two 50 mm x 360 mm joists		150	.053		3.04	1.90		4.94	6.30	
	2300	5 mm thick, 150 mm x 200 mm joist		145	.055		6.60	1.96		8.56	10.30	
	2400	150 mm x 250 mm joist		140	.057		7.80	2.03		9.83	11.70	
	2500	150 mm x 300 mm joist		135	.059		9.40	2.11		11.51	13.60	
	2700	6 mm thick, 150 mm x 360 mm joist		130	.062		11.65	2.19		13.84	16.20	
	2900	Plywood clips, extruded aluminum H clip, for 19 mm panels					.18			.18	.20	
	3000	Galvanized 18 ga. back-up clip					.17			.17	.19	
	3200	Post framing, 16 ga. galv. for 100 mm x 100 mm base, 2 piece	1 Carp	130	.062		6.55	2.19		8.74	10.60	
	3300	Cap		130	.062		3.21	2.19		5.40	6.95	

06090	Wood & Plastic Fastenings	CREW	DAILY OUTPUT	LABOR-HOURS	UNIT	2006 BARE COSTS				TOTAL INCL O&P	
						MAT.	LABOR	EQUIP.	TOTAL		
800											800
3500	Rafter anchors, 18 ga. galv., 38 mm wide, 131 mm long	1 Carp	145	.055	Ea.	.55	1.96		2.51	3.66	
3600	269 mm long		145	.055		1.09	1.96		3.05	4.25	
3800	Shear plates, 66 mm diameter		120	.067		1.96	2.37		4.33	5.85	
3900	100 mm diameter		115	.070		4.50	2.47		6.97	8.80	
4000	Sill anchors, embedded in concrete or block, 466 mm long		115	.070		1.32	2.47		3.79	5.30	
4100	Spike grids, 100 mm x 100 mm, flat or curved		120	.067		.49	2.37		2.86	4.23	
4400	Split rings, 63 mm diameter		120	.067		1.58	2.37		3.95	5.45	
4500	100 mm diameter		110	.073		2.45	2.59		5.04	6.75	
4550	Tie plate, 20 gauge, 175 mm x 79 mm		110	.073		2.45	2.59		5.04	6.75	
4560	Tie plate, 20 gauge, 127 mm x 105 mm		110	.073		2.45	2.59		5.04	6.75	
4575	Twist straps, 18 gauge, 300 mm x 32 mm		110	.073		2.45	2.59		5.04	6.75	
4580	Twist straps, 18 gauge, 400 mm x 32 mm		110	.073		2.45	2.59		5.04	6.75	
4600	Strap ties, 20 ga., 52 mm wide, 325 mm long		180	.044		1.43	1.58		3.01	4.03	
4700	Strap ties, 16 ga., 34 mm wide, 300 mm long		180	.044		1.43	1.58		3.01	4.03	
4800	600 mm long		160	.050		1.97	1.78		3.75	4.94	
5000	Toothed rings, 66 mm or 100 mm diameter		90	.089		1.35	3.16		4.51	6.40	
5200	Truss plates, nailed, 20 gauge, up to 10 m span		17	.471	Truss	9.75	16.75		26.50	37	
5400	Washers, 50 mm x 50 mm x 3 mm				Ea.	.31			.31	.34	
5500	75 mm x 75 mm x 5 mm				"	.80			.80	.88	
6101	Beam hangers, polymer painted										
6102	Bolted, 3 ga., (W x H x L)										
6104	83 mm x 229 mm x 305 mm top flange	1 Carp	1	8	h	7,650	284		7,934	8,875	
6106	133 mm x 229 mm x 305 mm top flange		1	8		7,975	284		8,259	9,200	
6108	133 mm x 279 mm x 298 mm top flange		1	8		17,900	284		18,184	20,100	
6110	175 mm x 229 mm x 305 mm top flange		1	8		8,225	284		8,509	9,500	
6112	175 mm x 279 mm x 343 mm top flange		1	8		18,800	284		19,084	21,100	
6114	225 mm x 279 mm x 394 mm top flange		1	8		20,100	284		20,384	22,600	
6116	Nailed, 3 ga., (W x H x L)										
6118	83 mm x 267 mm x 254 mm top flange	1 Carp	1.80	4.444	h	5,300	158		5,458	6,075	
6120	83 mm x 267 mm x 305 mm top flange		1.80	4.444		6,125	158		6,283	7,000	
6122	133 mm x 241 mm x 254 mm top flange		1.80	4.444		5,700	158		5,858	6,525	
6124	133 mm x 241 mm x 305 mm top flange		1.80	4.444		6,475	158		6,633	7,375	
6126	140 mm x 241 mm x 305 mm top flange		1.80	4.444		5,700	158		5,858	6,525	
6128	175 mm x 216 mm x 305 mm top flange		1.80	4.444		5,900	158		6,058	6,725	
6130	191 mm x 216 mm x 305 mm top flange		1.80	4.444		6,100	158		6,258	6,950	
6132	225 mm x 191 mm x 356 mm top flange		1.80	4.444		6,750	158		6,908	7,650	
6201	Beam and purlin hangers, galvanized, 12 ga.										
6202	Purlin or joist size, 76 mm x 203 mm	1 Carp	1.70	4.706	h	965	167		1,132	1,325	
6204	76 mm x 254 mm		1.70	4.706		1,075	167		1,242	1,450	
6206	76 mm x 305 mm		1.65	4.848		1,275	172		1,447	1,700	
6208	76 mm x 356 mm		1.65	4.848		1,500	172		1,672	1,925	
6210	76 mm x 406 mm		1.65	4.848		1,725	172		1,897	2,175	
6212	102 mm x 203 mm		1.65	4.848		965	172		1,137	1,350	
6214	102 mm x 254 mm		1.65	4.848		1,125	172		1,297	1,525	
6216	102 mm x 305 mm		1.60	5		1,200	178		1,378	1,600	
6218	102 mm x 356 mm		1.60	5		1,325	178		1,503	1,725	
6220	102 mm x 406 mm		1.60	5		1,500	178		1,678	1,925	
6224	152 mm x 254 mm		1.55	5.161		1,650	183		1,833	2,100	
6226	152 mm x 305 mm		1.55	5.161		1,775	183		1,958	2,225	
6228	152 mm x 356 mm		1.50	5.333		1,900	190		2,090	2,400	
6230	152 mm x 406 mm		1.50	5.333		2,175	190		2,365	2,700	
6300	Column bases										
6302	102 mm x 102 mm, 16 ga.	1 Carp	1.80	4.444	h	1,350	158		1,508	1,725	
6306	7 ga.		1.80	4.444		2,500	158		2,658	3,000	
6314	152 mm x 152 mm, 16 ga.		1.75	4.571		1,725	163		1,888	2,150	
6318	7 ga.		1.75	4.571		3,500	163		3,663	4,100	

WOOD & PLASTICS **6**

06090	Wood & Plastic Fastenings	CREW	DAILY OUTPUT	LABOR-HOURS	UNIT	2006 BARE COSTS				TOTAL INCL O&P	
						MAT.	LABOR	EQUIP.	TOTAL		
800	6326	203 mm x 203 mm, 7 ga.	1 Carp	1.65	4.848	h	5,850	172		6,022	6,725	**800**
	6330	203 mm x 254 mm, 7 ga.	↓	1.65	4.848	↓	6,250	172		6,422	7,150	
	6590	Joist hangers, heavy duty 12 ga., galvanized										
	6592	51 mm x 102 mm	1 Carp	1.75	4.571	h	1,100	163		1,263	1,450	
	6594	51 mm x 152 mm		1.65	4.848		1,175	172		1,347	1,575	
	6595	51 mm x 152 mm		1.65	4.848		1,175	172		1,347	1,575	
	6596	51 mm x 203 mm		1.65	4.848		1,250	172		1,422	1,650	
	6597	51 mm x 203 mm		1.65	4.848		1,250	172		1,422	1,650	
	6598	51 mm x 254 mm		1.65	4.848		1,350	172		1,522	1,750	
	6600	51 mm x 305 mm		1.65	4.848		1,500	172		1,672	1,925	
	6622	(2) 51 mm x 152 mm		1.60	5		1,475	178		1,653	1,875	
	6624	(2) 51 mm x 203 mm		1.60	5		1,600	178		1,778	2,025	
	6626	(2) 51 mm x 254 mm		1.55	5.161		1,775	183		1,958	2,250	
	6628	(2) 51 mm x 305 mm	↓	1.55	5.161	↓	2,175	183		2,358	2,650	
	6890	Purlin hangers, painted										
	6892	12 ga., 51 mm x 152 mm	1 Carp	1.80	4.444	h	1,350	158		1,508	1,725	
	6894	51 mm x 203 mm		1.80	4.444		1,425	158		1,583	1,800	
	6896	51 mm x 254 mm		1.80	4.444		1,450	158		1,608	1,850	
	6898	51 mm x 305 mm		1.75	4.571		1,575	163		1,738	1,975	
	6934	(2) 51 mm x 152 mm		1.70	4.706		1,350	167		1,517	1,750	
	6936	(2) 51 mm x 203 mm		1.70	4.706		1,500	167		1,667	1,900	
	6938	(2) 51 mm x 254 mm		1.70	4.706		1,650	167		1,817	2,050	
	6940	(2) 51 mm x 305 mm	↓	1.65	4.848	↓	1,775	172		1,947	2,225	
825	0010	**ROUGH HARDWARE**, average % of carpentry material										**825**
	0020	Minimum					.50%					
	0200	Maximum					1.50%					
	0210	In seismic or hurricane areas, up to					10%					
850	0010	**BRACING**										**850**
	0300	Let-in, "T" shaped, 22 ga. galv. steel, studs at 410 mm O.C.	1 Carp	177	.045	m	1.74	1.61		3.35	4.41	
	0400	Studs at 610 mm O.C.		183	.044		1.74	1.55		3.29	4.33	
	0500	16 ga. galv. steel straps, studs at 410 mm O.C.		183	.044		2.62	1.55		4.17	5.30	
	0600	Studs at 610 mm O.C.	↓	189	.042	↓	2.62	1.50		4.12	5.25	

06110	Wood Framing	CREW	DAILY OUTPUT	LABOR-HOURS	UNIT	2006 BARE COSTS				TOTAL INCL O&P	
						MAT.	LABOR	EQUIP.	TOTAL		
100	0010	**BLOCKING**										**100**
	2600	Miscellaneous, to wood construction										
	2620	50 mm x 100 mm	1 Carp	.40	19.943	m³	236	710		946	1,350	
	2625	Pneumatic nailed		.50	16.144		236	575		811	1,150	
	2660	50 mm x 200 mm		.64	12.557		271	445		716	995	
	2665	Pneumatic nailed	↓	.78	10.274	↓	271	365		636	870	
	2720	To steel construction										
	2740	50 mm x 100 mm	1 Carp	.33	24.216	m³	236	860		1,096	1,600	
	2780	50 mm x 200 mm	"	.50	16.144	"	271	575		846	1,200	
150	0010	**BRACING**										**150**
	0011	Let-in, with 25 mm x 150 mm boards, studs @ 400 mm O.C.	1 Carp	45.72	.175	m	1.80	6.20		8	11.70	

WOOD & PLASTICS | **6**

			06110	Wood Framing		CREW	DAILY OUTPUT	LABOR-HOURS	UNIT	2006 BARE COSTS				TOTAL INCL O&P	
										MAT.	LABOR	EQUIP.	TOTAL		
150	0200			Studs @ 600 mm O.C.		1 Carp	70.10	.114	m	1.80	4.06		5.86	8.30	150
200	0010	**BRIDGING**													200
	0011		Wood, for joists 400 mm O.C., 25 mm x 75 mm			1 Carp	1.30	6.154	h.Pr.	42.50	219		261.50	385	
	0015		Pneumatic nailed				1.70	4.706		42.50	167		209.50	305	
	0100		50 mm x 75 mm bridging				1.30	6.154		57	219		276	405	
	0105		Pneumatic nailed				1.70	4.706		57	167		224	325	
	0300		Steel, galv., 18 ga., for 50 mm x 250 mm joists at 300 mm O.C.				1.30	6.154		118	219		337	470	
	0400		600 mm O.C.				1.40	5.714		142	203		345	470	
	0600		For 50 mm x 350 mm joists at 400 mm O.C.				1.30	6.154		139	219		358	490	
	0700		600 mm O.C.				1.40	5.714		228	203		431	565	
	0900		Compression type, 400 mm O.C., 50 mm x 200 mm joists				2	4		138	142		280	375	
	1000		50 mm x 300 mm joists			▼	2	4	▼	138	142		280	375	
505	0010	**FRAMING, BEAMS & GIRDERS**			R061110 -30										505
	1000		Single, 50 mm x 150 mm			2 Carp	213	.075	m	1.94	2.67		4.61	6.30	
	1005		Pneumatic nailed				247	.065		1.94	2.30		4.24	5.70	
	1020		50 mm x 200 mm				198	.081		2.79	2.87		5.66	7.55	
	1025		Pneumatic nailed				230	.070		2.79	2.47		5.26	6.95	
	1040		50 mm x 250 mm				183	.087		4.23	3.11		7.34	9.50	
	1045		Pneumatic nailed				212	.075		4.23	2.68		6.91	8.85	
	1060		50 mm x 300 mm				168	.095		5.25	3.39		8.64	11	
	1065		Pneumatic nailed				194	.082		5.25	2.93		8.18	10.30	
	1080		50 mm x 350 mm				152	.105		7.15	3.74		10.89	13.70	
	1085		Pneumatic nailed				177	.090		7.15	3.21		10.36	12.85	
	1100		75 mm x 200 mm				168	.095		8.50	3.39		11.89	14.60	
	1120		75 mm x 250 mm				152	.105		10.70	3.74		14.44	17.65	
	1140		75 mm x 300 mm				137	.117		12.85	4.15		17	20.50	
	1160		75 mm x 350 mm			▼	122	.131		15.50	4.66		20.16	24.50	
	1180		100 mm x 200 mm			F-3	305	.131		8.50	4.73	2.09	15.32	18.95	
	1200		100 mm x 250 mm				290	.138		10.70	4.97	2.20	17.87	22	
	1220		100 mm x 300 mm				274	.146		13.25	5.25	2.32	20.82	25.50	
	1240		100 mm x 350 mm			▼	259	.154		16.25	5.55	2.46	24.26	29	
	2000		Double, 50 mm x 150 mm			2 Carp	191	.084		3.87	2.98		6.85	8.90	
	2005		Pneumatic nailed				221	.072		3.87	2.57		6.44	8.30	
	2020		50 mm x 200 mm				175	.091		5.60	3.25		8.85	11.20	
	2025		Pneumatic nailed				203	.079		5.60	2.80		8.40	10.50	
	2040		50 mm x 250 mm				168	.095		8.45	3.39		11.84	14.55	
	2045		Pneumatic nailed				194	.082		8.45	2.93		11.38	13.85	
	2060		50 mm x 300 mm				160	.100		10.50	3.56		14.06	17.10	
	2065		Pneumatic nailed				186	.086		10.50	3.06		13.56	16.30	
	2080		50 mm x 350 mm				145	.110		14.30	3.92		18.22	22	
	2085		Pneumatic nailed				168	.095		14.30	3.39		17.69	21	
	3000		Triple, 50 mm x 150 mm				168	.095		5.85	3.39		9.24	11.65	
	3005		Pneumatic nailed				194	.082		5.85	2.93		8.78	10.95	
	3020		50 mm x 200 mm				160	.100		8.40	3.56		11.96	14.80	
	3025		Pneumatic nailed				186	.086		8.40	3.06		11.46	14	
	3040		50 mm x 250 mm				152	.105		12.70	3.74		16.44	19.85	
	3045		Pneumatic nailed				177	.090		12.70	3.21		15.91	19	
	3060		50 mm x 300 mm				145	.110		15.75	3.92		19.67	23.50	
	3065		Pneumatic nailed				168	.095		15.75	3.39		19.14	22.50	
	3080		50 mm x 350 mm				137	.117		21.50	4.15		25.65	30	
	3085		Pneumatic nailed				159	.101	▼	21.50	3.58		25.08	29	
	3500		Single, 50 mm x 150 mm				1.65	9.686	m³	251	345		596	810	
	3505		Pneumatic nailed				1.92	8.350		251	297		548	735	
	3520		50 mm x 200 mm			▼ ▼	2.03	7.884	▼	271	280		551	735	

06110 | Wood Framing

			CREW	DAILY OUTPUT	LABOR-HOURS	UNIT	2006 BARE COSTS MAT.	LABOR	EQUIP.	TOTAL	TOTAL INCL O&P	
505	3525	Pneumatic nailed	2 Carp	2.35	6.801	m³	271	242		513	675	505
	3540	50 mm x 250 mm	R061110-30	2.36	6.781		330	241		571	735	
	3545	Pneumatic nailed		2.74	5.845		330	208		538	685	
	3560	50 mm x 300 mm		2.60	6.164		340	219		559	715	
	3565	Pneumatic nailed		3.01	5.314		340	189		529	670	
	3580	50 mm x 350 mm		2.76	5.795		395	206		601	755	
	3585	Pneumatic nailed		3.20	4.997		395	178		573	710	
	3600	75 mm x 200 mm		2.60	6.164		550	219		769	945	
	3620	75 mm x 250 mm		2.95	5.424		555	193		748	910	
	3640	75 mm x 300 mm		3.19	5.023		555	179		734	890	
	3660	75 mm x 350 mm		3.30	4.843		570	172		742	900	
	3680	100 mm x 200 mm	F-3	6.28	6.373		410	230	101	741	920	
	3700	100 mm x 250 mm		7.46	5.364		415	193	85.50	693.50	850	
	3720	100 mm x 300 mm		8.49	4.709		430	170	75	675	815	
	3740	100 mm x 350 mm		9.34	4.281		450	154	68	672	810	
	4000	Double, 50 mm x 150 mm	2 Carp	2.95	5.424		251	193		444	575	
	4005	Pneumatic nailed		3.42	4.676		251	166		417	535	
	4020	50 mm x 200 mm		3.78	4.238		271	151		422	535	
	4025	Pneumatic nailed		4.38	3.653		271	130		401	500	
	4040	50 mm x 250 mm		4.53	3.532		330	126		456	555	
	4045	Pneumatic nailed		5.26	3.045		330	108		438	530	
	4060	50 mm x 300 mm		5.19	3.082		340	110		450	545	
	4065	Pneumatic nailed		6.02	2.659		340	94.50		434.50	520	
	4080	50 mm x 350 mm		5.78	2.768		395	98.50		493.50	590	
	4085	Pneumatic nailed		6.70	2.388		395	85		480	565	
	5000	Triple, 50 mm x 150 mm		3.89	4.109		251	146		397	505	
	5005	Pneumatic nailed		4.51	3.550		251	126		377	470	
	5020	50 mm x 200 mm		4.96	3.229		271	115		386	475	
	5025	Pneumatic nailed		5.75	2.783		271	99		370	450	
	5040	50 mm x 250 mm		5.90	2.712		330	96.50		426.50	510	
	5045	Pneumatic nailed		6.84	2.338		330	83		413	490	
	5060	50 mm x 300 mm		6.73	2.379		340	84.50		424.50	505	
	5065	Pneumatic nailed		7.80	2.051		340	73		413	490	
	5080	50 mm x 350 mm		7.43	2.153		395	76.50		471.50	555	
	5085	Pneumatic nailed		7.91	2.022		395	72		467	545	
510	0010	**FRAMING, CEILINGS**										510
	6400	Suspended, 50 mm x 75 mm	2 Carp	1.18	13.561	m³	320	480		800	1,100	
	6450	50 mm x 100 mm		1.39	11.492		236	410		646	895	
	6500	50 mm x 150 mm		1.89	8.476		251	300		551	745	
	6550	50 mm x 200 mm		2.03	7.884		271	280		551	735	
515	0010	**FRAMING, COLUMNS**										515
	0100	100 mm x 100 mm	2 Carp	119	.134	m	4.10	4.78		8.88	12	
	0150	100 mm x 150 mm		83.82	.191		6.15	6.80		12.95	17.30	
	0200	100 mm x 200 mm		67.06	.239		8.50	8.50		17	22.50	
	0250	150 mm x 150 mm		65.53	.244		14.90	8.70		23.60	30	
	0300	150 mm x 200 mm		53.34	.300		19.80	10.65		30.45	38.50	
	0350	150 mm x 250 mm		45.72	.350		35.50	12.45		47.95	59	
	0400	100 mm x 100 mm		1.23	13.040	m³	400	465		865	1,150	
	0420	100 mm x 150 mm		1.30	12.328		395	440		835	1,125	
	0440	100 mm x 200 mm		1.39	11.492		410	410		820	1,100	
	0460	150 mm x 150 mm		1.53	10.432		640	370		1,010	1,275	
	0480	150 mm x 200 mm		1.65	9.686		640	345		985	1,225	
	0500	150 mm x 250 mm		1.77	9.041		925	320		1,245	1,525	

6 WOOD & PLASTICS

06110 | Wood Framing

		CREW	DAILY OUTPUT	LABOR-HOURS	UNIT	MAT.	LABOR	EQUIP.	TOTAL	TOTAL INCL O&P
520 0010	**HEAVY MILL TIMBER FRAMING**									**520**
0020	Beams, single 150 mm x 250 mm	2 Carp	2.60	6.164	m³	910	219		1,129	1,350
0100	Single 200 mm x 400 mm		2.83	5.650		1,150	201		1,351	1,575
0200	Built from 50 mm lumber, multiple 50 mm x 350 mm		2.12	7.534		395	268		663	850
0210	Built from 75 mm lumber, multiple 75 mm x 150 mm		1.65	9.686		540	345		885	1,125
0220	Multiple 75 mm x 200 mm		1.89	8.476		550	300		850	1,075
0230	Multiple 75 mm x 250 mm		2.12	7.534		555	268		823	1,025
0240	Multiple 75 mm x 300 mm		2.36	6.781		555	241		796	985
0250	Built from 100 mm lumber, multiple 100 mm x 150 mm		1.89	8.476		395	300		695	905
0260	Multiple 100 mm x 200 mm		2.12	7.534		410	268		678	870
0270	Multiple 100 mm x 250 mm		2.36	6.781		415	241		656	830
0280	Multiple 100 mm x 300 mm		2.60	6.164		430	219		649	810
0290	Columns, structural grade, 10 MPa, 100 mm x 100 mm		1.42	11.301		745	400		1,145	1,450
0300	150 mm x 150 mm		1.53	10.432		905	370		1,275	1,575
0400	200 mm x 200 mm		1.65	9.686		990	345		1,335	1,625
0500	250 mm x 250 mm		1.77	9.041		1,000	320		1,320	1,600
0600	300 mm x 300 mm		1.89	8.476		1,025	300		1,325	1,600
0800	Floor planks, 50 mm thick, T & G, 50 mm x 150 mm		2.48	6.458		795	230		1,025	1,225
0900	50 mm x 250 mm		2.60	6.164		795	219		1,014	1,225
1100	75 mm thick, 75 mm x 150 mm		2.48	6.458		570	230		800	985
1200	75 mm x 250 mm		2.60	6.164		570	219		789	970
1400	Girders, structural grade, 300 mm x 300 mm		1.89	8.476		820	300		1,120	1,375
1500	250 mm x 400 mm	▼	2.36	6.781	▼	800	241		1,041	1,250
2050	Roof planks, see division 06150-600									
2300	Roof purlins, 100 mm thick, structural grade	2 Carp	2.48	6.458	m³	560	230		790	970
2500	Roof trusses, add timber connectors, division 06090-800	"	1.06	15.068	"	545	535		1,080	1,425
530 0010	**FRAMING, JOISTS** R061110-30									**530**
2000	Joists, 50 mm x 100 mm	2 Carp	381	.042	m	1.21	1.49		2.70	3.67
2005	Pneumatic nailed		438	.037		1.21	1.30		2.51	3.37
2100	50 mm x 150 mm		381	.042		1.94	1.49		3.43	4.45
2105	Pneumatic nailed		438	.037		1.94	1.30		3.24	4.15
2150	50 mm x 200 mm		335	.048		2.79	1.70		4.49	5.70
2155	Pneumatic nailed		386	.041		2.79	1.47		4.26	5.35
2200	50 mm x 250 mm		274	.058		4.23	2.08		6.31	7.90
2205	Pneumatic nailed		315	.051		4.23	1.81		6.04	7.45
2250	50 mm x 300 mm		267	.060		5.25	2.13		7.38	9.05
2255	Pneumatic nailed		307	.052		5.25	1.85		7.10	8.65
2300	50 mm x 350 mm		235	.068		7.15	2.42		9.57	11.60
2305	Pneumatic nailed		270	.059		7.15	2.11		9.26	11.15
2350	75 mm x 150 mm		282	.057		6.30	2.02		8.32	10.05
2400	75 mm x 250 mm		238	.067		10.70	2.39		13.09	15.50
2450	75 mm x 300 mm		183	.087		12.85	3.11		15.96	19
2500	100 mm x 150 mm		244	.066		6.15	2.33		8.48	10.40
2550	100 mm x 250 mm		183	.087		10.70	3.11		13.81	16.65
2600	100 mm x 300 mm		137	.117		13.25	4.15		17.40	21
2605	Sister joist, 50 mm x 150 mm		244	.066		1.94	2.33		4.27	5.75
2606	Pneumatic nailed		293	.055		1.94	1.94		3.88	5.15
2610	50 mm x 200 mm		195	.082		2.79	2.92		5.71	7.60
2611	Pneumatic nailed		234	.068		2.79	2.43		5.22	6.85
2615	50 mm x 250 mm		163	.098		4.23	3.49		7.72	10.10
2616	Pneumatic nailed		196	.082		4.23	2.90		7.13	9.20
2620	50 mm x 300 mm		139	.115		5.25	4.09		9.34	12.10
2625	Pneumatic nailed		166	.096	▼	5.25	3.43		8.68	11.10
2650	Joists, 50 mm x 100 mm		1.96	8.169	m³	236	290		526	710
2655	Pneumatic nailed		2.27	7.063		236	251		487	650
2680	50 mm x 150 mm	▼	2.95	5.424	▼	251	193		444	575

WOOD & PLASTICS 5

06110 | Wood Framing

		CREW	DAILY OUTPUT	LABOR-HOURS	UNIT	2006 BARE COSTS MAT.	LABOR	EQUIP.	TOTAL	TOTAL INCL O&P	
530	2685	Pneumatic nailed	2 Carp	3.40	4.709	m³	251	167		418	535	**530**
	2700	50 mm x 200 mm		3.45	4.644		271	165		436	555	
	2705	Pneumatic nailed		3.96	4.036		271	143		414	520	
	2720	50 mm x 250 mm		3.52	4.551		330	162		492	610	
	2725	Pneumatic nailed		4.04	3.965		330	141		471	580	
	2740	50 mm x 300 mm		4.13	3.875		340	138		478	590	
	2745	Pneumatic nailed		4.74	3.373		340	120		460	560	
	2760	50 mm x 350 mm		4.22	3.788		395	135		530	645	
	2765	Pneumatic nailed		4.86	3.292		395	117		512	615	
	2780	75 mm x 150 mm		3.28	4.878		540	173		713	865	
	2790	75 mm x 200 mm		4.48	3.569		550	127		677	805	
	2800	75 mm x 250 mm		4.60	3.477		555	124		679	800	
	2820	75 mm x 300 mm		4.25	3.767		555	134		689	820	
	2840	100 mm x 150 mm		3.78	4.238		395	151		546	670	
	2850	100 mm x 200 mm		9.79	1.634		410	58		468	545	
	2860	100 mm x 250 mm		4.72	3.390		415	121		536	645	
	2880	100 mm x 300 mm		4.25	3.767	▼	430	134		564	680	
	3000	Composite wood joist 250 mm deep		.27	58.326	km	5,950	2,075		8,025	9,775	
	3010	300 mm deep		.27	59.653		6,425	2,125		8,550	10,400	
	3020	350 mm deep		.25	64.015		7,375	2,275		9,650	11,700	
	3030	400 mm deep		.24	67.300		8,350	2,400		10,750	12,900	
	4000	Open web joist 300 mm deep		.27	59.653		5,975	2,125		8,100	9,875	
	4010	350 mm deep		.25	64.015		6,975	2,275		9,250	11,200	
	4020	400 mm deep		.24	67.300		7,200	2,400		9,600	11,700	
	4030	450 mm deep		.23	70.938		7,350	2,525		9,875	12,000	
	6000	Composite rim joist, 32 mm x 240 mm		27.43	.583		6,525	20.50		6,545.50	7,200	
	6010	32 mm x 290 mm		.27	59.653		7,325	2,125		9,450	11,400	
	6020	32 mm x 365 mm		.25	64.015		9,200	2,275		11,475	13,700	
	6030	32 mm x 415 mm	▼	▼	.24	67.300	▼	10,500	2,400	12,900	15,200	
540	0010	**FRAMING MEMBERS**										**540**
	0100	25 mm x 25 mm				m	.32			.32	.35	
	0150	25 mm x 50 mm					.64			.64	.70	
	0200	25 mm x 75 mm					.93			.93	1.02	
	0250	25 mm x 100 mm					1.15			1.15	1.26	
	0300	50 mm x 75 mm					1.25			1.25	1.37	
	0350	50 mm x 100 mm					1.22			1.22	1.34	
	0400	50 mm x 150 mm					1.94			1.94	2.14	
	0450	50 mm x 200 mm					2.80			2.80	3.08	
	0500	50 mm x 250 mm					4.23			4.23	4.66	
	0550	50 mm x 300 mm					5.25			5.25	5.75	
	0600	50 mm x 350 mm					7.15			7.15	7.85	
	0650	75 mm x 150 mm					6.30			6.30	6.95	
	0700	75 mm x 200 mm					8.50			8.50	9.35	
	0750	75 mm x 250 mm					10.70			10.70	11.75	
	0800	75 mm x 300 mm					12.85			12.85	14.15	
	0850	75 mm x 350 mm					15.50			15.50	17.05	
	0900	100 mm x 150 mm					6.15			6.15	6.75	
	0950	100 mm x 200 mm					8.50			8.50	9.35	
	1000	100 mm x 250 mm					10.70			10.70	11.80	
	1100	100 mm x 300 mm					13.25			13.25	14.60	
	1150	100 mm x 350 mm				▼	16.25			16.25	17.85	
545	0010	**FRAMING, MISCELLANEOUS**										**545**
	8500	Firestops, 50 mm x 100 mm	2 Carp	1.20	13.295	m³	236	475		711	995	
	8505	Pneumatic nailed		1.46	10.936		236	390		626	865	
	8520	50 mm x 150 mm	▼	1.42	11.301	▼	251	400		651	900	

R061110 -30

Important: See the Reference Section for supporting data - Crews, Rental Equipment, City Cost Indexes and Reference Data

6

WOOD & PLASTICS

06110	Wood Framing	CREW	DAILY OUTPUT	LABOR-HOURS	UNIT	MAT.	LABOR	EQUIP.	TOTAL	TOTAL INCL O&P	
							2006 BARE COSTS				
545	8525	Pneumatic nailed	2 Carp	1.73	9.263	m³	251	330		581	790	**545**
	8540	50 mm x 200 mm		1.42	11.301		271	400		671	925	
	8560	50 mm x 305 mm		1.65	9.686		340	345		685	910	
	8600	Nailers, treated, wood construction, 50 mm x 100 mm		1.25	12.793		310	455		765	1,050	
	8605	Pneumatic nailed		1.50	10.661		310	380		690	930	
	8620	50 mm x 150 mm		1.77	9.041		315	320		635	850	
	8625	Pneumatic nailed		2.12	7.534		315	268		583	765	
	8640	50 mm x 200 mm		2.19	7.291		310	259		569	745	
	8645	Pneumatic nailed		2.63	6.076		310	216		526	675	
	8660	Steel construction, 50 mm x 100 mm		1.18	13.561		310	480		790	1,100	
	8680	50 mm x 150 mm		1.65	9.686		315	345		660	885	
	8700	50 mm x 200 mm		2.05	7.794		310	277		587	770	
	8760	Rough bucks, treated, for doors or windows, 50 mm x 150 mm		.94	16.951		315	605		920	1,300	
	8765	Pneumatic nailed		1.13	14.126		315	500		815	1,125	
	8780	50 mm x 200 mm		1.20	13.295		310	475		785	1,075	
	8785	Pneumatic nailed		1.44	11.079		310	395		705	955	
	8800	Stair stringers, 50 mm x 250 mm		.52	30.821		330	1,100		1,430	2,050	
	8820	50 mm x 300 mm		.61	26.079		340	925		1,265	1,825	
	8840	75 mm x 250 mm		.73	21.873		555	780		1,335	1,800	
	8860	75 mm x 300 mm		.90	17.843		555	635		1,190	1,600	
	8870	Composite LSL, 32 mm x 290 mm		39.62	.404	m	7.30	14.35		21.65	30.50	
	8880	32 mm x 365 mm		39.62	.404	"	9.20	14.35		23.55	32.50	
550	0010	**PARTITIONS** Wood stud with single bottom plate and										**550**
	0020	double top plate, no waste, std. & better lumber										
	0180	50 mm x 100 mm studs, 2450 mm high, studs 300 mm O.C.	2 Carp	24.38	.656	m	13.40	23.50		36.90	51.50	
	0185	300 mm O.C., pneumatic nailed		29.26	.547		13.40	19.45		32.85	45.50	
	0200	400 mm O.C.		30.48	.525		11	18.65		29.65	41	
	0205	400 mm O.C., pneumatic nailed		36.58	.437		11	15.55		26.55	36	
	0300	600 mm O.C.		38.10	.420		8.55	14.95		23.50	32.50	
	0305	600 mm O.C., pneumatic nailed		45.72	.350		8.55	12.45		21	29	
	0380	3050 mm high, studs 300 mm O.C.		24.38	.656		15.90	23.50		39.40	54	
	0385	300 mm O.C., pneumatic nailed		29.26	.547		15.90	19.45		35.35	48	
	0400	400 mm O.C.		30.48	.525		12.85	18.65		31.50	43	
	0405	400 mm O.C., pneumatic nailed		36.58	.437		12.85	15.55		28.40	38	
	0500	600 mm O.C.		38.10	.420		9.80	14.95		24.75	34	
	0505	600 mm O.C., pneumatic nailed		45.72	.350		9.80	12.45		22.25	30	
	0580	3650 mm high, studs 300 mm O.C.		19.81	.808		18.30	28.50		46.80	64.50	
	0585	300 mm O.C., pneumatic nailed		23.77	.673		18.30	24		42.30	57.50	
	0600	400 mm O.C.		24.38	.656		14.65	23.50		38.15	52.50	
	0605	400 mm O.C., pneumatic nailed		29.26	.547		14.65	19.45		34.10	46.50	
	0700	600 mm O.C.		30.48	.525		11	18.65		29.65	41	
	0705	600 mm O.C., pneumatic nailed		36.58	.437		11	15.55		26.55	36	
	0780	50 mm x 150 mm studs, 2450 mm high, studs 300 mm O.C.		21.34	.750		21.50	26.50		48	65	
	0785	300 mm O.C., pneumatic nailed		25.60	.625		21.50	22		43.50	58	
	0800	400 mm O.C.		27.43	.583		17.50	20.50		38	52	
	0805	400 mm O.C., pneumatic nailed		32.92	.486		17.50	17.30		34.80	46.50	
	0900	600 mm O.C.		35.05	.456		13.60	16.25		29.85	40.50	
	0905	600 mm O.C., pneumatic nailed		42.06	.380		13.60	13.50		27.10	36	
	0980	3050 mm high, studs 300 mm O.C.		21.34	.750		25.50	26.50		52	69.50	
	0985	300 mm O.C., pneumatic nailed		25.60	.625		25.50	22		47.50	62.50	
	1000	400 mm O.C.		27.43	.583		20.50	20.50		41	55	
	1005	400 mm O.C., pneumatic nailed		32.92	.486		20.50	17.30		37.80	49.50	
	1100	600 mm O.C.		35.05	.456		15.55	16.25		31.80	42.50	
	1105	600 mm O.C., pneumatic nailed		42.06	.380		15.55	13.50		29.05	38	
	1180	3650 mm high, studs 300 mm O.C.		16.76	.954		29	34		63	85	
	1185	300 mm O.C., pneumatic nailed		20.12	.795		29	28.50		57.50	76	

WOOD & PLASTICS 6

06110	Wood Framing	CREW	DAILY OUTPUT	LABOR-HOURS	UNIT	2006 BARE COSTS				TOTAL INCL O&P
						MAT.	LABOR	EQUIP.	TOTAL	
550 1200	400 mm O.C.	2 Carp	21.34	.750	m	23.50	26.50		50	67	**550**
1205	400 mm O.C., pneumatic nailed		25.60	.625		23.50	22		45.50	60
1300	600 mm O.C.		27.43	.583		17.50	20.50		38	52
1305	600 mm O.C., pneumatic nailed		32.92	.486		17.50	17.30		34.80	46.50
1400	For horizontal blocking, 50 mm x 100 mm, add		183	.087		1.21	3.11		4.32	6.20
1500	50 mm x 150 mm, add		183	.087		1.94	3.11		5.05	6.95
1600	For openings, add	▼	76.20	.210	▼		7.45		7.45	11.60
1700	Headers for above openings, material only, add				m³	271			271	298
552 0010	**FRAMING, PORCH OR DECK**										**552**
0100	Treated lumber, posts or columns, 100 mm x 100 mm	2 Carp	119	.134	m	4.63	4.78		9.41	12.55
0110	100 mm x 150 mm		83.82	.191		8.15	6.80		14.95	19.55
0120	100 mm x 200 mm		67.06	.239		10.90	8.50		19.40	25
0130	Girder, single, 100 mm x 100 mm		206	.078		4.63	2.76		7.39	9.40
0140	100 mm x 150 mm		183	.087		8.15	3.11		11.26	13.85
0150	100 mm x 200 mm		160	.100		10.90	3.56		14.46	17.50
0160	Double, 50 mm x 100 mm		191	.084		3.25	2.98		6.23	8.20
0170	50 mm x 150 mm		183	.087		5	3.11		8.11	10.35
0180	50 mm x 200 mm		175	.091		6.55	3.25		9.80	12.25
0190	50 mm x 250 mm		168	.095		9.95	3.39		13.34	16.20
0200	50 mm x 300 mm		160	.100		13	3.56		16.56	19.85
0210	Triple, 50 mm x 100 mm		175	.091		4.89	3.25		8.14	10.45
0220	50 mm x 150 mm		168	.095		7.50	3.39		10.89	13.50
0230	50 mm x 200 mm		160	.100		9.80	3.56		13.36	16.35
0240	50 mm x 250 mm		152	.105		14.95	3.74		18.69	22.50
0250	50 mm x 300 mm		145	.110		19.50	3.92		23.42	27.50
0260	Ledger, bolted 1200 mm O.C., 50 x 100 mm		122	.131		1.94	4.66		6.60	9.40
0270	50 x 150 mm		120	.133		2.79	4.74		7.53	10.50
0280	50 x 200 mm		119	.134		3.54	4.78		8.32	11.35
0300	50 x 300 mm		116	.138		6.75	4.90		11.65	15.05
0310	Joists, 50 mm x 100 mm		381	.042		1.64	1.49		3.13	4.12
0320	50 mm x 150 mm		381	.042		2.53	1.49		4.02	5.10
0330	50 mm x 200 mm		335	.048		3.28	1.70		4.98	6.25
0340	50 mm x 250 mm		274	.058		5	2.08		7.08	8.75
0350	50 mm x 300 mm	▼	267	.060		6	2.13		8.13	9.90
0360	Railings and trim , 25 mm x 102 mm	1 Carp	91.44	.087		1.97	3.11		5.08	7
0370	50 mm x 50 mm		91.44	.087		.98	3.11		4.09	5.90
0380	51 mm x 102 mm		91.44	.087		1.61	3.11		4.72	6.60
0390	51 mm x 152 mm		91.44	.087	▼	2.46	3.11		5.57	7.55
0400	Decking, 25 mm x 102 mm		25.55	.313	m²	20.50	11.15		31.65	40
0410	51 mm x 102 mm		27.87	.287		17.85	10.20		28.05	35.50
0420	51 mm x 152 mm		29.73	.269		17.55	9.55		27.10	34
0430	32 mm x 152 mm	▼	29.73	.269	▼	13.90	9.55		23.45	30
0440	Balusters, square, 50 mm x 50 mm	2 Carp	67.06	.239	m	1.02	8.50		9.52	14.30
0450	Turned, 50 mm x 50 mm		42.67	.375		1.35	13.35		14.70	22.50
0460	Stair stringer, 50 mm x 250 mm		39.62	.404		5	14.35		19.35	28
0470	50 mm x 300 mm		39.62	.404		6	14.35		20.35	29
0480	Stair treads, 25 mm x 100 mm		42.67	.375		2.10	13.35		15.45	23.50
0490	50 mm x 100 mm		42.67	.375		1.64	13.35		14.99	23
0500	50 mm x 150 mm		48.77	.328		2.46	11.65		14.11	21
0510	32 mm x 150 mm		48.77	.328	▼	1.97	11.65		13.62	20.50
0520	Turned handrail post, 100 x 100 mm		64	.250	Ea.	13.05	8.90		21.95	28
0530	Lattice panel, 1200 mm x 2400 mm		149	.107	m²	7.55	3.82		11.37	14.25
0540	Cedar, posts or columns, 100 mm x 100 mm		119	.134	m	10.45	4.78		15.23	18.95
0550	100 mm x 150 mm		83.82	.191		15.85	6.80		22.65	28
0560	100 mm x 200 mm		67.06	.239		24	8.50		32.50	39
0800	Decking, 25 mm x 100 Cam	▼	168	.095	▼	4.46	3.39		7.85	10.15

	06110	Wood Framing	CREW	DAILY OUTPUT	LABOR-HOURS	UNIT	MAT.	LABOR	EQUIP.	TOTAL	TOTAL INCL O&P	
							2006 BARE COSTS					
552	0810	50 mm x 100 mm	2 Carp	183	.087	m	8.65	3.11		11.76	14.35	**552**
	0820	50 mm x 150 mm		195	.082		13.90	2.92		16.82	19.85	
	0830	32 mm x 150 mm		195	.082		12.45	2.92		15.37	18.25	
	0840	Railings and trim, 25 mm x 100 mm		183	.087		4.46	3.11		7.57	9.75	
	0860	50 mm x 100 mm		183	.087		8.65	3.11		11.76	14.35	
	0870	50 mm x 150 mm		183	.087		13.90	3.11		17.01	20	
	0920	Stair treads, 25 mm x 100 mm		42.67	.375		4.46	13.35		17.81	26	
	0930	50 mm x 100 mm		42.67	.375		8.65	13.35		22	30.50	
	0940	50 mm x 150 mm		48.77	.328		13.90	11.65		25.55	33.50	
	0950	32 mm x 150 mm		48.77	.328		12.45	11.65		24.10	32	
	0980	Redwood, posts or columns, 100 mm x 100 mm		119	.134		21	4.78		25.78	30.50	
	0990	100 mm x 150 mm		83.82	.191		35	6.80		41.80	49	
	1000	100 mm x 200 mm	▼	67.06	.239	▼	61	8.50		69.50	80	
	1240	Redwood decking, 25 mm x 102 mm	1 Carp	25.55	.313	m²	54.50	11.15		65.65	77.50	
	1260	51 mm x 152 mm		31.59	.253		135	9		144	163	
	1270	32 mm x 152 mm	▼	29.73	.269	▼	87	9.55		96.55	111	
	1280	Railings and trim, 25 mm x 100 mm	2 Carp	183	.087	m	4.86	3.11		7.97	10.20	
	1310	50 mm x 150 mm		183	.087		18.95	3.11		22.06	26	
	1420	Alternative decking, wood / plastic composite, 38 x 150 mm		195	.082		6.35	2.92		9.27	11.50	
	1430	Vinyl, 38 mm x 140 mm		195	.082		11.40	2.92		14.32	17.10	
	1440	25 mm x 100 mm square edge fir		168	.095		4.27	3.39		7.66	9.95	
	1450	25 mm x 100 mm tongue and groove fir		137	.117		4.27	4.15		8.42	11.15	
	1460	25 mm x 100 mm mahogany	▼	168	.095	▼	4.27	3.39		7.66	9.95	
	1470	Accessories, joist hangers, 50 mm x 100 mm	1 Carp	160	.050	Ea.	.70	1.78		2.48	3.54	
	1480	50 mm x 100 mm - 50 mm x 300 mm	"	150	.053		.80	1.90		2.70	3.83	
	1530	Post ftg, incl excav, backfill, tube form & conc, 1200 x 200 mm dia	F-7	12	2.667		10.40	84		94.40	142	
	1540	250 mm diameter		11	2.909		14.70	91.50		106.20	159	
	1550	300 mm diameter	▼	10	3.200	▼	19.70	101		120.70	179	
555	0010	**FRAMING, ROOFS**										**555**
	5250	Composite rafter, 240 mm deep	2 Carp	175	.091	m	5.95	3.25		9.20	11.60	
	5260	290 mm deep		175	.091	"	6.45	3.25		9.70	12.10	
	6070	Fascia boards, 50 mm x 200 mm		.71	22.602	m³	271	805		1,076	1,550	
	6080	50 mm x 250 mm		.71	22.602		330	805		1,135	1,600	
	7000	Rafters, to 4 in 12 pitch, 50 mm x 150 mm		2.36	6.781		251	241		492	650	
	7060	50 mm x 200 mm		2.97	5.381		271	191		462	595	
	7300	Hip and valley rafters, 50 mm x 150 mm		1.79	8.922		251	315		566	770	
	7360	50 mm x 200 mm		2.27	7.063		271	251		522	690	
	7540	Hip and valley jacks, 50 mm x 150 mm		1.42	11.301		251	400		651	900	
	7600	50 mm x 200 mm	▼	1.53	10.432	▼	271	370		641	875	
	7780	For slopes steeper than 4 in 12, add						30%				
	7790	For dormers or complex roofs, add						50%				
	7800	Rafter tie, 25 mm x 100 mm, #3	2 Carp	.64	25.113	m³	445	895		1,340	1,900	
	7820	Ridge board, #2 or better, 25 mm x 150 mm		.71	22.602		590	805		1,395	1,900	
	7840	25 mm x 200 mm		.87	18.326		565	650		1,215	1,650	
	7860	25 mm x 250 mm		.99	16.144		660	575		1,235	1,625	
	7880	50 mm x 150 mm		1.18	13.561		251	480		731	1,025	
	7900	50 mm x 200 mm		1.42	11.301		271	400		671	925	
	7920	50 mm x 250 mm		1.56	10.274		330	365		695	930	
	7940	Roof cants, split, 100 mm x 100 mm		2.03	7.884		400	280		680	875	
	7960	150 mm x 150 mm		4.25	3.767		640	134		774	915	
	7980	Roof curbs, untreated, 50 mm x 150 mm		1.23	13.040		251	465		716	995	
	8000	50 mm x 300 mm		1.89	8.476		340	300		640	845	
	8020	Sister rafters, 50 mm x 150 mm		1.89	8.476		251	300		551	745	
	8040	50 mm x 200 mm		2.01	7.977		271	284		555	740	
	8060	50 mm x 250 mm		2.10	7.619		330	271		601	780	
	8080	50 mm x 300 mm	▼	2.15	7.451	▼	340	265		605	785	

WOOD & PLASTICS | 6

For expanded coverage of these items see *Means Interior Cost Data 2006*

06110 | Wood Framing

		CREW	DAILY OUTPUT	LABOR-HOURS	UNIT	2006 BARE COSTS MAT.	LABOR	EQUIP.	TOTAL	TOTAL INCL O&P	
560	**0010**	**FRAMING, SILLS**									**560**
	2000	Ledgers, nailed, 50 mm x 100 mm	2 Carp	230	.070	m	1.21	2.47		3.68	5.20
	2050	50 mm x 150 mm		183	.087		1.94	3.11		5.05	6.95
	2100	Bolted, not including bolts, 75 mm x 150 mm		99.06	.162		6.30	5.75		12.05	15.85
	2150	75 mm x 300 mm		71.02	.225		12.85	8		20.85	26.50
	2600	Mud sills, redwood, construction grade, 50 mm x 100 mm		273	.059		7.20	2.08		9.28	11.20
	2620	50 mm x 150 mm		238	.067		10.85	2.39		13.24	15.60
	4000	Sills, 50 mm x 100 mm		183	.087		1.21	3.11		4.32	6.20
	4050	50 mm x 150 mm		168	.095		1.94	3.39		5.33	7.40
	4080	50 mm x 200 mm		152	.105		2.79	3.74		6.53	8.95
	4200	Treated, 50 mm x 100 mm		168	.095		1.61	3.39		5	7
	4220	50 mm x 150 mm		152	.105		2.46	3.74		6.20	8.55
	4240	50 mm x 200 mm		137	.117		3.22	4.15		7.37	9.95
	4400	100 mm x 100 mm		137	.117		4.56	4.15		8.71	11.45
	4420	100 mm x 150 mm		107	.150		8.05	5.30		13.35	17.20
	4460	100 mm x 200 mm		91.44	.175		10.75	6.20		16.95	21.50
	4480	100 mm x 250 mm		79.25	.202	m³	13.45	7.20		20.65	26
	4482	Ledgers, nailed, 50 mm x 100 mm		1.18	13.561		236	480		716	1,000
	4484	50 mm x 150 mm		1.42	11.301		251	400		651	900
	4486	Bolted, not including bolts, 75 mm x 200 mm		1.53	10.432		550	370		920	1,175
	4488	75 mm x 300 mm		1.65	9.686		555	345		900	1,150
	4490	Mud sills, redwood, construction grade, 50 mm x 100 mm		1.39	11.492		1,400	410		1,810	2,175
	4492	50 mm x 150 mm		1.84	8.693		1,400	310		1,710	2,025
	4500	Sills, 50 mm x 100 mm		.94	16.951		236	605		841	1,200
	4520	50 mm x 150 mm		1.30	12.328		251	440		691	955
	4540	50 mm x 200 mm		1.58	10.120		271	360		631	860
	4600	Treated, 50 mm x 100 mm		.85	18.835		310	670		980	1,400
	4620	50 mm x 150 mm		1.18	13.561		315	480		795	1,100
	4640	50 mm x 200 mm		1.42	11.301		310	400		710	965
	4700	100 mm x 100 mm		1.42	11.301		440	400		840	1,100
	4720	100 mm x 150 mm		1.65	9.686		520	345		865	1,100
	4740	100 mm x 200 mm		1.89	8.476		520	300		820	1,050
	4760	100 mm x 250 mm		2.05	7.794		520	277		797	1,000
565	**0010**	**FRAMING, SLEEPERS**									**565**
	0100	On concrete, treated, 25 mm x 50 mm	2 Carp	716	.022	m	.33	.79		1.12	1.60
	0150	25 mm x 75 mm		610	.026		.62	.93		1.55	2.14
	0200	50 mm x 100 mm		457	.035		1.61	1.24		2.85	3.68
	0250	50 mm x 150 mm		396	.040		2.46	1.44		3.90	4.93
	0300	On concrete, treated, 25 mm x 50 mm		.92	17.386	m³	255	620		875	1,250
	0320	25 mm x 75 mm		1.18	13.561		320	480		800	1,100
	0340	50 mm x 100 mm		2.34	6.849		310	243		553	720
	0360	50 mm x 150 mm		3.07	5.216		315	185		500	640
570	**0010**	**FRAMING, SOFFITS & CANOPIES**									**570**
	1300	Canopy or soffit framing, 25 mm x 100 mm	2 Carp	.71	22.602	m³	590	805		1,395	1,900
	1340	25 mm x 200 mm		1.18	13.561		565	480		1,045	1,375
	1360	50 mm x 100 mm		.97	16.538		236	590		826	1,175
	1400	50 mm x 200 mm		1.58	10.120		271	360		631	860
	1420	75 mm x 100 mm		1.18	13.561		460	480		940	1,250
	1460	75 mm x 200 mm		1.42	11.301		550	400		950	1,225
575	**0010**	**FRAMING, TREATED LUMBER**									**575**
	0100	50 mm x 100 mm				m³	310			310	340
	0110	50 mm x 150 mm					315			315	350
	0120	50 mm x 200 mm					310			310	340

6

WOOD & PLASTICS

06110	Wood Framing		DAILY	LABOR-		2006 BARE COSTS				TOTAL	
		CREW	OUTPUT	HOURS	UNIT	MAT.	LABOR	EQUIP.	TOTAL	INCL O&P	
575	0130	50 mm x 250 mm				m³	380			380	415	**575**
	0140	50 mm x 300 mm					415			415	455	
	0200	100 mm x 100 mm					440			440	485	
	0210	100 mm x 150 mm					520			520	575	
	0220	100 mm x 200 mm					520			520	570	
590	0010	**FRAMING, WALLS**										**590**
	2000	Headers over openings, 50 mm x 150 mm R061110 -30	2 Carp	110	.145	m	1.94	5.15		7.09	10.20	
	2005	50 mm x 150 mm, pneumatic nailed		132	.121		1.94	4.31		6.25	8.85	
	2050	50 mm x 200 mm		104	.154		2.79	5.45		8.24	11.60	
	2055	50 mm x 200 mm, pneumatic nailed		124	.129		2.79	4.59		7.38	10.25	
	2100	50 mm x 250 mm		97.54	.164		4.23	5.85		10.08	13.75	
	2105	50 mm x 250 mm, pneumatic nailed		117	.137		4.23	4.86		9.09	12.20	
	2150	50 mm x 300 mm		91.44	.175		5.25	6.20		11.45	15.45	
	2155	50 mm x 300 mm, pneumatic nailed		110	.145		5.25	5.15		10.40	13.80	
	2200	100 mm x 300 mm		57.91	.276		13.25	9.80		23.05	30	
	2205	100 mm x 300 mm, pneumatic nailed		69.49	.230		13.25	8.20		21.45	27.50	
	2250	150 mm x 300 mm		42.67	.375		31.50	13.35		44.85	55.50	
	2255	150 mm x 300 mm, pneumatic nailed		51.21	.312		31.50	11.10		42.60	52	
	5000	Plates, untreated, 50 mm x 75 mm		259	.062		1.25	2.20		3.45	4.80	
	5005	50 mm x 75 mm, pneumatic nailed		311	.051		1.25	1.83		3.08	4.23	
	5020	50 mm x 100 mm		244	.066		1.21	2.33		3.54	4.98	
	5025	50 mm x 100 mm, pneumatic nailed		293	.055		1.21	1.94		3.15	4.37	
	5040	50 mm x 150 mm		229	.070		1.94	2.48		4.42	6	
	5045	50 mm x 150 mm, pneumatic nailed		274	.058		1.94	2.08		4.02	5.35	
	5120	Studs, 2450 mm high wall, 50 mm x 75 mm		366	.044		1.25	1.55		2.80	3.80	
	5125	50 mm x 75 mm, pneumatic nailed		439	.036		1.25	1.30		2.55	3.40	
	5140	50 mm x 100 mm		335	.048		1.21	1.70		2.91	3.99	
	5146	50 mm x 100 mm, pneumatic nailed		402	.040		1.21	1.41		2.62	3.55	
	5160	50 mm x 150 mm		305	.052		1.94	1.87		3.81	5.05	
	5166	50 mm x 150 mm, pneumatic nailed		366	.044		1.94	1.55		3.49	4.55	
	5180	75 mm x 100 mm		244	.066		3.58	2.33		5.91	7.55	
	5185	75 mm x 100 mm, pneumatic nailed		293	.055		3.58	1.94		5.52	6.95	
	5860	Headers over openings, 50 mm x 150 mm		.85	18.835	m³	251	670		921	1,325	
	5865	51 mm x 152 mm, pneumatic nailed		1.01	15.769		251	560		811	1,150	
	5880	50 mm x 200 mm		1.06	15.068		271	535		806	1,125	
	5885	51 mm x 203 mm, pneumatic nailed		1.27	12.557		271	445		716	995	
	5900	50 mm x 250 mm		1.25	12.793		330	455		785	1,075	
	5905	51 mm x 254 mm, pneumatic nailed		1.58	10.120		330	360		690	920	
	5920	50 mm x 300 mm		1.42	11.301		340	400		740	1,000	
	5925	51 mm x 305 mm, pneumatic nailed		1.70	9.417		340	335		675	895	
	5940	100 mm x 300 mm		1.79	8.922		430	315		745	965	
	5945	102 mm x 305 mm, pneumatic nailed		2.17	7.370		430	262		692	880	
	5960	150 mm x 300 mm		1.98	8.072		680	287		967	1,200	
	5965	152 mm x 305 mm, pneumatic nailed		2.38	6.727		680	239		919	1,125	
	6000	Plates, untreated, 50 mm x 75 mm		1.01	15.769		320	560		880	1,225	
	6005	51 mm x 76 mm, pneumatic nailed		1.23	13.040		320	465		785	1,075	
	6020	50 mm x 100 mm		1.25	12.793		236	455		691	970	
	6025	51 mm x 102 mm, pneumatic nailed		1.58	10.120		236	360		596	820	
	6040	50 mm x 150 mm		1.77	9.041		251	320		571	775	
	6045	51 mm x 152 mm, pneumatic nailed		2.12	7.534		251	268		519	690	
	6120	Studs, 2450 mm high wall, 50 mm x 75 mm		1.42	11.301		320	400		720	980	
	6125	51 mm x 76 mm, pneumatic nailed		1.70	9.417		320	335		655	875	
	6140	50 mm x 100 mm		2.17	7.370		236	262		498	670	
	6145	51 mm x 102 mm, pneumatic nailed		2.61	6.142		236	218		454	600	
	6160	50 mm x 150 mm		2.36	6.781		251	241		492	650	

WOOD & PLASTICS | **6**

06110 | Wood Framing

		CREW	DAILY OUTPUT	LABOR-HOURS	UNIT	2006 BARE COSTS				TOTAL INCL O&P
						MAT.	LABOR	EQUIP.	TOTAL	
590	6165	51 mm x 152 mm, pneumatic nailed R061110-30	2 Carp	2.83	5.650	m³	251	201		452	590
	6180	75 mm x 100 mm		1.89	8.476		460	300		760	980
	6185	76 mm x 102 mm, pneumatic nailed		2.27	7.063		460	251		711	900
	8200	For 3600 mm high walls, deduct						5%			
	8220	For stub wall, 1800 mm high, add						20%			
	8240	900 mm high, add						40%			
	8250	For second story & above, add						5%			
	8300	For dormer & gable, add						15%			
600	0010	**FURRING**									
	0012	Wood strips, 25 mm x 50 mm, on walls, on wood	1 Carp	168	.048	m	.62	1.69		2.31	3.33
	0015	On wood, pneumatic nailed		216	.037		.62	1.32		1.94	2.74
	0300	On masonry		151	.053		.62	1.88		2.50	3.62
	0400	On concrete		79.25	.101		.62	3.59		4.21	6.30
	0600	25 mm x 75 mm, on walls, on wood		168	.048		.92	1.69		2.61	3.66
	0605	On wood, pneumatic nailed		216	.037		.92	1.32		2.24	3.07
	0700	On masonry		151	.053		.92	1.88		2.80	3.95
	0800	On concrete		79.25	.101		.92	3.59		4.51	6.60
	0850	On ceilings, on wood		107	.075		.92	2.66		3.58	5.15
	0855	On wood, pneumatic nailed		137	.058		.92	2.08		3	4.25
	0900	On masonry		97.54	.082		.92	2.92		3.84	5.55
	0950	On concrete		64.01	.125		.92	4.44		5.36	7.90
700	0010	**GROUNDS**									
	0020	For casework, 25 mm x 50 mm wood strips, on wood	1 Carp	101	.079	m	.62	2.82		3.44	5.05
	0100	On masonry		86.87	.092		.62	3.27		3.89	5.80
	0200	On concrete		76.20	.105		.62	3.73		4.35	6.50
	0400	For plaster, 19 mm deep, on wood		137	.058		.62	2.08		2.70	3.92
	0500	On masonry		68.58	.117		.62	4.15		4.77	7.15
	0600	On concrete		53.34	.150		.62	5.35		5.97	9
	0700	On metal lath		60.96	.131		.62	4.67		5.29	7.95

06120 | Structural Panels

		CREW	DAILY OUTPUT	LABOR-HOURS	UNIT	MAT.	LABOR	EQUIP.	TOTAL	TOTAL INCL O&P
900	0010	**STRUCTURAL INSULATED PANELS**									
	0100	Structural insul. panels, 11 mm OSB both faces, EPS insul, 92 mm T	F-3	193	.207	m²	32.50	7.45	3.30	43.25	50.50
	0110	143 mm thick		160	.250		35	9	3.98	47.98	57
	0120	187 mm thick		132	.303		37.50	10.95	4.82	53.27	63.50
	0130	238 mm thick		105	.381		40.50	13.75	6.05	60.30	72.50
	0140	11 mm OSB one face, EPS insul, 92 mm thick		202	.198		16.80	7.15	3.15	27.10	33
	0150	143 mm thick		170	.235		19.80	8.50	3.75	32.05	38.50
	0160	187 mm thick		142	.282		22.50	10.15	4.48	37.13	45.50
	0170	238 mm thick		114	.351		26	12.65	5.60	44.25	54
	0180	289 mm thick		85.93	.465		27.50	16.80	7.40	51.70	64.50
	0190	11 mm OSB - 13 mm GWB faces, EPS insul, 92 mm T		193	.207		27	7.45	3.30	37.75	44.50
	0200	143 mm thick		160	.250		30	9	3.98	42.98	51.50
	0210	187 mm thick		132	.303		33.50	10.95	4.82	49.27	58.50
	0220	238 mm thick		105	.381		37	13.75	6.05	56.80	68.50
	0230	289 mm thick		76.64	.522		39.50	18.80	8.30	66.60	81.50
	0240	11 mm OSB - 13 mm MRGWB faces, EPS insul, 92 mm T		193	.207		27	7.45	3.30	37.75	44.50
	0250	143 mm thick		160	.250		30	9	3.98	42.98	51.50
	0300	For 13 mm GWB added to OSB skin, add					6.45			6.45	7.10
	0310	For 13 mm MRGWB added to OSB skin, add					8.05			8.05	8.95
	0320	For one T1-11 skin, add to OSB-OSB					8.05			8.05	8.95
	0330	For one 15 mm CDX skin, add to OSB-OSB					7.75			7.75	8.50

6 WOOD & PLASTICS

06150 | Wood Decking

		CREW	DAILY OUTPUT	LABOR-HOURS	UNIT	2006 BARE COSTS MAT.	LABOR	EQUIP.	TOTAL	TOTAL INCL O&P	
600	**0010**	**ROOF DECKS**									**600**
	0020	For laminated decks, see division 06170-550									
	0200	For cementitious decks, see division 03510-200									
	0400	Cedar planks, 75 mm thick	2 Carp	29.73	.538	m²	69	19.15		88.15	106
	0500	100 mm thick		23.23	.689		93	24.50		117.50	140
	0700	Douglas fir, 75 mm thick		29.73	.538		26	19.15		45.15	58.50
	0800	100 mm thick		23.23	.689		34.50	24.50		59	76
	1000	Hemlock, 75 mm thick		29.73	.538		26	19.15		45.15	58.50
	1100	100 mm thick		23.23	.689		34.50	24.50		59	76
	1300	Western white spruce, 75 mm thick		29.73	.538		25	19.15		44.15	57.50
	1400	100 mm thick	▼	23.23	.689	▼	33	24.50		57.50	74.50

06160 | Sheathing

		CREW	DAILY OUTPUT	LABOR-HOURS	UNIT	2006 BARE COSTS MAT.	LABOR	EQUIP.	TOTAL	TOTAL INCL O&P	
800	**0010**	**SHEATHING** Plywood on roof, CDX	2 Carp	149	.107	m²	6.65	3.82		10.47	13.25	**800**
	0030	8 mm thick [R061110-30]									
	0035	Pneumatic nailed		181	.088		6.65	3.14		9.79	12.20
	0050	10 mm thick [R061636-20]		142	.113		4.95	4.01		8.96	11.65
	0055	Pneumatic nailed		173	.092		4.95	3.29		8.24	10.50
	0100	13 mm thick		130	.123		6.55	4.38		10.93	14.10
	0105	Pneumatic nailed		159	.101		6.55	3.58		10.13	12.85
	0200	16 mm thick		121	.132		7.10	4.70		11.80	15.15
	0205	Pneumatic nailed		147	.109		7.10	3.87		10.97	13.85
	0300	19 mm thick		111	.144		8.85	5.10		13.95	17.70
	0305	Pneumatic nailed		136	.118		8.85	4.18		13.03	16.20
	0500	Plywood on walls with exterior CDX, 9 mm thick		111	.144		4.95	5.10		10.05	13.40
	0505	Pneumatic nailed		138	.116		4.95	4.12		9.07	11.80
	0600	13 mm thick		105	.152		6.55	5.40		11.95	15.75
	0605	Pneumatic nailed		130	.123		6.55	4.38		10.93	14.10
	0700	16 mm thick		97.55	.164		7.10	5.85		12.95	16.95
	0705	Pneumatic nailed		121	.132		7.10	4.70		11.80	15.15
	0800	19 mm thick		90.58	.177		8.85	6.30		15.15	19.50
	0805	Pneumatic nailed	▼	112	.143	▼	8.85	5.10		13.95	17.60
	1000	For shear wall construction, add						20%			
	1200	For structural 1 exterior plywood, add				m²	10%				
	1400	With boards, on roof 25 mm x 150 mm boards, laid horizontal	2 Carp	67.35	.238		12.90	8.45		21.35	27.50
	1500	Laid diagonal		60.39	.265		12.90	9.40		22.30	29
	1700	25 mm x 200 mm boards, laid horizontal		81.29	.197		11.95	7		18.95	24
	1800	Laid diagonal	▼	67.35	.238		11.95	8.45		20.40	26.50
	2000	For steep roofs, add						40%			
	2200	For dormers, hips and valleys, add					5%	50%			
	2400	Boards on walls, 25 mm x 150 mm boards, laid regular	2 Carp	60.39	.265		12.90	9.40		22.30	29
	2500	Laid diagonal		54.35	.294		12.90	10.45		23.35	30.50
	2700	25 mm x 200 mm boards, laid regular		71.07	.225		11.95	8		19.95	25.50
	2800	Laid diagonal		60.39	.265		11.95	9.40		21.35	28
	2850	Gypsum, weatherproof, 13 mm thick		105	.152		6.65	5.40		12.05	15.75
	2900	Sealed, 10 mm thick		102	.157		4.84	5.60		10.44	14.10
	3000	Wood fiber, regular, no vapor barrier, 13 mm thick		111	.144		5.90	5.10		11	14.55
	3100	16 mm thick		111	.144		7.75	5.10		12.85	16.50
	3300	No vapor barrier, in colors, 13 mm thick		111	.144		8.40	5.10		13.50	17.25
	3400	16 mm thick		111	.144		10.35	5.10		15.45	19.40
	3600	With vapor barrier one side, white, 13 mm thick		111	.144		5.90	5.10		11	14.55
	3700	Vapor barrier 2 sides, 13 mm thick		111	.144		8.30	5.10		13.40	17.15
	3800	Asphalt impregnated, 20 mm thick		111	.144		3.01	5.10		8.11	11.35
	3850	Intermediate, 13 mm thick	▼	111	.144	▼	1.94	5.10		7.04	10.15

WOOD & PLASTICS 6

06160 | Sheathing

			DAILY	LABOR-		2006 BARE COSTS				TOTAL	
		CREW	OUTPUT	HOURS	UNIT	MAT.	LABOR	EQUIP.	TOTAL	INCL O&P	
850	**0010**	**SUBFLOOR**									**850**
	0011	Plywood, CDX, 13 mm thick R061636-20	2 Carp	139	.115	m² Flr.	6.55	4.09		10.64	13.65
	0015	Pneumatic nailed		173	.092		6.55	3.29		9.84	12.40
	0017	Pneumatic nailed		173	.092		6.55	3.29		9.84	12.40
	0100	16 mm thick		125	.128		7.10	4.55		11.65	14.95
	0105	Pneumatic nailed		156	.103		7.10	3.65		10.75	13.55
	0200	19 mm thick		116	.138		8.85	4.90		13.75	17.35
	0205	Pneumatic nailed		144	.111		8.85	3.95		12.80	15.85
	0300	25 mm thick, 2-4-1 including underlayment		97.55	.164		12.80	5.85		18.65	23
	0450	25 mm x 200 mm S4S, laid regular		92.90	.172		11.95	6.10		18.05	23
	0460	Laid diagonal		78.97	.203		11.95	7.20		19.15	24.50
	0500	25 mm x 250 mm S4S, laid regular		102	.157		14.65	5.60		20.25	25
	0600	Laid diagonal		83.61	.191		14.65	6.80		21.45	27
900	**0010**	**UNDERLAYMENT**									**900**
	0030	Plywood, underlayment grade, 10 mm thick R061636-20	2 Carp	139	.115	m² Flr.	8.40	4.09		12.49	15.60
	0070	Pneumatic nailed		173	.092		8.40	3.29		11.69	14.35
	0100	13 mm thick		135	.119		9.35	4.21		13.56	16.90
	0105	Pneumatic nailed		167	.096		9.35	3.41		12.76	15.65
	0200	16 mm thick		130	.123		13.15	4.38		17.53	21
	0205	Pneumatic nailed		161	.099		13.15	3.53		16.68	19.90
	0300	19 mm thick		121	.132		14.20	4.70		18.90	23
	0305	Pneumatic nailed		150	.107		14.20	3.79		17.99	21.50
	0500	Particle board, 10 mm thick		139	.115		3.55	4.09		7.64	10.25
	0505	Pneumatic nailed		173	.092		3.55	3.29		6.84	9
	0600	13 mm thick		135	.119		3.98	4.21		8.19	10.95
	0605	Pneumatic nailed		167	.096		3.98	3.41		7.39	9.70
	0800	16 mm thick		130	.123		4.41	4.38		8.79	11.65
	0805	Pneumatic nailed		161	.099		4.41	3.53		7.94	10.35
	0900	19 mm thick		121	.132		6.05	4.70		10.75	13.95
	0905	Pneumatic nailed		150	.107		6.05	3.79		9.84	12.55
	1100	Hardboard, underlayment grade,1200 mm x 1200 mm, 5 mm thick		139	.115		4.20	4.09		8.29	11

06170 | Prefabricated Structural Wood

			DAILY	LABOR-		2006 BARE COSTS				TOTAL	
		CREW	OUTPUT	HOURS	UNIT	MAT.	LABOR	EQUIP.	TOTAL	INCL O&P	
550	**0010**	**LAMINATED ROOF DECK**									**550**
	0020	Pine or hemlock, 75 mm thick	2 Carp	39.48	.405	m²	32.50	14.40		46.90	58
	0100	100 mm thick		30.19	.530		43.50	18.85		62.35	77
	0300	Cedar, 75 mm thick		39.48	.405		39.50	14.40		53.90	66
	0400	100 mm thick		30.19	.530		50.50	18.85		69.35	85
	0600	Fir, 75 mm thick		39.48	.405		33	14.40		47.40	59
	0700	100 mm thick		30.19	.530		41.50	18.85		60.35	75
600	**0010**	**STRUCTURAL JOISTS** Fabricated "I" joists with wood flanges,									**600**
	0100	Plywood webs, incl. bridging & blocking, panels 600 mm O.C.									
	1200	4.6 m to 7.3 m span, 245 kg/m² live load	F-5	223	.144	m² Flr.	22.50	5.15		27.65	32.50
	1300	270 kg/m² live load		209	.153		24	5.50		29.50	35
	1400	7.3 m to 9.2 m span, 220 kg/m² live load		242	.132		27.50	4.77		32.27	38
	1500	270 kg/m² live load		223	.144		31.50	5.15		36.65	42.50
	1600	Tubular steel open webs, 220 kg/m², 600 mm O.C., 12 m span	F-3	581	.069		22.50	2.48	1.10	26.08	29.50
	1700	17 m span		720	.056		21.50	2	.88	24.38	28
	1800	21 m span		859	.047		28	1.68	.74	30.42	34.50
	1900	415 kg/m² live load, 7.9 m span		214	.187		26	6.75	2.98	35.73	42
980	**0010**	**ROOF TRUSSES**									**980**
	0020	For timber connectors, see div. 06090-800									
	0100	Fink (W) or King post type, 600 mm O.C.									
	0200	Metal plate connected, 4 in 12 slope									

06170	Prefabricated Structural Wood	CREW	DAILY OUTPUT	LABOR-HOURS	UNIT	2006 BARE COSTS				TOTAL INCL O&P	
						MAT.	LABOR	EQUIP.	TOTAL		
980 0210	7.3 m to 8.9 m span	F-3	279	.143	m² Flr.	19.80	5.15	2.28	27.23	32	**980**
0300	9.2 m to 13 m span		279	.143		22	5.15	2.28	29.43	34.50	
0400	13 m to 18 m span	↓	279	.143	↓	24	5.15	2.28	31.43	37	
0700	Glued and nailed, add					50%					
0800	Flat wood truss 4.9 m to 8.8 m span	F-3	279	.143	↓	23.50	5.15	2.28	30.93	36.50	

06180	Glued-Laminated Construction										
400 0010	**LAMINATED FRAMING** Not including decking										**400**
0020	13.6 kg, short term live load, 6.8 kg dead load										
0200	Straight roof beams, 6.1 m clear span, beams 2400 mm O.C.	F-3	238	.168	m² Flr.	19.15	6.05	2.68	27.88	33.50	
0300	Beams 4800 mm O.C.		297	.135		13.80	4.86	2.14	20.80	25	
0500	12 m clear span, beams 2400 mm O.C.		297	.135		36.50	4.86	2.14	43.50	50.50	
0600	Beams 4800 mm O.C.	↓	357	.112		30	4.04	1.78	35.82	41	
0800	18 m clear span, beams 2400 mm O.C.	F-4	268	.179		63	6.35	3.55	72.90	82.50	
0900	Beams 4800 mm O.C.	"	357	.134		47	4.77	2.67	54.44	62	
1100	Tudor arches, 9.2 m to 12 m clear span, frames 2400 mm O.C.	F-3	156	.256		82	9.25	4.08	95.33	109	
1200	Frames 4800 mm O.C.	"	208	.192		64	6.95	3.06	74.01	84.50	
1400	15 m to 18 m clear span, frames 2400 mm O.C.	F-4	204	.235		88.50	8.35	4.67	101.52	116	
1500	Frames 4800 mm O.C.		245	.196		75	6.95	3.89	85.84	98	
1700	Radial arches, 18 m clear span, frames 2400 mm O.C.		178	.270		82.50	9.55	5.35	97.40	112	
1800	Frames 4800 mm O.C.		268	.179		63.50	6.35	3.55	73.40	83.50	
2000	30 m clear span, frames 2400 mm O.C.		149	.322		85.50	11.40	6.40	103.30	119	
2100	Frames 4800 mm O.C.		223	.215		75	7.65	4.27	86.92	99.50	
2300	37 m clear span, frames 2400 mm O.C.		134	.358		114	12.70	7.10	133.80	152	
2400	Frames 4800 mm O.C.	↓	178	.270		104	9.55	5.35	118.90	135	
2600	Bowstring trusses, 6000 mm O.C., 12 m clear span	F-3	223	.179		51	6.45	2.86	60.31	69.50	
2700	18 m clear span	F-4	334	.144		46	5.10	2.85	53.95	61.50	
2800	30 m clear span	↓	372	.129		65	4.58	2.56	72.14	82	
2900	37 m clear span	↓	334	.144		70	5.10	2.85	77.95	88	
3000	For less than 2.36 m³, add					20%					
3050	For over 11.80 m³, deduct					10%					
3100	For premium appearance, add to m² prices					5%					
3300	For industrial type, deduct					15%					
3500	For stain and varnish, add					5%					
3900	For 19 mm laminations, add to straight					25%					
4100	Add to curved				↓	15%					
4300	Alternate pricing method: (use nominal footage of										
4310	components). Straight beams, camber less than 150 mm	F-3	8.26	4.843	m³	1,125	175	77	1,377	1,575	
4400	Columns, including hardware		4.72	8.476		1,200	305	135	1,640	1,950	
4600	Curved members, radius over 10 m		5.90	6.781		1,225	245	108	1,578	1,850	
4700	Radius 3 m to 10 m	↓	7.08	5.650		1,225	204	90	1,519	1,775	
4900	For complicated shapes, add maximum					100%					
5100	For pressure treating, add to straight					35%					
5200	Add to curved				↓	45%					
6000	Laminated veneer members, southern pine or western species										
6050	44 mm wide x 140 mm deep	2 Carp	146	.110	m	9.90	3.90		13.80	16.95	
6100	241 mm deep		146	.110		12.35	3.90		16.25	19.65	
6150	356 mm deep		137	.117		18.35	4.15		22.50	26.50	
6200	457 mm deep	↓	137	.117	↓	25	4.15		29.15	34	
6300	Parallel strand members, southern pine or western species										
6350	44 mm wide x 235 mm deep	2 Carp	146	.110	m	11.70	3.90		15.60	18.90	
6400	286 mm deep		137	.117		14.35	4.15		18.50	22.50	
6450	356 mm deep		122	.131		17.15	4.66		21.81	26	
6500	89 mm wide x 235 mm deep		146	.110		28.50	3.90		32.40	37	
6550	286 mm deep	↓	137	.117	↓	35	4.15		39.15	45	

WOOD & PLASTICS **6**

For expanded coverage of these items see Means Interior Cost Data 2006

06180	Glued-Laminated Construction	CREW	DAILY OUTPUT	LABOR-HOURS	UNIT	2006 BARE COSTS MAT.	LABOR	EQUIP.	TOTAL	TOTAL INCL O&P	
400 6600	356 mm deep	2 Carp	122	.131	m	41.50	4.66		46.16	53.50	**400**
6650	178 mm wide x 235 mm deep		137	.117		59.50	4.15		63.65	71.50	
6700	286 mm deep		128	.125		74	4.44		78.44	88.50	
6750	356 mm deep		122	.131		88.50	4.66		93.16	105	

06200 | Finish Carpentry

06220	Millwork	CREW	DAILY OUTPUT	LABOR-HOURS	UNIT	2006 BARE COSTS MAT.	LABOR	EQUIP.	TOTAL	TOTAL INCL O&P	
200 0010	**MOLDINGS, BASE**										**200**
0500	Base, stock pine, 14 mm x 88 mm	1 Carp	73.15	.109	m	7.25	3.89		11.14	14	
0550	14 mm x 113 mm		60.96	.131		7.85	4.67		12.52	15.90	
0561	Base shoe, oak, 19 mm x 25 mm		73.15	.109		3.71	3.89		7.60	10.10	
400 0010	**MOLDINGS, CASINGS**										**400**
0090	Apron, stock pine, 16 mm x 50 mm	1 Carp	76.20	.105	m	3.41	3.73		7.14	9.55	
0110	16 mm x 89 mm		67.06	.119		4.10	4.24		8.34	11.15	
0300	Band, stock pine, 17 mm x 29 mm		82.30	.097		1.87	3.46		5.33	7.45	
0350	17 mm x 44 mm		76.20	.105		3.67	3.73		7.40	9.85	
0700	Casing, stock pine, 17 mm x 64 mm		73.15	.109		4.20	3.89		8.09	10.70	
0750	17 mm x 89 mm		65.53	.122		6	4.34		10.34	13.35	
450 0010	**MOLDINGS, CEILINGS**										**450**
0600	Bed, stock pine, 14 mm x 44 mm	1 Carp	82.30	.097	m	2.89	3.46		6.35	8.60	
0650	14 mm x 50 mm		73.15	.109		3.90	3.89		7.79	10.35	
1200	Cornice molding, stock pine, 14 mm x 44 mm		101	.079		2.69	2.82		5.51	7.35	
1300	14 mm x 57 mm		91.44	.087		3.35	3.11		6.46	8.50	
2400	Cove scotia, stock pine, 14 mm x 44 mm		82.30	.097		2.13	3.46		5.59	7.75	
2500	17 mm x 70 mm		77.72	.103		4.10	3.66		7.76	10.25	
2600	Crown, stock pine, 14 mm x 92 mm		76.20	.105		6.10	3.73		9.83	12.55	
2700	17 mm x 117 mm		67.06	.119		10.20	4.24		14.44	17.80	
500 0010	**MOLDINGS, EXTERIOR**										**500**
1500	Cornice, boards, pine, 25 mm x 50 mm	1 Carp	101	.079	m	1.18	2.82		4	5.70	
1700	25 mm x 150 mm		76.20	.105		4.07	3.73		7.80	10.25	
2000	25 mm x 300 mm		54.86	.146		6.20	5.20		11.40	14.85	
2200	Three piece, built-up, pine, minimum		24.38	.328		7.60	11.65		19.25	26.50	
2300	Maximum		19.81	.404		16.45	14.35		30.80	40.50	
3000	Corner board, sterling pine, 25 mm x 100 mm		60.96	.131		2.26	4.67		6.93	9.75	
3100	25 mm x 150 mm		60.96	.131		4.92	4.67		9.59	12.65	
3350	Fascia, sterling pine, 25 mm x 150 mm		76.20	.105		4.92	3.73		8.65	11.20	
3370	25 mm x 200 mm		68.58	.117		6.25	4.15		10.40	13.30	
3400	Trim, exterior, sterling pine, back band		76.20	.105		2.43	3.73		6.16	8.45	
3500	Casing		76.20	.105		6.45	3.73		10.18	12.90	
3600	Crown		76.20	.105		6.20	3.73		9.93	12.60	
3700	Porch rail with balusters		6.71	1.193		50.50	42.50		93	122	
3800	Screen		120	.067		3.90	2.37		6.27	8	
4100	Verge board, sterling pine, 25 mm x 100 mm		60.96	.131		2.56	4.67		7.23	10.05	
4200	25 mm x 150 mm		60.96	.131		3.54	4.67		8.21	11.15	
4300	50 mm x 150 mm		50.29	.159		5.75	5.65		11.40	15.15	
4400	50 mm x 200 mm		50.29	.159		7.60	5.65		13.25	17.15	
4700	For redwood trim, add					200%					

06220	Millwork	CREW	DAILY OUTPUT	LABOR-HOURS	UNIT	2006 BARE COSTS				TOTAL INCL O&P
						MAT.	LABOR	EQUIP.	TOTAL	
700	**0010**	**MOLDINGS, TRIM**									**700**
0200	Astragal, stock pine, 17 mm x 44 mm	1 Carp	77.72	.103	m	3.81	3.66		7.47	9.90
0250	33 mm x 56 mm		73.15	.109		8.90	3.89		12.79	15.85
0800	Chair rail, stock pine, 16 mm x 63 mm		82.30	.097		4.72	3.46		8.18	10.60
0900	16 mm x 88 mm		73.15	.109		7.20	3.89		11.09	14
1000	Closet pole, stock pine, 28 mm diameter		60.96	.131		2.85	4.67		7.52	10.40
1100	Fir, 41 mm diameter		60.96	.131		5.25	4.67		9.92	13
3300	Half round, stock pine, 6 mm x 13 mm		82.30	.097		.75	3.46		4.21	6.20
3350	13 mm x 25 mm		77.72	.103		1.97	3.66		5.63	7.85
3400	Handrail, fir, single piece, stock, hardware not included									
3450	38 mm x 44 mm	1 Carp	24.38	.328	m	5.80	11.65		17.45	24.50
3470	Pine, 38 mm x 44 mm		24.38	.328		5.80	11.65		17.45	24.50
3500	38 mm x 64 mm		23.16	.345		5.35	12.30		17.65	25
3600	Lattice, stock pine, 6 mm x 29 mm		82.30	.097		1.35	3.46		4.81	6.90
3700	6 mm x 44 mm		76.20	.105		1.48	3.73		5.21	7.45
3800	Miscellaneous, custom, pine, 25 mm x 25 mm		82.30	.097		1.12	3.46		4.58	6.60
3900	25 mm x 75 mm		73.15	.109		2.26	3.89		6.15	8.55
4100	Birch or oak, nominal 25 mm x 25 mm		73.15	.109		1.74	3.89		5.63	7.95
4200	Nominal 25 mm x 75 mm		65.53	.122		5.95	4.34		10.29	13.30
4400	Walnut, nominal 25 mm x 25 mm		65.53	.122		2.85	4.34		7.19	9.90
4500	Nominal 25 mm x 75 mm		60.96	.131		8.55	4.67		13.22	16.65
4700	Teak, nominal 25 mm x 25 mm		65.53	.122		4.04	4.34		8.38	11.20
4800	Nominal 25 mm x 75 mm		60.96	.131		11.55	4.67		16.22	19.95
4900	Quarter round, stock pine, 6 mm x 6 mm		83.82	.095		.72	3.39		4.11	6.10
4950	19 mm x 19 mm		77.72	.103		2.10	3.66		5.76	8
5600	Wainscot moldings, 29 mm x 14 mm, 600 mm high, minimum		7.06	1.133	m²	113	40.50		153.50	187
5700	Maximum		6.04	1.325	"	211	47		258	305
800	**0010**	**MOLDINGS, WINDOW AND DOOR**									**800**
2800	Door moldings, stock, decorative, 29 mm wide, plain	1 Carp	17	.471	Set	43.50	16.75		60.25	74
2900	Detailed		17	.471	"	87.50	16.75		104.25	123
2960	Clear pine door jamb, no stops, 17 mm x 116 mm		73.15	.109	m	7.55	3.89		11.44	14.35
3150	Door trim set, 1 head and 2 sides, pine, 64 mm wide		5.90	1.356	Opng.	22	48		70	99
3170	89 mm wide		5.30	1.509	"	31	53.50		84.50	118
3250	Glass beads, stock pine, 10 mm x 13 mm		83.82	.095	m	1.02	3.39		4.41	6.40
3270	10 mm x 22 mm		82.30	.097		1.35	3.46		4.81	6.90
4850	Parting bead, stock pine, 10 mm x 19 mm		83.82	.095		1.15	3.39		4.54	6.60
4870	13 mm x 19 mm		77.72	.103		1.44	3.66		5.10	7.25
5000	Stool caps, stock pine, 17 mm x 89 mm		60.96	.131		7.20	4.67		11.87	15.20
5100	27 mm x 83 mm		45.72	.175		11.40	6.20		17.60	22.50
5300	Threshold, oak, 914 mm long, inside, 16 mm x 92 mm		32	.250	Ea.	7.75	8.90		16.65	22.50
5400	Outside, 38 mm x 194 mm		16	.500	"	35	17.80		52.80	66
5900	Window trim sets , including casings, header, stops,									
5910	stool and apron, 64 mm wide, minimum	1 Carp	13	.615	Opng.	30	22		52	66.50
5950	Average		10	.800		35	28.50		63.50	83
6000	Maximum		6	1.333		62.50	47.50		110	143
900	**0010**	**SOFFITS**									**900**
0200	Soffits, pine, 25 mm x 100 mm	2 Carp	128	.125	m	1.15	4.44		5.59	8.20
0210	25 mm x 150 mm		128	.125		1.80	4.44		6.24	8.90
0220	25 mm x 200 mm		128	.125		2.30	4.44		6.74	9.45
0230	25 mm x 250 mm		122	.131		3.54	4.66		8.20	11.15
0240	25 mm x 300 mm		122	.131		3.61	4.66		8.27	11.20
0250	STK cedar, 25 mm x 100 mm		128	.125		1.84	4.44		6.28	8.95
0260	25 mm x 150 mm		128	.125		2.72	4.44		7.16	9.90
0270	25 mm x 200 mm		128	.125		3.58	4.44		8.02	10.85
0280	25 mm x 250 mm		122	.131		7	4.66		11.66	14.95

WOOD & PLASTICS **6**

06220 | Millwork

		CREW	DAILY OUTPUT	LABOR-HOURS	UNIT	2006 BARE COSTS				TOTAL INCL O&P	
						MAT.	LABOR	EQUIP.	TOTAL		
900	0290	25 mm x 300 mm	2 Carp	122	.131	m	9.50	4.66		14.16	17.70	900
	1000	Exterior AC plywood, 6 mm thick		39.02	.410	m²	8.05	14.60		22.65	31.50	
	1050	10 mm thick		39.02	.410		8.40	14.60		23	32	
	1100	13 mm thick		39.02	.410		9.35	14.60		23.95	33	
	1150	Polyvinyl chloride, white, solid	1 Carp	21.37	.374		8.85	13.30		22.15	30	
	1160	Perforated	"	21.37	.374		8.85	13.30		22.15	30	

06250 | Prefinished Paneling

		CREW	DAILY OUTPUT	LABOR-HOURS	UNIT	MAT.	LABOR	EQUIP.	TOTAL	TOTAL INCL O&P	
200	0010	**PANELING, HARDBOARD**										200
	0050	Not incl. furring or trim, hardboard, tempered, 3 mm thick	2 Carp	46.45	.344	m²	3.44	12.25		15.69	23	
	0100	6 mm thick		46.45	.344		5.15	12.25		17.40	25	
	0300	Tempered pegboard, 3 mm thick		46.45	.344		4.20	12.25		16.45	23.50	
	0400	6 mm thick		46.45	.344		5.50	12.25		17.75	25	
	0600	Untempered hardboard, natural finish, 3 mm thick		46.45	.344		3.77	12.25		16.02	23.50	
	0700	6 mm thick		46.45	.344		3.66	12.25		15.91	23	
	0900	Untempered pegboard, 3 mm thick		46.45	.344		3.77	12.25		16.02	23.50	
	1000	6 mm thick		46.45	.344		4.20	12.25		16.45	23.50	
	1200	Plastic faced hardboard, 3 mm thick		46.45	.344		6.25	12.25		18.50	26	
	1300	6 mm thick		46.45	.344		8.30	12.25		20.55	28	
	1500	Plastic faced pegboard, 3 mm thick		46.45	.344		5.90	12.25		18.15	25.50	
	1600	6 mm thick		46.45	.344		7.30	12.25		19.55	27	
	1800	Wood grained, plain or grooved, 6 mm thick, minimum		46.45	.344		5.50	12.25		17.75	25	
	1900	Maximum		39.48	.405		11.65	14.40		26.05	35.50	
	2100	Moldings for hardboard, wood or aluminum, minimum		152	.105	m	1.15	3.74		4.89	7.15	
	2200	Maximum		130	.123	"	3.22	4.38		7.60	10.35	
500	0010	**PANELING, PLYWOOD**										500
	2400	Plywood, prefinished, 6 mm thick, 1219 x 2438 mm sheets [R061636-20]										
	2410	with vertical grooves. Birch faced, minimum	2 Carp	46.45	.344	m²	9.05	12.25		21.30	29	
	2420	Average		39.02	.410		13.80	14.60		28.40	37.50	
	2430	Maximum		32.52	.492		20	17.50		37.50	49	
	2600	Mahogany, African		37.16	.431		25.50	15.30		40.80	52.50	
	2700	Philippine (Lauan)		46.45	.344		11.10	12.25		23.35	31	
	2900	Oak or Cherry, minimum		46.45	.344		21.50	12.25		33.75	42.50	
	3000	Maximum		37.16	.431		33	15.30		48.30	60.50	
	3200	Rosewood		29.73	.538		47	19.15		66.15	81.50	
	3400	Teak		37.16	.431		33	15.30		48.30	60.50	
	3600	Chestnut		34.84	.459		49	16.35		65.35	79	
	3800	Pecan		37.16	.431		21	15.30		36.30	47.50	
	3900	Walnut, minimum		46.45	.344		28	12.25		40.25	50	
	3950	Maximum		37.16	.431		53.50	15.30		68.80	83	
	4000	Plywood, prefinished, 19 mm thick, stock grades, minimum		29.73	.538		12.80	19.15		31.95	44	
	4100	Maximum		20.81	.769		55	27.50		82.50	103	
	4300	Architectural grade, minimum		20.81	.769		40.50	27.50		68	87.50	
	4400	Maximum		14.86	1.076		62	38.50		100.50	128	
	4600	Plywood, "A" face, birch, V.C., 13 mm thick, natural		41.81	.383		19.25	13.60		32.85	42	
	4700	Select		41.81	.383		21	13.60		34.60	44.50	
	4900	Veneer core, 19 mm thick, natural		29.73	.538		20.50	19.15		39.65	52.50	
	5000	Select		29.73	.538		23	19.15		42.15	55	
	5200	Lumber core, 19 mm thick, natural		29.73	.538		30.50	19.15		49.65	63.50	
	5500	Plywood, knotty pine, 6 mm thick, A2 grade		41.81	.383		16.70	13.60		30.30	39.50	
	5600	A3 grade		41.81	.383		21	13.60		34.60	44.50	
	5800	19 mm thick, veneer core, A2 grade		29.73	.538		21.50	19.15		40.65	54	
	5900	A3 grade		29.73	.538		24.50	19.15		43.65	57	
	6100	Aromatic cedar, 6 mm thick, plywood		37.16	.431		21.50	15.30		36.80	47.50	
	6200	6 mm thick, particle board		37.16	.431		10.35	15.30		25.65	35.50	

6 WOOD & PLASTICS

06260 | Board Paneling

			CREW	DAILY OUTPUT	LABOR-HOURS	UNIT	2006 BARE COSTS				TOTAL INCL O&P
							MAT.	LABOR	EQUIP.	TOTAL	
400	0010	**PANELING, BOARDS**										400
	6400	Wood board paneling, 19 mm thick, knotty pine	2 Carp	27.87	.574	m²	15.05	20.50		35.55	48.50
	6500	Rough sawn cedar		27.87	.574		19.25	20.50		39.75	53
	6700	Redwood, clear, 25 mm x 100 mm boards		27.87	.574		45	20.50		65.50	81.50
	6900	Aromatic cedar, closet lining, boards	▼	25.55	.626	▼	34.50	22.50		57	72.50

06270 | Closet/Utility Wood Shelving

			CREW	DAILY OUTPUT	LABOR-HOURS	UNIT	2006 BARE COSTS				TOTAL INCL O&P
							MAT.	LABOR	EQUIP.	TOTAL	
200	0010	**SHELVING**										200
	0020	Pine, clear grade, no edge band, 25 mm x 200 mm	1 Carp	35.05	.228	m	5.95	8.10		14.05	19.20
	0100	25 mm x 250 mm		33.53	.239		7.60	8.50		16.10	21.50
	0200	25 mm x 300 mm		32	.250		10.15	8.90		19.05	25
	0600	Plywood, 19 mm thick with lumber edge, 300 mm wide		22.86	.350		5.45	12.45		17.90	25.50
	0700	600 mm wide		21.34	.375	▼	9.80	13.35		23.15	32
	0900	Bookcase, clear grade pine, shelves 300 mm O.C., 200 mm D shelf		6.50	1.230	m²	63.50	43.50		107	138
	1000	300 mm deep shelves		6.04	1.325	"	109	47		156	193
	1200	Adjustable closet rod and shelf, 300 mm wide, 900 mm long		20	.400	Ea.	8.70	14.20		22.90	31.50
	1300	2400 mm long		15	.533	"	22	18.95		40.95	53.50
	1500	Prefinished shelves with supports, stock, 200 mm wide		22.86	.350	m	12.55	12.45		25	33
	1600	250 mm wide	▼	21.34	.375	"	14	13.35		27.35	36.50

06400 | Architectural Woodwork

06410 | Custom Cabinets

			CREW	DAILY OUTPUT	LABOR-HOURS	UNIT	2006 BARE COSTS				TOTAL INCL O&P
							MAT.	LABOR	EQUIP.	TOTAL	
100	0010	**CABINETS** Corner china cabinets, stock pine,										100
	0020	2032 mm high, unfinished, minimum	2 Carp	6.60	2.424	Ea.	455	86		541	635
	0100	Maximum	"	4.40	3.636	"	1,000	129		1,129	1,300
	0300	Built-in drawer units, pine, 457 mm deep, 813 mm high, unfinished									
	0400	Minimum	2 Carp	16.15	.990	m	395	35		430	490
	0500	Maximum	"	12.19	1.312	"	480	46.50		526.50	605
	0700	Kitchen base cabinets, hardwood, not incl. counter tops,									
	0710	600 mm deep, 875 mm high, prefinished									
	0800	One top drawer, one door below, 300 mm wide	2 Carp	24.80	.645	Ea.	138	23		161	188
	0840	450 mm wide		23.30	.687		205	24.50		229.50	263
	0880	600 mm wide		22.30	.717		245	25.50		270.50	310
	1000	Four drawers, 305 mm wide		24.80	.645		325	23		348	395
	1040	457 mm wide		23.30	.687		280	24.50		304.50	350
	1060	610 mm wide		22.30	.717		305	25.50		330.50	375
	1200	Two top drawers, two doors below, 675 mm wide		22	.727		272	26		298	340
	1260	914 mm wide		20.30	.788		315	28		343	390
	1300	1219 mm wide		18.90	.847		360	30		390	440
	1500	Range or sink base, two doors below, 762 mm wide		21.40	.748		240	26.50		266.50	305
	1540	914 mm wide		20.30	.788		270	28		298	340
	1580	1219 mm wide	▼	18.90	.847		300	30		330	375
	1800	For sink front units, deduct					52			52	57.50
	2000	Corner base cabinets, 900 mm wide, standard	2 Carp	18	.889		395	31.50		426.50	485
	2100	Lazy Susan with revolving door	"	16.50	.970	▼	385	34.50		419.50	475
	4000	Kitchen wall cabinets, hardwood, 300 mm deep with two doors									
	4050	300 mm high, 750 mm wide	2 Carp	24.80	.645	Ea.	151	23		174	202
	4100	900 mm wide	▼	24	.667	▼	176	23.50		199.50	230

WOOD & PLASTICS   **6**

6 WOOD & PLASTICS

06410		Custom Cabinets	CREW	DAILY OUTPUT	LABOR-HOURS	UNIT	MAT.	LABOR	EQUIP.	TOTAL	TOTAL INCL O&P	
100	4400	375 mm high, 750 mm wide	2 Carp	24	.667	Ea.	158	23.50		181.50	211	**100**
	4440	900 mm wide		22.70	.705		179	25		204	236	
	4700	600 mm high, 750 mm wide		23.30	.687		197	24.50		221.50	254	
	4720	900 mm wide		22.70	.705		218	25		243	278	
	5000	750 mm high, one door, 300 mm wide		22	.727		133	26		159	187	
	5040	450 mm wide		20.90	.766		164	27		191	224	
	5060	600 mm wide		20.30	.788		185	28		213	248	
	5300	Two doors, 686 mm wide		19.80	.808		229	28.50		257.50	297	
	5340	900 mm wide		18.80	.851		253	30.50		283.50	325	
	5380	1200 mm wide		18.40	.870		310	31		341	390	
	6000	Corner wall, 750 mm high, 600 mm wide		18	.889		152	31.50		183.50	216	
	6050	762 mm wide		17.20	.930		180	33		213	250	
	6100	914 mm wide		16.50	.970		195	34.50		229.50	269	
	6500	Revolving Lazy Susan		15.20	1.053		298	37.50		335.50	385	
	7000	Broom cabinet, 2100 mm high, 600 mm deep, 450 mm wide		10	1.600		405	57		462	535	
	7500	Oven cabinets, 2134 mm high, 610 mm deep, 686 mm wide		8	2		590	71		661	760	
	7750	Valance board trim		121	.132	m	27.50	4.70		32.20	38	
	9000	For deluxe models of all cabinets, add					40%					
	9500	For custom built in place, add					25%	10%				
	9550	Rule of thumb, kitchen cabinets not including										
	9560	appliances & counter top, minimum	2 Carp	9.14	1.750	m	315	62		377	440	
	9600	Maximum	"	7.62	2.100	"	820	74.50		894.50	1,025	
	9610	For metal cabinets, see division 12310-750										
210	0010	**CASEWORK, FRAMES**										**210**
	0050	Base cabinets, counter storage, 914 mm high, one bay										
	0100	457 mm wide	1 Carp	2.70	2.963	Ea.	107	105		212	282	
	0400	Two bay, 914 mm wide		2.20	3.636		163	129		292	380	
	1100	Three bay, 1372 mm wide		1.50	5.333		194	190		384	510	
	2800	Book cases, one bay, 2134 mm high, 457 mm wide		2.40	3.333		126	119		245	325	
	3500	Two bay, 914 mm wide		1.60	5		183	178		361	480	
	4100	Three bay, 1372 mm wide		1.20	6.667		300	237		537	700	
	5100	Coat racks, one bay, 2134 mm high, 610 mm wide		4.50	1.778		126	63		189	237	
	5300	Two bay, 1219 mm wide		2.75	2.909		175	103		278	355	
	5800	Three bay, 1829 mm wide		2.10	3.810		258	135		393	495	
	6100	Wall mounted cabinet, one bay, 610 mm high, 457 mm wide		3.60	2.222		69.50	79		148.50	199	
	6800	Two bay, 914 mm wide		2.20	3.636		101	129		230	310	
	7400	Three bay, 1372 mm wide		1.70	4.706		126	167		293	400	
	8400	762 mm high, one bay, 457 mm wide		3.60	2.222		75.50	79		154.50	206	
	9000	Two bay, 914 mm wide		2.15	3.721		100	132		232	315	
	9400	Three bay, 1372 mm wide		1.60	5		125	178		303	415	
	9800	Wardrobe, 2134 mm high, single, 610 mm wide		2.70	2.963		139	105		244	315	
	9880	Partition & adjustable shelves, 1219 mm wide		1.70	4.706		176	167		343	455	
	9950	Partition, adjustable shelves & drawers, 1219 mm wide		1.40	5.714		265	203		468	605	
220	0010	**CABINET DOORS**										**220**
	2000	Glass panel, hardwood frame										
	2200	305 mm wide, 457 mm high	1 Carp	34	.235	Ea.	22	8.35		30.35	37.50	
	2600	762 mm high		32	.250		25.50	8.90		34.40	42	
	4450	457 mm wide, 457 mm high		32	.250		24	8.90		32.90	40	
	4550	762 mm high		29	.276		26.50	9.80		36.30	44.50	
	5000	Hardwood, raised panel										
	5100	305 mm wide, 457 mm high	1 Carp	16	.500	Ea.	33.50	17.80		51.30	64.50	
	5200	762 mm high		15	.533		50	18.95		68.95	84.50	
	5500	457 mm wide, 457 mm high		15	.533		37	18.95		55.95	70	
	5600	762 mm high		14	.571		45	20.50		65.50	81	
	6000	Plastic laminate on particle board										

			CREW	DAILY OUTPUT	LABOR-HOURS	UNIT	2006 BARE COSTS				TOTAL INCL O&P	
	06410	**Custom Cabinets**					MAT.	LABOR	EQUIP.	TOTAL		
220	6100	305 mm wide, 457 mm high	1 Carp	25	.320	Ea.	10.35	11.40		21.75	29	220
	6140	762 mm high		23	.348		17.25	12.35		29.60	38.50	
	6500	457 mm wide, 457 mm high		24	.333		15.50	11.85		27.35	35.50	
	6600	762 mm high		22	.364		26	12.95		38.95	48.50	
230	0010	**CABINET HARDWARE**										230
	1000	Catches, minimum	1 Carp	235	.034	Ea.	.84	1.21		2.05	2.80	
	1040	Maximum	"	80	.100	"	5.20	3.56		8.76	11.30	
	2000	Door/drawer pulls, handles										
	2200	Handles and pulls, projecting, metal, minimum	1 Carp	160	.050	Ea.	3.90	1.78		5.68	7.05	
	2240	Maximum		68	.118		7.15	4.18		11.33	14.40	
	2300	Wood, minimum		160	.050		3.90	1.78		5.68	7.05	
	2340	Maximum		68	.118		7.15	4.18		11.33	14.40	
	2600	Flush, metal, minimum		160	.050		3.90	1.78		5.68	7.05	
	2640	Maximum		68	.118		7.15	4.18		11.33	14.40	
	3000	Drawer tracks/glides, minimum		48	.167	Pr.	6.60	5.95		12.55	16.55	
	3040	Maximum		24	.333		19.25	11.85		31.10	39.50	
	4000	Cabinet hinges, minimum		160	.050		2.25	1.78		4.03	5.25	
	4040	Maximum		68	.118		7.85	4.18		12.03	15.15	
240	0010	**DRAWERS**										240
	0100	Solid hardwood front										
	1000	102 mm high, 305 mm wide	1 Carp	17	.471	Ea.	3.50	16.75		20.25	30	
	1200	457 mm wide	"	16	.500	"	4.95	17.80		22.75	33	
	2800	Plastic laminate on particle board front										
	3000	102 mm high, 305 mm wide	1 Carp	17	.471	Ea.	4.39	16.75		21.14	31	
	3200	457 mm wide	"	16	.500	"	6.60	17.80		24.40	35	
	5400	Plywood, flush panel front										
	6000	102 mm high, 305 mm wide	1 Carp	17	.471	Ea.	5.95	16.75		22.70	32.50	
	6200	457 mm wide	"	16	.500	"	8.90	17.80		26.70	37.50	
400	0010	**VANITIES**										400
	8000	Vanity bases, 2 doors, 762 mm high, 533 mm deep, 610 mm wide	2 Carp	20	.800	Ea.	184	28.50		212.50	247	
	8050	762 mm wide		16	1		211	35.50		246.50	288	
	8100	914 mm wide		13.33	1.200		282	42.50		324.50	375	
	8150	1219 mm wide		11.43	1.400		335	50		385	450	
	9000	For deluxe models of all vanities, add to above					40%					
	9500	For custom built in place, add to above					25%	10%				

	06415	**Countertops**										
100	0010	**COUNTER TOP**										100
	0020	Stock, plastic laminate, 600 mm wide w/backsplash, min.	1 Carp	9.14	.875	m	29	31		60	80.50	
	0100	Maximum		7.62	1.050		52.50	37.50		90	116	
	0300	Custom plastic, 22 mm thick, aluminum molding, no splash		9.14	.875		58	31		89	113	
	0400	Cove splash		9.14	.875		76	31		107	133	
	0600	32 mm thick, no splash		8.53	.937		66.50	33.50		100	125	
	0700	Square splash		8.53	.937		83	33.50		116.50	143	
	0900	Square edge, plastic face, 22 mm thick, no splash		9.14	.875		73	31		104	129	
	1000	With splash		9.14	.875		93	31		124	151	
	1200	For stainless channel edge, 22 mm thick, add					7.60			7.60	8.35	
	1300	32 mm thick, add					8.95			8.95	9.85	
	1500	For solid color suede finish, add					7.45			7.45	8.20	
	1700	For end splash, add				Ea.	15.15			15.15	16.65	
	1900	For cut outs, standard, add, minimum	1 Carp	32	.250		3.03	8.90		11.93	17.20	
	2000	Maximum		8	1		5.05	35.50		40.55	61	
	2100	Postformed, including backsplash and front edge		9.14	.875	m	30	31		61	81.50	
	2110	Mitred, add		12	.667	Ea.		23.50		23.50	37	
	2200	Built-in place, 635 mm wide, plastic laminate		7.62	1.050	m	40	37.50		77.50	102	

WOOD & PLASTICS 6

06415 | Countertops

			DAILY OUTPUT	LABOR-HOURS	UNIT	2006 BARE COSTS				TOTAL INCL O&P	
						MAT.	LABOR	EQUIP.	TOTAL		
			CREW								
100	2300	Ceramic tile mosaic	1 Carp	7.62	1.050	m	86	37.50		123.50	153	100
	2500	Marble, stock, with splash, 13 mm thick, minimum	1 Bric	5.18	1.544		106	56.50		162.50	203	
	2700	19 mm thick, maximum	"	3.96	2.019		265	74		339	405	
	2900	Maple, solid, laminated, 38 mm thick, no splash	1 Carp	8.53	.937		173	33.50		206.50	243	
	3000	With square splash		8.53	.937		206	33.50		239.50	278	
	3200	Stainless steel		2.23	3.588	m²	1,300	128		1,428	1,625	
	3400	Recessed cutting block with trim, 406 mm x 508 mm x 25 mm		8	1	Ea.	61.50	35.50		97	123	

06430 | Stairs & Railings

			DAILY OUTPUT	LABOR-HOURS	UNIT	MAT.	LABOR	EQUIP.	TOTAL	INCL O&P	
500	0010	**RAILINGS**									500
	0020	Custom design, architectural grade, hardwood, minimum	1 Carp	11.58	.691	m	18.40	24.50		42.90	58	
	0100	Maximum		9.14	.875		153	31		184	217	
	0300	Stock interior railing with spindles 150 mm O.C., 1219 mm long		12.19	.656		138	23.50		161.50	189	
	0400	2438 mm long		14.63	.547		69	19.45		88.45	107	
620	0010	**STAIRS, PREFABRICATED**									620
	0100	Box stairs, prefabricated, 914 mm wide										
	0110	Oak treads, up to 14 risers	2 Carp	39	.410	Riser	71.50	14.60		86.10	101	
	0600	With pine treads for carpet, up to 14 risers	"	39	.410	"	46	14.60		60.60	73	
	1100	For 1219 mm wide stairs, add				Flight	25%					
	1550	Stairs, prefabricated stair handrail with balusters	1 Carp	9.14	.875	m	209	31		240	279	
	1700	Basement stairs, prefabricated, pine treads										
	1710	Pine risers, 914 mm wide, up to 14 risers	2 Carp	52	.308	Riser	46	10.95		56.95	67.50	
	4000	Residential, wood, oak treads, prefabricated		1.50	10.667	Flight	930	380		1,310	1,625	
	4200	Built in place		.44	36.364	"	1,925	1,300		3,225	4,150	
	4400	Spiral, oak, 1372 mm diameter, unfinished, prefabricated,										
	4500	incl. railing, 2743 mm high	2 Carp	1.50	10.667	Flight	4,400	380		4,780	5,450	
630	0010	**STAIR PARTS**									630
	0020	Balusters, turned, 762 mm high, pine, minimum	1 Carp	28	.286	Ea.	3.70	10.15		13.85	19.85	
	0100	Maximum		26	.308		19	10.95		29.95	38	
	0300	762 mm high birch balusters, minimum		28	.286		6.30	10.15		16.45	23	
	0400	Maximum		26	.308		27.50	10.95		38.45	47	
	0600	1067 mm high, pine balusters, minimum		27	.296		4.90	10.55		15.45	22	
	0700	Maximum		25	.320		27.50	11.40		38.90	47.50	
	0900	1067 mm high birch balusters, minimum		27	.296		10.70	10.55		21.25	28	
	1000	Maximum		25	.320		38.50	11.40		49.90	60	
	1050	Baluster, stock pine, 32 mm x 32 mm		73.15	.109	m	10.25	3.89		14.14	17.30	
	1100	45 mm x 45 mm		67.06	.119	"	29	4.24		33.24	38.50	
	1200	Newels, 83 mm wide, starting, minimum		7	1.143	Ea.	38.50	40.50		79	106	
	1300	Maximum		6	1.333		320	47.50		367.50	425	
	1500	Landing, minimum		5	1.600		104	57		161	203	
	1600	Maximum		4	2		350	71		421	495	
	1800	Railings, oak, built-up, minimum		18.29	.437	m	104	15.55		119.55	138	
	1900	Maximum		16.76	.477		143	16.95		159.95	184	
	2100	Add for sub rail		33.53	.239		17.40	8.50		25.90	32.50	
	2300	Risers, beech, 19 mm x 191 mm high		19.51	.410		19	14.60		33.60	43.50	
	2400	Fir, 19 mm x 191 mm high		19.51	.410		5.25	14.60		19.85	28.50	
	2600	Oak, 19 mm x 191 mm high		19.51	.410		20	14.60		34.60	44.50	
	2800	Pine, 19 mm x 191 mm high		20.12	.398		10	14.15		24.15	33	
	2850	Skirt board, pine, 25 mm x 254 mm		16.76	.477		9.50	16.95		26.45	37	
	2900	25 mm x 305 mm		15.85	.505		11.30	17.95		29.25	40.50	
	3000	Treads, oak, 32 mm x 254 mm wide 914 mm long		18	.444	Ea.	29	15.80		44.80	56.50	
	3100	1219 mm long, oak		17	.471		116	16.75		132.75	154	
	3300	32 mm x 292 mm wide, 914 mm long, oak		18	.444		31	15.80		46.80	58.50	
	3400	1829 mm long, oak		14	.571		186	20.50		206.50	237	

		06430	Stairs & Railings	CREW	DAILY OUTPUT	LABOR-HOURS	UNIT	2006 BARE COSTS				TOTAL INCL O&P	
								MAT.	LABOR	EQUIP.	TOTAL		
630	3600		Beech treads, add				Ea.	40%					630
	3800		For mitered return nosings, add				m	12.80			12.80	14.05	

06440 | Wood Ornaments

150	0010	**BEAMS, DECORATIVE**										150
	0020	Rough sawn cedar, non-load bearing, 100 mm x 100 mm	2 Carp	54.86	.292	m	4.43	10.35		14.78	21	
	0100	100 mm x 150 mm		51.82	.309		8.55	11		19.55	26.50	
	0200	100 mm x 200 mm		48.77	.328		10.95	11.65		22.60	30	
	0300	100 mm x 250 mm		45.72	.350		15.20	12.45		27.65	36	
	0400	100 mm x 300 mm		42.67	.375		18.45	13.35		31.80	41.50	
	0500	200 mm x 200 mm	▼	39.62	.404	▼	26	14.35		40.35	51	
	1100	Beam connector plates see div 06090-800										
350	0010	**GRILLES** and panels, hardwood, sanded										350
	0020	0.6 m x 1.2 m to 1.2 m x 2.4 m, custom, unfinished, min.	1 Carp	3.53	2.266	m²	130	80.50		210.50	268	
	0050	Average		2.79	2.870		283	102		385	470	
	0100	Maximum		1.77	4.532		435	161		596	730	
	0300	As above, but prefinished, minimum		3.53	2.266		130	80.50		210.50	268	
	0400	Maximum	▼	1.77	4.532	▼	490	161		651	790	
400	0010	**LOUVERS**										400
	0020	Redwood, 610 mm diameter, full circle	1 Carp	16	.500	Ea.	136	17.80		153.80	178	
	0100	Half circle		16	.500		130	17.80		147.80	171	
	0200	Octagonal		16	.500		104	17.80		121.80	142	
	0300	Triangular, 5/12 pitch, 1524 mm at base	▼	16	.500	▼	220	17.80		237.80	270	
500	0010	**FIREPLACE MANTELS**										500
	0015	150 mm molding, 1.8 m x 1.1 m opening, min.	1 Carp	5	1.600	Opng.	139	57		196	242	
	0100	Maximum		5	1.600		173	57		230	279	
	0300	Prefabricated pine, colonial type, stock, deluxe		2	4		2,175	142		2,317	2,625	
	0400	Economy	▼	3	2.667	▼	375	95		470	560	
550	0010	**FIREPLACE MANTEL BEAMS**										550
	0020	Rough texture wood, 100 mm x 200 mm	1 Carp	10.97	.729	m	16.90	26		42.90	59	
	0100	100 mm x 250 mm		10.67	.750	"	20.50	26.50		47	64	
	0300	Laminated hardwood, 50 mm x 275 mm wide, 1829 mm long		5	1.600	Ea.	98.50	57		155.50	197	
	0400	2438 mm long		5	1.600	"	137	57		194	239	
	0600	Brackets for above, rough sawn		12	.667	Pr.	9.05	23.50		32.55	47	
	0700	Laminated	▼	12	.667	"	13.70	23.50		37.20	52	
700	0010	**COLUMNS**										700
	0050	Aluminum, round colonial, 152 mm diameter [6″]	2 Carp	24.38	.656	m	59.50	23.50		83	102	
	0100	203 mm diameter [8″]		18.97	.843		82	30		112	137	
	0200	254 mm diameter [10″]		16.76	.954		98.50	34		132.50	161	
	0250	Fir, stock units, hollow round, 152 mm diameter [6″]		24.38	.656		67.50	23.50		91	111	
	0300	203 mm diameter [8″]		24.38	.656		79.50	23.50		103	124	
	0350	254 mm diameter [10″]		21.34	.750		101	26.50		127.50	153	
	0400	Solid turned, to 2438 mm high, 89 mm diameter [3-1/2″]		24.38	.656		29.50	23.50		53	68.50	
	0500	114 mm diameter [4-1/2″]		22.86	.700		42	25		67	84.50	
	0600	140 mm diameter [5-1/2″]		21.34	.750		58.50	26.50		85	106	
	0800	Square columns, built-up, 127 mm x 127 mm		19.81	.808		54.50	28.50		83	104	
	0900	Solid, 89 mm x 89 mm		39.62	.404		25	14.35		39.35	50	
	1600	Hemlock, tapered, T & G, 305 mm diam, 3048 mm high		30.48	.525		125	18.65		143.65	167	
	1700	4877 mm high		19.81	.808		221	28.50		249.50	289	
	1900	3048 mm high, 356 mm diameter [14″]		30.48	.525		320	18.65		338.65	385	
	2000	5486 mm high		19.81	.808		305	28.50		333.50	380	
	2200	457 mm diameter, 3658 mm high [18″]		19.81	.808		430	28.50		458.50	520	
	2300	6096 mm high	▼	15.24	1.050	▼	415	37.50		452.50	515	

WOOD & PLASTICS 6

06440 | Wood Ornaments

		CREW	DAILY OUTPUT	LABOR-HOURS	UNIT	2006 BARE COSTS MAT.	LABOR	EQUIP.	TOTAL	TOTAL INCL O&P	
700											**700**
2500	508 mm diameter, 4267 mm high [20"]	2 Carp	12.19	1.312	m	510	46.50		556.50	635	
2600	6096 mm high	↓	10.67	1.500	↓	525	53.50		578.50	665	
2800	For flat pilasters, deduct				↓	33%					
3000	For splitting into halves, add				Ea.	114			114	125	
4000	Rough sawn cedar posts, 100 mm x 100 mm	2 Carp	76.20	.210	m	12.15	7.45		19.60	25	
4100	100 mm x 150 mm		71.63	.223		22	7.95		29.95	36.50	
4200	150 mm x 150 mm		67.06	.239		41	8.50		49.50	58	
4300	200 mm x 200 mm	↓	60.96	.262	↓	94	9.35		103.35	118	

06445 | Simulated Wood Ornaments

		CREW	DAILY OUTPUT	LABOR-HOURS	UNIT	MAT.	LABOR	EQUIP.	TOTAL	TOTAL INCL O&P	
100	0010	**MILLWORK, HIGH DENSITY POLYMER**									**100**
0100	Base, 14 mm x 81 mm	1 Carp	70.10	.114	m	4.40	4.06		8.46	11.10	
0200	Casing, fluted, 16 mm x 83 mm		65.53	.122		10.25	4.34		14.59	18	
0300	Chair rail, 14 mm x 57 mm		79.25	.101		6.70	3.59		10.29	12.95	
0600	Cove, 21 mm x 95 mm		79.25	.101		12.80	3.59		16.39	19.65	
0700	Crown, 19 mm x 97 mm		79.25	.101		19.25	3.59		22.84	26.50	
0800	Half round, 24 mm x 51 mm	↓	73.15	.109	↓	57.50	3.89		61.39	69	

06470 | Screen, Blinds & Shutters

		CREW	DAILY OUTPUT	LABOR-HOURS	UNIT	MAT.	LABOR	EQUIP.	TOTAL	TOTAL INCL O&P	
100	0010	**SHUTTERS, EXTERIOR**									**100**
0012	Aluminum, louvered, 406 mm wide, 914 mm long	1 Carp	10	.800	Pr.	45.50	28.50		74	94.50	
0400	2032 mm long		9	.889		91.50	31.50		123	150	
1000	Pine, louvered, primed, each 356 mm wide, 991 mm long		10	.800		86.50	28.50		115	140	
1100	1397 mm long		10	.800		117	28.50		145.50	174	
1500	Each 457 mm wide, 991 mm long		10	.800		92	28.50		120.50	146	
1600	1397 mm long		10	.800		129	28.50		157.50	187	
1620	Hemlock, louvered, 356 mm wide, 1702 mm long		10	.800		142	28.50		170.50	201	
1630	Each 406 mm wide, 660 mm long		10	.800		88.50	28.50		117	142	
1670	1295 mm long		10	.800		106	28.50		134.50	161	
1690	1800 mm long		10	.800		149	28.50		177.50	209	
1700	Door blinds, 2057 mm long, each 381 mm wide		9	.889		150	31.50		181.50	214	
1710	457 mm wide		9	.889		161	31.50		192.50	226	
1720	Hemlock, solid raised panel, each 406 mm wide, 991 mm long		10	.800		143	28.50		171.50	202	
1740	1295 mm long		10	.800		181	28.50		209.50	244	
1770	1800 mm long		10	.800		242	28.50		270.50	310	
1800	Door blinds, 2057 mm long, each 381 mm wide		9	.889		272	31.50		303.50	350	
1900	457 mm wide		9	.889		297	31.50		328.50	375	
2500	Polystyrene, solid raised panel, each 406 mm wide, 991 mm long		10	.800		38.50	28.50		67	86.50	
2700	1397 mm long		10	.800		48	28.50		76.50	97.50	
4500	Polystyrene, louvered, each 356 mm wide, 991 mm long		10	.800		26	28.50		54.50	73	
4750	1600 mm long		10	.800		37	28.50		65.50	85	
6000	Vinyl, louvered, each 356 mm x 1397 mm long		10	.800		32	28.50		60.50	79.50	
6200	Each 406 mm x 2032 mm long	↓	9	.889	↓	45	31.50		76.50	98.50	
200	0010	**SHUTTERS, INTERIOR** Wood, louvered,									**200**
0200	Two panel, 686 mm wide, 914 mm high	1 Carp	5	1.600	Set	108	57		165	208	
0300	838 mm wide, 914 mm high		5	1.600		142	57		199	245	
0500	1194 mm wide, 914 mm high		5	1.600		189	57		246	297	
1000	Four panel, 686 mm wide, 914 mm high		5	1.600		128	57		185	230	
1100	838 mm wide, 914 mm high		5	1.600		157	57		214	262	
1300	1194 mm wide, 914 mm high	↓	5	1.600	↓	223	57		280	335	

6 WOOD & PLASTICS

06510 | Struct Plastic Shapes & Plates

		CREW	DAILY OUTPUT	LABOR-HOURS	UNIT	MAT.	LABOR	EQUIP.	TOTAL	TOTAL INCL O&P	
400	**0010**	**CASTINGS, FIBERGLASS**									**400**
	0100	Angle, 25 mm x 25 mm x 3 mm thick	2 Sswk	73.15	.219	m	4.04	8.75		12.79	20	
	0120	75 mm x 75 mm x 6 mm thick		60.96	.262		16.75	10.50		27.25	37.50	
	0140	100 mm x 100 mm x 6 mm thick		60.96	.262		21.50	10.50		32	43	
	0160	100 mm x 100 mm x 9 mm thick		60.96	.262		33	10.50		43.50	55.50	
	0180	150 mm x 150 mm x 13 mm thick		48.77	.328		66.50	13.10		79.60	96.50	
	1000	Flat sheet, 3 mm thick		13.01	1.230	m²	46.50	49		95.50	140	
	1020	6 mm thick		11.15	1.435		91	57.50		148.50	203	
	1040	9 mm thick		9.29	1.722		155	69		224	294	
	1060	13 mm thick		7.43	2.153		203	86		289	380	
	2000	Handrail, 1050 mm high, 50 mm diam. rails pickets 1500 mm OC		9.75	1.640	m	143	65.50		208.50	275	
	3000	Round bar, 6 mm diam.		73.15	.219		1.61	8.75		10.36	17.50	
	3020	13 mm diam.		60.96	.262		2.85	10.50		13.35	22	
	3040	19 mm diam.		60.96	.262		5.75	10.50		16.25	25	
	3060	25 mm diam.		48.77	.328		10.15	13.10		23.25	34.50	
	3080	31 mm diam.		48.77	.328		11.55	13.10		24.65	36	
	3100	38 mm diam.		42.67	.375		14.15	15		29.15	42.50	
	3500	Round tube, 25 mm diam. x 3 mm thick		73.15	.219		5.50	8.75		14.25	22	
	3520	50 mm diam. x 6 mm thick		60.96	.262		17.80	10.50		28.30	38.50	
	3540	75 mm diam. x 6 mm thick		48.77	.328		28.50	13.10		41.60	54.50	
	4000	Square bar, 13 mm square		73.15	.219		11.15	8.75		19.90	28	
	4020	25 mm square		60.96	.262		14.25	10.50		24.75	34.50	
	4040	38 mm square		48.77	.328		29	13.10		42.10	55.50	
	4500	Square tube, 25 mm x 25 mm x 3 mm thick		73.15	.219		6.25	8.75		15	22.50	
	4520	50 mm x 50 mm x 3 mm thick		60.96	.262		12.15	10.50		22.65	32.50	
	4540	75 mm x 75 mm x 6 mm thick		48.77	.328		33	13.10		46.10	60	
	5000	Threaded rod, 9 mm diam.		97.54	.164		9.65	6.55		16.20	22.50	
	5020	13 mm diam.		97.54	.164		11.35	6.55		17.90	24.50	
	5040	16 mm diam.		85.34	.187		12.55	7.50		20.05	27.50	
	5060	19 mm diam.		85.34	.187		14.30	7.50		21.80	29.50	
	6000	Wide flange beam, 100 mm x 100 mm x 6 mm thick		36.58	.437		29.50	17.50		47	64	
	6020	150 mm x 150 mm x 6 mm thick		30.48	.525		56	21		77	99	
	6040	200 mm x 200 mm x 9 mm thick	▼	24.38	.656	▼	98	26		124	155	

06520 | Plastic Struct Assemblies

		CREW	DAILY OUTPUT	LABOR-HOURS	UNIT	MAT.	LABOR	EQUIP.	TOTAL	TOTAL INCL O&P	
100	**0010**	**STAIR TREAD, FIBERGLASS** Isophthalic Resin, 263 mm deep									**100**
	0100	600 mm wide	2 Sswk	52	.308	Ea.	34	12.30		46.30	59.50	
	0140	750 mm wide		52	.308		40.50	12.30		52.80	66.50	
	0180	900 mm wide		52	.308		47.50	12.30		59.80	74	
	0220	1050 mm wide	▼	52	.308	▼	56	12.30		68.30	83.50	
150	**0010**	**GRATING, FIBERGLASS**									**150**
	0100	Molded, green (for mod. corrosive environment)										
	0140	25 mm x 100 mm mesh, 25 mm thick	2 Sswk	37.16	.431	m²	113	17.20		130.20	155	
	0180	38 mm square mesh, 25 mm thick		37.16	.431		168	17.20		185.20	215	
	0220	31 mm thick		37.16	.431		134	17.20		151.20	178	
	0260	38 mm thick		37.16	.431		244	17.20		261.20	299	
	0300	50 mm square mesh, 50 mm thick	▼	29.73	.538	▼	229	21.50		250.50	291	
	1000	Orange (for highly corrosive environment)										
	1040	25 mm x 100 mm mesh, 25 mm thick	2 Sswk	37.16	.431	m²	142	17.20		159.20	187	
	1080	38 mm square mesh, 25 mm thick		37.16	.431		188	17.20		205.20	238	
	1120	31 mm thick		37.16	.431		195	17.20		212.20	246	
	1160	38 mm thick		37.16	.431		211	17.20		228.20	263	
	1200	50 mm square mesh, 50 mm thick	▼	29.73	.538	▼	252	21.50		273.50	315	
	3000	Pultruded, green (for mod. corrosive environment)										

WOOD & PLASTICS **6**

		06520	Plastic Struct Assemblies	CREW	DAILY OUTPUT	LABOR-HOURS	UNIT	2006 BARE COSTS MAT.	LABOR	EQUIP.	TOTAL	TOTAL INCL O&P	
150	3040		25 mm O.C. bar spacing, 25 mm thick	2 Sswk	37.16	.431	m²	168	17.20		185.20	216	150
	3080		38 mm thick		29.73	.538		187	21.50		208.50	245	
	3120		38 mm O.C. bar spacing, 25 mm thick		37.16	.431		137	17.20		154.20	181	
	3160		38 mm thick		37.16	.431		153	17.20		170.20	199	
	4000		Grating support legs, fixed height, no base				Ea.	41			41	45	
	4040		With base					36.50			36.50	40	
	4080		Adjustable to 1500 mm					55.50			55.50	61	
175	0010	**FLOOR GRATING, FIBERGLASS**											175
	0100		Reinforced polyester, fire retardant, 25 mm x 100 mm grid, 25 mm T	E-4	47.38	.675	m²	144	27.50	1.88	173.38	209	
	0200		38 x 150 mm mesh, 38 mm thick		46.45	.689		168	28	1.92	197.92	237	
	0300		With grit surface, 38 mm x 150 mm grid, 38 mm thick		46.45	.689		172	28	1.92	201.92	241	

06600 | Plastic Fabrications

		06620	Non-Structural Plastics	CREW	DAILY OUTPUT	LABOR-HOURS	UNIT	2006 BARE COSTS MAT.	LABOR	EQUIP.	TOTAL	TOTAL INCL O&P	
600	0010	**NETTING, FLEXIBLE PLASTIC**											600
	0020		3 mm square mesh	4 Clab	372	.086	m²	.43	2.36		2.79	4.10	
	0100		6 mm square mesh		372	.086		.54	2.36		2.90	4.32	
	0120		13 mm square mesh		372	.086		.54	2.36		2.90	4.32	
	0140		16 mm x 19 mm mesh		372	.086		.54	2.36		2.90	4.32	
	0160		31 mm x 38 mm mesh		372	.086		.65	2.36		3.01	4.42	
	0200		100 mm square mesh		372	.086		.75	2.36		3.11	4.53	
	1000		Poly clips				Ea.	.01			.01	.01	
810	0010	**SOLID SURFACE COUNTERTOPS**, Acrylic polymer											810
	0020		Pricing for orders of 30 m or greater										
	0100		635 mm wide, solid colors	2 Carp	8.53	1.875	m	145	66.50		211.50	263	
	0200		Patterned colors		8.53	1.875		184	66.50		250.50	305	
	0300		Premium patterned colors		8.53	1.875		230	66.50		296.50	355	
	0400		With silicone attached 100 mm backsplash, solid colors		8.23	1.944		159	69		228	283	
	0500		Patterned colors		8.23	1.944		201	69		270	330	
	0600		Premium patterned colors		8.23	1.944		251	69		320	385	
	0700		With hard seam attached 100 mm backsplash, solid colors		7.01	2.282		159	81		240	300	
	0800		Patterned colors		7.01	2.282		201	81		282	350	
	0900		Premium patterned colors		7.01	2.282		251	81		332	400	
	1000		Pricing for order of 15 - 30 m										
	1100		635 mm wide, solid colors	2 Carp	7.32	2.187	m	167	78		245	305	
	1200		Patterned colors		7.32	2.187		211	78		289	355	
	1300		Premium patterned colors		7.32	2.187		264	78		342	410	
	1400		With silicone attached 100 mm backsplash, solid colors		7.01	2.282		183	81		264	325	
	1500		Patterned colors		7.01	2.282		232	81		313	380	
	1600		Premium patterned colors		7.01	2.282		288	81		369	440	
	1700		With hard seam attached 100 mm backsplash, solid colors		6.10	2.625		183	93.50		276.50	345	
	1800		Patterned colors		6.10	2.625		232	93.50		325.50	400	
	1900		Premium patterned colors		6.10	2.625		288	93.50		381.50	460	
	2000		Pricing for order of .3 - 15 m										
	2100		635 mm wide, solid colors	2 Carp	6.10	2.625	m	196	93.50		289.50	360	
	2200		Patterned colors		6.10	2.625		248	93.50		341.50	420	
	2300		Premium patterned colors		6.10	2.625		310	93.50		403.50	485	
	2400		With silicone attached 100 mm backsplash, solid colors		5.79	2.763		215	98		313	390	

6 WOOD & PLASTICS

06620	Non-Structural Plastics	CREW	DAILY OUTPUT	LABOR-HOURS	UNIT	2006 BARE COSTS				TOTAL INCL O&P	
						MAT.	LABOR	EQUIP.	TOTAL		
810	2500	Patterned colors	2 Carp	5.79	2.763	m	272	98		370	450	810
	2600	Premium patterned colors		5.79	2.763		340	98		438	530	
	2700	With hard seam attached 100 mm backsplash, solid colors		4.57	3.500		215	124		339	430	
	2800	Patterned colors		4.57	3.500		272	124		396	495	
	2900	Premium patterned colors	▼	4.57	3.500	▼	340	124		464	570	
	3000	Sinks, pricing for order of 100 or greater units										
	3100	Single bowl, hard seamed, solid colors, 325 mm x 425 mm	1 Carp	3	2.667	Ea.	298	95		393	480	
	3200	250 mm x 375 mm		7	1.143		138	40.50		178.50	216	
	3300	Cutouts for sinks	▼	8	1	▼		35.50		35.50	55.50	
	3400	Sinks, pricing for order of 51 - 99 units										
	3500	Single bowl, hard seamed, solid colors, 325 mm x 425 mm	1 Carp	2.55	3.137	Ea.	345	112		457	550	
	3600	250 mm x 375 mm		6	1.333		159	47.50		206.50	248	
	3700	Cutouts for sinks	▼	7	1.143	▼		40.50		40.50	63.50	
	3800	Sinks, pricing for order of 1 - 50 units										
	3900	Single bowl, hard seamed, solid colors, 325 mm x 425 mm	1 Carp	2	4	Ea.	405	142		547	665	
	4000	250 mm x 375 mm		4.55	1.758		186	62.50		248.50	305	
	4100	Cutouts for sinks		5.25	1.524			54		54	84.50	
	4200	Cooktop cutouts, pricing for 100 or greater units		4	2		22	71		93	135	
	4300	51 - 99 units		3.40	2.353		25.50	83.50		109	158	
	4400	1 - 50 units	▼	3	2.667	▼	29.50	95		124.50	181	
850	0010	**VANITY TOPS**										850
	0015	Solid surface, center bowl, 432 mm x 483 mm	1 Carp	12	.667	Ea.	182	23.50		205.50	237	
	0020	483 mm x 635 mm		12	.667		220	23.50		243.50	279	
	0030	483 mm x 787 mm		12	.667		267	23.50		290.50	330	
	0040	483 mm x 930 mm		12	.667		310	23.50		333.50	375	
	0050	559 mm x 635 mm		10	.800		194	28.50		222.50	258	
	0060	559 mm x 787 mm		10	.800		227	28.50		255.50	294	
	0070	559 mm x 930 mm		10	.800		264	28.50		292.50	335	
	0080	559 mm x 1092 mm		10	.800		300	28.50		328.50	375	
	0090	559 mm x 1245 mm		10	.800		335	28.50		363.50	410	
	0110	559 mm x 1397 mm		8	1		380	35.50		415.50	470	
	0120	559 mm x 1549 mm		8	1		430	35.50		465.50	530	
	0220	Double bowl, 559 mm x 1549 mm		8	1		490	35.50		525.50	590	
	0230	Double bowl, 559 mm x 1855 mm	▼	8	1	▼	675	35.50		710.50	800	
	0240	For aggregate colors, add					35%					
	0250	For faucets and fittings see 15410-300										

For information about Means Estimating Seminars, see yellow pages 12 and 13 in back of book

WOOD & PLASTICS 6

For expanded coverage of these items see _Means Interior Cost Data 2006_

Division Notes

		CREW	DAILY OUTPUT	LABOR-HOURS	UNIT	2006 BARE COSTS				TOTAL INCL O&P
						MAT.	LABOR	EQUIP.	TOTAL	

Division 7
Thermal & Moisture Protection

Estimating Tips

07100 Dampproofing & Waterproofing

- Be sure of the job specifications before pricing this subdivision. The difference in cost between waterproofing and dampproofing can be great. Waterproofing will hold back standing water. Dampproofing prevents the transmission of water vapor. Also included in this section are vapor retarding membranes.

07200 Thermal Protection

- Insulation and fireproofing products are measured by area, thickness, volume or thermal resistance. Specifications may only give what the specific thermal resistance should be in a certain situation. The estimator may need to choose the type of insulation to meet that R value.

07300 Shingles, Roof Tiles & Roof Coverings
07400 Roofing & Siding Panels

- Many roofing and siding products are bought and sold by the square meter. Accessories necessary for a complete installation must be figured into any calculations for both material and labor.

07500 Membrane Roofing
07600 Flashing & Sheet Metal
07700 Roof Specialties & Accessories

- The items in these subdivisions compose a roofing system. No one component completes the installation and all must be estimated. Built-up or single ply membrane roofing systems are made up of many products and installation trades. Wood blocking at roof perimeters or penetrations, parapet coverings, reglets, roof drains, gutters, downspouts, sheet metal flashing, skylights, smoke vents and roof hatches all need to be considered along with the roofing material. Several different installation trades will need to work together on the roofing system. Inherent difficulties in the scheduling and coordination of various trades must be accounted for when estimating labor costs.

07900 Joint Sealers

- To complete the weather-tight shell the sealants and caulkings must be estimated. Where different materials meet—at expansion joints, at flashing penetrations, and at hundreds of other locations throughout a construction project—they provide another line of defense against water penetration. Often, an entire system is based on the proper location and placement of caulking or sealants. The detailed drawings that are included as part of a set of architectural plans, show typical locations for these materials. When caulking or sealants are shown at typical locations, this means the estimator must include them for all the locations where this detail is applicable. Be careful to keep different types of sealants separate, and remember to consider backer rods and primers if necessary.

Reference Numbers

Reference numbers are shown in bold squares at the beginning of some major classifications. These numbers refer to related items in the Reference Section. The reference information may be an estimating procedure, an alternate pricing method, or technical information.

Note: Not all subdivisions listed here necessarily appear in this publication.

			CREW	DAILY OUTPUT	LABOR-HOURS	UNIT	2006 BARE COSTS				TOTAL INCL O&P
	07060	**Selective Demolition**					MAT.	LABOR	EQUIP.	TOTAL	
110	0010	**SELECTIVE DEMO, THERMAL & MOISTURE PROTECTION** R024119-10										110
	0200	Waterproofing demo, scrape off, to 50 mm thick	2 Clab	186	.086	m²		2.36		2.36	3.67
	0210	Over 50 mm thick		163	.098	"		2.69		2.69	4.19
	0250	Protection / drain board		9.36	1.709	m³		47		47	73
	1000	Deck, roof, concrete plank	B-13	156	.359	m²		10.75	4.17	14.92	21
	1100	Gypsum plank		362	.155			4.63	1.80	6.43	9.15
	1150	Metal decking		325	.172			5.15	2	7.15	10.15
	1200	Wood, boards, tongue and groove, 50 mm x 150 mm	2 Clab	89.18	.179			4.92		4.92	7.65
	1220	50 mm x 250 mm		96.62	.166			4.54		4.54	7.05
	1280	Standard planks, 25 mm x 150 mm		100	.160			4.38		4.38	6.80
	1320	25 mm x 200 mm		108	.148			4.06		4.06	6.30
	1340	25 mm x 300 mm		111	.144			3.95		3.95	6.15
	1350	Plywood, to 25 mm thick		186	.086			2.36		2.36	3.67
	1360	Flashing, aluminum	1 Clab	26.94	.297			8.15		8.15	12.65
	2000	Gutters, aluminum or wood, edge hung		73.15	.109	m		3		3	4.66
	2100	Built-in		30.48	.262	"		7.20		7.20	11.20
	2200	Insulation removal, loose fitting		84.90	.094	m³		2.58		2.58	4.02
	2250	Air barrier		325	.025	m²		.67		.67	1.05
	2300	Batts or blankets		39.62	.202	m³		5.55		5.55	8.60
	2350	Rigid board		8.28	.966	"		26.50		26.50	41
	2500	Roof accessories, plumbing vent flashing		14	.571	Ea.		15.65		15.65	24.50
	2600	Adjustable metal chimney flashing		9	.889	"		24.50		24.50	38
	2650	Coping, sheet metal, up to 300 mm wide		73.15	.109	m		3		3	4.66
	2660	Concrete, up to 305 mm wide	2 Clab	48.77	.328	"		9		9	14
	3000	Roofing, built-up, 5 ply roof, no gravel	B-2	149	.268	m²		7.45		7.45	11.60
	3001	Including gravel		82.68	.484			13.45		13.45	21
	3100	Gravel removal, minimum		465	.086			2.39		2.39	3.72
	3120	Maximum		186	.215			6		6	9.30
	3400	Roof insulation board, up to 50 mm thick		362	.111			3.07		3.07	4.78
	3405	Over 50 mm thick		18.72	2.137	m³		59.50		59.50	92.50
	3450	Roll roofing, cold adhesive	1 Clab	111	.072	m²		1.97		1.97	3.07
	4000	Shingles, asphalt strip, 1 layer	B-2	325	.123			3.42		3.42	5.35
	4100	Slate		232	.172			4.79		4.79	7.45
	4300	Wood		204	.196			5.45		5.45	8.50
	4500	Skylight to 0.93 m²	1 Clab	8	1	Ea.		27.50		27.50	42.50
	5000	Siding, metal, horizontal		41.25	.194	m²		5.30		5.30	8.25
	5020	Vertical		37.16	.215			5.90		5.90	9.20
	5200	Wood, boards, vertical		37.16	.215			5.90		5.90	9.20
	5220	Clapboards, horizontal		35.30	.227			6.20		6.20	9.65
	5240	Shingles		32.52	.246			6.75		6.75	10.50
	5260	Textured plywood		67.35	.119			3.25		3.25	5.05

07100 | Dampproofing and Waterproofing

			CREW	DAILY OUTPUT	LABOR-HOURS	UNIT	2006 BARE COSTS				TOTAL INCL O&P
	07110	**Dampproofing**					MAT.	LABOR	EQUIP.	TOTAL	
100	0010	**BITUMINOUS ASPHALT COATING** For foundation										100
	0030	Brushed on, below grade, 1 coat	1 Rofc	61.78	.129	m²	.97	3.96		4.93	7.85
	0100	2 coat		46.45	.172		1.94	5.25		7.19	11
	0300	Sprayed on, below grade, 1 coat, 0.63 m²/L		77.11	.104		.97	3.17		4.14	6.50

			CREW	DAILY OUTPUT	LABOR-HOURS	UNIT	2006 BARE COSTS				TOTAL INCL O&P	
		07110 | Dampproofing					MAT.	LABOR	EQUIP.	TOTAL		
100	0400	2 coat, 0.50 m²/L	1 Rofc	46.45	.172	m²	1.83	5.25		7.08	11	**100**
	0500	Asphalt coating, with fibers				liter	1.10			1.10	1.22	
	0600	Troweled on, asphalt with fibers, 1.6 mm thick	1 Rofc	46.45	.172	m²	1.94	5.25		7.19	11.10	
	0700	3 mm thick	↓	37.16	.215		3.44	6.60		10.04	14.95	
	1000	13 mm thick	▼	32.52	.246	▼	11.30	7.55		18.85	25	
200	0010	**CEMENT PARGING**										**200**
	0020	Portland cement, 2 coats, 13 mm thick, regular P.C.	D-1	23.23	.689	m²	1.94	22		23.94	35.50	
	0100	Waterproofed Portland cement	"	23.23	.689	"	19.90	22		41.90	55.50	

07130 | Sheet Waterproofing

			CREW	DAILY OUTPUT	LABOR-HOURS	UNIT	MAT.	LABOR	EQUIP.	TOTAL	INCL O&P	
200	0010	**ELASTOMERIC WATERPROOFING**										**200**
	0090	EPDM, plain, 1.1 mm thick	2 Rofc	53.88	.297	m²	10.65	9.10		19.75	27	
	0100	1.5 mm thick		52.95	.302		11.50	9.25		20.75	28.50	
	0300	Nylon reinforced sheets, 1.1 mm thick		53.88	.297		16.15	9.10		25.25	33	
	0400	1.5 mm thick	▼	52.95	.302	▼	20.50	9.25		29.75	38	
	0600	Vulcanizing splicing tape for above, 50 mm wide				m	1.35			1.35	1.48	
	0700	100 mm wide				"	1.80			1.80	1.98	
	0900	Adhesive, bonding, 1.5 m²/L				liter	3.86			3.86	4.24	
	1000	Splicing, 1.84 m²/L				"	8.25			8.25	9.05	
	1200	Neoprene sheets, plain, 1.1 mm thick	2 Rofc	53.88	.297	m²	15.20	9.10		24.30	32	
	1300	1.5 mm thick		52.95	.302		24	9.25		33.25	41.50	
	1500	Nylon reinforced, 1.1 mm thick		53.88	.297		19.40	9.10		28.50	37	
	1600	1.5 mm thick		52.95	.302		22	9.25		31.25	40	
	1800	3 mm thick	▼	46.45	.344	▼	36.50	10.55		47.05	58	
	1900	Adhesive, splicing, 3.68 m²/L/coat				liter	4.80			4.80	5.30	
	2100	Fiberglass reinforced, fluid applied, 3 mm thick	2 Rofc	46.45	.344	m²	17.45	10.55		28	37	
	2200	Polyethylene and rubberized asphalt sheets, 3 mm thick		51.10	.313		5.80	9.60		15.40	22.50	
	2210	Asphaltic hardboard protection board, 3mm thick		46.45	.344		3.34	10.55		13.89	21.50	
	2220	6 mm thick		41.81	.383		5.90	11.70		17.60	26.50	
	2400	Polyvinyl chloride sheets, plain, 0.25 mm thick		53.88	.297		1.72	9.10		10.82	17.40	
	2500	0.51 mm thick		52.95	.302		2.69	9.25		11.94	18.70	
	2700	0.76 mm thick	▼	52.02	.308	▼	3.77	9.40		13.17	20	
	3000	Adhesives, trowel grade, 0.98 m²/L				liter	6.10			6.10	6.70	
	3100	Brush grade, 2.45 m²/L				"	6.10			6.10	6.70	
	3300	Bitumen modified polyurethane, fluid applied, 1 mm thick	2 Rofc	61.78	.259	m²	7.20	7.95		15.15	21.50	
	3600	Vinyl plastic, sprayed on, 0.6 to 1 mm thick	"	44.13	.363	"	10.75	11.10		21.85	30.50	
500	0010	**MEMBRANE WATERPROOFING**										**500**
	0012	On slabs, 1 ply, felt, mopped	G-1	279	.201	m²	2.05	5.75	1.21	9.01	13.35	
	0100	On slabs, 1 ply, glass fiber fabric, mopped		195	.287		2.26	8.20	1.73	12.19	18.35	
	0300	On slabs, 2 ply, felt, mopped		232	.241		4.20	6.90	1.45	12.55	17.95	
	0400	On slabs, 2 ply, glass fiber fabric, mopped		153	.366		4.95	10.45	2.20	17.60	25.50	
	0600	On slabs, 3 ply, felt, mopped		195	.287		6.25	8.20	1.73	16.18	23	
	0700	On slabs, 3 ply, glass fiber fabric, mopped	▼	144	.389		6.65	11.10	2.34	20.09	28.50	
	0710	Asphaltic hardboard protection board, 1/8" thick	2 Rofc	46.45	.344		3.34	10.55		13.89	21.50	
	1000	For adhered 6 mm EPS protection board, add		325	.049		1.72	1.51		3.23	4.50	
	1050	10 mm thick, add		325	.049		1.94	1.51		3.45	4.71	
	1060	13 mm thick, add		325	.049		2.15	1.51		3.66	4.93	
	1070	Fiberglass fabric, black, 780/390 mesh		1078	.015		1.07	.45		1.52	1.95	
	1080	White, 780/390 mesh		1078	.015		1.10	.45		1.55	1.98	
	1100	1.6 mm urethane, troweled		18.58	.861		7.10	26.50		33.60	52.50	
	1200	Roller applied	▼	11.15	1.435	▼	6.45	44		50.45	81.50	

THERMAL & MOISTURE PROTECTION 7

07160	Cement. & Reactive Waterproofing	CREW	DAILY OUTPUT	LABOR-HOURS	UNIT	2006 BARE COSTS				TOTAL INCL O&P
						MAT.	LABOR	EQUIP.	TOTAL	
150	**0010**	**CEMENTITIOUS WATERPROOFING** One coat cement base									150
0020	3 mm application, sprayed on	G-2	92.90	.258	m²	16.70	7.55	1.26	25.51	31.50
0030	Two coat, cementitious/metallic slurry, troweled, 6 mm thick	1 Cefi	23.04	.347		3.89	11.95		15.84	22
0040	3 coats, 9 mm thick		17.09	.468		5.85	16.10		21.95	30.50
0050	4 coats, 13 mm thick	↓	11.15	.718	↓	7.75	24.50		32.25	45

07170	Bentonite Waterproofing									
700	**0010**	**BENTONITE**									700
0020	Panels, 1200 mm x 1200 mm, 5 mm thick	1 Rofc	58.06	.138	m²	8.20	4.22		12.42	16.20
0100	Rolls, 9 mm thick, with geotextile fabric both sides	"	51.10	.157	"	10.45	4.79		15.24	19.65
0300	Granular bentonite, 23 kg bags (0.02 m³)				Bag	19.20			19.20	21
0400	9 mm thick, troweled on	1 Rofc	44.13	.181	m²	10.35	5.55		15.90	21
0500	Drain board, expanded polystyrene, binder encapsulated, 38 mm thick	1 Rohe	149	.054		11.50	1.21		12.71	14.75
0510	50 mm thick		149	.054		15.30	1.21		16.51	18.85
0520	75 mm thick		149	.054		23	1.21		24.21	27
0530	100 mm thick		149	.054		30.50	1.21		31.71	35.50
0600	With filter fabric, 38 mm thick		149	.054		13.90	1.21		15.11	17.35
0625	50 mm thick		149	.054		18.30	1.21		19.51	22
0650	75 mm thick		149	.054		26	1.21		27.21	30.50
0675	100 mm thick	↓	149	.054	↓	33.50	1.21		34.71	39
0700	Vapor retarder, see polyethelene, Div. 07260-100									

07190	Water Repellents									
700	**0010**	**RUBBER COATING**									700
0020	Water base liquid, roller applied	2 Rofc	650	.025	m²	6.15	.75		6.90	8.10
0200	Silicone or stearate, sprayed on CMU, 1 coat	1 Rofc	372	.022		3.44	.66		4.10	5
0300	2 coats	"	279	.029	↓	7	.88		7.88	9.15

07200 | Thermal Protection

07210	Building Insulation	CREW	DAILY OUTPUT	LABOR-HOURS	UNIT	2006 BARE COSTS				TOTAL INCL O&P
						MAT.	LABOR	EQUIP.	TOTAL	
150	**0010**	**BLOWN-IN INSULATION** Ceilings, with open access									150
0020	Cellulose, 89 mm thick, 2.3 m²·K/W	G-4	465	.052	m²	2.05	1.45	.49	3.99	5.05
0030	132 mm thick, 3.4 m²·K/W		353	.068		3.01	1.91	.64	5.56	6.90
0050	165 mm thick, 3.9 m²·K/W		279	.086		3.77	2.41	.81	6.99	8.85
0100	221 mm thick, 5.3 m²·K/W		242	.099		5.15	2.78	.94	8.87	10.95
0120	276 mm thick, 6.7 m²·K/W		167	.144		6.55	4.03	1.36	11.94	15
1000	Fiberglass, 125 mm thick, 1.9 m²·K/W		353	.068		1.83	1.91	.64	4.38	5.60
1050	150 mm thick, 2.3 m²·K/W		279	.086		2.26	2.41	.81	5.48	7.15
1100	213 mm thick, 3.4 m²·K/W		204	.118		3.12	3.30	1.11	7.53	9.80
1200	250 mm thick, 3.9 m²·K/W		167	.144		3.66	4.03	1.36	9.05	11.75
1300	300 mm thick, 4.6 m²·K/W		139	.173		4.41	4.85	1.63	10.89	14.20
2000	Mineral wool, 100 mm thick, 2.1 m²·K/W		325	.074		2.05	2.07	.70	4.82	6.25
2050	150 mm thick, 3.0 m²·K/W		232	.103		2.26	2.90	.98	6.14	8.05
2100	229 mm thick, 4.1 m²·K/W	↓	163	.147	↓	3.23	4.13	1.39	8.75	11.55
2500	Wall installation, incl. drilling & patching from outside, two 25 mm									
2510	diam. holes @ 400 mm O.C., top & mid-point of wall, add to above									
2700	For masonry	G-4	38.55	.623	m²	.65	17.45	5.85	23.95	34
2800	For wood siding	↓	78.04	.308	↓	.65	8.65	2.90	12.20	17.40

		07210	Building Insulation	CREW	DAILY OUTPUT	LABOR-HOURS	UNIT	2006 BARE COSTS MAT.	LABOR	EQUIP.	TOTAL	TOTAL INCL O&P	
150	2900		For stucco/plaster	G-4	61.78	.388	m²	.65	10.90	3.66	15.21	22	150
500	0010	**POURED INSULATION**											500
	0020		Cellulose fiber, 0.7 m²·K/W per 25 mm	1 Carp	5.66	1.413	m³	20.50	50.50		71	101	
	0040		Ceramic type (perlite), 0.6 m²·K/W per 25 mm		5.66	1.413		60	50.50		110.50	144	
	0080		Fiberglass wool, 0.7 m²·K/W per 25 mm		5.66	1.413		14.50	50.50		65	94	
	0100		Mineral wool, 0.5 m²·K/W per 25 mm		5.66	1.413		12.35	50.50		62.85	92	
	0300		Polystyrene, 0.7 m²·K/W per 25 mm		5.66	1.413		102	50.50		152.50	190	
	0400		Vermiculite or perlite, 0.5 m²·K/W per 25 mm		5.66	1.413		60	50.50		110.50	144	
	0700		Wood fiber, 0.7 m²·K/W per 25 mm		5.66	1.413		23	50.50		73.50	104	
550	0010	**MASONRY INSULATION** Vermiculite or perlite, poured											550
	0100		In cores of concrete block, 100 mm thick wall, 0.035 m³/m²	D-1	446	.036	m²	2.15	1.15		3.30	4.13	
	0200		150 mm thick wall, 0.053 m³/m²		279	.057		3.23	1.84		5.07	6.35	
	0300		200 mm thick wall, 0.078 m³/m²		223	.072		4.74	2.31		7.05	8.65	
	0400		250 mm thick wall, 0.102 m³/m²		172	.093		6.25	2.99		9.24	11.45	
	0500		300 mm thick wall, 0.127 m³/m²		111	.144		7.75	4.63		12.38	15.55	
	0550		For sand fill, deduct from above					70%					
	0600		Poured cavity wall, vermiculite or perlite, water repellant	D-1	7.08	2.261	m³	60	72.50		132.50	177	
	0700		Foamed in place, urethane in 67 mm cavity	G-2	96.15	.250	m²	4.41	7.30	1.22	12.93	17.35	
	0800		For each 25 mm added thickness, add	"	220	.109	"	1.29	3.19	.53	5.01	6.85	
600	0010	**PERIMETER INSULATION**											600
	0600		Polystyrene, expanded, 25 mm thick, 0.7 m²·K/W	1 Carp	63.17	.127	m²	2.69	4.50		7.19	10	
	0700		50 mm thick, 1.4 m²·K/W	"	62.71	.128	"	5.70	4.54		10.24	13.30	
700	0010	**REFLECTIVE INSULATION**											700
	0020		Aluminum foil on reinforced scrim	1 Carp	177	.045	m²	1.54	1.61		3.15	4.19	
	0100		Reinforced with woven polyolefin		177	.045		1.87	1.61		3.48	4.56	
	0500		With single bubble air space, 1.55 m²·K/W		139	.058		2.99	2.05		5.04	6.50	
	0600		With double bubble air space, 1.72 m²·K/W		139	.058		3.21	2.05		5.26	6.75	
800	0010	**SPRAYED INSULATION**											800
	0020		Fibrous/cementitious, finished wall, 25 mm tk., 0.7 m²·K/W	G-2	190	.126	m²	2.48	3.69	.62	6.79	9	
	0100		Attic, 132 mm thick, 3.4 m²·K/W	"	144	.167	"	3.98	4.88	.82	9.68	12.75	
	0300		Foam type, incl. preparation										
	0600		48 kg/m³, 25 mm thick, .7 m²·K/W	G-2	71.53	.336	m²	5.40	9.80	1.64	16.84	22.50	
	0700		50 mm thick, 1.3 m²·K/W	"	44.13	.544	"	11	15.90	2.66	29.56	39.50	
900	0010	**WALL INSULATION, RIGID**											900
	0040		Fiberglass, 24 kg/m³, unfaced, 25 mm thick, 0.7 m²·K/W	1 Carp	92.90	.086	m²	4.09	3.06		7.15	9.30	
	0060		38 mm thick, 1.1 m²·K/W		92.90	.086		5.25	3.06		8.31	10.55	
	0080		50 mm thick, 1.5 m²·K/W		92.90	.086		6.35	3.06		9.41	11.75	
	0120		75 mm thick, 2.2 m²·K/W		74.32	.108		7.30	3.83		11.13	14	
	0370		48 kg/m³, unfaced, 25 mm thick, 0.8 m²·K/W		92.90	.086		4.63	3.06		7.69	9.80	
	0390		38 mm thick, 1.1 m²·K/W		92.90	.086		8.85	3.06		11.91	14.45	
	0400		50 mm thick, 1.5 m²·K/W		82.68	.097		10.65	3.44		14.09	17.10	
	0420		64 mm thick, 1.9 m²·K/W		74.32	.108		13	3.83		16.83	20.50	
	0440		75 mm thick, 2.3 m²·K/W		74.32	.108		15.50	3.83		19.33	23	
	0520		Foil faced, 25 mm thick, 0.8 m²·K/W		92.90	.086		10.25	3.06		13.31	16.05	
	0540		38 mm thick, 1.1 m²·K/W		92.90	.086		13.90	3.06		16.96	20	
	0560		50 mm thick, 1.5 m²·K/W		82.68	.097		17.35	3.44		20.79	24.50	
	0580		64 mm thick, 1.9 m²·K/W		74.32	.108		20.50	3.83		24.33	28.50	
	0600		75 mm thick, 2.3 m²·K/W		74.32	.108		22.50	3.83		26.33	30.50	
	0670		96 kg/m³, unfaced, 25 mm thick, 0.8 m²·K/W		92.90	.086		9.90	3.06		12.96	15.60	
	0690		38 mm thick, 1.1 m²·K/W		82.68	.097		15.30	3.44		18.74	22	
	0700		50 mm thick, 1.5 m²·K/W		74.32	.108		21.50	3.83		25.33	29.50	
	0721		64 mm thick, 1.9 m²·K/W		74.32	.108		23.50	3.83		27.33	32	
	0741		75 thick, 2.3 m²·K/W		67.82	.118		28	4.19		32.19	37.50	

THERMAL & MOISTURE PROTECTION 7

07210 | Building Insulation

		CREW	DAILY OUTPUT	LABOR-HOURS	UNIT	2006 BARE COSTS				TOTAL INCL O&P
						MAT.	LABOR	EQUIP.	TOTAL	
900	0821	Foil faced, 25 mm thick, 0.8 m²·K/W	1 Carp	92.90	.086	m²	14	3.06		17.06	20	**900**
	0840	38 mm thick, 1.1 m²·K/W		82.68	.097		20	3.44		23.44	27.50
	0850	50 mm thick, 1.5 m²·K/W		74.32	.108		26.50	3.83		30.33	35
	0880	64 mm thick, 1.9 m²·K/W		74.32	.108		31.50	3.83		35.33	40.50
	0900	75 mm thick, 2.3 m²·K/W		67.82	.118		37.50	4.19		41.69	48
	1500	Foamglass, 38 mm thick, .7 m²·K/W		74.32	.108		13.45	3.83		17.28	21
	1550	75 mm thick, 1.6 m²·K/W	↓	67.82	.118	↓	32.50	4.19		36.69	42
	1600	Isocyanurate, 1200 mm x 2400 mm sheet, foil faced both sides									
	1610	13 mm thick, 0.7 m²·K/W	1 Carp	74.32	.108	m²	3.12	3.83		6.95	9.40
	1620	16 mm thick, 0.8 m²·K/W		74.32	.108		5.15	3.83		8.98	11.65
	1630	19 mm thick, 1.0 m²·K/W		74.32	.108		4.09	3.83		7.92	10.45
	1640	25 mm thick, 1.3 m²·K/W		74.32	.108		5.60	3.83		9.43	12.10
	1650	38 mm thick, 1.9 m²·K/W		67.82	.118		6.05	4.19		10.24	13.20
	1660	50 mm thick, 2.5 m²·K/W		67.82	.118		7.95	4.19		12.14	15.25
	1670	75 mm thick, 3.8 m²·K/W		67.82	.118		19.15	4.19		23.34	27.50
	1680	100 mm thick, 5.1 m²·K/W		67.82	.118		23.50	4.19		27.69	32.50
	1700	Perlite, 25 mm thick, 0.5 m²·K/W		74.32	.108		3.01	3.83		6.84	9.30
	1750	50 mm thick, 1.0 m²·K/W		67.82	.118		5.80	4.19		9.99	12.90
	1900	Extruded polystyrene, 172 kPa, 25mm thick, 0.9 m²·K/W		74.32	.108		5.05	3.83		8.88	11.55
	1940	50 mm thick 1.8 m²·K/W		67.82	.118		10	4.19		14.19	17.55
	1960	75 mm thick, 2.64 m²·K/W		67.82	.118		14	4.19		18.19	22
	2100	Expanded polystyrene, 25 mm thick, 0.68 m²·K/W		74.32	.108		2.37	3.83		6.20	8.55
	2120	50 mm thick, 1.35 m²·K/W		67.82	.118		5.90	4.19		10.09	13.10
	2140	75 mm thick, 2.02 m²·K/W	↓	67.82	.118	↓	7.30	4.19		11.49	14.60
950	0010	**WALL OR CEILING INSUL., NON-RIGID**										**950**
	0040	Fiberglass, kraft faced, batts or blankets									
	0060	88 mm thick, 1.9 m²·K/W, 275 mm wide	1 Carp	107	.075	m²	3.12	2.66		5.78	7.60
	0080	375 mm wide		149	.054		3.12	1.91		5.03	6.40
	0100	600 mm wide		149	.054		3.12	1.91		5.03	6.40
	0140	150 mm thick, 3.3 m²·K/W, 275 mm wide		92.90	.086		4.52	3.06		7.58	9.70
	0160	375 mm wide		125	.064		4.52	2.28		6.80	8.50
	0180	600 mm wide		149	.054		4.52	1.91		6.43	7.90
	0200	225 mm thick, 5.3 m²·K/W, 380 mm wide		107	.075		7.65	2.66		10.31	12.55
	0220	600 mm wide		125	.064		7.65	2.28		9.93	11.95
	0240	300 mm thick, 6.7 m²·K/W, 380 mm wide		92.90	.086		9.15	3.06		12.21	14.85
	0260	600 mm wide	↓	125	.064	↓	9.15	2.28		11.43	13.65
	0400	Fiberglass, foil faced, batts or blankets									
	0420	88 mm thick, 1.9 m²·K/W, 375 mm wide	1 Carp	149	.054	m²	5.05	1.91		6.96	8.55
	0440	600 mm wide		149	.054		5.05	1.91		6.96	8.55
	0460	150 mm thick, 3.3 m²·K/W, 375 mm wide		125	.064		5.50	2.28		7.78	9.60
	0480	600 mm wide		149	.054		5.50	1.91		7.41	9
	0500	225 mm thick, 5.3 m²·K/W, 380 mm wide		107	.075		8.40	2.66		11.06	13.40
	0550	600 mm wide	↓	125	.064	↓	8.40	2.28		10.68	12.80
	0800	Fiberglass, unfaced, batts or blankets									
	0820	88 mm thick, 1.9 m²·K/W, 375 mm wide	1 Carp	125	.064	m²	3.55	2.28		5.83	7.40
	0830	600 mm wide		149	.054		3.55	1.91		5.46	6.85
	0860	150 mm thick, 3.3 m²·K/W, 375 mm wide		107	.075		6.05	2.66		8.71	10.80
	0880	600 mm wide		125	.064		6.05	2.28		8.33	10.20
	0900	225 mm thick, 5.3 m²·K/W, 380 mm wide		92.90	.086		7.10	3.06		10.16	12.60
	0920	600 mm wide		107	.075		7.10	2.66		9.76	12
	0940	300 mm thick, 6.7 m²·K/W, 380 mm wide		92.90	.086		8.95	3.06		12.01	14.55
	0960	600 mm wide	↓	107	.075	↓	8.95	2.66		11.61	13.95
	1300	Mineral fiber batts, kraft faced									
	1320	88 mm thick, 2.1 m²·K/W	1 Carp	149	.054	m²	3.23	1.91		5.14	6.50
	1340	150 mm thick, 3.4 m²·K/W		149	.054		4.31	1.91		6.22	7.70
	1380	250 mm thick, 5.3 m²·K/W	↓	125	.064	↓	6.35	2.28		8.63	10.55

Important: See the Reference Section for supporting data - Crews, Rental Equipment, City Cost Indexes and Reference Data

07200 | Thermal Protection

		07210	Building Insulation	CREW	DAILY OUTPUT	LABOR-HOURS	UNIT	2006 BARE COSTS				TOTAL INCL O&P	
								MAT.	LABOR	EQUIP.	TOTAL		
950	1850		Friction fit wire insulation supports, 406 mm O.C.	1 Carp	960	.008	Ea.	.07	.30		.37	.54	950
	1900		For foil backing, add				m²	.43			.43	.43	

07220 | Roof and Deck Insulation

				CREW	DAILY OUTPUT	LABOR-HOURS	UNIT	MAT.	LABOR	EQUIP.	TOTAL	TOTAL INCL O&P	
700	0010	**ROOF DECK INSULATION**											700
	0020	Fiberboard low density, 13 mm thick 0.24 m²·K/W		1 Rofc	92.90	.086	m²	2.05	2.64		4.69	6.75	
	0030	25 mm thick 0.49 m²·K/W			74.32	.108		4.09	3.29		7.38	10.10	
	0080	38 mm thick 0.73 m²·K/W			74.32	.108		6.15	3.29		9.44	12.40	
	0100	50 mm thick 0.98 m²·K/W			74.32	.108		8.20	3.29		11.49	14.65	
	0110	Fiberboard high density, 13mm thick 0.23 m²·K/W			92.90	.086		2.15	2.64		4.79	6.85	
	0120	25 mm thick 0.44 m²·K/W			74.32	.108		4.31	3.29		7.60	10.35	
	0130	75 mm thick 0.67 m²·K/W			74.32	.108		6.45	3.29		9.74	12.70	
	0200	Fiberglass, 19 mm thick 0.5 m²·K/W			92.90	.086		5.40	2.64		8.04	10.35	
	0400	24 mm thick 0.65 m²·K/W			92.90	.086		7.20	2.64		9.84	12.40	
	0460	27 mm thick 0.73 m²·K/W			92.90	.086		9.05	2.64		11.69	14.35	
	0600	33 mm thick 0.92 m²·K/W			92.90	.086		12.40	2.64		15.04	18.10	
	0650	52 mm thick 1.5 m²·K/W			74.32	.108		13.25	3.29		16.54	20	
	0700	61 mm thick 1.76 m²·K/W			74.32	.108		15.05	3.29		18.34	22	
	1500	Foamglass, 38 mm thick 0.8 m²·K/W			74.32	.108		14	3.29		17.29	21	
	1530	75 mm thick 1.3 m²·K/W			65.03	.123	↓	28.50	3.76		32.26	38	
	1600	Tapered for drainage			1.44	5.556	m³	480	170		650	815	
	1650	Perlite, 13 mm thick 0.23 m²·K/W			97.55	.082	m²	3.01	2.51		5.52	7.60	
	1655	19 mm thick 0.4 m²·K/W			74.32	.108		3.23	3.29		6.52	9.15	
	1660	25 mm thick 0.5 m²·K/W			74.32	.108		3.44	3.29		6.73	9.35	
	1670	38 mm thick 0.7 m²·K/W			74.32	.108		4.31	3.29		7.60	10.35	
	1680	50 mm thick 0.9 m²·K/W			65.03	.123		6.90	3.76		10.66	13.95	
	1685	67 mm thick 1.17 m²·K/W			65.03	.123	↓	8.50	3.76		12.26	15.75	
	1690	Tapered for drainage			1.92	4.167	m³	259	128		387	500	
	1700	Polyisocyanurate, 32 kg/m³ density, 19 mm thick, 0.9 m²·K/W			139	.058	m²	6.45	1.76		8.21	10.10	
	1705	25 mm thick 1.3 m²·K/W			130	.062		7	1.88		8.88	10.95	
	1715	38 mm thick 1.9 m²·K/W			116	.069		7.45	2.11		9.56	11.80	
	1725	50 mm thick 2.5 m²·K/W			102	.078		9.80	2.40		12.20	14.80	
	1735	64 mm thick 2.93 m²·K/W			97.55	.082		10.75	2.51		13.26	16.10	
	1745	75 mm thick 3.9 m²·K/W			92.90	.086		15.70	2.64		18.34	22	
	1755	89 mm thick 4.4 m²·K/W			92.90	.086	↓	15.95	2.64		18.59	22	
	1765	Tapered for drainage		↓	3.36	2.381	m³	320	73		393	480	
	1900	Extruded Polystyrene											
	1910	103 kPa compressive strength, 25 mm thick, 0.9 m²·K/W		1 Rofc	139	.058	m²	4.63	1.76		6.39	8.05	
	1920	50 mm thick, 1.8 m²·K/W			116	.069	↓	7.10	2.11		9.21	11.45	
	1930	75 mm thick 2.6 m²·K/W			92.90	.086		9.25	2.64		11.89	14.70	
	1932	100 mm thick 3.5 m²·K/W			92.90	.086	↓	14.40	2.64		17.04	20.50	
	1934	Tapered for drainage			3.60	2.222	m³	186	68		254	320	
	1940	172 kPa compressive strength, 25 mm thick 0.88 m²·K/W			139	.058	m²	6.80	1.76		8.56	10.45	
	1942	51mm thick 1.76 m²·K/W			116	.069		13	2.11		15.11	17.90	
	1944	75 mm thick 2.64 m²·K/W			92.90	.086		19.80	2.64		22.44	26	
	1946	100 mm thick 3.5 m²·K/W			92.90	.086	↓	26.50	2.64		29.14	33.50	
	1948	Tapered for drainage			3.60	2.222	m³	212	68		280	350	
	1950	276 kPa compressive strength, 25 mm thick 0.88 m²·K/W			139	.058	m²	4.74	1.76		6.50	8.15	
	1952	50 mm thick 1.76 m²·K/W			116	.069		9.25	2.11		11.36	13.85	
	1954	75 mm thick 2.64 m²·K/W			92.90	.086		13.55	2.64		16.19	19.40	
	1956	100 mm thick 3.52 m²·K/W			92.90	.086	↓	18.20	2.64		20.84	24.50	
	1958	Tapered for drainage			3.36	2.381	m³	267	73		340	415	
	1960	414 kPa compressive strength, 25mm thick 0.88 m²·K/W			135	.059	m²	5.70	1.81		7.51	9.35	
	1962	50mm thick 1.76 m²·K/W			111	.072		10.10	2.21		12.31	14.85	
	1964	75mm thick 2.64 m²·K/W			90.58	.088		15.05	2.70		17.75	21	
	1966	100mm thick 3.52 m²·K/W		↓	88.26	.091	↓	21	2.77		23.77	27.50	

07220 | Roof and Deck Insulation

		CREW	DAILY OUTPUT	LABOR-HOURS	UNIT	MAT.	LABOR	EQUIP.	TOTAL	TOTAL INCL O&P	
700	**1968**	Tapered for drainage	1 Rofc	3.36	2.381	m³	320	73		393	480	**700**
	2010	Expanded polystyrene, 16 kg/m³ density, 19 mm thick 0.5 m²·K/W		139	.058	m²	2.69	1.76		4.45	6	
	2020	25 mm thick 0.68 m²·K/W		139	.058		2.69	1.76		4.45	6	
	2100	50 mm thick 1.35 m²·K/W		116	.069		5.70	2.11		7.81	9.85	
	2110	75 mm thick 2.02 m²·K/W		116	.069		6.55	2.11		8.66	10.80	
	2120	100 mm thick 2.7 m²·K/W		111	.072		7.95	2.21		10.16	12.45	
	2130	125 mm thick 3.38 m²·K/W		107	.075		9.90	2.29		12.19	14.75	
	2140	150 mm thick 4.09 m²·K/W		107	.075		11.65	2.29		13.94	16.70	
	2150	Tapered for drainage		3.60	2.222	m³	186	68		254	320	
	2400	Composites with 50 mm EPS										
	2410	25 mm fiberboard	1 Rofc	88.26	.091	m²	10.55	2.77		13.32	16.35	
	2420	11 mm oriented strand board		74.32	.108		12.50	3.29		15.79	19.40	
	2430	13 mm plywood		74.32	.108		13.45	3.29		16.74	20.50	
	2440	25 mm perlite		74.32	.108		11.10	3.29		14.39	17.75	
	2450	Composites with 37 mm polyisocyanurate										
	2460	25 mm fiberboard	1 Rofc	74.32	.108	m²	14.40	3.29		17.69	21.50	
	2470	25 mm perlite		78.97	.101		15.20	3.10		18.30	22	
	2480	11 mm oriented strand board		74.32	.108		17.45	3.29		20.74	25	

07240 | Ext. Insulation Finish Systems (EIFS)

		CREW	DAILY OUTPUT	LABOR-HOURS	UNIT	MAT.	LABOR	EQUIP.	TOTAL	TOTAL INCL O&P	
100	**0010**	**EXTERIOR INSULATION FINISH SYSTEM**										**100**
	0095	Field applied, 25 mm EPS insulation	J-1	27.41	1.460	m²	22.50	44.50	3.85	70.85	96	
	0100	With 13 mm cement board sheathing		20.44	1.957		30.50	60	5.15	95.65	130	
	0105	50 mm EPS insulation		27.41	1.460		26	44.50	3.85	74.35	100	
	0110	With 13 mm cement board sheathing		20.44	1.957		34	60	5.15	99.15	134	
	0115	75 mm EPS insulation		27.41	1.460		27	44.50	3.85	75.35	102	
	0120	With 13 mm cement board sheathing		20.44	1.957		35.50	60	5.15	100.65	135	
	0125	100 mm EPS insulation		27.41	1.460		32	44.50	3.85	80.35	107	
	0130	With 13 mm cement board sheathing		20.44	1.957		48	60	5.15	113.15	149	
	0140	premium finish add		118	.339		3.12	10.40	.89	14.41	20	
	0150	heavy duty reinforcement add		84.91	.471		18.60	14.45	1.24	34.29	44	
	0160	1.35 kg/m² metal lath substrate add	1 Lath	62.71	.128		2.66	4.18		6.84	9.10	
	0170	1.84 kg/m² metal lath substrate add	"	62.71	.128		2.77	4.18		6.95	9.25	
	0180	color or texture change,	J-1	118	.339		8.05	10.40	.89	19.34	25.50	
	0190	with substrate leveling base coat	1 Plas	49.24	.162		8.05	5.25		13.30	16.95	
	0210	with substrate sealing base coat	1 Pord	114	.070		.75	2.22		2.97	4.21	
	0370	V groove shape in panel face				m	1.77			1.77	1.94	
	0380	U groove shape in panel face				"	2.33			2.33	2.56	
	0440	For higher than one story, add						25%				

07260 | Vapor Retarders

		CREW	DAILY OUTPUT	LABOR-HOURS	UNIT	MAT.	LABOR	EQUIP.	TOTAL	TOTAL INCL O&P	
100	**0010**	**BUILDING PAPER**										**100**
	0020	Aluminum and kraft laminated, foil 1 side	1 Carp	344	.023	m²	.51	.83		1.34	1.86	
	0100	Foil 2 sides		344	.023		.86	.83		1.69	2.23	
	0400	Asphalt felt sheathing paper, 6.80 kg		344	.023		.37	.83		1.20	1.69	
	0450	Housewrap, exterior, spun bonded polypropylene										
	0470	Small roll	1 Carp	353	.023	m²	1.94	.81		2.75	3.40	
	0480	Large roll	"	372	.022	"	1.18	.76		1.94	2.48	
	0500	Material only, 1 m x 34 m roll				Ea.	61			61	67	
	0520	2.74 m x 34 m roll				"	110			110	121	
	0600	Polyethylene vapor barrier, standard, 0.05 mm thick	1 Carp	344	.023	m²	.10	.83		.93	1.40	
	0700	0.10 mm thick		344	.023		.29	.83		1.12	1.61	
	0900	0.15 mm thick		344	.023		.44	.83		1.27	1.77	
	1200	0.25 mm thick		344	.023		.55	.83		1.38	1.89	
	1300	Clear reinforced, fire retardant, 0.20 mm thick		344	.023		.92	.83		1.75	2.30	

07260	Vapor Retarders	CREW	DAILY OUTPUT	LABOR-HOURS	UNIT	2006 BARE COSTS				TOTAL INCL O&P	
						MAT.	LABOR	EQUIP.	TOTAL		
100	1350	Cross laminated type, 0.08 mm thick	1 Carp	344	.023	m²	.72	.83		1.55	2.08	100
	1400	0.10 mm thick		344	.023		.80	.83		1.63	2.17	
	1500	Red rosin paper, 46 m² rolls, 0.195 kg/m²		344	.023		.19	.83		1.02	1.49	
	1600	0.244 kg/m²		344	.023		.24	.83		1.07	1.55	
	1800	Reinf. waterproof, 0.05 mm polyethylene backing, 1 side		344	.023		.54	.83		1.37	1.88	
	1900	2 sides		344	.023		.71	.83		1.54	2.07	
	2100	Roof deck vapor barrier, class 1 metal decks	1 Rofc	344	.023		1.26	.71		1.97	2.60	
	2200	For all other decks	"	344	.023		.90	.71		1.61	2.20	
	2400	Waterproofed kraft with sisal or fiberglass fibers, minimum	1 Carp	344	.023		.58	.83		1.41	1.93	
	2500	Maximum	"	344	.023		1.45	.83		2.28	2.89	

07300 | Shingles, Roof Tiles and Roof Coverings

07310	Shingles	CREW	DAILY OUTPUT	LABOR-HOURS	UNIT	2006 BARE COSTS				TOTAL INCL O&P	
						MAT.	LABOR	EQUIP.	TOTAL		
050	0010	**ALUMINUM SHINGLES**										050
	0020	Mill finish 0.48 mm thick	1 Carp	46.45	.172	m²	17	6.10		23.10	28.50	
	0100	0.50 mm thick	"	46.45	.172		19	6.10		25.10	30.50	
	0300	For colors, add					1.63			1.63	1.79	
	0600	Ridge cap, 0.61 mm thick	1 Carp	51.82	.154	m	5.95	5.50		11.45	15.10	
	0700	End wall flashing, 0.61 mm thick		51.82	.154		5.30	5.50		10.80	14.35	
	0900	Valley section, 0.61 mm thick		51.82	.154		8.15	5.50		13.65	17.50	
	1000	Starter strip, 0.61 mm thick		122	.066		4.43	2.33		6.76	8.50	
	1200	Side wall flashing, 61 mm thick		51.82	.154		5.30	5.50		10.80	14.40	
	1500	Gable flashing, 61 mm thick		122	.066		4.92	2.33		7.25	9.05	
100	0010	**ASPHALT SHINGLES**										100
	0100	Standard strip shingles										
	0150	Inorganic, class A, 10.25 to 11.50 kg/m²	1 Rofc	51.10	.157	m²	4.12	4.79		8.91	12.70	
	0155	Pneumatic nailed		65.03	.123		4.12	3.76		7.88	10.95	
	0200	Organic, class C, 11.50 to 11.70 kg/m²		46.45	.172		4.63	5.25		9.88	14.05	
	0205	Pneumatic nailed		58.06	.138		4.63	4.22		8.85	12.25	
	0250	Standard, laminated multi-layered shingles										
	0300	Class A, 11.70 to 12.70 kg/m²	1 Rofc	41.81	.191	m²	5.10	5.85		10.95	15.55	
	0305	Pneumatic nailed		52.26	.153		5.10	4.68		9.78	13.55	
	0350	Class C, 12.70 to 14.65 kg/m²		37.16	.215		5.40	6.60		12	17.10	
	0355	Pneumatic nailed		46.45	.172		5.40	5.25		10.65	14.85	
	0400	Premium, laminated multi-layered shingles										
	0450	Class A, 12.70 to 14.65 kg/m²	1 Rofc	32.52	.246	m²	6.75	7.55		14.30	20.50	
	0455	Pneumatic nailed		40.60	.197		6.75	6.05		12.80	17.70	
	0500	Class C, 14.65 to 18.80 kg/m²		27.87	.287		8.10	8.80		16.90	24	
	0505	Pneumatic nailed		34.84	.230		8.10	7.05		15.15	21	
	0800	#15 felt underlayment		595	.013		.37	.41		.78	1.10	
	0825	#30 felt underlayment		539	.015		.77	.45		1.22	1.62	
	0850	Self adhering polyethylene and rubberized asphalt underlayment		204	.039		4.95	1.20		6.15	7.50	
	0900	Ridge shingles		101	.079	m	4.36	2.42		6.78	8.90	
	0905	Pneumatic nailed		126	.063	"	4.36	1.94		6.30	8.10	
	1000	For steep roofs (7 to 12 pitch or greater), add						50%				
500	0010	**FIBER CEMENT SHINGLES**										500
	0012	Field shingles, 406 mm x 237 mm, 24 kg/m²	1 Rofc	20.44	.391	m²	31.50	12		43.50	55	
	0110	Starters, 406 mm x 237 mm		91.44	.087	m	3.26	2.68		5.94	8.15	
	0120	Hip & ridge, 121 mm x 356 mm		30.48	.262	"	23.50	8.05		31.55	39.50	

07310	Shingles	CREW	DAILY OUTPUT	LABOR-HOURS	UNIT	2006 BARE COSTS				TOTAL INCL O&P
						MAT.	LABOR	EQUIP.	TOTAL	
500 0200	Shakes, 406 mm x 237 mm, 27 kg/m²	1 Rofc	20.44	.391	m²	28.50	12		40.50	52
0300	Hip & ridge, 121 mm x 356 mm		30.48	.262	m	23.50	8.05		31.55	39.50
0400	Hexagonal shape, 406 mm x 406 mm		27.87	.287	m²	21.50	8.80		30.30	38.50
0500	Square, 406 mm x 406 mm		27.87	.287	"	19.10	8.80		27.90	36
800 0010	SLATE, Buckingham, Virginia, black R073126-20									
0100	5 mm - 6 mm thick	1 Rots	16.26	.492	m²	62.50	15.05		77.55	94.50
0200	6 mm thick		16.26	.492		66	15.05		81.05	98.50
0900	Pennsylvania black, Bangor, #1 clear		16.26	.492		51.50	15.05		66.55	82
1200	Vermont, unfading, green, mottled green		16.26	.492		39.50	15.05		54.55	69
1300	Semi-weathering green & gray		16.26	.492		37.50	15.05		52.55	67
1400	Purple		16.26	.492		42	15.05		57.05	72
1500	Black or gray		16.26	.492		39	15.05		54.05	68
1600	Red		16.26	.492		126	15.05		141.05	165
1700	Variegated purple		16.26	.492		44	15.05		59.05	74
900 0010	STEEL SHINGLES									
0012	Galvanized, 26 gauge	1 Rots	20.44	.391	m²	19.65	12		31.65	42
0200	24 gauge	"	20.44	.391		20.50	12		32.50	43
0300	For colored galvanized shingles, add					5.40			5.40	5.90
0500	For 25 mm factory applied polystyrene insulation, add					3.95			3.95	4.34
980 0010	WOOD SHINGLES									
0012	400 mm No. 1 red cedar shingles, 125 mm exposure, on roof	1 Carp	23.23	.344	m²	20	12.25		32.25	41
0015	Pneumatic nailed		30.19	.265		20	9.40		29.40	36.50
0200	191 mm exposure, on walls		19.05	.420		13.40	14.95		28.35	38.50
0205	Pneumatic nailed		24.81	.323		13.40	11.45		24.85	32.50
0300	457 mm No. 1 red cedar perfections, 140 mm exposure, on roof		25.55	.313		18.05	11.15		29.20	37
0305	Pneumatic nailed		33.17	.241		18.05	8.60		26.65	33
0500	191 mm exposure, on walls		20.90	.383		13.30	13.60		26.90	35.50
0505	Pneumatic nailed		27.13	.295		13.30	10.50		23.80	31
0600	Resquared, and rebutted, 140 mm exposure, on roof		27.87	.287		22.50	10.20		32.70	41
0605	Pneumatic nailed		36.23	.221		22.50	7.85		30.35	37
0900	191 mm exposure, on walls		22.76	.351		16.60	12.50		29.10	37.50
0905	Pneumatic nailed		29.54	.271		16.60	9.65		26.25	33.50
1000	Add to above for fire retardant shingles, 406 mm long					3.77			3.77	4.14
1050	457 mm long					3.77			3.77	4.14
1060	Preformed ridge shingles	1 Carp	122	.066	m	5.40	2.33		7.73	9.60
1100	Hand-split red cedar shake, 13 mm x 610 mm, 254 mm exp.		23.23	.344	m²	12.45	12.25		24.70	32.50
1105	Pneumatic nailed		30.19	.265		12.45	9.40		21.85	28.50
1110	19 mm thick x 610 mm long, 255 mm exp. on roof		20.90	.383		12.45	13.60		26.05	34.50
1115	Pneumatic nailed		27.13	.295		12.45	10.50		22.95	30
1200	457 mm long, 216 mm exposure, 216 mm exp. on roof		18.58	.431		9.60	15.30		24.90	34.50
1205	Pneumatic nailed		24.15	.331		9.60	11.75		21.35	29
1210	19 mm thick x 457 mm long, 216 mm exp. on roof		16.72	.478		9.60	17		26.60	37
1215	Pneumatic nailed		21.74	.368		9.60	13.10		22.70	31
1255	254 mm exp. on walls		18.58	.431		10.85	15.30		26.15	36
1260	254 mm exposure on walls, pneumatic nailed		24.15	.331		10.85	11.75		22.60	30.50
1700	Add to above for fire retardant shakes, 610 mm long					5.40			5.40	5.90
1800	457 mm long					4.84			4.84	5.35
1810	Ridge shakes	1 Carp	107	.075	m	7.70	2.66		10.36	12.65
2000	White cedar shingles, 406 mm long, extras, 127 mm exposure		22.30	.359	m²	17.55	12.75		30.30	39
2005	Pneumatic nailed		28.99	.276		17.55	9.80		27.35	34.50
2050	127 mm exposure on walls		18.58	.431		17.55	15.30		32.85	43.50
2055	Pneumatic nailed		24.15	.331		17.55	11.75		29.30	37.50
2100	191 mm exposure, on walls		18.58	.431		12.55	15.30		27.85	38
2105	Pneumatic nailed		24.15	.331		12.55	11.75		24.30	32
2150	"B" grade, 127 mm exposure on walls		18.58	.431		15.70	15.30		31	41.50

7

THERMAL & MOISTURE PROTECTION

07310 | Shingles

			DAILY OUTPUT	LABOR-HOURS	UNIT	2006 BARE COSTS				TOTAL INCL O&P	
		CREW				MAT.	LABOR	EQUIP.	TOTAL		
980	2155	Pneumatic nailed	1 Carp	24.15	.331	m²	15.70	11.75		27.45	35.50	980
	2300	For 6.80 kg organic felt underlayment on roof, 1 layer, add		595	.013		.37	.48		.85	1.14	
	2400	2 layers, add		297	.027		.73	.96		1.69	2.30	
	2600	For steep roofs (6/10 pitch or greater), add to above						50%				
	2700	Panelized systems, No.1 cedar shingles on 8 mm CDX plywood										
	2800	On walls, 2.4 m strips, 178 mm or 356 mm exposure	2 Carp	65.03	.246	m²	34.50	8.75		43.25	51.50	
	3500	On roofs, 2.4 m strips, 178 mm or 356 mm exposure	1 Carp	27.87	.287		34.50	10.20		44.70	54	
	3505	Pneumatic nailed	"	37.16	.215		34.50	7.65		42.15	50	

07320 | Roof Tiles

			DAILY OUTPUT	LABOR-HOURS	UNIT	MAT.	LABOR	EQUIP.	TOTAL	INCL O&P	
100	0010	**ALUMINUM ROOF TILES**										100
	0020	Accessories included, 0.8 mm thick, mission tile	1 Carp	23.23	.344	m²	61	12.25		73.25	86	
	0200	Spanish tiles	"	27.87	.287	"	42	10.20		52.20	62	
200	0010	**CLAY TILE** ASTM C1167, GR 1, severe weathering, acces. incl.										200
	0200	Lanai tile or Classic tile, 17 pc/m²	1 Rots	15.33	.522	m²	44	15.95		59.95	75	
	0300	Americana, 17 pc/m², most colors		15.33	.522		56.50	15.95		72.45	89	
	0350	Green, gray or brown		15.33	.522		54.50	15.95		70.45	87	
	0400	Blue		15.33	.522		54.50	15.95		70.45	87	
	0600	Spanish tile, 18.4 pcs./m², red		16.72	.478		29	14.65		43.65	56.50	
	0800	Buff, green, gray, brown		16.72	.478		46.50	14.65		61.15	76	
	0900	Glazed white		16.72	.478		55.50	14.65		70.15	86	
	1100	Mission tile, 17.85 pcs./m², machine scored finish, red		10.68	.749		70	23		93	116	
	1700	French tile, 14.3 pcs./m², smooth finish, red		12.54	.638		63.50	19.50		83	103	
	1750	Blue or green		12.54	.638		76	19.50		95.50	117	
	1800	Norman black tile, 34.1 pcs./m²		9.29	.861		89	26.50		115.50	143	
	2200	Williamsburg tile, 17 pcs./m², aged cedar		12.54	.638		54.50	19.50		74	92.50	
	2250	Gray or green		12.54	.638		54.50	19.50		74	92.50	
	2510	One piece mission tile, natural red, 6.96 pc. per m²		15.33	.522		14.75	15.95		30.70	43	
	2530	Mission Tile, 12.4 pcs. per m²		10.68	.749		19.75	23		42.75	60.50	
300	0010	**CONCRETE TILE** Including installation of accessories										300
	0020	Corrugated, 325 mm x 413 mm, 9.7 pcs/m², 46 kg/m²										
	0050	Earthtone colors, nailed to wood deck	1 Rots	12.54	.638	m²	10.05	19.50		29.55	44	
	0150	Blues		12.54	.638		10.15	19.50		29.65	44	
	0200	Greens		12.54	.638		10.15	19.50		29.65	44	
	0250	Premium colors		12.54	.638		16.95	19.50		36.45	51.50	
	0500	Shakes, 330 mm x 419 mm, 9.7/m², 46 kg/m²										
	0600	All colors, nailed to wood deck	1 Rots	13.94	.574	m²	20.50	17.55		38.05	52.50	
	1500	Accessory pieces, ridge & hip, 254 mm x 419 mm, 3.63 kg each				Ea.	2.32			2.32	2.55	
	1700	Rake, 165 mm x 425 mm, 4.08 kg each					2.32			2.32	2.55	
	1800	Mansard hip, 254 mm x 419 mm, 4.17 kg each					2.32			2.32	2.55	
	1900	Hip starter, 254 mm x 419 mm, 4.76 kg each					9.80			9.80	10.75	
	2000	3 or 4 way apex, 254 mm each side, 5.22 kg each					10.55			10.55	11.60	

07400 | Roofing and Siding Panels

07410 | Metal Roof and Wall Panels

			DAILY OUTPUT	LABOR-HOURS	UNIT	2006 BARE COSTS				TOTAL INCL O&P	
		CREW				MAT.	LABOR	EQUIP.	TOTAL		
100	0010	**ALUMINUM ROOFING PANELS**										100
	0020	Corrugated or ribbed, 0.4 mm thick, natural	G-3	111	.288	m²	7.45	10.05		17.50	23.50	

7 THERMAL & MOISTURE PROTECTION

		07410	Metal Roof and Wall Panels	CREW	DAILY OUTPUT	LABOR-HOURS	UNIT	MAT.	LABOR	EQUIP.	TOTAL	TOTAL INCL O&P	
								\multicolumn{4}{c}{2006 BARE COSTS}			
100	0300		Painted	G-3	111	.288	m²	10.75	10.05		20.80	27.50	100
	0400		Corrugated, 0.46 mm thick, on steel frame, natural finish		111	.288		9.70	10.05		19.75	26	
	0600		Painted		111	.288		11.95	10.05		22	28.50	
	0700		Corrugated, on steel frame, natural, 0.6 mm thick		111	.288		14	10.05		24.05	31	
	0800		Painted		111	.288		16.90	10.05		26.95	34	
	0900		0.8 mm thick, natural		111	.288		16.60	10.05		26.65	33.50	
	1200		Painted		111	.288		23	10.05		33.05	40.50	
	1300		V-Beam, on steel frame construction, 0.8 mm thick, natural		111	.288		18.60	10.05		28.65	36	
	1500		Painted		111	.288		23.50	10.05		33.55	41	
	1600		1.0 mm thick, natural		111	.288		23	10.05		33.05	40.50	
	1800		Painted		111	.288		28	10.05		38.05	46	
	1900		1.3 mm thick, natural		111	.288		27.50	10.05		37.55	46	
	2100		Painted		111	.288		33.50	10.05		43.55	52	
	2200		For roofing on wood frame, deduct		427	.075	▼	.65	2.61		3.26	4.78	
	2400		Ridge cap, 0.8 mm thick, natural	▼	244	.131	m	6.65	4.56		11.21	14.35	
500	0010		MANSARD Colored aluminum, with battens, 0.81 mm thick										500
	0600		Stock units, straight surfaces	1 Shee	10.68	.749	m²	25.50	31.50		57	77	
	0700		Concave or convex surfaces		6.97	1.148	"	28.50	48.50		77	106	
	0800		For framing, to 1524 mm high, add		35.05	.228	m	8.65	9.60		18.25	24.50	
	0900		Soffits, to 305 mm wide	▼	11.61	.689	m²	14	29		43	60	
690	0010		METAL FACING PANELS										690
	0400		Textured alum, 1200 mm x 2400 mm x 8 mm ply backing, single face	2 Shee	34.84	.459	m²	28.50	19.35		47.85	61	
	0600		Double face		34.84	.459		41.50	19.35		60.85	76	
	0700		1200 mm x 3000 mm x 8 mm plywood backing, single face		34.84	.459		29.50	19.35		48.85	62.50	
	0900		Double face		34.84	.459		43	19.35		62.35	77.50	
	1000		1200 mm x 3600 mm x 8 mm plywood backing, single face		34.84	.459		38	19.35		57.35	71.50	
	1300		Smooth al, 6 mm panel, fluoropolymer finish, double face		34.84	.459		67.50	19.35		86.85	105	
	1350		Clear anodized finish, double face		34.84	.459		75.50	19.35		94.85	113	
	1400		Double face textured aluminum, structural panel, 25 mm EPS insul	▼	34.84	.459	▼	45.50	19.35		64.85	80	
	1500		Accessories, outside corner	1 Shee	53.34	.150	m	6.35	6.30		12.65	16.75	
	1600		Inside corner		53.34	.150		6.90	6.30		13.20	17.35	
	1800		Batten mounting clip		60.96	.131		1.28	5.55		6.83	9.90	
	1900		Low profile batten		146	.055		2.62	2.31		4.93	6.45	
	2100		High profile batten		146	.055		4.46	2.31		6.77	8.50	
	2200		Water table		60.96	.131		6.45	5.55		12	15.60	
	2400		Horizontal joint connector		60.96	.131		6	5.55		11.55	15.10	
	2500		Corner cap		60.96	.131		5.55	5.55		11.10	14.60	
	2700		H - moulding	▼	146	.055	▼	3.84	2.31		6.15	7.80	
700	0010		STEEL ROOFING PANELS										700
	0012		On steel frame, corrugated or ribbed, 0.3 mm galv	G-3	102	.314	m²	11.30	10.90		22.20	29.50	
	0100		28 ga		97.55	.328		11.85	11.40		23.25	30.50	
	0300		26 ga		92.90	.344		16.45	12		28.45	36.50	
	0400		24 ga		88.26	.363		20.50	12.60		33.10	42	
	0600		Colored, 28 ga.		97.55	.328		11.65	11.40		23.05	30.50	
	0700		26 ga.		92.90	.344		17.75	12		29.75	38	
	0710		Flat profile, 44 mm standing seams, 255 mm W, standard finish. 26 ga		92.90	.344		34.50	12		46.50	56.50	
	0715		24 ga.		88.26	.363		40.50	12.60		53.10	63.50	
	0720		22 ga.		83.61	.383		49.50	13.30		62.80	75	
	0725		Zinc aluminum alloy finish, 26 ga.		92.90	.344		27	12		39	48.50	
	0730		24 ga.		88.26	.363		32.50	12.60		45.10	55	
	0735		22 ga.		83.61	.383		37	13.30		50.30	61.50	
	0740		306 mm wide, standard finish, 26 ga.		92.90	.344		34.50	12		46.50	56.50	
	0745		24 ga.		88.26	.363		45.50	12.60		58.10	69.50	
	0750		Zinc aluminum alloy finish, 26 ga.	▼	92.90	.344	▼	39.50	12		51.50	62	

07410 | Metal Roof and Wall Panels

		CREW	DAILY OUTPUT	LABOR-HOURS	UNIT	2006 BARE COSTS				TOTAL INCL O&P	
						MAT.	LABOR	EQUIP.	TOTAL		
700	0755	24 ga.	G-3	88.26	.363	m²	32.50	12.60		45.10	55	700
	0840	Flat profile, 25 mm x 9.5 mm batt., 306 mm wide, stand. fin., 26 ga.		92.90	.344		30.50	12		42.50	52	
	0845	24 ga.		88.26	.363		35.50	12.60		48.10	59	
	0850	22 ga.		83.61	.383		43	13.30		56.30	67.50	
	0855	Zinc aluminum alloy finish, 26 ga.		92.90	.344		29.50	12		41.50	50.50	
	0860	24 ga.		88.26	.363		32.50	12.60		45.10	55.50	
	0865	22 ga.		83.61	.383		37.50	13.30		50.80	62	
	0870	420 mm wide, standard finish, 24 ga.		88.26	.363		35	12.60		47.60	58.50	
	0875	22 ga.		83.61	.383		39.50	13.30		52.80	64	
	0880	Zinc aluminum alloy finish, 24 ga.		88.26	.363		30.50	12.60		43.10	53.50	
	0885	22 ga.		83.61	.383		34.50	13.30		47.80	58.50	
	0890	Flat profile, 51 mm x 51 mm batt., 306 mm wide, stand. fin., 26 ga.		92.90	.344		35	12		47	57	
	0895	24 ga.		88.26	.363		42	12.60		54.60	65.50	
	0900	22 ga.		83.61	.383		51	13.30		64.30	77	
	0905	Zinc aluminum alloy finish, 26 ga.		92.90	.344		32.50	12		44.50	54.50	
	0910	24 ga.		88.26	.363		37	12.60		49.60	60.50	
	0915	22 ga.		83.61	.383		43.50	13.30		56.80	68	
	0920	420 mm wide, standard finish, 24 ga.		88.26	.363		38.50	12.60		51.10	62	
	0925	22 ga.		83.61	.383		45	13.30		58.30	70	
	0930	Zinc aluminum alloy finish, 24 ga.		88.26	.363		35	12.60		47.60	58	
	0935	22 ga.		83.61	.383		39.50	13.30		52.80	64	
	1200	Ridge, galvanized, 254 mm wide		244	.131	m	7.70	4.56		12.26	15.55	
	1210	508 mm wide	▼	229	.140	"	13.55	4.86		18.41	22.50	

07420 | Plastic Roof and Wall Panels

		CREW	DAILY OUTPUT	LABOR-HOURS	UNIT	MAT.	LABOR	EQUIP.	TOTAL	TOTAL INCL O&P	
770	0010	**FIBERGLASS CORRUGATED ROOFING PANELS**									770
	0012	Corrugated panels, roofing, 2.44 kg/m²	G-3	92.90	.344	m²	24	12		36	45	
	0100	3.66 kg/m²		92.90	.344		35	12		47	56.50	
	0300	Corrugated siding, 1.83 kg/m²		81.75	.391		21	13.60		34.60	44	
	0400	2.44 kg/m²		81.75	.391		24	13.60		37.60	47.50	
	0500	Fire retardant		81.75	.391		33.50	13.60		47.10	57.50	
	0600	3.66 kg/m² siding, textured		81.75	.391		34	13.60		47.60	58	
	0700	Fire retardant		81.75	.391		44.50	13.60		58.10	70	
	0900	Flat panels, 1.83 kg/m², clear or colors		81.75	.391		18.60	13.60		32.20	41.50	
	1100	Fire retardant, class A		81.75	.391		32.50	13.60		46.10	56.50	
	1300	2.44 kg/m², clear or colors		81.75	.391		24	13.60		37.60	47.50	
	1700	Sandwich panels, fiberglass, 40 mm thick, panels to 2 m²		16.72	1.914		233	66.50		299.50	360	
	1900	As above, but 70 mm thick, panels to 10 m²	▼	24.62	1.300	▼	173	45		218	261	

07440 | Faced Panels

		CREW	DAILY OUTPUT	LABOR-HOURS	UNIT	MAT.	LABOR	EQUIP.	TOTAL	TOTAL INCL O&P	
200	0010	**EXPOSED AGGREGATE PANELS**									200
	1400	Fiberglass polymer back-up,										
	1500	Small size aggregate	F-3	41.34	.968	m²	52.50	35	15.40	102.90	129	
	1600	Medium size aggregate		41.34	.968		59.50	35	15.40	109.90	136	
	1700	Large size aggregate	▼	41.34	.968	▼	61.50	35	15.40	111.90	138	

07460 | Siding

		CREW	DAILY OUTPUT	LABOR-HOURS	UNIT	MAT.	LABOR	EQUIP.	TOTAL	TOTAL INCL O&P	
100	0010	**ALUMINUM SIDING PANELS**									100
	0012	0.5 mm T, on steel construction, natural	G-3	72	.444	m²	7.95	15.45		23.40	32.50	
	0100	Painted		72	.444		10.35	15.45		25.80	35.50	
	0400	Farm type, 0.53 mm thick on steel frame, natural		72	.444		9.25	15.45		24.70	34.50	
	0600	Painted		72	.444		11.30	15.45		26.75	36.50	
	0700	Industrial type, corrugated, on steel, 0.6 mm thick, natural	▼	72	.444	▼	11.75	15.45		27.20	37	

THERMAL & MOISTURE PROTECTION **7**

				DAILY	LABOR-			2006 BARE COSTS				TOTAL	
07460		**Siding**	CREW	OUTPUT	HOURS	UNIT	MAT.	LABOR	EQUIP.	TOTAL	INCL O&P		
100	0900	Painted	G-3	72	.444	m²	14.10	15.45		29.55	39.50	**100**
	1000	0.8 mm thick, natural		72	.444		17.10	15.45		32.55	43	
	1200	Painted		72	.444		21.50	15.45		36.95	48	
	1300	V-Beam, on steel frame, 0.8 mm thick, natural		72	.444		20	15.45		35.45	46	
	1500	Painted		72	.444		24	15.45		39.45	50	
	1600	1.0 mm thick, natural		72	.444		24.50	15.45		39.95	51	
	1800	Painted		72	.444		28.50	15.45		43.95	55.50	
	1900	1.3 mm thick, natural		72	.444		28.50	15.45		43.95	55.50	
	2100	Painted		72	.444		34.50	15.45		49.95	62	
	2200	Ribbed, 76mm profile, on steel frame, 0.81 mm thick, natural		72	.444		16.70	15.45		32.15	42.50	
	2400	Painted		72	.444		22	15.45		37.45	48	
	2500	1 mm thick, natural		72	.444		20.50	15.45		35.95	46.50	
	2700	Painted		72	.444		24	15.45		39.45	50.50	
	2750	1.27 mm thick, natural		72	.444		23.50	15.45		38.95	50	
	2760	Painted		72	.444		27	15.45		42.45	53.50	
	3300	For siding on wood frame, deduct from above	▼	260	.123	▼	.75	4.28		5.03	7.45	
	3400	Screw fasteners, aluminum, self tapping, neoprene washer, 25 mm				k	141			141	155	
	3600	Stitch screws, self tapping, with neoprene washer, 16 mm				"	105			105	116	
	3630	Flashing, sidewall, 0.81 mm thick	G-3	244	.131	m	6.55	4.56		11.11	14.25	
	3650	End wall, 1 mm thick		244	.131		7.70	4.56		12.26	15.50	
	3670	Closure strips, corrugated, 0.81 mm thick		244	.131		1.94	4.56		6.50	9.20	
	3680	Ribbed, 102 mm or 203 mm, 0.81 mm thick		244	.131		1.94	4.56		6.50	9.20	
	3690	V-beam, 1 mm thick	▼	244	.131	▼	2.66	4.56		7.22	9.95	
	3800	Horizontal, colored clapboard, 203 mm wide, plain	2 Carp	47.84	.334	m²	13	11.90		24.90	33	
	3900	Insulated		47.84	.334		16.80	11.90		28.70	37	
	4000	Vertical board & batten, colored, non-insulated	▼	47.84	.334		13	11.90		24.90	33	
	4200	For simulated wood design, add				▼	.97			.97	1.08	
	4300	Corners for above, outside	2 Carp	157	.102	m	6.10	3.62		9.72	12.40	
	4500	Inside corners	"	157	.102	"	3.61	3.62		7.23	9.60	
	4600	Sandwich panels, 25 mm insulation, single story	G-3	36.70	.872	m²	55.50	30.50		86	108	
	4900	Multi-story	"	32.05	.998		78	34.50		112.50	140	
	5100	For baked enamel finish 1 side, add				▼	2.91			2.91	3.23	
	5200	See also Metal Facing Panels, division 07410-690										
300	0010	**FASCIA** Aluminum, reverse board and batten,										**300**
	0100	0.81 mm thick, colored, no furring included	1 Shee	13.47	.594	m²	32	25		57	73.50	
	0300	Steel, galv and enameled, stock, no furring, long panels		13.47	.594		32.50	25		57.50	74	
	0600	Short panels	▼	10.68	.749	▼	43	31.50		74.50	96	
500	0010	**FIBER CEMENT SIDING**										**500**
	0020	Lap siding, 8 mm thick, 150 mm wide, smooth texture	2 Carp	38.55	.415	m²	13.25	14.75		28	37.50	
	0025	Woodgrain texture		38.55	.415		13.25	14.75		28	37.50	
	0030	190 mm wide, smooth texture		39.48	.405		11.65	14.40		26.05	35	
	0035	Woodgrain texture		39.48	.405		11.65	14.40		26.05	35	
	0040	200 mm wide, smooth texture		39.48	.405		11.50	14.40		25.90	35	
	0045	Roughsawn texture		39.48	.405		11.50	14.40		25.90	35	
	0050	240 mm wide, smooth texture		40.88	.391		11.10	13.90		25	34	
	0055	Woodgrain texture		40.88	.391		11.10	13.90		25	34	
	0060	305 mm wide, smooth texture		42.27	.379		10.65	13.45		24.10	33	
	0065	Woodgrain texture		42.27	.379		10.65	13.45		24.10	33	
	0070	Panel siding, 8 mm thick, smooth texture		69.68	.230		9.60	8.15		17.75	23.50	
	0075	Stucco texture		69.68	.230		9.60	8.15		17.75	23.50	
	0080	Grooved woodgrain texture		69.68	.230		9.60	8.15		17.75	23.50	
	0085	V - grooved woodgrain texture		69.68	.230	▼	9.60	8.15		17.75	23.50	
	0090	Wood starter strip	▼	122	.131	m	.62	4.66		5.28	7.95	
600	0010	**VINYL SIDING**										**600**
	0020	Solid PVC panels, 200 mm to 250 mm wide, plain	1 Carp	23.69	.338	m²	7.95	12		19.95	27.50	

7

THERMAL & MOISTURE PROTECTION

07460 | Siding

		CREW	DAILY OUTPUT	LABOR-HOURS	UNIT	2006 BARE COSTS				TOTAL INCL O&P	
						MAT.	LABOR	EQUIP.	TOTAL		
600	0100	with 10 mm insulation	1 Carp	23.69	.338	m²	9.05	12		21.05	28.50	**600**
	0200	Soffit and fascia		19.04	.420	↓	17.20	14.95		32.15	42.50	
	0300	Window and door trim moldings		56.39	.142	m	1.31	5.05		6.36	9.30	
	0500	Corner posts, outside corner		62.48	.128		4.72	4.55		9.27	12.30	
	0600	Inside corner	↓	62.48	.128	↓	2.20	4.55		6.75	9.55	
750	0010	**SOFFIT**										**750**
	0012	Aluminum, residential, stock units, 0.51 mm thick	1 Carp	19.51	.410	m²	13.35	14.60		27.95	37	
	0100	Baked enamel on steel, 16 or 18 gauge		9.75	.820		49.50	29		78.50	100	
	0300	Polyvinyl chloride, white, solid		21.37	.374		8.85	13.30		22.15	30	
	0400	Perforated	↓	21.37	.374		8.85	13.30		22.15	30	
	0500	For colors, add				↓	.97			.97	1.08	
800	0010	**STEEL SIDING**										**800**
	0020	Beveled, vinyl coated, 200 mm wide, including fasteners	1 Carp	24.62	.325	m²	16.15	11.55		27.70	36	
	0050	254 mm wide	"	25.55	.313		17.20	11.15		28.35	36.50	
	0080	Galv, corrugated or ribbed, on steel frame, 30 gauge	G-3	74.32	.431		10.75	15		25.75	35	
	0100	28 gauge		73.86	.433		11.30	15.05		26.35	36	
	0300	26 gauge		73.39	.436		15.80	15.15		30.95	41	
	0400	24 gauge		72.93	.439		15.95	15.25		31.20	41	
	0600	22 gauge		71.53	.447		18.30	15.55		33.85	44	
	0700	Colored, corrugated/ribbed, on steel frame, 10 yr fnsh, 0.4 mm		74.32	.431		16.80	15		31.80	41.50	
	0900	0.5 mm		73.86	.433		17.55	15.05		32.60	43	
	1000	0.6 mm		73.39	.436		20.50	15.15		35.65	46	
	1020	0.9 mm		72.93	.439		26	15.25		41.25	52	
	1200	Factory sandwich panel, 26 ga., 25 mm insulation, galvanized		35.30	.906		43	31.50		74.50	96.50	
	1300	Colored 1 side		35.30	.906		52.50	31.50		84	107	
	1500	Galvanized 2 sides		35.30	.906		65	31.50		96.50	121	
	1600	Colored 2 sides		35.30	.906		67.50	31.50		99	124	
	1800	Acrylic paint face, regular paint liner	↓	35.30	.906		51	31.50		82.50	105	
	1900	For 51 mm thick polystyrene, add					7.85			7.85	8.60	
	2000	22 ga, galv, 51 mm insulation, baked enamel exterior	G-3	33.44	.957		99.50	33.50		133	162	
	2100	P.V.F. exterior finish	"	33.44	.957	↓	105	33.50		138.50	168	
900	0010	**WOOD SIDING, BOARDS**										**900**
	3200	Wood, cedar bevel, A grade, 13 mm x 150 mm	1 Carp	23.23	.344	m²	34.50	12.25		46.75	57	
	3300	13 mm x 200 mm		25.55	.313		33.50	11.15		44.65	54.50	
	3500	19 mm x 250 mm, clear grade		27.87	.287		39	10.20		49.20	59	
	3600	"B" grade		27.87	.287		30	10.20		40.20	49	
	3800	Cedar, rough sawn, 25 mm x 102 mm, A grade, natural		22.30	.359		31.50	12.75		44.25	54.50	
	3900	Stained		22.30	.359		35.50	12.75		48.25	59	
	4100	25 mm x 305 mm, board & batten, #3 & Btr., natural		24.15	.331		23.50	11.75		35.25	44.50	
	4200	Stained		24.15	.331		27.50	11.75		39.25	49	
	4400	25 mm x 203 mm channel siding, #3 & Btr., natural		23.23	.344		23	12.25		35.25	44.50	
	4500	Stained		23.23	.344		26	12.25		38.25	47.50	
	4700	Redwood, clear, beveled, vertical grain, 13 mm x 102 mm		18.58	.431		36.50	15.30		51.80	64.50	
	4750	13 mm x 152 mm		20.90	.383		31	13.60		44.60	55	
	4800	13 mm x 203 mm		23.23	.344		25	12.25		37.25	46.50	
	5000	19 mm x 254 mm		27.87	.287		40.50	10.20		50.70	60.50	
	5200	Channel siding, 25 mm x 254 mm, B grade	↓	26.48	.302		26	10.75		36.75	45	
	5250	Redwood, T&G boards, B grade, 25 mm x 102 mm	2 Carp	27.87	.574		31	20.50		51.50	66.50	
	5270	25 mm x 203 mm	"	34.84	.459		27	16.35		43.35	55	
	5400	White pine, rough sawn, 25 mm x 203 mm, natural	1 Carp	25.55	.313		7.85	11.15		19	26	
	5500	Stained	"	25.55	.313	↓	11.65	11.15		22.80	30	
950	0010	**WOOD PRODUCT SIDING**										**950**
	0030	Lap siding, hardboard, 11 mm x 200 mm, primed										
	0051	Wood grain texture finish	L-2	51.28	.312	m²	12.15	9.65		21.80	28.50	
	0100	Panels, 11 mm thick, smooth, textured or grooved, primed	2 Carp	65.03	.246	↓	9.05	8.75		17.80	23.50	

07460 | Siding

			CREW	DAILY OUTPUT	LABOR-HOURS	UNIT	2006 BARE COSTS				TOTAL INCL O&P	
							MAT.	LABOR	EQUIP.	TOTAL		
950	0200	Stained	2 Carp	65.03	.246	m²	10.35	8.75		19.10	25	950
	0700	Particle board, overlaid, 10 mm thick		69.68	.230		7.45	8.15		15.60	21	
	0900	Plywood, medium density overlaid, 9 mm thick		69.68	.230		9.05	8.15		17.20	22.50	
	1000	13 mm thick		65.03	.246		10.25	8.75		19	25	
	1100	19 mm thick		60.39	.265		13.80	9.40		23.20	30	
	1600	Texture 1-11, cedar, 16 mm thick, natural		62.71	.255		25.50	9.05		34.55	42	
	1700	Factory stained		62.71	.255		24	9.05		33.05	40	
	1900	Texture 1-11, fir, 16 mm thick, natural		62.71	.255		10.85	9.05		19.90	26	
	2000	Factory stained		62.71	.255		11.65	9.05		20.70	27	
	2050	Texture 1-11, S.Y.P., 16 mm thick, natural		62.71	.255		8.70	9.05		17.75	23.50	
	2100	Factory stained		62.71	.255		9.45	9.05		18.50	24.50	
	2200	Rough sawn cedar, 10 mm thick, natural		62.71	.255		11.65	9.05		20.70	27	
	2300	Factory stained		62.71	.255		12.90	9.05		21.95	28.50	
	2500	Rough sawn fir, 10 mm thick, natural		62.71	.255		6.25	9.05		15.30	21	
	2600	Factory stained		62.71	.255		7	9.05		16.05	22	
	2800	Redwood, textured siding, 16 mm thick		62.71	.255		19.40	9.05		28.45	35.50	
	3000	Polyvinyl chloride coated, 10 mm thick	▼	69.68	.230	▼	9.45	8.15		17.60	23	

07500 | Membrane Roofing

07510 | Built-Up Bituminous Roofing

			CREW	DAILY OUTPUT	LABOR-HOURS	UNIT	2006 BARE COSTS				TOTAL INCL O&P	
							MAT.	LABOR	EQUIP.	TOTAL		
050	0010	**BUILT-UP ROOFING COMPONENTS**										050
	0012	Asphalt coated felt, #30, 18.6 m²/roll, not mopped	1 Rofc	539	.015	m²	.77	.45		1.22	1.62	
	0200	#15, 37.2 m²/roll, plain or perforated, not mopped		539	.015		.37	.45		.82	1.17	
	0300	Roll roofing, smooth, #65		139	.058		.74	1.76		2.50	3.80	
	0500	#90		139	.058		2.05	1.76		3.81	5.25	
	0520	Mineralized		139	.058		1.78	1.76		3.54	4.95	
	0540	D.C. (Double coverage), 483 mm selvage edge	▼	92.90	.086	▼	3.44	2.64		6.08	8.25	
	0580	Adhesive (lap cement)				liter	1.05			1.05	1.15	
	0600	Steep, flat or dead level asphalt, 9.1 metric ton lots, bulk				Met. Ton	310			310	340	
	0800	Packaged				"	450			450	495	
300	0010	**BUILT-UP ROOFING** R075113-20										300
	0120	Asphalt flood coat with gravel/slag surfacing, not including										
	0140	Insulation, flashing or wood nailers										
	0200	Asphalt base sheet, 3 plies #15 asphalt felt, mopped	G-1	204	.275	m²	5.35	7.85	1.65	14.85	21	
	0350	On nailable decks		195	.287		6	8.20	1.73	15.93	22.50	
	0500	4 plies #15 asphalt felt, mopped		186	.301		7.70	8.60	1.81	18.11	25	
	0550	On nailable decks		177	.316		6.95	9.05	1.90	17.90	25	
	0700	Coated glass base sheet, 2 plies glass (type IV), mopped		204	.275		5.70	7.85	1.65	15.20	21.50	
	0850	3 plies glass, mopped		186	.301		6.80	8.60	1.81	17.21	24	
	0950	On nailable decks		177	.316		6.45	9.05	1.90	17.40	24.50	
	1100	4 plies glass fiber felt (type IV), mopped		186	.301		8.30	8.60	1.81	18.71	25.50	
	1150	On nailable decks		177	.316		7.55	9.05	1.90	18.50	25.50	
	1200	Coated & saturated base sheet, 3 plies #15 asph. felt, mopped		186	.301		6.15	8.60	1.81	16.56	23.50	
	1250	On nailable decks		177	.316		5.80	9.05	1.90	16.75	24	
	1300	4 plies #15 asphalt felt, mopped	▼	204	.275	▼	7.15	7.85	1.65	16.65	23	
	2000	Asphalt flood coat, smooth surface										
	2200	Asphalt base sheet & 3 plies #15 asphalt felt, mopped	G-1	223	.251	m²	5.90	7.20	1.51	14.61	20.50	
	2400	On nailable decks	▼	214	.262	▼	5.50	7.50	1.58	14.58	20.50	

THERMAL & MOISTURE PROTECTION 7

07510 | Built-Up Bituminous Roofing

		CREW	DAILY OUTPUT	LABOR-HOURS	UNIT	MAT.	LABOR	EQUIP.	TOTAL	TOTAL INCL O&P	
300	2600	4 plies #15 asphalt felt, mopped R075113-20	G-1	223	.251	m²	6.85	7.20	1.51	15.56	21.50	300
	2700	On nailable decks	↓	214	.262	↓	6.50	7.50	1.58	15.58	21.50	
	2900	Coated glass fiber base sheet, mopped, and 2 plies of										
	2910	glass fiber felt (type IV)	G-1	232	.241	m²	5.25	6.90	1.45	13.60	19.05	
	3100	On nailable decks		223	.251		4.96	7.20	1.51	13.67	19.30	
	3200	3 plies, mopped		214	.262		6.35	7.50	1.58	15.43	21.50	
	3300	On nailable decks		204	.275		5.95	7.85	1.65	15.45	21.50	
	3800	4 plies glass fiber felt (type IV), mopped		214	.262		7.40	7.50	1.58	16.48	22.50	
	3900	On nailable decks		204	.275		7.05	7.85	1.65	16.55	23	
	4000	Coated & saturated base sheet, 3 plies #15 asph. felt, mopped		223	.251		5.65	7.20	1.51	14.36	20	
	4200	On nailable decks		214	.262		5.30	7.50	1.58	14.38	20.50	
	4300	4 plies #15 organic felt, mopped	↓	204	.275	↓	6.65	7.85	1.65	16.15	22.50	
	4500	Coal tar pitch with gravel/slag surfacing										
	4600	4 plies #15 tarred felt, mopped	G-1	195	.287	m²	12.65	8.20	1.73	22.58	30	
	4800	3 plies glass fiber felt (type IV), mopped	"	177	.316	"	10.35	9.05	1.90	21.30	29	
	5000	Coated glass fiber base sheet, and 2 plies of										
	5010	glass fiber felt, (type IV), mopped	G-1	177	.316	m²	10.30	9.05	1.90	21.25	28.50	
	5300	On nailable decks		167	.335		9.15	9.60	2.02	20.77	28.50	
	5600	4 plies glass fiber felt (type IV), mopped		195	.287		14.35	8.20	1.73	24.28	31.50	
	5800	On nailable decks	↓	186	.301	↓	13.20	8.60	1.81	23.61	31	
400	0010	**CANTS**										400
	0012	100 mm x 100 mm treated timber, cut diagonally	1 Rofc	99.06	.081	m	4.53	2.47		7	9.20	
	0100	Foamglass		99.06	.081		7.05	2.47		9.52	12	
	0300	Mineral or fiber, trapezoidal, 25 mm x 100 mm x 1.2 m		99.06	.081		.56	2.47		3.03	4.82	
	0400	38 mm x 143 mm x 1.2 m	↓	99.06	.081	↓	.95	2.47		3.42	5.25	
700	0010	**ROOFING FELTS**										700
	0012	Glass fibered, #15, no mopping	1 Rofc	539	.015	m²	.48	.45		.93	1.30	
	0300	Base sheet, #45, channel vented		539	.015		2.06	.45		2.51	3.03	
	0400	#50, coated		539	.015		1.11	.45		1.56	2	
	0500	Cap, mineral surfaced		539	.015		2.06	.45		2.51	3.03	
	0600	Flashing membrane, #65		149	.054		3.15	1.64		4.79	6.25	
	0800	Coal tar fibered, #15, no mopping		539	.015		.94	.45		1.39	1.80	
	0900	Asphalt felt, #15, 37.2 m²/roll, no mopping		539	.015		.37	.45		.82	1.17	
	1100	#30, 18.6 m²/roll		539	.015		.77	.45		1.22	1.62	
	1200	Double coated, #33		539	.015		.74	.45		1.19	1.58	
	1400	#40, base sheet		539	.015		1.01	.45		1.46	1.88	
	1450	Coated and saturated		539	.015		.80	.45		1.25	1.66	
	1500	Tarred felt, organic, #15, 37.2 m² rolls		539	.015		1.17	.45		1.62	2.05	
	1550	#30, 18.6 m² roll	↓	539	.015		2	.45		2.45	2.97	
	1700	Add for mopping above felts, per ply, asphalt, 1.17 kg/m²	G-1	1784	.031		.53	.90	.19	1.62	2.31	
	1800	Coal tar mopping, 1.46 kg/m²		1728	.032		1	.93	.20	2.13	2.88	
	1900	Flood coat, with asphalt, 2.93 kg/m²		557	.101		1.32	2.87	.61	4.80	7	
	2000	With coal tar, 3.66 kg/m²	↓	520	.108	↓	2.49	3.08	.65	6.22	8.65	

07520 | Cold Applied Bituminous Roofing

		CREW	DAILY OUTPUT	LABOR-HOURS	UNIT	MAT.	LABOR	EQUIP.	TOTAL	TOTAL INCL O&P	
200	0010	**COLD APPLIED BUILT UP ROOF**									200
	0020	3-ply system (components listed below)	G-5	465	.086	m²		2.39	.31	2.70	4.41	
	0100	Spunbond poly. fabric, 45 g/m², 0.91 m W, 100 m²/roll				Ea.	132			132	146	
	0200	1.24 m wide, 136 m²/roll					183			183	201	
	0300	71 gram/m², 0.91 m wide, 100 m²/roll					199			199	219	
	0400	1.24 m wide, 136 m²/roll	↓				270			270	297	
	0500	Base & finish coat, 11 L/m², 19 L/can				liter	.93			.93	1.02	
	0600	Coating, ceramic granules, 4.6 m²/bag				Ea.	12.40			12.40	13.65	
	0700	Aluminum, 7.6 L/m²				liter	2.64			2.64	2.90	
	0800	Emulsion, fibered or non-fibered, 15.1 L/m²				"	1.21			1.21	1.33	

THERMAL & MOISTURE PROTECTION 7

THERMAL & MOISTURE PROTECTION

7

07530	Elastomeric Membrane Roofing	CREW	DAILY OUTPUT	LABOR-HOURS	UNIT	2006 BARE COSTS				TOTAL INCL O&P
						MAT.	LABOR	EQUIP.	TOTAL	
350 0010	**ELASTOMERIC ROOFING**									**350**
0100	For Elastomeric waterproofing, see Division 07130-200									
0110	Acrylic rubber, fluid applied, .5 mm thick	G-5	186	.215	m²	19.40	5.95	.79	26.14	32.50
0120	1.3 mm, reinforced		111	.360		30	10	1.32	41.32	51.50
0130	For walking surface, add	↓	83.61	.478		9.15	13.30	1.75	24.20	34.50
0300	Hypalon neoprene, fluid applied, 0.51 mm thick, not-reinforced	G-1	105	.533		22	15.25	3.21	40.46	54
0600	Non-woven polyester, reinforced		89.18	.628		22.50	17.95	3.78	44.23	59
0700	5 coat neoprene deck, 1.5 mm thick, under 900 m²		30.19	1.855		49	53	11.15	113.15	156
0900	Over 900 m²		58.06	.964		46	27.50	5.80	79.30	104
1300	Vinyl plastic traffic deck, sprayed, 0.05 to 0.10 mm thick		58.06	.964		14.30	27.50	5.80	47.60	69
1500	Vinyl and neoprene membrane traffic deck	↓	144	.389	↓	15.20	11.10	2.34	28.64	38
800 0010	**SINGLE-PLY MEMBRANE**									**800**
0800	Chlorosulfonated polyethylene-hypalon (CSPE), 0.89 mm,									
0900	1.22 kg/m², fully adhered with neoprene latex	G-5	242	.165	m²	13.25	4.59	.61	18.45	23
1100	Loose-laid & ballasted with stone (49 kg/m²)		474	.084		14	2.34	.31	16.65	19.70
1200	Mechanically attached		325	.123		13.50	3.42	.45	17.37	21
1300	Plates with adhesive attachment	↓	325	.123	↓	13.25	3.42	.45	17.12	21
3500	Ethylene propylene diene monomer (EPDM), 1.1 mm, 1.37 kg/m²									
3600	Loose-laid & ballasted with stone (49 kg/m²)	G-5	474	.084	m²	6.55	2.34	.31	9.20	11.55
3700	Mechanically attached		325	.123		5.75	3.42	.45	9.62	12.60
3800	Fully adhered with adhesive	↓	242	.165	↓	8.25	4.59	.61	13.45	17.55
4500	1.5 mm, 1.95 kg/m²									
4600	Loose-laid & ballasted with stone (49 kg/m²)	G-5	474	.084	m²	8.15	2.34	.31	10.80	13.25
4700	Mechanically attached		325	.123		7.20	3.42	.45	11.07	14.20
4800	Fully adhered with adhesive	↓	242	.165		9.70	4.59	.61	14.90	19.15
4810	1 mm, 1.37 kg/m², membrane only					3.37			3.37	3.71
4820	2 mm, 1.95 kg/m², membrane only				↓	4.68			4.68	5.15
4850	Seam tape for membrane, 75 mm x 30 m roll				Ea.	41			41	45
4900	Batten strips, 3 m sections					2.95			2.95	3.25
4910	Cover tape for batten strips, 152 mm x 30 m roll				↓	144			144	158
4930	Plate anchors				k	73.50			73.50	80.50
4970	Adhesive for fully adhered systems, 1.47 m²/L				liter	3.86			3.86	4.24
7500	Polyisobutylene (PIB), 3 mm, 2.78 kg/m²									
7600	Loose-laid & ballasted with stone/gravel (49 kg/m²)	G-5	474	.084	m²	13.75	2.34	.31	16.40	19.45
7700	Partially adhered with adhesive		325	.123		17.25	3.42	.45	21.12	25
7800	Hot asphalt attachment		325	.123		16.45	3.42	.45	20.32	24.50
7900	Fully adhered with contact cement	↓	242	.165	↓	17.80	4.59	.61	23	28

07540	PVC Single Ply Membrane									
800 0010	**POLYVINYL CHLORIDE ROOFING**									**800**
8200	Heat welded seams									
8700	Reinforced, 1.2 mm, 1.61 kg/m²									
8750	Loose-laid & ballasted with stone/gravel (58.6 kg/m²)	G-5	474	.084	m²	10.50	2.34	.31	13.15	15.85
8800	Mechanically attached		325	.123		9.50	3.42	.45	13.37	16.75
8850	Fully adhered with adhesive	↓	242	.165	↓	12.75	4.59	.61	17.95	22.50
8860	Reinforced, 1.5 mm, 1.95 kg/m²									
8870	Loose-laid & ballasted with stone/gravel (59 kg/m²)	G-5	474	.084	m²	9.70	2.34	.31	12.35	14.95
8880	Mechanically attached		325	.123		8.75	3.42	.45	12.62	15.90
8890	Fully adhered with adhesive	↓	242	.165	↓	12	4.59	.61	17.20	21.50

07550	Modified Bit. Membrane Roofing									
500 0010	**MODIFIED BITUMEN ROOFING** R075213-30									**500**
0020	Base sheet, #15 glass fiber felt, nailed to deck	1 Rofc	539	.015	m²	.56	.45		1.01	1.39
0030	Spot mopped to deck	G-1	2741	.020		.75	.58	.12	1.45	1.95
0040	Fully mopped to deck	"	1784	.031	↓	1.01	.90	.19	2.10	2.84

07550	Modified Bit. Membrane Roofing		CREW	DAILY OUTPUT	LABOR-HOURS	UNIT	2006 BARE COSTS				TOTAL INCL O&P
							MAT.	LABOR	EQUIP.	TOTAL	
500 0050	#15 organic felt, nailed to deck	R075213 -30	1 Rofc	539	.015	m²	.45	.45		.90	1.26	**500**
0060	Spot mopped to deck		G-1	2741	.020		.63	.58	.12	1.33	1.83
0070	Fully mopped to deck		"	1784	.031	↓	.90	.90	.19	1.99	2.72
0080	SBS modified, granule surf cap sheet, poly rein., mopped										
0600	3.8 mm		G-1	186	.301	m²	5.25	8.60	1.81	15.66	22.50
1100	4 mm			186	.301		7.65	8.60	1.81	18.06	25
1500	Glass fiber reinforced, mopped, 4 mm			186	.301		4.84	8.60	1.81	15.25	22
1600	Smooth surface cap sheet, mopped, 3.7 mm			195	.287		4.84	8.20	1.73	14.77	21.50
1700	Smooth surface flashing, 3.7 m			117	.479		4.84	13.70	2.88	21.42	31.50
1800	3.8 mm			117	.479		4.74	13.70	2.88	21.32	31.50
1900	Granular surface flashing, 3.8 mm			117	.479		5.25	13.70	2.88	21.83	32
2000	4 mm		↓	117	.479		7.65	13.70	2.88	24.23	34.50
2100	APP mod., smooth surf. cap sheet, poly. reinf., torched, 4 mm		G-5	195	.205		4.74	5.70	.75	11.19	15.60
2150	4.3 mm			195	.205		5.40	5.70	.75	11.85	16.35
2200	Granule surface cap sheet, poly. reinf., torched, 4.6 mm			186	.215		5.90	5.95	.79	12.64	17.55
2250	Smooth surface flashing, torched, 4 mm			117	.342		4.74	9.50	1.25	15.49	22.50
2300	2.3 mm			117	.342		5.40	9.50	1.25	16.15	23.50
2350	Granule surface flashing, torched, 4.6 mm		↓	117	.342		5.90	9.50	1.25	16.65	24
2400	Fibrated aluminum coating		1 Rofc	353	.023	↓	1.08	.69		1.77	2.36

07570	Foamed Coated Roofing	CREW	DAILY OUTPUT	LABOR-HOURS	UNIT	MAT.	LABOR	EQUIP.	TOTAL	TOTAL INCL O&P
350 0010	**FOAMED COATED ROOFING**										**350**
1600	Polyurethane spray-on with 0.51 mm silicone rubber coating applied									
1700	25 mm thick, 1.2 m²·K/W, minimum	G-2A	81.29	.295	m²	14.40	7.95	1.44	23.79	30.50
1800	Maximim		74.78	.321		19.40	8.60	1.57	29.57	37.50
1900	51 mm thick, 2.5 m²·K/W, minimum		63.64	.377		20.50	10.15	1.84	32.49	41
2000	Maximum		53.42	.449		26	12.05	2.20	40.25	51
2100	76 mm thick, 3.7 m²·K/W, minimum		46.45	.517		35	13.85	2.53	51.38	64.50
2200	Maximum	↓	40.88	.587	↓	45	15.75	2.87	63.62	78.50

07580	Roll Roofing	CREW	DAILY OUTPUT	LABOR-HOURS	UNIT	MAT.	LABOR	EQUIP.	TOTAL	TOTAL INCL O&P
200 0010	**ROLL ROOFING**										**200**
0100	Asphalt, mineral surface									
0200	1 ply #15 organic felt, 1 ply mineral surfaced									
0300	selvage roofing, lap 475 mm, nailed & mopped	G-1	251	.223	m²	4.50	6.40	1.34	12.24	17.25
0400	3 plies glass fiber felt (type IV), 1 ply mineral surfaced									
0500	selvage roofing, lapped 483 mm, mopped	G-1	232	.241	m²	7	6.90	1.45	15.35	21
0600	Coated glass fiber base sheet, 2 plies of glass fiber									
0700	felt (type IV), 1 ply mineral surfaced selvage									
0800	roofing, lapped 483 mm, mopped	G-1	232	.241	m²	7.65	6.90	1.45	16	21.50
0900	On nailable decks	"	223	.251	"	7.10	7.20	1.51	15.81	21.50
1000	3 plies glass fiber felt (type III), 1 ply mineral surfaced									
1100	selvage roofing, lapped 483 mm, mopped	G-1	232	.241	m²	7	6.90	1.45	15.35	21

07590	Roof Maintenance and Repairs	CREW	DAILY OUTPUT	LABOR-HOURS	UNIT	MAT.	LABOR	EQUIP.	TOTAL	TOTAL INCL O&P
300 0010	**ROOF COATINGS**										**300**
0012	Asphalt, brush grade, material only			liter	.78			.78	.86
0200	Asphalt base, fibered aluminum coating				2.59			2.59	2.85
0300	Asphalt primer, 19 L			↓	1.11			1.11	1.22
0600	Coal tar pitch, 91 kg barrels			Met. Ton	680			680	750
0700	Tar roof cement, 19 L lots			liter	1.71			1.71	1.88
0800	Glass fibered roof & patching cement, 19 L			"	1.23			1.23	1.35
0900	Reinforcing glass membrane, 42 m²/roll			Ea.	45.50			45.50	50
1000	Neoprene roof coating, 19 L, 0.82 L/m²			liter	5.80			5.80	6.35
1100	Roof patch & flashing cement, 19 L			↓	5.05			5.05	5.55

THERMAL & MOISTURE PROTECTION 7

07500 | Membrane Roofing

07590 | Roof Maintenance and Repairs

			CREW	DAILY OUTPUT	LABOR-HOURS	UNIT	2006 BARE COSTS MAT.	LABOR	EQUIP.	TOTAL	TOTAL INCL O&P	
300	1200	Roof resturant, glass fibered, 1.22 L/m²				liter	1.93			1.93	2.12	300
	1300	Mineral rubber, 1.22 L/m²			↓		1.24			1.24	1.37	

07600 | Flashing and Sheet Metal

07610 | Sheet Metal Roofing

			CREW	DAILY OUTPUT	LABOR-HOURS	UNIT	2006 BARE COSTS MAT.	LABOR	EQUIP.	TOTAL	TOTAL INCL O&P	
300	0010	**COPPER ROOFING**										300
	0012	Batten seam, over 90 m², 454 gram, 6.34 kg/m²	1 Shee	10.22	.783	m²	47.50	33		80.50	103	
	0200	510 gram, 7.08 kg/m²		9.29	.861		53	36.50		89.50	114	
	0300	567 gram, 7.81 kg/m²		9.29	.861		58.50	36.50		95	120	
	0400	Standing seam, over 90 m², 454 gram, 6.1 kg/m²		12.08	.662		45.50	28		73.50	93	
	0600	510 gram, 6.83 kg/m²		11.15	.718		51	30.50		81.50	103	
	0700	567 gram, 7.32 kg/m²		10.22	.783		55	33		88	112	
	0900	Flat seam, over 90 m², 454 gram, 5.61 kg/m²		11.15	.718		42	30.50		72.50	92.50	
	1000	567 gram, 7.08 kg/m²	↓	10.22	.783		53	33		86	109	
	1200	For abnormal conditions or small areas, add					25%	100%				
	1300	For lead-coated copper, add				↓	25%					
500	0010	**LEAD ROOFING**										500
	0020	24.40 kg/m², batten seam	1 Shee	11.15	.718	m²	42.50	30.50		73	93	
	0100	Flat seam	"	12.08	.662	"	42.50	28		70.50	89.50	
700	0010	**STAINLESS STEEL ROOFING**										700
	0020	Type 304, batten seam, 28 gauge	1 Shee	11.15	.718	m²	43	30.50		73.50	93.50	
	0100	26 gauge	"	10.68	.749		53.50	31.50		85	108	
	0200	For standing seam construction, deduct					2%					
	0500	For flat seam construction, deduct					3%					
	0800	For lead or teme coated stainless, 28 gauge, add					9.80			9.80	10.80	
	0900	For 26 gauge, add				↓	12.95			12.95	14.95	
900	0010	**ZINC/COPPER ALLOY ROOFING,** field fabricated										900
	0012	Batten seam, 0.51 mm thick	1 Shee	11.15	.718	m²	60.50	30.50		91	113	
	0100	0.69 mm thick		10.68	.749		73	31.50		104.50	129	
	0300	0.81 mm thick		10.22	.783		82.50	33		115.50	142	
	0400	1 mm thick	↓	9.75	.820		97	34.50		131.50	161	
	0600	For standing seam construction, deduct					2%					
	0700	For flat seam construction, deduct				↓	3%					

07650 | Flexible Flashing

			CREW	DAILY OUTPUT	LABOR-HOURS	UNIT	2006 BARE COSTS MAT.	LABOR	EQUIP.	TOTAL	TOTAL INCL O&P	
600	0010	**FLASHING**										600
	0011	Including up to 4 bends										
	0020	Aluminum, mill finish, 0.3 mm thick	1 Rofc	13.47	.594	m²	4.09	18.15		22.24	35.50	
	0030	0.41 mm thick		13.47	.594		6.05	18.15		24.20	37.50	
	0060	0.5 mm thick		13.47	.594		7.65	18.15		25.80	39.50	
	0100	0.8 mm thick		13.47	.594		12.60	18.15		30.75	45	
	0200	1 mm thick		13.47	.594		17	18.15		35.15	50	
	0300	1.3 mm thick	↓	13.47	.594		21.50	18.15		39.65	55	
	0325	Mill finish 127 mm x 178 mm step flashing, 0.41 mm thick		1920	.004	Ea.	.12	.13		.25	.35	
	0350	Mill finish 305 mm x 305 mm step flashing, 0.41 mm thick	↓	1600	.005	"	.48	.15		.63	.79	

Important: See the Reference Section for supporting data - Crews, Rental Equipment, City Cost Indexes and Reference Data

THERMAL & MOISTURE PROTECTION 7

07650 | Flexible Flashing

		CREW	DAILY OUTPUT	LABOR-HOURS	UNIT	MAT.	LABOR	EQUIP.	TOTAL	TOTAL INCL O&P	
						2006 BARE COSTS					
600	0400	Painted finish, add				m²	2.80			2.80	3.12	600
	0500	Fabric-backed 2 sides, 0.10 mm thick	1 Rofc	30.66	.261		10.85	8		18.85	25.50	
	0700	0.13 mm thick		30.66	.261		12.80	8		20.80	27.50	
	0750	Mastic-backed, self adhesive		42.73	.187		26.50	5.75		32.25	39	
	0800	Mastic-coated 2 sides, 0.10 mm thick		30.66	.261		10.85	8		18.85	25.50	
	1000	0.13 mm thick		30.66	.261		12.80	8		20.80	27.50	
	1100	0.41 mm thick		30.66	.261		14.10	8		22.10	29	
	1300	Asphalt flashing cement, 19 L				liter	1.26			1.26	1.39	
	1600	Copper, 5 kg/m², sheets, under 454 kg	1 Rofc	10.68	.749	m²	29.50	23		52.50	71.50	
	1700	Over 1814 kg		14.40	.556		33	17		50	65.50	
	1900	567 gram sheets, under 454 kg		10.22	.783		44	24		68	89	
	2000	Over 1814 kg		13.47	.594		41	18.15		59.15	76	
	2200	680 gram sheets, under 454 kg		9.75	.820		53	25		78	101	
	2300	Over 1814 kg		12.54	.638		49	19.50		68.50	87	
	2500	907 gram sheets, under 454 kg		9.29	.861		70.50	26.50		97	122	
	2600	Over 1814 kg		12.08	.662		65.50	20.50		86	107	
	2700	W shape for valleys, 0.45 kg, 610 mm wide		30.48	.262	m	20.50	8.05		28.55	36	
	2800	Copper, paperbacked 1 side, 0.6 kg/m²		30.66	.261	m²	9.70	8		17.70	24	
	2900	0.9 kg/m²		30.66	.261		12.70	8		20.70	27.50	
	3100	Paperbacked 2 sides, 0.6 kg/m²		30.66	.261		9.80	8		17.80	24.50	
	3150	0.9 kg/m²		30.66	.261		12.60	8		20.60	27.50	
	3200	1.5 kg/m²		30.66	.261		18.85	8		26.85	34.50	
	3250	2.1 kg/m²		30.66	.261		31	8		39	47.50	
	3400	Mastic-backed 2 sides, copper, 0.6 kg/m²		30.66	.261		11.50	8		19.50	26.50	
	3500	0.9 kg/m²		30.66	.261		14.20	8		22.20	29	
	3700	1.5 kg/m²		30.66	.261		21	8		29	36.50	
	3800	Fabric-backed 2 sides, copper, 0.6 kg/m²		30.66	.261		12.25	8		20.25	27	
	4000	0.9 kg/m²		30.66	.261		15.80	8		23.80	31	
	4100	1.5 kg/m²		30.66	.261		21.50	8		29.50	37	
	4300	Copper-clad stainless steel, 0.38 mm thick, under 225 kg		10.68	.749		35.50	23		58.50	78.50	
	4400	Over 900 kg		14.40	.556		34.50	17		51.50	67	
	4600	0.46 mm thick, under 225 kg		9.29	.861		47.50	26.50		74	96.50	
	4700	Over 900 kg		13.47	.594		34.50	18.15		52.65	69	
	4900	Fabric, asphalt-saturated cotton, specification grade		29.26	.273		2.31	8.35		10.66	16.75	
	5000	Utility grade		29.26	.273		1.46	8.35		9.81	15.80	
	5200	Open-mesh fabric, saturated, 1.36 kg/m²		29.26	.273		1.61	8.35		9.96	16	
	5300	Close-mesh fabric, saturated, 0.58 kg/m²		29.26	.273		1.70	8.35		10.05	16.05	
	5500	Fiberglass, resin-coated		29.26	.273		1.39	8.35		9.74	15.75	
	5600	Asphalt-coated, 1.36 kg/m²		29.26	.273		9.45	8.35		17.80	24.50	
	5800	Lead, 12.21 kg/m², up to 305 mm wide		12.54	.638		32.50	19.50		52	69	
	5900	Over 305 mm wide		12.54	.638		39.50	19.50		59	76.50	
	6100	Lead-coated copper, fabric-backed, 0.6 kg/m²		30.66	.261		16.60	8		24.60	32	
	6200	1.5 kg/m²		30.66	.261		19.05	8		27.05	34.50	
	6400	Mastic-backed 2 sides, 0.6 kg/m²		30.66	.261		12.90	8		20.90	28	
	6500	1.5 kg/m²		30.66	.261		16.05	8		24.05	31	
	6700	Paperbacked 1 side, 56.7 gram		30.66	.261		11.20	8		19.20	26	
	6800	85.05 gram		30.66	.261		13.15	8		21.15	28	
	7000	Paperbacked 2 sides, 0.6 kg/m²		30.66	.261		11.50	8		19.50	26.50	
	7100	1.5 kg/m²		30.66	.261		18.85	8		26.85	34.50	
	7300	Polyvinyl chloride, black, 0.25 mm thick		26.48	.302		1.83	9.25		11.08	17.75	
	7400	0.50 mm thick		26.48	.302		2.58	9.25		11.83	18.50	
	7600	0.75 mm thick		26.48	.302		3.44	9.25		12.69	19.45	
	7700	1.4 mm thick		26.48	.302		8.20	9.25		17.45	25	
	7900	Black or white for exposed roofs, 2 mm thick		26.48	.302		17.85	9.25		27.10	35.50	
	8060	PVC tape, 127 mm x 1 mm, for joint covers, 30 m/roll				Ea.	85			85	93.50	
	8100	Rubber, butyl, 0.79 mm thick	1 Rofc	26.48	.302	m²	8.05	9.25		17.30	24.50	

07650 | Flexible Flashing

		CREW	DAILY OUTPUT	LABOR-HOURS	UNIT	MAT.	LABOR	EQUIP.	TOTAL	TOTAL INCL O&P	
600	**8200**	2 mm thick	1 Rofc	26.48	.302	m²	12.05	9.25		21.30	29	**600**
	8300	Neoprene, cured, 2 mm thick		26.48	.302		17	9.25		26.25	34.50	
	8400	3 mm thick		26.48	.302		34.50	9.25		43.75	53.50	
	8500	Shower pan, bituminous membrane, 2.2 kg/m²		14.40	.556		12.50	17		29.50	43	
	8550	3 ply copper and fabric, 85 gram		14.40	.556		18.75	17		35.75	49.50	
	8600	2.2 kg/m²		14.40	.556		39	17		56	71.50	
	8650	Copper, 454 gram		9.29	.861		36	26.50		62.50	84	
	8700	Lead on copper and fabric, 1.5 kg/m²		14.40	.556		19.05	17		36.05	50	
	8800	198 gram		14.40	.556		34	17		51	66	
	8850	Polyvinyl chloride, 0.76 mm thick		14.86	.538		3.44	16.45		19.89	32	
	8900	Stainless steel sheets, 0.25 mm thick		14.40	.556		29	17		46	61	
	9000	0.4 mm thick		14.40	.556		36	17		53	68.50	
	9100	26 ga., 0.46 mm thick		14.40	.556		43.50	17		60.50	77	
	9200	24 ga., 0.64 mm thick	↓	14.40	.556	↓	57	17		74	91.50	
	9290	For mechanically keyed flashing, add					40%					
	9300	Stainless steel, paperbacked 2 sides, 0.13 mm thick	1 Rofc	30.66	.261	m²	25.50	8		33.50	41.50	
	9400	Terne coated stainless steel, 0.38 mm thick, 28 ga.		14.40	.556		54.50	17		71.50	89	
	9500	0.46 mm thick, 26 ga.		14.40	.556		61.50	17		78.50	96.50	
	9600	Zinc and copper alloy (brass), 0.51 mm thick		14.40	.556		38	17		55	71	
	9700	0.69 mm thick		14.40	.556		51	17		68	85.50	
	9800	0.81 mm thick		14.40	.556		60	17		77	95	
	9900	1 mm thick	↓	14.40	.556	↓	73	17		90	110	

07700 | Roof Specialties and Accessories

07710 | Manufactured Roof Specialties

		CREW	DAILY OUTPUT	LABOR-HOURS	UNIT	MAT.	LABOR	EQUIP.	TOTAL	TOTAL INCL O&P	
400	**0010**	**DOWNSPOUTS**										**400**
	0020	Aluminum 50 mm x 75 mm, 0.5 mm thick, embossed	1 Shee	57.91	.138	m	2.79	5.80		8.59	12.05	
	0100	Enameled		57.91	.138		4.13	5.80		9.93	13.50	
	0300	Enameled, 0.6 mm thick, 50 mm x 75 mm		54.86	.146		5.80	6.15		11.95	15.85	
	0400	75 mm x 100 mm		42.67	.187		8.20	7.90		16.10	21	
	0600	Round, corrugated alum., 75 mm diam., 0.5 mm T		57.91	.138		4.46	5.80		10.26	13.85	
	0700	100 mm diameter, 0.6 mm thick		42.67	.187	↓	6.65	7.90		14.55	19.50	
	0900	Wire strainer, round, 51 mm diameter [2"]		155	.052	Ea.	1.85	2.18		4.03	5.40	
	1000	102 mm diameter [4"]		155	.052		2	2.18		4.18	5.55	
	1200	Rectangular, perforated, 51 mm x 76 mm		145	.055		2.25	2.33		4.58	6.05	
	1300	76 mm x 102 mm		145	.055	↓	3.25	2.33		5.58	7.15	
	1500	Copper, round, 5 kg/m², stock, 50 mm diameter		57.91	.138	m	14.95	5.80		20.75	25.50	
	1600	75 mm diameter		57.91	.138		14.45	5.80		20.25	25	
	1800	100 mm diameter		44.20	.181		17.05	7.65		24.70	30.50	
	1900	125 mm diameter		39.62	.202		25.50	8.50		34	41	
	2100	Rectangular, corrugated copper, stock, 50 mm x 75 mm		57.91	.138		12.75	5.80		18.55	23	
	2200	75 mm x 100 mm		44.20	.181		15.10	7.65		22.75	28.50	
	2400	Rectangular, plain copper, stock, 51 mm x 76 mm		57.91	.138		17.20	5.80		23	28	
	2500	76 mm x 102 mm		44.20	.181	↓	19.50	7.65		27.15	33.50	
	2700	Wire strainers, rectangular, 51 mm x 76 mm		145	.055	Ea.	2.85	2.33		5.18	6.70	
	2800	76 mm x 102 mm		145	.055		4.45	2.33		6.78	8.50	
	3000	Round, 51 mm diameter [2"]		145	.055		2.65	2.33		4.98	6.50	
	3100	76 mm diameter [3"]		145	.055		3.70	2.33		6.03	7.65	
	3300	102 mm diameter [4"]		145	.055	↓	5.75	2.33		8.08	9.95	

Important: See the Reference Section for supporting data - Crews, Rental Equipment, City Cost Indexes and Reference Data

		07710	Manufactured Roof Specialties	CREW	DAILY OUTPUT	LABOR-HOURS	UNIT	MAT.	LABOR	EQUIP.	TOTAL	TOTAL INCL O&P	
400	3400		127 mm diameter [5″]	1 Shee	115	.070	Ea.	8.30	2.93		11.23	13.65	**400**
	3600		Lead-coated copper, round, stock, 50 mm diameter		57.91	.138	m	18.05	5.80		23.85	29	
	3700		75 mm diameter		57.91	.138		24	5.80		29.80	35.50	
	3900		100 mm diameter		44.20	.181		32	7.65		39.65	47	
	4200		150 mm diameter, corrugated		32	.250		49	10.55		59.55	70	
	4300		Rectangular, corrugated, stock, 50 mm x 75 mm		57.91	.138		25.50	5.80		31.30	37	
	4500		Plain, stock, 51 mm x 76 mm		57.91	.138		24	5.80		29.80	35.50	
	4600		76 mm x 102 mm		44.20	.181		26	7.65		33.65	41	
	4800		Steel, galvanized, round, corrugated, 50 mm or 75 mm diam, 0.4 mm		57.91	.138		4.40	5.80		10.20	13.75	
	4900		100 mm diameter, 0.4 mm		44.20	.181		4.59	7.65		12.24	16.80	
	5100		125 mm diameter, 0.4 mm		39.62	.202		7.70	8.50		16.20	21.50	
	5200		26 gauge		39.62	.202		7.70	8.50		16.20	21.50	
	5400		150 mm diameter, 0.4 mm		32	.250		10.90	10.55		21.45	28	
	5500		26 gauge		32	.250		10.65	10.55		21.20	28	
	5700		Rectangular, corrugated, 28 gauge, 51 mm x 76 mm		57.91	.138		4.13	5.80		9.93	13.50	
	5800		76 mm x 102 mm		44.20	.181		4.13	7.65		11.78	16.30	
	6000		Rectangular, plain, 0.4 mm, galvanized, 50 mm x 75 mm		57.91	.138		5.85	5.80		11.65	15.40	
	6100		75 mm x 100 mm		44.20	.181		5.20	7.65		12.85	17.45	
	6300		Epoxy painted, 0.6 mm, corrugated, 50 mm x 75 mm		57.91	.138		5.10	5.80		10.90	14.55	
	6400		75 mm x 100 mm		44.20	.181	▼	5.85	7.65		13.50	18.20	
	6600		Wire strainers, rectangular, 51 mm x 76 mm		145	.055	Ea.	1.57	2.33		3.90	5.30	
	6700		76 mm x 102 mm		145	.055		3.13	2.33		5.46	7	
	6900		Round strainers, 51 mm or 76 mm diameter [2″,3″]		145	.055		1.57	2.33		3.90	5.30	
	7000		102 mm diameter [4″]		145	.055		3.13	2.33		5.46	7	
	7200		127 mm diameter [5″]		145	.055		2.20	2.33		4.53	6	
	7300		152 mm diameter [6″]		115	.070	▼	2.63	2.93		5.56	7.40	
	7500		Steel pipe, black, extra heavy, 102 mm diameter [4″]		6.10	1.312	m	59.50	55.50		115	151	
	7600		152 mm diameter [6″]		5.49	1.458		126	61.50		187.50	234	
	7800		Stainless steel tubing, schedule 5, 51 mm x 76 mm or 76 mm diam.		57.91	.138		67	5.80		72.80	82.50	
	7900		76 mm x 102 mm or 102 mm diameter [3″,4″,4″]		44.20	.181		85	7.65		92.65	105	
	8100		102 mm x 127 mm or 127 mm diameter [4″,5″,5″]		41.15	.194		175	8.20		183.20	206	
	8200		Vinyl, rectangular, 51 mm x 76 mm		64.01	.125		2.36	5.25		7.61	10.70	
	8300		Round, 64 mm	▼	67.06	.119	▼	2.36	5.05		7.41	10.35	
450	0010		**DRIP EDGE**										**450**
	0020		Aluminum, 0.4 mm thick, 125 mm wide, mill finish	1 Carp	122	.066	m	.98	2.33		3.31	4.71	
	0100		White finish		122	.066		1.15	2.33		3.48	4.91	
	0200		203 mm wide, mill finish		122	.066		1.38	2.33		3.71	5.15	
	0300		Ice belt, 711 mm wide, mill finish		30.48	.262		12.35	9.35		21.70	28	
	0310		Vented, mill finish		122	.066		4.33	2.33		6.66	8.40	
	0320		Painted finish		122	.066		4.69	2.33		7.02	8.80	
	0400		Galvanized, 127 mm wide, mill finish		122	.066		1.31	2.33		3.64	5.05	
	0500		203 mm wide, mill finish		122	.066		1.61	2.33		3.94	5.40	
	0510		Rake edge, aluminum, 38 mm x 38 mm		122	.066		.49	2.33		2.82	4.19	
	0520		90 mm x 38 mm	▼	122	.066	▼	.62	2.33		2.95	4.32	
500	0010		**DOWNSPOUT ELBOWS**										**500**
	0020		Aluminum, 51 mm x 76 mm, embossed	1 Shee	100	.080	Ea.	3.30	3.37		6.67	8.85	
	0100		Enameled		100	.080		3.30	3.37		6.67	8.85	
	0200		76 mm x 102 mm, 0.64 mm thick, embossed		100	.080		3.30	3.37		6.67	8.85	
	0300		Enameled		100	.080		3.20	3.37		6.57	8.70	
	0400		Round corrugated, 76 mm, embossed, 0.51 mm thick		100	.080		1.98	3.37		5.35	7.40	
	0500		102 mm, 0.64 mm thick		100	.080		4.29	3.37		7.66	9.90	
	0600		Copper, 454 gram round, 51 mm diameter [2″]		100	.080		11.15	3.37		14.52	17.50	
	0700		76 mm diameter [3″]		100	.080		8.15	3.37		11.52	14.20	
	0800		102 mm diameter [4″]	▼	100	.080	▼	9.50	3.37		12.87	15.65	

		07710	Manufactured Roof Specialties	CREW	DAILY OUTPUT	LABOR-HOURS	UNIT	2006 BARE COSTS MAT.	LABOR	EQUIP.	TOTAL	TOTAL INCL O&P	
500	1000		51 mm x 76 mm corrugated	1 Shee	100	.080	Ea.	6.15	3.37		9.52	11.95	**500**
	1100		76 mm x 102 mm corrugated		100	.080		7.20	3.37		10.57	13.15	
	1300		Vinyl, 64 mm diameter, 45° or 75° [2-1/2"]		100	.080		2	3.37		5.37	7.40	
	1400		Tee Y junction		75	.107		8.50	4.50		13	16.30	
550	0010		**GRAVEL STOP**										**550**
	0020		Aluminum, 1.3 mm thick, 100 mm face height, mill finish	1 Shee	44.20	.181	m	14.05	7.65		21.70	27	
	0080		Duranodic finish		44.20	.181		13.55	7.65		21.20	26.50	
	0100		Painted		44.20	.181		15.65	7.65		23.30	29	
	0300		150 mm face height		41.15	.194		15.05	8.20		23.25	29	
	0350		Duranodic finish		41.15	.194		16	8.20		24.20	30.50	
	0400		Painted		41.15	.194		18.55	8.20		26.75	33	
	0600		200 mm face height		38.10	.210		18.85	8.85		27.70	34	
	0650		Duranodic finish		38.10	.210		18.30	8.85		27.15	33.50	
	0700		Painted		38.10	.210		18.65	8.85		27.50	34	
	0900		300 mm face height, 2 mm thick, 2 piece		30.48	.262		25	11.05		36.05	44.50	
	0950		Duranodic finish		30.48	.262		23	11.05		34.05	42.50	
	1000		Painted		30.48	.262		27.50	11.05		38.55	47	
	1350		Galv steel, 24 ga, 100 mm leg, plain, w/ continuous cleat, 100 mm		44.20	.181		5.50	7.65		13.15	17.80	
	1360		150 mm face height		44.20	.181		8.35	7.65		16	21	
	1500		Polyvinyl chloride, 150 mm face height		41.15	.194		12.05	8.20		20.25	26	
	1600		230 mm face height		38.10	.210		14.20	8.85		23.05	29.50	
	1800		Stainless steel, 24 ga., 150 mm face height		41.15	.194		26.50	8.20		34.70	41.50	
	1900		305 mm face height		30.48	.262		55	11.05		66.05	77.50	
	2100		20 ga., 150 mm face height		41.15	.194		30	8.20		38.20	45	
	2200		305 mm face height		30.48	.262		63	11.05		74.05	86	
650	0010		**GUTTERS**										**650**
	0012		Aluminum, stock units, 125 mm K type, 0.7 mm thick, plain	1 Shee	36.58	.219	m	4.36	9.20		13.56	19	
	0100		Enameled		36.58	.219		4.10	9.20		13.30	18.75	
	0300		125 mm K type type, 0.8 mm thick, plain		36.58	.219		7.70	9.20		16.90	22.50	
	0400		Enameled		36.58	.219		7.75	9.20		16.95	23	
	0700		Copper, half round, 5 kg/m², stock units, 100 mm wide		36.58	.219		13.45	9.20		22.65	29	
	0900		125 mm wide		36.58	.219		17.15	9.20		26.35	33	
	1000		150 mm wide		35.05	.228		18.65	9.60		28.25	35.50	
	1200		K type, 5 kg/m² gram, stock, 100 mm wide		36.58	.219		16.55	9.20		25.75	32.50	
	1300		125 mm wide		36.58	.219		18	9.20		27.20	34	
	1500		Lead coated copper, half round, stock, 100 mm wide		36.58	.219		24.50	9.20		33.70	41	
	1600		150 mm wide		35.05	.228		37.50	9.60		47.10	56	
	1800		K type, stock, 100 mm wide		36.58	.219		26.50	9.20		35.70	43.50	
	1900		125 mm wide		36.58	.219		29	9.20		38.20	46	
	2100		Stainless steel, half round or box, stock, 102 mm wide		36.58	.219		15.25	9.20		24.45	31	
	2200		127 mm wide		36.58	.219		16.40	9.20		25.60	32.50	
	2400		Steel, galv, half round or box, .4 mm, 125 mm wide, plain		36.58	.219		3.87	9.20		13.07	18.45	
	2500		Enameled		36.58	.219		3.67	9.20		12.87	18.25	
	2700		.5 mm, stock, 125 mm wide		36.58	.219		3.12	9.20		12.32	17.65	
	2800		150 mm wide		36.58	.219		5.25	9.20		14.45	19.95	
	3000		Vinyl, O.G., 102 mm wide	1 Carp	33.53	.239		2.79	8.50		11.29	16.30	
	3100		127 mm wide		33.53	.239		3.28	8.50		11.78	16.80	
	3200		102 mm half round, stock units		33.53	.239		2.23	8.50		10.73	15.65	
	3250		Joint connectors				Ea.	1.36			1.36	1.50	
	3300		Wood, clear treated cedar, fir or hemlock, 76 mm x 102 mm	1 Carp	30.48	.262	m	20.50	9.35		29.85	37	
	3400		102 mm x 127 mm	"	30.48	.262	"	24	9.35		33.35	41	
700	0010		**GUTTER GUARD**										**700**
	0020		152 mm wide strip, aluminum mesh	1 Carp	152	.053	m	1.21	1.87		3.08	4.26	
	0100		Vinyl mesh	"	152	.053	"	1.31	1.87		3.18	4.35	

Important: See the Reference Section for supporting data - Crews, Rental Equipment, City Cost Indexes and Reference Data

THERMAL & MOISTURE PROTECTION 7

07710	Manufactured Roof Specialties	CREW	DAILY OUTPUT	LABOR-HOURS	UNIT	2006 BARE COSTS				TOTAL INCL O&P	
						MAT.	LABOR	EQUIP.	TOTAL		
750	**0010**	**REGLET**									**750**
	0020	Aluminum, 0.6 mm thick, in concrete parapet	1 Carp	68.58	.117	m	3.35	4.15		7.50	10.10
	0100	Copper, 283.5 gram		68.58	.117		5.80	4.15		9.95	12.85
	0300	453.6 gram		68.58	.117		7.70	4.15		11.85	14.95
	0400	Galvanized steel, 0.6 mm		68.58	.117		2.46	4.15		6.61	9.15
	0600	Stainless steel, 0.51 mm thick		68.58	.117		5.70	4.15		9.85	12.70
	0700	Zinc and copper alloy, 567 gram		68.58	.117		6.40	4.15		10.55	13.50
	0900	Counter flashing for above, 305 mm wide, 0.81 mm aluminum	1 Shee	45.72	.175		4.33	7.40		11.73	16.10
	1000	Copper, 283.5 gram		45.72	.175		12.15	7.40		19.55	24.50
	1200	453.6 gram		45.72	.175		13.50	7.40		20.90	26
	1300	Galvanized steel, 0.51 mm thick		45.72	.175		2.26	7.40		9.66	13.85
	1500	Stainless steel, 0.51 mm thick		45.72	.175		10.05	7.40		17.45	22.50
	1600	Zinc and copper alloy, 567 gram		45.72	.175		11.35	7.40		18.75	24
800	**0010**	**EXPANSION JOINT**									**800**
	0300	Butyl or neoprene center with foam insulation, metal flanges									
	0400	Aluminum, 0.81 mm thick for openings to 65 mm	1 Rofc	50.29	.159	m	26.50	4.87		31.37	38
	0600	For joint openings to 90 mm		50.29	.159		31.50	4.87		36.37	43
	0610	For joint openibgs to 125 mm		50.29	.159		37.50	4.87		42.37	50
	0620	For joint openings to 200 mm		50.29	.159		59.50	4.87		64.37	73.50
	0700	Copper, 455 grams for openings to 65 mm		50.29	.159		37.50	4.87		42.37	50
	0900	For joint openings to 90 mm		50.29	.159		43	4.87		47.87	56
	0910	For joint openings to 125 mm		50.29	.159		50.50	4.87		55.37	64
	0920	For joint openings to 200 mm		50.29	.159		76	4.87		80.87	92
	1000	Galvanized steel, 26 ga. for openings to 65 mm		50.29	.159		22.50	4.87		27.37	33.50
	1200	For joint openings to 90 mm		50.29	.159		26.50	4.87		31.37	38
	1210	For joint openings to 125 mm		50.29	.159		33.50	4.87		38.37	45.50
	1220	For joint openings to 200 mm		50.29	.159		57	4.87		61.87	71
	1300	Lead-coated copper, 455 grams for openings to 65 mm		50.29	.159		67	4.87		71.87	82
	1500	For joint openings to 90 mm		50.29	.159		79	4.87		83.87	95.50
	1600	Stainless steel, 0.46 mm, for openings to 65 mm		50.29	.159		33.50	4.87		38.37	45.50
	1800	For joint openings to 90 mm		50.29	.159		38	4.87		42.87	50.50
	1810	For joint openings to 125 mm		50.29	.159		47	4.87		51.87	60.50
	1820	For joint openings to 200 mm		50.29	.159		72	4.87		76.87	87.50
	1900	Neoprene, double-seal type with thick center, 115 mm wide		38.10	.210		32	6.45		38.45	46.50
	1950	Polyethylene bellows, with galv steel flat flanges		30.48	.262		12.45	8.05		20.50	27.50
	1960	With galvanized angle flanges		30.48	.262		13.75	8.05		21.80	29
	2000	Roof joint with extruded aluminum cover, 50 mm	1 Shee	35.05	.228		86	9.60		95.60	109
	2100	Roof joint, plastic curbs, foam center, standard	1 Rofc	30.48	.262		31.50	8.05		39.55	48
	2200	Large		30.48	.262		42	8.05		50.05	59.50
	2300	Transitions, regular, minimum		10	.800	Ea.	82.50	24.50		107	133
	2350	Maximum		4	2		105	61		166	220
	2400	Large, minimum		9	.889		121	27		148	179
	2450	Maximum		3	2.667		127	81.50		208.50	278
	2500	Roof to wall joint with extruded aluminum cover	1 Shee	35.05	.228	m	73	9.60		82.60	95.50
	2600	See also Divisions 03150-250 & 05810-350									
	2700	Wall joint, closed cell foam on PVC cover, 230 mm wide	1 Rofc	38.10	.210	m	11.20	6.45		17.65	23
	2800	305 mm wide	"	35.05	.228	"	12.65	7		19.65	26

07720 | Roof Accessories

			CREW	DAILY OUTPUT	LABOR-HOURS	UNIT	MAT.	LABOR	EQUIP.	TOTAL	TOTAL INCL O&P
480	**0010**	**PITCH POCKETS**									**480**
	0100	Adjustable, 25 mm to 179 mm, welded corners, 25 mm deep	1 Rofc	48	.167	Ea.	10.75	5.10		15.85	20.50
	0200	Side extenders, 152 mm	"	240	.033	"	1.80	1.02		2.82	3.71
500	**0010**	**ROOF VENTS**									**500**
	0020	Mushroom for built-up roofs, aluminum	1 Rofc	30	.267	Ea.	24	8.15		32.15	40.50

THERMAL & MOISTURE PROTECTION 7

07720	Roof Accessories	CREW	DAILY OUTPUT	LABOR-HOURS	UNIT	2006 BARE COSTS				TOTAL INCL O&P	
						MAT.	LABOR	EQUIP.	TOTAL		
500 0100	PVC, 150 mm high	1 Rofc	30	.267	Ea.	27.50	8.15		35.65	44.50	**500**
550 0010	**RIDGE VENT**										**550**
0100	Aluminum strips, mill finish	1 Rofc	48.77	.164	m	3.97	5		8.97	12.85	
0150	Painted finish		48.77	.164	"	6.80	5		11.80	16	
0200	Connectors		48	.167	Ea.	1.99	5.10		7.09	10.85	
0300	End caps		48	.167	"	.96	5.10		6.06	9.70	
0400	Galvanized strips, with damper and bird screen		48.77	.164	m	6.80	5		11.80	16	
0430	Molded polyethylene, shingles not included		48.77	.164	"	8.85	5		13.85	18.25	
0440	End plugs		48	.167	Ea.	.96	5.10		6.06	9.70	
0450	Flexible roll, shingles not included	▼	48.77	.164	m	6.55	5		11.55	15.65	
560 0010	**SNOW GUARDS**										**560**
0100	Slate & asphalt shingle roofs	1 Rofc	160	.050	Ea.	7.75	1.53		9.28	11.10	
0200	Standing seam metal roofs		48	.167		12.25	5.10		17.35	22	
0300	Surface mount for metal roofs		48	.167	▼	6.75	5.10		11.85	16.10	
0400	Double rail pipe type, including pipe	▼	39.62	.202	m	58	6.20		64.20	74.50	
700 0010	**ROOF HATCHES** With curb, 25 mm fibrgls insul., 750 mm x 900 mm										**700**
0500	Aluminum curb and cover	G-3	10	3.200	Ea.	550	111		661	775	
0520	Galvanized steel curb and aluminum cover		10	3.200		380	111		491	590	
0540	Galvanized steel curb and cover		10	3.200		445	111		556	660	
0600	750 mm x 1350 mm, aluminum curb and cover		9	3.556		650	124		774	905	
0800	Galvanized steel curb and aluminum cover		9	3.556		540	124		664	785	
0900	Galvanized steel curb and cover		9	3.556		685	124		809	940	
1100	1200 mm x 1200 mm, aluminum curb and cover		8	4		670	139		809	950	
1120	Galvanized steel curb and aluminum cover		8	4		540	139		679	810	
1140	Galvanized steel curb and cover		8	4		625	139		764	905	
1200	750 mm x 2400 mm, aluminum curb and cover		6.60	4.848		1,375	169		1,544	1,750	
1400	Galvanized steel curb and aluminum cover		6.60	4.848		1,050	169		1,219	1,400	
1500	Galvanized steel curb and cover	▼	6.60	4.848		940	169		1,109	1,275	
1800	For plexiglass panels, 760 mm to 910 mm, add to above				▼	375			375	410	
800 0010	**ROOF WALKWAY**										**800**
0020	Asphalt impregnated, 900 mm x 1800 mm x 13 mm	1 Rofc	37.16	.215	m²	12.90	6.60		19.50	25.50	
0100	900 mm x 900 mm x 19 mm thick	"	37.16	.215		25	6.60		31.60	38	
0300	Concrete patio blocks, 51 mm thick, natural	1 Clab	10.68	.749		15.80	20.50		36.30	49.50	
0400	Colors	"	10.68	.749	▼	20	20.50		40.50	54	
850 0010	**SMOKE HATCHES** Unlabeled, not including hand winch operator										**850**
0200	For 915 mm long, add to roof hatches from Division 07720-700				Ea.	25%	5%				
0250	For 1200 mm long, add to roof hatches from Division 07720-700					20%	5%				
0300	For 2440 mm long, add to roof hatches from Division 07720-700				▼	10%	5%				
860 0010	**SMOKE VENT**, insulated, 1220 mm x 1220 mm										**860**
0100	Aluminum cover and frame	G-3	13	2.462	Ea.	1,150	85.50		1,235.50	1,375	
0200	Galvanized steel cover and frame		13	2.462		1,050	85.50		1,135.50	1,275	
0300	1220 mm x 2440 mm alum. cover and frame		8	4		1,550	139		1,689	1,925	
0400	Galvanized steel cover and frame	▼	8	4	▼	1,350	139		1,489	1,725	
870 0010	**VENTS, ONE-WAY**										**870**
0020	Plastic, for insulated decks, 1/100 m², minimum	1 Rofc	40	.200	Ea.	12.90	6.10		19	24.50	
0100	Maximum		20	.400		29.50	12.25		41.75	53.50	
0300	Aluminum	▼	30	.267		12.90	8.15		21.05	28	
0800	Polystyrene baffles, 305 mm wide for 405 mm O.C. rafter spacing	1 Carp	90	.089		.31	3.16		3.47	5.25	
0900	For 610 mm O.C. rafter spacing	"	110	.073	▼	.52	2.59		3.11	4.60	

07812	Cementitious Fireproofing	CREW	DAILY OUTPUT	LABOR-HOURS	UNIT	2006 BARE COSTS MAT.	LABOR	EQUIP.	TOTAL	TOTAL INCL O&P
600	0010 **SPRAYED** Mineral fiber or cementitious for fireproofing,										600
	0050 not incl tamping or canvas protection									
	0100 25 mm thick, on flat plate steel	G-2	279	.086	m²	4.63	2.52	.42	7.57	9.35
	0200 Flat decking		223	.108		4.63	3.15	.53	8.31	10.45
	0400 Beams		139	.173		4.63	5.05	.84	10.52	13.70
	0500 Corrugated or fluted decks		116	.207		6.90	6.05	1.01	13.96	17.90
	0700 Columns, 29 mm thick		102	.235		5.15	6.90	1.15	13.20	17.45
	0800 56 mm thick		65.03	.369		9.80	10.80	1.80	22.40	29
	0850 For tamping, add						10%			
	0900 For canvas protection, add	G-2	465	.052	m²	.65	1.51	.25	2.41	3.33
	1000 Acoustical sprayed, 25 mm thick, finished, straight work, minimum		48.31	.497		4.84	14.55	2.43	21.82	30
	1100 Maximum		18.58	1.292		5.15	38	6.30	49.45	70
	1300 Difficult access, minimum		20.90	1.148		5.15	33.50	5.60	44.25	63.50
	1400 Maximum		12.08	1.987		5.70	58	9.70	73.40	105
	1500 Intumescent epoxy fireproofing on wire mesh, 5 mm thick									
	1550 1 hour rating, exterior use	G-2	12.63	1.900	m²	60	55.50	9.30	124.80	161
	1600 Magnesium oxychloride, 15.9 kg to 18.1 kg density, 6 mm thick		279	.086		12.60	2.52	.42	15.54	18.20
	1650 13 mm thick		186	.129		25.50	3.77	.63	29.90	34.50
	1700 27.2 kg to 31.8 kg density, 6 mm thick		279	.086		16.70	2.52	.42	19.64	22.50
	1750 13 mm thick		186	.129		33.50	3.77	.63	37.90	43.50
	2000 Vermiculite cement, troweled or sprayed, 6 mm thick		279	.086		11.40	2.52	.42	14.34	16.90
	2050 13 mm thick		186	.129		22.50	3.77	.63	26.90	31.50

07840	Firestopping									
100	0010 **FIRESTOPPING** R078413-30										100
	0100 Metallic piping, non insulated									
	0110 Through walls, 50 mm diameter	1 Carp	16	.500	Ea.	9.95	17.80		27.75	38.50
	0120 100 mm diameter		14	.571		15.20	20.50		35.70	48.50
	0130 150 mm diameter		12	.667		20.50	23.50		44	59.50
	0140 300 mm diameter		10	.800		36.50	28.50		65	84.50
	0150 Through floors, 50 mm diameter		32	.250		6.05	8.90		14.95	20.50
	0160 100 mm diameter		28	.286		8.70	10.15		18.85	25.50
	0170 150 mm diameter		24	.333		11.40	11.85		23.25	31
	0180 300 mm diameter		20	.400		19.25	14.20		33.45	43
	0190 Metallic piping, insulated									
	0200 Through walls, 50 mm diameter	1 Carp	16	.500	Ea.	14.15	17.80		31.95	43
	0210 100 mm diameter		14	.571		19.40	20.50		39.90	53
	0220 150 mm diameter		12	.667		24.50	23.50		48	64
	0230 300 mm diameter		10	.800		40.50	28.50		69	89
	0240 Through floors, 50 mm diameter		32	.250		10.20	8.90		19.10	25
	0250 100 mm diameter		28	.286		12.85	10.15		23	30
	0260 150 mm diameter		24	.333		15.60	11.85		27.45	35.50
	0270 300 mm diameter		20	.400		19.25	14.20		33.45	43
	0280 Non metallic piping, non insulated									
	0290 Through walls, 50 mm diameter	1 Carp	12	.667	Ea.	41	23.50		64.50	82.50
	0300 100 mm diameter		10	.800		51.50	28.50		80	102
	0310 150 mm diameter		8	1		72	35.50		107.50	135
	0330 Through floors, 50 mm diameter		16	.500		32	17.80		49.80	63
	0340 100 mm diameter		6	1.333		40	47.50		87.50	118
	0350 150 mm diameter		6	1.333		48	47.50		95.50	127
	0370 Ductwork, insulated & non insulated, round									
	0380 Through walls, 150 mm diameter	1 Carp	12	.667	Ea.	21	23.50		44.50	60
	0390 300 mm diameter		10	.800		41.50	28.50		70	90.50
	0400 450 mm diameter		8	1		67.50	35.50		103	130
	0410 Through floors, 150 mm diameter		16	.500		11.45	17.80		29.25	40
	0420 300 mm diameter		14	.571		21	20.50		41.50	54.50

	07840	Firestopping	CREW	DAILY OUTPUT	LABOR-HOURS	UNIT	2006 BARE COSTS				TOTAL INCL O&P
							MAT.	LABOR	EQUIP.	TOTAL	
100	0430	450 mm diameter R078413-30	1 Carp	12	.667	Ea.	36.50	23.50		60	77
	0440	Ductwork, insulated & non insulated, rectangular									
	0450	W/ stiffener/closure angle, through walls, 150 x 300 mm	1 Carp	8	1	Ea.	17.35	35.50		52.85	74.50
	0460	300 mm x 600 mm		6	1.333		23	47.50		70.50	99.50
	0470	600 mm x 1200 mm		4	2		65.50	71		136.50	183
	0480	W/ stiffener.closure angle, through floors, 150 x 300 mm		10	.800		9.35	28.50		37.85	55
	0490	300 mm x 600 mm		8	1		16.85	35.50		52.35	74
	0500	600 mm x 1200 mm		6	1.333		33	47.50		80.50	111
	0510	Multi trade openings									
	0520	Through walls, 150 mm x 300 mm	1 Carp	2	4	Ea.	36.50	142		178.50	261
	0530	300 mm x 600 mm	"	1	8		147	284		431	605
	0540	600 mm x 1200 mm	2 Carp	1	16		585	570		1,155	1,525
	0550	1200 mm x 2400 mm	"	.75	21.333		2,350	760		3,110	3,775
	0560	Through floors, 150 mm x 300 mm	1 Carp	2	4		36.50	142		178.50	261
	0570	300 mm x 600 mm	"	1	8		147	284		431	605
	0580	600 mm x 1200 mm	2 Carp	.75	21.333		585	760		1,345	1,825
	0590	1200 mm x 2400 mm	"	.50	32		2,350	1,150		3,500	4,375
	0600	Structrual penetrations, through walls									
	0610	Steel beams, W200 x 15	1 Carp	8	1	Ea.	23	35.50		58.50	80.50
	0620	W310 x 36		6	1.333		36.50	47.50		84	114
	0630	W530 x 66		5	1.600		73	57		130	169
	0640	W920 x 201		3	2.667		177	95		272	345
	0650	Bar joists, 450 mm deep		6	1.333		33.50	47.50		81	111
	0660	600 mm deep		6	1.333		41.50	47.50		89	120
	0670	900 mm deep		5	1.600		62.50	57		119.50	157
	0680	1200 mm deep		4	2		73	71		144	191
	0690	Construction joints, floor slab at exterior wall									
	0700	Precast, brick, block or drywall exterior									
	0710	50 mm wide joint	1 Carp	38.10	.210	m	17.05	7.45		24.50	30.50
	0720	100 mm wide joint	"	22.86	.350	"	34	12.45		46.45	57
	0730	Metal panel, glass or curtain wall exterior									
	0740	50 mm wide joint	1 Carp	12.19	.656	m	40.50	23.50		64	81
	0750	100 mm wide joint	"	7.62	1.050	"	55	37.50		92.50	119
	0760	Floor slab to drywall partition									
	0770	Flat joint	1 Carp	30.48	.262	m	16.75	9.35		26.10	33
	0780	Fluted joint		15.24	.525		34	18.65		52.65	66.50
	0790	Etched fluted joint		22.86	.350		22	12.45		34.45	44
	0800	Floor slab to concrete/masonry partition									
	0810	Flat joint	1 Carp	22.86	.350	m	37.50	12.45		49.95	61
	0820	Fluted joint	"	15.24	.525	"	44.50	18.65		63.15	78
	0830	Concrete/CMU wall joints									
	0840	25 mm wide	1 Carp	30.48	.262	m	20.50	9.35		29.85	37
	0850	50 mm wide		22.86	.350		37.50	12.45		49.95	61
	0860	100 mm wide		15.24	.525		71.50	18.65		90.15	108
	0870	Concrete/CMU floor joints									
	0880	25 mm wide	1 Carp	60.96	.131	m	10.25	4.67		14.92	18.50
	0890	50 mm wide		45.72	.175		18.75	6.20		24.95	30
	0900	100 mm wide		30.48	.262		36	9.35		45.35	54

07920	Joint Sealants	CREW	DAILY OUTPUT	LABOR-HOURS	UNIT	2006 BARE COSTS				TOTAL INCL O&P
						MAT.	LABOR	EQUIP.	TOTAL	
800	0010	**CAULKING AND SEALANTS**										800
	0020	Acoustical sealant, elastomeric, cartridges				Ea.	2.28			2.28	2.51
	0030	Backer rod, polyethylene, 6 mm diameter [1/4"]	1 Bric	140	.057	m	.07	2.09		2.16	3.26
	0050	13 mm diameter [1/2"]		140	.057		.12	2.09		2.21	3.31
	0070	19 mm diameter [3/4"]		140	.057		.22	2.09		2.31	3.43
	0090	25 mm diameter [1"]	▼	140	.057	▼	.35	2.09		2.44	3.57
	0100	Acrylic latex caulk, white									
	0200	0.33 L cartridge				Ea.	1.88			1.88	2.07
	0500	6 mm x 13 mm	1 Bric	75.59	.106	m	.49	3.87		4.36	6.45
	0600	13 mm x 13 mm		76.20	.105		1.02	3.84		4.86	6.95
	0800	19 mm x 19 mm		70.10	.114		2.26	4.17		6.43	8.85
	0900	19 mm x 25 mm		60.96	.131		3.02	4.80		7.82	10.60
	1000	25 mm x 25 mm	▼	54.86	.146	▼	3.77	5.35		9.12	12.25
	1400	Butyl based, bulk				liter	6.05			6.05	6.65
	1500	Cartridges				"	7.35			7.35	8.10
	1700	Bulk, in place 6 mm x 13 mm, 12.4 m/L	1 Bric	70.10	.114	m	.49	4.17		4.66	6.85
	1800	13 mm x 13 mm, 6.2 m/L	"	54.86	.146	"	.98	5.35		6.33	9.20
	2000	Latex acrylic based, bulk				liter	6.30			6.30	6.95
	2100	Cartridges					7.75			7.75	8.55
	2300	Polysulfide compounds, 1 component, bulk				▼	11.90			11.90	13.10
	2600	1 or 2 component, in place, 6 mm x 6 mm, 24.8 m/L	1 Bric	44.20	.181	m	.49	6.60		7.09	10.55
	2700	13 mm x 6 mm, 12.4 m/L		41.15	.194		.95	7.10		8.05	11.85
	2900	19 mm x 10 mm, 5.5 m/L		39.62	.202		2.17	7.40		9.57	13.65
	3000	25 mm x 13 mm, 3.1 m/L	▼	39.62	.202	▼	3.90	7.40		11.30	15.50
	3200	Polyurethane, 1 or 2 component				liter	12.95			12.95	14.25
	3500	Bulk, in place, 6 mm x 6 mm, 24.8 m/L	1 Bric	45.72	.175	m	.52	6.40		6.92	10.35
	3600	13 mm x 6 mm		44.20	.181		1.05	6.60		7.65	11.20
	3800	19 mm x 10 mm		39.62	.202		2.36	7.40		9.76	13.85
	3900	25 mm x 13 mm	▼	33.53	.239	▼	4.17	8.70		12.87	17.90
	4100	Silicone rubber, bulk				liter	9.40			9.40	10.30
	4200	Cartridges				"	10.05			10.05	11.05
	4400	Neoprene gaskets, closed cell, adhesive, 3 mm x 9 mm	1 Bric	73.15	.109	m	.69	4		4.69	6.85
	4500	6 mm x 19 mm		65.53	.122		1.61	4.46		6.07	8.55
	4700	13 mm x 25 mm		60.96	.131		4.72	4.80		9.52	12.50
	4800	19 mm x 38 mm	▼	50.29	.159	▼	9.85	5.80		15.65	19.70
	5500	Resin epoxy coating, 2 component, heavy duty				liter	7			7	7.70
	5800	Tapes, sealant, P.V.C. foam adhesive, 2 mm x 6 mm				m	.16			.16	.17
	5900	2 mm x 13 mm					.23			.23	.25
	5950	2 mm x 25 mm					.38			.38	.42
	6000	3 mm x 13 mm				▼	.26			.26	.28
	6200	Urethane foam, 2 component, handy pack, 0.028 m³				Ea.	28.50			28.50	31
	6300	1.4 m³ pack				m³	510			510	560

For information about Means Estimating Seminars, see yellow pages 12 and 13 in back of book

THERMAL & MOISTURE PROTECTION **7**

Division Notes

		CREW	DAILY OUTPUT	LABOR-HOURS	UNIT	2006 BARE COSTS				TOTAL INCL O&P
						MAT.	LABOR	EQUIP.	TOTAL	

Division 8
Doors & Windows

Estimating Tips

08100 Metal Doors & Frames
- Most metal doors and frames look alike, but there may be significant differences among them. When estimating these items be sure to choose the line item that most closely compares to the specification or door schedule requirements regarding:
  - type of metal
  - metal gauge
  - door core material
  - fire rating
  - finish

08200 Wood & Plastic Doors
- Wood and plastic doors vary considerably in price. The primary determinant is the veneer material. Lauan, birch and oak are the most common veneers. Other variables include the following:
  - hollow or solid core
  - fire rating
  - flush or raised panel
  - finish
- If the specifications require compliance with AWI (Architectural Woodwork Institute) standards or acoustical standards, the cost of the door may increase substantially. All wood doors are priced pre-mortised for hinges and predrilled for cylindrical locksets.

- Frequently, doors, frames, and windows are unique in old buildings. Specified replacement units could be stock, custom (similar to the original) or exact reproduction. The estimator should work closely with a window consultant to determine any extra costs that may be associated with the unusual installation requirements.

08300 Specialty Doors
- There are many varieties of special doors, and they are usually priced per each. Add frames, hardware or operators required for a complete installation.

08510 Steel Windows
- Most metal windows are delivered preglazed. However, some metal windows are priced without glass. Refer to 08800 Glazing for glass pricing. The grade C indicates commercial grade windows, usually ASTM C-35.

08550 Wood Windows
- All wood windows are priced preglazed. The two glazing options priced are single pane float glass and insulating glass 13 mm thick. Add the cost of screens and grills if required.

08700 Hardware
- Hardware costs add considerably to the cost of a door. The most efficient method to determine the hardware requirements for a project is to review the door schedule. This schedule, in conjunction with the specifications, is all you should need to take off the door hardware.

- Door hinges are priced by the pair, with most doors requiring 1-1/2 pairs per door. The hinge prices do not include installation labor because it is included in door installation. Hinges are classified according to the frequency of use.

08800 Glazing
- Different openings require different types of glass. The three most common types are:
  - float
  - tempered
  - insulating
- Most exterior windows are glazed with insulating glass. Entrance doors and window walls, where the glass is less than 460 mm from the floor, are generally glazed with tempered glass. Interior windows and some residential windows are glazed with float glass.
- Energy efficient coatings are also available.

08900 Glazed Curtain Wall
- Glazed curtain walls consist of the metal tube framing and the glazing material. The cost data in this subdivision is presented for the metal tube framing alone or the composite wall. If your estimate requires a detailed takeoff of the framing, be sure to add the glazing cost.

Reference Numbers
Reference numbers are shown in bold squares at the beginning of some major classifications. These numbers refer to related items in the Reference Section. The reference information may be an estimating procedure, an alternate pricing method, or technical information.

Note: Not all subdivisions listed here necessarily appear in this publication.

08060	Selective Demolition	CREW	DAILY OUTPUT	LABOR-HOURS	UNIT	MAT.	LABOR	EQUIP.	TOTAL	TOTAL INCL O&P
110	0010 **SELECTIVE DEMOLITION, DOORS**									110
	0200 Doors, ext., 44 mm thick, single, 900 mm x 2100 mm high [R024119-10]	1 Clab	16	.500	Ea.		13.70		13.70	21.50
	0220 Double, 1800 mm x 2100 mm high		12	.667			18.25		18.25	28.50
	0500 Interior, 35 mm thick, single, 915 mm x 2135 mm high		20	.400			10.95		10.95	17.05
	0520 Double, 1830 mm x 2135 mm high		16	.500			13.70		13.70	21.50
	0700 Bi-folding, 900 mm x 2000 mm high		20	.400			10.95		10.95	17.05
	0720 1800 mm x 2000 mm high		18	.444			12.20		12.20	18.95
	0900 Bi-passing, 900 mm x 2000 mm high		16	.500			13.70		13.70	21.50
	0940 1800 mm x 2000 mm high	▼	14	.571			15.65		15.65	24.50
	1500 Remove and reset, minimum	1 Carp	8	1			35.50		35.50	55.50
	1520 Maximum		6	1.333			47.50		47.50	74
	2000 Frames, including trim, metal	▼	8	1			35.50		35.50	55.50
	2200 Wood	2 Carp	32	.500			17.80		17.80	27.50
	3000 Special doors, counter doors		6	2.667			95		95	148
	3100 Double acting		10	1.600			57		57	88.50
	3200 Floor door (trap type)		8	2			71		71	111
	3300 Glass, sliding, including frames		12	1.333			47.50		47.50	74
	3400 Overhead, commercial, 3700 mm x 3700 mm high		4	4			142		142	221
	3440 up to 6100 mm x 4900 mm high		3	5.333			190		190	295
	3445 up to 10.67m x 9.14m high		1	16			570		570	885
	3500 Residential, 2700 mm x 2100 mm high		8	2			71		71	111
	3540 4900 mm x 2100 mm high		7	2.286			81.50		81.50	127
	3600 Remove and reset, minimum		4	4			142		142	221
	3620 Maximum		2.50	6.400			228		228	355
	3700 Roll-up grille		5	3.200			114		114	177
	3800 Revolving door		2	8			284		284	445
	3900 Storefront swing door		3	5.333			190		190	295
	5020 Remove bay/bow window	▼	6	2.667	▼		95		95	148
	5032 Remove skylight, plstc domes, flush/curb mtd	G-3	36.70	.872	m²		30.50		30.50	47
	6600 Demo flexible transparent strip entrance	3 Shee	10.68	2.246	m² Surf		94.50		94.50	146
	7100 Remove double swing pneumatic doors, openers and sensors	2 Skwk	.50	32	Opng.		1,175		1,175	1,825
	7110 Remove automatic operators, industrial, sliding doors, to 3660 mm W	"	.40	40	"		1,450		1,450	2,275
	7550 Hangar door demo	2 Sswk	20.44	.783	m²		31.50		31.50	56.50
	7570 Remove shock absorbing door	"	1.90	8.421	Opng.		335		335	605
120	0010 **SELECTIVE DEMOLITION, WINDOWS**									120
	0200 Aluminum, including trim, to 1.12 m² [R024119-10]	1 Clab	16	.500	Ea.		13.70		13.70	21.50
	0240 To 2.32 m²		11	.727			19.95		19.95	31
	0280 To 4.65 m²		5	1.600			44		44	68
	0320 Storm windows, to 1.12 m²		27	.296			8.10		8.10	12.65
	0360 To 2.32 m²		21	.381			10.45		10.45	16.25
	0400 To 4.65 m²		16	.500	▼		13.70		13.70	21.50
	0600 Glass, minimum		18.58	.431	m²		11.80		11.80	18.35
	0620 Maximum		13.94	.574	"		15.75		15.75	24.50
	1000 Steel, including trim, to 1.12 m²		13	.615	Ea.		16.85		16.85	26.50
	1020 To 2.32 m²		9	.889			24.50		24.50	38
	1040 To 4.65 m²		4	2			55		55	85.50
	2000 Wood, including trim, to 1.15 m²		22	.364			9.95		9.95	15.50
	2020 To 2.32 m²		18	.444			12.20		12.20	18.95
	2060 To 4.65 m²		13	.615			16.85		16.85	26.50
	2065 To 13.94 m²	▼	8	1			27.50		27.50	42.50
	5020 Remove and reset window, minimum	1 Carp	6	1.333			47.50		47.50	74
	5040 Average		4	2			71		71	111
	5080 Maximum	▼	2	4	▼		142		142	221

8

DOORS & WINDOWS

08110	Steel Doors and Frames		CREW	DAILY OUTPUT	LABOR-HOURS	UNIT	2006 BARE COSTS				TOTAL INCL O&P
							MAT.	LABOR	EQUIP.	TOTAL	
200	0010	**COMMERCIAL STEEL DOORS** R081313 -20									**200**
	0015	Flush, full panel, hollow core									
	0020	35 mm thick, 20 ga., 610 mm x 2035 mm	2 Carp	20	.800	Ea.	246	28.50		274.50	315
	0040	815 mm x 2035 mm		18	.889		254	31.50		285.50	330
	0060	915 mm x 2035 mm		17	.941		259	33.50		292.50	335
	0100	915 mm x 2135 mm	▼	17	.941		241	33.50		274.50	315
	0120	For vision lite, add					84			84	92.50
	0140	For narrow lite, add					85.50			85.50	94
	0320	Half glass, 20 ga., 610 mm x 2035 mm	2 Carp	20	.800		365	28.50		393.50	450
	0340	815 mm x 2035 mm		18	.889		375	31.50		406.50	465
	0360	915 mm x 2035 mm		17	.941		380	33.50		413.50	470
	0400	915 mm x 2135 mm		17	.941		380	33.50		413.50	470
	0410	35 mm thick, 18 ga., 610 mm x 2035 mm		20	.800		295	28.50		323.50	370
	0420	915 mm x 2035 mm		17	.941		299	33.50		332.50	380
	0425	915 mm x 2135 mm	▼	17	.941		305	33.50		338.50	385
	0450	For vision lite, add					84			84	92.50
	0452	For narrow lite, add					85.50			85.50	94
	0460	Half glass, 18 ga., 610 mm x 2035 mm	2 Carp	20	.800		415	28.50		443.50	505
	0465	815 mm x 2035 mm		18	.889		430	31.50		461.50	525
	0470	915 mm x 2035 mm		17	.941		420	33.50		453.50	515
	0475	915 mm x 2135 mm		17	.941		425	33.50		458.50	520
	0500	Hollow core, 44 mm thick, full panel, 20 ga., 815 mm x 2035 mm		18	.889		277	31.50		308.50	355
	0520	915 mm x 2035 mm		17	.941		259	33.50		292.50	335
	0640	900 mm x 2100 mm		17	.941		305	33.50		338.50	390
	0680	1220 mm x 2135 mm		15	1.067		430	38		468	530
	0700	1220 mm x 2440 mm		13	1.231		490	44		534	610
	1000	18 ga., 815mm x 2035mm		17	.941		204	33.50		237.50	277
	1020	915 mm x 2035 mm		16	1		199	35.50		234.50	275
	1120	900 mm x 2100 mm		17	.941		345	33.50		378.50	425
	1180	1220 mm x 2135 mm		14	1.143		415	40.50		455.50	520
	1200	1220 mm x 2440 mm	▼	17	.941		490	33.50		523.50	590
	1212	For vision lite, add					84			84	92.50
	1214	For narrow lite, add					85.50			85.50	94
	1230	Half glass, 20 ga., 815 mm x 2035 mm	2 Carp	20	.800		380	28.50		408.50	465
	1240	915 mm x 2035 mm		18	.889		380	31.50		411.50	470
	1260	915 mm x 2135 mm		18	.889		390	31.50		421.50	480
	1320	18 ga., 815 mm x 2035 mm		18	.889		425	31.50		456.50	520
	1340	915 mm x 2035 mm		17	.941		420	33.50		453.50	515
	1360	915 mm x 2135 mm		17	.941		430	33.50		463.50	525
	1380	1220 mm x 2135 mm		15	1.067		535	38		573	650
	1400	1220 mm x 2440 mm		14	1.143		615	40.50		655.50	740
	1720	Insulated, 45 mm thick, full panel, 18 ga., 915 mm x 2035 mm		15	1.067		370	38		408	470
	1740	813 mm x 2134 mm		16	1		390	35.50		425.50	480
	1760	915 mm x 2135 mm		15	1.067		380	38		418	480
	1800	1220 mm x 2440 mm	▼	13	1.231		565	44		609	690
	1805	For vision lite, add					84			84	92.50
	1810	For narrow lite, add					85.50			85.50	94
	1820	Half glass, 18 ga., 915 mm x 2035 mm	2 Carp	16	1		495	35.50		530.50	600
	1840	815 mm x 2135 mm		17	.941		510	33.50		543.50	610
	1860	915 mm x 2135 mm		16	1		565	35.50		600.50	675
	1900	1220 mm x 2440 mm	▼	14	1.143		685	40.50		725.50	820
	2000	For bottom louver, add	▼			▼	138			138	152
	2020	For baked enamel finish, add					30%	15%			
	2040	For galvanizing, add	▼				15%				
250	0010	**DOOR FRAMES**									**250**
	0020	Steel channels with anchors and bar stops									

DOORS & WINDOWS | 8

08110	Steel Doors and Frames	CREW	DAILY OUTPUT	LABOR-HOURS	UNIT	2006 BARE COSTS				TOTAL INCL O&P	
						MAT.	LABOR	EQUIP.	TOTAL		
250											**250**
0100	150 mm channel @ 12 kg/m, 915 mm x 2135 mm door, 68 kg	E-4	13	2.462	Ea.	171	99.50	6.85	277.35	375	
0200	200 mm channel @ 17 kg/m, 1830 mm x 2440 mm door, 125 kg		9	3.556		315	144	9.90	468.90	615	
0300	2440 mm x 3660 mm door, weighs 181 kg		6.50	4.923		455	199	13.75	667.75	875	
0400	250 mm channel @ 22 kg/m, 3050 mm x 3050 mm door, 227 kg		6	5.333		570	216	14.90	800.90	1,025	
0500	3660 mm x 3660 mm door, weighs 272 kg		5.50	5.818		685	235	16.25	936.25	1,200	
0600	310 mm channel @ 31 kg/m, 3660 mm x 3660 mm door, 374 kg		4.50	7.111		940	288	19.85	1,247.85	1,575	
0700	3660 mm x 4880 mm door, weighs 454 kg		4	8		1,150	325	22.50	1,497.50	1,850	
0800	For frames without bar stops, light sections, deduct					15%					
0900	Heavy sections, deduct					10%					
300	0010 **FIRE DOOR**										**300**
0015	Steel, flush, "B" label, 90 minute										
0020	Full panel, 20 ga., 610 mm x 2035 mm	2 Carp	20	.800	Ea.	310	28.50		338.50	385	
0040	815 mm x 2035 mm		18	.889		320	31.50		351.50	400	
0060	915 mm x 2035 mm		17	.941		325	33.50		358.50	405	
0080	915 mm x 2135 mm		17	.941		335	33.50		368.50	420	
0140	18 ga., 915 mm x 2035 mm		16	1		365	35.50		400.50	455	
0160	815 mm x 2135 mm		17	.941		385	33.50		418.50	470	
0180	915 mm x 2135 mm		16	1		365	35.50		400.50	460	
0200	1220 mm x 2135 mm		15	1.067		480	38		518	585	
0220	For "A" label, 3 hour, 18 ga., use same price as "B" label										
0240	For vision lite, add				Ea.	104			104	115	
0520	Flush, "B" label 90 min., composite, 20 ga., 610 mm x 2035 mm	2 Carp	18	.889		400	31.50		431.50	490	
0540	815 mm x 2035 mm		17	.941		410	33.50		443.50	500	
0560	915 mm x 2035 mm		16	1		410	35.50		445.50	510	
0580	915 mm x 2135 mm		16	1		425	35.50		460.50	520	
0640	Flush, "A" label 3 hour, composite, 18 ga., 915 mm x 2035 mm		15	1.067		345	38		383	440	
0660	815 mm x 2135 mm		16	1		365	35.50		400.50	455	
0680	915 mm x 2135mm		15	1.067		360	38		398	455	
0700	1220 mm x 2135 mm		14	1.143		465	40.50		505.50	575	
600	0010 **RESIDENTIAL STEEL DOOR**										**600**
0020	Prehung, insulated, exterior										
0030	Embossed, full panel, 800 mm x 2000 mm	2 Carp	17	.941	Ea.	216	33.50		249.50	289	
0040	900 mm x 2000 mm		15	1.067		217	38		255	298	
0060	915 mm x 2135 mm		15	1.067		280	38		318	370	
0070	1625 mm x 2030 mm, double		8	2		440	71		511	595	
0220	Half glass, 815 mm x 2035mm		17	.941		261	33.50		294.50	340	
0240	915 mm x 2035 mm		16	1		261	35.50		296.50	345	
0260	915 mm x 2135 mm		16	1		315	35.50		350.50	405	
0270	1625 mm x 2035 mm, double		8	2		540	71		611	705	
0720	Raised plastic face, full panel, 815 mm x 2035 mm		16	1		256	35.50		291.50	340	
0740	915 mm x 2035 mm		15	1.067		258	38		296	345	
0760	915 mm x 2135 mm		15	1.067		261	38		299	345	
0780	1625 mm x 2035 mm, double		8	2		485	71		556	640	
0820	Half glass, 815 mm x 2035 mm		17	.941		287	33.50		320.50	365	
0840	915 mm x 2035 mm		16	1		290	35.50		325.50	375	
0860	915 mm x 2135 mm		16	1		320	35.50		355.50	405	
0880	1625 mm x 2035 mm, double		8	2		630	71		701	805	
1320	Flush face, full panel, 765 mm x 2035 mm		16	1		216	35.50		251.50	294	
1340	915 mm x 2035 mm		15	1.067		216	38		254	297	
1360	915 mm x 2135 mm		15	1.067		278	38		316	365	
1380	1625 mm x 2035 mm, double		8	2		405	71		476	555	
1420	Half glass, 815 mm x 2035 mm		17	.941		270	33.50		303.50	350	
1440	915 mm x 2035 mm		16	1		274	35.50		309.50	355	
1460	915 mm x 2135 mm		16	1		315	35.50		350.50	400	
1480	1625 mm x 2035 mm, double		8	2		530	71		601	690	

Important: See the Reference Section for supporting data - Crews, Rental Equipment, City Cost Indexes and Reference Data

08100 | Metal Doors and Frames

08110 | Steel Doors and Frames

		CREW	DAILY OUTPUT	LABOR-HOURS	UNIT	2006 BARE COSTS MAT.	LABOR	EQUIP.	TOTAL	TOTAL INCL O&P	
600	1500	Sidelight, full lite, 305 mm x 2032 mm				Ea.	211			211	232	**600**
	1510	305 mm x 2032 mm					242			242	267	
	1520	305 mm x 2032 mm					189			189	208	
	1530	305 mm x 2032 mm					230			230	253	
	2300	Interior, residential, closet, bi-fold, 2035 mm x 610 mm wide	2 Carp	16	1		140	35.50		175.50	210	
	2330	915 mm wide		16	1		157	35.50		192.50	229	
	2360	1220 mm wide		15	1.067		238	38		276	320	
	2400	1525 mm wide		14	1.143		276	40.50		316.50	370	
	2420	1830 mm wide	▼	13	1.231	▼	310	44		354	410	
820	0010	**STEEL FRAMES, KNOCK DOWN**										**820**
	0020	16 ga., up to 144 mm jamb depth										
	0025	2000 mm high, 900 mm wide, single	2 Carp	16	1	Ea.	121	35.50		156.50	189	
	0028	1067 mm wide, single		16	1		107	35.50		142.50	174	
	0030	1219 mm wide, single		16	1		107	35.50		142.50	174	
	0040	1830 mm wide, double		14	1.143		143	40.50		183.50	222	
	0045	2438 mm wide, double		14	1.143		149	40.50		189.50	228	
	0100	2135 mm high, 915 mm wide, single		16	1		124	35.50		159.50	192	
	0110	1067 mm wide, single		16	1		110	35.50		145.50	177	
	0112	1219 mm wide, single		16	1		188	35.50		223.50	263	
	0140	1830 mm wide, double		14	1.143		150	40.50		190.50	229	
	0145	2438 mm wide, double		14	1.143		140	40.50		180.50	218	
	1000	16 ga., up to 124 mm D, 2135 mm H, 915 mm W, single		16	1		121	35.50		156.50	189	
	1140	1830 mm wide, double		14	1.143		148	40.50		188.50	227	
	2800	14 ga., drywall, up to 98 mm D, 2134 mm H, 914 mm W, single		16	1		124	35.50		159.50	192	
	2840	1830 mm wide, single		14	1.143		152	40.50		192.50	231	
	3000	14 ga., drywall, up to 146 mm D, 2032 mm H, 914 mm W, single		16	1		134	35.50		169.50	204	
	3002	1067 mm wide, single		16	1		131	35.50		166.50	200	
	3005	1219 mm wide, single		16	1		131	35.50		166.50	200	
	3600	146 mm deep, 2135 mm high, 1220 mm W, single		15	1.067		103	38		141	173	
	3620	1829 mm wide, double		12	1.333		162	47.50		209.50	252	
	3640	2440 mm wide, double		12	1.333		136	47.50		183.50	224	
	3700	2440 mm high, 1220 mm wide, single		15	1.067		136	38		174	209	
	3740	2438 mm wide, double		12	1.333		169	47.50		216.50	260	
	4000	169 mm deep, 2100 mm high, 1200 mm wide, single		15	1.067		144	38		182	218	
	4020	1800 mm wide, double		12	1.333		165	47.50		212.50	256	
	4040	2400 mm		12	1.333		191	47.50		238.50	284	
	4100	2440 mm high, 1220 mm wide, single		15	1.067		162	38		200	237	
	4140	2440 mm wide, double		12	1.333		189	47.50		236.50	282	
	4400	222 mm deep, 2135 mm high, 1220 mm wide, single		15	1.067		152	38		190	226	
	4440	2440 mm wide, double		12	1.333		200	47.50		247.50	294	
	4500	2440 mm high, 1220 mm wide, single		15	1.067		178	38		216	254	
	4540	2440 mm wide, double	▼	12	1.333		208	47.50		255.50	305	
	4900	For welded frames, add					44.50			44.50	49	
	5400	14 ga., "B" label, to 144 mm D, 2100 mm H, 1200 mm W, single	2 Carp	15	1.067		160	38		198	235	
	5440	2400 mm wide, double		12	1.333		187	47.50		234.50	280	
	5800	171 mm deep, 2135 mm high, 1220 mm wide, single		15	1.067		140	38		178	213	
	5840	2440 mm wide, double		12	1.333		207	47.50		254.50	300	
	6200	219 mm deep, 2100 mm high, 1200 mm wide, single		15	1.067		170	38		208	246	
	6240	2400 mm wide, double	▼	12	1.333	▼	220	47.50		267.50	315	
	6300	For "A" label use same price as "B" label										
	6400	For baked enamel finish, add					30%	15%				
	6500	For galvanizing, add					15%					
	6600	For hospital stop, add				Ea.	263			263	289	
	7900	Transom lite frames, fixed, add	2 Carp	14.40	1.111	m²	460	39.50		499.50	565	
	8000	Movable, add	"	12.08	1.325	"	550	47		597	685	

DOORS & WINDOWS 8

08100 | Metal Doors and Frames

08160 | Sliding Metal Doors and Grilles

		CREW	DAILY OUTPUT	LABOR-HOURS	UNIT	2006 BARE COSTS MAT.	LABOR	EQUIP.	TOTAL	TOTAL INCL O&P	
300	0010	**STEEL, SLIDING**									**300**
	0020	Up to 15 240 mm x 5485 mm, electric, standard duty, minimum	L-5	33.44	1.674	m²	237	67	19.45	323.45	400	
	0100	Maximum		31.59	1.773		400	71	20.50	491.50	585	
	0500	Heavy duty, minimum		27.59	2.030		320	81	23.50	424.50	525	
	0600	Maximum	↓	25.73	2.176	↓	850	87	25.50	962.50	1,125	

08180 | Metal Screen and Storm Doors

		CREW	DAILY OUTPUT	LABOR-HOURS	UNIT	2006 BARE COSTS MAT.	LABOR	EQUIP.	TOTAL	TOTAL INCL O&P	
100	0010	**STORM DOORS & FRAMES** Aluminum, residential,									**100**
	0020	combination storm and screen										
	0400	Clear anodic coating, 2000 mm x 750 mm wide	2 Carp	15	1.067	Ea.	162	38		200	238	
	0420	800 mm wide		14	1.143		186	40.50		226.50	269	
	0440	900 mm wide	↓	14	1.143	↓	186	40.50		226.50	269	
	0500	For 2100 mm door height, add					5%					
	1000	Mill finish, 2000 mm x 750 mm wide	2 Carp	15	1.067	Ea.	216	38		254	296	
	1020	800 mm wide		14	1.143		216	40.50		256.50	300	
	1040	900 mm wide	↓	14	1.143		234	40.50		274.50	320	
	1100	For 2100 mm door, add					5%					
	1500	White painted, 2000 mm x 750 mm wide	2 Carp	15	1.067	Ea.	216	38		254	297	
	1520	800 mm wide		14	1.143		220	40.50		260.50	305	
	1540	900 mm wide	↓	14	1.143		229	40.50		269.50	315	
	1600	For 2100 mm door, add					5%					
	2000	Wood door & screen, see Division 08210-930										

08200 | Wood and Plastic Doors

08210 | Wood Doors

		CREW	DAILY OUTPUT	LABOR-HOURS	UNIT	2006 BARE COSTS MAT.	LABOR	EQUIP.	TOTAL	TOTAL INCL O&P	
450	0010	**KALAMEIN**										**450**
	0020	Interior, flush type, 915 mm x 2135 mm	2 Carp	4.30	3.721	Opng.	176	132		308	400	
720	0010	**PRE-HUNG DOORS**										**720**
	0300	Ext., wood, comb. storm & screen, 2055 mm x 765 mm W	2 Carp	15	1.067	Ea.	283	38		321	370	
	0320	815 mm wide		15	1.067		283	38		321	370	
	0340	915 mm wide	↓	15	1.067		291	38		329	380	
	0360	For 2135 mm high door, add				↓	24.50			24.50	27	
	0370	For aluminum storm doors, see Division 08180-100										
	1600	Entrance door, flush, birch, solid core										
	1620	117 mm solid jamb, 44 mm x 2032 mm x 815 mm wide	2 Carp	16	1	Ea.	288	35.50		323.50	370	
	1640	915 mm wide	"	16	1		296	35.50		331.50	380	
	1680	For 2135 mm high door, add				↓	17.50			17.50	19.25	
	2000	Entrance door, colonial, 6 panel pine										
	2020	117 mm solid jamb, 44 mm x 2035 mm x 815 mm wide	2 Carp	16	1	Ea.	520	35.50		555.50	625	
	2040	915 mm wide	"	16	1		520	35.50		555.50	625	
	2060	For 2135 mm high door, add					48			48	52.50	
	2200	For 143 mm solid jamb, add				↓	38.50			38.50	42	
	4000	Interior, passage door, 117 mm solid jamb										
	4400	Lauan, flush, solid core, 35 mm x 2035 mm x 765 mm wide	2 Carp	20	.800	Ea.	178	28.50		206.50	240	
	4420	815 mm wide		20	.800		178	28.50		206.50	240	
	4440	915 mm wide		19	.842		191	30		221	257	
	4600	Hollow core, 35 mm x 2035 mm x 765 mm wide	↓	20	.800	↓	120	28.50		148.50	177	

8 — Doors & Windows

Important: See the Reference Section for supporting data - Crews, Rental Equipment, City Cost Indexes and Reference Data

08210	Wood Doors	CREW	DAILY OUTPUT	LABOR-HOURS	UNIT	2006 BARE COSTS				TOTAL INCL O&P	
						MAT.	LABOR	EQUIP.	TOTAL		
720	4620	815 mm wide	2 Carp	20	.800	Ea.	118	28.50		146.50	175	**720**
	4640	915 mm wide	↓	19	.842		121	30		151	180	
	4700	For 2135 mm high door, add					23			23	25.50	
	5000	Birch, flush, solid core, 35 mm x 2035 mm x 765 mm wide	2 Carp	20	.800		166	28.50		194.50	227	
	5020	815 mm wide		20	.800		187	28.50		215.50	251	
	5040	915 mm wide		19	.842		200	30		230	267	
	5200	Hollow core, 35 mm x 2035 mm x 765 mm wide		20	.800		137	28.50		165.50	196	
	5220	815 mm wide		20	.800		144	28.50		172.50	203	
	5240	915 mm wide	↓	19	.842		144	30		174	205	
	5280	For 2135 mm high door, add					19.95			19.95	22	
	5500	Hardboard paneled, 35 mm x 2035 mm x 765 mm wide	2 Carp	20	.800		138	28.50		166.50	197	
	5520	815 mm wide		20	.800		145	28.50		173.50	204	
	5540	915 mm wide		19	.842		143	30		173	204	
	6000	Pine paneled, 35 mm x 2035 mm x 765 mm wide		20	.800		240	28.50		268.50	310	
	6020	815 mm wide		20	.800		261	28.50		289.50	330	
	6040	915 mm wide	↓	19	.842		266	30		296	340	
	6500	For 143 mm solid jamb, add					11.80			11.80	12.95	
	6520	For split jamb, deduct				↓	13.95			13.95	15.35	
850	0010	**TIN CLAD**										**850**
	0020	3 ply, 1829 mm x 2135 mm, dbl sliding, doors only	2 Carp	1	16	Opng.	1,375	570		1,945	2,400	
	1000	For electric operator, add	1 Elec	2	4	"	2,450	168		2,618	2,950	
900	0010	**WOOD DOOR, ARCHITECTURAL**										**900**
	0015	Flush, int., 35 mm, 7 ply, hollow core,										
	0020	Lauan face, 600 mm x 2000 mm	2 Carp	17	.941	Ea.	29	33.50		62.50	84	
	0040	765 mm x 2035 mm		17	.941		33	33.50		66.50	88.50	
	0080	915 mm x 2035 mm		17	.941		39	33.50		72.50	95	
	0100	1220 mm x 2035 mm		16	1		69.50	35.50		105	132	
	0120	Birch face, 600 mm x 2000 mm		17	.941		46	33.50		79.50	103	
	0140	750 mm x 2000 mm		17	.941		44.50	33.50		78	101	
	0180	900 mm x 2000 mm		17	.941		50	33.50		83.50	107	
	0200	1220 mm x 2035 mm		16	1		95	35.50		130.50	160	
	0220	Oak face, 610 mm x 2035 mm		17	.941		72.50	33.50		106	132	
	0240	765 mm x 2035 mm		17	.941		77.50	33.50		111	137	
	0280	915 mm x 2035 mm		17	.941		83	33.50		116.50	143	
	0300	1219 mm x 2032 mm		16	1		105	35.50		140.50	172	
	0320	Walnut face, 610 mm x 2032 mm		17	.941		147	33.50		180.50	213	
	0340	765 mm x 2035 mm		17	.941		150	33.50		183.50	216	
	0380	915 mm x 2035 mm		17	.941		155	33.50		188.50	223	
	0400	1220 mm x 2035 mm	↓	16	1		176	35.50		211.50	250	
	0430	For 2135 mm high, add					14.45			14.45	15.90	
	0440	For 2440 mm high, add					20.50			20.50	22.50	
	0480	For prefinishing, clear, add					32			32	35	
	0500	For prefinishing, stain, add					43.50			43.50	47.50	
	1320	M.D. overlay on hardboard, 610 mm x 2035 mm	2 Carp	17	.941		89.50	33.50		123	151	
	1340	765 mm x 2035 mm		17	.941		89.50	33.50		123	151	
	1380	915 mm x 2035 mm		17	.941		106	33.50		139.50	169	
	1400	1220 mm x 2035 mm	↓	16	1		146	35.50		181.50	217	
	1420	For 2135 mm high, add					7.95			7.95	8.75	
	1440	For 2440 mm high, add					21			21	23.50	
	1720	H.P. plastic laminate, 610 mm x 2035 mm	2 Carp	16	1		213	35.50		248.50	291	
	1740	765 mm x 2035 mm		16	1		213	35.50		248.50	291	
	1780	915 mm x 2035 mm		15	1.067		246	38		284	330	
	1800	1220 mm x 2035 mm	↓	14	1.143		340	40.50		380.50	440	
	1820	For 2135 mm high, add					8.35			8.35	9.20	
	1840	For 2440 mm high, add				↓	21.50			21.50	23.50	

DOORS & WINDOWS 8

For expanded coverage of these items see *Means Interior Cost Data 2006*

08210	Wood Doors	CREW	DAILY OUTPUT	LABOR-HOURS	UNIT	2006 BARE COSTS				TOTAL INCL O&P	
						MAT.	LABOR	EQUIP.	TOTAL		
900	2020	5 ply particle core, lauan face, 765 mm x 2035 mm	2 Carp	15	1.067	Ea.	73	38		111	140	**900**
	2040	915 mm x 2035 mm		14	1.143		76	40.50		116.50	147
	2080	915 mm x 2135 mm		13	1.231		85	44		129	162
	2100	1220 mm x 2135 mm		12	1.333		98	47.50		145.50	182
	2120	Birch face, 765 mm x 2035 mm		15	1.067		83	38		121	151
	2140	915 mm x 2035 mm		14	1.143		91	40.50		131.50	164
	2180	915 mm x 2135 mm		13	1.231		93	44		137	170
	2200	1220 mm x 2135 mm		12	1.333		113	47.50		160.50	199
	2220	Oak face, 765 mm x 2035 mm		15	1.067		92	38		130	160
	2240	915 mm x 2035 mm		14	1.143		101	40.50		141.50	175
	2280	915 mm x 2135 mm		13	1.231		104	44		148	182
	2300	1220 mm x 2135 mm		12	1.333		127	47.50		174.50	214
	2320	Walnut face, 610 mm x 2035 mm		15	1.067		102	38		140	171
	2340	765 mm x 2035 mm		14	1.143		116	40.50		156.50	192
	2380	915 mm x 2035 mm		13	1.231		131	44		175	212
	2400	1220 mm x 2035 mm	▼	12	1.333		171	47.50		218.50	262
	2440	For 2440 mm high, add					25			25	27.50
	2460	For 2440 mm high walnut, add					13.65			13.65	15.05
	2480	For solid wood core, add					30.50			30.50	33.50
	2720	For prefinishing, clear, add					19.95			19.95	22
	2740	For prefinishing, stain, add					44.50			44.50	49
	3320	M.D. overlay on hardboard, 765 mm x 2035 mm	2 Carp	14	1.143		89.50	40.50		130	162
	3340	915 mm x 2035 mm		13	1.231		93.50	44		137.50	171
	3380	915 mm x 2135 mm		12	1.333		95.50	47.50		143	179
	3400	1220 mm x 2135 mm	▼	10	1.600		117	57		174	217
	3440	For 2440 mm height, add					26			26	28.50
	3460	For solid wood core, add					34			34	37.50
	3720	W plastic laminate, 765 mm x 2035 mm	2 Carp	13	1.231		131	44		175	212
	3740	915 mm x 2035 mm		12	1.333		148	47.50		195.50	237
	3780	915 mm x 2135 mm		11	1.455		154	51.50		205.50	250
	3800	1220 mm x 2135 mm	▼	8	2		187	71		258	315
	3840	For 2440 mm height, add					26			26	28.50
	3860	For solid wood core, add					32			32	35
	4000	Ext., flush, solid wood core, birch, 44 mm x 2135 mm x 765 mm	2 Carp	15	1.067		158	38		196	232
	4020	815 mm wide		15	1.067		165	38		203	240
	4040	915 mm wide		14	1.143		176	40.50		216.50	258
	4100	Oak faced 44 mm x 2135 mm x 765 mm wide		15	1.067		174	38		212	250
	4120	815 mm wide		15	1.067		186	38		224	263
	4140	915 mm wide		14	1.143		198	40.50		238.50	282
	4200	Walnut faced, 44 mm x 2135 mm x 765 mm wide		15	1.067		255	38		293	340
	4220	815 mm wide		15	1.067		267	38		305	350
	4240	915 mm wide	▼	14	1.143		278	40.50		318.50	370
	4300	For 2035 mm high door, deduct from 2135 mm door				▼	14.20			14.20	15.60
910	0010	**WOOD DOORS, DECORATOR**										**910**
	3000	Solid wood, 44 mm stile and rail									
	3020	Mahogany, 915 mm x 2135 mm, minimum	2 Carp	14	1.143	Ea.	945	40.50		985.50	1,125
	3030	Maximum		10	1.600		1,300	57		1,357	1,550
	3040	1067 mm x 2440 mm, minimum		10	1.600		895	57		952	1,075
	3050	Maximum		8	2		1,625	71		1,696	1,900
	3100	Pine, 915 mm x 2135 mm, minimum		14	1.143		425	40.50		465.50	535
	3110	Maximum		10	1.600		705	57		762	865
	3120	1067 mm x 2440 mm, minimum		10	1.600		655	57		712	810
	3130	Maximum		8	2		1,075	71		1,146	1,300
	3200	Red oak, 915 mm x 2135 mm, minimum		14	1.143		1,475	40.50		1,515.50	1,700
	3210	Maximum	▼	10	1.600	▼	1,750	57		1,807	2,025

08210	Wood Doors	CREW	DAILY OUTPUT	LABOR-HOURS	UNIT	2006 BARE COSTS				TOTAL INCL O&P	
						MAT.	LABOR	EQUIP.	TOTAL		
910 3220	1067 mm x 2440 mm, minimum	2 Carp	10	1.600	Ea.	1,600	57		1,657	1,875	**910**
3230	Maximum	↓	8	2	↓	2,850	71		2,921	3,250	
4000	Hand carved door, mahogany										
4020	915 mm x 2135 mm, minimum	2 Carp	14	1.143	Ea.	1,450	40.50		1,490.50	1,650	
4030	Maximum		11	1.455		2,825	51.50		2,876.50	3,200	
4040	1067 mm x 2440 mm, minimum		10	1.600		1,800	57		1,857	2,100	
4050	Maximum		8	2		2,775	71		2,846	3,175	
4200	Red oak, 915 mm x 2135 mm, minimum		14	1.143		4,400	40.50		4,440.50	4,925	
4210	Maximum		11	1.455		12,100	51.50		12,151.50	13,400	
4220	1067 mm x 2440 mm, minimum	↓	10	1.600		4,950	57		5,007	5,550	
4280	For 2035 mm high door, deduct from 2135 mm door					31.50			31.50	34.50	
4400	For custom finish, add					335			335	370	
4600	Side light, mahogany, 2134 mm x 457 mm wide, minimum	2 Carp	18	.889		810	31.50		841.50	940	
4610	Maximum		14	1.143		2,375	40.50		2,415.50	2,700	
4620	2438 mm x 457 mm wide, minimum		14	1.143		1,525	40.50		1,565.50	1,750	
4630	Maximum		10	1.600		1,725	57		1,782	2,000	
4640	Side light, oak, 2135 mm x 457 mm wide, minimum		18	.889		935	31.50		966.50	1,075	
4650	Maximum		14	1.143		1,675	40.50		1,715.50	1,900	
4660	2440 mm x 457 mm wide, minimum		14	1.143		880	40.50		920.50	1,025	
4670	Maximum		10	1.600		1,675	57		1,732	1,925	
6520	Interior cafe doors, 765 mm opening, stock, panel pine		16	1		188	35.50		223.50	262	
6540	915 mm opening	↓	16	1	↓	196	35.50		231.50	271	
6550	Louvered pine										
6560	765 mm opening	2 Carp	16	1	Ea.	164	35.50		199.50	236	
8000	915 mm opening		16	1		175	35.50		210.50	248	
8010	765 mm opening, hardwood		16	1		282	35.50		317.50	365	
8020	915 mm opening	↓	16	1	↓	310	35.50		345.50	400	
8800	Pre-hung doors see Division 08210-720										
920 0010	**WOOD DOORS, PANELED**										**920**
0020	Interior, six panel, hollow core, 35 mm thick										
0040	Molded hardboard, 610 mm x 2035 mm	2 Carp	17	.941	Ea.	47.50	33.50		81	104	
0060	765 mm x 2035 mm		17	.941		51	33.50		84.50	109	
0070	790 mm x 2035 mm		17	.941		53.50	33.50		87	111	
0080	915 mm x 2035 mm		17	.941		56.50	33.50		90	114	
0140	Embossed print, molded hardboard, 610 mm x 2035 mm		17	.941		51	33.50		84.50	109	
0160	765 mm x 2035 mm		17	.941		51	33.50		84.50	109	
0180	915 mm x 2035 mm		17	.941		56.50	33.50		90	114	
0540	Six panel, solid, 35 mm thick, pine, 610 mm x 2035 mm		15	1.067		124	38		162	195	
0560	765 mm x 2035 mm		14	1.143		139	40.50		179.50	217	
0580	915 mm x 2035 mm		13	1.231		160	44		204	244	
1020	Two panel, bored rail, solid, 35 mm thick, pine, 455 mm x 2035 mm		16	1		226	35.50		261.50	305	
1040	610 mm x 2035 mm		15	1.067		297	38		335	385	
1060	765 mm x 2035 mm		14	1.143		340	40.50		380.50	440	
1340	Two panel, solid, 35 mm thick, fir, 610 mm x 2035 mm		15	1.067		124	38		162	195	
1360	765 mm x 2035 mm		14	1.143		139	40.50		179.50	217	
1380	915 mm x 2035 mm		13	1.231		340	44		384	445	
1740	Five panel, solid, 35 mm thick, fir, 610 mm x 2035 mm		15	1.067		222	38		260	305	
1760	765 mm x 2035 mm		14	1.143		355	40.50		395.50	455	
1780	915 mm x 2035 mm	↓	13	1.231	↓	355	44		399	460	
930 0010	**WOOD DOORS, RESIDENTIAL**										**930**
0200	Exterior, combination storm & screen, pine										
0260	815 mm wide	2 Carp	10	1.600	Ea.	272	57		329	390	
0280	915 mm wide		9	1.778		278	63		341	405	
0300	2100 mm x 900 mm wide		9	1.778		298	63		361	430	
0400	Full lite, 2055 mm x 765 mm wide	↓	11	1.455		290	51.50		341.50	400	

DOORS & WINDOWS 8

08210	Wood Doors	CREW	DAILY OUTPUT	LABOR-HOURS	UNIT	2006 BARE COSTS				TOTAL INCL O&P
						MAT.	LABOR	EQUIP.	TOTAL	
930 0420	815 mm wide	2 Carp	10	1.600	Ea.	290	57		347	410	930
0440	915 mm wide		9	1.778		299	63		362	430
0500	240 mm x 915 mm wide		9	1.778		320	63		383	455
0700	Dutch door, pine, 44 mm x 2035 mm x 815 mm wide, minimum		12	1.333		675	47.50		722.50	815
0720	Maximum		10	1.600		710	57		767	875
0800	915 mm wide, minimum		12	1.333		685	47.50		732.50	830
0820	Maximum		10	1.600		750	57		807	915
1000	Entrance door, colonial, 44 mm x 2035 mm x 815 mm wide		16	1		355	35.50		390.50	450
1020	6 panel pine, 915 mm wide		15	1.067		400	38		438	500
1100	8 panel pine, 815 mm wide		16	1		595	35.50		630.50	710
1120	915 mm wide	▼	15	1.067		535	38		573	650
1200	For tempered safety glass lites, (min of 2)add					62.50			62.50	68.50
1300	Flush, birch, solid core, 45 mm x 2000 mm x 800 mm wide	2 Carp	16	1		94.50	35.50		130	160
1320	915 mm wide		15	1.067		98	38		136	167
1350	2100 mm x 800 mm wide		16	1		104	35.50		139.50	170
1360	900 mm wide	▼	15	1.067		111	38		149	181
1380	For tempered safety glass lites, add				▼	93			93	102
2700	Interior, closet, bi-fold, w/hardware, no frame or trim incl.									
2720	Flush, birch, 1980 mm or 2035 mm x 765 mm wide	2 Carp	13	1.231	Ea.	48.50	44		92.50	122
2740	915 mm wide		13	1.231		52.50	44		96.50	126
2760	1220 mm wide		12	1.333		96	47.50		143.50	179
2780	1525 mm wide		11	1.455		97	51.50		148.50	188
2800	1830 mm wide		10	1.600		103	57		160	203
3000	Raised panel pine, 1950 mm or 2000 mm x 750 mm wide		13	1.231		154	44		198	237
3020	915 mm wide		13	1.231		196	44		240	283
3040	1220 mm wide		12	1.333		300	47.50		347.50	405
3060	1525 mm wide		11	1.455		360	51.50		411.50	475
3080	1830 mm wide		10	1.600		395	57		452	525
3200	Louvered, pine, 1950 mm or 2000 mm x 750 mm wide		13	1.231		104	44		148	182
3220	915 mm wide		13	1.231		157	44		201	241
3240	1220 mm wide		12	1.333		194	47.50		241.50	287
3260	1525 mm wide		11	1.455		220	51.50		271.50	325
3280	1830 mm wide	▼	10	1.600	▼	243	57		300	355
4400	Bi-passing closet, incl. hardware and frame, no trim incl.									
4420	Flush, lauan, 2035 mm x 1220 mm wide	2 Carp	12	1.333	Opng.	164	47.50		211.50	255
4440	1525 mm wide		11	1.455		180	51.50		231.50	279
4460	1830 mm wide		10	1.600		193	57		250	300
4600	Flush, birch, 2000 mm x 1200 mm wide		12	1.333		201	47.50		248.50	295
4620	1525 mm wide		11	1.455		203	51.50		254.50	305
4640	1800 mm wide		10	1.600		241	57		298	355
4800	Louvered, pine, 2000 mm x 1200 mm wide		12	1.333		400	47.50		447.50	515
4820	1500 mm wide		11	1.455		385	51.50		436.50	500
4840	1800 mm wide		10	1.600		490	57		547	630
5000	Paneled, pine, 2000 mm x 1200 mm wide		12	1.333		380	47.50		427.50	495
5020	1500 mm wide		11	1.455		400	51.50		451.50	520
5040	1800 mm wide	▼	10	1.600	▼	470	57		527	610
6100	Folding accordion, closet, including track and frame									
6120	Vinyl, 2 layer, stock (see also Division 10651-100)	2 Carp	37.16	.431	m²	33.50	15.30		48.80	60.50
6140	Woven mahogany and vinyl, stock		37.16	.431		21	15.30		36.30	47.50
6160	Wood slats with vinyl overlay, stock		37.16	.431		107	15.30		122.30	141
6180	Economy vinyl, stock		37.16	.431		18.75	15.30		34.05	44.50
6200	Rigid PVC	▼	37.16	.431	▼	50	15.30		65.30	79
6220	For custom partition, add					25%	10%			
7310	Passage doors, flush, no frame included									
7320	Hardboard, hollow core, 34 mm x 2000 mm x 450 mm W	2 Carp	18	.889	Ea.	40	31.50		71.50	93
7330	610 mm wide	▼	18	.889	▼	40.50	31.50		72	93.50

		08210	Wood Doors	CREW	DAILY OUTPUT	LABOR-HOURS	UNIT	2006 BARE COSTS				TOTAL INCL O&P	
								MAT.	LABOR	EQUIP.	TOTAL		
930	7340		765 mm wide	2 Carp	18	.889	Ea.	44.50	31.50		76	98	930
	7350		815 mm wide		18	.889		47	31.50		78.50	101	
	7360		915 mm wide		17	.941		49.50	33.50		83	107	
	7420		Lauan, hollow core, 34 mm x 2000 mm x 450 mm W		18	.889		28	31.50		59.50	80	
	7440		600 mm wide		18	.889		27	31.50		58.50	79	
	7450		711 mm wide		18	.889		30.50	31.50		62	82.50	
	7460		750 mm wide		18	.889		30.50	31.50		62	82.50	
	7480		800 mm wide		18	.889		32	31.50		63.50	84.50	
	7500		900 mm wide		17	.941		33.50	33.50		67	89	
	7700		Birch, hollow core, 34 mm x 2000 mm x 450 mm wide		18	.889		35.50	31.50		67	88	
	7720		600 mm wide		18	.889		40	31.50		71.50	93	
	7740		750 mm wide		18	.889		44.50	31.50		76	98	
	7760		800 mm wide		18	.889		46	31.50		77.50	99.50	
	7780		900 mm wide		17	.941		50	33.50		83.50	107	
	8000		Pine louvered, 34 mm x 2000 mm x 450 mm wide		19	.842		100	30		130	157	
	8020		600 mm wide		18	.889		125	31.50		156.50	187	
	8040		750 mm wide		18	.889		137	31.50		168.50	199	
	8060		815 mm wide		18	.889		144	31.50		175.50	207	
	8080		915 mm wide		17	.941		154	33.50		187.50	222	
	8300		Pine paneled, 35 mm x 2035 mm x 457 mm wide		19	.842		108	30		138	166	
	8320		610 mm wide		18	.889		125	31.50		156.50	187	
	8330		711 mm wide		18	.889		137	31.50		168.50	199	
	8340		765 mm wide		18	.889		140	31.50		171.50	203	
	8360		815 mm wide		18	.889		151	31.50		182.50	216	
	8380		915 mm wide		17	.941		158	33.50		191.50	226	
	8550		For over 20 doors, deduct					15%					
950	0010	**WOOD FIRE DOORS**											950
	0020		Particle core, 7 face plys, "B" label,										
	0040		1 hour, birch face, 45 mm x 750 mm x 2000 mm	2 Carp	14	1.143	Ea.	298	40.50		338.50	395	
	0080		915 mm x 2035 mm		13	1.231		310	44		354	410	
	0090		915 mm x 2135 mm		12	1.333		320	47.50		367.50	430	
	0100		1220 mm x 2135 mm		12	1.333		430	47.50		477.50	545	
	0140		Oak face, 765 mm x 2035 mm		14	1.143		298	40.50		338.50	395	
	0180		915 mm x 2035 mm		13	1.231		310	44		354	410	
	0190		915 mm x 2135 mm		12	1.333		325	47.50		372.50	430	
	0200		1220 mm x 2135 mm		12	1.333		420	47.50		467.50	540	
	0240		Walnut face, 765 mm x 2035 mm		14	1.143		390	40.50		430.50	495	
	0280		915 mm x 2035 mm		13	1.231		400	44		444	510	
	0290		915 mm x 2135 mm		12	1.333		420	47.50		467.50	535	
	0300		1220 mm x 2135 mm		12	1.333		565	47.50		612.50	695	
	0440		M.D. overlay on hardboard, 765 mm x 2035 mm		15	1.067		262	38		300	345	
	0480		915 mm x 2035 mm		14	1.143		272	40.50		312.50	365	
	0490		914 mm x 2134 mm		13	1.231		287	44		331	385	
	0500		1220 mm x 2135 mm		12	1.333		350	47.50		397.50	460	
	0540		H.P. plastic laminate, 765 mm x 2035 mm		13	1.231		350	44		394	455	
	0580		915 mm x 2035 mm		12	1.333		370	47.50		417.50	480	
	0590		915 mm x 2135 mm		11	1.455		370	51.50		421.50	490	
	0600		1220 mm x 2135 mm		10	1.600		475	57		532	610	
	0740		90 minutes, birch face, 45 mm x 750 mm x 2000 mm		14	1.143		256	40.50		296.50	345	
	0780		900 mm x 2000 mm		13	1.231		262	44		306	355	
	0790		900 mm x 2100 mm		12	1.333		315	47.50		362.50	425	
	0800		1200 mm x 2100 mm		12	1.333		380	47.50		427.50	490	
	0840		Oak face, 765 mm x 2035 mm		14	1.143		269	40.50		309.50	360	
	0880		915 mm x 2035 mm		13	1.231		279	44		323	375	
	0890		915 mm x 2135 mm		12	1.333		293	47.50		340.50	395	
	0900		1220 mm x 2135 mm		12	1.333		405	47.50		452.50	520	

DOORS & WINDOWS 8

For expanded coverage of these items see *Means Interior Cost Data 2006*

		CREW	DAILY OUTPUT	LABOR-HOURS	UNIT	MAT.	LABOR	EQUIP.	TOTAL	TOTAL INCL O&P	
08210	**Wood Doors**					2006 BARE COSTS					
950 0940	Walnut face, 765 mm x 2035 mm	2 Carp	14	1.143	Ea.	370	40.50		410.50	470	**950**
0980	915 mm x 2035 mm		13	1.231		380	44		424	485	
0990	915 mm x 2135 mm		12	1.333		395	47.50		442.50	510	
1000	1220 mm x 2135 mm		12	1.333		570	47.50		617.50	700	
1140	M.D. overlay on hardboard, 765 mm x 2035 mm		15	1.067		292	38		330	380	
1180	915 mm x 2035 mm		14	1.143		300	40.50		340.50	395	
1190	915 mm x 2135 mm		13	1.231		310	44		354	415	
1200	1220 mm x 2135 mm		12	1.333		425	47.50		472.50	540	
1240	For 2440 mm height, add					53.50			53.50	59	
1260	For 2440 mm height walnut, add					71			71	78	
1340	W plastic laminate, 765 mm x 2035 mm	2 Carp	13	1.231		365	44		409	470	
1380	915 mm x 2035 mm		12	1.333		380	47.50		427.50	495	
1390	915 mm x 2135 mm		11	1.455		385	51.50		436.50	505	
1400	1220 mm x 2135 mm		10	1.600		500	57		557	640	
2200	Custom architectural "B" label, flush, 44 mm thick, birch,										
2210	Solid core										
2220	765 mm x 2135 mm	2 Carp	15	1.067	Ea.	248	38		286	330	
2260	915 mm x 2135 mm		14	1.143		257	40.50		297.50	345	
2300	1220 mm x 2135 mm		13	1.231		360	44		404	465	
2420	1220 mm x 2440 mm		11	1.455		450	51.50		501.50	575	
2480	For oak veneer, add					50%					
2500	For walnut veneer, add					75%					
960 0010	**WOOD FRAMES**										**960**
0400	Exterior frame, incl. ext. trim, pine, 30 mm x 116 mm deep	2 Carp	114	.140	m	16.70	4.99		21.69	26	
0420	132 mm deep		114	.140		26	4.99		30.99	36.50	
0440	167 mm deep		114	.140		26	4.99		30.99	36.50	
0600	Oak, 30 mm x 116 mm deep		107	.150		31	5.30		36.30	42.50	
0620	132 mm deep		107	.150		35	5.30		40.30	47	
0640	167 mm deep		107	.150		39	5.30		44.30	51	
0800	Walnut, 30 mm x 116 mm deep		107	.150		36.50	5.30		41.80	48.50	
0820	132 mm deep		107	.150		52.50	5.30		57.80	66	
0840	167 mm deep		107	.150		62	5.30		67.30	77	
1000	Sills, 50 mm x 203 mm deep oak, no horns		30.48	.525		57.50	18.65		76.15	92.50	
1020	50 mm horns		30.48	.525		50.50	18.65		69.15	84.50	
1040	76 mm horns		30.48	.525		52.50	18.65		71.15	86.50	
1100	50 mm x 254 mm deep, oak, no horns		27.43	.583		73.50	20.50		94	113	
1120	51 mm horns		27.43	.583		69	20.50		89.50	108	
1140	76 mm horns		27.43	.583		69	20.50		89.50	108	
2000	Exterior, colonial, frame & trim, 915 mm opng., in-swing, minimum		22	.727	Ea.	310	26		336	380	
2010	Average		21	.762		460	27		487	550	
2020	Maximum		20	.800		1,050	28.50		1,078.50	1,200	
2100	1626 mm opening, in-swing, minimum		17	.941		355	33.50		388.50	440	
2120	Maximum		15	1.067		1,050	38		1,088	1,200	
2140	Out-swing, minimum		17	.941		360	33.50		393.50	445	
2160	Maximum		15	1.067		1,100	38		1,138	1,250	
2400	1830 mm opening, in-swing, minimum		16	1		340	35.50		375.50	425	
2420	Maximum		10	1.600		1,100	57		1,157	1,300	
2460	Out-swing, minimum		16	1		365	35.50		400.50	455	
2480	Maximum		10	1.600		1,275	57		1,332	1,525	
2600	For two sidelights, add, minimum		30	.533	Opng.	345	18.95		363.95	410	
2620	Maximum		20	.800	"	1,125	28.50		1,153.50	1,275	
2700	Custom birch frame, 915 mm opening		16	1	Ea.	199	35.50		234.50	275	
2750	1830 mm opening		16	1		300	35.50		335.50	385	
2900	Exterior, modern, plain trim, 914 mm opng., in-swing, minimum		26	.615		33.50	22		55.50	71	
2920	Average		24	.667		40	23.50		63.50	81	
2940	Maximum		22	.727		48.50	26		74.50	94	

8 DOORS & WINDOWS

08200 | Wood and Plastic Doors

08210 | Wood Doors

			CREW	DAILY OUTPUT	LABOR-HOURS	UNIT	2006 BARE COSTS MAT.	LABOR	EQUIP.	TOTAL	TOTAL INCL O&P	
960	3000	Interior frame, pine, 17 mm x 92 mm deep	2 Carp	114	.140	m	14.15	4.99		19.14	23.50	**960**
	3020	114 mm deep		114	.140		19.10	4.99		24.09	29	
	3200	Oak, 17 mm x 92 mm deep		107	.150		13.10	5.30		18.40	23	
	3220	116 mm deep		107	.150		14.15	5.30		19.45	24	
	3240	132 mm deep		107	.150		14.65	5.30		19.95	24.50	
	3400	Walnut, 17 mm x 92 mm deep		107	.150		22	5.30		27.30	33	
	3420	116 mm deep		107	.150		23.50	5.30		28.80	34	
	3440	132 mm deep		107	.150		24.50	5.30		29.80	35	
	3800	Threshold, oak, 16 mm x 92 mm deep		60.96	.262		8.55	9.35		17.90	24	
	3820	117 mm deep		57.91	.276		10.50	9.80		20.30	27	
	3840	143 mm deep	▼	54.86	.292	▼	18.55	10.35		28.90	36.50	
	4000	For casing see division 06220-400 & 06220-800										

08220 | Plastic Doors

			CREW	DAILY OUTPUT	LABOR-HOURS	UNIT	2006 BARE COSTS MAT.	LABOR	EQUIP.	TOTAL	TOTAL INCL O&P	
100	0010	**FIBERGLASS DOORS**										**100**
	0020	Pre-hung doors, ext, fiberglass, 813 mm wide x 2032 mm high	2 Carp	15	1.067	Ea.	305	38		343	395	
	0040	914 mm wide x 2032 mm high		15	1.067		305	38		343	395	
	0060	914 mm wide x 2134 mm high		15	1.067		375	38		413	475	
	0080	914 mm wide x 2032 mm high, with two lites		15	1.067		370	38		408	465	
	0100	914 mm wide x 2134 mm high, with two lites		15	1.067		440	38		478	545	
	0110	Half glass, 914 mm wide x 2032 mm high		15	1.067		415	38		453	515	
	0120	914 mm wide x 2032 mm high, low e		15	1.067		440	38		478	540	
	0130	914 mm wide x 2134 mm high		15	1.067		485	38		523	590	
	0140	914 mm wide x 2134 mm high	▼	15	1.067		510	38		548	620	
	0150	Side lights, 305 mm wide x 2032 mm high					281			281	310	
	0160	305 mm wide x 2032 mm high, low e					293			293	325	
	0180	305 mm wide x 2032 mm high full glass					320			320	355	
	0190	305 mm wide x 2032 mm high, low e				▼	345			345	380	

08260 | Sliding Wood and Plastic Doors

			CREW	DAILY OUTPUT	LABOR-HOURS	UNIT	2006 BARE COSTS MAT.	LABOR	EQUIP.	TOTAL	TOTAL INCL O&P	
700	0010	**GLASS, SLIDING**										**700**
	0012	Vinyl clad, 25 mm insul. glass, 1830 x 2085 mm high	2 Carp	4	4	Opng.	1,250	142		1,392	1,625	
	0030	1836 mm x 2448 mm high		4	4	Ea.	1,900	142		2,042	2,325	
	0100	2440 mm x 2085 mm high		4	4	Opng.	1,950	142		2,092	2,375	
	0500	3 leaf, 2745 mm x 2085 mm high		3	5.333		1,775	190		1,965	2,250	
	0600	3660 mm x 2085 mm high	▼	3	5.333	▼	2,200	190		2,390	2,725	

08300 | Specialty Doors

08310 | Access Doors and Panels

			CREW	DAILY OUTPUT	LABOR-HOURS	UNIT	2006 BARE COSTS MAT.	LABOR	EQUIP.	TOTAL	TOTAL INCL O&P	
100	0010	**ACCESS DOORS**										**100**
	1000	Fire rated door with lock										
	1100	Metal, 305 mm x 305 mm	1 Carp	10	.800	Ea.	143	28.50		171.50	202	
	1150	455 mm x 455 mm		9	.889		186	31.50		217.50	254	
	1200	610 mm x 610 mm		9	.889		225	31.50		256.50	296	
	1250	610 mm x 915 mm		8	1		300	35.50		335.50	385	
	1300	610 mm x 1220 mm		8	1		375	35.50		410.50	465	
	1350	915 mm x 915 mm		7.50	1.067		450	38		488	555	
	1400	1220 mm x 1220 mm		7.50	1.067		575	38		613	695	
	1600	Stainless steel, 305 mm x 305 mm	▼	10	.800	▼	255	28.50		283.50	325	

For expanded coverage of these items see *Means Interior Cost Data 2006*

08310 | Access Doors and Panels

		CREW	DAILY OUTPUT	LABOR-HOURS	UNIT	MAT.	LABOR	EQUIP.	TOTAL	TOTAL INCL O&P	
100	1650	455 mm x 455 mm	1 Carp	9	.889	Ea.	370	31.50		401.50	455	**100**
	1700	610 mm x 610 mm		9	.889		455	31.50		486.50	550	
	1750	610 mm x 915 mm		8	1		580	35.50		615.50	695	
	2000	Flush door for finishing										
	2100	Metal 203 mm x 203 mm	1 Carp	10	.800	Ea.	41	28.50		69.50	89.50	
	2150	305 mm x 305 mm	"	10	.800	"	46	28.50		74.50	95	
	3000	Recessed door for acoustic tile										
	3100	Metal, 305 mm x 305 mm	1 Carp	4.50	1.778	Ea.	63	63		126	168	
	3150	305 mm x 610 mm		4.50	1.778		82	63		145	189	
	3200	610 mm x 610 mm		4	2		110	71		181	232	
	3250	610 mm x 915 mm		4	2		141	71		212	267	
	4000	Recessed door for drywall										
	4100	Metal 305 mm x 305 mm	1 Carp	6	1.333	Ea.	70	47.50		117.50	151	
	4150	305 mm x 610 mm		5.50	1.455		104	51.50		155.50	195	
	4200	610 mm x 915 mm		5	1.600		166	57		223	271	
	6000	Standard door										
	6100	Metal, 205 mm x 205 mm	1 Carp	10	.800	Ea.	36.50	28.50		65	84.50	
	6150	305 mm x 305 mm		10	.800		41	28.50		69.50	89.50	
	6200	455 mm x 455 mm		9	.889		57	31.50		88.50	112	
	6250	610 mm x 610 mm		9	.889		73.50	31.50		105	130	
	6300	610 mm x 915 mm		8	1		109	35.50		144.50	176	
	6350	915 mm x 915 mm		8	1		133	35.50		168.50	202	
	6500	Stainless steel, 205 mm x 205 mm		10	.800		72.50	28.50		101	125	
	6550	305 mm x 305 mm		10	.800		96.50	28.50		125	151	
	6600	455 mm x 455 mm		9	.889		179	31.50		210.50	246	
	6650	610 mm x 610 mm		9	.889		234	31.50		265.50	305	
150	0010	**BULKHEAD CELLAR DOORS**										**150**
	0020	Steel, not incl. sides, 1120 mm x 1575 mm	1 Carp	5.50	1.455	Ea.	214	51.50		265.50	315	
	0100	1320 mm x 1855 mm		5.10	1.569		238	56		294	350	
	0500	W/ sides and foundation plates, 1450 mm x 1145 mm x 610 mm		4.70	1.702		280	60.50		340.50	405	
	0600	1065 mm x 1245 mm x 1295 mm		4.30	1.860		335	66		401	475	
300	0010	**FLOOR, COMMERCIAL**										**300**
	0020	Aluminum tile, steel frame, one leaf, 600 mm x 600 mm opng.	2 Sswk	3.50	4.571	Opng.	385	183		568	750	
	0050	1065 mm x 1065 mm opening		3.50	4.571		695	183		878	1,100	
	0500	Double leaf, 1220 mm x 1220 mm opening		3	5.333		1,025	213		1,238	1,500	
	0550	1525 mm x 1525 mm opening		3	5.333		1,500	213		1,713	2,025	
350	0010	**FLOOR, INDUSTRIAL**										**350**
	0020	Steel 1465 kg/m² L.L., single leaf, 600 mm x 600 mm, 79 kg	2 Sswk	6	2.667	Opng.	565	107		672	810	
	0050	915 mm x 915 mm opening, 136 kg		5.50	2.909		780	116		896	1,075	
	0300	Double leaf, 1220 mm x 1220 mm opening, 206 kg		5	3.200		1,175	128		1,303	1,525	
	0350	1525 mm x 1525 mm opening, 293 kg		4.50	3.556		1,525	142		1,667	1,950	
	1000	Alum., 1465 kg/m² L.L., single leaf, 600 mm x 600 mm, 27 kg		6	2.667		555	107		662	800	
	1050	915 mm x 915 mm opening, 45 kg		5.50	2.909		865	116		981	1,150	
	1500	Double leaf, 1220 mm x 1220 mm opening, 73 kg		5	3.200		1,375	128		1,503	1,725	
	1550	1525 mm x 1525 mm opening, 107 kg		4.50	3.556		1,800	142		1,942	2,250	
	2000	Alum., 732 kg/m² L.L., single leaf, 610 mm x 610 mm, 27 kg		6	2.667		510	107		617	750	
	2050	915 mm x 915 mm opening, 43 kg		5.50	2.909		755	116		871	1,050	
	2500	Double leaf, 1220 mm x 1220 mm opening, 68 kg		5	3.200		1,200	128		1,328	1,550	
	2550	1525 mm x 1525 mm opening, 104 kg		4.50	3.556		1,625	142		1,767	2,050	
400	0010	**KENNEL DOORS**										**400**
	0020	2 way, swinging type, 330 mm x 485 mm opening	2 Carp	11	1.455	Opng.	207	51.50		258.50	310	
	0100	430 mm x 735 mm opening	"	11	1.455	"	223	51.50		274.50	325	

08320	Detention Doors and Frames	CREW	DAILY OUTPUT	LABOR-HOURS	UNIT	2006 BARE COSTS				TOTAL INCL O&P	
						MAT.	LABOR	EQUIP.	TOTAL		
900	0010	**SECURITY GATES** See Division 10610-100										**900**
950	0010	**VAULT FRONT** See Division 11021-800										**950**

08330 | Coiling Doors and Grilles

			CREW	DAILY OUTPUT	LABOR-HOURS	UNIT	MAT.	LABOR	EQUIP.	TOTAL	TOTAL INCL O&P	
130	0010	**COUNTER DOORS**										**130**
	0020	Manual, incl. frm and hdwe, galv stl, 1200 mm roll-up, 1800 mm long	2 Carp	2	8	Opng.	950	284		1,234	1,500	
	0300	Galvanized steel, UL label		1.80	8.889		1,050	315		1,365	1,650	
	0600	Stainless steel, 1200 mm high roll-up, 1800 mm long		2	8		1,450	284		1,734	2,050	
	0700	3000 mm long		1.80	8.889		2,100	315		2,415	2,800	
	2000	Aluminum, 1200 mm high, 1200 mm long		2.20	7.273		920	259		1,179	1,400	
	2020	1800 mm long		2	8		1,025	284		1,309	1,575	
	2040	2400 mm long		1.90	8.421		1,175	299		1,474	1,775	
	2060	3000 mm long		1.80	8.889		1,400	315		1,715	2,050	
	2080	4200 mm long		1.40	11.429		2,025	405		2,430	2,875	
	2100	1800 mm, 1200 mm long		2	8		1,025	284		1,309	1,575	
	2120	1800 mm long		1.60	10		1,200	355		1,555	1,875	
	2140	3000 mm long	▼	1.40	11.429	▼	1,625	405		2,030	2,425	
640	0010	**COILING GRILLE**										**640**
	0015	Aluminum, manual, incl. frame, mill finish										
	0020	Top coiling, 1200 mm high, 1200 mm long	2 Sswk	3.20	5	Opng.	1,175	200		1,375	1,650	
	0030	1800 mm long		3.20	5		1,225	200		1,425	1,700	
	0040	2400 mm long		2.40	6.667		1,400	266		1,666	2,025	
	0050	3600 mm long		2.40	6.667		1,650	266		1,916	2,275	
	0060	4800 mm long		1.60	10		2,050	400		2,450	2,975	
	0070	1800 mm high, 1200 mm long		3.20	5		1,250	200		1,450	1,725	
	0080	1800 mm long		3.20	5		1,300	200		1,500	1,800	
	0090	2400 mm long		2.40	6.667		1,475	266		1,741	2,100	
	0100	3600 mm long		1.60	10		1,725	400		2,125	2,625	
	0110	4800 mm long		1.20	13.333		2,200	535		2,735	3,375	
	0200	Side coiling, 2400 mm high, 3600 long		.60	26.667		1,850	1,075		2,925	3,950	
	0220	5400 mm long		.50	32		2,850	1,275		4,125	5,425	
	0240	7300 mm long		.40	40		2,825	1,600		4,425	6,000	
	0260	3600 mm high, 3600 mm long		.50	32		2,675	1,275		3,950	5,250	
	0280	5400 mm long		.40	40		3,950	1,600		5,550	7,225	
	0300	7300 mm long	▼	.28	57.143	▼	5,275	2,275		7,550	9,900	
700	0010	**ROLLING GRILLE SUPPORTS**										**700**
	0020	Rolling grille supports, overhead framed	E-4	10.97	2.916	m	60.50	118	8.15	186.65	287	
720	0010	**ROLLING SERVICE DOORS** Stl, manual, 20 ga., incl. hardware										**720**
	0050	2400 mm x 2400 mm high	2 Sswk	1.60	10	Ea.	810	400		1,210	1,600	
	0100	3000 mm x 3000 mm high		1.40	11.429		1,075	455		1,530	2,000	
	0200	6000 mm x 3000 mm high		1	16		2,250	640		2,890	3,625	
	0300	3600 mm x 3600 mm high		1.20	13.333		1,375	535		1,910	2,475	
	0400	6000 mm x 3600 mm high		.90	17.778		2,300	710		3,010	3,825	
	0500	4200 mm x 4200 mm high		.80	20		1,800	800		2,600	3,450	
	0600	6000 mm x 4800 mm high		.60	26.667		2,700	1,075		3,775	4,900	
	0700	3000 mm x 6000 mm high		.50	32		1,625	1,275		2,900	4,100	
	1000	3600 mm x 3600 mm, crank operated, crank on door side		.80	20		1,275	800		2,075	2,850	
	1100	Crank thru wall	▼	.70	22.857		1,425	915		2,340	3,200	
	1300	For vision panel, add				▼	232			232	255	
	1400	For 22 ga., deduct				m²	7.65			7.65	8.40	
	1600	900 mm x 2 m pass door within rolling stl door, new construction				Ea.	1,250			1,250	1,375	
	1700	Existing construction	2 Sswk	2	8		1,250	320		1,570	1,950	
	2000	Class A fire doors, manual, 20 ga., 2400 mm x 2400 mm high	▼	1.40	11.429	▼	1,100	455		1,555	2,025	

For expanded coverage of these items see *Means Interior Cost Data 2006*

08330 | Coiling Doors and Grilles

		CREW	DAILY OUTPUT	LABOR-HOURS	UNIT	2006 BARE COSTS MAT.	LABOR	EQUIP.	TOTAL	TOTAL INCL O&P	
720	2100	3000 mm x 3000 mm high	2 Sswk	1.10	14.545	Ea.	1,450	580		2,030	2,650	720
	2200	6000 mm x 3000 mm high		.80	20		2,900	800		3,700	4,650	
	2300	3600 mm x 3600 mm high		1	16		1,925	640		2,565	3,275	
	2400	6000 mm x 3600 mm high		.80	20		3,350	800		4,150	5,125	
	2500	4200 mm x 4200 mm high		.60	26.667		2,500	1,075		3,575	4,675	
	2600	6000 mm x 4800 mm high		.50	32		4,175	1,275		5,450	6,900	
	2700	3000 mm x 6000 mm high	↓	.40	40	↓	2,650	1,600		4,250	5,775	
	3000	For 18 ga. doors, add				m²	8.85			8.85	9.70	
	3300	For enamel finish, add				"	10.55			10.55	11.65	
	3600	For safety edge bottom bar, pneumatic, add				m	42			42	46.50	
	3700	Electric, add					79.50			79.50	87.50	
	4000	For weatherstripping, extruded rubber, jambs, add					27.50			27.50	30	
	4100	Hood, add					17.40			17.40	19.15	
	4200	Sill, add				↓	9.95			9.95	10.95	
	4500	Motor operators, to 4265 mm x 4265 mm opening	2 Sswk	5	3.200	Ea.	780	128		908	1,100	
	4600	Over 4265 mm x 4265 mm, jack shaft type	"	5	3.200		770	128		898	1,075	
	4700	For fire door, additional fusible link, add				↓	15.15			15.15	16.65	

08340 | Special Function Doors

		CREW	DAILY OUTPUT	LABOR-HOURS	UNIT	MAT.	LABOR	EQUIP.	TOTAL	TOTAL INCL O&P	
100	0010	**COLD STORAGE**										100
	0020	Single, 20 ga. galvanized steel										
	0300	Horizontal sliding, 1500 mm x 2100 mm, manual, 88.9 mm thk	2 Carp	2	8	Ea.	2,600	284		2,884	3,325	
	0400	100 mm thick		2	8		3,175	284		3,459	3,925	
	0500	150 mm thick		2	8		2,850	284		3,134	3,600	
	0800	1500 mm x 2100 mm, power operation, 50 mm thick		1.90	8.421		4,750	299		5,049	5,700	
	0900	100 mm thick		1.90	8.421		4,825	299		5,124	5,775	
	1000	150 mm thick		1.90	8.421		5,500	299		5,799	6,525	
	1300	2700 mm x 3050 mm, manual operation, 50 mm insulation		1.70	9.412		3,825	335		4,160	4,750	
	1400	100 mm insulation		1.70	9.412		3,950	335		4,285	4,875	
	1500	150 mm insulation		1.70	9.412		4,775	335		5,110	5,775	
	1800	Power operation, 50 mm insulation		1.60	10		6,600	355		6,955	7,800	
	1900	100 mm insulation		1.60	10		6,750	355		7,105	7,975	
	2000	150 mm insulation	↓	1.70	9.412	↓	7,650	335		7,985	8,950	
	2300	For stainless steel face, add					20%					
	3000	Hinged, lightweight, 900 mm x 2100 mm, galv. 1 face, 50 mm thick	2 Carp	2	8	Ea.	1,200	284		1,484	1,775	
	3050	100 mm thick		1.90	8.421		1,400	299		1,699	2,000	
	3300	Aluminum doors, 900 mm x 2100 mm, 100 mm thick		1.90	8.421		1,125	299		1,424	1,725	
	3350	150 mm thick		1.40	11.429		2,025	405		2,430	2,850	
	3600	Stainless steel, 900 mm x 2100 mm, 100 mm thick		1.90	8.421		1,450	299		1,749	2,075	
	3650	150 mm thick		1.40	11.429		2,425	405		2,830	3,300	
	3900	Painted, 900 mm x 2100 mm, 100 mm thick		1.90	8.421		1,050	299		1,349	1,625	
	3950	150 mm thick	↓	1.40	11.429	↓	1,975	405		2,380	2,800	
	5000	Bi-parting, electric operated										
	5010	1800 mm x 2400 mm opening, galv. faces, 100 mm T for cooler	2 Carp	.80	20	Opng.	6,200	710		6,910	7,925	
	5050	For freezer, 100 mm thick		.80	20		6,825	710		7,535	8,625	
	5300	For door buck framing and door protection, add		2.50	6.400		505	228		733	915	
	6000	Galvanized batten door, galvanized hinges, 1200 mm x 2100 mm		2	8		1,525	284		1,809	2,125	
	6050	1800 mm x 2400 mm		1.80	8.889		2,075	315		2,390	2,800	
	6500	Fire door, 3 hr., 1800 mm x 2400 mm, single slide		.80	20		7,175	710		7,885	8,975	
	6550	Double, bi-parting	↓	.70	22.857	↓	10,700	815		11,515	13,100	
200	0010	**HANGAR DOOR**										200
	0020	Bi-fold, ovhd., 20 kg/m² wind load, incl. elec. oper.										
	0100	Without sheeting, 3660 mm high x 12 m	2 Sswk	22.30	.718	m²	146	28.50		174.50	213	
	0200	4875 mm high x 18 m	↓	21.37	.749	↓	186	30		216	258	

Important: See the Reference Section for supporting data - Crews, Rental Equipment, City Cost Indexes and Reference Data

08340	Special Function Doors	CREW	DAILY OUTPUT	LABOR-HOURS	UNIT	2006 BARE COSTS				TOTAL INCL O&P	
						MAT.	LABOR	EQUIP.	TOTAL		
200	0300	6095 mm high x 24 m	2 Sswk	20.44	.783	m²	189	31.50		220.50	265	**200**
610	0010	**ACOUSTICAL DOORS**										**610**
	0020	Including framed seals, 915 mm x 2135 mm, wood, 27 STC rating	2 Carp	1.50	10.667	Ea.	385	380		765	1,000	
	0100	Steel, 40 STC rating		1.50	10.667		1,450	380		1,830	2,200	
	0200	45 STC rating		1.50	10.667		1,925	380		2,305	2,725	
	0300	48 STC rating		1.50	10.667		2,425	380		2,805	3,275	
	0400	52 STC rating	↓	1.50	10.667	↓	2,925	380		3,305	3,800	
700	0010	**DOUBLE ACTING, SWING**										**700**
	1000	2 mm aluminum, 2135 mm high, 1220 mm wide	2 Carp	4.20	3.810	Pr.	1,825	135		1,960	2,200	
	1050	2030 mm wide	"	4	4	"	2,400	142		2,542	2,875	
	2000	Solid core wood, 19 mm thick, metal frame, stainless steel										
	2010	base plate, 2135 mm high opening, 1220 mm wide	2 Carp	4	4	Pr.	2,175	142		2,317	2,625	
	2050	2135 mm wide	"	3.80	4.211	"	2,400	150		2,550	2,875	
710	0010	**GLASS DOOR, SWING**										**710**
	0020	Including hdwe, 13 mm thick, tempered, 900 mm x 2100 mm opening	2 Glaz	2	8	Opng.	1,800	274		2,074	2,400	
	0100	1800 mm x 2100 mm opening	"	1.40	11.429	"	3,500	390		3,890	4,450	
720	0010	**SHOCK ABSORBING DOORS**										**720**
	0020	Rigid, no frame, 38 mm thick, 1525 mm x 2135 mm	2 Sswk	1.90	8.421	Opng.	1,275	335		1,610	2,000	
	0100	2440 mm x 2440 mm		1.80	8.889		1,800	355		2,155	2,625	
	0500	Flexible, no frame, insul., 4 mm tk, econo. 1525 mm x 2135 mm		2	8		1,575	320		1,895	2,300	
	0600	Deluxe		1.90	8.421		2,375	335		2,710	3,200	
	1000	2440 mm x 2440 mm opening, economy		2	8		2,450	320		2,770	3,250	
	1100	Deluxe	↓	1.90	8.421	↓	3,150	335		3,485	4,050	
740	0010	**TUBULAR STEEL SWING DOORS**										**740**
	0020	Tubular steel, 2135 mm high, single, 1015 mm opening	2 Sswk	2.50	6.400	Ea.	580	256		836	1,100	
	0100	Double, 1830 mm opening	"	2	8	Pr.	735	320		1,055	1,375	

08360 | Overhead Doors

			CREW	DAILY OUTPUT	LABOR-HOURS	UNIT	MAT.	LABOR	EQUIP.	TOTAL	INCL O&P	
550	0010	**OVERHEAD, COMMERCIAL** Frames not included										**550**
	1000	Stock, sectional, H.D., wood, 45 mm T, 2400 mm x 2400 mm H	2 Carp	2	8	Ea.	605	284		889	1,100	
	1100	3000 mm x 3000 mm high		1.80	8.889		910	315		1,225	1,500	
	1200	3600 mm x 3600 mm high		1.50	10.667		1,325	380		1,705	2,050	
	1300	Chain hoist, 4200 mm x 4200 mm high		1.30	12.308		1,925	440		2,365	2,800	
	1400	3600 mm x 4800 mm high		1	16		1,700	570		2,270	2,750	
	1500	6000 mm x 2400 mm high		1.30	12.270		1,575	435		2,010	2,400	
	1600	6000 mm x 4800 mm high		.65	24.615		3,975	875		4,850	5,700	
	1800	Center mullion openings, 2400 mm high		4	4		690	142		832	980	
	1900	6000 mm high	↓	2	8	↓	1,300	284		1,584	1,875	
	2100	For medium duty custom door, deduct					5%	5%				
	2150	For medium duty stock doors, deduct					10%	5%				
	2300	Fiberglass and alum., heavy duty, sectional, 3600 mm x 3600 mm H	2 Carp	1.50	10.667	Ea.	1,925	380		2,305	2,700	
	2450	Chain hoist, 6000 mm x 6000 mm high		.50	32		4,775	1,150		5,925	7,025	
	2600	Steel, 24 ga. sectional, manual, 2400 mm x 2400 mm high		2	8		530	284		814	1,025	
	2650	3000 mm x 3000 mm high		1.80	8.889		735	315		1,050	1,300	
	2700	3600 mm x 3600 mm high		1.50	10.667		1,000	380		1,380	1,700	
	2800	Chain hoist, 6000 mm x 4200 mm high	↓	.70	22.857	↓	2,975	815		3,790	4,525	
	2850	For 32 mm rigid insulation and 26 ga. galv.										
	2860	back panel, add				m²	27			27	29.50	
	2900	For electric trolley operator, 250 W, to 3660 mm x 3660 mm, add	1 Carp	2	4	Ea.	700	142		842	990	
	2950	Over 3600 mm x 3600 mm, 373 W, add	"	1	8	"	785	284		1,069	1,300	
600	0010	**RESIDENTIAL GARAGE DOORS** Including hardware, no frame										**600**
	0050	Hinged, wood, custom, double door, 2700 mm x 2100 mm	2 Carp	4	4	Ea.	390	142		532	645	

For expanded coverage of these items see *Means Interior Cost Data 2006*

DOORS & WINDOWS **8**

08360 | Overhead Doors

		CREW	DAILY OUTPUT	LABOR-HOURS	UNIT	2006 BARE COSTS				TOTAL INCL O&P	
						MAT.	LABOR	EQUIP.	TOTAL		
600	0070	4800 mm x 2100 mm	2 Carp	3	5.333	Ea.	660	190		850	1,025	**600**
	0200	Overhead, sectional, fiberglass, 2700 mm x 2100 mm, standard		5.28	3.030		610	108		718	840	
	0220	Deluxe		5.28	3.030		775	108		883	1,025	
	0300	4800 mm x 2100 mm, standard		6	2.667		1,100	95		1,195	1,350	
	0320	Deluxe		6	2.667		1,350	95		1,445	1,650	
	0500	Hardboard, 2700 mm x 2100 mm, standard		8	2		405	71		476	555	
	0520	Deluxe		8	2		540	71		611	705	
	0600	4800 mm x 2100 mm, standard		6	2.667		790	95		885	1,025	
	0620	Deluxe		6	2.667		920	95		1,015	1,175	
	0700	Metal, 2700 mm x 2100 mm, standard		5.28	3.030		475	108		583	690	
	0720	Deluxe		8	2		635	71		706	810	
	0800	4800 mm x 2100 mm, standard		3	5.333		605	190		795	960	
	0820	Deluxe		6	2.667		965	95		1,060	1,225	
	0900	Wood, 2700 mm x 2100 mm, standard		8	2		500	71		571	660	
	0920	Deluxe		8	2		1,425	71		1,496	1,675	
	1000	4800 mm x 2100 mm, standard		6	2.667		1,000	95		1,095	1,250	
	1020	Deluxe		6	2.667		2,100	95		2,195	2,450	
	1800	Door hardware, sectional	1 Carp	4	2		220	71		291	355	
	1810	Door tracks only		4	2		102	71		173	223	
	1820	One side only		7	1.143		71	40.50		111.50	142	
	3000	Swing-up, fiberglass, 2700 mm x 2100 mm, standard	2 Carp	8	2		690	71		761	865	
	3020	Deluxe		8	2		720	71		791	905	
	3100	4800 mm x 2100 mm, standard		6	2.667		870	95		965	1,100	
	3120	Deluxe		6	2.667		895	95		990	1,125	
	3200	Hardboard, 2700 mm x 2100 mm, standard		8	2		315	71		386	455	
	3220	Deluxe		8	2		420	71		491	570	
	3300	4800 mm x 2100 mm, standard		6	2.667		440	95		535	635	
	3320	Deluxe		6	2.667		655	95		750	870	
	3400	Metal, 2700 mm x 2100 mm, standard		8	2		345	71		416	490	
	3420	Deluxe		8	2		610	71		681	780	
	3500	4800 mm x 2100 mm, standard		6	2.667		540	95		635	745	
	3520	Deluxe		6	2.667		870	95		965	1,100	
	3600	Wood, 2700 mm x 2100 mm, standard		8	2		380	71		451	525	
	3620	Deluxe		8	2		665	71		736	840	
	3700	4800 mm x 2100 mm, standard		6	2.667		655	95		750	870	
	3720	Deluxe		6	2.667		925	95		1,020	1,175	
	3900	Door hardware only, swing up	1 Carp	4	2		109	71		180	231	
	3920	One side only		7	1.143		61.50	40.50		102	131	
	4000	For electric operator, economy, add		8	1		287	35.50		322.50	370	
	4100	Deluxe, including remote control		8	1		420	35.50		455.50	515	
	4500	For transmitter/receiver control , add to operator				Total	86.50			86.50	95.50	
	4600	Transmitters, additional				"	31.50			31.50	35	
800	0010	**TELESCOPING STEEL DOORS** Baked enamel finish										**800**
	1000	Overhead, 0.75 mm thick, electric operated, 3000 mm x 3000 mm	E-3	.80	30	Ea.	10,100	1,225	112	11,437	13,400	
	2000	6000 mm x 3000 mm		.60	40		13,400	1,625	149	15,174	17,800	
	3000	6000 mm x 4800 mm		.40	60		16,300	2,425	223	18,948	22,600	

08370 | Vertical Lift Doors

		CREW	DAILY OUTPUT	LABOR-HOURS	UNIT	MAT.	LABOR	EQUIP.	TOTAL	TOTAL INCL O&P	
950	0010	**VERTICAL LIFT DOORS**									**950**
	0020	Motorized, 14 ga. stl, incl. frm and ctrl pnl										
	0050	4800 mm x 4800 mm high	L-10	.50	48	Ea.	17,600	1,925	1,275	20,800	24,100	
	0100	3000 mm x 6000 mm high		1.30	18.462		23,900	740	490	25,130	28,100	
	0120	4500 mm x 6000 mm high		1.30	18.462		29,300	740	490	30,530	34,000	
	0140	6000 mm x 6000 mm high		1	24		34,400	960	635	35,995	40,200	
	0160	7500 mm x 6000 mm high		1	24		38,100	960	635	39,695	44,300	
	0170	9700 mm x 7300 mm high		.75	32		35,100	1,275	850	37,225	41,700	

Important: See the Reference Section for supporting data - Crews, Rental Equipment, City Cost Indexes and Reference Data

08300 | Specialty Doors

08370 | Vertical Lift Doors

			CREW	DAILY OUTPUT	LABOR-HOURS	UNIT	MAT.	LABOR	EQUIP.	TOTAL	TOTAL INCL O&P	
950	0180	6000 mm x 7500 mm high	L-10	1	24	Ea.	39,400	960	635	40,995	45,700	**950**
	0200	7500 mm x 7500 mm high		.70	34.286		44,400	1,375	910	46,685	52,000	
	0220	7500 mm x 9100 mm high		.70	34.286		52,000	1,375	910	54,285	60,500	
	0240	9100 mm x 9100 mm high		.70	34.286		59,500	1,375	910	61,785	69,000	
	0260	10.5 m x 9.1 m high	▼	.70	34.286	▼	65,500	1,375	910	67,785	75,500	

08380 | Traffic Doors

			CREW	DAILY OUTPUT	LABOR-HOURS	UNIT	MAT.	LABOR	EQUIP.	TOTAL	TOTAL INCL O&P	
480	0010	**FLEXIBLE TRANSPARENT STRIP ENTRANCE**										**480**
	0100	305 mm strip width, 2/3 overlap	3 Shee	12.54	1.914	m² Surf	56	80.50		136.50	186	
	0200	Full overlap		10.68	2.246		58.50	94.50		153	211	
	0220	203 mm strip width, 1/2 overlap		13.01	1.845		54.50	78		132.50	180	
	0240	Full overlap	▼	11.15	2.153	▼	71.50	90.50		162	219	
	0300	Add for suspension system, header mount				m	25			25	27.50	
	0400	Wall mount				"	25.50			25.50	28	

08400 | Entrances and Storefronts

08410 | Metal-Framed Storefronts

			CREW	DAILY OUTPUT	LABOR-HOURS	UNIT	MAT.	LABOR	EQUIP.	TOTAL	TOTAL INCL O&P	
110	0010	**ALUMINUM-FRAMED STOREFRONTS,** No glazing										**110**
	0020	Entrance, 915 mm x 2135 mm opening, clear anodized finish	2 Sswk	7	2.286	Opng.	254	91.50		345.50	445	
	0040	Bronze finish		7	2.286		310	91.50		401.50	505	
	0060	Black finish		7	2.286		350	91.50		441.50	550	
	0500	1830 mm x 2135 mm opening, clear finish		6	2.667		310	107		417	530	
	0520	Bronze finish		6	2.667		350	107		457	575	
	0540	Black finish		6	2.667		415	107		522	645	
	1000	With 915 mm H transoms, 915 mm x 3050 mm opng, clear finish		6.50	2.462		370	98.50		468.50	585	
	1050	Bronze finish		6.50	2.462		410	98.50		508.50	625	
	1100	Black finish		6.50	2.462		485	98.50		583.50	705	
	1500	With 915 mm H transoms, 1830 x 3050 mm opening, clr finish		5.50	2.909		450	116		566	705	
	1550	Bronze finish		5.50	2.909		485	116		601	740	
	1600	Black finish	▼	5.50	2.909	▼	575	116		691	840	
120	0010	**ALUMINUM DOORS** Commercial entrance, no glazing										**120**
	0020	Incl. hinges, push/pull, deadlock, cyl., threshold										
	0800	Narrow stile, no glazing, std hardware, pair of 750 mm x 2100 mm	2 Carp	1.70	9.412	Pr.	770	335		1,105	1,375	
	1000	900 mm x 2100 mm, single		3	5.333	Ea.	510	190		700	855	
	1200	Pair of 900 mm x 2100 mm		1.70	9.412	Pr.	1,025	335		1,360	1,650	
	1500	1050 mm x 2100 mm, single		3	5.333	Ea.	545	190		735	895	
	2000	Medium stile, pair of 750 mm x 2100 mm		1.70	9.412	Pr.	955	335		1,290	1,575	
	2100	900 mm x 2100 mm, single		3	5.333	Ea.	670	190		860	1,025	
	2200	Pair of 900 mm x 2100 mm		1.70	9.412	Pr.	1,300	335		1,635	1,975	
	2300	1050 mm x 2100 mm	▼	5.33	3.002	Ea.	790	107		897	1,025	
	5000	Flush panel doors, pair of 750 mm x 2100 mm	2 Sswk	2	8	Pr.	945	320		1,265	1,625	
	5050	900 mm x 2100 mm, single		2.50	6.400	Ea.	470	256		726	980	
	5100	Pair of 900 mm x 2100 mm		2	8	Pr.	945	320		1,265	1,625	
	5150	1050 mm x 2100 mm, single	▼	2.50	6.400	Ea.	560	256		816	1,075	
130	0010	**ALUMINUM DOORS & FRAMES** Entrance, narrow stile, including										**130**
	0015	Standard hardware, clr fin, not incl. glass, 750 mm x 2100 mm opng	2 Sswk	2	8	Ea.	695	320		1,015	1,325	

For expanded coverage of these items see *Means Interior Cost Data 2006*

		08410	Metal-Framed Storefronts	CREW	DAILY OUTPUT	LABOR-HOURS	UNIT	2006 BARE COSTS				TOTAL INCL O&P	
								MAT.	LABOR	EQUIP.	TOTAL		
130	0020		900 mm x 2100 mm opening	2 Sswk	2	8	Ea.	560	320		880	1,200	130
	0030		1050 mm x 2100 mm opening		2	8		500	320		820	1,125	
	0100		900 mm x 3000 mm opening, 900 mm high transom		1.80	8.889		785	355		1,140	1,500	
	0200		1050 mm x 3000 mm opening, 900 mm high transom		1.80	8.889		770	355		1,125	1,500	
	0280		1500 mm x 2100 opening		2	8	▼	850	320		1,170	1,525	
	0300		1800 mm x 2100 mm opening		1.30	12.308	Pr.	865	490		1,355	1,825	
	0400		1800 mm x 3000 mm opening, 900 mm high transom		1.10	14.545		730	580		1,310	1,850	
	0420		2100 mm x 2100 mm opening		1	16		830	640		1,470	2,075	
	0500		Wide stile, 750 mm x 2100 mm opening		2	8	Ea.	680	320		1,000	1,325	
	0520		900 mm x 2100 mm opening		2	8		695	320		1,015	1,350	
	0540		1050 mm x 2100 mm opening		2	8		740	320		1,060	1,400	
	0560		1500 mm x 2100 mm opening		2	8	▼	1,200	320		1,520	1,900	
	0580		1800 mm x 2100 mm opening		1.30	12.308	Pr.	1,150	490		1,640	2,150	
	0600		2100 mm x 2100 mm opening	▼	1	16	"	1,250	640		1,890	2,525	
	1100		For full vision doors, with 13 mm glass, add				Leaf	55%					
	1200		For non-standard size, add					67%					
	1300		Light bronze finish, add					36%					
	1400		Dark bronze finish, add					18%					
	1500		For black finish, add					36%					
	1600		Concealed panic device, add				▼	1,025			1,025	1,125	
	1700		Electric striker release, add				Opng.	262			262	288	
	1800		Floor check, add				Leaf	775			775	855	
	1900		Concealed closer, add				"	515			515	570	
	2000		Flush 900 X 2100 mm Insulated, 300 x 300 mm lite, clear finish	2 Sswk	2	8	Ea.	975	320		1,295	1,650	
140	0010		**STOREFRONT SYSTEMS** Aluminum frame, clear 10 mm plate glass,										140
	0020		incl. 915 mm x 2135 mm door with hardware (3715 m² max. wall)										
	0500		Wall height to 3660 mm high, commercial grade	2 Glaz	13.94	1.148	m²	149	39.50		188.50	224	
	0600		Institutional grade		12.08	1.325		187	45.50		232.50	275	
	0700		Monumental grade		10.68	1.498		283	51.50		334.50	390	
	1000		1830 mm x 2135 mm door with hardware, commercial grade		12.54	1.276		152	43.50		195.50	233	
	1100		Institutional grade		10.68	1.498		208	51.50		259.50	305	
	1200		Monumental grade	▼	9.29	1.722		385	59		444	515	
	1500		For bronze anodized finish, add					15%					
	1600		For black anodized finish, add					30%					
	1700		For stainless steel framing, add to monumental				▼	75%					
300	0010		**STAINLESS STEEL AND GLASS** Entrance unit, narrow stiles										300
	0020		915 mm x 2135 mm opening, including hardware, minimum	2 Sswk	1.60	10	Opng.	4,825	400		5,225	6,025	
	0050		Average		1.40	11.429		5,225	455		5,680	6,575	
	0100		Maximum	▼	1.20	13.333		5,600	535		6,135	7,100	
	1000		For solid bronze entrance units, statuary finish, add					60%					
	1100		Without statuary finish, add				▼	45%					
	2000		Balanced doors, 915 mm x 2135 mm, economy	2 Sswk	.90	17.778	Ea.	6,550	710		7,260	8,475	
	2100		Premium	"	.70	22.857	"	11,300	915		12,215	14,100	

		08460	Automatic Entrance Doors										
600	0010		**SLIDING ENTRANCE** 3.7 m x 2.3 m, 1.5 m x 2.1 m door, 2 way										600
	0020		mat activated, panic pushout, incl. operator & hardware,										
	0030		not including glass or glazing	2 Glaz	.70	22.857	Opng.	6,200	785		6,985	8,000	

		08470	Revolving Entrance Doors										
600	0010		**REVOLVING DOORS** Aluminum, 1950 mm to 2100 mm diameter,										600
	0020		2050 mm to 2100 mm high, stock units, minimum	4 Sswk	.75	42.667	Opng.	16,400	1,700		18,100	21,100	
	0050		Average		.60	53.333		20,200	2,125		22,325	26,000	
	0100		Maximum	▼	.45	71.111	▼	26,600	2,850		29,450	34,400	

08400 | Entrances and Storefronts

08470 | Revolving Entrance Doors

			CREW	DAILY OUTPUT	LABOR-HOURS	UNIT	2006 BARE COSTS MAT.	LABOR	EQUIP.	TOTAL	TOTAL INCL O&P	
600	1000	Stainless steel	4 Sswk	.30	105	Opng.	30,200	4,225		34,425	40,800	600
	1100	Solid bronze	↓	.15	213		35,300	8,525		43,825	54,000	
	1500	For automatic controls, add	2 Elec	2	8	↓	12,300	335		12,635	14,100	

08480 | Balanced Entrance Doors

			CREW	DAILY OUTPUT	LABOR-HOURS	UNIT	MAT.	LABOR	EQUIP.	TOTAL	TOTAL INCL O&P	
150	0010	**BALANCED DOORS**										150
	0020	Hardware & frame, alum. & glass, 0.9 m x 2 m, econ.	2 Sswk	.90	17.778	Ea.	5,300	710		6,010	7,100	
	0150	Premium	"	.70	22.857	"	6,600	915		7,515	8,900	

08490 | Sliding Storefronts

			CREW	DAILY OUTPUT	LABOR-HOURS	UNIT	MAT.	LABOR	EQUIP.	TOTAL	TOTAL INCL O&P	
100	0010	**SLIDING PANELS**										100
	0020	Mall fronts, aluminum & glass, 4570 mm x 2745 mm high	2 Glaz	1.30	12.308	Opng.	2,425	420		2,845	3,325	
	0100	7315 mm x 2745 mm high		.70	22.857		3,525	785		4,310	5,050	
	0200	14 630 mm x 2745 mm high, with fixed panels	↓	.90	17.778		6,575	610		7,185	8,150	
	0500	For bronze finish, add				↓	17%					

08500 | Windows

08510 | Steel Windows

			CREW	DAILY OUTPUT	LABOR-HOURS	UNIT	2006 BARE COSTS MAT.	LABOR	EQUIP.	TOTAL	TOTAL INCL O&P	
700	0010	**SCREENS**										700
	0020	For metal sash, aluminum or bronze mesh, flat screen	2 Sswk	111	.144	m²	39	5.75		44.75	53	
	0500	Wicket screen, inside window		92.90	.172		59.50	6.90		66.40	77.50	
	0800	Security screen, aluminum frame with stainless steel cloth		111	.144		209	5.75		214.75	240	
	0900	Steel grate, painted, on steel frame		149	.107		112	4.29		116.29	132	
	1000	For solar louvers, add	↓	14.86	1.076	↓	217	43		260	315	
	4000	See also Division 05580-900										
750	0010	**STEEL SASH** Custom units, glazing and trim not included, [R085123-10]										750
	0100	Casement, 100% vented	2 Sswk	18.58	.861	m²	450	34.50		484.50	555	
	0200	50% vented		18.58	.861		410	34.50		444.50	510	
	0300	Fixed		18.58	.861		282	34.50		316.50	370	
	1000	Projected, commercial, 40% vented		18.58	.861		470	34.50		504.50	580	
	1100	Intermediate, 50% vented		18.58	.861		515	34.50		549.50	625	
	1500	Industrial, horizontally pivoted		18.58	.861		490	34.50		524.50	600	
	1600	Fixed		18.58	.861		279	34.50		313.50	365	
	2000	Industrial security sash, 50% vented		18.58	.861		525	34.50		559.50	640	
	2100	Fixed		18.58	.861		430	34.50		464.50	530	
	2500	Picture window		18.58	.861		265	34.50		299.50	355	
	3000	Double hung		18.58	.861	↓	505	34.50		539.50	620	
	5000	Mullions for above, open interior face		73.15	.219	m	27.50	8.75		36.25	46.50	
	5100	With interior cover		73.15	.219	"	45.50	8.75		54.25	66.50	
770	0010	**STEEL WINDOWS** Stock, including frame, trim and insul. glass [R085123-10]										770
	0020	See also Division 13128-700										
	1000	Custom units, double hung, 815 mm x 1370 mm opening	2 Sswk	12	1.333	Ea.	595	53.50		648.50	750	
	1100	710 mm x 1145 mm opening		12	1.333		490	53.50		543.50	635	
	1500	Commercial projected, 1145 mm x 1650 mm opening		10	1.600		1,050	64		1,114	1,275	
	1600	2055 mm x 145 mm opening		7	2.286		1,375	91.50		1,466.50	1,700	

DOORS & WINDOWS | 8

		08510	Steel Windows		CREW	DAILY OUTPUT	LABOR-HOURS	UNIT	2006 BARE COSTS				TOTAL INCL O&P	
									MAT.	LABOR	EQUIP.	TOTAL		
770	2000		Intermediate projected, 840 mm x 145 mm opening	R085123 -10	2 Sswk	12	1.333	Ea.	585	53.50		638.50	735	770
	2100		145 mm x 1650 mm opening		↓	10	1.600	↓	1,175	64		1,239	1,425	

		08520	Aluminum Windows											
100	0010	**ALUMINUM SASH**												100
	0020		Stock, grade C, glaze & trim not incl., casement		2 Sswk	18.58	.861	m²	315	34.50		349.50	405	
	0050		Double hung			18.58	.861		320	34.50		354.50	410	
	0100		Fixed casement			18.58	.861		126	34.50		160.50	201	
	0150		Picture window			18.58	.861		135	34.50		169.50	211	
	0200		Projected window			18.58	.861		290	34.50		324.50	380	
	0250		Single hung			18.58	.861		149	34.50		183.50	226	
	0300		Sliding			18.58	.861	↓	192	34.50		226.50	274	
	1000		Mullions for above, tubular			73.15	.219	m	15.10	8.75		23.85	32.50	
	2000		Custom aluminum sash, grade hc, glazing not included, minimum			18.58	.861	m²	360	34.50		394.50	455	
	2100		Maximum		↓	7.90	2.026	"	465	81		546	660	
120	0010	**ALUMINUM WINDOWS** Incl. frame and glazing, Commercial grade												120
	0020		See also Division 05120-440											
	1000		Stock units, casement, 115 mm x 965 mm opening		2 Sswk	10	1.600	Ea.	310	64		374	455	
	1050		Add for storms						62.50			62.50	68.50	
	1600		Projected, with screen, 115 mm x 965 mm opening		2 Sswk	10	1.600		222	64		286	360	
	1700		Add for storms						59			59	64.50	
	2000		1345 mm x 1600 mm opening		2 Sswk	8	2		310	80		390	490	
	2100		Add for storms						81.50			81.50	90	
	2500		Enamel finish windows, 115 mm x 965 mm		2 Sswk	10	1.600		199	64		263	335	
	2600		1345 mm x 1600 mm			8	2		298	80		378	475	
	3000		Single hung, 610 x 915 mm opening, enameled, stnd glazed			10	1.600		149	64		213	279	
	3100		Insulating glass			10	1.600		181	64		245	315	
	3300		815 mm x 2030 mm opening, standard glazed			8	2		315	80		395	495	
	3400		Insulating glass			8	2		405	80		485	595	
	3700		1015 mm x 1525 mm opening, standard glazed			9	1.778		204	71		275	350	
	3800		Insulating glass			9	1.778		287	71		358	445	
	3890		Awning type, 915 mm x 915 mm opening standard glass			14	1.143		365	45.50		410.50	485	
	3900		Insulating glass			14	1.143		390	45.50		435.50	510	
	3910		915 mm x 1220 mm opening, standard glass			10	1.600		430	64		494	585	
	3920		Insulating glass			10	1.600		495	64		559	655	
	3930		915 mm x 1625 mm opening, standard glass			10	1.600		515	64		579	685	
	3940		Insulating glass			10	1.600		610	64		674	785	
	3950		1220 mm x 1625 mm opening, standard glass			9	1.778		565	71		636	755	
	3960		Insulating glass			9	1.778		680	71		751	875	
	4000		Sliding aluminum, 915 mm x 610 mm opening, standard glazed			10	1.600		168	64		232	300	
	4100		Insulating glass			10	1.600		186	64		250	320	
	4300		1525 mm x 915 mm opening, standard glazed			9	1.778		213	71		284	360	
	4400		Insulating glass			9	1.778		298	71		369	460	
	4600		2440 mm x 1220 mm opening, standard glazed			6	2.667		305	107		412	525	
	4700		Insulating glass			6	2.667		495	107		602	730	
	5000		2745 mm x 1525 mm opening, standard glazed			4	4		465	160		625	800	
	5100		Insulating glass			4	4		740	160		900	1,100	
	5500		Sliding, w/ thermal barrier and screen, 1830 x 1220 mm, 2 track			8	2		630	80		710	840	
	5700		4 track		↓	8	2		770	80		850	995	
	6000		For above units with bronze finish, add						12%					
	6200		For installation in concrete openings, add					↓	5%					
500	0010	**JALOUSIES**												500
	0020		Aluminum incl. glazing & screens, stock, 485 mm x 965 mm		2 Sswk	10	1.600	Ea.	136	64		200	265	
	0100		685 mm x 1220 mm			10	1.600		194	64		258	330	
	0200		115 mm x 610 mm		↓	10	1.600	↓	153	64		217	284	

08520 | Aluminum Windows

		DAILY	LABOR-			2006 BARE COSTS			TOTAL	
		CREW	OUTPUT	HOURS	UNIT	MAT.	LABOR	EQUIP.	TOTAL	INCL O&P	
500	0300	115 mm x 1600 mm	2 Sswk	10	1.600	Ea.	277	64		341	420	**500**
	1000	Mullions for above, 610 mm long		80	.200		10.45	8		18.45	26	
	1100	1600 mm long	▼	80	.200	▼	17.90	8		25.90	34	

08550 | Wood Windows

			CREW	DAILY OUTPUT	LABOR-HOURS	UNIT	MAT.	LABOR	EQUIP.	TOTAL	INCL O&P	
100	0010	**AWNING WINDOW** Including frame, screens and grills										**100**
	0100	Average quality,builders model,850 mm x 550 mm,dbl insulated glass	1 Carp	10	.800	Ea.	233	28.50		261.50	300	
	0200	Low E glass		10	.800		246	28.50		274.50	315	
	0300	1000 mm x 700 mm, double insulated glass		9	.889		294	31.50		325.50	375	
	0400	Low E Glass		9	.889		310	31.50		341.50	395	
	0500	1200 mm x 900 mm, double insulated glass		8	1		430	35.50		465.50	525	
	0600	Low E glass		8	1		450	35.50		485.50	550	
	1000	850 mm x 550 mm		10	.800		247	28.50		275.50	315	
	1100	1000 mm x 550 mm		10	.800		270	28.50		298.50	340	
	1200	900 mm x 700 mm		9	.889		287	31.50		318.50	365	
	1300	900 mm x 900 mm		9	.889		320	31.50		351.50	405	
	1400	1200 mm x 700 mm		8	1		345	35.50		380.50	435	
	1500	1500 mm x 900 mm		8	1		500	35.50		535.50	605	
	2000	Metal clad, deluxe, double insulated glass, 850 mm x 550 mm		10	.800		231	28.50		259.50	299	
	2100	1000 mm x 550 mm		10	.800		271	28.50		299.50	345	
	2200	900 mm x 625 mm		9	.889		251	31.50		282.50	325	
	2300	1000 mm x 750 mm		9	.889		315	31.50		346.50	395	
	2400	1200 mm x 700 mm		8	1		320	35.50		355.50	405	
	2500	1500 mm x 900 mm	▼	8	1	▼	340	35.50		375.50	430	
150	0010	**BOW-BAY WINDOW** Including frame, screens and grills,										**150**
	0020	end panels operable										
	1000	Bow,casement,wood,bldrs mdl,2400 x 1500mm dbl insul gls, 4 panel	2 Carp	10	1.600	Ea.	1,500	57		1,557	1,750	
	1050	Low E glass		10	1.600		1,200	57		1,257	1,425	
	1100	3000 mm x 1500 mm, double insulated glass, 6 panels		6	2.667		1,250	95		1,345	1,525	
	1200	Low E glass, 6 panels		6	2.667		1,325	95		1,420	1,600	
	1300	Vinyl clad,bldrs model,dbl insulated glass,1800 x 1200mm,3 panel		10	1.600		1,375	57		1,432	1,600	
	1340	2700 mm x 1200 mm, 4 panel		8	2		1,375	71		1,446	1,600	
	1380	3000 mm x 1800 mm, 5 panels		7	2.286		2,200	81.50		2,281.50	2,550	
	1420	3600 mm x 1800 mm, 6 panels		6	2.667		2,775	95		2,870	3,200	
	1600	Metal clad,casement,bldrs mdl,1800 x 1200mm,dbl insltd gls,3 panels		10	1.600		855	57		912	1,025	
	1640	2700 mm x 1200 mm , 4 panels		8	2		1,200	71		1,271	1,425	
	1680	3050 mm x 1525 mm , 5 panels		7	2.286		1,650	81.50		1,731.50	1,950	
	1720	3000 mm x 1800 mm , 6 panels		6	2.667		2,300	95		2,395	2,675	
	2000	Bay window,casement,bldrs mdl,2400 x 1525 mm,dbl insul gls,4 panels		10	1.600		1,675	57		1,732	1,925	
	2050	Low E glass,		10	1.600		2,025	57		2,082	2,325	
	2100	3600 mm x 1800 mm, Double insulated glass, 6 panels		6	2.667		2,075	95		2,170	2,450	
	2200	Low E glass		6	2.667		2,050	95		2,145	2,400	
	2280	1829 mm x 1219 mm		11	1.455		1,075	51.50		1,126.50	1,275	
	2300	Vinyl clad, premium, double insulated glass, 2440 mm x 1525 mm		10	1.600		1,275	57		1,332	1,500	
	2340	3050 mm x 1525 mm		8	2		1,800	71		1,871	2,100	
	2380	3050 mm x 1830 mm		7	2.286		1,900	81.50		1,981.50	2,200	
	2420	3660 mm x 1830 mm		6	2.667		2,250	95		2,345	2,625	
	2600	Metal cld, deluxe, dbl insul. glass, 2440 x 1525 mm H, 4 panels		10	1.600		1,475	57		1,532	1,725	
	2640	3050 mm x 1525 mm high, 5 panels		8	2		1,575	71		1,646	1,850	
	2680	3050 mm x 1830 mm high, 5 panels		7	2.286		1,875	81.50		1,956.50	2,175	
	2720	3660 mm x 1830 mm high, 6 panels		6	2.667		2,600	95		2,695	3,000	
	3000	Dbl hung, bldrs. model, bay, 2440 x 1220 mm H, dbl insulated glass		10	1.600		1,175	57		1,232	1,375	
	3050	Low E glass		10	1.600		1,250	57		1,307	1,475	
	3100	2745 mm x 1525 mm high, double insulated glass	▼	6	2.667		1,250	95		1,345	1,550	

DOORS & WINDOWS 8

08550 | Wood Windows

			CREW	DAILY OUTPUT	LABOR-HOURS	UNIT	2006 BARE COSTS				TOTAL INCL O&P	
							MAT.	LABOR	EQUIP.	TOTAL		
150	3200	Low E glass	2 Carp	6	2.667	Ea.	1,325	95		1,420	1,625	**150**
	3300	Vinyl clad, premium, double insulated glass, 2135 mm x 1370 mm		10	1.600		1,225	57		1,282	1,425	
	3340	2440 mm x 1370 mm		8	2		1,250	71		1,321	1,475	
	3380	2440 mm x 1525 mm		7	2.286		1,300	81.50		1,381.50	1,550	
	3420	2745 mm x 1525 mm		6	2.667		1,325	95		1,420	1,625	
	3600	Metal clad, deluxe, dbl insul. glass, 2135 mm x 1220 mm high		10	1.600		1,125	57		1,182	1,325	
	3640	2440 mm x 1220 mm high		8	2		1,175	71		1,246	1,375	
	3680	2440 mm x 1525 mm high		7	2.286		1,200	81.50		1,281.50	1,450	
	3720	2745 mm x 1525 mm high	▼	6	2.667	▼	1,275	95		1,370	1,550	
200	0010	**CASEMENT WINDOW** Including frame, screen, and grills										**200**
	0100	Avg. quality, bldrs. model, 600 mm x 900 mm H, dbl. insulated glass	1 Carp	10	.800	Ea.	186	28.50		214.50	250	
	0150	Low E glass		10	.800		225	28.50		253.50	293	
	0200	600 mm x 1350 mm high, double insulated glass		9	.889		242	31.50		273.50	315	
	0250	Low E glass		9	.889		320	31.50		351.50	405	
	0300	711 mm x 1800 mm high, double insulated glass		8	1		355	35.50		390.50	445	
	0350	Low E glass		8	1		385	35.50		420.50	480	
	0522	Vinyl clad, premium, double insulated glass, 600 mm x 900 mm		10	.800		264	28.50		292.50	335	
	0524	600 mm x 1200 mm		9	.889		310	31.50		341.50	390	
	0525	600 mm x 1500 mm		8	1		355	35.50		390.50	445	
	0528	600 mm x 1800 mm		8	1		400	35.50		435.50	495	
	8100	Metal clad, deluxe, dbl. insul. glass, 600 mm x 900 mm high		10	.800		206	28.50		234.50	272	
	8120	600 mm x 1200 mm high		9	.889		248	31.50		279.50	320	
	8140	600 mm x 1500 mm high		8	1		282	35.50		317.50	365	
	8160	600 mm x 1800 mm high	▼	8	1	▼	325	35.50		360.50	410	
	8200	For multiple leaf units, deduct for stationary sash											
	8220	600 mm high				Ea.	21.50			21.50	23.50	
	8240	1300 mm high					24.50			24.50	27	
	8260	1800 mm high					33			33	36.50	
	8300	For installation, add/leaf				▼		15%				
250	0010	**DOUBLE HUNG** Including frame, screens, and grills										**250**
	0100	Avg. qual, bldrs. model, 610 mm x 915 mm H, dbl insul. glass	1 Carp	10	.800	Ea.	204	28.50		232.50	270	
	0150	Low E glass		10	.800		186	28.50		214.50	250	
	0200	915 mm x 1220 mm high, double insulated glass		9	.889		266	31.50		297.50	340	
	0250	Low E glass		9	.889		249	31.50		280.50	325	
	0300	1220 mm x 1370 mm high, double insulated glass		8	1		310	35.50		345.50	395	
	0350	Low E glass		8	1		335	35.50		370.50	420	
	1000	Vinyl clad, premium, double insulated glass, 760 mm x 915 mm		10	.800		218	28.50		246.50	285	
	1100	915 mm x 1065 mm		10	.800		256	28.50		284.50	325	
	1200	915 mm x 1220 mm		9	.889		350	31.50		381.50	435	
	1300	915 mm x 1370 mm		9	.889		296	31.50		327.50	375	
	1400	915 mm x 1525 mm		8	1		310	35.50		345.50	395	
	1500	1065 mm x 1830 mm		8	1		360	35.50		395.50	455	
	2000	Metal clad, deluxe, dbl. insul. glass, 762 mm x 914 mm high		10	.800		219	28.50		247.50	285	
	2100	914 mm x 1067 mm high		10	.800		259	28.50		287.50	330	
	2200	915 mm x 1220 mm high		9	.889		275	31.50		306.50	355	
	2300	915 mm x 1370 mm high		9	.889		298	31.50		329.50	375	
	2400	915 mm x 1525 mm high		8	1		320	35.50		355.50	405	
	2500	1065 mm x 1830 mm high	▼	8	1	▼	385	35.50		420.50	480	
650	0010	**PALLADIAN WINDOWS**										**650**
	0020	Vinyl clad, double insulated glass, including frame and grills										
	0040	966 mm x 762 mm high	2 Carp	11	1.455	Ea.	1,200	51.50		1,251.50	1,400	
	0060	966 mm x 1472 mm		11	1.455		1,450	51.50		1,501.50	1,675	
	0080	966 mm x 1930 mm		10	1.600		1,625	57		1,682	1,900	
	0100	1220 mm x 1220 mm	▼	10	1.600	▼	1,450	57		1,507	1,700	

Note at row 0200/0250 (Double Hung): R085216 -10

Side tab: 8 DOORS & WINDOWS

08550	Wood Windows	CREW	DAILY OUTPUT	LABOR-HOURS	UNIT	2006 BARE COSTS				TOTAL INCL O&P	
						MAT.	LABOR	EQUIP.	TOTAL		
650											**650**
0120	1220 mm x 1625 mm	3 Carp	10	2.400	Ea.	1,750	85.50		1,835.50	2,050	
0140	1220 mm x 1830 mm		9	2.667		1,950	95		2,045	2,300	
0160	1220 mm x 2234 mm		9	2.667		1,950	95		2,045	2,300	
0180	1707 mm x 1472 mm		9	2.667		2,050	95		2,145	2,425	
0200	1707 mm x 2082 mm		9	2.667		2,375	95		2,470	2,750	
0220	1707 mm x 2362 mm		9	2.667		2,575	95		2,670	2,975	
0240	1830 mm x 2414 mm		8	3		3,225	107		3,332	3,725	
0260	2438 mm x 1830 mm	▼	8	3	▼	2,750	107		2,857	3,200	
670	0010	**PICTURE WINDOW** Including frame and grills									**670**
0100	Ave. quality, bldrs. model, 1065 mm x 1220 mm H, dbl insulated glass	2 Carp	12	1.333	Ea.	284	47.50		331.50	385	
0150	Low E glass		12	1.333		310	47.50		357.50	415	
0200	1220 mm x 1370 mm high, double insulated glass		11	1.455		315	51.50		366.50	425	
0250	Low E glass		11	1.455		345	51.50		396.50	460	
0300	1525 mm x 1220 mm high, double insulated glass		11	1.455		390	51.50		441.50	505	
0350	Low E glass		11	1.455		435	51.50		486.50	560	
0400	1830 mm x 1370 mm high, double insulated glass		10	1.600		495	57		552	635	
0450	Low E glass		10	1.600		560	57		617	705	
1000	Vinyl clad, premium, dbl. insul. glass, 1220 mm x 1220 mm		12	1.333		455	47.50		502.50	575	
1100	1220 mm x 1830 mm		11	1.455		675	51.50		726.50	820	
1200	1525 mm x 1830 mm		10	1.600		970	57		1,027	1,175	
1300	1830 mm x 1830 mm		10	1.600		990	57		1,047	1,175	
2000	Metal clad, deluxe, dbl. insul. glass, 1220 mm x 1220 mm high		12	1.333		325	47.50		372.50	430	
2100	1220 mm x 1830 mm high		11	1.455		480	51.50		531.50	605	
2200	1525 mm x 1830 mm high		10	1.600		525	57		582	670	
2300	1830 mm x 1830 mm high	▼	10	1.600	▼	605	57		662	755	
750	0010	**SLIDING WINDOW** Including frame, screen, and grills									**750**
0100	Average quality, 900 mm x 900 mm high, double insulated	1 Carp	10	.800	Ea.	153	28.50		181.50	213	
0120	Low E glass		10	.800		193	28.50		221.50	257	
0200	1200 mm x 1050 mm high, double insulated		9	.889		182	31.50		213.50	249	
0220	Low E glass		9	.889		227	31.50		258.50	299	
0300	1800 mm x 1500 mm high, double insulated		8	1		335	35.50		370.50	420	
0320	Low E glass		8	1		400	35.50		435.50	495	
1000	Vinyl clad, premium, dbl. insulated glass, 900 mm x 900 mm		10	.800		510	28.50		538.50	605	
1050	1200 mm x 1050 mm		9	.889		630	31.50		661.50	740	
1100	1500 mm x 1200 mm		9	.889		760	31.50		791.50	885	
1150	1830 mm x 1525 mm		8	1		965	35.50		1,000.50	1,100	
2000	Metal clad, deluxe, double insulated glass, 915 mm x 915 mm high		10	.800		290	28.50		318.50	365	
2050	1220 mm x 1065 mm high		9	.889		355	31.50		386.50	440	
2100	1525 mm x 1220 mm high		9	.889		430	31.50		461.50	520	
2150	1830 mm x 1525 mm high	▼	8	1	▼	655	35.50		690.50	775	
800	0010	**WINDOW GRILLE OR MUNTIN** Snap-in type									**800**
0020	Standard pattern interior grills										
2000	Wood, awning window, glass size 710 mm x 405 mm high	1 Carp	30	.267	Ea.	20	9.50		29.50	37.50	
2060	1120 mm x 610 mm high		32	.250		29.50	8.90		38.40	46.50	
2100	Casement, glass size, 510 mm x 915 mm high		30	.267		25	9.50		34.50	42.50	
2180	510 mm x 1420 mm high		32	.250	▼	36	8.90		44.90	53.50	
2200	Double hung, glass size, 405 mm x 610 mm high		24	.333	Set	44	11.85		55.85	67	
2280	815 mm x 815 mm high		34	.235	"	123	8.35		131.35	149	
2500	Picture, glass size, 1220 mm x 1220 mm high		30	.267	Ea.	141	9.50		150.50	170	
2580	1525 mm x 1725 mm high		28	.286	"	108	10.15		118.15	135	
2600	Sliding, glass size, 355 mm x 915 mm high		24	.333	Set	25	11.85		36.85	46	
2680	915 mm x 915 mm high	▼	22	.364	"	38.50	12.95		51.45	62	
820	0010	**WOOD SASH** Including glazing but not including trim									**820**
0050	Custom, 1525 mm x 1220 mm, 25 mm dbl. glazed, 5 mm thick lites	2 Carp	3.20	5	Ea.	164	178		342	455	

DOORS & WINDOWS 8

08550 | Wood Windows

			CREW	DAILY OUTPUT	LABOR-HOURS	UNIT	2006 BARE COSTS				TOTAL INCL O&P	
							MAT.	LABOR	EQUIP.	TOTAL		
820	0100	6 mm thick lites	2 Carp	5	3.200	Ea.	169	114		283	360	**820**
	0200	25 mm thick, triple glazed		5	3.200		385	114		499	600	
	0300	2135 mm x 1370 mm, 25 mm double glazed, 5 mm lites		4.30	3.721		395	132		527	635	
	0400	6 mm thick lites		4.30	3.721		445	132		577	690	
	0500	25 mm thick, triple glazed		4.30	3.721		505	132		637	760	
	0600	2590 mm x 1525 mm, 25 mm double glazed, 5 mm lites		3.50	4.571		530	163		693	840	
	0700	6 mm thick lites		3.50	4.571		580	163		743	895	
	0800	25 mm thick, triple glazed		3.50	4.571		585	163		748	900	
	0900	Window frames only, based on perimeter length				m	10.50			10.50	11.55	
	1200	Window sill, stock, per lineal meter					23.50			23.50	26	
	1250	Casing, stock					8.45			8.45	9.30	
	3000	Replacement sash, double hung, double glazing, to 1 m²	1 Carp	5.95	1.346	m²	185	48		233	278	
	3100	1 m² to 2 m²		8.73	.916		187	32.50		219.50	256	
	3200	2 m² and over		9.85	.812		163	29		192	224	
	3800	Triple glazing for above, add					26			26	28.50	
	7000	Sash, single lite, 610 mm x 610 mm high	1 Carp	20	.400	Ea.	34	14.20		48.20	59	
	7050	760 mm x 610 mm high		19	.421		39	14.95		53.95	66.50	
	7100	760 mm x 760 mm high		18	.444		45	15.80		60.80	74	
	7150	915 mm x 610 mm high		17	.471		48	16.75		64.75	79	
840	0010	**WOOD SCREENS**										**840**
	0020	Over 0.28 m², 20 mm frames	2 Carp	34.84	.459	m²	39.50	16.35		55.85	69	
	0100	30 mm frames	"	34.84	.459	"	69	16.35		85.35	102	

08560 | Plastic Windows

			CREW	DAILY OUTPUT	LABOR-HOURS	UNIT	2006 BARE COSTS				TOTAL INCL O&P	
							MAT.	LABOR	EQUIP.	TOTAL		
100	0010	**VINYL SINGLE HUNG WINDOWS**										**100**
	0100	Grids, low E, J fin, ext. jambs, 533 x 1346 mm	2 Carp	18	.889	Ea.	146	31.50		177.50	210	
	0110	533 x 1448 mm		17	.941		150	33.50		183.50	217	
	0120	533 x 1651 mm		16	1		156	35.50		191.50	227	
	0130	635 x 1014 mm		20	.800		138	28.50		166.50	197	
	0140	635 x 1346 mm		18	.889		152	31.50		183.50	217	
	0150	635 x 1448 mm		17	.941		156	33.50		189.50	223	
	0160	635 x 1651 mm		16	1		162	35.50		197.50	234	
	0170	737 x 1014 mm		18	.889		147	31.50		178.50	210	
	0180	737 x 1346 mm		18	.889		157	31.50		188.50	222	
	0190	737 x 1448 mm		17	.941		161	33.50		194.50	229	
	0200	737 x 1651 mm		16	1		167	35.50		202.50	240	
	0210	838 x 1014 mm		20	.800		152	28.50		180.50	212	
	0220	838 x 1346 mm		18	.889		163	31.50		194.50	229	
	0230	838 x 1448 mm		17	.941		167	33.50		200.50	236	
	0240	838 x 1651 mm		16	1		174	35.50		209.50	247	
	0250	940 x 1014 mm		20	.800		160	28.50		188.50	221	
	0260	940 x 1346 mm		18	.889		172	31.50		203.50	238	
	0270	940 x 1448 mm		17	.941		175	33.50		208.50	245	
	0280	940 x 1651 mm		16	1		182	35.50		217.50	257	
200	0010	**VINYL DOUBLE HUNG WINDOWS**										**200**
	0100	Grids, low E, J fin, ext. jambs, 533 x 1346 mm	2 Carp	18	.889	Ea.	167	31.50		198.50	233	
	0102	533 x 940 mm		18	.889		149	31.50		180.50	213	
	0104	533 x 1041 mm		18	.889		153	31.50		184.50	217	
	0106	533 x 1245 mm		18	.889		160	31.50		191.50	225	
	0110	533 x 1448 mm		17	.941		171	33.50		204.50	240	
	0120	533 x 1651 mm		16	1		177	35.50		212.50	251	
	0128	635 x 940 mm		20	.800		157	28.50		185.50	218	
	0130	635 x 1014 mm		20	.800		161	28.50		189.50	222	
	0140	635 x 1245 mm		18	.889		166	31.50		197.50	231	

08560 | Plastic Windows

		CREW	DAILY OUTPUT	LABOR-HOURS	UNIT	2006 BARE COSTS MAT.	LABOR	EQUIP.	TOTAL	TOTAL INCL O&P	
200											**200**
0145	635 x 1346 mm	2 Carp	18	.889	Ea.	172	31.50		203.50	239	
0150	635 x 1448 mm		17	.941		173	33.50		206.50	242	
0160	635 x 1651 mm		16	1		184	35.50		219.50	258	
0162	635 x 1753 mm		16	1		191	35.50		226.50	266	
0164	635 x 1956 mm		16	1		202	35.50		237.50	278	
0168	737 x 940 mm		18	.889		162	31.50		193.50	227	
0170	737 x 1014 mm		18	.889		166	31.50		197.50	231	
0172	737 x 1245 mm		18	.889		174	31.50		205.50	240	
0180	737 x 1346 mm		18	.889		178	31.50		209.50	244	
0190	737 x 1448 mm		17	.941		181	33.50		214.50	251	
0200	737 x 1651 mm		16	1		188	35.50		223.50	263	
0202	737 x 1753 mm		16	1		195	35.50		230.50	271	
0205	737 x 1956 mm		16	1		207	35.50		242.50	283	
0208	838 x 940 mm		20	.800		167	28.50		195.50	228	
0210	838 x 1014 mm		20	.800		170	28.50		198.50	232	
0215	838 x 1245 mm		20	.800		179	28.50		207.50	242	
0220	838 x 1346 mm		18	.889		183	31.50		214.50	250	
0230	838 x 1448 mm		17	.941		187	33.50		220.50	258	
0240	838 x 1651 mm		16	1		192	35.50		227.50	267	
0242	838 x 1753 mm		16	1		204	35.50		239.50	280	
0246	838 x 1956 mm		16	1		214	35.50		249.50	291	
0250	940 x 1014 mm		20	.800		174	28.50		202.50	237	
0255	940 x 1245 mm		20	.800		183	28.50		211.50	247	
0260	940 x 1346 mm		18	.889		191	31.50		222.50	259	
0270	940 x 1448 mm		17	.941		195	33.50		228.50	266	
0280	940 x 1651 mm		16	1		200	35.50		235.50	276	
0282	940 x 1753 mm		16	1		262	35.50		297.50	345	
0286	940 x 1956 mm		16	1		274	35.50		309.50	355	
0300	Solid vinyl, average quality, double insulated glass, 610 mm x 915 mm	1 Carp	10	.800		137	28.50		165.50	196	
0310	915 mm x 1220 mm		9	.889		165	31.50		196.50	230	
0320	1220 mm x 1370 mm		8	1		197	35.50		232.50	273	
0330	Premium, double insulated glass, 760 mm x 915 mm		10	.800		153	28.50		181.50	213	
0340	915 mm x 1065 mm		9	.889		177	31.50		208.50	244	
0350	915 mm x 2590 mm		9	.889		188	31.50		219.50	256	
0360	915 mm x 1370 mm		9	.889		193	31.50		224.50	261	
0370	915 mm x 1525 mm		8	1		198	35.50		233.50	274	
0380	1065 mm x 3860 mm		8	1		218	35.50		253.50	295	
300	0010 **VINYL CASEMENT WINDOWS**										**300**
0100	Grids, low E, J fin, ext. jambs, 1 lt, 533 x 1041 mm	2 Carp	20	.800	Ea.	216	28.50		244.50	283	
0110	533 x 1194 mm		20	.800		236	28.50		264.50	305	
0120	533 x 1346 mm		20	.800		255	28.50		283.50	325	
0128	610 x 889 mm		19	.842		208	30		238	275	
0130	610 x 1041 mm		19	.842		226	30		256	295	
0140	610 x 1194 mm		19	.842		244	30		274	315	
0150	610 x 1346 mm		19	.842		263	30		293	335	
0158	711 x 889 mm		19	.842		221	30		251	291	
0160	711 x 1041 mm		19	.842		239	30		269	310	
0170	711 x 1194 mm		19	.842		258	30		288	330	
0180	711 x 1346 mm		19	.842		284	30		314	360	
0184	711 x 1499 mm		19	.842		290	30		320	365	
0188	Two lites, 838 x 889 mm		18	.889		355	31.50		386.50	440	
0190	838 x 1041 mm		18	.889		380	31.50		411.50	470	
0200	838 x 1194 mm		18	.889		410	31.50		441.50	500	
0210	838 x 1346 mm		18	.889		440	31.50		471.50	530	
0212	838 x 1499 mm		18	.889		465	31.50		496.50	560	

DOORS & WINDOWS 8

08560 | Plastic Windows

			DAILY OUTPUT	LABOR-HOURS	UNIT	2006 BARE COSTS				TOTAL INCL O&P	
			CREW			MAT.	LABOR	EQUIP.	TOTAL		
300	0215	838 x 1829 mm	2 Carp	18	.889	Ea.	480	31.50		511.50	580	**300**
	0220	1041 x 1041 mm		18	.889		415	31.50		446.50	510
	0230	1041 x 1194 mm		18	.889		445	31.50		476.50	540
	0240	1041 x 1346 mm		17	.941		475	33.50		508.50	570
	0242	1041 x 1499 mm		17	.941		500	33.50		533.50	600
	0246	1041 x 1829 mm		17	.941		520	33.50		553.50	625
	0250	1194 x 1041 mm		17	.941		420	33.50		453.50	515
	0260	1194 x 1194 mm		17	.941		450	33.50		483.50	545
	0270	1194 x 1346 mm		17	.941		475	33.50		508.50	570
	0272	1194 x 1499 mm		17	.941		520	33.50		553.50	620
	0280	1422 x 1041 mm		15	1.067		450	38		488	555
	0290	1422 x 1194 mm		15	1.067		475	38		513	580
	0300	1422 x 1194 mm		15	1.067		520	38		558	630
	0302	1422 x 1499 mm		15	1.067		540	38		578	655
	0310	1422 x 1829 mm	↓	15	1.067		590	38		628	705
	0340	Solid vinyl, premium, double insulated glass, 600 mm x 900 mm high	1 Carp	10	.800		185	28.50		213.50	248
	0360	600 mm x 1200 mm high		9	.889		227	31.50		258.50	298
	0380	600 mm x 1500 mm high	↓	8	1	↓	258	35.50		293.50	340
400	0010	**VINYL PICTURE WINDOWS**										**400**
	0100	Grids, low E, J fin, ext. jambs, 838 x 1194 mm	2 Carp	12	1.333	Ea.	217	47.50		264.50	315
	0110	889 x 1803 mm		12	1.333		230	47.50		277.50	325
	0120	1041 x 1194 mm		12	1.333		252	47.50		299.50	350
	0130	1041x 1803 mm		12	1.333		273	47.50		320.50	375
	0140	1194x 1194 mm		12	1.333		285	47.50		332.50	390
	0150	1194 x 1803 mm		11	1.455		299	51.50		350.50	410
	0160	1346 x 1194 mm		11	1.455		280	51.50		331.50	390
	0170	1346 x 1803 mm		11	1.455		293	51.50		344.50	405
	0180	1499 x 1194 mm		11	1.455		320	51.50		371.50	430
	0190	1499 x 1803 mm		11	1.455		340	51.50		391.50	455
	0200	1803 x 1194 mm		10	1.600		350	57		407	480
	0210	1803 x 1803 mm	↓	10	1.600	↓	370	57		427	500

08580 | Special Function Windows

			CREW	DAILY OUTPUT	LABOR-HOURS	UNIT	MAT.	LABOR	EQUIP.	TOTAL	INCL O&P	
900	0010	**STORM WINDOWS** Aluminum, residential										**900**
	0300	Basement, mill finish, incl. fiberglass screen									
	0320	560 mm x 305 mm high	2 Carp	30	.533	Ea.	29.50	18.95		48.45	62
	0340	840 mm x 455 mm high	"	30	.533	"	32	18.95		50.95	65
	1600	Double-hung, combination, storm & screen									
	2000	Average quality, clear anodic coating, 610 mm x 1040 mm high	2 Carp	30	.533	Ea.	77.50	18.95		96.45	115
	2020	760 mm x 1525 mm high		28	.571		95	20.50		115.50	136
	2040	1220 mm x 1830 mm high		25	.640		115	23		138	163
	2400	White painted, 610 mm x 1040 mm high		30	.533		76.50	18.95		95.45	114
	2420	760 mm x 1525 mm high		28	.571		84.50	20.50		105	125
	2440	1220 mm x 1830 mm high		25	.640		92.50	23		115.50	138
	2600	Mill finish, 610 mm x 1040 mm high		30	.533		69.50	18.95		88.45	106
	2620	760 mm x 1525 mm high		28	.571		77.50	20.50		98	117
	2640	1220 mm x 150 mm high	↓	25	.640	↓	87	23		110	132

08620	Unit Skylights	CREW	DAILY OUTPUT	LABOR-HOURS	UNIT	2006 BARE COSTS				TOTAL INCL O&P	
						MAT.	LABOR	EQUIP.	TOTAL		
400	**0010**	**PREFABRICATED**, Glass block with metal frame										**400**
	0020	Minimum	G-3	24.62	1.300	m²	465	45		510	585	
	0100	Maximum	"	14.86	2.153	"	935	75		1,010	1,150	
800	**0010**	**SKYLIGHT** Plastic domes, flush or curb mounted, ten or										**800**
	0100	more units, curb not included										
	0300	Nominal size under 1 m², double	G-3	12.08	2.650	m²	206	92		298	370	
	0400	Single		14.86	2.153		150	75		225	281	
	0600	1 m² to 2 m², double		29.26	1.094		182	38		220	259	
	0700	Single		36.70	.872		93.50	30.50		124	150	
	0900	2 m² to 3 m², double		36.70	.872		166	30.50		196.50	230	
	1000	Single		43.20	.741		118	26		144	170	
	1200	3 m² to 6 m², double		43.20	.741		126	26		152	179	
	1300	Single	↓	56.67	.565		161	19.65		180.65	209	
	1500	For insulated 100 mm curbs, double, add					25%					
	1600	Single, add				↓	30%					
	1800	For integral insulated 225 mm curbs, double, add					30%					
	1900	Single, add					40%					
	2120	Ventilating insulated plexiglass dome with										
	2130	curb mounting, 915 mm x 915 mm	G-3	12	2.667	Ea.	380	93		473	565	
	2150	1320 mm x 1320 mm		12	2.667		570	93		663	770	
	2160	710 mm x 1320 mm		10	3.200		445	111		556	660	
	2170	915 mm x 1320 mm	↓	10	3.200		480	111		591	700	
	2180	For electric opening system, add				↓	285			285	315	
	2200	Field fabricated, factory type, aluminum and wire glass	G-3	11.15	2.870	m²	158	100		258	330	
	2300	Insulated safety glass with aluminum frame		14.86	2.153		920	75		995	1,150	
	2400	Sandwich panels, fiberglass, for walls, 40 mm thick, to 23 m²		18.58	1.722		167	60		227	277	
	2500	23 m² and up		24.62	1.300		150	45		195	235	
	2700	As above, but for roofs, 70 mm thick, to 23 m²		27.41	1.168		241	40.50		281.50	330	
	2800	23 m² and up	↓	30.66	1.044	↓	198	36.50		234.50	273	

08710	Door Hardware	CREW	DAILY OUTPUT	LABOR-HOURS	UNIT	2006 BARE COSTS				TOTAL INCL O&P	
						MAT.	LABOR	EQUIP.	TOTAL		
100	**0010**	**AUTOMATIC OPENERS COMMERCIAL**										**100**
	0020	Pneumatic, incl opener, motion sens, control box, tubing, compressor										
	0050	For single swing door, per opening	2 Skwk	.80	20	Ea.	3,725	730		4,455	5,225	
	0100	Pair, per opening		.50	32	Opng.	6,200	1,175		7,375	8,650	
	1000	For single sliding door, per opening		.60	26.667		4,125	975		5,100	6,050	
	1300	Bi-parting pair	↓	.50	32	↓	6,225	1,175		7,400	8,675	
	1420	Electronic door opener incl motion sens, 12V control box, motor										
	1450	For single swing door, per opening	2 Skwk	.80	20	Opng.	3,250	730		3,980	4,700	
	1500	Pair, per opening		.50	32		5,325	1,175		6,500	7,675	
	1600	For single sliding door, per opening		.60	26.667		3,525	975		4,500	5,400	
	1700	Bi-parting pair	↓	.50	32	↓	5,400	1,175		6,575	7,775	
	1750	Handicap actuator buttons, 2, including 12V DC wiring, add	1 Carp	1.50	5.333	Pr.	375	190		565	705	
120	**0010**	**AUTOMATIC OPENERS INDUSTRIAL**										**120**
	0015	Sliding doors, to 1830 mm wide	2 Skwk	.60	26.667	Opng.	4,875	975		5,850	6,875	
	0200	To 3660 mm wide	"	.40	40	"	5,925	1,450		7,375	8,800	
	0400	Over 3660 mm wide, add/mm of excess				m	2,125			2,125	2,350	

DOORS & WINDOWS | **8**

08710	Door Hardware		CREW	DAILY OUTPUT	LABOR-HOURS	UNIT	MAT.	LABOR	EQUIP.	TOTAL	TOTAL INCL O&P	
120	1000	Swing doors, to 1525 mm wide ♿	2 Skwk	.80	20	Ea.	2,850	730		3,580	4,250	**120**
	1860	Add for controls, wall pushbutton, 3 button		4	4		174	146		320	420	
	1870	Ceiling pull cord	↓	4.30	3.721	↓	149	136		285	375	
150	0010	**AVERAGE** Percentage for hardware, total job cost										**150**
	0025	Minimum									.75%	
	0050	Maximum									3.50%	
	0500	Total hardware for building, average distribution					85%	15%				
	1000	Door hardware, apartment, interior				Door	127			127	140	
	1500	Hospital bedroom, minimum					284			284	315	
	2000	Maximum				↓	625			625	685	
	2100	Pocket door				Ea.	127			127	140	
	2250	School, single exterior, incl. lever, not incl. panic device				Door	420			420	460	
	2500	Single interior, regular use, no lever included					280			280	310	
	2550	Including handicap lever					380			380	420	
	2600	Heavy use, incl. lever and closer					490			490	540	
	2850	Stairway, single interior				↓	700			700	770	
	3100	Double exterior, with panic device				Pr.	985			985	1,075	
	3600	Toilet, public, single interior				Door	154			154	170	
200	0010	**BOLTS, FLUSH** (security item)										**200**
	0020	Standard, concealed	1 Carp	7	1.143	Ea.	20	40.50		60.50	85.50	
	0800	Automatic fire exit	"	5	1.600		271	57		328	385	
	1600	For electric release, add	1 Elec	3	2.667		110	112		222	288	
	3000	Barrel, brass, 50 mm long	1 Carp	40	.200		4.30	7.10		11.40	15.80	
	3020	100 mm long		40	.200		2.92	7.10		10.02	14.25	
	3060	150 mm long	↓	40	.200	↓	8.70	7.10		15.80	20.50	
220	0010	**DOOR PROTECTORS**										**220**
	0020	44 mm x 19 mm U channel	2 Carp	24.38	.656	m	59	23.50		82.50	102	
	0021	44 mm x 32 mm U channel		24.38	.656	"	27.50	23.50		51	67	
	1000	Tear drop, spring-stl, 203 mm high x 483 mm long		15	1.067	Ea.	78	38		116	145	
	1010	Tear drop, spring-stl, 203 mm high x 813 mm long		15	1.067		96.50	38		134.50	165	
	1100	203 mm high x 483 mm long		15	1.067		199	38		237	278	
	1200	203 mm high x 813 mm long	↓	15	1.067	↓	248	38		286	330	
300	0010	**DOOR CLOSERS**										**300**
	0015	Door closer rack and pinion	1 Carp	6.50	1.231	Ea.	134	44		178	215	
	0020	Adjustable backcheck, 3 way mount, all sizes, regular arm		6	1.333		137	47.50		184.50	225	
	0040	Hold open arm		6	1.333		157	47.50		204.50	247	
	0100	Fusible link		6.50	1.231		122	44		166	202	
	0200	Non sized, regular arm		6	1.333		136	47.50		183.50	224	
	0240	Hold open arm		6	1.333		170	47.50		217.50	261	
	0400	4 way mount, non sized, regular arm		6	1.333		187	47.50		234.50	280	
	0440	Hold open arm	↓	6	1.333	↓	201	47.50		248.50	295	
	2000	Backcheck and adjustable power, hinge face mount										
	2010	All sizes, regular arm	1 Carp	6.50	1.231	Ea.	172	44		216	257	
	2040	Hold open arm		6.50	1.231		185	44		229	272	
	2400	Top jamb mount, all sizes, regular arm		6	1.333		172	47.50		219.50	263	
	2440	Hold open arm		6	1.333		185	47.50		232.50	278	
	2800	Top face mount, all sizes, regular arm		6.50	1.231		172	44		216	257	
	2840	Hold open arm		6.50	1.231		185	44		229	271	
	4000	Backcheck, overhead concealed, all sizes, regular arm		5.50	1.455		182	51.50		233.50	281	
	4040	Concealed arm		5	1.600		194	57		251	305	
	4400	Compact overhead, concealed, all sizes, regular arm		5.50	1.455		330	51.50		381.50	445	
	4440	Concealed arm	↓	5	1.600	↓	345	57		402	470	

8 DOORS & WINDOWS

08710	Door Hardware	CREW	DAILY OUTPUT	LABOR-HOURS	UNIT	2006 BARE COSTS				TOTAL INCL O&P	
						MAT.	LABOR	EQUIP.	TOTAL		
300											**300**
4800	Concealed in door, all sizes, regular arm	1 Carp	5.50	1.455	Ea.	123	51.50		174.50	216	
4840	Concealed arm		5	1.600		132	57		189	234	
4900	Floor concealed, all sizes, single acting		2.20	3.636		156	129		285	375	
4940	Double acting		2.20	3.636		201	129		330	420	
5000	For cast aluminum cylinder, deduct					16.50			16.50	18.15	
5040	For delayed action, add					29			29	32	
5080	For fusible link arm, add					11.95			11.95	13.15	
5120	For shock absorbing arm, add					36			36	39.50	
5160	For spring power adjustment, add					27.50			27.50	30.50	
6000	Closer-holder, hinge face mount, all sizes, exposed arm	1 Carp	6.50	1.231		127	44		171	207	
7000	Electronic closer-holder, hinge facemount, concealed arm		5	1.600		193	57		250	300	
7400	With built-in detector		5	1.600		580	57		637	730	
320	0010	**DEADLOCKS**									**320**
0011	Mortise, heavy duty, outside key	1 Carp	9	.889	Ea.	131	31.50		162.50	193	
0020	Double cylinder		9	.889		145	31.50		176.50	208	
0100	Medium duty, outside key		10	.800		102	28.50		130.50	157	
0110	Double cylinder		10	.800		127	28.50		155.50	185	
1000	Tubular, standard duty, outside key		10	.800		55	28.50		83.50	105	
1010	Double cylinder		10	.800		70.50	28.50		99	122	
1200	Night latch, outside key		10	.800		69	28.50		97.50	121	
340	0010	**DOORSTOPS**									**340**
0020	Holder and bumper, floor or wall	1 Carp	32	.250	Ea.	31.50	8.90		40.40	48.50	
1300	Wall bumper, 100 mm diameter, with rubber pad, aluminum		32	.250		9.35	8.90		18.25	24	
1600	Door bumper, floor type, aluminum		32	.250		4.82	8.90		13.72	19.15	
1900	Plunger type, door mounted		32	.250		25.50	8.90		34.40	42	
400	0010	**ENTRANCE LOCKS**									**400**
0015	Cylinder, grip handle, deadlocking latch	1 Carp	9	.889	Ea.	120	31.50		151.50	181	
0020	Deadbolt		8	1		146	35.50		181.50	217	
0100	Push and pull plate, dead bolt		8	1		139	35.50		174.50	209	
0900	For handicapped lever, add					152			152	167	
450	0010	**FLOOR CHECKS**									**450**
0020	For over 915 mm wide doors, single acting	1 Carp	2.50	3.200	Ea.	460	114		574	680	
0500	Double acting	"	2.50	3.200	"	585	114		699	820	
500	0010	**HASP**									**500**
0015	Steel, ssembly, 75 mm	1 Carp	26	.308	Ea.	2.68	10.95		13.63	20	
0020	113 mm		13	.615		3.43	22		25.43	38	
0040	150 mm		12.50	.640		5.20	23		28.20	41	
520	0010	**HINGES**									**520**
0012	Full mortise, avg. freq., stl base, USP, 114 mm x 114 mm [R087120-10]				Pr.	21			21	23.50	
0100	127 mm x 127 mm, USP					35.50			35.50	39	
0200	152 mm x 152 mm, USP					75.50			75.50	83	
0400	Brass base, 114 mm x 114 mm, US10					43.50			43.50	48	
0500	127 mm x 127 mm, US10					63.50			63.50	70	
0600	152 mm x 152 mm, US10					108			108	119	
0800	Stainless steel base, 114 mm x 114 mm, US32					65.50			65.50	72	
0900	For non removable pin, add				Ea.	2.38			2.38	2.62	
0910	For floating pin, driven tips, add					2.73			2.73	3	
0930	For hospital type tip on pin, add					11.80			11.80	12.95	
0940	For steeple type tip on pin, add					10.30			10.30	11.35	
0950	Full mortise, high frequency, steel base, 88 mm x 88 mm , US26D				Pr.	19.30			19.30	21	
1000	113 mm x 113 mm, USP					51.50			51.50	56.50	
1100	125 mm x 125 mm, USP					48			48	53	
1200	150 mm x 150 mm, USP					117			117	129	

8

DOORS & WINDOWS

For expanded coverage of these items see *Means Interior Cost Data 2006*

08710 | Door Hardware

			CREW	DAILY OUTPUT	LABOR-HOURS	UNIT	MAT.	LABOR	EQUIP.	TOTAL	TOTAL INCL O&P	
520	1400	Brass base, 88 mm x 88 mm, US4	R087120 -10			Pr.	40.50			40.50	44.50	**520**
	1430	113 mm x 113 mm, US10					69			69	76	
	1500	125 mm x 125 mm, US10					103			103	113	
	1600	150 mm x 150 mm, US10					149			149	164	
	1800	Stainless steel base, 113 mm x 113 mm, US32					111			111	122	
	1810	127mm x 113 mm, US32					154			154	170	
	1930	For hospital type tip on pin, add				Ea.	6.95			6.95	7.65	
	1950	Full mortise, low frequency, steel base, 88 mm x 88 mm, US26D				Pr.	10.10			10.10	11.10	
	2000	113 mm x 113 mm, USP					9.40			9.40	10.35	
	2100	125 mm x 125 mm, USP					26			26	28.50	
	2200	150 mm x 150 mm, USP					52			52	57	
	2300	113 mm x 113 mm, US3					15.20			15.20	16.70	
	2310	115 mm x 115 mm, US3					37.50			37.50	41.50	
	2400	Brass bass, 113 mm x 113 mm, US10					36.50			36.50	40	
	2500	125 mm x 125 mm, US10					55.50			55.50	61	
	2800	Stainless steel base, 114 mm x 114 mm, US32					63			63	69	
550	0010	**KICK PLATE**										**550**
	0020	Stainless steel	1 Carp	15	.533	Ea.	27.50	18.95		46.45	60	
	0500	Bronze		15	.533		35	18.95		53.95	67.50	
	2000	Aluminum, .050, with 3 beveled edges, 250 mm x 700 mm		15	.533		19.80	18.95		38.75	51.50	
	2010	250 mm x 750 mm		15	.533		21.50	18.95		40.45	53	
	2020	250 mm x 850 mm		15	.533		22.50	18.95		41.45	54	
	2040	250 mm x 950 mm		15	.533		25	18.95		43.95	57	
650	0010	**LOCKSET** Standard duty, cylindrical, with sectional trim										**650**
	0020	Non-keyed, passage	1 Carp	12	.667	Ea.	43	23.50		66.50	84	
	0100	Privacy		12	.667		53.50	23.50		77	95.50	
	0400	Keyed, single cylinder function		10	.800		74.50	28.50		103	127	
	0420	Hotel		8	1		106	35.50		141.50	173	
	0500	Lever handled, keyed, single cylinder function		10	.800		132	28.50		160.50	190	
	1000	Heavy duty with sectional trim, non-keyed, passages		12	.667		123	23.50		146.50	173	
	1100	Privacy		12	.667		156	23.50		179.50	209	
	1400	Keyed, single cylinder function		10	.800		184	28.50		212.50	248	
	1420	Hotel		8	1		275	35.50		310.50	355	
	1600	Communicating		10	.800		218	28.50		246.50	285	
	1690	For re-core cylinder, add					31			31	34	
	1700	Residential, interior door, minimum	1 Carp	16	.500		14.90	17.80		32.70	44	
	1720	Maximum		8	1		39.50	35.50		75	99	
	1800	Exterior, minimum		14	.571		33	20.50		53.50	68	
	1810	Average		8	1		65	35.50		100.50	127	
	1820	Maximum		8	1		140	35.50		175.50	209	
700	0010	**MORTISE LOCKSET** Comm., wrought knobs & full escutcheon trim										**700**
	0020	Non-keyed, passage, minimum	1 Carp	9	.889	Ea.	160	31.50		191.50	224	
	0030	Maximum		8	1		258	35.50		293.50	340	
	0040	Privacy, minimum		9	.889		170	31.50		201.50	236	
	0050	Maximum		8	1		278	35.50		313.50	360	
	0100	Keyed, office/entrance/apartment, minimum		8	1		195	35.50		230.50	270	
	0110	Maximum		7	1.143		335	40.50		375.50	435	
	0120	Single cylinder, typical, minimum		8	1		167	35.50		202.50	240	
	0130	Maximum		7	1.143		310	40.50		350.50	405	
	0200	Hotel, minimum		7	1.143		201	40.50		241.50	285	
	0210	Maximum		6	1.333		325	47.50		372.50	435	
	0300	Communication, double cylinder, minimum		8	1		201	35.50		236.50	277	
	0310	Maximum		7	1.143		260	40.50		300.50	350	
	1000	Wrought knobs and sectional trim, non-keyed, passage, minimum		10	.800		106	28.50		134.50	161	

8 DOORS & WINDOWS

08710	Door Hardware	CREW	DAILY OUTPUT	LABOR-HOURS	UNIT	2006 BARE COSTS				TOTAL INCL O&P	
						MAT.	LABOR	EQUIP.	TOTAL		
700											**700**
1010	Maximum	1 Carp	9	.889	Ea.	209	31.50		240.50	279	
1040	Privacy, minimum		10	.800		124	28.50		152.50	181	
1050	Maximum		9	.889		223	31.50		254.50	294	
1100	Keyed, entrance, office/apartment, minimum		9	.889		184	31.50		215.50	251	
1110	Maximum		8	1		265	35.50		300.50	350	
1120	Single cylinder, typical, minimum		9	.889		177	31.50		208.50	243	
1130	Maximum	▼	8	1	▼	257	35.50		292.50	340	
2000	Cast knobs and full escutcheon trim										
2010	Non-keyed, passage, minimum	1 Carp	9	.889	Ea.	225	31.50		256.50	296	
2020	Maximum		8	1		365	35.50		400.50	455	
2040	Privacy, minimum		9	.889		271	31.50		302.50	345	
2050	Maximum		8	1		390	35.50		425.50	485	
2120	Keyed, single cylinder, typical, minimum		8	1		271	35.50		306.50	355	
2130	Maximum		7	1.143		425	40.50		465.50	535	
2200	Hotel, minimum		7	1.143		300	40.50		340.50	395	
2210	Maximum		6	1.333		540	47.50		587.50	665	
3000	Cast knob and sectional trim, non-keyed, passage, minimum		10	.800		175	28.50		203.50	238	
3010	Maximum		10	.800		350	28.50		378.50	430	
3040	Privacy, minimum		10	.800		199	28.50		227.50	264	
3050	Maximum		10	.800		350	28.50		378.50	430	
3100	Keyed, office/entrance/apartment, minimum		9	.889		226	31.50		257.50	297	
3110	Maximum		9	.889		355	31.50		386.50	440	
3120	Single cylinder, typical, minimum		9	.889		226	31.50		257.50	297	
3130	Maximum	▼	9	.889		440	31.50		471.50	530	
3190	For re-core cylinder, add				▼	30.50			30.50	33.50	
3800	Cipher lockset	1 Carp	13	.615		715	22		737	820	
3900	Keyless, pushbutton type										
4000	Residential/light commercial, deadbolt, standard	1 Carp	9	.889	Ea.	102	31.50		133.50	161	
4010	Heavy duty		9	.889		121	31.50		152.50	182	
4020	Industrial, heavy duty, with deadbolt		9	.889		238	31.50		269.50	310	
4030	Key override		9	.889		264	31.50		295.50	340	
4040	Lever activated handle		9	.889		290	31.50		321.50	370	
4050	Key override		9	.889		320	31.50		351.50	405	
4060	Double sided pushbutton type		8	1		530	35.50		565.50	640	
4070	Key override	▼	8	1	▼	570	35.50		605.50	685	
750	**0010 PANIC DEVICE**										**750**
0015	For rim locks, single door, exit only	1 Carp	6	1.333	Ea.	375	47.50		422.50	485	
0020	Outside key and pull		5	1.600		425	57		482	555	
0200	Bar and vertical rod, exit only		5	1.600		540	57		597	685	
0210	Outside key and pull		4	2		645	71		716	820	
0400	Bar and concealed rod		4	2		545	71		616	705	
0600	Touch bar, exit only		6	1.333		430	47.50		477.50	545	
0610	Outside key and pull		5	1.600		515	57		572	660	
0700	Touch bar and vertical rod, exit only		5	1.600		585	57		642	735	
0710	Outside key and pull		4	2		685	71		756	860	
1000	Mortise, bar, exit only		4	2		480	71		551	640	
1600	Touch bar, exit only		4	2		550	71		621	715	
2000	Narrow stile, rim mounted, bar, exit only		6	1.333		580	47.50		627.50	710	
2010	Outside key and pull		5	1.600		630	57		687	780	
2200	Bar and vertical rod, exit only		5	1.600		595	57		652	745	
2210	Outside key and pull		4	2		595	71		666	765	
2400	Bar and concealed rod, exit only		3	2.667		695	95		790	915	
3000	Mortise, bar, exit only		4	2		505	71		576	665	
3600	Touch bar, exit only	▼	4	2	▼	735	71		806	920	

DOORS & WINDOWS **8**

08710	Door Hardware	CREW	DAILY OUTPUT	LABOR-HOURS	UNIT	2006 BARE COSTS MAT.	LABOR	EQUIP.	TOTAL	TOTAL INCL O&P	
770	**0010**	**DOOR AND WINDOW ACCESSORIES**										**770**
	0050	Door closing coord.,914 mm (for paired openings up to 1422 mm)	1 Carp	8	1	Ea.	87.50	35.50		123	152	
	0060	1219 mm (for paired openings up to 2134 mm)		8	1		93.50	35.50		129	159	
	0070	1422 mm (for paired openings up to 2438 mm)		8	1		103	35.50		138.50	169	
780	**0010**	**PUSH-PULL PLATE**										**780**
	0100	Push plate, 1 mm, 100 mm x 400 mm, aluminum	1 Carp	12	.667	Ea.	6.85	23.50		30.35	44.50	
	0500	Bronze		12	.667		17.35	23.50		40.85	56	
	1500	Pull handle and push bar, aluminum		11	.727		121	26		147	174	
	2000	Bronze		10	.800		157	28.50		185.50	217	
	3000	Push plate both sides, aluminum		14	.571		14.80	20.50		35.30	48	
	3500	Bronze		13	.615		37	22		59	74.50	
	4000	Door pull, designer style, cast aluminum, minimum		12	.667		64.50	23.50		88	108	
	5000	Maximum		8	1		325	35.50		360.50	415	
	6000	Cast bronze, minimum		12	.667		76	23.50		99.50	121	
	7000	Maximum		8	1		355	35.50		390.50	445	
	8000	Walnut, minimum		12	.667		58	23.50		81.50	101	
	9000	Maximum		8	1		325	35.50		360.50	410	
800	**0010**	**SPECIAL HINGES**										**800**
	0015	Paumelle, high frequency										
	0020	Steel base, 152 mm x 114 mm, US10				Pr.	134			134	147	
	0100	Bronze base, 127 mm x 114 mm, US10					170			170	187	
	0200	Paumelle, average frequency, steel base, 114 mm x 89 mm, US10					90.50			90.50	99.50	
	0400	Olive knuckle, low frequency, brass base, 152 mm x 114 mm, US10					155			155	171	
	1000	Electric hinge with concealed conductor, average frequency										
	1010	Steel base, 114 mm x 114 mm, US26D				Pr.	286			286	315	
	1100	Bronze base, 114 mm x 114 mm, US26D				"	300			300	330	
	1200	Electric hinge with concealed conductor, high frequency										
	1210	Steel base, 114 mm x 114 mm, US26D				Pr.	213			213	235	
	1600	Dbl. weight, 362 kg, steel base, removable pin, 127 x 152 mm, USP					125			125	137	
	1700	Steel base-welded pin, 127 mm x 152 mm, USP					141			141	155	
	1800	Triple weight, 907 kg, stl base, welded pin, 127 x 152 mm, USP					143			143	158	
	2000	Pivot reinf., high frequency, steel base, 197 mm door plate, USP					166			166	183	
	2200	Bronze base, 197 mm door plate, US10					198			198	218	
	3000	Swing clear, full mortise, full or half surface, high frequency,										
	3010	Steel base, 127 mm high, USP				Pr.	143			143	157	
	3200	Swing clear, full mortise, average frequency										
	3210	Steel base, 114 mm high, USP				Pr.	113			113	125	
	4000	Wide throw, average frequency, steel base, 114 mm x 152 mm, USP					86.50			86.50	95.50	
	4200	High frequency, steel base, 114 mm x 152 mm, USP					133			133	146	
	4600	Spring hinge, single acting, 152 mm flange, steel				Ea.	49			49	54	
	4700	Brass					86			86	94.50	
	4900	Double acting, 152 mm flange, steel					86			86	95	
	4950	Brass					141			141	155	
	9000	Continuous hinge, steel, full mortise, heavy duty	2 Carp	19.51	.820	m	36.50	29		65.50	85.50	

08720	Weatherstripping & Seals											
100	**0010**	**ASTRAGALS** One piece overlapping										**100**
	0400	Cadmium plated steel, flat, 5 mm x 50 mm	1 Carp	27.43	.292	m	10.05	10.35		20.40	27	
	0600	Prime coated steel, flat, 3 mm x 75 mm		27.43	.292		13.95	10.35		24.30	31.50	
	0800	Stainless steel, flat, 2 mm x 40 mm		27.43	.292		52.50	10.35		62.85	73.50	
	1000	Aluminum, flat, 3 mm x 50 mm		27.43	.292		10.05	10.35		20.40	27.50	
	1200	Nail on, "T" extrusion		36.58	.219		2.23	7.80		10.03	14.55	
	1300	Vinyl bulb insert		32	.250		3.61	8.90		12.51	17.80	
	1600	Screw on, "T" extrusion		27.43	.292		14.65	10.35		25	32.50	

8 DOORS & WINDOWS

08720 | Weatherstripping & Seals

		CREW	DAILY OUTPUT	LABOR-HOURS	UNIT	2006 BARE COSTS				TOTAL INCL O&P	
						MAT.	LABOR	EQUIP.	TOTAL		
100	1700	Vinyl insert	1 Carp	22.86	.350	m	9.90	12.45		22.35	30.50	100
	2000	"L" extrusion, neoprene bulbs		22.86	.350		5.70	12.45		18.15	25.50	
	2100	Neoprene sponge insert		22.86	.350		17.40	12.45		29.85	38.50	
	2200	Magnetic		22.86	.350		28.50	12.45		40.95	51	
	2400	Spring hinged security seal, with cam		22.86	.350		18.40	12.45		30.85	39.50	
	2600	Spring loaded locking bolt, vinyl insert		13.72	.583		25	20.50		45.50	60	
	2800	Neoprene sponge strip, "Z" shaped, aluminum		18.29	.437		11.90	15.55		27.45	37	
	2900	Solid neoprene strip, nail on aluminum strip	▼	27.43	.292	▼	10	10.35		20.35	27	
	3000	One piece stile protection										
	3020	Neoprene fabric loop, nail on aluminum strips	1 Carp	18.29	.437	m	1.77	15.55		17.32	26	
	3110	Flush mounted aluminum extrusion, 13 mm x 30 mm		18.29	.437		9.50	15.55		25.05	34.50	
	3140	20 mm x 35 mm		18.29	.437		12.05	15.55		27.60	37.50	
	3160	30 mm x 45 mm		18.29	.437		19.25	15.55		34.80	45	
	3300	Mortise, 15 mm x 20 mm		18.29	.437		10.20	15.55		25.75	35	
	3320	20 mm x 35 mm		18.29	.437		11.05	15.55		26.60	36	
	3600	Spring bronze strip, nail on type		32	.250		8.70	8.90		17.60	23.50	
	3620	Screw on, with retainer		22.86	.350		7.10	12.45		19.55	27	
	3800	Flexible stainless steel housing, pile insert, 13 mm door		32	.250		19.70	8.90		28.60	35.50	
	3820	20 mm door		32	.250		22	8.90		30.90	38.50	
	4000	Extruded aluminum retainer, flush mount, pile insert		32	.250		6.90	8.90		15.80	21.50	
	4080	Mortise, felt insert		27.43	.292		12.25	10.35		22.60	29.50	
	4160	Mortise with spring, pile insert		27.43	.292		9.60	10.35		19.95	26.50	
	4400	Rigid vinyl retainer, mortise, pile insert		32	.250		6.85	8.90		15.75	21.50	
	4600	Wool pile filler strip, aluminum backing	▼	32	.250	▼	6.90	8.90		15.80	21.50	
	5000	Two piece overlapping astragal, extruded aluminum retainer										
	5010	Pile insert	1 Carp	18.29	.437	m	8.90	15.55		24.45	34	
	5020	Vinyl bulb insert		18.29	.437		5.75	15.55		21.30	30.50	
	5040	Vinyl flap insert		18.29	.437		17.90	15.55		33.45	43.50	
	5060	Solid neoprene flap insert		18.29	.437		17.70	15.55		33.25	43.50	
	5080	Hypalon rubber flap insert		18.29	.437		18.15	15.55		33.70	44	
	5090	Snap on cover, pile insert		18.29	.437		20.50	15.55		36.05	47	
	5400	Magnetic aluminum, surface mounted		18.29	.437		70	15.55		85.55	101	
	5500	Interlocking aluminum, 16 mm x 25 mm neoprene bulb insert		13.72	.583		11.10	20.50		31.60	45	
	5600	Adjustable aluminum, 14 mm x 17 mm, pile insert	▼	13.72	.583		53	20.50		73.50	90.50	
	5790	For vinyl bulb, deduct					1.35			1.35	1.48	
	5800	Magnetic, adjustable, 14 mm x 17 mm	1 Carp	13.72	.583	▼	68	20.50		88.50	107	
	6000	Two piece stile protection										
	6010	Cloth backed rubber loop, 25 mm gap, nail on aluminum strips	1 Carp	13.72	.583	m	11.40	20.50		31.90	45	
	6040	Screw on aluminum strips		13.72	.583		17.70	20.50		38.20	52	
	6100	38 mm gap, screw on aluminum extrusion		13.72	.583		15.95	20.50		36.45	50	
	6240	Vinyl fabric loop, slotted aluminum extrusion, 25 mm gap		13.72	.583		5.65	20.50		26.15	38.50	
	6300	32 mm gap	▼	13.72	.583	▼	16.80	20.50		37.30	51	
300	0010	**WEATHERSTRIPPING**, Window, double hung, 900 mm x 1500 mm										300
	0020	Zinc	1 Carp	7.20	1.111	Opng.	12.05	39.50		51.55	75	
	0100	Bronze		7.20	1.111		23.50	39.50		63	87.50	
	0500	As above but heavy duty, zinc		4.60	1.739		15.35	62		77.35	113	
	0600	Bronze		4.60	1.739		26.50	62		88.50	126	
	1000	Doors, wood frame, interlocking, for 900 mm x 2100 mm door, zinc		3	2.667		13.75	95		108.75	163	
	1100	Bronze		3	2.667		21.50	95		116.50	172	
	1300	1800 mm x 2100 mm opening, zinc		2	4		15.05	142		157.05	238	
	1400	Bronze	▼	2	4	▼	28.50	142		170.50	252	
	1700	Wood frame, spring type, bronze										
	1800	900 mm x 2100 mm door	1 Carp	7.60	1.053	Opng.	17.70	37.50		55.20	78	
	1900	1800 mm x 2100 mm door	"	7	1.143	"	21	40.50		61.50	86.50	
	2200	Metal frame, spring type, bronze										
	2300	900 mm x 2100 mm door	1 Carp	3	2.667	Opng.	30	95		125	181	

DOORS & WINDOWS | 8

08720 | Weatherstripping & Seals

		CREW	DAILY OUTPUT	LABOR-HOURS	UNIT	MAT.	LABOR	EQUIP.	TOTAL	TOTAL INCL O&P	
300	2400	1800 mm x 2100 mm door	1 Carp	2.50	3.200	Opng.	41	114		155	222	300
	2500	For stainless steel, spring type, add					133%					
	2700	Metal frame, extruded sections, 900 mm x 2100 mm door, alum.	1 Carp	2	4	Opng.	40	142		182	265	
	2800	Bronze		2	4		101	142		243	330	
	3100	1800 mm x 2100 mm door, aluminum		1.20	6.667		51	237		288	425	
	3200	Bronze		1.20	6.667		119	237		356	500	
	3500	Threshold weatherstripping										
	3650	Door sweep, flush mounted, aluminum	1 Carp	25	.320	Ea.	11.80	11.40		23.20	30.50	
	3700	Vinyl		25	.320		14	11.40		25.40	33	
	5000	Garage door bottom weatherstrip, 3650 mm aluminum, clear		14	.571		18.90	20.50		39.40	52.50	
	5010	Bronze		14	.571		72	20.50		92.50	111	
	5050	Bottom protection, 3650 mm aluminum, clear		14	.571		21	20.50		41.50	54.50	
	5100	Bronze		14	.571		89.50	20.50		110	130	
800	0010	**THRESHOLDS**										800
	0011	900 mm long door saddles, aluminum	1 Carp	14.63	.547	m	11.90	19.45		31.35	43.50	
	0100	Aluminum, 200 mm wide, 13 mm thick		12	.667	Ea.	30.50	23.50		54	70.50	
	0500	Bronze		18.29	.437	m	105	15.55		120.55	139	
	0600	Bronze, panic threshold, 127 mm wide, 13 mm thick		12	.667	Ea.	60	23.50		83.50	103	
	0700	Rubber, 13 mm thick, 140 mm wide		20	.400		33.50	14.20		47.70	59	
	0800	70 mm wide		20	.400		14.05	14.20		28.25	37.50	

08770 | Door/Window Accessories

		CREW	DAILY OUTPUT	LABOR-HOURS	UNIT	MAT.	LABOR	EQUIP.	TOTAL	TOTAL INCL O&P	
100	0010	**AREA WINDOW WELL**, Galvanized steel 20 ga.										100
	0020	1 m wide, 300 mm deep	1 Sswk	29	.276	Ea.	12.95	11		23.95	34	
	0100	600 mm deep		23	.348		23	13.90		36.90	50.50	
	0300	16 ga., galv., 1 m wide, 300 mm deep		29	.276		18.25	11		29.25	40	
	0400	900 mm deep		23	.348		37	13.90		50.90	65.50	
	0600	Welded grating for above, 6.8 kg, painted		45	.178		33.50	7.10		40.60	50	
	0700	Galvanized		45	.178		81	7.10		88.10	102	
	0900	Translucent plastic cap for above		60	.133		13.40	5.35		18.75	24.50	
200	0010	**DOOR PROTECTION** Acrylic and neoprene cover for door										200
	0020	frame, high impact type	1 Carp	30.48	.262	m	24.50	9.35		33.85	41.50	
550	0010	**DETECTION SYSTEMS** See Division 13720										550

08810 | Glass

		CREW	DAILY OUTPUT	LABOR-HOURS	UNIT	MAT.	LABOR	EQUIP.	TOTAL	TOTAL INCL O&P	
100	0010	**ACOUSTICAL GLASS UNITS**, one lite										100
	0020	For 25 mm thick	2 Glaz	9.29	1.722	m²	255	59		314	370	
	0100	For 102 mm thick	"	7.43	2.153	"	435	73.50		508.50	590	
160	0010	**BEVELED GLASS**, with design patterns										160
	0020	Minimum	2 Glaz	13.94	1.148	m²	435	39.50		474.50	540	
	0050	Average		11.61	1.378		1,025	47		1,072	1,200	
	0100	Maximum		9.29	1.722		1,800	59		1,859	2,100	
250	0010	**FACETED**, Color tinted glass										250
	0020	Minimum	2 Glaz	8.83	1.813	m²	370	62		432	500	

8

DOORS & WINDOWS

		08810	Glass	CREW	DAILY OUTPUT	LABOR-HOURS	UNIT	2006 BARE COSTS				TOTAL INCL O&P	
								MAT.	LABOR	EQUIP.	TOTAL		
250	0100		Maximum	2 Glaz	6.97	2.296	m²	645	78.50		723.50	830	250
260	0010	**FLOAT GLASS**											260
	0020		5 mm thick, plain R088110-10	2 Glaz	12.08	1.325	m²	50.50	45.50		96	124	
	0200		Tempered, clear		12.08	1.325		59	45.50		104.50	133	
	0300		Tinted		12.08	1.325		75	45.50		120.50	151	
	0600		6 mm thick, clear, plain		11.15	1.435		59.50	49		108.50	141	
	0700		Tinted		11.15	1.435		58.50	49		107.50	139	
	0800		Tempered, clear		11.15	1.435		73	49		122	155	
	0900		Tinted		11.15	1.435		101	49		150	186	
	1600		9 mm thick, clear, plain		6.97	2.296		95	78.50		173.50	223	
	1700		Tinted		6.97	2.296		117	78.50		195.50	248	
	1800		Tempered, clear		6.97	2.296		147	78.50		225.50	281	
	1900		Tinted		6.97	2.296		182	78.50		260.50	320	
	2200		13 mm thick, clear, plain		5.11	3.131		190	107		297	370	
	2300		Tinted		5.11	3.131		205	107		312	390	
	2400		Tempered, clear		5.11	3.131		221	107		328	405	
	2500		Tinted		5.11	3.131		274	107		381	460	
	2800		16 mm thick, clear, plain		4.18	3.827		205	131		336	425	
	2900		Tempered, clear		4.18	3.827		235	131		366	455	
	3200		19 mm thick, clear, plain		3.25	4.921		264	169		433	545	
	3300		Tempered, clear		3.25	4.921		310	169		479	595	
	3600		25 mm thick, clear, plain	▼	2.79	5.741		440	197		637	780	
	8900		For low emissivity coating for 5 mm & 6 mm only, add to above				▼	15%					
270	0010	**FULL VISION**											270
	0020		Up to 3000 mm high	H-2	12.08	1.987	m²	590	63.50		653.50	745	
	0100		3000 mm to 6000 mm high, minimum		10.22	2.349		625	75		700	805	
	0150		Average		9.29	2.583		675	82.50		757.50	870	
	0200		Maximum	▼	7.43	3.229	▼	760	103		863	990	
300	0010	**GLAZING VARIABLES**											300
	0500		For high rise glazing, exterior, add per m² per story R088110-10				m²					.97	
	0600		For glass replacement, add				"		100%				
	0700		For gasket settings, add				m	13.40			13.40	14.75	
	0900		For sloped glazing, add				m²		25%				
	2000		Fabrication, polished edges, 6 mm thick				mm	.01			.01	.02	
	2100		13 mm thick					.04			.04	.04	
	2500		Mitered edges, 6 mm thick					.04			.04	.04	
	2600		13 mm thick	▼			▼	.06			.06	.06	
460	0010	**INSULATING GLASS** 2 lites 3 mm float, 13 mm thk, under 1.39 m²											460
	0020		Clear	2 Glaz	8.83	1.813	m²	84.50	62		146.50	187	
	0100		Tinted R088110-10		8.83	1.813		124	62		186	231	
	0200		2 lites 5 mm float,16 mm thk unit,1.39 to 2.79 m²,clr		8.36	1.914		103	65.50		168.50	212	
	0300		Tinted		8.36	1.914		105	65.50		170.50	215	
	0400		25 mm thk, dbl. glazed, 6 mm float, 2.79 to 6.50 m², clear		6.97	2.296		145	78.50		223.50	278	
	0500		Tinted		6.97	2.296		176	78.50		254.50	310	
	0600		25 mm thick double glazed, 6 mm float, 6 mm wire		6.97	2.296		215	78.50		293.50	355	
	0700		6 mm float, 6 mm tempered		6.97	2.296		215	78.50		293.50	355	
	0800		6 mm wire, 6 mm tempered		6.97	2.296		296	78.50		374.50	445	
	0900		Both lites, 6 mm wire		6.97	2.296		278	78.50		356.50	425	
	2000		Both lites, light & heat reflective		7.90	2.026		235	69.50		304.50	365	
	2500		Heat reflective, film inside, 25 mm thick unit, clear		7.90	2.026		205	69.50		274.50	330	
	2600		Tinted		7.90	2.026		221	69.50		290.50	350	
	3000		Film on weatherside, clear, 13 mm thick unit		8.83	1.813		147	62		209	255	
	3100		16 mm thick unit	▼	8.36	1.914	▼	179	65.50		244.50	295	

DOORS & WINDOWS 8

08810 | Glass

		Description		CREW	DAILY OUTPUT	LABOR-HOURS	UNIT	MAT.	LABOR	EQUIP.	TOTAL	TOTAL INCL O&P	
460	3200	25 mm thick unit	R088110 -10	2 Glaz	7.90	2.026	m²	202	69.50		271.50	325	460
500	0010	**LAMINATED GLASS** (security item)											500
	0020	Clear float, 0.75 mm vinyl, 6 mm thick		2 Glaz	8.36	1.914	m²	102	65.50		167.50	211	
	0100	10 mm thick			7.25	2.208		160	75.50		235.50	290	
	0200	2 mm vinyl, 13 mm thick			6.04	2.650		187	91		278	345	
	1000	16 mm thick			8.36	1.914		220	65.50		285.50	340	
	2000	Bullet-resisting, 30 mm thick, to 1.39 m²			1.49	10.764		605	370		975	1,225	
	2100	Over 1.39 m²			1.49	10.764		545	370		915	1,150	
	2500	55 mm thick, to 1.39 m²			1.11	14.352		720	490		1,210	1,550	
	2600	Over 1.39 m²		▼	1.11	14.352	▼	640	490		1,130	1,450	
550	0010	**LEAD GLASS** For X-ray, see Division 13091-600											550
600	0010	**OBSCURE GLASS**											600
	0020	3 mm thick, minimum		2 Glaz	13.01	1.230	m²	83	42		125	155	
	0100	Maximum			11.61	1.378		99	47		146	181	
	0300	6 mm thick, minimum			11.15	1.435		92	49		141	176	
	0400	Maximum		▼	9.75	1.640	▼	117	56		173	214	
650	0010	**PATTERNED GLASS**											650
	0020	Colored, 3 mm thick minimum		2 Glaz	13.01	1.230	m²	92.50	42		134.50	166	
	0100	Maximum			11.61	1.378		114	47		161	197	
	0300	5 mm thick, minimum			11.15	1.435		109	49		158	195	
	0400	Maximum		▼	9.75	1.640	▼	122	56		178	219	
675	0010	**REFLECTIVE GLASS**											675
	0100	6 mm float with fused metallic oxide, tinted		2 Glaz	10.68	1.498	m²	122	51.50		173.50	212	
	0500	6 mm float glass with reflective applied coating			10.68	1.498		102	51.50		153.50	190	
	2000	Solar film on glass, not including glass, minimum			16.72	.957		49	33		82	104	
	2010	Window protection film, controls blast damage			7.43	2.153		49	73.50		122.50	166	
	2050	Solar film on glass, not including glass, maximum		▼	20.90	.765	▼	114	26		140	165	
740	0010	**SANDBLASTED GLASS**, plate glass											740
	0020	3 mm thick		2 Glaz	14.86	1.076	m²	81.50	37		118.50	146	
	0100	5 mm thick			12.08	1.325		90	45.50		135.50	168	
	0500	6 mm thick			11.15	1.435		93.50	49		142.50	178	
	0600	10 mm thick		▼	6.97	2.296	▼	102	78.50		180.50	231	
760	0010	**SHEET GLASS**											760
	0020	3 mm thick		2 Glaz	14.86	1.076	m²	42.50	37		79.50	103	
	0200	6 mm thick		"	12.08	1.325	"	55.50	45.50		101	130	
780	0010	**SPANDREL GLASS**											780
	0020	Up to 92.9 m², 6 mm thick, standard colors		2 Glaz	10.22	1.566	m²	143	53.50		196.50	238	
	0200	92.9 m² to 185.8 m²		"	11.15	1.435	"	134	49		183	222	
	0300	For custom colors, add					Total	10%					
	0500	For 10 mm thick, add					m²	84.50			84.50	92.50	
	1000	For double coated, 6 mm thick, add						31.50			31.50	34.50	
	1200	For insulation on panels, add						52			52	57	
	2000	Panels, insulated, with aluminum backed fiberglass, 25 mm thick		2 Glaz	11.15	1.435		125	49		174	213	
	2100	50 mm thick		"	11.15	1.435		146	49		195	235	
	2500	With galvanized steel backing, add					▼	43.50			43.50	47.50	
850	0010	**WINDOW GLASS**											850
	0015	3 mm thick, clear float		2 Glaz	44.59	.359	m²	40.50	12.30		52.80	63.50	
	0500	5 mm thick, clear			44.59	.359		50	12.30		62.30	73.50	
	0600	Tinted		▼	44.59	.359	▼	56.50	12.30		68.80	80.50	

8

DOORS & WINDOWS

			CREW	DAILY OUTPUT	LABOR-HOURS	UNIT	2006 BARE COSTS				TOTAL INCL O&P	
	08810	**Glass**					MAT.	LABOR	EQUIP.	TOTAL		
850	0700	Tempered	2 Glaz	44.59	.359	m²	68.50	12.30		80.80	93.50	**850**
900	0010	**WIRE GLASS** (chicken wire) (security item)										**900**
	0012	6 mm thick, rough obscure	2 Glaz	12.54	1.276	m²	134	43.50		177.50	214	
	1000	Polished wire, 6 mm thick, diamond, clear		12.54	1.276		167	43.50		210.50	249	
	1500	Pinstripe, obscure		12.54	1.276		167	43.50		210.50	249	
	08830	**Mirrors**										
100	0010	**MIRRORS** No frames, wall type, 6 mm plate glass, polished edge										**100**
	0100	Up to 0.5 m²	2 Glaz	11.61	1.378	m²	72	47		119	151	
	0200	Over 0.5 m²		14.86	1.076		69.50	37		106.50	133	
	0500	Door type, 6 mm plate glass, up to 1.12 m²		14.86	1.076		75	37		112	139	
	1000	Float glass, up to 0.93 m², 3 mm thick		14.86	1.076		42	37		79	103	
	1100	5 mm thick		13.94	1.148		49	39.50		88.50	113	
	1500	300 mm x 300 mm wall tiles, m² edge, clear		18.12	.883		17.10	30.50		47.60	65	
	1600	Veined		18.12	.883		46.50	30.50		77	97	
	2000	6 mm thick, stock sizes, one way transparent		11.61	1.378		163	47		210	251	
	2010	Bathroom, unframed, laminated		14.86	1.076		120	37		157	188	
	08840	**Plastic Glazing**										
600	0010	**PLEXIGLASS ACRYLIC**, Masked										**600**
	0020	3 mm thick, cut sheets	2 Glaz	15.79	1.013	m²	37	34.50		71.50	93.50	
	0200	Full sheets		18.12	.883		19.40	30.50		49.90	67.50	
	0500	6 mm thick, cut sheets		15.33	1.044		65.50	36		101.50	126	
	0600	Full sheets		17.19	.931		35.50	32		67.50	87	
	0900	10 mm thick, cut sheets		14.40	1.111		120	38		158	190	
	1000	Full sheets		16.72	.957		64.50	33		97.50	121	
	1300	13 mm thick, cut sheets		12.54	1.276		138	43.50		181.50	218	
	1400	Full sheets		13.94	1.148		134	39.50		173.50	207	
	1700	19 mm thick, cut sheets		10.68	1.498		490	51.50		541.50	620	
	1800	Full sheets		12.08	1.325		283	45.50		328.50	380	
	2100	25 mm thick, cut sheets		9.75	1.640		555	56		611	695	
	2200	Full sheets		11.61	1.378		340	47		387	445	
	3000	Colored, 3 mm thick, cut sheets		15.79	1.013		114	34.50		148.50	179	
	3200	Full sheets		18.12	.883		73.50	30.50		104	127	
	3500	6 mm thick, cut sheets		15.33	1.044		128	36		164	194	
	3600	Full sheets		17.19	.931		87	32		119	144	
	4000	Mirrors, untinted, cut sheets, 3 mm thick		17.19	.931		53.50	32		85.50	107	
	4200	6 mm thick		16.72	.957		86	33		119	145	
650	0010	**POLYCARBONATE** (security item)										**650**
	0020	3 mm thick	2 Glaz	15.79	1.013	m²	64.50	34.50		99	124	
	0500	5 mm thick		15.33	1.044		78	36		114	140	
	1000	6 mm thick		14.40	1.111		86	38		124	152	
	1500	10 mm thick		13.94	1.148		159	39.50		198.50	234	
700	0010	**SECURITY FILM** clear, 22000 Kpa tensile strength, adh to glass [R088110 -10]										**700**
	0100	2 mils thick, daylight installation	H-2	88.26	.272	m²	7	8.70		15.70	21	
	0150	4 mils thick, daylight installation		74.32	.323		8.30	10.30		18.60	25	
	0200	6 mils thick, daylight installation		65.03	.369		8.85	11.80		20.65	27.50	
	0210	Install for anchorage		55.74	.431		9.80	13.75		23.55	32	
	0400	7 mils thick, daylight istallation		55.74	.431		9.35	13.75		23.10	31.50	
	0410	Install for anchorage		46.45	.517		10.45	16.50		26.95	36.50	
	0500	8 mils thick, daylight installation		46.45	.517		12.90	16.50		29.40	39	
	0510	Install for anchorage		46.45	.517		14.30	16.50		30.80	41	
	0600	15 mils thick, daylight installation		37.16	.646		27	20.50		47.50	61	

DOORS & WINDOWS 8

08800 | Glazing

08840 | Plastic Glazing

			CREW	DAILY OUTPUT	LABOR-HOURS	UNIT	MAT.	LABOR	EQUIP.	TOTAL	TOTAL INCL O&P	
700	0610	Install for anchorage	H-2	37.16	.646	m²	14.30	20.50		34.80	47.50	700
	0900	Security Film Anchorage, mech. attachment and cover plate R088110 -10	H-3	113	.142	m	21.50	4.29		25.79	30	
	0950	Security film anchorage, wet glaze structural caulking	1 Glaz	68.58	.117	"	3.94	4		7.94	10.40	
	1000	Adhered security film removal	1 Clab	25.55	.313	m²		8.60		8.60	13.35	
900	0010	**VINYL GLASS**										900
	0020	2 mm thick	2 Glaz	15.79	1.013	m²	87.50	34.50		122	149	
	0500	3 mm thick		15.79	1.013		110	34.50		144.50	175	
	1000	6 mm thick		14.40	1.111		154	38		192	228	
	1500	For non-standard sizes, add					15%					

08900 | Glazed Curtain Wall

08910 | Glazed Alumimum Curtain Wall

			CREW	DAILY OUTPUT	LABOR-HOURS	UNIT	MAT.	LABOR	EQUIP.	TOTAL	TOTAL INCL O&P	
200	0010	**CURTAIN WALLS**										200
	0020	Minimum	H-1	19.04	1.680	m²	251	62.50		313.50	380	
	0050	Average, single glazed		18.12	1.766		320	65.50		385.50	465	
	0150	Average, double glazed		16.72	1.914		450	71		521	615	
	0200	Maximum		14.86	2.153		1,200	80		1,280	1,450	
700	0010	**TUBE FRAMING** For window walls and store fronts, aluminum, stock										700
	0050	Plain tube frame, mill finish, 45 mm x 45 mm	2 Glaz	31.39	.510	m	21.50	17.45		38.95	50	
	0150	45 mm x 100 mm		29.87	.536		28.50	18.35		46.85	59.50	
	0200	45 mm x 115 mm		28.96	.553		33.50	18.95		52.45	65.50	
	0250	50 mm x 150 mm		27.13	.590		48	20		68	83.50	
	0350	100 mm x 100 mm		26.52	.603		47	20.50		67.50	83.50	
	0400	115 mm x 115 mm		25.91	.618		54.50	21		75.50	92	
	0450	Glass bead		73.15	.219		6.10	7.50		13.60	18.10	
	1000	Flush tube frm, mill finish, 6 mm glass, 45 x 100 mm, open header		24.38	.656		28	22.50		50.50	65	
	1050	Open sill		24.99	.640		24.50	22		46.50	60	
	1100	Closed back header		25.30	.632		39.50	21.50		61	76.50	
	1150	Closed back sill		25.91	.618		37.50	21		58.50	73	
	1200	Vertical mullion, one piece		22.86	.700		41.50	24		65.50	82.50	
	1250	Two piece		22.25	.719		44.50	24.50		69	86.50	
	1300	90° or 180° vertical corner post		22.86	.700		70.50	24		94.50	114	
	1400	45 mm x 115 mm, open header		24.38	.656		34.50	22.50		57	72	
	1450	Open sill		24.99	.640		28.50	22		50.50	64	
	1500	Closed back header		25.30	.632		42	21.50		63.50	79	
	1550	Closed back sill		25.91	.618		40.50	21		61.50	76.50	
	1600	Vertical mullion, one piece		22.86	.700		45	24		69	86	
	1650	Two piece		22.25	.719		47.50	24.50		72	90	
	1700	90° or 180° vertical corner post		22.86	.700		49	24		73	90	
	2000	Flush tube frame, mill fin. for glass, 50 mm x 115 mm, open header		22.86	.700		37.50	24		61.50	78	
	2050	Open sill		23.47	.682		33	23.50		56.50	72	
	2100	Closed back header		23.77	.673		41	23		64	80	
	2150	Closed back sill		24.38	.656		40.50	22.50		63	78.50	
	2200	Vertical mullion, one piece		21.34	.750		45.50	25.50		71	89	
	2250	Two piece		20.73	.772		48.50	26.50		75	93.50	
	2300	90° or 180° vertical corner post		21.34	.750		46	25.50		71.50	89.50	
	5000	Flush tube frm, mill fin., therm brk. 55 mmx 115 mm, open hdr.		22.56	.709		41.50	24.50		66	82.50	

08910 | Glazed Alumimum Curtain Wall

			CREW	DAILY OUTPUT	LABOR-HOURS	UNIT	2006 BARE COSTS				TOTAL INCL O&P	
							MAT.	LABOR	EQUIP.	TOTAL		
700	5050	Open sill	2 Glaz	22.86	.700	m	36	24		60	76.50	700
	5100	Vertical mullion, one piece		21.03	.761		50	26		76	94.50	
	5150	Two piece		20.42	.783		53.50	27		80.50	99.50	
	5200	90° or 180° vertical corner post		21.03	.761		48	26		74	92.50	
	6980	Door stop (snap in)	▼	116	.138	▼	8.55	4.72		13.27	16.55	
	7000	For joints, 90°, clip type, add				Ea.	20			20	22	
	7050	Screw spline joint, add					15.35			15.35	16.85	
	7100	For joint other than 90°, add				▼	32			32	35.50	
	8000	For bronze anodized aluminum, add					15%					
	8020	For black finish, add					27%					
	8050	For stainless steel materials, add					350%					
	8100	For monumental grade, add					50%					
	8150	For steel stiffener, add	2 Glaz	60.96	.262	m	25.50	9		34.50	42	
	8200	For 2 to 5 stories, add/story				Story		5%				
900	0010	**WINDOW WALLS,** Aluminum stock including glazing	H-2	14.86	1.615	m²	305	51.50		356.50	415	900
	0020	Minimum										
	0050	Average		13.01	1.845	▼	370	59		429	495	
	0100	Maximum	▼	10.22	2.349		1,150	75		1,225	1,400	
	0500	For translucent sandwich wall systems, see Div. 07420-770										
	0850	Cost of the above walls depends on material,										
	0860	finish, repetition, and size of units.										
	0870	The larger the opening, the lower the m² cost										
	1200	Double glazed acoustical window wall for airports,										
	1220	including 25 mm thick glass with 50 mm x 115 mm tube frame	H-2	3.72	6.459	m²	670	206		876	1,050	

08950 | Translucent Wall & Roof Assemblies

			CREW	DAILY OUTPUT	LABOR-HOURS	UNIT	MAT.	LABOR	EQUIP.	TOTAL	TOTAL INCL O&P	
100	0010	**SKYROOFS,** Translucent panels, 2-3/4" thick										100
	0020	Under 460 m²	G-3	36.70	.872	m² Hor.	218	30.50		248.50	287	
	0100	Over 460 m²		43.20	.741		195	26		221	255	
	0300	Continuous vaulted, semi-circular, to 2400 mm wide, double glazed		13.47	2.376		490	82.50		572.50	670	
	0400	Single glazed		14.86	2.153		330	75		405	480	
	0600	To 6000 mm wide, single glazed		16.26	1.968		370	68.50		438.50	510	
	0700	Over 6000 mm wide, single glazed		18.58	1.722		425	60		485	560	
	0900	Motorized opening type, single glazed, 1/3 opening		13.47	2.376		455	82.50		537.50	630	
	1000	Full opening	▼	12.08	2.650	▼	520	92		612	715	
	1200	Pyramid type units, self-supporting, to 9000 mm clear opening,										
	1300	square or circular, single glazed, minimum	G-3	18.58	1.722	m² Hor.	238	60		298	355	
	1310	Average		15.33	2.088		335	72.50		407.50	480	
	1400	Maximum		12.08	2.650		485	92		577	675	
	1500	Grid type, 1200 mm to 3000 mm modules, single glass glazed, min.		18.58	1.722		315	60		375	440	
	1550	Maximum		11.89	2.691		510	93.50		603.50	705	
	1600	Preformed acrylic, minimum		27.87	1.148		375	40		415	470	
	1650	Maximum	▼	16.26	1.968	▼	525	68.50		593.50	680	
	1800	Skyroofs, dome type, self-supporting, to 30.5 m clear opening, circular										
	1900	Rise to span ratio = 0.20										
	1920	Minimum	G-3	18.30	1.749	m² Hor.	156	61		217	265	
	1950	Maximum		10.50	3.048		520	106		626	735	
	2100	Rise to span ratio = 0.33, minimum		15.70	2.038		310	71		381	450	
	2150	Maximum		9.38	3.410		615	119		734	860	
	2200	Rise to span ratio = 0.50, minimum		13.75	2.327		470	81		551	640	
	2250	Maximum		8.08	3.959		680	138		818	965	
	2400	Ridge units, continuous, to 2440 mm wide, double		12.08	2.650		1,200	92		1,292	1,450	
	2500	Single		18.58	1.722		820	60		880	995	
	2700	Ridge and furrow units, over 1220 mm O.C., double, minimum		18.58	1.722		257	60		317	375	
	2750	Maximum		11.15	2.870		475	100		575	680	
	2800	Single, minimum	▼	19.88	1.610	▼	219	56		275	325	

08950	Translucent Wall & Roof Assemblies	CREW	DAILY OUTPUT	LABOR-HOURS	UNIT	2006 BARE COSTS				TOTAL INCL O&P	
						MAT.	LABOR	EQUIP.	TOTAL		
100 2850	Maximum	G-3	14.21	2.251	m² Hor.	465	78.50		543.50	630	100
3000	Rolling roof, translucent panels, flat roof, mininum		23.50	1.361	m²	198	47.50		245.50	292	
3030	Maximum		14.86	2.153		395	75		470	550	
3100	Lean-to skyroof, long span, double, minimum		18.30	1.749		265	61		326	385	
3150	Maximum		9.38	3.410		530	119		649	765	
3300	Single, minimum		29.82	1.073		199	37.50		236.50	277	
3350	Maximum	▼	14.86	2.153	▼	330	75		405	480	

For information about Means Estimating Seminars, see yellow pages 12 and 13 in back of book

Division 9
Finishes

Estimating Tips
General
- Room Finish Schedule: A complete set of plans should contain a room finish schedule. If one is not available, it would be well worth the time and effort to put one together. A room finish schedule should contain the room number, room name (for clarity), floor materials, base materials, wainscot materials, wainscot height, wall materials (for each wall), ceiling materials, ceiling height and special instructions.
- Surplus Finishes: Review the specifications to determine if there is any requirement to provide certain amounts of extra materials for the owner's maintenance department. In some cases the owner may require a substantial amount of materials, especially when it is a special order item or long lead time item.

09200 Plaster & Gypsum Board
- Lath is estimated by the square meter for both gypsum and metal lath, plus usually 5% allowance for waste. Furring, channels and accessories are measured by the linear meter. An extra 0.3 meters should be allowed for each accessory miter or stop.
- Plaster is also estimated by the square meter. Deductions for openings vary by preference, from zero deduction to 50% of all openings over 0.6 meters in width. Some estimators deduct a percentage of the total square meters for openings. The estimator should allow one square meter for each linear meter of horizontal interior or exterior angle located below the ceiling level. Also, double the areas of small radius work.
- Each room should be measured, perimeter times maximum wall height. Floors and ceiling areas are equal to length times width.
- Drywall accessories, studs, track, and acoustical caulking are all measured by the linear meter. Drywall taping is figured by the square meter. Gypsum wallboard is estimated by the square meter. No material deductions should be made for door or window openings under 2.97 square meters. Coreboard can be obtained in a 25 mm

thickness for solid wall and shaft work. Additions should be made to price out the inside or outside corners.
- Different types of partition construction should be listed separately on the quantity sheets. There may be walls with studs of various widths, double studded, and similar or dissimilar surface materials. Shaft work is usually different construction from surrounding partitions requiring separate quantities and pricing of the work.

09300 Tile
09400 Terrazzo
- Tile and terrazzo areas are taken off on a square meter basis. Trim and base materials are measured by the linear meter. Accent tiles are listed per each. Two basic methods of installation are used. Mud set is approximately 30% more expensive than the thin set. In terrazzo work, be sure to include the linear meter of embedded decorative strips, grounds, machine rubbing and power cleanup.

09600 Flooring
- Wood flooring is available in strip, parquet, or block configuration. The latter two types are set in adhesives with quantities estimated by the square meter. The laying pattern will influence labor costs and material waste. In addition to the material and labor for laying wood floors, the estimator must make allowances for sanding and finishing these areas unless the flooring is prefinished.
- Most of the various types of flooring are all measured on a square meter basis. Base is measured by the linear meter. If adhesive materials are to be quantified, they are estimated at a specified coverage rate by 3.78 liters depending upon the specified type and the manufacturer's recommendations.
- Sheet flooring is measured by the square meter. Roll widths vary, so consideration should be given to use the most economical width, as waste must be figured into the total quantity. Consider also the installation methods available, direct glue down or stretched.

09700 Wall Finishes
- Wall coverings are estimated by the square meter. The area to be covered is measured, length by height of wall above baseboards, to calculate the square meter of each wall. This figure is divided by the number of square meters in the single roll which is being used. Deduct, in full, the areas of openings such as doors and windows. Where a pattern match is required allow 25%-30% waste. 3.78 liters of paste should be sufficient to hang 12 single rolls of light to medium weight paper

09800 Acoustical Treatment
- Acoustical systems fall into several categories. The takeoff of these materials should be by the square meter of area with a 5% allowance for waste. Do not forget about scaffolding, if applicable, when estimating these systems.

09900 Paints & Coatings
- A major portion of the work in painting involves surface preparation. Be sure to include cleaning, sanding, filling and masking costs in the estimate.
- Painting is one area where bids vary to a greater extent than almost any other section of a project. This arises from the many methods of measuring surfaces to be painted. The estimator should check the plans and specifications carefully to be sure of the required number of coats.
- Protection of adjacent surfaces is not included in painting costs. When considering the method of paint application, an important factor is the amount of protection and masking required. These must be estimated separately and may be the determining factor in choosing the method of application.

Reference Numbers
Reference numbers are shown in bold squares at the beginning of some major classifications. These numbers refer to related items in the Reference Section. The reference information may be an estimating procedure, an alternate pricing method, or technical information.

Note: Not all subdivisions listed here necessarily appear in this publication.

09060	Selective Demolition		CREW	DAILY OUTPUT	LABOR-HOURS	UNIT	MAT.	2006 BARE COSTS LABOR	EQUIP.	TOTAL	TOTAL INCL O&P	
110	**0010**	**SELECTIVE DEMOLITION, CEILINGS**	R024119 -10										**110**
	0200	Ceiling, drywall, furred and nailed or screwed		2 Clab	74.32	.215	m²		5.90		5.90	9.20	
	0220	On metal frame			70.60	.227			6.20		6.20	9.65	
	0240	On suspension system, including system			66.89	.239			6.55		6.55	10.20	
	1000	Plaster, lime and horse hair, on wood lath, incl. lath			65.03	.246			6.75		6.75	10.50	
	1020	On metal lath			52.95	.302			8.30		8.30	12.90	
	1100	Gypsum, on gypsum lath			66.89	.239			6.55		6.55	10.20	
	1120	On metal lath			46.45	.344			9.45		9.45	14.70	
	1200	Suspended ceiling, mineral fiber, 600 x 600 mm or 600 x 1200 mm			139	.115			3.15		3.15	4.91	
	1250	On suspension system, incl. system			111	.144			3.95		3.95	6.15	
	1500	Tile, wood fiber, 300 mm x 300 mm, glued			83.61	.191			5.25		5.25	8.15	
	1540	Stapled			139	.115			3.15		3.15	4.91	
	1580	On suspension system, incl. system			70.60	.227			6.20		6.20	9.65	
	2000	Wood, tongue and groove, 25 mm x 100 mm			92.90	.172			4.72		4.72	7.35	
	2040	25 mm x 200 mm			102	.157			4.30		4.30	6.70	
	2400	Plywood or wood fiberboard, 1200 mm x 2400 mm sheets			111	.144			3.95		3.95	6.15	
120	**0010**	**SELECTIVE DEMOLITION, FLOORING**	R024119 -10										**120**
	0200	Brick with mortar		2 Clab	44.13	.363	m²		9.95		9.95	15.45	
	0400	Carpet, bonded, including surface scraping			186	.086			2.36		2.36	3.67	
	0440	Scrim applied			743	.022			.59		.59	.92	
	0480	Tackless			836	.019			.52		.52	.82	
	0600	Composition, acrylic or epoxy			37.16	.431			11.80		11.80	18.35	
	0700	Concrete, scarify skin		A-1A	20.90	.383			13.95	14.90	28.85	38	
	0800	Resilient, sheet goods		2 Clab	130	.123			3.37		3.37	5.25	
	0815	Cove base		1 Clab	305	.026	m		.72		.72	1.12	
	0820	For gym floors		2 Clab	83.61	.191	m²		5.25		5.25	8.15	
	0900	Vinyl composition tile, 300 mm x 300 mm			92.90	.172			4.72		4.72	7.35	
	2000	Tile, ceramic, thin set			62.71	.255			7		7	10.90	
	2020	Mud set			58.06	.276			7.55		7.55	11.75	
	2200	Marble, slate, thin set			62.71	.255			7		7	10.90	
	2220	Mud set			58.06	.276			7.55		7.55	11.75	
	2600	Terrazzo, thin set			41.81	.383			10.50		10.50	16.30	
	2620	Mud set			39.48	.405			11.10		11.10	17.30	
	2640	Cast in place			27.87	.574			15.75		15.75	24.50	
	3000	Wood, block, on end		1 Carp	37.16	.215			7.65		7.65	11.90	
	3200	Parquet			41.81	.191			6.80		6.80	10.60	
	3400	Strip flooring, interior, 56 mm x 20 mm thick			30.19	.265			9.40		9.40	14.65	
	3500	Exterior, porch flooring, 25 mm x 100 mm			20.44	.391			13.90		13.90	21.50	
	3800	Subfloor, tongue and groove, 25 mm x 150 mm			30.19	.265			9.40		9.40	14.65	
	3820	25 mm x 200 mm			39.95	.200			7.10		7.10	11.10	
	3840	25 mm x 250 mm			48.31	.166			5.90		5.90	9.15	
	4000	Plywood, nailed			55.74	.144			5.10		5.10	7.95	
	4100	Glued and nailed			37.16	.215			7.65		7.65	11.90	
	8000	Remove flooring, bead blast, minimum		A-1A	92.90	.086			3.14	3.35	6.49	8.60	
	8100	Maximum			37.16	.215			7.85	8.40	16.25	21.50	
	8150	Mastic only			139	.058			2.10	2.24	4.34	5.75	
130	**0010**	**SELECTIVE DEMOLITION, WALLS AND PARTITIONS**	R024119 -10										**130**
	0100	Brick, 100 mm to 300 mm thick		B-9C	6.23	6.425	m³		179	23.50	202.50	305	
	0200	Concrete block, 100 mm thick			92.90	.431	m²		11.95	1.58	13.53	20.50	
	0280	200 mm thick			75.25	.532			14.80	1.96	16.76	25	
	0300	Exterior stucco 25 mm thick over mesh		B-9	297	.135			3.74	.50	4.24	6.40	
	1000	Drywall, nailed or screwed		1 Clab	92.90	.086			2.36		2.36	3.67	
	1020	Glued and nailed			83.61	.096			2.62		2.62	4.08	
	1500	Fiberboard, nailed			83.61	.096			2.62		2.62	4.08	
	1520	Glued and nailed			74.32	.108			2.95		2.95	4.59	
	1568	Plenum barrier, sheet lead			27.87	.287			7.85		7.85	12.25	

Important: See the Reference Section for supporting data - Crews, Rental Equipment, City Cost Indexes and Reference Data

		09060	Selective Demolition	CREW	DAILY OUTPUT	LABOR-HOURS	UNIT	2006 BARE COSTS				TOTAL INCL O&P	
								MAT.	LABOR	EQUIP.	TOTAL		
130	2000	Movable walls, metal, 1500 mm high	R024119-10	1 Clab	27.87	.287	m²		7.85		7.85	12.25	130
	2020	2400 mm high		↓	37.16	.215			5.90		5.90	9.20	
	2200	Metal or wood studs, finish 2 sides, fiberboard		B-1	48.31	.497			13.95		13.95	21.50	
	2250	Lath and plaster			24.15	.994			28		28	43.50	
	2300	Plasterboard (drywall)			48.31	.497			13.95		13.95	21.50	
	2350	Plywood		↓	41.81	.574			16.10		16.10	25	
	3000	Plaster, lime and horsehair, on wood lath		1 Clab	37.16	.215			5.90		5.90	9.20	
	3020	On metal lath			31.12	.257			7.05		7.05	10.95	
	3400	Gypsum or perlite, on gypsum lath			38.09	.210			5.75		5.75	8.95	
	3420	On metal lath		↓	27.87	.287			7.85		7.85	12.25	
	3600	Plywood, one side		B-1	139	.173			4.85		4.85	7.55	
	3750	Terra cotta block and plaster, to 150 mm thick		"	16.26	1.476	↓		41.50		41.50	64.50	
	3800	Toilet partitions, slate or marble		1 Clab	5	1.600	Ea.		44		44	68	
	3820	Hollow metal		"	8	1	"		27.50		27.50	42.50	
	5000	Wallcovering, vinyl		1 Pape	65.03	.123	m²		3.92		3.92	5.90	
	5040	Designer		"	44.59	.179	"		5.70		5.70	8.60	

FINISHES 9

		09110	N.L.B. Wall Framing	CREW	DAILY OUTPUT	LABOR-HOURS	UNIT	2006 BARE COSTS				TOTAL INCL O&P	
								MAT.	LABOR	EQUIP.	TOTAL		
100	0010	**METAL STUDS AND TRACK**											100
	1600	Non-load bearing, galv, 2450 mm H, 25 ga. 41 mm W, 400 mm O.C.	1 Carp	57.51	.139	m²	2.58	4.95		7.53	10.50		
	1610	610 mm O.C.		88.26	.091		1.94	3.22		5.16	7.15		
	1620	64 mm wide, 400 mm O.C.		56.95	.140		3.12	4.99		8.11	11.15		
	1630	610 mm O.C.		87.14	.092		2.37	3.26		5.63	7.70		
	1640	92 mm wide, 400 mm O.C.		55.74	.144		3.55	5.10		8.65	11.85		
	1650	610 mm O.C.		85.93	.093		2.69	3.31		6	8.05		
	1660	102 mm wide, 400 mm O.C.		55.18	.145		3.77	5.15		8.92	12.20		
	1670	610 mm O.C.		85.93	.093		2.80	3.31		6.11	8.25		
	1680	152 mm wide, 400 mm O.C.		54.63	.146		5.50	5.20		10.70	14.15		
	1690	610 mm O.C.		84.17	.095		4.09	3.38		7.47	9.75		
	1700	20 ga. studs, 41 mm wide, 400 mm O.C.		45.89	.174		4.20	6.20		10.40	14.30		
	1710	610 mm O.C.		70.88	.113		3.12	4.01		7.13	9.70		
	1720	64 mm wide, 400 mm O.C.		45.34	.176		4.95	6.25		11.20	15.25		
	1730	610 mm O.C.		69.68	.115		3.77	4.08		7.85	10.45		
	1740	92 mm wide, 400 mm O.C.		44.68	.179		5.90	6.35		12.25	16.35		
	1750	610 mm O.C.		68.56	.117		4.41	4.15		8.56	11.30		
	1760	102 mm wide, 400 mm O.C.		44.13	.181		6.45	6.45		12.90	17.15		
	1770	610 mm O.C.		68.56	.117		4.84	4.15		8.99	11.85		
	1780	152 mm wide, 400 mm O.C.		43.57	.184		8.20	6.55		14.75	19.20		
	1790	610 mm O.C.		67.35	.119		6.15	4.22		10.37	13.35		
	2000	Non-load bearing, galv, 3000 mm H, 25 ga. 41 mm W, 400 mm O.C.		45.99	.174		2.37	6.20		8.57	12.35		
	2100	610 mm O.C.		70.60	.113		1.83	4.03		5.86	8.20		
	2200	64 mm wide, 400 mm O.C.		45.52	.176		2.91	6.25		9.16	13		
	2250	610 mm O.C.		69.68	.115		2.15	4.08		6.23	8.70		
	2300	92 mm wide, 400 mm O.C.		44.59	.179		3.34	6.40		9.74	13.60		
	2350	610 mm O.C.		68.75	.116		2.48	4.14		6.62	9.15		
	2400	102 mm wide, 400 mm O.C.		44.13	.181		3.55	6.45		10	13.95		
	2450	610 mm O.C.		68.75	.116		2.69	4.14		6.83	9.35		
	2500	152 mm wide, 400 mm O.C.		43.66	.183		5.15	6.50		11.65	15.85		

09110 | N.L.B. Wall Framing

		CREW	DAILY OUTPUT	LABOR-HOURS	UNIT	MAT.	LABOR	EQUIP.	TOTAL	TOTAL INCL O&P	
100	2550	610 mm O.C.	1 Carp	67.35	.119	m²	3.88	4.22		8.10	10.75	100
	2600	20 ga. studs, 41 mm wide, 400 mm O.C.		36.70	.218		3.98	7.75		11.73	16.35	
	2650	610 mm O.C.		56.67	.141		2.91	5		7.91	11.05	
	2700	64 mm wide, 400 mm O.C.		36.23	.221		4.74	7.85		12.59	17.35	
	2750	610 mm O.C.		55.74	.144		3.44	5.10		8.54	11.70	
	2800	92 mm wide, 400 mm O.C.		35.77	.224		5.60	7.95		13.55	18.55	
	2850	610 mm O.C.		54.81	.146		4.09	5.20		9.29	12.60	
	2900	102 mm wide, 400 mm O.C.		35.30	.227		6.15	8.05		14.20	19.35	
	2950	610 mm O.C.		54.81	.146		4.52	5.20		9.72	13.05	
	3000	152 mm wide, 400 mm O.C.		34.84	.230		7.75	8.15		15.90	21	
	3050	610 mm O.C.		53.88	.148		5.70	5.30		11	14.55	
	3060	Non-load bearing, galv, 3660 mm H, 25 ga. 41 mm W, 400 mm O.C.		38.37	.209		2.26	7.40		9.66	14.05	
	3070	610 mm O.C.		58.81	.136		1.72	4.84		6.56	9.40	
	3080	64 mm wide, 400 mm O.C.		37.90	.211		2.80	7.50		10.30	14.70	
	3090	610 mm O.C.		58.06	.138		2.05	4.90		6.95	9.90	
	3100	92 mm wide, 400 mm O.C.		37.16	.215		3.23	7.65		10.88	15.35	
	3110	610 mm O.C.		57.32	.140		2.37	4.96		7.33	10.35	
	3120	102 mm wide, 400 mm O.C.		36.79	.217		3.44	7.75		11.19	15.80	
	3130	610 mm O.C.		57.32	.140		2.48	4.96		7.44	10.45	
	3140	152 mm wide, 400 mm O.C.		36.42	.220		4.95	7.80		12.75	17.65	
	3150	610 mm O.C.		56.11	.143		3.66	5.05		8.71	11.90	
	3160	20 ga. studs, 41 mm wide, 400 mm O.C.		30.56	.262		3.77	9.30		13.07	18.60	
	3170	610 mm O.C.		47.19	.170		2.80	6.05		8.85	12.40	
	3180	64 mm wide, 400 mm O.C.		30.19	.265		4.52	9.40		13.92	19.60	
	3190	610 mm O.C.		46.45	.172		3.23	6.10		9.33	13.10	
	3200	92 mm wide, 400 mm O.C.		29.82	.268		5.40	9.55		14.95	21	
	3210	610 mm O.C.		45.71	.175		3.88	6.20		10.08	14	
	3220	102 mm wide, 400 mm O.C.		29.45	.272		5.80	9.65		15.45	21.50	
	3230	610 mm O.C.		45.71	.175		4.31	6.20		10.51	14.45	
	3240	152 mm wide, 400 mm O.C.		29.08	.275		7.45	9.80		17.25	23.50	
	3250	610 mm O.C.		44.87	.178		5.40	6.35		11.75	15.75	
	5000	Load bearing studs, see division 05410-400										

09120 | Ceiling Suspension

		CREW	DAILY OUTPUT	LABOR-HOURS	UNIT	MAT.	LABOR	EQUIP.	TOTAL	TOTAL INCL O&P	
100	0010	**CEILING SUSPENSION SYSTEMS** For gypsum board or plaster										100
	8000	Suspended ceilings, including carriers										
	8200	38 mm carriers, 610 mm O.C. with:										
	8300	22 mm channels, 406 mm O.C.	1 Lath	15.33	.522	m²	4.20	17.10		21.30	30	
	8320	610 mm O.C.		18.58	.431		3.44	14.10		17.54	25	
	8400	41 mm channels, 406 mm O.C.		14.40	.556		5.05	18.20		23.25	32.50	
	8420	610 mm O.C.		17.65	.453		3.98	14.85		18.83	26.50	
	8600	51 mm carriers, 610 mm O.C. with:										
	8700	22 mm channels, 406 mm O.C.	1 Lath	14.40	.556	m²	4.74	18.20		22.94	32	
	8720	610 mm O.C.		17.65	.453		3.98	14.85		18.83	26.50	
	8800	41 mm channels, 406 mm O.C.		13.47	.594		5.60	19.50		25.10	35	
	8820	610 mm O.C.		16.72	.478		4.52	15.70		20.22	28.50	

09130 | Acoustical Suspension

		CREW	DAILY OUTPUT	LABOR-HOURS	UNIT	MAT.	LABOR	EQUIP.	TOTAL	TOTAL INCL O&P	
100	0010	**CEILING SUSPENSION SYSTEMS** For boards and tile										100
	0050	Class A suspension system, 23 mm T bar, 600 mm x 1200 mm grid	1 Carp	74.32	.108	m²	5.60	3.83		9.43	12.20	
	0300	600 mm x 600 mm grid	"	60.39	.132		7.10	4.71		11.81	15.10	
	0350	For 14 mm grid, add					1.61			1.61	1.83	
	0360	For fire rated grid, add					.97			.97	1.08	
	0370	For colored grid, add					2.05			2.05	2.26	
	0400	Concealed Z bar suspension system, 300 mm module	1 Carp	48.31	.166		5.05	5.90		10.95	14.75	
	0600	38 mm carrier channels, 1219 mm O.C., add	"	43.66	.183		1.08	6.50		7.58	11.35	

09100 | Metal Support Assemblies

		09130	Acoustical Suspension	CREW	DAILY OUTPUT	LABOR-HOURS	UNIT	2006 BARE COSTS MAT.	LABOR	EQUIP.	TOTAL	TOTAL INCL O&P	
100	0700		Carrier channels for ceilings with										100
	0900		recessed lighting fixtures, add	1 Carp	42.73	.187	m²	1.94	6.65		8.59	12.50	
	1040		Hanging wire, 12 ga., 1200 mm long		604	.013		.61	.47		1.08	1.40	
	1080		2438 mm long		604	.013		1.21	.47		1.68	2.07	
	1200		Seismic bracing, 64 mm compression post with four bracing wires		50	.160	Ea.	2.61	5.70		8.31	11.70	

09200 | Plaster & Gypsum Board

		09205	Furring & Lathing	CREW	DAILY OUTPUT	LABOR-HOURS	UNIT	2006 BARE COSTS MAT.	LABOR	EQUIP.	TOTAL	TOTAL INCL O&P	
530	0010	FURRING	Beams & columns, 22 mm channels,galvanized										530
	0030		305 mm O.C.	1 Lath	14.40	.556	m²	2.69	18.20		20.89	30	
	0050		406 mm O.C.		15.79	.507		2.26	16.60		18.86	27	
	0070		610 mm O.C.		17.19	.465		1.51	15.25		16.76	24	
	0100		Ceilings, on steel, 22 mm channels, galvanized, 300 mm O.C.		19.51	.410		2.48	13.45		15.93	22.50	
	0300		400 mm O.C.		26.94	.297		2.26	9.75		12.01	17	
	0400		600 mm O.C.		39.02	.205		1.51	6.75		8.26	11.60	
	0600		41 mm channels, galvanized, 300 mm O.C.		17.65	.453		3.44	14.85		18.29	26	
	0700		406 mm O.C.		24.15	.331		3.12	10.85		13.97	19.60	
	0900		600 mm O.C.		36.23	.221		2.05	7.25		9.30	13	
	1000		Walls, 22 mm channels, galvanized, 300 mm O.C.		21.83	.366		2.48	12		14.48	20.50	
	1200		400 mm O.C.		24.62	.325		2.26	10.65		12.91	18.35	
	1300		610 mm O.C.		32.52	.246		1.51	8.05		9.56	13.60	
	1500		41 mm channels, galvanized, 300 mm O.C.		19.51	.410		3.44	13.45		16.89	24	
	1600		400 mm O.C.		22.30	.359		3.12	11.75		14.87	21	
	1800		610 mm O.C.		28.33	.282		2.05	9.25		11.30	16	
540	0010	GYPSUM LATH											540
	0012		Plain or perforated, nailed, 9 mm thick	1 Lath	71.07	.113	m²	6.45	3.69		10.14	12.60	
	0100		13 mm thick, nailed		66.89	.120		5.05	3.92		8.97	11.40	
	0300		Clipped to steel studs, 9 mm thick		62.71	.128		6.45	4.18		10.63	13.30	
	0400		13 mm thick		58.53	.137		5.05	4.48		9.53	12.20	
	0600		Firestop gypsum base, to steel studs, 10 mm thick		58.53	.137		4.41	4.48		8.89	11.50	
	0700		13 mm thick		54.35	.147		5.40	4.83		10.23	13.10	
	0900		Foil back, to steel studs, 10 mm thick		62.71	.128		4.74	4.18		8.92	11.40	
	1000		13 mm thick		58.53	.137		4.95	4.48		9.43	12.10	
	1500		For ceiling installations, add		181	.044			1.45		1.45	2.15	
	1600		For columns and beams, add		142	.056			1.85		1.85	2.75	
560	0010	METAL LATH											560
	0020		Diamond, expanded, 1.36 kg/m², painted				m²	3.41			3.41	3.76	
	0100		Galvanized, 1.36 kg/m²					2.83			2.83	3.12	
	0300		1.85 kg/m², painted					4.44			4.44	4.88	
	0400		Galvanized					3.72			3.72	4.09	
	0600		For 6.80 kg asphalt sheathing paper, add					.37			.37	.41	
	0900		Flat rib, 3 mm high, 1.25 kg, painted					3.64			3.64	3.99	
	1000		Foil backed					4.37			4.37	4.81	
	1200		1.85 kg/m², painted					3.92			3.92	4.32	
	1300		Galvanized					4.94			4.94	5.45	
	1500		For 6.80 kg asphalt sheating paper, add					.37			.37	.41	
	1800		High rib, 10 mm high, 1.85 kg/m², painted					5.25			5.25	5.80	

FINISHES 9

09205		Furring & Lathing	CREW	DAILY OUTPUT	LABOR-HOURS	UNIT	MAT.	LABOR	EQUIP.	TOTAL	TOTAL INCL O&P	
560	1900	Galvanized				m²	4.39			4.39	4.83	**560**
	2400	High rib, 19 mm high, painted, 2.93 kg/m²					4.52			4.52	4.95	
	2500	3.66 kg/m²					9.70			9.70	10.65	
	2800	Stucco mesh, painted, 1.63 kg					3.97			3.97	4.37	
	3000	K-lath, perforated, absorbent paper, regular					3.84			3.84	4.22	
	3100	Heavy duty					4.53			4.53	4.99	
	3300	Waterproof, heavy duty, grade B backing					4.44			4.44	4.88	
	3400	Fire resistant backing					4.90			4.90	5.40	
	3600	1.13 kg diamond painted, on wood framing, on walls	1 Lath	71.07	.113		3.41	3.69		7.10	9.25	
	3700	On ceilings		62.71	.128		3.41	4.18		7.59	9.95	
	3900	1.54 kg diamond painted, on wood framing, on walls		66.89	.120		3.92	3.92		7.84	10.15	
	4000	On ceilings		58.53	.137		3.92	4.48		8.40	10.95	
	4200	1.54 kg diamond painted, wired to steel framing		62.71	.128		3.92	4.18		8.10	10.50	
	4300	On ceilings		50.17	.159		3.92	5.25		9.17	12.05	
	4500	Columns and beams, wired to steel		33.44	.239		3.92	7.85		11.77	15.95	
	4600	Cornices, wired to steel		29.26	.273		3.92	8.95		12.87	17.65	
	4800	Screwed to steel studs, 1.13 kg		66.89	.120		3.41	3.92		7.33	9.60	
	4900	1.54 kg		62.71	.128		4.44	4.18		8.62	11.10	
	5100	Rib lath, painted, wired to steel, on walls, 1.13 kg		62.71	.128		3.64	4.18		7.82	10.20	
	5200	1.54 kg		58.53	.137		5.25	4.48		9.73	12.45	
	5400	1.81 kg	↓	54.35	.147		5.40	4.83		10.23	13.15	
	5500	For self-furring lath, add					.10			.10	.11	
	5700	Suspended ceiling system, incl. 1.54 kg diamond lath, painted	1 Lath	12.54	.638		14.20	21		35.20	46.50	
	5800	Galvanized	"	12.54	.638	↓	14.65	21		35.65	47	
	6000	Hollow metal stud partitions, 1.54 kg painted lath both sides										
	6010	Non-load bearing, 25 ga., w/rib lath 64 mm studs, 305 mm O.C.	1 Lath	16.97	.471	m²	14.20	15.45		29.65	38.50	
	6300	406 mm O.C.		17.64	.453		13.35	14.85		28.20	36.50	
	6350	610 mm O.C.		18.98	.422		12.70	13.85		26.55	34.50	
	6400	92 mm studs, 406 mm O.C.		16.30	.491		13.85	16.10		29.95	39.50	
	6600	610 mm O.C.		17.06	.469		13	15.40		28.40	37.50	
	6700	102 mm studs, 406 mm O.C.		17.06	.469		14.10	15.40		29.50	38.50	
	6900	610 mm O.C.		18.06	.443		13.15	14.55		27.70	36	
	7000	152 mm studs, 406 mm O.C.		16.30	.491		15.75	16.10		31.85	41.50	
	7100	610 mm O.C.		17.64	.453		14.35	14.85		29.20	38	
	7200	L.B. partitions, 16 ga., w/rib lath, 64 mm studs, 406 mm O.C.		16.72	.478		13.55	15.70		29.25	38.50	
	7300	92 mm studs, 16 ga.		16.47	.486		15.35	15.95		31.30	40.50	
	7500	102 mm studs, 16 ga.		16.30	.491		16	16.10		32.10	41.50	
	7600	152 mm studs, 16 ga.	↓	15.64	.512	↓	18.90	16.80		35.70	46	
600	0010	**SECURITY MESH**, expanded metal, flat, screwed to framing										**600**
	0100	On walls, 19 mm	2 Carp	139	.115	m²	16.80	4.09		20.89	25	
	0110	38 mm		149	.107		13.90	3.82		17.72	21.50	
	0200	On ceilings, 19 mm		125	.128		16.80	4.55		21.35	25.50	
	0210	38 mm	↓	135	.119	↓	13.90	4.21		18.11	22	
700	0010	**ACCESSORIES, PLASTER**										**700**
	0020	Casing bead, expanded flange, galvanized	1 Lath	82.30	.097	m	1.20	3.19		4.39	6.05	
	0900	Channels, cold rolled, 16 ga., 19 mm deep, galvanized					.76			.76	.84	
	1200	38 mm deep, 16 ga., galvanized					1.04			1.04	1.15	
	1620	Corner bead, expanded bullnose, 19 mm radius, #10 galvanized	1 Lath	79.25	.101		.98	3.31		4.29	6	
	1650	#1, galvanized		77.72	.103		1.55	3.38		4.93	6.70	
	1670	Expanded wing, 70 mm wide, galv. #1		80.77	.099		.99	3.25		4.24	5.90	
	1700	Inside corner, (corner rite) 76 mm x 76 mm, painted		79.25	.101		.87	3.31		4.18	5.85	
	1750	Strip-ex, 102 mm wide, painted		77.72	.103		.75	3.38		4.13	5.80	
	1800	Expansion joint, 19 mm grounds, limited expansion, galv., 1 piece		82.30	.097		3.12	3.19		6.31	8.15	
	2100	Extreme expansion, galvanized, 2 piece	↓	79.25	.101	↓	5.30	3.31		8.61	10.75	

Important: See the Reference Section for supporting data - Crews, Rental Equipment, City Cost Indexes and Reference Data

9 FINISHES

09210	Gypsum Plaster	CREW	DAILY OUTPUT	LABOR-HOURS	UNIT	2006 BARE COSTS				TOTAL INCL O&P	
						MAT.	LABOR	EQUIP.	TOTAL		
100	**0010**	**GYPSUM PLASTER**									**100**
	0020	36 kg bag, less than 0.9 metric ton				Bag	14.45			14.45	15.85	
	0100	Over 0.9 metric ton				"	13.20			13.20	14.50	
	0300	2 coats, no lath included, on walls	J-1	87.79	.456	m²	3.93	13.95	1.20	19.08	26.50	
	0400	On ceilings	"	76.92	.520		3.93	15.95	1.37	21.25	30	
	0600	2 coats on and incl. 10 mm gypsum lath on steel, on walls	J-2	81.10	.592		10.40	18.35	1.30	30.05	40.50	
	0700	On ceilings	"	69.40	.692		10.40	21.50	1.52	33.42	45.50	
	0900	3 coats, no lath included, on walls	J-1	72.74	.550		5.55	16.85	1.45	23.85	33	
	1000	On ceilings	"	65.22	.613		5.55	18.80	1.62	25.97	36.50	
	1200	3 coats on and including painted metal lath, on wood studs	J-2	71.90	.668		10.60	20.50	1.47	32.57	44.50	
	1300	On ceilings	"	63.96	.750	↓	10.60	23.50	1.65	35.75	48.50	
	1600	For irregular or curved surfaces, add						30%				
	1800	For columns & beams, add						50%				
200	**0010**	**GAUGING PLASTER**										**200**
	0020	45 kg bags, less than 0.9 metric ton				Bag	22.50			22.50	25	
	0100	Over 0.9 metric ton				"	18.75			18.75	20.50	
300	**0010**	**KEENES CEMENT**										**300**
	0020	In 45 kg bags, less than 0.9 metric ton				Bag	23.50			23.50	26	
	0100	Over 0.9 metric ton				"	23			23	25.50	
	0300	Finish only, add to plaster prices, standard	J-1	180	.222	m²	2.67	6.80	.59	10.06	13.85	
	0400	High quality	"	120	.333	"	2.69	10.20	.88	13.77	19.40	
500	**0010**	**PERLITE OR VERMICULITE PLASTER**										**500**
	0020	In 45 kg bags Under 200 bags				Bag	14.55			14.55	16	
	0100	Over 200 bags				"	13.10			13.10	14.40	
	0300	2 coats, no lath included, on walls	J-1	76.92	.520	m²	3.88	15.95	1.37	21.20	30	
	0400	On ceilings	"	66.05	.606		3.88	18.55	1.60	24.03	34	
	0600	2 coats, on and incl. 10 mm gypsum lath, on metal studs	J-2	70.23	.683		8.25	21	1.50	30.75	43	
	0700	On ceilings	"	58.53	.820		8.25	25.50	1.80	35.55	49.50	
	0900	3 coats, no lath included, on walls	J-1	61.87	.647		7	19.80	1.71	28.51	39.50	
	1000	On ceilings	"	52.67	.759		7	23.50	2	32.50	45	
	1200	3 coats, on and incl. painted metal lath, on metal studs	J-2	60.20	.797		12.25	24.50	1.75	38.50	53	
	1300	On ceilings		51	.941		12.25	29	2.07	43.32	60	
	1500	3 coats, on and incl. suspended metal lath ceiling	↓	30.94	1.552		21	48	3.41	72.41	100	
	1700	For irregular or curved surfaces, add to above						30%				
	1800	For columns and beams, add to above						50%				
	1900	For soffits, add to ceiling prices				↓		40%				
650	**0010**	**PLASTER PARTITION WALL**										**650**
	0400	Stud walls, 1.54 kg metal lath, 3 coat gypsum plaster, 2 sides										
	0600	50 mm x 100 mm wood studs, 400 mm O.C.	J-2	29.26	1.640	m²	34.50	51	3.61	89.11	118	
	0700	64 mm metal studs, 25 ga., 305 mm O.C.	↓	30.19	1.590	↓	32	49.50	3.50	85	113	
	0800	92 mm metal studs, 25 ga., 400 mm O.C.	↓	29.73	1.615	↓	32	50	3.55	85.55	115	
	0900	Gypsum lath, 2 coat vermiculite plaster, 2 sides										
	1000	50 mm x 100 mm wood studs, 400 mm O.C.	J-2	32.98	1.455	m²	38.50	45	3.20	86.70	114	
	1200	64 mm metal studs, 25 ga., 305 mm O.C.	↓	33.91	1.416		34	44	3.11	81.11	107	
	1300	92 mm metal studs, 25 ga., 400 mm O.C.	↓	33.44	1.435	↓	35	44.50	3.16	82.66	109	
900	**0010**	**THIN COAT**										**900**
	0012	1 coat veneer, not incl. lath	J-1	334	.120	m²	.75	3.67	.32	4.74	6.75	
	1000	In 23 kg bags				Bag	9.85			9.85	10.80	

FINISHES **9**

09220	Portland Cement Plaster	CREW	DAILY OUTPUT	LABOR-HOURS	UNIT	2006 BARE COSTS				TOTAL INCL O&P
						MAT.	LABOR	EQUIP.	TOTAL	
200	**0010** **STUCCO**									**200**
0015	3 coats 25 mm thk, float finish, w/ mesh, on wood frame	J-2	52.67	.911	m²	6.40	28	2	36.40	52
0100	On masonry construction, no mesh incl.	J-1	56.02	.714		2.45	22	1.89	26.34	38
0300	For trowel finish, add	1 Plas	142	.056			1.83		1.83	2.77
0400	For 19 mm thick, on masonry, deduct	J-1	736	.054		.68	1.66	.14	2.48	3.43
0600	For coloring and special finish, add, minimum	↓	573	.070		.44	2.14	.18	2.76	3.93
0700	Maximum	↓	167	.240		1.54	7.35	.63	9.52	13.50
0900	For soffits, add	J-2	130	.369		2.38	11.45	.81	14.64	21
1000	Exterior stucco, with bonding agent, 3 coats, on walls, no mesh incl.	J-1	167	.240		3.98	7.35	.63	11.96	16.15
1200	Ceilings		150	.267		3.98	8.15	.70	12.83	17.50
1300	Beams		66.89	.598		3.98	18.30	1.58	23.86	33.50
1500	Columns	↓	83.61	.478		3.98	14.65	1.26	19.89	28
1600	Mesh, painted, nailed to wood, 0.82 kg	1 Lath	50.17	.159		4.99	5.25		10.24	13.25
1800	1.63 kg		45.99	.174		3.97	5.70		9.67	12.85
1900	Wired to steel, painted, 0.82 kg		44.31	.181		4.99	5.90		10.89	14.30
2100	1.63 kg	↓	41.81	.191	↓	3.97	6.30		10.27	13.70

09250	Gypsum Board									
200	**0010** **CEMENTITIOUS BACKERBOARD**									**200**
0070	On floor 0.9 m x 1.2 m x 12.7 mm, sheets	2 Carp	48.77	.328	m²	11.20	11.65		22.85	30.50
0080	0.9 m x 1.52 m x 12.7 mm sheets		48.77	.328		11.20	11.65		22.85	30.50
0090	0.9 m x 1.83 m x 12.7 mm sheets		48.77	.328		8.20	11.65		19.85	27
0100	0.9 m x 1.2 m x 15.9 mm sheets		48.77	.328		11.65	11.65		23.30	31
0110	0.9 m x 1.52 m x 15.9 mm sheets		48.77	.328		11.40	11.65		23.05	31
0120	0.9 m x 1.83 m x 15.9 mm sheets		48.77	.328		11.50	11.65		23.15	31
0150	On wall, 0.9 m x 1.2 m x 12.7 mm sheets		32.52	.492		11.20	17.50		28.70	39.50
0160	0.9 m x 1.52 m x 12.7 mm sheets		32.52	.492		11.20	17.50		28.70	39.50
0170	0.9 m x 1.83 m x 12.7 mm sheets		32.52	.492		8.20	17.50		25.70	36
0180	0.9 m x 1.2 m x 15.9 mm sheets		32.52	.492		11.65	17.50		29.15	40
0190	0.9 m x 1.52 m x 15.9 mm sheets		32.52	.492		11.40	17.50		28.90	39.50
0200	0.9 m x 1.83 m x 15.9 mm sheets		32.52	.492		11.50	17.50		29	39.50
0250	On counter, 0.9 m x 1.2 m x 12.7 mm		16.72	.957		11.20	34		45.20	65.50
0260	0.9 m x 1.52 m x 12.7 mm sheets		16.72	.957		11.20	34		45.20	65.50
0270	0.9 m x 1.83 m x 12.7 mm sheets		16.72	.957		8.20	34		42.20	62
0300	0.9 m x 1.2 m x 15.9 mm sheets		16.72	.957		11.65	34		45.65	66
0310	0.9 m x 1.52 m x 15.9 mm sheets		16.72	.957		11.40	34		45.40	65.50
0320	0.9 m x 1.83 m x 15.9 mm sheets	↓	16.72	.957	↓	11.50	34		45.50	65.50
300	**0010** **BLUEBOARD** For use with thin coat									**300**
0100	plaster application (see division 09210-900)									
1000	10 mm thick, on walls or ceilings, standard, no finish included	2 Carp	177	.090	m²	2.48	3.21		5.69	7.70
1100	With thin coat plaster finish		81.29	.197		3.23	7		10.23	14.45
1400	On beams, columns, or soffits, standard, no finish included		62.71	.255		2.80	9.05		11.85	17.20
1450	With thin coat plaster finish		44.13	.363		3.66	12.90		16.56	24
3000	13 mm thick, on walls or ceilings, standard, no finish included		177	.090		2.91	3.21		6.12	8.25
3100	With thin coat plaster finish		81.29	.197		3.66	7		10.66	15
3300	Fire resistant, no finish included		177	.090		2.91	3.21		6.12	8.25
3400	With thin coat plaster finish		81.29	.197		3.66	7		10.66	15
3450	On beams, columns, or soffits, standard, no finish included		62.71	.255		3.34	9.05		12.39	17.75
3500	With thin coat plaster finish		44.13	.363		4.09	12.90		16.99	24.50
3700	Fire resistant, no finish included		62.71	.255		3.34	9.05		12.39	17.75
3800	With thin coat plaster finish		44.13	.363		4.09	12.90		16.99	24.50
5000	16 mm thick, on walls or ceilings, fire resistant, no finish included		177	.090		3.12	3.21		6.33	8.45
5100	With thin coat plaster finish		81.29	.197		3.88	7		10.88	15.20
5500	On beams, columns, or soffits, no finish included		62.71	.255		3.55	9.05		12.60	18.10
5600	With thin coat plaster finish	↓	44.13	.363	↓	4.41	12.90		17.31	25

9 FINISHES

		09250 | Gypsum Board	CREW	DAILY OUTPUT	LABOR-HOURS	UNIT	2006 BARE COSTS				TOTAL INCL O&P	
							MAT.	LABOR	EQUIP.	TOTAL		
300	6000	For high ceilings, over 2.4 m high, add	2 Carp	284	.056	m²		2		2	3.12	**300**
	6500	For over 3 stories high, add/story	↓	567	.028	↓		1		1	1.56	
500	0010	**CEILINGS** Gypsum drywall, fire rated, finished										**500**
	0100	Screwed to grid, channel or joists, 13 mm thick	2 Carp	71.07	.225	m²	3.23	8		11.23	16	
	0200	16 mm thick		71.07	.225		3.23	8		11.23	16	
	0300	Over 2.4 m high, 13 mm thick		57.13	.280		3.23	9.95		13.18	19.05	
	0400	16 mm thick	↓	57.13	.280	↓	3.23	9.95		13.18	19.05	
	0600	Grid suspension system, direct hung										
	0700	38 mm C.R.C., with 22 mm hi hat furring channel, 406 mm O.C.	2 Carp	55.74	.287	m²	10.75	10.20		20.95	28	
	0800	610 mm O.C.		83.61	.191		9.80	6.80		16.60	21.50	
	0900	92 mm C.R.C., with 22 mm hi hat furring channel, 406 mm O.C.		55.74	.287		11.40	10.20		21.60	28.50	
	1000	610 mm O.C.	↓	83.61	.191	↓	9.90	6.80		16.70	21.50	
700	0010	**DRYWALL** Gypsum plasterboard, nailed or screwed										**700**
	0100	to studs unless otherwise noted [R092910-10]										
	0150	10 mm thick, on walls, standard, no finish included	2 Carp	186	.086	m²	2.58	3.06		5.64	7.55	
	0200	On ceilings, standard, no finish included		167	.096		2.58	3.41		5.99	8.10	
	0250	On beams, columns, or soffits, no finish included		62.71	.255		2.58	9.05		11.63	16.90	
	0300	13 mm thick, on walls, standard, no finish included		186	.086		2.91	3.06		5.97	8	
	0350	Taped and finished (level 4 finish)		89.65	.178		3.23	6.35		9.58	13.55	
	0390	With compound skim coat (level 5 finish)		72	.222		3.66	7.90		11.56	16.40	
	0400	Fire resistant, no finish included		186	.086		3.23	3.06		6.29	8.30	
	0450	Taped and finished (level 4 finish)		89.65	.178		3.55	6.35		9.90	13.90	
	0490	With compound skim coat (level 5 finish)		72	.222		3.98	7.90		11.88	16.70	
	0500	Water resistant, no finish included		186	.086		2.69	3.06		5.75	7.75	
	0550	Taped and finished (level 4 finish)		89.65	.178		3.01	6.35		9.36	13.25	
	0590	With compound skim coat (level 5 finish)		72	.222		3.44	7.90		11.34	16.20	
	0600	Prefinished, vinyl, clipped to studs		83.61	.191		6.05	6.80		12.85	17.25	
	1000	On ceilings, standard, no finish included		167	.096		2.91	3.41		6.32	8.55	
	1050	Taped and finished (level 4 finish)		71.07	.225		3.23	8		11.23	16.10	
	1090	With compound skim coat (level 5 finish)		56.67	.282		3.66	10.05		13.71	19.75	
	1100	Fire resistant, no finish included		167	.096		3.23	3.41		6.64	8.85	
	1150	Taped and finished (level 4 finish)		71.07	.225		3.55	8		11.55	16.45	
	1195	With compound skim coat (level 5 finish)		56.67	.282		3.98	10.05		14.03	20	
	1200	Water resistant, no finish included		167	.096		2.69	3.41		6.10	8.30	
	1250	Taped and finished (level 4 finish)		71.07	.225		3.01	8		11.01	15.80	
	1290	With compound skim coat (level 5 finish)		56.67	.282		3.44	10.05		13.49	19.55	
	1500	On beams, columns, or soffits, standard, no finish included		62.71	.255		3.34	9.05		12.39	17.75	
	1550	Taped and finished (level 4 finish)		44.13	.363		3.23	12.90		16.13	23.50	
	1590	With compound skim coat (level 5 finish)		50.17	.319		3.66	11.35		15.01	21.50	
	1600	Fire resistant, no finish included		62.71	.255		3.77	9.05		12.82	18.20	
	1650	Taped and finished (level 4 finish)		44.13	.363		3.55	12.90		16.45	24	
	1690	With compound skim coat (level 5 finish)		50.17	.319		3.98	11.35		15.33	22	
	1700	Water resistant, no finish included		62.71	.255		3.12	9.05		12.17	17.55	
	1750	Taped and finished (level 4 finish)		44.13	.363		3.01	12.90		15.91	23.50	
	1790	With compound skim coat (level 5 finish)		50.17	.319		3.44	11.35		14.79	21.50	
	2000	16 mm thick, on walls, standard, no finish included		186	.086		3.23	3.06		6.29	8.30	
	2050	Taped and finished (level 4 finish)		89.65	.178		3.55	6.35		9.90	13.90	
	2090	With compound skim coat (level 5 finish)		72	.222		3.98	7.90		11.88	16.70	
	2100	Fire resistant, no finish included		186	.086		3.23	3.06		6.29	8.30	
	2150	Taped and finished (level 4 finish)		89.65	.178		3.55	6.35		9.90	13.90	
	2195	With compound skim coat (level 5 finish)		72	.222		3.98	7.90		11.88	16.70	
	2200	Water resistant, no finish included		186	.086		2.91	3.06		5.97	8	
	2250	Taped and finished (level 4 finish)		89.65	.178		3.23	6.35		9.58	13.55	
	2290	With compound skim coat (level 5 finish)		72	.222		3.66	7.90		11.56	16.40	
	2300	Prefinished, vinyl, clipped to studs	↓	83.61	.191		7	6.80		13.80	18.35	
	3000	On ceilings, standard, no finish included	↓	167	.096	↓	3.23	3.41		6.64	8.85	

09250 | Gypsum Board

				CREW	DAILY OUTPUT	LABOR-HOURS	UNIT	2006 BARE COSTS MAT.	LABOR	EQUIP.	TOTAL	TOTAL INCL O&P	
700	3050	Taped and finished (level 4 finish)	R092910-10	2 Carp	71.07	.225	m²	3.55	8		11.55	16.45	**700**
	3090	With compound skim coat (level 5 finish)			57.13	.280		3.98	9.95		13.93	19.90	
	3100	Fire resistant, no finish included			167	.096		3.23	3.41		6.64	8.85	
	3150	Taped and finished (level 4 finish)			71.07	.225		3.55	8		11.55	16.45	
	3190	With compound skim coat (level 5 finish)			57.13	.280		3.98	9.95		13.93	19.90	
	3200	Water resistant, no finish included			167	.096		2.91	3.41		6.32	8.55	
	3250	Taped and finished (level 4 finish)			71.07	.225		3.23	8		11.23	16.10	
	3290	With compound skim coat (level 5 finish)			57.13	.280		3.66	9.95		13.61	19.60	
	3500	On beams, columns, or soffits, no finish included			62.71	.255		3.77	9.05		12.82	18.20	
	3550	Taped and finished (level 4 finish)			44.13	.363		4.09	12.90		16.99	24.50	
	3590	With compound skim coat (level 5 finish)			35.30	.453		4.63	16.10		20.73	30	
	3600	Fire resistant, no finish included			62.71	.255		3.77	9.05		12.82	18.20	
	3650	Taped and finished (level 4 finish)			44.13	.363		4.09	12.90		16.99	24.50	
	3690	With compound skim coat (level 5 finish)			35.30	.453		3.98	16.10		20.08	29.50	
	3700	Water resistant, no finish included			62.71	.255		3.34	9.05		12.39	17.75	
	3750	Taped and finished (level 4 finish)			44.13	.363		3.77	12.90		16.67	24	
	3790	With compound skim coat (level 5 finish)			35.30	.453		3.66	16.10		19.76	29	
	4000	Fireproofing, beams or columns, 2 layers, 13 mm thick, incl. finish			30.66	.522		6.80	18.55		25.35	36.50	
	4050	16 mm thick			27.87	.574		7.20	20.50		27.70	40	
	4100	3 layers, 13 mm thick			20.90	.765		10	27		37	53.50	
	4150	16 mm thick			19.51	.820		10.75	29		39.75	57.50	
	5050	For 25 mm thick coreboard on columns		▼	44.59	.359		5.15	12.75		17.90	25.50	
	5100	For foil-backed board, add						1.08			1.08	1.18	
	5200	For work over 2.4 m high, add		2 Carp	284	.056			2		2	3.12	
	5270	For textured spray, add		2 Lath	149	.107		.54	3.52		4.06	5.90	
	5300	For over 3 stories high, add/story		2 Carp	567	.028	▼		1		1	1.56	
	5350	For finishing inner corners, add			290	.055	m	.26	1.96		2.22	3.31	
	5355	For finishing outer corners, add		▼	381	.042		.66	1.49		2.15	3.04	
	5500	For acoustical sealant, add/bead		1 Carp	152	.053	▼	.10	1.87		1.97	3.01	
	5550	Sealant, 1 quart tube					Ea.	4.89			4.89	5.40	
	5600	Sound deadening board, 6 mm gypsum		2 Carp	167	.096	m²	2.91	3.41		6.32	8.55	
	5650	13 mm wood fiber		"	167	.096	"	4.09	3.41		7.50	9.80	
800	0010	**HIGH ABUSE GYPSUM BOARD**, fiber reinforced, nailed or											**800**
	0100	screwed to studs unless otherwise noted											
	0110	13 mm thick, on walls, no finish included		2 Carp	167	.096	m²	5.70	3.41		9.11	11.55	
	0120	Taped and finished (level 4 finish)			80.82	.198		6.05	7.05		13.10	17.60	
	0130	With compound skim coat (level 5 finish)			65.03	.246		6.45	8.75		15.20	20.50	
	0150	On ceilings, no finish included			150	.107		5.70	3.79		9.49	12.15	
	0160	Taped and finished (level 4 finish)			64.10	.250		6.05	8.85		14.90	20.50	
	0170	With compound skim coat (level 5 finish)			51.10	.313		6.45	11.15		17.60	24.50	
	0210	16 mm thick, on walls, no finish included			167	.096		6.80	3.41		10.21	12.75	
	0220	Taped and finished (level 4 finish)			80.82	.198		7.10	7.05		14.15	18.80	
	0230	With compound skim coat (level 5 finish)			65.03	.246		7.55	8.75		16.30	22	
	0250	On ceilings, no finish included			150	.107		6.80	3.79		10.59	13.35	
	0260	Taped and finished (level 4 finish)			64.10	.250		7.10	8.85		15.95	21.50	
	0270	With compound skim coat (level 5 finish)			51.10	.313		7.55	11.15		18.70	25.50	
	0310	16 mm thick, on walls, very high impact, no finish included			167	.096		7.85	3.41		11.26	13.90	
	0320	Taped and finished (level 4 finish)			80.82	.198		8.20	7.05		15.25	20	
	0330	With compound skim coat (level 5 finish)			65.03	.246		8.60	8.75		17.35	23	
	0350	On ceilings, no finish included			150	.107		7.85	3.79		11.64	14.50	
	0360	Taped and finished (level 4 finish)			64.10	.250		8.20	8.85		17.05	23	
	0370	With compound skim coat (level 5 finish)		▼	51.10	.313	▼	8.60	11.15		19.75	27	
	0400	High abuse, gypsum core, paper face											
	0410	13 mm thick, on walls, no finish included		2 Carp	167	.096	m²	6.15	3.41		9.56	12.10	
	0420	Taped and finished (level 4 finish)			80.82	.198		6.45	7.05		13.50	18.15	
	0430	With compound skim coat (level 5 finish)		▼	65.03	.246	▼	6.90	8.75		15.65	21.50	

Important: See the Reference Section for supporting data - Crews, Rental Equipment, City Cost Indexes and Reference Data

09250 | Gypsum Board

		CREW	DAILY OUTPUT	LABOR-HOURS	UNIT	MAT.	LABOR	EQUIP.	TOTAL	TOTAL INCL O&P	
800	**0450** On ceilings, no finish included	2 Carp	150	.107	m²	6.15	3.79		9.94	12.70	**800**
	0460 Taped and finished (level 4 finish)		64.10	.250		6.45	8.85		15.30	21	
	0470 With compound skim coat (level 5 finish)		51.10	.313		6.90	11.15		18.05	25	
	0510 16 mm thick, on walls, no finish included		167	.096		6.80	3.41		10.21	12.75	
	0520 Taped and finished (level 4 finish)		80.82	.198		7.10	7.05		14.15	18.80	
	0530 With compound skim coat (level 5 finish)		65.03	.246		7.55	8.75		16.30	22	
	0550 On ceilings, no finish included		150	.107		6.80	3.79		10.59	13.35	
	0560 Taped and finished (level 4 finish)		64.10	.250		7.10	8.85		15.95	21.50	
	0570 With compound skim coat (level 5 finish)		51.10	.313		7.55	11.15		18.70	25.50	
	1000 For high ceilings, over 2.4 m high, add		255	.063			2.23		2.23	3.47	
	1010 For over 3 stories high, add/story	▼	511	.031	▼		1.11		1.11	1.73	

09260 | Gypsum Board Systems

		CREW	DAILY OUTPUT	LABOR-HOURS	UNIT	MAT.	LABOR	EQUIP.	TOTAL	TOTAL INCL O&P	
100	**0010** **PARTITION WALL** Stud wall, 2.4 m to 3.7 m high										**100**
	0050 13 mm, int., gypsum board, std, tape & finish 2 sides										
	0500 Inst. on and incl. 50 mm x 100 mm wood studs, 400 mm O.C.	2 Carp	28.80	.556	m²	11.10	19.75		30.85	43	
	1000 Metal studs, NLB, 25 ga., 400 mm O.C., 92 mm wide		32.52	.492		9.90	17.50		27.40	38	
	1200 152 mm wide		30.66	.522		11.75	18.55		30.30	42	
	1400 Water resistant, on 50 mm x 100 mm wood studs, 400 mm O.C.		28.80	.556		10.65	19.75		30.40	43	
	1600 Metal studs, NLB, 25 ga., 400 mm O.C., 92 mm wide		32.52	.492		9.45	17.50		26.95	37.50	
	1800 152 mm wide		30.66	.522		11.30	18.55		29.85	41.50	
	2000 Fire res., 2 layers, 1-1/2 hr.,on 50 mm x 100 mm wood studs		19.51	.820		18.20	29		47.20	65.50	
	2200 Metal studs, NLB, 25 ga., 400 mm O.C., 92 mm wide		23.23	.689		17	24.50		41.50	57	
	2400 152 mm wide		21.37	.749		18.85	26.50		45.35	62.50	
	2600 Fire & water res., 2 layers, 1-1/2 hr., 50 mm x 100 mm studs		19.51	.820		18.20	29		47.20	65.50	
	2800 Metal studs, NLB, 25 ga., 400 mm O.C., 92 mm wide		23.23	.689		17	24.50		41.50	57	
	3000 152 mm wide	▼	21.37	.749	▼	18.85	26.50		45.35	62.50	
	3200 15.9 mm, interior, gypsum board, std, tape & finish 2 sides										
	3400 Inst. on and incl. 50 mm x 100 mm wood studs, 400 mm O.C.	2 Carp	27.87	.574	m²	11.75	20.50		32.25	45	
	3600 610 mm O.C.		30.66	.522		10.65	18.55		29.20	41	
	3800 Metal studs, NLB, 25 ga., 400 mm O.C., 92 mm wide		31.59	.507		10.55	18		28.55	39.50	
	4000 152 mm wide		29.73	.538		12.40	19.15		31.55	43.50	
	4200 610 mm O.C., 92 mm wide		33.44	.478		9.70	17		26.70	37	
	4400 152 mm wide		31.59	.507		11.10	18		29.10	40	
	4800 Water resistant, on 50 mm x 100 mm wood studs, 400 mm O.C.		27.87	.574		11.10	20.50		31.60	44	
	5000 610 mm O.C.		30.66	.522		10	18.55		28.55	40	
	5200 Metal studs, NLB, 25 ga. 400 mm O.C., 92 mm wide		31.59	.507		9.90	18		27.90	39	
	5400 152 mm wide		29.73	.538		11.75	19.15		30.90	43	
	5600 610 mm O.C., 92 mm wide		33.44	.478		9.05	17		26.05	36.50	
	5800 152 mm wide		31.59	.507		10.45	18		28.45	39.50	
	6000 Fire res., 2 layers, 2 hr., on 50 x 100 mm wd stds, 400 mm O.C.		19.04	.840		17.10	30		47.10	65.50	
	6200 610 mm O.C.		21.83	.733		17.10	26		43.10	59.50	
	6400 Metal studs, NLB, 25 ga., 400 mm O.C., 92 mm wide		22.76	.703		17.20	25		42.20	58	
	6600 152 mm wide		20.90	.765		18.85	27		45.85	63.50	
	6800 610 mm O.C., 92 mm wide		24.62	.650		16.15	23		39.15	54	
	7000 152 mm wide		22.76	.703		17.55	25		42.55	58.50	
	7200 Fire & wat.res., 2 lyr, 2 hr., 50 x 100 mm stds, 400 mm OC		19.04	.840		18.20	30		48.20	66.50	
	7400 610 mm O.C.		21.83	.733		17.10	26		43.10	59.50	
	7600 Metal studs, NLB, 25 ga., 400 mm O.C., 92 mm wide		22.76	.703		17	25		42	58	
	7800 152 mm wide		20.90	.765		18.85	27		45.85	63.50	
	8000 610 mm O.C., 92 mm wide		24.62	.650		16.15	23		39.15	54	
	8200 152 mm wide	▼	22.76	.703	▼	17.55	25		42.55	58.50	
	8600 12.7 mm blueboard, mesh tape both sides										
	8620 Installed on and incl. 50 mm x 100 mm wd stds, 400 mm O.C.	2 Carp	27.87	.574	m²	11.75	20.50		32.25	45	
	8640 Metal studs, NLB, 25 ga., 400 mm O.C., 92 mm wide	▼	31.59	.507	▼	10.55	18		28.55	39.50	

FINISHES **9**

For expanded coverage of these items see *Means Interior Cost Data 2006*

09260 | Gypsum Board Systems

		CREW	DAILY OUTPUT	LABOR-HOURS	UNIT	2006 BARE COSTS				TOTAL INCL O&P	
						MAT.	LABOR	EQUIP.	TOTAL		
100	8660	152 mm wide	2 Carp	29.73	.538	m²	12.40	19.15		31.55	43.50	100
	9000	Exterior, 13 mm gypsum sheathing, 13 mm gypsum fnshd, interior,										
	9100	including foil faced insulation, metal studs, 20 ga.										
	9200	400 mm O.C., 92 mm wide	2 Carp	26.94	.594	m²	20.50	21		41.50	55.50	
	9400	152 mm wide	"	25.08	.638	"	23	22.50		45.50	60.50	
800	0010	**SHAFT WALL** Cavity type on 25 ga. J track & C-H studs, 610 mm O.C.										800
	0030	25 mm thick coreboard wall liner on shaft side										
	0040	2-hour assembly with double layer										
	0060	16 mm fire rated gypsum board on room side	2 Carp	20.44	.783	m²	11.65	28		39.65	56.50	
	0100	3-hour assembly with triple layer										
	0300	16 mm fire rated gypsum board on room side	2 Carp	16.72	.957	m²	14.85	34		48.85	69.50	
	0400	4-hr assembly, 25 mm coreboard, 16 mm fire rated gyp. board										
	0600	and 19 mm galv. metal furring channels, 610 mm O.C., with										
	0700	Double layer 16 mm fire rated gyp. board, room side	2 Carp	10.22	1.566	m²	13.15	55.50		68.65	101	
	0900	For taping & finishing, add/side	1 Carp	97.55	.082	"	.32	2.92		3.24	4.97	
	1000	For insulation, see div. 07210										

09270 | Drywall Accessories

		CREW	DAILY OUTPUT	LABOR-HOURS	UNIT	2006 BARE COSTS				TOTAL INCL O&P	
						MAT.	LABOR	EQUIP.	TOTAL		
100	0010	**ACCESSORIES, DRYWALL**										100
	0020	Casing bead, galvanized steel	1 Carp	88.39	.091	m	.59	3.22		3.81	5.65	
	0100	Vinyl		91.44	.087		.60	3.11		3.71	5.50	
	0300	Corner bead, galvanized steel, 25 mm x 25 mm		122	.066		.46	2.33		2.79	4.13	
	0400	Corner bead, galvanized steel, 32 mm x 32 mm		107	.075		.67	2.66		3.33	4.88	
	0600	Vinyl corner bead		122	.066		.61	2.33		2.94	4.30	
	0900	Furring channel, galv. steel, 22 mm deep, standard		79.25	.101		.78	3.59		4.37	6.45	
	1000	Resilient		77.72	.103		.77	3.66		4.43	6.55	
	1100	J trim, galvanized steel, 12.9 mm wide		91.44	.087		.61	3.11		3.72	5.50	
	1120	15.9 mm wide		89.92	.089		.55	3.16		3.71	5.50	
	1140	L trim, galvanized		91.44	.087		.52	3.11		3.63	5.40	
	1150	U trim, galvanized	▼	89.92	.089	▼	.54	3.16		3.70	5.50	
	1160	Screws #6 x 25 mm A				k	6.80			6.80	7.50	
	1170	#6 x 41 mm A				"	9.50			9.50	10.45	
	1200	For stud partitions, see Divisions 05410-400 and 09110-100										
	1500	Z stud, galvanized steel, 38 mm wide	1 Carp	79.25	.101	m	1.13	3.59		4.72	6.85	
	1600	51 mm wide	"	77.72	.103	"	1.45	3.66		5.11	7.30	

09280 | Gypsum Wallboard Repairs

		CREW	DAILY OUTPUT	LABOR-HOURS	UNIT	2006 BARE COSTS				TOTAL INCL O&P	
						MAT.	LABOR	EQUIP.	TOTAL		
100	0010	**GYPSUM WALLBOARD REPAIRS**										100
	0100	Fill and sand, pin / nail holes	1 Carp	960	.008	Ea.		.30		.30	.46	
	0110	Screw head pops		480	.017			.59		.59	.92	
	0120	Dents, up to 50 mm square		48	.167		.01	5.95		5.96	9.25	
	0130	50 to 100 mm square		24	.333		.03	11.85		11.88	18.50	
	0140	Cut square, patch, sand and finish, holes, up to 50 mm square		12	.667		.03	23.50		23.53	37	
	0150	50 to 100 mm square		11	.727		.07	26		26.07	40.50	
	0160	100 to 200 mm square		10	.800		.19	28.50		28.69	44.50	
	0170	204 to 305 mm square	▼	8	1	▼	.37	35.50		35.87	56	

9
FINISHES

09310	Ceramic Tile	CREW	DAILY OUTPUT	LABOR-HOURS	UNIT	2006 BARE COSTS				TOTAL INCL O&P
						MAT.	LABOR	EQUIP.	TOTAL	
100 0010	**CERAMIC TILE**									100
0050	Base, using 305 x 102 mm high pc. with 25 x 25 mm tiles, mud set	D-7	24.99	.640	m	13.35	19.45		32.80	43
0100	Thin set	"	39.01	.410		12.70	12.45		25.15	32.50
0300	For 152 mm high base, 25 mm x 25 mm tile face, add					2.10			2.10	2.30
0400	For 51 mm x 51 mm tile face, add to above					1.12			1.12	1.21
0600	Cove base, 108 mm x 108 mm high, mud set	D-7	27.74	.577		10.45	17.50		27.95	37
0700	Thin set		39.01	.410		10.55	12.45		23	30
0900	152 mm x 108 mm high, mud set		30.48	.525		9.65	15.95		25.60	34
1000	Thin set		41.76	.383		9.65	11.65		21.30	27.50
1200	Sanitary cove base, 152 mm x 108 mm high, mud set		28.35	.564		10.70	17.15		27.85	37
1300	Thin set		37.80	.423		12.15	12.85		25	32.50
1500	152 mm x 152 mm high, mud set		25.60	.625		13.30	19		32.30	42.50
1600	Thin set	▼	35.66	.449	▼	13.30	13.65		26.95	34.50
1800	Bathroom accessories, average		82	.195	Ea.	9.70	5.95		15.65	19.35
1900	Bathtub, 1.5 m, 108 x 108 mm tile wainscot, adhesive set 1.8 m H		2.90	5.517		140	168		308	400
2100	2134 mm high wainscot		2.50	6.400		160	194		354	460
2200	2438 mm high wainscot	▼	2.20	7.273	▼	170	221		391	510
2400	Bullnose trim, 108 mm x 108 mm, mud set		24.99	.640	m	9.25	19.45		28.70	38.50
2500	Thin set		39.01	.410		8.55	12.45		21	27.50
2700	152 mm x 108 mm bullnose trim, mud set		25.60	.625		7.60	19		26.60	36.50
2800	Thin set	▼	37.80	.423	▼	7.60	12.85		20.45	27.50
3000	Floors, natural clay, random or uniform, thin set, color group 1		17	.941	m²	39.50	28.50		68	85.50
3100	Color group 2		17	.941		43	28.50		71.50	89
3255	Floors, glazed, thin set, 150 mm x 150 mm, color group 1		18.58	.861		34	26		60	75.50
3260	200 mm x 200 mm tile		23.23	.689		34	21		55	67.50
3270	300 mm x 300 mm tile		30.19	.530		42.50	16.10		58.60	70
3280	400 mm x 400 mm tile		51.10	.313		58	9.50		67.50	78
3285	Border, 150 mm x 300 mm tile		25.55	.626		110	19.05		129.05	149
3290	76 mm x 300 mm tile		18.58	.861		325	26		351	395
3300	Porcelain type, 1 color, color group 2, 25 mm x 25 mm		17	.941		46	28.50		74.50	92.50
3310	51 mm x 51 mm or 51 mm x 25 mm, thin set	▼	17.65	.906		48	27.50		75.50	93.50
3350	For random blend, 2 colors, add					8.50			8.50	9.35
3360	4 colors, add					12.05			12.05	13.25
3370	For color group 3, add					4.95			4.95	5.50
3380	For abrasive non-slip tile, add					4.84			4.84	5.40
4300	Specialty tile, 108 mm x 108 mm x 13 mm, decorator finish	D-7	17	.941		96.50	28.50		125	148
4500	Add for epoxy grout, 2 mm joint, 25 mm x 25 mm tile		74.32	.215		5.90	6.55		12.45	16.15
4600	51 mm x 51 mm tile	▼	76.18	.210	▼	5.40	6.40		11.80	15.25
4800	Pregrouted sheets, walls, 108 mm x 108 mm, 152 mm x 108 mm									
4810	and 216 mm x 108 mm, 0.372 m² sheets, silicone grout	D-7	22.30	.718	m²	45.50	22		67.50	82
5100	Floors, unglazed, 0.186 m² sheets,									
5110	urethane adhesive	D-7	16.72	.957	m²	45.50	29		74.50	92.50
5400	Walls, interior, thin set, 108 mm x 108 mm tile		17.65	.906		21.50	27.50		49	64
5500	152 mm x 108 mm tile		17.65	.906		25.50	27.50		53	68.50
5700	216 mm x 108 mm tile		17.65	.906		35.50	27.50		63	80
5800	152 mm x 152 mm tile		18.58	.861		29.50	26		55.50	71
5810	203.2mm 203.2 mm tile		20.90	.765		39	23.50		62.50	77
5820	304.8mm x 304.8mm tile		27.87	.574		32.50	17.45		49.95	61.50
5830	406.4mm x 404.6mm tile		46.45	.344		35.50	10.45		45.95	54.50
6000	Decorated wall tile, 108 mm x 108 mm, minimum		25.08	.638		39.50	19.40		58.90	72
6100	Maximum		16.72	.957		420	29		449	505
6300	Exterior walls, frostproof, mud set, 108 mm x 108 mm		9.48	1.689		43.50	51.50		95	123
6400	35 mm x 35 mm		8.64	1.852		41.50	56.50		98	129
6600	Crystalline glazed, 108 mm x 108 mm, mud set, plain		9.29	1.722		36	52.50		88.50	117
6700	108 mm x 108 mm, scored tile		9.29	1.722		44.50	52.50		97	126
6900	152mm x 152mm plain	▼	8.64	1.852	▼	47.50	56.50		104	135

9

FINISHES

For expanded coverage of these items see *Means Interior Cost Data 2006*

09310 | Ceramic Tile

		CREW	DAILY OUTPUT	LABOR-HOURS	UNIT	MAT.	LABOR	EQUIP.	TOTAL	TOTAL INCL O&P	
						2006 BARE COSTS					
100	7000	For epoxy grout, 2 mm joints, 108 mm tile, add	D-7	74.32	.215	m²	3.55	6.55		10.10	13.50	100
	7200	For tile set in dry mortar, add		161	.099			3.02		3.02	4.43	
	7300	For tile set in portland cement mortar, add	↓	26.94	.594			18.05		18.05	26.50	
	9000	Regrout tile 108 mm x 108 mm, or larger, wall	1 Tilf	9.29	.861		1.40	29.50		30.90	45	
	9220	Floor	"	11.61	.689	↓	1.51	23.50		25.01	36	
300	0010	**CERAMIC TILE PANELS**										300
	0020	Insulated, over 92.9 m², 38 mm thick	D-7	20.44	.783	m²	95	24		119	140	
	0100	64 mm thick	"	20.44	.783	"	102	24		126	147	

09330 | Quarry Tile

		CREW	DAILY OUTPUT	LABOR-HOURS	UNIT	MAT.	LABOR	EQUIP.	TOTAL	TOTAL INCL O&P	
100	0010	**QUARRY TILE** Base, cove or sanitary, 50 mm or 127 mm high, mud set										100
	0100	13 mm thick	D-7	33.53	.477	m	15.05	14.50		29.55	38	
	0300	Bullnose trim, red, mud set, 152 mm x 152 mm x 13 mm thick		36.58	.437		12.85	13.30		26.15	33.50	
	0400	102 mm x 102 mm x 13 mm thick		33.53	.477		14.55	14.50		29.05	37.50	
	0600	102 mm x 203 mm x 13 mm thick, using 203 mm as edge		39.62	.404	↓	12.75	12.25		25	32	
	0700	Floors, mud set, 93 m² lots, red, 102 mm x 102 mm x 13 mm thick		11.15	1.435	m²	42	43.50		85.50	110	
	0900	152 mm x 152 mm x 13 mm thick		13.01	1.230		34	37.50		71.50	92	
	1000	102 mm x 203 mm x 13 mm thick	↓	12.08	1.325		42	40.50		82.50	105	
	1300	For waxed coating, add					6.65			6.65	7.30	
	1500	For colors other than green, add					3.98			3.98	4.41	
	1600	For abrasive surface, add					4.74			4.74	5.15	
	1800	Brown tile, imported, 152 mm x 152 mm x 19 mm	D-7	11.15	1.435		49.50	43.50		93	119	
	1900	203 mm x 203 mm x 25 mm		10.22	1.566		56.50	47.50		104	132	
	2100	For thin set mortar application, deduct		65.03	.246			7.45		7.45	10.95	
	2200	For epoxy grout & mortar, 152 mm x 152 mm x 13 mm, add		32.52	.492		18.20	14.95		33.15	42	
	2700	Stair tread, 152 mm x 152 mm x 19 mm, plain		4.65	3.445		52	105		157	211	
	2800	Abrasive		4.37	3.664		53	111		164	221	
	3000	Wainscot, 152 mm x 152 mm x 13 mm, thin set, red		9.75	1.640		39.50	50		89.50	117	
	3100	Colors other than green		9.75	1.640	↓	44	50		94	122	
	3300	Window sill, 152 mm wide, 19 mm thick		27.43	.583	m	15.10	17.70		32.80	42.50	
	3400	Corners	↓	80	.200	Ea.	5.15	6.10		11.25	14.60	

09350 | Glass Mosaics

		CREW	DAILY OUTPUT	LABOR-HOURS	UNIT	MAT.	LABOR	EQUIP.	TOTAL	TOTAL INCL O&P	
100	0010	**GLASS MOSAICS** 19 mm tile on 305 mm sheets, standard grout										100
	0300	Color group 1 & 2	D-7	6.78	2.359	m²	169	71.50		240.50	291	
	0350	Color group 3		6.78	2.359		178	71.50		249.50	300	
	0400	Color group 4		6.78	2.359		243	71.50		314.50	375	
	0450	Color group 5		6.78	2.359		273	71.50		344.50	405	
	0500	Color group 6		6.78	2.359		330	71.50		401.50	470	
	0600	Color group 7		6.78	2.359		385	71.50		456.50	525	
	0700	Color group 8, golds, silvers & specialties	↓	5.95	2.691	↓	555	82		637	730	

09370 | Metal Tile

		CREW	DAILY OUTPUT	LABOR-HOURS	UNIT	MAT.	LABOR	EQUIP.	TOTAL	TOTAL INCL O&P	
100	0010	**METAL TILE** 1219 mm x 1219 mm sheet, 24 ga., tile pattern, nailed										100
	0200	Stainless steel	2 Carp	47.56	.336	m²	250	11.95		261.95	294	
	0400	Aluminized steel	"	47.56	.336	"	135	11.95		146.95	167	

9 FINISHES

09410 | Portland Cement Terrazzo

		CREW	DAILY OUTPUT	LABOR-HOURS	UNIT	2006 BARE COSTS				TOTAL INCL O&P
						MAT.	LABOR	EQUIP.	TOTAL	
100	0010	**PORTLAND CEMENT TERRAZZO**, cast-in-place R096613-10									100
	0020	Cove base, 152 mm H, 16 ga. zinc	1 Mstz	6.10	1.312	m	9.85	45		54.85	76.50
	0100	Curb, 152 mm high and 152 mm wide		1.83	4.374		15.70	149		164.70	236
	0300	Divider strip for floors, 14 ga., 32 mm deep, zinc		114	.070		3.54	2.39		5.93	7.40
	0400	Brass		114	.070		6.35	2.39		8.74	10.50
	0600	Heavy top strip 6 mm thick, 32 mm deep, zinc		91.44	.087		5	2.98		7.98	9.90
	0900	Galv. bottoms, brass		91.44	.087		8.65	2.98		11.63	13.90
	1200	For thin set floors, 16 ga., 13 mm x 13 mm, zinc		107	.075		2.17	2.55		4.72	6.15
	1300	Brass	▼	107	.075	▼	4.33	2.55		6.88	8.50
	1500	Floor, bonded to concrete, 44 mm thick, gray cement	J-3	12.08	1.325	m²	27.50	41	19.50	88	112
	1600	White cement, mud set		12.08	1.325		31	41	19.50	91.50	116
	1800	Not bonded, 76 mm total thick., gray cement		10.68	1.498		34.50	46	22	102.50	130
	1900	White cement, mud set	▼	10.68	1.498	▼	37.50	46	22	105.50	133
	2100	For Venetian terrazzo, 25 mm topping, add					50%	50%			
	2200	For heavy duty abrasive terrazzo, add					50%	50%			
	2700	Monolithic terrazzo, 13 mm thick									
	2710	3048 mm panels	J-3	11.61	1.378	m²	25	42.50	20.50	88	113
	3000	Stairs, cast in place, pan filled treads		9.14	1.750	m	7.65	54	26	87.65	116
	3100	Treads and risers	▼	4.27	3.750	"	15.85	115	55	185.85	247
	3300	For stair landings, add to floor prices						50%	25%		
	3400	Stair stringers and fascia	J-3	2.79	5.741	m²	42	177	84.50	303.50	400
	3600	For abrasive metal nosings on stairs, add		45.72	.350	m	22	10.80	5.15	37.95	45.50
	3700	For abrasive surface finish, add		55.74	.287	m²	9.90	8.85	4.23	22.98	28.50
	3900	For raised abrasive strips, add		45.72	.350	m	2.89	10.80	5.15	18.84	24.50
	4000	Wainscot, bonded, 38 mm thick		2.79	5.741	m²	32	177	84.50	293.50	385
	4200	Epoxy terrazzo, 6 mm thick	▼	3.72	4.306	"	46.50	133	63.50	243	315
	4300	Stone chips, onyx gemstone, per 23 kg bag				Bag	13.50			13.50	14.85

09420 | Precast Terrazzo

		CREW	DAILY OUTPUT	LABOR-HOURS	UNIT	MAT.	LABOR	EQUIP.	TOTAL	TOTAL INCL O&P
900	0010	**TERRAZZO, PRECAST**									900
	0020	Base, 150 mm high, straight	1 Mstz	10.67	.750	m	31.50	25.50		57	72
	0100	Cove		9.14	.875		36	30		66	83.50
	0300	200 mm high base, straight		9.14	.875		32	30		62	79
	0400	Cove	▼	7.62	1.050		47	36		83	104
	0600	For white cement, add					1.28			1.28	1.41
	0700	For 16 ga. zinc toe strip, add					4.69			4.69	5.15
	0900	Curbs, 102 mm x 102 mm high	1 Mstz	5.79	1.381		90.50	47		137.50	169
	1000	200 mm x 200 mm high	"	4.57	1.750	▼	104	59.50		163.50	202
	1200	Floor tiles, non-slip, 25 mm thick, 300 mm x 300 mm	D-1	2.69	5.939	m²	174	191		365	485
	1300	32 mm thick, 300 mm x 300 mm		2.69	5.939		195	191		386	505
	1500	406 mm x 406 mm		2.14	7.488		212	241		453	600
	1600	38 mm thick, 406 mm x 406 mm	▼	1.95	8.201		194	264		458	615
	1800	For Venetian terrazzo, add					58			58	64
	1900	For white cement, add	▼			▼	5.40			5.40	5.90
	2400	Stair treads, 38 mm thick, non-slip, three line pattern	2 Mstz	21.34	.750	m	119	25.50		144.50	169
	2500	Nosing and two lines		21.34	.750		119	25.50		144.50	169
	2700	50 mm thick treads, straight		18.29	.875		127	30		157	183
	2800	Curved		15.24	1.050		165	36		201	234
	3000	Stair risers, 25 mm thick, to 155 mm high, straight sections		18.29	.875		29	30		59	75.50
	3100	Cove		15.24	1.050		42	36		78	99
	3300	Curved, 25 mm thick, to 155 mm high, vertical		14.63	1.094		57.50	37.50		95	118
	3400	Cove		11.58	1.381		110	47		157	190
	3600	Stair tread and riser, single piece, straight, minimum		18.29	.875		152	30		182	211
	3700	Maximum		12.19	1.312		196	45		241	282
	3900	Curved tread and riser, minimum	▼	12.19	1.312	▼	213	45		258	300

FINISHES 9

09420 | Precast Terrazzo

		CREW	DAILY OUTPUT	LABOR-HOURS	UNIT	2006 BARE COSTS				TOTAL INCL O&P	
						MAT.	LABOR	EQUIP.	TOTAL		
900	4000	Maximum	2 Mstz	9.75	1.640	m	268	56		324	375	**900**
	4200	Stair stringers, notched, 25 mm thick		7.62	2.100		87	71.50		158.50	201	
	4300	50 mm thick		6.71	2.386		103	81.50		184.50	232	
	4500	Stair landings, structural, non-slip, 38 mm thick		7.90	2.026	m²	310	69		379	440	
	4600	75 mm thick		6.97	2.296		440	78.50		518.50	600	
	4800	Wainscot, 305 mm x 305 mm x 25 mm tiles	1 Mstz	1.11	7.176		63	245		308	430	
	4900	406 mm x 406 mm x 38 mm tiles	"	.74	10.764		136	365		501	690	

09430 | Conductive Terrazzo

		CREW	DAILY OUTPUT	LABOR-HOURS	UNIT	2006 BARE COSTS				TOTAL INCL O&P	
						MAT.	LABOR	EQUIP.	TOTAL		
100	0010	**CONDUCTIVE TERRAZZO**									**100**
	2400	Bonded conductive floor for hospitals	J-3	8.36	1.914	m²	37.50	59	28	124.50	159	
	2500	Epoxy terrazzo, 6 mm thick, minimum		9.29	1.722		42.50	53	25.50	121	153	
	2550	Average		6.97	2.296		46.50	70.50	34	151	193	
	2600	Maximum		5.57	2.870		51	88.50	42.50	182	233	

09450 | Cast-in-Place Terrazzo

		CREW	DAILY OUTPUT	LABOR-HOURS	UNIT	2006 BARE COSTS				TOTAL INCL O&P	
						MAT.	LABOR	EQUIP.	TOTAL		
200	0010	**TILE OR TERRAZZO BASE**									**200**
	0020	Scratch coat only	1 Mstz	13.94	.574	m²	4.09	19.60		23.69	33	
	0500	Scratch and brown coat only	"	6.97	1.148	"	7.75	39		46.75	66	

09500 | Ceilings

09510 | Acoustical Ceilings

		CREW	DAILY OUTPUT	LABOR-HOURS	UNIT	2006 BARE COSTS				TOTAL INCL O&P	
						MAT.	LABOR	EQUIP.	TOTAL		
700	0010	**SUSPENDED ACOUSTIC CEILING TILES**, Not including									**700**
	0100	suspension system									
	0300	Fbgs board, film fcd, 600mm x600mm or 600mm x1200mm, 16mm T	1 Carp	58.06	.138	m²	5.60	4.90		10.50	13.80	
	0400	19 mm thick		55.74	.144		12.80	5.10		17.90	22	
	0500	76 mm thick, thermal, 1.9 m² K/W		41.81	.191		14.10	6.80		20.90	26	
	0600	Glass cloth faced fiberglass, 19 mm thick		46.45	.172		18.50	6.10		24.60	30	
	0700	25 mm thick		45.06	.178		20.50	6.30		26.80	32.50	
	0820	38 mm thick, nubby face		44.13	.181		25.50	6.45		31.95	38	
	1110	Mineral fbr tile, 600 x 600mm or 600 x 1200mm, 16mm T, fine text		58.06	.138		4.84	4.90		9.74	13.05	
	1115	Rough textured		58.06	.138		11.85	4.90		16.75	20.50	
	1125	19 mm thick, fine textured		55.74	.144		13.15	5.10		18.25	22.50	
	1130	Rough textured		55.74	.144		16.45	5.10		21.55	26	
	1135	Fissured		55.74	.144		19.50	5.10		24.60	29.50	
	1150	Tegular, 16 mm thick, fine textured		43.66	.183		11.65	6.50		18.15	23	
	1155	Rough textured		43.66	.183		15.05	6.50		21.55	27	
	1165	19 mm thick, fine textured		41.81	.191		16.45	6.80		23.25	28.50	
	1170	Rough textured		41.81	.191		18.60	6.80		25.40	31	
	1175	Fissured		41.81	.191		29	6.80		35.80	42.50	
	1180	For aluminum face, add					53			53	58	
	1185	For plastic film face, add					8.70			8.70	9.60	
	1190	For fire rating, add					3.88			3.88	4.31	
	1300	Mirror faced panels, 24 mm thick, 610 mm x 610 mm	1 Carp	46.45	.172		110	6.10		116.10	131	
	1900	Eggcrate, acrylic, 13 mm x 13 mm x 13 mm cubes		46.45	.172		15.95	6.10		22.05	27	
	2100	Polystyrene eggcrate, 10 mm x 10 mm x 13 mm cubes		47.38	.169		13.35	6		19.35	24	

09510	Acoustical Ceilings	CREW	DAILY OUTPUT	LABOR-HOURS	UNIT	2006 BARE COSTS				TOTAL INCL O&P	
						MAT.	LABOR	EQUIP.	TOTAL		
700	2200	13 mm x 13 mm x 13 mm cubes	1 Carp	46.45	.172	m²	17.85	6.10		23.95	29.50	**700**
	2400	Luminous panels, prismatic, acrylic		37.16	.215		19.40	7.65		27.05	33.50	
	2500	Polystyrene		37.16	.215		9.90	7.65		17.55	23	
	2700	Flat white acrylic		37.16	.215		33.50	7.65		41.15	49	
	2800	Polystyrene		37.16	.215		23	7.65		30.65	37.50	
	3000	Drop pan, white, acrylic		37.16	.215		49.50	7.65		57.15	66.50	
	3100	Polystyrene		37.16	.215		41.50	7.65		49.15	57.50	
	3600	Perforated aluminum sheets, 0.61 mm thick, corrugated, painted		45.52	.176		19.70	6.25		25.95	31.50	
	3700	Plain		46.45	.172		33.50	6.10		39.60	46	
	3720	Mineral fbr, 610 x 610 mm or 1219 mm, reveal edge pntd, 16 mm T		55.74	.144		11	5.10		16.10	20	
	3740	19 mm thick	↓	53.42	.150	↓	17.85	5.30		23.15	28	
760	0010	**SUSPENDED CEILINGS, COMPLETE** Including standard										**760**
	0100	suspension system but not incl. 38 mm carrier channels										
	0600	Fbrgls clng brd, 610 x 1219 x 16 mm, pln faced, supermarkets	1 Carp	46.45	.172	m²	11.20	6.10		17.30	22	
	0700	Offices, 610 mm x 1219 mm x 19 mm		35.30	.227		18.40	8.05		26.45	32.50	
	0800	Mineral fiber, on 23mm T bar susp. 600 x 600 x 19 mm board		32.05	.250		20	8.85		28.85	36	
	0810	600 mm x 1200 mm x 16 mm tile		35.30	.227		10.45	8.05		18.50	24	
	0820	Tegular, 600 mm x 600 mm x 16 mm tile on 14 mm grid		23.23	.344		20.50	12.25		32.75	41.50	
	0830	600mm x 1200mm x 19 mm tile		25.55	.313		23.50	11.15		34.65	43.50	
	0900	Luminous panels, prismatic, acrylic		23.69	.338		25	12		37	46	
	1200	Metal pan with acoustic pad, steel		6.97	1.148		37	41		78	105	
	1300	Painted aluminum		6.97	1.148		25.50	41		66.50	91.50	
	1500	Aluminum, degreased finish		6.97	1.148		43	41		84	111	
	1600	Stainless steel		6.97	1.148		80.50	41		121.50	152	
	1800	Tile, Z bar suspension, 16 mm mineral fiber tile		13.94	.574		17.35	20.50		37.85	51	
	1900	19 mm mineral fiber tile	↓	13.94	.574	↓	18.50	20.50		39	52.50	
	2400	For strip lighting, see division 16510-440										
	2500	For rooms under 45 m², add				m²		25%				
900	0010	**CEILING TILE**, Stapled or cemented										**900**
	0100	300 x 300 mm or 300 x 600 mm, not including furring										
	0600	Mineral fiber, vinyl coated, 16 mm thick	1 Carp	92.90	.086	m²	15.80	3.06		18.86	22	
	0700	19 mm thick		92.90	.086		15.20	3.06		18.26	21.50	
	0900	Fire rated, 19 mm thick, plain faced		92.90	.086		13.65	3.06		16.71	19.80	
	1000	Plastic coated face		92.90	.086		14.65	3.06		17.71	21	
	1200	Aluminum faced, 16 mm thick, plain		92.90	.086		12.05	3.06		15.11	18	
	3700	Wall application of above, add	↓	288	.028			.99		.99	1.54	
	3900	For ceiling primer, add					1.29			1.29	1.40	
	4000	For ceiling cement, add				↓	3.66			3.66	3.98	

09600 | Flooring

09620	Specialty Flooring	CREW	DAILY OUTPUT	LABOR-HOURS	UNIT	2006 BARE COSTS				TOTAL INCL O&P	
						MAT.	LABOR	EQUIP.	TOTAL		
100	0010	**ATHLETIC FLOORING**										**100**
	3700	Polyethylene, in rolls, no base incl., landscape surfaces	1 Tilf	25.55	.313	m²	27.50	10.75		38.25	46.50	
	3800	Nylon action surface, 3 mm thick		25.55	.313		29.50	10.75		40.25	48.50	
	3900	6 mm thick		25.55	.313		43	10.75		53.75	63	
	4000	10 mm thick		25.55	.313		54	10.75		64.75	75	
	4100	Golf tee surface with foam back	↓	21.83	.366		53.50	12.55		66.05	77	

For expanded coverage of these items see Means Interior Cost Data 2006

			CREW	DAILY OUTPUT	LABOR-HOURS	UNIT	2006 BARE COSTS				TOTAL INCL O&P	
		09620 | **Specialty Flooring**					MAT.	LABOR	EQUIP.	TOTAL		
100	4200	Practice putting, knitted nylon surface	1 Tilf	21.83	.366	m²	45	12.55		57.55	68	100
	4400	Polyurethane, thermoset, prefabricated in place, indoor										
	4500	10 mm thick for basketball, gyms, etc.	1 Tilf	9.29	.861	m²	42.50	29.50		72	90.50	
	4600	13 mm thick for professional sports		8.83	.906		57	31		88	109	
	4700	Outdoor, 6 mm thick, smooth, for tennis		9.29	.861		44	29.50		73.50	92	
	4800	Rough, for track, 10 mm thick		8.83	.906		51	31		82	102	
	5000	Poured in place, indoor, with finish, 6 mm thick		7.43	1.076		31.50	37		68.50	88.50	
	5050	10 mm thick		6.04	1.325		38	45.50		83.50	109	
	5100	13 mm thick		4.65	1.722		50.50	59		109.50	142	
	5500	Polyvinyl chloride, sheet goods for gyms, 6 mm thick		7.43	1.076		54	37		91	113	
	5600	10 mm thick	▼	5.57	1.435	▼	61	49		110	139	

			CREW	DAILY OUTPUT	LABOR-HOURS	UNIT	MAT.	LABOR	EQUIP.	TOTAL	INCL O&P	
		09631 | **Brick Flooring**										
100	0010	**BRICK FLOORING**										100
	0020	Acid proof shales, red, 203 mm x 95 mm x 32 mm thick	D-7	.43	37.209	k	760	1,125		1,885	2,475	
	0050	57 mm thick	D-1	.40	40		825	1,275		2,100	2,850	
	0200	Acid proof clay brick, 203 mm x 95 mm x 57 mm thick	"	.40	40	▼	785	1,275		2,060	2,825	
	0260	Cast ceramic, pressed, 102 mm x 203 mm x 13 mm, unglazed	D-7	9.29	1.722	m²	55.50	52.50		108	138	
	0270	Glazed		9.29	1.722		73.50	52.50		126	158	
	0280	Hand molded flooring, 102 mm x 203 mm x 19 mm, unglazed		8.83	1.813		73	55		128	162	
	0290	Glazed		8.83	1.813		91.50	55		146.50	182	
	0300	203 mm hexagonal, 19 mm thick, unglazed		7.90	2.026		80	61.50		141.50	179	
	0310	Glazed	▼	7.90	2.026		145	61.50		206.50	250	
	0400	Heavy duty indust., cement mortar bed, 50 mm thick, not incl. brick	D-1	7.43	2.153		7.55	69		76.55	113	
	0450	Acid proof joints, 6 mm wide	"	6.04	2.650		12.70	85		97.70	144	
	0500	Pavers, 200 mm x 100 mm, 25 mm to 32 mm thick, red	D-7	8.83	1.813		32.50	55		87.50	117	
	0510	Ironspot	"	8.83	1.813		45.50	55		100.50	131	
	0540	35 mm to 44 mm thick, red	D-1	8.83	1.813		31	58.50		89.50	123	
	0560	Ironspot		8.83	1.813		45	58.50		103.50	138	
	0580	57 mm thick, red		8.36	1.914		31.50	61.50		93	129	
	0590	Ironspot	▼	8.36	1.914	▼	49	61.50		110.50	148	
	0800	For sidewalks and patios with pavers, see division 02780-200										
	0870	For epoxy joints, add	D-1	55.74	.287	m²	24	9.25		33.25	40.50	
	0880	For Furan underlayment, add	"	55.74	.287		19.90	9.25		29.15	36	
	0890	For waxed surface, steam cleaned, add	A-1H	92.90	.086	▼	1.72	2.36	.62	4.70	6.30	

			CREW	DAILY OUTPUT	LABOR-HOURS	UNIT	MAT.	LABOR	EQUIP.	TOTAL	INCL O&P	
		09635 | **Marble Flooring**										
100	0010	**MARBLE**										100
	0020	Thin gauge tile, 305 mm x 152 mm, 10 mm, White Carara	D-7	5.57	2.870	m²	98	87		185	236	
	0100	Travertine		5.57	2.870		108	87		195	247	
	0200	305 mm x 305 mm x 10 mm, thin set, floors		5.57	2.870		68	87		155	203	
	0300	On walls	▼	4.83	3.312	▼	96	101		197	253	

			CREW	DAILY OUTPUT	LABOR-HOURS	UNIT	MAT.	LABOR	EQUIP.	TOTAL	INCL O&P	
		09637 | **Stone Flooring**										
100	0010	**SLATE TILE**										100
	0020	Vermont, 152 mm x 152 mm x 6 mm thick, thin set	D-7	16.72	.957	m²	46	29		75	93	
	0200	See also division 02780-600										
200	0010	**SLATE & STONE FLOORS** See division 02780-600										200
300	0010	**CONCRETE FLOORS** And toppings, see division 03350-300										300

9

FINISHES

09642 | Wood Athletic Flooring

		CREW	DAILY OUTPUT	LABOR-HOURS	UNIT	2006 BARE COSTS				TOTAL INCL O&P
						MAT.	LABOR	EQUIP.	TOTAL	
100	0010	**WOOD ATHLETIC FLOORING**									100
	0600	Gym floor, in mastic, over 2 ply felt, #2 & better									
	0700	20 mm thick maple	1 Carp	9.29	.861	m²	38	30.50		68.50	89.50
	0900	26 mm thick maple		9.10	.879		44	31		75	97
	1000	For 13 mm corkboard underlayment, add	↓	69.68	.115		8.50	4.08		12.58	15.70
	1300	For #1 grade maple, add				↓	4.52			4.52	4.95
	1600	Maple flooring, over sleepers, #2 & better									
	1700	20 mm thick	1 Carp	7.90	1.013	m²	41.50	36		77.50	102
	1900	26 mm thick	"	7.71	1.038		48	37		85	111
	2000	For #1 grade, add					4.84			4.84	5.40
	2200	For 19 mm subfloor, add	1 Carp	32.52	.246		10.75	8.75		19.50	25.50
	2300	With two 13 mm subfloors, 20 mm thick	"	6.41	1.248	↓	52	44.50		96.50	126
	2500	Maple, incl. finish, #2 & btr., 20 mm thick, on rubber									
	2600	Sleepers, with two 13 mm subfloors	1 Carp	7.06	1.133	m²	55.50	40.50		96	124
	2800	With steel spline, double connection to channels	"	6.78	1.180		59	42		101	131
	2900	For 26 mm maple, add					6.45			6.45	7.10
	3100	For #1 grade maple, add					4.84			4.84	5.40
	3500	For termite proofing all of the above, add					2.48			2.48	2.69
	3700	Portable hardwood, prefinished panels	1 Carp	7.71	1.038		74	37		111	139
	3720	Insulated with polystyrene, 25 mm thick, add		15.33	.522		6.45	18.55		25	36
	3750	Running tracks, Sitka spruce surface, 20 mm x 57 mm		5.76	1.389		136	49.50		185.50	226
	3770	19 mm plywood surface, finished	↓	9.29	.861	↓	32	30.50		62.50	83
	3800	See also resilient gym floors, division 09620-100									

09643 | Wood Block Flooring

		CREW	DAILY OUTPUT	LABOR-HOURS	UNIT	MAT.	LABOR	EQUIP.	TOTAL	INCL O&P
100	0010	**WOOD BLOCK FLOORING**									100
	0020	End grain flooring, coated, 51 mm thick	1 Carp	27.41	.292	m²	32	10.40		42.40	51.50
	0400	Natural finish, 25 mm thick, fir		11.61	.689		33	24.50		57.50	74.50
	0600	38 mm thick, pine		11.61	.689		32.50	24.50		57	74
	0700	50 mm thick, pine	↓	11.61	.689	↓	38	24.50		62.50	80

09644 | Wood Comp. Flooring

		CREW	DAILY OUTPUT	LABOR-HOURS	UNIT	MAT.	LABOR	EQUIP.	TOTAL	INCL O&P
100	0010	**WOOD COMPOSITION** Gym floors									100
	0100	57 mm x 175 mm x 10 mm, on 51 mm grout setting bed	D-7	13.94	1.148	m²	56	35		91	113
	0200	Thin set, on concrete	"	23.23	.689		51.50	21		72.50	87
	0300	Sanding and finishing, add	1 Carp	18.58	.431	↓	7.55	15.30		22.85	32.50

09647 | Wood Parquet Flooring

		CREW	DAILY OUTPUT	LABOR-HOURS	UNIT	MAT.	LABOR	EQUIP.	TOTAL	INCL O&P
100	0010	**WOOD PARQUET** flooring									100
	5200	Parquetry, standard, 8 mm thick, not incl. finish, oak, minimum	1 Carp	14.86	.538	m²	37	19.15		56.15	70.50
	5300	Maximum		9.29	.861		57	30.50		87.50	111
	5500	Teak, minimum		14.86	.538		49	19.15		68.15	83.50
	5600	Maximum		9.29	.861		85.50	30.50		116	142
	5650	21 mm thick, select grade oak, minimum		14.86	.538		94	19.15		113.15	133
	5700	Maximum		9.29	.861		143	30.50		173.50	205
	5800	Custom parquetry, including finish, minimum		9.29	.861		157	30.50		187.50	221
	5900	Maximum		4.65	1.722		209	61		270	325
	6700	Parquetry, prefinished white oak, 8 mm thk, min.		14.86	.538		37.50	19.15		56.65	71.50
	6800	Maximum		9.29	.861		77	30.50		107.50	133
	7000	Walnut or teak, parquetry, minimum		14.86	.538		53	19.15		72.15	88
	7100	Maximum	↓	9.29	.861	↓	92.50	30.50		123	150
	7200	Acrylic wood parquet blocks, 305 mm x 305 mm x 8 mm,									
	7210	irradiated, set in epoxy	1 Carp	14.86	.538	m²	77	19.15		96.15	115

FINISHES 9

9

FINISHES

09648 | Wood Strip Flooring

		CREW	DAILY OUTPUT	LABOR-HOURS	UNIT	2006 BARE COSTS				TOTAL INCL O&P
						MAT.	LABOR	EQUIP.	TOTAL	
100	0010 **WOOD**									100
	0020 Fir, vertical grain, 25 mm x 102 mm, not incl. finish, B & better	1 Carp	23.69	.338	m²	27	12		39	48
	0100 C grade & better		23.69	.338		25.50	12		37.50	46.50
	4000 Maple, strip, 20 mm x 57 mm, not incl. finish, select		15.79	.507		51.50	18		69.50	84.50
	4100 #2 & better		15.79	.507		32	18		50	63
	4300 26 mm x 83 mm, not incl. finish, #1 grade		15.79	.507		40	18		58	72
	4400 #2 & better	▼	15.79	.507	▼	35.50	18		53.50	67.50
	4600 Oak, white or red, 20 mm x 56 mm, not incl. finish									
	4700 #1 common	1 Carp	15.79	.507	m²	32.50	18		50.50	63.50
	4900 Select quartered, 56 mm wide		15.79	.507		30	18		48	61
	5000 Clear		15.79	.507		40	18		58	72
	6100 Prefinished, white oak, prime grade, 57 mm wide		15.79	.507		66.50	18		84.50	102
	6200 83 mm wide		17.19	.465		85.50	16.55		102.05	120
	6400 Ranch plank		13.47	.594		83	21		104	124
	6500 Hardwood blocks, 229 mm x 229 mm, 20 mm thick		14.86	.538		57	19.15		76.15	92.50
	7400 Yellow pine, 19 mm x 79 mm, T & G, C & better, not incl. finish	▼	18.58	.431		24.50	15.30		39.80	51
	7500 Refinish wood floor, sand, 2 cts poly, wax, soft wood, min	1 Clab	37.16	.215		7.65	5.90		13.55	17.60
	7600 Hard wood, max		12.08	.662		11.50	18.15		29.65	41
	7800 Sanding and finishing, 2 coats polyurethane	▼	27.41	.292	▼	7.65	8		15.65	21
	7900 Subfloor and underlayment, see division 06160									
	8015 Transition molding, 57.1 mm wide, 1524 mm long	1 Carp	19.20	.417	Ea.	12.90	14.80		27.70	37
	8300 Floating floor, wood composition strip, complete.	1 Clab	12.36	.647	m²	43	17.75		60.75	74.50
	8310 Floating floor components, T & G wood composite strips					37			37	41
	8320 Film					1.51			1.51	1.61
	8330 Foam					2.48			2.48	2.69
	8340 Adhesive					2.26			2.26	2.48
	8350 Installation kit				▼	1.72			1.72	1.94
	8360 Trim, 50 mm wide x 900 mm long				m	7.60			7.60	8.35
	8370 Reducer moulding				"	13.10			13.10	14.40

09651 | Resilient Base & Access.

		CREW	DAILY OUTPUT	LABOR-HOURS	UNIT	MAT.	LABOR	EQUIP.	TOTAL	TOTAL INCL O&P
100	0010 **STAIR TREADS AND RISERS** See index for materials other									100
	0100 than rubber and vinyl									
	0300 Rubber, molded tread, 305 mm wide, 8 mm thick, black	1 Tilf	35.05	.228	m	31	7.80		38.80	45.50
	0400 Colors		35.05	.228		29.50	7.80		37.30	44
	0600 6 mm thick, black		35.05	.228		28.50	7.80		36.30	43
	0700 Colors		35.05	.228		28.50	7.80		36.30	43
	0900 Grip strip safety tread, colors, 8 mm thick		35.05	.228		42.50	7.80		50.30	58.50
	1000 5 mm thick		36.58	.219	▼	30	7.50		37.50	44
	1200 Landings, smooth sheet rubber, 3 mm thick		11.15	.718	m²	45	24.50		69.50	85.50
	1300 5 mm thick		11.15	.718	"	57.50	24.50		82	99.50
	1500 Nosings, 75 mm wide, 5 mm thick, black		42.67	.187	m	9.50	6.40		15.90	19.90
	1600 Colors		42.67	.187		9.15	6.40		15.55	19.50
	1800 Risers, 175 mm high, 3 mm thick, flat		76.20	.105		11.10	3.60		14.70	17.55
	1900 Coved		76.20	.105		9.95	3.60		13.55	16.25
	2100 Vinyl, molded tread, 305 mm wide, colors, 3 mm thick		35.05	.228		12.65	7.80		20.45	25.50
	2200 6 mm thick		35.05	.228	▼	18.55	7.80		26.35	32
	2300 Landing material, 3 mm thick		18.58	.431	m²	42.50	14.75		57.25	68.50
	2400 Riser, 178 mm high, 3 mm thick, coved		53.34	.150	m	7.40	5.15		12.55	15.70
	2500 Tread and riser combined, 3 mm thick	▼	24.38	.328	"	21.50	11.25		32.75	40
200	0010 **RESILIENT BASE**									200
	0800 Base, cove, rubber or vinyl, 2 mm thick									
	1100 Standard colors, 64 mm high	1 Tilf	96.01	.083	m	1.71	2.85		4.56	6.05
	1150 102 mm high	▼	96.01	.083	▼	1.74	2.85		4.59	6.10

09651 | Resilient Base & Access.

		CREW	DAILY OUTPUT	LABOR-HOURS	UNIT	2006 BARE COSTS				TOTAL INCL O&P	
						MAT.	LABOR	EQUIP.	TOTAL		
200	1200	152 mm high	1 Tilf	96.01	.083	m	3.02	2.85		5.87	7.50	**200**
	1450	3 mm thick, standard colors, 64 mm high		96.01	.083		1.97	2.85		4.82	6.35	
	1500	102 mm high		96.01	.083	↓	1.97	2.85		4.82	6.35	
	1600	Corners, 64 mm high		315	.025	Ea.	1.17	.87		2.04	2.57	
	1630	102 mm high		315	.025		1.75	.87		2.62	3.21	
	1660	152 mm high	↓	315	.025	↓	2	.87		2.87	3.48	

09653 | Resilient Sheet Flooring

		CREW	DAILY OUTPUT	LABOR-HOURS	UNIT	MAT.	LABOR	EQUIP.	TOTAL	TOTAL INCL O&P	
100	0010	**RESILIENT SHEET FLOORING**										**100**
	5900	Rubber, sheet goods, 914 mm wide, 3 mm thick	1 Tilf	11.15	.718	m²	46	24.50		70.50	86.50	
	5950	5 mm thick		9.29	.861		64.50	29.50		94	115	
	6000	6 mm thick		8.36	.957		75.50	33		108.50	131	
	8000	Vinyl sheet goods, backed, 2 mm thick, minimum		23.23	.344		24	11.80		35.80	43.50	
	8050	Maximum		18.58	.431		29	14.75		43.75	53.50	
	8100	2 mm thick, minimum		21.37	.374		25.50	12.80		38.30	47.50	
	8150	Maximum		18.58	.431		36.50	14.75		51.25	62	
	8200	3 mm thick, minimum		21.37	.374		29	12.80		41.80	51	
	8250	Maximum	↓	18.58	.431	↓	46	14.75		60.75	72	
	8700	Adhesive cement, 3.75 L does 20 to 30 m²				liter	4.42			4.42	4.86	
	8800	Asphalt primer, 3.75 L/300 m²					2.72			2.72	2.99	
	8900	Emulsion, 3.75 L/140 m²					3.46			3.46	3.81	
	8950	Latex underlayment, liquid, fortified				↓	8.75			8.75	9.65	

09658 | Resilient Tile Flooring

		CREW	DAILY OUTPUT	LABOR-HOURS	UNIT	MAT.	LABOR	EQUIP.	TOTAL	TOTAL INCL O&P	
100	0010	**RESILIENT TILE FLOORING**										**100**
	2200	Cork tile, standard finish, 3 mm thick	1 Tilf	29.26	.273	m²	45.50	9.35		54.85	64	
	2250	5 mm thick		29.26	.273		53	9.35		62.35	72	
	2300	8 mm thick		29.26	.273		59	9.35		68.35	78.50	
	2350	13 mm thick		29.26	.273		69.50	9.35		78.85	90.50	
	2500	Urethane finish, 3 mm thick		29.26	.273		56	9.35		65.35	75.50	
	2550	5 mm thick		29.26	.273		60.50	9.35		69.85	80.50	
	2600	8 mm thick		29.26	.273		75.50	9.35		84.85	97	
	2650	13 mm thick		29.26	.273		106	9.35		115.35	131	
	6050	Tile, marbleized colors, 305 mm x 305 mm, 3 mm thick		37.16	.215		52.50	7.35		59.85	68.50	
	6100	5 mm thick		37.16	.215		70	7.35		77.35	88	
	6300	Special tile, plain colors, 3 mm thick		37.16	.215		54	7.35		61.35	70	
	6350	5 mm thick		37.16	.215		73	7.35		80.35	91.50	
	6410	Raised, radial or m², minimum		37.16	.215		85	7.35		92.35	104	
	6430	Maximum		37.16	.215		85	7.35		92.35	104	
	6450	For golf course, skating rink, etc., 6 mm thick		25.55	.313		85	10.75		95.75	109	
	6700	Synthetic turf, 10 mm thick	↓	8.36	.957	↓	39	33		72	91	
	6750	Interlocking 610 mm x 610 mm squares, 13 mm thick, not										
	6810	cemented, for playgrounds, minimum	1 Tilf	19.51	.410	m²	32	14.05		46.05	56	
	6850	Maximum		17.65	.453		83	15.50		98.50	114	
	7000	Vinyl composition tile, 305 mm x 305 mm, 2 mm thick		46.45	.172		9.25	5.90		15.15	18.90	
	7050	Embossed		46.45	.172		14.55	5.90		20.45	24.50	
	7100	Marbleized		46.45	.172		14.55	5.90		20.45	24.50	
	7150	Solid		46.45	.172		16.15	5.90		22.05	26.50	
	7200	2 mm thick, embossed		46.45	.172		11.10	5.90		17	21	
	7250	Marbleized		46.45	.172		16.15	5.90		22.05	26.50	
	7300	Solid		46.45	.172		21	5.90		26.90	32	
	7350	3 mm thick, marbleized		46.45	.172		12.60	5.90		18.50	22.50	
	7400	Solid		46.45	.172		22	5.90		27.90	33	
	7450	Conductive		46.45	.172		42	5.90		47.90	54.50	
	7500	Vinyl tile, 305 mm x 305 mm, 1 mm thick, minimum		46.45	.172		22.50	5.90		28.40	33.50	
	7550	Maximum	↓	46.45	.172	↓	42.50	5.90		48.40	55.50	

9

FINISHES

09600 | Flooring

09658 | Resilient Tile Flooring

			CREW	DAILY OUTPUT	LABOR-HOURS	UNIT	MAT.	LABOR	EQUIP.	TOTAL	TOTAL INCL O&P	
							2006 BARE COSTS					
100	7600	3 mm thick, minimum	1 Tilf	46.45	.172	m²	29.50	5.90		35.40	41	100
	7650	Solid colors		46.45	.172		54.50	5.90		60.40	68	
	7700	Marbleized or Travertine pattern		46.45	.172		42.50	5.90		48.40	55.50	
	7750	Florentine pattern		46.45	.172		47.50	5.90		53.40	61	
	7800	Maximum	↓	46.45	.172	↓	98	5.90		103.90	117	

09662 | Static Control Flooring

			CREW	DAILY OUTPUT	LABOR-HOURS	UNIT	MAT.	LABOR	EQUIP.	TOTAL	TOTAL INCL O&P	
100	0010	**CONDUCTIVE RESILIENT FLOORING**										100
	1700	Conductive flooring, rubber tile, 3 mm thick	1 Tilf	29.26	.273	m²	37.50	9.35		46.85	55.50	
	1800	Homogeneous vinyl tile, 3 mm thick	"	29.26	.273	"	51	9.35		60.35	70.50	

09670 | Fluid Applied Flooring

			CREW	DAILY OUTPUT	LABOR-HOURS	UNIT	MAT.	LABOR	EQUIP.	TOTAL	TOTAL INCL O&P	
100	0010	**RESILIENT TILE UNDERLAYMENT**										100
	3600	Latex underlayment, 3 mm thk., cementitious for resilient flooring	1 Tilf	14.86	.538	m²	16.70	18.45		35.15	45.50	

09673 | Composition Flooring

			CREW	DAILY OUTPUT	LABOR-HOURS	UNIT	MAT.	LABOR	EQUIP.	TOTAL	TOTAL INCL O&P	
100	0010	**COMPOSITION FLOORING**										100
	0020	Cementitous acrylic, 6 mm thick	C-6	48.31	.994	m²	14	28.50	.88	43.38	61	
	0100	10 mm thick	"	41.81	1.148		18.40	33	1.02	52.42	72	
	0200	Methyl methachrylate, 6 mm thick	C-8A	279	.172		54	5.15		59.15	67	
	0210	3mm thick	"	279	.172		29.50	5.15		34.65	40.50	
	0300	Cupric oxychloride, on bond coat, minimum	C-6	44.59	1.076		31.50	31	.96	63.46	83.50	
	0400	Maximum		39.02	1.230		52.50	35.50	1.09	89.09	114	
	0600	Epoxy, with colored quartz chips, broadcast, minimum		62.71	.765		25	22	.68	47.68	62	
	0700	Maximum		45.52	1.054		30	30.50	.94	61.44	81	
	0900	Trowelled, minimum		52.02	.923		32	26.50	.82	59.32	77	
	1000	Maximum	↓	44.59	1.076	↓	46.50	31	.96	78.46	100	
	1200	Heavy duty epoxy topping, 6 mm thick,										
	1300	45 to 90 m²	C-6	39.02	1.230	m²	55.50	35.50	1.09	92.09	117	
	1500	90 to 200 m²		41.81	1.148		43.50	33	1.02	77.52	100	
	1600	Over 900 m²	↓	44.59	1.076		40.50	31	.96	72.46	93.50	
	1800	Epoxy terrazzo, 6 mm thick, chemical resistant, minimum	J-3	18.58	.861		58.50	26.50	12.70	97.70	117	
	1900	Maximum		13.94	1.148		90	35.50	16.90	142.40	170	
	2100	Conductive, minimum		19.51	.820		72.50	25.50	12.05	110.05	130	
	2200	Maximum	↓	14.86	1.076		98.50	33	15.85	147.35	174	
	2400	Mastic, hot laid, 2 coat, 38 mm thick, standard, minimum	C-6	64.10	.749		36.50	21.50	.67	58.67	74	
	2500	Maximum		48.31	.994		46.50	28.50	.88	75.88	96.50	
	2700	Acidproof, minimum		56.20	.854		46.50	24.50	.76	71.76	90	
	2800	Maximum		32.52	1.476		64.50	42.50	1.31	108.31	138	
	3000	Neoprene, trowelled on, 6 mm thick, minimum		50.63	.948		35.50	27.50	.84	63.84	82.50	
	3100	Maximum		39.95	1.202		49	34.50	1.07	84.57	109	
	3150	Polyacrylate terrazzo, 6 mm thick, minimum		68.28	.703		32.50	20.50	.63	53.63	68	
	3170	Maximum		44.59	1.076		42.50	31	.96	74.46	95.50	
	3200	10 mm thick, minimum		57.60	.833		43	24	.74	67.74	85.50	
	3220	Maximum		44.59	1.076		63.50	31	.96	95.46	119	
	3300	Conductive terrazzo, 6 mm thick, minimum		41.81	1.148		76.50	33	1.02	110.52	137	
	3330	Maximum		28.33	1.694		98	49	1.51	148.51	185	
	3350	10 mm thick, minimum		33.91	1.416		102	41	1.26	144.26	176	
	3370	Maximum		23.69	2.026		126	58.50	1.80	186.30	231	
	3450	Granite, conductive, 6 mm thick, minimum		64.57	.743		80	21.50	.66	102.16	122	
	3470	Maximum		39.02	1.230		103	35.50	1.09	139.59	169	
	3500	10 mm thick, minimum		64.57	.743		117	21.50	.66	139.16	162	
	3520	Maximum		35.30	1.360		141	39.50	1.21	181.71	217	
	3600	Polyester, with colored quartz chips, 2 mm thick, minimum	↓	98.94	.485	↓	32	14	.43	46.43	57	

Important: See the Reference Section for supporting data - Crews, Rental Equipment, City Cost Indexes and Reference Data

09673	Composition Flooring	CREW	DAILY OUTPUT	LABOR-HOURS	UNIT	2006 BARE COSTS				TOTAL INCL O&P	
						MAT.	LABOR	EQUIP.	TOTAL		
100	3700	Maximum	C-6	52.02	.923	m²	44	26.50	.82	71.32	90.50	100
	3900	3 mm thick, minimum		75.25	.638		35	18.45	.57	54.02	67.50	
	4000	Maximum		62.71	.765		49	22	.68	71.68	89	
	4200	Polyester, heavy duty, compared to epoxy, add		241	.199		13	5.75	.18	18.93	23.50	
	4300	Polyurethane, with suspended vinyl chips, minimum		98.94	.485		69.50	14	.43	83.93	98.50	
	4500	Maximum		79.89	.601		100	17.35	.53	117.88	138	

09680	Carpet										
600	0010	**CARPET PAD** commercial grade										600
	9000	Sponge rubber pad, minimum	1 Tilf	125	.064	m²	4.28	2.19		6.47	7.95	
	9100	Maximum		125	.064		9.85	2.19		12.04	14.05	
	9200	Felt pad, minimum		125	.064		4.27	2.19		6.46	7.90	
	9300	Maximum		125	.064		8.05	2.19		10.24	12.05	
	9400	Bonded urethane pad, minimum		125	.064		4.72	2.19		6.91	8.40	
	9500	Maximum		125	.064		8.10	2.19		10.29	12.10	
	9600	Prime urethane pad, minimum		125	.064		2.71	2.19		4.90	6.20	
	9700	Maximum		125	.064		5	2.19		7.19	8.70	
800	0010	**CARPET** Commercial grades, direct cement										800
	0700	Nylon, level loop, 737 gram, light to medium traffic	1 Tilf	62.71	.128	m²	21	4.37		25.37	29.50	
	0720	795 gram, light to medium traffic		62.71	.128		19.80	4.37		24.17	28.50	
	0900	907 gram, medium traffic		62.71	.128		30	4.37		34.37	39.50	
	1100	1134 gram, medium to heavy traffic		62.71	.128		44.50	4.37		48.87	55	
	2920	Nylon plush, 851 gram, medium traffic		47.66	.168		22	5.75		27.75	32.50	
	3000	1021 gram, medium traffic		62.71	.128		29	4.37		33.37	38	
	3100	1191 gram, medium to heavy traffic		58.53	.137		26.50	4.68		31.18	36	
	3200	1304 gram, medium to heavy traffic		58.53	.137		39.50	4.68		44.18	50.50	
	3300	1531 gram, heavy traffic		58.53	.137		45	4.68		49.68	56.50	
	3340	1701 gram, heavy traffic		58.53	.137		63	4.68		67.68	76.50	
	3665	Olefin, 680 gram, light to medium traffic		62.71	.128		13.30	4.37		17.67	21	
	3670	737 gram, medium traffic		62.71	.128		13.05	4.37		17.42	21	
	3680	794 gram, medium to heavy traffic		62.71	.128		16.15	4.37		20.52	24	
	3700	907 gram, medium to heavy traffic		62.71	.128		19.85	4.37		24.22	28.50	
	3730	1191 gram, heavy traffic		58.53	.137		29	4.68		33.68	39	
	4110	Wool, level loop, 1134 gram, medium traffic		62.71	.128		99	4.37		103.37	115	
	4500	1418 gram, medium to heavy traffic		62.71	.128		101	4.37		105.37	117	
	4700	Patterned, 907 gram, medium to heavy traffic		58.53	.137		100	4.68		104.68	117	
	4900	1361 gram, heavy traffic		58.53	.137		102	4.68		106.68	119	
	5000	For less than full roll, add					25%					
	5100	For small rooms, less than 3600 mm wide, add						25%				
	5200	For large open areas (no cuts), deduct						25%				
	5600	For bound carpet baseboard, add	1 Tilf	91.44	.087	m	4.43	3		7.43	9.30	
	5610	For stairs, not incl. price of carpet, add	"	30	.267	Riser		9.15		9.15	13.40	
	5620	For borders and patterns, add to labor						18%				
	8950	For tackless, stretched installation, add padding to above										
	9850	For "branded" fiber, add				m²	25%					
900	0010	**CARPET TILE**										900
	0100	Tufted nylon, 450 mm x 450 mm, hard back, 565 gram	1 Tilf	125	.064	m²	24.50	2.19		26.69	30	
	0110	735 gram		125	.064		42	2.19		44.19	49.50	
	0200	Cushion back, 565 gram		125	.064		31	2.19		33.19	37	
	0210	735 gram		125	.064		48.50	2.19		50.69	56.50	
910	0010	**CARPET MAINTENANCE**										910
	0020	Steam clean, per cleaning, minimum	1 Clab	279	.029	m²	.75	.79		1.54	2.08	
	0500	Maximum	"	186	.043	"	1.18	1.18		2.36	3.12	

FINISHES 9

09720 | Wall Coverings

		CREW	DAILY OUTPUT	LABOR-HOURS	UNIT	2006 BARE COSTS				TOTAL INCL O&P
						MAT.	LABOR	EQUIP.	TOTAL	
100	**0010**	**WALL COVERING** Including sizing, add 10-30% waste @ takeoff	R097223-10								**100**
	0050	Aluminum foil	1 Pape	25.55	.313	m²	9.35	9.95		19.30	25.50
	0100	Copper sheets, 0.6 mm thick, vinyl backing		22.30	.359		50	11.45		61.45	72
	0300	Phenolic backing		22.30	.359		65	11.45		76.45	88.50
	0600	Cork tiles, light or dark, 305 mm x 305 mm x 5 mm		22.30	.359		40.50	11.45		51.95	61
	0700	8 mm thick		21.83	.366		34.50	11.65		46.15	55.50
	0900	6 mm basketweave		22.30	.359		52.50	11.45		63.95	75
	1000	13 mm natural, non-directional pattern		22.30	.359		67.50	11.45		78.95	91
	1100	19mm natural, non-directional pattern		22.30	.359		110	11.45		121.45	138
	1200	Granular surface, 305 mm x 914 mm, 13 mm thick		35.77	.224		11.40	7.10		18.50	23.50
	1300	25 mm thick		34.37	.233		14.75	7.40		22.15	27.50
	1500	Polyurethane coated, 305 mm x 305 mm x 5 mm thick		22.30	.359		35.50	11.45		46.95	56
	1600	8 mm thick		21.83	.366		50.50	11.65		62.15	73.50
	1800	Cork wallpaper, paperbacked, natural		44.59	.179		20	5.70		25.70	31
	1900	Colors		44.59	.179		25	5.70		30.70	36
	2100	Flexible wood veneer, 0.8 mm thick, plain woods		9.29	.861		21.50	27.50		49	65
	2200	Exotic woods	▼	8.83	.906	▼	32.50	29		61.50	79
	2400	Gypsum-based, fabric-backed, fire									
	2500	resistant for masonry walls, minimum	1 Pape	74.32	.108	m²	7.30	3.43		10.73	13.20
	2700	Maximum, (small quantities)	"	59.46	.135		12.25	4.29		16.54	19.90
	2750	Acrylic, modified, semi-rigid PVC, 0.71 mm thick	2 Carp	30.66	.522		10.25	18.55		28.80	40.50
	2800	1 mm thick	"	29.73	.538		13.45	19.15		32.60	45
	3000	Vinyl wall covering, fabric-backed, lightweight	1 Pape	59.46	.135		6.35	4.29		10.64	13.45
	3300	Medium weight, type 2		44.59	.179		7.85	5.70		13.55	17.20
	3400	Heavy weight, type 3	▼	40.41	.198	▼	12.60	6.30		18.90	23.50
	3600	Adhesive, 20 L lots				liter	2.36			2.36	2.59
	3700	Wallpaper, average workmanship, solid pattern, low cost paper	1 Pape	59.46	.135	m²	3.23	4.29		7.52	10
	3900	basic patterns (matching required), avg. cost paper		49.70	.161		7	5.15		12.15	15.45
	4000	Paper at $85 per double roll, quality workmanship	▼	40.41	.198	▼	16.45	6.30		22.75	27.50
	4100	Linen wall covering, paper backed									
	4150	Flame treatment, minimum				m²	7.55			7.55	8.30
	4180	Maximum					14			14	15.40
	4200	Grass cloths with lining paper, minimum	1 Pape	37.16	.215		7	6.85		13.85	18.05
	4300	Maximum	"	32.52	.246	▼	22.50	7.85		30.35	37

09770 | Special Wall Surfaces

		CREW	DAILY OUTPUT	LABOR-HOURS	UNIT	MAT.	LABOR	EQUIP.	TOTAL	TOTAL INCL O&P
400	**0010**	**FIBERGLASS REINFORCED PLASTIC** panels, 2 mm thick, on walls									**400**
	0020	Adhesive mounted, embossed surface	2 Carp	59.46	.269	m²	16.35	9.55		25.90	33
	0030	Smooth surface		59.46	.269		19.15	9.55		28.70	36
	0040	Fire rated, embossed surface		59.46	.269		21.50	9.55		31.05	38.50
	0050	Nylon rivet mounted, on drywall, embossed surface		44.59	.359		15.20	12.75		27.95	36.50
	0060	Smooth surface		44.59	.359		17.85	12.75		30.60	39.50
	0070	Fire rated, embossed surface		44.59	.359		19.90	12.75		32.65	42
	0080	On masonry, embossed surface		29.73	.538		15.20	19.15		34.35	46.50
	0090	Smooth surface		29.73	.538		19.15	19.15		38.30	51
	0100	Fire rated, embossed surface		29.73	.538		20	19.15		39.15	52.50
	0110	Nylon rivet and adhesive mounted, on drywall, embossed surface		22.30	.718		17.85	25.50		43.35	59
	0120	Smooth surface		22.30	.718		20	25.50		45.50	62
	0130	Fire rated, embossed surface		22.30	.718		22.50	25.50		48	64.50
	0140	On masonry, embossed surface		17.65	.906		17.85	32		49.85	69.50
	0150	Smooth surface		17.65	.906		20	32		52	72.50
	0160	Fire rated, embossed surface	▼	17.65	.906	▼	22.50	32		54.50	75
	0170	For moldings add	1 Carp	76.20	.105	m	.92	3.73		4.65	6.80
	0180	On ceilings, for lay in grid system, embossed surface		37.16	.215	m²	16.35	7.65		24	30
	0190	Smooth surface		37.16	.215		19.15	7.65		26.80	33
	0200	Fire rated, embossed surface	▼	37.16	.215	▼	21.50	7.65		29.15	35.50

09770 | Special Wall Surfaces

		CREW	DAILY OUTPUT	LABOR-HOURS	UNIT	2006 BARE COSTS MAT.	LABOR	EQUIP.	TOTAL	TOTAL INCL O&P	
700	**0010** **PANEL SYSTEM**										**700**
	0100 Raised panel, eng. wood core w/ wood veneer, std., paint grade	2 Carp	27.87	.574	m²	110	20.50		130.50	153	
	0110 Oak veneer		27.87	.574		182	20.50		202.50	232	
	0120 Maple veneer		27.87	.574		233	20.50		253.50	288	
	0130 Cherry veneer		27.87	.574		294	20.50		314.50	355	
	0300 Class I fire rated, paint grade		27.87	.574		131	20.50		151.50	176	
	0310 Oak veneer		27.87	.574		219	20.50		239.50	272	
	0320 Maple veneer		27.87	.574		278	20.50		298.50	335	
	0330 Cherry veneer		27.87	.574		355	20.50		375.50	420	
	0510 Beadboard, 16 mm MDF, standard, primed		27.87	.574		77.50	20.50		98	118	
	0520 Oak veneer, unfinished		27.87	.574		135	20.50		155.50	180	
	0530 Maple veneer, unfinished		27.87	.574		159	20.50		179.50	207	
	0610 Rustic paneling, 16 mm MDF, standard, maple veneer, unfinished	▼	27.87	.574	▼	205	20.50		225.50	257	
	5000 For prefinished paneling, see division 06250-500 & 06250-200										
750	**0010** **SLATWALL PANELS AND ACCESSORIES**										**750**
	0100 Slatwall panel, 1.2 m x 2.4 m x19 mm T, MDF, paint grade	1 Carp	46.45	.172	m²	15.20	6.10		21.30	26.50	
	0110 Melamine finish		46.45	.172		23	6.10		29.10	35	
	0120 High pressure plastic laminate finish	▼	46.45	.172		37	6.10		43.10	50	
	0130 Aluminum channel inserts, add				▼	29			29	32	
	0200 Accessories, corner forms, 2.4 m L				m	12.55			12.55	13.80	
	0210 T-connector, 2.4 m L					17.35			17.35	19.10	
	0220 J-mold, 2.4 m L					4.43			4.43	4.89	
	0230 Edge cap, 2.4 m L					2.89			2.89	3.18	
	0240 Finish end cap, 2.4 m L				▼	10.60			10.60	11.65	
	0300 Display hook, 100 mm L				Ea.	1			1	1.10	
	0310 150 mm L					1.11			1.11	1.22	
	0320 200 mm L					1.18			1.18	1.30	
	0330 250 mm L					1.30			1.30	1.43	
	0340 300 mm L					1.41			1.41	1.55	
	0350 Acrylic, 100 mm L					.82			.82	.90	
	0360 150 mm L					.95			.95	1.05	
	0370 200 mm L					1			1	1.10	
	0380 250 mm L					1.12			1.12	1.23	
	0400 Waterfall hanger, metal, 300 mm - 400 mm					6.65			6.65	7.30	
	0410 Acrylic					9.50			9.50	10.45	
	0500 Shelf bracket, metal, 200 mm					5.15			5.15	5.70	
	0510 250 mm					5.55			5.55	6.10	
	0520 300 mm					5.90			5.90	6.50	
	0530 350 mm					6.65			6.65	7.30	
	0540 400 mm					7.40			7.40	8.15	
	0550 Acrylic, 200 mm					3.08			3.08	3.39	
	0560 250 mm					3.45			3.45	3.80	
	0570 300 mm					3.82			3.82	4.20	
	0580 350 mm					4.31			4.31	4.74	
	0600 Shelf, acrylic, 300 mm x 400 mm x 6 mm					26			26	28.50	
	0610 300 mm x 600 mm x 6 mm				▼	43			43	47.50	

FINISHES 9

09800 | Acoustical Treatment

09820 | Acoustical Insul/Sealants

			CREW	DAILY OUTPUT	LABOR-HOURS	UNIT	2006 BARE COSTS				TOTAL INCL O&P
							MAT.	LABOR	EQUIP.	TOTAL	
500	0010	**SOUND ATTENUATION**										**500**
	0020	Blanket, 25 mm thick	1 Carp	85.93	.093	m²	2.69	3.31		6	8.15
	0500	38 mm thick		85.47	.094		2.69	3.33		6.02	8.20
	1000	50 mm thick		85	.094		3.34	3.35		6.69	8.85
	1500	75 mm thick		84.54	.095		4.52	3.36		7.88	10.20
	3000	Thermal or acoustical batt above ceiling, 51 mm thick		83.61	.096		4.95	3.40		8.35	10.80
	3100	76 mm thick		83.61	.096		7.45	3.40		10.85	13.50
	3200	102 mm thick		83.61	.096		9.15	3.40		12.55	15.40
	3400	Urethane plastic foam, open cell, on wall, 51 mm thick	2 Carp	190	.084		30.50	2.99		33.49	38
	3500	76 mm thick		144	.111		40.50	3.95		44.45	50
	3600	102 mm thick		97.55	.164		56.50	5.85		62.35	71
	3700	On ceiling, 51 mm thick		158	.101		30.50	3.60		34.10	39
	3800	76 mm thick		121	.132		40.50	4.70		45.20	51.50
	3900	102 mm thick		83.61	.191		56.50	6.80		63.30	72.50
	4000	Nylon matting 10 mm thick, with carbon black spinerette									
	4010	plus polyester fabric, on floor	J-4	372	.043	m²	22.50	1.31		23.81	27
	4200	Fiberglass reinf. backer board underlayment, 11 mm thick, on floor	"	74.32	.215	"	18.85	6.55		25.40	30.50
100	0011	**BARRIERS** Plenum										**100**
	0600	Aluminum foil, fiberglass reinf., parallel with joists	1 Carp	25.55	.313	m²	7.75	11.15		18.90	26
	0700	Perpendicular to joists		16.72	.478		9.60	17		26.60	37
	0900	Aluminum mesh, kraft paperbacked		25.55	.313		7.65	11.15		18.80	26
	0970	Fiberglass batts, kraft faced, 89 mm thick		130	.062		2.80	2.19		4.99	6.55
	0980	152 mm thick		121	.066		4.63	2.35		6.98	8.70
	1000	Sheet lead, 0.45 kg, 0.39 mm thick, perpendicular to joists		13.94	.574		29.50	20.50		50	64.50
	1100	Vinyl foam reinforced, 3 mm thick, 4.88 kg/m²		13.94	.574		43	20.50		63.50	79

09840 | Acoustical Wall Treatment

			CREW	DAILY OUTPUT	LABOR-HOURS	UNIT	MAT.	LABOR	EQUIP.	TOTAL	TOTAL INCL O&P
100	0010	**SOUND ABSORBING PANELS** Perforated steel facing, painted with										**100**
	0100	fiberglass or mineral filler, no backs, 56 mm thick, modular									
	0200	space units, ceiling or wall hung, white or colored	1 Carp	9.29	.861	m²	98.50	30.50		129	157
	0300	Fiberboard sound deadening panels, 13 mm thick	"	55.74	.144	"	3.23	5.10		8.33	11.50
	0500	Fiberglass panels, 1200 mm x 2400 mm x 25 mm thick, with									
	0600	glass cloth face for walls, cemented	1 Carp	14.40	.556	m²	66.50	19.75		86.25	105
	0700	38 mm thick, dacron covered, inner aluminum frame,									
	0710	wall mounted	1 Carp	27.87	.287	m²	81.50	10.20		91.70	106
	0900	Mineral fiberboard panels, fabric covered, 762 mm x 2743 mm,									
	1000	19 mm thick, concealed spline, wall mounted	1 Carp	13.94	.574	m²	59	20.50		79.50	96.50

09900 | Paints & Coatings

09910 | Paints

			CREW	DAILY OUTPUT	LABOR-HOURS	UNIT	2006 BARE COSTS				TOTAL INCL O&P
							MAT.	LABOR	EQUIP.	TOTAL	
100	0010	**CABINETS AND CASEWORK**										**100**
	1000	Primer coat, oil base, brushwork	1 Pord	60.39	.132	m²	.54	4.20		4.74	6.90
	2000	Paint, oil base, brushwork, 1 coat	[R099100 -20]	60.39	.132		.65	4.20		4.85	7.10
	3000	Stain, brushwork, wipe off		60.39	.132		.54	4.20		4.74	7
	4000	Shellac, 1 coat, brushwork		60.39	.132		.65	4.20		4.85	7
	4500	Varnish, 3 coats, brushwork, sand after 1st coat		30.19	.265		1.83	8.40		10.23	14.70
	5000	For latex paint, deduct					10%				

Important: See the Reference Section for supporting data - Crews, Rental Equipment, City Cost Indexes and Reference Data

09910	Paints		CREW	DAILY OUTPUT	LABOR-HOURS	UNIT	2006 BARE COSTS				TOTAL INCL O&P
							MAT.	LABOR	EQUIP.	TOTAL	
300	0010	**DOORS AND WINDOWS, EXTERIOR** R099100-20									**300**
	0100	Door frames & trim, only									
	0110	Brushwork, primer	1 Pord	156	.051	m	.16	1.63		1.79	2.65
	0120	Finish coat, exterior latex		156	.051		.20	1.63		1.83	2.65
	0130	Primer & 1 coat, exterior latex		91.44	.087		.36	2.77		3.13	4.57
	0140	Primer & 2 coats, exterior latex	↓	80.77	.099	↓	.56	3.14		3.70	5.35
	0150	Doors, flush, both sides, incl. frame & trim									
	0160	Roll & brush, primer	1 Pord	10	.800	Ea.	4.01	25.50		29.51	42.50
	0170	Finish coat, exterior latex		10	.800		4.42	25.50		29.92	43
	0180	Primer & 1 coat, exterior latex		7	1.143		8.45	36		44.45	64
	0190	Primer & 2 coats, exterior latex		5	1.600		12.85	50.50		63.35	90.50
	0200	Brushwork, stain, sealer & 2 coats polyurethane	↓	4	2	↓	17	63.50		80.50	114
	0210	Doors, French, both sides, 10-15 lite, incl. frame & trim									
	0220	Brushwork, primer	1 Pord	6	1.333	Ea.	2	42.50		44.50	65.50
	0230	Finish coat, exterior latex		6	1.333		2.21	42.50		44.71	66
	0240	Primer & 1 coat, exterior latex		3	2.667		4.21	84.50		88.71	132
	0250	Primer & 2 coats, exterior latex		2	4		6.30	127		133.30	198
	0260	Brushwork, stain, sealer & 2 coats polyurethane	↓	2.50	3.200	↓	6.10	101		107.10	160
	0270	Doors, louvered, both sides, incl. frame & trim									
	0280	Brushwork, primer	1 Pord	7	1.143	Ea.	4.01	36		40.01	59
	0290	Finish coat, exterior latex		7	1.143		4.42	36		40.42	59.50
	0300	Primer & 1 coat, exterior latex		4	2		8.45	63.50		71.95	105
	0310	Primer & 2 coats, exterior latex		3	2.667		12.60	84.50		97.10	141
	0320	Brushwork, stain, sealer & 2 coats polyurethane	↓	4.50	1.778	↓	17	56.50		73.50	104
	0330	Doors, panel, both sides, incl. frame & trim									
	0340	Roll & brush, primer	1 Pord	6	1.333	Ea.	4.01	42.50		46.51	68
	0350	Finish coat, exterior latex		6	1.333		4.42	42.50		46.92	68.50
	0360	Primer & 1 coat, exterior latex		3	2.667		8.45	84.50		92.95	136
	0370	Primer & 2 coats, exterior latex		2.50	3.200		12.60	101		113.60	167
	0380	Brushwork, stain, sealer & 2 coats polyurethane	↓	3	2.667	↓	17	84.50		101.50	146
	0400	Windows, per ext. side, based on 1.4 m²									
	0410	1 to 6 lite									
	0420	Brushwork, primer	1 Pord	13	.615	Ea.	.79	19.50		20.29	30.50
	0430	Finish coat, exterior latex		13	.615		.87	19.50		20.37	30.50
	0440	Primer & 1 coat, exterior latex		8	1		1.66	31.50		33.16	50
	0450	Primer & 2 coats, exterior latex		6	1.333		2.49	42.50		44.99	66
	0460	Stain, sealer & 1 coat varnish	↓	7	1.143	↓	2.42	36		38.42	57
	0470	7 to 10 lite									
	0480	Brushwork, primer	1 Pord	11	.727	Ea.	.79	23		23.79	35.50
	0490	Finish coat, exterior latex		11	.727		.87	23		23.87	35.50
	0500	Primer & 1 coat, exterior latex		7	1.143		1.66	36		37.66	56.50
	0510	Primer & 2 coats, exterior latex		5	1.600		2.49	50.50		52.99	79
	0520	Stain, sealer & 1 coat varnish	↓	6	1.333	↓	2.42	42.50		44.92	66
	0530	12 lite									
	0540	Brushwork, primer	1 Pord	10	.800	Ea.	.79	25.50		26.29	39
	0550	Finish coat, exterior latex		10	.800		.87	25.50		26.37	39
	0560	Primer & 1 coat, exterior latex		6	1.333		1.66	42.50		44.16	65.50
	0570	Primer & 2 coats, exterior latex		5	1.600		2.49	50.50		52.99	79
	0580	Stain, sealer & 1 coat varnish	↓	6	1.333	↓	2.40	42.50		44.90	66
	0590	For oil base paint, add	↓			↓	10%				
310	0010	**DOORS & WINDOWS, INTERIOR LATEX** R099100-20									**310**
	0100	Doors flush, both sides, incl. frame & trim									
	0110	Roll & brush, primer	1 Pord	10	.800	Ea.	3.63	25.50		29.13	42
	0120	Finish coat, latex		10	.800		3.66	25.50		29.16	42
	0130	Primer & 1 coat latex		7	1.143		7.30	36		43.30	62.50
	0140	Primer & 2 coats latex	↓	5	1.600	↓	10.75	50.50		61.25	88.50

FINISHES 9

09910 | Paints

		CREW	DAILY OUTPUT	LABOR-HOURS	UNIT	2006 BARE COSTS				TOTAL INCL O&P	
						MAT.	LABOR	EQUIP.	TOTAL		
310 0160	Spray, both sides, primer [R099100-20]	1 Pord	20	.400	Ea.	3.82	12.70		16.52	23.50	**310**
0170	Finish coat, latex		20	.400		3.84	12.70		16.54	23.50	
0180	Primer & 1 coat latex		11	.727		7.70	23		30.70	43	
0190	Primer & 2 coats latex		8	1		11.35	31.50		42.85	60.50	
0200	Doors, French, both sides, 10-15 lite, incl. frame & trim										
0210	Roll & bursh, primer	1 Pord	6	1.333	Ea.	1.81	42.50		44.31	65.50	
0220	Finish coat, latex		6	1.333		1.83	42.50		44.33	65.50	
0230	Primer & 1 coat latex		3	2.667		3.64	84.50		88.14	131	
0240	Primer & 2 coats latex		2	4		5.35	127		132.35	197	
0260	Doors, louvered, both sides, incl. frame & trim										
0270	Roll & brush, primer	1 Pord	7	1.143	Ea.	3.63	36		39.63	58.50	
0280	Finish coat, latex		7	1.143		3.66	36		39.66	58.50	
0290	Primer & 1 coat, latex		4	2		7.05	63.50		70.55	103	
0300	Primer & 2 coats, latex		3	2.667		10.95	84.50		95.45	139	
0320	Spray, both sides, primer		20	.400		3.82	12.70		16.52	23.50	
0330	Finish coat, latex		20	.400		3.84	12.70		16.54	23.50	
0340	Primer & 1 coat, latex		11	.727		7.70	23		30.70	43	
0350	Primer & 2 coats, latex		8	1		11.60	31.50		43.10	61	
0360	Doors, panel, both sides, incl. frame & trim										
0370	Roll & brush, primer	1 Pord	6	1.333	Ea.	3.82	42.50		46.32	67.50	
0380	Finish coat, latex		6	1.333		3.66	42.50		46.16	67.50	
0390	Primer & 1 coat, latex		3	2.667		7.30	84.50		91.80	135	
0400	Primer & 2 coats, latex		2.50	3.200		10.95	101		111.95	165	
0420	Spray, both sides, primer		10	.800		3.82	25.50		29.32	42	
0430	Finish coat, latex		10	.800		3.84	25.50		29.34	42	
0440	Primer & 1 coat, latex		5	1.600		7.70	50.50		58.20	85	
0450	Primer & 2 coats, latex		4	2		11.60	63.50		75.10	108	
0460	Windows, per interior side, based on 1.4 m²										
0470	1 to 6 lite										
0480	Brushwork, primer	1 Pord	13	.615	Ea.	.72	19.50		20.22	30.50	
0490	Finish coat, enamel		13	.615		.72	19.50		20.22	30.50	
0500	Primer & 1 coat enamel		8	1		1.44	31.50		32.94	49.50	
0510	Primer & 2 coats enamel		6	1.333		2.16	42.50		44.66	66	
0530	7 to 10 lite										
0540	Brushwork, primer	1 Pord	11	.727	Ea.	.72	23		23.72	35.50	
0550	Finish coat, enamel		11	.727		.72	23		23.72	35.50	
0560	Primer & 1 coat enamel		7	1.143		1.44	36		37.44	56	
0570	Primer & 2 coats enamel		5	1.600		2.16	50.50		52.66	79	
0590	12 lite										
0600	Brushwork, primer	1 Pord	10	.800	Ea.	.72	25.50		26.22	39	
0610	Finish coat, enamel		10	.800		.72	25.50		26.22	39	
0620	Primer & 1 coat enamel		6	1.333		1.44	42.50		43.94	65	
0630	Primer & 2 coats enamel		5	1.600		2.16	50.50		52.66	79	
0650	For oil base paint, add					10%					
320 0010	**DOORS AND WINDOWS, INTERIOR ALKYD (OIL BASE)** [R099100-20]										**320**
0500	Flush dr & frm, 914 x 2134 mm, oil, primer, brshwrk	1 Pord	10	.800	Ea.	2.20	25.50		27.70	40.50	
1000	Paint, 1 coat		10	.800		2.04	25.50		27.54	40	
1400	Stain, brushwork, wipe off		18	.444		1.06	14.10		15.16	22	
1600	Shellac, 1 coat, brushwork		25	.320		1.18	10.15		11.33	16.60	
1800	Varnish, 3 coats, brushwork, sand after 1st coat		9	.889		3.61	28		31.61	46.50	
2000	Panel door & frame, 914 mm x 2134 mm, oil, primer, brushwork		6	1.333		1.82	42.50		44.32	65.50	
2200	Paint, 1 coat		6	1.333		2.04	42.50		44.54	65.50	
2600	Stain, brshwrk, panel dr, 914 x 2134 mm, not incl. frm		16	.500		1.06	15.85		16.91	25	
2800	Shellac, 1 coat, brushwork		22	.364		1.18	11.55		12.73	18.65	
3000	Varnish, 3 coats, brushwork, sand after 1st coat		7.50	1.067		3.61	34		37.61	55	
4400	Windows, including frame and trim, per side										

09910	Paints	CREW	DAILY OUTPUT	LABOR-HOURS	UNIT	2006 BARE COSTS				TOTAL INCL O&P
						MAT.	LABOR	EQUIP.	TOTAL	
320 4600	Colonial type, 6/6 lites, 610 mm x 914 mm, oil primer, brush	1 Pord	14	.571	Ea.	.29	18.10		18.39	28	**320**
5800	Paint, 1 coat		14	.571		.32	18.10		18.42	28
6200	914 mm x 1524 mm opg, 6/6 lites, prm coat, brush R099100-20		12	.667		.72	21		21.72	33
6400	Paint, 1 coat		12	.667		.80	21		21.80	33
6800	1219 mm x 2438 mm opening, 6/6 lites, primer coat, brshwrk		8	1		1.54	31.50		33.04	49.50
7000	Paint, 1 coat		8	1		1.72	31.50		33.22	50
8000	Single lite type, 610 x 914 mm, oil, primer coat, brshwrk		33	.242		.29	7.70		7.99	11.90
8200	Paint, 1 coat		33	.242		.32	7.70		8.02	11.95
8600	914 mm x 1524 mm opening, primer coat, brushwork		20	.400		.72	12.70		13.42	19.90
8800	Paint, 1 coat		20	.400		.80	12.70		13.50	20
9200	1219 mm x 2438 mm opening, primer coat, brushwork		14	.571		1.54	18.10		19.64	29
9400	Paint, 1 coat		14	.571		1.72	18.10		19.82	29.50
400 0010	**FENCES** R099100-20										**400**
0100	Chain link or wire metal, one side, water base									
0110	Roll & brush, first coat	1 Pord	89.18	.090	m²	.65	2.84		3.49	4.93
0120	Second coat		119	.067		.54	2.13		2.67	3.86
0130	Spray, first coat		211	.038		.65	1.20		1.85	2.46
0140	Second coat		242	.033		.65	1.05		1.70	2.23
0150	Picket, water base									
0160	Roll & brush, first coat	1 Pord	80.36	.100	m²	.65	3.16		3.81	5.50
0170	Second coat		97.55	.082		.65	2.60		3.25	4.67
0180	Spray, first coat		211	.038		.65	1.20		1.85	2.56
0190	Second coat		242	.033		.65	1.05		1.70	2.33
0200	Stockade, water base									
0210	Roll & brush, first coat	1 Pord	96.62	.083	m²	.65	2.62		3.27	4.70
0220	Second coat		111	.072		.65	2.28		2.93	4.19
0230	Spray, first coat		211	.038		.65	1.20		1.85	2.56
0240	Second coat		242	.033		.65	1.05		1.70	2.33
500 0010	**FLOORS, INTERIOR**										**500**
0100	Concrete									
0110	Brushwork, latex, block filler									
0120	1st coat	1 Pord	90.58	.088	m²	1.29	2.80		4.09	5.75
0130	2nd coat		107	.075		.86	2.37		3.23	4.54
0140	3rd coat		121	.066		.75	2.10		2.85	3.91
0150	Roll, latex, block filler									
0160	1st coat	1 Pord	242	.033	m²	1.83	1.05		2.88	3.52
0170	2nd coat		302	.026		1.08	.84		1.92	2.44
0180	3rd coat		362	.022		.75	.70		1.45	1.92
0190	Spray, latex, block filler									
0200	1st coat	1 Pord	242	.033	m²	1.51	1.05		2.56	3.30
0210	2nd coat		302	.026		.86	.84		1.70	2.23
0220	3rd coat		362	.022		.65	.70		1.35	1.81
620 0010	**MISCELLANEOUS, EXTERIOR** R099100-20										**620**
0015	For painting metals, see Div. 05950-650									
0100	Railing, ext., decorative wood, incl. cap & baluster									
0110	newels & spindles @ 305 mm O.C.									
0120	Brushwork, stain, sand, seal & varnish									
0130	First coat	1 Pord	27.43	.292	m	1.57	9.25		10.82	15.70
0140	Second coat	"	36.58	.219	"	1.57	6.95		8.52	12.20
0150	Rough sawn wood, 1.1 m h, 51 mm sq posts, 152 mm O.C.									
0160	Brushwork, stain, each coat	1 Pord	27.43	.292	m	.49	9.25		9.74	14.50
0170	Wrought iron, 25.4 mm rail, 12.7 mm sq. verticals									
0180	Brushwork, zinc chromate, 1.5 m high, bars 152 mm O.C.									
0190	Primer	1 Pord	39.62	.202	m	1.67	6.40		8.07	11.50
0200	Finish coat		39.62	.202		.52	6.40		6.92	10.20
0210	Additional coat		57.91	.138		.62	4.38		5	7.25

FINISHES 9

For expanded coverage of these items see *Means Interior Cost Data 2006*

09910 | Paints

		CREW	DAILY OUTPUT	LABOR-HOURS	UNIT	2006 BARE COSTS MAT.	LABOR	EQUIP.	TOTAL	TOTAL INCL O&P
620	0220	Shutters or blinds, single panel, 610 mm x 1219 mm R099100-20									**620**
	0230	Brushwork, primer	1 Pord	20	.400	Ea.	.58	12.70		13.28	19.75
	0240	Finish coat, exterior latex		20	.400		.47	12.70		13.17	19.60
	0250	Primer & 1 coat, exterior latex		13	.615		.92	19.50		20.42	30.50
	0260	Spray, primer		35	.229		.84	7.25		8.09	11.85
	0270	Finish coat, exterior latex		35	.229		.99	7.25		8.24	12
	0280	Primer & 1 coat, exterior latex		20	.400		.91	12.70		13.61	20
	0290	For louvered shutters, add				m²	10%				
	0300	Stair stringers, exterior, metal									
	0310	Roll & brush, zinc chromate, to 356 mm, each coat	1 Pord	97.54	.082	m	.16	2.60		2.76	4.12
	0320	Rough sawn wood, 102 mm x 305 mm									
	0330	Roll & brush, exterior latex, each coat	1 Pord	65.53	.122	m	.23	3.87		4.10	6.10
	0340	Lattice, 51 mm x 51 mm, 76 mm O.C.									
	0350	Spray, latex, per side, each coat	1 Pord	44.13	.181	m²	.75	5.75	·	6.50	9.50
	0450	Decking, Ext., sealer, alkyd, brushwork, sealer coat		106	.075		.65	2.39		3.04	4.25
	0460	1st coat		106	.075		.65	2.39		3.04	4.25
	0470	2nd coat		121	.066		.43	2.10		2.53	3.70
	0500	Paint, alkyd, brushwork, primer coat		106	.075		.75	2.39		3.14	4.46
	0510	1st coat		106	.075		.75	2.39		3.14	4.46
	0520	2nd coat		121	.066		.54	2.10		2.64	3.81
	0600	Sand paint, alkyd, brushwork, 1 coat		13.94	.574		.97	18.20		19.17	28.50
630	0010	**MISCELLANEOUS, INTERIOR**									**630**
	2400	Floors, conc./wood, oil base, primer/sealer coat, brushwork	2 Pord	181	.088	m²	.65	2.80		3.45	4.97
	2450	Roller R099100-20		483	.033		.65	1.05		1.70	2.33
	2600	Spray		557	.029		.65	.91		1.56	2.12
	2650	Paint 1 coat, brushwork		181	.088		.65	2.80		3.45	4.87
	2800	Roller		483	.033		.65	1.05		1.70	2.23
	2850	Spray		557	.029		.65	.91		1.56	2.12
	3000	Stain, wood floor, brushwork, 1 coat		423	.038		.54	1.20		1.74	2.46
	3200	Roller		483	.033		.54	1.05		1.59	2.23
	3250	Spray		557	.029		.54	.91		1.45	2.02
	3400	Varnish, wood floor, brushwork		423	.038		.65	1.20		1.85	2.46
	3450	Roller		483	.033		.65	1.05		1.70	2.33
	3600	Spray		557	.029		.65	.91		1.56	2.12
	3650	For dust proofing or anti skid, see division 03350-300									
	3800	Grilles, per side, oil base, primer coat, brushwork	1 Pord	48.31	.166	m²	1.08	5.25		6.33	9.10
	3850	Spray		106	.075		1.08	2.39		3.47·	4.78
	3920	Paint 2 coats, brushwork		30.19	.265		2.26	8.40		10.66	15.15
	3940	Spray		60.39	.132		2.58	4.20		6.78	9.15
	5000	Pipe, thru 100 mm diam., primer or sealer, oil base, brshwrk [4″]	2 Pord	381	.042	m	.20	1.33		1.53	2.24
	5100	Spray		660	.024		.20	.77		.97	1.39
	5200	Paint 1 coat, brushwork		381	.042		.20	1.33		1.53	2.24
	5300	Spray		660	.024		.20	.77		.97	1.36
	5350	Paint 2 coats, brushwork		236	.068		.36	2.15		2.51	3.63
	5400	Spray		378	.042		.43	1.34		1.77	2.48
	5450	Thru 200 mm diameter, primer or sealer coat, brushwork [8″]		189	.085		.43	2.68		3.11	4.50
	5500	Spray		331	.048		.69	1.53		2.22	3.06
	5550	Paint 1 coat, brushwork		189	.085		.56	2.68		3.24	4.66
	5600	Spray		331	.048		.62	1.53		2.15	3
	5650	Paint 2 coats, brushwork		117	.137		.75	4.34		5.09	7.35
	5700	Spray		189	.085		.82	2.68		3.50	4.96
	5750	Thru 300 mm diameter, primer or sealer coat, brushwork [12″]		126	.127		.62	4.03		4.65	6.75
	5800	Spray		221	.072		.69	2.30		2.99	4.21
	5850	Paint 1 coat, brushwork		126	.127		.56	4.03		4.59	6.65
	6000	Spray		221	.072		.62	2.30		2.92	4.15

9 FINISHES

09910	Paints			CREW	DAILY OUTPUT	LABOR-HOURS	UNIT	2006 BARE COSTS				TOTAL INCL O&P	
								MAT.	LABOR	EQUIP.	TOTAL		
630	6200	Paint 2 coats, brushwork	R099100-20	2 Pord	79.25	.202	m	1.12	6.40		7.52	10.85	630
	6250	Spray			126	.127		1.25	4.03		5.28	7.40	
	6300	Thru 400 mm diameter, primer or sealer coat, brushwork [16"]			94.49	.169		.82	5.35		6.17	9	
	6350	Spray			165	.097		.92	3.07		3.99	5.65	
	6400	Paint 1 coat, brushwork			94.49	.169		.75	5.35		6.10	8.95	
	6450	Spray			165	.097		.85	3.07		3.92	5.60	
	6500	Paint 2 coats, brushwork			59.44	.269		1.48	8.55		10.03	14.50	
	6550	Spray			94.49	.169		1.64	5.35		6.99	9.90	
	7000	Trim, wood, incl. puttying, under 150 mm wide											
	7200	Primer coat, oil base, brushwork		1 Pord	198	.040	m	.07	1.28		1.35	2.03	
	7250	Paint, 1 coat, brushwork			198	.040		.10	1.28		1.38	2.03	
	7450	3 coats			99.06	.081		.26	2.56		2.82	4.16	
	7500	Over 150 mm wide, primer coat, brushwork			198	.040		.16	1.28		1.44	2.09	
	7550	Paint, 1 coat, brushwork			198	.040		.16	1.28		1.44	2.13	
	7650	3 coats			99.06	.081		.49	2.56		3.05	4.42	
	8000	Cornice, simple design, primer coat, oil base, brushwork			60.39	.132	m²	.54	4.20		4.74	6.90	
	8250	Paint, 1 coat			60.39	.132		.54	4.20		4.74	7	
	8350	Ornate design, primer coat			32.52	.246		.54	7.80		8.34	12.30	
	8400	Paint, 1 coat			32.52	.246		.54	7.80		8.34	12.40	
	8600	Balustrades, primer coat, oil base, brushwork			48.31	.166		.54	5.25		5.79	8.45	
	8650	Paint, 1 coat			48.31	.166		.54	5.25		5.79	8.55	
	8900	Trusses and wood frames, primer coat, oil base, brushwork			74.32	.108		.54	3.41		3.95	5.70	
	8950	Spray			111	.072		.54	2.28		2.82	4.09	
	9000	Paint 1 coat, brushwork			69.68	.115		.54	3.64		4.18	6.15	
	9200	Spray			111	.072		.65	2.28		2.93	4.19	
	9220	Paint 2 coats, brushwork			46.45	.172		1.08	5.45		6.53	9.40	
	9240	Spray			55.74	.144		1.29	4.55		5.84	8.25	
	9260	Stain, brushwork, wipe off			55.74	.144		.54	4.55		5.09	7.50	
	9280	Varnish, 3 coats, brushwork			25.55	.313		1.83	9.95		11.78	17	
	9350	For latex paint, deduct						10%					
650	0010	**COATINGS & PAINTS** In 19 L lots											650
	0050	For 380 L or more, deduct						10%				10%	
	0100	Paint, Exterior alkyd (oil base)											
	0200	Flat					liter	6.40			6.40	7.05	
	0300	Gloss						6.25			6.25	6.90	
	0400	Primer						6.40			6.40	7	
	0500	Latex (water base)											
	0600	Acrylic stain					liter	5.45			5.45	6	
	0700	Gloss enamel						6.55			6.55	7.20	
	0800	Flat						5.25			5.25	5.75	
	0900	Primer						5.55			5.55	6.15	
	1000	Semi-gloss						7.10			7.10	7.85	
	2400	Masonry, Exterior											
	2500	Alkali resistant primer					liter	5.30			5.30	5.85	
	2600	Block filler, epoxy						5.75			5.75	6.35	
	2700	Latex						2.80			2.80	3.08	
	2800	Latex, flat						5.50			5.50	6.05	
	2900	Semi-gloss						5.75			5.75	6.30	
	4000	Metal											
	4100	Galvanizing paint					liter	6.90			6.90	7.60	
	4200	High heat						12.30			12.30	13.50	
	4300	Heat resistant						7.45			7.45	8.20	
	4400	Machinery enamel, alkyd						7.25			7.25	7.95	
	4500	Metal pretreatment (polyvinyl butyral)						6.60			6.60	7.25	
	4600	Rust inhibitor, ferrous metal						6.40			6.40	7.05	
	4700	Zinc chromate						5.60			5.60	6.15	

FINISHES **9**

For expanded coverage of these items see *Means Interior Cost Data 2006*

09910 | Paints

		CREW	DAILY OUTPUT	LABOR-HOURS	UNIT	MAT.	LABOR	EQUIP.	TOTAL	TOTAL INCL O&P	
						2006 BARE COSTS					
650	4800	Zinc rich primer				liter	27			27	29.50	650
	5500	Coatings										
	5600	Heavy duty										
	5700	Acrylic urethane				liter	12.55			12.55	13.80	
	5800	Chlorinated rubber					8.45			8.45	9.30	
	5900	Coal tar epoxy					10.65			10.65	11.75	
	6000	Polyamide epoxy, finish					9.05			9.05	9.95	
	6100	Primer					8.90			8.90	9.80	
	6200	Silicone alkyd					9.05			9.05	9.95	
	6300	2 component solvent based acrylic epoxy					13			13	14.30	
	6400	Polyester epoxy					18.20			18.20	20	
	6500	Vinyl					7.60			7.60	8.35	
	6600	Special/Miscellaneous										
	6700	Aluminum				liter	6.45			6.45	7.10	
	6900	Dry fall out, flat					3.45			3.45	3.79	
	7000	Fire retardant, intumescent					10.35			10.35	11.35	
	7100	Linseed oil					3.22			3.22	3.54	
	7200	Shellac					5.95			5.95	6.55	
	7300	Swimming pool, epoxy or urethane base					10.25			10.25	11.30	
	7400	Rubber base					7.70			7.70	8.45	
	7500	Texture paint					3.78			3.78	4.15	
	7600	Turpentine					3.02			3.02	3.32	
	7700	Water repellent, 5% silicone					5.05			5.05	5.55	
700	0010	**SIDING EXTERIOR**, Alkyd (oil base) R099100-20				m²						700
	0450	Steel siding, oil base, paint 1 coat, brushwork	2 Pord	187	.086	m²	.65	2.71		3.36	4.74	
	0500	Spray		423	.038		.86	1.20		2.06	2.78	
	0800	Paint 2 coats, brushwork		121	.132		1.18	4.19		5.37	7.60	
	1000	Spray		423	.038		1.51	1.20		2.71	3.42	
	1200	Stucco, rough, oil base, paint 2 coats, brushwork		121	.132		1.18	4.19		5.37	7.60	
	1400	Roller		151	.106		1.29	3.36		4.65	6.45	
	1600	Spray		272	.059		1.29	1.86		3.15	4.32	
	1800	Texture 1-11 or clapboard, oil base, primer coat, brushwork		121	.132		1.08	4.19		5.27	7.50	
	2000	Spray		423	.038		1.08	1.20		2.28	2.99	
	2400	Paint 2 coats, brushwork		75.25	.213		1.72	6.75		8.47	12.10	
	2600	Spray		242	.066		1.94	2.10		4.04	5.30	
	3400	Stain 2 coats, brushwork		88.26	.181		1.08	5.75		6.83	9.85	
	4000	Spray		283	.057		1.18	1.79		2.97	3.99	
	4200	Wood shingles, oil base primer coat, brushwork		121	.132		.97	4.19		5.16	7.40	
	4400	Spray		362	.044		.86	1.40		2.26	3.08	
	5000	Paint 2 coats, brushwork		75.25	.213		1.40	6.75		8.15	11.75	
	5200	Spray		211	.076		1.40	2.40		3.80	5.15	
	6500	Stain 2 coats, brushwork		88.26	.181		1.08	5.75		6.83	9.85	
	7000	Spray		247	.065		1.51	2.05		3.56	4.70	
	8000	For latex paint, deduct					10%					
	8100	For work over 3.6 m H, from pipe scaffolding, add						15%				
	8200	For work over 3.6 m H, from extension ladder, add						25%				
	8300	For work over 3.6 m H, from swing staging, add						35%				
710	0010	**SIDING, MISC.**, latex paint R099100-20										710
	0100	Aluminum siding										
	0110	Brushwork, primer	2 Pord	211	.076	m²	.54	2.40		2.94	4.27	
	0120	Finish coat, exterior latex		211	.076		.43	2.40		2.83	4.16	
	0130	Primer & 1 coat exterior latex		121	.132		1.08	4.19		5.27	7.50	
	0140	Primer & 2 coats exterior latex		90.58	.177		1.51	5.60		7.11	10.15	
	0150	Mineral Fiber shingles										
	0160	Brushwork, primer	2 Pord	139	.115	m²	1.08	3.65		4.73	6.70	

Important: See the Reference Section for supporting data - Crews, Rental Equipment, City Cost Indexes and Reference Data

09910 | Paints

		CREW	DAILY OUTPUT	LABOR-HOURS	UNIT	MAT.	LABOR	EQUIP.	TOTAL	TOTAL INCL O&P	
710						2006 BARE COSTS					**710**
0170	Finish coat, industrial enamel R099100-20	2 Pord	139	.115	m²	1.08	3.65		4.73	6.70	
0180	Primer & 1 coat enamel		75.25	.213		2.15	6.75		8.90	12.50	
0190	Primer & 2 coats enamel		50.17	.319		3.23	10.10		13.33	18.80	
0200	Roll, primer		151	.106		1.18	3.36		4.54	6.35	
0210	Finish coat, industrial enamel		151	.106		1.18	3.36		4.54	6.35	
0220	Primer & 1 coat enamel		90.58	.177		2.37	5.60		7.97	11.05	
0230	Primer & 2 coats enamel		60.39	.265		3.55	8.40		11.95	16.55	
0240	Spray, primer		362	.044		.86	1.40		2.26	3.08	
0250	Finish coat, industrial enamel		362	.044		.97	1.40		2.37	3.19	
0260	Primer & 1 coat enamel		211	.076		1.83	2.40		4.23	5.65	
0270	Primer & 2 coats enamel		151	.106		2.91	3.36		6.27	8.15	
0280	Waterproof sealer, first coat		417	.038		.75	1.22		1.97	2.58	
0290	Second coat	▼	486	.033	▼	.65	1.04		1.69	2.32	
0300	Rough wood incl. shingles, shakes or rough sawn siding										
0310	Brushwork, primer	2 Pord	119	.134	m²	1.29	4.26		5.55	7.80	
0320	Finish coat, exterior latex		119	.134		.75	4.26		5.01	7.25	
0330	Primer & 1 coat exterior latex		89.18	.179		2.05	5.70		7.75	10.80	
0340	Primer & 2 coats exterior latex		65.03	.246		2.80	7.80		10.60	14.85	
0350	Roll, primer		272	.059		1.72	1.86		3.58	4.64	
0360	Finish coat, exterior latex		272	.059		.97	1.86		2.83	3.89	
0370	Primer & 1 coat exterior latex		166	.096		2.58	3.06		5.64	7.50	
0380	Primer & 2 coats exterior latex		121	.132		3.55	4.19		7.74	10.30	
0390	Spray, primer		362	.044		1.40	1.40		2.80	3.72	
0400	Finish coat, exterior latex		362	.044		.75	1.40		2.15	2.97	
0410	Primer & 1 coat exterior latex		242	.066		2.15	2.10		4.25	5.55	
0420	Primer & 2 coats exterior latex		193	.083		2.91	2.63		5.54	7.20	
0430	Waterproof sealer, first coat		417	.038		1.29	1.22		2.51	3.23	
0440	Second coat	▼	417	.038	▼	.75	1.22		1.97	2.58	
0450	Smooth wood incl. butt, T&G, beveled, drop or B&B siding										
0460	Brushwork, primer	2 Pord	216	.074	m²	.86	2.35		3.21	4.51	
0470	Finish coat, exterior latex		119	.134		.75	4.26		5.01	7.25	
0480	Primer & 1 coat exterior latex		74.32	.215		1.72	6.80		8.52	12.15	
0490	Primer & 2 coats exterior latex		58.53	.273		2.48	8.65		11.13	15.75	
0500	Roll, primer		211	.076		.97	2.40		3.37	4.70	
0510	Finish coat, exterior latex		211	.076		.86	2.40		3.26	4.59	
0520	Primer & 1 coat exterior latex		121	.132		1.83	4.19		6.02	8.35	
0530	Primer & 2 coats exterior latex		90.58	.177		2.69	5.60		8.29	11.45	
0540	Spray, primer		423	.038		.75	1.20		1.95	2.67	
0550	Finish coat, exterior latex		423	.038		.75	1.20		1.95	2.67	
0560	Primer & 1 coat exterior latex		242	.066		1.51	2.10		3.61	4.88	
0570	Primer & 2 coats exterior latex		181	.088		2.26	2.80		5.06	6.70	
0580	Waterproof sealer, first coat		486	.033		.75	1.04		1.79	2.32	
0590	Second coat	▼	556	.029	▼	.75	.91		1.66	2.12	
0600	For oil base paint, add	▼			▼	10%					
800	0010 **TRIM, EXTERIOR** R099100-20										**800**
0100	Door frames & trim (see Doors, interior or exterior)										
0110	Fascia, latex paint, one coat coverage										
0120	25 mm x 102 mm, brushwork	1 Pord	195	.041	m	.07	1.30		1.37	2.03	
0130	Roll		390	.021		.07	.65		.72	1.05	
0140	Spray		634	.013		.03	.40		.43	.67	
0150	25 mm x 152 mm to 25 mm x 254 mm, brushwork		195	.041		.20	1.30		1.50	2.19	
0160	Roll		375	.021		.23	.68		.91	1.25	
0170	Spray		640	.013		.16	.40		.56	.76	
0180	25 mm x 305 mm, brushwork		195	.041		.20	1.30		1.50	2.19	
0190	Roll		320	.025		.23	.79		1.02	1.42	
0200	Spray	▼	671	.012	▼	.16	.38		.54	.73	

For expanded coverage of these items see *Means Interior Cost Data 2006*

09910 | Paints

		CREW	DAILY OUTPUT	LABOR-HOURS	UNIT	2006 BARE COSTS				TOTAL INCL O&P
						MAT.	LABOR	EQUIP.	TOTAL	
800	0210	Gutters & downspouts, metal, zinc chromate paint	R099100-20								**800**
	0220	Brushwork, gutters, 127 mm, first coat	1 Pord	195	.041	m	.20	1.30		1.50	2.16
	0230	Second coat		293	.027		.16	.87		1.03	1.50
	0240	Third coat		390	.021		.13	.65		.78	1.14
	0250	Downspouts, 102 mm, first coat		195	.041		.20	1.30		1.50	2.16
	0260	Second coat		293	.027		.16	.87		1.03	1.50
	0270	Third coat		390	.021		.13	.65		.78	1.14
	0280	Gutters & downspouts, wood									
	0290	Brushwork, gutters, 127 mm, primer	1 Pord	195	.041	m	.16	1.30		1.46	2.16
	0300	Finish coat, exterior latex		195	.041		.16	1.30		1.46	2.16
	0310	Primer & 1 coat exterior latex		122	.066		.36	2.08		2.44	3.52
	0320	Primer & 2 coats exterior latex		99.06	.081		.56	2.56		3.12	4.48
	0330	Downspouts, 102 mm, primer		195	.041		.16	1.30		1.46	2.16
	0340	Finish coat, exterior latex		195	.041		.16	1.30		1.46	2.16
	0350	Primer & 1 coat exterior latex		122	.066		.36	2.08		2.44	3.52
	0360	Primer & 2 coats exterior latex		99.06	.081		.26	2.56		2.82	4.16
	0370	Molding, exterior, up to 355.6mm wide									
	0380	Brushwork, primer	1 Pord	195	.041	m	.20	1.30		1.50	2.19
	0390	Finish coat, exterior latex		195	.041		.20	1.30		1.50	2.19
	0400	Primer & 1 coat exterior latex		122	.066		.43	2.08		2.51	3.62
	0410	Primer & 2 coats exterior latex		96.01	.083		.43	2.64		3.07	4.47
	0420	Stain & fill		320	.025		.20	.79		.99	1.42
	0430	Shellac		564	.014		.23	.45		.68	.91
	0440	Varnish		389	.021		.23	.65		.88	1.24
910	0350	**WALLS, MASONRY (CMU), EXTERIOR**									**910**
	0360	Concrete masonry units (CMU), smooth surface									
	0370	Brushwork, latex, first coat	1 Pord	59.46	.135	m²	.43	4.27		4.70	6.85
	0380	Second coat		89.18	.090		.32	2.84		3.16	4.60
	0390	Waterproof sealer, first coat		68.37	.117		2.58	3.71		6.29	8.40
	0400	Second coat		103	.078		2.58	2.46		5.04	6.50
	0410	Roll, latex, paint, first coat		136	.059		.43	1.86		2.29	3.35
	0420	Second coat		166	.048		.32	1.53		1.85	2.62
	0430	Waterproof sealer, first coat		156	.051		2.58	1.63		4.21	5.25
	0440	Second coat		191	.042		2.58	1.33		3.91	4.80
	0450	Spray, latex, paint, first coat		181	.044		.32	1.40		1.72	2.54
	0460	Second coat		242	.033		.32	1.05		1.37	1.90
	0470	Waterproof sealer, first coat		209	.038		2.58	1.21		3.79	4.63
	0480	Second coat		278	.029		2.58	.91		3.49	4.17
	0490	Concrete masonry unit (CMU), porous									
	0500	Brushwork, latex, first coat	1 Pord	59.46	.135	m²	.75	4.27		5.02	7.25
	0510	Second coat		89.18	.090		.43	2.84		3.27	4.71
	0520	Waterproof sealer, first coat		68.37	.117		2.58	3.71		6.29	8.40
	0530	Second coat		103	.078		2.58	2.46		5.04	6.50
	0540	Roll latex, first coat		136	.059		.54	1.86		2.40	3.46
	0550	Second coat		166	.048		.32	1.53		1.85	2.73
	0560	Waterproof sealer, first coat		156	.051		2.58	1.63		4.21	5.25
	0570	Second coat		191	.042		2.58	1.33		3.91	4.80
	0580	Spray latex, first coat		181	.044		.43	1.40		1.83	2.54
	0590	Second coat		242	.033		.32	1.05		1.37	1.90
	0600	Waterproof sealer, first coat		209	.038		2.58	1.21		3.79	4.63
	0610	Second coat		278	.029		2.58	.91		3.49	4.17
920	0010	**WALLS AND CEILINGS, INTERIOR**									**920**
	0100	Concrete, dry wall or plaster, oil base, primer or sealer coat									
	0200	Smooth finish, brushwork	R099100-20 1 Pord	107	.075	m²	.54	2.37		2.91	4.11
	0240	Roller		125	.064		.54	2.03		2.57	3.60

9 FINISHES

			CREW	DAILY OUTPUT	LABOR-HOURS	UNIT	2006 BARE COSTS				TOTAL INCL O&P	
	09910	**Paints**					MAT.	LABOR	EQUIP.	TOTAL		
920	0280	Spray	1 Pord	255	.031	m²	.32	.99		1.31	1.93	**920**
	0300	Sand finish, brushwork		90.58	.088		.54	2.80		3.34	4.76	
	0340	Roller		107	.075		.54	2.37		2.91	4.22	
	0380	Spray		211	.038		.43	1.20		1.63	2.24	
	0800	Paint 2 coats, smooth finish, brushwork		63.17	.127		1.08	4.01		5.09	7.25	
	0840	Roller		74.32	.108		1.18	3.41		4.59	6.45	
	0880	Spray		151	.053		.97	1.68		2.65	3.61	
	0900	Sand finish, brushwork		56.20	.142		1.08	4.51		5.59	8	
	0940	Roller		94.76	.084		1.18	2.68		3.86	5.30	
	0980	Spray		158	.051		.97	1.61		2.58	3.50	
	1200	Paint 3 coats, smooth finish, brushwork		47.38	.169		1.61	5.35		6.96	9.90	
	1240	Roller		60.39	.132		1.72	4.20		5.92	8.30	
	1280	Spray		151	.053		1.51	1.68		3.19	4.14	
	1600	Glaze coating, 5 coats, spray, clear		83.61	.096		6.45	3.03		9.48	11.65	
	1640	Multicolor	▼	83.61	.096		9.15	3.03		12.18	14.65	
	1700	For latex paint, deduct					10%					
	1800	For ceiling installations, add				▼		25%				
	2000	Masonry or concrete block, oil base, primer or sealer coat										
	2100	Smooth finish, brushwork	1 Pord	114	.070	m²	.54	2.22		2.76	3.89	
	2180	Spray		223	.036		.75	1.14		1.89	2.46	
	2200	Sand finish, brushwork		101	.079		.75	2.51		3.26	4.64	
	2280	Spray		223	.036		.75	1.14		1.89	2.46	
	2800	Paint 2 coats, smooth finish, brushwork		70.23	.114		1.61	3.61		5.22	7.15	
	2880	Spray		126	.063		1.51	2.01		3.52	4.64	
	2900	Sand finish, brushwork		62.43	.128		1.61	4.06		5.67	7.80	
	2980	Spray		126	.063		1.51	2.01		3.52	4.64	
	3600	Glaze coating, 5 coats, spray, clear		83.61	.096		6.45	3.03		9.48	11.65	
	3620	Multicolor		83.61	.096		9.15	3.03		12.18	14.65	
	4000	Block filler, 1 coat, brushwork		39.48	.203		1.29	6.40		7.69	11.20	
	4100	Silicone, water repellent, 2 coats, spray	▼	186	.043		2.69	1.36		4.05	5.05	
	4120	For latex paint, deduct					10%					
	8200	For work 2.4 m - 4.6 m H, add						10%				
	8300	For work over 4.6 m H, add	▼			▼		20%				
940	0010	**DRY FALL PAINTING**										**940**
	0100	Walls										
	0200	Wallboard and smooth plaster, one coat, brush	1 Pord	84.54	.095	m²	.43	3		3.43	5.05	
	0210	Roll		145	.055		.43	1.75		2.18	3.17	
	0220	Spray		242	.033		.43	1.05		1.48	2.12	
	0230	Two coats, brush		48.31	.166		.97	5.25		6.22	9	
	0240	Roll		81.47	.098		.97	3.11		4.08	5.75	
	0250	Spray		145	.055		.97	1.75		2.72	3.71	
	0260	Concrete or textured plaster, one coat, brush		69.40	.115		.43	3.65		4.08	6.05	
	0270	Roll		121	.066		.43	2.10		2.53	3.70	
	0280	Spray		145	.055		.43	1.75		2.18	3.17	
	0290	Two coats, brush		39.20	.204		.97	6.45		7.42	10.85	
	0300	Roll		69.40	.115		.97	3.65		4.62	6.60	
	0310	Spray		121	.066		.97	2.10		3.07	4.24	
	0320	Concrete block, one coat, brush		69.40	.115		.43	3.65		4.08	6.05	
	0330	Roll		121	.066		.43	2.10		2.53	3.70	
	0340	Spray		145	.055		.43	1.75		2.18	3.17	
	0350	Two coats, brush		39.20	.204		.97	6.45		7.42	10.85	
	0360	Roll		69.40	.115		.97	3.65		4.62	6.60	
	0370	Spray		121	.066		.97	2.10		3.07	4.24	
	0380	Wood, one coat, brush		69.40	.115		.43	3.65		4.08	6.05	
	0390	Roll	▼	121	.066	▼	.43	2.10		2.53	3.70	

R099100 -20

09910 | Paints

		CREW	DAILY OUTPUT	LABOR-HOURS	UNIT	2006 BARE COSTS				TOTAL INCL O&P
						MAT.	LABOR	EQUIP.	TOTAL	
940	0400	Spray	1 Pord	81.47	.098	m²	.43	3.11		3.54	5.25	**940**
	0410	Two coats, brush		45.24	.177		.97	5.60		6.57	9.55
	0420	Roll		69.40	.115		.97	3.65		4.62	6.60
	0430	Spray	▼	60.39	.132	▼	.97	4.20		5.17	7.45
	0440	Ceilings									
	0450	Wallboard and smooth plaster, one coat, brush	1 Pord	55.74	.144	m²	.43	4.55		4.98	7.40
	0460	Roll		96.62	.083		.43	2.62		3.05	4.49
	0470	Spray		145	.055		.43	1.75		2.18	3.17
	0480	Two coats, brush		31.68	.253		.97	8		8.97	13.15
	0490	Roll		60.39	.132		.97	4.20		5.17	7.45
	0500	Spray		121	.066		.97	2.10		3.07	4.24
	0510	Concrete or textured plaster, one coat, brush		45.24	.177		.43	5.60		6.03	9
	0520	Roll		81.47	.098		.43	3.11		3.54	5.25
	0530	Spray		145	.055		.43	1.75		2.18	3.17
	0540	Two coats, brush		25.64	.312		.97	9.90		10.87	16
	0550	Roll		48.31	.166		.97	5.25		6.22	9
	0560	Spray		121	.066		.97	2.10		3.07	4.24
	0570	Structural steel, bar joists or metal deck, one coat, spray		145	.055		.43	1.75		2.18	3.17
	0580	Two coats, spray	▼	96.62	.083	▼	.97	2.62		3.59	5.05

09930 | Stains/Transp. Finishes

		CREW	DAILY OUTPUT	LABOR-HOURS	UNIT	MAT.	LABOR	EQUIP.	TOTAL	TOTAL INCL O&P
100	0010	**VARNISH**									**100**
	0012	1 coat + sealer, on wood trim, no sanding included	1 Pord	37.16	.215	m²	.75	6.80		7.55	11.15
	0100	Hardwood floors, 2 coats, no sanding included, roller	"	176	.045	"	1.40	1.44		2.84	3.68

09963 | Glazed Coatings

		CREW	DAILY OUTPUT	LABOR-HOURS	UNIT	MAT.	LABOR	EQUIP.	TOTAL	TOTAL INCL O&P
200	0010	**WALL COATINGS**									**200**
	0100	Acrylic glazed coatings, minimum	1 Pord	48.77	.164	m²	2.80	5.20		8	10.95
	0200	Maximum		28.33	.282		5.80	8.95		14.75	19.85
	0300	Epoxy coatings, minimum		48.77	.164		3.55	5.20		8.75	11.75
	0400	Maximum		15.79	.507		11	16.05		27.05	36
	0600	Exposed aggregate, troweled on, 2 mm to 6 mm, minimum		21.83	.366		5.40	11.60		17	23.50
	0700	Maximum (epoxy or polyacrylate)		12.08	.662		11.75	21		32.75	44.50
	0900	13 mm to 16 mm aggregate, minimum		12.08	.662		10.85	21		31.85	43.50
	1000	Maximum		7.43	1.076		18.50	34		52.50	72
	1200	25 mm aggregate size, minimum		8.36	.957		18.85	30.50		49.35	66.50
	1300	Maximum		5.11	1.566		29	49.50		78.50	107
	1500	Exposed aggregate, sprayed on, 3 mm aggregate, minimum		27.41	.292		5.05	9.25		14.30	19.55
	1600	Maximum		13.47	.594		9.35	18.85		28.20	39
	1800	High build epoxy, 1 mm, minimum		36.23	.221		6.05	7		13.05	17.20
	1900	Maximum		8.83	.906		10.35	28.50		38.85	55
	2100	Laminated epoxy with fiberglass, minimum		27.41	.292		6.55	9.25		15.80	21
	2200	Maximum		13.47	.594		11.65	18.85		30.50	41.50
	2400	Sprayed perlite or vermiculite, 2 mm thick, minimum		273	.029		2.37	.93		3.30	3.98
	2500	Maximum		59.46	.135		6.65	4.27		10.92	13.70
	2700	Vinyl plastic wall coating, minimum		68.28	.117		3.01	3.71		6.72	8.95
	2800	Maximum		22.30	.359		7.30	11.35		18.65	25
	3000	Urethane on smooth surface, 2 coats, minimum		105	.076		2.26	2.42		4.68	6.10
	3100	Maximum		61.78	.129		4.95	4.10		9.05	11.70
	3600	Ceramic-like glazed coating, cementitious, minimum		40.88	.196		4.31	6.20		10.51	14.10
	3700	Maximum		32.05	.250		7.20	7.90		15.10	19.85
	3900	Resin base, minimum		59.46	.135		3.01	4.27		7.28	9.75
	4000	Maximum	▼	30.66	.261	▼	4.84	8.25		13.09	17.85

9 FINISHES

09990	Paint Restoration	CREW	DAILY OUTPUT	LABOR-HOURS	UNIT	2006 BARE COSTS				TOTAL INCL O&P
						MAT.	LABOR	EQUIP.	TOTAL	
800	0010	**SANDING** and puttying interior trim, compared to									**800**
	0100	Painting 1 coat, on quality work				m		100%			
	0300	Medium work						50%			
	0400	Industrial grade				↓		25%			
	0500	Surface protection, placement and removal									
	0510	Basic drop cloths	1 Pord	595	.013	m²		.43		.43	.64
	0520	Masking with paper		74.32	.108		.32	3.41		3.73	5.45
	0530		↓	1486	.005	↓		.17		.17	.26
900	0010	**SURFACE PREPARATION, EXTERIOR**									**900**
	0015	Doors, per side, not incl. frames or trim									
	0020	Scrape & sand									
	0030	Wood, flush	1 Pord	57.23	.140	m²		4.43		4.43	6.70
	0040	Wood, detail		46.08	.174			5.50		5.50	8.30
	0050	Wood, louvered		26.01	.308			9.75		9.75	14.70
	0060	Wood, overhead	↓	57.23	.140	↓		4.43		4.43	6.70
	0070	Wire brush									
	0080	Metal, flush	1 Pord	59.46	.135	m²		4.27		4.27	6.40
	0090	Metal, detail		48.31	.166			5.25		5.25	7.90
	0100	Metal, louvered		33.44	.239			7.60		7.60	11.40
	0110	Metal or fibr., overhead		59.46	.135			4.27		4.27	6.40
	0120	Metal, roll up		52.02	.154			4.87		4.87	7.35
	0130	Metal, bulkhead	↓	59.46	.135	↓		4.27		4.27	6.40
	0140	Power wash, based on 17240 kPa operating pressure									
	0150	Metal, flush	B-9	208	.192	m²		5.35	.71	6.06	9.10
	0160	Metal, detail		197	.203			5.65	.75	6.40	9.60
	0170	Metal, louvered		186	.215			6	.79	6.79	10.15
	0180	Metal or fibr., overhead		223	.179			4.99	.66	5.65	8.50
	0190	Metal, roll up		223	.179			4.99	.66	5.65	8.50
	0200	Metal, bulkhead	↓	204	.196	↓		5.45	.72	6.17	9.30
	0400	Windows, per side, not incl. trim									
	0410	Scrape & sand									
	0420	Wood, 1-2 lite	1 Pord	29.73	.269	m²		8.55		8.55	12.85
	0430	Wood, 3-6 lite		26.01	.308			9.75		9.75	14.70
	0440	Wood, 7-10 lite		22.30	.359			11.35		11.35	17.15
	0450	Wood, 12 lite		18.58	.431			13.65		13.65	20.50
	0460	Wood, Bay / Bow	↓	29.73	.269	↓		8.55		8.55	12.85
	0470	Wire brush									
	0480	Metal, 1-2 lite	1 Pord	44.59	.179	m²		5.70		5.70	8.55
	0490	Metal, 3-6 lite		37.16	.215			6.80		6.80	10.30
	0500	Metal, Bay / Bow	↓	44.59	.179	↓		5.70		5.70	8.55
	0510	Power wash, based on 17240 kPa operating pressure									
	0520	1-2 lite	B-9	409	.098	m²		2.72	.36	3.08	4.63
	0530	3-6 lite		401	.100			2.77	.37	3.14	4.72
	0540	7-10 lite		394	.102			2.82	.37	3.19	4.80
	0550	12 lite		386	.104			2.88	.38	3.26	4.91
	0560	Bay / Bow	↓	409	.098	↓		2.72	.36	3.08	4.63
	0600	Siding, scrape and sand, light=10-30%, med.=30-70%									
	0610	Heavy=70-100%, % of surface to sand									
	0650	Texture 1-11, light	1 Pord	44.59	.179	m²		5.70		5.70	8.55
	0660	Med.		40.88	.196			6.20		6.20	9.35
	0670	Heavy		33.44	.239			7.60		7.60	11.40
	0680	Wood shingles, shakes, light		40.88	.196			6.20		6.20	9.35
	0690	Med.		33.44	.239			7.60		7.60	11.40
	0700	Heavy		26.01	.308			9.75		9.75	14.70
	0710	Clapboard, light		48.31	.166			5.25		5.25	7.90
	0720	Med.	↓	44.59	.179	↓		5.70		5.70	8.55

For expanded coverage of these items see *Means Interior Cost Data 2006*

FINISHES 9

09990	Paint Restoration	CREW	DAILY OUTPUT	LABOR-HOURS	UNIT	2006 BARE COSTS				TOTAL INCL O&P	
						MAT.	LABOR	EQUIP.	TOTAL		
900	0730	Heavy	1 Pord	37.16	.215	m²		6.80		6.80	10.30	**900**
	0740	Wire brush										
	0750	Aluminum, light	1 Pord	55.74	.144	m²		4.55		4.55	6.85	
	0760	Med.		48.31	.166			5.25		5.25	7.90	
	0770	Heavy	↓	40.88	.196	↓		6.20		6.20	9.35	
	0780	Pressure wash, based on 17240 kPa operating pressure										
	0790	Stucco	B-9	286	.140	m²		3.89	.51	4.40	6.60	
	0800	Aluminum or vinyl		297	.135			3.74	.50	4.24	6.40	
	0810	Siding, masonry, brick & block	↓	223	.179	↓		4.99	.66	5.65	8.50	
	1300	Miscellaneous, wire brush										
	1310	Metal, pedestrian gate	1 Pord	9.29	.861	m²		27.50		27.50	41	
	8000	For Chemical Washing, see Division 04930										
	8010	For Steam Cleaning, see Division 04930										
	8020	For Sand Blasting, see Division 05910-500 and 03350-350										
910	0010	**SURFACE PREPARATION, INTERIOR**										**910**
	0020	Doors										
	0030	Scrape & sand										
	0040	Wood, flush	1 Pord	57.23	.140	m²		4.43		4.43	6.70	
	0050	Wood, detail		46.08	.174			5.50		5.50	8.30	
	0060	Wood, louvered	↓	26.01	.308			9.75		9.75	14.70	
	0070	Wire brush										
	0080	Metal, flush	1 Pord	59.46	.135	m²		4.27		4.27	6.40	
	0090	Metal, detail		48.31	.166			5.25		5.25	7.90	
	0100	Metal, louvered	↓	33.44	.239	↓		7.60		7.60	11.40	
	0110	Hand wash										
	0120	Wood, flush	1 Pord	201	.040	m²		1.26		1.26	1.90	
	0130	Wood, detailed		186	.043			1.36		1.36	2.05	
	0140	Wood, louvered		126	.063			2.01		2.01	3.03	
	0150	Metal, flush		201	.040			1.26		1.26	1.90	
	0160	Metal, detail		186	.043			1.36		1.36	2.05	
	0170	Metal, louvered	↓	126	.063	↓		2.01		2.01	3.03	
	0400	Windows, per side, not incl. trim										
	0410	Scrape & sand										
	0420	Wood, 1-2 lite	1 Pord	33.44	.239	m²		7.60		7.60	11.40	
	0430	Wood, 3-6 lite		29.73	.269			8.55		8.55	12.85	
	0440	Wood, 7-10 lite		26.01	.308			9.75		9.75	14.70	
	0450	Wood, 12 lite		22.30	.359			11.35		11.35	17.15	
	0460	Wood, Bay / Bow	↓	33.44	.239	↓		7.60		7.60	11.40	
	0470	Wire brush										
	0480	Metal, 1-2 lite	1 Pord	48.31	.166	m²		5.25		5.25	7.90	
	0490	Metal, 3-6 lite		40.88	.196			6.20		6.20	9.35	
	0500	Metal, Bay / Bow	↓	48.31	.166	↓		5.25		5.25	7.90	
	0600	Walls, sanding, light=10-30%										
	0610	Med.=30-70%, heavy=70-100%, % of surface to sand										
	0650	Walls, sand										
	0660	Drywall, gypsum, plaster, light	1 Pord	286	.028	m²		.89		.89	1.34	
	0670	Drywall, gypsum, plaster, med.		201	.040			1.26		1.26	1.90	
	0680	Drywall, gypsum, plaster, heavy		85.75	.093			2.96		2.96	4.46	
	0690	Wood, T&G, light		223	.036			1.14		1.14	1.71	
	0700	Wood, T&G, med.		149	.054			1.70		1.70	2.56	
	0710	Wood, T&G, heavy	↓	74.32	.108	↓		3.41		3.41	5.15	
	0720	Walls, wash										
	0730	Drywall, gypsum, plaster	1 Pord	297	.027	m²		.85		.85	1.29	
	0740	Wood, T&G		297	.027			.85		.85	1.29	
	0750	Masonry, brick & block, smooth		260	.031			.98		.98	1.47	
	0760	Masonry, brick & block, coarse	↓	186	.043	↓		1.36		1.36	2.05	

			DAILY	LABOR-		2006 BARE COSTS				TOTAL	
09990	**Paint Restoration**	CREW	OUTPUT	HOURS	UNIT	MAT.	LABOR	EQUIP.	TOTAL	INCL O&P	
910	8000	For Chemical Washing, see Division 04930										910
	8010	For Steam Cleaning, see Division 04930										
	8020	For Sand Blasting, see Division 03350-350 and 05910-500										

For information about Means Estimating Seminars, see yellow pages 12 and 13 in back of book

FINISHES 9

Division Notes

	CREW	DAILY OUTPUT	LABOR-HOURS	UNIT	2006 BARE COSTS				TOTAL INCL O&P
					MAT.	LABOR	EQUIP.	TOTAL	

Division 10
Specialties

Estimating Tips

General

- The items in this division are usually priced per m² or each.
- Many items in Division 10 require some type of support system or special anchors that are not usually furnished with the item. The required anchors must be added to the estimate in the appropriate division.
- Some items in Division 10, such as lockers, may require assembly before installation. Verify the amount of assembly required. Assembly can often exceed installation time.

10150 Compartments & Cubicles

- Support angles and blocking are not included in the installation of toilet compartments, shower/dressing compartments, or cubicles. Appropriate line items from Divisions 5 or 6 may need to be added to support the installations.
- Toilet partitions are priced by the stall. A stall consists of a side wall, pilaster, and door with hardware. Toilet tissue holders and grab bars are extra.

10600 Partitions

- The required acoustical rating of a folding partition can have a significant impact on costs. Verify the sound transmission coefficient rating of the panel priced to the specification requirements.

10800 Toilet/Bath/Laundry Accessories

- Grab bar installation does not include supplemental blocking or backing to support the required load. When grab bars are installed at an existing facility, provisions must be made to attach the grab bars to solid structure.

Reference Numbers

Reference numbers are shown in bold squares at the beginning of some major classifications. These numbers refer to related items in the Reference Section. The reference information may be an estimating procedure, an alternate pricing method, or technical information.

Note: Not all subdivisions listed here necessarily appear in this publication.

10 SPECIALTIES

10110	Chalkboards	CREW	DAILY OUTPUT	LABOR-HOURS	UNIT	2006 BARE COSTS				TOTAL INCL O&P
						MAT.	LABOR	EQUIP.	TOTAL	
240	0010	**CHALKBOARDS** Porcelain enamel steel									240
0100	Freestanding, reversible									
0120	Economy, wood frame, 1219 mm x 1829 mm									
0140	Chalkboard both sides				Ea.	545			545	600
0160	Chalkboard one side, cork other side				"	455			455	500
0200	Standard, lightweight satin finished alum., 1219 mm x 1829 mm									
0220	Chalkboard both sides				Ea.	565			565	620
0240	Chalkboard one side, cork other side				"	490			490	540
0300	Deluxe, heavy duty extruded aluminum, 1219 mm x 1829 mm									
0320	Chalkboard both sides				Ea.	1,175			1,175	1,300
0340	Chalkboard one side, cork other side				"	1,075			1,075	1,175
0400	Sliding chalkboards									
0450	Vertical, one sliding board with back panel, wall mounted									
0500	2438 mm x 1219 mm	2 Carp	8	2	Ea.	1,525	71		1,596	1,775
0520	2438 mm x 2438 mm		7.50	2.133		2,125	76		2,201	2,475
0540	2438 mm x 3658 mm	↓	7	2.286	↓	2,650	81.50		2,731.50	3,025
0600	Two sliding boards, with back panel									
0620	2438 mm x 1219 mm	2 Carp	8	2	Ea.	2,325	71		2,396	2,650
0640	2438 mm x 2438 mm		7.50	2.133		3,200	76		3,276	3,650
0660	2438 mm x 3658 mm	↓	7	2.286	↓	3,975	81.50		4,056.50	4,500
0700	Horizontal, two track									
0800	1219 mm x 2438 mm, 2 sliding panels	2 Carp	8	2	Ea.	1,575	71		1,646	1,825
0820	1219 mm x 3658 mm, 2 sliding panels		7.50	2.133		1,975	76		2,051	2,300
0840	1219 mm x 4877 mm, 4 sliding panels	↓	7	2.286	↓	2,675	81.50		2,756.50	3,075
0900	Four track, four sliding panels									
0920	1219 mm x 2438 mm	2 Carp	8	2	Ea.	2,525	71		2,596	2,875
0940	1219 mm x 3658 mm		7.50	2.133		3,200	76		3,276	3,625
0960	1219 mm x 4877 mm	↓	7	2.286	↓	4,050	81.50		4,131.50	4,575
1200	Vertical, motor operated									
1400	One sliding panel with back panel									
1450	3048 mm x 1219 mm	2 Carp	4	4	Ea.	5,775	142		5,917	6,575
1500	3048 mm x 3048 mm	↓	3.75	4.267		6,700	152		6,852	7,600
1550	3048 mm x 4877 mm	↓	3.50	4.571	↓	7,650	163		7,813	8,650
1700	Two sliding panels with back panel									
1750	3048 mm x 1219 mm	2 Carp	4	4	Ea.	9,250	142		9,392	10,400
1800	3048 mm x 3048 mm		3.75	4.267		10,500	152		10,652	11,700
1850	3048 mm x 4877 mm	↓	3.50	4.571	↓	11,800	163		11,963	13,300
2000	Three sliding panels with back panel									
2100	3048 mm x 1219 mm	2 Carp	4	4	Ea.	12,100	142		12,242	13,500
2150	3048 mm x 3048 mm		3.75	4.267		14,000	152		14,152	15,600
2200	3048 mm x 4877 mm	↓	3.50	4.571	↓	15,800	163		15,963	17,700
2400	For projection screen, glass beaded, add				m²	116			116	128
2500	For remote control, 1 panel control, add				Ea.	455			455	500
2600	2 panel control, add				"	785			785	865
2800	For units without back panels, deduct				m²	53.50			53.50	59
2850	For liquid chalk porcelain panels, add				"	45.50			45.50	50
3000	Swing leaf, any comb. of chalkboard & cork, aluminum frame									
3100	Floor style, 6 panels									
3150	762 mm x 1016 mm panels				Ea.	880			880	965
3200	1200 mm x 1000 mm panels				"	2,275			2,275	2,525
3300	Wall mounted, 6 panels									
3400	762 mm x 1016 mm panels	2 Carp	16	1	Ea.	810	35.50		845.50	950
3450	1219 mm x 1016 mm panels	"	16	1	"	1,700	35.50		1,735.50	1,900
3600	Extra panels for swing leaf units									
3700	762 mm x 1016 mm panels				Ea.	263			263	290
3750	1219 mm x 1016 mm panels				"	315			315	345

Important: See the Reference Section for supporting data - Crews, Rental Equipment, City Cost Indexes and Reference Data

10110	Chalkboards	CREW	DAILY OUTPUT	LABOR-HOURS	UNIT	2006 BARE COSTS				TOTAL INCL O&P
						MAT.	LABOR	EQUIP.	TOTAL	
240	3900	Wall hung									240
	4000	Aluminum frame and chalktrough									
	4200	914 mm x 1219 mm	2 Carp	16	1	Ea.	150	35.50		185.50	221
	4300	914 mm x 1524 mm		15	1.067		218	38		256	299
	4500	1219 mm x 2438 mm		14	1.143		335	40.50		375.50	430
	4600	1219 mm x 3658 mm	▼	13	1.231	▼	480	44		524	600
	4700	Wood frame and chalktrough									
	4800	914 mm x 1219 mm	2 Carp	16	1	Ea.	155	35.50		190.50	227
	5000	914 mm x 1524 mm		15	1.067		217	38		255	297
	5100	1219 mm x 1524 mm		14	1.143		219	40.50		259.50	305
	5300	1219 mm x 2438 mm	▼	13	1.231	▼	310	44		354	410
	5400	Liquid chalk, white porcelain enamel, wall hung									
	5420	Deluxe units, aluminum trim and chalktrough									
	5450	1219 mm x 1219 mm	2 Carp	16	1	Ea.	191	35.50		226.50	266
	5500	1219 mm x 2438 mm		14	1.143		335	40.50		375.50	430
	5550	1219 mm x 3658 mm	▼	12	1.333	▼	455	47.50		502.50	575
	5700	Wood trim and chalktrough									
	5900	1219 mm x 1219 mm	2 Carp	16	1	Ea.	440	35.50		475.50	540
	6000	1219 mm x 1829 mm		15	1.067		530	38		568	640
	6200	1219 mm x 2438 mm	▼	14	1.143		615	40.50		655.50	740
	6300	Liquid chalk, felt tip markers					1.50			1.50	1.65
	6500	Erasers					1.53			1.53	1.68
	6600	Board cleaner, 226.8 gram bottle				▼	4.13			4.13	4.54

10120	Tack & Visual Aid Boards	CREW	DAILY OUTPUT	LABOR-HOURS	UNIT	MAT.	LABOR	EQUIP.	TOTAL	TOTAL INCL O&P
350	0010	**CONTROL BOARDS**										350
	0020	Magnetic, porcelain finish, 457 mm x 610 mm, fr.	2 Carp	8	2	Ea.	169	71		240	297
	0100	610 mm x 914 mm		7.50	2.133		235	76		311	375
	0200	914 mm x 1219 mm		7	2.286		295	81.50		376.50	450
	0300	1219 mm x 1829 mm		6	2.667		625	95		720	840
	0400	1219 mm x 2438 mm	▼	5	3.200	▼	745	114		859	995
940	0010	**BULLETIN BOARD**										940
	0020	Cork sheets, unbacked, no frame, 6 mm thick	2 Carp	26.94	.594	m²	39	21		60	75.50
	0100	13 mm thick		26.94	.594		66	21		87	106
	0300	Fabric-face, no frame, on 6 mm cork underlay		26.94	.594		55	21		76	93.50
	0400	On 6 mm cork on 6 mm hardboard		26.94	.594		66.50	21		87.50	106
	0600	With edges wrapped		26.94	.594		71	21		92	111
	0700	On 11 mm fire retardant core		26.94	.594		48	21		69	85.50
	0900	With edges wrapped	▼	26.94	.594		71	21		92	111
	1000	Designer fabric only, cut to size					20.50			20.50	22.50
	1200	6 mm vinyl cork, on 6 mm hardboard, no frame	2 Carp	26.94	.594		67	21		88	107
	1300	On 6 mm coreboard		26.94	.594	▼	66.50	21		87.50	107
	2000	For map and display rail, economy, add		117	.137	m	6.80	4.86		11.66	15.05
	2100	Deluxe, add		107	.150	"	9.95	5.30		15.25	19.25
	2120	Prefabricated, 6 mm cork, 914 mm x 1524 mm with alum. frame		16	1	Ea.	114	35.50		149.50	181
	2140	Wood frame		16	1		147	35.50		182.50	218
	2160	1219 mm x 1219 mm with aluminum frame		16	1		94.50	35.50		130	160
	2180	Wood frame		16	1		130	35.50		165.50	199
	2200	1219 mm x 2438 mm with aluminum frame		14	1.143		182	40.50		222.50	264
	2210	With wood frame		14	1.143		94.50	40.50		135	168
	2220	1219 mm x 3658 mm with aluminum frame	▼	12	1.333	▼	262	47.50		309.50	365
	2230	Bulletin board case, single glass door, with lock									
	2240	914 mm x 610 mm, economy	2 Carp	12	1.333	Ea.	198	47.50		245.50	291
	2250	Deluxe		12	1.333		450	47.50		497.50	570
	2260	1067 mm x 762 mm, economy	▼	12	1.333	▼	365	47.50		412.50	475

SPECIALTIES 10

10120	Tack & Visual Aid Boards	CREW	DAILY OUTPUT	LABOR-HOURS	UNIT	2006 BARE COSTS				TOTAL INCL O&P	
						MAT.	LABOR	EQUIP.	TOTAL		
940	2270	Deluxe	2 Carp	12	1.333	Ea.	465	47.50		512.50	585	**940**
	2300	Glass enclosed cabinets, alum., cork panel, hinged doors										
	2400	914 mm x 914 mm, 1 door	2 Carp	12	1.333	Ea.	510	47.50		557.50	635	
	2500	1219 mm x 1219 mm, 2 door		11	1.455		850	51.50		901.50	1,025	
	2600	1219 mm x 2134 mm, 3 door		10	1.600		1,400	57		1,457	1,650	
	2800	1219 mm x 3048 mm, 4 door		8	2		2,050	71		2,121	2,350	
	2900	For lights, add/door opening	1 Elec	13	.615		154	26		180	209	
	3100	Horizontal sliding units, 4 doors, 1219 mm x 2438 mm	2 Carp	9	1.778		1,575	63		1,638	1,825	
	3200	1219 mm x 3658 mm		7	2.286		1,975	81.50		2,056.50	2,300	
	3400	8 doors, 1219 mm x 4877 mm		5	3.200		2,675	114		2,789	3,125	
	3500	1219 mm x 7315 mm	▼	4	4	▼	3,600	142		3,742	4,200	

10160	Metal Toilet Compartments	CREW	DAILY OUTPUT	LABOR-HOURS	UNIT	2006 BARE COSTS				TOTAL INCL O&P	
						MAT.	LABOR	EQUIP.	TOTAL		
100	0010	**TOILET PARTITIONS, METAL**										**100**
	0110	Cubicles, ceiling hung										
	0200	Painted metal	2 Carp	4	4	Ea.	475	142		617	740	
	0500	Stainless steel	"	4	4		1,250	142		1,392	1,600	
	0600	For handicap units, incl. 1321 mm grab bars, add				▼	415			415	455	
	0900	Floor and ceiling anchored										
	1000	Painted metal	2 Carp	5	3.200	Ea.	470	114		584	690	
	1300	Stainless steel	"	5	3.200		1,450	114		1,564	1,775	
	1400	For handicap units, incl. 1321 mm grab bars, add				▼	287			287	315	
	1610	Floor mounted										
	1700	Painted metal	2 Carp	7	2.286	Ea.	530	81.50		611.50	705	
	2000	Stainless steel	"	7	2.286		1,550	81.50		1,631.50	1,825	
	2100	For handicap units, incl. 1321 mm grab bars, add					287			287	315	
	2200	For juvenile units, deduct				▼	39.50			39.50	43.50	
	2450	Floor mounted, headrail braced										
	2500	Painted metal	2 Carp	6	2.667	Ea.	470	95		565	665	
	2800	Stainless steel	"	6	2.667		1,450	95		1,545	1,725	
	2900	For handicap units, incl. 1321 mm grab bars, add					305			305	335	
	3000	Wall hung partitions, painted metal	2 Carp	7	2.286		595	81.50		676.50	780	
	3300	Stainless steel	"	7	2.286		1,425	81.50		1,506.50	1,675	
	3400	For handicap units, incl. 1321 mm grab bars, add				▼	305			305	335	
	4000	Screens, entrance, flr mounted, 1473 mm high, 1219 mm wide										
	4200	Painted metal	2 Carp	15	1.067	Ea.	212	38		250	293	
	4500	Stainless steel	"	15	1.067	"	765	38		803	905	
	4650	Urinal screen, 450 mm wide										
	4700	Painted metal	2 Carp	8	2	Ea.	204	71		275	335	
	5000	Stainless steel	"	8	2	"	550	71		621	715	
	5100	Floor mounted, head rail braced										
	5300	Painted metal	2 Carp	8	2	Ea.	198	71		269	330	
	5600	Stainless steel	"	8	2	"	640	71		711	815	
	5750	Pilaster, flush										
	5800	Painted metal	2 Carp	10	1.600	Ea.	269	57		326	385	
	6100	Stainless steel		10	1.600		485	57		542	620	
	6300	Urinal screen, post braced, painted metal	▼	10	1.600		294	57		351	415	

10160 | Metal Toilet Compartments

		CREW	DAILY OUTPUT	LABOR-HOURS	UNIT	MAT.	LABOR	EQUIP.	TOTAL	TOTAL INCL O&P	
100	6600	Stainless steel	2 Carp	10	1.600	Ea.	480	57		537	620	100
	6700	Wall hung, bracket supported										
	6800	Painted metal	2 Carp	10	1.600	Ea.	281	57		338	400	
	7100	Stainless steel		10	1.600		440	57		497	575	
	7400	Flange supported, painted metal		10	1.600		218	57		275	330	
	7700	Stainless steel		10	1.600		510	57		567	650	
	7800	Wedge type, painted metal		10	1.600		257	57		314	370	
	8100	Stainless steel		10	1.600		525	57		582	665	

10165 | Plastic Laminate Toilet Compartment

		CREW	DAILY OUTPUT	LABOR-HOURS	UNIT	MAT.	LABOR	EQUIP.	TOTAL	TOTAL INCL O&P	
100	0010	**TOILET PARTITIONS, PLASTIC LAMINATE**										100
	0110	Cubicles, ceiling hung										
	0300	Plastic laminate on particle board ♿	2 Carp	4	4	Ea.	575	142		717	850	
	0600	For handicap units, incl. 1321 mm grab bars, add				"	415			415	455	
	0900	Floor and ceiling anchored										
	1100	Plastic laminate on particle board	2 Carp	5	3.200	Ea.	665	114		779	905	
	1400	For handicap units, incl. 1321 mm grab bars, add				"	287			287	315	
	1610	Floor mounted										
	1800	Plastic laminate on particle board	2 Carp	7	2.286	Ea.	570	81.50		651.50	750	
	2450	Floor mounted, headrail braced										
	2600	Plastic laminate on particle board	2 Carp	6	2.667	Ea.	770	95		865	1,000	
	3400	For handicap units, incl. 1321 mm grab bars, add					305			305	335	
	4300	Entrance screen, floor mtd, 11450 mm high, 1200 mm wide	2 Carp	15	1.067		435	38		473	535	
	4800	Urinal screen, 450 mm wide, ceiling braced, plastic laminate		8	2		293	71		364	430	
	5400	Floor mounted, headrail braced		8	2		355	71		426	500	
	5900	Pilaster, flush, plastic laminate		10	1.600		360	57		417	485	
	6400	Post braced, plastic laminate		10	1.600		360	57		417	485	
	6700	Wall hung, bracket supported										
	6900	Plastic laminate on particle board	2 Carp	10	1.600	Ea.	167	57		224	272	
	7450	Flange supported										
	7500	Plastic laminate on particle board	2 Carp	10	1.600	Ea.	395	57		452	525	

10170 | Plastic Toilet Compartments

		CREW	DAILY OUTPUT	LABOR-HOURS	UNIT	MAT.	LABOR	EQUIP.	TOTAL	TOTAL INCL O&P	
100	0010	**TOILET PARTITIONS, PLASTIC**										100
	0110	Cubicles, ceiling hung										
	0250	Phenolic ♿	2 Carp	4	4	Ea.	955	142		1,097	1,275	
	0600	For handicap units, incl. 1321 mm grab bars, add				"	415			415	455	
	0900	Floor and ceiling anchored										
	1050	Phenolic	2 Carp	5	3.200	Ea.	985	114		1,099	1,250	
	1400	For handicap units, incl. 1321 mm grab bars, add				"	287			287	315	
	1610	Floor mounted										
	1750	Phenolic ♿	2 Carp	7	2.286	Ea.	1,075	81.50		1,156.50	1,325	
	2100	For handicap units, incl. 1321 mm grab bars, add					287			287	315	
	2200	For juvenile units, deduct					39.50			39.50	43.50	
	2450	Floor mounted, headrail braced										
	2550	Phenolic	2 Carp	6	2.667	Ea.	950	95		1,045	1,200	

10180 | Stone Toilet Compartments

		CREW	DAILY OUTPUT	LABOR-HOURS	UNIT	MAT.	LABOR	EQUIP.	TOTAL	TOTAL INCL O&P	
100	0010	**TOILET PARTITIONS, STONE**										100
	0100	Cubicles, ceiling hung, marble	2 Marb	2	8	Ea.	1,550	279		1,829	2,125	
	0600	For handicap units, incl. 1321 mm grab bars, add					415			415	455	
	0800	Floor & ceiling anchored, marble	2 Marb	2.50	6.400		1,675	223		1,898	2,200	
	1400	For handicap units, incl. 1321 mm grab bars, add					287			287	315	
	1600	Floor mounted, marble	2 Marb	3	5.333		995	186		1,181	1,375	

SPECIALTIES 10

10180 | Stone Toilet Compartments

100			CREW	DAILY OUTPUT	LABOR-HOURS	UNIT	2006 BARE COSTS MAT.	LABOR	EQUIP.	TOTAL	TOTAL INCL O&P	100
100	2400	Floor mounted, headrail braced, marble	2 Marb	3	5.333	Ea.	950	186		1,136	1,325	100
	2900	For handicap units, incl. 1321 mm grab bars, add					305			305	335	
	4100	Entrance screen, floor mounted marble, 1450 mm high, 1200 mm wide	2 Marb	9	1.778		625	62		687	780	
	4600	Urinal screen, 457 mm wide, ceiling braced, marble	D-1	6	2.667		625	85.50		710.50	815	
	5100	Floor mounted, head rail braced										
	5200	Marble	D-1	6	2.667	Ea.	545	85.50		630.50	730	
	5700	Pilaster, flush, marble		9	1.778		695	57		752	850	
	6200	Post braced, marble		9	1.778		690	57		747	840	

10185 | Shower/Dressing Compartments

100			CREW	DAILY OUTPUT	LABOR-HOURS	UNIT	MAT.	LABOR	EQUIP.	TOTAL	TOTAL INCL O&P	100
100	0010	**PARTITIONS, SHOWER** Floor mounted, no plumbing										100
	0100	Cabinet, incl. base, no door, painted steel, 25 mm thick walls	2 Shee	5	3.200	Ea.	765	135		900	1,050	
	0300	With door, fiberglass		4.50	3.556		620	150		770	910	
	0600	Galvanized and painted steel, 25 mm thick walls		5	3.200		805	135		940	1,100	
	0800	Stall, 25 mm thick wall, no base, enameled steel		5	3.200		875	135		1,010	1,175	
	1200	Stainless steel		5	3.200		1,400	135		1,535	1,750	
	1400	For double entry type, no doors, deduct					10%					
	1500	Circular fiberglass, cabinet 914 mm diameter, [36"]	2 Shee	4	4		635	169		804	960	
	1700	One piece, 914 mm diameter, less door [36"]		4	4		535	169		704	850	
	1800	With door		3.50	4.571		880	193		1,073	1,275	
	2000	Curved shell shower, no door needed		3	5.333		770	225		995	1,200	
	2300	For fiberglass seat, add to both above					109			109	120	
	2400	Glass stalls, with doors, no receptors, chrome on brass	2 Shee	3	5.333		1,250	225		1,475	1,725	
	2700	Anodized aluminum	"	4	4		870	169		1,039	1,225	
	2900	Marble shower stall, stock design, with shower door	2 Marb	1.20	13.333		1,925	465		2,390	2,825	
	3000	With curtain		1.30	12.308		1,700	430		2,130	2,525	
	3200	Receptors, precast terrazzo, 800 mm x 800 mm		14	1.143		236	40		276	320	
	3300	1200 mm x 850 mm		9.50	1.684		405	59		464	535	
	3500	Plastic, simulated terrazzo receptor, 813 mm x 813 mm		14	1.143		100	40		140	171	
	3600	800 mm x 1200 mm		12	1.333		147	46.50		193.50	233	
	3800	Precast concrete, colors, 800 mm x 800 mm		14	1.143		197	40		237	278	
	3900	1200 mm x 1200 mm		8	2		210	70		280	335	
	4100	Shower doors, economy plastic, 600 mm wide	1 Shee	9	.889		109	37.50		146.50	178	
	4200	Tempered glass door, economy		8	1		184	42		226	267	
	4400	Folding, tempered glass, aluminum frame		6	1.333		330	56		386	445	
	4500	Sliding, tempered glass, 1219 mm opening		6	1.333		217	56		273	325	
	4700	Deluxe, tempered glass, chrome on brass frame, minimum		8	1		269	42		311	360	
	4800	Maximum		1	8		735	335		1,070	1,325	
	4850	On anodized aluminum frame, minimum		2	4		128	169		297	400	
	4900	Maximum		1	8		430	335		765	995	
	5100	Shower enclosure, tempered glass, anodized alum. frame										
	5120	2 panel & door, corner unit, 813 mm x 813 mm	1 Shee	2	4	Ea.	430	169		599	735	
	5140	Neo-angle corner unit, 406 mm x 610 mm x 406 mm	"	2	4		795	169		964	1,125	
	5200	Shower surround, 3 wall, polypropylene, 800 mm x 800 mm	1 Carp	4	2		300	71		371	440	
	5220	PVC, 813 mm x 813 mm		4	2		272	71		343	410	
	5240	Fiberglass		4	2		315	71		386	455	
	5250	2 wall, polypropylene, 813 mm x 813 mm		4	2		225	71		296	360	
	5270	PVC		4	2		280	71		351	420	
	5290	Fiberglass		4	2		310	71		381	455	
	5300	Tub doors, tempered glass & frame, minimum	1 Shee	8	1		184	42		226	267	
	5400	Maximum		6	1.333		425	56		481	550	
	5600	Chrome plated, brass frame, minimum		8	1		240	42		282	330	
	5700	Maximum		6	1.333		475	56		531	610	
	5900	Tub/shower enclosure, temp. glass, alum. frame, minimum		2	4		325	169		494	615	
	6200	Maximum		1.50	5.333		635	225		860	1,050	
	6500	On chrome-plated brass frame, minimum		2	4		450	169		619	755	

10150 | Compartments & Cubicles

10185 | Shower/Dressing Compartments

			CREW	DAILY OUTPUT	LABOR-HOURS	UNIT	2006 BARE COSTS				TOTAL INCL O&P	
							MAT.	LABOR	EQUIP.	TOTAL		
100	6600	Maximum	1 Shee	1.50	5.333	Ea.	925	225		1,150	1,375	100
	6800	Tub surround, 3 wall, polypropylene	1 Carp	4	2		182	71		253	310	
	6900	PVC		4	2		277	71		348	415	
	7000	Fiberglass, minimum		4	2		310	71		381	450	
	7100	Maximum	▼	3	2.667	▼	530	95		625	735	

10190 | Cubicles

			CREW	DAILY OUTPUT	LABOR-HOURS	UNIT	2006 BARE COSTS				TOTAL INCL O&P	
							MAT.	LABOR	EQUIP.	TOTAL		
200	0010	**PARTITIONS, HOSPITAL**										200
	0020	Curtain track, box channel, ceiling mounted	1 Carp	41.15	.194	m	15.50	6.90		22.40	28	
	0100	Suspended	"	30.48	.262	"	21	9.35		30.35	38	
	0300	Curtains, nylon mesh tops, fire resistant, 373 gram/m²										
	0310	Polyester oxford cloth, 2743 mm ceiling height	1 Carp	130	.062	m	34.50	2.19		36.69	41	
	0500	2438 mm ceiling height		130	.062		29	2.19		31.19	35.50	
	0700	Designer oxford cloth	▼	130	.062	▼	73.50	2.19		75.69	84.50	
	0800	I.V. track systems										
	0820	I.V. track, oval	1 Carp	41.15	.194	m	15.05	6.90		21.95	27.50	
	0830	I.V. trolley		32	.250	Ea.	31.50	8.90		40.40	48.50	
	0840	I.V. pendent, (tree, 5 hook)	▼	32	.250	"	112	8.90		120.90	138	

10200 | Louvers & Vents

10210 | Wall Louvers

			CREW	DAILY OUTPUT	LABOR-HOURS	UNIT	2006 BARE COSTS				TOTAL INCL O&P	
							MAT.	LABOR	EQUIP.	TOTAL		
800	0010	**LOUVERS**										800
	0020	Aluminum with screen, residential, 200 mm x 200 mm	1 Carp	38	.211	Ea.	10.05	7.50		17.55	22.50	
	0100	300 mm x 300 mm		38	.211		11.80	7.50		19.30	24.50	
	0200	300 mm x 450 mm		35	.229		15.35	8.15		23.50	29.50	
	0250	350 mm x 600 mm		30	.267		20	9.50		29.50	37	
	0300	450 mm x 600 mm		27	.296		24.50	10.55		35.05	43.50	
	0500	600 mm x 750 mm		24	.333		30.50	11.85		42.35	52	
	0700	Triangle, adjustable, small		20	.400		28	14.20		42.20	53	
	0800	Large	▼	15	.533	▼	45.50	18.95		64.45	79.50	
	1200	Extruded aluminum, see division 15850-600										
	2100	Midget, aluminum, 19 mm deep, 25 mm diameter [3/4",1"]	1 Carp	85	.094	Ea.	.91	3.35		4.26	6.20	
	2150	76 mm diameter [3"]		60	.133		1.98	4.74		6.72	9.60	
	2200	102 mm diameter [4"]		50	.160		2.98	5.70		8.68	12.15	
	2250	152 mm diameter [6"]	▼	30	.267	▼	3.52	9.50		13.02	18.60	
	2300	Ridge vent strip, mill finish	1 Shee	47.24	.169	m	8.75	7.15		15.90	20.50	
	2330	Soffit vent, continuous, 76 mm wide, aluminum, mill finish	1 Carp	60.96	.131		1.67	4.67		6.34	9.10	
	2340	Baked enamel finish		60.96	.131	▼	1.51	4.67		6.18	8.90	
	2400	Under eaves vent, aluminum, mill finish, 406 mm x 102 mm		48	.167	Ea.	1.96	5.95		7.91	11.40	
	2500	406 mm x 203 mm		48	.167		2.17	5.95		8.12	11.65	
	7000	Vinyl gable vent, 203 mm x 203 mm		38	.211		10.90	7.50		18.40	23.50	
	7020	305 mm x 305 mm		38	.211		25	7.50		32.50	39	
	7080	305 mm x 457 mm		35	.229		30	8.15		38.15	45.50	
	7200	457 mm x 610 mm	▼	30	.267	▼	34	9.50		43.50	52.50	

10265 | Wall & Corner Guards

			CREW	DAILY OUTPUT	LABOR-HOURS	UNIT	2006 BARE COSTS				TOTAL INCL O&P
							MAT.	LABOR	EQUIP.	TOTAL	
200	0010	**CORNER GUARDS**										200
	0020	Steel angle, 25 mm x 25 mm x 6 mm, 2.2 kg/m	2 Carp	48.77	.328	m	18.05	11.65		29.70	38
	0100	50 mm x 50 mm x 6 mm angles, 4.8 kg/m		45.72	.350		32.50	12.45		44.95	55.50
	0200	75 mm x 75 mm x 8 mm angles, 9.1 kg/m		42.67	.375		49.50	13.35		62.85	75.50
	0300	100 mm x 100 mm x 8 mm angles, 12.2 kg/m		36.58	.437		59.50	15.55		75.05	89
	0350	For angles drilled and anchored to masonry, add					15%	120%			
	0370	Drilled and anchored to concrete, add					20%	170%			
	0400	For galvanized angles, add					35%				
	0450	For stainless steel angles, add				m	100%				
	0500	Steel door track/wheel guard, 1200 mm high	E-4	22	1.455	Ea.	122	59	4.06	185.06	244
	0800	Pipe bumper for truck doors, 2438 mm long, 152 mm dia., filled [6"]		20	1.600		320	64.50	4.46	388.96	475
	0900	203 mm diameter [8"]		20	1.600		490	64.50	4.46	558.96	660
250	0010	**CORNER PROTECTION**										250
	0100	Stainless steel, 16 ga., adhesive mount, 89 mm leg	1 Sswk	24.38	.328	m	62.50	13.10		75.60	92
	0200	12 ga. stainless, adhesive mount	"	24.38	.328		107	13.10		120.10	141
	0300	For screw mount, add						10%			
	0500	Vinyl acrylic, adhesive mount, 75 mm leg	1 Carp	39.01	.205		21.50	7.30		28.80	35
	0550	38 mm leg		48.77	.164		14.75	5.85		20.60	25.50
	0600	Screw mounted, 76 mm leg		24.38	.328		22.50	11.65		34.15	42.50
	0650	38 mm leg		30.48	.262		14.45	9.35		23.80	30.50
	0700	Clear plastic, screw mounted, 64 mm		18.29	.437		9.75	15.55		25.30	35
	1000	Vinyl cover, alum. retainer, surface mount, 76 mm x 76 mm		14.63	.547		20.50	19.45		39.95	53
	1050	51 mm x 51 mm		14.63	.547		16.25	19.45		35.70	48.50
	1100	Flush mounted, 76 mm x 76 mm		9.75	.820		45	29		74	95
	1150	51 mm x 51 mm		9.75	.820		33	29		62	82
500	0010	**WALLGUARD**										500
	0400	Rub rail, vinyl, adhesive mounted	1 Carp	56.39	.142	m	13.10	5.05		18.15	22.50
	0500	Neoprene, aluminum backing, 38 mm x 51 mm		33.53	.239		22.50	8.50		31	37.50
	1000	Trolley rail, PVC, clipped to wall, 127 mm high		56.39	.142		26	5.05		31.05	36.50
	1050	203 mm high		54.86	.146		32	5.20		37.20	43
	1200	Bed bumper, vinyl acrylic, alum. retainer, 533 mm long		10	.800	Ea.	37.50	28.50		66	86
	1300	1346 mm long with aligner		9	.889	"	111	31.50		142.50	171
	1400	Bumper, vinyl cover, alum. retain., cush. mnt., 38 mm x 70 mm		24.38	.328	m	37	11.65		48.65	59
	1500	51 mm x 108 mm		24.38	.328		52	11.65		63.65	75.50
	1600	Surface mounted, 44 mm x 92 mm		24.38	.328		35	11.65		46.65	57
	2000	Crash rail, vinyl cover, alum. retainer, 25 mm x 102 mm		33.53	.239		39	8.50		47.50	55.50
	2100	25 mm x 203 mm		27.43	.292		48.50	10.35		58.85	69.50
	2150	Vinyl inserts, aluminum plate, 25 mm x 64 mm		33.53	.239		39	8.50		47.50	56
	2200	25 mm x 127 mm		27.43	.292		53.50	10.35		63.85	75
	3000	Handrail/bumper, vinyl cover, alum. retainer									
	3010	Bracket mounted, flat rail, 140 mm	1 Carp	24.38	.328	m	47	11.65		58.65	69.50
	3100	165 mm		24.38	.328		74	11.65		85.65	99
	3200	Bronze bracket, 44 mm diam. rail [1-3/4"]		24.38	.328		60	11.65		71.65	84

10275 | Access Flooring

			CREW	DAILY OUTPUT	LABOR-HOURS	UNIT	2006 BARE COSTS				TOTAL INCL O&P
							MAT.	LABOR	EQUIP.	TOTAL	
150	0010	**PEDESTAL ACCESS FLOORS** Computer room application, metal										150
	0020	Particle board or steel panels, no covering, under 600 m²	2 Carp	37.16	.431	m²	156	15.30		171.30	195

10 SPECIALTIES

10275	Access Flooring	CREW	DAILY OUTPUT	LABOR-HOURS	UNIT	2006 BARE COSTS				TOTAL INCL O&P	
						MAT.	LABOR	EQUIP.	TOTAL		
150 0300	Metal covered, over 600 m²	2 Carp	41.81	.383	m²	90.50	13.60		104.10	121	150
0400	Aluminum, 610 mm panels	↓	46.45	.344		335	12.25		347.25	385	
0600	For carpet covering, add					64.50			64.50	71	
0700	For vinyl floor covering, add					65.50			65.50	72	
0900	For high pressure laminate covering, add					43.50			43.50	48	
0910	For snap on stringer system, add	2 Carp	92.90	.172	↓	14.85	6.10		20.95	26	
0950	Office applications, to 203 mm high, steel panels,										
0960	no covering, over 600 m²	2 Carp	46.45	.344	m²	104	12.25		116.25	134	
1000	Machine cutouts after initial installation	1 Carp	10	.800	Ea.	4.42	28.50		32.92	49.50	
1050	Pedestals, 152mm to 305mm	2 Carp	85	.188		8.50	6.70		15.20	19.75	
1100	Air conditioning grilles, 102 mm x 305 mm	1 Carp	17	.471		61	16.75		77.75	93	
1150	102 mm x 457 mm	"	14	.571	↓	83.50	20.50		104	124	
1200	Approach ramps, minimum	2 Carp	7.90	2.026	m²	248	72		320	385	
1300	Maximum	"	5.57	2.870	"	330	102		432	525	
1500	Handrail, 2 rail, aluminum	1 Carp	4.57	1.750	m	293	62		355	420	

10305	Manufactured Fireplaces	CREW	DAILY OUTPUT	LABOR-HOURS	UNIT	2006 BARE COSTS				TOTAL INCL O&P	
						MAT.	LABOR	EQUIP.	TOTAL		
100 0010	**FIREPLACE, PREFABRICATED** Free standing or wall hung										100
0100	with hood & screen, minimum	1 Carp	1.30	6.154	Ea.	980	219		1,199	1,425	
0150	Average		1	8		1,375	284		1,659	1,950	
0200	Maximum		.90	8.889	↓	3,350	315		3,665	4,200	
0500	Chimney dbl. wall, stainless, over 2.6 m, 178 mm diam., add [7"]		10.06	.795	m	177	28.50		205.50	239	
0600	254 mm diameter, add [10"]		9.75	.820		180	29		209	244	
0700	305 mm diameter, add [12"]		9.45	.847		246	30		276	320	
0800	356 mm diameter, add [14"]		9.14	.875	↓	325	31		356	405	
1000	Simulated brick chimney top, 1219 mm high, 406 mm x 406 mm		10	.800	Ea.	202	28.50		230.50	267	
1100	610 mm x 610 mm		7	1.143	"	375	40.50		415.50	480	
1500	Simulated logs, gas fired, 42 200 kJ, 610 mm long, minimum		7	1.143	Set	510	40.50		550.50	625	
1600	Maximum		6	1.333		680	47.50		727.50	825	
1700	Electric, 1580 kJ, 457 mm long, minimum		7	1.143		140	40.50		180.50	218	
1800	12 130 kJ, maximum		6	1.333	↓	300	47.50		347.50	405	
2000	Fireplace, built-in, 914 mm hearth, radiant		1.30	6.154	Ea.	540	219		759	935	
2100	Recirculating, small fan		1	8		860	284		1,144	1,400	
2150	Large fan		.90	8.889		1,500	315		1,815	2,150	
2200	1067 mm hearth, radiant		1.20	6.667		730	237		967	1,175	
2300	Recirculating, small fan		.90	8.889		925	315		1,240	1,525	
2350	Large fan		.80	10		1,500	355		1,855	2,200	
2400	1219 mm hearth, radiant		1.10	7.273		1,400	259		1,659	1,950	
2500	Recirculating, small fan		.80	10		1,750	355		2,105	2,475	
2550	Large fan		.70	11.429		2,700	405		3,105	3,600	
3000	See through, including doors		.80	10		2,300	355		2,655	3,075	
3200	Corner (2 wall)	↓	1	8	↓	2,300	284		2,584	2,975	

10310	Fireplace Specialties & Accessories										
100 0010	**FIREPLACE ACCESSORIES**										100
0020	Chimney screen, galv, 330 x 330 mm flue	1 Bric	8	1	Ea.	43	36.50		79.50	103	

10300 | Fireplaces & Stoves

10310 | Fireplace Specialties & Accessories

			CREW	DAILY OUTPUT	LABOR-HOURS	UNIT	MAT.	LABOR	EQUIP.	TOTAL	TOTAL INCL O&P	
100	0050	Galv., 610 mm x 610 mm flue	1 Bric	5	1.600	Ea.	100	58.50		158.50	199	100
	0200	Stainless steel, 330 mm x 330 mm flue		8	1		264	36.50		300.50	345	
	0250	508 mm x 508 mm flue		5	1.600		360	58.50		418.50	485	
	0400	Cleanout doors and frames, cast iron, 203 mm x 203 mm		12	.667		30	24.50		54.50	70	
	0450	305 mm x 305 mm		10	.800		36.50	29		65.50	84.50	
	0500	457 mm x 610 mm		8	1		105	36.50		141.50	171	
	0550	Cast iron frame, steel door, 610 mm x 762 mm		5	1.600		227	58.50		285.50	340	
	0800	Damper, rotary control, steel, 762 mm opening		6	1.333		64	48.50		112.50	144	
	0850	Cast iron, 762 mm opening		6	1.333		71	48.50		119.50	152	
	0880	914 mm opening		6	1.333		77	48.50		125.50	159	
	0900	1219 mm opening		6	1.333		110	48.50		158.50	195	
	0920	1524 mm opening		6	1.333		247	48.50		295.50	345	
	0950	1829 mm opening		5	1.600		297	58.50		355.50	415	
	1000	2134 mm opening, special order		5	1.600		635	58.50		693.50	790	
	1050	2438 mm opening, special order		4	2		645	73		718	820	
	1200	Steel plate, poker control, 1524 mm opening		8	1		225	36.50		261.50	305	
	1250	2134 mm opening, special opening		5	1.600		410	58.50		468.50	540	
	1400	"Universal" type, chain operated, 813 mm x 508 mm opening		8	1		175	36.50		211.50	249	
	1450	1219 mm x 610 mm opening		5	1.600		261	58.50		319.50	375	
	1600	Dutch Oven door and frame, cast iron, 305 mm x 381 mm opening		13	.615		92	22.50		114.50	136	
	1650	Copper plated, 305 mm x 381 mm opening		13	.615		177	22.50		199.50	229	
	1800	Fireplace forms, no accessories, 813 mm opening		3	2.667		510	97.50		607.50	710	
	1900	914 mm opening		2.50	3.200		650	117		767	895	
	2000	1016 mm opening		2	4		860	146		1,006	1,175	
	2100	1981 mm opening		1.50	5.333		1,250	195		1,445	1,675	
	2400	Squirrel and bird screens, galvanized, 203 mm x 203 mm flue		16	.500		37.50	18.30		55.80	69	
	2450	330 mm x 330 mm flue		12	.667		40.50	24.50		65	81.50	

10320 | Stoves

			CREW	DAILY OUTPUT	LABOR-HOURS	UNIT	MAT.	LABOR	EQUIP.	TOTAL	TOTAL INCL O&P	
100	0010	**WOODBURNING STOVES**										100
	0015	Cast iron, minimum	2 Carp	1.30	12.308	Ea.	910	440		1,350	1,675	
	0020	Average		1	16		1,400	570		1,970	2,425	
	0030	Maximum		.80	20		2,350	710		3,060	3,675	
	0050	For gas log lighter, add					40			40	44	

10340 | Manufactured Exterior Specialties

10342 | Cupolas

			CREW	DAILY OUTPUT	LABOR-HOURS	UNIT	MAT.	LABOR	EQUIP.	TOTAL	TOTAL INCL O&P	
100	0010	**CUPOLA**										100
	0020	Stock units, pine, painted, 457 mm sq., 0.7 m high, alum. roof	1 Carp	4.10	1.951	Ea.	150	69.50		219.50	273	
	0100	Copper roof		3.80	2.105		175	75		250	310	
	0300	584 mm square, 838 mm high, aluminum roof		3.70	2.162		325	77		402	480	
	0400	Copper roof		3.30	2.424		440	86		526	620	
	0600	762 mm square, 940 mm high, aluminum roof		3.70	2.162		460	77		537	625	
	0700	Copper roof		3.30	2.424		495	86		581	680	
	0900	Hexagonal, 787 mm wide, 1168 mm high, copper roof		4	2		620	71		691	790	
	1000	914 mm wide, 1270 mm high, copper roof		3.50	2.286		680	81.50		761.50	875	
	1200	For deluxe stock units, add to above					25%					

Important: See the Reference Section for supporting data - Crews, Rental Equipment, City Cost Indexes and Reference Data

10340 | Manufactured Exterior Specialties

10342	Cupolas	CREW	DAILY OUTPUT	LABOR-HOURS	UNIT	2006 BARE COSTS				TOTAL INCL O&P	
						MAT.	LABOR	EQUIP.	TOTAL		
100	1400	For custom built units, add to above				Ea.	50%	50%				100

10350 | Flagpoles

10355	Flagpoles	CREW	DAILY OUTPUT	LABOR-HOURS	UNIT	2006 BARE COSTS				TOTAL INCL O&P	
						MAT.	LABOR	EQUIP.	TOTAL		
400	0010	**FLAGPOLE**, Ground set										400
	0050	Not including base or foundation										
	0100	Aluminum, tapered, ground set 6.1 m high	K-1	2	8	Ea.	790	253	88	1,131	1,350	
	0200	7.5 m high		1.70	9.412		895	297	104	1,296	1,550	
	0300	9.1 m high		1.50	10.667		890	335	117	1,342	1,625	
	0400	10.7 m high		1.40	11.429		1,450	360	126	1,936	2,300	
	0500	12 m high		1.20	13.333		2,050	420	147	2,617	3,050	
	0600	15.2 m high		1	16		2,700	505	176	3,381	3,950	
	0700	18.3 m high		.90	17.778		4,350	560	196	5,106	5,875	
	0800	21.3 m high		.80	20		7,800	630	220	8,650	9,800	
	1100	Counterbalanced, internal halyard, 6.1 m high		1.80	8.889		2,050	281	98	2,429	2,800	
	1200	9.1 m high		1.50	10.667		2,275	335	117	2,727	3,150	
	1300	12.2 m high		1.30	12.308		3,800	390	136	4,326	4,925	
	1400	15.2 m high		1	16		7,000	505	176	7,681	8,675	
	2820	Aluminum, electronically operated, 9.1 m high		1.40	11.429		3,975	360	126	4,461	5,075	
	2840	10.7 m high		1.30	12.308		4,275	390	136	4,801	5,475	
	2860	11.9 m high		1.10	14.545		4,675	460	160	5,295	6,000	
	2880	13.7 m high		1	16		5,975	505	176	6,656	7,550	
	2900	15.2 m high		.90	17.778		6,350	560	196	7,106	8,050	
	3000	Fiberglass, tapered, ground set, 7.0 m high		2	8		1,050	253	88	1,391	1,650	
	3100	18.7 m high		1.50	10.667		1,275	335	117	1,727	2,050	
	3200	1.1 m high		1.40	11.429		1,650	360	126	2,136	2,525	
	3300	12.0 m high		1.20	13.333		2,150	420	147	2,717	3,175	
	3400	15.0 m high		1	16		4,375	505	176	5,056	5,775	
	3500	18 m high	▼	.90	17.778	▼	6,275	560	196	7,031	7,975	
	4300	Steel, direct imbedded installation										
	4400	Internal halyard, 6.1 m high	K-1	2.50	6.400	Ea.	1,350	202	70.50	1,622.50	1,875	
	4500	7.6 m high		2.50	6.400		1,500	202	70.50	1,772.50	2,050	
	4600	9.1 m high		2.30	6.957		1,775	220	76.50	2,071.50	2,375	
	4700	12.2 m high		2.10	7.619		2,625	241	84	2,950	3,375	
	4800	15.2 m high		1.90	8.421		3,900	266	92.50	4,258.50	4,775	
	5000	18.3 m high		1.80	8.889		5,875	281	98	6,254	7,025	
	5100	21.3 m high		1.60	10		7,175	315	110	7,600	8,500	
	5200	24.4 m high		1.40	11.429		9,475	360	126	9,961	11,100	
	5300	27.4 m high		1.20	13.333		13,000	420	147	13,567	15,100	
	5500	30.5 m high	▼	1	16	▼	13,400	505	176	14,081	15,800	
	6400	Wood poles, tapered, clear vertical grain fir with tilting										
	6410	base, not incl. foundation, 102 mm butt, 7.6 m high	K-1	1.90	8.421	Ea.	625	266	92.50	983.50	1,200	
	6800	152 mm butt, 9.1 m high	"	1.30	12.308	"	775	390	136	1,301	1,600	
	7300	Foundations for flagpoles, including										
	7400	excavation and concrete, to 10.7 m high poles	C-1	10	3.200	Ea.	500	107		607	715	
	7600	12.2 m to 15.2 m high		3.50	9.143		910	305		1,215	1,475	
	7700	Over 18.3 m high	▼	2	16	▼	1,025	535		1,560	1,950	

SPECIALTIES 10

10350 | Flagpoles

10355 | Flagpoles

		CREW	DAILY OUTPUT	LABOR-HOURS	UNIT	2006 BARE COSTS MAT.	LABOR	EQUIP.	TOTAL	TOTAL INCL O&P	
900	0010	**FLAGPOLE**, Structure mounted										900
	0100	Fiberglass, vertical wall set, 6 m long	K-1	1.50	10.667	Ea.	1,225	335	117	1,677	2,000	
	0200	7 m long		1.40	11.429		1,300	360	126	1,786	2,125	
	0300	8 m long		1.30	12.308		1,900	390	136	2,426	2,825	
	0800	6 m long outrigger		1.30	12.308		1,250	390	136	1,776	2,125	
	1300	Aluminum, vertical wall set, tapered, with base, 6 m high		1.20	13.333		925	420	147	1,492	1,825	
	1400	9 m high		1	16		2,175	505	176	2,856	3,375	
	2400	Outrigger poles with base, 3.7 m long		1.30	12.308		860	390	136	1,386	1,700	
	2500	4.3 m long	▼	1	16	▼	1,375	505	176	2,056	2,500	

10400 | Identification Devices

10410 | Directories

		CREW	DAILY OUTPUT	LABOR-HOURS	UNIT	2006 BARE COSTS MAT.	LABOR	EQUIP.	TOTAL	TOTAL INCL O&P	
100	0010	**DIRECTORY BOARDS**										100
	0050	Plastic, glass covered, 762 mm x 508 mm	2 Carp	3	5.333	Ea.	242	190		432	560	
	0100	914 mm x 1219 mm		2	8		770	284		1,054	1,300	
	0300	Grooved cork, 762 mm x 508 mm		3	5.333		355	190		545	685	
	0400	914 mm x 1219 mm		2	8		300	284		584	775	
	0600	Black felt, 762 mm x 508 mm		3	5.333		192	190		382	505	
	0700	914 mm x 1219 mm		2	8		330	284		614	805	
	0900	Outdoor, weatherproof, black plastic, 914 mm x 610 mm		2	8		685	284		969	1,200	
	1000	914 mm x 914 mm		1.50	10.667		790	380		1,170	1,450	
	1800	Indoor, economy, open face, 457 mm x 610 mm		7	2.286		113	81.50		194.50	252	
	1900	610 mm x 914 mm		7	2.286		137	81.50		218.50	277	
	2000	914 mm x 610 mm		6	2.667		156	95		251	320	
	2100	914 mm x 1219 mm		6	2.667		234	95		329	405	
	2400	Bldg. directory, alum., black felt panels, 1 door, 610 mm x 457 mm		4	4		281	142		423	530	
	2500	914 mm x 610 mm		3.50	4.571		320	163		483	610	
	2600	1219 mm x 813 mm		3	5.333		480	190		670	820	
	2700	914 mm x 1219 mm, 2 door		2.50	6.400		590	228		818	1,000	
	2800	914 mm x 1524 mm		2	8		845	284		1,129	1,375	
	2900	1219 mm x 1524 mm	▼	1	16		925	570		1,495	1,900	
	3100	For bronze enamel finish, add					15%					
	3200	For bronze anodized finish, add					25%					
	3400	For illuminated directory, single door unit, add				▼	154			154	170	
	3500	For 152 mm header panel, 6 letters/305 mm, add				m	89			89	98	

10430 | Exterior Signage

		CREW	DAILY OUTPUT	LABOR-HOURS	UNIT	2006 BARE COSTS MAT.	LABOR	EQUIP.	TOTAL	TOTAL INCL O&P	
200	0010	**SIGNS**										200
	0020	Letters, 50 mm high, 9 mm deep, cast bronze	1 Carp	24	.333	Ea.	20.50	11.85		32.35	41	
	0140	13 mm deep, cast aluminum		18	.444		20.50	15.80		36.30	47	
	0160	Cast bronze		32	.250		25.50	8.90		34.40	42.50	
	0300	152 mm high, 16 mm deep, cast aluminum		24	.333		28.50	11.85		40.35	50	
	0400	Cast bronze		24	.333		42.50	11.85		54.35	65.50	
	0600	200 mm high, 19 mm deep, cast aluminum		14	.571		32.50	20.50		53	67	
	0700	Cast bronze		20	.400		72.50	14.20		86.70	102	
	0900	250 mm high, 25 mm deep, cast aluminum		18	.444		34	15.80		49.80	62	
	1000	Bronze	▼	18	.444	▼	87.50	15.80		103.30	121	

10 SPECIALTIES

10400 | Identification Devices

10430 | Exterior Signage

			CREW	DAILY OUTPUT	LABOR-HOURS	UNIT	2006 BARE COSTS				TOTAL INCL O&P
							MAT.	LABOR	EQUIP.	TOTAL	
200	1200	305 mm high, 32 mm deep, cast aluminum	1 Carp	12	.667	Ea.	43.50	23.50		67	85	200
	1500	Cast bronze		18	.444		114	15.80		129.80	150
	1600	356 mm high, 59 mm deep, cast aluminum		12	.667		69.50	23.50		93	114
	1800	Fabricated stainless steel, 152 mm high, 51 mm deep		20	.400		113	14.20		127.20	146
	1900	305 mm high, 76 mm deep		18	.444		134	15.80		149.80	172
	2100	457 mm high, 76 mm deep		12	.667		196	23.50		219.50	252
	2200	610 mm high, 102 mm deep		10	.800		273	28.50		301.50	345
	2700	Acrylic, on high density foam, 300 mm high, 50 mm deep		20	.400		17.20	14.20		31.40	41
	2800	305 mm high, 51 mm deep		18	.444		42	15.80		57.80	70.50
	3900	Plaques, custom, 508 mm x 762 mm, to 450 letters, cast alum.	2 Carp	4	4		855	142		997	1,150
	4000	Cast bronze		4	4		1,125	142		1,267	1,475
	4200	762 mm x 914 mm, up to 900 letters cast aluminum		3	5.333		1,725	190		1,915	2,200
	4300	Cast bronze		3	5.333		2,225	190		2,415	2,750
	4500	914 mm x 1219 mm, for up to 1300 letters, cast bronze		2	8		3,375	284		3,659	4,150
	4800	Signs, reflective alum. street signs, dbl. face, 2-way, w/bracket		30	.533		71	18.95		89.95	108
	4900	4-way		30	.533		111	18.95		129.95	152
	5100	Exit signs, 24 ga. alum., 356 mm x 305 mm surface mounted	1 Carp	30	.267		33.50	9.50		43	52
	5200	254 mm x 178 mm		20	.400		19.55	14.20		33.75	43.50
	5400	Bracket mounted, double face, 305 mm x 254 mm		30	.267		33	9.50		42.50	51.50
	5500	Sticky back, stock decals, 356 mm x 254 mm	1 Clab	50	.160		8.30	4.38		12.68	15.95
	6000	Int. elec., wall mount, fiberglass panels, 2 lamps, 152 mm	1 Elec	8	1		104	42		146	177
	6100	203 mm	"	8	1		46	42		88	113
	6400	Replacement sign faces, 152 mm or 203 mm	1 Clab	50	.160		25	4.38		29.38	34.50

10440 | Interior Signage

			CREW	DAILY OUTPUT	LABOR-HOURS	UNIT	2006 BARE COSTS				TOTAL INCL O&P
							MAT.	LABOR	EQUIP.	TOTAL	
200	0010	**INTERIOR SIGNAGE**										200
	1010	Flexible door sign, adhesive, w/Braille, 16 mm letters, 100 x 100 mm	1 Clab	32	.250	Ea.	24	6.85		30.85	36.50
	1050	150 x 150 mm		32	.250		34.50	6.85		41.35	48
	1100	200 x 50 mm		32	.250		25	6.85		31.85	37.50
	1150	200 x 100 mm		32	.250		31	6.85		37.85	44.50
	1200	200 x 200 mm		32	.250		46	6.85		52.85	61
	1250	300 x 50 mm		32	.250		32	6.85		38.85	45.50
	1300	300 x 150 mm		32	.250		45	6.85		51.85	60
	1350	300 x 300 mm		32	.250		92	6.85		98.85	112
	1500	Graphic symbols, 50 mm x 50 mm		32	.250		11	6.85		17.85	23
	1550	150 mm x 150 mm		32	.250		28	6.85		34.85	41.50
	1600	200 mm x 200 mm		32	.250		28	6.85		34.85	41.50

10450 | Pedestrian Control Devices

10455 | Turnstiles

			CREW	DAILY OUTPUT	LABOR-HOURS	UNIT	2006 BARE COSTS				TOTAL INCL O&P
							MAT.	LABOR	EQUIP.	TOTAL	
900	0010	**TURNSTILES**										900
	0020	One way, 4 arm, 1168 mm diameter, economy, manual	2 Carp	5	3.200	Ea.	550	114		664	780
	0100	Electric		1.20	13.333		900	475		1,375	1,725
	0300	High security, galv., 1651 mm diameter, 2134 mm high, manual		1	16		2,325	570		2,895	3,450
	0350	Electric		.60	26.667		3,025	950		3,975	4,800
	0420	Three arm, 610 mm opening, light duty, manual		2	8		865	284		1,149	1,400
	0450	Heavy duty		1.50	10.667		2,150	380		2,530	2,950
	0460	Manual, with registering & controls, light duty		2	8		1,875	284		2,159	2,500

10450 | Pedestrian Control Devices

10455 | Turnstiles

			CREW	DAILY OUTPUT	LABOR-HOURS	UNIT	2006 BARE COSTS				TOTAL INCL O&P	
							MAT.	LABOR	EQUIP.	TOTAL		
900	0470	Heavy duty	2 Carp	1.50	10.667	Ea.	2,250	380		2,630	3,075	900
	0480	Electric, heavy duty	↓	1.10	14.545		2,625	515		3,140	3,675	
	0500	For coin or token operating, add				↓	420			420	460	
	1200	One way gate with horizontal bars, 1651 mm diameter										
	1300	2134 mm high, recreation or transit type	2 Carp	.80	20	Ea.	3,025	710		3,735	4,425	
	1500	For electronic counter, add				"	185			185	203	

10500 | Lockers

10505 | Metal Lockers

			CREW	DAILY OUTPUT	LABOR-HOURS	UNIT	2006 BARE COSTS				TOTAL INCL O&P	
							MAT.	LABOR	EQUIP.	TOTAL		
500	0011	**LOCKERS** Steel, baked enamel										500
	0110	Single tier box locker, 300 mm x 375 mm x 1800 mm	1 Shee	8	1	Ea.	165	42		207	247	
	0120	450 mm x 375 mm x 1800 mm		8	1		178	42		220	261	
	0130	300 mm x 450 mm x 1800 mm		8	1		170	42		212	252	
	0140	450 mm x 450 mm x 1800 mm		8	1		193	42		235	277	
	0410	Double tier, 300 mm x 375 mm x 900 mm		21	.381		175	16.05		191.05	218	
	0420	450 mm x 375 mm x 900 mm		21	.381		237	16.05		253.05	285	
	0430	300 mm x 450 mm x 900 mm		21	.381		184	16.05		200.05	227	
	0440	450 mm x 450 mm x 900 mm		21	.381		197	16.05		213.05	242	
	0500	Two person, 450 mm x 375 mm x 1800 mm		8	1		263	42		305	355	
	0510	450 mm x 450 mm x 1800 mm		8	1		270	42		312	360	
	0520	Duplex, 375 mm x 375 mm x 1800 mm		8	1		258	42		300	350	
	0530	375 mm x 525 mm x 1800 mm		8	1	↓	300	42		342	400	
	0600	5 tier box lockers, minimum		30	.267	Opng.	37	11.25		48.25	58.50	
	0700	Maximum		24	.333		46.50	14.05		60.55	72.50	
	0900	6 tier box lockers, minimum		36	.222		36.50	9.35		45.85	54.50	
	1000	Maximum		30	.267	↓	43.50	11.25		54.75	65	
	1100	Wire meshed wardrobe, floor. mtd., open front varsity type		7.50	1.067	Ea.	166	45		211	252	
	2100	Locker bench, laminated maple, top only		30.48	.262	m	43.50	11.05		54.55	64.50	
	2200	Pedestals, steel pipe	↓	25	.320	Ea.	35.50	13.50		49	60	
	2400	16-person locker unit with clothing rack										
	2500	72 wide x 381 mm deep x 1829 mm high	1 Shee	8	1	Ea.	470	42		512	585	
	2550	457 mm deep	"	8	1	"	400	42		442	505	
	3000	Wall mounted lockers, 4 person, with coat bar										
	3100	1219 mm wide x 457 mm deep x 305 mm high	1 Shee	8	1	Ea.	210	42		252	296	
	3250	Rack w/ 24 wire mesh baskets		1.50	5.333	Set	325	225		550	700	
	3260	30 baskets		1.25	6.400		256	270		526	695	
	3270	36 baskets		.95	8.421		325	355		680	900	
	3280	42 baskets	↓	.80	10	↓	560	420		980	1,275	
	3300	For built-in lock with 2 keys, add				Ea.	7			7	7.70	
	3600	For hanger rods, add				"	1.79			1.79	1.97	

10 SPECIALTIES

Important: See the Reference Section for supporting data - Crews, Rental Equipment, City Cost Indexes and Reference Data

			DAILY	LABOR-		2006 BARE COSTS				TOTAL	
10525	**Fire Prot. Specialties**	CREW	OUTPUT	HOURS	UNIT	MAT.	LABOR	EQUIP.	TOTAL	INCL O&P	
200	0010	**FIRE EQUIPMENT CABINETS** Not equipped, 20 ga. steel box,										**200**
	0040	recessed, D.S. glass in door, box size given										
	1000	Port. extinguisher, single, 200 mm x 300 mm x 675 mm, alum. door	Q-12	8	2	Ea.	96	75.50		171.50	219	
	1100	Steel door and frame	"	8	2	"	70	75.50		145.50	191	
	3000	Hose rack assy., 38 mm valve & 30 m hose, 610 mm x 1016 mm	•									
	3100	Aluminum door and frame	Q-12	6	2.667	Ea.	235	100		335	410	
	3200	Steel door and frame	↓	6	2.667		154	100		254	320	
	3300	Stainless steel door and frame	↓	6	2.667	↓	310	100		410	490	
	4000	Hose rack assy., 64 mm x 38 mm valve, 30 m hose										
	4100	Aluminum door and frame	Q-12	6	2.667	Ea.	237	100		337	410	
	4200	Steel door and frame		6	2.667		160	100		260	325	
	4300	Stainless steel door and frame	↓	6	2.667	↓	315	100		415	495	
	5000	Hose rack assy., 63 mm x 38 mm valve, 30 m hose										
	5100	Aluminum door and frame	Q-12	5	3.200	Ea.	300	120		420	510	
	5200	Steel door and frame		5	3.200		176	120		296	375	
	5300	Stainless steel door and frame	↓	5	3.200	↓	335	120		455	550	
	8000	Valve cabinet for 64 mm FD angle valve, 457 x 457 x 203 mm										
	8100	Aluminum door and frame	Q-12	12	1.333	Ea.	104	50		154	191	
	8200	Steel door and frame		12	1.333		86	50		136	170	
	8300	Stainless steel door and frame	↓	12	1.333	↓	140	50		190	230	
300	0010	**FIRE EXTINGUISHERS**										**300**
	0120	CO_2, portable with swivel horn, 2.3 kg				Ea.	130			130	143	
	0140	With hose and "H" horn, 4.5 kg					160			160	176	
	0160	6.8 kg					190			190	209	
	0180	9.1 kg					220			220	242	
	0360	Wheeled type, cart mounted, 22.7 kg					1,200			1,200	1,325	
	0400	45 kg				↓	1,625			1,625	1,775	
	1000	Dry chemical, pressurized										
	1040	Standard type, portable, painted, 1.1 kg				Ea.	27.50			27.50	30.50	
	1060	2.3 kg					60			60	66	
	1080	4.5 kg					80			80	88	
	1100	9.1 kg					110			110	121	
	1120	13.6 kg					157			157	173	
	1300	Standard type, wheeled, 68 kg					1,600			1,600	1,750	
	2000	ABC all purpose type, portable, 1.1 kg					30			30	33	
	2060	2.3 kg					45			45	49.50	
	2080	4.3 kg					68			68	75	
	2100	9.1 kg					98			98	108	
	2300	Wheeled, 20 kg					650			650	715	
	2360	68 kg					1,600			1,600	1,750	
	3000	Dry chemical, outside cartridge to 220 K, painted, 4.1 kg					200			200	220	
	3060	11.8 kg					250			250	275	
	5000	Pressurized water, 9.5 L, stainless steel					85			85	93.50	
	5060	With anti-freeze					135			135	149	
	9400	Installation of extinguishers, 12 or more, on wood	1 Carp	30	.267			9.50		9.50	14.75	
	9420	On masonry or concrete	"	15	.533	↓		18.95		18.95	29.50	

10535 | Awnings & Canopies

		CREW	DAILY OUTPUT	LABOR-HOURS	UNIT	2006 BARE COSTS MAT.	LABOR	EQUIP.	TOTAL	TOTAL INCL O&P
050	**0010 AWNINGS, FABRIC**									050
0020	Including acrylic canvas and frame, standard design									
0100	Door and window, slope, 900 mm high, 1200 mm wide	1 Carp	4.50	1.778	Ea.	585	63		648	745
0110	1800 mm wide		3.50	2.286		755	81.50		836.50	955
0120	2400 mm wide		3	2.667		925	95		1,020	1,175
0200	Quarter round convex, 1200 mm wide		3	2.667		910	95		1,005	1,150
0210	1800 mm wide		2.25	3.556		1,175	126		1,301	1,500
0220	2400 mm wide		1.80	4.444		1,450	158		1,608	1,850
0300	Dome, 1200 mm wide		7.50	1.067		350	38		388	450
0310	1800 mm wide		3.50	2.286		790	81.50		871.50	995
0320	2400 mm wide		2	4		1,400	142		1,542	1,775
0350	Elongated dome, 1200 mm wide		1.33	6.015		1,325	214		1,539	1,775
0360	1800 mm wide		1.11	7.207		1,575	256		1,831	2,150
0370	2400 mm wide	▼	1	8		1,850	284		2,134	2,475
1000	Entry or walkway, peak, 3.6 m long, 1200 mm wide	2 Carp	.90	17.778		4,175	630		4,805	5,550
1010	1800 mm wide		.60	26.667		6,425	950		7,375	8,550
1020	2400 mm wide		.40	40		8,875	1,425		10,300	12,000
1100	Radius with dome end, 1200 mm wide		1.10	14.545		3,175	515		3,690	4,275
1110	1800 mm wide		.70	22.857		5,075	815		5,890	6,875
1120	2400 mm wide	▼	.50	32	▼	7,225	1,150		8,375	9,725
2000	Retractable lateral arm awning, manual									
2010	To 3650 mm wide, 2600 mm projection	2 Carp	1.70	9.412	Ea.	940	335		1,275	1,550
2020	To 4270 mm wide, 2600 mm projection		1.10	14.545		1,100	515		1,615	2,000
2030	To 5790 mm wide, 2600 mm projection		.85	18.824		1,500	670		2,170	2,700
2040	To 7320 mm wide, 2600 mm projection	▼	.67	23.881		1,875	850		2,725	3,400
2050	Motor for above, add	1 Carp	2.67	3	▼	810	107		917	1,050
3000	Patio/deck canopy with frame									
3010	3.65 m wide, 3.65 m projection	2 Carp	2	8	Ea.	1,325	284		1,609	1,925
3020	4.88 m wide, 4.27 m projection	"	1.20	13.333		2,075	475		2,550	3,025
9000	For fire retardant canvas, add					7%				
9010	For lettering or graphics, add					35%				
9020	For painted or coated acrylic canvas, deduct					8%				
9030	For translucent or opaque vinyl canvas, add					10%				
9040	For 6 or more units, deduct				▼	20%	15%			
200	**0010 CANOPIES**									200
0020	Wall hung, alum., 0.8 mm, prefinished, 2438 mm x 3048 mm	K-2	1.30	18.462	Ea.	1,900	675	136	2,711	3,400
0300	2438 mm x 6096 mm		1.10	21.818		3,775	795	160	4,730	5,700
0500	3048 mm x 3048 mm		1.30	18.462		2,125	675	136	2,936	3,650
0700	3048 mm x 6096 mm		1.10	21.818		3,925	795	160	4,880	5,875
1000	3658 mm x 6096 mm		1	24		4,825	875	176	5,876	7,050
1360	3658 mm x 9144 mm		.80	30		7,100	1,100	220	8,420	9,950
1700	3658 mm x 12 192 mm	▼	.60	40		8,650	1,450	294	10,394	12,400
1900	For free standing units, add				▼	20%	10%			
2300	Aluminum entrance canopies, flat soffit, 0.8 mm									
2500	1067 mm x 1219 mm, clear anodized	2 Carp	4	4	Ea.	795	142		937	1,100
2700	Bronze anodized		4	4		1,400	142		1,542	1,750
3000	Polyurethane painted		4	4		1,125	142		1,267	1,475
3300	1372 mm x 3048 mm, clear anodized		2	8		2,200	284		2,484	2,850
3500	Bronze anodized		2	8		2,800	284		3,084	3,525
3700	Polyurethane painted	▼	2	8		2,350	284		2,634	3,025
4000	Wall downspout, 3048 mm, clear anodized	1 Carp	7	1.143		136	40.50		176.50	214
4300	Bronze anodized		7	1.143		237	40.50		277.50	325
4500	Polyurethane painted	▼	7	1.143	▼	204	40.50		244.50	288
7000	Carport, baked vinyl finish, 0.8 mm, 6.1 m x 3.0 m, no fndtns., min.	K-2	4	6	Car	3,400	219	44	3,663	4,175
7250	Maximum		2	12	"	6,800	440	88	7,328	8,325
7500	Walkway cover, to 3.66 m W, stl. vinyl fin. 0.8 mm no fndtns. min.	▼	23.23	1.033	m²	215	37.50	7.60	260.10	310

10530 | Protective Covers

10535 | Awnings & Canopies

			CREW	DAILY OUTPUT	LABOR-HOURS	UNIT	2006 BARE COSTS MAT.	LABOR	EQUIP.	TOTAL	TOTAL INCL O&P	
200	7750	Maximum	K-2	18.58	1.292	m²	233	47	9.50	289.50	350	200

10550 | Postal Specialties

10555 | Mail Delivery Systems

			CREW	DAILY OUTPUT	LABOR-HOURS	UNIT	2006 BARE COSTS MAT.	LABOR	EQUIP.	TOTAL	TOTAL INCL O&P	
600	0010	**MAIL BOXES**										600
	0020	Horiz., key lock, 125 mmH x 150 mmW x 375 mmD, rear	1 Carp	34	.235	Ea.	33	8.35		41.35	49	
	0100	Front loading		34	.235		37	8.35		45.35	53.50	
	0200	Double, 127 mmH x 305 mmW x 381 mmD, rear loading		26	.308		60	10.95		70.95	83	
	0300	Front loading		26	.308		63	10.95		73.95	86.50	
	0500	Quadruple, 254 mmH x 305 mmW x 381 mmD, rear loading		20	.400		112	14.20		126.20	145	
	0600	Front loading		20	.400		114	14.20		128.20	147	
	0800	Vert., front load, 381 mmH x 127 mmW x 152 mmD, alum.		34	.235		26	8.35		34.35	41.50	
	0900	Bronze, duranodic finish		34	.235		44	8.35		52.35	61.50	
	1000	Steel, enameled		34	.235		32	8.35		40.35	48	
	1700	Alphabetical directories, 120 names		10	.800		114	28.50		142.50	171	
	1800	Letter collection box		6	1.333		600	47.50		647.50	735	
	1900	Letter slot, residential		20	.400		60	14.20		74.20	88	
	2000	Post office type		8	1		220	35.50		255.50	298	
	2200	Post office counter window, with grille		2	4		515	142		657	785	
	2250	Key keeper, single key, aluminum		26	.308		37	10.95		47.95	58	
	2300	Steel, enameled	▼	26	.308	▼	100	10.95		110.95	127	
700	0010	**MAIL CHUTES**										700
	0020	Aluminum & glass, 356 mm wide, 116 mm deep	2 Shee	4	4	Floor	650	169		819	975	
	0100	219 mm deep		3.80	4.211		790	177		967	1,150	
	0300	219 mm x 88 mm, aluminum		5	3.200		600	135		735	870	
	0400	Bronze or stainless		4.50	3.556	▼	930	150		1,080	1,250	
	0600	Lobby collection boxes, aluminum		5	3.200	Ea.	1,800	135		1,935	2,175	
	0700	Bronze or stainless	▼	4.50	3.556	"	2,300	150		2,450	2,750	

10600 | Partitions

10605 | Wire Mesh Partitions

			CREW	DAILY OUTPUT	LABOR-HOURS	UNIT	2006 BARE COSTS MAT.	LABOR	EQUIP.	TOTAL	TOTAL INCL O&P	
100	0010	**PARTITIONS, WOVEN WIRE** For tool or stockroom enclosures										100
	0100	Channel frame, 38 mm diamond mesh, 10 ga. wire, painted										
	0300	Wall panels, 1200 mm wide, 2100 mm high	2 Carp	25	.640	Ea.	115	23		138	162	
	0400	2400 mm high		23	.696		123	24.50		147.50	175	
	0600	3000 mm high	▼	18	.889		145	31.50		176.50	209	
	0700	For 1500 mm wide panels, add					5%					
	0900	Ceiling panels, 3000 mm long, 600 mm wide	2 Carp	25	.640		92.50	23		115.50	138	
	1000	1200 mm wide	▼	15	1.067	▼	131	38		169	203	

SPECIALTIES 10

411

10605 | Wire Mesh Partitions

		CREW	DAILY OUTPUT	LABOR-HOURS	UNIT	MAT.	LABOR	EQUIP.	TOTAL	TOTAL INCL O&P	
100	1200	Panel with service window & shelf, 1500 mm wide, 2100 mm high	2 Carp	20	.800	Ea.	300	28.50		328.50	375	100
	1300	2400 mm high		15	1.067		310	38		348	405	
	1500	Sliding doors, full height, 900 mm wide, 2100 mm high		6	2.667		305	95		400	485	
	1600	3000 mm high		5	3.200		330	114		444	540	
	1800	1829 mm wide sliding door, 2134 mm full height		5	3.200		400	114		514	620	
	1900	3048 mm high		4	4		520	142		662	795	
	2100	Swinging doors, 914 mm wide, 2134 mm high, no transom		6	2.667		222	95		317	390	
	2200	2134 mm high, 914 mm transom		5	3.200		305	114		419	510	

10610 | Folding Gates

		CREW	DAILY OUTPUT	LABOR-HOURS	UNIT	MAT.	LABOR	EQUIP.	TOTAL	TOTAL INCL O&P	
100	0010	**SECURITY GATES** For roll up type, see division 08330-130										100
	0300	Scissors type folding gate, ptd. stl., single, 1950 mm high, 1676 mm	2 Sswk	4	4	Opng.	142	160		302	445	
	0350	1950 mm wide		4	4		138	160		298	440	
	0400	2286 mm wide		4	4		166	160		326	470	
	0600	Double gate, 2400 mm high, 2400 mm wide		2.50	6.400		217	256		473	700	
	0650	3000 mm wide		2.50	6.400		246	256		502	730	
	0700	3600 mm wide		2	8		340	320		660	950	
	0750	4200 mm wide		2	8		390	320		710	1,000	
	0900	Door gate, folding steel, 1219 mm wide, 1549 mm high		4	4		72.50	160		232.50	370	
	1000	1803 mm high		4	4		75	160		235	370	
	1200	2057 mm high		4	4		81.50	160		241.50	380	
	1300	Window gates, 610 mm to 1219 mm wide, 787 mm high		4	4		43	160		203	335	
	1500	1397 mm high		3.75	4.267		86	170		256	400	
	1600	2007 mm high		3.50	4.571		99.50	183		282.50	440	

10615 | Demountable Partitions

		CREW	DAILY OUTPUT	LABOR-HOURS	UNIT	MAT.	LABOR	EQUIP.	TOTAL	TOTAL INCL O&P	
100	0010	**PARTITIONS, MOVABLE OFFICE** Demountable, add for doors										100
	0100	Do not deduct door openings from total mm										
	0900	Demountable gypsum system on 51 mm to 64 mm										
	1000	steel studs, 2743 mm high, 76 mm to 95 mm thick										
	1200	Vinyl clad gypsum	2 Carp	14.63	1.094	m	157	39		196	234	
	1300	Fabric clad gypsum		13.41	1.193		390	42.50		432.50	495	
	1500	Steel clad gypsum		12.19	1.312		425	46.50		471.50	540	
	1600	1.75 system, aluminum framing, vinyl clad hardboard,										
	1800	paper honeycomb core panel, 44 mm to 64 mm thick										
	1900	2743 mm high	2 Carp	14.63	1.094	m	263	39		302	350	
	2100	2134 mm high		18.29	.875		236	31		267	310	
	2200	1524 mm high		24.38	.656		200	23.50		223.50	257	
	2250	Unitized gypsum system										
	2300	Unitized panel, 2743 mm high, 51 mm to 64 mm thick										
	2350	Vinyl clad gypsum	2 Carp	14.63	1.094	m	340	39		379	435	
	2400	Fabric clad gypsum	"	13.41	1.193	"	560	42.50		602.50	680	
	2500	Unitized mineral fiber system										
	2510	Unitized panel, 2743 mm high, 57 mm thick, aluminum frame										
	2550	Vinyl clad mineral fiber	2 Carp	14.63	1.094	m	335	39		374	430	
	2600	Fabric clad mineral fiber	"	13.41	1.193	"	505	42.50		547.50	620	
	2800	Movable steel walls, modular system										
	2900	Unitized panels, 2743 mm high, 1219 mm wide										
	3100	Baked enamel, pre-finished	2 Carp	18.29	.875	m	380	31		411	470	
	3200	Fabric clad steel		17.07	.937	"	550	33.50		583.50	655	
	5310	Trackless wall, cork finish, semi-acoustic, 41 mm thick, minimum		30.19	.530	m²	350	18.85		368.85	415	
	5320	Maximum		17.65	.906		330	32		362	415	
	5330	Acoustic, 51 mm thick, minimum		28.33	.565		283	20		303	340	
	5340	Maximum		20.90	.765		485	27		512	580	
	5500	For acoustical partitions, add, minimum					20			20	22.50	
	5550	Maximum					94.50			94.50	104	

Important: See the Reference Section for supporting data - Crews, Rental Equipment, City Cost Indexes and Reference Data

10 SPECIALTIES

10615 | Demountable Partitions

		CREW	DAILY OUTPUT	LABOR-HOURS	UNIT	2006 BARE COSTS MAT.	LABOR	EQUIP.	TOTAL	TOTAL INCL O&P	
100	5700	For doors, see Div. 08100 & 08200										100
	5800	For door hardware, see Div. 08700										
	6100	In-plant modular office system, w/prehung hollow core door										
	6200	76 mm thick polystyrene core panels										
	6250	3658 mm x 3658 mm, 2 wall	2 Clab	3.80	4.211	Ea.	2,675	115		2,790	3,125	
	6300	4 wall		1.90	8.421		3,725	231		3,956	4,450	
	6350	4877 mm x 4877 mm, 2 wall		3.60	4.444		4,175	122		4,297	4,800	
	6400	4 wall		1.80	8.889		5,375	244		5,619	6,275	

10630 | Port. Partitions/Screens/Panels

		CREW	DAILY OUTPUT	LABOR-HOURS	UNIT	2006 BARE COSTS MAT.	LABOR	EQUIP.	TOTAL	TOTAL INCL O&P	
100	0010	**PARTITIONS, PORTABLE** Divider panels, free standing, fiber core										100
	0020	Fabric face straight										
	0100	914 mm long, 1219 mm high	2 Carp	30.48	.525	m	350	18.65		368.65	415	
	0200	1524 mm high		27.43	.583		350	20.50		370.50	420	
	0500	1829 mm high		22.86	.700		400	25		425	480	
	0900	1524 mm long, 1219 mm high		53.34	.300		264	10.65		274.65	305	
	1000	1524 mm high		45.72	.350		325	12.45		337.45	375	
	1500	3861 mm high		38.10	.420		310	14.95		324.95	365	
	1600	1829 mm long, 1524 mm high		49.38	.324		259	11.50		270.50	305	
	3100	Curved, 914 mm long, 1524 mm high		27.43	.583		272	20.50		292.50	335	
	3150	1829 mm high		22.86	.700		315	25		340	385	
	3200	Economical panels, fabric face, 1219 mm long, 1524 mm high		40.23	.398		129	14.15		143.15	164	
	3250	1829 mm high		34.14	.469		139	16.65		155.65	179	
	3300	1524 mm long, 1524 mm high		45.72	.350		113	12.45		125.45	143	
	3350	1829 mm high		38.10	.420		118	14.95		132.95	152	
	3380	914 mm curved, 1524 mm high		27.43	.583		272	20.50		292.50	335	
	3390	1829 mm high		22.86	.700		315	25		340	385	
	3450	Acoustical panels, 60 to 90 NRC, 914 mm long, 1524 mm high		27.43	.583		315	20.50		335.50	380	
	3550	1829 mm high		22.86	.700		300	25		325	370	
	3600	1524 mm long, 1524 mm high		45.72	.350		218	12.45		230.45	259	
	3650	1829 mm high		38.10	.420		242	14.95		256.95	289	
	3700	1829 mm long, 1524 mm high		49.38	.324		211	11.50		222.50	250	
	3750	1829 mm high		42.06	.380		212	13.50		225.50	254	
	3800	Economy acoustical panels, 40 NRC, 1219 mm long, 1524 mm high		40.23	.398		129	14.15		143.15	164	
	3850	1829 mm high		34.14	.469		138	16.65		154.65	178	
	3900	1524 mm long, 1829 mm high		38.10	.420		118	14.95		132.95	152	
	3950	1829 mm long, 1524 mm high		49.38	.324		96.50	11.50		108	124	
	4000	Metal chalkboard, 1981 mm high, chalkboard, 1 side		38.10	.420		297	14.95		311.95	350	
	4100	Metal chalkboard, 2 sides		36.58	.437		340	15.55		355.55	400	
	4300	Tackboard, both sides		37.49	.427		268	15.15		283.15	320	

10651 | Accordion Folding Partitions

		CREW	DAILY OUTPUT	LABOR-HOURS	UNIT	2006 BARE COSTS MAT.	LABOR	EQUIP.	TOTAL	TOTAL INCL O&P	
100	0010	**PARTITIONS, FOLDING ACCORDION**										100
	0100	Vinyl covered, over 13.94 m², frame not included										
	0300	Residential, 6.10 kg/m², 2.4 m maximum height	2 Carp	27.87	.574	m²	187	20.50		207.50	238	
	0400	Commercial, 8.55 kg/m², 2.4 m maximum height		20.90	.765		214	27		241	278	
	0600	9.77 kg/m², 5.2 m maximum height		13.94	1.148		221	41		262	305	
	0700	Industrial, 19.53 kg/m², 6.1 m maximum height		6.97	2.296		315	81.50		396.50	475	
	0900	Acoustical, 14.65 kg/m², 5.2 m maximum height		9.29	1.722		248	61		309	370	
	1200	24.41 kg/m², 6.1 m maximum height		8.83	1.813		345	64.50		409.50	475	
	1300	26.86 kg/m², 5.2 m maximum height		8.36	1.914		405	68		473	550	
	1400	Fire rated, 21.97 kg/m², 6.1 m maximum height		14.86	1.076		405	38.50		443.50	505	
	1500	Vinyl clad wood or steel, electric operation, 24.41 kg/m²		14.86	1.076		465	38.50		503.50	575	
	1900	Wood, non-acoustic, birch or mahogany, to 3.0 m high		27.87	.574		251	20.50		271.50	310	

SPECIALTIES 10

10653 | Folding Panel Partitions

		CREW	DAILY OUTPUT	LABOR-HOURS	UNIT	2006 BARE COSTS MAT.	LABOR	EQUIP.	TOTAL	TOTAL INCL O&P
200 0010	**PARTITIONS, FOLDING LEAF** Acoustic, wood									**200**
0100	Vinyl faced, to 5.5 m high, 29 kg/m², minimum	2 Carp	5.57	2.870	m²	475	102		577	680
0150	Average		4.18	3.827		565	136		701	835
0200	Maximum		2.79	5.741		730	204		934	1,125
0400	Formica or hardwood finish, minimum		5.57	2.870		490	102		592	695
0500	Maximum		2.79	5.741		520	204		724	895
0600	Wood, low acoustical type, 22 kg/m², to 4.3 m high		4.65	3.445		355	122		477	580
1100	Steel, acoustical, 44 to 60 kg/m², vinyl faced, minimum		5.57	2.870		505	102		607	715
1200	Maximum		2.79	5.741		615	204		819	1,000
1700	Aluminum framed, acoustical, to 3.7 m high, 27 kg/m², minimum		5.57	2.870		340	102		442	535
1800	Maximum		2.79	5.741		410	204		614	770
2000	32 kg/m², minimum		5.57	2.870		360	102		462	555
2100	Maximum		2.79	5.741		440	204		644	805

10658 | Acoustic Air Wall

		CREW	DAILY OUTPUT	LABOR-HOURS	UNIT	2006 BARE COSTS MAT.	LABOR	EQUIP.	TOTAL	TOTAL INCL O&P
100 0010	**PARTITIONS, OPERABLE**									**100**
0020	Acoustic air wall, 41 mm thick, minimum	2 Carp	34.84	.459	m²	289	16.35		305.35	345
0100	Maximum		33.91	.472		495	16.75		511.75	570
0300	57 mm thick, minimum		33.44	.478		325	17		342	385
0400	Maximum		30.66	.522		570	18.55		588.55	655
0600	For track type, add to above				m	325			325	355
0700	Overhead track type, acoustical, 76 mm thick, 54 kg/m², minimum	2 Carp	32.52	.492	m²	730	17.50		747.50	830
0800	Maximum	"	27.87	.574	"	880	20.50		900.50	1,000

10674 | Storage Shelving

		CREW	DAILY OUTPUT	LABOR-HOURS	UNIT	2006 BARE COSTS MAT.	LABOR	EQUIP.	TOTAL	TOTAL INCL O&P
500 0010	**SHELVING**									**500**
0020	Metal, industrial, cross-braced, 914 mm wide, 305 mm deep	1 Sswk	16.26	.492	m² Shlf	104	19.65		123.65	150
0100	610 mm deep		30.66	.261		74.50	10.45		84.95	100
0300	1219 mm wide, 305 mm deep		17.19	.465		98	18.60		116.60	142
0400	610 mm deep		35.30	.227		71.50	9.05		80.55	95.50
1200	Enclosed sides, cross-braced back, 914 mm wide, 305 mm deep		16.26	.492		177	19.65		196.65	231
1300	610 mm deep		26.94	.297		115	11.85		126.85	148
1500	Fully enclosed, sides and back, 914 mm wide, 305 mm deep		13.94	.574		160	23		183	218
1600	610 mm deep		23.69	.338		98	13.50		111.50	133
1800	1219 mm wide, 305 mm deep		13.94	.574		110	23		133	163
1900	610 mm deep		26.94	.297		70.50	11.85		82.35	99
2200	Wide span, 726 kg capacity/shelf, 1829 mm wide, 610 mm deep		35.30	.227		78	9.05		87.05	102
2400	914 mm deep		40.88	.196		70.50	7.80		78.30	91.50
2600	2438 mm wide, 610 mm deep		40.88	.196		72.50	7.80		80.30	94
2800	914 mm deep		48.31	.166		61.50	6.60		68.10	80
4000	Pallet racks, steel frame 2268 kg capacity, 2438 mm L, 914 mm D	2 Sswk	41.81	.383		116	15.30		131.30	156
4200	1067 mm deep		46.45	.344		102	13.75		115.75	137
4400	1219 mm deep		48.31	.331		99.50	13.25		112.75	133
600 0010	**PARTS BINS** Metal, gray baked enamel finish									**600**
0100	1905 mm high, 914 mm wide									
0300	12 bins, 457 mm wide x 305 mm high, 305 mm deep	2 Clab	10	1.600	Ea.	243	44		287	335
0400	610 mm deep		10	1.600		320	44		364	420

10 SPECIALTIES

10670 | Storage Shelving

10674	Storage Shelving	CREW	DAILY OUTPUT	LABOR-HOURS	UNIT	2006 BARE COSTS				TOTAL INCL O&P	
						MAT.	LABOR	EQUIP.	TOTAL		
600	0600	72 bins, 152 mm wide x 152 mm high, 305 mm deep	2 Clab	8	2	Ea.	380	55		435	505	**600**
	0700	450 mm deep	↓	8	2	↓	850	55		905	1,025	
	1000	2210 mm high, 914 mm wide										
	1200	14 bins, 457 mm wide x 305 mm high, 305 mm deep	2 Clab	10	1.600	Ea.	255	44		299	350	
	1300	610 mm deep		10	1.600		335	44		379	440	
	1500	84 bins, 152 mm wide x 152 mm high, 305 mm deep		8	2		570	55		625	710	
	1600	610 mm deep	↓	8	2	↓	745	55		800	905	

10750 | Telephone Specialties

10755	Telephone Enclosures	CREW	DAILY OUTPUT	LABOR-HOURS	UNIT	2006 BARE COSTS				TOTAL INCL O&P	
						MAT.	LABOR	EQUIP.	TOTAL		
400	0010	**TELEPHONE ENCLOSURE**										**400**
	0300	Shelf type, wall hung, minimum	2 Carp	5	3.200	Ea.	1,025	114		1,139	1,300	
	0400	Maximum		5	3.200		2,625	114		2,739	3,075	
	0600	Booth type, painted steel, indoor or outdoor, minimum		1.50	10.667		3,200	380		3,580	4,125	
	0700	Maximum (stainless steel)		1.50	10.667		10,700	380		11,080	12,400	
	1300	Outdoor, acoustical, on post		3	5.333		1,400	190		1,590	1,850	
	1400	Phone carousel, pedestal mounted with dividers		.60	26.667		5,375	950		6,325	7,375	
	1900	Outdoor, drive-up type, wall mounted		4	4		860	142		1,002	1,175	
	2000	Post mounted, stainless steel posts	↓	3	5.333	↓	1,325	190		1,515	1,750	
	2200	Directory shelf, wall mounted, stainless steel										
	2300	3 binders	2 Carp	8	2	Ea.	1,025	71		1,096	1,225	
	2500	4 binders		7	2.286		1,175	81.50		1,256.50	1,400	
	2600	5 binders		6	2.667		1,500	95		1,595	1,800	
	2800	Table type, stainless steel, 4 binders		8	2		1,200	71		1,271	1,425	
	2900	7 binders	↓	7	2.286	↓	1,500	81.50		1,581.50	1,775	

10800 | Toilet/Bath/Laundry Accessories

10810	Toilet Accessories	CREW	DAILY OUTPUT	LABOR-HOURS	UNIT	2006 BARE COSTS				TOTAL INCL O&P	
						MAT.	LABOR	EQUIP.	TOTAL		
100	0010	**COMMERCIAL TOILET ACCESSORIES**										**100**
	0200	Curtain rod, stainless steel, 1500 mm long, 25 mm diameter	1 Carp	13	.615	Ea.	30.50	22		52.50	67.50	
	0300	32 mm diameter [1-1/4"]		13	.615		30	22		52	67.50	
	0400	Diaper changing station, horizontal, wall mounted, plastic	↓	10	.800	↓	172	28.50		200.50	234	
	0500	Dispenser units, combined soap & towel dispensers,										
	0510	mirror and shelf, flush mounted	1 Carp	10	.800	Ea.	350	28.50		378.50	430	
	0600	Towel dispenser and waste receptacle,										
	0610	68.130 L capacity	1 Carp	10	.800	Ea.	269	28.50		297.50	340	
	0800	Grab bar, straight, 32 mm dia., stainless steel, 457 mm L [1-1/4"]		24	.333		20	11.85		31.85	40.50	
	0900	600 mm long		23	.348		19.80	12.35		32.15	41.50	
	1000	762 mm long		22	.364		22	12.95		34.95	44.50	
	1100	914 mm long	↓	20	.400	↓	21	14.20		35.20	45	

SPECIALTIES 10

	10810	Toilet Accessories	CREW	DAILY OUTPUT	LABOR-HOURS	UNIT	2006 BARE COSTS				TOTAL INCL O&P	
							MAT.	LABOR	EQUIP.	TOTAL		
100	1105	1050 mm long	1 Carp	20	.400	Ea.	21.50	14.20		35.70	45.50	100
	1200	38 mm diameter, 610 mm long [1-1/2",24"]		23	.348		52	12.35		64.35	76.50	
	1300	914 mm long		20	.400		57	14.20		71.20	84.50	
	1310	1050 mm long		18	.444		48	15.80		63.80	77.50	
	1500	Tub bar, 32 mm diameter, 610 mm x 914 mm [1-1/4",24",36"]		14	.571		89.50	20.50		110	130	
	1600	Plus vertical arm		12	.667		90.50	23.50		114	137	
	1900	End tub bar, 25 mm dia., 90° angle, 406 x 813 mm		12	.667		124	23.50		147.50	174	
	2010	Tub/shower/toilet, 2-wall, 900 mm x 600 mm		12	.667		80	23.50		103.50	125	
	2300	Hand dryer, surface mounted, electric, 115 V, 20 A		4	2		510	71		581	670	
	2400	230 V, 10 A		4	2		510	71		581	670	
	2600	Hat and coat strip, stainless steel, 4 hook, 914 mm long		24	.333		50.50	11.85		62.35	74	
	2700	6 hook, 1524 mm long		20	.400		88	14.20		102.20	119	
	3000	Mirror, with stainless steel 19 mm square frame, 450 mm x 600 mm		20	.400		66	14.20		80.20	94.50	
	3100	914 mm x 610 mm		15	.533		147	18.95		165.95	192	
	3200	1219 mm x 610 mm		10	.800		193	28.50		221.50	257	
	3300	1829 mm x 610 mm		6	1.333		196	47.50		243.50	290	
	3500	With 125 mm stainless steel shelf, 450 mm x 600 mm		20	.400		157	14.20		171.20	194	
	3600	914 mm x 610 mm		15	.533		199	18.95		217.95	249	
	3700	1219 mm x 610 mm		10	.800		229	28.50		257.50	297	
	3800	1829 mm x 610 mm		6	1.333		340	47.50		387.50	450	
	4100	Mop holder strip, stainless steel, 5 holders, 1200 mm long		20	.400		86.50	14.20		100.70	117	
	4200	Napkin/tampon dispenser, recessed		15	.533		330	18.95		348.95	390	
	4300	Robe hook, single, regular		36	.222		5.30	7.90		13.20	18.15	
	4400	Heavy duty, concealed mounting		36	.222		12.15	7.90		20.05	25.50	
	4600	Soap dispenser, chrome, surface mounted, liquid		20	.400		47	14.20		61.20	73.50	
	4700	Powder		20	.400		44	14.20		58.20	70.50	
	5000	Recessed stainless steel, liquid		10	.800		140	28.50		168.50	199	
	5100	Powder		10	.800		183	28.50		211.50	246	
	5300	Soap tank, stainless steel, 3.785 L		10	.800		167	28.50		195.50	229	
	5400	18.925 L		5	1.600		230	57		287	340	
	5600	Shelf, stainless steel, 127 mm wide, 18 ga., 610 mm long		24	.333		45.50	11.85		57.35	68.50	
	5700	1219 mm long		16	.500		89.50	17.80		107.30	126	
	5800	203 mm wide shelf, 18 ga., 610 mm long		22	.364		53.50	12.95		66.45	78.50	
	5900	1219mm long		14	.571		114	20.50		134.50	157	
	6000	Toilet seat cover dispenser, stainless steel, recessed		20	.400		111	14.20		125.20	144	
	6050	Surface mounted		15	.533		27.50	18.95		46.45	59.50	
	6100	Toilet tissue dispenser, surface mounted, SS, single roll		30	.267		11.70	9.50		21.20	27.50	
	6200	Double roll		24	.333		16.30	11.85		28.15	36.50	
	6400	Towel bar, stainless steel, 457 mm long		23	.348		35	12.35		47.35	58	
	6500	762 mm long		21	.381		71.50	13.55		85.05	99.50	
	6700	Towel dispenser, stainless steel, surface mounted		16	.500		37.50	17.80		55.30	69	
	6800	Flush mounted, recessed		10	.800		209	28.50		237.50	275	
	7000	Towel holder, hotel type, 2 guest size		20	.400		16.15	14.20		30.35	40	
	7200	Towel shelf, stainless steel, 610 mm long, 203 mm wide		20	.400		62	14.20		76.20	90	
	7400	Tumbler holder, tumbler only		30	.267		26	9.50		35.50	43.50	
	7500	Soap, tumbler & toothbrush		30	.267		23.50	9.50		33	41	
	7700	Wall urn ash receiver, surface mount, 279 mm long		12	.667		117	23.50		140.50	166	
	7800	191 mm, long		18	.444		76	15.80		91.80	108	
	8000	Waste receptacles, stainless steel, with top, 49.205 L		10	.800		231	28.50		259.50	300	
	8100	136.26 L	▼	8	1	▼	425	35.50		460.50	520	

	10820	Bath Accessories										
400	0010	**MEDICINE CABINETS**										400
	0020	With mirror, st. st. fr., 406 mm x 559 mm, unlight	1 Carp	14	.571	Ea.	73	20.50		93.50	112	
	0100	Wood frame		14	.571		101	20.50		121.50	143	
	0300	Sliding mirror doors, 508 mm x 406 mm x 121 mm, unlighted	▼	7	1.143	▼	90.50	40.50		131	163	

Important: See the Reference Section for supporting data - Crews, Rental Equipment, City Cost Indexes and Reference Data

10 SPECIALTIES

10820		Bath Accessories	CREW	DAILY OUTPUT	LABOR-HOURS	UNIT	2006 BARE COSTS				TOTAL INCL O&P	
							MAT.	LABOR	EQUIP.	TOTAL		
400	0400	610 mm x 483 mm x 216 mm, lighted	1 Carp	5	1.600	Ea.	143	57		200	246	**400**
	0600	Triple door, 762 mm x 813 mm, unlighted, plywood body		7	1.143		214	40.50		254.50	300	
	0700	Steel body		7	1.143		282	40.50		322.50	375	
	0900	Oak door, wood body, beveled mirror, single door		7	1.143		138	40.50		178.50	216	
	1000	Double door		6	1.333		335	47.50		382.50	445	
	1200	Hotel cabinets, stainless, with lower shelf, unlighted		10	.800		181	28.50		209.50	244	
	1300	Lighted	↓	5	1.600	↓	269	57		326	385	

10885		Scales	CREW	DAILY OUTPUT	LABOR-HOURS	UNIT	2006 BARE COSTS				TOTAL INCL O&P	
							MAT.	LABOR	EQUIP.	TOTAL		
100	0010	**SCALES** Built-in floor scale, not incl. foundations										**100**
	0100	Dial type, 4.5 metric ton capacity, 2.4 m x 1.8 m platform	3 Carp	.50	48	Ea.	6,525	1,700		8,225	9,825	
	0300	2.7 m x 2.1 m platform		.40	60		8,475	2,125		10,600	12,700	
	0400	9 metric ton capacity, steel platform, 2.4 m x 1.8 m platform		.40	60		10,300	2,125		12,425	14,600	
	0600	2.7 m x 2.1 m platform	↓	.35	68.571	↓	9,650	2,450		12,100	14,400	
	0700	Truck scales, incl. steel weigh bridge,										
	0800	not including foundation, pits										
	1550	Digital, electronic, 90 metric ton capacity, 3.7 m x 3 m pl	3 Carp	.20	120	Ea.	10,900	4,275		15,175	18,600	
	1600	12 m x 3 m platform		.14	171		22,100	6,100		28,200	33,800	
	1640	18 m x 3 m platform		.13	184		27,300	6,575		33,875	40,300	
	1680	21.5 m x 3 m platform	↓	.12	200		30,500	7,100		37,600	44,700	
	2000	For standard automatic printing device, add					2,200			2,200	2,425	
	2100	For remote reading electronic system, add					2,000			2,000	2,200	
	2300	Concrete foundation pits for above, 2.4 m x 1.8 m, 3.82 m³ required	C-1	.50	64		830	2,150		2,980	4,275	
	2400	4.3 m x 1.8 m platform, 7.64 m³ required		.35	91.429		1,225	3,075		4,300	6,125	
	2600	15 m x 3 m platform, 22.92 m³ required		.25	128		1,650	4,300		5,950	8,475	
	2700	21 m x 3 m platform, 30.56 m³ required	↓	.15	213		3,575	7,150		10,725	15,100	
	2750	Crane scales, dial, 1 metric ton capacity					885			885	970	
	2780	4.5 metric ton capacity					1,025			1,025	1,150	
	2800	Digital, 1 metric ton capacity					1,850			1,850	2,025	
	2850	9.2 metric ton capacity				↓	3,725			3,725	4,100	
	2900	Low profile electronic warehouse scale,										
	3000	not incl. printer, 1.2 m x 1.2 m platform, 4536 kg capacity	2 Carp	.30	53.333	Ea.	1,825	1,900		3,725	4,950	
	3300	1.5 m x 2.1 m platform, 4536 kg capacity		.25	64		3,575	2,275		5,850	7,500	
	3400	9072 kg capacity	↓	.20	80		4,625	2,850		7,475	9,525	
	3500	For printers, incl. time, date & numbering, add					1,150			1,150	1,275	
	3800	Portable, beam type, capacity 454 kg, platform 457 mm x 610 mm					670			670	735	
	3900	Dial type, capacity 907 kg, platform 610 mm x 610 mm					1,150			1,150	1,275	
	4000	Digital type, capacity 454 kg, platform 610 mm x 762 mm					1,850			1,850	2,025	
	4100	Portable contractor truck scales, 45 metric ton cap., 12 m x 3 m					28,700			28,700	31,600	
	4200	18 m x 3 m platform				↓	25,700			25,700	28,300	

SPECIALTIES 10

10905	Coat Racks/Wardrobes	CREW	DAILY OUTPUT	LABOR-HOURS	UNIT	2006 BARE COSTS				TOTAL INCL O&P
						MAT.	LABOR	EQUIP.	TOTAL	
500 0010	**COAT RACKS & WARDROBES**									500
0020	Floor model hat & coat racks, 6 hangers									
0050	Standing, beech wood, 533 x 533 mm x 1.8 m, chrome				Ea.	223			223	245
0100	18 ga. tubular stl, 533 mm x 533 mm x 1.8 m, wood				"	237			237	261
0500	16 gauge steel frame, 22 gauge steel shelves									
0650	Single pedestal, 762 mm x 457 mm x 1600 mm				Ea.	170			170	187
0800	Single face rack, 737 mm x 470 mm x 1575 mm					173			173	190
0900	1295 mm x 470 mm x 1778 mm					305			305	335
0910	Double face rack, 991 mm x 660 mm x 1778 mm					201			201	221
0920	1600 mm x 660 mm x 1778 mm				↓	355			355	390
0940	For 51 mm ball casters, add				Set	80.50			80.50	89
1400	Utility hook strips, 10 mm x 64 mm x 457 mm, 6 hooks	1 Carp	48	.167	Ea.	51.50	5.95		57.45	66
1500	864 mm long, 12 hooks	"	48	.167	"	51	5.95		56.95	65.50
1650	Wall mounted racks, 16 gauge steel frame, 22 gauge steel shelves									
1850	305 mm x 381 mm x 660 mm, 6 hangers	1 Carp	32	.250	Ea.	125	8.90		133.90	152
2000	305 mm x 381 mm x 1270 mm, 12 hangers	"	32	.250	"	130	8.90		138.90	157
2150	Wardrobe cabinet, steel, baked enamel finish									
2300	914 mm x 533 mm x 1981 mm, incl. top shelf & hanger rod				Ea.	261			261	287

For information about Means Estimating Seminars, see yellow pages 12 and 13 in back of book

10 SPECIALTIES

Division 11
Equipment

Estimating Tips

General
- The items in this division are usually priced per m² or each. Many of these items are purchased by the owner for installation by the contractor. Check the specifications for responsibilities, and include time for receiving, installation, and mechanical and electrical hook-ups in the appropriate divisions.

- Many items in Division 11 require some type of support system that is not usually furnished with the item. Examples of these systems include blocking for the attachment casework and support angles for ceiling-hung projection screens. The required blocking or supports must be added to the estimate in the appropriate division.

- Some items in Division 11 may require assembly or electrical hook-ups. Verify the amount of assembly required or the need for a hard electrical connection and add the appropriate costs.

Reference Numbers
Reference numbers are shown in bold squares at the beginning of some major classifications. These numbers refer to related items in the Reference Section. The reference information may be an estimating procedure, an alternate pricing method, or technical information.

Note: Not all subdivisions listed here necessarily appear in this publication.

11012	Contractor Equipment		DAILY OUTPUT	LABOR-HOURS	UNIT	2006 BARE COSTS				TOTAL INCL O&P	
		CREW				MAT.	LABOR	EQUIP.	TOTAL		
100	0010	**CONTRACTOR EQUIPMENT** See Reference Section	R015433 -10									100

11013	Floor/Wall Cleaning Equipment										
800	0010	**VACUUM CLEANING**										800
	0020	Central, 3 inlet, residential	1 Skwk	.90	8.889	Total	640	325		965	1,200	
	0200	Commercial		.70	11.429		1,200	415		1,615	1,975	
	0400	5 inlet system, residential		.50	16		970	585		1,555	1,975	
	0600	7 inlet system, commercial		.40	20		1,075	730		1,805	2,325	
	0800	9 inlet system, residential		.30	26.667		1,375	975		2,350	3,025	
	4010	Rule of thumb: First 111.48 m², installed									1,180	
	4020	For each additional m², add				m²					2.02	

11021	Safes and Vault Doors		DAILY OUTPUT	LABOR-HOURS	UNIT	2006 BARE COSTS				TOTAL INCL O&P	
		CREW				MAT.	LABOR	EQUIP.	TOTAL		
600	0010	**SAFE**										600
	0015	Office, 4 hr. rating, 750 mm x 457 mm x 457 mm inside				Ea.	3,275			3,275	3,600	
	0100	1575 mm x 838 mm x 508 mm					7,100			7,100	7,825	
	0200	1 hr. rating, 864 mm x 508 mm x 508 mm					1,725			1,725	1,900	
	0250	1016 mm x 457 mm x 457 mm					3,450			3,450	3,775	
	0300	1575 mm x 838 mm x 508 mm, double door					4,450			4,450	4,900	
	0400	Data, 4 hr. rating, 597 mm x 495 mm x 432 mm inside					3,175			3,175	3,500	
	0450	1321 mm x 483 mm x 432 mm					6,150			6,150	6,775	
	0500	1695 mm x 933 mm x 432 mm, double door					9,550			9,550	10,500	
	0550	1819 mm x 933 mm x 1372 mm					8,975			8,975	9,875	
	0600	1 hr. rating, 686 mm x 483 mm x 406 mm					4,250			4,250	4,675	
	0700	1600 mm x 864 mm x 406 mm					7,900			7,900	8,700	
	0750	Diskette, 1 hr., 229 mm x 305 mm x 330 mm (inside)					2,925			2,925	3,225	
	0800	Money, "B" label, 229 mm x 356 mm x 356 mm					415			415	455	
	0900	Tool resistive, 610 mm x 610 mm x 508 mm					2,800			2,800	3,075	
	1050	Tool and torch resistive, 610 mm x 610 mm x 508 mm					6,875			6,875	7,550	
	1150	Jewelers, 584 mm x 508 mm x 457 mm					8,775			8,775	9,650	
	1200	1600 mm x 635 mm x 457 mm					15,300			15,300	16,800	
	1300	For handling into building, add, minimum	A-2	8.50	2.824			77.50	15.15	92.65	137	
	1400	Maximum	"	.78	30.769			845	165	1,010	1,475	
800	0010	**VAULT DOOR**										800
	0020	Door and frame, 800 mm x 1950 mm, clear opening										
	0100	1 hour test, 800 mm door, weighs 340 kg	2 Sswk	1.50	10.667	Opng.	3,550	425		3,975	4,675	
	0200	2 hour test, 800 mm door, weighs 430 kg		1.30	12.308		3,750	490		4,240	5,000	
	0250	1000 mm door, weighs 512 kg		1	16		4,275	640		4,915	5,875	
	0300	4 hour test, 800 mm door, weighs 465 kg		1.20	13.333		4,100	535		4,635	5,450	
	0350	1000 mm door, weighs 517 kg		.90	17.778		4,775	710		5,485	6,525	
	0500	For stainless steel front, including frame, add to above					1,750			1,750	1,925	
	0550	Back, add					1,750			1,750	1,925	
	0600	For time lock, two movement, add	1 Elec	2	4	Ea.	1,400	168		1,568	1,800	
	0650	Three movement, add	"	2	4		1,825	168		1,993	2,275	
	0800	Day gate, painted, steel, 815 mm wide	2 Sswk	1.50	10.667		1,675	425		2,100	2,625	
	0850	1000 mm wide		1.40	11.429		1,875	455		2,330	2,875	
	0900	Aluminum, 800 mm wide		1.50	10.667		2,550	425		2,975	3,575	

EQUIPMENT 11

11020 | Security & Vault Equipment

11021	Safes and Vault Doors	CREW	DAILY OUTPUT	LABOR-HOURS	UNIT	2006 BARE COSTS MAT.	LABOR	EQUIP.	TOTAL	TOTAL INCL O&P	
800	0950	1000 mm wide	2 Sswk	1.40	11.429	Ea.	2,825	455		3,280	3,950	800
	2050	Security vault door, class I, 900 mm wide, 90 mm thick	E-24	.19	166	Opng.	13,400	6,525	3,400	23,325	29,700	
	2100	Class II, 900 mm wide, 180 mm thick		.19	166		15,500	6,525	3,400	25,425	32,100	
	2150	Class III, 9R, 900 mm wide, 254 mm thick, minimum		.13	250		19,400	9,775	5,075	34,250	43,900	
	2160	Class V, type 1, 1000 mm door		2.48	12.903	Ea.	5,575	505	262	6,342	7,300	
	2170	Class V, type 2, 1000 mm door		2.48	12.903		5,525	505	262	6,292	7,250	
	2180	Day gate for class V vault	2 Sswk	2	8		1,150	320		1,470	1,825	

11030 | Teller & Service Equipment

11038	Bank Equipment	CREW	DAILY OUTPUT	LABOR-HOURS	UNIT	2006 BARE COSTS MAT.	LABOR	EQUIP.	TOTAL	TOTAL INCL O&P	
150	0010	**BANK EQUIPMENT**										150
	0020	Alarm system, police	2 Elec	1.60	10	Ea.	4,250	420		4,670	5,300	
	0100	With vault alarm	"	.40	40		16,800	1,675		18,475	21,000	
	0400	Bullet resistant teller window, 1100 mm x 1500 mm	1 Glaz	.60	13.333		2,825	455		3,280	3,800	
	0500	1200 mm x 1500 mm	"	.60	13.333		3,500	455		3,955	4,575	
	3000	Counters for banks, frontal only	2 Carp	1	16	Station	1,575	570		2,145	2,625	
	3100	Complete with steel undercounter	"	.50	32	"	3,100	1,150		4,250	5,175	
	4600	Door and frame, bullet-resistant, with vision panel, minimum	2 Sswk	1.10	14.545	Ea.	3,500	580		4,080	4,900	
	4700	Maximum		1.10	14.545		4,750	580		5,330	6,275	
	4800	Drive-up window, drawer & mike, not incl. glass, minimum		1	16		4,725	640		5,365	6,325	
	4900	Maximum		.50	32		7,950	1,275		9,225	11,000	
	5000	Night depository, with chest, minimum		1	16		6,500	640		7,140	8,300	
	5100	Maximum		.50	32		9,225	1,275		10,500	12,400	
	5200	Package receiver, painted		3.20	5		1,175	200		1,375	1,650	
	5300	Stainless steel		3.20	5		2,000	200		2,200	2,550	
	5400	Partitions, bullet-resistant, 30 mm glass, 2438 mm high	2 Carp	3.05	5.249	m	560	187		747	910	
	5450	Acrylic	"	3.05	5.249	"	1,075	187		1,262	1,475	
	5500	Pneumatic tube systems, 2 lane drive-up, complete	L-3	.25	64	Total	22,300	2,475		24,775	28,300	
	5550	With T.V. viewer	"	.20	80	"	43,200	3,100		46,300	52,500	
	5570	Safety deposit boxes, minimum	1 Sswk	44	.182	Opng.	49.50	7.25		56.75	67.50	
	5580	Maximum, 254 mm x 381 mm opening		19	.421		105	16.80		121.80	147	
	5590	Teller locker, average		15	.533		1,350	21.50		1,371.50	1,550	
	5600	Pass thru, bullet-res. window, painted steel, 600 mm x 900 mm	2 Sswk	1.60	10	Ea.	1,925	400		2,325	2,850	
	5700	1200 mm x 1200 mm		1.20	13.333		2,300	535		2,835	3,475	
	5800	1800 mm x 1000 mm		.80	20		2,875	800		3,675	4,600	
	5900	For stainless steel frames, add					20%					
	6100	Surveillance system, video camera, complete	2 Elec	1	16	Ea.	13,100	670		13,770	15,400	
	6110	For each additional camera, add				"	855			855	940	
	6120	CCTV system, see Div. 16850-600										
	6200	Twenty-four hour teller, single unit,										
	6300	automated deposit, cash and memo	L-3	.25	64	Ea.	40,000	2,475		42,475	47,800	
	7000	Vault front, see Div. 11021-800										

11041	Ecclesiastical Equipment	CREW	DAILY OUTPUT	LABOR-HOURS	UNIT	2006 BARE COSTS				TOTAL INCL O&P
						MAT.	LABOR	EQUIP.	TOTAL	
250	0010	**CHURCH EQUIPMENT**									250
	0020	Altar, wood, custom design, plain	1 Carp	1.40	5.714	Ea.	1,925	203		2,128	2,425
	0050	Deluxe	"	.20	40		9,225	1,425		10,650	12,400
	0070	Granite or marble, average	2 Marb	.50	32		7,400	1,125		8,525	9,850
	0090	Deluxe	"	.20	80		24,300	2,800		27,100	31,000
	0100	Arks, prefabricated, plain	2 Carp	.80	20		7,300	710		8,010	9,125
	0130	Deluxe, maximum	"	.20	80	▼	91,000	2,850		93,850	104,500
	0150	Baptistry, fiberglass, 1067 mm deep, x 4140 mm long,									
	0160	steps at both ends, incl. plumbing, minimum	L-8	1	20	Ea.	2,900	740		3,640	4,350
	0200	Maximum	"	.70	28.571		5,675	1,050		6,725	7,875
	0250	Add for filter, heater and lights				▼	1,250			1,250	1,375
	0300	Carillon, 4 octave (48 bells), with keyboard				System	661,500			661,500	727,500
	0320	2 octave (24 bells)					275,500			275,500	303,000
	0340	3 to 4 bell peal, minimum					60,500			60,500	66,500
	0360	Maximum				▼	441,000			441,000	485,000
	0380	Cast bronze bell, average				Ea.	77,000			77,000	85,000
	0400	Electronic, digital, minimum					13,800			13,800	15,200
	0410	With keyboard, maximum					66,000			66,000	73,000
	0500	Reconciliation room, wood, prefabricated, single, plain	1 Carp	.60	13.333		2,425	475		2,900	3,425
	0550	Deluxe		.40	20		6,675	710		7,385	8,450
	0650	Double, plain		.40	20		4,850	710		5,560	6,450
	0700	Deluxe		.20	40		15,200	1,425		16,625	18,900
	1000	Lecterns, wood, plain		5	1.600		690	57		747	850
	1100	Deluxe		2	4	▼	6,925	142		7,067	7,850
	1500	Pews, bench type, hardwood, minimum		6.10	1.312	m	220	46.50		266.50	315
	1550	Maximum	▼	4.57	1.750		435	62		497	575
	1570	For kneeler, add				▼	52.50			52.50	58
	2000	Pulpits, hardwood, prefabricated, plain	1 Carp	2	4	Ea.	1,225	142		1,367	1,575
	2100	Deluxe		1.60	5	"	8,325	178		8,503	9,425
	2500	Railing, hardwood, average	▼	7.62	1.050	m	495	37.50		532.50	605
	2700	Safes, see division 11021-600									
	3000	Seating, individual, oak, contour, laminated	1 Carp	21	.381	Person	133	13.55		146.55	168
	3100	Cushion seat		21	.381		121	13.55		134.55	154
	3200	Fully upholstered		21	.381		108	13.55		121.55	140
	3300	Combination, self-rising	▼	21	.381		335	13.55		348.55	385
	3500	For cherry, add				▼	30%				
	4000	Steeples, translucent fiberglass, 762 mm square, 4.6 m high	F-3	2	20	Ea.	3,775	720	320	4,815	5,625
	4150	7.6 m high		1.80	22.222		4,425	800	355	5,580	6,500
	4350	Opaque fiberglass, 610 mm square, 4.3 m high		2	20		3,150	720	320	4,190	4,950
	4500	8.5 m high	▼	1.80	22.222		3,550	800	355	4,705	5,550
	4600	Aluminum, baked finish, 4.3 m high, 406 mm square					1,875			1,875	2,050
	4620	6.1 m high, 1067 mm base					4,575			4,575	5,050
	4640	11 m high, 2438 mm base					17,800			17,800	19,600
	4660	18 m high, 4267 mm base					39,400			39,400	43,300
	4680	46 m high, custom					406,500			406,500	447,000
	4700	Porcelain enamel steeples, custom, 12 m high	F-3	.50	80		11,400	2,875	1,275	15,550	18,500
	4800	18 m high	"	.30	133	▼	19,800	4,800	2,125	26,725	31,600
	5000	Wall cross, aluminum, extruded, 51 mm x 51 mm section	1 Carp	10.36	.772	m	120	27.50		147.50	175
	5150	102 mm x 102 mm section		8.84	.905		173	32		205	240
	5300	Bronze, extruded, 25 mm x 51 mm section		9.45	.847		237	30		267	310
	5350	64 mm x 64 mm section		10.36	.772		360	27.50		387.50	440
	5450	Solid bar stock, 13 mm x 76 mm section		8.84	.905		470	32		502	570
	5600	Fiberglass, stock		10.36	.772		97.50	27.50		125	150
	5700	Stainless steel, 102 mm deep, channel section		8.84	.905		380	32		412	470
	5800	102 mm deep box section	▼	8.84	.905	▼	520	32		552	620

EQUIPMENT · 11

11050 | Library Equipment

11051	Library Equipment	CREW	DAILY OUTPUT	LABOR-HOURS	UNIT	2006 BARE COSTS				TOTAL INCL O&P
						MAT.	LABOR	EQUIP.	TOTAL	
0010	**LIBRARY EQUIPMENT**									
0020	Bookshelf, mtl, 2.3 m high, 254 mm shelf, dbl face	1 Carp	3.51	2.282	m	470	81		551	645
0300	Single face	"	3.66	2.187	"	525	78		603	700
0600	For 203 mm shelving, subtract from above					10%				
0700	For 305 mm shelving, add to above					10%				
0800	For 1067 mm high with countertop, subtract from above					20%				
1100	Magazine shelving, 2050 mm high, 305 mm deep, single face	1 Carp	3.51	2.282	m	580	81		661	765
1400	Double face		3.51	2.282	"	660	81		741	855
2500	Carrels, hardwood, 914 mm x 610 mm, minimum		5	1.600	Ea.	665	57		722	820
2650	Maximum	↓	4	2		855	71		926	1,050
2700	Card catalog file, 60 trays, complete					5,325			5,325	5,850
2720	Alternate method: each tray					88.50			88.50	97.50
3100	Chairs, wood				↓	145			145	159
3500	Charging desk, built-in, with counter, plastic laminated top	1 Carp	2.13	3.750	m	1,375	133		1,508	1,725
3700	Reading table, laminated top, 1524 mm x 914 mm				Ea.	555			555	610
3800	Mobile compacted shelving, hand crank, 2743 mm high									
3820	Double face, including track, 914 mm section				Ea.	1,025			1,025	1,125
3840	For electrical operation, add					25%				

11060 | Theater & Stage Equipment

11063	Stage Equipment	CREW	DAILY OUTPUT	LABOR-HOURS	UNIT	2006 BARE COSTS				TOTAL INCL O&P
						MAT.	LABOR	EQUIP.	TOTAL	
0010	**STAGE EQUIPMENT**									
0050	Control boards with dimmers and breakers, minimum	1 Elec	1	8	Ea.	11,500	335		11,835	13,200
0100	Average		.50	16		32,000	670		32,670	36,200
0150	Maximum	↓	.20	40	↓	106,000	1,675		107,675	119,000
0500	Curtain track, straight, light duty	2 Carp	6.10	2.625	m	69.50	93.50		163	222
0600	Heavy duty		5.49	2.916		149	104		253	325
0700	Curved sections		3.66	4.374	↓	460	156		616	750
1000	Curtains, velour, medium weight		55.74	.287	m²	73.50	10.20		83.70	97
1150	Silica based yarn, inherently fire retardant	↓	4.65	3.445		141	122		263	345
1500	Flooring, portable oak parquet, 914 mm x 914 mm sections				↓	124			124	137
1600	Cart to carry 20.9 m² of flooring				Ea.	345			345	380
2000	Lights, border, quartz, reflector, vented,									
2100	colored or white	1 Elec	6.10	1.312	m	490	55		545	620
2500	Spotlight, follow spot, with transformer, 2100 W	"	4	2	Ea.	1,200	84		1,284	1,425
2600	For no transformer, deduct					785			785	865
3000	Stationary spot, fresnel quartz, 152 mm lens	1 Elec	4	2		94	84		178	229
3100	203 mm lens		4	2		188	84		272	330
3500	Ellipsoidal quartz, 1000 W, 152 mm lens		4	2		285	84		369	440
3600	305 mm lens		4	2		470	84		554	645
4000	Strobe light, 1 to 15 flashes/second, quartz		3	2.667		650	112		762	880
4500	Color wheel, portable, five hole, motorized	↓	4	2	↓	150	84		234	290
5000	Stages, portable with steps, folding legs, stock, 200 mm high				m² Stg.	244			244	268
5100	406 mm high					236			236	260
5200	800 mm high					335			335	370
5300	1000 mm high					550			550	605
6000	Telescoping platforms, extruded alum., straight, minimum	4 Carp	14.59	2.194		264	78		342	410
6100	Maximum		7.15	4.473		355	159		514	645
6500	Pie-shaped, minimum	↓	13.94	2.296	↓	570	81.50		651.50	755

11060 | Theater & Stage Equipment

11063	Stage Equipment	CREW	DAILY OUTPUT	LABOR-HOURS	UNIT	2006 BARE COSTS				TOTAL INCL O&P	
						MAT.	LABOR	EQUIP.	TOTAL		
600	6600	Maximum	4 Carp	6.50	4.921	m² Stg.	640	175		815	975	**600**
	6800	For 19 mm plywood covered deck, deduct					36			36	40
	7000	Band risers, steel frame, plywood deck, minimum	4 Carp	25.55	1.253		283	44.50		327.50	380
	7100	Maximum	"	12.82	2.496		595	88.50		683.50	795
	7500	Chairs for above, self-storing, minimum	2 Carp	43	.372	Ea.	86.50	13.25		99.75	116
	7600	Maximum	"	40	.400	"	145	14.20		159.20	181
	8000	Rule of thumb: total stage equipment, minimum	4 Carp	9.29	3.445	m² Stg.	905	122		1,027	1,175
	8100	Maximum	"	2.32	13.778	"	5,075	490		5,565	6,375

11100 | Mercantile Equipment

11102	Barber Shop Equipment	CREW	DAILY OUTPUT	LABOR-HOURS	UNIT	2006 BARE COSTS				TOTAL INCL O&P	
						MAT.	LABOR	EQUIP.	TOTAL		
150	0010	**BARBER EQUIPMENT**										**150**
	0020	Chair, hydraulic, movable, minimum	1 Carp	24	.333	Ea.	430	11.85		441.85	495
	0050	Maximum	"	16	.500		2,750	17.80		2,767.80	3,050
	0200	Wall hung styling station with mirrors, minimum	L-2	8	2		415	62		477	550
	0300	Maximum	"	4	4		1,825	124		1,949	2,225
	0500	Sink, hair washing basin, rough plumbing not incl.	1 Plum	8	1		283	42.50		325.50	375
	1000	Sterilizer, liquid solution for tools					128			128	140
	1100	Total equipment, rule of thumb, per chair, minimum	L-8	1	20		1,475	740		2,215	2,775
	1150	Maximum	"	1	20		4,075	740		4,815	5,650

11103	Cash Register/Checking	CREW	DAILY OUTPUT	LABOR-HOURS	UNIT	2006 BARE COSTS				TOTAL INCL O&P	
						MAT.	LABOR	EQUIP.	TOTAL		
200	0010	**CHECKOUT COUNTER**										**200**
	0020	Supermarket conveyor, single belt	2 Clab	10	1.600	Ea.	2,300	44		2,344	2,600
	0100	Double belt, power take-away		9	1.778		3,850	48.50		3,898.50	4,325
	0400	Double belt, power take-away, incl. side scanning		7	2.286		4,800	62.50		4,862.50	5,375
	0800	Warehouse or bulk type		6	2.667		5,625	73		5,698	6,300
	1000	Scanning system, 2 lanes, w/registers, scan gun & memory				System	14,000			14,000	15,400
	1100	10 lanes, single processor, full scan, with scales				"	133,500			133,500	147,000
	2000	Register, restaurant, minimum				Ea.	570			570	625
	2100	Maximum					2,500			2,500	2,750
	2150	Store, minimum					570			570	625
	2200	Maximum					2,500			2,500	2,750

11104	Display Cases & Systems	CREW	DAILY OUTPUT	LABOR-HOURS	UNIT	2006 BARE COSTS				TOTAL INCL O&P	
						MAT.	LABOR	EQUIP.	TOTAL		
700	0010	**REFRIGERATED FOOD CASES**										**700**
	0030	Dairy, multi-deck, 3658 mm long	Q-5	3	5.333	Ea.	8,600	207		8,807	9,750
	0100	For rear sliding doors, add					1,225			1,225	1,350
	0200	Delicatessen case, service deli, 3658 mm long, single deck	Q-5	3.90	4.103		5,775	159		5,934	6,625
	0300	Multi-deck, 1.67 m² shelf display		3	5.333		7,075	207		7,282	8,100
	0400	Freezer, self-contained, chest-type, 0.84 m³		3.90	4.103		4,225	159		4,384	4,900
	0500	Glass door, upright, 2.18 m³		3.30	4.848		8,025	188		8,213	9,125
	0600	Frozen food, chest type, 3658 mm long		3.30	4.848		5,800	188		5,988	6,675
	0700	Glass door, reach-in, 5 door		3	5.333		11,100	207		11,307	12,500
	0800	Island case, 3658 mm long, single deck		3.30	4.848		6,575	188		6,763	7,500
	0900	Multi-deck		3	5.333		13,900	207		14,107	15,600
	1000	Meat case, 3658 mm long, single deck		3.30	4.848		4,775	188		4,963	5,525

Important: See the Reference Section for supporting data - Crews, Rental Equipment, City Cost Indexes and Reference Data

11100 | Mercantile Equipment

11104 | Display Cases & Systems

		CREW	DAILY OUTPUT	LABOR-HOURS	UNIT	2006 BARE COSTS				TOTAL INCL O&P
						MAT.	LABOR	EQUIP.	TOTAL	
700	1050	Multi-deck	Q-5	3.10	5.161	Ea.	8,225	200		8,425	9,350	**700**
	1100	Produce, 3658 mm long, single deck		3.30	4.848		6,325	188		6,513	7,225
	1200	Multi-deck	↓	3.10	5.161	↓	7,050	200		7,250	8,050

11110 | Commercial Laundry & Dry Cleaning Equipment

11119 | Laundry Cleaning

		CREW	DAILY OUTPUT	LABOR-HOURS	UNIT	2006 BARE COSTS				TOTAL INCL O&P
						MAT.	LABOR	EQUIP.	TOTAL	
450	0010	**LAUNDRY EQUIPMENT** Not incl. rough-in										**450**
	0500	Dryers, gas fired residential, 7.3 kg capacity, average	1 Plum	3	2.667	Ea.	650	114		764	885
	1000	Commercial, 13.6 kg capacity, coin operated, single		3	2.667		2,750	114		2,864	3,200
	1100	Double stacked		2	4		5,825	171		5,996	6,650
	1500	Industrial, 13.6 kg capacity		2	4		2,575	171		2,746	3,075
	1600	22.7 kg capacity	↓	1.70	4.706		3,125	201		3,326	3,725
	2000	Dry cleaners, electric, 9.1 kg capacity	L-1	.20	80		32,800	3,400		36,200	41,200
	2050	11.3 kg capacity		.17	94.118		42,700	3,975		46,675	53,000
	2100	13.6 kg capacity		.15	106		45,000	4,525		49,525	56,500
	2150	27.2 kg capacity	↓	.09	177		70,500	7,525		78,025	89,000
	3500	Folders, blankets & sheets, minimum	1 Elec	.17	47.059		29,900	1,975		31,875	35,900
	3700	King size with automatic stacker		.10	80		51,000	3,350		54,350	61,000
	3800	For conveyor delivery, add		.45	17.778		6,100	745		6,845	7,825
	4500	Ironers, institutional, 2794 mm, single roll	↓	.20	40		27,500	1,675		29,175	32,800
	4700	Lint collector, ductwork not included, 3.78 m³/s to 4.72 m³/s	Q-10	.30	80		8,025	3,150		11,175	13,700
	4800	Pressers, low capacity air operated	L-6	1.75	6.857		8,375	291		8,666	9,650
	4820	Hand operated		1.75	6.857		7,475	291		7,766	8,650
	4830	Extractor, low capacity		1.75	6.857		6,375	291		6,666	7,425
	4840	Ironer 1200 mm, 240 V		3.50	3.429		99,500	146		99,646	109,000
	4860	Coin dry cleaner, 9 kg		1.75	6.857		24,800	291		25,091	27,700
	4900	Spreader feeders, 240 V, 2 station		.70	17.143		48,400	730		49,130	54,500
	4920	4 station	↓	.35	34.286		59,500	1,450		60,950	67,500
	5000	Washers, residential, 4 cycle, average	1 Plum	3	2.667		695	114		809	935
	5300	Commercial, coin operated, average	"	3	2.667		1,125	114		1,239	1,400
	6000	Combination washer/extractor, 9.1 kg capacity	L-6	1.50	8		4,000	340		4,340	4,900
	6100	13.6 kg capacity		.80	15		7,775	635		8,410	9,500
	6200	22.7 kg capacity		.68	17.647		9,650	750		10,400	11,700
	6300	34 kg capacity		.30	40		19,800	1,700		21,500	24,400
	6350	56.7 kg capacity		.16	75		24,100	3,175		27,275	31,300
	6380	Washer extractor/dryer, 50 kg, 240 V		1	12		7,725	510		8,235	9,275
	6400	Washer extractor, 61 kg, 240 V		1	12		23,100	510		23,610	26,300
	6450	Pass through		1	12		62,500	510		63,010	70,000
	6500	91 kg		1	12		60,000	510		60,510	67,000
	6550	Pass through		1	12		65,000	510		65,510	72,000
	6600	Hand operated presser		.70	17.143		5,600	730		6,330	7,250
	6620	Mushroom press 115 V	↓	.70	17.143	↓	6,625	730		7,355	8,375

EQUIPMENT 11

11136	Projection Screens	CREW	DAILY OUTPUT	LABOR-HOURS	UNIT	2006 BARE COSTS				TOTAL INCL O&P
						MAT.	LABOR	EQUIP.	TOTAL	
500	**0010 PROJECTION SCREENS** Wall or ceiling hung, matte white									**500**
0100	Manually operated, economy	2 Carp	46.45	.344	m²	55.50	12.25		67.75	80
0300	Intermediate		41.81	.383		64.50	13.60		78.10	92
0400	Deluxe		37.16	.431		89.50	15.30		104.80	123
0600	Electric operated, matte white, 2.3 m², economy		5	3.200	Ea.	750	114		864	1,000
0700	Deluxe		4	4		1,575	142		1,717	1,950
0900	4.6 m², economy		3	5.333		735	190		925	1,100
1000	Deluxe		2	8		1,750	284		2,034	2,375
1200	Heavy duty, electric operated, 18.6 m²		1.50	10.667		3,425	380		3,805	4,375
1300	37.2 m²		1	16		4,225	570		4,795	5,525
1500	Rigid acrylic in wall, for rear projection, 6 mm thick	2 Glaz	2.79	5.741	m²	445	197		642	785
1600	13 mm thick (maximum size 3048 mm x 6096 mm)	"	2.32	6.889	"	790	236		1,026	1,225
600	**0010 MOVIE EQUIPMENT**									**600**
0020	Changeover, minimum				Ea.	410			410	450
0100	Maximum					795			795	875
0400	Film transport, incl. platters and autowind, minimum					4,425			4,425	4,875
0500	Maximum					12,600			12,600	13,800
0800	Lamphouses, incl. rectifiers, xenon, 1000 W	1 Elec	2	4		5,850	168		6,018	6,700
0900	1600 W		2	4		6,250	168		6,418	7,125
1000	2000 W		1.50	5.333		6,700	224		6,924	7,700
1100	4000 W		1.50	5.333		8,300	224		8,524	9,450
1400	Lenses, anamorphic, minimum					1,125			1,125	1,250
1500	Maximum					2,525			2,525	2,775
1800	Flat 35 mm, minimum					975			975	1,075
1900	Maximum					1,525			1,525	1,675
2200	Pedestals, for projectors, minimum					1,300			1,300	1,425
2300	Console type, maximum					9,425			9,425	10,400
2600	Projector mechanisms, 35 mm, minimum					9,750			9,750	10,700
2700	Maximum					13,400			13,400	14,700
3000	Projection screens, rigid, in wall, acrylic, 6 mm thick	2 Glaz	18.12	.883	m²	395	30.50		425.50	480
3100	13 mm thick	"	12.08	1.325	"	460	45.50		505.50	575
3300	Electric operated, heavy duty, 37 m²	2 Carp	1	16	Ea.	2,600	570		3,170	3,725
3320	Theatre projection screens, matte white, including frames	"	18.58	.861	m²	63	30.50		93.50	117
3400	Also see division 11136-500									
3700	Sound systems, incl. amplifier, single system, minimum	1 Elec	.90	8.889	Ea.	2,900	375		3,275	3,750
3800	Dolby/Super Sound, maximum		.40	20		15,900	840		16,740	18,800
4100	Dual system, minimum		.70	11.429		4,075	480		4,555	5,200
4200	Dolby/Super Sound, maximum		.40	20		14,600	840		15,440	17,300
4500	Sound heads, 35 mm					4,650			4,650	5,125
4900	Splicer, tape system, minimum					650			650	715
5000	Tape type, maximum					1,175			1,175	1,275
5300	Speakers, recessed behind screen, minimum	1 Elec	2	4		935	168		1,103	1,275
5400	Maximum	"	1	8		2,725	335		3,060	3,500
5700	Seating, painted steel, upholstered, minimum	2 Carp	35	.457		116	16.25		132.25	154
5800	Maximum	"	28	.571		370	20.50		390.50	440
6100	Rewind tables, minimum					2,325			2,325	2,550
6200	Maximum					4,150			4,150	4,575
7000	For automation, varying sophistication, minimum	1 Elec	1	8	System	2,100	335		2,435	2,800
7100	Maximum	2 Elec	.30	53.333	"	4,900	2,250		7,150	8,750

11140 | Vehicle Service Equipment

		11141	Service Station Equipment	CREW	DAILY OUTPUT	LABOR-HOURS	UNIT	2006 BARE COSTS MAT.	LABOR	EQUIP.	TOTAL	TOTAL INCL O&P	
100	0010		COMPRESSED AIR EQUIPMENT										100
	0030		Compressors, electric, 1 kW, standard controls	L-4	1.50	16	Ea.	281	530		811	1,125	
	0550		Dual controls		1.50	16		475	530		1,005	1,350	
	0600		4 kW, 115/230 V, standard controls		1	24		1,650	795		2,445	3,050	
	0650		Dual controls	▼	1	24	▼	1,775	795		2,570	3,175	
200	0010		FUEL DISPENSING EQUIPMENT										200
	1100		Product dispenser with vapor recovery for 6 nozzles, installed, not										
	1110		including piping to storage tanks				Ea.	18,900			18,900	20,800	
300	0010		LUBRICATION EQUIPMENT										300
	3000		Lube equipment, 3 reel type, with pumps, not including piping	L-4	.50	48	Set	7,175	1,600		8,775	10,400	
400	0010		SPRAY PAINTING EQUIPMENT										400
	4000		Spray painting booth, 7925 mm long, complete	L-4	.40	60	Ea.	13,300	2,000		15,300	17,700	

11150 | Parking Control Equipment

		11156	Parking Equipment	CREW	DAILY OUTPUT	LABOR-HOURS	UNIT	2006 BARE COSTS MAT.	LABOR	EQUIP.	TOTAL	TOTAL INCL O&P	
600	0010		PARKING EQUIPMENT										600
	5000		Barrier gate with programmable controller	2 Elec	3	5.333	Ea.	3,200	224		3,424	3,850	
	5020		Industrial	"	3	5.333		4,350	224		4,574	5,125	
	5100		Card reader	1 Elec	2	4		1,750	168		1,918	2,175	
	5120		Proximity with customer display	2 Elec	1	16		5,300	670		5,970	6,825	
	5200		Cashier booth, average	B-22	1	30		9,225	975	229	10,429	12,000	
	5300		Collector station, pay on foot	2 Elec	.20	80		105,500	3,350		108,850	121,000	
	5320		Credit card only		.50	32		19,500	1,350		20,850	23,400	
	5500		Exit verifier	▼	1	16		16,800	670		17,470	19,400	
	5600		Fee computer	1 Elec	1.50	5.333		12,800	224		13,024	14,400	
	5700		Full sign, 100 mm letters	"	2	4		1,175	168		1,343	1,525	
	5800		Inductive loop	2 Elec	4	4		162	168		330	430	
	5900		Ticket spitter with time/date stamp, standard		2	8		6,075	335		6,410	7,200	
	5920		Mag stripe encoding	▼	2	8		17,800	335		18,135	20,100	
	5950		Vehicle detector, microprocessor based	1 Elec	3	2.667		375	112		487	575	
	6000		Parking control software, minimum		.50	16		21,100	670		21,770	24,200	
	6020		Maximum	▼	.20	40	▼	87,500	1,675		89,175	99,000	

11160 | Loading Dock Equipment

		11161	Loading Dock Equipment	CREW	DAILY OUTPUT	LABOR-HOURS	UNIT	2006 BARE COSTS MAT.	LABOR	EQUIP.	TOTAL	TOTAL INCL O&P	
200	0010		DOCK BUMPERS Bolts not included										200
	0020		50mm x 150 mm to 100 mm x 200 mm, average	1 Carp	.71	11.301	m³	490	400		890	1,150	
400	0010		LOADING DOCK										400
	0020		Bumpers, rubber blocks 114 mm T, 250 mm H, 355 mm L	1 Carp	26	.308	Ea.	44.50	10.95		55.45	66	

11161 | Loading Dock Equipment

		CREW	DAILY OUTPUT	LABOR-HOURS	UNIT	MAT.	LABOR	EQUIP.	TOTAL	TOTAL INCL O&P	
400							2006 BARE COSTS					**400**
0200	610 mm long	1 Carp	22	.364	Ea.	84	12.95		96.95	112	
0300	900 mm long		17	.471		85.50	16.75		102.25	121	
0500	305 mm high, 356 mm long		25	.320		88.50	11.40		99.90	115	
0550	610 mm long		20	.400		97.50	14.20		111.70	130	
0600	914 mm long		15	.533		109	18.95		127.95	150	
0800	Rubber blocks 150 mm thick, 250 mm high, 350 mm long		22	.364		77.50	12.95		90.45	105	
0850	600 mm long		18	.444		107	15.80		122.80	143	
0900	900 mm long		13	.615		125	22		147	172	
0910	500 mm high, 279 mm long		13	.615		129	22		151	176	
0920	Extruded rubber bmprs, T section, 550 mm x 550 mm x 75 mm		41	.195		47.50	6.95		54.45	63.50	
0940	Molded rubber bumpers, 600 mm x 300 mm x 75 mm thick		20	.400		43.50	14.20		57.70	70	
1000	Welded installation of above bumpers	E-14	8	1		3.30	42	11.15	56.45	91.50	
1100	For drilled anchors, add/anchor	1 Carp	36	.222		5.90	7.90		13.80	18.80	
1300	Steel bumpers, see Div. 10265-200										
1350	Wood bumpers, see Div. 11161-200										
2200	Dock boards, heavy duty, 1500 x 1500 mm, alum, 2268 kg cap.				Ea.	1,250			1,250	1,375	
2700	4082 kg capacity					1,175			1,175	1,300	
3200	6804 kg capacity					1,300			1,300	1,425	
3600	Door seal for door perimeter, 300 mm x 300 mm, vinyl covered	1 Carp	7.92	1.009	m	76.50	36		112.50	140	
3900	Folding gates, see Div. 10610-100										
4200	Platform lifter, 1829 mm x 1829 mm, portable, 1361 kg capacity				Ea.	8,075			8,075	8,875	
4250	1814 kg capacity					9,925			9,925	10,900	
4400	Fixed, 1829 mm x 2438 mm, 2268 kg capacity	E-16	.70	22.857		8,700	935	128	9,763	11,400	
4500	Levelers, hinged for trucks, 9.1 metric ton cap., 1800 mm x 2400 mm		1.08	14.815		4,400	605	82.50	5,087.50	6,025	
4650	2100 mm x 2400 mm		1.08	14.815		4,175	605	82.50	4,862.50	5,775	
4670	Air bag power operated, 9.1 metric ton cap., 1.8 m x 2.4 m		1.08	14.815		5,100	605	82.50	5,787.50	6,800	
4680	2134 mm x 2438 mm		1.08	14.815		5,125	605	82.50	5,812.50	6,825	
4700	Hydraulic, 9.1 metric ton capacity, 1829 mm x 2438 mm		1.08	14.815		7,650	605	82.50	8,337.50	9,600	
4800	2134 mm x 2438 mm		1.08	14.815		8,225	605	82.50	8,912.50	10,200	
5000	Lights for loading docks, single arm, 600 mm long	1 Elec	3.80	2.105		127	88.50		215.50	272	
5700	Double arm, 1500 mm long	"	3.80	2.105		153	88.50		241.50	300	
5800	Loading dock safety restraints, manual style	E-16	1.08	14.815		2,800	605	82.50	3,487.50	4,275	
5900	Automatic style	"	1.08	14.815		4,700	605	82.50	5,387.50	6,375	
6200	Shelters, fabric, for truck or train, scissor arms, minimum	1 Carp	1	8		1,300	284		1,584	1,875	
6300	Maximum	"	.50	16		1,900	570		2,470	2,950	

11179 | Waste Handling Equipment

		CREW	DAILY OUTPUT	LABOR-HOURS	UNIT	MAT.	LABOR	EQUIP.	TOTAL	TOTAL INCL O&P	
150							2006 BARE COSTS					**150**
0010	**WASTE HANDLING**										
0020	Compactors, 115 V, 113 kg/hr., chute fed	L-4	1	24	Ea.	9,950	795		10,745	12,100	
0100	Hand fed		2.40	10		7,150	330		7,480	8,400	
0300	Multi-bag 230 V, 272 kg/hr, chute fed		1	24		8,975	795		9,770	11,100	
0400	Hand fed		1	24		7,700	795		8,495	9,700	
0500	Containerized, hand fed, 1.53 to 4.58 m³ containers, 113 kg/hr		1	24		9,425	795		10,220	11,600	
0550	For chute fed, add/floor		1	24		1,075	795		1,870	2,425	
1000	Heavy duty industrial compactor, 0.38 m³ capacity		1	24		6,375	795		7,170	8,225	
1050	0.76 m³ capacity		1	24		9,800	795		10,595	12,000	
1100	2.29 m³ capacity		.50	48		13,400	1,600		15,000	17,200	

Important: See the Reference Section for supporting data - Crews, Rental Equipment, City Cost Indexes and Reference Data

11170 | Solid Waste Handling Equipment

11179 | Waste Handling Equipment

			CREW	DAILY OUTPUT	LABOR-HOURS	UNIT	2006 BARE COSTS				TOTAL INCL O&P	
							MAT.	LABOR	EQUIP.	TOTAL		
150	1150	3.82 m³ capacity	L-4	.50	48	Ea.	16,600	1,600		18,200	20,700	150
	1200	Combination shredder/compactor (2268 kg/hr.)	↓	.50	48		32,400	1,600		34,000	38,200	
	1400	For handling hazardous waste materials, 208 L drum packer, std.					15,100			15,100	16,600	
	1410	208 L drum packer w/HEPA filter					18,800			18,800	20,700	
	1420	208 L drum packer w/charcoal & HEPA filter					25,100			25,100	27,600	
	1430	All of the above made explosion proof, add					11,500			11,500	12,600	
	1500	Crematory, not including building, 1 place	Q-3	.20	160		53,500	6,500		60,000	69,000	
	1750	2 place		.10	320		76,500	13,000		89,500	103,500	
	4400	Incinerator, gas, not incl. chimney, elec. or pipe, 23 kg/hr, minimum		.80	40		20,800	1,625		22,425	25,400	
	4420	Maximum		.70	45.714		27,200	1,850		29,050	32,700	
	4440	91 kg/hr, minimum (batch type)		.60	53.333		27,200	2,175		29,375	33,200	
	4460	Maximum (with feeder)		.50	64		53,000	2,600		55,600	62,000	
	4480	181 kg/hr, minimum (batch type)		.30	106		32,200	4,350		36,550	41,900	
	4500	Maximum (with feeder)		.25	128		60,500	5,200		65,700	75,000	
	4520	363 kg/hr, with feeder, minimum		.20	160		79,000	6,500		85,500	97,000	
	4540	Maximum		.17	188		108,000	7,650		115,650	130,500	
	4560	544 kg/hr, with feeder, minimum		.15	213		114,500	8,675		123,175	139,000	
	4580	Maximum		.11	290		137,500	11,800		149,300	169,500	
	4600	907 kg/hr, with feeder, minimum		.10	320		200,000	13,000		213,000	239,500	
	4620	Maximum		.05	640		336,500	26,000		362,500	409,000	
	4700	For heat recovery system, add, minimum		.25	128		65,000	5,200		70,200	79,500	
	4710	Add, maximum		.11	290		208,000	11,800		219,800	246,500	
	4720	For automatic ash conveyer, add		.50	64	↓	27,300	2,600		29,900	33,900	
	4750	Large municipal incinerators, incl. stack, minimum		.23	141	Met. Ton	18,300	5,750		24,050	28,800	
	4850	Maximum	↓	.09	352	"	48,700	14,400		63,100	75,000	
	5500	Shredder, municipal use, 0.882 kg/s				Ea.	247,000			247,000	272,000	
	5600	1.51 kg/s					526,500			526,500	579,000	
	5750	Shredder & baler, 45 metric ton/day					493,500			493,500	543,000	
	5800	Shredder, industrial, minimum					19,400			19,400	21,400	
	5850	Maximum					104,000			104,000	114,500	
	5900	Baler, industrial, minimum					7,775			7,775	8,575	
	5950	Maximum				↓	455,000			455,000	500,500	
	6000	Transfer station compactor, with power unit										
	6050	and pedestal, not including pit, 1.26 kg/s				Ea.	155,500			155,500	171,500	

11190 | Detention Equipment

11191 | Detention Equipment

			CREW	DAILY OUTPUT	LABOR-HOURS	UNIT	2006 BARE COSTS				TOTAL INCL O&P	
							MAT.	LABOR	EQUIP.	TOTAL		
150	0010	**DETENTION EQUIPMENT**										150
	0500	Bar front, rolling, 22 mm, 102 mm O.C., 2.1 m H, 1.5 m W, w/hdwe	E-4	2	16	Ea.	4,825	645	44.50	5,514.50	6,550	
	1000	Doors & frames, 914 x 2134 mm, complete, w/ hardware, single plate	↓	4	8		3,625	325	22.50	3,972.50	4,600	
	1650	Double plate		4	8	↓	4,500	325	22.50	4,847.50	5,525	
	2000	Cells, prefab., 1.5 to 2 m wide, 2 to 2.5 m high, 2 to 2.5 m deep										
	2010	bar front, cot, not incl. plumbing	E-4	1.50	21.333	Ea.	8,000	865	59.50	8,924.50	10,400	
	2500	Cot, bolted, single, painted steel		20	1.600		300	64.50	4.46	368.96	450	
	2700	Stainless steel	↓	20	1.600		835	64.50	4.46	903.96	1,025	
	3000	Toilet apparatus including wash basin, average	L-8	1.50	13.333		3,000	495		3,495	4,050	
	4000	Visitor cubicle, vision panel, no intercom	E-4	2	16	↓	2,625	645	44.50	3,314.50	4,125	

11285 | Hydraulic Gates

		CREW	DAILY OUTPUT	LABOR-HOURS	UNIT	2006 BARE COSTS				TOTAL INCL O&P
						MAT.	LABOR	EQUIP.	TOTAL	
150	**0010** **CANAL GATES** Cast iron body, fabricated frame									**150**
	0100 300 mm diameter	L-5A	4.60	6.957	Ea.	560	278	152	990	1,275
	0110 450 mm diameter		4	8		925	320	174	1,419	1,775
	0120 600 mm diameter		3.50	9.143		1,600	365	199	2,164	2,625
	0130 750 mm diameter		2.80	11.429		2,150	455	249	2,854	3,450
	0140 900 mm diameter		2.30	13.913		2,650	555	305	3,510	4,225
	0150 1050 mm diameter		1.70	18.824		3,875	755	410	5,040	6,025
	0160 1200 mm diameter		1.20	26.667		5,050	1,075	580	6,705	8,050
	0170 1350 mm diameter		.90	35.556		9,150	1,425	775	11,350	13,400
	0180 1500 mm diameter		.50	64		10,900	2,550	1,400	14,850	18,000
	0190 1650 mm diameter		.50	64		9,075	2,550	1,400	13,025	15,900
	0200 1800 mm diameter	↓	.40	80	↓	14,700	3,200	1,750	19,650	23,700
190	**0010** **FLAP GATES**									**190**
	0100 Aluminum, 450 mm diameter	L-5A	5	6.400	Ea.	2,200	256	139	2,595	3,025
	0110 600 mm diameter		4	8		2,350	320	174	2,844	3,325
	0120 750 mm diameter		3.50	9.143		2,825	365	199	3,389	3,975
	0130 900 mm diameter		2.80	11.429		3,550	455	249	4,254	4,975
	0140 1050 mm diameter		2.30	13.913		4,875	555	305	5,735	6,650
	0150 1200 mm diameter		1.70	18.824		5,875	755	410	7,040	8,225
	0160 1350 mm diameter		1.20	26.667		7,200	1,075	580	8,855	10,400
	0170 1500 mm diameter		.80	40		9,025	1,600	870	11,495	13,700
	0180 1650 mm diameter		.50	64		10,800	2,550	1,400	14,750	17,900
	0190 1800 mm diameter	↓	.40	80	↓	12,800	3,200	1,750	17,750	21,600
400	**0010** **KNIFE GATES**									**400**
	0100 Incl. handwheel operator for hub, 150 mm diameter	Q-23	7.70	3.117	Ea.	705	129	124	958	1,100
	0110 200 mm diameter		7.20	3.333		945	138	132	1,215	1,400
	0120 250 mm diameter		4.80	5		1,200	207	198	1,605	1,850
	0130 300 mm diameter		3.60	6.667		1,800	276	265	2,341	2,675
	0140 350 mm diameter		3.40	7.059		3,125	292	280	3,697	4,200
	0150 400 mm diameter		3.20	7.500		3,500	310	298	4,108	4,650
	0160 450 mm diameter		3	8		4,625	330	315	5,270	5,950
	0170 500 mm diameter		2.70	8.889		5,650	370	355	6,375	7,175
	0180 600 mm diameter		2.40	10		6,450	415	395	7,260	8,150
	0190 750 mm diameter		1.80	13.333		13,500	550	530	14,580	16,200
	0200 900 mm diameter	↓	1.20	20	↓	18,500	825	795	20,120	22,400
600	**0010** **SLIDE GATES**									**600**
	0100 Steel, self contained, incl. anchor bolts and grout, 300 x 300 mm	L-5A	4.60	6.957	Ea.	2,600	278	152	3,030	3,500
	0110 450 x 450 mm		4	8		2,525	320	174	3,019	3,525
	0120 600 x 600 mm		3.50	9.143		2,725	365	199	3,289	3,850
	0130 750 x 750 mm		2.80	11.429		3,300	455	249	4,004	4,725
	0140 900 x 900 mm		2.30	13.913		3,300	555	305	4,160	4,925
	0150 1050 x 1050 mm		1.70	18.824		3,850	755	410	5,015	5,975
	0160 1200 x 1200 mm		1.20	26.667		4,025	1,075	580	5,680	6,925
	0170 1350 x 1350 mm		.90	35.556		5,125	1,425	775	7,325	8,950
	0180 1500 x 1500 mm		.55	58.182		5,800	2,325	1,275	9,400	11,800
	0190 1800 x 1800 mm	↓	.36	88.889	↓	8,375	3,550	1,925	13,850	17,500
700	**0010** **SLUICE GATES** Cast iron, AWWA C501									**700**
	0100 Heavy duty, self contained w/crank oper. gate, 450 x 450 mm	L-5A	1.70	18.824	Ea.	5,075	755	410	6,240	7,325
	0110 600 x 600 mm		1.20	26.667		6,500	1,075	580	8,155	9,650
	0120 750 x 750 mm		1	32		7,850	1,275	695	9,820	11,600
	0130 900 x 900 mm		.90	35.556		9,450	1,425	775	11,650	13,700
	0140 1050 x 1050 mm		.80	40		11,200	1,600	870	13,670	16,000
	0150 1200 x 1200 mm		.50	64		11,400	2,550	1,400	15,350	18,500
	0160 1350 x 1350 mm	↓	.40	80	↓	14,500	3,200	1,750	19,450	23,500

11 EQUIPMENT

11280 | Hydraulic Gates & Valves

11285	Hydraulic Gates	CREW	DAILY OUTPUT	LABOR-HOURS	UNIT	2006 BARE COSTS				TOTAL INCL O&P	
						MAT.	LABOR	EQUIP.	TOTAL		
700	**0170**	1500 x 1500 mm	L-5A	.30	106	Ea.	16,700	4,275	2,325	23,300	28,300	**700**
	0180	1650 x 1650 mm		.30	106		20,700	4,275	2,325	27,300	32,600	
	0190	1800 x 1800 mm		.20	160		23,400	6,400	3,475	33,275	40,700	
	0200	1950 x 1950 mm		.20	160		31,600	6,400	3,475	41,475	49,600	
	0210	2100 x 2100 mm	▼	.10	320		29,500	12,800	6,975	49,275	62,500	
	0220	2250 x 2250 mm	E-20	.30	213		35,900	8,325	3,775	48,000	58,000	
	0230	2400 x 2400 mm		.30	213		40,100	8,325	3,775	52,200	62,500	
	0240	2700 x 2700 mm		.20	320		46,200	12,500	5,675	64,375	79,000	
	0250	3000 x 3000 mm		.10	640		54,500	25,000	11,400	90,900	116,000	
	0260	3300 x 3300 mm	▼	.10	640		96,000	25,000	11,400	132,400	161,500	

11300 | Fluid Waste Treatment & Disposal Equipment

11310	Sewage & Sludge Pumps	CREW	DAILY OUTPUT	LABOR-HOURS	UNIT	2006 BARE COSTS				TOTAL INCL O&P	
						MAT.	LABOR	EQUIP.	TOTAL		
100	**0010**	**OIL/WATER SEPARATORS**										**100**
	0020	Complete system, not including excavation and backfill										
	0100	Treated capacity 5.66 liters per second	B-22	1	30	Ea.	7,000	975	229	8,204	9,450	
	0200	14 - 17 liters per second	B-13	.75	74.667		10,700	2,225	870	13,795	16,100	
	0300	28 liters per second		.60	93.333		15,300	2,800	1,075	19,175	22,300	
	0400	68 - 85 liters per second		.30	186		22,700	5,600	2,175	30,475	36,000	
	0500	312 liters per second		.17	329		25,600	9,875	3,825	39,300	47,600	
	0600	623 liters per second	▼	.10	560	▼	34,700	16,800	6,500	58,000	71,000	
700	**0010**	**SEWAGE PUMPING STATIONS** Prefabricated steel, concrete										**700**
	0020	or fiberglass, 12.62 L/s	C-17D	.17	494	Total	37,500	18,300	3,125	58,925	73,000	
	0200	63.1 L/s		.07	1200		53,000	44,400	7,550	104,950	135,500	
	0500	Add for generator unit, 12.62 L/s, steel		.34	247		27,100	9,125	1,550	37,775	45,700	
	0600	Concrete		.51	164		17,200	6,100	1,050	24,350	29,500	
	1000	Add for generator unit, 63.1 L/s, steel		.30	280		28,800	10,300	1,775	40,875	49,800	
	1200	Concrete	▼	.38	221		24,300	8,175	1,400	33,875	41,000	
	1500	For drilled water well, if required, add	B-23	.50	80	▼	7,000	2,225	7,100	16,325	19,000	

11390	Pkg Sewage Treat Plants										
200	**0010**	**SEWAGE TREATMENT PLANTS,** not incl fencing or external piping										**200**
	0020	Steel packaged, blown air aeration plants										
	0100	4 kL/day				liter				4.60	5.35	
	0200	19 kL/day								3.11	3.59	
	0300	57 kL/day								1.71	1.96	
	0400	114 kL/day								1.62	1.89	
	0500	190 kL/day								1.26	1.39	
	0600	380 kL/day								1.08	1.19	
	0700	760 kL/day								.79	.87	
	0800	1900 kL/day				▼				.78	.87	
	1000	Concrete, extended aeration, primary and secondary treatment										
	1010	38 kL/day				liter				3.59	3.95	
	1100	114 kL/day								1.71	1.89	
	1200	190 kL/day								1.39	1.54	
	1400	380 kL/day								1.08	1.20	
	1500	1900 kL/day				▼				.80	.86	
	1700	Municipal wastewater treatment facility										
	1720	4 kL/day				liter				1.35	1.50	

11390	Pkg Sewage Treat Plants	CREW	DAILY OUTPUT	LABOR-HOURS	UNIT	2006 BARE COSTS				TOTAL INCL O&P	
						MAT.	LABOR	EQUIP.	TOTAL		
200	1740	5.5 kL/day				liter				1.34	1.48	200
	1760	7.5 kL/day								1.13	1.25
	1780	11.5 kL/day								.88	.98
	1800	19 kL/day								.82	.90
	2000	Holding tank system, not incl. excavation or backfill									
	2010	Recirculating chemical water closet	2 Plum	4	4	Ea.	810	171		981	1,150
	2100	For voltage converter, add	"	16	1		215	42.50		257.50	300
	2200	For high level alarm, add	1 Plum	7.80	1.026		123	44		167	201
900	0010	**WASTEWATER TREATMENT SYSTEM**										900
	0020	Fiberglass, 3785 L	B-21	1.29	21.705	Ea.	3,075	695	118	3,888	4,575
	0100	5.5 kL	"	1.03	27.184	"	6,725	875	148	7,748	8,925

11405	Food Storage Equipment	CREW	DAILY OUTPUT	LABOR-HOURS	UNIT	2006 BARE COSTS				TOTAL INCL O&P	
						MAT.	LABOR	EQUIP.	TOTAL		
110	0010	**FOOD STORAGE EQUIPMENT**										110
	2350	Cooler, reach-in, beverage, 1829 mm long	Q-1	6	2.667	Ea.	3,775	102		3,877	4,300
	4300	Freezers, reach-in, 1.23 m³		4	4		8,625	154		8,779	9,700
	4500	1.90 m³		3	5.333		9,825	205		10,030	11,100
	4600	Freezer, pre-fab, 2400 mm x 2400 mm w/refrigeration	2 Carp	.45	35.556		7,950	1,275		9,225	10,700
	4620	2400 mm x 3600 mm		.35	45.714		9,775	1,625		11,400	13,300
	4640	2400 mm x 4800 mm		.25	64		12,300	2,275		14,575	17,100
	4660	2400 mm x 6000 mm		.17	94.118		15,000	3,350		18,350	21,800
	4680	Reach-in, 1 compartment	Q-1	4	4		2,125	154		2,279	2,575
	4700	2 compartment		3	5.333		4,325	205		4,530	5,050
	8300	Refrigerators, reach-in type, 1.23 m³		5	3.200		5,200	123		5,323	5,900
	8310	With glass doors, 1.90 m³		4	4		7,575	154		7,729	8,575
	8320	Refrigerator, reach-in, 1 compartment	R-18	7.80	3.333		1,675	108		1,783	2,025
	8330	2 compartment		6.20	4.194		2,200	136		2,336	2,625
	8340	3 compartment		5.60	4.643		3,375	151		3,526	3,950
	8350	Pre-fab, with refrigeration, 2400 mm x 2400 mm	2 Carp	.45	35.556		5,275	1,275		6,550	7,775
	8360	2400 mm x 3600 mm		.35	45.714		6,800	1,625		8,425	10,000
	8370	2400 mm x 4800 mm		.25	64		8,650	2,275		10,925	13,100
	8380	2400 mm x 6000 mm		.17	94.118		10,600	3,350		13,950	16,900
	8390	Pass-thru/roll-in, 1 compartment	R-18	7.80	3.333		3,400	108		3,508	3,925
	8400	2 compartment		6.24	4.167		4,875	135		5,010	5,550
	8410	3 compartment		5.60	4.643		6,175	151		6,326	7,025
	8420	Walk-in, alum, door & flr, no refrig, 1800 x 1800 x 2250 mm	2 Carp	1.40	11.429		6,700	405		7,105	8,000
	8430	3000 mm x 1800 mm x 2250 mm		.55	29.091		9,600	1,025		10,625	12,200
	8440	3600 mm x 4200 mm x 2250 mm		.25	64		13,300	2,275		15,575	18,300
	8450	3600 mm x 6000 mm x 2250 mm		.17	94.118		16,400	3,350		19,750	23,200
	8460	Refrigerated cabinets, mobile					2,825			2,825	3,100
	8470	Refrigerator/freezer, reach-in, 1 compartment	R-18	5.60	4.643		4,350	151		4,501	5,000
	8480	2 compartment	"	4.80	5.417		6,275	176		6,451	7,175
	8600	Stainless steel shelving, louvered 4-tier, 500 mm x 900 mm	1 Clab	6	1.333		975	36.50		1,011.50	1,125
	8605	500 mm x 1200 mm		6	1.333		1,375	36.50		1,411.50	1,575
	8610	500 mm x 1800 mm		6	1.333		2,000	36.50		2,036.50	2,250
	8615	600 mm x 900 mm		6	1.333		1,275	36.50		1,311.50	1,450
	8620	600 mm x 1200 mm		6	1.333		1,475	36.50		1,511.50	1,675

11405 | Food Storage Equipment

		CREW	DAILY OUTPUT	LABOR-HOURS	UNIT	MAT.	LABOR	EQUIP.	TOTAL	TOTAL INCL O&P	
110											**110**
8625	600 mm x 1800 mm	1 Clab	6	1.333	Ea.	2,150	36.50		2,186.50	2,400	
8630	Flat 4-tier, 500 mm x 900 mm		6	1.333		1,125	36.50		1,161.50	1,300	
8635	500 mm x 1200 mm		6	1.333		1,350	36.50		1,386.50	1,525	
8640	500 mm x 1500 mm		6	1.333		1,550	36.50		1,586.50	1,750	
8645	600 mm x 900 mm		6	1.333		1,175	36.50		1,211.50	1,350	
8650	600 mm x 1200 mm		6	1.333		1,400	36.50		1,436.50	1,575	
8655	600 mm x 1800 mm		6	1.333		2,900	36.50		2,936.50	3,250	
8700	Galvanized shelving, louvered 4-tier, 500 mm x 900 mm		6	1.333		495	36.50		531.50	595	
8705	500 mm x 1200 mm		6	1.333		550	36.50		586.50	660	
8710	500 mm x 1800 mm		6	1.333		570	36.50		606.50	685	
8715	600 mm x 900 mm		6	1.333		450	36.50		486.50	550	
8720	600 mm x 1200 mm		6	1.333		590	36.50		626.50	705	
8725	600 mm x 1800 mm		6	1.333		845	36.50		881.50	985	
8730	Flat 4-tier, 500 mm x 900 mm		6	1.333		350	36.50		386.50	440	
8735	500 mm x 1200 mm		6	1.333		395	36.50		431.50	490	
8740	500 mm x 1800 mm		6	1.333		710	36.50		746.50	835	
8745	600 mm x 900 mm		6	1.333		340	36.50		376.50	430	
8750	600 mm x 1200 mm		6	1.333		535	36.50		571.50	640	
8755	600 mm x 1800 mm		6	1.333		755	36.50		791.50	885	
8760	Stainless steel dunnage rack, 600 mm x 900 mm		8	1		680	27.50		707.50	795	
8765	600 mm x 1200 mm		8	1		705	27.50		732.50	820	
8770	Galvanized dunnage rack, 600 mm x 900 mm		8	1		111	27.50		138.50	166	
8775	600 mm x 1200 mm	▼	8	1	▼	144	27.50		171.50	201	
800	0010 **WINE CELLAR**, refrigerated, Redwood interior, carpeted, walk-in type										**800**
	0020 2032 mm high, including racks										
	0200 2000 mm W x 1200 mm D for 900 bottles	2 Carp	1.50	10.667	Ea.	3,125	380		3,505	4,025	
	0250 2000 mm W x 1800 mm D for 1300 bottles		1.33	12.030		4,125	430		4,555	5,225	
	0300 2000 mm W x 2350 mm D for 1900 bottles		1.17	13.675		5,350	485		5,835	6,650	
	0400 2000 mm W x 3100 mm D for 2500 bottles	▼	1	16	▼	6,475	570		7,045	8,000	
	0600 Portable cabinets, red oak, reach-in temp. & humidity controlled										
	0650 676 mm W x 673 mm D x 1727 mm H for 235 bottles				Ea.	2,825			2,825	3,100	
	0660 813mm Wx 546mm Dx 1867mm H for 144 bottles					3,275			3,275	3,600	
	0670 813mm Wx 749mm Dx 1867mm H for 288 bottles					3,300			3,300	3,625	
	0680 1003mm W x 749mm D x 2197mm H for 440 bottles					3,750			3,750	4,125	
	0690 1334mm W x 749mm D x 1867mm H for 468 bottles					4,300			4,300	4,725	
	0700 1334mm W x 749mm D x 2197mm H for 572 bottles				▼	4,525			4,525	4,975	
	0730 Portable, red oak, can be built-in with glass door										
	0750 606mm W x 610mm D x 876mm H for 50 bottles				Ea.	1,025			1,025	1,125	

11410 | Food Preparation Equipment

		CREW	DAILY OUTPUT	LABOR-HOURS	UNIT	MAT.	LABOR	EQUIP.	TOTAL	TOTAL INCL O&P	
110	0010 **FOOD PREPARATION EQUIPMENT**										**110**
	1700 Choppers, 2.3 kg	R-18	7	3.714	Ea.	2,400	120		2,520	2,825	
	1720 7.3 kg		5	5.200		2,700	169		2,869	3,225	
	1740 15.9 to 18.1 kg	▼	4	6.500		3,000	211		3,211	3,625	
	1840 Coffee brewer, 5 burners	1 Plum	3	2.667		1,150	114		1,264	1,425	
	1850 Coffee urn, twin 22.7 L urns		2	4		2,650	171		2,821	3,150	
	1860 Single, 11.4 L	▼	3	2.667		1,675	114		1,789	2,025	
	3000 Fast food equipment, total package, minimum	6 Skwk	.08	600		139,500	21,900		161,400	187,500	
	3100 Maximum	"	.07	685		190,000	25,000		215,000	248,000	
	3800 Food mixers, 18.9 L	L-7	7	4		3,150	137		3,287	3,675	
	3850 38 L		5.40	5.185		6,775	177		6,952	7,725	
	3900 57 L		5	5.600		9,875	191		10,066	11,200	
	4040 76 L		3.90	7.179		11,200	245		11,445	12,800	
	4080 123 L	▼	2.20	12.727	▼	17,200	435		17,635	19,600	

EQUIPMENT 11

11410 | Food Preparation Equipment

		CREW	DAILY OUTPUT	LABOR-HOURS	UNIT	2006 BARE COSTS MAT.	LABOR	EQUIP.	TOTAL	TOTAL INCL O&P	
110	4100	Floor type, 19 L	L-7	15	1.867	Ea.	2,550	63.50		2,613.50	2,900	110
	4120	57 L		14	2		7,225	68.50		7,293.50	8,025	
	4140	76 L		12	2.333		9,875	79.50		9,954.50	10,900	
	4160	132 L	↓	8.60	3.256		22,200	111		22,311	24,600	
	6700	Peelers, small	R-18	8	3.250		1,150	105		1,255	1,425	
	6720	Large	"	6	4.333		3,650	141		3,791	4,250	
	6800	Pulper/extractor, close coupled, 4 kW	1 Plum	1.90	4.211		3,250	180		3,430	3,850	
	8580	Slicer with table	R-18	9	2.889	↓	4,175	93.50		4,268.50	4,750	

11415 | Food Delivery Carts and Conveyors

		CREW	DAILY OUTPUT	LABOR-HOURS	UNIT	MAT.	LABOR	EQUIP.	TOTAL	TOTAL INCL O&P	
110	0010	**FOOD DELIVERY CARTS**										110
	1650	Cabinet, heated, 1 compartment, reach-in	R-18	5.60	4.643	Ea.	2,325	151		2,476	2,800	
	1655	Pass-thru roll-in		5.60	4.643		3,975	151		4,126	4,575	
	1660	2 compartment, reach-in	↓	4.80	5.417		5,925	176		6,101	6,775	
	1670	Mobile					2,625			2,625	2,875	
	6850	Mobile rack w/pan slide					985			985	1,075	
	9180	Tray and silver dispenser, mobile	1 Clab	16	.500	↓	770	13.70		783.70	865	

11420 | Food Cooking Equipment

		CREW	DAILY OUTPUT	LABOR-HOURS	UNIT	MAT.	LABOR	EQUIP.	TOTAL	TOTAL INCL O&P	
110	0010	**COOKING EQUIPMENT**										110
	0020	Bake oven, gas, one section	Q-1	8	2	Ea.	4,700	77		4,777	5,275	
	0300	Two sections		7	2.286		9,450	88		9,538	10,500	
	0600	Three sections	↓	6	2.667		14,000	102		14,102	15,600	
	0900	Electric convection, single deck	L-7	4	7		4,875	239		5,114	5,725	
	1300	Broiler, without oven, standard	Q-1	8	2		2,950	77		3,027	3,375	
	1550	Infra-red	L-7	4	7		6,475	239		6,714	7,500	
	4750	Fryer, with twin baskets, modular model	Q-1	7	2.286		1,675	88		1,763	1,975	
	5000	Floor model on 152 mm legs	"	5	3.200		1,575	123		1,698	1,925	
	5100	Extra single basket, large					94			94	103	
	5300	914mm L SST griddle,600mm plate dp w/100mm legs,elec,208V 3Ph	Q-1	7	2.286		915	88		1,003	1,125	
	5550	1219 mm long	"	6	2.667		1,225	102		1,327	1,500	
	6200	Iced tea brewer	1 Plum	3.44	2.326		625	99.50		724.50	840	
	6350	75.7 L kettle w/steam jacket, tilting w/positive lock, S/S	L-7	7	4		5,100	137		5,237	5,825	
	6600	227.1 L	"	6	4.667		7,250	159		7,409	8,225	
	6900	Range, restaurant type, 6 burners and 1 standard oven, 914 mm W	Q-1	7	2.286		1,875	88		1,963	2,200	
	6950	Convection		7	2.286		2,650	88		2,738	3,050	
	7150	2 standard ovens, 610 mm griddle, 1524 mm wide		6	2.667		4,650	102		4,752	5,250	
	7200	1 standard, 1 convection oven		6	2.667		5,625	102		5,727	6,350	
	7450	Heavy duty, single 864 mm standard oven, open top		5	3.200		4,125	123		4,248	4,725	
	7500	Convection oven		5	3.200		5,675	123		5,798	6,425	
	7700	Griddle top		6	2.667		4,925	102		5,027	5,575	
	7750	Convection oven	↓	6	2.667		6,500	102		6,602	7,300	
	8850	Steamer, electric 27 kW	L-7	7	4		8,975	137		9,112	10,100	
	9100	Electric, 10 kW or gas 29 kW	"	5	5.600		4,325	191		4,516	5,050	
	9150	Toaster, conveyor type, 16-22 slices/minute					1,050			1,050	1,175	
	9160	Pop-up, 2 slot				↓	575			575	630	
	9200	For deluxe models of above equipment, add					75%					
	9400	Rule of thumb: Equipment cost based										
	9410	on kitchen work area										
	9420	Office buildings, minimum	L-7	7.15	3.914	m²	670	134		804	940	
	9450	Maximum		5.39	5.197		1,125	177		1,302	1,525	
	9550	Public eating facilities, minimum		7.15	3.914		875	134		1,009	1,175	
	9600	Maximum		4.27	6.552		1,425	224		1,649	1,925	
	9750	Hospitals, minimum		5.39	5.197		900	177		1,077	1,275	
	9800	Maximum	↓	3.62	7.728	↓	1,500	264		1,764	2,050	

11 EQUIPMENT

11425 | Hood and Ventilation Equipment

			CREW	DAILY OUTPUT	LABOR-HOURS	UNIT	2006 BARE COSTS				TOTAL INCL O&P	
							MAT.	LABOR	EQUIP.	TOTAL		
110	0010	**HOOD & VENTILATION EQUIPMENT**										110
	2970	Exhaust hood, sst, gutter all sides, 1200mm x 1200mm x 600mm	1 Carp	1.80	4.444	Ea.	3,250	158		3,408	3,825	
	2980	1200 mm x 1200 mm x 2100 mm	"	1.60	5		5,050	178		5,228	5,825	
	7950	Hood fire protection system, minimum	Q-1	3	5.333		3,450	205		3,655	4,075	
	8050	Maximum	"	1	16		25,300	615		25,915	28,800	

11430 | Food Dispensing Equipment

			CREW	DAILY OUTPUT	LABOR-HOURS	UNIT	MAT.	LABOR	EQUIP.	TOTAL	INCL O&P	
110	0010	**FOOD DISPENSING EQUIPMENT**										110
	1050	Butter pat dispenser	1 Clab	13	.615	Ea.	780	16.85		796.85	880	
	1100	Bread dispenser, counter top		13	.615		740	16.85		756.85	840	
	1900	Cup and glass dispenser, drop in		4	2		1,050	55		1,105	1,250	
	1920	Disposable cup, drop in		16	.500		320	13.70		333.70	375	
	2650	Dish dispenser, drop in, 300 mm		11	.727		1,200	19.95		1,219.95	1,350	
	2660	Mobile		10	.800		2,025	22		2,047	2,250	
	3300	Food warmer, counter, 1.2 kW					525			525	580	
	3550	1.6 kW					1,550			1,550	1,700	
	3600	Well, hot food, built-in, rectangular, 305 mm x 508 mm	R-30	10	2.600		360	86		446	525	
	3610	Circular, 6.6 L		10	2.600		325	86		411	490	
	3620	Refrigerated, 2 compartments		10	2.600		1,850	86		1,936	2,150	
	3630	3 compartments		9	2.889		1,850	96		1,946	2,200	
	3640	4 compartments		8	3.250		2,575	108		2,683	3,000	
	4720	Frost cold plate		9	2.889		12,700	96		12,796	14,100	
	5700	Hot chocolate dispenser	1 Plum	4	2		855	85.50		940.50	1,075	
	6250	Jet spray dispenser	R-18	4.50	5.778		2,650	187		2,837	3,175	
	6300	Juice dispenser, concentrate	"	4.50	5.778		925	187		1,112	1,300	
	6690	Milk dispenser, bulk, 2 flavor	R-30	8	3.250		1,175	108		1,283	1,475	
	6695	3 flavor	"	8	3.250		1,625	108		1,733	1,950	
	8800	Serving counter, straight	1 Carp	12.19	.656	m	2,050	23.50		2,073.50	2,275	
	8820	Curved section	"	9.14	.875	"	2,550	31		2,581	2,850	
	8825	Solid surface, see section 06620-810										
	8830	Soft serve ice cream machine, medium	R-18	11	2.364	Ea.	8,300	76.50		8,376.50	9,275	
	8840	Large	"	9	2.889	"	12,300	93.50		12,393.50	13,600	

11435 | Ice Machines

			CREW	DAILY OUTPUT	LABOR-HOURS	UNIT	MAT.	LABOR	EQUIP.	TOTAL	INCL O&P	
110	0010	**ICE MACHINES**										110
	5800	Ice cube maker, 23 kg/day	Q-1	6	2.667	Ea.	1,400	102		1,502	1,675	
	5900	113 kg/day		1.20	13.333		1,850	510		2,360	2,800	
	6050	227 kg/day		4	4		2,700	154		2,854	3,200	
	6060	With bin		1.20	13.333		2,200	510		2,710	3,200	
	6090	454 kg/day		1	16		6,625	615		7,240	8,200	
	6100	Ice flakers, 136 kg/day		1.60	10		3,025	385		3,410	3,900	
	6120	272 kg/day		.95	16.842		3,325	645		3,970	4,625	
	6130	454 kg/day		.75	21.333		4,950	820		5,770	6,650	
	6140	907 kg/day		.65	24.615		14,800	945		15,745	17,600	
	6160	Ice storage bin, 227 kg capacity	Q-5	1	16		1,800	620		2,420	2,900	
	6180	454 kg	"	.56	28.571		3,150	1,100		4,250	5,125	

11440 | Cleaning and Disposal Equipment

			CREW	DAILY OUTPUT	LABOR-HOURS	UNIT	MAT.	LABOR	EQUIP.	TOTAL	INCL O&P	
110	0010	**CLEANING & DISPOSAL EQUIPMENT**										110
	2700	Dishwasher, commercial, rack type										
	2720	10 to 12 racks/hour	Q-1	3.20	5	Ea.	3,875	192		4,067	4,575	
	2750	Semi-automatic 38 to 50 racks/hour	"	1.30	12.308		6,950	475		7,425	8,325	
	2800	Automatic, 190 to 230 racks/hour	L-6	.35	34.286		8,625	1,450		10,075	11,700	
	2820	235 to 275 racks/hour		.25	48		19,900	2,050		21,950	25,000	

EQUIPMENT 11

11440 | Cleaning and Disposal Equipment

		CREW	DAILY OUTPUT	LABOR-HOURS	UNIT	2006 BARE COSTS				TOTAL INCL O&P	
						MAT.	LABOR	EQUIP.	TOTAL		
110	2840	8750 to 12 500 dishes/hour	L-6	.10	120	Ea.	43,200	5,100		48,300	55,000	110
	2950	Dishwasher hood, canopy type	L-3A	3.05	3.937	m	1,525	154		1,679	1,925	
	2960	Pant leg type	"	2.50	4.800	Ea.	6,525	188		6,713	7,475	
	5200	Garbage disposal, 1 kW, 375 LPH	L-1	4.80	3.333		1,400	141		1,541	1,750	
	5210	2 kW, 450 LPH		4.60	3.478		2,250	147		2,397	2,700	
	5220	4 kW, 950 LPH		4.50	3.556		3,225	151		3,376	3,775	
	6750	Pot sink, 3 compartment	1 Plum	2.21	3.620	m	2,000	155		2,155	2,425	
	6760	Pot washer, small		1.60	5	Ea.	16,400	214		16,614	18,400	
	6770	Large		1.20	6.667		36,900	285		37,185	41,000	
	9170	Trash compactor, small, up to 57 kg compacted weight	L-4	4	6		17,500	199		17,699	19,500	
	9175	Large, up to 79.5 kg compacted weight	"	3	8		21,300	265		21,565	23,900	

11450 | Residential Equipment

11454 | Residential Appliances

		CREW	DAILY OUTPUT	LABOR-HOURS	UNIT	2006 BARE COSTS				TOTAL INCL O&P	
						MAT.	LABOR	EQUIP.	TOTAL		
500	0010	**RESIDENTIAL APPLIANCES**										500
	0020	Cooking range, 762 mm free standing, 1 oven, minimum	2 Clab	10	1.600	Ea.	249	44		293	340	
	0050	Maximum		4	4		1,550	110		1,660	1,875	
	0150	2 oven, minimum		10	1.600		1,525	44		1,569	1,750	
	0200	Maximum		10	1.600		1,600	44		1,644	1,850	
	0350	Built-in, 762 mm wide, 1 oven, minimum	1 Elec	6	1.333		460	56		516	590	
	0400	Maximum	2 Carp	2	8		1,350	284		1,634	1,950	
	0500	2 oven, conventional, minimum		4	4		960	142		1,102	1,275	
	0550	1 conventional, 1 microwave, maximum		2	8		1,525	284		1,809	2,125	
	0700	Free-standing, 533 mm wide range, 1 oven, minimum	2 Clab	10	1.600		256	44		300	350	
	0750	533 mm wide, maximum	"	4	4		283	110		393	480	
	0900	Counter top cook tops, 4 burner, standard, minimum	1 Elec	6	1.333		192	56		248	295	
	0950	Maximum		3	2.667		535	112		647	755	
	1050	As above, but with grille and griddle attachment, minimum		6	1.333		480	56		536	615	
	1100	Maximum		3	2.667		660	112		772	890	
	1200	Induction cooktop, 762 mm wide		3	2.667		535	112		647	755	
	1250	Microwave oven, minimum		4	2		85.50	84		169.50	219	
	1300	Maximum		2	4		405	168		573	695	
	1750	Compactor, residential size, 4 to 1 compaction, minimum	1 Carp	5	1.600		430	57		487	560	
	1800	Maximum	"	3	2.667		485	95		580	680	
	2000	Deep freeze, 0.43 to 0.64 m³, minimum	2 Clab	10	1.600		405	44		449	515	
	2050	Maximum		5	3.200		535	87.50		622.50	720	
	2200	0.84 m³, minimum		8	2		775	55		830	940	
	2250	Maximum		3	5.333		880	146		1,026	1,200	
	2450	Dehumidifier, portable, automatic, 7 L					154			154	170	
	2550	19 L					172			172	189	
	2750	Dishwasher, built-in, 2 cycles, minimum	L-1	4	4		256	169		425	535	
	2800	Maximum		2	8		299	340		639	835	
	2950	4 or more cycles, minimum		4	4		271	169		440	550	
	2960	Average		4	4		360	169		529	650	
	3000	Maximum		2	8		560	340		900	1,125	
	3200	Dryer, automatic, minimum	L-2	3	5.333		287	165		452	570	
	3250	Maximum	"	2	8		810	248		1,058	1,275	
	3300	Garbage disposal, sink type, minimum	L-1	10	1.600		44	68		112	149	

			CREW	DAILY OUTPUT	LABOR-HOURS	UNIT	2006 BARE COSTS				TOTAL INCL O&P	
11454		**Residential Appliances**					MAT.	LABOR	EQUIP.	TOTAL		
500	3350	Maximum	L-1	10	1.600	Ea.	150	68		218	266	**500**
	3550	Heater, electric, built-in, 1250 W, ceiling type, minimum	1 Elec	4	2		72	84		156	205	
	3600	Maximum		3	2.667		122	112		234	300	
	3700	Wall type, minimum		4	2		106	84		190	241	
	3750	Maximum		3	2.667		140	112		252	320	
	3900	1500 W wall type, with blower		4	2		131	84		215	269	
	3950	3000 W		3	2.667		267	112		379	460	
	4150	Hood for range, 2 speed, vented, 762 mm wide, minimum	L-3	5	3.200		38.50	124		162.50	234	
	4200	Maximum		3	5.333		615	207		822	1,000	
	4300	1067 mm wide, minimum		5	3.200		232	124		356	445	
	4330	Custom		5	3.200		645	124		769	900	
	4350	Maximum		3	5.333		785	207		992	1,175	
	4500	For ventless hood, 2 speed, add					15.55			15.55	17.10	
	4650	For vented 1 speed, deduct from maximum					40.50			40.50	44.50	
	4850	Humidifier, portable, 30.28 L/day					154			154	170	
	5000	56.78 L/day					185			185	204	
	5200	Icemaker, automatic, 9 kg/day	1 Plum	7	1.143		375	49		424	485	
	5350	23 kg/day	"	2	4		1,050	171		1,221	1,400	
	5500	Refrigerator, no frost, 0.28 m³ to 0.34 m³ minimum	2 Clab	10	1.600		455	44		499	570	
	5600	Maximum		6	2.667		705	73		778	890	
	5750	0.39 m³ to 0.45 m³, minimum		9	1.778		460	48.50		508.50	580	
	5800	Maximum		5	3.200		500	87.50		587.50	685	
	5950	0.50 m³ to 0.56 m³, minimum		8	2		525	55		580	665	
	6000	Maximum		4	4		870	110		980	1,125	
	6150	0.59 m³ to 0.81 m³, minimum		7	2.286		670	62.50		732.50	835	
	6200	Maximum		3	5.333		2,150	146		2,296	2,575	
	6400	Sump pump cellar drainer, pedestal, 249 W, molded PVC base	1 Plum	3	2.667		88.50	114		202.50	269	
	6450	Solid brass	"	2	4		185	171		356	460	
	6460	Sump pump, see also division 15440-940										
	6650	Washing machine, automatic, minimum	1 Plum	3	2.667	Ea.	288	114		402	485	
	6700	Maximum	"	1	8		1,050	340		1,390	1,675	
	6900	Water heater, electric, glass lined, 113.6 L, minimum	L-1	5	3.200		261	136		397	490	
	6950	Maximum		3	5.333		360	226		586	740	
	7100	302.8 L, minimum		2	8		500	340		840	1,050	
	7150	Maximum		1	16		695	680		1,375	1,800	
	7180	Water heater, gas, glass lined, 113.6 L, minimum	2 Plum	5	3.200		565	137		702	825	
	7220	Maximum		3	5.333		785	228		1,013	1,200	
	7260	189.3 L, minimum		2.50	6.400		595	273		868	1,075	
	7300	Maximum		1.50	10.667		830	455		1,285	1,600	
	7310	Water heater, see also division 15480-200										
	7350	Water softener, automatic, to 8 grains/L	2 Plum	5	3.200	Ea.	445	137		582	695	
	7400	To 26 grains/L	"	4	4		620	171		791	935	
	7450	Vent kits for dryers	1 Carp	10	.800		12.80	28.50		41.30	58.50	
550	0010	**DISAPPEARING STAIRWAY** No trim included										**550**
	0100	Custom grade, pine, 2591 mm ceiling, minimum	1 Carp	4	2	Ea.	117	71		188	240	
	0150	Average		3.50	2.286		118	81.50		199.50	257	
	0200	Maximum		3	2.667		195	95		290	365	
	0500	Heavy duty, pivoted, from 2311 mm to 3912 mm floor to floor		3	2.667		385	95		480	570	
	0600	4877 mm ceiling		2	4		1,275	142		1,417	1,625	
	0800	Economy folding, pine, 2591 mm ceiling		4	2		107	71		178	228	
	0900	2896 mm ceiling		4	2		116	71		187	239	
	1000	Fire escape, galvanized steel, 2438 mm to 3150 mm ceiling	2 Carp	1	16		1,350	570		1,920	2,375	
	1010	3200 mm to 4115 mm ceiling		1	16		1,700	570		2,270	2,750	
	1100	Automatic electric, aluminum, floor to floor height, 2438 to 2743 mm		1	16		7,000	570		7,570	8,575	
	1400	3353 mm to 3658 mm		.90	17.778		7,550	630		8,180	9,275	

11450 | Residential Equipment

11454 | Residential Appliances

			CREW	DAILY OUTPUT	LABOR-HOURS	UNIT	2006 BARE COSTS MAT.	LABOR	EQUIP.	TOTAL	TOTAL INCL O&P	
550	1700	4267 mm to 4572 mm	2 Carp	.70	22.857	Ea.	8,025	815		8,840	10,100	550

11460 | Unit Kitchens

			CREW	DAILY OUTPUT	LABOR-HOURS	UNIT	MAT.	LABOR	EQUIP.	TOTAL	TOTAL INCL O&P	
100	0010	**UNIT KITCHENS**										100
	1500	Combination range, refrigerator and sink, 762 mm wide, minimum	L-1	2	8	Ea.	755	340		1,095	1,325	
	1550	Maximum		1	16		1,500	680		2,180	2,675	
	1570	1524 mm wide, average		1.40	11.429		2,500	485		2,985	3,475	
	1590	1829 mm wide, average		1.20	13.333		2,825	565		3,390	3,950	
	1600	Office model, 1219 mm wide		2	8		2,125	340		2,465	2,850	
	1620	Refrigerator and sink only	↓	2.40	6.667	↓	2,125	282		2,407	2,775	
	1640	Combination range, refrigerator, sink, microwave										
	1660	oven and ice maker	L-1	.80	20	Ea.	4,150	845		4,995	5,850	

11470 | Darkroom Equipment

11471 | Darkroom Processing

			CREW	DAILY OUTPUT	LABOR-HOURS	UNIT	2006 BARE COSTS MAT.	LABOR	EQUIP.	TOTAL	TOTAL INCL O&P	
700	0010	**DARKROOM EQUIPMENT**										700
	0020	Developing sink, 125 mm deep, 600 mm x 1200 mm	Q-1	2	8	Ea.	4,450	305		4,755	5,350	
	0050	1200 mm x 1300 mm		1.70	9.412		4,500	360		4,860	5,500	
	0200	250 mm deep, 600 mm x 1200 mm		1.70	9.412		5,550	360		5,910	6,650	
	0250	600 mm x 2743 mm	↓	1.50	10.667		2,000	410		2,410	2,825	
	0500	Dryers, dehumidified filtered air, 900 mm x 625 mm x 1700 mm H	L-7	6	4.667		4,325	159		4,484	5,025	
	0550	1200 mm x 625 mm x 1700 mm high		5	5.600		7,975	191		8,166	9,075	
	2000	Processors, automatic, color print, minimum		4	7		11,200	239		11,439	12,700	
	2050	Maximum		.60	46.667		11,800	1,600		13,400	15,500	
	2300	Black and white print, minimum		2	14		8,500	480		8,980	10,100	
	2350	Maximum		.80	35		55,500	1,200		56,700	63,000	
	2600	Manual processor, 406 mm x 508 mm maximum print size		2	14		8,575	480		9,055	10,200	
	2650	508 mm x 610 mm maximum print size		1	28		8,075	955		9,030	10,400	
	3000	Viewing lites, 508 mm x 610 mm		6	4.667		350	159		509	630	
	3100	508 mm x 610 mm with color correction	↓	6	4.667		495	159		654	790	
	3500	Washers, round, minimum sheet 279 mm x 356 mm	Q-1	2	8		2,725	305		3,030	3,475	
	3550	Maximum sheet 508 mm x 610 mm		1	16		3,225	615		3,840	4,450	
	3800	Square, minimum sheet 508 mm x 610 mm		1	16		2,875	615		3,490	4,100	
	3900	Maximum sheet 1270 mm x 1422 mm	↓	.80	20	↓	4,475	770		5,245	6,075	
	4500	Combination tank sink, tray sink, washers, with										
	4510	dry side tables, average	Q-1	.45	35.556	Ea.	8,525	1,375		9,900	11,400	

11472 | Revolving Darkroom Doors

			CREW	DAILY OUTPUT	LABOR-HOURS	UNIT	MAT.	LABOR	EQUIP.	TOTAL	TOTAL INCL O&P	
370	0010	**DARKROOM DOORS**										370
	0015	Revolving, standard, 2 way, 900 mm diameter	2 Carp	3.10	5.161	Opng.	1,850	183		2,033	2,325	
	0020	1041 mm diameter		3.10	5.161		2,000	183		2,183	2,475	
	0050	3 way, 1295 mm diameter		1.40	11.429		2,475	405		2,880	3,350	
	1000	4 way, 1245 mm diameter		1.40	11.429		2,950	405		3,355	3,850	
	2000	Hinged safety, 2 way, 1041 mm diameter		2.30	6.957		2,375	247		2,622	2,975	
	2500	3 way, 1295 mm diameter		1.40	11.429		2,975	405		3,380	3,900	
	3000	Pop out safety, 2 way, 1041 mm diameter	↓	3.10	5.161	↓	2,900	183		3,083	3,475	

Important: See the Reference Section for supporting data - Crews, Rental Equipment, City Cost Indexes and Reference Data

11 EQUIPMENT

11470 | Darkroom Equipment

11472 | Revolving Darkroom Doors

		CREW	DAILY OUTPUT	LABOR-HOURS	UNIT	2006 BARE COSTS MAT.	LABOR	EQUIP.	TOTAL	TOTAL INCL O&P	
370	4000	3 way, 1295 mm diameter	2 Carp	1.40	11.429	Opng.	2,900	405		3,305	3,825	370
	5000	Wheelchair-type, pop out, 2195 mm diameter		1.40	11.429		3,000	405		3,405	3,925	
	5020	1800 mm diameter	↓	.90	17.778	↓	6,125	630		6,755	7,725	
	9300	For complete dark rooms, see division 13035-300										

11480 | Athletic Recreational & Therapeutic Equipment

11483 | Bowling Alleys

		CREW	DAILY OUTPUT	LABOR-HOURS	UNIT	2006 BARE COSTS MAT.	LABOR	EQUIP.	TOTAL	TOTAL INCL O&P	
150	0010	**BOWLING ALLEYS** Including alley, pinsetter, scorer,										150
	0020	counters and misc. supplies, minimum	4 Carp	.20	160	Lane	35,100	5,700		40,800	47,500	
	0150	Average	↓	.19	168		40,000	5,975		45,975	53,500	
	0300	Maximum	↓	.18	177		46,000	6,325		52,325	60,500	
	0400	Combo table ball rack, add					995			995	1,100	
	0600	For automatic scorer, add, minimum					6,350			6,350	7,000	
	0700	Maximum				↓	7,200			7,200	7,925	

11484 | Exercise Equipment

		CREW	DAILY OUTPUT	LABOR-HOURS	UNIT	2006 BARE COSTS MAT.	LABOR	EQUIP.	TOTAL	TOTAL INCL O&P	
400	0010	**HEALTH CLUB EQUIPMENT**										400
	0020	Abdominal rack, 2 board capacity				Ea.	435			435	480	
	0050	Abdominal board, upholstered					490			490	540	
	0200	Bicycle trainer, minimum					760			760	835	
	0300	Deluxe, electric					3,875			3,875	4,250	
	0400	Bar bell set, chrome plated steel, 11.34 kg					228			228	251	
	0420	45.36 kg					345			345	380	
	0450	90.72 kg				↓	665			665	735	
	0500	Weight plates, cast iron, per kg				kg	10.30			10.30	11.35	
	0520	Storage rack, 10 station				Ea.	735			735	810	
	0600	Circuit training apparatus, 12 machines minimum	2 Clab	1.25	12.800	Set	25,700	350		26,050	28,800	
	0700	Average	↓	1	16		31,600	440		32,040	35,400	
	0800	Maximum	↓	.75	21.333	↓	37,400	585		37,985	42,100	
	0820	Dumbbell set, cast iron, with rack and 5 pair					720			720	790	
	0900	Squat racks	2 Clab	5	3.200	Ea.	715	87.50		802.50	920	
	1200	Multi-station gym machine, 5 station					7,425			7,425	8,175	
	1250	9 station					10,400			10,400	11,500	
	1280	Rowing machine, hydraulic					1,375			1,375	1,500	
	1300	Treadmill, manual					1,350			1,350	1,475	
	1320	Motorized					3,800			3,800	4,175	
	1340	Electronic					5,100			5,100	5,600	
	1360	Cardio-testing					7,250			7,250	7,975	
	1400	Treatment/massage tables, minimum					485			485	535	
	1420	Deluxe, with accessories				↓	640			640	705	
	1600	For saunas, see Div. 13035-800										
	1640	For steam baths, see Div. 13035-940										
	1680	For whirlpool baths, see Div. 15418-100										
	1720	For additional exercise equipment, see Div. 11486-700										

11486 | Gymnasium Equipment

		CREW	DAILY OUTPUT	LABOR-HOURS	UNIT	2006 BARE COSTS MAT.	LABOR	EQUIP.	TOTAL	TOTAL INCL O&P	
700	0010	**SCHOOL EQUIPMENT**										700
	0200	For exterior equipment, see Div. 02880										

EQUIPMENT

439

11486 | Gymnasium Equipment

		CREW	DAILY OUTPUT	LABOR-HOURS	UNIT	2006 BARE COSTS				TOTAL INCL O&P
						MAT.	LABOR	EQUIP.	TOTAL	
700	0201	For chairs & desks, see Div. 12560-200										700
	0300	For chalkboards & bulletin boards, see Div. 10120-940 & 10110-240									
	0400	For lockers, see Div. 10505-500									
	1000	Basketball backstops, wall mtd., 1829 mm extended, fixed, minimum	L-2	1	16	Ea.	1,075	495		1,570	1,950
	1100	Maximum		1	16		1,575	495		2,070	2,500
	1200	Swing up, minimum		1	16		1,275	495		1,770	2,175
	1250	Maximum		1	16		5,000	495		5,495	6,275
	1300	Portable, manual, heavy duty, hydraulic		1.90	8.421		10,600	261		10,861	12,100
	1400	Ceiling suspended, stationary, minimum		.78	20.513		1,900	635		2,535	3,075
	1450	Fold up, with accessories, maximum		.40	40		5,700	1,250		6,950	8,200
	1600	For electrically operated, add	1 Elec	1	8		1,875	335		2,210	2,550
	2000	Benches, folding, in wall, 4267 mm table, 2 benches	L-4	2	12	Set	600	400		1,000	1,275
	3000	Bleachers, telescoping, manual to 15 tier, minimum	F-5	65	.492	Seat	71	17.75		88.75	106
	3100	Maximum		60	.533		107	19.25		126.25	147
	3300	16 to 20 tier, minimum		60	.533		170	19.25		189.25	217
	3400	Maximum		55	.582		213	21		234	267
	3600	21 to 30 tier, minimum		50	.640		177	23		200	231
	3700	Maximum		40	.800		213	29		242	279
	3900	For integral power operation, add, minimum	2 Elec	300	.053		35.50	2.24		37.74	42.50
	4000	Maximum	"	250	.064		57	2.69		59.69	66.50
	4100	Boxing ring, elevated, 6.6 m x 6.6 m	L-4	.10	240	Ea.	8,075	7,950		16,025	21,300
	4110	For cellular plastic foam padding, add		.10	240		2,550	7,950		10,500	15,200
	4120	Floor level, including posts and ropes only, 6 m x 6 m		.80	30		1,500	995		2,495	3,200
	4130	Canvas, 9 m x 9 m		5	4.800		965	159		1,124	1,325
	4150	Exercise equipment, bicycle trainer					405			405	445
	4180	Chinning bar, adjustable, wall mounted	1 Carp	5	1.600		287	57		344	405
	4200	Exercise ladder, 4877 mm x 483 mm, suspended	L-2	3	5.333		1,250	165		1,415	1,625
	4210	High bar, floor plate attached	1 Carp	4	2		1,075	71		1,146	1,300
	4240	Parallel bars, adjustable		4	2		2,375	71		2,446	2,700
	4270	Uneven parallel bars, adjustable		4	2		2,825	71		2,896	3,200
	4280	Wall mounted, adjustable	L-2	1.50	10.667	Set	815	330		1,145	1,400
	4300	Rope, ceiling mounted, 5486 mm long	1 Carp	3.66	2.186	Ea.	212	77.50		289.50	355
	4330	Side horse, vaulting		5	1.600		1,700	57		1,757	1,975
	4360	Treadmill, motorized, deluxe, training type		5	1.600		5,450	57		5,507	6,075
	4390	Weight lifting multi-station, minimum	2 Clab	1	16		2,575	440		3,015	3,500
	4450	Maximum	"	.50	32		14,200	875		15,075	17,000
	4480	For additional exercise equipment, see Div. 11484-400									
	4500	Gym divider curtain, mesh top, vinyl bottom, manual	L-4	46.45	.517	m²	83.50	17.10		100.60	118
	4700	Electric roll up	L-7	37.16	.754		122	25.50		147.50	174
	5500	Gym mats, 50 mm thick, naugahyde covered					38			38	42
	5600	Vinyl/nylon covered					57			57	63
	5800	Wall pads, 38 mm thick	2 Carp	59.46	.269		62.50	9.55		72.05	84
	6000	Wrestling mats, 25 mm thick, heavy duty					51			51	56
	7000	Scoreboards, baseball, minimum	R-3	1.30	15.385	Ea.	2,975	635	117	3,727	4,325
	7200	Maximum		.05	400		14,600	16,600	3,050	34,250	44,300
	7300	Football, minimum		.86	23.256		3,950	965	177	5,092	6,000
	7400	Maximum		.20	100		12,600	4,150	765	17,515	20,800
	7500	Basketball (one side), minimum		2.07	9.662		2,050	400	73.50	2,523.50	2,925
	7600	Maximum		.30	66.667		14,300	2,750	510	17,560	20,400
	7700	Hockey-basketball (four sides), minimum		.25	80		5,675	3,325	610	9,610	11,900
	7800	Maximum		.15	133		23,600	5,525	1,025	30,150	35,400

11488 | Shooting Ranges

700	0010	SHOOTING RANGE Incl. bullet traps, target provisions, controls,										700
	0100	separators, ceiling system, etc. Not incl. structural shell									

11 EQUIPMENT

		11488	Shooting Ranges	CREW	DAILY OUTPUT	LABOR-HOURS	UNIT	2006 BARE COSTS MAT.	LABOR	EQUIP.	TOTAL	TOTAL INCL O&P	
700	0200		Commercial	L-9	.64	56.250	Point	13,000	1,800		14,800	17,200	700
	0300		Law enforcement		.28	128		26,000	4,100		30,100	35,200	
	0400		National Guard armories		.71	50.704		12,500	1,625		14,125	16,400	
	0500		Reserve training centers		.71	50.704		10,500	1,625		12,125	14,200	
	0600		Schools and colleges		.32	112		22,000	3,600		25,600	30,000	
	0700		Major acadamies		.19	189		40,000	6,050		46,050	54,000	
	0800		For acoustical treatment, add					10%	10%				
	0900		For lighting, add					28%	25%				
	1000		For plumbing, add					5%	5%				
	1100		For ventilating system, add, minimum					40%	40%				
	1200		Add, average					25%	25%				
	1300		Add, maximum					35%	35%				

		11520	Industrial Equipment	CREW	DAILY OUTPUT	LABOR-HOURS	UNIT	2006 BARE COSTS MAT.	LABOR	EQUIP.	TOTAL	TOTAL INCL O&P	
250	0010	**DUST COLLECTION SYSTEMS** Commercial / industrial											250
	0120		Central vacuum units										
	0130		Includes stand, filters and motorized shaker										
	0200		0.236 m³/s, 254 mm inlet, 1.49 kW	Q-20	2.40	8.333	Ea.	3,650	325		3,975	4,500	
	0220		0.472 m³/s, 254 mm inlet, 2.24 kW		2.20	9.091		3,800	350		4,150	4,750	
	0240		0.708 m³/s, 254 mm inlet, 3.73 kW		2	10		3,900	385		4,285	4,875	
	0260		1.42 m³/s, 330 mm inlet, 7.46 kW		1.50	13.333		10,200	515		10,715	12,000	
	0280		2.36 m³/s, 406 mm inlet, 2 @ 7.46 kW		1	20		10,800	775		11,575	13,000	
	1000		Vacuum tubing, galvanized										
	1100		54 mm OD, 16 ga.	Q-9	134	.119	m	6.35	4.53		10.88	13.95	
	1110		64 mm OD, 16 ga.		128	.125		7.30	4.74		12.04	15.35	
	1120		76 mm OD, 16 ga.		122	.131		9.30	4.97		14.27	17.90	
	1130		89 mm OD, 16 ga.		116	.138		13.20	5.25		18.45	22.50	
	1140		102 mm OD, 16 ga.		110	.145		13.50	5.50		19	23.50	
	1150		127 mm OD, 14 ga.		97.54	.164		25	6.20		31.20	37	
	1160		152 mm OD, 14 ga.		85.34	.187		28	7.10		35.10	42	
	1170		203 mm OD, 14 ga.		60.96	.262		42.50	9.95		52.45	62	
	1180		254 mm OD, 12 ga.		48.77	.328		42.50	12.45		54.95	65.50	
	1190		305 mm OD, 12 ga.		36.58	.437		122	16.60		138.60	160	
	1200		356 mm OD, 12 ga.		24.38	.656		134	25		159	187	
	1940		Hose, flexible wire reinforced rubber										
	1956		76 mm dia.	Q-9	122	.131	m	18.15	4.97		23.12	27.50	
	1960		102 mm dia.		110	.145		20.50	5.50		26	31	
	1970		127 dia.		97.54	.164		23.50	6.20		29.70	35.50	
	1980		152 mm dia.		85.34	.187		25.50	7.10		32.60	39	
	2000		90° Elbow, slip fit										
	2110		54 mm dia.	Q-9	70	.229	Ea.	8.05	8.65		16.70	22	
	2120		64 mm dia.		65	.246		11.30	9.35		20.65	27	
	2130		76 mm dia.		60	.267		15.30	10.10		25.40	32.50	
	2140		89 mm dia.		55	.291		18.90	11.05		29.95	38	
	2150		102 mm dia.		50	.320		23.50	12.15		35.65	44.50	
	2160		127 mm dia.		45	.356		44.50	13.50		58	70	
	2170		152 mm dia.		40	.400		61.50	15.15		76.65	91	
	2180		203 mm dia.		30	.533		117	20		137	159	

11520	Industrial Equipment	CREW	DAILY OUTPUT	LABOR-HOURS	UNIT	2006 BARE COSTS				TOTAL INCL O&P
						MAT.	LABOR	EQUIP.	TOTAL	
250 2400	45° Elbow, slip fit									**250**
2410	54 mm dia.	Q-9	70	.229	Ea.	7.05	8.65		15.70	21
2420	64 mm dia.		65	.246		10.40	9.35		19.75	26
2430	76 mm dia.		60	.267		12.55	10.10		22.65	29.50
2440	89 mm dia.		55	.291		15.80	11.05		26.85	34.50
2450	102 mm dia.		50	.320		20.50	12.15		32.65	41
2460	127 mm dia.		45	.356		34.50	13.50		48	59
2470	152 mm dia.		40	.400		47	15.15		62.15	75
2480	203 mm dia.	▼	35	.457	▼	91.50	17.35		108.85	128
2800	90° TY, slip fit thru 152 mm dia.									
2810	54 mm dia.	Q-9	42	.381	Ea.	15.80	14.45		30.25	40
2820	64 mm dia.		39	.410		20.50	15.55		36.05	46.50
2830	76 mm dia.		36	.444		27	16.85		43.85	56
2840	89 mm dia.		33	.485		34.50	18.40		52.90	66.50
2850	102 mm dia.		30	.533		49.50	20		69.50	85.50
2860	127 mm dia.		27	.593		94.50	22.50		117	139
2870	152 mm dia.	▼	24	.667		129	25.50		154.50	180
2880	203 mm dia., butt end					273			273	300
2890	254 mm dia., butt end					495			495	540
2900	305 mm dia., butt end					495			495	540
2910	356 mm dia., butt end					895			895	985
2920	152 mm x 102 mm dia., butt end					78.50			78.50	86.50
2930	203 mm x 102 mm dia., butt end					149			149	163
2940	254 mm x 102 mm dia.; butt end					195			195	214
2950	305 mm x 102 mm dia., butt end				▼	223			223	246
3100	90° Elbow, butt end segmented									
3110	203 mm dia., butt end, segmented				Ea.	176			176	194
3120	254 mm dia., butt end, segmented					290			290	320
3130	305 mm dia., butt end, segmented					365			365	405
3140	356 mm dia., butt end, segmented				▼	460			460	505
3200	45° Elbow, butt end segmented									
3210	203 mm dia., butt end, segmented				Ea.	150			150	165
3220	254 mm dia., butt end, segmented					209			209	230
3230	305 mm dia., butt end, segmented					214			214	236
3240	356 mm dia., butt end, segmented				▼	264			264	290
3400	All butt end fittings require one coupling per joint									
3410	Labor for fitting included with couplings.									
3460	Compression coupling, galvanized, neoprene gasket									
3470	54 mm dia.	Q-9	44	.364	Ea.	12.30	13.80		26.10	35
3480	64 mm dia.		44	.364		12.30	13.80		26.10	35
3490	76 mm dia.		38	.421		16.15	15.95		32.10	42.50
3500	89 mm dia.		35	.457		18.05	17.35		35.40	46.50
3510	102 mm dia.		33	.485		19.60	18.40		38	50
3520	127 mm dia.		29	.552		22.50	21		43.50	57
3530	152 mm dia.		26	.615		26	23.50		49.50	64.50
3540	203 mm dia.		22	.727		47.50	27.50		75	95
3550	254 mm dia.		20	.800		69.50	30.50		100	124
3560	305 mm dia.		18	.889		87	33.50		120.50	148
3570	356 mm dia.	▼	16	1	▼	123	38		161	194
3800	Air gate valves, galvanized									
3810	54 mm dia.	Q-9	30	.533	Ea.	93.50	20		113.50	134
3820	64 mm dia.		28	.571		98	21.50		119.50	142
3830	76 mm dia.		26	.615		114	23.50		137.50	161
3840	102 mm dia.		23	.696		149	26.50		175.50	205
3850	152 mm dia.	▼	18	.889	▼	206	33.50		239.50	279

11520	Industrial Equipment	CREW	DAILY OUTPUT	LABOR-HOURS	UNIT	2006 BARE COSTS				TOTAL INCL O&P	
						MAT.	LABOR	EQUIP.	TOTAL		
300	0010	**EQUIPMENT INSTALLATION**									300
	0020	Industrial equipment, minimum	E-2	10.89	5.144	Met. Ton		200	131	331	490
	0200	Maximum	"	1.81	30.864	"		1,200	785	1,985	2,950
850	0010	**VOCATIONAL SHOP EQUIPMENT**									850
	0020	Benches, work, wood, average	2 Carp	5	3.200	Ea.	450	114		564	670
	0100	Metal, average		5	3.200		385	114		499	600
	0400	Combination belt & disc sander, 152 mm		4	4		710	142		852	1,000
	0700	Drill press, floor mounted, 305 mm, 373 W	↓	4	4		350	142		492	605
	0800	Dust collector, not incl. ductwork, 152 mm diameter [6″]	1 Shee	1.10	7.273		2,725	305		3,030	3,475
	1000	Grinders, double wheel, 373 W	2 Carp	5	3.200		192	114		306	390
	1300	Jointer, 102 mm, 559 W		4	4		1,000	142		1,142	1,325
	1600	Kilns, 0.45 m³, to 866 K		4	4		2,150	142		2,292	2,600
	1900	Lathe, woodworking, 254 mm, 373 W		4	4		915	142		1,057	1,225
	2200	Planer, 330 mm x 152 mm		4	4		1,725	142		1,867	2,125
	2500	Potter's wheel, motorized		4	4		980	142		1,122	1,300
	2800	Saws, band, 356 mm, 559 W		4	4		750	142		892	1,050
	3100	Metal cutting band saw, 356 mm		4	4		1,975	142		2,117	2,400
	3400	Radial arm saw, 254 mm, 1491 W		4	4		695	142		837	985
	3700	Scroll saw, 610 mm		4	4		1,450	142		1,592	1,825
	4000	Table saw, 254 mm, 2237 W		4	4		1,700	142		1,842	2,100
	4300	Welder AC arc, 30 A capacity	↓	4	4	↓	2,125	142		2,267	2,575

11620	Laboratory Equipment	CREW	DAILY OUTPUT	LABOR-HOURS	UNIT	2006 BARE COSTS				TOTAL INCL O&P	
						MAT.	LABOR	EQUIP.	TOTAL		
110	0010	**LABORATORY EQUIPMENT**									110
	0600	Fume hood, with countertop & base, not including HVAC									
	0610	Simple, minimum	2 Carp	1.65	9.721	m	1,800	345		2,145	2,525
	0620	Complex, including fixtures		.73	21.872		4,225	780		5,005	5,850
	0630	Special, maximum	↓	.52	30.878	↓	4,425	1,100		5,525	6,550
	0650	Ductwork, minimum	2 Shee	1	16	Hood	905	675		1,580	2,050
	0660	Maximum	"	.50	32	"	4,575	1,350		5,925	7,100
	0670	Service fixtures, average				Ea.	46.50			46.50	51
	0680	For sink assembly with hot and cold water, add	1 Plum	1.40	5.714		620	244		864	1,050
	0700	Glassware washer, undercounter, minimum	L-1	1.80	8.889		6,325	375		6,700	7,525
	0710	Maximum	"	1	16		8,625	680		9,305	10,500
	0750	Glove box, fiberglass, bacteriological					13,900			13,900	15,300
	0760	Controlled atmosphere					9,200			9,200	10,100
	0770	Radioisotope					11,900			11,900	13,100
	0780	Carcinogenic					11,900			11,900	13,100
	1000	Incubators, minimum					3,125			3,125	3,425
	1010	Maximum					17,400			17,400	19,100
	1200	Refrigerator, blood bank, 0.8 m³ emergency signal					6,625			6,625	7,275
	1210	Reach-in, 0.47 m³					6,650			6,650	7,325
	1400	Safety equipment, eye wash, hand held					310			310	340
	1450	Deluge shower					595			595	655
	1600	Sink, one piece plastic, flask wash, hose, free standing	1 Plum	1.60	5		1,575	214		1,789	2,050
	1610	Epoxy resin sink, 635 mm x 406 mm x 254 mm	"	2	4		175	171		346	450
	1700	Thermometer, electric, portable				↓	325			325	360

11600 | Laboratory Equipment

11620 | Laboratory Equipment

		CREW	DAILY OUTPUT	LABOR-HOURS	UNIT	2006 BARE COSTS				TOTAL INCL O&P	
						MAT.	LABOR	EQUIP.	TOTAL		
110	1800	Titration unit, four 2000 ml reservoirs				Ea.	9,025			9,025	9,950	110
	1850	Utensil washer-sanitizer	1 Plum	2	4	▼	9,425	171		9,596	10,700	
	1950	Utility table, acid resistant top with drawers	2 Carp	9.14	1.750	m	425	62		487	560	
	8000	Alternate pricing method: as percent of lab furniture										
	8050	Installation, not incl. plumbing & duct work				% Furn.					22%	
	8100	Plumbing, final connections, simple system									10%	
	8110	Moderately complex system									15%	
	8120	Complex system									20%	
	8150	Electrical, simple system									10%	
	8160	Moderately complex system									20%	
	8170	Complex system				▼					35%	

11700 | Medical Equipment

11710 | Medical Sterilizing Equipment

		CREW	DAILY OUTPUT	LABOR-HOURS	UNIT	2006 BARE COSTS				TOTAL INCL O&P	
						MAT.	LABOR	EQUIP.	TOTAL		
110	0010	**MEDICAL STERILIZING EQUIPMENT**										110
	0700	Distiller, water, steam heated, 189.25 L capacity	1 Plum	1.40	5.714	Ea.	15,900	244		16,144	17,900	
	5600	Sterilizers, floor loading, 711 mm x 1702 mm x 1321 mm, 1 door,					137,000			137,000	150,500	
	5650	Double door, steam					176,000			176,000	193,500	
	5800	General purpose, 508 mm x 508 mm x 965 mm, single door					13,200			13,200	14,500	
	6000	Portable, counter top, steam, minimum					3,275			3,275	3,600	
	6020	Maximum					5,125			5,125	5,650	
	6050	Portable, counter top, gas, 432 mm x 381 mm x 826 mm					33,900			33,900	37,300	
	6150	Manual washer/sterilizer	1 Plum	2	4	▼	46,500	171		46,671	51,500	
	6200	Steam generators, electric 10 kW to 180 kW, freestanding										
	6250	Minimum	1 Elec	3	2.667	Ea.	6,900	112		7,012	7,775	
	6300	Maximum	"	.70	11.429		24,900	480		25,380	28,100	
	8200	Bed pan washer-sanitizer	1 Plum	2	4	▼	6,150	171		6,321	7,025	

11720 | Examination & Treatment Equip.

		CREW	DAILY OUTPUT	LABOR-HOURS	UNIT	2006 BARE COSTS				TOTAL INCL O&P	
						MAT.	LABOR	EQUIP.	TOTAL		
110	0010	**EXAMINATION & TREATMENT EQUIPMENT**										110
	0300	Blood pressure unit, mercurial, wall				Ea.	130			130	143	
	0400	Diagnostic set, wall					560			560	615	
	4400	Scale, physician's, with height rod					254			254	279	
	6500	Surgery table, minor minimum	1 Sswk	.70	11.429		12,000	455		12,455	14,000	
	6520	Maximum	"	.50	16		22,500	640		23,140	26,000	
	6700	Surgical lights, doctor's office, single arm	2 Elec	2	8		1,125	335		1,460	1,725	
	6750	Dual arm	"	1	16	▼	5,025	670		5,695	6,525	

11730 | Patient Care Equipment

		CREW	DAILY OUTPUT	LABOR-HOURS	UNIT	2006 BARE COSTS				TOTAL INCL O&P	
						MAT.	LABOR	EQUIP.	TOTAL		
110	0010	**PATIENT CARE EQUIPMENT**										110
	0750	Exam room furnishings, average/room				Ea.	5,575			5,575	6,125	
	1800	Heat therapy unit, humidified, 660 mm x 1981 mm x 711 mm				"	2,925			2,925	3,225	
	2100	Hubbard tank with accessories, stainless steel,										
	2110	7.89 L/s at 310 kPa water pressure				Ea.	22,700			22,700	25,000	
	2150	For electric overhead hoist, add					2,475			2,475	2,725	
	2900	K-Module for heat therapy, 567 gram capacity, 297 K to 316 K					350			350	385	
	3600	Paraffin bath, 325 K, auto controlled				▼	1,300			1,300	1,425	

11730 | Patient Care Equipment

		CREW	DAILY OUTPUT	LABOR-HOURS	UNIT	2006 BARE COSTS				TOTAL INCL O&P	
						MAT.	LABOR	EQUIP.	TOTAL		
110	3900	Parallel bars for walking training, 3658 mm				Ea.	1,050			1,050	1,175	110
	4600	Station, dietary, medium, with ice					13,800			13,800	15,200	
	4700	Medicine					6,250			6,250	6,875	
	7000	Tables, physical therapy, walk off, electric	2 Carp	3	5.333		2,775	190		2,965	3,350	
	7150	Standard, vinyl top with base cabinets, minimum		3	5.333		1,125	190		1,315	1,525	
	7200	Maximum	↓	2	8		3,325	284		3,609	4,100	
	8400	Whirlpool bath, mobile, sst, 457 mm x 610 mm x 1524 mm					3,950			3,950	4,350	
	8450	Fixed, incl. mixing valves	1 Plum	2	4	↓	7,950	171		8,121	9,000	

11740 | Dental Equipment

		CREW	DAILY OUTPUT	LABOR-HOURS	UNIT	MAT.	LABOR	EQUIP.	TOTAL	TOTAL INCL O&P	
200	0010	**DENTAL EQUIPMENT**										200
	0020	Central suction system, minimum	1 Plum	1.20	6.667	Ea.	3,025	285		3,310	3,750	
	0100	Maximum	"	.90	8.889		23,400	380		23,780	26,300	
	0300	Air compressor, minimum	1 Skwk	.80	10		3,325	365		3,690	4,225	
	0400	Maximum		.50	16		23,100	585		23,685	26,300	
	0600	Chair, electric or hydraulic, minimum		.50	16		3,800	585		4,385	5,075	
	0700	Maximum	↓	.25	32		8,050	1,175		9,225	10,700	
	0800	Doctor's/assistant's stool, minimum					430			430	475	
	0850	Maximum					845			845	930	
	1000	Drill console with accessories, minimum	1 Skwk	1.60	5		3,425	183		3,608	4,050	
	1100	Maximum		1.60	5		9,800	183		9,983	11,100	
	2000	Light, ceiling mounted, minimum		8	1		1,975	36.50		2,011.50	2,225	
	2100	Maximum	↓	8	1		3,975	36.50		4,011.50	4,425	
	2200	Unit light, minimum	2 Skwk	5.33	3.002		1,475	110		1,585	1,800	
	2210	Maximum		5.33	3.002		4,825	110		4,935	5,500	
	2220	Track light, minimum		3.20	5		2,625	183		2,808	3,150	
	2230	Maximum	↓	3.20	5		5,750	183		5,933	6,600	
	2300	Sterilizers, steam portable, minimum					3,350			3,350	3,675	
	2350	Maximum					11,500			11,500	12,700	
	2600	Steam, institutional					14,400			14,400	15,800	
	2650	Dry heat, electric, portable, 3 trays					1,125			1,125	1,225	
	2700	Ultra-sonic cleaner, portable, minimum					905			905	995	
	2750	Maximum (institutional)					2,025			2,025	2,225	
	3000	X-ray unit, wall, minimum	1 Skwk	4	2		4,575	73		4,648	5,150	
	3010	Maximum		4	2		8,625	73		8,698	9,625	
	3100	Panoramic unit	↓	.60	13.333		20,700	485		21,185	23,600	
	3105	Deluxe, minimum	2 Skwk	1.60	10		29,900	365		30,265	33,500	
	3110	Maximum	"	1.60	10		69,000	365		69,365	76,500	
	3500	Developers, X-ray, average	1 Plum	5.33	1.501		4,025	64		4,089	4,525	
	3600	Maximum	"	5.33	1.501	↓	7,475	64		7,539	8,325	

11760 | Operating Room Equipment

		CREW	DAILY OUTPUT	LABOR-HOURS	UNIT	MAT.	LABOR	EQUIP.	TOTAL	TOTAL INCL O&P	
110	0010	**OPERATING ROOM EQUIPMENT**										110
	5000	Scrub, surgical, stainless steel, single station, minimum	1 Plum	3	2.667	Ea.	2,050	114		2,164	2,425	
	5100	Maximum					7,225			7,225	7,950	
	6550	Major surgery table, minimum	1 Sswk	.50	16		27,100	640		27,740	31,100	
	6570	Maximum	"	.50	16		89,500	640		90,140	99,500	
	6800	Surgical lights, major operating room, dual head, minimum	2 Elec	1	16		15,100	670		15,770	17,600	
	6850	Maximum	"	1	16	↓	25,100	670		25,770	28,600	

11770 | Radiology Equipment

		CREW	DAILY OUTPUT	LABOR-HOURS	UNIT	MAT.	LABOR	EQUIP.	TOTAL	TOTAL INCL O&P	
110	0010	**RADIOLOGY EQUIPMENT**										110
	8700	X-ray, mobile, minimum				Ea.	11,600			11,600	12,700	
	8750	Maximum					65,000			65,000	71,500	
	8900	Stationary, minimum				↓	36,100			36,100	39,700	

EQUIPMENT 11

		Radiology Equipment	CREW	DAILY OUTPUT	LABOR-HOURS	UNIT	2006 BARE COSTS				TOTAL INCL O&P	
		11770					MAT.	LABOR	EQUIP.	TOTAL		
110	8950	Maximum				Ea.	188,000			188,000	206,500	110
	9150	Developing processors, minimum					10,100			10,100	11,100	
	9200	Maximum				▼	26,400			26,400	29,000	

		**11781	Mortuary Equipment**	CREW	DAILY OUTPUT	LABOR-HOURS	UNIT	MAT.	LABOR	EQUIP.	TOTAL	TOTAL INCL O&P	
110	0010	**MORTUARY EQUIPMENT**										110
	0015	Autopsy table, standard	1 Plum	1	8	Ea.	7,650	340		7,990	8,950	
	0020	Deluxe	"	.60	13.333		13,800	570		14,370	16,100	
	3200	Mortuary refrigerator, end operated, 2 capacity					11,900			11,900	13,100	
	3300	6 capacity				▼	21,900			21,900	24,100	

For information about Means Estimating Seminars, see yellow pages 12 and 13 in back of book

11

EQUIPMENT

Division 12
Furnishings

Estimating Tips

General

- The items in this division are usually priced per m² or each. Most of these items are purchased by the owner and placed by the supplier. Do not assume the items in Division 12 will be purchased and installed by the supplier. Check the specifications for responsibilities and include receiving, storage, installation, and mechanical and electrical hook-ups in the appropriate divisions.

- Some items in this division require some type of support system that is not usually furnished with the item. Examples of these systems include blocking for the attachment casework and heavy drapery rods. The required blocking must be added to the estimate in the appropriate division.

Reference Numbers

Reference numbers are shown in bold squares at the beginning of some major classifications. These numbers refer to related items in the Reference Section. The reference information may be an estimating procedure, an alternate pricing method, or technical information.

Note: Not all subdivisions listed here necessarily appear in this publication.

12310	Metal Casework	CREW	DAILY OUTPUT	LABOR-HOURS	UNIT	2006 BARE COSTS				TOTAL INCL O&P	
						MAT.	LABOR	EQUIP.	TOTAL		
100	**0010**	**KEY CABINETS**									**100**
	0020	Wall mounted, 60 key capacity	1 Carp	20	.400	Ea.	74	14.20		88.20	104
	0200	Drawer type, 600 key capacity	1 Clab	15	.533		725	14.60		739.60	825
	0300	2,400 key capacity		20	.400		3,600	10.95		3,610.95	3,975
	0400	Tray type, 20 key capacity		50	.160		38.50	4.38		42.88	49.50
	0500	50 key capacity		40	.200		70	5.50		75.50	85.50
200	**0010**	**DISPLAY CASES** Free standing, all glass									**200**
	0020	Aluminum frame, 1067 mm high x 914 mm x 305 mm deep	2 Carp	8	2	Ea.	1,075	71		1,146	1,275
	0100	1778 mm high x 1219 mm x 457 mm deep	"	6	2.667		1,650	95		1,745	1,950
	0500	For wood bases, add					9%				
	0600	For hardwood frames, deduct					8%				
	0700	For bronze, baked enamel finish, add					10%				
	2000	Wall mounted, glass front, aluminum frame									
	2010	Non-illuminated, one section 914 mm x 1219 mm x 406 mm	2 Carp	5	3.200	Ea.	1,150	114		1,264	1,425
	2100	1524 mm x 1219 mm x 406 mm		5	3.200		1,250	114		1,364	1,550
	2200	1829 mm x 1219 mm x 406 mm		4	4		1,425	142		1,567	1,800
	2500	Two sections, 2438 mm x 1219 mm x 406 mm		2	8		1,725	284		2,009	2,350
	2600	3048 mm x 1219 mm x 406 mm		2	8		1,900	284		2,184	2,525
	3000	Three sections, 4877 mm x 1219 mm x 406 mm		1.50	10.667		3,275	380		3,655	4,200
	3500	For fluorescent lights, add				Section	260			260	286
	4000	Table exhibit cases, 610 mm W, 914 mm H, 1219 mm L, flat top	2 Carp	5	3.200	Ea.	740	114		854	990
	4100	914 mm wide, 914 mm high, 1219 mm long, sloping top	"	3	5.333	"	1,125	190		1,315	1,550
560	**0010**	**IRONING CENTER**									**560**
	0020	Including cabinet, board & light, minimum	1 Carp	2	4	Ea.	291	142		433	540
	0100	Maximum, see also Div. 11119-450	"	1.50	5.333	"	515	190		705	860
750	**0010**	**CASEWORK**									**750**
	0500	Hospital, base cabinets, laminated plastic	2 Carp	3.05	5.249	m	725	187		912	1,100
	1000	Stainless steel	"	3.05	5.249		995	187		1,182	1,400
	1200	For all drawers, add					37			37	40.50
	1300	Cabinet base trim, 102 mm high, enameled steel	2 Carp	60.96	.262		116	9.35		125.35	142
	1400	Stainless steel		60.96	.262		182	9.35		191.35	215
	1450	Counter top, laminated plastic, no backsplash		12.19	1.312		125	46.50		171.50	211
	1650	With backsplash		12.19	1.312		156	46.50		202.50	244
	1800	For sink cutout, add		12.20	1.311	Ea.		46.50		46.50	72.50
	1900	Stainless steel counter top		12.19	1.312	m	345	46.50		391.50	455
	2000	For drop-in stainless 1092 mm x 533 mm sink, add				Ea.	590			590	650
	2050	For solid surface countertops see Div. 06620-810									
	2100	Nurses station, door type, laminated plastic	2 Carp	3.05	5.249	m	840	187		1,027	1,225
	2200	Enameled steel		3.05	5.249		765	187		952	1,125
	2300	Stainless steel		3.05	5.249		1,075	187		1,262	1,475
	2400	For drawer type, add					340			340	375
	2500	Wall cabinets, laminated plastic	2 Carp	4.57	3.500		545	124		669	795
	2600	Enameled steel		4.57	3.500		635	124		759	895
	2700	Stainless steel		4.57	3.500		1,025	124		1,149	1,325
	2800	For glass doors, add					86.50			86.50	95
	3500	Kitchen, base cabinets, metal, minimum	2 Carp	9.14	1.750		186	62		248	300
	3600	Maximum		7.62	2.100		475	74.50		549.50	635
	3700	Wall cabinets, metal, minimum		9.14	1.750		186	62		248	300
	3800	Maximum		7.62	2.100		430	74.50		504.50	585
	5000	School, 610 mm deep, metal, 2134 mm high units		4.57	3.500		930	124		1,054	1,225
	5150	Counter height units		6.10	2.625		725	93.50		818.50	945
	5450	Wood, custom fabricated, 813 mm high counter		6.10	2.625		605	93.50		698.50	810
	5600	Add for counter top		17.07	.937		67.50	33.50		101	126
	5800	2134 mm high wall units		4.57	3.500		630	124		754	890
	6000	Laminated plastic finish is same price as wood									

12 FURNISHINGS

12300 | Manufactured Casework

12350 | Specialty Casework

		CREW	DAILY OUTPUT	LABOR-HOURS	UNIT	2006 BARE COSTS MAT.	LABOR	EQUIP.	TOTAL	TOTAL INCL O&P	
500	**0010**	**LABORATORY CASEWORK**										**500**
	0020	Cabinets, base, door units, metal	2 Carp	5.49	2.916	m	545	104		649	760	
	0300	Drawer units		5.49	2.916		1,175	104		1,279	1,450	
	0700	Tall storage cabinets, open, 2100 mm high		6.10	2.625		1,150	93.50		1,243.50	1,400	
	0900	With glazed doors		6.10	2.625		1,425	93.50		1,518.50	1,725	
	1300	Wall cabinets, metal, 318 mm deep, open		6.10	2.625		360	93.50		453.50	540	
	1500	With doors		6.10	2.625	▼	745	93.50		838.50	965	
	1550	Counter tops, not incl. base cabinets, acidproof, minimum		7.62	2.100	m²	286	74.50		360.50	430	
	1600	Maximum		6.50	2.460		385	87.50		472.50	560	
	1650	Stainless steel	▼	7.62	2.100	▼	840	74.50		914.50	1,025	
	6300	Rule of thumb: lab furniture including installation & connection										
	6320	High school				m²					345	
	6340	College									510	
	6360	Clinical, health care									435	
	6380	Industrial				▼					705	

12400 | Furnishings & Accessories

12460 | Furnishing Accessories

		CREW	DAILY OUTPUT	LABOR-HOURS	UNIT	2006 BARE COSTS MAT.	LABOR	EQUIP.	TOTAL	TOTAL INCL O&P	
900	**0010**	**ASH/TRASH RECEIVERS**										**900**
	1000	Ash urn, cylindrical metal										
	1020	203 mm diameter, 508 mm high [8",20"]	1 Clab	60	.133	Ea.	116	3.65		119.65	134	
	1060	254 mm diameter, 660 mm high [10",26"]	"	60	.133	"	163	3.65		166.65	185	
	2000	Combination ash/trash urn, metal										
	2020	203 mm diameter, 508 mm high [8",20"]	1 Clab	60	.133	Ea.	108	3.65		111.65	125	
	2050	254 mm diameter, 660 mm high [10",26"]	"	60	.133	"	182	3.65		185.65	206	
	4000	Trash receptacle, metal										
	4020	203 mm diameter, 381 mm high [8",15"]	1 Clab	60	.133	Ea.	62	3.65		65.65	73.50	
	4040	254 mm diameter, 457 mm high [10",18"]		60	.133		105	3.65		108.65	121	
	5040	406 mm x 203 mm x 356 mm high	▼	60	.133	▼	16	3.65		19.65	23.50	
	5500	Trash receptacle, plastic, with lid										
	5520	132 L	1 Clab	60	.133	Ea.	185	3.65		188.65	209	
	5540	170 L		60	.133		220	3.65		223.65	248	
	5550	Plastic recycling barrel, w/lid & wheels, 121L		60	.133		105	3.65		108.65	122	
	5560	246 L		60	.133		125	3.65		128.65	144	
	5570	360 L	▼	60	.133	▼	310	3.65		313.65	345	

12483 | Floor Mats & Frames

		CREW	DAILY OUTPUT	LABOR-HOURS	UNIT	2006 BARE COSTS MAT.	LABOR	EQUIP.	TOTAL	TOTAL INCL O&P	
200	**0010**	**FLOOR MATS**										**200**
	0020	Recessed, in-laid black rubber, 10 mm thick, solid	1 Clab	14.40	.556	m²	223	15.20		238.20	270	
	0050	Perforated		14.40	.556		188	15.20		203.20	231	
	0100	13 mm thick, solid		14.40	.556		211	15.20		226.20	256	
	0150	Perforated		14.40	.556		244	15.20		259.20	293	
	0200	In colors, 10 mm thick, solid		14.40	.556		215	15.20		230.20	261	
	0250	Perforated		14.40	.556		215	15.20		230.20	261	
	0300	13 mm thick, solid		14.40	.556		320	15.20		335.20	375	
	0350	Perforated		14.40	.556		320	15.20		335.20	375	
	0500	Link mats, including nosings, aluminum, 10 mm thick	▼	14.40	.556	▼	199	15.20		214.20	243	

FURNISHINGS 12

12483 | Floor Mats & Frames

			CREW	DAILY OUTPUT	LABOR-HOURS	UNIT	MAT.	LABOR	EQUIP.	TOTAL	TOTAL INCL O&P	
200	0550	Black rubber with galvanized tie rods	1 Clab	14.40	.556	m²	155	15.20		170.20	195	200
	0600	Steel, galvanized, 10 mm thick		14.40	.556		86	15.20		101.20	118	
	0650	Vinyl, in colors	↓	14.40	.556	↓	200	15.20		215.20	243	
	0750	Add for nosings, rubber				m	11.40			11.40	12.55	
	0850	Recess frames for above mats, aluminum	1 Carp	30.48	.262	↓	11.40	9.35		20.75	27	
	0870	Bronze	"	30.48	.262	↓	16.50	9.35		25.85	32.50	
	0900	Skate lock tile, 610 mm x 610 mm x 13 mm thick, rubber, black	1 Clab	11.61	.689	m²	157	18.90		175.90	203	
	0950	Color		11.61	.689	"	200	18.90		218.90	250	
	1000	305 mm x 610 mm border, black		22.86	.350	m	53.50	9.60		63.10	74	
	1100	Color		22.86	.350	"	72	9.60		81.60	94	
	1150	305 mm x 305 mm outside corner, black		9.29	.861	m²	149	23.50		172.50	200	
	1200	Color		9.29	.861		193	23.50		216.50	249	
	1500	Duckboard, aluminum slats		14.40	.556		163	15.20		178.20	204	
	1700	Hardwood strips on rubber base, to 1372 mm wide		14.40	.556		137	15.20		152.20	175	
	1800	Assembled with brass rods and vinyl spacers, to 1219 mm wide		14.40	.556		182	15.20		197.20	224	
	1850	Tire fabric, 19 mm thick		14.40	.556		119	15.20		134.20	155	
	1900	Vinyl, 914 mm wide, in colors, hollow top & bottoms		14.40	.556		93	15.20		108.20	126	
	1950	Solid top & bottom members	↓	14.40	.556	↓	93	15.20		108.20	126	

12492 | Blinds and Shades

			CREW	DAILY OUTPUT	LABOR-HOURS	UNIT	MAT.	LABOR	EQUIP.	TOTAL	TOTAL INCL O&P	
100	0010	**BLINDS, INTERIOR**										100
	0020	Horizontal, 25 mm aluminum slats, solid color, stock	1 Carp	54.81	.146	m²	31	5.20		36.20	42	
	0090	Custom, minimum		54.81	.146		28	5.20		33.20	38.50	
	0100	Maximum		40.88	.196		70	6.95		76.95	88	
	0250	51 mm aluminum slats, custom, minimum		54.81	.146		41	5.20		46.20	53	
	0350	Maximum		40.88	.196		140	6.95		146.95	165	
	0450	Stock, minimum		54.81	.146		45.50	5.20		50.70	58	
	0500	Maximum		40.88	.196		73.50	6.95		80.45	92	
	0600	51 mm steel slats, stock, minimum		54.81	.146		17.20	5.20		22.40	27	
	0630	Maximum		40.88	.196		48	6.95		54.95	64	
	0750	Custom, minimum		54.81	.146		15.50	5.20		20.70	25	
	0850	Maximum		37.16	.215		77	7.65		84.65	96.50	
	1500	Vertical, 76 mm to 127 mm PVC or cloth strips, minimum		42.73	.187		84.50	6.65		91.15	103	
	1600	Maximum		37.16	.215		103	7.65		110.65	125	
	1800	102 mm aluminum slats, minimum		42.73	.187		33.50	6.65		40.15	47.50	
	1900	Maximum		37.16	.215		80.50	7.65		88.15	101	
	1950	Mylar mirror-finish strips, to 203 mm wide, minimum		42.73	.187	↓	135	6.65		141.65	158	
	1970	Maximum		37.16	.215	↓	203	7.65		210.65	235	
	3000	Wood folding panels with movable louvers, 178 mm x 508 mm each		17	.471	Pr.	42	16.75		58.75	72.50	
	3300	203 mm x 711 mm each		17	.471		61	16.75		77.75	93	
	3450	229 mm x 914 mm each		17	.471		72.50	16.75		89.25	106	
	3600	254 mm x 1016 mm each		17	.471		82	16.75		98.75	116	
	4000	Fixed louver type, stock units, 203 mm x 508 mm each		17	.471		63	16.75		79.75	95.50	
	4150	254 mm x 711 mm each		17	.471		86.50	16.75		103.25	121	
	4300	305 mm x 914 mm each		17	.471		110	16.75		126.75	147	
	4450	457 mm x 1016 mm each		17	.471		131	16.75		147.75	170	
	5000	Insert panel type, stock, 178 mm x 508 mm each		17	.471		15.60	16.75		32.35	43	
	5150	203 mm x 711 mm each		17	.471		28.50	16.75		45.25	57.50	
	5300	229 mm x 914 mm each		17	.471		36.50	16.75		53.25	66	
	5450	254 mm x 1016 mm each		17	.471		39	16.75		55.75	69	
	5600	Raised panel type, stock, 254 mm x 610 mm each		17	.471		109	16.75		125.75	146	
	5650	305 mm x 660 mm each		17	.471		126	16.75		142.75	165	
	5700	356 mm x 762 mm each		17	.471		143	16.75		159.75	183	
	5750	406 mm x 914 mm each	↓	17	.471		160	16.75		176.75	202	
	6000	For custom built pine, add				↓	22%					
	6500	For custom built hardwood blinds, add				↓	42%					

12492 | Blinds and Shades

		CREW	DAILY OUTPUT	LABOR-HOURS	UNIT	MAT.	LABOR	EQUIP.	TOTAL	TOTAL INCL O&P	
600	**0010**	**SHADES**										**600**
	0020	Basswood, roll-up, stain finish, 10 mm slats	1 Carp	27.87	.287	m²	106	10.20		116.20	132	
	0200	22 mm slats		27.87	.287		100	10.20		110.20	126	
	0300	Vertical side slide, stain finish, 10 mm slats		27.87	.287		160	10.20		170.20	192	
	0400	22 mm slats	▼	27.87	.287		160	10.20		170.20	192	
	0500	For fire retardant finishes, add					16%					
	0600	For "B" rated finishes, add					20%					
	0900	Mylar, single layer, non-heat reflective	1 Carp	63.64	.126		62	4.47		66.47	75.50	
	1000	Double layered, heat reflective		63.64	.126		85.50	4.47		89.97	101	
	1100	Triple layered, heat reflective	▼	63.64	.126	▼	99.50	4.47		103.97	116	
	1200	For metal roller instead of wood, add/				Shade	3.42			3.42	3.76	
	1300	Vinyl coated cotton, standard	1 Carp	63.64	.126	m²	21	4.47		25.47	30	
	1400	Lightproof decorator shades		63.64	.126		18.85	4.47		23.32	28	
	1500	Vinyl, lightweight, 4 gauge		63.64	.126		4.63	4.47		9.10	12	
	1600	Heavyweight, 6 gauge		63.64	.126		14.55	4.47		19.02	23	
	1700	Vinyl laminated fiberglass, 6 ga., translucent		63.64	.126		21	4.47		25.47	30	
	1800	Lightproof		63.64	.126		33.50	4.47		37.97	43.50	
	3000	Woven aluminum, 10 mm thick, lightproof and fireproof	▼	32.52	.246	▼	47	8.75		55.75	65	

12493 | Curtains and Drapes

		CREW	DAILY OUTPUT	LABOR-HOURS	UNIT	MAT.	LABOR	EQUIP.	TOTAL	TOTAL INCL O&P	
200	**0010**	**DRAPERY HARDWARE**										**200**
	0030	Standard traverse, per m, minimum	1 Carp	17.98	.445	m	8.35	15.80		24.15	33.50	
	0100	Maximum		15.54	.515	"	34	18.30		52.30	66	
	4000	Traverse rods, adjustable, 700 mm to 1200 mm		22	.364	Ea.	21	12.95		33.95	43	
	4020	1200 mm to 2100 mm		20	.400		24.50	14.20		38.70	49	
	4040	1676 mm to 3048 mm		18	.444		28.50	15.80		44.30	56	
	4060	2100 mm to 3900 mm		16	.500		32	17.80		49.80	62.50	
	4080	2500 mm to 4500 mm		14	.571		34.50	20.50		55	69.50	
	4090	3900 mm to 5700 mm		13	.615		46	22		68	84.50	
	4100	5791 mm to 7925 mm		13	.615		55.50	22		77.50	95	
	4200	Double rods, adjustable, 750 mm to 1200 mm		9	.889		44	31.50		75.50	97.50	
	4220	1200 mm to 2150 mm		9	.889		58.50	31.50		90	113	
	4240	2150 mm to 3750 mm		8	1		67.50	35.50		103	130	
	4260	2500 mm to 4500 mm		7	1.143		72.50	40.50		113	143	
	4300	Tray and curtain rod, adjustable, 750 mm to 1200 mm		9	.889		26	31.50		57.50	77.50	
	4320	1200 mm to 2150 mm		9	.889		36.50	31.50		68	89	
	4340	2150 mm to 3750 mm		8	1		48	35.50		83.50	109	
	4360	2500 mm to 4500 mm	▼	7	1.143		51	40.50		91.50	120	
	4600	Valance, pinch pleated fabric, 305 mm D, up to 1.4 m long, min.					34			34	37.50	
	4610	Maximum					85.50			85.50	94	
	4620	Up to 1956 mm long, minimum					52.50			52.50	58	
	4630	Maximum					138			138	152	
	5000	Stationary rods, first 610 mm				▼	9.25			9.25	10.20	
	5020	Each additional m, add				m	11.15			11.15	12.25	
400	**0010**	**BLAST CURTAINS,** per m horizontal opening width, off-white or gray fabric										**400**
	0100	Blast curtains, drapery system, complete, including hardware, minimum				m					410	
	0120	Average									445	
	0140	Maximum				▼					475	

FURNISHINGS 12

12510	Office Furniture		CREW	DAILY OUTPUT	LABOR-HOURS	UNIT	2006 BARE COSTS				TOTAL INCL O&P	
							MAT.	LABOR	EQUIP.	TOTAL		
580	0010	**MULTI-MEDIA BOARDS** See Div. 10110-240										580
600	0010	**OFFICE CASE GOODS**										600
	0020	Desks, 737 mm H, dbl. pedestal, 762 mm x 1524 mm, metal, min.				Ea.	330			330	360	
	0030	Maximum					1,050			1,050	1,150	
	0160	Wood, minimum					350			350	385	
	0180	Maximum					2,050			2,050	2,250	
	0600	Desks, single pedestal, 762 mm x 1524 mm, metal, minimum					280			280	305	
	0620	Maximum					855			855	940	
	0640	Wood, minimum					455			455	500	
	0650	Maximum					640			640	705	
	0670	Executive return, 600 x 1050 mm, with box, file, wood, minimum					355			355	390	
	0680	Maximum					640			640	705	
	0720	Desks, secretarial, 750 mm x 1500 mm, metal, minimum					219			219	241	
	0730	Maximum					490			490	540	
	0740	Return, 500 mm x 1050 mm, minimum					282			282	310	
	0750	Maximum					390			390	430	
	0800	Wood, minimum					360			360	395	
	0810	Maximum					2,075			2,075	2,275	
	0820	Return, 500 mm x 1050 mm, minimum					350			350	385	
	0830	Maximum					1,125			1,125	1,225	
	0900	Desktop organizer, 1800 mm x 350 mm x 900 mm H, wood, min.					152			152	167	
	0920	Maximum					294			294	325	
	0940	1475mm x 300 mm x 575 mm high, steel, minimum					170			170	187	
	0960	Maximum					228			228	251	
	2000	Chairs, office type, executive, minimum					226			226	248	
	2150	Maximum					1,800			1,800	1,975	
	2200	Management, minimum					171			171	188	
	2250	Maximum					1,700			1,700	1,850	
	2280	Task, minimum					102			102	112	
	2290	Maximum					635			635	700	
	2300	Arm kit, minimum					50.50			50.50	55.50	
	2320	Maximum				▼	78			78	85.50	
	6000	Table, conference										
	6050	Boat, 2438 mm x 1067 mm, minimum				Ea.	610			610	670	
	6150	Maximum					4,125			4,125	4,525	
	6720	Rectangle, 2438 mm x 1067 mm, minimum					720			720	790	
	6740	Maximum				▼	3,400			3,400	3,725	
675	0010	**POSTS**										675
	0020	Portable for pedestrian traffic control, standard, minimum				Ea.	88			88	97	
	0100	Maximum					153			153	168	
	0300	Deluxe posts, minimum					176			176	193	
	0400	Maximum				▼	285			285	315	
	0600	Ropes for above posts, plastic covered, 38 mm diameter [1-1/2"]				m	30.50			30.50	34	
	0700	Chain core				"	31.50			31.50	34.50	
	1500	Portable security or safety barrier, black with 2150 mm yellow strap				Ea.	200			200	220	
	1510	3650 mm yellow strap					233			233	256	
	1550	Sign holder, standard design				▼	50			50	55	

12520	Seating		CREW	DAILY OUTPUT	LABOR-HOURS	UNIT	MAT.	LABOR	EQUIP.	TOTAL	TOTAL INCL O&P	
550	0010	**SEATING**										550
	0100	Bleachers, See Div. 02870 & 11486										
	0101	Benches, See Div. 02870, 10505, 11486										
	0500	Classroom, movable chair & desk type, minimum				Set					73	
	0600	Maximum				"					134.50	
	1000	Lecture hall, pedestal type, minimum	2 Carp	22	.727	Ea.	127	26		153	181	

Important: See the Reference Section for supporting data - Crews, Rental Equipment, City Cost Indexes and Reference Data

12 FURNISHINGS

12500 | Furniture

12520 | Seating

			CREW	DAILY OUTPUT	LABOR-HOURS	UNIT	2006 BARE COSTS MAT.	LABOR	EQUIP.	TOTAL	TOTAL INCL O&P	
550	1200	Maximum	2 Carp	14.50	1.103	Ea.	440	39		479	545	550
	2000	Auditorium chair, all veneer construction		22	.727		140	26		166	195	
	2200	Veneer back, padded seat		22	.727		176	26		202	235	
	2350	Fully upholstered, spring seat		22	.727		178	26		204	236	
	2450	For tablet arms, add					46			46	51	
	2500	For fire retardancy, CATB-133, add					19			19	21	

12540 | Hospitality Furniture

			CREW	DAILY OUTPUT	LABOR-HOURS	UNIT	MAT.	LABOR	EQUIP.	TOTAL	TOTAL INCL O&P	
100	0010	TABLES, FOLDING Laminated plastic tops										100
	1000	Tubular steel legs with glides										
	1020	457 mm x 1524 mm, minimum				Ea.	189			189	207	
	1040	Maximum					800			800	880	
	1840	914 mm x 2438 mm, minimum					213			213	234	
	1860	Maximum					1,575			1,575	1,725	
	2000	Round, wood stained, plywood top, 1524 mm dia., minimum [60″]					315			315	345	
	2020	Maximum					1,950			1,950	2,150	
500	0010	FURNITURE, HOTEL										500
	0020	Standard quality set, minimum				Room	1,500			1,500	1,650	
	0200	Maximum				″	7,900			7,900	8,675	
700	0010	FURNITURE, RESTAURANT										700
	0020	Bars, built-in, front bar	1 Carp	1.52	5.249	m	720	187		907	1,075	
	0200	Back bar	″	1.52	5.249	″	525	187		712	865	
	0300	Booth seating see Div. 12640-150										
	2000	Chair, bentwood side chair, metal, minimum				Ea.	51			51	56	
	2020	Maximum					82			82	90	
	2600	Upholstered seat & back, arms, minimum					164			164	180	
	2620	Maximum					350			350	385	

12560 | Institutional Furniture

			CREW	DAILY OUTPUT	LABOR-HOURS	UNIT	MAT.	LABOR	EQUIP.	TOTAL	TOTAL INCL O&P	
100	0010	BANK FURNITURE See Div. 12510-600										100
150	0010	CHURCH FURNITURE See Div. 11041-250										150
200	0010	FURNITURE, SCHOOL										200
	1000	Chair, molded plastic,										
	1100	Integral tablet arm, minimum				Ea.	71			71	78.50	
	1150	Maximum					235			235	258	
	2000	Desk, single pedestal, top book compartment, minimum					53.50			53.50	59	
	2020	Maximum					84			84	92.50	
	2200	Flip top, minimum					86			86	94.50	
	2220	Maximum					102			102	112	
300	0010	FURNITURE, DORMITORY										300
	1000	Chest, four drawer, minimum				Ea.	400			400	440	
	1020	Maximum				″	525			525	575	
	1050	Built-in, minimum	2 Carp	3.96	4.038	m	395	144		539	660	
	1150	Maximum		3.05	5.249		730	187		917	1,100	
	1200	Desk top, built-in, laminated plastic, 610 mm deep, minimum		15.24	1.050		83	37.50		120.50	149	
	1300	Maximum		12.19	1.312		249	46.50		295.50	345	
	1450	762 mm deep, minimum		15.24	1.050		106	37.50		143.50	175	
	1550	Maximum		12.19	1.312		465	46.50		511.50	585	
	1750	Dressing unit, built-in, minimum		3.66	4.374		590	156		746	890	
	1850	Maximum		2.44	6.562		1,775	233		2,008	2,325	
	8000	Rule of thumb: total cost for furniture, minimum				Student					2,125	
	8050	Maximum				″					4,050	

12500 | Furniture

12560 | Institutional Furniture

		CREW	DAILY OUTPUT	LABOR-HOURS	UNIT	MAT.	LABOR	EQUIP.	TOTAL	TOTAL INCL O&P	
400	**0010**	**FURNITURE, HOSPITAL**										**400**
	0020	Beds, manual, minimum				Ea.	775			775	855	
	0100	Maximum					2,375			2,375	2,600	
	0300	Manual and electric beds, minimum					990			990	1,100	
	0400	Maximum					2,650			2,650	2,900	
	0600	All electric hospital beds, minimum					1,250			1,250	1,375	
	0700	Maximum					3,825			3,825	4,225	
	0900	Manual, nursing home beds, minimum					755			755	830	
	1000	Maximum					1,575			1,575	1,725	
	1020	Overbed table, laminated top, minimum					320			320	355	
	1040	Maximum				▼	800			800	880	
	1100	Patient wall systems, not incl. plumbing, minimum				Room	960			960	1,050	
	1200	Maximum				"	1,775			1,775	1,950	
	2000	Geriatric chairs, minimum				Ea.	410			410	450	
	2020	Maximum				"	915			915	1,000	
700	**0010**	**FURNITURE, LIBRARY**										**700**
	0100	Attendant desk, 914 mm x 1575 mm x 737 mm high	1 Carp	16	.500	Ea.	2,400	17.80		2,417.80	2,650	
	0200	Book display,"A" frame,both sides,1050 mm x 1050 mm x 1.5 m high		16	.500		2,075	17.80		2,092.80	2,325	
	0220	Table with bulletin board, 1050 mm x 600 mm x 1225 mm high		16	.500		1,475	17.80		1,492.80	1,650	
	0800	Card catalogue, 30 tray unit		16	.500		2,575	17.80		2,592.80	2,875	
	0840	60 tray unit	▼	16	.500		4,525	17.80		4,542.80	5,025	
	0880	72 tray unit	2 Carp	16	1		5,075	35.50		5,110.50	5,625	
	1000	Carrels, single face, initial unit	1 Carp	16	.500		690	17.80		707.80	790	
	1500	Double face, initial unit	2 Carp	16	1		1,050	35.50		1,085.50	1,200	
	4000	Dictionary stand, stationary	1 Carp	16	.500		715	17.80		732.80	815	
	4020	Revolving		16	.500		286	17.80		303.80	345	
	4200	Exhibit case, table style, 1524 mm x 711 mm x 914 mm		11	.727		4,075	26		4,101	4,525	
	7000	Tables, card catalog reference, 610 mm x 1524 mm x 1067 mm	▼	16	.500	▼	920	17.80		937.80	1,050	

12600 | Multiple Seating

12640 | Booths & Tables

		CREW	DAILY OUTPUT	LABOR-HOURS	UNIT	MAT.	LABOR	EQUIP.	TOTAL	TOTAL INCL O&P	
150	**0010**	**BOOTHS**										**150**
	1000	Banquet, upholstered seat and back, custom										
	1500	Straight, minimum	2 Carp	12.19	1.312	m	450	46.50		496.50	570	
	1520	Maximum		10.97	1.458		880	52		932	1,050	
	1600	"L" or "U" shape, minimum		10.67	1.500		460	53.50		513.50	590	
	1620	Maximum	▼	9.14	1.750	▼	820	62		882	995	
	1800	Upholstered outside finished backs for										
	1810	single booths and custom banquets										
	1820	Minimum	2 Carp	13.41	1.193	m	52	42.50		94.50	124	
	1840	Maximum	"	12.19	1.312	"	156	46.50		202.50	245	
	3000	Fixed seating, one piece plastic chair and										
	3010	plastic laminate table top										
	3100	Two seat, 610 mm x 610 mm table, minimum	F-7	30	1.067	Ea.	555	33.50		588.50	665	
	3120	Maximum		26	1.231		790	38.50		828.50	930	
	3200	Four seat, 610 mm x 1220 mm table, minimum		28	1.143		550	36		586	660	
	3220	Maximum	▼	24	1.333	▼	940	42		982	1,100	

Important: See the Reference Section for supporting data - Crews, Rental Equipment, City Cost Indexes and Reference Data

12640	Booths & Tables	CREW	DAILY OUTPUT	LABOR-HOURS	UNIT	2006 BARE COSTS				TOTAL INCL O&P	
						MAT.	LABOR	EQUIP.	TOTAL		
150 5000	Mount in floor, wood fiber core with										**150**
5010	plastic laminate face, single booth										
5050	610 mm wide	F-7	30	1.067	Ea.	213	33.50		246.50	287	
5100	1220 mm wide	"	28	1.143	"	270	36		306	355	

12830	Interior Planters	CREW	DAILY OUTPUT	LABOR-HOURS	UNIT	2006 BARE COSTS				TOTAL INCL O&P	
						MAT.	LABOR	EQUIP.	TOTAL		
600 0010	**PLANTERS**										**600**
1000	Fiberglass, hanging, 305 mm diameter, 178 mm high [12",7"]				Ea.	43.50			43.50	48	
1500	Rectangular, 1219 mm long, 406 mm high x 381 mm wide					256			256	281	
1650	1524 mm long, 762 mm high, 711 mm wide					650			650	715	
2000	Round, 305 mm diameter, 330 mm high [12",13"]					85.50			85.50	94	
2050	635 mm high					94.50			94.50	104	
5000	Square, 254 mm side, 508 mm high					115			115	127	
5100	356 mm side, 381 mm high					120			120	132	
6000	Metal bowl, 813 mm diameter, 203 mm high, min. [32",8"]					345			345	380	
6050	Maximum					440			440	485	
8750	Wood, fiberglass liner, square										
8780	360 mm square, 381 mm high, minimum				Ea.	221			221	243	
8800	Maximum					272			272	299	
9400	Plastic cylinder, molded, 254 mm dia., 254 mm high [10", 10"]					8.95			8.95	9.85	
9500	279 mm diameter, 279 mm high [11",11"]					28.50			28.50	31	

FURNISHINGS 12

For information about Means Estimating Seminars, see yellow pages 12 and 13 in back of book

For expanded coverage of these items see *Means Interior Cost Data 2006*

Division Notes

	CREW	DAILY OUTPUT	LABOR-HOURS	UNIT	2006 BARE COSTS				TOTAL INCL O&P
					MAT.	LABOR	EQUIP.	TOTAL	

Division 13
Special Construction

Estimating Tips

General

- The items and systems in this division are usually estimated, purchased, supplied and installed as a unit by one or more subcontractors. The estimator must ensure that all parties are operating from the same set of specifications and assumptions and that all necessary items are estimated and will be provided. Many times the complex items and systems are covered but the more common ones such as excavation or a crane are overlooked for the very reason that everyone assumes nobody could miss them. The estimator should be the central focus and be able to ensure that all systems are complete.

- Another area where problems can develop in this division is at the interface between systems. The estimator must ensure, for instance, that anchor bolts, nuts and washers are estimated and included for the air-supported structures and pre-engineered buildings to be bolted to their foundations.

Utility supply is a common area where essential items or pieces of equipment can be missed or overlooked due to the fact that each subcontractor may feel it is another's responsibility. The estimator should also be aware of certain items which may be supplied as part of a package but installed by others, and ensure that the installing contractor's estimate includes the cost of installation. Conversely, the estimator must also ensure that items are not costed by two different subcontractors, resulting in an inflated overall estimate.

13120 Pre-Engineered Structures

- The foundations and floor slab, as well as rough mechanical and electrical, should be estimated, as this work is required for the assembly and erection of the structure. Generally, as noted in the book, the pre-engineered building comes as a shell and additional features, such as windows and doors, must be included by the estimator. Here again, the estimator must have a clear understanding of the scope of each portion of the work and all the necessary interfaces.

13200 Storage Tanks

- The prices in this subdivision for above- and below-ground storage tanks do not include foundations or hold-down slabs. The estimator should refer to Divisions 2 and 3 for foundation system pricing. In addition to the foundations, required tank accessories such as tank gauges, leak detection devices, and additional manholes and piping must be added to the tank prices.

Reference Numbers

Reference numbers are shown in bold squares at the beginning of some major classifications. These numbers refer to related items in the Reference Section. The reference information may be an estimating procedure, an alternate pricing method, or technical information.

Note: Not all subdivisions listed here necessarily appear in this publication.

		13005	Selective Demolition	CREW	DAILY OUTPUT	LABOR-HOURS	UNIT	2006 BARE COSTS				TOTAL INCL O&P	
								MAT.	LABOR	EQUIP.	TOTAL		
011	0010	**SELECTIVE DEMOLITION, AIR SUPPORTED STRUCTURES**											011
	0020	Tank covers, scrim, dbl layer, vinyl poly w/ hdw, blower & controls											
	0050	Round and rectangular [R024119 -10]	B-2	836	.048	m²		1.33		1.33	2.07	
	0100	Warehouse structures											
	0120	Poly/vinyl fabric, 794 g, incl tension cables & inflation system	4 Clab	836	.038	m² Flr.		1.05		1.05	1.63	
	0150	Reinforced vinyl, 340 g., 278 m²	"	465	.069			1.89		1.89	2.94	
	0200	1112 to 2224 m²	8 Clab	1858	.034			.94		.94	1.47	
	0250	Tedlar vinyl fabric, 794 g w/liner, to 278 m²	4 Clab	465	.069			1.89		1.89	2.94	
	0300	1112 to 2224 m²	8 Clab	1858	.034	▼		.94		.94	1.47	
	0350	Greenhouse/shelter, woven polyethylene with liner											
	0400	278 m²	4 Clab	465	.069	m² Flr.		1.89		1.89	2.94	
	0450	1112 to 2224 m²	8 Clab	1858	.034			.94		.94	1.47	
	0500	Tennis/gymnasium, poly/vinyl fabric, 794 g, incl thermal liner	4 Clab	836	.038			1.05		1.05	1.63	
	0600	Stadium/Convention Ctr, teflon coated fiberglass, incl thermal liner	9 Clab	3716	.019	▼		.53		.53	.83	
	0700	Doors, air lock, 4.6 m long, 3 m x 3 m	2 Carp	1.50	10.667	Ea.		380		380	590	
	0720	4.6 m x 4.6 m		.80	20			710		710	1,100	
	0750	Revolving personnel door, 1.8 m dia. x 2 m high	▼	1.50	10.667	▼		380		380	590	
012	0010	**SELECTIVE DEMOLITION, GARDEN HOUSES** [R024119 -10]											012
	0020	Garden house, prefab, wood, excl foundation, average	2 Clab	37.16	.431	m² Flr.		11.80		11.80	18.35	
013	0010	**SELECTIVE DEMOLITION, GREENHOUSES**, excl foundation											013
	0020	Greenhouse, resi-type, free standing, 2591mm long x 2286mm wide	2 Clab	14.86	1.076	m² Flr.		29.50		29.50	46	
	0030	3200mm wide [R024119 -10]		15.79	1.013			28		28	43	
	0040	4115mm wide		20.44	.783			21.50		21.50	33.50	
	0050	5182mm wide		29.73	.538			14.75		14.75	23	
	0060	Lean-to type, 1168mm wide		5.95	2.691			73.50		73.50	115	
	0070	2083mm wide	▼	11.15	1.435	▼		39.50		39.50	61	
	0080	Geodesic hemisphere, 3mm plexiglass glazing, 2438mm dia.		4	4	Ea.		110		110	171	
	0090	7315mm dia.		.80	20			550		550	855	
	0100	14 630mm dia.	▼	.40	40	▼		1,100		1,100	1,700	
035	0010	**SELECTIVE DEMOLITION, SPECIAL PURPOSE ROOMS** [R024119 -10]											035
	0100	Audiometric rooms, under 46.5 m² surface	4 Carp	18.58	1.722	m² Surf		61		61	95.50	
	0110	Over 46.5 m² surface	"	22.30	1.435	"		51		51	79.50	
	0200	Clean rooms, 3.66 m x 3.66 m soft wall, class 100	1 Carp	.30	26.667	Ea.		950		950	1,475	
	0210	Class 1000		.30	26.667			950		950	1,475	
	0220	Class 10,000		.35	22.857			815		815	1,275	
	0230	Class 100,000	▼	.35	22.857	▼		815		815	1,275	
	0300	Darkrooms, shell complete, 2.44 m high	2 Carp	20.44	.783	m² Flr.		28		28	43.50	
	0310	3.66 m high		10.22	1.566	"		55.50		55.50	86.50	
	0350	Darkrooms doors, mini-cylindrical, revolving		4	4	Ea.		142		142	221	
	0400	Music room, practice modular	▼	13.01	1.230	m² Surf		43.50		43.50	68	
	0500	Refrigeration structures and finishes											
	0510	Wall finish, 2 coat portland cement plaster, 13 mm thick	1 Clab	18.58	.431	m²		11.80		11.80	18.35	
	0520	Fiberglass panels, 3 mm thick		37.16	.215			5.90		5.90	9.20	
	0530	Ceiling finish, polystyrene plastic, 25 mm to 51 mm thick		46.45	.172			4.72		4.72	7.35	
	0540	102 mm thick	▼	41.81	.191	▼		5.25		5.25	8.15	
	0550	Refrigerator, prefab aluminum walk-in, 2.3 m H 1.8 m x 1.8 m OD	2 Carp	9.29	1.722	m² Flr.		61		61	95.50	
	0560	3.05 m x 3.05 m OD		14.86	1.076			38.50		38.50	59.50	
	0570	Over 13.9 m²		18.58	.861			30.50		30.50	47.50	
	0600	Sauna, prefabricated, including heater & controls, 2.1 m H, to 2.8 m		11.15	1.435			51		51	79.50	
	0610	To 3.7 m²		13.01	1.230			43.50		43.50	68	
	0620	To 5.6 m²		16.26	.984			35		35	54.50	
	0630	To 9.3 m²		20.44	.783			28		28	43.50	
	0640	To 12.1 m²	▼	23.23	.689	▼		24.50		24.50	38	
	0650	Steam bath, heater, timer, head, single, to 3.9 m³	1 Plum	2.20	3.636	Ea.		155		155	234	
	0660	To 8.4 m³	▼	2.20	3.636	▼		155		155	234	

13 SPECIAL CONSTRUCTION

		13005 Selective Demolition	CREW	DAILY OUTPUT	LABOR-HOURS	UNIT	MAT.	2006 BARE COSTS LABOR	EQUIP.	TOTAL	TOTAL INCL O&P	
035	0670	Comm. size, w/blow-down assembly, to 22.4 m³ R024119-10	1 Plum	1.80	4.444	Ea.		190		190	286	035
	0680	To 70 m³		1.60	5			214		214	320	
	0690	Multiple, for motels, apts, 14.2 m³, 2 baths		2	4			171		171	257	
	0700	28.4 m³, 4 baths		1.40	5.714			244		244	365	
036	0010	**SELECTIVE DEMOLITION, SOUND CONTROL** R024119-10										036
	0120	Acoustical enclosure, 102mm T walls & clg pnls, 39kg/m²	3 Carp	13.38	1.794	m² Surf		64		64	99.50	
	0130	51kg/m²		11.89	2.018			72		72	112	
	0140	Reverb chamber, parallel walls, 102mm thick		11.15	2.153			76.50		76.50	119	
	0150	Skewed walls, parallel roof, 102mm thick		10.22	2.349			83.50		83.50	130	
	0160	Skewed walls/roof, 102mm layer/air space		8.92	2.691			95.50		95.50	149	
	0170	Sound-absorbing pnls, painted metal, 762 x 2438 mm, under 93 m²		39.95	.601			21.50		21.50	33.50	
	0180	Over 93 m²		44.59	.538			19.15		19.15	30	
	0190	Flexible transparent curtain, clear	3 Shee	39.95	.601			25.50		25.50	39	
	0192	50% clear, 50% foam		39.95	.601			25.50		25.50	39	
	0194	25% clear, 75% foam		39.95	.601			25.50		25.50	39	
	0196	100% foam		39.95	.601			25.50		25.50	39	
	0200	Audio-masking system, incl speakers, amplfr, signal gnrtr										
	0205	Ceiling mounted, 464.5 m²	2 Elec	446	.036	m²		1.51		1.51	2.25	
	0210	929 m²		520	.031			1.29		1.29	1.93	
	0230	929 m²		818	.020			.82		.82	1.22	
038	0010	**SELECTIVE DEMOLITION, X-RAY/RADIO FREQ PROTECTION**										038
	0020	Shielding lead, lined door frame, excl hdwe, 2 mm thick	1 Clab	4.80	1.667	Ea.		45.50		45.50	71	
	0030	Lead sheets, 2 mm thick	2 Clab	25.08	.638	m²		17.50		17.50	27	
	0050	Lead shielding, 6 mm thick		25.08	.638			17.50		17.50	27	
	0060	6 mm thick		22.30	.718			19.65		19.65	30.50	
	0070	Lead glass, 6 mm thick, 2.0 mm LE, 305 mm x 406 mm	2 Glaz	26	.615	Ea.		21		21	32	
	0080	610 mm x 914 mm		16	1			34.50		34.50	52	
	0090	914 mm x 1524 mm		4	4			137		137	207	
	0100	Lead glass window Fr., w/2mm lead & voice passage, 914 x 1524 mm		4	4			137		137	207	
	0110	Lead glass window frame, 610 mm x 914 mm		16	1			34.50		34.50	52	
	0120	Lead gypsum board, 16 mm thick with 2 mm lead	2 Clab	29.73	.538	m²		14.75		14.75	23	
	0130	3 mm lead		26.01	.615			16.85		16.85	26	
	0140	.794 mm lead		37.16	.431			11.80		11.80	18.35	
	0150	Butt joints, 3 mm lead or thicker, 51 x 2134 mm L batten strip		480	.033	Ea.		.91		.91	1.42	
	0160	X-ray protection, avg. radiography room, to 28 m², 2 mm lead, min.		.50	32	Total		875		875	1,375	
	0170	Maximum		.30	53.333			1,450		1,450	2,275	
	0180	Deep therapy X-ray room, 250 KV cap, to 28 m², 2 mm lead, min		.20	80			2,200		2,200	3,400	
	0190	Maximum		.12	133			3,650		3,650	5,675	
	0880	Radio frequency shielding, prefab or screen-type Cu or stl, min.		33.44	.478	m² Surf		13.10		13.10	20.50	
	0890	Average		28.80	.556			15.20		15.20	23.50	
	0895	Maximum		26.94	.594			16.25		16.25	25.50	
039	0010	**SELECTIVE DEMOLITION, GEODESIC DOMES**										039
	0050	Shell only, interlocking plywood panels, 9 m diameter	F-5	3.20	10	Ea.		360		360	560	
	0060	10 m diameter		2.30	13.913			500		500	780	
	0070	12 m diameter		2	16			575		575	900	
	0080	14 m diameter		2.20	14.545			525		525	815	
	0090	17 m diameter		2	16			575		575	900	
	0100	18 m diameter		2	16			575		575	900	
	0110	19.8 m diameter		1.60	20			720		720	1,125	
040	0010	**SELECTIVE DEMOLITION, TENSION STRUCTURES**										040
	0020	Steel/alum frame, fabric shell, 18m clear span, 557 m²	B-41	186	.237	m² Flr.		6.75	1.26	8.01	11.85	
	0030	1115 m²		204	.216			6.15	1.15	7.30	10.80	
	0040	24 m clear span, 1932 m²		227	.194			5.50	1.03	6.53	9.70	
	0050	30 m clear span, 929 m²	L-5	404	.139			5.55	1.61	7.16	11.50	
	0060	2415 m²	"	427	.131			5.25	1.52	6.77	10.95	

SPECIAL CONSTRUCTION 13

459

		13005	Selective Demolition	CREW	DAILY OUTPUT	LABOR-HOURS	UNIT	2006 BARE COSTS				TOTAL INCL O&P	
								MAT.	LABOR	EQUIP.	TOTAL		
041	0010		**SELECTIVE DEMOLITION, HANGARS**										041
	0020		Prefab, steel T type, galv roof & walls, incl doors, excl fndtn	E-2	237	.236	m² Flr.		9.20	6	15.20	22.50	
	0030		Circular type, prefab, steel frame, plastic skin, incl foundation, 24 m dia	"	.50	112	Total		4,350	2,850	7,200	10,700	
042	0010		**SELECTIVE DEMOLITION, SILOS**										042
	0020		Conc stave, indstrl, conical/sloping bott, excl fndtn, 3.5m dia, 10.5mH	D-8	.22	181	Ea.		6,000		6,000	9,150	
	0030		5 m dia, 14 m h		.16	250			8,250		8,250	12,600	
	0040		7.5 m dia, 23 m h	▼	.10	400			13,200		13,200	20,100	
	0050		Steel, factory fabricated, 115 kL cap, painted or epoxy lined	L-5	2	28	▼		1,125	325	1,450	2,325	
046	0010		**SELECTIVE DEMOLITION, SWIMMING POOL EQUIPMENT**										046
	0020		Diving stand, stainless steel, 3 meter	2 Clab	3	5.333	Ea.		146		146	227	
	0030		1 meter		5	3.200			87.50		87.50	136	
	0040		Diving board, 4800 mm long, aluminum		5.40	2.963			81		81	126	
	0050		Fiberglass		5.40	2.963			81		81	126	
	0070		Ladders, heavy duty, stainless steel, 2 tread		14	1.143			31.50		31.50	48.50	
	0080		4 tread		12	1.333			36.50		36.50	57	
	0090		Lifeguard chair, stainless steel, fixed		5	3.200			87.50		87.50	136	
	0100		Slide, tubular, fiberglass, alum. handrails & ladder, 1500 mm, straight		4	4			110		110	171	
	0110		2.5 m, curved		6	2.667			73		73	114	
	0120		2.5 m, curved		3	5.333			146		146	227	
	0130		3600 mm straight, with platform		2.50	6.400			175		175	273	
	0140		Removable access ramp, stainless steel		4	4			110		110	171	
	0150		Removable stairs, stainless steel, collapsible	▼	4	4	▼		110		110	171	
047	0010		**SELECTIVE DEMOLITION, LIGHTNING PROTECTION**										047
	0020		Air terminal & base, copper, 10 mm dia x 254 mm, to 23 m h	1 Clab	16	.500	Ea.		13.70		13.70	21.50	
	0030		13 mm dia x 305 mm, over 23 m h		16	.500			13.70		13.70	21.50	
	0050		Aluminum, 13 mm dia x 305 mm, to 23 m h		16	.500			13.70		13.70	21.50	
	0060		16 mm dia x 305 mm, over 23 m h		16	.500	▼		13.70		13.70	21.50	
	0070		Cable, copper, 33 kg per 100 m, to 23 m high		195	.041	m		1.12		1.12	1.75	
	0080		56 kg per 100 m, over 23 m high		140	.057			1.57		1.57	2.44	
	0090		Aluminum, 15 kg per 100 m, to 23 m high		171	.047			1.28		1.28	2	
	0100		30 kg per 100 mg, over 23 m high		146	.055	▼		1.50		1.50	2.34	
	0110		Arrester, 175 V AC, to ground		16	.500	Ea.		13.70		13.70	21.50	
	0120		650 V AC, to ground	▼	13	.615	"		16.85		16.85	26.50	
128	0010		**SELECTIVE DEMOLITION, PRE-ENGINEERED STEEL BUILDINGS**										128
	0500		Pre-engd steel bldgs, rigid frame, clear span & multi post, excl salvage										
	0550		325 to 697 m²	L-11	125	.256	m² Flr.		8.20	11.35	19.55	26	
	0600		697 to 1162 m²		186	.172			5.50	7.65	13.15	17.35	
	0650		1162 m² or greater	▼	204	.157	▼		5	6.95	11.95	15.80	
	0700		Pre-engd steel building components										
	0710		Entrance canopy, including frame 1.2 m x 1.2 m	E-24	8	4	Ea.		157	81.50	238.50	360	
	0720		1.2 m x 2.4 m	"	7	4.571			179	93	272	410	
	0730		H.M doors, self framing, single leaf	2 Skwk	8	2			73		73	114	
	0740		Double leaf		5	3.200	▼		117		117	182	
	0760		Gutter, eave type		183	.087	m		3.19		3.19	4.97	
	0770		Sash, single slide, double slide or fixed		24	.667	Ea.		24.50		24.50	38	
	0780		Skylight, fiberglass, to 3 m²		16	1			36.50		36.50	57	
	0785		Roof vents, circular, 300 mm to 600 mm diameter		12	1.333			48.50		48.50	75.50	
	0790		Continuous, 3 m long	▼	8	2	▼		73		73	114	
	0900		Shelters, aluminum frame										
	0910		Aluminum frame, acrylic glazing, 0.9 x 2.7 x 2.4 m H	2 Skwk	2	8	Ea.		292		292	455	
	0920		2.7 x 3.7 x 2.4 m H	"	1.50	10.667	"		390		390	605	
	0930		Silos, concrete stave industrial, not incl foundations										

13005	Selective Demolition		CREW	DAILY OUTPUT	LABOR-HOURS	UNIT	2006 BARE COSTS				TOTAL INCL O&P	
							MAT.	LABOR	EQUIP.	TOTAL		
201	0010	**SELECTIVE DEMOLITION, STORAGE TANKS**										201
	0500	Steel tank, single wall, above ground, not incl fdn, pumps or piping										
	0510	Single wall, 1 mL	R024119 -10	Q-1	3	5.333	Ea.		205		205	310
	0520	2 thru 8 mL		B-34P	2	12			430	201	631	870
	0530	19 thru 38 mL		B-34Q	2	12			435	535	970	1,250
	0540	57 thru 114 mL		B-34N	2	4			113	223	336	420
	0600	Steel tank, double wall, above ground not incl fdn, pumps & piping										
	0620	2 thru 8 mL		B-34P	2	12	Ea.		430	201	631	870

13011	Air Supported Structures		CREW	DAILY OUTPUT	LABOR-HOURS	UNIT	2006 BARE COSTS				TOTAL INCL O&P	
							MAT.	LABOR	EQUIP.	TOTAL		
100	0010	**AIR SUPPORTED TANK COVERS**, vinyl polyester										100
	0100	scrim, double layer, with hardware, blower, standby & controls										
	0200	Round, 23 m diameter		B-2	418	.096	m²	72	2.66		74.66	83.50
	0300	30 m diameter			465	.086		66.50	2.39		68.89	76.50
	0400	46 m diameter			465	.086		50	2.39		52.39	58.50
	0500	Rectangular, 6 m x 6 m			418	.096		163	2.66		165.66	183
	0600	9 m x 12 m			418	.096		163	2.66		165.66	183
	0700	15 m x 18 m			418	.096		163	2.66		165.66	183
	0800	For single wall construction, deduct, minimum						4.74			4.74	5.15
	0900	Maximum						15.40			15.40	16.90
	1000	For maximum resistance to atmosphere or cold, add						6.45			6.45	7.10
	1100	For average shipping charges, add					Total	1,200			1,200	1,325
200	0010	**AIR SUPPORTED STRUCTURES**	R133113 -10									200
	0020	Site preparation, incl. anchor placement and utilities		B-11B	92.90	.172	m² Flr.	8.50	5.40	2.11	16.01	19.95
	0030	For concrete curb, see division 03310-240										
	0050	Warehouse, polyester/vinyl fabric, 794 gram, over 10 yr. life, welded										
	0060	Seams, tension cables, primary & auxiliary inflation system,										
	0070	airlock, personnel doors and liner										
	0100	464 m²		4 Clab	465	.069	m² Flr.	180	1.89		181.89	201
	0250	1115 m²		"	557	.057		130	1.57		131.57	145
	0400	2230 m²		8 Clab	1115	.057		95.50	1.57		97.07	107
	0500	4645 m²		"	1161	.055		82.50	1.51		84.01	93
	0700	340 gram reinforced vinyl fabric, 5 yr. life, sewn seams,										
	0710	accordian door, including liner										
	0750	278 m²		4 Clab	279	.115	m² Flr.	94	3.14		97.14	109
	0800	1115 m²		"	557	.057		70	1.57		71.57	79.50
	0850	2230 m²		8 Clab	1115	.057		69	1.57		70.57	78.50
	0950	Deduct for single layer						7.55			7.55	8.30
	1000	Add for welded seams						10.85			10.85	11.95
	1050	Add for double layer, welded seams included						22.50			22.50	24.50
	1250	Tedlar/vinyl fabric, 794 gram, with liner, over 10 yr. life,										
	1260	incl. overhead and personnel doors										
	1300	279 m²		4 Clab	279	.115	m² Flr.	175	3.14		178.14	198
	1450	1115 m²		"	557	.057		123	1.57		124.57	137
	1550	2230 m²		8 Clab	1115	.057		95.50	1.57		97.07	107
	1700	Deduct for single layer						15.50			15.50	17
	2250	Greenhouse/shelter, woven polyethylene with liner, 2 yr. life,										
	2260	sewn seams, including doors										

SPECIAL CONSTRUCTION 13

13011	Air Supported Structures		CREW	DAILY OUTPUT	LABOR-HOURS	UNIT	2006 BARE COSTS				TOTAL INCL O&P	
							MAT.	LABOR	EQUIP.	TOTAL		
200	2300	279 m²	R133113 -10	4 Clab	279	.115	m² Flr.	71	3.14		74.14	83.50	200
	2350	1115 m²		"	557	.057		72.50	1.57		74.07	82	
	2450	2230 m²		8 Clab	1115	.057		60.50	1.57		62.07	69	
	2550	Deduct for single layer					↓	6.80			6.80	7.45	
	2600	Tennis/gymnasium, polyester/vinyl fabric, 794 gram, over 10 yr. life											
	2610	including thermal liner, heat and lights											
	2650	669 m²		4 Clab	557	.057	m² Flr.	168	1.57		169.57	187	
	2750	1208 m²		"	604	.053		126	1.45		127.45	141	
	2850	Over 2230 m²		8 Clab	1115	.057		114	1.57		115.57	128	
	2860	For low temperature conditions, add					↓	7.65			7.65	8.40	
	2870	For average shipping charges, add					Total	3,100			3,100	3,425	
	2900	Thermal liner, translucent reinforced vinyl					m² Flr.	10.45			10.45	11.50	
	2950	Metalized mylar fabric and mesh, double liner					"	16.80			16.80	18.50	
	3050	Stadium/convention center, teflon coated fiberglass, heavy weight,											
	3060	over 20 yr. life, incl. thermal liner and heating system											
	3100	Minimum		9 Clab	2415	.030	m² Flr.	430	.82		430.82	475	
	3110	Maximum		"	1765	.041	"	500	1.12		501.12	555	
	3400	Doors, air lock, 4.6 m long, 3 m x 3 m		2 Carp	.80	20	Ea.	14,700	710		15,410	17,200	
	3600	4.6 m x 4.6 m		"	.50	32		21,200	1,150		22,350	25,100	
	3700	For each added 1.5 m length, add						3,050			3,050	3,350	
	3900	Revolving personnel door, 1.8 m diameter, 2 m high		2 Carp	.80	20	↓	10,700	710		11,410	12,900	
	4200	Double wall, self supporting, shell only, minimum					m² Flr.					226	
	4300	Maximum					"					420	

13035	Special Purpose Rooms		CREW	DAILY OUTPUT	LABOR-HOURS	UNIT	2006 BARE COSTS				TOTAL INCL O&P	
							MAT.	LABOR	EQUIP.	TOTAL		
150	0010	ANECHOIC CHAMBERS Standard units, 2134 mm ceiling heights											150
	0100	Area for pricing is net inside dimensions											
	0300	200 cycles/second cutoff, 2.32 m² floor area					m² Flr.	15,400			15,400	17,000	
	0400	4.65 m²										12,430	
	0600	6.97 m²										11,865	
	0700	9.29 m²					↓	11,600			11,600	12,800	
	0900	For 150 cycles/second cutoff, add to 9.29 m² room										30%	
	1000	For 100 cycles/second cutoff, add to 9.29 m² room										45%	
180	0010	AUDIOMETRIC ROOMS											180
	0020	Under 46 m² surface		4 Carp	9.10	3.515	m² Surf	495	125		620	740	
	0100	Over 46 m² surface		"	11.15	2.870	"	470	102		572	680	
200	0010	CLEAN ROOMS											200
	1100	Clean room, soft wall, 3.66 m x 3.66 m, Class 100		1 Carp	.18	44.444	Ea.	13,000	1,575		14,575	16,800	
	1110	Class 1,000			.18	44.444		10,100	1,575		11,675	13,600	
	1120	Class 10,000			.21	38.095		8,575	1,350		9,925	11,500	
	1130	Class 100,000		↓	.21	38.095	↓	7,925	1,350		9,275	10,800	
	2800	Ceiling grid support, slotted channel struts 1219 mm O.C., ea. way					m²					70	
	3000	Ceiling panel, vinyl coated foil on mineral substrate											
	3020	Sealed, non-perforated					m²					15	
	4000	Ceiling panel seal, silicone sealant, 121 m/L		1 Carp	45.72	.175	m	.75	6.20		6.95	10.50	
	4100	Two sided adhesive tape		"	73.15	.109	"	.33	3.89		4.22	6.40	

13035	Special Purpose Rooms	CREW	DAILY OUTPUT	LABOR-HOURS	UNIT	2006 BARE COSTS MAT.	LABOR	EQUIP.	TOTAL	TOTAL INCL O&P	
200	4200	Clips, one/panel				Ea.	.90			.90	.99	**200**
	6000	HEPA 0.6 m x 1.2 m,99.97% 76 mm dp beveled frame (silicone seal)					277			277	305	
	6040	152 mm deep skirted frame (channel seal)					320			320	350	
	6100	99.99% efficient, 76 mm deep beveled frame (silicone seal)					298			298	330	
	6140	152 mm deep skirted frame (channel seal)					340			340	375	
	6200	99.999% efficient, 76 mm deep beveled frame (silicone seal)					340			340	375	
	6240	152 mm deep skirted frame (channel seal)				▼	385			385	420	
	7000	Wall panel systems, including channel strut framing										
	7020	Polyester coated aluminum, particle board				m²					215	
	7100	Porcelain coated aluminum, particle board									375	
	7400	Wall panel support, slotted channel struts, to 3658 mm high				▼					194	
300	0010	**DARKROOMS**										**300**
	0020	Shell, complete except for door, 5.946 m², 2438 mm high	2 Carp	11.89	1.346	m² Flr.	775	48		823	930	
	0100	3658 mm high		5.95	2.691		915	95.50		1,010.50	1,150	
	0500	11.15 m² floor, 2438 mm high		11.15	1.435		570	51		621	710	
	0600	3658 mm high		5.57	2.870		690	102		792	920	
	0800	22.3 m² floor, 2438 mm high		11.15	1.435		395	51		446	515	
	0900	3658 mm high		5.57	2.870	▼	550	102		652	765	
	1200	Mini-cylindrical, revolving, unlined, 1219 mm diameter		3.50	4.571	Ea.	3,525	163		3,688	4,125	
	1400	1676 mm diameter	▼	2.50	6.400		5,575	228		5,803	6,475	
	1600	Add for lead lining, inner cylinder, 0.794 mm thick					1,425			1,425	1,575	
	1700	2 mm thick					1,675			1,675	1,825	
	1800	Add for lead lining, inner and outer cylinder, 0.794 mm thick					2,625			2,625	2,900	
	1900	2 mm thick				▼	2,750			2,750	3,025	
	2000	For darkroom door, see division 11472-370										
500	0010	**MUSIC ROOM**										**500**
	0020	Practice room, modular, perforated steel, under 46 m²	2 Carp	6.50	2.460	m² Surf	305	87.50		392.50	470	
	0100	Over 46 m²	"	7.43	2.153	"	258	76.50		334.50	400	
700	0010	**REFRIGERATION**										**700**
	0020	Curbs, 305 mm high, 102 mm thick, concrete	2 Carp	17.68	.905	m	11.55	32		43.55	62.50	
	1000	Doors, see division 08340-100										
	2400	Finishes, 2 coat portland cement plaster, 13 mm thick	1 Plas	4.46	1.794	m²	10.45	58		68.45	99.50	
	2500	For galvanized reinforcing mesh, add	1 Lath	31.12	.257		7	8.45		15.45	20.50	
	2700	5 mm thick latex cement	1 Plas	8.18	.979		17.85	32		49.85	67.50	
	2900	For glass cloth reinforced ceilings, add	"	41.81	.191		4.31	6.20		10.51	14.15	
	3100	Fiberglass panels, 3 mm thick	1 Carp	13.88	.576		23.50	20.50		44	58	
	3200	Polystyrene, plastic finish ceiling, 25 mm thick		25.45	.314		21.50	11.15		32.65	41.50	
	3400	51 mm thick		25.45	.314		25	11.15		36.15	44.50	
	3500	102 mm thick	▼	20.35	.393		27.50	14		41.50	52.50	
	3800	Floors, concrete, 102 mm thick	1 Cefi	8.64	.926		10.35	32		42.35	58.50	
	3900	152 mm thick	"	7.90	1.013	▼	15.50	35		50.50	68.50	
	4000	Insulation, 25 mm to 152 mm thick, cork				m³	400			400	435	
	4100	Urethane					390			390	430	
	4300	Polystyrene, regular					267			267	292	
	4400	Bead board				▼	199			199	220	
	4600	Installation of above, add/layer	2 Carp	61.09	.262	m²	3.23	9.30		12.53	18.05	
	4700	Wall and ceiling juncture		91.10	.176	m	4.92	6.25		11.17	15.10	
	4900	Partitions, galvanized sandwich panels, 102 mm thick, stock		20.36	.786	m²	68	28		96	119	
	5000	Aluminum or fiberglass	▼	20.36	.786	"	74.50	28		102.50	126	
	5200	Prefab walk-in, 2286 mm high, aluminum, incl. door & floors,										
	5210	not incl. partitions or refrig., 1829 mm x 1829 mm O.D. nominal	2 Carp	5.09	3.143	m² Flr.	1,225	112		1,337	1,525	
	5500	3048 mm x 3048 mm O.D. nominal		7.64	2.095		980	74.50		1,054.50	1,200	
	5700	3658 mm x 4267 mm O.D. nominal		10.18	1.571		880	56		936	1,050	
	5800	3658 mm x 6096 mm O.D. nominal	▼	10.18	1.571	▼	765	56		821	925	

SPECIAL CONSTRUCTION 13

13035	Special Purpose Rooms	CREW	DAILY OUTPUT	LABOR-HOURS	UNIT	2006 BARE COSTS				TOTAL INCL O&P	
						MAT.	LABOR	EQUIP.	TOTAL		
700	6100	For 2591 mm high, add				m² Flr.	5%					**700**
	6300	Rule of thumb for complete units, w/o doors & refrigeration, cooler	2 Carp	13.56	1.180		1,100	42		1,142	1,300	
	6400	Freezer		10.18	1.571	↓	1,300	56		1,356	1,500	
	6600	Shelving, plated or galvanized, steel wire type		33.44	.478	m² Hor.	96.50	17		113.50	133	
	6700	Slat shelf type	↓	34.84	.459	↓	119	16.35		135.35	157	
	6900	For stainless steel shelving, add					300%					
	7000	Vapor barrier, on wood walls	2 Carp	153	.105	m²	1.40	3.72		5.12	7.30	
	7200	On masonry walls	"	122	.131	"	3.66	4.66		8.32	11.25	
	7500	For air curtain doors, see division 15840-200										
800	0010	**SAUNA**										**800**
	0020	Prefab, incl. heatr & cntrls, 2.1 m H, 1829 mm x 1219 mm, C/C	L-7	2.20	12.727	Ea.	3,675	435		4,110	4,725	
	0050	1829 mm x 1219 mm, C/P		2	14		3,400	480		3,880	4,500	
	0400	1829 mm x 1524 mm, C/C		2	14		4,125	480		4,605	5,275	
	0450	1829 mm x 1524 mm, C/P		2	14		3,850	480		4,330	4,975	
	0600	1829 mm x 1829 mm, C/C		1.80	15.556		4,400	530		4,930	5,650	
	0650	1829 mm x 1829 mm, C/P		1.80	15.556		4,100	530		4,630	5,325	
	0800	1829 mm x 2743 mm, C/C		1.60	17.500		5,500	595		6,095	6,975	
	0850	1829 mm x 2743 mm, C/P		1.60	17.500		5,225	595		5,820	6,650	
	1000	2438 mm x 3658 mm, C/C		1.10	25.455		8,525	870		9,395	10,700	
	1050	2438 mm x 3658 mm, C/P		1.10	25.455		7,800	870		8,670	9,950	
	1200	2438 mm x 2438 mm, C/C		1.40	20		6,475	685		7,160	8,175	
	1250	2438 mm x 2438 mm, C/P		1.40	20		6,100	685		6,785	7,750	
	1400	2438 mm x 3048 mm, C/C		1.20	23.333		7,200	795		7,995	9,150	
	1450	2438 mm x 3048 mm, C/P		1.20	23.333		6,700	795		7,495	8,575	
	1600	3048 mm x 3658 mm, C/C		1	28		9,025	955		9,980	11,400	
	1650	3048 mm x 3658 mm, C/P	↓	1	28		8,175	955		9,130	10,500	
	1700	Door only, cedar, 610 mm x 1829 mm w/temp. insul. glass window	2 Carp	3.40	4.706		480	167		647	790	
	1800	Prehung, incl. jambs, pulls & hardware	"	12	1.333		490	47.50		537.50	615	
	2500	Heaters only (incl. above), wall mounted, to 5.6 m³					475			475	520	
	2750	To 8.4 m³					585			585	640	
	3000	Floor standing, to 20.16 m³, 10 kW, w/controls	1 Elec	3	2.667		1,500	112		1,612	1,825	
	3250	To 28 m³, 16 kW	"	3	2.667	↓	1,650	112		1,762	2,000	
900	0010	**SPORT COURT**										**900**
	0020	Floors, No. 2 & better maple, 20 mm thick				m² Flr.					71.75	
	0100	Walls, laminated plastic bonded to galv. steel studs				m² Wall					91	
	0150	Laminated fiberglass surfacing, minimum									22.50	
	0180	Maximum				↓					25.50	
	0300	Squash, regulation court in existing building, minimum				Court	15,300			15,300	16,800	
	0400	Maximum				"	28,500			28,500	31,400	
	0450	Rule of thumb for components:										
	0470	Walls	3 Carp	.15	160	Court	11,700	5,700		17,400	21,800	
	0500	Floor	"	.25	96		5,225	3,425		8,650	11,100	
	0550	Lighting	2 Elec	.60	26.667		1,650	1,125		2,775	3,500	
	0600	Handball, racquetball court in existing building, minimum	C-1	.20	160		2,025	5,350		7,375	10,600	
	0800	Maximum	"	.10	320		27,100	10,700		37,800	46,500	
	0900	Rule of thumb for components: walls	3 Carp	.12	200		10,700	7,100		17,800	22,800	
	1000	Floor		.25	96		5,550	3,425		8,975	11,400	
	1100	Ceiling	↓	.33	72.727		2,325	2,575		4,900	6,575	
	1200	Lighting	2 Elec	.60	26.667	↓	2,875	1,125		4,000	4,825	
940	0010	**STEAM BATH**										**940**
	0020	Heater, timer & head, single, to 3.9 m³	1 Plum	1.20	6.667	Ea.	1,075	285		1,360	1,600	
	0500	To 8.4 m³		1.10	7.273		1,200	310		1,510	1,800	
	1000	Commercial size, with blow-down assembly, to 22.4 m³	↓	.90	8.889	↓	4,175	380		4,555	5,150	

T3 SPECIAL CONSTRUCTION

13030 | Special Purpose Rooms

		13035	Special Purpose Rooms	CREW	DAILY OUTPUT	LABOR-HOURS	UNIT	2006 BARE COSTS				TOTAL INCL O&P	
								MAT.	LABOR	EQUIP.	TOTAL		
940	1500		To 70 m³	1 Plum	.80	10	Ea.	6,575	425		7,000	7,875	940
	2000		Multiple, motels, apts., 2 baths, w/ blow-down assm., 14.2 m	Q-1	1.30	12.308		4,125	475		4,600	5,225	
	2500		4 baths	"	.70	22.857		4,500	880		5,380	6,275	
	2700		Conversion unit for residential tub, including door					3,375			3,375	3,725	

13039 | Vaults

		13039	Vaults										
900	0010		VAULT FRONT See division 11021-800										900

13080 | Sound, Vibration & Seismic Control

		13081	Sound Control	CREW	DAILY OUTPUT	LABOR-HOURS	UNIT	2006 BARE COSTS				TOTAL INCL O&P	
								MAT.	LABOR	EQUIP.	TOTAL		
100	0010		AUDIO MASKING Acoustical enclosure, 102 mm thk wall & ceiling panels										100
	0020		39 kg/m², up to 3658 mm span	3 Carp	6.69	3.588	m² Surf	310	128		438	540	
	0300		Better quality panels, 51 kg/m²		5.95	4.037		355	144		499	615	
	0400		Reverb-chamber, 102 mm thick, parallel walls		5.57	4.306		440	153		593	725	
	0600		Skewed wall, parallel roof, 102 mm thick panels		5.11	4.697		500	167		667	815	
	0700		Skewed walls, skewed roof, 102 mm layers, 102 mm air space		4.46	5.382		565	191		756	920	
	0900		Sound-absorbing panels, pntd mtl, 762 mm x 2438 mm, 93 m²		19.97	1.202		117	42.50		159.50	196	
	1100		Over 93 m²		22.30	1.076		113	38.50		151.50	184	
	1200		Fabric faced		22.30	1.076		91	38.50		129.50	160	
	1500		Flexible transparent curtain, clear	3 Shee	19.97	1.202		70.50	50.50		121	156	
	1600		50% foam		19.97	1.202		98.50	50.50		149	186	
	1700		75% foam		19.97	1.202		98.50	50.50		149	186	
	1800		100% foam		19.97	1.202		98.50	50.50		149	186	
	3100		Audio masking system, including speakers, amplification										
	3110		and signal generator										
	3200		Ceiling mounted, 464.5 m²	2 Elec	223	.072	m²	12.80	3.01		15.81	18.60	
	3300		929 m²		260	.062		10.35	2.58		12.93	15.25	
	3400		Plenum mounted, 464.5 m²		353	.045		10.25	1.90		12.15	14.15	
	3500		929 m²		409	.039		6.90	1.64		8.54	10	

13090 | Radiation Protection

		13091	X-Ray/Radio Freq Protection	CREW	DAILY OUTPUT	LABOR-HOURS	UNIT	2006 BARE COSTS				TOTAL INCL O&P	
								MAT.	LABOR	EQUIP.	TOTAL		
600	0010		SHIELDING LEAD										600
	0100		Laminated lead in wood doors, 2 mm thick, no hardware				m²	325			325	360	
	0200		Lead lined door frame, not incl. hardware,										
	0210		2 mm thick lead, butt prepared for hardware	1 Lath	2.40	3.333	Ea.	390	109		499	590	
	0300		Lead sheets, 2 mm thick	2 Lath	12.54	1.276	m²	58.50	42		100.50	126	
	0400		3 mm thick		11.15	1.435		119	47		166	201	
	0500		Lead shielding, 6 mm thick		12.54	1.276		212	42		254	295	
	0550		13mm thick		11.15	1.435		430	47		477	540	

SPECIAL CONSTRUCTION 13

13091	X-Ray/Radio Freq Protection	CREW	DAILY OUTPUT	LABOR-HOURS	UNIT	2006 BARE COSTS				TOTAL INCL O&P	
						MAT.	LABOR	EQUIP.	TOTAL		
600	0600	Lead glass, 6 mm thick, 2.0 mm LE, 305 mm x 406 mm	2 Glaz	13	1.231	Ea.	269	42		311	360	**600**
	0700	610 mm x 914 mm		8	2		1,050	68.50		1,118.50	1,275	
	0800	914 mm x 1524 mm		2	8		3,075	274		3,349	3,825	
	0850	Window frame w/2 mm lead & voice passage, 914 x 1524 mm		2	8		1,125	274		1,399	1,675	
	0870	610 mm x 914 mm frame		8	2		650	68.50		718.50	820	
	0900	Lead gypsum board, 16 mm thick with 2 mm lead		14.86	1.076	m²	60.50	37		97.50	123	
	0910	3 mm lead		13.01	1.230		119	42		161	195	
	0930	0.794 mm lead	2 Lath	18.58	.861		37.50	28.50		66	83.50	
	0950	Lead headed nails (average 0.454 kg/sheet)				kg	13.65			13.65	15.05	
	1000	Butt joints in 3 mm lead or thicker, 51 mm batten strip x 2134 mm L	2 Lath	240	.067	Ea.	14.90	2.19		17.09	19.65	
	1200	X-ray protection, average radiography or fluoroscopy										
	1210	room, up to 28 m² floor, 2 mm lead, minimum	2 Lath	.25	64	Total	6,000	2,100		8,100	9,725	
	1500	Maximum, 2134 mm walls	"	.15	106	"	7,500	3,500		11,000	13,500	
	1600	Deep therapy X-ray room, 250 KV capacity,										
	1800	up to 28 m² floor, 6 mm lead, minimum	2 Lath	.08	200	Total	18,500	6,550		25,050	30,100	
	1900	Maximum, 2134 mm walls	"	.06	266	"	23,900	8,750		32,650	39,300	
	2000	X-ray viewing panels, clear lead plastic										
	2010	7 mm thick, 0.3 mm LE, 11 kg/m²	H-3	12.91	1.239	m²	1,575	37.50		1,612.50	1,800	
	2020	12 mm thick, 0.5 mm LE, 19 kg/m²		7.62	2.100		2,150	63.50		2,213.50	2,450	
	2030	18 mm thick, 0.8 mm LE, 28.8 kg/m²		5.02	3.189		2,325	96.50		2,421.50	2,700	
	2040	22 mm thick, 1.0 mm LE, 34.56 kg/m²		4.09	3.914		2,375	119		2,494	2,800	
	2050	35 mm thick, 1.5 mm LE, 55.2 kg/m²		2.60	6.151		2,675	187		2,862	3,200	
	2060	46 mm thick, 2.0 mm LE, 72.0 kg/m²		1.95	8.201		3,500	249		3,749	4,225	
	2090	For panels 1.12 m² to 4.46 m², add crating charge				Ea.					50	
	4000	X-ray barriers, modular, panels mounted within framework for										
	4002	attaching to floor, wall or ceiling, upper portion is clear lead										
	4005	plastic window panels 1219 mmH, lower portion is opaque leaded										
	4008	steel panels 914 mmH, structural supports not incl.										
	4010	1-section barrier, 914 mmW x 2134 mmH overall										
	4020	0.5 mm LE panels	H-3	6.40	2.500	Ea.	3,325	76		3,401	3,775	
	4030	0.8 mm LE panels		6.40	2.500		3,525	76		3,601	4,000	
	4040	1.0 mm LE panels		5.33	3.002		3,650	91		3,741	4,175	
	4050	1.5 mm LE panels		5.33	3.002		3,900	91		3,991	4,425	
	4060	2-section barrier, 1829 mmW x 2134 mmH overall										
	4070	0.5 mm LE panels	H-3	4	4	Ea.	6,875	121		6,996	7,750	
	4080	0.8 mm LE panels		4	4		7,275	121		7,396	8,175	
	4090	1.0 mm LE panels		3.56	4.494		7,425	136		7,561	8,375	
	5000	1.5 mm LE panels		3.20	5		8,025	152		8,177	9,050	
	5010	3-section barrier, 2743 mmW x 2134 mmH overall										
	5020	0.5 mm LE panels	H-3	3.20	5	Ea.	9,850	152		10,002	11,000	
	5030	0.8 mm LE panels		3.20	5		10,300	152		10,452	11,600	
	5040	1.0 mm LE panels		2.67	5.993		10,600	182		10,782	11,900	
	5050	1.5 mm LE panels		2.46	6.504		11,500	197		11,697	12,900	
	7000	X-ray barriers, mobile, mounted within framework w/casters on										
	7005	bottom, clear lead plastic window panels on upper portion,										
	7010	opaque on lower, 762 mmW x 1905 mmH overall, incl. frmwrk										
	7020	0.6 m upper w/0.5 mm LE, 1.2 m lower w/0.8 mm LE	1 Carp	16	.500	Ea.	2,250	17.80		2,267.80	2,500	
	7030	1219 mmW x 1905 mmH overall, incl. framework										
	7040	0.9 m upper w/0.5 mm LE, 0.9 m lower w/0.8 mm LE	1 Carp	16	.500	Ea.	4,000	17.80		4,017.80	4,425	
	7050	0.9 m upper w/1.0 mm LE, 0.9 m lower w/1.5 mm LE	"	16	.500	"	4,825	17.80		4,842.80	5,325	
	7060	1829 mmW x 1905 mmH overall, incl. framework										
	7070	0.9 m upper w/0.5 mm LE, 0.9 m lower w/0.8 mm LE	1 Carp	16	.500	Ea.	4,725	17.80		4,742.80	5,225	
	7080	0.9 m upper w/1.0 mm LE, 0.9 m lower w/1.5 mm LE	"	16	.500	"	5,975	17.80		5,992.80	6,600	
700	0010	**SHIELDING, RADIO FREQUENCY**										**700**
	0020	Prefabricated or screen-type copper or steel, minimum	2 Carp	16.72	.957	m² Surf	258	34		292	335	

13090 | Radiation Protection

13091 | X-Ray/Radio Freq Protection

			CREW	DAILY OUTPUT	LABOR-HOURS	UNIT	MAT.	LABOR	EQUIP.	TOTAL	TOTAL INCL O&P	
							2006 BARE COSTS					
700	0100	Average	2 Carp	14.40	1.111	m² Surf	280	39.50		319.50	370	700
	0150	Maximum	↓	13.47	1.188	↓	335	42		377	430	

13120 | Pre-Engineered Structures

13128 | Pre-Engineered Structures

			CREW	DAILY OUTPUT	LABOR-HOURS	UNIT	MAT.	LABOR	EQUIP.	TOTAL	TOTAL INCL O&P	
							2006 BARE COSTS					
045	0010	**GUARD HOUSE**										045
	0100	Prefab conc w/bullet resistant doors & windows, roof and wiring										
	0110	2400 x 2400 mm, Level III	L-10	1	24	Ea.	20,000	960	635	21,595	24,400	
	0120	2400 x 2400 mm, Level IV	"	1	24	"	25,000	960	635	26,595	29,900	
060	0010	**PORTABLE BOOTHS** Prefab aluminum with doors, windows, ext. roof										060
	0100	lights wiring & insulation, 1.4 m² building, O.D., painted, minimum				m²	2,875			2,875	3,175	
	0300	2.8 m² building, minimum					1,975			1,975	2,175	
	0400	4.6 m² building, minimum					1,475			1,475	1,625	
	0600	7.4 m² building, minimum					1,175			1,175	1,300	
	0700	9.3 m² building, minimum				↓	1,100			1,100	1,200	
	0900	Acoustical booth, 27 Db @ 1000 Hz, 1.4 m² floor				Ea.	3,025			3,025	3,325	
	1000	2134 mm x 2286 mm, including light & ventilation					6,200			6,200	6,825	
	1200	Ticket booth, galv. steel, not incl. foundations., 1219 mm x 1219 mm					5,125			5,125	5,650	
	1300	1219 mm x 1829 mm				↓	6,000			6,000	6,600	
070	0010	**CONTROL TOWERS**										070
	0020	Modular, 3.7 x 3 m, incl. instruments, min.				Ea.	491,500			491,500	540,500	
	0100	Maximum									1,020,000	
	0500	With standard 12 m tower, average				↓	685,000			685,000	753,500	
	1000	Temporary portable control towers, 2.4 m x 3.7 m,										
	1010	complete with one position communications				Ea.					280,000	
	2000	For fixed facilities, depending on height, minimum									60,000	
	2010	Maximum				↓					120,000	
160	0010	**TENSION STRUCTURES** Rigid steel or alum. frame, vinyl coated polyest										160
	0100	fabric shell, 18 m clear span, not incl. foundations or floors										
	0200	557 m²	B-41	92.90	.474	m² Flr.	117	13.50	2.52	133.02	153	
	0300	1115 m²		102	.431		103	12.30	2.29	117.59	135	
	0400	24 m clear span, 1932 m²	↓	113	.389		105	11.10	2.07	118.17	135	
	0410	30 m clear span, 929 m²	L-5	202	.277		117	11.10	3.22	131.32	151	
	0430	2415 m²		214	.262		103	10.45	3.04	116.49	135	
	0450	3345 m²		232	.241		101	9.65	2.80	113.45	131	
	0460	37 m clear span, 2230 m²		279	.201		122	8	2.33	132.33	152	
	0470	46 m clear span, 2787 m²	↓	557	.101		126	4.02	1.17	131.19	146	
	0480	61 m clear span, 3716 m²	E-6	743	.172	↓	153	6.80	2.21	162.01	182	
	0500	For roll-up door, 3.7 m x 4.3 m, add	L-2	1	16	Ea.	4,275	495		4,770	5,475	
	0600	For personnel doors, add, minimum				m² Flr.	5%					
	0700	Add, maximum					15%					
	0800	For site work, simple foundation, etc., add, minimum								13.45	21	
	0900	Add, maximum				↓				29.60	33	
200	0010	**COMFORT STATIONS** Prefab., stock, w/doors, windows & fixt.										200
	0100	Not incl. interior finish or electrical										
	0300	Mobile, on steel frame, minimum				m²	390			390	430	
	0350	Maximum					620			620	680	
	0400	Permanent, including concrete slab, minimum	B-12J	4.65	3.445	↓	1,525	113	181	1,819	2,050	
	0500	Maximum	"	3.99	4.005		2,225	131	211	2,567	2,875	

SPECIAL CONSTRUCTION — T13

		13128	Pre-Engineered Structures	CREW	DAILY OUTPUT	LABOR-HOURS	UNIT	2006 BARE COSTS				TOTAL INCL O&P	
								MAT.	LABOR	EQUIP.	TOTAL		
200	0600		Alternate pricing method, mobile, minimum				Fixture	1,775			1,775	1,950	200
	0650		Maximum					2,650			2,650	2,925	
	0700		Permanent, minimum	B-12J	.70	22.857		10,200	750	1,200	12,150	13,700	
	0750		Maximum	"	.50	32		17,000	1,050	1,675	19,725	22,200	
300	0010	**DOMES**											300
	0020		Revolv alum, elec drv for astronomy observation, shell only, stk units										
	0600		3048 mm diameter, 363 kg, dome	2 Carp	.25	64	Ea.	10,200	2,275		12,475	14,800	
	0700		Base		.67	23.881		3,525	850		4,375	5,200	
	0900		5486 mm diameter, 1134 kg, dome		.17	94.118		29,700	3,350		33,050	37,900	
	1000		Base		.33	48.485		9,850	1,725		11,575	13,500	
	1200		7315 mm diameter, 2041 kg, dome		.08	200		55,000	7,100		62,100	71,500	
	1300		Base		.25	64		17,700	2,275		19,975	23,100	
	1500		Bulk storage, shell only, dual radius hemispher. arch, steel										
	1600		framing, corrugated steel covering, 46 m diameter	E-2	51.10	1.096	m² Flr.	299	42.50	28	369.50	435	
	1700		122 m diameter	"	66.89	.837		244	32.50	21.50	298	350	
	1800		Wood framing, wood decking, to 122 m diameter	F-4	37.16	1.292		222	46	25.50	293.50	345	
	1900		Radial framed wood (50 mm x 150 mm), 13 mm thick										
	2000		plywood, asphalt shingles, 15 m diameter	F-3	186	.215	m² Flr.	415	7.75	3.42	426.17	470	
	2100		18 m diameter		177	.226		315	8.15	3.60	326.75	365	
	2200		22 m diameter		167	.240		283	8.65	3.81	295.46	330	
	2300		35 m diameter		161	.248		250	8.95	3.96	262.91	292	
	2400		46 m diameter		139	.288		216	10.40	4.58	230.98	259	
340	0010	**GEODESIC DOME** Shell only, interlocking plywood panels R133423 -30											340
	0400		9 m diameter	F-5	1.60	20	Ea.	19,300	720		20,020	22,300	
	0500		10 m diameter		1.14	28.070		22,800	1,000		23,800	26,600	
	0600		12 m diameter		1	32		25,600	1,150		26,750	30,000	
	0700		14 m diameter	F-3	1.13	35.556		31,500	1,275	565	33,340	37,300	
	0750		17 m diameter		1	40		45,200	1,450	635	47,285	52,500	
	0800		18 m diameter		1	40		58,500	1,450	635	60,585	67,500	
	0850		19.8 m diameter		.80	50		71,500	1,800	795	74,095	82,500	
	1100		Aluminum panel, with 150 mm insulation										
	1200		30 m diameter				m² Flr.	375			375	415	
	1300		150 m diameter				"	325			325	355	
	1600		Aluminum framed, plexiglass closure panels										
	1700		12 m diameter				m² Flr.	1,000			1,000	1,125	
	1800		61 m diameter				"	805			805	890	
	2100		Aluminum framed, aluminum closure panels										
	2200		12 m diameter				m² Flr.	300			300	330	
	2300		30.5 m diameter					226			226	249	
	2400		61 m diameter					215			215	237	
	2500		For VRP faced bonded fiberglass insulation, add									108	
	2700		Aluminum framed, fiberglass sandwich panel closure										
	2800		2 m diameter	2 Carp	13.94	1.148	m² Flr.	355	41		396	455	
	2900		9 m diameter	"	32.52	.492	"	325	17.50		342.50	380	
360	0010	**GARAGE COSTS**											360
	0020		Public parking, average				Car					15,400	
	0100		See Square Meter Costs in Reference Section										
	0300		Residential, prefab shell, stock, wood, single car, minimum	2 Carp	1	16	Total	3,175	570		3,745	4,375	
	0350		Maximum		.67	23.881		7,250	850		8,100	9,300	
	0400		Two car, minimum		.67	23.881		4,800	850		5,650	6,600	
	0450		Maximum		.50	32		9,575	1,150		10,725	12,300	
380	0010	**GARDEN HOUSE** Prefab wood, no floors or foundations											380
	0100		3 to 19 m², minimum	2 Carp	18.58	.861	m² Flr.	205	30.50		235.50	274	
	0300		Maximum	"	4.46	3.588	"	370	128		498	605	

13128	Pre-Engineered Structures	CREW	DAILY OUTPUT	LABOR-HOURS	UNIT	2006 BARE COSTS				TOTAL INCL O&P	
						MAT.	LABOR	EQUIP.	TOTAL		
500	**0010** **GRANDSTANDS** Permanent, municipal, including foundation										**500**
	0050 Steel understructure w/aluminum closed deck, minimum				Seat					130	
	0100 Maximum									225	
	0300 Steel, minimum					30			30	33	
	0400 Maximum					80			80	88	
	0600 Aluminum, extruded, stock design, minimum									70	
	0700 Maximum									115	
	0900 Composite, steel, wood and plastic, stock design, minimum					75			75	82.50	
	1000 Maximum					140			140	154	
510	**0010** **BLEACHERS**										**510**
	0020 Bleachers, outdoor, portable, 3 to 5 tiers, to 90 m long, min	2 Sswk	120	.133	Seat	36.50	5.35		41.85	49.50	
	0100 Maximum, less than 5 m long, prefabricated		80	.200		47.50	8		55.50	67	
	0200 6 to 20 tiers, minimum, up to 90 m long		120	.133		43.50	5.35		48.85	57.50	
	0300 Max., under 5 m, (highly prefabricated, on wheels)		80	.200		63.50	8		71.50	84	
	0500 Permanent grandstands, wood seat, steel frame, 600 mm row										
	0600 3 to 15 tiers, minimum	2 Sswk	60	.267	Seat	108	10.65		118.65	138	
	0700 Maximum		48	.333		119	13.30		132.30	155	
	0900 16 to 30 tiers, minimum		60	.267		125	10.65		135.65	156	
	0950 Average		55	.291		156	11.60		167.60	193	
	1000 Maximum		48	.333		187	13.30		200.30	230	
	1200 Seat backs only, 760 mm row, fiberglass		160	.100		22	4		26	31.50	
	1300 Steel and wood		160	.100		27.50	4		31.50	37	
	1400 NOTE: average seating is 457 mm in width										
540	**0010** **GREENHOUSE** Shell only, stock units, not incl. 610 mm stub walls,										**540**
	0020 foundation, floors, heat or compartments										
	0300 Residential type, free standing, 2591 mm long x 2286 mm wide	2 Carp	5.48	2.919	m² Flr.	470	104		574	675	
	0400 3200 mm wide		7.90	2.026		365	72		437	510	
	0600 4115 mm wide		10.03	1.595		320	56.50		376.50	445	
	0700 5182 mm wide		14.86	1.076		365	38.50		403.50	460	
	0900 Lean-to type, 1168 mm wide		3.16	5.066		415	180		595	740	
	1000 2083 mm wide		5.39	2.969		320	106		426	520	
	1500 Commercial, custom, truss frame, incl. equip., plumbing, elec.,										
	1510 benches and controls, under 186 m², minimum				m² Flr.					320	
	1550 Maximum					465			465	510	
	1700 Over 465 m², minimum					281			281	310	
	1750 Maximum									320	
	2000 Institutional, custom, rigid frame, including compartments and										
	2010 multi-controls, under 46 m², minimum				m² Flr.					830	
	2050 Maximum					1,200			1,200	1,325	
	2150 Over 186 m², minimum					475			475	525	
	2200 Maximum									700	
	2400 Concealed rigid frame, under 46 m², minimum									970	
	2450 Maximum									1,175	
	2550 Over 186 m², minimum									730	
	2600 Maximum									850	
	2800 Lean-to type, under 46 m², minimum									860	
	2850 Maximum									1,290	
	3000 Over 186 m², minimum									475	
	3050 Maximum									785	
	3600 For 6 mm clear plate glass, add				m² Surf	18.75			18.75	20.50	
	3700 For 6 mm tempered glass, add				"	42.50			42.50	46.50	
	3900 For cooling, add, minimum				m² Flr.	28.50			28.50	31	
	4000 Maximum					70			70	77	
	4200 For heaters, 4 kW, add					53.50			53.50	59	
	4300 18 kW, add					20			20	22	

SPECIAL CONSTRUCTION 13

	13128	Pre-Engineered Structures	CREW	DAILY OUTPUT	LABOR-HOURS	UNIT	2006 BARE COSTS MAT.	LABOR	EQUIP.	TOTAL	TOTAL INCL O&P	
540	4500	For benches, 610 mm x 1067 mm, add				m² Hor.	240			240	264	**540**
	4600	914 mm x 3048 mm, add				m²	130			130	143	
	4800	For controls, add, minimum				Total	2,250			2,250	2,475	
	4900	Maximum				"	13,300			13,300	14,700	
	5100	For humidification equipment, add				m³	.20			.20	.22	
	5200	For vinyl shading, add				m²	13			13	14.30	
	6000	Geodesic hemisphere, 3 mm plexiglass glazing										
	6050	2438 mm diameter	2 Carp	2	8	Ea.	2,525	284		2,809	3,225	
	6150	7315 mm diameter		.35	45.714		13,000	1,625		14,625	16,800	
	6250	14 630 mm diameter	↓	.20	80	↓	35,200	2,850		38,050	43,100	
580	0010	**HANGARS** Prefabricated steel T hangars, Galv. steel roof &										**580**
	0100	walls, incl. electric bi-folding doors, 4 or more units,										
	0110	not including floors or foundations, minimum	E-2	118	.475	m² Flr.	110	18.45	12.10	140.55	167	
	0130	Maximum		98.75	.567		115	22	14.45	151.45	180	
	0900	With bottom rolling doors, minimum		129	.434		105	16.90	11.05	132.95	157	
	1000	Maximum	↓	89.74	.624	↓	115	24.50	15.90	155.40	186	
	1200	Alternate pricing method:										
	1300	Galv. roof and walls, electric bi-folding doors, minimum	E-2	1.06	52.830	Plane	12,700	2,050	1,350	16,100	18,900	
	1500	Maximum		.91	61.538		14,100	2,400	1,575	18,075	21,400	
	1600	With bottom rolling doors, minimum		1.25	44.800		10,100	1,750	1,150	13,000	15,400	
	1800	Maximum	↓	.97	57.732	↓	12,000	2,250	1,475	15,725	18,800	
	2000	Circular type, prefab., steel frame, plastic skin, electric										
	2010	door, including foundations, 24 m diameter,										
	2020	for up to 5 light planes, minimum	E-2	.50	112	Total	63,000	4,350	2,850	70,200	80,000	
	2200	Maximum	"	.25	224	"	70,500	8,725	5,700	84,925	99,000	
600	0010	**KIOSKS**										**600**
	0020	Round, 1.5 m diameter, 2.4 m high, 6 mm fbgl. wall				Ea.	5,600			5,600	6,175	
	0100	25 mm insulated double wall, fiberglass					6,375			6,375	7,025	
	0500	Rectangular, 1.5 m x 2.7 m, 2.3 m high, 6 mm fiberglass wall					8,150			8,150	8,975	
	0600	25 mm insulated double wall, fiberglass				↓	9,700			9,700	10,700	
700	0010	**PRE-ENGINEERED STEEL BUILDINGS** R133419-10										**700**
	0100	Clear span rigid frame, 0.5 mm colored roofing and siding										
	0150	6.1 m wide, 3 m eave height	E-2	39.48	1.418	m² Flr.	78	55	36	169	221	
	0160	4.2 m eave height		32.52	1.722		82.50	67	44	193.50	256	
	0170	4.8 m eave height		29.73	1.884		87	73.50	48	208.50	275	
	0180	6 m eave height		25.55	2.192		96	85.50	56	237.50	315	
	0190	7.2 m eave height		22.30	2.512		109	97.50	64	270.50	360	
	0200	9 m to 12 m wide, 3 m eave height		49.70	1.127		64.50	44	28.50	137	178	
	0300	4.2 m eave height		41.81	1.340		68	52	34	154	203	
	0400	4.8 m eave height		38.55	1.453		71.50	56.50	37	165	217	
	0500	6 m eave height		33.44	1.674		78.50	65	42.50	186	246	
	0600	7.2 m eave height		29.73	1.884		88.50	73.50	48	210	276	
	0700	15 m to 30 m wide, 3 m eave height		80.36	.697		55.50	27	17.75	100.25	128	
	0800	4.2 m eave height		71.53	.783		59	30.50	19.95	109.45	140	
	0900	4.8 m eave height		67.82	.826		62.50	32	21	115.50	148	
	1000	6 m eave height		61.31	.913		67.50	35.50	23.50	126.50	161	
	1100	7.3 m eave height	↓	56.20	.996	↓	74	39	25.50	138.50	177	
	1200	Clear span tapered beam frame, 26 ga. colored roofing/siding										
	1300	9.14 m wide, 3 m eave height	E-2	49.70	1.127	m² Flr.	71.50	44	28.50	144	186	
	1400	4.25 m eave height		41.81	1.340		78.50	52	34	164.50	214	
	1500	4.85 m eave height		38.55	1.453		84	56.50	37	177.50	231	
	1600	6 m eave height		33.44	1.674		93	65	42.50	200.50	261	
	1700	12m wide, 3 m eave height		55.74	1.005		65	39	25.50	129.50	168	
	1800	4.25 m eave height	↓	47.38	1.182	↓	71	46	30	147	191	

13 SPECIAL CONSTRUCTION

13128	Pre-Engineered Structures	CREW	DAILY OUTPUT	LABOR-HOURS	UNIT	2006 BARE COSTS				TOTAL INCL O&P
						MAT.	LABOR	EQUIP.	TOTAL	
700 1900	4.85 m eave height R133419-10	E-2	44.13	1.269	m² Flr.	74.50	49.50	32.50	156.50	202 **700**
2000	6 m eave height		38.55	1.453		81.50	56.50	37	175	228
2100	15 m to 24 m wide, 3 m eave height		71.53	.783		62	30.50	19.95	112.45	143
2200	4.2 m eave height		62.71	.893		66	34.50	23	123.50	158
2300	4.8 m eave height		58.99	.949		69	37	24	130	166
2400	6 m eave height		52.49	1.067		77	41.50	27	145.50	186
2500	Single post 2-span frame, 0.5 mm colored roofing and siding									
2600	24 m wide, 4.2 m eave height	E-2	68.75	.815	m² Flr.	53.50	31.50	21	106	136
2700	4.8 m eave height		64.57	.867		56.50	33.50	22	112	145
2800	6 m eave height		58.06	.964		61.50	37.50	24.50	123.50	160
2900	7.2 m eave height		52.95	1.058		66.50	41	27	134.50	174
3000	30 m wide, 4.2 m eave height		77.57	.722		51.50	28	18.40	97.90	126
3100	4.8 m eave height		73.86	.758		51	29.50	19.30	99.80	129
3200	6 m eave height		67.82	.826		59.50	32	21	112.50	144
3300	7.2 m eave height		62.24	.900		64.50	35	23	122.50	156
3400	36 m wide, 4.2 m eave height		80.82	.693		51	27	17.65	95.65	122
3500	4.8 m eave height		77.11	.726		53.50	28.50	18.50	100.50	128
3600	6 m eave height		71.07	.788		57.50	30.50	20	108	139
3700	7.2 m eave height		65.49	.855		63	33.50	22	118.50	151
3800	Double post 3-span frame, 0.5 mm colored roofing and siding									
3900	45 m wide, 4.2 m eave height	E-2	85.93	.652	m² Flr.	47.50	25.50	16.60	89.60	114
4000	4.8 m eave height		82.68	.677		49	26.50	17.25	92.75	119
4100	6 m eave height		76.18	.735		52	28.50	18.75	99.25	128
4200	7.2 m eave height		71.07	.788		61.50	30.50	20	112	143
4300	Triple post 4-span frame, 0.5 mm colored roofing and siding									
4400	48 m wide, 4.2 m eave height	E-2	90.11	.621	m² Flr.	46	24	15.85	85.85	110
4500	4.8 m eave height		86.40	.648		48.50	25	16.50	90	115
4600	6 m eave height		80.82	.693		51	27	17.65	95.65	122
4700	7.2 m eave height		75.71	.740		60.50	29	18.85	108.35	137
4800	60 m wide, 4.2 m eave height		95.69	.585		46.50	23	14.90	84.40	107
4900	4.8 m eave height		92.44	.606		48	23.50	15.45	86.95	110
5000	6 m eave height		86.86	.645		52	25	16.45	93.45	119
5100	7.2 m eave height		82.22	.681		60.50	26.50	17.35	104.35	131
5200	Accessory items: add to the basic building cost above									
5250	Eave overhang, 0.6 m wide, 26 ga., with soffit	E-2	110	.509	m	70.50	19.80	12.95	103.25	126
5300	1.2 m wide, without soffit		91.44	.612		68	24	15.60	107.60	133
5350	With soffit		76.20	.735		94.50	28.50	18.75	141.75	174
5400	1.8 m wide, without soffit		76.20	.735		91.50	28.50	18.75	138.75	171
5450	With soffit		60.96	.919		119	35.50	23.50	178	219
5500	Entrance canopy, incl. frame, 1.2 m x 1.2 m		25	2.240	Ea.	277	87	57	421	520
5550	1.2 m x 2.4 m		19	2.947	"	375	115	75	565	695
5600	End wall roof overhang, 1.2 m wide, without soffit		259	.216	m	45	8.40	5.50	58.90	70.50
5650	With soffit		152	.368	"	66.50	14.35	9.40	90.25	108
5700	Doors, H.M. self-framing, incl. butts, lockset and trim									
5750	Single leaf, 900 mm x 2100 mm, economy	2 Sswk	5	3.200	Opng.	430	128		558	705
5800	Deluxe		4	4		505	160		665	845
5825	Glazed		4	4		515	160		675	860
5850	1050 mm x 2100 mm		4	4		530	160		690	875
5900	1200 mm x 2100 mm		3	5.333		565	213		778	1,000
5950	Double leaf, 1800 mm x 2100 mm		2	8		765	320		1,085	1,425
6000	Glazed		2	8		935	320		1,255	1,600
6050	Framing only, for openings, 900 mm x 2100 mm	E-2	25	2.240		132	87	57	276	360
6100	3000 mm x 3000 mm		21	2.667		435	104	68	607	735
6150	For windows below, 600 mm x 600 mm		25	2.240		140	87	57	284	365
6200	1200 mm x 900 mm		22	2.545		171	99	65	335	430
6250	Flashings, 26 ga., corner or eave, painted	2 Sswk	73.15	.219	m	11.05	8.75		19.80	28

		13128	Pre-Engineered Structures		CREW	DAILY OUTPUT	LABOR-HOURS	UNIT	2006 BARE COSTS				TOTAL INCL O&P	
									MAT.	LABOR	EQUIP.	TOTAL		
700	6300	Galvanized		R133419 -10	2 Sswk	73.15	.219	m	9.05	8.75		17.80	25.50	700
	6350	Rake flashing, painted				73.15	.219		11.85	8.75		20.60	29	
	6400	Galvanized				73.15	.219		9.95	8.75		18.70	26.50	
	6450	Ridge flashing, 450 mm wide, painted				73.15	.219		14.65	8.75		23.40	32	
	6500	Galvanized				73.15	.219		12.95	8.75		21.70	30	
	6550	Gutter, eave type, 26 ga., painted				97.54	.164		15.30	6.55		21.85	28.50	
	6600	Galvanized				97.54	.164		8.75	6.55		15.30	21.50	
	6650	Valley type, between buildings, painted				36.58	.437		24.50	17.50		42	58	
	6700	Galvanized			▼	36.58	.437	▼	26	17.50		43.50	60	
	6750	Insulation, rated 9.6 kg/m³ density, vinyl faced												
	6800	38 mm thick, 0.9 m²·K/W			2 Carp	214	.075	m²	2.37	2.66		5.03	6.70	
	6850	75 mm thick, 1.8 m²·K/W				214	.075		3.12	2.66		5.78	7.60	
	6900	100 mm thick, 1.9 m²·K/W				214	.075		4.20	2.66		6.86	8.75	
	6950	Foil/scrim/kraft (FSK) faced, 38 mm thick, 0.9 m²·K/W				214	.075		2.48	2.66		5.14	6.85	
	7000	50 mm thick, 1.1 m²·K/W				214	.075		3.12	2.66		5.78	7.60	
	7050	75 mm thick, 1.8 m²·K/W				214	.075		3.88	2.66		6.54	8.45	
	7100	100 mm thick, 2.3 m²·K/W				214	.075		4.84	2.66		7.50	9.55	
	7150	Metalized polyester (PSK) facing, 38 mm thick, 0.9 m²·K/W				214	.075		4.09	2.66		6.75	8.65	
	7200	50 mm thick, 1.1 m²·K/W				214	.075		4.63	2.66		7.29	9.20	
	7250	75 mm thick, 1.9 m²·K/W				214	.075		4.95	2.66		7.61	9.65	
	7300	100 mm thick, 2.3 m²·K/W				214	.075		5.80	2.66		8.46	10.50	
	7350	Vinyl/scrim/foil (VSF), 38 mm thick, 0.9 m²·K/W				214	.075		3.55	2.66		6.21	8	
	7400	50 mm thick, 1.1 m²·K/W				214	.075		4.41	2.66		7.07	9	
	7450	75 mm thick, 1.8 m²·K/W				214	.075		5.60	2.66		8.26	10.30	
	7500	100 mm thick, 2.3 m²·K/W			▼	214	.075	▼	6.25	2.66		8.91	11.05	
	7650	Sash, single slide, glazed, with screens, 600 mm x 600 mm			E-1	22	1.091	Opng.	81	42.50	4.06	127.56	166	
	7700	900 mm x 900 mm				14	1.714		182	67	6.40	255.40	325	
	7750	1200 mm x 900 mm				13	1.846		243	72	6.85	321.85	400	
	7800	1800 mm x 1200 mm				12	2		485	78	7.45	570.45	675	
	7850	Double slide sash, 900 mm x 900 mm				14	1.714		154	67	6.40	227.40	291	
	7900	1800 mm x 1200 mm				12	2		410	78	7.45	495.45	590	
	7950	Fixed glass, no screens, 900 mm x 900 mm				14	1.714		198	67	6.40	271.40	340	
	8000	1800 mm x 1200 mm				12	2		525	78	7.45	610.45	720	
	8050	Prefinished storm sash, 900 mm x 900 mm			▼	70	.343	▼	65	13.40	1.28	79.68	96	
	8100	Siding and roofing, see division 07300 & 07400												
	8200	Skylight, fiberglass panels, to 3 m²			E-1	10	2.400	Ea.	97	93.50	8.95	199.45	277	
	8250	Larger sizes, add for excess over 3 m²			"	27.87	.861	m²	35	33.50	3.20	71.70	99	
	8300	Roof vents, circular with damper, birdscreen												
	8350	and operator hardware, painted												
	8400	26 ga., 300 mm diameter [12"]			1 Sswk	4	2	Ea.	72	80		152	223	
	8450	500 mm diameter [20"]				3	2.667		148	107		255	355	
	8500	24 ga., 600 mm diameter [24"]				2	4		293	160		453	610	
	8550	Galvanized			▼	2	4		245	160		405	560	
	8600	Continuous, 26 ga., 3 m long, 225 mm wide			2 Sswk	4	4		20.50	160		180.50	310	
	8650	300 mm wide			"	4	4	▼	25	160		185	315	
800	0010	**SHELTERS**												800
	0020	Alum. frame, acrylic glazing, 914 mm x 2743 mm x 2438 mm			2 Sswk	1.14	14.035	Ea.	2,550	560		3,110	3,800	
	0100	2743 mm x 3658 mm x 2438 mm high			"	.73	21.918	"	4,500	875		5,375	6,525	
840	0010	**SILOS** Concrete stave industrial, not incl. foundations, conical or												840
	0100	sloping bottoms, 3.5 m diameter, 10.5 m high			D-8	.11	363	Ea.	19,500	12,000		31,500	39,700	
	0200	5 m diameter, 14 m high				.08	500		26,000	16,500		42,500	53,500	
	0400	7.5 m diameter, 23 m high			▼	.05	800		64,000	26,400		90,400	110,000	
	0500	Steel, factory fab., 115 kL cap., painted, minimum			L-5	1	56		17,200	2,250	650	20,100	23,700	
	0700	Maximum			▼	.50	112	▼	27,400	4,475	1,300	33,175	39,500	

Important: See the Reference Section for supporting data - Crews, Rental Equipment, City Cost Indexes and Reference Data

13128	Pre-Engineered Structures	CREW	DAILY OUTPUT	LABOR-HOURS	UNIT	2006 BARE COSTS				TOTAL INCL O&P	
						MAT.	LABOR	EQUIP.	TOTAL		
840 0800	Epoxy lined, minimum	L-5	1	56	Ea.	28,200	2,250	650	31,100	35,700	**840**
1000	Maximum	↓	.50	112	↓	35,700	4,475	1,300	41,475	48,600	
880 0010	**SWIMMING POOL ENCLOSURE** Translucent, free standing,										**880**
0020	not including foundations, heat or light										
0200	Economy, minimum	2 Carp	18.58	.861	m² Hor.	128	30.50		158.50	189	
0300	Maximum	↓	9.29	1.722	↓	375	61		436	510	
0400	Deluxe, minimum		9.29	1.722		445	61		506	585	
0600	Maximum	↓	6.50	2.460		1,975	87.50		2,062.50	2,300	
0700	For motorized roof, 40% opening, solid roof, add					73			73	80.50	
0800	Skylight type roof, add					77.50			77.50	85	
0900	Air-inflated, including blowers and heaters, minimum									37.50	
1000	Maximum				↓					72	

13151	Swimming Pools	CREW	DAILY OUTPUT	LABOR-HOURS	UNIT	2006 BARE COSTS				TOTAL INCL O&P	
						MAT.	LABOR	EQUIP.	TOTAL		
200 0010	**SWIMMING POOLS** Residential in-ground, vinyl lined, concrete sides										**200**
0020	Sides including equipment, sand bottom	B-52	27.87	2.009	m² Surf	126	65.50	14.25	205.75	255	
0100	Metal or polystyrene sides	B-14	38.09	1.260		105	36.50	5.95	147.45	179	
0200	Add for vermiculite bottom	[R131113-20]			↓	8.05			8.05	8.95	
0500	Gunite bottom and sides, white plaster finish										
0600	3.7 m x 9.1 m pool	B-52	13.47	4.157	m² Surf	208	135	29.50	372.50	470	
0720	4.9 m x 9.8 m pool	↓	14.40	3.889	↓	187	127	27.50	341.50	435	
0750	6.1 m x 12 m pool		23.23	2.411		167	78.50	17.10	262.60	325	
0810	Concrete bottom and sides, tile finish										
0820	3.7 m x 9.1 m pool	B-52	7.43	7.535	m² Surf	210	245	53.50	508.50	670	
0830	4.9 m x 9.8 m pool		8.83	6.345		174	206	45	425	560	
0840	6.1 m x 12 m pool	↓	12.08	4.637	↓	138	151	33	322	420	
1100	Motel, gunite with plaster finish, incl. medium										
1150	capacity filtration & chlorination	B-52	10.68	5.242	m² Surf	256	171	37	464	585	
1200	Municipal, gunite with plaster finish, incl. high										
1250	capacity filtration & chlorination	B-52	9.29	6.028	m² Surf	330	196	42.50	568.50	715	
1350	Add for formed gutters				m	175			175	193	
1360	Add for stainless steel gutters				"	520			520	570	
1600	For water heating system, see division 15510-880										
1700	Filtration and deck equipment only, as % of total				Total				20%	20%	
1800	Deck equipment, rule of thumb, 6.1 m x 12 m pool				m² Pool					14	
1900	465 m² pool				"					20.45	
3000	Painting pools, preparation + 3 coats, 6.1 m x 12 m pool, epoxy	2 Pord	.33	48.485	Total	670	1,525		2,195	3,050	
3100	Rubber base paint, 68 L	"	.33	48.485		505	1,525		2,030	2,875	
3500	12.8 m x 25 m pool, 284 L, epoxy paint	3 Pord	.14	171		2,825	5,425		8,250	11,300	
3600	Rubber base paint	"	.14	171	↓	2,175	5,425		7,600	10,600	
700 0010	**SWIMMING POOL EQUIPMENT**										**700**
0020	Diving stand, stainless steel, 3 meter	2 Carp	.40	40	Ea.	5,800	1,425		7,225	8,600	
0300	1 meter		2.70	5.926		5,175	211		5,386	6,000	
0600	Diving boards, 4800 mm long, aluminum		2.70	5.926		2,525	211		2,736	3,100	
0700	Fiberglass	↓	2.70	5.926	↓	2,025	211		2,236	2,550	
0900	Filter system, sand or diatomite type, incl. pump, 22.7 kL/hr.	2 Plum	1.80	8.889	Total	1,125	380		1,505	1,800	
1020	Add for chlorination system, 74 m² pool		3	5.333	Ea.	203	228		431	570	
1040	465 m² pool	↓	3	5.333	"	1,125	228		1,353	1,600	

SPECIAL CONSTRUCTION 13

473

	13151	Swimming Pools	CREW	DAILY OUTPUT	LABOR-HOURS	UNIT	2006 BARE COSTS				TOTAL INCL O&P	
							MAT.	LABOR	EQUIP.	TOTAL		
700	1100	Gutter system, stainless steel, with grating, stock,										**700**
	1110	contains supply and drainage system	E-1	6.10	3.937	m	555	154	14.65	723.65	890	
	1120	Integral gutter and 1524 mm high wall system, stainless steel	"	3.05	7.874	"	835	305	29.50	1,169.50	1,475	
	1200	Ladders, heavy duty, stainless steel, 2 tread	2 Carp	7	2.286	Ea.	465	81.50		546.50	640	
	1500	4 tread		6	2.667		575	95		670	785	
	1800	Lifeguard chair, stainless steel, fixed	↓	2.70	5.926		1,325	211		1,536	1,775	
	1900	Portable					1,000			1,000	1,100	
	2100	Lights, underwater, 12 volt, with transformer, 300 W	1 Elec	.40	20		145	840		985	1,400	
	2200	110 V, 500 W, standard	↓	.40	20		108	840		948	1,375	
	2400	Low water cutoff type	↓	.40	20	↓	135	840		975	1,400	
	2800	Heaters, see division 15510-880										
	3000	Pool covers, reinforced vinyl	3 Clab	167	.144	m²	3.23	3.94		7.17	9.70	
	3050	Automatic, electric									104	
	3100	Vinyl water tube	3 Clab	297	.081		2.48	2.21		4.69	6.15	
	3200	Maximum	"	279	.086		5.05	2.36		7.41	9.25	
	3250	Sealed air bubble polyethylene solar blanket				↓	2.15			2.15	2.37	
	3300	Slides, tubular, fiberglass, alum handrails & ladder, 1500 mm, straight	2 Carp	1.60	10	Ea.	2,325	355		2,680	3,100	
	3320	2.5m, curved		3	5.333		9,075	190		9,265	10,300	
	3400	3 m, curved		1	16		16,500	570		17,070	19,000	
	3420	3600 mm, straight with platform	↓	1.20	13.333	↓	1,725	475		2,200	2,650	
	4500	Hydraulic lift, movable pool bottom, single ram										
	4520	Under 93 m² area	L-9	.03	1200	Ea.	80,000	38,300		118,300	150,000	
	4600	Four ram lift, over 93 m²	"	.02	1800		96,500	57,500		154,000	199,000	
	5000	Removable access ramp, stainless steel	2 Clab	2	8		7,075	219		7,294	8,150	
	5500	Removable stairs, stainless steel, collapsible	"	2	8	↓	2,600	219		2,819	3,200	

	13171	Therapeutic Pools	CREW	DAILY OUTPUT	LABOR-HOURS	UNIT	2006 BARE COSTS				TOTAL INCL O&P	
							MAT.	LABOR	EQUIP.	TOTAL		
800	0010	**THERAPEUTIC POOLS** See division 15418										**800**

	13176	Ice Rinks	CREW	DAILY OUTPUT	LABOR-HOURS	UNIT	2006 BARE COSTS				TOTAL INCL O&P	
							MAT.	LABOR	EQUIP.	TOTAL		
500	0010	**ICE SKATING** Equipment incl. refrigeration, plumbing & cooling										**500**
	0020	coils & concrete slab, 26 m x 61 m rink										
	0300	55° system, 5 mos., 91 metric ton				Total	496,000			496,000	545,500	
	0700	90° system, 12 mos., 122 metric ton				"	550,500			550,500	605,500	
	1000	Dasher boards, 13 mm H.D. P.E. faced steel frame, 914 mm acrylic										
	1020	screen at sides, 1524 mm acrylic ends, 26 m x 61 m	F-5	.06	533	Ea.	115,000	19,200		134,200	156,500	
	1100	Fiberglass & aluminum construction, same sides and ends	"	.06	533	↓	160,000	19,200		179,200	206,000	
	1200	Subsoil heating system (recycled from compressor), 26 m x 61 m	Q-7	.27	118		25,000	4,875		29,875	34,800	

13175 | Ice Rinks

		13176 | Ice Rinks	CREW	DAILY OUTPUT	LABOR-HOURS	UNIT	2006 BARE COSTS MAT.	LABOR	EQUIP.	TOTAL	TOTAL INCL O&P	
500	1300	Subsoil insulation, 0.907 kg polystyrene with vapor barrier, 26 m	2 Carp	.14	114	Ea.	30,000	4,075		34,075	39,300	500

13200 | Storage Tanks

		13201 | Storage Tanks	CREW	DAILY OUTPUT	LABOR-HOURS	UNIT	2006 BARE COSTS MAT.	LABOR	EQUIP.	TOTAL	TOTAL INCL O&P	
200	0010	**ELEVATED STORAGE TANKS**, not incl pipe, pumps or foundation										200
	3000	Elevated water tanks, 30 m to bottom capacity line, incl painting										
	3010	189 kL				Ea.					193,764	
	3300	379 kL									267,300	
	3400	946 kL									368,851	
	3600	1893 kL									595,298	
	3700	2839 kL									820,577	
	3900	3785 kL									824,180	
300	0010	**PRESTRESSED CONCRETE WATER STORAGE TANKS**										300
	0020	Not incl. pipe or pumps, prestress conc., 946 kL				Ea.					284,200	
	0100	1893 kL									385,700	
	0300	3785 kL									548,100	
	0400	7570 kL									828,240	
	0600	15 140 kL									1,309,350	
	0700	22 710 kL									1,786,400	
	0750	30 280 kL									2,273,600	
	0800	37 850 kL									2,740,500	
	0910	Steel, ground level, ht/dia less than 1, not incl. fdn, 379 kL					137,000			137,000	150,500	
	1000	946 kL									159,910	
	1200	1893 kL									243,955	
	1250	2839 kL									290,645	
	1300	3785 kL									444,360	
	1500	7570 kL									717,858	
	1600	15 140 kL									1,114,723	
	1800	22 710 kL									1,546,606	
	1850	30 280 kL									2,025,179	
	1910	37 850 kL					2,745,000			2,745,000	3,020,000	
	2100	Steel standpipes, hgt/dia more than 1, 30 m to overflow, no fdn										
	2200	1893 kL				Ea.					320,993	
	2400	2839 kL									391,029	
	2500	3785 kL									476,238	
	2700	5678 kL									665,333	
	2800	7570 kL									817,075	
	3000	Steel, storage, above ground, including cradles, coating,										
	3020	fittings, not including fdn, pumps or piping										
	3040	Single wall, interior, 1 kL	Q-5	5	3.200	Ea.	325	124		449	540	
	3060	2 kL	"	2.70	5.926		1,600	230		1,830	2,100	
	3080	4 kL	Q-7	5	6.400		2,500	263		2,763	3,150	
	3100	6 kL		4.75	6.737		3,750	276		4,026	4,550	
	3120	8 kL		4.60	6.957		4,300	285		4,585	5,175	
	3140	19 kL		3.20	10		5,825	410		6,235	7,025	
	3150	38 kL		2	16		17,200	655		17,855	19,900	
	3160	57 kL		1.70	18.824		18,600	770		19,370	21,600	
	3170	76 kL		1.45	22.069		23,200	905		24,105	27,000	

SPECIAL CONSTRUCTION 13

			DAILY	LABOR-		2006 BARE COSTS				TOTAL
13201		**Storage Tanks**	CREW	OUTPUT	HOURS	UNIT	MAT.	LABOR	EQUIP.	TOTAL	INCL O&P
300	3180	95 kL	Q-7	1.30	24.615	Ea.	26,100	1,000		27,100	30,200	**300**
	3190	114 kL		1.10	29.091		32,100	1,200		33,300	37,100	
	3320	Double wall, 2 kL	Q-5	2.40	6.667		2,550	258		2,808	3,200	
	3330	8 kL	Q-7	4.15	7.711		6,250	315		6,565	7,350	
	3340	15 kL		3.60	8.889		11,100	365		11,465	12,800	
	3350	23 kL		2.40	13.333		13,100	545		13,645	15,300	
	3360	30 kL		2	16		16,800	655		17,455	19,500	
	3370	38 kL		1.80	17.778		18,600	730		19,330	21,500	
	3380	57 kL		1.50	21.333		28,200	875		29,075	32,400	
	3390	76 kL		1.30	24.615		32,100	1,000		33,100	36,900	
	3400	95 kL		1.15	27.826		39,100	1,150		40,250	44,700	
	3410	114 kL		1	32		42,800	1,325		44,125	49,100	
	4000	Fixed roof oil storage tanks, steel with fdn 900 mm D x 300 mm W										
	4200	795 kL				Ea.					1,730,000	
	4300	3816 kL									296,000	
	4500	8904 kL									588,500	
	4600	17 490 kL									855,000	
	4800	22 737 kL									1,096,500	
	4900	35 616 kL									1,645,500	
	5100	Floating roof gasoline tanks, steel, 795 kL									190,000	
	5200	3975 kL									325,000	
	5400	8745 kL									650,000	
	5500	15 900 kL									940,000	
	5700	23 850 kL									1,205,000	
	5800	35 775 kL									1,810,000	
	6000	Wood tanks, ground level, 50 mm cypress, 11 kL	C-1	.19	168		7,475	5,650		13,125	17,000	
	6100	64 mm cypress, 38 kL		.12	266		19,800	8,925		28,725	35,700	
	6300	75 mm redwood or 75 mm fir, 76 kL		.10	320		30,300	10,700		41,000	50,000	
	6400	114 kL		.08	400		37,400	13,400		50,800	62,000	
	6600	170 kL		.07	457		56,500	15,300		71,800	86,500	
	6700	Larger sizes, minimum				liter					.17	
	6900	Maximum				"					.21	
	7000	Vinyl coated fabric pillow tanks, freestanding, 19 kL	4 Clab	4	8	Ea.	3,975	219		4,194	4,725	
	7100	Supporting embankment not included, 95 kL	6 Clab	2	24		11,600	660		12,260	13,700	
	7200	189 kL	8 Clab	1.50	42.667		25,200	1,175		26,375	29,500	
	7300	379 kL	9 Clab	.90	80		43,700	2,200		45,900	51,500	
	7400	568 kL		.50	144		54,500	3,950		58,450	66,000	
	7500	757 kL		.40	180		54,500	4,925		59,425	67,500	
	7600	946 kL		.30	240		72,500	6,575		79,075	90,000	
800	0010	**UNDERGROUND STORAGE TANKS**										**800**
	0210	Fiberglass, underground, single wall, U.L. listed, not including										
	0220	manway or hold-down strap										
	0240	8 kL capacity	Q-7	4.57	7.002	Ea.	4,575	287		4,862	5,450	
	0250	15 kL capacity		3.55	9.014		6,500	370		6,870	7,700	
	0260	23 kL capacity		2.67	11.985		7,200	490		7,690	8,675	
	0280	38 kL capacity		2	16		9,700	655		10,355	11,700	
	0284	57 kL capacity		1.68	19.048		15,200	780		15,980	17,900	
	0290	76 kL capacity		1.45	22.069		19,400	905		20,305	22,700	
	0500	For manway, fittings and hold-downs, add					20%	15%				
	1020	Fiberglass, underground, double wall, U.L. listed										
	1030	includes manways, not incl. hold-down straps										
	1040	2.3 kL capacity	Q-5	2.42	6.612	Ea.	5,200	256		5,456	6,100	
	1050	4 kL capacity	"	2.25	7.111		6,800	276		7,076	7,925	
	1060	9 kL capacity	Q-7	4.16	7.692		9,950	315		10,265	11,400	
	1070	11 kL capacity		3.90	8.205		10,700	335		11,035	12,300	

13

SPECIAL CONSTRUCTION

13200 | Storage Tanks

13201 | Storage Tanks

			CREW	DAILY OUTPUT	LABOR-HOURS	UNIT	MAT.	LABOR	EQUIP.	TOTAL	TOTAL INCL O&P	
							2006 BARE COSTS					
800	1080	15 kL capacity	Q-7	3.64	8.791	Ea.	12,700	360		13,060	14,400	800
	1090	23 kL capacity		2.42	13.223		14,100	545		14,645	16,300	
	1100	30 kL capacity		2.08	15.385		16,100	630		16,730	18,700	
	1110	38 kL capacity		1.82	17.582		18,100	720		18,820	21,000	
	1120	45 kL capacity	▼	1.70	18.824	▼	22,300	770		23,070	25,800	
	2210	Fiberglass, underground, single wall, U.L. listed, including										
	2220	hold-down straps, no manways										
	2240	8 kL	Q-7	3.55	9.014	Ea.	4,850	370		5,220	5,900	
	2250	15 kL		2.90	11.034		6,775	455		7,230	8,125	
	2260	23 kL		2	16		7,775	655		8,430	9,550	
	2280	38 kL		1.60	20		10,300	820		11,120	12,500	
	2284	57 kL		1.39	23.022		15,800	945		16,745	18,700	
	2290	76 kL	▼	1.14	28.070	▼	20,200	1,150		21,350	24,000	
	3020	Fiberglass, underground, double wall, U.L. listed										
	3030	includes manways and hold-down straps										
	3040	2.3 kL	Q-5	1.86	8.602	Ea.	5,500	335		5,835	6,550	
	3050	4 kL	"	1.70	9.412		7,100	365		7,465	8,350	
	3060	9 kL	Q-7	3.29	9.726		10,200	400		10,600	11,900	
	3070	11 kL		3.13	10.224		11,000	420		11,420	12,700	
	3080	15 kL		2.93	10.922		13,000	450		13,450	15,000	
	3090	23 kL		1.86	17.204		14,300	705		15,005	16,900	
	3100	30 kL		1.65	19.394		16,200	795		16,995	19,000	
	3110	38 kL		1.48	21.622		18,700	885		19,585	21,900	
	3120	45 kL	▼	1.40	22.857	▼	22,900	940		23,840	26,600	
	5000	Steel underground, sti-P3, set in place, not incl. hold-down bars.										
	5500	Excavation, pad, pumps and piping not included										
	5510	Single wall, 2 kL capacity, 4.6 mm thick shell	Q-5	2.70	5.926	Ea.	1,000	230		1,230	1,450	
	5520	4 kL capacity, 4.6 mm thick shell	"	2.50	6.400		1,650	248		1,898	2,175	
	5530	8 kL capacity, 6 mm thick shell	Q-7	4.60	6.957		3,025	285		3,310	3,750	
	5535	10 kL capacity, 4.6 mm shell	Q-5	3	5.333		3,675	207		3,882	4,350	
	5540	19 kL capacity, 6 mm thick shell	Q-7	3.20	10		6,850	410		7,260	8,150	
	5580	57 kL capacity, 8 mm thick shell		1.70	18.824		13,200	770		13,970	15,800	
	5600	76 kL capacity, 8 mm thick shell		1.50	21.333		17,800	875		18,675	20,800	
	5610	95 kL capacity, 10 mm thick shell		1.30	24.615		21,500	1,000		22,500	25,200	
	5620	114 kL capacity, 10 mm thick shell		1.10	29.091		28,600	1,200		29,800	33,300	
	5630	151 kL capacity, 10 mm thick shell		.90	35.556		37,400	1,450		38,850	43,300	
	5640	189 kL capacity, 10 mm thick shell	▼	.80	40	▼	46,800	1,650		48,450	54,000	

13280 | Hazardous Material Remediation

13281 | Hazardous Material Remediation

			CREW	DAILY OUTPUT	LABOR-HOURS	UNIT	MAT.	LABOR	EQUIP.	TOTAL	TOTAL INCL O&P	
							2006 BARE COSTS					
120	0010	**BULK ASBESTOS REMOVAL**										120
	0020	Includes disposable tools and 2 suits and 1 respirator filter/day/worker										
	0100	Beams, W 250 x 28 [W 10 x 19]	A-9	71.63	.894	m	2.49	35		37.49	58.50	
	0110	W 310 x 33 [W 12 x 22]		64.01	1		2.79	39		41.79	65	
	0120	W 360 x 39 [W14 x 26]		54.86	1.167		3.25	45.50		48.75	75.50	
	0130	W 410 x 46 [W 16 x 31]		48.77	1.312		3.67	51.50		55.17	85	
	0140	W 460 x 60 [W 18 x 40]		42.67	1.500		4.17	58.50		62.67	97.50	
	0150	W 610 x 82 [W 24 x 55]	▼	33.53	1.909	▼	5.30	74.50		79.80	124	

	13281	Hazardous Material Remediation	CREW	DAILY OUTPUT	LABOR-HOURS	UNIT	MAT.	LABOR	EQUIP.	TOTAL	TOTAL INCL O&P	
							2006 BARE COSTS					
120	0160	W 760 x 161 [W 30 x 108]	A-9	25.91	2.470	m	6.90	96.50		103.40	161	120
	0170	W 920 x 223 [W 36 x 150]		21.95	2.916	↓	8.15	114		122.15	189	
	0200	Boiler insulation	↓	44.59	1.435	m²	4.63	56		60.63	94	
	0210	With metal lath add				%					50%	
	0300	Boiler breeching or flue insulation	A-9	48.31	1.325	m²	3.66	52		55.66	86	
	0310	For active boiler, add				%					100%	
	0400	Duct or AHU insulation	A-10B	40.88	.783	m²	2.15	30.50		32.65	51	
	0500	Duct vibration isolation joints, up to 16000 mm² duct	A-9	56	1.143	Ea.	3.19	44.50		47.69	74	
	0520	16 000 mm² to 32 000 mm² duct		48	1.333		3.72	52		55.72	86.50	
	0530	32 000 mm² to 48 000 mm² duct		40	1.600	↓	4.46	62.50		66.96	104	
	0600	Pipe insulation, air cell type, up to 100 mm diameter pipe		274	.234	m	.66	9.15		9.81	15.15	
	0610	100 mm to 200 mm diameter pipe		244	.262		.72	10.25		10.97	17	
	0620	250 mm to 300 mm diameter pipe		213	.300		.82	11.75		12.57	19.50	
	0630	350 mm to 400 mm diameter pipe		168	.381	↓	1.05	14.90		15.95	24.50	
	0650	Over 400 mm diameter pipe		60.39	1.060	m²	2.91	41.50		44.41	68.50	
	0700	With glove bag up to 75 mm diameter pipe		91.44	.700	m	10.35	27.50		37.85	55	
	1000	Pipe fitting insulation up to 100 mm diameter pipe		320	.200	Ea.	.56	7.80		8.36	12.95	
	1100	150 mm to 200 mm diameter pipe		304	.211		.59	8.25		8.84	13.65	
	1110	250 mm to 300 mm diameter pipe		192	.333		.93	13.05		13.98	21.50	
	1120	350 mm to 400 mm diameter pipe		128	.500	↓	1.39	19.55		20.94	32.50	
	1130	Over 400 mm diameter pipe		16.35	3.914	m²	10.85	153		163.85	254	
	1200	With glove bag, up to 200 mm diameter pipe		30.48	2.100	m	21.50	82		103.50	154	
	2000	Scrape foam fireproofing from flat surface		223	.287	m²	.75	11.20		11.95	18.60	
	2100	Irregular surfaces		111	.577		1.61	22.50		24.11	37	
	3000	Remove cementitious material from flat surface		167	.383		1.08	15		16.08	24.50	
	3100	Irregular surface		130	.492		1.40	19.25		20.65	32	
	4000	Scrape acoustical coating/fireproofing, from ceiling		297	.215		.65	8.45		9.10	14	
	5000	Remove VAT from floor by hand	↓	223	.287		.75	11.20		11.95	18.60	
	5100	By machine	A-11	446	.144	↓	.43	5.60	.07	6.10	9.40	
	5150	For 2 layers, add				%					50%	
	6000	Remove contaminated soil from crawl space by hand	A-9	11.32	5.654	m³	15.90	221		236.90	365	
	6100	With large production vacuum loader	A-12	19.81	3.231	"	8.85	126	25	159.85	238	
	7000	Radiator backing, not including radiator removal	A-9	111	.577	m²	1.61	22.50		24.11	37	
	8000	Cement-asbestos transite board	2 Asbe	92.90	.172		1.29	6.75		8.04	12.15	
	8100	Transite shingle siding	A-10B	69.68	.459		2.15	18		20.15	31	
	8200	Shingle roofing	"	186	.172		.75	6.75		7.50	11.50	
	8250	Built-up, no gravel, non-friable	B-2	130	.308		.75	8.55		9.30	14.15	
	8300	Asbestos millboard	2 Asbe	92.90	.172	↓	.86	6.75		7.61	11.60	
	9000	For type C respirator equipment, add				%					10%	
125	0010	**ASBESTOS ABATEMENT WORK AREA** Containment and preparation.										125
	0100	Pre-cleaning, HEPA vacuum and wet wipe, flat surfaces	A-10	1115	.057	m²	.11	2.24		2.35	3.77	
	0200	Protect carpeted area, 2 layers 0.152 mm poly on 19 mm plywood	"	92.90	.689		16.15	27		43.15	60.50	
	0300	Barrier, 50 mm x 100 mm , 13 mm plywood ea. side, 2.4 m high	2 Carp	37.16	.431		13.45	15.30		28.75	39	
	0310	3.6 m high		29.73	.538		15.05	19.15		34.20	46.50	
	0320	4.8 m high		18.58	.861		16.15	30.50		46.65	65.50	
	0400	Personnel decontam. chamber 50 mm x 100 mm, 19 mm ply ea. side		26.01	.615		27	22		49	63.50	
	0450	Waste decontam. chamber, 50 mm x 100 mm, 19 mm ply ea. side	↓	33.44	.478	↓	32.50	17		49.50	62	
	0500	Cover surfaces with polyethelene sheeting										
	0501	Including glue and tape										
	0550	Floors, each layer, 0.2 mm	A-10	743	.086	m²	1.18	3.37		4.55	6.65	
	0551	0.1 mm		836	.077		.65	2.99		3.64	5.50	
	0560	Walls, each layer, 0.2 mm		557	.115		1.18	4.49		5.67	8.40	
	0561	0.1 mm	↓	650	.098	↓	.65	3.85		4.50	6.85	
	0570	For heights above 3.6 m, add						20%				
	0575	For heights above 6 m, add						30%				

T3 SPECIAL CONSTRUCTION

			DAILY	LABOR-			2006 BARE COSTS				TOTAL	
13281	**Hazardous Material Remediation**	CREW	OUTPUT	HOURS	UNIT	MAT.	LABOR	EQUIP.	TOTAL	INCL O&P	
125	0580	For fire retardant poly, add					100%						**125**
	0590	For large open areas, deduct					10%	20%				
	0600	Seal floor penetrations with foam firestop, to 24 000 mm²	2 Carp	200	.080	Ea.	6.25	2.84		9.09	11.35	
	0610	24 000 mm² to 48 000 mm²		125	.128		12.50	4.55		17.05	21	
	0615	48 000 mm² to 96 000 mm²		80	.200		25	7.10		32.10	38.50	
	0620	Wall penetrations, to 24 000 mm²		180	.089		6.25	3.16		9.41	11.80	
	0630	24 000 mm² to 48 000 mm²		100	.160		12.50	5.70		18.20	22.50	
	0640	48 000 mm² to 96 000 mm²	▼	60	.267	▼	25	9.50		34.50	42.50	
	0800	Caulk seams with latex	1 Carp	70.10	.114	m	.49	4.06		4.55	6.85	
	0900	Set up neg. air machine, 4.7 - 9.4 m³/s per 700 m³ room volume	1 Asbe	4.30	1.860	Ea.		72.50		72.50	115	
130	0010	**DEMOLITION IN ASBESTOS CONTAMINATED AREA**										**130**
	0200	Ceiling, including suspension system, plaster and lath	A-9	195	.328	m²	.86	12.85		13.71	21.50	
	0210	Finished plaster, leaving wire lath		54.35	1.178		3.23	46		49.23	76.50	
	0220	Suspended acoustical tile		325	.197		.54	7.70		8.24	12.85	
	0230	Concealed tile grid system		279	.229		.65	8.95		9.60	14.95	
	0240	Metal pan grid system		139	.460		1.29	18		19.29	30	
	0250	Gypsum board		232	.276	▼	.75	10.80		11.55	17.90	
	0260	Lighting fixtures up to 600 mm x 1200 mm	▼	72	.889	Ea.	2.48	35		37.48	57.50	
	0400	Partitions, non load bearing										
	0410	Plaster, lath, and studs	A-9	64.10	.998	m²	8.70	39		47.70	71.50	
	0450	Gypsum board and studs	"	129	.496	"	1.40	19.40		20.80	32	
	9000	For type C respirator equipment, add				%					10%	
135	0010	**ASBESTOS ABATEMENT EQUIPMENT** and supplies, buy R028213 -20										**135**
	0200	Air filtration device, 1 m³/s				Ea.	2,500			2,500	2,750	
	0250	Large volume air sampling pump, minimum					365			365	400	
	0260	Maximum					680			680	745	
	0300	Airless sprayer unit, 2 gun					2,000			2,000	2,200	
	0350	Light stand, 500 W				▼	250			250	275	
	0400	Personal respirators										
	0410	Negative pressure, 1/2 face, dual operation, min.				Ea.	22.50			22.50	25	
	0420	Maximum					24			24	26.50	
	0450	P.A.P.R., full face, minimum					400			400	440	
	0460	Maximum					700			700	770	
	0470	Supplied air, full face, incl. air line, minimum					450			450	495	
	0480	Maximum					600			600	660	
	0500	Personnel sampling pump, minimum					450			450	495	
	0510	Maximum					750			750	825	
	1500	Power panel, 20 unit, incl. G.F.I.					1,800			1,800	1,975	
	1600	Shower unit, including pump and filters					1,125			1,125	1,250	
	1700	Supplied air system (type C)					10,000			10,000	11,000	
	1750	Vacuum cleaner, HEPA, 60 L, stainless steel, wet/dry					1,000			1,000	1,100	
	1760	210 L					2,200			2,200	2,425	
	1800	Vacuum loader, 0.227 kg/s					90,000			90,000	99,000	
	1900	Water atomizer unit, including 210 L drum					230			230	253	
	2000	Worker protection, whole body, foot, head cover & gloves, plastic					5.60			5.60	6.15	
	2500	Respirator, single use					10.30			10.30	11.35	
	2550	Cartridge for respirator					11.10			11.10	12.20	
	2570	Glove bag, 0.178 mm, 1270 mm x 1626 mm					8.50			8.50	9.35	
	2580	0.254 mm, 1118 mm x 1524 mm					8.40			8.40	9.25	
	3000	HEPA vacuum for work area, minimun					1,050			1,050	1,150	
	3050	Maximum					2,875			2,875	3,175	
	6000	Disposable polyethelene bags, 0.152 mm, 0.084 m³					1.15			1.15	1.27	
	6300	Disposable fiber drums, 0.084 m³					6.50			6.50	7.15	
	6400	Pressure sensitive caution lables, 76 mm x 127 mm					1.10			1.10	1.21	
	6450	279 mm x 432 mm					6.50			6.50	7.15	
	6500	Negative air machine, 5 m³/s					775			775	855	

SPECIAL CONSTRUCTION 13

13281	Hazardous Material Remediation	CREW	DAILY OUTPUT	LABOR-HOURS	UNIT	2006 BARE COSTS				TOTAL INCL O&P
						MAT.	LABOR	EQUIP.	TOTAL	
140	0010 **DECONTAMINATION CONTAINMENT AREA DEMOLITION** and clean-up									**140**
	0100　Spray exposed substrate with surfactant (bridging)									
	0200　　Flat surfaces	A-9	557	.115	m²	3.88	4.49		8.37	11.40
	0250　　Irregular surfaces		372	.172	"	3.34	6.75		10.09	14.30
	0300　　Pipes, beams, and columns		610	.105	m	1.84	4.10		5.94	8.55
	1000　Spray encapsulate polyethelene sheeting		743	.086	m²	3.34	3.37		6.71	9
	1100　Roll down polyethelene sheeting		743	.086	"		3.37		3.37	5.35
	1500　Bag polyethelene sheeting		400	.160	Ea.	.77	6.25		7.02	10.75
	2000　Fine clean exposed substrate, with nylon brush		223	.287	m²		11.20		11.20	17.75
	2500　　Wet wipe substrate		446	.144			5.60		5.60	8.90
	2600　　Vacuum surfaces, fine brush	▼	595	.108	▼		4.21		4.21	6.65
	3000　Structural demolition									
	3100　　Wood stud walls	A-9	260	.246	m²		9.65		9.65	15.20
	3500　　Window manifolds, not incl. window replacement		390	.164			6.40		6.40	10.15
	3600　　Plywood carpet protection	▼	186	.344	▼		13.45		13.45	21.50
	4000　Remove custom decontamination facility	A-10A	8	3	Ea.	15	118		133	203
	4100　Remove portable decontamination facility	3 Asbe	12	2	"	12.50	78		90.50	138
	5000　HEPA vacuum, shampoo carpeting	A-9	446	.144	m²	.54	5.60		6.14	9.55
	9000　Final cleaning of protected surfaces	A-10A	743	.032	"		1.27		1.27	2
145	0010 **OSHA TESTING**									**145**
	0100　Certified technician, minimum				Day					300
	0110　　Maximum				"					500
	0200　Personal sampling, PCM analysis, NIOSH 7400, minimum	1 Asbe	8	1	Ea.	2.75	39		41.75	65
	0210　　Maximum	"	4	2	"	3	78		81	127
	0300　Industrial hygenist, minimum				Day					400
	0310　　Maximum				"					550
	1000　Cleaned area samples	1 Asbe	8	1	Ea.	2.63	39		41.63	65
	1100　　PCM air sample analysis, NIOSH 7400, minimum		8	1		30	39		69	95
	1110　　　Maximum	▼	4	2		3.09	78		81.09	127
	1200　　TEM air sample analysis, NIOSH 7402, minimum									125
	1210　　　Maximum				▼					500
150	0010 **ENCAPSULATION WITH SEALANTS**									**150**
	0100　Ceilings and walls, minimum	A-9	1951	.033	m²	2.91	1.28		4.19	5.25
	0110　　Maximum		985	.065		4.52	2.54		7.06	8.95
	0200　Columns and beams, minimum		1236	.052		2.91	2.03		4.94	6.45
	0210　　Maximum		495	.129	▼	5.05	5.05		10.10	13.60
	0300　Pipes to 300 mm diameter including minor repairs, minimum [12"]		244	.262	m	1.21	10.25		11.46	17.55
	0310　　Maximum	▼	122	.525	"	3.41	20.50		23.91	36
155	0010 **WASTE PACKAGING, HANDLING, & DISPOSAL**									**155**
	0100　Collect and bag bulk material, 0.084 m³ bags, by hand	A-9	400	.160	Ea.	1.15	6.25		7.40	11.15
	0200　　Large production vacuum loader	A-12	880	.073		.80	2.84	.57	4.21	6
	1000　Double bag and decontaminate	A-9	960	.067		2.30	2.61		4.91	6.65
	2000　Containerize bagged material in drums, per 0.084 m³ drum	"	800	.080		6.50	3.13		9.63	12.10
	3000　Cart bags 15 m to dumpster	2 Asbe	400	.040	▼		1.56		1.56	2.47
	5000　Disposal charges, not including haul, minimum				m³					65
	5020　　Maximum				"					230
	5100　Remove refrigerant from system	1 Plum	18.14	.441	kg		18.85		18.85	28.50
	9000　For type C respirator equipment, add				%					10%
440	0010 **REMOVAL** Existing lead paint, by chemicals, per application　R028319 -60									**440**
	0050　Baseboard, to 150 mm wide	1 Pord	19.51	.410	m	5.15	13		18.15	25.50
	0070　　To 300 mm wide		9.75	.820	"	10.15	26		36.15	50
	0200　Balustrades, one side		2.60	3.076	m²	37.50	97.50		135	189
	1400　Cabinets, simple design		2.97	2.691		33	85.50		118.50	166
	1420　　Ornate design	▼	2.32	3.445		42.50	109		151.50	211

SPECIAL CONSTRUCTION　13

13281	Hazardous Material Remediation	CREW	DAILY OUTPUT	LABOR-HOURS	UNIT	2006 BARE COSTS				TOTAL INCL O&P
						MAT.	LABOR	EQUIP.	TOTAL	
440 1600	Cornice, simple design R028319-60	1 Pord	5.57	1.435	m²	17.75	45.50		63.25	88	**440**
1620	Ornate design		1.86	4.306		52.50	136		188.50	264
2800	Doors, one side, flush		7.80	1.025		12.80	32.50		45.30	63
2820	Two panel		7.43	1.076		13.25	34		47.25	66
2840	Four panel		4.18	1.914		23.50	60.50		84	117
2880	For trim, one side, add		19.51	.410	m	5.15	13		18.15	25.50
3000	Fence, picket, one side		2.79	2.870	m²	35.50	91		126.50	176
3200	Grilles, one side, simple design		2.79	2.870		35.50	91		126.50	176
3220	Ornate design		2.32	3.445		42.50	109		151.50	211
4400	Pipes, to 100 mm diameter [4"]		27.43	.292	m	3.71	9.25		12.96	18
4420	To 200 mm diameter [8"]		15.24	.525		6.45	16.65		23.10	32
4440	To 300 mm diameter [12"]		10.97	.729		9	23		32	45
4460	To 400 mm diameter [16"]		6.10	1.312		16.05	41.50		57.55	80
4500	For hangers, add		40	.200	Ea.	2.45	6.35		8.80	12.25
4800	Siding		8.36	.957	m²	12.15	30.50		42.65	59
5000	Trusses, open		5.11	1.566	m² Face	19.25	49.50		68.75	96
6200	Windows, one side only, dbl. hung, 1/1 light, 610 mm x 1.2 m high		4	2	Ea.	24.50	63.50		88	123
6220	762 mm x 1.5 m high		3	2.667		33	84.50		117.50	163
6240	914 mm x 1.8 m high		2.50	3.200		39.50	101		140.50	197
6280	1.0 m x 2.0 m high		2	4		49.50	127		176.50	245
6400	Colonial window, 6/6 light, 610 mm x 1.2 m high		2	4		49.50	127		176.50	246
6420	762 mm x 1.5 m high		1.50	5.333		65.50	169		234.50	330
6440	914 mm x 1.8 m high		1	8		98.50	254		352.50	490
6480	1.0 m x 2.0 m high		1	8		98.50	254		352.50	490
6600	8/8 light, 610 mm x 1.2 m high		2	4		49.50	127		176.50	246
6620	1.6 m x 2.0 m high		1	8		98.50	254		352.50	490
6800	12/12 light, 610 mm x 1.2 m high		1	8		98.50	254		352.50	490
6820	1.0 m x 2.0 m high		.75	10.667		131	340		471	655
6840	Window frame & trim items, include in pricing above									
460 0010	**LEAD PAINT ENCAPSULATION**, water based polymer coating, 35 mm DFT										**460**
0020	Interior, brushwork, trim, under 150 mm	1 Pord	73.15	.109	m	7.80	3.47		11.27	13.75
0030	150 mm to 300 mm wide		54.86	.146		10.35	4.62		14.97	18.35
0040	Balustrades		91.44	.087		6.25	2.77		9.02	11.05
0050	Pipe to 100 mm diameter [4"]		152	.053		3.74	1.67		5.41	6.60
0060	To 200 mm diameter [8"]		114	.070		4.95	2.22		7.17	8.80
0070	To 300 mm diameter [12"]		76.20	.105		7.45	3.33		10.78	13.20
0080	To 400 mm diameter [16"]		51.82	.154		10.95	4.89		15.84	19.40
0090	Cabinets, ornate design		18.58	.431	m²	31	13.65		44.65	54.50
0100	Simple design		23.23	.344	"	24.50	10.90		35.40	43.50
0110	Doors, 915 mm x 2135 mm, both sides, incl. frame & trim									
0120	Flush	1 Pord	6	1.333	Ea.	29	42.50		71.50	95.50
0130	French, 10-15 lite		3	2.667		5.80	84.50		90.30	133
0140	Panel		4	2		35	63.50		98.50	134
0150	Louvered		2.75	2.909		32	92		124	174
0160	Windows, per interior side, per 1.4 m²									
0170	1 to 6 lite	1 Pord	14	.571	Ea.	20	18.10		38.10	49.50
0180	7 to 10 lite		7.50	1.067		22	34		56	75.50
0190	12 lite		5.75	1.391		30	44		74	99.50
0200	Radiators		8	1		71	31.50		102.50	126
0210	Grilles, vents		25.55	.313	m²	22.50	9.95		32.45	39.50
0220	Walls, roller, drywall or plaster		92.90	.086		6.15	2.73		8.88	10.90
0230	With spunbonded reinforcing fabric		66.89	.120		7	3.79		10.79	13.45
0240	Wood		74.32	.108		7.65	3.41		11.06	13.55
0250	Ceilings, roller, drywall or plaster		83.61	.096		7	3.03		10.03	12.30
0260	Wood		65.03	.123		8.70	3.90		12.60	15.45

SPECIAL CONSTRUCTION 13

13281 | Hazardous Material Remediation

		CREW	DAILY OUTPUT	LABOR-HOURS	UNIT	MAT.	LABOR	EQUIP.	TOTAL	TOTAL INCL O&P	
460											**460**
0270	Exterior, brushwork, gutters and downspouts	1 Pord	91.44	.087	m	6.25	2.77		9.02	11.05	
0280	Columns		37.16	.215	m²	15.20	6.80		22	27	
0290	Spray, siding	↓	55.74	.144	"	10.25	4.55		14.80	18.15	
0300	Miscellaneous										
0310	Electrical conduit, brushwork, to 50 mm diameter [2"]	1 Pord	152	.053	m	3.74	1.67		5.41	6.60	
0320	Brick, block or concrete, spray		46.45	.172	m²	12.25	5.45		17.70	21.50	
0330	Steel, flat surfaces and tanks to 300 mm		46.45	.172		12.25	5.45		17.70	21.50	
0340	Beams, brushwork		37.16	.215		15.20	6.80		22	27	
0350	Trusses	↓	37.16	.215	↓	15.20	6.80		22	27	
600	**MOLD ABATEMENT WORK AREA** Containment and preparation										**600**
0010											
0100	Pre-cleaning, HEPA vacuum and wet wipe, flat surfaces	A-10	1115	.057	m²	.11	2.24		2.35	3.77	
0300	Barrier, 50 mm x 100 mm, 13 mm plywood ea. side, 2.4 m high	2 Carp	37.16	.431		13.45	15.30		28.75	39	
0310	3.6 m high		29.73	.538		15.05	19.15		34.20	46.50	
0320	4.8 m high		18.58	.861		16.15	30.50		46.65	65.50	
0400	Personnel decontam. chamber 50 mm x 100 mm, 19 mm ply ea. side		26.01	.615		27	22		49	63.50	
0450	Waste decontam. chamber, 50 mm x 100 mm, 19 mm ply each side	↓	33.44	.478	↓	32.50	17		49.50	62	
0500	Cover surfaces with polyethelene sheeting										
0501	Including glue and tape										
0550	Floors, each layer, 0.2 mm	A-10	743	.086	m²	1.18	3.37		4.55	6.65	
0551	0.1 mm		836	.077		.65	2.99		3.64	5.50	
0560	Walls, each layer, 0.2 mm		557	.115		.97	4.49		5.46	8.20	
0561	0.1 mm	↓	650	.098	↓	.75	3.85		4.60	6.95	
0570	For heights above 3.6 m, add						20%				
0575	For heights above 6 m, add						30%				
0580	For fire retardant poly, add					100%					
0590	For large open areas, deduct					10%	20%				
0600	Seal floor penetrations with foam firestop, to 24 000 mm²	2 Carp	200	.080	Ea.	6.25	2.84		9.09	11.35	
0610	24 000 mm² to 48 000 mm²		125	.128		12.50	4.55		17.05	21	
0615	48 000 mm² to 96 000 mm 2		80	.200		25	7.10		32.10	38.50	
0620	Wall penetrations, to 24 000 mm²		180	.089		6.25	3.16		9.41	11.80	
0630	24 000 mm² to 48 000 mm²		100	.160		12.50	5.70		18.20	22.50	
0640	48 000 mm² to 96 000 mm²	↓	60	.267	↓	25	9.50		34.50	42.50	
0800	Caulk seams with latex	1 Carp	70.10	.114	m	.49	4.06		4.55	6.85	
0900	Set up neg. air machine, 4.7 - 9.4 m³/s per 700 m³ room volume	1 Asbe	4.30	1.860	Ea.		72.50		72.50	115	
610	**DEMOLITION IN MOLD CONTAMINATED AREA**										**610**
0010											
0200	Ceiling, including suspension system, plaster and lath	A-9	195	.328	m²	.86	12.85		13.71	21.50	
0210	Finished plaster, leaving wire lath		54.35	1.178		3.23	46		49.23	76.50	
0220	Suspended acoustical tile		325	.197		.54	7.70		8.24	12.85	
0230	Concealed tile grid system		279	.229		.65	8.95		9.60	14.95	
0240	Metal pan grid system		139	.460		1.29	18		19.29	30	
0250	Gypsum board		232	.276		.75	10.80		11.55	17.90	
0255	Plywood		232	.276	↓	.75	10.80		11.55	17.90	
0260	Lighting fixtures up to 600 mm x 1200 mm	↓	72	.889	Ea.	2.48	35		37.48	57.50	
0400	Partitions, non load bearing										
0410	Plaster, lath, and studs	A-9	64.10	.998	m²	8.70	39		47.70	71.50	
0450	Gypsum board and studs		129	.496		1.40	19.40		20.80	32	
0465	carpet & pad		129	.496	↓	1.40	19.40		20.80	32	
0600	Pipe insulation, air cell type, up to 100 mm diameter pipe		274	.234	m	.66	9.15		9.81	15.15	
0610	100 mm to 200 mm diameter pipe		244	.262		.72	10.25		10.97	17	
0620	250 mm to 300 mm diameter pipe		213	.300		.82	11.75		12.57	19.50	
0630	350 mm to 400 mm diameter pipe		168	.381	↓	1.05	14.90		15.95	24.50	
0650	Over 400 mm diameter pipe	↓	60.39	1.060	m²	2.91	41.50		44.41	68.50	
9000	For type C respirator equipment, add				%					10%	

Important: See the Reference Section for supporting data - Crews, Rental Equipment, City Cost Indexes and Reference Data

13280 | Hazardous Material Remediation

13281	Hazardous Material Remediation	CREW	DAILY OUTPUT	LABOR-HOURS	UNIT	2006 BARE COSTS				TOTAL INCL O&P
						MAT.	LABOR	EQUIP.	TOTAL	
750	**0010** **MOLD ABATEMENT METHODS**								—	**750**
0020	Initial Inspection, average 3 bedroom home				Total					260
0030	Initial Inspection, average 5 bedroom home									360
0040	Air sample test, each									260
0050	Swab sample test, each									155
0060	Tape sample test, each									155
0070	After remediation air test, each									260
0080	Mold abatement plan, average 3 bedroom home									1,275
0090	Mold abatement plan, average 5 bedroom home									1,550
0100	Packup & removal of contents, average 3 bedroom home, excl storage									10,300
0110	Packup & removal of contents, average 5 bedroom home, excl storage									25,800
0120	For demolition in mold contaminated areas, see Div. 13281-130									
0130	For personal protection equipment, see Div. 13281-135									

13600 | Solar and Wind Energy Equipment

13630	Solar Collector Components	CREW	DAILY OUTPUT	LABOR-HOURS	UNIT	2006 BARE COSTS				TOTAL INCL O&P
						MAT.	LABOR	EQUIP.	TOTAL	
200	**0010** **SOLAR ENERGY** R235616-60									**200**
0020	System/Package prices, not including connecting									
0030	pipe, insulation, or special heating/plumbing fixtures									
0500	Hot water, standard package, low temperature									
0540	1 collector, circulator, fittings, 246 L tank	Q-1	.50	32	Ea.	1,625	1,225		2,850	3,625
0580	2 collectors, circulator, fittings, 454 L tank		.40	40		2,275	1,525		3,800	4,825
0620	3 collectors, circulator, fittings, 454 L tank		.34	47.059		2,950	1,800		4,750	5,975
0700	Medium temperature package									
0720	1 collector, circulator, fittings, 303 L tank	Q-1	.50	32	Ea.	1,725	1,225		2,950	3,750
0740	2 collectors, circulator, fittings, 454 L tank		.40	40		2,525	1,525		4,050	5,100
0780	3 collectors, circulator, fittings, 454 L tank		.30	53.333		3,550	2,050		5,600	6,975
0980	For each additional 454 L tank, add					880			880	970
2300	Circulators, air									
2310	Blowers									
2400	Reversible fan, 508 mm diameter, 2 speed [20"]	Q-9	18	.889	Ea.	108	33.50		141.50	171
2520	Space & DHW system, less duct work	"	.50	32		1,475	1,225		2,700	3,500
2870	62 W, 1.893 L/s	Q-1	10	1.600		330	61.50		391.50	460
3000	Collector panels, air with aluminum absorber plate									
3010	Wall or roof mount									
3040	Flat black, plastic glazing									
3080	1219 mm x 2438 mm	Q-9	6	2.667	Ea.	635	101		736	850
3200	Flush roof mount, 3048 mm to 4877 mm x 559 mm wide	"	29.26	.547	m	180	20.50		200.50	230
3300	Collector panels, liquid with copper absorber plate									
3330	Alum. fr., 1218 mm x 2438 mm, 4 mm single glazing	Q-1	9.50	1.684	Ea.	630	64.50		694.50	790
3390	Alum. fr., 1219 mm x 3047 mm, 4 mm single glazing		6	2.667		745	102		847	975
3450	Flat black, alum. frame, 1050 mm x 2250 mm		9	1.778		535	68.50		603.50	695
3500	1200 mm x 2400 mm		5.50	2.909		605	112		717	835
3520	Alum. frame, 1200 mm x 3750 mm		10	1.600		720	61.50		781.50	885
3540	Alum. frame, 1218 mm x 3809 mm		5	3.200		735	123		858	995
3600	Liquid, full wetted, plastic, alum. frame, 914 mm x 3048 mm		5	3.200		195	123		318	400
3650	Collector panel mounting, flat roof or ground rack		7	2.286		48	88		136	185
3670	Roof clamps		70	.229	Set	2.35	8.80		11.15	15.80

		13630	Solar Collector Components		CREW	DAILY OUTPUT	LABOR-HOURS	UNIT	2006 BARE COSTS MAT.	LABOR	EQUIP.	TOTAL	TOTAL INCL O&P	
200	3700		Roof strap, teflon	R235616 -60	1 Plum	62.48	.128	m	31	5.45		36.45	43	200
	3900		Differential controller with two sensors											
	3930		Thermostat, hard wired		1 Plum	8	1	Ea.	95	42.50		137.50	169	
	4100		Five station with digital read-out		"	3	2.667	"	218	114		332	410	
	4300		Heat exchanger											
	4580		Fluid to fluid package includes two circulating pumps											
	4590		expansion tank, check valve, relief valve											
	4600		controller, high temperature cutoff and sensors		Q-1	2.50	6.400	Ea.	695	246		941	1,125	
	4650		Heat transfer fluid											
	4700		Propylene glycol, inhibited anti-freeze		1 Plum	106	.075	liter	2.33	3.22		5.55	7.40	
	8250		Water storage tank with heat exchanger and electric element											
	8300		303 L with 51 mm x 0.907 kg density insulation		1 Plum	1.60	5	Ea.	835	214		1,049	1,250	
	8380		454 L with 51 mm x 0.907 kg density insulation			1.40	5.714		940	244		1,184	1,400	
	8400		454 L with 51 mm x 0.907 kg density insul., 3.7 m² heat coil			1.40	5.714		1,075	244		1,319	1,575	

		13710	Security Access	CREW	DAILY OUTPUT	LABOR-HOURS	UNIT	2006 BARE COSTS MAT.	LABOR	EQUIP.	TOTAL	TOTAL INCL O&P	
300	0010		**ACCESS CONTROL**										300
	0020		Card type, 1 time zone, minimum				Ea.	320			320	350	
	0040		Maximum					1,075			1,075	1,175	
	0060		3 time zones, minimum					785			785	865	
	0080		Maximum					1,800			1,800	2,000	
	0100		System with printer, and control console, 3 zones				Total	8,800			8,800	9,700	
	0120		6 zones				"	11,600			11,600	12,800	
	0140		For each door, minimum, add				Ea.	1,300			1,300	1,425	
	0160		Maximum, add					1,925			1,925	2,125	
	0220		Hand geometry scanner, 512 user memory, excl striker/power	1 Elec	3	2.667		1,600	112		1,712	1,925	
	0230		Memory upgrade, add for 9,700 user profiles		8	1		225	42		267	310	
	0240		Add for 32,500 user profiles		8	1		450	42		492	560	
	0250		Prison type, 256 user profiles, excl strike, power		3	2.667		2,150	112		2,262	2,550	
	0260		Memory upgrade, add for 3,300 user profiles		8	1		160	42		202	239	
	0270		Add for 9,700 user profiles		8	1		350	42		392	450	
	0280		Add for 27,900 user profiles		8	1		450	42		492	560	
	0290		All weather, 512 user memory, excl. striker/power		3	2.667		3,225	112		3,337	3,725	
	0300		Facial & fingerprint scanner, combination unit, excl striker/power		3	2.667		4,200	112		4,312	4,800	
	0310		Access for, for initial setup, excl striker/power		3	2.667		100	112		212	277	
400	0010		**ACCESS CONTROL**										400
	0200		Video cameras, wireless, hidden in exit signs, clocks, etc, incl receiver	1 Elec	3	2.667	Ea.	300	112		412	495	
	0210		Video camera accessories, single camera VCR		3	2.667		500	112		612	715	
	0220		multiple cameras VCR		3	2.667		1,500	112		1,612	1,825	
	0230		Video cameras, wireless, for under vehicle searching, complete		2	4		3,500	168		3,668	4,100	
	0240		Metal detector, hand-held, wand type unit									90	
	0250		Metal detector, walk through portal type, single zone	1 Elec	2	4		4,000	168		4,168	4,650	
	0260		Metal detector, walk through portal type, multi-zone		2	4		5,000	168		5,168	5,750	
	0270		Explosives detector, walk through portal type		2	4		3,500	168		3,668	4,100	
	0280		Explosives detector, hand-held, battery operated									28,000	
	0290		X-ray machine, desk top, for mail/small packages/letters	1 Elec	4	2		3,000	84		3,084	3,425	
	0300		X-ray machine, conveyor type, incl monitor, min		2	4		14,000	168		14,168	15,700	

			CREW	DAILY OUTPUT	LABOR-HOURS	UNIT	2006 BARE COSTS				TOTAL INCL O&P	
		**13710	Security Access**					MAT.	LABOR	EQUIP.	TOTAL		
400	0310	X-ray machine, conveyor type, incl monitor, max	1 Elec	2	4	Ea.	25,000	168		25,168	27,800	**400**
	0320	X-ray machine, large unit, for airports, incl monitor, min	2 Elec	1	16		35,000	670		35,670	39,500	
	0330	X-ray machine, large unit, for airports, incl monitor, max	"	.50	32	▼	60,000	1,350		61,350	68,000	

		**13720	Detection & Alarm**										
065	0010	**DETECTION SYSTEMS**, not including wires & conduits										**065**
	0100	Burglar alarm, battery operated, mechanical trigger	1 Elec	4	2	Ea.	254	84		338	405	
	0200	Electrical trigger		4	2		305	84		389	460	
	0400	For outside key control, add		8	1		72	42		114	142	
	0600	For remote signaling circuitry, add		8	1		114	42		156	189	
	0800	Card reader, flush type, standard		2.70	2.963		850	124		974	1,125	
	1000	Multi-code		2.70	2.963		1,100	124		1,224	1,375	
	1200	Door switches, hinge switch		5.30	1.509		53.50	63.50		117	154	
	1400	Magnetic switch		5.30	1.509		63	63.50		126.50	164	
	1600	Exit control locks, horn alarm		4	2		315	84		399	475	
	1800	Flashing light alarm		4	2		355	84		439	520	
	2000	Indicating panels, 1 channel	▼	2.70	2.963		335	124		459	555	
	2200	10 channel	2 Elec	3.20	5		1,150	210		1,360	1,575	
	2400	20 channel		2	8		2,250	335		2,585	2,975	
	2600	40 channel	▼	1.14	14.035		4,075	590		4,665	5,375	
	2800	Ultrasonic motion detector, 12 V	1 Elec	2.30	3.478		210	146		356	450	
	3000	Infrared photoelectric detector	"	2.30	3.478	▼	173	146		319	410	
	3594	Fire, alarm control panel										
	3600	4 zone	2 Elec	2	8	Ea.	940	335		1,275	1,525	
	3800	8 zone		1	16		1,400	670		2,070	2,550	
	4000	12 zone	▼	.67	23.988		1,825	1,000		2,825	3,500	
	4020	Alarm device	1 Elec	8	1		122	42		164	197	
	4050	Actuating device		8	1		292	42		334	385	
	4200	Battery and rack		4	2		690	84		774	885	
	4400	Automatic charger		8	1		445	42		487	555	
	4600	Signal bell		8	1		49.50	42		91.50	117	
	4800	Trouble buzzer or manual station		8	1		37	42		79	103	
	5000	Detector, rate of rise		8	1		33.50	42		75.50	99.50	
	5200	Smoke detector, ceiling type		6.20	1.290		75	54		129	164	
	5400	Duct type		3.20	2.500		250	105		355	430	
	5600	Strobe and horn		5.30	1.509		95	63.50		158.50	200	
	5800	Fire alarm horn		6.70	1.194		36.50	50		86.50	115	
	6000	Door holder, electro-magnetic		4	2		77.50	84		161.50	211	
	6200	Combination holder and closer		3.20	2.500		430	105		535	630	
	6400	Code transmitter		4	2		690	84		774	885	
	6600	Drill switch		8	1		86.50	42		128.50	158	
	6800	Master box		2.70	2.963		3,100	124		3,224	3,575	
	7000	Break glass station		8	1		50	42		92	118	
	7800	Remote annunciator, 8 zone lamp	▼	1.80	4.444		175	187		362	470	
	8000	12 zone lamp	2 Elec	2.60	6.154		300	258		558	715	
	8200	16 zone lamp	"	2.20	7.273	▼	300	305		605	785	

SPECIAL CONSTRUCTION 13

13834	Electric/Electronic Control	CREW	DAILY OUTPUT	LABOR-HOURS	UNIT	2006 BARE COSTS				TOTAL INCL O&P	
						MAT.	LABOR	EQUIP.	TOTAL		
200	**0010**	**CONTROL SYSTEMS, ELECTRONIC**										**200**
	0020	For electronic costs, add to division 13836-200				Ea.					15%	

13836	Pneumatic Controls	CREW	DAILY OUTPUT	LABOR-HOURS	UNIT	2006 BARE COSTS				TOTAL INCL O&P	
200	**0010**	**CONTROL SYSTEMS, PNEUMATIC** (Sub's quote incl. mat. & labor)										**200**
	0011	and nominal 15 m of tubing. add control panelboard if req'd.									
	0100	Heating and Ventilating, split system									
	0200	Mixed air control, economizer cycle, panel readout, tubing									
	0220	Up to 35 kW	Q-19	.68	35.294	Ea.	2,975	1,400		4,375	5,375
	0240	For 35 to 70 kW, add		.63	37.915		3,175	1,500		4,675	5,775
	0260	For over 70 kW, add	↓	.58	41.096	↓	3,450	1,625		5,075	6,250
	0300	Heating coil, hot water, 3 way valve,									
	0320	Freezestat, limit control on discharge readout	Q-5	.69	23.088	Ea.	2,200	895		3,095	3,775
	0500	Cooling coil, chilled water, room									
	0520	Thermostat, 3 way valve	Q-5	2	8	Ea.	985	310		1,295	1,550
	0600	Cooling tower, fan cycle, damper control,									
	0620	Control system including water readout in/out at panel	Q-19	.67	35.821	Ea.	3,900	1,425		5,325	6,450
	1000	Unit ventilator, day/night operation,									
	1100	freezestat, ASHRAE, cycle 2	Q-19	.91	26.374	Ea.	2,175	1,050		3,225	3,950
	2000	Compensated hot water from boiler, valve control,									
	2100	readout and reset at panel, up to 3.8 L/s	Q-19	.55	43.956	Ea.	4,050	1,750		5,800	7,075
	2120	For 7.6 L/s, add		.51	47.059		4,325	1,875		6,200	7,575
	2140	For 15 L/s, add		.49	49.180		4,525	1,950		6,475	7,950
	3000	Boiler room combustion air, damper to .46 m², controls		1.37	17.582		1,950	700		2,650	3,200
	3500	Fan coil, heating and cooling valves, 4 pipe control system		3	8		880	320		1,200	1,450
	3600	Heat exchanger system controls	↓	.86	27.907		1,900	1,100		3,000	3,775
	4000	Pneumatic thermostat, including controlling room radiator valve	Q-5	2.43	6.593		590	255		845	1,025
	4060	Pump control system	Q-19	3	8	↓	905	320		1,225	1,475
	4500	Air supply for pneumatic control system									
	4600	Tank mounted duplex compressor, starter, alternator,									
	4620	piping, dryer, PRV station and filter									
	4630	373 W	Q-19	.68	35.139	Ea.	7,275	1,400		8,675	10,100
	4660	1 kW		.58	41.739		8,875	1,650		10,525	12,300
	4690	4 kW	↓	.42	57.143	↓	21,100	2,275		23,375	26,600

13851	Detection & Alarm	CREW	DAILY OUTPUT	LABOR-HOURS	UNIT	2006 BARE COSTS				TOTAL INCL O&P	
						MAT.	LABOR	EQUIP.	TOTAL		
050	**0010**	**CLOCKS**										**050**
	0080	300 mm diameter, single face [12"]	1 Elec	8	1	Ea.	75.50	42		117.50	146
	0100	Double face	"	6.20	1.290	"	144	54		198	239
055	**0010**	**CLOCK SYSTEMS**, not including wires & conduits										**055**
	0100	Time system components, master controller	1 Elec	.33	24.242	Ea.	1,625	1,025		2,650	3,325
	0200	Program bell		8	1		53.50	42		95.50	122
	0400	Combination clock & speaker		3.20	2.500		189	105		294	365
	0600	Frequency generator		2	4		6,350	168		6,518	7,225
	0800	Job time automatic stamp recorder, minimum		4	2		445	84		529	615
	1000	Maximum	↓	4	2		680	84		764	875
	1200	Time stamp for correspondence, hand operated					340			340	375

13 **SPECIAL CONSTRUCTION**

13850 | Detection & Alarm

13851	Detection & Alarm	CREW	DAILY OUTPUT	LABOR-HOURS	UNIT	2006 BARE COSTS				TOTAL INCL O&P	
						MAT.	LABOR	EQUIP.	TOTAL		
1400	Fully automatic				Ea.	495			495	545	055
1600	Master time clock system, clocks & bells, 20 room	4 Elec	.20	160		4,075	6,725		10,800	14,500	
1800	50 room	"	.08	400		9,400	16,800		26,200	35,300	
2000	Time clock, 100 cards in & out, 1 color	1 Elec	3.20	2.500		950	105		1,055	1,200	
2200	2 colors		3.20	2.500		1,025	105		1,130	1,275	
2400	With 3 circuit program device, minimum		2	4		375	168		543	665	
2600	Maximum		2	4		545	168		713	850	
2800	Metal rack for 25 cards		7	1.143		60	48		108	138	
3000	Watchman's tour station		8	1		63	42		105	132	
3200	Annunciator with zone indication		1	8		230	335		565	755	
3400	Time clock with tape	▼	1	8	▼	610	335		945	1,175	

13900 | Fire Suppression

13910	Basic Fire Protection Matl/Methd	CREW	DAILY OUTPUT	LABOR-HOURS	UNIT	2006 BARE COSTS				TOTAL INCL O&P	
						MAT.	LABOR	EQUIP.	TOTAL		
0010	**FIRE HOSE AND EQUIPMENT**										400
0200	Adapters, rough brass, straight hose threads										
0220	One piece, female to male, rocker lugs										
0240	25 mm x 25 mm				Ea.	30.50			30.50	33.50	
2200	Hose, less couplings										
2260	Synthetic jacket, lined, 2069 kPa test, 40 mm diameter [1-1/2"]	Q-12	792	.020	m	6.05	.76		6.81	7.85	
2280	65 mm diameter [2-1/2"]		671	.024		10.05	.90		10.95	12.45	
2360	High strength, 3450 kPa test, 40 mm diameter [1-1/2"]		792	.020		6.25	.76		7.01	8	
2380	65 mm diameter [2-1/2"]	▼	671	.024	▼	10.85	.90		11.75	13.25	
2600	Hose rack, swinging, for 40 mm diameter hose, [1-1/2"]										
2620	Enameled steel, 15 m & 23 m lengths of hose	Q-12	20	.800	Ea.	33	30		63	82	
2640	30 m and 38 m lengths of hose	"	20	.800	"	33	30		63	82	
3750	Hydrants, wall, w/caps, single, flush, polished brass										
3800	64 mm x 64 mm	Q-12	5	3.200	Ea.	135	120		255	330	
3840	64 mm x 76 mm	"	5	3.200	"	180	120		300	380	
3950	Double, flush, polished brass										
4000	64 mm x 64 mm x 102 mm	Q-12	5	3.200	Ea.	360	120		480	580	
4040	64 mm x 64 mm x 152 mm	"	4.60	3.478		605	131		736	865	
4200	For polished chrome, add				▼	10%					
4350	Double, projecting, polished brass										
4400	64 mm x 64 mm x 102 mm	Q-12	5	3.200	Ea.	148	120		268	345	
4450	64 mm x 64 mm x 152 mm	"	4.60	3.478	"	286	131		417	515	
4460	Valve control, dbl. flush/projecting hydrant, cap &										
4470	chain, ext. rod & cplg., escutcheon, polished brass	Q-12	8	2	Ea.	192	75.50		267.50	325	
5600	Nozzles, brass										
5620	Adjustable fog, 20 mm booster line				Ea.	70.50			70.50	77.50	
5630	25 mm booster line					87			87	95.50	
5640	38 mm leader line					91			91	100	
5660	65 mm direct connection				▼	178			178	196	
5780	For chrome plated, add					8%					
5850	Electrical fire, adjustable fog, no shock										
5900	38 mm				Ea.	268			268	295	
5920	64 mm					365			365	400	
5980	For polished chrome, add				▼	6%					

SPECIAL CONSTRUCTION 13

		13910	**Basic Fire Protection Matl/Methd**	CREW	DAILY OUTPUT	LABOR-HOURS	UNIT	2006 BARE COSTS				TOTAL INCL O&P	
								MAT.	LABOR	EQUIP.	TOTAL		
400	6200		Heavy duty, comb. adj. fog and str. stream, with handle										400
	6210		25 mm booster line				Ea.	256			256	282	
	7140		Standpipe connections, wall, w/plugs & chains										
	7160		Single, flush, brass, 65 mm x 65 mm	Q-12	5	3.200	Ea.	101	120		221	293	
	7180		64 mm x 76 mm	"	5	3.200	"	104	120		224	297	
	7240		For polished chrome, add					15%					
	7280		Double, flush, polished brass										
	7300		65 mm x 65 mm x 100 mm	Q-12	5	3.200	Ea.	330	120		450	540	
	7330		64 mm x 64 mm x 152 mm	"	4.60	3.478	"	460	131		591	705	
	7400		For polished chrome, add					15%					
	7440		For sill cock combination, add				Ea.	50			50	55.50	
	7900		Three way, flush, polished brass										
	7920		64 mm (3) x 102 mm	Q-12	4.80	3.333	Ea.	930	125		1,055	1,225	
	7930		64 mm (3) x 152 mm	"	4.80	3.333	↓	940	125		1,065	1,225	
	8000		For polished chrome, add					9%					
	8020		Three way, projecting, polished brass										
	8040		64 mm(3) x 102 mm	Q-12	4.80	3.333	Ea.	620	125		745	870	
800	0010	**FIRE VALVES**											800
	0080		Wheel handle, 2069 kPa, 38 mm	1 Spri	12	.667	Ea.	30	28		58	75	
	0090		65 mm	"	7	1.143	"	51.50	48		99.50	129	
	0100		For polished brass, add					35%					
	0110		For polished chrome, add					50%					

		13920	**Fire Pumps**	CREW	DAILY OUTPUT	LABOR-HOURS	UNIT	MAT.	LABOR	EQUIP.	TOTAL	INCL O&P	
400	0010	**FIRE PUMPS** Including controller, fittings and relief valve											400
	0030		Diesel										
	0050		32 L/s, 345 kPa, 20.1 kW, 102 mm pump [4"]	Q-13	.64	50	Ea.	53,500	2,000		55,500	62,000	
	0200		47 L/s, 345 kPa, 32.8 kW, 127 mm pump [5"]		.60	53.333		57,000	2,125		59,125	66,000	
	0400		63 L/s, 690 kPa, 66 kW, 100 mm pump		.56	57.143		61,500	2,275		63,775	71,000	
	0700		126 L/s, 689 kPa, 125 kW, 152 mm pump [6"]		.34	94.118		65,000	3,750		68,750	77,000	
	0950		221 L/s, 689 kPa, 224 kW, 254 mm pump [10"]	↓	.24	133	↓	114,500	5,300		119,800	133,500	
	3000		Electric										
	3100		16 L/s, 379 kPa, 11 kW, 3,550 RPM, 50 mm pump	Q-13	.70	45.714	Ea.	12,600	1,825		14,425	16,600	
	3200		32 L/s, 345 kPa, 20 kW, 1770 RPM, 100 mm pump		.68	47.059		15,600	1,875		17,475	19,900	
	3350		47 L/s, 345 kPa, 33 kW, 1,770 RPM, 125 mm pump		.64	50		22,800	2,000		24,800	28,000	
	3400		47 L/s, 689 kPa, 49.2 kW, 3550 RPM, 100 mm pump	↓	.58	55.172	↓	20,200	2,200		22,400	25,500	
	5000		For jockey pump 25 mm, 2.2 kW, with control, add	Q-12	2	8	↓	2,800	300		3,100	3,525	

		13930	**Wet-Pipe Fire Supp. Sprinklers**	CREW	DAILY OUTPUT	LABOR-HOURS	UNIT	MAT.	LABOR	EQUIP.	TOTAL	INCL O&P	
400	0010	**SPRINKLER SYSTEM COMPONENTS**											400
	0600		Accelerator	1 Spri	8	1	Ea.	380	42		422	485	
	2600		Sprinkler heads, not including supply piping										
	2640		Dry, pendent, 13 mm orifice, 19 mm or 25 mm NPT										
	2700		375 mm to 450 mm length	1 Spri	14	.571	Ea.	58.50	24		82.50	101	
	2710		450 mm to 525 mm length		13	.615		61.50	25.50		87	107	
	2720		525 mm to 600 mm length		13	.615		64.50	25.50		90	110	
	2730		600 mm to 675 mm length		13	.615		68	25.50		93.50	114	
	3600		Foam-water, pendent or upright, 13 mm NPT	↓	12	.667	↓	61	28		89	110	
	3700		Standard spray, pendent or upright, brass, 330 K to 415 K										
	3730		15 mm NPT, 12mm orifice	1 Spri	16	.500	Ea.	6.75	21		27.75	39	
	3740		13 mm NPT, 13 mm orifice	"	16	.500	↓	5.10	21		26.10	37	
	3860		For wax and lead coating, add					6.25			6.25	6.85	
	3880		For wax coating, add					3.07			3.07	3.38	

13 SPECIAL CONSTRUCTION

13930	Wet-Pipe Fire Supp. Sprinklers	CREW	DAILY OUTPUT	LABOR-HOURS	UNIT	2006 BARE COSTS				TOTAL INCL O&P	
						MAT.	LABOR	EQUIP.	TOTAL		
400	3900	For lead coating, add				Ea.	3.69			3.69	4.06	400
	3920	For 455 K, same cost										
	3930	For 480 K, add				Ea.	6.25			6.25	6.85	
	3940	For 535 K, add				"	6.25			6.25	6.85	
	4500	Sidewall, horizontal, brass, 330 to 415 K										
	4520	13 mm NPT, 13 mm orifice	1 Spri	16	.500	Ea.	7.35	21		28.35	39.50	
	4540	For 455 K, same cost										
	4800	Recessed pendent, brass, 330 to 415 K										
	4820	13 mm NPT, 10 mm orifice	1 Spri	10	.800	Ea.	17.55	33.50		51.05	70	
	4830	13 mm NPT, 11 mm orifice		10	.800		12.45	33.50		45.95	64	
	4840	13 mm NPT, 13 mm orifice	↓	10	.800	↓	14.75	33.50		48.25	66.50	
	6200	Valves										
	6500	Check, swing, C.I. body, brass fittings, auto. ball drip										
	6520	102 mm size	Q-12	3	5.333	Ea.	163	201		364	485	
	6800	Check, wafer, butterfly type, C.I. body, bronze fittings										
	6820	102 mm size	Q-12	4	4	Ea.	207	151		358	455	
	7000	Deluge, assembly, incl. trim, pressure										
	7020	operated relief, emergency release, gauges										
	7040	51 mm size	Q-12	2	8	Ea.	1,025	300		1,325	1,575	
	7060	80 mm size	"	1.50	10.667	"	1,225	400		1,625	1,950	

13960	CO2 Fire Extinguishing										
200	0010	AUTOMATIC FIRE SUPPRESSION SYSTEMS										200
	0040	For detectors and control stations, see division 13850										
	1000	Dispersion nozzle, CO$_2$, 80 mm x 125 mm	1 Plum	18	.444	Ea.	50	19		69	83.50	
	1100	FM200, 38 mm	"	14	.571		50	24.50		74.50	91.50	
	2000	Extinguisher, CO$_2$ system, high pressure, 34 kg cylinder	Q-1	6	2.667	↓	950	102		1,052	1,200	
	2400	FM200 system, filled, with mounting bracket										
	2540	89 kg container	Q-1	4	4	Ea.	5,800	154		5,954	6,600	
	3000	Electro/mechanical release	L-1	4	4	↓	125	169		294	390	
	3400	Manual pull station	1 Plum	6	1.333	↓	45	57		102	135	
	6000	Average FM200 system, minimum				m^3	44			44	48.50	
	6020	Maximum				"	88.50			88.50	97	

For information about Means Estimating Seminars, see yellow pages 12 and 13 in back of book

SPECIAL CONSTRUCTION 13

Division Notes

		CREW	DAILY OUTPUT	LABOR-HOURS	UNIT	2006 BARE COSTS				TOTAL INCL O&P
						MAT.	LABOR	EQUIP.	TOTAL	

Division 14
Conveying Systems

Estimating Tips

General

- Many products in Division 14 will require some type of support or blocking for installation not included with the item itself. Examples are supports for conveyors or tube systems, attachment points for lifts, and footings for hoists or cranes. Add these supports in the appropriate division.

14100 Dumbwaiters
14200 Elevators

- Dumbwaiters and elevators are estimated and purchased in a method similar to buying a car. The manufacturer has a base unit with standard features. Added to this base unit price will be whatever options the owner or specifications require. Increased load capacity, additional vertical travel, additional stops, higher speed, and cab finish options are items to be considered. When developing an estimate for dumbwaiters and elevators, remember that some items needed by the installers may have to be included as part of the general contract.

Examples are:
  - shaftway
  - rail support brackets
  - machine room
  - electrical supply
  - sill angles
  - electrical connections
  - pits
  - roof penthouses
  - pit ladders

Check the job specifications and drawings before pricing.

- Installation of elevators and handicapped lifts in historic structures can require significant additional costs. The associated structural requirements may involve cutting into and repairing finishes, mouldings, flooring, etc. The estimator must account for these special conditions.

14300 Escalators & Moving Walks

- Escalators and moving walks are specialty items installed by specialty contractors. There are numerous options associated with these items. For specific options contact a manufacturer or contractor. In a method similar to estimating dumbwaiters and elevators, you should verify the extent of general contract work and add items as necessary.

14400 Lifts
14500 Material Handling
14600 Hoists & Cranes

- Products such as correspondence lifts, conveyors, chutes, pneumatic tube systems, material handling cranes and hoists, as well as other items specified in this subdivision, may require trained installers. The general contractor might not have any choice as to who will perform the installation or when it will be performed. Long lead times are often required for these products, making early decisions in scheduling necessary.

Reference Numbers

Reference numbers are shown in bold squares at the beginning of some major classifications. These numbers refer to related items in the Reference Section. The reference information may be an estimating procedure, an alternate pricing method, or technical information.

Note: Not all subdivisions listed here necessarily appear in this publication.

14100 | Dumbwaiters

14110 | Manual Dumbwaiters

			CREW	DAILY OUTPUT	LABOR-HOURS	UNIT	2006 BARE COSTS				TOTAL INCL O&P	
							MAT.	LABOR	EQUIP.	TOTAL		
400	0010	**DUMBWAITERS,** manual										400
	0020	2 stop, hand, minimum	2 Elev	.75	21.333	Ea.	2,450	1,075		3,525	4,300	
	0100	Maximum		.50	32	"	5,600	1,600		7,200	8,525	
	0300	For each additional stop, add	↓	.75	21.333	Stop	890	1,075		1,965	2,575	

14120 | Electric Dumbwaiters

			CREW	DAILY OUTPUT	LABOR-HOURS	UNIT	MAT.	LABOR	EQUIP.	TOTAL	TOTAL INCL O&P	
400	0010	**DUMBWAITERS,** electric										400
	0020	2 stop, electric, minimum	2 Elev	.13	123	Ea.	6,200	6,150		12,350	16,000	
	0100	Maximum		.11	145	"	18,600	7,275		25,875	31,300	
	0600	For each additional stop, add	↓	.54	29.630	Stop	2,725	1,475		4,200	5,200	
	0750	Correspondence lift,1 floor 2 stop, 20.45 kg capacity	2 Elec	.20	80	Ea.	7,025	3,350		10,375	12,700	

14200 | Elevators

14210 | Electric Traction Elevators

			CREW	DAILY OUTPUT	LABOR-HOURS	UNIT	2006 BARE COSTS				TOTAL INCL O&P	
							MAT.	LABOR	EQUIP.	TOTAL		
100	0010	**ELEVATOR SYSTEMS**										100
	7000	Residential, cab type, 1 floor, 2 stop, minimum	2 Elev	.20	80	Ea.	8,500	4,000		12,500	15,300	
	7100	Maximum		.10	160		14,300	8,000		22,300	27,700	
	7200	2 floor, 3 stop, minimum		.12	133		12,600	6,650		19,250	23,900	
	7300	Maximum	↓	.06	266	↓	20,600	13,300		33,900	42,500	
200	0010	**ELEVATORS**										200
	0020	For multi-story buildings, housing project, minimum	R142000 -10			% total					2.50%	
	0100	Maximum									4.50%	
	0300	Office building, minimum	R142000 -20								2.50%	
	0400	Maximum			↓						10%	
	0410	Holeless hydraulic passenger, 907 kg capacity, 2 story	R142000 -30	2 Elev	.10	160	Ea.	30,700	8,000		38,700	45,700	
	0420	1134 kg capacity		.10	160		33,800	8,000		41,800	49,100	
	0425	Electric freight, base unit, 1814 kg, 1.01 m/s, 4 stop, std fin	↓	.05	320		73,500	16,000		89,500	104,500	
	0450	For 2268 kg capacity, add	R142000 -40				5,250			5,250	5,775	
	0500	For 2722 kg capacity, add					9,275			9,275	10,200	
	0525	For 3175 kg capacity, add					12,400			12,400	13,700	
	0550	For 3629 kg capacity, add					17,000			17,000	18,700	
	0575	For 4536 kg capacity, add					20,300			20,300	22,300	
	0600	For 5443 kg capacity, add					24,800			24,800	27,300	
	0625	For 7257 kg capacity, add					29,800			29,800	32,800	
	0650	For 90720 kg capacity, add					32,900			32,900	36,200	
	0675	For increased speed, 1.27 m/s, add					9,875			9,875	10,900	
	0700	1.52 m/s, geared electric, add					12,300			12,300	13,500	
	0725	1.77 m/s, geared electric, add					14,500			14,500	16,000	
	0750	2.03 m/s, geared electric, add					16,400			16,400	18,100	
	0775	2.54 m/s, gearless electric, add					21,000			21,000	23,100	
	0800	3.04 m/s, gearless electric, add					23,200			23,200	25,500	
	0825	3.55 m/s, gearless electric, add					27,000			27,000	29,700	
	0850	4.06 m/s, gearless electric, add					30,000			30,000	33,000	
	0875	For class "B" loading, add					1,775			1,775	1,950	
	0900	For class "C-1" loading, add					4,400			4,400	4,850	
	0925	For class "C-2" loading, add					5,250			5,250	5,775	
	0950	For class "C-3" loading, add	↓			↓	7,225			7,225	7,950	

Important: See the Reference Section for supporting data - Crews, Rental Equipment, City Cost Indexes and Reference Data

14 CONVEYING SYSTEMS

14210 | Electric Traction Elevators

			CREW	DAILY OUTPUT	LABOR-HOURS	UNIT	2006 BARE COSTS				TOTAL INCL O&P	
							MAT.	LABOR	EQUIP.	TOTAL		
200	0975	For verticle travel over 12 m, add	2 Elev	2.21	7.240	m	345	360		705	920	200
	1000	For number of stops over 4, add	R142000 -10	.27	59.259	Stop	1,875	2,950		4,825	6,475	
	1625	Electric pass., base unit, 907 kg, 1.01 m/s, 4 stop, std. fin.		.05	320	Ea.	68,500	16,000		84,500	99,500	
	1650	For 1134 kg capacity, add	R142000 -20				2,800			2,800	3,075	
	1675	For 1361 kg capacity, add					4,275			4,275	4,700	
	1700	For 1588 kg capacity, add	R142000 -30				5,950			5,950	6,525	
	1725	For 1814 kg capacity, add					6,350			6,350	7,000	
	1750	For 2041 kg capacity, add	R142000 -40				8,350			8,350	9,200	
	1775	For 2268 kg capacity, add					10,600			10,600	11,700	
	1800	For increased speed, 1.27 m/s, geared electric, add					2,350			2,350	2,575	
	1825	1.52 m/s, geared electric, add					4,850			4,850	5,325	
	1850	1.77 m/s, geared electric, add					5,750			5,750	6,325	
	1875	2.03 m/s, geared electric, add					8,150			8,150	8,975	
	1900	2.54 m/s, gearless electric, add					38,300			38,300	42,100	
	1925	3.04 m/s, gearless electric, add					40,500			40,500	44,500	
	1950	3.55 m/s, gearless electric, add					44,300			44,300	48,700	
	1975	4.06 m/s, gearless electric, add					48,800			48,800	53,500	
	2000	For verticle travel over 12 m, add	2 Elev	2.21	7.240	m	350	360		710	925	
	2025	For number of stops over 4, add		.27	59.259	Stop	2,400	2,950		5,350	7,050	
	2400	Electric hospital, base unit, 1814 kg, 1.01 m/s, 4 stop, std fin		.05	320	Ea.	75,000	16,000		91,000	106,500	
	2425	For 2041 kg capacity, add					5,175			5,175	5,700	
	2450	For 2268 kg capacity, add					6,775			6,775	7,450	
	2475	For increased speed, 1.27 m/s, geared electric, add					2,525			2,525	2,775	
	2500	1.52 m/s, geared electric, add					4,800			4,800	5,275	
	2525	1.77 m/s, geared electric, add					5,850			5,850	6,450	
	2550	2.03 m/s, geared electric, add					8,175			8,175	8,975	
	2575	2.54 m/s, gearless electric, add					36,600			36,600	40,300	
	2600	3.04 m/s, gearless electric, add					40,500			40,500	44,500	
	2625	3.55 m/s, gearless electric, add					44,200			44,200	48,600	
	2650	4.06 m/s, gearless electric, add					48,800			48,800	53,500	
	2675	For verticle travel over 12 m, add	2 Elev	2.21	7.240	m	350	360		710	925	
	2700	For number of stops over 4, add	"	.27	59.259	Stop	2,975	2,950		5,925	7,700	

14240 | Hydraulic Elevators

			CREW	DAILY OUTPUT	LABOR-HOURS	UNIT	MAT.	LABOR	EQUIP.	TOTAL	TOTAL INCL O&P	
200	0010	**HYDRAULIC ELEVATORS**										200
	1025	Hydraulic freight, base unit, 907 kg, 0.25 m/s, 2 stop, std fin	2 Elev	.10	160	Ea.	36,100	8,000		44,100	51,500	
	1050	For 1134 kg capacity, add					2,625			2,625	2,875	
	1075	For 1361 kg capacity, add					4,150			4,150	4,550	
	1100	For 1588 kg capacity, add					6,275			6,275	6,900	
	1125	For 1814 kg capacity, add					6,700			6,700	7,375	
	1150	For 2041 kg capacity, add					7,725			7,725	8,500	
	1175	For 2268 kg capacity, add					10,500			10,500	11,600	
	1200	For 2722 kg capacity, add					11,100			11,100	12,200	
	1225	For 3175 kg capacity, add					17,000			17,000	18,700	
	1250	For 3629 kg capacity, add					19,000			19,000	20,900	
	1275	For 4536 kg capacity, add					20,200			20,200	22,200	
	1300	For 5443 kg capacity, add					24,400			24,400	26,800	
	1325	For 7257 kg capacity, add					31,800			31,800	35,000	
	1350	For 9072 kg capacity, add					35,300			35,300	38,800	
	1375	For increased speed, 0.50 m/s, add					755			755	830	
	1400	0.63 m/s, add					1,450			1,450	1,600	
	1425	0.76 m/s, add					2,725			2,725	3,000	
	1450	0.88 m/s, add					4,150			4,150	4,575	
	1475	For class "B" loading, add					1,725			1,725	1,900	
	1500	For class "C-1" loading, add					4,350			4,350	4,800	
	1525	For class "C-2" loading, add					5,225			5,225	5,725	

14240 | Hydraulic Elevators

		CREW	DAILY OUTPUT	LABOR-HOURS	UNIT	2006 BARE COSTS				TOTAL INCL O&P	
						MAT.	LABOR	EQUIP.	TOTAL		
200	1550	For class "C-3" loading, add				Ea.	7,175			7,175	7,875	200
	1575	For verticle travel over 6 m, add	2 Elev	2.21	7.240	m	1,150	360		1,510	1,800	
	1600	For number of stops over 2, add	↓	.27	59.259	Stop	510	2,950		3,460	4,975	
	2050	Hyd. pass., base unit, 680 kg, .508 m/s, 2 stop, std. fin.	↓	.10	160	Ea.	30,600	8,000		38,600	45,600	
	2075	For 907 kg capacity, add					975			975	1,075	
	2100	For 1134 kg capacity, add					2,000			2,000	2,200	
	2125	For 1361 kg capacity, add					3,625			3,625	4,000	
	2150	For 1588 kg capacity, add					5,375			5,375	5,925	
	2175	For 1814 kg capacity, add					6,475			6,475	7,125	
	2200	For 2041 kg capacity, add					7,800			7,800	8,600	
	2225	For 2268 kg capacity, add					11,800			11,800	13,000	
	2250	For increased speed, 0.63 m/s, add					1,150			1,150	1,275	
	2275	0.76 m/s, add					2,300			2,300	2,525	
	2300	0.88 m/s, add					3,725			3,725	4,100	
	2325	1.01 m/s, add				↓	5,425			5,425	5,975	
	2350	For verticle travel over 3.66 m, add	2 Elev	2.21	7.240	m	1,150	360		1,510	1,825	
	2375	For number of stops over 2, add	↓	.27	59.259	Stop	3,050	2,950		6,000	7,775	
	2725	Hydraulic hospital, base unit, 1814 kg, 0.50 m/s, 2 stop, std. fin.	↓	.10	160	Ea.	42,500	8,000		50,500	58,500	
	2750	For 1814 kg capacity, add					5,125			5,125	5,650	
	2775	For 2041 kg capacity, add					6,025			6,025	6,625	
	2800	For 2268 kg capacity, add					8,775			8,775	9,650	
	2825	For increased speed, 0.63 m/s, add					1,425			1,425	1,575	
	2850	0.76 m/s, add					2,400			2,400	2,650	
	2875	0.88 m/s, add					3,875			3,875	4,250	
	2900	1.01 m/s, add				↓	5,650			5,650	6,200	
	2925	For verticle travel over 3.66 m, add	2 Elev	2.21	7.240	m	770	360		1,130	1,375	
	2950	For number of stops over 2, add	"	.27	59.259	Stop	2,950	2,950		5,900	7,675	

14270 | Custom Elevator Cabs

		CREW	DAILY OUTPUT	LABOR-HOURS	UNIT	MAT.	LABOR	EQUIP.	TOTAL	TOTAL INCL O&P	
200	0010	**CAB FINISHES**										200
	3325	Passenger elevator cab finishes (based on 1588 kg cab size)										
	3350	Acrylic panel ceiling				Ea.	440			440	485	
	3375	Aluminum eggcrate ceiling					510			510	560	
	3400	Stainless steel doors					1,975			1,975	2,175	
	3425	Carpet flooring					370			370	405	
	3450	Epoxy flooring					305			305	335	
	3475	Quarry tile flooring					470			470	515	
	3500	Slate flooring					655			655	720	
	3525	Textured rubber flooring					119			119	131	
	3550	Stainless steel walls					2,650			2,650	2,925	
	3575	Stainless steel returns at door				↓	560			560	615	
	4450	Hospital elevator cab finishes (based on 1588 kg cab size)										
	4475	Aluminum eggcrate ceiling				Ea.	510			510	560	
	4500	Stainless steel doors					1,625			1,625	1,800	
	4525	Epoxy flooring					305			305	335	
	4550	Quarry tile flooring					465			465	515	
	4575	Textured rubber flooring					119			119	131	
	4600	Stainless steel walls					2,650			2,650	2,925	
	4625	Stainless steel returns at door				↓	560			560	615	

14280 | Elevator Equipment and Controls

		CREW	DAILY OUTPUT	LABOR-HOURS	UNIT	MAT.	LABOR	EQUIP.	TOTAL	TOTAL INCL O&P	
200	0010	**ELEVATOR CONTROLS AND DOORS**										200
	2975	Passenger elevator options										
	3000	2 car group automatic controls	2 Elev	.66	24.242	Ea.	2,450	1,200		3,650	4,475	
	3025	3 car group automatic controls	↓	.44	36.364	↓	3,675	1,825		5,500	6,725	

14280	Elevator Equipment and Controls	CREW	DAILY OUTPUT	LABOR-HOURS	UNIT	2006 BARE COSTS				TOTAL INCL O&P
						MAT.	LABOR	EQUIP.	TOTAL	
200	3050	4 car group automatic controls	2 Elev	.33	48.485	Ea.	6,075	2,425		8,500	10,300	200
	3075	5 car group automatic controls		.26	61.538		8,275	3,075		11,350	13,700
	3100	6 car group automatic controls		.22	72.727		12,500	3,625		16,125	19,200
	3125	Intercom service		3	5.333		325	266		591	755
	3150	Duplex car selective collective		.66	24.242		2,700	1,200		3,900	4,750
	3175	Center opening 1 speed doors		2	8		1,400	400		1,800	2,150
	3200	Center opening 2 speed doors		2	8		1,675	400		2,075	2,450
	3225	Rear opening doors (opposite front)		2	8		3,900	400		4,300	4,875
	3250	Side opening 2 speed doors		2	8		3,700	400		4,100	4,675
	3275	Automatic emergency power switching		.66	24.242		875	1,200		2,075	2,750
	3300	Manual emergency power switching	▼	8	2		335	100		435	520
	3625	Hall finishes, stainless steel doors					925			925	1,025
	3650	Stainless steel frames					925			925	1,025
	3675	12 month maintenance contract									2,640
	3700	Signal devices, hall lanterns	2 Elev	8	2		345	100		445	530
	3725	Position indicators, up to 3		9.40	1.702		239	85		324	390
	3750	Position indicators, per each over 3	▼	32	.500		66	25		91	110
	3775	High speed heavy duty door opener					1,675			1,675	1,850
	3800	Variable voltage, O.H. gearless machine, min.	2 Elev	.16	100		24,600	5,000		29,600	34,600
	3815	Maximum		.07	228		54,500	11,400		65,900	77,000
	3825	Basement installed geared machine	▼	.33	48.485	▼	12,200	2,425		14,625	17,000
	3850	Freight elevator options									
	3875	Doors, biparting	2 Elev	.66	24.242	Ea.	4,175	1,200		5,375	6,400
	3900	Power operated door and gate	"	.66	24.242		16,900	1,200		18,100	20,400
	3925	Finishes, steel plate floor					735			735	810
	3950	14 ga. 6 mm x 1224 mm steel plate walls					1,800			1,800	1,975
	3975	12 month maintenance contract									1,992
	4000	Signal devices, hall lanterns	2 Elev	8	2		345	100		445	530
	4025	Position indicators, up to 3		9.40	1.702		239	85		324	390
	4050	Position indicators, per each over 3		32	.500		66	25		91	110
	4075	Variable voltage basement installed geared machine	▼	.66	24.242	▼	13,600	1,200		14,800	16,700
	4100	Hospital elevator options									
	4125	2 car group automatic controls	2 Elev	.66	24.242	Ea.	2,450	1,200		3,650	4,475
	4150	3 car group automatic controls		.44	36.364		3,675	1,825		5,500	6,725
	4175	4 car group automatic controls		.33	48.485		6,075	2,425		8,500	10,300
	4200	5 car group automatic controls		.26	61.538		8,300	3,075		11,375	13,700
	4225	6 car group automatic controls		.22	72.727		12,500	3,625		16,125	19,200
	4250	Intercom service		3	5.333		325	266		591	755
	4275	Duplex car selective collective		.66	24.242		2,700	1,200		3,900	4,750
	4300	Center opening 1 speed doors		2	8		1,400	400		1,800	2,150
	4325	Center opening 2 speed doors		2	8		1,850	400		2,250	2,625
	4350	Rear opening doors (opposite front)		2	8		3,900	400		4,300	4,875
	4375	Side opening 2 speed doors		2	8		5,925	400		6,325	7,100
	4400	Automatic emergency power switching		.66	24.242		875	1,200		2,075	2,750
	4425	Manual emergency power switching	▼	8	2		335	100		435	520
	4675	Hall finishes, stainless steel doors					925			925	1,025
	4700	Stainless steel frames					925			925	1,025
	4725	12 month maintenance contract									3,985
	4750	Signal devices, hall lanterns	2 Elev	8	2		345	100		445	530
	4775	Position indicators, up to 3		9.40	1.702		239	85		324	390
	4800	Position indicators, per each over 3	▼	32	.500		66	25		91	110
	4825	High speed heavy duty door opener					1,675			1,675	1,850
	4850	Variable voltage, O.H. gearless machine, min.	2 Elev	.16	100		24,600	5,000		29,600	34,600
	4865	Maximum		.07	228		54,500	11,400		65,900	77,000
	4875	Basement installed geared machine	▼	.33	48.485	▼	12,200	2,425		14,625	17,000
	5000	Drilling for piston, casing included, 457 mm diameter	B-48	24.38	2.297	m	112	71.50	149	332.50	395

CONVEYING SYSTEMS **14**

495

14320 | Escalators

		CREW	DAILY OUTPUT	LABOR-HOURS	UNIT	2006 BARE COSTS				TOTAL INCL O&P
						MAT.	LABOR	EQUIP.	TOTAL	
300	0010 **ESCALATORS**										300
1000	Glass, 800 mm wide x 3050 mm floor to floor height [R143110 -10]	M-1	.07	457	Ea.	64,000	21,700	775	86,475	104,000
1010	1200 mm wide x 3050 mm floor to floor height		.07	457		70,500	21,700	775	92,975	111,000
1020	800 mm wide x 4500 mm floor to floor height		.06	533		67,000	25,300	900	93,200	113,000
1030	1200 mm wide x 4500 mm floor to floor height		.06	533		71,000	25,300	900	97,200	117,000
1040	800 mm wide x 6100 mm floor to floor height		.05	653		69,000	31,000	1,100	101,100	123,500
1050	1200 mm wide x 6100 mm floor to floor height		.05	653		74,000	31,000	1,100	106,100	129,000
1060	800 mm wide x 7600 mm floor to floor height		.04	800		74,500	38,000	1,350	113,850	140,000
1070	1200 mm wide x 7600 mm floor to floor height		.04	800		80,000	38,000	1,350	119,350	146,000
1080	Enameled steel, 800 mm wide x 3050 floor to floor height		.07	457		65,500	21,700	775	87,975	105,500
1090	1200 mm wide x 3050 mm floor to floor height		.07	457		70,500	21,700	775	92,975	111,000
1110	800 mm wide x 4500 mm floor to floor height		.06	533		67,000	25,300	900	93,200	113,000
1120	1200 mm wide x 4500 mm floor to floor height		.06	533		71,000	25,300	900	97,200	117,000
1130	800 mm wide x 6100 mm floor to floor height		.05	653		69,000	31,000	1,100	101,100	123,500
1140	1200 mm wide x 6100 mm floor to floor height		.05	653		74,000	31,000	1,100	106,100	129,000
1150	800 mm wide x 7600 mm floor to floor height		.04	800		74,500	38,000	1,350	113,850	140,000
1160	1200 mm wide x 7600 mm floor to floor height		.04	800		80,000	38,000	1,350	119,350	146,000
1170	Stainless steel, 800 mm wide x 3050 mm floor to floor height		.07	457		72,000	21,700	775	94,475	112,500
1180	1200 mm wide x 3050 mm floor to floor height		.07	457		75,500	21,700	775	97,975	116,500
1500	800 mm wide x 4500 mm floor to floor height		.06	533		73,500	25,300	900	99,700	119,500
1700	1200 mm wide x 4500 mm floor to floor height		.06	533		76,500	25,300	900	102,700	123,500
2300	800 mm wide x 7500 mm floor to floor height		.04	800		83,000	38,000	1,350	122,350	149,500
2500	1200 mm wide x 7500 mm floor to floor height		.04	800		88,000	38,000	1,350	127,350	155,000

14360 | Moving Walks

		CREW	DAILY OUTPUT	LABOR-HOURS	UNIT	2006 BARE COSTS				TOTAL INCL O&P
						MAT.	LABOR	EQUIP.	TOTAL	
500	0010 **MOVING RAMPS AND WALKS**										500
0020	Walk, 700 mm tread width, minimum [R143210 -20]	M-1	1.98	16.152	m	2,350	765	27.50	3,142.50	3,750
0100	90 m to 150 m, maximum		1.35	23.699		3,250	1,125	40	4,415	5,300
0300	1200 mm tread width walk, minimum		1.35	23.699		5,300	1,125	40	6,465	7,550
0400	30 m to 107 m, maximum		1.16	27.483		6,225	1,300	46.50	7,571.50	8,850
0600	Ramp, 12° incline, 900 mm tread width, minimum		1.61	19.922		4,350	945	33.50	5,328.50	6,200
0700	21 m to 27 m maximum		1.16	27.483		6,200	1,300	46.50	7,546.50	8,825
0900	1200 mm tread width, minimum		1.09	29.408		6,350	1,400	49.50	7,799.50	9,100
1000	12 m to 21 m, maximum		.89	36.078		8,000	1,700	61	9,761	11,400

14400 | Lifts

14420 | Wheelchair Lifts

		CREW	DAILY OUTPUT	LABOR-HOURS	UNIT	2006 BARE COSTS				TOTAL INCL O&P
						MAT.	LABOR	EQUIP.	TOTAL	
100	0010 **CHAIR / WHEELCHAIR LIFT**										100
7700	Stair climber (chair lift), single seat, minimum	2 Elev	1	16	Ea.	4,100	800		4,900	5,725
7800	Maximum		.20	80		5,650	4,000		9,650	12,200
8000	Wheelchair lift, minimum		1	16		5,625	800		6,425	7,400
8500	Maximum		.50	32		13,300	1,600		14,900	17,100
8700	Stair lift, minimum		1	16		11,100	800		11,900	13,500
8900	Maximum		.20	80		17,600	4,000		21,600	25,400

Important: See the Reference Section for supporting data - Crews, Rental Equipment, City Cost Indexes and Reference Data

14 CONVEYING SYSTEMS

14400 | Lifts

14450 | Vehicle Lifts

			CREW	DAILY OUTPUT	LABOR-HOURS	UNIT	2006 BARE COSTS MAT.	LABOR	EQUIP.	TOTAL	TOTAL INCL O&P	
700	0010	**HYDRAULIC LIFT AUTO/TRUCK**										**700**
	0200	Double post, 4080 kg frame	L-4	1.15	20.870	Ea.	5,150	690		5,840	6,750	
	0300	5450 kg frame		.85	28.235		7,300	935		8,235	9,475	
	0400	11 800 kg frame		.22	109		22,900	3,625		26,525	30,800	
	2200	Hoists, single post, 3629 kg capacity, swivel arms		.40	60		4,525	2,000		6,525	8,075	
	2400	Two posts, adjustable frames, 4990 kg capacity		.25	96		5,800	3,175		8,975	11,400	
	2500	10 886 kg capacity		.15	160		7,750	5,300		13,050	16,800	
	2700	3402 kg capacity, frame supports		.50	48		6,450	1,600		8,050	9,575	
	2800	Four post, roll on ramp		.50	48		5,800	1,600		7,400	8,875	
	2810	Hydraulic lifts, above ground, 2 post, clear floor, 2722 kg cap		2.67	8.989		6,075	298		6,373	7,150	
	2815	4082 kg capacity		2.29	10.480		14,400	345		14,745	16,300	
	2820	6804 kg capacity		2	12		16,300	400		16,700	18,500	
	2825	13 608 kg capacity		1.60	15		36,000	495		36,495	40,400	
	2830	4 post, ramp style, 11 340 kg capacity		2	12		14,100	400		14,500	16,100	
	2835	15 876 kg capacity		1	24		65,500	795		66,295	73,000	
	2840	22 680 kg capacity		1	24		73,500	795		74,295	81,500	
	2845	34 019 kg capacity	↓	1	24		85,000	795		85,795	95,000	
	2850	For drive thru tracks, add, minimum					905			905	995	
	2855	Maximum					1,550			1,550	1,700	
	2860	Ramp extensions, 900 mm (set of 2)					745			745	815	
	2865	Rolling jack platform					2,575			2,575	2,850	
	2870	Elec/hyd jacking beam					6,900			6,900	7,600	
	2880	Scissor lift, portable, 2722 kg capacity				↓	6,775			6,775	7,450	

14460 | Correspondence & Parcel Lifts

			CREW	DAILY OUTPUT	LABOR-HOURS	UNIT	MAT.	LABOR	EQUIP.	TOTAL	TOTAL INCL O&P	
200	0010	**CORRESPONDENCE LIFT**										**200**
	0020	1 floor 2 stop 11.3 kg capacity	2 Elev	.20	80	Ea.	5,025	4,000		9,025	11,500	
	0100	Hand, 2.3 kg capacity	"	.20	80	"	1,900	4,000		5,900	8,075	
600	0010	**PARCEL LIFT**										**600**
	0020	508 mm x 508 mm, 45.4 kg capacity, electric, per floor	2 Mill	.25	64	Ea.	7,850	2,375		10,225	12,200	

14500 | Material Transport

14510 | Material Handling

			CREW	DAILY OUTPUT	LABOR-HOURS	UNIT	2006 BARE COSTS MAT.	LABOR	EQUIP.	TOTAL	TOTAL INCL O&P	
900	0010	**MOTORIZED CAR** Distribution systems, single track,										**900**
	0100	9.1 kg/car capacity, material handling										
	0200	Minimum, 15 m to 30 m	4 Mill	.19	168	Station	25,300	6,250		31,550	37,000	
	0300	Maximum, 30 m to 61 m	"	.15	213	"	34,200	7,925		42,125	49,300	
	0400	Larger system, incl. hospital transport, track type,										
	0500	fully automated material handling system										
	0600	Minimum, complete system	4 Mill	.05	640	Ea.	112,000	23,700		135,700	158,500	
	0700	Maximum, with several stations	E-6	.05	2560	"	654,500	101,000	32,800	788,300	933,000	

14550 | Conveyors

			CREW	DAILY OUTPUT	LABOR-HOURS	UNIT	MAT.	LABOR	EQUIP.	TOTAL	TOTAL INCL O&P	
350	0010	**MATERIAL HANDLING** Conveyors, gravity 50 mm rollers, 75 mm O.C.										**350**
	0050	3 m sections with 2 supports, 272 kg capacity, 450 mm wide				Ea.	375			375	415	
	0100	612 mm wide					435			435	480	
	0150	623 kg capacity, 450 mm wide				↓	345			345	380	

14550	Conveyors	CREW	DAILY OUTPUT	LABOR-HOURS	UNIT	2006 BARE COSTS				TOTAL INCL O&P	
						MAT.	LABOR	EQUIP.	TOTAL		
350	0200	600 mm wide				Ea.	385			385	425	**350**
	0350	Horizontal belt, center drive and takeup, 0.30 m/s									
	0400	400 mm belt, 8 m length	2 Mill	.50	32	Ea.	2,650	1,175		3,825	4,675
	0450	600 mm belt, 12 450 mm length		.40	40		4,125	1,475		5,600	6,725
	0500	18 450 mm length	↓	.30	53.333	↓	5,500	1,975		7,475	8,975
	0600	Inclined belt, 3 m rise with horizontal loader and									
	0620	End idler assembly, 8.4 m length, 450 mm belt	2 Mill	.30	53.333	Ea.	7,750	1,975		9,725	11,500
	0700	600 mm belt	"	.15	106	"	8,700	3,950		12,650	15,400
	3600	Monorail, overhead, manual, channel type									
	3700	186 kg/m	1 Mill	7.92	1.009	m	28	37.50		65.50	86.50
	3900	744 kg/m	"	6.40	1.250	"	25.50	46.50		72	96.50
	4000	Trolleys for above, 2 wheel, 58 kg capacity				Ea.	54			54	59.50
	4200	4 wheel, 113 kg capacity					164			164	181
	4300	8 wheel, 454 kg capacity				↓	315			315	350
900	0010	**VERTICAL CONVEYOR** Automatic selective										**900**
	0100	to 10 floors, base price	2 Mill	.13	120	Floor	23,100	4,475		27,575	32,100
	0200	Add for electrical hook-up and testing	2 Elec	.50	32	"		1,350		1,350	2,000

14560	Chutes										
250	0010	**CHUTES** Linen or refuse, incl. sprinklers, 3658 mm floor height										**250**
	0050	Aluminized steel, 16 ga., 450 mm diameter	2 Shee	3.50	4.571	Floor	870	193		1,063	1,250
	0100	600 mm diameter		3.20	5		945	211		1,156	1,375
	0200	750 mm diameter		3	5.333		1,025	225		1,250	1,500
	0300	900 mm diameter		2.80	5.714		1,175	241		1,416	1,675
	0400	Galvanized steel, 16 ga., 450 mm diameter		3.50	4.571		770	193		963	1,150
	0500	600 mm diameter		3.20	5		870	211		1,081	1,275
	0600	750 mm diameter		3	5.333		975	225		1,200	1,425
	0700	900 mm diameter		2.80	5.714		1,150	241		1,391	1,650
	0800	Stainless steel, 450 mm diameter		3.50	4.571		1,800	193		1,993	2,275
	0900	600 mm diameter		3.20	5		1,900	211		2,111	2,425
	1000	750 mm diameter		3	5.333		2,250	225		2,475	2,825
	1005	900 mm diameter		2.80	5.714	↓	2,500	241		2,741	3,125
	1200	Linen chute bottom collector, aluminized steel		4	4	Ea.	1,075	169		1,244	1,425
	1300	Stainless steel		4	4		1,375	169		1,544	1,775
	1500	Refuse, bottom hopper, aluminized steel, 450 mm diameter		3	5.333		705	225		930	1,125
	1600	600 mm diameter		3	5.333		775	225		1,000	1,200
	1800	900 mm diameter		3	5.333	↓	1,325	225		1,550	1,800
	2900	Package chutes, spiral type, minimum		4.50	3.556	Floor	1,875	150		2,025	2,275
	3000	Maximum	↓	1.50	10.667	"	4,875	450		5,325	6,075

14580	Pneumatic Tube Systems										
800	0010	**PNEUMATIC TUBE SYSTEM** Single tube, 2 stations, blower										**800**
	0020	30 m long, stock									
	0100	75 mm diameter	2 Stpi	.12	133	Total	5,075	5,750		10,825	14,300
	0300	102 mm diameter	"	.09	177	"	5,725	7,650		13,375	17,800
	0400	Twin tube, two stations or more, conventional system									
	0600	65 mm round	2 Stpi	19.05	.840	m	33.50	36		69.50	91.50
	0700	75 mm round		14.02	1.141		38.50	49		87.50	117
	0900	100 mm round		15.12	1.058		42.50	45.50		88	116
	1000	100 mm x 175 mm oval	↓	11.46	1.396	↓	62.50	60		122.50	159
	1050	Add for blower		2	8	System	4,025	345		4,370	4,950
	1110	Plus for each round station, add		7.50	2.133	Ea.	450	92		542	635
	1150	Plus for each oval station, add		7.50	2.133	"	450	92		542	635
	1200	Alternate pricing method: base cost, minimum		.75	21.333	Total	4,925	920		5,845	6,775
	1300	Maximum	↓	.25	64	"	9,775	2,750		12,525	14,900

Important: See the Reference Section for supporting data - Crews, Rental Equipment, City Cost Indexes and Reference Data

14 CONVEYING SYSTEMS

14500 | Material Transport

14580 | Pneumatic Tube Systems

			CREW	DAILY OUTPUT	LABOR-HOURS	UNIT	2006 BARE COSTS				TOTAL INCL O&P	
							MAT.	LABOR	EQUIP.	TOTAL		
800	1500	Plus total system length, add, minimum	2 Stpi	28.47	.562	m	20.50	24		44.50	59.50	800
	1600	Maximum		11.46	1.396	"	61.50	60		121.50	158	
	1800	Completely automatic system, 102 mm round, 15 to 50 stations		.29	55.172	Station	15,500	2,375		17,875	20,600	
	2200	51 to 144 stations		.32	50		12,000	2,150		14,150	16,500	
	2400	152 mm round or 102 mm x 178 mm oval, 15 to 50 stations		.24	66.667		19,400	2,875		22,275	25,700	
	2800	51 to 144 stations	▼	.23	69.565	▼	16,300	3,000		19,300	22,400	

14600 | Hoists and Cranes

14610 | Fixed Hoists

			CREW	DAILY OUTPUT	LABOR-HOURS	UNIT	2006 BARE COSTS				TOTAL INCL O&P	
							MAT.	LABOR	EQUIP.	TOTAL		
500	0010	**MATERIAL HANDLING**										500
	1500	Cranes, portable hydraulic, floor type, 900 kg capacity				Ea.	1,900			1,900	2,100	
	1600	1800 kg capacity					3,200			3,200	3,525	
	1800	Movable gantry type, 3.7 m to 4.6 m range, 900 kg capacity					2,775			2,775	3,050	
	1900	2700 kg capacity					4,025			4,025	4,425	
	2100	Hoists, electric overhead, chain hook hung, 4500 mm lift, 9072 kg cap					1,850			1,850	2,025	
	2200	2722 kg capacity					2,300			2,300	2,525	
	2500	4536 kg capacity				▼	5,100			5,100	5,600	
	2600	For hand-pushed trolley, add					15%					
	2700	For geared trolley, add					30%					
	2800	For motor trolley, add					75%					
	3000	For lifts over 4600 mm, 907 kg, add				m	59			59	65	
	3100	4536 kg, add				"	131			131	144	
	3300	Lifts, scissor type, portable, electric, 900 mm high, 900 kg				Ea.	3,300			3,300	3,625	
	3400	1200 mm high, 1800 kg				"	3,600			3,600	3,950	

14630 | Bridge Cranes

			CREW	DAILY OUTPUT	LABOR-HOURS	UNIT	2006 BARE COSTS				TOTAL INCL O&P	
							MAT.	LABOR	EQUIP.	TOTAL		
300	0010	**CRANE RAIL**										300
	0020	Box beam bridge, no equipment included	E-4	1542	.021	kg	2.01	.84	.06	2.91	3.77	
	0200	Running track only, 0.052 kg/mm		2540	.013	"	1.06	.51	.04	1.61	2.13	
	0210	Running track only, 52 kg per m, 6 m piece	▼	48.77	.656	m	55	26.50	1.83	83.33	111	
700	0010	**OVERHEAD BRIDGE CRANES**										700
	0100	1 girder, 6000 mm span, 2720 kg	M-3	1	34	Ea.	19,100	1,375	151	20,626	23,200	
	0125	4540 kg		1	34		20,900	1,375	151	22,426	25,200	
	0150	6800 kg		1	34		24,900	1,375	151	26,426	29,600	
	0175	9070 kg		.80	42.500		32,900	1,700	189	34,789	39,000	
	0200	13 600 kg		.80	42.500		42,300	1,700	189	44,189	49,300	
	0225	9000 mm span, 2720 kg		1	34		19,900	1,375	151	21,426	24,100	
	0250	4540 kg		1	34		21,900	1,375	151	23,426	26,200	
	0275	6800 kg		1	34		26,200	1,375	151	27,726	31,000	
	0300	9070 kg		.80	42.500		34,100	1,700	189	35,989	40,300	
	0325	13 600 kg	▼	.80	42.500		44,200	1,700	189	46,089	51,500	
	0350	2 girder, 12 000 mm span, 2720 kg	M-4	.50	72		32,900	2,875	460	36,235	40,900	
	0375	4540 kg		.50	72		34,400	2,875	460	37,735	42,600	
	0400	6800 kg		.50	72		37,700	2,875	460	41,035	46,300	
	0425	9070 kg		.40	90		44,100	3,575	570	48,245	54,500	
	0450	13 600 kg		.40	90		60,000	3,575	570	64,145	72,000	
	0475	22 700 kg		.30	120		71,000	4,775	765	76,540	86,000	
	0500	15 000 mm span, 2720 kg	▼	.50	72	▼	37,400	2,875	460	40,735	46,000	

14630	Bridge Cranes	CREW	DAILY OUTPUT	LABOR-HOURS	UNIT	2006 BARE COSTS				TOTAL INCL O&P	
						MAT.	LABOR	EQUIP.	TOTAL		
700 0525	4540 kg	M-4	.50	72	Ea.	38,900	2,875	460	42,235	47,600	**700**
0550	6800 kg		.50	72		41,600	2,875	460	44,935	50,500	
0575	9070 kg		.40	90		48,000	3,575	570	52,145	59,000	
0600	13 600 kg		.40	90		63,000	3,575	570	67,145	75,000	
0625	22 700 kg	▼	.30	120	▼	74,500	4,775	765	80,040	89,500	

For information about Means Estimating Seminars, see yellow pages 12 and 13 in back of book

Important: See the Reference Section for supporting data - Crews, Rental Equipment, City Cost Indexes and Reference Data

Division 15
Mechanical

Estimating Tips

15100 Building Services Piping
This subdivision is primarily basic pipe and related materials. The pipe may be used by any of the mechanical disciplines, i.e., plumbing, fire protection, heating, and air conditioning.

- The piping section lists the add to labor for elevated pipe installation. These adds apply to all elevated pipe, fittings, valves, insulation, etc., that are placed above 3 m high. CAUTION: the correct percentage may vary for the same pipe. For example, the percentage add for the basic pipe installation should be based on the maximum height that the craftsman must install for that particular section. If the pipe is to be located 4 m above the floor but it is suspended on threaded rod from beams, the bottom flange of which is 5.5 m high (1.2 m rods), then the height is actually 5.5 m and the add is 20%. The pipe coverer, however, does not have to go above the 4 m and so his add should be 10%.

- Most pipe is priced first as straight pipe with a joint (coupling, weld, etc.) every 3 m and a hanger usually every 3 m. There are exceptions with hanger spacing such as: for cast iron pipe (1.5 m) and plastic pipe (3 per 3 m). Following each type of pipe there are several lines listing sizes and the amount to be subtracted to delete couplings and hangers. This is for pipe that is to be buried or supported together on trapeze hangers. The reason that the couplings are deleted is that these runs are usually long and frequently longer lengths of pipe are used. By deleting the couplings the estimator is expected to look up and add back the correct reduced number of couplings.

- When preparing an estimate it may be necessary to approximate the fittings. Fittings usually run between 25% and 50% of the cost of the pipe. The lower percentage is for simpler runs, and the higher number is for complex areas like mechanical rooms.

- For historic restoration projects, the systems must be as invisible as possible, and pathways must be sought for pipes, conduits, and ductwork. While installations in accessible spaces (such as basements and attics) are relatively straightforward to estimate, labor costs may be more difficult to determine when delivery systems must be concealed.

15400 Plumbing Fixtures & Equipment
- Plumbing fixture costs usually require two lines, the fixture itself and its "rough-in, supply and waste".

- In the Assemblies Section (Plumbing D2010) for the desired fixture, the System Components Group in the center of the page shows the fixture on the first line. The rest of the list (fittings, pipe, tubing, etc.) will total up to what we refer to in the Unit Price section as "Rough-in, supply, waste and vent". Note that for most fixtures we allow a nominal 2 meters of tubing to reach from the fixture to a main or riser.

- Remember that gas- and oil-fired units need venting.

15500 Heat Generation Equipment
- When estimating the cost of an HVAC system, check to see who is responsible for providing and installing the temperature control system. It is possible to overlook controls, assuming that they would be included in the electrical estimate.

- When looking up a boiler be careful on specified capacity. Some manufacturers rate their products on output while others use input.

- Include HVAC insulation for pipe, boiler and duct (wrap and liner).

- Be careful when looking up mechanical items to get the correct pressure rating and connection type (thread, weld, flange).

15700 Heating/Ventilation/ Air Conditioning Equipment
- Combination heating and cooling units are sized by the air conditioning requirements. (See Reference No. R236000-20 for preliminary sizing guide.)

- 3.5 kW of air conditioning is nominally 0.2 m³/s.

- Rectangular duct is taken off by the meter for each size, but its cost is usually estimated by the kilogram. Remember that SMACNA standards now base duct on internal pressure.

- Prefabricated duct is estimated and purchased like pipe: straight sections and fittings.

- Note that cranes or other lifting equipment are not included on any lines in Division 15. For example, if a crane is required to lift a heavy piece of pipe into place high above a gym floor, or to put a rooftop unit on the roof of a four-story building, etc., it must be added. Due to the potential for extreme variation—from nothing additional required, to a major crane or helicopter—we feel that including a nominal amount for "lifting contingency" would be useless and detract from the accuracy of the estimate. When using equipment rental from Means do not forget to include the cost of the operator(s).

Reference Numbers
Reference numbers are shown in bold squares at the beginning of some major classifications. These numbers refer to related items in the Reference Section. The reference information may be an estimating procedure, an alternate pricing method, or technical information.

Note: Not all subdivisions listed here necessarily appear in this publication.

Note: **i2 Trade Service,** *in part, has been used as a reference source for some of the material prices used in Division 15.*

15051 | Mechanical General

		CREW	DAILY OUTPUT	LABOR-HOURS	UNIT	2006 BARE COSTS				TOTAL INCL O&P
						MAT.	LABOR	EQUIP.	TOTAL	
100	0010	**BOILERS, GENERAL** Prices do not include flue piping, elec. wiring,									**100**
	0020	gas or oil piping, boiler base, pad, or tankless unless noted									
	0100	Boiler W: 10 KW = 34 lbs/steam/hr = 33,475 BTU/hr. R235000 -50									
	0120										
	0150	To convert SFR to BTU rating: Hot water, 150 x SFR;									
	0160	Forced hot water, 180 x SFR; steam, 240 x SFR									
700	0010	**PIPING** See also Divisions 02500 & 02600 for site work									**700**
	1000	Add to labor for elevated installation									
	1080	3.0 m to 4.6 m high						10%			
	1100	4.5 m to 6.0 m high						20%			
	1120	6.0 m to 7.5 m high						25%			
	1140	7.5 m to 9.0 m high						35%			
	1160	9.0 m to 10.5 m high						40%			
	1180	10.5 m to 12.0 m high						50%			
	1200	Over 12.0 m high						55%			

15055 | Selective Mech Demolition

		CREW	DAILY OUTPUT	LABOR-HOURS	UNIT	MAT.	LABOR	EQUIP.	TOTAL	TOTAL INCL O&P
300	0010	**HVAC DEMOLITION**									**300**
	0100	Air conditioner, split unit, 10 kW	Q-5	2	8	Ea.		310		310	465
	0150	Package unit, 10 kW	Q-6	3	8	"		320		320	485
	0298	Boilers									
	0300	Electric, up thru 148 kW	Q-19	2	12	Ea.		480		480	715
	0310	150 thru 518 kW	"	1	24			955		955	1,425
	0320	550 thru 2000 kW	Q-21	.40	80			3,250		3,250	4,875
	0330	2070 kW and up	"	.30	106			4,325		4,325	6,500
	0340	Gas and/or oil, steel, up thru 44 kW	Q-7	2.20	14.545			595		595	900
	0350	46.9 thru 586 kW		.80	40			1,650		1,650	2,475
	0360	615 thru 1319 kW		.50	64			2,625		2,625	3,950
	0370	1348 thru 2051 kW		.30	106			4,375		4,375	6,575
	0380	2080 thru 3516 kW		.16	200			8,200		8,200	12,400
	0390	3575 thru 7325 kW		.12	266			10,900		10,900	16,500
	1000	Ductwork, 100 mm high, 200 mm wide	1 Clab	60.96	.131	m		3.60		3.60	5.60
	1100	150 mm high, 200 mm wide		50.29	.159			4.36		4.36	6.80
	1200	250 mm high, 300 mm wide		38.10	.210			5.75		5.75	8.95
	1300	300 mm to 350 mm high, 400 mm to 450 mm wide		25.91	.309			8.45		8.45	13.15
	1400	450 mm high, 600 mm wide		20.42	.392			10.75		10.75	16.70
	1500	750 mm high, 900 mm wide		17.07	.469			12.85		12.85	20
	1540	1800 mm wide		15.24	.525			14.40		14.40	22.50
	3000	Mechanical equipment, light items. Unit is weight, not cooling.	Q-5	.82	19.596	Met. Ton		760		760	1,150
	3600	Heavy items	"	1	16.033	"		620		620	935
	3700	Deduct for salvage (when applicable), minimum				Job					55
	3710	Maximum				"					420
600	0010	**PLUMBING DEMOLITION**									**600**
	1020	Fixtures, including 3000 mm piping									
	1100	Bath tubs, cast iron	1 Plum	4	2	Ea.		85.50		85.50	129
	1120	Fiberglass		6	1.333			57		57	85.50
	1140	Steel		5	1.600			68.50		68.50	103
	1200	Lavatory, wall hung		10	.800			34		34	51.50
	1220	Counter top		8	1			42.50		42.50	64.50
	1250	Shower pan, minimum		24	.333			14.25		14.25	21.50
	1260	Shower pan, maximum		16	.500			21.50		21.50	32
	1300	Sink, single compartment		8	1			42.50		42.50	64.50
	1320	Double compartment		7	1.143			49		49	73.50
	1400	Water closet, floor mounted		8	1			42.50		42.50	64.50

15055 | Selective Mech Demolition

		CREW	DAILY OUTPUT	LABOR-HOURS	UNIT	MAT.	2006 BARE COSTS LABOR	EQUIP.	TOTAL	TOTAL INCL O&P	
600	1420	Wall mounted	1 Plum	7	1.143	Ea.		49		49	73.50	**600**
	1500	Urinal, floor mounted		4	2			85.50		85.50	129	
	1520	Wall mounted		7	1.143			49		49	73.50	
	1600	Water fountains, free standing		8	1			42.50		42.50	64.50	
	1620	Wall or deck mounted		6	1.333	▼		57		57	85.50	
	2000	Piping, metal, up thru 40 mm diameter		60.96	.131	m		5.60		5.60	8.45	
	2050	50 mm thru 90 mm diameter	▼	45.72	.175			7.45		7.45	11.25	
	2100	100 mm thru 150 mm diameter	2 Plum	30.48	.525			22.50		22.50	33.50	
	2150	200 mm thru 350 mm dia	"	18.29	.875			37.50		37.50	56	
	2153	400 mm thru 500 mm diameter	Q-18	21.34	1.125			45	3.24	48.24	71.50	
	2155	600 mm thru 650 mm diameter		16.76	1.432			57.50	4.12	61.62	91	
	2156	750 mm thru 900 mm diameter	▼	12.19	1.969	▼		79	5.65	84.65	125	
	2212	Deduct for salvage, aluminum scrap				Met. Ton					580	
	2214	Brass scrap									460	
	2216	Copper scrap									1,125	
	2218	Lead scrap									260	
	2220	Steel scrap				▼					72	
	2250	Water heater, 150 L	1 Plum	6	1.333	Ea.		57		57	85.50	

15080 | Mechanical Insulation

		CREW	DAILY OUTPUT	LABOR-HOURS	UNIT	MAT.	2006 BARE COSTS LABOR	EQUIP.	TOTAL	TOTAL INCL O&P	
200	0010	**DUCT INSULATION**										**200**
	0100	Rule of thumb, as a percentage of total mechanical costs				Job				10%		
	0110	Insulation req'd is based on the surface size/area to be covered										
	3000	Ductwork										
	3020	Blanket type, fiberglass, flexible										
	3030	Fire resistant liner, black coating one side										
	3050	13 mm thick, 0.907 kg density	Q-14	35.30	.453	m²	4.63	15.95		20.58	30	
	3060	25 mm thick, 0.68 kg density		32.52	.492		6.05	17.30		23.35	34	
	3070	40 mm thick, 0.68 kg density		29.73	.538		10.65	18.90		29.55	42	
	3080	50 mm thick, 0.68 kg density	▼	27.87	.574	▼	12.50	20		32.50	46	
	3140	FRK vapor barrier wrap, 0.34 kg density										
	3160	25 mm thick	Q-14	32.52	.492	m²	3.01	17.30		20.31	31	
	3170	40 mm thick		29.73	.538		3.23	18.90		22.13	33.50	
	3180	50 mm thick		27.87	.574		4.63	20		24.63	37	
	3190	76 mm thick		24.15	.662		5.70	23.50		29.20	43.50	
	3200	102 mm thick	▼	22.48	.712	▼	8.95	25		33.95	49.50	
	3210	Vinyl jacket, same as FRK										
	3280	Unfaced, 0.454 kg density										
	3310	25 mm thick	Q-14	33.44	.478	m²	3.88	16.80		20.68	31	
	3320	38 mm thick		30.66	.522		5.40	18.35		23.75	35	
	3330	51 mm thick	▼	28.80	.556	▼	6.05	19.55		25.60	37.50	
	3490	Board type, fiberglass liner, 1.4 kg density										
	3500	Fire resistant, black pigmented, 1 side										
	3520	25 mm thick	Q-14	13.94	1.148	m²	17.35	40.50		57.85	83	
	3540	38 mm thick		12.08	1.325		21	46.50		67.50	97	
	3560	51 mm thick	▼	11.15	1.435	▼	25.50	50.50		76	108	
	3600	FSK vapor barrier										
	3620	25 mm thick	Q-14	13.94	1.148	m²	18.95	40.50		59.45	85	
	3630	40 mm thick		12.08	1.325		24.50	46.50		71	101	
	3640	50 mm thick	▼	11.15	1.435	▼	29.50	50.50		80	113	
	3680	No finish										
	3700	25 mm thick	Q-14	15.79	1.013	m²	10.10	35.50		45.60	67.50	
	3710	38 mm thick		13.01	1.230		15.05	43		58.05	85	
	3720	51 mm thick	▼	12.08	1.325		18.95	46.50		65.45	94.50	
	3795	Finishes										
	3800	15 mm cement over 25 mm wire mesh, incl. corner bead	Q-14	9.29	1.722	m²	17.10	60.50		77.60	114	

MECHANICAL 15

15080	Mechanical Insulation	CREW	DAILY OUTPUT	LABOR-HOURS	UNIT	2006 BARE COSTS				TOTAL INCL O&P	
						MAT.	LABOR	EQUIP.	TOTAL		
200	3820	Glass cloth, pasted on	Q-14	15.79	1.013	m²	27	35.50		62.50	86	**200**
	3900	Weatherproof, non-metallic, 45 mil thick EPDM rubber sheeting	↓	16.72	.957	↓	26.50	33.50		60	82	
	9600	Minimum labor/equipment charge	1 Stpi	4	2	Job		86		86	130	
400	0010	**EQUIPMENT INSULATION**										**400**
	0100	Rule of thumb, as a percentage of total mechanical costs				Job				10%		
	0110	Insulation req'd is based on the surface size/area to be covered										
	1000	Boiler, 38 mm calcium silicate, 13 mm cement finish	Q-14	4.65	3.445	m²	38	121		159	233	
	1020	51 mm fiberglass	"	7.43	2.153	"	26.50	75.50		102	149	
	2000	Breeching, 51 mm calcium silicate with 13 mm cement finish, no lath										
	2020	Rectangular	Q-14	3.90	4.101	m²	50	144		194	283	
	2040	Round	"	3.60	4.450	"	52	156		208	305	
600	0010	**PIPING INSULATION**										**600**
	0100	Rule of thumb, as a percentage of total mechanical costs				Job				10%		
	0110	Insulation req'd is based on the surface size/area to be covered										
	4000	Pipe covering (price copper tube one size less than IPS)										
	6600	Fiberglass, with all service jacket										
	6840	25 mm wall, 15 mm iron pipe size	Q-14	73.15	.219	m	2.92	7.70		10.62	15.35	
	6870	25 mm iron pipe size		67.06	.239		3.41	8.40		11.81	17	
	6900	50 mm iron pipe size		60.96	.262		4.30	9.25		13.55	19.30	
	6920	80 mm iron pipe size		54.86	.292		5.20	10.25		15.45	22	
	6940	100 mm iron pipe size		45.72	.350		6.90	12.30		19.20	27	
	7320	50 mm wall, 15 mm iron pipe size		67.06	.239		8.55	8.40		16.95	22.50	
	7440	150 mm iron pipe size		30.48	.525		17.10	18.45		35.55	48	
	7460	200 mm iron pipe size		24.38	.656		21	23		44	59.50	
	7480	250 mm iron pipe size		21.34	.750		25	26.50		51.50	69	
	7490	300 mm iron pipe size	↓	19.81	.808	↓	28	28.50		56.50	75.50	
	7600	For fiberglass with standard canvas jacket, deduct					5%					
	7660	For fittings, add 914 mm for each fitting										
	7680	plus 1219 mm for each flange of the fitting										
	7700	For equipment or congested areas, add						20%				
	7720	Finishes, for 0.254 mm aluminum jacket, add	Q-14	18.58	.861	m²	6.35	30.50		36.85	55	
	7740	For 0.406 mm aluminum jacket, add		18.58	.861		9.25	30.50		39.75	58.50	
	7760	For 0.254 mm stainless steel, add	↓	14.86	1.076	↓	28	38		66	90.50	
	7780	For single layer of felt, add					10%	10%				
	7879	Rubber tubing, flexible closed cell foam										
	7880	10 mm wall, 6 mm iron pipe size	1 Asbe	36.58	.219	m	.79	8.55		9.34	14.35	
	7910	13 mm iron pipe size	↓	35.05	.228	↓	.98	8.90		9.88	15.20	
	7920	19 mm iron pipe size		35.05	.228		1.12	8.90		10.02	15.30	
	7930	25 mm iron pipe size		33.53	.239		1.28	9.30		10.58	16.15	
	7950	38 mm iron pipe size		33.53	.239		1.74	9.30		11.04	16.65	
	8100	15 mm wall, 8 mm iron pipe size		27.43	.292		1.18	11.40		12.58	19.30	
	8130	15 mm iron pipe size		27.13	.295		1.44	11.50		12.94	19.75	
	8140	20 mm iron pipe size		27.13	.295		1.61	11.50		13.11	19.95	
	8150	25 mm iron pipe size		26.82	.298		1.77	11.65		13.42	20.50	
	8170	40 mm iron pipe size		26.52	.302		2.49	11.80		14.29	21.50	
	8180	50 mm iron pipe size		26.21	.305		3.18	11.90		15.08	22.50	
	8200	76 mm iron pipe size		25.91	.309		4.49	12.05		16.54	24	
	8300	19 mm wall, 6 mm iron pipe size		27.43	.292		1.84	11.40		13.24	20	
	8330	13 mm iron pipe size		27.13	.295		2.39	11.50		13.89	21	
	8340	19 mm iron pipe size		27.13	.295		2.92	11.50		14.42	21.50	
	8350	25 mm iron pipe size		26.82	.298		3.31	11.65		14.96	22	
	8370	38 mm iron pipe size		26.52	.302		5.05	11.80		16.85	24	
	8380	51 mm iron pipe size		26.21	.305		5.95	11.90		17.85	25.50	
	8400	76 mm iron pipe size		25.91	.309		9.05	12.05		21.10	29	
	8444	25 mm wall, 13 mm iron pipe size	↓	26.21	.305	↓	4.63	11.90		16.53	24	

15050 | Basic Materials & Methods

			CREW	DAILY OUTPUT	LABOR-HOURS	UNIT	2006 BARE COSTS				TOTAL INCL O&P	
							MAT.	LABOR	EQUIP.	TOTAL		
600	8445	19 mm iron pipe size	1 Asbe	25.60	.312	m	5.60	12.20		17.80	25.50	600
	8446	25 mm iron pipe size		25.60	.312		6.55	12.20		18.75	26.50	
	8447	32 mm iron pipe size		24.99	.320		7.40	12.50		19.90	28	
	8448	38 mm iron pipe size		24.99	.320		8.55	12.50		21.05	29	
	8449	51 mm iron pipe size		24.38	.328		11.45	12.80		24.25	33	
	8450	64 mm iron pipe size	↓	24.38	.328	↓	14.95	12.80		27.75	37	
	8456	Rubber insulation tape, 3 mm x 51 mm x 9144 mm				Ea.	11.40			11.40	12.50	

15100 | Building Services Piping

15106 | Glass Pipe & Fittings

			CREW	DAILY OUTPUT	LABOR-HOURS	UNIT	2006 BARE COSTS				TOTAL INCL O&P	
							MAT.	LABOR	EQUIP.	TOTAL		
120	0010	**PIPE, GLASS** Borosilicate, couplings & hangers 3 m O.C.										120
	0020	Drainage										
	1100	40 mm diameter [1-1/2"]	Q-1	15.85	1.009	m	30.50	39		69.50	92	
	1120	50 mm diameter [2"]		13.41	1.193		40.50	46		86.50	114	
	1140	80 mm diameter [3"]		11.89	1.346		54.50	51.50		106	138	
	1160	100 mm diameter [4"]		9.14	1.750		97.50	67		164.50	208	
	1180	150 mm diameter [6"]	↓	7.92	2.019	↓	178	77.50		255.50	310	
	1870	To delete coupling & hanger, subtract										
	1880	40 mm diam. to 50 mm diam. [1-1/2"-2"]					19%	22%				
	1890	80 mm diam. to 150 mm diam. [3"-6"]					20%	17%				
	2000	Process supply (pressure), beaded joints										
	2040	15 mm diameter [1/2"]	1 Plum	10.97	.729	m	11	31		42	59	
	2060	20 mm diameter [3/4"]		9.45	.847		12.40	36		48.40	68	
	2080	25 mm diameter [1"]	↓	8.23	.972		44	41.50		85.50	111	
	2100	40 mm diameter [1-1/2"]	Q-1	14.33	1.117		53	43		96	123	
	2120	50 mm diameter [2"]		11.89	1.346		70.50	51.50		122	156	
	2140	80 mm diameter [3"]		10.36	1.544		102	59.50		161.50	202	
	2160	100 mm diameter [4"]		7.62	2.100		143	80.50		223.50	278	
	2180	150 mm diameter [6"]	↓	6.40	2.500	↓	283	96		379	455	
	2860	To delete coupling & hanger, subtract										
	2870	15 mm diam. to 25 mm diam. [1/2"-1"]					25%	33%				
	2880	40 mm diam. to 80 mm diam. [1-1/2"-3"]					22%	21%				
	2890	100 mm diam. to 150 mm diam. [4"-6"]					23%	15%				
	3800	Conical joint, transparent										
	3980	150 mm diameter [6"]	Q-1	6.40	2.500	m	1,875	96		1,971	2,225	
	4500	To delete couplings & hangers, subtract										
	4530	150 mm diam. [6"]					22%	26%				
160	0010	**PIPE, GLASS, FITTINGS**										160
	0020	Drainage, beaded ends										
	0040	Coupling & labor required at joints not incl. in fitting										
	0050	price. Add 1 per joint for installed price										
	0070	90° Bend or sweep, 40 mm [1-1/2"]				Ea.	20.50			20.50	22.50	
	0090	50 mm [2"]					26			26	28.50	
	0100	80 mm [3"]					42.50			42.50	47	
	0110	100 mm [4"]					68.50			68.50	75	
	0120	150 mm (sweep only) [6"]				↓	208			208	229	
	0200	45° Bend or sweep same as 90°										

MECHANICAL **15**

			DAILY	LABOR-		2006 BARE COSTS				TOTAL	
15106 | Glass Pipe & Fittings		CREW	OUTPUT	HOURS	UNIT	MAT.	LABOR	EQUIP.	TOTAL	INCL O&P	
160	0350	Tee, single sanitary, 40 mm [1-1/2"]				Ea.	33			33	36.50	160
	0370	50 mm [2"]					33			33	36.50	
	0380	80 mm [3"]					49.50			49.50	54.50	
	0390	100 mm [4"]					88.50			88.50	97.50	
	0400	150 mm [6"]					238			238	262	
	0410	Tee, straight, 40 mm [1-1/2"]					41			41	45	
	0430	50 mm [2"]					41			41	45	
	0440	80 mm [3"]					59			59	65	
	0450	100 mm [4"]					104			104	115	
	0460	150 mm [6"]					256			256	282	
	0500	Coupling, stainless steel, TFE seal ring										
	0520	40 mm [1-1/2"]	Q-1	32	.500	Ea.	14.30	19.20		33.50	45	
	0530	50 mm [2"]		30	.533		18.10	20.50		38.60	51	
	0540	80 mm [3"]		25	.640		24.50	24.50		49	64	
	0550	100 mm [4"]		23	.696		42	26.50		68.50	86	
	0560	150 mm [6"]		20	.800		94.50	30.50		125	151	
	0600	Coupling, stainless steel, bead to plain end										
	0610	40 mm [1-1/2"]	Q-1	36	.444	Ea.	18.45	17.10		35.55	46	
	0620	50 mm [2"]		34	.471		23	18.10		41.10	52.50	
	0630	80 mm [3"]		29	.552		39	21		60	75	
	0640	100 mm [4"]		27	.593		58.50	23		81.50	99	
	0650	150 mm [6"]		24	.667		198	25.50		223.50	257	
	2000	Process supply (pressure), beaded ends										
	2050	90° Sweep elbow, 15 mm [1/2"]				Ea.	20.50			20.50	22.50	
	2070	20 mm [3/4"]					24.50			24.50	27	
	2080	25 mm [1"]					32.50			32.50	35.50	
	2090	40 mm [1-1/2"]					34			34	37.50	
	2100	50 mm [2"]					45.50			45.50	50	
	2120	80 mm [3"]					72			72	79	
	2130	100 mm [4"]					101			101	111	
	2140	150 mm [6"]					253			253	278	
	2150	45° Sweep elbow, same as 90°										
	2250	Tee, straight, 15 mm [1/2"]				Ea.	35			35	38.50	
	2270	20 mm [3/4"]					35			35	38.50	
	2280	25 mm [1"]					39			39	43	
	2290	40 mm [1-1/2"]					58			58	63.50	
	2300	50 mm [2"]					68.50			68.50	75.50	
	2310	80 mm [3"]					83.50			83.50	92	
	2320	100 mm [4"]					140			140	154	
	2330	150 mm [6"]					261			261	287	
	2350	Coupling, Viton liner, for temperatures to 475°K										
	2370	15 mm [1/2"]	Q-1	40	.400	Ea.	29.50	15.35		44.85	55.50	
	2380	20 mm [3/4"]		37	.432		33.50	16.60		50.10	62	
	2390	25 mm [1"]		35	.457		40	17.55		57.55	70.50	
	2400	40 mm [1-1/2"]		32	.500		45	19.20		64.20	79	
	2410	50 mm [2"]		30	.533		54.50	20.50		75	91	
	2420	80 mm [3"]		25	.640		100	24.50		124.50	147	
	2430	100 mm [4"]		23	.696		120	26.50		146.50	173	
	2440	150 mm [6"]		20	.800		255	30.50		285.50	330	
	2550	For beaded joint armored fittings, add					200%					
	2600	Conical ends. Flange set, gasket & labor not incl. in fitting										
	2620	price. Add 1 per joint for installed price.										
	2650	90° Sweep elbow, 25 mm [1"]				Ea.	53			53	58.50	
	2670	40 mm [1-1/2"]					77			77	84.50	
	2680	50 mm [2"]					89			89	98	
	2690	80 mm [3"]					178			178	196	

			DAILY	LABOR-		2006 BARE COSTS				TOTAL	
15106	**Glass Pipe & Fittings**	CREW	OUTPUT	HOURS	UNIT	MAT.	LABOR	EQUIP.	TOTAL	INCL O&P	
160	2700	100 mm [4"]				Ea.	340			340	375	160
	2710	150 mm [6"]					645			645	710	
	2750	Cross (straight), add					55%					
	2850	Tee, add					20%					

			DAILY	LABOR-		2006 BARE COSTS				TOTAL	
15107	**Metal Pipe & Fittings**	CREW	OUTPUT	HOURS	UNIT	MAT.	LABOR	EQUIP.	TOTAL	INCL O&P	
220	0010	**PIPE, BRASS** Plain end,										220
	0900	Field threaded, coupling & clevis hanger 3048 mm O.C.										
	0920	Regular weight										
	1120	15 mm diameter [1/2"]	1 Plum	14.63	.547	m	14.40	23.50		37.90	51	
	1140	20 mm diameter [3/4"]		14.02	.571		19.45	24.50		43.95	58	
	1160	25 mm diameter [1"]		13.11	.610		28	26		54	70	
	1180	32 mm diameter [1-1/4"]	Q-1	21.95	.729		43	28		71	89	
	1200	40 mm diameter [1-1/2"]		19.81	.808		51	31		82	103	
	1220	50 mm diameter [2"]		16.15	.990		70	38		108	135	
320	0010	**PIPE, CAST IRON** Soil, on hangers 1.5 m O.C.			R221113 -50							320
	0020	Single hub, service wt., lead & oakum joints 3 m O.C.										
	2120	50 mm diameter [2"]	Q-1	19.20	.833	m	16.45	32		48.45	66	
	2140	80 mm diameter [3"]		18.29	.875		23	33.50		56.50	75.50	
	2160	100 mm diameter [4"]		16.76	.954		29.50	36.50		66	87	
	2180	125 mm diameter [5"]	Q-2	23.16	1.036		40	41.50		81.50	106	
	2200	150 mm diameter [6"]	"	22.25	1.079		49.50	43		92.50	119	
	2220	200 mm diameter [8"]	Q-3	17.98	1.779		76	72.50		148.50	193	
	2240	250 mm diameter [10"]		16.46	1.944		138	79		217	271	
	2260	300 mm diameter [12"]		14.63	2.187		193	89		282	345	
	2320	For service weight, double hub, add					10%					
	2340	For extra heavy, single hub, add					48%	4%				
	2360	For extra heavy, double hub, add					71%	4%				
	2400	Lead for caulking, (180g/dia mm)	Q-1	72.58	.220	kg	2.29	8.45		10.74	15.25	
	2420	Oakum for caulking, (22g/dia mm)	"	18.14	.882	"	3.97	34		37.97	55.50	
	2960	To delete hangers, subtract										
	2970	50 mm diam. to 100 mm diam. [2"-4"]					16%	19%				
	2980	125 mm to 200 mm dia [5"-8"]					14%	14%				
	2990	250 mm diam. to 375 mm diam. [10"-15"]					13%	19%				
	3000	Single hub, service wt, push-on gasket joints 3 m O.C.										
	3010	50 mm diameter [2"]	Q-1	20.12	.795	m	17.60	30.50		48.10	65.50	
	3020	80 mm diameter [3"]		19.20	.833		24.50	32		56.50	75	
	3030	100 mm diameter [4"]		17.37	.921		31	35.50		66.50	87.50	
	3040	125 mm diameter [5"]	Q-2	24.08	.997		43	39.50		82.50	108	
	3050	150 mm diameter [6"]	"	22.86	1.050		52.50	42		94.50	121	
	3060	200 mm diameter [8"]	Q-3	18.90	1.693		82.50	69		151.50	195	
	3070	250 mm diameter [10"]		17.07	1.875		148	76.50		224.50	278	
	3080	300 mm diameter [12"]		14.94	2.143		206	87		293	360	
	3100	For service weight, double hub, add					65%					
	3110	For extra heavy, single hub, add					48%	4%				
	3120	For extra heavy, double hub, add					29%	4%				
	3130	To delete hangers, subtract										
	3140	50 mm diam. to 100 mm diam. [2"-4"]					12%	21%				
	3150	125 mm diam. to 200 mm diam. [5"-8"]					10%	16%				
	3160	250 mm diam. to 375 mm diam. [10"-15"]					9%	21%				
	4000	No hub, couplings 3 m O.C.										
	4100	40 mm diameter [1-1/2"]	Q-1	21.64	.739	m	17.20	28.50		45.70	62	
	4120	50 mm diameter [2"]		20.42	.783		17.70	30		47.70	65	
	4140	80 mm diameter [3"]		19.51	.820		23.50	31.50		55	73.50	
	4160	100 mm diameter [4"]		17.68	.905		30	35		65	85.50	

MECHANICAL 15

For expanded coverage of these items see *Means Mechanical or Plumbing Cost Data 2006*

15107 | Metal Pipe & Fittings

		CREW	DAILY OUTPUT	LABOR-HOURS	UNIT	2006 BARE COSTS MAT.	LABOR	EQUIP.	TOTAL	TOTAL INCL O&P	
320 4180	125 mm diameter [5"] R221113-50	Q-2	25.30	.949	m	42.50	38		80.50	104	**320**
4200	150 mm diameter [6"]	"	24.08	.997		53.50	39.50		93	119	
4220	200 mm diameter [8"]	Q-3	21.03	1.522		83.50	62		145.50	185	
4240	250 mm diameter [10"]	"	18.59	1.721		149	70		219	269	
4280	To delete hangers, subtract										
4290	40 mm diam. to 150 mm diam. [1-1/2"-6"]					22%	47%				
4300	200 mm diam. to 250 mm diam. [8"-10"]					21%	44%				
360 0010	**PIPE, CAST IRON, FITTINGS** Soil R221113-50										**360**
0040	Hub and spigot, service weight, lead & oakum joints										
0080	1/4 bend, 50 mm [2"]	Q-1	16	1	Ea.	10.35	38.50		48.85	69.50	
0120	80 mm [3"]		14	1.143		13.80	44		57.80	81	
0140	100 mm [4"]		13	1.231		21.50	47.50		69	94.50	
0160	125 mm [5"]	Q-2	18	1.333		30	53		83	113	
0180	150 mm [6"]	"	17	1.412		37.50	56.50		94	126	
0200	200 mm [8"]	Q-3	11	2.909		113	118		231	300	
0220	250 mm [10"]		10	3.200		165	130		295	380	
0224	300 mm [12"]		9	3.556		224	145		369	465	
0266	Closet bend, 80 mm diam. with flange 250 mm x 400 mm	Q-1	14	1.143		67	44		111	140	
0268	400 x 400 mm [16"-16"]		12	1.333		92.50	51		143.50	178	
0270	Closet bend, 100mm diam., 25x100mm ring,150x400mm		13	1.231		74	47.50		121.50	153	
0280	200 x 400 mm [8"-16"]		13	1.231		66	47.50		113.50	144	
0290	250 x 300 mm [10"-12"]		12	1.333		62.50	51		113.50	146	
0300	250 x 450 mm [10"-18"]		11	1.455		87	56		143	180	
0310	300 x 400 mm [12"-16"]		11	1.455		75	56		131	167	
0330	400 x 400 mm [16"-16"]		10	1.600		94.50	61.50		156	197	
0340	1/8 bend, 50 mm [2"]		16	1		7.35	38.50		45.85	66	
0350	80 mm [3"]		14	1.143		11.55	44		55.55	78.50	
0360	100 mm [4"]		13	1.231		16.90	47.50		64.40	89.50	
0380	125 mm [5"]	Q-2	18	1.333		23.50	53		76.50	106	
0400	150 mm [6"]	"	17	1.412		28.50	56.50		85	116	
0420	200 mm [8"]	Q-3	11	2.909		85	118		203	272	
0440	250 mm [10"]		10	3.200		123	130		253	330	
0460	300 mm [12"]		9	3.556		233	145		378	475	
0500	Sanitary tee, 50 mm [2"]	Q-1	10	1.600		14.50	61.50		76	108	
0540	80 mm [3"]		9	1.778		23.50	68.50		92	129	
0620	100 mm [4"]		8	2		28.50	77		105.50	148	
0700	125 mm [5"]	Q-2	12	2		57	79.50		136.50	183	
0800	150 mm [6"]	"	11	2.182		64.50	87		151.50	202	
0880	200 mm [8"]	Q-3	7	4.571		171	186		357	470	
1000	Tee, 50 mm [2"]	Q-1	10	1.600		19.15	61.50		80.65	114	
1060	80 mm [3"]		9	1.778		31	68.50		99.50	137	
1120	100 mm [4"]		8	2		40	77		117	160	
1200	125 mm [5"]	Q-2	12	2		61	79.50		140.50	188	
1300	150 mm [6"]	"	11	2.182		83	87		170	223	
1380	200 mm [8"]	Q-3	7	4.571		183	186		369	480	
1400	Combination Y and 1/8 bend										
1420	50 mm [2"]	Q-1	10	1.600	Ea.	18	61.50		79.50	112	
1460	80 mm [3"]		9	1.778		27.50	68.50		96	133	
1520	100 mm [4"]		8	2		38	77		115	158	
1540	125 mm [5"]	Q-2	12	2		72	79.50		151.50	199	
1560	150 mm [6"]		11	2.182		91	87		178	231	
1580	200 mm [8"]		7	3.429		224	137		361	455	
1582	300 mm [12"]	Q-3	6	5.333		505	217		722	880	
1600	Double Y, 50 mm [2"]	Q-1	8	2		23	77		100	142	
1610	80 mm [3"]		7	2.286		40	88		128	176	

15 MECHANICAL

			CREW	DAILY OUTPUT	LABOR-HOURS	UNIT	2006 BARE COSTS				TOTAL INCL O&P	
							MAT.	LABOR	EQUIP.	TOTAL		
360	1620	100 mm [4"]		Q-1	6.50	2.462	Ea.	52.50	94.50		147	200	360
	1630	125 mm [5"]	R221113 -50	Q-2	9	2.667		89.50	106		195.50	259
	1640	150 mm [6"]		"	8	3		134	120		254	330
	1650	200 mm [8"]		Q-3	5.50	5.818		330	237		567	715
	1660	250 mm [10"]			5	6.400		465	260		725	900
	1670	300 mm [12"]		▼	4.50	7.111		665	289		954	1,175
	1740	Reducer, 80 mm x 50 mm [3" x 2"]		Q-1	15	1.067		10.10	41		51.10	72.50
	1750	100 mm x 50 mm [4" x 2"]			14.50	1.103		10.60	42.50		53.10	75.50
	1760	100 mm x 80 mm [4" x 3"]			14	1.143		13.15	44		57.15	80.50
	1770	125 mm x 50 mm [5" x 2"]			14	1.143		18.40	44		62.40	86
	1780	125 mm x 80 mm [5" x 3"]			13.50	1.185		17.90	45.50		63.40	88
	1790	125 mm x 100 mm [5" x 4"]			13	1.231		16.95	47.50		64.45	89.50
	1800	150 mm x 50 mm [6" x 2"]			13.50	1.185		23.50	45.50		69	94.50
	1810	150 mm x 80 mm [6" x 3"]			13	1.231		23.50	47.50		71	97
	1840	150 mm x 125 mm [6" x 5"]		▼	11	1.455		28.50	56		84.50	116
	1880	200 mm x 80 mm [8" x 3"]		Q-2	13.50	1.778		48.50	71		119.50	160
	1900	200 mm x 100 mm [8" x 4"]			13	1.846		44	73.50		117.50	160
	1920	200 mm x 125 mm [8" x 5"]			12	2		46	79.50		125.50	171
	1940	200 mm x 150 mm [8" x 6"]		▼	12	2		45.50	79.50		125	170
	1960	Increaser, 50 mm x 80 mm [2"-3"]		Q-1	15	1.067		24	41		65	88
	1980	50 mm x 100 mm [2"-4"]			14	1.143		24	44		68	92.50
	2000	50 mm x 125 mm [2"-5"]			13	1.231		29	47.50		76.50	103
	2020	80 mm x 100 mm [3"-4"]			13	1.231		26.50	47.50		74	100
	2040	80 mm x 125 mm [3"-5"]			13	1.231		29	47.50		76.50	103
	2060	80 mm x 150 mm [3"-6"]			12	1.333		36	51		87	117
	2070	100 mm x 125 mm [4"-5"]			13	1.231		29	47.50		76.50	103
	2080	100 mm x 150 mm [4"-6"]		▼	12	1.333		36	51		87	117
	2090	100 mm x 200 mm [4"-8"]		Q-2	13	1.846		73.50	73.50		147	192
	2100	125 mm x 150 mm [5"-6"]		Q-1	11	1.455		53.50	56		109.50	143
	2110	125 mm x 200 mm [5"-8"]		Q-2	12	2		86.50	79.50		166	215
	2120	150 mm x 200 mm [6"-8"]			12	2		86.50	79.50		166	215
	2130	150 mm x 250 mm [6"-10"]			8	3		156	120		276	350
	2140	200 mm x 250 mm [8"-10"]			6.50	3.692		156	147		303	395
	2150	250 mm x 300 mm [10"-12"]		▼	5.50	4.364		279	174		453	565
	2500	Y, 50 mm [2"]		Q-1	10	1.600		13.15	61.50		74.65	107
	2510	80 mm [3"]			9	1.778		24.50	68.50		93	130
	2520	100 mm [4"]		▼	8	2		32.50	77		109.50	152
	2530	125 mm [5"]		Q-2	12	2		55	79.50		134.50	181
	2540	150 mm [6"]		"	11	2.182		75	87		162	214
	2550	200 mm [8"]		Q-3	7	4.571		184	186		370	480
	2560	250 mm [10"]			6	5.333		297	217		514	650
	2570	300 mm [12"]			5	6.400		510	260		770	950
	2580	375 mm [15"]		▼	4	8		1,075	325		1,400	1,700
	3000	For extra heavy, add					▼	44%	4%			
	3600	Hub and spigot, service weight gasket joint										
	3605	Note: gaskets and joint labor have										
	3606	been included with all listed fittings.										
	3610	1/4 bend, 50 mm [2"]		Q-1	20	.800	Ea.	13.70	30.50		44.20	61.50
	3620	80 mm [3"]			17	.941		18.25	36		54.25	74.50
	3630	100 mm [4"]		▼	15	1.067		27	41		68	91.50
	3640	125 mm [5"]		Q-2	21	1.143		39	45.50		84.50	111
	3650	150 mm [6"]		"	19	1.263		46.50	50.50		97	127
	3660	200 mm [8"]		Q-3	12	2.667		133	109		242	310
	3670	250 mm [10"]			11	2.909		196	118		314	395
	3680	300 mm [12"]		▼	10	3.200		263	130		393	485
	3700	Closet bend-250 x400 mm, 80 mm pipe/ring		Q-1	17	.941	▼	71	36		107	133

15100 | Building Services Piping

15107 | Metal Pipe & Fittings

			CREW	DAILY OUTPUT	LABOR-HOURS	UNIT	2006 BARE COSTS				TOTAL INCL O&P
							MAT.	LABOR	EQUIP.	TOTAL	
3710	400 mm x 400 mm [16"-16"]		Q-1	15	1.067	Ea.	96.50	41		137.50	168
3730	Closet bend-150x400mm,100mm pipe,/25x100mm ring			15	1.067		79.50	41		120.50	149
3740	200 mm x 400 mm,[8" x16"]	R221113 -50		15	1.067		72	41		113	141
3750	250 mm x 300 mm [10"-12"]			14	1.143		68	44		112	141
3760	250 mm x 450 mm [10"-18"]			13	1.231		92.50	47.50		140	173
3770	300 mm x 400 mm [12"-16"]			13	1.231		80.50	47.50		128	160
3780	400 mm x 400 mm [16"-16"]			12	1.333		100	51		151	187
3800	1/8 bend, 50 mm [2"]			20	.800		10.75	30.50		41.25	58.50
3810	80 mm [3"]			17	.941		15.95	36		51.95	72
3820	100 mm [4"]			15	1.067		22.50	41		63.50	86.50
3830	125 mm [5"]		Q-2	21	1.143		32	45.50		77.50	104
3840	150 mm [6"]		"	19	1.263		37.50	50.50		88	118
3850	200 mm [8"]		Q-3	12	2.667		105	109		214	279
3860	250 mm [10"]			11	2.909		153	118		271	345
3870	300 mm [12"]			10	3.200		271	130		401	495
3900	Sanitary Tee, 50 mm [2"]		Q-1	12	1.333		21.50	51		72.50	101
3910	80 mm [3"]			10	1.600		32	61.50		93.50	128
3920	100 mm [4"]			9	1.778		40	68.50		108.50	147
3930	125 mm [5"]		Q-2	13	1.846		74	73.50		147.50	193
3940	150 mm [6"]		"	11	2.182		82.50	87		169.50	222
3950	200 mm [8"]		Q-3	8	4		211	163		374	475
3980	Tee, 50 mm [2"]		Q-1	12	1.333		26	51		77	106
3990	80 mm [3"]			10	1.600		40	61.50		101.50	137
4000	100 mm [4"]			9	1.778		51	68.50		119.50	159
4010	127 mm [5"]		Q-2	13	1.846		78.50	73.50		152	198
4020	150 mm [6"]		"	11	2.182		101	87		188	242
4030	200 mm [8"]		Q-3	8	4		222	163		385	490
4060	Combination Y and 1/8 bend										
4070	50 mm [2"]		Q-1	12	1.333	Ea.	25	51		76	105
4080	80 mm [3"]			10	1.600		36.50	61.50		98	133
4090	100 mm [4"]			9	1.778		49	68.50		117.50	157
4100	125 mm [5"]		Q-2	13	1.846		89	73.50		162.50	209
4110	150 mm [6"]		"	11	2.182		109	87		196	251
4120	200 mm [8"]		Q-3	8	4		264	163		427	535
4160	Double Y, 50 mm [2"]		Q-1	10	1.600		33	61.50		94.50	129
4170	80 mm [3"]			8	2		53	77		130	175
4180	100 mm [4"]			7	2.286		69	88		157	208
4190	125 mm [5"]		Q-2	10	2.400		116	95.50		211.50	271
4200	150 mm [6"]		"	9	2.667		161	106		267	340
4210	200 mm [8"]		Q-3	6	5.333		390	217		607	755
4220	250 mm [10"]			5	6.400		555	260		815	1,000
4230	300 mm [12"]			4.50	7.111		785	289		1,074	1,300
4260	Reducer, 80 mm x 50 mm [3"-2"]		Q-1	17	.941		17.90	36		53.90	74
4270	100 mm x 50 mm [4"-2"]			16.50	.970		19.60	37.50		57.10	77.50
4280	100 mm x 80 mm [4"-3"]			16	1		23	38.50		61.50	83.50
4290	125 mm x 50 mm [5"-2"]			16	1		30.50	38.50		69	91.50
4300	125 mm x 80 mm [5"-3"]			15.50	1.032		31	39.50		70.50	93.50
4310	125 mm x 100 mm [5"-4"]			15	1.067		31.50	41		72.50	96
4320	150 mm x 50 mm [6"-2"]			15.50	1.032		36	39.50		75.50	99
4330	150 mm x 80 mm [6"-3"]			15	1.067		37	41		78	102
4340	150 mm x 125 mm [6"-5"]			13	1.231		46.50	47.50		94	122
4360	200 mm x 80 mm [8"-3"]		Q-2	15	1.600		72.50	64		136.50	176
4370	200 mm x 100 mm [8"-4"]			15	1.600		70	64		134	173
4380	200 mm x 125 mm [8"-5"]			14	1.714		74.50	68.50		143	185
4390	200 mm x 150 mm [8"-6"]			14	1.714		74.50	68.50		143	185
4430	Increaser, 50 mm x 80 mm [2"-3"]		Q-1	17	.941		28.50	36		64.50	86

Important: See the Reference Section for supporting data - Crews, Rental Equipment, City Cost Indexes and Reference Data

15107	Metal Pipe & Fittings		CREW	DAILY OUTPUT	LABOR-HOURS	UNIT	2006 BARE COSTS				TOTAL INCL O&P
							MAT.	LABOR	EQUIP.	TOTAL	
360	4440	50 mm x 100 mm (2" x 4")	R221113 -50	Q-1	16	1	Ea.	30	38.50		68.50	91	360
	4450	50 mm x 125 mm (2"-5")			15	1.067		37.50	41		78.50	103
	4460	80 mm x 100 mm (3"-4")			15	1.067		32	41		73	97
	4470	80 mm x 125 mm (3"-5")			15	1.067		37.50	41		78.50	103
	4480	80 mm x 150 mm (3"-6")			14	1.143		45	44		89	116
	4490	100 mm x 125 mm [4"-5"]			15	1.067		37.50	41		78.50	103
	4500	100 mm x 150 mm [4"-6"]			14	1.143		45	44		89	116
	4510	100 mm x 200 mm [4"-8"]		Q-2	15	1.600		93.50	64		157.50	199
	4520	125 mm x 150 mm [5"-6"]		Q-1	13	1.231		62.50	47.50		110	140
	4530	125 mm x 200 mm [5"-8"]		Q-2	14	1.714		106	68.50		174.50	220
	4540	150 mm x 200 mm [6"-8"]			14	1.714		106	68.50		174.50	220
	4550	150 mm x 250 mm [6"-10"]			10	2.400		187	95.50		282.50	350
	4560	200 mm x 250 mm [8"-10"]			8.50	2.824		187	113		300	375
	4570	250 mm x 300 mm [10"-12"]			7.50	3.200		320	128		448	540
	4600	Y, 50 mm [2"]		Q-1	12	1.333		19.95	51		70.95	99
	4610	80 mm [3"]			10	1.600		33	61.50		94.50	129
	4620	100 mm [4"]			9	1.778		44	68.50		112.50	152
	4630	125 mm [5"]		Q-2	13	1.846		72.50	73.50		146	191
	4640	150 mm [6"]		"	11	2.182		93	87		180	233
	4650	200 mm [8"]		Q-3	8	4		223	163		386	490
	4660	250 mm [10"]			7	4.571		355	186		541	675
	4670	300 mm [12"]			6	5.333		585	217		802	970
	4900	For extra heavy, add						44%	4%			
	4940	Gasket and making push-on joint										
	4950	50 mm [2"]		Q-1	40	.400	Ea.	3.38	15.35		18.73	26.50
	4960	80 mm [3"]			35	.457		4.43	17.55		21.98	31.50
	4970	100 mm [4"]			32	.500		5.65	19.20		24.85	35
	4980	125 mm [5"]		Q-2	43	.558		8.70	22		30.70	43
	4990	150 mm [6"]		"	40	.600		9	24		33	46
	5000	200 mm [8"]		Q-3	32	1		19.95	40.50		60.45	83
	5010	250 mm [10"]			29	1.103		30.50	45		75.50	101
	5020	300 mm [12"]			25	1.280		39	52		91	121
	5022	375 mm [15"]			21	1.524		46.50	62		108.50	145
	5030	Note: gaskets and joint labor have										
	5040	Been included with all listed fittings.										
	5990	No hub										
	6000	Cplg. & labor required at joints not incl. in fitting										
	6010	price. Add 1 coupling per joint for installed price										
	6020	1/4 Bend, 40 mm [1-1/2"]					Ea.	5.80			5.80	6.40
	6060	50 mm [2"]						6.35			6.35	7
	6080	80 mm [3"]						8.75			8.75	9.65
	6120	100 mm [4"]						12.65			12.65	13.95
	6140	125 mm [5"]						29			29	32
	6160	150 mm [6"]						32			32	35
	6180	200 mm [8"]						88			88	96.50
	6184	1/4 Bend, long sweep, 40 mm [1-1/2"]						13.65			13.65	15.05
	6186	50 mm [2"]						13.65			13.65	15.05
	6188	80 mm [3"]						16.20			16.20	17.80
	6189	100 mm [4"]						26			26	28.50
	6190	125 mm [5"]						47.50			47.50	52.50
	6191	150 mm [6"]						58			58	64
	6192	200 mm [8"]						141			141	155
	6193	250 mm [10"]						253			253	279
	6200	1/8 Bend, 40 mm [1-1/2"]						4.85			4.85	5.35
	6210	50 mm [2"]						5.40			5.40	5.95
	6212	80 mm [3"]						7.30			7.30	8.05

MECHANICAL 15

15107 | Metal Pipe & Fittings

		CREW	DAILY OUTPUT	LABOR-HOURS	UNIT	2006 BARE COSTS				TOTAL INCL O&P
						MAT.	LABOR	EQUIP.	TOTAL	
360											360
6214	100 mm [4"]				Ea.	9.25			9.25	10.20
6216	125 mm [5"]	R221113 -50				19.70			19.70	21.50
6218	150 mm [6"]					21.50			21.50	23.50
6220	200 mm [8"]					61.50			61.50	68
6222	250 mm [10"]					115			115	127
6380	Sanitary Tee, tapped, 38 mm [1-1/2"]					10.70			10.70	11.75
6382	50 mm x 40 mm [2"-1-1/2"]					9.65			9.65	10.60
6384	50 mm [2"]					10.75			10.75	11.85
6386	80 mm x 51 mm [3"-2"]					15			15	16.50
6388	80 mm [3"]					28			28	30.50
6390	100 mm x 40 mm [4"-1-1/2"]					13.35			13.35	14.70
6392	100 mm x 50 mm [4"-2"]					15.05			15.05	16.55
6394	150x40 mm [6"-1-1/2"]					31			31	34
6396	150x 50mm [6"-2"]					32			32	35
6459	Sanitary Tee, 40 mm [1-1/2"]					8.05			8.05	8.85
6460	50 mm [2"]					8.05			8.05	8.85
6470	80 mm [3"]					10.65			10.65	11.70
6472	100 mm [4"]					16.50			16.50	18.15
6474	125 mm [5"]					47			47	52
6476	150 mm [6"]					48			48	53
6478	200 mm [8"]					195			195	214
6730	Y, 40 mm [1-1/2"]					8.20			8.20	9
6740	50 mm [2"]					8.20			8.20	9
6750	80 mm [3"]					11.60			11.60	12.75
6760	100 mm [4"]					18.95			18.95	21
6762	125 mm [5"]					44			44	48.50
6764	150 mm [6"]					50.50			50.50	55.50
6768	200 mm [8"]					119			119	131
6791	Y, reducing, 80 mm x 50 mm [3"-2"]					8.75			8.75	9.65
6792	100 mm x 50 mm [4"-2"]					12.15			12.15	13.40
6793	125 mm x 50 mm [5"-2"]					27.50			27.50	30
6794	150 mm x 50 mm [6"-2"]					30.50			30.50	33.50
6795	150 mm x 100 mm [6"-4"]					39.50			39.50	43.50
6796	200 mm x 100 mm [8"-4"]					68			68	74.50
6797	200 mm x 150 mm [8"-6"]					83.50			83.50	92
6798	250 mm x 150 mm [10"-6"]					188			188	207
6799	250 mm x 200 mm [10"-8"]					226			226	248
6800	Double Y, 50 mm [2"]					12.65			12.65	13.95
6920	80 mm [3"]					23.50			23.50	25.50
7000	100 mm [4"]					50			50	54.50
7100	150 mm [6"]					88			88	96.50
7120	200 mm [8"]					251			251	276
7200	Combination Y and 1/8 Bend									
7220	40 mm [1-1/2"]				Ea.	9.10			9.10	10
7260	50 mm [2"]					8.70			8.70	9.60
7320	80 mm [3"]					13.25			13.25	14.60
7400	100 mm [4"]					24.50			24.50	26.50
7480	125 mm [5"]					53.50			53.50	58.50
7500	150mm [6"]					67.50			67.50	74.50
7520	200 mm [8"]					167			167	184
7800	Reducer, 80 mm x 50 mm [3"-2"]					4.62			4.62	5.10
7820	100 mm x 50 mm [4"-2"]					7.20			7.20	7.90
7840	100 mm x 80 mm [4"-3"]					7.20			7.20	7.90
8000	Coupling, standard (by CISPI Mfrs.)									
8020	40 mm [1-1/2"]	Q-1	48	.333	Ea.	6	12.80		18.80	26
8040	50 mm [2"]		44	.364		6	13.95		19.95	27.50

Important: See the Reference Section for supporting data - Crews, Rental Equipment, City Cost Indexes and Reference Data

15 MECHANICAL

15107	Metal Pipe & Fittings		CREW	DAILY OUTPUT	LABOR-HOURS	UNIT	2006 BARE COSTS				TOTAL INCL O&P	
							MAT.	LABOR	EQUIP.	TOTAL		
360 8080	80 mm [3"]	R221113 -50	Q-1	38	.421	Ea.	7.15	16.20		23.35	32.50	**360**
8120	100 mm [4"]		↓	33	.485	↓	8.45	18.65		27.10	37.50	
8160	125 mm [5"]		Q-2	44	.545		20.50	21.50		42	55	
8180	150 mm [6"]		"	40	.600		21.50	24		45.50	59.50	
8200	200 mm [8"]		Q-3	33	.970		40.50	39.50		80	104	
8220	250 mm [10"]		"	26	1.231	↓	56	50		106	137	
8300	Coupling, cast iron clamp & neoprene gasket (by MG)											
8310	40 mm [1-1/2"]		Q-1	48	.333	Ea.	7.75	12.80		20.55	28	
8320	50 mm [2"]			44	.364		8	13.95		21.95	30	
8330	80 mm [3"]			38	.421		8.35	16.20		24.55	33.50	
8340	100 mm [4"]		↓	33	.485		9.80	18.65		28.45	39	
8350	125 mm [5"]		Q-2	44	.545		18.55	21.50		40.05	53	
8360	150 mm [6"]		"	40	.600		25.50	24		49.50	64	
8380	200 mm [8"]		Q-3	33	.970		44.50	39.50		84	108	
8400	250 mm [10"]		"	26	1.231	↓	68	50		118	151	
8600	Coupling, Stainless steel, heavy duty											
8620	40nmm [1-1/2"]		Q-1	48	.333	Ea.	8.55	12.80		21.35	28.50	
8630	50 mm [2"]			44	.364		8.90	13.95		22.85	31	
8640	50 mm x 40 mm [2"-1-1/2"]			44	.364		13.70	13.95		27.65	36	
8650	80 mm [3"]			38	.421		9.65	16.20		25.85	35	
8660	100 mm [4"]			33	.485		10.90	18.65		29.55	40	
8670	100 mm x 80 mm [4"-3"]		↓	33	.485		23.50	18.65		42.15	54	
8680	125 mm [5"]		Q-2	44	.545		23.50	21.50		45	58.50	
8690	150 mm [6"]		"	40	.600		26.50	24		50.50	65	
8700	200 mm [8"]		Q-3	33	.970		44.50	39.50		84	109	
8710	250 mm [10"]		"	26	1.231	↓	56	50		106	137	
420 0010	**PIPE, COPPER** Solder joints											**420**
1000	Type K tubing, couplings & clevis hangers 3 m O.C.											
1100	8 mm diameter [1/4"]		1 Plum	25.60	.312	m	5.45	13.35		18.80	26	
1120	10 mm diameter [3/8"]			24.99	.320		6.50	13.65		20.15	27.50	
1140	15 mm diameter [1/2"]			23.77	.337		7.45	14.35		21.80	29.50	
1160	18 mm diameter [5/8"]			23.47	.341		9.30	14.55		23.85	32.50	
1180	20 mm diameter [3/4"]			22.56	.355		11.80	15.15		26.95	36	
1200	25 mm diameter [1"]			20.12	.398		16.55	17		33.55	43.50	
1220	32 mm diameter [1-1/4"]			17.07	.469		20.50	20		40.50	52.50	
1240	40 mm diameter [1-1/2"]			15.24	.525		25.50	22.50		48	61.50	
1260	50 mm diameter [2"]		↓	12.19	.656		39	28		67	85	
1280	65 mm diameter [2-1/2"]		Q-1	18.29	.875		59.50	33.50		93	116	
1300	80 mm diameter [3"]			16.46	.972		83	37.50		120.50	147	
1320	90 mm diameter [3-1/2"]			12.80	1.250		111	48		159	195	
1330	100 mm diameter [4"]			11.58	1.381		140	53		193	234	
1340	125 mm diameter [5"]		↓	9.75	1.640		365	63		428	495	
1360	150 mm diameter [6"]		Q-2	11.58	2.072		525	82.50		607.50	705	
1380	200 mm diameter [8"]		"	10.36	2.316		935	92.50		1,027.50	1,175	
1390	For other than full hard temper, add					↓	13%					
1440	For silver solder, add							15%				
1800	For medical clean, (oxygen class), add						12%					
1950	To delete cplgs. & hngrs., 8-25 mm pipe, subtract						27%	60%				
1960	32-80 mm pipe, subtract [1-1/4"-3"]						14%	52%				
1970	90-125 mm pipe, subtract [3-1/2"-5"]						10%	60%				
1980	150-200 mm pipe, subtract [6"-8"]						19%	53%				
2000	Type L cu tubing,cplngs & hangers 3 m O.C.											
2100	8 mm diameter [1/4"]		1 Plum	26.82	.298	m	3.61	12.75		16.36	23	
2120	10 mm diameter [3/8"]			25.60	.312		5	13.35		18.35	25.50	
2140	15 mm diameter [1/2"]			24.69	.324		5.80	13.85		19.65	27.50	
2160	18 mm diameter [5/8"]		↓	24.08	.332	↓	8.15	14.20		22.35	30.50	

For expanded coverage of these items see *Means Mechanical or Plumbing Cost Data 2006*

		15107 Metal Pipe & Fittings	CREW	DAILY OUTPUT	LABOR-HOURS	UNIT	MAT.	LABOR	EQUIP.	TOTAL	TOTAL INCL O&P	
420	2180	20 mm diameter [3/4"]	1 Plum	23.16	.345	m	8.55	14.75		23.30	31.50	420
	2200	25 mm diameter [1"]		20.73	.386		11.95	16.50		28.45	38	
	2220	32 mm diameter [1-1/4"]		17.68	.453		16.65	19.30		35.95	47.50	
	2240	40 mm diameter [1-1/2"]		15.85	.505		21.50	21.50		43	56	
	2260	50 mm diameter [2"]	▼	12.80	.625		33	26.50		59.50	76.50	
	2280	65 mm diameter [2-1/2"]	Q-1	18.90	.847		51	32.50		83.50	105	
	2300	80 mm diameter [3"]		17.07	.937		69	36		105	130	
	2320	90 mm diameter [3-1/2"]		13.11	1.221		93.50	47		140.50	174	
	2340	100 mm diameter [4"]		11.89	1.346		117	51.50		168.50	207	
	2360	125 mm diameter [5"]	▼	10.36	1.544		330	59.50		389.50	450	
	2380	150 mm diameter [6"]	Q-2	12.19	1.969		405	78.50		483.50	570	
	2400	200 mm diameter [8"]	"	10.97	2.187		720	87		807	920	
	2410	For other than full hard temper, add				▼	21%					
	2590	For silver solder, add						15%				
	4000	Type DWV tubing, cplings & hangers 3 m OC										
	4100	32 mm diameter [1-1/4"]	1 Plum	18.29	.437	m	14.60	18.70		33.30	44	
	4120	40 mm diameter [1-1/2"]		16.46	.486		18.25	21		39.25	51	
	4140	50 mm diameter [2"]	▼	13.41	.597		24.50	25.50		50	65	
	4160	80 mm diameter [3"]	Q-1	17.68	.905		43	35		78	100	
	4180	100 mm diameter [4"]		12.19	1.312		75.50	50.50		126	159	
	4200	125 mm diameter [5"]	▼	10.97	1.458		225	56		281	330	
	4220	150 mm diameter [6"]	Q-2	12.80	1.875	▼	320	74.50		394.50	465	
460	0010	**PIPE, COPPER, FITTINGS** Wrought unless otherwise noted										460
	0040	Solder joints, copper x copper										
	0070	90° elbow, 8 mm [1/4"]	1 Plum	22	.364	Ea.	1.70	15.55		17.25	25.50	
	0090	10 mm [3/8"]		22	.364		1.61	15.55		17.16	25.50	
	0100	15 mm [1/2"]		20	.400		.51	17.10		17.61	26	
	0110	18 mm [5/8"]		19	.421		3.28	18		21.28	30.50	
	0120	20 mm [3/4"]		19	.421		1.14	18		19.14	28.50	
	0130	25 mm [1"]		16	.500		2.82	21.50		24.32	35	
	0140	32 mm [1-1/4"]		15	.533		4.26	23		27.26	39	
	0150	40 mm [1-1/2"]		13	.615		6.65	26.50		33.15	47	
	0160	50 mm [2"]	▼	11	.727		12.10	31		43.10	60	
	0170	65 mm [2-1/2"]	Q-1	13	1.231		25.50	47.50		73	99	
	0180	80 mm [3"]		11	1.455		34	56		90	122	
	0190	90 mm [3-1/2"]		10	1.600		119	61.50		180.50	224	
	0200	100 mm [4"]		9	1.778		87	68.50		155.50	199	
	0210	125 mm [5"]	▼	6	2.667		365	102		467	555	
	0220	150 mm [6"]	Q-2	9	2.667		485	106		591	695	
	0230	200 mm [8"]	"	8	3		1,800	120		1,920	2,150	
	0250	45° elbow, 8 mm [1/4"]	1 Plum	22	.364		3.18	15.55		18.73	27	
	0270	10 mm [3/8"]		22	.364		2.59	15.55		18.14	26.50	
	0280	15 mm [1/2"]		20	.400		.94	17.10		18.04	26.50	
	0290	18 mm [5/8"]		19	.421		4.93	18		22.93	32.50	
	0300	20 mm [3/4"]		19	.421		1.64	18		19.64	29	
	0310	25 mm [1"]		16	.500		4.13	21.50		25.63	36.50	
	0320	32 mm [1-1/4"]		15	.533		5.90	23		28.90	41	
	0330	40 mm [1-1/2"]		13	.615		7.10	26.50		33.60	47.50	
	0340	50 mm [2"]	▼	11	.727		11.80	31		42.80	59.50	
	0350	65 mm [2-1/2"]	Q-1	13	1.231		25	47.50		72.50	98.50	
	0360	80 mm [3"]		13	1.231		37.50	47.50		85	112	
	0370	90 mm [3-1/2"]		10	1.600		66	61.50		127.50	165	
	0380	100 mm [4"]		9	1.778		79.50	68.50		148	191	
	0390	125 mm [5"]	▼	6	2.667		310	102		412	495	
	0400	150 mm [6"]	Q-2	9	2.667		485	106		591	695	
	0410	200 mm [8"]	"	8	3	▼	1,650	120		1,770	2,000	

15107 | Metal Pipe & Fittings

		CREW	DAILY OUTPUT	LABOR-HOURS	UNIT	2006 BARE COSTS				TOTAL INCL O&P	
						MAT.	LABOR	EQUIP.	TOTAL		
460	0450	Tee, 8 mm [1/4"] R221113-50	1 Plum	14	.571	Ea.	3.57	24.50		28.07	40.50	**460**
	0470	10 mm [3/8"]		14	.571		2.73	24.50		27.23	39.50	
	0480	15 mm [1/2"]		13	.615		.87	26.50		27.37	40.50	
	0490	18 mm [5/8"]		12	.667		5.95	28.50		34.45	49.50	
	0500	20 mm [3/4"]		12	.667		2.11	28.50		30.61	45.50	
	0510	25 mm [1"]		10	.800		6.50	34		40.50	58.50	
	0520	32 mm [1-1/4"]		9	.889		9.35	38		47.35	67.50	
	0530	40 mm [1-1/2"]		8	1		14.45	42.50		56.95	80.50	
	0540	50 mm [2"]		7	1.143		22.50	49		71.50	98.50	
	0550	65 mm [2-1/2"]	Q-1	8	2		45.50	77		122.50	166	
	0560	80 mm [3"]		7	2.286		69.50	88		157.50	209	
	0570	90 mm [3-1/2"]		6	2.667		201	102		303	375	
	0580	100 mm [4"]		5	3.200		168	123		291	370	
	0590	125 mm [5"]		4	4		550	154		704	835	
	0600	150 mm [6"]	Q-2	6	4		750	159		909	1,075	
	0610	200 mm [8"]	"	5	4.800		2,900	191		3,091	3,500	
	0612	Tee, reducing on the outlet, 8 mm [1/4"]	1 Plum	15	.533		6.15	23		29.15	41.50	
	0613	10 mm [3/8"]		15	.533		5.40	23		28.40	40.50	
	0614	15 mm [1/2"]		14	.571		4.74	24.50		29.24	41.50	
	0615	18 mm [5/8"]		13	.615		9.55	26.50		36.05	50	
	0616	20 mm [3/4"]		12	.667		2.97	28.50		31.47	46.50	
	0617	25 mm [1"]		11	.727		6.90	31		37.90	54	
	0618	32 mm [1-1/4"]		10	.800		10.10	34		44.10	62.50	
	0619	40 mm [1-1/2"]		9	.889		10.75	38		48.75	69	
	0620	50 mm [2"]		8	1		17	42.50		59.50	83	
	0621	65 mm [2-1/2"]	Q-1	9	1.778		53	68.50		121.50	162	
	0622	80 mm [3"]		8	2		58.50	77		135.50	180	
	0623	100 mm [4"]		6	2.667		113	102		215	279	
	0624	125 mm [5"]		5	3.200		520	123		643	760	
	0625	150 mm [6"]	Q-2	7	3.429		715	137		852	990	
	0626	200 mm [8"]	"	6	4		2,900	159		3,059	3,450	
	0630	Tee, reducing on the run, 8 mm [1/4"]	1 Plum	15	.533		7.20	23		30.20	42.50	
	0631	10 mm [3/8"]		15	.533		9.65	23		32.65	45	
	0632	15 mm [1/2"]		14	.571		6.50	24.50		31	43.50	
	0633	18 mm [5/8"]		13	.615		10.15	26.50		36.65	50.50	
	0634	20 mm [3/4"]		12	.667		5.25	28.50		33.75	49	
	0635	25 mm [1"]		11	.727		8.20	31		39.20	55.50	
	0636	32 mm [1-1/4"]		10	.800		13.30	34		47.30	66	
	0637	40 mm [1-1/2"]		9	.889		23	38		61	82	
	0638	50 mm [2"]		8	1		29.50	42.50		72	96.50	
	0639	65 mm [2-1/2"]	Q-1	9	1.778		68.50	68.50		137	179	
	0640	80mm [3"]		8	2		101	77		178	227	
	0641	100 mm [4"]		6	2.667		224	102		326	400	
	0642	125 mm [5"]		5	3.200		520	123		643	760	
	0643	150 mm [6"]	Q-2	7	3.429		790	137		927	1,075	
	0644	200 mm [8"]	"	6	4		3,050	159		3,209	3,600	
	0650	Coupling, 8 mm [1/4"]	1 Plum	24	.333		.40	14.25		14.65	22	
	0670	10 mm [3/8"]		24	.333		.52	14.25		14.77	22	
	0680	15 mm [1/2"]		22	.364		.40	15.55		15.95	24	
	0690	18 mm [5/8"]		21	.381		1.22	16.25		17.47	26	
	0700	20 mm [3/4"]		21	.381		.76	16.25		17.01	25.50	
	0710	25 mm [1"]		18	.444		1.56	19		20.56	30	
	0715	32 mm [1-1/4"]		17	.471		2.90	20		22.90	33	
	0716	40 mm [1-1/2"]		15	.533		3.84	23		26.84	38.50	
	0718	50 mm [2"]		13	.615		6.40	26.50		32.90	46.50	
	0721	65 mm [2-1/2"]	Q-1	15	1.067		13.60	41		54.60	76.50	

MECHANICAL 15

For expanded coverage of these items see *Means Mechanical or Plumbing Cost Data 2006*

15107 | Metal Pipe & Fittings

			CREW	DAILY OUTPUT	LABOR-HOURS	UNIT	MAT.	LABOR	EQUIP.	TOTAL	TOTAL INCL O&P	
460	0722	80 mm [3"]	R221113 -50	Q-1	13	1.231	Ea.	20.50	47.50		68	93.50	460
	0724	90 mm [3-1/2"]			8	2		39	77		116	159	
	0726	100 mm [4"]			7	2.286		43	88		131	180	
	0728	125 mm [5"]			6	2.667		106	102		208	270	
	0731	150 mm [6"]		Q-2	8	3		175	120		295	370	
	0732	200 mm [8"]		"	7	3.429		565	137		702	830	
	0850	Unions, 8 mm [1/4"]		1 Plum	21	.381		10.30	16.25		26.55	36	
	0870	10 mm [3/8"]			21	.381		10.35	16.25		26.60	36	
	0880	15 mm [1/2"]			19	.421		5.50	18		23.50	33	
	0890	18 mm [5/8"]			18	.444		23.50	19		42.50	54.50	
	0900	20 mm [3/4"]			18	.444		6.90	19		25.90	36	
	0910	25 mm [1"]			15	.533		11.90	23		34.90	47.50	
	0920	32 mm [1-1/4"]			14	.571		20.50	24.50		45	59	
	0930	40 mm [1-1/2"]			12	.667		27	28.50		55.50	73	
	0940	50 mm [2"]			10	.800		46.50	34		80.50	103	
	0950	65 mm [2-1/2"]		Q-1	12	1.333		101	51		152	188	
	0960	80 mm [3"]		"	10	1.600		262	61.50		323.50	380	
	0980	Adapter, copper x male IPS, 8 mm [1/4"]		1 Plum	20	.400		5.60	17.10		22.70	31.50	
	0990	10 mm [3/8"]			20	.400		2.79	17.10		19.89	28.50	
	1000	15 mm [1/2"]			18	.444		1.07	19		20.07	29.50	
	1010	20 mm [3/4"]			17	.471		1.79	20		21.79	32	
	1020	25 mm [1"]			15	.533		4.64	23		27.64	39.50	
	1030	32 mm [1-1/4"]			13	.615		7.25	26.50		33.75	47.50	
	1040	40 mm [1-1/2"]			12	.667		8.30	28.50		36.80	52	
	1050	50 mm [2"]			11	.727		14.05	31		45.05	62	
	1060	65 mm [2-1/2"]		Q-1	10.50	1.524		54	58.50		112.50	148	
	1070	80 mm [3"]			10	1.600		71.50	61.50		133	171	
	1080	90 mm [3-1/2"]			9	1.778		104	68.50		172.50	217	
	1090	100 mm [4"]			8	2		104	77		181	230	
	1200	125 mm [5"]			6	2.667		490	102		592	695	
	1210	150 mm [6"]		Q-2	8.50	2.824		535	113		648	760	
	1250	Cross, 15 mm [1/2"]		1 Plum	10	.800		7.45	34		41.45	59.50	
	1260	20 mm [3/4"]			9.50	.842		14.50	36		50.50	70	
	1270	25 mm [1"]			8	1		24.50	42.50		67	91.50	
	1280	32 mm [1-1/4"]			7.50	1.067		35.50	45.50		81	108	
	1290	40 mm [1-1/2"]			6.50	1.231		50.50	52.50		103	135	
	1300	50 mm [2"]			5.50	1.455		95.50	62		157.50	199	
	1310	65mm [2-1/2"]		Q-1	6.50	2.462		220	94.50		314.50	385	
	1320	80 mm [3"]		"	5.50	2.909		276	112		388	475	
	1500	Tee fitting, mechanically formed, (Type 1).											
	1520	15 mm run [1/2"], 10 to 15 mm branch size		1 Plum	80	.100	Ea.		4.27		4.27	6.45	
	1530	20 mm run [3/4"], 10 to 20 mm branch size			60	.133			5.70		5.70	8.55	
	1540	25 mm run [1"], 10 to 25 mm branch size			54	.148			6.35		6.35	9.50	
	1550	32 mm run [1-1/4"], 10 to 32 mm branch size			48	.167			7.10		7.10	10.70	
	1560	40 mm run [1-1/2"], 10 to 40 mm branch size			40	.200			8.55		8.55	12.85	
	1570	50 mm run [2"], 10 to 50 mm branch size			35	.229			9.75		9.75	14.70	
	1580	65 mm run size, 15 mm to 50 mm branch size			32	.250			10.70		10.70	16.05	
	1590	80 mm run size, 25-50 mm branch size			26	.308			13.15		13.15	19.75	
	1600	100 mm run size, 25 to 50 mm branch size			24	.333			14.25		14.25	21.50	
	1640	Tee fitting, mechanically formed, (Type 2)											
	1650	65 mm run size, 65 mm branch size		1 Plum	12.50	.640	Ea.		27.50		27.50	41	
	1660	80 mm run size, 65 mm to 80 mm branch size			12	.667			28.50		28.50	43	
	1670	90 mm run size, 65 mm to 90 mm branch size			11	.727			31		31	46.50	
	1680	100 mm run size, 65-100 mm branch size			10.50	.762			32.50		32.50	49	
	1698	125 mm run size, 50 to 100 mm branch size			9.50	.842			36		36	54	
	1700	150 mm run size, 50 to 100 mm branch size			8.50	.941			40		40	60.50	

Important: See the Reference Section for supporting data - Crews, Rental Equipment, City Cost Indexes and Reference Data

MECHANICAL 15

15107 | Metal Pipe & Fittings

		CREW	DAILY OUTPUT	LABOR-HOURS	UNIT	2006 BARE COSTS				TOTAL INCL O&P	
						MAT.	LABOR	EQUIP.	TOTAL		
460	1710	200 mm run size, 50 to 100 mm branch size R221113 -50	1 Plum	7	1.143	Ea.		49		49	73.50	**460**
	2000	DWV, solder joints, copper x copper										
	2030	90° Elbow, 32 mm [1-1/4"]	1 Plum	13	.615	Ea.	5.20	26.50		31.70	45.50	
	2050	40 mm [1-1/2"]		12	.667		7.80	28.50		36.30	51.50	
	2070	50 mm [2"]	▼	10	.800		10.20	34		44.20	62.50	
	2090	80 mm [3"]	Q-1	10	1.600		25	61.50		86.50	120	
	2100	100 mm [4"]	"	9	1.778		120	68.50		188.50	235	
	2150	45° Elbow, 32 mm [1-1/4"]	1 Plum	13	.615		4.80	26.50		31.30	45	
	2170	40 mm [1-1/2"]		12	.667		3.97	28.50		32.47	47.50	
	2180	50 mm [2"]	▼	10	.800		9.10	34		43.10	61.50	
	2190	80 mm [3"]	Q-1	10	1.600		19.35	61.50		80.85	114	
	2200	100 mm [4"]	"	9	1.778		92	68.50		160.50	204	
	2250	Tee, Sanitary, 32 mm [1-1/4"]	1 Plum	9	.889		9.50	38		47.50	67.50	
	2270	40 mm [1-1/2"]		8	1		12.80	42.50		55.30	78.50	
	2290	50 mm [2"]	▼	7	1.143		14.95	49		63.95	90	
	2310	80 mm [3"]	Q-1	7	2.286		54	88		142	192	
	2330	100 mm [4"]	"	6	2.667		138	102		240	305	
	2400	Coupling, 32 mm [1-1/4"]	1 Plum	14	.571		2.43	24.50		26.93	39	
	2420	40 mm [1-1/2"]		13	.615		3.04	26.50		29.54	43	
	2440	50 mm [2"]	▼	11	.727		4.20	31		35.20	51	
	2460	80 mm [3"]	Q-1	11	1.455		8.15	56		64.15	93	
	2480	100 mm [4"]	"	10	1.600	▼	26	61.50		87.50	121	
500	0010	**PIPE, CORROSION RESISTANT** No couplings or hangers										**500**
	0020	Iron alloy, drain, mechanical joint										
	1000	40 mm diameter [1-1/2"]	Q-1	21.34	.750	m	74.50	29		103.50	126	
	1100	50 mm diameter [2"]		20.12	.795		85	30.50		115.50	140	
	1120	80 mm diameter [3"]		18.29	.875		119	33.50		152.50	182	
	1140	100 mm diameter [4"]	▼	15.85	1.009	▼	152	39		191	226	
	1980	Iron alloy, drain, B&S joint										
	2000	50 mm diameter [2"]	Q-1	16.46	.972	m	92.50	37.50		130	158	
	2100	80 mm diameter [3"]		15.85	1.009		133	39		172	205	
	2120	100 mm diameter [4"]	▼	14.63	1.094		184	42		226	265	
	2140	150 mm diameter [6"]	Q-2	17.98	1.335		305	53		358	420	
	2160	200 mm diameter [8"]	"	16.46	1.458	▼	555	58		613	700	
	2980	Plastic, epoxy, fiberglass filament wound, B&S joint										
	3000	50 mm diameter [2"]	Q-1	18.90	.847	m	23.50	32.50		56	75	
	3100	80 mm diameter [3"]		15.54	1.029		27	39.50		66.50	89.50	
	3120	100 mm diameter [4"]		13.72	1.167		40	45		85	112	
	3140	150 mm diameter [6"]	▼	9.75	1.640		56.50	63		119.50	157	
	3160	200 mm diameter [8"]	Q-2	11.58	2.072		99.50	82.50		182	233	
	3180	250 mm diameter [10"]	▼	9.75	2.461		138	98		236	300	
	3200	300 mm diameter [12"]		8.53	2.812	▼	165	112		277	350	
	3980	Polyester, fiberglass filament wound, B&S joint										
	4000	50 mm diameter [2"]	Q-1	18.90	.847	m	32.50	32.50		65	85	
	4100	80 mm diameter [3"]		15.54	1.029		45	39.50		84.50	109	
	4120	100 mm diameter [4"]		13.72	1.167		69.50	45		114.50	144	
	4140	150 mm diameter [6"]	▼	9.75	1.640		115	63		178	222	
	4160	200 mm diameter [8"]	Q-2	11.58	2.072		117	82.50		199.50	252	
	4180	250 mm diameter [10"]	▼	9.75	2.461		179	98		277	345	
	4200	300 mm diameter [12"]		8.53	2.812	▼	216	112		328	405	
	4980	Polypropylene, acid resistant, fire retardant, schedule 40										
	5000	40 mm diameter [1-1/2"]	Q-1	20.73	.772	m	3.71	29.50		33.21	48.50	
	5100	50 mm diameter [2"]		18.90	.847		5	32.50		37.50	54.50	
	5120	80 mm diameter [3"]		15.54	1.029		14.05	39.50		53.55	75	
	5140	100 mm diameter [4"]		13.72	1.167		14.95	45		59.95	84	
	5160	150 mm diameter [6"]	▼	9.75	1.640	▼	25	63		88	123	

MECHANICAL 15

15107 | Metal Pipe & Fittings

		CREW	DAILY OUTPUT	LABOR-HOURS	UNIT	MAT.	LABOR	EQUIP.	TOTAL	TOTAL INCL O&P	
500	5980	Proxylene, fire retardant, Schedule 40										**500**
	6000	40 mm diameter [1-1/2"]	Q-1	20.73	.772	m	16.55	29.50		46.05	62.50	
	6100	50 mm diameter [2"]		18.90	.847		22.50	32.50		55	74	
	6120	80 mm diameter [3"]		15.54	1.029		41	39.50		80.50	105	
	6140	100 mm diameter [4"]		13.72	1.167		58	45		103	132	
	6160	150 mm diameter [6"]	▼	9.75	1.640		104	63		167	210	
	6820	For Schedule 80, add				▼	35%	2%				
560	0010	**PIPE, CORROSION RESISTANT, FITTINGS**										**560**
	0030	Iron alloy										
	0050	Mechanical joint										
	0060	1/4 Bend, 40 mm [1-1/2"]	Q-1	12	1.333	Ea.	47	51		98	129	
	0080	50 mm [2"]		10	1.600		76.50	61.50		138	177	
	0090	80 mm [3"]		9	1.778		92	68.50		160.50	204	
	0100	100 mm [4"]		8	2		105	77		182	232	
	0110	1/8 Bend, 40 mm [1-1/2"]		12	1.333		30	51		81	110	
	0130	50 mm [2"]		10	1.600		51	61.50		112.50	149	
	0140	80 mm [3"]		9	1.778		68	68.50		136.50	178	
	0150	100 mm [4"]	▼	8	2	▼	91	77		168	216	
	0160	Tee and Y, sanitary, straight										
	0170	40 mm [1-1/2"]	Q-1	8	2	Ea.	51	77		128	172	
	0180	50 mm [2"]		7	2.286		68	88		156	207	
	0190	80 mm [3"]		6	2.667		105	102		207	270	
	0200	100 mm [4"]		5	3.200		194	123		317	400	
	0360	Coupling, 40 mm [1-1/2"]		14	1.143		35.50	44		79.50	106	
	0380	50 mm [2"]		12	1.333		41	51		92	122	
	0390	80 mm [3"]		11	1.455		42.50	56		98.50	131	
	0400	100 mm [4"]	▼	10	1.600	▼	48.50	61.50		110	146	
	0500	Bell & Spigot										
	0510	1/4 and 1/16 bend, 50 mm [2"]	Q-1	16	1	Ea.	53.50	38.50		92	117	
	0520	80 mm [3"]		14	1.143		125	44		169	203	
	0530	100 mm [4"]	▼	13	1.231		128	47.50		175.50	211	
	0540	150 mm [6"]	Q-2	17	1.412		253	56.50		309.50	365	
	0550	200 mm [8"]	"	12	2		1,050	79.50		1,129.50	1,275	
	0620	1/8 bend, 50 mm [2"]	Q-1	16	1		60	38.50		98.50	124	
	0640	80 mm [3"]		14	1.143		111	44		155	188	
	0650	100 mm [4"]	▼	13	1.231		117	47.50		164.50	200	
	0660	150 mm [6"]	Q-2	17	1.412		212	56.50		268.50	320	
	0680	200 mm [8"]	"	12	2		825	79.50		904.50	1,025	
	0700	Tee, sanitary, 50 mm [2"]	Q-1	10	1.600		114	61.50		175.50	218	
	0710	80 mm [3"]		9	1.778		370	68.50		438.50	510	
	0720	100 mm [4"]	▼	8	2		305	77		382	450	
	0730	150 mm [6"]	Q-2	11	2.182		385	87		472	550	
	0740	200 mm [8"]	"	8	3		1,050	120		1,170	1,325	
	1800	Y, sanitary, 50 mm [2"]	Q-1	10	1.600		120	61.50		181.50	225	
	1820	80 mm [3"]		9	1.778		215	68.50		283.50	340	
	1830	100 mm [4"]	▼	8	2		193	77		270	330	
	1840	150 mm [6"]	Q-2	11	2.182		640	87		727	835	
	1850	200 mm [8"]	"	8	3	▼	1,775	120		1,895	2,125	
	3000	Epoxy, filament wound										
	3030	Quick-lock joint										
	3040	90° Elbow, 50 mm [2"]	Q-1	28	.571	Ea.	67.50	22		89.50	108	
	3060	80 mm [3"]		16	1		78	38.50		116.50	144	
	3070	100 mm [4"]		13	1.231		106	47.50		153.50	188	
	3080	150 mm [6"]	▼	8	2		155	77		232	286	
	3090	200 mm [8"]	Q-2	9	2.667	▼	286	106		392	475	

15 MECHANICAL

15107 | Metal Pipe & Fittings

		CREW	DAILY OUTPUT	LABOR-HOURS	UNIT	2006 BARE COSTS				TOTAL INCL O&P	
						MAT.	LABOR	EQUIP.	TOTAL		
560	3100	250 mm [10"]	Q-2	7	3.429	Ea.	360	137		497	600	560
	3110	300 mm [12"]		6	4		515	159		674	805	
	3120	45° Elbow, 50 mm [2"]	Q-1	28	.571		52.50	22		74.50	90.50	
	3130	80 mm [3"]		16	1		74.50	38.50		113	140	
	3140	100 mm [4"]		13	1.231		103	47.50		150.50	184	
	3150	150 mm [6"]		8	2		155	77		232	286	
	3160	200 mm [8"]	Q-2	9	2.667		286	106		392	475	
	3170	250 mm [10"]		7	3.429		360	137		497	600	
	3180	300 mm [12"]		6	4		515	159		674	805	
	3190	Tee, 50 mm [2"]	Q-1	19	.842		161	32.50		193.50	226	
	3200	80 mm [3"]		11	1.455		195	56		251	299	
	3210	100 mm [4"]		9	1.778		234	68.50		302.50	360	
	3220	150 mm [6"]		5	3.200		360	123		483	580	
	3230	200 mm [8"]	Q-2	6	4		445	159		604	730	
	3240	250 mm [10"]		5	4.800		600	191		791	950	
	3250	300 mm [12"]		4	6		875	239		1,114	1,325	
	4000	Polypropylene, acid resistant										
	4020	Non-pressure, electrofusion joints										
	4050	1/4 bend, 40 mm [1-1/2"]	1 Plum	16	.500	Ea.	17	21.50		38.50	50.50	
	4060	50 mm [2"]	Q-1	28	.571		21	22		43	56	
	4080	80 mm [3"]		17	.941		36.50	36		72.50	94.50	
	4090	100 mm [4"]		14	1.143		58.50	44		102.50	130	
	4110	150 mm [6"]		8	2		140	77		217	270	
	4150	1/4 Bend, long sweep										
	4170	40 mm [1-1/2"]	1 Plum	16	.500	Ea.	18.35	21.50		39.85	52	
	4180	50 mm [2"]	Q-1	28	.571		24.50	22		46.50	60	
	4200	80 mm [3"]		17	.941		41	36		77	99.50	
	4210	100 mm [4"]		14	1.143		59.50	44		103.50	132	
	4250	1/8 bend, 40 mm [1-1/2"]	1 Plum	16	.500		17	21.50		38.50	50.50	
	4260	50 mm [2"]	Q-1	28	.571		20.50	22		42.50	55.50	
	4280	80 mm [3"]		17	.941		37.50	36		73.50	96	
	4290	100 mm [4"]		14	1.143		42.50	44		86.50	113	
	4310	150 mm [6"]		8	2		117	77		194	245	
	4400	Tee, sanitary										
	4420	40 mm [1-1/2"]	1 Plum	10	.800	Ea.	21	34		55	74.50	
	4430	50 mm [2"]	Q-1	17	.941		25	36		61	82	
	4450	80 mm [3"]		11	1.455		50.50	56		106.50	140	
	4460	100 mm [4"]		9	1.778		75.50	68.50		144	186	
	4480	150 mm [6"]		5	3.200		128	123		251	325	
	4490	Tee, sanitary reducing, 50 x 50 x 40 mm [2"x2"x1-1/2"]		17	.941		25	36		61	82	
	4492	80 mm x 80 mm x 50 mm [3"x3"x2"]		11	1.455		49	56		105	138	
	4494	100 mm x 100 mm x 80 mm [4"x4"x3"]		9	1.778		71	68.50		139.50	181	
	4496	150 mm x 150 mm x 100 mm [6"x6"x4"]		5	3.200		244	123		367	455	
	4500	Tee/wye, long turn										
	4520	40 mm [1-1/2"]	1 Plum	10	.800	Ea.	28.50	34		62.50	82.50	
	4530	50 mm [2"]	Q-1	17	.941		35.50	36		71.50	93.50	
	4550	80 mm [3"]		11	1.455		60.50	56		116.50	151	
	4570	100 mm [4"]		9	1.778		85	68.50		153.50	197	
	4650	Wye 45°, 40 mm [1-1/2"]	1 Plum	10	.800		21.50	34		55.50	75.50	
	4652	50 mm [2"]	Q-1	17	.941		31	36		67	88.50	
	4653	80 mm [3"]		11	1.455		53.50	56		109.50	143	
	4654	100 mm [4"]		9	1.778		78.50	68.50		147	189	
	4656	150 mm [6"]		5	3.200		200	123		323	405	
620	0010	**PIPE, STEEL**										620
	0012	The steel pipe in this section does not include fittings such as ells, tees,										

MECHANICAL 15

15107 | Metal Pipe & Fittings

		CREW	DAILY OUTPUT	LABOR-HOURS	UNIT	2006 BARE COSTS				TOTAL INCL O&P
						MAT.	LABOR	EQUIP.	TOTAL	
620	0014	For fittings either add a % (usually 25 to 35%) or see									620
	0015	the Mechanical or Plumbing Cost Data [R221113 -50]									
	0020	All pipe sizes are to Spec. A-53 unless noted otherwise									
	0050	Schedule 40, threaded, with couplings, and clevis type									
	0060	hangers sized for covering, 3 m O.C.									
	0540	Black, 6 mm diameter [1/4"]	1 Plum	20.12	.398	m	6.70	17·		23.70	33
	0550	10 mm diameter [3/8"]		19.81	.404		6.65	17.25		23.90	33.50
	0560	15 mm diameter [1/2"]		19.20	.417		6.75	17.80		24.55	34.50
	0570	20 mm diameter [3/4"]		18.59	.430		7.85	18.35		26.20	36
	0580	25 mm diameter [1"]		16.15	.495		11.40	21		32.40	44.50
	0590	32 mm diameter [1-1/4"]	Q-1	27.13	.590		14.95	22.50		37.45	50.50
	0600	40 mm diameter [1-1/2"]		24.38	.656		17.55	25		42.55	57.50
	0610	50 mm diameter [2"]		19.51	.820		23.50	31.50		55	73
	0620	65 mm diameter [2-1/2"]		15.24	1.050		36	40.50		76.50	100
	0630	80 mm diameter [3"]		13.11	1.221		47	47		94	122
	0640	90 mm diameter [3-1/2"]		12.19	1.312		63.50	50.50		114	146
	0650	100 mm diameter [4"]		10.97	1.458		69	56		125	161
	0660	125 mm diameter [5"]		7.92	2.019		103	77.50		180.50	230
	0670	150 mm diameter [6"]	Q-2	9.45	2.540		152	101		253	320
	0680	200 mm diameter [8"]		8.23	2.916		232	116		348	430
	0690	250 mm diameter [10"]		7.01	3.423		325	136		461	565
	0700	300 mm diameter [12"]		5.49	4.374		360	174		534	660
	0809	A-106, gr. A/B, seamless w/cplgs. & hangers									
	0811	8 mm diameter [1/4"]	1 Plum	20.12	.398	m	22	17		39	49.50
	0812	10 mm diameter [3/8"]		19.81	.404		20	17.25		37.25	48
	0813	15 mm diameter [1/2"]		19.20	.417		12.75	17.80		30.55	41
	0814	20 mm diameter [3/4"]		18.59	.430		17.65	18.35		36	47
	0815	25 mm diameter [1"]		16.15	.495		22	21		43	56
	0816	32 mm diameter [1-1/4"]	Q-1	27.13	.590		27.50	22.50		50	64.50
	0817	40 mm diameter [1-1/2"]		24.38	.656		33.50	25		58.50	75
	0819	A-53, 50 mm diameter [2"]		19.51	.820		37.50	31.50		69	89
	0821	65 mm diameter [2-1/2"]		15.24	1.050		47	40.50		87.50	112
	0822	80 mm diameter [3"]		13.11	1.221		61	47		108	138
	0823	100 mm diameter [4"]		10.97	1.458		93.50	56		149.50	188
	1220	To delete coupling & hanger, subtract									
	1230	8 mm diam. to 20 mm diam. [1/4"-3/4"]					31%	56%			
	1240	25 mm diam. to 40 mm diam. [1"-1-1/2"]					23%	51%			
	1250	50 mm diam. to 100 mm diam. [2"-4"]					23%	41%			
	1260	125 mm diam. to 300mm dia					21%	45%			
	1280	All pipe sizes are to Spec. A-53 unless noted otherwise									
	1281	Schedule 40, threaded, with couplings and clevis type									
	1282	hangers sized for covering, 3 m O. C.									
	1290	Galvanized, 6 mm diameter [1/4"]	1 Plum	20.12	.398	m	7.95	17		24.95	34.50
	1300	10 mm diameter [3/8"]		19.81	.404		8.80	17.25		26.05	35.50
	1310	15 mm diameter [1/2"]		19.20	.417		9.10	17.80		26.90	37
	1320	20 mm diameter [3/4"]		18.59	.430		10.90	18.35		29.25	39.50
	1330	25 mm diameter [1"]		16.15	.495		14.80	21		35.80	48.50
	1340	32 mm diameter [1-1/4"]	Q-1	27.13	.590		19.75	22.50		42.25	55.50
	1350	40 mm diameter [1-1/2"]		24.38	.656		23.50	25		48.50	63.50
	1360	50 mm diameter [2"]		19.51	.820		31	31.50		62.50	81.50
	1370	65 mm diameter [2-1/2"]		15.24	1.050		51	40.50		91.50	117
	1380	80 mm diameter [3"]		13.11	1.221		65.50	47		112.50	143
	1390	90 mm diameter [3-1/2"]		12.19	1.312		83	50.50		133.50	167
	1400	100 mm diameter [4"]		10.97	1.458		94.50	56		150.50	189
	1410	125 mm diameter [5"]		7.92	2.019		129	77.50		206.50	259
	1420	150 mm diameter [6"]	Q-2	9.45	2.540		173	101		274	345

15
MECHANICAL

15107 | Metal Pipe & Fittings

		CREW	DAILY OUTPUT	LABOR-HOURS	UNIT	2006 BARE COSTS MAT.	LABOR	EQUIP.	TOTAL	TOTAL INCL O&P	
620	1430	200 mm diameter [8"] R221113 -50	Q-2	8.23	2.916	m	267	116		383	470	**620**
	1440	250 mm diameter [10"]		7.01	3.423		298	136		434	535	
	1450	300 mm diameter [12"]	▼	5.49	4.374	▼	415	174		589	715	
	1750	To delete coupling & hanger, subtract										
	1760	8 mm diam. to 20 mm diam. [1/4"-3/4"]					31%	56%				
	1770	25 mm diam. to 40 mm diam. [1"-1-1/2"]					23%	51%				
	1780	50 mm diam. to 100 mm diam. [2"-4"]					23%	41%				
	1790	125 mm diam. to 300 mm diam. [5"-12"]					21%	45%				
	2000	Welded, sch. 40, on yoke & roll hangers, sized for covering,										
	2010	3 m O.C.										
	2040	Black, 25 mm diameter [1"]	Q-15	28.35	.564	m	16.55	21.50	2.44	40.49	53.50	
	2070	50 mm diameter [2"]		18.59	.861		29	33	3.72	65.72	86	
	2090	80 mm diameter [3"]		13.11	1.221		46	47	5.25	98.25	127	
	2110	100 mm diameter [4"]		11.28	1.419		63	54.50	6.15	123.65	158	
	2120	125 mm diameter [5"]	▼	9.75	1.640		83	63	7.10	153.10	194	
	2130	150 mm diameter [6"]	Q-16	10.97	2.187		113	87	6.30	206.30	262	
	2140	200 mm diameter [8"]		8.84	2.715		169	108	7.80	284.80	360	
	2150	250 mm diameter [10"]		7.32	3.281		305	131	9.45	445.45	540	
	2160	300 mm diameter [12"]	▼	5.79	4.144	▼	355	165	11.95	531.95	655	
640	0010	**PIPE, STEEL, FITTINGS** Threaded R221113 -50										**640**
	0020	Cast Iron										
	1300	Extra heavy weight, black										
	1310	Couplings, steel straight										
	1320	8 mm [1/4"]	1 Plum	19	.421	Ea.	1.86	18		19.86	29	
	1330	10 mm [3/8"]		19	.421		2.04	18		20.04	29	
	1340	15 mm [1/2"]		19	.421		2.75	18		20.75	30	
	1350	20 mm [3/4"]		18	.444		2.94	19		21.94	31.50	
	1360	25 mm [1"]	▼	15	.533		3.74	23		26.74	38.50	
	1370	32 mm [1-1/4"]	Q-1	26	.615		6	23.50		29.50	42	
	1380	40 mm [1-1/2"]		24	.667		6	25.50		31.50	45	
	1390	50 mm [2"]		21	.762		9.15	29.50		38.65	54	
	1400	65 mm [2-1/2"]		18	.889		13.60	34		47.60	66.50	
	1410	80 mm [3"]		14	1.143		16.15	44		60.15	84	
	1420	90 mm [3-1/2"]		12	1.333		21.50	51		72.50	101	
	1430	100 mm [4"]		10	1.600		25.50	61.50		87	121	
	1440	125 mm [5"]	▼	6	2.667		36.50	102		138.50	194	
	1450	150 mm [6"]	Q-2	8	3		44.50	120		164.50	229	
	1460	200 mm [8"]		7	3.429		81	137		218	295	
	1470	250 mm [10"]		6	4		128	159		287	380	
	1480	300 mm [12"]	▼	4	6	▼	171	239		410	550	
	1510	90° Elbow, straight										
	1520	15 mm [1/2"]	1 Plum	15	.533	Ea.	14.05	23		37.05	50	
	1530	20 mm [3/4"]		14	.571		14.85	24.50		39.35	53	
	1540	25 mm [1"]	▼	13	.615		17.95	26.50		44.45	59.50	
	1550	32 mm [1-1/4"]	Q-1	22	.727		26.50	28		54.50	71.50	
	1560	40 mm [1-1/2"]		20	.800		33	30.50		63.50	83	
	1580	50 mm [2"]		18	.889		41	34		75	96.50	
	1590	60 mm [2-1/2"]		14	1.143		95.50	44		139.50	171	
	1600	80 mm [3"]		10	1.600		126	61.50		187.50	232	
	1610	100 mm [4"]	▼	6	2.667		265	102		367	445	
	1620	150 mm [6"]	Q-2	7	3.429	▼	680	137		817	955	
	1650	45° Elbow, straight										
	1660	15 mm [1/2"]	1 Plum	15	.533	Ea.	20	23		43	56.50	
	1670	20 mm [3/4"]		14	.571		19.35	24.50		43.85	58	
	1680	25 mm [1"]	▼	13	.615	▼	23	26.50		49.50	65	

MECHANICAL 15

			CREW	DAILY OUTPUT	LABOR-HOURS	UNIT	MAT.	LABOR	EQUIP.	TOTAL	TOTAL INCL O&P	
		**15107	Metal Pipe & Fittings**					**2006 BARE COSTS**					
640	1690	32 mm [1-1/4"] R221113 -50	Q-1	22	.727	Ea.	42	28		70	88.50	**640**
	1700	40 mm [1-1/2"]		20	.800		60	30.50		90.50	113	
	1710	50 mm [2"]		18	.889		60	34		94	118	
	1720	65 mm [2-1/2"]		14	1.143		103	44		147	180	
	1800	Tee, straight										
	1810	15 mm [1/2"]	1 Plum	9	.889	Ea.	22	38		60	81.50	
	1820	20 mm [3/4"]		9	.889		22.50	38		60.50	82	
	1830	25 mm [1"]		8	1		27	42.50		69.50	94	
	1840	32 mm [1-1/4"]	Q-1	14	1.143		40	44		84	110	
	1850	40 mm [1-1/2"]		13	1.231		51.50	47.50		99	128	
	1860	50 mm [2"]		11	1.455		63.50	56		119.50	154	
	1870	65 mm [2-1/2"]		9	1.778		136	68.50		204.50	252	
	1880	80 mm [3"]		6	2.667		185	102		287	360	
	1890	100 mm [4"]		4	4		360	154		514	625	
	1900	150 mm [6"]	Q-2	4	6		785	239		1,024	1,225	
	4000	Standard weight, black										
	4010	Couplings, steel straight, merchants										
	4030	8 mm [1/4"]	1 Plum	19	.421	Ea.	.50	18		18.50	27.50	
	4040	10 mm [3/8"]		19	.421		.60	18		18.60	27.50	
	4050	15 mm [1/2"]		19	.421		.64	18		18.64	27.50	
	4060	20 mm [3/4"]		18	.444		.82	19		19.82	29.50	
	4070	25 mm [1"]		15	.533		1.15	23		24.15	36	
	4080	32 mm [1-1/4"]	Q-1	26	.615		1.46	23.50		24.96	37	
	4090	40 mm [1-1/2"]		24	.667		1.85	25.50		27.35	40.50	
	4100	50 mm [2"]		21	.762		2.65	29.50		32.15	47	
	4110	65 mm [2-1/2"]		18	.889		7.55	34		41.55	60	
	4120	80 mm [3"]		14	1.143		10.60	44		54.60	77.50	
	4130	90 mm [3-1/2"]		12	1.333		18.80	51		69.80	97.50	
	4140	100 mm [4"]		10	1.600		20	61.50		81.50	115	
	4150	125 mm [5"]		6	2.667		34.50	102		136.50	192	
	4160	150 mm [6"]	Q-2	8	3		41	120		161	226	
	4200	Standard weight, galvanized										
	4210	Couplings, steel straight, merchants										
	4230	8 mm [1/4"]	1 Plum	19	.421	Ea.	.58	18		18.58	27.50	
	4240	10 mm [3/8"]		19	.421		.74	18		18.74	28	
	4250	15 mm [1/2"]		19	.421		.78	18		18.78	28	
	4260	20 mm [3/4"]		18	.444		.99	19		19.99	29.50	
	4270	25 mm [1"]		15	.533		1.38	23		24.38	36	
	4280	32 mm [1-1/4"]	Q-1	26	.615		1.76	23.50		25.26	37.50	
	4290	40 mm [1-1/2"]		24	.667		2.19	25.50		27.69	41	
	4300	50 mm [2"]		21	.762		3.27	29.50		32.77	47.50	
	4310	65 mm [2-1/2"]		18	.889		9.25	34		43.25	61.50	
	4320	80 mm [3"]		14	1.143		12.30	44		56.30	79.50	
	4330	90 mm [3-1/2"]		12	1.333		21.50	51		72.50	101	
	4340	100 mm [4"]		10	1.600		21.50	61.50		83	117	
	4350	125 mm [5"]		6	2.667		40.50	102		142.50	199	
	4360	150 mm [6"]	Q-2	8	3		49.50	120		169.50	235	
690	0010	**PIPE, GROOVED-JOINT STEEL FITTINGS & VALVES**										**690**
	0012	Fittings are ductile iron. Steel fittings noted.										
	0020	Pipe includes coupling & clevis type hanger 3 m O.C.										
	1000	Schedule 40, black										
	1040	20 mm diameter [3/4"]	1 Plum	21.64	.370	m	9.80	15.80		25.60	35	
	1050	25 mm diameter [1"]		19.20	.417		11.90	17.80		29.70	40	
	1060	32 mm diameter [1-1/4"]		17.68	.453		15.70	19.30		35	46.50	
	1070	40 mm diameter [1-1/2"]		15.54	.515		18.20	22		40.20	53	

Important: See the Reference Section for supporting data - Crews, Rental Equipment, City Cost Indexes and Reference Data

15 MECHANICAL

		CREW	DAILY OUTPUT	LABOR-HOURS	UNIT	2006 BARE COSTS				TOTAL INCL O&P
						MAT.	LABOR	EQUIP.	TOTAL	
690 1080	50 mm diameter [2"]	1 Plum	12.19	.656	m	23.50	28		51.50	68	690
1090	65 mm diameter [2-1/2"]	Q-1	17.37	.921		33.50	35.50		69	90
1100	80 mm diameter [3"]		15.24	1.050		42	40.50		82.50	107
1110	100 mm diameter [4"]		13.72	1.167		56.50	45		101.50	130
1120	125 mm diameter [5"]		11.28	1.419		73.50	54.50		128	163
1130	150 mm diameter [6"]	Q-2	12.80	1.875		102	74.50		176.50	225
1140	200 mm diameter [8"]		11.28	2.128		155	85		240	298
1150	250 mm diameter [10"]		9.45	2.540		249	101		350	425
1160	305 mm diameter [12"]		8.23	2.916		310	116		426	515
1170	356 mm diameter [14"]		6.10	3.937		395	157		552	670
1180	406 mm diameter [16"]		5.18	4.632		430	185		615	755
1190	457 mm diameter [18"]		4.27	5.624		510	224		734	895
1200	508 mm diameter [20"]		3.66	6.562		770	261		1,031	1,250
1210	610 mm diameter [24"]		3.05	7.874		905	315		1,220	1,475
1740	To delete coupling & hanger, subtract									
1750	19 mm diam. to 51 mm diam. [3/4"-2"]					65%	27%			
1760	64 mm diam. to 127 mm diam. [2-1/2"-5"]					41%	18%			
1770	152 mm diam. to 305 mm diam. [6"-12"]					31%	13%			
1780	356 mm diam. to 610 mm diam. [14"-24"]					35%	10%			
1800	Galvanized									
1840	20 mm diameter [3/4"]	1 Plum	21.64	.370	m	12.75	15.80		28.55	38
1850	25 mm diameter [1"]		19.20	.417		14.80	17.80		32.60	43.50
1860	32 mm diameter [1-1/4"]		17.68	.453		19.65	19.30		38.95	50.50
1870	40 mm diameter [1-1/2"]		15.54	.515		23	22		45	58.50
1880	50 mm diameter [2"]		12.19	.656		29.50	28		57.50	74.50
1890	65 mm diameter [2-1/2"]	Q-1	17.37	.921		45	35.50		80.50	103
1900	80 mm diameter [3"]		15.24	1.050		57.50	40.50		98	124
1910	100 mm diameter [4"]		13.72	1.167		80	45		125	156
1920	125 mm diameter [5"]		11.28	1.419		95	54.50		149.50	187
1930	150 mm diameter [6"]	Q-2	12.80	1.875		129	74.50		203.50	253
1940	203 mm diameter [8"]		11.28	2.128		201	85		286	350
1950	254 mm diameter [10"]		9.45	2.540		320	101		421	505
1960	305 mm diameter [12"]		8.23	2.916		400	116		516	610
2540	To delete coupling & hanger, subtract									
2550	19 mm diam. to 51 mm diam. [3/4"-2"]					36%	27%			
2560	64 mm diam. to 127 mm diam. [2-1/2"-5"]					19%	18%			
2570	152 mm diam. to 305 mm diam. [6"-12"]					14%	13%			
3990	Fittings: cplg. & labor required at joints not incl. in fitting									
3994	price. Add 1 per joint for installed price.									
4000	Elbow, 90° or 45°, painted									
4030	19 mm diameter [3/4"]	1 Plum	50	.160	Ea.	24.50	6.85		31.35	37.50
4040	25 mm diameter [1"]		50	.160		13.15	6.85		20	25
4050	32 mm diameter [1-1/4"]		40	.200		13.15	8.55		21.70	27.50
4060	38 mm diameter [1-1/2"]		33	.242		13.15	10.35		23.50	30
4070	51 mm diameter [2"]		25	.320		13.15	13.65		26.80	35
4080	64 mm diameter [2-1/2"]	Q-1	40	.400		13.15	15.35		28.50	37.50
4090	76 mm diameter [3"]		33	.485		23.50	18.65		42.15	53.50
4100	102 mm diameter [4"]		25	.640		25.50	24.50		50	65
4110	127 mm diameter [5"]		20	.800		61	30.50		91.50	114
4120	152 mm diameter [6"]	Q-2	25	.960		72	38.50		110.50	137
4250	For galvanized elbows, add					26%				
4690	Tee, painted									
4700	19 mm diameter [3/4"]	1 Plum	38	.211	Ea.	26.50	9		35.50	42.50
4740	25 mm diameter [1"]		33	.242		20	10.35		30.35	37.50
4750	32 mm diameter [1-1/4"]		27	.296		20	12.65		32.65	41
4760	38 mm diameter [1-1/2"]		22	.364		20	15.55		35.55	45.50

MECHANICAL 15

15107	Metal Pipe & Fittings	CREW	DAILY OUTPUT	LABOR-HOURS	UNIT	2006 BARE COSTS				TOTAL INCL O&P	
						MAT.	LABOR	EQUIP.	TOTAL		
690	4770	51 mm diameter [2"]	1 Plum	17	.471	Ea.	20	20		40	52	**690**
	4780	64 mm diameter [2-1/2"]	Q-1	27	.593		20	23		43	56.50	
	4790	76 mm diameter [3"]		22	.727		28	28		56	73	
	4800	102 mm diameter [4"]		17	.941		43	36		79	102	
	4810	127 mm diameter [5"]		13	1.231		101	47.50		148.50	182	
	4820	152 mm diameter [6"]	Q-2	17	1.412		117	56.50		173.50	213	
	4830	203 mm diameter [8"]		14	1.714		256	68.50		324.50	385	
	4840	254 mm diameter [10"]		12	2		535	79.50		614.50	705	
	4850	305 mm diameter [12"]		10	2.400		745	95.50		840.50	960	
	4851	356 mm diameter [14"]		9	2.667		785	106		891	1,025	
	4852	406 mm diameter [16"]		8	3		890	120		1,010	1,150	
	4853	457 mm diameter [18"]	Q-3	11	2.909		1,100	118		1,218	1,400	
	4854	508 mm diameter [20"]		10	3.200		1,600	130		1,730	1,950	
	4855	610 mm diameter [24"]		8	4		2,425	163		2,588	2,925	
	4900	For galvanized tees, add					24%					
	4906	Couplings, rigid style, painted										
	4908	25 mm diameter [1"]	1 Plum	100	.080	Ea.	9.70	3.42		13.12	15.80	
	4909	32 mm diameter [1-1/4"]		100	.080		9.70	3.42		13.12	15.80	
	4910	38 mm diameter [1-1/2"]		67	.119		9.70	5.10		14.80	18.30	
	4912	51 mm diameter [2"]		50	.160		9.95	6.85		16.80	21.50	
	4914	64 mm diameter [2-1/2"]	Q-1	80	.200		11.40	7.70		19.10	24	
	4916	76 mm diameter [3"]		67	.239		13.30	9.20		22.50	28.50	
	4918	102 mm diameter [4"]		50	.320		18.90	12.30		31.20	39.50	
	4920	127 mm diameter [5"]		40	.400		24.50	15.35		39.85	50	
	4922	152 mm diameter [6"]	Q-2	50	.480		32.50	19.15		51.65	65	
	4939	Couplings										
	4940	Flexible, standard, painted										
	4950	19 mm diameter [3/4"]	1 Plum	100	.080	Ea.	7	3.42		10.42	12.85	
	4960	25 mm diameter [1"]		100	.080		7	3.42		10.42	12.85	
	4970	32 mm diameter [1-1/4"]		80	.100		9.25	4.27		13.52	16.65	
	4980	38 mm diameter [1-1/2"]		67	.119		10.15	5.10		15.25	18.80	
	4990	51 mm diameter [2"]		50	.160		10.70	6.85		17.55	22	
	5000	64 mm diameter [2-1/2"]	Q-1	80	.200		12.75	7.70		20.45	25.50	
	5010	76 mm diameter [3"]		67	.239		14.10	9.20		23.30	29.50	
	5020	89 mm diameter [3-1/2"]		57	.281		20.50	10.80		31.30	39	
	5030	102 mm diameter [4"]		50	.320		20.50	12.30		32.80	41	
	5040	127 mm diameter [5"]		40	.400		31.50	15.35		46.85	57.50	
	5050	152 mm diameter [6"]	Q-2	50	.480		37	19.15		56.15	70	
	5070	203 mm diameter [8"]		42	.571		60.50	23		83.50	101	
	5090	254 mm diameter [10"]		35	.686		100	27.50		127.50	152	
	5110	305 mm diameter [12"]		32	.750		114	30		144	171	
	5120	356 mm diameter [14"]		24	1		139	40		179	213	
	5130	406 mm diameter [16"]		20	1.200		182	48		230	272	
	5140	457 mm diameter [18"]		18	1.333		213	53		266	315	
	5150	508 mm diameter [20"]		16	1.500		335	60		395	460	
	5160	610 mm diameter [24"]		13	1.846		370	73.50		443.50	515	
	5200	For galvanized couplings, add					33%					
	5750	Flange, w/groove gasket, black steel										
	5760	ANSI class 125 and 150, painted										
	5780	51 mm pipe size [2"]	1 Plum	23	.348	Ea.	44	14.85		58.85	71	
	5790	64 mm pipe size [2-1/2"]	Q-1	37	.432		55	16.60		71.60	85.50	
	5800	76 mm pipe size [3"]		31	.516		59	19.85		78.85	95	
	5820	102 mm pipe size [4"]		23	.696		79.50	26.50		106	128	
	5830	127 mm pipe size [5"]		19	.842		92	32.50		124.50	150	
	5840	152 mm pipe size [6"]	Q-2	23	1.043		100	41.50		141.50	174	
	5850	203 mm pipe size [8"]		17	1.412		113	56.50		169.50	210	

15 MECHANICAL

		DAILY OUTPUT	LABOR-HOURS	UNIT	2006 BARE COSTS				TOTAL INCL O&P	
		CREW			MAT.	LABOR	EQUIP.	TOTAL		
690 5860	254 mm pipe size [10"]	Q-2	14	1.714	Ea.	180	68.50		248.50	300	**690**
5870	305 mm pipe size [12"]		12	2		235	79.50		314.50	380	
5880	356 mm pipe size [14"]		10	2.400		540	95.50		635.50	740	
5890	406 mm pipe size [16"]		9	2.667		625	106		731	850	
5900	457 mm pipe size [18"]		6	4		770	159		929	1,100	
5910	508 mm pipe size [20"]		5	4.800		930	191		1,121	1,325	
5920	610 mm pipe size [24"]	↓	4.50	5.333	↓	1,200	213		1,413	1,625	
8000	Butterfly valve, 2 position handle, with standard trim										
8010	38 mm pipe size [1-1/2"]	1 Plum	50	.160	Ea.	124	6.85		130.85	147	
8020	51 mm pipe size [2"]	"	38	.211		124	9		133	151	
8030	76 mm pipe size [3"]	Q-1	50	.320		178	12.30		190.30	215	
8050	102 mm pipe size [4"]	"	38	.421		196	16.20		212.20	241	
8070	152 mm pipe size [6"]	Q-2	38	.632		395	25		420	475	
8080	203 mm pipe size [8"]		27	.889		570	35.50		605.50	680	
8090	254 mm pipe size [10"]	↓	20	1.200	↓	1,000	48		1,048	1,175	
8200	With stainless steel trim										
8240	38 mm pipe size [1-1/2"]	1 Plum	50	.160	Ea.	158	6.85		164.85	183	
8250	51 mm pipe size [2"]	"	38	.211		158	9		167	187	
8270	76 mm pipe size [3"]	Q-1	50	.320		212	12.30		224.30	252	
8280	102 mm pipe size [4"]	"	38	.421		230	16.20		246.20	278	
8300	152 mm pipe size [6"]	Q-2	38	.632		450	25		475	535	
8310	203 mm pipe size [8"]		27	.889		690	35.50		725.50	815	
8320	254 mm pipe size [10"]	↓	20	1.200	↓	985	48		1,033	1,150	
9000	Cut one groove, labor										
9010	19 mm pipe size [3/4"]	Q-1	152	.105	Ea.		4.05		4.05	6.10	
9020	25 mm pipe size [1"]		140	.114			4.39		4.39	6.60	
9030	32 mm pipe size [1-1/4"]		124	.129			4.96		4.96	7.45	
9040	38 mm pipe size [1-1/2"]		114	.140			5.40		5.40	8.10	
9050	51 mm pipe size [2"]		104	.154			5.90		5.90	8.90	
9060	64 mm pipe size [2-1/2"]		96	.167			6.40		6.40	9.65	
9070	76 mm pipe size [3"]		88	.182			7		7	10.50	
9080	89 mm pipe size [3-1/2"]		83	.193			7.40		7.40	11.15	
9090	102 mm pipe size [4"]		78	.205			7.90		7.90	11.85	
9100	127 mm pipe size [5"]		72	.222			8.55		8.55	12.85	
9110	152 mm pipe size [6"]		70	.229			8.80		8.80	13.20	
9120	203 mm pipe size [8"]		54	.296			11.40		11.40	17.15	
9130	254 mm pipe size [10"]		38	.421			16.20		16.20	24.50	
9140	305 mm pipe size [12"]		30	.533			20.50		20.50	31	
9150	356 mm pipe size [14"]		20	.800			30.50		30.50	46.50	
9160	406 mm pipe size [16"]		19	.842			32.50		32.50	48.50	
9170	457 mm pipe size [18"]		18	.889			34		34	51.50	
9180	508 mm pipe size [20"]		17	.941			36		36	54.50	
9190	610 mm pipe size [24"]	↓	15	1.067	↓		41		41	61.50	
9210	Roll one groove										
9220	19 mm pipe size [3/4"]	Q-1	266	.060	Ea.		2.31		2.31	3.48	
9230	25 mm pipe size [1"]		228	.070			2.70		2.70	4.06	
9240	32 mm pipe size [1-1/4"]		200	.080			3.07		3.07	4.63	
9250	38 mm pipe size [1-1/2"]		178	.090			3.45		3.45	5.20	
9260	51 mm pipe size [2"]		116	.138			5.30		5.30	8	
9270	64 mm pipe size [2-1/2"]		110	.145			5.60		5.60	8.40	
9280	76 mm pipe size [3"]		100	.160			6.15		6.15	9.25	
9290	89 mm pipe size [3-1/2"]		94	.170			6.55		6.55	9.85	
9300	102 mm pipe size [4"]		86	.186			7.15		7.15	10.75	
9310	127 mm pipe size [5"]		84	.190			7.30		7.30	11	
9320	152 mm pipe size [6"]		80	.200			7.70		7.70	11.55	
9330	203 mm pipe size [8"]	↓	66	.242	↓		9.30		9.30	14	

MECHANICAL 15

15107 | Metal Pipe & Fittings

		CREW	DAILY OUTPUT	LABOR-HOURS	UNIT	2006 BARE COSTS				TOTAL INCL O&P
						MAT.	LABOR	EQUIP.	TOTAL	
690 9340	254 mm pipe size [10″]	Q-1	58	.276	Ea.		10.60		10.60	15.95 **690**
9350	305 mm pipe size [12″]		46	.348			13.35		13.35	20
9360	356 mm pipe size [14″]		30	.533			20.50		20.50	31
9370	406 mm pipe size [16″]		28	.571			22		22	33
9380	457 mm pipe size [18″]		27	.593			23		23	34.50
9390	508 mm pipe size [20″]		25	.640			24.50		24.50	37
9400	610 mm pipe size [24″]	↓	23	.696	↓		26.50		26.50	40
920 0010	**PIPE, STAINLESS STEEL**									**920**
3500	Threaded, couplings and hangers 3 m O.C.									
3520	Schedule 40, type 304									
3540	6 mm diameter [1/4″]	1 Plum	16.46	.486	m	18.80	21		39.80	51.50
3550	10 mm diameter [3/8″]		16.15	.495		22	21		43	56
3560	15 mm diameter [1/2″]		15.85	.505		26	21.50		47.50	61
3580	25 mm diameter [1″]	↓	13.72	.583		43.50	25		68.50	85
3610	50 mm diameter [2″]	Q-1	17.37	.921		88.50	35.50		124	151
3640	100 mm diameter [4″]	Q-2	15.54	1.544		274	61.50		335.50	395
3740	For small quantities, add				↓	10%				
4250	Schedule 40, type 316									
4290	6 mm diameter [1/4″]	1 Plum	16.46	.486	m	22	21		43	55
4300	10 mm diameter [3/8″]		16.15	.495		26.50	21		47.50	61
4310	15 mm diameter [1/2″]		15.85	.505		34.50	21.50		56	70.50
4320	20 mm diameter [3/4″]		15.54	.515		42.50	22		64.50	79.50
4330	25 mm diameter [1″]	↓	13.72	.583		55.50	25		80.50	98.50
4360	50 mm diameter [2″]	Q-1	17.37	.921		113	35.50		148.50	179
4390	100 mm diameter [4″]	Q-2	15.54	1.544		360	61.50		421.50	490
4490	For small quantities, add				↓	10%				

15108 | Plastic Pipe & Fittings

		CREW	DAILY OUTPUT	LABOR-HOURS	UNIT	MAT.	LABOR	EQUIP.	TOTAL	TOTAL INCL O&P
520 0010	**PIPE, PLASTIC**									**520**
0020	FRP,cplngs 3 m OC hngrs 1/m									
0200	High strength									
0240	50 mm diameter [2″]	Q-1	17.68	.905	m	30	35		65	85.50
0260	80 mm diameter [3″]		15.54	1.029		43	39.50		82.50	107
0280	100 mm diameter [4″]		14.33	1.117		61	43		104	132
0300	150 mm diameter [6″]	↓	11.58	1.381		101	53		154	191
0320	200 mm diameter [8″]	Q-2	14.63	1.640		145	65.50		210.50	259
0340	250 mm diameter [10″]		12.19	1.969		241	78.50		319.50	385
0360	300 mm diameter [12″]	↓	10.97	2.187	↓	279	87		366	435
0550	To delete coupling & hangers, subtract									
0560	50 mm diam. to 150 mm diam. [2″-6″]					33%	56%			
0570	200 mm diam. to 300 mm diam. [8″-12″]					31%	52%			
0600	PVC, high impact/pressure, cplgs. 3 m O.C., hangers 1/m									
1800	PVC, couplings 3 m OC, hangers 1/m									
1820	Schedule 40									
1860	15 mm diameter [1/2″]	1 Plum	16.46	.486	m	3.74	21		24.74	35
1870	20 mm diameter [3/4″]		15.54	.515		4.04	22		26.04	37.50
1880	25 mm diameter [1″]		14.02	.571		4.53	24.50		29.03	41.50
1890	32 mm diameter [1-1/4″]		12.80	.625		5.15	26.50		31.65	45.50
1900	40mm diameter [1-1/2″]	↓	10.97	.729		5.55	31		36.55	53
1910	50 mm diameter [2″]	Q-1	17.98	.890		6.50	34		40.50	58.50
1920	65 mm diameter [2-1/2″]		17.07	.937		8.85	36		44.85	64
1930	80 mm diameter [3″]		16.15	.990		11.40	38		49.40	70
1940	100 mm diameter [4″]		14.63	1.094		14.40	42		56.40	79
1950	125 mm diameter [5″]	↓	13.11	1.221	↓	18.85	47		65.85	91

15108 | Plastic Pipe & Fittings

		CREW	DAILY OUTPUT	LABOR-HOURS	UNIT	2006 BARE COSTS				TOTAL INCL O&P	
						MAT.	LABOR	EQUIP.	TOTAL		
520	1960	150 mm diameter [6"]	Q-1	11.89	1.346	m	25	51.50		76.50	106	**520**
	4100	DWV type, schedule 40, couplings 3 m O.C., hangers 3 per 3 m										
	4120	ABS										
	4140	32 mm diameter [1-1/4"]	1 Plum	12.80	.625	m	4.13	26.50		30.63	44.50	
	4150	40 mm diameter [1-1/2"]	"	10.97	.729		4.23	31		35.23	51.50	
	4160	50 mm diameter [2"]	Q-1	17.98	.890		4.66	34		38.66	56.50	
	4170	80 mm diameter [3"]		16.15	.990		8.65	38		46.65	67	
	4180	100 mm diameter [4"]		14.63	1.094		11.45	42		53.45	75.50	
	4190	150 mm diameter [6"]		11.89	1.346		26	51.50		77.50	107	
	4360	To delete coupling & hangers, subtract										
	4370	32 mm diam. [1-1/4"]					64%	68%				
	4380	40 mm diam. to 150 mm diam. [1-1/2"-6"]					54%	57%				
	4400	PVC										
	4410	32 mm diameter [1-1/4"]	1 Plum	12.80	.625	m	4.49	26.50		30.99	45	
	4420	40 mm diameter [1-1/2"]	"	10.97	.729		4.59	31		35.59	52	
	4460	50 mm diameter [2"]	Q-1	17.98	.890		5.20	34		39.20	57	
	4470	76 mm diameter [3"]		16.15	.990		9.70	38		47.70	68	
	4480	100 mm diameter [4"]		14.63	1.094		12.60	42		54.60	77	
	4490	150 mm diameter [6"]		11.89	1.346		23.50	51.50		75	104	
	4500	200 mm diameter [8"]	Q-2	14.63	1.640		55	65.50		120.50	159	
	4750	To delete coupling & hangers, subtract										
	4760	32 mm diam. to 40 mm diam. [1-1/4"-1-1/2"]					71%	64%				
	4770	50 mm diam. to 200 mm diam. [2"-8"]					60%	57%				
	5360	CPVC, couplings 3 m O.C., hangers 3 per 3 m										
	5380	Schedule 40										
	5460	15 mm diameter [1/2"]	1 Plum	16.46	.486	m	8.30	21		29.30	40	
	5470	20 mm diameter [3/4"]		15.54	.515		10.90	22		32.90	45	
	5480	25 mm diameter [1"]		14.02	.571		13.35	24.50		37.85	51	
	5490	32 mm diameter [1-1/4"]		12.80	.625		15.45	26.50		41.95	57	
	5500	40 mm diameter [1-1/2"]		10.97	.729		17.30	31		48.30	66	
	5510	50 mm diameter [2"]	Q-1	17.98	.890		21.50	34		55.50	75	
	5520	65 mm diameter [2-1/2"]		17.07	.937		34	36		70	91.50	
	5530	80 mm diameter [3"]		16.15	.990		42	38		80	104	
	5540	100 mm diameter [4"]		14.63	1.094		67.50	42		109.50	137	
	5550	150 mm diameter [6"]		13.11	1.221		108	47		155	190	
	5730	To delete coupling & hangers, subtract										
	5740	15 mm diam. to 20 mm diam. [1/2"-3/4"]					37%	77%				
	5750	25 mm diam. to 32 mm diam. [1"-1-1/4"]					27%	70%				
	5760	40 mm diam. to 80 mm diam. [1-1/2"-3"]					21%	57%				
	5770	100 mm diam. to 150 mm diam. [4"-6"]					16%	57%				
	7280	Polyethylene, flexible, no couplings or hangers										
	7300	SDR 15, 690 kPa										
	7310	20 mm diameter [3/4"]				m	.79			.79	.85	
	7350	25 mm diameter [1"]					1.05			1.05	1.15	
	7360	32 mm diameter [1-1/4"]					1.87			1.87	2.07	
	7370	40 mm diameter [1-1/2"]					2.26			2.26	2.49	
	7380	50 mm diameter [2"]					3.71			3.71	4.07	
	8120	SDR 9, 1380 kPa										
	8150	20 mm diameter [3/4"]				m	.92			.92	1.02	
	8160	25 mm diameter [1"]					1.51			1.51	1.67	
	8170	32 mm diameter [1-1/4"]					2.26			2.26	2.49	
560	0010	**PIPE, PLASTIC, FITTINGS**										**560**
	0030	Epoxy resin, fiberglass reinforced, general service										
	0090	Elbow, 90°, 50 mm [2"]	Q-1	33.10	.483	Ea.	67.50	18.60		86.10	103	
	0100	80 mm [3"]		20.80	.769		78	29.50		107.50	131	

MECHANICAL **15**

15108	Plastic Pipe & Fittings		CREW	DAILY OUTPUT	LABOR-HOURS	UNIT	2006 BARE COSTS				TOTAL INCL O&P
							MAT.	LABOR	EQUIP.	TOTAL	
560	0110	100 mm [4"]	Q-1	16.50	.970	Ea.	106	37.50		143.50	173
	0120	150 mm [6"]		10.10	1.584		155	61		216	262
	0130	200 mm [8"]	Q-2	9.30	2.581		286	103		389	470
	0140	250 mm [10"]		8.50	2.824		360	113		473	565
	0150	300 mm [12"]		7.60	3.158		515	126		641	755
	0160	45° Elbow, same as 90°									
	0290	Tee, 50 mm [2"]	Q-1	20	.800	Ea.	37	30.50		67.50	87
	0300	80 mm [3"]		13.90	1.151		44	44		88	115
	0310	100 mm [4"]		11	1.455		61.50	56		117.50	152
	0320	150 mm [6"]		6.70	2.388		172	92		264	325
	0330	200 mm [8"]	Q-2	6.20	3.871		1,100	154		1,254	1,425
	0340	250 mm [10"]		5.70	4.211		1,250	168		1,418	1,625
	0350	300 mm [12"]		5.10	4.706		1,525	188		1,713	1,950
	0380	Couplings									
	0410	50 mm [2"]	Q-1	33.10	.483	Ea.	11.85	18.60		30.45	41
	0420	80 mm [3"]		20.80	.769		13.30	29.50		42.80	59
	0430	100 mm [4"]		16.50	.970		18.35	37.50		55.85	76
	0440	150 mm [6"]		10.10	1.584		43.50	61		104.50	139
	0450	200 mm [8"]	Q-2	9.30	2.581		74	103		177	237
	0460	250 mm [10"]		8.50	2.824		108	113		221	288
	0470	300 mm [12"]		7.60	3.158		146	126		272	350
	0473	High corrosion resistant couplings, add					30%				
	2100	PVC schedule 80, socket joint									
	2110	90° elbow, 15 mm [1/2"]	1 Plum	30.30	.264	Ea.	.71	11.25		11.96	17.75
	2130	20 mm [3/4"]		26	.308		.90	13.15		14.05	20.50
	2140	25 mm [1"]		22.70	.352		1.46	15.05		16.51	24
	2150	32 mm [1-1/4"]		20.20	.396		1.94	16.90		18.84	27.50
	2160	40 mm [1-1/2"]		18.20	.440		2.08	18.75		20.83	30.50
	2170	50 mm [2"]	Q-1	33.10	.483		2.50	18.60		21.10	31
	2180	80 mm [3"]		20.80	.769		6.65	29.50		36.15	52
	2190	100 mm [4"]		16.50	.970		7.50	37.50		45	64.50
	2200	150 mm [6"]		10.10	1.584		67	61		128	165
	2210	200 mm [8"]	Q-2	9.30	2.581		185	103		288	360
	2250	45° elbow, 15 mm [1/2"]	1 Plum	30.30	.264		1.34	11.25		12.59	18.40
	2270	20 mm [3/4"]		26	.308		2.03	13.15		15.18	22
	2280	25 mm [1"]		22.70	.352		3.01	15.05		18.06	26
	2290	32 mm [1-1/4"]		20.20	.396		3.87	16.90		20.77	30
	2300	40 mm [1-1/2"]		18.20	.440		4.57	18.75		23.32	33
	2310	50 mm [2"]	Q-1	33.10	.483		5.95	18.60		24.55	34.50
	2320	80 mm [3"]		20.80	.769		15.10	29.50		44.60	61
	2330	100 mm [4"]		16.50	.970		27	37.50		64.50	86
	2340	150 mm [6"]		10.10	1.584		34.50	61		95.50	130
	2350	200 mm [8"]	Q-2	9.30	2.581		175	103		278	345
	2400	Tee, 15 mm [1/2"]	1 Plum	20.20	.396		1.98	16.90		18.88	27.50
	2420	20 mm [3/4"]		17.30	.462		2.08	19.75		21.83	32
	2430	25 mm [1"]		15.20	.526		2.59	22.50		25.09	37
	2440	32 mm [1-1/4"]		13.50	.593		7.15	25.50		32.65	46
	2450	40 mm [1-1/2"]		12.10	.661		7.15	28		35.15	50.50
	2460	50 mm [2"]	Q-1	20	.800		8.90	30.50		39.40	56.50
	2470	80 mm [3"]		13.90	1.151		12.15	44		56.15	80
	2480	100 mm [4"]		11	1.455		14.05	56		70.05	99.50
	2490	150 mm [6"]		6.70	2.388		47.50	92		139.50	191
	2500	200 mm [8"]	Q-2	6.20	3.871		215	154		369	470
	2510	Flange, socket, 1034 kPa, 15 mm [1/2"]	1 Plum	55.60	.144		3.86	6.15		10.01	13.50
	2514	20 mm [3/4"]		47.60	.168		4.13	7.20		11.33	15.35
	2518	25 mm [1"]		41.70	.192		4.59	8.20		12.79	17.40

15 MECHANICAL

			DAILY	LABOR-			2006 BARE COSTS				TOTAL	
15108	**Plastic Pipe & Fittings**	CREW	OUTPUT	HOURS	UNIT	MAT.	LABOR	EQUIP.	TOTAL		INCL O&P	
560	2522	40 mm [1-1/2"]	1 Plum	33.30	.240	Ea.	4.82	10.25		15.07	21	560
	2526	50 mm [2"]	Q-1	60.60	.264		6.45	10.15		16.60	22.50	
	2530	100 mm [4"]		30.30	.528		13.85	20.50		34.35	45.50	
	2534	150 mm [6"]		18.50	.865		21.50	33		54.50	74	
	2538	200 mm [8"]	Q-2	17.10	1.404		107	56		163	201	
	2550	Coupling, 15 mm [1/2"]	1 Plum	30.30	.264		1.29	11.25		12.54	18.35	
	2570	20 mm [3/4"]		26	.308		1.73	13.15		14.88	21.50	
	2580	25 mm [1"]		22.70	.352		1.77	15.05		16.82	24.50	
	2590	32 mm [1-1/4"]		20.20	.396		2.69	16.90		19.59	28.50	
	2600	40 mm [1-1/2"]		18.20	.440		2.89	18.75		21.64	31	
	2610	50 mm [2"]	Q-1	33.10	.483		3.10	18.60		21.70	31.50	
	2620	80 mm [3"]		20.80	.769		8.80	29.50		38.30	54	
	2630	100 mm [4"]		16.50	.970		11.10	37.50		48.60	68	
	2640	150 mm [6"]		10.10	1.584		23.50	61		84.50	118	
	2650	200 mm [8"]	Q-2	9.30	2.581		37.50	103		140.50	196	
	2660	250 mm [10"]		8.50	2.824		40.50	113		153.50	214	
	2670	300 mm [12"]		7.60	3.158		47	126		173	241	
	4500	DWV, ABS, non pressure, socket joints										
	4540	1/4 Bend, 32 mm [1-1/4"]	1 Plum	20.20	.396	Ea.	3.15	16.90		20.05	29	
	4560	40 mm [1-1/2"]	"	18.20	.440		2.41	18.75		21.16	30.50	
	4570	50 mm [2"]	Q-1	33.10	.483		3.73	18.60		22.33	32	
	4580	80 mm [3"]		20.80	.769		8.95	29.50		38.45	54.50	
	4590	100 mm [4"]		16.50	.970		15.10	37.50		52.60	72.50	
	4600	150 mm [6"]		10.10	1.584		71.50	61		132.50	171	
	4650	1/8 Bend, same as 1/4 Bend										
	4800	Tee, sanitary										
	4820	32 mm [1-1/4"]	1 Plum	13.50	.593	Ea.	3.57	25.50		29.07	42	
	4830	40 mm [1-1/2"]	"	12.10	.661		3.29	28		31.29	46	
	4840	50 mm [2"]	Q-1	20	.800		4.81	30.50		35.31	52	
	4850	80 mm [3"]		13.90	1.151		12.15	44		56.15	80	
	4860	100 mm [4"]		11	1.455		22.50	56		78.50	109	
	4862	Tee, sanitary, reducing, 50 mm x 40 mm [2"-1-1/2"]		22	.727		4.64	28		32.64	47	
	4864	80 mm x 50 mm [3"-2"]		15.30	1.046		8.85	40		48.85	70	
	4868	100 mm x 80 mm [4"-3"]		12.10	1.322		22	51		73	101	
	4870	Combination Y and 1/8 bend										
	4872	40 mm [1-1/2"]	1 Plum	12.10	.661	Ea.	6.70	28		34.70	50	
	4874	50 mm [2"]	Q-1	20	.800		7.75	30.50		38.25	55	
	4876	80 mm [3"]		13.90	1.151		19.05	44		63.05	87.50	
	4878	100 mm [4"]		11	1.455		38.50	56		94.50	126	
	4880	80 mm x 40 mm [3"-1-1/2"]		15.50	1.032		16.70	39.50		56.20	78	
	4882	100 mm x 80 mm [4"-3"]		12.10	1.322		28	51		79	107	
	4900	Wye, 32 mm [1-1/4"]	1 Plum	13.50	.593		3.93	25.50		29.43	42.50	
	4902	40 mm [1-1/2"]	"	12.10	.661		4.21	28		32.21	47	
	4904	50 mm [2"]	Q-1	20	.800		5.95	30.50		36.45	53	
	4906	80 mm [3"]		13.90	1.151		14.70	44		58.70	82.50	
	4908	100 mm [4"]		11	1.455		29	56		85	116	
	4910	150 mm [6"]		6.70	2.388		89	92		181	236	
	4918	80 mm x 40 mm [3"-1-1/2"]		15.50	1.032		9.80	39.50		49.30	70.50	
	4920	100 mm x 80 mm [4"-3"]		12.10	1.322		21	51		72	99.50	
	4922	150 mm x 100 mm [6"-4"]		6.90	2.319		73.50	89		162.50	215	
	4930	Double Wye, 40 mm [1-1/2"]	1 Plum	9.10	.879		9.60	37.50		47.10	67	
	4932	50 mm [2"]	Q-1	16.60	.964		11.45	37		48.45	68	
	4934	80 mm [3"]		10.40	1.538		29.50	59		88.50	122	
	4936	100 mm [4"]		8.25	1.939		60	74.50		134.50	178	
	4940	50 mm x 40 mm [2"-1-1/2"]		16.80	.952		11.45	36.50		47.95	67.50	
	4942	80 mm x 51 mm [3"-2"]		10.60	1.509		22	58		80	112	

MECHANICAL 15

15108	Plastic Pipe & Fittings	CREW	DAILY OUTPUT	LABOR-HOURS	UNIT	2006 BARE COSTS				TOTAL INCL O&P
						MAT.	LABOR	EQUIP.	TOTAL	
560 4944	100 mm x 80 mm [4"-3"]	Q-1	8.45	1.893	Ea.	47.50	73		120.50	163 **560**
4946	150 mm x 100 mm [6"-4"]		7.25	2.207		94.50	85		179.50	232
4950	Reducer bushing, 50 mm x 40 mm [2"-1-1/2"]		36.40	.440		1.58	16.90		18.48	27
4952	80 mm x 40 mm [3"-1-1/2"]		27.30	.586		5.95	22.50		28.45	40.50
4954	100 mm x 50 mm [4"-2"]		18.20	.879		12.20	34		46.20	64.50
4956	150 mm x 100 mm [6"-4"]		11.10	1.441		32.50	55.50		88	120
4960	Couplings, 40 mm [1-1/2"]	1 Plum	18.20	.440		1.01	18.75		19.76	29
4962	50 mm [2"]	Q-1	33.10	.483		1.50	18.60		20.10	29.50
4963	80 mm [3"]		20.80	.769		4.35	29.50		33.85	49.50
4964	100 mm [4"]		16.50	.970		6.80	37.50		44.30	63.50
4966	150 mm [6"]		10.10	1.584		26.50	61		87.50	121
4970	50 mm x 40 mm [2"-1-1/2"]		33.30	.480		2.48	18.45		20.93	30.50
4972	80 mm x 40 mm [3"-1-1/2"]		21	.762		6.95	29.50		36.45	51.50
4974	100 mm x 80 mm [4"-3"]		16.70	.958		12.35	37		49.35	69
4978	Closet flange, 100 mm [4"]	1 Plum	32	.250		6.40	10.70		17.10	23
4980	100 mm x 80 mm [4"-3"]	"	34	.235		6.60	10.05		16.65	22.50
5000	DWV, PVC, schedule 40, socket joints									
5040	1/4 bend, 32 mm [1-1/4"]	1 Plum	20.20	.396	Ea.	4.50	16.90		21.40	30.50
5060	40 mm [1-1/2"]	"	18.20	.440		1.71	18.75		20.46	30
5070	50 mm [2"]	Q-1	33.10	.483		2.52	18.60		21.12	31
5080	80 mm [3"]		20.80	.769		7.20	29.50		36.70	52.50
5090	100 mm [4"]		16.50	.970		12.50	37.50		50	70
5100	150 mm [6"]		10.10	1.584		47.50	61		108.50	144
5105	200 mm [8"]	Q-2	9.30	2.581		96.50	103		199.50	261
5110	1/4 bend, long sweep, 40 mm [1-1/2"]	1 Plum	18.20	.440		3.74	18.75		22.49	32
5112	50 mm [2"]	Q-1	33.10	.483		4.03	18.60		22.63	32.50
5114	80 mm [3"]		20.80	.769		9.50	29.50		39	55
5116	100 mm [4"]		16.50	.970		17.70	37.50		55.20	75.50
5150	1/8 bend, 32 mm [1-1/4"]	1 Plum	20.20	.396		2.69	16.90		19.59	28.50
5170	40 mm [1-1/2"]	"	18.20	.440		1.57	18.75		20.32	29.50
5180	50 mm [2"]	Q-1	33.10	.483		2.43	18.60		21.03	30.50
5190	80 mm [3"]		20.80	.769		6.65	29.50		36.15	52
5200	100 mm [4"]		16.50	.970		11.10	37.50		48.60	68.50
5210	150 mm [6"]		10.10	1.584		46	61		107	142
5215	200 mm [8"]	Q-2	9.30	2.581		96	103		199	261
5250	Tee, sanitary 32 mm [1-1/4"]	1 Plum	13.50	.593		4.12	25.50		29.62	42.50
5254	40 mm [1-1/2"]	"	12.10	.661		2.95	28		30.95	46
5255	50 mm [2"]	Q-1	20	.800		4.14	30.50		34.64	51
5256	80 mm [3"]		13.90	1.151		9.75	44		53.75	77.50
5257	100 mm [4"]		11	1.455		16.95	56		72.95	103
5259	150 mm [6"]		6.70	2.388		91	92		183	238
5261	200 mm [8"]	Q-2	6.20	3.871		190	154		344	440
5264	50 mm x 40 mm [2"-1-1/2"]	Q-1	22	.727		4.67	28		32.67	47
5266	80 mm x 40 mm [3"-1-1/2"]		15.50	1.032		5.90	39.50		45.40	66
5268	100 mm x 80 mm [4"-3"]		12.10	1.322		22.50	51		73.50	101
5271	150 mm x 100 mm [6"-4"]		6.90	2.319		76	89		165	218
5314	Combination Y & 1/8 bend, 40 mm [1-1/2"]	1 Plum	12.10	.661		5.35	28		33.35	48.50
5315	50 mm [2"]	Q-1	20	.800		7.15	30.50		37.65	54.50
5317	80 mm [3"]		13.90	1.151		13	44		57	81
5318	100 mm [4"]		11	1.455		22	56		78	109
5324	Combination Y & 1/8 bend reducing									
5325	50 mm x 50 mm x 40 mm [2"x2"x1-1/2"]	Q-1	22	.727	Ea.	8.05	28		36.05	51
5327	80 mm x 80 mm x 40 mm [3"x3"x1-1/2"]		15.50	1.032		14	39.50		53.50	75
5328	80 mm x 80 mm x 50 mm [3" x 3" x 2"]		15.30	1.046		11.65	40		51.65	73.50
5329	100 mm x 100 mm x 50 mm (4"x4"x2")		12.20	1.311		19.50	50.50		70	97.50
5331	Wye, 32 mm [1-1/4"]	1 Plum	13.50	.593		4	25.50		29.50	42.50

15108 | Plastic Pipe & Fittings

		CREW	DAILY OUTPUT	LABOR-HOURS	UNIT	2006 BARE COSTS				TOTAL INCL O&P
						MAT.	LABOR	EQUIP.	TOTAL	
5332	40 mm [1-1/2"]	1 Plum	12.10	.661	Ea.	3.95	28		31.95	47
5333	50 mm [2"]	Q-1	20	.800		5.05	30.50		35.55	52
5334	80 mm [3"]		13.90	1.151		14.45	44		58.45	82.50
5335	100 mm [4"]		11	1.455		21	56		77	107
5336	150 mm [6"]		6.70	2.388		44	92		136	187
5337	200 mm [8"]	Q-2	6.20	3.871		73.50	154		227.50	315
5341	50 mm x 40 mm [2"-1-1/2"]	Q-1	22	.727		5.90	28		33.90	48.50
5342	80 mm x 40 mm [3"-1-1/2"]		15.50	1.032		8.30	39.50		47.80	68.50
5343	100 mm x 80 mm [4"-3"]		12.10	1.322		16.75	51		67.75	95
5344	150 mm x 100 mm [6"-4"]		6.90	2.319		52	89		141	192
5345	200 mm x 150 mm [8"-6"]	Q-2	6.40	3.750		56	149		205	287
5347	Double wye, 40 mm [1-1/2"]	1 Plum	9.10	.879		8	37.50		45.50	65.50
5348	50 mm [2"]	Q-1	16.60	.964		9.15	37		46.15	65.50
5349	80 mm [3"]		10.40	1.538		23.50	59		82.50	115
5350	100 mm [4"]		8.25	1.939		48	74.50		122.50	165
5354	50 mm x 40 mm [2"-1-1/2"]		16.80	.952		8.40	36.50		44.90	64
5355	80 mm x 50 mm [3"-2"]		10.60	1.509		17.60	58		75.60	107
5356	100 mm x 80 mm [4"-3"]		8.45	1.893		38	73		111	152
5357	150 mm x 100 mm [6"-4"]		7.25	2.207		79	85		164	215
5374	Coupling, 32 mm [1-1/4"]	1 Plum	20.20	.396		2.50	16.90		19.40	28.50
5376	40 mm [1-1/2"]	"	18.20	.440		.76	18.75		19.51	29
5378	50 mm [2"]	Q-1	33.10	.483		1.03	18.60		19.63	29
5380	80 mm [3"]		20.80	.769		3.75	29.50		33.25	48.50
5390	100 mm [4"]		16.50	.970		6.15	37.50		43.65	63
5400	150 mm [6"]		10.10	1.584		20	61		81	114
5402	200 mm [8"]	Q-2	9.30	2.581		43	103		146	202
5404	50 mm x 40 mm [2"-1-1/2"]	Q-1	33.30	.480		1.97	18.45		20.42	30
5406	80 mm x 40 mm [3"-1-1/2"]		21	.762		6.25	29.50		35.75	51
5408	100 mm x 80 mm [4"-3"]		16.70	.958		11.05	37		48.05	67.50
5410	Reducer bushing, 50 mm x 32 mm [2"-1-1/4"]		36.50	.438		2.56	16.85		19.41	28.50
5412	80 mm x 40 mm [3"-1-1/2"]		27.30	.586		5.85	22.50		28.35	40.50
5414	100 mm x 50 mm [4"-2"]		18.20	.879		10.80	34		44.80	63
5416	150 mm x 100 mm [6"-4"]		11.10	1.441		31	55.50		86.50	118
5418	200 mm x 150 mm [8"-6"]	Q-2	10.20	2.353		62	94		156	210
5425	Closet flange 100 mm [4"]	Q-1	32	.500		6.30	19.20		25.50	36
5426	100 mm x 80 mm [4"-3"]	"	34	.471		10.85	18.10		28.95	39
5450	Solvent cement for PVC, industrial grade				liter	13.80			13.80	15.20
7340	PVC flange, slip-on, Sch 80 std., 15 mm [1/2"]	1 Plum	22	.364	Ea.	8.20	15.55		23.75	32.50
7350	20 mm [3/4"]		21	.381		8.75	16.25		25	34
7360	25 mm [1"]		18	.444		9.75	19		28.75	39
7370	32 mm [1-1/4"]		17	.471		10.05	20		30.05	41
7380	40 mm [1-1/2"]		16	.500		10.25	21.50		31.75	43.50
7390	50 mm [2"]	Q-1	26	.615		13.65	23.50		37.15	50.50
7400	65 mm [2-1/2"]		24	.667		21	25.50		46.50	61.50
7410	80 mm [3"]		18	.889		23.50	34		57.50	77
7420	100 mm [4"]		15	1.067		29.50	41		70.50	94
7430	150 mm [6"]		10	1.600		49.50	61.50		111	147
7440	200 mm [8"]	Q-2	11	2.182		93.50	87		180.50	234
7550	Union, schedule 40, socket joints, 15 mm [1/2"]	1 Plum	19	.421		3	18		21	30.50
7560	20 mm [3/4"]		18	.444		3.25	19		22.25	32
7570	25 mm [1"]		15	.533		3.34	23		26.34	38
7580	32 mm [1-1/4"]		14	.571		10.75	24.50		35.25	48.50
7590	40 mm [1-1/2"]		13	.615		11.20	26.50		37.70	52
7600	50 mm [2"]	Q-1	20	.800		15.10	30.50		45.60	63

560

			DAILY	LABOR-		2006 BARE COSTS				TOTAL	
15110	**Valves**	CREW	OUTPUT	HOURS	UNIT	MAT.	LABOR	EQUIP.	TOTAL	INCL O&P	
100	0010	**VALVES, BRASS**										100
	0500	Gas stops, threaded									
	0530	15 mm	1 Plum	24	.333	Ea.	5.75	14.25		20	28
	0540	20 mm		22	.364		14.25	15.55		29.80	39
	0550	25 mm		19	.421		18.20	18		36.20	47
	0560	32 mm		15	.533		22.50	23		45.50	59.50
160	0010	**VALVES, BRONZE**										160
	1020	Angle, 1034 kPa, rising stem, threaded									
	1030	6 mm	1 Plum	24	.333	Ea.	57	14.25		71.25	84
	1040	8 mm		24	.333		57	14.25		71.25	84
	1050	10 mm		24	.333		57	14.25		71.25	84
	1060	15 mm		22	.364		57	15.55		72.55	86
	1070	20 mm		20	.400		77.50	17.10		94.60	111
	1080	25 mm		19	.421		110	18		128	148
	1100	40 mm		13	.615		218	26.50		244.50	280
	1110	50 mm		11	.727		345	31		376	425
	1380	Ball, 1034 kPa, threaded									
	1400	8 mm	1 Plum	24	.333	Ea.	8.10	14.25		22.35	30.50
	1430	10 mm		24	.333		8.10	14.25		22.35	30.50
	1450	15 mm		22	.364		8.10	15.55		23.65	32.50
	1460	20 mm		20	.400		13.35	17.10		30.45	40
	1470	25 mm		19	.421		16.85	18		34.85	45.50
	1480	32 mm		15	.533		19.05	23		42.05	55.50
	1490	40 mm		13	.615		25	26.50		51.50	67
	1500	50 mm		11	.727		30	31		61	79.50
	1510	65 mm		9	.889		100	38		138	168
	1520	80 mm		8	1		153	42.50		195.50	233
	1750	Check, swing, 1034 kPa, regrinding disc, threaded									
	1800	6 mm	1 Plum	24	.333	Ea.	29	14.25		43.25	53.50
	1830	8 mm		24	.333		29	14.25		43.25	53.50
	1840	10 mm		24	.333		30	14.25		44.25	54.50
	1850	15 mm		24	.333		30	14.25		44.25	54.50
	1860	20 mm		20	.400		40.50	17.10		57.60	70
	1870	25 mm		19	.421		60	18		78	93
	1880	32 mm		15	.533		84	23		107	127
	1890	40 mm		13	.615		99	26.50		125.50	149
	1900	50 mm		11	.727		146	31		177	207
	1910	65 mm	Q-1	15	1.067		242	41		283	330
	2000	For 1379 kPa, add					5%	10%			
	2040	For 2068 kPa, add					15%	15%			
	2850	Gate, N.R.S., soldered, 862 kPa									
	2900	10 mm	1 Plum	24	.333	Ea.	23	14.25		37.25	46.50
	2920	15 mm		24	.333		19.50	14.25		33.75	43
	2940	20 mm		20	.400		22.50	17.10		39.60	50
	2950	25 mm		19	.421		32	18		50	62
	2960	32 mm		15	.533		50	23		73	89.50
	2970	40 mm		13	.615		56.50	26.50		83	102
	2980	50 mm		11	.727		76.50	31		107.50	131
	2990	65 mm	Q-1	15	1.067		191	41		232	272
	3000	76 mm	"	13	1.231		250	47.50		297.50	345
	3850	Rising stem, soldered, 2068 kPa									
	3950	25 mm	1 Plum	19	.421	Ea.	83.50	18		101.50	119
	3980	50 mm	"	11	.727		222	31		253	291
	4000	80 mm	Q-1	13	1.231		700	47.50		747.50	840
	4250	Threaded, 1034 kPa									
	4310	8 mm	1 Plum	24	.333	Ea.	29	14.25		43.25	53.50

Important: See the Reference Section for supporting data - Crews, Rental Equipment, City Cost Indexes and Reference Data

	15110 | Valves	CREW	DAILY OUTPUT	LABOR-HOURS	UNIT	2006 BARE COSTS				TOTAL INCL O&P
						MAT.	LABOR	EQUIP.	TOTAL	
160 4320	10 mm	1 Plum	24	.333	Ea.	29	14.25		43.25	53.50
4330	15 mm		24	.333		27	14.25		41.25	51.50
4340	20 mm		20	.400		32	17.10		49.10	61
4350	25 mm		19	.421		43	18		61	74.50
4360	32 mm		15	.533		57.50	23		80.50	98
4370	40 mm		13	.615		72.50	26.50		99	120
4380	50 mm		11	.727		98.50	31		129.50	156
4390	65 mm	Q-1	15	1.067		230	41		271	315
4400	80 mm	"	13	1.231		320	47.50		367.50	425
4500	For 2068 kPa, threaded, add					100%	15%			
4540	For chain operated type, add					15%				
4850	Globe, 1034 kPa, rising stem, threaded									
4920	8 mm	1 Plum	24	.333	Ea.	42	14.25		56.25	67.50
4940	10 mm		24	.333		42	14.25		56.25	67.50
4950	15 mm		24	.333		42	14.25		56.25	67.50
4960	20 mm		20	.400		57	17.10		74.10	88
4970	25 mm		19	.421		89.50	18		107.50	125
4980	32 mm		15	.533		141	23		164	190
4990	40 mm		13	.615		172	26.50		198.50	229
5000	50 mm		11	.727		257	31		288	330
5010	65 mm	Q-1	15	1.067		515	41		556	630
5020	80 mm	"	13	1.231		735	47.50		782.50	880
5120	For 2068 kPa threaded, add					50%	15%			
5600	Relief, pressure & temperature, self-closing, ASME, threaded									
5640	20 mm	1 Plum	28	.286	Ea.	85.50	12.20		97.70	112
5650	25 mm		24	.333		124	14.25		138.25	159
5660	32 mm		20	.400		249	17.10		266.10	300
5670	40 mm		18	.444		475	19		494	555
5680	50 mm		16	.500		520	21.50		541.50	600
5950	Pressure, poppet type, threaded									
6000	15 mm	1 Plum	30	.267	Ea.	15.35	11.40		26.75	34
6040	20 mm	"	28	.286	"	42	12.20		54.20	64.50
6400	Pressure, water, ASME, threaded									
6440	20 mm	1 Plum	28	.286	Ea.	49	12.20		61.20	72.50
6450	25 mm		24	.333		53.50	14.25		67.75	80.50
6460	32 mm		20	.400		157	17.10		174.10	199
6470	40 mm		18	.444		232	19		251	285
6480	50 mm		16	.500		360	21.50		381.50	425
6490	65 mm		15	.533		1,175	23		1,198	1,325
6900	Reducing, water pressure									
6920	2068 kPa to 172-517 kPa, threaded or sweat									
6940	15 mm	1 Plum	24	.333	Ea.	145	14.25		159.25	182
6950	20 mm		20	.400		145	17.10		162.10	186
6960	25 mm		19	.421		225	18		243	274
6970	32 mm		15	.533		400	23		423	480
6980	40 mm		13	.615		605	26.50		631.50	710
8350	Tempering, water, sweat connections									
8400	15 mm	1 Plum	24	.333	Ea.	52	14.25		66.25	78.50
8440	20 mm	"	20	.400	"	63.50	17.10		80.60	95
8650	Threaded connections									
8700	15 mm	1 Plum	24	.333	Ea.	63.50	14.25		77.75	91
8740	20 mm		20	.400		242	17.10		259.10	292
8750	25 mm		19	.421		267	18		285	320
8760	32 mm		15	.533		425	23		448	500
8770	40 mm		13	.615		460	26.50		486.50	550
8780	50 mm		11	.727		690	31		721	805

MECHANICAL 15

For expanded coverage of these items see *Means Mechanical or Plumbing Cost Data 2006*

	15110	Valves	CREW	DAILY OUTPUT	LABOR-HOURS	UNIT	2006 BARE COSTS				TOTAL INCL O&P	
							MAT.	LABOR	EQUIP.	TOTAL		
200	0010	VALVES, IRON BODY										200
	1020	Butterfly, wafer type, gear actuator, 1380 kPa										
	1030	50 mm	1 Plum	14	.571	Ea.	120	24.50		144.50	169	
	1040	65 mm	Q-1	9	1.778		123	68.50		191.50	239	
	1050	80 mm		8	2		128	77		205	257	
	1060	100 mm		5	3.200		160	123		283	360	
	1070	125 mm	Q-2	5	4.800		193	191		384	500	
	1080	150 mm	"	5	4.800		218	191		409	530	
	1650	Gate, 862 kPa, N.R.S.										
	2150	Flanged										
	2200	50 mm	1 Plum	5	1.600	Ea.	345	68.50		413.50	485	
	2240	65 mm	Q-1	5	3.200		355	123		478	575	
	2260	80 mm		4.50	3.556		400	137		537	645	
	2280	100 mm		3	5.333		570	205		775	940	
	2300	150 mm	Q-2	3	8		970	320		1,290	1,550	
	3550	OS&Y, 862 kPa, flanged										
	3600	50 mm	1 Plum	5	1.600	Ea.	251	68.50		319.50	380	
	3660	80 mm	Q-1	4.50	3.556		290	137		427	525	
	3680	100 mm	"	3	5.333		415	205		620	765	
	3700	150 mm	Q-2	3	8		685	320		1,005	1,225	
	3900	For 1448 kPa, flanged, add					200%	10%				
	4350	Globe, OS&Y										
	5450	Swing check, 862 kPa, threaded										
	5470	25 mm	1 Plum	13	.615	Ea.	355	26.50		381.50	430	
	5500	50 mm	"	11	.727		445	31		476	535	
	5540	65 mm	Q-1	15	1.067		685	41		726	810	
	5550	80 mm		13	1.231		765	47.50		812.50	910	
	5560	100 mm		10	1.600		1,225	61.50		1,286.50	1,450	
	5950	Flanged										
	6000	50 mm	1 Plum	5	1.600	Ea.	172	68.50		240.50	292	
	6040	65 mm	Q-1	5	3.200		210	123		333	415	
	6050	80 mm		4.50	3.556		225	137		362	455	
	6060	100 mm		3	5.333		350	205		555	695	
	6070	150 mm	Q-2	3	8		605	320		925	1,150	
500	0010	VALVES, PLASTIC										500
	1100	Angle, PVC, threaded										
	1110	6 mm	1 Plum	26	.308	Ea.	49.50	13.15		62.65	74.50	
	1120	13 mm		26	.308		49.50	13.15		62.65	74.50	
	1130	19 mm		25	.320		58.50	13.65		72.15	85	
	1140	25 mm		23	.348		70	14.85		84.85	99.50	
	1150	Ball, PVC, socket or threaded, single union										
	1230	13 mm	1 Plum	26	.308	Ea.	12.65	13.15		25.80	33.50	
	1240	19 mm		25	.320		15.40	13.65		29.05	37.50	
	1250	25 mm		23	.348		18.75	14.85		33.60	43	
	1260	32 mm		21	.381		27.50	16.25		43.75	55	
	1270	38 mm		20	.400		32	17.10		49.10	61	
	1280	51 mm		17	.471		42	20		62	76.50	
	1290	64 mm	Q-1	26	.615		196	23.50		219.50	251	
	1300	76 mm		24	.667		238	25.50		263.50	300	
	1310	102 mm		20	.800		400	30.50		430.50	485	
	1360	For PVC, flanged, add					100%	15%				
	1650	CPVC, socket or threaded, single union										
	1700	13 mm	1 Plum	26	.308	Ea.	35	13.15		48.15	58.50	
	1720	19 mm		25	.320		44	13.65		57.65	69	
	1730	25 mm		23	.348		52.50	14.85		67.35	80.50	
	1750	32 mm		21	.381		88	16.25		104.25	122	

15110 | Valves

		CREW	DAILY OUTPUT	LABOR-HOURS	UNIT	2006 BARE COSTS				TOTAL INCL O&P
						MAT.	LABOR	EQUIP.	TOTAL	
500										
1760	38 mm	1 Plum	20	.400	Ea.	88	17.10		105.10	122
1840	For CPVC, flanged, add					65%	15%			
1880	For true union, socket or threaded, add					50%	5%			
2050	Polypropylene, threaded									
2100	6 mm	1 Plum	26	.308	Ea.	36	13.15		49.15	59.50
2120	10 mm		26	.308		36	13.15		49.15	59.50
2130	13 mm		26	.308		36	13.15		49.15	59.50
2140	19 mm		25	.320		45.50	13.65		59.15	70.50
2150	25 mm		23	.348		53.50	14.85		68.35	81.50
2160	32 mm		21	.381		77.50	16.25		93.75	110
2170	38 mm		20	.400		89.50	17.10		106.60	124
2180	51 mm		17	.471		122	20		142	165
3150	Ball check, PVC, socket or threaded									
3200	6 mm	1 Plum	26	.308	Ea.	30.50	13.15		43.65	53.50
3220	10 mm		26	.308		30.50	13.15		43.65	53.50
3240	13 mm		26	.308		30.50	13.15		43.65	53.50
3250	19 mm		25	.320		34	13.65		47.65	58
3260	25 mm		23	.348		43	14.85		57.85	69.50
3270	32 mm		21	.381		71.50	16.25		87.75	104
3280	38 mm		20	.400		71.50	17.10		88.60	105
3290	51 mm		17	.471		98	20		118	138
3310	76 mm	Q-1	24	.667		271	25.50		296.50	335
3320	102 mm	"	20	.800		385	30.50		415.50	465
3360	For PVC, flanged, add					50%	15%			
4850	Foot valve, PVC, socket or threaded									
4900	13 mm	1 Plum	34	.235	Ea.	49.50	10.05		59.55	69.50
4930	19 mm		32	.250		56	10.70		66.70	77.50
4940	25 mm		28	.286		73	12.20		85.20	98.50
4950	32 mm		27	.296		140	12.65		152.65	173
4960	38 mm		26	.308		140	13.15		153.15	174
6350	Y sediment strainer, PVC, socket or threaded									
6400	13 mm	1 Plum	26	.308	Ea.	36	13.15		49.15	59.50
6440	19 mm		24	.333		39	14.25		53.25	64.50
6450	25 mm		23	.348		47	14.85		61.85	74.50
6460	32 mm		21	.381		77.50	16.25		93.75	110
6470	38 mm		20	.400		77.50	17.10		94.60	111
700										
0010	**VALVES, STEEL**									
0800	Cast									
1350	Check valve, swing type, 1034 kPa, flanged									
1370	25 mm	1 Plum	10	.800	Ea.	380	34		414	465
1400	50 mm	"	8	1		610	42.50		652.50	735
1440	65 mm	Q-1	5	3.200		550	123		673	790
1450	80 mm		4.50	3.556		720	137		857	995
1460	100 mm		3	5.333		1,025	205		1,230	1,425
1540	For 2068 kPa, flanged, add					50%	15%			
1548	For 4137 kPa, flanged, add					110%	20%			
1950	Gate valve, 1034 kPa, flanged									
2000	50 mm	1 Plum	8	1	Ea.	650	42.50		692.50	780
2040	65 mm	Q-1	5	3.200		920	123		1,043	1,200
2050	80 mm		4.50	3.556		920	137		1,057	1,225
2060	100 mm		3	5.333		1,150	205		1,355	1,550
2070	150 mm	Q-2	3	8		1,775	320		2,095	2,450
3650	Globe valve, 1034 kPa, flanged									
3700	50 mm	1 Plum	8	1	Ea.	820	42.50		862.50	965
3740	65 mm	Q-1	5	3.200		1,050	123		1,173	1,325
3750	80 mm		4.50	3.556		1,050	137		1,187	1,350

MECHANICAL 15

15110 | Valves

		CREW	DAILY OUTPUT	LABOR-HOURS	UNIT	2006 BARE COSTS				TOTAL INCL O&P	
						MAT.	LABOR	EQUIP.	TOTAL		
700	3760	100 mm	Q-1	3	5.333	Ea.	1,525	205		1,730	1,975	**700**
	3770	150 mm	Q-2	3	8	↓	2,400	320		2,720	3,100	
	5150	Forged										
	5650	Check valve, class 800, horizontal, socket or threaded										
	5698	Threaded										
	5700	8 mm	1 Plum	24	.333	Ea.	67.50	14.25		81.75	96	
	5720	10 mm		24	.333		67.50	14.25		81.75	96	
	5730	15 mm		24	.333		67.50	14.25		81.75	96	
	5740	20 mm		20	.400		73.50	17.10		90.60	107	
	5750	25 mm		19	.421		85.50	18		103.50	122	
	5760	32 mm	↓	15	.533	↓	168	23		191	220	

15120 | Piping Specialties

		CREW	DAILY OUTPUT	LABOR-HOURS	UNIT	2006 BARE COSTS				TOTAL INCL O&P	
						MAT.	LABOR	EQUIP.	TOTAL		
120	0010	**AIR CONTROL** With strainer										**120**
	0030	Air separator, with strainer										
	0040	50 mm diameter [2"]	Q-5	6	2.667	Ea.	655	103		758	875	
	0080	65 mm diameter [2-1/2"]		5	3.200		745	124		869	1,000	
	0100	80 mm diameter [3"]		4	4		1,125	155		1,280	1,475	
	0120	100 mm diameter [4"]	↓	3	5.333		1,650	207		1,857	2,100	
	0130	125 mm diameter [5"]	Q-6	3.60	6.667		2,100	268		2,368	2,700	
	0140	150 mm diameter [6"]	"	3.40	7.059	↓	2,500	284		2,784	3,175	
160	0010	**AUTOMATIC AIR VENT**										**160**
	0020	Cast iron body, stainless steel internals, float type										
	0060	15 mm NPT inlet, 2068 kPa [1/2"]	1 Stpi	12	.667	Ea.	76.50	28.50		105	128	
	0220	20 mm NPT inlet, 1725 kPa	"	10	.800		246	34.50		280.50	320	
	0340	40 mm NPT inlet, 1725 kPa	Q-5	12	1.333	↓	765	51.50		816.50	920	
300	0010	**EXPANSION JOINTS**										**300**
	0100	Bellows type, neoprene cover, flanged spool										
	0140	152 mm face to face, 32 mm diameter [1-1/4"]	1 Stpi	11	.727	Ea.	210	31.50		241.50	278	
	0160	38 mm diameter [1-1/2"]	"	10.60	.755		210	32.50		242.50	280	
	0180	51 mm diameter [2"]	Q-5	13.30	1.203		214	46.50		260.50	305	
	0190	64 mm diameter [2-1/2"]		12.40	1.290		222	50		272	320	
	0200	76 mm diameter [3"]		11.40	1.404		252	54.50		306.50	360	
	0480	254 mm face to face, 51 mm diameter [10",2"]		13	1.231		315	47.50		362.50	415	
	0500	64 mm diameter [2-1/2"]		12	1.333		330	51.50		381.50	445	
	0520	76 mm diameter [3"]		11	1.455		335	56.50		391.50	455	
	0540	102 mm diameter [4"]		8	2		375	77.50		452.50	530	
	0560	127 mm diameter [5"]		7	2.286		455	88.50		543.50	635	
	0580	152 mm diameter [6"]	↓	6	2.667	↓	465	103		568	670	
320	0010	**EXPANSION TANKS**										**320**
	1505	Fiberglass and steel single / double wall storage, see Div 13201										
	1510	Tank leak detection systems, see Div 13851-350										
	2000	Steel, liquid expansion, ASME, painted, 57 L capacity	Q-5	17	.941	Ea.	365	36.50		401.50	455	
	2020	91 L capacity		14	1.143		380	44.50		424.50	480	
	2040	114 L capacity		12	1.333		420	51.50		471.50	540	
	2060	151 L capacity		10	1.600		460	62		522	600	
	2080	227 L capacity		8	2		550	77.50		627.50	720	
	2100	303 L capacity		7	2.286		615	88.50		703.50	810	
	2120	379 L capacity		6	2.667		805	103		908	1,050	
	3000	Steel ASME expansion, rubber diaphragm, 71.9 L cap. accept.		12	1.333		1,700	51.50		1,751.50	1,950	
	3020	117 L capacity		8	2		1,875	77.50		1,952.50	2,200	
	3040	231 L capacity		6	2.667		1,975	103		2,078	2,325	
	3080	450 L capacity	↓	4	4	↓	2,850	155		3,005	3,350	

		15120 | **Piping Specialties**	CREW	DAILY OUTPUT	LABOR-HOURS	UNIT	2006 BARE COSTS				TOTAL INCL O&P	
							MAT.	LABOR	EQUIP.	TOTAL		
320	3100	598 L capacity	Q-5	3.80	4.211	Ea.	3,950	163		4,113	4,600	**320**
	3140	1200 L capacity		2.80	5.714		5,975	221		6,196	6,900	
	3180	1998 L capacity	↓	2.40	6.667	↓	9,700	258		9,958	11,100	
	5950											
350	0010	**FLEXIBLE CONNECTORS**, Corrugated, 24mm O.D., 15 mm I.D.										**350**
	0050	Gas, seamless brass, steel fittings										
	0200	300 mm long [12"]	1 Plum	36	.222	Ea.	12.35	9.50		21.85	28	
	0220	450 mm long [18"]		36	.222		15.30	9.50		24.80	31	
	0240	600 mm long [24"]		34	.235		18.05	10.05		28.10	35	
	0260	750 mm long [30"]		34	.235		19.55	10.05		29.60	36.50	
	0280	900 mm long [36"]		32	.250		21.50	10.70		32.20	40	
	0320	1200 mm long [48"]		30	.267		27.50	11.40		38.90	47	
	0340	1500 mm long [60"]		30	.267		32.50	11.40		43.90	53	
	0360	1800 mm long [72"]	↓	30	.267	↓	37.50	11.40		48.90	58.50	
	2000	Water, copper tubing, dielectric separators										
	2100	300 mm long [12"]	1 Plum	36	.222	Ea.	9.70	9.50		19.20	25	
	2220	375 mm long [15"]		36	.222		10.80	9.50		20.30	26	
	2240	450 mm long [18"]		36	.222		11.75	9.50		21.25	27	
	2260	600 mm long [24"]	↓	34	.235	↓	14.40	10.05		24.45	31	
370	0010	**FLEXIBLE METAL HOSE** Connectors, standard lengths										**370**
	0100	Bronze braided, bronze ends										
	0120	10 mm diameter x 300 mm [3/8",12"]	1 Stpi	26	.308	Ea.	16.95	13.25		30.20	38.50	
	0160	20 mm diameter x 300 mm [3/4",12"]		20	.400		23.50	17.20		40.70	52	
	0180	25 mm diameter x 450 mm [1",18"]		19	.421		30.50	18.15		48.65	61	
	0200	40 mm diameter x 450 mm [1-1/2",18"]		13	.615		48.50	26.50		75	93	
	0220	50 mm diameter x 450 mm [2",18"]	↓	11	.727	↓	58	31.50		89.50	111	
520	0010	**HYDRONIC HEATING CONTROL VALVES**										**520**
	0100	Radiator supply, 15 mm diameter [1/2"]	1 Stpi	24	.333	Ea.	41.50	14.35		55.85	67.50	
	0120	20 mm diameter [3/4"]		20	.400		42.50	17.20		59.70	72.50	
	0140	25 mm diameter [1"]		19	.421		53	18.15		71.15	85.50	
	0160	32 mm diameter [1-1/4"]	↓	15	.533	↓	74	23		97	116	
	0500	For low pressure steam, add					25%					
670	0010	**PRESSURE REGULATOR**										**670**
	3000	Steam, high capacity, bronze body, stainless steel trim										
	3020	Threaded, 13 mm diameter [1/2"]	1 Stpi	24	.333	Ea.	985	14.35		999.35	1,100	
	3030	19 mm diameter [3/4"]		24	.333		985	14.35		999.35	1,100	
	3040	25 mm diameter [1"]		19	.421		1,100	18.15		1,118.15	1,225	
	3060	32 mm diameter [1-1/4"]		15	.533		1,225	23		1,248	1,350	
	3080	38 mm diameter [1-1/2"]		13	.615		1,400	26.50		1,426.50	1,575	
	3100	51 mm diameter [2"]	↓	11	.727		1,700	31.50		1,731.50	1,925	
	3120	64 mm diameter [2-1/2"]	Q-5	12	1.333		2,125	51.50		2,176.50	2,425	
	3140	76 mm diameter [3"]	"	11	1.455	↓	2,425	56.50		2,481.50	2,750	
	3500	Flanged connection, iron body, 861 kPa W.S.P.										
	3520	76 mm diameter [3"]	Q-5	11	1.455	Ea.	2,675	56.50		2,731.50	3,000	
	3540	102 mm diameter [4"]	"	5	3.200	"	3,350	124		3,474	3,875	
760	0010	**STEAM TRAP**										**760**
	0030	Cast iron body, threaded										
	0040	Inverted bucket										
	0050	15 mm pipe size [1/2"]	1 Stpi	12	.667	Ea.	112	28.50		140.50	166	
	0070	20 mm pipe size [3/4"]		10	.800		194	34.50		228.50	266	
	0100	25 mm pipe size [1"]		9	.889		300	38.50		338.50	390	
	0120	32 mm pipe size [1-1/4"]	↓	8	1	↓	450	43		493	560	
	1000	Float & thermostatic, 103 kPa										

15120 | Piping Specialties

		CREW	DAILY OUTPUT	LABOR-HOURS	UNIT	2006 BARE COSTS				TOTAL INCL O&P	
						MAT.	LABOR	EQUIP.	TOTAL		
760	1010	20 mm pipe size [3/4"]	1 Stpi	16	.500	Ea.	89.50	21.50		111	131	**760**
	1020	25 mm pipe size [1"]		15	.533		107	23		130	153	
	1040	40 mm pipe size [1-1/2"]		9	.889		191	38.50		229.50	268	
	1060	50 mm pipe size [2"]		6	1.333		350	57.50		407.50	470	
820	0010	**STRAINERS, Y TYPE** Bronze body										**820**
	0050	Screwed, 1034 kPa, 8 mm pipe size [1/4"]	1 Stpi	24	.333	Ea.	12.80	14.35		27.15	35.50	
	0070	10 mm pipe size [3/8"]		24	.333		16.85	14.35		31.20	40	
	0100	15 mm pipe size [1/2"]		20	.400		16.85	17.20		34.05	44.50	
	0140	25 mm pipe size [1"]		17	.471		19.85	20.50		40.35	52.50	
	0160	40 mm pipe size [1-1/2"]		14	.571		42.50	24.50		67	84	
	0180	50 mm pipe size [2"]		13	.615		57	26.50		83.50	103	
	0182	80 mm pipe size [3"]		12	.667		360	28.50		388.50	440	
	0200	2069 kPa, 65 mm pipe size [2-1/2"]	Q-5	17	.941		305	36.50		341.50	390	
	0220	80 mm pipe size [3"]		16	1		600	39		639	720	
	0240	100 mm pipe size [4"]		15	1.067		1,375	41.50		1,416.50	1,550	
	0500	For 2068 kPa rating 8 mm thru 50 mm dia, add					15%					
	1000	Flanged, 1034 kPa, 40 mm pipe size [1-1/2"]	1 Stpi	11	.727	Ea.	315	31.50		346.50	390	
	1020	50 mm pipe size [2"]	"	8	1		365	43		408	470	
	1030	Flanged, 1034 kPa, 65 mm [2-1/2"]	Q-5	5	3.200		560	124		684	805	
	1040	80 mm pipe size [3"]		4.50	3.556		695	138		833	965	
	1060	100 mm pipe size [4"]		3	5.333		1,050	207		1,257	1,450	
	1100	150 mm pipe size [6"]	Q-6	3	8		2,800	320		3,120	3,550	
	1106	200 mm pipe size [8"]	"	2.60	9.231		3,425	370		3,795	4,325	
	1500	For 2068 kPa rating, add					40%					
840	0010	**STRAINERS, Y TYPE** Iron body										**840**
	0050	Screwed, 1725 kPa, 8 mm pipe size [1/4"]	1 Stpi	20	.400	Ea.	7.05	17.20		24.25	34	
	0070	10 mm pipe size [3/8"]		20	.400		7.05	17.20		24.25	34	
	0100	15 mm pipe size [1/2"]		20	.400		7.05	17.20		24.25	34	
	0140	25 mm pipe size [1"]		16	.500		11.40	21.50		32.90	45	
	0160	40 mm pipe size [1-1/2"]		12	.667		18.80	28.50		47.30	63.50	
	0180	50 mm pipe size [2"]		8	1		29	43		72	97	
	0220	80 mm pipe size [3"]	Q-5	11	1.455		160	56.50		216.50	260	
	0240	100 mm pipe size [4"]	"	5	3.200		270	124		394	485	
	0500	For galvanized body, add					50%					
	1000	Flanged, 862 kPa, 40 mm pipe size [1-1/2"]	1 Stpi	11	.727	Ea.	91	31.50		122.50	147	
	1020	50 mm pipe size [2"]	"	8	1		68	43		111	140	
	1040	80 mm pipe size [3"]	Q-5	4.50	3.556		90	138		228	305	
	1060	100 mm pipe size [4"]	"	3	5.333		164	207		371	490	
	1080	125 mm pipe size [5"]	Q-6	3.40	7.059		256	284		540	705	
	1100	150 mm pipe size [6"]	"	3	8		315	320		635	830	
	1500	For 1723 kPa rating, add					20%					
	2000	For galvanized body, add					50%					
	2500	For steel body, add					40%					
920	0010	**VENTURI FLOW** Measuring device										**920**
	0050	15 mm diameter [1/2"]	1 Stpi	24	.333	Ea.	175	14.35		189.35	215	
	0120	25 mm diameter [1"]		19	.421		169	18.15		187.15	214	
	0140	32 mm diameter [1-1/4"]		15	.533		203	23		226	258	
	0160	40 mm diameter [1-1/2"]		13	.615		206	26.50		232.50	267	
	0180	50 mm diameter [2"]		11	.727		222	31.50		253.50	291	
	0220	80 mm diameter [3"]	Q-5	14	1.143		380	44.50		424.50	485	
	0240	100 mm diameter [4"]	"	11	1.455		595	56.50		651.50	740	
	0280	150 mm diameter [6"]	Q-6	3.50	6.857		780	276		1,056	1,275	
	0500	For meter, add					1,175			1,175	1,300	
940	0010	**WATER SUPPLY METERS**										**940**
	1000	Detector, serves dual systems such as fire and domestic or										

15120	Piping Specialties	CREW	DAILY OUTPUT	LABOR-HOURS	UNIT	2006 BARE COSTS				TOTAL INCL O&P
						MAT.	LABOR	EQUIP.	TOTAL	
940	1020	process water, wide range cap., UL and FM approved									940
1100	76 mm mainline x 51 mm by-pass, 25.2 L/s	Q-1	3.60	4.444	Ea.	5,150	171		5,321	5,900
1140	102 mm mainline x 51 mm by-pass, 44.2 L/s	"	2.50	6.400		5,150	246		5,396	6,025
1180	150 mm mainline x 80 mm by-pass, 101 L/s	Q-2	2.60	9.231		7,775	370		8,145	9,100
1220	200 mm mainline x 100 mm by-pass, 177 L/s		2.10	11.429		11,500	455		11,955	13,300
1260	254 mm mainline x 152 mm by-pass, 278 L/s		2	12		15,700	480		16,180	18,000
1300	254mm x 305mm mainlines x 152mm by-pass, 341 L/s		1.70	14.118		20,700	565		21,265	23,600
2000	Domestic/commercial, bronze									
2020	Threaded									
2060	18 mm diameter, to 1.3 L/s [5/8"]	1 Plum	16	.500	Ea.	40	21.50		61.50	76
2080	20 mm diameter, to 2.0 L/s [3/4"]		14	.571		67.50	24.50		92	111
2100	25 mm diameter, to 3.2 L/s [1"]		12	.667		94	28.50		122.50	146
2300	Threaded/flanged									
2340	40 mm diameter, to 6.3 L/s [1-1/2"]	1 Plum	8	1	Ea.	330	42.50		372.50	430
2360	50 mm diameter, to 10 L/s [2"]	"	6	1.333	"	415	57		472	540
2600	Flanged, compound									
2640	76 mm diameter, 20.2 L/s [3"]	Q-1	3	5.333	Ea.	1,950	205		2,155	2,425
2660	100 mm diameter, to 32 L/s [4"]		1.50	10.667		3,025	410		3,435	3,950
2680	150 mm diameter, to 63 L/s [6"]		1	16		4,350	615		4,965	5,700
2700	200 mm diameter, to 114 L/s [8"]		.80	20		8,600	770		9,370	10,600
7000	Turbine									
7260	Flanged									
7300	51 mm diameter, to 10.1 L/s [2"]	1 Plum	7	1.143	Ea.	455	49		504	575
7320	76 mm diameter, to 28.4 L/s [3"]	Q-1	3.60	4.444		745	171		916	1,075
7340	102 mm diameter, to 41 L/s [4"]	"	2.50	6.400		1,350	246		1,596	1,850
7360	152 mm diameter, to 113.6 L/s [6"]	Q-2	2.60	9.231		2,550	370		2,920	3,350
7380	203 mm diameter, to 158 L/s [8"]		2.10	11.429		4,050	455		4,505	5,125
7400	254 mm diameter, to 347 L/s [10"]		1.70	14.118		5,400	565		5,965	6,800

15140	Domestic Water Piping	CREW	DAILY OUTPUT	LABOR-HOURS	UNIT	MAT.	LABOR	EQUIP.	TOTAL	TOTAL INCL O&P
100	0010	BACKFLOW PREVENTER Includes valves									100
0020	and four test cocks, corrosion resistant, automatic operation									
1000	Double check principle									
1010	Threaded, with ball valves									
1020	20 mm pipe size [3/4"]	1 Plum	16	.500	Ea.	148	21.50		169.50	195
1030	25 mm pipe size [1"]		14	.571		164	24.50		188.50	217
1040	40mm pipe size [1-1/2"]		10	.800		295	34		329	375
1050	50mm pipe size [2"]		7	1.143		365	49		414	475
1080	Threaded, with gate valves									
1100	20 mm pipe size [3/4"]	1 Plum	16	.500	Ea.	580	21.50		601.50	670
1120	25 mm pipe size [1"]		14	.571		620	24.50		644.50	715
1140	40mm pipe size [1-1/2"]		10	.800		755	34		789	880
1160	50mm pipe size [2"]		7	1.143		925	49		974	1,100
1200	Flanged, valves are gate									
1210	80mm pipe size [3"]	Q-1	4.50	3.556	Ea.	1,600	137		1,737	1,950
1220	100mm pipe size [4"]	"	3	5.333		1,975	205		2,180	2,450
1230	150mm pipe size [6"]	Q-2	3	8		2,900	320		3,220	3,675
1240	200 mm pipe size [8"]		2	12		5,275	480		5,755	6,525
1250	250 mm pipe size [10"]		1	24		7,800	955		8,755	10,100
1300	Flanged, valves are OS&Y									
1380	80mm pipe size [3"]	Q-1	4.50	3.556	Ea.	2,300	137		2,437	2,725
1400	100mm pipe size [4"]	"	3	5.333		3,750	205		3,955	4,425
1420	150mm pipe size [6"]	Q-2	3	8		5,975	320		6,295	7,025
4000	Reduced pressure principle									
4100	Threaded, bronze, valves are ball									
4120	20 mm pipe size [3/4"]	1 Plum	16	.500	Ea.	220	21.50		241.50	274

For expanded coverage of these items see *Means Mechanical or Plumbing Cost Data 2006*

MECHANICAL 15

15140 | Domestic Water Piping

		CREW	DAILY OUTPUT	LABOR-HOURS	UNIT	2006 BARE COSTS				TOTAL INCL O&P	
						MAT.	LABOR	EQUIP.	TOTAL		
100	4140	25 mm pipe size [1"]	1 Plum	14	.571	Ea.	236	24.50		260.50	297	100
	4150	32 mm pipe size [1-1/4"]		12	.667		405	28.50		433.50	490	
	4160	40 mm pipe size [1-1/2"]		10	.800		445	34		479	540	
	4180	50 mm pipe size [2"]		7	1.143		500	49		549	625	
	5000	Flanged, valves are OS&Y										
	5060	65 mm pipe size [2-1/2"]	Q-1	5	3.200	Ea.	2,550	123		2,673	2,975	
	5080	80 mm pipe size [3"]		4.50	3.556		2,675	137		2,812	3,150	
	5100	100 mm pipe size [4"]		3	5.333		3,375	205		3,580	4,000	
	5120	150 mm pipe size [6"]	Q-2	3	8		4,875	320		5,195	5,850	
	5600	Flanged, iron, valves are OS&Y										
	5660	65 mm pipe size [2-1/2"]	Q-1	5	3.200	Ea.	2,050	123		2,173	2,425	
	5680	80 mm pipe size [3"]		4.50	3.556		2,150	137		2,287	2,575	
	5700	100 mm pipe size [4"]		3	5.333		2,700	205		2,905	3,275	
	5720	150 mm pipe size [6"]	Q-2	3	8		3,900	320		4,220	4,775	
	5740	200 mm pipe size [8"]		2	12		6,875	480		7,355	8,275	
	5760	250 mm pipe size [10"]		1	24		9,200	955		10,155	11,600	
600	0010	**VACUUM BREAKERS** Hot or cold water										600
	1030	Anti-siphon, brass										
	1040	8 mm size [1/4"]	1 Plum	24	.333	Ea.	24	14.25		38.25	47.50	
	1050	10 mm size [3/8"]		24	.333		24	14.25		38.25	47.50	
	1060	15 mm size [1/2"]		24	.333		27	14.25		41.25	51	
	1080	20 mm size [3/4"]		20	.400		32	17.10		49.10	60.50	
	1100	25 mm size [1"]		19	.421		50	18		68	81.50	
	1120	32 mm size [1-1/4"]		15	.533		87	23		110	131	
	1140	40 mm size [1-1/2"]		13	.615		103	26.50		129.50	153	
	1160	50 mm size [2"]		11	.727		160	31		191	223	
	1300	For polished chrome, (6 mm thru 25 mm), add					50%					
	1900	Vacuum relief, water service, bronze										
	2000	15 mm size [1/2"]	1 Plum	30	.267	Ea.	23.50	11.40		34.90	42.50	
800	0010	**WATER HAMMER ARRESTORS / SHOCK ABSORBERS**										800
	0490	Copper										
	0500	20 mm male I.P.S. For 1 to 11 fixtures [3/4"]	1 Plum	12	.667	Ea.	15.55	28.50		44.05	60	
	0600	25 mm male I.P.S., For 12 to 32 fixtures [1"]		8	1		39.50	42.50		82	108	
	0700	32 mm male I.P.S. For 33 to 60 fixtures		8	1		46	42.50		88.50	115	
	0800	40 mm male I.P.S. For 61 to 113 fixtures		8	1		62.50	42.50		105	134	
	0900	50 mm male I.P.S. For 114 to 154 fixtures		8	1		100	42.50		142.50	175	
	1000	65 mm male I.P.S. For 155 to 330 fixtures		4	2		285	85.50		370.50	445	

15150 | Sanitary Waste and Vent Piping

		CREW	DAILY OUTPUT	LABOR-HOURS	UNIT	MAT.	LABOR	EQUIP.	TOTAL	INCL O&P	
100	0010	**BACKWATER VALVES**, C.I. body										100
	6980	Bronze gate and automatic flapper valves										
	7000	80 mm and 100 mm pipe size [3",4"]	Q-1	13	1.231	Ea.	1,050	47.50		1,097.50	1,250	
	7100	125 mm and 150 mm pipe size [5",6"]	"	13	1.231	"	1,625	47.50		1,672.50	1,850	
	7240	Bronze flapper valve, bolted cover										
	7260	50 mm pipe size [2"]	Q-1	16	1	Ea.	310	38.50		348.50	400	
	7300	100 mm pipe size [4"]	"	13	1.231		595	47.50		642.50	725	
	7340	150 mm pipe size [6"]	Q-2	17	1.412		855	56.50		911.50	1,025	
200	0010	**CLEANOUTS**										200
	0060	Floor type										
	0080	Round or square, scoriated nickel bronze top										
	0100	50 mm pipe size [2"]	1 Plum	10	.800	Ea.	102	34		136	164	
	0120	80 mm pipe size [3"]		8	1		153	42.50		195.50	233	
	0140	100 mm pipe size [4"]		6	1.333		153	57		210	254	
	0160	125 mm pipe size [5"]		4	2		194	85.50		279.50	340	
	0180	150 mm pipe size [6"]	Q-1	6	2.667		194	102		296	365	

15 MECHANICAL

15150	Sanitary Waste and Vent Piping	CREW	DAILY OUTPUT	LABOR-HOURS	UNIT	2006 BARE COSTS				TOTAL INCL O&P	
						MAT.	LABOR	EQUIP.	TOTAL		
200	**0200**	200 mm pipe size [8"]	Q-1	4	4	Ea.	345	154		499	605	**200**
	0340	Recessed for tile, same price									
	0980	Round top, recessed for terrazzo									
	1000	50 mm pipe size [2"]	1 Plum	9	.889	Ea.	102	38		140	169
	1080	80 mm pipe size [3"]		6	1.333		153	57		210	254
	1100	100 mm pipe size [4"]	↓	4	2		153	85.50		238.50	297
	1120	125 mm pipe size [5"]	Q-1	6	2.667		194	102		296	365
	1140	150 mm pipe size [6"]		5	3.200		194	123		317	400
	1160	200 mm pipe size [8"]	↓	4	4	↓	345	154		499	605
	2000	Round scoriated nickel bronze top, extra heavy duty									
	2060	50 mm pipe size [2"]	1 Plum	9	.889	Ea.	139	38		177	209
	2080	80 mm pipe size [3"]		6	1.333		190	57		247	295
	2100	100 mm pipe size [4"]	↓	4	2		190	85.50		275.50	340
	2120	125 mm pipe size [5"]	Q-1	6	2.667		231	102		333	410
	2140	150 mm pipe size [6"]		5	3.200		231	123		354	440
	2160	200 mm pipe size [8"]	↓	4	4	↓	380	154		534	650
	4000	Wall type, square smooth cover, over wall frame									
	4060	50 mm pipe size [2"]	1 Plum	14	.571	Ea.	167	24.50		191.50	220
	4080	80 mm pipe size [3"]		12	.667		179	28.50		207.50	240
	4100	100 mm pipe size [4"]		10	.800		193	34		227	264
	4120	125 mm pipe size [5"]		9	.889		279	38		317	360
	4140	150 mm pipe size [6"]	↓	8	1		310	42.50		352.50	405
	4160	200 mm pipe size [8"]	Q-1	11	1.455	↓	415	56		471	540
	5000	Extension, C.I.;bronze countersunk plug, 200 mm long [8"]									
	5040	50 mm pipe size [2"]	1 Plum	16	.500	Ea.	87	21.50		108.50	128
	5060	80 mm pipe size [3"]		14	.571		101	24.50		125.50	148
	5080	100 mm pipe size [4"]		13	.615		104	26.50		130.50	154
	5100	125 mm pipe size [5"]		12	.667		162	28.50		190.50	221
	5120	150 mm pipe size [6"]	↓	11	.727	↓	196	31		227	263
250	**0010**	**CLEANOUT TEE**										**250**
	0100	Cast iron, B&S, with countersunk plug									
	0200	50 mm pipe size [2"]	1 Plum	4	2	Ea.	139	85.50		224.50	281
	0220	80 mm pipe size [3"]	↓	3.60	2.222		151	95		246	310
	0240	100 mm pipe size [4"]	↓	3.30	2.424		188	104		292	365
	0260	125 mm pipe size [5"]	Q-1	5.50	2.909		410	112		522	620
	0280	150 mm pipe size [6"]	"	5	3.200		510	123		633	745
	0300	200 mm pipe size [8"]	Q-3	5	6.400		585	260		845	1,025
	0500	For round smooth access cover, same price									
	0600	For round scoriated access cover, same price									
	0700	For square smooth access cover, add				↓	60%				
	4000	Plastic, tees and adapters, Add plugs									
	4010	ABS, DWV									
	4020	Cleanout tee, 40 mm pipe size [1-1/2"]	1 Plum	15	.533	Ea.	6.05	23		29.05	41
	4030	50 mm pipe size [2"]	Q-1	27	.593		7.15	23		30.15	42.50
	4040	80 mm pipe size [3"]	↓	21	.762		14.30	29.50		43.80	59.50
	4050	100 mm pipe size [4"]	↓	16	1		26.50	38.50		65	87
	4100	Cleanout plug, 40 mm pipe size [1-1/2"]	1 Plum	32	.250		1.11	10.70		11.81	17.25
	4110	50 mm pipe size [2"]	Q-1	56	.286		1.46	11		12.46	18.10
	4120	80 mm pipe size [3"]		36	.444		2.22	17.10		19.32	28
	4130	100 mm pipe size [4"]	↓	30	.533		4.57	20.50		25.07	36
	4180	Cleanout adapter fitting, 40 mm pipe size [1-1/2"]	1 Plum	32	.250		1.75	10.70		12.45	18
	4190	50 mm pipe size [2"]	Q-1	56	.286		2.46	11		13.46	19.20
	4200	80 mm pipe size [3"]		36	.444		6.95	17.10		24.05	33
	4210	100 mm pipe size [4"]	↓	30	.533	↓	11.10	20.50		31.60	43.50
	5000	PVC, DWV									

MECHANICAL 15

				DAILY	LABOR-			2006 BARE COSTS				TOTAL	
15150		**Sanitary Waste and Vent Piping**	CREW	OUTPUT	HOURS	UNIT	MAT.	LABOR	EQUIP.	TOTAL		INCL O&P	
250	5010	Cleanout tee, 40 mm pipe size [1-1/2"]	1 Plum	15	.533	Ea.	6.05	23		29.05		41	**250**
	5020	50 mm pipe size [2"]	Q-1	27	.593		7.15	23		30.15		42.50	
	5030	80 mm pipe size [3"]		21	.762		14.30	29.50		43.80		59.50	
	5040	100 mm pipe size [4"]	↓	16	1		26.50	38.50		65		87	
	5090	Cleanout plug, 40 mm pipe size [1-1/2"]	1 Plum	32	.250		1.11	10.70		11.81		17.25	
	5100	50 mm pipe size [2"]	Q-1	56	.286		1.46	11		12.46		18.10	
	5110	80 mm pipe size [3"]		36	.444		2.22	17.10		19.32		28	
	5120	100 mm pipe size [4"]		30	.533		4.58	20.50		25.08		36	
	5130	150 mm pipe size [6"]	↓	24	.667		13.95	25.50		39.45		54	
	5170	Cleanout adapter fitting, 40 mm pipe size [1-1/2"]	1 Plum	32	.250		1.75	10.70		12.45		18	
	5180	50 mm pipe size [2"]	Q-1	56	.286		2.46	11		13.46		19.20	
	5190	80 mm pipe size [3"]		36	.444		6.95	17.10		24.05		33	
	5200	100 mm pipe size [4"]		30	.533		11.10	20.50		31.60		43.50	
	5210	150 mm pipe size [6"]	↓	24	.667		25.50	25.50		51		66.50	
300	0010	**FLOOR AND AREA DRAINS**											**300**
	0400	Deck, auto park, C.I., 325 mm top [13"]											
	0440	80, 100, 125, and 150 mm pipe	Q-1	8	2	Ea.	745	77		822		935	
	0480	For galvanized body, add				"	355			355		390	
	2000	Floor, medium duty, C.I., deep flange, 175 mm dia top [7"]											
	2040	50 mm and 80 mm pipe size [2",3"]	Q-1	12	1.333	Ea.	104	51		155		192	
	2080	For galvanized body, add					44			44		48.50	
	2120	For polished bronze top, add				↓	58			58		64	
	2400	Heavy duty, with sediment bucket, C.I., 300 mm dia. loose grate											
	2420	50, 80, 100, 125, and 150 mm pipe	Q-1	9	1.778	Ea.	360	68.50		428.50		500	
	2460	For polished bronze top, add				"	148			148		162	
	2500	Heavy duty, cleanout & trap w/bucket, C.I., 375 mm top [15"]											
	2540	50, 80, and 100 mm pipe size [2",3",4"]	Q-1	6	2.667	Ea.	3,325	102		3,427		3,800	
	2560	For galvanized body, add					770			770		845	
	2580	For polished bronze top, add				↓	270			270		297	
400	0010	**INTERCEPTORS**											**400**
	0150	Grease, cast iron, 0.25 L/s, 3.6 kg fat capacity	1 Plum	4	2	Ea.	615	85.50		700.50		810	
	0200	0.44 L/s, 6.35 kg fat capacity		4	2		855	85.50		940.50		1,075	
	1000	0.63 L/s, 9 kg fat capacity		4	2		1,000	85.50		1,085.50		1,225	
	1040	0.95 L/s, 13.6 kg fat capacity		4	2		1,500	85.50		1,585.50		1,775	
	1060	1.3 L/s, 18 kg fat capacity	↓	3	2.667		1,825	114		1,939		2,175	
	1120	Fabricated steel, 3.2 L/s, 45 kg fat capacity	Q-1	2	8		3,350	305		3,655		4,150	
	1160	6.3 L/s, 91 kg fat capacity	"	2	8	↓	7,600	305		7,905		8,825	
	1580	For seepage pan, add					7%						
	3000	Hair, cast iron, 32 mm and 38 mm pipe connection	1 Plum	8	1	Ea.	231	42.50		273.50		320	
	3100	For chrome-plated cast iron, add					142			142		156	
	4000	Oil, fabricated steel, 631 mL/s, 51 mm pipe size	1 Plum	4	2		1,375	85.50		1,460.50		1,625	
	4100	947 mL/s, 51 mm or 76 mm pipe size		4	2		1,875	85.50		1,960.50		2,200	
	4120	1262 mL/s, 51 mm or 76 mm pipe size	↓	3	2.667		2,275	114		2,389		2,675	
	4220	6310 mL/s, 76 mm pipe size	Q-1	2	8		7,600	305		7,905		8,825	
	6000	Solids, precious metals recovery, C.I., 32 mm to 51 mm pipe	1 Plum	4	2		350	85.50		435.50		515	
	6100	Dental Lab., large, C.I., 40 mm to 50 mm pipe	"	3	2.667	↓	1,225	114		1,339		1,525	
600	0010	**SHOWER DRAINS**											**600**
	2780	Shower, with strainer, uniform diam. trap, bronze top											
	2800	50 and 80 mm pipe size [2",3"]	Q-1	8	2	Ea.	220	77		297		360	
	2820	100 mm pipe size [4"]	"	7	2.286		243	88		331		400	
	2840	For galvanized body, add				↓	86			86		94.50	
800	0010	**TRAPS**											**800**
	0030	Cast iron, service weight											

Important: See the Reference Section for supporting data - Crews, Rental Equipment, City Cost Indexes and Reference Data

15150 | Sanitary Waste and Vent Piping

			CREW	DAILY OUTPUT	LABOR-HOURS	UNIT	MAT.	LABOR	EQUIP.	TOTAL	TOTAL INCL O&P	
800	0050	Running P trap, without vent										800
	1100	50 mm [2"]	Q-1	16	1	Ea.	28.50	38.50		67	89.50	
	1140	80 mm [3"]		14	1.143		47.50	44		91.50	119	
	1150	100 mm [4"]		13	1.231		82.50	47.50		130	162	
	1160	150 mm [6"]	Q-2	17	1.412		360	56.50		416.50	480	
	1180	Running trap, single hub, with vent										
	2080	80 mm pipe size, 80 mm vent [3",3"]	Q-1	14	1.143	Ea.	65.50	44		109.50	138	
	2120	100 mm pipe size, 100 mm vent [4",4"]	"	13	1.231		89	47.50		136.50	169	
	2300	For double hub, vent, add					10%	20%				
	3000	P trap, B&S, 50 mm pipe size [2"]	Q-1	16	1		19.60	38.50		58.10	79.50	
	3040	80 mm pipe size [3"]	"	14	1.143		29.50	44		73.50	98	
	3350	Deep seal trap, B&S										
	3400	32 mm pipe size [1-1/4"]	Q-1	14	1.143	Ea.	34	44		78	103	
	3410	40mm pipe size [1-1/2"]		14	1.143		34	44		78	103	
	3420	50 mm pipe size [2"]		14	1.143		22	44		66	90.50	
	3440	80 mm pipe size [3"]		12	1.333		36	51		87	117	
	4700	Copper, drainage, drum trap										
	4800	80 mm x 125 mm solid, 40 mm pipe size [3",5",1-1/2"]	1 Plum	16	.500	Ea.	43.50	21.50		65	79.50	
	4840	80 mm x 150 mm swivel, 40 mm pipe size [3",6",1-1/2"]	"	16	.500	"	69	21.50		90.50	108	
	5100	P trap, standard pattern										
	5200	32 mm pipe size [1-1/4"]	1 Plum	18	.444	Ea.	32	19		51	63.50	
	5240	40 mm pipe size [1-1/2"]		17	.471		31	20		51	64	
	5260	50 mm pipe size [2"]		15	.533		48	23		71	87	
	5280	80 mm pipe size [3"]		11	.727		115	31		146	173	
	5340	With cleanout and slip joint										
	5360	32 mm pipe size [1-1/4"]	1 Plum	18	.444	Ea.	30.50	19		49.50	62.50	
	5400	40 mm pipe size [1-1/2"]	"	17	.471	"	47	20		67	82	
900	0010	**VENT FLASHING, CAPS**										900
	0120	Vent caps										
	0140	Cast iron										
	0180	64 mm - 92 mm pipe [2-1/2" - 3-5/8"]	1 Plum	21	.381	Ea.	35	16.25		51.25	62.50	
	0190	102 mm - 105 mm pipe [4" - 4-1/8"]	"	19	.421	"	41.50	18		59.50	73	
	0900	Vent flashing										
	1000	Aluminum with lead ring										
	1020	32 mm pipe [1-1/4"]	1 Plum	20	.400	Ea.	6.65	17.10		23.75	33	
	1030	40 mm pipe [1-1/2"]		20	.400		7.15	17.10		24.25	33.50	
	1040	50 mm pipe [2"]		18	.444		7.15	19		26.15	36.50	
	1050	80 mm pipe [3"]		17	.471		7.90	20		27.90	38.50	
	1060	100 mm pipe [4"]		16	.500		9.55	21.50		31.05	42.50	
	1350	Copper with neoprene ring										
	1400	32 mm pipe [1-1/4"]	1 Plum	20	.400	Ea.	14.05	17.10		31.15	41	
	1430	40 mm pipe [1-1/2"]		20	.400		14.05	17.10		31.15	41	
	1440	50 mm pipe [2"]		18	.444		14.85	19		33.85	45	
	1450	80 mm pipe [3"]		17	.471		17.40	20		37.40	49	
	1460	100 mm pipe [4"]		16	.500		19.20	21.50		40.70	53	

15160 | Storm Drainage Piping

			CREW	DAILY OUTPUT	LABOR-HOURS	UNIT	MAT.	LABOR	EQUIP.	TOTAL	TOTAL INCL O&P	
500	0010	**STORM AREA DRAINS**										500
	0140	Cornice, C.I., 45° or 90° outlet										
	0200	80 mm and 100 mm pipe size [3",4"]	Q-1	12	1.333	Ea.	171	51		222	265	
	0260	For galvanized body, add					34.50			34.50	37.50	
	0280	For polished bronze dome, add					29.50			29.50	32.50	
	3860	Roof, flat metal deck, C.I. body, 300 mm C.I. dome [12"]										
	3890	80 mm pipe size [3"]	Q-1	14	1.143	Ea.	204	44		248	291	
	3920	150 mm pipe size [6"]	"	10	1.600	"	355	61.50		416.50	485	

MECHANICAL 15

15160 | Storm Drainage Piping

		CREW	DAILY OUTPUT	LABOR-HOURS	UNIT	2006 BARE COSTS MAT.	LABOR	EQUIP.	TOTAL	TOTAL INCL O&P	
500	4620	Main, all aluminum, 300 mm low profile dome [12"]									500
	4640	50, 80 & 100 mm pipe size	Q-1	14	1.143	Ea.	226	44		270	315
	4980	Scupper floor, oblique strainer, C.I.									
	5000	150,175 mm top, 50, 80 & 100 mm pipe	Q-1	16	1	Ea.	147	38.50		185.50	220
	5100	200 x 300 mm top, 125 and 150 mm pipe	"	14	1.143		286	44		330	380
	5160	For galvanized body, add					40%				
	5200	For polished bronze strainer, add					85%				
	5980	Trench, floor, heavy duty, modular, C.I., 300 x 300 mm top									
	6000	50,80,100,125, & 150 mm pipe	Q-1	8	2	Ea.	465	77		542	625
	6100	For polished bronze top, add				"	228			228	250

15180 | Heating and Cooling Piping

		CREW	DAILY OUTPUT	LABOR-HOURS	UNIT	2006 BARE COSTS MAT.	LABOR	EQUIP.	TOTAL	TOTAL INCL O&P	
200	0010	**PUMPS, CIRCULATING** Heated or chilled water application									200
	0600	Bronze, sweat connections, 19 W, in line									
	0640	19 mm size [3/4"]	Q-1	16	1	Ea.	134	38.50		172.50	206
	1000	Flange connection, 20 mm to 40 mm size [3/4" to 1-1/2"]									
	1040	62 W	Q-1	6	2.667	Ea.	365	102		467	560
	1060	93 W		6	2.667		630	102		732	850
	1100	249 W		6	2.667		705	102		807	930
	1140	50 mm size, 124 W [2"]		5	3.200		905	123		1,028	1,175
	1180	64 mm size, 186 W [2-1/2"]		5	3.200		1,150	123		1,273	1,450
	2000	Cast iron, flange connection									
	2040	19 mm to 38 mm size, in line, 62 W [3/4" to 1-1/2"]	Q-1	6	2.667	Ea.	239	102		341	415
	2100	249 W		6	2.667		445	102		547	645
	2140	51 mm size, 124 W [2"]		5	3.200		490	123		613	720
	2180	64 mm size, 186 W [2-1/2"]		5	3.200		625	123		748	875
	2220	80 mm size, 186 W [3"]		4	4		640	154		794	930
	2600	For non-ferrous impeller, add					3%				
300	0010	**PUMPS, CONDENSATE RETURN SYSTEM**									300
	0200	Simplex, 560 W motor, float sw, controls, 38 L C.I. receiver, .4-.9 L/s	Q-1	1	16	Ea.	1,850	615		2,465	2,950
	1000	Duplex, 2 pumps, 560 W motors, float switch,									
	1060	alternator asssembly, 55 L C.I. receiver	Q-1	.50	32	Ea.	5,050	1,225		6,275	7,400
800	0010	**STEAM CONDENSATE METER**									800
	0100	226.8 kg/hr	1 Stpi	14	.571	Ea.	2,300	24.50		2,324.50	2,550
	0140	680.4 kg/hr	"	7	1.143	"	2,475	49		2,524	2,800

15200 | Process Piping

15230 | Industrial Process Piping

		CREW	DAILY OUTPUT	LABOR-HOURS	UNIT	2006 BARE COSTS MAT.	LABOR	EQUIP.	TOTAL	TOTAL INCL O&P	
500	0010	**PUMPS, GENERAL UTILITY** With motor									500
	2000	Single stage									
	3000	Double suction,									
	3190	56 kW to 157 L/s [75 HP, to 2500 GPM]	Q-3	.28	114	Ea.	12,100	4,650		16,750	20,300
	3220	75 kW, to 189 L/s [100 HP, to 3000 GPM]		.26	123		15,400	5,000		20,400	24,500
	3240	112 kW, to 252 L/s [150 HP, to 4000 GPM]		.24	133		20,700	5,425		26,125	31,000

15 MECHANICAL

			DAILY	LABOR-		2006 BARE COSTS				TOTAL	
15410		**Plumbing Fixtures**	CREW	OUTPUT	HOURS	UNIT	MAT.	LABOR	EQUIP.	TOTAL	INCL O&P
040	0010	**FIXTURES** Includes trim fittings unless otherwise noted	R224000 -40									040
	0080	For rough-in, supply, waste, and vent, see add for each type									
	0120	For electric water coolers, see division 15413									
	0160	For color, unless otherwise noted, add				Ea.	20%				
200	0010	**CARRIERS/SUPPORTS** For plumbing fixtures										200
	0500	Drinking fountain, wall mounted									
	0600	Plate type with studs, top back plate	1 Plum	7	1.143	Ea.	46	49		95	125
	0700	Top front and back plate		7	1.143		57.50	49		106.50	137
	0800	Top & bottom, front & back plates, w/bearing jacks		7	1.143		83.50	49		132.50	165
	3000	Lavatory, concealed arm									
	3050	Floor mounted, single									
	3100	High back fixture	1 Plum	6	1.333	Ea.	177	57		234	281
	3200	Flat slab fixture		6	1.333		223	57		280	330
	3220	Paraplegic		6	1.333		290	57		347	405
	3250	Floor mounted, back to back									
	3300	High back fixtures	1 Plum	5	1.600	Ea.	370	68.50		438.50	510
	3400	Flat slab fixtures		5	1.600		455	68.50		523.50	605
	3430	Paraplegic		5	1.600		320	68.50		388.50	455
	3500	Wall mounted, in stud or masonry									
	3600	High back fixture	1 Plum	6	1.333	Ea.	153	57		210	254
	3700	Flat slab fixture	"	6	1.333	"	196	57		253	300
	4600	Sink, floor mounted									
	4650	Exposed arm system									
	4700	Single heavy fixture	1 Plum	5	1.600	Ea.	435	68.50		503.50	580
	4750	Single heavy sink with slab		5	1.600		350	68.50		418.50	490
	4800	Back to back, standard fixtures		5	1.600		320	68.50		388.50	455
	4850	Back to back, heavy fixtures		5	1.600		350	68.50		418.50	490
	4900	Back to back, heavy sink with slab		5	1.600		350	68.50		418.50	490
	4950	Exposed offset arm system									
	5000	Single heavy deep fixture	1 Plum	5	1.600	Ea.	410	68.50		478.50	555
	5100	Plate type system									
	5200	With bearing jacks, single fixture	1 Plum	5	1.600	Ea.	475	68.50		543.50	625
	5300	With exposed arms, single heavy fixture		5	1.600		605	68.50		673.50	775
	5400	Wall mounted, exposed arms, single heavy fixture		5	1.600		221	68.50		289.50	345
	6000	Urinal, floor mounted, 50 mm or 80 mm coupling, blowout type		6	1.333		275	57		332	385
	6100	With fixture or hanger bolts, blowout or washout		6	1.333		196	57		253	300
	6200	With bearing plate		6	1.333		196	57		253	300
	6300	Wall mounted, plate type system		6	1.333		170	57		227	273
	6980	Water closet, siphon jet									
	7000	Horizontal, adjustable, caulk									
	7040	Single,100 mm pipe size [4"]	1 Plum	5.33	1.501	Ea.	360	64		424	490
	7050	100 mm pipe size, paraplegic [4"]		5.33	1.501		360	64		424	490
	7060	125 mm pipe size [5"]		5.33	1.501		475	64		539	615
	7100	Double,100 mm pipe size [4"]		5	1.600		675	68.50		743.50	850
	7110	100 mm pipe size, paraplegic [4"]		5	1.600		675	68.50		743.50	850
	7120	125 mm pipe size [5"]		5	1.600		830	68.50		898.50	1,025
	7160	Horizontal, adjustable, extended, caulk									
	7180	Single, 100 mm pipe size [4"]	1 Plum	5.33	1.501	Ea.	360	64		424	490
	7200	125 mm pipe size [5"]		5.33	1.501		475	64		539	615
	7240	Double, 100 mm pipe size [4"]		5	1.600		670	68.50		738.50	840
	7260	125 mm pipe size [5"]		5	1.600		830	68.50		898.50	1,025
	7400	Vertical, adjustable, caulk or thread									
	7440	Single, 100 mm pipe size [4"]	1 Plum	5.33	1.501	Ea.	495	64		559	635
	7460	125 mm pipe size [5"]		5.33	1.501		605	64		669	765
	7480	150 mm pipe size [6"]		5	1.600		695	68.50		763.50	870
	7520	Double, 100 mm pipe size [4"]		5	1.600		735	68.50		803.50	915

For expanded coverage of these items see *Means Mechanical or Plumbing Cost Data 2006*

MECHANICAL 15

	15410	Plumbing Fixtures	CREW	DAILY OUTPUT	LABOR-HOURS	UNIT	2006 BARE COSTS				TOTAL INCL O&P	
							MAT.	LABOR	EQUIP.	TOTAL		
200	7540	125 mm pipe size [5"]	1 Plum	5	1.600	Ea.	850	68.50		918.50	1,050	**200**
	7560	150 mm pipe size [6"]	↓	4	2	↓	940	85.50		1,025.50	1,150	
	7600	Vertical, adjustable, extended, caulk										
	7620	Single, 100 mm pipe size [4"]	1 Plum	5.33	1.501	Ea.	495	64		559	635	
	7640	125 mm pipe size [5"]		5.33	1.501		605	64		669	765	
	7680	150 mm pipe size [6"]		5	1.600		695	68.50		763.50	870	
	7720	Double, 100 mm pipe size [4"]		5	1.600		735	68.50		803.50	915	
	7740	125 mm pipe size [5"]		5	1.600		850	68.50		918.50	1,050	
	7760	150 mm pipe size [6"]	↓	4	2	↓	940	85.50		1,025.50	1,150	
	7780	Water closet, blow out										
	7800	Vertical offset, caulk or thread										
	7820	Single, 100 mm pipe size [4"]	1 Plum	5.33	1.501	Ea.	370	64		434	505	
	7840	Double, 100 mm pipe size [4"]	"	5	1.600	"	635	68.50		703.50	805	
	7880	Vertical offset, extended, caulk										
	7900	Single, 100 mm pipe size [4"]	1 Plum	5.33	1.501	Ea.	465	64		529	610	
	7920	Double, 100 mm pipe size [4"]	"	5	1.600	"	730	68.50		798.50	910	
	7960	Vertical, for floor mounted back-outlet										
	7980	Single, 100 mm thread, 50 mm vent [4",2"]	1 Plum	5.33	1.501	Ea.	330	64		394	460	
	8000	Double, 100 mm thread, 50 mm vent [4",2"]	"	6	1.333	"	625	57		682	770	
	8040	Vertical, for floor mounted back-outlet, extended										
	8060	Single, 100 mm caulk, 50 mm vent [4",2"]	1 Plum	6	1.333	Ea.	330	57		387	450	
	8080	Double, 100 mm caulk, 50 mm vent [4",2"]	"	6	1.333	"	625	57		682	770	
	8200	Water closet, residential										
	8220	Vertical centerline, floor mount										
	8240	Single, 80 mm caulk, 50 mm or 80 mm vent [3",2",3"]	1 Plum	6	1.333	Ea.	297	57		354	410	
	8260	100 mm caulk, 50 mm or 100 mm vent [4",2",4"]		6	1.333		385	57		442	505	
	8280	80 mm copper sweat, 80 mm vent [3",3"]		6	1.333		267	57		324	380	
	8300	100 mm copper sweat, 100 mm vent [4",4"]	↓	6	1.333	↓	325	57		382	440	
	8400	Vertical offset, floor mount										
	8420	Single, 80 mm or 100 mm caulk, vent [3",4"]	1 Plum	4	2	Ea.	370	85.50		455.50	540	
	8440	80 mm or 100 mm copper sweat, vent [3",4"]		5	1.600		370	68.50		438.50	515	
	8460	Double, 80 mm or 100 mm caulk, vent [3",4"]		4	2		635	85.50		720.50	830	
	8480	80 mm or 100 mm copper sweat, vent [3",4"]	↓	5	1.600	↓	635	68.50		703.50	805	
	9000	Water cooler (electric), floor mounted										
	9100	Plate type with bearing plate, single	1 Plum	6	1.333	Ea.	188	57		245	293	
300	0010	**FAUCETS/FITTINGS**										**300**
	0150	Bath, faucets, diverter spout combination, sweat	1 Plum	8	1	Ea.	69.50	42.50		112	141	
	0200	For integral stops, IPS unions, add				↓	73			73	80.50	
	0420	Bath, press-bal mix valve w/diverter, spout, shower hd, arm/flange	1 Plum	8	1	↓	116	42.50		158.50	193	
	0810	Bidet										
	0812	Fitting, over the rim, swivel spray/pop-up drain	1 Plum	8	1	Ea.	138	42.50		180.50	217	
	0840	Flush valves, with vacuum breaker										
	0850	Water closet										
	0860	Exposed, rear spud	1 Plum	8	1	Ea.	118	42.50		160.50	194	
	0920	Urinal										
	0930	Exposed, stall	1 Plum	8	1	Ea.	108	42.50		150.50	184	
	0940	Wall, (washout)		8	1		108	42.50		150.50	184	
	0950	Pedestal, top spud		8	1		110	42.50		152.50	186	
	0960	Concealed, stall		8	1		124	42.50		166.50	201	
	0970	Wall (washout) ♿	↓	8	1	↓	123	42.50		165.50	201	
	0971	Automatic flush sensor and operator for										
	0972	urinals or water closets	1 Plum	5.33	1.501	Ea.	355	64		419	485	
	1000	Kitchen sink faucets, top mount, cast spout		10	.800		51.50	34		85.50	108	
	1100	For spray, add	↓	24	.333	↓	11.05	14.25		25.30	33.50	
	1300	Single control lever handle										

Important: See the Reference Section for supporting data - Crews, Rental Equipment, City Cost Indexes and Reference Data

15410	Plumbing Fixtures	CREW	DAILY OUTPUT	LABOR-HOURS	UNIT	2006 BARE COSTS				TOTAL INCL O&P
						MAT.	LABOR	EQUIP.	TOTAL	
300 1310	With pull out spray									**300**
1320	Polished chrome	1 Plum	10	.800	Ea.	167	34		201	235
1330	Polished brass		10	.800		200	34		234	272
1340	White	↓	10	.800	↓	184	34		218	254
1348	With spray thru escutcheon									
1350	Polished chrome	1 Plum	10	.800	Ea.	117	34		151	180
1360	Polished brass		10	.800		140	34		174	206
1370	White		10	.800		128	34		162	193
2000	Laundry faucets, shelf type, IPS or copper unions		12	.667		42.50	28.50		71	89.50
2100	Lavatory faucet, centerset, without drain	↓	10	.800	↓	37.50	34		71.50	92.50
2210	Porcelain cross handles and pop-up drain									
2220	Polished chrome	1 Plum	6.66	1.201	Ea.	110	51.50		161.50	198
2230	Polished brass	"	6.66	1.201	"	165	51.50		216.50	259
2260	Single lever handle and pop-up drain									
2270	Black nickel	1 Plum	6.66	1.201	Ea.	212	51.50		263.50	310
2280	Polished brass		6.66	1.201		212	51.50		263.50	310
2290	Polished chrome		6.66	1.201		141	51.50		192.50	232
2810	Automatic sensor and operator, with faucet head		6.15	1.301		300	55.50		355.50	415
3000	Service sink faucet, cast spout, pail hook, hose end		14	.571		72	24.50		96.50	116
4000	Shower by-pass valve with union		18	.444		50.50	19		69.50	84
4200	Shower thermostatic mixing valve, concealed	↓	8	1	↓	250	42.50		292.50	340
4220	Shower pressure balancing mixing valve,									
4230	With shower head, arm, flange and diverter tub spout									
4240	Chrome	1 Plum	6.14	1.303	Ea.	138	55.50		193.50	235
4250	Polished brass		6.14	1.303		193	55.50		248.50	296
4260	Satin		6.14	1.303		193	55.50		248.50	296
4270	Polished chrome/brass		6.14	1.303		158	55.50		213.50	258
5000	Sillcock, compact, brass, IPS or copper to hose	↓	24	.333	↓	5.50	14.25		19.75	27.50

15411	Commercial/Indust Fixtures									
400 0010	**HOT WATER DISPENSERS**									**400**
0160	Commercial, 100 cup, 11.3 A	1 Plum	14	.571	Ea.	330	24.50		354.50	400
3180	Household, 60 cup	"	14	.571	"	166	24.50		190.50	220
500 0010	**HYDRANTS**									**500**
0050	Wall type, moderate climate, bronze, encased									
0200	20 mm IPS connection [3/4"]	1 Plum	16	.500	Ea.	395	21.50		416.50	465
0300	25 mm IPS connection [1"]	"	14	.571		395	24.50		419.50	470
0500	Anti-siphon type				↓	340			340	375
1000	Non-freeze, bronze, exposed									
1100	20 mm IPS connection, 100 mm to 225 mm wall	1 Plum	14	.571	Ea.	268	24.50		292.50	330
1120	250 mm to 350 mm thick wall [10",14"]		12	.667		290	28.50		318.50	365
1140	375 mm to 475 mm thick wall [15",19"]		12	.667		325	28.50		353.50	400
1160	500 to 600 mm thick wall [20",24"]	↓	10	.800	↓	350	34		384	435
1200	For 25 mm IPS connection, add					15%	10%			
1240	For 20 mm adapter type vacuum breaker, add [3/4"]				Ea.	33.50			33.50	37
1280	For anti-siphon type, add				"	70			70	77
2000	Non-freeze bronze, encased, anti-siphon type									
2100	20 mm IPS conn, 125 mm to 225 mm wall	1 Plum	14	.571	Ea.	460	24.50		484.50	545
2140	375 mm to 475 mm thick wall [15",19"]	"	12	.667	"	520	28.50		548.50	615
3000	Ground box type, bronze frame, 20 mm IPS connection [3/4"]									
3080	Non-freeze, all bronze, polished face, set flush									
3100	600 mm depth of bury	1 Plum	8	1	Ea.	495	42.50		537.50	605
3180	1800 mm depth of bury		7	1.143		645	49		694	785
3220	2400 mm depth of bury	↓	5	1.600		720	68.50		788.50	895
3400	For 25 mm IPS connection, add [1"]				↓	15%	10%			

MECHANICAL 15

15411 | Commercial/Indust Fixtures

			CREW	DAILY OUTPUT	LABOR-HOURS	UNIT	MAT.	LABOR	EQUIP.	TOTAL	TOTAL INCL O&P	
500	3550	For 50 mm connection, add [2"]				Ea.	445%	24%				500
	3600	For tapped drain port in box, add			↓		45.50			45.50	50	
	5000	Moderate climate, all bronze, polished face										
	5020	and scoriated cover, set flush										
	5100	20 mm IPS connection [3/4"]	1 Plum	16	.500	Ea.	350	21.50		371.50	415	
	5120	25 mm IPS connection [1"]	"	14	.571		350	24.50		374.50	420	
	5200	For tapped drain port in box, add			↓		45.50			45.50	50	
700	0010	**URINALS**										700
	3000	Wall hung, vitreous china, with hanger & self-closing valve										
	3100	Siphon jet type	Q-1	3	5.333	Ea.	315	205		520	655	
	3120	Blowout type		3	5.333		350	205		555	695	
	3300	Rough-in, supply, waste & vent		2.83	5.654		136	217		353	475	
	5000	Stall type, vitreous china, includes valve		2.50	6.400		520	246		766	945	
	5100	76 mm seam cover, add		12	1.333		154	51		205	246	
	5200	152 mm seam cover, add		12	1.333		170	51		221	264	
	6980	Rough-in, supply, waste and vent	↓	1.99	8.040	↓	184	310		494	670	
840	0010	**WASH FOUNTAINS**										840
	1900	Group, foot control										
	2000	Precast terrazzo, circular, 900 mm diam., 5 or 6 persons [36"]	Q-2	3	8	Ea.	2,975	320		3,295	3,725	
	2100	1350 mm diameter for 8 or 10 persons [54"]		2.50	9.600		3,600	385		3,985	4,525	
	2400	Semi-circular, 900 mm diam. for 3 persons [36"]		3	8		2,650	320		2,970	3,400	
	2500	1350 mm diam. for 4 or 5 persons [54"]		2.50	9.600		3,275	385		3,660	4,175	
	2700	Quarter circle (corner), 1372 mm for 3 persons		3.50	6.857		3,275	273		3,548	4,000	
	3000	Stainless steel, circular, 900 mm diameter [36"]		3.50	6.857		3,400	273		3,673	4,125	
	3100	1350 mm diameter [54"]		2.80	8.571		4,475	340		4,815	5,450	
	3400	Semi-circular, 900 mm diameter [36"]		3.50	6.857		2,925	273		3,198	3,625	
	3500	1350 mm diameter [54"]	↓	2.80	8.571	↓	3,850	340		4,190	4,775	
	5610	Group, infrared control, barrier free ♿										
	5614	Precast terrazzo										
	5620	Semi-circular 914 mm diam. for 3 persons [36"]	Q-2	3	8	Ea.	3,900	320		4,220	4,775	
	5630	1168 mm diam. for 4 persons [46"]		2.80	8.571		4,375	340		4,715	5,325	
	5640	Circular, 1372 mm diam. for 8 persons, button control [54"]	↓	2.50	9.600		6,400	385		6,785	7,600	
	5700	Rough-in, supply, waste and vent for above wash fountains	Q-1	1.82	8.791		176	340		516	705	
	6200	Duo for small washrooms, stainless steel		2	8		2,000	305		2,305	2,675	
	6500	Rough-in, supply, waste & vent for duo fountains	↓	2.02	7.921	↓	98	305		403	570	

15412 | Drinking Fountains

			CREW	DAILY OUTPUT	LABOR-HOURS	UNIT	MAT.	LABOR	EQUIP.	TOTAL	TOTAL INCL O&P	
200	0010	**DRINKING FOUNTAIN** For connection to cold water supply [R224000-40]										200
	1000	Wall mounted, non-recessed										
	1400	Bronze, with no back	1 Plum	4	2	Ea.	1,075	85.50		1,160.50	1,300	
	1800	Cast aluminum, enameled, for correctional institutions		4	2		660	85.50		745.50	860	
	2000	Fiberglass, 305 mm back, single bubbler unit		4	2		590	85.50		675.50	780	
	2040	Dual bubbler		3.20	2.500		855	107		962	1,100	
	2400	Precast stone, no back		4	2		535	85.50		620.50	720	
	2700	Stainless steel, single bubbler, no back		4	2		855	85.50		940.50	1,075	
	2740	With back		4	2		975	85.50		1,060.50	1,200	
	2780	Dual handle & wheelchair projection type ♿		4	2		540	85.50		625.50	725	
	2820	Dual level for handicapped type ♿	↓	3.20	2.500	↓	1,025	107		1,132	1,275	
	3300	Vitreous china										
	3340	175 mm back	1 Plum	4	2	Ea.	420	85.50		505.50	590	
	3940	For vandal-resistant bottom plate, add					52.50			52.50	57.50	
	3960	For freeze-proof valve system, add	1 Plum	2	4		410	171		581	705	
	3980	For rough-in, supply and waste, add	↓ "	2.21	3.620	↓	75.50	155		230.50	315	

15412 | Drinking Fountains

		CREW	DAILY OUTPUT	LABOR-HOURS	UNIT	2006 BARE COSTS				TOTAL INCL O&P
						MAT.	LABOR	EQUIP.	TOTAL	
200	4000	Wall mounted, semi-recessed									**200**
	4200	Poly-marble, single bubbler	1 Plum	4	2	Ea.	680	85.50		765.50	875
	4600	Stainless steel, satin finish, single bubbler		4	2		870	85.50		955.50	1,100
	4900	Vitreous china, single bubbler		4	2		565	85.50		650.50	750
	5980	For rough-in, supply and waste, add		1.83	4.372		75.50	187		262.50	365
	6000	Wall mounted, fully recessed									
	6400	Poly-marble, single bubbler	1 Plum	4	2	Ea.	765	85.50		850.50	970
	6800	Stainless steel, single bubbler		4	2		700	85.50		785.50	895
	7560	For freeze-proof valve system, add		2	4		500	171		671	810
	7580	For rough-in, supply and waste, add		1.83	4.372		75.50	187		262.50	365
	7600	Floor mounted, pedestal type									
	7700	Aluminum, architectural style, C.I. base	1 Plum	2	4	Ea.	460	171		631	760
	7780	Wheelchair handicap unit		2	4		1,050	171		1,221	1,425
	8400	Stainless steel, architectural style		2	4		1,200	171		1,371	1,550
	8600	Enameled iron, heavy duty service, 2 bubblers		2	4		1,100	171		1,271	1,450
	8660	4 bubblers		2	4		1,175	171		1,346	1,550
	8880	For freeze-proof valve system, add		2	4		335	171		506	625
	8900	For rough-in, supply and waste, add		1.83	4.372		75.50	187		262.50	365
	9100	Deck mounted									
	9500	Stainless steel, circular receptor	1 Plum	4	2	Ea.	280	85.50		365.50	440
	9760	White enameled steel, 356 mm x 229 mm receptor		4	2		253	85.50		338.50	405
	9860	White enameled cast iron, 610 mm x 406 mm receptor		3	2.667		242	114		356	435
	9980	For rough-in, supply and waste, add		1.83	4.372		75.50	187		262.50	365

R224000-40

15413 | Electric Water Coolers

		CREW	DAILY OUTPUT	LABOR-HOURS	UNIT	MAT.	LABOR	EQUIP.	TOTAL	TOTAL INCL O&P
900	0010	**WATER COOLER**									**900**
	0030	See line 15413-900-9800 for rough-in, waste & vent									
	0040	for all water coolers									
	0100	Wall mounted, non-recessed									
	0140	15.1 L/hr	Q-1	4	4	Ea.	500	154		654	780
	0160	30.2 L/hr, barrier free, sensor operated		4	4		630	154		784	925
	0180	31 L/hr		4	4		560	154		714	845
	0600	For hot and cold water, add					144			144	158
	0640	For stainless steel cabinet, add					73.50			73.50	81
	1000	Dual height, 31 L/hr	Q-1	3.80	4.211		805	162		967	1,125
	1040	54.1 L/hr	"	3.80	4.211		855	162		1,017	1,175
	1240	For stainless steel cabinet, add					112			112	123
	2600	Wheelchair type, 30 L/hr	Q-1	4	4		1,425	154		1,579	1,800
	3300	Semi-recessed, 30.7 L/hr		4	4		840	154		994	1,150
	3320	45.4 L/hr		4	4		885	154		1,039	1,200
	4600	Floor mounted, flush-to-wall									
	4640	15.1 L/hr	1 Plum	3	2.667	Ea.	520	114		634	745
	4680	31 L/hr		3	2.667		575	114		689	800
	4720	54 L/hr		3	2.667		620	114		734	855
	4960	50 L/hr hot and cold water		3	2.667		865	114		979	1,125
	4980	For stainless steel cabinet, add					112			112	123
	5000	Dual height, 31 L/hr	1 Plum	2	4		870	171		1,041	1,200
	5040	54.1 L/hr	"	2	4		900	171		1,071	1,250
	5120	For stainless steel cabinet, add					147			147	162
	9800	For supply, waste & vent, all coolers	1 Plum	2.21	3.620		75.50	155		230.50	315

R224000-50

15414 | Emergency Fixtures

		CREW	DAILY OUTPUT	LABOR-HOURS	UNIT	MAT.	LABOR	EQUIP.	TOTAL	TOTAL INCL O&P
200	0010	**INDUSTRIAL SAFETY FIXTURES** Rough-in not included										**200**
	1000	Eye wash fountain									

MECHANICAL 15

15414	Emergency Fixtures	CREW	DAILY OUTPUT	LABOR-HOURS	UNIT	2006 BARE COSTS				TOTAL INCL O&P	
						MAT.	LABOR	EQUIP.	TOTAL		
200 1400	Plastic bowl, pedestal mounted	Q-1	4	4	Ea.	196	154		350	445	**200**
1600	Unmounted		4	4		138	154		292	385	
1800	Wall mounted		4	4		141	154		295	385	
2000	Stainless steel, pedestal mounted		4	4		293	154		447	550	
2200	Unmounted		4	4		192	154		346	440	
2400	Wall mounted		4	4		185	154		339	435	
3000	Eye wash, portable, self-contained					395			395	435	
4000	Eye and face wash, combination fountain										
4200	Stainless steel, pedestal mounted	Q-1	4	4	Ea.	415	154		569	685	
4400	Unmounted		4	4		315	154		469	580	
4600	Wall mounted		4	4		320	154		474	585	
5000	Shower, single head, drench, ball valve, pull, freestanding		4	4		273	154		427	530	
5200	Horizontal or vertical supply		4	4		249	154		403	505	
6000	Multi-nozzle, eye/face wash combination		4	4		505	154		659	790	
6400	Multi-nozzle, 12 spray, shower only		4	4		1,300	154		1,454	1,675	
6600	For freeze-proof, add		6	2.667		287	102		389	470	

15418	Resi/Comm/Industrial Fixtures	CREW	DAILY OUTPUT	LABOR-HOURS	UNIT	MAT.	LABOR	EQUIP.	TOTAL	TOTAL INCL O&P	
100 0010	**BATHS**										**100**
0100	Tubs, recessed porcelain enamel on cast iron, with trim										
0180	1219 mm x 1067 mm	Q-1	4	4	Ea.	1,625	154		1,779	2,000	
0220	1800 mm x 900 mm		3	5.333		1,675	205		1,880	2,125	
2000	Enameled formed steel, 1372 mm long		5.80	2.759		335	106		441	525	
4000	Soaking, acrylic w/ pop-up drain, 1524 mm x 813 mm x 457 mm		5.50	2.909		830	112		942	1,075	
4100	1500 mm x 1200 mm x 475 mm deep		5	3.200		635	123		758	880	
6000	Whirlpool, bath with vented overflow, molded fiberglass										
6100	1676 mm x 1219 mm x 610 mm	Q-1	1	16	Ea.	2,700	615		3,315	3,900	
9600	Rough-in, supply, waste and vent, for all above tubs, add	"	2.07	7.729	"	166	297		463	625	
400 0010	**LAUNDRY SINKS** With trim										**400**
0020	Porcelain enamel on cast iron, black iron frame										
0050	610 mm x 525 mm, single compartment	Q-1	6	2.667	Ea.	345	102		447	535	
0100	650 mm x 525 mm, single compartment	"	6	2.667	"	340	102		442	530	
2000	Molded stone, on wall hanger or legs										
2020	550 mm x 575 mm, single compartment	Q-1	6	2.667	Ea.	124	102		226	290	
2100	1143 mm x 533 mm, double compartment	"	5	3.200	"	257	123		380	470	
3000	Plastic, on wall hanger or legs										
3020	457 mm x 584 mm, single compartment	Q-1	6.50	2.462	Ea.	94	94.50		188.50	246	
3300	1016 mm x 610 mm, double compartment		5.50	2.909		224	112		336	415	
5000	Stainless steel, counter top, 559 mm x 432 mm single compartment		6	2.667		54.50	102		156.50	214	
5200	838 mm x 559 mm, double compartment		5	3.200		66.50	123		189.50	259	
9600	Rough-in, supply, waste and vent, for all laundry sinks		2.14	7.477		130	287		417	575	
450 0010	**LAVATORIES** With trim, white unless noted otherwise	R224000 -40									**450**
0500	Vanity top, porcelain enamel on cast iron										
0600	500 mm x 450 mm	Q-1	6.40	2.500	Ea.	212	96		308	380	
0640	836 mm x 482 mm oval		6.40	2.500		445	96		541	630	
0720	475 mm round		6.40	2.500		199	96		295	365	
0860	For color, add					25%					
1000	Cultured marble, 483 mm x 432 mm, single bowl	Q-1	6.40	2.500	Ea.	163	96		259	325	
1040	635 mm x 483 mm, single bowl	"	6.40	2.500	"	175	96		271	335	
1580	For color, same price										
1900	Stainless st, self-rimming, 635 mm x 559 mm, single bowl, ledge	Q-1	6.40	2.500	Ea.	247	96		343	415	
1960	432 mm x 559 mm, single bowl		6.40	2.500		240	96		336	410	
2600	Steel, enameled, 500 mm x 425 mm, single bowl		5.80	2.759		138	106		244	310	
2660	483 mm round		5.80	2.759		135	106		241	310	
2900	Vitreous china, 500 mm x 400 mm, single bowl		5.40	2.963		222	114		336	415	

		Description	CREW	DAILY OUTPUT	LABOR-HOURS	UNIT	MAT.	LABOR	EQUIP.	TOTAL	TOTAL INCL O&P	
15418		**Resi/Comm/Industrial Fixtures**										
450	2960	508 mm x 432 mm, single bowl R224000-40	Q-1	5.40	2.963	Ea.	147	114		261	330	**450**
	3020	483 mm round, single bowl		5.40	2.963		145	114		259	330	
	3200	559 mm x 330 mm, single bowl	↓	5.40	2.963	↓	234	114		348	430	
	3560	For color, add					50%					
	3580	Rough-in, supply, waste and vent for all above lavatories	Q-1	2.30	6.957	Ea.	113	267		380	525	
	4000	Wall hung										
	4040	Porcelain enamel on cast iron, 400 mm x 350 mm, single bowl	Q-1	8	2	Ea.	345	77		422	495	
	4180	500 mm x 450 mm, single bowl		8	2		241	77		318	380	
	4240	550 mm x 475 mm, single bowl	↓	8	2	↓	410	77		487	570	
	4580	For color, add					30%					
	6000	Vitreous china, 450 mm x 375 mm, single bowl with backsplash	Q-1	7	2.286	Ea.	204	88		292	355	
	6180	483 mm x 483 mm, corner style	"	8	2	"	435	77		512	595	
	6500	For color, add					30%					
	6960	Rough-in, supply, waste and vent for above lavatories	Q-1	1.66	9.639	Ea.	310	370		680	895	
500	0010	**SHOWERS**										**500**
	1500	Stall, with drain only. Add for valve and door/curtain										
	3000	Fiberglass, one piece, with 3 walls, 800 mm x 800 mm	Q-1	5.50	2.909	Ea.	420	112		532	630	
	3100	900 mm x 900 mm		5.50	2.909		465	112		577	680	
	3250	1626 mm x 1670 mm x 2070 mm fold. seat, whlchr.		3.80	4.211		2,900	162		3,062	3,450	
	4000	Polypropylene, stall only, w/ molded-stone flr, 750 mm x 750 mm		2	8		330	305		635	825	
	4200	Rough-in, supply, waste and vent for above showers	↓	2.05	7.805		244	300		544	720	
	4520	Showers, fiberglass receptor only	1 Plum	8	1		187	42.50		229.50	270	
	5000	Built-in, head, arm, 250 mL/s valve		4	2		123	85.50		208.50	265	
	5200	Head, arm, by-pass, integral stops, handles	↓	3.60	2.222	↓	196	95		291	360	
	6000	Group, w/pressure balancing valve, rough-in and rigging not included										
	6800	Column, 6 heads, no receptors, less partitions	Q-1	3	5.333	Ea.	2,125	205		2,330	2,650	
	6900	With enameled partitions		1	16		4,550	615		5,165	5,925	
	7600	5 heads, no receptors, less partitions		3	5.333		1,625	205		1,830	2,075	
	7620	4 heads (1 handicap) no receptors, less partitions		3	5.333		2,275	205		2,480	2,800	
	7700	With enameled partitions		1	16		1,625	615		2,240	2,700	
	8000	Wall, 2 heads, no receptors, less partitions		4	4		935	154		1,089	1,250	
	8100	With partitions	↓	2	8	↓	2,275	305		2,580	2,975	
600	0010	**SINKS** With faucets and drain										**600**
	2000	Kitchen, countertop, P.E. on C.I., 600 mm x 525 mm single bowl	Q-1	5.60	2.857	Ea.	232	110		342	420	
	2100	775 mm x 550 mm single bowl R224000-40		5.60	2.857		217	110		327	405	
	2200	800 mm x 525 mm double bowl		4.80	3.333		315	128		443	540	
	2300	1067 mm x 533 mm double bowl		4.80	3.333		480	128		608	725	
	3000	Stainless stl, self rimming, 483 mm x 457 mm sgl bowl		5.60	2.857		380	110		490	580	
	3100	635 mm x 559 mm single bowl		5.60	2.857		420	110		530	630	
	4000	Steel, enameled, with ledge, 610 mm x 533 mm single bowl		5.60	2.857		114	110		224	290	
	4100	813 mm x 533 mm double bowl	↓	4.80	3.333		150	128		278	360	
	4960	For color sinks except stainless steel, add					10%					
	4980	For rough-in, supply, waste and vent, counter top sinks	Q-1	2.14	7.477	↓	130	287		417	575	
	5000	Kitchen, raised deck, P.E. on C.I.										
	5100	813 x 533 mm, dual level, dbl bowl	Q-1	2.60	6.154	Ea.	265	236		501	645	
	5700	For color, add					20%					
	5790	For rough-in, supply, waste & vent, sinks	Q-1	1.85	8.649		130	330		460	645	
	6650	Service, floor, corner, P.E. on C.I., 700 mm x 700 mm		4.40	3.636		585	140		725	855	
	6790	For rough-in, supply, waste & vent, floor service sinks		1.64	9.756		305	375		680	900	
	7000	Service, wall, P.E. on C.I., roll rim, 559 mm x 457 mm		4	4		480	154		634	760	
	7100	600 mm x 500 mm		4	4		530	154		684	815	
	8600	Vitreous china, 559 mm x 508 mm	↓	4	4		420	154		574	695	
	8960	For stainless steel rim guard, front or side, add					37			37	40.50	
	8980	For rough-in, supply, waste & vent, wall service sinks	Q-1	1.30	12.308	↓	500	475		975	1,250	
900	0010	**WATER CLOSETS** R224000-40										**900**
	0030	For automatic flush, see 15410-300-0972										

MECHANICAL 15

For expanded coverage of these items see *Means Mechanical or Plumbing Cost Data 2006*

15418 | Resi/Comm/Industrial Fixtures

		CREW	DAILY OUTPUT	LABOR-HOURS	UNIT	2006 BARE COSTS				TOTAL INCL O&P	
						MAT.	LABOR	EQUIP.	TOTAL		
900	0150	Tank type, vitreous china, incl. seat, supply pipe w/stop R224000 -40									900
	0200	Wall hung, one piece	Q-1	5.30	3.019	Ea.	390	116		506	605
	0400	Two piece, close coupled		5.30	3.019		480	116		596	700
	0960	For rough-in, supply, waste, vent and carrier		2.73	5.861		445	225		670	825
	1000	Floor mounted, one piece		5.30	3.019		495	116		611	720
	1100	Two piece, close coupled		5.30	3.019		169	116		285	360
	1960	For color, add					30%				
	1980	For rough-in, supply, waste and vent	Q-1	3.05	5.246	Ea.	191	202		393	515
	3000	Bowl only, with flush valve, seat									
	3100	Wall hung	Q-1	5.80	2.759	Ea.	350	106		456	545
	3200	For rough-in, supply, waste and vent, single WC		2.56	6.250		460	240		700	865
	3300	Floor mounted		5.80	2.759		291	106		397	480
	3350	With wall outlet		5.80	2.759		510	106		616	720
	3400	For rough-in, supply, waste and vent, single WC		2.84	5.634		209	217		426	555

15440 | Plumbing Pumps

		CREW	DAILY OUTPUT	LABOR-HOURS	UNIT	MAT.	LABOR	EQUIP.	TOTAL	TOTAL INCL O&P	
400	0010	**PUMPS, GRINDER SYSTEM** Compl., incl. check valve, tank, std.									400
	0020	controls incl. alarm/disconnect panel w/wire. Excavation not included									
	0260	Simplex, 568 mL/s at 414 kPa, 265 L tank				Ea.	2,525			2,525	2,775
	0300	For manway, 660 mm I.D., 457 mm high, add					320			320	355
	0340	660 mm I.D., 914 mm high, add					445			445	485
	0380	1092 mm I.D., 1219 mm high, add					495			495	545
	0600	Simplex, 0.567 L/s at 276 kPaG, 568 L tank					2,725			2,725	3,000
	0660	For manway, add									
	0700	660 mm I.D., 914 mm high, add				Ea.	590			590	650
	0740	660 mm I.D., 1219 mm high, add					670			670	740
	2000	Duplex, 1.136 L/s at 414 kPaG, 568 L tank					1,800			1,800	1,975
	2060	For manway 1092 mm I.D., 1219 mm high, add					1,475			1,475	1,625
	2400	For core only					1,350			1,350	1,475
800	0010	**PUMPS, SEWAGE EJECTOR** With operating and level controls									800
	0100	Simplex system incl. tank, cover, pump 4.5 m head									
	0500	140 L PE tank, 757 mL/s, 373 W, 51 mm discharge [2"]	Q-1	3.20	5	Ea.	385	192		577	715
	0510	76 mm discharge [3"]		3.10	5.161		415	198		613	755
	0530	5.49 L/s, 522 W, 51 mm discharge [2"]		3.20	5		585	192		777	930
	0540	76 mm discharge [3"]		3.10	5.161		635	198		833	995
	0600	170 L coated stl tank, 757 mL/s, 373 W, 51 mm discharge		3	5.333		680	205		885	1,050
	0610	76 mm discharge [3"]		2.90	5.517		710	212		922	1,100
	0630	5.49 L/s, 522 W, 51 mm discharge [2"]		3	5.333		870	205		1,075	1,275
	0640	76 mm discharge [3"]		2.90	5.517		920	212		1,132	1,350
	0660	8 L/s, 746 W, 51 mm discharge [2"]		2.80	5.714		945	220		1,165	1,375
	0680	76 mm discharge [3"]		2.70	5.926		995	228		1,223	1,450
	0700	265 L PE tank, 760 mL/s, 373 W, 50 mm discharge [2"]		2.60	6.154		750	236		986	1,175
	0710	76 mm discharge [3"]		2.40	6.667		800	256		1,056	1,275
	0730	5 L/s, 520 W, 50 mm discharge [2"]		2.50	6.400		960	246		1,206	1,425
	0740	76 mm discharge [3"]		2.30	6.957		1,025	267		1,292	1,525
	0760	8.5 L/s, 750 W, 50 mm discharge [2"]		2.20	7.273		1,050	279		1,329	1,575
	0770	76 mm discharge [3"]		2	8		1,125	305		1,430	1,700
	0800	284 L coated stl. tank, 0.757 L/s, 373 W, 51 mm discharge [2"]		2.40	6.667		775	256		1,031	1,225
	0810	76 mm discharge		2.20	7.273		805	279		1,084	1,300
	0830	5.49 L/s, 522 W, 51 mm discharge		2.30	6.957		970	267		1,237	1,475
	0840	76 mm discharge		2.10	7.619		1,025	293		1,318	1,575
	0860	8.455 L/s, 746 W, 51 mm discharge		2	8		1,050	305		1,355	1,625
	0880	76 mm discharge		1.80	8.889		1,100	340		1,440	1,725
	1040	Duplex system incl. tank, covers, pumps									
	1060	416 L fiberglass tank, 1.514 L/s, 373 W, 51 mm discharge	Q-1	1.60	10	Ea.	1,475	385		1,860	2,175

15
MECHANICAL

		15440	Plumbing Pumps	CREW	DAILY OUTPUT	LABOR-HOURS	UNIT	2006 BARE COSTS MAT.	LABOR	EQUIP.	TOTAL	TOTAL INCL O&P	
800	1080		76 mm discharge	Q-1	1.40	11.429	Ea.	1,525	440		1,965	2,350	800
	1100		10.979 L/s, 522 W, 51 mm discharge		1.50	10.667		1,900	410		2,310	2,725	
	1120		76 mm discharge		1.30	12.308		2,000	475		2,475	2,900	
	1140		16.911 L/s, 746 W, 51 mm discharge		1.20	13.333		2,075	510		2,585	3,075	
	1160		76 mm discharge	▼	1	16		2,150	615		2,765	3,300	
	1260		511 L coated stl. tank, 1.514 L/s, 373 W, 51 mm discharge	Q-2	1.70	14.118		1,525	565		2,090	2,525	
	2000		76 mm discharge		1.60	15		1,600	600		2,200	2,675	
	2640		10.979 L/s, 522 W, 51 mm discharge		1.60	15		1,975	600		2,575	3,075	
	2660		76 mm discharge		1.50	16		2,100	640		2,740	3,250	
	2700		16.911 L/s, 746 W, 51 mm discharge		1.30	18.462		2,150	735		2,885	3,475	
	3040		76 mm discharge		1.10	21.818		2,300	870		3,170	3,825	
	3060		1041 L coated stl. tank, 1.514 L/s, 373 W, 51 mm discharge		1.50	16		1,900	640		2,540	3,025	
	3080		76 mm discharge		1.40	17.143		1,925	685		2,610	3,150	
	3100		10.979 L/s, 522 W, 51 mm discharge		1.40	17.143		2,450	685		3,135	3,725	
	3120		76 mm discharge		1.30	18.462		2,475	735		3,210	3,825	
	3140		16.911 L/s, 746 W, 51 mm discharge		1.10	21.818		2,675	870		3,545	4,250	
	3160		76 mm discharge	▼	.90	26.667	▼	2,825	1,075		3,900	4,700	
	3260		Pump system accessories, add										
	3300		Alarm horn and lights, 115V mercury switch	Q-1	8	2	Ea.	86.50	77		163.50	211	
	3340		Switch, mag. contactor, alarm bell, light, 3 level control		5	3.200		395	123		518	620	
	3380		Alternator, mercury switch activated	▼	4	4	▼	750	154		904	1,050	
900	0010		**PUMPS, PEDESTAL SUMP** With float control										900
	0400		Molded PVC base, 1.325 L/s at 4572 mm head, 249 W	1 Plum	5	1.600	Ea.	88.50	68.50		157	201	
	0800		Iron base, 1.325 L/s at 4572 mm head, 249 W		5	1.600		113	68.50		181.50	228	
	1200		Solid brass, 1.325 L/s at 4572 mm head, 249 W	▼	5	1.600	▼	185	68.50		253.50	305	
940	0010		**PUMPS, SUBMERSIBLE** Sump										940
	7000		Sump pump, automatic										
	7100		Plastic, 32 mm discharge, 186 W	1 Plum	6	1.333	Ea.	114	57		171	211	
	7140		249 W		5	1.600		138	68.50		206.50	254	
	7160		373 W		5	1.600		168	68.50		236.50	288	
	7180		38mm discharge, 373 W		4	2		188	85.50		273.50	335	
	7500		Cast iron, 32 mm discharge, 186 W		6	1.333		131	57		188	230	
	7540		249 W		6	1.333		155	57		212	256	
	7560		373 W	▼	5	1.600	▼	187	68.50		255.50	310	

		15460	Domestic Water Cond Equipment										
900	0010		**WATER SOFTENER**										900
	5800		Softener systems, automatic, intermediate sizes										
	5820		available, may be used in multiples.										
	6000		Hardness capacity between regenerations and flow										
	6100		150 000 grains 2.3 L/s cont., 3.2 L/s peak	Q-1	1.20	13.333	Ea.	4,250	510		4,760	5,450	
	6200		300 000 grains, 5.1 L/s cont., 7.1 L/s peak		1	16		6,300	615		6,915	7,850	
	6300		750 000 grains, 10.1 L/s cont., 14.5 L/s peak		.80	20		8,200	770		8,970	10,200	
	6400		900 000 grains, 11.7 L/s cont., 17.0 L/s peak	▼	.70	22.857	▼	13,200	880		14,080	15,800	

		15480	Domestic Water Heaters										
200	0010		**WATER HEATERS**										200
	0020		For solar, see division 13600										
	1000		Residential, electric, glass lined tank, 5 yr, 38 L, single element	1 Plum	2.30	3.478	Ea.	227	149		376	470	
	1040		76 L, single element		2.20	3.636		259	155		414	520	
	1060		114 L, double element		2.20	3.636		290	155		445	555	
	1080		151 L, double element		2	4		310	171		481	600	
	1100		197 L, double element		2	4		370	171		541	660	
	1180		454 L, double element	▼	1.40	5.714	▼	800	244		1,044	1,250	

MECHANICAL 15

15480 | Domestic Water Heaters

		CREW	DAILY OUTPUT	LABOR-HOURS	UNIT	MAT.	LABOR	EQUIP.	TOTAL	TOTAL INCL O&P	
200	**2000** Gas fired, foam lined tank, 10 yr, vent not incl.,										**200**
2040	114 L	1 Plum	2	4	Ea.	630	171		801	945	
2100	284 L		1.50	5.333		820	228		1,048	1,250	
2120	379 L		1.30	6.154		1,250	263		1,513	1,775	
3000	Oil fired, glass lined tank, 5 yr, vent not included, 114 L		2	4		795	171		966	1,125	
3040	189 L		1.80	4.444		1,375	190		1,565	1,775	
3060	265 L		1.50	5.333		1,525	228		1,753	2,025	
3080	322 L		1.40	5.714		4,750	244		4,994	5,600	
4000	Commercial, 55° rise. NOTE: for each size tank, a range of										
4010	heaters between the ones shown are available										
4020	Electric										
4100	19 L, 3 kW, 45.4 L/hr, 208V	1 Plum	2	4	Ea.	1,800	171		1,971	2,225	
4120	38 L, 6 kW, 94.6 L/hr, 208V		2	4		1,975	171		2,146	2,425	
4140	189 L, 9 kW, 140 L/hr, 208V		1.80	4.444		2,725	190		2,915	3,275	
4160	189 L, 36 kW, 560 L/hr, 208V		1.80	4.444		4,250	190		4,440	4,950	
4300	757 L, 15 kW, 231 L/hr, 480V	Q-1	1.70	9.412		13,100	360		13,460	14,900	
4320	757 L, 120 kW , 1855 L/hr, 480V		1.70	9.412		17,900	360		18,260	20,200	
4460	1514 L, 30 kW, 466 L/hr, 480V		1	16		18,000	615		18,615	20,700	
5400	Modulating step control, 2-5 steps	1 Elec	5.30	1.509		1,675	63.50		1,738.50	1,950	
5440	6-10 steps		3.20	2.500		2,175	105		2,280	2,550	
5460	11-15 steps		2.70	2.963		2,600	124		2,724	3,025	
5480	16-20 steps		1.60	5		3,050	210		3,260	3,700	
6000	Gas fired, flush jacket, std. controls, vent not incl.										
6040	22 kW input, 275 L/hr	1 Plum	1.40	5.714	Ea.	1,575	244		1,819	2,100	
6060	29 kW input, 360 L/hr		1.40	5.714		2,775	244		3,019	3,425	
6080	35 kW input, 465 L/hr		1.20	6.667		3,000	285		3,285	3,725	
6180	59 kW input, 727 L/hr		.60	13.333		4,300	570		4,870	5,600	
6200	70 kW input, 871 L/hr		.50	16		4,675	685		5,360	6,175	
6900	For low water cutoff, add		8	1		242	42.50		284.50	330	
6960	For bronze body hot water circulator, add		4	2		1,150	85.50		1,235.50	1,400	
8000	Oil fired, glass lined, UL listed, std. controls, vent not incl.										
8060	41 kW input, .141 L/s	Q-1	2.13	7.512	Ea.	1,150	289		1,439	1,675	
8080	58 kW input, .20 L/s		2	8		11,800	305		12,105	13,500	
8100	74.7 kW input, .259 L/s		1.60	10		12,100	385		12,485	13,900	
8160	158 kW input, .545 L/s		.96	16.667		16,100	640		16,740	18,700	
8180	211 kW input, .726 L/s		.92	17.391		16,400	670		17,070	19,000	
8280	366 kW input, 1.26 L/s	Q-2	1.22	19.672		25,100	785		25,885	28,900	
8300	440 kW input, 1.51 L/s	"	1.16	20.690		27,300	825		28,125	31,400	
8900	For low water cutoff, add	1 Plum	8	1		242	42.50		284.50	330	
8960	For bronze body hot water circulator, add	"	4	2		905	85.50		990.50	1,125	
900	**0010 HEAT TRANSFER PACKAGES** Complete, controls,										**900**
0020	expansion tank, converter, air separator										
1000	Hot water, 355 K enter, 366 K leaving, 103 kPa steam										
1010	One pump system, 1.767 L/s	Q-6	.75	32	Ea.	13,400	1,275		14,675	16,600	
1020	2.209 L/s		.70	34.286		14,700	1,375		16,075	18,300	
1040	3.471 L/s		.65	36.923		16,700	1,475		18,175	20,600	
1060	8.203 L/s		.55	43.636		20,700	1,750		22,450	25,500	
1080	16.091 L/s		.40	60		28,200	2,400		30,600	34,600	
1100	34.705 L/s		.30	80		35,600	3,225		38,825	44,000	

15490 | Pool and Fountain Equipment

		CREW	DAILY OUTPUT	LABOR-HOURS	UNIT	MAT.	LABOR	EQUIP.	TOTAL	TOTAL INCL O&P	
100	**0010 FOUNTAINS/AERATORS**										**100**
0100	Pump w/controls										
0200	Single phase, 30 m cord, 373 W pump	2 Skwk	4.40	3.636	Ea.	2,900	133		3,033	3,400	
0300	559 W pump		4.30	3.721		3,300	136		3,436	3,850	

15 MECHANICAL

15400 | Plumbing Fixtures & Equipment

		15490	Pool and Fountain Equipment	CREW	DAILY OUTPUT	LABOR-HOURS	UNIT	2006 BARE COSTS				TOTAL INCL O&P	
								MAT.	LABOR	EQUIP.	TOTAL		
100	0400		746 W pump	2 Skwk	4.20	3.810	Ea.	3,350	139		3,489	3,900	**100**
	0500		1100 W pump		4.10	3.902		3,450	142		3,592	4,025	
	0600		1491 W pump		4	4		3,500	146		3,646	4,075	
	0700		Three phase, 60 m cord, 3.7 kW pump		3.90	4.103		8,500	150		8,650	9,575	
	0800		5.6 kW pump		3.80	4.211		9,575	154		9,729	10,700	
	0900		7.5 kW pump		3.70	4.324		10,700	158		10,858	12,000	
	1000		11.2 kW pump		3.60	4.444		12,700	162		12,862	14,200	
	1100		Nozzles, minimum		8	2		212	73		285	350	
	1200		Maximum		8	2		279	73		352	420	
	1300		Lights w/mounting kits, 200 W		18	.889		350	32.50		382.50	435	
	1400		300 W		18	.889		390	32.50		422.50	475	
	1500		500 W		18	.889		425	32.50		457.50	520	
	1600		Color blender	▼	12	1.333	▼	330	48.50		378.50	440	

15500 | Heat Generation Equipment

		15510	Heating Boilers and Accessories	CREW	DAILY OUTPUT	LABOR-HOURS	UNIT	2006 BARE COSTS				TOTAL INCL O&P	
								MAT.	LABOR	EQUIP.	TOTAL		
300	0010		**BOILERS, ELECTRIC, ASME** Standard controls and trim										**300**
	1000		Steam, 6 kW	Q-19	1.20	20	Ea.	2,900	795		3,695	4,400	
	1160		60 kW		1	24		4,925	955		5,880	6,825	
	1220		112 kW		.75	32		6,725	1,275		8,000	9,300	
	1280		222 kW	▼	.55	43.636		11,100	1,750		12,850	14,900	
	1380		518 kW	Q-21	.36	88.889		18,800	3,600		22,400	26,100	
	1480		1080 kW		.25	128		26,300	5,200		31,500	36,800	
	1600		2340 kW	▼	.16	200		54,000	8,125		62,125	71,500	
	2000		Hot water, 7.5 kW	Q-19	1.30	18.462		3,050	735		3,785	4,475	
	2100		90 kW		1.10	21.818		4,475	870		5,345	6,225	
	2220		296 kW	▼	.55	43.636		9,850	1,750		11,600	13,400	
	2500		1036 kW	Q-21	.34	94.118		22,900	3,825		26,725	31,000	
	2680		2401 kW		.25	128		47,700	5,200		52,900	60,500	
	2820		3600 kW	▼	.16	200	▼	67,000	8,125		75,125	86,000	
400	0010		**BOILERS, GAS FIRED** Natural or propane, standard controls										**400**
	1000		Cast iron, with insulated jacket										
	2000		Steam, gross output, 24 kW	Q-7	1.40	22.857	Ea.	1,575	940		2,515	3,125	
	2080		59 kW		.90	35.556		2,550	1,450		4,000	5,000	
	2180		117 kW		.56	56.838		3,925	2,325		6,250	7,825	
	2240		224 kW		.43	74.419		8,525	3,050		11,575	14,000	
	2320		550 kW		.30	106		12,900	4,375		17,275	20,800	
	2440		1383 kW		.15	207		24,700	8,525		33,225	40,000	
	2480		1788 kW		.13	246		61,000	10,100		71,100	82,000	
	2540		2043 kW		.10	320		68,500	13,100		81,600	95,500	
	3000		Hot water, gross output, 23 kW		1.46	21.918		1,500	900		2,400	3,000	
	3140		94 kW		.80	40		3,050	1,650		4,700	5,825	
	3260		319 kW		.40	80		10,300	3,275		13,575	16,300	
	3380		957 kW		.18	179		19,400	7,375		26,775	32,400	
	3480		1788 kW		.13	250		61,500	10,300		71,800	83,000	
	3540		2 043 kW	▼	.09	359	▼	69,000	14,800		83,800	98,000	
	7000		For tankless water heater, add					10%					

MECHANICAL 15

15510	Heating Boilers and Accessories	CREW	DAILY OUTPUT	LABOR-HOURS	UNIT	MAT.	LABOR	EQUIP.	TOTAL	TOTAL INCL O&P	
						2006 BARE COSTS					
460	0010	**BOILERS, GAS/OIL** Combination with burners and controls										**460**
	1000	Cast iron with insulated jacket									
	2000	Steam, gross output, 211 kW	Q-7	.43	74.074	Ea.	7,600	3,050		10,650	12,900
	2080	469 kW		.30	107		12,500	4,400		16,900	20,300
	2140	791 kW		.19	165		17,800	6,800		24,600	29,700
	2280	1618 kW		.14	235		67,500	9,650		77,150	88,500
	2340	1873 kW		.11	296		72,000	12,200		84,200	98,000
	2380	2043 kW	↓	.09	372	↓	76,500	15,300		91,800	107,000
	2900	Hot water, gross output									
	2910	59 kW	Q-6	.62	39.024	Ea.	6,150	1,575		7,725	9,125
	2920	88 kW		.49	49.080		6,150	1,975		8,125	9,750
	2930	117 kW		.41	57.971		7,200	2,325		9,525	11,400
	2940	147 kW	↓	.36	67.039		7,775	2,700		10,475	12,600
	3000	171 kW	Q-7	.44	72.072		10,400	2,950		13,350	15,900
	3060	428 kW		.28	113		15,300	4,650		19,950	23,800
	3160	1198 kW		.16	195		36,000	8,000		44,000	51,500
	3300	3956 kW	↓	.04	727	↓	111,500	29,800		141,300	167,500
500	0010	**BOILERS, OIL FIRED** Standard controls, flame retention burner										**500**
	1000	Cast iron, with insulated flush jacket									
	2000	Steam, gross output, 32 kW	Q-7	1.20	26.667	Ea.	2,250	1,100		3,350	4,125
	2060	61 kW		.90	35.556		3,050	1,450		4,500	5,575
	2180	318 kW		.38	85.106		7,725	3,500		11,225	13,800
	2280	879 kW		.19	170		17,300	6,975		24,275	29,500
	2380	1618 kW		.14	235		57,000	9,650		66,650	77,500
	2460	2043 kW	↓	.09	363	↓	73,000	14,900		87,900	103,000
	3000	Hot water, same price as steam									
	4000	For tankless coil in smaller sizes, add					15%				
	5000	Steel, insulated jacket, burner									
	6000	Steam, full water leg construction, gross output									
	6020	42 kW	Q-6	1.33	18.045	Ea.	4,250	725		4,975	5,775
	6120	137 kW		.70	34.483		7,825	1,375		9,200	10,700
	6180	295 kW		.39	61.381		12,100	2,475		14,575	17,100
	6280	703 kW	↓	.19	125	↓	23,200	5,025		28,225	33,100
	6400	Larger sizes are same as steel, gas fired									
	7000	Hot water, gross output, 30 kW	Q-6	1.60	15	Ea.	1,400	605		2,005	2,425
	7120	123 kW		.70	34.483		5,225	1,375		6,600	7,800
	7200	246 kW		.44	55.172		10,600	2,225		12,825	14,900
	7260	492 kW		.29	83.624		14,500	3,350		17,850	21,100
	7320	923 kW	↓	.13	184	↓	21,200	7,425		28,625	34,500
	7340	For tankless coil in steam or hot water, add					7%				
880	0010	**SWIMMING POOL HEATERS** Not including wiring, external										**880**
	0020	piping, base or pad,									
	0160	Gas fired, input, 44 kW	Q-6	1.50	16	Ea.	1,925	645		2,570	3,100
	0200	58 kW		1	24		2,050	965		3,015	3,725
	0280	147 kW		.40	60		8,325	2,400		10,725	12,800
	0400	469 kW	↓	.14	171		18,100	6,900		25,000	30,300
	2000	Electric, 12 kW, 18 168 L pool	Q-19	3	8		1,925	320		2,245	2,600
	2020	15 kW, 27 252 L pool		2.80	8.571		1,950	340		2,290	2,650
	2040	24 kW, 36 336 L pool		2.40	10		2,625	400		3,025	3,500
	2100	55 kW, 90 840 L pool	↓	1.20	20	↓	3,725	795		4,520	5,300

15530 | Furnaces

			CREW	DAILY OUTPUT	LABOR-HOURS	UNIT	MAT.	LABOR	EQUIP.	TOTAL	TOTAL INCL O&P	
400	0010	**FURNACES** Hot air heating, blowers, standard controls										**400**
	0020	not including gas, oil or flue piping									

15530 | Furnaces

			CREW	DAILY OUTPUT	LABOR-HOURS	UNIT	MAT.	LABOR	EQUIP.	TOTAL	TOTAL INCL O&P	
							2006 BARE COSTS					
400	1000	Electric, UL listed, heat staging, 240 V										**400**
	1020	3 kW	Q-20	5	4	Ea.	325	155		480	595	
	1040	5 kW		4.80	4.167		340	161		501	620	
	1060	8 kW		4.60	4.348		410	168		578	715	
	1100	10 kW		4.40	4.545		425	176		601	735	
	3000	Gas, AGA certified, upflow, direct drive models										
	3020	13 kW input	Q-9	4	4	Ea.	455	152		607	735	
	3040	18 kW input		3.80	4.211		605	160		765	910	
	3060	22 kW input		3.60	4.444		640	169		809	965	
	3100	29 kW input		3.20	5		680	190		870	1,025	
	6000	Oil, UL listed, atomizing gun type burner										
	6020	16 kW output	Q-9	3.60	4.444	Ea.	745	169		914	1,075	
	6030	25 kW output		3.50	4.571		775	173		948	1,125	
	6040	28 kW output		3.40	4.706		790	178		968	1,150	
	6060	39 kW output		3.20	5		1,100	190		1,290	1,500	
	6080	45 kW output		3	5.333		1,225	202		1,427	1,625	

15540 | Fuel-Fired Heaters

			CREW	DAILY OUTPUT	LABOR-HOURS	UNIT	MAT.	LABOR	EQUIP.	TOTAL	TOTAL INCL O&P	
300	0010	**DUCT FURNACES** Includes burner, controls, stainless steel										**300**
	0020	heat exchanger. Gas fired, electric ignition										
	0030	Indoor installation										
	0100	35 kW output	Q-5	4	4	Ea.	1,525	155		1,680	1,900	
	0130	59 kW output		2.70	5.926		2,050	230		2,280	2,600	
	0140	70 kW output		2.30	6.957		2,150	270		2,420	2,750	
	0180	94 kW output		1.60	10		2,575	390		2,965	3,400	
	0300	For powered venter and adapter, add					355			355	390	
	0500	For required flue pipe, see 15550										
	1000	Outdoor installation, with vent cap										
	1020	22 kW output	Q-5	4	4	Ea.	2,000	155		2,155	2,425	
	1060	35 kW output		4	4		2,550	155		2,705	3,050	
	1100	55 kW output		3	5.333		3,650	207		3,857	4,325	
	1140	88 kW output		1.80	8.889		4,750	345		5,095	5,750	
	1180	132 kW output		1.40	11.429		5,275	445		5,720	6,475	
	1400	For powered venter, add					10%					
900	0010	**SPACE HEATERS** Cabinet, grilles, fan, controls, burner,										**900**
	0020	thermostat, no piping. For flue see 15550										
	1000	Gas fired, floor mounted										
	1100	18 kW output	Q-5	10	1.600	Ea.	620	62		682	780	
	1140	29 kW output		8	2		680	77.50		757.50	860	
	1180	53 kW output		6	2.667		925	103		1,028	1,175	
	2000	Suspension mounted, propeller fan, 6 kW output		8.50	1.882		480	73		553	635	
	2040	18 kW output		7	2.286		590	88.50		678.50	785	
	2060	23 kW output		6	2.667		675	103		778	895	
	2100	38 kW output		5	3.200		875	124		999	1,150	
	2240	94 kW output		2	8		1,800	310		2,110	2,450	
	2500	For powered venter and adapter, add					305			305	335	
	5000	Wall furnace, 5 kW output	Q-5	6	2.667		545	103		648	755	
	5020	7 kW output		5	3.200		575	124		699	815	
	5040	10 kW output		4	4		770	155		925	1,075	
	6000	Oil fired, suspension mounted, 28 kW output		4	4		2,325	155		2,480	2,800	
	6040	41 kW output		3	5.333		2,500	207		2,707	3,050	
	6060	54 kW output		3	5.333		2,675	207		2,882	3,250	

15550 | Breechings, Chimneys & Stacks

			CREW	DAILY OUTPUT	LABOR-HOURS	UNIT	MAT.	LABOR	EQUIP.	TOTAL	TOTAL INCL O&P	
440	0010	**VENT CHIMNEY** Prefab metal, U.L. listed										**440**
	0020	Gas, double wall, galvanized steel										

MECHANICAL 15

15550	Breechings, Chimneys & Stacks	CREW	DAILY OUTPUT	LABOR-HOURS	UNIT	2006 BARE COSTS				TOTAL INCL O&P
						MAT.	LABOR	EQUIP.	TOTAL	
0080	80 mm diameter [3"]	Q-9	21.95	.729	m	13	27.50		40.50	57
0100	100 mm diameter [4"]		20.73	.772		16.25	29.50		45.75	63
0120	125 mm diameter [5"]		19.51	.820		18.85	31		49.85	69
0140	150 mm diameter [6"]		18.29	.875		22	33		55	75
0160	175 mm diameter [7"]		17.07	.937		32.50	35.50		68	90.50
0180	200 mm diameter [8"]		15.85	1.009		36	38.50		74.50	98.50
0200	250 mm diameter [10"]		14.63	1.094		76	41.50		117.50	148
0220	300 mm diameter [12"]		13.41	1.193		101	45.50		146.50	181
0260	400 mm diameter [16"]		12.19	1.312		231	50		281	330
0300	508 mm diameter [20"]	Q-10	10.97	2.187		350	86		436	520
0480	965 mm diameter [38"]	"	7.32	3.281		1,125	129		1,254	1,425
7800	All fuel, double wall, stainless steel, 150 mm diameter [6"]	Q-9	18.29	.875		101	33		134	162
7802	175 mm diameter [7"]		17.07	.937		130	35.50		165.50	198
7804	200 mm diameter [8"]		15.85	1.009		152	38.50		190.50	226
7806	250 mm diameter [10"]		14.63	1.094		221	41.50		262.50	310
7808	300 mm diameter [12"]		13.41	1.193		297	45.50		342.50	395
7810	350 mm diameter [14"]		12.80	1.250		390	47.50		437.50	505
9000	High temp. (1366 K), steel jacket, acid resist. refractory lining									
9010	11 ga. galvanized jacket, U.L. listed									
9020	Straight section, 1219 mm long, 254 mm diameter [48",10"]	Q-10	13.30	1.805	Ea.	365	71		436	515
9040	457 mm diameter [18"]		7.40	3.243		560	128		688	810
9070	914 mm diameter [36"]		2.70	8.889		1,200	350		1,550	1,875
9090	1219 mm diameter [48"]	Q-11	2.70	11.852		1,750	475		2,225	2,650
9120	Tee section, 254 mm diameter [10"]	Q-10	4.40	5.455		775	215		990	1,175
9140	457 mm diameter [18"]		2.40	10		1,100	395		1,495	1,825
9170	914 mm diameter [36"]		.80	30		3,625	1,175		4,800	5,825
9190	1219 mm diameter [48"]	Q-11	.90	35.556		6,175	1,425		7,600	9,000
9220	Cleanout pier section, 254 mm diameter [10"]	Q-10	3.50	6.857		600	270		870	1,075
9240	457 mm diameter [18"]		1.90	12.632		805	495		1,300	1,650
9270	914 mm diameter [36"]		.75	32		1,475	1,250		2,725	3,575
9290	1219 mm diameter [48"]	Q-11	.75	42.667		2,250	1,725		3,975	5,125
9320	For drain, add					53%				
9330	Elbow, 30° and 45°, 254 mm diameter [10"]	Q-10	6.60	3.636		510	143		653	780
9350	457 mm diameter [18"]		3.70	6.486		820	255		1,075	1,300
9380	914 mm diameter [36"]		1.30	18.462		2,025	725		2,750	3,375
9400	1219 mm diameter [48"]	Q-11	1.35	23.704		3,550	950		4,500	5,375
9430	For 60° and 90° elbow, add					112%				
9440	End cap, 254 mm diameter [10"]	Q-10	26	.923		870	36.50		906.50	1,000
9460	457 mm diameter [18"]		15	1.600		1,225	63		1,288	1,450
9490	914 mm diameter [36"]		5	4.800		2,150	189		2,339	2,675
9510	1219 mm diameter [48"]	Q-11	5.30	6.038		3,150	242		3,392	3,850
9540	Increaser (1 diameter), 254 mm diameter [10"]	Q-10	6.60	3.636		430	143		573	695
9560	457 mm diameter [18"]		3.70	6.486		675	255		930	1,125
9590	914 mm diameter [36"]		1.30	18.462		1,325	725		2,050	2,575
9592	1067 mm diameter (42")	Q-11	1.60	20		1,450	805		2,255	2,850
9594	1219 mm diameter (48")		1.35	23.704		1,750	950		2,700	3,400
9596	1372 mm diameter (54")		1.10	29.091		2,125	1,175		3,300	4,125
9600	For expansion joints, add to straight section					7%				
9610	For 6 mm hot rolled steel jacket, add					157%				
9620	For 1894 K very high temperature, add					89%				
9630	26 ga. aluminized jacket, straight section, 1219 mm long									
9640	254 mm diameter [10"]	Q-10	4.66	5.146	m	710	202		912	1,100
9660	457 mm diameter [18"]		2.59	9.264		1,100	365		1,465	1,750
9690	914 mm diameter [36"]		.94	25.400		2,525	1,000		3,525	4,325
9810	Draw band, galv. stl., 11 gauge, 254 mm diameter [10"]		32	.750	Ea.	58.50	29.50		88	110
9820	305 mm diameter [12"]		30	.800		63	31.50		94.50	118

15500 | Heat Generation Equipment

15550 | Breechings, Chimneys & Stacks

			CREW	DAILY OUTPUT	LABOR-HOURS	UNIT	2006 BARE COSTS				TOTAL INCL O&P	
							MAT.	LABOR	EQUIP.	TOTAL		
440	9830	457 mm diameter [18"]	Q-10	26	.923	Ea.	78.50	36.50		115	143	**440**
	9860	914 mm diameter [36"]	↓	18	1.333		133	52.50		185.50	227	
	9880	1219 mm diameter [48"]	Q-11	20	1.600	↓	228	64.50		292.50	350	
600	0010	**INDUCED DRAFT FANS**										**600**
	1000	Breeching installation										
	1800	Hot gas, 588K, variable pitch pulley and motor										
	1860	203 mm diam. inlet, 186 W, 1phase, 0.529 m³/s [8"]	Q-9	4	4	Ea.	1,600	152		1,752	1,975	
	1900	305 mm diam. inlet, 559 W, 3 phase, 1.397 m³/s [12"]		3	5.333		2,175	202		2,377	2,700	
	1940	457 mm diam. inlet, 2237 W, 3 phase, 4.305 m³/s [18"]		2	8		3,375	305		3,680	4,200	
	1980	610 mm diam. inlet, 5593 W, 3 phase, 8.383 m³/s [24"]	↓	.80	20	↓	5,300	760		6,060	7,025	
	2300	For multi-blade damper at fan inlet, add					20%					

15600 | Refrigeration Equipment

15620 | Packaged Water Chillers

			CREW	DAILY OUTPUT	LABOR-HOURS	UNIT	2006 BARE COSTS				TOTAL INCL O&P	
							MAT.	LABOR	EQUIP.	TOTAL		
100	0010	**ABSORPTION WATER CHILLERS**										**100**
	0020	Steam or hot water, water cooled										
	0050	360 kW	Q-7	.13	240	Ea.	122,500	9,850		132,350	150,000	
	0400	1477 kW	"	.10	323	"	212,000	13,300		225,300	253,500	
600	0010	**CENTRIFUGAL/SCREW/RECIP. WATER CHILLERS,** W/ standard controls										**600**
	0020	Centrifugal liquid chiller, water cooled										
	0030	not including water tower										
	0100	Open drive, 7034 kW	Q-7	.07	477	Ea.	617,500	19,600		637,100	708,500	
	0110	Screw, liquid chiller, air cooled, insulated evaporator										
	0124	563 kW	Q-7	.13	246	Ea.	78,000	10,100		88,100	100,500	
	0128	633 kW		.13	250		87,500	10,300		97,800	112,000	
	0132	739 kW		.12	258		96,500	10,600		107,100	122,000	
	0136	950 kW		.12	266		110,500	10,900		121,400	138,000	
	0140	1125 kW		.12	275		138,500	11,300		149,800	169,500	
	0490	Packaged w/integral air cooled condenser, 53 kW cool		.37	86.486		15,300	3,550		18,850	22,200	
	0500	70 kW cooling		.34	94.955		20,300	3,900		24,200	28,200	
	0520	140 kW cooling		.30	108		27,500	4,425		31,925	36,900	
	0600	352 kW cooling	↓	.25	129	↓	59,000	5,300		64,300	73,000	
	0680	Scroll water cooled, single compressor, hermetic, tower not incl.										
	0700	7 to 17 kW cooling	Q-5	.57	28.021	Ea.	3,175	1,075		4,250	5,100	
	0740	28 kW cooling	"	.31	52.117		4,675	2,025		6,700	8,175	
	0760	35 kW cooling	Q-6	.36	67.039		6,300	2,700		9,000	11,000	
	0800	70 kW cooling	Q-7	.38	83.990		9,275	3,450		12,725	15,400	
	0820	105 kW cooling	"	.33	96.096	↓	10,700	3,950		14,650	17,600	
	0980	Water cooled, multiple compr., semi-hermetic, tower not incl.										
	1000	53 kW cooling	Q-6	.36	65.934	Ea.	12,600	2,650		15,250	17,900	
	1020	70 kW cooling	Q-7	.41	78.049		13,800	3,200		17,000	19,900	
	1060	105 kW cooling		.31	101		16,300	4,175		20,475	24,300	
	1100	176 kW cooling		.28	113		29,200	4,675		33,875	39,200	
	1160	352 kW cooling		.18	179		52,000	7,375		59,375	68,500	
	1180	422 kW cooling		.16	196		59,000	8,050		67,050	77,000	
	1200	492 kW cooling	↓	.16	202	↓	69,500	8,300		77,800	89,000	
	1450	Water cooled, dual compressors, semi-hermetic, tower not incl.										
	1500	281 kW cooling	Q-7	.14	222	Ea.	25,600	9,125		34,725	41,900	

MECHANICAL 15

15620 | Packaged Water Chillers

		CREW	DAILY OUTPUT	LABOR-HOURS	UNIT	2006 BARE COSTS				TOTAL INCL O&P
						MAT.	LABOR	EQUIP.	TOTAL	
600	**1520** 352 kW cooling	Q-7	.14	228	Ea.	32,000	9,375		41,375	49,300	**600**
	1540 422 kW cooling		.14	231		36,100	9,525		45,625	54,000	
	1580 528 kW cooling, screw compressors		.13	240		55,500	9,875		65,375	76,000	
	1620 703 kW cooling, screw compressors		.13	250		67,500	10,300		77,800	90,000	
	1660 1024 kW cooling, screw compressors		.12	260		76,500	10,700		87,200	100,000	

15640 | Packaged Cooling Towers

		CREW	DAILY OUTPUT	LABOR-HOURS	UNIT	MAT.	LABOR	EQUIP.	TOTAL	TOTAL INCL O&P	
400	**0010** **COOLING TOWERS** Packaged units										**400**
	0070 Galvanized steel										
	0080 Induced draft, crossflow										
	0100 Vertical, belt drive, 215 kW	Q-6	317	.076	kW	24.50	3.04		27.54	31.50	
	0150 352 kW		352	.068		19.60	2.74		22.34	25.50	
	0200 405 kW		383	.063		19.05	2.52		21.57	25	
	0250 461kW		422	.057		16.90	2.29		19.19	22	
	0260 570 kW		464	.052		14.20	2.08		16.28	18.75	
	1500 Induced air, double flow										
	1900 Vertical, gear drive, 588 kW	Q-6	443	.054	kW	26.50	2.18		28.68	32.50	
	2000 1045 kW		454	.053		18.80	2.12		20.92	23.50	
	2100 2050 kW		464	.052		15.55	2.08		17.63	20	
	2150 2990 kW		499	.048		15.35	1.93		17.28	19.80	
	2200 3575 kW		528	.045		14.65	1.83		16.48	18.90	
	3000 For higher capacities, use multiples										
	3500 For pumps and piping, add	Q-6	134	.179	kW	12.35	7.20		19.55	24.50	
	4000 For absorption systems, add				"	75%	75%				
	5000 Fiberglass										
	5010 Draw thru										
	5100 211 kW	Q-6	1.50	16	Ea.	3,275	645		3,920	4,575	
	5120 440 kW		.99	24.242		6,700	975		7,675	8,850	
	5140 1055 kW		.43	55.814		15,700	2,250		17,950	20,700	
	5160 2110 kW		.22	109		28,200	4,375		32,575	37,600	
	5180 3517 kW		.15	160		48,400	6,425		54,825	62,500	
	6000 Stainless steel										
	6010 Induced draft, crossflow, horizontal, belt drive										
	6100 200 kW	Q-6	1.50	16	Ea.	8,775	645		9,420	10,600	
	6120 320 kW		.99	24.242		12,600	975		13,575	15,400	
	6140 390 kW		.43	55.814		14,900	2,250		17,150	19,800	
	6160 445 kW		.22	109		16,100	4,375		20,475	24,300	

15660 | Liquid Coolers/Evap Condensers

		CREW	DAILY OUTPUT	LABOR-HOURS	UNIT	MAT.	LABOR	EQUIP.	TOTAL	TOTAL INCL O&P	
100	**0010** **CONDENSERS** Ratings are for 17° K TD, R-22										**100**
	0080 Air cooled, belt drive, propeller fan										
	0240 169 kW	Q-6	.69	34.985	Ea.	10,500	1,400		11,900	13,700	
	0280 211 kW		.58	41.308		11,700	1,650		13,350	15,400	
	0320 253 kW		.47	51.173		13,800	2,050		15,850	18,300	
	0360 288 kW		.40	60.302		16,400	2,425		18,825	21,700	
	0380 317 kW		.39	61.697		17,500	2,475		19,975	23,000	
	1550 Air cooled, direct drive, propeller fan										
	1590 3.5 kW	Q-5	3.80	4.211	Ea.	560	163		723	860	
	1600 5.3 kW		3.60	4.444		665	172		837	990	
	1620 7 kW		3.20	5		735	194		929	1,100	
	1640 18 kW		2	8		1,500	310		1,810	2,125	
	1660 35 kW		1.40	11.429		2,300	445		2,745	3,200	
	1690 56 kW		1.10	14.545		3,250	565		3,815	4,425	
	1720 91 kW		.84	19.002		4,200	735		4,935	5,725	
	1760 144 kW	Q-6	.77	31.008		7,700	1,250		8,950	10,300	

15600 | Refrigeration Equipment

		15660	Liquid Coolers/Evap Condensers	CREW	DAILY OUTPUT	LABOR-HOURS	UNIT	2006 BARE COSTS				TOTAL INCL O&P	
								MAT.	LABOR	EQUIP.	TOTAL		
100	1800		222 kW	Q-6	.55	44.037	Ea.	12,100	1,775		13,875	16,100	100

15700 | Heating/Ventilating/Air Conditioning Equipment

		15710	Heat Exchangers	CREW	DAILY OUTPUT	LABOR-HOURS	UNIT	2006 BARE COSTS				TOTAL INCL O&P	
								MAT.	LABOR	EQUIP.	TOTAL		
900	0010	**HEAT EXCHANGERS**											900
	0016	Shell & tube type, 2 or 4 pass, 20 mm O.D. copper tubes,											
	0020	C.I. heads, C.I. tube sheet, steel shell											
	0100	Hot water 277 K to 355 K, by steam at 69 kPa											
	0120	0.5 L/s	Q-5	6	2.667	Ea.	1,175	103		1,278	1,425	
	0140	0.6 L/s		5	3.200		1,750	124		1,874	2,100	
	0160	2.5 L/s		4	4		2,725	155		2,880	3,225	
	0180	4.0 L/s		2	8		4,175	310		4,485	5,075	
	0200	6.0 L/s		1	16		5,575	620		6,195	7,075	
	0220	7.6 L/s	Q-6	1.50	16		7,325	645		7,970	9,025	
	1000	Hot water 277 K to 333 K, by water at 365 K											
	1020	0.44 L/s	Q-5	6	2.667	Ea.	1,425	103		1,528	1,725	
	1040	1.0 L/s		5	3.200		2,025	124		2,149	2,400	
	1060	2.1 L/s		4	4		3,075	155		3,230	3,600	
	1100	4.7 L/s		1.50	10.667		5,575	415		5,990	6,750	
	3000	Plate type,											
	3100	25 200 mL/S	Q-6	.80	30	Ea.	25,400	1,200		26,600	29,700	
	3120	50 400 mL/S	"	.50	48		43,700	1,925		45,625	51,000	
	3140	75 600 mL/S	Q-7	.34	94.118		65,000	3,850		68,850	77,500	
	3160	113 400 mL/S	"	.24	133		86,500	5,475		91,975	103,000	

		15720	Air Handling Units	CREW	DAILY OUTPUT	LABOR-HOURS	UNIT	MAT.	LABOR	EQUIP.	TOTAL	TOTAL INCL O&P	
500	0010	**MAKE-UP AIR UNIT**											500
	0020	Indoor suspension, natural/LP gas, direct fired,											
	0030	standard control. For flue see division 15550-440											
	0040	294 K temperature rise, kW is input											
	0100	0.944 m³/s, 49 kW	Q-6	3	8	Ea.	6,950	320		7,270	8,100	
	0160	2.832 m³/s, 147 kW		1.50	16		8,750	645		9,395	10,600	
	0220	5.664 m³/s, 295 kW		1	24		10,900	965		11,865	13,500	
	0300	11.328 m³/s, 588 kW	Q-7	1	32		14,200	1,325		15,525	17,600	
	0400	23.6 m³/s, 1225 kW	"	.80	40		19,500	1,650		21,150	23,900	
	0600	For discharge louver assembly, add					10%					
	0700	For filters, add					20%					
	0800	For air shut-off damper section, add					10%					
	0900	For vertical unit, add					10%					

		15730	Unitary Air Conditioning Equip	CREW	DAILY OUTPUT	LABOR-HOURS	UNIT	MAT.	LABOR	EQUIP.	TOTAL	TOTAL INCL O&P	
200	0010	**COMPUTER ROOM UNITS**											200
	1000	Air cooled, includes remote condenser but not											
	1020	interconnecting tubing or refrigerant											
	1080	10 kW	Q-5	.50	32	Ea.	9,000	1,250		10,250	11,800	
	1120	17 kW		.45	35.556		12,000	1,375		13,375	15,300	
	1160	21 kW		.30	53.333		22,200	2,075		24,275	27,500	

MECHANICAL 15

For expanded coverage of these items see *Means Mechanical or Plumbing Cost Data 2006*

		15730	Unitary Air Conditioning Equip		CREW	DAILY OUTPUT	LABOR-HOURS	UNIT	2006 BARE COSTS				TOTAL INCL O&P	
									MAT.	LABOR	EQUIP.	TOTAL		
200	1200		28 kW		Q-5	.27	59.259	Ea.	25,100	2,300		27,400	31,100	**200**
	1240		35 kW			.25	64		26,100	2,475		28,575	32,400	
	1280		54 kW			.22	72.727		29,200	2,825		32,025	36,400	
	1290		62 kW		↓	.20	80		33,200	3,100		36,300	41,200	
	1320		70 kW		Q-6	.29	82.759		34,800	3,325		38,125	43,300	
	1360		81 kW		"	.28	85.714	↓	36,600	3,450		40,050	45,400	
500	0010	**PACKAGED TERMINAL AIR CONDITIONER** Cabinet, wall sleeve,												**500**
	0100	louver, electric heat, thermostat, manual changeover, 208 V												
	0200		1.8 kW cooling, 2.6 kW heat		Q-5	6	2.667	Ea.	955	103		1,058	1,200	
	0220		2.6 kW cooling, 4 kW heat			5	3.200		1,000	124		1,124	1,275	
	0240		3.5 kW cooling, 4.1 kW heat			4	4		1,100	155		1,255	1,450	
	0260		4.4 kW cooling, 4.1 kW heat		↓	3	5.333		1,300	207		1,507	1,750	
	0500	For hot water coil, increase heat by 10%, add								5%	10%			
	1000	For steam, increase heat output by 30%, add						↓		8%	10%			
600	0010	**ROOF TOP AIR CONDITIONERS** Standard controls, curb, economizer												**600**
	1000	Single zone, electric cool, gas heat												
	1090		7 kW cooling, 16 kW heating	R236000 -20	Q-5	.93	17.204	Ea.	2,900	665		3,565	4,200	
	1100		10 kW cooling, 18 kW heating			.70	22.857		2,925	885		3,810	4,550	
	1120		14 kW cooling, 28 kW heating			.61	26.403		4,250	1,025		5,275	6,225	
	1140		17 kW cooling, 33 kW heating			.56	28.521		4,650	1,100		5,750	6,800	
	1145		21 kW cooling, 41 kW heating			.52	30.769		5,450	1,200		6,650	7,800	
	1150		26 kW cooling, 50 kW heating		↓	.50	32.258		6,800	1,250		8,050	9,350	
	1160		35 kW cooling, 59 kW heating		Q-6	.67	35.982		8,725	1,450		10,175	11,800	
	1170		43 kW cooling, 67 kW heating			.63	37.975		9,950	1,525		11,475	13,200	
	1190		62 kW cooling, 97 kW heating		↓	.52	45.889		16,300	1,850		18,150	20,700	
	1200		70 kW cooling, 106 kW heating		Q-7	.67	47.976		18,200	1,975		20,175	23,000	
	1210		89 kW cooling, 132 kW heating			.56	57.554		22,400	2,350		24,750	28,300	
	1220		105 kW cooling, 158 kW heating		↓	.47	68.376	↓	28,300	2,800		31,100	35,300	
	2000	Multizone, electric cool, gas heat, economizer												
	2100		54 kW cooling, 106 kW heating		Q-7	.61	52.545	Ea.	76,000	2,150		78,150	87,500	
	2120		70 kW cooling, 106 kW heating			.53	60.038		87,500	2,475		89,975	99,500	
	2200		140 kW cooling, 158 kW heating			.28	113		118,000	4,675		122,675	137,000	
	2210		50 ton cooling, 540 MBH heating			.23	142		129,500	5,825		135,325	151,000	
	2220		244 kW cooling, 440 kW heating			.16	198		140,000	8,150		148,150	166,500	
	2240		283 kW cooling, 440 kW heating			.14	228		160,000	9,375		169,375	190,000	
	2260		318 kW cooling, 440 kW heating			.13	256		164,000	10,500		174,500	196,000	
	2280		368 kW cooling, 440 kW heating		↓	.11	290		185,500	11,900		197,400	222,000	
	2400	For hot water heat coil, deduct								5%				
	2500	For steam heat coil, deduct								2%				
	2600	For electric heat, deduct			↓			↓		3%	5%			
840	0010	**SELF-CONTAINED SINGLE PACKAGE**												**840**
	0100	Air cooled, for free blow or duct, not incl. remote condenser												
	0200		10.5 kW cooling		Q-5	1	16	Ea.	2,975	620		3,595	4,200	
	0220		17.6 kW cooling		Q-6	1.20	20		3,525	805		4,330	5,075	
	0240		35 kW cooling		Q-7	1	32		6,025	1,325		7,350	8,600	
	0260		70 kW cooling			.90	35.556		14,100	1,450		15,550	17,700	
	0280		105 kW cooling		↓	.80	40		14,700	1,650		16,350	18,700	
	0340		210 kW cooling		Q-8	.40	80	↓	29,300	3,275	173	32,748	37,300	
	0490	For duct mounting no price change												
	0500	For steam heating coils, add						Ea.	10%	10%				
	1000	Water cooled for free blow or duct, not including tower												
	1010	Constant volume												
	1100		10.5 kW cooling		Q-6	1	24	Ea.	2,825	965		3,790	4,550	
	1120		17.6 kW cooling		"	1	24		3,650	965		4,615	5,475	
	1140		35 kW cooling		Q-7	.90	35.556		7,225	1,450		8,675	10,200	
	1160		70 kW cooling		↓	.80	40	↓	24,200	1,650		25,850	29,200	

Important: See the Reference Section for supporting data - Crews, Rental Equipment, City Cost Indexes and Reference Data

		15730	Unitary Air Conditioning Equip	CREW	DAILY OUTPUT	LABOR-HOURS	UNIT	2006 BARE COSTS				TOTAL INCL O&P	
								MAT.	LABOR	EQUIP.	TOTAL		
840	1180		105 kW cooling	Q-7	.70	45.714	Ea.	32,100	1,875		33,975	38,200	840
		15740	**Heat Pumps**										
100	0010		**AIR-SOURCE HEAT PUMPS** (Not including interconnecting tubing)										100
	1000		Air to air, split system, not including curbs, pads, or ductwork										
	1020		7 kW cooling, 2 kW heat @ 255 K	Q-5	1.20	13.333	Ea.	1,625	515		2,140	2,575	
	1060		17 kW cooling, 8 kW heat @ 255 K		.50	32		2,825	1,250		4,075	5,000	
	1080		26 kW cooling, 10 kW heat @ 255 K	↓	.30	53.333		6,250	2,075		8,325	9,975	
	1100		35 kW cooling, 15 kW heat @ 255 K	Q-6	.38	63.158		8,375	2,550		10,925	13,100	
	1120		54 kW cooling, 19 kW heat @ 255 K		.26	92.308		11,900	3,700		15,600	18,700	
	1130		70 kW cooling, 25 kW heat @ 255 K		.20	120		15,600	4,825		20,425	24,500	
	1140		88 kW cooling, 35 kW heat @ 255 K	↓	.20	120	↓	18,800	4,825		23,625	28,000	
	1300		Supplementary electric heat coil, included										
	1500		Single package, not including curbs, pads, or plenums										
	1520		7 kW cooling, 2 kW heat @ 255 K	Q-5	1.50	10.667	Ea.	2,525	415		2,940	3,400	
	1580		14 kW cooling, 4 kW heat @ 255 K		.96	16.667		3,550	645		4,195	4,875	
	1640		26 kW cooling, 10 kW heat @ 255 K	↓	.40	40	↓	5,850	1,550		7,400	8,750	
800	0010		**WATER-SOURCE HEAT PUMPS** (Not including interconnecting tubing)										800
	2000		Water source to air, single package										
	2100		3 kW cooling, 4 kW heat @ 297 K	Q-5	2	8	Ea.	1,125	310		1,435	1,725	
	2140		7 kW cooling, 6 kW heat @ 297 K		1.70	9.412		1,300	365		1,665	1,975	
	2220		17 kW cooling, 8 kW heat @ 297 K	↓	.90	17.778		2,000	690		2,690	3,200	
	3960		For supplementary heat coil, add				↓	10%					
	4000		For increase in capacity thru use										
	4020		of solar collector, size boiler at 60%										
		15750	**Humidity Control Equipment**										
500	0010		**HUMIDIFIERS**										500
	0520		Steam, room or duct, filter, regulators, auto. controls, 220 V										
	0540		5 kg/hr	Q-5	6	2.667	Ea.	2,175	103		2,278	2,550	
	0560		9 kg/hr		5	3.200		2,400	124		2,524	2,800	
	0580		15 kg/hr		4	4		2,450	155		2,605	2,925	
	0600		22.7 kg/hr		4	4		3,025	155		3,180	3,550	
	0620		45.4 kg/hr	↓	3	5.333	↓	3,600	207		3,807	4,275	
		15760	**Terminal Heating & Cooling Units**										
100	0010		**COILS, FLANGED**										100
	0500		Chilled water cooling, 6 rows, 600 mm x 1200 mm	Q-5	3.20	5	Ea.	2,875	194		3,069	3,450	
	1000		Direct expansion cooling, 6 rows, 600 mm x 1200 mm		2.80	5.714		3,125	221		3,346	3,750	
	1500		Hot water heating, 1 row, 600 mm x 1200 mm		4	4		1,150	155		1,305	1,475	
	2000		Steam heating, 1 row, 600 mm x 1200 mm	↓	3.06	5.229	↓	1,625	203		1,828	2,075	
200	0010		**DUCT HEATERS** Electric, 480 V, 3 Ph										200
	0020		Finned tubular insert, 533 K										
	0100		200 mm wide x 150 mm high, 4.0 kW	Q-20	16	1.250	Ea.	570	48.50		618.50	700	
	0120		300 mm high, 8.0 kW		15	1.333		940	51.50		991.50	1,100	
	0140		450 mm high, 12.0 kW		14	1.429		1,325	55.50		1,380.50	1,525	
	0160		600 mm high, 16.0 kW		13	1.538		1,700	59.50		1,759.50	1,975	
	0180		750 mm high, 20.0 kW		12	1.667		2,075	64.50		2,139.50	2,375	
	0300		300 mm wide x 150 mm high, 6.7 kW		15	1.333		600	51.50		651.50	745	
	0360		600 mm high, 26.7 kW		12	1.667		1,750	64.50		1,814.50	2,025	
	0700		600 mm wide x 150 mm high, 17.8 kW		13	1.538		705	59.50		764.50	865	
	0760		600 mm high, 71.1 kW	↓	10	2	↓	2,200	77.50		2,277.50	2,525	
	8000		To obtain BTU multiply kW by 3413										

For expanded coverage of these items see *Means Mechanical or Plumbing Cost Data 2006*

MECHANICAL 15

15760 | Terminal Heating & Cooling Units

		CREW	DAILY OUTPUT	LABOR-HOURS	UNIT	MAT.	LABOR	EQUIP.	TOTAL	TOTAL INCL O&P
250	**0010 ELECTRIC HEATING**, not incl. conduit or feed wiring									**250**
1100	Rule of thumb: Baseboard units, including control	1 Elec	4.40	1.818	kW	79	76.50		155.50	201
1300	Baseboard heaters, 0.6 m long, 375 W		8	1	Ea.	33	42		75	98.50
1400	0.9 m long, 500 W		8	1		39	42		81	106
1600	1.2 m long, 750 W		6.70	1.194		46	50		96	126
1800	1.5 m long, 935 W		5.70	1.404		55	59		114	149
2000	1.8 m long, 1.13 kW		5	1.600		61	67		128	167
2400	2.4 m long, 1.5 kW		4	2		76.50	84		160.50	210
2950	Wall heaters with fan, 120 to 277 V									
3600	Thermostats, integral	1 Elec	16	.500	Ea.	18	21		39	51.50
3800	Line voltage, 1 pole	"	8	1	"	22.50	42		64.50	87
300	**0010 FAN COIL AIR CONDITIONING** Cabinet mounted, filters, controls									**300**
0100	Chilled water, 1.8 kW cooling	Q-5	8	2	Ea.	600	77.50		677.50	775
0120	3.5 kW cooling		6	2.667		720	103		823	950
0140	5 kW cooling		5.50	2.909		825	113		938	1,075
0150	7 kW cooling		5.25	3.048		1,050	118		1,168	1,325
0180	10.5 kW cooling		4	4		1,700	155		1,855	2,100
0190	26 kW cooling		2.70	5.926		1,825	230		2,055	2,350
0262	For hot water coil, add					40%	10%			
0940	Direct expansion, for use w/air cooled condensing, 5.3 kW cooling	Q-5	5	3.200	Ea.	470	124		594	705
1000	17.6 kW cooling	"	3	5.333		875	207		1,082	1,275
1040	35 kW cooling	Q-6	2.60	9.231		2,450	370		2,820	3,250
1060	70 kW cooling	"	.70	34.286		4,300	1,375		5,675	6,825
500	**0010 HEATING & VENTILATING UNITS** Classroom									**500**
0020	Includes filter, heating/cooling coils, standard controls									
0080	0.354 m³/s, 7 kW cooling	Q-6	2	12	Ea.	3,275	480		3,755	4,325
0120	0.59 m³/s, 10.5 kW cooling		1.40	17.143		3,975	690		4,665	5,400
0140	0.708 m³/s, 14 kW cooling		.80	30		4,275	1,200		5,475	6,525
0500	For electric heat, add					35%				
1000	For no cooling, deduct					25%	10%			
600	**0010 HYDRONIC HEATING** Terminal units, not incl. main supply pipe									**600**
1000	Radiation									
1100	Panel, baseboard, C.I., including supports, no covers	Q-5	14.02	1.141	m	88.50	44		132.50	164
1150	Fin tube, wall hung, 356 mm slope top cover, with damper									
1200	32 mm copper tube, 108 mm alum. fin	Q-5	11.58	1.381	m	90.50	53.50		144	180
1250	32 mm steel tube, 108 mm steel fin	"	10.97	1.458	"	81.50	56.50		138	175
1500	Note: fin tube may also require corners, caps, etc.									
1990	Convector unit, floor recessed, flush, with trim									
2000	for under large glass wall areas, no damper	Q-5	6.10	2.625	m	90.50	102		192.50	253
2100	For unit with damper	"				129			129	142
2210	425 mm H x 600 mm L	Q-5	10	1.600	Ea.	66.50	62		128.50	167
2214	425 mm H x 900 mm L		8.60	1.860		99.50	72		171.50	219
2218	425 mm H x 1200 mm L		7.40	2.162		133	84		217	272
2222	525 mm H x 600 mm L		9	1.778		66.50	69		135.50	177
2226	525 mm H x 900 mm L		8.20	1.951		100	75.50		175.50	224
2228	525 mm H x 1200 mm L		6.80	2.353		133	91		224	283
2240	For knob operated damper, add					140%				
2241	For metal trim strips, add	Q-5	64	.250	Ea.	4.12	9.70		13.82	19.15
2243	For snap-on inlet grille, add					10%	10%			
2245	For hinged access door, add	Q-5	64	.250	Ea.	14.70	9.70		24.40	31
2246	For air chamber, auto-venting, add	"	58	.276	"	4.59	10.70		15.29	21
3000	Radiators, cast iron									
3100	Free standing or wall hung, 6 tube, 635 mm high	Q-5	96	.167	Section	24	6.45		30.45	36
3200	4 tube, 483 mm high	"	96	.167	"	16.65	6.45		23.10	28

15 MECHANICAL

15700 | Heating/Ventilating/Air Conditioning Equipment

MECHANICAL 15

		15760	Terminal Heating & Cooling Units	CREW	DAILY OUTPUT	LABOR-HOURS	UNIT	2006 BARE COSTS MAT.	LABOR	EQUIP.	TOTAL	TOTAL INCL O&P	
600	3250		Adj. brackets, 2 per wall radiator up to 30 sections	1 Stpi	32	.250	Ea.	19.25	10.75		30	37	600
	9500		To convert SFR to BTU rating: Hot water, 150 x SFR										
	9510		Forced hot water, 180 x SFR; steam, 240 x SFR										
700	0010	**INFRARED UNIT**											700
	0020		Gas fired, unvented, electric ignition, 100% shutoff.										
	0030		Piping and wiring not included										
	0060		Input, 4 kW	Q-5	7	2.286	Ea.	380	88.50		468.50	555	
	0120		13 kW		5	3.200		430	124		554	660	
	0160		18 kW		4	4		465	155		620	745	
	0240		35 kW		2	8		895	310		1,205	1,450	
800	0010	**UNIT HEATERS**											800
	3950		Unit heater, propeller 115 V 14 kPa steam, 289 K ent. air										
	4000		Horizontal, 3.5 W	Q-5	12	1.333	Ea.	278	51.50		329.50	385	
	4060		12.9 kW		8	2		425	77.50		502.50	585	
	4140		28.4 kW		6	2.667		620	103		723	835	
	4180		46.2 kW		4	4		830	155		985	1,150	
	4240		84.1 kW		2	8		1,325	310		1,635	1,925	
	4250		95.5 kW		1.90	8.421		1,625	325		1,950	2,275	
	4260		106.6 kW		1.80	8.889		1,700	345		2,045	2,400	
	4270		118.4 kW		1.60	10		2,200	390		2,590	3,000	
	4300		For vertical diffuser, add					154			154	169	
	4310		Vertical flow, 11.7 kW	Q-5	11	1.455		415	56.50		471.50	545	
	4314		17.1 kW		8	2		470	77.50		547.50	630	
	4326		38.4 kW		4	4		655	155		810	955	
	4346		87 kW		1.80	8.889		1,125	345		1,470	1,750	
	4354		123 kW, (460 V)	Q-6	1.80	13.333		1,575	535		2,110	2,525	
	4358		146.5 kW, (460 V)		1.71	14.035		2,225	565		2,790	3,300	
	4362		167 kW, (460 V)		1.40	17.143		3,275	690		3,965	4,625	
	4366		181.7 kW, (460 V)		1.30	18.462		3,750	740		4,490	5,250	
	4370		281.3 kW, (460 V)		1.10	21.818		6,500	875		7,375	8,475	

15800 | Air Distribution

		15810	Ducts	CREW	DAILY OUTPUT	LABOR-HOURS	UNIT	2006 BARE COSTS MAT.	LABOR	EQUIP.	TOTAL	TOTAL INCL O&P	
100	0010	**METAL DUCTWORK**	R233100 -20										100
	0020		Fabricated rectangular, includes fittings, joints, supports,										
	0030		allowance for flexible connections, no insulation										
	0031		NOTE: Fabrication and installation are combined										
	0040		as LABOR cost. Approx. 25% fittings assumed.										
	0100		Aluminum, alloy 3003-H14, under 45 kg	Q-10	34.02	.705	kg	5.50	28		33.50	49	
	0110		45 to 226 kg		36.29	.661		4.41	26		30.41	45	
	0120		227 to 454 kg		43.09	.557		3.64	22		25.64	38	
	0140		454 to 907 kg		54.43	.441		3.53	17.35		20.88	30.50	
	0150		908 to 2268 kg		58.97	.407		3.48	16		19.48	28.50	
	0160		Over 2268 kg		65.77	.365		3.42	14.35		17.77	26	
	0500		Galvanized steel, under 90 kg		107	.224		2.09	8.80		10.89	15.90	
	0520		91 to 227 kg		111	.216		1.79	8.50		10.29	15.05	
	0540		227 to 454 kg		116	.207		1.57	8.15		9.72	14.25	

15810 | Ducts

		CREW	DAILY OUTPUT	LABOR-HOURS	UNIT	2006 BARE COSTS				TOTAL INCL O&P
						MAT.	LABOR	EQUIP.	TOTAL	
100	0560	454 to 907 kg R233100-20	Q-10	120	.200	kg	1.19	7.85		9.04	13.40
	0570	908 to 2268 kg		125	.192		1.10	7.55		8.65	12.85
	0580	Over 2268 kg		129	.186		1.06	7.30		8.36	12.45
	1000	Stainless steel, type 304, under 45 kg		74.84	.321		5.50	12.60		18.10	25.50
	1020	45 to 227 kg		79.38	.302		5.05	11.90		16.95	24
	1030	227 to 454 kg		86.18	.278		4.85	10.95		15.80	22.50
	1040	454 to 907 kg		90.72	.265		4.19	10.40		14.59	20.50
	1050	908 to 2268 kg		102	.235		4.19	9.25		13.44	18.85
	1060	Over 2268 kg		107	.224		4.19	8.80		12.99	18.20
	1100	For medium pressure ductwork, add						15%			
	1200	For high pressure ductwork, add						40%			
300	0010	**FIBEROUS GLASS DUCTWORK** R233100-20									
	3490	Rigid fiberglass duct board, foil reinf. kraft facing									
	3500	Rectangular, 25 mm thick, alum. faced, (FRK), std. weight	Q-10	32.52	.738	m² Surf	6.65	29		35.65	52
500	0010	**FLEXIBLE DUCTS** R233100-20									
	1300	Flexible, coated fiberglass fabric on corr. resist. metal helix									
	1400	pressure to 300 mm (WG) UL-181									
	1500	Non-insulated, 76 mm diameter [3"]	Q-9	122	.131	m	3.54	4.97		8.51	11.55
	1540	127 mm diameter [5"]		97.54	.164		4.43	6.20		10.63	14.50
	1560	152 mm diameter [6"]		85.34	.187		5.25	7.10		12.35	16.70
	1580	178 mm diameter [7"]		73.15	.219		6.25	8.30		14.55	19.65
	1600	203 mm diameter [8"]		60.96	.262		7.20	9.95		17.15	23.50
	1640	254 mm diameter [10"]		48.77	.328		9.05	12.45		21.50	29
	1660	305 mm diameter [12"]		36.58	.437		10.95	16.60		27.55	37.50
	1900	Insulated, 25 mm Th., PE jacket, 80 mm dia [1",3"]		116	.138		7.30	5.25		12.55	16.05
	1910	100 mm diameter [4"]		104	.154		7.30	5.85		13.15	17
	1920	125 mm diameter [5"]		91.44	.175		7.85	6.65		14.50	18.90
	1940	150 mm diameter [6"]		79.25	.202		8.35	7.65		16	21
	1960	175 mm diameter [7"]		67.06	.239		9.70	9.05		18.75	24.50
	1980	200 mm diameter [8"]		54.86	.292		10.35	11.05		21.40	28.50
	2020	250 mm diameter [10"]		42.67	.375		12.15	14.20		26.35	35.50
	2040	300 mm diameter [12"]		30.48	.525		15.15	19.90		35.05	47

15820 | Duct Accessories

		CREW	DAILY OUTPUT	LABOR-HOURS	UNIT	MAT.	LABOR	EQUIP.	TOTAL	TOTAL INCL O&P
300	0010	**DUCT ACCESSORIES**								
	0050	Air extractors, 305 mm x 102 mm	1 Shee	24	.333	Ea.	16.50	14.05		30.55	39.50
	0100	203 mm x 152 mm		22	.364		16.50	15.35		31.85	41.50
	0200	508 mm x 203 mm		16	.500		37.50	21		58.50	73.50
	0280	610 mm x 305 mm		10	.800		50.50	33.50		84	108
	1000	Duct access door, insulated, 150 mm x 150 mm		14	.571		12.90	24		36.90	51
	1020	250 mm x 250 mm		11	.727		19.80	30.50		50.30	69
	1040	300 mm x 300 mm		10	.800		21	33.50		54.50	75.50
	1050	300 mm x 450 mm		9	.889		33	37.50		70.50	93.50
	1070	450 mm x 450 mm		8	1		37.50	42		79.50	107
	1074	600 mm x 450 mm		8	1		45	42		87	115
	2000	Fabrics for flexible connections, with metal edge		30.48	.262	m	6.05	11.05		17.10	23.50
	2100	Without metal edge		48.77	.164	"	4.36	6.90		11.26	15.45
	3000	Fire damper, curtain type, 1-1/2 hr rated, vertical, 150 mm x 150 mm		24	.333	Ea.	14.75	14.05		28.80	38
	3020	203 mm x 152 mm		22	.364		15.10	15.35		30.45	40
	3240	400 mm x 350 mm		18	.444		32.50	18.75		51.25	65
	3400	610 mm x 508 mm		8	1		45.50	42		87.50	115
	5990	Multi-blade dampers, opposed blade, 200mm x 150mm		24	.333		20	14.05		34.05	43.50
	5994	200 mm x 200 mm		22	.364		21	15.35		36.35	46.50
	5996	250 mm x 250 mm		21	.381		24	16.05		40.05	51

15 MECHANICAL

15820	Duct Accessories	CREW	DAILY OUTPUT	LABOR-HOURS	UNIT	2006 BARE COSTS				TOTAL INCL O&P
						MAT.	LABOR	EQUIP.	TOTAL	
300 6000	300 mm x 300 mm	1 Shee	21	.381	Ea.	27	16.05		43.05	54.50	**300**
6020	300 mm x 450 mm		18	.444		37	18.75		55.75	69.50
6030	350 mm x 250 mm		20	.400		26.50	16.85		43.35	55
6031	350 mm x 350 mm		17	.471		33	19.85		52.85	66.50
6033	400 mm x 300 mm		17	.471		33	19.85		52.85	66.50
6035	400 mm x 400 mm		16	.500		41	21		62	77.50
6037	450 mm x 400 mm		15	.533		45	22.50		67.50	84
6038	450 mm x 450 mm		15	.533		49	22.50		71.50	88
6070	500 mm x 400 mm		14	.571		49	24		73	90.50
6072	500 mm x 500 mm		13	.615		58.50	26		84.50	104
6074	550 mm x 450 mm		14	.571		58.50	24		82.50	101
6076	600 mm x 400 mm		11	.727		57.50	30.50		88	111
6078	600 mm x 500 mm		8	1		68	42		110	140
6080	600 mm x 600 mm		8	1		79	42		121	152
6110	650 mm x 650 mm		6	1.333		85.50	56		141.50	181
6133	750 mm x 750 mm	Q-9	6.60	2.424		124	92		216	278
6135	800 mm x 800 mm		6.40	2.500		144	95		239	305
6180	1200 mm x 900 mm		5.60	2.857		236	108		344	425
7000	Splitter damper assembly, self-locking, 300 mm rod	1 Shee	24	.333		17.40	14.05		31.45	40.50
7020	914 mm rod		22	.364		22.50	15.35		37.85	48.50
7040	1219 mm rod		20	.400		25	16.85		41.85	53.50
7060	1829 mm rod		18	.444		30.50	18.75		49.25	62.50
8000	Multi-blade dampers, parallel blade									
8100	200 mm x 200 mm	1 Shee	24	.333	Ea.	59	14.05		73.05	86.50
8140	400 mm x 250 mm		20	.400		78	16.85		94.85	112
8200	600 mm x 400 mm		11	.727		100	30.50		130.50	157
8260	750 mm x 450 mm		7	1.143		127	48		175	214
9000	Silencers, noise control for air flow, duct				m³/s	104			104	114

15830	Fans									
100 0010	**FANS** [R233400-10]										**100**
0020	Air conditioning and process air handling									
0030	Axial flow, compact, low sound, 64 mm S.P.									
0050	1.794 m³/s, 3.7 kW	Q-20	3.40	5.882	Ea.	3,925	228		4,153	4,650
0080	3.021 m³/s, 3.7 kW		2.80	7.143		4,375	277		4,652	5,250
0100	4.956 m³/s, 5.6 kW		2.40	8.333		5,450	325		5,775	6,500
0120	7.363 m³/s, 7.5 kW		1.60	12.500		6,875	485		7,360	8,300
0200	In-line centrifugal, supply/exhaust booster									
0220	aluminum wheel/hub, disc. switch, 6 mm S.P.									
0240	0.236 m³/s, 254 mm dia conn [10"]	Q-20	3	6.667	Ea.	940	258		1,198	1,425
0260	0.651 m³/s, 305 mm dia conn [12"]		2	10		995	385		1,380	1,700
0280	0.717 m³/s, 406 mm dia connection [16"]		2	10		1,075	385		1,460	1,800
0300	1.208 m³/s, 457 mm dia connection [18"]		1	20		1,175	775		1,950	2,475
0320	1.643 m³/s, 508 mm dia connection [20"]		.80	25		1,400	970		2,370	3,025
0326	2.36 m³/s, 500 mm dia connection		.75	26.667		1,525	1,025		2,550	3,250
1500	Vaneaxial, low pressure, 0.944 m³/s, 372 W		3.60	5.556		1,350	215		1,565	1,825
1520	1.888 m³/s, 746 W		3.20	6.250		1,475	242		1,717	2,000
1540	3.776 m³/s, 1.5 kW		2.80	7.143		2,000	277		2,277	2,600
2500	Ceiling fan, right angle, extra quiet, 3 mm S.P.									
2520	0.045 m³/s	Q-20	20	1	Ea.	160	38.50		198.50	236
2540	0.099 m³/s		19	1.053		189	41		230	271
2560	0.182 m³/s		18	1.111		240	43		283	330
2580	0.418 m³/s		16	1.250		470	48.50		518.50	595
2600	0.779 m³/s		13	1.538		655	59.50		714.50	810
2620	1.397 m³/s		11	1.818		870	70.50		940.50	1,075
2640	For wall or roof cap, add	1 Shee	16	.500		160	21		181	209

MECHANICAL 15

15830	Fans	CREW	DAILY OUTPUT	LABOR-HOURS	UNIT	2006 BARE COSTS				TOTAL INCL O&P
						MAT.	LABOR	EQUIP.	TOTAL	
2660	For straight thru fan, add				Ea.	10%				
2680	For speed control switch, add [R233400-10]	1 Elec	16	.500	↓	87	21		108	128
3000	Paddle blade air circulator, 3 speed switch									
3020	1067 mm, 2.36 m³/s high, 1.416 m³/s low	1 Elec	2.40	3.333	Ea.	73.50	140		213.50	290
3040	1321 mm, 3.068 m³/s high, 1.888 m³/s low	"	2.20	3.636	"	74.50	153		227.50	310
3100	For antique white motor, same cost									
3200	For brass plated motor, same cost									
3300	For light adaptor kit, add				Ea.	27.50			27.50	30
3500	Centrifugal, airfoil, motor and drive, complete									
3520	0.472 m³/s, 373 W	Q-20	2.50	8	Ea.	1,100	310		1,410	1,675
3540	0.944 m³/s, 746 W		2	10		1,175	385		1,560	1,900
3560	1.888 m³/s, 2.2 kW		1.80	11.111		1,475	430		1,905	2,275
3580	3.776 m³/s, 5.6 kW		1.40	14.286		2,375	555		2,930	3,450
3600	5.664 m³/s, 7.5 kW		1	20		2,925	775		3,700	4,400
4500	Corrosive fume resistant, plastic									
4600	roof ventilators, centrifugal, V belt drive, motor									
4620	6 mm S.P., 0.118 m³/s, 186 W	Q-20	6	3.333	Ea.	2,525	129		2,654	3,000
4640	0.422 m³/s, 248 W		5	4		2,750	155		2,905	3,250
4660	0.769 m³/s, 373 W		4	5		3,250	194		3,444	3,875
4680	1.057 m³/s, 746 W		3	6.667		3,400	258		3,658	4,125
5000	Utility set, centrifugal, V belt drive, motor									
5020	8 mm S.P., 0.57 m³/s, 186 W	Q-20	6	3.333	Ea.	2,850	129		2,979	3,350
5040	1.024 m³/s, 248 W		5	4		2,850	155		3,005	3,375
5060	1.265 m³/s, 373 W		4	5		2,875	194		3,069	3,450
5080	1.425 m³/s, 560 W		3	6.667		2,900	258		3,158	3,600
5100	15 mm S.P., 1.7 m³/s, 746 W		2	10		4,200	385		4,585	5,225
5120	1.704 m³/s, 1.1 kW		1.60	12.500		4,250	485		4,735	5,425
5140	1.945 m³/s, 1.5 kW		1.40	14.286		4,325	555		4,880	5,600
6000	Propeller exhaust, wall shutter, 8 mm S.P.									
6020	Direct drive, two speed									
6100	0.177 m³/s, 75 W	Q-20	10	2	Ea.	287	77.50		364.50	435
6120	0.345 m³/s, 107 W		9	2.222		320	86		406	480
6140	0.472 m³/s, 93 W		8	2.500		445	97		542	640
6200	2.228 m³/s, 746 W		5	4		710	155		865	1,025
6300	V-belt drive, 3 phase									
6320	2.915 m³/s, 559 W	Q-20	5	4	Ea.	595	155		750	890
6340	3.54 m³/s, 559 W		5	4		630	155		785	925
6360	4.767 m³/s, 746 W		4.50	4.444		770	172		942	1,100
6380	6.75 m³/s, 1.1 kW		4	5		905	194		1,099	1,300
6650	Residential, bath exhaust, grille, back draft damper									
6660	0.024 m³/s	Q-20	24	.833	Ea.	30	32.50		62.50	82.50
6670	0.052 m³/s		22	.909		50	35		85	109
6680	Light combination, squirrel cage, 100 W, 0.033 m³/s		24	.833		63	32.50		95.50	119
6700	Light/heater combination, ceiling mounted									
6710	0.033 m³/s, 1450 W	Q-20	24	.833	Ea.	74.50	32.50		107	132
6800	Heater combination, recessed, 0.033 m³/s		24	.833		36	32.50		68.50	89
6820	With 2 infrared bulbs		23	.870		53.50	33.50		87	111
6900	Kitchen exhaust, grille, complete, 0.076 m³/s		22	.909		63.50	35		98.50	124
6910	0.085 m³/s		20	1		54	38.50		92.50	119
6920	0.127 m³/s		18	1.111		97.50	43		140.50	173
6930	0.165 m³/s		16	1.250		74.50	48.50		123	156
6940	Residential roof jacks and wall caps									
6944	Wall cap with back draft damper									
6946	80 mm and 100 mm dia. round duct (3" & 4")	1 Shee	11	.727	Ea.	13.05	30.50		43.55	61.50
6948	152 mm dia. round duct (6")	"	11	.727	"	31	30.50		61.50	81
6958	Roof jack with bird screen and back draft damper									

15 MECHANICAL

100

15830 | Fans

			CREW	DAILY OUTPUT	LABOR-HOURS	UNIT	MAT.	LABOR	EQUIP.	TOTAL	TOTAL INCL O&P	
100	6960	80 mm & 100 mm dia. round duct (3″ & 4″)	1 Shee	11	.727	Ea.	12.55	30.50		43.05	61	100
	6962	82 mm x 254 mm rect. duct (3-1/4″ x 10″) [R233400 -10]	″	10	.800	″	22.50	33.50		56	77	
	6980	Transition										
	6982	82 mm x 254 mm to 152 mm dia. (3-1/4″ x 10″ to 6″)	1 Shee	20	.400	Ea.	13.85	16.85		30.70	41.50	
	7000	Roof exhauster, centrifugal, aluminum housing, 300 mm galvanized										
	7020	curb, bird screen, back draft damper, 8 mm S.P.										
	7100	Direct drive, 0.15 m³/s, 0.08 m² damper	Q-20	7	2.857	Ea.	360	111		471	565	
	7120	0.28 m³/s, 0.08 m² damper		6	3.333		365	129		494	600	
	7140	0.38 m³/s, 0.11 m² damper		5	4		365	155		520	635	
	7160	0.68 m³/s, 0.11 m² damper		4.20	4.762		465	184		649	790	
	7180	0.97 m³/s, 0.16 m² damper		4	5		465	194		659	805	
	7200	V-belt drive, 0.78 m³/s, 0.09 m² damper		6	3.333		790	129		919	1,075	
	7220	1.3 m³/s, 0.28 m² damper		5	4		890	155		1,045	1,225	
	7230	1.6 m³/s, 0.5 m² sq. damper [21″]		4.50	4.444		985	172		1,157	1,350	
	7240	2.3 m³/s, 0.34 m² damper		4	5		1,225	194		1,419	1,650	
	7260	4 m³/s, 0.51 m² damper		3	6.667		1,525	258		1,783	2,075	
	7280	6.5 m³/s, 0.8 m² damper		2	10		2,100	385		2,485	2,925	
	7300	9.7 m³/s, 1.2 m² damper		1	20		4,475	775		5,250	6,100	
	7320	For 2 speed winding, add					15%					
	7340	For explosionproof motor, add				Ea.	330			330	360	
	7360	For belt driven, top discharge, add					15%					
	7500	Utility set, steel construction, pedestal, 8 mm S.P.										
	7520	Direct drive, 0.071 m³/s, 93 W	Q-20	6.40	3.125	Ea.	660	121		781	910	
	7540	0.23 m³/s, 124 W		5.80	3.448		830	134		964	1,125	
	7560	0.92 m³/s, 373 W		4.80	4.167		970	161		1,131	1,325	
	7580	1.1 m³/s, 559 W		4.40	4.545		1,800	176		1,976	2,250	
	7600	1.6 m³/s, 1.1 kW		3	6.667		2,000	258		2,258	2,600	
	7680	V-belt drive, drive cover, 3 phase										
	7700	0.38 m³/s, 186 W	Q-20	6	3.333	Ea.	520	129		649	770	
	7720	0.6 m³/s, 249 W		5	4		545	155		700	830	
	7740	0.94 m³/s, 746 W		4.60	4.348		640	168		808	965	
	7760	1.4 m³/s, 559 W		4.20	4.762		865	184		1,049	1,225	
	8500	Wall exhausters, centrifugal, auto damper, 6 mm S.P.										
	8520	Direct drive, 0.29 m³/s, 37 W	Q-20	14	1.429	Ea.	232	55.50		287.50	340	
	8540	0.38 m³/s, 62 W		13	1.538		240	59.50		299.50	355	
	8560	0.39 m³/s, 124 W		12	1.667		380	64.50		444.50	515	
	8580	0.62 m³/s, 186 W		12	1.667		385	64.50		449.50	520	
	9500	V-belt drive, 3 phase										
	9520	1.322 m³/s, 186 W	Q-20	9	2.222	Ea.	1,050	86		1,136	1,275	
	9540	1.765 m³/s, 373 W	″	8	2.500	″	1,100	97		1,197	1,350	

15840 | Air Terminal Units

			CREW	DAILY OUTPUT	LABOR-HOURS	UNIT	MAT.	LABOR	EQUIP.	TOTAL	TOTAL INCL O&P	
200	0010	**AIR CURTAINS** Incl. motor starters, transformers,										200
	0050	door switches & temperature controls										
	0100	Shipping and receiving doors, unheated, minimal wind stoppage										
	0150	2400 mm high, multiples of 900 mm wide	2 Shee	1.83	8.749	m	1,050	370		1,420	1,725	
	0160	1500 mm wide		3.05	5.249		1,025	221		1,246	1,475	
	0210	3000 mm high, multiples of 1200 mm wide		2.44	6.562		1,000	277		1,277	1,525	
	0250	3600 mm high, 1000 mm wide		2.13	7.499		1,100	315		1,415	1,675	
	0260	3600 mm wide		1.83	8.749		1,000	370		1,370	1,675	
	0350	4800 mm high, 1000 mm wide		2.13	7.499		1,050	315		1,365	1,625	
	0360	3600 mm wide		1.83	8.749		1,175	370		1,545	1,850	
	0500	Maximum wind stoppage										
	0550	3000 mm high, multiples of 1200 mm wide	2 Shee	2.44	6.562	m	805	277		1,082	1,300	
	0650	4200 mm high, multiples of 1200 mm wide		2.44	6.562		805	277		1,082	1,300	
	0750	6000 mm high, multiples of 2400 mm wide		1.83	8.749		2,800	370		3,170	3,675	

MECHANICAL 15

For expanded coverage of these items see *Means Mechanical or Plumbing Cost Data 2006*

15840 | Air Terminal Units

		CREW	DAILY OUTPUT	LABOR-HOURS	UNIT	MAT.	LABOR	EQUIP.	TOTAL	TOTAL INCL O&P	
200	1100	Heated, maximum wind stoppage, steam heat										**200**
	1150	3000 mm high, multiples of 1200 mm wide	2 Shee	1.83	8.749	m	1,375	370		1,745	2,100	
	1250	4200 mm high, multiples of 1200 mm wide		1.83	8.749		1,525	370		1,895	2,250	
	1350	6000 mm high, multiples of 2400 mm wide	↓	1.22	13.123	↓	4,175	555		4,730	5,450	
	1500	Customer entrance doors, unheated, minimal wind stoppage										
	1550	3000 mm high, multiples of 900 mm wide	2 Shee	1.83	8.749	m	925	370		1,295	1,600	
	1560	1500 mm wide		3.05	5.249		1,000	221		1,221	1,450	
	1650	Max. wind stoppage, 3600 mm H, multiples of 1200 mm W	↓	2.44	6.562	↓	740	277		1,017	1,225	
	1700	Heated, minimal wind stoppage, electric heat										
	1750	2400 mm high, multiples of 900 mm wide	2 Shee	1.83	8.749	m	1,850	370		2,220	2,600	
	1760	Multiples of 1500 mm wide		3.05	5.249		1,325	221		1,546	1,800	
	1850	3000 mm high, multiples of 900 mm wide		1.83	8.749		1,575	370		1,945	2,300	
	1860	Multiples of 1525 mm wide	↓	3.05	5.249		1,775	221		1,996	2,300	
	1950	Maximum wind stoppage, steam heat										
	1960	3660 mm high, multiples of 1220 mm wide	2 Shee	2.44	6.562	m	1,675	277		1,952	2,275	
	2000	Walk-in coolers and freezers, ambient air, minimal wind stoppage										
	2050	2440 mm high, multiples of 915 mm wide	2 Shee	1.83	8.749	m	690	370		1,060	1,325	
	2060	Multiples of 1525 mm wide		3.05	5.249		750	221		971	1,175	
	2250	Max. wind stoppage, 3660 mm H, multiples of 915 mm W		1.83	8.749		790	370		1,160	1,450	
	2450	Conveyor openings or service windows, unheated, 1525 mm high		1.52	10.499		880	445		1,325	1,650	
	2460	Heated, electric, 1525 mm high, 760 mm wide	↓	1.52	10.499	↓	830	445		1,275	1,600	

15850 | Air Outlets & Inlets

		CREW	DAILY OUTPUT	LABOR-HOURS	UNIT	MAT.	LABOR	EQUIP.	TOTAL	TOTAL INCL O&P	
300	0010	**DIFFUSERS** Aluminum, opposed blade damper unless noted										**300**
	0100	Ceiling, linear, also for sidewall										
	0500	Perf, 600 mm x 600 mm lay-in panel size, 150 mm x 150 mm	1 Shee	16	.500	Ea.	72	21		93	112	
	0520	200 mm x 200 mm		15	.533		74	22.50		96.50	116	
	0530	225 mm x 225 mm		14	.571		77	24		101	122	
	0540	250 mm x 250 mm		14	.571		79.50	24		103.50	124	
	0560	300 mm x 300 mm		12	.667		82	28		110	134	
	0590	400 mm x 400 mm		11	.727		112	30.50		142.50	171	
	0600	450 mm x 450 mm		10	.800		120	33.50		153.50	184	
	0610	500 mm x 500 mm		10	.800		137	33.50		170.50	203	
	0620	600 mm x 600 mm		9	.889		159	37.50		196.50	233	
	1000	Rectangular, 1 to 4 way blow, 150 mm x 150 mm		16	.500		44.50	21		65.50	81.50	
	1010	200 mm x 200 mm		15	.533		53	22.50		75.50	93	
	1014	225 mm x 225 mm		15	.533		57	22.50		79.50	97.50	
	1016	250 mm x 250 mm		15	.533		67.50	22.50		90	109	
	1020	300 mm x 150 mm		15	.533		77.50	22.50		100	120	
	1040	300 mm x 225 mm		14	.571		84.50	24		108.50	130	
	1060	300 mm x 300 mm		12	.667		78.50	28		106.50	130	
	1070	350 mm x 150 mm		13	.615		84	26		110	132	
	1074	350 mm x 350 mm		12	.667		103	28		131	157	
	1150	450 mm x 450 mm		9	.889		145	37.50		182.50	217	
	1160	525 mm x 525 mm		8	1		183	42		225	266	
	1170	600 mm x 300 mm		10	.800		136	33.50		169.50	202	
	1500	Round, butterfly damper, steel, 150 mm diameter		18	.444		17.95	18.75		36.70	48.50	
	1520	200 mm diameter		16	.500		19.35	21		40.35	54	
	1540	250 mm diameter		14	.571		24	24		48	63.50	
	1560	300 mm diameter		12	.667		31.50	28		59.50	78	
	1580	350 mm diameter		10	.800		41	33.50		74.50	97	
	2000	T bar mount, 600mm x 600mm lay-in frame, 150mm x 150mm		16	.500		94.50	21		115.50	137	
	2020	225 mm x 225 mm		14	.571		105	24		129	152	
	2040	300 mm x 300 mm		12	.667		135	28		163	192	
	2060	375 mm x 375 mm	↓	11	.727	↓	173	30.50		203.50	237	

15 MECHANICAL

			DAILY	LABOR-		2006 BARE COSTS				TOTAL	
15850		**Air Outlets & Inlets**								**INCL O&P**	
			CREW	OUTPUT	HOURS	UNIT	MAT.	LABOR	EQUIP.	TOTAL	
300	2080	450 mm x 450 mm	1 Shee	10	.800	Ea.	190	33.50		223.50	262	**300**
	6000	For steel diffusers instead of aluminum, deduct					10%				
500	0010	**GRILLES**										**500**
	0020	Aluminum									
	1000	Air return, 152 mm x 152 mm	1 Shee	26	.308	Ea.	14.30	12.95		27.25	36
	1020	254 mm x 152 mm		24	.333		16.90	14.05		30.95	40
	1080	406 mm x 203 mm		22	.364		25.50	15.35		40.85	51.50
	1100	305 mm x 305 mm		22	.364		25.50	15.35		40.85	51.50
	1120	610 mm x 305 mm		18	.444		45	18.75		63.75	78.50
	1220	610 mm x 457 mm		16	.500		49.50	21		70.50	87
	1280	914 mm x 610 mm		14	.571		99	24		123	146
	3000	Filter grille with filter, 305 mm x 305 mm		24	.333		42.50	14.05		56.55	68
	3020	457 mm x 305 mm		20	.400		62	16.85		78.85	94
	3040	610 mm x 457 mm		18	.444		75.50	18.75		94.25	112
	3060	610 mm x 610 mm		16	.500		118	21		139	162
	6000	For steel grilles instead of aluminum in above, deduct					10%				
600	0010	**LOUVERS**										**600**
	0100	Aluminum, extruded, with screen, mill finish									
	1000	Brick vent, (see also division 04090-860)									
	1100	Standard, 102 mm deep, 203 mm wide, 127 mm high	1 Shee	24	.333	Ea.	25	14.05		39.05	49
	1200	Modular, 102 mm deep, 197 mm wide, 127 mm high		24	.333		26	14.05		40.05	50
	1300	Speed brick, 102 mm deep, 295 mm wide, 98 mm high		24	.333		26	14.05		40.05	50
	1400	Fuel oil brick, 102 mm deep, 203 mm wide, 127 mm high		24	.333		44.50	14.05		58.55	70.50
	2000	Cooling tower and mechanical equip., screens, light weight		3.72	2.153	m²	120	90.50		210.50	272
	2020	Standard weight		3.25	2.460		320	104		424	510
	2500	Dual combination, automatic, intake or exhaust		1.86	4.306		435	181		616	760
	2520	Manual operation		1.86	4.306		325	181		506	635
	2540	Electric or pneumatic operation		1.86	4.306		325	181		506	635
	2560	Motor, for electric or pneumatic		14	.571	Ea.	350	24		374	420
	3000	Fixed blade, continuous line									
	3100	Mullion type, stormproof	1 Shee	2.60	3.076	m²	325	130		455	555
	3200	Stormproof		2.60	3.076		325	130		455	555
	3300	Vertical line		2.60	3.076		430	130		560	670
	3500	For damper to use with above, add					50%	30%			
	3520	Motor, for damper, electric or pneumatic	1 Shee	14	.571	Ea.	350	24		374	420
	4000	Operating, 45°, manual, electric or pneumatic		2.23	3.588	m²	325	151		476	590
	4100	Motor, for electric or pneumatic		14	.571	Ea.	350	24		374	420
	4200	Penthouse, roof		5.20	1.538	m²	192	65		257	310
	4300	Walls		3.72	2.153		450	90.50		540.50	635
	5000	Thinline, under 102 mm thick, fixed blade		3.72	2.153		187	90.50		277.50	345
	5010	Finishes, applied by mfr. at additional cost, available in colors									
	5020	Prime coat only, add				m²	26			26	28.50
	5040	Baked enamel finish coating, add					48			48	52.50
	5060	Anodized finish, add					52			52	57
	5080	Duranodic finish, add					94			94	103
	5100	Fluoropolymer finish coating, add					147			147	162
	9980	For small orders (under 10 pieces), add					25%				
700	0010	**REGISTERS**										**700**
	0980	Air supply									
	1000	Ceiling/wall, O.B. damper, anodized aluminum									
	1010	One or two way deflection, adj. curved face bars									
	1020	200 mm x 100 mm	1 Shee	26	.308	Ea.	21.50	12.95		34.45	44
	1120	300 mm x 300 mm		18	.444		35.50	18.75		54.25	68
	1240	500 mm x 150 mm		18	.444		35.50	18.75		54.25	68
	1340	610 mm x 203 mm		13	.615		48	26		74	92.50

MECHANICAL 15

For expanded coverage of these items see *Means Mechanical or Plumbing Cost Data 2006*

15850 | Air Outlets & Inlets

			CREW	DAILY OUTPUT	LABOR-HOURS	UNIT	MAT.	LABOR	EQUIP.	TOTAL	TOTAL INCL O&P	
700	1350	610 mm x 457 mm	1 Shee	12	.667	Ea.	89	28		117	142	700
	2700	Above registers in steel instead of aluminum, deduct				↓	10%					
	4000	Floor, toe operated damper, enameled steel										
	4020	100 mm x 200 mm	1 Shee	32	.250	Ea.	18.65	10.55		29.20	37	
	4100	203 mm x 254 mm		22	.364		23	15.35		38.35	48.50	
	4140	254 mm x 254 mm		20	.400		27	16.85		43.85	55.50	
	4220	356 mm x 356 mm		16	.500		95	21		116	138	
	4240	356 mm x 508 mm	↓	15	.533	↓	127	22.50		149.50	175	
	4980	Air return										
	5000	Ceiling or wall, fixed 45° face blades										
	5010	Adjustable O.B. damper, anodized aluminum										
	5020	100 mm x 200 mm	1 Shee	26	.308	Ea.	25.50	12.95		38.45	48	
	5060	150 mm x 250 mm		19	.421		28	17.75		45.75	58.50	
	5280	600 mm x 600 mm		11	.727		120	30.50		150.50	179	
	5300	600 mm x 900 mm	↓	8	1	↓	187	42		229	271	
	6000	For steel construction instead of aluminum, deduct						10%				
800	0010	**VENTILATORS** Base & damper										800
	1280	Rotary ventilators, wind driven, galvanized										
	1300	100 mm neck diameter [4"]	Q-9	20	.800	Ea.	52	30.50		82.50	105	
	1340	150 mm neck diameter [6"]		16	1		58	38		96	123	
	1400	300 mm neck diameter [12"]		10	1.600		84.50	60.50		145	186	
	1500	600 mm neck diameter, 1.5 m³/s [24"]		8	2		245	76		321	385	
	1540	900 mm neck diameter, 2.6 m³/s [36"]	↓	6	2.667	↓	720	101		821	945	
	2000	Stationary, gravity, syphon, galvanized										
	2160	150 mm neck diameter, 0.031 m³/s [6"]	Q-9	16	1	Ea.	45	38		83	108	
	2240	300 mm neck diameter, 0.076 m³/s [12"]		10	1.600		94.50	60.50		155	198	
	2340	600 mm neck diameter, 0.43 m³/s [24"]		8	2		257	76		333	400	
	2380	900 mm neck diameter, 0.94 m³/s [36"]		6	2.667		805	101		906	1,050	
	4200	Stationary mushroom, aluminum, 400 mm orifice diameter [16"]		10	1.600		360	60.50		420.50	490	
	4220	650 mm orifice diameter [26"]		6.15	2.602		530	98.50		628.50	730	
	4230	750 mm orifice diameter [30"]		5.71	2.802		775	106		881	1,025	
	4240	950 mm orifice diameter [38"]		5	3.200		1,100	121		1,221	1,400	
	4250	1050 mm orifice diameter [42"]		4.70	3.404		1,475	129		1,604	1,800	
	4260	1250 mm orifice diameter [50"]	↓	4.44	3.604	↓	1,750	137		1,887	2,125	
	5000	Relief vent										
	5500	Rectangular, aluminum, galvanized curb										
	5510	intake/exhaust, 1 mm SP										
	5600	0.236 m³/s, 305 mm x 406 mm	Q-9	8	2	Ea.	410	76		486	565	
	5640	0.472 m³/s, 305 mm x 610 mm		6.60	2.424		485	92		577	675	
	5680	1.416 m³/s, 508 mm x 1067 mm	↓	4	4	↓	845	152		997	1,175	

15860 | Air Cleaning Devices

			CREW	DAILY OUTPUT	LABOR-HOURS	UNIT	MAT.	LABOR	EQUIP.	TOTAL	TOTAL INCL O&P	
100	0010	**AIR FILTERS**										100
	0050	Activated charcoal type, full flow				m³/s	1,275			1,275	1,400	
	0060	Activated charcoal type, full flow, impreg. media 300 mm deep					370			370	410	
	0070	Activated charcoal type, HEPA filter & frame for field erection					370			370	410	
	0080	Activated charcoal type, HEPA filter-diffuser, ceiling install.				↓	530			530	585	
	2000	Electronic air cleaner, self-contained										
	2150	0.24 m³/s	1 Shee	2.30	3.478	Ea.	695	147		842	990	
	2200	0.472 m³/s		2.20	3.636		725	153		878	1,025	
	2250	0.57 m³/s	↓	2.10	3.810	↓	800	161		961	1,125	
	2950	Mechanical media filtration units										
	3000	High efficiency type, with frame, non-supported				m³/s	95.50			95.50	105	
	3100	Supported type					117			117	128	
	4000	Medium efficiency, extended surface					11.65			11.65	12.80	
	4500	Permanent washable				↓	42.50			42.50	46.50	

15860	Air Cleaning Devices	CREW	DAILY OUTPUT	LABOR-HOURS	UNIT	2006 BARE COSTS				TOTAL INCL O&P	
						MAT.	LABOR	EQUIP.	TOTAL		
100	5000	Renewable disposable roll				m³/s	254			254	280	**100**
	5500	Throwaway glass or paper media type				Ea.	4.60			4.60	5.05	

15950 | Testing/Adjusting/Balancing

15955	HVAC Test/Adjust/Balance	CREW	DAILY OUTPUT	LABOR-HOURS	UNIT	2006 BARE COSTS				TOTAL INCL O&P	
						MAT.	LABOR	EQUIP.	TOTAL		
100	0010	**BALANCING, AIR** (Subcontractor's quote incl. material & labor)										**100**
	0900	Heating and ventilating equipment										
	1000	Centrifugal fans, utility sets				Ea.					311.76	
	1100	Heating and ventilating unit									467.64	
	1200	In-line fan									467.64	
	1300	Propeller and wall fan									88.33	
	1400	Roof exhaust fan									207.84	
	2000	Air conditioning equipment, central station									675.48	
	2100	Built-up low pressure unit									623.52	
	2200	Built-up high pressure unit									727.44	
	2500	Multi-zone A.C. and heating unit									467.64	
	2600	For each zone over one, add									103.92	
	2700	Package A.C. unit									259.80	
	2800	Rooftop heating and cooling unit									363.72	
	3000	Supply, return, exhaust, registers & diffusers, avg. height ceiling									62.35	
	3100	High ceiling									93.53	
	3200	Floor height				↓					51.96	
700	0010	**PIPING, TESTING**										**700**
	0100	Nondestructive Testing										
	0110	Nondestructive hydraulic pressure test, isolate & 1 hr. hold										
	0120	25 mm - 100 mm pipe [1" - 4"]										
	0140	0 - 75 m [0 - 250 L.F.]	1 Stpi	1.33	6.015	Ea.		259		259	390	
	0160	75 m - 150 m [250 - 500 L.F.]	"	.80	10			430		430	650	
	0180	150 m - 300 m [500 - 1000 L.F.]	Q-5	1.14	14.035	↓		545		545	820	
	0200	300 m - 600 m [1000 - 2000 L.F.]	"	.80	20	↓		775		775	1,175	
	0300	150 mm - 250 mm pipe [6" - 10"]										
	0320	0 - 75 m [0 - 250 L.F.]	Q-5	1	16	Ea.		620		620	935	
	0340	75 m - 150 m [250 - 500 L.F.]	↓	.73	21.918			850		850	1,275	
	0360	150 m - 300 m [500 - 1000 L.F.]		.53	30.189			1,175		1,175	1,750	
	0380	300 m - 600 m [1000 - 2000 L.F.]	↓	.38	42.105	↓		1,625		1,625	2,450	
	1000	Pneumatic pressure test, includes soaping joints										
	1120	25 mm - 100 mm pipe [1" - 4"]										
	1140	0 - 75 m [0 - 250 L.F.]	Q-5	2.67	5.993	Ea.	7.05	232		239.05	360	
	1160	75 m - 150 m [250 - 500 L.F.]	↓	1.33	12.030		14.10	465		479.10	715	
	1180	150 m - 300 m [500 - 1000 L.F.]		.80	20		21	775		796	1,200	
	1200	300 m - 600 m [1000 - 2000 L.F.]	↓	.50	32	↓	28	1,250		1,278	1,900	
	1300	150 mm - 250 mm pipe [6" - 10"]										
	1320	0 - 75 m [0 - 250 L.F.]	Q-5	1.33	12.030	Ea.	7.05	465		472.05	710	
	1340	75 m - 150 m [250 - 500 L.F.]		.67	23.881		14.10	925		939.10	1,425	
	1360	150 m - 300 m [500 - 1000 L.F.]		.40	40		28	1,550		1,578	2,350	
	1380	300 m - 600 m [1000 - 2000 L.F.]	↓	.25	64		35	2,475		2,510	3,775	
	2000	X-Ray of welds										
	2110	50 mm diam. [2"]	1 Stpi	8	1	Ea.	8.85	43		51.85	75	

MECHANICAL 15

15955	HVAC Test/Adjust/Balance	CREW	DAILY OUTPUT	LABOR-HOURS	UNIT	2006 BARE COSTS				TOTAL INCL O&P	
						MAT.	LABOR	EQUIP.	TOTAL		
700 2120	80 mm diam. [3"]	1 Stpi	8	1	Ea.	8.85	43		51.85	75	**700**
2130	100 mm diam. [4"]		8	1		13.30	43		56.30	79.50	
2140	150 mm diam. [6"]		8	1		13.30	43		56.30	79.50	
2150	200 mm diam. [8"]		6.60	1.212		13.30	52		65.30	93	
2160	250 mm diam. [10"]	↓	6	1.333	↓	17.70	57.50		75.20	106	
3000	Liquid penetration of welds										
3110	50 mm diam. [2"]	1 Stpi	14	.571	Ea.	2.48	24.50		26.98	39.50	
3120	80 mm diam. [3"]		13.60	.588		2.48	25.50		27.98	40.50	
3130	100 mm diam. [4"]		13.40	.597		2.48	25.50		27.98	41	
3140	150 mm diam. [6"]		13.20	.606		2.48	26		28.48	42	
3150	200 mm diam. [8"]		13	.615		3.72	26.50		30.22	44	
3160	250 mm diam. [10"]	↓	12.80	.625	↓	3.72	27		30.72	44.50	
900 0010	**BALANCING, WATER** (Subcontractor's quote incl. material & labor)										**900**
0050	Air cooled condenser				Ea.					181.44	
0100	Cabinet unit heater									62.21	
0200	Chiller									440.64	
0300	Convector									51.84	
0500	Cooling tower									336.96	
0600	Fan coil unit, unit ventilator									93.31	
0700	Fin tube and radiant panels									103.68	
0800	Main and duct re-heat coils									95.90	
0810	Heat exchanger									95.90	
1000	Pumps									228.10	
1100	Unit heater				↓					72.58	

For information about Means Estimating Seminars, see yellow pages 12 and 13 in back of book

15 MECHANICAL

Division 16
Electrical

Estimating Tips

16060 Grounding & Bonding

- When taking off grounding system, identify separately the type and size of wire and list each unique type of ground connection.

16100 Wiring Methods

- Conduit should be taken off in three main categories: power distribution, branch power, and branch lighting, so the estimator can concentrate on systems and components, therefore making it easier to ensure all items have been accounted for.

- For cost modifications for elevated conduit installation, add the percentages to labor according to the height of installation and only the quantities exceeding the different height levels, not to the total conduit quantities.

- Remember that aluminum wiring of equal ampacity is larger in diameter than copper and may require larger conduit.

- If more than three wires at a time are being pulled, deduct percentages from the labor hours of that grouping of wires.

- The estimator should take the weights of materials into consideration when completing a takeoff. Topics to consider include: How will the materials be supported? What methods of support are available? How high will the support structure have to reach? Will the final support structure be able to withstand the total burden? Is the support material included or separate from the fixture, equipment, and material specified?

16200 Electrical Power

- Do not overlook the costs for equipment used in the installation. If scaffolding or highlifts are available in the field, contractors may use them in lieu of the proposed ladders and rolling staging.

16400 Low-Voltage Distribution

- Supports and concrete pads may be shown on drawings for the larger equipment, or the support system may be just a piece of plywood for the back of a panelboard. In either case, it must be included in the costs.

16500 Lighting

- Fixtures should be taken off room by room, using the fixture schedule, specifications, and the ceiling plan. For large concentrations of lighting fixtures in the same area, deduct the percentages from labor hours.

16700 Communications
16800 Sound & Video

- When estimating material costs for special systems, it is always prudent to obtain manufacturers' quotations for equipment prices and special installation requirements which will affect the total costs.

Reference Numbers

Reference numbers are shown in bold squares at the beginning of some major classifications. These numbers refer to related items in the Reference Section. The reference information may be an estimating procedure, an alternate pricing method, or technical information.

Note: Not all subdivisions listed here necessarily appear in this publication.

Note: **i2 Trade Service,** *in part, has been used as a reference source for some of the material prices used in Division 16.*

16055	Selective Demolition	CREW	DAILY OUTPUT	LABOR-HOURS	UNIT	2006 BARE COSTS				TOTAL INCL O&P
						MAT.	LABOR	EQUIP.	TOTAL	
300	**0010 ELECTRICAL DEMOLITION**									300
0020	Conduit to 4500 mm high, including fittings & hangers									
0100	Rigid galvanized steel, 13 mm to 25 mm diameter	1 Elec	73.76	.108	m		4.56		4.56	6.80
0120	31 mm to 50 mm	"	60.96	.131			5.50		5.50	8.20
0140	65 mm to 90 mm	2 Elec	92.05	.174			7.30		7.30	10.90
0160	100 mm to 150 mm	"	48.77	.328			13.80		13.80	20.50
0200	Electric metallic tubing (EMT), 13 mm to 25 mm	1 Elec	120	.067			2.80		2.80	4.17
0220	31 mm to 38 mm		99.36	.081			3.38		3.38	5.05
0240	50 mm to 75 mm	↓	71.93	.111			4.67		4.67	6.95
0260	88 mm to 100 mm	2 Elec	94.49	.169	↓		7.10		7.10	10.60
0270	Armored cable, (BX) avg. 15 m runs									
0280	#14, 2 wire	1 Elec	210	.038	m		1.60		1.60	2.39
0290	#14, 3 wire		174	.046			1.93		1.93	2.88
0300	#12, 2 wire		184	.043			1.83		1.83	2.72
0310	#12, 3 wire		157	.051			2.14		2.14	3.19
0320	#10, 2 wire		157	.051			2.14		2.14	3.19
0330	#10, 3 wire		130	.062			2.58		2.58	3.85
0340	#8, 3 wire	↓	104	.077	↓		3.23		3.23	4.82
0350	Non metallic sheathed cable (Romex)									
0360	#14, 2 wire	1 Elec	219	.037	m		1.53		1.53	2.29
0370	#14, 3 wire		200	.040			1.68		1.68	2.50
0380	#12, 2 wire		192	.042			1.75		1.75	2.61
0390	#10, 3 wire	↓	137	.058	↓		2.45		2.45	3.66
0400	Wiremold raceway, including fittings & hangers									
0420	No. 3000	1 Elec	76.20	.105	m		4.41		4.41	6.55
0440	No. 4000		66.14	.121			5.10		5.10	7.55
0460	No. 6000	↓	50.60	.158	↓		6.65		6.65	9.90
0465	Telephone/power pole		12	.667	Ea.		28		28	41.50
0470	Non-metallic, straight section	↓	146	.055	m		2.30		2.30	3.43
0500	Channels, steel, including fittings & hangers									
0520	19 mm x 38 mm	1 Elec	93.88	.085	m		3.58		3.58	5.35
0540	38 mm x 38 mm		81.99	.098			4.10		4.10	6.10
0560	38 mm x 47 mm	↓	69.80	.115	↓		4.81		4.81	7.15
0600	Copper bus duct, indoor, 3 phase									
0610	Including hangers & supports									
0620	225 A	2 Elec	41.15	.389	m		16.35		16.35	24.50
0640	400 A		32.31	.495			21		21	31
0660	600 A		26.21	.610			25.50		25.50	38
0680	1000 A		18.29	.875			37		37	55
0700	1600 A		12.19	1.312			55		55	82
0720	3000 A	↓	3.05	5.249	↓		220		220	330
1300	Transformer, dry type, 1 ph, incl. removal of									
1320	supports, wire & conduit terminations									
1340	1 kVA	1 Elec	7.70	1.039	Ea.		43.50		43.50	65
1420	75 kVA	2 Elec	2.50	6.400	"		269		269	400
1440	3 Phase to 600 V, primary									
1460	3 kVA	1 Elec	3.85	2.078	Ea.		87.50		87.50	130
1520	75 kVA	2 Elec	2.70	5.926			249		249	370
1550	300 kVA	R-3	1.80	11.111			460	85	545	785
1570	750 kVA	"	1.10	18.182	↓		755	139	894	1,275
1800	Wire, THW-THWN-THHN, removed from									
1810	in place conduit, to 4.6 m high									
1830	#14	1 Elec	1981	.004	m		.17		.17	.25
1840	#12		1676	.005			.20		.20	.30
1850	#10		1387	.006			.24		.24	.36
1860	#8	↓	1231	.007	↓		.27		.27	.41

16 ELECTRICAL

16055 | Selective Demolition

		CREW	DAILY OUTPUT	LABOR-HOURS	UNIT	2006 BARE COSTS				TOTAL INCL O&P
						MAT.	LABOR	EQUIP.	TOTAL	
300	1870	#6	1 Elec	994	.008	m		.34		.34	.50	**300**
	1880	#4	2 Elec	1615	.010			.42		.42	.62
	1890	#3		1524	.011			.44		.44	.66
	1900	#2		1359	.012			.49		.49	.74
	1910	1/0		1012	.016			.66		.66	.99
	1920	2/0		890	.018			.76		.76	1.13
	1930	3/0		762	.021			.88		.88	1.31
	1940	4/0		671	.024			1		1	1.49
	1950	250 kcmil		610	.026			1.10		1.10	1.64
	1960	300 kcmil		579	.028			1.16		1.16	1.73
	1970	350 kcmil		549	.029			1.22		1.22	1.82
	1980	400 kcmil		518	.031			1.30		1.30	1.93
	1990	500 kcmil	▼	494	.032	▼		1.36		1.36	2.03
	2000	Interior fluorescent fixtures, incl. supports									
	2010	& whips, to 4500 mm high									
	2100	Recessed drop-in 600 mm x 600 mm, 2 lamp	2 Elec	35	.457	Ea.		19.20		19.20	28.50
	2120	600 mm x 1200 mm, 2 lamp		33	.485			20.50		20.50	30.50
	2140	600 mm x 1200 mm, 4 lamp		30	.533			22.50		22.50	33.50
	2160	1200 mm x 1200 mm, 4 lamp	▼	20	.800	▼		33.50		33.50	50
	2180	Surface mount, acrylic lens & hinged frame									
	2200	300 mm x 1200 mm, 2 lamp	2 Elec	44	.364	Ea.		15.25		15.25	23
	2220	600 mm x 600 mm, 2 lamp		44	.364			15.25		15.25	23
	2260	600 mm x 1200 mm, 4 lamp		33	.485			20.50		20.50	30.50
	2280	1200 mm x 1200 mm, 4 lamp	▼	23	.696	▼		29		29	43.50
	2300	Strip fixtures, surface mount									
	2320	1200 mm long, 1 lamp	2 Elec	53	.302	Ea.		12.70		12.70	18.90
	2340	1200 mm long, 2 lamp		50	.320			13.45		13.45	20
	2360	2400 mm long, 1 lamp		42	.381			16		16	24
	2380	2400 mm long, 2 lamp	▼	40	.400	▼		16.80		16.80	25
	2400	Pendant mount, industrial, incl. removal									
	2410	of chain or rod hangers, to 4.6 m high									
	2420	1200 mm long, 2 lamp	2 Elec	35	.457	Ea.		19.20		19.20	28.50
	2440	2400 mm long, 2 lamp	"	27	.593	"		25		25	37

16060 | Grounding & Bonding

		CREW	DAILY OUTPUT	LABOR-HOURS	UNIT	2006 BARE COSTS				TOTAL INCL O&P
						MAT.	LABOR	EQUIP.	TOTAL	
800	0010	**GROUNDING**									**800**
	0030	Rod, copper clad, 2.4 m long, 15 mm diameter [1/2"]	1 Elec	5.50	1.455	Ea.	13.70	61		74.70	106
	0050	20 mm diameter [3/4"]		5.30	1.509		28.50	63.50		92	126
	0080	3.0 m long, 15 mm diameter [1/2"]		4.80	1.667		17.70	70		87.70	123
	0100	20 mm diameter [3/4"]		4.40	1.818		31.50	76.50		108	149
	0130	4.6 m long, 20 mm diameter [3/4"]		4	2	▼	87	84		171	221
	0390	Bare copper wire, #8 stranded		335	.024	m	.52	1		1.52	2.06
	0400	#6	▼	305	.026		.94	1.10		2.04	2.67
	0600	#2	2 Elec	305	.052		2.17	2.20		4.37	5.65
	0800	3/0		201	.080		4.83	3.34		8.17	10.30
	1000	4/0	▼	174	.092		6.05	3.86		9.91	12.40
	1200	250 kcmil	3 Elec	219	.110	▼	7.15	4.60		11.75	14.70
	1800	Water pipe ground clamps, heavy duty									
	2000	Bronze, 15 mm to 25 mm diameter [1/2" - 1"]	1 Elec	8	1	Ea.	14.40	42		56.40	78.50
	2100	32 mm to 50 mm diameter [1-1/4" - 2"]		8	1		18.95	42		60.95	83.50
	2200	65 mm to 80 mm diameter [2-1/2" - 3"]		6	1.333		41	56		97	129
	2800	Brazed connections, #6 wire		12	.667		12.35	28		40.35	55
	3000	#2 wire		10	.800		16.55	33.50		50.05	68
	3100	3/0 wire		8	1		25	42		67	90
	3200	4/0 wire	▼	7	1.143	▼	28.50	48		76.50	103

ELECTRICAL 16

16050 | Basic Electrical Materials & Methods

		16060	Grounding & Bonding	CREW	DAILY OUTPUT	LABOR-HOURS	UNIT	MAT.	LABOR	EQUIP.	TOTAL	TOTAL INCL O&P	
									2006 BARE COSTS				
800	3400		250 kcmil wire	1 Elec	5	1.600	Ea.	33	67		100	137	**800**
	3600		500 kcmil wire	↓	4	2	↓	41	84		125	170	

16100 | Wiring Methods

		16120	Conductors & Cables	CREW	DAILY OUTPUT	LABOR-HOURS	UNIT	MAT.	LABOR	EQUIP.	TOTAL	TOTAL INCL O&P	
									2006 BARE COSTS				
120	0010	**ARMORED CABLE**											**120**
	0050		600 V, copper (BX), #14, 2 conductor, solid	1 Elec	73.15	.109	m	1.92	4.59		6.51	8.95	
	0100		3 conductor, solid		67.06	.119		3.05	5		8.05	10.80	
	0150		#12, 2 conductor, solid		70.10	.114		1.95	4.79		6.74	9.30	
	0200		3 conductor, solid		60.96	.131		3.11	5.50		8.61	11.60	
	0250		#10, 2 conductor, solid		60.96	.131		3.54	5.50		9.04	12.10	
	0300		3 conductor, solid		48.77	.164		4.89	6.90		11.79	15.65	
	0350		#8, 3 conductor, solid		39.62	.202		8.80	8.50		17.30	22.50	
	0400		3 conductor with PVC jacket, in cable tray, #6	↓	94.49	.085		10.50	3.56		14.06	16.90	
	0450		#4	2 Elec	165	.097		13.60	4.07		17.67	21	
	0500		#2		140	.114		18.05	4.80		22.85	27	
	0550		#1		122	.131		24.50	5.50		30	35	
	0600		1/0		110	.145		29	6.10		35.10	40.50	
	0650		2/0		104	.154		35	6.45		41.45	48	
	0700		3/0		97.54	.164		41	6.90		47.90	55.50	
	0750		4/0	↓	91.44	.175		47	7.35		54.35	62.50	
	0800		250 kcmil	3 Elec	110	.218		52.50	9.15		61.65	71.50	
	0850		350 kcmil		101	.238		70	10		80	92	
	0900		500 kcmil	↓	91.44	.262	↓	91.50	11		102.50	116	
	1050		5 kV, copper, 3 conductor with PVC jacket,										
	1060		non-shielded, in cable tray, #4	2 Elec	116	.138	m	21	5.80		26.80	31.50	
	1100		#2		110	.145		27	6.10		33.10	38.50	
	1200		#1		91.44	.175		34.50	7.35		41.85	49	
	1400		1/0		88.39	.181		39.50	7.60		47.10	55	
	1600		2/0		79.25	.202		46	8.50		54.50	63	
	2000		4/0	↓	73.15	.219		61	9.20		70.20	80.50	
	2100		250 kcmil	3 Elec	101	.238		84	10		94	107	
	2150		350 kcmil		96.01	.250		103	10.50		113.50	130	
	2200		500 kcmil	↓	82.30	.292	↓	126	12.25		138.25	157	
	2400		15 kV, copper, 3 conductor with PVC jacket galv steel armored										
	2500		grounded neutral, in cable tray, #2	2 Elec	91.44	.175	m	43.50	7.35		50.85	59	
	2600		#1		85.34	.187		46.50	7.85		54.35	63	
	2800		1/0		79.25	.202		53.50	8.50		62	71.50	
	2900		2/0		67.06	.239		70.50	10		80.50	92.50	
	3000		4/0	↓	57.91	.276		80	11.60		91.60	105	
	3100		250 kcmil	3 Elec	82.30	.292		89	12.25		101.25	116	
	3150		350 kcmil		73.15	.328		105	13.80		118.80	136	
	3200		500 kcmil	↓	64.01	.375	↓	141	15.75		156.75	179	
	3400		15 kV, copper, 3 conductor with PVC jacket,										
	3450		ungrounded neutral, in cable tray, #2	2 Elec	79.25	.202	m	47	8.50		55.50	64	
	3500		#1		70.10	.228		52	9.60		61.60	72	
	3600		1/0		60.96	.262		59.50	11		70.50	82	
	3700		2/0		57.91	.276		73	11.60		84.60	98	
	3800		4/0	↓	48.77	.328	↓	88.50	13.80		102.30	118	

Important: See the Reference Section for supporting data - Crews, Rental Equipment, City Cost Indexes and Reference Data

16120	Conductors & Cables	CREW	DAILY OUTPUT	LABOR-HOURS	UNIT	2006 BARE COSTS				TOTAL INCL O&P
						MAT.	LABOR	EQUIP.	TOTAL	
120	4000	250 kcmil	3 Elec	64.01	.375	m	103	15.75		118.75	137	**120**
	4050	350 kcmil		59.44	.404		135	16.95		151.95	175
	4100	500 kcmil	▼	54.86	.437		165	18.35		183.35	209
	9010	600 V, copper (MC) steel clad, #14, 2 wire	1 Elec	73.15	.109		2.01	4.59		6.60	9.05
	9020	3 wire		67.06	.119		3.13	5		8.13	10.90
	9040	#12, 2 wire		70.10	.114		2.05	4.79		6.84	9.40
	9050	3 wire		60.96	.131		3.16	5.50		8.66	11.65
	9070	#10, 2 wire		60.96	.131		3.63	5.50		9.13	12.20
	9080	3 wire		48.77	.164		5.60	6.90		12.50	16.40
	9100	#8, 2 wire, stranded		54.86	.146		6.90	6.10		13	16.70
	9110	3 wire, stranded		39.62	.202		10.60	8.50		19.10	24.50
	9200	600 V, copper (MC) aluminum clad, #14, 2 wire		80.77	.099		2.01	4.16		6.17	8.40
	9210	3 wire		74.68	.107		3.13	4.50		7.63	10.15
	9220	4 wire		67.06	.119		4.19	5		9.19	12.05
	9230	#12, 2 wire		77.72	.103		2.05	4.32		6.37	8.70
	9240	3 wire		67.06	.119		3.16	5		8.16	10.90
	9250	4 wire		60.96	.131		4.28	5.50		9.78	12.90
	9260	#10, 2 wire		67.06	.119		3.72	5		8.72	11.55
	9270	3 wire		54.86	.146		5.75	6.10		11.85	15.50
	9280	4 wire	▼	47.24	.169	▼	9.10	7.10		16.20	20.50
230	0010	**CABLE TERMINATIONS**										**230**
	0015	Wire connectors, screw type, #22 to #14	1 Elec	260	.031	Ea.	.07	1.29		1.36	2.01
	0020	#18 to #12		240	.033		.08	1.40		1.48	2.18
	0025	#18 to #10		240	.033		.13	1.40		1.53	2.23
	0030	Screw-on connectors, insulated, #18 to #12		240	.033		.09	1.40		1.49	2.19
	0035	#16 to #10		230	.035		.12	1.46		1.58	2.31
	0040	#14 to #8		210	.038		.25	1.60		1.85	2.67
	0045	#12 to #6		180	.044		.42	1.87		2.29	3.24
	0050	Terminal lugs, solderless, #16 to #10		50	.160		.46	6.70		7.16	10.50
	0100	#8 to #4		30	.267		.64	11.20		11.84	17.40
	0150	#2 to #1		22	.364		.86	15.25		16.11	24
	0200	1/0 to 2/0		16	.500		2.22	21		23.22	34
	0250	3/0		12	.667		2.95	28		30.95	45
	0300	4/0		11	.727		2.95	30.50		33.45	49
	0350	250 kcmil		9	.889		2.95	37.50		40.45	59
	0400	350 kcmil		7	1.143		3.80	48		51.80	75.50
	0450	500 kcmil	▼	6	1.333	▼	7.40	56		63.40	91.50
	1600	Crimp, 1 hole lugs, copper or aluminum, 600 V									
	1620	#14	1 Elec	60	.133	Ea.	.32	5.60		5.92	8.70
	1630	#12		50	.160		.45	6.70		7.15	10.50
	1640	#10		45	.178		.45	7.45		7.90	11.65
	1780	#8		36	.222		1.72	9.35		11.07	15.80
	1800	#6		30	.267		2.78	11.20		13.98	19.75
	2000	#4		27	.296		3	12.45		15.45	22
	2200	#2		24	.333		4.07	14		18.07	25.50
	2400	#1		20	.400		4.90	16.80		21.70	30.50
	2600	2/0		15	.533		6.60	22.50		29.10	41
	2800	3/0		12	.667		8.20	28		36.20	50.50
	3000	4/0		11	.727		8.90	30.50		39.40	55.50
	3200	250 kcmil		9	.889		10.60	37.50		48.10	67
	3400	300 kcmil		8	1		11.90	42		53.90	75.50
	3500	350 kcmil		7	1.143		12.30	48		60.30	85
	3600	400 kcmil		6.50	1.231		15.30	51.50		66.80	94
	3800	500 kcmil	▼	6	1.333	▼	18.60	56		74.60	104
280	0010	**CONTROL CABLE**										**280**
	0020	600 V, copper, #14 THWN wire with PVC jacket, 2 wires	1 Elec	274	.029	m	.51	1.23		1.74	2.39

For expanded coverage of these items see *Means Electrical Cost Data 2006*

16120	Conductors & Cables	CREW	DAILY OUTPUT	LABOR-HOURS	UNIT	2006 BARE COSTS MAT.	LABOR	EQUIP.	TOTAL	TOTAL INCL O&P	
280	0100	4 wires	1 Elec	213	.038	m	.84	1.58		2.42	3.27	**280**
	0200	6 wires		183	.044		1.50	1.84		3.34	4.39	
	0300	8 wires		162	.049		1.84	2.07		3.91	5.10	
	0400	10 wires		146	.055		2.16	2.30		4.46	5.80	
	0500	12 wires		131	.061		2.47	2.56		5.03	6.55	
	0600	14 wires		116	.069		3.01	2.90		5.91	7.65	
	0700	16 wires		107	.075		3.10	3.14		6.24	8.10	
	0800	18 wires		101	.079		3.68	3.33		7.01	9	
	0900	20 wires		91.44	.087		4.27	3.67		7.94	10.20	
	1000	22 wires		85.34	.094		4.37	3.94		8.31	10.65	
500	0010	**MINERAL INSULATED CABLE** 600 V										**500**
	0100	1 conductor, #12	1 Elec	48.77	.164	m	7.75	6.90		14.65	18.75	
	0200	#10		48.77	.164		10.05	6.90		16.95	21.50	
	0400	#8		45.72	.175		11.85	7.35		19.20	24	
	0500	#6		42.67	.187		13.30	7.85		21.15	26.50	
	0600	#4	2 Elec	73.15	.219		18.40	9.20		27.60	33.50	
	0800	#2		67.06	.239		24.50	10		34.50	42	
	0900	#1		64.01	.250		28	10.50		38.50	46.50	
	1000	1/0		60.96	.262		32.50	11		43.50	52	
	1100	2/0		57.91	.276		38.50	11.60		50.10	60	
	1200	3/0		54.86	.292		45.50	12.25		57.75	68.50	
	1400	4/0		48.77	.328		52.50	13.80		66.30	78	
	1410	250 kcmil	3 Elec	73.15	.328		59.50	13.80		73.30	86	
	1420	350 kcmil		59.44	.404		72.50	16.95		89.45	105	
	1430	500 kcmil		59.44	.404		91	16.95		107.95	126	
550	0010	**NON-METALLIC SHEATHED CABLE** 600 V										**550**
	0100	Copper with ground wire, (Romex)										
	0150	#14, 2 conductor	1 Elec	82.30	.097	m	.60	4.08		4.68	6.75	
	0200	3 conductor		73.15	.109		.96	4.59		5.55	7.90	
	0250	#12, 2 conductor		76.20	.105		.88	4.41		5.29	7.50	
	0300	3 conductor		67.06	.119		1.59	5		6.59	9.20	
	0350	#10, 2 conductor		67.06	.119		1.58	5		6.58	9.20	
	0400	3 conductor		54.86	.146		1.99	6.10		8.09	11.35	
	0450	#8, 3 conductor		45.72	.175		4.51	7.35		11.86	15.90	
	0500	#6, 3 conductor		42.67	.187		7.05	7.85		14.90	19.50	
	0550	SE type SER aluminum cable, 3 RHW and										
	0600	1 bare neutral, 3 #8 & 1 #8	1 Elec	48.77	.164	m	4.35	6.90		11.25	15.05	
	0650	3 #6 & 1 #6	"	42.67	.187		4.92	7.85		12.77	17.15	
	0700	3 #4 & 1 #6	2 Elec	73.15	.219		5.50	9.20		14.70	19.75	
	0750	3 #2 & 1 #4		67.06	.239		8.10	10		18.10	24	
	0800	3 #1/0 & 1 #2		60.96	.262		12.30	11		23.30	30	
	0850	3 #2/0 & 1 #1		54.86	.292		14.50	12.25		26.75	34	
	0900	3 #4/0 & 1 #2/0		48.77	.328		20.50	13.80		34.30	43	
	1450	UF underground feeder cable, copper with ground, #14, 2 conductor	1 Elec	122	.066		.81	2.75		3.56	4.99	
	1500	#12, 2 conductor		107	.075		1.16	3.14		4.30	5.95	
	1550	#10, 2 conductor		91.44	.087		1.85	3.67		5.52	7.55	
	1600	#14, 3 conductor		107	.075		1.12	3.14		4.26	5.90	
	1650	#12, 3 conductor		91.44	.087		1.61	3.67		5.28	7.30	
	1700	#10, 3 conductor		76.20	.105		2.50	4.41		6.91	9.30	
700	0010	**SHIELDED CABLE** Splicing & terminations not included										**700**
	0040	Copper, XLP shielding, 5 kV, #6	2 Elec	134	.119	m	3.72	5		8.72	11.55	
	0050	#4		134	.119		4.83	5		9.83	12.75	
	0100	#2		122	.131		5.70	5.50		11.20	14.50	
	0200	#1		122	.131		6.55	5.50		12.05	15.40	
	0400	1/0		116	.138		7.20	5.80		13	16.60	

16 ELECTRICAL

16120	Conductors & Cables	CREW	DAILY OUTPUT	LABOR-HOURS	UNIT	2006 BARE COSTS				TOTAL INCL O&P	
						MAT.	LABOR	EQUIP.	TOTAL		
700	0600	2/0	2 Elec	110	.145	m	9.10	6.10		15.20	19.10	**700**
	0800	4/0	↓	97.54	.164		11.80	6.90		18.70	23.50	
	1000	250 kcmil	3 Elec	137	.175		14.60	7.35		21.95	27	
	1200	350 kcmil		119	.202		18.80	8.45		27.25	33	
	1400	500 kcmil	↓	110	.218		23	9.15		32.15	39	
	1600	15 kV, ungrounded neutral, #1	2 Elec	122	.131		7.95	5.50		13.45	16.95	
	1800	1/0		116	.138		9.50	5.80		15.30	19.10	
	2000	2/0		110	.145		10.85	6.10		16.95	21	
	2200	4/0	↓	97.54	.164		14.40	6.90		21.30	26	
	2400	250 kcmil	3 Elec	137	.175		15.95	7.35		23.30	28.50	
	2600	350 kcmil		119	.202		20	8.45		28.45	34.50	
	2800	500 kcmil	↓	110	.218		25	9.15		34.15	41	
	3000	25 kV, grounded neutral, #1/0	2 Elec	110	.145		13.15	6.10		19.25	23.50	
	3200	2/0		104	.154		14.50	6.45		20.95	25.50	
	3400	4/0	↓	91.44	.175		18.15	7.35		25.50	31	
	3600	250 kcmil	3 Elec	128	.188		22.50	7.90		30.40	37	
	3800	350 kcmil		110	.218		26.50	9.15		35.65	42.50	
	3900	500 kcmil	↓	101	.238		31	10		41	49	
	4000	35 kV, grounded neutral, #1/0	2 Elec	104	.154		13.95	6.45		20.40	25	
	4200	2/0		97.54	.164		16.40	6.90		23.30	28.50	
	4400	4/0	↓	85.34	.187		20.50	7.85		28.35	34.50	
	4600	250 kcmil	3 Elec	119	.202		24	8.45		32.45	39	
	4800	350 kcmil		101	.238		29	10		39	47	
	5000	500 kcmil	↓	91.44	.262		34	11		45	54	
	5050	Aluminum, XLP shielding, 5 kV, #2	2 Elec	152	.105		4.61	4.42		9.03	11.65	
	5070	#1		134	.119		4.76	5		9.76	12.70	
	5090	1/0		122	.131		5.50	5.50		11	14.25	
	5100	2/0		116	.138		6.20	5.80		12	15.45	
	5150	4/0	↓	110	.145		7.30	6.10		13.40	17.15	
	5200	250 kcmil	3 Elec	146	.164		8.85	6.90		15.75	20	
	5220	350 kcmil		137	.175		10.45	7.35		17.80	22.50	
	5240	500 kcmil		119	.202		13.40	8.45		21.85	27.50	
	5260	750 kcmil	↓	110	.218		17.70	9.15		26.85	33	
	5300	15 kV aluminum, XLP, #1	2 Elec	134	.119		5.90	5		10.90	13.95	
	5320	1/0		122	.131		6.10	5.50		11.60	14.90	
	5340	2/0		116	.138		7.30	5.80		13.10	16.65	
	5360	4/0	↓	110	.145		8.05	6.10		14.15	18	
	5380	250 kcmil	3 Elec	146	.164		9.65	6.90		16.55	21	
	5400	350 kcmil		137	.175		10.80	7.35		18.15	23	
	5420	500 kcmil		119	.202		14.95	8.45		23.40	29	
	5440	750 kcmil	↓	110	.218	↓	22.50	9.15		31.65	38	
900	0010	**WIRE** R260519 -92										**900**
	0020	600 V, type THW, copper, solid, #14	1 Elec	396	.020	m	.14	.85		.99	1.41	
	0030	#12		335	.024		.21	1		1.21	1.72	
	0040	#10		305	.026		.32	1.10		1.42	1.99	
	0050	Stranded, #14 R260533 -22		396	.020		.16	.85		1.01	1.44	
	0100	#12		335	.024		.24	1		1.24	1.75	
	0120	#10		305	.026		.36	1.10		1.46	2.04	
	0140	#8		244	.033		.65	1.38		2.03	2.76	
	0160	#6	↓	198	.040		1.07	1.70		2.77	3.70	
	0180	#4	2 Elec	323	.050		1.64	2.08		3.72	4.90	
	0200	#3		305	.052		2.05	2.20		4.25	5.55	
	0220	#2		274	.058		2.59	2.45		5.04	6.50	
	0240	#1		244	.066		3.28	2.75		6.03	7.70	
	0260	1/0	↓	201	.080	↓	3.64	3.34		6.98	9	

ELECTRICAL 16

16120 | Conductors & Cables

			CREW	DAILY OUTPUT	LABOR-HOURS	UNIT	2006 BARE COSTS				TOTAL INCL O&P	
							MAT.	LABOR	EQUIP.	TOTAL		
900	0280	2/0	2 Elec	177	.090	m	4.26	3.80		8.06	10.35	900
	0300	3/0	R260519 -92	152	.105		5.45	4.42		9.87	12.55	
	0350	4/0	▼	134	.119		6.75	5		11.75	14.85	
	0400	250 kcmil	R260533 -22	183	.131		8.05	5.50		13.55	17.05	
	0420	300 kcmil	3 Elec	174	.138		9.55	5.80		15.35	19.15	
	0450	350 kcmil		165	.145		11.20	6.10		17.30	21.50	
	0480	400 kcmil		155	.155		12.80	6.50		19.30	24	
	0490	500 kcmil	▼	146	.164		15.80	6.90		22.70	27.50	
	0530	Aluminum, stranded, #8	1 Elec	274	.029		.59	1.23		1.82	2.48	
	0540	#6	"	244	.033		.80	1.38		2.18	2.93	
	0560	#4	2 Elec	396	.040		1	1.70		2.70	3.63	
	0580	#2		323	.050		1.35	2.08		3.43	4.59	
	0600	#1		274	.058		1.98	2.45		4.43	5.85	
	0620	1/0		244	.066		2.37	2.75		5.12	6.70	
	0640	2/0		219	.073		2.81	3.07		5.88	7.65	
	0680	3/0		201	.080		3.48	3.34		6.82	8.80	
	0700	4/0	▼	189	.085		3.87	3.56		7.43	9.55	
	0720	250 kcmil	3 Elec	265	.091		4.72	3.80		8.52	10.85	
	0740	300 kcmil		247	.097		6.50	4.08		10.58	13.25	
	0760	350 kcmil		229	.105		6.65	4.40		11.05	13.85	
	0780	400 kcmil		210	.114		7.75	4.80		12.55	15.70	
	0800	500 kcmil		183	.131		9.10	5.50		14.60	18.20	
	0850	600 kcmil		174	.138		10.85	5.80		16.65	20.50	
	0880	700 kcmil		155	.155		12.50	6.50		19	23.50	
	0900	750 kcmil	▼	146	.164		12.65	6.90		19.55	24	
	0920	Type THWN-THHN, copper, solid, #14	1 Elec	396	.020		.14	.85		.99	1.41	
	0940	#12		335	.024		.21	1		1.21	1.72	
	0960	#10		305	.026		.32	1.10		1.42	1.99	
	1000	Stranded, #14		396	.020		.16	.85		1.01	1.44	
	1200	#12		335	.024		.24	1		1.24	1.75	
	1250	#10		305	.026		.36	1.10		1.46	2.04	
	1300	#8		244	.033		.65	1.38		2.03	2.76	
	1350	#6	▼	198	.040		1.07	1.70		2.77	3.70	
	1400	#4	2 Elec	323	.050	▼	1.64	2.08		3.72	4.90	

16131 | Cable Trays

			CREW	DAILY OUTPUT	LABOR-HOURS	UNIT	MAT.	LABOR	EQUIP.	TOTAL	INCL O&P	
105	0010	**CABLE TRAY LADDER TYPE** w/ftngs & supports,102 mm dp,to 4.6 m										105
	0160	Galvanized steel tray										
	0170	102 mm rung spacing, 152 mm wide	2 Elec	29.87	.536	m	35	22.50		57.50	72	
	0200	305 mm wide		26.21	.610		42	25.50		67.50	84	
	0400	457 mm wide		24.99	.640		48.50	27		75.50	93.50	
	0600	610 mm wide		23.77	.673		56	28.50		84.50	104	
	3200	Aluminum tray, 102 mm deep, 152 mm rung spacing, 152 mm wide		40.84	.392		42	16.45		58.45	70.50	
	3220	305 mm wide		37.80	.423		47	17.80		64.80	78	
	3230	457 mm wide		34.75	.460		52.50	19.35		71.85	86.50	
	3240	610 mm wide	▼	32.31	.495	▼	61	21		82	98	
	9980	Allow. for tray ftngs., 5% min.-20% max.										

16132 | Conduit & Tubing

			CREW	DAILY OUTPUT	LABOR-HOURS	UNIT	MAT.	LABOR	EQUIP.	TOTAL	INCL O&P	
205	0010	**CONDUIT** To 4.6 m high, includes 2 terminations, 2 elbows and										205
	0020	11 beam clamps & couplings per 30.5 m	R260533 -22									
	0300	Aluminum, 15 mm diameter [1/2"]	1 Elec	30.48	.262	m	6.35	11		17.35	23.50	
	0500	20 mm diameter [3/4"]		27.43	.292		8.50	12.25		20.75	27.50	
	0700	25 mm diameter [1"]		24.38	.328		11.25	13.80		25.05	33	
	1000	32 mm diameter [1-1/4"]	▼	21.34	.375	▼	15.10	15.75		30.85	40	

			CREW	DAILY OUTPUT	LABOR-HOURS	UNIT	2006 BARE COSTS				TOTAL INCL O&P	
16132		**Conduit & Tubing**					MAT.	LABOR	EQUIP.	TOTAL		
205	1030	40 mm diameter [1-1/2"]	1 Elec	19.81	.404	m	18.15	16.95		35.10	45.50	205
	1050	50 mm diameter [2"]		18.29	.437		24.50	18.35		42.85	54.50	
	1070	65 mm diameter [2-1/2"]	▼	15.24	.525		39.50	22		61.50	76.50	
	1100	80 mm diameter [3"]	2 Elec	27.43	.583		53.50	24.50		78	95.50	
	1130	90 mm diameter [3-1/2"]		24.38	.656		70	27.50		97.50	118	
	1140	100 mm diameter [4"]	▼	21.34	.750		84	31.50		115.50	140	
	1750	Rigid galvanized steel, 15 mm diameter [1/2"]	1 Elec	27.43	.292		8	12.25		20.25	27	
	1770	20 mm diameter [3/4"]		24.38	.328		9.10	13.80		22.90	30.50	
	1800	25 mm diameter [1"]		19.81	.404		13.40	16.95		30.35	40.50	
	1830	32 mm diameter [1-1/4"]		18.29	.437		18.30	18.35		36.65	47.50	
	1850	40 mm diameter [1-1/2"]		16.76	.477		21	20		41	53	
	1870	50 mm diameter [2"]		13.72	.583		27	24.50		51.50	66.50	
	1900	65 mm diameter [2-1/2"]	▼	10.67	.750		51	31.50		82.50	103	
	1930	80 mm diameter [3"]	2 Elec	15.24	1.050		62	44		106	134	
	1950	90 mm diameter [3-1/2"]		13.41	1.193		76	50		126	158	
	1970	100 mm diameter [4"]	▼	12.19	1.312		87	55		142	178	
	2500	Steel, intermediate conduit (IMC), 15 mm diameter [1/2"]	1 Elec	30.48	.262		7	11		18	24	
	2530	20 mm diameter [3/4"]		27.43	.292		8.30	12.25		20.55	27.50	
	2550	25 mm diameter [1"]		21.34	.375		11.95	15.75		27.70	36.50	
	2570	32 mm diameter [1-1/4"]		19.81	.404		15.75	16.95		32.70	43	
	2600	40 mm diameter [1-1/2"]		18.29	.437		18.35	18.35		36.70	47.50	
	2630	50 mm diameter [2"]		15.24	.525		23.50	22		45.50	58.50	
	2650	65 mm diameter [2-1/2"]	▼	12.19	.656		48	27.50		75.50	94	
	2670	80 mm diameter [3"]	2 Elec	18.29	.875		62.50	37		99.50	124	
	2700	90 mm diameter [3-1/2"]		16.46	.972		77	41		118	146	
	2730	100 mm diameter [4"]	▼	15.24	1.050		88	44		132	163	
	5000	Electric metallic tubing (EMT), 15 mm diameter [1/2"]	1 Elec	51.82	.154		1.97	6.50		8.47	11.80	
	5020	20 mm diameter [3/4"]		39.62	.202		3.22	8.50		11.72	16.20	
	5040	25 mm diameter [1"]		35.05	.228		5.70	9.60		15.30	20.50	
	5060	32 mm diameter [1-1/4"]		30.48	.262		8.75	11		19.75	26	
	5080	40 mm diameter [1-1/2"]		27.43	.292		11.05	12.25		23.30	30.50	
	5100	50 mm diameter [2"]		24.38	.328		14	13.80		27.80	36	
	5120	65 mm diameter [2-1/2"]	▼	18.29	.437		26.50	18.35		44.85	57	
	5140	80 mm diameter [3"]	2 Elec	30.48	.525		36	22		58	72.50	
	5160	90 mm diameter [3-1/2"]		27.43	.583		46	24.50		70.50	87	
	5180	100 mm diameter [4"]	▼	24.38	.656	▼	52.50	27.50		80	98.50	
	9900	Add to labor for higher elevated installation										
	9910	4.6 m to 6.1 m high, add						10%				
	9920	6.1 m to 7.6 m high, add						20%				
	9930	7.6 m to 9.1 m high, add						25%				
	9940	9.1 m to 10.7 m high, add						30%				
	9950	10.7 m to 12.2 m high, add						35%				
	9960	Over 12.2 m high, add						40%				
	9980	Allow. for cond. ftngs., 5% min.-20% max.										
230	0010	**CONDUIT IN CONCRETE SLAB** Including terminations,										230
	0020	fittings and supports										
	3230	PVC, schedule 40, 15 mm diameter [1/2"]	1 Elec	82.30	.097	m	1.87	4.08		5.95	8.15	
	3250	20 mm diameter [3/4"]		70.10	.114		2.20	4.79		6.99	9.60	
	3270	25 mm diameter [1"]		60.96	.131		2.89	5.50		8.39	11.40	
	3300	32 mm diameter [1-1/4"]		51.82	.154		4.13	6.50		10.63	14.20	
	3330	40 mm diameter [1-1/2"]		42.67	.187		4.99	7.85		12.84	17.25	
	3350	50 mm diameter [2"]		36.58	.219		6.40	9.20		15.60	21	
	3370	65 mm diameter [2-1/2"]	▼	27.43	.292		10.65	12.25		22.90	30	
	3400	80 mm diameter [3"]	2 Elec	48.77	.328		13.50	13.80		27.30	35.50	
	3430	90 mm diameter [3-1/2"]		36.58	.437		18	18.35		36.35	47.50	
	3440	100 mm diameter [4"]	▼	30.48	.525	▼	20	22		42	55	

R260533
-22

ELECTRICAL 16

			DAILY	LABOR-		2006 BARE COSTS				TOTAL	
16132	**Conduit & Tubing**	CREW	OUTPUT	HOURS	UNIT	MAT.	LABOR	EQUIP.	TOTAL	INCL O&P	
230 3450	125 mm diameter [5"]	2 Elec	24.38	.656	m	30	27.50		57.50	74	**230**
3460	150 mm diameter [6"]	↓	18.29	.875	↓	43	37		80	102	
3530	Sweeps, 25 mm diameter, 762 mm radius [1"]	1 Elec	32	.250	Ea.	15.70	10.50		26.20	33	
3550	32 mm diameter [1-1/4"]		24	.333		19.85	14		33.85	43	
3570	40 mm diameter [1-1/2"]		21	.381		20.50	16		36.50	46.50	
3600	50 mm diameter [2"]		18	.444		21.50	18.65		40.15	51.50	
3630	65 mm diameter [2-1/2"]		14	.571		36.50	24		60.50	76.50	
3650	80 mm diameter [3"]		10	.800		42	33.50		75.50	96	
3670	90 mm diameter [3-1/2"]		8	1		62.50	42		104.50	132	
3700	100 mm diameter [4"]		7	1.143		60	48		108	137	
3710	125 mm diameter [5"]	↓	6	1.333		91	56		147	184	
3730	Couplings, 15 mm diameter [1/2"]					.41			.41	.45	
3750	20 mm diameter [3/4"]					.50			.50	.55	
3770	25 mm diameter [1"]					.77			.77	.85	
3800	32 mm diameter [1-1/4"]					1.02			1.02	1.12	
3830	40 mm diameter [1-1/2"]					1.40			1.40	1.54	
3850	50 mm diameter [2"]					1.86			1.86	2.05	
3870	65 mm diameter [2-1/2"]					3.30			3.30	3.63	
3900	80 mm diameter [3"]					5.40			5.40	5.95	
3930	90 mm diameter [3-1/2"]					6			6	6.60	
3950	100 mm diameter [4"]					8.35			8.35	9.20	
3960	125 mm diameter [5"]					21			21	23.50	
3970	150 mm diameter [6"]					27			27	30	
4030	End bells 25 mm diameter, PVC [1"]	1 Elec	60	.133		4	5.60		9.60	12.75	
4050	32 mm diameter [1-1/4"]		53	.151		4.95	6.35		11.30	14.90	
4100	40 mm diameter [1-1/2"]		48	.167		4.95	7		11.95	15.90	
4150	50 mm diameter [2"]		34	.235		7.30	9.90		17.20	23	
4170	65 mm diameter [2-1/2"]		27	.296		8.10	12.45		20.55	27.50	
4200	80 mm diameter [3"]		20	.400		8.60	16.80		25.40	34.50	
4250	90 mm diameter [3-1/2"]		16	.500		9.40	21		30.40	42	
4300	100 mm diameter [4"]		14	.571		10.20	24		34.20	47	
4310	125 mm diameter [5"]		12	.667		16.05	28		44.05	59	
4320	150 mm diameter [6"]		9	.889	↓	17.60	37.50		55.10	75	
4350	Rigid galvanized steel, 15 mm diameter [1/2"]		60.96	.131	m	7.30	5.50		12.80	16.25	
4400	20 mm diameter [3/4"]		51.82	.154		8.40	6.50		14.90	18.90	
4450	25 mm diameter [1"]		39.62	.202		12.75	8.50		21.25	26.50	
4500	32 mm diameter [1-1/4"]		33.53	.239		16.85	10		26.85	33.50	
4600	40 mm diameter [1-1/2"]		30.48	.262		19.60	11		30.60	38	
4800	50 mm diameter [2"]	↓	27.43	.292	↓	25	12.25		37.25	46	
240 0010	**CONDUIT IN TRENCH** Includes terminations and fittings										**240**
0020	Does not include excavation or backfill, see div. 02315										
0200	Rigid galvanized steel, 50 mm diameter [2"]	1 Elec	45.72	.175	m	24	7.35		31.35	37.50	
0400	65 mm diameter [2-1/2"]	"	30.48	.262		47	11		58	68	
0600	80 mm diameter [3"]	2 Elec	48.77	.328		58	13.80		71.80	84	
0800	90 mm diameter [3-1/2"]		42.67	.375		73.50	15.75		89.25	104	
1000	100 mm diameter [4"]		30.48	.525		81	22		103	122	
1200	125 mm diameter [5"]		24.38	.656		176	27.50		203.50	234	
1400	150 mm diameter [6"]	↓	18.29	.875	↓	257	37		294	340	
16133	**Multi-outlet Assemblies**										
540 0010	**TRENCH DUCT** Steel with cover										**540**
0020	Standard adjustable, depths to 102 mm										
0100	Straight, single compartment, 229 mm wide	2 Elec	12.19	1.312	m	262	55		317	370	
0200	305 mm wide	↓	9.75	1.640		299	69		368	435	

			DAILY	LABOR-		2006 BARE COSTS				TOTAL	
16133	**Multi-outlet Assemblies**	CREW	OUTPUT	HOURS	UNIT	MAT.	LABOR	EQUIP.	TOTAL	INCL O&P	
540	0400	457 mm wide	2 Elec	7.92	2.019	m	395	85		480	560	**540**
	0600	610 mm wide		6.71	2.386		510	100		610	710	
	0800	762 mm wide		6.10	2.625		615	110		725	840	
	1000	914 mm wide		4.88	3.281	↓	740	138		878	1,025	
	1200	Horizontal elbow, 229 mm wide		5.40	2.963	Ea.	296	124		420	510	
	1400	305 mm wide		4.60	3.478		340	146		486	595	
	1600	457 mm wide		4	4		440	168		608	730	
	1800	610 mm wide		3.20	5		615	210		825	990	
	2000	762 mm wide		2.60	6.154		820	258		1,078	1,275	
	2200	914 mm wide		2.40	6.667		1,075	280		1,355	1,600	
	2400	Vertical elbow, 229 mm wide		5.40	2.963		103	124		227	298	
	2600	305 mm wide		4.60	3.478		112	146		258	340	
	2800	457 mm wide		4	4		129	168		297	390	
	3000	610 mm wide		3.20	5		160	210		370	490	
	3200	762 mm wide		2.60	6.154		177	258		435	580	
	3400	914 mm wide		2.40	6.667		194	280		474	630	
	3600	Cross, 229 mm wide		4	4		485	168		653	785	
	3800	305 mm wide		3.20	5		515	210		725	880	
	4000	457 mm wide		2.60	6.154		615	258		873	1,050	
	4200	610 mm wide		2.20	7.273		780	305		1,085	1,300	
	4400	762 mm wide		2	8		1,000	335		1,335	1,625	
	4600	914 mm wide		1.80	8.889		1,275	375		1,650	1,950	
	4800	End closure, 229 mm wide		14.40	1.111		30.50	46.50		77	103	
	5000	305 mm wide		12	1.333		34.50	56		90.50	122	
	5200	457 mm wide		10	1.600		53	67		120	159	
	5400	610 mm wide		8	2		70	84		154	202	
	5600	762 mm wide		6.60	2.424		87.50	102		189.50	249	
	5800	914 mm wide		5.80	2.759		104	116		220	287	
	6000	Tees, 229 mm wide		4	4		295	168		463	575	
	6200	305 mm wide		3.60	4.444		340	187		527	655	
	6400	457 mm wide		3.20	5		440	210		650	795	
	6600	610 mm wide		3	5.333		630	224		854	1,025	
	6800	762 mm wide		2.60	6.154		820	258		1,078	1,275	
	7000	914 mm wide		2	8		1,075	335		1,410	1,675	
	7200	Riser and cabinet connector, 229 mm wide		5.40	2.963		129	124		253	325	
	7400	305 mm wide		4.60	3.478		150	146		296	385	
	7600	457 mm wide		4	4		184	168		352	450	
	7800	610 mm wide		3.20	5		223	210		433	560	
	8000	762 mm wide	↓	2.60	6.154		257	258		515	670	
	8200	914 mm wide		2	8	↓	299	335		634	830	
	8400	Insert assembly, cell to conduit adapter, 32 mm	1 Elec	16	.500		51	21		72	87.50	
560	0010	**UNDERFLOOR DUCT**										**560**
	0100	Duct, 35 mm x 79 mm blank, standard	2 Elec	48.77	.328	m	32	13.80		45.80	55.50	
	0200	35 mm x 184 mm blank, super duct		36.58	.437		64	18.35		82.35	98	
	0400	22 mm or 35 mm insert , 610 mm O.C., 35 mm, x 79 mm, std.		42.67	.375		43	15.75		58.75	71	
	0600	35 mm x 184 mm, super duct	↓	30.48	.525	↓	75	22		97	116	
	0800	Junction box, single duct, 1 level, 79 mm	1 Elec	4	2	Ea.	294	84		378	450	
	1000	Junction box, single duct, 1 level, 184 mm		2.70	2.963		340	124		464	560	
	1200	1 level, 2 duct, 79 mm		3.20	2.500		390	105		495	585	
	1400	Junction box, 1 level, 2 duct, 184 mm		2.30	3.478		1,000	146		1,146	1,325	
	1600	Triple duct, 79 mm		2.30	3.478		665	146		811	950	
	1800	Insert to conduit adapter, 19 mm & 25 mm		32	.250		24	10.50		34.50	42	
	2000	Support, single cell		27	.296		36	12.45		48.45	58	
	2200	Super duct		16	.500		36	21		57	71	
	2400	Double cell	↓	16	.500	↓	36	21		57	71	

ELECTRICAL 16

16133 | Multi-outlet Assemblies

		CREW	DAILY OUTPUT	LABOR-HOURS	UNIT	2006 BARE COSTS				TOTAL INCL O&P	
						MAT.	LABOR	EQUIP.	TOTAL		
560	2600	Triple cell	1 Elec	11	.727	Ea.	36	30.50		66.50	85	560
	2800	Vertical elbow, standard duct		10	.800		64.50	33.50		98	121	
	3000	Super duct		8	1		64.50	42		106.50	134	
	3200	Cabinet connector, standard duct		32	.250		49	10.50		59.50	69	
	3400	Super duct		27	.296		49	12.45		61.45	72	
	3600	Conduit adapter, 25 mm to 32 mm		32	.250		49	10.50		59.50	69	
	3800	51 mm to 32 mm		27	.296		58.50	12.45		70.95	82.50	
	4000	Outlet, low tension (telephone, computer, etc.)		8	1		68.50	42		110.50	138	
	4200	High tension, receptacle (120 V)	▼	8	1	▼	68.50	42		110.50	138	
800	0010	**SURFACE RACEWAY**										800
	0090	Metal, straight section										
	0100	No. 500	1 Elec	30.48	.262	m	2.66	11		13.66	19.35	
	0110	No. 700		30.48	.262		2.99	11		13.99	19.75	
	0400	No. 1500, small pancake		27.43	.292		5.40	12.25		17.65	24	
	0600	No. 2000, base & cover, blank		27.43	.292		5.30	12.25		17.55	24	
	0800	No. 3000, base & cover, blank		22.86	.350		10.65	14.70		25.35	34	
	1000	No. 4000, base & cover, blank		19.81	.404		17.20	16.95		34.15	44.50	
	1200	No. 6000, base & cover, blank		15.24	.525	▼	29	22		51	65	
	2400	Fittings, elbows, No. 500		40	.200	Ea.	1.47	8.40		9.87	14.10	
	2800	Elbow cover, No. 2000		40	.200		2.80	8.40		11.20	15.60	
	2880	Tee, No. 500		42	.190		2.82	8		10.82	15	
	2900	No. 2000		27	.296		8.80	12.45		21.25	28.50	
	3000	Switch box, No. 500		16	.500		9.55	21		30.55	42	
	3400	Telephone outlet, No. 1500		16	.500		10.60	21		31.60	43	
	3600	Junction box, No. 1500	▼	16	.500	▼	7.40	21		28.40	39.50	
	3800	Plugmold wired sections, No. 2000										
	4000	1 circuit, 6 outlets, 0.9 m long	1 Elec	8	1	Ea.	27	42		69	92	
	4100	2 circuits, 8 outlets, 1.8 m long	"	5.30	1.509	"	45	63.50		108.50	144	

16134 | Wireway & Aux Gutters

		CREW	DAILY OUTPUT	LABOR-HOURS	UNIT	MAT.	LABOR	EQUIP.	TOTAL	TOTAL INCL O&P	
150	0010	**WIREWAY** to 4.6 m high										150
	0100	Screw cover, NEMA 1 w/ fittings and supports, 64 mm x 64 mm	1 Elec	13.72	.583	m	36	24.50		60.50	76	
	0200	102 mm x 102 mm	"	12.19	.656		38	27.50		65.50	83	
	0400	152 mm x 152 mm	2 Elec	18.29	.875		64.50	37		101.50	126	
	0600	203 mm x 203 mm	"	12.19	1.312		108	55		163	201	
	4475	Screw cover, NEMA 3R w/ fittings and supports, 64 mm x 64 mm	1 Elec	10.97	.729		80.50	30.50		111	135	
	4480	152 mm x 152 mm	2 Elec	16.76	.954		108	40		148	178	
	4485	203 mm x 203 mm		10.97	1.458		173	61		234	282	
	4490	305 mm x 305 mm	▼	5.49	2.916	▼	241	122		363	450	
	9980	Allow. for wireway ftngs., 5% min.-20% max.										

16136 | Boxes

		CREW	DAILY OUTPUT	LABOR-HOURS	UNIT	MAT.	LABOR	EQUIP.	TOTAL	TOTAL INCL O&P	
600	0010	**OUTLET BOXES**										600
	0020	Pressed steel, octagon, 102 mm	1 Elec	20	.400	Ea.	2.21	16.80		19.01	27.50	
	0060	Covers, blank		64	.125		.92	5.25		6.17	8.85	
	0100	Extension rings		40	.200		3.64	8.40		12.04	16.50	
	0150	Square, 102 mm		20	.400		2.24	16.80		19.04	27.50	
	0200	Extension rings		40	.200		3.71	8.40		12.11	16.60	
	0250	Covers, blank		64	.125		1.04	5.25		6.29	9	
	0300	Plaster rings		64	.125		2.03	5.25		7.28	10.10	
	0650	Switchbox		27	.296		3.54	12.45		15.99	22.50	
	1100	Concrete, floor, 1 gang		5.30	1.509		67.50	63.50		131	169	
	2000	Poke-thru fitting, fire rated, for 95 mm floor		6.80	1.176		100	49.50		149.50	184	
	2040	For 178 mm floor	▼	6.80	1.176	▼	100	49.50		149.50	184	

16136	Boxes		CREW	DAILY OUTPUT	LABOR-HOURS	UNIT	2006 BARE COSTS				TOTAL INCL O&P	
							MAT.	LABOR	EQUIP.	TOTAL		
600	2100	Pedestal, 15 A, duplex receptacle & blank plate	1 Elec	5.25	1.524	Ea.	103	64		167	210	**600**
	2120	Duplex receptacle and telephone plate		5.25	1.524		104	64		168	210	
	2140	Pedestal, 20 A, duplex recept. & phone plate		5	1.600		105	67		172	215	
	2200	Abandonment plate		32	.250		30.50	10.50		41	49	
700	0010	**PULL BOXES & CABINETS**										**700**
	0100	Sheet met, box, NEMA 1, SC, 152 mm W x 152 mm H x 102 mm D	1 Elec	8	1	Ea.	10.80	42		52.80	74.50	
	0200	203 mm W x 203 mm H x 102 mm D R260533 -70		8	1		14.80	42		56.80	79	
	0300	254 mm W x 305 mm H x 152 mm D		5.30	1.509		26	63.50		89.50	124	
	0400	406 mm W x 508 mm H x 203 mm D		4	2		98.50	84		182.50	233	
	0500	508 mm W x 610 mm H x 203 mm D		3.20	2.500		115	105		220	284	
	0600	610 mm W x 915 mm H x 203 mm D		2.70	2.963		163	124		287	365	
	0650	Hinged cabinets, NEMA 1, 152 mm W x 152 mm H x 102 mm D		8	1		11	42		53	74.50	
	0800	305 mm W x 406 mm H x 152 mm D		4.70	1.702		36	71.50		107.50	147	
	1000	508 mm W x 508 mm H x 152 mm D		3.60	2.222		72.50	93.50		166	219	
	1200	508 mm W x 508 mm H x 203 mm D		3.20	2.500		145	105		250	315	
	1400	610 mm W x 914 mm H x 203 mm D		2.70	2.963		233	124		357	440	
	1600	610 mm W x 1067 mm H x 203 mm D		2	4		355	168		523	640	
	2100	Pull box, NEMA 3R, type SC, raintight & weatherproof										
	2150	152 mm L x 152 mm W x 152 mm D	1 Elec	10	.800	Ea.	18.50	33.50		52	70.50	
	2200	203 mm L x 152 mm W x 152 mm D		8	1		23	42		65	88	
	2250	254 mm L x 152 mm W x 152 mm D		7	1.143		31	48		79	106	
	2300	305 mm L x 305 mm W x 152 mm D		5	1.600		43.50	67		110.50	148	
	2350	406 mm L x 406 mm W x 152 mm D		4.50	1.778		86.50	74.50		161	207	
	2400	508 mm L x 508 mm W x 152 mm D		4	2		119	84		203	256	
	2450	610 mm L x 457 mm W x 203 mm D		3	2.667		130	112		242	310	
	2500	610 mm L x 610 mm W x 254 mm D		2.50	3.200		174	134		308	390	
	2550	760 mm L x 600 mm W x 300 mm D		2	4		315	168		483	600	
	2600	900 mm L x 900 mm W x 300 mm D		1.50	5.333		415	224		639	795	
	2800	Cast iron, pull boxes for surface mounting										
	3000	NEMA 4, watertight & dust tight										
	3050	152 mm L x 152 mm W x 152 mm D	1 Elec	4	2	Ea.	164	84		248	305	
	3100	203 mm L x 152 mm W x 152 mm D		3.20	2.500		222	105		327	400	
	3150	254 mm L x 152 mm W x 152 mm D		2.50	3.200		277	134		411	505	
	3200	305 mm L x 305 mm W x 152 mm D		2.30	3.478		465	146		611	735	
	3250	406 mm L x 406 mm W x 152 mm D		1.30	6.154		955	258		1,213	1,425	
	3300	508 mm L x 508 mm W x 152 mm D		.80	10		1,775	420		2,195	2,575	
	3350	610 mm L x 457 mm W x 203 mm D		.70	11.429		1,875	480		2,355	2,775	
	3400	610 mm L x 610 mm W x 254 mm D		.50	16		2,875	670		3,545	4,150	
	3450	760 mm L x 610 mm W x 305 mm D		.40	20		4,800	840		5,640	6,550	
	3500	915 mm L x 915 mm W x 305 mm D		.20	40		5,950	1,675		7,625	9,050	
	6000	J.I.C. wiring boxes, NEMA 12, dust tight & drip tight										
	6050	152 mm L x 203 mm W x 102 mm D	1 Elec	10	.800	Ea.	43	33.50		76.50	97.50	
	6100	203 mm L x 254 mm W x 102 mm D		8	1		54	42		96	122	
	6150	305 mm L x 356 mm W x 152 mm D		5.30	1.509		88	63.50		151.50	191	
	6200	356 mm L x 406 mm W x 152 mm D		4.70	1.702		105	71.50		176.50	222	
	6250	406 mm L x 508 mm W x 152 mm D		4.40	1.818		223	76.50		299.50	360	
	6300	610 mm L x 762 mm W x 152 mm D		3.20	2.500		330	105		435	515	
	6350	610 mm L x 762 mm W x 203 mm D		2.90	2.759		350	116		466	560	
	6400	610 mm L x 914 mm W x 203 mm D		2.70	2.963		390	124		514	615	
	6450	610 mm L x 1067 mm W x 203 mm D		2.30	3.478		425	146		571	690	
	6500	610 mm L x 1219 mm W x 203 mm D		2	4		465	168		633	760	
	7000	Cabinets, current transformer										
	7050	Single door, 610 mm H x 610 mm W x 254 mm D	1 Elec	1.60	5	Ea.	139	210		349	470	
	7100	762 mm H x 610 mm W x 254 mm D		1.30	6.154		153	258		411	555	
	7150	914 mm H x 610 mm W x 254 mm D		1.10	7.273		161	305		466	630	
	7200	762 mm H x 762 mm W x 254 mm D		1	8		167	335		502	685	

ELECTRICAL 16

For expanded coverage of these items see *Means Electrical Cost Data 2006*

16136 | Boxes

		CREW	DAILY OUTPUT	LABOR-HOURS	UNIT	2006 BARE COSTS				TOTAL INCL O&P
						MAT.	LABOR	EQUIP.	TOTAL	
700 7250	914 mm H x 762 mm W x 254 mm D R260533-70	1 Elec	.90	8.889	Ea.	230	375		605	810	**700**
7300	914 mm H x 914 mm W x 254 mm D		.80	10		244	420		664	895
7500	Double door, 1220 mm H x 915 mm W x 254 mm D		.60	13.333		440	560		1,000	1,325
7550	610 mm H x 610 mm W x 305 mm D		1	8		230	335		565	755

16139 | Residential Wiring

		CREW	DAILY OUTPUT	LABOR-HOURS	UNIT	MAT.	LABOR	EQUIP.	TOTAL	TOTAL INCL O&P
700 0010	**RESIDENTIAL WIRING**										**700**
0020	6 m avg. runs and #14/2 wiring incl. unless otherwise noted									
1000	Service & panel, includes 7.3 m SE-AL cable, service eye, meter,									
1010	Socket, panel board, main bkr., ground rod, 15 or 20 A									
1020	1-pole circuit breakers, and misc. hardware									
1100	100 A, with 10 branch breakers	1 Elec	1.19	6.723	Ea.	490	282		772	960
1110	With PVC conduit and wire		.92	8.696		535	365		900	1,125
1120	With RGS conduit and wire		.73	10.959		710	460		1,170	1,475
1150	150 A, with 14 branch breakers		1.03	7.767		760	325		1,085	1,325
1170	With PVC conduit and wire		.82	9.756		850	410		1,260	1,550
1180	With RGS conduit and wire		.67	11.940		1,200	500		1,700	2,050
1200	200 A, with 18 branch breakers	2 Elec	1.80	8.889		995	375		1,370	1,650
1220	With PVC conduit and wire		1.46	10.959		1,100	460		1,560	1,875
1230	With RGS conduit and wire		1.24	12.903		1,525	540		2,065	2,500
1800	Lightning surge suppressor for above services, add	1 Elec	32	.250		43.50	10.50		54	63.50
2000	Switch devices									
2100	Single pole, 15 A, Ivory, with a 1-gang box, cover plate,									
2110	Type NM (Romex) cable	1 Elec	17.10	.468	Ea.	8.45	19.65		28.10	39
2120	Type MC (BX) cable		14.30	.559		22	23.50		45.50	59
2130	EMT & wire		5.71	1.401		28	59		87	119
2150	3-way, #14/3, type NM cable		14.55	.550		12.10	23		35.10	48
2170	Type MC cable		12.31	.650		30	27.50		57.50	74
2180	EMT & wire		5	1.600		30.50	67		97.50	134
2200	4-way, #14/3, type NM cable		14.55	.550		25.50	23		48.50	63
2220	Type MC cable		12.31	.650		44	27.50		71.50	88.50
2230	EMT & wire		5	1.600		44	67		111	149
2250	S.P., 20 A, #12/2, type NM cable		13.33	.600		14.50	25		39.50	53.50
2270	Type MC cable		11.43	.700		26.50	29.50		56	73
2280	EMT & wire		4.85	1.649		34.50	69.50		104	141
2290	S.P. rotary dimmer, 600W, no wiring		17	.471		16.90	19.75		36.65	48
2300	S.P. rotary dimmer, 600 W, type NM cable		14.55	.550		20.50	23		43.50	57
2320	Type MC cable		12.31	.650		34	27.50		61.50	78
2330	EMT & wire		5	1.600		41	67		108	145
2350	3-way rotary dimmer, type NM cable		13.33	.600		18.30	25		43.30	57.50
2370	Type MC cable		11.43	.700		31.50	29.50		61	79
2380	EMT & wire		4.85	1.649		39	69.50		108.50	146
2400	Interval timer wall switch, 20 A, 1-30 min., #12/2									
2410	Type NM cable	1 Elec	14.55	.550	Ea.	40	23		63	78
2420	Type MC cable		12.31	.650		47.50	27.50		75	92.50
2430	EMT & wire		5	1.600		60	67		127	166
2500	Decorator style									
2510	S.P., 15 A, type NM cable	1 Elec	17.10	.468	Ea.	12	19.65		31.65	42.50
2520	Type MC cable		14.30	.559		25.50	23.50		49	63
2530	EMT & wire		5.71	1.401		31.50	59		90.50	123
2550	3-way, #14/3, type NM cable		14.55	.550		15.65	23		38.65	52
2570	Type MC cable		12.31	.650		34	27.50		61.50	77.50
2580	EMT & wire		5	1.600		34	67		101	138
2600	4-way, #14/3, type NM cable		14.55	.550		29.50	23		52.50	66.50
2620	Type MC cable		12.31	.650		47.50	27.50		75	92.50
2630	EMT & wire		5	1.600		47.50	67		114.50	153

Important: See the Reference Section for supporting data - Crews, Rental Equipment, City Cost Indexes and Reference Data

16139 | Residential Wiring

		16139 Residential Wiring	CREW	DAILY OUTPUT	LABOR-HOURS	UNIT	2006 BARE COSTS MAT.	LABOR	EQUIP.	TOTAL	TOTAL INCL O&P	
700	2650	S.P., 20 A, #12/2, type NM cable	1 Elec	13.33	.600	Ea.	18.05	25		43.05	57.50	700
	2670	Type MC cable		11.43	.700		30	29.50		59.50	77	
	2680	EMT & wire		4.85	1.649		38	69.50		107.50	145	
	2700	S.P., slide dimmer, type NM cable		17.10	.468		27	19.65		46.65	59.50	
	2720	Type MC cable		14.30	.559		40.50	23.50		64	79.50	
	2730	EMT & wire		5.71	1.401		47.50	59		106.50	140	
	2750	S.P., touch dimmer, type NM cable		17.10	.468		24	19.65		43.65	56	
	2770	Type MC cable		14.30	.559		37.50	23.50		61	76	
	2780	EMT & wire		5.71	1.401		44.50	59		103.50	137	
	2800	3-way touch dimmer, type NM cable		13.33	.600		44	25		69	86	
	2820	Type MC cable		11.43	.700		57.50	29.50		87	107	
	2830	EMT & wire		4.85	1.649		64.50	69.50		134	174	
	3000	Combination devices										
	3100	S.P. switch/15 A recpt., Ivory, 1-gang box, plate										
	3110	Type NM cable	1 Elec	11.43	.700	Ea.	17.50	29.50		47	63.50	
	3120	Type MC cable		10	.800		31	33.50		64.50	84	
	3130	EMT & wire		4.40	1.818		38	76.50		114.50	156	
	3150	S.P. switch/pilot light, type NM cable		11.43	.700		18.20	29.50		47.70	64	
	3170	Type MC cable		10	.800		31.50	33.50		65	85	
	3180	EMT & wire		4.43	1.806		38.50	76		114.50	156	
	3190	2-S.P. switches, 2-#14/2, no wiring		14	.571		7	24		31	43.50	
	3200	2-S.P. switches, 2-#14/2, type NM cables		10	.800		20.50	33.50		54	72.50	
	3220	Type MC cable		8.89	.900		42	38		80	103	
	3230	EMT & wire		4.10	1.951		40	82		122	166	
	3250	3-way switch/15 A recpt., #14/3, type NM cable		10	.800		24.50	33.50		58	77	
	3270	Type MC cable		8.89	.900		42.50	38		80.50	104	
	3280	EMT & wire		4.10	1.951		43	82		125	169	
	3300	2-3 way switches, 2-#14/3, type NM cables		8.89	.900		32.50	38		70.50	92.50	
	3320	Type MC cable		8	1		63.50	42		105.50	133	
	3330	EMT & wire		4	2		47.50	84		131.50	178	
	3350	S.P. switch/20 A recpt., #12/2, type NM cable		10	.800		28	33.50		61.50	80.50	
	3370	Type MC cable		8.89	.900		35.50	38		73.50	95.50	
	3380	EMT & wire		4.10	1.951		48	82		130	175	
	3400	Decorator style										
	3410	S.P. switch/15 A recpt., type NM cable	1 Elec	11.43	.700	Ea.	21	29.50		50.50	67	
	3420	Type MC cable		10	.800		34.50	33.50		68	88	
	3430	EMT & wire		4.40	1.818		41.50	76.50		118	160	
	3450	S.P. switch/pilot light, type NM cable		11.43	.700		21.50	29.50		51	68	
	3470	Type MC cable		10	.800		35	33.50		68.50	88.50	
	3480	EMT & wire		4.40	1.818		42	76.50		118.50	161	
	3500	2-S.P. switches, 2-#14/2, type NM cables		10	.800		24	33.50		57.50	76.50	
	3520	Type MC cable		8.89	.900		45.50	38		83.50	107	
	3530	EMT & wire		4.10	1.951		43.50	82		125.50	170	
	3550	3-way/15 A recpt., #14/3, type NM cable		10	.800		28	33.50		61.50	81	
	3570	Type MC cable		8.89	.900		46	38		84	108	
	3580	EMT & wire		4.10	1.951		46.50	82		128.50	173	
	3650	2-3 way switches, 2-#14/3, type NM cables		8.89	.900		36	38		74	96.50	
	3670	Type MC cable		8	1		67	42		109	137	
	3680	EMT & wire		4	2		51.50	84		135.50	182	
	3700	S.P. switch/20 A recpt., #12/2, type NM cable		10	.800		31.50	33.50		65	84.50	
	3720	Type MC cable		8.89	.900		39	38		77	99.50	
	3730	EMT & wire		4.10	1.951		51.50	82		133.50	179	
	4000	Receptacle devices										
	4010	Duplex outlet, 15 A recpt., Ivory, 1-gang box, plate										
	4015	Type NM cable	1 Elec	14.55	.550	Ea.	6.80	23		29.80	42	
	4020	Type MC cable		12.31	.650		20	27.50		47.50	63	

16139 | Residential Wiring

		CREW	DAILY OUTPUT	LABOR-HOURS	UNIT	2006 BARE COSTS				TOTAL INCL O&P
						MAT.	LABOR	EQUIP.	TOTAL	
700	4030	EMT & wire	1 Elec	5.33	1.501	Ea.	26.50	63		89.50	123	700
	4050	With #12/2, type NM cable		12.31	.650		8.55	27.50		36.05	50
	4070	Type MC cable		10.67	.750		20.50	31.50		52	69.50
	4080	EMT & wire		4.71	1.699		28.50	71.50		100	138
	4100	20 A recpt., #12/2, type NM cable		12.31	.650		16.20	27.50		43.70	58.50
	4120	Type MC cable		10.67	.750		28	31.50		59.50	78
	4130	EMT & wire	▼	4.71	1.699	▼	36.50	71.50		108	146
	4140	For GFI see line 4300 below									
	4150	Decorator style, 15 A recpt., type NM cable	1 Elec	14.55	.550	Ea.	10.35	23		33.35	46
	4170	Type MC cable		12.31	.650		24	27.50		51.50	66.50
	4180	EMT & wire		5.33	1.501		30	63		93	127
	4200	With #12/2 type NM cable		12.31	.650		12.10	27.50		39.60	54
	4220	Type MC cable		10.67	.750		24	31.50		55.50	73.50
	4230	EMT & wire		4.71	1.699		32	71.50		103.50	142
	4250	20 A recpt. #12/2 type NM cable		12.31	.650		19.75	27.50		47.25	62
	4270	Type MC cable		10.67	.750		31.50	31.50		63	82
	4280	EMT & wire		4.71	1.699		40	71.50		111.50	150
	4300	GFI, 15 A recpt., type NM cable		12.31	.650		35.50	27.50		63	79.50
	4320	Type MC cable		10.67	.750		48.50	31.50		80	101
	4330	EMT & wire		4.71	1.699		55	71.50		126.50	167
	4350	GFI with #12/2 type NM cable		10.67	.750		37	31.50		68.50	87.50
	4370	Type MC cable		9.20	.870		49	36.50		85.50	109
	4380	EMT & wire		4.21	1.900		57	80		137	182
	4400	20 A recpt., #12/2 type NM cable		10.67	.750		38.50	31.50		70	89.50
	4420	Type MC cable		9.20	.870		50.50	36.50		87	110
	4430	EMT & wire		4.21	1.900		59	80		139	184
	4500	Weather-proof cover for above receptacles, add	▼	32	.250	▼	4.40	10.50		14.90	20.50
	4550	Air conditioner outlet, 20 A, 240 V recpt.									
	4560	9 m of #12/2, 2 pole circuit breaker									
	4570	Type NM cable	1 Elec	10	.800	Ea.	48.50	33.50		82	104
	4580	Type MC cable		9	.889		63.50	37.50		101	126
	4590	EMT & wire		4	2		67.50	84		151.50	200
	4600	Decorator style, type NM cable		10	.800		52.50	33.50		86	108
	4620	Type MC cable		9	.889		67.50	37.50		105	130
	4630	EMT & wire	▼	4	2	▼	72	84		156	204
	4650	Dryer outlet, 30 A, 240 V recpt., 6 m of #10/3									
	4660	2 pole circuit breaker									
	4670	Type NM cable	1 Elec	6.41	1.248	Ea.	57.50	52.50		110	142
	4680	Type MC cable		5.71	1.401		70.50	59		129.50	165
	4690	EMT & wire	▼	3.48	2.299	▼	71.50	96.50		168	223
	4700	Range outlet, 50 A, 240 V recpt., 9 m of #8/3									
	4710	Type NM cable	1 Elec	4.21	1.900	Ea.	89	80		169	217
	4720	Type MC cable		4	2		133	84		217	271
	4730	EMT & wire		2.96	2.703		91	114		205	269
	4750	Central vacuum outlet, Type NM cable		6.40	1.250		53.50	52.50		106	138
	4770	Type MC cable		5.71	1.401		73.50	59		132.50	168
	4780	EMT & wire	▼	3.48	2.299	▼	75	96.50		171.50	227
	4800	30 A, 110 V locking recpt., #10/2 circ. bkr.									
	4810	Type NM cable	1 Elec	6.20	1.290	Ea.	60	54		114	148
	4820	Type MC cable		5.40	1.481		85.50	62		147.50	187
	4830	EMT & wire	▼	3.20	2.500	▼	82.50	105		187.50	248
	4900	Low voltage outlets									
	4910	Telephone recpt., 6 m of 4/C phone wire	1 Elec	26	.308	Ea.	7.95	12.90		20.85	28
	4920	TV recpt., 6 m of RG59U coax wire, F type connector	"	16	.500	"	13.20	21		34.20	46
	4950	Door bell chime, transformer, 2 buttons, 18 m of bellwire									
	4970	Economy model	1 Elec	11.50	.696	Ea.	55	29		84	104

16 ELECTRICAL

16139 | Residential Wiring

		CREW	DAILY OUTPUT	LABOR-HOURS	UNIT	2006 BARE COSTS				TOTAL INCL O&P
						MAT.	LABOR	EQUIP.	TOTAL	
700	4980	Custom model	1 Elec	11.50	.696	Ea.	86	29		115	138	700
	4990	Luxury model, 3 buttons	↓	9.50	.842	↓	232	35.50		267.50	310
	6000	Lighting outlets									
	6050	Wire only (for fixture) type NM cable	1 Elec	32	.250	Ea.	4.71	10.50		15.21	21
	6070	Type MC cable		24	.333		14.30	14		28.30	36.50
	6080	EMT & wire		10	.800		19.35	33.50		52.85	71.50
	6100	Box (102 mm) and wire (for fixture), type NM cable		25	.320		10.45	13.45		23.90	31.50
	6120	Type MC cable		20	.400		20	16.80		36.80	47
	6130	EMT & wire	↓	11	.727	↓	25	30.50		55.50	73
	6200	Fixtures (use with lines 6050 or 6100 above)									
	6210	Canopy style, economy grade	1 Elec	40	.200	Ea.	26	8.40		34.40	41
	6220	Custom grade		40	.200		48	8.40		56.40	65.50
	6250	Dining room chandelier, economy grade		19	.421		78	17.70		95.70	113
	6260	Custom grade		19	.421		230	17.70		247.70	280
	6270	Luxury grade		15	.533		510	22.50		532.50	595
	6310	Kitchen fixture (fluorescent), economy grade		30	.267		57	11.20		68.20	79
	6320	Custom grade		25	.320		165	13.45		178.45	202
	6350	Outdoor, wall mounted, economy grade		30	.267		27.50	11.20		38.70	47
	6360	Custom grade		30	.267		104	11.20		115.20	131
	6370	Luxury grade		25	.320		235	13.45		248.45	279
	6410	Outdoor PAR floodlights, 1 lamp, 150 W		20	.400		26.50	16.80		43.30	54
	6420	2 lamp, 150 W each		20	.400		43	16.80		59.80	72.50
	6430	For infrared security sensor, add		32	.250		90	10.50		100.50	115
	6450	Outdoor, quartz-halogen, 300 W flood		20	.400		39	16.80		55.80	68
	6600	Recessed downlight, round, pre-wired, 50 or 75 W trim		30	.267		36	11.20		47.20	56
	6610	With shower light trim		30	.267		45	11.20		56.20	66
	6620	With wall washer trim		28	.286		54	12		66	77.50
	6630	With eye-ball trim	↓	28	.286		54	12		66	77.50
	6640	For direct contact with insulation, add					1.80			1.80	1.98
	6700	Porcelain lamp holder	1 Elec	40	.200		3.80	8.40		12.20	16.70
	6710	With pull switch		40	.200		4.13	8.40		12.53	17.05
	6750	Fluor. strip, 1-20 W tube, wrap around diffuser, 600 mm		24	.333		53	14		67	79.50
	6760	1-40 W tube, 1200 mm		24	.333		65	14		79	92.50
	6770	2-40 W tubes, 1200 mm		20	.400		79	16.80		95.80	112
	6780	With residential ballast		20	.400		89.50	16.80		106.30	124
	6800	Bathroom heat lamp, 1-250 W		28	.286		38.50	12		50.50	60.50
	6810	2-250 W lamps	↓	28	.286	↓	63	12		75	87.50
	6820	For timer switch, see line 2400									
	6900	Outdoor post lamp, incl. post, fixture, 10.7 m of #14/2									
	6910	Type NMC cable	1 Elec	3.50	2.286	Ea.	186	96		282	345
	6920	Photo-eye, add		27	.296		30	12.45		42.45	51.50
	6950	Clock dial time switch, 24 hr., w/enclosure, type NM cable		11.43	.700		56.50	29.50		86	106
	6970	Type MC cable		11	.727		70	30.50		100.50	123
	6980	EMT & wire	↓	4.85	1.649	↓	76	69.50		145.50	187
	7000	Alarm systems									
	7050	Smoke detectors, box, #14/3, type NM cable	1 Elec	14.55	.550	Ea.	30	23		53	67.50
	7070	Type MC cable		12.31	.650		44	27.50		71.50	89
	7080	EMT & wire	↓	5	1.600		44	67		111	149
	7090	For relay output to security system, add				↓	12.15			12.15	13.35
	8000	Residential equipment									
	8050	Disposal hook-up, incl. switch, outlet box, 0.9 m of flex									
	8060	20 A, 1 pole circ. bkr., and 7.6 m of #12/2									
	8070	Type NM cable	1 Elec	10	.800	Ea.	24.50	33.50		58	77
	8080	Type MC cable	↓	8	1	↓	38	42		80	105
	8090	EMT & wire	↓	5	1.600	↓	47.50	67		114.50	152
	8100	Trash compactor or dishwasher hook-up, incl. outlet box,									

							2006 BARE COSTS				TOTAL	
	16139	**Residential Wiring**	CREW	DAILY OUTPUT	LABOR-HOURS	UNIT	MAT.	LABOR	EQUIP.	TOTAL	INCL O&P	
700	8110	0.9 m of flex, 15 A, 1 pole circ. bkr., and 7.6 m of #14/2										700
	8120	Type NM cable	1 Elec	10	.800	Ea.	17.85	33.50		51.35	69.50	
	8130	Type MC cable		8	1		33.50	42		75.50	99.50	
	8140	EMT & wire	↓	5	1.600	↓	41	67		108	146	
	8150	Hot water sink dispensor hook-up, use line 8100										
	8200	Vent/exhaust fan hook-up, type NM cable	1 Elec	32	.250	Ea.	4.71	10.50		15.21	21	
	8220	Type MC cable		24	.333		14.30	14		28.30	36.50	
	8230	EMT & wire	↓	10	.800	↓	19.35	33.50		52.85	71.50	
	8250	Bathroom vent fan, 0.024 m³/s (use with above hook-up)										
	8260	Economy model	1 Elec	15	.533	Ea.	22.50	22.50		45	58.50	
	8270	Low noise model		15	.533		32	22.50		54.50	68.50	
	8280	Custom model	↓	12	.667	↓	117	28		145	171	
	8300	Bathroom or kitchen vent fan, 0.052 m³/s										
	8310	Economy model	1 Elec	15	.533	Ea.	60.50	22.50		83	100	
	8320	Low noise model	"	15	.533	"	79	22.50		101.50	121	
	8350	Paddle fan, variable speed (w/o lights)										
	8360	Economy model (AC motor)	1 Elec	10	.800	Ea.	105	33.50		138.50	166	
	8370	Custom model (AC motor)		10	.800		182	33.50		215.50	250	
	8380	Luxury model (DC motor)		8	1		360	42		402	460	
	8390	Remote speed switch for above, add	↓	12	.667	↓	26	28		54	70	
	8500	Whole house exhaust fan, ceiling mount, 920 mm, variable speed										
	8510	Remote switch, incl. shutters, 20 A, 1 pole circ. bkr.										
	8520	9 m of #12/2, type NM cable	1 Elec	4	2	Ea.	685	84		769	880	
	8530	Type MC cable		3.50	2.286		705	96		801	920	
	8540	EMT & wire	↓	3	2.667	↓	715	112		827	950	
	8600	Whirlpool tub hook-up, incl. timer switch, outlet box										
	8610	0.9 m of flex, 20 A, 1 pole GFI circ. bkr.										
	8620	9 m of #12/2, type NM cable	1 Elec	5	1.600	Ea.	100	67		167	210	
	8630	Type MC cable		4.20	1.905		109	80		189	239	
	8640	EMT & wire	↓	3.40	2.353	↓	118	99		217	276	
	8650	Hot water heater hook-up, incl. 1-2 pole circ. bkr., box;										
	8660	0.9 m of flex, 6 m of #10/2, type NM cable	1 Elec	5	1.600	Ea.	26.50	67		93.50	129	
	8670	Type MC cable		4.20	1.905		44	80		124	167	
	8680	EMT & wire	↓	3.40	2.353	↓	42.50	99		141.50	194	
	9000	Heating/air conditioning										
	9050	Furnace/boiler hook-up, incl. firestat, local on-off switch										
	9060	Emergency switch, and 12 m of type NM cable	1 Elec	4	2	Ea.	46	84		130	176	
	9070	Type MC cable		3.50	2.286		67.50	96		163.50	218	
	9080	EMT & wire	↓	1.50	5.333	↓	79.50	224		303.50	425	
	9100	Air conditioner hook-up, incl. local 60 A disc. switch										
	9110	0.9 m sealtite, 40 A, 2 pole circuit breaker										
	9130	12 m of #8/2, type NM cable	1 Elec	3.50	2.286	Ea.	169	96		265	330	
	9140	Type MC cable		3	2.667		240	112		352	430	
	9150	EMT & wire	↓	1.30	6.154	↓	190	258		448	595	
	9200	Heat pump hook-up, 1-40 & 1-100 A 2 pole circ. bkr.										
	9210	Local disconnect switch, 0.9 m sealtite										
	9220	12 m of #8/2 & 9 m of #3/2										
	9230	Type NM cable	1 Elec	1.30	6.154	Ea.	425	258		683	855	
	9240	Type MC cable		1.08	7.407		585	310		895	1,100	
	9250	EMT & wire	↓	.94	8.511	↓	460	355		815	1,050	
	9500	Thermostat hook-up, using low voltage wire										
	9520	Heating only	1 Elec	24	.333	Ea.	6	14		20	27.50	
	9530	Heating/cooling	"	20	.400	"	7.15	16.80		23.95	33	

16 ELECTRICAL

16100 | Wiring Methods

16140 | Wiring Devices

		CREW	DAILY OUTPUT	LABOR-HOURS	UNIT	2006 BARE COSTS				TOTAL INCL O&P
						MAT.	LABOR	EQUIP.	TOTAL	
500	**0010**	**LOW VOLTAGE SWITCHING**									**500**
	3600	Relays, 120 V or 277 V standard	1 Elec	12	.667	Ea.	26	28		54	70
	3800	Flush switch, standard		40	.200		9.05	8.40		17.45	22.50
	4000	Interchangeable		40	.200		11.80	8.40		20.20	25.50
	4100	Surface switch, standard		40	.200		6.60	8.40		15	19.80
	4200	Transformer 115 V to 25 V		12	.667		93	28		121	144
	4400	Master control, 12 circuit, manual		4	2		94	84		178	228
	4500	25 circuit, motorized		4	2		102	84		186	237
	4600	Rectifier, silicon		12	.667		30.50	28		58.50	75
	4800	Switchplates, 1 gang, 1, 2 or 3 switch, plastic		80	.100		3	4.20		7.20	9.55
	5000	Stainless steel		80	.100		8.10	4.20		12.30	15.15
	5400	2 gang, 3 switch, stainless steel		53	.151		15.65	6.35		22	26.50
	5500	4 switch, plastic		53	.151		6.70	6.35		13.05	16.80
	5800	3 gang, 9 switch, stainless steel		32	.250		50	10.50		60.50	70.50
910	**0010**	**WIRING DEVICES**									**910**
	0200	Toggle switch, quiet type, single pole, 15 A	1 Elec	40	.200	Ea.	4.88	8.40		13.28	17.85
	0600	3 way, 15 A		23	.348		7	14.60		21.60	29.50
	0900	4 way, 15 A		15	.533		21	22.50		43.50	56.50
	1650	Dimmer switch, 120 V, incandescent, 600 W, 1 pole		16	.500		11.35	21		32.35	44
	2460	Receptacle, duplex, 120 V, grounded, 15 A		40	.200		1.17	8.40		9.57	13.80
	2470	20 A		27	.296		8.85	12.45		21.30	28.50
	2490	Dryer, 30 A		15	.533		10.80	22.50		33.30	45.50
	2500	Range, 50 A		11	.727		11.40	30.50		41.90	58
	2600	Wall plates, stainless steel, 1 gang		80	.100		1.80	4.20		6	8.25
	2800	2 gang		53	.151		4.10	6.35		10.45	13.95
	3200	Lampholder, keyless		26	.308		9.70	12.90		22.60	30
	3400	Pullchain with receptacle		22	.364		9.20	15.25		24.45	33

16150 | Wiring Connections

		CREW	DAILY OUTPUT	LABOR-HOURS	UNIT	2006 BARE COSTS				TOTAL INCL O&P
						MAT.	LABOR	EQUIP.	TOTAL	
275	**0010**	**MOTOR CONNECTIONS**									**275**
	0020	Flexible conduit and fittings, 115 V, 1 phase, up to 0.75 kW motor	1 Elec	8	1	Ea.	4.97	42		46.97	68
	0120	230 V, 7.46 W motor, 3 phase		4.20	1.905		9.15	80		89.15	129
	0200	19 kW motor		2.70	2.963		19.80	124		143.80	207
	0400	37 kW motor		2.20	3.636		55	153		208	289
	0600	75 kW motor		1.50	5.333		141	224		365	490

16200 | Electrical Power

16210 | Electrical Utility Services

		CREW	DAILY OUTPUT	LABOR-HOURS	UNIT	2006 BARE COSTS				TOTAL INCL O&P
						MAT.	LABOR	EQUIP.	TOTAL	
600	**0010**	**METER CENTERS AND SOCKETS**									**600**
	0100	Sockets, single position, 4 terminal, 100 A	1 Elec	3.20	2.500	Ea.	36	105		141	197
	0200	150 A		2.30	3.478		40	146		186	263
	0300	200 A		1.90	4.211		53.50	177		230.50	325
	2000	Meter center, main fusible switch, 1P 3W 120/240 V									
	2030	400 A	2 Elec	1.60	10	Ea.	1,400	420		1,820	2,150
	2040	600 A		1.10	14.545		2,425	610		3,035	3,575
	2050	800 A		.90	17.778		3,800	745		4,545	5,325
	2060	Rainproof 1P 3W 120/240 V, 400 A		1.60	10		1,400	420		1,820	2,150
	2070	600 A		1.10	14.545		2,425	610		3,035	3,575

16210 | Electrical Utility Services

		CREW	DAILY OUTPUT	LABOR-HOURS	UNIT	2006 BARE COSTS MAT.	LABOR	EQUIP.	TOTAL	TOTAL INCL O&P	
600	2080	800 A	2 Elec	.90	17.778	Ea.	3,800	745		4,545	5,325	600
	2100	3P 4W 120/208 V, 400 A		1.60	10		1,600	420		2,020	2,375	
	2110	600 A		1.10	14.545		2,975	610		3,585	4,175	
	2120	800 A		.90	17.778		5,550	745		6,295	7,225	
	2130	Rainproof 3P 4W 120/208 V, 400 A		1.60	10		1,600	420		2,020	2,375	
	2140	600 A		1.10	14.545		2,975	610		3,585	4,175	
	2150	800 A	▼	.90	17.778	▼	5,550	745		6,295	7,225	
	2170	Main circuit breaker, 1P 3W 120/240 V										
	2180	400 A	2 Elec	1.60	10	Ea.	2,500	420		2,920	3,375	
	2190	600 A		1.10	14.545		3,375	610		3,985	4,600	
	2200	800 A		.90	17.778		3,925	745		4,670	5,450	
	2210	1000 A		.80	20		5,400	840		6,240	7,200	
	2220	1200 A		.76	21.053		7,300	885		8,185	9,350	
	2230	1600 A		.68	23.529		12,500	990		13,490	15,200	
	2240	Rainproof 1P 3W 120/240 V, 400 A		1.60	10		2,500	420		2,920	3,375	
	2250	600 A		1.10	14.545		3,375	610		3,985	4,600	
	2260	800 A		.90	17.778		3,925	745		4,670	5,450	
	2270	1000 A		.80	20		5,400	840		6,240	7,200	
	2280	1200 A		.76	21.053		7,300	885		8,185	9,350	
	2300	3P 4W 120/208 V, 400 A		1.60	10		2,850	420		3,270	3,750	
	2310	600 A		1.10	14.545		4,000	610		4,610	5,300	
	2320	800 A		.90	17.778		4,775	745		5,520	6,375	
	2330	1000 A		.80	20		6,250	840		7,090	8,125	
	2340	1200 A		.76	21.053		8,000	885		8,885	10,100	
	2350	1600 A		.68	23.529		12,500	990		13,490	15,200	
	2360	Rainproof 3P 4W 120/208 V, 400 A		1.60	10		2,850	420		3,270	3,750	
	2370	600 A		1.10	14.545		4,000	610		4,610	5,300	
	2380	800 A		.90	17.778		4,775	745		5,520	6,375	
	2390	1000 A		.76	21.053		6,250	885		7,135	8,200	
	2400	1200 A	▼	.68	23.529	▼	8,000	990		8,990	10,300	

16220 | Motors & Generators

		CREW	DAILY OUTPUT	LABOR-HOURS	UNIT	MAT.	LABOR	EQUIP.	TOTAL	TOTAL INCL O&P	
610	0010	**MOTORS** 230/460 V, 60 HZ										610
	0050	Dripproof, premium efficiency, 1.15 service factor										
	0060	1800 RPM, 0.19 kW	1 Elec	5.33	1.501	Ea.	134	63		197	241	
	0070	0.25 kW		5.33	1.501		146	63		209	255	
	0080	0.37 kW		5.33	1.501		172	63		235	283	
	0090	0.56 kW		5.33	1.501		192	63		255	305	
	0100	0.75 kW		4.50	1.778		201	74.50		275.50	330	
	0250	4 kW		4.50	1.778		305	74.50		379.50	450	
	0350	7.46 kW	▼	4	2		550	84		634	730	
	0450	15 kW	2 Elec	5.20	3.077	▼	875	129		1,004	1,150	

16230 | Generator Assemblies

		CREW	DAILY OUTPUT	LABOR-HOURS	UNIT	MAT.	LABOR	EQUIP.	TOTAL	TOTAL INCL O&P	
450	0010	**GENERATOR SET**										450
	0020	Gas or gasoline operated, includes battery,										
	0050	charger, muffler & transfer switch										
	0200	3 phase, 4 wire, 277/480 V, 7.5 kW	R-3	.83	24.096	Ea.	6,000	1,000	184	7,184	8,300	
	0300	11.5 kW		.71	28.169		8,500	1,175	215	9,890	11,300	
	0400	20 kW		.63	31.746		10,000	1,325	242	11,567	13,200	
	0500	35 kW		.55	36.364		12,000	1,500	277	13,777	15,800	
	0520	60 kW		.50	40		16,500	1,650	305	18,455	21,000	
	0600	80 kW		.40	50		20,500	2,075	380	22,955	26,100	
	0700	100 kW		.33	60.606		22,500	2,500	460	25,460	29,100	
	0800	125 kW		.28	71.429		46,000	2,950	545	49,495	55,500	
	0900	185 kW	▼	.25	80	▼	61,000	3,325	610	64,935	72,500	

16 ELECTRICAL

16230 | Generator Assemblies

		CREW	DAILY OUTPUT	LABOR-HOURS	UNIT	MAT.	LABOR	EQUIP.	TOTAL	TOTAL INCL O&P	
450	2000	Diesel engine, including battery, charger,										450
	2010	muffler, automatic transfer switch & day tank, 30 kW	R-3	.55	36.364	Ea.	16,000	1,500	277	17,777	20,200	
	2100	50 kW		.42	47.619		19,700	1,975	365	22,040	25,100	
	2200	75 kW		.35	57.143		25,700	2,375	435	28,510	32,300	
	2300	100 kW		.31	64.516		28,500	2,675	490	31,665	35,900	
	2400	125 kW		.29	68.966		30,000	2,850	525	33,375	37,900	
	2500	150 kW		.26	76.923		34,400	3,175	585	38,160	43,200	
	2600	175 kW		.25	80		37,500	3,325	610	41,435	46,900	
	2700	200 kW		.24	83.333		38,700	3,450	635	42,785	48,500	
	2800	250 kW		.23	86.957		45,600	3,600	665	49,865	56,000	
	2900	300 kW		.22	90.909		49,300	3,775	695	53,770	60,500	
	3000	350 kW		.20	100		55,500	4,150	765	60,415	68,500	
	3100	400 kW		.19	105		69,000	4,350	805	74,155	83,000	
	3200	500 kW		.18	111		86,500	4,600	850	91,950	103,000	
	3220	600 kW	▼	.17	117		113,500	4,875	900	119,275	133,500	
	3240	750 kW	R-13	.38	110	▼	142,500	4,425	445	147,370	164,000	

16270 | Transformers

		CREW	DAILY OUTPUT	LABOR-HOURS	UNIT	MAT.	LABOR	EQUIP.	TOTAL	TOTAL INCL O&P	
200	0010	**DRY TYPE TRANSFORMER**										200
	0050	Single phase, 240/480 V primary 120/240 V secondary										
	0100	1 kVA	1 Elec	2	4	Ea.	219	168		387	490	
	0300	2 kVA		1.60	5		330	210		540	675	
	0500	3 kVA		1.40	5.714		405	240		645	810	
	0700	5 kVA	▼	1.20	6.667		560	280		840	1,025	
	0900	7.5 kVA	2 Elec	2.20	7.273		780	305		1,085	1,300	
	1100	10 KVA		1.60	10		955	420		1,375	1,675	
	1300	15 kVA		1.20	13.333		1,300	560		1,860	2,250	
	1500	25 kVA		1	16		1,775	670		2,445	2,950	
	1700	37.5 kVA		.80	20		2,300	840		3,140	3,775	
	1900	50 kVA		.70	22.857		2,750	960		3,710	4,425	
	2100	75 kVA		.65	24.615		3,625	1,025		4,650	5,525	
	2190	480 V primary 120/240 V secondary, nonvent., 15 kVA		1.20	13.333		1,375	560		1,935	2,350	
	2200	25 kVA		.90	17.778		2,025	745		2,770	3,350	
	2210	37 kVA		.75	21.333		2,400	895		3,295	3,975	
	2220	50 kVA	▼	.65	24.615	▼	2,850	1,025		3,875	4,700	
	2300	3 phase, 480 V primary, 120/208 V secondary										
	2310	Ventilated, 3 kVA	1 Elec	1	8	Ea.	725	335		1,060	1,300	
	2700	6 kVA		.80	10		1,000	420		1,420	1,725	
	2900	9 kVA	▼	.70	11.429		1,125	480		1,605	1,975	
	3100	15 kVA	2 Elec	1.10	14.545		1,500	610		2,110	2,550	
	3300	30 kVA		.90	17.778		1,775	745		2,520	3,075	
	3500	45 kVA		.80	20		2,125	840		2,965	3,575	
	3700	75 kVA	▼	.70	22.857		3,200	960		4,160	4,950	
	3900	112.5 kVA	R-3	.90	22.222		4,250	920	170	5,340	6,225	
	4100	150 kVA		.85	23.529		5,550	975	180	6,705	7,750	
	4300	225 kVA		.65	30.769		7,525	1,275	235	9,035	10,400	
	4500	300 kVA		.55	36.364		9,500	1,500	277	11,277	13,000	
	4700	500 kVA		.45	44.444		15,700	1,850	340	17,890	20,400	
	4800	750 kVA	▼	.35	57.143	▼	27,600	2,375	435	30,410	34,300	
600	0010	**OIL-FILLED TRANSFORMER** Pad mounted, primary delta or Y, [R262213-60]										600
	0050	5 kV or 15 kV, with taps, 277/480 V secondary, 3 phase										
	0100	150 kVA	R-3	.65	30.769	Ea.	7,275	1,275	235	8,785	10,200	
	0110	225 kVA		.55	36.364		8,250	1,500	277	10,027	11,600	
	0200	300 kVA		.45	44.444		10,300	1,850	340	12,490	14,500	
	0300	500 kVA	▼	.40	50		14,500	2,075	380	16,955	19,500	

For expanded coverage of these items see *Means Electrical Cost Data 2006*

ELECTRICAL T6

16270 | Transformers

			DAILY OUTPUT	LABOR-HOURS	UNIT	2006 BARE COSTS				TOTAL INCL O&P	
			CREW			MAT.	LABOR	EQUIP.	TOTAL		
600	0400	750 kVA	R-3	.38	52.632	Ea.	18,600	2,175	400	21,175	24,100	600
	0500	1000 kVA		.26	76.923		22,000	3,175	585	25,760	29,600	
	0600	1500 kVA		.23	86.957		26,100	3,600	665	30,365	34,900	
	0700	2000 kVA		.20	100		33,000	4,150	765	37,915	43,300	
	0800	3750 kVA		.16	125		62,000	5,175	955	68,130	77,000	
610	0010	**TRANSFORMER, LIQUID-FILLED** Pad mounted										610
	0020	5 kV or 15 kV primary, 277/480 V secondary, 3 phase										
	0050	225 kVA	R-3	.55	36.364	Ea.	11,200	1,500	277	12,977	14,900	
	0100	300 kVA		.45	44.444		13,300	1,850	340	15,490	17,700	
	0200	500 kVA		.40	50		16,800	2,075	380	19,255	22,000	
	0250	750 kVA		.38	52.632		21,700	2,175	400	24,275	27,500	
	0300	1000 kVA		.26	76.923		25,200	3,175	585	28,960	33,100	

R262213 -60

16280 | Power Filters & Conditioners

			DAILY OUTPUT	LABOR-HOURS	UNIT	MAT.	LABOR	EQUIP.	TOTAL	TOTAL INCL O&P	
300	0010	**CAPACITORS** Indoor									300
	0020	240 V, single & 3 phase, 0.5 kVAR	1 Elec	2.70	2.963	Ea.	300	124		424	515	
	0100	1.0 kVAR		2.70	2.963		365	124		489	585	
	0150	2.5 kVAR		2	4		410	168		578	700	
	0200	5.0 kVAR		1.80	4.444		490	187		677	820	
	0250	7.5 kVAR		1.60	5		570	210		780	940	
	0300	10 kVAR		1.50	5.333		685	224		909	1,100	
	0350	15 kVAR		1.30	6.154		945	258		1,203	1,425	
	0400	20 kVAR		1.10	7.273		1,175	305		1,480	1,750	
	0450	25 kVAR		1	8		1,375	335		1,710	2,025	
	1000	480 V, single & 3 phase, 1 kVAR		2.70	2.963		275	124		399	485	
	1050	2 kVAR		2.70	2.963		315	124		439	535	
	1100	5 kVAR		2	4		400	168		568	690	
	1150	7.5 kVAR		2	4		430	168		598	725	
	1200	10 kVAR		2	4		480	168		648	780	
	1250	15 kVAR		2	4		595	168		763	905	
	1300	20 kVAR		1.60	5		660	210		870	1,050	
	1350	30 kVAR		1.50	5.333		825	224		1,049	1,250	
	1400	40 kVAR		1.20	6.667		1,050	280		1,330	1,575	
	1450	50 kVAR		1.10	7.273		1,200	305		1,505	1,775	

16290 | Power Measurement & Control

			DAILY OUTPUT	LABOR-HOURS	UNIT	MAT.	LABOR	EQUIP.	TOTAL	TOTAL INCL O&P	
800	0010	**SWITCHBOARD INSTRUMENTS** 3 phase, 4 wire										800
	0100	AC indicating, ammeter & switch	1 Elec	8	1	Ea.	1,675	42		1,717	1,900	
	0200	Voltmeter & switch		8	1		1,675	42		1,717	1,900	
	0300	Wattmeter		8	1		3,300	42		3,342	3,700	
	0400	AC recording, ammeter		4	2		5,875	84		5,959	6,600	
	0500	Voltmeter		4	2		5,875	84		5,959	6,600	
	0600	Ground fault protection, zero sequence		2.70	2.963		5,200	124		5,324	5,875	
	0700	Ground return path		2.70	2.963		5,200	124		5,324	5,875	
	0800	3 current transformers, 5 to 800 A		2	4		2,425	168		2,593	2,900	
	0900	1000 to 1500 A		1.30	6.154		3,475	258		3,733	4,200	
	1200	2000 to 4000 A		1	8		4,100	335		4,435	5,000	
	1300	Fused potential transformer, maximum 600 V		8	1		910	42		952	1,075	

16 ELECTRICAL

16310	Transmission & Dist Accessories	CREW	DAILY OUTPUT	LABOR-HOURS	UNIT	2006 BARE COSTS				TOTAL INCL O&P
						MAT.	LABOR	EQUIP.	TOTAL	
600	**0010** **LINE POLES & FIXTURES**									**600**
0100	Digging holes in earth, average	R-5	25.14	3.500	Ea.		127	63.50	190.50	262
0105	In rock, average	"	4.51	19.512			710	355	1,065	1,475
0200	Wood poles, material handling and spotting	R-7	6.49	7.396			215	20.50	235.50	355
0220	Erect wood poles & backfill holes, in earth	R-5	6.77	12.999		1,100	475	236	1,811	2,175
0250	In rock	"	5.87	14.991		1,100	545	272	1,917	2,325
0260	Disposal of surplus material	R-7	33.59	1.429	km		41.50	3.99	45.49	68.50
0300	Crossarms for wood pole structure									
0310	Material handling and spotting	R-7	14.55	3.299	Ea.		96	9.20	105.20	157
0320	Install crossarms	R-5	11	8	"	360	291	145	796	995
0330	Disposal of surplus material	R-7	64.37	.746	km		21.50	2.08	23.58	36
0400	Formed plate pole structure									
0410	Material handling and spotting	R-7	2.40	20	Ea.		580	56	636	955
0420	Erect steel plate pole	R-5	1.95	45.128		6,775	1,650	820	9,245	10,800
0500	Guys, anchors and hardware for pole, in earth		7.04	12.500		410	455	227	1,092	1,375
0510	In rock		17.96	4.900		490	178	89	757	905
0900	Foundations for line poles									
0920	Excavation, in earth	R-5	104	.846	m³		31	15.35	46.35	63.50
0940	In rock	"	15.29	5.755	"		209	104	313	430
0950	See also Division 02400									
0960	Concrete foundations	R-5	8.41	10.463	m³	148	380	190	718	945
0970	See also Division 03300									
610	**0010** **LINE TOWERS & FIXTURES**									**610**
0100	Excavation and backfill, earth	R-5	104	.846	m³		31	15.35	46.35	63.50
0105	Rock		16.41	5.363	"		195	97.50	292.50	400
0200	Steel footings (grillage) in earth		3.55	24.809	Met. Ton	1,675	900	450	3,025	3,700
0205	In rock		2.90	30.313	"	1,675	1,100	550	3,325	4,100
0290	See also Division 02400									
0300	Rock anchors	R-5	5.87	14.991	Ea.	360	545	272	1,177	1,525
0400	Concrete foundations	"	9.83	8.957	m³	148	325	163	636	830
0490	See also Division 03300									
0500	Towers-material handling and spotting	R-7	20.47	2.345	Met. Ton		68	6.55	74.55	112
0540	Steel tower erection	R-5	6.94	12.680		1,625	460	230	2,315	2,725
0550	Lace and box		6.44	13.662		1,625	495	248	2,368	2,800
0560	Painting total structure		1.47	59.864	Ea.	289	2,175	1,075	3,539	4,800
0570	Disposal of surplus material	R-7	33.59	1.429	km		41.50	3.99	45.49	68.50
0600	Special towers-material handling and spotting	"	11.17	4.298	Met. Ton		125	12	137	205
0640	Special steel structure erection	R-6	5.91	14.878		2,000	540	545	3,085	3,625
0650	Special steel lace and box	"	5.71	15.422		2,000	560	565	3,125	3,675
0670	Disposal of surplus material	R-7	12.67	3.790	km		110	10.55	120.55	181
700	**0010** **OVERHEAD LINE CONDUCTORS & DEVICES**									**700**
0100	Conductors, primary circuits									
0110	Material handling and spotting	R-5	15.89	5.537	km		201	101	302	415
0120	For river crossing, add		17.88	4.923			179	89.50	268.50	370
0150	Conductors, per wire, 210 to 636 kcmil		3.19	27.630		4,900	1,000	500	6,400	7,450
0160	795 to 954 kcmil		3.04	28.959		7,225	1,050	525	8,800	10,100
0170	1000 to 1600 kcmil		2.39	36.839		11,800	1,350	670	13,820	15,800
0180	Over 1600 kcmil		2.19	40.114		15,700	1,450	730	17,880	20,300
0200	For river crossing, add, 210 to 636 kcmil		2.02	43.672			1,600	795	2,395	3,275
0220	795 to 954 kcmil		1.77	49.682			1,800	900	2,700	3,725
0230	1000 to 1600 kcmil		1.58	55.829			2,025	1,025	3,050	4,175
0240	Over 1600 kcmil		1.41	62.246			2,275	1,125	3,400	4,650
0300	Joints and dead ends	R-8	6	8	Ea.	1,050	295	35.50	1,380.50	1,650
0400	Sagging	R-5	11.92	7.385	km		269	134	403	550
0500	Clipping, per structure, 69 kV	R-10	9.60	5	Ea.		197	50.50	247.50	350
0510	161 kV		5.33	9.006			355	91	446	630

ELECTRICAL 16

		Transmission & Dist Accessories	CREW	DAILY OUTPUT	LABOR-HOURS	UNIT	2006 BARE COSTS				TOTAL INCL O&P	
							MAT.	LABOR	EQUIP.	TOTAL		
700	0520	345 to 500 kV	R-10	2.53	18.972	Ea.		750	192	942	1,325	**700**
	0600	Make and install jumpers, per structure, 69 kV	R-8	3.20	15		287	555	66	908	1,225	
	0620	161 kV		1.20	40		575	1,475	176	2,226	3,050	
	0640	345 to 500 kV	↓	.32	150		965	5,525	660	7,150	10,100	
	0700	Spacers	R-10	68.57	.700		57	27.50	7.05	91.55	112	
	0720	For river crossings, add	"	60	.800	↓		31.50	8.10	39.60	56.50	
	0800	Installing pulling line (500 kV only)	R-9	2.36	27.162	km	360	930	90	1,380	1,900	
	0810	Disposal of surplus material	R-7	11.20	4.285			125	11.95	136.95	204	
	0820	With trailer mounted reel stands	"	22.06	2.176	↓		63.50	6.05	69.55	104	
	0900	Insulators and hardware, primary circuits										
	0920	Material handling and spotting, 69 kV	R-7	480	.100	Ea.		2.91	.28	3.19	4.78	
	0930	161 kV		686	.070			2.03	.20	2.23	3.34	
	0950	345 to 500 kV	↓	960	.050			1.45	.14	1.59	2.38	
	1000	Disk insulators, 69 kV	R-5	880	.100		58.50	3.64	1.82	63.96	72	
	1020	161 kV		978	.090		67	3.27	1.63	71.90	80	
	1040	345 to 500 kV	↓	1100	.080	↓	67	2.91	1.45	71.36	79.50	
	1060	See line 16360-800-7400 for pin or pedestal insulator										
	1100	Install disk insulator at river crossing, add										
	1110	69 kV	R-5	587	.150	Ea.		5.45	2.72	8.17	11.20	
	1120	161 kV		880	.100			3.64	1.82	5.46	7.50	
	1140	345 to 500 kV	↓	880	.100	↓		3.64	1.82	5.46	7.50	
	1150	Disposal of surplus material	R-7	67.17	.715	km		21	1.99	22.99	34	
	1300	Overhead ground wire installation										
	1320	Material handling and spotting	R-7	9.18	5.228	km		152	14.60	166.60	250	
	1340	Overhead ground wire	R-5	2.86	30.769		2,150	1,125	560	3,835	4,650	
	1350	At river crossing, add		1.90	46.285			1,675	840	2,515	3,450	
	1360	Disposal of surplus material	↓	67.17	1.310	↓		47.50	24	71.50	98	
	1400	Installing conductors, underbuilt circuits										
	1420	Material handling and spotting	R-7	9.18	5.228	km		152	14.60	166.60	250	
	1440	Conductors, per wire, 210 to 636 kcmil	R-5	3.19	27.630		4,900	1,000	500	6,400	7,450	
	1450	795 to 954 kcmil		3.04	28.959		7,225	1,050	525	8,800	10,100	
	1460	1000 to 1600 kcmil		2.39	36.839		11,800	1,350	670	13,820	15,800	
	1470	Over 1600 kcmil	↓	2.19	40.114	↓	15,700	1,450	730	17,880	20,300	
	1500	Joints and dead ends	R-8	6	8	Ea.	1,050	295	35.50	1,380.50	1,650	
	1550	Sagging	R-5	14.30	6.154	km		224	112	336	460	
	1600	Clipping, per structure, 69 kV	R-10	9.60	5	Ea.		197	50.50	247.50	350	
	1620	161 kV		5.33	9.006			355	91	446	630	
	1640	345 to 500 kV	↓	2.53	18.972			750	192	942	1,325	
	1700	Making and installing jumpers, per structure, 69 kV	R-8	5.87	8.177		287	300	36	623	810	
	1720	161 kV		.96	50		575	1,850	221	2,646	3,650	
	1740	345 to 500 kV	↓	.32	150		965	5,525	660	7,150	10,100	
	1800	Spacers	R-10	96	.500	↓	57	19.75	5.05	81.80	98	
	1810	Disposal of surplus material	R-7	11.20	4.285	km		125	11.95	136.95	204	
	2000	Insulators and hardware for underbuilt circuits										
	2100	Material handling and spotting	R-7	1200	.040	Ea.		1.16	.11	1.27	1.91	
	2150	Disk insulators, 69 kV	R-8	600	.080		58.50	2.95	.35	61.80	69.50	
	2160	161 kV		686	.070		67	2.58	.31	69.89	77.50	
	2170	345 to 500 kV	↓	800	.060	↓	67	2.21	.26	69.47	77	
	2180	Disposal of surplus material	R-7	67.17	.715	km		21	1.99	22.99	34	
	2300	Sectionalizing switches, 69 kV	R-5	1.26	69.841	Ea.	15,100	2,550	1,275	18,925	21,800	
	2310	161 kV		.80	110		17,000	4,000	2,000	23,000	26,900	
	2500	Protective devices		5.50	16	↓	4,925	580	290	5,795	6,625	
	2600	Clearance poles, 5 poles per km										
	2650	In earth, 69 kV	R-5	1.87	47.140	km	2,525	1,725	855	5,105	6,300	
	2660	161 kV	"	1.03	85.441	↓	4,150	3,100	1,550	8,800	10,900	
	2670	345 to 500 kV	R-6	.77	113		4,975	4,150	4,175	13,300	16,300	

			DAILY	LABOR-		2006 BARE COSTS				TOTAL	
	16310	**Transmission & Dist Accessories**				MAT.	LABOR	EQUIP.	TOTAL	INCL O&P	
			CREW	OUTPUT	HOURS	UNIT						
700	2800	In rock, 69 kV	R-5	1.11	79.249	km	2,525	2,875	1,450	6,850	8,700	**700**
	2820	161 kV	"	.56	156		4,150	5,675	2,825	12,650	16,200	
	2840	345 to 500 kV	R-6	.39	227	↓	4,975	8,275	8,350	21,600	27,200	
850	0010	**TRANSMISSION LINE RIGHT OF WAY**										**850**
	0100	Clearing right of way	B-87	2.70	14.818	Hectare		515	920	1,435	1,775	
	0200	Restoration & seeding	B-10D	1.62	7.413	"	2,325	249	865	3,439	3,875	

	16330	**Med-Voltage Switching & Protection**

			CREW	DAILY OUTPUT	LABOR-HOURS	UNIT	MAT.	LABOR	EQUIP.	TOTAL	INCL O&P	
760	0010	**SWITCHGEAR**, Incorporate switch with cable connections, transformer,										**760**
	0100	& Low Voltage section										
	0200	Load interrupter switch, 600 A, 2 position										
	0300	NEMA 1, 4.8 kV, 300 kVA & below w/CLF fuses	R-3	.40	50	Ea.	15,900	2,075	380	18,355	21,000	
	0400	400 kVA & above w/CLF fuses		.38	52.632		17,800	2,175	400	20,375	23,200	
	0500	Non fusible		.41	48.780		12,900	2,025	370	15,295	17,600	
	0600	13.8 kV, 300 kVA & below		.38	52.632		20,000	2,175	400	22,575	25,700	
	0700	400 kVA & above		.36	55.556		20,000	2,300	425	22,725	25,900	
	0800	Non fusible	↓	.40	50		15,200	2,075	380	17,655	20,200	
	0900	Cable lugs for 2 feeders 4.8 kV or 13.8 kV	1 Elec	8	1		485	42		527	600	
	1000	Pothead, one 3 conductor or three 1 conductor		4	2		2,325	84		2,409	2,675	
	1100	Two 3 conductor or six 1 conductor		2	4		4,600	168		4,768	5,300	
	1200	Key interlocks	↓	8	1	↓	540	42		582	655	
	1300	Lightning arresters, Distribution class (no charge)										
	1400	Intermediate class or line type 4.8 kV	1 Elec	2.70	2.963	Ea.	2,625	124		2,749	3,050	
	1500	13.8 kV		2	4		3,475	168		3,643	4,050	
	1600	Station class, 4.8 kV		2.70	2.963		4,475	124		4,599	5,100	
	1700	13.8 kV	↓	2	4		7,700	168		7,868	8,725	
	1800	Transformers, 4.8 kV to 480/277 V, 75 kVA	R-3	.68	29.412		13,600	1,225	224	15,049	17,100	
	1900	112.5 kVA		.65	30.769		16,600	1,275	235	18,110	20,500	
	2000	150 kVA		.57	35.088		18,900	1,450	268	20,618	23,300	
	2100	225 kVA		.48	41.667		21,700	1,725	320	23,745	26,800	
	2200	300 kVA		.41	48.780		24,300	2,025	370	26,695	30,100	
	2300	500 kVA		.36	55.556		32,000	2,300	425	34,725	39,100	
	2400	750 kVA		.29	68.966		36,300	2,850	525	39,675	44,800	
	2500	13.8 kV to 480/277 V, 75 kVA		.61	32.787	.	19,200	1,350	250	20,800	23,400	
	2600	112.5 kVA		.55	36.364		25,500	1,500	277	27,277	30,600	
	2700	150 kVA		.49	40.816		25,700	1,700	310	27,710	31,200	
	2800	225 kVA		.41	48.780		29,700	2,025	370	32,095	36,000	
	2900	300 kVA		.37	54.054		30,300	2,250	410	32,960	37,100	
	3000	500 kVA		.31	64.516		33,500	2,675	490	36,665	41,400	
	3100	750 kVA	↓	.42	47.733		36,900	1,975	365	39,240	44,000	
	3200	Forced air cooling & temperature alarm	1 Elec	1	8	↓	2,975	335		3,310	3,750	
	3300	Low voltage components										
	3400	Maximum panel height 1250 mm, single or twin row										
	3500	Breaker heights, type FA or FH, 152 mm										
	3600	type KA or KH, 203 mm										
	3700	type LA, 279 mm										
	3800	type MA, 356 mm										
	3900	Breakers, 2 pole, 15 to 60 A, type FA	1 Elec	5.60	1.429	Ea.	305	60		365	425	
	4000	70 to 100 A, type FA		4.20	1.905		375	80		455	535	
	4100	15 to 60 A, type FH		5.60	1.429		485	60		545	625	
	4200	70 to 100 A, type FH		4.20	1.905		570	80		650	745	
	4300	125 to 225 A, type KA		3.40	2.353		870	99		969	1,100	
	4400	125 to 225 A, type KH		3.40	2.353		2,025	99		2,124	2,375	
	4500	125 to 400 A, type LA	↓	2.50	3.200	↓	1,600	134		1,734	1,950	

ELECTRICAL 16

		16330	Med-Voltage Switching & Protection	CREW	DAILY OUTPUT	LABOR-HOURS	UNIT	2006 BARE COSTS				TOTAL INCL O&P	
								MAT.	LABOR	EQUIP.	TOTAL		
760	4600		125 to 600 A, type MA	1 Elec	1.80	4.444	Ea.	2,575	187		2,762	3,100	760
	4700		700 & 800 A, type MA		1.50	5.333		3,250	224		3,474	3,900	
	4800		3 pole, 15 to 60 A, type FA		5.30	1.509		380	63.50		443.50	515	
	4900		70 to 100 A, type FA		4	2		465	84		549	635	
	5000		15 to 60 A, type FH		5.30	1.509		570	63.50		633.50	720	
	5100		70 to 100 A, type FH		4	2		645	84		729	835	
	5200		125 to 225 A, type KA		3.20	2.500		1,100	105		1,205	1,350	
	5300		125 to 225 A, type KH		3.20	2.500		2,450	105		2,555	2,850	
	5400		125 to 400 A, type LA		2.30	3.478		1,925	146		2,071	2,350	
	5500		125 to 600 A, type MA		1.60	5		3,175	210		3,385	3,825	
	5600		700 & 800 A, type MA	▼	1.30	6.154	▼	4,150	258		4,408	4,950	

		16360	Unit Substations										
800	0010	**SUBSTATION EQUIPMENT**											800
	1000		Main conversion equipment										
	1050		Power transformers, 13 to 26 kV	R-11	1.72	32.558	MVA	17,600	1,275	450	19,325	21,700	
	1060		46 kV		3.50	16		16,500	630	220	17,350	19,300	
	1070		69 kV		3.11	18.006		14,100	710	248	15,058	16,800	
	1080		110 kV		3.29	17.021		13,400	670	234	14,304	16,000	
	1090		161 kV		4.31	12.993		12,400	510	179	13,089	14,600	
	1100		500 kV		7	8	▼	12,300	315	110	12,725	14,200	
	1200		Grounding transformers	▼	3.11	18.006	Ea.	78,000	710	248	78,958	87,000	
	1300		Station capacitors										
	1350		Synchronous, 13 to 26 kV	R-11	3.11	18.006	MVAR	4,900	710	248	5,858	6,750	
	1360		46 kV		3.33	16.817		6,250	665	231	7,146	8,125	
	1370		69 kV		3.81	14.698		6,150	580	202	6,932	7,875	
	1380		161 kV		6.51	8.602		5,750	340	118	6,208	6,975	
	1390		500 kV		10.37	5.400		5,000	213	74.50	5,287.50	5,900	
	1450		Static, 13 to 26 kV		3.11	18.006		4,150	710	248	5,108	5,925	
	1460		46 kV		3.01	18.605		5,250	735	256	6,241	7,150	
	1470		69 kV		3.81	14.698		5,100	580	202	5,882	6,725	
	1480		161 kV		6.51	8.602		4,725	340	118	5,183	5,850	
	1490		500 kV		10.37	5.400	▼	4,325	213	74.50	4,612.50	5,150	
	1600		Voltage regulators, 13 to 26 kV	▼	.75	74.667	Ea.	186,000	2,950	1,025	189,975	210,000	
	2000		Power circuit breakers										
	2050		Oil circuit breakers, 13 to 26 kV	R-11	1.12	50	Ea.	41,500	1,975	690	44,165	49,400	
	2060		46 kV		.75	74.667		60,500	2,950	1,025	64,475	72,000	
	2070		69 kV		.45	124		141,000	4,900	1,700	147,600	164,000	
	2080		161 kV		.16	350		215,500	13,800	4,825	234,125	263,000	
	2090		500 kV		.06	933		806,000	36,800	12,800	855,600	956,000	
	2100		Air circuit breakers, 13 to 26 kV		.56	100		43,400	3,950	1,375	48,725	55,000	
	2110		161 kV		.24	233		186,500	9,200	3,200	198,900	223,000	
	2150		Gas circuit breakers, 13 to 26 kV		.56	100		163,000	3,950	1,375	168,325	187,000	
	2160		161 kV		.08	700		214,000	27,600	9,625	251,225	287,500	
	2170		500 kV		.04	1400		754,500	55,000	19,300	828,800	934,000	
	2200		Vacuum circuit breakers, 13 to 26 kV	▼	.56	100	▼	35,600	3,950	1,375	40,925	46,600	
	3000		Disconnecting switches										
	3050		Gang operated switches										
	3060		Manual operation, 13 to 26 kV	R-11	1.65	33.939	Ea.	10,000	1,350	465	11,815	13,500	
	3070		46 kV		1.12	50		16,600	1,975	690	19,265	22,000	
	3080		69 kV		.80	70		18,700	2,750	965	22,415	25,700	
	3090		161 kV		.56	100		22,400	3,950	1,375	27,725	32,200	
	3100		500 kV		.14	400		61,000	15,800	5,500	82,300	97,000	
	3110		Motor operation, 161 kV		.51	109		33,800	4,325	1,500	39,625	45,400	
	3120		500 kV	▼	.28	200	▼	93,500	7,875	2,750	104,125	118,000	

16360	Unit Substations	CREW	DAILY OUTPUT	LABOR-HOURS	UNIT	2006 BARE COSTS				TOTAL INCL O&P	
						MAT.	LABOR	EQUIP.	TOTAL		
800 3250	Circuit switches, 161 kV	R-11	.41	136	Ea.	67,500	5,375	1,875	74,750	84,000	800
3300	Single pole switches										
3350	Disconnecting switches, 13 to 26 kV	R-11	28	2	Ea.	8,800	79	27.50	8,906.50	9,825	
3360	46 kV		8	7		15,500	276	96.50	15,872.50	17,500	
3370	69 kV		5.60	10		17,200	395	138	17,733	19,600	
3380	161 kV		2.80	20		65,500	790	275	66,565	73,500	
3390	500 kV		.22	254		186,500	10,000	3,500	200,000	224,000	
3450	Grounding switches, 46 kV		5.60	10		23,600	395	138	24,133	26,700	
3460	69 kV		3.73	15.013		24,500	590	207	25,297	28,100	
3470	161 kV		2.24	25		26,000	985	345	27,330	30,500	
3480	500 kV	▼	.62	90.323	▼	32,300	3,550	1,250	37,100	42,200	
4000	Instrument transformers										
4050	Current transformers, 13 to 26 kV	R-11	14	4	Ea.	2,275	158	55	2,488	2,825	
4060	46 kV		9.33	6.002		6,675	237	82.50	6,994.50	7,800	
4070	69 kV		7	8		6,925	315	110	7,350	8,225	
4080	161 kV		1.87	29.947		22,400	1,175	410	23,985	26,900	
4100	Potential transformers, 13 to 26 kV		11.20	5		3,275	197	69	3,541	3,975	
4110	46 kV		8	7		6,725	276	96.50	7,097.50	7,925	
4120	69 kV		6.22	9.003		7,125	355	124	7,604	8,525	
4130	161 kV		2.24	25		15,400	985	345	16,730	18,800	
4140	500 kV	▼	1.40	40	▼	46,000	1,575	550	48,125	53,500	
7000	Conduit, conductors, and insulators										
7100	Conduit, metallic	R-11	254	.220	kg	5.10	8.70	3.03	16.83	22	
7110	Non-metallic		363	.154		4.63	6.10	2.12	12.85	16.55	
7200	Wire and cable	▼	318	.176	▼	4.32	6.95	2.42	13.69	17.80	
7290	See also Division 16100										
7300	Bus	R-11	268	.209	kg	6.20	8.25	2.88	17.33	22.50	
7400	Insulators, pedestal type	"	112	.500	Ea.		19.70	6.90	26.60	37	
7490	See also Line 16310-700-1000										
7500	Grounding systems	R-11	127	.441	kg	20.50	17.40	6.05	43.95	55	
7600	Manholes	"	4.15	13.494	Ea.	3,225	530	186	3,941	4,550	
7690	See also Division 02080										
7700	Cable tray	R-11	12.19	4.593	m	43.50	181	63	287.50	390	
8000	Protective equipment										
8050	Lightning arresters, 13 to 26 kV	R-11	18.67	2.999	Ea.	1,125	118	41.50	1,284.50	1,450	
8060	46 kV		14	4		3,050	158	55	3,263	3,650	
8070	69 kV		11.20	5		3,900	197	69	4,166	4,675	
8080	161 kV		5.60	10		5,375	395	138	5,908	6,675	
8090	500 kV		1.40	40		18,200	1,575	550	20,325	23,100	
8150	Reactors and resistors, 13 to 26 kV		28	2		2,550	79	27.50	2,656.50	2,950	
8160	46 kV		4.31	12.993		7,675	510	179	8,364	9,425	
8170	69 kV		2.80	20		12,500	790	275	13,565	15,300	
8180	161 kV		2.24	25		14,300	985	345	15,630	17,600	
8190	500 kV		.08	700		57,500	27,600	9,625	94,725	115,000	
8250	Fuses, 13 to 26 kV		18.67	2.999		1,500	118	41.50	1,659.50	1,900	
8260	46 kV		11.20	5		1,725	197	69	1,991	2,275	
8270	69 kV		8	7		1,800	276	96.50	2,172.50	2,500	
8280	161 kV	▼	4.67	11.991	▼	2,250	475	165	2,890	3,400	
9000	Station service equipment										
9100	Conversion equipment										
9110	Station service transformers	R-11	5.60	10	Ea.	67,500	395	138	68,033	74,500	
9120	Battery chargers		11.20	5	"	2,800	197	69	3,066	3,450	
9200	Control batteries	▼	14	4	K.A.H.	64	158	55	277	370	
9210											

16410 | Encl. Switches & Circuit Breakers

			CREW	DAILY OUTPUT	LABOR-HOURS	UNIT	MAT.	LABOR	EQUIP.	TOTAL	TOTAL INCL O&P	
200	0010	**CIRCUIT BREAKERS** (in enclosure)										200
	0100	Enclosed (NEMA 1), 600 V, 3 pole, 30 A	1 Elec	3.20	2.500	Ea.	500	105		605	705	
	0200	60 A		2.80	2.857		500	120		620	730	
	0400	100 A		2.30	3.478		570	146		716	845	
	0600	225 A		1.50	5.333		1,325	224		1,549	1,775	
	0700	400 A	2 Elec	1.60	10		2,250	420		2,670	3,100	
	0800	600 A		1.20	13.333		3,250	560		3,810	4,400	
	1000	800 A		.94	17.021		4,250	715		4,965	5,750	
800	0010	**SAFETY SWITCHES**										800
	0100	General duty, 240 V, 3 pole NEMA 1, fusible, 30 A	1 Elec	3.20	2.500	Ea.	82	105		187	247	
	0200	60 A		2.30	3.478		138	146		284	370	
	0300	100 A		1.90	4.211		238	177		415	525	
	0400	200 A		1.30	6.154		510	258		768	950	
	0500	400 A	2 Elec	1.80	8.889		1,300	375		1,675	1,975	
	0600	600 A	"	1.20	13.333		2,400	560		2,960	3,475	
	2900	Heavy duty, 240 V, 3 pole NEMA 1 fusible										
	2910	30 A	1 Elec	3.20	2.500	Ea.	132	105		237	300	
	3000	60 A		2.30	3.478		224	146		370	465	
	3300	100 A		1.90	4.211		350	177		527	655	
	3500	200 A		1.30	6.154		605	258		863	1,050	
	3700	400 A	2 Elec	1.80	8.889		1,550	375		1,925	2,250	
	3900	600 A	"	1.20	13.333		2,675	560		3,235	3,775	
	5510	Heavy duty, 600 V, 3 pole 3ph. NEMA 3R fusible, 30 A	1 Elec	3.10	2.581		375	108		483	575	
	5520	60 A		2.20	3.636		440	153		593	715	
	5530	100 A		1.80	4.444		690	187		877	1,025	
	5540	200 A		1.20	6.667		950	280		1,230	1,475	
	5550	400 A	2 Elec	1.60	10		2,225	420		2,645	3,075	

16420 | Enclosed Controllers

			CREW	DAILY OUTPUT	LABOR-HOURS	UNIT	MAT.	LABOR	EQUIP.	TOTAL	TOTAL INCL O&P	
220	0010	**CONTROL STATIONS**										220
	0050	NEMA 1, heavy duty, stop/start	1 Elec	8	1	Ea.	127	42		169	203	
	0100	Stop/start, pilot light		6.20	1.290		173	54		227	271	
	0200	Hand/off/automatic		6.20	1.290		94	54		148	184	
	0400	Stop/start/reverse		5.30	1.509		171	63.50		234.50	283	

16440 | Swbds, Panels & Control Centers

			CREW	DAILY OUTPUT	LABOR-HOURS	UNIT	MAT.	LABOR	EQUIP.	TOTAL	TOTAL INCL O&P	
660	0010	**MOTOR STARTERS & CONTROLS**										660
	0050	Magnetic, FVNR, with enclosure and heaters, 480 V										
	0100	3.7 kW, size 0	1 Elec	2.30	3.478	Ea.	224	146		370	465	
	0200	7.5 kW, size 1	"	1.60	5		251	210		461	590	
	0300	19 kW, size 2	2 Elec	2.20	7.273		470	305		775	975	
	0400	37 kW, size 3		1.80	8.889		770	375		1,145	1,400	
	0500	75 kW, size 4		1.20	13.333		1,700	560		2,260	2,700	
	0600	149 kW, size 5		.90	17.778		4,000	745		4,745	5,525	
	0700	Combination, with motor circuit protectors, 4 kW, size 0	1 Elec	1.80	4.444		725	187		912	1,075	
	0800	7.46 kW, size 1	"	1.30	6.154		755	258		1,013	1,225	
	0900	19 kW, size 2	2 Elec	2	8		1,050	335		1,385	1,675	
	1000	37 kW, size 3		1.32	12.121		1,525	510		2,035	2,425	
	1200	75 kW, size 4		.80	20		3,325	840		4,165	4,900	
	1400	Combination, with fused switch, 4 kW, size 0	1 Elec	1.80	4.444		555	187		742	890	
	1600	7.46 kW, size 1	"	1.30	6.154		595	258		853	1,050	
	1800	19 kW, size 2	2 Elec	2	8		965	335		1,300	1,550	
	2000	37 kW, size 3		1.32	12.121		1,625	510		2,135	2,550	
	2200	75 kW, size 4		.80	20		2,850	840		3,690	4,375	

Important: See the Reference Section for supporting data - Crews, Rental Equipment, City Cost Indexes and Reference Data

16 ELECTRICAL

16440	Swbds, Panels & Control Centers	CREW	DAILY OUTPUT	LABOR-HOURS	UNIT	2006 BARE COSTS MAT.	LABOR	EQUIP.	TOTAL	TOTAL INCL O&P
700	**0010 PANELBOARD & LOAD CENTER CIRCUIT BREAKERS**									**700**
0050	Bolt-on, 10 k AIC., 120 V, 1 pole									
0100	15 to 50 A	1 Elec	10	.800	Ea.	13	33.50		46.50	64.50
0200	60 A		8	1		13	42		55	77
0300	70 A	↓	8	1	↓	24.50	42		66.50	89.50
0350	240 V, 2 pole									
0400	15 to 50 A	1 Elec	8	1	Ea.	28.50	42		70.50	94
0500	60 A		7.50	1.067		28.50	45		73.50	98.50
0600	80 to 100 A		5	1.600		73.50	67		140.50	181
0700	3 pole, bolt-on, 15 to 60 A		6.20	1.290		90.50	54		144.50	181
0800	70 A		5	1.600		114	67		181	225
0900	80 to 100 A		3.60	2.222		129	93.50		222.50	281
1000	22 k AIC, 240 V, 2 pole, 70 - 225 A		2.70	2.963		555	124		679	795
1100	3 pole, 70 - 225 A		2.30	3.478		620	146		766	900
1200	14 k AIC, 277 V, 1 pole, 15 - 30 A		8	1		34.50	42		76.50	101
1300	22 k AIC, 480 V, 2 pole, 70 - 225 A		2.70	2.963		555	124		679	795
1400	3 pole, 70 - 225 A	↓	2.30	3.478	↓	685	146		831	975
720	**0010 PANELBOARDS** (Commercial use)									**720**
0050	NQOD, w/20 A 1 pole bolt-on circuit breakers									
0100	3 wire, 120/240 V, 100 A main lugs									
0150	10 circuits	1 Elec	1	8	Ea.	420	335		755	965
0200	14 circuits		.88	9.091		495	380		875	1,125
0250	18 circuits		.75	10.667		540	450		990	1,275
0300	20 circuits	↓	.65	12.308		610	515		1,125	1,450
0350	225 A main lugs, 24 circuits	2 Elec	1.20	13.333		690	560		1,250	1,600
0400	30 circuits		.90	17.778		805	745		1,550	2,000
0450	36 circuits		.80	20		920	840		1,760	2,250
0500	38 circuits		.72	22.222		985	935		1,920	2,475
0550	42 circuits	↓	.66	24.242		1,025	1,025		2,050	2,650
0600	4 wire, 120/208 V, 100 A main lugs, 12 circuits	1 Elec	1	8		475	335		810	1,025
0650	16 circuits		.75	10.667		550	450		1,000	1,275
0700	20 circuits		.65	12.308		640	515		1,155	1,475
0750	24 circuits		.60	13.333		695	560		1,255	1,600
0800	30 circuits	↓	.53	15.094		800	635		1,435	1,825
0850	225 A main lugs, 32 circuits	2 Elec	.90	17.778		900	745		1,645	2,125
0900	34 circuits		.84	19.048		925	800		1,725	2,225
0950	36 circuits		.80	20		945	840		1,785	2,300
1000	42 circuits	↓	.68	23.529	↓	1,050	990		2,040	2,650
1200	NEHB, w/20 A, 1 pole bolt-on circuit breakers									
1250	4 wire, 277/480 V, 100 A main lugs, 12 circuits	1 Elec	.88	9.091	Ea.	915	380		1,295	1,575
1300	20 circuits	"	.60	13.333		1,350	560		1,910	2,300
1350	225 A main lugs, 24 circuits	2 Elec	.90	17.778		1,550	745		2,295	2,850
1400	30 circuits		.80	20		1,875	840		2,715	3,300
1450	36 circuits	↓	.72	22.222	↓	2,175	935		3,110	3,800
1600	NQOD panel, w/20 A, 1 pole, circuit breakers									
1650	3 wire, 120/240 V with main circuit breaker									
1700	100 A main, 12 circuits	1 Elec	.80	10	Ea.	585	420		1,005	1,275
1750	20 circuits	"	.60	13.333		750	560		1,310	1,650
1800	225 A main, 30 circuits	2 Elec	.68	23.529		1,425	990		2,415	3,050
1850	42 circuits		.52	30.769		1,675	1,300		2,975	3,750
1900	400 A main, 30 circuits		.54	29.630		2,000	1,250		3,250	4,050
1950	42 circuits	↓	.50	32	↓	2,225	1,350		3,575	4,450
2000	4 wire, 120/208 V with main circuit breaker									
2050	100 A main, 24 circuits	1 Elec	.47	17.021	Ea.	875	715		1,590	2,025
2100	30 circuits	"	.40	20	↓	990	840		1,830	2,350

For expanded coverage of these items see *Means Electrical Cost Data 2006*

16440	Swbds, Panels & Control Centers	CREW	DAILY OUTPUT	LABOR-HOURS	UNIT	2006 BARE COSTS				TOTAL INCL O&P	
						MAT.	LABOR	EQUIP.	TOTAL		
720 2200	225 A main, 32 circuits	2 Elec	.72	22.222	Ea.	1,675	935		2,610	3,225	**720**
2250	42 circuits		.56	28.571		1,825	1,200		3,025	3,800	
2300	400 A main, 42 circuits		.48	33.333		2,475	1,400		3,875	4,800	
2350	600 A main, 42 circuits		.40	40		3,675	1,675		5,350	6,525	
2400	NEHB, with 20 A, 1 pole circuit breaker										
2450	4 wire, 277/480 V with main circuit breaker										
2500	100 A main, 24 circuits	1 Elec	.42	19.048	Ea.	1,800	800		2,600	3,175	
2550	30 circuits	"	.38	21.053		2,100	885		2,985	3,650	
2600	225 A main, 30 circuits	2 Elec	.72	22.222		2,650	935		3,585	4,325	
2650	42 circuits	"	.56	28.571		3,275	1,200		4,475	5,400	
800 0010	**SWITCHBOARDS** distribution section										**800**
0100	Aluminum bus bars, not including breakers										
0200	120/208 or 277/480 V, 4 wire, 600 A	2 Elec	1	16	Ea.	1,475	670		2,145	2,625	
0300	800 A		.88	18.182		1,900	765		2,665	3,225	
0400	1000 A		.80	20		2,000	840		2,840	3,450	
0500	1200 A		.72	22.222		2,700	935		3,635	4,375	
0600	1600 A		.66	24.242		3,150	1,025		4,175	5,000	
0700	2000 A		.62	25.806		3,700	1,075		4,775	5,700	
0800	2500 A		.60	26.667		4,175	1,125		5,300	6,275	
0900	3000 A		.56	28.571		5,050	1,200		6,250	7,375	
0950	4000 A		.52	30.769		7,375	1,300		8,675	10,100	
820 0010	**SWITCHBOARDS** feeder section group mounted devices										**820**
0030	Circuit breakers										
0160	FA frame, 15 to 60 A, 240 V, 1 pole	1 Elec	8	1	Ea.	83	42		125	154	
0280	FA frame, 70 to 100 A, 240 V, 1 pole		7	1.143		109	48		157	192	
0420	KA frame, 70 to 225 A		3.20	2.500		915	105		1,020	1,150	
0430	LA frame, 125 to 400 A		2.30	3.478		2,075	146		2,221	2,500	
0460	MA frame, 450 to 600 A		1.60	5		3,425	210		3,635	4,075	
0470	700 to 800 A		1.30	6.154		4,425	258		4,683	5,250	
0480	MAL frame, 1000 A		1	8		4,600	335		4,935	5,575	
0490	PA frame, 1200 A		.80	10		9,350	420		9,770	10,900	
0500	Branch circuit, fusible switch, 600 V, double 30/30 A		4	2		935	84		1,019	1,150	
0550	60/60 A		3.20	2.500		950	105		1,055	1,200	
0600	100/100 A		2.70	2.963		1,325	124		1,449	1,625	
0650	Single, 30 A		5.30	1.509		630	63.50		693.50	790	
0700	60 A		4.70	1.702		630	71.50		701.50	800	
0750	100 A		4	2		885	84		969	1,100	
0800	200 A		2.70	2.963		1,325	124		1,449	1,625	
0850	400 A		2.30	3.478		2,500	146		2,646	2,975	
0900	600 A		1.80	4.444		2,950	187		3,137	3,525	
0950	800 A		1.30	6.154		4,850	258		5,108	5,700	
1000	1200 A		.80	10		5,775	420		6,195	6,975	
840 0010	**SWITCHBOARDS** Incoming main service section										**840**
0100	Aluminum bus bars, not including CT's or PT's										
0200	No main disconnect, includes CT compartment										
0300	120/208 V, 4 wire, 600 A	2 Elec	1	16	Ea.	3,450	670		4,120	4,800	
0400	800 A		.88	18.182		3,450	765		4,215	4,950	
0500	1000 A		.80	20		4,150	840		4,990	5,800	
0600	1200 A		.72	22.222		4,150	935		5,085	5,950	
0700	1600 A		.66	24.242		4,150	1,025		5,175	6,075	
0800	2000 A		.62	25.806		4,450	1,075		5,525	6,525	
1000	3000 A		.56	28.571		5,900	1,200		7,100	8,275	
2000	Fused switch & CT compartment										
2100	120/208 V, 4 wire, 400 A	2 Elec	1.12	14.286	Ea.	3,775	600		4,375	5,050	

16 ELECTRICAL

16440	Swbds, Panels & Control Centers	CREW	DAILY OUTPUT	LABOR-HOURS	UNIT	2006 BARE COSTS				TOTAL INCL O&P	
						MAT.	LABOR	EQUIP.	TOTAL		
840											**840**
2200	600 A	2 Elec	.94	17.021	Ea.	4,475	715		5,190	5,975	
2300	800 A		.84	19.048		9,325	800		10,125	11,500	
2400	1200 A	▼	.68	23.529	▼	12,100	990		13,090	14,800	
2900	Pressure switch & CT compartment										
3000	120/208 V, 4 wire, 800 A	2 Elec	.80	20	Ea.	8,375	840		9,215	10,500	
3100	1200 A		.66	24.242		16,200	1,025		17,225	19,300	
3200	1600 A		.62	25.806		17,200	1,075		18,275	20,600	
3300	2000 A	▼	.56	28.571		18,300	1,200		19,500	21,900	
4400	Circuit breaker, molded case & CT compartment										
4600	3 pole, 4 wire, 600 A	2 Elec	.94	17.021	Ea.	7,225	715		7,940	9,000	
4800	800 A		.84	19.048		8,625	800		9,425	10,700	
5000	1200 A	▼	.68	23.529	▼	11,800	990		12,790	14,400	
5100	Copper bus bars, not incl. CT's or PT's, add, minimum						15%				

16450	Enclosed Bus Assemblies										
320											**320**
0010	**COPPER BUS DUCT** 3.05 m long										
0050	3 pole 4 wire, plug-in/indoor, straight section, 225 A	2 Elec	12.19	1.312	m	485	55		540	610	
1000	400 A		9.75	1.640		485	69		554	635	
1500	600 A		7.92	2.019		485	85		570	655	
2400	800 A		6.10	2.625		575	110		685	795	
2450	1000 A		5.49	2.916		635	122		757	880	
2500	1350 A		4.88	3.281		905	138		1,043	1,200	
2510	1600 A		3.66	4.374		1,025	184		1,209	1,400	
2520	2000 A		3.05	5.249		1,300	220		1,520	1,750	
2550	Feeder, 600 A		8.53	1.875		425	78.50		503.50	580	
2600	800 A		6.71	2.386		515	100		615	715	
2700	1000 A		6.10	2.625		575	110		685	795	
2800	1350 A		5.49	2.916		845	122		967	1,125	
2900	1600 A		4.27	3.750		965	157		1,122	1,275	
3000	2000 A	▼	3.66	4.374		1,250	184		1,434	1,625	
3100	Elbows, 225 A		4	4	Ea.	870	168		1,038	1,200	
3200	400 A		3.60	4.444		870	187		1,057	1,250	
3300	600 A		3.20	5		870	210		1,080	1,275	
3400	800 A		2.80	5.714		945	240		1,185	1,400	
3500	1000 A		2.60	6.154		1,050	258		1,308	1,525	
3600	1350 A		2.40	6.667		1,250	280		1,530	1,800	
3700	1600 A		2.20	7.273		1,350	305		1,655	1,925	
3800	2000 A		1.80	8.889		1,675	375		2,050	2,375	
4000	End box, 225 A		34	.471		118	19.75		137.75	160	
4100	400 A		32	.500		118	21		139	162	
4200	600 A		28	.571		118	24		142	166	
4300	800 A		26	.615		118	26		144	169	
4400	1000 A		24	.667		118	28		146	172	
4500	1350 A		22	.727		118	30.50		148.50	176	
4600	1600 A		20	.800		118	33.50		151.50	180	
4700	2000 A		18	.889		144	37.50		181.50	215	
4800	Cable tap box, end, 225 A		3.20	5		875	210		1,085	1,275	
5000	400 A		2.60	6.154		875	258		1,133	1,350	
5100	600 A		2.20	7.273		880	305		1,185	1,425	
5200	800 A		2	8		985	335		1,320	1,575	
5300	1000 A		1.60	10		1,075	420		1,495	1,825	
5400	1350 A		1.40	11.429		1,300	480		1,780	2,150	
5500	1600 A		1.20	13.333		1,450	560		2,010	2,425	
5600	2000 A		1	16		1,675	670		2,345	2,825	
5700	Switchboard stub, 225 A	▼	5.40	2.963	▼	885	124		1,009	1,150	

ELECTRICAL 16

For expanded coverage of these items see *Means Electrical Cost Data 2006*

16450	Enclosed Bus Assemblies	CREW	DAILY OUTPUT	LABOR-HOURS	UNIT	2006 BARE COSTS				TOTAL INCL O&P	
						MAT.	LABOR	EQUIP.	TOTAL		
320	5800	400 A	2 Elec	4.60	3.478	Ea.	885	146		1,031	1,200	320
5900	600 A		4	4		885	168		1,053	1,225
6000	800 A		3.20	5		1,075	210		1,285	1,500
6100	1000 A		3	5.333		1,250	224		1,474	1,675
6200	1350 A		2.60	6.154		1,600	258		1,858	2,125
6300	1600 A		2.40	6.667		1,800	280		2,080	2,400
6400	2000 A		2	8		2,175	335		2,510	2,900
6490	Tee fittings, 225 A		2.40	6.667		1,200	280		1,480	1,750
6500	400 A		2	8		1,200	335		1,535	1,825
6600	600 A		1.80	8.889		1,200	375		1,575	1,875
6700	800 A		1.60	10		1,375	420		1,795	2,150
6800	1350 A		1.20	13.333		2,000	560		2,560	3,025
7000	1600 A		1	16		2,275	670		2,945	3,500
7100	2000 A		.80	20		2,700	840		3,540	4,200
7200	Plug-in fusible switches w/3 fuses, 600 V, 3 pole, 30 A	1 Elec	4	2		525	84		609	700
7300	60 A		3.60	2.222		590	93.50		683.50	790
7400	100 A		2.70	2.963		900	124		1,024	1,175
7500	200 A	2 Elec	3.20	5		1,600	210		1,810	2,100
7600	400 A		1.40	11.429		4,700	480		5,180	5,900
7700	600 A		.90	17.778		5,325	745		6,070	7,000
7800	800 A		.66	24.242		7,525	1,025		8,550	9,800
7900	1200 A		.50	32		14,100	1,350		15,450	17,500
8000	Plug-in circuit breakers, molded case, 15 to 50 A	1 Elec	4.40	1.818		495	76.50		571.50	660
8100	70 to 100 A	"	3.10	2.581		550	108		658	765
8200	150 to 225 A	2 Elec	3.40	4.706		1,425	198		1,623	1,875
8300	250 to 400 A		1.40	11.429		2,625	480		3,105	3,600
8400	500 to 600 A		1	16		3,550	670		4,220	4,900
8500	700 to 800 A		.64	25		4,375	1,050		5,425	6,375
8600	900 to 1000 A		.56	28.571		6,250	1,200		7,450	8,675
8700	1200 A		.44	36.364		7,500	1,525		9,025	10,600

16510	Interior Luminaires	CREW	DAILY OUTPUT	LABOR-HOURS	UNIT	2006 BARE COSTS				TOTAL INCL O&P	
						MAT.	LABOR	EQUIP.	TOTAL		
440	0010	INTERIOR LIGHTING FIXTURES Including lamps, mounting										440
0030	hardware and connections									
0100	Fluorescent, C.W. lamps, troffer, recess mounted in grid, RS									
0200	Acrylic lens, 300 mm W x 1200 mm L, two 40 W	1 Elec	5.70	1.404	Ea.	48.50	59		107.50	142
0210	300 mm W x 1200 mm L, three 40 W		5.40	1.481		56.50	62		118.50	155
0300	600 mm W x 600 mm L, two U40 W		5.70	1.404		52	59		111	145
0400	600 mm W x 1200 mm L, two 40 W		5.30	1.509		52	63.50		115.50	152
0500	600 mm W x 1200 mm L, three 40 W		5	1.600		55.50	67		122.50	161
0600	600 mm W x 1200 mm L, four 40 W		4.70	1.702		59	71.50		130.50	172
0700	1200 mmW x 1200 mmL, four 40 W	2 Elec	6.40	2.500		310	105		415	495
0800	1200 mm W x 1200 mm L, six 40 W		6.20	2.581		320	108		428	515
0900	1200 mm W x 1200 mm L, eight 40 W		5.80	2.759		325	116		441	535
0910	Acrylic lens, 300 mm W x 600 mm L, two 32 W	1 Elec	5.70	1.404		56	59		115	150
0930	600 mm W x 600 mm L, two U32 W		5.70	1.404		80	59		139	176
0940	600 mm W x 600 mm L, two 32 W		5.30	1.509		72	63.50		135.50	174
0950	600 mm W x 1200 mm L, three 32 W		5	1.600		76.50	67		143.50	184

Important: See the Reference Section for supporting data - Crews, Rental Equipment, City Cost Indexes and Reference Data

16 ELECTRICAL

16510	Interior Luminaires	CREW	DAILY OUTPUT	LABOR-HOURS	UNIT	2006 BARE COSTS				TOTAL INCL O&P	
						MAT.	LABOR	EQUIP.	TOTAL		
440 0960	2'W x 4'L, four 32 watt	1 Elec	4.70	1.702	Ea.	78.50	71.50		150	194	**440**
1000	Surface mounted, RS										
1030	Acrylic lens with hinged & latched door frame										
1100	300 mm W x 1200 mm L, two 40 W	1 Elec	7	1.143	Ea.	73.50	48		121.50	153	
1110	300 mm W x 1200 mm L, three 40 W		6.70	1.194		75.50	50		125.50	158	
1200	600 mm W x 600 mm L, two U40 W		7	1.143		79	48		127	158	
1300	600 mm W x 1200 mm L, two 40 W		6.20	1.290		90.50	54		144.50	181	
1400	600 mm W x 1200 mm L, three 40 W		5.70	1.404		91.50	59		150.50	188	
1500	600 mm W x 1200 mm L, four 40 W		5.30	1.509		93.50	63.50		157	198	
1600	1200 mm W x 1200 mm L, four 40 W	2 Elec	7.20	2.222		430	93.50		523.50	615	
1700	1200 mm W x 1200 mm L, six 40 W		6.60	2.424		470	102		572	665	
1800	1200 mm W x 1200 mm L, eight 40 W		6.20	2.581		485	108		593	695	
1900	600 mm W x 2400 mm L, four 40 W		6.40	2.500		163	105		268	335	
2000	600 mm W x 2400 mm L, eight 40 W		6.20	2.581		187	108		295	370	
2100	Strip fixture										
2130	Surface mounted										
2200	1200 mm long, one 40 W RS	1 Elec	8.50	.941	Ea.	28.50	39.50		68	90	
2300	1200 mm long, two 40 W RS		8	1		30.50	42		72.50	96	
2400	1200 mm long, one 40 W, SL		8	1		41.50	42		83.50	108	
2500	1200 mm long, two 40 W, SL		7	1.143		56.50	48		104.50	134	
2600	2400 mm long, one 75 W, SL	2 Elec	13.40	1.194		42.50	50		92.50	122	
2700	2400 mm long, two 75 W, SL	"	12.40	1.290		51.50	54		105.50	138	
2800	1200 mm long, two 60 W, HO	1 Elec	6.70	1.194		83	50		133	167	
2900	2400 mm long, two 110 W, HO	2 Elec	10.60	1.509		87.50	63.50		151	191	
3000	Pendent mounted, industrial, white porcelain enamel										
3100	1200 mm long, two 40 W, RS	1 Elec	5.70	1.404	Ea.	47	59		106	140	
3200	1200 mm long, two 60 W, HO	"	5	1.600		74.50	67		141.50	182	
3300	2400 mm long, two 75 W, SL	2 Elec	8.80	1.818		88.50	76.50		165	212	
3400	2400 mm long, two 110 W, HO	"	8	2		113	84		197	250	
3470	Troffer, air handling, 600 mm W x 1200 mm L 4-40 W, RS	1 Elec	4	2		83	84		167	217	
3480	600 mm W x 600 mm L with two U40 W RS		5.50	1.455		72.50	61		133.50	171	
3490	Air connector insulated, 125 mm diameter		20	.400		56.50	16.80		73.30	87.50	
3500	150 mm diameter		20	.400		58	16.80		74.80	88.50	
4450	Incandescent, high hat can, round alzak reflector, prewired										
4470	100 W	1 Elec	8	1	Ea.	60.50	42		102.50	129	
4480	150 W		8	1		87.50	42		129.50	159	
4500	300 W		6.70	1.194		204	50		254	300	
4600	Square glass lens with metal trim, prewired										
4630	100 W	1 Elec	6.70	1.194	Ea.	47	50		97	127	
4700	200 W		6.70	1.194		82	50		132	166	
4800	300 W		5.70	1.404		122	59		181	222	
4900	Ceiling/wall, surface mounted, metal cylinder, 75 W		10	.800		48	33.50		81.50	103	
4920	150 W		10	.800		69	33.50		102.50	126	
5200	Ceiling, surface mounted, opal glass drum										
5300	203 mm, one 60 W lamp	1 Elec	10	.800	Ea.	37.50	33.50		71	91.50	
5400	254 mm, two 60 W lamps		8	1		42	42		84	109	
5500	305 mm, four 60 W lamps		6.70	1.194		60	50		110	141	
6010	Vapor tight, incandescent, ceiling mounted, 200 W		6.20	1.290		53.50	54		107.50	140	
6100	Fluorescent, surface mounted, 2 lamps, 1200 mm L, RS, 40 W		3.20	2.500		97	105		202	264	
6300	Explosionproof										
6510	Incandescent, ceiling mounted, 200 W	1 Elec	4	2	Ea.	700	84		784	895	
6600	Fluorescent, RS, 1200 mm long, ceiling mounted, two 40 W		2.70	2.963		1,775	124		1,899	2,125	
6850	Vandalproof, surface mounted, fluorescent, two 40 W		3.20	2.500		211	105		316	390	
6860	Incandescent, one 150 W		8	1		54.50	42		96.50	123	
7500	Ballast replacement, by weight of ballast, to 4.6 m high										
7520	Indoor fluorescent, less than 0.9 kg	1 Elec	10	.800	Ea.		33.50		33.50	50	

ELECTRICAL 16

16510 | Interior Luminaires

		CREW	DAILY OUTPUT	LABOR-HOURS	UNIT	2006 BARE COSTS				TOTAL INCL O&P	
						MAT.	LABOR	EQUIP.	TOTAL		
440	7540	Two 40 W, watt reducer, 0.9 to 2.3 kg	1 Elec	9.40	.851	Ea.	23	35.50		58.50	79	**440**
	7560	Two F96 slimline, over 2.3 kg		8	1		39	42		81	105	
	7580	Vaportite ballast, less than 0.9 kg		9.40	.851			35.50		35.50	53.50	
	7600	0.9 kg to 2.3 kg		8.90	.899			38		38	56.50	
	7620	Over 2.3 kg		7.60	1.053			44		44	66	
	7630	Electronic ballast for two tubes		8	1		34.50	42		76.50	101	
	7640	Dimmable ballast one lamp		8	1		49.50	42		91.50	117	
	7650	Dimmable ballast two-lamp		7.60	1.053		80	44		124	154	

16520 | Exterior Luminaires

		CREW	DAILY OUTPUT	LABOR-HOURS	UNIT	2006 BARE COSTS				TOTAL INCL O&P	
						MAT.	LABOR	EQUIP.	TOTAL		
300	0010	**EXTERIOR FIXTURES** With lamps										**300**
	0200	Wall mounted, incandescent, 100 W	1 Elec	8	1	Ea.	28	42		70	93.50	
	0400	Quartz, 500 W		5.30	1.509		54	63.50		117.50	154	
	0420	1500 W		4.20	1.905		102	80		182	232	
	1100	Wall pack, low pressure sodium, 35 W		4	2		214	84		298	360	
	1150	55 W		4	2		255	84		339	405	
	1160	High pressure sodium, 70 W		4	2		200	84		284	345	
	1170	150 W		4	2		230	84		314	380	
	1180	Metal Halide, 175 W		4	2		255	84		339	405	
	1190	250 W		4	2		265	84		349	415	
	1200	Floodlights with ballast and lamp,										
	1400	pole mounted, pole not included										
	1950	Metal halide, 175 W	1 Elec	2.70	2.963	Ea.	300	124		424	515	
	2000	400 W	2 Elec	4.40	3.636		360	153		513	625	
	2200	1000 W		4	4		505	168		673	805	
	2210	1500 W		3.70	4.324		530	182		712	855	
	2250	Low pressure sodium, 55 W	1 Elec	2.70	2.963		485	124		609	720	
	2270	90 W		2	4		535	168		703	840	
	2290	180 W		2	4		680	168		848	1,000	
	2340	High pressure sodium, 70 W		2.70	2.963		207	124		331	410	
	2360	100 W		2.70	2.963		212	124		336	420	
	2380	150 W		2.70	2.963		249	124		373	460	
	2400	400 W	2 Elec	4.40	3.636		365	153		518	630	
	2600	1000 W	"	4	4		530	168		698	835	
	2610	Incandescent, 300 W	1 Elec	4	2		76.50	84		160.50	209	
	2620	500 W	"	4	2		122	84		206	259	
	2630	1000 W	2 Elec	6	2.667		131	112		243	310	
	2640	1500 W	"	6	2.667		144	112		256	325	
	2650	Roadway area luminaire, low pressure sodium, 135 W	1 Elec	2	4		535	168		703	840	
	2700	180 W	"	2	4		565	168		733	870	
	2750	Metal halide, 400 W	2 Elec	4.40	3.636		460	153		613	735	
	2760	1000 W		4	4		520	168		688	820	
	2780	High pressure sodium, 400 W		4.40	3.636		475	153		628	755	
	2790	1000 W		4	4		545	168		713	850	
	2800	Light poles, anchor base										
	2820	not including concrete bases										
	2840	Aluminum pole, 2.4 m high	1 Elec	4	2	Ea.	585	84		669	770	
	2850	3 m high		4	2		610	84		694	795	
	2860	3.7 m high		3.80	2.105		635	88.50		723.50	830	
	2870	4.3 m high		3.40	2.353		665	99		764	875	
	2880	4.9 m high		3	2.667		730	112		842	965	
	3000	6.1 m high	R-3	2.90	6.897		760	286	52.50	1,098.50	1,325	
	3200	9.1 m high		2.60	7.692		1,475	320	58.50	1,853.50	2,175	
	3400	10.7 m high		2.30	8.696		1,600	360	66.50	2,026.50	2,400	
	3600	12.2 m high		2	10		1,825	415	76.50	2,316.50	2,725	
	3800	Bracket arms, 1 arm	1 Elec	8	1		100	42		142	173	

16 ELECTRICAL

16520 | Exterior Luminaires

		CREW	DAILY OUTPUT	LABOR-HOURS	UNIT	2006 BARE COSTS				TOTAL INCL O&P	
						MAT.	LABOR	EQUIP.	TOTAL		
300	4000	2 arms	1 Elec	8	1	Ea.	201	42		243	284	300
	4200	3 arms		5.30	1.509		300	63.50		363.50	425
	4400	4 arms		5.30	1.509		400	63.50		463.50	535
	4500	Steel pole, galvanized, 2.4 m high		3.80	2.105		510	88.50		598.50	695
	4510	3 m high		3.70	2.162		535	91		626	725
	4520	3.7 m high		3.40	2.353		580	99		679	780
	4530	4.3 m high		3.10	2.581		615	108		723	840
	4540	4.9 m high		2.90	2.759		655	116		771	895
	4550	5.5 m high	▼	2.70	2.963		690	124		814	945
	4600	6.1 m high	R-3	2.60	7.692		910	320	58.50	1,288.50	1,550
	4800	9.1 m high		2.30	8.696		1,075	360	66.50	1,501.50	1,800
	5000	10.7 m high		2.20	9.091		1,175	375	69.50	1,619.50	1,950
	5200	12.2 m high	▼	1.70	11.765		1,450	485	90	2,025	2,425
	5400	Bracket arms, 1 arm	1 Elec	8	1		149	42		191	227
	5600	2 arms		8	1		231	42		273	315
	5800	3 arms		5.30	1.509		250	63.50		313.50	370
	6000	4 arms	▼	5.30	1.509		350	63.50		413.50	475
	6100	Fiberglass pole, 1 or 2 fixtures, 6.1 m high	R-3	4	5		465	207	38	710	860
	6200	9.1 m high		3.60	5.556		575	230	42.50	847.50	1,025
	6300	10.7 m high		3.20	6.250		900	259	47.50	1,206.50	1,425
	6400	12.2 m high	▼	2.80	7.143		1,050	296	54.50	1,400.50	1,650
	6420	Wood pole, 114 mm x 130 mm, 2.4 m high	1 Elec	6	1.333		269	56		325	380
	6430	3 m high		6	1.333		305	56		361	420
	6440	3.7 m high		5.70	1.404		385	59		444	515
	6450	4.6 m high		5	1.600		450	67		517	595
	6460	6.1 m high	▼	4	2	▼	545	84		629	725
	6500	Bollard light, lamp & ballast, 1.07 m high with polycarbonate lens									
	6800	Metal halide, 175 W	1 Elec	3	2.667	Ea.	645	112		757	870
	6900	High pressure sodium, 70 W		3	2.667		660	112		772	890
	7000	100 W		3	2.667		660	112		772	890
	7100	150 W		3	2.667		645	112		757	870
	7200	Incandescent, 150 W	▼	3	2.667	▼	465	112		577	675
	7300	Transformer bases, not including concrete bases									
	7320	Maximum pole size, steel, 12.2 m high	1 Elec	2	4	Ea.	1,100	168		1,268	1,475
	7340	Cast aluminum, 9.1 m high		3	2.667		600	112		712	825
	7350	12.2 m high	▼	2.50	3.200	▼	905	134		1,039	1,200
	7380	Landscape recessed uplight, incl. housing, ballast, transformer									
	7390	& reflector									
	7420	Incandescent, 250 W	1 Elec	5	1.600	Ea.	465	67		532	615
	7440	Quartz, 250 W		5	1.600		445	67		512	585
	7460	500 W	▼	4	2	▼	455	84		539	625
	7500	Replacement (H.I.D.) ballasts,									
	7510	Multi-tap 120/208/240/277 V									
	7550	High pressure sodium, 70 W	1 Elec	10	.800	Ea.	142	33.50		175.50	206
	7560	100 W		9.40	.851		148	35.50		183.50	217
	7570	150 W		9	.889		160	37.50		197.50	232
	7580	250 W		8.50	.941		239	39.50		278.50	320
	7590	400 W		7	1.143		269	48		317	370
	7600	1000 W		6	1.333		370	56		426	495
	7610	Metal halide, 175 W		8	1		87	42		129	159
	7620	250 W		8	1		113	42		155	187
	7630	400 W		7	1.143		141	48		189	227
	7640	1000 W		6	1.333		241	56		297	350
	7650	1500 W		5	1.600		299	67		366	430
	7810	Walkway luminaire, square 400 mm, metal halide 250 W		2.70	2.963		520	124		644	755
	7820	High pressure sodium, 70 W	▼	3	2.667	▼	595	112		707	820

ELECTRICAL 16

16520	Exterior Luminaires	CREW	DAILY OUTPUT	LABOR-HOURS	UNIT	2006 BARE COSTS				TOTAL INCL O&P	
						MAT.	LABOR	EQUIP.	TOTAL		
300 7830	100 W	1 Elec	3	2.667	Ea.	605	112		717	830	**300**
7840	150 W		3	2.667		605	112		717	830	
7850	200 W		3	2.667		610	112		722	835	
7910	Round 480 mm, metal halide, 250 W		2.70	2.963		755	124		879	1,025	
7920	High pressure sodium, 70 W		3	2.667		830	112		942	1,075	
7930	100 W		3	2.667		830	112		942	1,075	
7940	150 W		3	2.667		835	112		947	1,075	
7950	250 W		2.70	2.963		875	124		999	1,150	
8000	Sphere 350 mm opal, incandescent, 200 W		4	2		244	84		328	395	
8020	Sphere 450 mm opal, incandescent, 300 W		3.50	2.286		295	96		391	470	
8040	Sphere 400 mm clear, high pressure sodium, 70 W		3	2.667		510	112		622	725	
8050	100 W		3	2.667		545	112		657	765	
8100	Cube 400 mm opal, incandescent, 300 W		3.50	2.286		325	96		421	505	
8120	High pressure sodium, 70 W		3	2.667		475	112		587	690	
8130	100 W		3	2.667		490	112		602	700	
8230	Lantern, high pressure sodium, 70 W		3	2.667		420	112		532	630	
8240	100 W		3	2.667		455	112		567	665	
8250	150 W		3	2.667		425	112		537	635	
8260	250 W		2.70	2.963		595	124		719	840	
8270	Incandescent, 300 W		3.50	2.286		315	96		411	490	
8330	Reflector 550 mm w/globe, high pressure sodium, 70 W		3	2.667		395	112		507	600	
8340	100 W		3	2.667		405	112		517	610	
8350	150 W		3	2.667		410	112		522	615	
8360	250 W		2.70	2.963		520	124		644	760	
500 0010	**LANDSCAPE FIXTURES** Incl. conduit, wire, trench										**500**
0030	Bollards										
0040	Incandescent, 600 mm	1 Elec	2.50	3.200	Ea.	305	134		439	535	
0050	900 mm		2	4		365	168		533	650	
0060	1050 mm		2	4		385	168		553	675	
0070	H.I.D., 600 mm		2.50	3.200		300	134		434	530	
0080	900 mm		2	4		335	168		503	620	
0090	1050 mm		2	4		435	168		603	730	
0100	Concrete, 450 mm dia. [18″]		1.20	6.667		1,150	280		1,430	1,700	
0110	600 mm dia. [24″]		.75	10.667		1,450	450		1,900	2,275	
0120	Dry niche										
0130	300 W 120 V	1 Elec	4	2	Ea.	745	84		829	945	
0140	1000 W 120 V		2	4		1,000	168		1,168	1,350	
0150	300 W 12 V		4	2		745	84		829	945	
0160	Low voltage										
0170	Recessed uplight	1 Elec	2.20	3.636	Ea.	243	153		396	495	
0180	Walkway		4	2		222	84		306	370	
0190	Malibu - 5 light set		3	2.667		178	112		290	365	
0200	Mushroom 600 mm pier		4	2		182	84		266	325	
0210	Recessed, adjustable										
0220	Incandescent, 150 W	1 Elec	2.50	3.200	Ea.	370	134		504	610	
0230	300 W	″	2.50	3.200	″	430	134		564	670	
0250	Recessed uplight										
0260	Incandescent, 50 W	1 Elec	2.50	3.200	Ea.	340	134		474	575	
0270	150 W		2.50	3.200		355	134		489	590	
0280	300 W		2.50	3.200		395	134		529	635	
0310	Quartz 500 W		2.50	3.200		380	134		514	615	
0400	Recessed wall light										
0410	Incandescent 100 W	1 Elec	4	2	Ea.	186	84		270	330	
0420	Fluorescent		4	2		169	84		253	310	
0430	H.I.D. 100 W		3	2.667		345	112		457	545	
0500	Step lights										

Important: See the Reference Section for supporting data - Crews, Rental Equipment, City Cost Indexes and Reference Data

16520 | Exterior Luminaires

		CREW	DAILY OUTPUT	LABOR-HOURS	UNIT	2006 BARE COSTS				TOTAL INCL O&P	
						MAT.	LABOR	EQUIP.	TOTAL		
500	0510	Incandescent	1 Elec	5	1.600	Ea.	171	67		238	288	500
	0520	Fluorescent	"	5	1.600	"	182	67		249	300	
	0600	Tree lights, surface adjustable										
	0610	Incandescent 50 W	1 Elec	3	2.667	Ea.	405	112		517	610	
	0620	Incandescent 100 W		3	2.667		425	112		537	630	
	0630	Incandescent 150 W		2	4		450	168		618	745	
	0700	Underwater lights										
	0710	150 W 120 V	1 Elec	6	1.333	Ea.	675	56		731	825	
	0720	300 W 120 V		6	1.333		805	56		861	970	
	0730	1000 W 120 V		4	2		1,025	84		1,109	1,250	
	0740	50 W 12 V		6	1.333		765	56		821	925	
	0750	300 W 12 V		6	1.333		690	56		746	845	
	0800	Walkway, adjustable										
	0810	Fluorescent, 600 mm	1 Elec	3	2.667	Ea.	289	112		401	485	
	0820	Fluorescent, 1200 mm		3	2.667		305	112		417	505	
	0830	Fluorescent 2400 mm		2	4		600	168		768	910	
	0840	Incandescent, 50 W		4	2		273	84		357	425	
	0850	150 W		4	2		294	84		378	450	
	0900	Wet niche										
	0910	300 W 120 V	1 Elec	2.50	3.200	Ea.	600	134		734	860	
	0920	1000 W 120 V		1.50	5.333		720	224		944	1,125	
	0930	300 W 12 V		2.50	3.200		685	134		819	955	

16525 | Aviation Lighting

		CREW	DAILY OUTPUT	LABOR-HOURS	UNIT	MAT.	LABOR	EQUIP.	TOTAL	INCL O&P	
100	0010	**AIRPORT LIGHTING**										100
	0100	Runway centerline, bidir., semi-flush, 200 W, w/shallow insert base	R-22	12.40	3.006	Ea.	1,375	106		1,481	1,650	
	0120	Flush, 200 W, w/shallow insert base		12.40	3.006		1,375	106		1,481	1,650	
	0130	for mounting in base housing		18.64	2		720	71		791	895	
	0150	Touchdown zone light, unidirectional, 200 W, w/shallow insert base		12.40	3.006		1,125	106		1,231	1,375	
	0160	115 W		12.40	3.006		1,200	106		1,306	1,450	
	0180	Unidirectional, 200 W, for mounting in base housing		18.60	2.004		545	71		616	705	
	0190	115 W		18.60	2.004		590	71		661	755	
	0210	Runway edge & threshold light, bidir., 200 W, for base housing		9.36	3.983		775	141		916	1,075	
	0240	Threshold & approach light, unidir., 200 W, for base housing		9.36	3.983		450	141		591	710	
	0260	Runway edge, bi-directional, 2-115 W, for base housing		12.40	3.006		1,200	106		1,306	1,475	
	0280	Runway threshold & end, bidir., 2-115 W, for base housing		12.40	3.006		1,325	106		1,431	1,600	
	0370	45 W, flush, for mounting in base housing		18.64	2		650	71		721	820	
	0380	115 W		18.64	2		665	71		736	835	
	1200	Wind cone, 3.7 m lighted assembly, rigid, w/obstruction light	R-21	1.36	24.118		7,600	1,025	51	8,676	9,900	
	1210	Without obstruction light		1.52	21.579		6,375	905	46	7,326	8,400	
	1220	Unlighted assembly, w/obstruction light		1.68	19.524		6,950	820	41.50	7,811.50	8,925	
	1230	Without obstruction light		1.84	17.826		6,375	750	38	7,163	8,175	
	1240	Wind cone slip fitter, 65 mm pipe		21.84	1.502		83.50	63	3.18	149.68	190	
	1250	Wind cone sock, 3.7 m x 0.9 m, cotton		6.56	5		460	210	10.60	680.60	830	
	1260	Nylon		6.56	5		470	210	10.60	690.60	840	

16530 | Emergency Lighting

		CREW	DAILY OUTPUT	LABOR-HOURS	UNIT	MAT.	LABOR	EQUIP.	TOTAL	INCL O&P	
320	0010	**EXIT AND EMERGENCY LIGHTING**										320
	0080	Exit light, ceiling or wall mount, incandescent, single face	1 Elec	8	1	Ea.	42	42		84	109	
	0100	Double face		6.70	1.194		46	50		96	126	
	0200	L.E.D. standard, single face		8	1		64.50	42		106.50	134	
	0220	Double face		6.70	1.194		67.50	50		117.50	150	
	0240	L.E.D. w/battery unit, single face		4.40	1.818		125	76.50		201.50	252	
	0260	Double face		4	2		127	84		211	265	
	0300	Emergency light units, battery operated										

ELECTRICAL 16

			DAILY	LABOR-			2006 BARE COSTS				TOTAL	
	16530	**Emergency Lighting**	CREW	OUTPUT	HOURS	UNIT	MAT.	LABOR	EQUIP.	TOTAL	**INCL O&P**	
320	0350	Twin sealed beam light, 25 W, 6 V each										**320**
	0500	Lead battery operated	1 Elec	4	2	Ea.	122	84		206	259	
	0700	Nickel cadmium battery operated	↓	4	2		580	84		664	765	
	0900	Self-contained fluorescent lamp pack	↓	10	.800	↓	135	33.50		168.50	199	

	16585	**Lamps**										
600	0010	**LAMPS**										**600**
	0080	Fluorescent, rapid start, cool white, 600 mm long, 20 W	1 Elec	1	8	h	310	335		645	840	
	0100	1200 mm long, 40 W		.90	8.889		305	375		680	890	
	0200	Slimline, 1200 mm long, 40 W		.90	8.889		700	375		1,075	1,325	
	0210	1200 mm long, 30 W energy saver		.90	8.889		700	375		1,075	1,325	
	0400	High output, 1200 mm long, 60 W		.90	8.889		890	375		1,265	1,525	
	0410	2400 mm long, 95 W energy saver		.80	10		855	420		1,275	1,575	
	0500	2400 mm long, 110 W		.80	10		855	420		1,275	1,575	
	0512	600 mm long, T5, 14 watt energy saver		1	8		1,225	335		1,560	1,850	
	0514	900 mm long, T5, 21 watt energy saver		.90	8.889		1,225	375		1,600	1,900	
	0516	1200 mm long, T5, 28 watt energy saver		.90	8.889		1,050	375		1,425	1,700	
	0560	Twin tube compact lamp		.90	8.889		535	375		910	1,150	
	0570	Double twin tube compact lamp		.80	10		1,300	420		1,720	2,050	
	0600	Mercury vapor, mogul base, deluxe white, 100 W		.30	26.667		2,725	1,125		3,850	4,650	
	0650	175 W		.30	26.667		2,000	1,125		3,125	3,875	
	0700	250 W		.30	26.667		3,600	1,125		4,725	5,625	
	0800	400 W		.30	26.667		2,875	1,125		4,000	4,850	
	0900	1000 W		.20	40		6,750	1,675		8,425	9,900	
	1000	Metal halide, mogul base, 175 W		.30	26.667		3,400	1,125		4,525	5,400	
	1100	250 W		.30	26.667		3,850	1,125		4,975	5,900	
	1200	400 W		.30	26.667		3,650	1,125		4,775	5,675	
	1300	1000 W		.20	40		9,100	1,675		10,775	12,500	
	1320	1000 W, 125 000 initial lumens		.20	40		12,700	1,675		14,375	16,400	
	1330	1500 W		.20	40		12,900	1,675		14,575	16,700	
	1350	High pressure sodium, 70 W		.30	26.667		3,950	1,125		5,075	6,025	
	1360	100 W		.30	26.667		4,275	1,125		5,400	6,375	
	1370	150 W		.30	26.667		4,225	1,125		5,350	6,325	
	1380	250 W		.30	26.667		4,500	1,125		5,625	6,625	
	1400	400 W		.30	26.667		4,625	1,125		5,750	6,750	
	1450	1000 W		.20	40		13,200	1,675		14,875	17,000	
	1500	Low pressure sodium, 35 W		.30	26.667		5,975	1,125		7,100	8,250	
	1550	55 W		.30	26.667		6,550	1,125		7,675	8,900	
	1600	90 W		.30	26.667		7,600	1,125		8,725	10,100	
	1650	135 W		.20	40		9,725	1,675		11,400	13,200	
	1700	180 W		.20	40		10,700	1,675		12,375	14,200	
	1750	Quartz line, clear, 500 W		1.10	7.273		930	305		1,235	1,475	
	1760	1500 W		.20	40		3,650	1,675		5,325	6,500	
	1762	Spot, MR 16, 50 W		1.30	6.154		785	258		1,043	1,250	
	3000	Guards, fluorescent lamp, 1200 mm long		1	8		905	335		1,240	1,500	
	3200	2400 mm long	↓	.90	8.889	↓	1,800	375		2,175	2,550	

16 **ELECTRICAL**

			DAILY	LABOR-		2006 BARE COSTS				TOTAL
16710	**Communication Circuits**	CREW	OUTPUT	HOURS	UNIT	MAT.	LABOR	EQUIP.	TOTAL	INCL O&P
400	0010	**FIBER OPTICS**									**400**
	0020	Fiber optics cable only, Added costs depend on the type of fiber									
	0030	special connectors, optical modems, and networking parts.									
	0040	Specialized tools & techniques cause installation costs to vary.									
	0070	Cable, minimum, bulk simplex	1 Elec	244	.033	m	.75	1.38		2.13	2.88
	0080	Cable, maximum, bulk plenum quad	"	69.80	.115	"	3.84	4.81		8.65	11.35
	0150	Fiber optic jumper				Ea.	55.50			55.50	61
	0200	Fiber optic pigtail					30			30	33
	0300	Fiber optic connector	1 Elec	24	.333		16.90	14		30.90	39.50
	0350	Fiber optic finger splice		32	.250		32	10.50		42.50	50.50
	0400	Transceiver (low cost bi-directional)		8	1		425	42		467	535
	0450	Multi-channel rack enclosure (10 modules)		2	4		475	168		643	775
	0500	Fiber optic patch panel (12 ports)	↓	6	1.333	↓	178	56		234	280
	1000	Cable, 62.5 microns, direct burial, 4 fiber	R-15	366	.131	m	2.43	5.35	.72	8.50	11.45
	1020	Indoor, 2 fiber	R-19	305	.066		1.12	2.76		3.88	5.30
	1040	Outdoor, aerial/duct	"	509	.039		6.20	1.65		7.85	9.25
	1060	50 microns, direct burial, 8 fiber	R-22	1219	.031		5.10	1.08		6.18	7.25
	1080	12 fiber		1219	.031		6.90	1.08		7.98	9.25
	1100	Indoor, 12 fiber	↓	231	.161	↓	5.40	5.70		11.10	14.55
	1120	Connectors, 62.5 micron cable, transmission	R-19	40	.500	Ea.	9.40	21		30.40	42
	1140	Cable splice		40	.500		6.45	21		27.45	38.50
	1160	125 micron cable, transmission		16	1.250		9.25	52.50		61.75	88.50
	1180	Receiver, 2 km range		20	1		256	42		298	345
	1200	3 km range		20	1		485	42		527	595
	1220	10 km range		5	4		685	168		853	1,000
	1240	Transmitter, 2 km range		20	1		255	42		297	345
	1260	3 km range		20	1		485	42		527	595
	1280	10 km range		5	4		685	168		853	1,000
	1300	Modem, 2 km range		5	4		325	168		493	610
	1320	10 km range		5	4		900	168		1,068	1,250
	1340	3 km range, 12 channel		5	4		1,800	168		1,968	2,225
	1360	Repeater, 2 km range		10	2		430	84		514	595
	1380	3 km range		10	2		430	84		514	595
	1400	10 km range		5	4		1,100	168		1,268	1,450
	1420	2 km range, digital		5	4		430	168		598	720
	1440	Transceiver, 3 km range		5	4		745	168		913	1,075
	1460	2 km range, digital		5	4		255	168		423	530
	1480	Cable enclosure, interior NEMA 13		7	2.857		143	120		263	335
	1500	Splice w/enclosure encapsulant	↓	16	1.250	↓	214	52.50		266.50	315

			DAILY	LABOR-		2006 BARE COSTS				TOTAL
16720	**Tel and Intercomm Equipment**	CREW	OUTPUT	HOURS	UNIT	MAT.	LABOR	EQUIP.	TOTAL	INCL O&P
600	0010	**NURSE CALL SYSTEMS**									**600**
	0100	Single bedside call station	1 Elec	8	1	Ea.	190	42		232	272
	0200	Ceiling speaker station		8	1		55.50	42		97.50	124
	0400	Emergency call station		8	1		94	42		136	167
	0600	Pillow speaker		8	1		174	42		216	255
	0800	Double bedside call station		4	2		180	84		264	325
	1000	Duty station		4	2		143	84		227	282
	1200	Standard call button		8	1		76	42		118	146
	1400	Lights, corridor, dome or zone indicator	↓	8	1	↓	42.50	42		84.50	109
	1600	Master control station for 20 stations	2 Elec	.65	24.615	Total	3,375	1,025		4,400	5,275

For expanded coverage of these items see *Means Electrical Cost Data 2006*

ELECTRICAL 16

16 ELECTRICAL

		16820	Sound Reinforcement	CREW	DAILY OUTPUT	LABOR-HOURS	UNIT	2006 BARE COSTS				TOTAL INCL O&P	
								MAT.	LABOR	EQUIP.	TOTAL		
300	0010	**DOORBELL SYSTEM** Incl. transformer, button & signal											**300**
	0100	152 mm bell		1 Elec	4	2	Ea.	86	84		170	220	
	0200	Buzzer		"	4	2	"	68.50	84		152.50	200	
800	0010	**PUBLIC ADDRESS SYSTEM**											**800**
	0100	Conventional, office		1 Elec	5.33	1.501	Speaker	107	63		170	212	
	0200	Industrial		"	2.70	2.963	"	207	124		331	410	
	0400	Explosionproof system is 3 times cost of central control											
	0600	Installation costs run about 120% of material cost											
840	0010	**SOUND SYSTEM** not including rough-in wires, cables & conduits											**840**
	0100	Components, outlet, projector		1 Elec	8	1	Ea.	49.50	42		91.50	117	
	0200	Microphone			4	2		55	84		139	186	
	0400	Speakers, ceiling or wall			8	1		93.50	42		135.50	166	
	0600	Trumpets			4	2		174	84		258	315	
	0800	Privacy switch			8	1		69.50	42		111.50	139	
	1000	Monitor panel			4	2		310	84		394	465	
	1200	Antenna, AM/FM			4	2		173	84		257	315	
	1400	Volume control			8	1		69.50	42		111.50	139	
	1600	Amplifier, 250 W			1	8		1,375	335		1,710	2,025	
	1800	Cabinets		↓	1	8		675	335		1,010	1,250	
	2000	Intercom, 25 station capacity, master station		2 Elec	2	8		1,625	335		1,960	2,275	
	2200	Remote station		1 Elec	8	1		130	42		172	206	
	2400	Intercom outlets			8	1		76	42		118	147	
	2600	Handset			4	2		252	84		336	405	
	2800	Emergency call system, 12 zones, annunciator			1.30	6.154		760	258		1,018	1,225	
	3000	Bell			5.30	1.509		78.50	63.50		142	181	
	3200	Light or relay			8	1		39.50	42		81.50	106	
	3400	Transformer			4	2		173	84		257	315	
	3600	House telephone, talking station			1.60	5		370	210		580	725	
	3800	Press to talk, release to listen		↓	5.30	1.509		86.50	63.50		150	190	
	4000	System-on button						51.50			51.50	57	
	4200	Door release		1 Elec	4	2		92.50	84		176.50	227	
	4400	Combination speaker and microphone			8	1		157	42		199	236	
	4600	Termination box			3.20	2.500		49.50	105		154.50	212	
	4800	Amplifier or power supply			5.30	1.509	↓	570	63.50		633.50	720	
	5000	Vestibule door unit			16	.500	Name	105	21		126	147	
	5200	Strip cabinet			27	.296	Ea.	197	12.45		209.45	236	
	5400	Directory		↓	16	.500	"	93	21		114	134	

		16850	Television Equipment										
600	0010	**T.V. SYSTEMS** not including rough-in wires, cables & conduits											**600**
	0100	Master TV antenna system											
	0200	VHF reception & distribution, 12 outlets		1 Elec	6	1.333	Outlet	171	56		227	272	
	0400	30 outlets			10	.800		112	33.50		145.50	173	
	0600	100 outlets			13	.615		114	26		140	165	
	0800	VHF & UHF reception & distribution, 12 outlets			6	1.333		169	56		225	270	
	1000	30 outlets			10	.800		112	33.50		145.50	173	
	1200	100 outlets			13	.615		114	26		140	165	
	1400	School and deluxe systems, 12 outlets			2.40	3.333		223	140		363	455	
	1600	30 outlets			4	2		196	84		280	340	
	1800	80 outlets		↓	5.30	1.509	↓	188	63.50		251.50	300	
	2000	Closed circuit, surveillance, one station (camera & monitor)		2 Elec	2.60	6.154	Total	1,075	258		1,333	1,575	
	2200	For additional camera stations, add		1 Elec	2.70	2.963	Ea.	605	124		729	850	
	2400	Industrial quality, one station (camera & monitor)		2 Elec	2.60	6.154	Total	2,250	258		2,508	2,850	

16850	Television Equipment	CREW	DAILY OUTPUT	LABOR-HOURS	UNIT	2006 BARE COSTS				TOTAL INCL O&P	
						MAT.	LABOR	EQUIP.	TOTAL		
600	2600	For additional camera stations, add	1 Elec	2.70	2.963	Ea.	1,375	124		1,499	1,675	600
	2610	For low light, add		2.70	2.963		1,100	124		1,224	1,400
	2620	For very low light, add		2.70	2.963		8,125	124		8,249	9,125
	2800	For weatherproof camera station, add		1.30	6.154		845	258		1,103	1,325
	3000	For pan and tilt, add		1.30	6.154		2,200	258		2,458	2,775
	3200	For zoom lens - remote control, add, minimum		2	4		2,025	168		2,193	2,475
	3400	Maximum		2	4		7,375	168		7,543	8,375
	3410	For automatic iris for low light, add	↓	2	4	↓	1,775	168		1,943	2,200
	3600	Educational T.V. studio, basic 3 camera system, black & white,									
	3800	electrical & electronic equip. only, minimum	4 Elec	.80	40	Total	10,500	1,675		12,175	14,000
	4000	Maximum (full console)		.28	114		44,700	4,800		49,500	56,500
	4100	As above, but color system, minimum		.28	114		59,000	4,800		63,800	72,000
	4120	Maximum	↓	.12	266	↓	257,000	11,200		268,200	299,000
	4200	For film chain, black & white, add	1 Elec	1	8	Ea.	12,000	335		12,335	13,700
	4250	Color, add		.25	32		14,600	1,350		15,950	18,000
	4400	For video tape recorders, add, minimum	↓	1	8		2,525	335		2,860	3,275
	4600	Maximum	4 Elec	.40	80	↓	21,000	3,350		24,350	28,100

For information about Means Estimating Seminars, see yellow pages 12 and 13 in back of book

ELECTRICAL 16

For expanded coverage of these items see *Means Electrical Cost Data 2006*

Division Notes

		CREW	DAILY OUTPUT	LABOR-HOURS	UNIT	2006 BARE COSTS				TOTAL INCL O&P
						MAT.	LABOR	EQUIP.	TOTAL	

Assemblies Section

Table of Contents

How to Use the Assemblies Cost Tables

The following is a detailed explanation of a sample Assemblies Cost Table. Most Assembly Tables are separated into three parts: 1) an illustration of the system to be estimated; 2) the components and related costs of a typical system; and 3) the costs for similar systems with dimensional and/or size variations. For costs of the components that comprise these systems, or assemblies, refer to the Unit Price Section. Next to each bold number below is the described item with the appropriate component of the sample entry following in parentheses. In most cases, if the work is to be subcontracted, the general contractor will need to add an additional markup (RSMeans suggests using 10%) to the "Total" figures.

System/Line Numbers (G2040 210 1000)

Each Assemblies Cost Line has been assigned a unique identification number based on the UNIFORMAT II classification system.

UNIFORMAT II Major Group

G2040 **210** **1000**

UNIFORMAT II Level 3

Means Major Classification

Means Individual Line Number

G20 Site Improvements
G2040 Site Development

1

2

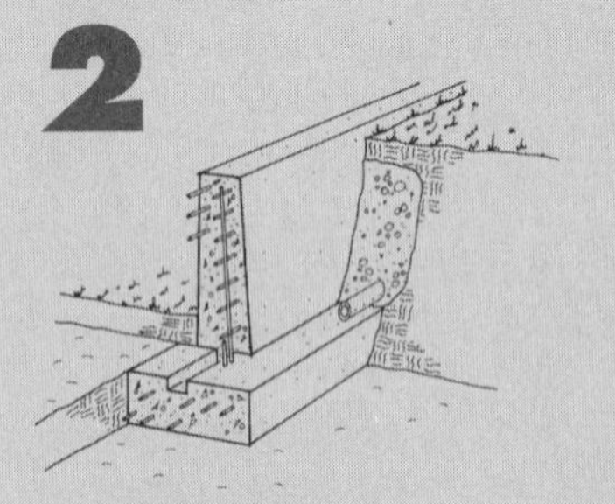

There are four basic types of Concrete Retaining Wall Systems: reinforced concrete with level backfill; reinforced concrete with sloped backfill or surcharge; unreinforced with level backfill; and unreinforced with sloped backfill or surcharge. System elements include: all necessary forms (4 uses); 21 MPa concrete with an 200 mm chute; all necessary reinforcing steel; and underdrain. Exposed concrete is patched and rubbed.

The Expanded System Listing shows walls that range in thickness from 250 mm to 4600 mm for reinforced concrete walls with level backfill and 300 mm to 600 mm for reinforced walls with sloped backfill. Walls range from a height of 1200 mm to 6100 mm. Unreinforced level and sloped backfill walls range from a height of 900 mm to 3000 mm.

4 **5** **6**

System Components **3**	QUANTITY	UNIT	MAT.	INST.	TOTAL
SYSTEM G2040 210 1000					
CONC.RETAIN. WALL REINFORCED, LEVEL BACKFILL 1200 mm HIGH					
Forms in place, cont. wall footing & keyway, 4 uses	.610	m²	6.16	22.56	28.72
Forms in place, retaining wall forms, battered to 2400 mm high, 4 uses	2.438	m²CA	15.48	173.13	188.61
Reinforcing in place, walls, #10M to #25M	.012	Met. Ton	11.73	9.11	20.84
Concrete ready mix, regular weight, 21 MPa	.512	m³	63.96		63.96
Placing concrete and vibrating footing con., shallow direct chute	.186	m³		4.45	4.45
Placing concrete and vibrating walls, 200 mm thick, direct chute	.326	m³		10.33	10.33
Pipe bedding, crushed or screened bank run gravel	1.000	m	**7** .74	3.26	0
Pipe, subdrainage, corrugated plastic, 100 mm diameter	1.000	m	.62	**8** .51	
Finish walls and break ties, patch walls	1.219	m²	.39	.88	**9** .27
TOTAL			106.48	233.23	339.71

7 **8** **9**

G2040 210	Concrete Retaining Walls	MAT.	INST.	TOTAL
1000	Conc.retain.wall,reinforced,level backfill,1200mm high x 650mm, 250mm thick	106	233	339
1200	1800 mm high x 975 mm base, 250 mm thick	158	340	498
1400	2400 mm high x 1300 mm base, 250 mm thick	207	445	652
1600	3100 mm high x 1600 mm base, 325 mm thick	274	655	929
1800	3700 mm high x 2000 mm base, 350 mm thick	340	785	1,125
2200	4900 mm high x 2600 mm base, 400 mm thick	515	1,050	1,565
2600	6100 mm high x 3200 mm base, 450 mm thick	760	1,375	2,135
3000	Sloped backfill, 1200 mm high x 950 mm base, 300 mm thick	132	244	376

2 Illustration

At the top of most assembly tables are an illustration, a brief description, and the design criteria used to develop the cost.

3 System Components

The components of a typical system are listed separately to show what has been included in the development of the total system price. The table below contains prices for other similar systems with dimensional and/or size variations.

4 Quantity

This is the number of line item units required for one system unit. For example, we assume that it will take .610 m² of footing forms per L.F. of wall.

5 Unit of Measure for Each Item

The abbreviated designation indicates the unit of measure, as defined by industry standards, upon which the price of the component is based. For example, reinforcing is priced per Met. Ton and concrete is priced per m³.

6 Unit of Measure for Each System (Cost per m)

Costs shown in the three right-hand columns have been adjusted by the component quantity and unit of measure for the entire system. In this example, "Cost per m" is the unit of measure for this system, or assembly.

7 Materials (106.48)

This column contains the Materials Cost of each component. These cost figures are bare costs plus 10% for profit.

8 Installation (233.23)

Installation includes labor and equipment plus the installing contractor's overhead and profit. Equipment costs are the bare rental costs plus 10% for profit. The labor overhead and profit are defined on the inside back cover of this book.

9 Total (339.71)

The figure in this column is the sum of the material and installation costs.

Material Cost	+	Installation Cost	=	Total
$106.48	+	$233.23	=	$339.71

A SUBSTRUCTURE

A1010 Standard Foundations

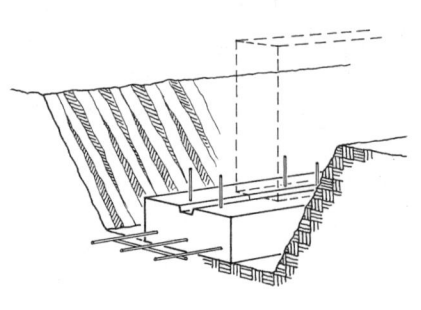

The Strip Footing System includes: excavation; hand trim; all forms needed for footing placement; forms for 50 mm x 150 mm keyway (four uses); dowels; and 21 MPa concrete.

The footing size required varies for different soils. Soil bearing capacities are listed for 145 Pa and 285 Pa. Depths of the system range from 200 mm to 600 mm. Widths range from 400 mm to 2400 mm. Smaller strip footings may not require reinforcement.

Please see the reference section for further design and cost information.

SUBSTRUCTURE A

System Components	QUANTITY	UNIT	COST PER m		
			MAT.	INST.	TOTAL
SYSTEM A1010 110 2500					
STRIP FOOTING,LOAD 75 kN/m,SOIL CAP. 145 kPa,600 mm W x 300 mm D,REINF.					
Trench excavation	.371	m³		3.32	3.32
Hand trim	.610	m²		4.76	4.76
Compacted backfill	.186	m³		.74	.74
Formwork, 4 uses	.610	m²	6.16	22.56	28.72
Keyway form, 4 uses	1.000	m	1.08	2.91	3.99
Reinforcing, fy = 400 MPa	4.464	kg	4.82	4.87	9.69
Dowels	6.562	Ea.	4.53	14.24	18.77
Concrete, f'c = 21 MPa	.186	m³	23.20		23.20
Place concrete, direct chute	.186	m³		4.45	4.45
Screed finish	.610	m²		3.69	3.69
TOTAL			39.79	61.54	101.33

A1010 110	Strip Footings	COST PER m		
		MAT.	INST.	TOTAL
2100	Strip footing,load 40 kN/m,soil capacity 145 kPa, 400 mm wide x 200 mm deep	19.45	37	56.45
2300	Load 55 kN/m soil capacity, 145 kPa, 600 mm wide x 200 mm deep, plain	25	41	66
2500	Load 75 kN/m, soil capacity 145 kPa, 600 mm wide x 300 mm deep, reinf.	40	61.50	101.50
2700	Load 160 kN/m, soil capacity 285 kPa, 600 mm wide x 300 mm deep, reinf.	40	61.50	101.50
2900	Load 100 kN/m, soil capacity 145 kPa, 800 mm wide x 300 mm deep, reinf.	49.50	67.50	117
3100	Load 215 kN/m, soil capacity 285 kPa, 800 mm wide x 300 mm deep, reinf.	49.50	67.50	117
3300	Load 135 kN/m, soil capacity 145 kPa, 1000 mm wide x 300 mm deep, reinf.	59	73.50	132.50
3500	Load 270 kN/m, soil capacity 285 kPa, 1000 mm wide x 300 mm deep, reinf.	59.50	74	133.50
3700	Load 145 kN/m, soil capacity 145 kPa, 1200 mm wide x 300 mm deep, reinf.	66	79.50	145.50
3900	Load 325 kN/m, soil capacity 285 kPa, 1200 mm wide x 300 mm deep, reinf.	71	84.50	155.50
4100	Load 170 kN/m, soil capacity 145 kPa, 1400 mm wide x 300 mm deep, reinf.	78.50	87.50	166
4300	Load 375 kN/m, soil capacity 285 kPa, 1400 mm wide x 300 mm deep, reinf.	85	94.50	179.50
4500	Load 145 kN/m, soil capacity 145 kPa, 1200 mm wide x 400 mm deep, reinf.	83.50	92	175.50
4700	Load 320 kN/m, soil capacity 285 kPa, 1200 mm wide x 400 mm deep, reinf.	85.50	94	179.50
4900	Load 170 kN/m, soil capacity 145 kPa, 1400 mm wide x 400 mm deep, reinf.	96	130	226
5100	Load 375 kN/m, soil capacity 285 kPa, 1400 mm wide x 400 mm deep, reinf.	101	135	236
5300	Load 195 kN/m, soil capacity 145 kPa, 1600 mm wide x 400 mm deep, reinf.	111	109	220
5500	Load 430 kN/m, soil capacity 285 kPa, 1600 mm wide x 400 mm deep, reinf.	119	117	236
5700	Load 220 kN/m, soil capacity 145 kPa, 1800 mm wide x 500 mm deep, reinf.	148	131	279
5900	Load 480 kN/m, soil capacity 285 kPa, 1800 mm wide x 500 mm deep, reinf.	157	140	297
6100	Load 265 kN/m, soil capacity 145 kPa, 2200 mm wide x 600 mm deep, reinf.	210	165	375
6300	Load 590 kN/m, soil capacity 285 kPa, 2200 mm wide x 600 mm deep, reinf.	229	184	413
6500	Load 290 kN/m, soil capacity 145 kPa, 2400 mm wide x 600 mm deep, reinf.	229	176	405
6700	Load 640 kN/m, soil capacity 285 kPa, 2400 mm wide x 600 mm deep, reinf.	244	191	435

A1010 Standard Foundations

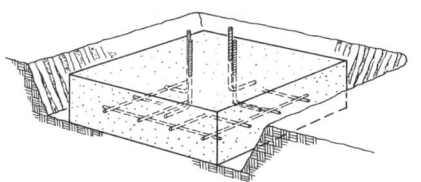

The Spread Footing System includes: excavation; backfill; forms (four uses); all reinforcement; 21 MPa concrete (chute placed); and screed finish.

Footing systems are priced per individual unit. The Expanded System Listing at the bottom shows footings that range from 900 mm square x 300 mm deep, to 6100 mm square x 1070 mm deep. It is assumed that excavation is done by a truck mounted hydraulic excavator with an operator and oiler.

Backfill is with a dozer, and compaction by air tamp. The excavation and backfill equipment is assumed to operate at 23 m³ per hour.

Please see the reference section for further design and cost information.

System Components	QUANTITY	UNIT	COST EACH MAT.	COST EACH INST.	COST EACH TOTAL
SYSTEM A1010 210 7100					
SPREAD FOOTINGS, LOAD 10 MET.TON,SOIL CAPACITY 145 kPa, 900 m sq. x 300 mm					
Bulk excavation	.451	m³		4.25	4.25
Hand trim	.836	m²		6.54	6.54
Compacted backfill	.199	m³		.79	.79
Formwork, 4 uses	1.115	m²	7.80	48.50	56.30
Reinforcing, fy = 400 MPa	.005	Met. Ton	5.36	5.98	11.34
Dowel or anchor bolt templates	1.829	m	5.52	20.12	25.64
Concrete, f'c = 21 MPa	.252	m³	31.54		31.54
Place concrete, direct chute	.252	m³		6.06	6.06
Screed finish	.836	m²		4.06	4.06
TOTAL			50.22	96.30	146.52

A1010 210	Spread Footings	COST EACH MAT.	COST EACH INST.	COST EACH TOTAL
7090	Spread footings, 21 MPa concrete, chute delivered			
7100	Load 10 met. ton, soil capacity 145 kPa, 900 mm sq. x 300 mm deep	50	96	146
7150	Load 25 met. ton, soil capacity 145 kPa, 1400 mm sq. x 300 mm deep	107	167	274
7200	Load 25 met. ton, soil capacity 285 kPa, 900 mm sq. x 300 mm deep	50	96	146
7250	Load 35 met. ton, soil capacity 145 kPa, 1700 mm sq. x 325 mm deep	169	236	405
7300	Load 35 met. ton, soil capacity 285 kPa, 1200 mm sq. x 300 mm deep	86.50	143	229.50
7350	Load 45 met. ton, soil capacity 145 kPa, 1800 mm sq. x 350 mm deep	214	283	497
7410	Load 45 met. ton, soil capacity 285 kPa, 1400 mm sq. x 375 mm deep	131	196	327
7450	Load 55 met. ton, soil capacity 145 kPa, 2100 mm sq. x 425 mm deep	340	405	745
7500	Load 55 met. ton, soil capacity 285 kPa, 1500 mm sq. x 400 mm deep	169	235	404
7550	Load 70 met. ton, soil capacity 145 kPa 2300 mm sq. x 450 mm deep	410	475	885
7610	Load 70 met. ton, soil capacity 285 kPa, 1700 mm sq. x 450 mm deep	225	296	521
7650	Load 90 met. ton, soil capacity 145 kPa, 2600 mm sq. x 500 mm deep	585	630	1,215
7700	Load 90 met. ton, soil capacity 285 kPa, 1800 mm sq. x 500 mm deep	295	365	660
7750	Load 135 met. ton, soil capacity 145 kPa, 3200 mm sq. x 625 mm deep	1,075	1,025	2,100
7810	Load 135 met. ton, soil capacity 285 kPa, 2300 mm sq. x 625 mm deep	560	615	1,175
7850	Load 180 met. ton, soil capacity 145 kPa, 3800 mm sq. x 700 mm deep	1,700	1,500	3,200
7900	Load 180 met. ton, soil capacity 285 kPa, 2600 mm sq. x 675 mm deep	775	800	1,575
7950	Load 225 met. ton, soil capacity 145 kPa, 4300 mm sq. x 775 mm deep	2,375	2,000	4,375
8010	Load 225 met. ton, soil capacity 285 kPa, 2900 mm sq. x 750 mm deep	1,075	1,050	2,125

SUBSTRUCTURE

A

A1010 Standard Foundations

A1010 210	Spread Footings	COST EACH		
		MAT.	INST.	TOTAL
8050	Load 270 met. ton, soil capacity 145 kPa, 4900 mm sq. x 875 mm deep	3,450	2,750	6,200
8100	Load 270 met. ton, soil capacity 285 kPa, 3200 mm sq. x 825 mm deep	1,450	1,350	2,800
8150	Load 320 met. ton, soil capacity 145 kPa, 5200 mm sq. x 925 mm deep	4,075	3,125	7,200
8200	Load 320 met. ton, soil capacity 285 kPa, 3500 mm sq. x 900 mm deep	1,850	1,650	3,500
8250	Load 365 met. ton, soil capacity 145 kPa, 5500 mm sq. x 975 mm deep	4,825	3,625	8,450
8300	Load 365 met. ton, soil capacity 285 kPa, 3700 mm sq. x 925 mm deep	2,075	1,800	3,875
8350	Load 410 met. ton, soil capacity 145 kPa, 5800 mm sq. x 1000 mm deep	5,600	4,150	9,750
8400	Load 410 met. ton, soil capacity 285 kPa, 4000 mm sq. x 975 mm deep	2,550	2,125	4,675
8450	Load 455 met. ton, soil capacity 145 kPa, 6100 mm sq. x 1050 mm deep	6,450	4,650	11,100
8500	Load 455 met. ton, soil capacity 285 kPa, 4100 mm sq. x 1025 mm deep	2,900	2,375	5,275
8550	Load 545 met. ton, soil capacity 285 kPa, 4600 mm sq. x 1200 mm deep	3,850	3,025	6,875
8600	Load 635 met. ton, soil capacity 285 kPa, 4900 mm sq. x 1175 mm deep	4,700	3,600	8,300
8650	Load 725 met. ton, soil capacity 285 kPa, 5500 mm sq. x 1300 mm deep	6,525	4,775	11,300

Important: See the Reference Section for critical supporting data - Reference Numbers and City Cost Indexes

A10 Foundations

A1010 Standard Foundations

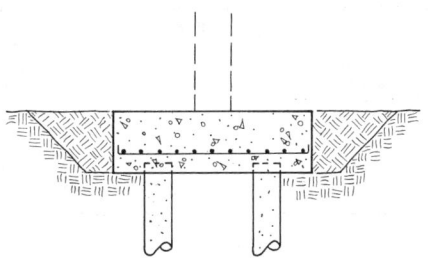

These pile cap systems include excavation with a truck mounted hydraulic excavator, hand trimming, compacted backfill, forms for concrete, templates for dowels or anchor bolts, reinforcing steel and concrete placed and screeded.

Pile embedment is assumed as 150 mm. Design is consistent with the Concrete Reinforcing Steel Institute Handbook, f'c = 21 MPa, fy = 400 MPa

Please see the reference section for further design and cost information.

System Components	QUANTITY	UNIT	COST EACH		
			MAT.	INST.	TOTAL
SYSTEM A1010 250 5100					
CAP FOR 2 PILES 1975mmX1075mmX15met. ton PILE 200mm COL,200kN COL. LD.					
Excavation, bulk, hyd excavator, truck mtd. 750 mm bucket 0.38 m³	2.210	m³		20.79	20.79
Trim sides and bottom of trench, regular soil	2.137	m²		16.71	16.71
Dozer backfill & roller compaction	1.147	m³		4.57	4.57
Forms in place pile cap, square or rectangular, 4 uses	3.066	m²CA	28.97	144.09	173.06
Templates for dowels or anchor bolts	8.000	Ea.	7.36	26.82	34.18
Reinforcing in place footings, #25M to #40M	.023	Met. Ton	21.21	14.40	35.61
Concrete ready mix, regular weight, 21 MPa	1.070	m³	133.80		133.80
Place and vibrate concrete for pile caps, under 3.82 m³, direct chute	1.070	m³		33.91	33.91
Monolithic screed finish	2.137	m²		10.38	10.38
TOTAL			191.34	271.67	463.01

A1010 250		Pile Caps						
	NO. PILES	SIZE mm X mm X mm	PILE CAPACITY (met. ton)	COLUMN SIZE (mm)	COLUMN LOAD (kn)	COST EACH		
						MAT.	INST.	TOTAL
5100	2	1975x1075x500	15	200	200	191	272	463
5150		650	35	200	700	235	330	565
5200		875	75	280	1400	320	425	745
5250		950	110	360	2100	340	455	795
5300	3	1675x1550x575	15	200	330	227	315	542
5350		700	35	250	1000	254	355	609
5400		825	75	360	2100	293	400	693
5450		975	110	430	3200	340	460	800
5500	4	1675x1675x450	15	250	210	256	315	571
5550		750	35	280	620	375	455	830
5600		900	75	410	1300	440	525	965
5650		975	110	480	1900	465	550	1,015
5700	6	2600x1675x450	15	300	320	425	460	885
5750		940	35	360	930	695	670	1,365
5800		1016	75	480	1900	790	750	1,540
5850		1143	110	610	2900	885	830	1,715
5900	8	2600x2350x475	15	300	410	645	620	1,265
5950		900	35	410	1200	930	810	1,740
6000		1125	75	560	2500	1,150	980	2,130
6050		1200	110	690	3800	1,250	1,050	2,300

A SUBSTRUCTURE

A1010 Standard Foundations

A1010 250	Pile Caps

	NO. PILES	SIZE mm X mm X mm	PILE CAPACITY (met. ton)	COLUMN SIZE (mm)	COLUMN LOAD (kn)	COST EACH		
						MAT.	INST.	TOTAL
6100	10	3500x2350x525	15	360	500	910	765	1,675
6150		1000	35	430	1500	1,400	1,075	2,475
6200		1200	75	640	3100	1,700	1,300	3,000
6250		1250	110	790	4700	1,800	1,375	3,175
6300	12	3500x2600x550	15	380	1400	1,150	925	2,075
6350		12505	35	480	4000	1,825	1,375	3,200
6400		1325	75	690	8300	2,025	1,500	3,525
6450		1400	110	860	12500	2,200	1,625	3,825
6500	14	3500x3275x600	15	410	1100	1,500	1,150	2,650
6550		1050	35	530	4700	1,975	1,400	3,375
6600		1400	75	740	9600	2,600	1,800	4,400
6700	16	3500x3500x650	15	460	1800	1,800	1,325	3,125
6750		1225	35	560	5300	2,425	1,675	4,100
6800		1525	75	790	10900	3,025	2,025	5,050
6900	18	3950x3500x700	15	510	2000	2,050	1,425	3,475
6950		1250	35	580	6000	2,825	1,875	4,700
7000		1425	75	840	12300	3,325	2,200	5,525
7100	20	4425x3500x750	15	510	2300	2,525	1,725	4,250
7150		1325	35	610	6600	3,400	2,200	5,600

SUBSTRUCTURE A

Important: See the Reference Section for critical supporting data - Reference Numbers and City Cost Indexes

A10 Foundations

A1010 Standard Foundations

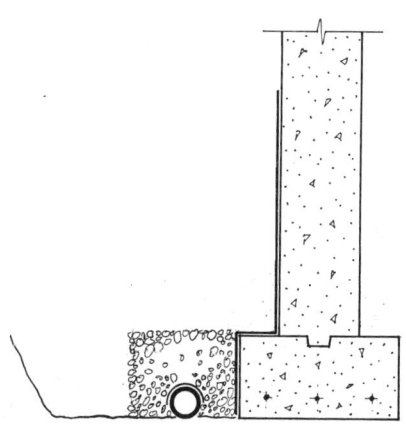

General: Footing drains can be placed either inside or outside of foundation walls depending upon the source of water to be intercepted. If the source of subsurface water is principally from grade or a subsurface stream above the bottom of the footing, outside drains should be used. For high water tables, use inside drains or both inside and outside.

The effectiveness of underdrains depends on good waterproofing. This must be carefully installed and protected during construction.

Costs below include the labor and materials for the pipe and 150 mm only of gravel or crushed stone around pipe. Excavation and backfill are not included.

System Components	QUANTITY	UNIT	COST PER m		
			MAT.	INST.	TOTAL
SYSTEM A1010 310 1000					
FOUNDATION UNDERDRAIN, OUTSIDE ONLY, PVC 100 mm DIAM.					
PVC pipe 100 mm diam. S.D.R. 35	1.000	m	9.10	10.20	19.30
Pipe bedding, graded gravel 19 mm to 13 mm	.176	m³	7.02	2.07	9.09
TOTAL			16.12	12.27	28.39

A1010 310	Foundation Underdrain	COST PER m		
		MAT.	INST.	TOTAL
1000	Foundation underdrain, outside only, PVC, 100 mm diameter	16.10	12.30	28.40
1100	150 mm diameter	28	13.55	41.55
1200	Bituminous fiber, 100 mm diameter	16.10	12.30	28.40
1300	150 mm diameter	28	13.55	41.55
1400	Porous concrete, 150 mm diameter	18.65	14.80	33.45
1450	200 mm diameter	23	19.80	42.80
1500	300 mm diameter	41	22.50	63.50
1600	Corrugated metal, 16 ga. asphalt coated, 150 mm diameter	20	23.50	43.50
1650	200 mm diameter	26.50	25	51.50
1700	250 mm diameter	32.50	26	58.50
2000	Vitrified clay, C-211, 100 mm diameter	14.80	22	36.80
2050	150 mm diameter	22	27.50	49.50
2100	300 mm diameter	54.50	33.50	88
3000	Outside and inside, PVC, 100 mm diameter	32.50	25	57.50
3100	150 mm diameter	56.50	27	83.50
3200	Bituminous fiber, 100 mm diameter	32.50	25	57.50
3300	150 mm diameter	56.50	27	83.50
3400	Porous concrete, 150 mm diameter	37.50	29.50	67
3450	200 mm diameter	46	40	86
3500	300 mm diameter	82	45.50	127.50
3600	Corrugated metal, 16 ga., asphalt coated, 150 mm diameter	40	47	87
3650	200 mm diameter	53.50	49	102.50
3700	250 mm diameter	65.50	51.50	117
4000	Vitrified clay, C-211, 100 mm diameter	29.50	43.50	73
4050	150 mm diameter	44	55.50	99.50
4100	300 mm diameter	109	66.50	175.50

A | **SUBSTRUCTURE**

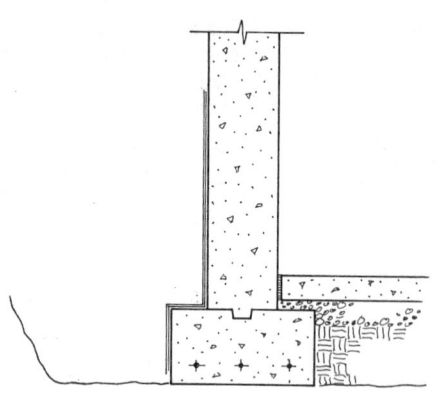

General: Apply foundation wall dampproofing over clean concrete giving particular attention to the joint between the wall and the footing. Use care in backfilling to prevent damage to the dampproofing.

Costs for four types of dampproofing are listed below.

System Components	QUANTITY	UNIT	COST PER m		
			MAT.	INST.	TOTAL
SYSTEM A1010 320 1000					
FOUNDATION DAMPPROOFING, BITUMINOUS, 1 COAT, 1200 mm HIGH					
Bituminous asphalt dampproofing brushed on below grade, 1 coat	1.219	m²	1.32	8.23	9.55
Labor for protection of dampproofing during backfilling	1.219	m²		3.35	3.35
TOTAL			1.32	11.58	12.90

A1010 320	Foundation Dampproofing	COST PER m		
		MAT.	INST.	TOTAL
1000	Foundation dampproofing, bituminous, 1 coat, 1200 mm high	1.32	11.60	12.92
1400	2400 mm high	2.63	23	25.63
1800	3700 mm high	3.95	36	39.95
2000	2 coats, 1200 mm high	2.50	14.25	16.75
2400	2400 mm high	5	28.50	33.50
2800	3700 mm high	7.50	44	51.50
3000	Asphalt with fibers, 2 mm thick, 1200 mm high	2.62	14.25	16.87
3400	2400 mm high	5.25	28.50	33.75
3800	3700 mm high	7.85	44	51.85
4000	3 mm thick, 1200 mm high	4.60	17	21.60
4400	2400 mm high	9.20	34	43.20
4800	3700 mm high	13.80	52	65.80
5000	Asphalt coated board and mastic, 6 mm thick, 1200 mm high	8.90	15.50	24.40
5400	2400 mm high	17.80	31	48.80
5800	3700 mm high	26.50	47.50	74
6000	13 mm thick, 1200 mm high	13.35	21.50	34.85
6400	2400 mm high	26.50	43	69.50
6800	3700 mm high	40	66	106
7000	Metallic coating, 5 coats, 16 mm thick, on walls, 1200 mm high	22.50	15.80	38.30
7400	2400 mm high	45	31.50	76.50
7800	3700 mm high	67.50	47.50	115
8000	25 mm thick, on slabs, 600 mm wide	2.60	10.75	13.35
8400	1200 mm wide	5.20	21.50	26.70
8800	1800 mm wide	7.80	32	39.80

SUBSTRUCTURE A

A1020 Special Foundations

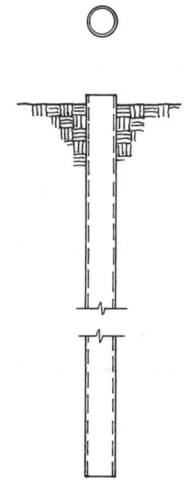

The Cast-in-Place Concrete Pile System includes: a defined number of 28 MPa concrete piles with thin-wall, straight-sided, steel shells that have a standard steel plate driving point. An allowance for cutoffs is included.

The Expanded System Listing shows costs per cluster of piles. Clusters range from one pile to twenty piles. Loads vary from 220 kN to 7000 kN. Both end-bearing and friction-type piles are shown.

Please see the reference section for cost of mobilization of the pile driving equipment and other design and cost information.

System Components	QUANTITY	UNIT	COST EACH		
			MAT.	INST.	TOTAL
SYSTEM A1020 110 2220					
CIP SHELL CONCRETE PILE, 8 m LONG, 220 kN LOAD, END BEARING, 1 PILE					
Cast in place piles, end bearing, no mobil, 7 Ga. shell, 300 mm diam.	8.230	m	567.84	240.72	808.56
Steel pipe pile standard point, 300 mm or 350 mm diameter pile	1.000	Ea.	53	71	124
Pile cutoff, conc. pile with thin steel shell	1.000	Ea.		12	12
TOTAL			620.84	323.72	944.56

A1020 110	C.I.P. Concrete Piles	COST EACH		
		MAT.	INST.	TOTAL
2220	CIP shell concrete pile, 8 m long, 220 kN load, end bearing, 1 pile	620	325	945
2240	440 kN load, end bearing, 2 pile cluster	1,250	645	1,895
2260	890 kN load, end bearing, 4 pile cluster	2,475	1,300	3,775
2280	1800 kN load, end bearing, 7 pile cluster	4,350	2,275	6,625
2300	10 pile cluster	6,200	3,225	9,425
2320	3600 kN load, end bearing, 13 pile cluster	12,400	5,950	18,350
2340	17 pile cluster	16,200	7,775	23,975
2360	5300 kN load, end bearing, 14 pile cluster	13,400	6,425	19,825
2380	19 pile cluster	18,200	8,700	26,900
2400	7100 kN load, end bearing, 19 pile cluster	18,200	8,700	26,900
2420	15 m long, 220 kN load, end bearing, 1 pile	1,175	560	1,735
2440	Friction type, 2 pile cluster	2,250	1,100	3,350
2460	3 pile cluster	3,350	1,650	5,000
2480	440 kN load, end bearing, 2 pile cluster	2,325	1,100	3,425
2500	Friction type, 4 pile cluster	4,475	2,200	6,675
2520	6 pile cluster	6,725	3,350	10,075
2540	890 kN load, end bearing, 4 pile cluster	4,675	2,200	6,875
2560	Friction type, 8 pile cluster	8,950	4,450	13,400
2580	10 pile cluster	11,200	5,550	16,750
2600	1800 kN load, end bearing, 7 pile cluster	8,175	3,900	12,075
2620	Friction type, 16 pile cluster	17,900	8,900	26,800
2640	19 pile cluster	21,300	10,600	31,900
2660	3600 kN load, end bearing, 14 pile cluster	25,700	10,500	36,200
2680	20 pile cluster	36,700	15,000	51,700
2700	5300 kN load, end bearing, 15 pile cluster	27,500	11,300	38,800
2720	7100 kN load, end bearing, 20 pile cluster	36,700	15,000	51,700

SUBSTRUCTURE

A

A1020 Special Foundations

A1020 110	C.I.P. Concrete Piles	COST EACH		
		MAT.	INST.	TOTAL
3740	23 m long, 220 kN load, end bearing, 1 pile	1,800	940	2,740
3760	Friction type, 2 pile cluster	3,475	1,875	5,350
3780	3 pile cluster	5,200	2,800	8,000
3800	440 kN load, end bearing, 2 pile cluster	3,600	1,875	5,475
3820	Friction type, 3 pile cluster	5,200	2,800	8,000
3840	5 pile cluster	8,650	4,700	13,350
3860	890 kN load, end bearing, 4 pile cluster	7,225	3,750	10,975
3880	6 pile cluster	10,800	5,650	16,450
3900	Friction type, 6 pile cluster	10,400	5,650	16,050
3910	7 pile cluster	12,100	6,575	18,675
3920	1800 kN load, end bearing, 7 pile cluster	12,600	6,575	19,175
3930	11 pile cluster	19,800	10,300	30,100
3940	Friction type, 12 pile cluster	20,800	11,300	32,100
3950	14 pile cluster	24,200	13,200	37,400
3960	3600 kN load, end bearing, 15 pile cluster	42,400	18,500	60,900
3970	20 pile cluster	56,500	24,600	81,100
3980	5300 kN load, end bearing, 17 pile cluster	48,000	21,000	69,000

Important: See the Reference Section for critical supporting data - Reference Numbers and City Cost Indexes

SUBSTRUCTURE A

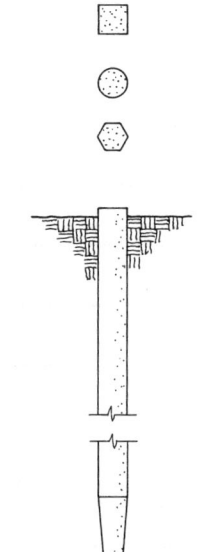

The Precast Concrete Pile System includes: pre-stressed concrete piles; standard steel driving point; and an allowance for cutoffs.

The Expanded System Listing shows costs per cluster of piles. Clusters range from one pile to twenty piles. Loads vary from 220 kN to 7000 kN. Both end-bearing and friction type piles are listed.

Please see the reference section for cost of mobilization of the pile driving equipment and other design and cost information.

System Components	QUANTITY	UNIT	COST EACH MAT.	COST EACH INST.	COST EACH TOTAL
SYSTEM A1020 120 2220					
PRECAST CONCRETE PILE, 15 m LONG, 220 kN LOAD, END BEARING, 1 PILE					
Precast, prestressed conc. piles, 250 mm square, no mobil.	16.154	m	533.10	405.47	938.57
Steel pipe pile standard point, 200 mm to 250 mm diameter	1.000	Ea.	38.25	62.50	100.75
Piling special costs cutoffs concrete piles plain	1.000	Ea.		83	83
TOTAL			571.35	550.97	1,122.32

A1020 120	Precast Concrete Piles	COST EACH MAT.	COST EACH INST.	COST EACH TOTAL
2220	Precast conc pile, 15 m long, 220 kN load, end bearing, 1 pile	570	550	1,120
2240	Friction type, 2 pile cluster	1,450	1,150	2,600
2260	4 pile cluster	2,900	2,300	5,200
2280	440 kN load, end bearing, 2 pile cluster	1,150	1,100	2,250
2300	Friction type, 2 pile cluster	1,450	1,150	2,600
2320	4 pile cluster	2,900	2,300	5,200
2340	7 pile cluster	5,075	4,000	9,075
2360	890 kN load, end bearing, 3 pile cluster	1,725	1,650	3,375
2380	4 pile cluster	2,275	2,200	4,475
2400	Friction type, 8 pile cluster	5,775	4,575	10,350
2420	9 pile cluster	6,500	5,150	11,650
2440	14 pile cluster	10,100	8,025	18,125
2460	1800 kN load, end bearing, 6 pile cluster	3,425	3,300	6,725
2480	8 pile cluster	4,575	4,400	8,975
2500	Friction type, 14 pile cluster	10,100	8,025	18,125
2520	16 pile cluster	11,600	9,125	20,725
2540	18 pile cluster	13,000	10,300	23,300
2560	3600 kN load, end bearing, 12 pile cluster	7,425	7,150	14,575
2580	16 pile cluster	9,150	8,825	17,975
2600	5300 kN load, end bearing, 19 pile cluster	30,700	12,800	43,500
2620	20 pile cluster	32,300	13,500	45,800
2640	7100 kN load, end bearing, 19 pile cluster	30,700	12,800	43,500
4660	30 m long, 220 kN load, end bearing, 1 pile	1,100	950	2,050
4680	Friction type, 1 pile	1,375	980	2,355

SUBSTRUCTURE

A

A1020 Special Foundations

A1020 120	Precast Concrete Piles	COST EACH		
		MAT.	INST.	TOTAL
4700	2 pile cluster	2,750	1,950	4,700
4720	440 kN load, end bearing, 2 pile cluster	2,200	1,900	4,100
4740	Friction type, 2 pile cluster	2,750	1,950	4,700
4760	3 pile cluster	4,150	2,950	7,100
4780	4 pile cluster	5,525	3,925	9,450
4800	890 kN load, end bearing, 3 pile cluster	3,275	2,850	6,125
4820	4 pile cluster	4,375	3,775	8,150
4840	Friction type, 3 pile cluster	4,150	2,950	7,100
4860	5 pile cluster	6,900	4,900	11,800
4880	1800 kN load, end bearing, 6 pile cluster	6,575	5,675	12,250
4900	8 pile cluster	8,750	7,600	16,350
4910	Friction type, 8 pile cluster	11,000	7,875	18,875
4920	10 pile cluster	13,800	9,825	23,625
4930	3600 kN load, end bearing, 13 pile cluster	14,200	12,400	26,600
4940	16 pile cluster	17,500	15,200	32,700
4950	5300 kN load, end bearing, 19 pile cluster	59,500	22,200	81,700
4960	20 pile cluster	10,100	6,225	16,325
4970	7100 kN load, end bearing, 19 pile cluster	59,500	22,200	81,700

Important: See the Reference Section for critical supporting data - Reference Numbers and City Cost Indexes

SUBSTRUCTURE A

A1020 Special Foundations

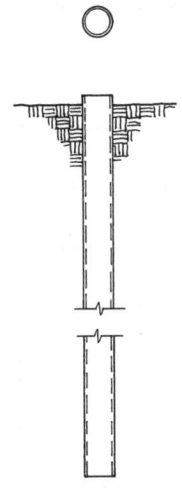

The Steel Pipe Pile System includes: steel pipe sections filled with 28 MPa concrete; a standard steel driving point; splices when required and an allowance for cutoffs.

The Expanded System Listing shows costs per cluster of piles. Clusters range from one pile to twenty piles. Loads vary from 220 kN to 7000 kN. Both end-bearing and friction-type piles are shown.

Please see the reference section for cost of mobilization of the pile driving equipment and other design and cost information.

System Components	QUANTITY	UNIT	COST EACH		
			MAT.	INST.	TOTAL
SYSTEM A1020 130 2220					
CONC. FILL STEEL PIPE PILE,15 m LONG,220 kN LOAD,END BEARING,1 PILE					
Piles, steel, pipe, conc. filled, 300 mm diameter	16.154	m	1,292.35	681.72	1,974.07
Steel pipe pile, standard point, for 300 mm or 350 mm diameter pipe	1.000	Ea.	106	142	248
Pile cut off, concrete pile, thin steel shell	1.000	Ea.		12	12
TOTAL			1,398.35	835.72	2,234.07

A1020 130	Steel Pipe Piles	COST EACH		
		MAT.	INST.	TOTAL
2220	Conc. fill steel pipe pile,15 m long,220 kN load,end bearing,1 pile	1,400	835	2,235
2240	Friction type, 2 pile cluster	2,800	1,675	4,475
2250	440 kN load, end bearing, 2 pile cluster	2,800	1,675	4,475
2260	3 pile cluster	4,200	2,525	6,725
2300	Friction type, 4 pile cluster	5,600	3,325	8,925
2320	5 pile cluster	7,000	4,175	11,175
2340	10 pile cluster	14,000	8,375	22,375
2360	890 kN load, end bearing, 3 pile cluster	4,200	2,525	6,725
2380	4 pile cluster	5,600	3,325	8,925
2400	Friction type, 4 pile cluster	5,600	3,325	8,925
2420	8 pile cluster	11,200	6,700	17,900
2440	9 pile cluster	12,600	7,525	20,125
2460	1800 kN load, end bearing, 6 pile cluster	8,400	5,000	13,400
2480	7 pile cluster	9,800	5,850	15,650
2500	Friction type, 9 pile cluster	12,600	7,525	20,125
2520	16 pile cluster	22,400	13,400	35,800
2540	19 pile cluster	26,600	15,900	42,500
2560	3600 kN load, end bearing, 11 pile cluster	15,400	9,175	24,575
2580	14 pile cluster	19,600	11,700	31,300
2600	15 pile cluster	21,000	12,600	33,600
2620	Friction type, 17 pile cluster	23,800	14,200	38,000
2640	5300 kN load, end bearing, 16 pile cluster	22,400	13,400	35,800
2660	20 pile cluster	28,000	16,700	44,700
2680	7100 kN load, end bearing, 17 pile cluster	23,800	14,200	38,000
3700	30 m long, 220 kN load, end bearing, 1 pile	2,750	1,675	4,425
3720	Friction type, 1 pile	2,750	1,675	4,425

A SUBSTRUCTURE

A1020 Special Foundations

A1020 130	Steel Pipe Piles	COST EACH		
		MAT.	INST.	TOTAL
3740	2 pile cluster	5,500	3,300	8,800
3760	440 kN load, end bearing, 2 pile cluster	5,500	3,300	8,800
3780	Friction type, 2 pile cluster	5,500	3,300	8,800
3800	3 pile cluster	8,275	4,975	13,250
3820	890 kN load, end bearing, 3 pile cluster	8,275	4,975	13,250
3840	4 pile cluster	11,000	6,625	17,625
3860	Friction type, 3 pile cluster	8,275	4,975	13,250
3880	4 pile cluster	11,000	6,625	17,625
3900	1800 kN load, end bearing, 6 pile cluster	16,500	9,925	26,425
3910	7 pile cluster	19,300	11,600	30,900
3920	Friction type, 5 pile cluster	13,800	8,300	22,100
3930	8 pile cluster	22,000	13,300	35,300
3940	3600 kN load, end bearing, 11 pile cluster	30,300	18,200	48,500
3950	14 pile cluster	38,600	23,300	61,900
3960	15 pile cluster	41,300	24,900	66,200
3970	5300 kN load, end bearing, 16 pile cluster	44,100	26,500	70,600
3980	20 pile cluster	11,200	9,975	21,175
3990	7100 kN load, end bearing, 17 pile cluster	46,800	28,200	75,000

Important: See the Reference Section for critical supporting data - Reference Numbers and City Cost Indexes

SUBSTRUCTURE A

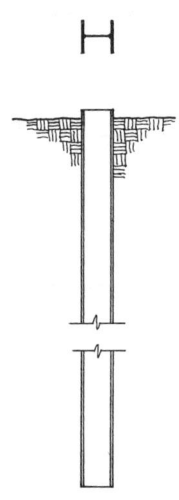

A Steel "H" Pile System includes: steel H sections; heavy duty driving point; splices where applicable and allowance for cutoffs.

The Expanded System Listing shows costs per cluster of piles. Clusters range from one pile to seventeen piles. Loads vary from 220 kN to 8900 kN. All loads for Steel H Pile systems are given in terms of end bearing capacity.

Steel sections range from 250 mm x 250 mm to 350 mm x 350 mm in the Expanded System Listing. The 350 mm x 350 mm steel section is used for all H piles used in applications requiring a working load over 3600 kN.

Please see the reference section for cost of mobilization of the pile driving equipment and other design and cost information.

System Components	QUANTITY	UNIT	COST EACH		
			MAT.	INST.	TOTAL
SYSTEM A1020 140 2220 **STEEL H PILES, 15 m LONG, 440 kN LOAD, END BEARING, 1 PILE**					
Steel H piles 250 mm x 250 mm, 63 kg/m	16.154	m	872.34	464.44	1,336.78
Heavy duty point, 250 mm	1.000	Ea.	142	144	286
Pile cut off, steel pipe or H piles	1.000	Ea.		24	24
TOTAL			1,014.34	632.44	1,646.78

A1020 140	Steel H Piles	COST EACH		
		MAT.	INST.	TOTAL
2220	Steel H piles, 15 m long, 440 kN load, end bearing, 1 pile	1,025	630	1,655
2260	2 pile cluster	2,025	1,275	3,300
2280	890 kN load, end bearing, 2 pile cluster	2,025	1,275	3,300
2300	3 pile cluster	3,050	1,900	4,950
2320	1800 kN load, end bearing, 3 pile cluster	3,050	1,900	4,950
2340	4 pile cluster	4,050	2,525	6,575
2360	6 pile cluster	6,075	3,800	9,875
2380	3600 kN load, end bearing, 5 pile cluster	5,075	3,150	8,225
2400	7 pile cluster	7,100	4,425	11,525
2420	12 pile cluster	12,200	7,575	19,775
2440	5300 kN load, end bearing, 8 pile cluster	8,125	5,050	13,175
2460	11 pile cluster	11,200	6,950	18,150
2480	17 pile cluster	17,200	10,800	28,000
2500	7100 kN load, end bearing, 10 pile cluster	12,700	6,600	19,300
2520	14 pile cluster	17,800	9,250	27,050
2540	8900 kN load, end bearing, 12 pile cluster	15,300	7,925	23,225
2560	18 pile cluster	22,900	11,900	34,800
3580	30 m long, 220 kN load, end bearing, 1 pile	3,375	1,450	4,825
3600	440 kN load, end bearing, 1 pile	3,375	1,450	4,825
3620	2 pile cluster	6,750	2,925	9,675
3640	890 kN load, end bearing, 2 pile cluster	6,750	2,925	9,675
3660	3 pile cluster	10,100	4,375	14,475
3680	1800 kN load, end bearing, 3 pile cluster	10,100	4,375	14,475
3700	4 pile cluster	13,500	5,875	19,375
3720	6 pile cluster	20,200	8,800	29,000
3740	3600 kN load, end bearing, 5 pile cluster	16,900	7,325	24,225

SUBSTRUCTURE

A

A1020 Special Foundations

A1020 140	Steel H Piles	COST EACH		
		MAT.	INST.	TOTAL
3760	7 pile cluster	23,600	10,300	33,900
3780	12 pile cluster	40,500	17,600	58,100
3800	5300 kN load, end bearing, 8 pile cluster	27,000	11,700	38,700
3820	11 pile cluster	37,100	16,100	53,200
3840	17 pile cluster	57,500	24,900	82,400
3860	7100 kN load, end bearing, 10 pile cluster	33,700	14,700	48,400
3880	14 pile cluster	47,200	20,500	67,700
3900	8900 kN load, end bearing, 12 pile cluster	40,500	17,600	58,100
3920	18 pile cluster	60,500	26,300	86,800

SUBSTRUCTURE

A

Important: See the Reference Section for critical supporting data - Reference Numbers and City Cost Indexes

A1020   Special Foundations

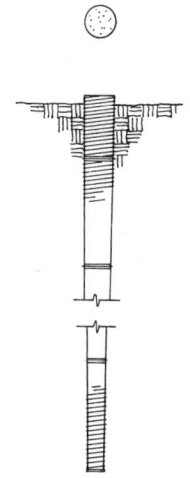

The Step Tapered Steel Pile System includes: step tapered piles filled with 28 MPa. concrete. The cost for splices and pile cutoffs is included.

The Expanded System Listing shows costs per cluster of piles. Clusters range from one pile to twenty-four piles. Both end bearing piles and friction piles are listed. Loads vary from 220 kN to 7000 kN.

Please see the reference section for cost of mobilization of the pile driving equipment and other design and cost information.

System Components	QUANTITY	UNIT	COST EACH		
			MAT.	INST.	TOTAL
SYSTEM A1020 150 1000					
STEEL PILE, STEP TAPERED, 15 m LONG, 220 kN LOAD, END BEARING, 1 PILE					
St. shell step tapered conc. fld. piles, 200mm tip 270 kN cap. to 18m	16.154	m	500.79	382.05	882.84
Pile cutoff, steel pipe or H piles	1.000	Ea.		24	24
TOTAL			500.79	406.05	906.84

A1020 150	Step-Tapered Steel Piles	COST EACH		
		MAT.	INST.	TOTAL
1000	Steel pile, step tapered, 15 m long, 220 kN load, end bearing, 1 pile	500	405	905
1200	Friction type, 3 pile cluster	1,500	1,225	2,725
1400	440 kN load, end bearing, 2 pile cluster	1,000	810	1,810
1600	Friction type, 4 pile cluster	2,000	1,625	3,625
1800	890 kN load, end bearing, 4 pile cluster	2,000	1,625	3,625
2000	Friction type, 6 pile cluster	3,000	2,450	5,450
2200	1800 kN load, end bearing, 7 pile cluster	3,500	2,850	6,350
2400	Friction type, 10 pile cluster	5,000	4,050	9,050
2600	3600 kN load, end bearing, 14 pile cluster	7,000	5,675	12,675
2800	Friction type, 18 pile cluster	9,025	7,300	16,325
3000	5300 kN load, end bearing, 16 pile cluster	8,925	6,975	15,900
3200	Friction type, 21 pile cluster	11,700	9,150	20,850
3400	7100 kN load, end bearing, 18 pile cluster	10,000	7,850	17,850
3600	Friction type, 24 pile cluster	13,400	10,500	23,900
5000	30 m long, 220 kN load, end bearing, 1 pile	1,000	815	1,815
5200	Friction type, 2 pile cluster	2,000	1,625	3,625
5400	440 kN load, end bearing, 2 pile cluster	2,000	1,625	3,625
5600	Friction type, 3 pile cluster	3,025	2,450	5,475
5800	890 kN load, end bearing, 4 pile cluster	4,025	3,250	7,275
6000	Friction type, 5 pile cluster	5,025	4,075	9,100
6200	1800 kN load, end bearing, 7 pile cluster	7,025	5,675	12,700
6400	Friction type, 8 pile cluster	8,025	6,475	14,500
6600	3600 kN load, end bearing, 15 pile cluster	15,100	12,200	27,300
6800	Friction type, 16 pile cluster	16,100	13,000	29,100
7000	5300 kN load, end bearing, 17 pile cluster	9,725	7,200	16,925
7200	Friction type, 19 pile cluster	22,400	16,100	38,500
7400	7100 kN load, end bearing, 20 pile cluster	23,500	16,900	40,400
7600	Friction type, 22 pile cluster	5,325	4,250	9,575

SUBSTRUCTURE

A

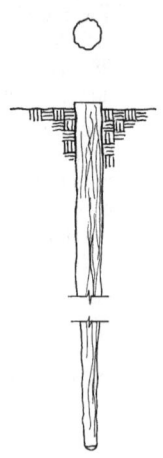

The Treated Wood Pile System includes: pressure treated wood piles; a standard steel driving point; and an allowance for cutoffs.

The Expanded System Listing shows costs per cluster of piles. Clusters range from three piles to twenty piles. Loads vary from 220 kN to 1800 kN. Both end-bearing and friction type piles are listed.

Please see the reference section for cost of mobilization of the pile driving equipment and other design and cost information.

SUBSTRUCTURE A

System Components	QUANTITY	UNIT	COST EACH		
			MAT.	INST.	TOTAL
SYSTEM A1020 160 2220					
WOOD PILES, 8 m LONG, 220 kN LOAD, END BEARING, 3 PILE CLUSTER					
Wood piles, treated, 300 mm butt, 200 mm tip, up to 9 m long	24.689	m	864.11	691.28	1,555.39
Point for driving wood piles	3.000	Ea.	66	69	135
Pile cutoff, wood piles	3.000	Ea.		36	36
TOTAL			930.11	796.28	1,726.39

A1020 160	Treated Wood Piles	COST EACH		
		MAT.	INST.	TOTAL
2220	Wood piles, 8 m long, 220 kN load, end bearing, 3 pile cluster	930	800	1,730
2240	Friction type, 3 pile cluster	865	730	1,595
2260	5 pile cluster	1,450	1,200	2,650
2280	440 kN load, end bearing, 4 pile cluster	1,250	1,075	2,325
2300	5 pile cluster	1,550	1,325	2,875
2320	6 pile cluster	1,850	1,600	3,450
2340	Friction type, 5 pile cluster	1,450	1,200	2,650
2360	6 pile cluster	1,725	1,450	3,175
2380	10 pile cluster	2,875	2,425	5,300
2400	890 kN load, end bearing, 8 pile cluster	2,475	2,125	4,600
2420	10 pile cluster	3,100	2,650	5,750
2440	12 pile cluster	3,725	3,200	6,925
2460	Friction type, 10 pile cluster	2,875	2,425	5,300
2480	1800 kN load, end bearing, 16 pile cluster	4,950	4,250	9,200
2500	20 pile cluster	6,200	5,325	11,525
4520	15 m long, 220 kN load, end bearing, 3 pile cluster	2,025	1,175	3,200
4540	4 pile cluster	2,700	1,575	4,275
4560	Friction type, 2 pile cluster	1,275	720	1,995
4580	3 pile cluster	1,925	1,075	3,000
4600	440 kN load, end bearing, 5 pile cluster	3,375	1,975	5,350
4620	8 pile cluster	5,400	3,150	8,550
4640	Friction type, 3 pile cluster	1,925	1,075	3,000
4660	5 pile cluster	3,200	1,800	5,000
4680	890 kN load, end bearing, 9 pile cluster	6,075	3,550	9,625
4700	10 pile cluster	6,750	3,950	10,700
4720	15 pile cluster	10,100	5,925	16,025
4740	Friction type, 5 pile cluster	3,200	1,800	5,000
4760	6 pile cluster	3,850	2,150	6,000

A10 Foundations

A1020 Special Foundations

A1020 160	Treated Wood Piles	COST EACH		
		MAT.	INST.	TOTAL
4780	10 pile cluster	6,425	3,600	10,025
4800	1800 kN load, end bearing, 18 pile cluster	12,100	7,100	19,200
4820	20 pile cluster	13,500	7,875	21,375
4840	Friction type, 9 pile cluster	5,775	3,250	9,025
4860	10 pile cluster	6,425	3,600	10,025
9000	Add for boot for driving tip, each pile	22	16.90	38.90

A1020 Special Foundations

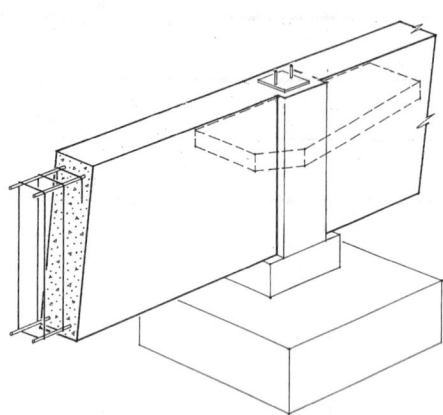

The Grade Beam System includes: excavation with a truck mounted backhoe; hand trim; backfill; forms (four uses); reinforcing steel; and 21 MPa concrete placed from chute.

Superimposed loads vary in the listing from 900 kg/m to 5700 kg/m. In the Expanded System Listing, the span of the beams varies from 4.5 m to 12 m. Depth varies from 700 mm to 1300 mm. Width varies from 300 mm to 1200mm.

Please see the reference section for further design and cost information.

System Components	QUANTITY	UNIT	COST PER m		
			MAT.	INST.	TOTAL
SYSTEM A1020 210 2220					
GRADE BEAM, 4.5 m SPAN, 700 mm DEEP, 300 mm WIDE, 115 kN/m LOAD					
Excavation, trench, hydraulic backhoe, 0.29 m³ bucket	.652	m³		5.84	5.84
Trim sides and bottom of trench, regular soil	.610	m²		4.76	4.76
Backfill, by hand, compaction in 150 mm layers, using vibrating plate	.426	m³		3.45	3.45
Forms in place, grade beam, 4 uses	1.433	m²CA	9.53	65.90	75.43
Reinforcing in place, beams & girders, #25M to #45M	.057	Met. Ton	57.96	48.07	106.03
Concrete ready mix, regular weight, 21 MPa	.226	m³	28.22		28.22
Place and vibrate conc. for grade beam, direct chute	.226	m³		4.29	4.29
TOTAL			95.71	132.31	228.02

A1020 210	Grade Beams	COST PER m		
		MAT.	INST.	TOTAL
2220	Grade beam, 4.5 m span, 700 mm deep, 300 mm wide, 115 kN/m load	95.50	132	227.50
2240	350 mm wide, 175 kN/m load	99	133	232
2260	1000 mm deep, 300 mm wide, 235 kN/m load	97	164	261
2280	290 kN/m load	112	176	288
2300	1300 mm deep, 300 mm wide, 440 kN/m load	129	220	349
2320	585 kN/m load	159	245	404
2340	730 kN/m load	190	268	458
3360	6 m span, 700 mm deep, 300 mm wide, 30 kN/m load	59	102	161
3380	400 mm wide, 60 kN/m load	79	114	193
3400	1000 mm deep, 300 mm wide, 115 kN/m load	97	164	261
3420	175 kN/m load	124	184	308
3440	350 mm wide, 235 kN/m load	152	205	357
3460	1300 mm deep, 300 mm wide, 290 kN/m load	162	248	410
3480	350 mm wide, 440 kN/m load	227	296	523
3500	500 mm wide, 585 kN/m load	266	310	576
3520	600 mm wide, 730 kN/m load	330	355	685
4540	9 m span, 700 mm deep, 300 mm wide, 15 kN/m load	62	105	167
4560	350 mm wide, 30 kN/m load	100	145	245
4580	1000 mm deep, 300 mm wide, 60 kN/m load	119	173	292
4600	450 mm wide, 115 kN/m load	171	211	382
4620	1300 mm deep, 350 mm wide, 175 kN/m load	218	289	507
4640	500 mm wide, 235 kN/m load	267	315	582
4660	600 mm wide, 290 kN/m load	330	355	685
4680	900 mm wide, 440 kN/m load	480	450	930
4700	1200 mm wide, 585 kN/m load	635	550	1,185
5720	12 m span, 1000 mm deep, 300 mm wide, 15 kN/m load	85	153	238

SUBSTRUCTURE

A10 Foundations

A1020 Special Foundations

A1020 210	Grade Beams	COST PER m		
		MAT.	INST.	TOTAL
5740	30 kN/m load	106	171	277
5760	1300 mm deep, 300 mm wide, 60 kN/m load	159	245	404
5780	500 mm wide, 115 kN/m load	258	305	563
5800	700 mm wide, 175 kN/m load	355	375	730
5820	950 mm wide, 235 kN/m load	490	455	945
5840	1150 mm wide, 290 kN/m load	645	560	1,205

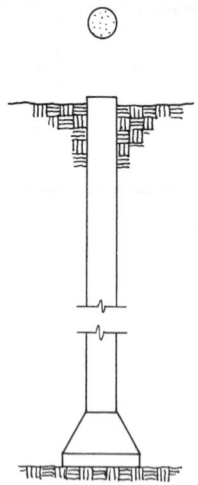

Caisson Systems are listed for three applications: stable ground, wet ground and soft rock. Concrete used is 21 MPa placed from chute. Included are a bell at the bottom of the caisson shaft (if applicable) along with required excavation and disposal of excess excavated material up to 3.2 km from job site.

The Expanded System lists cost per caisson. End-bearing loads vary from 890 kN to 1400 kN. The dimensions of the caissons range from 600 mm x 15 m to 2100 mm x 60 m.

Please see the reference section for further design and cost information.

SUBSTRUCTURE A

System Components	QUANTITY	UNIT	COST EACH		
			MAT.	INST.	TOTAL
SYSTEM A1020 310 2200					
CAISSON, STABLE GRND., 21MPa CONC, 480kPa BRNG, 890 kN LOAD, 600mmx15m					
Caissons, drilled, to 15 m, 600 mm shaft diameter, 0.291 m³/m	15.240	m	693.42	1,562.10	2,255.52
Reinforcing in place, columns, #10M to #25M	.054	Met. Ton	55.79	83.01	138.80
Caisson bell excavation and concrete, 1200 mm diameter 0.34 m³	1.000	Ea.	42.50	298	340.50
Load & haul excess excavation, 3.2 km	4.771	m³		30.54	30.54
TOTAL			791.71	1,973.65	2,765.36

A1020 310	Caissons	COST EACH		
		MAT.	INST.	TOTAL
2200	Caisson, stable grnd., 21 MPa conc, 480kPa brng, 890 kN load, 600mm x 15m	790	1,975	2,765
2400	1800 kN load, 750 mm x 15 m	1,300	3,175	4,475
2600	3600 kN load, 925 mm x 30 m	3,575	7,550	11,125
2800	5300 kN load, 1225 mm x 30 m	6,175	9,500	15,675
3000	7100 kN load, 1525 mm x 46 m	13,900	14,600	28,500
3200	10700 kN load, 1825 mm x 46 m	20,200	19,000	39,200
3400	14200 kN load, 2125 mm x 61 m	36,500	27,400	63,900
5000	Wet ground, 21 MPa conc., 480 kPa brng, 890 kN load, 600 mm x 15m	690	2,700	3,390
5200	1800 kN load, 750 mm x 15 m	1,150	4,650	5,800
5400	3600 kN load, 925 mm x 30 m	3,100	12,500	15,600
5600	5300 kN load, 1225 mm x 30 m	5,350	17,500	22,850
5800	7100 kN load, 1525 mm x 46 m	11,900	37,600	49,500
6000	10700 kN load, 1825 mm x 46 m	17,400	46,400	63,800
6200	14200 kN load, 2125 mm x 61 m	31,300	74,500	105,800
7800	Soft rock, 21 MPa conc., 480 kPa brng, 890 kN load, 600mm x 15m	690	13,900	14,590
8000	1800 kN load, 750 mm x 15 m	1,150	22,500	23,650
8200	3600 kN load, 925 mm x 30 m	3,100	59,000	62,100
8400	5300 kN load, 1225 mm x 30 m	5,350	86,500	91,850
8600	7100 kN load, 1525 mm x 46 m	11,900	177,500	189,400
8800	10700 kN load, 1825 mm x 46 m	17,400	211,500	228,900
9000	14200 kN load, 2125 mm x 61 m	31,300	337,000	368,300

Important: See the Reference Section for critical supporting data - Reference Numbers and City Cost Indexes

A1020 Special Foundations

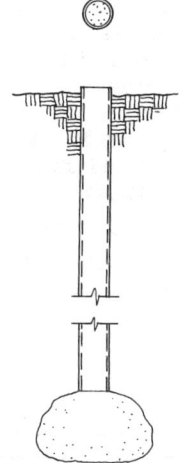

Pressure Injected Piles are usually uncased up to 8 m and cased over 8 m depending on soil conditions.

These costs include excavation and hauling of excess materials; steel casing over 8 m; reinforcement; 21 MPa concrete; plus mobilization and demobilization of equipment for a distance of up to 80 km to and from the job site.

The Expanded System lists cost per cluster of piles. Clusters range from one pile to eight piles. End-bearing loads range from 220 kN to 7100 kN.

Please see the reference section for further design and cost information.

System Components	QUANTITY	UNIT	COST EACH		
			MAT.	INST.	TOTAL
SYSTEM A1020 710 4200					
PRESSURE INJECTED FOOTING, END BEARING,15m LONG,220 kN LOAD, 1 PILE					
Pressure injected footings, cased, 130-270 kN cap., 300 mm diam.	15.240	m	739.14	1,012.70	1,751.84
Pile cutoff, concrete pile with thin steel shell	1.000	Ea.		12	12
TOTAL			739.14	1,024.70	1,763.84

A1020 710	Pressure Injected Footings	COST EACH		
		MAT.	INST.	TOTAL
2200	Pressure injected footing, end bearing,8 m long,220 kN load, 1 pile	345	720	1,065
2400	440 kN load, 1 pile	345	720	1,065
2600	2 pile cluster	685	1,425	2,110
2800	890 kN load, 2 pile cluster	685	1,425	2,110
3200	1800 kN load, 4 pile cluster	1,375	2,850	4,225
3400	7 pile cluster	2,400	5,025	7,425
3800	5300 kN load, 6 pile cluster	2,875	5,425	8,300
4000	7100 kN load, 7 pile cluster	3,350	6,350	9,700
4200	15 m long, 220 kN load, 1 pile	740	1,025	1,765
4400	440 kN load, 1 pile	1,275	1,025	2,300
4600	2 pile cluster	2,550	2,050	4,600
4800	890 kN load, 2 pile cluster	2,550	2,050	4,600
5000	4 pile cluster	5,100	4,100	9,200
5200	1800 kN load, 4 pile cluster	5,100	4,100	9,200
5400	8 pile cluster	10,200	8,200	18,400
5600	3600 kN load, 7 pile cluster	8,900	7,175	16,075
5800	5300 kN load, 6 pile cluster	8,175	6,150	14,325
6000	7100 kN load, 7 pile cluster	9,550	7,175	16,725

SUBSTRUCTURE

A

A2020 Basement Walls

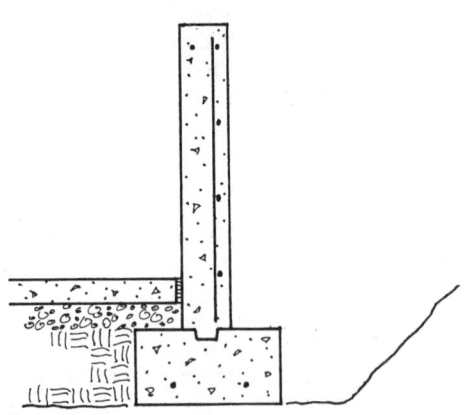

The Foundation Bearing Wall System includes: forms up to 4900 mm high (four uses); 21 MPa concrete placed and vibrated; and form removal with breaking form ties and patching walls. The wall systems list walls from 150 mm to 400 mm thick and are designed with minimum reinforcement.

Excavation and backfill are not included.

Please see the reference section for further design and cost information.

SUBSTRUCTURE A

System Components	QUANTITY	UNIT	COST PER m		
			MAT.	INST.	TOTAL
SYSTEM A2020 110 1500					
FOUNDATION WALL, CAST IN PLACE, DIRECT CHUTE, 1200 mm HIGH, 150 mm THICK					
Formwork	2.438	m²CA	11.80	113.39	125.19
Reinforcing	4.911	kg	4.84	3.76	8.60
Unloading & sorting reinforcing	4.911	kg		.23	.23
Concrete, 21 MPa	.186	m³	23.20		23.20
Place concrete, direct chute	.186	m³		5.88	5.88
Finish walls, break ties and patch voids, one side	1.219	m²	.39	9.88	10.27
TOTAL			40.23	133.14	173.37

A2020 110	Walls, Cast in Place							
	WALL HEIGHT (mm)	PLACING METHOD	CONCRETE (m³ per m)	REINFORCING (kg per m)	WALL THICKNESS (mm)	COST PER m		
						MAT.	INST.	TOTAL
1500	1200mm	direct chute	.186	4.911	150	40	133	173
1520			.248	7.143	200	50.50	137	187.50
1540			.309	8.929	250	59.50	139	198.50
1560			.371	10.715	300	69	142	211
1580			.434	12.054	350	78.50	144	222.50
1600			.494	14.049	400	88	148	236
1700	1200mm	pumped	.186	4.911	150	40	136	176
1720			.248	7.143	200	50.50	141	191.50
1740			.309	8.929	250	59.50	144	203.50
1760			.371	10.715	300	69	149	218
1780			.434	12.054	350	78.50	151	229.50
1800			.494	14.049	400	88	155	243
3000	1800mm	direct chute	.278	7.367	150	60.50	199	259.50
3020			.374	10.715	200	75.50	205	280.50
3040			.462	13.394	250	89	209	298
3060			.557	16.073	300	104	213	317
3080			.652	18.082	350	118	218	336
3100			.753	21.415	400	133	223	356

Important: See the Reference Section for critical supporting data - Reference Numbers and City Cost Indexes

A20 Basement Construction

A2020 Basement Walls

A2020 110				Walls, Cast in Place				

	WALL HEIGHT (mm)	PLACING METHOD	CONCRETE (m³ per m)	REINFORCING (kg per m)	WALL THICKNESS (mm)	COST PER m		
						MAT.	INST.	TOTAL
3200	1800mm	pumped	.278	7.367	150	60.50	204	264.50
3220			.374	10.715	200	75.50	212	287.50
3240			.462	13.394	250	89	217	306
3260			.557	16.073	300	104	223	327
3280			.652	18.082	350	118	226	344
3300			.753	21.415	400	133	233	366
5000	2400mm	direct chute	.371	9.821999	150	80.50	266	346.50
5020			.499	14.287	200	101	273	374
5040			.627	17.858	250	120	279	399
5060			.742	21.415	300	138	286	424
5080			.87	24.094	350	144	287	431
5100			.988	28.559	400	176	297	473
5200	2400mm	pumped	.371	9.821999	150	80.50	273	353.50
5220			4.992	14.287	200	101	282	383
5240			.627	17.858	250	120	289	409
5260			.742	21.415	300	138	296	434
5280			.87	24.094	350	144	298	442
5300			.988	28.559	400	176	310	486
6020	3000mm	direct chute	.622	17.858	200	126	340	466
6040			.77	22.308	250	149	350	499
6060			.928	26.773	300	173	355	528
6080			1.086	30.121	350	196	360	556
6100			1.237	35.702	400	220	370	590
6220	3000mm	pumped	.622	17.858	200	126	350	476
6240			.77	22.308	250	149	365	514
6260			.928	26.772	300	173	370	543
6280			1.086	30.122	350	196	375	571
6300			1.237	35.702	400	220	385	605
7220	3700mm	pumped	.748	21.416	200	151	425	576
7240			.926	26.772	250	179	435	614
7260			1.114	32.13	300	207	445	652
7280			1.304	36.148	350	235	450	685
7300			1.482	42.846	400	264	465	729
7420	3700mm	crane & bucket	.748	21.416	200	151	440	591
7440			.926	26.772	250	179	450	629
7460			1.114	32.13	300	207	470	677
7480			1.304	36.148	350	235	480	715
7500			1.482	42.846	400	264	500	764
8220	4300mm	pumped	.87	24.986	200	176	495	671
8240			1.079	31.238	250	208	505	713
8260			1.302	37.488	300	242	520	762
8280			1.523	42.16	350	275	530	805
8300			1.731	49.988	400	310	540	850
8420	4300mm	crane & bucket	.87	24.986	200	176	515	691
8440			1.079	31.238	250	208	530	738
8460			1.302	37.488	300	242	545	787
8480			1.523	42.16	350	275	560	835
8500			1.731	49.988	400	310	575	885
9220	4900mm	pumped	.996	28.558	200	201	565	766
9240			1.234	35.702	250	238	575	813
9260			1.487	42.846	300	277	595	872
9280			1.738	48.202	350	315	600	915
9300			1.977	57.118	400	350	620	970

SUBSTRUCTURE

A

A2020 Basement Walls

A2020 110		Walls, Cast in Place						
	WALL HEIGHT (mm)	PLACING METHOD	CONCRETE (m³ per m)	REINFORCING (kg per m)	WALL THICKNESS (mm)	COST PER m		
						MAT.	INST.	TOTAL
9420	4900mm	crane & bucket	.996	28.558	200	201	590	791
9440			1.234	35.702	250	238	600	838
9460			1.487	42.846	300	277	625	902
9480			1.738	48.202	350	315	635	950
9500			1.977	57.118	400	350	660	1,010

For information about Means Estimating Seminars, see yellow pages 12 and 13 in back of book

Important: See the Reference Section for critical supporting data - Reference Numbers and City Cost Indexes

SUBSTRUCTURE

A

G BUILDING SITEWORK

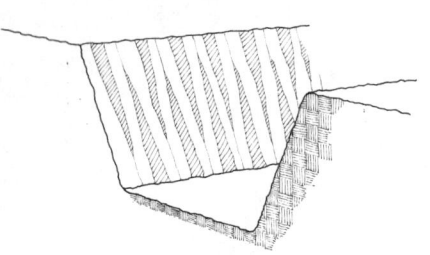

Trenching Systems are shown on a cost per meter basis. The systems include: excavation; backfill and removal of spoil; and compaction for various depths and trench bottom widths. The backfill has been reduced to accommodate a pipe of suitable diameter and bedding. See System

12.3-310 for bedding costs. The slope for trench sides varies from 0:1 to 2:1.

The Expanded System Listing shows Trenching Systems that range from 600 mm to 3600 mm in width. Depths range from 600 mm to 7300 mm.

System Components

	QUANTITY	UNIT	COST PER m		
			EQUIP.	LABOR	TOTAL
SYSTEM G1030 805 1310					
TRENCHING, BACKHOE, 0 TO 1 SLOPE, 600 mm WIDE, 600 mm DP, 0.29 m³ BUCKET					
Excavation, trench, hyd. backhoe, track mtd., 0.29 m³ bucket	.436	m³	.94	2.97	3.91
Backfill and load spoil, from stockpile	.436	m³	.31	.88	1.19
Compaction by rammer tamper, 200 mm lifts, 4 passes	.035	m³	.01	.12	.13
Remove excess spoil, 4.58 m³ dump truck, 3 km roundtrip	.401	m³	1.89	1.82	3.71
TOTAL			3.15	5.79	8.94

G1030 805 — Trenching

		COST PER m		
		EQUIP.	LABOR	TOTAL
1310	Trenching, backhoe, 0 to 1 slope, 600 mm wide, 600 mm deep, 0.29 m³ bucket	3.15	5.80	8.95
1320	900 mm deep, 0.29 m³ bucket	3.86	8.45	12.30
1330	1200 mm deep, 0.29 m³ bucket	4.59	11.15	15.75
1340	1800 mm deep, 0.29 m³ bucket	5.80	14.30	20
1350	2400 mm deep, 0.38 m³ bucket	6.80	18.15	25
1360	3100 mm deep, 0.76 m³ bucket	10	21.50	31.50
1400	1200 mm wide, 600 mm deep, 0.29 m³ bucket	6.60	11.65	18.25
1410	900 mm deep, 0.29 m³ bucket	9.35	17.30	26.50
1420	1200 mm deep, 0.38 m³ bucket	9.45	17.45	27
1430	1800 mm deep, 0.38 m³ bucket	12.15	26.50	38.50
1440	2400 mm deep, 0.38 m³ bucket	18.55	34	52.50
1450	3100 mm deep, 0.76 m³ bucket	21	42.50	63.50
1460	3700 mm deep, 0.76 m³ bucket	26.50	54.50	81
1470	4600 mm deep, 1.15 m³ bucket	22.50	46.50	69
1480	5500 mm deep, 1.91 m³ bucket	31	66.50	97.50
1520	1800 mm wide, 1800 mm deep, 0.48 m³ bucket	24	37	61
1530	2400 mm deep, 0.57 m³ bucket	29	44	73
1540	3100 mm deep, 0.76 m³ bucket	31.50	53.50	85
1550	3700 mm deep, 0.96 m³ bucket	27.50	42	69.50
1560	4900 mm deep, 1.53 m³ bucket	36	48	84
1570	6100 mm deep, 2.67 m³ bucket	49	85.50	135
1580	7300 mm deep, 2.67 m³ bucket	79	143	222
1640	2400 mm wide, 3700 mm deep, 0.96 m³ bucket	41	57.50	98.50
1650	4600 mm deep, 1.15 m³ bucket	48	71	119
1660	5500 mm deep, 1.91 m³ bucket	56	72	128
1680	7300 mm deep, 2.67 m³ bucket	109	191	300
1730	3100 mm wide, 6100 mm deep, 2.67 m³ bucket	88.50	144	233
1740	7300 mm deep, 2.67 m³ bucket	138	240	380
1800	1/2 to 1 slope, 600 mm wide, 600 mm deep, 0.29 m³ bucket	4.21	7.45	11.65
1810	900 mm deep, 0.29 m³ bucket	5.85	12.25	18.10
1820	1200 mm deep, 0.29 m³ bucket	7.45	17.90	25.50
1840	1800 mm deep, 0.29 m³ bucket	11.05	28	39
1860	2400 mm deep, 0.38 m³ bucket	15.45	42.50	58
1880	3100 mm deep, 0.76 m³ bucket	27.50	58	85.50

Important: See the Reference Section for critical supporting data - Reference Numbers and City Cost Indexes

G1030 Site Earthwork

G1030 805	Trenching	COST PER m		
		EQUIP.	LABOR	TOTAL
2300	1200 mm wide, 600 mm deep, 0.29 m³ bucket	7	12.40	19.40
2310	900 mm deep, 0.29 m³ bucket	11.55	20.50	32
2320	1200 mm deep, 0.38 m³ bucket	12.40	21.50	34
2340	1800 mm deep, 0.38 m³ bucket	17.35	37	54.50
2360	2400 mm deep, 0.38 m³ bucket	28.50	50.50	79
2380	3100 mm deep, 0.76 m³ bucket	37.50	74.50	112
2400	3700 mm deep, 0.76 m³ bucket	39.50	80	120
2430	4600 mm deep, 1.15 m³ bucket	50.50	99.50	150
2460	5500 mm deep, 1.91 m³ bucket	69	119	188
2840	1800 mm wide, 1800 mm deep, 0.48 m³ bucket	28.50	40.50	69
2860	2400 mm deep, 0.57 m³ bucket	40.50	61	102
2880	3100 mm deep, 0.76 m³ bucket	47.50	80	128
2900	3700 mm deep, 0.96 m³ bucket	44.50	67	112
2940	4900 mm deep, 1.53 m³ bucket	69.50	107	177
2980	6100 mm deep, 2.67 m³ bucket	106	184	290
3020	7300 mm deep, 2.67 m³ bucket	193	340	535
3100	2400 mm wide, 3650 mm deep, 0.96 m³ bucket	59.50	82	142
3120	4600 mm deep, 1.15 m³ bucket	76	110	186
3140	5500 mm deep, 1.91 m³ bucket	96	120	216
3180	7300 mm deep, 2.67 m³ bucket	222	385	605
3270	3100 mm wide, 6100 mm deep, 2.67 m³ bucket	138	229	365
3280	7300 mm deep, 2.67 m³ bucket	253	435	690
3500	1 to 1 slope, 600 mm wide, 600 mm deep, 0.29 m³ bucket	4.92	8.70	13.60
3520	900 mm deep, 0.29 m³ bucket	7.55	15.20	23
3540	1200 mm deep, 0.29 m³ bucket	9.80	23	33
3560	1800 mm deep, 0.29 m³ bucket	15.10	36.50	51.50
3580	2400 mm deep, 0.38 m³ bucket	22	57.50	79.50
3600	3100 mm deep, 0.76 m³ bucket	46	92.50	139
3800	1200 mm wide, 600 mm deep, 0.29 m³ bucket	8.40	14.80	23
3820	900 mm deep, 0.29 m³ bucket	14.75	26	41
3840	1200 mm deep, 0.38 m³ bucket	15.05	24	39
3860	1800 mm deep, 0.38 m³ bucket	21.50	44.50	66
3880	2400 mm deep, 0.38 m³ bucket	38.50	66.50	105
3900	3100 mm deep, 0.76 m³ bucket	50	95.50	146
3920	3700 mm deep, 0.76 m³ bucket	72.50	139	212
3940	4600 mm deep, 1.15 m³ bucket	70	129	199
3960	5500 mm deep, 1.91 m³ bucket	99	156	255
4030	1800 mm wide, 1800 mm deep, 0.48 m³ bucket	34.50	49.50	84
4040	2400 mm deep, 0.57 m³ bucket	50.50	79	130
4050	3100 mm deep, 0.76 m³ bucket	61	110	171
4060	3700 mm deep, 0.96 m³ bucket	59	99	158
4070	4900 mm deep, 1.53 m³ bucket	94	140	234
4080	6100 mm deep, 2.67 m³ bucket	152	291	445
4090	7300 mm deep, 2.67 m³ bucket	287	565	850
4500	2400 mm wide, 3700 mm deep, 0.96 m³ bucket	76	112	188
4550	4600 mm deep, 1.15 m³ bucket	100	159	259
4600	5500 mm deep, 1.91 m³ bucket	130	185	315
4650	7300 mm deep, 2.67 m³ bucket	315	600	915
4800	3100 mm wide, 6100 mm deep, 2.67 m³ bucket	192	335	525
4850	7300 mm deep, 2.67 m³ bucket	340	640	980
5000	1-1/2 to 1 slope, 600 mm wide, 600 mm deep, 0.29 m³ bucket	5.70	10.05	15.75
5020	900 mm deep, 0.29 m³ bucket	9.25	17.95	27
5040	1200 mm deep, 0.29 m³ bucket	12	27	39
5060	1800 mm deep, 0.29 m³ bucket	18.65	43	61.50
5080	2400 mm deep, 0.38 m³ bucket	27	68	95
5100	3100 mm deep, 0.76 m³ bucket	51.50	95.50	147
5300	1200 mm wide, 600 mm deep, 0.29 m³ bucket	7.95	13.95	22
5320	900 mm deep, 0.29 m³ bucket	14.40	25.50	40

G1030 805	Trenching	COST PER m		
		EQUIP.	LABOR	TOTAL
5340	1200 mm deep, 0.38 m³ bucket	17.90	28	46
5360	1800 mm deep, 0.38 m³ bucket	26	50.50	76.50
5380	2400 mm deep, 0.38 m³ bucket	46	75.50	122
5400	3100 mm deep, 0.76 m³ bucket	61	109	170
5420	3700 mm deep, 0.76 m³ bucket	89.50	161	251
5450	4.5 m deep, 1.15 m³ bucket	87	143	230
5480	5500 mm deep, 1.91 m³ bucket	124	170	294
5660	1800 mm wide, 1800 mm deep, 0.48 m³ bucket	40	55	95
5680	2400 mm deep, 0.57 m³ bucket	59	88	147
5700	3100 mm deep, 0.76 m³ bucket	72	122	194
5720	3700 mm deep, 0.96 m³ bucket	69	105	174
5760	4900 mm deep, 1.53 m³ bucket	114	150	264
5800	6100 mm deep, 2.67 m³ bucket	188	330	520
6050	4600 mm deep, 1.15 m³ bucket	119	174	293
6080	5500 mm deep, 1.91 m³ bucket	157	199	355
6140	7300 mm deep, 2.67 m³ bucket	390	680	1,075
6300	3100 mm wide, 6100 mm deep, 2.67 m³ bucket	230	375	605
6350	7300 mm deep, 2.67 m³ bucket	415	710	1,125
6600	2 to 1 slope, 600 mm wide, 600 mm deep, 0.29 m³ bucket	8.90	9.20	18.10
6620	900 mm deep, 0.29 m³ bucket	10.60	19.90	30.50
6640	1200 mm deep, 0.29 m³ bucket	13.70	30	43.50
6660	1800 mm deep, 0.29 m³ bucket	20	45.50	65.50
6680	2400 mm deep, 0.38 m³ bucket	31	77	108
6700	3100 mm deep, 0.76 m³ bucket	58.50	108	167
6900	1200 mm wide, 600 mm deep, 0.29 m³ bucket	7.90	13.90	22
6920	900 mm deep, 0.29 m³ bucket	14.75	26	41
6940	1200 mm deep, 0.38 m³ bucket	19.95	30	50
6960	1800 mm deep, 0.38 m³ bucket	28.50	55	83.50
6980	2400 mm deep, 0.38 m³ bucket	51.50	83	135
7000	3100 mm deep, 0.76 m³ bucket	67.50	120	188
7020	3700 mm deep, 0.76 m³ bucket	100	180	280
7050	4600 mm deep, 1.15 m³ bucket	98	161	259
7080	5500 mm deep, 1.91 m³ bucket	140	192	330
7260	1800 mm wide, 1800 mm deep, 0.48 m³ bucket	44	59	103
7280	2400 mm deep, 0.57 m³ bucket	65	95	160
7300	3100 mm deep, 0.57 m³ bucket	86	136	222
7320	3700 mm deep, 0.96 m³ bucket	77.50	118	196
7360	4900 mm deep, 1.53 m³ bucket	127	167	294
7400	6100 mm deep, 2.67 m³ bucket	211	370	580
7440	7300 mm deep, 2.67 m³ bucket	405	725	1,125
7620	2400 mm wide, 3700 mm deep, 0.96 m³ bucket	99.50	134	234
7650	4600 mm deep, 1.15 m³ bucket	133	191	325
7680	5500 mm deep, 1.91 m³ bucket	175	220	395
7740	7300 mm deep, 2.67 m³ bucket	435	755	1,200
7920	3100 mm wide, 6100 mm deep, 2.67 m³ bucket	254	410	665
7940	7300 mm deep, 2.67 m³ bucket	460	785	1,250

BUILDING SITEWORK G

Important: See the Reference Section for critical supporting data - Reference Numbers and City Cost Indexes

G1030 Site Earthwork

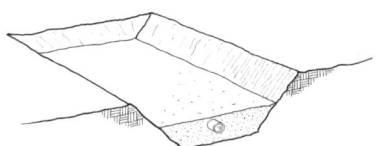

The Pipe Bedding System is shown for various pipe diameters. Compacted bank sand is used for pipe bedding and to fill 300 mm over the pipe. No backfill is included. Various side slopes are shown to accommodate different soil conditions. Pipe sizes vary from 150 mm to 2125 mm diameter.

System Components			COST PER m		
	QUANTITY	UNIT	MAT.	INST.	TOTAL
SYSTEM G1030 815 1440					
PIPE BEDDING, SIDE SLOPE 0 TO 1, 300 mm WIDE, PIPE SIZE 150 mm DIAMETER					
Borrow, bank sand, 3 km haul, machine spread	.168	m³	1.02	1.23	2.25
Compaction, vibrating plate	.168	m³		.40	.40
TOTAL			1.02	1.63	2.65

G1030 815	Pipe Bedding	COST PER m		
		MAT.	INST.	TOTAL
1440	Pipe bedding, side slope 0 to 1, 300 mm wide, pipe size 150 mm diameter	1.02	1.63	2.65
1460	600 mm wide, pipe size 200 mm diameter	2.25	3.63	5.88
1480	Pipe size 250 mm diameter	2.31	3.72	6.03
1500	Pipe size 300 mm diameter	2.37	3.82	6.19
1520	900 mm wide, pipe size 350 mm diameter	3.90	6.30	10.20
1540	Pipe size 375 mm diameter	3.95	6.35	10.30
1560	Pipe size 400 mm diameter	3.99	6.45	10.44
1580	Pipe size 450 mm diameter	4.08	6.60	10.68
1600	1200 mm wide, pipe size 500 mm diameter	5.90	9.50	15.40
1620	Pipe size 525 mm diameter	5.95	9.60	15.55
1640	Pipe size 600 mm diameter	6.10	9.80	15.90
1660	Pipe size 750 mm diameter	6.25	10.10	16.35
1680	1800 mm wide, pipe size 800 mm diameter	10.90	17.60	28.50
1700	Pipe size 900 mm diameter	11.20	18.05	29.25
1720	2100 mm wide, pipe size 1200 mm diameter	14.60	23.50	38.10
1740	2400 mm wide, pipe size 1500 mm diameter	18.20	29.50	47.70
1760	3100 mm wide, pipe size 1825 mm diameter	26.50	42.50	69
1780	3700 mm wide, pipe size 2125 mm diameter	35.50	57.50	93
2140	Side slope 1/2 to 1, 300 mm wide, pipe size 150 mm diameter	2.14	3.46	5.60
2160	600 mm wide, pipe size 200 mm diameter	3.55	5.75	9.30
2180	Pipe size 250 mm diameter	3.85	6.20	10.05
2200	Pipe size 300 mm diameter	4.11	6.65	10.76
2220	900 mm wide, pipe size 350 mm diameter	5.90	9.55	15.45
2240	Pipe size 375 mm diameter	6.05	9.80	15.85
2260	Pipe size 400 mm diameter	6.25	10.10	16.35
2280	Pipe size 450 mm diameter	6.60	10.65	17.25
2300	1200 mm wide, pipe size 500 mm diameter	8.70	14.05	22.75
2320	Pipe size 525 mm diameter	8.90	14.40	23.30
2340	Pipe size 600 mm diameter	9.55	15.40	24.95
2360	Pipe size 750 mm diameter	10.75	17.30	28.05
2380	1800 mm wide, pipe size 800 mm diameter	15.80	25.50	41.30
2400	Pipe size 900 mm diameter	16.90	27.50	44.40
2420	2100 mm wide, pipe size 1200 mm diameter	23	37.50	60.50
2440	2400 mm wide, pipe size 1500 mm diameter	30	48.50	78.50
2460	3100 mm wide, pipe size 1825 mm diameter	42	68	110
2480	3700 mm wide, pipe size 2125 mm diameter	56	90.50	146.50
2620	Side slope 1 to 1, 300 mm wide, pipe size 150 mm diameter	3.26	5.25	8.51
2640	600 mm wide, pipe size 200 mm diameter	4.90	7.90	12.80

G1030 815	Pipe Bedding	COST PER m		
		MAT.	INST.	TOTAL
2660	Pipe size 250 mm diameter	5.35	8.65	14
2680	Pipe size 300 mm diameter	5.90	9.55	15.45
2700	900 mm wide, pipe size 350 mm diameter	7.90	12.75	20.65
2720	Pipe size 375 mm diameter	8.20	13.20	21.40
2740	Pipe size 400 mm diameter	8.50	13.70	22.20
2760	Pipe size 450 mm diameter	9.10	14.70	23.80
2780	1200 mm wide, pipe size 500 mm diameter	11.55	18.60	30.15
2800	Pipe size 525 mm diameter	11.90	19.20	31.10
2820	Pipe size 600 mm diameter	13	21	34
2840	Pipe size 750 mm diameter	15.20	24.50	39.70
2860	1800 mm wide, pipe size 800 mm diameter	20.50	33.50	54
2880	Pipe size 900 mm diameter	22.50	36.50	59
2900	2100 mm wide, pipe size 1200 mm diameter	31.50	51	82.50
2920	2400 mm wide, pipe size 1500 mm diameter	42	67.50	109.50
2940	3100 mm wide, pipe size 1825 mm diameter	58	93.50	151.50
2960	3700 mm wide, pipe size 2125 mm diameter	76.50	123	199.50
3000	Side slope 1-1/2 to 1, 300 mm wide, pipe size 150 mm diameter	4.39	7.10	11.49
3020	600 mm wide, pipe size 200 mm diameter	6.20	10	16.20
3040	Pipe size 250 mm diameter	6.90	11.15	18.05
3060	Pipe size 300 mm diameter	7.65	12.30	19.95
3080	900 mm wide, pipe size 350 mm diameter	9.90	15.95	25.85
3100	Pipe size 375 mm diameter	10.30	16.60	26.90
3120	Pipe size 400 mm diameter	10.75	17.30	28.05
3140	Pipe size 450 mm diameter	11.65	18.85	30.50
3160	1200 mm wide, pipe size 500 mm diameter	14.40	23.50	37.90
3180	Pipe size 525 mm diameter	14.85	24	38.85
3200	Pipe size 600 mm diameter	16.45	26.50	42.95
3220	Pipe size 750 mm diameter	19.75	32	51.75
3240	1800 mm wide, pipe size 800 mm diameter	25.50	41.50	67
3260	Pipe size 900 mm diameter	28.50	45.50	74
3280	2100 mm wide, pipe size 1200 mm diameter	40	64.50	104.50
3300	2400 mm wide, pipe size 1500 mm diameter	54	86.50	140.50
3320	3100 mm wide, pipe size 1825 mm diameter	73.50	119	192.50
3340	3700 mm wide, pipe size 2125 mm diameter	96.50	156	252.50
3400	Side slope 2 to 1, 300 mm wide, pipe size 150 mm diameter	5.60	9.05	14.65
3420	600 mm wide, pipe size 200 mm diameter	7.55	12.20	19.75
3440	Pipe size 250 mm diameter	8.40	13.55	21.95
3460	Pipe size 300 mm diameter	9.40	15.15	24.55
3480	900 mm wide, pipe size 350 mm diameter	11.90	19.20	31.10
3500	Pipe size 375 mm diameter	12.35	19.95	32.30
3520	Pipe size 400 mm diameter	13	21	34
3540	Pipe size 450 mm diameter	14.20	23	37.20
3560	1200 mm wide, pipe size 500 mm diameter	17.20	28	45.20
3580	Pipe size 525 mm diameter	17.80	28.50	46.30
3600	Pipe size 600 mm diameter	19.85	32	51.85
3620	Pipe size 750 mm diameter	24	39.50	63.50
3640	1800 mm wide, pipe size 800 mm diameter	30.50	49	79.50
3660	Pipe size 900 mm diameter	34	55	89
3680	2100 mm wide, pipe size 1200 mm diameter	48.50	78.50	127
3700	2400 mm wide, pipe size 1500 mm diameter	65.50	106	171.50
3720	3100 mm wide, pipe size 1825 mm diameter	89.50	144	233.50
3740	3700 mm wide, pipe size 2125 mm diameter	117	189	306

G2030 Pedestrian Paving

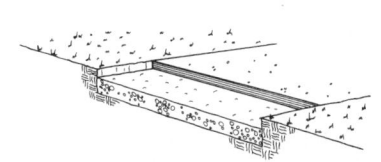

The Bituminous Sidewalk System includes: excavation; compacted gravel base (hand graded), bituminous surface; and hand grading along the edge of the completed walk.

The Expanded System Listing shows Bituminous Sidewalk systems with wearing course depths ranging from 25 mm to 60 mm of bituminous material. The gravel base ranges from 100 mm to 200 mm. Sidewalk widths are shown ranging from 900 mm to 1500 mm. Costs are on a per meter basis.

System Components	QUANTITY	UNIT	COST PER m		
			MAT.	INST.	TOTAL
SYSTEM G2030 110 1580					
BITUMINOUS SIDEWALK, 25 mm THICK PAVING, 100 mm GRAVEL BASE, 900 mm WIDTH					
Excavation, bulk, dozer, push 15 m	.115	m³		.24	.24
Borrow, bank run gravel, haul 3.2 km, spread w/dozer, no compaction	.093	m³	2.41	.68	3.09
Compact w/ vib plate, 200 mm lifts	.093	m³		.23	.23
Fine grade, area to be paved, small area	.913	m²		7.31	7.31
Sidewalk, bituminous, no base, 25 mm thick	.913	m²	2.41	1.74	4.15
Backfill by hand, no compaction, light soil	.015	m³		.48	.48
TOTAL			4.82	10.68	15.50

G2030 110	Bituminous Sidewalks	COST PER m		
		MAT.	INST.	TOTAL
1580	Bituminous sidewalk, 25 mm thick paving, 100 mm gravel base, 900 mm width	4.82	10.65	15.47
1600	1200 mm width	6.40	11.60	18
1620	1500 mm width	8.15	12.20	20.35
1640	150 mm gravel base, 900 mm width	6.05	11.20	17.25
1660	1200 mm width	8.05	12.35	20.40
1680	1500 mm width	10.05	13.50	23.55
1700	200 mm gravel base, 900 mm width	7.25	11.80	19.05
1720	1200 mm width	9.65	13.10	22.75
1740	1500 mm width	12.10	14.45	26.55
1800	40 mm thick paving, 100 mm gravel base, 900 mm width	6.15	11.15	17.30
1820	1200 mm width	8.15	12.45	20.60
1840	1500 mm width	10.05	14.35	24.40
1860	150 mm gravel base, 900 mm width	7.40	11.75	19.15
1880	1200 mm width	9.65	13.80	23.45
1900	1500 mm width	12.20	14.65	26.85
1920	200 mm gravel base, 900 mm width	8.45	12.90	21.35
1940	1200 mm width	11.40	13.90	25.30
1960	1500 mm width	14.10	16.15	30.25
2120	50 mm thick paving, 100 mm gravel base, 900 mm width	7.20	12.90	20.10
2140	1200 mm width	9.60	14.45	24.05
2160	1500 mm width	12.05	16.10	28.15
2180	150 mm gravel base, 900 mm width	8.45	13.50	21.95
2200	1200 mm width	11.20	15.25	26.45
2220	1500 mm width	14.10	16.95	31.05
2240	200 mm gravel base, 900 mm width	9.65	14.05	23.70
2260	1200 mm width	12.85	16	28.85
2280	1500 mm width	16.10	17.90	34
2400	60 mm thick paving, 100 mm gravel base, 900 mm width	10.05	14.45	24.50
2420	1200 mm width	13.40	16.40	29.80
2440	1500 mm width	16.90	18.50	35.40
2460	150 mm gravel base, 900 mm width	11.30	15	26.30
2480	1200 mm width	15.05	17.20	32.25
2500	1500 mm width	18.95	19.40	38.35
2520	200 mm gravel base, 900 mm width	12.50	15.55	28.05
2540	1200 mm width	16.65	17.95	34.60
2560	1500 mm width	21	20.50	41.50

BUILDING SITEWORK

G

G2030 Pedestrian Paving

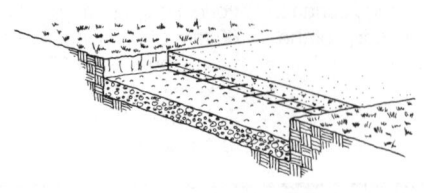

The Concrete Sidewalk System includes: excavation; compacted gravel base (hand graded); forms; welded wire fabric; and 21 MPa air-entrained concrete (broom finish).

The Expanded System Listing shows Concrete Sidewalk systems with wearing course depths ranging from 100 mm to 150 mm. The gravel base ranges from 100 mm to 200 mm. Sidewalk widths are shown ranging from 900 mm to 1500 mm. Costs are on a per meter basis.

System Components	QUANTITY	UNIT	COST PER m		
			MAT.	INST.	TOTAL
SYSTEM G2030 120 1580					
CONCRETE, SIDEWALK 100 mm THICK, 100 mm GRAVEL BASE, 900 mm WIDE					
Excavation, box out with dozer	.251	m³		.53	.53
Gravel base, haul 3.2 km, spread with dozer	.093	m³	2.41	.68	3.09
Compaction with vibrating plate	.093	m³		.23	.23
Fine grade by hand	.913	m²		7.41	7.41
Concrete in place including forms and reinforcing	.093	m³	15.73	19.66	35.39
Backfill edges by hand	.025	m³		.80	.80
TOTAL			18.14	29.31	47.45

G2030 120	Concrete Sidewalks	COST PER m		
		MAT.	INST.	TOTAL
1580	Concrete sidewalk, 100 mm thick, 100 mm gravel base, 900 mm wide	18.15	29	47.15
1600	1200 mm wide	24	36.50	60.50
1620	1500 mm wide	30.50	43.50	74
1640	150 mm gravel base, 900 mm wide	19.40	30	49.40
1660	1200 mm wide	26	37	63
1680	1500 mm wide	32.50	44.50	77
1700	200 mm gravel base, 900 mm wide	20.50	30.50	51
1720	1200 mm wide	27.50	38	65.50
1740	1500 mm wide	34	45.50	79.50
1800	130 mm thick concrete, 100 mm gravel base, 900 mm wide	23.50	31.50	55
1820	1200 mm wide	31	39.50	70.50
1840	1500 mm wide	39	47	86
1860	150 mm gravel base, 900 mm wide	24.50	32.50	57
1900	1500 mm wide	41	46.50	87.50
1920	200 mm gravel base, 900 mm wide	26	32.50	58.50
1940	1200 mm wide	34.50	43	77.50
1960	1500 mm wide	43	47.50	90.50
2120	150 mm thick concrete, 100 mm gravel base, 900 mm wide	26.50	33	59.50
2140	1200 mm wide	35.50	41.50	77
2160	1500 mm wide	44.50	49.50	94
2180	150 mm gravel base, 900 mm wide	28	34	62
2200	1200 mm wide	37	42	79
2220	1500 mm wide	46.50	50.50	97
2240	200 mm gravel base, 900 mm wide	29	34	63
2260	1200 mm wide	39	43	82
2280	1500 mm wide	48.50	51.50	100

Important: See the Reference Section for critical supporting data - Reference Numbers and City Cost Indexes

G2040 Site Development

There are four basic types of Concrete Retaining Wall Systems: reinforced concrete with level backfill; reinforced concrete with sloped backfill or surcharge; unreinforced with level backfill; and unreinforced with sloped backfill or surcharge. System elements include: all necessary forms (4 uses); 21 MPa concrete with an 200 mm chute; all necessary reinforcing steel; and underdrain. Exposed concrete is patched and rubbed.

The Expanded System Listing shows walls that range in thickness from 250 mm to 4600 mm for reinforced concrete walls with level backfill and 300 mm to 600 mm for reinforced walls with sloped backfill. Walls range from a height of 1200 mm to 6100 mm. Unreinforced level and sloped backfill walls range from a height of 900 mm to 3000 mm.

System Components	QUANTITY	UNIT	COST PER m		
			MAT.	INST.	TOTAL
SYSTEM G2040 210 1000					
CONC.RETAIN. WALL REINFORCED, LEVEL BACKFILL 1200 mm HIGH					
Forms in place, cont. wall footing & keyway, 4 uses	.610	m²	6.16	22.56	28.72
Forms in place, retaining wall forms, battered to 2400 mm high, 4 uses	2.438	m²CA	15.48	173.13	188.61
Reinforcing in place, walls, #10M to #25M	.012	Met. Ton	11.73	9.11	20.84
Concrete ready mix, regular weight, 21 MPa	.512	m³	63.96		63.96
Placing concrete and vibrating footing con., shallow direct chute	.186	m³		4.45	4.45
Placing concrete and vibrating walls, 200 mm thick, direct chute	.326	m³		10.33	10.33
Pipe bedding, crushed or screened bank run gravel	1.000	m	8.14	3.26	11.40
Pipe, subdrainage, corrugated plastic, 100 mm diameter	1.000	m	.62	.51	1.13
Finish walls and break ties, patch walls	1.219	m²	.39	9.88	10.27
TOTAL			106.48	233.23	339.71

G2040 210	Concrete Retaining Walls	COST PER m		
		MAT.	INST.	TOTAL
1000	Conc.retain.wall,reinforced,level backfill,1200mm high x 650mm, 250mm thick	106	233	339
1200	1800 mm high x 975 mm base, 250 mm thick	158	340	498
1400	2400 mm high x 1300 mm base, 250 mm thick	207	445	652
1600	3100 mm high x 1600 mm base, 325 mm thick	274	655	929
1800	3700 mm high x 2000 mm base, 350 mm thick	340	785	1,125
2200	4900 mm high x 2600 mm base, 400 mm thick	515	1,050	1,565
2600	6100 mm high x 3200 mm base, 450 mm thick	760	1,375	2,135
3000	Sloped backfill, 1200 mm high x 950 mm base, 300 mm thick	132	244	376
3200	1800 mm high x 1400 mm base, 300 mm thick	188	350	538
3400	2400 mm high x 1800 mm base, 300 mm thick	252	460	712
3600	3100 mm high x 2700 mm base, 400 mm thick	365	690	1,055
3800	3700 mm high x 2700 mm base, 450 mm thick	475	835	1,310
4200	4900 mm high x 3600 mm base, 525 mm thick	795	1,175	1,970
4600	6000 mm high x 4600 mm base, 600 mm thick	1,250	1,600	2,850
5000	Unreinforced, level backfill, 900 mm high x 450 mm base	60	158	218
5200	1200 mm high x 600 mm base	93	210	303
5400	1800 mm high x 900 mm base	174	325	499
5600	2400 mm high x 1200 mm base	277	430	707
5800	3100 mm high x 1500 mm base	415	675	1,090
7000	Sloped backfill, 900 mm high x 600 mm base	72	163	235
7200	1200 mm high x 900 mm base	120	218	338
7400	1800 mm high x 1500 mm base	251	345	596
7600	2400 mm high x 2100 mm base	425	470	895
7800	3100 mm high x 2700 mm base	655	730	1,385

BUILDING SITEWORK

G

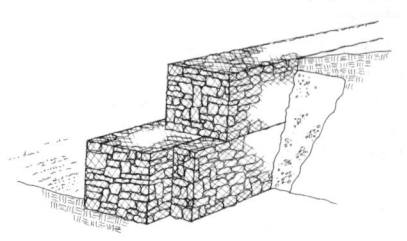

The Gabion Retaining Wall Systems list three types of surcharge conditions: level, sloped and highway, for two different facing configurations: stepped and straight with batter. Costs are expressed per meter for heights ranging from 1800 mm to 5500 mm for retaining sandy or clay soil. For protection against sloughing in wet clay materials counterforts have been added for these systems. Drainage stone has been added behind the walls to avoid any additional lateral pressures.

System Components	QUANTITY	UNIT	COST PER m		
			MAT.	INST.	TOTAL
SYSTEM G2040 270 1000					
GABION RET. WALL, LEVEL BACKFILL, STEPPED FACE, 1200 mm BASE, 1800 mm HIGH					
900 mm x 900 mm cross section gabion	6.562	Ea.	162.58	400.26	562.84
900 mm x 300 mm cross section gabion	3.281	Ea.	36.27	23.98	60.25
Crushed stone drainage	.552	m³	25.08	4.61	29.69
TOTAL			223.93	428.85	652.78

G2040 270	Gabion Retaining Walls	COST PER m		
		MAT.	INST.	TOTAL
1000	Gabion ret. wall, level backfill, stepped face, 1200 mm base, 1800 mm high,	224	430	654
1040	Clay soil with counterforts @ 4900 mm O.C.	345	880	1,225
1100	1500 mm base, 2700 mm high, sandy soil	390	680	1,070
1140	Clay soil with counterforts @ 4900 mm O.C.	660	1,350	2,010
1200	1800 mm base, 3700 mm high, sandy soil	565	1,075	1,640
1240	Clay soil with counterforts @ 4900 mm O.C.	1,050	2,875	3,925
1300	2300 mm base, 4600 mm high, sandy soil	790	1,500	2,290
1340	Clay soil with counterforts @ 4900 mm O.C.	1,900	4,025	5,925
1400	2700 mm base, 5500 mm high, sandy soil	1,050	2,125	3,175
1440	Clay soil with counterforts @ 4900 mm O.C.	2,150	4,825	6,975
2000	Straight face w/1:6 batter, 900 mm base, 1800 mm high, sandy soil	213	410	623
2040	Clay soil with counterforts @ 4900 mm O.C.	335	860	1,195
2100	1400 mm base, 2700 mm high, sandy soil	360	670	1,030
2140	Clay soil with counterforts @ 4900 mm O.C.	635	1,350	1,985
2200	1800 mm base, 3700 mm high, sandy soil	545	1,075	1,620
2240	Clay soil with counterforts @ 4900 mm O.C.	1,025	2,875	3,900
2300	2300 mm base, 4600 mm high, sandy soil	775	1,550	2,325
2340	Clay soil with counterforts @ 4900 mm O.C.	1,525	3,175	4,700
2400	5500 mm high, sandy soil	1,000	2,000	3,000
2440	Clay soil with counterforts @ 4900 mm O.C.	1,500	2,300	3,800
3000	Backfill sloped 1-1/2:1, stepped face, 1400mm base, 1800mm high, sandy soil	235	460	695
3040	Clay soil with counterforts @ 4900 mm O.C.	355	910	1,265
3100	1800 mm base, 2700 mm high, sandy soil	410	860	1,270
3140	Clay soil with counterforts @ 4900 mm O.C.	680	1,525	2,205
3200	2300 mm base, 3700 mm high, sandy soil	635	1,325	1,960
3240	Clay soil with counterforts @ 4900 mm O.C.	1,125	3,125	4,250
3300	2700 mm base, 4600 mm high, sandy soil	885	1,925	2,810
3340	Clay soil with counterforts @ 4900 mm O.C.	2,000	4,450	6,450
3400	3200 mm base, 5500 mm high, sandy soil	1,200	2,600	3,800
3440	Clay soil with counterforts @ 4900 mm O.C.	2,300	5,275	7,575
4000	Straight face with 1:6 batter, 1400 mm base, 1800 mm high, sandy soil	261	465	726
4040	Clay soil with counterforts @ 4900 mm O.C.	380	910	1,290
4100	1800 mm base, 2700 mm high, sandy soil	445	870	1,315
4140	Clay soil with counterforts @ 4900 mm O.C.	720	1,550	2,270

Important: See the Reference Section for critical supporting data - Reference Numbers and City Cost Indexes

BUILDING SITEWORK G

G20 Site Improvements

G2040 Site Development

G2040 270	Gabion Retaining Walls	COST PER m		
		MAT.	INST.	TOTAL
4200	2300 mm base, 3700 mm high, sandy soil	675	1,325	2,000
4240	Clay soil with counterforts @ 4900 mm O.C.	1,150	3,125	4,275
4300	2700 mm base, 4600 mm high, sandy soil	940	1,925	2,865
4340	Clay soil with counterforts @ 4900 mm O.C.	2,050	4,450	6,500
4400	5500 mm high, sandy soil	1,200	2,525	3,725
4440	Clay soil with counterforts @ 4900 mm O.C.	2,300	5,250	7,550
5000	Highway surcharge, straight face, 1800 mm base, 1800 mm high, sandy soil	350	805	1,155
5040	Clay soil with counterforts @ 4900 mm O.C.	470	1,250	1,720
5100	2700 mm base, 2700 mm high, sandy soil	605	1,400	2,005
5140	Clay soil with counterforts @ 4900 mm O.C.	880	2,075	2,955
5200	3700 mm high, sandy soil	1,325	2,100	3,425
5240	Clay soil with counterforts @ 4900 mm O.C.	1,725	5,200	6,925
5300	3700 mm base, 4600 mm high, sandy soil	1,200	2,825	4,025
5340	Clay soil with counterforts @ 4900 mm O.C.	1,950	4,475	6,425
5400	5500 mm high, sandy soil	1,550	3,625	5,175
5440	Clay soil with counterforts @ 4900 mm O.C.	2,625	6,325	8,950

G3030 Storm Sewer

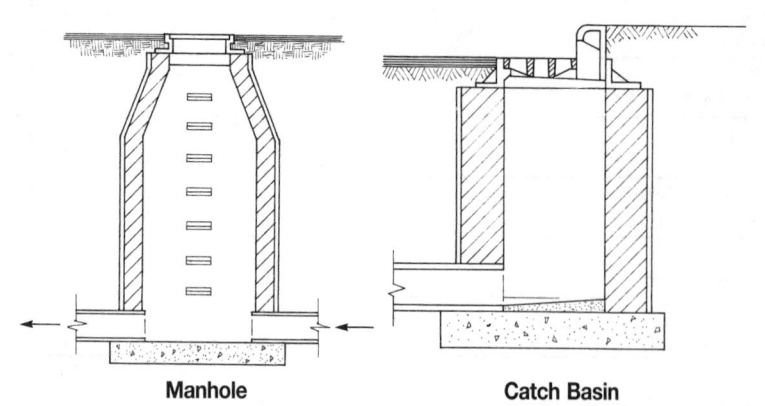

Manhole **Catch Basin**

The Manhole and Catch Basin System includes: excavation with a backhoe; a formed concrete footing; frame and cover; cast iron steps and compacted backfill.

The Expanded System Listing shows manholes that have a 1200 mm, 1500 mm and 1800 mm inside diameter riser. Depths range from 1200 mm to 4300 mm. Construction material shown is either concrete, concrete block, precast concrete, or brick.

BUILDING SITEWORK G

System Components	QUANTITY	UNIT	COST PER EACH		
			MAT.	INST.	TOTAL
SYSTEM G3030 210 1920					
MANHOLE/CATCH BASIN, BRICK, 1200 mm I.D. RISER, 1200 mm DEEP					
Excavation, hydraulic backhoe, 0.29 m³ bucket	11.327	m³		87.44	87.44
Trim sides and bottom of excavation	5.946	m²		46.50	46.50
Forms in place, manhole base, 4 uses	1.858	m²CA	13.01	80.83	93.84
Reinforcing in place footings, #10M to #25M	.017	Met. Ton	16.98	18.96	35.94
Concrete, 3000 psi	.707	m³	88.40		88.40
Place and vibrate concrete, footing, direct chute	.707	m³		36.86	36.86
Catch basin or MH, brick, 1200 mm ID, 1200 mm deep	1.000	Ea.	395	785	1,180
Catch basin or MH steps; heavy galvanized cast iron	1.000	Ea.	15.35	11.15	26.50
Catch basin or MH frame and cover	1.000	Ea.	246	174	420
Fill, granular	9.904	m³	114.39		114.39
Backfill, spread with wheeled front end loader	9.904	m³		23.37	23.37
Backfill compaction, 300 mm lifts, air tamp	9.904	m³		86.07	86.07
TOTAL			889.13	1,350.18	2,239.31

G3030 210	Manholes & Catch Basins	COST PER EACH		
		MAT.	INST.	TOTAL
1920	Manhole/catch basin, brick, 1200 mm I.D. riser, 1200 mm deep	890	1,350	2,240
1940	1800 mm deep	1,175	1,900	3,075
1960	2400 mm deep	1,500	2,575	4,075
1980	3100 mm deep	2,000	3,175	5,175
3000	3700 mm deep	2,625	3,450	6,075
3020	4300 mm deep	3,300	4,800	8,100
3200	Block, 1200 mm I.D. riser, 1200 mm deep	825	1,075	1,900
3220	1800 mm deep	1,075	1,550	2,625
3240	2400 mm deep	1,325	2,125	3,450
3260	3100 mm deep	1,550	2,600	4,150
3280	3700 mm deep	1,900	3,300	5,200
3300	4300 mm deep	2,275	4,050	6,325
4620	Concrete, cast-in-place, 1200 mm I.D. riser, 1200 mm deep	1,025	1,875	2,900
4640	1800 mm deep	1,375	2,525	3,900
4660	2400 mm deep	1,850	3,650	5,500
4680	3100 mm deep	2,225	4,525	6,750
4700	3700 mm deep	2,675	5,600	8,275
4720	4300 mm deep	3,200	6,700	9,900
5820	Concrete, precast, 1200 mm I.D. riser, 1200 mm deep	1,275	995	2,270
5840	1800 mm deep	1,625	1,350	2,975

Important: See the Reference Section for critical supporting data - Reference Numbers and City Cost Indexes

G3030 Storm Sewer

G3030 210	Manholes & Catch Basins	COST PER EACH		
		MAT.	INST.	TOTAL
5860	2400 mm deep	1,950	1,875	3,825
5880	3100 mm deep	2,400	2,325	4,725
5900	3700 mm deep	2,950	2,850	5,800
5920	4300 mm deep	3,550	3,600	7,150
6000	1500 mm I.D. riser, 1200 mm deep	1,375	1,100	2,475
6020	1800 mm deep	1,800	1,525	3,325
6040	2400 mm deep	2,250	2,025	4,275
6060	3100 mm deep	2,825	2,575	5,400
6080	3700 mm deep	3,450	3,275	6,725
6100	4300 mm deep	4,125	4,000	8,125
6200	1800 mm I.D. riser, 1200 mm deep	1,975	1,425	3,400
6220	1800 mm deep	2,575	1,900	4,475
6240	2400 mm deep	3,150	2,625	5,775
6260	3100 mm deep	3,950	3,350	7,300
6280	3700 mm deep	4,800	4,200	9,000
6300	4350 mm deep	5,700	5,100	10,800

G3030 Storm Sewer

The Headwall Systems are listed in concrete and two different stone wall materials for two different backfill slope conditions. The backfill slope directly affects the length of the wing walls. Walls are listed for eight different culvert sizes from 760 mm to 2100 mm diameter. For pipes less than 760 mm diameter refer to the Unit Price section for reinforced concrete culvert with flared ends. Excavation and backfill are included in the system components, and are figured from an elevation 600 mm below the bottom of the pipe.

System Components	QUANTITY	UNIT	COST PER EACH		
			MAT.	INST.	TOTAL
SYSTEM G3030 310 2000					
HEADWALL C.I.P. CONCRETE FOR 750 mm PIPE, 900 mm LONG WING WALLS					
Excavation, hydraulic backhoe, 0.29 m³ bucket	1.911	m³		165.27	165.27
Formwork, 2 uses	14.586	m²CA	313.09	1,456.26	1,769.35
Reinforcing in place including dowels	20.412	kg	23.47	64.50	87.97
Concrete, 21 MPa, in place	1.988	m³	248.48		248.48
Place concrete, spread footings, direct chute	1.988	m³		103.61	103.61
Backfill, dozer	1.911	m³		37.89	37.89
TOTAL			585.04	1,827.53	2,412.57

G3030 310	Headwalls	COST PER EACH		
		MAT.	INST.	TOTAL
2000	Headwall,1-1/2 to 1 slope soil,C.I.P. conc,750mm pipe,900mm long wing walls	585	1,825	2,410
2020	Pipe size 900 mm, 1100 mm long wing walls	730	2,200	2,930
2040	Pipe size 1050 mm, 1200 mm long wing walls	885	2,575	3,460
2060	Pipe size 1200 mm, 1400 mm long wing walls	1,075	3,000	4,075
2080	Pipe size 1350 mm, 1500 mm long wing walls	1,275	3,500	4,775
2100	Pipe size 1500 mm, 1700 mm long wing walls	1,475	3,975	5,450
2120	Pipe size 1825 mm, 2000 mm long wing walls	1,975	5,100	7,075
2140	Pipe size 2125 mm, 2300 mm long wing walls	2,500	6,300	8,800
3000	$320/met. ton stone, pipe size 750 mm, 900 mm long wing walls	1,850	665	2,515
3020	Pipe size 900 mm, 1100 mm long wing walls	2,375	795	3,170
3040	Pipe size 1050 mm, 1200 mm long wing walls	2,975	945	3,920
3060	Pipe size 1200 mm, 1400 mm long wing walls	3,675	1,125	4,800
3080	Pipe size 1350 mm, 1500 mm long wing walls	4,425	1,325	5,750
3100	Pipe size 1500 mm, 1700 mm long wing walls	5,250	1,525	6,775
3120	Pipe size 1825 mm, 2000 mm long wing walls	7,250	2,025	9,275
3140	Pipe size 2125 mm, 2300 mm long wing walls	9,550	2,600	12,150
4500	2 to 1 slope soil,C.I.P. concrete,pipe size 750 mm,1300 mm long wing walls	700	2,125	2,825
4520	Pipe size 900 mm, 1500 mm long wing walls	885	2,575	3,460
4540	Pipe size 1050 mm, 1800 mm long wing walls	1,100	3,125	4,225
4560	Pipe size 1200 mm, 2000 mm long wing walls	1,325	3,600	4,925
4580	Pipe size 1350 mm, 2200 mm long wing walls	1,575	4,200	5,775
4600	Pipe size 1500 mm, 2400 mm long wing walls	1,825	4,825	6,650
4620	Pipe size 1825 mm, 2900 mm long wing walls	2,450	6,250	8,700
4640	Pipe size 2125 mm, 3400 mm long wing walls	3,175	7,800	10,975
5500	$320/met. ton stone, pipe size 750 mm, 1300 mm long wing walls	2,225	760	2,985
5520	Pipe size 900 mm, 1500 mm long wing walls	2,925	935	3,860
5540	Pipe size 1050 mm, 1800 mm long wing walls	3,725	1,150	4,875
5560	Pipe size 1200 mm, 2000 mm long wing walls	4,600	1,350	5,950

Important: See the Reference Section for critical supporting data - Reference Numbers and City Cost Indexes

G3030 Storm Sewer

G3030 310	Headwalls	COST PER EACH		
		MAT.	INST.	TOTAL
5580	Pipe size 1350 mm, 2200 mm long wing walls	5,575	1,600	7,175
5600	Pipe size 1500 mm, 2400 mm long wing walls	6,625	1,875	8,500
5620	Pipe size 1825 mm, 2900 mm long wing walls	9,175	2,500	11,675
5640	Pipe size 2125 mm, 3400 mm long wing walls	12,200	3,275	15,475

For information about Means Estimating Seminars, see yellow pages 12 and 13 in back of book

BUILDING SITEWORK

G

Reference Section

All the reference information is in one section, making it easy to find what you need to know . . . and easy to use the book on a daily basis. This section is visually identified by a vertical gray bar on the page edges.

In this Reference Section, we've included Equipment Rental Costs, a listing of rental and operating costs; Crew Listings, a full listing of all crews and equipment, and their costs; Historical Cost Indexes for cost comparisons over time; City Cost Indexes and Location Factors for adjusting costs to the region you are in; Reference Tables, where you will find explanations, estimating information and procedures, or technical data; Change Orders, information on pricing changes to contract documents; Square Meter Costs that allow you to make a rough estimate for the overall cost of a project; Metric Units used in the construction trades; and an explanation of all the Abbreviations in the book.

Table of Contents

Construction Equipment Rental Costs

Estimating Tips

- This section contains the average costs to rent and operate hundreds of pieces of construction equipment. This is useful information when estimating the time and material requirements of any particular operation in order to establish a unit or total cost. Equipment costs include not only rental, but also operating costs for equipment under normal use.

Rental Costs

- Equipment rental rates are obtained from industry sources throughout North America - contractors, suppliers, dealers, manufacturers, and distributors.
- Rental rates vary throughout the country with larger cities generally having lower rates. Lease plans for new equipment are available for periods in excess of six months with a percentage of payments applying toward purchase.
- Monthly rental rates vary from 2% to 5% of the purchase price of the equipment depending on the anticipated life of the equipment and its wearing parts.
- Weekly rental rates are about 1/3 the monthly rates, and daily rental rates are about 1/3 the weekly rate.

Operating Costs

- The operating costs include parts and labor for routine servicing such as repair and replacement of pumps, filters and worn lines. Normal operating expendables such as fuel, lubricants, tires and electricity (where applicable) are also included.
- Extraordinary operating expendables with highly variable wear patterns such as diamond bits and blades are excluded. These costs can be found as material costs in the unit price section.
- The hourly operating costs listed do not include the operator's wages.

Crew Equipment Cost/Day

- Any power equipment required by a crew is shown in the Crew Listings with a daily cost.
- The daily cost of crew equipment is based on dividing the weekly rental rate by 5 (number of working days in the week), and then adding the hourly operating cost times 8 (the number of hours in a day). This "Crew Equipment Cost/Day" is shown in the far right column of the Equipment Rental pages.
- If equipment is needed for only one or two days, it is best to develop your own cost by including components for daily rent and hourly operating cost. This is important when the listed Crew for a task does not contain the equipment needed, such as a crane for lifting mechanical heating/cooling equipment up onto a roof.

Mobilization/ Demobilization

- The cost to move construction equipment from an equipment yard or rental company to the jobsite and back again is not included in equipment rental costs listed in the Reference section, nor in the bare equipment cost of any Unit Cost line item, nor in any equipment costs shown in the Crew listings.
- Mobilization (to the site) and demobilization (from the site) costs can be found in the Unit Cost section.
- If a piece of equipment is already at the jobsite, it is not appropriate to utilize mob/demob costs again in an estimate.

01 54 33 | Rental Equipment

		UNIT	HOURLY OPER. COST	RENT PER DAY	RENT PER WEEK	RENT PER MONTH	CREW EQUIPMENT COST/DAY	
10	0010	**CONCRETE EQUIPMENT RENTAL** without operators	R015433 -10						**10**
	0150	For batch plant, see div. 01590-500							
	0200	Bucket, concrete lightweight, 0.382 m³	Ea.	.55	16	48	144	14	
	0300	0.764 m³	R033105 -70	.60	19	57	171	16.20	
	0400	1.146 m³		.75	26	78	234	21.60	
	0500	1.528 m³		.85	30.50	92	276	25.20	
	0580	6 m³		4.55	205	615	1,850	159.40	
	0600	Cart, concrete, operator walking, 0.28 m³		2.15	58.50	175	525	52.20	
	0700	Operator riding, 0.504 m³		3.30	80	240	720	74.40	
	0800	Conveyer for concrete, portable, gas, 400 mm wide, 8 m long		7.30	117	350	1,050	128.40	
	0900	14 m long		7.65	142	425	1,275	146.20	
	1000	17 m long		7.80	150	450	1,350	152.40	
	1100	Core drill, electric, 1864 W, 25 mm to 200 mm bit diameter		2.00	81	243	730	64.60	
	1150	3.7 kW, 200 mm to 450 mm cores		6.61	104	312.80	940	115.45	
	1200	Finisher, concrete floor, gas, riding trowel, 1200 mm diameter		4.80	86.50	260	780	90.40	
	1300	Gas, manual, 3 blade, 900 mm trowel		1.00	16.65	50	150	18	
	1400	4 blade, 1200 mm trowel		1.45	21	63	189	24.20	
	1500	Float, hand-operated (Bull float) 1200 mm wide		.08	13.35	40	120	8.65	
	1570	Curb builder, 10.4 kw, gas, single screw		9.75	210	630	1,900	204	
	1590	Double screw		10.40	242	725	2,175	228.20	
	1600	Grinder, concrete and terrazzo, electric, floor		2.68	116	349	1,050	91.25	
	1700	Wall grinder		1.35	58.50	175	525	45.80	
	1800	Mixer, powered, mortar and concrete, gas, 0.17 m³, 13.4 kW		5.20	107	320	960	105.60	
	1900	0.28 m³, 18.6 kW		6.30	128	385	1,150	127.40	
	2000	0.45 m³		6.65	152	455	1,375	144.20	
	2100	Concrete, stationary, tilt drum, 1.5 m³		5.35	200	600	1,800	162.80	
	2120	Pump, concrete, truck mounted 100 mm line, 24 m boom		21.70	880	2,635	7,900	700.60	
	2140	125 mm line, 34 m boom		28.40	1,175	3,495	10,500	926.20	
	2160	Mud jack, 1.5 m³/hr		5.79	118	355	1,075	117.30	
	2180	6.3 m³/hr		7.56	158	473.20	1,425	155.10	
	2190	Shotcrete pump rig, 10 m³/Hr		11.80	233	700	2,100	234.40	
	2600	Saw, concrete, manual, gas, 13.4 kW		3.50	35	105	315	49	
	2650	Self-propelled, gas, 22.4 kW		6.75	88.50	265	795	107	
	2700	Vibrators, concrete, electric, 60 cycle, 1.4 kW		.37	8.35	25	75	7.95	
	2800	2.2 kW		.56	12.35	37	111	11.90	
	2900	Gas engine, 3.7 kW		.90	15	45	135	16.20	
	3000	6.0 kW		1.30	18.35	55	165	21.40	
	3050	Vibrating screed, gas engine, 6.0 kW		2.27	66.50	199	595	57.95	
	3100	Concrete transit mixer, hydraulic drive							
	3120	6 x 4, 185 kW, 6.1 m³, rear discharge		34.85	540	1,625	4,875	603.80	
	3200	Front discharge		40.95	675	2,020	6,050	731.60	
	3300	6 x 6, 200 kW, 9.2 m³, rear discharge		40.05	635	1,900	5,700	700.40	
	3400	Front discharge		41.85	680	2,045	6,125	743.80	
20	0010	**EARTHWORK EQUIPMENT RENTAL** without operators	R015433 -10						**20**
	0040	Aggregate spreader, push type 2.4 m to 3.7 m wide	Ea.	1.80	29.50	88	264	32	
	0045	Tailgate type, 7.4 m wide	"	1.75	31.50	95	285	33	
	0050	Augers for vertical drilling	R312323 -30						
	0055	Earth auger, truck-mounted, for fence & sign posts	Ea.	8.40	480	1,445	4,325	356.20	
	0060	For borings and monitoring wells	R312316 -40	30.95	640	1,920	5,750	631.60	
	0070	Earth auger, portable, trailer mounted		1.75	23	69	207	27.80	
	0075	Earth auger, truck-mounted, for caissons, water wells, utility poles	R312316 -45	160.90	3,475	10,460	31,400	3,379	
	0080	Auger, horizontal boring machine, 300 mm to 900 mm diameter, 34 kW		16.70	190	570	1,700	247.60	
	0090	300 mm to 1200 mm diameter, 48 kW		23.50	340	1,020	3,050	392	
	0095	Auger, for fence posts, gas engine, hand held		.35	4.33	13	39	5.40	
	0100	Excavator, diesel hydraulic, crawler mounted, 0.382 m³ cap.		16.05	345	1,035	3,100	335.40	
	0120	0.478 m³ capacity		19.55	465	1,395	4,175	435.40	
	0140	0.573 m³ capacity		22.85	490	1,475	4,425	477.80	

01 54 33 | Rental Equipment

			UNIT	HOURLY OPER. COST	RENT PER DAY	RENT PER WEEK	RENT PER MONTH	CREW EQUIPMENT COST/DAY
0150	0.764 m³ capacity		Ea.	27.40	570	1,705	5,125	560.20
0200	1.146 m³ capacity	R015433 -10		33.85	755	2,260	6,775	722.80
0300	1.528 m³ capacity			43.70	950	2,850	8,550	919.60
0320	1.91 m³ capacity	R312323 -30		56.20	1,275	3,840	11,500	1,218
0340	2.674 m³ capacity			95.20	2,100	6,305	18,900	2,023
0341	Attachments	R312316 -40						
0342	Bucket thumbs			2.55	208	625	1,875	145.40
0345	Grapples	R312316 -45		1.00	219	656	1,975	139.20
0350	Gradall type, truck mounted, 2.7 met. ton @ 4.5 m radius, 0.478 m³			38.75	885	2,655	7,975	841
0370	0.764 m³ capacity			44.65	1,025	3,085	9,250	974.20
0400	Backhoe-loader, 30 kW to 34 kW, 0.478 m³ capacity			8.60	183	550	1,650	178.80
0450	34 kW to 45 kW, 0.573 m³ capacity			10.90	232	695	2,075	226.20
0460	60 kW, 0.955 m³ capacity			13.95	277	830	2,500	277.60
0470	84 kW, 1.146 m³ backhoe			18.55	400	1,200	3,600	388.40
0480	Attachments							
0482	Compactor, 9000 kg			4.35	118	355	1,075	105.80
0485	Hydraulic hammer, 1 kJ			2.00	70	210	630	58
0486	Hydraulic hammer 1.45 kJ			4.20	133	400	1,200	113.60
0500	Brush chipper, gas engine, 150 mm cutter head, 26 kW			6.60	102	305	915	113.80
0550	300 mm cutter head, 97 kW			10.25	152	455	1,375	173
0600	380 mm cutter head, 123 kW			14.30	162	485	1,450	211.40
0750	Bucket, clamshell, general purpose, 0.287 m³			1.00	35	105	315	29
0800	0.382 m³			1.10	41.50	125	375	33.80
0850	0.573 m³			1.25	51.50	155	465	41
0900	0.764 m³			1.30	55	165	495	43.40
0950	1.146 m³			2.05	75	225	675	61.40
1000	1.528 m³			2.20	85	255	765	68.60
1010	Bucket, dragline, medium duty, 0.382 m³			.60	22.50	67	201	18.20
1020	0.573 m³			.60	23.50	71	213	19
1030	0.764 m³			.65	25.50	76	228	20.40
1040	1.146 m³			1.00	38.50	115	345	31
1050	1.528 m³			1.05	43.50	130	390	34.40
1070	2.292 m³			1.60	58.50	175	525	47.80
1200	Compactor, manually guided 2-drum vibratory smooth roller, 5.6 kW			4.85	145	435	1,300	125.80
1250	Rammer compactor, gas, 454 kg blow			1.75	38.50	115	345	37
1300	Vibratory plate, gas, 330 mm plate, 454 kg blow			1.65	22	66	198	26.40
1350	600 mm plate, 2268 kg blow			2.00	31	93	279	34.60
1370	Curb builder/extruder, 10.4 kW gas single screw			9.75	212	635	1,900	205
1390	Double screw			10.40	248	745	2,225	232.20
1500	Disc harrow attachment, for tractor			.37	61.50	185	555	39.95
1750	Extractor, piling, see lines 2500 to 2750							
1810	Feller buncher, shearing & accumulating trees, 75 kW		Ea.	22.60	520	1,555	4,675	491.80
1860	Grader, self-propelled, 11 000 kg			18.25	400	1,200	3,600	386
1910	14 000 kg			21.05	480	1,445	4,325	457.40
1920	18 000 kg			30.55	745	2,230	6,700	690.40
1930	25 000 kg			40.80	1,050	3,130	9,400	952.40
1950	Hammer, pavement demo., hyd., gas, self-prop., 450 to 570 kg			18.30	305	915	2,750	329.40
2000	Diesel 590 to 680 kg			28.20	610	1,830	5,500	591.60
2050	Pile driving hammer, steam or air, 5.6 kJ @ 225 BPM			6.60	272	815	2,450	215.80
2100	12 kJ @ 145 BPM			8.50	440	1,325	3,975	333
2150	20 kJ @ 60 BPM			8.85	475	1,425	4,275	355.80
2200	30 kJ @ 111 BPM			11.75	525	1,570	4,700	408
2250	Leads, 20 kJ hammers		m	.10	4.76	14.27	43	3.65
2300	30 kJ hammers and heavier		"	.16	8.10	24.28	73	6.15
2350	Diesel type hammer, 30 kJ		Ea.	25.50	605	1,820	5,450	568
2400	60 kJ			34.75	655	1,970	5,900	672
2450	190 kJ			56.45	1,125	3,380	10,100	1,128
2500	Vib. elec. hammer/extractor, 200 kW diesel generator, 25 kW			28.70	635	1,910	5,725	611.60

RENTAL EQUIPMENT

01 54 33 | Rental Equipment

			UNIT	HOURLY OPER. COST	RENT PER DAY	RENT PER WEEK	RENT PER MONTH	CREW EQUIPMENT COST/DAY	
20	2550	60 kW	R015433 -10	Ea.	49.50	930	2,785	8,350	953	20
	2600	110 kW			91.90	1,825	5,440	16,300	1,823	
	2700	Extractor, steam or air, 950 J	R312323 -30		14.85	435	1,310	3,925	380.80	
	2750	1350 J			16.95	540	1,620	4,850	459.60	
	2800	Log chipper, up to 560 mm diam., 450 kW	R312316 -40		41.71	1,100	3,331.60	10,000	1,000	
	2850	Logger, for skidding & stacking logs, 110 kW			36.70	825	2,470	7,400	787.60	
	2900	Rake, spring tooth, with tractor	R312316 -45		8.04	214	643	1,925	192.90	
	3000	Roller, vibratory, tandem, smooth drum, 15 kW			5.40	118	355	1,075	114.20	
	3050	26 kW			7.75	225	675	2,025	197	
	3100	Towed type vibratory compactor, smooth drum, 37 kW			24.70	435	1,300	3,900	457.60	
	3150	Sheepsfoot, 37 kW			25.50	455	1,370	4,100	478	
	3170	Landfill compactor, 164 kW			50.30	1,200	3,595	10,800	1,121	
	3200	Pneumatic tire roller, 60 kW			9.15	310	930	2,800	259.20	
	3250	60 kW			14.75	540	1,615	4,850	441	
	3300	Sheepsfoot vibratory roller, 150 kW			38.75	895	2,690	8,075	848	
	3320	254 kW			53.55	1,300	3,870	11,600	1,202	
	3350	Smooth drum vibratory roller, 56 kW			15.90	485	1,460	4,375	419.20	
	3400	93 kW			20.25	600	1,800	5,400	522	
	3410	Rotary mower, brush, 1500 mm, with tractor			11.50	232	695	2,075	231	
	3450	Scrapers, towed type, 8 to 10 m³ capacity			4.42	211	632	1,900	161.75	
	3500	9.175 to 12.988 m³ capacity			1.50	281	842	2,525	180.40	
	3550	Scrapers, self-propelled, 4 x 4 drive, 2 engine, 11 m³ capacity			84.60	1,500	4,500	13,500	1,577	
	3600	2 engine, 18 m³ capacity			121.45	2,300	6,890	20,700	2,350	
	3640	24 - 34 m³ capacity			143.20	2,675	8,060	24,200	2,758	
	3650	Self-loading, 8.4 m³ capacity			40.95	825	2,475	7,425	822.60	
	3700	17 m³ capacity			76.90	1,725	5,190	15,600	1,653	
	3710	Screening plant 82 kW w / 1.5 m x 3 m screen			23.00	380	1,135	3,400	411	
	3720	1.5 m x 5 m screen			25.05	475	1,430	4,300	486.40	
	3850	Shovels, see Cranes division 01590-600								
	3860	Shovel/backhoe bucket, 0.4 m³		Ea.	1.85	56.50	170	510	48.80	
	3870	0.6 m³			1.90	63.50	190	570	53.20	
	3880	0.8 m³			2.00	71.50	215	645	59	
	3890	1.1 m³			2.15	86.50	260	780	69.20	
	3910	2.3 m³			2.45	120	360	1,075	91.60	
	3950	Stump chipper, 460 mm deep, 22 kW			4.56	45	135	405	63.50	
	4110	Tractor, crawler, with bulldozer, torque converter, diesel 60 kW			16.35	315	940	2,825	318.80	
	4150	78 kW			22.35	475	1,420	4,250	462.80	
	4200	104 kW			27.50	590	1,765	5,300	573	
	4260	149 kW			41.10	985	2,960	8,875	920.80	
	4310	224 kW			53.40	1,275	3,850	11,600	1,197	
	4360	306 kW			72.30	1,600	4,790	14,400	1,536	
	4370	375 kW			95.55	2,125	6,395	19,200	2,043	
	4380	522 kW			142.40	3,350	10,045	30,100	3,148	
	4400	Loader, crawler, torque conv., diesel, 1.1 m³, 60 kW			15.15	325	970	2,900	315.20	
	4450	1.1 to 1.3 m³, 71 kW			17.55	385	1,160	3,475	372.40	
	4510	1.3 to 1.7 m³, 97 kW			23.75	610	1,835	5,500	557	
	4530	1.9 to 2.5 m³, 142 kW			35.75	845	2,540	7,625	794	
	4560	2.7 to 3.8 m³, 205 kW			48.55	1,200	3,610	10,800	1,110	
	4610	Tractor loader, wheel, torque conv., 4 x 4, 0.8 to 1 m³, 48 kW			9.95	192	575	1,725	194.60	
	4620	1.1 to 1.3 m³, 60 kW			12.55	235	705	2,125	241.40	
	4650	1.3 to 1.5 m³, 75 kW			14.25	280	840	2,525	282	
	4710	1.9 to 2.7 m³, 97 kW			15.30	305	915	2,750	305.40	
	4730	2.3 to 3.4 m³, 127 kW			21.00	460	1,380	4,150	444	
	4760	4 to 4.4 m³, 201 kW			34.75	705	2,115	6,350	701	
	4810	5.4 to 6 m³, 280 kW			59.05	1,250	3,740	11,200	1,220	
	4870	9.6 m³, 515 kW			82.40	1,925	5,810	17,400	1,821	
	4880	Wheeled, skid steer, 0.28 m³, 22 kW gas			8.75	140	420	1,250	154	
	4890	0.76 m³, 58 kW, diesel			10.50	197	590	1,775	202	

01 54 | Construction Aids

01 54 33 | Rental Equipment

		01 54 33 Rental Equipment		UNIT	HOURLY OPER. COST	RENT PER DAY	RENT PER WEEK	RENT PER MONTH	CREW EQUIPMENT COST/DAY	
20	4891	Attachments for all skid steer loaders	R015433-10							20
	4892	Auger		Ea.	.40	66.50	200	600	43.20	
	4893	Backhoe	R312323-30		.67	112	336	1,000	72.55	
	4894	Broom			.61	101	303	910	65.50	
	4895	Forks	R312316-40		.22	37	111	335	23.95	
	4896	Grapple			.50	83.50	250	750	54	
	4897	Concrete hammer	R312316-45		.96	159	478	1,425	103.30	
	4898	Tree spade			.95	158	475	1,425	102.60	
	4899	Trencher			.64	106	318	955	68.70	
	4900	Trencher, chain, boom type, gas, operator walking, 9 kW			2.70	43.50	130	390	47.60	
	4910	Operator riding, 30 kW			8.85	243	730	2,200	216.80	
	5000	Wheel type, diesel, 1200 mm deep, 300 mm wide 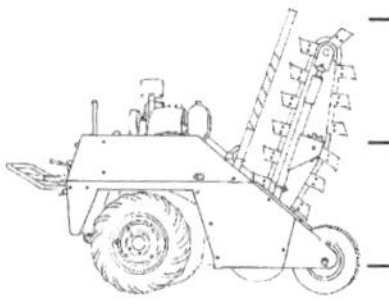			47.60	745	2,230	6,700	826.80	
	5100	Diesel, 1800 mm deep, 500 mm wide			64.90	1,775	5,295	15,900	1,578	
	5150	Ladder type, diesel, 1500 mm deep, 200 mm wide			23.25	705	2,115	6,350	609	
	5200	Diesel, 2400 mm deep, 400 mm wide			57.25	1,675	5,010	15,000	1,460	
	5210	Tree spade, self-propelled			11.10	267	800	2,400	248.80	
	5250	Truck, dump, tandem, 11 metric tons payload			20.45	272	815	2,450	326.60	
	5300	Three axle dump, 14.5 metric tons payload			27.95	420	1,265	3,800	476.60	
	5350	Dump trailer only, rear dump, 12.6 m^3			4.25	115	345	1,025	103	
	5400	15.3 m^3 			4.60	132	395	1,175	115.80	
	5450	Flatbed, single axle, 1.4 metric tons rating			11.85	56.50	170	510	128.80	
	5500	2.7 metric tons rating			15.15	91.50	275	825	176.20	
	5550	Off highway rear dump, 22.7 metric tons capacity			40.40	980	2,945	8,825	912.20	
	5600	31.7 metric tons capacity			41.30	1,000	3,015	9,050	933.40	
	5610	45.4 metric tons capacity			53.15	1,250	3,780	11,300	1,181	
	5620	59 metric tons capacity			56.85	1,375	4,105	12,300	1,276	
	5630	90.7 metric tons capacity			73.00	1,775	5,325	16,000	1,649	
	6000	Vibratory plow, 19 kW, walking		↓	4.65	58.50	175	525	72.20	
40	0010	**GENERAL EQUIPMENT RENTAL** Without operators	R314116-45							40
	0150	Aerial lift, scissor type, to 4.6 m high, 450 kg cap., electric		Ea.	2.35	48.50	145	435	47.80	
	0160	To 7.6 m high, 900 kg capacity	R314116-40		2.85	70	210	630	64.80	
	0170	Telescoping boom to 12 m high, 227 kg capacity, gas			12.25	278	835	2,500	265	
	0180	13.5 m high, 227 kg cap	R312323-30		13.05	320	960	2,875	296.40	
	0190	To 18 m high, 272 kg capacity			15.00	425	1,280	3,850	376	
	0195	Air compressor, portable, 3.06 L/s, electric			.40	12.35	37	111	10.60	
	0196	Gasoline			.57	18.65	56	168	15.75	
	0200	Air compressor, portable, gas engine, 0.028 m^3/s			4.90	33	99	297	59	
	0300	0.076 m^3/s			8.70	46.50	140	420	97.60	
	0400	Diesel engine, rotary screw, 0.118 m^3/s			8.60	85	255	765	119.80	
	0500	0.172 m^3/s			11.50	112	335	1,000	159	
	0550	0.212 m^3/s			13.90	130	390	1,175	189.20	
	0600	0.283 m^3/s			23.65	218	655	1,975	320.20	
	0700	0.354 m^3/s			25.55	228	685	2,050	341.40	
	0800	For silenced models, small sizes, add								
	0900	Large sizes, add		↓						
	0920	Air tools and accessories								
	0930	Breaker, pavement, 30 kg		Ea.	.40	9.65	29	87	9	
	0940	36 kg 			.40	10.65	32	96	9.60	
	0950	Drills, hand (jackhammer) 29 kg			.45	15.35	46	138	12.80	
	0960	Track or wagon, swing boom, 100 mm drifter			38.85	695	2,090	6,275	728.80	
	0970	125 mm drifter			45.75	760	2,280	6,850	822	
	0975	Track mounted quarry drill, 150 mm diameter drill			55.65	965	2,900	8,700	1,025	
	0980	Dust control/drill			.86	14.65	44	132	15.70	
	0990	Hammer, chipping, 5 kg			.40	20	60	180	15.20	
	1000	Hose, air with couplings, 15 m long, 19 mm (3/4") diameter			.04	6	18	54	3.90	
	1100	25 mm (1") diameter			.04	6	18	54	3.90	
	1200	40 mm (1-1/2") diameter			.04	7.35	22	66	4.70	
	1300	50 mm (2") diameter		↓	.10	17.35	52	156	11.20	

01 54 33 | Rental Equipment

			UNIT	HOURLY OPER. COST	RENT PER DAY	RENT PER WEEK	RENT PER MONTH	CREW EQUIPMENT COST/DAY	
40	1400	65 mm (2-1/2") diameter	Ea.	.12	19.35	58	174	12.55	40
	1410	80 mm (3") diameter		.17	28.50	86	258	18.55	
	1450	Drill, steel, 22 mm (7/8") x 600 mm		.05	8.35	25	75	5.40	
	1460	22 mm (7/8") x 1800 mm		.06	9.35	28	84	6.10	
	1520	Moil points		.03	4.33	13	39	2.85	
	1525	Pneumatic nailer w/accessories		.41	27.50	82	246	19.70	
	1530	Sheeting driver for 27 kg breaker		.10	7.35	22	66	5.20	
	1540	For 41 kg breaker		.15	10	30	90	7.20	
	1550	Spade, 11 kg		.35	6.65	20	60	6.80	
	1560	Tamper, single, 16 kg		.58	38.50	116	350	27.85	
	1570	Triple, 64 kg		.87	58	174	520	41.75	
	1580	Wrenches, impact, air powered, up to 20 mm bolt		.25	7.65	23	69	6.60	
	1590	Up to 32 mm bolt		.35	16.35	49	147	12.60	
	1600	Barricades, barrels, reflectorized, 1 to 50 barrels		.02	3.10	9.30	28	2	
	1610	100 to 200 barrels		.01	2.33	7	21	1.50	
	1620	Barrels with flashers, 1 to 50 barrels		.02	3.77	11.30	34	2.40	
	1630	100 to 200 barrels		.02	3	9	27	1.95	
	1640	Barrels with steady burn type C lights		.03	5	15	45	3.25	
	1650	Illuminated board, trailer mounted, with generator		.65	115	345	1,025	74.20	
	1670	Portable barricade, stock, with flashers, 1 to 6 units		.02	3.77	11.30	34	2.40	
	1680	25 to 50 units		.02	3.50	10.50	31.50	2.25	
	1690	Butt fusion machine, electric		18.65	330	990	2,975	347.20	
	1695	Electro fusion machine		8.55	133	400	1,200	148.40	
	1700	Carts, brick, hand powered, 450 kg capacity		.30	49.50	148	445	32	
	1800	Gas engine, 680 kg, 2.3 m lift		3.09	94.50	284	850	81.50	
	1822	Dehumidifier, medium, 0.504 g/s, 0.071 m³/s		.75	45.50	137	410	33.40	
	1824	Large, 1.512 g/s, 0.283 m³/s		1.48	90.50	272	815	66.25	
	1830	Distributor, asphalt, trailer mtd, 7570 L, 28 kW diesel		7.25	252	755	2,275	209	
	1840	11 355 L, 28 kW diesel		8.55	290	870	2,600	242.40	
	1850	Drill, rotary hammer, electric, 40 mm diameter		.40	24.50	74	222	18	
	1860	Carbide bit for above		.04	6	18	54	3.90	
	1865	Rotary, crawler, 186 kW		95.45	1,800	5,410	16,200	1,846	
	1870	Emulsion sprayer, 246 L, 3.73 kW gas engine		2.00	75	225	675	61	
	1880	757 L, 3.73 kW engine		5.15	125	375	1,125	116.20	
	1920	Floodlight, mercury vapor, or quartz, on tripod							
	1930	1000 W	Ea.	.30	12	36	108	9.60	
	1940	2000 W		.52	22	66	198	17.35	
	1950	Floodlights, trailer mounted w/generator, 1 - 300 W light		2.55	65	195	585	59.40	
	1960	2 - 1000 W lights		3.60	110	330	990	94.80	
	2000	4 - 300 W lights		2.95	76.50	230	690	69.60	
	2020	Forklift, wheeled, for brick, 5.5 m, 1360 kg, 2 wheel drive, gas		15.25	182	545	1,625	231	
	2040	8.5 m, 1815 kg, 4 wheel drive, diesel		12.55	237	710	2,125	242.40	
	2050	For rough terrain, 3630 kg, 5 m lift, 50 kW		16.90	375	1,125	3,375	360.20	
	2060	For plant, 3.6 met. ton capacity, 60 kW, 2 wheel drive gas		8.85	88.50	265	795	123.80	
	2080	9.1 met. ton capacity, 90 kW, 2 wheel drive, diesel		13.30	162	485	1,450	203.40	
	2100	Generator, electric, gas engine, 1.5 kW to 3 kW		1.90	12	36	108	22.40	
	2200	5 kW		2.60	17.35	52	156	31.20	
	2300	10 kW		4.30	33.50	100	300	54.40	
	2400	25 kW		8.35	61.50	185	555	103.80	
	2500	Diesel engine, 20 kW		6.40	66.50	200	600	91.20	
	2600	50 kW		12.25	71.50	215	645	141	
	2700	100 kW		18.15	80	240	720	193.20	
	2800	250 kW		51.65	148	445	1,325	502.20	
	2850	Hammer, hydraulic, for mounting on boom, to 680 J		1.85	63.50	190	570	52.80	
	2860	680 to 1600 J		3.25	103	310	930	88	
	2900	Heaters, space, oil or electric, 15 kW		1.25	7.35	22	66	14.40	
	3000	29 kW		2.26	10.65	32	96	24.50	
	3100	88 kW		7.24	33.50	100	300	77.90	

Reference codes: R314116-45, R314116-40, R312323-30

01 54 33 | Rental Equipment

		UNIT	HOURLY OPER. COST	RENT PER DAY	RENT PER WEEK	RENT PER MONTH	CREW EQUIPMENT COST/DAY	
40	3150	147 kW	R314116 -45	Ea.	14.60	50	150	450	146.80	40
	3200	Hose, water, suction with coupling, 6 m long, 50 mm diameter			.02	3.33	10	30	2.15	
	3210	75 mm diameter	R314116 -40		.03	5.65	17	51	3.65	
	3220	100 mm diameter			.03	6.35	19	57	4.05	
	3230	150 mm diameter	R312323 -30		.11	18.65	56	168	12.10	
	3240	200 mm diameter			.27	45.50	137	410	29.55	
	3250	Discharge hose with coupling, 15 m long, 50 mm diameter			.01	2	6	18	1.30	
	3260	75 mm diameter			.02	3.33	10	30	2.15	
	3270	100 mm diameter			.02	4.67	14	42	2.95	
	3280	150 mm diameter			.06	11.65	35	105	7.50	
	3290	200 mm diameter			.33	54.50	164	490	35.45	
	3295	Insulation blower			.11	7.35	22	66	5.30	
	3300	Ladders, extension type, 5 m to 11 m long			.20	33	99	297	21.40	
	3400	12 m to 18 m long			.30	49.50	149	445	32.20	
	3405	Lance for cutting concrete			2.36	77	231	695	65.10	
	3407	Lawn mower, rotary, 600 mm, 3.7 kW			1.05	23.50	71	213	22.60	
	3408	1200 mm, self propelled			3.52	126	377	1,125	103.55	
	3410	Level, laser type, for pipe and sewer leveling			1.22	81	243	730	58.35	
	3430	Electronic			.81	54	162	485	38.90	
	3440	Laser type, rotating beam for grade control			1.20	79.50	239	715	57.40	
	3460	Builders level with tripod and rod			.08	12.65	38	114	8.25	
	3500	Light towers, towable, with diesel generator, 2000 W			2.95	76.50	230	690	69.60	
	3600	4000 W			3.60	110	330	990	94.80	
	3700	Mixer, powered, plaster and mortar, 0.17 m³, 5.2 kW			1.30	21.50	65	195	23.40	
	3800	0.28 m³, 6.7 kW			1.55	35	105	315	33.40	
	3850	Nailer, pneumatic			.41	27.50	82	246	19.70	
	3900	Paint sprayers complete, 0.004 m³/s			.69	45.50	137	410	32.90	
	4000	0.008 m³/s			1.12	74.50	223	670	53.55	
	4020	Pavers, bituminous, rubber tires, 2.4 m wide, 37 kW, gas			28.05	800	2,405	7,225	705.40	
	4030	3 m wide, 112 kW			62.65	1,525	4,595	13,800	1,420	
	4050	Crawler, 2.4 m wide, 75 kW, diesel			58.30	1,650	4,915	14,700	1,449	
	4060	3 m wide, 112 kW			70.45	1,925	5,810	17,400	1,726	
	4070	Concrete paver, 3.6 m to 7.3 m wide, 186 kW			58.75	1,350	4,035	12,100	1,277	
	4080	Placer-spreader-trimmer, 7.3 m wide, 224 kW			92.80	2,500	7,475	22,400	2,237	
	4100	Pump, centrifugal gas pump, 38 mm, 15 L/s			2.75	40	120	360	46	
	4200	50 mm, 30 L/s			3.55	43.50	130	390	54.40	
	4300	75 mm, 57 L/s			3.70	45	135	405	56.60	
	4400	150 mm, 341 L/s			17.75	163	490	1,475	240	
	4500	Submersible electric pump, 32 mm, 3.5 L/s			.36	15	45	135	11.90	
	4600	38 mm, 5.2 L/s			.39	17	51	153	13.30	
	4700	50 mm, 7.6 L/s			.45	21.50	64	192	16.40	
	4800	75 mm, 18.9 L/s			.95	36.50	110	330	29.60	
	4900	100 mm, 35 L/s			6.34	145	435	1,300	137.70	
	5000	150 mm, 100 L/s			9.47	213	640	1,925	203.75	
	5100	Diaphragm pump, gas, single, 38 mm diameter			.88	40	120	360	31.05	
	5200	50 mm diameter			2.90	50	150	450	53.20	
	5300	75 mm diameter			2.90	50	150	450	53.20	
	5400	Double, 100 mm diameter			3.90	70	210	630	73.20	
	5500	Trash pump, self-priming, gas, 50 mm diameter			2.80	19.65	59	177	34.20	
	5600	Diesel, 100 mm diameter			6.85	55	165	495	87.80	
	5650	Diesel, 150 mm diameter			21.40	122	365	1,100	244.20	
	5655	Grout Pump	▼		8.05	70	210	630	106.40	
	5660	Rollers, see division 01590-200								
	5700	Salamanders, L.P. gas fired, 29 kW		Ea.	2.27	11.35	34	102	24.95	
	5705	14.7 kW			1.70	8	24	72	18.40	
	5720	Sandblaster, portable, open top, 84 L capacity			.40	20.50	62	186	15.60	
	5730	0.17 m³ capacity			.65	29.50	88	264	22.80	
	5740	Accessories for above	▼	▼	.11	18.65	56	168	12.10	

01 54 33 | Rental Equipment

			UNIT	HOURLY OPER. COST	RENT PER DAY	RENT PER WEEK	RENT PER MONTH	CREW EQUIPMENT COST/DAY
5750	Sander, floor	R314116 -45	Ea.	.75	18.35	55	165	17
5760	Edger			.55	17.65	53	159	15
5800	Saw, chain, gas engine, 450 mm long	R314116 -40		1.35	16	48	144	20.40
5900	900 mm long			.55	48.50	145	435	33.40
5950	1500 mm long	R312323 -30		.55	50	150	450	34.40
6000	Masonry, table mounted, 350 mm diameter, 3.73 kW			1.30	56	168	505	44
6050	Portable cut-off, 6 kW			1.45	29.50	88	264	29.20
6100	Circular, hand held, electric, 184 mm diameter			.20	5	15	45	4.60
6200	300 mm diameter			.27	8.35	25	75	7.15
6250	Wall saw, w/hydraulic power, 7.5 kW			2.04	108	323.20	970	80.95
6275	Shot blaster, walk behind, 500 mm wide			1.70	495	1,490	4,475	311.60
6300	Steam cleaner, 379 L/hr			2.40	63.50	190	570	57.20
6310	757 L/hr			3.25	76.50	230	690	72
6340	Tar Kettle/Pot 1514 L			3.29	51.50	155	465	57.30
6350	Torch, cutting, acetylene-oxygen, 45 m hose			1.50	7	21	63	16.20
6360	Hourly operating cost includes tips and gas			8.10				64.80
6410	Toilet, portable chemical			.11	18	54	162	11.70
6420	Recycle flush type			.13	22	66	198	14.25
6430	Toilet, fresh water flush, garden hose,			.15	24.50	74	222	16
6440	Hoisted, non-flush, for high rise			.13	21.50	65	195	14.05
6450	Toilet, trailers, minimum			.22	37.50	112	335	24.15
6460	Maximum			.67	112	337	1,000	72.75
6465	Tractor, farm with attachment			10.30	223	670	2,000	216.40
6470	Trailer, office, see division 01520-500							
6500	Trailers, platform, flush deck, 2 axle, 22 metric tons		Ea.	4.25	90	270	810	88
6600	36 metric tons capacity			5.55	125	375	1,125	119.40
6700	3 axle, 45 metric tons capacity			6.00	138	415	1,250	131
6800	68 metric tons capacity			7.45	182	545	1,625	168.60
6810	Trailer mounted cable reel for H.V. line work			4.56	217	652	1,950	166.90
6820	Trailer mounted cable tensioning rig			8.96	425	1,280	3,850	327.70
6830	Cable pulling rig			58.52	2,425	7,260	21,800	1,920
6850	Trailer, storage, see division 01520-500							
6900	Water tank, engine driven discharge, 19 kL		Ea.	5.60	120	360	1,075	116.80
6925	38 kL			7.70	168	505	1,525	162.60
6950	Water truck, off highway, 23 kL			49.25	700	2,095	6,275	813
7010	Tram car for H.V. line work, powered, 2 conductor			5.78	118	354	1,050	117.05
7020	Transit (builder's level) with tripod			.08	12.65	38	114	8.25
7030	Trench box, 1100 kg, 1.30 m x 2.44 m			.62	104	311	935	67.15
7040	2700 kg 1.83 m x 6.10 m			.71	118	354	1,050	76.50
7050	Trench box, 3600 kg 2.4 m x 4.9 m			.93	156	467	1,400	100.85
7060	3500 kg, 2.44 m x 6.10 m			1.20	200	599	1,800	129.40
7065	4100 kg, 2.44 m x 7.32 m			1.35	226	677	2,025	146.20
7070	5400 kg, 3.0 m x 6.1 m			1.77	294	883	2,650	190.75
7100	Truck, pickup, 0.68 metric ton capacity, 2 wheel drive			5.75	53.50	160	480	78
7200	4 wheel drive			5.90	63.50	190	570	85.20
7250	Crew carrier, 9 passenger			9.40	97.50	292.80	880	133.75
7290	Tool van, 11 000 kg G.V.W.			8.80	103	310	930	132.40
7300	Tractor, 4 x 2, 27 metric tons capacity, 145 kW			13.65	165	495	1,475	208.20
7410	186 kW			18.20	243	730	2,200	291.60
7500	6 x 2, 36 metric tons capacity, 179 kW			17.45	270	810	2,425	301.60
7600	6 x 4, 41 metric tons capacity, 179 kW			22.20	293	880	2,650	353.60
7620	Vacuum truck, hazardous material, 9500 liter			7.38	310	926	2,775	244.25
7625	19 000 liter			11.37	410	1,234.80	3,700	337.90
7640	Tractor, with A frame, boom and winch, 168 kW			15.75	223	670	2,000	260
7650	Vacuum, H.E.P.A., 61 L, wet/dry		Ea.	.27	24	72	216	16.55
7655	208 L, wet/dry			.60	36	108	325	26.40
7660	Water tank, portable			1.00	9.65	29	87	13.80
7690	Large production vacuum loader, 1.5 m³/s			16.25	615	1,850	5,550	500

40

RENTAL EQUIPMENT

01 54 33 | Rental Equipment

			UNIT	HOURLY OPER. COST	RENT PER DAY	RENT PER WEEK	RENT PER MONTH	CREW EQUIPMENT COST/DAY	
40	7700	Welder, electric, 200 A	Ea.	3.52	42.50	127	380	53.55	40
	7800	300 A		5.04	48	144	430	69.10	
	7900	Gas engine, 200 A		6.65	25.50	76	228	68.40	
	8000	300 A		9.00	28.50	86	258	89.20	
	8100	Wheelbarrow, any size		.06	10.65	32	96	6.90	
	8200	Wrecking ball, 1800 kg		1.95	70	210	630	57.60	
50	0010	**HIGHWAY EQUIPMENT RENTAL** without operators							50
	0050	Asphalt batch plant, portable drum mixer, 90 metric tons/hr.	Ea.	57.45	1,350	4,025	12,100	1,265	
	0060	180 metric tons/hr.		63.80	1,400	4,230	12,700	1,356	
	0070	270 metric tons/hr.		74.25	1,675	5,005	15,000	1,595	
	0100	Backhoe attachment, long stick, up to 138 kW, 3.2 m long		.31	20.50	61	183	14.70	
	0140	Up to 186 kW, 3.7 m long		.33	22	66	198	15.85	
	0180	Over 186 kW, 4.6 m long		.44	29	87	261	20.90	
	0200	Special dipper arm, up to 75 kW, 9.8 m long		.90	59.50	179	535	43	
	0240	Over 75 kW, 10.1 m long		1.12	74.50	224	670	53.75	
	0300	Concrete batch plant, portable, electric, 150 m³/Hr		10.42	495	1,490	4,475	381.35	
	0500	Grader attachment, ripper/scarifier, rear mounted							
	0520	Up to 100 kW	Ea.	2.85	60	180	540	58.80	
	0540	Up to 135 kW		3.40	76.50	230	690	73.20	
	0580	Up to 185 kW		3.75	86.50	260	780	82	
	0700	Pvmt. removal bucket, for hyd. excavator, up to 67 kW		1.45	45	135	405	38.60	
	0740	Up to 150 kW		1.65	65	195	585	52.20	
	0780	Over 150 kW		1.75	78.50	235	705	61	
	0900	Aggregate spreader, self-propelled, 139 kW		37.80	790	2,375	7,125	777.40	
	1000	Chemical spreader, 2.3 m³		2.25	51.50	155	465	49	
	1900	Hammermill, traveling, 186 kW		47.35	1,750	5,220	15,700	1,423	
	2000	Horizontal borer, 75 mm diam., 7 kW gas driven		4.25	53.50	160	480	66	
	2200	Hydromulchers, gas power, 11 000 liters, for truck mounting		10.65	187	560	1,675	197.20	
	2400	Joint & crack cleaner, walk behind, 19 kW		2.10	48.50	145	435	45.80	
	2500	Filler, trailer mounted, 1500 liter, 15 kW		6.05	182	545	1,625	157.40	
	3000	Paint striper, self propelled, double line, 22 kW		5.20	155	465	1,400	134.60	
	3200	Post drivers, 150 mm I-Beam frame, for truck mounting		8.25	415	1,240	3,725	314	
	3400	Road sweeper, self propelled, 2.4 m wide, 67 kW		25.40	485	1,450	4,350	493.20	
	4000	Road mixer, self-propelled, 97 kW		30.65	595	1,780	5,350	601.20	
	4100	231 kW		58.55	2,050	6,150	18,500	1,698	
	4200	Cold mix paver, incl pug mill and bitumen tank,							
	4220	123 kW	Ea.	70.45	1,950	5,825	17,500	1,729	
	4250	Paver, asphalt, wheel or crawler, 97 kW, diesel		69.15	1,850	5,540	16,600	1,661	
	4300	Paver, road widener, gas 300 mm to 1800 mm, 50 kW		31.70	680	2,035	6,100	660.60	
	4400	Diesel, 600 mm to 4200 mm, 66 kW		42.35	995	2,980	8,950	934.80	
	4600	Slipform pavers, curb and gutter, 2 track, 56 kW		25.20	655	1,960	5,875	593.60	
	4700	4 track, 123 kW		35.25	760	2,280	6,850	738	
	4800	Median barrier, 160 kW		35.85	790	2,370	7,100	760.80	
	4901	Trailer, low bed, 68 metric ton capacity		7.90	178	535	1,600	170.20	
	5000	Road planer, walk behind, 250 mm cutting width, 7 kW		2.05	27	81	243	32.60	
	5100	Self propelled, 300 mm cutting width, 48 kW		5.45	150	450	1,350	133.60	
	5200	Pavement profiler, 1200 mm to 1800 mm wide, 336 kW		161.05	2,925	8,755	26,300	3,039	
	5300	2400 mm to 3000 mm wide, 559 kW		256.35	4,350	13,040	39,100	4,659	
	5400	Roadway plate, steel 25 mm x 2.4 m x 6 m		.06	10.35	31	93	6.70	
	5600	Stabilizer, self propelled, 112 kW		29.65	565	1,690	5,075	575.20	
	5700	231 kW		48.65	1,200	3,570	10,700	1,103	
	5800	Striper, thermal, truck mounted 450 liter paint, 112 kW		33.15	485	1,450	4,350	555.20	
	6000	Tar kettle, 1200 liter, trailer mounted		2.97	36.50	110	330	45.75	
	7000	Tunnel locomotive, diesel, 7 to 11 metric ton		22.10	555	1,670	5,000	510.80	
	7005	Electric, 9 metric ton		21.20	630	1,895	5,675	548.60	
	7010	Muck cars, 0.38 m³ capacity		1.55	21	63	189	25	
	7020	0.76 m³ capacity		1.75	29	87	261	31.40	
	7030	1.53 m³ capacity		1.90	33.50	100	300	35.20	

Reference boxes: R314116-45, R314116-40, R312323-30

RENTAL EQUIPMENT

01 54 33 | Rental Equipment

		UNIT	HOURLY OPER. COST	RENT PER DAY	RENT PER WEEK	RENT PER MONTH	CREW EQUIPMENT COST/DAY	
50	7040	Side dump, 1.53 m³ capacity	Ea.	2.10	41.50	125	375	41.80	**50**
	7050	2.29 m³ capacity		2.80	48.50	145	435	51.40	
	7060	3.82 m³ capacity		3.95	60	180	540	67.60	
	7100	Ventilating blower for tunnel, 5.6 kW		1.25	47	141	425	38.20	
	7110	7.5 kW		1.44	48.50	146	440	40.70	
	7120	15 kW		2.33	62	186	560	55.85	
	7140	30 kW		4.04	88.50	265	795	85.30	
	7160	45 kW		6.16	137	410	1,225	131.30	
	7175	56 kW		7.89	182	545	1,625	172.10	
	7180	149 kW		17.60	270	810	2,425	302.80	
	7800	Windrow loader, elevating		35.00	905	2,715	8,150	823	
60	0010	**LIFTING AND HOISTING EQUIPMENT RENTAL** without operators							**60**
	0120	Aerial lift truck, 2 person, to 24 m	Ea.	19.35	600	1,805	5,425	515.80	
	0140	Boom work platform, 12 m snorkel		10.60	222	665	2,000	217.80	
	0150	Crane, flatbed mounted, 2.7 metric ton cap.		14.65	197	590	1,775	235.20	
	0200	Crane, climbing, 30 m jib, 2700 kg capacity, 2 m/s		54.75	1,375	4,150	12,500	1,268	
	0300	30 m jib, 4600 kg capacity, 1.4 m/s		60.25	1,750	5,250	15,800	1,532	
	0400	Tower, static, 40 m high, 30 m jib,							
	0500	2800 kg capacity at 2 m/s	Ea.	57.95	1,600	4,790	14,400	1,422	
	0600	Crawler mounted, lattice boom, 0.382 m³, 14 metric tons at 3658 mm radius		22.08	595	1,790	5,375	534.65	
	0700	0.573 m³, 18 metric tons at 3658 radius		29.44	715	2,140	6,425	663.50	
	0800	0.764 m³, 23 metric tons at 3658 radiu		39.25	945	2,830	8,500	880	
	0900	146 m³, 36 metric tons at 3.6 m radius		43.65	975	2,925	8,775	934.20	
	1000	1.528 m³, 45 metric tons at 3.6 m radius		54.25	1,375	4,155	12,500	1,265	
	1100	2.292 m³, 68 metric tons at 3.6 m radius		51.75	1,350	4,060	12,200	1,226	
	1200	90 metric tons capacity, 18 m boom		66.30	1,775	5,300	15,900	1,590	
	1300	150 metric tons capacity, 18 m boom		96.50	2,300	6,890	20,700	2,150	
	1400	180 metric tons capacity, 20 m boom		102.55	2,450	7,375	22,100	2,295	
	1500	315 metic ton capacity, 24 m boom		149.45	3,475	10,450	31,400	3,286	
	1600	Truck mounted, lattice boom, 6 x 4, 18 metric tons at 3.0 m radius		30.88	1,175	3,500	10,500	947.05	
	1700	23 metric tons at 3.0 m radius		32.98	1,250	3,740	11,200	1,012	
	1800	8 x 4, 27 metric tons at 3.0 m radius		42.67	1,325	3,970	11,900	1,135	
	1900	36 metric tons at 3.6 m radius		36.20	1,400	4,200	12,600	1,130	
	2000	8 x 4, 54 metric tons at 4.6 m radius		45.96	1,475	4,440	13,300	1,256	
	2050	74 metric tons at 4.6 m radius		48.65	1,550	4,670	14,000	1,323	
	2100	82 metric tons at 4.6 m radius		50.65	1,700	5,110	15,300	1,427	
	2200	100 metric tons at 4.6 m radius		51.45	1,900	5,675	17,000	1,547	
	2300	136 metric tons at 5.5 m radius		47.40	1,925	5,800	17,400	1,539	
	2350	150 metric tons at 5.5 m radius		80.90	2,275	6,825	20,500	2,012	
	2400	Truck mounted, hydraulic, 11 metric tons capacity		35.20	590	1,775	5,325	636.60	
	2500	23 metric tons capacity		35.35	615	1,840	5,525	650.80	
	2550	30 metric tons capacity		36.20	645	1,935	5,800	676.60	
	2560	36 metric tons capacity		34.65	620	1,855	5,575	648.20	
	2600	50 metric tons capacity		51.45	900	2,700	8,100	951.60	
	2700	73 metric tons capacity		61.10	950	2,850	8,550	1,059	
	2720	90 metric tons capacity		83.70	2,425	7,250	21,800	2,120	
	2740	110 metric tons capacity		90.35	2,625	7,900	23,700	2,303	
	2760	136 metric tons capacity		110.25	3,325	9,990	30,000	2,880	
	2800	Self-propelled, 4 x 4, with telescoping boom, 4.5 metric tons		16.15	293	880	2,650	305.20	
	2900	11 metric tons capacity		27.05	535	1,600	4,800	536.40	
	3000	14 metric tons capacity		28.20	595	1,785	5,350	582.60	
	3050	18 metric tons capacity		29.05	620	1,855	5,575	603.40	
	3100	23 metric tons capacity		33.35	715	2,150	6,450	696.80	
	3150	36 metric tons capacity		48.00	885	2,660	7,975	916	
	3200	Derricks, guy, 18 metric ton capacity, 18 m boom, 23 m mast		15.61	340	1,016	3,050	328.10	
	3300	30 m boom, 35 m mast		24.93	580	1,740	5,225	547.45	
	3400	Stiffleg, 18 metric ton capacity, 21 m boom, 11 m mast		17.67	435	1,310	3,925	403.35	

R015433 -10
R015433 -15
R312316 -45

01 54 33 | Rental Equipment

			UNIT	HOURLY OPER. COST	RENT PER DAY	RENT PER WEEK	RENT PER MONTH	CREW EQUIPMENT COST/DAY	
60	3500	30 m boom, 14 m mast	Ea.	27.59	705	2,120	6,350	644.70	60
	3550	Helicopter, small, lift to 565 kg maximum, w/pilot	R015433 -10	74.47	2,725	8,210	24,600	2,238	
	3600	Hoists, chain type, overhead, manual, 0.68 metric tons	R015433 -15	.10	1.33	4	12	1.60	
	3900	9.1 metric tons		.60	10	30	90	10.80	
	4000	Hoist and tower, 2268 kg cap., portable electric, 12 m high	R312316 -45	4.25	195	585	1,750	151	
	4100	For each added 3 m section, add		.09	15.35	46	138	9.90	
	4200	Hoist and single tubular tower, 2268 kg electric, 30 m high		5.74	272	817	2,450	209.30	
	4300	For each added 2 m section, add		.15	25.50	77	231	16.60	
	4400	Hoist and double tubular tower, 2268 kg, 30 m high		6.15	300	900	2,700	229.20	
	4500	For each added 2 m section, add		.17	29	87	261	18.75	
	4550	Hoist and tower, mast type, 2722 kg, 30 m high		6.65	310	933	2,800	239.80	
	4570	For each added 3 m section, add		.11	18.65	56	168	12.10	
	4600	Hoist and tower, personnel, electric, 907 kg, 30 m @ 635 mm/s		13.77	830	2,490	7,475	608.15	
	4700	1361 kg, 30 m @ 1 m/s		15.70	935	2,810	8,425	687.60	
	4800	1361 kg, 46 m @ 1.5 m/s		17.40	1,050	3,150	9,450	769.20	
	4900	1814 kg, 30 m @ 1.5 m/s		18.03	1,075	3,210	9,625	786.25	
	5000	2722 kg, 30 m @ 1.4 m/s		19.49	1,125	3,370	10,100	829.90	
	5100	For added heights up to 150 m, add	m	.03	5.45	16.40	49	3.55	
	5200	Jacks, hydraulic, 18 metric tons	Ea.	.05	4.67	14	42	3.20	
	5500	90 metric tons	"	.30	13.35	40	120	10.40	
	6000	Jacks, hydraulic, climbing with 15 m jackrods							
	6010	and control consoles, minimum 3 mo. rental							
	6100	27 metric tons capacity	Ea.	1.68	112	336	1,000	80.65	
	6150	For each added 3 m jackrod section, add		.05	3.33	10	30	2.40	
	6300	45 metric tons capacity		2.71	180	541	1,625	129.90	
	6350	For each added 3 m jackrod section, add		.06	4	12	36	2.90	
	6500	113 metric tons capacity		7.10	475	1,420	4,250	340.80	
	6550	For each added 3 m jackrod section, add		.49	32.50	97	291	23.30	
	6600	Cable jack, 9.1 metric tons capacity with 60 m cable		1.41	93.50	281	845	67.50	
	6650	For each added 15 m of cable, add		.16	10.35	31	93	7.50	
70	0010	**WELLPOINT EQUIPMENT RENTAL** without operators	R312319 -90						70
	0020	Based on 2 months rental							
	0100	Combination jetting & wellpoint pump, 45 kW diesel	Ea.	10.93	278	833	2,500	254.05	
	0200	High pressure gas jet pump, 150 kW, 2000 kPa	"	21.98	237	712	2,125	318.25	
	0300	Discharge pipe, 200 mm diameter	m	.03	1.48	4.43	13.30	1.15	
	0350	300 mm diameter		.03	2.19	6.56	19.70	1.55	
	0400	Header pipe, flows up to 9.5 L/s, 100 mm diameter		.03	1.33	4	12	1.05	
	0500	25 L/s, 150 mm diameter		.03	1.59	4.76	14.25	1.20	
	0600	50 L/s, 200 mm diameter		.03	2.19	6.56	19.70	1.55	
	0700	95 L/s, 250 mm diameter		.03	2.30	6.89	20.50	1.65	
	0800	158 L/s, 300 mm diameter		.07	4.34	13.02	39	3.15	
	0900	284 L/s, 400 mm diameter		.10	5.55	16.67	50	4.10	
	0950	For quick coupling aluminum and plastic pipe, add		.10	5.75	17.22	51.50	4.25	
	1100	Wellpoint, 7.6 m long, with fittings & riser pipe, 40 or 50 mm diam	Ea.	.05	3.50	10.49	31.50	2.50	
	1200	Wellpoint pump, diesel powered, 100 mm diameter, 15 kW		5.06	160	480	1,450	136.50	
	1300	150 mm diameter, 22 kW		6.72	199	596	1,800	172.95	
	1400	200 mm suction, 30 kW		9.12	272	817	2,450	236.35	
	1500	250 mm suction, 56 kW		13.06	320	955	2,875	295.50	
	1600	300 mm suction, 75 kW		19.21	510	1,530	4,600	459.70	
	1700	300 mm suction, 130 kW		26.64	560	1,680	5,050	549.10	
80	0010	**MARINE EQUIPMENT RENTAL** without operators							80
	0200	Barge, 360 metric ton, 9 m wide x 27 m long	Ea.	17.00	240	720	2,150	280	
	0240	725 metric ton, 14 m wide x 27 m long		28.10	345	1,030	3,100	430.80	
	2000	Tugboat, diesel, 75 kW		18.80	170	510	1,525	252.40	
	2040	186 kW		35.50	315	945	2,825	473	
	2080	283 kW	Ea.	80.55	935	2,805	8,425	1,205	

Crew No.	Bare Costs		Incl. Subs O & P		Cost Per Labor-Hour	
	Hr.	Daily	Hr.	Daily	Bare Costs	Incl. O&P
Crew A-1	Hr.	Daily	Hr.	Daily	Bare Costs	Incl. O&P
1 Building Laborer	$27.40	$219.20	$42.65	$341.20	$27.40	$42.65
1 Concrete saw, gas manual		49.00		53.90	6.13	6.74
8 L.H., Daily Totals		$268.20		$395.10	$33.53	$49.39
Crew A-1A	Hr.	Daily	Hr.	Daily	Bare Costs	Incl. O&P
1 Skilled Worker	$36.50	$292.00	$56.80	$454.40	$36.50	$56.80
1 Shot Blaster, 500 mm		311.60		342.75	38.95	42.85
8 L.H., Daily Totals		$603.60		$797.15	$75.45	$99.65
Crew A-1B	Hr.	Daily	Hr.	Daily	Bare Costs	Incl. O&P
1 Building Laborer	$27.40	$219.20	$42.65	$341.20	$27.40	$42.65
1 Concr. saw, gas, self-prop.		107.00		117.70	13.38	14.71
8 L.H., Daily Totals		$326.20		$458.90	$40.78	$57.36
Crew A-1C	Hr.	Daily	Hr.	Daily	Bare Costs	Incl. O&P
1 Building Laborer	$27.40	$219.20	$42.65	$341.20	$27.40	$42.65
1 Brush saw		20.40		22.45	2.55	2.81
8 L.H., Daily Totals		$239.60		$363.65	$29.95	$45.46
Crew A-1D	Hr.	Daily	Hr.	Daily	Bare Costs	Incl. O&P
1 Building Laborer	$27.40	$219.20	$42.65	$341.20	$27.40	$42.65
1 Vibrating plate, gas, 450 mm		26.40		29.05	3.30	3.63
8 L.H., Daily Totals		$245.60		$370.25	$30.70	$46.28
Crew A-1E	Hr.	Daily	Hr.	Daily	Bare Costs	Incl. O&P
1 Building Laborer	$27.40	$219.20	$42.65	$341.20	$27.40	$42.65
1 Vibrating plate, gas, 525 mm		34.60		38.05	4.33	4.76
8 L.H., Daily Totals		$253.80		$379.25	$31.73	$47.41
Crew A-1F	Hr.	Daily	Hr.	Daily	Bare Costs	Incl. O&P
1 Building Laborer	$27.40	$219.20	$42.65	$341.20	$27.40	$42.65
1 Rammer/tamper, gas, 200 mm		37.00		40.70	4.63	5.09
8 L.H., Daily Totals		$256.20		$381.90	$32.03	$47.74
Crew A-1G	Hr.	Daily	Hr.	Daily	Bare Costs	Incl. O&P
1 Building Laborer	$27.40	$219.20	$42.65	$341.20	$27.40	$42.65
1 Rammer/tamper, gas, 200 mm		37.00		40.70	4.63	5.09
8 L.H., Daily Totals		$256.20		$381.90	$32.03	$47.74
Crew A-1H	Hr.	Daily	Hr.	Daily	Bare Costs	Incl. O&P
1 Building Laborer	$27.40	$219.20	$42.65	$341.20	$27.40	$42.65
1 Pressure washer		57.20		62.90	7.15	7.87
8 L.H., Daily Totals		$276.40		$404.10	$34.55	$50.52
Crew A-1J	Hr.	Daily	Hr.	Daily	Bare Costs	Incl. O&P
1 Building Laborer	$27.40	$219.20	$42.65	$341.20	$27.40	$42.65
1 Rototiller		103.55		113.90	12.94	14.24
8 L.H., Daily Totals		$322.75		$455.10	$40.34	$56.89
Crew A-1K	Hr.	Daily	Hr.	Daily	Bare Costs	Incl. O&P
1 Building Laborer	$27.40	$219.20	$42.65	$341.20	$27.40	$42.65
1 Lawn aerator		22.60		24.85	2.83	3.11
8 L.H., Daily Totals		$241.80		$366.05	$30.23	$45.76
Crew A-1L	Hr.	Daily	Hr.	Daily	Bare Costs	Incl. O&P
1 Building Laborer	$27.40	$219.20	$42.65	$341.20	$27.40	$42.65
1 Power blower/vacuum		22.60		24.85	2.83	3.11
8 L.H., Daily Totals		$241.80		$366.05	$30.23	$45.76

Crew No.	Bare Costs		Incl. Subs O & P		Cost Per Labor-Hour	
	Hr.	Daily	Hr.	Daily	Bare Costs	Incl. O&P
Crew A-1M	Hr.	Daily	Hr.	Daily	Bare Costs	Incl. O&P
1 Building Laborer	$27.40	$219.20	$42.65	$341.20	$27.40	$42.65
1 Snow blower		103.55		113.90	12.94	14.24
8 L.H., Daily Totals		$322.75		$455.10	$40.34	$56.89
Crew A-2	Hr.	Daily	Hr.	Daily	Bare Costs	Incl. O&P
2 Laborers	$27.40	$438.40	$42.65	$682.40	$27.47	$42.48
1 Truck Driver (light)	27.60	220.80	42.15	337.20		
1 Light Truck, 1.4 metric ton		128.80		141.70	5.37	5.90
24 L.H., Daily Totals		$788.00		$1161.30	$32.84	$48.38
Crew A-2A	Hr.	Daily	Hr.	Daily	Bare Costs	Incl. O&P
2 Laborers	$27.40	$438.40	$42.65	$682.40	$27.47	$42.48
1 Truck Driver (light)	27.60	220.80	42.15	337.20		
1 Light Truck, 1.4 metric ton		128.80		141.70		
1 Concrete Saw		107.00		117.70	9.83	10.81
24 L.H., Daily Totals		$895.00		$1279.00	$37.30	$53.29
Crew A-3	Hr.	Daily	Hr.	Daily	Bare Costs	Incl. O&P
1 Truck Driver (heavy)	$28.35	$226.80	$43.30	$346.40	$28.35	$43.30
1 Dump Truck, 11 metric ton		326.60		359.25	40.83	44.91
8 L.H., Daily Totals		$553.40		$705.65	$69.18	$88.21
Crew A-3A	Hr.	Daily	Hr.	Daily	Bare Costs	Incl. O&P
1 Truck Driver (light)	$27.60	$220.80	$42.15	$337.20	$27.60	$42.15
1 Pickup Truck (4x4)		85.20		93.70	10.65	11.72
8 L.H., Daily Totals		$306.00		$430.90	$38.25	$53.87
Crew A-3B	Hr.	Daily	Hr.	Daily	Bare Costs	Incl. O&P
1 Equip. Oper. (medium)	$36.70	$293.60	$55.40	$443.20	$32.53	$49.35
1 Truck Driver (heavy)	28.35	226.80	43.30	346.40		
1 Dump Truck, 18 metric tons		476.60		524.25		
1 F.E. Loader, 2.3 m³		305.40		335.95	48.88	53.76
16 L.H., Daily Totals		$1302.40		$1649.80	$81.41	$103.11
Crew A-3C	Hr.	Daily	Hr.	Daily	Bare Costs	Incl. O&P
1 Equip. Oper. (light)	$35.20	$281.60	$53.10	$424.80	$35.20	$53.10
1 Wheeled Skid Steer Loader		202.00		222.20	25.25	27.78
8 L.H., Daily Totals		$483.60		$647.00	$60.45	$80.88
Crew A-3D	Hr.	Daily	Hr.	Daily	Bare Costs	Incl. O&P
1 Truck Driver, Light	$27.60	$220.80	$42.15	$337.20	$27.60	$42.15
1 Pickup Truck (4x4)		85.20		93.70		
1 Flatbed Trailer, 22 metric ton		88.00		96.80	21.65	23.82
8 L.H., Daily Totals		$394.00		$527.70	$49.25	$65.97
Crew A-3E	Hr.	Daily	Hr.	Daily	Bare Costs	Incl. O&P
1 Equip. Oper. (crane)	$38.10	$304.80	$57.50	$460.00	$33.23	$50.40
1 Truck Driver (heavy)	28.35	226.80	43.30	346.40		
1 Pickup Truck (4x4)		85.20		93.70	5.33	5.86
16 L.H., Daily Totals		$616.80		$900.10	$38.56	$56.26
Crew A-3F	Hr.	Daily	Hr.	Daily	Bare Costs	Incl. O&P
1 Equip. Oper. (crane)	$38.10	$304.80	$57.50	$460.00	$33.23	$50.40
1 Truck Driver (heavy)	28.35	226.80	43.30	346.40		
1 Pickup Truck (4x4)		85.20		93.70		
1 Tractor, 6x2, 36 tonne Cap		301.60		331.75		
1 Lowbed Trailer, 68 metric ton		170.20		187.20	34.81	38.29
16 L.H., Daily Totals		$1088.60		$1419.05	$68.04	$88.69

Crew No.	Bare Costs		Incl. Subs O & P		Cost Per Labor-Hour	

Crew A-3G

	Hr.	Daily	Hr.	Daily	Bare Costs	Incl. O&P
1 Equip. Oper. (crane)	$38.10	$304.80	$57.50	$460.00	$33.23	$50.40
1 Truck Driver (heavy)	28.35	226.80	43.30	346.40		
1 Pickup Truck (4x4)		85.20		93.70		
1 Tractor, 6x4, 40 tonne Cap.		353.60		388.95		
1 Lowbed Trailer, 68 metric ton		170.20		187.20	38.06	41.87
16 L.H., Daily Totals		$1140.60		$1476.25	$71.29	$92.27

Crew A-4

	Hr.	Daily	Hr.	Daily	Bare Costs	Incl. O&P
2 Carpenters	$35.55	$568.80	$55.35	$885.60	$34.27	$52.82
1 Painter, Ordinary	31.70	253.60	47.75	382.00		
24 L.H., Daily Totals		$822.40		$1267.60	$34.27	$52.82

Crew A-5

	Hr.	Daily	Hr.	Daily	Bare Costs	Incl. O&P
2 Laborers	$27.40	$438.40	$42.65	$682.40	$27.42	$42.59
.25 Truck Driver (light)	27.60	55.20	42.15	84.30		
.25 Light Truck, 1.4 metric ton		32.20		35.40	1.79	1.97
18 L.H., Daily Totals		$525.80		$802.10	$29.21	$44.56

Crew A-6

	Hr.	Daily	Hr.	Daily	Bare Costs	Incl. O&P
1 Instrument Man	$36.50	$292.00	$56.80	$454.40	$35.38	$54.30
1 Rodman/Chainman	34.25	274.00	51.80	414.40		
1 Laser Transit/Level		58.35		64.20	3.64	4.01
16 L.H., Daily Totals		$624.35		$933.00	$39.02	$58.31

Crew A-7

	Hr.	Daily	Hr.	Daily	Bare Costs	Incl. O&P
1 Chief Of Party	$45.80	$366.40	$71.30	$570.40	$38.85	$59.97
1 Instrument Man	36.50	292.00	56.80	454.40		
1 Rodman/Chainman	34.25	274.00	51.80	414.40		
1 Laser Transit/Level		58.35		64.20	2.43	2.67
24 L.H., Daily Totals		$990.75		$1503.40	$41.28	$62.64

Crew A-8

	Hr.	Daily	Hr.	Daily	Bare Costs	Incl. O&P
1 Chief Of Party	$45.80	$366.40	$71.30	$570.40	$37.70	$57.93
1 Instrument Man	36.50	292.00	56.80	454.40		
2 Rodmen/Chainmen	34.25	548.00	51.80	828.80		
1 Laser Transit/Level		58.35		64.20	1.82	2.00
32 L.H., Daily Totals		$1264.75		$1917.80	$39.52	$59.93

Crew A-9

	Hr.	Daily	Hr.	Daily	Bare Costs	Incl. O&P
1 Asbestos Foreman	$39.55	$316.40	$62.55	$500.40	$39.11	$61.85
7 Asbestos Workers	39.05	2186.80	61.75	3458.00		
64 L.H., Daily Totals		$2503.20		$3958.40	$39.11	$61.85

Crew A-10

	Hr.	Daily	Hr.	Daily	Bare Costs	Incl. O&P
1 Asbestos Foreman	$39.55	$316.40	$62.55	$500.40	$39.11	$61.85
7 Asbestos Workers	39.05	2186.80	61.75	3458.00		
64 L.H., Daily Totals		$2503.20		$3958.40	$39.11	$61.85

Crew A-10A

	Hr.	Daily	Hr.	Daily	Bare Costs	Incl. O&P
1 Asbestos Foreman	$39.55	$316.40	$62.55	$500.40	$39.22	$62.02
2 Asbestos Workers	39.05	624.80	61.75	988.00		
24 L.H., Daily Totals		$941.20		$1488.40	$39.22	$62.02

Crew A-10B

	Hr.	Daily	Hr.	Daily	Bare Costs	Incl. O&P
1 Asbestos Foreman	$39.55	$316.40	$62.55	$500.40	$39.18	$61.95
3 Asbestos Workers	39.05	937.20	61.75	1482.00		
32 L.H., Daily Totals		$1253.60		$1982.40	$39.18	$61.95

Crew A-10C

	Hr.	Daily	Hr.	Daily	Bare Costs	Incl. O&P
3 Asbestos Workers	$39.05	$937.20	$61.75	$1482.00	$39.05	$61.75
1 Flatbed truck		128.80		141.70	5.37	5.90
24 L.H., Daily Totals		$1066.00		$1623.70	$44.42	$67.65

Crew A-10D

	Hr.	Daily	Hr.	Daily	Bare Costs	Incl. O&P
2 Asbestos Workers	$39.05	$624.80	$61.75	$988.00	$37.16	$57.49
1 Equip. Oper. (crane)	38.10	304.80	57.50	460.00		
1 Equip. Oper. Oiler	32.45	259.60	48.95	391.60		
1 Hydraulic crane, 30 metric tons		676.60		744.25	21.14	23.26
32 L.H., Daily Totals		$1865.80		$2583.85	$58.30	$80.75

Crew A-11

	Hr.	Daily	Hr.	Daily	Bare Costs	Incl. O&P
1 Asbestos Foreman	$39.55	$316.40	$62.55	$500.40	$39.11	$61.85
7 Asbestos Workers	39.05	2186.80	61.75	3458.00		
2 Chipping Hammers		30.40		33.45	.48	.52
64 L.H., Daily Totals		$2533.60		$3991.85	$39.59	$62.37

Crew A-12

	Hr.	Daily	Hr.	Daily	Bare Costs	Incl. O&P
1 Asbestos Foreman	$39.55	$316.40	$62.55	$500.40	$39.11	$61.85
7 Asbestos Workers	39.05	2186.80	61.75	3458.00		
1 Large Prod. Vac. Loader		500.00		550.00	7.81	8.59
64 L.H., Daily Totals		$3003.20		$4508.40	$46.92	$70.44

Crew A-13

	Hr.	Daily	Hr.	Daily	Bare Costs	Incl. O&P
1 Equip. Oper. (light)	$35.20	$281.60	$53.10	$424.80	$35.20	$53.10
1 Large Prod. Vac. Loader		500.00		550.00	62.50	68.75
8 L.H., Daily Totals		$781.60		$974.80	$97.70	$121.85

Crew B-1

	Hr.	Daily	Hr.	Daily	Bare Costs	Incl. O&P
1 Laborer Foreman (outside)	$29.40	$235.20	$45.80	$366.40	$28.07	$43.70
2 Laborers	27.40	438.40	42.65	682.40		
24 L.H., Daily Totals		$673.60		$1048.80	$28.07	$43.70

Crew B-1A

	Hr.	Daily	Hr.	Daily	Bare Costs	Incl. O&P
1 Laborer Foreman	$29.40	$235.20	$45.80	$366.40	$28.07	$43.70
2 Laborers	27.40	438.40	42.65	682.40		
2 Cutting Torches		32.40		35.65		
2 Gases		129.60		142.55	6.75	7.43
24 L.H., Daily Totals		$835.60		$1227.00	$34.82	$51.13

Crew B-1B

	Hr.	Daily	Hr.	Daily	Bare Costs	Incl. O&P
1 Laborer Foreman	$29.40	$235.20	$45.80	$366.40	$30.58	$47.15
2 Laborers	27.40	438.40	42.65	682.40		
1 Equip. Oper. (crane)	38.10	304.80	57.50	460.00		
2 Cutting Torches		32.40		35.65		
2 Gases		129.60		142.55		
1 Hyd. Crane, 11 Metric Ton		636.60		700.25	24.96	27.45
32 L.H., Daily Totals		$1777.00		$2387.25	$55.54	$74.60

Crew B-2

	Hr.	Daily	Hr.	Daily	Bare Costs	Incl. O&P
1 Laborer Foreman (outside)	$29.40	$235.20	$45.80	$366.40	$27.80	$43.28
4 Laborers	27.40	876.80	42.65	1364.80		
40 L.H., Daily Totals		$1112.00		$1731.20	$27.80	$43.28

CREWS

Crew B-3

	Bare Costs Hr.	Daily	Incl. Subs O & P Hr.	Daily	Bare Costs	Incl. O&P
1 Laborer Foreman (outside)	$29.40	$235.20	$45.80	$366.40	$29.60	$45.52
2 Laborers	27.40	438.40	42.65	682.40		
1 Equip. Oper. (med.)	36.70	293.60	55.40	443.20		
2 Truck Drivers (heavy)	28.35	453.60	43.30	692.80		
1 F.E. Loader, T.M., 1.91 m³		794.00		873.40		
2 Dump Trucks, 15 metric ton		953.20		1048.50	36.40	40.04
48 L.H., Daily Totals		$3168.00		$4106.70	$66.00	$85.56

Crew B-3A

	Bare Costs Hr.	Daily	Incl. Subs O & P Hr.	Daily	Bare Costs	Incl. O&P
4 Laborers	$27.40	$876.80	$42.65	$1364.80	$29.26	$45.20
1 Equip. Oper. (med.)	36.70	293.60	55.40	443.20		
1 Hyd. Excavator, 1.15 m³		722.80		795.10	18.07	19.88
40 L.H., Daily Totals		$1893.20	·	$2603.10	$47.33	$65.08

Crew B-3B

	Bare Costs Hr.	Daily	Incl. Subs O & P Hr.	Daily	Bare Costs	Incl. O&P
2 Laborers	$27.40	$438.40	$42.65	$682.40	$29.96	$46.00
1 Equip. Oper. (med.)	36.70	293.60	55.40	443.20		
1 Truck Driver (heavy)	28.35	226.80	43.30	346.40		
1 Backhoe Loader, 60 kW		277.60		305.35		
1 Dump Truck, 15 metric ton		476.60		524.25	23.57	25.93
32 L.H., Daily Totals		$1713.00		$2301.60	$53.53	$71.93

Crew B-3C

	Bare Costs Hr.	Daily	Incl. Subs O & P Hr.	Daily	Bare Costs	Incl. O&P
3 Laborers	$27.40	$657.60	$42.65	$1023.60	$29.73	$45.84
1 Equip. Oper. (med.)	36.70	293.60	55.40	443.20		
1 F.E. Crawler Ldr., 3 m³		1110.00		1221.00	34.69	38.16
32 L.H., Daily Totals		$2061.20		$2687.80	$64.42	$84.00

Crew B-4

	Bare Costs Hr.	Daily	Incl. Subs O & P Hr.	Daily	Bare Costs	Incl. O&P
1 Laborer Foreman (outside)	$29.40	$235.20	$45.80	$366.40	$27.89	$43.28
4 Laborers	27.40	876.80	42.65	1364.80		
1 Truck Driver (heavy)	28.35	226.80	43.30	346.40		
1 Tractor, 4 x 2, 145 kW		208.20		229.00		
1 Platform Trailer		119.40		131.35	6.83	7.51
48 L.H., Daily Totals		$1666.40		$2437.95	$34.72	$50.79

Crew B-5

	Bare Costs Hr.	Daily	Incl. Subs O & P Hr.	Daily	Bare Costs	Incl. O&P
1 Laborer Foreman (outside)	$29.40	$235.20	$45.80	$366.40	$30.34	$46.74
4 Laborers	27.40	876.80	42.65	1364.80		
2 Equip. Oper. (med.)	36.70	587.20	55.40	886.40		
1 Air Compr., 118 L/s		119.80		131.80		
2 Air Tools & Accessories		18.00		19.80		
2-15 m Hoses, 40 mm Dia. (1.5")		9.40		10.35		
1 F.E. Loader, T.M., 1.91 m³		794.00		873.40	16.81	18.49
56 L.H., Daily Totals		$2640.40		$3652.95	$47.15	$65.23

Crew B-5A

	Bare Costs Hr.	Daily	Incl. Subs O & P Hr.	Daily	Bare Costs	Incl. O&P
1 Laborer Foreman	$29.40	$235.20	$45.80	$366.40	$29.93	$46.02
6 Laborers	27.40	1315.20	42.65	2047.20		
2 Equip. Oper. (med.)	36.70	587.20	55.40	886.40		
1 Equip. Oper. (light)	35.20	281.60	53.10	424.80		
2 Truck Drivers (heavy)	28.35	453.60	43.30	692.80		
1 Air Compr. 172 L/s		159.00		174.90		
2 Pavement Breakers		18.00		19.80		
8 Air Hoses w/Coup., 25 mm		31.20		34.30		
2 Dump Trucks, 11 metric ton		653.20		718.50	8.97	9.87
96 L.H., Daily Totals		$3734.20		$5365.10	$38.90	$55.89

Crew B-5B

	Bare Costs Hr.	Daily	Incl. Subs O & P Hr.	Daily	Bare Costs	Incl. O&P
1 Powderman	$36.50	$292.00	$56.80	$454.40	$32.49	$49.58
2 Equip. Oper. (med.)	36.70	587.20	55.40	886.40		
3 Truck Drivers (heavy)	28.35	680.40	43.30	1039.20		
1 F.E. Ldr. 1.91 m³		305.40		335.95		
3 Dump Trucks, 15 metric ton		1429.80		1572.80		
1 Air Compr. 172 L/s		159.00		174.90	39.46	43.41
48 L.H., Daily Totals		$3453.80		$4463.65	$71.95	$92.99

Crew B-5C

	Bare Costs Hr.	Daily	Incl. Subs O & P Hr.	Daily	Bare Costs	Incl. O&P
3 Laborers	$27.40	$657.60	$42.65	$1023.60	$30.77	$47.05
1 Equip. Oper. (medium)	36.70	293.60	55.40	443.20		
2 Truck Drivers (heavy)	28.35	453.60	43.30	692.80		
1 Equip. Oper. (crane)	38.10	304.80	57.50	460.00		
1 Equip. Oper. Oiler	32.45	259.60	48.95	391.60		
2 Dump Trucks, 15 metric ton		953.20		1048.50		
1 F.E. Crawler Ldr., 3 m³		1110.00		1221.00		
1 Hyd. Crane, 23 metric ton		696.80		766.50	43.13	47.44
64 L.H., Daily Totals		$4729.20		$6047.20	$73.90	$94.49

Crew B-6

	Bare Costs Hr.	Daily	Incl. Subs O & P Hr.	Daily	Bare Costs	Incl. O&P
2 Laborers	$27.40	$438.40	$42.65	$682.40	$30.00	$46.13
1 Equip. Oper. (light)	35.20	281.60	53.10	424.80		
1 Backhoe Loader, 36 kW		226.20		248.80	9.43	10.37
24 L.H., Daily Totals		$946.20		$1356.00	$39.43	$56.50

Crew B-6A

	Bare Costs Hr.	Daily	Incl. Subs O & P Hr.	Daily	Bare Costs	Incl. O&P
.5 Laborer Foremen (outside)	$29.40	$117.60	$45.80	$183.20	$31.52	$48.38
1 Laborer	27.40	219.20	42.65	341.20		
1 Equip. Oper. (med.)	36.70	293.60	55.40	443.20		
1 Vacuum Trk., 18925 L		337.90		371.70	16.90	18.58
20 L.H., Daily Totals		$968.30		$1339.30	$48.42	$66.96

Crew B-6B

	Bare Costs Hr.	Daily	Incl. Subs O & P Hr.	Daily	Bare Costs	Incl. O&P
2 Laborer Foremen (out)	$29.40	$470.40	$45.80	$732.80	$28.07	$43.70
4 Laborers	27.40	876.80	42.65	1364.80		
1 Winch Truck		305.20		335.70		
1 Flatbed Truck		128.80		141.70		
1 Butt Fusion Machine		347.20		381.90	16.28	17.90
48 L.H., Daily Totals		$2128.40		$2956.90	$44.35	$61.60

Crew B-7

	Bare Costs Hr.	Daily	Incl. Subs O & P Hr.	Daily	Bare Costs	Incl. O&P
1 Laborer Foreman (outside)	$29.40	$235.20	$45.80	$366.40	$29.28	$45.30
4 Laborers	27.40	876.80	42.65	1364.80		
1 Equip. Oper. (med.)	36.70	293.60	55.40	443.20		
1 Chipping Machine		173.00		190.30		
1 F.E. Loader, T.M., 1.91 m³		794.00		873.40		
2 Chain Saws		66.80		73.50	21.54	23.69
48 L.H., Daily Totals		$2439.40		$3311.60	$50.82	$68.99

Crew B-7A

	Bare Costs Hr.	Daily	Incl. Subs O & P Hr.	Daily	Bare Costs	Incl. O&P
2 Laborers	$27.40	$438.40	$42.65	$682.40	$30.00	$46.13
1 Equip. Oper. (light)	35.20	281.60	53.10	424.80		
1 Rake w/Tractor		192.90		212.20		
2 Chain Saws		40.80		44.90	9.74	10.71
24 L.H., Daily Totals		$953.70		$1364.30	$39.74	$56.84

Crew No.	Bare Costs		Incl. Subs O & P		Cost Per Labor-Hour	
Crew B-8	Hr.	Daily	Hr.	Daily	Bare Costs	Incl. O&P
1 Laborer Foreman (outside)	$29.40	$235.20	$45.80	$366.40	$30.84	$47.18
2 Laborers	27.40	438.40	42.65	682.40		
2 Equip. Oper. (med.)	36.70	587.20	55.40	886.40		
1 Equip. Oper. Oiler	32.45	259.60	48.95	391.60		
2 Truck Drivers (heavy)	28.35	453.60	43.30	692.80		
1 Hyd. Crane, 22 metric ton		650.80		715.90		
1 F.E. Loader, T.M., 1.91 m³		794.00		873.40		
2 Dump Trucks, 15 metric ton		953.20		1048.50	37.47	41.22
64 L.H., Daily Totals		$4372.00		$5657.40	$68.31	$88.40

Crew No.	Bare Costs		Incl. Subs O & P		Cost Per Labor-Hour	
Crew B-9	Hr.	Daily	Hr.	Daily	Bare Costs	Incl. O&P
1 Laborer Foreman (outside)	$29.40	$235.20	$45.80	$366.40	$27.80	$43.28
4 Laborers	27.40	876.80	42.65	1364.80		
1 Air Compr., 118 L/s		119.80		131.80		
2 Air Tools & Accessories		18.00		19.80		
2-15 m Hoses, 40 mm (1.5")		9.40		10.35	3.68	4.05
40 L.H., Daily Totals		$1259.20		$1893.15	$31.48	$47.33

Crew No.	Bare Costs		Incl. Subs O & P		Cost Per Labor-Hour	
Crew B-9A	Hr.	Daily	Hr.	Daily	Bare Costs	Incl. O&P
2 Laborers	$27.40	$438.40	$42.65	$682.40	$27.72	$42.87
1 Truck Driver (heavy)	28.35	226.80	43.30	346.40		
1 Water Tanker		116.80		128.50		
1 Tractor		208.20		229.00		
2-15 m Disch. Hoses, 80mm (3")		4.30		4.75	13.72	15.09
24 L.H., Daily Totals		$994.50		$1391.05	$41.44	$57.96

Crew No.	Bare Costs		Incl. Subs O & P		Cost Per Labor-Hour	
Crew B-9B	Hr.	Daily	Hr.	Daily	Bare Costs	Incl. O&P
2 Laborers	$27.40	$438.40	$42.65	$682.40	$27.72	$42.87
1 Truck Driver (heavy)	28.35	226.80	43.30	346.40		
2-15 m Disch. Hoses, 80mm (3")		4.30		4.75		
1 Water Tanker		116.80		128.50		
1 Tractor		208.20		229.00		
1 Pressure Washer		50.60		55.65	15.83	17.41
24 L.H., Daily Totals		$1045.10		$1446.70	$43.55	$60.28

Crew No.	Bare Costs		Incl. Subs O & P		Cost Per Labor-Hour	
Crew B-9C	Hr.	Daily	Hr.	Daily	Bare Costs	Incl. O&P
1 Laborer Foreman (outside)	$29.40	$235.20	$45.80	$366.40	$27.80	$43.28
4 Laborers	27.40	876.80	42.65	1364.80		
1 Air Compr., 118 L/s		119.80		131.80		
2-15 m Hoses, 40mm Dia.(1.5")		9.40		10.35		
2 Breakers, Pavement, 30 kg.		18.00		19.80	3.68	4.05
40 L.H., Daily Totals		$1259.20		$1893.15	$31.48	$47.33

Crew No.	Bare Costs		Incl. Subs O & P		Cost Per Labor-Hour	
Crew B-9D	Hr.	Daily	Hr.	Daily	Bare Costs	Incl. O&P
1 Laborer Foreman (Outside)	$29.40	$235.20	$45.80	$366.40	$27.80	$43.28
4 Common Laborers	27.40	876.80	42.65	1364.80		
1 Air Compressor, 118 L/s		119.80		131.80		
2 Air hoses, 38 mm x 15 m		9.40		10.35		
2 Air tampers		55.70		61.25	4.63	5.09
40 L.H., Daily Totals		$1296.90		$1934.60	$32.43	$48.37

Crew No.	Bare Costs		Incl. Subs O & P		Cost Per Labor-Hour	
Crew B-10	Hr.	Daily	Hr.	Daily	Bare Costs	Incl. O&P
1 Equip. Oper. (med.)	$36.70	$293.60	$55.40	$443.20	$33.60	$51.15
.5 Laborer	27.40	109.60	42.65	170.60		
12 L.H., Daily Totals		$403.20		$613.80	$33.60	$51.15

Crew No.	Bare Costs		Incl. Subs O & P		Cost Per Labor-Hour	
Crew B-10A	Hr.	Daily	Hr.	Daily	Bare Costs	Incl. O&P
1 Equip. Oper. (med.)	$36.70	$293.60	$55.40	$443.20	$33.60	$51.15
.5 Laborer	27.40	109.60	42.65	170.60		
1 Walk behind compactor, 5.6 kW		125.80		138.40	10.48	11.53
12 L.H., Daily Totals		$529.00		$752.20	$44.08	$62.68

Crew No.	Bare Costs		Incl. Subs O & P		Cost Per Labor-Hour	
Crew B-10B	Hr.	Daily	Hr.	Daily	Bare Costs	Incl. O&P
1 Equip. Oper. (med.)	$36.70	$293.60	$55.40	$443.20	$33.60	$51.15
.5 Laborer	27.40	109.60	42.65	170.60		
1 Dozer, 150 kW		920.80		1012.90	76.73	84.41
12 L.H., Daily Totals		$1324.00		$1626.70	$110.33	$135.56

Crew No.	Bare Costs		Incl. Subs O & P		Cost Per Labor-Hour	
Crew B-10C	Hr.	Daily	Hr.	Daily	Bare Costs	Incl. O&P
1 Equip. Oper. (med.)	$36.70	$293.60	$55.40	$443.20	$33.60	$51.15
.5 Laborer	27.40	109.60	42.65	170.60		
1 Dozer, 150 kW		920.80		1012.90		
1 Vibratory Roller, Towed		457.60		503.35	114.87	126.35
12 L.H., Daily Totals		$1781.60		$2130.05	$148.47	$177.50

Crew No.	Bare Costs		Incl. Subs O & P		Cost Per Labor-Hour	
Crew B-10D	Hr.	Daily	Hr.	Daily	Bare Costs	Incl. O&P
1 Equip. Oper. (med.)	$36.70	$293.60	$55.40	$443.20	$33.60	$51.15
.5 Laborer	27.40	109.60	42.65	170.60		
1 Dozer, 150 kW		920.80		1012.90		
1 Sheepsft. Roller, Towed		478.00		525.80	116.57	128.22
12 L.H., Daily Totals		$1802.00		$2152.50	$150.17	$179.37

Crew No.	Bare Costs		Incl. Subs O & P		Cost Per Labor-Hour	
Crew B-10E	Hr.	Daily	Hr.	Daily	Bare Costs	Incl. O&P
1 Equip. Oper. (med.)	$36.70	$293.60	$55.40	$443.20	$33.60	$51.15
.5 Laborer	27.40	109.60	42.65	170.60		
1 Tandem Roller, 4.5 met. ton		114.20		125.60	9.52	10.47
12 L.H., Daily Totals		$517.40		$739.40	$43.12	$61.62

Crew No.	Bare Costs		Incl. Subs O & P		Cost Per Labor-Hour	
Crew B-10F	Hr.	Daily	Hr.	Daily	Bare Costs	Incl. O&P
1 Equip. Oper. (med.)	$36.70	$293.60	$55.40	$443.20	$33.60	$51.15
.5 Laborer	27.40	109.60	42.65	170.60		
1 Tandem Roller, 9 met. ton		197.00		216.70	16.42	18.06
12 L.H., Daily Totals		$600.20		$830.50	$50.02	$69.21

Crew No.	Bare Costs		Incl. Subs O & P		Cost Per Labor-Hour	
Crew B-10G	Hr.	Daily	Hr.	Daily	Bare Costs	Incl. O&P
1 Equip. Oper. (med.)	$36.70	$293.60	$55.40	$443.20	$33.60	$51.15
.5 Laborer	27.40	109.60	42.65	170.60		
1 Sheepsft. Roll., 100 kW		848.00		932.80	70.67	77.73
12 L.H., Daily Totals		$1251.20		$1546.60	$104.27	$128.88

Crew No.	Bare Costs		Incl. Subs O & P		Cost Per Labor-Hour	
Crew B-10H	Hr.	Daily	Hr.	Daily	Bare Costs	Incl. O&P
1 Equip. Oper. (med.)	$36.70	$293.60	$55.40	$443.20	$33.60	$51.15
.5 Laborer	27.40	109.60	42.65	170.60		
1 Diaphr. Water Pump, 51 mm		53.20		58.50		
1-6 m Suct., 50 mm (2")		2.15		2.35		
2-15 m Disch., 50 mm (2")		2.60		2.85	4.83	5.31
12 L.H., Daily Totals		$461.15		$677.50	$38.43	$56.46

Crew No.	Bare Costs		Incl. Subs O & P		Cost Per Labor-Hour	
Crew B-10I	Hr.	Daily	Hr.	Daily	Bare Costs	Incl. O&P
1 Equip. Oper. (med.)	$36.70	$293.60	$55.40	$443.20	$33.60	$51.15
.5 Laborer	27.40	109.60	42.65	170.60		
1 Diaphr. Water Pump, 102 mm		73.20		80.50		
1-6 m Suction, 100 mm (4")		4.05		4.45		
2-15 m Disch., 100 mm (4")		5.90		6.50	6.93	7.62
12 L.H., Daily Totals		$486.35		$705.25	$40.53	$58.77

Crew No.	Bare Costs		Incl. Subs O & P		Cost Per Labor-Hour	
Crew B-10J	Hr.	Daily	Hr.	Daily	Bare Costs	Incl. O&P
1 Equip. Oper. (med.)	$36.70	$293.60	$55.40	$443.20	$33.60	$51.15
.5 Laborer	27.40	109.60	42.65	170.60		
1 Centr. Pump, 80 mm (3")		56.60		62.25		
1-6 m Suction, 80 mm (3")		3.65		4.00		
2-15 m Disch., 80 mm (3")		4.30		4.75	5.38	5.92
12 L.H., Daily Totals		$467.75		$684.80	$38.98	$57.07

Crew B-10K

	Bare Costs Hr.	Daily	Incl. Subs O&P Hr.	Daily	Cost Per Labor-Hour Bare Costs	Incl. O&P
1 Equip. Oper. (med.)	$36.70	$293.60	$55.40	$443.20	$33.60	$51.15
.5 Laborer	27.40	109.60	42.65	170.60		
1 Centr. Pump, 150 mm (6")		240.00		264.00		
1-6 m Suction, 150 mm (6")		12.10		13.30		
2-15 m Disch., 150 mm (6")		15.00		16.50	22.26	24.48
12 L.H., Daily Totals		$670.30		$907.60	$55.86	$75.63

Crew B-10L

	Bare Costs Hr.	Daily	Incl. Subs O&P Hr.	Daily	Cost Per Labor-Hour Bare Costs	Incl. O&P
1 Equip. Oper. (med.)	$36.70	$293.60	$55.40	$443.20	$33.60	$51.15
.5 Laborer	27.40	109.60	42.65	170.60		
1 Dozer, 55 kW		318.80		350.70	26.57	29.22
12 L.H., Daily Totals		$722.00		$964.50	$60.17	$80.37

Crew B-10M

	Bare Costs Hr.	Daily	Incl. Subs O&P Hr.	Daily	Cost Per Labor-Hour Bare Costs	Incl. O&P
1 Equip. Oper. (med.)	$36.70	$293.60	$55.40	$443.20	$33.60	$51.15
.5 Laborer	27.40	109.60	42.65	170.60		
1 Dozer, 225 kW		1197.00		1316.70	99.75	109.73
12 L.H., Daily Totals		$1600.20		$1930.50	$133.35	$160.88

Crew B-10N

	Bare Costs Hr.	Daily	Incl. Subs O&P Hr.	Daily	Cost Per Labor-Hour Bare Costs	Incl. O&P
1 Equip. Oper. (med.)	$36.70	$293.60	$55.40	$443.20	$33.60	$51.15
.5 Laborer	27.40	109.60	42.65	170.60		
1 F.E. Loader, T.M., 1.15 m³		315.20		346.70	26.27	28.89
12 L.H., Daily Totals		$718.40		$960.50	$59.87	$80.04

Crew B-10O

	Bare Costs Hr.	Daily	Incl. Subs O&P Hr.	Daily	Cost Per Labor-Hour Bare Costs	Incl. O&P
1 Equip. Oper. (med.)	$36.70	$293.60	$55.40	$443.20	$33.60	$51.15
.5 Laborer	27.40	109.60	42.65	170.60		
1 F.E. Loader, T.M., 1.72 m³		557.00		612.70	46.42	51.06
12 L.H., Daily Totals		$960.20		$1226.50	$80.02	$102.21

Crew B-10P

	Bare Costs Hr.	Daily	Incl. Subs O&P Hr.	Daily	Cost Per Labor-Hour Bare Costs	Incl. O&P
1 Equip. Oper. (med.)	$36.70	$293.60	$55.40	$443.20	$33.60	$51.15
.5 Laborer	27.40	109.60	42.65	170.60		
1 F.E. Loader, T.M., 1.91 m³		794.00		873.40	66.17	72.78
12 L.H., Daily Totals		$1197.20		$1487.20	$99.77	$123.93

Crew B-10Q

	Bare Costs Hr.	Daily	Incl. Subs O&P Hr.	Daily	Cost Per Labor-Hour Bare Costs	Incl. O&P
1 Equip. Oper. (med.)	$36.70	$293.60	$55.40	$443.20	$33.60	$51.15
.5 Laborer	27.40	109.60	42.65	170.60		
1 F.E. Loader, T.M., 3.82 m³		1110.00		1221.00	92.50	101.75
12 L.H., Daily Totals		$1513.20		$1834.80	$126.10	$152.90

Crew B-10R

	Bare Costs Hr.	Daily	Incl. Subs O&P Hr.	Daily	Cost Per Labor-Hour Bare Costs	Incl. O&P
1 Equip. Oper. (med.)	$36.70	$293.60	$55.40	$443.20	$33.60	$51.15
.5 Laborer	27.40	109.60	42.65	170.60		
1 F.E. Loader, W.M., 0.76 m³		194.60		214.05	16.22	17.84
12 L.H., Daily Totals		$597.80		$827.85	$49.82	$68.99

Crew B-10S

	Bare Costs Hr.	Daily	Incl. Subs O&P Hr.	Daily	Cost Per Labor-Hour Bare Costs	Incl. O&P
1 Equip. Oper. (med.)	$36.70	$293.60	$55.40	$443.20	$33.60	$51.15
.5 Laborer	27.40	109.60	42.65	170.60		
1 F.E. Loader, W.M., 1.15 m³		241.40		265.55	20.12	22.13
12 L.H., Daily Totals		$644.60		$879.35	$53.72	$73.28

Crew B-10T

	Bare Costs Hr.	Daily	Incl. Subs O&P Hr.	Daily	Cost Per Labor-Hour Bare Costs	Incl. O&P
1 Equip. Oper. (med.)	$36.70	$293.60	$55.40	$443.20	$33.60	$51.15
.5 Laborer	27.40	109.60	42.65	170.60		
1 F.E. Loader, W.M., 1.91 m³		305.40		335.95	25.45	28.00
12 L.H., Daily Totals		$708.60		$949.75	$59.05	$79.15

Crew B-10U

	Bare Costs Hr.	Daily	Incl. Subs O&P Hr.	Daily	Cost Per Labor-Hour Bare Costs	Incl. O&P
1 Equip. Oper. (med.)	$36.70	$293.60	$55.40	$443.20	$33.60	$51.15
.5 Laborer	27.40	109.60	42.65	170.60		
1 F.E. Loader, W.M., 4.20 m³		701.00		771.10	58.42	64.26
12 L.H., Daily Totals		$1104.20		$1384.90	$92.02	$115.41

Crew B-10V

	Bare Costs Hr.	Daily	Incl. Subs O&P Hr.	Daily	Cost Per Labor-Hour Bare Costs	Incl. O&P
1 Equip. Oper. (med.)	$36.70	$293.60	$55.40	$443.20	$33.60	$51.15
.5 Laborer	27.40	109.60	42.65	170.60		
1 Dozer, 525 kW		3148.00		3462.80	262.33	288.57
12 L.H., Daily Totals		$3551.20		$4076.60	$295.93	$339.72

Crew B-10W

	Bare Costs Hr.	Daily	Incl. Subs O&P Hr.	Daily	Cost Per Labor-Hour Bare Costs	Incl. O&P
1 Equip. Oper. (med.)	$36.70	$293.60	$55.40	$443.20	$33.60	$51.15
.5 Laborer	27.40	109.60	42.65	170.60		
1 Dozer, 80 kW		462.80		509.10	38.57	42.42
12 L.H., Daily Totals		$866.00		$1122.90	$72.17	$93.57

Crew B-10X

	Bare Costs Hr.	Daily	Incl. Subs O&P Hr.	Daily	Cost Per Labor-Hour Bare Costs	Incl. O&P
1 Equip. Oper. (med.)	$36.70	$293.60	$55.40	$443.20	$33.60	$51.15
.5 Laborer	27.40	109.60	42.65	170.60		
1 Dozer, 305 kW		1536.00		1689.60	128.00	140.80
12 L.H., Daily Totals		$1939.20		$2303.40	$161.60	$191.95

Crew B-10Y

	Bare Costs Hr.	Daily	Incl. Subs O&P Hr.	Daily	Cost Per Labor-Hour Bare Costs	Incl. O&P
1 Equip. Oper. (med.)	$36.70	$293.60	$55.40	$443.20	$33.60	$51.15
.5 Laborer	27.40	109.60	42.65	170.60		
1 Vibratory Drum Roller		419.20		461.10	34.93	38.43
12 L.H., Daily Totals		$822.40		$1074.90	$68.53	$89.58

Crew B-11A

	Bare Costs Hr.	Daily	Incl. Subs O&P Hr.	Daily	Cost Per Labor-Hour Bare Costs	Incl. O&P
1 Equipment Oper. (med.)	$36.70	$293.60	$55.40	$443.20	$32.05	$49.03
1 Laborer	27.40	219.20	42.65	341.20		
1 Dozer, 150 kW		920.80		1012.90	57.55	63.31
16 L.H., Daily Totals		$1433.60		$1797.30	$89.60	$112.34

Crew B-11B

	Bare Costs Hr.	Daily	Incl. Subs O&P Hr.	Daily	Cost Per Labor-Hour Bare Costs	Incl. O&P
1 Equipment Oper. (light)	$35.20	$281.60	$53.10	$424.80	$31.30	$47.88
1 Laborer	27.40	219.20	42.65	341.20		
1 Air Powered Tamper		27.85		30.65		
1 Air Compr. 172 L/s		159.00		174.90		
2-15 m Hoses, 40mm Dia.(1.5")		9.40		10.35	12.27	13.50
16 L.H., Daily Totals		$697.05		$981.90	$43.57	$61.38

Crew B-11C

	Bare Costs Hr.	Daily	Incl. Subs O&P Hr.	Daily	Cost Per Labor-Hour Bare Costs	Incl. O&P
1 Equipment Oper. (med.)	$36.70	$293.60	$55.40	$443.20	$32.05	$49.03
1 Laborer	27.40	219.20	42.65	341.20		
1 Backhoe Loader, 35 kW		226.20		248.80	14.14	15.55
16 L.H., Daily Totals		$739.00		$1033.20	$46.19	$64.58

Crew B-11J

	Bare Costs Hr.	Daily	Incl. Subs O&P Hr.	Daily	Cost Per Labor-Hour Bare Costs	Incl. O&P
1 Equipment Oper. (med.)	$36.70	$293.60	$55.40	$443.20	$32.05	$49.03
1 Laborer	27.40	219.20	42.65	341.20		
1 Grader, 13 608 kg		457.40		503.15		
1 Ripper, beam & 1 shank		73.20		80.50	33.16	36.48
16 L.H., Daily Totals		$1043.40		$1368.05	$65.21	$85.51

Crews

Crew B-11K

Crew No.	Bare Costs Hr.	Daily	Incl. Subs O & P Hr.	Daily	Cost Per L.H. Bare Costs	Incl. O&P
1 Equipment Oper. (med.)	$36.70	$293.60	$55.40	$443.20	$32.05	$49.03
1 Laborer	27.40	219.20	42.65	341.20		
1 Trencher, 2 438 mm D.		1460.00		1606.00	91.25	100.38
16 L.H., Daily Totals		$1972.80		$2390.40	$123.30	$149.41

Crew B-11L

Crew No.	Bare Costs Hr.	Daily	Incl. Subs O & P Hr.	Daily	Cost Per L.H. Bare Costs	Incl. O&P
1 Equipment Oper. (med.)	$36.70	$293.60	$55.40	$443.20	$32.05	$49.03
1 Laborer	27.40	219.20	42.65	341.20		
1 Grader, 13 608 kg		457.40		503.15	28.59	31.45
16 L.H., Daily Totals		$970.20		$1287.55	$60.64	$80.48

Crew B-11M

Crew No.	Bare Costs Hr.	Daily	Incl. Subs O & P Hr.	Daily	Cost Per L.H. Bare Costs	Incl. O&P
1 Equipment Oper. (med.)	$36.70	$293.60	$55.40	$443.20	$32.05	$49.03
1 Laborer	27.40	219.20	42.65	341.20		
1 Backhoe Loader, 60 kW		277.60		305.35	17.35	19.09
16 L.H., Daily Totals		$790.40		$1089.75	$49.40	$68.12

Crew B-11N

Crew No.	Bare Costs Hr.	Daily	Incl. Subs O & P Hr.	Daily	Cost Per L.H. Bare Costs	Incl. O&P
1 Laborer Foreman	$29.40	$235.20	$45.80	$366.40	$30.32	$46.27
2 Equipment Operators (med.)	36.70	587.20	55.40	886.40		
6 Truck Drivers (hvy.)	28.35	1360.80	43.30	2078.40		
1 F.E. Loader, 4.20 m³		701.00		771.10		
1 Dozer, 298 kW		1536.00		1689.60		
6 Off Hwy. Tks. 45 metric tons		7086.00		7794.60	129.49	142.43
72 L.H., Daily Totals		$11506.20		$13586.50	$159.81	$188.70

Crew B-11Q

Crew No.	Bare Costs Hr.	Daily	Incl. Subs O & P Hr.	Daily	Cost Per L.H. Bare Costs	Incl. O&P
1 Equipment Operator (med.)	$36.70	$293.60	$55.40	$443.20	$33.60	$51.15
.5 Laborer	27.40	109.60	42.65	170.60		
1 Dozer, 104 kW		573.00		630.30	47.75	52.53
12 L.H., Daily Totals		$976.20		$1244.10	$81.35	$103.68

Crew B-11R

Crew No.	Bare Costs Hr.	Daily	Incl. Subs O & P Hr.	Daily	Cost Per L.H. Bare Costs	Incl. O&P
1 Equipment Operator (med.)	$36.70	$293.60	$55.40	$443.20	$33.60	$51.15
.5 Laborer	27.40	109.60	42.65	170.60		
1 Dozer, 149 kW		920.80		1012.90	76.73	84.41
12 L.H., Daily Totals		$1324.00		$1626.70	$110.33	$135.56

Crew B-11S

Crew No.	Bare Costs Hr.	Daily	Incl. Subs O & P Hr.	Daily	Cost Per L.H. Bare Costs	Incl. O&P
1 Equipment Operator (med.)	$36.70	$293.60	$55.40	$443.20	$33.60	$51.15
.5 Laborer	27.40	109.60	42.65	170.60		
1 Dozer, 224 kW		1197.00		1316.70		
1 Ripper, Beam & 1 Shank		73.20		80.50	105.85	116.44
12 L.H., Daily Totals		$1673.40		$2011.00	$139.45	$167.59

Crew B-11T

Crew No.	Bare Costs Hr.	Daily	Incl. Subs O & P Hr.	Daily	Cost Per L.H. Bare Costs	Incl. O&P
1 Equipment Operator (med.)	$36.70	$293.60	$55.40	$443.20	$33.60	$51.15
.5 Laborer	27.40	109.60	42.65	170.60		
1 Dozer, 306 kW		1536.00		1689.60		
1 Ripper, Beam & 2 Shanks		82.00		90.20	134.83	148.32
12 L.H., Daily Totals		$2021.20		$2393.60	$168.43	$199.47

Crew B-11U

Crew No.	Bare Costs Hr.	Daily	Incl. Subs O & P Hr.	Daily	Cost Per L.H. Bare Costs	Incl. O&P
1 Equipment Operator (med.)	$36.70	$293.60	$55.40	$443.20	$33.60	$51.15
.5 Laborer	27.40	109.60	42.65	170.60		
1 Dozer, 388 kW		2043.00		2247.30	170.25	187.28
12 L.H., Daily Totals		$2446.20		$2861.10	$203.85	$238.43

Crew B-11V

Crew No.	Bare Costs Hr.	Daily	Incl. Subs O & P Hr.	Daily	Cost Per L.H. Bare Costs	Incl. O&P
3 Laborers	$27.40	$657.60	$42.65	$1023.60	$27.40	$42.65
1 Walk behind compactor, 5.6 kW		125.80		138.40	5.24	5.77
24 L.H., Daily Totals		$783.40		$1162.00	$32.64	$48.42

Crew B-11W

Crew No.	Bare Costs Hr.	Daily	Incl. Subs O & P Hr.	Daily	Cost Per L.H. Bare Costs	Incl. O&P
1 Equipment Operator (med.)	$36.70	$293.60	$55.40	$443.20	$28.97	$44.25
1 Common Laborer	27.40	219.20	42.65	341.20		
10 Truck Drivers, Heavy	28.35	2268.00	43.30	3464.00		
1 Dozer, 150 kW		920.80		1012.90		
1 Vib. roller, smth., towed, 21 MT		457.60		503.35		
10 Dump Truck, 9 Metric Ton		3266.00		3592.60	48.38	53.22
96 L.H., Daily Totals		$7425.20		$9357.25	$77.35	$97.47

Crew B-11Y

Crew No.	Bare Costs Hr.	Daily	Incl. Subs O & P Hr.	Daily	Cost Per L.H. Bare Costs	Incl. O&P
1 Laborer Foreman (Outside)	$29.40	$235.20	$45.80	$366.40	$30.72	$47.25
5 Common Laborers	27.40	1096.00	42.65	1706.00		
3 Equipment Operators (med.)	36.70	880.80	55.40	1329.60		
1 Dozer, 60 kW		318.80		350.70		
2 Walk behind compactor, 5.6 kW		251.60		276.75		
4 Vibratory plate, gas, 525 mm		138.40		152.25	9.84	10.83
72 L.H., Daily Totals		$2920.80		$4181.70	$40.56	$58.08

Crew B-12A

Crew No.	Bare Costs Hr.	Daily	Incl. Subs O & P Hr.	Daily	Cost Per L.H. Bare Costs	Incl. O&P
1 Equip. Oper. (crane)	$38.10	$304.80	$57.50	$460.00	$32.75	$50.08
1 Laborer	27.40	219.20	42.65	341.20		
1 Hyd. Excavator, 0.76 m³		560.20		616.20	35.01	38.51
16 L.H., Daily Totals		$1084.20		$1417.40	$67.76	$88.59

Crew B-12B

Crew No.	Bare Costs Hr.	Daily	Incl. Subs O & P Hr.	Daily	Cost Per L.H. Bare Costs	Incl. O&P
1 Equip. Oper. (crane)	$38.10	$304.80	$57.50	$460.00	$32.75	$50.08
1 Laborer	27.40	219.20	42.65	341.20		
1 Hyd. Excavator, 1.15 m³		722.80		795.10	45.18	49.69
16 L.H., Daily Totals		$1246.80		$1596.30	$77.93	$99.77

Crew B-12C

Crew No.	Bare Costs Hr.	Daily	Incl. Subs O & P Hr.	Daily	Cost Per L.H. Bare Costs	Incl. O&P
1 Equip. Oper. (crane)	$38.10	$304.80	$57.50	$460.00	$32.75	$50.08
1 Laborer	27.40	219.20	42.65	341.20		
1 Hyd. Excavator, 1.53 m³		919.60		1011.55	57.48	63.22
16 L.H., Daily Totals		$1443.60		$1812.75	$90.23	$113.30

Crew B-12D

Crew No.	Bare Costs Hr.	Daily	Incl. Subs O & P Hr.	Daily	Cost Per L.H. Bare Costs	Incl. O&P
1 Equip. Oper. (crane)	$38.10	$304.80	$57.50	$460.00	$32.75	$50.08
1 Laborer	27.40	219.20	42.65	341.20		
1 Hyd. Excavator, 2.67 m³		2023.00		2225.30	126.44	139.08
16 L.H., Daily Totals		$2547.00		$3026.50	$159.19	$189.16

Crew B-12E

Crew No.	Bare Costs Hr.	Daily	Incl. Subs O & P Hr.	Daily	Cost Per L.H. Bare Costs	Incl. O&P
1 Equip. Oper. (crane)	$38.10	$304.80	$57.50	$460.00	$32.75	$50.08
1 Laborer	27.40	219.20	42.65	341.20		
1 Hyd. Excavator, 0.38 m³		335.40		368.95	20.96	23.06
16 L.H., Daily Totals		$859.40		$1170.15	$53.71	$73.14

Crew B-12F

Crew No.	Bare Costs Hr.	Daily	Incl. Subs O & P Hr.	Daily	Cost Per L.H. Bare Costs	Incl. O&P
1 Equip. Oper. (crane)	$38.10	$304.80	$57.50	$460.00	$32.75	$50.08
1 Laborer	27.40	219.20	42.65	341.20		
1 Hyd. Excavator, 0.57 m³		477.80		525.60	29.86	32.85
16 L.H., Daily Totals		$1001.80		$1326.80	$62.61	$82.93

CREWS

Crew B-12G

Crew No.	Bare Costs Hr.	Daily	Incl. Subs O&P Hr.	Daily	Cost Per Labor-Hour Bare Costs	Incl. O&P
1 Equip. Oper. (crane)	$38.10	$304.80	$57.50	$460.00	$32.75	$50.08
1 Laborer	27.40	219.20	42.65	341.20		
1 Power Shovel, 0.38 m³		534.65		588.10		
1 Clamshell Bucket, 0.38 m³		33.80		37.20	35.53	39.08
16 L.H., Daily Totals		$1092.45		$1426.50	$68.28	$89.16

Crew B-12H

Crew No.	Bare Costs Hr.	Daily	Incl. Subs O&P Hr.	Daily	Cost Per Labor-Hour Bare Costs	Incl. O&P
1 Equip. Oper. (crane)	$38.10	$304.80	$57.50	$460.00	$32.75	$50.08
1 Laborer	27.40	219.20	42.65	341.20		
1 Power Shovel, 0.76 m³		880.00		968.00		
1 Clamshell Bucket, 0.76 m³		43.40		47.75	57.71	63.48
16 L.H., Daily Totals		$1447.40		$1816.95	$90.46	$113.56

Crew B-12I

Crew No.	Bare Costs Hr.	Daily	Incl. Subs O&P Hr.	Daily	Cost Per Labor-Hour Bare Costs	Incl. O&P
1 Equip. Oper. (crane)	$38.10	$304.80	$57.50	$460.00	$32.75	$50.08
1 Laborer	27.40	219.20	42.65	341.20		
1 Power Shovel, 0.57 m³		663.50		729.85		
1 Dragline Bucket, 0.57 m³		19.00		20.90	42.66	46.92
16 L.H., Daily Totals		$1206.50		$1551.95	$75.41	$97.00

Crew B-12J

Crew No.	Bare Costs Hr.	Daily	Incl. Subs O&P Hr.	Daily	Cost Per Labor-Hour Bare Costs	Incl. O&P
1 Equip. Oper. (crane)	$38.10	$304.80	$57.50	$460.00	$32.75	$50.08
1 Laborer	27.40	219.20	42.65	341.20		
1 Gradall, 2.72 metric ton		841.00		925.10	52.56	57.82
16 L.H., Daily Totals		$1365.00		$1726.30	$85.31	$107.90

Crew B-12K

Crew No.	Bare Costs Hr.	Daily	Incl. Subs O&P Hr.	Daily	Cost Per Labor-Hour Bare Costs	Incl. O&P
1 Equip. Oper. (crane)	$38.10	$304.80	$57.50	$460.00	$32.75	$50.08
1 Laborer	27.40	219.20	42.65	341.20		
1 Gradall, 2.72 metric ton		974.20		1071.60	60.89	66.98
16 L.H., Daily Totals		$1498.20		$1872.80	$93.64	$117.06

Crew B-12L

Crew No.	Bare Costs Hr.	Daily	Incl. Subs O&P Hr.	Daily	Cost Per Labor-Hour Bare Costs	Incl. O&P
1 Equip. Oper. (crane)	$38.10	$304.80	$57.50	$460.00	$32.75	$50.08
1 Laborer	27.40	219.20	42.65	341.20		
1 Power Shovel, 0.38 m³		534.65		588.10		
1 F.E. Attachment, 0.38 m³		48.80		53.70	36.46	40.11
16 L.H., Daily Totals		$1107.45		$1443.00	$69.21	$90.19

Crew B-12M

Crew No.	Bare Costs Hr.	Daily	Incl. Subs O&P Hr.	Daily	Cost Per Labor-Hour Bare Costs	Incl. O&P
1 Equip. Oper. (crane)	$38.10	$304.80	$57.50	$460.00	$32.75	$50.08
1 Laborer	27.40	219.20	42.65	341.20		
1 Power Shovel, 0.57 m³		663.50		729.85		
1 F.E. Attachment, 0.57 m³		53.20		58.50	44.79	49.27
16 L.H., Daily Totals		$1240.70		$1589.55	$77.54	$99.35

Crew B-12N

Crew No.	Bare Costs Hr.	Daily	Incl. Subs O&P Hr.	Daily	Cost Per Labor-Hour Bare Costs	Incl. O&P
1 Equip. Oper. (crane)	$38.10	$304.80	$57.50	$460.00	$32.75	$50.08
1 Laborer	27.40	219.20	42.65	341.20		
1 Power Shovel, 0.76 m³		880.00		968.00		
1 F.E. Attachment, 0.76 m³		59.00		64.90	58.69	64.56
16 L.H., Daily Totals		$1463.00		$1834.10	$91.44	$114.64

Crew B-12O

Crew No.	Bare Costs Hr.	Daily	Incl. Subs O&P Hr.	Daily	Cost Per Labor-Hour Bare Costs	Incl. O&P
1 Equip. Oper. (crane)	$38.10	$304.80	$57.50	$460.00	$32.75	$50.08
1 Laborer	27.40	219.20	42.65	341.20		
1 Power Shovel, 1.15 m³		934.20		1027.60		
1 F.E. Attachment, 1.15 m³		69.20		76.10	62.71	68.98
16 L.H., Daily Totals		$1527.40		$1904.90	$95.46	$119.06

Crew B-12P

Crew No.	Bare Costs Hr.	Daily	Incl. Subs O&P Hr.	Daily	Cost Per Labor-Hour Bare Costs	Incl. O&P
1 Equip. Oper. (crane)	$38.10	$304.80	$57.50	$460.00	$32.75	$50.08
1 Laborer	27.40	219.20	42.65	341.20		
1 Crawler Crane, 36 metric ton		934.20		1027.60		
1 Dragline Bucket, 1.15 m³		31.00		34.10	60.33	66.36
16 L.H., Daily Totals		$1489.20		$1862.90	$93.08	$116.44

Crew B-12Q

Crew No.	Bare Costs Hr.	Daily	Incl. Subs O&P Hr.	Daily	Cost Per Labor-Hour Bare Costs	Incl. O&P
1 Equip. Oper. (crane)	$38.10	$304.80	$57.50	$460.00	$32.75	$50.08
1 Laborer	27.40	219.20	42.65	341.20		
1 Hyd. Excavator, 0.48 m³		435.40		478.95	27.21	29.93
16 L.H., Daily Totals		$959.40		$1280.15	$59.96	$80.01

Crew B-12R

Crew No.	Bare Costs Hr.	Daily	Incl. Subs O&P Hr.	Daily	Cost Per Labor-Hour Bare Costs	Incl. O&P
1 Equip. Oper. (crane)	$38.10	$304.80	$57.50	$460.00	$32.75	$50.08
1 Laborer	27.40	219.20	42.65	341.20		
1 Hyd. Excavator, 1.15 m³		722.80		795.10	45.18	49.69
16 L.H., Daily Totals		$1246.80		$1596.30	$77.93	$99.77

Crew B-12S

Crew No.	Bare Costs Hr.	Daily	Incl. Subs O&P Hr.	Daily	Cost Per Labor-Hour Bare Costs	Incl. O&P
1 Equip. Oper. (crane)	$38.10	$304.80	$57.50	$460.00	$32.75	$50.08
1 Laborer	27.40	219.20	42.65	341.20		
1 Hyd. Excavator, 1.91 m³		1218.00		1339.80	76.13	83.74
16 L.H., Daily Totals		$1742.00		$2141.00	$108.88	$133.82

Crew B-12T

Crew No.	Bare Costs Hr.	Daily	Incl. Subs O&P Hr.	Daily	Cost Per Labor-Hour Bare Costs	Incl. O&P
1 Equip. Oper. (crane)	$38.10	$304.80	$57.50	$460.00	$32.75	$50.08
1 Laborer	27.40	219.20	42.65	341.20		
1 Crawler Crane, 68 metric ton		1226.00		1348.60		
1 F.E. Attachment, 2.29 m³		91.60		100.75	82.35	90.59
16 L.H., Daily Totals		$1841.60		$2250.55	$115.10	$140.67

Crew B-12V

Crew No.	Bare Costs Hr.	Daily	Incl. Subs O&P Hr.	Daily	Cost Per Labor-Hour Bare Costs	Incl. O&P
1 Equip. Oper. (crane)	$38.10	$304.80	$57.50	$460.00	$32.75	$50.08
1 Laborer	27.40	219.20	42.65	341.20		
1 Crawler Crane, 68 metric ton		1226.00		1348.60		
1 Dragline Bucket, 2.29 m³		47.80		52.60	79.61	87.57
16 L.H., Daily Totals		$1797.80		$2202.40	$112.36	$137.65

Crew B-13

Crew No.	Bare Costs Hr.	Daily	Incl. Subs O&P Hr.	Daily	Cost Per Labor-Hour Bare Costs	Incl. O&P
1 Laborer Foreman (outside)	$29.40	$235.20	$45.80	$366.40	$29.94	$46.12
4 Laborers	27.40	876.80	42.65	1364.80		
1 Equip. Oper. (crane)	38.10	304.80	57.50	460.00		
1 Equip. Oper. Oiler	32.45	259.60	48.95	391.60		
1 Hyd. Crane, 23 metric ton		650.80		715.90	11.62	12.78
56 L.H., Daily Totals		$2327.20		$3298.70	$41.56	$58.90

Crew B-13A

Crew No.	Bare Costs Hr.	Daily	Incl. Subs O&P Hr.	Daily	Cost Per Labor-Hour Bare Costs	Incl. O&P
1 Laborer Foreman	$29.40	$235.20	$45.80	$366.40	$30.61	$46.93
2 Laborers	27.40	438.40	42.65	682.40		
2 Equipment Operators	36.70	587.20	55.40	886.40		
2 Truck Drivers (heavy)	28.35	453.60	43.30	692.80		
1 Crane, 68 metric ton		1226.00		1348.60		
1 F.E. Lder, 2.87 m³		1110.00		1221.00		
2 Dump Trucks, 11 metric ton		653.20		718.50	53.38	58.72
56 L.H., Daily Totals		$4703.60		$5916.10	$83.99	$105.65

Crew No.	Bare Costs		Incl. Subs O & P		Cost Per Labor-Hour	

Crew B-13B

	Hr.	Daily	Hr.	Daily	Bare Costs	Incl. O&P
1 Laborer Foreman (outside)	$29.40	$235.20	$45.80	$366.40	$29.94	$46.12
4 Laborers	27.40	876.80	42.65	1364.80		
1 Equip. Oper. (crane)	38.10	304.80	57.50	460.00		
1 Equip. Oper. Oiler	32.45	259.60	48.95	391.60		
1 Hyd. Crane, 23 metric ton		951.60		1046.75	16.99	18.69
56 L.H., Daily Totals		$2628.00		$3629.55	$46.93	$64.81

Crew B-13C

	Hr.	Daily	Hr.	Daily	Bare Costs	Incl. O&P
1 Laborer Foreman (outside)	$29.40	$235.20	$45.80	$366.40	$29.94	$46.12
4 Laborers	27.40	876.80	42.65	1364.80		
1 Equip. Oper. (crane)	38.10	304.80	57.50	460.00		
1 Equip. Oper. Oiler	32.45	259.60	48.95	391.60		
1 Crawler Crane, 91 metric ton		1590.00		1749.00	28.39	31.23
56 L.H., Daily Totals		$3266.40		$4331.80	$58.33	$77.35

Crew B-14

	Hr.	Daily	Hr.	Daily	Bare Costs	Incl. O&P
1 Laborer Foreman (outside)	$29.40	$235.20	$45.80	$366.40	$29.03	$44.92
4 Laborers	27.40	876.80	42.65	1364.80		
1 Equip. Oper. (light)	35.20	281.60	53.10	424.80		
1 Backhoe Loader, 36 kW		226.20		248.80	4.71	5.18
48 L.H., Daily Totals		$1619.80		$2404.80	$33.74	$50.10

Crew B-15

	Hr.	Daily	Hr.	Daily	Bare Costs	Incl. O&P
1 Equipment Oper. (med)	$36.70	$293.60	$55.40	$443.20	$30.60	$46.66
.5 Laborer	27.40	109.60	42.65	170.60		
2 Truck Drivers (heavy)	28.35	453.60	43.30	692.80		
2 Dump Trucks, 14 metric ton		953.20		1048.50		
1 Dozer, 149 kW		920.80		1012.90	66.93	73.62
28 L.H., Daily Totals		$2730.80		$3368.00	$97.53	$120.28

Crew B-16

	Hr.	Daily	Hr.	Daily	Bare Costs	Incl. O&P
1 Laborer Foreman (outside)	$29.40	$235.20	$45.80	$366.40	$28.14	$43.60
2 Laborers	27.40	438.40	42.65	682.40		
1 Truck Driver (heavy)	28.35	226.80	43.30	346.40		
1 Dump Truck, 14 metric ton		476.60		524.25	14.89	16.38
32 L.H., Daily Totals		$1377.00		$1919.45	$43.03	$59.98

Crew B-17

	Hr.	Daily	Hr.	Daily	Bare Costs	Incl. O&P
2 Laborers	$27.40	$438.40	$42.65	$682.40	$29.59	$45.43
1 Equip. Oper. (light)	35.20	281.60	53.10	424.80		
1 Truck Driver (heavy)	28.35	226.80	43.30	346.40		
1 Backhoe Loader, 36 kW		226.20		248.80		
1 Dump Truck, 10 metric ton		326.60		359.25	17.28	19.00
32 L.H., Daily Totals		$1499.60		$2061.65	$46.87	$64.43

Crew B-17A

	Hr.	Daily	Hr.	Daily	Bare Costs	Incl. O&P
2 Laborer Foremen	$29.40	$470.40	$45.80	$732.80	$29.82	$46.42
6 Laborers	27.40	1315.20	42.65	2047.20		
1 Skilled Worker Foreman	38.50	308.00	59.90	479.20		
1 Skilled Worker	36.50	292.00	56.80	454.40		
80 L.H., Daily Totals		$2385.60		$3713.60	$29.82	$46.42

Crew B-18

	Hr.	Daily	Hr.	Daily	Bare Costs	Incl. O&P
1 Laborer Foreman (outside)	$29.40	$235.20	$45.80	$366.40	$28.07	$43.70
2 Laborers	27.40	438.40	42.65	682.40		
1 Vibrating Compactor		34.60		38.05	1.44	1.59
24 L.H., Daily Totals		$708.20		$1086.85	$29.51	$45.29

Crew B-19

	Hr.	Daily	Hr.	Daily	Bare Costs	Incl. O&P
1 Pile Driver Foreman	$36.70	$293.60	$60.35	$482.80	$35.52	$56.56
4 Pile Drivers	34.70	1110.40	57.05	1825.60		
2 Equip. Oper. (crane)	38.10	609.60	57.50	920.00		
1 Equip. Oper. Oiler	32.45	259.60	48.95	391.60		
1 Crane, 36 metric ton		934.20		1027.60		
18 m Pile Leads		66.00		72.60		
1 Hammer, Diesel, 30 kJ		568.00		624.80	24.55	27.01
64 L.H., Daily Totals		$3841.40		$5345.00	$60.07	$83.57

Crew B-19A

	Hr.	Daily	Hr.	Daily	Bare Costs	Incl. O&P
1 Pile Driver Foreman	$36.70	$293.60	$60.35	$482.80	$35.52	$56.56
4 Pile Drivers	34.70	1110.40	57.05	1825.60		
2 Equip. Oper. (crane)	38.10	609.60	57.50	920.00		
1 Equip. Oper. Oiler	32.45	259.60	48.95	391.60		
1 Crawler Crane, 68 metric ton		1226.00		1348.60		
18 m Leads, 33 150 J		114.00		125.40		
1 Hammer, Diesel, 60 kJ		672.00		739.20	31.39	34.53
64 L.H., Daily Totals		$4285.20		$5833.20	$66.91	$91.09

Crew B-20

	Hr.	Daily	Hr.	Daily	Bare Costs	Incl. O&P
1 Laborer Foreman (out)	$29.40	$235.20	$45.80	$366.40	$31.10	$48.42
1 Skilled Worker	36.50	292.00	56.80	454.40		
1 Laborer	27.40	219.20	42.65	341.20		
24 L.H., Daily Totals		$746.40		$1162.00	$31.10	$48.42

Crew B-20A

	Hr.	Daily	Hr.	Daily	Bare Costs	Incl. O&P
1 Laborer Foreman	$29.40	$235.20	$45.80	$366.40	$33.41	$51.03
1 Laborer	27.40	219.20	42.65	341.20		
1 Plumber	42.70	341.60	64.25	514.00		
1 Plumber Apprentice	34.15	273.20	51.40	411.20		
32 L.H., Daily Totals		$1069.20		$1632.80	$33.41	$51.03

Crew B-21

	Hr.	Daily	Hr.	Daily	Bare Costs	Incl. O&P
1 Laborer Foreman (out)	$29.40	$235.20	$45.80	$366.40	$32.10	$49.71
1 Skilled Worker	36.50	292.00	56.80	454.40		
1 Laborer	27.40	219.20	42.65	341.20		
.5 Equip. Oper. (crane)	38.10	152.40	57.50	230.00		
.5 S.P. Crane, 4.5 metric ton		152.60		167.85	5.45	6.00
28 L.H., Daily Totals		$1051.40		$1559.85	$37.55	$55.71

Crew B-21A

	Hr.	Daily	Hr.	Daily	Bare Costs	Incl. O&P
1 Laborer Foreman	$29.40	$235.20	$45.80	$366.40	$34.35	$52.32
1 Laborer	27.40	219.20	42.65	341.20		
1 Plumber	42.70	341.60	64.25	514.00		
1 Plumber Apprentice	34.15	273.20	51.40	411.20		
1 Equip. Oper. (crane)	38.10	304.80	57.50	460.00		
1 S.P. Crane, 11 metric ton		536.40		590.05	13.41	14.75
40 L.H., Daily Totals		$1910.40		$2682.85	$47.76	$67.07

Crew B-21B

	Hr.	Daily	Hr.	Daily	Bare Costs	Incl. O&P
1 Laborer Foreman	$29.40	$235.20	$45.80	$366.40	$29.94	$46.25
3 Laborers	27.40	657.60	42.65	1023.60		
1 Equip. Oper. (crane)	38.10	304.80	57.50	460.00		
1 Hyd. Crane, 11 Metric Ton		636.60		700.25	15.92	17.51
40 L.H., Daily Totals		$1834.20		$2550.25	$45.86	$63.76

Crew No.	Bare Costs Hr.	Daily	Incl. Subs O & P Hr.	Daily	Cost Per Labor-Hour Bare Costs	Incl. O&P
Crew B-21C	Hr.	Daily	Hr.	Daily	Bare Costs	Incl. O&P
1 Laborer Foreman	$29.40	$235.20	$45.80	$366.40	$29.94	$46.12
4 Laborers	27.40	876.80	42.65	1364.80		
1 Equip. Oper. (crane)	38.10	304.80	57.50	460.00		
1 Equip. Oper. Oiler	32.45	259.60	48.95	391.60		
2 Cutting Torches		32.40		35.65		
2 Gases		129.60		142.55		
1 Crane, 82 Metric Ton		1427.00		1569.70	28.38	31.21
56 L.H., Daily Totals		$3265.40		$4330.70	$58.32	$77.33
Crew B-22	Hr.	Daily	Hr.	Daily	Bare Costs	Incl. O&P
1 Laborer Foreman (out)	$29.40	$235.20	$45.80	$366.40	$32.50	$50.23
1 Skilled Worker	36.50	292.00	56.80	454.40		
1 Laborer	27.40	219.20	42.65	341.20		
.75 Equip. Oper. (crane)	38.10	228.60	57.50	345.00		
.75 S.P. Crane, 4.5 metric ton		228.90		251.80	7.63	8.39
30 L.H., Daily Totals		$1203.90		$1758.80	$40.13	$58.62
Crew B-22A	Hr.	Daily	Hr.	Daily	Bare Costs	Incl. O&P
1 Laborer Foreman (out)	$29.40	$235.20	$45.80	$366.40	$31.43	$48.64
1 Skilled Worker	36.50	292.00	56.80	454.40		
2 Laborers	27.40	438.40	42.65	682.40		
.75 Equipment Oper. (crane)	38.10	228.60	57.50	345.00		
.75 Crane, 4.5 Metric Tons		228.90		251.80		
1 Generator, 5 kW		31.20		34.30		
1 Butt Fusion Machine		347.20		381.90	15.98	17.58
38 L.H., Daily Totals		$1801.50		$2516.20	$47.41	$66.22
Crew B-22B	Hr.	Daily	Hr.	Daily	Bare Costs	Incl. O&P
1 Skilled Worker	$36.50	$292.00	$56.80	$454.40	$31.95	$49.73
1 Laborer	27.40	219.20	42.65	341.20		
1 Electro Fusion Machine		148.40		163.25	9.28	10.20
16 L.H., Daily Totals		$659.60		$958.85	$41.23	$59.93
Crew B-23	Hr.	Daily	Hr.	Daily	Bare Costs	Incl. O&P
1 Laborer Foreman (outside)	$29.40	$235.20	$45.80	$366.40	$27.80	$43.28
4 Laborers	27.40	876.80	42.65	1364.80		
1 Drill Rig		3379.00		3716.90		
1 Light Truck, 2.7 metric ton		176.20		193.80	88.88	97.77
40 L.H., Daily Totals		$4667.20		$5641.90	$116.68	$141.05
Crew B-23A	Hr.	Daily	Hr.	Daily	Bare Costs	Incl. O&P
1 Laborer Foreman (outside)	$29.40	$235.20	$45.80	$366.40	$31.17	$47.95
1 Laborer	27.40	219.20	42.65	341.20		
1 Equip. Operator (medium)	36.70	293.60	55.40	443.20		
1 Drill Rig, Wells		3379.00		3716.90		
1 Pickup Truck, 0.7 metric ton		78.00		85.80	144.04	158.45
24 L.H., Daily Totals		$4205.00		$4953.50	$175.21	$206.40
Crew B-23B	Hr.	Daily	Hr.	Daily	Bare Costs	Incl. O&P
1 Laborer Foreman (outside)	$29.40	$235.20	$45.80	$366.40	$31.17	$47.95
1 Laborer	27.40	219.20	42.65	341.20		
1 Equip. Operator (medium)	36.70	293.60	55.40	443.20		
1 Drill Rig, Wells		3379.00		3716.90		
1 Pickup Truck, 0.7 metric ton		78.00		85.80		
1 Pump, Cntfgl., 150 mm (6")		240.00		264.00	154.04	169.45
24 L.H., Daily Totals		$4445.00		$5217.50	$185.21	$217.40

Crew No.	Bare Costs Hr.	Daily	Incl. Subs O & P Hr.	Daily	Cost Per Labor-Hour Bare Costs	Incl. O&P
Crew B-24	Hr.	Daily	Hr.	Daily	Bare Costs	Incl. O&P
1 Cement Finisher	$34.40	$275.20	$50.75	$406.00	$32.45	$49.58
1 Laborer	27.40	219.20	42.65	341.20		
1 Carpenter	35.55	284.40	55.35	442.80		
24 L.H., Daily Totals		$778.80		$1190.00	$32.45	$49.58
Crew B-25	Hr.	Daily	Hr.	Daily	Bare Costs	Incl. O&P
1 Laborer Foreman	$29.40	$235.20	$45.80	$366.40	$30.12	$46.41
7 Laborers	27.40	1534.40	42.65	2388.40		
3 Equip. Oper. (med.)	36.70	880.80	55.40	1329.60		
1 Asphalt Paver, 97 kW		1661.00		1827.10		
1 Tandem Roller, 9 metric ton		197.00		216.70		
1 Roller, Pneumatic Wheel		259.20		285.10	24.06	26.47
88 L.H., Daily Totals		$4767.60		$6413.30	$54.18	$72.88
Crew B-25B	Hr.	Daily	Hr.	Daily	Bare Costs	Incl. O&P
1 Laborer Foreman	$29.40	$235.20	$45.80	$366.40	$30.67	$47.16
7 Laborers	27.40	1534.40	42.65	2388.40		
4 Equip. Oper. (medium)	36.70	1174.40	55.40	1772.80		
1 Asphalt Paver, 97 kW		1661.00		1827.10		
2 Rollers, Steel Wheel		394.00		433.40		
1 Roller, Pneumatic Wheel		259.20		285.10	24.11	26.52
96 L.H., Daily Totals		$5258.20		$7073.20	$54.78	$73.68
Crew B-25C	Hr.	Daily	Hr.	Daily	Bare Costs	Incl. O&P
1 Laborer Foreman	$29.40	$235.20	$45.80	$366.40	$30.83	$47.43
3 Laborers	27.40	657.60	42.65	1023.60		
2 Equip. Oper. (medium)	36.70	587.20	55.40	886.40		
1 Asphalt Paver, 97 kW		1661.00		1827.10		
1 Roller, Steel Wheel		197.00		216.70	38.71	42.58
48 L.H., Daily Totals		$3338.00		$4320.20	$69.54	$90.01
Crew B-26	Hr.	Daily	Hr.	Daily	Bare Costs	Incl. O&P
1 Laborer Foreman (outside)	$29.40	$235.20	$45.80	$366.40	$31.01	$48.03
6 Laborers	27.40	1315.20	42.65	2047.20		
2 Equip. Oper. (med.)	36.70	587.20	55.40	886.40		
1 Rodman (reinf.)	39.50	316.00	65.05	520.40		
1 Cement Finisher	34.40	275.20	50.75	406.00		
1 Grader, 13 608 kg		457.40		503.15		
1 Paving Mach. & Equip.		2237.00		2460.70	30.62	33.68
88 L.H., Daily Totals		$5423.20		$7190.25	$61.63	$81.71
Crew B-27	Hr.	Daily	Hr.	Daily	Bare Costs	Incl. O&P
1 Laborer Foreman (outside)	$29.40	$235.20	$45.80	$366.40	$27.90	$43.44
3 Laborers	27.40	657.60	42.65	1023.60		
1 Berm Machine		232.20		255.40	7.26	7.98
32 L.H., Daily Totals		$1125.00		$1645.40	$35.16	$51.42
Crew B-28	Hr.	Daily	Hr.	Daily	Bare Costs	Incl. O&P
2 Carpenters	$35.55	$568.80	$55.35	$885.60	$32.83	$51.12
1 Laborer	27.40	219.20	42.65	341.20		
24 L.H., Daily Totals		$788.00		$1226.80	$32.83	$51.12
Crew B-29	Hr.	Daily	Hr.	Daily	Bare Costs	Incl. O&P
1 Laborer Foreman (outside)	$29.40	$235.20	$45.80	$366.40	$29.94	$46.12
4 Laborers	27.40	876.80	42.65	1364.80		
1 Equip. Oper. (crane)	38.10	304.80	57.50	460.00		
1 Equip. Oper. Oiler	32.45	259.60	48.95	391.60		
1 Gradall, 2.7 metric ton		841.00		925.10	15.02	16.52
56 L.H., Daily Totals		$2517.40		$3507.90	$44.96	$62.64

Crews

Crew No.	Bare Costs Hr.	Daily	Incl. Subs O & P Hr.	Daily	Cost Per Labor-Hour Bare Costs	Incl. O&P
Crew B-30	Hr.	Daily	Hr.	Daily	Bare Costs	Incl. O&P
1 Equip. Oper. (med.)	$36.70	$293.60	$55.40	$443.20	$31.13	$47.33
2 Truck Drivers (heavy)	28.35	453.60	43.30	692.80		
1 Hyd. Excavator, 1.15 m³		722.80		795.10		
2 Dump Trucks, 14 metric ton		953.20		1048.50	69.83	76.82
24 L.H., Daily Totals		$2423.20		$2979.60	$100.96	$124.15

Crew B-31	Hr.	Daily	Hr.	Daily	Bare Costs	Incl. O&P
1 Laborer Foreman (outside)	$29.40	$235.20	$45.80	$366.40	$29.43	$45.82
3 Laborers	27.40	657.60	42.65	1023.60		
1 Carpenter	35.55	284.40	55.35	442.80		
1 Air Compr., 118 L/s		119.80		131.80		
1 Sheeting Driver		7.20		7.90		
2-15 m Hoses, 40mm Dia.(1.5")		9.40		10.35	3.41	3.75
40 L.H., Daily Totals		$1313.60		$1982.85	$32.84	$49.57

Crew B-32	Hr.	Daily	Hr.	Daily	Bare Costs	Incl. O&P
1 Laborer	$27.40	$219.20	$42.65	$341.20	$34.38	$52.21
3 Equip. Oper. (med.)	36.70	880.80	55.40	1329.60		
1 Grader, 13 608 kg		457.40		503.15		
1 Tandem Roller, 9 metric ton		197.00		216.70		
1 Dozer, 149 kW		920.80		1012.90	49.23	54.15
32 L.H., Daily Totals		$2675.20		$3403.55	$83.61	$106.36

Crew B-32A	Hr.	Daily	Hr.	Daily	Bare Costs	Incl. O&P
1 Laborer	$27.40	$219.20	$42.65	$341.20	$33.60	$51.15
2 Equip. Oper. (medium)	36.70	587.20	55.40	886.40		
1 Grader, 13 608 kg		457.40		503.15		
1 Roller, Vibr., 13 154 kg³		522.00		574.20	40.81	44.89
24 L.H., Daily Totals		$1785.80		$2304.95	$74.41	$96.04

Crew B-32B	Hr.	Daily	Hr.	Daily	Bare Costs	Incl. O&P
1 Laborer	$27.40	$219.20	$42.65	$341.20	$33.60	$51.15
2 Equip. Oper. (medium)	36.70	587.20	55.40	886.40		
1 Dozer, 149 kW		920.80		1012.90		
1 Roller, Vibr., 13 154 kg		522.00		574.20	60.12	66.13
24 L.H., Daily Totals		$2249.20		$2814.70	$93.72	$117.28

Crew B-32C	Hr.	Daily	Hr.	Daily	Bare Costs	Incl. O&P
1 Laborer Foreman	$29.40	$235.20	$45.80	$366.40	$32.38	$49.55
2 Laborers	27.40	438.40	42.65	682.40		
3 Equip. Oper. (medium)	36.70	880.80	55.40	1329.60		
1 Grader, 13 608 kg		457.40		503.15		
1 Roller, Steel Wheel		197.00		216.70		
1 Dozer, 149 kW		920.80		1012.90	32.82	36.10
48 L.H., Daily Totals		$3129.60		$4111.15	$65.20	$85.65

Crew B-33A	Hr.	Daily	Hr.	Daily	Bare Costs	Incl. O&P
1 Equip. Oper. (med.)	$36.70	$293.60	$55.40	$443.20	$34.04	$51.76
.5 Laborer	27.40	109.60	42.65	170.60		
.25 Equip. Oper. (med.)	36.70	73.40	55.40	110.80		
1 Scraper, Towed, 5.35 m³		161.75		177.95		
1.25 Dozer, 224 kW		1496.25		1645.90	118.43	130.28
14 L.H., Daily Totals		$2134.60		$2548.45	$152.47	$182.04

Crew B-33B	Hr.	Daily	Hr.	Daily	Bare Costs	Incl. O&P
1 Equip. Oper. (med.)	$36.70	$293.60	$55.40	$443.20	$34.04	$51.76
.5 Laborer	27.40	109.60	42.65	170.60		
.25 Equip. Oper. (med.)	36.70	73.40	55.40	110.80		
1 Scraper, Towed, 7.64 m³		180.40		198.45		
1.25 Dozer, 224 kW		1496.25		1645.90	119.76	131.74
14 L.H., Daily Totals		$2153.25		$2568.95	$153.80	$183.50

Crew B-33C	Hr.	Daily	Hr.	Daily	Bare Costs	Incl. O&P
1 Equip. Oper. (med.)	$36.70	$293.60	$55.40	$443.20	$34.04	$51.76
.5 Laborer	27.40	109.60	42.65	170.60		
.25 Equip. Oper. (med.)	36.70	73.40	55.40	110.80		
1 Scraper, Towed, 9.17 m³		180.40		198.45		
1.25 Dozer, 224 kW		1496.25		1645.90	119.76	131.74
14 L.H., Daily Totals		$2153.25		$2568.95	$153.80	$183.50

Crew B-33D	Hr.	Daily	Hr.	Daily	Bare Costs	Incl. O&P
1 Equip. Oper. (med.)	$36.70	$293.60	$55.40	$443.20	$34.04	$51.76
.5 Laborer	27.40	109.60	42.65	170.60		
.25 Equip. Oper. (med.)	36.70	73.40	55.40	110.80		
1 S.P. Scraper, 10.7 m³		1577.00		1734.70		
.25 Dozer, 224 kW		299.25		329.20	134.02	147.42
14 L.H., Daily Totals		$2352.85		$2788.50	$168.06	$199.18

Crew B-33E	Hr.	Daily	Hr.	Daily	Bare Costs	Incl. O&P
1 Equip. Oper. (med.)	$36.70	$293.60	$55.40	$443.20	$34.04	$51.76
.5 Laborer	27.40	109.60	42.65	170.60		
.25 Equip. Oper. (med.)	36.70	73.40	55.40	110.80		
1 S.P. Scraper, 18.34 m³		2350.00		2585.00		
.25 Dozer, 224 kW		299.25		329.20	189.23	208.16
14 L.H., Daily Totals		$3125.85		$3638.80	$223.27	$259.92

Crew B-33F	Hr.	Daily	Hr.	Daily	Bare Costs	Incl. O&P
1 Equip. Oper. (med.)	$36.70	$293.60	$55.40	$443.20	$34.04	$51.76
.5 Laborer	27.40	109.60	42.65	170.60		
.25 Equip. Oper. (med.)	36.70	73.40	55.40	110.80		
1 Elev. Scraper, 8.4 m³		822.60		904.85		
.25 Dozer, 224 kW		299.25		329.20	80.13	88.15
14 L.H., Daily Totals		$1598.45		$1958.65	$114.17	$139.91

Crew B-33G	Hr.	Daily	Hr.	Daily	Bare Costs	Incl. O&P
1 Equip. Oper. (med.)	$36.70	$293.60	$55.40	$443.20	$34.04	$51.76
.5 Laborer	27.40	109.60	42.65	170.60		
.25 Equip. Oper. (med.)	36.70	73.40	55.40	110.80		
1 Elev. Scraper, 15.28 m³		1653.00		1818.30		
.25 Dozer, 224 kW		299.25		329.20	139.45	153.39
14 L.H., Daily Totals		$2428.85		$2872.10	$173.49	$205.15

Crew B-33H	Hr.	Daily	Hr.	Daily	Bare Costs	Incl. O&P
.25 Laborer	$27.40	$54.80	$42.65	$85.30	$35.10	$53.20
1 Equipment Operator (med.)	36.70	293.60	55.40	443.20		
.2 Equipment Operator (med.)	36.70	58.72	55.40	88.64		
1 Scraper, 24-34 m³		2758.00		3033.80		
.2 Dozer, 298 kW		307.20		337.90	264.24	290.67
11. L.H., Daily Totals		$3472.32		$3988.84	$299.34	$343.87

Crew B-33J	Hr.	Daily	Hr.	Daily	Bare Costs	Incl. O&P
1 Equipment Operator (med.)	$36.70	$293.60	$55.40	$443.20	$36.70	$55.40
1 Scraper 13 m³		1577.00		1734.70	197.13	216.84
8 L.H., Daily Totals		$1870.60		$2177.90	$233.83	$272.24

Crew B-34A	Hr.	Daily	Hr.	Daily	Bare Costs	Incl. O&P
1 Truck Driver (heavy)	$28.35	$226.80	$43.30	$346.40	$28.35	$43.30
1 Dump Truck, 11 metric ton		326.60		359.25	40.83	44.91
8 L.H., Daily Totals		$553.40		$705.65	$69.18	$88.21

Crew No.	Bare Costs Hr.	Daily	Incl. Subs O & P Hr.	Daily	Cost Per Labor-Hour Bare Costs	Incl. O&P
Crew B-34B	Hr.	Daily	Hr.	Daily	Bare Costs	Incl. O&P
1 Truck Driver (heavy)	$28.35	$226.80	$43.30	$346.40	$28.35	$43.30
1 Dump Truck, 15 metric ton		476.60		524.25	59.58	65.53
8 L.H., Daily Totals		$703.40		$870.65	$87.93	$108.83
Crew B-34C	Hr.	Daily	Hr.	Daily	Bare Costs	Incl. O&P
1 Truck Driver (heavy)	$28.35	$226.80	$43.30	$346.40	$28.35	$43.30
1 Truck Tractor, 36 metric ton		301.60		331.75		
1 Dump Trailer, 12.61 m³		103.00		113.30	50.58	55.63
8 L.H., Daily Totals		$631.40		$791.45	$78.93	$98.93
Crew B-34D	Hr.	Daily	Hr.	Daily	Bare Costs	Incl. O&P
1 Truck Driver (heavy)	$28.35	$226.80	$43.30	$346.40	$28.35	$43.30
1 Truck Tractor, 36 metric ton		301.60		331.75		
1 Dump Trailer, 15.28 m³		115.80		127.40	52.18	57.39
8 L.H., Daily Totals		$644.20		$805.55	$80.53	$100.69
Crew B-34E	Hr.	Daily	Hr.	Daily	Bare Costs	Incl. O&P
1 Truck Driver (heavy)	$28.35	$226.80	$43.30	$346.40	$28.35	$43.30
1 Truck, Off Hwy., 23 met. ton		912.20		1003.40	114.03	125.43
8 L.H., Daily Totals		$1139.00		$1349.80	$142.38	$168.73
Crew B-34F	Hr.	Daily	Hr.	Daily	Bare Costs	Incl. O&P
1 Truck Driver (heavy)	$28.35	$226.80	$43.30	$346.40	$28.35	$43.30
1 Truck, Off Hwy., 16.81 m³		933.40		1026.75	116.68	128.34
8 L.H., Daily Totals		$1160.20		$1373.15	$145.03	$171.64
Crew B-34G	Hr.	Daily	Hr.	Daily	Bare Costs	Incl. O&P
1 Truck Driver (heavy)	$28.35	$226.80	$43.30	$346.40	$28.35	$43.30
1 Truck, Off Hwy., 25.98 m³		1181.00		1299.10	147.63	162.39
8 L.H., Daily Totals		$1407.80		$1645.50	$175.98	$205.69
Crew B-34H	Hr.	Daily	Hr.	Daily	Bare Costs	Incl. O&P
1 Truck Driver (heavy)	$28.35	$226.80	$43.30	$346.40	$28.35	$43.30
1 Truck, Off Hwy., 32.09 m³		1276.00		1403.60	159.50	175.45
8 L.H., Daily Totals		$1502.80		$1750.00	$187.85	$218.75
Crew B-34J	Hr.	Daily	Hr.	Daily	Bare Costs	Incl. O&P
1 Truck Driver (heavy)	$28.35	$226.80	$43.30	$346.40	$28.35	$43.30
1 Truck, Off Hwy., 45.84 m³		1649.00		1813.90	206.13	226.74
8 L.H., Daily Totals		$1875.80		$2160.30	$234.48	$270.04
Crew B-34K	Hr.	Daily	Hr.	Daily	Bare Costs	Incl. O&P
1 Truck Driver (heavy)	$28.35	$226.80	$43.30	$346.40	$28.35	$43.30
1 Truck Tractor, 179 kW		353.60		388.95		
1 Low Bed Trailer		170.20		187.20	65.48	72.02
8 L.H., Daily Totals		$750.60		$922.55	$93.83	$115.32
Crew B-34N	Hr.	Daily	Hr.	Daily	Bare Costs	Incl. O&P
1 Truck Driver (heavy)	$28.35	$226.80	$43.30	$346.40	$28.35	$43.30
1 Dump Truck, 11 metric ton		326.60		359.25		
1 Flatbed trailer, 36 metric ton		119.40		131.35	55.75	61.33
8 L.H., Daily Totals		$672.80		$837.00	$84.10	$104.63

Crew No.	Bare Costs Hr.	Daily	Incl. Subs O & P Hr.	Daily	Cost Per Labor-Hour Bare Costs	Incl. O&P
Crew B-34P	Hr.	Daily	Hr.	Daily	Bare Costs	Incl. O&P
1 Pipe Fitter	$43.05	$344.40	$64.80	$518.40	$35.78	$54.12
1 Truck Driver (light)	27.60	220.80	42.15	337.20		
1 Equip. Oper. (medium)	36.70	293.60	55.40	443.20		
1 Flatbed Truck, 3 metric ton		176.20		193.80		
1 Backhoe Loader, 36 kW		226.20		248.80	16.77	18.44
24 L.H., Daily Totals		$1261.20		$1741.40	$52.55	$72.56
Crew B-34Q	Hr.	Daily	Hr.	Daily	Bare Costs	Incl. O&P
1 Pipe Fitter	$43.05	$344.40	$64.80	$518.40	$36.25	$54.82
1 Truck Driver (light)	27.60	220.80	42.15	337.20		
1 Equip. Oper. (crane)	38.10	304.80	57.50	460.00		
1 Flatbed Trailer, 22 metric ton		88.00		96.80		
1 Dump Truck, 11 metric ton		326.60		359.25		
1 Hyd. Crane, 23 metric ton		650.80		715.90	44.39	48.83
24 L.H., Daily Totals		$1935.40		$2487.55	$80.64	$103.65
Crew B-34R	Hr.	Daily	Hr.	Daily	Bare Costs	Incl. O&P
1 Pipe Fitter	$43.05	$344.40	$64.80	$518.40	$36.25	$54.82
1 Truck Driver (light)	27.60	220.80	42.15	337.20		
1 Equip. Oper. (crane)	38.10	304.80	57.50	460.00		
1 Flatbed Trailer, 22 metric ton		88.00		96.80		
1 Dump Truck, 11 metric ton		326.60		359.25		
1 Hyd. Crane, 23 metric ton		650.80		715.90		
1 Hyd. Excavator, 0.76 m³		560.20		616.20	67.73	74.51
24 L.H., Daily Totals		$2495.60		$3103.75	$103.98	$129.33
Crew B-34S	Hr.	Daily	Hr.	Daily	Bare Costs	Incl. O&P
2 Pipe Fitters	$43.05	$688.80	$64.80	$1036.80	$38.14	$57.60
1 Truck Driver (heavy)	28.35	226.80	43.30	346.40		
1 Equip. Oper. (crane)	38.10	304.80	57.50	460.00		
1 Flatbed trailer, 36 metric ton		119.40		131.35		
1 Truck Tractor, 36 metric ton		301.60		331.75		
1 Truck Crane, 73 metric ton		1059.00		1164.90		
1 Hyd. Excavator, 1.53 m³		919.60		1011.55	74.99	82.49
32 L.H., Daily Totals		$3620.00		$4482.75	$113.13	$140.09
Crew B-34T	Hr.	Daily	Hr.	Daily	Bare Costs	Incl. O&P
2 Pipe Fitters	$43.05	$688.80	$64.80	$1036.80	$38.14	$57.60
1 Truck Driver (heavy)	28.35	226.80	43.30	346.40		
1 Equip. Oper. (crane)	38.10	304.80	57.50	460.00		
1 Flatbed trailer, 36 metric ton		119.40		131.35		
1 Truck Tractor, 36 metric ton		301.60		331.75		
1 Truck Crane, 73 metric ton		1059.00		1164.90	46.25	50.88
32 L.H., Daily Totals		$2700.40		$3471.20	$84.39	$108.48
Crew B-35	Hr.	Daily	Hr.	Daily	Bare Costs	Incl. O&P
1 Laborer Foreman (out)	$29.40	$235.20	$45.80	$366.40	$34.43	$52.66
1 Skilled Worker	36.50	292.00	56.80	454.40		
1 Welder (plumber)	42.70	341.60	64.25	514.00		
1 Laborer	27.40	219.20	42.65	341.20		
1 Equip. Oper. (crane)	38.10	304.80	57.50	460.00		
1 Equip. Oper. Oiler	32.45	259.60	48.95	391.60		
1 Electric Welding Mach.		69.10		76.00		
1 Hyd. Excavator, 0.57 m³		477.80		525.60	11.39	12.53
48 L.H., Daily Totals		$2199.30		$3129.20	$45.82	$65.19

Crew B-35A

Crew No.	Bare Costs Hr.	Daily	Incl. Subs O&P Hr.	Daily	Bare Costs	Incl. O&P
1 Laborer Foreman (out)	$29.40	$235.20	$45.80	$366.40	$33.42	$51.23
2 Laborers	27.40	438.40	42.65	682.40		
1 Skilled Worker	36.50	292.00	56.80	454.40		
1 Welder (plumber)	42.70	341.60	64.25	514.00		
1 Equip. Oper. (crane)	38.10	304.80	57.50	460.00		
1 Equip. Oper. Oiler	32.45	259.60	48.95	391.60		
1 Welder, 300 A		89.20		98.10		
1 Crane, 68 metric ton		1226.00		1348.60	23.49	25.83
56 L.H., Daily Totals		$3186.80		$4315.50	$56.91	$77.06

Crew B-36

Crew No.	Bare Costs Hr.	Daily	Incl. Subs O&P Hr.	Daily	Bare Costs	Incl. O&P
1 Laborer Foreman (outside)	$29.40	$235.20	$45.80	$366.40	$31.52	$48.38
2 Laborers	27.40	438.40	42.65	682.40		
2 Equip. Oper. (med.)	36.70	587.20	55.40	886.40		
1 Dozer, 149 kW		920.80		1012.90		
1 Aggregate Spreader		32.00		35.20		
1 Tandem Roller, 9 metric ton		197.00		216.70	28.75	31.62
40 L.H., Daily Totals		$2410.60		$3200.00	$60.27	$80.00

Crew B-36A

Crew No.	Bare Costs Hr.	Daily	Incl. Subs O&P Hr.	Daily	Bare Costs	Incl. O&P
1 Laborer Foreman (outside)	$29.40	$235.20	$45.80	$366.40	$33.00	$50.39
2 Laborers	27.40	438.40	42.65	682.40		
4 Equip. Oper. (med.)	36.70	1174.40	55.40	1772.80		
1 Dozer, 149 kW		920.80		1012.90		
1 Aggregate Spreader		32.00		35.20		
1 Roller, Steel Wheel		197.00		216.70		
1 Roller, Pneumatic Wheel		259.20		285.10	25.16	27.68
56 L.H., Daily Totals		$3257.00		$4371.50	$58.16	$78.07

Crew B-36B

Crew No.	Bare Costs Hr.	Daily	Incl. Subs O&P Hr.	Daily	Bare Costs	Incl. O&P
1 Laborer Foreman (outside)	$29.40	$235.20	$45.80	$366.40	$32.42	$49.50
2 Laborers	27.40	438.40	42.65	682.40		
4 Equip. Oper. (medium)	36.70	1174.40	55.40	1772.80		
1 Truck Driver, Heavy	28.35	226.80	43.30	346.40		
1 Grader, 14 000 kg		457.40		503.15		
1 F.E. Loader, crl, 1.15 m³		372.40		409.65		
1 Dozer, 224 kW		1197.00		1316.70		
1 Roller, vibratory		522.00		574.20		
1 Truck, Tractor, 179 kW		353.60		388.95		
1 Water Tanker, 19 kL		116.80		128.50	47.18	51.89
64 L.H., Daily Totals		$5094.00		$6489.15	$79.60	$101.39

Crew B-36C

Crew No.	Bare Costs Hr.	Daily	Incl. Subs O&P Hr.	Daily	Bare Costs	Incl. O&P
1 Laborer Foreman (outside)	$29.40	$235.20	$45.80	$366.40	$33.57	$51.06
3 Equip. Oper. (medium)	36.70	880.80	55.40	1329.60		
1 Truck Driver, Heavy	28.35	226.80	43.30	346.40		
1 Grader, 14 000 kg		457.40		503.15		
1 Dozer, 224 kW		1197.00		1316.70		
1 Roller, vibratory		522.00		574.20		
1 Truck, Tractor, 179 kW		353.60		388.95		
1 Water Tanker, 19 kL		116.80		128.50	66.17	72.79
40 L.H., Daily Totals		$3989.60		$4953.90	$99.74	$123.85

Crew B-37

Crew No.	Bare Costs Hr.	Daily	Incl. Subs O&P Hr.	Daily	Bare Costs	Incl. O&P
1 Laborer Foreman (outside)	$29.40	$235.20	$45.80	$366.40	$29.03	$44.92
4 Laborers	27.40	876.80	42.65	1364.80		
1 Equip. Oper. (light)	35.20	281.60	53.10	424.80		
1 Tandem Roller, 4.5 met. ton		114.20		125.60	2.38	2.62
48 L.H., Daily Totals		$1507.80		$2281.60	$31.41	$47.54

Crew B-38

Crew No.	Bare Costs Hr.	Daily	Incl. Subs O&P Hr.	Daily	Bare Costs	Incl. O&P
1 Laborer Foreman (outside)	$29.40	$235.20	$45.80	$366.40	$31.22	$47.92
2 Laborers	27.40	438.40	42.65	682.40		
1 Equip. Oper. (light)	35.20	281.60	53.10	424.80		
1 Equip. Oper. (medium)	36.70	293.60	55.40	443.20		
1 Backhoe Loader, 36 kW		226.20		248.80		
1 Demol. Hammer, (453.6 kg)		113.60		124.95		
1 F.E. Loader (127 kW)		444.00		488.40		
1 Pavt. Rem. Bucket		52.20		57.40	20.90	22.99
40 L.H., Daily Totals		$2084.80		$2836.35	$52.12	$70.91

Crew B-39

Crew No.	Bare Costs Hr.	Daily	Incl. Subs O&P Hr.	Daily	Bare Costs	Incl. O&P
1 Laborer Foreman (outside)	$29.40	$235.20	$45.80	$366.40	$29.03	$44.92
4 Laborers	27.40	876.80	42.65	1364.80		
1 Equip. Oper. (light)	35.20	281.60	53.10	424.80		
1 Air Compr., 118 L/s		119.80		131.80		
2 Air Tools & Accessories		18.00		19.80		
2-15 m Hoses, 40mm Dia.(1.5")		9.40		10.35	3.07	3.38
48 L.H., Daily Totals		$1540.80		$2317.95	$32.10	$48.30

Crew B-40

Crew No.	Bare Costs Hr.	Daily	Incl. Subs O&P Hr.	Daily	Bare Costs	Incl. O&P
1 Pile Driver Foreman (out)	$36.70	$293.60	$60.35	$482.80	$35.52	$56.56
4 Pile Drivers	34.70	1110.40	57.05	1825.60		
2 Equip. Oper. (crane)	38.10	609.60	57.50	920.00		
1 Equip. Oper. Oiler	32.45	259.60	48.95	391.60		
1 Crane, 36 metric ton		934.20		1027.60		
1 Vibratory Hammer & Gen.		1823.00		2005.30	43.08	47.39
64 L.H., Daily Totals		$5030.40		$6652.90	$78.60	$103.95

Crew B-40B

Crew No.	Bare Costs Hr.	Daily	Incl. Subs O&P Hr.	Daily	Bare Costs	Incl. O&P
1 Laborer Foreman	$29.40	$235.20	$45.80	$366.40	$30.36	$46.70
3 Laborers	27.40	657.60	42.65	1023.60		
1 Equip. Oper. (crane)	38.10	304.80	57.50	460.00		
1 Equip. Oper. Oiler	32.45	259.60	48.95	391.60		
1 Crane, 36 Metric Ton		1130.00		1243.00	23.54	25.90
48 L.H., Daily Totals		$2587.20		$3484.60	$53.90	$72.60

Crew B-41

Crew No.	Bare Costs Hr.	Daily	Incl. Subs O&P Hr.	Daily	Bare Costs	Incl. O&P
1 Laborer Foreman (outside)	$29.40	$235.20	$45.80	$366.40	$28.48	$44.18
4 Laborers	27.40	876.80	42.65	1364.80		
.25 Equip. Oper. (crane)	38.10	76.20	57.50	115.00		
.25 Equip. Oper. Oiler	32.45	64.90	48.95	97.90		
.25 Crawler Crane, 36 met. ton		233.55		256.90	5.31	5.84
44 L.H., Daily Totals		$1486.65		$2201.00	$33.79	$50.02

Crew B-42

Crew No.	Bare Costs Hr.	Daily	Incl. Subs O&P Hr.	Daily	Bare Costs	Incl. O&P
1 Laborer Foreman (outside)	$29.40	$235.20	$45.80	$366.40	$31.19	$49.34
4 Laborers	27.40	876.80	42.65	1364.80		
1 Equip. Oper. (crane)	38.10	304.80	57.50	460.00		
1 Equip. Oper. Oiler	32.45	259.60	48.95	391.60		
1 Welder	39.95	319.60	71.90	575.20		
1 Hyd. Crane, 23 metric ton		650.80		715.90		
1 Gas Welding Machine		89.20		98.10		
1 Horz. Boring Csg. Mch.		392.00		431.20	17.69	19.46
64 L.H., Daily Totals		$3128.00		$4403.20	$48.88	$68.80

Crew No.	Bare Costs		Incl. Subs O & P		Cost Per Labor-Hour	
Crew B-43	Hr.	Daily	Hr.	Daily	Bare Costs	Incl. O&P
1 Laborer Foreman (outside)	$29.40	$235.20	$45.80	$366.40	$30.36	$46.70
3 Laborers	27.40	657.60	42.65	1023.60		
1 Equip. Oper. (crane)	38.10	304.80	57.50	460.00		
1 Equip. Oper. Oiler	32.45	259.60	48.95	391.60		
1 Drill Rig & Augers		3379.00		3716.90	70.40	77.44
48 L.H., Daily Totals		$4836.20		$5958.50	$100.76	$124.14
Crew B-44	Hr.	Daily	Hr.	Daily	Bare Costs	Incl. O&P
1 Pile Driver Foreman	$36.70	$293.60	$60.35	$482.80	$34.89	$55.78
4 Pile Drivers	34.70	1110.40	57.05	1825.60		
2 Equip. Oper. (crane)	38.10	609.60	57.50	920.00		
1 Laborer	27.40	219.20	42.65	341.20		
1 Crane, 36 metric ton		934.20		1027.60		
14 m Leads, 20 337 J		49.50		54.45	15.41	16.95
64 L.H., Daily Totals		$3216.50		$4651.65	$50.30	$72.73
Crew B-45	Hr.	Daily	Hr.	Daily	Bare Costs	Incl. O&P
1 Equip. Oper. (med.)	$36.70	$293.60	$55.40	$443.20	$32.53	$49.35
1 Truck Driver (heavy)	28.35	226.80	43.30	346.40		
1 Dist. Tank Truck, 11.355 L		242.40		266.65		
1 Tractor, 4 x 2, 186 kW		291.60		320.75	33.38	36.71
16 L.H., Daily Totals		$1054.40		$1377.00	$65.91	$86.06
Crew B-46	Hr.	Daily	Hr.	Daily	Bare Costs	Incl. O&P
1 Pile Driver Foreman	$36.70	$293.60	$60.35	$482.80	$31.38	$50.40
2 Pile Drivers	34.70	555.20	57.05	912.80		
3 Laborers	27.40	657.60	42.65	1023.60		
1 Chain Saw, 914 mm Long		33.40		36.75	.70	.77
48 L.H., Daily Totals		$1539.80		$2455.95	$32.08	$51.17
Crew B-47	Hr.	Daily	Hr.	Daily	Bare Costs	Incl. O&P
1 Blast Foreman	$29.40	$235.20	$45.80	$366.40	$30.67	$47.18
1 Driller	27.40	219.20	42.65	341.20		
1 Equip. Oper. (light)	35.20	281.60	53.10	424.80		
1 Crawler Type Drill, 102 mm		728.80		801.70		
1 Air Compr., 283.2 L/s		320.20		352.20		
2-15 m Hoses, 80 mm Dia.(3")		37.10		40.80	45.26	49.78
24 L.H., Daily Totals		$1822.10		$2327.10	$75.93	$96.96
Crew B-47A	Hr.	Daily	Hr.	Daily	Bare Costs	Incl. O&P
1 Drilling Foreman	$29.40	$235.20	$45.80	$366.40	$33.32	$50.75
1 Equip. Oper. (heavy)	38.10	304.80	57.50	460.00		
1 Oiler	32.45	259.60	48.95	391.60		
1 Quarry Drill		822.00		904.20	34.25	37.68
24 L.H., Daily Totals		$1621.60		$2122.20	$67.57	$88.43
Crew B-47C	Hr.	Daily	Hr.	Daily	Bare Costs	Incl. O&P
1 Laborer	$27.40	$219.20	$42.65	$341.20	$31.30	$47.88
1 Equip. Oper. (light)	35.20	281.60	53.10	424.80		
1 Air Compressor, 354 L/s		341.40		375.55		
2-15 m Air Hoses, 80 mm		37.10		40.80		
1 Air Track Drill, 100 mm		728.80		801.70	69.21	76.13
16 L.H., Daily Totals		$1608.10		$1984.05	$100.51	$124.01
Crew B-47E	Hr.	Daily	Hr.	Daily	Bare Costs	Incl. O&P
1 Laborer Foreman	$29.40	$235.20	$45.80	$366.40	$27.90	$43.44
3 Laborers	27.40	657.60	42.65	1023.60		
1 Truck, Flatbed, 2.7 metric ton		176.20		193.80	5.51	6.06
32 L.H., Daily Totals		$1069.00		$1583.80	$33.41	$49.50

Crew No.	Bare Costs		Incl. Subs O & P		Cost Per Labor-Hour	
Crew B-47G	Hr.	Daily	Hr.	Daily	Bare Costs	Incl. O&P
1 Laborer Foreman	$29.40	$235.20	$45.80	$366.40	$29.85	$46.05
2 Laborers	27.40	438.40	42.65	682.40		
1 Equip. Oper. (light)	35.20	281.60	53.10	424.80		
1 Air Track Drill, 100 mm		728.80		801.70		
1 Air Compr., 283 L/s		320.20		352.20		
2-15 m Hoses, 75 mm Dia.		37.10		40.80		
1 Grout Pump		106.40		117.05	37.27	41.00
32 L.H., Daily Totals		$2147.70		$2785.35	$67.12	$87.05
Crew B-48	Hr.	Daily	Hr.	Daily	Bare Costs	Incl. O&P
1 Laborer Foreman (outside)	$29.40	$235.20	$45.80	$366.40	$31.05	$47.61
3 Laborers	27.40	657.60	42.65	1023.60		
1 Equip. Oper. (crane)	38.10	304.80	57.50	460.00		
1 Equip. Oper. Oiler	32.45	259.60	48.95	391.60		
1 Equip. Oper. (light)	35.20	281.60	53.10	424.80		
1 Centr. Pump, 150 mm (6")		240.00		264.00		
1-6 m Suct. Hose, 150 mm (6")		12.10		13.30		
1-15 m Disch., 150 mm (6")		7.50		8.25		
1 Drill Rig & Augers		3379.00		3716.90	64.98	71.47
56 L.H., Daily Totals		$5377.40		$6668.85	$96.03	$119.08
Crew B-49	Hr.	Daily	Hr.	Daily	Bare Costs	Incl. O&P
1 Laborer Foreman (outside)	$29.40	$235.20	$45.80	$366.40	$32.48	$50.35
3 Laborers	27.40	657.60	42.65	1023.60		
2 Equip. Oper. (crane)	38.10	609.60	57.50	920.00		
2 Equip. Oper. Oilers	32.45	519.20	48.95	783.20		
1 Equip. Oper. (light)	35.20	281.60	53.10	424.80		
2 Pile Drivers	34.70	555.20	57.05	912.80		
1 Hyd. Crane, 23 metric ton		650.80		715.90		
1 Centr. Pump, 150 mm (6")		240.00		264.00		
1-6 m Suct. Hose, 150 mm (6")		12.10		13.30		
1-15 m Disch., 150 mm (6")		7.50		8.25		
1 Drill Rig & Augers		3379.00		3716.90	48.74	53.62
88 L.H., Daily Totals		$7147.80		$9149.15	$81.22	$103.97
Crew B-50	Hr.	Daily	Hr.	Daily	Bare Costs	Incl. O&P
2 Pile Driver Foremen	$36.70	$587.20	$60.35	$965.60	$33.75	$53.92
6 Pile Drivers	34.70	1665.60	57.05	2738.40		
2 Equip. Oper. (crane)	38.10	609.60	57.50	920.00		
1 Equip. Oper. Oiler	32.45	259.60	48.95	391.60		
3 Laborers	27.40	657.60	42.65	1023.60		
1 Crane, 36 metric ton		934.20		1027.60		
18 m Leads, 20 337 J		66.00		72.60		
1 Hammer, 20 337 J		355.80		391.40		
1 Air Compr., 283 L/s		320.20		352.20		
2-15 m Hoses, 80 mm Dia.(3")		37.10		40.80		
1 Chain Saw, 914 mm Long		33.40		36.75	15.62	17.19
112 L.H., Daily Totals		$5526.30		$7960.55	$49.37	$71.11
Crew B-51	Hr.	Daily	Hr.	Daily	Bare Costs	Incl. O&P
1 Laborer Foreman (outside)	$29.40	$235.20	$45.80	$366.40	$27.77	$43.09
4 Laborers	27.40	876.80	42.65	1364.80		
1 Truck Driver (light)	27.60	220.80	42.15	337.20		
1 Light Truck, 1.4 metric ton		128.80		141.70	2.68	2.95
48 L.H., Daily Totals		$1461.60		$2210.10	$30.45	$46.04

Crew B-52

Crew B-52	Bare Costs Hr.	Bare Costs Daily	Incl. Subs O&P Hr.	Incl. Subs O&P Daily	Cost Per Labor-Hour Bare Costs	Cost Per Labor-Hour Incl. O&P
1 Carpenter Foreman	$37.55	$300.40	$58.45	$467.60	$32.54	$50.39
1 Carpenter	35.55	284.40	55.35	442.80		
3 Laborers	27.40	657.60	42.65	1023.60		
1 Cement Finisher	34.40	275.20	50.75	406.00		
.5 Rodman (reinf.)	39.50	158.00	65.05	260.20		
.5 Equip. Oper. (med.)	36.70	146.80	55.40	221.60		
.5 F.E. Ldr., T.M., 1.91 m³		397.00		436.70	7.09	7.80
56 L.H., Daily Totals		$2219.40		$3258.50	$39.63	$58.19

Crew B-53

Crew B-53	Bare Costs Hr.	Bare Costs Daily	Incl. Subs O&P Hr.	Incl. Subs O&P Daily	Cost Per Labor-Hour Bare Costs	Cost Per Labor-Hour Incl. O&P
1 Equip. Oper. (light)	$35.20	$281.60	$53.10	$424.80	$35.20	$53.10
1 Trencher, Chain, 9 kW		47.60		52.35	5.95	6.55
8 L.H., Daily Totals		$329.20		$477.15	$41.15	$59.65

Crew B-54

Crew B-54	Bare Costs Hr.	Bare Costs Daily	Incl. Subs O&P Hr.	Incl. Subs O&P Daily	Cost Per Labor-Hour Bare Costs	Cost Per Labor-Hour Incl. O&P
1 Equip. Oper. (light)	$35.20	$281.60	$53.10	$424.80	$35.20	$53.10
1 Trencher, Chain, 30 kW		216.80		238.50	27.10	29.81
8 L.H., Daily Totals		$498.40		$663.30	$62.30	$82.91

Crew B-54A

Crew B-54A	Bare Costs Hr.	Bare Costs Daily	Incl. Subs O&P Hr.	Incl. Subs O&P Daily	Cost Per Labor-Hour Bare Costs	Cost Per Labor-Hour Incl. O&P
.17 Laborer Foreman (outside)	$29.40	$39.98	$45.80	$62.29	$35.64	$54.01
1 Equipment Operator (med.)	36.70	293.60	55.40	443.20		
1 Wheel Trencher, 50 kW		826.80		909.50	88.33	97.17
9.36 L.H., Daily Totals		$1160.38		$1414.99	$123.97	$151.18

Crew B-54B

Crew B-54B	Bare Costs Hr.	Bare Costs Daily	Incl. Subs O&P Hr.	Incl. Subs O&P Daily	Cost Per Labor-Hour Bare Costs	Cost Per Labor-Hour Incl. O&P
.25 Laborer Foreman (outside)	$29.40	$58.80	$45.80	$91.60	$35.24	$53.48
1 Equipment Operator (med.)	36.70	293.60	55.40	443.20		
1 Wheel Trencher, 112 kW		1578.00		1735.80	157.80	173.58
10 L.H., Daily Totals		$1930.40		$2270.60	$193.04	$227.06

Crew B-55

Crew B-55	Bare Costs Hr.	Bare Costs Daily	Incl. Subs O&P Hr.	Incl. Subs O&P Daily	Cost Per Labor-Hour Bare Costs	Cost Per Labor-Hour Incl. O&P
2 Laborers	$27.40	$438.40	$42.65	$682.40	$27.47	$42.48
1 Truck Driver (light)	27.60	220.80	42.15	337.20		
1 Flatbed Truck w/Auger		631.60		694.75		
1 Truck, 2.7 metric ton		176.20		193.80	33.66	37.02
24 L.H., Daily Totals		$1467.00		$1908.15	$61.13	$79.50

Crew B-56

Crew B-56	Bare Costs Hr.	Bare Costs Daily	Incl. Subs O&P Hr.	Incl. Subs O&P Daily	Cost Per Labor-Hour Bare Costs	Cost Per Labor-Hour Incl. O&P
1 Laborer	$27.40	$219.20	$42.65	$341.20	$31.30	$47.88
1 Equip. Oper. (light)	35.20	281.60	53.10	424.80		
1 Crawler Type Drill, 100 mm		728.80		801.70		
1 Air Compr., 283 L/s		320.20		352.20		
1-15 m Air Hose, 80 mm Dia.		18.55		20.40	66.73	73.40
16 L.H., Daily Totals		$1568.35		$1940.30	$98.03	$121.28

Crew B-57

Crew B-57	Bare Costs Hr.	Bare Costs Daily	Incl. Subs O&P Hr.	Incl. Subs O&P Daily	Cost Per Labor-Hour Bare Costs	Cost Per Labor-Hour Incl. O&P
1 Laborer Foreman (outside)	$29.40	$235.20	$45.80	$366.40	$31.66	$48.44
2 Laborers	27.40	438.40	42.65	682.40		
1 Equip. Oper. (crane)	38.10	304.80	57.50	460.00		
1 Equip. Oper. (light)	35.20	281.60	53.10	424.80		
1 Equip. Oper. Oiler	32.45	259.60	48.95	391.60		
1 Power Shovel, 0.76 m³		880.00		968.00		
1 Clamshell Bucket, 0.76 m³		43.40		47.75		
1 Centr. Pump, 150 mm (6")		240.00		264.00		
1-6 m Suct. Hose, 150 mm (6")		12.10		13.30		
20-15 m Disch., 150 mm (6")		150.00		165.00	27.61	30.38
48 L.H., Daily Totals		$2845.10		$3783.25	$59.27	$78.82

Crew B-58

Crew B-58	Bare Costs Hr.	Bare Costs Daily	Incl. Subs O&P Hr.	Incl. Subs O&P Daily	Cost Per Labor-Hour Bare Costs	Cost Per Labor-Hour Incl. O&P
2 Laborers	$27.40	$438.40	$42.65	$682.40	$30.00	$46.13
1 Equip. Oper. (light)	35.20	281.60	53.10	424.80		
1 Backhoe Loader, 36 kW		226.20		248.80		
1 Small Helicopter, w/pilot		2238.00		2461.80	102.68	112.94
24 L.H., Daily Totals		$3184.20		$3817.80	$132.68	$159.07

Crew B-59

Crew B-59	Bare Costs Hr.	Bare Costs Daily	Incl. Subs O&P Hr.	Incl. Subs O&P Daily	Cost Per Labor-Hour Bare Costs	Cost Per Labor-Hour Incl. O&P
1 Truck Driver (heavy)	$28.35	$226.80	$43.30	$346.40	$28.35	$43.30
1 Truck, 27 metric ton		208.20		229.00		
1 Water tank, 22 712 L		116.80		128.50	40.63	44.69
8 L.H., Daily Totals		$551.80		$703.90	$68.98	$87.99

Crew B-59A

Crew B-59A	Bare Costs Hr.	Bare Costs Daily	Incl. Subs O&P Hr.	Incl. Subs O&P Daily	Cost Per Labor-Hour Bare Costs	Cost Per Labor-Hour Incl. O&P
2 Laborers	$27.40	$438.40	$42.65	$682.40	$27.72	$42.87
1 Truck Driver (heavy)	28.35	226.80	43.30	346.40		
1 Water tank, 18 925 L		116.80		128.50		
1 Truck, 27 metric ton		208.20		229.00	13.54	14.90
24 L.H., Daily Totals		$990.20		$1386.30	$41.26	$57.77

Crew B-60

Crew B-60	Bare Costs Hr.	Bare Costs Daily	Incl. Subs O&P Hr.	Incl. Subs O&P Daily	Cost Per Labor-Hour Bare Costs	Cost Per Labor-Hour Incl. O&P
1 Laborer Foreman (outside)	$29.40	$235.20	$45.80	$366.40	$32.16	$49.11
2 Laborers	27.40	438.40	42.65	682.40		
1 Equip. Oper. (crane)	38.10	304.80	57.50	460.00		
2 Equip. Oper. (light)	35.20	563.20	53.10	849.60		
1 Equip. Oper. Oiler	32.45	259.60	48.95	391.60		
1 Crawler Crane, 36 metric ton		934.20		1027.60		
14 m Leads, 20 337 J		49.50		54.45		
1 Backhoe Loader, 36 kW		226.20		248.80	21.65	23.81
56 L.H., Daily Totals		$3011.10		$4080.85	$53.81	$72.92

Crew B-61

Crew B-61	Bare Costs Hr.	Bare Costs Daily	Incl. Subs O&P Hr.	Incl. Subs O&P Daily	Cost Per Labor-Hour Bare Costs	Cost Per Labor-Hour Incl. O&P
1 Laborer Foreman (outside)	$29.40	$235.20	$45.80	$366.40	$29.36	$45.37
3 Laborers	27.40	657.60	42.65	1023.60		
1 Equip. Oper. (light)	35.20	281.60	53.10	424.80		
1 Cement Mixer, 1.53 m³		162.80		179.10		
1 Air Compr., 76 L/s		97.60		107.35	6.51	7.16
40 L.H., Daily Totals		$1434.80		$2101.25	$35.87	$52.53

Crew B-62

Crew B-62	Bare Costs Hr.	Bare Costs Daily	Incl. Subs O&P Hr.	Incl. Subs O&P Daily	Cost Per Labor-Hour Bare Costs	Cost Per Labor-Hour Incl. O&P
2 Laborers	$27.40	$438.40	$42.65	$682.40	$30.00	$46.13
1 Equip. Oper. (light)	35.20	281.60	53.10	424.80		
1 Loader, Skid Steer		154.00		169.40	6.42	7.06
24 L.H., Daily Totals		$874.00		$1276.60	$36.42	$53.19

Crew B-63

Crew B-63	Bare Costs Hr.	Bare Costs Daily	Incl. Subs O&P Hr.	Incl. Subs O&P Daily	Cost Per Labor-Hour Bare Costs	Cost Per Labor-Hour Incl. O&P
4 Laborers	$27.40	$876.80	$42.65	$1364.80	$28.96	$44.74
1 Equip. Oper. (light)	35.20	281.60	53.10	424.80		
1 Loader, Skid Steer		154.00		169.40	3.85	4.24
40 L.H., Daily Totals		$1312.40		$1959.00	$32.81	$48.98

Crew B-64

Crew B-64	Bare Costs Hr.	Bare Costs Daily	Incl. Subs O&P Hr.	Incl. Subs O&P Daily	Cost Per Labor-Hour Bare Costs	Cost Per Labor-Hour Incl. O&P
1 Laborer	$27.40	$219.20	$42.65	$341.20	$27.50	$42.40
1 Truck Driver (light)	27.60	220.80	42.15	337.20		
1 Power Mulcher (small)		113.80		125.20		
1 Light Truck, 1.4 metric ton		128.80		141.70	15.16	16.68
16 L.H., Daily Totals		$682.60		$945.30	$42.66	$59.08

Left Column

Crew No.	Bare Costs Hr.	Daily	Incl. Subs O&P Hr.	Daily	Cost Per Labor-Hour Bare Costs	Incl. O&P
Crew B-65	Hr.	Daily	Hr.	Daily	Bare Costs	Incl. O&P
1 Laborer	$27.40	$219.20	$42.65	$341.20	$27.50	$42.40
1 Truck Driver (light)	27.60	220.80	42.15	337.20		
1 Power Mulcher (large)		185.20		203.70		
1 Light Truck, 1.4 metric ton		128.80		141.70	19.63	21.59
16 L.H., Daily Totals		$754.00		$1023.80	$47.13	$63.99
Crew B-66	Hr.	Daily	Hr.	Daily	Bare Costs	Incl. O&P
1 Equip. Oper. (light)	$35.20	$281.60	$53.10	$424.80	$35.20	$53.10
1 Backhoe Ldr. w/Attchmt.		178.80		196.70	22.35	24.59
8 L.H., Daily Totals		$460.40		$621.50	$57.55	$77.69
Crew B-67	Hr.	Daily	Hr.	Daily	Bare Costs	Incl. O&P
1 Millwright	$37.10	$296.80	$54.80	$438.40	$36.15	$53.95
1 Equip. Oper. (light)	35.20	281.60	53.10	424.80		
1 Forklift		242.40		266.65	15.15	16.67
16 L.H., Daily Totals		$820.80		$1129.85	$51.30	$70.62
Crew B-68	Hr.	Daily	Hr.	Daily	Bare Costs	Incl. O&P
2 Millwrights	$37.10	$593.60	$54.80	$876.80	$36.47	$54.23
1 Equip. Oper. (light)	35.20	281.60	53.10	424.80		
1 Forklift		242.40		266.65	10.10	11.11
24 L.H., Daily Totals		$1117.60		$1568.25	$46.57	$65.34
Crew B-69	Hr.	Daily	Hr.	Daily	Bare Costs	Incl. O&P
1 Laborer Foreman (outside)	$29.40	$235.20	$45.80	$366.40	$30.36	$46.70
3 Laborers	27.40	657.60	42.65	1023.60		
1 Equip Oper. (crane)	38.10	304.80	57.50	460.00		
1 Equip Oper. Oiler	32.45	259.60	48.95	391.60		
1 Truck Crane, 73 metric ton		1059.00		1164.90	22.06	24.27
48 L.H., Daily Totals		$2516.20		$3406.50	$52.42	$70.97
Crew B-69A	Hr.	Daily	Hr.	Daily	Bare Costs	Incl. O&P
1 Laborer Foreman	$29.40	$235.20	$45.80	$366.40	$30.45	$46.65
3 Laborers	27.40	657.60	42.65	1023.60		
1 Equip. Oper. (medium)	36.70	293.60	55.40	443.20		
1 Concrete Finisher	34.40	275.20	50.75	406.00		
1 Curb Paver		593.60		652.95	12.37	13.60
48 L.H., Daily Totals		$2055.20		$2892.15	$42.82	$60.25
Crew B-69B	Hr.	Daily	Hr.	Daily	Bare Costs	Incl. O&P
1 Laborer Foreman	$29.40	$235.20	$45.80	$366.40	$30.45	$46.65
3 Laborers	27.40	657.60	42.65	1023.60		
1 Equip. Oper. (medium)	36.70	293.60	55.40	443.20		
1 Cement Finisher	34.40	275.20	50.75	406.00		
1 Curb/Gutter Paver		738.00		811.80	15.38	16.91
48 L.H., Daily Totals		$2199.60		$3051.00	$45.83	$63.56
Crew B-70	Hr.	Daily	Hr.	Daily	Bare Costs	Incl. O&P
1 Laborer Foreman (outside)	$29.40	$235.20	$45.80	$366.40	$31.67	$48.56
3 Laborers	27.40	657.60	42.65	1023.60		
3 Equip. Oper. (med.)	36.70	880.80	55.40	1329.60		
1 Motor Grader, 13 608 kg		457.40		503.15		
1 Grader Attach., Ripper		73.20		80.50		
1 Road Sweeper, S.P.		493.20		542.50		
1 F.E. Loader, 1.34 m³		241.40		265.55	22.59	24.85
56 L.H., Daily Totals		$3038.80		$4111.30	$54.26	$73.41

Right Column

Crew No.	Bare Costs Hr.	Daily	Incl. Subs O&P Hr.	Daily	Cost Per Labor-Hour Bare Costs	Incl. O&P
Crew B-71	Hr.	Daily	Hr.	Daily	Bare Costs	Incl. O&P
1 Laborer Foreman (outside)	$29.40	$235.20	$45.80	$366.40	$31.67	$48.56
3 Laborers	27.40	657.60	42.65	1023.60		
3 Equip. Oper. (med.)	36.70	880.80	55.40	1329.60		
1 Pvmt. Profiler, 559 kW		4659.00		5124.90		
1 Road Sweeper, S.P.		493.20		542.50		
1 F.E. Loader, 1.34 m³		241.40		265.55	96.31	105.95
56 L.H., Daily Totals		$7167.20		$8652.55	$127.98	$154.51
Crew B-72	Hr.	Daily	Hr.	Daily	Bare Costs	Incl. O&P
1 Laborer Foreman (outside)	$29.40	$235.20	$45.80	$366.40	$32.30	$49.42
3 Laborers	27.40	657.60	42.65	1023.60		
4 Equip. Oper. (med.)	36.70	1174.40	55.40	1772.80		
1 Pvmt. Profiler, 559 kW		4659.00		5124.90		
1 Hammermill, 186 kW		1423.00		1565.30		
1 Windrow Loader		823.00		905.30		
1 Mix Paver 123 kW		1729.00		1901.90		
1 Roller, Pneu., 11 metric ton		259.20		285.10	138.96	152.85
64 L.H., Daily Totals		$10960.40		$12945.30	$171.26	$202.27
Crew B-73	Hr.	Daily	Hr.	Daily	Bare Costs	Incl. O&P
1 Laborer Foreman (outside)	$29.40	$235.20	$45.80	$366.40	$33.46	$51.01
2 Laborers	27.40	438.40	42.65	682.40		
5 Equip. Oper. (med.)	36.70	1468.00	55.40	2216.00		
1 Road Mixer, 231 kW		1698.00		1867.80		
1 Roller, Tandem, 11 met. ton		197.00		216.70		
1 Hammermill, 186 kW		1423.00		1565.30		
1 Motor Grader, 13 608 kg		457.40		503.15		
.5 F.E. Loader, 1.34 m³		120.70		132.75		
.5 Truck, 27 metric ton		104.10		114.50		
.5 Water Tank 18 925 L		58.40		64.25	63.42	69.76
64 L.H., Daily Totals		$6200.20		$7729.25	$96.88	$120.77
Crew B-74	Hr.	Daily	Hr.	Daily	Bare Costs	Incl. O&P
1 Laborer Foreman (outside)	$29.40	$235.20	$45.80	$366.40	$32.54	$49.58
1 Laborer	27.40	219.20	42.65	341.20		
4 Equip. Oper. (med.)	36.70	1174.40	55.40	1772.80		
2 Truck Drivers (heavy)	28.35	453.60	43.30	692.80		
1 Motor Grader, 13 608 kg		457.40		503.15		
1 Grader Attach., Ripper		73.20		80.50		
2 Stabilizers, 231 kW		2206.00		2426.60		
1 Flatbed Truck, 3 metric ton		176.20		193.80		
1 Chem. Spreader, Towed		49.00		53.90		
1 Vibr. Roller, 13 154 kg		522.00		574.20		
1 Water Tank 18 925 L		116.80		128.50		
1 Truck, 27 metric ton		208.20		229.00	59.51	65.46
64 L.H., Daily Totals		$5891.20		$7362.85	$92.05	$115.04
Crew B-75	Hr.	Daily	Hr.	Daily	Bare Costs	Incl. O&P
1 Laborer Foreman (outside)	$29.40	$235.20	$45.80	$366.40	$33.14	$50.48
1 Laborer	27.40	219.20	42.65	341.20		
4 Equip. Oper. (med.)	36.70	1174.40	55.40	1772.80		
1 Truck Driver (heavy)	28.35	226.80	43.30	346.40		
1 Motor Grader, 13 608 kg		457.40		503.15		
1 Grader Attach., Ripper		73.20		80.50		
2 Stabilizers, 231 kW		2206.00		2426.60		
1 Dist. Truck, 11 355 L		242.40		266.65		
1 Vibr. Roller, 13 154 kg		522.00		574.20	62.52	68.77
56 L.H., Daily Totals		$5356.60		$6677.90	$95.66	$119.25

CREWS

Crew No.	Bare Costs		Incl. Subs O & P		Cost Per Labor-Hour	

Crew B-76

	Hr.	Daily	Hr.	Daily	Bare Costs	Incl. O&P
1 Dock Builder Foreman	$36.70	$293.60	$60.35	$482.80	$35.43	$56.62
5 Dock Builders	34.70	1388.00	57.05	2282.00		
2 Equip. Oper. (crane)	38.10	609.60	57.50	920.00		
1 Equip. Oper. Oiler	32.45	259.60	48.95	391.60		
1 Crawler Crane, 45 metric ton		1265.00		1391.50		
1 Barge, 363 metric ton		280.00		308.00		
1 Hammer, 20 337 J		355.80		391.40		
18 m Leads, 20 337 J		66.00		72.60		
1 Air Compr., 283 L/s		320.20		352.20		
2-15m Air Hoses, 80mm Dia.(3")		37.10		40.80	32.32	35.55
72 L.H., Daily Totals		$4874.90		$6632.90	$67.75	$92.17

Crew B-76A

	Hr.	Daily	Hr.	Daily	Bare Costs	Incl. O&P
1 Laborer Foreman	$29.40	$235.20	$45.80	$366.40	$29.62	$45.69
5 Laborers	27.40	1096.00	42.65	1706.00		
1 Equip. Oper. (crane)	38.10	304.80	57.50	460.00		
1 Equip. Oper. Oiler	32.45	259.60	48.95	391.60		
1 Crawler Crane, 45 metric ton		1265.00		1391.50		
1 Barge, 363 metric ton		280.00		308.00	24.14	26.55
64 L.H., Daily Totals		$3440.60		$4623.50	$53.76	$72.24

Crew B-77

	Hr.	Daily	Hr.	Daily	Bare Costs	Incl. O&P
1 Laborer Foreman	$29.40	$235.20	$45.80	$366.40	$27.84	$43.18
3 Laborers	27.40	657.60	42.65	1023.60		
1 Truck Driver (light)	27.60	220.80	42.15	337.20		
1 Crack Cleaner, 19 kW		45.80		50.40		
1 Crack Filler, Trailer Mtd.		157.40		173.15		
1 Flatbed Truck, 3 metric ton		176.20		193.80	9.49	10.43
40 L.H., Daily Totals		$1493.00		$2144.55	$37.33	$53.61

Crew B-78

	Hr.	Daily	Hr.	Daily	Bare Costs	Incl. O&P
1 Laborer Foreman	$29.40	$235.20	$45.80	$366.40	$27.77	$43.09
4 Laborers	27.40	876.80	42.65	1364.80		
1 Truck Driver (light)	27.60	220.80	42.15	337.20		
1 Paint Striper, S.P.		134.60		148.05		
1 Flatbed Truck, 3 metric ton		176.20		193.80		
1 Pickup Truck, 0.7 metric ton		78.00		85.80	8.10	8.91
48 L.H., Daily Totals		$1721.60		$2496.05	$35.87	$52.00

Crew B-78A

	Hr.	Daily	Hr.	Daily	Bare Costs	Incl. O&P
1 Equip. Oper. (light)	$35.20	$281.60	$53.10	$424.80	$35.20	$53.10
1 Line Remov. (metal balls) 86 kW		795.80		875.40	99.48	109.42
8 L.H., Daily Totals		$1077.40		$1300.20	$134.68	$162.52

Crew B-78B

	Hr.	Daily	Hr.	Daily	Bare Costs	Incl. O&P
2 Laborers	$27.40	$438.40	$42.65	$682.40	$28.27	$43.81
.25 Equip. Oper. (light)	35.20	70.40	53.10	106.20		
1 Pickup Truck, 0.7 metric ton		78.00		85.80		
1 Line Remov. (walk behind) 8 kW		44.80		49.30		
.25 Road Sweeper, SP		123.30		135.65	13.67	15.04
18 L.H., Daily Totals		$754.90		$1059.35	$41.94	$58.85

Crew B-79

	Hr.	Daily	Hr.	Daily	Bare Costs	Incl. O&P
1 Laborer Foreman	$29.40	$235.20	$45.80	$366.40	$27.84	$43.18
3 Laborers	27.40	657.60	42.65	1023.60		
1 Truck Driver (light)	27.60	220.80	42.15	337.20		
1 Thermo. Striper, T.M.		555.00		610.70		
1 Flatbed Trucks, 3 metric ton		176.20		193.80		
2 Pickup Trucks, 0.7 metric ton		156.00		171.60	22.19	24.40
40 L.H., Daily Totals		$2001.00		$2703.30	$50.03	$67.58

Crew B-79A

	Hr.	Daily	Hr.	Daily	Bare Costs	Incl. O&P
1.5 Equip. Oper. (light)	$35.20	$422.40	$53.10	$637.20	$35.20	$53.10
.5 Line Remov. (grinder) 86 kW		332.70		365.95		
1 Line Remov. (metal balls) 86 kW		795.80		875.40	94.04	103.45
12 L.H., Daily Totals		$1550.90		$1878.55	$129.24	$156.55

Crew B-80

	Hr.	Daily	Hr.	Daily	Bare Costs	Incl. O&P
1 Laborer Foreman	$29.40	$235.20	$45.80	$366.40	$29.90	$45.93
1 Laborer	27.40	219.20	42.65	341.20		
1 Truck Driver (light)	27.60	220.80	42.15	337.20		
1 Equip. Oper. (light)	35.20	281.60	53.10	424.80		
1 Flatbed Truck, 3 metric ton		176.20		193.80		
1 Fence Post Auger, TM		356.20		391.80	16.64	18.30
32 L.H., Daily Totals		$1489.20		$2055.20	$46.54	$64.23

Crew B-80A

	Hr.	Daily	Hr.	Daily	Bare Costs	Incl. O&P
3 Laborers	$27.40	$657.60	$42.65	$1023.60	$27.40	$42.65
1 Flatbed Truck, 3 metric ton		176.20		193.80	7.34	8.08
24 L.H., Daily Totals		$833.80		$1217.40	$34.74	$50.73

Crew B-80B

	Hr.	Daily	Hr.	Daily	Bare Costs	Incl. O&P
3 Laborers	$27.40	$657.60	$42.65	$1023.60	$29.35	$45.26
1 Equip. Oper. (light)	35.20	281.60	53.10	424.80		
1 Crane, Flatbed Mtd., 7 met. ton		235.20		258.70	7.35	8.09
32 L.H., Daily Totals		$1174.40		$1707.10	$36.70	$53.35

Crew B-80C

	Hr.	Daily	Hr.	Daily	Bare Costs	Incl. O&P
2 Laborers	$27.40	$438.40	$42.65	$682.40	$27.47	$42.48
1 Truck Driver (light)	27.60	220.80	42.15	337.20		
1 Light Truck, 1.4 metric ton		128.80		141.70		
1 Manual fence post auger, gas		5.40		5.95	5.59	6.15
24 L.H., Daily Totals		$793.40		$1167.25	$33.06	$48.63

Crew B-81

	Hr.	Daily	Hr.	Daily	Bare Costs	Incl. O&P
1 Laborer	$27.40	$219.20	$42.65	$341.20	$30.82	$47.12
1 Equip. Oper. (med.)	36.70	293.60	55.40	443.20		
1 Truck Driver (heavy)	28.35	226.80	43.30	346.40		
1 Hydromulcher, T.M.		197.20		216.90		
1 Tractor Truck, 4 x 2		208.20		229.00	16.89	18.58
24 L.H., Daily Totals		$1145.00		$1576.70	$47.71	$65.70

Crew B-82

	Hr.	Daily	Hr.	Daily	Bare Costs	Incl. O&P
1 Laborer	$27.40	$219.20	$42.65	$341.20	$31.30	$47.88
1 Equip. Oper. (light)	35.20	281.60	53.10	424.80		
1 Horiz. Borer, 4 kW		66.00		72.60	4.13	4.54
16 L.H., Daily Totals		$566.80		$838.60	$35.43	$52.42

Crew B-83

	Hr.	Daily	Hr.	Daily	Bare Costs	Incl. O&P
1 Tugboat Captain	$36.70	$293.60	$55.40	$443.20	$32.05	$49.03
1 Tugboat Hand	27.40	219.20	42.65	341.20		
1 Tugboat, 185 kW		473.00		520.30	29.56	32.52
16 L.H., Daily Totals		$985.80		$1304.70	$61.61	$81.55

Crew B-84

	Hr.	Daily	Hr.	Daily	Bare Costs	Incl. O&P
1 Equip. Oper. (med.)	$36.70	$293.60	$55.40	$443.20	$36.70	$55.40
1 Rotary Mower/Tractor		231.00		254.10	28.88	31.76
8 L.H., Daily Totals		$524.60		$697.30	$65.58	$87.16

CREWS

Crews

Crew No.	Bare Costs		Incl. Subs O & P		Cost Per Labor-Hour	

Crew B-85	Hr.	Daily	Hr.	Daily	Bare Costs	Incl. O&P
3 Laborers	$27.40	$657.60	$42.65	$1023.60	$29.45	$45.33
1 Equip. Oper. (med.)	36.70	293.60	55.40	443.20		
1 Truck Driver (heavy)	28.35	226.80	43.30	346.40		
1 Aerial Lift Truck, 24 m		515.80		567.40		
1 Brush Chipper, 97 kW		173.00		190.30		
1 Pruning Saw, Rotary		7.15		7.85	17.40	19.14
40 L.H., Daily Totals		$1873.95		$2578.75	$46.85	$64.47

Crew B-86	Hr.	Daily	Hr.	Daily	Bare Costs	Incl. O&P
1 Equip. Oper. (med.)	$36.70	$293.60	$55.40	$443.20	$36.70	$55.40
1 Stump Chipper, S.P.		63.50		69.85	7.94	8.73
8 L.H., Daily Totals		$357.10		$513.05	$44.64	$64.13

Crew B-86A	Hr.	Daily	Hr.	Daily	Bare Costs	Incl. O&P
1 Equip. Oper. (medium)	$36.70	$293.60	$55.40	$443.20	$36.70	$55.40
1 Grader, 13 608 kg		457.40		503.15	57.18	62.89
8 L.H., Daily Totals		$751.00		$946.35	$93.88	$118.29

Crew B-86B	Hr.	Daily	Hr.	Daily	Bare Costs	Incl. O&P
1 Equip. Oper. (medium)	$36.70	$293.60	$55.40	$443.20	$36.70	$55.40
1 Dozer, 149 kW		920.80		1012.90	115.10	126.61
8 L.H., Daily Totals		$1214.40		$1456.10	$151.80	$182.01

Crew B-87	Hr.	Daily	Hr.	Daily	Bare Costs	Incl. O&P
1 Laborer	$27.40	$219.20	$42.65	$341.20	$34.84	$52.85
4 Equip. Oper. (med.)	36.70	1174.40	55.40	1772.80		
2 Feller Bunchers, 37 kW		983.60		1081.95		
1 Log Chipper, 559 mm Tree		1000.00		1100.00		
1 Dozer, 78 kW		462.80		509.10		
1 Chainsaw, Gas, 914 mm Long		33.40		36.75	62.00	68.19
40 L.H., Daily Totals		$3873.40		$4841.80	$96.84	$121.04

Crew B-88	Hr.	Daily	Hr.	Daily	Bare Costs	Incl. O&P
1 Laborer	$27.40	$219.20	$42.65	$341.20	$35.37	$53.58
6 Equip. Oper. (med.)	36.70	1761.60	55.40	2659.20		
2 Feller Bunchers, 37 kW		983.60		1081.95		
1 Log Chipper, 559 mm Tree		1000.00		1100.00		
2 Log Skidders, 37 kW		1575.20		1732.70		
1 Dozer, 78 kW		462.80		509.10		
1 Chainsaw, Gas, 914 mm Long		33.40		36.75	72.41	79.65
56 L.H., Daily Totals		$6035.80		$7460.90	$107.78	$133.23

Crew B-89	Hr.	Daily	Hr.	Daily	Bare Costs	Incl. O&P
1 Equip. Oper. (light)	$35.20	$281.60	$53.10	$424.80	$31.40	$47.63
1 Truck Driver (light)	27.60	220.80	42.15	337.20		
1 Truck, Stake Body, 3 met. ton		176.20		193.80		
1 Concrete Saw		107.00		117.70		
1 Water Tank, 246 L		13.80		15.20	18.56	20.42
16 L.H., Daily Totals		$799.40		$1088.70	$49.96	$68.05

Crew B-89A	Hr.	Daily	Hr.	Daily	Bare Costs	Incl. O&P
1 Skilled Worker	$36.50	$292.00	$56.80	$454.40	$31.95	$49.73
1 Laborer	27.40	219.20	42.65	341.20		
1 Core Drill (large)		115.45		127.00	7.22	7.94
16 L.H., Daily Totals		$626.65		$922.60	$39.17	$57.67

Crew No.	Bare Costs		Incl. Subs O & P		Cost Per Labor-Hour	

Crew B-89B	Hr.	Daily	Hr.	Daily	Bare Costs	Incl. O&P
1 Equip. Oper. (light)	$35.20	$281.60	$53.10	$424.80	$31.40	$47.63
1 Truck Driver, Light	27.60	220.80	42.15	337.20		
1 Wall Saw, Hydraulic, 7.5 kW		80.95		89.05		
1 Generator, Diesel, 100 kW		193.20		212.50		
1 Water Tank, 246 L		13.80		15.20		
1 Flatbed Truck, 2.7 metric ton		176.20		193.80	29.01	31.91
16 L.H., Daily Totals		$966.55		$1272.55	$60.41	$79.54

Crew B-90	Hr.	Daily	Hr.	Daily	Bare Costs	Incl. O&P
1 Laborer Foreman (outside)	$29.40	$235.20	$45.80	$366.40	$29.84	$45.82
3 Laborers	27.40	657.60	42.65	1023.60		
2 Equip. Oper. (light)	35.20	563.20	53.10	849.60		
2 Truck Drivers (heavy)	28.35	453.60	43.30	692.80		
1 Road Mixer, 231 kW		1698.00		1867.80		
1 Dist. Truck, 7570 L		209.00		229.90	29.80	32.78
64 L.H., Daily Totals		$3816.60		$5030.10	$59.64	$78.60

Crew B-90A	Hr.	Daily	Hr.	Daily	Bare Costs	Incl. O&P
1 Laborer Foreman	$29.40	$235.20	$45.80	$366.40	$33.00	$50.39
2 Laborers	27.40	438.40	42.65	682.40		
4 Equip. Oper. (medium)	36.70	1174.40	55.40	1772.80		
2 Graders, 13 608 kg		914.80		1006.30		
1 Roller, Steel Wheel		197.00		216.70		
1 Roller, Pneumatic Wheel		259.20		285.10	24.48	26.93
56 L.H., Daily Totals		$3219.00		$4329.70	$57.48	$77.32

Crew B-90B	Hr.	Daily	Hr.	Daily	Bare Costs	Incl. O&P
1 Laborer Foreman	$29.40	$235.20	$45.80	$366.40	$32.38	$49.55
2 Laborers	27.40	438.40	42.65	682.40		
3 Equip. Oper. (medium)	36.70	880.80	55.40	1329.60		
1 Roller, Steel Wheel		197.00		216.70		
1 Roller, Pneumatic Wheel		259.20		285.10		
1 Road Mixer, 231 kW		1698.00		1867.80	44.88	49.37
48 L.H., Daily Totals		$3708.60		$4748.00	$77.26	$98.92

Crew B-91	Hr.	Daily	Hr.	Daily	Bare Costs	Incl. O&P
1 Laborer Foreman (outside)	$29.40	$235.20	$45.80	$366.40	$32.42	$49.50
2 Laborers	27.40	438.40	42.65	682.40		
4 Equip. Oper. (med.)	36.70	1174.40	55.40	1772.80		
1 Truck Driver (heavy)	28.35	226.80	43.30	346.40		
1 Dist. Truck, 11 355 L		242.40		266.65		
1 Aggreg. Spreader, S.P.		777.40		855.15		
1 Roller, Pneu., 11 metric ton		259.20		285.10		
1 Roller, Steel, 9.1 metric ton		197.00		216.70	23.06	25.37
64 L.H., Daily Totals		$3550.80		$4791.60	$55.48	$74.87

Crew B-92	Hr.	Daily	Hr.	Daily	Bare Costs	Incl. O&P
1 Laborer Foreman (outside)	$29.40	$235.20	$45.80	$366.40	$27.90	$43.44
3 Laborers	27.40	657.60	42.65	1023.60		
1 Crack Cleaner, 19 kW		45.80		50.40		
1 Air Compressor		59.00		64.90		
1 Tar Kettle, T.M.		45.75		50.35		
1 Flatbed Truck, 3 metric ton		176.20		193.80	10.21	11.23
32 L.H., Daily Totals		$1219.55		$1749.45	$38.11	$54.67

Crew B-93	Hr.	Daily	Hr.	Daily	Bare Costs	Incl. O&P
1 Equip. Oper. (med.)	$36.70	$293.60	$55.40	$443.20	$36.70	$55.40
1 Feller Buncher, 37 kW		491.80		541.00	61.48	67.62
8 L.H., Daily Totals		$785.40		$984.20	$98.18	$123.02

Crew B-94A	Hr.	Daily	Hr.	Daily	Bare Costs	Incl. O&P
1 Laborer	$27.40	$219.20	$42.65	$341.20	$27.40	$42.65
1 Diaph. Pump, 50 mm (2")		53.20		58.50		
1-6 m Suction, 50 mm (2")		2.15		2.35		
2-15 m Disch., 50 mm (2")		2.60		2.85	7.24	7.97
8 L.H., Daily Totals		$277.15		$404.90	$34.64	$50.62

Crew B-94B	Hr.	Daily	Hr.	Daily	Bare Costs	Incl. O&P
1 Laborer	$27.40	$219.20	$42.65	$341.20	$27.40	$42.65
1 Diaph. Water Pump, 100 mm		73.20		80.50		
1-6 m Suct. Hose, 100 mm (4")		4.05		4.45		
2-15 m Disch., 100 mm (4")		5.90		6.50	10.39	11.43
8 L.H., Daily Totals		$302.35		$432.65	$37.79	$54.08

Crew B-94C	Hr.	Daily	Hr.	Daily	Bare Costs	Incl. O&P
1 Laborer	$27.40	$219.20	$42.65	$341.20	$27.40	$42.65
1 Centr. Pump, 80 mm (3")		56.60		62.25		
1-6 m Suct. Hose, 80 mm (3")		3.65		4.00		
2-15 m Disch., 80 mm (3")		4.30		4.75	8.07	8.88
8 L.H., Daily Totals		$283.75		$412.20	$35.47	$51.53

Crew B-94D	Hr.	Daily	Hr.	Daily	Bare Costs	Incl. O&P
1 Laborer	$27.40	$219.20	$42.65	$341.20	$27.40	$42.65
1 Centr. Water Pump, 150 mm		240.00		264.00		
1-6 m Suction, 150 mm (6")		12.10		13.30		
2-15 m Disch., 150 mm (6")		15.00		16.50	33.39	36.73
8 L.H., Daily Totals		$486.30		$635.00	$60.79	$79.38

Crew B-95A	Hr.	Daily	Hr.	Daily	Bare Costs	Incl. O&P
1 Equip. Oper. (crane)	$38.10	$304.80	$57.50	$460.00	$32.75	$50.08
1 Laborer	27.40	219.20	42.65	341.20		
1 Hyd. Excavator, 0.48 m³		435.40		478.95	27.21	29.93
16 L.H., Daily Totals		$959.40		$1280.15	$59.96	$80.01

Crew B-95B	Hr.	Daily	Hr.	Daily	Bare Costs	Incl. O&P
1 Equip. Oper. (crane)	$38.10	$304.80	$57.50	$460.00	$32.75	$50.08
1 Laborer	27.40	219.20	42.65	341.20		
1 Hyd. Excavator, 1.15 m³		722.80		795.10	45.18	49.69
16 L.H., Daily Totals		$1246.80		$1596.30	$77.93	$99.77

Crew B-95C	Hr.	Daily	Hr.	Daily	Bare Costs	Incl. O&P
1 Equip. Oper. (crane)	$38.10	$304.80	$57.50	$460.00	$32.75	$50.08
1 Laborer	27.40	219.20	42.65	341.20		
1 Hyd. Excavator, 1.91 m³		1218.00		1339.80	76.13	83.74
16 L.H., Daily Totals		$1742.00		$2141.00	$108.88	$133.82

Crew C-1	Hr.	Daily	Hr.	Daily	Bare Costs	Incl. O&P
3 Carpenters	$35.55	$853.20	$55.35	$1328.40	$33.51	$52.18
1 Laborer	27.40	219.20	42.65	341.20		
32 L.H., Daily Totals		$1072.40		$1669.60	$33.51	$52.18

Crew C-2	Hr.	Daily	Hr.	Daily	Bare Costs	Incl. O&P
1 Carpenter Foreman (out)	$37.55	$300.40	$58.45	$467.60	$34.53	$53.75
4 Carpenters	35.55	1137.60	55.35	1771.20		
1 Laborer	27.40	219.20	42.65	341.20		
48 L.H., Daily Totals		$1657.20		$2580.00	$34.53	$53.75

Crew C-2A	Hr.	Daily	Hr.	Daily	Bare Costs	Incl. O&P
1 Carpenter Foreman (out)	$37.55	$300.40	$58.45	$467.60	$34.33	$52.98
3 Carpenters	35.55	853.20	55.35	1328.40		
1 Cement Finisher	34.40	275.20	50.75	406.00		
1 Laborer	27.40	219.20	42.65	341.20		
48 L.H., Daily Totals		$1648.00		$2543.20	$34.33	$52.98

Crew C-3	Hr.	Daily	Hr.	Daily	Bare Costs	Incl. O&P
1 Rodman Foreman	$41.50	$332.00	$68.35	$546.80	$36.19	$58.37
4 Rodmen (reinf.)	39.50	1264.00	65.05	2081.60		
1 Equip. Oper. (light)	35.20	281.60	53.10	424.80		
2 Laborers	27.40	438.40	42.65	682.40		
3 Stressing Equipment		31.20		34.30		
.5 Grouting Equipment		77.55		85.30	1.70	1.87
64 L.H., Daily Totals		$2424.75		$3855.20	$37.89	$60.24

Crew C-4	Hr.	Daily	Hr.	Daily	Bare Costs	Incl. O&P
1 Rodman Foreman	$41.50	$332.00	$68.35	$546.80	$40.00	$65.88
3 Rodmen (reinf.)	39.50	948.00	65.05	1561.20		
3 Stressing Equipment		31.20		34.30	.98	1.07
32 L.H., Daily Totals		$1311.20		$2142.30	$40.98	$66.95

Crew C-5	Hr.	Daily	Hr.	Daily	Bare Costs	Incl. O&P
1 Rodman Foreman	$41.50	$332.00	$68.35	$546.80	$38.58	$62.14
4 Rodmen (reinf.)	39.50	1264.00	65.05	2081.60		
1 Equip. Oper. (crane)	38.10	304.80	57.50	460.00		
1 Equip. Oper. Oiler	32.45	259.60	48.95	391.60		
1 Hyd. Crane, 23 metric ton		650.80		715.90	11.62	12.78
56 L.H., Daily Totals		$2811.20		$4195.90	$50.20	$74.92

Crew C-6	Hr.	Daily	Hr.	Daily	Bare Costs	Incl. O&P
1 Laborer Foreman (outside)	$29.40	$235.20	$45.80	$366.40	$28.90	$44.53
4 Laborers	27.40	876.80	42.65	1364.80		
1 Cement Finisher	34.40	275.20	50.75	406.00		
2 Gas Engine Vibrators		42.80		47.10	.89	.98
48 L.H., Daily Totals		$1430.00		$2184.30	$29.79	$45.51

Crew C-7	Hr.	Daily	Hr.	Daily	Bare Costs	Incl. O&P
1 Laborer Foreman (outside)	$29.40	$235.20	$45.80	$366.40	$29.99	$46.02
5 Laborers	27.40	1096.00	42.65	1706.00		
1 Cement Finisher	34.40	275.20	50.75	406.00		
1 Equip. Oper. (med.)	36.70	293.60	55.40	443.20		
1 Equip. Oper. (oiler)	32.45	259.60	48.95	391.60		
2 Gas Engine Vibrators		42.80		47.10		
1 Concrete Bucket, 0.76 m³		16.20		17.80		
1 Hyd. Crane, 50 metric ton		951.60		1046.75	14.04	15.44
72 L.H., Daily Totals		$3170.20		$4424.85	$44.03	$61.46

Crew C-7A	Hr.	Daily	Hr.	Daily	Bare Costs	Incl. O&P
1 Laborer Foreman (outside)	$29.40	$235.20	$45.80	$366.40	$27.89	$43.21
5 Laborers	27.40	1096.00	42.65	1706.00		
2 Truck Drivers (Heavy)	28.35	453.60	43.30	692.80		
2 Conc. Transit Mixers		1487.60		1636.35	23.24	25.57
64 L.H., Daily Totals		$3272.40		$4401.55	$51.13	$68.78

Crews

Crew No.	Bare Costs		Incl. Subs O & P		Cost Per Labor-Hour	

Crew C-7B

	Hr.	Daily	Hr.	Daily	Bare Costs	Incl. O&P
1 Laborer Foreman (outside)	$29.40	$235.20	$45.80	$366.40	$29.62	$45.69
5 Laborers	27.40	1096.00	42.65	1706.00		
1 Equipment Operator (heavy)	38.10	304.80	57.50	460.00		
1 Equipment Oiler	32.45	259.60	48.95	391.60		
1 Conc. Bucket, 1.5 m³		25.20		27.70		
1 Truck Crane, 150 metric ton		2012.00		2213.20	31.83	35.01
64 L.H., Daily Totals		$3932.80		$5164.90	$61.45	$80.70

Crew C-7C

	Hr.	Daily	Hr.	Daily	Bare Costs	Incl. O&P
1 Laborer Foreman (outside)	$29.40	$235.20	$45.80	$366.40	$29.98	$46.23
5 Laborers	27.40	1096.00	42.65	1706.00		
2 Equipment Operators (medium)	36.70	587.20	55.40	886.40		
2 Wheel Loaders, 3.06 m³		888.00		976.80	13.88	15.26
64 L.H., Daily Totals		$2806.40		$3935.60	$43.86	$61.49

Crew C-7D

	Hr.	Daily	Hr.	Daily	Bare Costs	Incl. O&P
1 Laborer Foreman (outside)	$29.40	$235.20	$45.80	$366.40	$29.01	$44.92
5 Laborers	27.40	1096.00	42.65	1706.00		
1 Equipment Operator (med.)	36.70	293.60	55.40	443.20		
1 Concrete Conveyer		152.40		167.65	2.72	2.99
56 L.H., Daily Totals		$1777.20		$2683.25	$31.73	$47.91

Crew C-8

	Hr.	Daily	Hr.	Daily	Bare Costs	Incl. O&P
1 Laborer Foreman (outside)	$29.40	$235.20	$45.80	$366.40	$31.01	$47.24
3 Laborers	27.40	657.60	42.65	1023.60		
2 Cement Finishers	34.40	550.40	50.75	812.00		
1 Equip. Oper. (med.)	36.70	293.60	55.40	443.20		
1 Concrete Pump (small)		700.60		770.65	12.51	13.76
56 L.H., Daily Totals		$2437.40		$3415.85	$43.52	$61.00

Crew C-8A

	Hr.	Daily	Hr.	Daily	Bare Costs	Incl. O&P
1 Laborer Foreman (outside)	$29.40	$235.20	$45.80	$366.40	$30.07	$45.88
3 Laborers	27.40	657.60	42.65	1023.60		
2 Cement Finishers	34.40	550.40	50.75	812.00		
48 L.H., Daily Totals		$1443.20		$2202.00	$30.07	$45.88

Crew C-8B

	Hr.	Daily	Hr.	Daily	Bare Costs	Incl. O&P
1 Laborer Foreman (outside)	$29.40	$235.20	$45.80	$366.40	$29.66	$45.83
3 Laborers	27.40	657.60	42.65	1023.60		
1 Equipment Operator	36.70	293.60	55.40	443.20		
1 Vibrating Screed		57.95		63.75		
1 Vibratory Roller		522.00		574.20		
1 Dozer, 149 kW		920.80		1012.90	37.52	41.27
40 L.H., Daily Totals		$2687.15		$3484.05	$67.18	$87.10

Crew C-8C

	Hr.	Daily	Hr.	Daily	Bare Costs	Incl. O&P
1 Laborer Foreman (outside)	$29.40	$235.20	$45.80	$366.40	$30.45	$46.65
3 Laborers	27.40	657.60	42.65	1023.60		
1 Cement Finisher	34.40	275.20	50.75	406.00		
1 Equipment Operator (med.)	36.70	293.60	55.40	443.20		
1 Shotcrete Rig, 9 m³/hr		234.40		257.85	4.88	5.37
48 L.H., Daily Totals		$1696.00		$2497.05	$35.33	$52.02

Crew C-8D

	Hr.	Daily	Hr.	Daily	Bare Costs	Incl. O&P
1 Laborer Foreman (outside)	$29.40	$235.20	$45.80	$366.40	$31.60	$48.08
1 Laborer	27.40	219.20	42.65	341.20		
1 Cement Finisher	34.40	275.20	50.75	406.00		
1 Equipment Operator (light)	35.20	281.60	53.10	424.80		
1 Compressor, 118 L/S		119.80		131.80		
2 hoses, 25 mm		7.80		8.60	3.99	4.39
32 L.H., Daily Totals		$1138.80		$1678.80	$35.59	$52.47

Crew C-8E

	Hr.	Daily	Hr.	Daily	Bare Costs	Incl. O&P
1 Laborer Foreman (outside)	$29.40	$235.20	$45.80	$366.40	$31.60	$48.08
1 Laborer	27.40	219.20	42.65	341.20		
1 Cement Finisher	34.40	275.20	50.75	406.00		
1 Equipment Operator (light)	35.20	281.60	53.10	424.80		
1 Compressor, 118 L/s		119.80		131.80		
2-15 m Hoses, 25 mm		7.80		8.60		
1 Concrete Pump (small)		700.60		770.65	25.88	28.47
32 L.H., Daily Totals		$1839.40		$2449.45	$57.48	$76.55

Crew C-10

	Hr.	Daily	Hr.	Daily	Bare Costs	Incl. O&P
1 Laborer	$27.40	$219.20	$42.65	$341.20	$32.07	$48.05
2 Cement Finishers	34.40	550.40	50.75	812.00		
24 L.H., Daily Totals		$769.60		$1153.20	$32.07	$48.05

Crew C-10B

	Hr.	Daily	Hr.	Daily	Bare Costs	Incl. O&P
3 Laborers	$27.40	$657.60	$42.65	$1023.60	$30.20	$45.89
2 Cement Finishers	34.40	550.40	50.75	812.00		
1 Concrete mixer, 0.28 m³		127.40		140.15		
2 Concrete finishers, 1200 mm dia		48.40		53.25	4.40	4.83
40 L.H., Daily Totals		$1383.80		$2029.00	$34.60	$50.72

Crew C-11

	Hr.	Daily	Hr.	Daily	Bare Costs	Incl. O&P
1 Struc. Steel Foreman	$41.95	$335.60	$75.50	$604.00	$39.13	$68.15
6 Struc. Steel Workers	39.95	1917.60	71.90	3451.20		
1 Equip. Oper. (crane)	38.10	304.80	57.50	460.00		
1 Equip. Oper. Oiler	32.45	259.60	48.95	391.60		
1 Truck Crane, 136 metric ton		1539.00		1692.90	21.38	23.51
72 L.H., Daily Totals		$4356.60		$6599.70	$60.51	$91.66

Crew C-12

	Hr.	Daily	Hr.	Daily	Bare Costs	Incl. O&P
1 Carpenter Foreman (out)	$37.55	$300.40	$58.45	$467.60	$34.95	$54.11
3 Carpenters	35.55	853.20	55.35	1328.40		
1 Laborer	27.40	219.20	42.65	341.20		
1 Equip. Oper. (crane)	38.10	304.80	57.50	460.00		
1 Hyd. Crane, 11 metric ton		636.60		700.25	13.26	14.59
48 L.H., Daily Totals		$2314.20		$3297.45	$48.21	$68.70

Crew C-13

	Hr.	Daily	Hr.	Daily	Bare Costs	Incl. O&P
1 Struc. Steel Worker	$39.95	$319.60	$71.90	$575.20	$38.48	$66.38
1 Welder	39.95	319.60	71.90	575.20		
1 Carpenter	35.55	284.40	55.35	442.80		
1 Gas Welding Machine		89.20		98.10	3.72	4.09
24 L.H., Daily Totals		$1012.80		$1691.30	$42.20	$70.47

Crew C-14

	Hr.	Daily	Hr.	Daily	Bare Costs	Incl. O&P
1 Carpenter Foreman (out)	$37.55	$300.40	$58.45	$467.60	$34.57	$54.11
5 Carpenters	35.55	1422.00	55.35	2214.00		
4 Laborers	27.40	876.80	42.65	1364.80		
4 Rodmen (reinf.)	39.50	1264.00	65.05	2081.60		
2 Cement Finishers	34.40	550.40	50.75	812.00		
1 Equip. Oper. (crane)	38.10	304.80	57.50	460.00		
1 Equip. Oper. Oiler	32.45	259.60	48.95	391.60		
1 Crane, 72 metric ton		1059.00		1164.90	7.35	8.09
144 L.H., Daily Totals		$6037.00		$8956.50	$41.92	$62.20

Crews

Crew C-14A

Crew No.	Bare Costs Hr.	Daily	Incl. Subs O & P Hr.	Daily	Cost Per Labor-Hour Bare Costs	Incl. O&P
1 Carpenter Foreman (out)	$37.55	$300.40	$58.45	$467.60	$35.61	$55.83
16 Carpenters	35.55	4550.40	55.35	7084.80		
4 Rodmen (reinf.)	39.50	1264.00	65.05	2081.60		
2 Laborers	27.40	438.40	42.65	682.40		
1 Cement Finisher	34.40	275.20	50.75	406.00		
1 Equip. Oper. (med.)	36.70	293.60	55.40	443.20		
1 Gas Engine Vibrator		21.40		23.55		
1 Concrete Pump (small)		700.60		770.65	3.61	3.97
200 L.H., Daily Totals		$7844.00		$11959.80	$39.22	$59.80

Crew C-14B

Crew No.	Bare Costs Hr.	Daily	Incl. Subs O & P Hr.	Daily	Cost Per Labor-Hour Bare Costs	Incl. O&P
1 Carpenter Foreman (out)	$37.55	$300.40	$58.45	$467.60	$35.56	$55.63
16 Carpenters	35.55	4550.40	55.35	7084.80		
4 Rodmen (reinf.)	39.50	1264.00	65.05	2081.60		
2 Laborers	27.40	438.40	42.65	682.40		
2 Cement Finishers	34.40	550.40	50.75	812.00		
1 Equip. Oper. (med.)	36.70	293.60	55.40	443.20		
1 Gas Engine Vibrator		21.40		23.55		
1 Concrete Pump (small)		700.60		770.65	3.47	3.82
208 L.H., Daily Totals		$8119.20		$12365.80	$39.03	$59.45

Crew C-14C

Crew No.	Bare Costs Hr.	Daily	Incl. Subs O & P Hr.	Daily	Cost Per Labor-Hour Bare Costs	Incl. O&P
1 Carpenter Foreman (out)	$37.55	$300.40	$58.45	$467.60	$33.85	$53.00
6 Carpenters	35.55	1706.40	55.35	2656.80		
2 Rodmen (reinf.)	39.50	632.00	65.05	1040.80		
4 Laborers	27.40	876.80	42.65	1364.80		
1 Cement Finisher	34.40	275.20	50.75	406.00		
1 Gas Engine Vibrator		21.40		23.55	.19	.21
112 L.H., Daily Totals		$3812.20		$5959.55	$34.04	$53.21

Crew C-14D

Crew No.	Bare Costs Hr.	Daily	Incl. Subs O & P Hr.	Daily	Cost Per Labor-Hour Bare Costs	Incl. O&P
1 Carpenter Foreman (out)	$37.55	$300.40	$58.45	$467.60	$35.29	$55.05
18 Carpenters	35.55	5119.20	55.35	7970.40		
2 Rodmen (reinf.)	39.50	632.00	65.05	1040.80		
2 Laborers	27.40	438.40	42.65	682.40		
1 Cement Finisher	34.40	275.20	50.75	406.00		
1 Equip. Oper. (med.)	36.70	293.60	55.40	443.20		
1 Gas Engine Vibrator		21.40		23.55		
1 Concrete Pump (small)		700.60		770.65	3.61	3.97
200 L.H., Daily Totals		$7780.80		$11804.60	$38.90	$59.02

Crew C-14E

Crew No.	Bare Costs Hr.	Daily	Incl. Subs O & P Hr.	Daily	Cost Per Labor-Hour Bare Costs	Incl. O&P
1 Carpenter Foreman (out)	$37.55	$300.40	$58.45	$467.60	$34.84	$55.28
2 Carpenters	35.55	568.80	55.35	885.60		
4 Rodmen (reinf.)	39.50	1264.00	65.05	2081.60		
3 Laborers	27.40	657.60	42.65	1023.60		
1 Cement Finisher	34.40	275.20	50.75	406.00		
1 Gas Engine Vibrator		21.40		23.55	.24	.27
88 L.H., Daily Totals		$3087.40		$4887.95	$35.08	$55.55

Crew C-14F

Crew No.	Bare Costs Hr.	Daily	Incl. Subs O & P Hr.	Daily	Cost Per Labor-Hour Bare Costs	Incl. O&P
1 Laborer Foreman (out)	$29.40	$235.20	$45.80	$366.40	$32.29	$48.40
2 Laborers	27.40	438.40	42.65	682.40		
6 Cement Finishers	34.40	1651.20	50.75	2436.00		
1 Gas Engine Vibrator		21.40		23.55	.30	.33
72 L.H., Daily Totals		$2346.20		$3508.35	$32.59	$48.73

Crew C-14G

Crew No.	Bare Costs Hr.	Daily	Incl. Subs O & P Hr.	Daily	Cost Per Labor-Hour Bare Costs	Incl. O&P
1 Laborer Foreman (out)	$29.40	$235.20	$45.80	$366.40	$31.69	$47.73
2 Laborers	27.40	438.40	42.65	682.40		
4 Cement Finishers	34.40	1100.80	50.75	1624.00		
1 Gas Engine Vibrator		21.40		23.55	.38	.42
56 L.H., Daily Totals		$1795.80		$2696.35	$32.07	$48.15

Crew C-14H

Crew No.	Bare Costs Hr.	Daily	Incl. Subs O & P Hr.	Daily	Cost Per Labor-Hour Bare Costs	Incl. O&P
1 Carpenter Foreman (out)	$37.55	$300.40	$58.45	$467.60	$34.99	$54.60
2 Carpenters	35.55	568.80	55.35	885.60		
1 Rodman (reinf.)	39.50	316.00	65.05	520.40		
1 Laborer	27.40	219.20	42.65	341.20		
1 Cement Finisher	34.40	275.20	50.75	406.00		
1 Gas Engine Vibrator		21.40		23.55	.45	.49
48 L.H., Daily Totals		$1701.00		$2644.35	$35.44	$55.09

Crew C-15

Crew No.	Bare Costs Hr.	Daily	Incl. Subs O & P Hr.	Daily	Cost Per Labor-Hour Bare Costs	Incl. O&P
1 Carpenter Foreman (out)	$37.55	$300.40	$58.45	$467.60	$33.24	$51.52
2 Carpenters	35.55	568.80	55.35	885.60		
3 Laborers	27.40	657.60	42.65	1023.60		
2 Cement Finishers	34.40	550.40	50.75	812.00		
1 Rodman (reinf.)	39.50	316.00	65.05	520.40		
72 L.H., Daily Totals		$2393.20		$3709.20	$33.24	$51.52

Crew C-16

Crew No.	Bare Costs Hr.	Daily	Incl. Subs O & P Hr.	Daily	Cost Per Labor-Hour Bare Costs	Incl. O&P
1 Laborer Foreman (outside)	$29.40	$235.20	$45.80	$366.40	$32.90	$51.19
3 Laborers	27.40	657.60	42.65	1023.60		
2 Cement Finishers	34.40	550.40	50.75	812.00		
1 Equip. Oper. (med.)	36.70	293.60	55.40	443.20		
2 Rodmen (reinf.)	39.50	632.00	65.05	1040.80		
1 Concrete Pump (small)		700.60		770.65	9.73	10.70
72 L.H., Daily Totals		$3069.40		$4456.65	$42.63	$61.89

Crew C-17

Crew No.	Bare Costs Hr.	Daily	Incl. Subs O & P Hr.	Daily	Cost Per Labor-Hour Bare Costs	Incl. O&P
2 Skilled Worker Foremen	$38.50	$616.00	$59.90	$958.40	$36.90	$57.42
8 Skilled Workers	36.50	2336.00	56.80	3635.20		
80 L.H., Daily Totals		$2952.00		$4593.60	$36.90	$57.42

Crew C-17A

Crew No.	Bare Costs Hr.	Daily	Incl. Subs O & P Hr.	Daily	Cost Per Labor-Hour Bare Costs	Incl. O&P
2 Skilled Worker Foremen	$38.50	$616.00	$59.90	$958.40	$36.91	$57.42
8 Skilled Workers	36.50	2336.00	56.80	3635.20		
.125 Equip. Oper. (crane)	38.10	38.10	57.50	57.50		
.125 Crane, 72 metric ton		132.38		145.60	1.63	1.80
81 L.H., Daily Totals		$3122.48		$4796.70	$38.54	$59.22

Crew C-17B

Crew No.	Bare Costs Hr.	Daily	Incl. Subs O & P Hr.	Daily	Cost Per Labor-Hour Bare Costs	Incl. O&P
2 Skilled Worker Foremen	$38.50	$616.00	$59.90	$958.40	$36.93	$57.42
8 Skilled Workers	36.50	2336.00	56.80	3635.20		
.25 Equip. Oper. (crane)	38.10	76.20	57.50	115.00		
.25 Crane, 72 metric ton		264.75		291.25		
.25 Walk Behind Power Tools		6.05		6.65	3.30	3.63
82 L.H., Daily Totals		$3299.00		$5006.50	$40.23	$61.05

Crew C-17C

Crew No.	Bare Costs Hr.	Daily	Incl. Subs O & P Hr.	Daily	Cost Per Labor-Hour Bare Costs	Incl. O&P
2 Skilled Worker Foremen	$38.50	$616.00	$59.90	$958.40	$36.94	$57.42
8 Skilled Workers	36.50	2336.00	56.80	3635.20		
.375 Equip. Oper. (crane)	38.10	114.30	57.50	172.50		
.375 Crane, 73 metric ton		397.13		436.85	4.78	5.26
83 L.H., Daily Totals		$3463.43		$5202.95	$41.72	$62.68

Crew C-17D

	Bare Costs		Incl. Subs O & P		Cost Per Labor-Hour	
	Hr.	Daily	Hr.	Daily	Bare Costs	Incl. O&P
2 Skilled Worker Foremen	$38.50	$616.00	$59.90	$958.40	$36.96	$57.42
8 Skilled Workers	36.50	2336.00	56.80	3635.20		
.5 Equip. Oper. (crane)	38.10	152.40	57.50	230.00		
.5 Crane, 72 metric ton		529.50		582.45	6.30	6.93
84 L.H., Daily Totals		$3633.90		$5406.05	$43.26	$64.35

Crew C-17E

	Bare Costs		Incl. Subs O & P		Cost Per Labor-Hour	
	Hr.	Daily	Hr.	Daily	Bare Costs	Incl. O&P
2 Skilled Worker Foremen	$38.50	$616.00	$59.90	$958.40	$36.90	$57.42
8 Skilled Workers	36.50	2336.00	56.80	3635.20		
1 Hyd. Jack with Rods		80.65		88.70	1.01	1.11
80 L.H., Daily Totals		$3032.65		$4682.30	$37.91	$58.53

Crew C-18

	Bare Costs		Incl. Subs O & P		Cost Per Labor-Hour	
	Hr.	Daily	Hr.	Daily	Bare Costs	Incl. O&P
.125 Laborer Foreman (out)	$29.40	$29.40	$45.80	$45.80	$27.62	$43.00
1 Laborer	27.40	219.20	42.65	341.20		
1 Concrete Cart, 0.28 m³		52.20		57.40	5.80	6.38
9 L.H., Daily Totals		$300.80		$444.40	$33.42	$49.38

Crew C-19

	Bare Costs		Incl. Subs O & P		Cost Per Labor-Hour	
	Hr.	Daily	Hr.	Daily	Bare Costs	Incl. O&P
.125 Laborer Foreman (out)	$29.40	$29.40	$45.80	$45.80	$27.62	$43.00
1 Laborer	27.40	219.20	42.65	341.20		
1 Concrete Cart, 0.5 m³		74.40		81.85	8.27	9.09
9 L.H., Daily Totals		$323.00		$468.85	$35.89	$52.09

Crew C-20

	Bare Costs		Incl. Subs O & P		Cost Per Labor-Hour	
	Hr.	Daily	Hr.	Daily	Bare Costs	Incl. O&P
1 Laborer Foreman (outside)	$29.40	$235.20	$45.80	$366.40	$29.69	$45.65
5 Laborers	27.40	1096.00	42.65	1706.00		
1 Cement Finisher	34.40	275.20	50.75	406.00		
1 Equip. Oper. (med.)	36.70	293.60	55.40	443.20		
2 Gas Engine Vibrators		42.80		47.10		
1 Concrete Pump (small)		700.60		770.65	11.62	12.78
64 L.H., Daily Totals		$2643.40		$3739.35	$41.31	$58.43

Crew C-21

	Bare Costs		Incl. Subs O & P		Cost Per Labor-Hour	
	Hr.	Daily	Hr.	Daily	Bare Costs	Incl. O&P
1 Laborer Foreman (outside)	$29.40	$235.20	$45.80	$366.40	$29.69	$45.65
5 Laborers	27.40	1096.00	42.65	1706.00		
1 Cement Finisher	34.40	275.20	50.75	406.00		
1 Equip. Oper. (med.)	36.70	293.60	55.40	443.20		
2 Gas Engine Vibrators		42.80		47.10		
1 Concrete Conveyer		152.40		167.65	3.05	3.36
64 L.H., Daily Totals		$2095.20		$3136.35	$32.74	$49.01

Crew C-22

	Bare Costs		Incl. Subs O & P		Cost Per Labor-Hour	
	Hr.	Daily	Hr.	Daily	Bare Costs	Incl. O&P
1 Rodman Foreman	$41.50	$332.00	$68.35	$546.80	$39.68	$65.12
4 Rodmen (reinf.)	39.50	1264.00	65.05	2081.60		
.125 Equip. Oper. (crane)	38.10	38.10	57.50	57.50		
.125 Equip. Oper. Oiler	32.45	32.45	48.95	48.95		
.125 Hyd. Crane, 22 metric ton		81.35		89.50	1.94	2.13
42 L.H., Daily Totals		$1747.90		$2824.35	$41.62	$67.25

Crew C-23

	Bare Costs		Incl. Subs O & P		Cost Per Labor-Hour	
	Hr.	Daily	Hr.	Daily	Bare Costs	Incl. O&P
2 Skilled Worker Foremen	$38.50	$616.00	$59.90	$958.40	$36.66	$56.71
6 Skilled Workers	36.50	1752.00	56.80	2726.40		
1 Equip. Oper. (crane)	38.10	304.80	57.50	460.00		
1 Equip. Oper. Oiler	32.45	259.60	48.95	391.60		
1 Crane, 82 metric ton		1427.00		1569.70	17.84	19.62
80 L.H., Daily Totals		$4359.40		$6106.10	$54.50	$76.33

Crew C-23A

	Bare Costs		Incl. Subs O & P		Cost Per Labor-Hour	
	Hr.	Daily	Hr.	Daily	Bare Costs	Incl. O&P
1 Laborer Foreman (outside)	$29.40	$235.20	$45.80	$366.40	$30.95	$47.51
2 Laborers	27.40	438.40	42.65	682.40		
1 Equip. Oper. (crane)	38.10	304.80	57.50	460.00		
1 Equip. Oper. Oiler	32.45	259.60	48.95	391.60		
1 Crane, 91 metric ton capacity		1590.00		1749.00		
3 Conc. bucket, 6 m³		478.20		526.00	51.71	56.88
40 L.H., Daily Totals		$3306.20		$4175.40	$82.66	$104.39

Crew C-24

	Bare Costs		Incl. Subs O & P		Cost Per Labor-Hour	
	Hr.	Daily	Hr.	Daily	Bare Costs	Incl. O&P
2 Skilled Worker Foremen	$38.50	$616.00	$59.90	$958.40	$36.66	$56.71
6 Skilled Workers	36.50	1752.00	56.80	2726.40		
1 Equip. Oper. (crane)	38.10	304.80	57.50	460.00		
1 Equip. Oper. Oiler	32.45	259.60	48.95	391.60		
1 Truck Crane, 136 metric ton		1539.00		1692.90	19.24	21.16
80 L.H., Daily Totals		$4471.40		$6229.30	$55.90	$77.87

Crew C-25

	Bare Costs		Incl. Subs O & P		Cost Per Labor-Hour	
	Hr.	Daily	Hr.	Daily	Bare Costs	Incl. O&P
2 Rodmen (reinf.)	$39.50	$632.00	$65.05	$1040.80	$31.03	$51.65
2 Rodman Helpers	22.55	360.80	38.25	612.00		
32 L.H., Daily Totals		$992.80		$1652.80	$31.03	$51.65

Crew C-27

	Bare Costs		Incl. Subs O & P		Cost Per Labor-Hour	
	Hr.	Daily	Hr.	Daily	Bare Costs	Incl. O&P
2 Cement Finishers	$34.40	$550.40	$50.75	$812.00	$34.40	$50.75
1 Concrete Saw		107.00		117.70	6.69	7.36
16 L.H., Daily Totals		$657.40		$929.70	$41.09	$58.11

Crew C-28

	Bare Costs		Incl. Subs O & P		Cost Per Labor-Hour	
	Hr.	Daily	Hr.	Daily	Bare Costs	Incl. O&P
1 Cement Finisher	$34.40	$275.20	$50.75	$406.00	$34.40	$50.75
1 Portable Air Compressor		15.75		17.35	1.96	2.16
8 L.H., Daily Totals		$290.95		$423.35	$36.36	$52.91

Crew D-1

	Bare Costs		Incl. Subs O & P		Cost Per Labor-Hour	
	Hr.	Daily	Hr.	Daily	Bare Costs	Incl. O&P
1 Bricklayer	$36.55	$292.40	$55.65	$445.20	$32.15	$48.95
1 Bricklayer Helper	27.75	222.00	42.25	338.00		
16 L.H., Daily Totals		$514.40		$783.20	$32.15	$48.95

Crew D-2

	Bare Costs		Incl. Subs O & P		Cost Per Labor-Hour	
	Hr.	Daily	Hr.	Daily	Bare Costs	Incl. O&P
3 Bricklayers	$36.55	$877.20	$55.65	$1335.60	$33.26	$50.75
2 Bricklayer Helpers	27.75	444.00	42.25	676.00		
.5 Carpenter	35.55	142.20	55.35	221.40		
44 L.H., Daily Totals		$1463.40		$2233.00	$33.26	$50.75

Crew D-3

	Bare Costs		Incl. Subs O & P		Cost Per Labor-Hour	
	Hr.	Daily	Hr.	Daily	Bare Costs	Incl. O&P
3 Bricklayers	$36.55	$877.20	$55.65	$1335.60	$33.15	$50.53
2 Bricklayer Helpers	27.75	444.00	42.25	676.00		
.25 Carpenter	35.55	71.10	55.35	110.70		
42 L.H., Daily Totals		$1392.30		$2122.30	$33.15	$50.53

Crew D-4

	Bare Costs		Incl. Subs O & P		Cost Per Labor-Hour	
	Hr.	Daily	Hr.	Daily	Bare Costs	Incl. O&P
1 Bricklayer	$36.55	$292.40	$55.65	$445.20	$31.81	$48.31
2 Bricklayer Helpers	27.75	444.00	42.25	676.00		
1 Equip. Oper. (light)	35.20	281.60	53.10	424.80		
1 Grout Pump, 1.5 m³/hr		117.30		129.05		
1 Hose & Hopper		15.60		17.15		
1 Accessory		12.10		13.30	4.53	4.98
32 L.H., Daily Totals		$1163.00		$1705.50	$36.34	$53.29

Crew D-5

	Bare Costs		Incl. Subs O & P		Cost Per Labor-Hour	
	Hr.	Daily	Hr.	Daily	Bare Costs	Incl. O&P
1 Bricklayer	$36.55	$292.40	$55.65	$445.20	$36.55	$55.65
8 L.H., Daily Totals		$292.40		$445.20	$36.55	$55.65

CREWS

Crew No.	Bare Costs Hr.	Daily	Incl. Subs O & P Hr.	Daily	Cost Per Labor-Hour Bare Costs	Incl. O&P
Crew D-6	Hr.	Daily	Hr.	Daily	Bare Costs	Incl. O&P
3 Bricklayers	$36.55	$877.20	$55.65	$1335.60	$32.29	$49.21
3 Bricklayer Helpers	27.75	666.00	42.25	1014.00		
.25 Carpenter	35.55	71.10	55.35	110.70		
50 L.H., Daily Totals		$1614.30		$2460.30	$32.29	$49.21
Crew D-7	Hr.	Daily	Hr.	Daily	Bare Costs	Incl. O&P
1 Tile Layer	$34.25	$274.00	$50.30	$402.40	$30.38	$44.60
1 Tile Layer Helper	26.50	212.00	38.90	311.20		
16 L.H., Daily Totals		$486.00		$713.60	$30.38	$44.60
Crew D-8	Hr.	Daily	Hr.	Daily	Bare Costs	Incl. O&P
3 Bricklayers	$36.55	$877.20	$55.65	$1335.60	$33.03	$50.29
2 Bricklayer Helpers	27.75	444.00	42.25	676.00		
40 L.H., Daily Totals		$1321.20		$2011.60	$33.03	$50.29
Crew D-9	Hr.	Daily	Hr.	Daily	Bare Costs	Incl. O&P
3 Bricklayers	$36.55	$877.20	$55.65	$1335.60	$32.15	$48.95
3 Bricklayer Helpers	27.75	666.00	42.25	1014.00		
48 L.H., Daily Totals		$1543.20		$2349.60	$32.15	$48.95
Crew D-10	Hr.	Daily	Hr.	Daily	Bare Costs	Incl. O&P
1 Bricklayer Foreman	$38.55	$308.40	$58.65	$469.20	$33.74	$51.26
1 Bricklayer	36.55	292.40	55.65	445.20		
2 Bricklayer Helpers	27.75	444.00	42.25	676.00		
1 Equip. Oper. (crane)	38.10	304.80	57.50	460.00		
1 Truck Crane, 11 metric ton		536.40		590.05	13.41	14.75
40 L.H., Daily Totals		$1886.00		$2640.45	$47.15	$66.01
Crew D-11	Hr.	Daily	Hr.	Daily	Bare Costs	Incl. O&P
1 Bricklayer Foreman	$38.55	$308.40	$58.65	$469.20	$34.28	$52.18
1 Bricklayer	36.55	292.40	55.65	445.20		
1 Bricklayer Helper	27.75	222.00	42.25	338.00		
24 L.H., Daily Totals		$822.80		$1252.40	$34.28	$52.18
Crew D-12	Hr.	Daily	Hr.	Daily	Bare Costs	Incl. O&P
1 Bricklayer Foreman	$38.55	$308.40	$58.65	$469.20	$32.65	$49.70
1 Bricklayer	36.55	292.40	55.65	445.20		
2 Bricklayer Helpers	27.75	444.00	42.25	676.00		
32 L.H., Daily Totals		$1044.80		$1590.40	$32.65	$49.70
Crew D-13	Hr.	Daily	Hr.	Daily	Bare Costs	Incl. O&P
1 Bricklayer Foreman	$38.55	$308.40	$58.65	$469.20	$34.04	$51.94
1 Bricklayer	36.55	292.40	55.65	445.20		
2 Bricklayer Helpers	27.75	444.00	42.25	676.00		
1 Carpenter	35.55	284.40	55.35	442.80		
1 Equip. Oper. (crane)	38.10	304.80	57.50	460.00		
1 Truck Crane, 11 metric ton		536.40		590.05	11.18	12.29
48 L.H., Daily Totals		$2170.40		$3083.25	$45.22	$64.23
Crew E-1	Hr.	Daily	Hr.	Daily	Bare Costs	Incl. O&P
1 Welder Foreman	$41.95	$335.60	$75.50	$604.00	$39.03	$66.83
1 Welder	39.95	319.60	71.90	575.20		
1 Equip. Oper. (light)	35.20	281.60	53.10	424.80		
1 Gas Welding Machine		89.20		98.10	3.72	4.09
24 L.H., Daily Totals		$1026.00		$1702.10	$42.75	$70.92

Crew No.	Bare Costs Hr.	Daily	Incl. Subs O & P Hr.	Daily	Cost Per Labor-Hour Bare Costs	Incl. O&P
Crew E-2	Hr.	Daily	Hr.	Daily	Bare Costs	Incl. O&P
1 Struc. Steel Foreman	$41.95	$335.60	$75.50	$604.00	$38.90	$67.08
4 Struc. Steel Workers	39.95	1278.40	71.90	2300.80		
1 Equip. Oper. (crane)	38.10	304.80	57.50	460.00		
1 Equip. Oper. Oiler	32.45	259.60	48.95	391.60		
1 Crane, 82 metric ton		1427.00		1569.70	25.48	28.03
56 L.H., Daily Totals		$3605.40		$5326.10	$64.38	$95.11
Crew E-3	Hr.	Daily	Hr.	Daily	Bare Costs	Incl. O&P
1 Struc. Steel Foreman	$41.95	$335.60	$75.50	$604.00	$40.62	$73.10
1 Struc. Steel Worker	39.95	319.60	71.90	575.20		
1 Welder	39.95	319.60	71.90	575.20		
1 Gas Welding Machine		89.20		98.10	3.72	4.09
24 L.H., Daily Totals		$1064.00		$1852.50	$44.34	$77.19
Crew E-4	Hr.	Daily	Hr.	Daily	Bare Costs	Incl. O&P
1 Struc. Steel Foreman	$41.95	$335.60	$75.50	$604.00	$40.45	$72.80
3 Struc. Steel Workers	39.95	958.80	71.90	1725.60		
1 Gas Welding Machine		89.20		98.10	2.79	3.07
32 L.H., Daily Totals		$1383.60		$2427.70	$43.24	$75.87
Crew E-5	Hr.	Daily	Hr.	Daily	Bare Costs	Incl. O&P
2 Struc. Steel Foremen	$41.95	$671.20	$75.50	$1208.00	$39.42	$68.89
5 Struc. Steel Workers	39.95	1598.00	71.90	2876.00		
1 Equip. Oper. (crane)	38.10	304.80	57.50	460.00		
1 Welder	39.95	319.60	71.90	575.20		
1 Equip. Oper. Oiler	32.45	259.60	48.95	391.60		
1 Crane, 82 metric ton		1427.00		1569.70		
1 Gas Welding Machine		89.20		98.10	18.95	20.85
80 L.H., Daily Totals		$4669.40		$7178.60	$58.37	$89.74
Crew E-6	Hr.	Daily	Hr.	Daily	Bare Costs	Incl. O&P
3 Struc. Steel Foremen	$41.95	$1006.80	$75.50	$1812.00	$39.44	$69.07
9 Struc. Steel Workers	39.95	2876.40	71.90	5176.80		
1 Equip. Oper. (crane)	38.10	304.80	57.50	460.00		
1 Welder	39.95	319.60	71.90	575.20		
1 Equip. Oper. Oiler	32.45	259.60	48.95	391.60		
1 Equip. Oper. (light)	35.20	281.60	53.10	424.80		
1 Crane, 82 metric ton		1427.00		1569.70		
1 Gas Welding Machine		89.20		98.10		
1 Air Compr., 76 L/s		97.60		107.35		
2 Impact Wrenches		25.20		27.70	12.80	14.09
128 L.H., Daily Totals		$6687.80		$10643.25	$52.24	$83.16
Crew E-7	Hr.	Daily	Hr.	Daily	Bare Costs	Incl. O&P
1 Struc. Steel Foreman	$41.95	$335.60	$75.50	$604.00	$39.42	$68.89
4 Struc. Steel Workers	39.95	1278.40	71.90	2300.80		
1 Equip. Oper. (crane)	38.10	304.80	57.50	460.00		
1 Equip. Oper. Oiler	32.45	259.60	48.95	391.60		
1 Welder Foreman	41.95	335.60	75.50	604.00		
2 Welders	39.95	639.20	71.90	1150.40		
1 Crane, 82 metric ton		1427.00		1569.70		
2 Gas Welding Machines		178.40		196.25	20.07	22.07
80 L.H., Daily Totals		$4758.60		$7276.75	$59.49	$90.96

Crew E-8

	Bare Costs Hr.	Daily	Incl. Subs O & P Hr.	Daily	Cost Per Labor-Hour Bare Costs	Incl. O&P
1 Struc. Steel Foreman	$41.95	$335.60	$75.50	$604.00	$39.17	$68.13
4 Struc. Steel Workers	39.95	1278.40	71.90	2300.80		
1 Welder Foreman	41.95	335.60	75.50	604.00		
4 Welders	39.95	1278.40	71.90	2300.80		
1 Equip. Oper. (crane)	38.10	304.80	57.50	460.00		
1 Equip. Oper. Oiler	32.45	259.60	48.95	391.60		
1 Equip. Oper. (light)	35.20	281.60	53.10	424.80		
1 Crane, 82 metric ton		1427.00		1569.70		
4 Gas Welding Machines		356.80		392.50	17.15	18.87
104 L.H., Daily Totals		$5857.80		$9048.20	$56.32	$87.00

Crew E-9

	Bare Costs Hr.	Daily	Incl. Subs O & P Hr.	Daily	Cost Per Labor-Hour Bare Costs	Incl. O&P
2 Struc. Steel Foremen	$41.95	$671.20	$75.50	$1208.00	$39.44	$69.07
5 Struc. Steel Workers	39.95	1598.00	71.90	2876.00		
1 Welder Foreman	41.95	335.60	75.50	604.00		
5 Welders	39.95	1598.00	71.90	2876.00		
1 Equip. Oper. (crane)	38.10	304.80	57.50	460.00		
1 Equip. Oper. Oiler	32.45	259.60	48.95	391.60		
1 Equip. Oper. (light)	35.20	281.60	53.10	424.80		
1 Crane, 82 metric ton		1427.00		1569.70		
5 Gas Welding Machines		446.00		490.60	14.63	16.10
128 L.H., Daily Totals		$6921.80		$10900.70	$54.07	$85.17

Crew E-10

	Bare Costs Hr.	Daily	Incl. Subs O & P Hr.	Daily	Cost Per Labor-Hour Bare Costs	Incl. O&P
1 Welder Foreman	$41.95	$335.60	$75.50	$604.00	$40.95	$73.70
1 Welder	39.95	319.60	71.90	575.20		
1 Gas Welding Machine		89.20		98.10		
1 Truck, 2.7 metric ton		176.20		193.80	16.59	18.25
16 L.H., Daily Totals		$920.60		$1471.10	$57.54	$91.95

Crew E-11

	Bare Costs Hr.	Daily	Incl. Subs O & P Hr.	Daily	Cost Per Labor-Hour Bare Costs	Incl. O&P
2 Painters, Struc. Steel	$32.45	$519.20	$60.45	$967.20	$31.88	$54.16
1 Building Laborer	27.40	219.20	42.65	341.20		
1 Equip. Oper. (light)	35.20	281.60	53.10	424.80		
1 Air Compressor 118 L/s		119.80		131.80		
1 Sand Blaster		15.60		17.15		
1 Sand Blasting Accessory		12.10		13.30	4.61	5.07
32 L.H., Daily Totals		$1167.50		$1895.45	$36.49	$59.23

Crew E-12

	Bare Costs Hr.	Daily	Incl. Subs O & P Hr.	Daily	Cost Per Labor-Hour Bare Costs	Incl. O&P
1 Welder Foreman	$41.95	$335.60	$75.50	$604.00	$38.58	$64.30
1 Equip. Oper. (light)	35.20	281.60	53.10	424.80		
1 Gas Welding Machine		89.20		98.10	5.58	6.13
16 L.H., Daily Totals		$706.40		$1126.90	$44.16	$70.43

Crew E-13

	Bare Costs Hr.	Daily	Incl. Subs O & P Hr.	Daily	Cost Per Labor-Hour Bare Costs	Incl. O&P
1 Welder Foreman	$41.95	$335.60	$75.50	$604.00	$39.70	$68.03
.5 Equip. Oper. (light)	35.20	140.80	53.10	212.40		
1 Gas Welding Machine		89.20		98.10	7.43	8.18
12 L.H., Daily Totals		$565.60		$914.50	$47.13	$76.21

Crew E-14

	Bare Costs Hr.	Daily	Incl. Subs O & P Hr.	Daily	Cost Per Labor-Hour Bare Costs	Incl. O&P
1 Welder Foreman	$41.95	$335.60	$75.50	$604.00	$41.95	$75.50
1 Gas Welding Machine		89.20		98.10	11.15	12.27
8 L.H., Daily Totals		$424.80		$702.10	$53.10	$87.77

Crew E-16

	Bare Costs Hr.	Daily	Incl. Subs O & P Hr.	Daily	Cost Per Labor-Hour Bare Costs	Incl. O&P
1 Welder Foreman	$41.95	$335.60	$75.50	$604.00	$40.95	$73.70
1 Welder	39.95	319.60	71.90	575.20		
1 Gas Welding Machine		89.20		98.10	5.58	6.13
16 L.H., Daily Totals		$744.40		$1277.30	$46.53	$79.83

Crew E-17

	Bare Costs Hr.	Daily	Incl. Subs O & P Hr.	Daily	Cost Per Labor-Hour Bare Costs	Incl. O&P
1 Structural Steel Foreman	$41.95	$335.60	$75.50	$604.00	$40.95	$73.70
1 Structural Steel Worker	39.95	319.60	71.90	575.20		
1 Power Tool		5.40		5.95	.34	.37
16 L.H., Daily Totals		$660.60		$1185.15	$41.29	$74.07

Crew E-18

	Bare Costs Hr.	Daily	Incl. Subs O & P Hr.	Daily	Cost Per Labor-Hour Bare Costs	Incl. O&P
1 Structural Steel Foreman	$41.95	$335.60	$75.50	$604.00	$39.70	$69.32
3 Structural Steel Workers	39.95	958.80	71.90	1725.60		
1 Equipment Operator (med.)	36.70	293.60	55.40	443.20		
1 Crane, 18 metric ton		947.05		1041.75	23.68	26.04
40 L.H., Daily Totals		$2535.05		$3814.55	$63.38	$95.36

Crew E-19

	Bare Costs Hr.	Daily	Incl. Subs O & P Hr.	Daily	Cost Per Labor-Hour Bare Costs	Incl. O&P
1 Structural Steel Worker	$39.95	$319.60	$71.90	$575.20	$39.03	$66.83
1 Structural Steel Foreman	41.95	335.60	75.50	604.00		
1 Equip. Oper. (light)	35.20	281.60	53.10	424.80		
1 Power Tool		5.40		5.95		
1 Crane, 18 metric ton		947.05		1041.75	39.68	43.65
24 L.H., Daily Totals		$1889.25		$2651.70	$78.71	$110.48

Crew E-20

	Bare Costs Hr.	Daily	Incl. Subs O & P Hr.	Daily	Cost Per Labor-Hour Bare Costs	Incl. O&P
1 Structural Steel Foreman	$41.95	$335.60	$75.50	$604.00	$39.03	$67.68
5 Structural Steel Workers	39.95	1598.00	71.90	2876.00		
1 Equip. Oper. (crane)	38.10	304.80	57.50	460.00		
1 Oiler	32.45	259.60	48.95	391.60		
1 Power Tool		5.40		5.95		
1 Crane, 36 metric ton		1130.00		1243.00	17.74	19.51
64 L.H., Daily Totals		$3633.40		$5580.55	$56.77	$87.19

Crew E-22

	Bare Costs Hr.	Daily	Incl. Subs O & P Hr.	Daily	Cost Per Labor-Hour Bare Costs	Incl. O&P
1 Skilled Worker Foreman	$38.50	$308.00	$59.90	$479.20	$37.17	$57.83
2 Skilled Workers	36.50	584.00	56.80	908.80		
24 L.H., Daily Totals		$892.00		$1388.00	$37.17	$57.83

Crew E-24

	Bare Costs Hr.	Daily	Incl. Subs O & P Hr.	Daily	Cost Per Labor-Hour Bare Costs	Incl. O&P
3 Structural Steel Workers	$39.95	$958.80	$71.90	$1725.60	$39.14	$67.78
1 Equipment Operator (medium)	36.70	293.60	55.40	443.20		
1-32 metric ton crane		650.80		715.90	20.34	22.37
32 L.H., Daily Totals		$1903.20		$2884.70	$59.48	$90.15

Crew E-25

	Bare Costs Hr.	Daily	Incl. Subs O & P Hr.	Daily	Cost Per Labor-Hour Bare Costs	Incl. O&P
1 Welder Foreman	$41.95	$335.60	$75.50	$604.00	$41.95	$75.50
1 Cutting Torch		16.20		17.80		
1 Gas		64.80		71.30	10.13	11.14
8 L.H., Daily Totals		$416.60		$693.10	$52.08	$86.64

Crew F-3

	Bare Costs Hr.	Daily	Incl. Subs O & P Hr.	Daily	Cost Per Labor-Hour Bare Costs	Incl. O&P
4 Carpenters	$35.55	$1137.60	$55.35	$1771.20	$36.06	$55.78
1 Equip. Oper. (crane)	38.10	304.80	57.50	460.00		
1 Hyd. Crane, 11 metric ton		636.60		700.25	15.92	17.51
40 L.H., Daily Totals		$2079.00		$2931.45	$51.98	$73.29

Crew F-4

	Bare Costs Hr.	Daily	Incl. Subs O & P Hr.	Daily	Cost Per Labor-Hour Bare Costs	Incl. O&P
4 Carpenters	$35.55	$1137.60	$55.35	$1771.20	$35.46	$54.64
1 Equip. Oper. (crane)	38.10	304.80	57.50	460.00		
1 Equip. Oper. Oiler	32.45	259.60	48.95	391.60		
1 Hyd. Crane, 50 metric ton		951.60		1046.75	19.83	21.81
48 L.H., Daily Totals		$2653.60		$3669.55	$55.29	$76.45

Crew No.	Bare Costs		Incl. Subs O & P		Cost Per Labor-Hour	
Crew F-5	Hr.	Daily	Hr.	Daily	Bare Costs	Incl. O&P
1 Carpenter Foreman	$37.55	$300.40	$58.45	$467.60	$36.05	$56.13
3 Carpenters	35.55	853.20	55.35	1328.40		
32 L.H., Daily Totals		$1153.60		$1796.00	$36.05	$56.13
Crew F-6	Hr.	Daily	Hr.	Daily	Bare Costs	Incl. O&P
2 Carpenters	$35.55	$568.80	$55.35	$885.60	$32.80	$50.70
2 Building Laborers	27.40	438.40	42.65	682.40		
1 Equip. Oper. (crane)	38.10	304.80	57.50	460.00		
1 Hyd. Crane, 11 metric ton		636.60		700.25	15.92	17.51
40 L.H., Daily Totals		$1948.60		$2728.25	$48.72	$68.21
Crew F-7	Hr.	Daily	Hr.	Daily	Bare Costs	Incl. O&P
2 Carpenters	$35.55	$568.80	$55.35	$885.60	$31.48	$49.00
2 Building Laborers	27.40	438.40	42.65	682.40		
32 L.H., Daily Totals		$1007.20		$1568.00	$31.48	$49.00
Crew G-1	Hr.	Daily	Hr.	Daily	Bare Costs	Incl. O&P
1 Roofer Foreman	$32.60	$260.80	$55.30	$442.40	$28.59	$48.51
4 Roofers, Composition	30.60	979.20	51.95	1662.40		
2 Roofer Helpers	22.55	360.80	38.25	612.00		
1 Application Equipment		146.20		160.80		
1 Tar Kettle/Pot		57.30		63.05		
1 Crew Truck		133.75		147.15	6.02	6.62
56 L.H., Daily Totals		$1938.05		$3087.80	$34.61	$55.13
Crew G-2	Hr.	Daily	Hr.	Daily	Bare Costs	Incl. O&P
1 Plasterer	$32.45	$259.60	$49.10	$392.80	$29.25	$44.65
1 Plasterer Helper	27.90	223.20	42.20	337.60		
1 Building Laborer	27.40	219.20	42.65	341.20		
1 Grouting Equipment		117.30		129.05	4.89	5.38
24 L.H., Daily Totals		$819.30		$1200.65	$34.14	$50.03
Crew G-2A	Hr.	Daily	Hr.	Daily	Bare Costs	Incl. O&P
1 Roofer, composition	$30.60	$244.80	$51.95	$415.60	$26.85	$44.28
1 Roofer Helper	22.55	180.40	38.25	306.00		
1 Building Laborer	27.40	219.20	42.65	341.20		
1 Spray Equipment		117.30		129.05	4.89	5.38
24 L.H., Daily Totals		$761.70		$1191.85	$31.74	$49.66
Crew G-3	Hr.	Daily	Hr.	Daily	Bare Costs	Incl. O&P
2 Sheet Metal Workers	$42.15	$674.40	$64.95	$1039.20	$34.78	$53.80
2 Building Laborers	27.40	438.40	42.65	682.40		
32 L.H., Daily Totals		$1112.80		$1721.60	$34.78	$53.80
Crew G-4	Hr.	Daily	Hr.	Daily	Bare Costs	Incl. O&P
1 Laborer Foreman (outside)	$29.40	$235.20	$45.80	$366.40	$28.07	$43.70
2 Building Laborers	27.40	438.40	42.65	682.40		
1 Light Truck, 1.4 metric ton		128.80		141.70		
1 Air Compr., 76 L/s		97.60		107.35	9.43	10.38
24 L.H., Daily Totals		$900.00		$1297.85	$37.50	$54.08
Crew G-5	Hr.	Daily	Hr.	Daily	Bare Costs	Incl. O&P
1 Roofer Foreman	$32.60	$260.80	$55.30	$442.40	$27.78	$47.14
2 Roofers, Composition	30.60	489.60	51.95	831.20		
2 Roofer Helpers	22.55	360.80	38.25	612.00		
1 Application Equipment		146.20		160.80	3.66	4.02
40 L.H., Daily Totals		$1257.40		$2046.40	$31.44	$51.16

Crew No.	Bare Costs		Incl. Subs O & P		Cost Per Labor-Hour	
Crew G-6A	Hr.	Daily	Hr.	Daily	Bare Costs	Incl. O&P
2 Roofers, Composition	$30.60	$489.60	$51.95	$831.20	$30.60	$51.95
1 Small Compressor		10.60		11.65		
2 Pneumatic Nailers		39.40		43.35	3.12	3.43
16 L.H., Daily Totals		$539.60		$886.20	$33.72	$55.38
Crew G-7	Hr.	Daily	Hr.	Daily	Bare Costs	Incl. O&P
1 Carpenter	$35.55	$284.40	$55.35	$442.80	$35.55	$55.35
1 Small Compressor		10.60		11.65		
1 Pneumatic Nailer		19.70		21.65	3.78	4.16
8 L.H., Daily Totals		$314.70		$476.10	$39.33	$59.51
Crew H-1	Hr.	Daily	Hr.	Daily	Bare Costs	Incl. O&P
2 Glaziers	$34.25	$548.00	$51.80	$828.80	$37.10	$61.85
2 Struc. Steel Workers	39.95	639.20	71.90	1150.40		
32 L.H., Daily Totals		$1187.20		$1979.20	$37.10	$61.85
Crew H-2	Hr.	Daily	Hr.	Daily	Bare Costs	Incl. O&P
2 Glaziers	$34.25	$548.00	$51.80	$828.80	$31.97	$48.75
1 Building Laborer	27.40	219.20	42.65	341.20		
24 L.H., Daily Totals		$767.20		$1170.00	$31.97	$48.75
Crew H-3	Hr.	Daily	Hr.	Daily	Bare Costs	Incl. O&P
1 Glazier	$34.25	$274.00	$51.80	$414.40	$30.33	$46.38
1 Helper	26.40	211.20	40.95	327.60		
16 L.H., Daily Totals		$485.20		$742.00	$30.33	$46.38
Crew J-1	Hr.	Daily	Hr.	Daily	Bare Costs	Incl. O&P
3 Plasterers	$32.45	$778.80	$49.10	$1178.40	$30.63	$46.34
2 Plasterer Helpers	27.90	446.40	42.20	675.20		
1 Mixing Machine, 0.17 m^3		105.60		116.15	2.64	2.90
40 L.H., Daily Totals		$1330.80		$1969.75	$33.27	$49.24
Crew J-2	Hr.	Daily	Hr.	Daily	Bare Costs	Incl. O&P
3 Plasterers	$32.45	$778.80	$49.10	$1178.40	$30.99	$46.74
2 Plasterer Helpers	27.90	446.40	42.20	675.20		
1 Lather	32.80	262.40	48.75	390.00		
1 Mixing Machine, 0.17 m^3		105.60		116.15	2.20	2.42
48 L.H., Daily Totals		$1593.20		$2359.75	$33.19	$49.16
Crew J-3	Hr.	Daily	Hr.	Daily	Bare Costs	Incl. O&P
1 Terrazzo Worker	$34.10	$272.80	$50.05	$400.40	$30.80	$45.20
1 Terrazzo Helper	27.50	220.00	40.35	322.80		
1 Terrazzo Grinder, Electric		91.25		100.40		
1 Terrazzo Mixer		144.20		158.60	14.72	16.19
16 L.H., Daily Totals		$728.25		$982.20	$45.52	$61.39
Crew J-4	Hr.	Daily	Hr.	Daily	Bare Costs	Incl. O&P
1 Tile Layer	$34.25	$274.00	$50.30	$402.40	$30.38	$44.60
1 Tile Layer Helper	26.50	212.00	38.90	311.20		
16 L.H., Daily Totals		$486.00		$713.60	$30.38	$44.60
Crew K-1	Hr.	Daily	Hr.	Daily	Bare Costs	Incl. O&P
1 Carpenter	$35.55	$284.40	$55.35	$442.80	$31.58	$48.75
1 Truck Driver (light)	27.60	220.80	42.15	337.20		
1 Truck w/Power Equip.		176.20		193.80	11.01	12.11
16 L.H., Daily Totals		$681.40		$973.80	$42.59	$60.86

CREWS

Crew K-2	Hr.	Daily	Hr.	Daily	Bare Costs	Incl. O&P
1 Struc. Steel Foreman	$41.95	$335.60	$75.50	$604.00	$36.50	$63.18
1 Struc. Steel Worker	39.95	319.60	71.90	575.20		
1 Truck Driver (light)	27.60	220.80	42.15	337.20		
1 Truck w/Power Equip.		176.20		193.80	7.34	8.08
24 L.H., Daily Totals		$1052.20		$1710.20	$43.84	$71.26

Crew L-1	Hr.	Daily	Hr.	Daily	Bare Costs	Incl. O&P
1 Electrician	$42.00	$336.00	$62.60	$500.80	$42.35	$63.43
1 Plumber	42.70	341.60	64.25	514.00		
16 L.H., Daily Totals		$677.60		$1014.80	$42.35	$63.43

Crew L-2	Hr.	Daily	Hr.	Daily	Bare Costs	Incl. O&P
1 Carpenter	$35.55	$284.40	$55.35	$442.80	$30.98	$48.15
1 Carpenter Helper	26.40	211.20	40.95	327.60		
16 L.H., Daily Totals		$495.60		$770.40	$30.98	$48.15

Crew L-3	Hr.	Daily	Hr.	Daily	Bare Costs	Incl. O&P
1 Carpenter	$35.55	$284.40	$55.35	$442.80	$38.81	$59.56
.5 Electrician	42.00	168.00	62.60	250.40		
.5 Sheet Metal Worker	42.15	168.60	64.95	259.80		
16 L.H., Daily Totals		$621.00		$953.00	$38.81	$59.56

Crew L-3A	Hr.	Daily	Hr.	Daily	Bare Costs	Incl. O&P
1 Carpenter Foreman (outside)	$37.55	$300.40	$58.45	$467.60	$39.08	$60.62
.5 Sheet Metal Worker	42.15	168.60	64.95	259.80		
12 L.H., Daily Totals		$469.00		$727.40	$39.08	$60.62

Crew L-4	Hr.	Daily	Hr.	Daily	Bare Costs	Incl. O&P
2 Skilled Workers	$36.50	$584.00	$56.80	$908.80	$33.13	$51.52
1 Helper	26.40	211.20	40.95	327.60		
24 L.H., Daily Totals		$795.20		$1236.40	$33.13	$51.52

Crew L-5	Hr.	Daily	Hr.	Daily	Bare Costs	Incl. O&P
1 Struc. Steel Foreman	$41.95	$335.60	$75.50	$604.00	$39.97	$70.36
5 Struc. Steel Workers	39.95	1598.00	71.90	2876.00		
1 Equip. Oper. (crane)	38.10	304.80	57.50	460.00		
1 Hyd. Crane, 23 metric ton		650.80		715.90	11.62	12.78
56 L.H., Daily Totals		$2889.20		$4655.90	$51.59	$83.14

Crew L-5A	Hr.	Daily	Hr.	Daily	Bare Costs	Incl. O&P
1 Structural Steel Foreman	$41.95	$335.60	$75.50	$604.00	$39.99	$69.20
2 Structural Steel Workers	39.95	639.20	71.90	1150.40		
1 Equip. Oper. (crane)	38.10	304.80	57.50	460.00		
1 Crane, SP, 23 metric ton		696.80		766.50	21.78	23.95
32 L.H., Daily Totals		$1976.40		$2980.90	$61.77	$93.15

Crew L-6	Hr.	Daily	Hr.	Daily	Bare Costs	Incl. O&P
1 Plumber	$42.70	$341.60	$64.25	$514.00	$42.47	$63.70
.5 Electrician	42.00	168.00	62.60	250.40		
12 L.H., Daily Totals		$509.60		$764.40	$42.47	$63.70

Crew L-7	Hr.	Daily	Hr.	Daily	Bare Costs	Incl. O&P
2 Carpenters	$35.55	$568.80	$55.35	$885.60	$34.14	$52.76
1 Building Laborer	27.40	219.20	42.65	341.20		
.5 Electrician	42.00	168.00	62.60	250.40		
28 L.H., Daily Totals		$956.00		$1477.20	$34.14	$52.76

Crew L-8	Hr.	Daily	Hr.	Daily	Bare Costs	Incl. O&P
2 Carpenters	$35.55	$568.80	$55.35	$885.60	$36.98	$57.13
.5 Plumber	42.70	170.80	64.25	257.00		
20 L.H., Daily Totals		$739.60		$1142.60	$36.98	$57.13

Crew L-9	Hr.	Daily	Hr.	Daily	Bare Costs	Incl. O&P
1 Laborer Foreman (inside)	$27.90	$223.20	$43.45	$347.60	$31.92	$51.54
2 Building Laborers	27.40	438.40	42.65	682.40		
1 Struc. Steel Worker	39.95	319.60	71.90	575.20		
.5 Electrician	42.00	168.00	62.60	250.40		
36 L.H., Daily Totals		$1149.20		$1855.60	$31.92	$51.54

Crew L-10	Hr.	Daily	Hr.	Daily	Bare Costs	Incl. O&P
1 Structural Steel Foreman	$41.95	$335.60	$75.50	$604.00	$40.00	$68.30
1 Structural Steel Worker	39.95	319.60	71.90	575.20		
1 Equip. Oper. (crane)	38.10	304.80	57.50	460.00		
1 Hyd. Crane, 11 metric ton		636.60		700.25	26.53	29.18
24 L.H., Daily Totals		$1596.60		$2339.45	$66.53	$97.48

Crew L-11	Hr.	Daily	Hr.	Daily	Bare Costs	Incl. O&P
2 Wreckers	$27.40	$438.40	$48.60	$777.60	$32.03	$51.95
1 Equip. Oper. (crane)	38.10	304.80	57.50	460.00		
1 Equip. Oper. (light)	35.20	281.60	53.10	424.80		
1 Hyd. Excavator, 1.9 m^3		1218.00		1339.80		
1 Skid steer loader		202.00		222.20	44.38	48.81
32 L.H., Daily Totals		$2444.80		$3224.40	$76.41	$100.76

Crew M-1	Hr.	Daily	Hr.	Daily	Bare Costs	Incl. O&P
3 Elevator Constructors	$49.95	$1198.80	$74.55	$1789.20	$47.45	$70.81
1 Elevator Apprentice	39.95	319.60	59.60	476.80		
5 Hand Tools		54.00		59.40	1.69	1.86
32 L.H., Daily Totals		$1572.40		$2325.40	$49.14	$72.67

Crew M-3	Hr.	Daily	Hr.	Daily	Bare Costs	Incl. O&P
1 Electrician Foreman (out)	$44.00	$352.00	$65.60	$524.80	$40.11	$60.29
1 Common Laborer	27.40	219.20	42.65	341.20		
.25 Equipment Operator, Medium	36.70	73.40	55.40	110.80		
1 Elevator Constructor	49.95	399.60	74.55	596.40		
1 Elevator Apprentice	39.95	319.60	59.60	476.80		
.25 Crane, SP, 4 x 4, 18 metric ton		150.85		165.95	4.44	4.88
34 L.H., Daily Totals		$1514.65		$2215.95	$44.55	$65.17

Crew M-4	Hr.	Daily	Hr.	Daily	Bare Costs	Incl. O&P
1 Electrician Foreman (out)	$44.00	$352.00	$65.60	$524.80	$39.76	$59.78
1 Common Laborer	27.40	219.20	42.65	341.20		
.25 Equipment Operator, Crane	38.10	76.20	57.50	115.00		
.25 Equipment Operator, Oiler	32.45	64.90	48.95	97.90		
1 Elevator Constructor	49.95	399.60	74.55	596.40		
1 Elevator Apprentice	39.95	319.60	59.60	476.80		
.25 Crane, hydr, SP, 36 met ton		229.00		251.90	6.36	7.00
36 L.H., Daily Totals		$1660.50		$2404.00	$46.12	$66.78

Crew Q-1	Hr.	Daily	Hr.	Daily	Bare Costs	Incl. O&P
1 Plumber	$42.70	$341.60	$64.25	$514.00	$38.43	$57.83
1 Plumber Apprentice	34.15	273.20	51.40	411.20		
16 L.H., Daily Totals		$614.80		$925.20	$38.43	$57.83

CREWS

Crews

Crew Q-1C	Hr.	Daily	Hr.	Daily	Bare Costs	Incl. O&P
1 Plumber	$42.70	$341.60	$64.25	$514.00	$37.85	$57.02
1 Plumber Apprentice	34.15	273.20	51.40	411.20		
1 Equip. Oper. (medium)	36.70	293.60	55.40	443.20		
1 Trencher, Chain		1460.00		1606.00	60.83	66.92
24 L.H., Daily Totals		$2368.40		$2974.40	$98.68	$123.94

Crew Q-2	Hr.	Daily	Hr.	Daily	Bare Costs	Incl. O&P
2 Plumbers	$42.70	$683.20	$64.25	$1028.00	$39.85	$59.97
1 Plumber Apprentice	34.15	273.20	51.40	411.20		
24 L.H., Daily Totals		$956.40		$1439.20	$39.85	$59.97

Crew Q-3	Hr.	Daily	Hr.	Daily	Bare Costs	Incl. O&P
1 Plumber Foreman (inside)	$43.20	$345.60	$65.00	$520.00	$40.69	$61.23
2 Plumbers	42.70	683.20	64.25	1028.00		
1 Plumber Apprentice	34.15	273.20	51.40	411.20		
32 L.H., Daily Totals		$1302.00		$1959.20	$40.69	$61.23

Crew Q-4	Hr.	Daily	Hr.	Daily	Bare Costs	Incl. O&P
1 Plumber Foreman (inside)	$43.20	$345.60	$65.00	$520.00	$40.69	$61.23
1 Plumber	42.70	341.60	64.25	514.00		
1 Welder (plumber)	42.70	341.60	64.25	514.00		
1 Plumber Apprentice	34.15	273.20	51.40	411.20		
1 Electric Welding Mach.		69.10		76.00	2.16	2.38
32 L.H., Daily Totals		$1371.10		$2035.20	$42.85	$63.61

Crew Q-5	Hr.	Daily	Hr.	Daily	Bare Costs	Incl. O&P
1 Steamfitter	$43.05	$344.40	$64.80	$518.40	$38.75	$58.33
1 Steamfitter Apprentice	34.45	275.60	51.85	414.80		
16 L.H., Daily Totals		$620.00		$933.20	$38.75	$58.33

Crew Q-6	Hr.	Daily	Hr.	Daily	Bare Costs	Incl. O&P
2 Steamfitters	$43.05	$688.80	$64.80	$1036.80	$40.18	$60.48
1 Steamfitter Apprentice	34.45	275.60	51.85	414.80		
24 L.H., Daily Totals		$964.40		$1451.60	$40.18	$60.48

Crew Q-7	Hr.	Daily	Hr.	Daily	Bare Costs	Incl. O&P
1 Steamfitter Foreman (inside)	$43.55	$348.40	$65.55	$524.40	$41.03	$61.75
2 Steamfitters	43.05	688.80	64.80	1036.80		
1 Steamfitter Apprentice	34.45	275.60	51.85	414.80		
32 L.H., Daily Totals		$1312.80		$1976.00	$41.03	$61.75

Crew Q-8	Hr.	Daily	Hr.	Daily	Bare Costs	Incl. O&P
1 Steamfitter Foreman (inside)	$43.55	$348.40	$65.55	$524.40	$41.03	$61.75
1 Steamfitter	43.05	344.40	64.80	518.40		
1 Welder (steamfitter)	43.05	344.40	64.80	518.40		
1 Steamfitter Apprentice	34.45	275.60	51.85	414.80		
1 Electric Welding Mach.		69.10		76.00	2.16	2.38
32 L.H., Daily Totals		$1381.90		$2052.00	$43.19	$64.13

Crew Q-9	Hr.	Daily	Hr.	Daily	Bare Costs	Incl. O&P
1 Sheet Metal Worker	$42.15	$337.20	$64.95	$519.60	$37.93	$58.45
1 Sheet Metal Apprentice	33.70	269.60	51.95	415.60		
16 L.H., Daily Totals		$606.80		$935.20	$37.93	$58.45

Crew Q-10	Hr.	Daily	Hr.	Daily	Bare Costs	Incl. O&P
2 Sheet Metal Workers	$42.15	$674.40	$64.95	$1039.20	$39.33	$60.62
1 Sheet Metal Apprentice	33.70	269.60	51.95	415.60		
24 L.H., Daily Totals		$944.00		$1454.80	$39.33	$60.62

Crew Q-11	Hr.	Daily	Hr.	Daily	Bare Costs	Incl. O&P
1 Sheet Metal Foreman (inside)	$42.65	$341.20	$65.70	$525.60	$40.16	$61.89
2 Sheet Metal Workers	42.15	674.40	64.95	1039.20		
1 Sheet Metal Apprentice	33.70	269.60	51.95	415.60		
32 L.H., Daily Totals		$1285.20		$1980.40	$40.16	$61.89

Crew Q-12	Hr.	Daily	Hr.	Daily	Bare Costs	Incl. O&P
1 Sprinkler Installer	$41.80	$334.40	$63.10	$504.80	$37.63	$56.80
1 Sprinkler Apprentice	33.45	267.60	50.50	404.00		
16 L.H., Daily Totals		$602.00		$908.80	$37.63	$56.80

Crew Q-13	Hr.	Daily	Hr.	Daily	Bare Costs	Incl. O&P
1 Sprinkler Foreman (inside)	$42.30	$338.40	$63.85	$510.80	$39.84	$60.14
2 Sprinkler Installers	41.80	668.80	63.10	1009.60		
1 Sprinkler Apprentice	33.45	267.60	50.50	404.00		
32 L.H., Daily Totals		$1274.80		$1924.40	$39.84	$60.14

Crew Q-14	Hr.	Daily	Hr.	Daily	Bare Costs	Incl. O&P
1 Asbestos Worker	$39.05	$312.40	$61.75	$494.00	$35.15	$55.58
1 Asbestos Apprentice	31.25	250.00	49.40	395.20		
16 L.H., Daily Totals		$562.40		$889.20	$35.15	$55.58

Crew Q-15	Hr.	Daily	Hr.	Daily	Bare Costs	Incl. O&P
1 Plumber	$42.70	$341.60	$64.25	$514.00	$38.43	$57.83
1 Plumber Apprentice	34.15	273.20	51.40	411.20		
1 Electric Welding Mach.		69.10		76.00	4.32	4.75
16 L.H., Daily Totals		$683.90		$1001.20	$42.75	$62.58

Crew Q-16	Hr.	Daily	Hr.	Daily	Bare Costs	Incl. O&P
2 Plumbers	$42.70	$683.20	$64.25	$1028.00	$39.85	$59.97
1 Plumber Apprentice	34.15	273.20	51.40	411.20		
1 Electric Welding Mach.		69.10		76.00	2.88	3.17
24 L.H., Daily Totals		$1025.50		$1515.20	$42.73	$63.14

Crew Q-17	Hr.	Daily	Hr.	Daily	Bare Costs	Incl. O&P
1 Steamfitter	$43.05	$344.40	$64.80	$518.40	$38.75	$58.33
1 Steamfitter Apprentice	34.45	275.60	51.85	414.80		
1 Electric Welding Mach.		69.10		76.00	4.32	4.75
16 L.H., Daily Totals		$689.10		$1009.20	$43.07	$63.08

Crew Q-17A	Hr.	Daily	Hr.	Daily	Bare Costs	Incl. O&P
1 Steamfitter	$43.05	$344.40	$64.80	$518.40	$38.53	$58.05
1 Steamfitter Apprentice	34.45	275.60	51.85	414.80		
1 Equip. Oper. (crane)	38.10	304.80	57.50	460.00		
1 Truck Crane, 11 metric ton		636.60		700.25		
1 Electric Welding Mach.		69.10		76.00	29.40	32.34
24 L.H., Daily Totals		$1630.50		$2169.45	$67.93	$90.39

Crew Q-18	Hr.	Daily	Hr.	Daily	Bare Costs	Incl. O&P
2 Steamfitters	$43.05	$688.80	$64.80	$1036.80	$40.18	$60.48
1 Steamfitter Apprentice	34.45	275.60	51.85	414.80		
1 Electric Welding Mach.		69.10		76.00	2.88	3.17
24 L.H., Daily Totals		$1033.50		$1527.60	$43.06	$63.65

Crew Q-19	Hr.	Daily	Hr.	Daily	Bare Costs	Incl. O&P
1 Steamfitter	$43.05	$344.40	$64.80	$518.40	$39.83	$59.75
1 Steamfitter Apprentice	34.45	275.60	51.85	414.80		
1 Electrician	42.00	336.00	62.60	500.80		
24 L.H., Daily Totals		$956.00		$1434.00	$39.83	$59.75

CREWS

Crews

Crew Q-20	Hr.	Daily	Hr.	Daily	Bare Costs	Incl. O&P
1 Sheet Metal Worker	$42.15	$337.20	$64.95	$519.60	$38.74	$59.28
1 Sheet Metal Apprentice	33.70	269.60	51.95	415.60		
.5 Electrician	42.00	168.00	62.60	250.40		
20 L.H., Daily Totals		$774.80		$1185.60	$38.74	$59.28

Crew Q-21	Hr.	Daily	Hr.	Daily	Bare Costs	Incl. O&P
2 Steamfitters	$43.05	$688.80	$64.80	$1036.80	$40.64	$61.01
1 Steamfitter Apprentice	34.45	275.60	51.85	414.80		
1 Electrician	42.00	336.00	62.60	500.80		
32 L.H., Daily Totals		$1300.40		$1952.40	$40.64	$61.01

Crew Q-22	Hr.	Daily	Hr.	Daily	Bare Costs	Incl. O&P
1 Plumber	$42.70	$341.60	$64.25	$514.00	$38.43	$57.83
1 Plumber Apprentice	34.15	273.20	51.40	411.20		
1 Truck Crane, 11 metric ton		636.60		700.25	39.79	43.77
16 L.H., Daily Totals		$1251.40		$1625.45	$78.22	$101.60

Crew Q-22A	Hr.	Daily	Hr.	Daily	Bare Costs	Incl. O&P
1 Plumber	$42.70	$341.60	$64.25	$514.00	$35.59	$53.95
1 Plumber Apprentice	34.15	273.20	51.40	411.20		
1 Laborer	27.40	219.20	42.65	341.20		
1 Equip. Oper. (crane)	38.10	304.80	57.50	460.00		
1 Truck Crane, 11 metric ton		636.60		700.25	19.89	21.88
32 L.H., Daily Totals		$1775.40		$2426.65	$55.48	$75.83

Crew Q-23	Hr.	Daily	Hr.	Daily	Bare Costs	Incl. O&P
1 Plumber Foreman	$44.70	$357.60	$67.25	$538.00	$41.37	$62.30
1 Plumber	42.70	341.60	64.25	514.00		
1 Equip. Oper. (medium)	36.70	293.60	55.40	443.20		
1 Power Tools		5.40		5.95		
1 Crane, 18 metric ton		947.05		1041.75	39.68	43.65
24 L.H., Daily Totals		$1945.25		$2542.90	$81.05	$105.95

Crew R-1	Hr.	Daily	Hr.	Daily	Bare Costs	Incl. O&P
1 Electrician Foreman	$42.50	$340.00	$63.35	$506.80	$36.88	$55.51
3 Electricians	42.00	1008.00	62.60	1502.40		
2 Helpers	26.40	422.40	40.95	655.20		
48 L.H., Daily Totals		$1770.40		$2664.40	$36.88	$55.51

Crew R-1A	Hr.	Daily	Hr.	Daily	Bare Costs	Incl. O&P
1 Electrician	$42.00	$336.00	$62.60	$500.80	$34.20	$51.78
1 Helper	26.40	211.20	40.95	327.60		
16 L.H., Daily Totals		$547.20		$828.40	$34.20	$51.78

Crew R-2	Hr.	Daily	Hr.	Daily	Bare Costs	Incl. O&P
1 Electrician Foreman	$42.50	$340.00	$63.35	$506.80	$37.06	$55.79
3 Electricians	42.00	1008.00	62.60	1502.40		
2 Helpers	26.40	422.40	40.95	655.20		
1 Equip. Oper. (crane)	38.10	304.80	57.50	460.00		
1 S.P. Crane, 4.5 metric ton		305.20		335.70	5.45	6.00
56 L.H., Daily Totals		$2380.40		$3460.10	$42.51	$61.79

Crew R-3	Hr.	Daily	Hr.	Daily	Bare Costs	Incl. O&P
1 Electrician Foreman	$42.50	$340.00	$63.35	$506.80	$41.42	$61.88
1 Electrician	42.00	336.00	62.60	500.80		
.5 Equip. Oper. (crane)	38.10	152.40	57.50	230.00		
.5 S.P. Crane, 4.5 metric ton		152.60		167.85	7.63	8.39
20 L.H., Daily Totals		$981.00		$1405.45	$49.05	$70.27

Crew R-4	Hr.	Daily	Hr.	Daily	Bare Costs	Incl. O&P
1 Struc. Steel Foreman	$41.95	$335.60	$75.50	$604.00	$40.76	$70.76
3 Struc. Steel Workers	39.95	958.80	71.90	1725.60		
1 Electrician	42.00	336.00	62.60	500.80		
1 Gas Welding Machine		89.20		98.10	2.23	2.45
40 L.H., Daily Totals		$1719.60		$2928.50	$42.99	$73.21

Crew R-5	Hr.	Daily	Hr.	Daily	Bare Costs	Incl. O&P
1 Electrician Foreman	$42.50	$340.00	$63.35	$506.80	$36.37	$54.80
4 Electrician Linemen	42.00	1344.00	62.60	2003.20		
2 Electrician Operators	42.00	672.00	62.60	1001.60		
4 Electrician Groundmen	26.40	844.80	40.95	1310.40		
1 Crew Truck		133.75		147.15		
1 Tool Van		132.40		145.65		
1 Pickup Truck, 0.7 metric ton		78.00		85.80		
.2 Crane, 50 metric ton		190.32		209.35		
.2 Crane, 11 metric ton		127.32		140.05		
.2 Auger, Truck Mtd.		675.80		743.40		
1 Tractor w/Winch		260.00		286.00	18.15	19.97
88 L.H., Daily Totals		$4798.39		$6579.40	$54.52	$74.77

Crew R-6	Hr.	Daily	Hr.	Daily	Bare Costs	Incl. O&P
1 Electrician Foreman	$42.50	$340.00	$63.35	$506.80	$36.37	$54.80
4 Electrician Linemen	42.00	1344.00	62.60	2003.20		
2 Electrician Operators	42.00	672.00	62.60	1001.60		
4 Electrician Groundmen	26.40	844.80	40.95	1310.40		
1 Crew Truck		133.75		147.15		
1 Tool Van		132.40		145.65		
1 Pickup Truck, 0.7 metric ton		78.00		85.80		
.2 Crane, 50 metric ton		190.32		209.35		
.2 Crane, 11 metric ton		127.32		140.05		
.2 Auger, Truck Mtd.		675.80		743.40		
1 Tractor w/Winch		260.00		286.00		
3 Cable Trailers		500.70		550.75		
.5 Tensioning Rig		163.85		180.25		
.5 Cable Pulling Rig		960.00		1056.00	36.62	40.28
88 L.H., Daily Totals		$6422.94		$8366.40	$72.99	$95.08

Crew R-7	Hr.	Daily	Hr.	Daily	Bare Costs	Incl. O&P
1 Electrician Foreman	$42.50	$340.00	$63.35	$506.80	$29.08	$44.68
5 Electrician Groundmen	26.40	1056.00	40.95	1638.00		
1 Crew Truck		133.75		147.15	2.79	3.07
48 L.H., Daily Totals		$1529.75		$2291.95	$31.87	$47.75

Crew R-8	Hr.	Daily	Hr.	Daily	Bare Costs	Incl. O&P
1 Electrician Foreman	$42.50	$340.00	$63.35	$506.80	$36.88	$55.51
3 Electrician Linemen	42.00	1008.00	62.60	1502.40		
2 Electrician Groundmen	26.40	422.40	40.95	655.20		
1 Pickup Truck, 0.7 metric ton		78.00		85.80		
1 Crew Truck		133.75		147.15	4.41	4.85
48 L.H., Daily Totals		$1982.15		$2897.35	$41.29	$60.36

Crew R-9	Hr.	Daily	Hr.	Daily	Bare Costs	Incl. O&P
1 Electrician Foreman	$42.50	$340.00	$63.35	$506.80	$34.26	$51.87
1 Electrician Lineman	42.00	336.00	62.60	500.80		
2 Electrician Operators	42.00	672.00	62.60	1001.60		
4 Electrician Groundmen	26.40	844.80	40.95	1310.40		
1 Pickup Truck, 0.7 metric ton		78.00		85.80		
1 Crew Truck		133.75		147.15	3.31	3.64
64 L.H., Daily Totals		$2404.55		$3552.55	$37.57	$55.51

Crews

Crew No.	Bare Costs		Incl. Subs O & P		Cost Per Labor-Hour	

Crew R-10	Hr.	Daily	Hr.	Daily	Bare Costs	Incl. O&P
1 Electrician Foreman	$42.50	$340.00	$63.35	$506.80	$39.48	$59.12
4 Electrician Linemen	42.00	1344.00	62.60	2003.20		
1 Electrician Groundman	26.40	211.20	40.95	327.60		
1 Crew Truck		133.75		147.15		
3 Tram Cars		351.15		386.25	10.10	11.11
48 L.H., Daily Totals		$2380.10		$3371.00	$49.58	$70.23

Crew R-11	Hr.	Daily	Hr.	Daily	Bare Costs	Incl. O&P
1 Electrician Foreman	$42.50	$340.00	$63.35	$506.80	$39.43	$59.13
4 Electricians	42.00	1344.00	62.60	2003.20		
1 Equip. Oper. (crane)	38.10	304.80	57.50	460.00		
1 Common Laborer	27.40	219.20	42.65	341.20		
1 Crew Truck		133.75		147.15		
1 Crane, 11 metric ton		636.60		700.25	13.76	15.13
56 L.H., Daily Totals		$2978.35		$4158.60	$53.19	$74.26

Crew R-12	Hr.	Daily	Hr.	Daily	Bare Costs	Incl. O&P
1 Carpenter Foreman	$36.05	$288.40	$56.15	$449.20	$33.14	$52.31
4 Carpenters	35.55	1137.60	55.35	1771.20		
4 Common Laborers	27.40	876.80	42.65	1364.80		
1 Equip. Oper. (med.)	36.70	293.60	55.40	443.20		
1 Steel Worker	39.95	319.60	71.90	575.20		
1 Dozer, 149 kW		920.80		1012.90		
1 Pickup Truck, 0.7 metric ton		78.00		85.80	11.35	12.49
88 L.H., Daily Totals		$3914.80		$5702.30	$44.49	$64.80

Crew R-13	Hr.	Daily	Hr.	Daily	Bare Costs	Incl. O&P
1 Electrician Foreman	$42.50	$340.00	$63.35	$506.80	$40.09	$59.90
3 Electricians	42.00	1008.00	62.60	1502.40		
.25 Equip. Oper. (crane)	38.10	76.20	57.50	115.00		
1 Equipment Oiler	32.45	259.60	48.95	391.60		
.25 Hyd Crane, 30 metric ton		169.15		186.05	4.03	4.43
42 L.H., Daily Totals		$1852.95		$2701.85	$44.12	$64.33

Crew R-15	Hr.	Daily	Hr.	Daily	Bare Costs	Incl. O&P
1 Electrician Foreman	$42.50	$340.00	$63.35	$506.80	$40.95	$61.14
4 Electricians	42.00	1344.00	62.60	2003.20		
1 Equipment Operator	35.20	281.60	53.10	424.80		
1 Aerial Lift Truck		265.00		291.50	5.52	6.07
48 L.H., Daily Totals		$2230.60		$3226.30	$46.47	$67.21

Crew R-18	Hr.	Daily	Hr.	Daily	Bare Costs	Incl. O&P
.25 Electrician Foreman	$42.50	$85.00	$63.35	$126.70	$32.44	$49.33
1 Electrician	42.00	336.00	62.60	500.80		
2 Helpers	26.40	422.40	40.95	655.20		
26 L.H., Daily Totals		$843.40		$1282.70	$32.44	$49.33

Crew R-19	Hr.	Daily	Hr.	Daily	Bare Costs	Incl. O&P
.5 Electrician Foreman	$42.50	$170.00	$63.35	$253.40	$42.10	$62.75
2 Electricians	42.00	672.00	62.60	1001.60		
20 L.H., Daily Totals		$842.00		$1255.00	$42.10	$62.75

Crew R-21	Hr.	Daily	Hr.	Daily	Bare Costs	Incl. O&P
1 Electrician Foreman	$42.50	$340.00	$63.35	$506.80	$41.99	$62.61
3 Electricians	42.00	1008.00	62.60	1502.40		
.1 Equip. Oper. (med.)	36.70	29.36	55.40	44.32		
.1 Hyd. Crane 23 metric ton		69.68		76.65	2.12	2.34
32. L.H., Daily Totals		$1447.04		$2130.17	$44.11	$64.95

Crew R-22	Hr.	Daily	Hr.	Daily	Bare Costs	Incl. O&P
.66 Electrician Foreman	$42.50	$224.40	$63.35	$334.49	$35.38	$53.41
2 Helpers	26.40	422.40	40.95	655.20		
2 Electricians	42.00	672.00	62.60	1001.60		
37.28 L.H., Daily Totals		$1318.80		$1991.29	$35.38	$53.41

Crew R-30	Hr.	Daily	Hr.	Daily	Bare Costs	Incl. O&P
.25 Electrician	$44.00	$88.00	$65.60	$131.20	$33.17	$50.55
1 Electrician	42.00	336.00	62.60	500.80		
2 Laborers (Semi-Skilled)	27.40	438.40	42.65	682.40		
26 L.H., Daily Totals		$862.40		$1314.40	$33.17	$50.55

Crew R-31	Hr.	Daily	Hr.	Daily	Bare Costs	Incl. O&P
1 Electrician	$42.00	$336.00	$62.60	$500.80	$42.00	$62.60
1 Core Drill, Elec, 1.86 kW		64.60		71.05	8.08	8.88
8 L.H., Daily Totals		$400.60		$571.85	$50.08	$71.48

Crew W-41E	Hr.	Daily	Hr.	Daily	Bare Costs	Incl. O&P
1 Laborer (Semi-Skilled)	$27.40	$219.20	$42.65	$341.20	$36.98	$56.21
1 Plumber	42.70	341.60	64.25	514.00		
.5 Plumber	44.70	178.80	67.25	269.00		
20 L.H., Daily Totals		$739.60		$1124.20	$36.98	$56.21

Historical Cost Indexes

The table below lists both the Means Historical Cost Index based on Jan. 1, 1993 = 100 as well as the computed value of an index based on Jan. 1, 2006 costs. Since the Jan. 1, 2006 figure is estimated, space is left to write in the actual index figures as they become available through either the quarterly "Means Construction Cost Indexes" or as printed in the "Engineering News-Record." To compute the actual index based on Jan. 1, 2006 = 100, divide the Historical Cost Index for a particular year by the actual Jan. 1, 2006 Construction Cost Index. Space has been left to advance the index figures as the year progresses.

Year	Historical Cost Index Jan. 1, 1993 = 100		Current Index Based on Jan. 1, 2006 = 100		Year	Historical Cost Index Jan. 1, 1993 = 100	Current Index Based on Jan. 1, 2006 = 100		Year	Historical Cost Index Jan. 1, 1993 = 100	Current Index Based on Jan. 1, 2006 = 100	
	Est.	Actual	Est.	Actual		Actual	Est.	Actual		Actual	Est.	Actual
Oct 2006					July 1991	96.8	62.1		July 1973	37.7	24.2	
July 2006					1990	94.3	60.5		1972	34.8	22.3	
April 2006					1989	92.1	59.1		1971	32.1	20.6	
Jan 2006	155.9		100.0	100.0	1988	89.9	57.6		1970	28.7	18.4	
July 2005		151.6	97.2		1987	87.7	56.2		1969	26.9	17.3	
2004		143.7	92.2		1986	84.2	54.0		1968	24.9	16.0	
2003		132.0	84.7		1985	82.6	53.0		1967	23.5	15.1	
2002		128.7	82.6		1984	82.0	52.6		1966	22.7	14.6	
2001		125.1	80.2		1983	80.2	51.4		1965	21.7	13.9	
2000		120.9	77.5		1982	76.1	48.8		1964	21.2	13.6	
1999		117.6	75.4		1981	70.0	44.9		1963	20.7	13.3	
1998		115.1	73.8		1980	62.9	40.3		1962	20.2	13.0	
1997		112.8	72.4		1979	57.8	37.1		1961	19.8	12.7	
1996		110.2	70.7		1978	53.5	34.3		1960	19.7	12.6	
1995		107.6	69.0		1977	49.5	31.8		1959	19.3	12.4	
1994		104.4	67.0		1976	46.9	30.1		1958	18.8	12.1	
1993		101.7	65.2		1975	44.8	28.7		1957	18.4	11.8	
1992		99.4	63.8		1974	41.4	26.6		1956	17.6	11.3	

Adjustments to Costs

The Historical Cost Index can be used to convert National Average building costs at a particular time to the approximate building costs for some other time.

Example:

Estimate and compare construction costs for different years in the same city.

To estimate the National Average construction cost of a building in 1970, knowing that it cost $900,000 in 2006:

INDEX in 1970 = 28.7

INDEX in 2006 = 155.9

Note: The City Cost Indexes for Canada can be used to convert U.S. National averages to local costs in Canadian dollars.

Time Adjustment using the Historical Cost Indexes:

$$\frac{\text{Index for Year A}}{\text{Index for Year B}} \times \text{Cost in Year B} = \text{Cost in Year A}$$

$$\frac{\text{INDEX 1970}}{\text{INDEX 2006}} \times \text{Cost 2006} = \text{Cost 1970}$$

$$\frac{28.7}{155.9} \times \$900,000 = .184 \times \$900,000 = \$165,600$$

The construction cost of the building in 1970 is $165,600.

How to Use the City Cost Indexes

What you should know before you begin

Means City Cost Indexes (CCI) are an extremely useful tool to use when you want to compare costs from city to city and region to region.

This publication contains average construction cost indexes for 719 U.S. and Canadian cities covering over 930 three-digit zip code locations, as listed directly under each city.

Keep in mind that a City Cost Index number is a *percentage ratio* of a specific city's cost to the national average cost of the same item at a stated time period.

In other words, these index figures represent relative construction *factors* (or, if you prefer, multipliers) for Material and Installation costs, as well as the weighted average for Total In Place costs for each CSI MasterFormat division. Installation costs include both labor and equipment rental costs. When estimating equipment rental rates only, for a specific location, use 01590 EQUIPMENT RENTAL index.

The 30 City Average Index is the average of 30 major U.S. cities and serves as a National Average.

Index figures for both material and installation are based on the 30 major city average of 100 and represent the cost relationship as of July 1, 2005. The index for each division is computed from representative material and labor quantities for that division. The weighted average for each city is a weighted total of the components listed above it, but does not include relative productivity between trades or cities.

As changes occur in local material prices, labor rates and equipment rental rates, the impact of these changes should be accurately measured by the change in the City Cost Index for each particular city (as compared to the 30 City Average).

> *Therefore, if you know (or have estimated) building costs in one city today, you can easily convert those costs to expected building costs in another city.*

> *In addition, by using the Historical Cost Index, you can easily convert National Average building costs at a particular time to the approximate building costs for some other time. The City Cost Indexes can then be applied to calculate the costs for a particular city.*

Quick Calculations

Location Adjustment Using the City Cost Indexes:

$$\frac{\text{Index for City A}}{\text{Index for City B}} \text{ x Cost in City B} = \text{Cost in City A}$$

Time Adjustment for the National Average Using the Historical Cost Index:

$$\frac{\text{Index for Year A}}{\text{Index for Year B}} \text{ x Cost in Year B} = \text{Cost in Year A}$$

Adjustment from the National Average:

$$\frac{\text{Index for City A}}{100} \text{ x National Average Cost} = \text{Cost in City A}$$

Since each of the other RSMeans publications contains many different items, any *one* item multiplied by the particular city index may give incorrect results. However, the larger the number of items compiled, the closer the results should be to actual costs for that particular city.

The City Cost Indexes for Canadian cities are calculated using Canadian material and equipment prices and labor rates, in Canadian dollars. Therefore, indexes for Canadian cities can be used to convert U.S. National Average prices to local costs in Canadian dollars.

How to use this section

1. Compare costs from city to city.

In using the Means Indexes, remember that an index number is not a fixed number but a *ratio:* It's a percentage ratio of a building component's cost at any stated time to the National Average cost of that same component at the same time period. Put in the form of an equation:

$$\frac{\text{Specific City Cost}}{\text{National Average Cost}} \text{ x 100} = \text{City Index Number}$$

Therefore, when making cost comparisons between cities, do not subtract one city's index number from the index number of another city and read the result as a percentage difference. Instead, divide one city's index number by that of the other city. The resulting number may then be used as a multiplier to calculate cost differences from city to city.

The formula used to find cost differences between cities for the purpose of comparison is as follows:

$$\frac{\text{City A Index}}{\text{City B Index}} \text{ x City B Cost (Known)} = \text{City A Cost (Unknown)}$$

In addition, you can use *Means CCI* to calculate and compare costs division by division between cities using the same basic formula. (Just be sure that you're comparing similar divisions.)

2. Compare a specific city's construction costs with the National Average.

When you're studying construction location feasibility, it's advisable to compare a prospective project's cost index with an index of the National Average cost.

For example, divide the weighted average index of construction costs of a specific city by that of the 30 City Average, which = 100.

$$\frac{\text{City Index}}{100} = \text{\% of National Average}$$

As a result, you get a ratio that indicates the relative cost of construction in that city in comparison with the National Average.

3. Covert U.S. National Average to actual costs in Canadian City.

$$\frac{\text{Index for Canadian City}}{100} \times \text{National Average Cost} = \text{Cost in Canadian City in \$ CAN}$$

4. Adjust construction cost data based on a National Average.

When you use a source of construction cost data which is based on a National Average (such as *Means cost data publications*), it is necessary to adjust those costs to a specific location.

$$\frac{\text{City Index}}{100} \times \frac{\text{"Book" Cost Based on}}{\text{National Average Costs}} = \frac{\text{City Cost}}{\text{(Unknown)}}$$

5. When applying the City Cost Indexes to demolition projects, use the appropriate division installation index. For example, for removal of existing doors and windows, use the Division 8 index.

What you might like to know about how we developed the Indexes

To create a reliable index, RSMeans researched the building type most often constructed in the United States and Canada. Because it was concluded that no one type of building completely represented the building construction industry, nine different types of buildings were combined to create a composite model.

The exact material, labor and equipment quantities are based on detailed analysis of these nine building types, then each quantity is weighted in proportion to expected usage. These various material items, labor hours, and equipment rental rates are thus combined to form a composite building representing as closely as possible the actual usage of materials, labor and equipment used in the North American Building Construction Industry.

The following structures were chosen to make up that composite model:

1. Factory, 1 story
2. Office, 2–4 story
3. Store, Retail
4. Town Hall, 2–3 story
5. High School, 2–3 story
6. Hospital, 4–8 story
7. Garage, Parking
8. Apartment, 1–3 story
9. Hotel/Motel, 2–3 story

For the purposes of ensuring the timeliness of the data, the components of the index for the composite model have been streamlined. They currently consist of:

- specific quantities of 66 commonly used construction materials;
- specific labor-hours for 21 building construction trades; and
- specific days of equipment rental for 6 types of construction equipment (normally used to install the 66 material items by the 21 trades.)

A sophisticated computer program handles the updating of all costs for each city on a quarterly basis. Material and equipment price quotations are gathered quarterly from 719 cities in the United States and Canada. These prices and the latest negotiated labor wage rates for 21 different building trades are used to compile the quarterly update of the City Cost Index.

The 30 major U.S. cities used to calculate the National Average are:

Atlanta, GA	Memphis, TN
Baltimore, MD	Milwaukee, WI
Boston, MA	Minneapolis, MN
Buffalo, NY	Nashville, TN
Chicago, IL	New Orleans, LA
Cincinnati, OH	New York, NY
Cleveland, OH	Philadelphia, PA
Columbus, OH	Phoenix, AZ
Dallas, TX	Pittsburgh, PA
Denver, CO	St. Louis, MO
Detroit, MI	San Antonio, TX
Houston, TX	San Diego, CA
Indianapolis, IN	San Francisco, CA
Kansas City, MO	Seattle, WA
Los Angeles, CA	Washington, DC

F.Y.I.: The CSI MasterFormat Divisions

1. General Requirements
2. Site Construction
3. Concrete
4. Masonry
5. Metals
6. Wood & Plastics
7. Thermal & Moisture Protection
8. Doors & Windows
9. Finishes
10. Specialties
11. Equipment
12. Furnishings
13. Special Construction
14. Conveying Systems
15. Mechanical
16. Electrical

The information presented in the CCI is organized according to the Construction Specifications Institute (CSI) MasterFormat.

What the CCI does not indicate

The weighted average for each city is a total of the components listed above weighted to reflect typical usage, but it does *not* include the productivity variations between trades or cities.

In addition, the CCI does not take into consideration factors such as the following:

- managerial efficiency
- competitive conditions
- automation
- restrictive union practices
- unique local requirements
- regional variations due to specific building codes

City Cost Indexes

DIVISION		UNITED STATES 30 CITY AVERAGE			ALABAMA ANNISTON 362			BIRMINGHAM 350 - 352			BUTLER 369			DECATUR 356			DOTHAN 363		
		MAT.	INST.	TOTAL	MAT.	INST.	TOTAL	MAT.	INST.	TOTAL	MAT.	INST.	TOTAL	MAT.	INST.	TOTAL	MAT.	INST.	TOTAL
01590	EQUIPMENT RENTAL	.0	100.0	100.0	.0	101.2	101.2	.0	101.3	101.3	.0	97.7	97.7	.0	101.2	101.2	.0	97.7	97.7
02	SITE CONSTRUCTION	100.0	100.0	100.0	88.9	91.5	90.8	83.2	92.9	90.3	101.8	85.3	89.7	81.6	91.3	88.7	99.4	85.3	89.1
03100	CONCRETE FORMS & ACCESSORIES	100.0	100.0	100.0	92.3	34.5	42.5	95.3	73.4	76.4	87.7	45.3	51.1	97.5	48.3	55.0	98.5	45.1	52.5
03200	CONCRETE REINFORCEMENT	100.0	100.0	100.0	92.1	86.2	89.1	92.1	87.9	90.0	97.3	55.3	76.0	92.1	53.0	72.3	97.3	55.3	76.0
03300	CAST-IN-PLACE CONCRETE	100.0	100.0	100.0	91.1	39.3	69.8	93.3	65.9	82.1	88.9	47.1	71.7	90.6	54.3	75.7	88.9	47.0	71.7
03	CONCRETE	100.0	100.0	100.0	94.8	47.9	71.5	91.7	74.6	83.2	95.8	49.5	72.8	90.5	52.9	71.8	95.3	49.4	72.5
04	MASONRY	100.0	100.0	100.0	86.0	34.2	53.9	88.5	76.2	80.9	90.4	35.5	56.3	86.9	50.4	64.3	91.2	35.5	56.6
05	METALS	100.0	100.0	100.0	92.4	89.3	91.4	92.8	94.2	93.3	91.3	76.5	86.6	94.2	77.7	88.9	91.3	76.3	86.5
06	WOOD & PLASTICS	100.0	100.0	100.0	90.9	33.9	61.2	96.3	73.7	84.5	85.5	46.8	65.3	96.3	47.5	70.9	97.7	46.8	71.2
07	THERMAL & MOISTURE PROTECTION	100.0	100.0	100.0	96.8	39.8	72.8	96.8	79.3	89.4	96.8	53.1	78.4	96.6	58.8	80.7	96.8	47.9	76.2
08	DOORS & WINDOWS	100.0	100.0	100.0	94.4	46.2	81.8	98.3	76.5	92.6	94.4	49.4	82.6	98.3	48.0	85.1	94.4	49.4	82.7
09200	PLASTER & GYPSUM BOARD	100.0	100.0	100.0	96.7	32.7	56.9	103.0	73.5	84.7	95.2	45.9	64.6	100.6	46.6	67.0	102.6	45.9	67.4
095,098	CEILINGS & ACOUSTICAL TREATMENT	100.0	100.0	100.0	97.3	32.7	58.7	104.3	73.5	85.9	97.3	45.9	66.6	102.4	46.6	69.1	97.3	45.9	66.6
09600	FLOORING	100.0	100.0	100.0	101.2	34.3	83.1	104.3	48.3	89.2	107.7	27.1	85.8	104.3	50.4	89.7	115.9	32.7	93.4
097,099	WALL FINISHES, PAINTS & COATINGS	100.0	100.0	100.0	94.8	31.9	56.4	94.8	70.9	80.2	94.8	54.0	69.9	94.8	50.5	67.8	94.8	54.0	69.9
09	FINISHES	100.0	100.0	100.0	97.8	33.6	64.3	101.1	68.1	83.9	101.0	42.5	70.6	100.1	48.5	73.2	104.4	43.5	72.7
10 - 14	TOTAL DIV. 10000 - 14000	100.0	100.0	100.0	100.0	45.7	88.4	100.0	85.3	96.9	100.0	48.6	89.0	100.0	50.5	89.5	100.0	48.6	89.0
15	MECHANICAL	100.0	100.0	100.0	99.9	37.1	73.8	99.9	67.0	86.3	97.2	39.7	73.3	99.9	48.1	78.4	97.2	39.6	73.3
16	ELECTRICAL	100.0	100.0	100.0	95.2	26.0	61.6	97.5	67.0	82.7	97.0	36.6	67.7	96.7	48.2	73.1	95.9	36.6	67.1
01 - 16	WEIGHTED AVERAGE	100.0	100.0	100.0	95.7	46.5	73.9	96.5	75.2	87.0	96.1	49.0	75.2	96.2	55.9	78.3	96.3	49.0	75.3

DIVISION		ALABAMA EVERGREEN 364			GADSDEN 359			HUNTSVILLE 357 - 358			JASPER 355			MOBILE 365 - 366			MONTGOMERY 360 - 361		
		MAT.	INST.	TOTAL	MAT.	INST.	TOTAL	MAT.	INST.	TOTAL	MAT.	INST.	TOTAL	MAT.	INST.	TOTAL	MAT.	INST.	TOTAL
01590	EQUIPMENT RENTAL	.0	97.7	97.7	.0	101.2	101.2	.0	101.2	101.2	.0	101.2	101.2	.0	97.7	97.7	.0	97.7	97.7
02	SITE CONSTRUCTION	102.3	85.6	90.7	87.6	91.9	90.7	81.3	92.9	89.8	87.3	91.7	90.5	94.2	85.8	88.0	92.5	86.4	88.0
03100	CONCRETE FORMS & ACCESSORIES	83.4	46.8	51.8	87.9	44.4	50.4	97.5	66.7	70.9	94.5	36.5	44.4	97.6	51.4	57.7	96.3	46.7	53.5
03200	CONCRETE REINFORCEMENT	97.4	55.3	76.0	97.5	86.6	92.0	92.1	74.8	83.3	92.1	86.3	89.2	95.1	51.8	73.1	95.1	86.3	90.6
03300	CAST-IN-PLACE CONCRETE	88.9	49.4	72.7	90.7	50.2	74.0	88.3	64.5	78.5	100.4	44.8	77.6	93.2	53.0	76.7	94.9	47.8	75.5
03	CONCRETE	95.8	50.9	73.5	94.6	56.1	75.5	89.4	68.7	79.1	97.9	50.7	74.4	92.3	53.6	73.0	93.0	56.2	74.7
04	MASONRY	90.4	39.8	59.0	85.4	42.2	58.6	88.1	68.4	75.9	82.1	36.9	54.1	88.8	50.8	65.3	87.9	35.2	55.3
05	METALS	91.3	76.3	86.5	92.4	91.4	92.1	94.2	89.8	92.8	92.4	90.1	91.7	93.0	77.4	88.0	92.8	90.8	92.1
06	WOOD & PLASTICS	80.5	46.8	62.9	86.2	42.7	63.6	96.3	66.3	80.7	93.3	34.7	62.8	96.5	51.1	72.9	94.9	46.8	69.8
07	THERMAL & MOISTURE PROTECTION	96.7	48.9	77.0	96.6	63.2	82.6	96.6	74.3	87.2	96.7	44.5	74.7	96.6	67.5	84.4	96.2	61.2	81.5
08	DOORS & WINDOWS	94.4	49.4	82.6	94.3	52.2	83.3	98.3	63.2	89.1	94.3	51.1	83.0	98.3	51.4	86.0	98.3	56.8	87.4
09200	PLASTER & GYPSUM BOARD	92.2	45.9	63.5	94.8	41.7	61.8	100.6	65.9	79.0	97.2	33.5	57.6	100.6	50.3	69.3	100.6	45.9	66.6
095,098	CEILINGS & ACOUSTICAL TREATMENT	97.3	45.9	66.6	97.3	41.7	64.1	104.3	65.9	81.4	96.4	33.5	58.9	104.3	50.3	72.1	104.3	45.9	69.5
09600	FLOORING	104.7	27.1	83.7	98.9	43.2	83.9	104.3	51.9	90.1	102.3	32.9	83.5	114.8	52.4	97.9	113.1	27.1	89.8
097,099	WALL FINISHES, PAINTS & COATINGS	94.8	54.0	69.9	94.8	61.5	74.5	94.8	61.9	74.8	94.8	39.1	60.8	99.6	54.9	72.3	94.8	54.0	69.9
09	FINISHES	99.7	43.6	70.5	96.8	44.6	69.6	100.6	62.8	80.9	97.8	35.1	65.1	105.3	51.3	77.2	104.3	42.4	72.1
10 - 14	TOTAL DIV. 10000 - 14000	100.0	50.3	89.4	100.0	76.5	95.0	100.0	83.4	96.5	100.0	46.5	88.6	100.0	64.3	92.4	100.0	76.5	95.0
15	MECHANICAL	97.2	42.0	74.3	101.9	42.6	77.3	99.9	67.2	86.3	101.9	58.1	83.7	99.9	68.2	86.7	99.9	40.2	75.1
16	ELECTRICAL	94.6	36.6	66.4	96.4	67.0	82.1	97.5	68.8	83.6	96.3	26.3	62.3	97.5	50.1	74.4	96.7	67.4	82.5
01 - 16	WEIGHTED AVERAGE	95.7	50.2	75.5	96.1	58.8	79.6	96.3	72.0	85.5	96.5	52.1	76.8	97.2	61.1	81.2	97.0	57.1	79.3

DIVISION		ALABAMA PHENIX CITY 368			SELMA 367			TUSCALOOSA 354			ALASKA ANCHORAGE 995 - 996			FAIRBANKS 997			JUNEAU 998		
		MAT.	INST.	TOTAL	MAT.	INST.	TOTAL	MAT.	INST.	TOTAL	MAT.	INST.	TOTAL	MAT.	INST.	TOTAL	MAT.	INST.	TOTAL
01590	EQUIPMENT RENTAL	.0	97.7	97.7	.0	97.7	97.7	.0	101.2	101.2	.0	118.4	118.4	.0	118.4	118.4	.0	118.4	118.4
02	SITE CONSTRUCTION	106.2	86.4	91.7	99.1	86.3	89.7	81.8	91.6	89.0	143.7	133.7	136.4	127.6	133.7	132.1	139.3	133.7	135.2
03100	CONCRETE FORMS & ACCESSORIES	92.1	41.1	48.1	89.2	46.0	51.9	97.4	38.5	46.6	133.2	115.6	118.0	136.3	119.9	122.1	134.7	115.6	118.2
03200	CONCRETE REINFORCEMENT	97.3	75.1	86.0	97.3	86.2	91.7	92.1	86.6	89.3	141.4	108.0	124.4	144.1	108.0	125.7	108.4	108.0	108.2
03300	CAST-IN-PLACE CONCRETE	88.9	48.3	72.2	88.9	47.7	71.9	91.9	45.7	72.9	197.4	115.2	163.6	164.6	115.7	144.5	198.3	115.2	164.1
03	CONCRETE	98.5	51.8	75.4	94.5	55.8	75.3	91.2	51.9	71.7	153.6	113.5	133.7	134.1	115.5	124.9	148.8	113.4	131.3
04	MASONRY	90.4	40.3	59.3	94.3	35.6	57.9	87.0	37.5	56.3	227.3	121.5	161.7	214.0	121.5	156.7	221.8	121.5	159.6
05	METALS	91.3	86.6	89.8	91.3	89.9	90.8	93.4	91.3	92.7	129.2	102.5	120.7	129.3	102.8	120.8	129.3	102.4	120.7
06	WOOD & PLASTICS	90.5	39.0	63.7	87.3	46.8	66.2	96.3	37.1	65.5	119.3	113.4	116.2	119.6	118.8	119.2	119.3	113.4	116.2
07	THERMAL & MOISTURE PROTECTION	97.1	61.9	82.3	96.6	53.1	78.3	96.6	60.0	81.2	172.8	115.6	148.7	170.2	118.7	148.6	170.8	115.6	147.5
08	DOORS & WINDOWS	94.4	49.9	82.7	94.4	56.8	84.6	98.3	56.7	87.4	125.9	111.0	122.0	123.1	113.4	120.6	123.1	111.0	119.9
09200	PLASTER & GYPSUM BOARD	97.8	37.9	60.6	96.3	45.9	65.0	100.6	36.0	60.5	132.8	113.6	120.9	142.5	119.1	128.0	132.8	113.6	120.9
095,098	CEILINGS & ACOUSTICAL TREATMENT	97.3	37.9	61.9	97.3	45.9	64.0	100.6	36.0	60.5	132.4	113.6	121.2	132.4	119.1	124.4	132.4	113.6	121.2
09600	FLOORING	110.7	27.1	88.1	108.5	27.1	86.5	104.3	39.7	86.8	165.0	125.3	154.2	165.0	125.3	154.2	165.0	125.3	154.2
097,099	WALL FINISHES, PAINTS & COATINGS	94.8	54.0	69.9	94.8	54.0	69.9	94.8	46.5	65.3	166.6	114.6	134.9	166.6	128.9	143.6	166.6	114.6	134.9
09	FINISHES	102.6	38.1	69.0	101.2	42.5	70.6	100.6	37.9	67.9	154.3	117.4	135.1	153.7	122.1	137.3	153.0	117.4	134.5
10 - 14	TOTAL DIV. 10000 - 14000	100.0	75.9	94.9	100.0	48.6	89.0	100.0	74.1	94.5	100.0	115.2	103.2	100.0	115.9	103.4	100.0	115.2	103.2
15	MECHANICAL	97.2	40.7	73.8	97.2	40.0	73.5	99.9	35.4	73.1	100.5	108.7	103.9	100.5	116.9	107.3	100.5	105.5	102.6
16	ELECTRICAL	96.5	61.3	79.4	95.7	36.6	67.0	97.0	67.0	82.4	148.4	114.2	131.8	150.2	114.2	132.7	150.2	114.2	132.7
01 - 16	WEIGHTED AVERAGE	96.6	54.9	78.1	95.9	51.6	76.2	96.3	55.4	78.1	133.7	114.7	125.3	130.3	117.6	124.6	132.5	114.0	124.3

COST INDEXES

Table 1

DIVISION		ALASKA KETCHIKAN 999			ARIZONA CHAMBERS 865			FLAGSTAFF 860			GLOBE 855			KINGMAN 864			MESA/TEMPE 852		
		MAT.	INST.	TOTAL	MAT.	INST.	TOTAL	MAT.	INST.	TOTAL	MAT.	INST.	TOTAL	MAT.	INST.	TOTAL	MAT.	INST.	TOTAL
01590	EQUIPMENT RENTAL	.0	118.4	118.4	.0	95.0	95.0	.0	95.0	95.0	.0	98.3	98.3	.0	95.0	95.0	.0	98.3	98.3
02	SITE CONSTRUCTION	190.6	133.7	148.9	63.5	100.1	90.3	81.4	100.8	95.6	95.0	103.6	101.3	63.4	100.8	90.8	86.0	103.9	99.1
03100	CONCRETE FORMS & ACCESSORIES	126.1	115.5	117.0	96.5	54.9	60.6	102.7	67.6	72.4	93.9	54.6	60.0	94.1	66.4	70.2	97.3	63.0	67.7
03200	CONCRETE REINFORCEMENT	113.0	108.0	110.5	105.8	73.7	89.5	105.6	74.5	89.8	104.3	73.7	88.8	105.9	70.2	87.7	105.0	74.5	89.5
03300	CAST-IN-PLACE CONCRETE	320.0	115.2	235.8	90.0	62.4	78.7	90.2	76.2	84.4	97.0	62.7	82.9	89.8	62.5	78.6	97.7	69.5	86.1
03	CONCRETE	224.0	113.4	169.1	97.6	61.1	79.5	116.9	71.7	94.5	108.7	61.1	85.0	97.2	65.6	81.5	101.1	67.4	84.4
04	MASONRY	222.9	121.5	160.0	100.5	50.7	69.6	100.1	65.1	78.4	105.0	50.6	71.3	100.5	63.1	77.3	105.1	52.5	72.5
05	METALS	129.4	102.4	120.8	93.3	65.7	84.5	93.8	69.5	86.0	94.1	66.1	85.1	93.9	65.2	84.7	94.3	70.3	86.7
06	WOOD & PLASTICS	109.5	113.4	111.5	101.8	52.9	76.3	108.2	68.2	87.4	95.5	53.0	73.4	97.9	68.2	82.4	99.1	68.4	83.1
07	THERMAL & MOISTURE PROTECTION	173.4	115.6	149.0	97.8	61.1	82.4	99.7	69.1	86.8	99.8	58.0	82.2	97.8	66.9	84.8	99.1	63.4	84.1
08	DOORS & WINDOWS	123.5	111.0	120.2	102.0	56.7	90.1	102.1	68.3	93.3	98.8	56.8	87.8	102.2	66.9	92.9	98.8	68.1	90.8
09200	PLASTER & GYPSUM BOARD	127.7	113.6	118.9	89.5	51.6	66.0	92.3	67.4	76.8	94.9	51.6	68.0	85.6	67.4	74.3	96.8	67.4	78.5
095,098	CEILINGS & ACOUSTICAL TREATMENT	127.6	113.6	119.2	103.9	51.6	72.7	104.8	67.4	82.5	97.4	51.6	70.1	104.8	67.4	82.5	97.4	67.4	79.5
09600	FLOORING	165.0	125.3	154.2	92.0	49.4	80.4	94.5	49.6	82.4	95.0	49.4	82.7	91.0	67.0	84.5	96.5	61.8	87.1
097,099	WALL FINISHES, PAINTS & COATINGS	166.6	114.6	134.9	93.2	47.1	65.1	93.2	56.3	70.7	102.8	47.1	68.8	93.2	56.3	70.7	102.8	56.3	74.5
09	FINISHES	154.1	117.4	135.0	94.5	52.0	72.4	97.4	62.6	79.3	98.0	52.2	74.1	93.8	65.2	78.9	97.8	61.5	78.9
10 - 14	TOTAL DIV. 10000 - 14000	100.0	115.2	103.2	100.0	78.6	95.4	100.0	81.2	96.0	100.0	79.0	95.5	100.0	80.4	95.8	100.0	76.8	95.0
15	MECHANICAL	98.6	103.3	100.6	97.0	75.8	88.2	100.2	76.8	90.5	95.3	69.1	84.4	97.0	71.3	86.3	100.2	70.0	83.6
16	ELECTRICAL	150.2	114.2	132.7	100.5	77.9	89.5	99.5	64.1	82.3	94.4	65.8	80.5	100.5	45.1	73.6	91.4	64.1	78.2
01 - 16	WEIGHTED AVERAGE	141.9	113.6	129.3	96.8	68.3	84.1	100.4	72.3	87.9	98.1	65.5	83.6	96.7	67.2	83.6	97.8	68.9	85.0

Table 2

DIVISION		ARIZONA PHOENIX 850,853			PRESCOTT 863			SHOW LOW 859			TUCSON 856 - 857			ARKANSAS BATESVILLE 725			CAMDEN 717		
		MAT.	INST.	TOTAL	MAT.	INST.	TOTAL	MAT.	INST.	TOTAL	MAT.	INST.	TOTAL	MAT.	INST.	TOTAL	MAT.	INST.	TOTAL
01590	EQUIPMENT RENTAL	.0	98.9	98.9	.0	95.0	95.0	.0	98.3	98.3	.0	98.3	98.3	.0	86.0	86.0	.0	86.0	86.0
02	SITE CONSTRUCTION	86.5	104.6	99.8	69.6	100.0	91.9	97.1	103.6	101.9	82.9	104.4	98.6	72.8	83.5	80.6	73.6	83.1	80.6
03100	CONCRETE FORMS & ACCESSORIES	98.4	68.9	73.0	98.1	54.6	60.5	101.8	55.0	61.5	97.9	68.6	72.6	80.5	43.3	48.4	80.2	30.5	37.3
03200	CONCRETE REINFORCEMENT	103.3	74.6	88.7	105.6	70.8	87.9	105.0	73.8	89.2	86.8	74.5	80.5	96.2	47.0	71.2	97.5	25.9	61.1
03300	CAST-IN-PLACE CONCRETE	97.8	76.7	89.1	90.1	62.3	78.7	97.0	62.9	83.0	100.5	76.5	90.7	77.7	44.8	64.2	79.6	33.2	60.6
03	CONCRETE	100.7	72.5	86.7	102.7	60.4	81.7	110.8	61.4	86.3	99.3	72.3	85.9	79.3	45.3	62.4	80.8	31.8	56.5
04	MASONRY	93.3	66.2	76.5	100.5	57.1	73.6	105.0	52.5	72.5	95.0	65.1	76.5	99.9	41.2	63.5	108.6	35.6	63.3
05	METALS	95.8	71.2	87.9	93.8	65.1	84.6	93.9	66.7	85.2	95.0	70.1	87.1	94.5	58.4	82.9	94.5	49.5	80.1
06	WOOD & PLASTICS	100.1	69.7	84.3	103.3	52.9	77.1	103.8	53.0	77.3	99.3	69.7	83.9	82.2	44.5	62.6	82.0	31.5	55.7
07	THERMAL & MOISTURE PROTECTION	99.0	69.2	86.5	98.3	60.0	82.2	100.0	61.4	83.8	100.7	65.6	85.9	97.9	44.5	75.4	97.7	33.5	70.7
08	DOORS & WINDOWS	100.0	69.1	91.9	102.1	56.2	90.1	97.9	56.8	87.1	95.9	69.1	88.9	96.1	41.6	81.9	92.4	27.5	75.4
09200	PLASTER & GYPSUM BOARD	98.9	68.7	80.1	89.6	51.6	66.0	98.6	51.6	69.4	97.1	68.7	79.5	79.0	43.6	57.0	79.0	30.3	48.8
095,098	CEILINGS & ACOUSTICAL TREATMENT	104.2	68.7	83.0	103.1	51.6	72.4	97.4	51.6	70.1	97.4	68.7	80.3	84.5	43.6	60.1	84.5	30.3	52.2
09600	FLOORING	96.8	64.6	88.1	92.8	49.4	81.1	98.5	56.2	87.0	96.0	49.6	83.4	106.1	48.2	90.5	106.0	16.0	81.6
097,099	WALL FINISHES, PAINTS & COATINGS	102.8	58.3	75.7	93.2	47.1	65.1	102.8	47.1	68.8	100.5	56.3	73.5	97.0	43.0	64.1	97.0	33.6	58.3
09	FINISHES	100.0	66.5	82.6	95.0	52.0	72.6	99.8	53.4	75.6	97.5	63.5	79.8	90.6	44.4	66.6	90.6	28.7	58.4
10 - 14	TOTAL DIV. 10000 - 14000	100.0	81.7	96.1	100.0	78.6	95.4	100.0	79.0	95.5	100.0	81.7	96.1	100.0	44.5	88.2	100.0	40.7	87.4
15	MECHANICAL	100.2	76.9	90.6	100.2	75.4	89.9	95.3	75.8	87.2	100.2	70.5	87.9	95.1	40.7	72.6	95.1	31.4	68.7
16	ELECTRICAL	100.5	64.1	82.8	99.2	64.1	82.1	91.8	69.2	80.8	93.4	62.0	78.1	97.7	46.6	72.8	94.4	38.8	67.4
01 - 16	WEIGHTED AVERAGE	98.8	73.6	87.6	98.3	66.8	84.3	98.2	68.0	84.8	97.1	71.2	85.6	93.3	48.4	73.3	93.2	38.9	69.1

Table 3

DIVISION		ARKANSAS FAYETTEVILLE 727			FORT SMITH 729			HARRISON 726			HOT SPRINGS 719			JONESBORO 724			LITTLE ROCK 720 - 722		
		MAT.	INST.	TOTAL	MAT.	INST.	TOTAL	MAT.	INST.	TOTAL	MAT.	INST.	TOTAL	MAT.	INST.	TOTAL	MAT.	INST.	TOTAL
01590	EQUIPMENT RENTAL	.0	86.0	86.0	.0	86.0	86.0	.0	86.0	86.0	.0	86.0	86.0	.0	107.8	107.8	.0	86.0	86.0
02	SITE CONSTRUCTION	72.3	83.5	80.5	77.2	83.6	81.9	77.5	83.5	81.9	77.1	82.8	81.3	99.8	98.9	99.1	77.0	83.6	81.8
03100	CONCRETE FORMS & ACCESSORIES	75.6	39.5	44.4	97.5	41.0	48.7	85.5	43.3	49.1	77.2	28.6	35.3	84.2	46.3	51.5	91.7	57.6	62.3
03200	CONCRETE REINFORCEMENT	96.2	45.4	70.4	97.3	74.4	85.7	95.7	47.0	71.0	95.6	24.6	59.6	93.0	47.7	70.0	97.5	69.6	83.3
03300	CAST-IN-PLACE CONCRETE	77.7	44.3	64.0	88.9	66.5	79.7	86.2	44.8	69.2	81.5	32.3	61.3	84.7	55.3	72.6	88.9	66.7	79.8
03	CONCRETE	78.9	43.2	61.2	86.9	56.8	72.0	85.8	45.3	65.7	83.7	30.4	57.2	83.5	51.3	67.5	86.5	63.4	75.0
04	MASONRY	89.9	41.7	60.0	96.0	55.2	70.7	100.2	41.2	63.6	80.8	27.6	47.8	91.4	45.5	62.9	94.1	55.2	70.0
05	METALS	94.5	57.7	82.7	96.3	70.7	88.2	95.3	58.2	83.5	94.5	48.2	79.7	90.6	73.8	85.3	94.5	62.9	84.9
06	WOOD & PLASTICS	77.3	39.8	57.8	100.0	39.3	68.4	87.9	44.5	65.3	78.6	28.5	52.5	85.7	47.3	65.7	96.9	61.3	78.4
07	THERMAL & MOISTURE PROTECTION	99.1	46.3	76.9	99.6	47.8	77.8	98.2	44.5	75.6	98.0	34.4	71.2	101.5	50.1	79.9	98.0	50.1	77.8
08	DOORS & WINDOWS	96.1	39.8	81.4	96.9	44.8	83.3	96.9	41.6	82.5	92.4	27.6	75.4	98.5	46.2	84.9	96.9	56.3	86.3
09200	PLASTER & GYPSUM BOARD	76.1	38.8	52.9	83.8	38.4	55.6	82.7	43.6	58.4	77.2	27.3	46.2	92.6	46.2	63.7	83.8	60.9	69.6
095,098	CEILINGS & ACOUSTICAL TREATMENT	84.5	38.8	57.2	89.7	38.4	59.1	88.7	43.6	61.8	84.5	27.3	50.4	87.6	46.2	62.9	89.7	60.9	72.5
09600	FLOORING	102.5	48.2	87.8	115.3	73.1	103.9	109.3	48.2	92.8	104.4	48.2	89.2	77.1	43.6	68.0	116.6	73.1	104.8
097,099	WALL FINISHES, PAINTS & COATINGS	97.0	34.9	59.1	97.0	56.3	72.1	97.0	43.0	64.1	97.0	35.9	59.7	85.5	50.6	64.2	97.0	57.7	73.0
09	FINISHES	89.2	40.5	63.8	95.5	48.0	70.8	93.4	44.4	66.0	90.1	33.0	60.4	87.4	46.4	66.0	87.4	61.1	77.7
10 - 14	TOTAL DIV. 10000 - 14000	100.0	58.9	91.2	100.0	71.9	94.0	100.0	44.5	88.2	100.0	40.5	87.3	100.0	51.2	89.6	100.0	74.6	94.6
15	MECHANICAL	95.1	37.4	71.2	100.1	41.6	75.8	95.1	40.7	72.6	95.1	32.4	69.1	100.2	41.5	75.8	100.1	57.9	82.6
16	ELECTRICAL	92.0	43.0	68.2	95.2	66.6	81.3	96.3	46.6	72.1	96.4	38.0	68.0	102.1	46.6	75.1	94.9	68.4	82.0
01 - 16	WEIGHTED AVERAGE	91.9	46.7	71.8	95.9	56.9	78.6	94.5	48.4	74.0	92.4	38.4	68.4	95.2	53.4	76.6	95.1	63.8	81.2

COST INDEXES

ARKANSAS / CALIFORNIA

DIVISION		PINE BLUFF 716			RUSSELLVILLE 728			TEXARKANA 718			WEST MEMPHIS 723			ALHAMBRA 917 - 918			ANAHEIM 928		
		MAT.	INST.	TOTAL	MAT.	INST.	TOTAL	MAT.	INST.	TOTAL	MAT.	INST.	TOTAL	MAT.	INST.	TOTAL	MAT.	INST.	TOTAL
01590	EQUIPMENT RENTAL	.0	86.0	86.0	.0	86.0	86.0	.0	86.7	86.7	.0	107.8	107.8	.0	97.0	97.0	.0	102.2	102.2
02	SITE CONSTRUCTION	79.2	83.6	82.4	74.2	83.5	81.0	94.6	84.4	87.1	107.2	98.9	101.1	98.8	107.3	105.1	103.2	109.0	107.5
03100	CONCRETE FORMS & ACCESSORIES	76.8	57.4	60.0	81.4	39.4	45.1	84.7	38.4	44.8	90.3	46.3	52.4	118.6	116.5	116.8	105.7	121.7	119.5
03200	CONCRETE REINFORCEMENT	97.4	69.6	83.3	96.8	48.9	72.5	97.0	46.2	71.2	93.0	47.7	70.0	107.3	112.7	110.0	95.4	112.6	104.1
03300	CAST-IN-PLACE CONCRETE	81.5	66.6	75.4	81.4	44.7	66.3	88.9	43.1	70.1	88.8	55.3	75.0	107.7	109.7	108.5	108.1	118.5	112.4
03	CONCRETE	84.3	63.2	73.9	82.1	43.9	63.1	83.8	42.5	63.3	89.7	51.3	70.6	111.9	112.4	112.2	107.3	118.0	112.6
04	MASONRY	115.9	55.2	78.3	95.4	41.2	61.8	96.0	31.9	56.2	78.3	45.5	58.0	125.6	111.8	117.0	93.2	111.0	104.2
05	METALS	95.0	69.1	86.8	94.5	59.0	83.1	87.6	57.7	78.1	90.0	73.8	84.8	91.0	96.4	92.8	110.5	99.8	107.1
06	WOOD & PLASTICS	78.1	61.3	69.4	83.7	39.3	60.6	87.6	41.0	63.3	91.9	47.3	68.7	103.0	116.5	110.0	96.3	121.9	109.6
07	THERMAL & MOISTURE PROTECTION	98.0	50.1	77.9	99.4	43.9	76.0	99.2	41.0	74.7	102.0	50.1	80.2	98.7	110.0	103.5	102.7	115.2	107.9
08	DOORS & WINDOWS	92.4	56.3	83.0	96.1	39.3	81.3	97.4	39.7	82.3	98.5	46.2	84.9	95.6	113.6	100.3	102.7	116.6	106.3
09200	PLASTER & GYPSUM BOARD	76.8	60.9	66.9	79.0	38.4	53.8	81.9	40.1	55.9	94.4	46.2	64.4	96.8	117.1	109.4	98.8	122.4	113.5
095,098	CEILINGS & ACOUSTICAL TREATMENT	85.4	60.9	70.8	84.5	38.4	57.0	92.3	40.1	61.1	86.6	46.2	62.5	97.4	117.1	109.1	120.4	122.4	121.6
09600	FLOORING	104.2	73.1	95.8	105.7	48.2	90.1	107.0	41.6	89.3	79.9	43.6	70.1	107.5	104.8	106.7	122.8	104.8	117.9
097,099	WALL FINISHES, PAINTS & COATINGS	97.0	57.7	73.0	97.0	43.0	64.1	97.0	35.5	59.4	85.5	50.6	64.2	111.1	109.0	109.8	110.9	109.0	109.8
09	FINISHES	90.3	61.1	75.1	90.7	41.4	65.0	94.1	38.8	65.3	88.9	46.4	66.7	104.5	113.8	109.3	114.3	117.5	116.0
10 - 14	TOTAL DIV. 10000 - 14000	100.0	74.6	94.6	100.0	43.9	88.0	100.0	36.6	86.5	100.0	51.2	89.6	100.0	98.0	99.6	100.0	113.0	102.8
15	MECHANICAL	100.1	43.9	76.8	95.1	35.6	70.4	100.1	36.7	73.8	95.3	41.5	73.0	95.3	106.0	99.7	100.2	107.3	103.1
16	ELECTRICAL	94.6	68.4	81.8	95.2	46.6	71.5	96.3	38.8	68.4	103.6	46.6	75.9	122.3	113.4	118.0	89.5	105.3	97.2
01 - 16	WEIGHTED AVERAGE	95.2	60.9	80.0	93.2	46.6	72.5	94.6	44.0	72.1	94.6	53.4	76.3	102.7	109.1	105.5	102.6	110.6	106.1

CALIFORNIA

DIVISION		BAKERSFIELD 932 - 933			BERKELEY 947			EUREKA 955			FRESNO 936 - 938			INGLEWOOD 903 - 905			LONG BEACH 906 - 908		
		MAT.	INST.	TOTAL	MAT.	INST.	TOTAL	MAT.	INST.	TOTAL	MAT.	INST.	TOTAL	MAT.	INST.	TOTAL	MAT.	INST.	TOTAL
01590	EQUIPMENT RENTAL	.0	99.6	99.6	.0	103.5	103.5	.0	99.2	99.2	.0	99.6	99.6	.0	96.7	96.7	.0	96.7	96.7
02	SITE CONSTRUCTION	108.0	106.1	106.6	145.4	104.6	115.5	118.6	103.1	107.2	109.5	106.2	107.1	93.4	104.6	101.6	100.8	104.6	103.6
03100	CONCRETE FORMS & ACCESSORIES	97.9	121.2	118.0	120.3	133.9	132.1	116.6	119.3	119.0	101.9	124.6	121.5	109.2	116.4	115.4	102.7	116.4	114.5
03200	CONCRETE REINFORCEMENT	108.6	112.5	110.6	100.6	113.6	107.2	105.3	112.5	109.0	91.9	112.8	102.6	112.8	112.7	112.7	111.8	112.7	112.2
03300	CAST-IN-PLACE CONCRETE	103.2	117.4	109.1	145.7	117.3	134.0	112.0	106.7	109.8	113.0	113.3	113.1	83.8	110.6	94.8	95.4	110.6	101.7
03	CONCRETE	106.5	117.3	111.9	125.8	123.3	124.6	119.1	112.7	115.9	108.9	117.5	113.2	97.7	112.8	105.2	107.3	112.8	110.0
04	MASONRY	109.1	110.3	109.9	145.8	124.0	132.3	115.4	103.8	108.2	112.7	113.5	113.2	80.1	111.9	99.8	89.0	111.9	103.2
05	METALS	105.1	98.9	103.1	110.3	105.1	108.6	110.1	95.8	105.5	110.7	100.4	107.4	104.9	97.6	102.6	104.8	97.6	102.5
06	WOOD & PLASTICS	86.8	122.0	105.1	118.0	135.5	127.1	110.0	125.8	118.2	100.2	125.8	113.5	96.3	116.4	106.8	89.2	116.4	103.4
07	THERMAL & MOISTURE PROTECTION	99.2	109.1	103.4	113.2	128.2	119.5	106.6	97.6	102.8	95.9	111.4	102.4	100.5	111.9	105.3	100.7	111.9	105.4
08	DOORS & WINDOWS	101.5	114.4	104.9	106.0	127.7	111.7	102.8	110.1	104.7	101.9	116.5	105.7	91.0	113.6	96.9	91.0	113.6	96.9
09200	PLASTER & GYPSUM BOARD	98.8	122.4	113.5	108.0	135.9	125.3	103.5	126.3	117.6	96.6	126.3	115.0	102.6	117.1	111.6	99.7	117.1	110.5
095,098	CEILINGS & ACOUSTICAL TREATMENT	120.4	122.4	121.6	113.1	135.9	122.1	122.5	126.3	124.7	120.4	126.3	123.9	110.8	117.1	114.5	110.8	117.1	114.5
09600	FLOORING	117.0	93.7	110.7	121.9	126.1	123.0	122.0	103.8	117.1	133.3	135.6	133.9	115.7	104.8	112.7	111.8	104.8	109.9
097,099	WALL FINISHES, PAINTS & COATINGS	110.5	95.5	101.4	115.4	137.8	129.0	109.9	59.0	78.8	133.7	96.7	111.1	100.3	109.0	105.6	100.3	109.0	105.6
09	FINISHES	114.5	114.1	114.3	118.4	133.7	126.3	118.0	112.0	114.9	120.7	124.5	122.7	108.9	113.7	111.4	107.7	113.7	110.8
10 - 14	TOTAL DIV. 10000 - 14000	100.0	111.9	102.5	100.0	128.3	106.0	100.0	121.8	104.6	100.0	125.3	105.4	100.0	97.8	99.5	100.0	97.8	99.5
15	MECHANICAL	100.2	96.2	98.6	95.4	135.4	112.0	95.3	95.4	95.4	100.2	109.1	103.9	95.2	105.9	99.7	95.2	105.9	99.7
16	ELECTRICAL	89.6	97.2	93.3	104.1	138.9	121.0	98.6	80.5	89.8	88.4	96.6	92.4	100.0	113.4	106.5	99.7	113.4	106.4
01 - 16	WEIGHTED AVERAGE	102.3	105.9	103.9	110.1	126.7	117.5	105.9	101.1	103.8	104.1	110.9	107.1	98.2	109.0	103.0	99.6	109.0	103.8

CALIFORNIA

DIVISION		LOS ANGELES 900 - 902			MARYSVILLE 959			MODESTO 953			MOJAVE 935			OAKLAND 946			OXNARD 930		
		MAT.	INST.	TOTAL	MAT.	INST.	TOTAL	MAT.	INST.	TOTAL	MAT.	INST.	TOTAL	MAT.	INST.	TOTAL	MAT.	INST.	TOTAL
01590	EQUIPMENT RENTAL	.0	98.6	98.6	.0	99.2	99.2	.0	99.2	99.2	.0	99.6	99.6	.0	103.5	103.5	.0	98.2	98.2
02	SITE CONSTRUCTION	100.5	107.8	105.8	114.3	105.0	107.5	107.8	105.6	106.2	101.6	106.0	104.8	154.4	104.7	117.9	109.5	103.9	105.4
03100	CONCRETE FORMS & ACCESSORIES	105.8	121.6	119.4	104.0	123.6	120.9	98.9	124.7	121.1	112.5	111.8	111.9	106.3	134.8	130.9	104.0	121.8	119.3
03200	CONCRETE REINFORCEMENT	112.8	112.9	112.9	105.3	112.8	109.1	109.2	112.8	111.1	110.6	112.4	111.5	102.9	113.6	108.3	108.6	112.5	110.6
03300	CAST-IN-PLACE CONCRETE	91.3	116.8	101.8	125.2	111.5	119.6	112.0	113.3	112.5	94.1	114.1	102.3	141.7	117.9	131.9	109.6	117.7	112.9
03	CONCRETE	104.0	117.3	110.6	120.9	116.4	118.7	111.0	117.5	114.2	100.9	112.0	106.4	126.7	123.9	125.3	110.0	117.7	113.8
04	MASONRY	95.6	117.3	109.1	116.4	110.8	112.9	113.1	113.5	113.3	109.4	108.0	108.6	154.1	125.1	136.1	113.7	109.6	111.1
05	METALS	111.7	98.7	107.6	109.6	99.7	106.4	106.6	100.6	104.7	106.7	98.6	104.1	104.6	105.1	104.8	104.9	99.5	103.1
06	WOOD & PLASTICS	92.3	121.6	107.6	95.6	125.8	111.3	90.6	125.8	108.9	101.7	110.7	106.4	102.8	136.2	120.2	95.1	122.0	109.1
07	THERMAL & MOISTURE PROTECTION	100.7	117.4	107.7	105.9	108.9	107.2	105.5	110.8	107.7	99.9	106.9	102.9	110.1	128.7	118.0	105.4	113.5	108.8
08	DOORS & WINDOWS	97.0	116.4	102.1	102.0	118.1	106.2	100.8	118.1	105.3	97.3	108.3	100.2	106.1	128.1	111.8	100.4	116.6	104.6
09200	PLASTER & GYPSUM BOARD	104.1	122.4	115.5	96.8	126.3	115.1	99.2	126.3	116.0	107.5	110.8	109.5	101.1	136.5	123.1	98.8	122.4	113.5
095,098	CEILINGS & ACOUSTICAL TREATMENT	123.5	122.4	122.9	120.8	126.3	124.1	115.7	126.3	122.0	120.8	110.8	114.8	116.7	136.5	128.5	120.4	122.4	121.6
09600	FLOORING	113.5	104.8	111.2	116.4	113.9	115.7	117.0	114.1	116.2	124.6	93.7	116.2	116.1	126.1	118.8	117.0	104.8	113.7
097,099	WALL FINISHES, PAINTS & COATINGS	100.3	109.0	105.6	109.9	115.8	113.5	109.9	110.4	110.2	109.9	87.8	96.4	115.4	137.8	129.0	109.9	102.7	105.5
09	FINISHES	112.0	117.3	114.7	114.4	122.1	118.4	113.2	122.3	118.0	117.8	106.1	111.7	117.0	134.3	126.0	114.3	116.9	115.7
10 - 14	TOTAL DIV. 10000 - 14000	100.0	112.3	102.8	100.0	124.4	105.2	100.0	125.3	105.4	100.0	109.6	102.0	100.0	128.8	106.1	100.0	113.3	102.8
15	MECHANICAL	100.1	107.2	103.1	95.3	105.6	99.6	100.2	109.1	103.9	95.3	94.9	95.1	100.3	136.0	115.1	100.2	107.3	103.2
16	ELECTRICAL	97.0	114.2	105.3	95.3	120.1	107.4	97.9	99.7	98.8	87.7	97.2	92.3	103.4	138.9	120.6	95.4	106.8	100.9
01 - 16	WEIGHTED AVERAGE	102.5	112.1	106.8	105.0	112.5	108.3	104.3	111.0	107.3	100.5	103.1	101.7	110.7	127.1	118.0	103.8	110.0	106.5

COST INDEXES

CALIFORNIA

DIVISION		PALM SPRINGS 922			PALO ALTO 943			PASADENA 910 - 912			REDDING 960			RICHMOND 948			RIVERSIDE 925		
		MAT.	INST.	TOTAL	MAT.	INST.	TOTAL	MAT.	INST.	TOTAL	MAT.	INST.	TOTAL	MAT.	INST.	TOTAL	MAT.	INST.	TOTAL
01590	EQUIPMENT RENTAL	.0	100.7	100.7	.0	103.5	103.5	.0	97.0	97.0	.0	99.2	99.2	.0	103.5	103.5	.0	100.7	100.7
02	SITE CONSTRUCTION	93.7	106.4	103.1	139.9	104.6	114.0	95.5	107.3	104.2	114.3	105.0	107.5	153.2	104.7	117.6	100.8	106.7	105.2
03100	CONCRETE FORMS & ACCESSORIES	101.4	116.6	114.5	103.8	133.9	129.7	104.7	117.6	115.8	103.7	123.5	120.8	123.8	134.7	133.2	106.2	121.6	119.5
03200	CONCRETE REINFORCEMENT	110.8	112.5	111.6	100.6	113.6	107.2	108.2	112.7	110.5	105.3	112.8	109.1	100.6	113.6	107.2	107.6	112.5	110.1
03300	CAST-IN-PLACE CONCRETE	98.4	112.6	104.3	123.1	117.2	120.7	102.2	109.7	105.3	124.1	111.5	118.9	141.7	117.9	131.9	107.1	118.5	111.8
03	CONCRETE	102.3	113.6	107.9	112.6	123.2	117.9	106.7	112.9	109.8	119.7	116.4	118.1	127.6	123.8	125.7	108.8	117.9	113.3
04	MASONRY	85.9	108.3	99.8	120.7	124.0	122.7	107.6	111.8	110.2	116.3	110.8	112.9	145.5	125.1	132.8	86.9	110.6	101.6
05	METALS	111.3	99.2	107.5	102.7	105.0	103.4	91.1	96.5	92.8	109.9	99.6	106.6	102.7	105.0	103.5	110.8	99.7	107.2
06	WOOD & PLASTICS	90.5	116.7	104.1	99.8	135.5	118.4	88.2	118.0	103.7	95.3	125.8	111.1	122.3	136.2	129.5	96.3	121.9	109.6
07	THERMAL & MOISTURE PROTECTION	102.7	112.1	106.6	109.7	128.4	117.6	98.3	110.2	103.3	105.9	109.5	107.4	110.4	128.1	117.8	102.9	114.7	107.9
08	DOORS & WINDOWS	99.6	113.7	103.3	106.1	125.0	111.0	95.6	114.4	100.5	103.2	118.1	107.1	106.1	128.1	111.8	102.7	116.6	106.3
09200	PLASTER & GYPSUM BOARD	94.0	117.1	108.3	99.6	135.9	122.1	92.0	118.6	108.5	97.2	126.3	115.3	109.6	136.5	126.3	97.7	122.4	113.0
095,098	CEILINGS & ACOUSTICAL TREATMENT	111.4	117.1	114.8	114.8	135.9	127.4	97.4	118.6	110.0	127.2	126.3	126.7	114.8	136.5	127.8	116.2	122.4	119.9
09600	FLOORING	119.1	104.8	115.2	115.1	126.1	118.1	102.0	104.8	102.7	116.2	113.9	115.6	124.0	126.1	124.6	121.5	104.8	117.0
097,099	WALL FINISHES, PAINTS & COATINGS	107.4	109.0	108.4	115.4	137.8	129.0	111.1	109.0	109.8	109.9	115.8	113.5	115.4	137.8	129.0	107.4	109.0	108.4
09	FINISHES	109.4	113.9	111.7	115.3	133.7	124.9	102.0	114.6	108.5	116.0	122.1	119.2	120.2	134.3	127.5	112.2	117.5	115.0
10 - 14	TOTAL DIV. 10000 - 14000	100.0	98.5	99.7	100.0	128.3	106.0	100.0	98.2	99.6	100.0	124.4	105.2	100.0	128.8	106.1	100.0	113.0	102.8
15	MECHANICAL	95.3	106.0	99.7	95.4	137.5	112.8	95.3	106.0	99.7	100.2	105.6	102.5	95.4	133.4	111.2	100.1	107.3	103.1
16	ELECTRICAL	93.0	99.7	96.3	103.3	140.8	121.5	119.3	113.4	116.5	98.1	104.2	101.1	103.9	136.0	119.5	89.6	99.7	94.5
01 - 16	WEIGHTED AVERAGE	100.1	107.2	103.2	105.5	127.2	115.1	100.4	109.3	104.4	106.6	110.3	108.3	109.5	126.1	116.9	102.2	109.5	105.5

CALIFORNIA

DIVISION		SACRAMENTO 942,956 - 958			SALINAS 939			SAN BERNARDINO 923 - 924			SAN DIEGO 919 - 921			SAN FRANCISCO 940 - 941			SAN JOSE 951		
		MAT.	INST.	TOTAL	MAT.	INST.	TOTAL	MAT.	INST.	TOTAL	MAT.	INST.	TOTAL	MAT.	INST.	TOTAL	MAT.	INST.	TOTAL
01590	EQUIPMENT RENTAL	.0	103.0	103.0	.0	99.6	99.6	.0	100.7	100.7	.0	97.1	97.1	.0	108.8	108.8	.0	100.0	100.0
02	SITE CONSTRUCTION	118.7	111.3	113.3	125.0	106.3	111.3	78.3	106.5	99.0	101.2	100.7	100.8	156.7	111.4	123.5	147.9	99.8	112.6
03100	CONCRETE FORMS & ACCESSORIES	104.8	125.2	122.4	108.2	125.4	123.0	110.6	117.2	116.3	108.1	108.7	108.6	106.7	135.6	131.7	106.3	134.7	130.8
03200	CONCRETE REINFORCEMENT	95.2	112.9	104.2	109.2	113.3	111.3	107.6	112.5	110.1	104.9	112.4	108.7	117.2	113.9	115.5	95.9	113.6	104.9
03300	CAST-IN-PLACE CONCRETE	122.1	114.1	118.8	108.3	113.6	110.4	73.9	113.7	90.2	112.0	106.1	109.6	141.6	119.5	132.5	129.7	117.5	124.7
03	CONCRETE	115.9	118.0	116.9	120.4	118.0	119.2	81.4	114.3	97.7	112.9	107.9	110.4	129.0	124.9	127.0	117.9	123.8	120.8
04	MASONRY	123.3	113.6	117.3	108.9	120.9	116.3	93.4	110.2	103.8	99.4	108.6	105.1	154.4	128.5	138.3	149.4	125.2	134.4
05	METALS	100.0	100.0	100.0	109.8	101.7	107.2	110.7	99.2	107.0	108.4	98.8	105.3	110.7	107.1	109.5	104.2	106.9	105.1
06	WOOD & PLASTICS	97.5	126.0	112.4	100.4	125.8	113.6	100.7	116.7	109.0	102.2	105.8	104.1	102.8	136.4	120.3	102.0	135.9	119.7
07	THERMAL & MOISTURE PROTECTION	110.6	113.2	111.7	101.0	122.5	110.1	101.8	112.6	106.4	107.8	103.4	105.9	113.3	130.6	120.6	101.9	129.5	113.5
08	DOORS & WINDOWS	117.6	118.2	117.7	101.8	119.7	106.4	99.7	113.7	103.4	103.8	106.6	104.5	110.5	128.4	115.2	92.7	127.9	101.9
09200	PLASTER & GYPSUM BOARD	98.7	126.3	115.8	100.1	126.3	116.4	99.9	117.1	110.6	97.4	105.7	102.6	103.9	136.5	124.2	100.4	136.5	122.8
095,098	CEILINGS & ACOUSTICAL TREATMENT	116.7	126.3	122.4	120.8	126.3	124.1	115.7	117.1	116.5	104.2	105.7	105.1	125.8	136.5	132.2	108.7	136.5	125.3
09600	FLOORING	118.9	113.9	117.6	117.9	126.1	120.2	123.8	104.8	118.7	110.7	104.8	109.1	116.1	126.1	118.8	113.6	126.1	117.2
097,099	WALL FINISHES, PAINTS & COATINGS	112.8	115.8	114.7	110.8	137.8	127.3	107.4	109.0	108.4	108.2	109.0	108.7	115.4	149.6	136.3	111.5	137.8	127.5
09	FINISHES	115.6	123.1	119.5	116.5	127.5	122.2	111.7	114.2	113.0	107.7	107.8	107.8	119.9	136.1	128.3	112.5	134.1	123.7
10 - 14	TOTAL DIV. 10000 - 14000	100.0	126.1	105.6	100.0	125.4	105.4	100.0	99.2	99.8	100.0	110.9	102.3	100.0	129.4	106.3	100.0	128.1	106.0
15	MECHANICAL	100.2	110.6	104.5	95.3	109.1	101.1	95.3	104.5	99.1	100.3	104.0	101.8	100.3	148.6	120.3	100.2	138.0	115.9
16	ELECTRICAL	97.7	106.1	101.8	88.4	123.2	105.3	93.0	105.2	98.9	98.9	96.4	97.7	103.4	150.5	126.3	103.3	140.8	121.5
01 - 16	WEIGHTED AVERAGE	106.6	113.0	109.4	104.1	116.4	109.6	97.9	108.1	102.4	104.0	103.9	104.0	112.8	132.9	121.7	107.3	127.5	116.3

CALIFORNIA

DIVISION		SAN LUIS OBISPO 934			SAN MATEO 944			SAN RAFAEL 949			SANTA ANA 926 - 927			SANTA BARBARA 931			SANTA CRUZ 950		
		MAT.	INST.	TOTAL	MAT.	INST.	TOTAL	MAT.	INST.	TOTAL	MAT.	INST.	TOTAL	MAT.	INST.	TOTAL	MAT.	INST.	TOTAL
01590	EQUIPMENT RENTAL	.0	99.6	99.6	.0	103.5	103.5	.0	103.6	103.6	.0	100.7	100.7	.0	99.6	99.6	.0	100.0	100.0
02	SITE CONSTRUCTION	115.8	106.1	108.7	149.8	104.7	116.7	132.8	111.1	116.9	91.9	106.5	102.6	109.4	106.1	107.0	148.3	99.6	112.6
03100	CONCRETE FORMS & ACCESSORIES	114.8	117.4	117.0	111.4	134.8	131.6	118.2	134.1	131.9	111.4	116.6	115.9	104.7	121.7	119.4	106.4	125.5	122.9
03200	CONCRETE REINFORCEMENT	110.6	112.5	111.6	100.6	113.8	107.3	101.2	113.7	107.5	111.3	112.5	111.9	108.6	112.6	110.6	118.9	113.4	116.1
03300	CAST-IN-PLACE CONCRETE	116.4	115.3	115.9	137.2	117.9	129.2	159.7	116.4	141.9	94.4	115.1	102.9	109.2	117.6	112.6	129.4	115.7	123.8
03	CONCRETE	118.7	114.9	116.8	123.3	123.9	123.6	148.1	122.8	135.5	99.8	114.5	107.1	109.9	117.6	113.7	121.4	119.1	120.3
04	MASONRY	111.1	110.3	110.6	145.2	128.9	135.1	119.1	128.3	124.8	82.2	108.7	98.6	109.6	110.7	110.3	151.2	121.1	132.5
05	METALS	107.4	99.1	104.8	102.6	105.4	103.5	103.4	101.1	102.6	110.8	99.3	107.1	105.3	99.5	103.5	112.6	105.9	110.5
06	WOOD & PLASTICS	104.2	116.8	110.7	108.5	136.2	122.9	108.9	135.9	122.9	100.8	116.7	109.9	95.1	122.0	109.1	102.0	125.9	114.5
07	THERMAL & MOISTURE PROTECTION	100.9	110.2	104.8	110.1	130.3	118.6	115.9	126.5	120.4	103.1	112.2	106.9	100.3	112.5	105.4	101.9	123.8	111.1
08	DOORS & WINDOWS	99.6	111.6	102.7	106.0	125.4	111.1	116.4	127.9	119.4	98.9	113.7	102.8	101.5	116.6	105.5	94.8	119.8	101.4
09200	PLASTER & GYPSUM BOARD	108.6	117.1	113.8	103.7	136.5	124.1	107.0	136.5	125.3	100.3	117.1	110.7	98.8	122.4	113.5	104.3	126.3	118.0
095,098	CEILINGS & ACOUSTICAL TREATMENT	120.8	117.1	118.6	114.8	136.5	127.8	124.1	136.5	131.5	115.7	117.1	116.5	120.4	122.4	121.6	117.4	126.3	122.7
09600	FLOORING	125.9	96.1	117.8	118.2	126.1	120.4	129.1	126.1	128.3	124.5	104.8	119.2	117.0	96.1	111.3	113.9	126.1	117.2
097,099	WALL FINISHES, PAINTS & COATINGS	109.9	102.7	105.5	115.4	137.8	129.0	111.2	137.8	127.4	107.4	109.0	108.4	109.9	102.7	105.5	111.5	137.8	127.5
09	FINISHES	119.3	111.8	115.4	117.4	134.3	126.2	120.8	134.0	127.7	113.0	113.9	113.5	114.5	115.4	115.0	115.2	127.6	121.7
10 - 14	TOTAL DIV. 10000 - 14000	100.0	102.4	100.6	100.0	128.8	106.1	100.0	127.9	105.9	100.0	98.5	99.7	100.0	113.3	102.8	100.0	125.8	105.5
15	MECHANICAL	95.3	107.0	100.2	95.4	134.8	111.7	95.3	133.7	111.2	95.3	106.0	99.8	100.2	107.3	103.2	100.2	139.0	113.9
16	ELECTRICAL	87.7	100.5	93.9	103.3	142.8	122.5	96.9	110.0	103.2	93.0	105.0	98.8	86.7	110.0	98.0	102.2	123.2	112.4
01 - 16	WEIGHTED AVERAGE	103.4	107.9	105.4	108.4	127.8	117.0	110.2	122.8	115.8	99.9	108.2	103.5	102.6	110.5	106.1	109.4	116.4	112.5

		CALIFORNIA																COLORADO		
	DIVISION	SANTA ROSA			STOCKTON			SUSANVILLE			VALLEJO			VAN NUYS			ALAMOSA		
		954			952			961			945			913 - 916			811		
		MAT.	INST.	TOTAL	MAT.	INST.	TOTAL	MAT.	INST.	TOTAL	MAT.	INST.	TOTAL	MAT.	INST.	TOTAL	MAT.	INST.	TOTAL
01590	EQUIPMENT RENTAL	.0	99.9	99.9	.0	99.2	99.2	.0	99.2	99.2	.0	103.6	103.6	.0	97.0	97.0	.0	96.9	96.9
02	SITE CONSTRUCTION	107.4	105.6	106.1	107.5	105.6	106.1	122.1	105.0	109.6	115.5	111.4	112.5	113.6	107.3	109.0	128.9	94.6	103.7
03100	CONCRETE FORMS & ACCESSORIES	102.8	133.2	129.0	104.4	124.8	122.0	105.4	123.6	121.1	106.1	133.4	129.6	112.8	116.5	116.0	101.1	79.4	82.4
03200	CONCRETE REINFORCEMENT	106.3	113.7	110.0	109.2	112.9	111.1	105.3	112.8	109.1	102.5	113.7	108.2	108.2	112.7	110.5	114.5	80.9	97.4
03300	CAST-IN-PLACE CONCRETE	123.0	114.7	119.6	109.1	113.3	110.8	112.9	111.5	112.3	127.1	115.3	122.2	107.8	109.7	108.5	100.5	83.6	93.5
03	CONCRETE	120.6	122.0	121.3	109.9	117.6	113.7	121.0	116.4	118.7	119.6	122.2	120.9	121.6	112.4	117.1	116.4	81.2	98.9
04	MASONRY	113.9	126.0	121.4	113.3	113.5	113.4	115.4	110.8	112.5	87.4	125.9	111.3	125.6	111.8	117.0	135.9	79.7	101.0
05	METALS	111.1	102.9	108.5	106.8	100.7	104.8	108.9	99.7	106.0	103.3	102.1	102.9	90.2	96.4	92.2	96.0	82.9	91.8
06	WOOD & PLASTICS	91.9	135.6	114.7	97.1	125.8	112.0	97.5	125.8	112.2	95.9	135.9	116.7	97.2	116.5	107.3	100.2	81.6	90.5
07	THERMAL & MOISTURE PROTECTION	103.6	126.5	113.2	105.6	110.8	107.8	106.6	109.4	107.8	113.1	126.0	118.5	99.5	110.0	103.9	100.9	83.9	93.8
08	DOORS & WINDOWS	100.5	128.0	107.7	100.8	118.1	105.3	103.1	118.1	107.0	118.9	128.1	121.3	95.4	113.6	100.2	93.4	85.8	91.4
09200	PLASTER & GYPSUM BOARD	95.5	136.5	121.0	99.2	126.3	116.0	97.9	126.3	115.5	101.4	136.5	123.2	94.8	117.1	108.6	83.1	80.8	81.6
095,098	CEILINGS & ACOUSTICAL TREATMENT	115.7	136.5	128.1	120.4	126.3	123.9	120.8	126.3	124.1	126.1	136.5	132.3	94.0	117.1	107.7	100.5	80.8	88.7
09600	FLOORING	119.8	126.1	121.5	117.0	114.1	116.2	117.0	113.9	116.1	123.7	126.1	124.4	105.2	104.8	105.1	110.6	68.4	99.2
097,099	WALL FINISHES, PAINTS & COATINGS	107.4	137.8	125.9	109.9	116.6	114.0	109.9	131.2	122.9	112.0	137.8	127.7	111.1	109.0	109.8	115.2	30.1	63.3
09	FINISHES	111.8	133.6	123.1	114.4	123.0	118.9	115.5	123.9	119.9	117.4	133.7	125.9	103.5	113.8	108.9	103.1	72.0	86.9
10 - 14	TOTAL DIV. 10000 - 14000	100.0	126.3	105.6	100.0	125.4	105.4	100.0	124.3	105.2	100.0	127.0	105.8	100.0	98.0	99.6	100.0	89.8	97.8
15	MECHANICAL	95.3	133.5	111.2	100.2	104.4	101.9	95.3	106.5	100.0	100.2	122.9	109.6	95.3	106.0	99.7	95.2	71.0	85.2
16	ELECTRICAL	93.3	110.0	101.4	97.9	123.1	110.1	98.5	104.2	101.3	93.0	123.5	107.9	119.3	113.4	116.4	91.8	80.5	86.3
01 - 16	WEIGHTED AVERAGE	104.1	122.0	112.1	104.4	113.5	108.4	105.7	110.7	107.9	105.5	122.2	112.9	103.5	109.1	106.0	101.3	79.5	91.7

		COLORADO																		
	DIVISION	BOULDER			COLORADO SPRINGS			DENVER			DURANGO			FORT COLLINS			FORT MORGAN		
		803			808 - 809			800 - 802			813			805			807		
		MAT.	INST.	TOTAL	MAT.	INST.	TOTAL	MAT.	INST.	TOTAL	MAT.	INST.	TOTAL	MAT.	INST.	TOTAL	MAT.	INST.	TOTAL
01590	EQUIPMENT RENTAL	.0	96.7	96.7	.0	95.2	95.2	.0	100.3	100.3	.0	96.9	96.9	.0	96.7	96.7	.0	96.7	96.7
02	SITE CONSTRUCTION	91.2	98.7	96.7	93.3	96.8	95.8	91.9	106.0	102.3	122.6	94.6	102.0	102.2	99.2	100.0	92.9	99.3	97.6
03100	CONCRETE FORMS & ACCESSORIES	104.0	82.8	85.7	92.5	85.1	86.1	101.8	85.9	88.1	107.9	79.5	83.4	101.2	81.4	84.1	104.3	82.7	85.7
03200	CONCRETE REINFORCEMENT	113.6	80.9	97.0	112.8	88.5	100.5	112.8	88.6	100.5	114.5	80.9	97.4	113.7	79.9	96.5	113.9	80.9	97.1
03300	CAST-IN-PLACE CONCRETE	91.4	83.8	88.3	95.1	86.5	91.5	88.9	86.9	88.1	115.5	84.9	102.9	103.2	82.2	94.6	89.7	83.7	87.2
03	CONCRETE	103.0	82.8	93.0	106.7	86.3	96.6	100.9	86.8	93.9	119.0	81.7	100.5	113.1	81.4	97.4	101.6	82.7	92.2
04	MASONRY	100.7	79.9	87.8	106.0	81.1	90.6	105.7	83.3	91.8	122.2	79.7	95.9	118.2	59.1	81.6	117.2	79.9	94.1
05	METALS	95.4	82.8	91.4	98.0	89.3	95.2	100.3	89.5	96.9	96.0	83.1	91.9	96.2	81.3	91.5	95.2	82.9	91.2
06	WOOD & PLASTICS	104.5	85.4	94.6	93.3	87.8	90.4	102.9	87.7	95.0	108.6	81.6	94.5	102.7	85.4	93.7	104.9	85.4	94.8
07	THERMAL & MOISTURE PROTECTION	105.4	82.0	95.6	105.9	85.4	97.3	105.4	82.1	95.6	100.8	83.9	93.7	105.8	70.9	91.1	105.3	81.3	95.2
08	DOORS & WINDOWS	95.9	87.9	93.8	98.5	91.3	96.6	99.8	91.2	97.6	100.6	85.8	96.7	95.8	87.9	93.8	95.8	87.9	93.7
09200	PLASTER & GYPSUM BOARD	99.6	85.3	90.7	87.8	87.6	87.7	97.4	87.6	91.3	92.3	80.8	85.1	95.9	85.3	89.3	100.7	85.3	91.1
095,098	CEILINGS & ACOUSTICAL TREATMENT	95.9	85.3	89.6	101.7	87.6	93.3	101.8	87.6	93.3	100.5	80.8	88.7	95.9	85.3	89.6	95.9	85.3	89.6
09600	FLOORING	112.0	96.6	107.8	106.9	85.2	101.0	107.8	96.6	104.8	115.8	68.4	103.0	107.7	68.4	97.1	112.5	68.4	100.6
097,099	WALL FINISHES, PAINTS & COATINGS	105.6	75.0	86.9	105.6	51.7	72.7	105.6	78.4	89.0	115.2	30.1	63.3	105.6	51.3	72.5	105.6	68.3	82.8
09	FINISHES	100.2	84.2	91.9	98.4	81.6	89.7	100.0	87.2	93.3	105.3	72.0	88.0	98.8	76.0	86.9	100.4	78.6	89.0
10 - 14	TOTAL DIV. 10000 - 14000	100.0	89.0	97.7	100.0	89.8	97.8	100.0	89.8	97.8	100.0	89.8	97.8	100.0	88.0	97.4	100.0	89.0	97.7
15	MECHANICAL	95.1	84.6	90.8	100.1	79.9	91.7	100.1	87.4	94.8	95.2	79.6	88.8	100.1	83.1	93.0	95.1	84.6	90.8
16	ELECTRICAL	96.4	94.2	95.3	99.6	88.6	94.3	101.5	94.9	98.3	91.2	80.3	85.9	96.4	94.2	95.3	96.7	94.2	95.4
01 - 16	WEIGHTED AVERAGE	97.7	86.5	92.7	100.4	85.7	93.9	100.6	89.9	95.8	101.7	81.4	92.7	101.1	82.4	92.8	98.4	85.8	92.8

		COLORADO																		
	DIVISION	GLENWOOD SPRINGS			GOLDEN			GRAND JUNCTION			GREELEY			MONTROSE			PUEBLO		
		816			804			815			806			814			810		
		MAT.	INST.	TOTAL	MAT.	INST.	TOTAL	MAT.	INST.	TOTAL	MAT.	INST.	TOTAL	MAT.	INST.	TOTAL	MAT.	INST.	TOTAL
01590	EQUIPMENT RENTAL	.0	100.0	100.0	.0	96.7	96.7	.0	100.0	100.0	.0	96.7	96.7	.0	98.4	98.4	.0	96.9	96.9
02	SITE CONSTRUCTION	137.7	102.3	111.7	104.1	99.6	100.8	121.8	101.8	107.1	90.1	97.9	95.8	131.3	98.0	106.9	114.1	96.2	101.0
03100	CONCRETE FORMS & ACCESSORIES	97.6	82.6	84.7	95.7	82.6	84.4	106.6	81.8	85.2	98.6	48.1	55.1	97.0	82.4	84.4	103.4	85.4	87.8
03200	CONCRETE REINFORCEMENT	113.4	80.9	96.9	113.9	80.9	97.1	113.7	79.8	96.5	113.6	79.3	96.2	113.2	80.8	96.8	109.5	88.5	98.8
03300	CAST-IN-PLACE CONCRETE	100.4	84.1	93.7	89.8	83.7	87.3	111.2	83.2	99.7	86.3	58.9	75.0	100.5	84.1	93.7	99.7	87.7	94.8
03	CONCRETE	121.4	82.9	102.3	112.3	82.7	97.6	115.4	81.9	98.8	98.4	58.6	78.6	112.5	82.7	97.7	105.3	86.9	96.2
04	MASONRY	109.1	79.9	91.0	120.7	80.0	95.4	144.2	61.5	92.9	112.2	39.2	66.9	114.6	79.7	93.0	105.2	80.2	89.7
05	METALS	95.7	83.7	91.9	95.3	83.0	91.4	96.9	80.5	91.7	96.2	79.6	90.9	94.8	82.5	90.9	98.4	90.3	95.8
06	WOOD & PLASTICS	95.0	85.6	90.1	97.0	85.4	91.0	105.9	85.6	95.3	99.8	46.8	72.2	95.1	85.8	90.2	102.7	88.2	95.1
07	THERMAL & MOISTURE PROTECTION	100.7	80.5	92.2	106.3	78.6	94.7	99.7	67.1	86.0	105.2	59.0	85.7	100.9	84.4	93.9	99.2	84.9	93.2
08	DOORS & WINDOWS	99.8	88.0	96.7	95.9	87.3	93.6	100.5	88.0	97.2	95.8	66.9	88.3	100.8	88.1	97.5	95.2	91.5	94.2
09200	PLASTER & GYPSUM BOARD	101.0	85.3	91.2	94.0	85.3	88.6	109.1	85.3	94.3	94.8	45.6	64.2	79.7	85.3	83.2	87.4	87.6	87.5
095,098	CEILINGS & ACOUSTICAL TREATMENT	98.8	85.3	90.7	95.9	85.3	89.6	98.8	85.3	90.7	95.9	45.6	65.9	100.5	85.3	91.4	111.6	87.6	97.3
09600	FLOORING	109.6	63.0	97.0	105.1	68.4	95.1	115.2	68.4	102.5	106.3	68.4	96.1	113.0	56.6	97.7	111.5	96.6	107.5
097,099	WALL FINISHES, PAINTS & COATINGS	115.2	68.3	86.6	105.6	75.0	86.9	105.6	75.0	86.9	105.6	31.9	60.7	115.2	30.1	63.3	115.2	47.9	74.1
09	FINISHES	105.4	77.7	91.0	98.3	79.3	88.4	106.8	79.4	92.5	97.6	48.5	72.0	103.2	72.1	87.0	105.2	83.8	94.1
10 - 14	TOTAL DIV. 10000 - 14000	100.0	89.4	97.7	100.0	89.0	97.7	100.0	89.3	97.7	100.0	79.3	95.6	100.0	89.8	97.8	100.0	90.7	98.0
15	MECHANICAL	95.2	79.7	88.8	95.1	84.4	90.7	100.1	72.7	88.7	100.1	77.5	90.7	95.2	79.7	88.8	100.1	71.2	88.1
16	ELECTRICAL	88.9	81.9	85.5	96.7	94.2	95.4	90.9	63.9	77.8	96.4	94.1	95.3	90.9	63.9	77.8	91.8	80.6	86.3
01 - 16	WEIGHTED AVERAGE	101.1	83.3	93.2	99.9	85.8	93.6	103.6	76.8	91.7	98.7	70.4	86.1	100.3	79.6	91.1	100.0	83.1	92.5

DIVISION		COLORADO SALIDA 812			CONNECTICUT BRIDGEPORT 066			CONNECTICUT BRISTOL 060			CONNECTICUT HARTFORD 061			CONNECTICUT MERIDEN 064			CONNECTICUT NEW BRITAIN 060		
		MAT.	INST.	TOTAL	MAT.	INST.	TOTAL	MAT.	INST.	TOTAL	MAT.	INST.	TOTAL	MAT.	INST.	TOTAL	MAT.	INST.	TOTAL
01590	EQUIPMENT RENTAL	.0	98.4	98.4	.0	101.5	101.5	.0	101.5	101.5	.0	101.5	101.5	.0	102.1	102.1	.0	101.5	101.5
02	SITE CONSTRUCTION	122.2	98.4	104.7	103.3	104.9	104.5	102.4	104.9	104.2	102.8	104.9	104.4	100.3	105.8	104.3	102.6	104.9	104.3
03100	CONCRETE FORMS & ACCESSORIES	106.4	82.4	85.7	102.2	123.2	120.3	102.2	122.7	119.9	101.1	122.8	119.8	102.0	123.1	120.2	102.5	122.8	120.0
03200	CONCRETE REINFORCEMENT	113.0	80.8	96.6	107.5	128.1	117.9	107.5	128.0	117.9	107.5	128.0	117.9	107.5	128.0	117.9	107.5	128.0	117.9
03300	CAST-IN-PLACE CONCRETE	115.1	84.0	102.3	108.1	117.0	111.7	101.2	116.9	107.7	102.4	116.9	108.3	97.4	117.0	105.4	102.9	116.9	108.6
03	CONCRETE	114.3	82.6	98.6	109.3	121.7	115.5	106.0	121.4	113.7	106.5	121.4	113.9	104.1	121.6	112.8	106.8	121.4	114.1
04	MASONRY	146.1	79.7	104.9	109.2	127.8	120.7	101.0	127.8	117.6	101.1	127.8	117.6	100.7	127.8	117.5	101.1	127.8	117.6
05	METALS	94.5	82.5	90.7	100.9	124.6	108.5	100.9	124.1	108.3	106.0	124.1	111.8	97.9	124.4	106.4	97.3	124.1	105.9
06	WOOD & PLASTICS	104.3	85.8	94.7	100.4	122.3	111.8	100.4	122.3	111.8	100.4	122.3	111.8	100.4	122.3	111.8	100.4	122.3	111.8
07	THERMAL & MOISTURE PROTECTION	99.8	84.4	93.3	101.8	126.2	112.1	102.0	122.4	110.6	100.8	122.4	109.9	102.0	122.4	110.6	102.0	123.4	111.0
08	DOORS & WINDOWS	93.5	88.1	92.1	106.9	131.8	113.4	106.9	121.8	110.8	106.9	121.8	110.8	109.6	131.8	115.4	106.9	121.8	110.8
09200	PLASTER & GYPSUM BOARD	82.7	85.3	84.3	105.5	122.0	115.8	105.5	122.0	115.8	105.5	122.0	115.8	107.2	122.0	116.4	105.5	122.0	115.8
095,098	CEILINGS & ACOUSTICAL TREATMENT	100.5	85.3	91.4	98.5	122.0	112.5	98.5	122.0	112.5	98.5	122.0	112.5	103.4	122.0	114.5	98.5	122.0	112.5
09600	FLOORING	118.6	56.6	101.9	99.6	120.6	105.3	99.6	120.6	105.3	99.6	120.6	105.3	99.6	120.6	105.3	99.6	120.6	105.3
097,099	WALL FINISHES, PAINTS & COATINGS	115.2	30.1	63.3	91.5	118.5	108.0	91.5	114.1	105.3	91.5	114.1	105.3	91.5	121.8	110.0	91.5	114.1	105.3
09	FINISHES	104.5	72.1	87.6	102.1	122.0	112.5	102.1	121.5	112.2	102.1	121.5	112.2	103.6	122.4	113.4	102.1	121.5	112.2
10 - 14	TOTAL DIV. 10000 - 14000	100.0	89.9	97.8	100.0	115.4	103.3	100.0	115.3	103.3	100.0	115.4	103.3	100.0	115.4	103.3	100.0	115.3	103.3
15	MECHANICAL	95.2	70.9	85.1	100.1	111.8	104.9	100.1	111.8	104.9	100.1	111.8	104.9	95.1	111.8	102.1	100.1	111.8	104.9
16	ELECTRICAL	91.1	80.5	86.0	97.7	109.6	103.5	97.7	105.8	101.6	97.3	110.7	103.8	97.6	109.7	103.5	97.7	105.8	101.7
01 - 16	WEIGHTED AVERAGE	101.3	80.2	91.9	102.4	117.8	109.2	101.6	116.6	108.2	102.3	117.3	109.0	100.1	117.8	108.0	101.1	116.6	108.0

DIVISION		CONNECTICUT NEW HAVEN 065			CONNECTICUT NEW LONDON 063			CONNECTICUT NORWALK 068			CONNECTICUT STAMFORD 069			CONNECTICUT WATERBURY 067			CONNECTICUT WILLIMANTIC 062		
		MAT.	INST.	TOTAL	MAT.	INST.	TOTAL	MAT.	INST.	TOTAL	MAT.	INST.	TOTAL	MAT.	INST.	TOTAL	MAT.	INST.	TOTAL
01590	EQUIPMENT RENTAL	.0	102.1	102.1	.0	102.1	102.1	.0	101.5	101.5	.0	101.5	101.5	.0	101.5	101.5	.0	101.5	101.5
02	SITE CONSTRUCTION	102.4	105.8	104.9	94.7	105.8	102.8	103.0	104.9	104.4	103.7	105.0	104.6	102.8	104.9	104.4	103.1	104.9	104.4
03100	CONCRETE FORMS & ACCESSORIES	102.0	123.1	120.2	102.0	122.8	119.9	102.2	123.4	120.5	102.2	123.6	120.7	102.2	123.1	120.2	102.2	122.8	120.0
03200	CONCRETE REINFORCEMENT	107.5	128.0	117.9	84.3	128.0	106.5	107.5	128.1	117.9	107.5	128.1	118.0	107.5	128.0	117.9	107.5	128.0	117.9
03300	CAST-IN-PLACE CONCRETE	104.7	117.0	109.7	89.1	116.9	100.5	106.3	125.5	114.2	108.1	125.6	115.3	108.1	117.0	111.7	108.1	117.0	111.7
03	CONCRETE	121.5	121.6	121.5	93.6	121.4	107.4	108.5	124.7	116.5	109.3	124.9	117.0	109.3	121.6	115.4	105.9	117.6	111.7
04	MASONRY	101.2	127.8	117.7	99.6	127.8	117.1	101.4	127.6	117.6	101.5	127.6	117.7	101.5	127.8	117.8	100.9	127.8	117.6
05	METALS	97.5	124.4	106.1	97.2	124.1	105.8	100.9	124.7	108.5	100.9	125.0	108.6	100.9	124.5	108.4	100.7	124.2	108.2
06	WOOD & PLASTICS	100.4	122.3	111.8	100.4	122.3	111.8	100.4	122.3	111.8	100.4	122.3	111.8	100.4	122.3	111.8	100.4	122.3	111.8
07	THERMAL & MOISTURE PROTECTION	102.1	123.4	111.0	101.9	122.4	110.6	102.0	127.4	112.7	102.0	127.4	112.7	102.0	123.4	111.0	102.2	120.7	110.0
08	DOORS & WINDOWS	106.9	131.8	113.4	110.3	121.2	113.1	106.9	131.8	113.4	106.9	131.8	113.4	106.9	131.8	113.4	110.3	131.8	115.9
09200	PLASTER & GYPSUM BOARD	105.5	122.0	115.8	105.5	122.0	115.8	105.5	122.0	115.8	105.5	122.0	115.8	105.5	122.0	115.8	105.5	122.0	115.8
095,098	CEILINGS & ACOUSTICAL TREATMENT	98.5	122.0	112.5	96.6	122.0	111.8	98.5	122.0	112.5	98.5	122.0	112.5	98.5	122.0	112.5	96.6	122.0	111.8
09600	FLOORING	99.6	120.6	105.3	99.6	120.6	105.3	99.6	120.6	105.3	99.6	120.6	105.3	99.6	120.6	105.3	99.6	120.6	105.3
097,099	WALL FINISHES, PAINTS & COATINGS	91.5	118.5	108.0	91.5	114.1	105.3	91.5	118.5	108.0	91.5	118.5	108.0	91.5	121.8	110.0	91.5	121.8	110.0
09	FINISHES	102.2	122.0	112.5	101.2	121.5	111.8	102.1	122.0	112.5	102.2	122.0	112.5	102.0	122.4	112.6	101.8	122.4	112.5
10 - 14	TOTAL DIV. 10000 - 14000	100.0	115.4	103.3	100.0	115.4	103.3	100.0	115.4	103.3	100.0	115.5	103.3	100.0	115.4	103.3	100.0	115.4	103.3
15	MECHANICAL	100.1	111.8	104.9	95.1	111.8	102.1	100.1	111.9	105.0	100.1	111.9	105.0	100.1	111.8	104.9	100.1	111.8	104.9
16	ELECTRICAL	97.6	109.7	103.5	94.8	110.7	102.5	97.7	109.6	103.5	97.7	150.4	123.3	97.2	109.6	103.2	97.7	110.6	104.0
01 - 16	WEIGHTED AVERAGE	102.8	117.8	109.4	98.1	117.3	106.7	101.9	118.2	109.1	102.0	124.0	111.8	101.9	117.7	108.9	101.8	117.2	108.7

DIVISION		D.C. WASHINGTON 200 - 205			DELAWARE DOVER 199			DELAWARE NEWARK 197			DELAWARE WILMINGTON 198			FLORIDA DAYTONA BEACH 321			FLORIDA FORT LAUDERDALE 333		
		MAT.	INST.	TOTAL	MAT.	INST.	TOTAL	MAT.	INST.	TOTAL	MAT.	INST.	TOTAL	MAT.	INST.	TOTAL	MAT.	INST.	TOTAL
01590	EQUIPMENT RENTAL	.0	102.3	102.3	.0	118.2	118.2	.0	118.2	118.2	.0	118.4	118.4	.0	97.7	97.7	.0	89.3	89.3
02	SITE CONSTRUCTION	102.8	89.7	93.2	101.7	110.8	108.4	101.7	110.8	108.4	87.4	111.1	104.8	113.4	86.5	93.7	99.5	73.6	80.5
03100	CONCRETE FORMS & ACCESSORIES	98.5	80.8	83.2	96.5	98.5	98.2	96.5	98.5	98.2	96.2	98.5	98.2	98.5	67.8	72.0	96.0	66.3	70.4
03200	CONCRETE REINFORCEMENT	105.5	92.8	99.0	101.4	95.6	98.5	102.3	95.6	98.9	102.3	95.6	98.9	95.1	84.3	89.6	95.1	70.1	82.4
03300	CAST-IN-PLACE CONCRETE	114.5	86.2	102.9	85.6	91.4	88.0	85.6	91.4	88.0	80.5	91.4	85.0	90.4	69.2	81.7	94.9	67.2	83.5
03	CONCRETE	111.9	86.3	99.2	100.9	96.5	98.7	101.1	96.5	98.8	98.6	96.5	97.5	91.1	72.6	81.9	93.1	68.6	80.9
04	MASONRY	95.0	81.4	86.6	107.0	91.8	97.6	105.0	91.8	96.8	111.1	91.8	99.1	87.6	63.7	72.8	87.8	65.6	74.1
05	METALS	104.9	110.0	106.5	105.5	112.6	107.8	105.5	112.6	107.8	104.2	112.6	105.6	94.6	94.3	94.5	94.5	87.4	92.2
06	WOOD & PLASTICS	96.4	81.1	88.4	94.5	100.0	97.4	94.5	100.0	97.4	94.5	100.0	97.4	97.5	70.3	83.3	92.9	63.8	77.7
07	THERMAL & MOISTURE PROTECTION	103.6	84.8	95.7	102.1	105.0	103.3	102.2	105.0	103.4	101.8	105.0	103.2	99.6	68.0	86.3	99.6	72.6	88.2
08	DOORS & WINDOWS	102.7	90.9	99.6	94.6	102.5	96.7	94.6	102.5	96.7	94.3	102.5	96.5	100.7	68.0	92.1	98.3	61.1	88.6
09200	PLASTER & GYPSUM BOARD	108.8	80.6	91.3	111.5	99.9	104.3	111.5	99.9	104.3	110.7	99.9	104.0	100.6	70.0	81.6	100.2	63.4	77.3
095,098	CEILINGS & ACOUSTICAL TREATMENT	103.7	80.6	90.0	104.3	99.9	101.6	104.3	99.9	101.6	99.9	99.9	99.9	104.3	70.0	83.9	104.3	63.4	79.9
09600	FLOORING	115.9	103.1	112.5	85.2	105.3	90.6	85.2	105.3	90.6	85.0	105.3	90.5	118.7	71.8	106.0	118.7	58.3	102.4
097,099	WALL FINISHES, PAINTS & COATINGS	117.2	84.9	97.4	92.8	93.6	93.3	92.8	93.6	93.3	92.8	93.6	93.3	110.6	69.5	85.6	106.7	48.1	71.0
09	FINISHES	104.6	85.0	94.4	102.8	99.2	100.9	102.8	99.2	100.9	101.6	99.2	100.3	108.7	68.8	87.9	106.7	62.3	83.6
10 - 14	TOTAL DIV. 10000 - 14000	100.0	96.1	99.2	100.0	93.2	98.6	100.0	93.2	98.6	100.0	93.2	98.6	100.0	81.6	96.1	100.0	87.9	97.4
15	MECHANICAL	99.9	91.4	96.4	100.3	113.6	105.8	100.3	113.6	105.8	100.3	113.6	105.8	99.9	70.3	87.6	99.9	67.2	86.3
16	ELECTRICAL	99.2	98.3	98.7	96.3	102.2	99.2	96.3	102.2	99.2	96.2	102.2	99.1	96.6	65.4	81.4	96.6	73.4	85.3
01 - 16	WEIGHTED AVERAGE	102.5	91.2	97.5	100.7	103.9	102.1	100.7	103.9	102.1	99.7	103.9	101.5	98.3	72.8	87.0	97.7	70.3	85.5

FLORIDA

DIVISION		FORT MYERS 339,341			GAINESVILLE 326,344			JACKSONVILLE 320,322			LAKELAND 338			MELBOURNE 329			MIAMI 330 - 332,340		
		MAT.	INST.	TOTAL	MAT.	INST.	TOTAL	MAT.	INST.	TOTAL	MAT.	INST.	TOTAL	MAT.	INST.	TOTAL	MAT.	INST.	TOTAL
01590	EQUIPMENT RENTAL	.0	97.7	97.7	.0	97.7	97.7	.0	97.7	97.7	.0	97.7	97.7	.0	97.7	97.7	.0	89.3	89.3
02	SITE CONSTRUCTION	110.7	86.4	92.9	123.4	86.0	96.0	113.5	87.0	94.1	112.8	86.0	93.2	121.5	87.1	96.2	98.8	73.5	80.2
03100	CONCRETE FORMS & ACCESSORIES	91.0	72.1	74.7	92.2	51.4	57.0	98.2	51.8	58.2	86.4	72.6	74.5	93.9	72.0	75.0	96.3	66.4	70.5
03200	CONCRETE REINFORCEMENT	96.2	85.9	90.9	100.9	48.8	74.4	95.1	49.2	71.8	98.5	87.1	92.7	96.2	84.4	90.2	95.1	70.1	82.4
03300	CAST-IN-PLACE CONCRETE	99.0	58.7	82.4	103.9	48.1	81.0	91.3	57.0	77.2	101.2	59.7	84.1	109.0	75.0	95.0	92.4	67.4	82.1
03	CONCRETE	93.9	71.2	82.7	102.0	51.5	76.9	91.5	54.8	73.3	95.7	71.9	83.9	102.4	76.5	89.5	91.9	68.7	80.4
04	MASONRY	81.5	50.9	62.5	101.4	42.4	64.8	87.2	49.9	64.1	98.0	73.5	82.8	86.6	73.5	78.5	86.2	66.0	73.7
05	METALS	97.1	91.8	95.4	93.8	77.3	88.5	94.1	78.2	89.1	97.0	93.0	95.7	103.2	95.0	100.6	99.6	87.7	95.8
06	WOOD & PLASTICS	89.6	75.0	82.0	90.8	50.5	69.8	97.5	50.5	73.0	84.7	75.0	79.7	92.7	70.3	81.0	92.9	63.8	77.7
07	THERMAL & MOISTURE PROTECTION	99.3	54.7	80.5	99.9	54.2	80.6	99.8	57.1	81.8	99.2	60.6	83.0	100.0	75.6	89.7	103.3	69.4	89.0
08	DOORS & WINDOWS	99.4	70.8	91.9	98.8	47.4	85.3	100.7	47.9	86.9	99.3	68.6	91.3	99.9	73.9	93.1	98.3	61.7	88.7
09200	PLASTER & GYPSUM BOARD	96.3	74.9	83.0	96.4	49.7	67.4	100.6	49.7	69.0	94.2	74.9	82.2	96.7	70.0	80.1	100.2	63.4	77.3
095,098	CEILINGS & ACOUSTICAL TREATMENT	97.3	74.9	83.9	96.4	49.7	68.6	104.3	49.7	71.7	96.4	74.9	83.6	99.2	70.0	81.8	104.3	63.4	79.9
09600	FLOORING	114.4	40.0	94.3	115.0	34.3	93.1	118.7	47.9	99.5	111.2	50.7	94.9	115.2	71.8	103.5	126.0	59.0	107.9
097,099	WALL FINISHES, PAINTS & COATINGS	110.6	44.9	70.5	110.6	43.2	69.5	110.6	46.4	71.4	110.6	44.9	70.5	110.6	94.7	100.9	106.7	48.1	71.0
09	FINISHES	104.8	63.9	83.5	105.8	46.7	75.0	108.7	49.8	78.0	103.5	66.1	84.0	106.3	74.0	89.5	108.8	62.4	84.7
10 - 14	TOTAL DIV. 10000 - 14000	100.0	82.1	96.2	100.0	79.3	95.6	100.0	76.7	95.0	100.0	82.1	96.2	100.0	85.3	96.9	100.0	87.9	97.4
15	MECHANICAL	97.2	51.7	78.3	98.7	66.9	85.5	99.9	49.5	79.0	97.2	71.3	86.5	99.9	75.9	89.9	99.9	73.8	89.1
16	ELECTRICAL	98.7	39.5	69.9	96.9	40.9	69.7	96.4	66.9	82.0	96.9	45.7	72.0	97.5	72.1	85.2	97.5	71.6	84.9
01 - 16	WEIGHTED AVERAGE	97.7	62.8	82.2	99.6	57.7	81.0	98.3	59.6	81.1	98.4	70.5	86.0	100.8	77.8	90.6	98.7	71.4	86.6

FLORIDA

DIVISION		ORLANDO 327 - 328,347			PANAMA CITY 324			PENSACOLA 325			SARASOTA 342			ST. PETERSBURG 337			TALLAHASSEE 323		
		MAT.	INST.	TOTAL	MAT.	INST.	TOTAL	MAT.	INST.	TOTAL	MAT.	INST.	TOTAL	MAT.	INST.	TOTAL	MAT.	INST.	TOTAL
01590	EQUIPMENT RENTAL	.0	97.7	97.7	.0	97.7	97.7	.0	97.7	97.7	.0	97.7	97.7	.0	97.7	97.7	.0	97.7	97.7
02	SITE CONSTRUCTION	114.1	86.2	93.6	127.7	83.6	95.4	125.4	86.1	96.6	114.9	86.1	93.8	114.4	85.5	93.2	114.8	85.6	93.4
03100	CONCRETE FORMS & ACCESSORIES	98.2	69.4	73.3	97.3	26.9	36.5	87.5	49.5	54.8	98.5	72.2	75.8	95.8	45.8	52.6	98.2	38.0	46.3
03200	CONCRETE REINFORCEMENT	95.1	80.8	87.8	99.3	48.1	73.3	101.8	48.5	74.7	95.1	87.1	91.0	98.5	55.4	76.6	95.1	48.7	71.5
03300	CAST-IN-PLACE CONCRETE	98.3	69.1	86.3	96.0	33.5	70.3	96.0	53.2	78.4	103.5	59.5	85.4	102.3	53.5	82.2	94.7	47.0	75.1
03	CONCRETE	92.7	72.6	82.8	99.7	34.9	67.5	98.7	52.3	75.7	97.4	71.7	84.6	97.2	52.0	74.8	93.1	45.1	69.3
04	MASONRY	91.3	63.7	74.2	92.0	26.8	51.6	89.1	48.4	63.9	88.5	73.5	79.2	136.3	47.8	81.4	90.2	37.7	57.6
05	METALS	100.5	92.7	98.0	94.4	63.8	84.6	94.3	77.5	89.0	97.4	92.6	95.9	97.7	78.3	91.5	87.8	76.6	84.2
06	WOOD & PLASTICS	97.5	72.8	84.7	96.3	26.9	60.2	85.8	49.8	67.1	97.5	75.0	85.8	94.6	45.7	69.2	95.9	36.2	64.8
07	THERMAL & MOISTURE PROTECTION	99.8	68.3	86.5	100.0	30.5	70.8	99.8	51.3	79.4	99.6	60.6	83.1	99.5	48.2	77.9	99.8	45.2	76.8
08	DOORS & WINDOWS	100.7	67.7	92.1	98.3	25.7	79.3	98.3	48.4	85.2	100.7	68.0	92.2	99.3	44.8	85.1	99.3	39.6	83.7
09200	PLASTER & GYPSUM BOARD	103.0	72.7	84.1	98.7	25.5	53.2	94.6	49.0	66.3	100.6	74.9	84.6	97.9	44.8	64.9	100.6	35.0	59.9
095,098	CEILINGS & ACOUSTICAL TREATMENT	104.3	72.7	85.4	98.3	25.5	54.9	98.3	49.0	68.9	102.4	74.9	86.0	98.3	44.8	66.4	104.3	35.0	63.0
09600	FLOORING	118.7	71.8	106.0	118.1	18.4	91.1	112.0	50.9	95.5	118.7	50.5	100.3	117.0	50.7	99.1	118.7	37.2	96.7
097,099	WALL FINISHES, PAINTS & COATINGS	110.6	53.3	75.6	110.6	23.7	57.6	110.6	53.9	76.0	110.6	44.9	70.5	110.6	44.9	70.5	110.6	38.2	66.4
09	FINISHES	109.0	68.4	87.9	107.8	24.5	64.4	105.2	50.1	76.5	108.2	66.0	86.2	106.3	46.4	75.1	108.8	36.9	71.4
10 - 14	TOTAL DIV. 10000 - 14000	100.0	81.8	96.1	100.0	42.9	87.8	100.0	48.0	88.9	100.0	82.1	96.2	100.0	50.2	89.4	100.0	72.9	94.2
15	MECHANICAL	99.9	65.8	85.8	99.9	24.3	68.5	99.9	48.2	78.5	99.8	52.1	80.0	99.9	48.0	78.4	99.9	38.5	74.4
16	ELECTRICAL	97.3	43.5	71.2	95.6	32.6	64.9	98.7	62.5	81.1	96.6	35.7	67.0	96.9	47.0	72.6	97.0	40.3	69.4
01 - 16	WEIGHTED AVERAGE	99.7	68.6	85.9	99.4	36.8	71.6	99.1	57.2	80.5	99.5	65.1	84.2	101.6	54.2	80.5	97.6	48.2	75.6

FLORIDA / GEORGIA

DIVISION		TAMPA 335 - 336,346			WEST PALM BEACH 334,349			ALBANY 317,398			ATHENS 306			ATLANTA 300 - 303,399			AUGUSTA 308 - 309		
		MAT.	INST.	TOTAL	MAT.	INST.	TOTAL	MAT.	INST.	TOTAL	MAT.	INST.	TOTAL	MAT.	INST.	TOTAL	MAT.	INST.	TOTAL
01590	EQUIPMENT RENTAL	.0	97.7	97.7	.0	89.3	89.3	.0	90.1	90.1	.0	92.2	92.2	.0	92.8	92.8	.0	92.2	92.2
02	SITE CONSTRUCTION	114.6	86.1	93.7	96.1	73.8	79.7	99.7	76.0	82.3	103.2	92.6	95.4	99.4	95.7	96.7	96.0	92.4	93.4
03100	CONCRETE FORMS & ACCESSORIES	99.7	72.7	76.4	100.1	65.9	70.6	97.5	44.6	51.9	93.8	40.9	48.2	98.1	76.6	79.6	95.1	50.2	56.4
03200	CONCRETE REINFORCEMENT	95.1	87.1	91.1	97.8	69.7	83.5	94.7	90.8	92.7	99.6	87.4	93.4	98.9	92.6	95.7	100.0	74.6	87.1
03300	CAST-IN-PLACE CONCRETE	100.0	59.7	83.4	90.2	63.3	79.1	96.6	44.8	75.3	99.0	47.8	78.0	99.0	72.8	88.2	93.5	47.3	74.5
03	CONCRETE	95.8	72.0	84.0	90.4	67.0	78.8	93.8	54.9	74.5	103.2	52.7	78.1	100.6	78.3	89.5	96.1	54.3	75.4
04	MASONRY	88.7	73.5	79.2	87.4	59.7	70.2	90.6	38.7	58.4	75.3	53.7	61.9	90.7	68.6	77.0	90.9	39.1	58.8
05	METALS	97.8	93.0	96.3	93.8	86.5	91.5	93.8	87.3	91.7	90.9	72.6	85.1	91.8	80.4	88.1	90.6	67.8	83.3
06	WOOD & PLASTICS	99.0	75.0	86.5	97.8	63.8	80.1	96.3	41.2	67.6	92.5	35.5	62.8	97.0	78.4	87.3	93.9	51.7	71.9
07	THERMAL & MOISTURE PROTECTION	99.8	60.6	83.3	99.3	64.9	84.8	96.6	53.4	78.4	101.8	51.7	80.7	101.8	74.9	90.5	101.3	48.4	79.0
08	DOORS & WINDOWS	100.7	68.7	92.3	97.4	61.1	87.9	98.3	48.9	85.4	93.8	45.4	81.1	99.6	75.9	93.4	93.8	52.8	83.1
09200	PLASTER & GYPSUM BOARD	100.6	74.9	84.6	102.2	63.4	78.1	100.3	40.2	63.0	112.5	34.2	63.9	114.0	78.3	91.8	112.9	50.8	74.3
095,098	CEILINGS & ACOUSTICAL TREATMENT	104.3	74.9	86.8	97.3	63.4	77.1	103.4	40.2	65.7	115.5	34.2	67.0	115.5	78.3	93.3	116.4	50.8	77.3
09600	FLOORING	118.7	50.7	100.3	121.5	48.7	101.8	118.7	31.1	95.0	82.3	56.1	75.2	83.9	74.6	81.4	82.7	39.6	71.1
097,099	WALL FINISHES, PAINTS & COATINGS	110.6	44.9	70.5	106.7	44.5	68.7	106.7	39.9	65.9	90.7	33.3	55.6	90.7	81.5	85.1	90.7	37.9	58.5
09	FINISHES	108.7	66.1	86.5	105.9	59.9	82.0	106.5	40.2	71.9	98.2	41.5	68.6	98.5	76.7	87.2	98.1	46.6	71.3
10 - 14	TOTAL DIV. 10000 - 14000	100.0	84.0	96.6	100.0	87.8	97.4	100.0	76.8	95.0	100.0	50.0	89.3	100.0	83.3	96.4	100.0	63.0	92.1
15	MECHANICAL	99.9	71.3	88.0	97.2	71.7	86.7	99.9	47.9	78.3	95.1	76.4	87.4	100.1	78.1	91.0	100.1	63.8	85.0
16	ELECTRICAL	96.5	47.0	72.4	97.5	50.7	74.8	92.2	65.0	78.9	95.1	61.8	78.9	95.5	80.1	88.0	95.7	48.1	72.6
01 - 16	WEIGHTED AVERAGE	99.4	70.8	86.7	96.6	66.6	83.3	97.2	56.3	79.1	95.3	62.2	80.6	97.7	79.0	89.4	96.3	57.5	79.0

GEORGIA

DIVISION	COLUMBUS 318 - 319			DALTON 307			GAINESVILLE 305			MACON 310 - 312			SAVANNAH 313 - 314			STATESBORO 304		
	MAT.	INST.	TOTAL	MAT.	INST.	TOTAL	MAT.	INST.	TOTAL	MAT.	INST.	TOTAL	MAT.	INST.	TOTAL	MAT.	INST.	TOTAL
01590 EQUIPMENT RENTAL	.0	90.1	90.1	.0	106.1	106.1	.0	92.2	92.2	.0	102.6	102.6	.0	91.2	91.2	.0	91.2	91.2
02 SITE CONSTRUCTION	99.6	76.8	82.9	104.0	95.6	97.8	102.4	91.9	94.7	100.4	93.9	95.6	100.2	76.9	83.1	104.5	74.4	82.5
03100 CONCRETE FORMS & ACCESSORIES	97.4	57.1	62.6	84.7	44.9	50.4	97.7	43.8	51.2	96.8	54.1	59.9	97.7	51.2	57.6	78.2	40.8	45.9
03200 CONCRETE REINFORCEMENT	95.1	91.9	93.5	99.1	53.2	75.8	99.4	65.9	82.4	96.3	91.1	93.7	96.0	74.8	85.3	98.7	32.0	64.8
03300 CAST-IN-PLACE CONCRETE	96.2	53.7	78.7	96.2	34.9	71.0	104.1	48.2	81.1	94.9	44.8	74.3	93.2	49.9	75.4	98.8	42.5	75.7
03 CONCRETE	93.7	63.8	78.9	102.3	44.4	73.5	104.7	50.1	77.6	93.2	59.3	76.4	92.4	56.6	74.6	102.3	41.6	72.1
04 MASONRY	89.8	55.6	68.6	78.5	29.2	47.9	85.9	31.7	52.3	103.8	37.4	62.6	92.6	52.3	67.6	77.9	39.4	54.0
05 METALS	93.4	92.0	92.9	92.4	65.6	83.8	90.2	61.7	81.1	89.2	90.6	89.7	89.5	82.1	87.2	95.6	67.1	86.5
06 WOOD & PLASTICS	96.3	56.8	75.7	75.9	48.7	61.7	96.7	41.1	67.8	103.8	55.8	78.8	109.5	48.7	77.8	68.2	42.7	54.9
07 THERMAL & MOISTURE PROTECTION	96.3	61.0	81.5	103.1	40.4	76.7	101.7	44.4	77.6	95.1	56.1	78.7	96.7	51.5	77.7	101.9	37.6	74.9
08 DOORS & WINDOWS	98.3	59.3	88.1	94.9	36.3	79.6	93.8	39.1	79.5	96.7	57.8	86.5	99.4	49.2	86.3	95.7	33.1	79.4
09200 PLASTER & GYPSUM BOARD	100.3	56.2	72.9	98.6	47.8	67.0	114.0	40.0	68.0	105.8	55.1	74.3	100.6	47.8	67.8	98.7	41.7	63.3
095,098 CEILINGS & ACOUSTICAL TREATMENT	103.4	56.2	75.3	120.3	47.8	77.0	115.5	40.0	70.4	99.5	55.1	73.0	104.3	47.8	70.6	121.3	41.7	73.8
09600 FLOORING	118.7	52.4	100.8	83.4	13.4	64.5	83.8	13.4	64.8	92.3	36.5	77.2	118.7	46.7	99.2	96.4	46.7	83.0
097,099 WALL FINISHES, PAINTS & COATINGS	106.7	47.4	70.5	81.8	39.9	56.2	90.7	34.9	56.6	108.5	46.7	70.8	107.5	47.4	70.8	89.2	41.4	60.0
09 FINISHES	106.4	54.4	79.3	104.0	39.2	70.2	98.7	36.2	66.2	94.6	49.4	71.0	107.0	49.6	77.1	107.0	42.6	73.4
10 - 14 TOTAL DIV. 10000 - 14000	100.0	78.6	95.4	100.0	24.9	84.0	100.0	50.7	89.5	100.0	76.8	95.1	100.0	65.9	92.7	100.0	47.4	88.8
15 MECHANICAL	99.9	48.3	78.5	95.2	32.8	69.3	95.1	67.5	83.7	99.9	40.4	75.2	99.9	46.1	77.6	95.7	35.9	70.9
16 ELECTRICAL	92.6	49.7	71.7	105.5	73.3	89.9	95.1	61.7	78.9	90.3	66.4	78.7	94.1	61.9	78.4	96.2	30.2	64.1
01 - 16 WEIGHTED AVERAGE	97.2	60.4	80.8	97.4	49.4	76.1	96.0	55.7	78.1	95.7	59.1	79.4	97.0	57.7	79.6	97.1	43.6	73.4

GEORGIA / HAWAII / IDAHO

DIVISION	VALDOSTA 316			WAYCROSS 315			HILO 967			HONOLULU 968			STATES & POSS. , GUAM 969			BOISE 836 - 837		
	MAT.	INST.	TOTAL	MAT.	INST.	TOTAL	MAT.	INST.	TOTAL	MAT.	INST.	TOTAL	MAT.	INST.	TOTAL	MAT.	INST.	TOTAL
01590 EQUIPMENT RENTAL	.0	90.1	90.1	.0	90.1	90.1	.0	99.4	99.4	.0	99.4	99.4	.0	176.0	176.0	.0	101.7	101.7
02 SITE CONSTRUCTION	109.7	76.2	85.1	106.5	74.4	83.0	134.3	106.4	113.8	139.5	106.4	115.2	175.1	105.3	123.9	78.2	102.4	95.9
03100 CONCRETE FORMS & ACCESSORIES	84.3	44.3	49.8	86.7	41.5	47.7	107.0	142.9	138.0	107.5	142.9	138.1	109.1	48.7	57.0	97.6	81.3	83.5
03200 CONCRETE REINFORCEMENT	97.0	42.3	69.2	97.0	64.3	80.4	109.6	126.2	118.0	108.6	126.2	117.5	1386.8	26.7	695.8	108.2	76.7	92.2
03300 CAST-IN-PLACE CONCRETE	94.5	50.5	76.4	106.8	42.9	80.5	209.1	127.9	175.7	201.5	127.9	171.3	181.1	107.1	150.7	94.2	88.1	91.7
03 CONCRETE	97.6	47.7	72.9	101.6	48.0	75.0	158.5	133.3	146.0	154.7	133.3	144.1	350.5	65.5	209.0	104.3	82.7	93.6
04 MASONRY	95.9	47.6	65.9	96.8	39.4	61.2	134.1	128.9	130.9	133.8	128.9	130.8	185.1	33.0	90.8	133.5	72.3	95.6
05 METALS	93.0	70.5	85.8	92.4	77.2	87.5	118.7	110.9	116.2	144.4	110.9	133.7	155.2	77.5	130.4	99.6	76.1	92.1
06 WOOD & PLASTICS	81.8	40.2	60.1	84.9	42.6	62.9	99.1	146.9	124.0	98.9	146.9	123.9	107.8	48.5	76.9	95.4	81.3	88.0
07 THERMAL & MOISTURE PROTECTION	96.5	56.7	79.7	96.3	39.7	72.5	105.6	128.9	115.4	107.5	128.9	116.5	115.3	58.7	91.5	92.7	78.0	86.5
08 DOORS & WINDOWS	93.7	37.3	79.0	93.7	42.5	80.3	100.6	137.9	110.3	105.8	137.9	114.2	100.4	41.6	85.0	94.7	76.6	89.9
09200 PLASTER & GYPSUM BOARD	92.0	39.1	59.2	94.6	41.7	61.7	99.2	148.0	129.5	120.6	148.0	137.6	3174.6	34.0	1223.5	85.5	80.6	82.5
095,098 CEILINGS & ACOUSTICAL TREATMENT	98.3	39.1	63.0	96.4	41.7	63.8	117.4	148.0	135.6	120.4	148.0	136.9	258.8	34.0	124.7	110.7	80.6	92.7
09600 FLOORING	110.3	37.3	90.5	111.8	31.1	89.9	145.4	132.6	141.9	168.2	132.6	158.6	117.8	32.7	94.8	99.0	53.8	86.7
097,099 WALL FINISHES, PAINTS & COATINGS	106.7	34.6	62.7	106.7	41.4	66.8	103.3	148.6	131.0	104.0	148.6	131.2	109.9	39.8	67.1	102.2	50.3	70.5
09 FINISHES	102.1	41.3	70.4	102.2	39.3	69.5	123.1	143.0	133.5	134.5	143.0	138.9	646.4	44.8	333.1	99.0	73.2	85.6
10 - 14 TOTAL DIV. 10000 - 14000	100.0	64.7	92.5	100.0	47.3	88.8	100.0	123.5	105.0	100.0	123.5	105.0	100.0	97.1	99.4	100.0	80.8	95.9
15 MECHANICAL	99.9	40.6	75.3	96.7	42.0	74.0	100.2	116.5	107.0	100.2	116.5	107.0	102.8	34.9	74.6	100.1	75.5	89.9
16 ELECTRICAL	90.8	29.5	61.0	94.7	53.7	74.8	106.1	123.3	114.5	108.9	123.3	115.9	118.8	31.9	76.5	96.0	77.0	86.8
01 - 16 WEIGHTED AVERAGE	96.9	47.6	75.0	97.0	50.0	76.1	115.0	124.8	119.3	120.6	124.8	122.5	196.4	52.7	132.5	100.2	78.9	90.7

IDAHO / ILLINOIS

DIVISION	COEUR D'ALENE 838			IDAHO FALLS 834			LEWISTON 835			POCATELLO 832			TWIN FALLS 833			BLOOMINGTON 617		
	MAT.	INST.	TOTAL	MAT.	INST.	TOTAL	MAT.	INST.	TOTAL	MAT.	INST.	TOTAL	MAT.	INST.	TOTAL	MAT.	INST.	TOTAL
01590 EQUIPMENT RENTAL	.0	94.6	94.6	.0	101.7	101.7	.0	94.6	94.6	.0	101.7	101.7	.0	101.7	101.7	.0	101.6	101.6
02 SITE CONSTRUCTION	76.7	96.5	91.2	76.1	100.9	94.3	82.1	96.7	92.8	79.3	102.4	96.2	85.8	100.3	96.5	95.0	95.0	95.0
03100 CONCRETE FORMS & ACCESSORIES	105.3	66.8	72.1	90.3	35.6	43.1	111.5	68.0	74.0	97.7	80.8	83.1	98.5	35.2	43.9	83.7	104.3	101.5
03200 CONCRETE REINFORCEMENT	115.7	97.1	106.3	110.1	75.5	92.5	115.7	97.3	106.4	108.5	76.2	92.1	110.4	75.9	92.9	100.5	96.4	98.4
03300 CAST-IN-PLACE CONCRETE	98.3	83.5	92.2	86.5	54.7	73.4	102.2	92.7	98.3	93.5	87.9	91.2	95.9	44.2	74.7	98.9	100.5	99.5
03 CONCRETE	108.3	78.5	93.5	93.8	50.7	72.4	112.1	82.3	97.3	100.9	82.3	91.7	108.5	46.9	78.0	97.2	101.9	99.5
04 MASONRY	134.4	85.0	103.8	127.6	31.0	67.7	134.7	85.0	103.9	129.6	64.5	89.2	132.8	37.8	73.9	111.5	110.3	110.8
05 METALS	93.0	85.9	90.7	107.6	73.0	96.6	92.6	87.3	90.9	107.7	75.0	97.3	107.7	72.6	96.5	96.4	105.9	99.4
06 WOOD & PLASTICS	97.1	61.8	78.7	87.7	33.0	59.2	103.3	61.8	81.7	95.4	81.3	88.0	96.0	34.1	63.7	83.1	101.0	92.4
07 THERMAL & MOISTURE PROTECTION	135.7	77.7	111.3	92.4	47.5	73.5	135.9	79.7	112.2	92.9	68.8	82.7	93.7	45.4	73.3	94.0	103.2	97.8
08 DOORS & WINDOWS	114.6	64.3	101.4	97.9	44.0	83.8	114.5	67.6	102.3	94.7	70.1	88.2	97.9	41.9	83.3	89.8	97.2	91.7
09200 PLASTER & GYPSUM BOARD	140.0	60.7	90.7	80.3	31.0	49.7	141.8	60.7	91.4	85.5	80.6	82.5	85.0	32.1	52.1	98.3	100.8	99.9
095,098 CEILINGS & ACOUSTICAL TREATMENT	131.5	60.7	89.3	101.4	31.0	59.4	131.5	60.7	89.3	111.6	80.6	93.1	103.9	32.1	61.1	88.6	100.8	95.9
09600 FLOORING	133.2	57.3	112.7	95.6	53.8	84.3	136.4	91.3	124.2	99.3	53.8	87.0	100.6	53.8	87.9	89.0	111.1	95.0
097,099 WALL FINISHES, PAINTS & COATINGS	127.5	74.7	95.3	102.2	41.4	65.1	127.5	74.7	95.3	102.2	52.4	71.8	102.2	30.9	58.7	90.8	96.6	94.3
09 FINISHES	156.2	64.9	108.6	94.5	38.3	65.3	157.5	71.9	112.9	99.4	73.4	85.9	98.3	37.2	66.5	90.7	105.1	98.2
10 - 14 TOTAL DIV. 10000 - 14000	100.0	65.9	92.7	100.0	46.3	88.5	100.0	82.8	96.3	100.0	80.8	95.9	100.0	45.5	88.4	100.0	88.3	97.5
15 MECHANICAL	100.0	62.1	84.3	101.0	62.3	85.0	101.2	89.1	96.2	100.1	75.4	89.8	100.1	62.1	84.3	95.1	96.4	95.6
16 ELECTRICAL	83.0	83.6	83.3	82.0	76.1	79.1	81.3	83.6	82.4	91.1	76.1	83.8	83.3	69.9	76.8	96.0	87.3	91.7
01 - 16 WEIGHTED AVERAGE	106.6	75.9	93.0	98.3	58.8	80.8	107.4	83.6	96.8	100.3	77.3	90.1	100.8	57.7	81.7	95.8	99.2	97.3

COST INDEXES

ILLINOIS

DIVISION		CARBONDALE 629 MAT.	INST.	TOTAL	CENTRALIA 628 MAT.	INST.	TOTAL	CHAMPAIGN 618-619 MAT.	INST.	TOTAL	CHICAGO 606-608 MAT.	INST.	TOTAL	DECATUR 625 MAT.	INST.	TOTAL	EAST ST. LOUIS 620-622 MAT.	INST.	TOTAL
01590	EQUIPMENT RENTAL	.0	109.1	109.1	.0	109.1	109.1	.0	102.4	102.4	.0	91.5	91.5	.0	102.4	102.4	.0	109.1	109.1
02	SITE CONSTRUCTION	102.9	95.7	97.6	103.2	96.3	98.2	104.3	96.0	98.2	99.1	90.8	93.0	89.3	96.0	94.2	105.4	96.4	98.8
03100	CONCRETE FORMS & ACCESSORIES	94.0	99.8	99.0	96.4	112.7	110.4	91.1	102.6	101.1	99.7	137.6	132.4	97.3	102.2	101.5	91.0	110.2	107.6
03200	CONCRETE REINFORCEMENT	99.3	86.2	92.6	99.3	99.1	99.2	100.5	92.8	96.6	100.5	146.9	124.1	97.4	100.7	99.1	99.2	99.0	99.1
03300	CAST-IN-PLACE CONCRETE	92.5	85.3	89.5	92.9	104.4	97.6	114.7	105.5	110.9	113.0	137.9	123.2	101.0	102.6	101.7	94.5	105.3	99.0
03	CONCRETE	88.4	93.3	90.8	88.8	108.0	98.4	109.7	102.1	106.0	110.4	138.5	124.4	98.7	102.4	100.5	89.7	107.3	98.4
04	MASONRY	74.4	100.2	90.4	74.4	109.7	96.3	134.1	93.4	108.9	93.7	136.8	120.4	73.9	106.8	94.3	74.6	111.5	97.5
05	METALS	95.6	109.5	100.0	95.6	117.4	102.6	96.4	102.9	98.4	98.5	124.9	106.9	98.7	106.9	101.3	96.4	117.1	103.0
06	WOOD & PLASTICS	98.3	99.6	99.0	101.2	115.0	108.4	91.1	99.0	95.2	101.6	135.2	119.1	99.7	99.0	99.3	95.3	111.2	103.6
07	THERMAL & MOISTURE PROTECTION	89.4	90.1	89.7	89.5	106.1	96.5	94.6	103.2	98.2	101.2	129.7	113.2	94.7	99.9	96.9	89.5	101.9	94.7
08	DOORS & WINDOWS	86.0	102.1	90.2	86.0	113.8	93.3	90.4	98.2	92.5	103.2	139.4	112.6	96.3	100.1	97.3	86.1	111.7	92.8
09200	PLASTER & GYPSUM BOARD	102.7	99.5	100.7	103.8	115.2	110.9	101.3	98.8	99.7	101.3	136.0	122.9	105.6	98.8	101.4	102.0	111.3	107.8
095,098	CEILINGS & ACOUSTICAL TREATMENT	88.6	99.5	95.1	88.6	115.2	104.5	88.6	98.8	94.7	104.2	136.0	123.2	95.5	98.8	97.4	88.6	111.3	102.2
09600	FLOORING	106.5	97.2	103.9	107.8	118.0	110.5	92.5	106.5	96.3	83.5	127.9	95.6	96.7	105.6	99.1	105.1	118.0	108.6
097,099	WALL FINISHES, PAINTS & COATINGS	98.8	93.0	95.3	98.8	101.9	100.6	90.8	106.2	100.2	78.9	134.3	112.7	90.8	101.7	97.4	98.8	101.9	100.6
09	FINISHES	94.9	99.1	97.1	95.5	113.3	104.7	92.7	103.9	98.5	92.6	135.7	115.0	95.8	102.9	99.5	94.4	110.0	102.6
10-14	TOTAL DIV. 10000-14000	100.0	87.3	97.9	100.0	90.0	97.9	100.0	89.0	97.7	100.0	123.8	105.1	100.0	88.6	97.6	100.0	90.2	97.9
15	MECHANICAL	95.0	98.7	96.5	95.0	85.0	90.9	95.1	108.4	100.6	100.0	126.1	110.8	100.0	106.2	102.6	99.9	92.3	96.8
16	ELECTRICAL	95.0	90.6	92.8	96.4	89.5	93.0	99.2	95.4	97.3	98.8	117.1	107.7	98.7	89.1	94.0	95.9	89.5	92.8
01-16	WEIGHTED AVERAGE	92.9	97.3	94.9	93.3	101.1	96.8	99.2	100.8	99.9	100.1	126.3	111.7	97.0	101.1	98.8	94.5	102.0	97.8

ILLINOIS

DIVISION		EFFINGHAM 624 MAT.	INST.	TOTAL	GALESBURG 614 MAT.	INST.	TOTAL	JOLIET 604 MAT.	INST.	TOTAL	KANKAKEE 609 MAT.	INST.	TOTAL	LA SALLE 613 MAT.	INST.	TOTAL	NORTH SUBURBAN 600-603 MAT.	INST.	TOTAL
01590	EQUIPMENT RENTAL	.0	102.4	102.4	.0	101.6	101.6	.0	89.3	89.3	.0	89.3	89.3	.0	101.6	101.6	.0	89.3	89.3
02	SITE CONSTRUCTION	96.2	95.8	95.9	97.3	95.0	95.6	99.0	89.6	92.1	92.7	88.8	89.9	96.8	95.6	95.9	98.2	90.1	92.3
03100	CONCRETE FORMS & ACCESSORIES	102.3	100.5	100.8	91.0	104.2	102.4	100.6	138.1	133.0	92.5	108.4	106.2	107.4	111.0	110.5	99.7	136.0	131.0
03200	CONCRETE REINFORCEMENT	100.5	95.8	98.1	100.0	96.4	98.1	100.5	123.7	112.3	101.4	112.5	107.1	100.1	104.9	102.5	100.5	146.9	124.1
03300	CAST-IN-PLACE CONCRETE	100.7	77.2	91.0	101.8	102.2	102.0	112.9	128.4	119.3	105.2	115.0	109.2	101.7	109.6	104.9	113.0	139.9	124.0
03	CONCRETE	99.7	92.1	95.9	100.1	102.4	101.2	110.4	131.0	120.6	103.9	111.1	107.5	101.2	109.7	105.4	110.4	138.4	124.3
04	MASONRY	78.2	86.5	83.3	111.6	110.2	110.8	94.7	127.6	115.1	93.2	112.5	105.2	111.6	109.0	110.0	93.7	134.1	118.8
05	METALS	96.3	103.1	98.5	96.4	105.7	99.4	96.2	112.5	101.4	96.2	106.0	99.4	96.5	112.8	101.7	97.5	122.7	105.5
06	WOOD & PLASTICS	102.3	99.0	100.6	90.9	101.2	96.3	103.2	141.5	123.1	94.5	108.2	101.6	107.6	111.6	109.7	101.6	134.7	118.9
07	THERMAL & MOISTURE PROTECTION	94.3	92.6	93.6	94.1	101.2	97.1	100.6	123.4	110.2	99.8	114.0	105.8	94.4	99.7	96.6	101.2	129.4	113.1
08	DOORS & WINDOWS	90.4	99.1	92.6	89.8	95.1	91.1	101.1	136.6	110.4	93.0	111.8	97.9	89.8	105.2	93.8	101.1	139.2	111.1
09200	PLASTER & GYPSUM BOARD	104.9	98.8	101.1	101.3	101.1	101.2	98.7	142.5	125.9	95.0	108.3	103.3	107.2	111.7	110.0	101.3	135.6	122.6
095,098	CEILINGS & ACOUSTICAL TREATMENT	88.6	98.8	94.7	88.6	101.1	96.1	104.2	142.5	127.1	104.2	108.3	106.7	88.6	111.7	102.4	104.2	135.6	122.9
09600	FLOORING	98.1	106.5	100.3	92.4	107.5	96.5	83.1	127.9	95.2	80.0	106.8	87.2	100.5	108.1	102.5	83.5	111.2	91.0
097,099	WALL FINISHES, PAINTS & COATINGS	90.8	102.0	97.6	90.8	97.9	95.1	76.7	132.0	110.5	76.7	79.9	78.7	90.8	96.6	94.3	78.9	132.0	111.3
09	FINISHES	94.4	102.4	98.6	92.3	104.6	98.7	91.9	136.9	115.3	90.1	103.8	97.3	95.5	109.4	102.8	92.5	131.0	112.6
10-14	TOTAL DIV. 10000-14000	100.0	87.5	97.3	100.0	86.4	97.1	100.0	120.9	104.5	100.0	99.5	99.9	100.0	86.0	97.0	100.0	117.9	103.8
15	MECHANICAL	95.1	106.4	99.8	95.1	98.5	96.5	100.1	113.4	105.6	95.2	101.3	97.7	95.1	92.1	93.9	100.0	121.1	108.7
16	ELECTRICAL	96.6	90.6	93.7	96.9	83.9	90.6	97.1	103.8	100.4	92.4	100.2	96.2	94.0	100.2	97.1	97.0	126.9	111.6
01-16	WEIGHTED AVERAGE	95.1	97.1	96.0	96.5	98.9	97.6	99.4	118.5	107.9	95.7	104.1	99.4	96.7	102.6	99.3	99.5	125.3	111.0

ILLINOIS

DIVISION		PEORIA 615-616 MAT.	INST.	TOTAL	QUINCY 623 MAT.	INST.	TOTAL	ROCKFORD 610-611 MAT.	INST.	TOTAL	ROCK ISLAND 612 MAT.	INST.	TOTAL	SOUTH SUBURBAN 605 MAT.	INST.	TOTAL	SPRINGFIELD 626-627 MAT.	INST.	TOTAL
01590	EQUIPMENT RENTAL	.0	101.6	101.6	.0	102.4	102.4	.0	101.6	101.6	.0	101.6	101.6	.0	89.3	89.3	.0	102.4	102.4
02	SITE CONSTRUCTION	97.9	95.1	95.8	94.9	95.5	95.4	97.1	95.7	96.1	95.6	94.6	94.8	98.2	90.1	92.3	95.3	96.0	95.8
03100	CONCRETE FORMS & ACCESSORIES	95.6	105.0	103.7	99.6	98.8	98.9	99.9	117.1	114.8	93.1	104.6	103.0	99.7	136.0	131.0	99.2	102.5	102.0
03200	CONCRETE REINFORCEMENT	97.4	96.4	96.9	100.1	96.6	98.3	97.4	125.2	111.5	100.0	96.1	98.0	100.5	146.9	124.1	97.4	96.8	97.1
03300	CAST-IN-PLACE CONCRETE	98.7	103.8	100.8	100.8	99.1	100.1	101.0	96.1	99.0	99.6	99.7	99.7	113.0	139.9	124.0	93.8	103.0	97.6
03	CONCRETE	97.5	103.3	100.4	99.2	98.9	99.1	98.9	111.6	105.2	98.3	101.7	100.0	110.4	138.4	124.3	95.4	102.0	98.6
04	MASONRY	115.2	111.9	113.1	97.3	96.6	96.9	89.2	107.7	100.7	111.5	101.2	105.1	93.7	134.1	118.8	75.5	107.7	95.4
05	METALS	98.7	106.0	101.0	96.4	104.5	99.0	98.7	121.1	105.8	96.4	104.8	99.1	97.5	122.7	105.5	101.3	105.2	102.5
06	WOOD & PLASTICS	99.7	101.1	100.4	99.5	99.0	99.2	99.7	117.4	108.9	93.1	105.0	99.3	101.6	134.7	118.9	102.3	99.0	100.6
07	THERMAL & MOISTURE PROTECTION	94.9	103.9	98.7	94.3	94.4	94.4	97.9	112.6	104.1	94.1	97.3	95.5	101.2	129.4	113.1	94.4	100.9	97.1
08	DOORS & WINDOWS	96.3	99.1	97.0	91.2	99.1	93.3	96.3	115.7	101.4	89.8	97.6	91.8	101.1	139.2	111.1	96.3	99.1	97.0
09200	PLASTER & GYPSUM BOARD	105.6	100.9	102.7	103.8	98.8	100.7	105.6	117.7	113.1	102.0	104.9	103.8	101.3	135.6	122.6	105.6	98.8	101.4
095,098	CEILINGS & ACOUSTICAL TREATMENT	95.5	100.9	98.7	88.6	98.8	94.7	95.5	117.7	108.7	88.6	104.9	98.4	104.2	135.6	122.9	95.5	98.8	97.4
09600	FLOORING	96.7	107.5	99.6	96.7	101.1	97.9	96.7	89.3	94.7	93.8	91.6	93.2	83.5	111.2	91.0	96.9	105.6	99.3
097,099	WALL FINISHES, PAINTS & COATINGS	90.8	102.1	97.4	92.1	92.1	92.1	90.8	104.9	99.4	90.8	97.9	95.1	78.9	132.0	111.3	90.8	101.7	97.4
09	FINISHES	95.9	104.3	100.3	93.8	100.1	97.1	95.9	110.9	103.7	92.7	102.2	97.6	92.5	131.0	112.6	95.9	102.3	99.7
10-14	TOTAL DIV. 10000-14000	100.0	88.8	97.6	100.0	89.0	97.7	100.0	98.0	99.6	100.0	85.5	96.9	100.0	117.9	103.8	100.0	88.9	97.6
15	MECHANICAL	100.0	98.4	99.3	95.1	93.5	94.4	100.1	101.9	100.8	95.1	95.6	95.3	100.0	121.1	108.7	100.0	101.9	100.8
16	ELECTRICAL	98.0	86.6	92.4	94.1	77.8	86.2	98.1	107.4	102.6	89.3	85.0	87.2	97.0	126.9	111.6	98.6	86.9	92.9
01-16	WEIGHTED AVERAGE	99.1	99.9	99.4	95.7	94.6	95.2	98.1	107.9	102.4	95.4	97.0	96.1	99.5	125.3	111.0	97.3	99.8	98.4

COST INDEXES

715

INDIANA

DIVISION		ANDERSON 460			BLOOMINGTON 474			COLUMBUS 472			EVANSVILLE 476 - 477			FORT WAYNE 467 - 468			GARY 463 - 464		
		MAT.	INST.	TOTAL	MAT.	INST.	TOTAL	MAT.	INST.	TOTAL	MAT.	INST.	TOTAL	MAT.	INST.	TOTAL	MAT.	INST.	TOTAL
01590	EQUIPMENT RENTAL	.0	96.7	96.7	.0	85.9	85.9	.0	85.9	85.9	.0	121.3	121.3	.0	96.7	96.7	.0	96.7	96.7
02	SITE CONSTRUCTION	88.2	96.4	94.2	77.0	95.8	90.8	74.1	95.5	89.8	82.0	130.1	117.2	89.2	96.5	94.6	89.2	96.5	96.4
03100	CONCRETE FORMS & ACCESSORIES	96.4	79.9	82.2	100.5	80.7	83.4	94.1	77.9	80.2	93.1	82.2	83.7	95.0	77.8	80.2	96.5	106.3	105.0
03200	CONCRETE REINFORCEMENT	94.9	80.7	87.7	84.1	80.4	82.2	84.4	80.4	82.4	92.2	80.4	86.2	94.9	77.5	86.0	94.9	100.3	97.6
03300	CAST-IN-PLACE CONCRETE	101.8	82.0	93.6	100.8	79.5	92.0	100.3	75.6	90.2	96.3	92.0	94.5	108.1	80.4	96.8	106.3	111.1	108.3
03	CONCRETE	96.2	81.3	88.8	102.7	79.9	91.4	102.0	77.3	89.7	102.9	85.4	94.2	99.1	79.2	89.3	98.4	106.9	102.6
04	MASONRY	92.3	84.9	87.7	90.3	80.3	84.1	90.2	80.2	84.0	86.6	86.6	86.6	92.4	84.7	87.6	93.7	107.8	102.5
05	METALS	91.6	88.7	90.7	94.4	75.7	88.4	94.4	75.1	88.2	87.1	86.3	86.8	91.6	87.2	90.2	91.6	103.7	95.5
06	WOOD & PLASTICS	113.1	79.2	95.4	118.2	79.7	98.2	112.1	76.1	93.4	97.3	79.8	88.2	112.9	76.1	93.7	111.1	106.1	108.5
07	THERMAL & MOISTURE PROTECTION	100.3	82.3	92.7	91.4	84.3	88.4	91.0	84.1	88.1	95.5	88.1	92.4	100.1	86.1	94.2	99.6	105.8	102.2
08	DOORS & WINDOWS	98.9	80.9	94.2	103.1	81.2	97.4	98.6	79.2	93.5	95.6	80.4	91.7	98.9	76.0	92.9	98.9	104.4	100.4
09200	PLASTER & GYPSUM BOARD	96.6	79.2	85.8	97.0	79.9	86.4	93.3	76.2	82.7	91.6	78.5	83.4	95.8	76.1	83.6	91.7	106.9	101.1
095,098	CEILINGS & ACOUSTICAL TREATMENT	91.5	79.2	84.2	81.6	79.9	80.6	81.6	76.2	78.4	86.2	78.5	81.6	91.5	76.1	82.3	91.5	106.9	100.7
09600	FLOORING	88.2	92.7	89.4	104.5	92.7	101.3	98.6	92.7	97.0	97.5	88.7	95.1	88.2	80.8	86.2	88.2	119.9	96.8
097,099	WALL FINISHES, PAINTS & COATINGS	91.5	77.0	82.6	90.2	89.0	89.5	90.2	89.0	89.5	96.8	88.0	91.4	91.5	80.1	84.5	91.5	107.9	101.5
09	FINISHES	89.7	81.9	85.7	94.7	83.7	89.0	92.3	81.6	86.7	92.8	83.7	88.1	89.5	78.2	83.6	88.9	109.2	99.5
10 - 14	TOTAL DIV. 10000 - 14000	100.0	90.7	98.0	100.0	90.2	97.9	100.0	89.8	97.8	100.0	95.3	99.0	100.0	89.3	97.7	100.0	94.3	98.8
15	MECHANICAL	99.9	78.8	91.2	99.7	84.9	93.6	94.8	86.9	91.5	99.9	86.8	94.5	99.9	81.2	92.2	99.9	101.3	100.5
16	ELECTRICAL	85.1	90.6	87.8	101.6	88.0	95.0	100.9	90.6	95.8	95.0	87.6	91.4	86.2	82.9	84.6	94.4	104.1	99.1
01 - 16	WEIGHTED AVERAGE	94.9	84.8	90.4	98.2	84.1	91.9	96.1	84.0	90.7	95.3	90.0	93.0	95.4	83.1	89.9	96.2	104.3	99.8

INDIANA

DIVISION		INDIANAPOLIS 461 - 462			KOKOMO 469			LAFAYETTE 479			LAWRENCEBURG 470			MUNCIE 473			NEW ALBANY 471		
		MAT.	INST.	TOTAL	MAT.	INST.	TOTAL	MAT.	INST.	TOTAL	MAT.	INST.	TOTAL	MAT.	INST.	TOTAL	MAT.	INST.	TOTAL
01590	EQUIPMENT RENTAL	.0	91.5	91.5	.0	96.7	96.7	.0	85.9	85.9	.0	108.1	108.1	.0	96.2	96.2	.0	96.0	96.0
02	SITE CONSTRUCTION	87.9	100.1	96.9	84.9	96.5	93.4	74.8	95.5	90.0	72.9	115.4	104.0	77.2	95.8	90.8	69.6	99.7	91.6
03100	CONCRETE FORMS & ACCESSORIES	97.0	84.4	86.1	99.7	78.7	81.6	91.4	81.9	83.2	89.9	76.5	78.3	90.9	79.7	81.2	88.6	74.3	76.3
03200	CONCRETE REINFORCEMENT	94.2	82.2	88.1	85.8	80.6	83.2	84.1	78.7	81.4	83.3	81.4	82.3	93.0	80.7	86.8	84.7	82.1	83.4
03300	CAST-IN-PLACE CONCRETE	97.9	85.0	92.6	100.7	83.0	93.5	100.9	82.5	93.3	94.3	77.2	87.3	106.0	81.4	95.9	97.4	77.1	89.0
03	CONCRETE	98.1	83.8	91.0	93.1	81.1	87.2	102.1	81.1	91.7	95.1	78.2	86.7	101.4	81.0	91.3	100.6	77.1	86.9
04	MASONRY	99.6	84.4	90.1	92.0	83.0	86.5	96.1	82.2	87.4	75.7	79.0	77.8	92.6	84.9	87.8	82.7	73.4	76.9
05	METALS	93.5	78.2	88.6	88.8	88.9	88.8	93.3	74.5	87.3	89.4	88.1	89.0	96.1	89.0	93.8	91.6	83.3	89.0
06	WOOD & PLASTICS	110.1	83.8	96.4	116.5	77.1	96.0	108.7	81.6	94.6	96.1	75.1	85.1	110.3	79.0	94.0	98.2	73.9	85.5
07	THERMAL & MOISTURE PROTECTION	98.5	86.4	93.4	100.0	81.3	92.1	91.0	83.0	87.6	90.8	81.5	86.9	93.7	82.3	88.9	85.9	70.1	79.3
08	DOORS & WINDOWS	106.7	83.8	100.8	93.2	79.8	89.7	97.2	81.7	93.1	97.5	78.0	92.4	98.0	80.8	93.5	95.3	77.9	90.7
09200	PLASTER & GYPSUM BOARD	94.8	83.7	87.9	100.8	77.1	86.1	88.9	81.8	84.5	80.2	75.0	77.0	91.3	79.2	83.8	90.5	73.6	80.0
095,098	CEILINGS & ACOUSTICAL TREATMENT	95.8	83.7	88.6	90.7	77.1	82.6	73.1	81.8	78.3	91.3	75.0	81.6	82.5	79.2	80.5	86.2	73.6	78.7
09600	FLOORING	88.0	92.7	89.3	92.8	88.1	91.5	97.3	94.4	96.5	72.6	92.7	78.0	96.6	92.7	95.5	95.6	68.8	87.3
097,099	WALL FINISHES, PAINTS & COATINGS	91.5	89.0	90.0	91.5	87.7	89.2	90.2	84.1	86.5	92.0	79.1	84.1	90.2	77.0	82.2	96.8	72.1	81.7
09	FINISHES	90.8	86.4	88.5	91.4	81.1	86.0	89.2	84.5	86.8	85.0	79.9	82.4	91.4	81.8	86.4	92.3	73.0	82.3
10 - 14	TOTAL DIV. 10000 - 14000	100.0	92.4	98.4	100.0	90.3	97.9	100.0	90.8	98.0	100.0	55.3	90.5	100.0	90.3	97.9	100.0	55.3	90.5
15	MECHANICAL	99.9	87.7	94.9	95.0	87.0	91.7	94.8	76.5	87.2	95.5	80.9	89.5	99.7	78.7	91.0	95.0	78.9	88.3
16	ELECTRICAL	104.4	90.6	97.7	89.4	88.2	88.8	100.4	85.7	93.3	92.4	71.8	82.4	89.7	74.6	82.4	93.7	71.6	83.0
01 - 16	WEIGHTED AVERAGE	98.8	87.2	93.7	93.0	85.7	89.8	95.7	82.4	89.8	92.0	81.8	87.5	96.2	82.4	90.1	93.6	77.5	86.5

DIVISION		INDIANA SOUTH BEND 465 - 466			TERRE HAUTE 478			WASHINGTON 475			IOWA BURLINGTON 526			CARROLL 514			CEDAR RAPIDS 522 - 524		
		MAT.	INST.	TOTAL	MAT.	INST.	TOTAL	MAT.	INST.	TOTAL	MAT.	INST.	TOTAL	MAT.	INST.	TOTAL	MAT.	INST.	TOTAL
01590	EQUIPMENT RENTAL	.0	106.8	106.8	.0	121.3	121.3	.0	121.3	121.3	.0	99.6	99.6	.0	99.6	99.6	.0	95.6	95.6
02	SITE CONSTRUCTION	88.2	96.2	94.1	83.3	130.0	117.6	83.3	128.5	116.5	87.6	95.5	93.4	78.2	93.3	89.3	89.7	94.6	93.3
03100	CONCRETE FORMS & ACCESSORIES	100.9	81.5	84.2	94.6	82.2	83.9	95.3	81.3	83.2	95.7	72.0	75.3	81.2	49.9	54.2	102.1	80.6	83.6
03200	CONCRETE REINFORCEMENT	94.9	76.7	85.6	92.2	80.9	86.4	85.1	63.7	74.2	92.6	81.3	86.9	93.3	51.9	72.3	93.3	84.5	88.8
03300	CAST-IN-PLACE CONCRETE	99.0	85.1	93.3	93.2	91.2	92.4	101.5	97.1	99.7	107.8	53.5	85.5	107.8	58.5	87.5	108.0	78.2	95.8
03	CONCRETE	91.1	83.2	87.2	105.7	85.2	95.5	111.3	83.4	97.4	100.1	68.3	84.3	98.7	54.4	76.7	100.3	81.1	90.8
04	MASONRY	90.6	84.6	86.9	93.4	85.7	88.6	86.7	88.4	87.7	100.0	62.0	76.4	102.7	46.5	67.9	106.7	79.6	89.9
05	METALS	91.6	101.2	94.7	87.8	87.0	87.6	83.2	76.0	80.9	85.7	92.2	87.8	85.8	72.4	81.5	87.8	92.8	89.4
06	WOOD & PLASTICS	113.0	80.3	96.0	99.6	81.2	90.0	100.0	79.6	89.4	102.1	77.5	89.3	86.1	51.5	68.1	109.0	80.0	93.8
07	THERMAL & MOISTURE PROTECTION	99.4	87.4	94.3	95.6	85.1	91.1	95.7	86.5	91.8	97.9	65.2	84.2	98.2	53.7	79.5	98.9	81.8	91.7
08	DOORS & WINDOWS	92.3	79.8	89.0	96.2	82.0	92.5	93.1	75.7	88.6	93.1	72.0	87.6	98.3	47.7	85.1	99.3	84.7	95.5
09200	PLASTER & GYPSUM BOARD	96.2	80.3	86.3	91.6	79.9	84.3	89.8	78.3	82.7	100.2	77.0	85.8	93.9	50.2	66.8	103.7	79.8	88.9
095,098	CEILINGS & ACOUSTICAL TREATMENT	91.5	80.3	84.9	86.2	79.9	82.5	75.1	78.3	77.0	109.6	77.0	90.1	109.6	50.2	74.2	114.7	79.8	93.9
09600	FLOORING	88.2	85.3	87.4	97.5	97.2	97.4	98.7	83.5	94.6	105.9	47.2	90.0	98.5	41.5	83.1	122.3	55.6	104.2
097,099	WALL FINISHES, PAINTS & COATINGS	91.5	83.8	86.8	96.8	93.3	94.7	96.8	91.9	93.8	101.4	84.7	91.2	101.4	39.7	63.7	105.5	76.5	87.8
09	FINISHES	89.7	82.4	85.9	92.8	86.1	89.4	90.4	83.4	86.8	105.1	69.0	86.3	100.9	46.9	72.8	111.9	74.6	92.5
10 - 14	TOTAL DIV. 10000 - 14000	100.0	90.2	97.9	100.0	94.4	98.8	100.0	95.3	99.0	100.0	85.1	96.8	100.0	90.4	98.0	100.0	94.3	98.4
15	MECHANICAL	99.9	82.8	92.8	99.9	76.9	90.4	95.0	86.2	91.4	95.4	74.2	86.6	95.4	51.7	77.3	100.3	85.9	94.3
16	ELECTRICAL	99.2	88.6	94.1	93.5	88.0	90.9	94.0	87.5	90.9	98.3	78.0	88.4	99.0	41.0	70.8	96.1	85.5	90.9
01 - 16	WEIGHTED AVERAGE	95.3	86.8	91.5	96.0	88.3	92.6	94.0	88.4	91.5	96.0	75.5	86.9	95.8	55.7	78.0	98.9	84.4	92.5

IOWA

DIVISION		COUNCIL BLUFFS 515			CRESTON 508			DAVENPORT 527 - 528			DECORAH 521			DES MOINES 500 - 503,509			DUBUQUE 520		
		MAT.	INST.	TOTAL	MAT.	INST.	TOTAL	MAT.	INST.	TOTAL	MAT.	INST.	TOTAL	MAT.	INST.	TOTAL	MAT.	INST.	TOTAL
01590	EQUIPMENT RENTAL	.0	95.3	95.3	.0	99.6	99.6	.0	99.6	99.6	.0	99.6	99.6	.0	101.6	101.6	.0	94.2	94.2
02	SITE CONSTRUCTION	95.4	91.4	92.4	78.2	94.4	90.0	87.8	99.1	96.1	86.3	93.9	91.8	80.9	94.9	94.9	87.7	91.6	90.5
03100	CONCRETE FORMS & ACCESSORIES	80.2	56.9	60.1	81.8	73.4	74.6	101.7	93.1	94.3	92.6	46.8	53.1	103.7	80.4	83.6	82.0	69.2	70.9
03200	CONCRETE REINFORCEMENT	95.2	78.9	86.9	92.8	78.4	85.4	93.3	95.5	94.4	92.6	52.0	72.0	93.3	79.6	86.3	92.0	84.2	88.1
03300	CAST-IN-PLACE CONCRETE	112.4	72.1	95.8	107.4	61.0	88.3	104.0	99.0	102.0	104.8	55.1	84.4	108.3	76.5	95.2	105.8	91.0	99.7
03	CONCRETE	102.1	67.5	84.9	98.5	70.8	84.7	98.3	96.0	97.2	97.9	51.9	75.1	99.4	79.6	89.6	96.6	80.4	88.6
04	MASONRY	106.4	76.0	87.6	102.7	50.5	70.3	102.7	91.8	96.0	121.1	43.3	72.8	100.5	77.9	86.5	107.6	73.9	86.7
05	METALS	93.0	89.3	91.8	85.7	85.0	85.4	87.8	101.9	92.3	85.8	71.9	81.4	87.9	91.9	89.2	86.4	91.9	88.2
06	WOOD & PLASTICS	84.8	52.8	68.1	86.7	80.3	83.4	109.0	91.5	99.9	98.8	47.8	72.2	110.5	80.3	94.8	86.8	66.9	76.5
07	THERMAL & MOISTURE PROTECTION	98.2	66.9	85.0	99.0	58.4	81.9	98.4	92.1	95.7	98.1	48.1	77.1	99.1	78.4	90.4	98.5	70.3	86.6
08	DOORS & WINDOWS	98.3	60.4	88.4	111.2	75.5	101.9	99.3	92.4	97.5	96.6	49.8	84.4	99.3	83.5	95.1	98.3	77.6	92.9
09200	PLASTER & GYPSUM BOARD	93.9	51.9	67.8	93.9	79.9	85.2	103.7	91.3	96.0	99.1	46.4	66.4	100.9	79.9	87.8	94.3	66.4	76.9
095,098	CEILINGS & ACOUSTICAL TREATMENT	109.6	51.9	75.1	109.6	79.9	91.9	114.7	91.3	100.8	109.6	46.4	71.9	113.0	79.9	93.2	109.6	66.4	83.8
09600	FLOORING	97.3	45.5	83.3	98.8	41.5	83.3	108.8	84.1	102.1	105.2	58.9	92.6	108.6	41.5	90.4	111.1	38.7	91.5
097,099	WALL FINISHES, PAINTS & COATINGS	97.2	69.5	80.3	101.4	49.0	69.4	101.4	93.3	96.5	101.4	42.0	65.1	101.4	78.7	87.6	104.4	76.5	87.4
09	FINISHES	101.2	54.5	76.9	100.9	65.9	82.7	107.7	91.1	99.1	104.6	48.7	75.5	105.9	72.4	88.4	105.7	63.0	83.5
10 - 14	TOTAL DIV. 10000 - 14000	100.0	69.3	93.4	100.0	86.0	97.0	100.0	95.1	98.9	100.0	51.1	89.6	100.0	90.9	98.1	100.0	87.6	97.4
15	MECHANICAL	100.3	79.0	91.5	95.4	53.8	78.1	100.3	95.3	98.2	95.4	44.8	74.4	100.3	80.2	92.0	100.3	78.9	91.4
16	ELECTRICAL	101.0	85.5	93.4	94.2	53.5	74.4	94.7	93.0	93.9	96.1	40.8	69.2	96.6	81.2	89.1	99.6	68.6	84.6
01 - 16	WEIGHTED AVERAGE	99.3	75.2	88.6	96.4	66.0	82.9	97.9	94.9	96.5	96.8	52.8	77.3	97.8	82.2	90.9	97.8	77.2	88.7

IOWA

DIVISION		FORT DODGE 505			MASON CITY 504			OTTUMWA 525			SHENANDOAH 516			SIBLEY 512			SIOUX CITY 510 - 511		
		MAT.	INST.	TOTAL	MAT.	INST.	TOTAL	MAT.	INST.	TOTAL	MAT.	INST.	TOTAL	MAT.	INST.	TOTAL	MAT.	INST.	TOTAL
01590	EQUIPMENT RENTAL	.0	99.6	99.6	.0	99.6	99.6	.0	94.2	94.2	.0	95.3	95.3	.0	99.6	99.6	.0	99.6	99.6
02	SITE CONSTRUCTION	83.7	93.0	90.5	83.7	93.8	91.1	88.0	90.0	89.4	93.3	89.0	90.2	95.8	93.3	94.0	97.3	95.5	95.9
03100	CONCRETE FORMS & ACCESSORIES	82.6	52.5	56.6	88.1	47.6	53.2	90.4	60.2	64.3	82.3	60.9	63.9	82.9	42.7	48.2	102.1	69.0	73.5
03200	CONCRETE REINFORCEMENT	92.8	51.5	71.8	92.6	66.7	79.5	92.6	75.6	84.0	95.2	77.1	86.0	95.2	66.1	80.4	93.3	66.8	79.9
03300	CAST-IN-PLACE CONCRETE	100.8	46.4	78.5	100.8	55.4	82.2	108.5	61.8	89.3	108.6	50.5	84.7	106.4	50.5	83.4	107.1	56.5	86.3
03	CONCRETE	94.0	51.4	72.8	94.3	55.2	74.9	99.5	64.8	82.3	99.7	61.4	80.7	98.7	51.2	75.1	99.6	65.2	82.6
04	MASONRY	100.1	27.3	55.2	114.4	49.0	73.8	104.2	51.9	71.8	106.1	35.2	62.2	127.1	39.6	72.9	99.7	59.4	74.7
05	METALS	85.8	71.0	81.0	85.8	79.5	83.8	85.7	86.8	86.0	92.3	82.0	89.0	85.9	77.9	83.3	87.8	82.9	86.2
06	WOOD & PLASTICS	87.4	58.5	72.3	93.3	47.8	69.6	96.0	61.9	78.2	86.8	68.6	77.3	87.8	42.5	64.2	109.0	69.3	88.3
07	THERMAL & MOISTURE PROTECTION	98.3	35.9	72.0	97.8	49.4	77.5	98.7	59.4	82.1	97.5	45.5	75.6	97.9	50.9	78.1	98.4	57.6	81.2
08	DOORS & WINDOWS	103.7	49.6	89.5	94.0	53.8	83.5	98.3	65.9	89.8	87.9	65.2	81.9	94.4	49.3	82.6	99.3	66.4	90.7
09200	PLASTER & GYPSUM BOARD	94.3	57.5	71.4	96.1	46.4	65.3	97.6	61.2	75.0	94.3	68.1	78.0	94.3	41.0	61.2	103.7	68.5	81.9
095,098	CEILINGS & ACOUSTICAL TREATMENT	109.6	57.5	78.5	109.6	46.4	71.9	109.6	61.2	80.7	109.6	68.1	84.8	109.6	41.0	68.7	114.7	68.5	87.1
09600	FLOORING	100.4	58.9	89.2	103.3	58.9	91.3	115.1	62.7	100.9	98.2	42.8	83.2	99.6	42.4	84.1	109.3	58.6	95.6
097,099	WALL FINISHES, PAINTS & COATINGS	101.4	18.0	50.5	101.4	38.4	62.9	104.4	82.4	90.9	97.2	27.1	54.5	101.4	39.7	63.7	102.6	66.0	80.3
09	FINISHES	102.4	50.8	75.5	103.5	48.3	74.7	107.5	62.4	84.0	101.5	55.0	77.3	103.3	41.9	71.3	109.0	66.8	87.0
10 - 14	TOTAL DIV. 10000 - 14000	100.0	66.4	92.8	100.0	52.0	89.8	100.0	82.4	96.2	100.0	81.2	96.0	100.0	63.0	92.1	100.0	87.6	97.4
15	MECHANICAL	95.4	67.9	84.0	95.4	69.6	84.7	95.4	72.3	85.8	95.4	34.1	70.0	95.4	40.7	72.7	100.3	82.4	92.9
16	ELECTRICAL	100.6	54.4	78.1	99.6	56.3	78.6	98.2	78.0	88.4	96.1	32.5	65.2	96.1	41.0	69.3	96.1	78.0	87.3
01 - 16	WEIGHTED AVERAGE	96.2	58.1	79.3	96.0	62.0	80.9	96.8	71.1	85.4	96.2	52.9	77.0	97.0	51.5	76.8	98.4	74.9	88.0

IOWA / KANSAS

DIVISION		IOWA SPENCER 513			IOWA WATERLOO 506 - 507			KANSAS BELLEVILLE 669			KANSAS COLBY 677			KANSAS DODGE CITY 678			KANSAS EMPORIA 668		
		MAT.	INST.	TOTAL	MAT.	INST.	TOTAL	MAT.	INST.	TOTAL	MAT.	INST.	TOTAL	MAT.	INST.	TOTAL	MAT.	INST.	TOTAL
01590	EQUIPMENT RENTAL	.0	99.6	99.6	.0	99.6	99.6	.0	103.6	103.6	.0	103.6	103.6	.0	103.6	103.6	.0	101.6	101.6
02	SITE CONSTRUCTION	95.9	93.3	94.0	88.3	94.5	92.9	112.1	92.5	97.7	109.5	92.4	97.0	112.0	92.9	98.0	103.1	89.9	93.4
03100	CONCRETE FORMS & ACCESSORIES	90.2	42.5	49.1	102.6	48.1	55.6	93.9	47.7	54.1	101.6	42.0	50.2	93.4	67.9	71.4	84.1	38.6	44.8
03200	CONCRETE REINFORCEMENT	95.2	41.8	68.1	93.3	84.0	88.6	105.4	56.9	80.8	108.0	56.3	81.7	105.4	57.3	81.0	103.9	49.5	76.3
03300	CAST-IN-PLACE CONCRETE	106.4	50.5	83.4	108.0	47.1	83.0	119.8	55.9	93.5	115.2	55.5	90.7	117.3	56.4	92.2	115.8	42.1	85.5
03	CONCRETE	99.2	46.5	73.1	100.3	55.9	78.3	117.5	53.7	85.8	114.8	50.9	83.1	115.8	63.0	89.6	109.9	43.5	76.9
04	MASONRY	127.1	39.6	72.9	101.3	60.8	76.2	95.2	50.6	67.6	96.7	40.9	62.1	104.1	57.6	75.3	101.0	40.4	63.5
05	METALS	85.9	66.9	79.8	87.8	90.5	88.7	94.7	77.7	89.3	95.1	76.0	89.0	96.1	79.8	90.9	83.3	38.5	60.0
06	WOOD & PLASTICS	95.9	42.5	68.1	109.7	47.8	77.5	93.0	48.6	69.9	100.9	42.1	70.3	92.3	75.4	83.5	83.3	38.5	60.0
07	THERMAL & MOISTURE PROTECTION	98.8	44.7	76.1	98.2	52.9	79.1	95.5	57.7	79.6	95.5	47.8	75.4	95.4	62.2	81.4	94.1	40.7	71.6
08	DOORS & WINDOWS	107.8	42.2	90.7	95.0	60.8	86.0	96.1	49.0	83.8	96.2	45.4	82.9	96.1	63.5	87.6	93.7	41.1	80.0
09200	PLASTER & GYPSUM BOARD	97.6	41.0	62.4	103.7	46.4	68.1	103.0	47.1	68.3	106.7	40.3	65.5	102.3	74.6	85.1	99.0	36.7	60.3
095,098	CEILINGS & ACOUSTICAL TREATMENT	109.6	41.0	68.7	114.7	46.4	74.0	86.9	47.1	63.2	86.9	40.3	59.1	86.9	74.6	79.6	86.9	36.7	57.0
09600	FLOORING	103.0	42.4	86.6	110.5	58.9	96.5	96.2	48.9	83.4	100.4	48.9	86.4	96.0	48.9	83.3	90.3	45.5	78.1
097,099	WALL FINISHES, PAINTS & COATINGS	101.4	39.7	63.7	102.6	35.5	61.7	90.8	48.2	64.8	90.8	48.2	64.8	90.8	48.2	64.8	90.8	48.2	64.8
09	FINISHES	104.9	41.9	72.0	108.3	47.9	76.8	94.2	47.4	69.8	95.8	43.6	68.6	94.0	63.1	77.9	91.0	40.2	64.6
10 - 14	TOTAL DIV. 10000 - 14000	100.0	79.7	95.7	100.0	80.5	95.9	100.0	47.7	88.9	100.0	46.8	88.7	100.0	51.1	89.6	100.0	42.8	87.8
15	MECHANICAL	95.4	40.8	72.7	100.3	45.2	77.4	95.1	71.3	85.2	95.1	46.1	74.8	100.0	64.3	85.2	95.1	64.1	82.2
16	ELECTRICAL	97.7	41.0	70.1	96.1	67.0	82.0	102.5	69.6	86.5	102.5	49.1	76.6	99.4	74.8	87.4	99.7	73.2	86.9
01 - 16	WEIGHTED AVERAGE	98.8	49.9	77.1	97.8	62.0	81.9	99.2	63.7	83.5	99.2	53.1	78.8	100.4	68.4	86.2	97.4	57.7	79.8

City Cost Indexes

KANSAS

DIVISION		FORT SCOTT 667 MAT.	INST.	TOTAL	HAYS 676 MAT.	INST.	TOTAL	HUTCHINSON 675 MAT.	INST.	TOTAL	INDEPENDENCE 673 MAT.	INST.	TOTAL	KANSAS CITY 660-662 MAT.	INST.	TOTAL	LIBERAL 679 MAT.	INST.	TOTAL
01590	EQUIPMENT RENTAL	.0	102.8	102.8	.0	103.6	103.6	.0	103.6	103.6	.0	103.6	103.6	.0	100.1	100.1	.0	103.6	103.6
02	SITE CONSTRUCTION	99.8	91.9	94.0	114.8	93.2	98.9	92.1	92.8	92.6	112.8	93.3	98.5	93.1	93.3	93.2	115.2	92.8	98.7
03100	CONCRETE FORMS & ACCESSORIES	103.2	72.6	76.8	98.7	67.9	72.2	87.0	29.1	37.1	112.1	60.0	67.2	99.3	94.4	95.1	93.9	28.7	37.6
03200	CONCRETE REINFORCEMENT	103.3	71.9	87.3	105.4	57.0	80.8	105.4	46.9	75.7	104.8	64.5	84.3	100.1	87.4	93.7	106.9	46.8	76.4
03300	CAST-IN-PLACE CONCRETE	107.4	60.5	88.1	90.8	56.7	76.7	84.1	40.9	66.3	117.7	42.5	86.8	91.9	96.1	93.6	90.8	40.7	70.2
03	CONCRETE	105.3	69.2	87.4	105.8	63.1	84.6	88.9	38.4	63.8	117.4	56.1	86.9	97.0	94.0	95.5	107.9	38.1	73.2
04	MASONRY	102.0	59.9	75.9	104.3	50.9	71.2	96.3	39.3	60.9	93.7	51.5	67.5	104.3	97.0	99.8	102.1	39.3	63.2
05	METALS	94.5	85.8	91.7	94.7	80.4	90.1	94.5	72.2	87.4	94.5	82.7	90.7	100.9	96.4	99.5	95.0	71.7	87.5
06	WOOD & PLASTICS	103.5	78.5	90.4	97.8	75.4	86.1	86.3	27.3	55.6	111.9	66.1	88.0	99.4	93.4	96.3	93.0	27.3	58.8
07	THERMAL & MOISTURE PROTECTION	94.9	70.2	84.5	95.8	60.5	80.9	94.4	41.9	72.3	95.6	64.4	82.4	93.2	98.6	95.5	95.6	36.9	71.1
08	DOORS & WINDOWS	93.7	73.1	88.3	96.1	63.5	87.6	96.1	35.0	80.1	93.7	60.3	85.0	95.1	90.3	93.9	96.2	35.0	80.2
09200	PLASTER & GYPSUM BOARD	105.2	77.7	88.1	105.2	74.6	86.2	100.8	25.2	53.8	111.9	65.0	82.8	98.8	93.1	95.2	103.0	25.2	54.7
095,098	CEILINGS & ACOUSTICAL TREATMENT	86.9	77.7	81.4	86.9	74.6	79.6	86.9	25.2	50.1	86.9	65.0	73.9	87.8	93.1	91.0	86.9	25.2	50.1
09600	FLOORING	106.7	49.2	91.2	99.0	48.9	85.5	92.4	48.9	80.7	105.7	48.9	90.4	87.2	103.1	91.5	96.2	48.9	83.4
097,099	WALL FINISHES, PAINTS & COATINGS	93.1	48.2	65.7	90.8	48.2	64.8	90.8	48.2	64.8	90.8	48.2	64.8	98.8	86.4	91.2	90.8	48.2	64.8
09	FINISHES	97.0	66.3	81.0	95.6	63.1	78.7	91.5	33.9	61.5	98.3	56.9	76.7	92.3	94.0	93.2	94.7	33.9	63.0
10-14	TOTAL DIV. 10000 - 14000	100.0	58.4	91.1	100.0	51.1	89.6	100.0	43.4	87.9	100.0	48.7	89.1	100.0	80.7	95.9	100.0	43.3	87.9
15	MECHANICAL	95.1	67.2	83.5	95.1	64.3	82.3	95.1	37.7	71.3	95.1	69.7	84.6	99.9	95.9	98.2	95.1	31.8	68.8
16	ELECTRICAL	99.0	78.7	89.1	101.4	69.6	85.9	96.4	66.4	81.8	98.6	78.7	88.9	104.3	106.8	105.5	99.4	38.8	69.9
01-16	WEIGHTED AVERAGE	97.6	72.3	86.4	98.5	67.0	84.5	94.5	49.5	74.5	99.0	67.7	85.1	98.9	96.1	97.7	98.3	44.2	74.3

KANSAS / KENTUCKY

DIVISION		SALINA 674 MAT.	INST.	TOTAL	TOPEKA 664-666 MAT.	INST.	TOTAL	WICHITA 670-672 MAT.	INST.	TOTAL	ASHLAND 411-412 MAT.	INST.	TOTAL	BOWLING GREEN 421-422 MAT.	INST.	TOTAL	CAMPTON 413-414 MAT.	INST.	TOTAL
01590	EQUIPMENT RENTAL	.0	103.6	103.6	.0	101.6	101.6	.0	103.6	103.6	.0	98.8	98.8	.0	96.0	96.0	.0	103.5	103.5
02	SITE CONSTRUCTION	101.9	93.3	95.6	94.9	90.6	91.8	96.0	93.2	94.0	103.5	86.2	90.8	69.6	100.2	92.0	77.9	99.3	93.6
03100	CONCRETE FORMS & ACCESSORIES	89.2	47.9	53.6	98.6	44.2	51.7	95.4	56.4	61.7	85.7	102.6	100.3	84.1	83.2	83.3	87.3	33.0	40.4
03200	CONCRETE REINFORCEMENT	104.8	79.5	92.0	97.4	95.0	96.2	97.4	80.0	88.5	86.2	100.5	93.5	83.5	71.7	77.5	84.3	61.4	72.7
03300	CAST-IN-PLACE CONCRETE	101.7	48.2	79.7	93.0	52.4	76.3	88.6	57.9	76.0	88.6	102.2	94.2	87.9	81.4	85.2	98.1	39.0	73.8
03	CONCRETE	102.8	55.5	79.4	94.9	58.2	76.7	92.6	62.6	77.7	96.2	102.7	99.4	94.4	80.5	87.5	98.4	41.5	70.2
04	MASONRY	118.6	50.5	76.4	98.6	62.3	76.1	92.5	65.4	75.7	92.4	105.1	100.3	93.9	78.7	84.5	92.0	29.8	53.5
05	METALS	95.9	89.4	93.8	101.3	97.4	100.0	101.3	88.8	97.3	90.7	110.0	96.9	92.1	78.1	87.6	91.6	67.2	83.8
06	WOOD & PLASTICS	88.4	48.6	67.7	96.4	40.0	67.0	94.3	55.1	73.9	79.3	101.3	90.8	92.9	82.8	87.6	91.3	31.9	60.4
07	THERMAL & MOISTURE PROTECTION	95.0	57.2	79.1	94.9	69.3	84.1	94.6	65.2	82.2	86.7	99.6	92.1	85.8	79.1	83.0	95.8	40.5	72.5
08	DOORS & WINDOWS	96.1	52.9	84.8	96.3	59.1	86.6	96.3	61.7	87.3	93.6	96.9	94.4	95.3	75.5	90.1	96.8	48.2	84.1
09200	PLASTER & GYPSUM BOARD	101.6	47.1	67.7	103.6	38.3	63.0	103.6	53.7	72.6	63.6	101.4	87.1	87.5	82.6	84.5	87.5	29.3	51.4
095,098	CEILINGS & ACOUSTICAL TREATMENT	86.9	47.1	63.2	87.8	38.3	58.2	87.8	53.7	67.5	81.5	101.4	93.4	86.2	82.6	84.1	86.2	29.3	52.3
09600	FLOORING	93.9	35.7	78.2	97.5	45.5	83.4	96.7	78.7	91.8	77.9	101.8	84.3	93.2	60.3	84.3	94.6	41.9	80.4
097,099	WALL FINISHES, PAINTS & COATINGS	90.8	38.0	58.5	90.8	67.0	76.3	90.8	61.2	72.7	99.9	99.9	99.9	96.8	74.4	83.1	96.8	32.4	57.5
09	FINISHES	92.7	43.4	67.0	93.8	44.5	68.1	93.7	60.7	76.5	80.5	102.5	92.0	91.1	78.1	84.3	91.6	33.7	61.4
10-14	TOTAL DIV. 10000 - 14000	100.0	62.0	91.9	100.0	65.3	92.6	100.0	65.9	92.7	100.0	84.6	96.7	100.0	60.8	91.6	100.0	52.4	89.8
15	MECHANICAL	100.0	64.2	85.2	100.0	71.6	88.2	100.0	65.4	85.7	95.0	95.0	95.0	99.9	86.6	94.4	95.1	34.4	69.9
16	ELECTRICAL	99.1	70.8	85.4	103.0	78.7	91.2	101.3	70.4	86.5	90.6	82.8	86.8	93.9	85.5	89.8	91.5	27.0	60.1
01-16	WEIGHTED AVERAGE	99.2	63.9	83.5	98.7	69.3	85.6	97.9	69.5	85.3	92.5	96.9	94.5	94.6	82.7	89.3	94.1	43.7	71.7

KENTUCKY

DIVISION		CORBIN 407-409 MAT.	INST.	TOTAL	COVINGTON 410 MAT.	INST.	TOTAL	ELIZABETHTOWN 427 MAT.	INST.	TOTAL	FRANKFORT 406 MAT.	INST.	TOTAL	HAZARD 417-418 MAT.	INST.	TOTAL	HENDERSON 424 MAT.	INST.	TOTAL
01590	EQUIPMENT RENTAL	.0	103.5	103.5	.0	108.1	108.1	.0	96.0	96.0	.0	103.5	103.5	.0	103.5	103.5	.0	121.3	121.3
02	SITE CONSTRUCTION	75.4	99.7	93.2	74.3	116.5	105.2	64.5	100.0	90.5	76.7	101.3	94.8	76.0	101.0	94.3	73.0	129.7	114.6
03100	CONCRETE FORMS & ACCESSORIES	84.0	32.6	39.7	82.9	86.9	86.4	78.7	82.6	82.0	94.8	57.4	62.5	84.0	32.8	39.8	91.0	82.7	83.8
03200	CONCRETE REINFORCEMENT	83.9	61.4	72.5	82.9	99.0	91.1	83.9	91.3	87.6	85.1	72.8	78.9	84.7	61.4	72.9	83.6	94.4	89.1
03300	CAST-IN-PLACE CONCRETE	92.8	39.4	70.9	93.9	85.0	90.2	79.5	79.7	79.6	92.8	66.2	81.9	94.3	39.5	71.8	77.7	94.4	84.6
03	CONCRETE	93.5	41.6	67.7	96.6	88.8	92.7	86.7	83.3	85.0	94.5	63.9	79.3	95.1	41.7	68.6	91.6	89.0	90.3
04	MASONRY	92.2	30.0	53.6	106.3	95.3	99.5	77.0	76.6	76.8	90.9	72.4	79.4	88.9	30.0	52.4	98.1	94.8	96.0
05	METALS	91.6	67.6	83.9	89.4	95.6	91.4	91.5	86.7	90.0	99.1	77.9	92.3	91.6	67.6	84.0	93.0	71.2	85.6
06	WOOD & PLASTICS	87.7	31.9	58.6	88.5	84.0	86.1	87.2	82.8	84.9	99.6	55.8	76.8	87.7	31.9	58.6	95.2	79.3	86.9
07	THERMAL & MOISTURE PROTECTION	95.6	40.5	72.4	91.0	93.0	91.8	85.4	79.5	82.9	95.7	66.8	83.5	95.7	40.5	72.5	95.1	94.3	94.8
08	DOORS & WINDOWS	96.1	37.8	80.9	98.5	87.0	95.5	95.3	84.4	92.5	95.4	63.6	87.1	97.2	37.8	81.7	93.6	83.7	91.0
09200	PLASTER & GYPSUM BOARD	86.0	29.3	50.8	76.6	84.1	81.3	85.6	82.6	83.8	91.6	53.9	68.2	86.0	29.3	50.8	87.6	78.0	81.6
095,098	CEILINGS & ACOUSTICAL TREATMENT	86.2	29.3	52.3	88.7	84.1	86.0	86.2	82.6	84.1	86.2	53.9	66.9	86.2	29.3	52.3	75.1	78.0	76.8
09600	FLOORING	92.6	41.9	78.9	69.8	97.0	77.2	90.1	73.3	85.6	98.3	51.2	85.6	92.6	41.9	78.9	96.6	91.4	95.2
097,099	WALL FINISHES, PAINTS & COATINGS	96.8	32.4	57.5	92.0	89.4	90.5	96.8	78.4	85.6	96.8	49.0	67.6	96.8	32.4	57.5	96.8	99.6	98.5
09	FINISHES	90.6	33.7	61.0	83.2	88.7	86.1	89.6	80.4	84.8	93.1	54.6	73.1	90.7	33.7	61.0	88.9	85.8	87.3
10-14	TOTAL DIV. 10000 - 14000	100.0	52.4	89.8	100.0	97.1	99.4	100.0	80.2	95.8	100.0	63.8	92.3	100.0	52.4	89.8	100.0	90.4	94.8
15	MECHANICAL	95.0	33.9	69.7	95.6	89.6	93.1	95.3	85.0	91.1	100.0	60.0	83.4	95.1	33.9	69.7	95.3	82.4	90.0
16	ELECTRICAL	91.7	27.0	60.2	94.2	81.4	88.0	91.3	85.5	88.4	94.0	59.5	77.2	91.5	27.0	60.1	93.4	84.1	88.9
01-16	WEIGHTED AVERAGE	93.3	43.3	71.1	93.8	91.9	93.0	91.1	84.5	88.2	96.3	66.7	83.1	93.5	43.4	71.3	92.0	90.4	91.3

KENTUCKY

DIVISION		LEXINGTON 403 - 405			LOUISVILLE 400 - 402			OWENSBORO 423			PADUCAH 420			PIKEVILLE 415 - 416			SOMERSET 425 - 426		
		MAT.	INST.	TOTAL	MAT.	INST.	TOTAL	MAT.	INST.	TOTAL	MAT.	INST.	TOTAL	MAT.	INST.	TOTAL	MAT.	INST.	TOTAL
01590	EQUIPMENT RENTAL	.0	103.5	103.5	.0	96.0	96.0	.0	121.3	121.3	.0	121.3	121.3	.0	98.8	98.8	.0	103.5	103.5
02	SITE CONSTRUCTION	77.3	102.2	95.6	67.1	100.0	91.2	81.9	129.3	116.6	75.4	129.7	115.3	113.3	84.2	91.9	69.4	99.3	91.3
03100	CONCRETE FORMS & ACCESSORIES	98.1	70.4	74.2	92.8	82.4	83.8	89.1	63.1	66.6	86.6	81.5	82.2	95.7	57.5	62.7	84.7	33.1	40.2
03200	CONCRETE REINFORCEMENT	92.2	95.4	93.8	92.2	96.7	94.5	83.6	92.6	88.2	84.1	84.7	84.4	86.7	53.8	70.0	83.9	61.5	72.5
03300	CAST-IN-PLACE CONCRETE	95.0	78.9	88.4	91.9	79.2	86.7	90.7	74.4	84.0	82.8	88.0	84.9	97.4	69.8	86.1	77.7	70.0	74.5
03	CONCRETE	96.9	78.4	87.7	95.0	84.1	89.6	103.6	73.0	88.4	96.0	84.4	90.3	109.8	62.7	86.4	81.7	52.2	67.1
04	MASONRY	90.6	46.7	63.4	92.2	81.4	85.5	90.7	50.5	65.8	93.6	93.8	93.7	90.7	54.0	67.9	84.8	29.8	50.7
05	METALS	93.6	88.1	91.8	101.0	89.3	97.3	84.0	86.2	84.7	81.9	86.3	83.3	90.6	82.7	88.1	91.6	67.3	83.8
06	WOOD & PLASTICS	104.0	73.9	88.3	102.5	82.8	92.2	93.1	61.7	76.7	90.6	79.5	84.8	89.7	54.8	71.5	88.7	31.9	59.1
07	THERMAL & MOISTURE PROTECTION	95.9	77.0	87.9	86.0	81.4	84.0	95.5	66.6	83.3	95.2	82.6	89.9	87.4	53.7	73.2	95.1	40.5	72.1
08	DOORS & WINDOWS	96.2	79.5	91.8	96.2	86.4	93.6	93.6	74.8	88.7	92.7	78.6	89.1	94.3	57.8	84.8	96.1	48.2	83.6
09200	PLASTER & GYPSUM BOARD	94.3	72.5	80.8	95.3	82.6	87.4	86.5	59.9	70.0	85.8	78.2	81.1	66.9	53.7	58.7	87.1	29.3	51.2
095,098	CEILINGS & ACOUSTICAL TREATMENT	91.3	72.5	80.1	91.3	82.6	86.1	75.1	59.9	66.0	75.1	78.2	77.0	81.5	53.7	64.9	86.2	29.3	52.3
09600	FLOORING	98.3	41.9	83.0	96.8	73.3	90.5	95.7	60.3	86.1	94.3	60.3	85.1	83.0	49.8	74.0	92.9	41.9	79.1
097,099	WALL FINISHES, PAINTS & COATINGS	96.8	57.6	72.9	96.8	81.7	87.5	96.8	99.6	98.5	96.8	77.3	84.9	99.9	64.2	78.1	96.8	32.4	57.5
09	FINISHES	94.9	63.3	78.4	94.3	80.5	87.1	88.8	65.6	76.7	88.1	76.7	82.2	83.1	56.3	69.2	90.4	33.7	60.8
10 - 14	TOTAL DIV. 10000 - 14000	100.0	76.7	95.0	100.0	79.9	95.7	100.0	94.6	98.9	100.0	70.1	93.6	100.0	56.2	90.7	100.0	52.4	89.8
15	MECHANICAL	99.9	47.0	78.0	99.9	84.6	93.6	99.9	51.7	79.9	95.3	86.2	91.5	95.0	54.7	78.3	95.3	34.0	69.9
16	ELECTRICAL	94.4	49.5	72.6	94.4	85.5	90.1	93.5	84.1	88.9	95.7	85.7	90.8	93.4	48.1	71.4	91.9	27.0	60.3
01 - 16	WEIGHTED AVERAGE	96.0	65.7	82.5	96.4	85.4	91.5	94.4	73.5	85.1	92.2	88.4	90.5	95.0	60.3	79.6	91.6	45.1	70.9

LOUISIANA

DIVISION		ALEXANDRIA 713 - 714			BATON ROUGE 707 - 708			HAMMOND 704			LAFAYETTE 705			LAKE CHARLES 706			MONROE 712		
		MAT.	INST.	TOTAL	MAT.	INST.	TOTAL	MAT.	INST.	TOTAL	MAT.	INST.	TOTAL	MAT.	INST.	TOTAL	MAT.	INST.	TOTAL
01590	EQUIPMENT RENTAL	.0	86.7	86.7	.0	86.2	86.2	.0	87.0	87.0	.0	87.0	87.0	.0	86.2	86.2	.0	86.7	86.7
02	SITE CONSTRUCTION	102.0	85.5	89.9	107.4	85.0	91.0	107.4	86.5	92.1	108.3	86.4	92.2	108.9	84.8	91.2	102.0	85.0	89.5
03100	CONCRETE FORMS & ACCESSORIES	79.6	43.5	48.5	98.1	59.7	65.0	77.9	58.8	61.4	97.5	53.5	59.5	98.6	59.2	64.6	78.9	43.7	48.5
03200	CONCRETE REINFORCEMENT	98.8	66.2	82.2	94.4	64.3	79.1	93.1	65.3	78.9	94.4	64.2	79.1	94.4	64.3	79.1	97.7	66.3	81.8
03300	CAST-IN-PLACE CONCRETE	92.9	43.0	72.4	83.4	51.0	70.1	84.3	49.9	70.2	83.9	50.3	70.1	88.3	52.3	73.5	92.9	46.6	73.8
03	CONCRETE	88.4	48.8	68.7	91.8	58.1	75.0	90.7	57.6	74.2	91.9	55.1	73.7	94.2	58.4	76.4	88.2	50.0	69.2
04	MASONRY	113.5	52.4	75.6	98.3	53.9	70.8	97.6	60.7	74.7	97.6	52.9	69.9	96.0	54.9	70.6	108.3	56.3	76.1
05	METALS	86.4	75.0	82.8	94.0	71.2	86.7	89.3	72.6	84.0	88.6	71.5	83.1	88.6	71.2	83.0	86.4	73.3	82.2
06	WOOD & PLASTICS	81.8	42.5	61.3	104.7	63.0	83.0	82.4	60.8	71.2	104.7	55.4	79.0	102.7	62.3	81.7	81.0	42.9	61.2
07	THERMAL & MOISTURE PROTECTION	99.6	53.6	80.2	98.9	58.9	82.1	101.0	64.2	85.5	101.7	57.0	82.9	101.1	58.6	83.2	99.6	56.3	81.4
08	DOORS & WINDOWS	99.0	49.6	86.1	103.8	60.0	92.4	99.1	63.8	89.9	103.8	55.9	91.3	103.8	61.2	92.7	99.0	54.5	87.4
09200	PLASTER & GYPSUM BOARD	78.1	41.6	55.4	97.9	62.6	75.9	89.1	60.3	71.2	97.9	54.7	71.1	97.9	61.8	75.5	77.7	42.1	55.6
095,098	CEILINGS & ACOUSTICAL TREATMENT	88.8	41.6	60.7	98.4	62.6	77.0	99.2	60.3	76.0	97.5	54.7	72.0	98.4	61.8	76.6	88.8	42.1	60.9
09600	FLOORING	104.4	65.9	94.0	108.0	64.6	96.2	96.5	57.8	86.0	108.2	43.3	90.6	108.2	50.5	92.6	104.0	41.7	87.1
097,099	WALL FINISHES, PAINTS & COATINGS	97.0	40.6	62.6	98.0	46.6	66.6	98.0	61.6	75.8	98.0	46.6	66.6	98.0	42.9	64.4	97.0	48.5	67.4
09	FINISHES	92.3	46.5	68.4	101.9	59.5	79.8	97.3	58.9	77.3	101.7	50.3	74.9	101.9	56.1	78.0	92.2	42.8	66.5
10 - 14	TOTAL DIV. 10000 - 14000	100.0	66.7	92.9	100.0	76.7	95.0	100.0	63.7	92.3	100.0	75.3	94.7	100.0	76.7	95.0	100.0	61.0	91.7
15	MECHANICAL	100.1	62.5	84.5	100.0	58.0	82.6	95.1	58.4	79.9	100.0	62.8	84.6	100.0	63.5	84.8	100.1	53.8	80.9
16	ELECTRICAL	92.0	60.4	76.6	94.0	63.7	79.3	93.2	56.9	75.5	94.4	66.2	80.7	94.0	66.2	80.4	93.7	61.3	78.0
01 - 16	WEIGHTED AVERAGE	95.4	59.5	79.5	98.1	62.9	82.5	95.0	62.7	80.7	97.4	62.3	81.8	97.5	64.1	82.7	95.3	57.9	78.7

LOUISIANA / MAINE

DIVISION		NEW ORLEANS 700 - 701			SHREVEPORT 710 - 711			THIBODAUX 703			AUGUSTA 043			BANGOR 044			BATH 045		
		MAT.	INST.	TOTAL	MAT.	INST.	TOTAL	MAT.	INST.	TOTAL	MAT.	INST.	TOTAL	MAT.	INST.	TOTAL	MAT.	INST.	TOTAL
01590	EQUIPMENT RENTAL	.0	87.3	87.3	.0	86.7	86.7	.0	87.0	87.0	.0	101.5	101.5	.0	101.5	101.5	.0	101.5	101.5
02	SITE CONSTRUCTION	111.5	87.5	93.9	100.7	84.8	89.1	109.9	86.5	92.7	84.3	102.4	97.6	84.2	100.4	96.1	82.3	102.5	97.1
03100	CONCRETE FORMS & ACCESSORIES	98.2	67.7	71.9	98.5	46.5	53.7	91.5	68.2	71.4	100.9	63.4	68.5	95.0	89.9	90.6	90.1	63.5	67.2
03200	CONCRETE REINFORCEMENT	94.4	67.3	80.6	97.3	66.0	81.4	93.1	62.5	77.5	87.5	111.6	99.7	87.5	111.8	99.8	86.5	111.6	99.2
03300	CAST-IN-PLACE CONCRETE	87.5	75.5	82.6	91.6	50.2	74.6	90.6	53.5	75.4	83.7	61.8	74.7	83.7	63.1	75.3	83.7	61.8	74.7
03	CONCRETE	93.7	70.7	82.3	88.3	52.4	70.5	95.6	62.4	79.2	98.6	72.0	85.4	97.7	84.1	90.9	97.7	72.1	85.0
04	MASONRY	97.8	62.4	75.8	101.0	49.6	69.1	122.3	59.6	83.5	97.1	50.9	68.5	114.9	58.4	79.9	121.8	50.5	77.6
05	METALS	101.2	75.5	93.0	82.7	72.7	79.5	89.3	71.1	83.5	87.7	85.4	87.0	87.3	82.1	85.7	86.2	85.5	86.0
06	WOOD & PLASTICS	99.0	70.4	84.1	101.9	46.7	73.1	91.9	73.2	82.1	99.5	61.4	79.7	93.0	96.0	94.5	87.4	61.4	73.8
07	THERMAL & MOISTURE PROTECTION	102.2	63.6	85.9	98.7	55.0	80.3	102.3	62.2	85.4	102.2	52.9	81.5	102.1	59.4	84.1	102.1	57.2	83.2
08	DOORS & WINDOWS	104.5	70.3	95.6	96.9	52.0	85.2	104.8	69.7	95.6	103.8	60.6	92.5	103.7	81.0	97.8	103.7	62.3	92.9
09200	PLASTER & GYPSUM BOARD	96.8	70.2	80.3	83.8	45.9	60.3	94.2	73.0	81.1	104.0	59.5	76.3	100.4	95.0	97.0	97.5	59.5	73.9
095,098	CEILINGS & ACOUSTICAL TREATMENT	97.5	70.2	81.2	89.7	45.9	63.6	99.2	73.0	83.6	90.0	59.5	71.8	87.5	95.0	91.9	85.5	59.5	70.0
09600	FLOORING	108.6	49.5	92.6	115.1	61.8	100.7	104.4	55.1	91.1	99.6	42.0	84.0	97.1	52.4	85.0	94.9	42.0	80.6
097,099	WALL FINISHES, PAINTS & COATINGS	99.7	67.2	79.9	97.0	40.6	62.6	99.7	61.6	76.5	91.5	35.1	57.1	91.5	32.4	55.4	91.5	35.1	57.1
09	FINISHES	102.1	64.2	82.4	96.5	48.0	71.3	100.7	65.7	82.5	98.6	55.2	76.0	96.7	77.9	86.9	95.1	55.2	74.3
10 - 14	TOTAL DIV. 10000 - 14000	100.0	79.3	95.6	100.0	70.0	93.6	100.0	78.7	95.5	100.0	59.6	91.4	100.0	79.4	95.6	100.0	59.6	91.4
15	MECHANICAL	100.0	65.4	85.6	100.1	62.4	84.4	95.1	62.2	81.4	100.1	76.0	90.1	100.1	77.1	90.5	95.1	76.0	87.2
16	ELECTRICAL	95.1	69.0	82.4	92.6	70.2	81.7	91.9	69.0	80.7	97.6	87.8	92.9	95.9	85.0	90.6	94.2	87.8	91.1
01 - 16	WEIGHTED AVERAGE	99.8	69.7	86.4	94.6	61.3	79.8	97.7	67.2	84.2	97.4	73.2	86.7	97.7	79.8	89.8	96.3	73.4	86.1

MAINE

DIVISION		HOULTON 047			KITTERY 039			LEWISTON 042			MACHIAS 046			PORTLAND 040 - 041			ROCKLAND 048		
		MAT.	INST.	TOTAL	MAT.	INST.	TOTAL	MAT.	INST.	TOTAL	MAT.	INST.	TOTAL	MAT.	INST.	TOTAL	MAT.	INST.	TOTAL
01590	EQUIPMENT RENTAL	.0	101.5	101.5	.0	101.5	101.5	.0	101.5	101.5	.0	101.5	101.5	.0	101.5	101.5	.0	101.5	101.5
02	SITE CONSTRUCTION	83.8	102.5	97.5	78.7	102.5	96.1	82.9	100.4	95.7	83.2	102.4	97.3	81.6	100.4	95.4	79.9	102.4	96.4
03100	CONCRETE FORMS & ACCESSORIES	98.8	77.4	80.3	90.7	63.5	67.3	101.0	89.9	91.4	95.7	63.4	67.8	100.1	89.9	91.3	96.8	63.5	68.1
03200	CONCRETE REINFORCEMENT	87.5	111.6	99.7	85.9	111.6	99.0	107.5	111.8	109.7	87.5	111.5	99.7	107.5	111.8	109.7	87.5	111.6	99.7
03300	CAST-IN-PLACE CONCRETE	83.7	61.9	74.7	83.7	61.8	74.7	95.3	63.1	82.1	83.7	61.8	74.7	86.8	63.1	77.0	85.5	61.8	75.7
03	CONCRETE	98.6	78.3	88.5	93.3	72.1	82.8	103.1	84.1	93.6	98.2	72.0	85.2	98.9	84.1	91.5	95.5	72.1	83.9
04	MASONRY	97.6	50.5	68.4	112.3	50.5	74.0	98.9	58.4	73.8	97.6	50.5	68.4	96.5	58.4	72.8	91.2	51.8	66.8
05	METALS	86.4	85.6	86.1	86.1	85.5	85.9	90.4	82.1	87.8	86.4	85.3	86.1	91.8	82.1	88.7	86.3	85.5	86.1
06	WOOD & PLASTICS	97.3	80.0	88.3	87.9	61.4	74.1	99.5	96.0	97.7	94.0	61.4	77.0	99.5	96.0	97.7	95.0	61.4	77.5
07	THERMAL & MOISTURE PROTECTION	102.2	59.1	84.1	101.7	57.2	83.0	102.0	59.4	84.1	102.2	57.1	83.2	101.8	59.4	84.0	101.9	57.5	83.2
08	DOORS & WINDOWS	103.8	72.4	95.6	103.7	62.3	92.9	106.9	81.0	100.2	103.8	60.1	92.4	106.9	81.0	100.2	103.7	60.1	92.3
09200	PLASTER & GYPSUM BOARD	102.9	78.6	87.8	97.5	59.5	73.9	106.2	95.0	99.2	101.8	59.5	75.5	106.2	95.0	99.2	101.1	59.5	75.3
095,098	CEILINGS & ACOUSTICAL TREATMENT	88.1	78.6	82.4	85.5	59.5	70.0	98.5	95.0	96.4	88.1	59.5	71.0	98.5	95.0	96.4	85.5	59.5	70.0
09600	FLOORING	98.5	42.0	83.2	95.2	42.0	80.8	99.6	52.4	86.8	97.4	20.6	76.6	99.6	52.4	86.8	97.7	42.0	82.6
097,099	WALL FINISHES, PAINTS & COATINGS	91.5	35.1	57.1	91.5	35.1	57.1	91.5	32.4	55.4	91.5	35.1	57.1	91.5	32.4	55.4	91.5	35.1	57.1
09	FINISHES	97.7	66.1	81.2	94.9	55.2	74.2	100.8	77.9	88.9	97.2	51.5	73.4	100.8	77.9	88.9	96.2	55.2	74.9
10 - 14	TOTAL DIV. 10000 - 14000	100.0	61.9	91.9	100.0	59.6	91.4	100.0	79.4	95.6	100.0	59.6	91.4	100.0	79.4	95.6	100.0	59.6	91.4
15	MECHANICAL	95.1	76.0	87.2	95.1	76.0	87.2	100.1	77.1	90.5	95.1	76.0	87.2	100.1	77.1	90.5	95.1	76.0	87.2
16	ELECTRICAL	97.7	87.8	92.9	94.2	87.8	91.1	97.7	85.0	91.5	97.7	87.8	92.9	97.8	85.0	91.6	97.6	87.8	92.9
01 - 16	WEIGHTED AVERAGE	96.0	76.4	87.3	95.2	73.4	85.5	98.9	79.8	90.4	95.9	72.8	85.6	98.5	79.8	90.2	95.1	73.4	85.4

DIVISION		MAINE WATERVILLE 049			MARYLAND ANNAPOLIS 214			BALTIMORE 210 - 212			COLLEGE PARK 207 - 208			CUMBERLAND 215			EASTON 216		
		MAT.	INST.	TOTAL	MAT.	INST.	TOTAL	MAT.	INST.	TOTAL	MAT.	INST.	TOTAL	MAT.	INST.	TOTAL	MAT.	INST.	TOTAL
01590	EQUIPMENT RENTAL	.0	101.5	101.5	.0	99.1	99.1	.0	103.3	103.3	.0	102.2	102.2	.0	99.1	99.1	.0	99.1	99.1
02	SITE CONSTRUCTION	84.0	102.4	97.5	96.1	88.6	90.6	97.1	93.1	94.1	100.0	87.9	91.2	88.4	88.6	88.5	94.9	85.3	87.9
03100	CONCRETE FORMS & ACCESSORIES	89.5	63.5	67.0	102.8	57.5	63.8	102.9	74.8	78.6	82.9	71.3	72.9	92.3	79.5	81.3	89.8	34.9	42.4
03200	CONCRETE REINFORCEMENT	87.5	111.6	99.7	93.3	88.0	90.6	93.3	88.1	90.6	105.5	81.7	93.4	82.4	76.0	79.1	81.7	20.9	50.8
03300	CAST-IN-PLACE CONCRETE	83.7	61.8	74.7	99.0	78.0	90.4	99.0	79.0	90.8	115.6	74.5	98.7	88.4	82.6	86.1	98.2	43.1	75.6
03	CONCRETE	99.1	72.1	85.7	100.5	71.8	86.3	100.5	79.8	90.3	111.5	75.6	93.7	89.7	80.8	85.3	97.2	37.0	67.3
04	MASONRY	107.7	50.9	72.5	93.2	72.5	80.4	92.7	72.5	80.2	108.3	69.2	84.0	90.9	77.0	82.3	104.2	37.4	62.8
05	METALS	86.4	85.4	86.1	100.3	97.8	99.5	101.3	98.7	100.5	91.1	95.6	92.5	97.4	91.0	95.4	97.6	61.0	85.9
06	WOOD & PLASTICS	86.6	61.4	73.5	100.0	53.0	75.5	100.0	76.0	87.5	79.7	74.4	77.0	88.9	78.9	83.7	85.9	35.1	59.5
07	THERMAL & MOISTURE PROTECTION	102.2	56.3	82.9	97.8	77.4	89.2	97.8	80.1	90.4	103.3	75.9	91.8	97.5	80.1	90.2	97.6	41.3	73.9
08	DOORS & WINDOWS	103.8	60.6	92.5	92.5	70.8	86.8	94.5	83.2	91.5	97.2	77.4	92.0	93.6	83.6	91.0	91.7	32.1	76.1
09200	PLASTER & GYPSUM BOARD	97.7	59.5	74.0	103.3	52.1	71.5	103.3	75.5	86.1	99.8	73.7	83.6	99.6	78.8	86.7	98.1	33.8	58.2
095,098	CEILINGS & ACOUSTICAL TREATMENT	88.1	59.5	71.0	103.1	52.1	72.7	103.1	75.5	86.7	94.4	73.7	82.1	103.1	78.8	88.6	103.1	33.8	61.8
09600	FLOORING	94.6	42.0	80.4	92.0	78.6	88.4	92.0	78.6	88.4	106.1	64.5	94.9	87.5	79.2	85.2	86.7	16.0	67.5
097,099	WALL FINISHES, PAINTS & COATINGS	91.5	35.1	57.1	95.9	87.4	90.7	95.9	87.4	90.7	117.2	81.7	95.5	95.9	61.8	75.1	95.9	25.9	53.2
09	FINISHES	95.9	55.2	74.7	96.0	62.8	78.7	96.1	76.3	85.8	98.0	71.0	83.9	93.8	77.4	85.3	93.6	31.3	61.2
10 - 14	TOTAL DIV. 10000 - 14000	100.0	59.6	91.4	100.0	82.5	96.3	100.0	85.8	97.0	100.0	67.2	93.0	100.0	89.0	97.7	100.0	49.0	89.1
15	MECHANICAL	95.1	76.0	87.2	100.0	86.0	94.2	100.0	86.1	94.2	95.0	82.7	89.9	95.1	74.8	86.7	95.1	27.9	67.2
16	ELECTRICAL	97.7	87.8	92.9	103.1	82.5	93.1	105.1	91.2	98.4	98.3	95.8	97.0	101.9	84.3	93.4	101.4	63.4	82.9
01 - 16	WEIGHTED AVERAGE	96.3	73.3	86.1	98.9	79.2	90.1	99.5	84.7	92.9	98.4	81.4	90.8	95.4	81.2	89.1	96.8	44.8	73.7

DIVISION		MARYLAND ELKTON 219			HAGERSTOWN 217			SALISBURY 218			SILVER SPRING 209			WALDORF 206			MASSACHUSETTS BOSTON 020 - 022, 024		
		MAT.	INST.	TOTAL	MAT.	INST.	TOTAL	MAT.	INST.	TOTAL	MAT.	INST.	TOTAL	MAT.	INST.	TOTAL	MAT.	INST.	TOTAL
01590	EQUIPMENT RENTAL	.0	99.1	99.1	.0	99.1	99.1	.0	99.1	99.1	.0	93.7	93.7	.0	93.7	93.7	.0	108.4	108.4
02	SITE CONSTRUCTION	82.9	86.4	85.5	86.9	88.7	88.2	94.9	85.5	88.0	88.8	80.5	82.7	95.3	80.7	84.6	89.4	109.1	103.8
03100	CONCRETE FORMS & ACCESSORIES	97.1	64.6	69.1	91.2	76.4	78.4	106.8	33.0	43.1	92.1	61.2	65.5	100.0	54.2	60.5	106.5	137.9	133.6
03200	CONCRETE REINFORCEMENT	81.7	61.6	71.5	82.4	75.8	79.1	81.7	61.4	71.4	104.4	81.6	92.8	105.1	61.4	82.9	106.3	145.1	126.0
03300	CAST-IN-PLACE CONCRETE	79.5	50.4	67.5	84.2	60.4	74.4	98.2	42.4	75.3	118.3	67.0	97.2	132.6	60.5	103.0	108.1	145.5	123.5
03	CONCRETE	82.7	60.0	71.4	86.1	71.8	79.0	98.3	42.8	70.8	109.5	68.3	89.1	120.6	59.1	90.1	112.2	141.0	126.5
04	MASONRY	91.0	41.1	60.1	96.6	77.0	84.4	103.7	35.0	61.1	106.9	69.6	83.7	91.2	65.9	75.5	118.0	148.1	136.7
05	METALS	97.7	72.6	89.7	97.5	90.2	95.2	97.7	63.7	86.8	95.3	91.1	94.0	95.3	80.1	90.4	103.0	127.4	110.8
06	WOOD & PLASTICS	93.8	71.3	82.1	87.5	75.8	81.4	104.3	31.1	66.2	87.1	61.3	73.6	94.9	54.8	74.0	103.5	138.9	121.9
07	THERMAL & MOISTURE PROTECTION	97.2	50.2	77.4	97.2	71.2	86.3	97.9	59.0	81.5	102.6	80.1	93.1	103.1	75.1	91.3	102.4	142.5	119.3
08	DOORS & WINDOWS	91.7	58.5	83.1	91.7	76.8	87.8	92.0	26.8	75.0	90.6	70.3	85.3	91.5	53.6	81.6	102.9	136.8	111.7
09200	PLASTER & GYPSUM BOARD	101.1	71.0	82.4	98.5	75.5	84.2	105.2	29.7	58.2	106.6	61.2	78.4	109.2	54.6	75.3	108.6	139.1	127.6
095,098	CEILINGS & ACOUSTICAL TREATMENT	103.1	71.0	83.9	104.0	75.5	87.0	103.1	29.7	59.3	103.7	61.2	78.4	103.7	54.6	74.4	97.6	139.1	122.3
09600	FLOORING	89.4	61.1	81.8	87.2	79.2	85.0	93.7	61.1	84.9	113.6	64.5	100.3	118.6	61.1	103.1	100.3	163.7	117.4
097,099	WALL FINISHES, PAINTS & COATINGS	95.9	79.2	85.7	95.9	39.7	61.6	95.9	29.3	55.2	122.8	81.7	97.7	122.8	68.3	89.5	94.6	153.1	130.3
09	FINISHES	94.3	66.6	79.9	93.6	73.0	82.9	96.8	36.8	65.5	100.3	63.2	81.0	102.4	56.2	78.4	101.1	145.0	124.0
10 - 14	TOTAL DIV. 10000 - 14000	100.0	54.8	90.4	100.0	88.6	97.6	100.0	47.9	88.9	100.0	63.3	92.2	100.0	51.4	89.6	100.0	120.4	104.4
15	MECHANICAL	95.1	79.7	88.7	100.0	90.0	95.9	95.1	61.7	81.2	95.0	82.4	89.8	95.0	79.9	88.7	100.1	128.8	112.0
16	ELECTRICAL	103.1	64.2	84.2	101.7	84.3	93.2	100.1	63.4	82.3	94.1	95.7	94.9	92.0	95.7	93.8	96.8	120.4	108.3
01 - 16	WEIGHTED AVERAGE	94.6	66.7	82.2	96.2	81.9	89.8	97.3	53.5	77.8	97.6	77.9	88.9	98.3	72.6	86.9	102.5	132.0	115.6

City Cost Indexes

MASSACHUSETTS

DIVISION		BROCKTON 023			BUZZARDS BAY 025			FALL RIVER 027			FITCHBURG 014			FRAMINGHAM 017			GREENFIELD 013		
		MAT.	INST.	TOTAL	MAT.	INST.	TOTAL	MAT.	INST.	TOTAL	MAT.	INST.	TOTAL	MAT.	INST.	TOTAL	MAT.	INST.	TOTAL
01590	EQUIPMENT RENTAL	.0	103.7	103.7	.0	103.7	103.7	.0	104.9	104.9	.0	101.5	101.5	.0	102.8	102.8	.0	101.5	101.5
02	SITE CONSTRUCTION	87.2	105.0	100.2	78.0	105.0	97.8	86.2	105.1	100.1	79.6	104.9	98.2	76.4	104.6	97.1	83.0	103.8	98.3
03100	CONCRETE FORMS & ACCESSORIES	106.1	124.2	121.7	103.2	122.8	120.1	106.1	123.1	120.8	96.6	122.8	119.2	104.4	122.5	120.0	94.6	105.1	103.6
03200	CONCRETE REINFORCEMENT	107.5	144.9	126.5	86.2	124.1	105.4	107.5	124.2	115.9	86.2	140.2	113.6	86.1	144.8	115.9	90.1	122.2	106.4
03300	CAST-IN-PLACE CONCRETE	102.9	141.7	118.8	85.4	141.5	108.5	99.5	142.2	117.1	89.6	141.3	110.9	89.6	106.3	96.5	92.1	119.4	103.3
03	CONCRETE	109.4	133.4	121.3	92.4	128.8	110.4	107.7	129.2	118.4	90.7	131.5	111.0	93.7	120.5	107.0	94.5	112.7	103.5
04	MASONRY	114.1	141.8	131.3	104.2	141.8	127.5	113.9	141.7	131.1	100.2	141.5	125.8	106.6	141.6	128.3	104.8	116.8	112.3
05	METALS	99.8	124.5	107.6	95.1	115.5	101.6	99.8	116.0	105.0	95.0	119.2	102.7	95.0	123.5	104.1	96.7	106.7	99.9
06	WOOD & PLASTICS	102.5	123.5	113.4	99.2	122.1	111.2	102.5	122.5	112.9	94.9	122.1	109.1	101.7	121.8	112.2	92.6	103.7	98.4
07	THERMAL & MOISTURE PROTECTION	102.3	135.2	116.1	101.6	131.7	114.2	102.2	130.7	114.2	101.6	129.2	113.2	101.7	131.9	114.4	101.7	112.5	106.2
08	DOORS & WINDOWS	100.8	127.7	107.8	96.3	120.0	102.5	100.8	120.1	105.8	105.8	126.5	111.2	96.8	127.6	104.8	105.7	109.8	106.7
09200	PLASTER & GYPSUM BOARD	103.8	123.2	115.8	99.1	121.8	113.2	103.8	121.8	115.0	100.4	121.8	113.7	102.6	121.8	114.6	102.1	102.9	102.6
095,098	CEILINGS & ACOUSTICAL TREATMENT	99.5	123.2	113.6	86.5	121.8	107.6	99.5	121.8	112.8	85.5	121.8	107.2	85.5	121.8	107.2	96.6	102.9	100.4
09600	FLOORING	100.7	163.7	117.8	99.2	163.7	116.6	100.5	163.7	117.6	97.1	163.7	115.1	99.1	163.7	116.6	96.2	128.6	104.9
097,099	WALL FINISHES, PAINTS & COATINGS	94.0	130.1	116.0	94.0	130.1	116.0	94.0	130.1	116.0	91.5	130.1	115.1	92.8	130.1	115.5	91.5	104.5	99.4
09	FINISHES	100.9	132.3	117.3	96.0	131.5	114.5	100.9	131.7	117.0	93.5	131.5	114.1	96.1	131.3	114.4	98.3	109.2	104.0
10 - 14	TOTAL DIV. 10000 - 14000	100.0	116.9	103.6	100.0	116.8	103.6	100.0	117.6	103.8	100.0	110.8	102.3	100.0	116.0	103.4	100.0	106.5	101.4
15	MECHANICAL	100.1	110.1	104.2	95.1	109.8	101.2	100.1	109.9	104.2	95.8	108.0	100.9	95.8	122.5	106.9	95.8	98.4	96.9
16	ELECTRICAL	95.8	103.4	99.5	93.4	103.4	98.3	95.4	103.4	99.3	98.1	106.8	102.3	95.1	120.4	107.4	98.1	93.7	96.0
01 - 16	WEIGHTED AVERAGE	101.1	121.0	110.0	95.5	119.0	105.9	100.9	119.1	109.0	96.6	119.8	106.9	96.1	123.8	108.4	97.9	105.3	101.2

MASSACHUSETTS

DIVISION		HYANNIS 026			LAWRENCE 019			LOWELL 018			NEW BEDFORD 027			PITTSFIELD 012			SPRINGFIELD 010 - 011		
		MAT.	INST.	TOTAL	MAT.	INST.	TOTAL	MAT.	INST.	TOTAL	MAT.	INST.	TOTAL	MAT.	INST.	TOTAL	MAT.	INST.	TOTAL
01590	EQUIPMENT RENTAL	.0	103.7	103.7	.0	103.7	103.7	.0	101.5	101.5	.0	104.9	104.9	.0	101.5	101.5	.0	101.5	101.5
02	SITE CONSTRUCTION	83.7	105.0	99.3	87.7	105.0	100.4	86.8	104.9	100.1	85.8	105.1	100.0	87.9	103.6	99.4	87.3	103.8	99.4
03100	CONCRETE FORMS & ACCESSORIES	96.5	122.8	119.2	105.9	123.2	120.9	102.6	123.4	120.6	106.1	123.1	120.8	102.6	99.2	99.7	102.8	107.7	107.0
03200	CONCRETE REINFORCEMENT	86.2	124.1	105.4	106.6	132.5	119.8	107.5	132.6	120.2	107.5	124.2	115.9	89.4	116.9	103.4	107.5	122.2	114.9
03300	CAST-IN-PLACE CONCRETE	93.9	141.5	113.5	103.7	141.7	119.3	94.3	141.8	113.8	96.6	142.2	115.3	102.9	117.8	109.0	98.1	119.4	106.9
03	CONCRETE	99.0	128.8	113.8	109.6	130.6	120.1	100.4	130.6	115.4	106.3	129.2	117.7	101.6	108.5	105.1	102.3	113.8	108.0
04	MASONRY	113.0	141.8	130.8	113.4	141.8	131.0	98.8	141.5	125.3	113.7	141.7	131.0	99.5	114.2	108.6	99.1	116.8	110.1
05	METALS	96.1	115.5	102.3	97.1	119.4	104.2	97.1	116.7	103.3	99.8	116.0	105.0	96.9	104.4	99.3	99.7	106.7	101.9
06	WOOD & PLASTICS	92.1	122.1	107.8	102.5	122.1	112.7	101.5	122.1	112.3	102.5	122.5	112.9	101.5	97.3	99.3	101.5	107.2	104.5
07	THERMAL & MOISTURE PROTECTION	101.8	131.7	114.4	102.2	135.0	116.0	102.0	134.2	115.6	102.2	130.7	114.2	102.0	110.7	105.7	102.0	112.9	106.6
08	DOORS & WINDOWS	97.0	120.0	103.0	108.0	124.5	107.0	106.9	124.5	111.5	100.8	120.1	105.8	106.9	104.9	106.4	106.9	111.7	108.2
09200	PLASTER & GYPSUM BOARD	96.1	121.8	112.1	106.2	121.8	115.9	106.2	121.8	115.9	103.8	121.8	115.0	106.2	96.3	100.0	106.2	106.5	106.4
095,098	CEILINGS & ACOUSTICAL TREATMENT	88.4	121.8	108.4	98.5	121.8	112.4	98.5	121.8	112.4	99.5	121.8	112.8	98.5	96.3	97.2	98.5	106.5	103.3
09600	FLOORING	96.6	163.7	114.7	99.6	163.7	117.0	99.6	163.7	117.0	100.5	163.7	117.6	99.8	128.6	107.6	99.6	128.6	107.4
097,099	WALL FINISHES, PAINTS & COATINGS	94.0	130.1	116.0	91.6	130.1	115.1	91.5	133.1	116.9	94.0	130.1	116.0	91.5	104.5	99.4	93.3	104.5	100.1
09	FINISHES	95.6	131.5	114.3	100.6	131.5	116.7	100.6	131.8	116.9	100.8	131.7	116.9	100.6	104.8	102.8	100.7	111.3	106.2
10 - 14	TOTAL DIV. 10000 - 14000	100.0	116.8	103.6	100.0	116.8	103.6	100.0	116.8	103.6	100.0	117.6	103.8	100.0	104.7	101.0	100.0	106.9	101.5
15	MECHANICAL	100.1	109.8	104.1	100.1	111.4	104.8	100.1	124.6	110.2	100.1	109.9	104.2	100.1	97.0	98.8	100.1	98.4	99.4
16	ELECTRICAL	94.0	103.4	98.6	97.3	120.4	108.5	97.7	120.4	108.7	96.7	103.4	100.0	97.7	93.7	95.8	97.7	93.7	95.8
01 - 16	WEIGHTED AVERAGE	98.1	119.0	107.4	100.9	122.6	110.5	99.7	125.0	111.0	100.8	119.1	109.0	99.9	103.0	101.3	100.4	105.9	102.8

MASSACHUSETTS / MICHIGAN

DIVISION		WORCESTER (MA) 015 - 016			ANN ARBOR 481			BATTLE CREEK 490			BAY CITY 487			DEARBORN 481			DETROIT 482		
		MAT.	INST.	TOTAL	MAT.	INST.	TOTAL	MAT.	INST.	TOTAL	MAT.	INST.	TOTAL	MAT.	INST.	TOTAL	MAT.	INST.	TOTAL
01590	EQUIPMENT RENTAL	.0	101.5	101.5	.0	112.2	112.2	.0	105.1	105.1	.0	112.2	112.2	.0	112.2	112.2	.0	97.7	97.7
02	SITE CONSTRUCTION	86.8	104.9	100.1	77.2	95.3	90.5	82.3	88.0	86.5	69.6	93.8	87.3	76.9	95.7	90.7	90.4	97.5	95.6
03100	CONCRETE FORMS & ACCESSORIES	103.2	122.8	120.1	97.4	119.4	116.4	97.2	87.2	88.5	97.5	95.5	95.8	97.3	123.9	120.2	98.8	124.0	120.5
03200	CONCRETE REINFORCEMENT	107.5	143.9	126.0	95.2	122.2	108.9	92.2	85.0	88.5	95.2	121.3	108.5	95.2	123.1	109.4	94.6	123.1	109.1
03300	CAST-IN-PLACE CONCRETE	94.3	141.5	113.7	92.3	107.9	98.7	94.9	98.8	96.5	88.3	92.1	89.9	90.2	121.6	103.1	99.1	121.7	108.4
03	CONCRETE	100.4	132.3	116.3	94.0	116.4	105.1	96.8	90.1	93.4	92.0	100.1	96.1	92.9	123.2	108.0	97.2	121.9	109.5
04	MASONRY	98.8	141.7	125.4	98.8	103.1	101.5	97.4	84.4	89.4	98.5	86.1	90.8	98.7	122.7	113.6	97.8	122.7	113.2
05	METALS	99.7	120.7	106.4	102.4	125.5	109.8	92.4	82.7	89.3	103.1	121.4	108.9	102.5	127.8	110.6	107.1	106.2	106.8
06	WOOD & PLASTICS	102.0	122.1	112.5	100.1	122.6	111.8	99.6	85.5	92.3	100.1	96.3	98.1	100.1	124.0	112.5	101.1	124.0	113.0
07	THERMAL & MOISTURE PROTECTION	102.0	129.3	113.5	102.8	108.6	105.2	91.0	83.1	87.7	100.7	88.2	95.4	101.9	124.8	111.5	99.5	124.8	110.2
08	DOORS & WINDOWS	106.9	127.5	112.3	96.6	117.8	102.1	91.1	80.2	88.2	96.6	101.1	97.8	96.6	118.5	102.3	98.2	121.1	104.2
09200	PLASTER & GYPSUM BOARD	106.2	121.8	115.9	98.1	122.0	112.9	92.9	80.4	85.1	98.1	94.9	96.1	98.1	123.4	113.8	98.1	123.4	113.8
095,098	CEILINGS & ACOUSTICAL TREATMENT	98.5	121.8	112.4	91.3	122.0	109.6	91.3	80.4	84.8	92.3	94.9	93.8	91.3	123.4	110.4	92.3	123.4	110.8
09600	FLOORING	99.6	166.4	116.1	88.0	105.9	92.6	87.8	57.0	79.5	87.6	122.8	97.2	87.7	122.8	97.2	88.4	113.8	103.9
097,099	WALL FINISHES, PAINTS & COATINGS	91.5	130.1	115.1	86.6	99.7	94.6	96.8	83.6	88.7	86.6	78.2	81.5	86.6	113.8	103.2	88.4	113.8	103.9
09	FINISHES	100.6	130.8	116.3	92.4	115.4	104.4	94.2	86.7	90.3	92.4	86.1	89.1	92.3	123.0	108.3	93.3	123.0	108.8
10 - 14	TOTAL DIV. 10000 - 14000	100.0	110.8	102.3	100.0	110.1	102.2	100.0	97.8	99.5	100.0	91.7	98.2	100.0	112.7	102.7	100.0	112.7	102.7
15	MECHANICAL	100.1	108.1	103.4	100.0	99.6	99.8	99.9	89.0	95.4	100.0	92.4	96.8	100.0	117.5	107.3	100.0	120.0	108.3
16	ELECTRICAL	97.7	106.8	102.1	94.1	85.8	90.0	94.0	92.5	93.3	93.2	81.2	87.4	94.1	119.2	106.3	95.8	119.1	107.1
01 - 16	WEIGHTED AVERAGE	100.1	120.0	109.0	97.4	105.9	101.2	95.5	87.9	92.1	96.9	93.5	95.4	97.3	118.9	106.9	99.1	117.6	107.3

City Cost Indexes

MICHIGAN

DIVISION		FLINT 484-485 MAT.	INST.	TOTAL	GAYLORD 497 MAT.	INST.	TOTAL	GRAND RAPIDS 493,495 MAT.	INST.	TOTAL	IRON MOUNTAIN 498-499 MAT.	INST.	TOTAL	JACKSON 492 MAT.	INST.	TOTAL	KALAMAZOO 491 MAT.	INST.	TOTAL
01590	EQUIPMENT RENTAL	.0	112.2	112.2	.0	107.6	107.6	.0	105.1	105.1	.0	94.2	94.2	.0	107.6	107.6	.0	105.1	105.1
02	SITE CONSTRUCTION	68.2	94.3	87.3	78.8	85.1	83.4	82.1	87.6	86.1	84.9	95.4	92.6	97.7	87.2	90.0	82.6	88.0	86.5
03100	CONCRETE FORMS & ACCESSORIES	101.0	94.9	95.7	94.2	62.6	66.9	97.7	76.6	79.5	88.3	86.9	87.1	91.7	87.3	87.9	97.2	86.9	88.3
03200	CONCRETE REINFORCEMENT	95.2	121.9	108.8	85.9	114.2	100.2	92.2	83.3	87.7	85.8	101.6	93.8	83.3	121.6	102.8	92.2	85.0	88.5
03300	CAST-IN-PLACE CONCRETE	93.0	96.4	94.4	94.6	75.1	86.6	95.1	96.7	95.8	112.1	91.6	103.7	94.5	92.7	93.8	96.8	98.7	97.6
03	CONCRETE	94.5	101.4	97.9	93.0	77.8	85.5	96.9	84.3	90.7	102.7	91.3	97.1	87.8	96.5	92.1	100.0	89.9	95.0
04	MASONRY	98.9	89.8	93.3	108.9	74.6	87.6	94.8	54.4	69.7	94.4	91.4	92.6	88.6	89.7	89.3	97.6	84.4	89.4
05	METALS	102.5	123.0	109.0	93.9	108.6	98.6	93.9	78.4	89.0	93.5	92.9	93.3	94.0	119.9	102.3	92.4	82.3	89.2
06	WOOD & PLASTICS	103.6	96.3	99.8	92.5	59.6	75.3	98.0	77.2	87.2	88.6	86.3	87.4	91.3	84.9	88.0	99.6	85.5	92.3
07	THERMAL & MOISTURE PROTECTION	100.8	91.0	96.6	89.8	67.0	80.2	91.4	60.6	78.4	92.5	82.0	88.1	89.1	92.4	90.5	91.0	83.1	87.7
08	DOORS & WINDOWS	96.6	101.1	97.8	92.5	66.4	85.7	94.3	68.6	87.6	97.2	76.4	91.8	91.6	91.9	91.7	91.1	80.2	88.2
09200	PLASTER & GYPSUM BOARD	99.6	94.9	96.7	94.6	56.9	71.2	92.9	71.9	79.8	63.2	86.5	77.7	94.0	82.9	87.1	92.9	80.4	85.1
095,098	CEILINGS & ACOUSTICAL TREATMENT	91.3	94.9	93.4	88.7	56.9	69.8	91.3	71.9	79.7	89.4	86.5	87.6	90.5	82.9	86.0	91.3	80.4	84.8
09600	FLOORING	87.8	71.6	83.5	90.6	60.1	82.3	97.0	42.3	82.2	109.3	96.9	105.9	88.8	82.8	87.2	97.0	67.1	88.9
097,099	WALL FINISHES, PAINTS & COATINGS	86.6	85.0	85.6	92.0	53.9	68.8	96.8	41.2	62.8	117.4	45.0	73.2	92.0	91.6	91.8	96.8	83.6	88.7
09	FINISHES	92.1	89.0	90.5	94.3	59.9	76.4	94.2	66.3	79.7	95.8	84.3	89.8	95.0	86.2	90.5	94.2	82.6	88.1
10-14	TOTAL DIV. 10000-14000	100.0	90.3	97.9	100.0	83.7	96.5	100.0	94.5	98.8	100.0	93.3	98.6	100.0	90.8	98.0	100.0	97.8	99.5
15	MECHANICAL	100.0	99.9	99.9	95.3	85.7	91.3	99.9	55.0	81.3	95.2	85.3	91.1	95.3	89.9	93.1	99.9	84.6	93.6
16	ELECTRICAL	94.1	108.0	100.9	92.0	59.9	76.4	94.8	59.8	77.8	98.5	83.6	91.2	95.6	85.7	90.8	93.7	88.6	91.2
01-16	WEIGHTED AVERAGE	97.2	99.9	98.4	94.5	76.9	86.7	96.0	68.1	83.6	96.3	87.7	92.5	93.8	92.3	93.2	95.8	85.8	91.4

MICHIGAN / MINNESOTA

DIVISION		LANSING 488-489 MAT.	INST.	TOTAL	MUSKEGON 494 MAT.	INST.	TOTAL	ROYAL OAK 480,483 MAT.	INST.	TOTAL	SAGINAW 486 MAT.	INST.	TOTAL	TRAVERSE CITY 496 MAT.	INST.	TOTAL	BEMIDJI 566 MAT.	INST.	TOTAL
01590	EQUIPMENT RENTAL	.0	112.2	112.2	.0	105.1	105.1	.0	94.6	94.6	.0	112.2	112.2	.0	94.2	94.2	.0	98.3	98.3
02	SITE CONSTRUCTION	83.2	93.8	91.0	80.4	87.8	85.8	80.0	94.6	90.7	70.4	93.8	87.5	72.3	94.0	88.2	85.3	98.2	94.7
03100	CONCRETE FORMS & ACCESSORIES	100.9	95.9	96.6	97.9	83.3	85.3	93.3	110.0	107.7	97.4	95.4	95.7	88.3	60.3	64.1	84.5	96.2	94.6
03200	CONCRETE REINFORCEMENT	95.2	121.7	108.7	92.7	84.8	88.7	86.3	116.0	101.4	95.2	121.3	108.5	87.1	84.0	85.5	95.3	109.7	102.6
03300	CAST-IN-PLACE CONCRETE	92.3	91.8	92.1	94.6	95.1	94.8	80.8	113.8	94.4	91.2	92.1	91.6	87.5	65.7	78.5	102.3	102.8	102.5
03	CONCRETE	94.2	100.3	97.2	95.0	87.0	91.1	81.3	111.3	96.2	93.4	100.1	96.7	84.7	67.3	76.0	95.1	102.1	98.6
04	MASONRY	92.6	91.8	92.1	94.9	77.8	84.3	93.0	120.3	110.0	100.3	86.1	91.5	92.6	74.8	81.6	101.0	106.4	104.4
05	METALS	100.8	122.1	107.6	90.1	81.9	87.5	106.1	95.5	102.7	102.5	121.3	108.5	93.4	82.8	90.0	88.8	123.7	100.0
06	WOOD & PLASTICS	103.0	96.4	99.6	96.5	82.5	89.2	95.6	107.6	101.8	96.2	96.3	96.2	88.6	57.5	72.4	77.2	93.7	85.8
07	THERMAL & MOISTURE PROTECTION	100.8	90.9	96.6	90.2	76.4	84.4	98.9	114.0	105.3	101.2	88.2	95.7	91.6	67.3	81.3	99.9	95.1	97.9
08	DOORS & WINDOWS	96.6	101.2	97.8	90.3	79.2	87.4	96.7	104.9	98.8	96.1	101.1	97.4	97.2	58.6	87.1	99.1	112.4	102.6
09200	PLASTER & GYPSUM BOARD	100.5	95.0	97.1	82.1	77.3	79.1	95.8	106.6	102.5	98.1	94.9	96.1	63.2	56.9	59.3	105.6	93.8	98.3
095,098	CEILINGS & ACOUSTICAL TREATMENT	91.3	95.0	93.5	93.0	77.3	83.7	90.6	106.6	100.1	91.3	94.9	93.4	89.4	56.9	70.0	126.6	93.8	107.1
09600	FLOORING	96.3	79.4	91.7	96.2	77.9	91.3	85.1	114.6	93.1	88.0	57.0	79.6	109.2	60.1	95.9	102.0	131.2	109.9
097,099	WALL FINISHES, PAINTS & COATINGS	100.8	91.1	94.9	96.0	59.7	73.9	88.4	109.8	101.4	86.6	78.2	81.5	117.4	55.1	79.4	97.2	98.5	97.9
09	FINISHES	96.7	92.3	94.4	92.3	79.4	85.6	91.2	110.8	101.4	92.2	86.1	89.1	95.1	58.6	76.1	108.5	102.4	105.4
10-14	TOTAL DIV. 10000-14000	100.0	91.7	98.2	100.0	96.1	99.2	100.0	96.7	99.3	100.0	91.7	98.2	100.0	78.4	95.4	100.0	87.5	97.3
15	MECHANICAL	100.0	90.0	95.8	99.8	86.5	94.3	95.4	111.8	102.2	100.0	92.1	96.7	95.2	85.1	91.0	95.6	91.0	93.7
16	ELECTRICAL	92.3	86.1	89.3	94.2	82.5	88.5	96.0	105.8	100.7	93.4	81.2	87.5	94.1	59.2	77.1	103.3	99.1	101.3
01-16	WEIGHTED AVERAGE	97.2	95.2	96.3	94.4	83.6	89.6	95.3	107.9	100.9	97.0	93.4	95.4	93.3	73.1	84.3	97.3	101.2	99.0

MINNESOTA

DIVISION		BRAINERD 564 MAT.	INST.	TOTAL	DETROIT LAKES 565 MAT.	INST.	TOTAL	DULUTH 556-558 MAT.	INST.	TOTAL	MANKATO 560 MAT.	INST.	TOTAL	MINNEAPOLIS 553-555 MAT.	INST.	TOTAL	ROCHESTER 559 MAT.	INST.	TOTAL
01590	EQUIPMENT RENTAL	.0	101.6	101.6	.0	98.3	98.3	.0	101.8	101.8	.0	101.6	101.6	.0	106.2	106.2	.0	101.8	101.8
02	SITE CONSTRUCTION	84.4	103.8	98.6	83.5	98.7	94.6	83.0	102.7	97.4	81.1	103.2	97.3	82.9	108.9	101.9	82.1	101.9	96.7
03100	CONCRETE FORMS & ACCESSORIES	86.5	99.6	97.8	81.1	104.6	101.3	99.7	122.4	119.3	95.9	104.0	102.9	100.4	138.3	133.1	100.2	111.1	109.6
03200	CONCRETE REINFORCEMENT	94.2	109.7	102.1	95.3	109.7	102.6	95.6	110.1	102.9	94.1	128.1	111.4	95.8	128.9	112.6	95.6	128.3	112.2
03300	CAST-IN-PLACE CONCRETE	111.3	108.7	110.2	99.3	107.7	102.8	109.7	108.8	109.3	102.4	113.2	106.8	108.0	126.4	115.6	105.9	103.4	104.9
03	CONCRETE	99.2	105.6	102.4	92.7	107.4	100.0	101.0	116.0	108.4	94.8	112.7	103.7	101.8	132.7	117.2	99.2	112.7	105.9
04	MASONRY	124.8	115.6	119.1	123.3	115.4	118.4	109.1	118.0	114.6	111.9	113.2	112.7	109.1	137.5	126.7	108.4	116.5	113.4
05	METALS	90.1	123.5	100.8	88.8	123.1	99.8	93.0	126.6	103.8	90.0	133.9	104.0	95.8	139.1	109.6	92.9	136.6	106.9
06	WOOD & PLASTICS	94.8	93.4	94.1	73.5	101.2	87.9	112.7	123.1	118.1	105.5	102.2	103.8	113.1	137.3	125.7	113.1	110.1	111.6
07	THERMAL & MOISTURE PROTECTION	98.2	110.1	103.2	99.7	87.1	94.4	100.4	121.0	109.0	98.5	100.6	99.4	100.3	136.5	115.5	100.2	107.6	103.4
08	DOORS & WINDOWS	86.7	112.3	93.4	99.1	116.5	103.6	96.7	124.4	104.0	91.6	123.7	100.0	99.9	142.8	111.1	96.7	128.1	104.9
09200	PLASTER & GYPSUM BOARD	88.4	93.8	91.8	103.7	101.5	102.4	105.9	124.2	117.3	93.2	102.8	99.2	106.1	138.6	126.3	105.7	110.9	108.9
095,098	CEILINGS & ACOUSTICAL TREATMENT	72.6	93.8	85.3	126.6	101.5	111.6	100.3	124.2	114.6	72.6	102.8	90.6	101.2	138.6	123.5	99.5	110.9	106.3
09600	FLOORING	100.7	131.2	109.0	100.5	131.2	108.8	105.0	122.6	109.8	103.3	120.9	108.1	102.4	131.2	110.2	104.8	78.5	97.7
097,099	WALL FINISHES, PAINTS & COATINGS	93.1	98.5	96.4	97.2	98.5	98.0	91.4	112.0	104.0	101.3	104.7	103.4	97.6	127.9	116.1	93.7	103.8	99.9
09	FINISHES	91.8	104.5	98.4	107.7	109.0	108.4	102.5	121.8	112.5	93.6	107.1	100.6	102.3	136.2	120.0	102.3	104.8	103.6
10-14	TOTAL DIV. 10000-14000	100.0	90.3	97.9	100.0	91.7	98.2	100.0	102.8	100.6	100.0	89.1	97.7	100.0	109.3	102.0	100.0	99.9	100.0
15	MECHANICAL	94.7	95.8	95.2	95.6	95.4	95.5	100.1	114.6	106.1	94.7	98.5	96.3	100.1	119.3	108.1	100.1	106.4	102.7
16	ELECTRICAL	101.4	100.6	101.0	103.2	70.1	87.1	103.5	100.6	102.1	107.5	87.2	97.6	105.3	114.3	109.7	103.5	87.2	95.6
01-16	WEIGHTED AVERAGE	96.0	105.1	100.0	97.9	100.6	99.1	99.5	114.5	106.2	96.3	105.9	100.6	100.5	126.7	112.2	99.2	108.5	103.4

City Cost Indexes

MINNESOTA / MISSISSIPPI

DIVISION		SAINT PAUL 550-551			ST. CLOUD 563			THIEF RIVER FALLS 567			WILLMAR 562			WINDOM 561			BILOXI 395		
		MAT.	INST.	TOTAL	MAT.	INST.	TOTAL	MAT.	INST.	TOTAL	MAT.	INST.	TOTAL	MAT.	INST.	TOTAL	MAT.	INST.	TOTAL
01590	EQUIPMENT RENTAL	.0	101.8	101.8	.0	101.6	101.6	.0	98.3	98.3	.0	101.6	101.6	.0	101.6	101.6	.0	98.3	98.3
02	SITE CONSTRUCTION	85.1	103.4	98.5	79.9	105.4	98.6	84.1	98.1	94.4	79.2	99.9	94.4	73.4	98.7	92.0	95.9	85.9	90.1
03100	CONCRETE FORMS & ACCESSORIES	91.1	134.3	128.4	83.1	133.0	126.2	85.4	95.5	94.2	82.8	59.0	62.3	88.1	55.0	59.5	95.9	41.2	48.8
03200	CONCRETE REINFORCEMENT	92.3	128.8	110.8	94.3	128.4	111.6	95.6	109.6	102.7	94.0	127.7	111.1	94.0	126.5	110.5	95.1	64.7	79.6
03300	CAST-IN-PLACE CONCRETE	109.0	125.6	115.8	98.0	123.8	108.6	101.4	93.9	98.3	99.7	81.4	92.2	86.2	58.2	74.7	105.1	44.8	80.3
03	CONCRETE	103.3	130.7	116.9	90.6	129.3	109.8	93.9	98.7	96.3	90.6	81.4	86.1	81.7	71.5	76.7	97.9	48.8	73.5
04	MASONRY	118.8	137.4	130.3	109.5	128.6	121.3	101.0	106.4	104.4	112.3	92.6	100.1	123.0	63.2	85.9	90.1	37.0	57.2
05	METALS	92.4	138.7	107.2	90.6	136.1	105.1	89.0	123.0	99.8	90.0	126.2	101.5	89.9	124.5	100.9	90.4	83.0	88.0
06	WOOD & PLASTICS	102.7	132.1	118.0	91.2	132.0	112.5	78.4	93.7	86.3	90.9	56.1	72.8	96.6	56.1	75.5	96.2	41.1	67.5
07	THERMAL & MOISTURE PROTECTION	100.2	135.8	115.2	98.2	121.3	108.0	101.1	81.2	92.7	98.1	62.0	82.9	98.1	51.2	78.4	96.4	45.9	75.2
08	DOORS & WINDOWS	94.1	140.0	106.1	91.6	139.9	104.2	99.1	112.4	102.6	89.1	81.4	87.1	93.4	81.4	90.3	98.3	48.0	85.1
09200	PLASTER & GYPSUM BOARD	100.4	133.5	121.0	87.3	133.5	116.0	105.2	93.8	98.2	87.3	55.5	67.6	89.1	55.5	68.3	103.5	40.0	64.1
095,098	CEILINGS & ACOUSTICAL TREATMENT	97.8	133.5	119.1	72.6	133.5	108.9	126.6	93.8	107.1	72.6	55.5	62.4	72.6	55.5	62.4	104.3	40.0	66.0
09600	FLOORING	97.8	131.2	106.8	96.7	131.2	106.0	101.6	131.2	109.6	98.5	120.9	104.5	101.4	120.9	106.7	118.7	39.6	97.3
097,099	WALL FINISHES, PAINTS & COATINGS	97.6	127.9	116.1	101.3	127.9	117.5	97.2	98.5	98.0	97.2	98.5	98.0	97.2	104.7	101.8	106.7	35.2	63.1
09	FINISHES	99.5	133.2	117.0	90.6	132.5	112.4	108.3	102.4	105.3	90.9	71.9	81.0	91.6	70.2	80.5	107.2	39.8	72.1
10 - 14	TOTAL DIV. 10000 - 14000	100.0	108.3	101.8	100.0	107.1	101.5	100.0	87.4	97.3	100.0	58.7	91.2	100.0	55.2	90.4	100.0	51.4	89.6
15	MECHANICAL	100.1	115.7	106.6	99.7	117.9	107.2	95.6	90.6	93.5	94.7	93.7	94.3	107.5	77.5	87.6	99.9	53.9	80.8
16	ELECTRICAL	104.1	114.2	109.0	101.4	114.2	107.7	100.5	70.1	85.7	101.4	76.7	89.4	107.5	87.1	97.6	97.0	61.0	79.5
01 - 16	WEIGHTED AVERAGE	99.7	124.5	110.7	95.7	123.3	108.0	96.8	96.1	96.5	94.4	87.4	91.3	95.1	80.3	88.5	97.8	55.5	79.0

MISSISSIPPI

DIVISION		CLARKSDALE 386			COLUMBUS 397			GREENVILLE 387			GREENWOOD 389			JACKSON 390 - 392			LAUREL 394		
		MAT.	INST.	TOTAL	MAT.	INST.	TOTAL	MAT.	INST.	TOTAL	MAT.	INST.	TOTAL	MAT.	INST.	TOTAL	MAT.	INST.	TOTAL
01590	EQUIPMENT RENTAL	.0	98.3	98.3	.0	98.3	98.3	.0	98.3	98.3	.0	98.3	98.3	.0	98.3	98.3	.0	98.3	98.3
02	SITE CONSTRUCTION	98.8	85.1	88.7	100.5	85.4	89.5	105.5	85.7	91.0	102.3	84.8	89.5	98.1	85.7	89.0	106.6	84.1	90.1
03100	CONCRETE FORMS & ACCESSORIES	84.5	18.3	27.4	82.2	29.5	36.8	80.4	34.0	40.4	95.8	20.1	30.5	95.6	38.8	46.6	82.5	23.3	31.4
03200	CONCRETE REINFORCEMENT	102.4	16.9	59.0	101.8	30.6	65.6	103.1	43.2	72.6	102.4	45.9	73.7	95.1	46.8	70.6	102.4	30.0	65.6
03300	CAST-IN-PLACE CONCRETE	102.5	26.7	71.3	107.2	36.9	78.3	105.5	39.2	78.3	110.2	26.6	75.8	103.1	41.7	77.8	104.8	29.7	73.9
03	CONCRETE	96.1	23.3	60.0	99.5	34.4	67.2	101.0	39.6	70.5	102.4	29.6	66.2	96.9	43.2	70.3	101.6	29.0	65.6
04	MASONRY	90.0	19.8	46.5	113.8	23.3	57.7	133.3	38.3	74.4	90.6	18.5	45.9	92.0	38.3	58.8	110.3	23.9	56.8
05	METALS	88.0	59.0	78.8	87.9	65.8	80.8	88.8	72.8	83.7	88.0	71.2	82.6	90.4	74.6	85.3	88.0	64.2	80.4
06	WOOD & PLASTICS	83.1	17.0	48.7	80.8	30.6	54.7	78.4	33.3	54.9	96.3	19.2	56.1	96.3	39.6	66.7	81.1	24.1	51.4
07	THERMAL & MOISTURE PROTECTION	96.2	27.6	67.3	96.3	27.5	67.4	96.5	41.2	73.2	96.6	23.2	65.7	96.3	42.4	73.6	96.5	26.2	66.9
08	DOORS & WINDOWS	97.6	21.1	77.6	97.6	28.7	79.6	97.7	37.8	82.0	97.6	30.4	80.1	98.7	42.3	83.9	94.2	25.0	76.1
09200	PLASTER & GYPSUM BOARD	94.6	15.3	45.4	94.2	29.3	53.9	92.0	32.1	54.8	102.0	17.6	49.5	103.5	38.5	63.1	92.0	22.6	48.9
095,098	CEILINGS & ACOUSTICAL TREATMENT	96.4	15.3	48.1	96.4	29.3	56.4	98.3	32.1	58.8	96.4	17.6	49.4	104.3	38.5	65.0	107.7	18.7	83.6
09600	FLOORING	111.3	23.0	87.4	109.9	19.9	85.6	108.9	33.8	88.6	118.7	20.7	92.2	118.7	39.1	97.2	106.7	21.8	54.9
097,099	WALL FINISHES, PAINTS & COATINGS	106.7	14.8	50.6	106.7	21.8	54.9	106.7	34.6	62.7	106.7	20.2	53.9	106.7	34.6	62.7	106.7	21.8	54.9
09	FINISHES	101.5	18.6	58.3	101.3	27.4	62.8	101.4	33.8	66.2	105.1	20.1	60.9	107.2	38.5	71.4	100.7	22.8	60.1
10 - 14	TOTAL DIV. 10000 - 14000	100.0	40.9	87.4	100.0	43.5	87.9	100.0	49.8	89.3	100.0	41.3	87.5	100.0	50.6	89.5	100.0	26.1	84.2
15	MECHANICAL	97.7	18.8	65.0	97.2	25.6	67.5	99.9	34.3	72.7	97.7	19.8	65.4	99.9	34.0	72.5	97.2	21.8	65.9
16	ELECTRICAL	96.3	24.6	61.4	94.6	27.0	61.7	96.3	37.1	67.6	96.3	25.1	61.7	97.8	37.1	68.3	96.0	25.0	61.5
01 - 16	WEIGHTED AVERAGE	95.9	30.8	67.0	97.1	36.7	70.3	99.3	44.6	75.0	97.2	33.3	68.8	97.8	46.1	74.8	97.1	33.4	68.8

MISSISSIPPI / MISSOURI

DIVISION		MCCOMB 396			MERIDIAN 393			TUPELO 388			BOWLING GREEN 633			CAPE GIRARDEAU 637			CHILLICOTHE 646		
		MAT.	INST.	TOTAL	MAT.	INST.	TOTAL	MAT.	INST.	TOTAL	MAT.	INST.	TOTAL	MAT.	INST.	TOTAL	MAT.	INST.	TOTAL
01590	EQUIPMENT RENTAL	.0	98.3	98.3	.0	98.3	98.3	.0	98.3	98.3	.0	108.3	108.3	.0	108.3	108.3	.0	103.4	103.4
02	SITE CONSTRUCTION	92.6	85.4	87.3	97.0	85.7	88.7	97.0	85.0	88.2	93.0	92.7	92.8	94.3	93.3	93.5	104.0	91.4	94.7
03100	CONCRETE FORMS & ACCESSORIES	82.2	57.6	61.0	78.9	33.0	39.3	81.1	23.6	31.5	91.4	90.9	91.0	83.4	77.2	78.1	91.2	79.8	81.3
03200	CONCRETE REINFORCEMENT	103.1	39.5	70.8	101.8	46.3	73.6	100.3	46.5	73.0	93.4	98.4	95.9	94.5	84.5	89.4	108.4	94.9	101.6
03300	CAST-IN-PLACE CONCRETE	93.0	40.2	71.3	99.6	41.7	75.7	102.5	38.3	76.1	88.4	84.1	86.6	87.4	83.3	85.7	97.0	71.8	86.6
03	CONCRETE	89.6	49.7	69.8	93.7	40.6	67.3	95.5	35.3	65.6	88.8	91.3	90.0	87.8	82.4	85.1	101.1	80.6	90.9
04	MASONRY	116.2	38.5	68.0	89.8	28.6	51.8	123.2	30.5	65.7	117.3	97.6	105.1	113.8	81.7	93.9	104.5	90.7	96.0
05	METALS	88.1	70.8	82.6	88.7	73.7	83.9	87.9	72.8	83.1	92.0	116.2	99.7	92.7	109.1	98.0	93.9	81.5	87.5
06	WOOD & PLASTICS	80.8	64.8	72.5	77.1	32.0	53.6	79.0	22.5	49.6	87.0	91.7	89.5	78.5	74.5	76.5	92.7	82.1	88.3
07	THERMAL & MOISTURE PROTECTION	95.9	42.9	73.6	96.1	39.1	72.1	96.2	34.3	70.1	92.8	100.9	96.2	92.7	80.8	87.7	92.7	82.1	88.3
08	DOORS & WINDOWS	97.7	53.1	86.0	97.6	36.6	81.7	97.6	28.9	79.7	93.9	88.3	92.5	93.9	75.5	89.1	90.9	76.6	87.2
09200	PLASTER & GYPSUM BOARD	94.2	64.4	75.7	92.0	30.7	53.9	92.0	20.9	47.9	104.5	91.3	96.3	100.8	73.7	84.0	97.7	80.7	87.2
095,098	CEILINGS & ACOUSTICAL TREATMENT	96.4	64.4	77.3	98.3	30.7	58.0	96.4	20.9	51.4	86.1	91.3	89.2	86.1	73.7	78.7	82.0	80.7	81.3
09600	FLOORING	109.9	23.0	86.4	107.6	27.5	85.9	109.3	25.4	86.6	97.1	74.8	91.0	92.7	74.8	87.9	88.9	53.7	79.3
097,099	WALL FINISHES, PAINTS & COATINGS	106.7	32.6	61.5	106.7	34.2	62.5	106.7	34.6	62.7	101.0	104.3	103.0	101.0	75.7	85.6	93.1	57.2	71.2
09	FINISHES	100.8	50.2	74.4	100.5	31.5	64.5	100.5	24.8	61.1	92.9	89.0	90.9	91.0	75.5	82.9	91.5	73.5	82.1
10 - 14	TOTAL DIV. 10000 - 14000	100.0	72.6	94.2	100.0	49.8	89.3	100.0	42.1	87.7	100.0	69.2	93.4	100.1	66.8	92.9	100.0	65.4	92.6
15	MECHANICAL	97.2	30.3	69.4	99.9	34.0	72.6	98.0	24.5	67.5	95.2	101.4	97.8	100.1	100.9	100.5	95.1	51.2	76.9
16	ELECTRICAL	93.3	61.3	77.7	96.0	61.3	79.1	96.1	24.6	61.3	101.4	92.5	97.1	101.4	114.6	107.8	98.7	42.8	71.5
01 - 16	WEIGHTED AVERAGE	95.7	52.1	76.4	96.0	46.7	74.1	97.3	37.3	70.6	95.7	95.8	95.7	96.4	92.3	94.6	96.7	71.2	85.4

COST INDEXES

MISSOURI

DIVISION		COLUMBIA 652			FLAT RIVER 636			HANNIBAL 634			HARRISONVILLE 647			JEFFERSON CITY 650-651			JOPLIN 648		
		MAT.	INST.	TOTAL	MAT.	INST.	TOTAL	MAT.	INST.	TOTAL	MAT.	INST.	TOTAL	MAT.	INST.	TOTAL	MAT.	INST.	TOTAL
01590	EQUIPMENT RENTAL	.0	109.1	109.1	.0	108.3	108.3	.0	108.3	108.3	.0	103.4	103.4	.0	109.1	109.1	.0	107.1	107.1
02	SITE CONSTRUCTION	103.4	94.6	96.9	95.7	93.4	94.0	90.2	91.6	91.2	95.8	95.6	95.7	105.5	94.0	97.0	103.6	98.8	100.1
03100	CONCRETE FORMS & ACCESSORIES	85.0	72.0	73.8	98.3	86.6	88.2	89.4	70.7	73.2	87.6	95.5	94.4	99.6	63.6	68.5	105.0	73.5	77.8
03200	CONCRETE REINFORCEMENT	101.1	113.3	107.3	94.5	109.5	102.2	92.9	83.2	88.0	108.0	103.6	105.8	99.7	110.3	105.1	111.9	75.7	93.5
03300	CAST-IN-PLACE CONCRETE	90.4	84.3	87.9	91.2	84.6	88.5	83.7	78.1	81.4	99.3	99.1	99.3	94.5	71.1	84.9	105.1	72.6	91.7
03	CONCRETE	86.8	85.8	86.3	91.5	91.7	91.6	85.3	77.3	81.3	97.8	98.9	98.3	90.4	77.0	83.7	101.6	74.6	88.2
04	MASONRY	131.0	86.2	103.2	115.1	83.6	95.6	108.3	88.5	96.0	99.0	103.6	101.9	94.4	86.2	89.3	96.5	61.7	74.9
05	METALS	93.5	121.2	102.4	91.9	120.2	100.9	92.0	107.2	96.8	97.0	109.1	100.9	93.3	119.6	101.7	99.4	89.8	96.3
06	WOOD & PLASTICS	89.1	66.2	77.2	94.6	85.4	89.8	84.8	70.1	77.1	89.2	92.1	90.7	105.1	55.3	79.2	106.9	73.3	89.4
07	THERMAL & MOISTURE PROTECTION	89.6	84.8	87.6	93.0	96.7	94.6	92.6	89.1	91.1	91.9	103.6	96.9	89.7	81.9	86.4	93.5	69.6	83.4
08	DOORS & WINDOWS	90.8	91.5	91.0	93.9	88.2	92.4	93.9	72.6	88.4	91.3	98.8	93.3	87.7	84.8	86.9	92.2	76.2	88.0
09200	PLASTER & GYPSUM BOARD	100.1	65.2	78.4	108.2	84.9	93.7	103.0	69.1	82.0	90.7	91.6	91.3	106.8	54.0	74.0	103.4	72.3	84.1
095,098	CEILINGS & ACOUSTICAL TREATMENT	88.6	65.2	74.6	86.1	84.9	85.4	86.1	69.1	76.0	82.0	91.6	87.8	88.6	54.0	68.0	85.4	72.3	77.6
09600	FLOORING	101.9	91.1	98.9	100.8	74.8	93.7	95.9	74.8	90.2	85.0	98.3	88.6	110.3	91.1	105.1	111.5	50.3	95.0
097,099	WALL FINISHES, PAINTS & COATINGS	98.8	72.2	82.6	101.0	75.7	85.6	101.0	81.6	89.2	97.5	90.4	93.2	98.8	72.2	82.6	92.5	44.7	63.4
09	FINISHES	93.1	72.3	82.3	94.7	82.3	88.2	92.2	71.6	81.5	89.0	94.8	92.0	96.7	65.9	80.7	98.3	66.2	81.6
10-14	TOTAL DIV. 10000-14000	100.0	90.6	98.0	100.0	68.8	93.4	100.0	63.1	92.1	100.0	86.8	97.2	100.0	89.2	97.7	100.0	77.0	95.1
15	MECHANICAL	99.9	101.8	100.7	95.2	101.9	98.0	95.2	96.5	95.8	95.0	102.1	97.9	99.9	101.8	100.7	100.1	57.5	82.4
16	ELECTRICAL	99.0	92.5	95.8	105.9	114.6	110.1	100.2	84.8	92.7	105.2	109.4	107.3	100.8	92.5	96.8	96.8	62.1	80.0
01-16	WEIGHTED AVERAGE	97.1	92.4	95.0	96.7	97.0	96.8	94.5	86.3	90.9	96.4	101.3	98.6	96.1	89.7	93.3	98.6	70.5	86.1

MISSOURI

DIVISION		KANSAS CITY 640-641			KIRKSVILLE 635			POPLAR BLUFF 639			ROLLA 654-655			SEDALIA 653			SIKESTON 638		
		MAT.	INST.	TOTAL	MAT.	INST.	TOTAL	MAT.	INST.	TOTAL	MAT.	INST.	TOTAL	MAT.	INST.	TOTAL	MAT.	INST.	TOTAL
01590	EQUIPMENT RENTAL	.0	104.9	104.9	.0	99.5	99.5	.0	102.0	102.0	.0	109.1	109.1	.0	100.1	100.1	.0	102.0	102.0
02	SITE CONSTRUCTION	97.4	98.1	97.9	89.1	87.7	88.1	75.6	92.2	87.7	102.8	92.9	95.6	94.4	90.3	91.4	79.0	92.3	88.7
03100	CONCRETE FORMS & ACCESSORIES	103.9	106.6	106.3	81.2	60.3	63.2	81.2	73.2	74.3	93.8	86.9	87.8	91.1	66.0	69.4	82.4	73.4	74.6
03200	CONCRETE REINFORCEMENT	106.3	114.2	110.3	93.7	82.4	88.0	96.6	85.2	90.8	101.6	58.8	79.9	100.1	93.8	96.9	95.9	75.7	85.7
03300	CAST-IN-PLACE CONCRETE	97.6	107.7	101.8	91.2	67.5	81.4	70.1	65.6	68.3	92.5	77.5	86.3	96.6	70.8	86.0	75.0	74.9	74.9
03	CONCRETE	97.4	108.7	103.0	100.9	68.3	84.7	78.8	74.0	76.4	88.8	79.5	84.2	100.4	74.1	87.4	82.3	75.4	78.9
04	MASONRY	101.3	106.8	104.7	120.8	75.1	92.5	112.9	63.0	82.0	107.7	50.3	72.1	113.7	57.7	79.0	112.8	68.8	85.5
05	METALS	106.9	114.9	109.5	91.7	94.7	92.6	92.2	97.2	93.8	93.3	90.2	92.3	91.8	100.9	94.7	92.5	92.6	92.5
06	WOOD & PLASTICS	106.2	106.1	106.2	72.8	58.9	65.5	71.9	74.6	73.3	98.1	97.9	98.0	91.4	59.5	74.8	73.2	74.6	73.9
07	THERMAL & MOISTURE PROTECTION	92.3	108.0	98.9	97.0	81.3	90.4	97.0	78.8	89.4	89.7	64.0	78.9	93.6	80.9	88.2	97.2	72.2	86.6
08	DOORS & WINDOWS	98.3	109.1	101.1	97.8	66.7	89.7	98.7	78.5	93.4	90.8	75.2	86.7	94.2	76.4	89.5	98.7	67.9	90.7
09200	PLASTER & GYPSUM BOARD	100.1	106.0	103.7	97.2	57.6	72.6	97.9	73.7	82.8	102.7	97.7	99.6	96.0	58.3	72.5	99.0	73.7	83.3
095,098	CEILINGS & ACOUSTICAL TREATMENT	91.4	106.0	100.1	83.5	57.6	68.1	86.1	73.7	78.7	88.6	97.7	94.1	86.9	58.3	69.8	86.1	73.7	78.7
09600	FLOORING	91.6	98.3	93.4	75.1	55.1	69.7	88.0	62.3	81.1	106.4	46.9	90.3	83.4	98.3	87.4	88.6	50.3	78.2
097,099	WALL FINISHES, PAINTS & COATINGS	97.5	114.6	107.9	96.6	44.7	64.9	96.0	87.3	90.7	98.8	86.7	91.4	98.8	90.4	93.7	96.0	64.3	76.7
09	FINISHES	94.7	105.7	100.4	88.8	56.2	71.8	90.4	72.6	81.1	94.9	81.0	87.6	90.7	73.0	81.5	91.0	67.8	78.9
10-14	TOTAL DIV. 10000-14000	100.0	97.4	99.4	100.0	58.3	91.1	100.0	61.1	91.7	100.0	63.8	92.3	100.0	78.5	95.4	100.0	62.1	91.9
15	MECHANICAL	99.9	105.4	102.2	95.3	94.8	95.1	95.3	82.5	90.0	95.0	73.3	86.0	95.0	91.4	93.5	95.3	83.5	90.4
16	ELECTRICAL	108.3	109.4	108.8	99.5	84.7	92.3	100.2	114.6	107.2	97.5	92.4	95.0	97.5	109.4	103.3	99.3	114.6	106.7
01-16	WEIGHTED AVERAGE	100.9	106.8	103.5	96.8	79.2	89.0	94.0	83.8	89.5	95.0	78.6	87.7	96.1	84.9	91.2	94.5	83.2	89.5

MISSOURI / MONTANA

DIVISION		SPRINGFIELD 656-658			ST. JOSEPH 644-645			ST. LOUIS 630-631			BILLINGS 590-591			BUTTE 597			GREAT FALLS 594		
		MAT.	INST.	TOTAL	MAT.	INST.	TOTAL	MAT.	INST.	TOTAL	MAT.	INST.	TOTAL	MAT.	INST.	TOTAL	MAT.	INST.	TOTAL
01590	EQUIPMENT RENTAL	.0	102.8	102.8	.0	103.4	103.4	.0	109.6	109.6	.0	98.6	98.6	.0	98.3	98.3	.0	98.3	98.3
02	SITE CONSTRUCTION	97.3	93.8	94.7	98.9	93.7	95.1	94.0	96.4	95.8	87.1	96.0	93.6	94.5	94.6	94.6	98.1	95.4	96.1
03100	CONCRETE FORMS & ACCESSORIES	100.4	73.7	77.4	103.8	85.8	88.2	98.3	105.4	104.4	96.8	63.6	68.2	84.1	61.2	64.3	98.0	64.5	69.1
03200	CONCRETE REINFORCEMENT	97.4	112.9	105.3	105.0	103.3	104.2	86.5	111.3	99.1	93.3	75.7	84.3	101.2	75.5	88.1	93.3	75.6	84.3
03300	CAST-IN-PLACE CONCRETE	103.6	66.9	88.5	97.6	102.8	99.7	87.4	107.4	95.6	119.3	69.6	98.9	122.8	71.1	101.5	129.8	56.8	99.8
03	CONCRETE	100.2	79.7	90.0	97.2	95.6	96.4	87.5	108.2	97.8	105.3	68.9	87.2	105.3	68.3	86.9	110.5	64.9	87.8
04	MASONRY	88.0	82.9	84.8	100.8	93.1	96.0	94.9	111.6	105.3	121.6	70.9	90.1	121.2	65.4	86.6	125.1	74.1	93.5
05	METALS	97.4	107.2	100.5	102.8	106.4	104.0	97.4	123.4	105.7	106.3	86.1	99.8	99.2	85.4	94.8	102.5	85.5	97.3
06	WOOD & PLASTICS	100.0	73.2	86.0	106.9	83.2	94.6	94.4	103.1	98.9	101.2	63.3	81.4	88.5	61.7	74.5	103.9	63.3	82.7
07	THERMAL & MOISTURE PROTECTION	94.1	78.7	87.6	92.7	94.5	93.5	92.6	106.7	98.5	98.8	68.2	85.9	98.3	67.3	85.2	98.9	67.8	85.8
08	DOORS & WINDOWS	96.3	82.2	92.6	97.0	94.0	96.2	93.0	110.7	97.6	99.0	62.7	89.5	95.9	61.7	86.9	99.3	62.7	89.7
09200	PLASTER & GYPSUM BOARD	103.8	72.3	84.3	104.7	82.5	90.9	109.1	103.0	105.3	103.7	62.6	78.2	95.0	61.0	73.8	103.7	62.6	78.2
095,098	CEILINGS & ACOUSTICAL TREATMENT	88.6	72.3	78.9	90.6	82.5	85.7	91.2	103.0	98.2	112.8	62.6	82.8	112.1	61.0	81.6	114.7	62.6	83.6
09600	FLOORING	105.4	50.3	90.5	93.9	80.6	90.3	100.5	97.1	99.6	107.3	49.0	91.5	98.4	39.0	82.3	107.3	52.9	92.6
097,099	WALL FINISHES, PAINTS & COATINGS	93.1	64.0	75.4	93.1	77.3	83.5	101.0	108.4	105.6	97.2	59.7	74.3	97.2	40.6	62.7	97.2	45.8	65.8
09	FINISHES	96.6	67.7	81.6	95.7	83.4	89.3	95.9	103.5	99.9	106.4	60.0	82.2	102.2	54.7	77.2	107.0	59.9	82.4
10-14	TOTAL DIV. 10000-14000	100.0	88.2	97.5	100.0	92.0	98.3	100.0	99.8	100.0	100.0	67.1	93.0	100.0	66.1	92.6	100.0	68.4	93.3
15	MECHANICAL	100.0	69.7	87.4	100.1	92.4	96.9	100.1	108.3	103.5	100.3	76.5	90.4	100.3	69.5	87.5	100.3	78.2	91.1
16	ELECTRICAL	102.4	67.1	85.2	105.3	93.2	99.4	104.3	114.6	109.3	94.2	78.7	86.6	100.9	75.6	88.6	94.2	73.3	84.0
01-16	WEIGHTED AVERAGE	98.4	78.7	89.6	99.9	93.3	97.0	97.0	109.0	102.3	102.1	74.5	89.8	101.2	70.9	87.7	102.7	73.8	89.9

MONTANA

DIVISION		HAVRE 595 MAT.	INST.	TOTAL	HELENA 596 MAT.	INST.	TOTAL	KALISPELL 599 MAT.	INST.	TOTAL	MILES CITY 593 MAT.	INST.	TOTAL	MISSOULA 598 MAT.	INST.	TOTAL	WOLF POINT 592 MAT.	INST.	TOTAL
01590	EQUIPMENT RENTAL	.0	98.3	98.3	.0	98.3	98.3	.0	98.3	98.3	.0	98.3	98.3	.0	98.3	98.3	.0	98.3	98.3
02	SITE CONSTRUCTION	102.1	95.1	97.0	100.1	95.4	96.6	84.7	94.5	91.9	91.2	95.1	94.1	77.3	94.5	89.9	108.9	95.1	98.8
03100	CONCRETE FORMS & ACCESSORIES	75.9	60.6	62.7	98.1	65.0	69.5	87.7	57.2	61.4	95.8	60.7	65.5	87.7	57.4	61.5	87.6	59.3	63.2
03200	CONCRETE REINFORCEMENT	102.1	63.6	82.5	96.6	63.3	79.7	104.0	80.8	92.2	101.6	63.8	82.4	103.0	81.1	91.9	103.1	63.6	83.0
03300	CAST-IN-PLACE CONCRETE	132.3	61.6	103.2	132.2	66.6	105.3	106.6	63.0	88.7	116.7	61.4	94.0	90.4	63.1	79.2	130.8	57.0	100.4
03	CONCRETE	113.2	62.5	88.0	112.2	66.1	89.3	95.3	64.7	80.1	102.4	62.5	82.6	84.1	64.9	74.5	116.9	60.3	88.8
04	MASONRY	122.1	67.6	88.3	120.3	66.7	87.1	120.3	53.3	78.8	127.2	72.1	93.0	146.1	58.7	91.9	128.2	67.6	90.6
05	METALS	95.7	79.0	90.3	102.0	78.9	94.6	95.5	86.4	92.6	95.0	79.6	90.0	95.9	86.9	93.0	95.0	79.0	89.9
06	WOOD & PLASTICS	79.5	59.8	69.2	104.0	63.3	82.8	92.9	58.0	74.7	100.6	59.8	79.3	92.9	58.0	74.7	91.8	58.0	74.2
07	THERMAL & MOISTURE PROTECTION	98.6	63.2	83.7	98.9	67.6	85.7	98.0	65.5	84.3	98.3	68.5	85.8	97.6	68.8	85.5	99.1	64.0	84.3
08	DOORS & WINDOWS	96.0	57.4	85.9	98.7	59.5	88.5	95.9	62.0	87.1	95.4	57.4	85.5	95.9	62.0	87.1	95.4	56.4	85.2
09200	PLASTER & GYPSUM BOARD	91.3	59.0	71.2	103.1	62.6	77.9	96.8	57.2	72.2	102.4	59.0	75.5	96.8	57.2	72.2	98.7	57.2	72.9
095,098	CEILINGS & ACOUSTICAL TREATMENT	112.1	59.0	80.4	112.1	62.6	82.6	112.1	57.2	79.3	109.6	59.0	79.4	112.1	57.2	79.3	109.6	57.2	78.3
09600	FLOORING	94.9	42.9	80.8	107.3	51.3	92.1	100.6	61.0	89.9	107.0	48.1	91.1	100.6	61.0	89.9	102.4	42.9	86.3
097,099	WALL FINISHES, PAINTS & COATINGS	97.2	45.2	65.5	97.2	50.8	68.9	97.2	45.8	65.8	97.2	42.7	63.9	97.2	45.8	65.8	97.2	42.7	63.9
09	FINISHES	101.0	55.3	77.2	106.3	60.7	82.5	102.7	56.2	78.5	105.1	56.0	79.5	102.3	56.2	78.3	104.1	53.9	78.0
10 - 14	TOTAL DIV. 10000 - 14000	100.0	54.3	90.3	100.0	56.4	90.7	100.0	51.5	89.7	100.0	54.3	90.3	100.0	51.5	89.7	100.0	54.1	90.2
15	MECHANICAL	95.4	77.1	87.8	100.3	73.9	89.4	95.4	69.3	84.6	95.4	77.0	87.8	100.3	69.3	87.4	95.4	77.1	87.8
16	ELECTRICAL	94.2	61.6	78.4	94.2	74.6	84.7	98.0	78.4	88.5	94.2	63.9	79.4	99.0	78.4	89.0	94.2	63.9	79.4
01 - 16	WEIGHTED AVERAGE	99.6	68.9	86.0	102.5	71.5	88.7	97.8	69.4	85.2	98.8	70.0	86.0	98.9	70.1	86.1	100.8	68.7	86.6

NEBRASKA

DIVISION		ALLIANCE 693 MAT.	INST.	TOTAL	COLUMBUS 686 MAT.	INST.	TOTAL	GRAND ISLAND 688 MAT.	INST.	TOTAL	HASTINGS 689 MAT.	INST.	TOTAL	LINCOLN 683-685 MAT.	INST.	TOTAL	MCCOOK 690 MAT.	INST.	TOTAL
01590	EQUIPMENT RENTAL	.0	96.6	96.6	.0	101.6	101.6	.0	101.6	101.6	.0	101.6	101.6	.0	101.6	101.6	.0	101.6	101.6
02	SITE CONSTRUCTION	97.1	95.3	95.8	95.6	89.0	90.7	100.3	91.5	93.9	99.1	89.6	92.2	91.6	91.5	91.5	99.8	89.0	91.9
03100	CONCRETE FORMS & ACCESSORIES	87.5	25.7	34.2	97.4	24.8	34.8	96.8	49.6	56.1	100.7	45.6	53.2	101.5	45.8	53.4	94.0	31.0	39.7
03200	CONCRETE REINFORCEMENT	116.3	44.5	79.8	106.6	49.7	77.7	106.0	74.7	90.1	106.0	49.6	77.3	97.4	75.6	86.3	108.0	74.3	90.9
03300	CAST-IN-PLACE CONCRETE	108.8	36.2	78.9	111.4	46.0	84.5	118.0	57.4	93.1	118.0	50.7	90.3	106.5	59.9	87.4	118.0	36.4	84.4
03	CONCRETE	119.3	34.0	76.9	105.2	38.7	72.2	109.7	58.4	84.2	110.0	49.5	80.0	101.7	57.7	79.8	109.9	42.7	76.5
04	MASONRY	108.1	29.0	59.1	113.6	30.0	61.7	105.9	48.6	70.4	114.9	39.3	68.0	96.1	64.9	76.7	103.7	28.9	57.4
05	METALS	100.5	55.4	86.1	92.3	70.7	85.4	93.6	83.9	90.5	95.0	70.9	87.3	98.7	85.4	94.5	95.2	79.4	90.1
06	WOOD & PLASTICS	85.4	24.6	53.7	97.0	22.0	57.9	96.3	44.1	69.1	100.5	44.1	71.1	101.0	38.5	68.5	93.9	30.6	60.9
07	THERMAL & MOISTURE PROTECTION	101.9	34.6	73.6	94.4	31.8	68.0	94.5	57.4	78.9	94.6	46.0	74.1	94.8	61.1	80.6	94.5	35.1	69.5
08	DOORS & WINDOWS	90.3	27.9	74.0	90.1	33.0	75.2	90.1	49.0	79.4	90.1	45.0	78.3	95.5	48.6	83.3	90.2	38.1	76.6
09200	PLASTER & GYPSUM BOARD	94.4	22.5	49.7	101.9	19.7	50.8	101.6	42.4	64.8	103.0	42.4	65.4	105.6	36.7	62.8	101.6	28.5	56.2
095,098	CEILINGS & ACOUSTICAL TREATMENT	89.2	22.5	49.4	86.9	19.7	46.8	86.9	42.4	60.4	86.9	42.4	60.4	95.5	36.7	60.4	86.9	28.5	52.1
09600	FLOORING	94.0	28.5	76.2	95.0	36.1	79.1	94.9	34.7	78.6	96.5	36.1	80.2	96.7	41.8	81.8	93.7	28.5	76.0
097,099	WALL FINISHES, PAINTS & COATINGS	145.1	21.8	69.9	90.8	50.0	65.9	90.8	41.3	60.6	90.8	50.0	65.9	90.8	41.7	60.8	90.8	23.7	49.9
09	FINISHES	93.8	25.3	58.1	93.0	28.0	59.2	93.1	44.2	67.6	93.8	43.8	67.7	96.2	42.5	68.2	92.8	28.8	59.5
10 - 14	TOTAL DIV. 10000 - 14000	100.0	46.4	88.6	100.0	59.2	91.3	100.0	68.9	93.4	100.0	66.2	92.8	100.0	68.2	93.2	100.0	45.5	88.4
15	MECHANICAL	95.3	26.2	66.7	95.1	64.5	82.4	100.0	74.8	89.6	95.1	70.4	84.9	100.0	74.9	89.6	95.1	67.9	83.8
16	ELECTRICAL	92.6	18.1	56.4	92.3	43.0	68.3	91.0	70.3	80.9	90.4	43.0	67.4	96.6	70.3	83.8	94.7	18.2	57.5
01 - 16	WEIGHTED AVERAGE	99.0	35.9	71.0	96.1	49.6	75.5	97.5	65.6	83.3	97.1	56.7	79.2	98.2	67.1	84.4	96.9	48.2	75.3

NEBRASKA / NEVADA

DIVISION		NORFOLK 687 MAT.	INST.	TOTAL	NORTH PLATTE 691 MAT.	INST.	TOTAL	OMAHA 680-681 MAT.	INST.	TOTAL	VALENTINE 692 MAT.	INST.	TOTAL	CARSON CITY 897 MAT.	INST.	TOTAL	ELKO 898 MAT.	INST.	TOTAL
01590	EQUIPMENT RENTAL	.0	90.9	90.9	.0	101.6	101.6	.0	90.9	90.9	.0	94.0	94.0	.0	101.7	101.7	.0	101.7	101.7
02	SITE CONSTRUCTION	79.0	88.3	85.8	101.0	89.7	92.7	80.3	90.3	87.6	84.2	92.1	90.0	61.9	104.0	92.8	56.6	103.8	91.3
03100	CONCRETE FORMS & ACCESSORIES	82.6	45.6	50.7	96.6	45.9	52.8	96.5	73.6	76.7	83.8	22.9	31.3	94.2	100.7	99.8	106.4	89.1	91.5
03200	CONCRETE REINFORCEMENT	106.8	57.1	81.5	107.4	74.4	90.7	102.0	76.4	89.0	108.1	40.0	73.5	113.7	114.1	113.9	115.5	113.8	114.7
03300	CAST-IN-PLACE CONCRETE	112.1	56.0	89.0	118.0	50.8	90.4	110.6	74.1	95.6	103.8	42.5	78.6	109.7	91.4	102.2	98.4	80.9	91.2
03	CONCRETE	103.7	52.1	78.1	110.0	54.3	82.3	103.1	74.5	88.9	106.4	34.3	70.6	109.3	99.9	104.6	101.6	91.0	96.4
04	MASONRY	119.9	61.6	83.7	91.3	39.3	59.1	101.4	78.0	86.9	102.9	30.0	57.7	131.3	80.6	99.9	131.8	79.0	99.0
05	METALS	95.6	65.4	86.0	93.8	81.0	89.7	98.0	77.4	91.5	106.7	56.0	90.5	93.3	102.3	96.2	92.6	100.8	95.2
06	WOOD & PLASTICS	80.2	43.4	61.1	96.2	44.1	69.0	95.9	74.9	84.9	80.7	21.3	49.8	90.7	103.2	97.2	105.0	89.9	97.2
07	THERMAL & MOISTURE PROTECTION	92.4	54.7	76.5	94.5	46.1	74.1	90.2	71.5	82.3	93.0	31.1	66.9	99.8	88.9	95.2	99.7	80.9	91.8
08	DOORS & WINDOWS	92.1	46.1	80.0	89.4	51.3	79.5	98.7	67.5	90.6	92.2	26.7	75.1	94.3	108.8	98.1	96.9	94.0	96.1
09200	PLASTER & GYPSUM BOARD	101.0	42.4	64.6	102.0	42.4	65.0	107.9	74.7	87.3	104.6	19.7	51.9	83.6	103.0	95.7	93.2	89.5	90.9
095,098	CEILINGS & ACOUSTICAL TREATMENT	100.6	42.4	65.9	88.6	42.4	61.1	105.7	74.7	87.2	104.6	19.7	54.0	103.9	103.0	103.4	103.9	89.5	95.3
09600	FLOORING	117.0	43.4	97.1	94.8	36.1	78.9	122.7	46.3	102.1	119.9	43.4	99.2	103.0	63.8	90.5	106.6	63.8	95.0
097,099	WALL FINISHES, PAINTS & COATINGS	148.6	50.0	88.4	90.8	50.0	65.9	148.6	71.0	101.2	148.6	33.5	78.4	102.2	78.7	87.9	102.2	83.5	90.8
09	FINISHES	110.1	44.8	76.1	93.6	43.8	67.6	114.1	67.9	90.0	112.9	27.2	68.3	96.8	91.3	93.9	99.8	83.9	91.5
10 - 14	TOTAL DIV. 10000 - 14000	100.0	64.6	92.4	100.0	51.4	89.6	100.0	71.2	93.9	100.0	37.6	86.7	100.0	107.8	101.7	100.0	85.3	96.9
15	MECHANICAL	94.9	70.4	84.7	100.0	73.8	89.1	99.8	81.1	92.0	94.7	23.4	65.2	100.1	93.0	97.2	97.7	81.3	90.9
16	ELECTRICAL	91.4	54.0	73.2	92.9	43.0	68.7	90.6	86.3	88.5	90.5	42.9	67.4	91.0	110.9	100.7	90.5	68.4	79.7
01 - 16	WEIGHTED AVERAGE	97.7	60.6	81.3	97.1	58.8	80.1	99.3	78.2	89.9	99.2	38.5	72.3	98.6	97.9	98.3	97.6	85.4	92.2

COST INDEXES

725

NEVADA / NEW HAMPSHIRE

DIVISION		ELY 893 MAT.	INST.	TOTAL	LAS VEGAS 889 - 891 MAT.	INST.	TOTAL	RENO 894 - 895 MAT.	INST.	TOTAL	CHARLESTON 036 MAT.	INST.	TOTAL	CLAREMONT 037 MAT.	INST.	TOTAL	CONCORD 032 - 033 MAT.	INST.	TOTAL
01590	EQUIPMENT RENTAL	.0	101.7	101.7	.0	101.7	101.7	.0	101.7	101.7	.0	101.5	101.5	.0	101.5	101.5	.0	101.5	101.5
02	SITE CONSTRUCTION	61.5	103.4	92.2	61.5	106.2	94.3	61.6	104.0	92.7	80.9	98.2	93.6	75.2	98.2	92.0	87.8	99.9	96.6
03100	CONCRETE FORMS & ACCESSORIES	97.8	71.9	75.5	93.1	108.7	106.6	94.3	100.8	99.9	88.0	37.5	44.4	95.2	37.5	45.4	89.6	62.2	66.0
03200	CONCRETE REINFORCEMENT	114.2	105.1	109.6	108.1	119.3	113.8	108.1	118.5	113.4	85.9	57.9	71.7	85.9	57.9	71.7	85.9	88.6	87.3
03300	CAST-IN-PLACE CONCRETE	105.7	78.2	94.4	103.5	108.8	105.6	111.7	91.4	103.4	100.2	53.0	80.8	92.1	53.0	76.0	97.4	96.3	96.9
03	CONCRETE	109.9	80.9	95.5	105.4	110.4	107.8	109.4	100.7	105.1	103.0	47.6	75.5	94.9	47.6	71.4	99.8	79.2	89.6
04	MASONRY	137.0	70.6	95.8	122.2	100.4	108.7	130.7	80.6	99.6	93.3	41.4	61.1	93.5	41.4	61.2	96.6	92.2	93.9
05	METALS	92.5	96.9	93.9	93.9	107.4	98.2	93.7	104.2	97.1	94.5	66.0	85.4	94.5	66.0	85.4	95.1	85.2	91.9
06	WOOD & PLASTICS	94.7	71.3	82.5	90.4	107.7	99.4	90.7	103.2	97.2	85.3	36.8	60.1	93.3	36.8	63.9	86.7	54.2	69.8
07	THERMAL & MOISTURE PROTECTION	100.3	75.4	89.8	103.2	101.9	102.7	99.8	88.9	95.2	101.3	45.3	77.7	101.2	45.3	77.6	101.8	87.3	95.7
08	DOORS & WINDOWS	96.8	81.5	92.8	94.7	112.4	99.3	94.7	109.8	98.6	103.6	40.2	87.0	104.9	40.2	88.0	104.9	63.1	94.0
09200	PLASTER & GYPSUM BOARD	89.1	70.3	77.5	84.4	107.7	98.8	85.5	103.0	96.4	96.0	34.3	57.6	99.3	34.3	58.9	96.0	52.1	68.7
095,098	CEILINGS & ACOUSTICAL TREATMENT	103.9	70.3	83.9	109.9	107.7	108.6	111.6	103.0	106.5	85.5	34.3	54.9	85.5	34.3	54.9	85.5	52.1	65.6
09600	FLOORING	104.1	71.4	95.2	100.3	99.2	100.0	100.3	63.8	90.5	93.3	40.1	78.9	96.3	40.1	81.1	93.9	102.9	96.3
097,099	WALL FINISHES, PAINTS & COATINGS	102.2	78.7	87.9	102.2	117.1	111.3	102.2	78.7	87.9	91.5	32.1	55.3	91.5	32.1	55.3	91.5	94.9	93.6
09	FINISHES	98.9	71.9	84.8	98.3	107.7	103.2	99.0	91.3	95.0	93.4	36.7	63.9	94.4	36.7	64.4	94.4	71.7	82.6
10 - 14	TOTAL DIV. 10000 - 14000	100.0	70.9	93.8	100.0	103.3	100.7	100.0	107.8	101.7	100.0	45.7	88.4	100.0	45.7	88.4	100.0	74.6	94.6
15	MECHANICAL	97.7	76.8	89.0	100.1	107.1	103.0	100.1	93.1	97.2	95.1	39.5	72.1	95.1	39.5	72.1	100.1	90.3	96.0
16	ELECTRICAL	90.6	68.4	79.8	93.0	108.8	100.6	91.0	110.9	100.7	95.5	35.7	66.4	95.5	35.7	66.4	97.6	72.7	85.5
01 - 16	WEIGHTED AVERAGE	98.7	79.0	90.0	98.3	107.1	102.2	98.9	98.2	98.6	96.6	47.9	75.0	95.9	47.9	74.6	98.4	82.4	91.3

NEW HAMPSHIRE / NEW JERSEY

DIVISION		KEENE 034 MAT.	INST.	TOTAL	LITTLETON 035 MAT.	INST.	TOTAL	MANCHESTER 031 MAT.	INST.	TOTAL	NASHUA 030 MAT.	INST.	TOTAL	PORTSMOUTH 038 MAT.	INST.	TOTAL	ATLANTIC CITY 082,084 MAT.	INST.	TOTAL
01590	EQUIPMENT RENTAL	.0	101.5	101.5	.0	101.5	101.5	.0	101.5	101.5	.0	101.5	101.5	.0	101.5	101.5	.0	99.4	99.4
02	SITE CONSTRUCTION	87.9	98.2	95.5	75.1	97.8	91.8	87.7	99.9	96.6	89.0	99.9	97.0	83.2	98.9	94.7	91.5	105.1	101.5
03100	CONCRETE FORMS & ACCESSORIES	93.6	38.6	46.1	105.3	43.8	52.3	102.3	62.2	67.7	102.8	62.2	67.8	89.6	57.8	62.2	112.9	125.4	123.7
03200	CONCRETE REINFORCEMENT	85.9	69.0	77.3	86.7	41.4	63.7	107.5	88.6	97.9	107.5	88.6	97.9	85.9	88.5	87.2	81.9	110.4	96.4
03300	CAST-IN-PLACE CONCRETE	100.7	53.4	81.2	90.3	47.5	72.7	103.8	96.3	100.7	95.3	96.3	95.7	90.4	90.0	90.2	85.4	129.9	103.7
03	CONCRETE	102.9	50.3	76.8	94.4	45.4	70.1	107.3	79.2	93.3	103.2	79.2	91.3	94.1	75.0	84.6	96.1	122.9	109.4
04	MASONRY	97.4	41.4	62.7	106.9	40.0	65.4	99.1	92.2	94.8	99.4	92.2	94.9	94.7	81.9	86.8	101.7	121.3	113.9
05	METALS	95.0	71.3	87.5	95.1	58.8	83.5	99.7	85.2	95.1	99.7	85.2	95.0	96.1	83.1	92.0	94.8	100.1	96.5
06	WOOD & PLASTICS	91.4	36.8	63.0	103.6	42.2	71.6	101.5	54.2	76.9	101.5	54.2	76.9	86.7	54.2	69.8	113.2	126.0	119.9
07	THERMAL & MOISTURE PROTECTION	101.8	41.7	78.8	101.3	49.7	79.6	101.8	87.3	95.7	102.2	87.3	95.9	101.8	90.3	97.0	101.9	121.5	110.2
08	DOORS & WINDOWS	102.1	49.3	88.3	106.0	39.8	88.7	106.9	63.1	95.5	106.9	63.1	95.5	107.8	57.3	94.6	104.5	119.1	108.3
09200	PLASTER & GYPSUM BOARD	98.6	34.3	58.6	111.5	39.8	66.9	106.2	52.1	72.6	106.2	52.1	72.6	96.0	52.1	68.7	108.5	125.8	119.3
095,098	CEILINGS & ACOUSTICAL TREATMENT	85.5	34.3	54.9	85.5	39.8	58.2	98.5	52.1	70.8	98.5	52.1	70.8	87.5	52.1	66.3	85.5	125.8	109.6
09600	FLOORING	95.7	65.7	87.6	106.6	40.1	88.6	99.8	102.9	100.7	99.6	102.9	100.5	93.9	102.9	96.3	104.0	126.8	110.2
097,099	WALL FINISHES, PAINTS & COATINGS	91.5	32.1	55.3	91.5	65.6	75.7	91.5	94.9	93.6	91.5	94.9	93.6	91.5	38.9	59.4	91.5	115.5	106.1
09	FINISHES	95.5	42.0	67.6	99.3	44.8	70.9	100.8	71.7	85.6	101.0	71.7	85.7	94.6	62.8	78.1	99.8	124.9	112.9
10 - 14	TOTAL DIV. 10000 - 14000	100.0	64.9	92.5	100.0	48.2	89.0	100.0	74.6	94.6	100.0	74.6	94.6	100.0	70.7	93.7	100.0	116.4	103.5
15	MECHANICAL	95.1	42.3	73.2	95.1	80.5	89.1	100.1	90.3	96.0	100.1	90.3	96.0	100.1	84.8	93.7	99.7	116.3	106.6
16	ELECTRICAL	95.5	35.7	66.4	96.5	72.7	84.9	97.8	72.7	85.6	97.6	72.7	85.5	95.9	72.7	84.6	93.6	114.8	103.9
01 - 16	WEIGHTED AVERAGE	97.2	50.9	76.6	97.4	61.6	81.5	100.9	82.4	92.7	100.6	82.4	92.5	97.8	78.0	89.0	98.4	116.5	106.4

NEW JERSEY

DIVISION		CAMDEN 081 MAT.	INST.	TOTAL	DOVER 078 MAT.	INST.	TOTAL	ELIZABETH 072 MAT.	INST.	TOTAL	HACKENSACK 076 MAT.	INST.	TOTAL	JERSEY CITY 073 MAT.	INST.	TOTAL	LONG BRANCH 077 MAT.	INST.	TOTAL
01590	EQUIPMENT RENTAL	.0	99.4	99.4	.0	101.5	101.5	.0	101.5	101.5	.0	101.5	101.5	.0	99.4	99.4	.0	98.9	98.9
02	SITE CONSTRUCTION	92.2	105.0	101.6	102.7	105.4	104.7	106.2	105.4	105.6	103.0	104.8	104.3	92.2	104.8	101.4	96.9	105.3	103.0
03100	CONCRETE FORMS & ACCESSORIES	102.8	125.6	122.4	99.0	130.2	125.9	113.1	130.1	127.8	99.0	130.0	125.7	102.8	130.2	126.4	103.4	133.3	129.2
03200	CONCRETE REINFORCEMENT	107.5	114.7	111.1	82.8	114.4	98.8	82.8	114.4	98.8	82.8	114.4	98.8	107.5	114.4	111.0	82.8	114.1	98.7
03300	CAST-IN-PLACE CONCRETE	82.8	125.3	99.3	106.3	131.2	116.5	91.3	131.1	107.7	103.9	131.1	115.1	82.8	131.1	102.7	92.1	130.1	107.7
03	CONCRETE	97.2	122.3	109.6	104.3	126.5	115.3	99.6	126.5	113.0	102.4	126.4	114.3	97.2	126.3	111.6	101.4	127.3	114.2
04	MASONRY	91.3	121.9	110.3	96.4	127.6	115.8	113.5	127.6	122.3	100.6	127.6	117.4	91.3	127.6	113.8	105.7	121.8	115.7
05	METALS	99.5	101.7	100.2	94.8	107.3	98.8	95.8	107.2	99.5	94.9	107.1	98.7	99.6	104.3	101.1	94.9	103.7	97.7
06	WOOD & PLASTICS	101.5	126.0	114.3	100.5	131.1	116.4	117.0	131.1	124.4	100.5	131.1	116.4	101.5	131.1	116.9	102.7	135.4	119.8
07	THERMAL & MOISTURE PROTECTION	101.7	122.5	110.5	102.2	133.5	115.4	102.5	133.5	115.6	102.0	125.9	112.1	101.7	133.5	115.1	101.9	131.5	114.3
08	DOORS & WINDOWS	106.9	120.2	110.4	110.3	124.1	113.9	108.1	124.1	112.3	107.3	124.1	111.7	106.9	124.1	111.4	102.9	126.4	109.1
09200	PLASTER & GYPSUM BOARD	106.2	125.8	118.4	102.3	131.0	120.1	109.6	131.0	122.9	102.3	131.0	120.1	106.2	131.0	121.6	104.1	135.6	123.7
095,098	CEILINGS & ACOUSTICAL TREATMENT	98.5	125.8	114.8	85.5	131.0	112.7	87.5	131.0	113.4	85.5	131.0	112.7	98.5	131.0	117.9	85.5	135.6	115.4
09600	FLOORING	99.6	126.8	107.0	98.4	126.8	106.1	104.5	126.8	110.5	98.4	126.8	106.1	99.6	126.8	107.0	99.9	126.8	107.2
097,099	WALL FINISHES, PAINTS & COATINGS	91.5	115.5	106.1	91.5	133.8	117.3	91.5	133.8	117.3	91.5	133.8	117.3	91.5	133.8	117.3	91.5	115.5	106.1
09	FINISHES	101.3	125.1	113.7	97.3	130.1	114.4	101.1	130.1	116.2	97.2	130.1	114.3	101.3	130.1	116.3	98.2	130.6	115.1
10 - 14	TOTAL DIV. 10000 - 14000	100.0	116.6	103.5	100.0	118.7	104.0	100.0	118.7	104.0	100.0	118.7	104.0	100.0	118.7	104.0	100.0	117.7	103.8
15	MECHANICAL	100.1	116.6	106.9	99.7	125.9	110.6	100.1	121.9	109.1	99.7	125.9	110.6	100.1	125.9	110.8	99.7	127.2	111.1
16	ELECTRICAL	98.0	114.8	106.2	95.3	135.0	114.6	95.9	135.0	114.9	95.3	137.6	115.9	99.5	137.6	118.0	95.0	144.2	118.9
01 - 16	WEIGHTED AVERAGE	99.6	116.8	107.2	99.7	124.5	110.7	100.8	123.6	110.9	99.5	124.5	110.6	99.8	124.5	110.8	99.1	125.3	110.8

City Cost Indexes

COST INDEXES

NEW JERSEY

DIVISION		NEWARK 070 - 071			NEW BRUNSWICK 088 - 089			PATERSON 074 - 075			POINT PLEASANT 087			SUMMIT 079			TRENTON 085 - 086		
		MAT.	INST.	TOTAL	MAT.	INST.	TOTAL	MAT.	INST.	TOTAL	MAT.	INST.	TOTAL	MAT.	INST.	TOTAL	MAT.	INST.	TOTAL
01590	EQUIPMENT RENTAL	.0	101.5	101.5	.0	98.9	98.9	.0	101.5	101.5	.0	98.9	98.9	.0	101.5	101.5	.0	98.9	98.9
02	SITE CONSTRUCTION	110.6	105.4	106.8	104.6	105.3	105.1	104.7	104.8	104.8	105.9	105.3	105.4	103.8	105.4	105.0	94.0	105.3	102.2
03100	CONCRETE FORMS & ACCESSORIES	101.0	130.2	126.2	106.5	130.0	126.8	101.6	130.0	126.1	100.0	122.7	119.6	102.6	130.1	126.3	101.7	129.7	125.8
03200	CONCRETE REINFORCEMENT	107.5	114.4	111.0	82.8	114.1	98.7	107.5	114.4	111.0	82.8	114.0	98.7	82.8	114.4	98.8	107.5	115.0	111.3
03300	CAST-IN-PLACE CONCRETE	96.7	131.2	110.9	105.6	125.2	113.7	105.7	131.1	116.1	105.6	130.0	115.6	88.2	131.1	105.9	98.1	125.1	109.2
03	CONCRETE	103.8	126.5	115.1	113.6	124.2	118.9	108.1	126.4	117.2	113.2	122.5	117.8	96.2	126.5	111.2	104.5	124.1	114.2
04	MASONRY	101.1	127.6	117.6	99.5	121.8	113.3	97.0	127.6	116.0	91.4	121.8	110.3	99.3	127.6	116.9	90.9	121.8	110.1
05	METALS	99.6	107.3	102.0	94.9	103.6	97.7	94.4	107.0	98.4	94.9	103.4	97.6	94.8	107.2	98.8	94.4	102.2	96.9
06	WOOD & PLASTICS	103.3	131.1	117.8	106.1	131.0	119.1	103.3	131.1	117.8	98.8	121.4	110.6	104.6	131.1	118.4	101.5	131.0	116.9
07	THERMAL & MOISTURE PROTECTION	101.9	133.5	115.2	102.1	131.0	114.3	102.3	125.9	112.2	102.1	130.0	113.8	102.7	133.5	115.7	100.6	122.1	109.6
08	DOORS & WINDOWS	113.1	124.1	116.0	98.8	124.0	105.4	113.1	124.1	116.0	100.9	118.5	105.5	115.7	124.1	117.9	106.9	123.3	111.2
09200	PLASTER & GYPSUM BOARD	106.2	131.0	121.6	105.6	131.0	121.4	106.2	131.0	121.6	102.3	121.2	114.0	104.1	131.0	120.8	106.2	131.0	121.6
095,098	CEILINGS & ACOUSTICAL TREATMENT	98.5	131.0	117.9	85.5	131.0	112.7	98.5	131.0	117.9	85.5	121.2	106.8	85.5	131.0	112.7	98.5	131.0	117.9
09600	FLOORING	99.8	126.8	107.1	101.3	126.6	108.2	99.6	126.8	107.0	98.4	126.8	106.1	99.9	126.8	107.2	99.8	126.8	107.1
097,099	WALL FINISHES, PAINTS & COATINGS	91.5	133.8	117.3	91.5	133.8	117.3	91.5	133.8	117.3	91.5	115.5	106.1	91.5	133.8	117.3	91.5	115.5	106.2
09	FINISHES	101.8	130.1	116.5	99.6	130.1	115.4	101.4	130.1	116.4	98.2	122.4	110.8	98.3	130.1	114.9	101.3	128.0	115.2
10 - 14	TOTAL DIV. 10000 - 14000	100.0	118.7	104.0	100.0	118.5	104.0	100.0	118.7	104.0	100.0	114.3	103.0	100.0	118.7	104.0	100.0	117.1	103.6
15	MECHANICAL	100.1	124.7	110.3	99.7	127.3	111.1	100.1	125.9	110.8	99.7	123.9	109.7	99.7	121.9	108.9	100.1	127.2	111.3
16	ELECTRICAL	99.4	137.6	118.0	94.2	135.0	114.1	99.5	135.0	116.8	93.6	144.2	118.2	95.9	135.0	114.9	98.1	134.7	115.9
01 - 16	WEIGHTED AVERAGE	102.1	124.6	112.1	100.0	123.4	110.4	101.4	124.2	111.5	99.5	122.3	109.6	99.8	123.6	110.4	99.6	122.6	109.8

DIVISION		NEW JERSEY VINELAND 080,083			NEW MEXICO ALBUQUERQUE 870 - 872			CARRIZOZO 883			CLOVIS 881			FARMINGTON 874			GALLUP 873		
		MAT.	INST.	TOTAL	MAT.	INST.	TOTAL	MAT.	INST.	TOTAL	MAT.	INST.	TOTAL	MAT.	INST.	TOTAL	MAT.	INST.	TOTAL
01590	EQUIPMENT RENTAL	.0	99.4	99.4	.0	116.3	116.3	.0	116.3	116.3	.0	116.3	116.3	.0	116.3	116.3	.0	116.3	116.3
02	SITE CONSTRUCTION	96.0	105.0	102.5	78.5	111.2	102.1	98.1	111.2	107.7	86.1	111.2	104.5	83.8	111.2	103.9	91.9	111.2	106.1
03100	CONCRETE FORMS & ACCESSORIES	97.1	125.6	121.7	96.0	68.6	72.4	96.0	68.6	72.4	96.0	68.4	72.1	96.0	68.6	72.4	96.0	68.6	72.4
03200	CONCRETE REINFORCEMENT	81.9	110.4	96.4	108.7	68.1	88.1	116.6	68.1	91.9	117.9	57.0	87.0	118.8	68.1	93.0	113.7	68.1	90.5
03300	CAST-IN-PLACE CONCRETE	92.1	130.2	107.8	102.6	74.2	90.9	95.2	74.2	86.5	95.1	74.0	86.4	103.2	74.2	91.3	97.1	74.2	87.7
03	CONCRETE	100.8	123.1	111.9	105.3	71.5	88.5	119.9	71.5	95.8	108.5	69.2	89.0	109.1	71.5	90.4	115.1	71.5	93.4
04	MASONRY	88.7	121.9	109.3	119.3	64.4	85.3	112.9	64.4	82.9	113.0	64.4	82.9	123.6	64.4	86.9	113.0	64.4	82.9
05	METALS	94.7	100.1	96.5	100.1	87.7	96.2	97.2	87.7	94.2	96.9	82.1	92.2	97.8	87.7	94.6	97.2	87.7	94.2
06	WOOD & PLASTICS	95.6	126.0	111.5	95.4	70.1	82.2	95.4	70.1	82.2	95.4	70.1	82.2	95.4	70.1	82.2	95.4	70.1	82.2
07	THERMAL & MOISTURE PROTECTION	101.6	121.7	110.1	99.0	73.4	88.2	101.1	73.4	89.4	99.8	73.4	88.7	99.6	73.4	88.6	100.7	73.4	89.2
08	DOORS & WINDOWS	100.4	119.1	105.3	94.6	72.2	88.8	93.5	72.2	88.0	93.7	68.8	87.2	97.5	72.2	90.9	97.5	72.2	90.9
09200	PLASTER & GYPSUM BOARD	101.1	125.8	116.5	87.5	68.8	75.9	82.7	68.8	74.0	82.7	68.8	74.0	82.7	68.8	74.0	82.7	68.8	74.0
095,098	CEILINGS & ACOUSTICAL TREATMENT	85.5	125.8	109.6	109.0	68.8	85.0	100.5	68.8	81.6	100.5	68.8	81.6	100.5	68.8	81.6	100.5	68.8	81.6
09600	FLOORING	97.3	126.8	105.3	100.3	62.2	90.0	100.3	62.2	90.0	100.3	62.2	90.0	100.3	62.2	90.0	100.3	62.2	90.0
097,099	WALL FINISHES, PAINTS & COATINGS	91.5	115.5	106.1	102.2	58.6	75.6	102.2	58.6	75.6	102.2	58.6	75.6	102.2	58.6	75.6	102.2	58.6	75.6
09	FINISHES	97.0	125.1	111.6	99.1	66.2	82.0	97.9	66.2	81.4	96.8	66.2	80.9	96.5	66.2	80.7	97.4	66.2	81.2
10 - 14	TOTAL DIV. 10000 - 14000	100.0	116.6	103.5	100.0	71.5	93.9	100.0	71.5	93.9	100.0	71.5	93.9	100.0	71.5	93.9	100.0	71.5	93.9
15	MECHANICAL	99.7	119.6	108.0	100.1	72.2	88.5	97.2	72.2	86.8	97.2	71.7	86.6	100.1	72.2	88.5	97.1	72.2	86.7
16	ELECTRICAL	93.6	114.8	103.9	84.0	78.1	81.1	84.0	78.1	81.1	81.9	78.1	80.0	82.4	78.1	80.3	81.7	78.1	80.0
01 - 16	WEIGHTED AVERAGE	97.6	117.3	106.3	98.5	76.2	88.6	99.0	76.2	88.9	97.0	75.2	87.3	98.7	76.2	88.7	98.3	76.2	88.5

NEW MEXICO

DIVISION		LAS CRUCES 880			LAS VEGAS 877			ROSWELL 882			SANTA FE 875			SOCORRO 878			TRUTH/CONSEQUENCES 879		
		MAT.	INST.	TOTAL	MAT.	INST.	TOTAL	MAT.	INST.	TOTAL	MAT.	INST.	TOTAL	MAT.	INST.	TOTAL	MAT.	INST.	TOTAL
01590	EQUIPMENT RENTAL	.0	86.8	86.8	.0	116.3	116.3	.0	116.3	116.3	.0	116.3	116.3	.0	116.3	116.3	.0	86.8	86.8
02	SITE CONSTRUCTION	89.2	86.6	87.3	82.7	111.2	103.6	88.2	111.2	105.1	78.2	111.2	102.4	80.0	111.2	102.9	99.4	86.6	90.0
03100	CONCRETE FORMS & ACCESSORIES	93.4	66.9	70.5	96.0	68.6	72.4	96.0	68.5	72.3	96.0	68.6	72.4	96.0	68.6	72.4	93.4	67.0	70.7
03200	CONCRETE REINFORCEMENT	112.7	56.6	84.2	115.6	68.0	91.4	117.9	57.0	87.0	116.6	68.1	91.9	117.9	68.1	92.6	111.3	56.7	83.5
03300	CAST-IN-PLACE CONCRETE	89.8	64.7	79.5	100.4	74.1	89.6	95.1	74.1	86.5	97.0	74.2	87.6	98.4	74.2	88.4	107.8	64.8	90.1
03	CONCRETE	87.2	64.9	76.1	106.3	71.4	89.0	109.4	69.3	89.5	103.8	71.5	87.7	105.1	71.5	88.4	99.0	65.0	82.1
04	MASONRY	108.7	58.9	77.9	113.3	63.8	82.6	124.4	64.4	87.2	116.8	64.4	84.3	113.2	64.4	83.0	111.1	60.1	79.5
05	METALS	97.7	73.6	90.0	97.0	87.7	94.0	97.8	82.3	92.8	97.8	87.7	94.6	97.2	87.7	94.2	96.9	73.8	89.5
06	WOOD & PLASTICS	87.2	68.7	77.6	95.4	70.1	82.2	95.4	70.1	82.2	95.4	70.1	82.2	95.4	70.1	82.2	87.2	68.7	77.6
07	THERMAL & MOISTURE PROTECTION	84.3	66.6	76.8	99.3	73.3	88.3	99.9	73.4	88.7	99.2	73.4	88.3	99.2	73.4	88.4	84.8	66.9	77.3
08	DOORS & WINDOWS	87.1	68.0	82.1	93.7	72.2	88.1	93.5	68.8	87.1	93.7	72.2	88.1	93.6	72.2	88.0	87.0	68.0	82.0
09200	PLASTER & GYPSUM BOARD	85.1	68.8	75.0	82.7	68.8	74.0	82.7	68.8	74.0	82.7	68.8	74.0	82.7	68.8	74.0	85.1	68.8	75.0
095,098	CEILINGS & ACOUSTICAL TREATMENT	98.5	68.8	80.7	100.5	68.8	81.6	100.5	68.8	81.6	100.5	68.8	81.6	100.5	68.8	81.6	98.5	68.8	80.7
09600	FLOORING	132.4	62.4	113.5	100.3	62.2	90.0	100.3	62.2	90.0	100.3	62.2	90.0	100.3	62.2	90.0	132.4	64.5	114.1
097,099	WALL FINISHES, PAINTS & COATINGS	95.6	58.6	73.0	102.2	58.6	75.6	102.2	58.6	75.6	102.2	58.6	75.6	102.2	58.6	75.6	95.6	58.6	73.0
09	FINISHES	110.8	65.2	87.1	96.4	66.2	80.7	96.9	66.2	80.9	96.3	66.2	80.6	96.3	66.2	80.7	111.3	65.7	87.5
10 - 14	TOTAL DIV. 10000 - 14000	100.0	67.7	93.1	100.0	71.5	93.9	100.0	71.5	93.9	100.0	71.5	93.9	100.0	71.5	93.9	100.0	67.8	93.1
15	MECHANICAL	100.3	71.6	88.4	97.1	72.2	86.7	100.1	72.0	88.4	100.1	72.2	88.5	97.1	72.2	86.7	96.8	71.6	86.4
16	ELECTRICAL	84.7	58.9	72.2	84.0	78.1	81.1	83.1	78.1	80.7	84.0	78.1	81.1	82.2	78.1	80.2	87.3	78.1	82.8
01 - 16	WEIGHTED AVERAGE	95.8	67.9	83.4	96.9	76.2	87.7	98.7	75.3	88.3	97.5	76.2	88.0	96.5	76.2	87.5	96.9	70.8	85.3

City Cost Indexes

		NEW MEXICO			NEW YORK													
		TUCUMCARI			ALBANY			BINGHAMTON			BRONX			BROOKLYN			BUFFALO	
	DIVISION	884			120 - 122			137 - 139			104			112			140 - 142	
		MAT.	INST.	TOTAL	MAT.	INST.	TOTAL	MAT.	INST.	TOTAL	MAT.	INST.	TOTAL	MAT.	INST.	TOTAL	MAT.	INST.	TOTAL
01590	EQUIPMENT RENTAL	.0	116.3	116.3	.0	115.6	115.6	.0	115.8	115.8	.0	117.2	117.2	.0	117.2	117.2	.0	93.1	93.1
02	SITE CONSTRUCTION	85.8	111.2	104.4	72.1	106.6	97.4	94.4	90.2	91.3	126.1	128.3	127.7	119.9	130.9	128.0	96.3	93.7	94.4
03100	CONCRETE FORMS & ACCESSORIES	96.0	68.4	72.1	97.4	93.1	93.7	102.5	80.5	83.5	105.9	169.5	160.7	109.7	168.9	160.7	98.9	115.2	113.0
03200	CONCRETE REINFORCEMENT	115.6	57.0	85.8	103.3	99.1	101.2	102.3	86.9	94.5	105.0	197.8	152.2	103.6	182.5	143.7	104.1	104.6	104.4
03300	CAST-IN-PLACE CONCRETE	95.1	74.0	86.4	87.7	100.8	93.1	104.8	95.5	101.0	105.1	160.6	127.9	106.7	158.6	128.0	115.7	119.1	117.1
03	CONCRETE	107.9	69.2	88.7	100.6	97.7	99.2	100.5	88.8	94.7	106.3	169.8	137.9	112.6	166.1	139.2	106.8	113.7	110.2
04	MASONRY	127.0	64.4	88.2	89.1	97.2	94.2	105.8	88.3	95.0	92.4	158.4	133.3	114.9	158.3	141.8	104.7	121.1	114.9
05	METALS	96.9	82.1	92.2	96.4	108.6	100.3	92.9	115.2	100.0	92.5	145.1	109.3	102.0	139.6	114.0	100.3	94.9	98.6
06	WOOD & PLASTICS	95.4	70.1	82.2	97.0	92.4	94.6	105.9	79.6	92.2	103.5	173.8	140.1	108.7	173.4	142.4	102.4	114.6	108.8
07	THERMAL & MOISTURE PROTECTION	99.7	73.4	88.7	93.8	92.8	93.4	102.5	87.3	96.1	113.9	156.6	131.9	110.3	156.1	129.6	97.7	110.5	103.1
08	DOORS & WINDOWS	93.5	68.8	87.0	95.6	88.7	93.8	90.8	77.6	87.3	92.1	169.8	112.4	89.9	166.2	109.8	92.9	103.7	95.7
09200	PLASTER & GYPSUM BOARD	82.7	68.8	74.0	109.8	91.9	98.7	114.6	78.5	92.2	104.7	175.3	148.6	108.5	175.3	150.0	99.6	114.7	109.0
095,098	CEILINGS & ACOUSTICAL TREATMENT	100.5	68.8	81.6	97.9	91.9	94.3	97.9	78.5	86.4	80.5	175.3	137.1	78.4	175.3	136.2	94.6	114.7	106.6
09600	FLOORING	100.3	62.2	90.0	87.1	96.5	89.7	97.7	91.2	95.9	95.1	173.9	116.4	104.4	173.9	123.2	97.9	119.6	103.8
097,099	WALL FINISHES, PAINTS & COATINGS	102.2	58.6	75.6	86.1	78.4	81.4	91.1	82.4	85.8	96.9	151.4	130.2	114.1	151.4	136.9	95.1	114.3	106.8
09	FINISHES	96.8	66.2	80.9	97.2	92.0	94.5	98.3	82.2	89.9	98.8	169.1	135.4	107.0	168.8	139.2	95.5	116.7	106.6
10 - 14	TOTAL DIV. 10000 - 14000	100.0	71.5	93.9	100.0	95.7	99.1	100.0	92.8	98.5	100.0	140.4	108.6	100.0	139.3	108.4	100.0	107.3	101.6
15	MECHANICAL	97.2	71.7	86.6	100.3	91.8	96.8	100.5	84.5	93.9	100.2	164.6	126.9	99.8	164.3	126.6	99.8	99.6	99.7
16	ELECTRICAL	84.0	78.1	81.1	103.6	94.1	99.0	101.6	87.4	94.7	102.4	166.1	133.4	102.3	167.9	134.2	100.2	99.2	99.7
01 - 16	WEIGHTED AVERAGE	97.9	75.2	87.8	97.8	96.4	97.2	98.5	88.8	94.2	99.8	159.9	126.5	103.3	159.1	128.1	99.8	105.7	102.4

		NEW YORK																
		ELMIRA			FAR ROCKAWAY			FLUSHING			GLENS FALLS			HICKSVILLE			JAMAICA	
	DIVISION	148 - 149			116			113			128			115,117,118			114	
		MAT.	INST.	TOTAL	MAT.	INST.	TOTAL	MAT.	INST.	TOTAL	MAT.	INST.	TOTAL	MAT.	INST.	TOTAL	MAT.	INST.	TOTAL
01590	EQUIPMENT RENTAL	.0	116.0	116.0	.0	117.2	117.2	.0	117.2	117.2	.0	115.6	115.6	.0	117.2	117.2	.0	117.2	117.2
02	SITE CONSTRUCTION	91.5	89.0	89.7	123.9	130.9	129.0	123.9	130.9	129.0	62.2	106.2	94.4	112.7	131.1	126.2	117.7	130.9	127.4
03100	CONCRETE FORMS & ACCESSORIES	81.1	74.2	75.1	93.5	168.9	158.5	98.0	168.9	159.1	82.8	87.6	86.9	89.3	154.4	145.5	98.0	168.8	159.1
03200	CONCRETE REINFORCEMENT	105.4	84.7	94.9	103.6	182.5	143.7	105.5	182.5	144.6	103.0	94.3	98.6	103.6	182.3	143.6	103.6	182.5	143.7
03300	CAST-IN-PLACE CONCRETE	108.1	79.0	96.2	115.5	158.6	133.2	115.5	158.6	133.2	79.6	93.7	85.4	98.0	159.1	123.1	106.7	158.5	128.0
03	CONCRETE	100.4	80.3	90.4	118.8	166.1	142.3	119.4	166.1	142.6	90.1	91.9	91.0	103.9	159.8	131.7	111.8	166.0	138.7
04	MASONRY	101.2	71.5	82.8	119.2	158.3	143.5	113.1	158.3	141.1	91.2	87.7	89.0	109.0	160.5	140.9	117.1	158.3	142.7
05	METALS	95.8	117.5	102.7	102.1	139.5	114.0	102.1	139.5	114.0	93.0	106.1	97.2	103.1	138.8	114.5	102.1	139.5	114.0
06	WOOD & PLASTICS	84.5	72.2	78.1	91.2	173.4	134.0	96.4	173.4	136.5	81.0	88.6	84.9	87.2	154.5	122.3	96.4	173.4	136.5
07	THERMAL & MOISTURE PROTECTION	98.1	79.2	90.2	110.1	156.1	129.5	110.2	156.1	129.5	92.8	88.4	91.0	109.7	149.3	126.4	110.0	154.8	128.9
08	DOORS & WINDOWS	93.6	73.4	88.3	88.3	166.2	108.7	88.3	166.2	108.7	89.4	85.6	88.4	88.3	156.0	106.0	88.3	166.2	108.7
09200	PLASTER & GYPSUM BOARD	94.1	71.2	79.9	102.2	175.3	147.6	104.4	175.3	148.5	99.7	88.0	92.4	101.4	156.0	135.3	104.4	175.3	148.5
095,098	CEILINGS & ACOUSTICAL TREATMENT	91.2	71.2	79.3	78.4	175.3	136.2	78.4	175.3	136.2	91.2	88.0	89.3	77.5	156.0	124.3	78.4	175.3	136.2
09600	FLOORING	87.5	74.1	83.9	98.6	173.9	119.0	100.3	173.9	120.2	78.7	88.3	81.3	97.3	86.2	94.3	100.3	173.9	120.2
097,099	WALL FINISHES, PAINTS & COATINGS	93.3	80.7	85.6	114.1	151.4	136.9	114.1	151.4	136.9	86.1	78.4	81.4	114.1	151.4	136.9	114.1	151.4	136.9
09	FINISHES	90.5	74.0	81.9	104.7	168.8	138.1	105.5	168.8	138.5	90.7	86.7	88.6	103.2	140.1	122.4	105.1	168.8	138.3
10 - 14	TOTAL DIV. 10000 - 14000	100.0	91.0	98.1	100.0	139.3	108.4	100.0	139.3	108.4	100.0	64.9	92.5	100.0	137.8	108.1	100.0	139.3	108.4
15	MECHANICAL	95.0	83.9	90.4	94.9	164.3	123.7	94.9	164.3	123.7	95.5	86.0	91.6	99.8	152.0	121.5	94.9	164.3	123.7
16	ELECTRICAL	98.9	95.1	97.0	109.3	166.1	136.9	109.3	166.1	136.9	96.8	94.1	95.5	102.0	155.1	127.8	101.0	166.1	132.7
01 - 16	WEIGHTED AVERAGE	96.3	85.4	91.4	103.4	158.8	128.0	103.3	158.8	128.0	92.7	91.3	92.1	101.2	149.5	122.7	101.5	158.8	126.9

		NEW YORK																
		JAMESTOWN			KINGSTON			LONG ISLAND CITY			MONTICELLO			MOUNT VERNON			NEW ROCHELLE	
	DIVISION	147			124			111			127			105			108	
		MAT.	INST.	TOTAL	MAT.	INST.	TOTAL	MAT.	INST.	TOTAL	MAT.	INST.	TOTAL	MAT.	INST.	TOTAL	MAT.	INST.	TOTAL
01590	EQUIPMENT RENTAL	.0	89.5	89.5	.0	117.2	117.2	.0	117.2	117.2	.0	117.2	117.2	.0	117.2	117.2	.0	117.2	117.2
02	SITE CONSTRUCTION	92.8	89.4	90.3	119.5	128.2	125.9	121.5	130.9	128.4	114.7	128.2	124.6	135.3	125.3	127.9	133.8	125.2	127.5
03100	CONCRETE FORMS & ACCESSORIES	81.2	85.0	84.5	84.5	112.0	108.2	103.2	168.8	159.8	92.9	112.1	109.5	93.9	139.7	133.4	112.0	139.5	135.7
03200	CONCRETE REINFORCEMENT	105.6	90.3	97.8	103.4	132.2	118.0	103.6	182.5	143.7	102.6	132.2	117.7	103.8	181.3	143.2	103.9	181.3	143.2
03300	CAST-IN-PLACE CONCRETE	112.2	110.5	111.5	107.5	122.9	113.9	110.2	158.5	130.0	100.7	123.0	109.8	117.6	132.3	123.6	117.6	132.2	123.6
03	CONCRETE	103.4	94.5	99.0	112.6	119.2	115.9	115.2	166.0	140.5	107.3	119.2	113.2	116.6	143.8	130.1	116.4	143.7	129.9
04	MASONRY	109.3	94.6	100.2	107.0	125.6	118.5	111.7	158.3	140.6	99.4	125.6	115.6	98.4	132.0	119.2	98.4	132.0	119.2
05	METALS	93.1	85.2	90.6	102.0	113.6	105.7	102.0	139.5	114.0	102.0	113.8	105.8	92.3	134.8	105.9	92.5	134.7	106.0
06	WOOD & PLASTICS	83.3	78.5	80.8	82.3	106.8	95.1	102.2	173.4	139.3	91.5	106.8	99.5	90.9	140.3	116.6	111.0	140.3	126.3
07	THERMAL & MOISTURE PROTECTION	97.5	92.9	95.6	112.1	133.4	121.1	110.2	154.8	129.0	111.8	133.4	120.9	115.0	140.0	125.5	115.1	140.0	125.6
08	DOORS & WINDOWS	93.4	80.6	90.1	93.4	119.2	100.2	88.3	166.2	108.7	88.4	119.2	96.4	92.1	148.2	106.8	92.2	148.2	106.8
09200	PLASTER & GYPSUM BOARD	89.5	77.7	82.2	97.3	107.0	103.3	107.0	175.3	149.5	100.6	107.0	104.6	98.9	140.9	125.0	110.3	140.9	129.3
095,098	CEILINGS & ACOUSTICAL TREATMENT	86.9	77.7	81.4	72.5	107.0	93.1	78.4	175.3	136.2	72.5	107.0	93.1	78.0	140.9	115.5	78.0	140.9	115.5
09600	FLOORING	90.4	88.7	89.9	91.1	90.5	90.9	102.0	173.9	121.5	94.2	90.5	93.2	87.2	168.7	109.2	94.7	168.7	114.7
097,099	WALL FINISHES, PAINTS & COATINGS	95.1	83.0	87.7	111.8	124.4	119.5	114.1	151.4	136.9	111.8	105.3	107.8	94.5	175.0	143.6	94.5	141.5	123.2
09	FINISHES	89.7	84.5	87.0	99.8	107.6	103.9	106.2	168.8	138.8	100.9	105.4	103.3	95.6	149.1	123.5	99.3	145.3	123.3
10 - 14	TOTAL DIV. 10000 - 14000	100.0	100.9	100.2	100.0	121.0	104.5	100.0	139.3	108.4	100.0	121.0	104.5	100.0	133.2	107.1	100.0	133.0	107.1
15	MECHANICAL	94.9	80.6	88.9	95.5	106.7	100.1	99.8	164.3	126.6	95.5	120.4	105.8	95.5	131.7	110.5	95.5	131.7	110.5
16	ELECTRICAL	97.7	90.6	94.3	98.5	125.7	111.7	101.5	166.1	132.9	98.5	125.7	111.7	99.0	147.1	122.4	99.0	147.1	122.4
01 - 16	WEIGHTED AVERAGE	96.3	87.9	92.6	100.9	117.3	108.2	103.1	158.8	127.8	99.5	119.9	108.5	99.5	138.5	116.8	100.0	138.0	116.9

City Cost Indexes

NEW YORK

DIVISION		NEW YORK 100-102			NIAGARA FALLS 143			PLATTSBURGH 129			POUGHKEEPSIE 125-126			QUEENS 110			RIVERHEAD 119		
		MAT.	INST.	TOTAL	MAT.	INST.	TOTAL	MAT.	INST.	TOTAL	MAT.	INST.	TOTAL	MAT.	INST.	TOTAL	MAT.	INST.	TOTAL
01590	EQUIPMENT RENTAL	.0	117.5	117.5	.0	89.5	89.5	.0	97.8	97.8	.0	117.2	117.2	.0	117.2	117.2	.0	117.2	117.2
02	SITE CONSTRUCTION	136.2	128.7	130.7	95.1	90.7	91.9	86.5	102.9	98.5	115.8	128.3	125.0	116.0	130.9	126.9	113.4	131.1	126.3
03100	CONCRETE FORMS & ACCESSORIES	111.1	181.7	172.0	81.1	115.8	111.0	87.9	84.4	84.9	84.5	119.2	114.5	89.6	168.8	157.9	94.9	154.4	146.2
03200	CONCRETE REINFORCEMENT	111.1	180.5	146.4	104.1	98.5	101.3	107.5	98.6	103.0	103.4	132.4	118.1	105.5	182.5	144.6	105.6	182.3	144.6
03300	CAST-IN-PLACE CONCRETE	118.4	160.9	135.8	116.1	115.4	115.8	97.4	94.2	96.1	104.1	122.4	111.7	101.4	158.5	124.9	99.7	159.1	124.1
03	CONCRETE	117.7	172.4	144.9	105.9	111.4	108.6	104.0	90.0	97.1	109.6	122.3	115.9	107.3	166.0	136.5	104.8	159.8	132.1
04	MASONRY	102.8	158.4	137.3	117.7	120.2	119.3	87.8	87.9	87.9	100.3	123.8	114.9	106.1	158.3	138.5	115.1	160.5	143.2
05	METALS	104.4	145.4	117.5	95.9	89.6	93.8	97.8	84.9	93.7	102.1	115.2	106.3	102.0	139.5	114.0	103.5	138.8	114.8
06	WOOD & PLASTICS	108.6	189.9	150.9	83.2	113.9	99.2	89.1	81.9	85.4	82.3	116.3	100.0	87.4	173.4	132.2	93.1	154.5	125.1
07	THERMAL & MOISTURE PROTECTION	114.3	161.4	134.1	97.6	109.3	102.5	101.3	86.5	95.1	112.0	133.6	121.1	109.8	154.8	128.7	110.3	149.3	126.7
08	DOORS & WINDOWS	97.7	178.7	118.8	93.5	101.4	95.5	98.5	83.0	94.5	93.4	124.3	101.5	88.3	166.2	108.7	88.3	156.0	106.0
09200	PLASTER & GYPSUM BOARD	113.9	191.9	162.4	89.5	114.1	104.7	119.1	80.6	95.2	97.3	116.7	109.4	101.4	175.3	147.4	104.2	156.0	136.4
095,098	CEILINGS & ACOUSTICAL TREATMENT	107.8	191.9	158.0	86.9	114.1	103.1	97.9	80.6	87.5	72.5	116.7	98.9	78.4	175.3	136.2	82.6	156.0	126.4
09600	FLOORING	96.6	173.9	117.5	90.2	112.6	96.2	101.1	96.5	99.8	91.1	125.2	100.3	97.3	173.9	118.0	98.6	86.2	95.2
097,099	WALL FINISHES, PAINTS & COATINGS	96.9	151.6	130.3	95.1	107.7	102.8	105.3	74.7	86.6	111.8	105.3	107.8	114.1	151.4	136.9	114.1	151.4	136.9
09	FINISHES	108.0	178.6	144.8	89.8	115.1	103.0	99.8	85.3	92.2	99.7	117.9	109.1	103.6	168.8	137.5	105.2	140.1	123.4
10-14	TOTAL DIV. 10000-14000	100.0	142.4	109.1	100.0	108.6	101.8	100.0	60.8	91.6	100.0	121.5	104.6	100.0	139.3	108.4	100.0	137.8	108.1
15	MECHANICAL	100.2	164.6	126.9	94.9	102.0	97.8	95.4	88.3	92.5	95.5	106.9	100.2	99.8	164.3	126.6	99.9	152.0	121.5
16	ELECTRICAL	109.9	181.3	144.6	96.4	98.9	97.6	91.3	91.3	91.3	98.5	125.7	111.7	102.0	166.1	133.2	103.5	155.1	128.6
01-16	WEIGHTED AVERAGE	106.0	164.4	131.9	97.3	104.7	100.6	96.8	88.4	93.1	100.1	119.4	108.7	101.4	158.8	126.9	102.2	149.5	123.2

NEW YORK

DIVISION		ROCHESTER 144-146			SCHENECTADY 123			STATEN ISLAND 103			SUFFERN 109			SYRACUSE 130-132			UTICA 133-135		
		MAT.	INST.	TOTAL	MAT.	INST.	TOTAL	MAT.	INST.	TOTAL	MAT.	INST.	TOTAL	MAT.	INST.	TOTAL	MAT.	INST.	TOTAL
01590	EQUIPMENT RENTAL	.0	116.5	116.5	.0	115.6	115.6	.0	117.2	117.2	.0	117.2	117.2	.0	115.6	115.6	.0	115.6	115.6
02	SITE CONSTRUCTION	74.3	107.4	98.6	71.8	106.6	97.3	140.6	128.3	131.6	130.2	125.3	126.6	93.2	107.0	103.3	69.9	105.4	96.0
03100	CONCRETE FORMS & ACCESSORIES	99.0	102.7	102.2	102.1	92.6	93.9	93.2	169.4	159.0	104.5	133.8	129.8	101.3	89.9	91.4	102.6	84.0	86.6
03200	CONCRETE REINFORCEMENT	105.1	85.9	95.3	102.0	99.1	100.5	105.0	197.8	152.2	103.9	132.7	118.5	103.3	96.1	99.6	103.3	84.5	93.8
03300	CAST-IN-PLACE CONCRETE	110.1	101.6	106.6	96.2	100.1	97.8	117.6	160.5	135.2	113.9	128.2	119.8	97.2	97.2	97.2	89.0	93.3	90.7
03	CONCRETE	112.1	100.0	106.1	104.8	97.2	101.0	118.4	169.8	143.9	113.0	130.7	121.8	103.6	94.4	99.1	101.5	88.4	95.0
04	MASONRY	102.7	100.2	101.2	90.0	95.9	93.7	105.7	158.4	138.4	98.2	126.6	115.8	96.4	99.3	98.2	88.7	91.2	90.2
05	METALS	96.7	107.1	100.0	96.4	108.6	100.3	90.5	145.0	107.9	90.5	114.9	98.3	96.3	106.6	99.6	94.4	102.1	96.9
06	WOOD & PLASTICS	99.3	103.7	101.6	102.5	92.4	97.2	89.4	173.8	133.4	102.8	138.2	121.2	102.5	87.5	94.7	102.5	83.6	92.7
07	THERMAL & MOISTURE PROTECTION	94.6	98.4	96.2	93.9	92.2	93.2	114.2	155.4	131.6	115.0	135.5	123.6	102.2	96.1	99.6	93.8	92.4	93.2
08	DOORS & WINDOWS	96.0	92.6	95.1	95.6	88.7	93.8	92.1	169.8	112.4	92.2	136.2	103.7	93.5	84.2	91.1	95.6	78.9	91.2
09200	PLASTER & GYPSUM BOARD	101.5	103.8	102.9	109.8	91.9	98.7	99.1	175.3	146.5	103.7	138.8	125.5	109.8	87.0	95.6	109.8	82.9	93.1
095,098	CEILINGS & ACOUSTICAL TREATMENT	99.0	103.8	101.9	97.9	91.9	94.3	80.5	175.3	137.1	78.0	138.8	114.3	97.9	87.0	91.4	97.9	82.9	89.0
09600	FLOORING	87.4	104.8	92.1	87.1	96.5	89.7	91.0	173.9	113.4	90.7	60.0	82.4	88.6	92.4	89.6	87.1	87.1	87.1
097,099	WALL FINISHES, PAINTS & COATINGS	93.0	103.4	99.4	86.1	78.4	81.4	96.9	151.4	130.2	94.5	109.7	103.8	91.6	86.2	88.3	86.1	84.0	84.8
09	FINISHES	96.7	103.8	100.4	96.9	91.7	94.2	97.8	169.1	134.9	96.9	116.8	107.3	98.2	89.6	93.7	96.9	84.3	90.4
10-14	TOTAL DIV. 10000-14000	100.0	101.1	100.2	100.0	95.2	99.0	100.0	140.8	108.6	100.0	130.8	106.6	100.0	96.3	99.2	100.0	92.7	98.4
15	MECHANICAL	99.8	94.7	97.7	100.3	91.1	96.5	100.2	164.6	126.9	95.5	123.0	106.9	100.3	89.0	95.6	100.3	87.2	94.8
16	ELECTRICAL	103.9	94.9	99.5	101.1	94.1	97.7	102.4	166.1	133.4	106.3	135.6	120.6	101.1	92.2	96.8	97.7	91.8	94.8
01-16	WEIGHTED AVERAGE	99.8	99.7	99.8	98.1	95.9	97.1	101.6	159.9	127.5	99.8	126.2	111.5	99.0	94.7	97.1	96.9	90.9	94.2

NEW YORK / NORTH CAROLINA

DIVISION		WATERTOWN 136			WHITE PLAINS 106			YONKERS 107			ASHEVILLE 287-288			CHARLOTTE 281-282			DURHAM 277		
		MAT.	INST.	TOTAL	MAT.	INST.	TOTAL	MAT.	INST.	TOTAL	MAT.	INST.	TOTAL	MAT.	INST.	TOTAL	MAT.	INST.	TOTAL
01590	EQUIPMENT RENTAL	.0	115.6	115.6	.0	117.2	117.2	.0	117.2	117.2	.0	93.3	93.3	.0	93.3	93.3	.0	99.6	99.6
02	SITE CONSTRUCTION	77.9	107.4	99.6	125.6	125.2	125.4	135.5	125.1	127.8	102.7	72.2	80.3	102.9	72.2	80.4	102.7	82.0	87.5
03100	CONCRETE FORMS & ACCESSORIES	84.7	90.5	89.7	110.7	139.5	135.6	110.9	139.6	135.7	95.1	42.7	49.9	102.7	42.8	51.0	97.0	43.0	50.4
03200	CONCRETE REINFORCEMENT	104.0	85.1	94.4	103.9	181.3	143.2	108.0	181.3	145.2	93.8	46.9	70.0	94.2	47.0	70.2	94.2	55.0	74.3
03300	CAST-IN-PLACE CONCRETE	103.6	83.1	95.2	104.3	132.3	115.8	116.8	132.3	123.1	96.5	49.3	77.1	99.0	49.3	78.5	96.8	49.4	77.3
03	CONCRETE	114.3	87.9	101.2	105.9	143.7	124.7	116.5	143.8	130.1	100.3	47.6	74.1	101.3	47.7	74.7	99.8	49.3	74.8
04	MASONRY	89.8	91.2	90.7	97.5	132.0	118.9	102.4	132.0	120.7	80.5	40.1	55.5	85.9	40.1	57.5	82.8	40.1	56.3
05	METALS	94.5	101.6	96.7	91.6	134.7	105.4	100.5	134.8	111.4	92.4	76.6	87.4	94.9	76.7	89.1	97.5	80.3	92.0
06	WOOD & PLASTICS	83.0	92.4	87.9	109.5	140.3	125.5	109.3	140.3	125.4	94.7	43.2	67.9	103.9	43.2	72.3	97.0	43.2	69.0
07	THERMAL & MOISTURE PROTECTION	94.0	92.1	93.2	114.8	140.0	125.4	115.1	140.0	125.6	101.3	45.7	77.9	101.3	47.3	78.6	101.8	46.2	78.4
08	DOORS & WINDOWS	95.6	81.0	91.8	92.2	148.2	106.8	95.8	148.3	109.5	92.6	43.4	79.7	96.6	43.4	82.7	96.6	46.1	83.4
09200	PLASTER & GYPSUM BOARD	102.0	92.0	95.8	106.6	140.9	127.9	113.5	140.9	130.5	103.4	41.5	64.9	109.6	41.5	67.3	109.6	41.5	67.3
095,098	CEILINGS & ACOUSTICAL TREATMENT	97.9	92.0	94.4	78.0	140.9	115.5	106.1	140.9	126.9	94.1	41.5	62.7	100.9	41.5	65.5	100.9	41.5	65.5
09600	FLOORING	79.3	87.1	81.4	92.8	168.7	113.3	92.4	168.7	113.1	101.8	48.1	87.3	105.8	48.1	90.2	106.0	48.1	90.3
097,099	WALL FINISHES, PAINTS & COATINGS	86.1	70.8	76.8	94.5	151.4	129.2	94.5	175.0	143.6	110.4	37.7	66.1	110.4	37.7	66.1	110.4	37.7	66.1
09	FINISHES	94.0	87.6	90.7	97.6	146.4	123.0	106.1	149.1	128.5	97.7	43.1	69.3	101.5	43.1	71.1	101.6	43.1	71.2
10-14	TOTAL DIV. 10000-14000	100.0	78.5	95.4	100.0	133.2	107.1	100.0	133.2	107.1	100.0	59.0	91.3	100.0	59.0	91.3	100.0	74.5	94.6
15	MECHANICAL	100.3	74.1	89.5	100.5	131.7	113.4	100.5	131.7	113.4	100.1	42.6	76.2	100.1	42.6	76.2	100.1	42.8	76.3
16	ELECTRICAL	101.1	91.4	96.4	99.0	147.1	122.4	106.4	147.1	126.2	97.8	39.4	69.4	99.6	38.5	69.9	97.3	37.7	68.3
01-16	WEIGHTED AVERAGE	98.5	88.5	94.1	99.5	138.2	116.7	104.5	138.5	119.6	96.8	48.9	75.5	98.6	48.9	76.5	98.4	50.7	77.2

NORTH CAROLINA

DIVISION		ELIZABETH CITY 279			FAYETTEVILLE 283			GASTONIA 280			GREENSBORO 270,272 - 274			HICKORY 286			KINSTON 285		
		MAT.	INST.	TOTAL	MAT.	INST.	TOTAL	MAT.	INST.	TOTAL	MAT.	INST.	TOTAL	MAT.	INST.	TOTAL	MAT.	INST.	TOTAL
01590	EQUIPMENT RENTAL	.0	104.7	104.7	.0	99.6	99.6	.0	93.3	93.3	.0	99.6	99.6	.0	99.6	99.6	.0	99.6	99.6
02	SITE CONSTRUCTION	107.3	82.8	89.3	100.7	82.0	87.0	102.3	72.2	80.2	102.6	82.0	87.5	101.7	80.8	86.4	100.3	81.5	86.5
03100	CONCRETE FORMS & ACCESSORIES	81.0	25.1	32.8	92.7	43.1	49.9	102.9	42.8	51.1	97.1	42.9	50.3	89.9	24.1	33.2	85.7	24.4	32.8
03200	CONCRETE REINFORCEMENT	93.3	45.6	69.0	93.3	55.0	73.8	94.2	47.0	70.2	94.2	55.0	74.3	93.8	20.5	56.5	93.3	20.6	56.3
03300	CAST-IN-PLACE CONCRETE	96.8	33.0	70.6	93.4	49.4	75.3	94.3	49.3	75.8	95.9	49.3	76.8	96.5	35.1	71.2	93.1	36.2	69.7
03	CONCRETE	100.0	33.9	67.2	97.4	49.4	73.6	99.1	47.7	73.6	99.4	49.2	74.5	99.9	29.6	65.0	96.8	30.2	63.7
04	MASONRY	96.5	25.9	52.7	86.6	40.1	57.8	83.9	40.1	56.8	82.6	40.1	56.3	70.3	25.7	42.7	76.4	26.1	45.2
05	METALS	92.0	74.2	86.3	97.4	80.3	91.9	92.9	76.7	87.8	98.0	80.1	92.3	92.6	65.8	84.0	91.4	67.4	83.8
06	WOOD & PLASTICS	79.2	25.3	51.1	91.8	43.2	66.5	103.9	43.2	72.3	97.0	43.2	69.0	88.5	24.5	55.2	84.2	24.5	53.1
07	THERMAL & MOISTURE PROTECTION	101.2	26.0	69.5	101.5	46.9	78.5	101.5	47.3	78.7	101.8	46.2	78.4	101.5	28.6	70.8	101.4	29.1	71.0
08	DOORS & WINDOWS	93.5	27.9	76.4	92.7	46.1	80.5	96.6	43.4	82.7	96.6	46.1	83.4	92.6	24.0	74.7	92.7	24.4	74.9
09200	PLASTER & GYPSUM BOARD	100.3	22.3	51.8	103.6	41.5	65.0	109.6	41.5	67.3	109.6	41.5	67.3	103.4	22.3	53.0	101.9	22.3	52.4
095,098	CEILINGS & ACOUSTICAL TREATMENT	100.9	22.3	54.0	94.9	41.5	63.1	100.9	41.5	65.5	100.9	41.5	65.5	94.1	22.3	51.2	98.3	22.3	53.0
09600	FLOORING	95.1	15.5	73.5	101.9	48.1	87.4	106.0	48.1	90.3	106.0	48.1	90.3	101.7	14.1	78.0	98.3	21.2	77.4
097,099	WALL FINISHES, PAINTS & COATINGS	110.4	12.8	50.9	110.4	37.7	66.1	110.4	37.7	66.1	110.4	37.7	66.1	110.4	15.1	52.3	110.4	16.6	53.2
09	FINISHES	97.2	21.3	57.7	98.0	43.1	71.1	100.5	43.1	71.1	101.6	43.1	71.2	97.8	20.8	57.7	97.5	22.6	58.5
10 - 14	TOTAL DIV. 10000 - 14000	100.0	72.0	94.0	100.0	74.5	94.6	100.0	59.0	91.3	100.0	59.0	91.3	100.0	53.9	90.2	100.0	69.4	93.5
15	MECHANICAL	95.1	26.9	66.8	100.1	42.8	76.3	100.1	42.6	76.2	100.1	42.7	76.3	95.1	18.5	63.3	95.1	19.2	63.7
16	ELECTRICAL	98.2	29.1	64.6	94.7	37.7	67.0	97.3	38.5	68.7	97.9	37.7	68.6	95.5	29.1	63.2	95.3	29.1	63.1
01 - 16	WEIGHTED AVERAGE	96.4	37.8	70.4	97.1	50.8	76.5	97.7	48.9	76.0	98.5	50.3	77.1	94.8	33.9	67.8	94.4	35.1	68.1

NORTH CAROLINA / **NORTH DAKOTA**

DIVISION		MURPHY 289			RALEIGH 275 - 276			ROCKY MOUNT 278			WILMINGTON 284			WINSTON-SALEM 271			BISMARCK 585		
		MAT.	INST.	TOTAL	MAT.	INST.	TOTAL	MAT.	INST.	TOTAL	MAT.	INST.	TOTAL	MAT.	INST.	TOTAL	MAT.	INST.	TOTAL
01590	EQUIPMENT RENTAL	.0	93.3	93.3	.0	99.6	99.6	.0	99.6	99.6	.0	93.3	93.3	.0	99.6	99.6	.0	98.3	98.3
02	SITE CONSTRUCTION	104.0	70.9	79.7	103.7	82.0	87.8	105.5	80.8	87.4	103.6	72.2	80.6	102.9	82.0	87.6	90.5	96.5	94.9
03100	CONCRETE FORMS & ACCESSORIES	103.6	24.1	35.1	99.9	43.0	50.8	88.6	29.0	37.2	96.6	43.1	50.4	98.5	42.8	50.4	95.1	49.8	56.0
03200	CONCRETE REINFORCEMENT	93.3	20.4	56.3	94.2	55.0	74.3	93.3	20.5	56.3	94.5	55.0	74.4	94.2	47.0	70.2	101.2	78.4	89.7
03300	CAST-IN-PLACE CONCRETE	99.8	31.3	71.6	102.4	49.4	80.6	94.7	30.5	68.3	96.1	49.4	76.9	99.0	49.3	78.5	105.1	55.2	84.6
03	CONCRETE	103.4	28.3	66.1	102.8	49.3	76.2	101.1	30.2	65.9	100.2	49.4	75.0	101.0	47.7	74.5	99.6	58.2	79.1
04	MASONRY	72.5	23.4	42.1	87.4	40.1	58.1	76.6	27.7	46.3	70.4	40.1	51.6	82.8	40.1	56.3	104.5	65.1	80.1
05	METALS	90.5	65.4	82.5	95.3	80.3	90.5	91.4	65.8	83.2	91.9	80.3	88.2	95.3	76.7	89.3	92.6	83.0	89.5
06	WOOD & PLASTICS	104.7	24.5	62.9	100.3	43.2	70.6	87.2	29.9	57.4	96.5	43.2	68.8	97.0	43.2	69.0	88.3	44.7	65.6
07	THERMAL & MOISTURE PROTECTION	101.5	30.5	71.6	101.6	46.3	78.3	101.7	27.8	70.6	101.3	46.9	78.4	101.8	46.2	78.4	99.2	55.6	80.8
08	DOORS & WINDOWS	92.6	27.7	75.6	93.4	46.1	81.1	92.7	26.2	75.3	92.7	46.1	80.5	96.6	43.9	82.8	99.4	50.4	86.6
09200	PLASTER & GYPSUM BOARD	107.8	22.3	54.7	109.6	41.5	67.3	101.8	27.8	55.8	104.3	41.5	65.3	109.6	41.5	67.3	113.5	43.6	70.0
095,098	CEILINGS & ACOUSTICAL TREATMENT	94.1	22.3	51.2	100.9	41.5	65.5	94.9	27.8	54.9	94.9	41.5	63.1	100.9	41.5	65.5	138.5	43.6	81.9
09600	FLOORING	106.5	22.6	83.8	106.0	48.1	90.3	99.8	15.0	76.9	102.8	48.1	88.0	106.0	48.1	90.3	107.8	75.0	98.9
097,099	WALL FINISHES, PAINTS & COATINGS	110.4	17.2	53.5	110.4	37.7	66.1	110.4	28.7	60.5	110.4	37.7	66.1	110.4	37.7	66.1	97.2	37.5	60.8
09	FINISHES	99.9	22.5	59.6	101.7	43.1	71.2	97.5	26.6	60.5	98.3	43.1	69.6	101.6	43.1	71.2	114.7	52.0	82.0
10 - 14	TOTAL DIV. 10000 - 14000	100.0	53.9	90.2	100.0	74.5	94.6	100.0	70.8	93.8	100.0	74.5	94.6	100.0	59.0	91.3	100.0	69.1	93.4
15	MECHANICAL	95.1	18.6	63.4	100.1	42.8	76.3	95.1	18.6	63.4	100.1	42.8	76.3	100.1	42.7	76.2	100.5	65.3	85.9
16	ELECTRICAL	98.6	29.1	64.8	97.9	37.7	68.6	100.1	29.1	65.6	98.3	37.7	68.8	97.9	37.7	68.6	93.6	69.7	82.0
01 - 16	WEIGHTED AVERAGE	95.7	33.0	67.9	98.4	50.7	77.2	95.7	35.5	69.0	96.4	49.9	75.7	98.2	49.6	76.6	99.3	66.6	84.8

NORTH DAKOTA

DIVISION		DEVILS LAKE 583			DICKINSON 586			FARGO 580 - 581			GRAND FORKS 582			JAMESTOWN 584			MINOT 587		
		MAT.	INST.	TOTAL	MAT.	INST.	TOTAL	MAT.	INST.	TOTAL	MAT.	INST.	TOTAL	MAT.	INST.	TOTAL	MAT.	INST.	TOTAL
01590	EQUIPMENT RENTAL	.0	98.3	98.3	.0	98.3	98.3	.0	98.3	98.3	.0	98.3	98.3	.0	98.3	98.3	.0	98.3	98.3
02	SITE CONSTRUCTION	94.9	95.0	95.0	102.8	94.0	96.4	89.9	96.5	94.8	98.7	94.0	95.3	94.0	94.0	94.0	96.2	96.5	96.4
03100	CONCRETE FORMS & ACCESSORIES	99.4	44.0	51.6	87.2	43.3	49.3	96.1	50.1	56.4	91.8	43.4	50.1	89.1	43.1	49.4	86.9	56.4	60.6
03200	CONCRETE REINFORCEMENT	101.2	78.7	89.8	102.2	51.3	76.4	93.3	77.6	85.3	99.7	78.6	89.0	101.8	61.7	81.5	103.1	78.5	90.6
03300	CAST-IN-PLACE CONCRETE	119.0	53.9	92.2	107.6	52.6	85.0	110.2	57.3	88.4	107.6	52.6	85.0	117.5	52.5	90.8	107.6	53.8	85.5
03	CONCRETE	109.3	55.2	82.5	108.4	49.2	79.0	110.0	58.9	84.7	105.5	54.3	80.1	107.7	51.0	79.6	104.2	60.7	82.6
04	MASONRY	112.2	66.9	84.1	115.5	59.0	80.5	106.9	48.7	70.8	106.1	66.6	81.6	125.9	42.6	74.3	107.1	66.4	81.9
05	METALS	92.6	82.7	89.4	92.5	68.4	84.8	95.0	81.6	90.7	92.6	79.0	88.2	92.6	71.1	85.7	92.9	83.1	89.8
06	WOOD & PLASTICS	92.7	41.1	65.9	80.1	41.1	59.8	88.3	45.4	66.0	84.7	41.1	62.0	82.2	41.1	60.8	79.8	53.7	66.2
07	THERMAL & MOISTURE PROTECTION	99.6	56.0	81.3	100.1	53.1	80.3	99.8	52.6	79.9	99.8	56.0	81.4	99.4	49.8	78.5	99.5	56.7	81.5
08	DOORS & WINDOWS	99.4	45.5	85.3	99.4	39.1	83.6	99.4	50.7	86.7	99.4	45.5	85.3	99.4	41.8	84.3	99.5	55.3	88.0
09200	PLASTER & GYPSUM BOARD	118.3	39.9	69.6	110.1	39.9	66.5	113.5	44.3	70.5	111.6	39.9	67.0	111.2	39.9	66.9	110.1	52.8	74.5
095,098	CEILINGS & ACOUSTICAL TREATMENT	138.5	39.9	79.7	138.5	39.9	79.7	138.5	44.3	82.3	138.5	39.9	79.7	138.5	39.9	79.7	138.5	52.8	87.4
09600	FLOORING	112.2	44.8	94.0	103.4	44.8	87.6	107.6	44.8	90.6	106.0	44.8	89.4	104.7	44.8	88.5	103.2	77.5	96.3
097,099	WALL FINISHES, PAINTS & COATINGS	97.2	28.7	55.4	97.2	28.7	55.4	97.2	71.0	81.2	97.2	28.7	55.4	97.2	28.7	55.4	97.2	31.4	57.1
09	FINISHES	117.0	41.2	77.5	113.8	41.2	76.0	114.5	49.9	80.9	114.4	41.2	76.3	113.6	41.2	75.9	113.2	57.0	84.0
10 - 14	TOTAL DIV. 10000 - 14000	100.0	44.9	88.2	100.0	45.0	88.3	100.0	69.2	93.4	100.0	45.0	88.3	100.0	66.4	92.8	100.0	70.3	93.7
15	MECHANICAL	95.6	66.0	83.3	95.6	63.2	82.1	100.5	69.3	87.6	100.5	45.2	77.5	95.6	45.0	74.6	100.5	63.4	85.1
16	ELECTRICAL	93.0	46.6	70.4	101.0	72.5	87.2	93.5	68.7	81.4	96.2	69.7	83.3	93.0	46.5	70.4	99.2	70.9	85.4
01 - 16	WEIGHTED AVERAGE	99.9	60.8	82.6	100.7	60.5	82.9	100.9	65.2	85.1	100.5	59.2	82.2	100.0	52.6	78.9	100.6	67.8	86.1

DIVISION		NORTH DAKOTA			OHIO														
		WILLISTON			AKRON			ATHENS			CANTON			CHILLICOTHE			CINCINNATI		
		588			442 - 443			457			446 - 447			456			451 - 452		
		MAT.	INST.	TOTAL	MAT.	INST.	TOTAL	MAT.	INST.	TOTAL	MAT.	INST.	TOTAL	MAT.	INST.	TOTAL	MAT.	INST.	TOTAL
01590	EQUIPMENT RENTAL	.0	98.3	98.3	.0	97.4	97.4	.0	90.5	90.5	.0	97.4	97.4	.0	102.8	102.8	.0	102.5	102.5
02	SITE CONSTRUCTION	96.7	94.0	94.7	101.7	106.7	105.4	87.0	94.0	92.2	101.8	106.8	105.5	76.5	108.6	100.1	74.4	108.5	99.4
03100	CONCRETE FORMS & ACCESSORIES	93.8	43.3	50.2	98.7	96.7	96.9	92.1	90.0	90.3	98.7	85.6	87.4	94.7	98.0	97.6	97.0	88.3	89.5
03200	CONCRETE REINFORCEMENT	104.2	51.3	77.3	94.8	96.8	95.9	96.8	88.2	92.4	94.8	83.6	89.1	93.3	83.1	88.1	98.9	86.0	92.4
03300	CAST-IN-PLACE CONCRETE	107.6	52.6	85.0	98.5	105.1	101.2	101.8	101.7	101.7	99.5	101.8	100.4	92.4	97.0	94.3	85.1	88.8	86.6
03	CONCRETE	105.6	49.2	77.6	98.6	98.6	98.6	102.8	92.8	97.9	99.1	90.1	94.6	97.5	94.6	96.1	92.4	88.1	90.3
04	MASONRY	101.9	59.0	75.3	89.7	102.5	97.7	74.9	98.7	89.6	90.5	92.7	91.9	82.7	97.9	92.1	82.4	94.2	89.7
05	METALS	92.7	68.4	85.0	89.8	81.8	87.2	93.7	74.4	87.5	89.8	75.5	85.2	86.0	85.7	86.0	88.0	88.0	88.0
06	WOOD & PLASTICS	86.3	41.1	62.8	92.3	94.8	93.6	82.7	88.0	85.5	92.6	84.0	88.1	94.3	97.1	95.8	96.8	86.7	91.5
07	THERMAL & MOISTURE PROTECTION	99.7	52.9	80.0	99.6	100.7	100.0	96.4	84.1	91.2	100.1	96.2	98.4	96.0	95.1	95.6	94.2	96.2	95.1
08	DOORS & WINDOWS	99.5	39.1	83.7	104.3	96.7	102.3	95.5	77.6	90.8	98.7	77.2	93.1	88.8	89.3	88.9	96.1	85.5	93.4
09200	PLASTER & GYPSUM BOARD	111.6	39.9	67.0	97.7	94.2	95.6	97.3	87.3	91.1	98.5	83.2	89.0	98.5	97.3	97.7	99.8	86.6	91.6
095,098	CEILINGS & ACOUSTICAL TREATMENT	138.5	39.9	79.7	90.4	94.2	92.7	103.5	87.3	93.8	90.4	83.2	86.1	98.0	97.3	97.6	99.8	86.6	91.6
09600	FLOORING	107.2	44.8	90.3	103.4	94.3	100.9	130.0	96.7	121.0	103.6	85.6	98.7	107.3	96.0	104.2	108.4	99.3	106.0
097,099	WALL FINISHES, PAINTS & COATINGS	97.2	28.7	55.4	107.9	118.5	114.4	111.3	53.2	75.9	107.9	90.4	97.2	107.8	98.5	102.1	107.8	92.8	98.6
09	FINISHES	114.7	41.2	76.4	99.4	98.5	98.9	102.9	88.1	95.2	99.6	85.1	92.1	98.7	98.2	98.5	99.3	90.6	94.8
10 - 14	TOTAL DIV. 10000 - 14000	100.0	45.0	88.3	100.0	86.8	97.2	100.0	70.6	93.7	100.0	75.0	94.7	100.0	81.4	96.0	100.0	92.7	98.4
15	MECHANICAL	95.6	61.9	81.6	99.9	102.0	100.8	94.9	69.8	84.5	99.9	86.3	94.2	95.4	93.0	94.4	99.8	88.5	95.1
16	ELECTRICAL	96.6	72.5	84.9	98.2	94.9	96.6	101.0	57.0	79.6	97.4	91.0	94.3	98.3	89.3	93.9	98.1	81.5	90.1
01 - 16	WEIGHTED AVERAGE	99.3	60.2	81.9	97.8	98.0	97.9	96.3	79.9	89.0	97.3	88.4	93.4	93.5	94.2	93.8	94.9	90.2	92.8

DIVISION		OHIO																	
		CLEVELAND			COLUMBUS			DAYTON			HAMILTON			LIMA			LORAIN		
		441			430 - 432			453 - 454			450			458			440		
		MAT.	INST.	TOTAL	MAT.	INST.	TOTAL	MAT.	INST.	TOTAL	MAT.	INST.	TOTAL	MAT.	INST.	TOTAL	MAT.	INST.	TOTAL
01590	EQUIPMENT RENTAL	.0	97.7	97.7	.0	96.4	96.4	.0	96.7	96.7	.0	102.8	102.8	.0	93.5	93.5	.0	97.4	97.4
02	SITE CONSTRUCTION	101.7	106.9	105.5	85.6	101.8	97.5	73.2	108.1	98.8	73.5	108.7	99.3	82.1	94.3	91.1	101.1	106.0	104.7
03100	CONCRETE FORMS & ACCESSORIES	98.8	102.7	102.1	99.2	86.3	88.1	96.9	79.2	81.6	98.9	88.3	89.8	92.1	86.9	87.6	98.7	103.7	103.0
03200	CONCRETE REINFORCEMENT	95.4	97.0	96.2	76.7	83.2	80.0	98.9	82.1	90.4	98.9	86.0	92.4	96.8	82.2	89.4	94.8	96.9	95.9
03300	CAST-IN-PLACE CONCRETE	96.7	112.5	103.2	88.9	90.9	89.7	77.5	89.1	82.3	84.8	87.4	85.9	93.5	99.2	95.8	93.8	107.5	99.4
03	CONCRETE	97.8	104.0	100.9	91.4	87.0	89.2	88.7	82.9	85.8	92.3	87.6	90.0	96.2	89.9	93.1	96.3	102.6	99.4
04	MASONRY	94.0	107.3	102.3	95.8	91.1	92.9	81.8	88.8	86.1	82.3	94.2	89.7	103.4	89.4	94.7	86.7	99.9	94.9
05	METALS	91.2	85.0	89.2	96.1	80.6	91.1	87.3	77.8	84.2	87.4	88.0	87.6	93.7	81.2	89.7	90.3	82.8	87.9
06	WOOD & PLASTICS	91.5	99.2	95.5	103.7	84.9	93.9	97.9	75.8	86.4	96.8	86.7	91.5	82.7	85.2	84.0	92.3	105.0	98.9
07	THERMAL & MOISTURE PROTECTION	98.3	112.4	104.2	98.1	93.0	96.0	99.4	89.7	95.3	96.1	96.0	96.1	96.0	100.2	97.9	100.0	107.2	103.1
08	DOORS & WINDOWS	95.2	99.1	96.2	99.5	83.5	95.3	96.3	78.6	91.7	94.0	85.5	91.7	95.5	82.6	92.1	98.7	102.2	99.6
09200	PLASTER & GYPSUM BOARD	96.9	99.8	98.1	95.2	84.3	88.5	99.8	75.4	84.7	96.8	86.6	91.6	97.3	84.3	89.2	97.7	104.4	102.0
095,098	CEILINGS & ACOUSTICAL TREATMENT	88.6	99.8	94.7	93.9	84.3	88.2	99.8	75.4	85.3	98.8	86.6	91.6	102.5	84.3	91.7	90.4	104.7	98.9
09600	FLOORING	103.2	108.0	104.5	97.5	91.3	95.8	111.2	86.3	104.4	108.4	99.3	106.0	129.1	93.3	119.5	103.6	106.7	104.4
097,099	WALL FINISHES, PAINTS & COATINGS	107.9	121.9	116.4	98.4	98.5	98.4	107.8	90.3	97.1	107.8	91.6	97.9	111.4	75.8	89.6	107.9	121.9	116.4
09	FINISHES	98.9	105.3	102.2	97.5	88.2	92.7	100.3	80.8	90.2	99.3	90.5	94.7	102.1	86.4	94.0	99.5	106.7	103.3
10 - 14	TOTAL DIV. 10000 - 14000	100.0	105.2	101.1	100.0	92.8	98.5	100.0	90.6	98.0	100.0	92.7	98.4	100.0	83.5	96.5	100.0	103.2	100.7
15	MECHANICAL	99.9	108.1	103.3	100.0	91.7	96.6	100.8	88.0	95.5	100.3	88.5	95.4	94.9	89.4	92.6	99.9	90.8	96.1
16	ELECTRICAL	98.2	107.1	102.6	98.1	88.4	93.4	96.1	88.3	92.3	96.3	77.7	87.3	101.2	84.4	93.1	97.5	88.6	93.2
01 - 16	WEIGHTED AVERAGE	97.2	104.3	100.4	97.3	89.6	93.9	94.6	87.0	91.2	94.5	89.6	92.3	96.8	87.9	92.8	96.9	97.1	97.0

DIVISION		OHIO																	
		MANSFIELD			MARION			SPRINGFIELD			STEUBENVILLE			TOLEDO			YOUNGSTOWN		
		448 - 449			433			455			439			434 - 436			444 - 445		
		MAT.	INST.	TOTAL	MAT.	INST.	TOTAL	MAT.	INST.	TOTAL	MAT.	INST.	TOTAL	MAT.	INST.	TOTAL	MAT.	INST.	TOTAL
01590	EQUIPMENT RENTAL	.0	97.4	97.4	.0	96.1	96.1	.0	96.7	96.7	.0	101.7	101.7	.0	99.5	99.5	.0	97.4	97.4
02	SITE CONSTRUCTION	96.9	106.8	104.1	81.1	102.0	96.4	73.7	107.2	98.3	121.5	112.2	114.6	84.9	102.9	98.1	101.6	107.1	105.6
03100	CONCRETE FORMS & ACCESSORIES	87.1	102.8	100.7	95.2	85.5	86.9	96.9	85.4	87.0	96.1	92.7	93.1	99.2	96.4	96.8	98.7	89.8	91.0
03200	CONCRETE REINFORCEMENT	86.3	83.9	85.1	70.8	83.4	77.2	98.9	82.1	90.4	68.8	93.4	81.3	76.7	95.3	86.2	94.8	96.7	95.8
03300	CAST-IN-PLACE CONCRETE	91.2	94.0	92.4	81.2	90.6	85.1	81.3	89.0	84.5	88.2	101.3	93.6	88.9	105.7	95.8	97.6	103.4	100.0
03	CONCRETE	90.8	95.2	93.0	83.8	86.4	85.1	90.6	85.6	88.1	87.6	94.9	91.3	91.4	98.9	95.1	98.2	95.0	96.6
04	MASONRY	88.9	101.2	96.5	98.3	97.4	97.8	82.1	88.6	86.2	84.3	95.9	91.5	105.7	102.8	103.9	89.9	95.2	93.2
05	METALS	90.5	77.2	86.2	95.0	77.8	89.5	87.3	77.6	84.2	91.2	81.3	88.0	95.9	89.2	93.8	89.8	82.1	87.3
06	WOOD & PLASTICS	79.8	105.0	92.9	99.5	84.9	91.9	99.1	84.5	91.5	95.3	90.4	92.8	103.7	94.2	98.7	92.3	87.4	89.7
07	THERMAL & MOISTURE PROTECTION	99.4	99.5	99.5	97.6	87.2	93.2	99.3	90.5	95.6	109.9	97.5	104.7	100.3	107.8	103.4	100.2	99.0	99.7
08	DOORS & WINDOWS	98.1	91.2	96.3	94.1	80.2	90.4	94.4	80.7	90.8	94.8	90.4	93.7	97.4	93.7	96.4	98.7	93.6	97.4
09200	PLASTER & GYPSUM BOARD	93.2	104.7	100.3	93.7	84.3	87.9	99.8	84.3	90.2	94.5	89.5	91.4	95.2	93.9	94.4	97.7	86.6	90.8
095,098	CEILINGS & ACOUSTICAL TREATMENT	92.9	104.7	99.9	93.9	84.3	88.2	99.8	84.3	90.6	120.9	89.5	90.0	96.6	96.4	96.6	90.4	86.6	88.1
09600	FLOORING	98.4	107.5	100.8	95.6	87.1	93.3	111.2	86.3	104.4	124.0	99.3	117.3	96.6	96.4	96.6	103.6	93.7	100.9
097,099	WALL FINISHES, PAINTS & COATINGS	107.9	70.7	85.2	98.4	60.6	75.3	107.8	90.3	97.1	115.6	97.4	104.5	98.4	106.5	103.4	107.9	100.0	103.1
09	FINISHES	97.7	101.3	99.6	96.5	83.2	89.5	100.3	85.9	92.8	113.2	94.0	103.2	97.2	97.0	97.1	99.6	90.9	95.0
10 - 14	TOTAL DIV. 10000 - 14000	100.0	82.7	96.3	100.0	71.3	93.9	100.0	91.6	98.2	100.0	99.8	100.0	100.0	87.6	97.3	100.0	85.7	97.0
15	MECHANICAL	95.0	90.8	93.2	95.1	91.7	93.7	100.8	88.0	95.4	95.4	86.7	91.8	100.0	102.9	101.2	99.9	92.5	96.8
16	ELECTRICAL	95.2	88.3	91.9	92.6	88.3	90.5	96.1	87.0	91.7	84.6	106.3	95.1	97.9	105.7	101.7	97.5	93.1	95.4
01 - 16	WEIGHTED AVERAGE	94.6	93.8	94.2	93.9	88.4	91.5	94.6	87.9	91.6	95.2	95.1	95.1	97.6	100.0	98.7	97.2	93.3	95.5

DIVISION		OHIO			OKLAHOMA													
		ZANESVILLE 437-438			ARDMORE 734			CLINTON 736			DURANT 747			ENID 737			GUYMON 739		
		MAT.	INST.	TOTAL	MAT.	INST.	TOTAL	MAT.	INST.	TOTAL	MAT.	INST.	TOTAL	MAT.	INST.	TOTAL	MAT.	INST.	TOTAL
01590	EQUIPMENT RENTAL	.0	96.1	96.1	.0	78.2	78.2	.0	77.1	77.1	.0	77.1	77.1	.0	77.1	77.1	.0	77.1	77.1
02	SITE CONSTRUCTION	83.6	102.5	97.5	104.8	90.5	94.3	106.1	88.9	93.5	99.0	88.6	91.4	108.0	88.9	94.0	111.4	87.7	94.0
03100	CONCRETE FORMS & ACCESSORIES	91.6	85.5	86.3	90.8	46.9	52.9	89.5	40.2	47.0	81.3	46.9	51.6	93.5	39.4	46.8	97.9	24.2	34.3
03200	CONCRETE REINFORCEMENT	70.3	92.9	81.8	97.2	82.4	89.6	97.7	82.4	89.9	98.1	66.9	82.2	97.0	82.4	89.6	97.7	33.7	65.2
03300	CAST-IN-PLACE CONCRETE	85.6	90.1	87.4	94.8	47.1	75.2	91.6	47.1	73.3	88.6	47.0	71.5	91.6	50.0	74.5	91.6	30.9	66.7
03	CONCRETE	87.3	88.0	87.6	90.2	53.7	72.1	89.4	50.7	70.2	85.2	50.8	68.1	90.0	51.4	70.8	92.4	28.8	60.8
04	MASONRY	94.7	86.8	89.8	92.8	61.4	73.3	119.7	61.4	83.5	91.6	62.0	73.3	101.6	61.4	76.6	96.1	23.2	50.9
05	METALS	96.6	81.1	91.6	88.8	67.1	81.9	88.9	67.0	81.9	88.9	60.4	79.8	89.9	67.1	82.6	89.3	39.1	73.3
06	WOOD & PLASTICS	95.5	84.9	89.9	93.8	46.4	69.1	92.7	37.4	63.9	83.7	46.4	64.3	96.9	36.4	65.4	101.1	23.8	60.8
07	THERMAL & MOISTURE PROTECTION	97.7	90.9	94.8	99.6	62.6	84.0	99.7	61.7	83.7	99.3	60.8	83.1	99.9	61.6	83.7	100.2	29.4	70.4
08	DOORS & WINDOWS	94.1	85.0	91.7	95.4	56.1	85.1	95.4	51.3	83.9	95.4	52.2	84.1	95.4	50.5	83.7	95.5	25.9	77.3
09200	PLASTER & GYPSUM BOARD	91.9	84.3	87.2	80.7	45.7	59.0	80.3	36.6	53.1	76.6	45.7	57.4	81.4	35.5	52.9	82.1	22.5	45.1
095,098	CEILINGS & ACOUSTICAL TREATMENT	93.9	84.3	88.2	79.4	45.7	59.3	79.4	36.6	53.8	79.4	45.7	59.3	80.3	35.5	53.6	81.9	22.5	46.5
09600	FLOORING	93.4	87.1	91.7	111.6	53.9	96.0	109.9	51.4	94.0	105.3	70.5	95.9	112.4	51.4	95.9	114.8	30.4	92.0
097,099	WALL FINISHES, PAINTS & COATINGS	98.4	64.9	77.9	97.0	65.4	77.7	97.0	65.4	77.7	97.0	65.4	77.7	97.0	65.4	77.7	97.0	19.4	49.7
09	FINISHES	95.7	83.4	89.3	93.0	48.9	70.0	92.6	43.2	66.9	90.2	52.5	70.6	93.8	42.5	67.1	95.3	24.7	58.6
10-14	TOTAL DIV. 10000-14000	100.0	77.6	95.2	100.0	60.9	91.7	100.0	59.7	91.4	100.0	61.1	91.7	100.0	59.6	91.4	100.0	55.5	90.5
15	MECHANICAL	95.1	83.8	90.4	95.1	65.3	82.8	95.1	65.3	82.8	95.1	65.5	82.8	100.1	65.3	85.6	95.1	27.2	67.0
16	ELECTRICAL	92.5	83.3	88.0	93.9	71.5	83.0	94.9	71.5	83.5	96.4	63.3	80.3	94.9	71.5	83.5	96.4	20.0	59.3
01-16	WEIGHTED AVERAGE	94.3	86.1	90.6	94.0	63.7	80.5	95.3	62.0	80.6	93.1	61.7	79.2	96.0	62.0	80.9	95.2	32.9	67.5

DIVISION		OKLAHOMA																
		LAWTON 735			MCALESTER 745			MIAMI 743			MUSKOGEE 744			OKLAHOMA CITY 730-731			PONCA CITY 746		
		MAT.	INST.	TOTAL	MAT.	INST.	TOTAL	MAT.	INST.	TOTAL	MAT.	INST.	TOTAL	MAT.	INST.	TOTAL	MAT.	INST.	TOTAL
01590	EQUIPMENT RENTAL	.0	78.2	78.2	.0	77.1	77.1	.0	86.7	86.7	.0	86.7	86.7	.0	78.5	78.5	.0	77.1	77.1
02	SITE CONSTRUCTION	103.4	90.5	94.0	91.3	88.2	89.0	93.4	85.9	87.9	94.1	84.9	87.4	104.7	91.1	94.7	99.9	88.9	91.8
03100	CONCRETE FORMS & ACCESSORIES	97.4	53.9	59.9	78.9	48.7	52.8	93.7	57.9	62.8	98.8	35.7	44.3	98.5	46.5	53.6	88.6	47.3	53.0
03200	CONCRETE REINFORCEMENT	97.3	82.4	89.7	97.7	50.2	73.6	96.2	82.4	89.2	97.0	38.5	67.3	97.3	82.4	89.7	97.0	82.3	89.6
03300	CAST-IN-PLACE CONCRETE	88.6	50.1	72.8	77.7	49.3	66.0	81.4	49.1	68.1	82.4	38.5	64.4	95.9	53.3	78.4	91.0	41.2	70.5
03	CONCRETE	86.8	57.9	72.4	76.9	49.1	63.1	81.1	60.2	70.7	82.6	38.3	60.6	90.4	55.7	73.1	87.2	51.8	69.6
04	MASONRY	96.0	61.4	74.5	109.7	61.8	80.0	94.6	62.8	74.8	111.2	54.9	76.3	97.8	63.5	76.5	86.3	62.0	71.2
05	METALS	93.3	67.2	84.9	88.8	50.0	76.4	88.8	81.9	86.6	89.8	58.6	79.9	95.2	67.1	86.2	88.8	66.8	81.8
06	WOOD & PLASTICS	100.0	55.8	77.0	81.0	50.4	65.1	96.8	60.4	77.9	102.1	36.0	67.7	101.9	44.8	72.1	92.0	47.1	68.6
07	THERMAL & MOISTURE PROTECTION	99.6	63.6	84.5	99.0	61.7	83.3	99.4	63.7	84.4	99.5	47.6	77.6	98.9	63.3	83.9	99.5	61.6	83.6
08	DOORS & WINDOWS	96.9	61.0	87.5	95.4	49.3	83.4	95.4	64.1	87.2	95.4	35.4	79.7	96.9	55.0	86.0	95.4	56.5	85.2
09200	PLASTER & GYPSUM BOARD	83.8	55.5	66.2	75.5	49.8	59.6	81.4	60.0	68.1	85.3	34.9	54.0	83.8	44.1	59.1	80.3	46.4	59.3
095,098	CEILINGS & ACOUSTICAL TREATMENT	89.7	55.5	69.3	79.4	49.8	61.8	79.4	60.0	67.8	89.7	34.9	57.0	89.7	44.1	62.5	79.4	46.4	59.7
09600	FLOORING	115.3	51.4	98.0	104.0	51.4	89.7	113.3	70.5	101.8	116.5	41.4	96.2	115.3	51.4	98.0	109.3	51.4	93.7
097,099	WALL FINISHES, PAINTS & COATINGS	97.0	65.4	77.7	97.0	48.2	67.2	97.0	80.4	86.9	97.0	35.0	59.2	97.0	65.4	77.7	97.0	65.4	77.7
09	FINISHES	96.9	53.9	74.5	89.1	48.9	68.2	92.7	62.6	77.0	96.9	37.0	65.7	97.0	48.0	71.4	92.0	49.0	69.6
10-14	TOTAL DIV. 10000-14000	100.0	62.0	91.9	100.0	61.6	91.8	100.0	63.8	92.3	100.0	59.0	91.3	100.0	61.5	91.8	100.0	61.2	91.7
15	MECHANICAL	100.1	65.3	85.6	95.1	38.3	71.6	95.1	65.1	82.7	100.1	29.2	70.6	100.1	66.4	86.1	95.1	64.6	82.5
16	ELECTRICAL	96.4	71.5	84.3	94.9	68.0	81.8	96.3	68.1	82.6	94.5	34.3	65.2	95.8	71.5	84.0	94.6	64.6	80.0
01-16	WEIGHTED AVERAGE	96.4	65.3	82.6	92.6	55.1	75.9	93.0	67.5	81.7	95.6	44.0	72.7	97.1	64.3	82.6	93.1	62.2	79.4

DIVISION		OKLAHOMA												OREGON					
		POTEAU 749			SHAWNEE 748			TULSA 740-741			WOODWARD 738			BEND 977			EUGENE 974		
		MAT.	INST.	TOTAL	MAT.	INST.	TOTAL	MAT.	INST.	TOTAL	MAT.	INST.	TOTAL	MAT.	INST.	TOTAL	MAT.	INST.	TOTAL
01590	EQUIPMENT RENTAL	.0	86.0	86.0	.0	77.1	77.1	.0	86.7	86.7	.0	77.1	77.1	.0	99.6	99.6	.0	99.6	99.6
02	SITE CONSTRUCTION	75.0	84.4	81.9	102.9	88.6	92.4	100.4	85.9	89.8	106.7	88.9	93.7	121.3	104.7	109.1	108.7	104.7	105.7
03100	CONCRETE FORMS & ACCESSORIES	85.9	47.7	52.9	81.1	45.7	50.6	98.5	46.8	53.9	89.6	40.1	46.9	108.2	105.8	106.1	104.0	105.7	105.5
03200	CONCRETE REINFORCEMENT	98.2	82.4	90.2	97.0	55.2	75.8	97.3	82.3	89.7	97.0	82.4	89.6	101.4	99.3	100.4	105.8	99.3	102.5
03300	CAST-IN-PLACE CONCRETE	81.4	48.6	68.0	93.9	46.2	74.3	89.8	48.8	72.9	91.6	47.1	73.3	110.4	105.4	108.4	106.9	105.4	106.3
03	CONCRETE	82.7	55.5	69.2	88.7	47.7	68.4	87.4	55.1	71.4	89.7	50.7	70.3	117.8	104.1	111.0	108.3	104.0	106.2
04	MASONRY	95.4	62.1	74.7	110.9	60.5	79.6	95.3	63.5	75.6	89.9	61.4	72.2	115.7	103.3	108.0	113.4	103.3	107.1
05	METALS	88.8	81.7	86.6	88.8	54.3	77.8	92.6	81.0	88.9	89.0	67.1	82.0	93.6	96.8	94.7	94.2	96.8	95.0
06	WOOD & PLASTICS	88.8	47.2	67.2	83.6	46.1	64.1	100.7	46.1	72.3	92.9	37.3	64.0	99.7	105.5	102.7	95.2	105.5	100.6
07	THERMAL & MOISTURE PROTECTION	99.5	61.7	83.6	99.5	62.3	83.9	99.5	60.2	82.9	99.8	61.7	83.7	106.0	97.6	102.5	105.2	95.3	101.0
08	DOORS & WINDOWS	95.4	56.9	85.3	95.4	45.9	82.4	96.9	55.3	86.0	95.4	51.2	83.9	98.5	106.2	100.5	98.9	106.2	100.8
09200	PLASTER & GYPSUM BOARD	79.6	46.4	59.0	76.6	45.5	57.3	83.8	45.3	59.9	81.0	36.4	53.3	96.9	105.5	102.2	95.1	105.5	101.5
095,098	CEILINGS & ACOUSTICAL TREATMENT	79.4	46.4	59.7	79.4	45.5	59.1	89.7	45.3	63.2	81.9	36.4	54.8	101.2	105.5	103.7	106.0	105.5	105.7
09600	FLOORING	108.5	70.5	98.2	105.2	40.9	87.8	115.1	53.5	98.4	109.9	53.9	94.8	117.5	98.6	112.3	115.4	98.6	110.9
097,099	WALL FINISHES, PAINTS & COATINGS	97.0	65.4	77.7	97.0	43.5	64.4	97.0	51.7	69.3	97.0	65.4	77.7	115.2	72.0	88.8	115.2	72.0	88.8
09	FINISHES	90.3	53.0	70.9	90.4	43.7	66.1	96.4	47.6	71.0	93.4	43.6	67.5	110.7	100.8	105.6	110.1	100.8	105.2
10-14	TOTAL DIV. 10000-14000	100.0	61.6	91.8	100.0	60.5	91.6	100.0	62.2	91.9	100.0	59.7	91.4	100.0	106.4	101.4	100.0	106.4	101.4
15	MECHANICAL	95.1	64.7	82.5	95.1	64.6	82.5	100.1	60.3	83.6	95.1	65.3	82.8	95.3	106.1	99.8	100.2	106.1	102.7
16	ELECTRICAL	94.7	68.1	81.7	96.5	71.5	84.3	96.4	44.7	71.3	96.3	71.5	84.2	99.5	100.7	100.0	98.2	100.7	99.4
01-16	WEIGHTED AVERAGE	92.3	64.7	80.1	94.6	60.2	79.3	96.2	59.9	80.1	94.2	62.1	79.9	102.2	102.9	102.5	101.7	102.8	102.2

		OREGON																
	DIVISION	KLAMATH FALLS			MEDFORD			PENDLETON			PORTLAND			SALEM			VALE		
		976			975			978			970 - 972			973			979		
		MAT.	INST.	TOTAL	MAT.	INST.	TOTAL	MAT.	INST.	TOTAL	MAT.	INST.	TOTAL	MAT.	INST.	TOTAL	MAT.	INST.	TOTAL
01590	EQUIPMENT RENTAL	.0	99.6	99.6	.0	99.6	99.6	.0	96.5	96.5	.0	99.6	99.6	.0	99.6	99.6	.0	96.5	96.5
02	SITE CONSTRUCTION	126.5	104.7	110.5	118.2	104.7	108.3	114.8	97.2	101.9	109.8	104.7	106.1	108.3	104.7	105.6	100.6	96.5	97.6
03100	CONCRETE FORMS & ACCESSORIES	99.9	105.5	104.7	98.8	105.5	104.6	101.6	105.8	105.3	105.6	106.0	105.9	105.4	105.8	105.8	109.3	104.7	105.3
03200	CONCRETE REINFORCEMENT	101.4	99.3	100.3	103.1	99.3	101.2	100.7	99.5	100.1	106.6	99.5	103.0	106.7	99.5	103.0	98.2	99.3	98.8
03300	CAST-IN-PLACE CONCRETE	110.5	105.3	108.3	110.4	105.3	108.3	111.2	106.8	109.4	114.8	105.5	111.0	110.9	105.4	108.7	87.5	101.3	93.2
03	CONCRETE	120.7	103.9	112.4	114.9	103.9	109.4	100.2	104.6	102.4	112.4	104.2	108.3	110.5	104.1	107.3	84.0	102.0	92.9
04	MASONRY	132.2	103.3	114.3	110.6	103.3	106.1	120.4	106.0	111.5	114.1	106.0	109.3	118.4	105.7	110.6	118.6	105.7	110.6
05	METALS	93.6	96.5	94.6	93.8	96.5	94.7	101.0	97.3	99.8	95.4	97.2	96.0	94.6	97.1	95.4	100.9	94.2	98.8
06	WOOD & PLASTICS	89.9	105.5	98.0	88.6	105.5	97.4	92.7	105.6	99.4	96.3	105.5	101.1	96.3	105.5	101.1	102.1	105.6	103.9
07	THERMAL & MOISTURE PROTECTION	106.2	92.6	100.5	105.8	92.6	100.3	93.9	97.4	95.4	105.1	101.5	103.6	105.1	98.2	102.2	93.3	95.9	94.4
08	DOORS & WINDOWS	98.5	106.2	100.5	101.5	106.2	102.7	94.5	106.3	97.6	96.4	106.2	99.0	98.3	106.2	100.3	94.4	99.4	95.7
09200	PLASTER & GYPSUM BOARD	94.0	105.5	101.1	93.3	105.5	100.8	77.7	105.5	94.9	92.9	105.5	100.7	92.9	105.5	100.7	83.2	105.5	97.0
095,098	CEILINGS & ACOUSTICAL TREATMENT	111.4	105.5	107.9	116.2	105.5	109.8	68.9	105.5	90.7	106.0	105.5	105.7	106.0	105.5	105.7	68.9	105.5	90.7
09600	FLOORING	113.7	98.6	109.6	112.9	98.6	109.1	80.1	98.6	85.1	115.4	98.6	110.9	115.4	98.6	110.9	82.3	98.6	86.7
097,099	WALL FINISHES, PAINTS & COATINGS	115.2	62.6	83.1	115.2	62.6	83.1	96.0	72.0	81.3	115.2	72.0	88.8	115.2	72.0	88.8	96.0	72.0	81.3
09	FINISHES	112.1	99.7	105.7	112.3	99.7	105.8	75.9	100.9	88.9	109.8	100.8	105.1	109.7	100.8	105.0	76.6	100.9	89.2
10 - 14	TOTAL DIV. 10000 - 14000	100.0	106.4	101.4	100.0	106.4	101.4	100.0	106.6	101.4	100.0	106.4	101.4	100.0	106.4	101.4	100.0	106.7	101.4
15	MECHANICAL	95.3	106.1	99.8	100.2	106.1	102.7	97.0	106.4	100.9	100.2	108.9	103.8	100.2	106.1	102.7	97.0	81.4	90.5
16	ELECTRICAL	98.3	90.2	94.3	101.5	90.2	96.0	89.1	99.8	94.3	98.6	105.4	101.9	98.3	100.7	99.5	89.1	77.1	83.2
01 - 16	WEIGHTED AVERAGE	103.4	101.1	102.4	103.3	101.1	102.3	96.5	102.6	99.2	102.2	104.6	103.3	102.2	103.2	102.6	94.4	93.2	93.9

		PENNSYLVANIA																
	DIVISION	ALLENTOWN			ALTOONA			BEDFORD			BRADFORD			BUTLER			CHAMBERSBURG		
		181			166			155			167			160			172		
		MAT.	INST.	TOTAL	MAT.	INST.	TOTAL	MAT.	INST.	TOTAL	MAT.	INST.	TOTAL	MAT.	INST.	TOTAL	MAT.	INST.	TOTAL
01590	EQUIPMENT RENTAL	.0	115.6	115.6	.0	115.6	115.6	.0	113.8	113.8	.0	115.6	115.6	.0	115.6	115.6	.0	114.8	114.8
02	SITE CONSTRUCTION	91.9	106.8	102.8	96.1	106.7	103.9	98.7	104.7	103.1	91.1	106.5	102.4	87.0	107.6	102.1	86.5	103.7	99.1
03100	CONCRETE FORMS & ACCESSORIES	100.5	111.9	110.3	82.7	85.8	85.4	83.7	84.5	84.4	85.6	86.8	86.6	84.6	98.8	96.8	84.7	83.6	83.8
03200	CONCRETE REINFORCEMENT	103.3	101.9	102.6	100.2	93.8	96.9	99.8	81.8	90.7	102.3	93.8	98.0	100.9	109.6	105.3	100.7	97.9	99.3
03300	CAST-IN-PLACE CONCRETE	88.1	101.9	93.7	98.2	85.6	93.0	102.9	78.9	93.0	93.8	91.4	92.8	86.7	95.7	90.4	95.2	78.3	88.3
03	CONCRETE	98.1	107.5	102.8	93.5	88.7	91.1	97.4	83.4	90.5	99.9	91.1	95.5	85.2	101.0	93.0	105.7	85.9	95.9
04	MASONRY	93.7	99.2	97.1	95.8	83.7	88.3	96.2	85.8	89.8	92.4	89.7	90.7	98.0	95.4	96.4	94.2	82.6	87.0
05	METALS	96.9	121.4	104.7	90.5	115.3	98.5	95.1	107.0	98.9	95.3	115.6	101.8	90.3	124.7	101.3	95.2	113.8	101.2
06	WOOD & PLASTICS	101.8	115.2	108.8	79.2	87.6	83.6	81.4	85.7	83.6	85.6	85.7	85.7	81.0	98.5	90.1	89.7	85.7	87.6
07	THERMAL & MOISTURE PROTECTION	102.2	116.2	108.1	101.1	93.0	97.7	101.7	89.1	96.4	102.1	95.2	99.2	100.8	99.4	100.2	101.1	76.8	90.8
08	DOORS & WINDOWS	93.5	110.8	98.0	88.0	93.1	89.4	90.9	89.0	90.4	93.9	96.6	94.6	88.0	108.4	93.3	90.4	86.5	89.4
09200	PLASTER & GYPSUM BOARD	107.4	115.3	112.3	100.3	87.0	92.1	91.4	85.0	87.4	101.2	85.0	91.2	100.7	98.2	99.2	105.2	85.0	92.7
095,098	CEILINGS & ACOUSTICAL TREATMENT	88.5	115.3	104.5	92.8	87.0	89.4	88.6	85.0	86.5	91.2	85.0	87.5	93.8	98.2	96.4	91.2	85.0	87.5
09600	FLOORING	88.6	96.4	90.7	82.6	57.0	75.6	88.2	57.8	79.9	81.8	102.6	87.5	83.5	101.1	88.3	86.0	57.6	78.3
097,099	WALL FINISHES, PAINTS & COATINGS	91.6	90.0	90.6	86.8	107.8	99.6	99.5	86.3	91.4	91.6	87.6	89.2	86.8	105.8	98.4	91.6	73.7	80.7
09	FINISHES	95.5	106.3	101.2	93.9	81.9	87.7	92.4	78.9	85.3	93.3	88.4	90.7	94.1	99.2	96.7	94.4	77.5	85.6
10 - 14	TOTAL DIV. 10000 - 14000	100.0	105.9	101.3	100.0	97.8	99.5	100.0	98.4	99.7	100.0	99.8	100.0	100.0	103.1	100.7	100.0	95.8	99.1
15	MECHANICAL	100.3	109.5	104.1	99.8	90.4	95.9	95.2	91.4	93.6	95.5	90.8	93.6	95.0	91.8	93.7	95.5	91.4	93.8
16	ELECTRICAL	100.7	99.5	100.1	89.8	105.0	97.2	93.4	105.0	99.0	94.9	105.0	99.8	90.3	105.1	97.5	93.2	86.8	90.1
01 - 16	WEIGHTED AVERAGE	98.0	107.4	102.2	94.4	94.5	94.5	95.2	92.6	94.0	95.7	96.6	96.1	92.3	102.3	96.7	95.9	89.9	93.3

		PENNSYLVANIA																
	DIVISION	DOYLESTOWN			DUBOIS			ERIE			GREENSBURG			HARRISBURG			HAZLETON		
		189			158			164 - 165			156			170 - 171			182		
		MAT.	INST.	TOTAL	MAT.	INST.	TOTAL	MAT.	INST.	TOTAL	MAT.	INST.	TOTAL	MAT.	INST.	TOTAL	MAT.	INST.	TOTAL
01590	EQUIPMENT RENTAL	.0	92.6	92.6	.0	113.8	113.8	.0	115.6	115.6	.0	113.8	113.8	.0	114.8	114.8	.0	115.6	115.6
02	SITE CONSTRUCTION	107.1	88.6	93.5	103.4	105.4	104.9	92.7	107.4	103.5	94.9	106.5	103.4	82.9	105.5	99.5	85.0	106.8	101.0
03100	CONCRETE FORMS & ACCESSORIES	81.6	131.9	125.0	82.9	87.0	86.4	99.8	95.6	96.1	91.3	98.8	97.8	93.2	88.2	88.9	79.1	89.3	87.9
03200	CONCRETE REINFORCEMENT	99.9	130.8	115.6	99.2	93.8	96.5	102.3	90.9	96.5	99.2	109.6	104.4	103.3	98.4	100.8	100.3	112.8	106.7
03300	CAST-IN-PLACE CONCRETE	83.2	96.4	88.7	99.2	92.4	96.4	96.5	84.2	91.5	95.4	95.4	95.4	97.2	93.4	95.6	83.2	91.5	86.6
03	CONCRETE	93.4	118.8	106.0	98.2	91.5	94.9	92.6	92.0	92.3	92.7	100.8	96.8	100.3	93.4	96.9	90.4	96.0	93.2
04	MASONRY	97.1	130.8	118.0	96.3	89.7	92.2	86.0	95.9	92.2	106.8	94.1	99.0	94.4	89.6	91.4	105.8	96.5	100.0
05	METALS	94.3	117.7	101.8	95.1	114.7	101.4	90.7	114.4	98.3	95.0	123.5	104.1	98.8	119.1	105.3	96.6	125.5	105.9
06	WOOD & PLASTICS	80.8	134.5	108.7	80.4	85.7	83.1	98.3	95.1	96.6	89.7	98.5	94.3	95.5	87.1	91.1	79.1	87.1	83.3
07	THERMAL & MOISTURE PROTECTION	100.8	130.8	113.4	102.0	95.5	99.2	100.8	98.2	99.7	101.6	98.7	100.4	105.2	108.6	106.6	101.6	106.0	103.5
08	DOORS & WINDOWS	95.3	137.4	106.3	90.9	96.6	92.4	88.2	93.2	89.5	90.9	108.3	95.5	93.5	94.2	93.7	94.2	98.7	95.4
09200	PLASTER & GYPSUM BOARD	98.8	135.2	121.4	90.6	85.0	87.2	107.4	94.7	99.5	94.1	98.2	96.7	107.4	86.5	94.4	99.3	86.5	91.3
095,098	CEILINGS & ACOUSTICAL TREATMENT	87.8	135.2	116.1	88.6	85.0	86.5	88.5	94.7	92.2	87.8	98.2	94.0	88.5	86.5	87.3	89.5	86.5	87.7
09600	FLOORING	72.6	103.5	81.0	87.8	102.6	91.8	90.8	84.7	89.1	92.0	84.9	90.1	88.8	87.5	88.4	78.9	94.8	83.2
097,099	WALL FINISHES, PAINTS & COATINGS	91.0	88.0	89.2	99.5	105.8	103.3	98.1	91.3	94.0	99.5	105.8	103.3	91.6	87.3	89.0	91.6	98.6	95.9
09	FINISHES	86.8	122.6	105.5	92.5	89.8	91.1	97.0	92.8	94.8	93.5	97.1	95.4	94.5	87.4	90.8	91.3	89.7	90.5
10 - 14	TOTAL DIV. 10000 - 14000	100.0	91.4	98.2	100.0	99.8	100.0	100.0	102.9	100.6	100.0	103.0	100.6	100.0	97.4	99.4	100.0	101.6	100.3
15	MECHANICAL	95.0	126.4	108.0	95.2	93.6	94.6	99.8	94.0	97.4	95.2	96.2	95.6	100.3	94.0	97.7	95.5	85.7	91.4
16	ELECTRICAL	93.9	117.7	105.5	93.9	105.0	99.3	91.3	90.6	91.0	93.9	105.0	99.3	99.7	86.8	93.4	95.6	91.7	93.7
01 - 16	WEIGHTED AVERAGE	94.7	119.7	105.8	95.4	97.3	96.2	94.4	96.7	95.4	95.3	102.2	98.3	98.2	95.3	96.9	95.3	96.6	95.9

PENNSYLVANIA

DIVISION		INDIANA 157			JOHNSTOWN 159			KITTANNING 162			LANCASTER 175 - 176			LEHIGH VALLEY 180			MONTROSE 188		
		MAT.	INST.	TOTAL	MAT.	INST.	TOTAL	MAT.	INST.	TOTAL	MAT.	INST.	TOTAL	MAT.	INST.	TOTAL	MAT.	INST.	TOTAL
01590	EQUIPMENT RENTAL	.0	113.8	113.8	.0	113.8	113.8	.0	115.6	115.6	.0	114.8	114.8	.0	115.6	115.6	.0	115.6	115.6
02	SITE CONSTRUCTION	93.2	106.0	102.6	99.3	106.0	104.2	89.6	107.7	102.9	78.3	105.5	98.3	89.2	106.8	102.1	87.9	105.4	100.7
03100	CONCRETE FORMS & ACCESSORIES	84.5	89.1	88.5	82.9	88.6	87.9	84.6	98.9	96.9	87.0	87.7	87.6	93.0	112.1	109.5	80.1	89.4	88.2
03200	CONCRETE REINFORCEMENT	98.4	109.7	104.1	99.8	109.4	104.7	100.9	109.7	105.4	100.3	98.4	99.3	100.3	102.3	101.3	105.1	112.7	109.0
03300	CAST-IN-PLACE CONCRETE	93.6	95.4	94.3	103.8	89.0	97.7	90.1	95.8	92.4	81.0	93.1	86.0	90.1	102.0	95.0	88.3	90.8	89.3
03	CONCRETE	90.5	96.6	93.5	98.1	94.1	96.1	87.8	101.1	94.4	93.1	93.0	93.1	97.1	107.7	102.4	95.6	95.6	95.6
04	MASONRY	92.9	97.1	95.5	93.7	90.1	91.5	101.6	97.1	98.9	99.6	89.6	93.4	93.8	99.2	97.2	93.8	97.2	95.9
05	METALS	95.2	123.7	104.3	95.1	121.9	103.7	90.4	125.1	101.5	95.2	118.5	102.7	96.6	121.7	104.6	95.4	121.7	103.8
06	WOOD & PLASTICS	82.3	85.7	84.1	80.4	87.6	84.2	81.0	98.5	90.1	92.5	87.1	89.7	93.7	115.2	104.9	80.0	87.9	84.1
07	THERMAL & MOISTURE PROTECTION	101.5	97.2	99.7	101.7	95.0	98.9	100.9	99.8	100.4	100.6	103.4	101.8	102.0	109.9	105.3	101.6	98.3	100.3
08	DOORS & WINDOWS	90.9	101.3	93.6	90.9	97.8	92.7	88.0	108.4	93.3	90.4	94.6	91.5	94.2	110.5	98.5	90.6	99.2	92.9
09200	PLASTER & GYPSUM BOARD	91.7	85.0	87.6	90.4	87.0	88.3	100.7	98.2	99.2	107.1	86.5	94.3	103.7	115.3	110.9	99.7	87.3	92.0
095,098	CEILINGS & ACOUSTICAL TREATMENT	88.6	85.0	86.5	87.8	87.0	87.3	93.8	98.2	96.4	91.2	86.5	88.4	89.5	115.3	104.9	91.2	87.3	88.9
09600	FLOORING	88.6	102.6	92.4	87.8	72.5	83.6	83.5	102.6	88.7	87.0	91.2	88.1	84.8	100.6	89.1	79.6	72.3	77.6
097,099	WALL FINISHES, PAINTS & COATINGS	99.5	105.8	103.3	99.5	107.8	104.6	86.8	105.8	98.4	91.6	72.9	80.2	91.6	84.8	87.4	91.6	98.6	95.9
09	FINISHES	92.3	92.0	92.1	91.9	86.7	89.2	94.2	99.6	97.0	94.6	84.9	89.6	94.0	106.5	100.5	92.2	86.9	89.4
10 - 14	TOTAL DIV. 10000 - 14000	100.0	101.4	100.3	100.0	102.0	100.0	100.0	103.1	100.7	100.0	97.3	99.4	100.0	106.1	101.3	100.0	102.0	100.4
15	MECHANICAL	95.2	96.2	95.6	95.2	94.1	94.7	95.0	98.5	96.4	95.5	93.9	94.9	95.5	109.5	101.3	95.5	85.8	91.5
16	ELECTRICAL	93.9	105.0	99.3	93.9	105.0	99.3	89.8	105.1	97.2	94.5	57.3	76.4	95.6	150.2	122.1	94.9	90.8	92.9
01 - 16	WEIGHTED AVERAGE	94.1	100.7	97.0	95.1	98.1	96.5	92.8	103.6	97.6	94.8	90.6	92.9	95.9	114.4	104.1	94.8	95.5	95.1

PENNSYLVANIA

DIVISION		NEW CASTLE 161			NORRISTOWN 194			OIL CITY 163			PHILADELPHIA 190 - 191			PITTSBURGH 150 - 152			POTTSVILLE 179		
		MAT.	INST.	TOTAL	MAT.	INST.	TOTAL	MAT.	INST.	TOTAL	MAT.	INST.	TOTAL	MAT.	INST.	TOTAL	MAT.	INST.	TOTAL
01590	EQUIPMENT RENTAL	.0	115.6	115.6	.0	95.3	95.3	.0	115.6	115.6	.0	94.7	94.7	.0	115.4	115.4	.0	114.8	114.8
02	SITE CONSTRUCTION	87.5	107.7	102.3	94.8	95.7	95.5	85.9	106.7	101.2	101.8	95.2	97.0	98.3	109.1	106.2	81.3	105.2	98.9
03100	CONCRETE FORMS & ACCESSORIES	84.6	98.4	96.5	80.9	129.5	122.8	84.6	85.2	85.1	99.3	136.4	131.3	100.6	99.0	99.2	77.4	88.0	86.6
03200	CONCRETE REINFORCEMENT	99.6	104.3	102.0	100.9	136.5	119.0	100.9	104.2	102.6	104.1	137.0	120.8	100.3	109.9	105.2	99.5	99.4	99.4
03300	CAST-IN-PLACE CONCRETE	87.5	95.9	91.0	80.8	127.1	99.8	85.0	96.5	89.7	99.2	129.3	111.6	99.1	95.5	97.6	86.3	85.0	85.8
03	CONCRETE	85.6	99.9	92.7	95.3	129.5	112.3	84.0	94.1	89.0	108.1	133.5	120.7	96.1	101.0	98.6	96.8	90.6	93.7
04	MASONRY	96.5	95.4	95.8	112.6	125.6	120.7	97.6	93.9	95.3	98.8	135.5	121.5	88.6	100.3	95.8	93.5	90.1	91.4
05	METALS	90.4	122.0	100.5	101.2	126.1	109.2	90.4	120.4	100.0	105.0	127.8	112.3	96.3	124.3	105.3	95.4	118.7	102.9
06	WOOD & PLASTICS	81.0	98.5	90.1	77.7	129.0	104.4	81.0	81.5	81.2	97.2	136.3	117.6	100.0	98.5	99.2	82.1	87.1	84.7
07	THERMAL & MOISTURE PROTECTION	100.8	97.4	99.4	102.1	133.0	115.1	100.7	96.7	99.0	102.9	136.8	117.1	101.9	100.3	101.2	100.7	105.1	102.5
08	DOORS & WINDOWS	88.0	106.5	92.8	88.7	136.1	101.1	88.0	87.6	87.9	96.7	140.2	108.1	93.5	108.3	97.4	90.5	94.5	91.5
09200	PLASTER & GYPSUM BOARD	100.7	98.2	99.2	96.3	129.7	117.0	100.7	80.7	88.3	104.1	137.2	124.6	99.1	98.2	98.6	102.3	86.5	92.4
095,098	CEILINGS & ACOUSTICAL TREATMENT	93.8	98.2	96.4	98.3	129.7	117.0	93.8	80.7	86.0	98.3	137.2	121.5	88.6	98.2	94.4	91.2	86.5	88.4
09600	FLOORING	83.5	71.8	80.3	78.4	131.8	92.9	83.5	102.6	88.7	87.0	138.6	101.0	96.6	104.3	98.7	82.7	91.2	85.0
097,099	WALL FINISHES, PAINTS & COATINGS	86.8	105.8	98.4	94.6	138.4	121.3	86.8	105.8	98.4	94.6	138.4	121.3	99.5	113.5	108.0	91.6	98.6	95.9
09	FINISHES	94.1	93.6	93.8	96.7	130.8	114.5	94.0	88.5	91.1	100.7	137.2	119.7	95.9	101.0	98.6	92.7	89.5	91.0
10 - 14	TOTAL DIV. 10000 - 14000	100.0	103.1	100.7	100.0	113.1	102.8	100.0	100.9	100.2	100.0	125.9	105.5	100.0	103.0	100.6	100.0	97.6	99.5
15	MECHANICAL	95.0	93.8	94.5	95.2	129.2	109.3	95.0	93.4	94.3	100.1	133.4	113.9	100.1	98.8	99.6	95.5	83.7	90.6
16	ELECTRICAL	90.3	104.5	97.2	92.0	130.9	110.9	91.9	105.0	98.3	96.1	138.2	116.6	95.9	105.1	100.3	92.8	93.4	93.1
01 - 16	WEIGHTED AVERAGE	92.3	101.0	96.2	96.6	126.0	109.7	92.3	98.1	94.9	101.0	131.1	114.4	97.0	104.2	100.2	94.5	94.0	94.3

PENNSYLVANIA

DIVISION		READING 195 - 196			SCRANTON 184 - 185			STATE COLLEGE 168			STROUDSBURG 183			SUNBURY 178			UNIONTOWN 154		
		MAT.	INST.	TOTAL	MAT.	INST.	TOTAL	MAT.	INST.	TOTAL	MAT.	INST.	TOTAL	MAT.	INST.	TOTAL	MAT.	INST.	TOTAL
01590	EQUIPMENT RENTAL	.0	118.2	118.2	.0	115.6	115.6	.0	114.8	114.8	.0	115.6	115.6	.0	115.6	115.6	.0	113.8	113.8
02	SITE CONSTRUCTION	99.2	111.1	107.9	92.4	106.8	103.0	83.4	105.5	99.6	86.8	105.3	100.4	90.9	106.5	102.3	93.7	106.5	103.1
03100	CONCRETE FORMS & ACCESSORIES	98.8	88.6	90.0	100.7	88.7	90.3	83.7	85.7	85.5	86.5	91.4	90.8	90.5	86.4	87.0	76.2	98.6	95.5
03200	CONCRETE REINFORCEMENT	102.3	99.6	100.9	103.3	112.7	108.1	101.5	93.6	97.5	103.7	102.5	103.1	102.3	98.4	100.3	99.2	109.2	104.3
03300	CAST-IN-PLACE CONCRETE	72.5	98.9	83.3	92.0	92.1	92.1	88.7	75.3	83.2	86.7	80.8	84.3	94.3	92.9	93.8	93.6	95.3	94.3
03	CONCRETE	94.9	95.7	95.3	100.0	95.9	98.0	99.9	85.1	92.6	94.4	91.1	92.8	100.7	92.4	96.5	90.0	100.6	95.3
04	MASONRY	100.4	92.1	95.3	94.1	94.5	94.4	99.4	81.4	88.2	90.7	95.6	93.8	92.4	89.7	90.7	108.4	88.2	95.9
05	METALS	101.5	120.2	107.5	98.8	125.3	107.3	95.1	115.1	101.5	96.7	116.0	102.9	95.2	118.0	102.5	94.9	122.8	103.8
06	WOOD & PLASTICS	97.3	86.2	91.6	101.8	85.6	93.3	88.5	87.6	88.1	87.0	91.1	89.1	90.8	85.3	87.9	73.0	98.5	86.3
07	THERMAL & MOISTURE PROTECTION	102.3	110.8	105.8	102.1	102.8	102.4	100.9	96.7	99.1	101.8	85.1	94.8	102.0	103.6	102.7	101.4	97.1	99.6
08	DOORS & WINDOWS	94.3	94.4	94.4	93.5	97.9	94.7	90.4	93.1	91.1	94.2	89.0	92.9	90.6	93.2	91.3	90.9	108.3	95.4
09200	PLASTER & GYPSUM BOARD	107.4	85.7	93.9	109.8	84.9	94.3	103.8	87.0	93.4	101.4	90.6	94.7	102.4	84.6	91.3	87.5	98.2	94.2
095,098	CEILINGS & ACOUSTICAL TREATMENT	87.1	85.7	86.3	97.9	84.9	90.2	88.6	87.0	87.7	87.8	90.6	89.5	88.6	84.6	86.2	87.8	98.2	94.0
09600	FLOORING	85.2	101.6	89.6	88.6	91.1	89.3	85.4	62.1	79.1	82.4	64.1	77.4	83.8	53.0	75.5	84.5	80.0	83.3
097,099	WALL FINISHES, PAINTS & COATINGS	92.8	94.9	94.1	94.6	98.6	97.0	91.6	107.8	101.5	91.6	82.4	86.0	91.6	98.6	95.9	99.5	105.8	103.3
09	FINISHES	97.8	91.1	94.3	98.2	89.4	93.6	93.2	83.0	87.9	92.3	85.4	88.7	93.4	81.3	87.1	90.2	95.3	92.8
10 - 14	TOTAL DIV. 10000 - 14000	100.0	100.2	100.0	100.0	101.8	100.4	100.0	95.3	99.0	100.0	80.4	95.8	100.0	97.2	99.4	100.0	103.0	100.6
15	MECHANICAL	100.3	107.9	103.5	100.3	86.2	94.5	95.5	90.7	93.5	95.5	86.2	91.6	95.5	93.9	94.8	95.2	96.2	95.6
16	ELECTRICAL	97.9	93.5	95.7	100.7	91.7	96.3	93.6	105.0	99.2	95.6	150.1	122.1	93.6	88.2	91.0	91.2	105.0	98.0
01 - 16	WEIGHTED AVERAGE	98.8	100.9	99.7	98.8	96.3	97.7	95.4	93.9	94.7	95.2	101.1	97.8	95.4	94.4	95.0	94.2	101.2	97.3

City Cost Indexes

PENNSYLVANIA

DIVISION		WASHINGTON 153			WELLSBORO 169			WESTCHESTER 193			WILKES-BARRE 186 - 187			WILLIAMSPORT 177			YORK 173 - 174		
		MAT.	INST.	TOTAL	MAT.	INST.	TOTAL	MAT.	INST.	TOTAL	MAT.	INST.	TOTAL	MAT.	INST.	TOTAL	MAT.	INST.	TOTAL
01590	EQUIPMENT RENTAL	.0	113.8	113.8	.0	115.6	115.6	.0	95.3	95.3	.0	115.6	115.6	.0	115.6	115.6	.0	114.8	114.8
02	SITE CONSTRUCTION	93.7	106.5	103.1	95.1	105.2	102.5	101.1	94.4	96.2	84.7	106.6	100.7	82.0	105.1	98.9	82.2	105.5	99.3
03100	CONCRETE FORMS & ACCESSORIES	84.6	98.9	96.9	84.9	83.1	83.3	88.5	133.1	127.0	89.5	87.8	88.1	86.5	68.2	70.7	81.3	88.1	87.2
03200	CONCRETE REINFORCEMENT	99.2	109.7	104.5	101.5	112.6	107.1	99.9	105.9	103.0	102.3	102.3	102.3	101.5	83.8	92.5	102.3	98.4	100.3
03300	CAST-IN-PLACE CONCRETE	93.6	95.4	94.3	93.0	87.4	90.7	89.5	128.6	105.6	83.2	91.1	86.5	79.8	77.3	78.8	86.7	93.3	89.4
03	CONCRETE	90.6	100.9	95.7	102.3	91.6	97.0	103.6	126.1	114.7	91.5	93.1	92.3	88.2	76.2	82.2	98.1	93.3	95.8
04	MASONRY	93.2	97.1	95.6	98.7	88.5	92.4	107.2	131.4	122.2	106.7	92.9	98.2	84.8	79.6	81.5	93.4	89.6	91.0
05	METALS	94.9	123.8	104.1	95.2	122.3	103.9	101.2	116.1	106.0	95.3	120.0	103.2	95.3	110.5	100.1	96.4	119.0	103.6
06	WOOD & PLASTICS	82.4	98.5	90.8	85.0	82.5	83.7	85.8	134.4	111.1	89.8	85.4	87.5	87.0	66.2	76.2	86.2	87.1	86.6
07	THERMAL & MOISTURE PROTECTION	101.5	99.6	100.7	102.3	94.4	99.0	102.5	130.8	114.4	101.6	104.9	103.0	101.5	96.5	99.4	100.8	108.6	104.0
08	DOORS & WINDOWS	90.8	108.3	95.4	93.9	96.2	94.5	88.7	115.0	95.6	90.6	95.0	91.8	90.6	69.9	85.2	90.4	94.2	91.4
09200	PLASTER & GYPSUM BOARD	91.5	98.2	95.7	100.5	81.8	88.9	99.2	135.2	121.6	102.6	84.8	91.5	102.3	65.1	79.2	103.8	86.5	93.0
095,098	CEILINGS & ACOUSTICAL TREATMENT	87.8	98.2	94.0	88.6	81.8	84.6	98.3	135.2	120.3	91.2	84.8	87.4	91.2	65.1	75.6	90.2	86.5	88.0
09600	FLOORING	88.7	102.6	92.5	81.6	63.4	76.7	82.2	138.6	97.5	83.5	91.1	85.5	82.4	62.1	76.9	84.3	91.2	86.2
097,099	WALL FINISHES, PAINTS & COATINGS	99.5	105.8	103.3	91.6	98.6	95.9	94.6	138.4	121.3	91.6	97.1	95.0	91.6	97.1	95.0	91.6	87.3	89.0
09	FINISHES	92.0	99.7	96.0	92.7	80.2	86.2	98.7	134.3	117.2	93.6	88.7	91.0	93.1	68.8	80.4	93.2	87.4	90.2
10 - 14	TOTAL DIV. 10000 - 14000	100.0	103.0	100.6	100.0	96.2	99.2	100.0	123.6	105.0	100.0	101.2	100.3	100.0	91.7	98.2	100.0	97.4	99.4
15	MECHANICAL	95.2	96.2	95.6	95.5	92.7	94.4	95.2	130.2	109.7	95.5	85.3	91.3	95.5	89.4	93.0	100.3	94.0	97.7
16	ELECTRICAL	93.3	105.1	99.0	94.9	90.3	92.6	91.9	108.2	99.8	95.6	91.7	93.7	93.9	60.6	77.7	94.5	86.8	90.7
01 - 16	WEIGHTED AVERAGE	94.0	102.8	98.0	96.4	94.2	95.4	97.6	122.0	108.5	95.2	94.9	95.1	93.4	82.4	88.5	96.2	95.3	95.8

DIVISION		PUERTO RICO SAN JUAN 009			RHODE ISLAND NEWPORT 028			PROVIDENCE 029			SOUTH CAROLINA AIKEN 298			BEAUFORT 299			CHARLESTON 294		
		MAT.	INST.	TOTAL	MAT.	INST.	TOTAL	MAT.	INST.	TOTAL	MAT.	INST.	TOTAL	MAT.	INST.	TOTAL	MAT.	INST.	TOTAL
01590	EQUIPMENT RENTAL	.0	89.1	89.1	.0	103.4	103.4	.0	103.4	103.4	.0	99.3	99.3	.0	99.3	99.3	.0	99.3	99.3
02	SITE CONSTRUCTION	116.4	91.8	98.4	82.5	103.7	98.1	81.9	104.0	98.1	119.1	83.5	93.0	113.9	80.2	89.2	96.5	80.9	85.1
03100	CONCRETE FORMS & ACCESSORIES	94.9	19.5	29.9	106.1	117.6	116.0	104.9	117.7	115.9	100.0	72.1	75.9	98.2	28.8	38.3	97.1	37.3	45.5
03200	CONCRETE REINFORCEMENT	193.0	12.0	101.1	107.5	115.8	111.7	107.5	115.8	111.7	95.3	71.1	83.0	94.4	29.5	61.4	94.2	67.2	80.5
03300	CAST-IN-PLACE CONCRETE	108.0	30.4	76.1	83.7	111.0	94.9	88.5	111.0	97.7	72.8	73.0	72.8	72.7	39.9	59.2	85.3	47.8	69.9
03	CONCRETE	114.5	22.5	68.8	100.1	114.5	107.2	102.3	114.5	108.4	103.8	73.3	88.7	100.8	34.8	68.0	94.3	48.3	71.5
04	MASONRY	220.6	16.8	94.3	106.5	128.4	120.1	108.9	128.4	121.0	74.5	64.2	68.1	89.0	25.0	49.3	90.3	35.1	56.1
05	METALS	114.8	30.4	87.8	99.7	109.9	103.0	99.7	109.9	103.0	91.2	88.0	90.1	91.2	64.0	82.5	92.7	80.6	88.8
06	WOOD & PLASTICS	91.1	19.8	53.9	102.4	116.5	109.8	102.4	116.5	109.8	100.5	74.0	86.7	98.4	29.0	62.3	97.0	36.5	65.5
07	THERMAL & MOISTURE PROTECTION	148.2	23.2	95.6	102.1	116.5	108.2	101.0	116.5	107.5	102.6	71.2	89.3	102.2	33.1	73.1	101.4	42.6	76.7
08	DOORS & WINDOWS	153.5	16.5	117.7	100.8	115.8	104.7	100.8	115.8	104.7	92.6	70.2	76.8	92.6	29.0	76.0	96.6	42.1	82.4
09200	PLASTER & GYPSUM BOARD	258.0	17.4	108.5	103.3	116.1	111.2	103.3	116.1	111.2	106.3	73.1	85.7	110.3	26.9	58.5	109.6	34.6	63.0
095,098	CEILINGS & ACOUSTICAL TREATMENT	334.4	17.4	145.3	95.9	116.1	107.9	97.8	116.1	108.7	94.1	73.1	81.6	100.9	26.9	56.8	100.9	34.6	61.3
09600	FLOORING	199.8	16.4	150.1	100.5	129.7	108.4	100.5	129.7	108.4	104.8	16.8	81.0	106.6	30.7	86.0	106.0	39.2	87.9
097,099	WALL FINISHES, PAINTS & COATINGS	199.0	15.2	86.8	94.0	120.2	110.0	94.0	120.2	110.0	110.4	74.7	88.6	110.4	26.0	58.9	110.4	38.2	66.3
09	FINISHES	250.3	18.7	129.7	100.0	120.7	110.8	100.3	120.7	110.9	101.3	62.0	80.8	103.8	28.8	64.7	101.9	37.2	68.2
10 - 14	TOTAL DIV. 10000 - 14000	100.0	19.6	82.9	100.0	97.4	99.4	100.0	97.4	99.4	100.0	74.0	94.4	100.0	37.7	86.7	100.0	64.6	92.4
15	MECHANICAL	103.7	15.3	67.0	100.1	109.6	104.0	100.1	109.6	104.0	95.1	68.8	84.2	95.1	25.4	66.2	100.1	43.0	76.4
16	ELECTRICAL	129.8	15.3	74.2	96.7	105.4	101.0	96.5	105.4	100.8	95.6	72.4	84.3	98.7	25.9	63.3	97.3	32.4	65.8
01 - 16	WEIGHTED AVERAGE	135.2	25.5	86.5	99.6	112.7	105.4	99.9	112.7	105.5	96.1	72.0	85.4	97.0	36.2	70.0	97.3	47.9	75.3

DIVISION		SOUTH CAROLINA COLUMBIA 290 - 292			FLORENCE 295			GREENVILLE 296			ROCK HILL 297			SPARTANBURG 293			SOUTH DAKOTA ABERDEEN 574		
		MAT.	INST.	TOTAL	MAT.	INST.	TOTAL	MAT.	INST.	TOTAL	MAT.	INST.	TOTAL	MAT.	INST.	TOTAL	MAT.	INST.	TOTAL
01590	EQUIPMENT RENTAL	.0	99.3	99.3	.0	99.3	99.3	.0	99.3	99.3	.0	99.3	99.3	.0	99.3	99.3	.0	98.3	98.3
02	SITE CONSTRUCTION	95.8	80.9	84.9	107.4	80.9	88.0	102.3	80.6	86.3	99.4	80.2	85.3	102.1	80.6	86.3	86.1	93.5	91.5
03100	CONCRETE FORMS & ACCESSORIES	101.9	39.5	48.1	82.4	39.4	45.3	96.9	39.2	47.2	95.1	27.8	37.0	100.8	39.2	47.7	95.6	38.8	46.6
03200	CONCRETE REINFORCEMENT	94.2	67.2	80.5	93.9	67.2	80.3	93.8	47.4	70.2	94.6	18.3	55.8	93.8	47.4	70.2	99.9	47.4	73.2
03300	CAST-IN-PLACE CONCRETE	80.5	47.4	66.9	72.7	47.7	62.4	93.5	45.5	69.7	72.7	35.2	57.3	72.7	47.6	62.4	102.6	49.5	80.8
03	CONCRETE	92.3	49.2	70.9	94.8	49.2	72.2	93.5	45.5	69.7	91.6	30.7	61.3	93.8	45.5	69.8	98.9	45.6	72.4
04	MASONRY	89.1	36.0	56.1	74.6	35.1	50.1	72.5	35.1	49.3	96.6	23.0	50.9	74.5	35.1	50.1	108.2	57.1	76.5
05	METALS	92.7	80.5	88.8	91.7	80.4	88.1	91.7	73.0	85.7	91.2	59.6	81.1	91.7	73.0	85.7	98.9	68.9	89.3
06	WOOD & PLASTICS	102.9	39.7	70.0	80.6	39.7	59.3	96.9	39.7	67.1	94.9	28.0	60.1	101.3	39.7	69.2	101.2	37.0	67.7
07	THERMAL & MOISTURE PROTECTION	101.4	42.4	76.6	101.6	42.9	76.9	101.6	42.9	76.9	101.5	28.8	70.9	101.7	42.9	77.0	98.1	50.9	78.3
08	DOORS & WINDOWS	96.6	43.8	82.8	92.7	43.8	79.9	92.6	39.2	78.6	92.6	25.9	75.2	92.6	39.2	78.7	95.1	39.7	80.6
09200	PLASTER & GYPSUM BOARD	109.6	37.9	65.0	98.8	37.9	60.9	104.9	37.9	63.2	104.1	25.9	55.5	106.7	37.9	63.9	105.5	35.6	62.1
095,098	CEILINGS & ACOUSTICAL TREATMENT	100.9	37.9	63.3	94.9	37.9	60.9	94.1	37.9	60.5	94.1	25.9	53.4	94.1	37.9	60.5	113.4	35.6	67.0
09600	FLOORING	105.8	39.2	87.8	95.7	39.2	80.4	103.3	40.0	86.1	102.1	30.3	82.7	105.0	40.0	87.4	108.1	56.1	94.1
097,099	WALL FINISHES, PAINTS & COATINGS	110.4	38.2	66.3	110.4	38.2	66.3	110.4	38.2	66.3	110.4	26.0	58.9	110.4	38.2	66.3	97.2	34.4	58.9
09	FINISHES	101.8	39.0	69.1	96.7	39.0	66.7	99.3	39.1	68.0	98.6	28.2	61.9	100.1	39.1	68.3	106.9	41.1	72.6
10 - 14	TOTAL DIV. 10000 - 14000	100.0	65.0	92.5	100.0	64.9	92.5	100.0	64.9	92.5	100.0	62.2	91.9	100.0	64.9	92.5	100.0	53.6	90.1
15	MECHANICAL	100.1	36.2	73.6	100.1	36.2	73.6	100.1	36.2	73.6	95.1	18.2	63.2	100.1	36.2	73.6	100.2	38.6	74.6
16	ELECTRICAL	97.9	34.2	66.9	95.5	19.5	58.6	97.4	31.1	65.2	97.4	19.2	59.4	97.4	31.1	65.2	96.7	59.3	78.5
01 - 16	WEIGHTED AVERAGE	97.1	47.3	75.0	95.4	45.2	73.1	95.7	45.4	73.4	95.3	33.0	67.6	95.9	45.4	73.5	99.5	53.1	78.9

COST INDEXES

735

SOUTH DAKOTA

DIVISION		MITCHELL 573			MOBRIDGE 576			PIERRE 575			RAPID CITY 577			SIOUX FALLS 570 - 571			WATERTOWN 572		
		MAT.	INST.	TOTAL	MAT.	INST.	TOTAL	MAT.	INST.	TOTAL	MAT.	INST.	TOTAL	MAT.	INST.	TOTAL	MAT.	INST.	TOTAL
01590	EQUIPMENT RENTAL	.0	98.3	98.3	.0	98.3	98.3	.0	98.3	98.3	.0	98.3	98.3	.0	99.5	99.5	.0	98.3	98.3
02	SITE CONSTRUCTION	83.1	93.5	90.7	83.1	93.5	90.7	84.4	93.5	91.1	84.6	93.3	90.9	85.6	95.4	92.8	83.0	93.5	90.7
03100	CONCRETE FORMS & ACCESSORIES	94.6	40.2	47.6	84.3	38.7	45.0	94.0	40.5	47.8	103.1	36.0	45.2	94.1	41.5	48.7	80.3	38.8	44.5
03200	CONCRETE REINFORCEMENT	99.2	47.6	73.0	101.8	47.4	74.2	99.4	61.1	79.9	93.3	61.0	76.9	93.3	61.1	77.0	96.6	47.7	71.7
03300	CAST-IN-PLACE CONCRETE	99.6	48.0	78.4	99.6	49.5	79.0	99.6	45.4	77.4	98.9	43.0	75.9	103.0	48.3	80.5	99.6	49.5	79.0
03	CONCRETE	96.6	45.6	71.3	96.3	45.5	71.1	96.6	47.4	72.2	95.9	44.6	70.4	96.1	48.9	72.7	95.2	45.5	70.6
04	MASONRY	96.5	60.5	74.2	104.8	57.1	75.2	104.8	60.5	77.3	104.7	53.7	73.1	102.2	60.5	76.3	130.4	61.6	87.8
05	METALS	98.2	68.9	88.9	98.2	68.8	88.8	98.9	75.1	91.3	100.7	75.3	92.6	100.8	75.7	92.8	98.2	69.2	88.9
06	WOOD & PLASTICS	100.0	39.1	68.3	88.7	37.0	61.8	99.2	39.1	67.9	105.4	34.9	68.7	99.2	39.7	68.2	84.2	37.0	59.6
07	THERMAL & MOISTURE PROTECTION	98.0	49.9	77.7	98.0	50.9	78.2	98.2	50.1	78.0	98.6	47.5	77.1	98.2	51.3	78.4	97.7	52.1	78.5
08	DOORS & WINDOWS	93.5	40.3	79.6	96.5	38.8	81.4	98.3	44.6	84.3	99.3	42.4	84.4	99.3	45.0	85.1	93.5	38.8	79.2
09200	PLASTER & GYPSUM BOARD	104.1	37.8	62.9	99.6	35.6	59.8	103.7	37.8	62.8	105.5	33.5	60.8	105.5	38.4	63.8	96.7	35.6	58.8
095,098	CEILINGS & ACOUSTICAL TREATMENT	110.9	37.8	67.3	113.4	35.6	67.0	110.9	37.8	67.3	117.7	33.5	67.5	117.7	38.4	70.4	110.9	35.6	66.0
09600	FLOORING	107.6	56.1	93.7	101.8	56.1	89.4	107.3	40.0	89.1	107.3	75.8	98.8	107.3	79.7	99.8	100.0	56.1	88.1
097,099	WALL FINISHES, PAINTS & COATINGS	97.2	38.9	61.6	97.2	34.4	58.9	97.2	39.2	61.8	97.2	39.2	61.8	97.2	39.2	61.8	97.2	34.4	58.9
09	FINISHES	105.8	42.9	73.0	104.0	41.1	71.3	105.6	39.6	71.2	107.7	43.4	74.2	107.6	48.1	76.6	102.4	41.1	70.5
10 - 14	TOTAL DIV. 10000 - 14000	100.0	46.7	86.8	100.0	53.5	90.1	100.0	63.8	92.3	100.0	61.5	91.8	100.0	64.0	92.3	100.0	53.5	90.1
15	MECHANICAL	95.3	35.6	70.5	95.3	39.7	72.2	100.2	37.0	73.9	100.2	34.5	72.9	100.2	36.0	73.6	95.3	39.7	72.2
16	ELECTRICAL	95.2	44.4	70.5	96.7	39.6	68.9	93.0	54.7	74.4	93.5	54.8	74.7	92.9	71.9	82.7	94.4	39.6	67.8
01 - 16	WEIGHTED AVERAGE	96.9	50.8	76.4	97.5	50.4	76.6	98.8	53.5	78.7	99.4	52.2	78.4	99.2	57.3	80.6	97.9	51.0	77.1

TENNESSEE

DIVISION		CHATTANOOGA 373 - 374			COLUMBIA 384			COOKEVILLE 385			JACKSON 383			JOHNSON CITY 376			KNOXVILLE 377 - 379		
		MAT.	INST.	TOTAL	MAT.	INST.	TOTAL	MAT.	INST.	TOTAL	MAT.	INST.	TOTAL	MAT.	INST.	TOTAL	MAT.	INST.	TOTAL
01590	EQUIPMENT RENTAL	.0	104.5	104.5	.0	97.7	97.7	.0	97.7	97.7	.0	104.9	104.9	.0	97.4	97.4	.0	97.4	97.4
02	SITE CONSTRUCTION	101.4	96.2	97.6	90.3	84.7	86.2	96.8	85.1	88.3	98.9	95.4	96.3	110.8	84.9	91.8	88.4	85.0	85.9
03100	CONCRETE FORMS & ACCESSORIES	98.3	45.6	52.8	81.0	50.6	54.8	81.2	39.8	45.5	89.0	37.2	44.3	82.9	46.4	51.4	97.2	46.4	53.4
03200	CONCRETE REINFORCEMENT	87.9	42.9	65.0	89.7	44.9	66.9	89.7	47.0	68.0	89.7	42.7	65.8	88.5	43.1	65.4	87.9	43.1	65.1
03300	CAST-IN-PLACE CONCRETE	102.0	47.8	79.7	91.4	51.4	75.0	103.6	49.0	81.1	101.1	41.3	76.5	82.0	52.4	69.8	95.8	48.0	76.1
03	CONCRETE	92.9	47.7	70.4	93.3	51.5	72.6	102.9	46.3	74.8	94.7	41.7	68.4	96.9	49.6	73.4	90.0	48.2	69.2
04	MASONRY	99.1	43.1	64.4	112.7	51.3	74.6	107.8	39.5	65.5	113.0	33.5	63.7	111.8	42.9	69.1	76.5	42.9	55.7
05	METALS	93.5	75.5	87.8	89.3	76.5	85.2	89.4	77.2	85.5	91.0	73.8	85.5	90.9	75.1	85.9	94.1	75.4	88.1
06	WOOD & PLASTICS	101.3	46.6	72.8	70.9	53.9	62.0	71.1	39.5	54.7	86.7	37.1	60.8	75.3	47.7	60.9	90.9	47.7	68.4
07	THERMAL & MOISTURE PROTECTION	97.9	50.2	77.8	93.0	56.4	77.6	93.5	46.9	73.9	95.1	39.5	71.7	92.2	47.9	73.5	90.3	48.1	72.6
08	DOORS & WINDOWS	101.2	46.5	86.9	92.6	48.2	81.0	92.7	43.0	79.7	101.8	39.2	85.5	97.0	47.6	84.1	93.7	47.6	81.6
09200	PLASTER & GYPSUM BOARD	84.6	45.7	60.4	89.5	53.2	66.9	89.5	38.4	57.7	90.1	35.9	56.4	90.0	46.9	63.2	97.3	46.9	66.0
095,098	CEILINGS & ACOUSTICAL TREATMENT	96.4	45.7	66.2	87.9	53.2	67.2	87.9	38.4	58.4	96.2	35.9	60.2	91.9	46.9	65.0	94.5	46.9	66.1
09600	FLOORING	104.4	49.4	89.5	91.3	25.6	73.6	91.5	23.9	73.2	92.8	23.9	74.2	97.7	50.8	85.0	103.5	50.8	89.2
097,099	WALL FINISHES, PAINTS & COATINGS	108.9	45.8	70.4	95.2	42.6	63.1	95.2	46.5	65.5	96.0	31.0	56.3	106.7	53.7	74.4	106.7	53.7	74.4
09	FINISHES	98.2	46.3	71.2	95.1	45.9	69.5	95.6	36.9	65.0	95.9	34.1	63.7	98.5	47.9	72.1	94.0	47.9	70.0
10 - 14	TOTAL DIV. 10000 - 14000	100.0	51.6	89.7	100.0	51.0	89.6	100.0	57.0	90.8	100.0	48.7	89.1	100.0	59.3	91.3	100.0	59.4	91.3
15	MECHANICAL	99.9	44.8	77.1	97.2	52.1	78.5	97.2	44.0	75.1	100.0	55.5	81.5	99.8	59.0	82.9	99.8	59.0	82.9
16	ELECTRICAL	104.2	52.8	79.2	93.4	44.8	69.8	94.9	43.7	70.0	100.9	51.1	76.7	92.5	51.1	72.4	97.9	59.7	79.3
01 - 16	WEIGHTED AVERAGE	98.6	54.1	78.8	94.9	55.2	77.3	96.2	49.9	75.6	98.3	51.7	77.6	97.3	56.4	79.1	94.7	57.5	78.2

TENNESSEE / TEXAS

DIVISION		MCKENZIE 382			MEMPHIS 375,380 - 381			NASHVILLE 370 - 372			ABILENE 795 - 796			AMARILLO 790 - 791			AUSTIN 786 - 787		
		MAT.	INST.	TOTAL	MAT.	INST.	TOTAL	MAT.	INST.	TOTAL	MAT.	INST.	TOTAL	MAT.	INST.	TOTAL	MAT.	INST.	TOTAL
01590	EQUIPMENT RENTAL	.0	97.7	97.7	.0	102.3	102.3	.0	105.4	105.4	.0	86.7	86.7	.0	86.7	86.7	.0	85.8	85.8
02	SITE CONSTRUCTION	96.1	85.7	88.5	93.8	91.7	92.3	93.5	98.7	97.3	100.7	85.2	89.3	101.1	86.0	90.1	89.4	84.9	86.1
03100	CONCRETE FORMS & ACCESSORIES	90.8	37.0	44.4	98.5	67.1	71.4	97.7	65.0	69.5	95.2	42.7	49.9	97.4	54.0	59.9	92.8	57.1	62.0
03200	CONCRETE REINFORCEMENT	89.9	35.3	62.1	102.3	69.3	85.5	86.9	68.2	77.4	94.5	52.6	73.2	94.5	54.0	73.9	98.8	52.0	75.0
03300	CAST-IN-PLACE CONCRETE	101.3	44.4	77.9	88.9	71.3	81.6	96.6	69.2	85.3	93.1	42.2	72.2	97.9	48.2	77.5	87.7	49.1	71.9
03	CONCRETE	101.7	41.5	71.8	89.2	70.5	79.9	90.6	68.5	79.6	88.4	45.4	67.0	90.8	52.8	71.9	80.5	54.0	67.3
04	MASONRY	110.8	53.6	75.3	89.1	76.4	81.2	84.4	65.7	72.8	98.8	53.8	70.9	100.7	51.3	70.1	92.3	54.7	69.0
05	METALS	89.4	76.6	85.3	94.8	94.4	94.7	95.1	91.5	94.0	96.3	66.6	86.8	96.3	67.1	87.0	95.4	65.0	85.7
06	WOOD & PLASTICS	81.7	34.8	57.3	96.7	67.9	81.7	97.6	66.2	81.2	96.6	42.0	68.2	98.0	57.1	76.7	93.4	60.9	76.5
07	THERMAL & MOISTURE PROTECTION	93.5	52.2	76.1	91.9	74.4	84.5	95.9	66.7	83.6	99.6	49.8	78.6	101.6	48.9	79.4	88.8	54.1	74.2
08	DOORS & WINDOWS	92.7	35.4	77.7	100.8	69.3	92.6	98.5	67.4	90.4	92.8	45.1	80.3	92.8	51.4	81.9	94.5	58.8	85.2
09200	PLASTER & GYPSUM BOARD	93.9	33.6	56.4	94.4	67.4	77.7	92.4	65.8	75.9	83.8	41.1	57.3	83.8	56.6	66.9	88.1	60.5	71.0
095,098	CEILINGS & ACOUSTICAL TREATMENT	87.9	33.6	55.5	104.2	67.4	82.3	94.8	65.8	77.5	89.7	41.1	60.7	89.7	56.6	70.0	87.7	60.5	71.5
09600	FLOORING	95.1	49.1	82.7	97.7	45.1	83.4	107.6	77.3	99.4	115.3	68.0	102.5	115.1	63.8	101.2	100.5	48.9	86.6
097,099	WALL FINISHES, PAINTS & COATINGS	95.2	54.2	70.2	99.8	59.4	75.2	110.5	68.1	84.6	95.8	54.6	70.7	95.8	37.8	60.4	92.5	42.5	62.0
09	FINISHES	97.3	39.9	67.4	96.5	61.6	78.3	103.4	67.7	84.8	96.4	48.3	71.4	96.4	54.4	74.5	91.4	54.1	72.0
10 - 14	TOTAL DIV. 10000 - 14000	100.0	34.3	86.0	100.0	80.4	95.8	100.0	79.4	95.6	100.0	73.3	94.3	100.0	61.9	91.9	100.0	62.1	91.9
15	MECHANICAL	97.2	72.7	87.0	99.9	75.5	89.8	99.8	71.0	87.9	100.1	45.1	77.3	100.1	51.5	79.9	99.8	59.5	83.1
16	ELECTRICAL	94.7	67.2	81.3	98.3	76.7	87.8	102.3	69.4	86.3	96.5	48.8	73.3	97.0	60.0	79.0	99.5	69.8	85.0
01 - 16	WEIGHTED AVERAGE	96.4	59.4	80.0	96.6	76.2	87.5	97.5	73.7	86.9	96.6	53.3	77.4	97.2	58.0	79.8	94.0	61.7	80.0

COST INDEXES

TEXAS

		BEAUMONT 776 - 777			BROWNWOOD 768			BRYAN 778			CHILDRESS 792			CORPUS CHRISTI 783 - 784			DALLAS 752 - 753		
DIVISION		MAT.	INST.	TOTAL	MAT.	INST.	TOTAL	MAT.	INST.	TOTAL	MAT.	INST.	TOTAL	MAT.	INST.	TOTAL	MAT.	INST.	TOTAL
01590	EQUIPMENT RENTAL	.0	88.6	88.6	.0	86.7	86.7	.0	88.6	88.6	.0	86.7	86.7	.0	94.6	94.6	.0	98.1	98.1
02	SITE CONSTRUCTION	96.9	87.1	89.7	107.4	83.9	90.2	86.5	88.4	87.9	112.9	84.5	92.1	124.2	80.2	92.0	126.2	86.3	97.0
03100	CONCRETE FORMS & ACCESSORIES	102.3	53.0	59.8	93.3	28.8	37.7	79.0	43.2	48.1	93.5	52.4	58.0	96.3	38.7	46.6	93.7	59.8	64.5
03200	CONCRETE REINFORCEMENT	92.9	44.1	68.1	94.9	26.5	60.1	95.3	52.0	73.3	94.7	41.3	67.5	97.9	50.1	73.6	96.8	56.2	76.2
03300	CAST-IN-PLACE CONCRETE	87.4	54.5	73.9	100.5	37.4	74.6	70.1	67.8	69.1	95.6	46.5	75.4	104.0	45.5	79.9	96.6	56.6	80.2
03	CONCRETE	86.4	52.7	69.7	95.0	32.6	64.0	72.7	54.4	63.6	95.5	49.0	72.4	90.1	45.2	67.8	88.4	59.9	74.0
04	MASONRY	100.4	58.8	74.6	126.2	49.7	78.7	139.9	63.6	92.6	102.9	49.6	69.8	84.0	50.0	62.9	102.3	61.1	76.8
05	METALS	99.8	64.2	88.4	94.5	51.0	80.6	100.2	68.4	90.0	94.3	59.2	83.1	94.9	77.9	89.5	93.4	82.3	89.9
06	WOOD & PLASTICS	110.9	53.4	81.0	95.1	27.6	59.9	77.1	35.6	55.5	96.0	57.1	75.7	105.9	38.1	70.6	95.8	60.4	77.4
07	THERMAL & MOISTURE PROTECTION	105.1	58.5	85.5	99.7	41.2	75.1	99.8	60.1	83.1	100.2	46.1	77.4	91.3	46.9	72.6	99.5	63.6	84.4
08	DOORS & WINDOWS	97.4	49.2	84.8	88.9	28.5	73.1	99.3	46.3	85.5	89.9	48.8	79.2	101.5	39.9	85.4	105.5	55.9	92.6
09200	PLASTER & GYPSUM BOARD	93.6	52.8	68.3	82.7	26.3	47.7	83.3	34.5	53.0	82.7	56.6	66.5	88.1	36.9	56.3	87.6	59.8	70.3
095,098	CEILINGS & ACOUSTICAL TREATMENT	91.3	52.8	68.3	84.5	26.3	49.8	85.1	34.5	54.9	84.5	56.6	67.9	87.7	36.9	57.4	92.9	59.8	73.2
09600	FLOORING	112.1	72.0	101.2	113.9	40.6	94.1	83.6	65.2	78.6	112.8	42.5	93.8	113.6	48.4	95.9	104.3	59.3	92.1
097,099	WALL FINISHES, PAINTS & COATINGS	94.0	51.0	67.8	95.8	30.4	55.9	89.3	61.8	72.5	95.8	40.2	61.9	105.9	41.9	66.8	104.7	54.1	73.8
09	FINISHES	92.8	56.1	73.6	95.0	30.8	61.6	80.5	47.5	63.3	95.3	49.9	71.6	97.1	40.1	67.4	98.4	59.0	77.8
10 - 14	TOTAL DIV. 10000 - 14000	100.0	77.3	95.2	100.0	44.3	88.1	100.0	63.2	92.1	100.0	61.2	91.7	100.0	74.0	94.5	100.0	78.8	95.5
15	MECHANICAL	100.0	66.1	85.9	95.1	29.7	68.0	95.1	72.2	85.6	95.1	50.2	76.5	99.8	46.0	77.5	99.8	67.8	86.5
16	ELECTRICAL	94.4	72.5	83.7	92.8	28.7	61.6	94.0	70.9	82.8	96.5	39.1	68.6	97.6	54.3	76.6	96.1	65.5	81.2
01 - 16	WEIGHTED AVERAGE	97.1	64.1	82.4	96.5	39.5	71.2	94.7	64.5	81.3	96.1	52.5	76.7	97.3	53.1	77.7	98.3	67.1	84.4

TEXAS

		DEL RIO 788			DENTON 762			EASTLAND 764			EL PASO 798 - 799,885			FORT WORTH 760 - 761			GALVESTON 775		
DIVISION		MAT.	INST.	TOTAL	MAT.	INST.	TOTAL	MAT.	INST.	TOTAL	MAT.	INST.	TOTAL	MAT.	INST.	TOTAL	MAT.	INST.	TOTAL
01590	EQUIPMENT RENTAL	.0	85.8	85.8	.0	93.9	93.9	.0	86.7	86.7	.0	86.7	86.7	.0	86.7	86.7	.0	99.1	99.1
02	SITE CONSTRUCTION	109.4	82.7	89.8	108.1	77.5	85.6	110.7	84.0	91.1	101.2	84.8	89.2	100.6	85.6	89.6	126.4	85.9	96.7
03100	CONCRETE FORMS & ACCESSORIES	90.5	25.3	34.3	101.1	38.1	46.7	94.4	33.9	42.2	94.7	47.2	53.7	96.2	59.0	64.1	88.4	68.1	70.9
03200	CONCRETE REINFORCEMENT	98.8	18.6	58.0	96.3	52.5	74.1	95.1	42.9	68.6	94.5	48.9	71.4	94.5	56.1	75.0	94.8	64.1	79.2
03300	CAST-IN-PLACE CONCRETE	111.9	32.3	79.2	77.7	49.4	66.1	106.2	38.2	78.2	95.4	37.0	71.4	91.6	51.6	75.2	93.0	69.1	83.2
03	CONCRETE	103.5	27.7	65.9	76.1	46.7	61.5	99.2	38.0	68.8	89.4	44.8	67.3	87.7	56.6	72.2	87.8	69.2	78.5
04	MASONRY	98.6	44.8	65.2	134.4	51.3	82.9	97.3	48.7	67.2	97.5	51.4	68.9	94.3	61.1	73.7	96.6	63.6	76.2
05	METALS	94.1	46.4	78.8	94.1	78.6	89.1	94.3	53.9	81.4	96.1	62.3	85.3	93.1	68.3	85.2	101.9	89.2	97.9
06	WOOD & PLASTICS	89.4	25.6	56.2	106.3	38.2	70.8	101.4	34.6	66.6	96.6	50.5	72.6	103.3	60.2	80.8	93.6	68.6	80.8
07	THERMAL & MOISTURE PROTECTION	88.8	34.2	65.8	99.6	47.8	77.8	100.1	41.5	75.4	99.2	53.9	80.1	99.9	54.8	80.9	99.5	66.1	85.4
08	DOORS & WINDOWS	95.3	21.9	76.1	104.7	42.9	88.6	77.5	31.9	65.5	92.8	45.8	80.5	86.8	55.8	78.7	104.8	67.4	95.0
09200	PLASTER & GYPSUM BOARD	84.3	24.3	47.0	87.2	37.1	56.1	82.7	33.5	52.1	83.8	49.8	62.7	83.8	59.8	68.9	89.7	68.3	76.4
095,098	CEILINGS & ACOUSTICAL TREATMENT	84.3	24.3	48.5	90.5	37.1	58.6	84.5	33.5	54.1	89.7	49.8	65.9	89.7	59.8	71.9	88.5	68.3	76.5
09600	FLOORING	97.2	19.5	76.2	105.9	47.4	90.1	148.1	29.3	115.9	115.3	69.5	102.9	149.4	47.4	121.8	98.4	65.2	89.4
097,099	WALL FINISHES, PAINTS & COATINGS	92.5	26.8	52.4	106.7	30.8	60.4	97.0	31.8	57.2	95.8	36.3	59.5	97.0	54.0	70.7	102.3	58.6	75.6
09	FINISHES	90.2	24.8	56.1	92.5	38.5	64.3	105.4	33.0	67.7	96.4	50.7	72.6	106.6	56.4	80.5	90.6	66.5	78.1
10 - 14	TOTAL DIV. 10000 - 14000	100.0	41.6	87.6	100.0	46.1	88.5	100.0	44.8	88.2	100.0	59.9	91.5	100.0	78.3	95.4	100.0	82.7	96.3
15	MECHANICAL	94.9	19.4	63.5	95.1	47.5	75.4	95.1	30.0	68.1	100.1	34.4	72.8	100.1	59.7	83.3	95.0	72.0	85.3
16	ELECTRICAL	99.2	20.4	60.9	95.7	46.5	71.8	92.8	35.7	65.0	95.7	56.2	76.5	95.7	63.8	80.2	95.2	56.4	76.3
01 - 16	WEIGHTED AVERAGE	96.5	33.2	68.4	96.5	51.6	76.6	95.6	42.0	71.8	96.6	51.6	76.6	96.2	62.8	81.4	97.3	70.6	85.4

TEXAS

		GIDDINGS 789			GREENVILLE 754			HOUSTON 770 - 772			HUNTSVILLE 773			LAREDO 780			LONGVIEW 756		
DIVISION		MAT.	INST.	TOTAL	MAT.	INST.	TOTAL	MAT.	INST.	TOTAL	MAT.	INST.	TOTAL	MAT.	INST.	TOTAL	MAT.	INST.	TOTAL
01590	EQUIPMENT RENTAL	.0	85.8	85.8	.0	95.3	95.3	.0	98.8	98.8	.0	88.6	88.6	.0	85.8	85.8	.0	88.9	88.9
02	SITE CONSTRUCTION	95.9	83.2	86.6	117.2	81.2	90.8	124.2	85.6	95.9	106.3	85.9	91.4	89.9	84.7	86.1	109.7	89.3	94.7
03100	CONCRETE FORMS & ACCESSORIES	88.6	33.9	41.4	84.5	28.6	36.3	90.8	68.2	71.3	87.3	35.3	42.4	90.5	39.0	46.1	80.0	29.6	36.5
03200	CONCRETE REINFORCEMENT	99.5	43.7	71.1	97.2	26.6	61.3	94.6	64.2	79.2	95.6	29.1	61.8	98.8	50.3	74.2	96.2	24.3	59.7
03300	CAST-IN-PLACE CONCRETE	94.8	38.1	71.5	88.0	37.7	67.3	90.2	69.2	81.5	96.3	45.4	75.4	80.0	61.3	72.3	102.3	34.9	74.6
03	CONCRETE	86.0	38.1	62.2	81.9	33.4	57.8	85.8	69.3	77.6	92.8	38.8	66.0	80.2	49.8	65.1	94.5	31.6	63.3
04	MASONRY	105.8	50.7	71.6	157.1	50.0	90.7	96.4	65.5	77.2	138.5	34.0	73.7	92.6	53.8	68.5	153.0	46.6	87.0
05	METALS	93.6	54.1	81.0	91.2	64.5	82.7	104.4	89.5	99.6	100.1	53.3	85.2	96.3	63.9	85.9	84.5	50.1	73.5
06	WOOD & PLASTICS	88.5	34.6	60.4	86.4	27.7	55.8	96.1	68.6	81.8	86.9	36.9	60.9	89.4	37.9	62.6	79.3	30.0	53.6
07	THERMAL & MOISTURE PROTECTION	89.3	42.0	69.4	99.3	42.5	75.4	99.4	66.9	85.7	100.9	39.0	74.8	87.9	51.1	72.4	98.9	36.9	72.8
08	DOORS & WINDOWS	94.5	31.8	78.1	104.3	28.6	84.5	107.1	67.4	96.7	99.3	31.7	81.6	95.1	40.6	80.9	93.9	27.5	76.6
09200	PLASTER & GYPSUM BOARD	82.9	33.5	52.2	83.7	26.3	48.1	91.9	68.3	77.2	87.7	35.9	55.5	85.1	36.9	55.2	79.7	28.8	48.1
095,098	CEILINGS & ACOUSTICAL TREATMENT	83.4	33.5	53.6	89.5	26.3	51.8	92.8	68.3	78.2	85.1	35.9	56.4	87.7	36.9	57.4	85.2	28.8	51.6
09600	FLOORING	96.9	29.3	78.6	99.0	47.4	85.0	99.6	65.2	90.3	87.3	29.3	71.6	97.1	48.4	83.9	105.6	41.8	88.3
097,099	WALL FINISHES, PAINTS & COATINGS	92.5	31.8	55.4	104.7	26.2	56.8	102.3	61.8	77.6	89.3	32.6	54.7	92.5	55.5	69.9	96.2	34.9	58.8
09	FINISHES	89.0	32.9	59.8	94.9	31.5	61.9	97.3	66.9	81.5	83.2	34.2	57.7	90.1	41.5	64.8	99.0	32.3	64.2
10 - 14	TOTAL DIV. 10000 - 14000	100.0	43.1	87.9	100.0	44.2	88.1	100.0	82.8	96.3	100.0	43.2	87.9	100.0	71.4	93.9	100.0	24.3	83.9
15	MECHANICAL	94.9	30.0	68.0	94.9	29.2	67.6	99.9	72.4	88.5	95.1	28.3	67.4	99.8	41.5	75.6	94.8	28.8	67.4
16	ELECTRICAL	96.3	35.7	66.9	92.8	28.7	61.7	97.0	71.0	84.3	94.0	35.0	65.4	99.3	62.8	81.5	92.4	55.9	74.6
01 - 16	WEIGHTED AVERAGE	94.0	42.1	71.0	97.7	40.7	72.4	99.6	73.0	87.7	97.7	40.3	72.2	94.6	53.7	76.5	96.9	42.6	72.8

TEXAS

DIVISION		LUBBOCK 793-794			LUFKIN 759			MCALLEN 785			MCKINNEY 750			MIDLAND 797			ODESSA 797		
		MAT.	INST.	TOTAL	MAT.	INST.	TOTAL	MAT.	INST.	TOTAL	MAT.	INST.	TOTAL	MAT.	INST.	TOTAL	MAT.	INST.	TOTAL
01590	EQUIPMENT RENTAL	.0	96.9	96.9	.0	88.9	88.9	.0	94.6	94.6	.0	95.3	95.3	.0	96.9	96.9	.0	86.7	86.7
02	SITE CONSTRUCTION	130.2	83.3	95.8	103.4	89.3	93.1	128.5	80.2	93.1	112.3	82.0	90.1	133.9	83.3	96.8	100.9	85.5	89.6
03100	CONCRETE FORMS & ACCESSORIES	94.8	41.9	49.2	83.6	41.4	47.2	95.1	36.5	44.6	83.3	43.7	49.1	99.1	40.0	48.1	95.1	39.9	47.5
03200	CONCRETE REINFORCEMENT	95.8	52.5	73.8	97.8	32.5	64.6	98.2	50.1	73.7	97.2	52.5	74.5	96.9	52.4	74.3	94.5	52.3	73.1
03300	CAST-IN-PLACE CONCRETE	93.3	48.3	74.8	91.6	40.4	70.5	113.0	43.8	84.6	82.5	51.1	69.6	99.2	45.5	77.1	93.1	44.3	73.1
03	CONCRETE	87.5	48.1	68.0	87.9	40.5	64.3	95.5	43.6	69.7	77.9	49.6	63.9	91.8	46.3	69.2	88.4	44.8	66.7
04	MASONRY	98.2	49.7	68.1	118.4	39.8	69.7	98.1	50.0	68.2	168.6	54.6	97.9	116.1	50.1	75.2	98.8	50.0	68.6
05	METALS	99.9	80.8	93.8	91.3	57.6	80.5	93.9	77.5	88.7	91.1	77.5	86.8	98.5	80.3	92.7	95.7	65.6	86.1
06	WOOD & PLASTICS	97.1	42.1	68.5	87.7	43.0	64.4	104.3	35.5	68.4	84.9	43.8	63.5	102.5	40.1	70.0	96.6	39.9	67.1
07	THERMAL & MOISTURE PROTECTION	91.9	50.8	74.6	98.7	42.1	74.9	91.3	45.1	71.8	99.1	51.7	79.2	92.2	47.1	73.2	99.6	46.2	77.1
08	DOORS & WINDOWS	104.1	43.3	88.2	75.7	39.7	66.3	100.1	39.0	84.1	104.3	46.0	89.0	103.0	41.8	87.0	92.8	41.7	79.4
09200	PLASTER & GYPSUM BOARD	84.5	41.1	57.5	80.0	42.1	56.5	88.1	34.3	54.6	82.6	42.8	57.9	86.6	39.0	57.0	83.8	39.0	56.0
095,098	CEILINGS & ACOUSTICAL TREATMENT	92.3	41.1	61.7	75.0	42.1	55.4	87.7	34.3	55.8	89.5	42.8	61.6	87.9	39.0	58.7	89.7	39.0	59.4
09600	FLOORING	106.6	40.2	88.6	142.2	39.0	114.3	112.9	64.5	99.8	98.5	47.4	84.7	108.4	39.7	89.8	115.3	39.7	94.9
097,099	WALL FINISHES, PAINTS & COATINGS	107.3	34.0	62.5	96.2	39.3	61.5	105.9	30.0	59.6	104.7	49.0	70.7	107.3	33.5	62.3	95.8	34.0	58.0
09	FINISHES	98.4	39.9	67.9	106.9	40.7	72.4	97.2	40.5	67.7	94.3	44.6	68.4	98.5	38.6	67.3	96.4	38.6	66.3
10 - 14	TOTAL DIV. 10000 - 14000	100.0	73.0	94.2	100.0	61.0	91.7	100.0	71.3	93.9	100.0	48.0	88.9	100.0	59.6	91.4	100.0	59.2	91.3
15	MECHANICAL	99.7	48.3	78.3	94.8	33.5	69.4	94.9	32.5	69.0	94.9	49.3	76.0	94.7	37.7	71.1	100.1	37.7	74.2
16	ELECTRICAL	95.1	45.9	71.2	93.6	49.0	71.9	96.9	36.3	67.4	92.9	55.9	74.9	95.1	42.8	69.6	96.5	42.7	70.4
01 - 16	WEIGHTED AVERAGE	98.6	53.5	78.6	94.5	46.7	73.3	97.2	47.4	75.1	97.6	55.5	78.9	98.7	49.9	77.0	96.6	48.5	75.2

TEXAS

DIVISION		PALESTINE 758			SAN ANGELO 769			SAN ANTONIO 781-782			TEMPLE 765			TEXARKANA 755			TYLER 757		
		MAT.	INST.	TOTAL	MAT.	INST.	TOTAL	MAT.	INST.	TOTAL	MAT.	INST.	TOTAL	MAT.	INST.	TOTAL	MAT.	INST.	TOTAL
01590	EQUIPMENT RENTAL	.0	88.9	88.9	.0	86.7	86.7	.0	88.7	88.7	.0	86.7	86.7	.0	88.9	88.9	.0	88.9	88.9
02	SITE CONSTRUCTION	112.3	88.8	95.1	103.2	85.9	90.5	89.6	89.5	89.5	90.5	84.6	86.2	96.8	89.8	91.6	109.3	89.8	95.0
03100	CONCRETE FORMS & ACCESSORIES	74.0	33.9	39.4	93.6	37.8	45.5	90.5	58.5	62.9	97.6	36.1	44.6	91.5	36.0	43.6	85.5	34.8	41.8
03200	CONCRETE REINFORCEMENT	95.5	43.6	69.1	94.7	52.4	73.2	105.5	52.5	78.6	94.9	51.2	72.7	95.4	55.9	75.3	96.2	56.0	75.8
03300	CAST-IN-PLACE CONCRETE	83.6	37.8	64.8	94.8	42.6	73.3	78.5	68.9	74.6	77.6	48.5	65.7	84.3	44.1	67.8	100.5	43.0	76.9
03	CONCRETE	87.3	37.9	62.8	90.7	43.2	67.1	80.6	61.6	71.1	78.3	44.3	61.4	82.2	43.7	63.1	94.1	42.8	68.7
04	MASONRY	113.1	48.1	72.8	122.9	49.6	77.5	92.4	64.3	75.0	133.6	52.7	83.4	172.8	50.2	96.8	161.8	51.2	93.3
05	METALS	91.1	53.0	78.9	94.6	63.3	84.6	97.0	68.0	87.7	94.3	63.0	84.3	84.4	66.8	78.8	91.0	67.2	83.4
06	WOOD & PLASTICS	77.0	34.6	54.9	95.4	38.2	65.6	89.4	57.3	72.7	104.6	34.6	68.2	91.7	34.1	61.7	89.4	32.2	59.6
07	THERMAL & MOISTURE PROTECTION	99.1	43.1	75.5	99.5	44.4	76.3	87.9	64.8	78.1	99.1	47.2	77.2	98.5	49.1	77.7	99.0	53.5	79.8
08	DOORS & WINDOWS	75.6	32.5	64.4	88.9	40.7	76.3	97.1	56.1	86.4	75.4	37.3	65.4	93.9	40.3	79.9	75.6	39.2	66.1
09200	PLASTER & GYPSUM BOARD	74.8	33.5	49.2	82.7	37.2	54.4	85.1	56.7	67.5	84.2	33.5	52.7	83.6	33.0	52.2	80.0	31.0	49.6
095,098	CEILINGS & ACOUSTICAL TREATMENT	75.0	33.5	50.3	84.5	37.2	56.3	87.7	56.7	69.2	84.5	33.5	54.1	81.8	33.0	52.7	75.0	31.0	48.8
09600	FLOORING	131.3	29.3	103.7	114.1	35.5	92.8	97.1	55.3	85.8	151.3	48.7	123.5	115.7	64.9	102.0	144.4	43.3	117.0
097,099	WALL FINISHES, PAINTS & COATINGS	96.2	31.8	56.9	95.8	35.0	58.7	92.5	55.5	69.9	97.0	35.0	59.2	96.2	41.0	62.5	96.2	41.0	62.5
09	FINISHES	103.6	32.7	66.7	94.8	36.4	64.4	90.1	56.9	72.8	105.5	37.4	70.0	101.0	41.0	69.8	107.9	35.4	70.1
10 - 14	TOTAL DIV. 10000 - 14000	100.0	44.8	88.2	100.0	58.1	91.1	100.0	77.9	95.3	100.0	44.7	88.2	100.0	71.6	93.9	100.0	71.3	93.9
15	MECHANICAL	94.8	30.0	67.9	95.1	43.5	73.7	99.8	72.4	88.4	95.1	35.7	70.5	94.8	39.4	71.8	94.8	63.0	81.6
16	ELECTRICAL	90.1	40.5	66.0	96.5	34.1	66.1	99.5	62.8	81.7	93.9	64.7	79.7	93.5	72.5	83.3	92.4	57.9	75.6
01 - 16	WEIGHTED AVERAGE	93.6	43.0	71.1	96.2	47.7	74.7	95.0	67.0	82.5	94.5	50.3	74.9	96.6	54.0	77.7	97.4	56.2	79.1

TEXAS / UTAH

DIVISION		VICTORIA 779			WACO 766-767			WAXAHACHIE 751			WHARTON 774			WICHITA FALLS 763			LOGAN 843		
		MAT.	INST.	TOTAL	MAT.	INST.	TOTAL	MAT.	INST.	TOTAL	MAT.	INST.	TOTAL	MAT.	INST.	TOTAL	MAT.	INST.	TOTAL
01590	EQUIPMENT RENTAL	.0	97.6	97.6	.0	86.7	86.7	.0	95.3	95.3	.0	99.1	99.1	.0	86.7	86.7	.0	100.7	100.7
02	SITE CONSTRUCTION	132.5	81.5	95.1	99.6	85.5	89.3	115.4	81.8	90.8	140.7	83.4	98.7	100.3	85.2	89.2	89.7	100.6	97.7
03100	CONCRETE FORMS & ACCESSORIES	89.0	41.5	48.0	95.8	41.4	48.8	83.3	48.6	53.4	82.4	37.9	44.0	95.8	43.0	50.3	100.9	59.6	65.3
03200	CONCRETE REINFORCEMENT	91.0	32.5	61.3	94.5	51.8	72.8	97.2	52.5	74.5	94.7	63.4	78.8	94.5	53.0	73.4	108.5	77.6	92.8
03300	CAST-IN-PLACE CONCRETE	104.5	41.9	78.7	84.1	53.2	71.4	87.1	41.8	68.5	107.3	49.7	83.7	89.8	47.9	72.6	86.7	70.5	80.0
03	CONCRETE	94.4	42.2	68.5	84.0	48.5	66.4	81.1	48.7	65.0	97.8	48.8	73.5	86.8	47.6	67.3	108.6	67.2	88.0
04	MASONRY	113.1	39.9	67.7	95.7	59.7	73.4	157.7	52.5	92.5	97.6	34.9	58.7	96.1	59.6	73.4	118.6	60.4	82.6
05	METALS	100.1	75.4	92.2	96.2	65.9	86.5	91.2	77.4	86.8	101.9	81.8	95.5	96.2	67.4	87.0	99.4	75.4	91.7
06	WOOD & PLASTICS	97.4	43.1	69.2	102.3	37.1	68.3	84.9	51.6	67.6	87.0	39.1	62.0	102.3	42.0	70.9	87.9	57.9	72.3
07	THERMAL & MOISTURE PROTECTION	101.9	43.5	77.3	100.1	50.6	79.2	99.2	48.8	78.0	99.8	49.6	78.7	100.1	54.2	80.8	99.4	67.7	86.1
08	DOORS & WINDOWS	104.7	41.5	88.2	86.8	38.8	74.3	104.3	50.2	90.2	104.8	43.5	88.8	86.8	45.0	75.9	89.3	59.7	81.6
09200	PLASTER & GYPSUM BOARD	88.4	42.1	59.7	83.8	36.1	54.2	82.8	50.9	63.0	86.7	37.9	56.4	83.8	41.1	57.3	82.9	56.5	66.5
095,098	CEILINGS & ACOUSTICAL TREATMENT	89.4	42.1	61.2	89.7	36.1	57.7	90.4	50.9	66.8	88.5	37.9	58.4	89.7	41.1	60.7	101.4	56.5	74.6
09600	FLOORING	97.3	39.0	81.5	149.6	36.9	119.1	98.5	39.0	82.4	95.5	29.3	77.6	150.0	77.7	130.5	100.3	59.6	89.3
097,099	WALL FINISHES, PAINTS & COATINGS	102.5	39.3	63.9	97.0	35.0	59.2	104.7	39.3	64.8	102.3	32.6	59.8	98.7	55.5	72.4	102.2	47.9	69.0
09	FINISHES	89.0	40.9	63.9	106.6	38.5	71.2	94.8	46.2	69.4	90.1	35.6	61.7	106.9	50.4	77.5	97.6	57.6	76.7
10 - 14	TOTAL DIV. 10000 - 14000	100.0	47.6	88.8	100.0	75.4	94.8	100.0	49.2	89.2	100.0	46.0	88.5	100.0	60.1	91.5	100.0	70.7	93.7
15	MECHANICAL	95.1	33.5	69.6	100.1	55.5	81.6	94.9	48.2	75.5	95.0	27.8	67.2	100.1	51.3	79.8	100.1	71.2	88.1
16	ELECTRICAL	99.8	49.0	75.1	96.5	67.1	82.2	92.8	63.1	78.4	98.8	35.1	67.8	99.6	65.6	83.1	91.0	74.5	83.0
01 - 16	WEIGHTED AVERAGE	99.2	47.6	76.3	96.4	57.5	79.1	97.5	56.3	79.2	99.1	45.1	75.1	97.1	57.9	79.7	99.1	70.5	86.4

DIVISION		UTAH OGDEN 842,844 MAT.	INST.	TOTAL	PRICE 845 MAT.	INST.	TOTAL	PROVO 846-847 MAT.	INST.	TOTAL	SALT LAKE CITY 840-841 MAT.	INST.	TOTAL	VERMONT BELLOWS FALLS 051 MAT.	INST.	TOTAL	BENNINGTON 052 MAT.	INST.	TOTAL
01590	EQUIPMENT RENTAL	.0	100.7	100.7	.0	99.5	99.5	.0	99.5	99.5	.0	100.7	100.7	.0	101.5	101.5	.0	101.5	101.5
02	SITE CONSTRUCTION	78.4	100.6	94.6	87.3	97.4	94.7	86.3	98.6	95.3	78.4	100.5	94.6	77.7	98.0	92.6	77.2	98.0	92.4
03100	CONCRETE FORMS & ACCESSORIES	100.9	59.6	65.3	103.5	32.8	42.5	102.5	59.7	65.6	99.4	59.7	65.2	102.6	39.6	48.3	99.7	39.4	47.7
03200	CONCRETE REINFORCEMENT	113.4	77.6	95.2	116.4	77.1	96.4	117.4	77.7	97.2	110.7	77.7	93.9	85.9	78.4	82.1	85.9	78.3	82.1
03300	CAST-IN-PLACE CONCRETE	88.0	70.5	80.8	86.8	46.5	70.2	86.8	70.5	80.1	95.5	70.5	85.2	97.2	54.6	79.7	97.2	54.5	79.6
03	CONCRETE	99.3	67.2	83.3	110.1	46.9	78.7	108.6	67.2	88.0	117.1	67.2	92.3	99.5	52.8	76.3	99.3	52.6	76.1
04	MASONRY	111.9	60.4	80.0	124.3	41.7	73.0	124.3	60.4	84.7	129.2	60.4	86.6	97.1	38.1	60.5	107.2	38.1	64.3
05	METALS	99.9	75.4	92.0	97.0	73.4	89.5	97.7	75.4	90.6	103.0	75.4	94.2	94.8	71.5	87.3	94.8	71.2	87.2
06	WOOD & PLASTICS	87.9	57.9	72.3	90.8	30.4	59.4	89.5	57.9	73.0	88.1	57.9	72.3	101.5	37.5	68.2	98.2	37.5	66.6
07	THERMAL & MOISTURE PROTECTION	98.2	67.7	85.4	101.2	54.5	81.6	101.2	67.7	87.1	100.9	67.7	86.9	101.3	56.2	82.3	101.3	47.8	78.8
08	DOORS & WINDOWS	89.3	59.7	81.6	93.5	43.5	80.4	93.5	59.7	84.7	89.3	59.7	81.6	103.7	45.5	88.5	103.7	43.0	87.8
09200	PLASTER & GYPSUM BOARD	82.9	56.5	66.5	84.4	28.4	49.6	83.3	56.5	66.7	82.9	56.5	66.5	103.4	34.9	60.9	101.5	34.9	60.2
095,098	CEILINGS & ACOUSTICAL TREATMENT	101.4	56.5	74.6	101.4	28.4	57.8	101.4	56.5	74.6	100.4	56.5	74.2	85.5	34.9	55.4	85.5	34.9	55.4
09600	FLOORING	100.3	59.6	89.3	101.5	44.3	86.0	101.1	59.6	89.9	99.9	59.6	89.0	99.6	38.1	83.0	98.2	38.1	82.0
097,099	WALL FINISHES, PAINTS & COATINGS	102.2	47.9	69.0	102.2	47.9	69.0	102.2	61.1	77.1	102.2	61.1	77.1	91.5	34.1	56.5	91.5	34.1	56.5
09	FINISHES	96.5	57.6	76.2	98.6	35.0	65.5	98.2	59.1	77.8	96.6	59.1	77.1	96.1	37.8	65.8	95.5	37.8	65.4
10-14	TOTAL DIV. 10000-14000	100.0	70.7	93.7	100.0	53.5	90.1	100.0	70.7	93.7	100.0	70.7	93.7	100.0	47.3	88.8	100.0	47.3	88.8
15	MECHANICAL	100.1	71.2	88.1	97.5	40.4	73.8	100.1	71.2	88.1	100.2	71.2	88.2	95.1	36.2	70.7	95.1	39.8	72.2
16	ELECTRICAL	91.5	74.5	83.3	95.6	44.2	70.6	91.2	74.5	83.1	95.0	74.5	85.0	97.6	36.0	67.6	97.6	36.0	67.6
01-16	WEIGHTED AVERAGE	97.4	70.5	85.5	99.6	50.1	77.6	99.6	70.5	86.7	101.3	70.7	87.7	97.1	48.8	75.6	97.4	49.1	76.0

DIVISION		VERMONT BRATTLEBORO 053 MAT.	INST.	TOTAL	BURLINGTON 054 MAT.	INST.	TOTAL	GUILDHALL 059 MAT.	INST.	TOTAL	MONTPELIER 056 MAT.	INST.	TOTAL	RUTLAND 057 MAT.	INST.	TOTAL	ST. JOHNSBURY 058 MAT.	INST.	TOTAL
01590	EQUIPMENT RENTAL	.0	101.5	101.5	.0	101.5	101.5	.0	101.5	101.5	.0	101.5	101.5	.0	101.5	101.5	.0	101.5	101.5
02	SITE CONSTRUCTION	78.5	98.0	92.8	80.6	98.8	93.9	76.8	97.6	92.0	79.2	98.8	93.5	80.6	98.8	93.9	76.8	97.6	92.0
03100	CONCRETE FORMS & ACCESSORIES	102.9	39.5	48.2	92.4	50.5	56.2	99.2	35.4	44.2	96.5	50.5	56.9	103.2	50.5	57.8	96.5	35.5	43.9
03200	CONCRETE REINFORCEMENT	85.1	78.4	81.7	107.5	53.9	80.2	86.7	34.3	60.1	85.5	53.9	69.4	107.5	53.9	80.2	85.1	52.8	68.7
03300	CAST-IN-PLACE CONCRETE	100.2	54.6	81.5	99.8	62.9	84.6	94.2	50.2	76.1	100.2	62.9	84.9	95.2	62.9	81.9	94.2	50.2	76.1
03	CONCRETE	102.0	52.7	77.5	104.7	56.0	80.5	97.1	41.2	69.4	101.6	56.0	79.0	103.2	56.0	79.7	96.7	44.4	70.7
04	MASONRY	106.9	38.1	64.2	99.8	60.2	75.2	107.2	39.9	65.5	90.8	60.2	71.8	87.5	60.2	70.6	134.2	39.9	75.7
05	METALS	94.8	71.4	87.3	101.4	65.9	90.0	94.8	54.7	82.0	94.8	66.0	85.6	99.7	66.0	88.9	94.8	54.8	82.0
06	WOOD & PLASTICS	101.9	37.5	68.4	89.0	48.5	67.9	97.0	34.0	64.2	92.8	48.5	69.8	102.0	48.5	74.2	92.8	34.0	62.2
07	THERMAL & MOISTURE PROTECTION	101.4	48.8	79.3	101.7	55.2	82.1	101.1	42.8	76.6	101.2	56.4	82.3	101.5	56.4	82.5	101.0	42.8	76.5
08	DOORS & WINDOWS	103.7	45.5	88.5	106.9	45.2	90.8	103.7	32.6	85.1	103.7	45.2	88.4	106.9	45.2	90.8	103.7	32.6	85.1
09200	PLASTER & GYPSUM BOARD	103.4	34.9	60.9	104.7	46.3	68.4	108.2	31.3	60.4	112.6	46.3	71.4	104.7	46.3	68.4	112.6	31.3	62.1
095,098	CEILINGS & ACOUSTICAL TREATMENT	85.5	34.9	55.4	92.6	46.3	64.9	85.5	31.3	53.2	85.5	46.3	62.1	92.6	46.3	64.9	85.5	31.3	53.2
09600	FLOORING	99.8	38.1	83.1	99.6	68.8	91.3	103.4	38.1	85.7	107.7	68.8	97.2	99.6	68.8	91.3	107.7	38.1	88.8
097,099	WALL FINISHES, PAINTS & COATINGS	91.5	34.1	56.5	91.5	39.6	59.9	91.5	26.4	51.8	91.5	39.6	59.9	91.5	39.6	59.9	91.5	27.7	52.6
09	FINISHES	96.3	37.8	65.8	98.3	52.1	74.2	97.9	34.1	64.7	100.0	52.1	75.0	98.2	52.1	74.2	99.9	34.3	65.7
10-14	TOTAL DIV. 10000-14000	100.0	47.3	88.8	100.0	92.0	98.3	100.0	45.7	88.4	100.0	92.0	98.3	100.0	92.0	98.3	100.0	45.7	88.4
15	MECHANICAL	95.1	39.8	72.2	100.1	66.1	86.0	95.1	62.4	81.6	95.1	66.1	83.1	100.1	66.1	86.0	95.1	62.4	81.6
16	ELECTRICAL	97.6	36.0	67.6	98.3	68.7	83.9	97.6	28.0	63.7	97.6	68.7	83.5	97.6	68.7	83.6	97.6	28.0	63.7
01-16	WEIGHTED AVERAGE	97.9	49.3	76.3	100.5	64.9	84.7	97.4	48.6	75.7	97.3	64.9	82.9	99.5	64.9	84.1	98.8	49.1	76.7

DIVISION		VERMONT WHITE RIVER JCT. 050 MAT.	INST.	TOTAL	VIRGINIA ALEXANDRIA 223 MAT.	INST.	TOTAL	ARLINGTON 222 MAT.	INST.	TOTAL	BRISTOL 242 MAT.	INST.	TOTAL	CHARLOTTESVILLE 229 MAT.	INST.	TOTAL	CULPEPER 227 MAT.	INST.	TOTAL
01590	EQUIPMENT RENTAL	.0	101.5	101.5	.0	101.2	101.2	.0	99.6	99.6	.0	99.6	99.6	.0	104.8	104.8	.0	99.6	99.6
02	SITE CONSTRUCTION	80.6	98.3	93.6	113.3	86.2	93.4	124.2	82.5	93.6	107.4	80.6	87.8	112.0	83.7	91.3	111.5	81.9	89.8
03100	CONCRETE FORMS & ACCESSORIES	96.6	40.7	48.4	92.4	75.2	77.5	91.1	71.7	74.4	85.8	33.1	40.4	83.9	40.6	46.5	80.9	66.2	68.2
03200	CONCRETE REINFORCEMENT	85.9	31.1	58.1	83.7	85.9	84.8	94.4	62.8	78.3	94.5	72.4	83.3	93.9	71.8	82.7	94.4	56.3	75.0
03300	CAST-IN-PLACE CONCRETE	100.2	51.5	80.2	99.9	80.8	92.0	97.2	76.4	88.6	96.8	41.3	74.0	100.7	52.8	81.0	99.5	42.1	75.9
03	CONCRETE	103.6	43.4	73.7	102.7	80.2	91.5	107.5	72.7	90.2	103.6	45.3	74.6	104.1	52.6	78.5	100.9	57.3	79.3
04	MASONRY	119.9	46.0	74.1	87.0	73.7	78.7	99.1	67.8	79.7	87.3	30.8	52.2	112.0	42.1	68.7	101.0	34.3	59.6
05	METALS	94.8	54.3	81.9	101.6	98.2	100.5	100.4	85.6	95.7	99.5	83.7	94.4	99.7	88.9	96.3	99.8	80.0	93.5
06	WOOD & PLASTICS	94.9	40.4	66.5	96.0	75.6	85.3	92.4	75.6	83.6	84.0	32.4	57.1	81.9	39.1	59.6	80.5	75.6	77.9
07	THERMAL & MOISTURE PROTECTION	101.4	40.2	75.7	100.6	81.5	92.6	102.2	72.3	89.6	101.7	39.9	75.7	101.4	49.8	79.7	101.5	47.3	78.6
08	DOORS & WINDOWS	103.7	38.0	86.5	96.6	78.0	91.7	94.6	71.7	88.6	97.9	43.2	83.6	95.9	45.4	82.7	96.3	67.0	88.6
09200	PLASTER & GYPSUM BOARD	100.4	37.9	61.6	109.6	74.7	87.9	105.8	74.7	86.5	101.5	30.4	57.3	100.4	36.4	60.7	101.0	74.7	84.7
095,098	CEILINGS & ACOUSTICAL TREATMENT	85.5	37.9	57.1	100.9	74.7	85.3	94.9	74.7	82.9	94.1	30.4	56.1	94.1	36.4	59.7	94.9	74.7	82.9
09600	FLOORING	97.1	38.1	81.1	106.0	91.4	102.0	104.2	67.5	94.2	99.6	33.7	81.8	97.9	39.8	82.2	97.9	66.7	89.5
097,099	WALL FINISHES, PAINTS & COATINGS	91.5	21.1	48.5	121.8	84.2	98.8	121.8	84.2	98.8	110.4	32.8	63.0	110.4	54.9	76.5	121.8	39.2	71.4
09	FINISHES	95.2	37.9	65.3	102.3	78.8	90.1	100.6	72.6	86.0	97.2	32.3	63.4	96.6	40.6	67.4	97.2	65.2	80.5
10-14	TOTAL DIV. 10000-14000	100.0	47.4	88.8	100.0	86.6	97.1	100.0	68.2	93.2	100.0	70.1	93.6	100.0	74.1	94.5	100.0	55.1	90.4
15	MECHANICAL	95.1	41.0	72.7	99.9	86.3	94.3	100.1	82.7	92.8	95.1	55.6	78.8	95.1	65.7	82.9	95.1	38.3	71.5
16	ELECTRICAL	97.6	32.1	65.7	98.3	95.1	96.8	95.8	95.2	95.5	97.7	29.9	64.7	97.7	71.5	84.9	100.0	48.1	74.8
01-16	WEIGHTED AVERAGE	98.6	46.5	75.4	99.9	84.9	93.3	100.5	79.3	91.1	97.9	49.0	76.2	99.1	61.5	82.4	98.5	55.3	79.3

VIRGINIA

DIVISION		FAIRFAX 220 - 221			FARMVILLE 239			FREDERICKSBURG 224 - 225			GRUNDY 246			HARRISONBURG 228			LYNCHBURG 245		
		MAT.	INST.	TOTAL	MAT.	INST.	TOTAL	MAT.	INST.	TOTAL	MAT.	INST.	TOTAL	MAT.	INST.	TOTAL	MAT.	INST.	TOTAL
01590	EQUIPMENT RENTAL	.0	99.6	99.6	.0	104.8	104.8	.0	99.6	99.6	.0	99.6	99.6	.0	99.6	99.6	.0	99.6	99.6
02	SITE CONSTRUCTION	122.9	82.5	93.3	108.1	82.9	89.6	111.0	82.0	89.7	105.2	80.3	86.9	119.6	82.0	92.0	105.9	80.8	87.5
03100	CONCRETE FORMS & ACCESSORIES	84.2	71.8	73.5	98.2	27.2	36.9	84.2	45.6	50.9	89.4	29.0	37.3	79.6	28.0	35.1	85.8	36.3	43.1
03200	CONCRETE REINFORCEMENT	94.4	85.8	90.0	94.5	45.5	69.6	95.2	84.2	89.6	93.2	45.9	69.2	94.4	56.2	75.0	93.9	72.5	83.0
03300	CAST-IN-PLACE CONCRETE	97.2	76.3	88.6	100.6	45.8	78.1	98.7	47.4	77.6	96.8	40.4	73.6	97.1	58.9	81.4	96.8	46.4	76.1
03	CONCRETE	107.0	77.0	92.1	103.3	38.7	71.2	100.6	55.2	78.1	102.3	37.9	70.3	104.6	46.1	75.6	102.1	48.5	75.5
04	MASONRY	99.0	67.3	79.3	96.1	32.3	56.5	99.4	35.3	59.6	90.7	33.3	55.1	98.2	32.8	57.6	104.9	33.4	60.6
05	METALS	99.9	94.4	98.1	99.7	63.8	88.2	99.8	91.7	97.2	99.5	67.0	89.1	99.7	79.9	93.4	99.7	84.0	94.6
06	WOOD & PLASTICS	84.0	75.6	79.6	98.3	26.1	60.7	84.0	45.9	64.2	87.7	28.0	56.6	79.2	22.9	49.9	84.0	35.3	58.6
07	THERMAL & MOISTURE PROTECTION	102.0	72.1	89.4	101.5	33.9	73.0	101.5	45.0	77.7	101.7	36.6	74.3	101.8	51.7	80.7	101.5	41.9	76.4
08	DOORS & WINDOWS	94.6	77.2	90.0	96.2	26.0	77.9	95.9	57.6	85.9	97.9	27.9	79.6	96.2	38.5	81.1	96.2	44.7	82.8
09200	PLASTER & GYPSUM BOARD	101.8	74.7	85.0	107.8	23.1	55.2	101.8	44.3	66.0	102.6	25.8	54.9	100.4	20.7	50.9	101.5	33.4	59.2
095,098	CEILINGS & ACOUSTICAL TREATMENT	94.9	74.7	82.9	94.1	23.1	51.7	94.9	44.3	64.7	94.1	25.8	53.4	94.1	20.7	50.3	94.1	33.4	57.9
09600	FLOORING	99.9	58.2	88.6	106.0	54.4	92.0	99.9	66.7	90.9	101.4	30.4	82.1	97.4	66.7	89.1	99.6	37.0	82.6
097,099	WALL FINISHES, PAINTS & COATINGS	121.8	84.2	99.8	110.4	34.7	64.2	121.8	39.2	71.4	110.4	32.8	63.0	121.8	42.7	73.5	110.4	32.8	63.0
09	FINISHES	98.8	70.9	84.2	99.8	32.9	64.5	97.9	47.9	71.8	97.8	29.2	62.0	97.3	35.2	65.0	97.1	35.1	64.8
10 - 14	TOTAL DIV. 10000 - 14000	100.0	68.0	93.2	100.0	44.2	88.1	100.0	72.3	94.1	100.0	43.5	88.0	100.0	69.8	93.6	100.0	71.2	93.9
15	MECHANICAL	95.1	82.6	89.9	95.1	26.5	66.6	95.1	78.1	88.0	95.1	34.5	70.0	95.1	38.9	71.8	95.1	65.1	82.7
16	ELECTRICAL	98.7	95.2	97.0	95.2	44.1	70.3	96.0	95.2	95.6	97.7	58.0	78.4	97.9	36.9	68.3	98.7	29.2	64.9
01 - 16	WEIGHTED AVERAGE	99.3	80.6	91.0	98.3	41.1	72.9	98.0	68.4	84.9	98.0	44.4	74.2	98.8	47.1	75.8	98.6	52.2	78.0

VIRGINIA

DIVISION		NEWPORT NEWS 236			NORFOLK 233 - 235			PETERSBURG 238			PORTSMOUTH 237			PULASKI 243			RICHMOND 230 - 232		
		MAT.	INST.	TOTAL	MAT.	INST.	TOTAL	MAT.	INST.	TOTAL	MAT.	INST.	TOTAL	MAT.	INST.	TOTAL	MAT.	INST.	TOTAL
01590	EQUIPMENT RENTAL	.0	104.8	104.8	.0	105.5	105.5	.0	104.8	104.8	.0	104.7	104.7	.0	99.6	99.6	.0	104.8	104.8
02	SITE CONSTRUCTION	106.7	84.5	90.4	106.0	85.6	91.1	110.4	84.9	91.7	105.3	83.9	89.6	104.7	80.1	86.7	107.3	84.9	90.9
03100	CONCRETE FORMS & ACCESSORIES	97.2	59.4	64.6	101.7	59.6	65.4	89.7	56.0	60.6	84.9	59.4	62.9	89.4	28.0	36.4	98.2	56.0	61.8
03200	CONCRETE REINFORCEMENT	94.2	74.6	84.2	94.2	74.6	84.2	93.9	72.1	82.8	93.9	74.6	84.1	93.2	45.9	69.1	94.2	72.1	83.0
03300	CAST-IN-PLACE CONCRETE	97.8	54.8	80.1	100.7	54.9	81.9	100.7	54.7	81.8	96.8	54.3	79.4	96.8	39.0	73.0	104.1	54.7	83.8
03	CONCRETE	100.4	62.1	81.4	102.1	62.2	82.3	104.2	60.1	82.3	99.0	61.9	80.6	102.3	37.0	69.8	103.4	60.1	81.9
04	MASONRY	91.7	53.8	68.2	98.0	53.8	70.6	103.9	49.4	70.1	96.5	53.6	69.9	84.6	30.8	51.3	90.3	49.4	65.0
05	METALS	101.7	90.5	98.1	100.7	90.7	97.5	99.8	89.8	96.6	100.7	89.0	97.0	99.5	66.9	89.1	103.6	89.8	99.2
06	WOOD & PLASTICS	97.0	63.3	79.4	102.6	63.3	82.1	88.2	57.7	72.3	83.4	63.3	72.9	87.7	28.0	56.6	98.2	57.7	77.1
07	THERMAL & MOISTURE PROTECTION	101.4	51.2	80.3	101.2	51.2	80.2	101.4	54.2	81.5	101.4	51.2	80.3	101.7	35.6	73.9	101.1	54.2	81.3
08	DOORS & WINDOWS	96.6	60.0	87.0	96.6	62.6	87.7	95.9	55.5	85.4	96.7	62.6	87.8	97.9	27.9	79.6	96.6	55.5	85.9
09200	PLASTER & GYPSUM BOARD	109.6	61.3	79.6	109.6	61.3	79.6	102.9	55.6	73.5	101.8	61.3	76.6	102.6	25.8	54.9	109.6	55.6	76.0
095,098	CEILINGS & ACOUSTICAL TREATMENT	100.9	61.3	77.3	100.9	61.3	77.3	94.9	55.6	71.4	100.9	61.3	77.3	94.1	25.8	53.4	100.9	55.6	73.8
09600	FLOORING	106.0	39.4	88.0	105.8	39.4	87.8	100.9	62.5	90.5	96.9	39.4	81.3	101.4	30.4	82.1	105.8	62.5	94.1
097,099	WALL FINISHES, PAINTS & COATINGS	110.4	42.0	68.7	110.4	63.3	81.7	110.4	54.9	76.5	110.4	63.3	81.7	110.4	32.8	63.0	110.4	54.9	76.5
09	FINISHES	101.7	53.8	76.8	101.6	56.3	78.0	98.0	56.9	76.6	97.8	56.3	76.2	97.8	28.5	61.7	101.5	56.9	78.3
10 - 14	TOTAL DIV. 10000 - 14000	100.0	78.1	95.3	100.0	78.1	95.3	100.0	77.5	95.2	100.0	78.4	95.4	100.0	42.6	87.8	100.0	77.5	95.2
15	MECHANICAL	100.1	63.4	84.8	100.1	63.6	84.9	95.1	67.4	83.6	100.1	63.6	84.9	95.1	27.1	66.9	100.1	67.4	86.5
16	ELECTRICAL	97.7	59.2	79.0	97.8	63.1	80.9	97.9	71.5	85.1	96.2	63.1	80.1	97.7	58.0	78.4	98.4	71.5	85.3
01 - 16	WEIGHTED AVERAGE	99.6	64.7	84.1	100.0	65.8	84.8	98.9	66.7	84.6	98.9	65.5	84.0	97.7	42.4	73.1	100.3	66.7	85.4

VIRGINIA / WASHINGTON

DIVISION		ROANOKE 240 - 241			STAUNTON 244			WINCHESTER 226			CLARKSTON 994			EVERETT 982			OLYMPIA 985		
		MAT.	INST.	TOTAL	MAT.	INST.	TOTAL	MAT.	INST.	TOTAL	MAT.	INST.	TOTAL	MAT.	INST.	TOTAL	MAT.	INST.	TOTAL
01590	EQUIPMENT RENTAL	.0	99.6	99.6	.0	104.8	104.8	.0	99.6	99.6	.0	90.0	90.0	.0	104.0	104.0	.0	104.0	104.0
02	SITE CONSTRUCTION	104.3	80.7	87.0	108.7	83.1	89.9	118.3	81.5	91.3	103.0	88.4	92.3	93.8	115.2	109.5	97.0	115.2	110.4
03100	CONCRETE FORMS & ACCESSORIES	97.0	36.0	44.4	89.1	35.3	42.7	82.4	37.7	43.9	113.2	79.0	83.7	108.0	101.8	102.7	98.3	102.1	101.6
03200	CONCRETE REINFORCEMENT	94.2	72.7	83.3	93.9	45.7	69.4	93.8	81.1	87.4	107.1	87.7	97.2	108.0	94.7	101.2	108.5	94.7	101.5
03300	CAST-IN-PLACE CONCRETE	109.9	46.3	83.7	100.7	46.6	78.4	97.1	57.9	81.0	118.2	85.0	104.5	94.5	106.3	99.3	99.2	108.3	102.9
03	CONCRETE	106.2	48.3	77.5	103.2	43.2	73.4	104.0	54.7	79.5	108.4	82.6	95.6	95.7	101.4	98.5	98.7	102.2	100.4
04	MASONRY	91.8	38.8	58.9	101.8	34.5	60.1	95.1	32.7	56.4	115.1	83.3	95.4	132.8	103.0	114.4	125.8	103.7	112.1
05	METALS	101.4	84.0	95.9	99.7	76.8	92.4	99.8	88.7	96.3	93.8	81.9	90.0	106.0	88.1	100.3	106.5	88.1	100.7
06	WOOD & PLASTICS	97.0	35.3	64.9	87.7	35.3	60.4	82.3	36.0	58.2	95.8	78.0	86.5	98.7	101.1	99.9	87.8	101.1	94.7
07	THERMAL & MOISTURE PROTECTION	101.4	44.6	77.5	101.3	40.3	75.6	102.0	53.0	81.3	138.8	78.6	113.4	104.6	98.4	102.0	104.2	97.3	101.3
08	DOORS & WINDOWS	96.6	44.7	83.1	96.2	38.6	81.2	98.0	51.5	85.8	111.1	80.8	103.2	98.6	98.8	98.7	98.6	98.8	98.6
09200	PLASTER & GYPSUM BOARD	109.6	33.4	62.2	102.6	32.6	59.1	101.8	34.1	59.7	124.7	77.2	95.2	104.7	101.1	102.4	98.9	101.1	100.3
095,098	CEILINGS & ACOUSTICAL TREATMENT	100.9	33.4	60.6	94.1	32.6	57.4	94.9	34.1	58.6	102.9	77.2	87.6	111.3	101.1	105.2	107.7	101.1	103.7
09600	FLOORING	106.0	35.3	86.9	100.6	35.6	83.0	99.1	66.7	90.4	109.9	58.9	96.1	119.1	105.1	115.3	111.6	105.1	109.9
097,099	WALL FINISHES, PAINTS & COATINGS	110.4	32.8	63.0	110.4	27.1	59.6	121.8	27.6	64.4	112.6	74.5	89.3	111.1	90.5	98.5	111.1	92.2	99.6
09	FINISHES	101.5	34.8	66.8	97.5	34.6	64.7	98.2	40.9	68.3	121.8	74.2	97.0	113.4	101.3	107.1	109.5	101.6	105.4
10 - 14	TOTAL DIV. 10000 - 14000	100.0	71.3	93.9	100.0	72.7	94.2	100.0	50.4	89.4	100.0	96.8	99.3	100.0	103.2	100.7	100.0	103.2	100.7
15	MECHANICAL	100.1	39.3	74.9	95.1	34.1	69.8	95.1	60.1	80.6	95.6	82.7	90.2	99.9	101.9	100.7	100.0	100.6	100.3
16	ELECTRICAL	97.7	34.2	66.8	96.7	37.4	67.9	96.4	33.4	65.8	98.8	97.6	98.2	103.6	97.7	100.7	103.4	100.6	102.0
01 - 16	WEIGHTED AVERAGE	100.2	48.1	77.1	98.4	45.6	75.0	98.6	53.8	78.7	103.9	84.3	95.2	103.6	101.0	102.4	102.3	101.3	101.9

WASHINGTON

DIVISION		RICHLAND 993 MAT.	INST.	TOTAL	SEATTLE 980-981,987 MAT.	INST.	TOTAL	SPOKANE 990-992 MAT.	INST.	TOTAL	TACOMA 983-984 MAT.	INST.	TOTAL	VANCOUVER 986 MAT.	INST.	TOTAL	WENATCHEE 988 MAT.	INST.	TOTAL
01590	EQUIPMENT RENTAL	.0	90.0	90.0	.0	103.7	103.7	.0	90.0	90.0	.0	104.0	104.0	.0	97.1	97.1	.0	104.0	104.0
02	SITE CONSTRUCTION	104.4	89.4	93.3	97.8	113.0	109.0	105.2	89.4	93.6	96.6	115.2	110.3	108.7	100.7	102.8	108.1	114.2	112.6
03100	CONCRETE FORMS & ACCESSORIES	113.4	80.3	84.8	98.4	102.5	101.9	120.4	80.3	85.8	98.4	102.2	101.7	99.2	96.9	97.2	99.9	80.7	83.3
03200	CONCRETE REINFORCEMENT	102.9	87.7	95.1	106.8	94.8	100.7	103.5	87.7	95.5	106.8	94.7	100.6	107.6	94.5	100.9	107.5	90.1	98.7
03300	CAST-IN-PLACE CONCRETE	118.5	83.6	104.1	99.2	108.5	103.0	122.7	83.6	106.6	97.2	108.3	101.8	108.6	100.1	105.1	99.3	86.1	93.9
03	CONCRETE	108.0	82.7	95.4	98.4	102.5	100.5	110.7	82.7	96.8	97.4	102.3	99.8	107.2	97.2	102.2	105.2	84.3	94.8
04	MASONRY	115.0	83.3	95.3	126.8	103.7	112.5	116.5	83.3	95.9	126.6	103.7	112.4	127.2	100.1	110.4	129.8	84.9	101.9
05	METALS	94.1	82.0	90.2	107.7	90.1	102.1	96.7	82.0	92.0	107.7	88.2	101.4	104.8	89.6	100.0	105.4	84.0	98.6
06	WOOD & PLASTICS	96.0	79.6	87.5	88.8	101.1	95.2	105.8	79.6	92.2	87.8	101.1	94.7	82.3	96.6	89.7	88.9	78.7	83.6
07	THERMAL & MOISTURE PROTECTION	139.9	81.2	115.2	104.6	100.6	102.9	137.7	81.1	113.8	104.2	99.2	102.1	106.8	90.9	100.1	104.9	79.5	94.2
08	DOORS & WINDOWS	111.7	76.2	102.4	100.8	98.8	100.3	111.5	76.2	102.2	99.3	98.8	99.1	96.2	95.3	96.0	98.8	76.2	92.9
09200	PLASTER & GYPSUM BOARD	124.7	78.9	96.3	98.2	101.1	100.0	129.5	78.9	98.1	100.2	101.1	100.7	98.7	96.8	97.5	101.5	78.1	86.9
095,098	CEILINGS & ACOUSTICAL TREATMENT	107.7	78.9	90.5	114.5	101.1	106.5	107.7	78.9	90.5	114.7	101.1	106.6	109.7	96.8	102.0	107.7	78.1	90.0
09600	FLOORING	110.2	43.2	92.0	111.0	105.1	109.4	113.9	74.2	103.1	111.6	105.1	109.9	117.5	79.5	107.2	115.0	74.2	103.9
097,099	WALL FINISHES, PAINTS & COATINGS	112.6	74.5	89.3	111.1	92.2	99.6	112.6	74.5	89.3	111.1	92.2	99.6	119.0	68.1	87.9	111.1	74.5	88.8
09	FINISHES	123.2	71.9	96.5	111.0	101.6	106.1	125.0	78.3	100.7	111.5	101.6	106.3	109.5	90.3	99.5	111.8	78.1	94.2
10 - 14	TOTAL DIV. 10000 - 14000	100.0	83.1	96.4	100.0	103.2	100.7	100.0	83.0	96.4	100.0	103.2	100.7	100.0	80.6	95.9	100.0	97.1	99.4
15	MECHANICAL	100.5	98.5	99.7	99.9	112.2	105.0	100.5	87.7	95.2	100.0	100.6	100.3	100.1	104.0	101.7	95.1	86.0	91.3
16	ELECTRICAL	95.8	97.6	96.7	103.4	106.5	104.9	94.1	81.8	88.1	103.4	100.6	102.0	114.0	103.1	108.7	104.2	85.9	95.3
01 - 16	WEIGHTED AVERAGE	105.0	86.9	96.9	103.8	104.7	104.2	105.7	83.3	95.8	103.5	101.4	102.6	105.4	97.7	102.0	103.5	86.6	96.0

WASHINGTON / WEST VIRGINIA

DIVISION		YAKIMA 989 MAT.	INST.	TOTAL	BECKLEY 258-259 MAT.	INST.	TOTAL	BLUEFIELD 247-248 MAT.	INST.	TOTAL	BUCKHANNON 262 MAT.	INST.	TOTAL	CHARLESTON 250-253 MAT.	INST.	TOTAL	CLARKSBURG 263-264 MAT.	INST.	TOTAL
01590	EQUIPMENT RENTAL	.0	104.0	104.0	.0	99.6	99.6	.0	99.6	99.6	.0	99.6	99.6	.0	99.6	99.6	.0	99.6	99.6
02	SITE CONSTRUCTION	99.8	113.8	110.1	99.1	84.9	88.7	98.9	84.8	88.6	105.1	84.6	90.0	102.1	86.0	90.3	105.6	84.6	90.2
03100	CONCRETE FORMS & ACCESSORIES	98.9	95.8	96.2	84.5	90.6	89.8	86.2	89.5	89.0	85.4	91.3	90.5	104.9	91.7	93.5	82.4	91.3	90.1
03200	CONCRETE REINFORCEMENT	107.2	93.4	100.2	92.8	87.1	89.9	92.8	69.6	81.0	93.4	89.0	91.2	94.2	87.3	90.7	93.4	91.6	92.4
03300	CAST-IN-PLACE CONCRETE	103.9	81.6	94.7	94.6	106.9	99.7	94.6	105.0	98.9	94.4	101.5	97.3	95.7	108.0	100.8	103.4	101.6	102.6
03	CONCRETE	102.1	89.9	96.0	97.7	96.2	97.0	97.8	91.8	94.8	101.2	95.0	98.1	99.9	97.1	98.5	104.9	95.5	100.2
04	MASONRY	118.8	71.1	89.2	90.6	92.1	91.6	86.2	92.1	89.9	97.0	90.3	92.9	89.0	93.4	91.8	98.5	90.3	93.4
05	METALS	105.9	82.8	98.5	99.7	98.4	99.3	99.7	91.1	96.9	99.9	99.8	99.8	101.7	99.4	100.9	99.9	100.8	100.2
06	WOOD & PLASTICS	88.2	101.1	95.0	83.7	90.9	87.5	85.6	90.9	88.4	84.7	90.6	87.8	105.6	90.5	97.7	81.3	90.6	86.1
07	THERMAL & MOISTURE PROTECTION	104.4	79.5	94.0	101.4	90.6	96.9	101.5	84.9	94.5	101.7	87.7	95.8	101.3	91.1	97.0	101.7	87.7	95.8
08	DOORS & WINDOWS	98.8	85.0	95.2	97.2	84.2	93.8	98.1	80.2	93.4	98.2	84.5	94.6	97.8	84.0	94.2	98.2	90.2	96.1
09200	PLASTER & GYPSUM BOARD	100.0	101.1	100.7	99.0	90.5	93.7	99.4	90.5	93.9	99.8	90.2	93.8	108.9	90.1	97.2	98.4	90.2	93.3
095,098	CEILINGS & ACOUSTICAL TREATMENT	109.6	101.1	104.5	91.5	90.5	90.9	91.5	90.5	90.9	93.2	90.2	91.4	98.3	90.1	93.4	93.2	90.2	91.4
09600	FLOORING	113.0	65.0	100.0	96.2	107.1	99.1	96.9	107.1	99.7	96.4	103.4	98.3	105.8	107.1	106.2	95.1	103.4	97.3
097,099	WALL FINISHES, PAINTS & COATINGS	111.1	74.5	88.8	110.4	44.7	70.3	110.4	44.7	70.3	110.4	101.3	104.9	110.4	95.9	101.6	110.4	101.3	104.9
09	FINISHES	110.7	88.3	99.0	94.7	89.3	91.9	95.0	89.3	92.0	95.7	94.8	95.3	100.9	95.2	97.9	95.1	94.8	94.9
10 - 14	TOTAL DIV. 10000 - 14000	100.0	98.5	99.7	100.0	71.2	93.9	100.0	71.1	93.8	100.0	94.6	98.8	100.0	95.0	98.9	100.0	94.6	98.8
15	MECHANICAL	100.0	97.6	99.0	95.1	83.4	90.3	95.1	67.1	83.5	95.1	94.7	95.0	100.1	84.9	93.8	95.1	92.7	94.1
16	ELECTRICAL	106.1	97.6	102.0	94.4	87.3	90.9	96.8	56.0	76.9	98.0	96.4	97.2	97.7	87.3	92.6	98.0	96.4	97.2
01 - 16	WEIGHTED AVERAGE	103.7	91.8	98.4	96.6	88.8	93.1	96.8	79.4	89.1	98.1	93.4	96.0	99.5	91.0	95.7	98.5	93.4	96.2

WEST VIRGINIA

DIVISION		GASSAWAY 266 MAT.	INST.	TOTAL	HUNTINGTON 255-257 MAT.	INST.	TOTAL	LEWISBURG 249 MAT.	INST.	TOTAL	MARTINSBURG 254 MAT.	INST.	TOTAL	MORGANTOWN 265 MAT.	INST.	TOTAL	PARKERSBURG 261 MAT.	INST.	TOTAL
01590	EQUIPMENT RENTAL	.0	99.6	99.6	.0	99.6	99.6	.0	99.6	99.6	.0	99.6	99.6	.0	99.6	99.6	.0	99.6	99.6
02	SITE CONSTRUCTION	102.4	85.1	89.7	103.5	86.8	91.2	115.1	84.9	92.9	102.7	82.9	88.2	99.7	84.9	88.8	108.3	85.9	91.9
03100	CONCRETE FORMS & ACCESSORIES	84.7	91.4	90.5	98.1	93.7	94.3	82.8	90.2	89.2	84.5	81.2	81.7	82.9	91.3	90.2	87.5	91.7	91.1
03200	CONCRETE REINFORCEMENT	93.4	87.3	90.3	94.2	85.5	89.8	93.4	69.7	81.4	92.8	76.0	84.3	93.4	91.6	92.4	92.8	89.0	90.9
03300	CAST-IN-PLACE CONCRETE	98.8	107.4	102.3	103.2	112.5	107.0	94.7	106.8	99.7	99.2	93.7	96.9	94.4	101.6	97.3	96.5	102.1	98.8
03	CONCRETE	101.2	96.8	99.0	103.0	99.2	101.1	108.0	92.7	100.4	101.3	85.2	93.3	97.6	95.5	96.5	103.1	95.4	99.2
04	MASONRY	100.1	92.3	95.3	91.2	94.4	93.2	90.2	92.1	91.4	93.1	80.0	85.0	118.1	90.3	100.9	77.0	91.0	85.7
05	METALS	99.8	99.4	99.7	101.7	98.8	100.8	99.8	91.8	97.2	100.0	88.1	96.2	99.9	100.8	100.2	100.3	99.8	100.2
06	WOOD & PLASTICS	83.9	90.6	87.4	97.0	91.8	94.3	81.6	90.9	86.5	83.7	83.1	83.4	81.6	90.6	86.3	85.8	90.6	88.3
07	THERMAL & MOISTURE PROTECTION	101.5	89.6	96.5	101.6	91.6	97.4	102.3	90.6	97.4	101.7	72.1	89.2	101.5	87.7	95.7	101.7	88.4	96.1
08	DOORS & WINDOWS	96.1	84.0	92.9	96.6	84.3	93.4	98.1	80.2	93.5	99.5	71.9	92.3	99.5	90.2	97.1	97.2	82.0	93.2
09200	PLASTER & GYPSUM BOARD	99.5	90.2	93.7	107.8	91.4	97.6	98.4	90.5	93.5	99.5	82.4	88.9	98.4	90.2	93.3	102.1	90.2	94.7
095,098	CEILINGS & ACOUSTICAL TREATMENT	93.2	90.2	91.4	94.1	91.4	92.5	93.2	90.5	91.6	93.2	82.4	86.8	93.2	90.2	91.4	93.2	90.2	91.4
09600	FLOORING	96.2	107.1	99.2	105.8	104.3	105.4	95.2	107.1	98.4	96.2	67.0	88.3	95.2	103.4	97.4	100.2	102.9	100.9
097,099	WALL FINISHES, PAINTS & COATINGS	110.4	95.9	101.6	110.4	93.1	99.9	110.4	44.7	70.3	110.4	52.8	75.2	110.4	101.3	104.9	110.4	96.6	102.0
09	FINISHES	95.4	95.0	95.2	99.5	95.7	97.5	95.9	89.3	92.5	95.4	76.3	85.4	94.7	94.8	94.8	97.2	94.4	95.8
10 - 14	TOTAL DIV. 10000 - 14000	100.0	94.6	98.8	100.0	96.2	99.2	100.0	71.1	93.8	100.0	68.5	93.3	100.0	84.5	96.7	100.0	95.0	98.9
15	MECHANICAL	95.1	92.4	94.0	100.1	88.4	95.2	95.1	83.3	90.2	95.1	80.3	89.0	95.1	94.5	94.9	100.1	89.7	95.8
16	ELECTRICAL	98.0	87.3	92.8	97.7	95.8	96.8	94.4	56.0	75.7	99.6	76.7	88.5	98.2	96.4	97.3	98.2	96.9	97.5
01 - 16	WEIGHTED AVERAGE	97.9	92.2	95.4	99.6	93.4	96.9	98.3	83.1	91.6	98.1	80.0	90.1	98.6	93.5	96.4	98.7	92.6	96.0

City Cost Indexes

WEST VIRGINIA / WISCONSIN

DIVISION		PETERSBURG 268 MAT.	INST.	TOTAL	ROMNEY 267 MAT.	INST.	TOTAL	WHEELING 260 MAT.	INST.	TOTAL	BELOIT 535 MAT.	INST.	TOTAL	EAU CLAIRE 547 MAT.	INST.	TOTAL	GREEN BAY 541-543 MAT.	INST.	TOTAL
01590	EQUIPMENT RENTAL	.0	99.6	99.6	.0	99.6	99.6	.0	99.6	99.6	.0	100.1	100.1	.0	100.6	100.6	.0	98.3	98.3
02	SITE CONSTRUCTION	99.1	85.3	89.0	102.0	85.3	89.7	108.9	85.6	91.8	91.4	105.3	101.6	86.5	102.9	98.5	89.2	98.8	96.2
03100	CONCRETE FORMS & ACCESSORIES	86.5	85.1	85.3	81.9	84.6	84.2	89.5	91.6	91.3	99.0	98.6	98.6	98.1	99.3	99.2	109.2	98.5	100.0
03200	CONCRETE REINFORCEMENT	92.8	80.7	86.7	93.4	72.5	82.8	92.2	91.6	91.9	89.2	107.0	98.3	91.8	105.6	98.8	90.1	93.3	91.7
03300	CAST-IN-PLACE CONCRETE	94.4	99.0	96.3	98.8	98.8	98.8	96.5	101.7	98.6	106.2	99.3	103.4	100.3	98.4	99.5	103.8	100.4	102.4
03	CONCRETE	97.7	89.8	93.8	101.0	88.0	94.6	103.1	95.7	99.4	98.5	100.5	99.5	96.9	100.4	98.7	98.2	98.4	98.3
04	MASONRY	93.1	90.3	91.4	90.5	75.7	81.3	100.1	90.3	94.1	102.8	101.1	101.7	92.8	101.5	98.2	124.3	100.5	109.5
05	METALS	99.9	96.2	98.8	100.0	92.0	97.4	100.4	101.1	100.6	94.7	104.0	97.7	91.9	105.2	96.2	94.3	99.5	96.0
06	WOOD & PLASTICS	85.8	83.1	84.4	80.5	83.1	81.8	87.7	90.6	89.2	110.1	98.3	104.0	112.2	98.4	105.0	118.1	98.4	107.8
07	THERMAL & MOISTURE PROTECTION	101.6	79.1	92.1	101.6	75.5	90.6	101.8	88.6	96.2	96.2	100.5	98.0	97.4	90.9	94.6	99.4	89.0	95.0
08	DOORS & WINDOWS	99.5	78.5	94.0	99.5	76.8	93.5	98.1	90.2	96.0	101.9	99.4	101.2	100.9	96.7	99.8	99.1	89.9	96.7
09200	PLASTER & GYPSUM BOARD	99.8	82.4	89.0	97.6	82.4	88.2	102.4	90.2	94.8	98.1	98.7	98.4	105.0	98.7	101.1	97.3	98.7	98.1
095,098	CEILINGS & ACOUSTICAL TREATMENT	93.2	82.4	86.8	93.2	82.4	86.8	93.2	90.2	91.4	86.1	98.7	93.6	98.5	98.7	98.6	91.8	98.7	95.9
09600	FLOORING	97.4	103.4	99.0	94.9	75.5	89.7	101.4	103.4	101.9	93.5	114.5	99.2	90.5	109.4	95.6	108.3	109.4	108.6
097,099	WALL FINISHES, PAINTS & COATINGS	110.4	80.6	92.2	110.4	80.6	92.2	110.4	101.3	104.9	91.5	96.3	94.4	87.7	94.2	91.7	97.2	77.1	84.9
09	FINISHES	95.6	88.1	91.7	94.7	82.3	88.3	97.6	94.8	96.2	93.9	101.6	97.9	97.3	100.9	99.2	100.7	98.7	99.7
10-14	TOTAL DIV. 10000-14000	100.0	93.6	98.6	100.0	93.6	98.6	100.0	102.5	100.5	100.0	92.1	98.3	100.0	94.8	98.9	100.0	93.1	98.5
15	MECHANICAL	95.1	95.5	95.3	95.1	95.2	95.2	100.1	94.6	97.8	99.8	92.6	96.8	100.2	90.9	96.3	100.5	90.1	96.2
16	ELECTRICAL	101.1	85.5	93.5	100.4	85.5	93.2	95.6	96.4	95.9	96.0	92.8	94.4	101.5	83.1	92.6	96.6	90.8	93.8
01-16	WEIGHTED AVERAGE	97.9	89.3	94.1	97.9	86.5	92.9	99.7	94.2	97.2	98.0	98.4	98.2	97.8	96.2	97.1	99.7	95.2	97.7

WISCONSIN

DIVISION		KENOSHA 531 MAT.	INST.	TOTAL	LA CROSSE 546 MAT.	INST.	TOTAL	LANCASTER 538 MAT.	INST.	TOTAL	MADISON 537 MAT.	INST.	TOTAL	MILWAUKEE 530,532 MAT.	INST.	TOTAL	NEW RICHMOND 540 MAT.	INST.	TOTAL
01590	EQUIPMENT RENTAL	.0	97.9	97.9	.0	100.6	100.6	.0	100.1	100.1	.0	100.1	100.1	.0	86.3	86.3	.0	101.6	101.6
02	SITE CONSTRUCTION	96.5	102.1	100.6	80.6	102.8	96.9	90.4	105.0	101.1	90.8	105.5	101.6	92.2	93.8	93.4	81.8	104.0	98.1
03100	CONCRETE FORMS & ACCESSORIES	109.2	108.7	108.7	82.6	98.8	96.6	98.0	96.9	97.1	99.8	99.4	99.5	102.1	114.2	112.6	91.6	109.6	107.2
03200	CONCRETE REINFORCEMENT	89.1	101.5	95.4	91.5	92.2	91.9	90.3	92.2	91.3	89.2	92.4	90.8	89.2	101.7	95.6	89.4	105.5	97.6
03300	CAST-IN-PLACE CONCRETE	115.6	102.6	110.2	90.2	97.8	93.3	105.5	98.6	102.7	104.2	100.9	102.8	104.2	108.5	105.9	104.4	95.0	100.5
03	CONCRETE	103.7	105.1	104.4	88.3	97.5	92.9	98.2	96.7	97.5	97.6	98.7	98.1	97.9	109.0	103.4	95.0	103.8	99.4
04	MASONRY	99.2	111.4	106.8	92.0	101.5	97.9	102.8	101.2	101.8	103.2	104.3	103.9	102.7	118.9	112.7	118.3	101.7	108.0
05	METALS	95.3	102.7	97.7	91.9	99.0	94.1	92.8	96.1	93.9	96.4	97.4	96.7	98.1	94.1	96.8	92.1	104.5	96.0
06	WOOD & PLASTICS	115.9	108.3	111.9	94.7	98.4	96.6	109.2	98.3	103.5	110.2	98.3	104.0	112.9	113.1	113.0	100.4	112.8	106.8
07	THERMAL & MOISTURE PROTECTION	96.4	103.7	99.5	96.8	90.0	93.9	96.0	84.2	91.1	95.7	97.3	96.4	94.9	111.9	102.1	98.3	102.3	100.0
08	DOORS & WINDOWS	96.9	108.0	99.8	100.9	86.7	97.2	97.3	87.7	94.8	101.9	96.3	100.4	104.2	110.6	105.8	88.8	104.5	92.9
09200	PLASTER & GYPSUM BOARD	90.0	108.9	101.8	98.3	98.7	98.5	96.4	98.7	97.8	98.1	98.7	98.4	99.0	113.7	108.1	90.0	113.7	104.7
095,098	CEILINGS & ACOUSTICAL TREATMENT	80.5	108.9	97.4	96.8	98.7	97.9	79.3	98.7	90.8	86.1	98.7	93.6	89.5	113.7	103.9	70.0	113.7	96.1
09600	FLOORING	109.7	110.7	109.9	82.9	109.5	90.1	92.8	114.6	98.7	93.5	107.2	97.2	96.2	121.1	102.2	100.8	109.4	103.1
097,099	WALL FINISHES, PAINTS & COATINGS	101.5	101.1	101.3	87.7	71.6	77.9	91.5	84.7	87.3	91.5	96.3	94.4	93.8	111.3	104.5	101.3	94.2	96.9
09	FINISHES	96.6	109.0	103.1	93.3	98.4	96.0	91.7	98.5	95.2	93.8	101.0	97.5	95.8	115.8	106.2	91.8	107.6	100.0
10-14	TOTAL DIV. 10000-14000	100.0	101.0	100.2	100.0	94.8	98.9	100.0	60.4	91.6	100.0	93.3	98.6	100.0	103.7	100.8	100.0	95.9	99.1
15	MECHANICAL	100.0	98.2	99.3	100.2	90.6	96.2	94.9	88.7	92.3	99.8	93.2	97.0	99.8	103.2	101.2	94.7	85.1	90.7
16	ELECTRICAL	96.0	96.0	96.0	101.8	91.7	96.9	95.5	91.6	93.6	97.5	95.3	96.4	97.1	103.1	100.0	99.6	91.7	95.7
01-16	WEIGHTED AVERAGE	98.5	103.1	100.5	96.1	95.7	95.9	95.8	94.0	95.0	98.3	98.1	98.2	99.0	106.2	102.2	95.5	98.5	96.8

WISCONSIN

DIVISION		OSHKOSH 549 MAT.	INST.	TOTAL	PORTAGE 539 MAT.	INST.	TOTAL	RACINE 534 MAT.	INST.	TOTAL	RHINELANDER 545 MAT.	INST.	TOTAL	SUPERIOR 548 MAT.	INST.	TOTAL	WAUSAU 544 MAT.	INST.	TOTAL
01590	EQUIPMENT RENTAL	.0	98.3	98.3	.0	100.1	100.1	.0	100.1	100.1	.0	98.3	98.3	.0	101.6	101.6	.0	98.3	98.3
02	SITE CONSTRUCTION	81.4	98.8	94.1	81.2	105.3	98.9	91.2	106.0	102.0	93.1	98.6	97.1	78.8	104.1	97.3	77.4	98.8	93.1
03100	CONCRETE FORMS & ACCESSORIES	89.8	98.2	97.1	89.7	99.1	97.8	99.3	108.6	107.3	86.9	97.9	96.4	89.8	99.5	98.1	89.1	98.1	96.9
03200	CONCRETE REINFORCEMENT	90.3	93.2	91.8	90.4	92.4	91.4	89.2	101.4	95.4	90.4	93.6	92.0	89.4	94.2	91.8	90.4	92.1	91.3
03300	CAST-IN-PLACE CONCRETE	96.2	100.3	97.9	90.6	103.1	95.7	104.2	102.2	103.3	109.1	100.2	105.4	98.4	100.9	99.4	89.7	100.3	94.0
03	CONCRETE	90.2	98.4	94.2	86.2	99.3	92.7	97.5	104.9	101.2	101.9	98.1	100.0	89.6	99.2	94.4	85.3	98.0	91.6
04	MASONRY	105.0	100.5	102.2	101.7	104.3	103.3	102.6	111.3	108.0	123.9	100.5	109.4	117.6	110.3	113.1	104.5	100.5	102.0
05	METALS	92.4	99.2	94.6	93.4	97.1	94.6	96.4	102.7	98.4	92.3	99.1	94.4	92.9	101.0	95.5	92.2	98.7	94.2
06	WOOD & PLASTICS	96.4	98.4	97.4	99.5	98.3	98.8	110.4	108.3	109.3	93.1	98.4	95.9	98.5	97.3	97.9	95.5	98.4	97.0
07	THERMAL & MOISTURE PROTECTION	98.5	90.6	95.1	95.4	97.1	96.1	96.3	103.2	99.2	99.2	85.3	93.4	98.0	103.0	100.1	98.3	84.2	92.3
08	DOORS & WINDOWS	94.4	89.9	93.2	97.5	96.3	97.2	101.9	108.0	103.5	94.4	90.0	93.2	88.7	95.2	90.4	94.6	89.6	93.3
09200	PLASTER & GYPSUM BOARD	87.7	98.7	94.5	92.2	98.7	96.2	98.1	108.9	104.8	86.6	98.7	94.1	89.8	97.8	94.8	87.0	98.7	94.2
095,098	CEILINGS & ACOUSTICAL TREATMENT	91.8	98.7	95.9	80.1	98.7	91.2	86.1	108.9	99.7	91.8	98.7	95.9	70.9	97.8	87.0	91.8	98.7	95.9
09600	FLOORING	98.5	109.4	101.4	88.8	114.5	95.8	93.5	110.7	98.1	97.5	109.4	100.7	102.1	123.2	107.8	98.3	109.4	101.3
097,099	WALL FINISHES, PAINTS & COATINGS	97.2	77.1	84.9	91.5	84.7	87.3	91.4	92.2	91.9	97.2	77.0	84.9	93.1	104.4	100.0	97.2	79.1	86.2
09	FINISHES	96.0	97.5	96.8	89.6	100.4	95.2	93.8	107.9	101.2	96.3	98.3	97.3	91.6	104.5	98.3	95.6	99.0	97.3
10-14	TOTAL DIV. 10000-14000	100.0	83.8	96.6	100.0	82.0	96.2	100.0	101.0	100.2	100.0	69.6	93.5	100.0	94.5	98.8	100.0	93.1	98.5
15	MECHANICAL	95.6	90.0	93.3	94.9	93.0	94.1	99.8	95.4	97.9	95.6	89.8	93.2	94.7	106.7	99.7	95.6	89.7	93.1
16	ELECTRICAL	100.1	93.4	96.8	99.1	95.3	97.2	95.8	98.0	96.9	99.5	86.4	93.1	104.1	98.5	101.4	101.2	86.4	94.0
01-16	WEIGHTED AVERAGE	95.5	95.2	95.4	94.4	97.7	95.9	98.1	102.9	100.2	98.0	93.7	96.1	95.4	102.8	98.7	94.9	94.3	94.6

742

City Cost Indexes

WYOMING

DIVISION		CASPER 826 MAT.	INST.	TOTAL	CHEYENNE 820 MAT.	INST.	TOTAL	NEWCASTLE 827 MAT.	INST.	TOTAL	RAWLINS 823 MAT.	INST.	TOTAL	RIVERTON 825 MAT.	INST.	TOTAL	ROCK SPRINGS 829-831 MAT.	INST.	TOTAL
01590	EQUIPMENT RENTAL	.0	101.7	101.7	.0	101.7	101.7	.0	101.7	101.7	.0	101.7	101.7	.0	101.7	101.7	.0	101.7	101.7
02	SITE CONSTRUCTION	79.2	100.0	94.4	79.0	100.0	94.4	79.3	99.6	94.2	94.3	99.6	98.2	88.0	99.7	96.6	82.0	99.6	94.9
03100	CONCRETE FORMS & ACCESSORIES	97.7	42.6	50.1	98.5	42.8	50.5	92.4	30.4	38.9	97.4	30.4	39.6	91.0	35.1	42.8	99.2	30.4	39.8
03200	CONCRETE REINFORCEMENT	115.6	51.7	83.2	108.5	51.9	79.7	117.3	51.5	83.9	116.8	51.5	83.6	117.9	51.6	84.2	117.9	51.5	84.2
03300	CAST-IN-PLACE CONCRETE	97.6	75.8	88.7	97.6	75.9	88.7	98.1	42.5	75.3	98.2	42.5	75.3	98.2	43.6	75.7	98.1	42.5	75.3
03	CONCRETE	104.1	56.5	80.5	103.0	56.7	80.0	104.2	39.7	72.2	117.6	39.7	78.9	112.2	42.1	77.4	104.8	39.6	72.4
04	MASONRY	107.0	38.1	64.3	104.4	40.2	64.6	105.4	31.1	59.3	105.4	31.1	59.3	105.4	33.7	60.9	170.7	31.1	84.2
05	METALS	97.2	62.2	86.1	98.0	62.7	86.7	94.4	61.2	83.8	94.4	61.2	83.8	94.5	61.3	83.9	95.0	61.2	84.2
06	WOOD & PLASTICS	95.1	40.5	66.7	95.1	40.5	66.7	89.9	28.7	58.0	95.1	28.7	60.5	88.5	34.1	60.2	97.9	28.7	61.9
07	THERMAL & MOISTURE PROTECTION	99.6	53.5	80.2	99.6	54.0	80.4	99.9	44.1	76.4	101.7	44.1	77.4	101.0	47.1	78.3	100.0	44.1	76.5
08	DOORS & WINDOWS	93.5	43.2	80.3	94.2	43.2	80.9	100.1	36.8	83.5	99.7	36.8	83.3	99.9	39.7	84.2	100.4	36.8	83.8
09200	PLASTER & GYPSUM BOARD	82.7	38.7	55.4	82.7	38.7	55.4	80.5	26.6	47.0	82.7	26.6	47.9	80.1	32.2	50.3	88.2	26.6	50.0
095,098	CEILINGS & ACOUSTICAL TREATMENT	100.5	38.7	63.7	99.6	38.7	63.3	100.5	26.6	56.4	100.5	26.6	56.4	100.5	32.2	59.7	100.5	26.6	56.4
09600	FLOORING	100.3	41.2	84.3	100.3	66.2	91.1	97.4	40.4	82.4	100.3	40.4	84.1	96.7	40.4	81.5	102.6	40.4	85.8
097,099	WALL FINISHES, PAINTS & COATINGS	102.2	54.9	73.3	102.0	54.9	73.3	102.2	33.3	60.2	102.2	33.3	60.2	102.2	30.6	58.5	102.2	33.3	60.2
09	FINISHES	96.5	42.4	68.3	96.2	47.6	70.9	95.2	31.3	61.9	97.8	31.3	63.1	95.8	34.6	63.9	98.0	31.3	63.2
10 - 14	TOTAL DIV. 10000 - 14000	100.0	79.2	95.6	100.0	79.3	95.6	100.0	47.3	88.8	100.0	47.3	88.8	100.0	48.7	89.1	100.0	47.3	88.8
15	MECHANICAL	100.1	65.9	85.9	100.1	60.0	83.4	97.5	54.9	79.9	97.5	54.9	79.9	97.5	55.9	80.3	100.1	54.9	81.3
16	ELECTRICAL	90.6	63.3	77.3	90.6	74.4	82.7	91.5	74.4	83.2	91.5	74.4	83.2	91.5	63.3	77.8	89.9	62.7	76.7
01 - 16	WEIGHTED AVERAGE	97.7	59.9	80.9	97.6	61.2	81.4	97.2	53.1	77.6	99.4	53.1	78.8	98.4	53.1	78.3	101.5	51.4	79.2

WYOMING / CANADA

DIVISION		SHERIDAN 828 MAT.	INST.	TOTAL	WHEATLAND 822 MAT.	INST.	TOTAL	WORLAND 824 MAT.	INST.	TOTAL	YELLOWSTONE NAT'L PA 821 MAT.	INST.	TOTAL	BARRIE, ONTARIO MAT.	INST.	TOTAL	BATHURST, NEW BRUNSWICK MAT.	INST.	TOTAL
01590	EQUIPMENT RENTAL	.0	101.7	101.7	.0	101.7	101.7	.0	101.7	101.7	.0	101.7	101.7	.0	100.4	100.4	.0	98.6	98.6
02	SITE CONSTRUCTION	85.4	100.0	96.1	84.7	99.7	95.7	80.5	99.6	94.5	80.6	99.9	94.7	116.6	98.7	103.5	96.0	94.3	94.8
03100	CONCRETE FORMS & ACCESSORIES	91.4	42.5	49.2	94.7	37.9	45.7	94.8	30.4	39.2	94.9	34.9	43.1	122.6	80.2	86.1	97.6	57.0	62.6
03200	CONCRETE REINFORCEMENT	117.9	51.8	84.3	117.3	51.6	83.9	117.9	51.5	84.2	120.0	51.6	85.2	179.9	85.7	132.0	148.3	57.8	102.3
03300	CAST-IN-PLACE CONCRETE	98.2	75.8	89.0	107.0	44.0	81.1	98.1	42.5	75.3	98.1	55.3	80.5	174.7	79.0	135.4	144.0	55.3	107.5
03	CONCRETE	110.2	56.5	83.5	111.4	43.5	77.7	104.5	39.6	72.3	104.8	46.0	75.6	156.2	81.1	118.9	139.7	57.2	97.8
04	MASONRY	105.4	40.4	65.1	106.1	34.5	61.7	105.4	31.1	59.3	105.5	36.9	62.9	170.2	88.9	119.8	158.7	59.0	96.9
05	METALS	94.5	62.4	84.2	94.4	61.6	83.9	94.5	61.2	83.9	95.0	61.4	84.2	107.8	86.2	100.9	105.3	67.6	93.3
06	WOOD & PLASTICS	88.8	40.5	63.6	92.1	37.2	63.5	92.1	28.7	59.1	92.1	31.8	60.7	120.2	78.9	98.7	93.4	57.2	74.5
07	THERMAL & MOISTURE PROTECTION	100.7	54.1	81.1	100.3	48.3	78.4	100.0	44.1	76.4	99.6	48.9	78.2	109.5	82.7	98.2	103.8	58.3	84.6
08	DOORS & WINDOWS	100.3	44.3	85.6	98.7	41.4	83.7	100.3	36.8	83.7	93.7	38.5	79.2	91.1	79.0	87.9	86.1	50.9	76.9
09200	PLASTER & GYPSUM BOARD	80.1	38.7	54.4	80.9	35.3	52.5	80.9	26.6	47.2	81.3	29.7	49.3	158.6	78.4	108.8	171.1	56.1	99.7
095,098	CEILINGS & ACOUSTICAL TREATMENT	100.5	38.7	63.7	100.5	35.3	61.6	100.5	26.6	56.4	102.2	29.7	59.0	95.8	78.4	85.4	96.6	56.1	72.4
09600	FLOORING	96.9	45.4	82.9	98.8	40.4	83.0	98.8	40.4	83.0	98.8	49.9	85.6	132.9	87.9	120.7	113.7	42.0	94.3
097,099	WALL FINISHES, PAINTS & COATINGS	102.2	54.9	73.3	102.2	33.3	60.2	102.2	33.3	60.2	102.2	33.3	60.2	109.8	81.6	92.6	109.1	46.9	71.1
09	FINISHES	95.6	43.3	68.4	96.0	36.8	65.2	95.7	31.3	62.2	96.2	36.0	64.9	120.4	82.1	100.5	115.4	53.2	83.0
10 - 14	TOTAL DIV. 10000 - 14000	100.0	79.2	95.6	100.0	68.7	93.3	100.0	47.3	88.8	100.0	53.4	90.1	140.0	65.4	124.1	140.0	56.6	122.2
15	MECHANICAL	97.5	58.5	81.4	97.5	56.2	80.4	97.5	54.9	79.9	97.5	57.8	81.1	103.7	78.4	93.2	103.7	55.0	83.5
16	ELECTRICAL	91.5	62.7	77.5	91.5	74.4	83.2	91.5	62.7	77.5	90.5	62.7	77.0	124.7	89.2	107.4	126.1	61.0	94.4
01 - 16	WEIGHTED AVERAGE	98.1	58.7	80.6	98.1	56.0	79.4	97.4	51.4	77.0	96.8	54.6	78.0	119.9	84.1	104.0	115.1	60.9	91.1

CANADA

DIVISION		BRANDON, MANITOBA MAT.	INST.	TOTAL	BRANTFORD, ONTARIO MAT.	INST.	TOTAL	BRIDGEWATER, NS MAT.	INST.	TOTAL	CALGARY, ALBERTA MAT.	INST.	TOTAL	CAP-DE-LA-MADELEINE, PQ MAT.	INST.	TOTAL	CHARLESBOURG, QUEBEC MAT.	INST.	TOTAL
01590	EQUIPMENT RENTAL	.0	102.9	102.9	.0	100.4	100.4	.0	98.6	98.6	.0	107.0	107.0	.0	99.2	99.2	.0	99.2	99.2
02	SITE CONSTRUCTION	111.8	98.1	101.8	114.3	99.2	103.2	98.7	95.5	96.4	116.6	104.8	107.9	93.7	96.5	95.8	93.7	96.5	95.8
03100	CONCRETE FORMS & ACCESSORIES	121.7	65.1	72.9	122.5	87.0	91.9	90.3	64.1	67.7	122.0	74.0	80.6	129.8	78.9	85.9	129.8	78.9	85.9
03200	CONCRETE REINFORCEMENT	161.5	54.3	107.0	174.7	84.3	128.8	148.3	47.6	97.1	161.5	64.2	112.0	148.3	80.6	113.9	148.3	80.6	113.9
03300	CAST-IN-PLACE CONCRETE	157.5	68.9	121.1	170.5	97.8	140.6	175.2	64.3	129.6	191.6	84.3	147.5	137.4	86.6	116.6	137.4	86.6	116.6
03	CONCRETE	138.2	65.1	101.9	153.3	90.3	122.0	150.4	61.8	106.4	161.3	76.4	119.1	133.6	82.1	108.0	133.6	82.1	108.0
04	MASONRY	164.6	61.4	100.6	166.7	93.0	121.0	160.9	65.2	101.6	181.1	71.9	113.4	161.1	79.1	110.2	161.1	79.1	110.2
05	METALS	107.9	73.5	96.9	107.2	87.4	100.8	106.3	70.3	94.8	135.6	83.2	118.9	105.6	83.1	98.4	105.6	83.1	98.4
06	WOOD & PLASTICS	116.1	65.8	89.9	118.4	85.7	101.4	84.5	63.5	73.6	116.2	73.6	94.0	131.5	78.7	104.0	131.5	78.7	104.0
07	THERMAL & MOISTURE PROTECTION	104.2	66.9	88.5	109.4	87.6	100.2	104.4	65.2	87.9	117.5	77.6	100.7	104.4	83.3	95.5	104.4	83.3	95.5
08	DOORS & WINDOWS	91.9	59.1	83.4	90.9	84.5	89.2	82.8	59.0	76.6	91.9	70.6	86.3	91.9	74.0	87.2	91.9	74.0	87.2
09200	PLASTER & GYPSUM BOARD	131.4	64.5	89.8	158.6	85.4	113.1	164.6	62.6	101.3	152.4	72.4	102.7	195.3	78.1	122.5	195.3	78.1	122.5
095,098	CEILINGS & ACOUSTICAL TREATMENT	95.8	64.5	77.1	95.8	85.4	89.6	95.8	62.6	76.0	110.2	72.4	87.7	95.8	78.1	85.2	95.8	78.1	85.2
09600	FLOORING	132.9	62.5	113.8	132.9	87.9	120.7	109.6	59.9	96.2	132.9	82.4	119.2	132.9	88.6	120.9	132.9	88.6	120.9
097,099	WALL FINISHES, PAINTS & COATINGS	109.2	53.3	75.1	109.1	89.6	97.2	109.1	58.2	78.1	109.1	73.8	87.5	109.1	83.3	93.3	109.1	83.3	93.3
09	FINISHES	115.3	63.6	88.4	119.3	87.8	102.9	112.8	62.9	86.8	122.3	75.4	97.9	124.2	81.5	102.0	124.2	81.5	102.0
10 - 14	TOTAL DIV. 10000 - 14000	140.0	59.3	122.8	140.0	67.6	124.6	140.0	58.1	122.5	140.0	79.8	127.2	140.0	73.8	125.9	140.0	73.8	125.9
15	MECHANICAL	103.7	65.9	88.0	103.7	81.0	94.3	103.7	66.4	88.2	101.4	74.7	90.3	104.1	71.8	90.7	104.1	71.8	90.7
16	ELECTRICAL	125.4	68.6	97.8	124.7	88.7	107.2	128.2	63.5	96.7	117.6	84.2	101.4	120.5	71.6	96.7	120.5	71.6	96.7
01 - 16	WEIGHTED AVERAGE	117.0	68.5	95.5	119.1	87.6	105.1	116.5	67.1	94.6	124.3	79.6	104.4	115.9	78.9	99.5	115.9	78.9	99.5

COST INDEXES

743

CANADA

DIVISION		CHARLOTTETOWN, PEI			CHICOUTIMI, QUEBEC			CORNER BROOK, NFLD			CORNWALL, ONTARIO			DALHOUSIE, NB			DARTMOUTH, NOVA SCOTIA		
		MAT.	INST.	TOTAL	MAT.	INST.	TOTAL	MAT.	INST.	TOTAL	MAT.	INST.	TOTAL	MAT.	INST.	TOTAL	MAT.	INST.	TOTAL
01590	EQUIPMENT RENTAL	.0	98.6	98.6	.0	99.2	99.2	.0	98.6	98.6	.0	100.4	100.4	.0	98.6	98.6	.0	98.6	98.6
02	SITE CONSTRUCTION	106.7	93.8	97.2	93.0	96.5	95.6	114.3	94.2	99.6	112.1	98.8	102.3	96.0	94.3	94.8	106.9	95.5	98.5
03100	CONCRETE FORMS & ACCESSORIES	90.6	52.1	57.4	129.8	78.7	85.7	99.5	55.1	61.2	119.8	80.3	85.8	97.6	57.0	62.6	90.4	64.1	67.8
03200	CONCRETE REINFORCEMENT	148.3	47.7	97.2	111.9	80.5	96.0	148.3	49.5	98.1	174.7	84.1	128.7	148.3	57.8	102.3	154.9	47.6	100.4
03300	CAST-IN-PLACE CONCRETE	166.1	54.6	120.3	135.6	86.6	115.5	179.6	62.6	131.5	153.3	89.3	127.0	144.0	55.3	107.5	177.1	64.3	130.7
03	CONCRETE	154.4	52.9	104.0	126.9	82.0	104.6	164.7	57.5	111.5	144.8	84.4	114.8	137.9	57.2	97.8	152.4	61.8	107.4
04	MASONRY	160.4	56.9	96.3	160.1	79.1	109.9	161.3	58.3	97.4	165.4	85.4	115.8	158.6	59.0	96.9	172.7	65.2	106.1
05	METALS	106.4	63.9	92.8	105.2	83.0	98.1	108.2	68.3	95.5	107.2	86.0	100.4	105.3	67.6	93.3	108.4	70.3	96.2
06	WOOD & PLASTICS	84.8	51.6	67.5	131.5	78.7	104.0	94.6	54.2	73.6	116.9	79.7	97.5	93.4	57.2	74.5	84.5	63.5	73.6
07	THERMAL & MOISTURE PROTECTION	104.8	56.0	84.2	104.3	83.3	95.5	106.5	57.7	86.0	109.2	81.7	97.6	106.1	58.3	85.9	104.4	65.2	87.9
08	DOORS & WINDOWS	94.1	45.5	81.4	91.5	68.2	85.4	98.0	51.9	86.0	91.9	78.7	88.5	86.1	50.9	76.9	82.8	59.0	76.6
09200	PLASTER & GYPSUM BOARD	168.9	50.4	95.3	195.3	78.1	122.5	171.4	53.1	97.9	230.5	79.2	136.5	171.1	56.1	99.7	167.9	62.6	102.5
095,098	CEILINGS & ACOUSTICAL TREATMENT	111.1	50.4	74.9	95.8	78.1	85.2	97.5	53.1	71.0	99.2	79.2	87.3	96.6	56.1	72.4	108.5	62.6	81.1
09600	FLOORING	109.7	56.0	95.2	132.9	88.6	120.9	114.6	50.9	97.4	132.9	86.7	120.4	113.7	42.0	94.3	109.6	59.9	96.2
097,099	WALL FINISHES, PAINTS & COATINGS	109.1	39.0	66.3	109.1	83.3	93.3	109.1	56.1	76.8	109.1	83.5	93.5	109.1	46.9	71.1	109.1	58.2	78.1
09	FINISHES	118.0	51.7	83.5	124.2	81.5	102.0	116.7	54.5	84.3	131.1	82.2	105.6	115.4	53.2	83.0	116.5	62.9	88.6
10 - 14	TOTAL DIV. 10000 - 14000	140.0	56.0	122.1	140.0	73.8	125.9	140.0	56.7	122.2	140.0	65.0	124.0	140.0	56.6	122.2	140.0	58.1	122.5
15	MECHANICAL	103.7	50.4	81.5	103.7	71.8	90.4	103.7	56.5	84.1	104.1	78.9	93.6	103.7	55.0	83.5	103.7	66.4	88.2
16	ELECTRICAL	120.0	51.6	86.8	120.5	71.6	96.7	120.4	57.4	89.8	123.0	89.6	106.8	125.4	61.0	94.1	125.1	63.5	95.2
01 - 16	WEIGHTED AVERAGE	117.7	56.9	90.7	115.0	78.6	98.8	119.8	60.9	93.7	119.1	84.3	103.6	115.1	60.9	91.1	117.8	67.1	95.3

CANADA

DIVISION		EDMONTON, ALBERTA			FORT MCMURRAY, ALBERTA			FREDERICTON, NB			GATINEAU, QUEBEC			GRANBY, QUEBEC			HALIFAX, NOVA SCOTIA		
		MAT.	INST.	TOTAL	MAT.	INST.	TOTAL	MAT.	INST.	TOTAL	MAT.	INST.	TOTAL	MAT.	INST.	TOTAL	MAT.	INST.	TOTAL
01590	EQUIPMENT RENTAL	.0	107.0	107.0	.0	103.7	103.7	.0	98.6	98.6	.0	99.2	99.2	.0	99.2	99.2	.0	98.6	98.6
02	SITE CONSTRUCTION	120.3	104.8	108.9	111.6	101.2	104.0	95.4	94.3	94.6	93.4	96.5	95.7	94.0	96.5	95.9	99.1	95.5	96.5
03100	CONCRETE FORMS & ACCESSORIES	119.7	74.0	80.3	119.2	73.9	80.1	120.1	57.3	65.9	129.8	78.7	85.8	129.8	78.7	85.8	90.5	64.2	67.8
03200	CONCRETE REINFORCEMENT	161.5	64.2	112.0	161.5	64.2	112.0	149.2	57.9	102.8	157.0	80.5	118.1	157.0	80.5	118.1	154.9	47.8	100.4
03300	CAST-IN-PLACE CONCRETE	207.0	84.3	156.5	157.5	83.3	127.0	137.1	55.3	103.4	135.6	86.6	115.5	140.1	86.6	118.1	178.5	66.1	132.3
03	CONCRETE	168.6	76.4	122.8	144.6	76.0	110.5	136.4	57.4	97.2	134.1	82.0	108.2	136.3	82.0	109.3	153.1	62.3	108.1
04	MASONRY	182.3	71.9	113.9	161.6	71.8	105.9	158.1	60.4	97.6	161.0	79.1	110.2	161.3	79.1	110.3	172.7	67.5	107.5
05	METALS	136.4	83.2	119.4	107.9	83.1	100.0	105.2	68.0	93.4	105.6	83.0	98.4	105.6	83.0	98.4	120.8	70.5	104.7
06	WOOD & PLASTICS	112.7	73.6	92.3	112.7	73.5	92.3	119.2	57.2	86.9	131.5	78.7	104.0	131.5	78.7	104.0	84.5	63.5	73.6
07	THERMAL & MOISTURE PROTECTION	117.7	77.6	100.8	114.9	76.8	98.8	106.4	59.2	86.5	104.4	83.3	95.5	104.3	83.3	95.5	104.8	65.8	88.3
08	DOORS & WINDOWS	91.9	70.6	86.3	91.9	70.5	86.3	86.0	49.9	76.6	91.9	69.8	86.1	91.9	69.8	86.1	82.8	59.0	76.6
09200	PLASTER & GYPSUM BOARD	154.5	72.4	103.5	137.2	72.4	96.9	189.2	56.1	106.5	146.7	78.1	104.1	149.2	78.1	105.0	167.9	62.6	102.5
095,098	CEILINGS & ACOUSTICAL TREATMENT	125.6	72.4	93.9	95.8	72.4	81.8	105.1	56.1	75.9	95.8	78.1	85.2	95.8	78.1	85.2	108.5	62.6	81.1
09600	FLOORING	132.9	82.4	119.2	132.9	82.4	119.2	128.0	42.0	104.7	132.9	88.6	120.9	132.9	88.6	120.9	109.6	59.9	96.2
097,099	WALL FINISHES, PAINTS & COATINGS	109.2	73.8	87.6	109.2	73.8	87.6	109.1	59.8	79.0	109.1	83.3	93.3	109.1	83.3	93.3	109.1	58.2	78.1
09	FINISHES	126.5	75.4	99.9	116.4	75.4	95.0	124.5	54.7	88.1	116.8	81.5	98.4	117.2	81.5	98.6	116.5	62.9	88.6
10 - 14	TOTAL DIV. 10000 - 14000	140.0	79.8	127.2	140.0	79.5	127.1	140.0	56.6	122.2	140.0	73.8	125.9	140.0	73.8	125.9	140.0	58.1	122.5
15	MECHANICAL	101.5	74.7	90.4	103.7	74.7	91.7	103.7	62.1	86.4	104.1	71.8	90.7	103.7	71.8	90.4	101.0	66.4	86.7
16	ELECTRICAL	118.3	84.2	101.8	114.4	84.2	99.7	133.6	76.0	105.6	120.5	71.6	96.7	121.7	71.6	97.3	127.4	68.3	98.7
01 - 16	WEIGHTED AVERAGE	125.9	79.6	105.3	116.6	79.1	100.0	117.0	64.9	93.8	115.3	78.7	99.1	115.7	78.7	99.3	119.3	68.2	96.6

CANADA

DIVISION		HAMILTON, ONTARIO			HULL, QUEBEC			JOLIETTE, QUEBEC			KAMLOOPS, BC			KINGSTON, ONTARIO			KITCHENER, ONTARIO		
		MAT.	INST.	TOTAL	MAT.	INST.	TOTAL	MAT.	INST.	TOTAL	MAT.	INST.	TOTAL	MAT.	INST.	TOTAL	MAT.	INST.	TOTAL
01590	EQUIPMENT RENTAL	.0	103.1	103.1	.0	99.2	99.2	.0	99.2	99.2	.0	103.5	103.5	.0	103.1	103.1	.0	103.0	103.0
02	SITE CONSTRUCTION	112.7	103.2	105.7	93.4	96.5	95.7	94.2	96.5	95.9	117.3	100.4	104.9	112.1	103.0	105.4	101.9	102.6	102.4
03100	CONCRETE FORMS & ACCESSORIES	121.4	84.1	89.2	129.8	78.7	85.8	129.8	78.9	85.9	117.3	81.5	86.4	119.9	80.5	85.9	115.6	77.1	82.4
03200	CONCRETE REINFORCEMENT	175.4	85.7	129.8	157.0	80.5	118.1	148.3	80.6	113.9	116.5	78.4	97.1	174.7	84.1	128.7	110.3	85.6	97.8
03300	CAST-IN-PLACE CONCRETE	157.2	91.8	130.3	135.6	86.6	115.5	141.2	86.6	118.8	122.2	90.7	109.2	153.3	89.3	127.0	146.7	74.0	116.8
03	CONCRETE	146.9	87.2	117.2	134.1	82.0	108.2	135.4	82.1	108.9	140.1	84.2	112.3	144.8	84.4	114.8	124.8	77.9	101.5
04	MASONRY	168.1	88.5	118.7	161.0	79.1	110.2	161.4	79.1	110.3	168.4	86.1	117.4	172.0	85.4	118.3	163.6	84.5	114.6
05	METALS	139.3	86.8	122.5	105.6	83.0	98.4	105.6	83.1	98.4	107.8	83.3	100.0	108.7	86.0	101.5	116.3	86.2	106.7
06	WOOD & PLASTICS	117.4	82.9	99.5	131.5	78.7	104.0	131.5	78.7	104.0	98.1	80.0	88.7	116.9	79.8	97.6	110.9	75.5	92.4
07	THERMAL & MOISTURE PROTECTION	110.0	85.8	99.8	104.3	83.3	95.5	104.3	83.3	95.5	121.5	80.5	104.2	109.2	82.8	98.1	108.6	81.5	97.2
08	DOORS & WINDOWS	91.9	82.7	89.5	91.9	69.8	86.1	91.9	74.0	87.2	87.8	78.1	85.3	91.9	78.5	88.4	83.4	77.1	81.8
09200	PLASTER & GYPSUM BOARD	191.0	82.5	123.6	146.7	78.1	104.1	195.3	78.1	122.5	140.7	79.1	102.4	234.6	79.3	138.1	157.6	74.9	106.2
095,098	CEILINGS & ACOUSTICAL TREATMENT	113.7	82.5	95.1	95.8	78.1	85.2	95.8	78.1	85.2	95.8	79.1	85.8	115.4	79.3	93.9	107.7	74.9	88.1
09600	FLOORING	132.9	87.9	120.7	132.9	88.6	120.9	132.9	88.6	120.9	130.0	49.5	108.2	132.9	86.7	120.4	128.4	87.9	117.5
097,099	WALL FINISHES, PAINTS & COATINGS	109.1	85.8	101.0	109.1	83.3	93.3	109.1	83.3	93.3	109.1	76.2	89.0	109.1	91.0	98.1	109.1	87.7	96.1
09	FINISHES	128.6	86.2	106.5	116.8	81.5	98.4	124.2	81.5	102.0	117.3	75.3	95.4	135.7	83.1	108.3	120.1	80.2	99.4
10 - 14	TOTAL DIV. 10000 - 14000	140.0	85.3	128.3	140.0	73.8	125.9	140.0	73.8	125.9	140.0	82.9	127.8	140.0	65.0	124.0	140.0	83.3	127.9
15	MECHANICAL	101.5	82.4	93.6	103.7	71.8	90.4	103.7	71.8	90.5	103.7	74.8	91.7	104.1	79.0	93.7	101.0	78.2	91.8
16	ELECTRICAL	130.4	91.7	111.6	124.2	71.6	98.6	121.7	71.6	97.3	126.8	80.8	104.4	123.0	88.4	106.1	125.2	88.5	107.4
01 - 16	WEIGHTED AVERAGE	124.5	87.9	108.2	115.6	78.7	99.2	116.2	78.9	99.7	117.8	81.8	101.8	120.1	84.7	104.4	115.6	83.7	101.4

CANADA

DIVISION		LAVAL, QUEBEC			LETHBRIDGE, ALBERTA			LLOYDMINSTER, ALBERTA			LONDON, ONTARIO			MEDICINE HAT, ALBERTA			MONCTON, NEW BRUNSWICK		
		MAT.	INST.	TOTAL	MAT.	INST.	TOTAL	MAT.	INST.	TOTAL	MAT.	INST.	TOTAL	MAT.	INST.	TOTAL	MAT.	INST.	TOTAL
01590	EQUIPMENT RENTAL	.0	99.2	99.2	.0	103.7	103.7	.0	103.7	103.7	.0	103.2	103.2	.0	103.7	103.7	.0	98.6	98.6
02	SITE CONSTRUCTION	94.0	96.5	95.9	112.2	101.2	104.1	111.6	101.2	104.0	112.6	103.0	105.5	110.3	101.2	103.6	95.3	94.3	94.6
03100	CONCRETE FORMS & ACCESSORIES	130.0	78.7	85.8	121.5	73.9	80.4	119.2	73.9	80.1	127.1	79.8	86.3	121.5	73.9	80.4	97.6	57.0	62.6
03200	CONCRETE REINFORCEMENT	157.0	80.5	118.1	161.5	64.2	112.0	161.5	64.2	112.0	135.2	84.2	109.3	161.5	64.2	112.0	148.3	57.8	102.3
03300	CAST-IN-PLACE CONCRETE	140.1	86.6	118.1	172.4	83.3	135.8	157.5	83.3	127.0	157.2	89.5	129.4	157.5	83.3	127.0	138.7	55.3	104.4
03	CONCRETE	136.3	82.0	109.3	152.0	75.9	114.2	144.6	76.0	110.5	140.9	84.2	112.7	144.8	75.9	110.6	135.3	57.2	96.5
04	MASONRY	161.3	79.1	110.3	162.7	71.8	106.4	161.6	71.8	105.9	168.0	85.8	117.0	161.6	71.8	105.9	158.3	59.0	96.7
05	METALS	105.6	83.0	98.4	107.9	83.0	100.0	107.9	83.1	100.0	124.5	86.3	112.3	107.9	83.0	100.0	105.3	67.6	93.3
06	WOOD & PLASTICS	131.5	78.7	104.0	116.2	73.5	94.0	112.7	73.5	92.3	117.4	78.1	97.0	116.2	73.5	94.0	93.4	57.2	74.5
07	THERMAL & MOISTURE PROTECTION	104.7	83.3	95.7	114.9	76.7	98.9	113.8	76.8	98.2	112.6	84.0	100.6	117.2	76.7	100.2	106.1	58.3	85.9
08	DOORS & WINDOWS	91.9	69.8	86.1	91.9	70.5	86.3	91.9	70.5	86.3	92.9	79.1	89.3	91.9	70.5	86.3	86.1	50.9	76.9
09200	PLASTER & GYPSUM BOARD	151.6	78.1	105.9	139.0	72.4	97.6	137.2	72.4	96.9	191.5	77.6	120.7	139.0	72.4	97.6	171.1	56.1	99.7
095,098	CEILINGS & ACOUSTICAL TREATMENT	95.8	78.1	85.2	95.8	72.4	81.8	95.8	72.4	81.8	115.4	77.6	92.8	95.8	72.4	81.8	96.6	56.1	72.4
09600	FLOORING	132.9	88.6	120.9	132.9	82.4	119.2	132.9	82.4	119.2	134.9	87.9	122.2	132.9	82.4	119.2	113.7	42.0	94.3
097,099	WALL FINISHES, PAINTS & COATINGS	109.1	83.3	93.3	109.1	73.8	87.5	109.2	73.8	87.6	109.1	95.8	101.0	109.1	73.8	87.5	109.1	46.9	71.1
09	FINISHES	117.5	81.5	98.8	116.6	75.4	95.1	116.4	75.4	95.0	129.8	83.0	105.4	116.6	75.4	95.1	115.4	53.2	83.0
10 - 14	TOTAL DIV. 10000 - 14000	140.0	73.8	125.9	140.0	79.5	127.1	140.0	79.5	127.1	140.0	84.1	128.1	140.0	79.5	127.1	140.0	56.6	122.2
15	MECHANICAL	101.0	71.8	88.9	103.7	72.2	90.6	104.1	74.7	91.9	101.6	79.0	92.2	103.7	72.2	90.6	103.7	55.0	83.5
16	ELECTRICAL	122.1	71.6	97.5	114.4	84.2	99.7	114.4	84.2	99.7	130.7	88.7	110.3	114.4	84.2	99.7	130.6	61.0	96.7
01 - 16	WEIGHTED AVERAGE	115.2	78.7	99.0	117.6	78.6	100.3	116.7	79.1	100.0	121.9	85.3	105.6	116.7	78.6	99.8	115.4	60.9	91.2

CANADA

DIVISION		MONTREAL, QUEBEC			MOOSE JAW, SASKATCHEWAN			NEWCASTLE, NB			NEW GLASGOW, NOVA SCOTIA			NORTH BAY, ONTARIO			OSHAWA, ONTARIO		
		MAT.	INST.	TOTAL	MAT.	INST.	TOTAL	MAT.	INST.	TOTAL	MAT.	INST.	TOTAL	MAT.	INST.	TOTAL	MAT.	INST.	TOTAL
01590	EQUIPMENT RENTAL	.0	101.0	101.0	.0	98.6	98.6	.0	98.6	98.6	.0	98.6	98.6	.0	100.4	100.4	.0	103.0	103.0
02	SITE CONSTRUCTION	93.7	96.8	96.0	111.4	94.5	99.0	96.0	94.3	94.8	99.0	95.5	96.4	114.3	98.5	102.7	114.2	102.8	105.8
03100	CONCRETE FORMS & ACCESSORIES	130.3	79.0	86.1	101.9	54.2	60.8	97.6	57.0	62.6	90.3	64.1	67.7	122.5	78.1	84.2	122.6	80.6	86.4
03200	CONCRETE REINFORCEMENT	159.6	80.6	119.4	113.9	63.3	88.2	148.3	57.8	102.3	148.3	47.6	97.1	174.7	83.6	128.4	174.7	86.2	129.8
03300	CAST-IN-PLACE CONCRETE	137.5	88.2	117.2	152.8	63.5	116.1	144.0	55.3	107.5	177.1	64.3	130.7	170.5	77.6	132.3	169.7	79.2	132.5
03	CONCRETE	135.5	82.7	109.3	127.2	59.9	93.8	137.9	57.2	97.8	151.3	61.8	106.9	153.3	79.3	116.5	152.9	81.4	117.4
04	MASONRY	164.5	79.1	111.5	160.1	58.5	97.1	158.7	59.0	96.9	161.1	65.2	101.7	166.7	81.8	114.0	166.7	87.8	117.7
05	METALS	125.0	83.5	111.7	104.7	70.8	93.9	105.3	67.6	93.3	106.3	70.3	94.8	107.2	85.8	100.3	107.2	86.5	100.6
06	WOOD & PLASTICS	131.5	79.0	104.2	95.2	52.8	73.2	93.4	57.2	74.5	84.5	63.5	73.6	118.4	78.4	97.6	118.4	78.9	97.9
07	THERMAL & MOISTURE PROTECTION	105.3	83.9	96.3	103.6	59.9	85.2	106.1	58.3	85.9	104.4	65.2	87.9	109.4	79.6	96.8	109.6	82.0	98.0
08	DOORS & WINDOWS	91.9	69.9	86.2	86.9	51.9	77.8	86.1	50.9	76.9	82.8	59.0	76.6	90.9	76.5	87.2	90.9	80.2	88.1
09200	PLASTER & GYPSUM BOARD	154.6	78.1	107.1	136.9	51.6	83.9	171.1	56.1	99.7	164.6	62.6	101.3	158.6	77.9	108.5	159.9	78.4	109.3
095,098	CEILINGS & ACOUSTICAL TREATMENT	107.7	78.1	90.0	95.8	51.6	69.4	96.6	56.1	72.4	95.8	62.6	76.0	95.8	77.9	85.1	100.9	78.4	87.5
09600	FLOORING	132.9	88.6	120.9	120.0	56.3	102.7	113.7	42.0	94.3	109.6	59.9	96.2	132.9	86.7	120.4	132.9	90.2	121.3
097,099	WALL FINISHES, PAINTS & COATINGS	109.1	83.3	93.3	109.1	60.7	79.6	109.1	46.9	71.1	109.1	58.2	78.1	109.1	82.8	93.1	109.1	100.6	103.9
09	FINISHES	121.0	81.7	100.5	112.5	55.1	82.6	115.4	53.2	83.0	112.8	62.9	86.8	119.3	80.6	99.1	120.7	84.7	102.0
10 - 14	TOTAL DIV. 10000 - 14000	140.0	74.6	126.0	140.0	56.0	122.1	140.0	56.6	122.2	140.0	58.1	122.5	140.0	63.8	123.7	140.0	84.5	128.2
15	MECHANICAL	101.6	71.9	89.3	104.1	59.7	85.7	103.7	55.0	83.5	103.7	66.4	88.2	103.7	77.1	92.6	101.0	80.4	92.5
16	ELECTRICAL	121.1	71.6	97.0	126.0	61.9	94.8	125.4	61.0	94.1	121.1	63.5	93.1	124.7	89.5	107.6	126.3	89.2	108.3
01 - 16	WEIGHTED AVERAGE	118.5	79.0	101.0	114.2	62.9	91.4	115.1	60.9	91.1	115.8	67.1	94.2	119.1	82.4	102.8	118.7	85.7	104.1

CANADA

DIVISION		OTTAWA, ONTARIO			OWEN SOUND, ONTARIO			PETERBOROUGH, ONTARIO			PORTAGE LA PRAIRIE, MB			PRINCE ALBERT, SS			PRINCE GEORGE, BC		
		MAT.	INST.	TOTAL	MAT.	INST.	TOTAL	MAT.	INST.	TOTAL	MAT.	INST.	TOTAL	MAT.	INST.	TOTAL	MAT.	INST.	TOTAL
01590	EQUIPMENT RENTAL	.0	103.0	103.0	.0	100.4	100.4	.0	100.4	100.4	.0	102.9	102.9	.0	98.6	98.6	.0	103.5	103.5
02	SITE CONSTRUCTION	112.9	102.8	105.5	116.6	98.6	103.4	114.3	98.8	102.9	112.2	98.1	101.9	107.6	94.6	98.1	121.3	100.4	106.0
03100	CONCRETE FORMS & ACCESSORIES	120.1	80.6	86.1	122.6	76.7	83.0	122.5	79.0	85.0	121.7	65.1	72.9	119.0	54.1	60.7	108.0	81.5	85.1
03200	CONCRETE REINFORCEMENT	173.8	84.1	128.2	179.9	85.6	132.0	174.7	84.1	128.7	161.5	54.3	107.0	119.0	63.2	90.7	116.5	78.4	97.1
03300	CAST-IN-PLACE CONCRETE	159.6	89.4	130.7	174.7	73.8	133.3	170.5	79.0	132.9	157.5	68.9	121.1	145.1	63.5	111.6	153.3	90.7	127.5
03	CONCRETE	147.7	84.6	116.3	156.2	77.7	117.2	153.3	80.3	117.0	138.2	65.1	101.9	124.3	59.8	92.3	154.4	84.2	119.6
04	MASONRY	168.1	85.4	116.8	170.2	86.8	118.5	166.7	87.8	117.8	164.6	61.4	100.6	159.7	58.5	97.0	170.7	86.1	118.3
05	METALS	116.4	86.3	106.8	107.8	86.0	100.9	107.2	86.2	100.5	107.9	73.5	96.9	104.8	70.7	93.9	107.8	83.3	100.0
06	WOOD & PLASTICS	116.9	79.7	97.5	120.2	75.5	96.9	118.4	77.2	97.0	116.1	65.8	89.9	95.2	52.8	73.2	98.1	80.0	88.7
07	THERMAL & MOISTURE PROTECTION	110.2	82.6	98.6	109.5	80.3	97.2	109.4	83.9	98.7	104.2	66.9	88.5	103.5	59.0	84.8	116.7	80.5	101.5
08	DOORS & WINDOWS	91.9	79.9	88.8	91.1	75.9	87.1	90.9	78.1	87.6	91.9	59.1	83.4	85.5	51.9	76.7	87.8	78.1	85.5
09200	PLASTER & GYPSUM BOARD	232.4	79.2	137.2	158.6	74.9	106.6	158.6	76.7	107.7	131.4	64.5	89.8	136.9	51.6	83.9	138.2	79.1	101.5
095,098	CEILINGS & ACOUSTICAL TREATMENT	106.8	79.2	90.3	95.8	74.9	83.3	95.8	76.7	84.4	95.8	64.5	77.1	95.8	51.6	69.4	95.8	79.1	85.8
09600	FLOORING	132.9	86.7	120.4	132.9	87.9	120.7	132.9	86.7	120.4	132.9	62.5	113.8	120.0	56.3	102.7	127.1	49.5	106.1
097,099	WALL FINISHES, PAINTS & COATINGS	109.1	89.6	97.2	109.8	81.6	92.6	109.1	84.9	94.3	109.2	53.3	75.1	109.1	51.8	74.1	109.1	76.2	89.0
09	FINISHES	133.3	82.9	107.0	120.4	79.5	99.1	119.3	81.2	99.4	115.4	63.6	88.4	112.5	54.1	82.0	116.1	75.3	94.8
10 - 14	TOTAL DIV. 10000 - 14000	140.0	82.3	127.7	140.0	64.2	123.8	140.0	65.2	124.1	140.0	59.3	122.8	140.0	56.0	122.1	140.0	82.9	127.8
15	MECHANICAL	101.6	79.3	92.4	103.7	77.2	92.7	103.7	80.2	93.9	103.7	65.9	88.0	104.1	59.7	85.7	103.7	74.8	91.7
16	ELECTRICAL	122.1	89.7	106.3	126.6	88.3	108.0	124.7	89.2	107.4	125.4	68.6	97.8	126.0	61.9	94.8	124.3	80.8	103.2
01 - 16	WEIGHTED AVERAGE	120.5	85.5	105.0	120.1	82.5	103.4	119.1	84.1	103.5	117.0	68.5	95.5	113.7	62.7	91.1	119.1	81.8	102.5

COST INDEXES

CANADA

DIVISION		QUEBEC, QUEBEC			RED DEER, ALBERTA			REGINA, SASKATCHEWAN			RIMOUSKI, QUEBEC			ROUYN-NORANDA, QUEBEC			SAINT HYACINTHE, QUEBEC		
		MAT.	INST.	TOTAL	MAT.	INST.	TOTAL	MAT.	INST.	TOTAL	MAT.	INST.	TOTAL	MAT.	INST.	TOTAL	MAT.	INST.	TOTAL
01590	EQUIPMENT RENTAL	.0	101.4	101.4	.0	103.7	103.7	.0	98.6	98.6	.0	99.2	99.2	.0	99.2	99.2	.0	99.2	99.2
02	SITE CONSTRUCTION	95.5	96.9	96.5	110.3	101.2	103.6	112.8	94.5	99.4	93.9	96.5	95.8	93.4	96.5	95.7	94.0	96.5	95.9
03100	CONCRETE FORMS & ACCESSORIES	130.2	79.3	86.3	135.1	73.9	82.3	102.0	54.2	60.8	129.8	78.7	85.7	129.8	78.7	85.8	129.8	78.7	85.8
03200	CONCRETE REINFORCEMENT	148.3	80.6	113.9	161.5	64.2	112.0	126.1	63.3	94.2	111.9	80.5	96.0	157.0	80.5	118.1	157.0	80.5	118.1
03300	CAST-IN-PLACE CONCRETE	151.5	88.6	125.7	157.5	83.3	127.0	162.8	63.5	122.0	142.7	86.6	119.6	135.6	86.6	115.5	140.1	86.6	118.1
03	CONCRETE	140.4	82.9	111.9	145.7	75.9	111.1	134.0	59.9	97.2	130.3	82.0	106.3	134.1	82.0	108.2	136.3	82.0	109.3
04	MASONRY	165.4	79.1	111.9	161.6	71.8	105.9	168.3	58.9	100.5	160.6	79.1	110.1	161.0	79.1	110.2	161.3	79.1	110.3
05	METALS	120.7	83.8	108.9	107.9	83.0	100.0	105.6	70.8	94.5	105.2	83.0	98.1	105.6	83.0	98.4	105.6	83.0	98.4
06	WOOD & PLASTICS	132.2	79.1	104.5	116.2	73.5	94.0	95.2	52.8	73.2	131.5	78.7	104.0	131.5	78.7	104.0	131.5	78.7	104.0
07	THERMAL & MOISTURE PROTECTION	104.9	84.1	96.1	124.6	76.7	104.5	105.5	60.0	86.4	104.3	83.3	95.5	104.3	83.3	95.5	104.7	83.3	95.7
08	DOORS & WINDOWS	91.9	76.8	88.0	91.9	70.5	86.3	86.9	51.9	77.8	91.5	68.2	85.4	91.9	69.8	86.1	91.9	69.8	86.1
09200	PLASTER & GYPSUM BOARD	197.2	78.1	123.2	139.0	72.4	97.6	168.0	51.6	95.7	194.7	78.1	122.2	146.1	78.1	103.8	150.9	78.1	105.7
095,098	CEILINGS & ACOUSTICAL TREATMENT	97.5	78.1	85.9	95.8	72.4	81.8	122.2	51.6	80.1	93.2	78.1	84.2	93.2	78.1	84.2	93.2	78.1	84.2
09600	FLOORING	132.9	88.6	120.9	134.9	82.4	120.7	120.0	56.3	102.7	132.9	88.6	120.9	132.9	88.6	120.9	132.9	88.6	120.9
097,099	WALL FINISHES, PAINTS & COATINGS	109.1	83.3	93.3	109.1	73.8	87.5	109.1	60.7	79.6	109.1	83.3	93.3	109.1	83.3	93.3	109.1	83.3	93.3
09	FINISHES	125.0	81.7	102.4	117.2	75.4	95.4	124.0	55.1	88.1	123.5	81.5	101.6	116.1	81.5	98.1	116.8	81.5	98.4
10 - 14	TOTAL DIV. 10000 - 14000	140.0	74.7	126.1	140.0	79.5	127.1	140.0	56.0	122.1	140.0	73.8	125.9	140.0	73.8	125.9	140.0	73.8	125.9
15	MECHANICAL	101.5	71.9	89.2	103.7	72.2	90.6	101.3	59.7	84.0	103.7	71.8	90.4	103.7	71.8	90.4	101.0	71.8	88.9
16	ELECTRICAL	125.9	71.6	99.5	114.4	84.2	99.7	126.4	61.9	95.0	121.7	71.6	97.3	121.7	71.6	97.3	122.1	71.6	97.5
01 - 16	WEIGHTED AVERAGE	119.4	79.4	101.6	117.1	78.6	100.0	116.1	62.9	92.5	115.5	78.6	99.1	115.3	78.7	99.0	115.1	78.7	98.9

CANADA

DIVISION		SAINT JOHN, N B			SARNIA, ONTARIO			SASKATOON, SASKATCHEWAN			SAULT STE MARIE, ON			SHERBROOKE, QUEBEC			SOREL, QUEBEC		
		MAT.	INST.	TOTAL	MAT.	INST.	TOTAL	MAT.	INST.	TOTAL	MAT.	INST.	TOTAL	MAT.	INST.	TOTAL	MAT.	INST.	TOTAL
01590	EQUIPMENT RENTAL	.0	98.6	98.6	.0	100.4	100.4	.0	98.6	98.6	.0	100.4	100.4	.0	99.2	99.2	.0	99.2	99.2
02	SITE CONSTRUCTION	96.1	94.3	94.8	112.6	98.8	102.5	107.9	94.6	98.1	102.3	98.5	99.5	94.0	96.5	95.9	94.2	96.5	95.9
03100	CONCRETE FORMS & ACCESSORIES	120.1	57.3	65.9	120.9	85.5	90.4	102.0	54.2	60.7	108.5	78.4	82.5	129.8	78.7	85.8	129.8	78.9	85.9
03200	CONCRETE REINFORCEMENT	148.3	57.9	102.3	123.5	85.5	104.2	119.0	63.3	90.7	112.2	83.7	97.7	157.0	80.5	118.1	148.3	80.6	113.9
03300	CAST-IN-PLACE CONCRETE	141.9	55.3	106.3	157.3	90.7	129.9	148.0	63.5	113.2	141.2	77.7	115.1	140.1	86.6	118.1	141.2	86.6	118.8
03	CONCRETE	138.6	57.4	98.3	138.6	87.4	113.1	125.7	59.8	93.0	122.0	79.5	100.9	136.3	82.0	109.3	135.4	82.1	108.9
04	MASONRY	176.5	60.4	104.5	177.4	90.3	123.4	168.0	58.9	100.3	163.2	81.8	112.7	161.3	79.1	110.3	161.4	79.1	110.3
05	METALS	105.2	68.0	93.4	107.1	86.5	100.5	105.7	70.7	94.5	106.5	86.1	100.0	105.6	83.0	98.4	105.6	83.1	98.4
06	WOOD & PLASTICS	119.2	57.2	86.9	117.4	84.8	100.4	93.8	52.8	72.5	103.4	78.4	90.4	131.5	78.7	104.0	131.5	78.7	104.0
07	THERMAL & MOISTURE PROTECTION	106.4	59.2	86.5	109.5	87.7	100.3	104.3	59.1	85.3	108.2	81.0	96.8	104.4	83.3	95.5	104.3	83.3	95.5
08	DOORS & WINDOWS	86.0	49.9	76.6	92.9	81.6	89.9	86.0	51.9	77.0	83.9	77.7	82.3	91.9	69.8	86.1	91.9	74.0	87.2
09200	PLASTER & GYPSUM BOARD	189.2	56.1	106.5	187.8	84.4	123.5	148.5	51.6	88.3	145.6	77.9	103.5	148.5	78.1	104.8	194.7	78.1	122.2
095,098	CEILINGS & ACOUSTICAL TREATMENT	105.1	56.1	75.9	100.9	84.4	91.1	122.2	51.6	80.1	95.8	77.9	85.1	93.2	78.1	84.2	93.2	78.1	84.2
09600	FLOORING	128.0	42.0	104.7	132.9	91.6	121.7	120.0	56.3	102.7	123.6	86.7	113.6	132.9	88.6	120.9	132.9	88.6	120.9
097,099	WALL FINISHES, PAINTS & COATINGS	109.1	59.8	79.0	109.1	96.0	101.1	109.1	51.8	74.1	109.1	89.0	96.8	109.1	83.3	93.3	109.1	83.3	93.3
09	FINISHES	124.5	54.7	88.1	125.0	88.2	105.9	120.8	54.1	86.1	113.9	81.3	96.9	116.4	81.5	98.2	123.5	81.5	101.6
10 - 14	TOTAL DIV. 10000 - 14000	140.0	56.6	122.2	140.0	66.7	124.4	140.0	56.0	122.1	140.0	82.8	127.8	140.0	73.8	125.9	140.0	73.8	125.9
15	MECHANICAL	103.7	62.1	86.4	103.7	84.9	95.9	101.2	59.7	84.0	103.7	74.6	91.6	104.1	71.8	90.7	103.7	71.8	90.5
16	ELECTRICAL	133.4	76.0	105.5	130.2	91.4	111.3	126.5	61.9	95.1	127.8	89.5	109.2	121.7	71.6	97.3	121.7	71.6	97.3
01 - 16	WEIGHTED AVERAGE	118.1	64.9	94.5	119.3	87.9	105.4	114.6	62.8	91.6	114.0	82.7	100.1	115.7	78.7	99.3	116.2	78.9	99.6

CANADA

DIVISION		ST CATHARINES, ONTARIO			ST JEROME, QUEBEC			ST JOHNS, NEWFOUNDLAND			SUDBURY, ONTARIO			SUMMERSIDE, PEI			SYDNEY, NOVA SCOTIA		
		MAT.	INST.	TOTAL	MAT.	INST.	TOTAL	MAT.	INST.	TOTAL	MAT.	INST.	TOTAL	MAT.	INST.	TOTAL	MAT.	INST.	TOTAL
01590	EQUIPMENT RENTAL	.0	100.4	100.4	.0	99.2	99.2	.0	98.6	98.6	.0	100.4	100.4	.0	98.6	98.6	.0	98.6	98.6
02	SITE CONSTRUCTION	102.2	98.6	99.5	93.4	96.5	95.7	114.8	94.2	99.7	102.3	98.5	99.5	106.9	93.8	97.3	93.8	95.5	95.1
03100	CONCRETE FORMS & ACCESSORIES	113.2	78.7	83.4	129.8	78.7	85.8	99.5	55.1	61.2	108.5	78.4	82.5	90.6	52.1	57.4	90.3	64.1	67.7
03200	CONCRETE REINFORCEMENT	111.3	85.6	98.3	157.0	80.5	118.1	164.7	49.5	106.2	112.2	83.7	97.7	148.3	47.7	97.2	148.3	47.6	97.1
03300	CAST-IN-PLACE CONCRETE	140.1	78.3	114.7	135.6	86.6	115.5	179.6	62.6	131.5	141.2	77.7	115.1	168.0	54.6	121.4	136.4	64.3	106.7
03	CONCRETE	121.6	80.1	101.0	134.1	82.0	108.2	167.3	57.5	112.8	122.0	79.5	100.9	155.4	52.9	104.5	131.7	61.8	97.0
04	MASONRY	163.1	82.0	112.8	161.0	79.1	110.2	163.0	58.3	98.1	163.2	81.8	112.7	160.4	56.9	96.3	158.1	65.2	100.5
05	METALS	106.6	86.3	100.1	105.6	83.0	98.4	108.2	68.3	95.5	106.5	86.1	100.0	106.4	63.9	92.8	106.3	70.3	94.8
06	WOOD & PLASTICS	108.3	79.5	93.3	131.5	78.7	104.0	94.6	54.2	73.6	103.4	78.4	90.4	84.8	51.6	67.5	84.5	63.5	73.6
07	THERMAL & MOISTURE PROTECTION	108.6	82.6	97.6	104.3	83.3	95.5	108.1	57.7	86.9	108.2	81.0	96.8	104.8	56.0	84.2	104.4	65.2	87.9
08	DOORS & WINDOWS	83.0	79.6	82.1	91.9	69.8	86.1	98.0	51.9	86.0	83.9	77.7	82.3	94.1	45.5	81.4	82.8	59.0	76.6
09200	PLASTER & GYPSUM BOARD	142.2	79.0	103.0	146.1	78.1	103.8	173.8	53.1	98.8	145.6	77.9	103.5	165.0	50.4	93.8	164.6	62.6	101.3
095,098	CEILINGS & ACOUSTICAL TREATMENT	100.9	79.0	87.8	93.2	78.1	84.2	106.8	53.1	74.8	95.8	77.9	85.1	95.8	50.4	68.7	95.8	62.6	76.0
09600	FLOORING	126.6	87.9	116.1	132.9	88.6	120.9	114.6	50.9	97.4	123.6	86.7	113.6	109.7	56.0	95.2	109.6	59.9	96.2
097,099	WALL FINISHES, PAINTS & COATINGS	109.1	89.6	97.2	109.1	83.3	93.3	109.1	56.1	76.8	109.1	89.0	96.8	109.1	39.0	66.3	109.1	54.2	75.3
09	FINISHES	115.6	81.9	98.0	116.1	81.5	98.1	119.4	54.5	85.6	113.9	81.3	96.9	113.6	51.7	81.4	112.8	62.9	86.8
10 - 14	TOTAL DIV. 10000 - 14000	140.0	63.4	123.7	140.0	73.8	125.9	140.0	56.7	122.2	140.0	82.8	127.8	140.0	56.0	122.1	140.0	58.1	122.5
15	MECHANICAL	101.0	77.3	91.2	103.7	71.8	90.4	101.4	56.5	82.8	103.7	74.6	91.7	103.7	50.4	81.5	103.7	66.4	88.2
16	ELECTRICAL	128.2	89.7	109.5	122.9	71.6	98.0	121.0	57.4	90.1	124.7	89.5	107.6	119.3	51.6	86.4	121.1	63.5	93.1
01 - 16	WEIGHTED AVERAGE	113.6	83.1	100.0	115.4	78.7	99.1	120.1	60.9	93.8	113.7	82.7	99.9	117.3	56.9	90.5	113.3	67.1	92.8

COST INDEXES

		CANADA																
	DIVISION	THUNDER BAY, ONTARIO			TIMMINS, ONTARIO			TORONTO, ONTARIO			TROIS RIVIERES, QUEBEC			TRURO, NOVA SCOTIA			VANCOUVER, B C		
		MAT.	INST.	TOTAL	MAT.	INST.	TOTAL	MAT.	INST.	TOTAL	MAT.	INST.	TOTAL	MAT.	INST.	TOTAL	MAT.	INST.	TOTAL
01590	EQUIPMENT RENTAL	.0	100.4	100.4	.0	100.4	100.4	.0	103.2	103.2	.0	99.2	99.2	.0	98.6	98.6	.0	111.3	111.3
02	SITE CONSTRUCTION	107.9	98.7	101.1	114.3	98.5	102.7	113.8	103.6	106.3	94.2	96.5	95.9	99.0	95.5	96.4	118.6	107.2	110.3
03100	CONCRETE FORMS & ACCESSORIES	122.6	79.7	85.6	122.5	78.1	84.2	123.0	90.4	94.8	129.8	78.9	85.9	90.3	64.1	67.7	114.5	82.2	86.7
03200	CONCRETE REINFORCEMENT	99.5	85.0	92.1	174.7	83.6	128.4	170.9	86.6	128.0	148.3	80.6	113.9	148.3	47.6	97.1	173.2	78.6	125.1
03300	CAST-IN-PLACE CONCRETE	154.2	89.6	127.6	170.5	77.6	132.3	163.3	98.8	136.7	141.2	86.6	118.8	177.1	64.3	130.7	153.2	91.5	127.9
03	CONCRETE	130.4	84.3	107.5	153.3	79.3	116.5	149.2	92.5	121.1	135.4	82.1	108.9	151.3	61.8	106.9	147.9	85.2	116.8
04	MASONRY	163.9	82.8	113.6	166.7	81.8	114.0	186.9	96.2	130.7	161.4	79.1	110.3	161.1	65.2	101.7	177.3	86.1	120.8
05	METALS	106.4	85.4	99.7	107.2	85.8	100.3	130.4	88.5	117.0	105.6	83.1	98.4	106.3	70.3	94.8	141.5	88.4	124.6
06	WOOD & PLASTICS	118.4	79.9	98.3	118.4	78.4	97.6	118.4	88.9	103.0	131.5	78.7	104.0	84.5	63.5	73.6	113.9	80.2	96.3
07	THERMAL & MOISTURE PROTECTION	108.8	81.9	97.5	109.4	79.6	96.8	110.2	92.3	102.7	104.3	83.3	95.5	104.4	65.2	87.9	116.6	86.9	104.1
08	DOORS & WINDOWS	82.0	79.2	81.2	90.9	76.5	87.2	90.9	87.7	90.1	91.9	74.0	87.2	82.8	59.0	76.6	94.4	78.4	90.2
09200	PLASTER & GYPSUM BOARD	170.8	79.4	114.0	158.6	77.9	108.5	174.9	88.6	121.3	194.7	78.1	122.2	164.6	62.6	101.3	149.2	79.1	105.7
095,098	CEILINGS & ACOUSTICAL TREATMENT	95.8	79.4	86.0	95.8	77.9	85.1	121.3	88.6	101.8	93.2	78.1	84.2	95.8	62.6	76.0	111.1	79.1	92.0
09600	FLOORING	132.9	50.6	110.6	132.9	86.7	120.4	132.9	93.2	122.1	132.9	88.6	120.9	109.6	59.9	96.2	132.9	85.3	120.0
097,099	WALL FINISHES, PAINTS & COATINGS	109.1	91.6	98.4	109.1	82.8	93.1	112.1	100.6	105.1	109.1	83.3	93.3	109.1	58.2	78.1	109.1	95.2	100.6
09	FINISHES	120.8	76.4	97.7	119.3	80.6	99.1	128.4	92.2	109.6	123.5	81.5	101.6	112.8	62.9	86.8	122.6	83.8	102.4
10 - 14	TOTAL DIV. 10000 - 14000	140.0	64.3	123.9	140.0	63.8	123.7	140.0	87.4	128.8	140.0	73.8	125.9	140.0	58.1	122.5	140.0	83.2	127.9
15	MECHANICAL	101.0	78.0	91.5	103.7	77.1	92.6	101.5	88.0	95.9	103.7	71.8	90.5	103.7	66.4	88.2	101.5	81.4	93.2
16	ELECTRICAL	125.2	87.9	107.1	127.8	89.5	109.2	129.8	92.3	111.6	121.7	71.6	97.3	121.1	63.5	93.1	132.9	80.8	107.5
01 - 16	WEIGHTED AVERAGE	114.8	82.8	100.6	119.4	82.4	103.0	124.2	92.1	109.9	116.2	78.9	99.6	115.8	67.1	94.2	125.7	85.6	107.9

		CANADA																
	DIVISION	VICTORIA, B C			WHITEHORSE, YUKON			WINDSOR, ONTARIO			WINNIPEG, MANITOBA			YARMOUTH, NOVA SCOTIA			YELLOWKNIFE, NWT		
		MAT.	INST.	TOTAL	MAT.	INST.	TOTAL	MAT.	INST.	TOTAL	MAT.	INST.	TOTAL	MAT.	INST.	TOTAL	MAT.	INST.	TOTAL
01590	EQUIPMENT RENTAL	.0	107.3	107.3	.0	99.2	99.2	.0	100.4	100.4	.0	105.3	105.3	.0	98.6	98.6	.0	98.7	98.7
02	SITE CONSTRUCTION	121.2	106.4	110.3	108.5	94.6	98.3	98.3	98.8	98.7	112.2	100.0	103.3	98.7	95.5	96.4	121.3	97.5	103.8
03100	CONCRETE FORMS & ACCESSORIES	108.0	81.5	85.1	102.0	53.4	60.0	122.6	80.3	86.1	122.2	65.2	73.0	90.3	64.1	67.7	102.0	73.2	77.2
03200	CONCRETE REINFORCEMENT	116.5	78.4	97.1	119.0	61.3	89.7	109.0	84.3	96.5	161.5	54.3	107.0	148.3	47.6	97.1	119.0	65.1	91.6
03300	CAST-IN-PLACE CONCRETE	153.3	90.7	127.5	152.8	62.8	115.8	143.5	90.7	121.8	157.5	70.2	121.6	175.2	64.3	129.6	253.7	79.2	182.0
03	CONCRETE	154.4	84.2	119.6	128.0	58.8	93.7	123.5	84.9	104.3	144.6	65.6	105.4	150.4	61.8	106.4	176.8	74.2	125.8
04	MASONRY	170.7	86.1	118.3	160.2	57.1	96.3	163.3	89.0	117.2	178.5	61.5	106.0	160.9	65.2	101.6	167.8	69.2	106.7
05	METALS	107.8	83.3	100.0	105.8	69.9	94.4	106.5	86.6	100.1	139.2	73.7	118.3	106.3	70.3	94.8	106.6	76.3	96.9
06	WOOD & PLASTICS	98.1	80.0	88.7	95.2	52.1	72.8	118.4	78.5	97.6	116.1	65.9	90.0	84.5	63.5	73.6	95.2	74.4	84.3
07	THERMAL & MOISTURE PROTECTION	116.7	80.5	101.5	103.5	58.1	84.4	108.6	85.5	98.9	105.3	68.2	89.6	104.4	65.2	87.9	104.7	73.9	91.7
08	DOORS & WINDOWS	87.8	78.1	85.3	85.5	50.7	76.4	81.7	79.0	81.0	91.9	59.2	83.4	82.8	59.0	76.6	85.5	64.3	80.0
09200	PLASTER & GYPSUM BOARD	138.2	79.1	101.5	119.9	50.8	77.0	163.5	78.0	110.4	153.9	64.5	98.4	164.6	62.6	101.3	122.3	73.8	92.2
095,098	CEILINGS & ACOUSTICAL TREATMENT	95.8	79.1	85.8	95.8	50.8	68.9	95.8	78.0	85.1	107.7	64.5	81.9	95.8	62.6	76.0	95.8	73.8	82.6
09600	FLOORING	127.1	49.5	106.1	120.0	54.6	102.3	132.9	88.6	120.9	132.9	62.5	113.8	109.6	59.9	96.2	120.0	83.8	110.2
097,099	WALL FINISHES, PAINTS & COATINGS	109.1	76.2	89.0	109.1	50.9	73.6	109.1	91.2	98.2	109.2	53.3	75.1	109.1	58.2	78.1	109.1	74.2	87.8
09	FINISHES	116.1	75.3	94.8	109.9	53.1	80.3	119.4	83.1	100.5	122.0	63.7	91.6	112.8	62.9	86.8	110.2	74.7	91.7
10 - 14	TOTAL DIV. 10000 - 14000	140.0	82.9	127.8	140.0	55.4	122.0	140.0	65.5	124.1	140.0	59.6	122.9	140.0	58.1	122.5	140.0	78.9	127.0
15	MECHANICAL	103.7	74.8	91.7	103.7	58.7	85.0	101.0	80.5	92.5	101.5	65.9	86.7	103.7	66.4	88.2	103.7	74.6	91.6
16	ELECTRICAL	124.9	80.8	103.5	126.0	61.3	94.6	133.4	89.7	112.1	131.9	68.6	101.1	121.1	63.5	93.1	127.2	85.5	106.9
01 - 16	WEIGHTED AVERAGE	119.1	82.3	102.8	114.0	61.9	90.8	114.6	85.4	101.6	124.1	68.9	99.6	115.6	67.1	94.1	120.4	77.4	101.3

Location Factors

Costs shown in *Means cost data publications* are based on National Averages for materials and installation. To adjust these costs to a specific location, simply multiply the base cost by the factor and divide by 100 for that city. The data is arranged alphabetically by state and postal zip code numbers. For a city not listed, use the factor for a nearby city with similar economic characteristics.

STATE/ZIP	CITY	MAT.	INST.	TOTAL
ALABAMA				
350-352	Birmingham	96.5	75.2	87.0
354	Tuscaloosa	96.3	55.4	78.1
355	Jasper	96.5	52.1	76.8
356	Decatur	96.2	55.9	78.3
357-358	Huntsville	96.3	72.0	85.5
359	Gadsden	96.1	58.8	79.6
360-361	Montgomery	97.0	57.1	79.3
362	Anniston	95.7	46.5	73.9
363	Dothan	96.3	49.0	75.3
364	Evergreen	95.7	50.2	75.5
365-366	Mobile	97.2	61.1	81.2
367	Selma	95.9	51.6	76.2
368	Phenix City	96.6	54.9	78.1
369	Butler	96.1	49.0	75.2
ALASKA				
995-996	Anchorage	133.7	114.7	125.3
997	Fairbanks	130.3	117.6	124.6
998	Juneau	132.5	114.0	124.3
999	Ketchikan	141.9	113.6	129.3
ARIZONA				
850,853	Phoenix	98.8	73.6	87.6
852	Mesa/Tempe	97.8	68.9	85.0
855	Globe	98.1	65.5	83.6
856-857	Tucson	97.1	71.2	85.6
859	Show Low	98.2	68.0	84.8
860	Flagstaff	100.4	72.3	87.9
863	Prescott	98.3	66.8	84.3
864	Kingman	96.7	67.2	83.6
865	Chambers	96.8	68.3	84.1
ARKANSAS				
716	Pine Bluff	95.2	60.9	80.0
717	Camden	93.2	38.9	69.1
718	Texarkana	94.6	44.0	72.1
719	Hot Springs	92.4	38.4	68.4
720-722	Little Rock	95.1	63.8	81.2
723	West Memphis	94.6	53.4	76.3
724	Jonesboro	95.2	53.4	76.6
725	Batesville	93.3	48.4	73.3
726	Harrison	94.5	48.4	74.0
727	Fayetteville	91.9	46.7	71.8
728	Russellville	93.2	46.6	72.5
729	Fort Smith	95.9	56.9	78.6
CALIFORNIA				
900-902	Los Angeles	102.5	112.1	106.8
903-905	Inglewood	98.2	109.0	103.0
906-908	Long Beach	99.6	109.0	103.8
910-912	Pasadena	100.4	109.3	104.4
913-916	Van Nuys	103.5	109.1	106.0
917-918	Alhambra	102.7	109.1	105.5
919-921	San Diego	104.0	103.9	104.0
922	Palm Springs	100.1	107.2	103.2
923-924	San Bernardino	97.9	108.1	102.4
925	Riverside	102.2	109.5	105.5
926-927	Santa Ana	99.9	108.2	103.5
928	Anaheim	102.6	110.6	106.1
930	Oxnard	103.8	110.0	106.5
931	Santa Barbara	102.6	110.5	106.1
932-933	Bakersfield	102.3	105.9	103.9
934	San Luis Obispo	103.4	107.9	105.4
935	Mojave	100.5	103.1	101.7
936-938	Fresno	104.1	110.9	107.1
939	Salinas	104.1	116.4	109.6
940-941	San Francisco	112.8	132.9	121.7
942,956-958	Sacramento	106.6	113.0	109.4
943	Palo Alto	105.5	127.2	115.1
944	San Mateo	108.4	127.8	117.0
945	Vallejo	105.5	122.2	112.9
946	Oakland	110.7	127.1	118.0
947	Berkeley	110.1	126.7	117.5
948	Richmond	109.5	126.1	116.9
949	San Rafael	110.2	122.8	115.8
950	Santa Cruz	109.4	116.4	112.5

STATE/ZIP	CITY	MAT.	INST.	TOTAL
CALIFORNIA (CONT'D)				
951	San Jose	107.3	127.5	116.3
952	Stockton	104.4	113.5	108.4
953	Modesto	104.3	111.0	107.3
954	Santa Rosa	104.1	122.0	112.1
955	Eureka	105.9	101.1	103.8
959	Marysville	105.0	112.5	108.3
960	Redding	106.6	110.3	108.3
961	Susanville	105.7	110.7	107.9
COLORADO				
800-802	Denver	100.6	89.9	95.8
803	Boulder	97.7	86.5	92.7
804	Golden	99.9	85.8	93.6
805	Fort Collins	101.1	82.4	92.8
806	Greeley	98.7	70.4	86.1
807	Fort Morgan	98.4	85.8	92.8
808-809	Colorado Springs	100.4	85.7	93.9
810	Pueblo	100.0	83.1	92.5
811	Alamosa	101.3	79.5	91.7
812	Salida	101.3	80.2	91.9
813	Durango	101.7	81.4	92.7
814	Montrose	100.3	79.6	91.1
815	Grand Junction	103.6	76.8	91.7
816	Glenwood Springs	101.1	83.3	93.2
CONNECTICUT				
060	New Britain	101.1	116.6	108.0
061	Hartford	102.3	117.3	109.0
062	Willimantic	101.8	117.2	108.7
063	New London	98.1	117.3	106.7
064	Meriden	100.1	117.8	108.0
065	New Haven	102.8	117.8	109.4
066	Bridgeport	102.4	117.8	109.2
067	Waterbury	101.9	117.7	108.9
068	Norwalk	101.9	118.2	109.1
069	Stamford	102.0	124.0	111.8
D.C.				
200-205	Washington	102.5	91.2	97.5
DELAWARE				
197	Newark	100.7	103.9	102.1
198	Wilmington	99.7	103.9	101.5
199	Dover	100.7	103.9	102.1
FLORIDA				
320,322	Jacksonville	98.3	59.6	81.1
321	Daytona Beach	98.3	72.8	87.0
323	Tallahassee	97.6	48.2	75.6
324	Panama City	99.4	36.8	71.6
325	Pensacola	99.1	57.2	80.5
326,344	Gainesville	99.6	57.7	81.0
327-328,347	Orlando	99.7	68.6	85.9
329	Melbourne	100.8	77.8	90.6
330-332,340	Miami	98.7	71.4	86.6
333	Fort Lauderdale	97.7	70.3	85.5
334,349	West Palm Beach	96.6	66.6	83.3
335-336,346	Tampa	99.4	70.8	86.7
337	St. Petersburg	101.6	54.2	80.5
338	Lakeland	98.4	70.5	86.0
339,341	Fort Myers	97.7	62.8	82.2
342	Sarasota	99.5	65.1	84.2
GEORGIA				
300-303,399	Atlanta	97.7	79.0	89.4
304	Statesboro	97.1	43.6	73.4
305	Gainesville	96.0	55.7	78.1
306	Athens	95.3	62.2	80.6
307	Dalton	97.4	49.4	76.1
308-309	Augusta	96.3	57.5	79.0
310-312	Macon	95.7	59.1	79.4
313-314	Savannah	97.0	57.7	79.6
315	Waycross	97.0	50.0	76.1
316	Valdosta	96.9	47.6	75.0
317,398	Albany	97.2	56.3	79.1
318-319	Columbus	97.2	60.4	80.8

STATE/ZIP	CITY	MAT.	INST.	TOTAL
HAWAII				
967	Hilo	115.0	124.8	119.3
968	Honolulu	120.6	124.8	122.5
STATES & POSS.				
969	Guam	196.4	52.7	132.5
IDAHO				
832	Pocatello	100.3	77.3	90.1
833	Twin Falls	100.8	57.7	81.7
834	Idaho Falls	98.3	58.8	80.8
835	Lewiston	107.4	83.6	96.8
836-837	Boise	100.2	78.9	90.7
838	Coeur d'Alene	106.6	75.9	93.0
ILLINOIS				
600-603	North Suburban	99.5	125.3	111.0
604	Joliet	99.4	118.5	107.9
605	South Suburban	99.5	125.3	111.0
606-608	Chicago	100.1	126.3	111.7
609	Kankakee	95.7	104.1	99.4
610-611	Rockford	98.1	107.9	102.4
612	Rock Island	95.4	97.0	96.1
613	La Salle	96.7	102.6	99.3
614	Galesburg	96.5	98.9	97.6
615-616	Peoria	99.1	99.9	99.4
617	Bloomington	95.8	99.2	97.3
618-619	Champaign	99.2	100.8	99.9
620-622	East St. Louis	94.5	102.0	97.8
623	Quincy	95.7	94.6	95.2
624	Effingham	95.1	97.1	96.0
625	Decatur	97.0	101.1	98.8
626-627	Springfield	97.3	99.8	98.4
628	Centralia	93.3	101.1	96.8
629	Carbondale	92.9	97.3	94.9
INDIANA				
460	Anderson	94.9	84.8	90.4
461-462	Indianapolis	98.8	87.2	93.7
463-464	Gary	96.2	104.3	99.8
465-466	South Bend	95.3	86.8	91.5
467-468	Fort Wayne	95.4	83.1	89.9
469	Kokomo	93.0	85.7	89.8
470	Lawrenceburg	92.0	81.8	87.5
471	New Albany	93.6	77.5	86.5
472	Columbus	96.1	84.0	90.7
473	Muncie	96.2	82.4	90.1
474	Bloomington	98.2	84.1	91.9
475	Washington	94.0	88.4	91.5
476-477	Evansville	95.3	90.0	93.0
478	Terre Haute	96.0	88.3	92.6
479	Lafayette	95.7	82.4	89.8
IOWA				
500-503,509	Des Moines	97.8	82.2	90.9
504	Mason City	96.0	62.0	80.9
505	Fort Dodge	96.2	58.1	79.3
506-507	Waterloo	97.8	62.0	81.9
508	Creston	96.4	66.0	82.9
510-511	Sioux City	98.4	74.9	88.0
512	Sibley	97.0	51.5	76.8
513	Spencer	98.8	49.9	77.1
514	Carroll	95.8	55.7	78.0
515	Council Bluffs	99.3	75.2	88.6
516	Shenandoah	96.2	52.9	77.0
520	Dubuque	97.8	77.2	88.7
521	Decorah	96.8	52.8	77.3
522-524	Cedar Rapids	98.9	84.4	92.5
525	Ottumwa	96.8	71.1	85.4
526	Burlington	96.0	75.5	86.9
527-528	Davenport	97.9	94.9	96.5
KANSAS				
660-662	Kansas City	98.9	96.1	97.7
664-666	Topeka	98.7	69.3	85.6
667	Fort Scott	97.6	72.3	86.4
668	Emporia	97.4	57.7	79.8
669	Belleville	99.2	63.7	83.5
670-672	Wichita	97.9	69.5	85.3
673	Independence	99.0	67.7	85.1
674	Salina	99.2	63.9	83.5
675	Hutchinson	94.5	49.5	74.5
676	Hays	98.5	67.0	84.5
677	Colby	99.2	53.1	78.8

STATE/ZIP	CITY	MAT.	INST.	TOTAL
KANSAS (CONT'D)				
678	Dodge City	100.4	68.4	86.2
679	Liberal	98.3	44.2	74.3
KENTUCKY				
400-402	Louisville	96.4	85.4	91.5
403-405	Lexington	96.0	65.7	82.5
406	Frankfort	96.3	66.7	83.1
407-409	Corbin	93.3	43.3	71.1
410	Covington	93.8	91.9	93.0
411-412	Ashland	92.5	96.9	94.5
413-414	Campton	94.1	43.7	71.7
415-416	Pikeville	95.0	60.3	79.6
417-418	Hazard	93.5	43.4	71.3
420	Paducah	92.2	88.4	90.5
421-422	Bowling Green	94.6	82.7	89.3
423	Owensboro	94.4	73.5	85.1
424	Henderson	92.0	90.4	91.3
425-426	Somerset	91.6	45.1	70.9
427	Elizabethtown	91.1	84.5	88.2
LOUISIANA				
700-701	New Orleans	99.8	69.7	86.4
703	Thibodaux	97.7	67.2	84.2
704	Hammond	95.0	62.7	80.7
705	Lafayette	97.4	62.3	81.8
706	Lake Charles	97.5	64.1	82.7
707-708	Baton Rouge	98.1	62.9	82.5
710-711	Shreveport	94.6	61.3	79.8
712	Monroe	95.3	57.9	78.7
713-714	Alexandria	95.4	59.5	79.5
MAINE				
039	Kittery	95.2	73.4	85.5
040-041	Portland	98.5	79.8	90.2
042	Lewiston	98.9	79.8	90.4
043	Augusta	97.4	73.2	86.7
044	Bangor	97.7	79.8	89.8
045	Bath	96.3	73.4	86.1
046	Machias	95.9	72.8	85.6
047	Houlton	96.0	76.4	87.3
048	Rockland	95.1	73.4	85.4
049	Waterville	96.3	73.3	86.1
MARYLAND				
206	Waldorf	98.3	72.6	86.9
207-208	College Park	98.4	81.4	90.8
209	Silver Spring	97.6	77.9	88.9
210-212	Baltimore	99.5	84.7	92.9
214	Annapolis	98.9	79.2	90.1
215	Cumberland	95.4	81.2	89.1
216	Easton	96.8	44.8	73.7
217	Hagerstown	96.2	81.9	89.8
218	Salisbury	97.3	53.5	77.8
219	Elkton	94.6	66.7	82.2
MASSACHUSETTS				
010-011	Springfield	100.4	105.9	102.8
012	Pittsfield	99.9	103.0	101.3
013	Greenfield	97.9	105.3	101.2
014	Fitchburg	96.6	119.8	106.9
015-016	Worcester	100.1	120.0	109.0
017	Framingham	96.1	123.8	108.4
018	Lowell	99.7	125.0	111.0
019	Lawrence	100.9	122.6	110.5
020-022, 024	Boston	102.5	132.0	115.6
023	Brockton	101.1	121.0	110.0
025	Buzzards Bay	95.5	119.0	105.9
026	Hyannis	98.1	119.0	107.4
027	New Bedford	100.8	119.1	109.0
MICHIGAN				
480,483	Royal Oak	95.3	107.9	100.9
481	Ann Arbor	97.4	105.9	101.2
482	Detroit	99.1	117.6	107.3
484-485	Flint	97.2	99.9	98.4
486	Saginaw	97.0	93.4	95.4
487	Bay City	96.9	93.5	95.4
488-489	Lansing	97.2	95.2	96.3
490	Battle Creek	95.5	87.9	92.1
491	Kalamazoo	95.8	85.8	91.4
492	Jackson	93.8	92.3	93.2
493,495	Grand Rapids	96.0	68.1	83.6
494	Muskegon	94.4	83.6	89.6

Location Factors

STATE/ZIP	CITY	MAT.	INST.	TOTAL
MICHIGAN (CONT'D)				
496	Traverse City	93.3	73.1	84.3
497	Gaylord	94.5	76.9	86.7
498-499	Iron Mountain	96.3	87.7	92.5
MINNESOTA				
550-551	Saint Paul	99.7	124.5	110.7
553-555	Minneapolis	100.5	126.7	112.2
556-558	Duluth	99.5	114.5	106.2
559	Rochester	99.2	108.5	103.4
560	Mankato	96.3	105.9	100.6
561	Windom	95.1	80.3	88.5
562	Willmar	94.4	87.4	91.3
563	St. Cloud	95.7	123.3	108.0
564	Brainerd	96.0	105.1	100.0
565	Detroit Lakes	97.9	100.6	99.1
566	Bemidji	97.3	101.2	99.0
567	Thief River Falls	96.8	96.1	96.5
MISSISSIPPI				
386	Clarksdale	95.9	30.8	67.0
387	Greenville	99.3	44.6	75.0
388	Tupelo	97.3	37.3	70.6
389	Greenwood	97.2	33.3	68.8
390-392	Jackson	97.8	46.1	74.8
393	Meridian	96.0	46.7	74.1
394	Laurel	97.1	33.4	68.8
395	Biloxi	97.8	55.5	79.0
396	McComb	95.7	52.1	76.4
397	Columbus	97.1	36.7	70.3
MISSOURI				
630-631	St. Louis	97.0	109.0	102.3
633	Bowling Green	95.7	95.8	95.7
634	Hannibal	94.5	86.3	90.9
635	Kirksville	96.8	79.2	89.0
636	Flat River	96.7	97.0	96.8
637	Cape Girardeau	96.4	92.3	94.6
638	Sikeston	94.5	83.2	89.5
639	Poplar Bluff	94.0	83.8	89.5
640-641	Kansas City	100.9	106.8	103.5
644-645	St. Joseph	99.9	93.3	97.0
646	Chillicothe	96.7	71.2	85.4
647	Harrisonville	96.4	101.3	98.6
648	Joplin	98.6	70.5	86.1
650-651	Jefferson City	96.1	89.7	93.3
652	Columbia	97.1	92.4	95.0
653	Sedalia	96.1	84.9	91.2
654-655	Rolla	95.0	78.6	87.7
656-658	Springfield	98.4	78.7	89.6
MONTANA				
590-591	Billings	102.1	74.5	89.8
592	Wolf Point	100.8	68.7	86.6
593	Miles City	98.8	70.0	86.0
594	Great Falls	102.7	73.8	89.9
595	Havre	99.6	68.9	86.0
596	Helena	102.5	71.5	88.7
597	Butte	101.2	70.9	87.7
598	Missoula	98.9	70.1	86.1
599	Kalispell	97.8	69.4	85.2
NEBRASKA				
680-681	Omaha	99.3	78.2	89.9
683-685	Lincoln	98.2	67.1	84.4
686	Columbus	96.1	49.6	75.5
687	Norfolk	97.7	60.6	81.3
688	Grand Island	97.5	65.6	83.3
689	Hastings	97.1	56.7	79.2
690	Mccook	96.9	48.2	75.3
691	North Platte	97.1	58.8	80.1
692	Valentine	99.2	38.5	72.3
693	Alliance	99.0	35.9	71.0
NEVADA				
889-891	Las Vegas	98.3	107.1	102.2
893	Ely	98.7	79.0	90.0
894-895	Reno	98.9	98.2	98.6
897	Carson City	98.6	97.9	98.3
898	Elko	97.6	85.4	92.2
NEW HAMPSHIRE				
030	Nashua	100.6	82.4	92.5
031	Manchester	100.9	82.4	92.7

STATE/ZIP	CITY	MAT.	INST.	TOTAL
NEW HAMPSHIRE (CONT'D)				
032-033	Concord	98.4	82.4	91.3
034	Keene	97.2	50.9	76.6
035	Littleton	97.4	61.6	81.5
036	Charleston	96.6	47.9	75.0
037	Claremont	95.9	47.9	74.6
038	Portsmouth	97.8	78.0	89.0
NEW JERSEY				
070-071	Newark	102.1	124.6	112.1
072	Elizabeth	100.8	123.6	110.9
073	Jersey City	99.8	124.5	110.8
074-075	Paterson	101.4	124.2	111.5
076	Hackensack	99.5	124.5	110.6
077	Long Branch	99.1	125.3	110.8
078	Dover	99.7	124.5	110.7
079	Summit	99.8	123.6	110.4
080,083	Vineland	97.6	117.3	106.3
081	Camden	99.6	116.8	107.2
082,084	Atlantic City	98.4	116.5	106.4
085-086	Trenton	99.6	122.6	109.8
087	Point Pleasant	99.5	122.3	109.6
088-089	New Brunswick	100.0	123.4	110.4
NEW MEXICO				
870-872	Albuquerque	98.5	76.2	88.6
873	Gallup	98.3	76.2	88.5
874	Farmington	98.7	76.2	88.7
875	Santa Fe	97.5	76.2	88.0
877	Las Vegas	96.9	76.2	87.7
878	Socorro	96.5	76.2	87.5
879	Truth/Consequences	96.9	70.8	85.3
880	Las Cruces	95.8	67.9	83.4
881	Clovis	97.0	75.2	87.3
882	Roswell	98.7	75.3	88.3
883	Carrizozo	99.0	76.2	88.9
884	Tucumcari	97.9	75.2	87.8
NEW YORK				
100-102	New York	106.0	164.4	131.9
103	Staten Island	101.6	159.9	127.5
104	Bronx	99.8	159.9	126.5
105	Mount Vernon	99.5	138.5	116.8
106	White Plains	99.5	138.2	116.7
107	Yonkers	104.5	138.5	119.6
108	New Rochelle	100.0	138.0	116.9
109	Suffern	99.8	126.2	111.5
110	Queens	101.4	158.8	126.9
111	Long Island City	103.1	158.8	127.8
112	Brooklyn	103.3	159.1	128.1
113	Flushing	103.3	158.8	128.0
114	Jamaica	101.5	158.8	126.9
115,117,118	Hicksville	101.2	149.5	122.7
116	Far Rockaway	103.4	158.8	128.0
119	Riverhead	102.2	149.5	123.2
120-122	Albany	97.8	96.4	97.2
123	Schenectady	98.1	95.9	97.1
124	Kingston	100.9	117.3	108.2
125-126	Poughkeepsie	100.1	119.4	108.7
127	Monticello	99.5	119.9	108.5
128	Glens Falls	92.7	91.3	92.1
129	Plattsburgh	96.8	88.4	93.1
130-132	Syracuse	99.0	94.7	97.1
133-135	Utica	96.9	90.9	94.2
136	Watertown	98.5	88.5	94.1
137-139	Binghamton	98.5	88.8	94.2
140-142	Buffalo	99.8	105.7	102.4
143	Niagara Falls	97.3	104.7	100.6
144-146	Rochester	99.8	99.7	99.8
147	Jamestown	96.3	87.9	92.6
148-149	Elmira	96.3	85.4	91.4
NORTH CAROLINA				
270,272-274	Greensboro	98.5	50.3	77.1
271	Winston-Salem	98.2	49.6	76.6
275-276	Raleigh	98.4	50.7	77.2
277	Durham	98.4	50.7	77.2
278	Rocky Mount	95.7	35.5	69.0
279	Elizabeth City	96.4	37.8	70.4
280	Gastonia	97.7	48.9	76.0
281-282	Charlotte	98.6	48.9	76.5
283	Fayetteville	97.1	50.8	76.5
284	Wilmington	96.4	49.9	75.7
285	Kinston	94.4	35.1	68.1

STATE/ZIP	CITY	MAT.	INST.	TOTAL
NORTH CAROLINA (CONT'D)				
286	Hickory	94.8	33.9	67.8
287-288	Asheville	96.8	48.9	75.5
289	Murphy	95.7	33.0	67.9
NORTH DAKOTA				
580-581	Fargo	100.9	65.2	85.1
582	Grand Forks	100.5	59.2	82.2
583	Devils Lake	99.9	60.8	82.6
584	Jamestown	100.0	52.6	78.9
585	Bismarck	99.3	66.6	84.8
586	Dickinson	100.7	60.5	82.9
587	Minot	100.6	67.8	86.1
588	Williston	99.3	60.2	81.9
OHIO				
430-432	Columbus	97.3	89.6	93.9
433	Marion	93.9	88.4	91.5
434-436	Toledo	97.6	100.0	98.7
437-438	Zanesville	94.3	86.1	90.6
439	Steubenville	95.2	95.1	95.1
440	Lorain	96.9	97.1	97.0
441	Cleveland	97.2	104.3	100.4
442-443	Akron	97.8	98.0	97.9
444-445	Youngstown	97.2	93.3	95.5
446-447	Canton	97.3	88.4	93.4
448-449	Mansfield	94.6	93.8	94.2
450	Hamilton	94.5	89.6	92.3
451-452	Cincinnati	94.9	90.2	92.8
453-454	Dayton	94.6	87.0	91.2
455	Springfield	94.6	87.9	91.6
456	Chillicothe	93.5	94.2	93.8
457	Athens	96.3	79.9	89.0
458	Lima	96.8	87.9	92.8
OKLAHOMA				
730-731	Oklahoma City	97.1	64.3	82.6
734	Ardmore	94.0	63.7	80.5
735	Lawton	96.4	65.3	82.6
736	Clinton	95.3	62.0	80.6
737	Enid	96.0	62.0	80.9
738	Woodward	94.2	62.1	79.9
739	Guymon	95.2	32.9	67.5
740-741	Tulsa	96.2	59.9	80.1
743	Miami	93.0	67.5	81.7
744	Muskogee	95.6	44.0	72.7
745	Mcalester	92.6	55.1	75.9
746	Ponca City	93.1	62.2	79.4
747	Durant	93.1	61.7	79.2
748	Shawnee	94.6	60.2	79.3
749	Poteau	92.3	64.7	80.1
OREGON				
970-972	Portland	102.2	104.6	103.3
973	Salem	102.2	103.2	102.6
974	Eugene	101.7	102.8	102.2
975	Medford	103.3	101.1	102.3
976	Klamath Falls	103.4	101.1	102.4
977	Bend	102.2	102.9	102.5
978	Pendleton	96.5	102.6	99.2
979	Vale	94.4	93.2	93.9
PENNSYLVANIA				
150-152	Pittsburgh	97.0	104.2	100.2
153	Washington	94.0	102.8	98.0
154	Uniontown	94.2	101.2	97.3
155	Bedford	95.2	92.6	94.0
156	Greensburg	95.3	102.2	98.3
157	Indiana	94.1	100.7	97.0
158	Dubois	95.4	97.3	96.2
159	Johnstown	95.1	98.1	96.5
160	Butler	92.3	102.3	96.7
161	New Castle	92.3	101.0	96.2
162	Kittanning	92.8	103.6	97.6
163	Oil City	92.3	98.1	94.9
164-165	Erie	94.4	96.7	95.4
166	Altoona	94.4	94.5	94.5
167	Bradford	95.7	96.6	96.1
168	State College	95.4	93.9	94.7
169	Wellsboro	96.4	94.2	95.4
170-171	Harrisburg	98.2	95.3	96.9
172	Chambersburg	95.9	89.9	93.3
173-174	York	96.2	95.3	95.8
175-176	Lancaster	94.8	90.6	92.9

STATE/ZIP	CITY	MAT.	INST.	TOTAL
PENNSYLVANIA (CONT'D)				
177	Williamsport	93.4	82.4	88.5
178	Sunbury	95.4	94.4	95.0
179	Pottsville	94.5	94.0	94.3
180	Lehigh Valley	95.9	114.4	104.1
181	Allentown	98.0	107.4	102.2
182	Hazleton	95.3	96.6	95.9
183	Stroudsburg	95.2	101.1	97.8
184-185	Scranton	98.8	96.3	97.7
186-187	Wilkes-Barre	95.2	94.9	95.1
188	Montrose	94.8	95.5	95.1
189	Doylestown	94.7	119.7	105.8
190-191	Philadelphia	101.0	131.1	114.4
193	Westchester	97.6	122.0	108.5
194	Norristown	96.6	126.0	109.7
195-196	Reading	98.8	100.9	99.7
PUERTO RICO				
009	San Juan	135.2	25.5	86.5
RHODE ISLAND				
028	Newport	99.6	112.7	105.4
029	Providence	99.9	112.7	105.6
SOUTH CAROLINA				
290-292	Columbia	97.1	47.3	75.0
293	Spartanburg	95.9	45.4	73.5
294	Charleston	97.3	47.9	75.3
295	Florence	95.4	45.2	73.1
296	Greenville	95.7	45.4	73.4
297	Rock Hill	95.3	33.0	67.6
298	Aiken	96.1	72.0	85.4
299	Beaufort	97.0	36.2	70.0
SOUTH DAKOTA				
570-571	Sioux Falls	99.2	57.3	80.6
572	Watertown	97.9	51.0	77.1
573	Mitchell	96.9	50.8	76.4
574	Aberdeen	99.5	53.1	78.9
575	Pierre	98.8	53.5	78.7
576	Mobridge	97.5	50.4	76.6
577	Rapid City	99.4	52.2	78.4
TENNESSEE				
370-372	Nashville	97.5	73.7	86.9
373-374	Chattanooga	98.6	54.1	78.8
375,380-381	Memphis	96.6	76.2	87.5
376	Johnson City	97.3	56.4	79.1
377-379	Knoxville	94.7	57.5	78.2
382	Mckenzie	96.4	59.4	80.0
383	Jackson	98.3	51.7	77.6
384	Columbia	94.9	55.2	77.3
385	Cookeville	96.2	49.9	75.6
TEXAS				
750	Mckinney	97.6	55.5	78.9
751	Waxahackie	97.5	56.3	79.2
752-753	Dallas	98.3	67.1	84.4
754	Greenville	97.7	40.7	72.4
755	Texarkana	96.6	54.0	77.7
756	Longview	96.9	42.6	72.8
757	Tyler	97.4	56.2	79.1
758	Palestine	93.6	43.0	71.1
759	Lufkin	94.5	46.7	73.3
760-761	Fort Worth	96.2	62.8	81.4
762	Denton	96.5	51.6	76.6
763	Wichita Falls	97.1	57.9	79.7
764	Eastland	95.6	42.0	71.8
765	Temple	94.5	50.3	74.9
766-767	Waco	96.4	57.5	79.1
768	Brownwood	96.5	39.5	71.2
769	San Angelo	96.2	47.7	74.7
770-772	Houston	99.6	73.0	87.7
773	Huntsville	97.7	40.3	72.2
774	Wharton	99.1	45.1	75.1
775	Galveston	97.3	70.6	85.4
776-777	Beaumont	97.1	64.1	82.4
778	Bryan	94.7	64.5	81.3
779	Victoria	99.2	47.6	76.3
780	Laredo	94.6	53.7	76.5
781-782	San Antonio	95.0	67.0	82.5
783-784	Corpus Christi	97.3	53.1	77.7
785	Mc Allen	97.2	47.4	75.1
786-787	Austin	94.6	61.7	80.0

STATE/ZIP	CITY	MAT.	INST.	TOTAL
TEXAS (CONT'D)				
788	Del Rio	96.5	33.2	68.4
789	Giddings	94.0	42.1	71.0
790-791	Amarillo	97.2	58.0	79.8
792	Childress	96.1	52.5	76.7
793-794	Lubbock	98.6	53.5	78.6
795-796	Abilene	96.6	53.3	77.4
797	Midland	98.7	49.9	77.0
798-799,885	El Paso	96.6	51.6	76.6
UTAH				
840-841	Salt Lake City	101.3	70.7	87.7
842,844	Ogden	97.4	70.5	85.5
843	Logan	99.1	70.5	86.4
845	Price	99.6	50.1	77.6
846-847	Provo	99.6	70.5	86.7
VERMONT				
050	White River Jct.	98.6	46.5	75.4
051	Bellows Falls	97.1	48.8	75.6
052	Bennington	97.4	49.1	76.0
053	Brattleboro	97.9	49.3	76.3
054	Burlington	100.5	64.9	84.7
056	Montpelier	97.3	64.9	82.9
057	Rutland	99.5	64.9	84.1
058	St. Johnsbury	98.8	49.1	76.7
059	Guildhall	97.4	48.6	75.7
VIRGINIA				
220-221	Fairfax	99.3	80.6	91.0
222	Arlington	100.5	79.3	91.1
223	Alexandria	99.9	84.9	93.3
224-225	Fredericksburg	98.0	68.4	84.9
226	Winchester	98.6	53.8	78.7
227	Culpeper	98.5	55.3	79.3
228	Harrisonburg	98.8	47.1	75.8
229	Charlottesville	99.1	61.5	82.4
230-232	Richmond	100.3	66.7	85.4
233-235	Norfolk	100.0	65.8	84.8
236	Newport News	99.6	64.7	84.1
237	Portsmouth	98.9	65.5	84.0
238	Petersburg	98.9	66.7	84.6
239	Farmville	98.3	41.1	72.9
240-241	Roanoke	100.2	48.1	77.1
242	Bristol	97.9	49.0	76.2
243	Pulaski	97.7	42.4	73.1
244	Staunton	98.4	45.6	75.0
245	Lynchburg	98.6	52.2	78.0
246	Grundy	98.0	44.4	74.2
WASHINGTON				
980-981,987	Seattle	103.8	104.7	104.2
982	Everett	103.6	101.0	102.4
983-984	Tacoma	103.5	101.4	102.6
985	Olympia	102.3	101.3	101.9
986	Vancouver	105.4	97.7	102.0
988	Wenatchee	103.5	86.6	96.0
989	Yakima	103.7	91.8	98.4
990-992	Spokane	105.7	83.3	95.8
993	Richland	105.0	86.9	96.9
994	Clarkston	103.9	84.3	95.2
WEST VIRGINIA				
247-248	Bluefield	96.8	79.4	89.1
249	Lewisburg	98.3	83.1	91.6
250-253	Charleston	99.5	91.0	95.7
254	Martinsburg	98.1	80.0	90.1
255-257	Huntington	99.6	93.4	96.9
258-259	Beckley	96.6	88.8	93.1
260	Wheeling	99.7	94.2	97.2
261	Parkersburg	98.7	92.6	96.0
262	Buckhannon	98.1	93.4	96.0
263-264	Clarksburg	98.5	93.4	96.2
265	Morgantown	98.6	93.5	96.4
266	Gassaway	97.9	92.2	95.4
267	Romney	97.9	86.5	92.9
268	Petersburg	97.9	89.3	94.1
WISCONSIN				
530,532	Milwaukee	99.0	106.2	102.2
531	Kenosha	98.5	103.1	100.5
534	Racine	98.1	102.9	100.2
535	Beloit	98.0	98.4	98.2
537	Madison	98.3	98.1	98.2

STATE/ZIP	CITY	MAT.	INST.	TOTAL
WISCONSIN (CONT'D)				
538	Lancaster	95.8	94.0	95.0
539	Portage	94.4	97.7	95.9
540	New Richmond	95.5	98.5	96.8
541-543	Green Bay	99.7	95.2	97.7
544	Wausau	94.9	94.3	94.6
545	Rhinelander	98.0	93.7	96.1
546	La Crosse	96.1	95.7	95.9
547	Eau Claire	97.8	96.2	97.1
548	Superior	95.4	102.8	98.7
549	Oshkosh	95.5	95.2	95.4
WYOMING				
820	Cheyenne	97.6	61.2	81.4
821	Yellowstone Nat'l Park	96.8	54.6	78.0
822	Wheatland	98.1	56.0	79.4
823	Rawlins	99.4	53.1	78.8
824	Worland	97.4	51.4	77.0
825	Riverton	98.4	53.1	78.3
826	Casper	97.7	59.9	80.9
827	Newcastle	97.2	53.1	77.6
828	Sheridan	98.1	58.7	80.6
829-831	Rock Springs	101.5	51.4	79.2
CANADIAN FACTORS (reflect Canadian currency)				
ALBERTA				
	Calgary	124.3	79.6	104.4
	Edmonton	125.9	79.6	105.3
	Fort McMurray	116.6	79.1	100.0
	Lethbridge	117.6	78.6	100.3
	Lloydminster	116.7	79.1	100.0
	Medicine Hat	116.7	78.6	99.8
	Red Deer	117.1	78.6	100.0
BRITISH COLUMBIA				
	Kamloops	117.8	81.8	101.8
	Prince George	119.1	81.8	102.5
	Vancouver	125.7	85.6	107.9
	Victoria	119.1	82.3	102.8
MANITOBA				
	Brandon	117.0	68.5	95.5
	Portage la Prairie	117.0	68.5	95.5
	Winnipeg	124.1	68.9	99.6
NEW BRUNSWICK				
	Bathurst	115.1	60.9	91.1
	Dalhousie	115.1	60.9	91.1
	Fredericton	117.0	64.9	93.8
	Moncton	115.4	60.9	91.2
	Newcastle	115.1	60.9	91.1
	Saint John	118.1	64.9	94.5
NEWFOUNDLAND				
	Corner Brook	119.8	60.9	93.7
	St. John's	120.1	60.9	93.8
NORTHWEST TERRITORIES				
	Yellowknife	120.4	77.4	101.3
NOVA SCOTIA				
	Dartmouth	117.8	67.1	95.3
	Halifax	119.3	68.2	96.6
	New Glasgow	115.8	67.1	94.2
	Sydney	113.3	67.1	92.8
	Yarmouth	115.6	67.1	94.1
ONTARIO				
	Barrie	119.9	84.1	104.0
	Brantford	119.1	87.6	105.1
	Cornwall	119.1	84.3	103.6
	Hamilton	124.5	87.9	108.2
	Kingston	120.1	84.7	104.4
	Kitchener	115.6	83.7	101.4
	London	121.9	85.3	105.6
	North Bay	119.1	82.4	102.8
	Oshawa	118.7	85.7	104.1
	Ottawa	120.5	85.5	105.0
	Owen Sound	120.1	82.5	103.4
	Peterborough	119.1	84.1	103.5
	Sarnia	119.3	87.9	105.4
	St. Catharines	113.6	83.1	100.0
	Sudbury	113.7	82.7	99.9

Location Factors

STATE/ZIP	CITY	MAT.	INST.	TOTAL
ONTARIO (CONT'D)	Thunder Bay	114.8	82.8	100.6
	Toronto	124.2	92.1	109.9
	Windsor	114.6	85.4	101.6
PRINCE EDWARD ISLAND	Charlottetown	117.7	56.9	90.7
	Summerside	117.3	56.9	90.5
QUEBEC	Cap-de-la-Madeleine	115.9	78.9	99.5
	Charlesbourg	115.9	78.9	99.5
	Chicoutimi	115.0	78.6	98.8
	Gatineau	115.3	78.7	99.1
	Laval	115.2	78.7	99.0
	Montreal	118.5	79.0	101.0
	Quebec	119.4	79.4	101.6
	Sherbrooke	115.7	78.7	99.3
	Trois Rivieres	116.2	78.9	99.6
SASKATCHEWAN	Moose Jaw	114.2	62.9	91.4
	Prince Albert	113.7	62.7	91.1
	Regina	116.1	62.9	92.5
	Saskatoon	114.6	62.8	91.6
YUKON	Whitehorse	114.0	61.9	90.8

R011105-05 Tips for Accurate Estimating

1. Use pre-printed or columnar forms for orderly sequence of dimensions and locations and for recording telephone quotations.

2. Use only the front side of each paper or form except for certain pre-printed summary forms.

3. Be consistent in listing dimensions: For example, length x width x height. This helps in rechecking to ensure that, the total length of partitions is appropriate for the building area.

4. Use printed (rather than measured) dimensions where given.

5. Add up multiple printed dimensions for a single entry where possible.

6. Measure all other dimensions carefully.

7. Use each set of dimensions to calculate multiple related quantities.

8. Preferred metric scales, all multiples of 1, 2 or 5, shall be used (1:100 equals approximately 1/8″ - 1′).

9. Do not "round off" quantities until the final summary.

10. Mark drawings with different colors as items are taken off.

11. Keep similar items together, different items separate.

12. Identify location and drawing numbers to aid in future checking for completeness.

13. Measure or list everything on the drawings or mentioned in the specifications.

14. It may be necessary to list items not called for to make the job complete.

15. Be alert for: Notes on plans such as N.T.S. (not to scale); changes in scale throughout the drawings; reduced size drawings; discrepancies between the specifications and the drawings.

16. Develop a consistent pattern of performing an estimate. For example:
    a. Start the quantity takeoff at the lower floor and move to the next higher floor.
    b. Proceed from the main section of the building to the wings.
    c. Proceed from south to north or vice versa, clockwise or counterclockwise.
    d. Take off floor plan quantities first, elevations next, then detail drawings.

17. List all gross dimensions that can be either used again for different quantities, or used as a rough check of other quantities for verification (exterior perimeter, gross floor area, individual floor areas, etc.).

18. Utilize design symmetry or repetition (repetitive floors, repetitive wings, symmetrical design around a center line, similar room layouts, etc.). Note: Extreme caution is needed here so as not to omit or duplicate an area.

19. When figuring alternatives, it is best to total all items involved in the basic system, then total all items involved in the alternates. Therefore you work with positive numbers in all cases. When adds and deducts are used, it is often confusing whether to add or subtract a portion of an item; especially on a complicated or involved alternate.

R011105-10 Unit Gross Area Requirements

The figures in the table below indicate typical ranges in square meters as a function of the "occupant" unit. This table is best used in the preliminary design stages to help determine the probable size requirement for the total project.

Building Type	Unit	Gross Area in Square Meters		
		1/4	Median	3/4
Apartments	Unit	6.1	80	102
Auditorium & Play Theaters	Seat	1.7	2.3	3.5
Bowling Alleys	Lane		87	
Churches & Synagogues	Seat	1.9	2.6	3.6
Dormitories	Bed	19	21	26
Fraternity & Sorority Houses	Bed	20	29	34
Garages, Parking	Car	30	33	36
Hospitals	Bed	64	79	100
Hotels	Rental Unit	44	56	66
Housing for the elderly	Unit	48	59	70
Housing, Public	Unit	65	81	96
Ice Skating Rinks	Total	2500	2800	3300
Motels	Rental Unit	33	43	58
Nursing Homes	Bed	27	33	42
Restaurants	Seat	2.1	2.7	3.6
Schools, Elementary	Pupil	6	7	8
Junior High & Middle	↓	8	10	12
Senior High		9	12	14
Vocational	↓	10	13	18
Shooting Ranges	Point		42	
Theaters & Movies	Seat		1.4	

R011105-20 Floor Area Ratios

Table below lists commonly used gross to net area and net to gross area ratios expressed in % for various building types.

Building Type	Gross to Net Ratio	Net to Gross Ratio	Building Type	Gross to Net Ratio	Net to Gross Ratio
Apartment	156	64	School Buildings (campus type)		
Bank	140	72	Administrative	150	67
Church	142	70	Auditorium	142	70
Courthouse	162	61	Biology	161	62
Department Store	123	81	Chemistry	170	59
Garage	118	85	Classroom	152	66
Hospital	183	55	Dining Hall	138	72
Hotel	158	63	Dormitory	154	65
Laboratory	171	58	Engineering	164	61
Library	132	76	Fraternity	160	63
Office	135	75	Gymnasium	142	70
Restaurant	141	70	Science	167	60
Warehouse	108	93	Service	120	83
			Student Union	172	59

The gross area of a building is the total floor area based on outside dimensions.

The net area of a building is the usable floor area for the function intended and excludes such items as stairways, corridors and mechanical rooms. In the case of a commercial building, it might be considered as the "leasable area."

R011105-30 Occupancy Determinations

Description		Square Meters Required per Person		
		BOCA	SBC	UBC
Assembly Areas	Fixed Seats	**	0.6	0.7
	Movable Seats		1.4	1.4
	Concentrated	0.7		
	Unconcentrated	1.4		
	Standing Space	0.3		
Educational	Unclassified			
	Classrooms	1.9	3.7	1.9
	Shop Areas	4.6	9.3	4.6
Institutional	Unclassified		11.6	
	In-Patient Areas	22.3		
	Sleeping Areas	11.1		
Mercantile	Basement	2.8	2.8	1.9
	Ground Floor	2.8	2.8	2.8
	Upper Floors	5.6	5.6	4.6
Office		9.3	9.3	9.3

BOCA=Building Officials & Code Administrators
SBC=Southern Building Code
UBC=Uniform Building Code

** The occupancy load for assembly area with fixed seats shall be determined by the number of fixed seats installed.

RO111105-40 Weather Data and Design Conditions

City	Latitude (1) 0	1'	Winter Temperatures (1) Med. of Annual Extremes	99%	97½%	Winter Degree Days (2)	Summer (Design Dry Bulb) Temperatures and Relative Humidity 1%	2½%	5%
UNITED STATES									
Albuquerque, NM	35	0	5.1	12	16	4,400	96/61	94/61	92/61
Atlanta, GA	33	4	11.9	17	22	3,000	94/74	92/74	90/73
Baltimore, MD	39	2	7	14	17	4,600	94/75	91/75	89/74
Birmingham, AL	33	3	13	17	21	2,600	96/74	94/75	92/74
Bismarck, ND	46	5	-32	-23	-19	8,800	95/68	91/68	88/67
Boise, ID	43	3	1	3	10	5,800	96/65	94/64	91/64
Boston, MA	42	2	-1	6	9	5,600	91/73	88/71	85/70
Burlington, VT	44	3	-17	-12	-7	8,200	88/72	85/70	82/69
Charleston, WV	38	2	3	7	11	4,400	92/74	90/73	87/72
Charlotte, NC	35	1	13	18	22	3,200	95/74	93/74	91/74
Casper, WY	42	5	-21	-11	-5	7,400	92/58	90/57	87/57
Chicago, IL	41	5	-8	-3	2	6,600	94/75	91/74	88/73
Cincinnati, OH	39	1	0	1	6	4,400	92/73	90/72	88/72
Cleveland, OH	41	2	-3	1	5	6,400	91/73	88/72	86/71
Columbia, SC	34	0	16	20	24	2,400	97/76	95/75	93/75
Dallas, TX	32	5	14	18	22	2,400	102/75	100/75	97/75
Denver, CO	39	5	-10	-5	1	6,200	93/59	91/59	89/59
Des Moines, IA	41	3	-14	-10	-5	6,600	94/75	91/74	88/73
Detroit, MI	42	2	-3	3	6	6,200	91/73	88/72	86/71
Great Falls, MT	47	3	-25	-21	-15	7,800	91/60	88/60	85/59
Hartford, CT	41	5	-4	3	7	6,200	91/74	88/73	85/72
Houston, TX	29	5	24	28	33	1,400	97/77	95/77	93/77
Indianapolis, IN	39	4	-7	-2	2	5,600	92/74	90/74	87/73
Jackson, MS	32	2	16	21	25	2,200	97/76	95/76	93/76
Kansas City, MO	39	1	-4	2	6	4,800	99/75	96/74	93/74
Las Vegas, NV	36	1	18	25	28	2,800	108/66	106/65	104/65
Lexington, KY	38	0	-1	3	8	4,600	93/73	91/73	88/72
Little Rock, AR	34	4	11	15	20	3,200	99/76	96/77	94/77
Los Angeles, CA	34	0	36	41	43	2,000	93/70	89/70	86/69
Memphis, TN	35	0	10	13	18	3,200	98/77	95/76	93/76
Miami, FL	25	5	39	44	47	200	91/77	90/77	89/77
Milwaukee, WI	43	0	-11	-8	-4	7,600	90/74	87/73	84/71
Minneapolis, MN	44	5	-22	-16	-12	8,400	92/75	89/73	86/71
New Orleans, LA	30	0	28	29	33	1,400	93/78	92/77	90/77
New York, NY	40	5	6	11	15	5,000	92/74	89/73	87/72
Norfolk, VA	36	5	15	20	22	3,400	93/77	91/76	89/76
Oklahoma City, OK	35	2	4	9	13	3,200	100/74	97/74	95/73
Omaha, NE	41	2	-13	-8	-3	6,600	94/76	91/75	88/74
Philadelphia, PA	39	5	6	10	14	4,400	93/75	90/74	87/72
Phoenix, AZ	33	3	27	31	34	1,800	109/71	107/71	105/71
Pittsburgh, PA	40	3	-1	3	7	6,000	91/72	88/71	86/70
Portland, ME	43	4	-10	-6	-1	7,600	87/72	84/71	81/69
Portland, OR	45	4	18	17	23	4,600	89/68	85/67	81/65
Portsmouth, NH	43	1	-8	-2	2	7,200	89/73	85/71	83/70
Providence, RI	41	4	-1	5	9	6,000	89/73	86/72	83/70
Rochester, NY	43	1	-5	1	5	6,800	91/73	88/71	85/70
Salt Lake City, UT	40	5	0	3	8	6,000	97/62	95/62	92/61
San Francisco, CA	37	5	36	38	40	3,000	74/63	71/62	69/61
Seattle, WA	47	4	22	22	27	5,200	85/68	82/66	78/65
Sioux Falls, SD	43	4	-21	-15	-11	7,800	94/73	91/72	88/71
St. Louis, MO	38	4	-3	3	8	5,000	98/75	94/75	91/75
Tampa, FL	28	0	32	36	40	680	92/77	91/77	90/76
Trenton, NJ	40	1	4	11	14	5,000	91/75	88/74	85/73
Washington, DC	38	5	7	14	17	4,200	93/75	91/74	89/74
Wichita, KS	37	4	-3	3	7	4,600	101/72	98/73	96/73
Wilmington, DE	39	4	5	10	14	5,000	92/74	89/74	87/73
ALASKA									
Anchorage	61	1	-29	-23	-18	10,800	71/59	68/58	66/56
Fairbanks	64	5	-59	-51	-47	14,280	82/62	78/60	75/59
CANADA									
Edmonton, Alta.	53	3	-30	-29	-25	11,000	85/66	82/65	79/63
Halifax, N.S.	44	4	-4	1	5	8,000	79/66	76/65	74/64
Montreal, Que.	45	3	-20	-16	-10	9,000	88/73	85/72	83/71
Saskatoon, Sask.	52	1	-35	-35	-31	11,000	89/68	86/66	83/65
St. John's, N.F.	47	4	1	3	7	8,600	77/66	75/65	73/64
Saint John, N.B.	45	2	-15	-12	-8	8,200	80/67	77/65	75/64
Toronto, Ont.	43	4	-10	-5	-1	7,000	90/73	87/72	85/71
Vancouver, B.C.	49	1	13	15	19	6,000	79/67	77/66	74/65
Winnipeg, Man.	49	5	-31	-30	-27	10,800	89/73	86/71	84/70

(1) Handbook of Fundamentals, ASHRAE, Inc., NY 1989
(2) Local Climatological Annual Survey, USDC Env. Science Services
Administration, Asheville, NC

R011105-50 Metric Conversion Factors

Description: This table is primarily for converting customary U.S. units in the left hand column to SI metric units in the right hand column. In addition, conversion factors for some commonly encountered Canadian and non-SI metric units are included.

If You Know		Multiply By		To Find
Length	Inches x	25.4[a]	=	Millimeters
	Feet x	0.3048[a]	=	Meters
	Yards x	0.9144[a]	=	Meters
	Miles (statute) x	1.609	=	Kilometers
Area	Square inches x	645.2	=	Square millimeters
	Square feet x	0.0929	=	Square meters
	Square yards x	0.8361	=	Square meters
Volume	Cubic inches x	16,387	=	Cubic millimeters
(Capacity)	Cubic feet x	0.02832	=	Cubic meters
	Cubic yards x	0.7646	=	Cubic meters
	Gallons (U.S. liquids)[b] x	0.003785	=	Cubic meters[c]
	Gallons (Canadian liquid)[b] x	0.004546	=	Cubic meters[c]
	Ounces (U.S. liquid)[b] x	29.57	=	Milliliters[c, d]
	Quarts (U.S. liquid)[b] x	0.9464	=	Liters[c, d]
	Gallons (U.S. liquid)[b] x	3.785	=	Liters[c, d]
Force	Kilograms force[d] x	9.807	=	Newtons
	Pounds force x	4.448	=	Newtons
	Pounds force x	0.4536	=	Kilograms force[d]
	Kips x	4448	=	Newtons
	Kips x	453.6	=	Kilograms force[d]
Pressure,	Kilograms force per square centimeter[d] x	0.09807	=	Megapascals
Stress,	Pounds force per square inch (psi) x	0.006895	=	Megapascals
Strength	Kips per square inch x	6.895	=	Megapascals
(Force per unit area)	Pounds force per square inch (psi) x	0.07031	=	Kilograms force per square centimeter[d]
	Pounds force per square foot x	47.88	=	Pascals
	Pounds force per square foot x	4.882	=	Kilograms force per square meter[d]
Bending	Inch-pounds force x	0.01152	=	Meter-kilograms force[d]
Moment	Inch-pounds force x	0.1130	=	Newton-meters
Or Torque	Foot-pounds force x	0.1383	=	Meter-kilograms force[d]
	Foot-pounds force x	1.356	=	Newton-meters
	Meter-kilograms force[d] x	9.807	=	Newton-meters
Mass	Ounces (avoirdupois) x	28.35	=	Grams
	Pounds (avoirdupois) x	0.4536	=	Kilograms
	Tons (metric) x	1000	=	Kilograms
	Tons, short (2000 pounds) x	907.2	=	Kilograms
	Tons, short (2000 pounds) x	0.9072	=	Megagrams[e]
Mass per	Pounds mass per cubic foot x	16.02	=	Kilograms per cubic meter
Unit	Pounds mass per cubic yard x	0.5933	=	Kilograms per cubic meter
Volume	Pounds mass per gallon (U.S. liquid)[b] x	119.8	=	Kilograms per cubic meter
	Pounds mass per gallon (Canadian liquid)[b] x	99.78	=	Kilograms per cubic meter
Temperature	Degrees Fahrenheit (F-32)/1.8		=	Degrees Celsius
	Degrees Fahrenheit (F+459.67)/1.8		=	Degrees Kelvin
	Degrees Celsius C+273.15		=	Degrees Kelvin

[a] The factor given is exact
[b] One U.S. gallon = 0.8327 Canadian gallon
[c] 1 liter = 1000 milliliters = 1000 cubic centimeters
 1 cubic decimeter = 0.001 cubic meter

[d] Metric but not SI unit
[e] Called "tonne" in England and "metric ton" in other metric countries

R011105-60 Weights and Measures

Measures of Length
1 Mile = 1760 Yards = 5280 Feet
1 Yard = 3 Feet = 36 inches
1 Foot = 12 Inches
1 Mil = 0.001 Inch
1 Fathom = 2 Yards = 6 Feet
1 Rod = 5.5 Yards = 16.5 Feet
1 Hand = 4 Inches
1 Span = 9 Inches
1 Micro-inch = One Millionth Inch or 0.000001 Inch
1 Micron = One Millionth Meter + 0.00003937 Inch

Surveyor's Measure
1 Mile = 8 Furlongs = 80 Chains
1 Furlong = 10 Chains = 220 Yards
1 Chain = 4 Rods = 22 Yards = 66 Feet = 100 Links
1 Link = 7.92 Inches

Square Measure
1 Square Mile = 640 Acres = 6400 Square Chains
1 Acre = 10 Square Chains = 4840 Square Yards = 43,560 Sq. Ft.
1 Square Chain = 16 Square Rods = 484 Square Yards = 4356 Sq. Ft.
1 Square Rod = 30.25 Square Yards = 272.25 Square Feet = 625 Square Lines
1 Square Yard = 9 Square Feet
1 Square Foot = 144 Square Inches
An Acre equals a Square 208.7 Feet per Side

Cubic Measure
1 Cubic Yard = 27 Cubic Feet
1 Cubic Foot = 1728 Cubic Inches
1 Cord of Wood = 4 x 4 x 8 Feet = 128 Cubic Feet
1 Perch of Masonry = 16½ x 1½ x 1 Foot = 24.75 Cubic Feet

Avoirdupois or Commercial Weight
1 Gross or Long Ton = 2240 Pounds
1 Net or Short ton = 2000 Pounds
1 Pound = 16 Ounces = 7000 Grains
1 Ounce = 16 Drachms = 437.5 Grains
1 Stone = 14 Pounds

Shipping Measure
For Measuring Internal Capacity of a Vessel:
  1 Register Ton = 100 Cubic Feet

For Measurement of Cargo:
  Approximately 40 Cubic Feet of Merchandise is considered a Shipping Ton, unless that bulk would weigh more than 2000 Pounds, in which case Freight Charge may be based upon weight.

40 Cubic Feet = 32.143 U.S. Bushels = 31.16 Imp. Bushels

Liquid Measure
1 Imperial Gallon = 1.2009 U.S. Gallon = 277.42 Cu. In.
1 Cubic Foot = 7.48 U.S. Gallons

R011110-10 Architectural Fees

Tabulated below are typical percentage fees by project size, for good professional architectural service. Fees may vary from those listed depending upon degree of design difficulty and economic conditions in any particular area.

Rates can be interpolated horizontally and vertically. Various portions of the same project requiring different rates should be adjusted proportionately. For alterations, add 50% to the fee for the first $500,000 of project cost and add 25% to the fee for project cost over $500,000.

Architectural fees tabulated below include Structural, Mechanical and Electrical Engineering Fees. They do not include the fees for special consultants such as kitchen planning, security, acoustical, interior design, etc.

Civil Engineering fees are included in the Architectural fee for project sites requiring minimal design such as city sites. However, separate Civil Engineering fees must be added when utility connections require design, drainage calculations are needed, stepped foundations are required, or provisions are required to protect adjacent wetlands.

Building Types	Total Project Size in Thousands of Dollars						
	100	250	500	1,000	5,000	10,000	50,000
Factories, garages, warehouses, repetitive housing	9.0%	8.0%	7.0%	6.2%	5.3%	4.9%	4.5%
Apartments, banks, schools, libraries, offices, municipal buildings	12.2	12.3	9.2	8.0	7.0	6.6	6.2
Churches, hospitals, homes, laboratories, museums, research	15.0	13.6	12.7	11.9	9.5	8.8	8.0
Memorials, monumental work, decorative furnishings	—	16.0	14.5	13.1	10.0	9.0	8.3

R011110-30   Engineering Fees

Typical **Structural Engineering Fees** based on type of construction and total project size. These fees are included in Architectural Fees.

Type of Construction	Total Project Size (in thousands of dollars)			
	$500	$500-$1,000	$1,000-$5,000	Over $5000
Industrial buildings, factories & warehouses	Technical payroll times 2.0 to 2.5	1.60%	1.25%	1.00%
Hotels, apartments, offices, dormitories, hospitals, public buildings, food stores		2.00%	1.70%	1.20%
Museums, banks, churches and cathedrals		2.00%	1.75%	1.25%
Thin shells, prestressed concrete, earthquake resistive		2.00%	1.75%	1.50%
Parking ramps, auditoriums, stadiums, convention halls, hangars & boiler houses		2.50%	2.00%	1.75%
Special buildings, major alterations, underpinning & future expansion		Add to above 0.5%	Add to above 0.5%	Add to above 0.5%

For complex reinforced concrete or unusually complicated structures, add 20% to 50%.

Typical **Mechanical and Electrical Engineering Fees** are based on the size of the subcontract. The fee structure for both are shown below. These fees are included in Architectural Fees.

Type of Construction	Subcontract Size							
	$25,000	$50,000	$100,000	$225,000	$350,000	$500,000	$750,000	$1,000,000
Simple structures	6.4%	5.7%	4.8%	4.5%	4.4%	4.3%	4.2%	4.1%
Intermediate structures	8.0	7.3	6.5	5.6	5.1	5.0	4.9	4.8
Complex structures	10.1	9.0	9.0	8.0	7.5	7.5	7.0	7.0

For renovations, add 15% to 25% to applicable fee.

R012153-10   Repair and Remodeling

Cost figures are based on new construction utilizing the most cost-effective combination of labor, equipment and material with the work scheduled in proper sequence to allow the various trades to accomplish their work in an efficient manner.

The costs for repair and remodeling work must be modified due to the following factors that may be present in any given repair and remodeling project.

1. Equipment usage curtailment due to the physical limitations of the project, with only hand-operated equipment being used.

2. Increased requirement for shoring and bracing to hold up the building while structural changes are being made and to allow for temporary storage of construction materials on above-grade floors.

3. Material handling becomes more costly due to having to move within the confines of an enclosed building. For multi-story construction, low capacity elevators and stairwells may be the only access to the upper floors.

4. Large amount of cutting and patching and attempting to match the existing construction is required. It is often more economical to remove entire walls rather than create many new door and window openings. This sort of trade-off has to be carefully analyzed.

5. Cost of protection of completed work is increased since the usual sequence of construction usually cannot be accomplished.

6. Economies of scale usually associated with new construction may not be present. If small quantities of components must be custom fabricated due to job requirements, unit costs will naturally increase. Also, if only small work areas are available at a given time, job scheduling between trades becomes difficult and subcontractor quotations may reflect the excessive start-up and shut-down phases of the job.

7. Work may have to be done on other than normal shifts and may have to be done around an existing production facility which has to stay in production during the course of the repair and remodeling.

8. Dust and noise protection of adjoining non-construction areas can involve substantial special protection and alter usual construction methods.

9. Job may be delayed due to unexpected conditions discovered during demolition or removal. These delays ultimately increase construction costs.

10. Piping and ductwork runs may not be as simple as for new construction. Wiring may have to be snaked through walls and floors.

11. Matching "existing construction" may be impossible because materials may no longer be manufactured. Substitutions may be expensive.

12. Weather protection of existing structure requires additional temporary structures to protect building at openings.

13. On small projects, because of local conditions, it may be necessary to pay a tradesman for a minimum of four hours for a task that is completed in one hour.

All of the above areas can contribute to increased costs for a repair and remodeling project. Each of the above factors should be considered in the planning, bidding and construction stage in order to minimize the increased costs associated with repair and remodeling jobs.

RO12157-20 Construction Time Requirements

Table at left is average construction time in months for different types of building projects. Table at right is the construction time in months for different size projects. Design time runs 25% to 40% of construction time.

Type Building	Construction Time	Project Value	Construction Time
Industrial Buildings	12 Months	Under $1,400,000	10 Months
Commercial Buildings	15 Months	Up to $3,800,000	15 Months
Research & Development	18 Months	Up to $19,000,000	21 Months
Institutional Buildings	20 Months	over $19,000,000	28 Months

RO12909-80 Sales Tax by State

State sales tax on materials is tabulated below (5 states have no sales tax). Many states allow local jurisdictions, such as a county or city, to levy additional sales tax.

Some projects may be sales tax exempt, particularly those constructed with public funds.

State	Tax (%)	State	Tax (%)	State	Tax (%)	State	Tax (%)
Alabama	4	Illinois	6.25	Montana	0	Rhode Island	7
Alaska	0	Indiana	6	Nebraska	5.5	South Carolina	5
Arizona	5.6	Iowa	5	Nevada	6.5	South Dakota	4
Arkansas	6	Kansas	5.3	New Hampshire	0	Tennessee	7
California	6.25	Kentucky	6	New Jersey	6	Texas	6.25
Colorado	2.9	Louisiana	4	New Mexico	5	Utah	4.75
Connecticut	6	Maine	5	New York	4.25	Vermont	6
Delaware	0	Maryland	5	North Carolina	4.5	Virginia	4
District of Columbia	5.75	Massachusetts	5	North Dakota	5	Washington	6.5
Florida	6	Michigan	6	Ohio	6	West Virginia	6
Georgia	4	Minnesota	6.5	Oklahoma	4.5	Wisconsin	5
Hawaii	4	Mississippi	7	Oregon	0	Wyoming	4
Idaho	6	Missouri	4.225	Pennsylvania	6	Average	4.83 %

Sales Tax by Province (Canada)

GST - a value-added tax, which the government imposes on most goods and services provided in or imported into Canada. PST - a retail sales tax, which five of the provinces impose on the price of most goods and some services. QST - a value-added tax, similar to the federal GST, which Quebec imposes. HST - Three provinces have combined their retail sales tax with the federal GST into one harmonized tax.

Province	PST (%)	QST (%)	GST(%)	HST(%)
Alberta	0	0	7	0
British Columbia	7.5	0	7	0
Manitoba	7	0	7	0
New Brunswick	0	0	0	15
Newfoundland	0	0	0	15
Northwest Territories	0	0	7	0
Nova Scotia	0	0	0	15
Ontario	8	0	7	0
Prince Edward Island	10	0	7	0
Quebec	0	7.5	7	0
Saskatchewan	6	0	7	0
Yukon	0	0	7	0

R012909-85   Unemployment Taxes and Social Security Taxes

State Unemployment Tax rates vary not only from state to state, but also with the experience rating of the contractor. The Federal Unemployment Tax rate is 6.2% of the first $7,000 of wages. This is reduced by a credit of up to 5.4% for timely payment to the state. The minimum Federal Unemployment Tax is 0.8% after all credits.

Social Security (FICA) for 2006 is estimated at time of publication to be 7.65% of wages up to $90,000.

R012909-90   Overtime

One way to improve the completion date of a project or eliminate negative float from a schedule is to compress activity duration times. This can be achieved by increasing the crew size or working overtime with the proposed crew.

To determine the costs of working overtime to compress activity duration times, consider the following examples. Below is an overtime efficiency and cost chart based on a five, six, or seven day week with an eight through twelve hour day. Payroll percentage increases for time and one half and double time are shown for the various working days.

Days per Week	Hours per Day	Production Efficiency					Payroll Cost Factors	
		1 Week	2 Weeks	3 Weeks	4 Weeks	Average 4 Weeks	@ 1-1/2 Times	@ 2 Times
	8	100%	100%	100%	100%	100 %	100 %	100 %
	9	100	100	95	90	96.25	105.6	111.1
5	10	100	95	90	85	91.25	110.0	120.0
	11	95	90	75	65	81.25	113.6	127.3
	12	90	85	70	60	76.25	116.7	133.3
	8	100	100	95	90	96.25	108.3	116.7
	9	100	95	90	85	92.50	113.0	125.9
6	10	95	90	85	80	87.50	116.7	133.3
	11	95	85	70	65	78.75	119.7	139.4
	12	90	80	65	60	73.75	122.2	144.4
	8	100	95	85	75	88.75	114.3	128.6
	9	95	90	80	70	83.75	118.3	136.5
7	10	90	85	75	65	78.75	121.4	142.9
	11	85	80	65	60	72.50	124.0	148.1
	12	85	75	60	55	68.75	126.2	152.4

R013113-40   Builder's Risk Insurance

Builder's Risk Insurance is insurance on a building during construction. Premiums are paid by the owner or the contractor. Blasting, collapse and underground insurance would raise total insurance costs above those listed. Floater policy for materials delivered to the job runs $.75 to $1.25 per $100 value. Contractor equipment insurance runs $.50 to $1.50 per $100 value. Insurance for miscellaneous tools to $1,500 value runs from $3.00 to $7.50 per $100 value.

Tabulated below are New England Builder's Risk insurance rates in dollars per $100 value for $1,000 deductible. For $25,000 deductible, rates can be reduced 13% to 34%. On contracts over $1,000,000, rates may be lower than those tabulated. Policies are written annually for the total completed value in place. For "all risk" insurance (excluding flood, earthquake and certain other perils) add $.025 to total rates below.

Coverage	Frame Construction (Class 1)			Brick Construction (Class 4)			Fire Resistive (Class 6)		
	Range		Average	Range		Average	Range		Average
Fire Insurance	$.350 to	$ .850	$.600	$.158 to	$.189	$.174	$.052 to	$.080	$.070
Extended Coverage	.115 to	.200	.158	.080 to	.105	.101	.081 to	.105	.100
Vandalism	.012 to	.016	.014	.008 to	.011	.011	.008 to	.011	.010
Total Annual Rate	$.477 to	$1.066	$.772	$.246 to	$.305	$.286	$.141 to	$.196	$.180

R013113-50   General Contractor's Overhead

There are two distinct types of overhead on a construction project: Project Overhead and Main Office Overhead. Project Overhead includes those costs at a construction site not directly associated with the installation of construction materials. Examples of Project Overhead costs include the following:

1. Superintendent
2. Construction office and storage trailers
3. Temporary sanitary facilities
4. Temporary utilities
5. Security fencing
6. Photographs
7. Clean up
8. Performance and payment bonds

The above Project Overhead items are also referred to as General Requirements and therefore are estimated in Division 1. Division 1 is the first division listed in the CSI MasterFormat but it is usually the last division estimated. The sum of the costs in Divisions 1 through 16 is referred to as the sum of the direct costs.

All construction projects also include indirect costs. The primary components of indirect costs are the contractor's Main Office Overhead and profit. The amount of the Main Office Overhead expense varies depending on the the following:

1. Owner's compensation
2. Project managers and estimator's wages
3. Clerical support wages
4. Office rent and utilities
5. Corporate legal and accounting costs
6. Advertising
7. Automobile expenses
8. Association dues
9. Travel and entertainment expenses

These costs are usually calculated as a percentage of annual sales volume. This percentage can range from 35% for a small contractor doing less than $500,000 to 5% for a large contractor with sales in excess of $100 million.

R013113-60   Workers' Compensation Insurance Rates by Trade

The table below tabulates the national averages for Workers' Compensation insurance rates by trade and type of building. The average "Insurance Rate" is multiplied by the "% of Building Cost" for each trade. This produces the "Workers' Compensation Cost" by % of total labor cost, to be added for each trade by building type to determine the weighted average Workers' Compensation rate for the building types analyzed.

Trade	Insurance Rate (% Labor Cost)			% of Building Cost			Workers' Compensation		
	Range		Average	Office Bldgs.	Schools & Apts.	Mfg.	Office Bldgs.	Schools & Apts.	Mfg.
Excavation, Grading, etc.	4.2 % to	19.5%	10.6%	4.8%	4.9%	4.5%	.51%	.52%	.48%
Piles & Foundations	7.7 to	76.7	22.1	7.1	5.2	8.7	1.57	1.15	1.92
Concrete	5.2 to	35.8	15.8	5.0	14.8	3.7	.79	2.34	.58
Masonry	5.1 to	28.3	14.9	6.9	7.5	1.9	1.03	1.12	.28
Structural Steel	7.8 to	103.8	39.7	10.7	3.9	17.6	4.25	1.55	6.99
Miscellaneous & Ornamental Metals	4.9 to	24.8	12.2	2.8	4.0	3.6	.34	.49	.44
Carpentry & Millwork	7.0 to	53.2	18.4	3.7	4.0	0.5	.68	.74	.09
Metal or Composition Siding	5.3 to	32.1	17.3	2.3	0.3	4.3	.40	.05	.74
Roofing	7.8 to	78.9	32.4	2.3	2.6	3.1	.75	.84	1.00
Doors & Hardware	4.5 to	24.9	11.8	0.9	1.4	0.4	.11	.17	.05
Sash & Glazing	4.6 to	33.7	13.9	3.5	4.0	1.0	.49	.56	.14
Lath & Plaster	4.3 to	40.2	14.0	3.3	6.9	0.8	.46	.97	.11
Tile, Marble & Floors	2.4 to	31.8	9.5	2.6	3.0	0.5	.25	.29	.05
Acoustical Ceilings	2.4 to	29.7	11.4	2.4	0.2	0.3	.27	.02	.03
Painting	4.5 to	33.7	13.3	1.5	1.6	1.6	.20	.21	.21
Interior Partitions	7.0 to	53.2	18.4	3.9	4.3	4.4	.72	.79	.81
Miscellaneous Items	2.6 to	137.6	17.3	5.2	3.7	9.7	.90	.64	1.68
Elevators	2.8 to	42.8	6.9	2.1	1.1	2.2	.14	.08	.15
Sprinklers	3.2 to	22.1	8.6	0.5	—	2.0	.04	—	.17
Plumbing	2.7 to	12.4	8.2	4.9	7.2	5.2	.40	.59	.43
Heat., Vent., Air Conditioning	4.0 to	27.2	11.8	13.5	11.0	12.9	1.59	1.30	1.52
Electrical	2.6 to	12.7	6.8	10.1	8.4	11.1	.69	.57	.75
Total	2.4 % to	137.6%	—	100.0%	100.0%	100.0%	16.58%	14.99%	18.62%
				Overall Weighted Average	16.73%				

Workers' Compensation Insurance Rates by States

The table below lists the weighted average Workers' Compensation base rate for each state with a factor comparing this with the national average of 16.3%.

State	Weighted Average	Factor	State	Weighted Average	Factor	State	Weighted Average	Factor
Alabama	25.7%	158	Kentucky	20.5%	126	North Dakota	13.1%	80
Alaska	24.0	147	Louisiana	28.2	173	Ohio	14.5	89
Arizona	7.5	46	Maine	20.9	128	Oklahoma	14.7	90
Arkansas	13.7	84	Maryland	19.1	117	Oregon	13.0	80
California	18.8	115	Massachusetts	13.4	82	Pennsylvania	14.3	88
Colorado	10.9	67	Michigan	17.4	107	Rhode Island	21.1	129
Connecticut	22.4	137	Minnesota	25.5	156	South Carolina	16.8	103
Delaware	17.3	106	Mississippi	17.9	110	South Dakota	17.1	105
District of Columbia	17.1	105	Missouri	17.7	109	Tennessee	18.1	111
Florida	23.3	143	Montana	20.4	125	Texas	12.7	78
Georgia	22.9	140	Nebraska	22.9	140	Utah	12.3	75
Hawaii	16.9	104	Nevada	13.6	83	Vermont	25.0	153
Idaho	11.7	72	New Hampshire	22.6	139	Virginia	13.5	83
Illinois	18.8	115	New Jersey	12.2	75	Washington	10.9	67
Indiana	7.1	44	New Mexico	17.7	109	West Virginia	13.2	81
Iowa	12.3	75	New York	14.4	88	Wisconsin	14.7	90
Kansas	8.9	55	North Carolina	15.4	94	Wyoming	9.1	56
			Weighted Average for U.S. is	16.7% of payroll = 100%				

Rates in the following table are the base or manual costs per $100 of payroll for Workers' Compensation in each state. Rates are usually applied to straight time wages only and not to premium time wages and bonuses.

The weighted average skilled worker rate for 35 trades is 16.3%. For bidding purposes, apply the full value of Workers' Compensation directly to total labor costs, or if labor is 38%, materials 42% and overhead and profit 20% of total cost, carry 38/80 x 16.3% =7.7% of cost (before overhead and profit) into overhead. Rates vary not only from state to state but also with the experience rating of the contractor.

Rates are the most current available at the time of publication.

R013113-60 Workers' Compensation Insurance Rates by Trade and State (cont.)

State	Carpentry — 3 stories or less 5651	Carpentry — interior cab. work 5437	Carpentry — general 5403	Concrete Work — NOC 5213	Concrete Work — flat (flr., sdwk.) 5221	Electrical Wiring — inside 5190	Excavation — earth NOC 6217	Excavation — rock 6217	Glaziers 5462	Insulation Work 5479	Lathing 5443	Masonry 5022	Painting & Decorating 5474	Pile Driving 6003	Plastering 5480	Plumbing 5183	Roofing 5551	Sheet Metal Work (HVAC) 5538	Steel Erection — door & sash 5102	Steel Erection — inter, ornam. 5102	Steel Erection — structure 5040	Steel Erection — NOC 5057	Tile Work — (interior ceramic) 5348	Waterproofing 9014	Wrecking 5701
AL	29.57	16.77	31.62	11.24	10.07	8.55	12.90	12.90	26.88	17.64	12.34	23.46	32.15	38.02	32.13	11.79	65.04	21.69	19.37	19.37	56.15	31.21	12.21	6.54	56.15
AK	21.71	15.70	15.84	13.63	12.88	13.24	20.07	20.07	26.84	31.05	11.29	31.31	21.16	51.97	26.78	12.29	41.90	14.82	12.85	12.85	50.64	25.20	9.35	9.63	50.64
AZ	6.88	5.05	11.62	5.56	3.49	3.85	4.74	4.74	7.40	7.69	4.27	5.37	4.87	9.83	5.20	3.90	12.19	5.65	9.50	9.50	18.98	8.11	2.04	2.34	50.31
AR	11.87	9.14	15.06	11.65	6.81	5.31	9.16	9.16	11.83	30.45	7.61	10.44	10.67	16.93	12.24	5.61	22.08	12.99	7.43	7.43	36.81	16.98	6.36	4.67	38.24
CA	28.28	9.07	28.28	13.18	13.18	9.93	7.88	7.88	16.37	23.34	10.90	14.55	19.45	21.18	18.12	11.59	39.79	15.33	13.55	13.55	23.93	21.91	7.74	19.45	21.91
CO	14.58	6.77	10.47	11.38	6.17	4.52	8.41	8.41	7.33	10.76	4.73	10.57	8.54	13.61	8.00	7.15	19.17	10.23	6.98	6.98	27.39	12.94	7.07	4.70	27.39
CT	18.69	17.64	27.54	26.02	12.53	8.70	10.59	10.59	16.86	19.70	19.27	24.01	18.88	25.92	24.21	10.86	43.51	13.48	16.02	16.02	61.01	29.58	9.39	6.44	50.86
DE	17.74	17.74	14.29	14.46	11.19	7.77	11.13	11.13	13.41	14.29	15.42	14.16	18.57	23.83	15.42	9.28	31.57	11.76	14.60	14.60	33.94	14.60	11.81	14.16	33.94
DC	12.26	11.99	12.28	14.08	16.31	7.27	9.99	9.99	23.64	10.47	9.64	19.37	10.29	17.95	12.01	14.00	18.86	8.55	17.12	17.12	46.07	20.00	28.91	4.70	46.07
FL	24.98	19.24	28.32	24.68	12.25	10.16	12.77	12.77	18.92	17.94	13.27	20.39	19.51	47.98	24.53	10.82	37.58	16.40	13.31	13.31	56.10	32.45	10.28	8.90	56.10
GA	32.05	17.69	24.16	15.16	10.44	8.97	16.86	16.86	15.69	22.02	17.16	20.30	17.13	31.16	19.94	10.77	46.74	17.52	13.97	13.97	44.46	48.01	9.64	8.55	44.46
HI	17.38	11.62	28.20	13.74	12.27	7.10	7.73	7.73	20.55	21.19	10.94	17.87	11.33	20.15	15.61	6.06	33.24	8.02	11.11	11.11	31.71	21.70	9.78	11.54	31.71
ID	11.92	6.31	13.47	10.64	5.99	5.25	6.90	6.90	9.58	8.14	6.31	9.91	8.88	15.41	9.45	5.99	29.54	9.11	7.94	7.94	33.81	14.13	7.27	4.09	33.81
IL	17.63	14.46	17.30	27.55	11.47	19.13	9.77	9.77	17.62	12.10	9.88	16.21	10.52	28.78	11.82	9.62	27.47	13.74	15.18	15.18	52.69	23.84	13.78	4.52	52.69
IN	5.29	14.84	6.93	5.29	13.20	2.58	4.34	4.34	4.67	4.81	2.49	5.20	5.25	8.82	4.02	2.73	11.74	4.35	5.20	5.20	18.17	8.49	3.31	2.82	18.17
IA	11.94	5.77	9.18	13.10	7.10	4.40	5.19	5.19	12.22	8.84	5.48	9.22	8.30	10.69	8.48	5.61	17.71	6.88	8.97	8.97	42.71	42.83	4.81	3.70	31.77
KS	9.72	8.65	10.24	7.86	5.92	3.32	4.15	4.15	7.75	7.72	4.21	8.13	6.79	10.74	7.35	5.65	18.03	6.03	7.26	7.26	22.40	11.92	5.40	3.31	23.40
KY	19.74	12.89	18.65	24.60	7.70	8.10	15.00	15.00	21.87	24.34	11.81	10.91	14.25	22.06	15.53	8.42	54.39	17.05	14.51	14.51	50.29	28.09	12.52	5.04	50.26
LA	23.14	24.93	53.17	26.43	15.61	9.88	17.49	17.49	20.14	21.43	24.51	28.33	29.58	31.25	22.37	8.64	77.12	21.20	18.98	18.98	51.76	24.21	13.77	13.78	66.41
ME	13.43	10.42	43.59	24.29	11.12	3.72	12.56	12.56	13.35	15.39	14.03	17.26	16.07	31.44	17.95	9.07	32.07	10.07	15.08	15.08	37.84	61.81	11.66	5.92	37.84
MD	13.71	5.95	10.55	18.82	8.70	5.15	12.91	12.91	25.52	20.07	11.37	13.31	9.37	26.22	14.21	8.58	48.72	12.74	13.66	13.66	56.25	37.81	8.03	7.14	26.80
MA	9.93	6.07	16.09	17.95	8.10	3.69	6.49	6.49	7.55	13.48	5.76	13.04	7.54	14.31	5.34	4.55	38.25	7.47	12.35	12.35	35.13	27.39	9.24	3.32	28.82
MI	19.68	11.72	21.53	19.00	8.83	5.94	9.96	9.96	11.76	13.31	13.59	17.06	13.34	36.96	15.69	7.35	33.24	9.51	9.92	9.92	36.96	27.43	10.97	5.38	39.06
MN	19.53	19.80	41.48	14.65	15.73	7.53	14.95	14.95	15.68	11.90	22.05	19.18	16.95	24.63	22.05	10.88	70.18	14.23	10.58	10.58	104.13	33.30	14.55	6.68	132.92
MS	16.68	14.87	24.71	12.67	9.14	5.85	12.50	12.50	13.10	15.77	8.50	11.88	13.12	19.91	28.30	9.82	38.94	20.61	10.87	10.87	32.50	30.23	10.01	6.49	32.50
MO	24.01	11.40	15.92	15.49	10.23	6.90	9.30	9.30	11.03	17.88	11.24	14.55	13.46	16.69	15.51	9.66	31.18	13.82	11.26	11.26	53.28	36.77	8.63	6.32	53.28
MT	21.18	10.79	19.52	12.88	10.39	6.82	19.69	19.69	13.67	20.72	22.29	15.50	10.31	60.73	12.40	10.22	45.28	12.09	12.45	12.45	40.26	22.93	6.26	5.14	40.26
NE	23.85	14.32	22.60	27.52	12.77	9.52	18.07	18.07	21.82	33.02	12.77	23.10	17.70	23.25	17.17	11.13	38.42	17.27	13.90	13.90	60.90	30.50	10.52	7.63	50.40
NV	20.81	8.10	10.79	9.56	7.96	6.16	10.76	10.76	12.49	10.61	9.45	9.87	9.00	16.26	9.36	7.27	18.09	21.51	10.59	10.59	24.87	29.06	6.55	5.43	4.09
NH	27.93	14.04	25.82	31.58	12.51	7.80	18.33	18.33	13.55	38.86	9.24	22.82	12.51	20.88	15.76	11.83	49.65	16.44	13.62	13.62	52.78	22.94	14.95	6.55	52.78
NJ	12.63	9.48	12.63	10.77	8.25	4.50	8.33	8.33	7.06	12.93	11.17	12.87	10.95	13.78	11.17	6.14	32.47	6.17	10.72	10.72	22.56	12.00	4.75	4.81	26.39
NM	29.00	8.32	20.65	18.60	11.34	7.16	10.46	10.46	14.57	12.61	8.59	12.43	13.86	21.32	11.32	8.92	30.32	10.81	19.90	19.90	40.88	27.34	8.55	7.65	40.88
NY	17.66	6.98	15.91	17.16	13.57	6.32	10.02	10.02	12.02	9.09	5.11	19.20	13.75	16.81	8.43	8.53	33.44	17.47	9.63	9.63	18.80	19.14	9.43	6.09	11.29
NC	16.52	12.25	14.99	13.22	7.32	8.85	9.45	9.45	11.16	11.72	10.68	9.98	10.17	15.13	14.39	8.39	24.40	12.94	7.40	7.40	68.39	20.27	6.05	4.83	68.39
ND	10.49	10.49	10.49	5.94	5.94	4.15	5.93	5.93	10.49	10.49	9.25	8.19	6.57	21.54	9.25	5.46	20.98	5.46	21.45	21.45	21.54	21.54	10.49	20.98	14.48
OH	7.48	9.85	9.98	11.81	11.78	6.16	9.02	9.02	8.65	14.76	38.58	12.85	13.30	23.94	3.76	7.06	24.80	9.48	10.09	10.09	31.53	15.16	9.91	5.99	31.53
OK	13.44	9.11	12.70	12.65	6.63	5.82	10.25	10.25	16.10	19.50	8.46	10.98	9.09	20.29	12.21	7.24	22.32	9.49	13.78	13.78	39.78	24.01	6.95	5.88	39.78
OR	16.72	9.16	15.83	14.23	8.06	4.50	9.48	9.48	14.13	8.79	7.41	14.91	11.59	15.81	11.05	5.30	22.92	9.68	9.53	9.53	29.02	11.69	10.63	4.52	29.02
PA	13.48	13.48	12.50	14.67	10.14	6.58	8.34	8.34	9.68	12.50	11.87	12.38	14.03	17.07	11.87	7.29	27.70	8.07	15.29	15.29	24.67	15.29	8.67	12.38	24.67
RI	19.53	11.65	18.07	18.23	16.24	4.43	10.38	10.38	12.85	22.78	11.97	25.11	24.13	37.66	17.25	8.52	33.92	10.29	14.07	14.07	59.49	37.50	14.36	7.70	78.79
SC	22.26	17.21	19.93	13.51	7.81	9.42	11.95	11.95	14.32	11.08	8.53	13.27	16.47	18.32	18.15	8.83	43.19	10.97	11.47	11.47	24.55	27.29	8.90	5.98	24.55
SD	21.00	17.87	17.14	19.81	7.46	5.25	15.58	15.58	11.54	12.77	8.94	7.73	12.16	23.97	11.77	10.03	22.29	11.34	11.11	11.11	62.70	23.72	7.24	4.06	62.70
TN	30.13	13.71	22.70	18.06	9.32	8.15	13.14	13.14	11.29	13.81	12.27	15.91	13.61	16.19	16.22	10.57	36.23	13.55	12.52	12.52	42.02	23.33	9.07	6.20	42.02
TX	13.47	9.95	13.47	11.11	8.00	7.76	8.79	8.79	9.83	12.47	6.78	12.04	8.63	16.18	8.63	7.35	19.58	16.38	9.69	9.69	31.53	13.20	6.81	7.37	14.13
UT	9.45	6.23	10.52	7.68	7.99	7.05	6.38	6.38	9.55	10.95	12.32	13.75	15.29	17.24	10.60	6.41	26.27	6.30	8.99	8.99	23.51	23.51	6.06	5.83	26.53
VT	24.65	11.60	22.01	36.23	15.66	8.18	13.75	13.75	25.11	29.18	12.53	23.26	13.86	23.35	18.10	11.77	42.54	15.20	15.81	15.81	72.09	57.39	9.47	12.14	72.09
VA	13.36	8.95	11.28	12.51	5.79	5.62	7.58	7.58	9.32	10.10	17.52	9.64	10.83	15.93	9.09	6.69	21.06	8.52	11.01	11.01	49.54	21.40	6.14	3.14	49.54
WA	10.29	10.29	10.29	7.89	7.89	3.28	8.11	8.11	12.02	10.01	10.29	9.82	9.08	18.17	11.05	4.63	18.54	4.64	12.96	12.96	8.65	8.65	10.03	10.23	8.65
WV	12.30	12.30	12.30	27.37	27.37	5.86	8.90	8.90	6.99	13.15	13.15	11.94	12.35	13.34	12.35	5.48	14.63	6.99	10.89	10.89	10.89	10.89	13.15	4.46	48.62
WI	12.40	11.36	16.86	12.29	9.90	4.24	7.44	7.44	13.31	11.62	8.09	18.99	14.32	13.64	12.11	5.42	34.72	7.16	9.18	9.18	37.88	24.06	13.99	5.52	37.88
WY	8.27	8.27	8.27	8.27	8.27	8.27	8.27	8.27	8.27	8.27	8.27	8.27	8.27	8.27	8.27	8.27	8.27	8.27	8.27	8.27	8.27	8.27	8.27	8.27	8.27
AVG.	17.27	11.81	18.39	15.82	10.29	6.84	10.63	10.63	13.87	15.76	11.44	14.92	13.29	22.07	14.00	8.22	32.40	11.76	12.19	12.19	39.61	24.35	9.52	7.04	40.06

RO13113-60 Workers' Compensation (cont.) (Canada in Canadian dollars)

Province		Alberta	British Columbia	Manitoba	Ontario	New Brunswick	Newfndld. & Labrador	Northwest Territories	Nova Scotia	Prince Edward Island	Quebec	Saskat-chewan	Yukon
Carpentry—3 stories or less	Rate	9.71	7.19	4.57	4.62	4.83	9.78	4.61	7.67	6.04	15.41	7.33	5.58
	Code	42143	721028	40102	723	422	4226	4-41	4226	401	80110	B1317	202
Carpentry—interior cab. work	Rate	1.60	6.16	4.57	4.62	4.31	6.37	4.61	5.74	4.05	15.41	3.68	5.58
	Code	42133	721021	40102	723	427	4270	4-41	4274	402	80110	B11-27	202
CARPENTRY—general	Rate	9.71	7.19	4.57	4.62	4.83	6.37	4.61	7.67	6.04	15.41	7.33	5.58
	Code	42143	721028	40102	723	422	4299	4-41	4226	401	80110	B1317	202
CONCRETE WORK—NOC	Rate	5.96	7.53	7.60	15.25	4.83	9.78	4.61	4.83	6.04	16.51	7.33	5.58
	Code	42104	721010	40110	748	422	4224	4-41	4224	401	80100	B13-14	203
CONCRETE WORK—flat (flr. sidewalk)	Rate	5.96	7.53	7.60	15.25	4.83	9.78	4.61	4.83	6.04	16.51	7.33	5.58
	Code	42104	721010	40110	748	422	4224	4-41	4224	401	80100	B13-14	203
ELECTRICAL Wiring—inside	Rate	2.50	2.17	2.52	2.94	2.35	3.25	4.13	2.23	4.05	7.64	3.68	5.58
	Code	42124	721019	40203	704	426	4261	4-46	4261	402	80170	B11-05	206
EXCAVATION—earth NOC	Rate	3.31	3.85	3.84	4.20	3.16	5.53	3.71	4.11	4.25	8.36	4.37	5.58
	Code	40604	721031	40706	711	421	4214	4-43	4214	404	80030	R11-06	207
EXCAVATION—rock	Rate	3.31	3.85	3.84	4.20	3.16	5.53	3.71	4.11	4.25	8.36	4.37	5.58
	Code	40604	721031	40706	711	421	4214	4-43	4214	404	80030	R11-06	207
GLAZIERS	Rate	4.17	4.07	4.57	8.12	5.58	7.07	4.61	7.67	4.05	14.36	7.33	5.58
	Code	42121	715020	40109	751	423	4233	4-41	4233	402	80150	B13-04	212
INSULATION WORK	Rate	3.33	5.88	4.57	8.12	5.58	7.07	4.61	7.67	6.04	15.41	6.15	5.58
	Code	42184	721029	40102	751	423	4234	4-41	4234	401	80110	B12-07	202
LATHING	Rate	6.71	8.99	4.57	4.62	4.31	6.37	4.61	5.74	4.05	15.41	7.33	5.58
	Code	42135	721033	40102	723	427	4279	4-41	4271	402	80110	B13-16	202
MASONRY	Rate	5.96	8.99	4.57	11.44	5.58	7.07	4.61	7.67	6.04	16.51	7.33	5.58
	Code	42102	721037	40102	741	423	4231	4-41	4231	401	80100	B13-18	202
PAINTING & DECORATING	Rate	5.20	5.88	3.45	6.41	4.31	6.37	4.61	5.74	4.05	15.41	6.15	5.58
	Code	42111	721041	40105	719	427	4275	4-41	4275	402	80110	B12-01	202
PILE DRIVING	Rate	5.96	5.63	3.84	5.84	4.83	9.78	3.71	4.83	6.04	8.36	7.33	5.58
	Code	42159	722004	40706	732	422	4221	4-43	4221	401	80030	B13-10	202
PLASTERING	Rate	6.71	8.99	5.13	6.41	4.31	6.37	4.61	5.74	4.05	15.41	6.15	5.58
	Code	42135	721042	40108	719	427	4271	4-41	4271	402	80110	B12-21	202
PLUMBING	Rate	2.50	4.67	2.88	3.67	3.09	3.87	4.13	2.23	4.05	7.61	3.68	5.58
	Code	42122	721043	40204	707	424	4241	4-46	4241	402	80160	B11-01	214
ROOFING	Rate	9.88	10.10	6.52	11.60	10.25	9.78	4.61	9.49	6.04	22.49	7.33	5.58
	Code	42118	721036	40403	728	430	4236	4-41	4236	401	80130	B13-20	202
SHEET METAL WORK (HVAC)	Rate	2.50	4.67	6.52	3.67	3.09	3.87	4.13	3.23	4.05	7.61	3.68	7.39
	Code	42117	721043	40402	707	424	4244	4-46	4244	402	80160	B11-07	208
STEEL ERECTION—door & sash	Rate	3.33	13.44	9.89	15.25	4.83	9.78	4.61	7.67	6.04	29.92	7.33	5.58
	Code	42106	722005	40502	748	422	4227	4-41	4227	401	80080	B13-22	202
STEEL ERECTION—inter., ornam.	Rate	3.33	13.44	9.89	15.25	4.83	9.78	4.61	7.67	6.04	29.92	7.33	5.58
	Code	42106	722005	40502	748	422	4227	4-41	4227	401	80080	B13-22	202
STEEL ERECTION—structure	Rate	3.33	13.44	9.89	15.25	4.83	9.78	4.61	7.67	6.04	29.92	7.33	5.58
	Code	42106	722005	40502	748	422	4227	4-41	4227	401	80080	B13-22	202
STEEL ERECTION—NOC	Rate	3.33	13.44	9.89	15.25	4.83	9.78	4.61	7.67	6.04	29.92	7.33	5.58
	Code	42106	722005	40502	748	422	4227	4-41	4227	401	80080	B13-22	202
TILE WORK—inter. (ceramic)	Rate	4.62	5.67	2.17	6.41	4.31	6.37	4.61	5.74	4.05	15.41	7.33	5.58
	Code	42113	721054	40103	719	427	4276	4-41	4276	402	80110	B13-01	202
WATERPROOFING	Rate	5.20	5.88	4.57	4.62	5.58	6.37	4.61	7.67	4.05	22.49	6.15	5.58
	Code	42139	721016	40102	723	423	4299	4-41	4239	402	80130	B12-17	202
WRECKING	Rate	3.31	5.91	6.46	15.25	3.16	5.53	3.71	4.11	6.04	15.41	7.33	5.58
	Code	40604	721005	40106	748	421	4211	4-43	4211	401	80110	B13-09	202

R013113-80 Performance Bond

This table shows the cost of a Performance Bond for a construction job scheduled to be completed in 12 months. Add 1% of the premium cost per month for jobs requiring more than 12 months to complete. The rates are "standard" rates offered to contractors that the bonding company considers financially sound and capable of doing the work. Preferred rates are offered by some bonding companies based upon financial strength of the contractor. Actual rates vary from contractor to contractor and from bonding company to bonding company. Contractors should prequalify through a bonding agency before submitting a bid on a contract that requires a bond.

Contract Amount	Building Construction Class B Projects			Highways & Bridges					
				Class A New Construction			Class A-1 Highway Resurfacing		
First $ 100,000 bid	$25.00 per M			$15.00 per M			$9.40 per M		
Next 400,000 bid	$ 2,500	plus	$15.00 per M	$ 1,500	plus	$10.00 per M	$ 940	plus	$7.20 per M
Next 2,000,000 bid	8,500	plus	10.00 per M	5,500	plus	7.00 per M	3,820	plus	5.00 per M
Next 2,500,000 bid	28,500	plus	7.50 per M	19,500	plus	5.50 per M	15,820	plus	4.50 per M
Next 2,500,000 bid	47,250	plus	7.00 per M	33,250	plus	5.00 per M	28,320	plus	4.50 per M
Over 7,500,000 bid	64,750	plus	6.00 per M	45,750	plus	4.50 per M	39,570	plus	4.00 per M

R015113-65 Temporary Power Equipment

Cost data for the temporary equipment was developed utilizing the following information.

1) Re-usable material-services, transformers, equipment and cords are based on new purchase and prorated to three projects.
2) PVC feeder includes trench and backfill.
3) Connections include disconnects and fuses.
4) Labor units include an allowance for removal.
5) No utility company charges or fees are included.
6) Concrete pads or vaults are not included.
7) Utility company conduits not included.

R015423-10   Steel Tubular Scaffolding

On new construction, tubular scaffolding is efficient up to 18 000 mm high or five stories. Above this it is usually better to use a hung scaffolding if construction permits. Swing scaffolding operations may interfere with tenants. In this case the tubular is more practical at all heights.

In repairing or cleaning the front of an existing building the cost of tubular scaffolding per m² of building front increases as the height increases above the first tier. The first tier cost is relatively high due to leveling and alignment.

The minimum efficient crew for erection is three workers. For heights over 15 000 mm, a crew of four is more efficient. Use two or more on top and two at the bottom for handing up or hoisting. Four workers can erect and dismantle about nine frames per hour up to five stories. From five to eight stories they will average six frames per hour. With 2100 mm horizontal spacing, this will run about 36 m² and 24 m² of wall surface, respectively. Time for placing planks must be added to the above. On heights above 15 000 mm, five planks can be placed per labor-hour.

The table below shows the number of pieces required to erect tubular steel scaffolding for 93 m² of building frontage. This area is made up of a scaffolding system that is 12 frames (11 bays) long by 2 frames high.

For jobs under twenty-five frames, add 50% to rental cost. Rental rates will be lower for jobs over three months duration. Large quantities for long periods can reduce rental rates by 20%.

Description of Component	CSI Line Item	Number of pieces for 93 m² of Building Front	Unit
1500 mm Wide Standard Frame, 900 mm High	01540-750-2200	24	Ea.
Leveling Jack & Plate	01540-750-2650	24	
Cross Brace	01540-750-2500	44	
Side Arm Bracket, 500 mm	01540-750-2700	12	
Guardrail Post	01540-750-2550	12	
Guardrail, 2100 mm section	01540-750-2600	22	
Stairway Section	01540-750-2900	2	
Stairway Starter Bar	01540-750-2910	1	
Stairway Inside Handrail	01540-750-2920	2	
Stairway Outside Handrail	01540-750-2930	2	
Walk-Thru Frame Guardrail	01540-750-2940	2	

Scaffolding is often used as falsework over 4500 mm high during construction of cast-in-place concrete beams and slabs. For heavy beam construction, 600 mm wide scaffolding is generally used. The span between frames depends upon the load to be carried with a maximum span of 1500 mm.

Heavy duty scaffolding with a capacity of 4500 kg/leg can be spaced up to 3000 mm O.C. depending upon form support design and loading.

Scaffolding used as horizontal shoring requires less than half the material required with conventional shoring.

On new construction, erection is done by carpenters.

Rolling towers supporting horizontal shores can reduce labor and speed the job. For maintenance work, catwalks with spans up to 21 000 mm can be supported by the rolling towers.

R015423-20   Pump Staging

Pump staging is generally not available for rent. The table below shows the number of pieces required to erect pump staging for 216 m² of building frontage. This area is made up of a pump jack system that is 3 poles (2 bays) wide by 2 poles high.

Item	CSI Line Item	Number of pieces for 216 m² of Building Front	Unit
Aluminum pole section, 7200 mm long	01540-550-0200	6	Ea.
Aluminum splice joint, 1800 mm long	01540-550-0600	3	
Aluminum foldable brace	01540-550-0900	3	
Aluminum pump jack	01540-550-0700	3	
Aluminum support for workbench/back safety rail	01540-550-1000	3	
Aluminum scaffold plank/workbench, 350 mm W x 7200 mm L	01540-550-1100	4	
Safety net, 6600 mm long	01540-550-1250	2	
Aluminum plank end safety rail	01540-550-1200	2	

The cost in place for this 216 m² will depend on how many uses are realized during the life of the equipment. Several options are given in Division 01540-550.

R015433-10   Contractor Equipment

Rental Rates shown in the front of the book pertain to late model high quality machines in excellent working condition, rented from equipment dealers. Rental rates from contractors may be substantially lower than the rental rates from equipment dealers depending upon economic conditions; for older, less productive machines, reduce rates by a maximum of 15%. Any overtime must be added to the base rates. For shift work, rates are lower. Usual rule of thumb is 150% of one shift rate for two shifts; 200% for three shifts.

For periods of less than one week, operated equipment is usually more economical to rent than renting bare equipment and hiring an operator.

Equipment moving and mobilization costs must be added to rental rates where applicable. A large crane, for instance, may take two days to erect and two days to dismantle.

Rental rates vary throughout the country with larger cities generally having lower rates. Lease plans for new equipment are available for periods in excess of six months with a percentage of payments applying toward purchase.

Monthly rental rates vary from 2% to 5% of the cost of the equipment depending on the anticipated life of the equipment and its wearing parts. Weekly rates are about 1/3 the monthly rates and daily rental rates about 1/3 the weekly rate.

The hourly operating costs for each piece of equipment include costs to the user such as fuel, oil, lubrication, normal expendables for the equipment, and a percentage of mechanic's wages chargeable to maintenance. The hourly operating costs listed do not include the operator's wages.

The daily cost for equipment used in the standard crews is figured by dividing the weekly rate by five, then adding eight times the hourly operating cost to give the total daily equipment cost, not including the operator. This figure is in the right hand column of the Equipment Listings under Crew Equipment Cost/Day.

Pile Driving rates shown for pile hammer and extractor do not include leads, crane, boiler or compressor. Vibratory pile driving requires an added field specialist during set-up and pile driving operation for the electric model. The hydraulic model requires a field specialist for set-up only. Up to 125 reuses of sheet piling are possible using vibratory drivers. For normal conditions, crane capacity for hammer type and size are as follows.

Crane Capacity	Hammer Type and Size		
	Air or Steam	Diesel	Vibratory
23 metric ton	to 11 863 J		52 kW
36 metric ton	20 337 J	to 43 386 J	127 kW
54 metric ton	33 895 J		224 kW
91 metric ton		151 852 J	

Cranes should be specified for the job by size, building and site characteristics, availability, performance characteristics, and duration of time required.

Backhoes & Shovels rent for about the same as equivalent size cranes but maintenance and operating expense is higher. Crane operators rate must be adjusted for high boom heights. Average adjustments: for 40 m boom add 2% per hour; over 56 m, add 4% per hour; over 63 m, add 6% per hour; over 75 m, add 8% per hour and over 89 m, add 12% per hour.

Tower Cranes of the climbing or static type have jibs from 15 m to 61 m and capacities at maximum reach range from 1,814 to 6,350 kilograms. Lifting capacities increase up to maximum load as the hook radius decreases.

Typical rental rates, based on purchase price are about 2% to 3% per month.

Erection and dismantling runs between 500 and 2000 labor hours. Climbing operation takes 10 labor hours per 6 m climb. Crane dead time is about 5 hours per 12 m climb. If crane is bolted to side of the building add cost of ties and extra mast sections. Climbing cranes have from 24 to 55 m of mast while static cranes have 24 to 244 m of mast.

Truck Cranes can be converted to tower cranes by using tower attachments. Mast heights over 122 m have been used.

A single 30 m high material **Hoist and Tower** can be erected and dismantled in about 400 labor hours; a double 30 m high hoist and tower in about 600 labor hours. Erection times for additional heights are 3 and 4 labor hours per vertical meter respectively up to 46 m, and 4 to 5 labor hours per vertical meter over 46 m high. A 12 m high portable Buck hoist takes about 160 labor hours to erect and dismantle. Additional heights take 2 labor hours per vertical meter to 24 m and 3 labor hours per vertical meter for the next 30 m. Most material hoists do not meet local code requirements for carrying personnel.

A 46 m high **Personnel Hoist** requires about 500 to 800 labor hours to erect and dismantle. Budget erection time at 5 labor hours per vertical meter for all trades. Local code requirements or labor scarcity requiring overtime can add up to 50% to any of the above erection costs.

Earthmoving Equipment: The selection of earthmoving equipment depends upon the type and quantity of material, moisture content, haul distance, haul road, time available, and equipment available. Short haul cut and fill operations may require dozers only, while another operation may require excavators, a fleet of trucks, and spreading and compaction equipment. Stockpiled material and granular material are easily excavated with front end loaders. Scrapers are most economically used with hauls between 90 m and 2.4 km if adequate haul roads can be maintained. Shovels are often used for blasted rock and any material where a vertical face of 2.4 m or more can be excavated. Special conditions may dictate the use of draglines, clamshells, or backhoes. Spreading and compaction equipment must be matched to the soil characteristics, the compaction required and the rate the fill is being supplied.

R015433-15   Heavy Lifting

Hydraulic Climbing Jacks

The use of hydraulic heavy lift systems is an alternative to conventional type crane equipment. The lifting, lowering, pushing, or pulling mechanism is a hydraulic climbing jack moving on a square steel jackrod from 40 mm to 100 mm square, or a steel cable. The jackrod or cable can be vertical or horizontal, stationary or movable, depending on the individual application. When the jackrod is stationary, the climbing jack will climb the rod and push or pull the load along with itself. When the climbing jack is stationary, the jackrod is movable with the load attached to the end and the climbing jack will lift or lower the jackrod with the attached load. The heavy lift system is normally operated by a single control lever located at the hydraulic pump.

The system is flexible in that one or more climbing jacks can be applied wherever a load support point is required, and the rate of lift synchronized.

Economic benefits have been demonstrated on projects such as: erection of ground assembled roofs and floors, complete bridge spans, girders and trusses, towers, chimney liners and steel vessels, storage tanks, and heavy machinery. Other uses are raising and lowering offshore work platforms, caissons, tunnel sections and pipelines.

R019313-10 Facility Maintenance - Frequency Table

The following table lists "average" frequency data for selected facility maintenance activities. The frequencies given are for a normal standard of maintenance under average conditions.

Activity	Average Frequency	Notes
Acoustical tile, cleaning		
Heavy smoking	2-3 years	
Non-smoking	10 years	
Carpet, cleaning		Frequency depends on the type of carpet and the occupancy in the building.
Heavy traffic	Weekly	
Light traffic	Every 6 weeks	
Carpet, vacuuming		6-8 passes by machine in key areas.
Heavy traffic	Daily	
Light traffic	Twice weekly	
Offices	Weekly	
Corridor and lobby, policing		Includes picking up loose trash, removing cigarette butts from in and around jardinieres and/or sand urns. Frequency depends on occupancy in the building.
Main	4 times daily	
Secondary	Daily	
Elevator, cleaning		The frequency of elevator cleaning is dependent upon the amount of traffic, size of the elevator car and the occupancy in the building.
Passenger	Daily	
Freight	Weekly	
Elevator lobby, cleaning	Daily	Includes sweeping, mopping and rinsing.
Escalator, cleaning	Daily	Frequency given is for a normal standard of cleaning under average conditions. Approximately 40% of escalator treads are exposed when escalator is stopped.
Floors, mopping		Frequency given is for a normal standard of cleaning under average conditions.
Main corridors	Daily	
Secondary	Weekly	
Floors, sweeping	Daily	Frequency given is for a normal standard of cleaning under average conditions.
Floors, waxing and polishing		Frequency given is for a normal standard of cleaning under average conditions.
Office area	Every 9 weeks	
Open area	Every 9 weeks	
Flower beds		Frequency depends on geographic location.
Fall clean-up	Every year	
Fertilize	2 times per year	
Mulch	Every year	
Police-up	30 times per year	
Weed		
With mulch	15 times per year	
No mulch	25 times per year	
Lawn, mowing	30 times per year	Frequency depends on geographic location.
Fertilize	2 times per year	
Sweep	3 times per year	
Weed control	2 times per year	
Edge-trim		
Walks	30 times per year	
Shrub	10 times per year	
Light fixtures		Frequency depends on occupancy in the building.
Dusting	Every month	
Washing	Every 3 months	
Shrub Areas		Frequency depends on geographic location.
Fertilize	Every year	
Mulch	2 times per year	
Police-up	30 times per year	
Prune	5 times per year	
Weed	10 times per year	
Stairway		Frequency is dependent upon weather conditions, type of building, type of occupancy, and amount of traffic.
Sweeping and dusting	Daily	
Mopping or scrubbing	Weekly	
Toilet, cleaning	Daily	Includes toilet cleaning, collect waste, cleaning wash basins, urinals, water closets, partitions, walls and floors.
Trash collection	Daily	
Trees		Frequency depends on geographic location.
Fertilize	Every year	
Prune	2 times per year	
Pest control		
Spray	3 times per year	
Systemic	Every year	
Urn and Jardiniere cleaning	Daily	Includes the removal of refuse and debris, cleaning and polishing.
Walks, sweeping	30 times per year	
Walls, periodic cleaning	Every 6 months	Damp wipe, spot removal.
Windows, washing	Every month	Frequency given is for a normal standard of cleaning under average conditions.

R019313-20 Facility Maintenance Labor-Hours

This section lists minimum and maximum cleaning times per unit.
For more information, see *Means Facilities Maintenance Standards* book.

	Unit			Unit		
	Minimum	Maximum			Minimum	Maximum
Blast Cleaning			Seamless Floor Repair		0.5 m²/Hr	1.4 m²/Hr
White-metal	9 m²/Hr					
Near-white	16 m²/Hr		Wood Floors			
Commercial	34 m²/Hr		Sanding		3.7 m²/Hr	5.6 m²/Hr
Brush-off	81 m²/Hr		Sealing		1.8 m²/Hr	2.8 m²/Hr
Paint Application			Waxing*			
Brushing	12 m²/Hr		Wood Floor Repair			
Rolling	12 m²/Hr		Loose boards or tiles		4.6 m²/Hr	23 m²/Hr
Spraying	46 m²/Hr		Wood strip floor			
Plaster Cleaning			replacement		2.8 m²/Hr	5.6 m²/Hr
Wall dusting	22 sec/m²	33 sec./m²	Window Washing		2.8 m²/Hr	42 m²/Hr
Vacuuming	43 sec/m²	54 sec./m²	Venetian Blinds, cleaning		15 min./set of blinds	
Spot washing	12 m²/Hr	17 m²/Hr.				
Thorough cleaning	26 m²/Hr		This section lists average cleaning times per each, or square foot			
Plaster Repair						Average
Gypsum and lime repair	4 m²/Hr	8 m²/Hr.	Floor Operations			
Ceramic Tile Repair			Sweeping			
General	0.7 m²/Hr	1.3 m²/Hr.	Halls and Corridors			16 min./100 m²
Adhesive tile setting	0.8 m²/Hr	1.5 m²/Hr.	General Rooms			32 min./100 m²
Pointing tile joints	0.9 m²/Hr.	1.9 m²/Hr.	Dust Mop (unobstructed)			11 min./100 m²
Floor Cleaning			Dust Mop (obstructed)			16 min./100 m²
Manual sweeping	11 min./100 m²	27 min./100 m²	Damp Mop (unobstructed)			22 min./100 m²
Dust mopping	5 min./100 m²	22 min./100 m²	Damp Mop (obstructed)			43 min./100 m²
Buffing	16 min./100 m²	43 min./100 m²	Wet Mop and Rinse			108 min./100 m²
Spray buffin	22 min./100 m²	54 min./100 m²	Hand Scrubbing 12" brush			323 min./100 m²
Damp mopping	16 min./100 m²	32 min./100 m²	Deck Scrubbing			108 min./100 m²
Wet mopping	32 min./100 m²	54 min./100 m²	Machine Scrubbing			
Scrubbing	54 min./100 m²	151 min./100 m²	300 mm			54 min./100 m²
Scrubbing using electric floor			350 mm			43 min./100 m²
machine			400 mm			38 min./100 m²
General	16 min./100 m²	32 min./100 m²	450 mm			33 min./100 m²
Stripping	108 min./100 m²	215 min./100 m²	475 mm			30 min./100 m²
Waxing and Buffing using power			525 mm			27 min./100 m²
maching			575 mm			25 min./100 m²
Rewaxing	16 min./100 m²	32 min./100 m²	600 mm			22 min./100 m²
Stripping and rewaxing	108 min./100 m	323 min./100 m²	800 mm			19 min./100 m²
(two coats)			900 mm			16 min./100 m²
Waxing and buffing	32 min./100 m²	75 min./100 m²	Automatic Scrub Machine (600 mm)			5 min./100 m²
(one coat)			Vacuum (unobstructed)			22 min./100 m²
Carpets			Vacuum (obstructed)			32 min./100 m²
Dry vacuuming	16 min./100 m²	43 min./100 m²	Waxing			32 min./100 m²
Wet vacuuming	32 min./100 m²	54 min./100 m²	Machine Polish (475 mm machine)			16 min./100 m²
Carpet mopping	22 min./100 m²	43 min./100 m²	Rectangular Machine (1200 mm plate)			5 min./100 m²
Shampooing	188 min./100 m²	269 min./100 m²	Buff with steel wool			22 min./100 m²
Resilient Floor Repair			Strip and Rewax			161 min./100 m²
Grinding	4.6 m²/Hr	80 m²/Hr	Dry Strip and Rewax			129 min./100 m²
Floor Replacement			Spray Buffing (unobstructed)			32 min./100 m²
Removal (by hand)			Spray Buffing (obstructed)			48 min./100 m²
Tiles	9.0 m²/Hr	12.0 m²/Hr	Carpeting			
Sheet	11.1 m²/Hr	14.8 m²/Hr	Vacuuming (unobstructed)			22 min./100 m²
Hardwood	3.7 m²/Hr	5.6 m²/Hr	Vacuuming (obstructed)			32 min./100 m²
Replacement			Spot Vacuuming			16 min./100 m²
Ceramic	0.9 m²/Hr	1.9 m²/Hr	Shampoo (dry foam)			65 min. /100 m²
Resilient	3.7 m²/Hr	6.5 m²/Hr	Pile Lift			32 min./100 m²
Hardwood	2.3 m²/Hr	3.3 m²/Hr				
Add for related items:			Lockers			.20 min
Replace wood subfloor	7.4 m²/Hr	9.3 m²/Hr	Radiators			.30 min.
Replace underlayment	7.0 m²/Hr	8.4 m²/Hr	Tables (medium)			.50 min.
Replace floor moulding	0.9 m²/Hr	2.8 m²/Hr	Telephones			.15 min.
			Towel dispensers			.12 min.

*See waxing and buffing using electric power machine.

	Average
Towel Disposal Cans	.40 min.
Typewriter and Stand	.50 min
Wash Basin (office)	.60 min
Waste Basin (office)	.50 min.
Window Sill	.20 min.
Venetian Blinds, std. size	3.50 min.
Washrooms	
Cleaning commode	4 min.
Door (spot wash both sides)	1 min.
Mirrors	1 min.
Sanitary napkin dispenser	.50 min.
Urinals	3 min.
Wash basin-soap dispenser	3 min.
General cleaning	129 min./100 m²
Wall Washing	
Painted walls (manual)	258 min./100 m²
Painted walls (machine)	161 min./100 m²
Marble walls (manual)	97 min./100 m²
Ceiling Washing	
Ceiling washing (manual)	323 min./100 m²
Ceiling washing (machine)	194 min./100 m²
Window Washing	
Single pane	135 min./100 m²
Multi-pane	183 min./100 m²
Frosted single pane	205 min./100 m²
Opaque glass	54 min./100 m²
Plate glass	38 min./100 m²
Office partitions (glass)	118 min./100 m²
Dusting Lamps and Lighting Fixtures	
Wall fluorescent fixtures	.15 min.
Desk fluorescent lamp	.30 min.
Table lamp and shade	.60 min.
Floor lamp and shade	.60 min.
Washing Fluorescent Light Fixtures	
Ceiling fixtures (egg crate) 1200 mm ea.	9 min.
Ceiling fixtures (egg crate) 2400 mm ea.	12 min.
Dusting	
Air conditioners	.30 min.
Ash trays (desk)	.25 min.
Book cases (3-tier set)	.30 min.
Chairs	.30 min.
Cigarette stands	.40 min.

	Average
Couch	.25 min.
Desks	.80 min.
Desk Trays	.15 min.
File cabinets (4 drawer)	.40 min.
Fabric Upholstery Cleaning	
Whisk or vacuum armless chair	.50 min.
Armchair	1 min.
Couch	2 min.
Shampooing armless chair	4 min.
Armchair	7 min.
Couch	20 min.
Stairway Cleaning	
Sweep and dust, 1 flight, 15 steps	6 min.
Damp mop, 1 flight, 15 steps	5 min.
Scrubbing (hand)	20 min.
Carpentry	
Repair door surface closer	1-1/2 Hrs.
Repair concealed door closer	4 Hrs.
Repair door damage at shop	4 Hrs.
Repair and replace screens	1 Hr.
Repair broken glass	1-1/2 Hrs.
Repair and replace ceiling tile	2 Hrs.
Repair small drywall damage	4 Hrs.
Repair and replace sash balance	4 Hrs.
Change lock on door	1-1/2 Hrs.
Painting	
Repaint 6 m x 4.5 m room, 1 coat	34 Hrs.
Repaint small bathroom, 1 coat	11-1/2 Hrs.
Repaint exterior window	2-1/2 Hrs.
Plumbing and Steamfitting	
Clear stopped water closet	3 Hrs.
Clear stopped basin	1-1/2 Hrs.
Replace leaking radiator valve	6 Hrs.
Clear external sewer stoppage	9 Hrs.
Install replacement valve, faucet, or trap	10 Hrs.
Electrical	
Replace fluorescent lamp ballast	2 Hrs.
Replace fractional HP motor	10 Hrs.
Replace blown fuse or reset	
Circuit breaker	1-1/2 Hrs.
Repair exterior light damage	4 Hrs.
Control circuit problems	5 Hrs.

R019313-30 A Review of Major Building Materials–Advantages and Disadvantages

Material	Advantages	Disadvantages
Aluminum	Lightweight High resistance to corrosion Low electrical resistance Good conductor	Softness Limited strength for structural uses Low stiffness High rate of thermal expansion Low rate of fire resistance Relatively high cost
Concrete	High compressive resistance Durable Resistance to moisture, rot, insects, fire and wear Watertight (depending on water-cement ratio)	Workability Lack of tensile strength Lack of resistance to many types of chemical exposure, such as salt Hard to remove stains
Copper	High resistance to corrosion Good electrical conductivity Workable Forms its own surface protection Thermal contraction and expansion not high	Costly Strength varies with treatment and mechanical working
Epoxy	Can have varied properties depending on composition Liquid use very applicable Controllable Excellent strength properties Small creep during curing Hard Tough Resistant to abrasion Resistant to corrosion, salts, acids, petroleum products, solvents and other chemicals Adhesion to surfaces good	Color Form
Glass	Considerably strong Non-corrosive	Brittle, subject to shattering under shock Transmits energy at a rapid rate Large sizes expensive
Lead	Good resistance to corrosion	Heavy High coefficient of thermal expansion Difficult to hold in place
Masonry	Available in small units Appearance (available in many textures, sizes and colors) Good insulator	Stains hard to remove Porous (absorption rate high) Shrinkage of mortar Thermal and expansion cracking Others — see concrete
Paper	Readily available Low in cost Used in conjunction with other materials	Susceptible to water damage Susceptible to rot Relatively weak Highly combustible
Plastics — general	Applicable to many uses Relatively low in cost Workable into many shapes High strength Lightweight Non-corrosive Others — see specific plastic entry	Lack of resistance to fire Low stiffness High rate of thermal expansion Low thermal conductivity Some cases of chemical or physical instability with time Non-salvageable Others — see specific plastic entry

R019313-30 A Review of Major Building Materials—Advantages and Disadvantages (cont.)

Material	Advantages	Disadvantages
Plastics — specific		
Acrylics	Transparent	Easily scratched
	Hard	
	Weather resistant	
	Shatter resistant	
Alkyds	Water resistant	
	Touch	
	Good adhesive properties	
Melamines	Hard	
	Durable	
	Abrasive resistant	
	Chemical and heat resistant	
Polyamides	Hard	Costly
(nylon)	Tough	
	Wear resistant	
Polyesters	Weather and chemical resistant	
	Stiff	
	Hard	
Polyethylene	Flexible	Easily scratched
	Tough	
	Translucent	
	Low cost	
Polystyrene	Hard	Brittle
	Clear	
	Water and chemical resistant	
	Low cost	
Vinyls	Tough	
	Wear resistant	
	Stain resistant	
Plywood	Stronger than standard lumber	Dimensional stability not as good as that of standard structural lumber
	Durable	Poor quality glues are possible
	High resistance to impact and load	Thermal expansion and contraction more of a problem
	Not as affected by moisture changes as standard wood	
	Workability	
	See Wood for others	
Steel	Strong	Costly
	Most resistant to aging	Resultant loss of strength when exposed to intense heat,
	Most reliable in quality, non-combustible,	heat, rapid heat gain and loss
	non-rotting, dimensionally stable with time and moisture change	Corrosive when exposed to moisture and air or other
	Resistant to staining	corrosive conditions
Tin	Extremely workable	Heavy
	Very resistant to corrosion	Loss of strength when exposed to heat
Wood	Readily available	Paintability changes with moisture content
	Relatively low in cost	Combustible
	Simple to work with	Susceptible to rot and insect infestation
	Available in many shapes and forms	Soft and easily damaged (porous)
	Good insulating properties	Dimensional changes due to changes in temperature and moisture
		Strength changes with changing moisture content
Zinc	Fairly high resistance to corrosion	Brittle
	Forms its own protective surface	
	Workable	

R024119-10   Demolition Defined

Whole Building Demolition - Demolition of the whole building with no concern for any particular building element, component, or material type being demolished. This type of demolition is accomplished with large pieces of construction equipment that break up the structure, load it into trucks and haul it to a disposal site, but disposal or dump fees are not included. Demolition of below-grade foundation elements, such as footings, foundation walls, grade beams, slabs on grade, etc., is not included. Certain mechanical equipment containing flammable liquids or ozone-depleting refrigerants, electric lighting elements, communication equipment components, and other building elements may contain hazardous waste, and must be removed, either selectively or carefully, as hazardous waste before the building can be demolished.

Foundation Demolition (Division) - Demolition of below-grade foundation footings, foundation walls, grade beams, and slabs on grade. This type of demolition is accomplished by hand or pneumatic hand tools, and does not include saw cutting, or handling, loading, hauling, or disposal of the debris.

Gutting - Removal of building interior finishes and electrical/mechanical systems down to the load-bearing and sub-floor elements of the rough building frame, with no concern for any particular building element, component, or material type being demolished. This type of demolition is accomplished by hand or pneumatic hand tools, and includes loading into trucks, but not hauling, disposal or dump fees; scaffolding; or shoring. Certain mechanical equipment containing flammable liquids or ozone-depleting refrigerants, electric lighting elements, communication equipment components, and other building elements may contain hazardous waste, and must be removed, either selectively or carefully, as hazardous waste, before the building is gutted.

Selective Demolition - Demolition of a selected building element, component, or finish, with some concern for surrounding or adjacent elements, components, or finishes (see the first Subdivision (s) at the beginning of appropriate Divisions). This type of demolition is accomplished by hand or pneumatic hand tools, and does not include handling, loading,

storing, hauling, or disposal of the debris ; scaffolding; or shoring . "Gutting" methods may be used in order to save time, but damage that is caused to surrounding or adjacent elements, components, or finishes may have to be repaired at a later time.

Careful Removal - Removal of a piece of service equipment, building element or component, or material type, with great concern for both the removed item and surrounding or adjacent elements, components or finishes. The purpose of careful removal may be to protect the removed item for later re-use, preserve a higher salvage value of the removed item, or replace an item while taking care to protect surrounding or adjacent elements, components, connections, or finishes from cosmetic and/or structural damage. An approximation of the time required to perform this type of removal is 1/3 to 1/2 the time it would take to install a new item of like kind (see Reference Numbers R220105-10 and R260105-30). This type of removal is accomplished by hand or pneumatic hand tools, and does not include loading, hauling, or storing the removed item; scaffolding; shoring; or lifting equipment.

Cutout Demolition - Demolition of a small quantity of floor, wall, roof, or other assembly, with concern for the appearance and structural integrity of the surrounding materials. This type of demolition is accomplished by hand or pneumatic hand tools, and does not include saw cutting; handling, loading, hauling, or disposal of debris; scaffolding; or shoring.

Rubbish Handling - Work activities that involve handling, loading or hauling of debris. Generally, the cost of rubbish handling must be added to the cost of all types of demolition, with the exception of whole building demolition.

Minor Site Demolition - Demolition of site elements outside the footprint of a building. This type of demolition is accomplished by hand or pneumatic hand tools, or with larger pieces of construction equipment, and may include loading a removed item onto a truck (check the Crew for equipment used). It does not include saw cutting , hauling or disposal of debris , and, sometimes, handling or loading.

R026510-20 Underground Storage Tank Removal

Underground Storage Tank Removal can be divided into two categories: Non-Leaking and Leaking. Prior to removing an underground storage tank, tests should be made, with the proper authorities present, to determine whether a tank has been leaking or the surrounding soil has been contaminated.

To safely remove Liquid Underground Storage Tanks:
1. Excavate to the top of the tank.
2. Disconnect all piping.
3. Open all tank vents and access ports.
4. Remove all liquids and/or sludge.
5. Purge the tank with an inert gas.
6. Provide access to the inside of the tank and clean out the interior using proper personal protective equipment (PPE).
7. Excavate soil surrounding the tank using proper PPE for on-site personnel.
8. Pull and properly dispose of the tank.
9. Clean up the site of all contaminated material.
10. Install new tanks or close the excavation.

R028213-20 Asbestos Removal Process

Asbestos removal is accomplished by a specialty contractor who understands the federal and state regulations regarding the handling and disposal of the material. The process of asbestos removal is divided into many individual steps. An accurate estimate can be calculated only after all the steps have been priced.

The steps are generally as follows:
1. Obtain an asbestos abatement plan from an industrial hygienist.
2. Monitor the air quality in and around the removal area and along the path of travel between the removal area and transport area. This establishes the background contamination.
3. Construct a two part decontamination chamber at entrance to removal area.
4. Install a HEPA filter to create a negative pressure in the removal area.
5. Install wall, floor and ceiling protection as required by the plan, usually 2 layers of fireproof 6 mil polyethylene.
6. Industrial hygienist visually inspects work area to verify compliance with plan.
7. Provide temporary supports for conduit and piping affected by the removal process.
8. Proceed with asbestos removal and bagging process. Monitor air quality as described in Step #2. Discontinue operations when contaminate levels exceed applicable standards.
9. Document the legal disposal of materials in accordance with EPA standards.
10. Thoroughly clean removal area including all ledges, crevices and surfaces.
11. Post abatement inspection by industrial hygienist to verify plan compliance.
12. Provide a certificate from a licensed industrial hygienist attesting that contaminate levels are within acceptable standards before returning area to regular use.

R028319-60   Lead Paint Remediation Methods

Lead paint remediation can be accomplished by the following methods.
1. Abrasive blast
2. Chemical stripping
3. Power tool cleaning with vacuum collection system
4. Encapsulation
5. Remove and replace
6. Enclosure

Each of these methods has strengths and weakness depending on the specific circumstances of the project. The following is an overview of each method.

1. **Abrasive blasting** is usually accomplished with sand or recyclable metallic blast. Before work can begin, the area must be contained to ensure the blast material with lead does not escape to the atmosphere. The use of vacuum blast greatly reduces the containment requirements. Lead abatement equipment that may be associated with this work includes a negative air machine. In addition, it is necessary to have an industrial hygienist monitor the project on a continual basis. When the work is complete, the spent blast sand with lead must be disposed of as a hazardous material. If metallic shot was used, the lead is separated from the shot and disposed of as hazardous material. Worker protection includes disposable clothing and respiratory protection.

2. **Chemical stripping** requires strong chemicals be applied to the surface to remove the lead paint. Before the work can begin, the area under/adjacent to the work area must be covered to catch the chemical and removed lead. After the chemical is applied to the painted surface it is usually covered with paper. The chemical is left in place for the specified period, then the paper with lead paint is pulled or scraped off. The process may require several chemical applications. The paper with chemicals and lead paint adhered to it, plus the containment and loose scrapings collected by a HEPA (High Efficiency Particulate Air Filter) vac, must be disposed of as a hazardous material. The chemical stripping process usually requires a neutralizing agent and several wash downs after the paint is removed. Worker protection includes a neoprene or other compatible protective clothing and respiratory protection with face shield. An industrial hygienist is required intermittently during the process.

3. **Power tool cleaning** is accomplished using shrouded needle blasting guns. The shrouding with different end configurations is held up against the surface to be cleaned. The area is blasted with hardened needles and the shroud captures the lead with a HEPA vac and deposits it in a holding tank. An industrial hygienist monitors the project, protective clothing and a respirator is required until air samples prove otherwise. When the work is complete the lead must be disposed of as a hazardous material.

4. **Encapsulation** is a method that leaves the well bonded lead paint in place after the peeling paint has been removed. Before the work can begin, the area under/adjacent to the work must be covered to catch the scrapings. The scraped surface is then washed with a detergent and rinsed. The prepared surface is covered with approximately 10 mils of paint. A reinforcing fabric can also be embedded in the paint covering. The scraped paint and containment must be disposed of as a hazardous material. Workers must wear protective clothing and respirators.

5. **Remove and replace** is an effective way to remove lead paint from windows, gypsum walls and concrete masonry surfaces. The painted materials are removed and new materials are installed. Workers should wear a respirator and tyvek suit. The demolished materials must be disposed of as hazardous waste if it fails the TCLP (Toxicity Characteristic Leachate Process) test.

6. **Enclosure** is the process that permanently seals lead painted materials in place. This process has many applications such as covering lead painted drywall with new drywall, covering exterior construction with tyvek paper then residing, or covering lead painted structural members with aluminum or plastic. The seams on all enclosing materials must be securely sealed. An industrial hygienist monitors the project, and protective clothing and a respirator is required until air samples prove otherwise.

All the processes require clearance monitoring and wipe testing as required by the hygienist.

R031113-10   Wall Form Materials

Aluminum Forms

Approximate weight is 15 kg per m² C.A. Standard widths are available from 100 mm to 920 mm with 920 mm most common. Standard lengths of 0.6 m, 1.2 m, 1.8 m to 2.4 m are available. Forms are lightweight and fewer ties are needed with the wider widths. The form face is either smooth or textured.

Metal Framed Plywood Forms

Manufacturers claim over 75 reuses of plywood and over 300 reuses of steel frames. Many specials such as corners, fillers, pilasters, etc. are available. Monthly rental is generally about 15% of purchase price for first month and 9% per month thereafter with 90% of rental applied to purchase for the first month and decreasing percentages thereafter. Aluminum framed forms cost 25% to 30% more than steel framed.

After the first month, extra days may be prorated from the monthly charge. Rental rates do not include ties, accessories, cleaning, loss of hardware or freight in and out. Approximate weight is 24 kg/m² for steel; 15 kg/m² for aluminum.

Forms can be rented with option to buy.

Plywood Forms, Job Fabricated

There are two types of plywood used for concrete forms.

1. Exterior plyform which is completely waterproof. This is face oiled to facilitate stripping. Ten reuses can be expected with this type with 25 reuses possible.
2. An overlaid type consists of a resin fiber fused to exterior plyform. No oiling is required except to facilitate cleaning. This is available in both high density (HDO) and medium density overlaid (MDO). Using HDO, 50 reuses can be expected with 200 possible.

Plyform is available in 16 mm and 19 mm thickness. High density overlaid is available in 10 mm, 13 mm, 16 mm and 19 mm thickness.

16 mm thick is sufficient for most building forms, while 19 mm is best on heavy construction.

Plywood Forms, Modular, Prefabricated

There are many plywood forming systems without frames. Most of these are manufactured from 29 mm (HDO) plywood and have some hardware attached. These are used principally for foundation walls 2.4 mm or less high. With care and maintenance, 100 reuses can be attained with decreasing quality of surface finish.

Steel Forms

Approximate weight is 32 kg/m² C.A. including accessories. Standard widths are available from 50 mm to 610 mm, with 610 mm most common. Standard lengths are from 0.6 m to 2.4 m, with 1.2 m the most common. Forms are easily ganged into modular units.

Forms are usually leased for 15% of the purchase price per month prorated daily over 30 days.

Rental may be applied to sale price and usually rental forms are bought. With careful handling and cleaning 200 to 400 reuses are possible.

Straight wall gang forms up to 3.6 m x 6.1 m or 2.4 m x 9.1 m can be fabricated. These crane handled forms usually lease for approx. 9% per month.

Individual job analysis is available from the manufacturer at no charge.

R031113-30 Slipforms

The slipform method of forming may be used for forming circular silo and multi-celled storage bin type structures over 9 m high, and building core shear walls over eight stories high. The shear walls, usually enclose elevator shafts, stairwells, mechanical spaces, and toilet rooms. Reuse of the form on duplicate structures will reduce the height necessary and spread the cost of building the form. Slipform systems can be used to cast chimneys, towers, piers, dams, underground shafts or other structures capable of being extruded.

Slipforms are usually 1.2 m high and are raised semi-continuously by jacks climbing on rods which are embedded in the concrete. The jacks are powered by a hydraulic, pneumatic, or electric source and are available in 3, 5, and 20 metric ton capacities. Interior work decks and exterior scaffolds must be provided for placing inserts, embedded items, reinforcing steel, and concrete. Scaffolds below the form for finishers may be required. The interior work decks are often used as roof slab forms on silos and bin work. Form raising rates will range from 150 mm to 510 mm per hour for silos; 150 mm to 760 mm per hour for buildings; and 150 mm to 1250 mm per hour for shaft work.

Reinforcing bars and stressing strands are usually hoisted by crane or gin pole and the concrete material can be hoisted by crane, winch-powered skip, or pumps. The slipform system is operated on a continuous 24 hour day when a monolithic structure is desired. For least cost, the system is operated only during normal working hours.

Placing concrete will range from 0.5 to 1.5 labor-hours per m^3. Bucks, blockouts, keyways, weldplates, etc. are extra.

R031113-40 Forms for Reinforced Concrete

Design Economy

Avoid many sizes in proportioning beams and columns.

From story to story avoid changing column dimensions. Gain strength by adding steel or using a richer mix. If a change in size of column is necessary, vary one dimension only to minimize form alterations. Keep beams and columns the same width.

From floor to floor in a multi-story building vary beam depth, not width, as that will leave slab panel form unchanged. It is cheaper to vary the strength of a beam from floor to floor by means of steel area than by 50 mm changes in either width or depth.

Cost Factors

Material includes the cost of lumber, cost of rent for metal pans or forms if used, nails, form ties, form oil, bolts and accessories.

Labor includes the cost of carpenters to make up, erect, remove and repair, plus common labor to clean and move. Having carpenters remove forms minimizes repairs.

Improper alignment and condition of forms will increase finishing cost. When forms are heavily oiled, concrete surfaces must be neutralized before finishing. Special curing compounds will cause spillages to spall off in first frost. Gang forming methods will reduce costs on large projects.

Materials Used

Boards are seldom used unless their architectural finish is required. Generally, steel, fiberglass and plywood are used for contact surfaces. Labor on plywood is 10% less than with boards. The plywood is backed up with 50 mm x 100 mm's at 310 mm to 820 mm O.C. Walers are generally 2 - 50 mm x 100 mm's. Column forms are held together with steel yokes or bands. Shoring is with adjustable shoring or scaffolding for high ceilings.

Reuse

Floor and column forms can be reused four or possibly five times without excessive repair. Remember to allow for 10% waste on each reuse.

When modular sized wall forms are made, up to twenty uses can be expected with exterior plyform.

When forms are reused, the cost to erect, strip, clean and move will not be affected. 10% replacement of lumber should be included and about one hour of carpenter time for repairs on each reuse per 9 m^2.

The reuse cost for certain accessory items normally rented on a monthly basis will be lower than the cost for the first use.

After fifth use, new material required plus time needed for repair prevent form cost from dropping further and it may go up. Much depends on care in stripping, the number of special bays, changes in beam or column sizes and other factors.

Costs for multiple use of formwork may be developed as follows:

2 Uses	3 Uses	4 Uses
$\dfrac{\text{(1st Use + Reuse)}}{2} = \text{avg. cost/2 uses}$	$\dfrac{\text{(1st Use + 2 Reuse)}}{3} = \text{avg. cost/3 uses}$	$\dfrac{\text{(1st use + 3 Reuse)}}{4} = \text{avg. cost/4 uses}$

R031113-60 Formwork Labor Hours

Item	Unit	Hours Required			Total Hours	Multiple Use		
		Fabricate	Erect & Strip	Clean & Move	1 Use	2 Use	3 Use	4 Use
Beam and Girder, interior beams, 310 mm wide	10 m²	6.9	8.9	1.4	17.2	14.3	13.3	12.9
Hung from steel beams		6.2	8.3	1.4	15.9	13.3	12.5	12.0
Beam sides only, 920 mm high		6.2	7.7	1.4	15.4	12.8	11.9	11.5
Beam bottoms only, 610 mm wide		7.1	14.0	1.4	22.5	19.5	18.5	18.0
Box out for openings		10.6	10.8	1.2	22.6	17.8	16.2	15.4
Buttress forms, to 2.4 m high		6.5	7.0	1.3	14.7	12.0	11.1	10.7
Centering, steel, 19 mm rib lath			1.1		1.1			
9.5 mm rib lath or slab form	▼		1.0		1.0			
Chamfer strip or keyway	100 m		4.9		4.9	4.9	4.9	4.9
Columns, fiber tube 200 mm diameter			67.5		67.5			
310 mm			69.8		69.8			
410 mm			75.1		75.1			
510 mm			77.7		77.7			
610 mm			80.7		80.7			
760 mm	▼		83.9		83.9			
Round Steel, 310 mm diameter			72.1		72.1	72.1	72.1	72.1
410 mm			83.9		83.9	83.9	83.9	83.9
510 mm			100.0		100.0	100.0	100.0	100.0
610 mm	▼		123.6		123.6	123.6	123.6	123.6
Plywood 200 mm x 200 mm	10 m²	7.5	11.8	1.3	20.6	17.4	16.3	15.8
310 mm x 310 mm		6.5	11.3	1.3	19.0	16.3	15.4	15.0
410 mm x 410 mm		6.3	10.8	1.3	18.4	15.8	14.9	14.4
610 mm x 610 mm		6.2	10.5	1.3	18.1	15.5	14.6	14.2
Steel framed plywood 200 mm x 200 mm			10.8	1.1	11.8	11.8	11.8	11.8
310 mm x 310 mm			10.0	1.1	11.1	11.1	11.1	11.1
410 mm x 410 mm			9.1	1.1	10.2	10.2	10.2	10.2
610 mm x 610 mm			8.4	1.1	9.5	9.5	9.5	9.5
Drop head forms, plywood		9.7	13.4	1.6	24.7	20.4	19.0	18.3
Coping forms		9.1	16.1	1.6	26.9	22.9	21.5	20.8
Culvert, box			15.6	4.6	20.2	20.2	20.2	20.2
Curb forms, 150 mm to 310 mm high, on grade		5.4	9.1	1.3	15.8	13.7	12.9	12.6
On elevated slabs	▼	6.5	11.6	1.3	19.4	16.7	15.8	15.3
Edge forms to 150 mm high, on grade	100 m	6.6	11.5	2.0	20.1	17.3	16.4	16.0
180 mm to 310 mm high	10 m²	2.7	5.4	1.1	9.1	8.3	8.1	7.9
Equipment foundations		10.8	19.4	2.2	32.3	27.4	25.8	25.0
Flat slabs, including drops		3.8	6.5	1.3	11.5	10.2	9.7	9.5
Hung from steel		3.2	5.9	1.3	10.4	9.4	9.0	8.8
Closed deck for domes		3.2	6.2	1.3	10.8	9.7	9.3	9.1
Open deck for pans		2.4	5.7	1.1	9.1	8.5	8.3	8.2
Footings, continuous, 310 mm high		3.8	3.8	1.6	9.1	7.8	7.3	7.1
Spread, 310 mm high		5.1	4.5	1.7	11.3	9.3	8.6	8.3
Pile caps, square or rectangular		4.8	5.4	1.6	11.8	9.9	9.3	9.0
Grade beams, 610 mm deep		2.7	5.7	1.3	9.7	8.9	8.6	8.5
Lintel or Sill forms		8.6	18.3	2.2	29.0	25.3	24.0	23.4
Spandrel beams, 310 mm wide		9.7	12.0	1.4	23.1	18.8	17.4	16.7
Stairs			26.9	4.3	31.2	31.2	31.2	31.2
Trench forms in floor		4.8	15.1	1.6	21.5	19.6	19.0	18.7
Walls, Plywood, at grade, to 2.4 m high		5.4	7.0	1.6	14.0	11.8	11.1	10.8
2.4 m to 4.9 m		8.1	8.6	1.6	18.3	14.8	13.6	13.0
4.9 m to 6.1 m		9.7	10.8	1.6	22.0	17.7	16.3	15.6
Foundation walls, to 2.4 m high		4.8	7.0	1.1	12.9	11.0	10.4	10.1
2.4 m to 4.9 m high		5.9	8.1	1.1	15.1	12.6	11.8	11.4
Retaining wall to 3.7 m high, battered		6.5	9.1	1.6	17.2	14.5	13.6	13.2
Radial walls to 3.7 m high, smooth		8.6	10.2	2.2	21.0	17.2	16.0	15.3
But in 0.6 m chords		7.5	8.6	1.6	17.7	14.5	13.4	12.9
Prefabricated modular, to 2.4 m high		—	4.6	1.1	5.7	5.7	5.7	5.7
Steel, to 2.4 m high		—	7.3	1.3	8.6	8.6	8.6	8.6
2.4 m to 4.9 m high		—	9.8	1.6	11.4	11.4	11.4	11.4
Steel framed plywood to 2.4 m high		—	7.3	1.3	8.6	8.6	8.6	8.6
2.4 m to 4.9 m high	▼	—	10.0	1.3	11.3	11.3	11.3	11.3

R032110-10 Reinforcing Steel Weights and Measures

Bar Designation No.**	Nominal Weight Lb./Ft.	U.S. Customary Units Nominal Dimensions*			SI Units Nominal Dimensions*			
		Diameter in.	Cross Sectional Area, in.²	Perimeter in.	Nominal Weight kg/m	Diameter mm	Cross Sectional Area, cm²	Perimeter mm
3	0.376	0.375	0.11	1.178	0.560	9.52	0.71	29.9
4	0.688	0.500	0.20	1.571	0.994	12.70	1.29	39.9
5	1.043	0.625	0.31	1.963	1.552	15.88	2.00	49.9
6	1.502	0.750	0.44	2.356	2.235	19.05	2.84	59.8
7	2.044	0.875	0.60	2.749	3.042	22.22	3.87	69.8
8	2.670	1.000	0.79	3.142	3.973	25.40	5.10	79.8
9	3.400	1.128	1.00	3.544	5.059	28.65	6.45	90.0
10	4.303	1.270	1.27	3.990	6.403	32.26	8.19	101.4
11	5.313	1.410	1.56	4.430	7.906	35.81	10.06	112.5
14	7.650	1.693	2.25	5.320	11.384	43.00	14.52	135.1
18	13.600	2.257	4.00	7.090	20.238	57.33	25.81	180.1

* The nominal dimensions of a deformed bar are equivalent to those of a plain round bar having the same weight per foot as the deformed bar.
** Bar numbers are based on the number of eighths of an inch included in the nominal diameter of the bars.

R032110-20 Metric Rebar Specification - ASTM A615-81

	Grade 300 (300 MPa* = 43,560 psi; +8.7% vs. Grade 40)			
	Grade 400 (400 MPa* = 58,000 psi; −3.4% vs. Grade 60)			
Bar No.	Diameter mm	Area mm²	Equivalent in.²	Comparison with U.S. Customary Bars
10M	11.3	100	.16	Between #3 & #4
15M	16.0	200	.31	#5 (.31 in.²)
20M	19.5	300	.47	#6 (.44 in.²)
25M	25.2	500	.78	#8 (.79 in.²)
30M	29.9	700	1.09	#9 (1.00 in.²)
35M	35.7	1000	1.55	#11 (1.56 in.²)
45M	43.7	1500	2.33	#14 (2.25 in.²)
55M	56.4	2500	3.88	#18 (4.00 in.²)

* MPa = megapascals

R032110-25 Comparison of U.S. Customary Units and SI Units for Reinforcing Bars

Bar Designation No.[b]	U.S. Customary Units						
	Nominal Weight, lb/ft	Nominal Dimensions[a]			Deformation Requirements, in.		
		Diameter in.	Cross Sectional Area, in.²	Perimeter in.	Maximum Average Spacing	Minimum Average Height	Maximum Gap (Chord of 12-1/2% of Nominal Perimeter)
3	0.376	0.375	0.11	1.178	0.262	0.015	0.143
4	0.668	0.500	0.20	1.571	0.350	0.020	0.191
5	1.043	0.625	0.31	1.963	0.437	0.028	0.239
6	1.502	0.750	0.44	2.356	0.525	0.038	0.286
7	2.044	0.875	0.60	2.749	0.612	0.044	0.334
8	2.670	1.000	0.79	3.142	0.700	0.050	0.383
9	3.400	1.128	1.00	3.544	0.790	0.056	0.431
10	4.303	1.270	1.27	3.990	0.889	0.064	0.487
11	5.313	1.410	1.56	4.430	0.987	0.071	0.540
14	7.65	1.693	2.25	5.32	1.185	0.085	0.648
18	13.60	2.257	4.00	7.09	1.58	0.102	0.864

Bar Designation No.[b]	SI UNITS						
	Nominal Weight kg/m	Nominal Dimensions[a]			Deformation Requirements, mm		
		Diameter, mm	Cross Sectional Area, cm²	Perimeter, mm	Maximum Average Spacing	Minimum Average Height	Maximum Gap (Chord of 12-1/2% of Nominal Perimeter)
3	0.560	9.52	0.71	29.9	6.7	0.38	3.5
4	0.994	12.70	1.29	39.9	8.9	0.51	4.9
5	1.552	15.88	2.00	49.9	11.1	0.71	6.1
6	2.235	19.05	2.84	59.8	13.3	0.96	7.3
7	3.042	22.22	3.87	69.8	15.5	1.11	8.5
8	3.973	25.40	5.10	79.8	17.8	1.27	9.7
9	5.059	28.65	6.45	90.0	20.1	1.42	10.9
10	6.403	32.26	8.19	101.4	22.6	1.62	11.4
11	7.906	35.81	10.06	112.5	25.1	1.80	13.6
14	11.384	43.00	14.52	135.1	30.1	2.16	16.5
18	20.238	57.33	25.81	180.1	40.1	2.59	21.9

[a]Nominal dimensions of a deformed bar are equivalent to those of a plain round bar having the same weight per foot as the deformed bar.

[b]Bar numbers are based on the number of eighths of an inch included in the nominal diameter of the bars.

R032110-40   Weight of Steel Reinforcing Per m² of Wall (kg/m²)

Reinforced Weights: The table below suggests the weights per m² for reinforcing steel in walls. Weights are approximate and will be the same for all grades of steel bars. For bars in two directions, add weights for each size and spacing.

C/C Spacing in mm	#10M Wt. (kg/m²)	#15M Wt. (kg/m²)	#20M Wt. (kg/m²)	#25M Wt. (kg/m²)	#30M Wt. (kg/m²)	#35M Wt. (kg/m²)	#45M Wt. (kg/m²)	#55M Wt. (kg/m²)
50 mm	15.70	31.40	47.10	78.50	109.90			
75 mm	10.47	20.93	31.40	52.33	73.27	157.00	235.50	392.50
100 mm	7.85	15.70	23.55	39.25	54.95	104.67	157.00	261.67
125 mm	6.28	12.56	18.84	31.40	43.96	78.50	117.75	196.25
150 mm	5.23	10.47	15.70	26.17	36.63	62.80	94.20	157.00
200 mm	3.93	7.85	11.78	19.63	27.48	52.33	78.50	130.83
250 mm	3.14	6.28	9.42	15.70	21.98	39.25	58.88	98.13
300 mm	2.62	5.23	7.85	13.08	18.32	31.40	47.10	78.50
450 mm	1.74	3.49	5.23	8.72	12.21	26.17	39.25	65.42
600 mm	1.31	2.62	3.93	6.54	9.16	17.44	26.17	43.61
750 mm	1.05	2.09	3.14	5.23	7.33	13.08	19.63	32.71
800 mm	0.98	1.96	2.94	4.91	6.87	10.47	15.70	26.17
1050 mm	0.75	1.50	2.24	3.74	5.23	9.81	14.72	24.53
1200 mm	0.65	1.31	1.96	3.27	4.58	7.48	11.21	18.69

R032110-50   Minimum Wall Reinforcement Weight (kg/m²)

This table lists the approximate minimum wall reinforcement weights per m² according to the specification of 0.12% of gross area for vertical bars and 0.20% of gross area for horizontal bars.

Location	Wall Thickness	Bar Size	Horizontal Steel Spacing C/C	cm² per m	Total Wt. per m²	Bar Size	Vertical Steel Spacing C/C	cm² per m	Total Wt. per m²	Horizontal & Vertical Steel Total Weight per m²
Both Faces	250 mm	#10M	460 mm	275	4.35	#10M	460 mm	1.48	2.45 kg	6.80 kg
	310 mm	#10M	410 mm	318	4.89	#10M	410 mm	1.91	2.94	7.83
	360 mm	#10M	360 mm	360	5.59	#10M	330 mm	2.12	3.39	8.98
	410 mm	#10M	310 mm	423	6.52	#10M	280 mm	2.54	4.00	10.52
	460 mm	#15M	430 mm	466	7.19	#10M	460 mm	2.75	4.35	11.54
One Face	150 mm	#10M	230 mm	318	2.45	#10M	460 mm	1.48	1.23	3.68
	200 mm	#10M	310 mm	423	3.26	#10M	280 mm	2.54	2.00	5.26
	250 mm	#15M	380 mm	529	4.07	#10M	410 mm	3.18	2.45	6.52

R032110-70   Bend, Place and Tie Reinforcing

Placing and tying by rodmen for footings and slabs runs from nine hrs. per metric ton for heavy bars to fifteen hrs. per metric ton for light bars. For beams, columns, and walls, production runs from eight hrs. per metric ton for heavy bars to twenty hrs. per metric ton for light bars. Overall average for typical reinforced concrete buildings is about fourteen hrs. per metric ton. These production figures include the time for placing of accessories and usual inserts, but not their material cost (allow 15% of the

cost of delivered bent rods). Equipment handling is necessary for the larger size bars so that installation costs for the very heavy bars will not decrease proportionately.

Installation costs for splicing reinforcing bars includes allowance for equipment to hold the bars in place while splicing as well as necessary scaffolding for iron workers.

R032110-80   Shop-Fabricated Reinforcing Steel

The material prices for reinforcing, shown in the unit cost sections of the book, are for 45 metric tons or more of shop-fabricated reinforcing steel and include:

1. Mill base price of reinforcing steel
2. Mill grade/size/length extras
3. Mill delivery to the fabrication shop
4. Shop storage and handling
5. Shop drafting/detailing

6. Shop shearing and bending
7. Shop listing
8. Shop delivery to the job site

Both material and installation costs can be considerably higher for small jobs consisting primarily of smaller bars, while material costs may be slightly lower for larger jobs

R032205-30   Common Stock Styles of Welded Wire Fabric

This table provides some of the basic specifications, sizes, and weights of welded wire fabric used for reinforcing concrete.

New Designation		Old Designation		Steel Area per Foot				Approximate Weight per 100 S.F.	
Spacing — Cross Sectional Area (in.) — (Sq. in. 100)		Spacing — Wire Gauge (in.) — (AS & W)		Longitudinal		Transverse			
				in.	cm	in.	cm	lbs	kg
Rolls	6 x 6 — W1.4 x W1.4	6 x 6 — 10 x 10		.028	.071	.028	.071	21	9.53
	6 x 6 — W2.0 x W2.0	6 x 6 — 8 x 8	1	.040	.102	.040	.102	29	13.15
	6 x 6 — W2.9 x W2.9	6 x 6 — 6 x 6		.058	.147	.058	.147	42	19.05
	6 x 6 — W4.0 x W4.0	6 x 6 — 4 x 4		.080	.203	.080	.203	58	26.91
	4 x 4 — W1.4 x W1.4	4 x 4 — 10 x 10		.042	.107	.042	.107	31	14.06
	4 x 4 — W2.0 x W2.0	4 x 4 — 8 x 8	1	.060	.152	.060	.152	43	19.50
	4 x 4 — W2.9 x W2.9	4 x 4 — 6 x 6		.087	.227	.087	.227	62	28.12
	4 x 4 — W4.0 x W4.0	4 x 4 — 4 x 4		.120	.305	.120	.305	85	38.56
Sheets	6 x 6 — W2.9 x W2.9	6 x 6 — 6 x 6		.058	.147	.058	.147	42	19.05
	6 x 6 — W4.0 x W4.0	6 x 6 — 4 x 4		.080	.203	.080	.203	58	26.31
	6 x 6 — W5.5 x W5.5	6 x 6 — 2 x 2	2	.110	.279	.110	.279	80	36.29
	4 x 4 — W1.4 x W1.4	4 x 4 — 4 x 4		.120	.305	.120	.305	85	38.56

NOTES: 1. Exact W—number size for 8 gauge is W2.1
2. Exact W—number size for 2 gauge is W5.4

R033053-10 Spread Footings

General: A spread footing is used to convert a concentrated load (from one superstructure column, or substructure grade beams) into an allowable area load on supporting soil.

Because of punching action from the column load, a spread footing is usually thicker than strip footings which support wall loads. One or two story commercial or residential buildings should have no less than 30 mm thick spread footings. Heavier loads require no less than 600 mm thick. Spread footings may be square, rectangular or octagonal in plan.

Spread footings tend to minimize excavation and foundation materials, as well as labor and equipment. Another advantage is that footings and soil conditions can be readily examined. They are the most widely used type of footing, especially in mild climates and for buildings of four stories or under. This is because they are usually more economical than other types, if suitable soil and site conditions exist.

They are used when suitable supporting soil is located within several meters of the surface or line of subsurface excavation. Suitable soil types include sands and gravels, gravels with a small amount of clay or silt, hardpan, chalk, and rock. Pedestals may be used to bring the column base load down to the top of footing. Alternately, undesirable soil between underside of footing and top of bearing level can be removed and replaced with lean concrete mix or compacted granular material.

Depth of footing should be below topsoil, uncompacted fill, muck, etc. It must be lower than frost penetration but should be above the water table. It must not be at the ground surface because of potential surface erosion. If the ground slopes, approximately one horizontal meter of edge protection must remain. Differential footing elevations may overlap soil stresses or cause excavation problems if clear spacing between footings is less than the difference in depth.

Other footing types are usually used for the following reasons:

A. Bearing capacity of soil is low.
B. Very large footings are required, at a cost disadvantage.
C. Soil under footing (shallow or deep) is very compressible, with probability of causing excessive or differential settlement.
D. Good bearing soil is deep.
E. Potential for scour action exists.
F. Varying subsoil conditions within building perimeter.

Cost of spread footings for a building is determined by:
1. The soil bearing capacity.
2. Typical bay size.
3. Total load (live plus dead) per m^2 for roof and elevated floor levels.
4. The size and shape of the building.
5. Footing configuration. Does the building utilize outer spread footings or are there continuous perimeter footings only or a combination of spread footings plus continuous footings?

Soil Bearing Capacity in kPa

Bearing Material	Typical Allowable Bearing Capacity
Hard sound rock	5746 kPa
Medium hard rock	3830
Hardpan overlaying rock	1149
Compact gravel and boulder-gravel; very compact sandy gravel	958
Soft rock	766
Loose gravel; sandy gravel; compact sand; very compact sand-inorganic silt	575
Hard dry consolidated clay	479
Loose coarse to medium sand; medium compact fine sand	383
Compact sand-clay	287
Loose fine sand; medium compact sand-inorganic silts	192
Firm or stiff clay	143
Loose saturated sand-clay; medium soft clay	95

R033053-50 Industrial Chimneys

Foundation requirements in m³ of concrete for various sized chimneys.

Size Chimney	1.81 tonne Soil	2.72 tonne Soil	Size Chimney	1.81 tonne Soil	2.72 tonne Soil	Size Chimney	1.81 tonne Soil	2.72 tonne Soil
22 860 mm x 914 mm	9.94 m³	8.41 m³	48 768 mm x 1981 mm	65.75 m³	58.11 m³	91 440 mm x 3048 mm	248.48 m³	187.32 m³
25 908 mm x 1676 mm	14.53	12.23	53 340 mm x 2134 mm	82.57	72.63	106 680 mm x 3658 mm	322.64	244.66
30 480 mm x 1524 mm	18.35	15.29	60 960 mm x 1829 mm	95.57	80.28	121 920 mm x 4267 mm	397.57	305.82
38 100 mm x 1676 mm	32.88	27.53	76 200 mm x 2438 mm	175.85	133.80	152 400 mm x 5486 mm	554.30	439.62

R033053-60 Maximum Depth of Frost Penetration in Inches

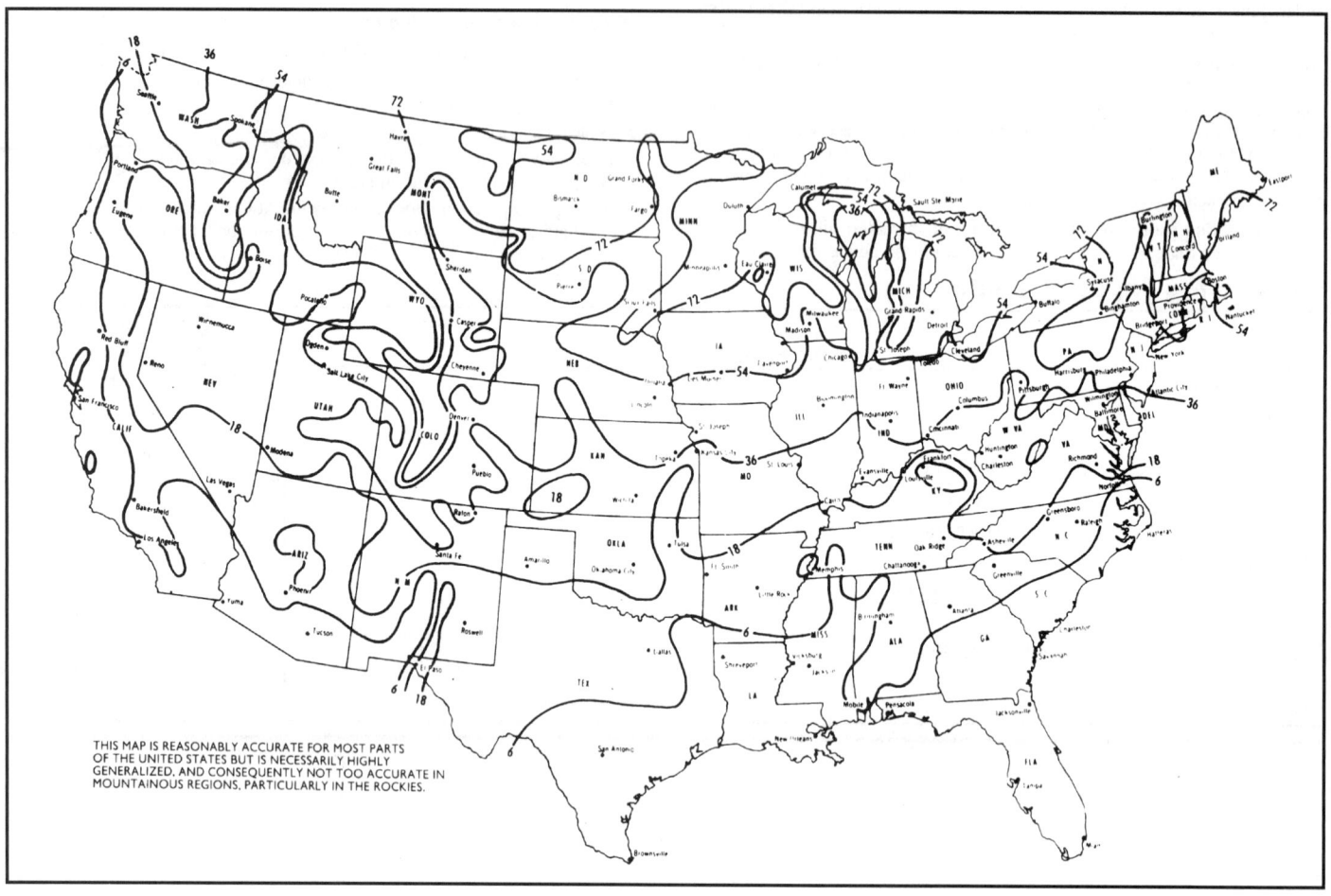

THIS MAP IS REASONABLY ACCURATE FOR MOST PARTS OF THE UNITED STATES BUT IS NECESSARILY HIGHLY GENERALIZED, AND CONSEQUENTLY NOT TOO ACCURATE IN MOUNTAINOUS REGIONS, PARTICULARLY IN THE ROCKIES.

R033105-10 Proportionate Quantities

The tables below show both quantities per m² of floor areas as well as form and reinforcing quantities per m³. Unusual structural requirements would increase the ratios below. High strength reinforcing would reduce the steel weights. Figures are for 21 MPa concrete and 414 MPa reinforcing unless specified otherwise.

Type of Construction	Live Load	Span	Per m² of Floor Area				Per m³ of Concrete		
			Concrete	Forms	Reinf.	Pans	Forms	Reinf.	Pans
Flat Plate	244 kg/m²	4.57 m	0.14 m³	1.06 m²	8.35 kg		7.53 m²	59.92 kg	
		6.10	0.19	1.02	11.72		5.35	61.70	
		7.62	0.24	1.02	14.79		4.25	61.70	
	488	4.57	0.14	1.04	10.45		7.41	74.75	
		6.10	0.22	1.02	13.28		4.74	61.70	
		7.62	0.25	1.01	16.94		4.01	67.04	
Flat Plate (waffle construction) 510 mm domes	244	6.10	0.13	1.00	10.25	0.84 m²	7.66	80.09	6.44 m²
		7.62	0.16	1.00	14.16	0.89	6.32	88.99	5.59
		9.14	0.20	1.00	18.06	0.87	5.10	91.96	4.50
	488	6.10	0.16	1.00	11.23	0.84	6.44	74.16	5.47
		7.62	0.20	1.00	15.62	0.83	5.10	80.09	4.25
		9.14	0.23	1.00	21.48	0.81	4.37	94.92	3.52
Waffle Construction 760 mm domes	244	7.62	0.21	1.06	8.93	0.68	5.10	42.72	4.86
		9.14	0.23	1.06	11.67	0.69	4.74	51.61	4.74
		10.67	0.26	1.05	13.23	0.69	4.01	50.43	4.74
		12.19	0.24	1.00	23.44	0.68	4.25	97.89	4.86
Flat Slab (two way with drop panels)	244	6.10	0.19	1.03	11.42		5.47	60.51	
		7.62	0.23	1.03	14.60		4.37	62.29	
		9.14	0.29	1.03	19.97		3.52	68.82	
	488	6.10	0.20	1.03	13.82		5.23	70.60	
		7.62	0.24	1.03	18.94		4.25	78.91	
		9.14	0.29	1.03	22.75		3.52	77.72	
	976	6.10	0.22	1.03	14.79		4.62	66.45	
		7.62	0.26	1.03	20.65		3.89	78.91	
		9.14	0.32	1.03	25.88		3.16	80.09	
One Way Joists 510 mm Pans	244	4.57	0.11	1.04	6.84	0.93	9.48	62.29	8.51
		6.10	0.13	1.05	8.79	0.94	8.14	71.19	7.29
		7.62	0.14	1.05	12.69	0.94	7.29	88.99	5.56
	488	4.57	0.12	1.07	9.28	0.93	9.36	83.06	8.02
		6.10	0.13	1.08	11.72	0.94	8.14	88.99	7.05
		7.62	0.16	1.07	17.09	0.94	6.68	109.76	5.95
One Way Joists 200 mm x 410 mm filler blocks	244	4.57	0.10	1.06	8.79	8.72 Ea.	10.21	86.02	84 Ea.
		6.10	0.12	1.08	10.74	8.83	8.87	86.02	72
		7.62	0.14	1.07	15.62	8.93	7.66	112.72	64
	488	4.57	0.12	1.07	9.28	8.72	8.99	77.13	73
		6.10	0.14	1.09	13.67	8.83	7.78	94.92	63
		7.62	0.16	1.10	17.58	8.93	6.80	112.72	55
One Way Beam & Slab	244	4.57	0.13	1.30	8.45		10.21	65.85	
		6.10	0.16	1.28	12.74		8.26	81.87	
		7.62	0.20	1.25	13.57		6.44	69.41	
	488	4.57	0.13	1.30	9.28		10.21	72.38	
		6.10	0.16	1.35	13.13		8.26	91.36	
		7.62	0.21	1.37	19.19		6.56	86.02	
	976	4.57	0.13	1.31	10.94		9.72	81.28	
		6.10	0.18	1.40	16.11		7.90	96.70	
		7.62	0.21	1.42	23.88		6.44	108.57	
Two Way Beam & Slab	488	4.57	0.14	1.20	11.03		8.38	77.13	
		6.10	0.19	1.29	14.94		6.68	77.72	
		7.62	0.25	1.33	18.50		5.23	72.97	
	976	4.57	0.15	1.25	13.18		4.98	88.40	
		6.10	0.20	1.32	19.72		6.56	97.89	
		7.62	0.27	1.32	29.69		4.98	110.95	

R033105-10 Proportionate Quantities (cont.)

			27 MPa Concrete and 414 MPa Reinforcing—Form and Reinforcing Quantities per m³		
Item	**Size**	**Forms**	**Reinforcing**	**Minimum**	**Maximum**
Columns (square tied)	250 mm x 250 mm	15.79 m² C.A.	#15M to #35M	130 kg	519 kg
	300 mm x 300 mm	13.12	#20M to #45M	119	567
	350 mm x 350 mm	11.18	#25M to #45M	113	534
	400 mm x 400 mm	9.84	#20M to #45M	111	642
	450 mm x 450 mm	8.75	#20M to #45M	101	538
	500 mm x 500 mm	7.90	#25M to #55M	89	641
	550 mm x 550 mm	7.17	#25M to #55M	91	535
	600 mm x 600 mm	6.56	#25M to #55M	97	524
	650 mm x 650 mm	6.08	#30M to #55M	100	590
	700 mm x 700 mm	5.59	#30M to #55M	87	513
	750 mm x 750 mm	5.23	#35M to #55M	87	583
	800 mm x 800 mm	4.86	#35M to #55M	104	514
	850 mm x 850 mm	4.62	#35M to #55M	93	458
	900 mm x 900 mm	4.37	#35M to #55M	104	505
	950 mm x 950 mm	4.13	#35M to #55M	94	454
	1000 mm x 1000 mm	3.89	#35M to #55M	85	411

Item	**Size**	**Form**	**Spiral**	**Reinforcing**	**Minimum**	**Maximum**
Columns (spirally reinforced)	310 mm dia.	14 m	113 kg	#10 to #35	98 kg	893 kg
		14	113	#45 & #55	—	1100
	360 mm	10	101	#10 to #35	89	970
		10	101	#45 & #55	475	1000
	410 mm	8	95	#10 to #35	95	950
		8	95	#45 & #55	359	1080
	460 mm	6	89	#10 to #35	95	915
		6	89	#45 & #55	285	1075
	510 mm	5	77	#10 to #35	92	865
		5	77	#45 & #55	228	1020
	560 mm	4	74	#10 to #35	98	775
		4	74	#45 & #55	136	995
	610 mm	4	71	#10 to #35	116	800
		4	71	#45 & #55	172	1150
	660 mm	3	59	#10 to #35	119	729
		3	59	#45 & #55	139	1035
	710 mm	3	56	#10 to #35	104	700
		3	56	#45 & #55	119	1075
	760 mm	2	53	#10 to #35	107	670
		2	53	#45 & #55	104	1015
	815 mm	2	50	#10 to #35	110	615
		2	50	#45 & #55	92	955
	865 mm	2	47	#10 to #35	107	600
		2	47	#45 & #55	101	855
	920 mm	2	44	#10 to #35	98	570
		2	44	#45 & #55	92	865
	1015 mm	1	42	#10 to #35	98	500
		1	42	#45 & #55	86	765

R033105-10 Proportionate Quantities (cont.)

		21 MPa Concrete and 414 MPa Reinforcing—Form and Reinforcing Quantities per m³				
Item	Type	Loading	Height	m³/m	Forms/m³	Reinf./m³
Retaining Walls	Cantilever	Level Backfill	1.2 m	0.5 m³	5.95 m²	21 kg
			2.4	1.3	5.10	27
			3.7	2.0	4.25	42
			4.9	2.8	3.89	50
			6.1	4.0	3.40	62
		Highway Surcharge	1.2	0.8	4.98	21
			2.4	1.3	4.37	33
			3.7	2.0	4.01	53
			4.9	3.0	3.65	71
			6.1	4.3	3.28	92
		Railroad Surcharge	1.2	1.0	3.40	27
			2.4	2.0	3.04	39
			3.7	3.3	2.67	53
			4.9	4.8	2.43	59
			6.1	6.5	2.19	71
	Gravity, with Vertical Face	Level Backfill	1.2	1.0	4.50	None
			2.1	1.5	3.28	
			3.0	3.0	2.43	
		Sloping Surcharge	1.2	0.8	3.77	
			2.1	2.0	2.55	
			3.0	4.0	1.82	↓

		Live Load in Metric Tons per Meter							
	Span	Under 1.49 Metric tons		2.98 to 4.46 Metric tons		5.95 to 7.44 Metric tons		8.93 to 10.42 Metric tons	
		Forms	Reinf.	Forms	Reinf.	Forms	Reinf.	Forms	Reinf.
Beams	3.0 m	—	—	10.94 m²	100.86 kg	10.33 m²	103.82 kg	9.11 m²	109.76 kg
	4.9	15.80 m²	97.89 kg	10.33	106.79	9.11	106.79	7.90	133.49
	6.1	13.37	100.86	9.11	109.76	7.53	118.66	6.20	118.66
	7.9	10.94	100.86	7.90	127.55	7.53	127.55	—	—
		10.33	103.82	7.29	118.66	—	—	—	—

Item	Size	Type	Forms per m³	Reinforcing per m³
Spread Footings	Under 1 m³	4 880 kg/m² soil	3.0 m²	26 kg
		24 400	3.0	25
		48 800	3.0	31
	1 m³ to 4 m³	4 880	1.7	29
		24 400	1.7	30
		48 800	1.7	30
	Over 4 m³	4 880	1.1	32
		24 400	1.1	31
		48 800	1.1	33
Pile Caps (30 Tonne Concrete Piles)	Under 4 m³	shallow caps	2.4	39
		medium	2.4	30
		deep	2.4	24
	4 m³ to 8 m³	shallow	1.7	33
		medium	1.8	27
		deep	1.8	24
	8 m³ to 16 m³	shallow	1.3	36
		medium	1.3	27
		deep	1.5	21
	Over 16 m³	shallow	1.1	36
		medium	1.1	27
		deep	1.2	24

R033105-10 Proportionate Quantities (cont.)

21 MPa Concrete and 414 MPa psi Reinforcing — Form and Reinforcing Quantities per m³							
Item	Pile Spacing		50 T Pile	100 T Pile	50 T Pile	100 T Pile
Pile Caps (Steel H Piles)	Under 4 m³	610 mm O.C.	2.9 m²	2.9 m²	44 kg	53 kg
		760 mm	3.0	3.0	47	59
		915 mm	2.9	2.9	47	65
	4 m³ to 8 m³	610 mm	1.8	1.8	47	65
		760 mm	1.8	1.8	50	65
		915 mm	1.8	1.8	44	53
	Over 8 m³	610 mm	1.6	1.6	50	53
		760 mm	1.3	1.3	50	56
		915 mm	1.2	1.2	50	53

		200 mm Thick		250 mm Thick		310 mm Thick		380 mm Thick	
Item	Height	Forms	Reinf.	Forms	Reinf.	Forms	Reinf.	Forms	Reinf.
Basement Walls	2.13 m	9.8 m²	26 kg	7.9 m²	27 kg	6.5 m²	26 kg	5 m²	25 kg
	2.44		26		27		26		25
	2.74		27		27		26		25
	3.05		34		27		26		25
	3.66		49		30		31		25
	4.27		69		38		38		30
	4.88				51		53		39
	5.49	↓		↓		↓	63	↓	42

R033105-20 Materials for One m³ of Concrete

This is an approximate method of figuring quantities of cement, sand and gravel for a field mix with waste allowance included.

With gravel as coarse aggregate for barrels of cement required, divide 10 by total mix; that is, for 1:2:4 mix, 10 divided by 7 = 1-3/7 barrels.

For metric tons of sand multiply barrels of cement by parts of sand and then by 0.2; that is, for the 1:2:4 mix, as above, 1-3/7 x 2 x .2 = .57 metric tons.

Metric tons of gravel are in the same ratio to metric tons of sand as parts in mix, or 4/2 x 0.57 = 1.14 metric tons.

If coarse aggregate is crushed stone, use 10-1/2 instead of 10 as given for gravel.

1 bag cement = 43 kg	1 m³ sand or gravel = 1600 kg	1 m³ crushed stone = 1525 kg
4 bags = 159 L	1 metric ton = 0.62 m³	1 metric ton = 0.66 m³

Average carload of cement is 692 bags; of sand or gravel is 51 metric tons.

Do not stack stored cement over 10 bags high.

R033105-30 Metric Equivalents of Cement Content for Concrete Mixes

94 Pound Bags per Cubic Yard	Kilograms per Cubic Meter	94 Pound Bags per Cubic Yard	Kilograms per Cubic Meter
1.0	55.77	7.0	390.4
1.5	83.65	7.5	418.3
2.0	111.5	8.0	446.2
2.5	139.4	8.5	474.0
3.0	167.3	9.0	501.9
3.5	195.2	9.5	529.8
4.0	223.1	10.0	557.7
4.5	251.0	10.5	585.6
5.0	278.8	11.0	613.5
5.5	306.7	11.5	641.3
6.0	334.6	12.0	669.2
6.5	362.5	12.5	697.1

a. If you know the cement content in pounds per cubic yard,
   multiply by .5933 to obtain kilograms per cubic meter.

b. If you know the cement content in 94 pound bags per cubic yard,
   multiply by 55.77 to obtain kilograms per cubic meter.

R033105-40 Metric Equivalents of Common Concrete Strengths
(to convert other psi values to megapascals, multiply by 0.006895)

U.S. Values psi	SI Value Megapascals	Non-SI Metric Value kgf/cm²*
2000	14	140
2500	17	175
3000	21	210
3500	24	245
4000	28	280
4500	31	315
5000	34	350
6000	41	420
7000	48	490
8000	55	560
9000	62	630
10,000	69	705

* kilograms force per square centimeter

R033105-50 Quantities of Cement, Sand and Stone for One m³ of Concrete per Various Mixes

This table can be used to determine the quantities of the ingredients for smaller quantities of site mixed concrete.

Concrete (m³)	Mix = 1:1:1-3/4			Mix = 1:2:2.25			Mix = 1:2.25:3			Mix = 1:2.25:3		
	Cement (sacks)	Sand (m³)	Stone (m³)	Cement (sacks)	Sand (m³)	Stone (m³)	Cement (sacks)	Sand (m³)	Stone (m³)	Cement (sacks)	Sand (m³)	Stone (m³)
1	13	0.37	0.63	10.14	0.56	0.65	8.17	0.52	0.70	7	0.56	0.74
2	26	0.74	1.26	20.27	1.12	1.30	16.35	1.04	1.40	13	1.12	1.48
3	39	1.11	1.89	30.41	1.68	1.95	24.52	1.56	2.10	20	1.68	2.22
4	52	1.48	2.52	40.55	2.24	2.60	32.70	2.08	2.80	26	2.24	2.96
5	65	1.85	3.15	50.68	2.80	3.25	40.87	2.60	3.50	33	2.80	3.70
6	78	2.22	3.78	60.82	3.36	3.90	49.05	3.12	4.20	39	3.36	4.44
7	92	2.59	4.41	70.96	3.92	4.55	57.22	3.64	4.90	46	3.92	5.18
8	105	2.96	5.04	81.09	4.48	5.20	65.40	4.16	5.60	52	4.48	5.92
9	118	3.33	5.67	91.23	5.04	5.85	73.57	4.68	6.30	59	5.04	6.66
10	131	3.70	6.30	101.37	5.60	6.50	81.75	5.20	7.00	65	5.60	7.40
11	144	4.07	6.93	111.50	6.16	7.15	89.92	5.72	7.70	72	6.16	8.14
12	157	4.44	7.56	121.64	6.72	7.80	98.10	6.24	8.40	78	6.72	8.88
13	170	4.82	8.20	131.79	7.28	8.46	106.28	6.76	9.10	85	7.28	9.62
14	183	5.18	8.82	141.91	7.84	9.10	114.45	7.28	9.80	92	7.84	10.36
15	196	5.56	9.46	152.06	8.40	9.76	122.63	7.80	10.50	98	8.40	11.10
16	209	5.92	10.08	162.19	8.96	10.40	130.80	8.32	11.20	105	8.96	11.84
17	222	6.30	10.72	172.34	9.52	11.06	138.98	8.84	11.90	111	9.52	12.58
18	235	6.66	11.34	182.46	10.08	11.70	147.14	9.36	12.60	118	10.08	13.32
19	249	7.04	11.98	192.61	10.64	12.36	155.33	9.84	13.30	124	10.64	14.06
20	262	7.40	12.60	202.73	11.20	13.00	163.49	10.40	14.00	131	11.20	14.80
21	275	7.77	13.23	212.87	11.76	13.65	171.67	10.92	14.70	137	11.76	15.54
22	288	8.14	13.86	222.42	12.32	14.30	179.84	11.44	15.40	144	12.32	16.28
23	301	8.51	14.49	233.14	12.88	14.95	188.02	11.96	16.10	150	12.88	17.02
24	314	8.88	15.12	243.28	13.44	15.60	196.19	12.48	16.80	157	13.44	17.76
25	327	9.25	15.75	253.42	14.00	16.25	204.37	13.00	17.50	163	14.00	18.50
26	340	9.64	16.40	263.58	14.56	16.92	212.57	13.52	18.20	170	14.56	19.24
27	353	10.00	17.00	273.70	15.12	17.56	220.73	14.04	18.90	177	15.02	20.00
28	366	10.36	17.64	283.83	15.68	18.20	228.89	14.56	19.60	183	15.68	20.72
29	379	10.74	18.28	293.97	16.24	18.86	237.08	15.08	20.30	190	16.24	21.46

R033105-65 Field Mix Concrete

Presently most building jobs are built with ready mix concrete except for isolated locations and some larger jobs requiring over 7650 m³ where land is readily available for setting up a temporary batch plant.

The most economical mix is a controlled mix using local aggregate proportioned by trial to give the required strength with the least cost of material.

R033105-70 Placing Ready Mixed Concrete

For ground pours allow for 5% waste when figuring quantities.

Prices in the front of the book assume normal deliveries. If deliveries are made before 8 A.M. or after 5 P.M. or on Saturday afternoons add 30%. Negotiated discounts for large volumes are not included in prices in front of book.

For the lower floors without truck access, concrete may be wheeled in rubber tired buggies, conveyer handled, crane handled or pumped. Pumping is economical if there is top steel. Conveyers are more efficient for thick slabs.

At higher floors the rubber tired buggies may be hoisted by a hoisting tower then wheeled to location. Placement by a conveyer is limited to three

floors and is best for high volume pours. Pumped concrete is best when building has no crane access. Concrete may be pumped directly as high as thirty-six stories using special pumping techniques. Normal maximum height is about fifteen stories.

Best pumping aggregate is screened and graded bank gravel rather than crushed stone.

Pumping downward is more difficult than pumping upwards. Horizontal distance from pump to pour may increase preparation time prior to pour. Placing by cranes, either mobile, climbing or tower types continues as the most efficient method for high rise concrete buildings.

REFERENCE TABLES

R033105-80 Slab on Grade

General: Ground slabs are classified on the basis of use. Thickness is generally controlled by the heaviest concentrated load supported. If load area is greater than 0.05 m² soil bearing may be important. The base granular fill must be a uniformly compacted material of limited capillarity, such as gravel or crushed rock. Concrete is placed on this surface of the vapor barrier on top of base.

Ground slabs are either single or two course floors. Single course are widely used. Two course floors have a subsequent wear resistant topping.

Reinforcement is provided to maintain tightly closed cracks.

Control joints limit crack locations and provide for differential horizontal movement only. Isolation joints allow both horizontal and vertical differential movement.

Use of Table: Determine appropriate type of slab (A, B, C, or D) by considering type of use or amount of abrasive wear of traffic type.

Determine thickness by maximum allowable wheel load or uniform load,

opposite 1st column, thickness. Increase the controlling thickness if details require, and select either plain or reinforced slab thickness and type.

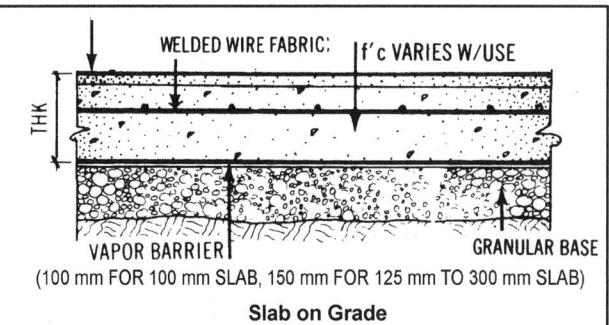

Slab on Grade

Thickness and Loading Assumptions by Type of Use

SLAB THICKNESS (IN.)	TYPE	A Non Little Foot Only Load* tonne	B Light Light Pneumatic Wheels Load* tonne	C Normal Moderate Solid Rubber Wheels Load* tonne	D Heavy Severe Steel Tires Load* tonne	◄ Slab I.D. ◄ Industrial ◄ Abrasion ◄ Type of Traffic Max. Uniform Load to Slab ▼ (kg/m²)
100 mm	Reinf. Plain	1.8 tonne				488
125 mm	Reinf. Plain	2.7 tonne	1.8 tonne			976
150 mm	Reinf. Plain		3.6 tonne	3.6 tonne	2.7 tonne	2400 to 3900
175 mm	Reinf. Plain			4.1 tonne	3.6 tonne	7300
200 mm	Reinf. Plain				5.0 tonne	
250 mm	Reinf. Plain				6.3 tonne	* Max. Wheel Load in tonnes (incl. impact)
300 mm	Reinf. Plain					
D E S I G N A S S U M P T I O N S	Concrete, Chuted	f'c = 3.5 KSI 24 mPa	28 mPa	31 mPa	Slab @ 24 mPa	ASSUMPTIONS BY SLAB TYPE
	Toppings			25 mm Integral	25 mm Bonded	
	Finish	Steel Trowel	Steel Trowel	Steel Trowel	Screed & Steel Trowel	
	Compacted Granular Base	100 mm deep for 100 mm slab thickness 150 mm deep for 125 mm slab thickness & greater				ASSUMPTIONS FOR ALL SLAB TYPES
	Vapor Barrier	0.5 mm thick polyethylene				
	Forms & Joints	Allowances included				
	Rein-forcement	WWF as required ≥ 413 mPa				

R033105-85   Lift Slabs

The cost advantage of the lift slab method is due to placing all concrete, reinforcing steel, inserts and electrical conduit at ground level and in reduction of formwork. Minimum economical project size is about 2700 m². Slabs may be tilted for parking garage ramps.

It is now used in all types of buildings and has gone up to 22 stories high in apartment buildings. Current trend is to use post-tensioned flat plate slabs with spans from 7 m to 11 m. Cylindrical void forms are used when deep slabs are required.

One half kilogram of prestressing steel is about equal to 3.2 kg of conventional reinforcing.

To be considered cured for stressing and lifting, a slab must have attained 75% of design strength. Seven days are usually sufficient with four to five days possible if high early strength cement is used. Slabs can be stacked using two coats of a non-bonding agent to insure that slabs do not stick to each other. Lifting is done by companies specializing in this work. Lift rate is 2 m to 5 m per hour with an average of 3 m per hour. Total areas up to 3000 m² have been lifted at one time. Use of 24 to 36 jacking columns is common. Most economical bay sizes are 7 m to 9 m with four to fourteen stories most efficient. Continuous design reduces reinforcing steel cost. Use of post-tensioned slabs allows larger bay sizes.

R034105-30 Prestressed Precast Concrete Structural Units

See also R03410-090 for post-tensioned prestressed concrete.

Type	Location	Depth	Span in m		Live Load kg per m²
Double Tee 2.4 m to 3.0 m	Floor	710 mm to 865 mm	18 to 25		245 to 390
	Roof	310 mm to 610 mm	9 to 16		195
	Wall	Width 2.4 m	Up to 17 m high		Wind
Multiple Tee 2.4 m	Roof	200 mm to 310 mm	4 to 12		195
	Floor	200 mm to 310 mm	4 to 9		490
Plank or	Roof or Floor		Roof	Floor	
		100 mm	4	4	195 for Roof
		150 mm	7	5	
		200 mm	8	8	
		250 mm	10	9	490 for Floor
		310 mm	13	10	
Single Tee 2.4 m to 3.0 m	Roof	710 mm	12		
		815 mm	24		
		920 mm	30		195
		1220 mm	37		
AASHO Girder	Bridges	Type 4	30		
		5	34		Highway
		6	38		
Box Beam 1.2 m	Bridges	380 mm	12		
		690 mm	to		Highway
		840 mm	30		

The majority of precast projects today utilize double tees rather than single tees because of speed and ease of installation. As a result casting beds at manufacturing plants are normally formed for double tees. Single tee projects will therefore require an initial set up charge to be spread over the individual single tee costs.

For floors, a 50 to 75 mm topping is field cast over the shapes. For roofs, insulating concrete or rigid insulation is placed over the shapes. Topping is not included in the above costs.

Member lengths up to 12 m are standard haul, 12 m to 18 m require special permits and lengths over 18 m must be escorted. Over width and/or over length can add up to 100% on hauling costs.

Large heavy members may require two cranes for lifting which would increase erection costs by about 45%. An eight man crew can install to 20 double tees, or 45 to 70 quad tees or planks per day.

Grouting of connections must also be included.

Several system buildings utilizing precast members are available. Heights can go up to 22 stories for apartment buildings. Optimum design ratio is 3 m² of surface to 1 m² of floor area.

R034136-90   Prestressed Concrete, Post-tensioned

In post-tensioned concrete the steel tendons are tensioned after the concrete has reached about 3/4 of its ultimate strength. The cableways are grouted after tensioning to provide bond between the steel and concrete. If bond is to be prevented, the tendons are coated with a corrosion-preventative grease and wrapped with waterproof paper or plastic. Bonded tendons are usually used when ultimate strength (beams & girders) is a controlling factor.

High strength concrete is used to fully utilize the steel, thereby reducing the size and weight of the member. A plasticizing agent may be added to reduce water content. Maximum size aggregate ranges from 12.7 to 38 mm depending on the spacing of the tendons.

The types of steel commonly used are bars and strands. Job conditions determine which is best suited. Bars are best for vertical prestresses since they are easy to support. The trend is for steel manufacturers to supply a finished package, cut to length, which reduces field preparation to a minimum.

Bars vary from 19 to 35 mm diameter. Table below gives time in labor-hours per tendon for placing, tensioning and grouting (if required) a 23 m beam. Tendons used in buildings are not usually grouted; tendons for bridges usually are grouted. For strands the table indicates the labor-hours per kg for typical prestressed units 30 m long. Simple span beams usually require one end stressing regardless of lengths. Continuous beams are usually stressed from two ends. Long slabs are poured from the center outward and stressed in 23 m increments after the initial 46 m center pour.

Labor Hours per Tendon and per kilogram of Prestress Steel						
Length	30 m Beam		23 m Beam		30 m Slab	
Type Steel	Strand		Bars		Strand	
Diameter	12.7 mm		19 mm	35 mm	12.7 mm	15 mm
Number	4	12	1	1	1	1
Force in kilonewtons	445	1335	187	636	111	156
Preparation & Placing Cables	3.6	7.4	0.9	2.9	0.9	1.1
Stressing Cables	2.0	2.4	0.8	1.6	0.5	0.5
Grouting, if required	2.5	3.0	0.6	1.3		
Total Labor Hours	8.1	12.8	2.3	5.8	1.4	1.6
Prestressing Steel Weights (kg)	98	290	52	172	24	34
Labor hours per kg, Bonded	0.083	0.044	0.044	0.034		
Non-bonded					0.059	0.048

Flat Slab construction — 28 MPa concrete with span to depth ratio between 36 and 44. Two way post-tensioned steel averages 5.0 kg per m^2 for 2.2 m to 2.6 m bays (usually strand) and additional reinforcing steel averages 2.5 kg per m^2.

Pan and Joist construction — 28 MPa concrete with span to depth ratio 28 to 30. Post-tensioned steel averages 3.9 kg per m^2 and reinforcing steel about 5.0 kg per m^2. Placing and stressing averages 40 hours per metric ton of total material.

Beam construction — 28 to 34 MPa concrete. Steel weights vary greatly.

Labor cost per kg goes down as the size and length of the tendon increases. The primary economic consideration is the cost per Newton for the member.

Post-tensioning becomes feasible for beams and girders over 9 m long; for continuous two-way slabs over 6 m clear; also in transferring upper building loads over longer spans at lower levels. Post-tension suppliers will provide engineering services at no cost to the user. Substantial economies are possible by using post-tensioned Lift Slabs.

| **Concrete** | **R0345** | **Precast Architectural Concrete** |

R034513-10   Precast Concrete Wall Panels

Panels are either solid or insulated with plain, colored or textured finishes. Transportation is an important cost factor. Prices shown in the unit cost section of the book are based on delivery within 80 kilometers of a plant including fabricators' overhead and profit. Engineering data is available from fabricators to assist with construction details. Usual minimum job size for economical use of panels is about 465 m^2. Small jobs can double the prices shown. For large, highly repetitive jobs, deduct up to 15% from the prices shown.

50 mm thick panels cost about the same as 75 mm thick panels and maximum panel size is less. For building panels faced with granite, marble or stone, add the material prices from those unit cost sections to the plain panel price shown. There is a growing trend toward aggregate facings and broken rib finish rather than plain gray concrete panels.

No allowance has been made, in the unit cost section, for supporting steel framework. On one story buildings, panels may rest on grade beams and require only wind bracing and fasteners. On multi-story buildings panels can span from column to column and floor to floor. Plastic designed steel framed structures may have large deflections which slow down erection and raise costs.

Large panels are more economical than small panels on a m^2 basis. When figuring areas include all protrusions, returns, etc. Overhangs can triple erection costs. Panels over 15 m have been produced. Larger flat units should be prestressed. Vacuum lifting of smooth finish panels eliminates inserts and can speed erection.

R034713-20   Tilt Up Concrete Panels

The advantage of tilt up construction is in the low cost of forms and the placing of concrete and reinforcing. Panels up to 20 m high and 140 mm thick have been tilted using strongbacks. Tilt up has been used for one to five story buildings and is well suited for warehouses, stores, offices, schools and residences.

The panels are cast in forms on the floor slab. Most jobs use 140 mm thick solid reinforced concrete panels. Sandwich panels with a layer of insulating materials are also used. Where dampness is a factor, lightweight aggregate is used. Optimum panel size is 28 m² to 46 m².

Slabs are usually poured with 21 MPa concrete which permits tilting seven days after pouring. Slabs may be stacked on top of each other and are separated from each other by either two coats of bond breaker or a film of polyethylene. Use of high early strength cement allows tilting two days after a pour. Tilting up is done with a roller outrigger crane with a capacity of at least 1-1/2 times the weight of the panel at the required reach. Exterior precast columns can be set at the same time as the panels; interior precast columns can be set first and the panels clipped directly to them. The use of cast-in-place concrete columns is diminishing due to shrinkage problems. Structural steel columns are sometimes used if crane rails are planned. Panels can be clipped to the columns or lowered between the flanges. Steel channels with anchors may be used as edge forms for the slab. When the panels are lifted the channels form an integral steel column to take structural loads. Roof loads can be carried directly by the panels for wall heights to 4.3 m.

Requirements of local building codes may be a limiting factor and should be checked. Building floor slabs should be poured first and should be a minimum of 125 mm thick with 100% compaction of soil or 150 mm thick with less than 100% compaction.

Setting times as fast as nine minutes per panel have been observed, but a safer expectation would be four panels per hour with a crane and a four man setting crew. If crane erects from inside building, some provision must be made to get crane out after walls are erected. Good yarding procedure is important to minimize delays. Equalizing three point lifting beams and self-releasing pick-up hooks speed erection. If panels must be carried to their final location, setting time per panel will be increased and erection costs may approach the erection cost range of architectural precast wall panels. Placing panels into slots formed in continuous footers will speed erection.

Reinforcing should be with no. 15 bars with vertical bars on the bottom. If surface is to be sandblasted, stainless steel chairs should be used to prevent rust staining.

Use of a broom finish is popular since the unavoidable surface blemishes are concealed.

Precast columns run from three to five times the m³ price of the panels only.

R035216-10   Lightweight Concrete

Lightweight aggregate concrete is usually purchased ready mixed, but it can also be field mixed.

Vermiculite or Perlite come in bags of 0.11 m³ under various trade names. Weight is about 128 kg per m³. For insulating roof fill use 1:6 mix. For structural deck use 1:4 mix over gypsum boards, steeltex, steel centering, etc. supported by closely spaced joists or bulb trees. For structural slabs use 1:3:2 vermiculite sand concrete over steeltex, metal lath, steel centering, etc. on joists spaced 0.6 m O.C. for maximum L.L. of 390 kg/m². Use same

mix for slab base fill over steel flooring or regular reinforced concrete slab when tile, terrazzo or other finish is to be laid over.

For slabs on grade use 1:3:2 mix when tile, etc. finish is to be laid over. If radiant heating units are installed use a 1:6 mix for a base. After coils are in place, cover with a regular granolithic finish (mix 1:3:2) to a minimum depth of 38 mm over top of units.

Reinforce all slabs with 150 x 150 or 200 x 200 welded wire mesh.

R040130-10   Cleaning Face Brick

On smooth brick a man can clean 6.5 m² an hour; on rough brick 4.6 m² per hour. Use one liter muriatic acid to 75 liters of water for 93 m². Do not use acid solution until wall is at least seven days old, but a mild soap solution may be used after two days. Commercial cleaners cost from $3.00 to $4.00 per liter.

Time has been allowed for clean-up in brick prices.

R040513-10   Cement Mortar (material only)

Type N - 1:1:6 mix by volume. Use everywhere above grade except as noted below. - 1:3 mix using conventional masonry cement which saves handling two separate bagged materials.

Type M - 1:1/4:3 mix by volume, or 1 part cement, 1/4 (10% by wt.) lime, 3 parts sand. Use for heavy loads and where earthquakes or hurricanes may occur. Also for reinforced brick, sewers, manholes and everywhere below grade.

Mix Proportions by Volume and Compressive Strength of Mortar

Where Used	Mortar Type	\multicolumn{4}{c}{Allowable Proportions by Volume}	Compressive Strength @ 28 days
		Portland Cement	Masonry Cement	Hydrated Lime	Masonry Sand	
Plain Masonry	M	1	1	—	6	
		1	—	1/4	3	17.2 MPa
	S	1/2	1	—	4	
		1	—	1/4 to 1/2	4	12.4 MPa
	N	—	1	—	3	
		1	—	1/2 to 1-1/4	6	5.2 MPa
	O	—	1	—	3	
		1	—	1-1/4 to 2-1/2	9	2.4 MPa
	K	1	—	2-1/2 to 4	12	0.5 MPa
Reinforced Masonry	PM	1	1	—	6	17.2 MPa
	PL	1	—	1/4 to 1/2	4	17.2 MPa

Note: The total aggregate should be between 2.25 to 3 times the sum of the cement and lime used.

The labor cost to mix the mortar is included in the productivity and labor cost of unit price lines in unit cost sections for brickwork, blockwork and stonework.

The material cost of mixed mortar is included in the material cost of those same unit price lines, and includes the cost of renting and operating a 0.28 m³ mixer at the rate of 5.66 m³ per day.

There are two types of mortar color used. One type is the inert additive type with about 45 kg per M brick as the typical quantity required. These colors are also available in smaller batch size bags (.45 kg to 6.8 kg) which can be placed directly into the mixer without measuring. The other type is premixed and replaces the masonry cement. Dark green color has the highest cost.

R040519-50   Masonry Reinforcing

Horizontal joint reinforcing helps prevent wall cracks where wall movement may occur and in many locations is required by code. Horizontal joint reinforcing is generally not considered to be structural reinforcing and an unreinforced wall may still contain joint reinforcing. Reinforcing strips come in 3 meter and 3.7 meter lengths and in truss and ladder shapes, with and without drips. Field labor runs between 2.7 to 5.3 hours per 30 meters for wall thicknesses up to 300 mm. The wire meets ASTM A82 for cold drawn steel wire and the typical size is 9 ga. sides and ties with 5 mm diameter also available. Typical finish is mill galvanized with zinc coating at 30 grams per m^2. Class I (122 grams per m^2) and Class III (244 grams per m^2.) are also available, as is hot dipped galvanizing at 457 grams per m^2.

R042110-10   Economy in Bricklaying

Have adequate supervision. Be sure bricklayers are always supplied with materials so there is no waiting. Place best bricklayers at corners and openings.

Use only screened sand for mortar. Otherwise, labor time will be wasted picking out pebbles. Use seamless metal tubs for mortar as they do not leak or catch the trowel. Locate stack and mortar for easy wheeling.

Have brick delivered for stacking. This makes for faster handling, reduces chipping and breakage, and requires less storage space. Many dealers will deliver select common in 610 mm x 914 mm x 1219 mm pallets or face brick packaged. This affords quick handling with a crane or forklift and easy tonging in units of ten, which reduces waste.

Use wider bricks for one wythe wall construction. Keep scaffolding away from wall to allow mortar to fall clear and not stain wall.

On large jobs develop specialized crews for each type of masonry unit.

Consider designing for prefabricated panel construction on high rise projects.

Avoid excessive corners or openings. Each opening adds about 50% to labor cost for area of opening.

Bolting stone panels and using window frames as stops reduces labor costs and speeds up erection.

R042110-20   Common and Face Brick

Common building brick manufactured according to ASTM C62 and facing brick manufactured according to ASTM C216 are the two standard bricks available for general building use.

Building brick is made in three grades; SW, where high resistance to damage caused by cyclic freezing is required; MW, where moderate resistance to cyclic freezing is needed; and NW, where little resistance to cyclic freezing is needed. Facing brick is made in only the two grades SW and MW. Additionally, facing brick is available in three types; FBS, for general use; FBX, for general use where a higher degree of precision and lower permissible variation in size than FBS is needed; and FBA, for general use to produce characteristic architectural effects resulting from non-uniformity in size and texture of the units.

In figuring the material cost of brickwork, an allowance of 25% mortar waste and 3% brick breakage was included. If bricks are delivered palletized with 280 to 300 per pallet, or packaged, allow only 1-1/2% for breakage. Packaged or palletized delivery is practical when a job is big enough to have a crane or other equipment available to handle a package of brick. This is so on all industrial work but not always true on small commercial buildings.

The use of buff and gray face is increasing, and there is a continuing trend to the Norman, Roman, Jumbo and SCR brick.

Common red clay brick for backup is not used that often. Concrete block is the most usual backup material with occasional use of sand lime or cement brick. Building brick is commonly used in solid walls for strength and as a fire stop.

Brick panels built on the ground and then crane erected to the upper floors have proven to be economical. This allows the work to be done under cover and without scaffolding.

R042110-50   Brick, Block & Mortar Quantities

Running Bond							For Other Bonds Standard Size Add to m² Quantities in Table to Left		
Number of Brick per m² of Wall - Single Wythe with 10 mm Joints					m³ of Mortar per k Bricks, Waste Included				
Type Brick	Nominal Size (incl. mortar) L H W		Modular Coursing	Number of Brick per m²	10 mm Joint	13 mm Joint	Bond Type	Description	Factor
Standard	203 mm x 68 mm x 102 mm		3C=203 mm	73.66	0.29	0.37	Common	full header every fifth course	+20%
Economy	203 mm x 102 mm x 102 mm		1C=102 mm	48.43	0.32	0.41		full header every sixth course	+16.7%
Engineer	203 mm x 81 mm x 102 mm		5C=406 mm	60.60	0.30	0.39	English	full header every second course	+50%
Fire	229 mm x 64 mm x 114 mm		2C=127 mm	68.89	250 kg Fireclay	—	Flemish	alternate headers every course	+33.3%
Jumbo	305 mm x 102 mm x 152 mm or 203 mm		1C=102 mm	32.29	0.67	0.87		every sixth course	+5.6%
Norman	305 mm x 68 mm x 102 mm		3C=203 mm	48.44	0.40	0.51	Header = W x H exposed		+100%
Norwegian	305 mm x 81 mm x 102 mm		5C=406 mm	40.36	0.42	0.53	Rowlock = H x W exposed		+100%
Roman	305 mm x 51 mm x 102 mm		2C=102 mm	64.58	0.38	0.48	Rowlock stretcher = L x W exposed		+33.3%
SCR	305 mm x 68 mm x 152 mm		3C=203 mm	48.43	0.61	0.79	Soldier = H x L exposed		—
Utility	305 mm x 102 mm x 102 mm		1C=102 mm	32.29	0.44	0.56	Sailor = W x L exposed		-33.3%

Concrete Blocks Nominal Size		Approximate Weight per m²		Blocks per m²	Mortar per k block, waste included	
		Standard	Lightweight		Partitions	Back up
51 mm	x 203 mm x 406 mm	98 kg/m²	73 kg/m²	12	0.77 m³	1.02 m³
102 mm		146	98		1.16	1.44
152 mm		205	146		1.59	1.87
203 mm		268	185		2.04	2.32
254 mm		341	229		2.47	2.75
305 mm	↓	415 ↓	268 ↓	↓	2.89 ↓	3.17 ↓

R042210-20   Concrete Block

The material cost of special block such as corner, jamb and head block can be figured at the same price as ordinary block of same size. Labor on specials is about the same as equal sized regular block.

Bond beam and 400 mm high lintel blocks are more expensive than regular units of equal size. Lintel blocks are 200 mm long and either 200 mm or 400 mm high.

Use of motorized mortar spreader box will speed construction of continuous walls.

Hollow non-load bearing units are made according to ASTM C129 and hollow load bearing units according to ASTM C90.

R042210-30 Fully Grouted Reinforced Masonry Wall Capacities Per Meter (KN & N.m/m)

	Earthquake Zones 1, 2 & 3				Allowable Vertical Wall Loads									Allowable Wall Moments Without Vertical Wall Loads	
					Eccentric Loads				Without Wind or Eccentric Loads		With Wind			Not Wind or Earthquake	
Thk.	Length Or Height		Type Wall	Rebar	2.38 N.m/m		1.19 N.m/m					Inspection			
					Inspection		Inspection		Inspection		No		Yes	Inspection	
T (Nom.) (mm)	h' (m)	h'/T (m/m)	Brick/Grout/C.M.U. (mm)	Size and Spacing (mm O.C.)	No (kN/m)	Yes (kN/m)	No (kN/m)	Yes (kN/m)	No (kN/m)	Yes (kN/m)	73 kg/m² (kN/m)	146 kg/m² (kN/m)	73 kg/m² & 146 kg/m² (kN/m)	No (N.m/m)	Yes (N.m/m)
200			100/0/100	15m@800	147	341	171	357	194	387	194	194	387	2.35	3.28
	2.4	12	Solid CMU		186	419	209	442	233	465	233	233	465	2.65	3.32
			100/0/100		136	315	158	336	179	358	179	179	358	2.35	3.28
	3.6	18	Solid CMU		243	387	265	409	214	430	214	214	430	2.65	3.32
			100/0/100		114	264	132	282	150	301	150	150	301	2.35	3.28
	4.8	24	Solid CMU		144	324	162	342	180	360	180	180	360	2.65	3.32
250			100/0/150	15m@600	211	460	230	479	249	498	249	249	498	3.31	5.51
			Solid CMU		260	562	279	578	298	597	298	298	597	4.30	5.56
	2.4	9.6	100/50/100		211	460	230	479	249	498	249	249	498	3.81	5.51
			100/0/150		203	443	221	461	239	479	239	239	479	3.81	5.51
			Solid CMU		251	538	269	557	287	575	287	287	575	4.30	5.56
	3.6	14.4	100/50/100		203	443	221	461	239	479	239	239	479	3.81	5.51
			100/0/150		187	409	204	426	221	443	221	221	443	3.81	5.51
			Solid CMU		232	498	249	514	266	531	266	266	531	4.30	5.56
	4.8	19.2	100/50/100		187	409	204	426	221	443	221	221	443	3.81	5.51
			100/0/150		163	354	177	368	191	382	191	191	382	3.81	5.51
			Solid CMU		201	450	215	445	230	460	230	230	460	4.30	5.56
	6.0	24	100/50/100		163	354	177	368	191	382	191	191	382	3.81	5.51
300			100/0/200	25m@1200	271	514	287	589	303	605	303	303	605	5.66	8.47
			Solid CMU		332	695	347	710	363	727	363	363	727	6.39	8.53
	2.4	8	100/50/150	15m@500	271	574	287	589	303	605	303	303	605	4.05	6.38
			100/0/200	25m@1200	266	562	281	577	296	592	296	296	592	5.66	8.47
			Solid CMU		325	681	340	696	355	711	355	355	711	6.39	8.53
	3.6	12	100/50/150	15m@500	266	562	281	577	296	572	296	296	592	4.05	6.38
			100/200	25m@1200	255	538	269	553	284	568	284	284	568	5.66	8.47
			Solid CMU		311	652	326	667	341	681	341	341	681	6.39	8.53
	4.8	16	100/50/150	15m@500	255	538	269	553	284	568	284	284	568	4.05	6.38
			100/0/200	25m@1200	209	441	220	454	233	465	233	233	465	5.66	8.47
			Solid CMU		250	535	267	546	279	559	279	279	559	6.39	8.53
	7.2	24	100/50/150	15m@500	209	441	220	454	233	465	233	233	465	4.05	6.38
400			100/0/300	25m@800	385	794	397	805	409	818	409	409	818	10.61	16.69
			Solid CMU		467	958	479	970	490	981	490	490	981	11.99	17.14
	2.4	6	100/50/250	15m@375	385	794	397	805	409	818	409	409	818	5.02	8.19
			100/0/300	25m@800	382	787	393	799	405	810	405	405	810	10.61	16.69
			Solid CMU		463	964	474	961	486	972	486	486	972	11.99	17.14
	3.6	9	100/50/250	15m@375	382	787	393	799	405	810	405	405	810	5.02	8.19
			100/0/300	25m@800	375	774	387	785	413	797	398	398	797	10.61	16.69
			Solid CMU		455	933	466	945	478	956	478	478	956	11.99	17.14
	4.8	12	100/50/250		375	774	387	785	398	797	398	398	797	5.02	8.19
			100/0/300	25m@800	297	612	306	621	315	630	294*	274*	630	10.61	16.69
			Solid CMU		360	738	369	747	378	756	358*	358*	756	11.99	17.14
	9.6	24	100/50/250	15m@375	297	612	306	621	315	680	875*	274*	630	5.02	8.19

*Zone 3 Only

RO42210-30 Unreinforced Masonry Wall Capacities Per Meter (kN & N.m/m)

Thk. T (Nom.) (mm)	Length Or Height h' (m)	h'/t (m/m)	Type of Wall	Eccentric Loads 2.38 N.m/m kN/m	Eccentric Loads 1.19 N.m/m kN/m	Without Wind or Eccentric Loads kN/m	With Wind 73 kg/m² kN/m	With Wind 146 kg/m² kN/m	Not Wind or Earthquake Inspection No N.m/m	Not Wind or Earthquake Inspection Yes N.m/m	Wind or Earthquake Inspection No Nom/m	Wind or Earthquake Inspection Yes Nom/m
200			Solid Brick	71	109	148	148	148	.39	.78	.53	1.05
			Solid CM Units	129	168	706	206	206	.24	.48	.32	.63
	2.4	12	Hollow CM Units	116	155	194	194	194	.24	.48	.20	.43
			Solid Brick	69	107	144	144	144	.39	.78	.53	1.05
			Solid CM Units	125	163	201	201	201	.24	.48	.31	.63
	3.0	15	Hollow CM Units	—	65	92	92	88	.15	.32	.20	.43
			Solid Brick	—	101	137	137	116	.39	.78	.53	1.05
			Solid CM Units	119	155	191	191	188	.24	.48	.31	.63
	3.6	18	Hollow CM Units	—	62	88	88	69	.15	.32	.20	.43
300			Solid Brick	179	206	232	232	232	.92	1.84	1.22	2.45
			Solid CM Units	270	296	322	322	332	.54	1.09	.73	1.46
			Hollow CM Units	115	133	152	152	152	.37	.77	.51	1.02
	2.4	8	Brick & Hollow CMU	80	114	147	147	147	.46	.92	.61	1.22
			Solid Brick	176	201	227	227	227	.92	1.84	1.22	2.45
			Solid CM Units	196	215	234	234	234	.54	1.09	.73	1.46
			Hollow CM Units	112	130	148	148	148	.37	.77	.51	1.02
	3.6	12	Brick & Hollow CMU	79	112	144	144	132	.46	.92	.61	1.22
			Solid Brick	169	193	217	217	209	.92	1.84	1.22	2.45
			Solid CM Units	253	278	302	302	302	.54	1.09	.73	1.46
			Hollow CM Units	107	125	142	142	132	.37	.77	.51	1.02
	4.8	16	Brick & Hollow CMU	76	107	138	133	—	.46	.92	.61	1.22
400			Solid Brick	271	291	311	311	311	1.65	3.32	2.21	4.42
			Solid CM Units	393	413	432	432	432	.99	1.99	1.33	2.65
			Hollow CM Units	158	177	194	194	194	.51	1.04	.68	1.38
	3.6	9	Brick & Hollow CMU	136	162	187	187	187	.75	1.50	.99	1.99
			Solid Brick	267	286	305	305	305	1.65	3.32	2.21	4.42
			Solid CM Units	387	406	425	425	425	.99	1.99	1.33	2.65
			Hollow CM Units	156	174	191	191	191	.51	1.04	.68	1.38
	4.8	12	Brick & Hollow CMU	133	159	185	185	162	.75	1.50	.99	1.99
			Solid Brick	260	278	296	296	296	1.65	3.32	2.21	4.42
			Solid CM Units	376	394	412	412	412	.99	1.99	1.33	2.65
			Hollow CM Units	152	169	185	185	160	.51	1.04	.68	1.38
	6.0	15	Brick & Hollow CMU	130	155	179	175	—	.75	1.50	.99	1.99
500			Solid Brick	362	377	392	392	392	2.62	5.24	3.49	6.97
			Solid CM Units	515	531	546	546	546	1.56	3.13	2.09	4.18
			Hollow CM Units	236	250	265	265	265	.85	1.72	1.14	2.30
	3.6	7.2	Brick & Hollow CMU	235	252	269	269	269	1.38	2.75	1.84	3.67
			Solid Brick	358	374	390	390	390	2.62	5.24	3.49	6.97
			Solid CM Units	511	527	542	542	542	1.56	3.13	2.09	4.18
			Hollow CM Units	233	248	263	263	263	.85	1.72	1.14	2.30
	4.8	9.6	Brick & Hollow CMU	233	249	267	267	267	1.38	2.75	1.84	3.67
			Solid Brick	346	360	376	376	376	2.62	5.24	3.49	6.97
			Solid CM Units	493	508	522	522	522	1.56	3.13	2.09	4.18
			Hollow CM Units	225	239	254	254	234	.85	1.72	1.14	2.30
	7.2	14.4	Brick & Hollow CMU	224	241	257	257	222	1.38	2.75	1.84	3.67

For information about Means Estimating Seminars, see yellow pages 12 and 13 in back of book

R050516-30   Coating Structural Steel

On field-welded jobs, the shop-applied primer coat is necessarily omitted. All painting must be done in the field and usually consists of red oxide rust inhibitive paint or an aluminum paint. The table below shows paint coverage and daily production for field painting.

See Division 05950-650 for hot-dipped galvanizing and for field-applied cold galvanizing and other paints and protective coatings.

See Division 05910-500 for steel surface preparation treatments such as wire brushing, pressure washing and sand blasting.

Type Construction	Surface Area per Metric Ton	Coat	One Liter Covers		In 8 Hrs. Person Covers		Average per Metric Ton Spray	
			Brush	Spray	Brush	Spray	Liters	Labor-hours
Light Structural	25 m² to 50 m²	1st	12.25 m²	11.15 m²	59.45 m²	185.80 m²	3.75 liters	1.76 L.H.
		2nd	11.05	10.05	74.30	223.00	4.15	1.43
		3rd	11.05	10.05	89.20	297.30	4.15	1.10
Medium	15 m² to 25 m²	All	9.80	8.95	148.65	297.30	2.50	0.66
Heavy Structural	5 m² to 10 m²	1st	9.80	8.95	178.35	371.60	0.85	0.22
		2nd	9.80	8.95	185.80	371.60	0.85	0.22
		3rd	9.80	8.95	185.80	371.60	0.85	0.22
Weighted Average	20 m²	All	9.80	8.95	125.40	278.70	2.50	0.66

R050521-20   Welded Structural Steel

Usual weight reductions with welded design run 10% to 20% compared with bolted or riveted connections. This amounts to about the same total cost compared with bolted structures since field welding is more expensive than bolts. For normal spans of 5400 mm to 7200 mm figure 7 to 8 connections per metric ton.

Trusses — For welded trusses add 4% to weight of main members for connections. Up to 15% less steel can be expected in a welded truss compared to one that is shop bolted. Cost of erection is the same whether shop bolted or welded.

General — Typical electrodes for structural steel welding are E6010, E6011, E60T and E70T. Typical buildings vary between 0.8 kg to 3.3

kg of weld rod per metric ton of steel. Buildings utilizing continuous design require about three times as much welding as conventional welded structures. In estimating field erection by welding, it is best to use the average meter of weld per metric ton to arrive at the welding cost per metric ton. The type, size and position of the weld will have a direct bearing on the cost per meter. A typical field welder will deposit 0.8 kg to 0.9 kg of weld rod per hour manually. Using semiautomatic methods can increase production by as much as 50% to 75%.

R050523-10   High Strength Bolts

Common bolts (A307) are usually used in secondary connections (see Division 05090-150).

High strength bolts (A325 and A490) are usually specified for primary connections such as column splices, beam and girder connections to columns, column bracing, connections for supports of operating equipment or of other live loads which produce impact or reversal of stress, and in structures carrying cranes of over 5 metric ton capacity.

Allow 20 field bolts per metric ton of steel for a 6 story office building, apartment house or light industrial building. For 6 to 12 stories allow 25 bolts per metric ton, and above 12 stories, 30 bolts per metric ton. On power stations, 20 to 25 bolts per metric ton are needed.

| Metals | R0512 | Structural Steel Framing |

R051223-10   Structural Steel

The bare material prices for structural steel, shown in the unit cost sections of the book, are for 90 metric tons of shop-fabricated structural steel and include:

1. Mill base price of structural steel
2. Mill scrap/grade/size/length extras
3. Mill delivery to a metals service center (warehouse)
4. Service center storage and handling
5. Service center delivery to a fabrication shop
6. Shop storage and handling
7. Shop drafting/detailing
8. Shop fabrication
9. Shop coat of primer paint
10. Shop listing
11. Shop delivery to the job site

In unit cost sections of the book that contain items for field fabrication of steel components, the bare material cost of steel includes:

1. Mill base price of structural steel
2. Mill scrap/grade/size/length extras
3. Mill delivery to a metals service center (warehouse)
4. Service center storage and handling
5. Service center delivery to the job site

R051223-15   Structural Steel Estimating for Repair and Remodeling Projects

The correct approach to estimating structural steel is dependent upon the amount of steel required for the particular project. If the project is a sizable addition to a building, the data can be used directly from the cost data book. This is not the case however if the project requires a small amount of steel, for instance to reinforce existing roof or floor structural systems. To better understand this, please refer to the unit price line for a W410x46 beam with bolted connections. Assume your project requires the reinforcement of the structural members of the roof system of a 3 story building required for the installation a new piece of HVAC equipment. The project will need 4 pieces of W410x46 that are 9 m long. After pricing this 36 m job using the unit prices for a W410x46, an analysis will reveal that the price is wholly inadequate.

The first problem is apparent if you examine the amount of steel that can be installed per day. The unit price line indicates 274 m per day can be installed by a 5 man crew with an 82 metric ton crane. This productivity is correct for new construction but certainly is not for repair and remodeling work. Installation of new structural steel considers that each member is installed from the foundation to the roof of the structure in a planned and systematic manner with a crane having unrestricted access to all parts of the project. Additionally each connection is planned and detailed with full field access for fit-up and final bolting. The erection is planned and progresses such that interferences and conflicts with other structural members are minimized, if not completely eliminated. All of these assumptions are clearly not the case with a repair and remodeling job, and a significant decrease in the stated productivity will be observed.

A crane will certainly be needed to lift the members into the general area of the project but in most cases will not be able to place the beams into their final position. An opening in the existing roof may not be large enough to permit the beams to pass through and it may be necessary to bring them into the building through existing windows or doors. Moving the beams to the actual area where they will be installed may involve hand labor and the use of dollies. Finally, hoists and/or jacks may be needed for final positioning.

The connection of new members to existing can often be accomplished by field bolting with accurate field measurements and good planning but in many cases access to both sides of existing members is not possible and field welding becomes the only alternative. In addition to the cost of the actual welding, protection of existing finishes, systems and structure and fire protection must be considered.

New beams can never be installed tight to the existing decks or floors which they must support and the use of shims and tack welding becomes necessary. Additionally, further planning and cost is involved in assuring that existing loads are minimized during the installation and shimming process.

It is apparent that installation of structural steel as part of a repair and remodeling project involves more than simply installing the members and estimating the cost in the same manner as new construction. The best procedure for estimating the total cost is adequate planning and coordination of each process and activity that will be needed. Unit costs for the materials, labor and equipment can then be attached to each needed activity and a final, complete price can be determined.

R051223-20   Steel Estimating Quantities

One estimate on erection is that a crane can handle 35 to 60 pieces per day. Say average is 45. With usual sizes of beams, girders, and columns, this would amount to about 18.15 metric tons. The type of connection greatly affects the speed of erection. Moment connections for continuous design slow down production and increase erection costs.

Short open web bar joists can be set at the rate of 75 to 80 per day, with 50 per day being the average for setting long span joists. After main members are calculated, add the following for usual allowances: base plates 2% to 3%; column splices 4% to 5%; and miscellaneous details 4% to 5%, for a total of 10% to 13% in addition to main members.

Ratio of column to beam tonnage varies depending on type of steels used, typical spans, story heights and live loads.

It is more economical to keep the column size constant and to vary the strength of the column by using high strength steels. This also saves floor space. Buildings have recently gone as high as ten stories with 200 mm high-strength columns. For light columns under W200 x 46.1 sections, concrete filled steel columns are economical.

High strength steels may be used in columns and beams to save floor space and to meet head room requirements. High strength steels in some sizes sometimes require long lead times.

Round, square and rectangular columns, both plain and concrete filled, are readily available and save floor area, but are higher in cost per pound than rolled columns. For high unbraced columns, tube columns may be less expensive.

Below are average minimum figures for the weights of the structural steel frame for different types of buildings using A36 M steel, rolled shapes and simple joints. For economy in domes, rise to span ratio = 0.13. Open web joist framing systems will reduce weights by 10% to 40%. Composite design can reduce figures up to 25% but additional concrete floor slab thickness may be required. Continuous design can reduce the weights up to 20%. There are many building codes with different live load requirements and different structural requirements, such as hurricane and earthquake loadings which can alter the figures.

Structural Steel Weights per m² of Floor Area									
Type of Building	No. of Stories	Avg. Spans	L.L. kg/m²	kg per m²	Type of Building	No. of Stories	Avg. Spans	L.L. kg/m²	kg per m²
Steel Frame Mfg.	1	6000 x 6000 mm	147	29.50	Apartments	2-8	6000 x 6000 mm	195	39.00
		9000 x 9000 mm		49.00		9-25			68.50
		12 000 x 12 000 mm		68.50	Office	to 10	Various	390	49.00
Parking garage	4	Various	390	41.50		20			88.00
Domes (Schwedler)*	1	60 000 mm	147	49.00		30			127.00
		90 000 mm		73.00		over 50			171.00

R051223-25 Common Structural Steel Specifications

ASTM 992 is the all-purpose carbon grade steel widely used in building and bridge construction.

The other high-strength steels listed below may each have certain advantages over ASTM A992 stuctural carbon steel, depending on the application. They have proven to be economical choices where, due to lighter members, the reduction of dead load and the associated savings in shipping cost can be significant.

ASTM A588 atmospheric weathering, high-strength low-alloy steels can be used in the bare (uncoated) condition, where exposure to normal atmosphere causes a tightly adherant oxide to form on the surface protecting the steel from further oxidation. ASTM A242 corrosion-resistant, high-strength low-alloy steels have enhanced atmospheric corrosion resistance of at least two times that of carbon structural steels with copper, or four times that of carbon structural steels without copper. The reduction or elimination of maintenance resulting from the use of these steels often offsets their higher initial cost.

Steel Type	ASTM Designation	Minimum Yield Stress in MPa	Shapes Available
Carbon	A36	250	All structural shape groups, and plates & bars up thru 200 mm thick
	A529	290	Structural shape group 1, and plates & bars up thru 13 mm thick
High-Strength Low-Alloy Manganese-Vanadium	A441	275	Plates & bars over 100 mm up thru 200 mm thick
		290	Structural shape groups 4 & 5, and plates & bars over 38 up thru 100 mm thick
		315	Structural shape group 3, and plates & bars over 19 up thru 38 mm thick
		345	Structural shape groups 1 & 2, and plates & bars up thru 19 mm thick
High-Strength Low-Alloy Columbium-Vanadium	A572	290	All structural shape groups, and plates & bars up thru 150 mm thick
		345	All structural shape groups, and plates & bars up thru 100 mm thick
		415	Structural shape groups 1 & 2, and plates & bars up thru 31 mm thick
		450	Structural shape group 1, and plates & bars up thru 31 mm thick
High-Strength Low-Alloy Columbium-Vanadium	A992	345	All structural shape groups
Corrosion-Resistant High-Strength Low-Alloy	A242	290	Structural shape groups 4 & 5, and plates & bars over 38 up thru 100 mm thick
		315	Structural shape group 3, and plates & bars over 19 up thru 38 mm thick
		345	Structural shape groups 1 & 2, and plates & bars up thru 19 mm thick
Weathering High-Strength Low-Alloy	A588	290	Plates & bars over 125 up thru 200 mm thick
		315	Plates & bars over 100 up thru 125 mm thick
		345	All structural shape groups, and plates & bars up thru 100 mm thick
Quenched and Tempered Low-Alloy	A852	485	Plates & bars up thru 100 mm thick
Quenched and Tempered Alloy	A514	620	Plates & bars over 63 mm up thru 150 mm thick
		690	Plates & bars up thru 63 mm thick

R051223-30 High Strength Steels

The mill price of these steels may be higher than A992 carbon steel but their proper use can achieve overall savings thru total reduced weights. For columns with L/r over 100, A992 steel is best; under 100, high strength steels are economical. For heavy columns high strength steels are economical when cover plates are eliminated. There is no economy using high strength steels for clip angles or supports or for beams where deflection governs. Thinner members are more economical than thick.

The per metric ton erection and fabricating costs of the high strength steels will be higher than for A992 since the same number of pieces, but less weight, will be installed.

RO51223-35   Common Steel Sections

The upper portion of this table shows the name, shape, common designation and basic characteristics of commonly used steel sections. The lower portion explains how to read the designations used for the above illustrated common sections.

Shape & Designation	Name & Characteristics	Shape & Designation	Name & Characteristics
W	Wide Flange Parallel flange surfaces	MC	Miscellaneous Channel Infrequently rolled by some producers
S	American Standard Beam (I Beam) Sloped inner flange	L	Angle Equal or unequal legs, constant thickness
M	Miscellaneous Beams Cannot be classified as W, HP or S; infrequently rolled by some producers	T	Structural Tee Cut from W, M or S on center of web
C	American Standard Channel Sloped inner flange	HP	Bearing Pile Parallel flanges and equal flange and web thickness

Common drawing designations follow:

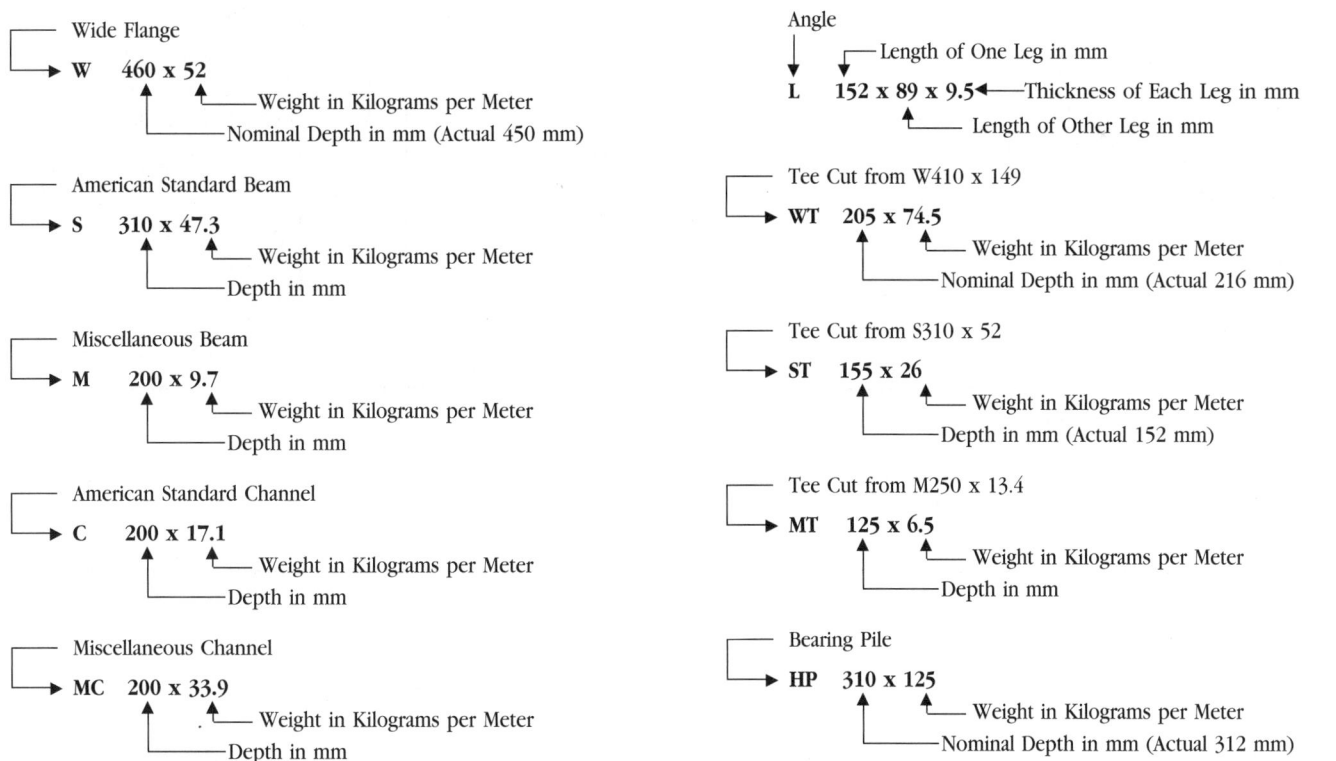

Wide Flange
W   460 x 52
Weight in Kilograms per Meter
Nominal Depth in mm (Actual 450 mm)

American Standard Beam
S   310 x 47.3
Weight in Kilograms per Meter
Depth in mm

Miscellaneous Beam
M   200 x 9.7
Weight in Kilograms per Meter
Depth in mm

American Standard Channel
C   200 x 17.1
Weight in Kilograms per Meter
Depth in mm

Miscellaneous Channel
MC   200 x 33.9
Weight in Kilograms per Meter
Depth in mm

Angle
L   152 x 89 x 9.5
Length of One Leg in mm
Thickness of Each Leg in mm
Length of Other Leg in mm

Tee Cut from W410 x 149
WT   205 x 74.5
Weight in Kilograms per Meter
Nominal Depth in mm (Actual 216 mm)

Tee Cut from S310 x 52
ST   155 x 26
Weight in Kilograms per Meter
Depth in mm (Actual 152 mm)

Tee Cut from M250 x 13.4
MT   125 x 6.5
Weight in Kilograms per Meter
Depth in mm

Bearing Pile
HP   310 x 125
Weight in Kilograms per Meter
Nominal Depth in mm (Actual 312 mm)

REFERENCE TABLES

R051223-45   Installation Time for Structural Steel Building Components

The following tables show the expected average installation times for various structural steel shapes. Table A presents installation times for column shapes, Table B for beams, Table C for light framing and bolts, and Table D for structural steel for various project types.

Table A		
Description	Labor-Hours	Unit
Columns		
Steel, Concrete Filled		
75 mm Diameter	0.933	Ea.
150 mm Diameter	1.120	Ea.
Steel Pipe		
75 mm Diameter	0.933	Ea.
200 mm Diameter	1.120	Ea.
300 mm Diameter	1.244	Ea.
Structural Tubing		
100 mm x 100 mm	0.966	Ea.
200 mm x 200 mm	1.120	Ea.
300 mm x 200 mm	1.167	Ea.
Wide Flange 2 Tier		
W 200 x 46	0.171	m
W 200 x 100	0.187	m
W 250 x 67	0.177	m
W 250 x 167	0.190	m
W 310 x 74	0.177	m
W 310 x 283	0.200	m
W 360 x 110	0.187	m
W 360 x 262	0.200	m

Table B				
Description	Labor-Hours	Unit	Labor-Hours	Unit
Beams, Wide Flange				
W 150 x 13	0.949	Ea.	0.305	m
W 250 x 32	1.037	Ea.	0.279	m
W 310 x 38	1.037	Ea.	0.210	m
W 360 x 51	1.333	Ea.	0.226	m
W 410 x 46	1.333	Ea.	0.203	m
W 460 x 74	2.162	Ea.	0.289	m
W 530 x 92	2.222	Ea.	0.253	m
W 610 x 113	2.353	Ea.	0.236	m
W 690 x 140	2.581	Ea.	0.220	m
W 760 x 161	2.857	Ea.	0.220	m
W 840 x 193	3.200	Ea.	0.233	m
W 920 x 446	3.810	Ea.	0.253	m

Table C		
Description	Labor-Hours	Unit
Light Framing		
Angles 100 mm and Larger	0.121	kg
Less than 100 mm	0.201	kg
Channels 200 mm and Larger	0.106	kg
Less than 200 mm	0.159	kg
Cross Bracing Angles	0.121	kg
Rods	0.075	kg
Hanging Lintels	0.152	kg
High Strength Bolts in Place		
19 mm Bolts	0.070	Ea.
22 mm Bolts	0.076	Ea.

Table D				
Description	Labor-Hours	Unit	Labor-Hours	Unit
Apartments, Nursing Homes, etc.				
1-2 Stories	4.211	Piece	8.562	Metric Ton
3-6 Stories	4.444	Piece	8.731	Metric Ton
7-15 Stories	4.923	Piece	9.936	Metric Ton
Over 15 Stories	5.333	Piece	10.151	Metric Ton
Offices, Hospitals, etc.				
1-2 Stories	4.211	Piece	8.562	Metric Ton
3-6 Stories	4.741	Piece	9.798	Metric Ton
7-15 Stories	4.923	Piece	9.936	Metric Ton
Over 15 Stories	5.120	Piece	10.151	Metric Ton
Industrial Buildings				
1 Story	3.478	Piece	6.837	Metric Ton

R051223-50   Subpurlins

Bulb tee subpurlins are structural members designed to support and reinforce a variety of roof deck systems such as precast cement fiber roof deck tiles, monolithic roof deck systems, and gypsum or lightweight concrete over formboard. Other uses include interstitial service ceiling systems, wall panel systems, and joist anchoring in bond beams.

See Unit Price section for pricing on a square meter basis @ 816 mm O.C. Maximum span is based on a 3-span condition with a total allowable vertical load of 195 kg/m².

R051223-80   Dimensions and Weights of Sheet Steel

Gauge No.	Approximate Thickness					Weight		
	Inches (in fractions)	Inches (in decimal parts)		Millimeters				per Square
	Wrought Iron	Wrought Iron	Steel	Steel		per S.F. in Ounces	per S.F. in Lbs.	Meter in Kg.
0000000	1/2"	.5	.4782	12.146		320	20.000	97.650
000000	15/32"	.46875	.4484	11.389		300	18.750	91.550
00000	7/16"	.4375	.4185	10.630		280	17.500	85.440
0000	13/32"	.40625	.3886	9.870		260	16.250	79.330
000	3/8"	.375	.3587	9.111		240	15.000	73.240
00	11/32"	.34375	.3288	8.352		220	13.750	67.130
0	5/16"	.3125	.2989	7.592		200	12.500	61.030
1	9/32"	.28125	.2690	6.833		180	11.250	54.930
2	17/64"	.265625	.2541	6.454		170	10.625	51.880
3	1/4"	.25	.2391	6.073		160	10.000	48.820
4	15/64"	.234375	.2242	5.695		150	9.375	45.770
5	7/32"	.21875	.2092	5.314		140	8.750	42.720
6	13/64"	.203125	.1943	4.935		130	8.125	39.670
7	3/16"	.1875	.1793	4.554		120	7.500	36.320
8	11/64"	.171875	.1644	4.176		110	6.875	33.570
9	5/32"	.15625	.1495	3.797		100	6.250	30.520
10	9/64"	.140625	.1345	3.416		90	5.625	27.460
11	1/8"	.125	.1196	3.038		80	5.000	24.410
12	7/64"	.109375	.1046	2.657		70	4.375	21.360
13	3/32"	.09375	.0897	2.278		60	3.750	18.310
14	5/64"	.078125	.0747	1.897		50	3.125	15.260
15	9/128"	.0713125	.0673	1.709		45	2.813	13.730
16	1/16"	.0625	.0598	1.519		40	2.500	12.210
17	9/160"	.05625	.0538	1.367		36	2.250	10.990
18	1/20"	.05	.0478	1.214		32	2.000	9.765
19	7/160"	.04375	.0418	1.062		28	1.750	8.544
20	3/80"	.0375	.0359	.912		24	1.500	7.324
21	11/320"	.034375	.0329	.836		22	1.375	6.713
22	1/32"	.03125	.0299	.759		20	1.250	6.103
23	9/320"	.028125	.0269	.683		18	1.125	5.490
24	1/40"	.025	.0239	.607		16	1.000	4.882
25	7/320"	.021875	.0209	.531		14	.875	4.272
26	3/160"	.01875	.0179	.455		12	.750	3.662
27	11/640"	.0171875	.0164	.417		11	.688	3.357
28	1/64"	.015625	.0149	.378		10	.625	3.052

R053100-10   Decking Descriptions

General - All Deck Products

Steel deck is made by cold forming structural grade sheet steel into a repeating pattern of parallel ribs. The strength and stiffness of the panels are the result of the ribs and the material properties of the steel. Deck lengths can be varied to suit job conditions, but because of shipping considerations, are usually less than 12 meters. Standard deck width varies with the product used but full sheets are usually 300 mm, 450 mm, 600 mm, 750 mm, or 900 mm. Deck is typically furnished in a standard width with the ends cut square. Any cutting for width, such as at openings or for angular fit, is done at the job site.

Deck is typically attached to the building frame with arc puddle welds, self-drilling screws, or powder or pneumatically driven pins. Sheet to sheet fastening is done with screws, button punching (crimping), or welds.

Composite Floor Deck After installation and adequate fastening, floor deck serves several purposes. It (a) acts as a working platform, (b) stabilizes the frame, (c) serves as a concrete form for the slab, and (d) reinforces the slab to carry the design loads applied during the life of the building. Composite decks are distinguished by the presence of shear connector devices as part of the deck. These devices are designed to mechanically lock the concrete and deck together so that the concrete and the deck work together to carry subsequent floor loads. These shear connector devices can be rolled-in embossments, lugs, holes, or wires welded to the panels. The deck profile can also be used to interlock concrete and steel.

Composite deck finishes are either galvanized (zinc coated) or phosphatized/painted. Galvanized deck has a zinc coating on both the top and bottom surfaces. The phosphatized/painted deck has a bare (phosphatized) top surface that will come into contact with the concrete. This bare top surface can be expected to develop rust before the concrete is placed. The bottom side of the deck has a primer coat of paint.

Composite floor deck is normally installed so the panel ends do not overlap on the supporting beams. Shear lugs or panel profile shape often prevent a tight metal to metal fit if the panel ends overlap; the air gap caused by overlapping will prevent proper fusion with the structural steel supports when the panel end laps are shear stud welded.

Adequate end bearing of the deck must be obtained as shown on the drawings. If bearing is actually less in the field than shown on the drawings, further investigation is required.

Roof Deck

Roof deck is not designed to act compositely with other materials. Roof deck acts alone in transferring horizontal and vertical loads into the building frame. Roof deck rib openings are usually narrower than floor deck rib openings. This provides adequate support of rigid thermal insulation board.

Roof deck is typically installed to endlap approximately 50 mm over supports. However, it can be butted (or lapped more than 50 mm) to solve field fit problems. Since designers frequently use the installed deck system as part of the horizontal bracing system (the deck as a diaphragm), any fastening substitution or change should be approved by the designer. Continuous perimeter support of the deck is necessary to limit edge deflection in the finished roof and may be required for diaphragm shear transfer.

Standard roof deck finishes are galvanized or primer painted. The standard factory applied paint for roof deck is a primer paint and is not intended to weather for extended periods of time. Field painting or touching up of abrasions and deterioration of the primer coat or other protective finishes is the responsibility of the contractor.

Cellular Deck

Cellular deck is made by attaching a bottom steel sheet to a roof deck or composite floor deck panel. Cellular deck can be used in the same manner as floor deck. Electrical, telephone, and data wires are easily run through the chase created between the deck panel and the bottom sheet.

When used as part of the electrical distribution system, the cellular deck must be installed so that the ribs line up and create a smooth cell transition at abutting ends. The joint that occurs at butting cell ends must be taped or otherwise sealed to prevent wet concrete from seeping into the cell. Cell interiors must be free of welding burrs, or other sharp intrusions, to prevent damage to wires.

When used as a roof deck, the bottom flat plate is usually left exposed to view. Care must be maintained during erection to keep good alignment and prevent damage.

Cellular deck is sometimes used with the flat plate on the top side to provide a flat working surface. Installation of the deck for this purpose requires special methods for attachment to the frame because the flat plate, now on the top, can prevent direct access to the deck material that is bearing on the structural steel. It may be advisable to treat the flat top surface to prevent slipping.

Cellular deck is always furnished galvanized or painted over galvanized.

Form Deck Form deck can be any floor or roof deck product used as a concrete form. Connections to the frame are by the same methods used to anchor floor and roof deck. Welding washers are recommended when welding deck that is less than 20 gauge thickness.

Form deck is furnished galvanized, prime painted, or uncoated. Galvanized deck must be used for those roof deck systems where form deck is used to carry a lightweight insulating concrete fill.

R061110-30 Lumber Product Material Prices

The price of forest products fluctuates widely from location to location and from season to season depending upon economic conditions. The bare material prices in the unit cost sections of the book show the National Average material prices in effect Jan. 1 of this book year. It must be noted that lumber prices in general may change significantly during the year.

Availability of certain items depends upon geographic location and must be checked prior to firm-price bidding.

Wood, Plastics & Composites	R0616	Sheathing

R061636-20 Plywood

There are two types of plywood used in construction: interior, which is moisture resistant but not waterproofed, and exterior, which is waterproofed.

The grade of the exterior surface of the plywood sheets is designated by the first letter: A, for smooth surface with patches allowed; B, for solid surface with patches and plugs allowed; C, which may be surface plugged or may have knot holes up to 25 mm wide; and D, which is used only for interior type plywood and may have knot holes up to 63 mm wide. "Structural Grade" is specifically designed for engineered applications such as box beams. All CC & DD grades have roof and floor spans marked on them.

Underlayment grade plywood runs from 6 mm to 31 mm thick. Thicknesses 16 mm and over have optional tongue and groove joints which eliminates the need for blocking the edges. Underlayment 15 mm and over may be referred to as Sturd-i-Floor.

The price of plywood can fluctuate widely due to geographic and economic conditions.

Typical uses for various plywood grades are as follows:

AA-AD Interior — cupboards, shelving, paneling, furniture

BB Plyform — concrete form plywood

CDX — wall and roof sheathing

Structural — box beams, girders, stressed skin panels

AA-AC Exterior — fences, signs, siding, soffits, etc.

Underlayment — base for resilient floor coverings

Overlaid HDO — high density for concrete forms & highway signs

Overlaid MDO — medium density for painting, siding, soffits & signs

303 Siding — exterior siding, textured, striated, embossed, etc.

Thermal & Moisture Protection | R0731 | Shingles & Shakes

R073126-20   Roof Slate

400, 450 and 500 mm are standard lengths, and slate usually comes in random widths. For standard 5 mm thickness use 38 mm copper nails. Allow for 3% breakage.

Thermal & Moisture Protection | R0751 | Built-Up Bituminous Roofing

R075113-20   Built-Up Roofing

Asphalt is available in kegs of 45 kg each; coal tar pitch in 254 kg kegs. Prepared roofing felts are available in a wide range of sizes, weights & characteristics. However, the most commonly used are #15 (40 m² per roll, 0.63 kg per m²) and #30 (20 m² per roll, 1.32 kg per m²).

Inter-ply bitumen varies from 1.17 kg per m² (asphalt) to 1.46 kg per m² (coal tar) per ply, ± 25%. Flood coat bitumen also varies from 2.93 kg per m² (asphalt) to 3.66 kg per m² (coal tar), ± 25%. Expendable equipment (mops, brooms, screeds, etc.) runs about 16% of the bitumen cost. For new, inexperienced crews this factor may be much higher.

Rigid insulation board is typically applied in two layers. The first is mechanically attached to nailable decks or spot or solid mopped to non-nailable decks; the second layer is then spot or solid mopped to the first layer. Membrane application follows the insulation, except in protected membrane roofs, where the membrane goes down first and the insulation on top, followed with ballast (stone or concrete pavers). Insulation and related labor costs are NOT included in prices for built-up roofing.

Thermal & Moisture Protection | R0752 | Modified Bituminous & Membrane Roofing

R075213-30   Modified Bitumen Roofing

The cost of modified bitumen roofing is highly dependent on the type of installation that is planned. Installation is based on the type of modifier used in the bitumen. The two most popular modifiers are atactic polypropylene (APP) and styrene butadiene styrene (SBS). The modifiers are added to heated bitumen during the manufacturing process to change its characteristics. A polyethylene, polyester or fiberglass reinforcing sheet is then sandwiched between layers of this bitumen. When completed, the result is a pre-assembled, built-up roof that has increased elasticity and weatherablility. Some manufacturers include a surfacing material such as ceramic or mineral granules, metal particles or sand.

The preferred method of adhering SBS-modified bitumen roofing to the substrate is with hot-mopped asphalt (much the same as built-up roofing). This installation method requires a tar kettle/pot to heat the asphalt, as well as the labor, tools and equipment necessary to distribute and spread the hot asphalt.

The alternative method for applying APP and SBS modified bitumen is as follows. A skilled installer uses a torch to melt a small pool of bitumen off the membrane. This pool must form across the entire roll for proper adhesion. The installer must unroll the roofing at a pace slow enough to melt the bitumen, but fast enough to prevent damage to the rest of the membrane.

Modified bitumen roofing provides the advantages of both built-up and single-ply roofing. Labor costs are reduced over those of built-up roofing because only a single ply is necessary. The elasticity of single-ply roofing is attained with the reinforcing sheet and polymer modifiers. Modifieds have some self-healing characteristics and because of their multi-layer construction, they offer the reliability and safety of built-up roofing.

Thermal & Moisture Protection | R0784 | Firestopping

R078413-30   Firestopping

Firestopping is the sealing of structural, mechanical, electrical and other penetrations through fire-rated assemblies. The basic components of firestop systems are safing insulation and firestop sealant on both sides of wall penetrations and the top side of floor penetrations.

Pipe penetrations are assumed to be through concrete, grout, or joint compound and can be sleeved or unsleeved. Costs for the penetrations and sleeves are not included. An annular space of 25 mm is assumed. Escutcheons are not included.

Metallic pipe is assumed to be copper, aluminum, cast iron or similar metallic material. Insulated metallic pipe is assumed to be covered with a thermal insulating jacket of varying thickness and materials.

Non-metallic pipe is assumed to be PVC, CPVC, FR Polypropylene or similar plastic piping material. Intumescent firestop sealant or wrap strips are included. Collars on both sides of wall penetrations and a sheet metal plate on the underside of floor penetrations are included.

Ductwork is assumed to be sheet metal, stainless steel or similar metallic material. Duct penetrations are assumed to be through concrete, grout or joint compound. Costs for penetrations and sleeves are not included. An annular space of 13 mm is assumed.

Multi-trade openings include costs for sheet metal forms, firestop mortar, wrap strips, collars and sealants as necessary.

Structural penetrations joints are assumed to be 13 mm or less. CMU walls are assumed to be within 38 mm of metal deck. Drywall walls are assumed to be tight to the underside of metal decking.

Metal panel, glass or curtain wall systems include a spandrel area of 1.5 m filled with mineral wool foil-faced insulation. Fasteners and stiffeners are included.

Openings | RO813 | Metal Doors

RO081313-20   Steel Door Selection Guide

Standard steel doors are classified into four levels, as recommended by the Steel Door Institute in the chart below. Each of the four levels offers a range of construction models and designs, to meet architectural requirements for preference and appearance, including full flush, seamless, and stile & rail. Recommended minimum gauge requirements are also included.

For complete standard steel door construction specifications and available sizes, refer to the Steel Door Institute Technical Data Series, ANSI A250.8-98 (SDI-100), and ANSI A250.4-94 Test Procedure and Acceptance Criteria for Physical Endurance of Steel Door and Hardware Reinforcements.

| | | | | For Full Flush or Seamless | | |
	Level	Model	Construction	Min. Gauge	Thickness (in)	Thickness (mm)
I	Standard Duty	1	Full Flush			
		2	Seamless	20	0.032	0.8
II	Heavy Duty	1	FullFlush			
		2	Seamless	18	0.042	1.0
III	Extra Heavy Duty	1	FullFlush			
		2	Seamless			
		3	*Stile&Rail	16	0.053	1.3
IV	Maximum Duty	1	FullFlush			
		2	Seamless	14	0.067	1.6

*Stiles & rails are 16 gauge; flush panels, when specified, are 18 gauge

Openings | RO851 | Metal Windows

RO085123-10   Steel Sash

Ironworker crew will erect 2.32 m² or 1.3 sash unit per hour, whichever is less.

Mechanic will point 9.14 m per hour.

Painter will paint 8.36 m² per coat per hour.

Glazier production depends on light size.

Allow 0.45 kg special steel sash putty per 406 mm x 508 mm light.

Openings | RO852 | Wood Windows

RO085216-10   Window Estimates

To ensure a complete window estimate, be sure to include the material and labor costs for each window, as well as the material and labor costs for an interior wood trim set.

Openings	**R0853**	**Plastic Windows**

R085313-20   Replacement Windows

Replacement windows are typically measured per United millimeter.

United millimeters are calculated by adding the width and height of the window opening in millimeters.

The labor cost for replacement windows includes removal of sash, existing sash balance or weights, parting bead where necessary and installation of new window.

Debris hauling and dump fees are not included.

Openings	**R0871**	**Door Hardware**

R087120-10   Hinges

All closer equipped doors should have ball bearing hinges. Lead lined or extremely heavy doors require special strength hinges. Usually 1-1/2 pair of hinges are used per door up to 2286 mm high openings. Table below shows typical hinge requirements.

Use Frequency	Type Hinge Required	Type of Opening	Type of Structure
High	Heavy weight	Entrances	Banks, Office buildings, Schools, Stores & Theaters
	ball bearing	Toilet Rooms	Office buildings and Schools
Average	Standard	Entrances	Dwellings
	weight	Corridors	Office buildings and Schools
	ball bearing	Toilet Rooms	Stores
Low	Plain bearing	Interior	Dwellings

Door Thickness	Weight of Doors in Pascals				
	White Pine	Oak	Hollow Core	Solid Core	Hollow Metal
35 mm	144 Pa	287 Pa	72 Pa	168 — 192 Pa	311 Pa
44 mm	168	335	120	215 — 251	311
57 mm	215	431	—	263 — 323	311

Openings	**R0881**	**Glass Glazing**

R088110-10   Glazing Productivity

Glass sizes are estimated by the "united millimeter" (height + width). Table below shows the number of lights glazed in an eight hour period by the crew size indicated, for glass up to 6 mm thick. Square or nearly square lights are more economical on a m² basis. Long slender lights will have a high m² installation cost. For insulated glass reduce production by 33%. For 13 mm plate glass reduce production by 50%. Production time for glazing with two glaziers per day averages; 6 mm plate glass 11.15 m²; 13 mm plate glass 5.11 m²; 13 mm insulated glass 8.83 m²; 19 mm insulated glass 6.97 m².

Glazing Method	United Millimeters per Light							
	1016 mm	1524 mm	2032 mm	2540 mm	3429 mm	4191 mm	5080 mm	6096 mm
Number of Men in Crew	1	1	1	1	2	3	3	4
Industrial sash, putty	60	45	24	15	18	—	—	—
With stops, putty bed	50	36	21	12	16	8	4	3
Wood stops, rubber	40	27	15	9	11	6	3	2
Metal stops, rubber	30	24	14	9	9	6	3	2
Structural glass	10	7	4	3	—	—	—	—
Corrugated glass	12	9	7	4	4	4	3	—
Storefronts	16	15	13	11	7	6	4	4
Skylights, puttyglass	60	36	21	12	16	—	—	—
Thiokol set	15	15	11	9	9	6	3	2
Vinyl set, snap on	18	18	13	12	12	7	5	4
Maximum area per light	0.26 m²	0.585 m²	1.0 m²	1.6 m²	2.9 m²	4.366 m²	6.41 m²	9.29 m²

R092000-50 Lath, Plaster and Gypsum Board

Gypsum board lath is available in 10 mm thick x 400 mm wide x 1200 mm long sheets as a base material for multi-layer plaster applications. It is also available as a base for either multi-layer or veneer plaster applications in 13 mm and 16 mm thick x 1200 mm wide x 2400, 3000 or 3600 mm long sheets. Fasteners are screws or blued ring shank nails for wood framing and screws for metal framing.

Metal lath is available in diamond mesh pattern with flat or self furring profiles. Paper backing is available for applications where excessive plaster waste needs to be avoided. A slotted mesh ribbed lath should be used in areas where the span between structural supports is greater than normal. Most metal lath comes in 675 x 2400 mm sheets. Diamond mesh weighs 0.95, 1.35 or 1.90 kilograms per square meter, slotted mesh lath weighs 1.50 or 1.85 kilograms per square meter. Metal lath can be nailed, screwed or tied in place.

Many **accessories** are available. Corner beads, flat reinforcing strips, casing beads, control and expansion joints, furring brackets and channels are some examples. Note that accessories are not included in plaster or stucco line items.

Plaster is defined as a material or combination of materials that when mixed with a suitable amount of water, forms a plastic mass or paste. When applied to a surface, the paste adheres to it and subsequently hardens, preserving in a rigid state the form or texture imposed during the period of elasticity.

Gypsum plaster is made from ground calcined gypsum. It is mixed with aggregates and water for use as a base coat plaster.

Vermiculite plaster is a fire-retardant plaster covering used on steel beams, concrete slabs and other heavy construction materials. Vermiculite is a group name for certain clay minerals, hydrous silicates or aluminum, magnesium and iron that have been expanded by heat.

Perlite plaster is a plaster using perlite as an aggregate instead of sand. Perlite is a volcanic glass that has been expanded by heat.

Gauging plaster is a mix of gypsum plaster and lime putty that when applied produces a quick drying finish coat.

Veneer plaster is a one or two component gypsum plaster used as a thin finish coat over special gypsum board.

Keenes cement is a white cementitious material manufactured from gypsum that has been burned at a high temperature and ground to a fine powder. Alum is added to accelerate the set. The resulting plaster is hard and strong and accepts and maintains a high polish, hence it is used as a finishing plaster.

Stucco is a Portland cement based plaster used primarily as an exterior finish.

Plaster is used on both interior and exterior surfaces. Generally it is applied in multiple-coat systems. A three-coat system uses the terms scratch, brown and finish to identify each coat. A two-coat system uses base and finish to describe each coat. Each type of plaster and application system has attributes that are chosen by the designer to best fit the intended use.

Gypsum Plaster Quantities per m²	2 Coat, 16 mm Thick		3 Coat, 19 mm Thick		
	Base	Finish	Scratch	Brown	Finish
	1:3 Mix	2:1 Mix	1:2 Mix	1:3 Mix	2:1 Mix
Gypsum plaster	7.05 kg		7.32 kg	3.52 kg	
Sand	0.016 m³		0.017 m³	0.012 m³	
Finish hydrated lime		1.84 kg			1.84 kg
Gauging plaster		0.92 kg			0.92 kg

Vermiculite or Perlite Plaster Quantities per m²	2 Coat, 16 mm Thick		3 Coat, 19 mm Thick		
	Base	Finish	Scratch	Brown	Finish
Gypsum plaster	6.78 kg		7.86 kg	4.34 kg	
Vermiculite or perlite	0.09 bags		0.10 bags	0.04 bags	
Finish hydrated lime		1.84 kg			1.84 kg
Gauging plaster		0.92 kg			0.92 kg

Stucco Three-Coat System Quantities per m²	On Wood	On
	Frame	Masonry
Portland cement	0.35 bags	0.25 bags
Sand	0.024 m³	0.018 m³
Hydrated lime	0.98 kg	0.65 kg

R092910-10   Levels of Gypsum Drywall Finish

In the past, contract documents often used phrases such as "industry standard" and "workmanlike finish" to specify the expected quality of gypsum board wall and ceiling installations. The vagueness of these descriptions led to unacceptable work and disputes.

In order to resolve this problem, four major trade associations concerned with the manufacture, erection, finish and decoration of gypsum board wall and ceiling systems have developed an industry-wide *Recommended Levels of Gypsum Board Finish*.

The finish of gypsum board walls and ceilings for specific final decoration is dependent on a number of factors. A primary consideration is the location of the surface and the degree of decorative treatment desired. Painted and unpainted surfaces in warehouses and other areas where appearance is normally not critical may simply require the taping of wallboard joints and 'spotting' of fastener heads. Blemish-free, smooth, monolithic surfaces often intended for painted and decorated walls and ceilings in habitated structures, ranging from single-family dwellings through monumental buildings, require additional finishing prior to the application of the final decoration.

Other factors to be considered in determining the level of finish of the gypsum board surface are (1) the type of angle of surface illumination (both natural and artificial lighting), and (2) the paint and method of application or the type and finish of wallcovering specified as the final decoration. Critical lighting conditions, gloss paints, and thin wallcoverings require a higher level of gypsum board finish than do heavily textured surfaces which are subsequently painted or surfaces which are to be decorated with heavy grade wallcoverings.

The following descriptions were developed jointly by the Association of the Wall and Ceiling Industries-International (AWCI), Ceiling & Interior Systems Construction Association (CISCA), Gypsum Association (GA), and Painting and Decorating Contractors of America (PDCA) as a guide.

Level 0: No taping, finishing, or accessories required. This level of finish may be useful in temporary construction or whenever the final decoration has not been determined.

Level 1: All joints and interior angles shall have tape set in joint compound. Surface shall be free of excess joint compound. Tool marks and ridges are acceptable. Frequently specified in plenum areas above ceilings, in attics, in areas where the assembly would generally be concealed or in building service corridors, and other areas not normally open to public view.

Level 2: All joints and interior angles shall have tape embedded in joint compound and wiped with a joint knife leaving a thin coating of joint compound over all joints and interior angles. Fastener heads and accessories shall be covered with a coat of joint compound. Surface shall be free of excess joint compound. Tool marks and ridges are acceptable. Joint compound applied over the body of the tape at the time of tape embedment shall be considered a separate coat of joint compound and shall satisfy the conditions of this level. Specified where water-resistant gypsum backing board is used as a substrate for tile; may be specified in garages, warehouse storage, or other similar areas where surface appearance is not of primary concern.

Level 3: All joints and interior angles shall have tape embedded in joint compound and one additional coat of joint compound applied over all joints and interior angles. Fastener heads and accessories shall be covered with two separate coats of joint compound. All joint compound shall be smooth and free of tool marks and ridges. Typically specified in appearance areas which are to receive heavy- or medium-texture (spray or hand applied) finishes before final painting, or where heavy-grade wallcoverings are to be applied as the final decoration. This level of finish is not recommended where smooth painted surfaces or light to medium wallcoverings are specified.

Level 4: All joints and interior angles shall have tape embedded in joint compound and two separate coats of joint compound applied over all flat joints and one separate coat of joint compound applied over interior angles. Fastener heads and accessories shall be covered with three separate coats of joint compound. All joint compound shall be smooth and free of tool marks and ridges. This level should be specified where flat paints, light textures, or wallcoverings are to be applied. In critical lighting areas, flat paints applied over light textures tend to reduce joint photographing. Gloss, semi-gloss, and enamel paints are not recommended over this level of finish. The weight, texture, and sheen level of wallcoverings applied over this level of finish should be carefully evaluated. Joints and fasteners must be adequately concealed if the wallcovering material is lightweight, contains limited pattern, has a gloss finish, or any combination of these finishes is present. Unbacked vinyl wallcoverings are not recommended over this level of finish.

Level 5: All joints and interior angles shall have tape embedded in joint compound and two separate coats of joint compound applied over all flat joints and one separate coat of joint compound applied over interior angles. Fastener heads and accessories shall be covered with three separate coats of joint compound. A thin skim coat of joint compound or a material manufactured especially for this purpose, shall be applied to the entire surface. The surface shall be smooth and free of tool marks and ridges. This level of finish is highly recommended where gloss, semi-gloss, enamel, or nontextured flat paints are specified or where severe lighting conditions occur. This highest quality finish is the most effective method to provide a uniform surface and minimize the possibility of joint photographing and of fasteners showing through the final decoration.

| **R0966** | **Terrazzo Flooring**

R096613-10   Terrazzo Floor

The table below lists quantities required per m² of 16 mm terrazzo topping, either bonded or not bonded.

Description	Bonded to Concrete 29 mm Bed, 1:4 Mix	Not Bonded 54 mm Bed and 6 mm Sand
Portland cement, 43 kg bag	0.65 bag	0.85 bag
Sand	0.03 m³	0.06 m³
Divider strips, 1.2 m squares	1.80 m³	1.80 m³
Terrazzo fill, 23 kg bag	1.3 bags	1.3 bags
15 # tarred felt		1 m²
Mesh 50 x 50 #14 galvanized		1 m²
Crew J-3	0.083 days	0.094 days

600 mm x 600 mmm panels require 3333 mm divider strip per m²
900 mm x 900 mmm panels require 2222 mm divider strip per m²
1200 mm x 1200 mmm panels require 1667 mm divider strip per m²
1500 mm x 1500 mmm panels require 1333 mm divider strip per m²
1800 mm x 1800 mmm panels require 1111 mm divider strip per m²

| **R0972** | **Wall Coverings**

R097223-10   Wall Covering

The table below lists the quantities required per m² of wall covering.

Description	Medium Price Paper	Expensive Paper
Paper	0.172 dbl. rolls	0.172 dbl. rolls
Wall sizing	0.102 liter	0.102 liter
Vinyl wall paste	0.245 liter	0.245 liter
Apply sizing	0.032 hour	0.032 hour
Apply paper	0.129 hour	0.161 hour

Most wallpapers now come in double rolls only.
To remove old paper, allow 0.14 hours per m²

R099100-20   Painting

Item	Coat	3.785 liters Cover			In 8 Hours Painter Covers			Labor Hours per m²		
		Brush	Roller	Spray	Brush	Roller	Spray	Brush	Roller	Spray
Paint wood siding	prime	23 m²	21 m²	27 m²	107 m²	121 m²	211 m²	0.075	0.066	0.038
	others	25	23	27	121	151	242	0.066	0.053	0.033
Paint exterior trim	prime	37	—	—	60	—	—	0.132	—	—
	1st	44	—	—	74	—	—	0.108	—	—
	2nd	48	—	—	91	—	—	0.088	—	—
Paint shingle siding	prime	25	24	28	60	91	181	0.133	0.088	0.044
	others	33	32	35	74	107	211	0.108	0.075	0.038
Stain shingle siding	1st	17	16	19	70	105	209	0.114	0.076	0.038
	2nd	25	23	27	84	123	242	0.095	0.065	0.033
Paint brick masonry	prime	17	13	15	70	74	167	0.114	0.108	0.048
	1st	25	21	27	76	91	211	0.105	0.088	0.038
	2nd	32	28	33	76	107	272	0.105	0.075	0.029
Paint interior plaster or drywall	prime	37	35	46	107	186	302	0.075	0.043	0.026
	others	42	39	46	121	214	372	0.066	0.037	0.022
Paint interior doors and windows	prime	37	—	—	60	—	—	0.133	—	—
	1st	39	—	—	74	—	—	0.108	—	—
	2nd	42	—	—	91	—	—	0.088	—	—

Special Construction | **R1311** | **Swimming Pools**

R131113-20   Swimming Pools

Pool prices given per square meter of surface area include pool structure, filter and chlorination equipment, pumps, related piping, ladders/steps, maintenance kit, skimmer and vacuum system. Decks and electrical service to equipment are not included.

Residential in-ground pool construction can be divided into two categories: vinyl lined and gunite. Vinyl lined pool walls are constructed of different materials including wood, concrete, plastic or metal. The bottom is often graded with sand over which the vinyl liner is installed. Vermiculite or soil cement bottoms may be substituted for an added cost.

Gunite pool construction is used both in residential and municipal installations. These structures are steel reinforced for strength and finished with a white cement limestone plaster.

Municipal pools will have a higher cost because plumbing codes require more expensive materials, chlorination equipment and higher filtration rates.

Municipal pools greater than 167 m² require gutter systems to control waves. This gutter may be formed into the concrete wall. Often a vinyl/stainless steel gutter or gutter/wall system is specified, which will raise the pool cost.

Competition pools usually require tile bottoms and sides with contrasting lane striping, which will also raise the pool cost.

R133113-10 Air Supported Structures

Air supported structures are made from fabrics that can be classified into two groups: temporary and permanent. Temporary fabrics include nylon, woven polyethylene, vinyl film, and vinyl coated dacron. These have lifespans that range from five to fifteen plus years. The cost per square foot includes a fabric shell, tension cables, primary and back-up inflation systems and doors. The lower cost structures are used for construction shelters, bulk storage and pond covers. The more expensive are used for recreational structures and warehouses.

Permanent fabrics are teflon coated fiberglass. The life of this structure is twenty plus years. The high cost limits its application to architectural designed structures which call for a clear span covered area, such as stadiums and convention centers. Both temporary and permanent structures are available in translucent fabrics which eliminates the need for daytime lighting.

Areas to be covered vary from 1000 m^2 to any area up to 300 m by any length. Height restrictions range from a maximum of 1/2 of width to a minimum of 1/6 of the width. Erection of even the largest of the temporary structures requires no more than a week.

Centrifugal fans provide the inflation necessary to support the structure during application of live loads. Airlocks are usually used at large entrances to prevent loss of static pressure. Some manufacturers employ propeller fans which generate sufficient airflow (14.2 m^3/s) to eliminate the need for airlocks. These fans may also be automatically controlled to resist high wind conditions, regulate humidity (air changes), and provide cooling and heat.

Insulation can be provided with the addition of a second or even third interior liner, creating a dead air space with an "R" value of 0.70 to 1.58 m^2 x 'K/W. Some structures allow for the liner to be collapsed into the outer shell to enable the internal heat to melt accumulated snow. For cooling or air conditioning, the exterior face of the liner can be aluminized to reflect the sun's heat.

R133113-90 Seismic Bracing

Sometimes referred to as anti-sway bracing, this support system is required in earthquake areas. The individual components must be assembled to make a required system.

Example				Additionally, height factors must be taken ainto account. Add the following percentages to labor for elevated installations:	
	C-Clamp 10 mm rod	2 ea.		4.5 m to 6.0 m high	10%
	Rod, continuous thread 10 mm	3 m		6.1 m to 7.5 m high	20%
	Field, weld 25 mm	2 ea.		7.6 m to 9.0 m high	30%
				9.1 m to 10.5 m high	40%
				10.6 m to 12.0 m high	50%
				12.1 m to 15.0 m high	60%

R133419-10 Pre-engineered Steel Buildings

These buildings are manufactured by many companies and normally erected by franchised dealers throughout the U.S. The four basic types are: Rigid Frames, Truss type, Post and Beam and the Sloped Beam type. Most popular roof slope is low pitch of 1 mm in 12 mm. The minimum economical area of these buildings is about 280 m² of floor area. Bay sizes are usually 6 m to 7.2 m but can go as high as 9 m with heavier girts and purlins. Eave heights are usually 3.6 m to 7.2 m with 5.4 m to 6 m most typical.

Material prices shown in the Unit Price section are bare costs for the building shell only and do not include floors, foundations, anchor bolts, interior finishes or utilities. Costs assume at least three bays of 7.2 m each, a 1 mm in 12 mm roof slope, and they are based on 1440 Pa roof load and 960

Pa wind load (wind load is a function of wind speed, building height, and terrain characteristics; this should be determined by a Registered Structural Engineer) and no unusual requirements. Costs include the structural frame (columns, beams, girts, purlins, bracing, etc.), 26 ga. non-insulated colored corrugated or ribbed roofing and siding panels, fasteners, closures, trim and flashing but no allowance for doors, windows, skylights, gutters or downspouts. Very large projects would generally cost less for materials than the prices shown. For roof panel substitutions and wall panel substitutions, see appropriate Unit Price sections.

Conditions at the site, weather, shape and size of the building, and labor availability will affect the erection cost of the building.

R133423-30 Dome Structures

Steel — The four types are Lamella, Schwedler, Arch and Geodesic. For maximum economy, rise should be about 15 to 20% of diameter. Most common diameters are in the 61 m to 91 m range. Lamella domes weigh about 2.41 kg/m² of floor area less than Schwedler domes. Schwedler dome weight in kg per m² approaches 0.046 times the diameter. Domes below 38 m diameter weigh .07 times diameter and the cost per metric ton of steel is higher. See R051223-20 for estimating weight.

Wood — Small domes are of sawn lumber, larger ones are laminated. In larger sizes, triaxial and triangular cost about the same; radial domes cost more. Radial domes are economical in the 18 m to 21 m diameter range. Most economical range of all types is 24 m to 61 m diameters. Diameters can

run over 400′. All costs are quoted above the foundation. Prices include 50 mm decking and a tension tie ring in place.

Plywood — Stock prefab geodesic domes are available with diameters from 7 m to 18m.

Fiberglass — Aluminum framed translucent sandwich panels with spans from 1.5 m to 14 m are commercially available.

Aluminum — Stressed skin aluminum panels form geodesic domes with spans ranging from 24 m to 70 m. An aluminum space truss, triangulated or nontriangulated, with aluminum or clear acrylic closure panels can be used for clear spans of 12 m to 125 m.

R142000-10   Freight Elevators

Capacities run from 680 kg to over 45 360 kg with 1360 kg to 4535 kg most common. Travel speeds are generally lower and control less intricate than on passenger elevators. Costs in the Unit Price sections are for hydraulic and geared elevators.

R142000-20   Elevator Selective Costs   See R142000-40 for cost development.

A. Base Unit	Passenger		Freight		Hospital	
	Hydraulic	**Electric**	**Hydraulic**	**Electric**	**Hydraulic**	**Electric**
Capacity	680 kg	907 kg	907 kg	1814 kg	1814 kg	1814 kg
Speed	0.508 m/s	1.016 m/s	0.254 m/s	1.016 m/s	0.508 m/s	1.016 m/s
#Stops/Travel m	2/3.66	4/12.19	2/6.10	4/12.19	2/6.10	4/12.19
Push Button Oper.	Yes	Yes	Yes	Yes	Yes	Yes
Telephone Box & Wire	"	"	"	"	"	"
Emergency Lighting	"	"	No	No	"	"
Cab	Plastic Lam. Walls	Plastic Lam. Walls	Painted Steel	Painted Steel	Plastic Lam. Walls	Plastic Lam. Walls
Cove Lighting	Yes	Yes	No	No	Yes	Yes
Floor	V.C.T.	V.C.T.	Wood w/Safety Treads	Wood w/Safety Treads	V.C.T.	V.C.T.
Doors, & Speedside Slide	Yes	Yes	Yes	Yes	Yes	Yes
Gates, Manual	No	No	No	No	No	No
Signals, Lighted Buttons	Car and Hall	Car and Hall	Car and Hall	Car and Hall	Car and Hall	Car and Hall
O.H. Geared Machine	N.A.	Yes	Yes	Yes	N.A.	Yes
Variable Voltage Contr.	"	"	N.A.	"	"	"
Emergency Alarm	Yes	"	Yes	"	Yes	"
Class "A" Loading	N.A.	N.A.	"	"	N.A.	N.A.

Conveying Equipment	R1420	Elevators

R142000-30   Passenger Elevators

Electric elevators are used generally but hydraulic elevators can be used for lifts up to 21.34 m and where large capacities are required. Hydraulic speeds are limited to 1.016 m/s but cars are self leveling at the stops. On low rises, hydraulic installation runs about 15% less than standard electric types but on higher rises this installation cost advantage is reduced. Maintenance of hydraulic elevators is about the same as electric type but underground portion is not included in the maintenance contract.

In electric elevators there are several control systems available, the choice of which will be based upon elevator use, size, speed and cost criteria. The two types of drives are geared for low speeds and gearless for 2.286 m/s and over.

The tables on the preceding pages illustrate typical installed costs of the various types of elevators available.

R142000-40   Elevator Cost Development

To price a new car or truck from the factory, you must start with the manufacturer's basic model, then add or exchange optional equipment and features. The same is true for pricing elevators.

Requirement: One passenger elevator, five story hydraulic, 1134 kg capacity, 3.6 m floor to floor, speed 0.762 m/s, emergency power switching and maintenance contract.

Example:

Description		Adjustment	
A. Base Elevator: Hydraulic Passenger, 680 kg Capacity, 0.508 m/s, 2 Stops, Standard Finish		1	Ea.
B. Capacity Adjustment (1134 kg)		1	Ea.
C. Excess Travel Adjustment: 14.4 m Total Travel (4 x 3.6 m) minus 3.6 m Base Unit Travel =		10.8	m
D. Stops Adjustment: 5 Total Stops minus 2 Stops (Base Unit) =		3	Stops
E. Speed Adjustment (0.762 m/s)		1	Ea.
F. Options:			
1. Intercom Service		1	Ea.
2. Emergency Power Switching, Automatic		1	Ea.
3. Stainless Steel Entrance Doors		5	Ea.
4. Maintenance Contract (12 Months)		1	Ea.
5. Position Indicator for main floor level (none indicated in Base Unit)		1	Ea.

Conveying Equipment	R1431	Escalators

R143110-10   Escalators

Moving stairs can be used for buildings where 600 or more people are to be carried to the second floor or beyond. Freight cannot be carried on escalators and at least one elevator must be available for this function.

Carrying capacity is 5,000 to 8,000 people per hour. Power requirement is 2 to 3 kW per hour and incline angle is 30°.

Conveying Equipment	R1432	Moving Walks

R143210-20   Moving Ramps and Walks

These are a specialized form of conveyor 0.91 m to 1.83 m wide with capacities of 3,600 to 18,000 persons per hour. Maximum speed is 0.711 m/s and normal incline is 0° to 15°.

Local codes will determine the maximum angle. Outdoor units would require additional weather protection.

R211226-10 Standpipe Systems

The basis for standpipe system design is National Fire Protection Association NFPA 14, however, the authority having jurisdiction should be consulted for special conditions, local requirements and approval.

Standpipe systems, properly designed and maintained, are an effective and valuable time saving aid for extinguishing fires, especially in the upper stories of tall buildings, the interior of large commercial or industrial malls, or other areas where construction features or access make the laying of temporary hose lines time consuming and/or hazardous. Standpipes are frequently installed with automatic sprinkler systems for maximum protection.

There are three general classes of service for standpipe systems:
Class I for use by fire departments and personnel with special training for heavy streams (65 mm hose connections).
Class II for use by building occupants until the arrival of the fire department (40 mm hose connector with hose).

Class III for use by either fire departments and trained personnel or by the building occupants (both 65 mm and 40 mm hose connections or one 65 mm hose valve with an easily removable 65 mm by 40 mm adapter).

Standpipe systems are also classified by the way water is supplied to the system. The four basic types are:
Type 1: Wet standpipe system having supply valve open and water pressure maintained at all times.
Type 2: Standpipe system so arranged through the use of approved devices as to admit water to the system automatically by opening a hose valve.
Type 3: Standpipe system arranged to admit water to the system through manual operation of approved remote control devices located at each hose station.
Type 4: Dry standpipe having no permanent water supply.

R211226-20 NFPA 14 Basic Standpipe Design

Class	Design-Use	Pipe Size Minimums	Water Supply Minimums
Class I	65 mm hose connection on each floor All areas within 45 m of an exit in every exit stairway Fire Department Trained Personnel	Height to 30 m, 100 mm dia. Heights above 30 m, 150 mm dia. (82 m max. except with pressure regulators 120 m max.)	For each standpipe riser 32 L/s flow For common supply pipe allow 32 L/s for first standpipe plus 16 L/s for each additional standpipe (160 L/s max. total) 30 min. duration 450 kPa at 32 L/s
Class II	40 mm hose connection with hose on each floor All areas within 39 m of hose connection measured along path of hose travel Occupant personnel	Height to 15 m, 50 mm dia. Height above 15 m, 65 mm dia.	For each standpipe riser 6 L/s flow For multiple riser common supply pipe 6 L/s 300 min. duration, 450 kPa at 6 L/s
Class III	Both of above. Class I valved connections will meet Class III with additional 65 mm by 40 mm adapter and 40 mm hose.	Same as Class I	Same as Class I

*Note: Where 2 or more standpipes are installed in the same building or section of building they shall be interconnected at the bottom.

Combined Systems

Combined systems are systems where the risers supply both automatic sprinklers and 65 mm hose connection outlets for fire department use. In such a system the sprinkler spacing pattern shall be in accordance with NFPA 13 while the risers and supply piping will be sized in accordance with NFPA 14. When the building is completely sprinklered the risers may be sized by hydraulic calculation. The minimum size riser for buildings not completely sprinklered is 150 mm.
The minimum water supply of a completely sprinklered, light hazard, high-rise occupancy building will be 32 L/s while the supply required for other types of completely sprinklered high-rise buildings is 64 L/s.

General System Requirements

1. Approved valves will be provided at the riser for controlling branch lines to hose outlets.
2. A hose valve will be provided at each outlet for attachment of hose.
3. Where pressure at any standpipe outlet exceeds 690 kPa a pressure reducer must be installed to limit the pressure to 690 kPa. Note that

the pressure head due to gravity in 30 m of riser is 300 kPa. This must be overcome by city pressure, fire pumps, or gravity tanks to provide adequate pressure at the top of the riser.
4. Each hose valve on a wet system having linen hose shall have an automatic drip connection to prevent valve leakage from entering the hose.
5. Each riser will have a valve to isolate it from the rest of the system.
6. One or more fire department connections as an auxiliary supply shall be provided for each Class I or Class III standpipe system. In buildings having two or more zones, a connection will be provided for each zone.
7. There will be no shutoff valve in the fire department connection, but a check valve will be located in the line before it joins the system.
8. All hose connections street side will be identified on a cast plate or fitting as to purpose.

R211313-10 Sprinkler Systems (Automatic)

Sprinkler systems may be classified by type as follows:

1. **Wet Pipe System.** A system employing automatic sprinklers attached to a piping system containing water and connected to a water supply so that water discharges immediately from sprinklers opened by a fire.

2. **Dry Pipe System.** A system employing automatic sprinklers attached to a piping system containing air under pressure, the release of which as from the opening of sprinklers permits the water pressure to open a valve known as a "dry pipe valve". The water then flows into the piping system and out the opened sprinklers.

3. **Pre-Action System.** A system employing automatic sprinklers attached to a piping system containing air that may or may not be under pressure, with a supplemental heat responsive system of generally more sensitive characteristics than the automatic sprinklers themselves, installed in the same areas as the sprinklers; actuation of the heat responsive system, as from a fire, opens a valve which permits water to flow into the sprinkler piping system and to be discharged from any sprinklers which may be open.

4. **Deluge System.** A system employing open sprinklers attached to a piping system connected to a water supply through a valve which is opened by the operation of a heat responsive system installed in the same areas as the sprinklers. When this valve opens, water flows into the piping system and discharges from all sprinklers attached thereto.

5. **Combined Dry Pipe and Pre-Action Sprinkler System.** A system employing automatic sprinklers attached to a piping system containing air under pressure with a supplemental heat responsive system of generally more sensitive characteristics than the automatic sprinklers themselves, installed in the same areas as the sprinklers; operation of the heat responsive system, as from a fire, actuates tripping devices which open dry pipe valves simultaneously and without loss of air pressure in the system. Operation of the heat responsive system also opens

approved air exhaust valves at the end of the feed main which facilitates the filling of the system with water which usually precedes the opening of sprinklers. The heat responsive system also serves as an automatic fire alarm system.

6. **Limited Water Supply System.** A system employing automatic sprinklers and conforming to these standards but supplied by a pressure tank of limited capacity.

7. **Chemical Systems.** Systems using halon, carbon dioxide, dry chemical or high expansion foam as selected for special requirements. Agent may extinguish flames by chemically inhibiting flame propagation, suffocate flames by excluding oxygen, interrupting chemical action of oxygen uniting with fuel or sealing and cooling the combustion center.

8. **Firecycle System.** Firecycle is a fixed fire protection sprinkler system utilizing water as its extinguishing agent. It is a time delayed, recycling, preaction type which automatically shuts the water off when heat is reduced below the detector operating temperature and turns the water back on when that temperature is exceeded. The system senses a fire condition through a closed circuit electrical detector system which controls water flow to the fire automatically. Batteries supply up to 90 hour emergency power supply for system operation. The piping system is dry (until water is required) and is monitored with pressurized air. Should any leak in the system piping occur, an alarm will sound, but water will not enter the system until heat is sensed by a firecycle detector.

Area coverage sprinkler systems may be laid out and fed from the supply in any one of several patterns as shown below. It is desirable, if possible, to utilize a central feed and achieve a shorter flow path from the riser to the furthest sprinkler. This permits use of the smallest sizes of pipe possible with resulting savings.

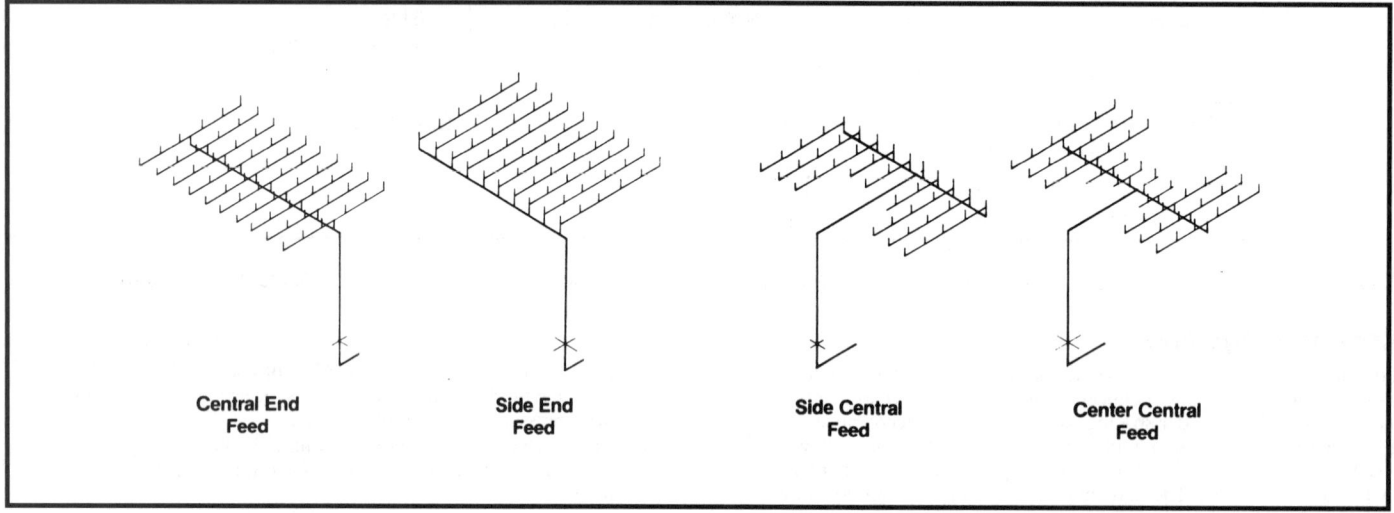

Central End Feed   Side End Feed   Side Central Feed   Center Central Feed

R211313-20   System Classification

System Classification
Rules for installation of sprinkler systems vary depending on the classification of occupancy falling into one of three categories as follows:

Light Hazard Occupancy
The protection area allotted per sprinkler should not exceed 20.9 m² with the maximum distance between lines and sprinklers on lines being 4.5 m. The sprinklers do not need to be staggered. Branch lines should not exceed eight sprinklers on either side of a cross main. Each large area requiring more than 100 sprinklers and without a sub-dividing partition should be supplied by feed mains or risers sized for ordinary hazard occupancy.
Maximum system area = 4830 m²

Included in this group are:

Churches	Nursing Homes
Clubs	Offices
Educational	Residential
Hospitals	Restaurants
Institutional	Theaters and Auditoriums
Libraries	(except stages and prosceniums)
(except large stack rooms)	Unused Attics
Museums	

Ordinary Hazard Occupancy
The protection area allotted per sprinkler shall not exceed 121 m² of noncombustible ceiling and 12.1 m² of combustible ceiling. The maximum allowable distance between sprinkler lines and sprinklers on line is 4.5 m. Sprinklers shall be staggered if the distance between heads exceeds 3.6 m. Branch lines should not exceed eight sprinklers on either side of a cross main.
Maximum system area = 4830 m².

Included in this group are:

Group 1	Group 2
Automotive Parking and Showrooms	Cereal Mills
Bakeries	Chemical Plants—Ordinary
Beverage manufacturing	Confectionery Products
Canneries	Distilleries
Dairy Products Manufacturing/Processing	Dry Cleaners
Electronic Plans	Feed Mills
Glass and Glass Products Manufacturing	Horse Stables
Laundries	Leather Goods Manufacturing
Restaurant Service Areas	Libraries—Large Stack Room Areas
	Machine Shops
	Metal Working
	Mercantile
	Paper and Pulp Mills
	Paper Process Plants
	Piers and Wharves
	Post Offices
	Printing and Publishing
	Repair Garages
	Stages
	Textile Manufacturing
	Tire Manufacturing
	Tobacco Products Manufacturing
	Wood Machining
	Wood Product Assembly

Extra Hazard Occupancy
The protection area allotted per sprinkler shall not exceed 9.3 m² of noncombustible ceiling and 9.3 m² of combustible ceiling. The maximum allowable distance between lines and between sprinklers on lines is 3.6 m. Sprinklers on alternate lines shall be staggered if the distance between sprinklers on lines exceeds 2.4 m. Branch lines should not exceed six sprinklers on either side of a cross main.

Maximum system area:
   Design by pipe schedule = 2323 m²
   Design by hydraulic calculation = 3716 m²

Included in this group are:

Group 1	Group 2
Aircraft hangars	Asphalt Saturating
Combustible Hydraulic Fluid Use Area	Flammable Liquids Spraying
Die Casting	Flow Coating
Metal Extruding	Manufactured/Modular Home Building Assemblies (where finished enclosure is present and has combustible interiors)
Plywood/Particle Board Manufacturing	
Printing (inkls with flash points < 38 degrees C	
Rubber Reclaiming, Compounding, Drying, Milling, Vulcanizing	Open Oil Quenching
Saw Mills	Plastics Processing
Textile Picking, Opening, Blending, Garnetting, Carding, Combing of Cotton, Synthetics, Wood Shoddy, or Burlap	Solvent Cleaning
Upholstering with Plastic Foams	Varnish and Paint Dipping

R211313-30 Sprinkler Quantities for Various Sizes and Types of Pipe

Sprinkler Quantities: The table below lists the usual maximum number of sprinkler heads for each size of copper and steel pipe for both wet and dry systems. These quantities may be adjusted to meet individual structural needs or local code requirements. Maximum area on any one floor for one system is: light hazard and ordinary hazard 4830 m², extra hazardous 2323 m².

| Pipe Size | Light Hazard Occupancy | | Ordinary Hazard Occupancy | | Extra Hazard Occupancy | |
Diameter	Steel Pipe	Copper Pipe	Steel Pipe	Copper Pipe	Steel Pipe	Copper Pipe
25 mm	2 sprinklers	2 sprinklers	2 sprinklers	2 sprinklers	1 sprinklers	1 sprinklers
32 mm	3	3	3	3	2	2
40 mm	5	5	5	5	5	5
50 mm	10	12	10	12	8	8
65 mm	30	40	20	25	15	20
80 mm	60	65	40	45	27	30
90 mm	100	115	65	75	40	45
100 mm			100	115	55	65
125 mm			160	180	90	100
150 mm			275	300	150	170

Dry Pipe Systems: A dry pipe system should be installed where a wet pipe system is impractical, as in rooms or buildings which cannot be properly heated.

The use of an approved dry pipe system is more desirable than shutting off the water supply during cold weather.

Not more than 2840 liters of system capacity should be controlled by one dry pipe valve. Where two or more dry pipe valves are used, systems should preferably be divided horizontally.

R211313-40 Adjustment for Sprinkler/Standpipe Installations

Quality/Complexity Multiplier (For all installations)

Economy installation, add . . . . 0 to 5%
Good quality, medium complexity, add . . . . 5 to 15%
Above average quality and complexity, add . . . . 15 to 25%

R220105-10 Demolition (Selective vs. Removal for Replacement)

Demolition can be divided into two basic categories.

One type of demolition involves the removal of material with no concern for its replacement. The labor-hours to estimate this work are found under "Selective Demolition" in the Fire Protection, Plumbing and HVAC Divisions. It is selective in that individual items or all the material installed as a system or trade grouping such as plumbing or heating systems are removed. This may be accomplished by the easiest way possible, such as sawing, torch cutting, or sledge hammer as well as simple unbolting.

The second type of demolition is the removal of some item for repair or replacement. This removal may involve careful draining, opening of unions,

disconnecting and tagging of electrical connections, capping of pipes/ducts to prevent entry of debris or leakage of the material contained as well as transport of the item away from its in-place location to a truck/dumpster. An approximation of the time required to accomplish this type of demolition is to use half of the time indicated as necessary to install a new unit. For example; installation of a new pump might be listed as requiring 6 labor-hours so if we had to estimate the removal of the old pump we would allow an additional 3 hours for a total of 9 hours. That is, the complete replacement of a defective pump with a new pump would be estimated to take 9 labor-hours.

R220523-80 Valve Materials

VALVE MATERIALS

Bronze:
Bronze is one of the oldest materials used to make valves. It is most commonly used in hot and cold water systems and other non-corrosive services. It is often used as a seating surface in larger iron body valves to ensure tight closure.

Carbon Steel:
Carbon steel is a high strength material. Therefore, valves made from this metal are used in higher pressure services, such as steam lines up to 4137 kPa at 454°C. Many steel valves are available with butt-weld ends for economy and are generally used in high pressure steam service as well as other higher pressure non-corrosive services.

Forged Steel:
Valves from tough carbon steel are used in service up to 13 790 kPa and temperatures up to 538°C in Gate, Globe and Check valves.

Iron:
Valves are normally used in medium to large pipe lines to control non-corrosive fluid and gases, where pressures do not exceed 1724 kPa at 232°C or 3448 kPa cold water, oil or gas.

Stainless Steel:
Developed steel alloys can be used in over 90% corrosive services.

Plastic PVC:
This is used in a great variety of valves generally in high corrosive service with lower temperatures and pressures.

VALVE SERVICE PRESSURES

Pressure ratings on valves provide an indication of the safe operating pressure for a valve at some elevated temperature. This temperature is dependent upon the materials used and the fabrication of the valve. When specific data is not available, a good "rule-of-thumb" to follow is the temperature of saturated steam on the primary rating indicated on the valve body. Example: The valve has the number 1034S printed on the side indicating 1034 kPa and hence, a maximum operating temperature of 186°C (temperature of saturated steam and 1034 kPa).

DEFINITIONS

1. "WOG" – Water, oil, gas (cold working pressures).
2. "SWP" – Steam working pressure.
3. 100% area (full port) – means the area through the valve is equal to or greater than the area of standard pipe.
4. "Standard Opening" – means that the area through the valve is less than the area of standard pipe and therefore these valves should be used only where restriction of flow is unimportant.
5. "Round Port" – means the valve has a full round opening through the plug and body, of the same size and area as standard pipe.
6. "Rectangular Port" – valves have rectangular shaped ports through the plug body. The area of the port is either equal to 100% of the area of standard pipe, or restricted (standard opening). In either case it is clearly marked.
7. "ANSI" – American National Standards Institute.

R220523-90 Valve Selection Considerations

INTRODUCTION: In any piping application, valve performance is critical. Valves should be selected to give the best performance at the lowest cost.

The following is a list of performance characteristics generally expected of valves.

1. Stopping flow or starting it.
2. Throttling flow (Modulation).
3. Flow direction changing.
4. Checking backflow (Permitting flow in only one direction).
5. Relieving or regulating pressure.

In order to properly select the right valve, some facts must be determined.

A. What liquid or gas will flow through the valve?
B. Does the fluid contain suspended particles?
C. Does the fluid remain in liquid form at all times?
D. Which metals does fluid corrode?
E. What are the pressure and temperature limits? (As temperature and pressure rise, so will the price of the valve.)
F. Is there constant line pressure?
G. Is the valve merely an on-off valve?
H. Will checking of backflow be required?
I. Will the valve operate frequently or infrequently?

Valves are classified by design type into such classifications as Gate, Globe, Angle, Check, Ball, Butterfly and Plug. They are also classified by end connection, stem, pressure restrictions and material such as bronze, cast iron, etc. Each valve has a specific use. A quality valve used correctly will provide a lifetime of trouble-free service, but a high quality valve installed in the wrong service may require frequent attention.

PLUMBING

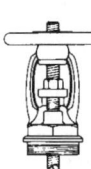

STEM TYPES
(OS & Y)—Rising Stem-Outside Screw and Yoke

Offers a visual indication of whether the valve is open or closed. Recommended where high temperatures, corrosives, and solids in the line might cause damage to inside-valve stem threads. The stem threads are engaged by the yoke bushing so the stem rises through the hand wheel as it is turned.

(R.S.)—Rising Stem-Inside Screw

Adequate clearance for operation must be provided because both the hand wheel and the stem rise.
The valve wedge position is indicated by the position of the stem and hand wheel.

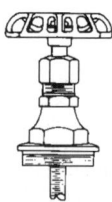

(N.R.S.)—Non-Rising Stem-Inside Screw

A minimum clearance is required for operating this type of valve. Excessive wear or damage to stem threads inside the valve may be caused by heat, corrosion, and solids. Because the hand wheel and stem do not rise, wedge position cannot be visually determined.

VALVE TYPES
Gate Valves

Provide full flow, minute pressure drop, minimum turbulence and minimum fluid trapped in the line.
They are normally used where operation is infrequent.

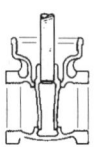

Globe Valves

Globe valves are designed for throttling and/or frequent operation with positive shut-off. Particular attention must be paid to the several types of seating materials available to avoid unnecessary wear. The seats must be compatible with the fluid in service and may be composition or metal. The configuration of the Globe valve opening causes turbulence which results in increased resistance. Most bronze Globe valves are rising stem-inside screw, but they are also available on O.S. & Y.

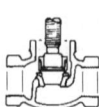

Angle Valves

The fundamental difference between the Angle valve and the Globe valve is the fluid flow through the Angle valve. It makes a 90° turn and offers less resistance to flow than the Globe valve while replacing an elbow. An Angle valve thus reduces the number of joints and installation time.

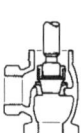

REFERENCE TABLES

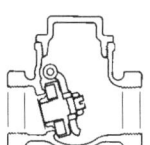
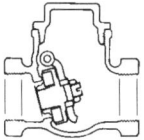

Check Valves

Check valves are designed to prevent backflow by automatically seating when the direction of fluid is reversed.

Swing Check valves are generally installed with Gate-valves, as they provide comparable full flow. Usually recommended for lines where flow velocities are low and should not be used on lines with pulsating flow. Recommended for horizontal installation, or in vertical lines only where flow is upward.

Lift Check Valves

These are commonly used with Globe and Angle valves since they have similar diaphragm seating arrangements and are recommended for preventing backflow of steam, air, gas and water, and on vapor lines with high flow velocities. For horizontal lines, horizontal lift checks should be used and vertical lift checks for vertical lines.

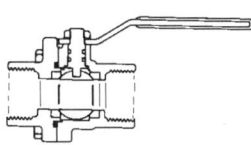

Ball Valves

Ball valves are light and easily installed, yet because of modern elastomeric seats, provide tight closure. Flow is controlled by rotating up to 90° a drilled ball which fits tightly against resilient seals. This ball seats with flow in either direction, and valve handle indicates the degree of opening. Recommended for frequent operation readily adaptable to automation, ideal for installation where space is limited.

Butterfly Valves

Butterfly valves provide bubble-tight closure with excellent throttling characteristics. They can be used for full-open, closed and for throttling applications.

The Butterfly valve consists of a disc within the valve body which is controlled by a shaft. In its closed position, the valve disc seals against a resilient seat. The disc position throughout the full 90° rotation is visually indicated by the position of the operator.

A Butterfly valve is only a fraction of the weight of a Gate valve and requires no gaskets between flanges in most cases. Recommended for frequent operation and adaptable to automation where space is limited.

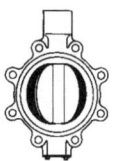

Wafer and Lug type bodies when installed between two pipe flanges, can be easily removed from the line. The pressure of the bolted flanges holds the valve in place.

Locating lugs makes installation easier.

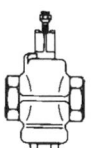

Plug Valves

Lubricated plug valves, because of the wide range of service to which they are adapted, may be classified as all purpose valves. They can be safely used at all pressure and vacuums, and at all temperatures up to the limits of available lubricants. They are the most satisfactory valves for the handling of gritty suspensions and many other destructive, erosive, corrosive and chemical solutions.

R221113-40 Plumbing Approximations for Quick Estimating

Water Control

Water Meter; Backflow Preventer, ... 10 to 15% of Fixtures
Shock Absorbers; Vacuum Breakers;
Mixer.

Pipe And Fittings .... 30 to 60% of Fixtures

> **Note:** Lower percentage for compact buildings or larger buildings with plumbing in one area.
> Larger percentage for large buildings with plumbing spread out.
> In extreme cases pipe may be more than 100% of fixtures.
> Percentages **do not** include special purpose or process piping.

Plumbing Labor

1 & 2 Story Residential .... Rough-in Labor = 80% of Materials
Apartment Buildings .... Rough-in Labor = 90 to 100% of Materials
Labor for handling and placing fixtures is approximately 25 to 30% of fixtures

Quality/Complexity Multiplier (for all installations)

Economy installation, add.. 0 to 5%
Good quality, medium complexity, add .... 5 to 15%
Above average quality and complexity, add .... 15 to 25%

R221113-50 Pipe Material Considerations

1. Malleable fittings should be used for gas service.
2. Malleable fittings are used where there are stresses/strains due to expansion and vibration.
3. Cast fittings may be broken as an aid to disassembling of heating lines frozen by long use, temperature and minerals.
4. Cast iron pipe is extensively used for underground and submerged service.
5. Type M (light wall) copper tubing is available in hard temper only and is used for nonpressure and less severe applications than K and L.

6. Type L (medium wall) copper tubing, available hard or soft for interior service.
7. Type K (heavy wall) copper tubing, available in hard or soft temper for use where conditions are severe. For underground and interior service.
8. Hard drawn tubing requires fewer hangers or supports but should not be bent. Silver brazed fittings are recommended, however soft solder is normally used.
9. Type DMV (very light wall) copper tubing designed for drainage, waste and vent plus other non-critical pressure services.

Domestic/Imported Pipe and Fittings Cost

The prices shown in this publication for steel/cast iron pipe and steel, cast iron, malleable iron fittings are based on domestic production sold at the normal trade discounts. The above listed items of foreign manufacture may be available at prices of 1/3 to 1/2 those shown. Some imported items after minor machining or finishing operations are being sold as domestic to further complicate the system.

Caution: Most pipe prices in this book also include a coupling and pipe hangers which for the larger sizes can add significantly to the per meter cost and should be taken into account when comparing "book cost" with quoted supplier's cost.

R221113-70 Piping to 3 m High

When taking off pipe, it is important to identify the different material types and joining procedures, as well as distances between supports and components required for proper support.

During the takeoff, measure through all fittings. Do not subtract the lengths of the fittings, valves, or strainers, etc. This added length plus the final rounding of the totals will compensate for nipples and waste.

When rounding off totals always increase the actual amount to correspond with manufacturer's shipping lengths.

A. Both red brass and yellow brass pipe are normally furnished in 3.6 m lengths, plain end. The Unit Price section includes in the per meter costs two field threads and one coupling per 3 m length. A carbon steel clevis type hanger assembly every 3 m is also prorated into the linear foot costs, including both material and labor.

B. Cast iron soil pipe is furnished in either 1.5 m or 3 m lengths. For pricing purposes, the Unit Price section features 3 m lengths with a joint and a carbon steel clevis hanger assembly every 1.5 m prorated into the per meter costs of both material and labor.

Three methods of joining are considered lead and oakum poured joints, or push-on gasket type joints for the bell and spigot pipe, and a joint clamp for the no-hub soil pipe. The labor and material costs for each of these individual joining procedures are also prorated into the per meter costs.

C. Copper tubing covers types K, L, M, and DWV which are furnished in 6 m lengths. Means pricing data is based on a tubing cut each length and a coupling and two soft soldered joints every 3 m. A carbon steel, clevis type hanger assembly every 3 m is also prorated into the per meter costs. The prices for refrigeration tubing are for materials only. Labor for full lengths may be based on the type L labor but short cut measures in tight areas can increase the installation labor-hours from 20 to 40%.

D. Corrosion-resistant piping does not lend itself to one particular standard of hanging or support assembly due to its diversity of application and placement. The several varieties of corrosion-resistant piping do not include any material or labor costs for hanger assemblies (See the Unit Price section for appropriate selection).

E. Glass pipe is furnished in standard lengths either 15 m or 3 m long, beaded on one end. Special orders for diverse lengths beaded on both ends are also available. For pricing purposes, R.S. Means features 3 m lengths with a coupling and a carbon steel band hanger assembly every 3 m prorated into the per meter costs.

Glass pipe is also available with conical ends and standard lengths ranging from 150 m through 900 m in 150 m increments, then up to 3 m in 300 mm increments. Special lengths can be customized for particular installation requirements.

For pricing purposes, Means has based the labor and material pricing on 3 m lengths. Included in these costs per meter are the prorated costs for a flanged assembly every 3 m consisting of two flanges, a gasket, two insertable seals, and the required number of bolts and nuts. A carbon steel band hanger assembly based on 3 m center lines has also been prorated into the costs per meter for labor and materials.

F. Plastic pipe of several compositions and joining methods are considered. Fiberglass reinforced pipe (FRP) is priced, based on 3 m lengths (6 m are also available), with coupling and epoxy joints every 3 m. FRP is furnished in both "General Service" and "High Strength." A carbon steel clevis hanger assembly, 3 for every 3 m, is built into the prorated labor and material costs on a per meter basis.

The PVC and CPVC pipe schedules 40, 80 and 120, plus SDR ratings are all based on 6 m lengths with a coupling installed every 3 m, as well as a carbon steel clevis hanger assembly every 1 m. The PVC and ABS type DWV piping is based on 3 m lengths with solvent weld couplings every 3 m, and with carbon steel clevis hanger assemblies,

3 for every 3 m. The rest of the plastic piping in this section is based on flexible 30 m coils and does not include any coupling or supports.

This section ends with PVC drain and sewer piping based on 3 m lengths with bell and spigot ends and 0-ring type, push-on joints.

G. Stainless steel piping, includes both weld end and threaded piping, both in the type 304 and 316 specification and in the following schedules, 5, 10, 40, 80, and 160. Although this piping is usually furnished in 6 m lengths, this cost grouping has a joint (either heli-arc butt-welded or threads and coupling) every 3 m. A carbon steel clevis type hanger assembly is also included at 3 m intervals and prorated into the per meter costs.

H. Carbon steel pipe includes both black and galvanized. This section encompasses schedules 40 (standard) and 80 (extra heavy).

Several common methods of joining steel pipe — such as thread and coupled, butt welded, and flanged (1034 kPa weld neck flanges) are also included.

For estimating purposes, it is assumed that the piping is purchased in 6 m lengths and that a compatible joint is made up every 3 m. These joints are prorated into the labor and material costs per meter. The following hanger and support assemblies every 3 m are also included: carbon steel clevis for the T & C pipe, and single rod roll type for both the welded and flanged piping. All of these hangers are oversized to accommodate pipe insulation 19 mm thick through 125 mm pipe size and 38 m thick from 150 mm through 300 mm pipe size.

I. Grooved joint steel pipe is prices both black and galvanized, in schedules 10, 40, and 80, furnished in 6 m lengths. This section describes two joining methods: cut groove and roll groove. The schedule 10 piping is roll-grooved, while the heavier schedules are cut-grooved. The labor and material costs are prorated into per meter prices, including a coupled joint every 3 m, as well as a carbon steel clevis hanger assembly.

Notes:

The pipe hanger assemblies mentioned in the preceding paragraphs include the described hanger; appropriately sized steel, box-type insert and nut; plus a threaded hanger rod. On average, the distance from the pipe center line to the insert face is 600 mm.

C clamps are used when the pipe is to be supported from steel shapes rather than anchored in the slab. C clamps are slightly less costly than inserts. However, to save time in estimating, it is advisable to use the given line number cost, rather than substituting a C clamp for the insert.

Add to piping labor for elevated installation:

3.0 m to 4.5 m high	10%	9.0 m to 10.5 m high	40%
4.5 m to 6.0 m high	20%	10.5 m to 12.0 m high	50%
6.0 m to 7.5 m high	25%	Over 12.0 m high	55%
7.5 m to 9.0 m high	35%		

When using the percentage adds for elevated piping installations as shown above, bear in mind that the given heights are for the pipe supports, even though the insert, anchor, or clamp may be a meter or more higher than the pipe itself.

An allowance has been included in the piping installation time for testing and minor tightening of leaking joints, fittings, stuffing boxes, packing glands, etc. For extraordinary test requirements such as x-rays, prolonged pressure or demonstration tests, a percentage of the piping labor, based on the estimator's experience, must be added to the labor total. A testing service specializing in weld x-rays should be consulted for pricing if it is an estimate requirement. Equipment installation time includes start-up with associated adjustments.

R221316-10 Drainage Requirements

Drainage lines must have a slope to maintain flow for proper operation. This slope should not be less than 20 mm per meter for 80 mm diameter or smaller pipe and not less than 10 mm per meter for 100 mm diameter or larger pipe. The capacity of building drainage systems is calculated on a basis of "drainage fixture units" (d.f.u.) as per the following chart.

Type of Fixture	d.f.u. Value	Type of Fixture	d.f.u. Value
Automatic clothes washer (50 mm standpipe)	3	Service sink (trap standard)	3
Bathroom group (water closet, lavatory and		Service sink (P trap)	2
bathtub or shower) tank type closet	6	Urinal, pedestal, syphon jet blowout	6
Bathtub (with or without overhead shower)	2	Urinal, wall hung	4
Clinic sink	6	Urinal, stall washout	4
Combination sink & tray with food disposal	4	Wash sink (cir. or mult.) per faucet set	2
Dental unit or cuspidor	1	Water closet, tank operated	4
Dental lavatory	1	Water closet, valve operated	6
Drinking fountain	1/2	Fixtures not listed above	
Dishwasher, domestic	2	Trap size 32 mm or smaller	1
Floor drains with 50 mm waste	3	Trap size 40 mm	2
Kitchen sink, domestic with one 40 mm trap	2	Trap size 50 mm	3
Kitchen sink, domestic with food disposal	2	Trap size 65 mm	4
Lavatory with 32 mm waste	1	Trap size 80 mm	5
Laundry tray (1 or 2 compartment)	2	Trap size 100 mm	6
Shower stall, domestic	2		

For continuous or nearly continuous flow into the sytem from a pump, air conditioning equipment or other item, allow 0.5 fixture units for each liter per minute of flow.

When the "drainage fixture units" (d.f.u.) for each horizontal branch or vertical stack is computed from the table above, the appropriate pipe size for each branch or stack is determined from the table below.

R221316-20 Allowable Fixture Units (d.f.u.) for Branches and Stacks

Pipe Diam.	Horiz. Branch (not incl. drains)	Stack Size for 3 Stories or 3 Levels	Stack size for Over 3 levels	Maximum for 1 Story building Stack
40 mm	3	4	8	2
50 mm	6	10	24	6
65 mm	12	20	42	9
80 mm	20*	48*	72*	20*
100 mm	160	240	500	90
125 mm	360	540	1100	200
150 mm	620	960	1900	350
200 mm	1400	2200	3600	600
250 mm	2500	3800	5600	1000
300 mm	3900	6000	8400	1500
375 mm	7000			

*Not more than two water closets or bathroom groups within each branch interval nor more than six water closets or bathroom groups on the stack.

Stacks sized for the total may be reduced as load decreases at each story to a minimum diameter of 1/2 the maximum diameter.

R224000-10 Hot Water Consumption Rates

Type of Building	Size Factor	Maximum Hourly Demand	Average Day Demand
Apartment Dwellings	No. of Apartments: Up to 20 21 to 50 51 to 75 76 to 100 101 to 200 201 up	45 L per apt. 38 L per apt. 32 L per apt. 26 L per apt. 23 L per apt. 19 L per apt.	159 L per apt. 157 L per apt. 144 L per apt. 140 L per apt. 136 L per apt. 132 L per apt.
Dormitories	Men Women	14 L per man 19 L per woman	50 L per man 47 L per woman
Hospitals	Per bed	87 L per patient	341 L per patient
Hotels	Single room with bath Double room with bath	64 L per unit 102 L per unit	189 L per unit 303 L per unit
Motels	No. of units: Up to 20 21 to 100 101 Up	23 L per unit 19 L per unit 15 L per unit	76 L per unit 53 L per unit 38 L per unit
Nursing Homes		17 L per bed	70 L per bed
Office buildings		1.5 L per person	3.8 L per person
Restaurants	Full meal type Drive-in snack type	6 L /max. meals/hr. 2.6 L /max. meals/hr.	9 L per meal 2.6 L per meal
Schools	Elementary Secondary & High	2.3 L per student 3.8 L per student	2.3 L per student 7 L per student

For evaluation purposes, recovery rate and storage capacity are inversely proportional. Water heaters should be sized so that the maximum hourly demand anticipated can be met in addition to allowance for the heat loss from the pipes and storage tank.

R224000-20 Fixture Demands in Liters Per Fixture Per Hour

Table below is based on 60°C final temperature except for dishwashers in public places (*) where 82°C water is mandatory.

Fixture	Apartment House	Club	Gym	Hospital	Hotel	Indust. Plant	Office	Private Home	School
Bathtubs	76	76	114	76	76			76	
Dishwashers, automatic	57	189-568*		189-568*	189-757*	76-379*		57	76-379*
Kitchen sink	38	76		76	114	76	76	38	76
Laundry, stationary tubs	76	106		106	106			76	
Laundry, automatic wash	284	284		378	568			284	
Private lavatory	8	8	8	8	8	8	8	8	8
Public lavatory	15	23	30	23	30	45	23		57
Showers	114	568	852	284	284	852	114	114	852
Service sink	76	76		76	114	76	76	57	76
Demand factor	0.30	0.30	0.40	0.25	0.25	0.40	0.30	0.30	0.40
Storage capacity factor	1.25	0.90	1.00	0.60	0.80	1.00	2.00	0.70	1.00

To obtain the probable maximum demand multiply the total demands for the fixtures (liters/fixture/hour) by the demand factor. The heater should have a heating capacity in liters per hour equal to this maximum. The storage tank should have a capacity in liters equal to the probable maximum demand multiplied by the storage capacity factor.

R224000-30 Minimum Plumbing Fixture Requirements

Minimum Plumbing Fixture Requirements

Type of Building or Occupancy (2)	Water Closets (14) (Fixtures per Person)		Urinals (5,10) (Fixtures per Person)		Lavatories (Fixtures per Person)		Bathtubs or Showers (Fixtures per Person)	Drinking Fountains (Fixtures per Person) (3, 13)
	Male	Female	Male	Female	Male	Female		
Assembly Places- Theatres, Auditoriums, Convention Halls, etc.-for permanent employee use	1: 1 - 15 2: 16 - 35 3: 36 - 55	1: 1 - 15 2: 16 - 35 3: 36 - 55	0: 1 - 9 1: 10 - 50		1 per 40	1 per 40		
	Over 55, add 1 fixture for each additional 40 persons		Add one fixture for each additional 50 males					
Assembly Places- Theatres, Auditoriums, Convention Halls, etc. - for public use	1: 1 - 100 2: 101 - 200 3: 201 - 400	3: 1 - 50 4: 51 - 100 8: 101 - 200 11: 201 - 400	1: 1 - 100 2: 101 - 200 3: 201 - 400 4: 401 - 600		1: 1 - 200 2: 201 - 400 3: 401 - 750	1: 1 - 200 2: 201 - 400 3: 401 - 750		1: 1 - 150 2: 151 - 400 3: 401 - 750
	Over 400, add 1 fixture for each additional 500 males and 1 for each additional 125 females		Over 600, add 1 fixture for each additional 300 males		Over 750, add 1 fixture for each additional 500 persons			Over 750, add one fixture for each additional 500 persons
Dormitories (9) School or Labor	1 per 10	1 per 8	1 per 25		1 per 12	1 per 12	1 per 8	1 per 150 (12)
	Add 1 fixture for each additional 25 males (over 10) and 1 for each additional 20 females (over 8)		Over 150, add 1 fixture for each additional 50 males		Over 12 add 1 fixture for each additional 20 males and 1 for each 15 additional females		For females add 1 bathtub per 30. Over 150, add 1 per 20	
Dormitories- for Staff Use	1: 1 - 15 2: 16 - 35 3: 36 - 55	1: 1 - 15 3: 16 - 35 4: 36 - 55	1 per 50		1 per 40	1 per 40	1 per 8	
	Over 55, add 1 fixture for each additional 40 persons							
Dwellings: Single Dwelling Multiple Dwelling or Apartment House	1 per dwelling 1 per dwelling or apartment unit				1 per dwelling 1 per dwelling or apartment unit		1 per dwelling 1 per dwelling or apartment unit	
Hospital Waiting rooms	1 per room				1 per room			1 per 150 (12)
Hospitals- for employee use	1: 1 - 15 2: 16 - 35 3: 36 - 55	1: 1 - 15 3: 16 - 35 4: 36 - 55	0: 1 - 9 1: 10 - 50		1 per 40	1 per 40		
	Over 55, add 1 fixture for each additional 40 persons		Add 1 fixture for each additional 50 males					
Hospitals: Individual Room Ward Room	1 per room 1 per 8 patients				1 per room 1 per 10 patients		1 per room 1 per 20 patients	1 per 150 (12)
Industrial (6) Warehouses Workshops, Foundries and similar establishments- for employee use	1: 1 -10 2: 11 - 25 3: 26 - 50 4: 51 - 75 5: 76 - 100	1: 1 -10 2: 11 - 25 3: 26 - 50 4: 51 - 75 5: 76 - 100			Up to 100, per 10 persons Over 100, 1 per 15 persons (7, 8)		1 shower for each 15 persons exposed to excessive heat or to skin contamination with poisonous, infectious or irritating material	1 per 150 (12)
	Over 100, add 1 fixture for each additional 30 persons							
Institutional - Other than Hospitals or Penal Institutions (on each occupied floor)	1 per 25	1 per 20	0: 1 - 9 1: 10 - 50		1 per 10	1 per 10	1 per 8	1 per 150 (12)
			Add 1 fixture for each additional 50 males					
Institutional - Other than Hospitals or Penal Institutions (on each occupied floor)- for employee use	1: 1 - 15 2: 16 - 35 3: 36 - 55	1: 1 - 15 3: 16 - 35 4: 36 - 55	0: 1 - 9 1: 10 - 50		1 per 40	1 per 40	1 per 8	1 per 150 (12)
	Over 55, add 1 fixture for each additional 40 persons		Add 1 fixture for each additional 50 males					
Office or Public Buildings	1: 1 - 100 2: 101 - 200 3: 201 - 400	3: 1 - 50 4: 51 - 100 8: 101 - 200 11: 201 - 400	1: 1 - 100 2: 101 - 200 3: 201 - 400 4: 401 - 600		1: 1 - 200 2: 201 - 400 3: 401 - 750	1: 1 - 200 2: 201 - 400 3: 401 - 750		1 per 150 (12)
	Over 400, add 1 fixture for each additional 500 males and 1 for each additional 150 females		Over 600, add 1 fixture for each additional 300 males		Over 750, add 1 fixture for each additional 500 persons			
Office or Public Buildings - for employee use	1: 1 - 15 2: 16 - 35 3: 36 - 55	1: 1 - 15 3: 16 - 35 4: 36 - 55	0: 1 - 9 1: 10 - 50		1 per 40	1 per 40		
	Over 55, add 1 fixture for each additional 40 persons		Add 1 fixture for each additional 50 males					

R224000-30 Minimum Plumbing Fixture Requirements (cont.)

Minimum Plumbing Fixture Requirements

Type of Building or Occupancy	Water Closets (14) (Fixtures per Person)		Urinals (5, 10) (Fixtures per Person)		Lavatories (Fixtures per Person)		Bathtubs or Showers (Fixtures per Person)	Drinking Fountains (Fixtures per Person) (3, 13)
	Male	Female	Male	Female	Male	Female		
Penal Institutions - for employee use	1: 1 - 15 2: 16 - 35 3: 36 - 55 Over 55, add 1 fixture for each additional 40 persons	1: 1 - 15 3: 16 - 35 4: 36 - 55	0: 1 - 9 1: 10 - 50 Add 1 fixture for each additional 50 males		1 per 40	1 per 40		1 per 150 (12)
Penal Institutions - for prison use Cell	1 per cell				1 per cell			1 per cellblock floor
Exercise room	1 per exercise room		1 per exercise room		1 per exercise room			1 per exercise room
Restaurants, Pubs and Lounges (11)	1: 1 - 50 2: 51 - 150 3: 151 - 300 Over 300, add 1 fixture for each additional 200 persons	1: 1 - 50 2: 51 - 150 4: 151 - 300	1: 1 - 150 Over 150, add 1 fixture for each additional 150 males		1: 1 - 150 2: 151 - 200 3: 201 - 400 Over 400, add 1 fixture for each additional 400 persons	1: 1 - 150 2: 151 - 200 3: 201 - 400		
Schools - for staff use All Schools	1: 1 - 15 2: 16 - 35 3: 36 - 55 Over 55, add 1 fixture for each additional 40 persons	1: 1 - 15 3: 16 - 35 4: 36 - 55	1 per 50		1 per 40	1 per 40		
Schools - for student use: Nursery	1: 1 - 20 2: 21 - 50 Over 50, add 1 fixture for each additional 50 persons	1: 1 - 20 2: 21 - 50			1: 1 - 25 2: 26 - 50 Over 50, add 1 fixture for each additional 50 persons	1: 1 - 25 2: 26 - 50		1 per 150 (12)
Elementary	1 per 30	1 per 25	1 per 75		1 per 35	1 per 35		1 per 150 (12)
Secondary	1 per 40	1 per 30	1 per 35		1 per 40	1 per 40		1 per 150 (12)
Others (Colleges, Universities, Adult Centers, etc.	1 per 40	1 per 30	1 per 35		1 per 40	1 per 40		1 per 150 (12)
Worship Places Educational and Activities Unit	1 per 150	1 per 75	1 per 150		1 per 2 water closets			1 per 150 (12)
Worship Places Principal Assembly Place	1 per 150	1 per 75	1 per 150		1 per 2 water closets			1 per 150 (12)

Notes:
1. The figures shown are based upon one (1) fixture being the minimum required for the number of persons indicated or any fraction thereof.
2. Building categories not shown on this table shall be considered separately by the Administrative Authority.
3. Drinking fountains shall not be installed in toilet rooms.
4. Laundry trays. One (1) laundry tray or one (1) automatic washer standpipe for each dwelling unit or one (1) laundry trays or one (1) automatic washer standpipes, or combination thereof, for each twelve (12) apartments. Kitchen sinks, one (1) for each dwelling or apartment unit.
5. For each urinal added in excess of the minimum required, one water closet may be deducted. The number of water closets shall not be reduced to less than two-thirds (2/3) of the minimum requirement.
6. As required by ANSI Z4.1-1968, Sanitation in Places of Employment.
7. Where there is exposure to skin contamination with poisonous, infectious, or irritating materials, provide one (1) lavatory for each five (5) persons.
8. Twenty-four (24) lineal inches of wash sink or eighteen (18) inches of a circular basin, when provided with water outlets for such space shall be considered equivalent to one (1) lavatory.
9. Laundry trays, one (1) for each fifty (50) persons. Service sinks, one (1) for each hundred (100) persons.
10. General. In applying this schedule of facilities, consideration shall be given to the accessibility of the fixtures. Conformity purely on a numerical basis may not result in an installation suited to the need of the individual establishment. For example, schools should be provided with toilet facilities on each floor having classrooms.
    a. Surrounding materials, wall and floor space to a point two (2) feet in front of urinal lip and four (4) feet above the floor, and at least two (2) feet to each side of the urinal shall be lined with non-absorbent materials.
    b. Trough urinals shall be prohibited.
11. A restaurant is defined as a business which sells food to be consumed on the premises.
    a. The number of occupants for a drive-in restaurant shall be considered as equal to the number of parking stalls.
    b. Employee toilet facilities shall not to be included in the above restaurant requirements. Hand washing facilities shall be available in the kitchen for employees.
12. Where food is consumed indoors, water stations may be substituted for drinking fountains. Offices, or public buildings for use by more than six (6) persons shall have one (1) drinking fountain for the first one hundred fifty (150) persons and one additional fountain for each three hundred (300) persons thereafter.
13. There shall be a minimum of one (1) drinking fountain per occupied floor in schools, theaters, auditoriums, dormitories, offices of public building.
14. The total number of water closets for females shall be at least equal to the total number of water closets and urinals required for males.

R224000-40 Plumbing Fixture Installation Time

Item	Rough-In	Set	Total Hours	Item	Rough-In	Set	Total Hours
Bathtub	5	5	10	Shower head only	2	1	3
Bathtub and shower, cast iron	6	6	12	Shower drain	3	1	4
Fire hose reel and cabinet	4	2	6	Shower stall, slate		15	15
Floor drain to 102 mm diameter	3	1	4	Slop sink	5	3	8
Grease trap, single, cast iron	5	3	8	Test 6 fixtures			14
Kitchen gas range		4	4	Urinal, wall	6	2	8
Kitchen sink, single	4	4	8	Urinal, pedestal or floor	6	4	10
Kitchen sink, double	6	6	12	Water closet and tank	4	3	7
Laundry tubs	4	2	6	Water closet and tank, wall hung	5	3	8
Lavatory wall hung	5	3	8	Water heater, 170 L gas, automatic	5	2	7
Lavatory pedestal	5	3	8	Water heaters, 245 L gas, automatic	5	2	7
Shower and stall	6	4	10	Water heaters, electric, plumbing only	4	2	6

Fixture prices in front of book are based on the cost per fixture set in place. The rough-in cost, which must be added for each fixture, includes carrier, if required, some supply, waste and vent pipe connecting fittings and stops. The lengths of rough-in pipe are nominal runs which would connect to the larger runs and stacks. The supply runs and DWV runs and stacks must be accounted for in separate entries. In the eastern half of the United States it is common for the plumber to carry these to a point 1.5 m outside the building.

R224000-50 Water Cooler Application

Type of Service	Requirement
Office, School or Hospital	12 persons per 1 mL/s
Office, Lobby or Department Store	4 or 5 mL/s per fountain
Light manufacturing	7 persons per 1 mL/s
Heavy manufacturing	5 persons per 1 mL/s
Hot heavy manufacturing	4 persons per 1 mL/s
Hotel	0.08 mL/s per room
Theatre	1 mL/s per 100 seats

R230500-10 Subcontractors

On the unit cost pages of the R.S. Means Cost Data books, the last column is entitled "Total Incl. O&P". This is normally the cost of the installing contractor. In the HVAC division, this is the cost of the mechanical contractor. If the particular work being estimated is to be performed by a sub to the mechanical contractor, the mechanical's profit and handling charge (usually 10%) is added to the total of the last column.

R233100-10 Loudness Levels for Moving Air thru Fans, Diffusers, Register, Etc. (Measured in Sones)

Recommended Loudness Levels (Sones)

Area	Very Quiet	Quiet	Noisy	Area	Very Quiet	Quiet	Noisy	Area	Very Quiet	Quiet	Noisy
Auditoriums				**Hotels**				**Offices**			
Auditorium lobbies	3	4	6	Banquet Rooms	1.5	3	6	Conference rooms	1	1.7	3
Concert and opera halls	0.8	1	1.5	Individual rooms, suites	1	2	4	Drafting	2	4	8
Courtrooms	2	3	4	Kitchens and laundries	4	7	10	General open offices	2	4	8
Lecture halls	1.5	2	3	Lobbies	2	4	8	Halls and corridors	2.5	5	10
Movie theaters	1.5	2	3	**Indoor Sports**				Professional offices	1.5	3	6
Churches and Schools				Bowling alleys	3	4	6	Reception room	1.5	3	6
Kitchens	4	6	8	Gymnasiums	2	4	6	Tabulation & computation	3	6	12
Laboratories	2	4	6	Swimming pools	4	7	10	**Public Buildings**			
Libraries	1.5	2	3	**Manufacturing Areas**				Banks	2	4	6
Recreation halls	2	4	8	Assembly lines	5	12	30	Court houses	2.5	4	6
Sanctuaries	1	1.7	3	Foreman's office	3	5	8	Museums	2	3	4
Schools and classrooms	1.5	2.5	4	Foundries	10	20	40	Planetariums	1.5	2	3
Hospital and Clinics				General storage	5	10	20	Post offices	2.5	4	6
Laboratories	2	4	6	Heavy machinery	10	25	60	Public libraries	1.5	2	4
Lobbies, waiting rooms	2	4	6	Light machinery	5	12	30	Waiting rooms	3	5	8
Halls and corridors	2	4	6	Tool maintenance	4	7	10	**Restaurants**			
Operating rooms	1.5	2.5	4	**Stores**				Cafeterias	3	6	10
Private rooms	1	1.7	3	Department stores	3	6	8	Night clubs	2.5	4	6
Wards	1.5	2.5	4	Supermarkets	4	7	10	Restaurants	2	4	8

R233100-20 Ductwork

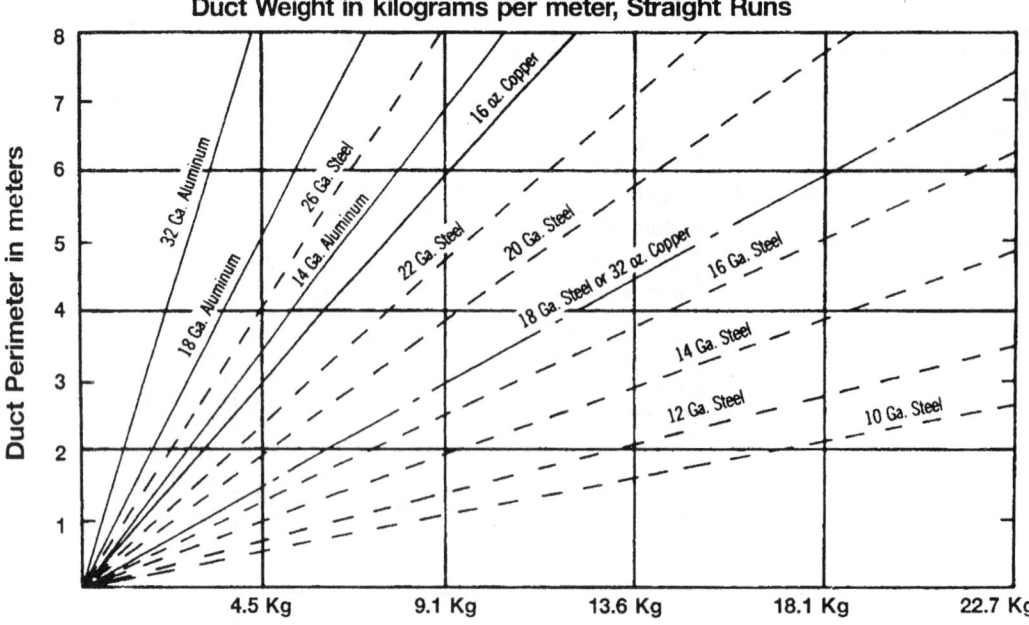

Duct Weight in kilograms per meter, Straight Runs

Add to the above for fittings; 90° elbow is 0.9 m L.F.; 45° elbow is 0.8 m ; offset is 1.2 m; transition offset is 1.8 m ; square-to-round transition is 1.2 m, 90° reducing elbow is 1.5 m. For bracing and waste, add 20% to aluminum and copper, 15% to steel.

R233100-30 Duct Fabrication/Installation

The labor cost for fabricated sheet metal duct includes both the cost of fabrication and installation of the duct. The split is approximately 40% for fabrication, 60% for installation. It is for this reason that the percentage add for elevated installation of fabricated duct is less than the percentage add for prefabricated/preformed duct.

Example: assume a piece of duct cost $100 installed (labor only)

$$\text{Sheet Metal Fabrication} = 40\% = \$40$$
$$\text{Installation} = 60\% = \$60$$

The add for elevated installation is:

$$\frac{\text{Based on total labor}}{\text{(fabrication \& installation)}} = \$100 \times 6\% = \$6.00$$

$$\frac{\text{Based on installation cost only}}{\text{(Material purchased prefabricated)}} = \$60 \times 10\% = \$6.00$$

The $6.00 markup (3 m to 4.5 m high) is the same.

R233100-40 Sheet Metal Calculator (Weight in Kg/m of Length)

Gauge	26	24	22	20	18	16
Wt. (kg/m²)	4.42	5.44	6.86	8.08	10.52	12.96
SMACNA Max. Dimension – Long Side (mm)		750	1350	2100	2125 Up	
Sum-2 sides (mm)						
50	.4	.6	.7	.9	1.2	1.3
75	.7	1.0	1.2	1.3	1.6	2.1
100	1.0	1.3	1.5	1.8	2.2	2.7
125	1.2	1.6	1.9	2.2	2.8	3.4
150	1.5	1.9	2.2	2.5	3.4	4.0
175	1.8	2.2	2.7	3.0	4.0	4.8
200	1.9	2.5	3.0	3.4	4.5	5.4
225	2.2	2.8	3.4	3.8	5.1	6.1
250	2.5	3.3	3.7	4.3	5.7	6.7
275	2.7	3.6	4.2	4.8	6.3	7.5
300	3.0	3.8	4.5	5.2	6.9	8.0
325	3.3	4.2	4.9	5.7	7.3	8.8
350	3.4	4.5	5.2	6.1	7.9	9.4
375	3.7	4.8	5.7	6.6	8.5	10.1
400	4.0	5.1	6.0	6.9	9.1	10.7
425	4.2	5.5	6.4	7.3	9.7	11.5
450	4.5	5.8	6.7	7.7	10.1	12.1
475	4.8	6.1	7.2	8.2	10.7	12.8
500	4.9	6.9	7.5	8.6	11.3	13.4
525	5.2	6.7	7.9	9.1	11.9	14.2
550	5.5	7.6	8.2	9.5	12.5	14.7
575	5.7	7.5	8.6	10.0	13.0	15.5
600	6.0	7.7	8.9	10.4	13.6	16.1
625	6.3	8.0	9.4	10.9	14.2	16.8
650	6.4	8.3	9.7	11.2	14.7	17.4
675	6.7	8.6	10.1	11.6	15.3	18.2
700	7.0	8.9	10.4	12.1	15.8	18.8
725	7.2	9.2	10.9	12.5	16.4	19.5
750	7.5	9.7	11.2	13.0	17.0	20.1
775	7.7	10.0	11.6	13.4	17.6	20.9
800	7.9	10.3	11.9	13.9	18.2	21.5
825	8.2	10.6	12.4	14.3	18.6	22.2
850	8.5	10.9	12.7	14.7	19.2	22.8
875	8.6	11.2	13.1	15.2	19.8	23.5
900	8.9	11.6	13.4	15.5	20.4	24.1
925	4.2	11.9	13.9	15.9	21.0	24.9
950	9.4	12.2	14.2	16.4	21.5	25.5
975	9.7	12.5	14.6	16.8	22.1	26.2
1000	10.0	12.8	14.9	17.3	22.6	26.8
1025	10.1	13.1	15.3	17.7	23.2	27.6
1050	10.4	13.4	15.6	18.2	23.8	28.2
1075	10.7	13.7	16.1	18.6	24.3	28.9
1100	10.9	14.2	16.4	19.1	24.9	29.5
1125	11.2	14.5	16.8	19.5	25.5	30.2
1150	11.5	14.7	17.1	19.8	26.1	30.8
1175	11.6	15.0	17.6	20.3	26.7	31.6
1200	11.9	15.3	17.9	20.7	27.1	32.2
1225	12.2	15.6	18.3	21.2	27.7	32.9
1250	12.4	15.9	18.6	21.6	28.3	33.5
1275	12.7	16.4	19.1	22.1	28.9	34.3
1300	13.0	16.7	19.4	22.5	29.5	34.9
1325	13.1	17.0	19.8	22.9	29.9	35.6
1350	13.4	17.3	20.1	23.4	30.5	36.2
1375	13.7	17.6	20.6	23.8	31.1	37.0

Gauge	26	24	22	20	18	16
Wt. (kg/m²)	.42	5.64	6.86	8.08	10.52	12.96
SMACNA Max. Dimension – Long Side (mm)		750	1350	2100	2125 Up	
Sum-2 Sides (mm)						
1400	13.9	17.9	20.9	24.1	31.7	37.5
1425	14.2	18.3	21.3	24.6	32.3	38.3
1450	14.5	18.6	21.6	25.0	32.8	38.9
1475	14.6	18.9	22.1	25.5	33.4	39.6
1500	14.9	19.2	22.4	25.9	34.0	40.2
1525	15.2	19.5	22.8	26.4	34.6	41.0
1550	15.3	19.8	23.1	26.8	35.2	41.6
1575	15.6	20.1	23.5	27.3	35.8	42.3
1600	15.9	20.4	23.8	27.7	36.2	42.9
1625	16.1	20.7	24.3	28.2	36.8	43.7
1650	16.4	21.0	24.6	28.5	37.4	44.3
1675	16.7	21.3	25.0	28.9	38.0	45.0
1700	16.8	21.2	25.3	29.4	38.4	45.6
1725	17.1	22.1	25.8	29.8	39.0	46.3
1750	17.4	22.4	26.1	30.2	39.6	46.9
1775	17.6	22.6	26.5	30.7	40.2	47.7
1800	17.9	22.9	26.8	31.1	40.8	48.3
1825	18.2	23.2	27.3	31.6	41.3	49.0
1850	18.3	23.5	27.6	32.0	41.9	49.6
1875	18.6	24.0	28.0	32.5	42.5	50.4
1900	18.9	24.3	28.3	32.8	43.1	51.0
1925	19.1	24.6	28.8	33.2	43.7	51.7
1950	19.4	24.9	29.1	33.7	44.1	52.3
1975	19.7	25.2	29.5	34.1	44.7	53.0
2000	19.8	25.5	29.8	34.6	45.3	53.6
2025	20.1	25.8	30.2	35.0	45.9	54.4
2050	20.4	26.1	30.5	35.5	46.5	55.0
2075	20.6	26.5	31.0	35.9	46.9	55.7
2100	20.9	26.8	31.5	36.4	47.5	56.3
2125	21.2	27.1	31.7	36.8	48.1	57.1
2150	21.3	27.4	32.0	37.1	48.7	57.7
2175	21.6	27.7	32.5	37.5	49.3	58.4
2200	21.9	28.0	32.8	38.0	49.8	59.0
2225	22.1	28.3	33.2	38.4	50.4	59.7
2250	22.4	28.8	33.5	38.9	51.0	60.3
2275	22.6	29.1	34.0	39.3	51.6	61.1
2300	22.8	29.4	34.3	39.8	52.2	61.7
2325	23.1	29.7	34.7	40.2	52.6	6.24
2350	23.4	29.9	35.0	40.7	53.2	63.0
2375	23.5	30.2	35.5	41.1	53.8	63.8
2400	23.8	30.5	35.8	41.4	54.4	64.4
2425	24.1	31.0	36.2	41.9	55.0	65.1
2450	24.3	30.3	36.5	42.3	55.4	65.7
2475	24.6	31.6	51.0	42.8	56.0	66.5
2500	29.9	31.9	37.3	43.2	56.6	67.1
2525	25.0	32.2	37.7	43.7	57.2	67.8
2550	25.3	32.5	38.0	44.1	57.8	68.4
2575	25.6	32.8	38.4	44.6	58.3	69.1
2600	25.8	33.2	38.7	45.0	58.9	69.7
2625	26.1	33.5	39.2	45.4	57.5	70.5
2650	26.4	33.8	39.5	45.7	60.0	71.1
2675	26.5	34.1	39.9	46.2	60.6	71.8
2700	26.8	34.4	40.2	46.6	61.1	72.4
2725	27.1	34.7	40.7	47.1	61.7	73.2
2750	27.3	35.0	41.0	47.5	62.3	73.8

Example: If duct is 850 mm x 500 mm x 4.5 m long, 850 mm is greater than 750 mm maximum, for 24 ga. so must be 22 ga. 850 + 500 = 1350 mm going across from 1350 mm find 20.1 kg/m. 20.1 kg/m x 4.5 m = 90.5 kg. For surface area, 90.5 kg ÷ 6.86 kg/m² = 13.2 m² **Note:** Figures include an allowance for scrap.

R233100-50 Ductwork Packages (per kW of Cooling)

System	Sheet Metal	Insulation	Diffusers	Return Register
Roof Top Unit Single Zone	15 kg	4.8 m²	1	1
Roof Top Unit Multizone	31 kg	9.7 m²	2	1
Self-contained Air or Water Cooled	14 kg	—	2	—
Split System Air Cooled	13 kg	—	2	—

Systems reflect most common usage.
Refer to system graphics for duct layout.

R233400-10 Recommended Ventilation Air Changes

Table below lists range of time in seconds per change for various types of facilities.

Assembly Halls 120-600	Dance Halls 120-600	Laundries 60-180
Auditoriums 120-600	Dining Rooms 180-600	Markets 120-600
Bakeries 120-180	Dry Cleaners 60-300	Offices 120-600
Banks 180-600	Factories 120-300	Pool Rooms 120-300
Bars 120-300	Garages 120-600	Recreation Rooms .120-600
Beauty Parlors 120-300	Generator Rooms 120-300	Sales Rooms 120-600
Boiler Rooms 60-300	Gymnasiums 120-600	Theaters 120-480
Bowling Alleys 120-600	Kitchens-Hospitals 120-300	Toilets 120-300
Churches 300-600	Kitchens-Restaurant 60-180	Transformer Rooms 60-300

m³/s air required for changes = Volume of room in m³ ÷ Seconds per change.

R233700-60 Diffuser Evaluation

L/s = V × An × K where V = Outlet velocity in meters per second. An = Neck area in square meters and K = Diffuser delivery factor. An undersized diffuser for a desired L/s will produce a high velocity and noise level. When air moves past people at a velocity in excess of 0.127 m/s, an annoying draft is felt. An oversized diffuser will result in low velocity with poor mixing. Consideration must be given to avoid vertical stratification or horizontal areas of stagnation.

R235000-10 Heating Systems

Heating Systems

The basic function of a heating system is to bring an enclosed volume up to a desired temperature and then maintain that temperature within a reasonable range. To accomplish this, the selected system must have sufficient capacity to offset transmission losses resulting from the temperature difference on the interior and exterior of the enclosing walls in addition to losses due to cold air infiltration through cracks, crevices and around doors and windows. The amount of heat to be furnished is dependent upon the building size, construction, temperature difference, air leakage, use, shape, orientation and exposure. Air circulation is also an important consideration. Circulation will prevent stratification which could result in heat losses through uneven temperatures at various levels. For example, the most

efficient use of unit heaters can usually be achieved by circulating the space volume through the total number of units once every 20 minutes or 3 times an hour. This general rule must, of course, be adapted for special cases such as large buildings with low ratios of heat transmitting surface to cubical volume. The type of occupancy of a building will have considerable bearing on the number of heat transmitting units and the location selected. It is axiomatic, however, that the basis of any successful heating system is to provide the maximum amount of heat at the points of maximum heat loss such as exposed walls, windows, and doors. Large roof areas, wind direction, and wide doorways create problems of excessive heat loss and require special consideration and treatment.

Heat Transmission

Heat transfer is an important parameter to be considered during selection of the exterior wall style, material and window area. A high rate of transfer will permit greater heat loss during the wintertime with the resultant increase in heating energy costs and a greater rate of heat gain in the summer with proportionally greater cooling cost. Several terms are used to describe various aspects of heat transfer. However, for general estimating purposes this book lists U values for systems of construction materials. U is the "overall heat transfer coefficient." It is defined as the heat flow per hour through one square foot when the temperature difference in the air on either side of the structure wall, roof, ceiling or floor is one degree Fahrenheit. The structural segment may be a single homogeneous material or a composite.

Total heat transfer is found using the following equation:

$Q = AU(T_2 - T_1)$ where
$\quad Q$ = Heat flow, kW
$\quad A$ = Area, m^2
$\quad U$ = Overall heat transfer coefficient
$(T_2 - T_1)$ = Difference in temperature of air on each side of the construction component. (Also abbreviated TD)

Note that heat can flow through all surfaces of any building and this flow is in addition to heat gain or loss due to ventilation, infiltration and generation (appliances, machinery, people).

R235000-20 Heating Approximations for Quick Estimating

Oil Piping & Boiler Room Piping

Small System . . . . 20 to 30% of Boiler

Complex System
with Pumps, Headers, Etc. . . . . 80 to 110% of Boiler

Breeching With Insulation:

Small . . . . 10 to 15% of Boiler

Large . . . . 15 to 25% of Boiler

Coils: . . . . 15 to 30% of Containing Unit

Balancing (Independent) . . . . 1/2% of H.V.A.C. Estimate

Quality/Complexity Adjustment: For all heating installations add these adjustments to the estimate to more closely allow for the equipment and conditions of the particular job under consideration.

Economy installation, add . . . . 0 to 5% of System

Good quality, medium complexity, add . . . . 5 to 15% of System

Above average quality and complexity, add . . . . 15 to 25% of System

R235000-30 The Basics of a Heating System

The function of a heating system is to achieve and maintain a desired temperature in a room or building by replacing the amount of heat being dissipated. There are four kinds of heating systems: hot-water, steam, warm-air and electric resistance. Each has certain essential and similar elements with the exception of electric resistance heating.

The basic elements of a heating system are:

A. A **combustion chamber** in which fuel is burned and heat transferred to a conveying medium.

B. The **"fluid"** used for conveying the heat (water, steam or air).

C. **Conductors** or pipes for transporting the fluid to specific desired locations.

D. A means of disseminating the heat, sometimes called **terminal units.**

A. The **combustion chamber** in a furnace heats air which is then distributed. This is called a warm-air system.

The combustion chamber in a boiler heats water which is either distributed as hot water or steam and this is termed a hydronic system.

The maximum allowable working pressures are limited by ASME "Code for Heating Boilers" to 103 kPa for steam and 1103 kPa for hot water heating boilers, with a maximum temperature limitation of 121°C. Hot water boilers are generally rated for a working pressure of 207 kPa. High pressure boilers are governed by the ASME "Code for Power Boilers" which is used almost universally for boilers operating over 103 kPa. High pressure boilers used for a combination of heating/process loads are usually designed for 1034 kPa.

Boiler ratings are usually indicated as either Gross or Net Output. The Gross Load is equal to the Net Load plus a piping and pickup allowance. When this allowance cannot be determined, divide the gross output rating by 1.25 for a value equal to or greater than the net heat loss requirement of the building.

B. Of the three **fluids** used, steam carries the greatest amount of heat per unit volume. This is due to the fact that it gives up its latent heat of vaporization at a temperature considerably above room temperature. Another advantage is that the pressure to produce a positive circulation is readily available. Piping conducts the steam to terminal units and returns condensate to the boiler.

The **steam system** is well adapted to large buildings because of its positive circulation, its comparatively economical installation and its ability to deliver large quantities of heat. Nearly all large office buildings, stores, hotels, and industrial buildings are so heated, in addition to many residences.

Hot water, when used as the heat carrying fluid, gives up a portion of its sensible heat and then returns to the boiler or heating apparatus for reheating. As the heat conveyed by each kg of water is about one-fiftieth of the heat conveyed by a kg of steam, it is necessary to circulate about fifty times as much water as steam by weight (although only one-thirtieth as much by volume). The hot water system is usually, although not necessarily, designed to operate at temperatures below that of the ordinary steam system and so the amount of heat transfer surface must be correspondingly greater. A temperature of 88°C to 93°C is normally the maximum. Circulation in small buildings may depend on the difference in density between hot water and the cool water returning to the boiler; circulating pumps are normally used to maintain a desired rate of flow. Pumps permit a greater degree of flexibility and better control.

In **warm-air** furnace systems, cool air is taken from one or more points in the building, passed over the combustion chamber and flue gas passages and then distributed through a duct system. A disadvantage of this system is that the ducts take up much more building volume than steam or hot water pipes. Advantages of this system are the relative ease with which humidification can be accomplished by the evaporation of water as the air circulates through the heater, and the lack of need for expensive disseminating units as the warm air simply becomes part of the interior atmosphere of the building.

C. Conductors (pipes and ducts) have been lightly treated in the discussion of conveying fluids. For more detailed information such as sizing and distribution methods, the reader is referred to technical publications such as the American Society of Heating, Refrigerating and Air-Conditioning Engineers "Handbook of Fundamentals."

D. Terminal units come in an almost infinite variety of sizes and styles, but the basic principles of operation are very limited. As previously mentioned, warm-air systems require only a simple register or diffuser to mix heated air with that present in the room. Special application items such as radiant coils and infrared heaters are available to meet particular conditions but are not usually considered for general heating needs. Most heating is accomplished by having air flow over coils or pipes containing the heat transporting medium (steam, hot-water, electricity). These units, while varied, may be separated into two general types, (1) radiator/convectors and (2) unit heaters.

Radiator/convectors may be cast, fin-tube or pipe assemblies. They may be direct, indirect, exposed, concealed or mounted within a cabinet enclosure, upright or baseboard style. These units are often collectively referred to as "radiatiors" or "radiation" although none gives off heat either entirely by radiation or by convection but rather a combination of both. The air flows over the units as a gravity "current." It is necessary to have one or more heat-emitting units in each room. The most efficient placement is low along an outside wall or under a window to counteract the cold coming into the room and achieve an even distribution.

In contrast to radiator/convectors which operate most effectively against the walls of smaller rooms, **unit heaters** utilize a fan to move air over heating coils and are very effective in locations of relatively large volume. Unit heaters, while usually suspended overhead, may be floor mounted. They also may take in fresh outside air for ventilation. The heat distributed by unit heaters may be from a remote source and conveyed by a fluid or it may be from the combustion of fuel in each individual heater. In the latter case the only piping required would be for fuel, however, a vent for the products of combustion would be necessary.

The following list gives may of the advantages of unit heaters for applications other than office or residential:

a. Large capacity so smaller number of units are required, **b.** Piping system simplified, **c.** Space saved where they are located overhead out of the way, **d.** Rapid heating directed where needed with effective wide distribution, **e.** Difference between floor and ceiling temperature reduced, **f.** Circulation of air obtained, and ventilation with introduction of fresh air possible, **g.** Heat output flexible and easily controlled.

R235000-35 Heating (42° degrees latitude)

$$\text{Approximate m}^2 \text{ radiation} = \frac{\text{m}^2 \text{ sash}}{2} + \frac{\text{m}^2 \text{ wall} + \text{roof} - \text{sash}}{20} + \frac{\text{m}^3 \text{ building}}{61}$$

R235000-50 Factor for Determining Heat Loss for Various Types of Buildings

General: While the most accurate estimates of heating requirements would naturally be based on detailed information about the building being considered, it is possible to arrive at a reasonable approximation using the following procedure:

1. Calculate the cubic volume of the room or building.
2. Select the appropriate factor from Table 1 below. Note that the factors apply only to inside temperatures listed in the first column and to 255 K outside temperature.
3. If the building has bad north and west exposures, multiply the heat loss factor by 1.1.
4. If the outside design temperature is other than 255 K, multiply the factor from Table 1 by the factor from Table 2.
5. Multiply the cubic volume by the factor selected from Table 1. This will give the estimated W heat loss which must be made up to maintain inside temperature.

Table 1 — Building Type	Conditions	Qualifications	Loss Factor*
Factories & Industrial Plants General Office Areas at 295 K	One Story	Skylight in Roof	64.2
		No Skylight in Roof	59.0
	Multiple Story	Two Story	47.6
		Three Story	44.5
		Four Story	42.4
		Five Story	40.4
		Six Story	37.3
	All Walls Exposed	Flat Roof	71.4
		Heated Space Above	53.8
	One Long Warm Common Wall	Flat Roof	65.2
		Heated Space Above	48.6
	Warm Common Walls on Both Long Sides	Flat Roof	60.0
		Heated Space Above	42.4
Warehouses at 290 K	All Walls Exposed	Skylights in Roof	56.9
		No Skylight in Roof	52.8
		Heated Space Above	41.4
	One Long Warm Common Wall	Skylight in Roof	51.7
		No Skylight in Roof	50.7
		Heated Space Above	35.2
	Warm Common Walls on Both Long Sides	Skylight in Roof	48.6
		No Skylight in Roof	45.5
		Heated Space Above	31.0

*Note: This table tends to be conservative particularly for new buildings designed for minimum energy consumption.

Table 2 — Outside Design Temperature Correction Factor (for Degrees Kelvin)									
Outside Design Temperature	283	278	272	266	261	255	250	244	239
Correction Factor	0.29	0.43	0.57	0.72	0.86	1.00	1.14	1.28	1.43

R235000-70 Transmission of Heat

R235000-70 and R235000-80 provide a way to calculate heat transmission of various construction materials from their U values and the TD (Temperature Difference).

1. From the Exterior Enclosure Division or elsewhere, determine U values for the construction desired.
2. Determine the coldest design temperature. The difference between this temperature and the desired interior temperature is the TD (temperature difference).

3. Enter R235000-70 or R235000-80 at correct U Value. Cross horizontally to the intersection with appropriate TD. Read transmission per square foot from bottom of figure.
4. Multiply this value of BTU per hour transmission per square foot of area by the total surface area of that type of construction.

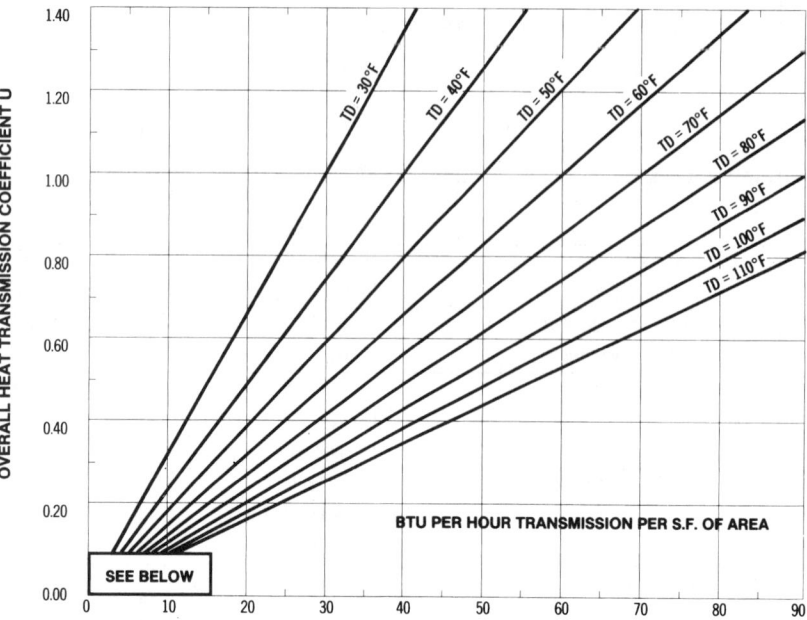

R235000-80 Transmission of Heat (Low Rate)

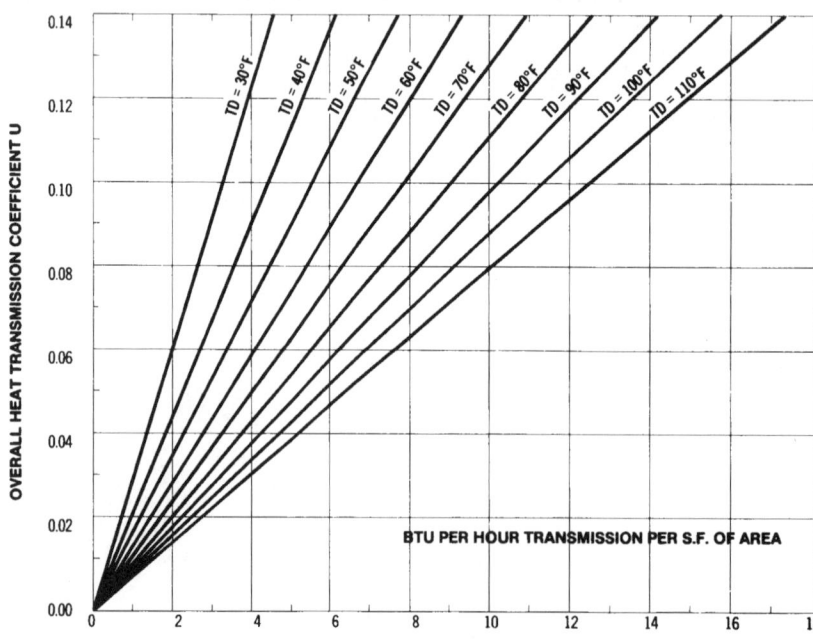

844

R235616-60 Solar Heating (Space and Hot Water)

Collectors should face as close to due South as possible, however, variations of up to 20 degrees on either side of true South are acceptable. Local climate and collector type may influence the choice between east or west deviations. Obviously they should be located so they are not shaded from the sun's rays. Incline collectors at a slope of latitude minus 5 degrees for domestic hot water and latitude plus 15 degrees for space heating.

Flat plate collectors consist of a number of components as follows: Insulation to reduce heat loss through the bottom and sides of the collector. The enclosure which contains all the components in this assembly is usually weatherproof and prevents dust, wind and water from coming in contact with the absorber plate. The cover plate usually consists of one or more layers of a variety of glass or plastic and reduces the reradiation by creating an air space which traps the heat between the cover and the absorber plates.

The absorber plate must have a good thermal bond with the fluid passages. The absorber plate is usually metallic and treated with a surface coating which improves absorptivity. Black or dark paints or selective coatings are used for this purpose, and the design of this passage and plate combination helps determine a solar system's effectiveness.

Heat transfer fluid passage tubes are attached above and below or integral with an absorber plate for the purpose of transferring thermal energy from the absorber plate to a heat transfer medium. The heat exchanger is a device for transferring thermal energy from one fluid to another.

Piping and storage tanks should be well insulated to minimize heat losses.

Size domestic water heating storage tanks to hold 0.08 m^3 of water per user, minimum, plus 0.04 m^3 per dishwasher or washing machine. For domestic water heating an optimum collector size is approximately 0.07 m^2 of area per 0.004 m^3 of water storage. For space heating of residences and small commercial applications the collector is commonly sized between 30% and 50% of the internal floor area. For space heating of large commercial applications, collector areas less than 30% of the internal floor area can still provide significant heat reductions.

A supplementary heat source is recommended for Northern states for December through February.

The solar energy transmission per m^2 of collector surface varies greatly with the material used. Initial cost, heat transmittance and useful life are obviously interrelated.

R236000-10 Air Conditioning

General: The purpose of air conditioning is to control the environment of a space so that comfort is provided for the occupants and/or conditions are suitable for the processes or equipment contained therein. The several items which should be evaluated to define system objectives are:

Temperature Control
Humidity Control
Cleanliness
Odor, smoke and fumes
Ventilation

Efforts to control the above parameters must also include consideration of the degree or tolerance of variation, the noise level introduced, the velocity of air motion and the energy requirements to accomplish the desired results.

The variation in **temperature** and **humidity** is a function of the sensor and the controller. The controller reacts to a signal from the sensor and produces the appropriate suitable response in either the terminal unit, the conductor of the transporting medium (air, steam, chilled water, etc.), or the source (boiler, evaporating coils, etc.).

The **noise level** is a by-product of the energy supplied to moving components of the system. Those items which usually contribute the most noise are pumps, blowers, fans, compressors and diffusers. The level of noise can be partially controlled through use of vibration pads, isolators, proper sizing, shields, baffles and sound absorbing liners.

Some **air motion** is necessary to prevent stagnation and stratification. The maximum acceptable velocity varies with the degree of heating or cooling which is taking place. Most people feel air moving past them at velocities in excess of 0.127 m/s as an annoying draft. However, velocities up to $.229$ m/s may be acceptable in certain cases. Ventilation, expressed as air changes per hour and percentage of fresh air, is usually an item regulated by local codes.

Selection of the system to be used for a particular application is usually a trade-off. In some cases the building size, style, or room available for mechanical use limits the range of possibilities. Prime factors influencing the decision are first cost and total life (operating, maintenance and replacement costs). The accuracy with which each parameter is determined will be an important measure of the reliability of the decision and subsequent satisfactory operation of the installed system.

Heat delivery may be desired from an air conditioning system. Heating capability usually is added as follows: A gas fired burner or hot water/steam/electric coils may be added to the air handling unit directly and heat all air equally. For limited or localized heat requirements the water/steam/electric coils may be inserted into the duct branch supplying the cold areas. Gas fired duct furnaces are also available.

Note: When water or steam coils are used the cost of the piping and boiler must also be added. For a rough estimate use the cost per square foot of the appropriate sized hydronic system with unit heaters. This will provide a cost for the boiler and piping, and the unit heaters of the system would equate to the approximate cost of the heating coils.

R236000-20 Air Conditioning Requirements

Watts per hour per m² of floor area and m² per kW of air conditioning.

Type of Building	W / m²	m² / kW	Type of Building	W / m²	m² / kW	Type of Building	W / m²	m² / kW
Apartments, Individual	82	11.9	Dormitory, Rooms	126	7.9	Libraries	158	6.3
Corridors	69	14.5	Corridors	95	10.6	Low Rise Office, Exterior	120	8.5
Auditoriums & Theaters	126	7.9/5*	Dress Shops	136	7.4	Interior	104	9.5
Banks	158	6.3	Drug Stores	252	4.0	Medical Centers	88	11.2
Barber Shops	151	6.6	Factories	126	7.9	Motels	88	11.2
Bars & Taverns	419	2.4	High Rise Office—Ext. Rms.	145	6.9	Office (small suite)	136	7.4
Beauty Parlors	208	4.7	Interior Rooms	117	8.6	Post Office, Individual Office	132	7.5
Bowling Alleys	214	4.6	Hospitals, Core	136	7.4	Central Area	145	6.9
Churches	113	8.7/6*	Perimeter	145	6.9	Residences	63	15.8
Cocktail Lounges	214	4.6	Hotel, Guest Rooms	139	7.3	Restaurants	189	5.3
Computer Rooms	444	2.2	Corridors	95	10.6	Schools & Colleges	145	6.9
Dental Offices	164	6.1	Public Spaces	173	5.8	Shoe Stores	173	5.8
Dept. Stores, Basement	107	9.2	Industrial Plants, Offices	120	8.5	Shop'g. Ctrs., Supermarkets	107	9.2
Main Floor	126	7.9	General Offices	107	9.2	Retail Stores	151	6.6
Upper Floor	95	10.6	Plant Areas	126	7.9	Specialty	189	5.3

*Persons per kW

1 ton of air conditioning = 3517 watts = 3.517 kW

R236000-030 Psychrometric Table

Dewpoint or Saturation Temperature (F)

Relative humidity (%)	100	32	35	40	45	50	55	60	65	70	75	80	85	90	95	100
	90	30	33	37	42	47	52	57	62	67	72	77	82	87	92	97
	80	27	30	34	39	44	49	54	58	64	68	73	78	83	88	93
	70	24	27	31	36	40	45	50	55	60	64	69	74	79	84	88
	60	20	24	28	32	36	41	46	51	55	60	65	69	74	79	83
	50	16	20	24	28	33	36	41	46	50	55	60	64	69	73	78
	40	12	15	18	23	27	31	35	40	45	49	53	58	62	67	71
	30	8	10	14	18	21	25	29	33	37	42	46	50	54	59	62
	20	6	7	8	9	13	16	20	24	28	31	35	40	43	48	52
	10	4	4	5	5	6	8	9	10	13	17	20	24	27	30	34

32 35 40 45 50 55 60 65 70 75 80 85 90 95 100
Dry bulb temperature (F)

This table shows the relationship between RELATIVE HUMIDITY, DRY BULB TEMPERATURE AND DEWPOINT.

As an example, assume that the thermometer in a room reads 75°F, and we know that the relative humidity is 50%. The chart shows the dewpoint temperature to be 55°. That is, any surface colder than 55°F will "sweat" or collect condensing moisture. This surface could be the outside of an uninsulated chilled water pipe in the summertime, or the inside surface of a wall or deck in the wintertime. After determining the extreme ambient parameters, the table at the left is useful in determining which surfaces need insulation or vapor barrier protection.

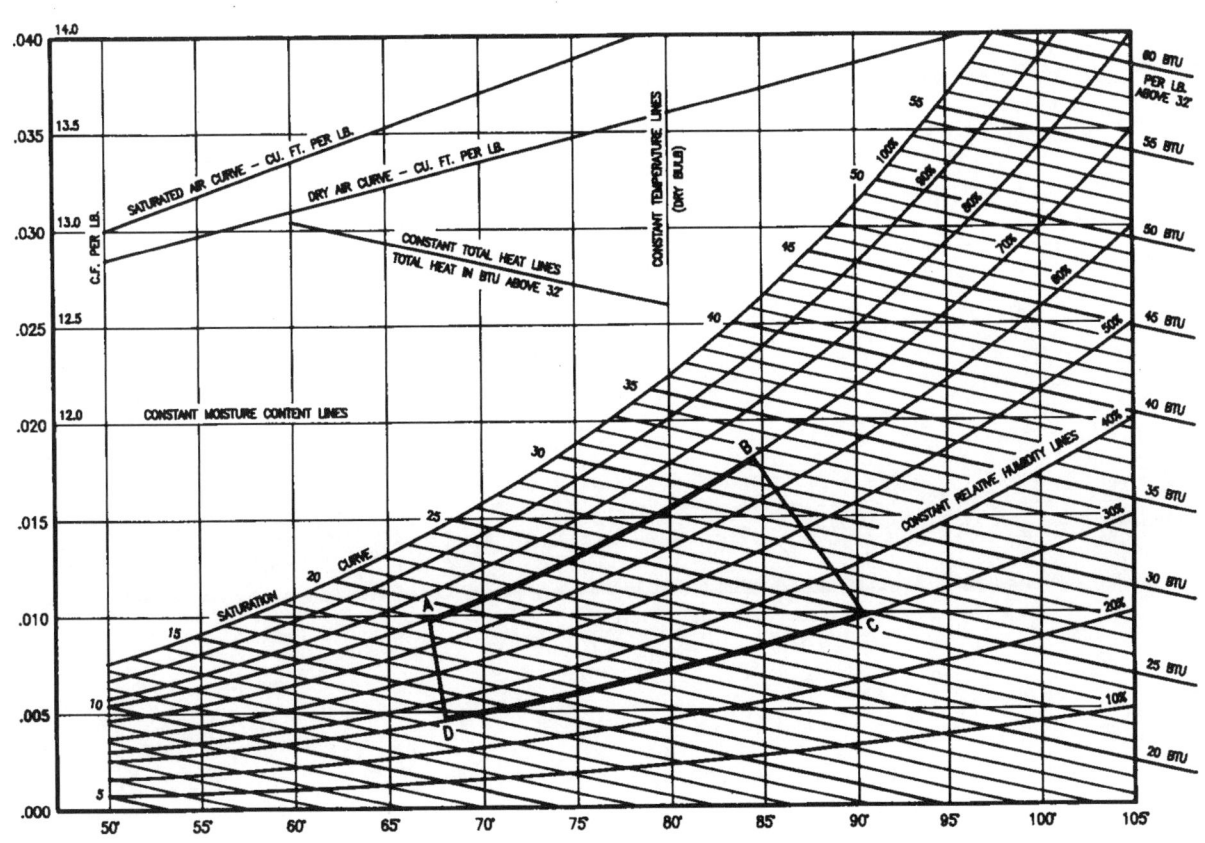

TEMPERATURE DEGREES FAHRENHEIT
TOTAL PRESSURE = 14.696 LB. PER SQ. IN. ABS.

Psychrometric chart showing different variables based on one pound of dry air. Space marked A B C D is temperature-humidity range which is most comfortable for majority of people.

R236000-90 Quality/Complexity Adjustment for Air Conditioning Systems

Economy installation, add . . . . 0 to 5%

Good quality, medium complexity, add . . . . 5 to 15%

Above average quality and complexity, add . . . . 15 to 25%

Add the above adjustments to the estimate to more closely allow for the equipment and conditions of the particular job under consideration.

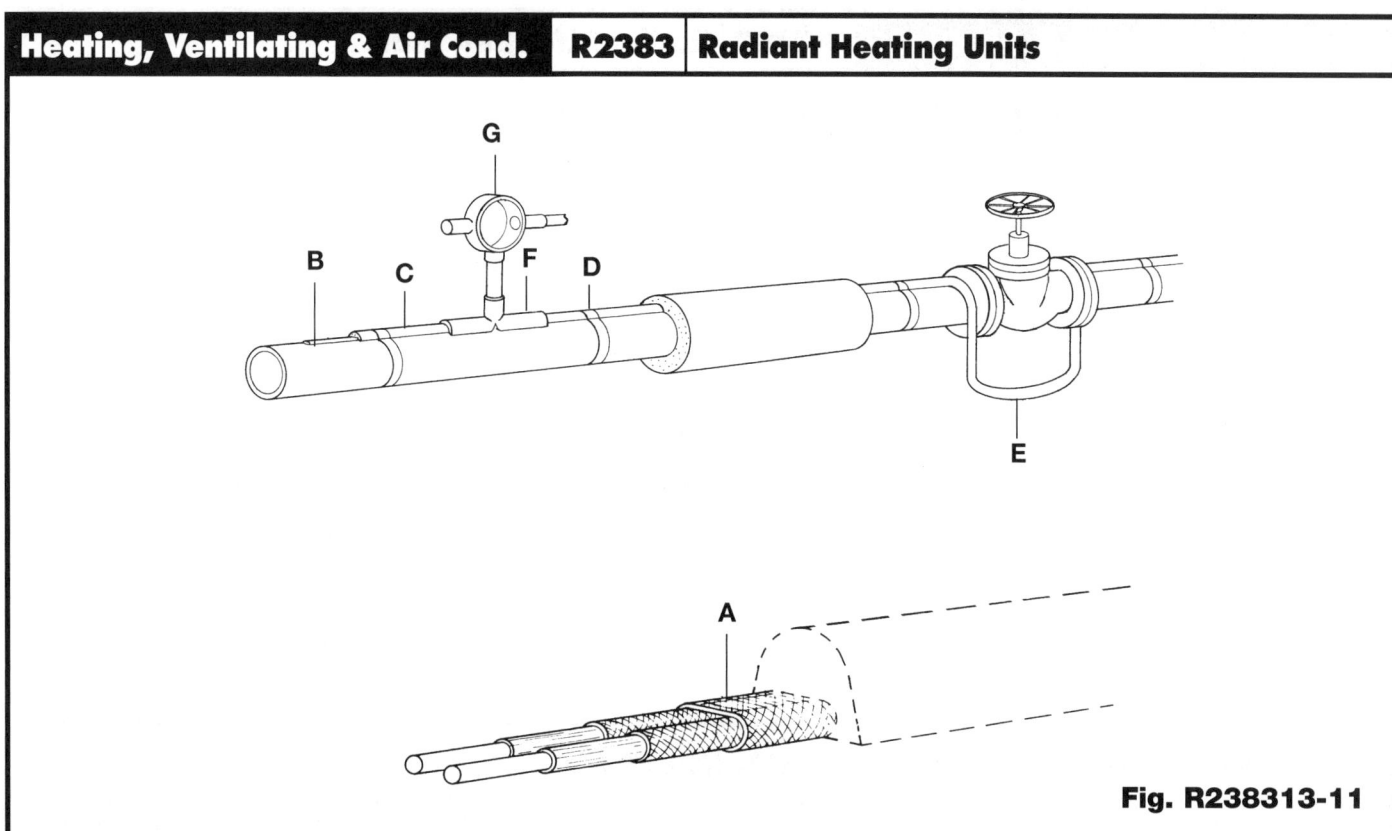

Fig. R238313-11

R238313-10   Heat Trace Systems

Before you can determine the cost of a HEAT TRACE installation the method of attachment must be established. There are (4) common methods:

1. Cable is simply attached to the pipe with polyester tape every 3.6 m.
2. Cable is attached with a continuous cover of 50 mm wide aluminum tape.
3. Cable is attached with factory extruded heat transfer cement and covered with metallic raceway with clips every 3 m.
4. Cable is attached between layers of pipe insulation using either clips or polyester tape.

Example: Components for method 3 must include:
  A. Heat trace cable by voltage and watts per meter.
  B. Heat transfer cement, 1 liter per 4.8 m of cover.
  C. Metallic raceway by size and type.
  D. Raceway clips by size of pipe.

When taking off distances of cable add the following for each valve in the system. (E)

In all of the above methods each component of the system must be priced individually.

SCREWED OR WELDED VALVE:			FLANGED VALVE:			BUTTERFLY VALVES:		
15 mm	=	150 mm	15 mm	=	300 mm	15 mm	=	0 mm
20 mm	=	225 mm	20 mm	=	450 mm	20 mm	=	0 mm
25 mm	=	300 mm	25 mm	=	600 mm	25 mm	=	300 mm
40 mm	=	450 mm	40 mm	=	750 mm	40 mm	=	450 mm
50 mm	=	600 mm	50 mm	=	750 mm	50 mm	=	600 mm
65 mm	=	750 mm	65 mm	=	900 mm	65 mm	=	750 mm
80 mm	=	750 mm	80 mm	=	1050 mm	80 mm	=	750 mm
100 mm	=	1200 mm	100 mm	=	1200 mm	100 mm	=	900 mm
150 mm	=	2100 mm	150 mm	=	2400 mm	150 mm	=	1050 mm
200 mm	=	2.85 mm	200 mm	=	3.30 mm	200 mm	=	1200 mm
250 mm	=	3.75 mm	250 mm	=	4.20 mm	250 mm	=	1200 mm
300 mm	=	4.50 mm	300 mm	=	4.98 mm	300 mm	=	1500 mm
350 mm	=	5.40 mm	350 mm	=	5.88 mm	350 mm	=	1650 mm
400 mm	=	6.48 mm	400 mm	=	6.90 mm	400 mm	=	1800 mm
450 mm	=	7.68 mm	450 mm	=	8.10 mm	450 mm	=	1950 mm
500 mm	=	8.58 mm	500 mm	=	9.00 mm	500 mm	=	2100 mm
600 mm	=	10.2 mm	600 mm	=	10.80 mm	600 mm	=	2400 mm
750 mm	=	12.0 mm	750 mm	=	12.60 mm	750 mm	=	3000 mm

R238313-10   Heat Trace Systems (cont.)

Add the following quantities of heat transfer cement for each valve:

Nominal Valve Size	Liters of Cement per Valve
15 mm	0.5
20 mm	0.8
25 mm	1.1
40 mm	1.4
50 mm	1.6
65 mm	2.6
80 mm	2.7
100 mm	3.8
150 mm	5.4
200 mm	5.6
250 mm	5.7
300 mm	6.1
350 mm	6.6
400 mm	7.6
450 mm	8.5
500 mm	9.5
600 mm	11.4
750 mm	14.2

The following must be added to the list of components to accurately price HEAT TRACE systems:

1. Expediter fitting and clamp fasteners (F)
2. Junction box and nipple connected to expediter fitting (G)
3. Field installed terminal blocks within junction box
4. Ground lugs
5. Piping from power source to expediter fitting
6. Controls
7. Thermostats
8. Branch wiring
9. Cable splices
10. End of cable terminations
11. Branch piping fittings and boxes

Deduct the following percentages from labor if cable lengths in the same area exceed:

45 m to 75 m	10%	106 m to 150 m	20%
76 m to 105 m	15%	Over 150 m	25%

Add the following percentages to labor for elevated installations:

4.5 m to 6.0 m high	10%	9.1 m to 10.5 m high	40%
6.1 m to 7.5 m high	20%	10.6 m to 12.0 m high	50%
7.6 m to 9.0 m high	30%	Over 12.0 m high	60%

R238313-20   Spiral-Wrapped Heat Trace Cable (Pitch Table)

In order to increase the amount of heat, occasionally heat trace cable is wrapped in a spiral fashion around a pipe; increasing the number of meters of heater cable per meter of pipe.

Engineers first determine the heat loss per meter of pipe (based on the insulating material, its thickness, and the temperature differential across it). A ratio is then calculated by the formula:

$$\text{Meters of Heat Trace per meter of Pipe} = \frac{\text{Watts/Meter of Heat Loss}}{\text{Watts/Meter of the Cable}}$$

The linear distance between wraps (pitch) is then taken from a chart or table. Generally, the pitch is listed on a drawing leaving the estimator to calculate the total length of heat tape required. An approximation may be taken from this table.

	Meters of Heat Trace Per Meter of Pipe															
	Nominal Pipe Size in mm															
Pitch In mm	25	32	40	50	65	80	100	150	200	250	300	350	400	450	500	600
88	1.80															
100	1.65															
125	1.46	1.60	1.80													
150	1.34	1.45	1.55	1.75												
175	1.25	1.35	1.43	1.57	1.75											
200	1.20	1.28	1.34	1.45	1.60	1.80										
225	1.16	1.23	1.28	1.37	1.51	1.68										
250	1.13	1.19	1.24	1.32	1.44	1.57	1.82									
375	1.06	1.08	1.10	1.15	1.21	1.29	1.42	1.78								
500	1.04	1.05	1.06	1.08	1.13	1.17	1.25	1.49	1.73							
625		1.04	1.04	1.06	1.08	1.11	1.17	1.33	1.51	1.72						
750				1.04	1.05	1.07	1.12	1.24	1.37	1.54	1.70	1.80				
875					1.06	1.06	1.09	1.17	1.28	1.42	1.54	1.64	1.78			
1000						1.05	1.07	1.14	1.22	1.33	1.44	1.52	1.64	1.75		
1250							1.05	1.09	1.15	1.22	1.29	1.35	1.44	1.53	1.64	1.83
1500								1.06	1.11	1.16	1.21	1.25	1.31	1.39	1.46	1.62
1750								1.05	1.08	1.12	1.17	1.19	1.24	1.30	1.35	1.47
2000									1.06	1.09	1.13	1.15	1.19	1.24	1.28	1.38
2250									1.04	1.06	1.10	1.13	1.16	1.19	1.23	1.32
2500										1.05	1.08	1.10	1.13	1.15	1.19	1.23

Note: Common practice would normally limit the lower end of the table to 5% of additional heat and above 80% an engineer would likely opt for two (2) parallel cables.

R260105-30   Electrical Demolition (Removal for Replacement)

The purpose of this reference number is to provide a guide to users for electrical "removal for replacement" by applying the rule of thumb: 1/3 of new installation time (typical range from 20% to 50%) for removal. Remember to use reasonable judgment when applying the suggested percentage factor. For example:

Contractors have been requested to remove an existing fluorescent lighting fixture and replace with a new fixture utilizing energy saver lamps and electronic ballast:

In order to fully understand the extent of the project, contractors should visit the job site and estimate the time to perform the renovation work in accordance with applicable national, state and local regulation and codes.

The contractor may need to add extra labor hours to his estimate if he discovers unknown concealed conditions such as: contaminated asbestos ceiling, broken acoustical ceiling tile and need to repair, patch and touch-up paint all the damaged or disturbed areas, tasks normally assigned to general contractors. In addition, the owner could request that the contractors salvage the materials removed and turn over the materials to the owner or dispose of the materials to a reclamation station. The normal removal item is 0.5 labor hour for a lighting fixture and 1.5 labor-hours for new installation time. Revise the estimate times from 2 labor-hours work up to a minimum 4 labor-hours work for just fluorescent lighting fixture.

For removal of large concentrations of lighting fixtures in the same area, apply an "economy of scale" to reduce estimating labor hours.

R260519-20   Armored Cable

Armored Cable – Quantities are taken off in the same manner as wire.

Bx Type Cable – Productivities are based on an average run of 15 m before terminating at a box fixture, etc. Each 15 m section includes field preparation of (2) ends with hacksaw, identification and tagging of wire. Set up is open coil type without reels attaching cable to snake and pulling across a suspended ceiling or open face wood or steel studding, price does not include drilling of studs.

Cable in Tray – Productivities are based on an average run of 30 m with set up of pulling equipment for (2) 90° bends, attaching cable to pull-in means, identification and tagging of wires, set up of reels. Wire termination to breakers equipment, etc., are not included.

Job Conditions – Productivities are based on new construction to a height of 4.5 m using rolling staging in an unobstructed area. Material staging is assumed to be within 30 m of work being performed.

R260519-80   Undercarpet Systems

Takeoff Procedure for Power Systems: List components for each fitting type, tap, splice, and bend on your quantity takeoff sheet. Each component must be priced separately. Start at the power supply transition fittings and survey each circuit for the components needed. List the quantities of each component under a specific circuit number. Use the floor plan layout scale to get cable footage.

Reading across the list, combine the totals of each component in each circuit and list the total quantity in the last column. Calculate approximately 5% for scrap for items such as cable, top shield, tape, and spray adhesive. Also provide for final variations that may occur on-site.

Suggested guidelines are:
1. Equal amounts of cable and top shield should be priced.
2. For each roll of cable, price a set of cable splices.
3. For every meter of cable, price 2.5 m of hold-down tape.
4. For every 3 rolls of hold-down tape, price 1 can of spray adhesive.

Adjust final figures wherever possible to accommodate standard packaging of the product. This information is available from the distributor.

Each transition fitting requires:
1. 1 base
2. 1 cover
3. 1 transition block

Each floor fitting requires:
1. 1 frame/base kit
2. 1 transition block
3. 2 covers (duplex/blank)

Each tap requires:
1. 1 tap connector for each conductor
2. 1 pair insulating patches
3. 2 top shield connectors

Each splice requires:
1. 1 splice connector for each conductor
2. 1 pair insulating patches
3. 3 top shield connectors

Each cable bend requires:
1. 2 top shield connectors

Each cable dead end (outside of transition block) requires:
1. 1 pair insulating patches

Labor does not include:
1. Patching or leveling uneven floors.
2. Filling in holes or removing projections from concrete slabs.
3. Sealing porous floors.
4. Sweeping and vacuuming floors.
5. Removal of existing carpeting.
6. Carpet square cut-outs.
7. Installation of carpet squares.

Takeoff Procedures for Telephone Systems: After reviewing floor plans identify each transition. Number or letter each cable run from that fitting.

Start at the transition fitting and survey each circuit for the components needed. List the cable type, terminations, cable length, and floor fitting type under the specific circuit number. Use the floor plan layout scale to get the cable footage. Add some extra length (next higher increment of 1.5 m) to preconnectorized cable.

Transition fittings require:
1. 1 base plate
2. 1 cover
3. 1 transition block

Floor fittings require:
1. 1 frame/base kit
2. 2 covers
3. Modular jacks

Reading across the list, combine the list of components in each circuit and list the total quantity in the last column. Calculate the necessary scrap factors for such items as tape, bottom shield and spray adhesive. Also provide for final variations that may occur on-site.

Adjust final figures whenever possible to accommodate standard packaging. Check that items such as transition fittings, floor boxes, and floor fittings that are to utilize both power and telephone have been priced as combination fittings, so as to avoid duplication.

Make sure to include marking of floors and drilling of fasteners if fittings specified are not the adhesive type.

Labor does not include:
1. Conduit or raceways before transition of floor boxes
2. Telephone cable before transition boxes
3. Terminations before transition boxes
4. Floor preparation as described in power section

Be sure to include all cable folds when pricing labor.

Takeoff Procedure for Data Systems: Start at the transition fittings and take off quantities in the same manner as the telephone system, keeping in mind that data cable does not require top or bottom shields.

The data cable is simply cross-taped on the cable run to the floor fitting.

Data cable can be purchased in either bulk form in which case coaxial connector material and labor must be priced, or in preconnectorized cut lengths.

Data cable cannot be folded and must be notched at 25 mm intervals. A count of all turns must be added to the labor portion of the estimate. (Note: Some manufacturers have prenotched cable.)

Notching required:
1. 90 degree turn requires 8 notches per side.
2. 180 degree turn requires 16 notches per side.

Floor boxes, transition boxes, and fittings are the same as described in the power and telephone procedures.

Since undercarpet systems require special hand tools, be sure to include this cost in proportion to number of crews involved in the installation.

Job Conditions: Productivity is based on new construction in an unobstructed area. Staging area is assumed to be within 60 m of work being performed.

R260519-90 Wire

Wire quantities are taken off by either measuring each cable run or by extending the conduit and raceway quantities times the number of conductors in the raceway. Ten percent should be added for waste and tie-ins.

Price per meter of wire includes:
1. Set up wire coils or spools on racks
2. Attaching wire to pull-in means
3. Measuring and cutting wire
4. Pulling wire into a raceway
5. Identifying and tagging

Price does not include:
1. Connections to breakers, panelboards or equipment
2. Splices

Job Conditions: Productivity is based on new construction to a height of 4.5 m using rolling staging in an unobstructed area. Material staging is assumed to be within 30 m of work being performed.

Economy of Scale: If more than three wires at a time are being pulled, deduct the following percentages from the labor of that grouping:

4-5 wires	25%
6-10 wires	30%
11-15 wires	35%
over 15	40%

If a wire pull is less than 30 m in length and is interrupted several times by boxes, lighting outlets, etc., it may be necessary to add the following lengths to each wire being pulled:

Junction box to junction box	0.6 m
Lighting panel to junction box	1.8 m
Distribution panel to sub panel	2.4 m
Switchboard to distribution panel	3.6 m
Switchboard to motor control center	6 m
Switchboard to cable tray	12 m

Measure of Drops and Riser: It is important when taking off wire quantities to include the wire for drops to electrical equipment. If heights of electrical equipment are not clearly stated, use the following guide:

	Bottom A.F.F.	Top A.F.F.	Inside Cabinet
Safety switch to 100A	1500 mm	1800 mm	600 mm
Safety switch 400 to 600A	1200 mm	1800 mm	900 mm
100A panel 12 to 30 circuit	1200 mm	1800 mm	900 mm
42 circuit panel	900 mm	1800 mm	1200 mm
Switch box	900 mm	1050 mm	300 mm
Switchgear	0 mm	2400 mm	2400 mm
Motor control centers	0 mm	2400 mm	2400 mm
Transformers - wall mount	1200 mm	2400 mm	600 mm
Transformers - floor mount	0 mm	3600 mm	1200 mm

ELECTRICAL

R260519-91 Maximum Circuit Length (approximate) for Various Power Requirements Assuming THW, Copper Wire @ 75° C, Based Upon a 4% Voltage Drop

Maximum Circuit Length: Table R260519-91 indicates typical maximum installed length a circuit can have and still maintain an adequate voltage level at the point of use. The circuit length is similar to the conduit length.

If the circuit length for an ampere load and a copper wire size exceeds the length obtained from Table R260519-91, use the next largest wire size to compensate voltage drop.

Example: A 130 ampere load at 480 volts, 3 phase, 3 wire with No. 1 wire can be run a maximum of 167 m and provide satisfactory operation. If the same load is to be wired at the end of a 188 m circuit, then a larger wire must be used.

		Maximum Circuit Length in Meters				
		2 Wire, 1 Phase		3 Wire, 3 Phase		
Amperes	Wire Size	120V	240V	240V	480V	600V
15	14*	15	32	36	72	90
	14	15	30	36	71	89
20	12*	18	38	44	87	108
	12	18	36	42	84	105
30	10*	20	39	47	92	114
	10	20	39	45	90	113
50	8	18	38	44	86	107
65	6	23	45	53	104	131
85	4	27	56	63	128	159
115	2	33	65	75	150	186
130	1	36	72	83	167	207
150	1/0	39	78	92	182	228
175	2/0	42	86	99	197	246
200	3/0	47	95	108	218	272
230	4/0	51	104	119	239	297
255	250	56	110	126	254	317
285	300	59	119	137	273	342
310	350	63	126	146	293	366
380	500	74	147	170	339	425

*Solid Conductor

Note: The circuit length is the one-way distance between the origin and the load.

REFERENCE TABLES

R260519-92   Minimum Copper and Aluminum Wire Size Allowed for Various Types of Insulation

Minimum Wire Sizes

Amperes	Copper		Aluminum		Amperes	Copper		Aluminum	
	THW THWN or XHHW	THHN XHHW *	THW XHHW	THHN XHHW *		THW THWN or XHHW	THHN XHHW *	THW XHHW	THHN XHHW *
15A	#14	#14	#12	#12	195	3/0	2/0	250kcmil	4/0
20	#12	#12	#10	#10	200	3/0	3/0	250kcmil	4/0
25	#10	#10	#10	#10	205	4/0	3/0	250kcmil	4/0
30	#10	#10	# 8	# 8	225	4/0	3/0	300kcmil	250kcmil
40	# 8	# 8	# 8	# 8	230	4/0	4/0	300kcmil	250kcmil
45	# 8	# 8	# 6	# 8	250	250kcmil	4/0	350kcmil	300kcmil
50	# 8	# 8	# 6	# 6	255	250kcmil	4/0	400kcmil	300kcmil
55	# 6	# 8	# 4	# 6	260	300kcmil	4/0	400kcmil	350kcmil
60	# 6	# 6	# 4	# 6	270	300kcmil	250kcmil	400kcmil	350kcmil
65	# 6	# 6	# 4	# 4	280	300kcmil	250kcmil	500kcmil	350kcmil
75	# 4	# 6	# 3	# 4	285	300kcmil	250kcmil	500kcmil	400kcmil
85	# 4	# 4	# 2	# 3	290	350kcmil	250kcmil	500kcmil	400kcmil
90	# 3	# 4	# 2	# 2	305	350kcmil	300kcmil	500kcmil	400kcmil
95	# 3	# 4	# 1	# 2	310	350kcmil	300kcmil	500kcmil	500kcmil
100	# 3	# 3	# 1	# 2	320	400kcmil	300kcmil	600kcmil	500kcmil
110	# 2	# 3	1/0	# 1	335	400kcmil	350kcmil	600kcmil	500kcmil
115	# 2	# 2	1/0	# 1	340	500kcmil	350kcmil	600kcmil	500kcmil
120	# 1	# 2	1/0	1/0	350	500kcmil	350kcmil	700kcmil	500kcmil
130	# 1	# 2	2/0	1/0	375	500kcmil	400kcmil	700kcmil	600kcmil
135	1/0	# 1	2/0	1/0	380	500kcmil	400kcmil	750kcmil	600kcmil
150	1/0	# 1	3/0	2/0	385	600kcmil	500kcmil	750kcmil	600kcmil
155	2/0	1/0	3/0	3/0	420	600kcmil	500kcmil		700kcmil
170	2/0	1/0	4/0	3/0	430		500kcmil		750kcmil
175	2/0	2/0	4/0	3/0	435		600kcmil		750kcmil
180	3/0	2/0	4/0	4/0	475		600kcmil		

*Dry Locations Only

Notes:
1. Size #14 to 4/0 is in AWG units (American Wire Gauge).
2. Size 250 to 750 is in kcmil units (Thousand Circular Mils).
3. Use next higher ampere value if exact value is not listed in table.
4. For loads that operate continuously increase ampere value by 25% to obtain proper wire size.
5. Refer to Table R260519-91 for the maximum circuit length for the various size wires.
6. Table R260519-92 has been written for estimating purpose only, based on ambient temperature of 30 °C (86° F); for ambient temperature other than 30 °C (86° F), ampacity correction factors will be applied.

R260519-93 Metric Equivalent, Wire

U.S. vs. European Wire – Approximate Equivalents			
United States		European	
Size AWG or kcmil	Area Cir. Mils.(cmil) mm²	Size mm²	Area Cir. Mils.
18	1620/.82	.75	1480
16	2580/1.30	1.0	1974
14	4110/2.08	1.5	2961
12	6530/3.30	2.5	4935
10	10,380/5.25	4	7896
8	16,510/8.36	6	11,844
6	26,240/13.29	10	19,740
4	41,740/21.14	16	31,584
3	52,620/26.65	25	49,350
2	66,360/33.61	–	–
1	83,690/42.39	35	69,090
1/0	105,600/53.49	50	98,700
2/0	133,100/67.42	–	–
3/0	167,800/85.00	70	138,180
4/0	211,600/107.19	95	187,530
250	250,000/126.64	120	236,880
300	300,000/151.97	150	296,100
350	350,000/177.30	–	–
400	400,000/202.63	185	365,190
500	500,000/253.29	240	473,760
600	600,000/303.95	300	592,200
700	700,000/354.60	–	–
750	750,000/379.93	–	–

R260519-94 Size Required and Weight (kg per 1000 m) of Aluminum and Copper THW Wire by Ampere Load

Amperes	Copper Size	Aluminum Size	Copper Weight	Aluminum Weight
15	14	12	36	16
20	12	10	49	25
30	10	8	71	58
45	8	6	115	77
65	6	4	167	107
85	4	2	249	150
100	3	1	305	202
115	2	1/0	375	241
130	1	2/0	482	289
150	1/0	3/0	591	347
175	2/0	4/0	731	420
200	3/0	250	905	516
230	4/0	300	1121	600
255	250	400	1338	762
285	300	500	1589	923
310	350	500	1835	923
335	400	600	2077	1149
380	500	750	2579	1415

R260526-80 Grounding

Grounding – When taking off grounding systems, identify separately the type and size of wire.

Example:

Bare copper & size

Bare aluminum & size

Insulated copper & size

Insulated aluminum & size

Count the number of ground rods and their size.

Example:

1. 2.4 m grounding rod – 18 mm dia. 20 Ea.
2. 3.0 m grounding rod – 18 mm dia. 12 Ea.
3. 4.5 m grounding rod – 20 mm dia. 4 Ea.

Count the number of connections; the size of the largest wire will determine the productivity.

Example:

Braze a #2 wire to a #4/0 cable

The 4/0 cable will determine the L.H. and cost to be used.

Include individual connections to:

1. Ground rods
2. Building steel
3. Equipment
4. Raceways

Price does not include:

1. Excavation
2. Backfill
3. Sleeves or raceways used to protect grounding wires
4. Wall penetrations
5. Floor cutting
6. Core drilling

Job Conditions: Productivity is based on a ground floor area, using cable reels in an unobstructed area. Material staging area assumed to be within 30 m of work being performed.

R260533-20 Conduit to 4.5 m High

List conduit by quantity, size, and type. Do not deduct for lengths occupied by fittings, since this will be allowance for scrap. Example:

  A. Aluminum — size
  B. Rigid Galvanized — size
  C. Steel Intermediate (IMC) — size
  D. Rigid Steel, plastic coated 20 Mil — size
  E. Rigid Steel, plastic coated 40 Mil — size
  F. Electric Metallic Tubing (EMT) — size
  G. PVC Schedule 40 — size

Types (A) thru (E) listed above contain the following per 30 m:
  1. (11) Threaded couplings
  2. (11) Beam type hangers
  3. (2) Factory sweeps
  4. (2) Fiber bushings
  5. (4) Locknuts
  6. (2) Field threaded pipe terminations
  7. (2) Removal of concentric knockouts

Type (F) contains per 30 m:
  1. (11) Set screw couplings
  2. (11) Beam clamps
  3. (2) Field bends on 15 mm and 20 mm diameter
  4. (2) Factory sweeps for 25 mm and above
  5. (2) Set screw steel connectors
  6. (2) Removal of concentric knockouts

Type (G) contains per 30 m:
  1. (11) Field cemented couplings
  2. (34) Beam clamps

  3. (2) Factory sweeps
  4. (2) Adapters
  5. (2) Locknuts
  6. (2) Removal of concentric knockouts

Labor-hours for all conduit to 4.5 m high include:
  1. Unloading by hand
  2. Hauling by hand to an area up to 60 m from loading dock
  3. Set up of rolling staging
  4. Installation of conduit and fittings, as described in Conduit models (A) thru (G)

Not included in the material and labor are:
  1. Staging rental or purchase
  2. Structural Modifications
  3. Wire
  4. Junction boxes
  5. Fittings in excess of those described in conduit models (A) thru (G)
  6. Painting of conduit

Fittings

Only those fittings listed above are included in the per m totals, although they should be listed separately from conduit lengths, without prices, to insure proper quantities for material procurement.

If the fittings required exceed the quantities included in the model conduit runs, then material and labor costs must be added to the difference. If actual needs per 30 m of conduit are: (2) sweeps, (4) LB's and (1) field bend.

Then, in this case (4) LB's and (1) field bend must be priced additionally.

R260533-21 Hangers

It is sometimes desirable to substitute an alternate style of hanger if the desired support is not the type described in the conduit models.

One approach is the substitution method:
  1. Find the cost of the type hanger described in the conduit model
  2. Calculate the cost of the desired type hanger (it may be necessary to calculate individual components such as drilling, expansion shields, etc.)
  3. Calculate the cost difference (delta) between the two types of hangers
  4. Multiply the cost delta by the number of hangers in the model
  5. Divide the total delta cost for hangers in the model by the length of the model to find the delta cost for that model, per m
  6. Modify the given unit costs per m for the model by the delta cost per m

Another approach to hanger configurations would be to start with the conduit only and add all the supports and any other items as separate lines. This procedure is most useful if the project involves racking many runs of conduit on a single hanger, for instance a trapeze type hanger.

Example: Five (5) 50 mm RGS conduits, 15 m each, are to be run on trapeze hangers from one pull box to another. The run includes one 90° bend.

  1. List the hanger's components to create an assembly cost for each 600 mm wide trapeze.
  2. List the components for the 15 m conduit run, noting that 6 trapeze supports will be required.

Job Conditions: Productivities are based on new construction to 4.5 m high, using scaffolding in an unobstructed area. Material storage is assumed to be within 30 m of work being performed.

Add to labor for elevated installations:

4.6 m to 6.1 m High 10%		9.1 m to 10.7 m High 30%
6.1 m to 7.6 m High 20%		10.7 m to 12.2 m High 35%
7.6 m to 9.1 m High 25%		Over 12.2 m High 40%

Add these percentages to the per m labor cost, but not to fittings. Add these percentages only to quantities exceeding the different height levels, not the total conduit quantities.

Linear foot price for labor does not include penetrations in walls or floors, and must be added to the estimate

R260533-22 Conductors in Conduit

Table below lists maximum number of conductors for various sized conduit using THW, TW or THWN insulations.

Copper Wire Size	15 mm (½")			20 mm (¾")			25 mm (1")			32 mm (1-¼")			40 mm (1-½")			50 mm (2")			65 mm (2-½")			80 mm (3")		90 mm (3-½")		100 mm (4")	
	TW	THW	THWN	TW	THW	THWN	TW	THW	THWN	TW	THW	THWN	TW	THW	THWN	TW	THW	THWN	TW	THW	THWN	THW	THWN	THW	THWN	THW	THWN
#14	9	6	13	15	10	24	25	16	39	44	29	69	60	40	94	99	65	154	142	93		143		192			
#12	7	4	10	12	8	18	19	13	29	35	24	51	47	32	70	78	53	114	111	76	164	117		157			
#10	5	4	6	9	6	11	15	11	18	26	19	32	36	26	44	60	43	73	85	61	104	95	160	127		163	
#8	2	1	3	4	3	5	7	5	9	12	10	16	17	13	22	28	22	36	40	32	51	49	79	66	106	85	136
#6		1	1		2	4		4	6		7	11		10	15		16	26		23	37	36	57	48	76	62	98
#4		1	1		1	2		3	4		5	7		7	9		12	16		17	22	27	35	36	47	47	60
#3		1	1		1	1		2	3		4	6		6	8		10	13		15	19	23	29	31	39	40	51
#2		1	1		1	1		2	3		4	5		5	7		9	11		13	16	20	25	27	33	34	43
#1					1	1		1	1		3	3		4	5		6	8		9	12	14	18	19	25	25	32
1/0					1	1		1	1		2	3		3	4		5	7		8	10	12	15	16	21	21	27
2/0					1	1		1	1		1	2		3	3		5	6		7	8	10	13	14	17	18	22
3/0					1	1		1	1		1	1		2	3		4	5		6	7	9	11	12	14	15	18
4/0						1		1	1		1	1		1	2		3	4		5	6	7	9	10	12	13	15
250 kcmil								1	1		1	1		1	1		2	3		4	4	6	7	8	10	10	12
300								1	1		1	1		1	1		2	3		3	4	5	6	7	8	9	11
350												1		1	1		1	2		3	3	4	5	6	7	8	9
400											1	1		1	1		1	1		2	3	4	5	5	6	7	8
500											1	1		1	1		1	1		1	2	3	4	4	5	6	7
600												1		1	1		1	1		1	1	3	3	4	4	5	5
700														1	1		1	1		1	1	2	3	3	4	4	5
750														1	1		1	1		1	1	2	2	3	3	4	4

Notes:
1. All conduit sizes are industry standard conduit designated by their nominal millimeter (mm) diameter equivalent.
2. kcmil = 0.5067 mm^2 for wire soft conversion approximate equilvalent.

R260533-23 Metric Equivalent, Conduit

U.S. vs. European Conduit – Approximate Equivalents			
United States		European	
Trade Size	Inside Diameter Inch/mm	Trade Size	Inside Diameter mm
½	.622/15.8	11	16.4
¾	.824/20.9	16	19.9
1	1.049/26.6	21	25.5
1¼	1.380/35.0	29	34.2
1½	1.610/40.9	36	44.0
2	2.067/52.5	42	51.0
2½	2.469/62.7		
3	3.068/77.9		
3½	3.548/90.12		
4	4.026/102.3		
5	5.047/128.2		
6	6.065/154.1		

REFERENCE TABLES

R260533-24 Conduit Weight Comparisons (kg per meter) Empty

Type	15 mm	20 mm	25 mm	32 mm	40 mm	50 mm	65 mm	80 mm	90 mm	100 mm	125 mm	150 mm
Rigid Aluminum	0.4	0.6	0.8	1.1	1.3	1.8	2.8	3.7	4.4	5.2	7.1	9.4
Rigid Steel	1.2	1.6	2.3	3.0	3.7	4.9	7.9	10.2	12.4	14.5	19.6	26.0
Intermediate Steel (IMC)	0.9	1.2	1.7	2.2	2.7	3.6	6.0	7.3	8.5	9.5		
Electrical Metallic Tubing (EMT)	0.4	0.7	1.0	1.4	1.7	2.1	3.2	3.9	5.4	5.8		
Polyvinyl Chloride, Schedule 40	0.2	0.3	0.5	0.6	0.8	1.0	1.6	2.1	2.5	3.0	4.0	5.2
Polyvinyl Chloride Encased Burial						0.6		1.0	1.3	1.6	2.2	3.0
Fibre Duct Encased Burial						1.9		2.44	2.7	3.1	6.0	7.6
Fibre Duct Direct Burial						2.2		3.7	4.5	5.3		
Transite Encased Burial						2.4		3.6	4.3	4.9	6.7	8.2
Transite Direct Burial						3.3		4.6		6.0	8.0	9.5

R260533-25 Conduit Weight Comparisons (kg per meter) with Maximum Cable Fill*

Type	15 mm	20 mm	25 mm	32 mm	40 mm	50 mm	65 mm	80 mm	90 mm	100 mm	125 mm	150 mm
Rigid Galvanized Steel (RGS)	7.5	2.1	3.5	5.3	6.8	10.7	15.2	21.6	26.1	32.0	45.9	64.7
Intermediate Steel (IMC)	1.3	1.7	2.8	4.4	5.6	9.1	13.2	18.8	22.4	27.3		
Electrical Metallic Tubing (EMT)	0.8	1.7	2.7	4.4	5.5	6.6	9.6	13.9	18.1	22.9		

*Conduit & Heaviest Conductor Combination

R260533-60 Wireway

When "taking off" Wireway, list by size and type.

Example:
1. Screw cover, unflanged + size
2. Screw cover, flanged + size
3. Hinged cover, flanged + size
4. Hinged cover, unflanged + size

Each 3 m length on Wireway contains:
1. 3 m of cover either screw or hinged type
2. (1) Coupling or flange gasket
3. (1) Wall type mount

All fittings must be priced separately.

Substitution of hanger types is done the same as described in R260533-21, "HANGERS".

Labor-hours for wireway include:
1. Unloading by hand
2. Hauling by hand up to 30 m from loading dock
3. Measuring and marking
4. Mounting wall bracket using (2) anchor type lead fasteners
5. Installing wireway on brackets, to 4.5 m high (For higher elevations use factors in R260533-20)

Job Conditions: Productivity is based on new construction, to a height of 4.5 m using rolling staging in an unobstructed area.

Material staging area is assumed to be within 30 m of work being performed.

R260533-65 Outlet Boxes

Outlet boxes should be included on the same takeoff sheet as branch piping or devices to better explain what is included in each circuit.

Each unit price in this section is a stand alone item and contains no other component unless specified. For example, to estimate a duplex outlet, components that must be added are:

1. 100 mm square box
2. 100 mm plaster ring
3. Duplex receptacle
4. Device cover

The method of mounting outlet boxes is (2) plastic shield fasteners.

Outlet boxes plastic, labor-hours include:
1. Marking box location on wood studding
2. Mounting box

Economy of Scale – For large concentrations of plastic boxes in the same area deduct the following percentages from labor-hour totals:

1	to	10	0%
11	to	25	20%
26	to	50	25%
51	to	100	30%
	over	100	35%

Note: It is important to understand that these percentages are not used on the total job quantities, but only areas where concentrations exceed the levels specified.

R260533-70 Pull Boxes and Cabinets

List cabinets and pull boxes by NEMA type and size.

Example:	**TYPE**	**SIZE**
	NEMA 1	150 mm x 150 mm x 100 mm
	NEMA 3R	155 mm x 150 mm x 100 m

Labor-hours for wall mount (indoor or outdoor) installations include:
1. Unloading and uncrating
2. Handling of enclosures up to 60 m from loading dock using a dolly or pipe rollers
3. Measuring and marking
4. Drilling (4) anchor type lead fasteners using a hammer drill
5. Mounting and leveling boxes

Note: A plywood backboard is not included.

Labor-hours for ceiling mounting include:
1. Unloading and uncrating
2. Handling boxes up to 30 mm from loading dock

3. Measuring and marking
4. Drilling (4) anchor type lead fasteners using a hammer drill
5. Installing and leveling boxes to a height of 4.5 m using rolling staging

Labor-hours for free standing cabinets include:
1. Unloading and uncrating
2. Handling of cabinets up to 60 m from loading dock using a dolly or pipe rollers
3. Marking of floor
4. Drilling (4) anchor type lead fasteners using a hammer drill
5. Leveling and shimming

Labor-hours for telephone cabinets include:
1. Unloading and uncrating
2. Handling cabinets up to 60 m using a dolly or pipe rollers
3. Measuring and marking
4. Mounting and leveling, using (4) lead anchor type fasteners

R260533-75 Weight Comparisons of Common Size Cast Boxes in Lbs.

Size NEMA 4 or 9	Cast Iron	Cast Aluminum	Size NEMA 7	Cast Iron	Cast Aluminum
152 mm x 152 mm x 152 mm	7.7	3.2	152 mm x 152 mm x 152 mm	18	6.8
203 mm x 152 mm x 152 mm	9.5	3.6	203 mm x 152 mm x 152 mm	23	8.6
254 mm x 152 mm x 152 mm	10.4	4.1	254 mm x 152 mm x 152 mm	25	9.5
305 mm x 305 mm x 152 mm	24	9.1	305 mm x 305 mm x 152 mm	45	17
406 mm x 406 mm x 152 mm	44	16	406 mm x 406 mm x 152 mm	64	24
508 mm x 508 mm x 152 mm	60	23	508 mm x 508 mm x 152 mm	82	30
610 mm x 457 mm x 203 mm	68	25	610 mm x 457 mm x 203 mm	113	42
610 mm x 610 mm x 254 mm	108	40	610 mm x 610 mm x 254 mm	162	60
762 mm x 610 mm x 305 mm	147	54	762 mm x 610 mm x 254 mm	215	80
914 mm x 914 mm x 305 mm	227	84	762 mm x 610 mm x 305 mm	231	86

REFERENCE TABLES

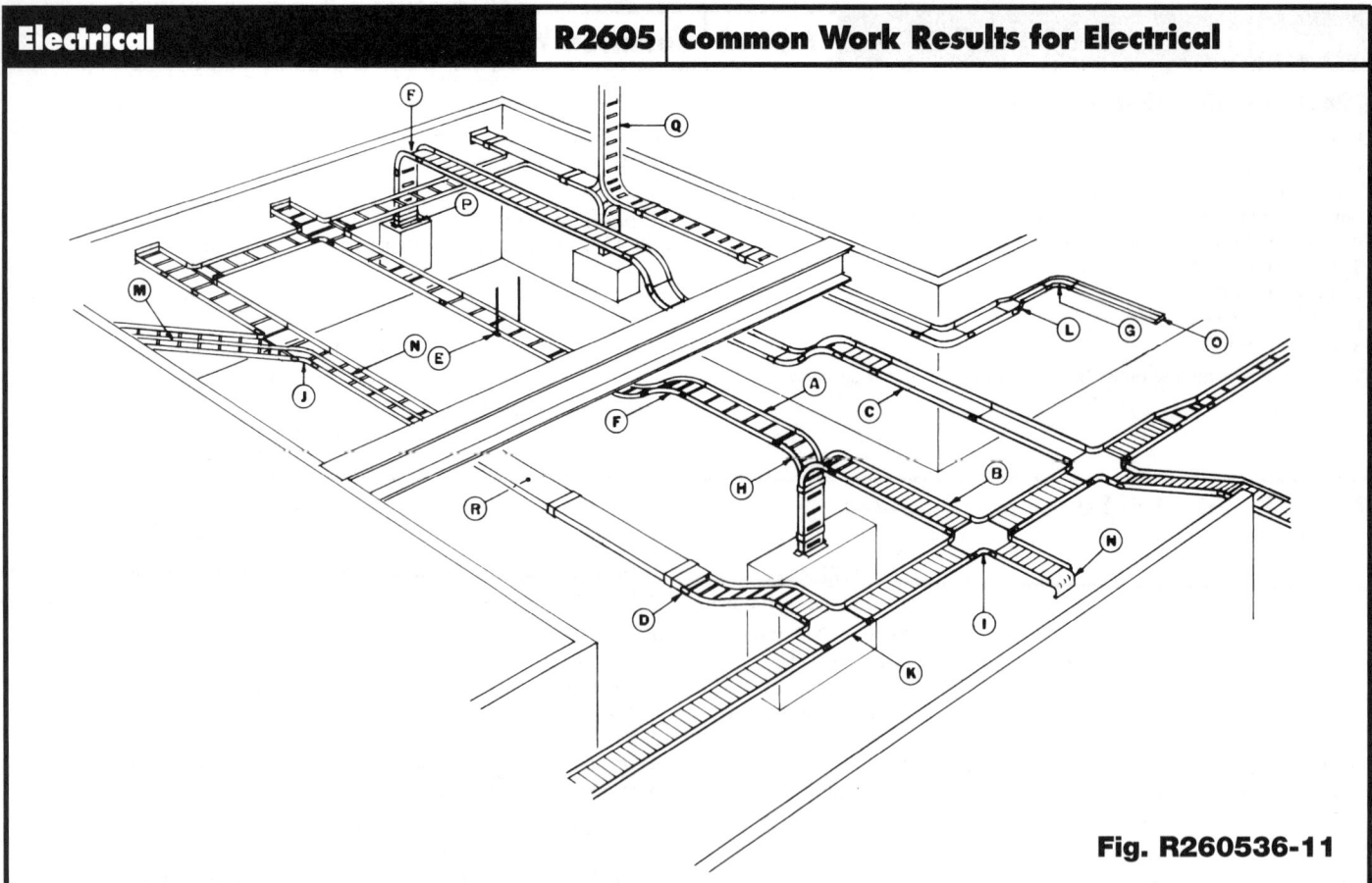

Fig. R260536-11

R260536-10   Cable Tray

Cable Tray - When taking off cable tray it is important to identify separately the different types and sizes involved in the system being estimated. (Fig. R260536-11)

    A. – Ladder Type, galvanized or aluminum
    B. – Trough Type, galvanized or aluminum
    C. – Solid Bottom, galvanized or aluminum

The unit of measure is calculated in meters do not deduct from this distance any length occupied by fittings, this will be the only allowance for scrap. Be sure to include all vertical drops to panels, switch gear, etc.

Hangers – Included in the length of cable tray is

    D. – 1 – Pair of connector plates per 3.6 m.
    E. – 1 – Pair clamp type hangers and 1.2 m of 10 mm threaded rod per
           3.6 m.

Not included are structural supports, which must be priced in addition to the hangers.

Fittings – Identify separately the different types of fittings
  1.) Ladder Type, galvanized or aluminum
  2.) Trough Type, galvanized or aluminum
  3.) Solid Bottom Type, galvanized or aluminum

The configuration, radius and rung spacing must also be listed. The unit of measure is "Ea."

    F. – Elbow, vertical Ea.
    G. – Elbow, horizontal Ea.
    H. – Tee, vertical Ea.
    I. – Cross, horizontal Ea.
    J. – Wye, horizontal Ea.
    K. – Tee, horizontal Ea.
    L. – Reducing fitting Ea.

Depending on the use of the system other examples of units which must be included are:

    M. – Divider strip Meter
    N. – Drop-outs Ea.
    O. – End caps Ea.

    P. – Panel connectors Ea.

Wire and cable are not included and should be taken off separately, see Unit Price sections.

Job Conditions – Unit prices are based on a new installation to a work plane of 4.5 m using rolling staging.

Add to labor for elevated installations

4.5 m to 6.0 m High	10%
6.1 m to 7.5 m High	20%
7.6 m to 9.0 m High	25%
9.1 m to 10.5 m High	30%
10.6 m to 12.0 m High	35%
Over 12.0 m High	40%

Add these percentages for cable tray length totals but not to fittings. Add percentages to only those quantities that fall in the different elevations, in other words, if the total quantity of cable tray is 60 m but only 20 m is above 4.5 m then the 10% is added to the 20 m only.

Cable tray costs do not include penetrations through walls and floors which must be added to the estimate.

Cable Tray Covers

Covers – Cable tray covers are taken off in the same manner as the tray itself, making distinctions as to the type of cover. (Fig. R260536-11)

    Q. – Vented, galvanized or aluminum
    R. – Solid, galvanized or aluminum

Cover configurations are taken off separately noting type, specific radius and widths.

Note: Care should be taken to identify from plans and specifications exactly what is being covered. In many systems only vertical fittings are covered to retain wire and cable.

R260539-30   Conduit In Concrete Slab

List conduit by quantity, size and type.

Example:
  A. Rigid galvanized steel + size
  B. P.V.C. + size

Rigid galvanized steel (A) contains per meter:
1. (20) Ties to slab reinforcing
2. (11) Threaded steel couplings
3. (2) Factory sweeps
4. (2) Field threaded conduit terminations

  5. (2) Fiber bushings + locknuts
  6. (2) Removal of concentric knockouts

P.V.C. (B) contains per meter:
1. (20) Ties to slab reinforcing
2. (11) Field cemented couplings
3. (2) Factory sweeps
4. (2) Adapters
5. (2) Removal of concentric knockouts

R260539-40   Conduit In Trench

Conduit in trench is galvanized steel and contains per meter:
1. (11) Threaded couplings
2. (2) Factory sweeps
3. (2) Fiber bushings + (4) locknuts
4. (2) Field threaded conduit terminations
5. (2) Removal of concentric knockouts

Note:

Conduit in Unit Price sections do not include:
1. Floor cutting
2. Excavation or backfill
3. Grouting or patching

Conduit fittings in excess of those listed in the above Conduit model must be added. (Refer to R260533-20 for Procedure example.)

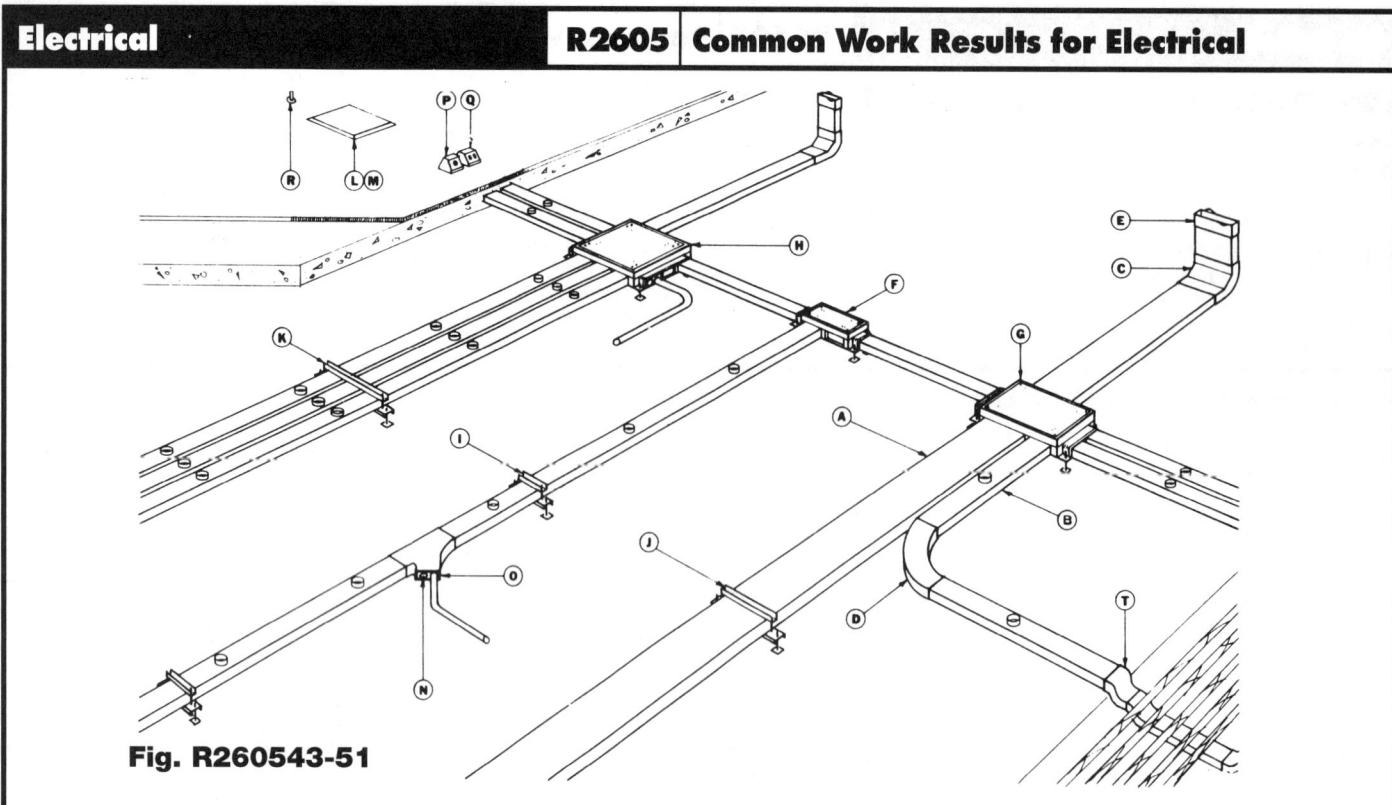

Fig. R260543-51

R260543-50   Underfloor Duct

When pricing Underfloor Duct it is important to identify and list each component, since costs vary significantly from one type of fitting to another. Do not deduct boxes or fittings from length of run totals; this will be your allowance for scrap.

The first step is to identify the system as either:

       FIG. R260543-51 Single Level

       FIG. R260543-52 Dual Level

Single Level System

Include on your ''takeoff sheet'' the following unit price items, making sure to distinquish between Standard and Super duct:

  A. Feeder duct (blank) in meters.
  B. Distribution duct (Inserts 600 mm on center) in meters.
  C. Elbows (Vertical) Ea.
  D. Elbows (Horizontal) Ea.
  E. Cabinet connector Ea.
  F. Single duct junction box Ea.
  G. Double duct junction box Ea.
  H. Triple duct junction box Ea.
  I. Support, single cell Ea.
  J. Support, double cell Ea.
  K. Support, triple cell Ea.
  L. Carpet pan Ea.
  M. Terrazzo pan Ea.
  N. Insert to conduit adapter Ea.
  O. Conduit adapter Ea.
  P. Low tension outlet Ea.
  Q. High tension outlet Ea.
  R. Galvanized nipple Ea.
  S. Wire per meter
  T. Offset (Duct type) Ea.

Dual Level System + Labor
see next page

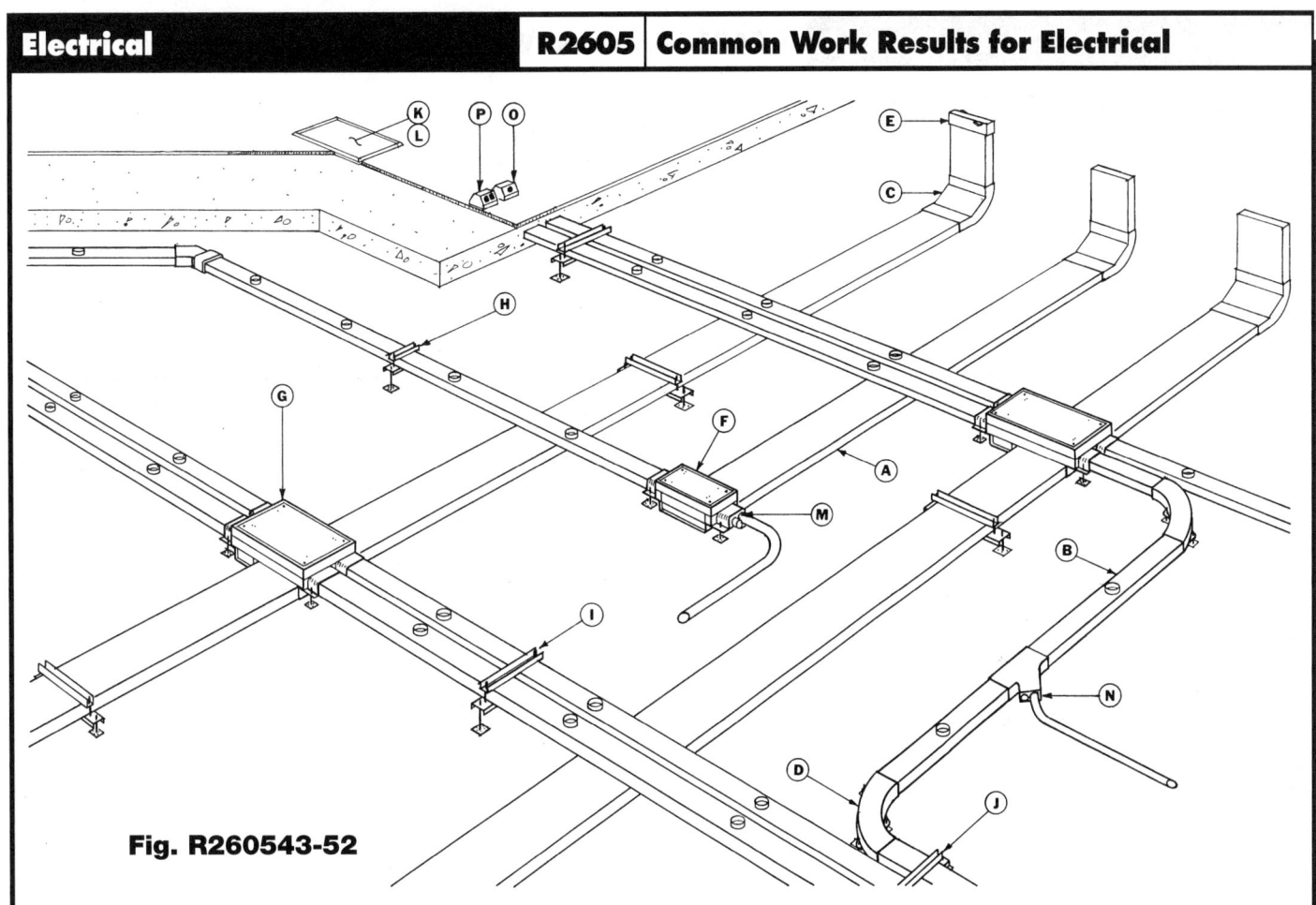

Fig. R260543-52

R260543-50 Underfloor Duct (cont.)
Dual Level

Include the following when "taking off" Dual Level systems:

Distinguish between Standard and Super duct.

- A. Feeder duct (blank) in meters
- B. Distribution duct (Inserts 600 mm on center) in meters
- C. Elbows (Vertical) Ea.
- D. Elbows (Horizontal) Ea.
- E. Cabinet connector Ea.
- F. Single duct, 2 level, junction box Ea.
- G. Double duct, 2 level, junction box Ea.
- H. Support, single cell Ea.
- I. Support, double cell Ea.
- J. Support, triple cell Ea.
- K. Carpet pan Ea.
- L. Terrazzo pan Ea.
- M. Insert to conduit adapter Ea.
- N. Conduit adapter Ea.
- O. Low tension outlet Ea.
- P. High tension outlet Ea.
- Q. Wire per meter

Note: Make sure to include risers in length of run totals. High tension outlets include box, receptacle, covers and related mounting hardware.

Labor-hours for both Single and Dual Level systems include:
1. Unloading and uncrating
2. Hauling up to 60 m from loading dock
3. Measuring and marking
4. Setting raceway and fittings in slab or on grade
5. Leveling raceway and fittings

Labor-hours do not include:
1. Floor cutting
2. Excavation or backfill
3. Concrete pour
4. Grouting or patching
5. Wire or wire pulls
6. Additional outlets after concrete is poured
7. Piping to or from Underfloor Duct

Note: Installation is based on installing up to 45 m of duct. If quantities exceed this, deduct the following percentages:
1. 46 m to 75 m - 10%
2. 76 m to 105 m - 15%
3. 106 m to 150 m - 20%
4. over 150 m - 25%

Deduct these percentages from labor only.

Deduct these percentages from straight sections only.

Do not deduct from fittings or junction boxes.

Job Conditions: Productivity is based on new construction.

Underfloor duct to be installed on first three floors.

Material staging area within 30 m of work being performed.

Area unobstructed and duct not subject to physical damage.

R260580-75   Motor Connections

Motor connections should be listed by size and type of motor.
Included in the material and labor cost is:
1. (2) Flex connectors
2. 450 mm of flexible metallic wireway
3. Wire identification and termination
4. Test for rotation
5. (2) or (3) Conductors

Price does not include:
1. Mounting of motor
2. Disconnect Switch
3. Motor Starter
4. Controls
5. Conduit or wire ahead of flex

Note: When "Taking off" Motor connections, it is advisable to list connections on the same quantity sheet as Motors, Motor Starters and controls.

R260590-05 Typical Overhead Service Entrance

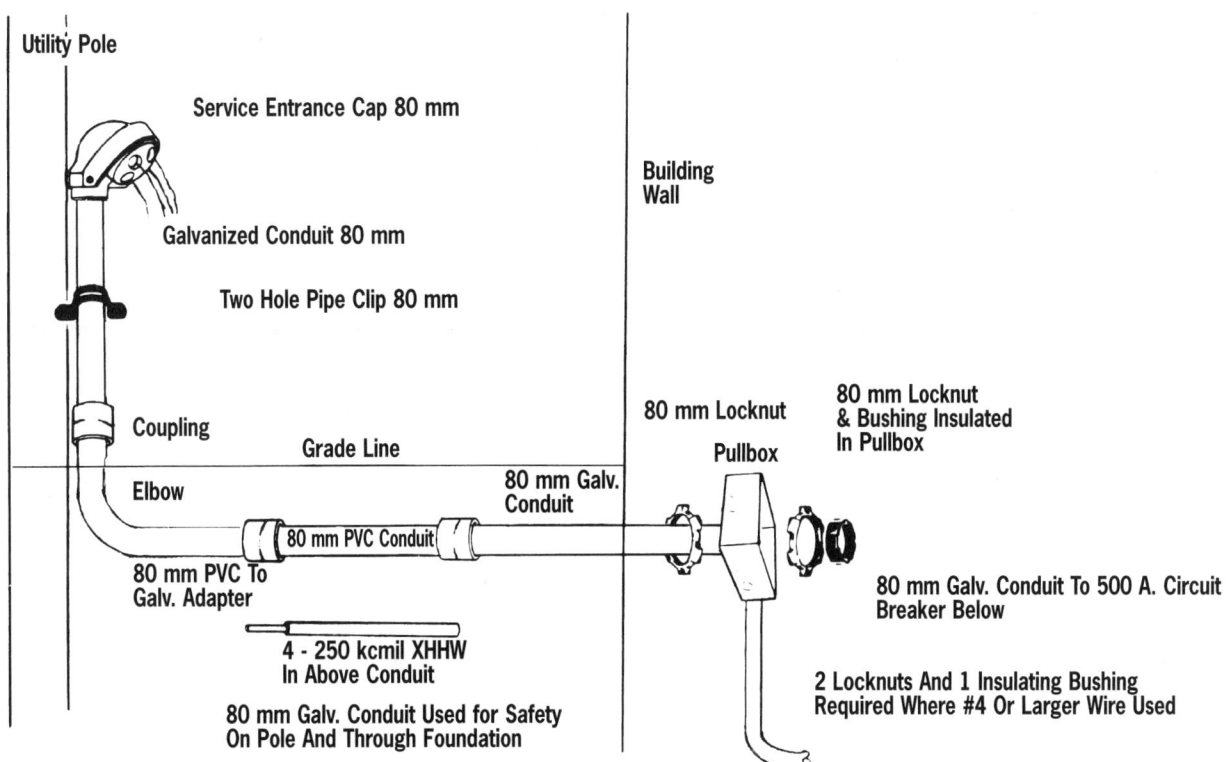

Utility Pole

Service Entrance Cap 80 mm

Building Wall

Galvanized Conduit 80 mm

Two Hole Pipe Clip 80 mm

Coupling

Grade Line

Elbow

80 mm PVC Conduit

80 mm Galv. Conduit

80 mm PVC To Galv. Adapter

4 - 250 kcmil XHHW In Above Conduit

80 mm Galv. Conduit Used for Safety On Pole And Through Foundation

80 mm Locknut

Pullbox

80 mm Locknut & Bushing Insulated In Pullbox

80 mm Galv. Conduit To 500 A. Circuit Breaker Below

2 Locknuts And 1 Insulating Bushing Required Where #4 Or Larger Wire Used

R260913-80 Switchboard Instruments

Switchboard instruments are added to the price of switchboards according to job specifications. This equipment is usually included when ordering "Gear" from the manufacturer and will arrive factory installed.

Included in the labor cost is:
1. Internal wiring connections
2. Wire identification
3. Wire tagging

Transition sections include:
1. Uncrating
2. Hauling sections up to 30 m from loading dock
3. Positioning sections
4. Leveling sections
5. Bolting enclosures
6. Bolting vertical bus bars

Price does not include:
1. Equipment pads
2. Steel channels embedded or grouted in concrete
3. Special knockouts
4. Rigging

Job Conditions: Productivity is based on new construction, equipment to be installed on the first floor within 60 m of the loading dock.

R262213-10   Electric Circuit Voltages

General: The following method provides the user with a simple non-technical means of obtaining comparative costs of wiring circuits. The circuits considered serve the electrical loads of motors, electric heating, lighting and transformers, for example, that require low voltage 60 Hertz alternating current.

The method used here is suitable only for obtaining estimated costs. It is **not** intended to be used as a substitute for electrical engineering design applications.

Conduit and wire circuits can represent from twenty to thirty percent of the total building electrical cost. By following the described steps and using the tables the user can translate the various types of electric circuits into estimated costs.

Wire Size: Wire size is a function of the electric load which is usually listed in one of the following units:

1. Amperes (A)
2. Watts (W)
3. Kilowatts (kW)
4. Volt amperes (VA)
5. Kilovolt amperes (kVA)

These units of electric load must be converted to amperes in order to obtain the size of wire necessary to carry the load. To convert electric load units to amperes one must have an understanding of the voltage classification of the power source and the voltage characteristics of the electrical equipment or load to be energized. The seven A.C. circuits commonly used are illustrated in Figures R262213-11 thru R262213-17 showing the tranformer load voltage and the point of use voltage at the point on the circuit where the load is connected. The difference between the source and point of use voltages is attributed to the circuit voltage drop and is considered to be approximately 4%.

Motor Voltages: Motor voltages are listed by their point of use voltage and not the power source voltage.

For example: 460 volts instead of 480 volts
200 instead of 208 volts
115 volts instead of 120 volts

Lighting and Heating Voltages: Lighting and heating equipment voltages are listed by the power source voltage and not the point of wire voltage.

For example: 480, 277, 120 volt lighting
480 volt heating or air conditioning unit
208 volt heating unit

Transformer Voltages: Transformer primary (input) and secondary (output) voltages are listed by the power source voltage.

For example: Single phase 10 kVA
Primary 240/480 volts
Secondary 120/240 volts

In this case, the primary voltage may be 240 volts with a 120 volts secondary or may be 480 volts with either a 120V or a 240V secondary.

For example: Three phase 10 kVA
Primary 480 volts
Secondary 208Y/120 volts

In this case the transformer is suitable for connection to a circuit with a 3 phase 3 wire or 3 phase 4 wire circuit with a 480 voltage. This application will provide a secondary circuit of 3 phase 4 wire with 208 volts between phase wires and 120 volts between any phase wire and the neutral (white) wire.

R262213-11

3 Wire, 1 Phase, 120/240 Volt System

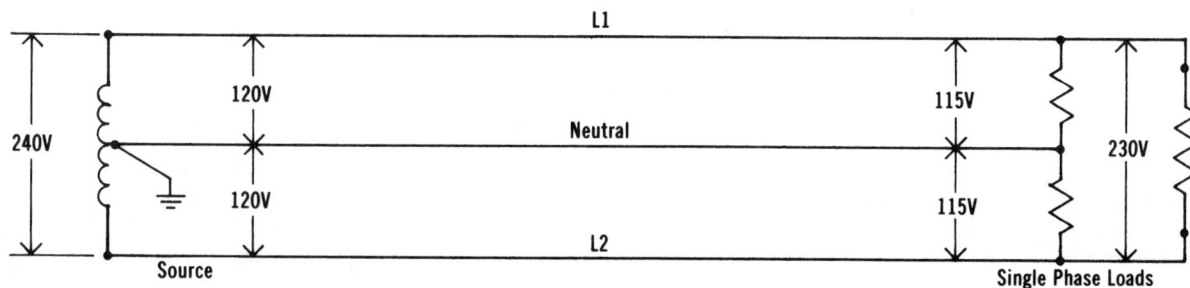

R262213-12

4 wire, 3 Phase, 208Y/120 Volt System

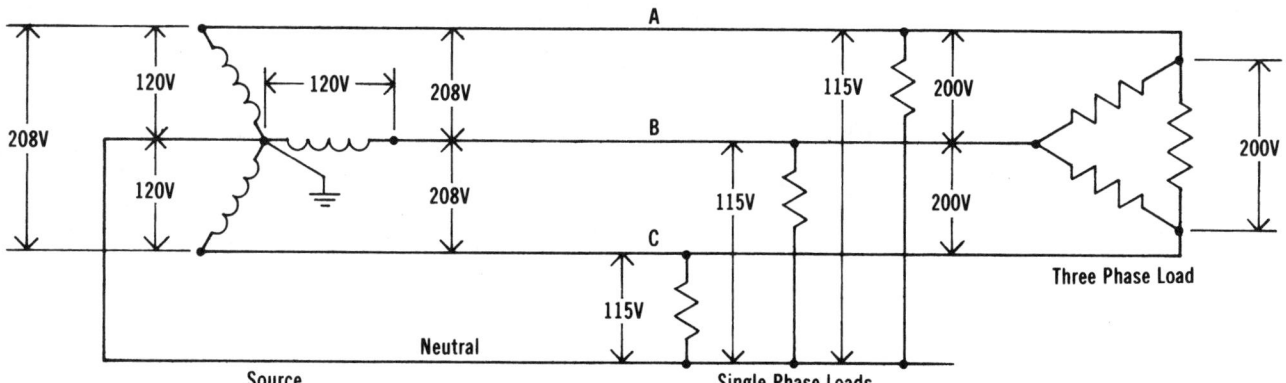

R262213-13   Electric Circuit Voltages (cont.)

3 Wire, 3 Phase 240 Volt System

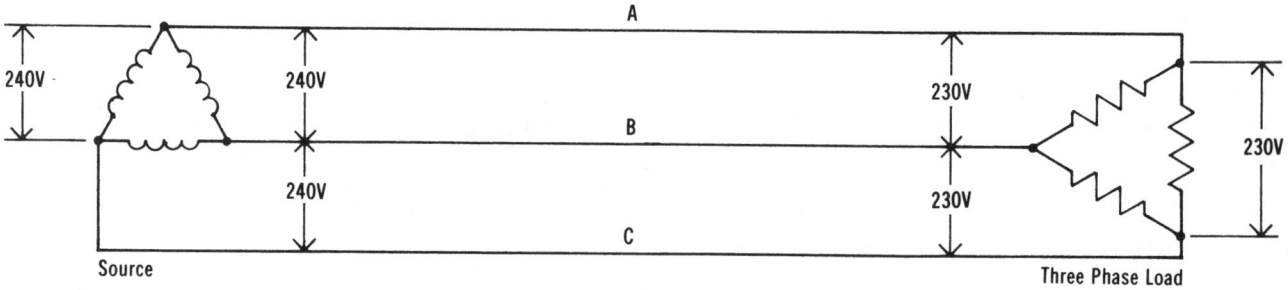

R262213-14

4 Wire, 3 Phase, 240/120 Volt System

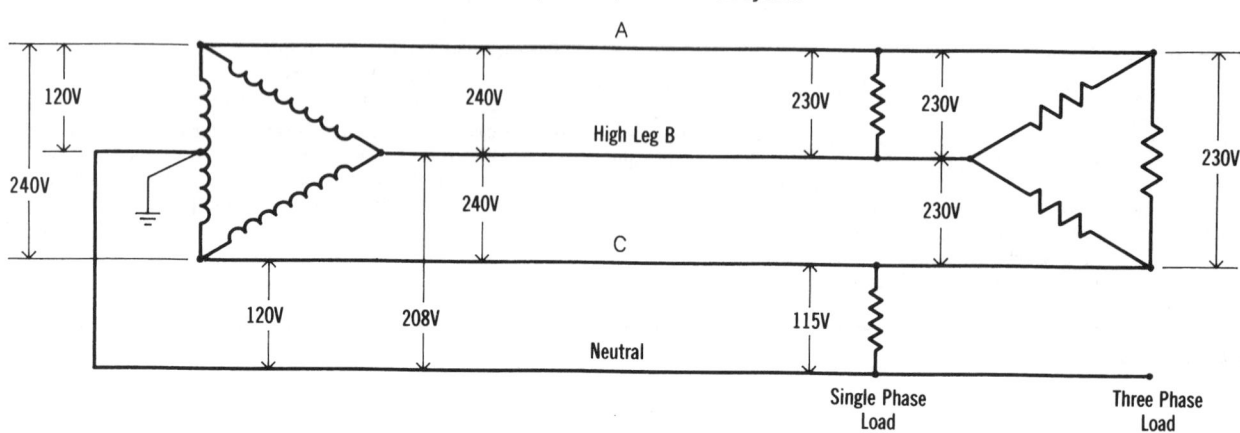

R262213-15

3 Wire, 3 Phase 480 Volt System

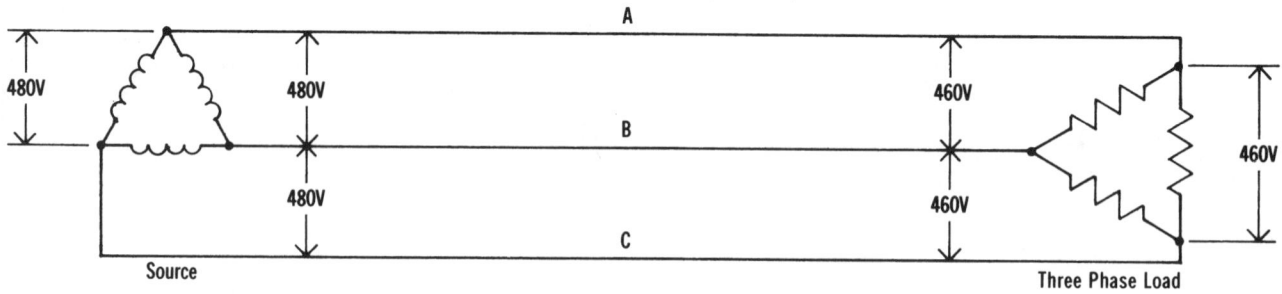

R262213-16

4 Wire, 3 Phase, 480Y/277 Volt System

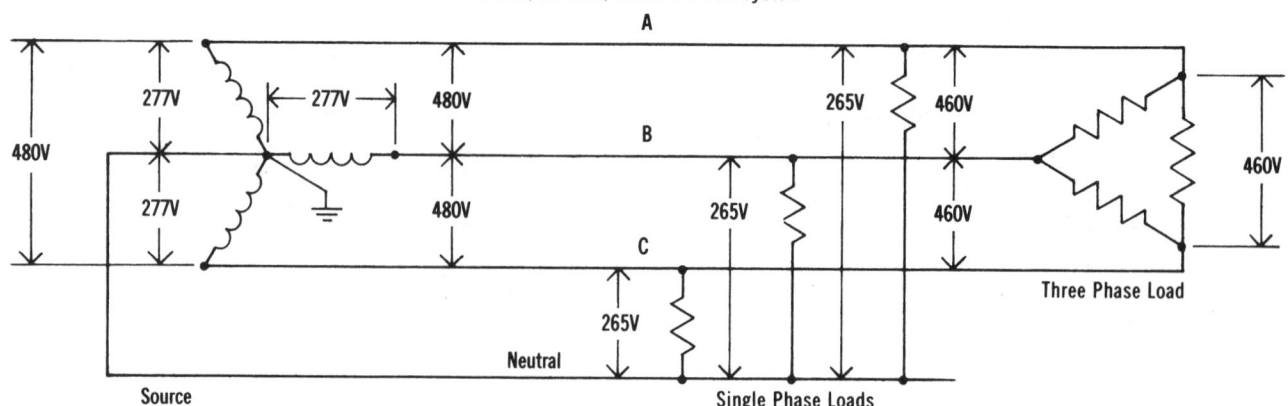

R262213-17   Electric Circuit Voltages (cont.)

3 Wire, 3 Phase, 600 Volt System

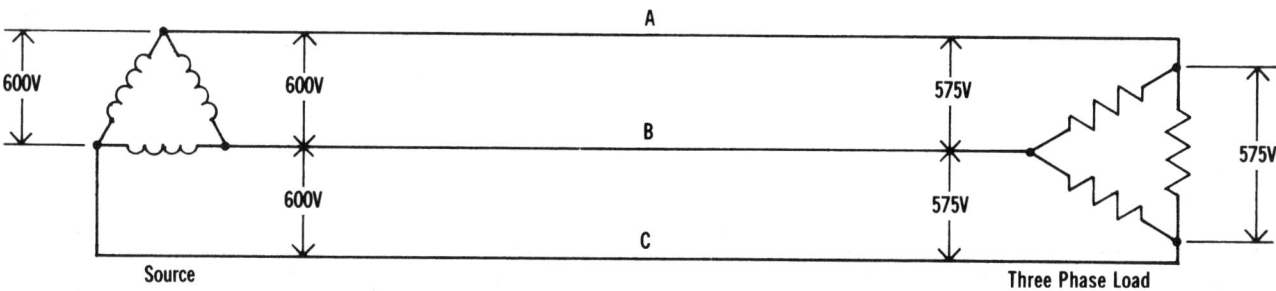

R262213-20   kW Value/Cost Determination

General: Lighting and electric heating loads are expressed in watts and kilowatts.

Cost Determination:

The proper ampere values can be obtained as follows:

1. Convert watts to kilowatts
   (watts 1000 ÷ kilowatts)
2. Determine voltage rating of equipment.
3. Determine whether equipment is single phase or three phase.
4. Refer to Table R262213-21 to find ampere value from kW, Ton and Btu/hr. values.
5. Determine type of wire insulation – TW, THW, THWN.
6. Determine if wire is copper or aluminum.
7. Refer to Table R260519-92 to obtain copper or aluminum wire size from ampere values.
8. Next refer to Table R260533-22 for the proper conduit size to accommodate the number and size of wires in each particular case.
9. Next refer to unit cost data for the per meter cost of the conduit.
10. Next refer to unit cost data for the per meter cost of the wire. Multiply cost of wire per meter x number of wires in the circuits to obtain total wire cost per meter.
11. Add values obtained in Step 9 and 10 for total cost per meter for conduit and wire x length of circuit = Total Cost.

Notes:

1. 1 Phase refers to single phase, 2 wire circuits.
2. 3 Phase refers to three phase, 3 wire circuits.
3. For circuits which operate continuously for 3 hours or more, multiply the ampere values by 1.25 for a given kw requirement.
4. For kW ratings not listed, add ampere values.
5. "Length of Circuit" refers to the one way distance of the run, not to the total sum of wire lengths.

   For example: Find the ampere value of 9 kW at 208 volt, single phase.

$$\begin{array}{r} 4\ \text{kW} = 19.2\text{A} \\ 5\ \text{kW} = 24.0\text{A} \\ \hline 9\ \text{kW} = 43.2\text{A} \end{array}$$

R262213-21   Ampere Values as Determined by kW Requirements, BTU/HR or Ton, Voltage and Phase Values

			Ampere Values						
			120V	208V		240V		277V	480V
kW	Ton	BTU/HR	1 Phase	1 Phase	3 Phase	1 Phase	3 Phase	1 Phase	3 Phase
0.5	.1422	1,707	4.2A	2.4A	1.4A	2.1A	1.2A	1.8A	0.6A
0.75	.2133	2,560	6.2	3.6	2.1	3.1	1.9	2.7	.9
1.0	.2844	3,413	8.3	4.9	2.8	4.2	2.4	3.6	1.2
1.25	.3555	4,266	10.4	6.0	3.5	5.2	3.0	4.5	1.5
1.5	.4266	5,120	12.5	7.2	4.2	6.3	3.1	5.4	1.8
2.0	.5688	6,826	16.6	9.7	5.6	8.3	4.8	7.2	2.4
2.5	.7110	8,533	20.8	12.0	7.0	10.4	6.1	9.1	3.1
3.0	.8532	10,239	25.0	14.4	8.4	12.5	7.2	10.8	3.6
4.0	1.1376	13,652	33.4	19.2	11.1	16.7	9.6	14.4	4.8
5.0	1.4220	17,065	41.6	24.0	13.9	20.8	12.1	18.1	6.1
7.5	2.1331	25,598	62.4	36.0	20.8	31.2	18.8	27.0	9.0
10.0	2.8441	34,130	83.2	48.0	27.7	41.6	24.0	36.5	12.0
12.5	3.5552	42,663	104.2	60.1	35.0	52.1	30.0	45.1	15.0
15.0	4.2662	51,195	124.8	72.0	41.6	62.4	37.6	54.0	18.0
20.0	5.6883	68,260	166.4	96.0	55.4	83.2	48.0	73.0	24.0
25.0	7.1104	85,325	208.4	120.2	70.0	104.2	60.0	90.2	30.0
30.0	8.5325	102,390		144.0	83.2	124.8	75.2	108.0	36.0
35.0	9.9545	119,455		168.0	97.1	145.6	87.3	126.0	42.1
40.0	11.3766	136,520		192.0	110.8	166.4	96.0	146.0	48.0
45.0	12.7987	153,585			124.8	187.5	112.8	162.0	54.0
50.0	14.2208	170,650			140.0	208.4	120.0	180.4	60.0
60.0	17.0650	204,780			166.4		150.4	216.0	72.0
70.0	19.9091	238,910			194.2		174.6		84.2
80.0	22.7533	273,040			221.6		192.0		96.0
90.0	25.5975	307,170					225.6		108.0
100.0	28.4416	341,300							120.0

R262213-25 kVA Value/Cost Determination

General: Control transformers are listed in VA. Step-down and power transformers are listed in kVA.

Cost Determination:

1. Convert VA to kVA. Volt amperes (VA) ÷ 1000 = Kilovolt amperes (kVA).
2. Determine voltage rating of equipment.
3. Determine whether equipment is single phase or three phase.
4. Refer to Table R262213-26 to find ampere value from kVA value.
5. Determine type of wire insulation – TW, THW, THWN.
6. Determine if wire is copper or aluminum.
7. Refer to Table R260519-92 to obtain copper or aluminum wire size from ampere values.

8. Next refer to Table R260533-22 for the proper conduit size to accommodate the number and size of wires in each particular case.
9. Next refer to unit price data for the per meter cost of the conduit.
10. Next refer to unit price data for the per meter cost of the wire. Multiply cost of wire per meter x number of wires in the circuits to obtain total wire cost.
11. Add values obtained in Step 9 and 10 for total cost per meter for conduit and wire x length of circuit = Total Cost.

Example: A transformer rated 10 kVA 480 volts primary, 240 volts secondary, 3 phase has the capacity to furnish the following:

1. Primary amperes = 10 kVA x 1.20 = 12 amperes
   (from Table R262213-26)
2. Secondary amperes = 10 kVA x 2.40 = 24 amperes
   (from Table R262213-26)

Note: Transformers can deliver generally 125% of their rated kVA. For instance, a 10 kVA rated transformer can safely deliver 12.5 kVA.

R262213-26 Multiplier Values for kVA to Amperes Determined by Voltage and Phase Values

Volts	Multiplier for Circuits	
	2 Wire, 1 Phase	3 Wire, 3 Phase
115	8.70	
120	8.30	
230	4.30	2.51
240	4.16	2.40
200	5.00	2.89
208	4.80	2.77
265	3.77	2.18
277	3.60	2.08
460	2.17	1.26
480	2.08	1.20
575	1.74	1.00
600	1.66	0.96

R262213-27 Central Air Conditioning Watts per m², of Floor Area and m² per kW of Air Conditioning

Type Building	Watts per m²	m² per kW	Type Building	Watts per m²	m² per kW	Type Building	Watts per m²	m² per kW
Apartments, Individual	82	12.2	Dormitory, Rooms	126	7.9	Libraries	158	6.3
Corridors	69	14.5	Corridors	95	10.5	Low Rise Office, Ext.	170	8.3
Auditoriums & Theaters	126	7.0/5*	Dress Shops	136	7.4	Interior	104	9.6
Banks	158	6.3	Drug Stores	252	4.0	Medical Centers	88	11.4
Barber Shops	151	6.6	Factories	126	7.9	Motels	88	11.4
Bars & Taverns	420	2.4	High Rise Off.-Ext. Rms.	145	6.9	Office (small suite)	136	7.4
Beauty Parlors	208	4.8	Interior Rooms	117	8.5	Post Office, Int. Office	133	7.5
Bowling Alleys	215	4.7	Hospitals, Core	136	7.4	Central Area	145	6.5
Churches	114	8.8/6*	Perimeter	145	6.9	Residences	63	15.9
Cocktail Lounges	215	4.7	Hotels, Guest Rooms	139	7.2	Restaurants	189	5.3
Computer Rooms	445	2.2	Public Spaces	174	5.7	Schools & Colleges	145	6.9
Dental Offices	164	6.1	Corridors	95	10.5	Shoe Stores	174	5.7
Dept. Stores, Basement	107	9.3	Industrial Plants, Offices	120	8.3	Shop'g. Ctrs., Sup. Mkts.	107	9.3
Main Floor	126	7.9	General Offices	107	9.3	Retail Stores	151	6.6
Upper Floor	95	10.5	Plant Areas	126	7.9	Specialty Shops	189	5.3

*Persons per ton

R262213-60 Oil Filled Transformers

Transformers in this section include:
1. Rigging (as required)
2. Rental of crane and operator
3. Setting of oil filled transformer
4. (4) Anchor bolts, nuts and washers in concrete pad

Price does not include:
1. Primary and secondary terminations
2. Transformer pad
3. Equipment grounding
4. Cable
5. Conduit locknuts or bushings

Transformers in Unit Price sections for dry type, back-boost and isolating transformers include:

Transformers in these sections include:
1. Unloading and uncrating
2. Hauling transformer to within 60 m of loading dock

3. Setting in place
4. Wall mounting hardware
5. Testing

Price does not include:
1. Structural supports
2. Suspension systems
3. Welding or fabrication
4. Primary & secondary terminations

Add the following percentages to the labor for ceiling mounted transformers:

3.0 m to 4.5 m	= + 15%
4.6 m to 7.5 m	= + 30%
Over 7.5 m	= + 35%

Job Conditions: Productivities are based on new construction. Installation is assumed to be on the first floor, in an obstructed area to a height of 3 m. Material staging area is within 30 m of final transformer location.

R262213-65 Transformer Weight (Lbs.) by kVA

Oil Filled 3 Phase 5/15 KV To 480/277			
kVA	kg.	kVA	kg.
150	816	1000	2810
300	1315	1500	3810
500	2130	2000	4400
750	2405	3000	6805

Dry 240/480 To 120/240 V			
1 Phase		3 Phase	
kVA	kg.	kVA	kg.
1	10.64	3	41
2	16	6	61
3	27	9	77
5	33	15	100
7.5	59	30	141
10	68	45	181
15	93	75	272
25	116	112.5	430
37.5	134	150	520
50	154	225	715
75	249	300	850
100	304	500	1295
167	408	750	1950

R262416-50   Load Centers and Panelboards

When pricing Load Centers list panels by size and type. List Breakers in a separate column of the "Quantity Sheet," and define by phase and ampere rating.

Material and Labor prices include breakers; for example: a 100A, 3-Wire, 102/240V, 18 circuit panel w/ main breaker, as described in the unit cost section, contains 18 single pole 20A breakers.

If you do not choose to include a full panel of single pole breakers, use the following method to adjust material and labor costs.

Example: In an 18 circuit panel only 16 single pole breakers are desired, requiring that the cost of 2 breakers be subtracted from the panel cost.

1. Go to the appropriate unit cost section of the book to find the unit prices of the given circuit breaker type
2. Modify those costs as follows: Bare material price x 0.50   Bare labor cost x 0.60
3. Multiply those modified bare costs by 2 (breakers in this example)
4. Subtract the modified costs for the 2 breakers from the given cost of the panel.

Labor-hours for Load Center installation includes:

1. Unloading, uncrating, and handling enclosures 60 m from unloading area.
2. Measuring and marking
3. Drilling (4) lead anchor type fasteners using a hammer drill
4. Mounting and leveling panel to a height of 1.8 m
5. Preparation and termination of feeder cable to lugs or main breaker
6. Branch circuit identification
7. Lacing using tie wraps
8. Testing and load balancing
9. Marking panel directory

Not included in the material and labor are:

1. Modifications to enclosure
2. Structural supports
3. Additional lugs
4. Plywood backboards
5. Painting or lettering

Note: Knockouts are included in the price of terminating pipe runs and need not be added to the Load Center costs.

Job Conditions: Productivity is based on new construction to a height of 1.8 m, in a unobstructed area. Material staging area is assumed to be within 30 m of work being performed.

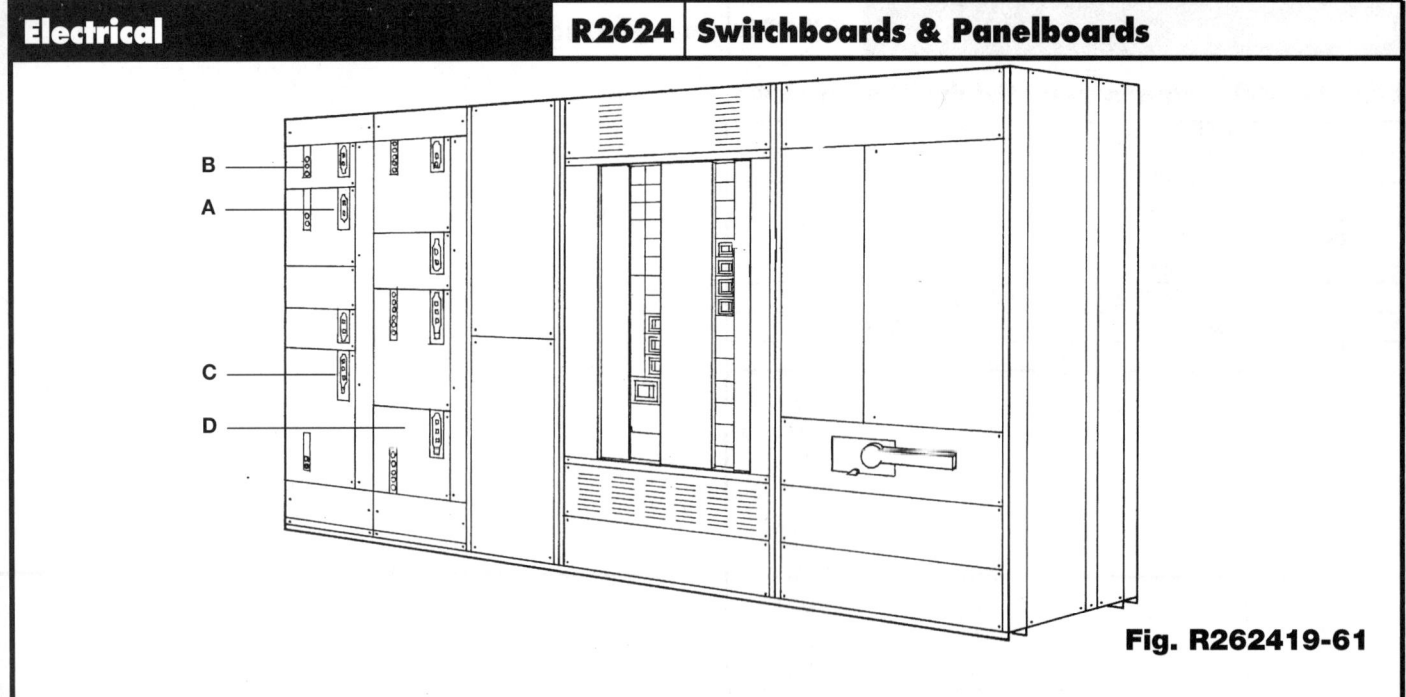

Fig. R262419-61

R262419-60   Motor Control Centers

When taking off Motor Control Centers, list the size, type and height of structures.

Example:
1. 600A, 22,000 RMS, 1800 mm high
2. 600A, back to back, 1800 mm high

Next take off individual starters; the number of structures can also be determined by adding the height in inches of starters divided by the height of the structure, and list on the same quantity sheet as the structures. Identify starters by Type, Horsepower rating, Size and Height in inches.

Example:
  A. Class l, Type B, FVNR starter, 19 kW, 450 mm high.
      Add to the list with Starters, factory installed controls.

Example:
  B. Pilot lights
  C. Push buttons
  D. Auxiliary contacts

Identify starters and structures as either copper or aluminum and by the NEMA type of enclosure.

When pricing starters and structures, be sure to add or deduct adjustments, using lines in the unit price section.

Included in the cost of Motor Control Structures are:
1. Uncrating
2. Hauling to location within 30 m of loading dock
3. Setting structures
4. Leveling
5. Aligning
6. Bolting together structure frames
7. Bolting horizontal bus bars

Labor-hours do not include:
1. Equipment pad
2. Steel channels embedded or grouted in concrete
3. Pull boxes
4. Special knockouts
5. Main switchboard section
6. Transition section
7. Instrumentation
8. External control wiring
9. Conduit or wire

Material for Starters includes:
1. Circuit breaker or fused disconnect
2. Magnetic motor starter
3. Control transformer
4. Control fuse and fuse block

Labor-hours for Starters include:
1. Handling
2. Installing starter within structure
3. Internal wiring connections
4. Lacing within enclosure
5. Testing
6. Phasing

Job Conditions: Productivity is based on new construction. Motor control location assumed to be on first floor in an unobstructed area. Material staging area within 30 m of final location.

Note: Additional labor-hours must be added if M.C.C. is to be installed on other than the first floor, or if rigging is required.

R262419-65 Motor Starters and Controls

Motor starters should be listed on the same "Quantity Sheet" as Motors and Motor Connections. Identify each starter by:

1. Size
2. Voltage
3. Type

Example:

A. FVNR, 480V, 2HP, Size 00
B. FVNR, 480V, 5HP, Size 0, Combination type
C. FVR, 480V, Size 2

The NEMA type of enclosure should also be identified.

Included in the labor-hours are:

1. Unloading, uncrating and handling of starters up to 60 m from loading area
2. Measuring and marking
3. Drilling (4) anchor type lead fasteners, using a hammer drill
4. Mounting and leveling starter
5. Connecting wire or cable to line and load sides of starter (when already lugged)
6. Installation of (3) thermal type heaters
7. Testing

The following is not included unless specified in the unit price description.

1. Control transformer
2. Controls, either factory or field installed
3. Conduit and wire to or from starter
4. Plywood backboard
5. Cable terminations

The following material and labor has been included for Combination type starters:

1. Unloading, uncrating and handling of starter up to 60 m of loading dock
2. Measuring and marking
3. Drilling (4) anchor type lead fasteners using a hammer drill
4. Mounting and leveling
5. Connecting prepared cable conductors
6. Installation of (3) dual element cartridge type fuses
7. Installation of (3) thermal type heaters
8. Test for rotation

MOTOR CONTROLS

When pricing motor controls make sure you consider the type of control system being utilized. If the controls are factory installed and located in the enclosure itself, then you would add to your starter price the items 5200 thru 5800 in the Unit Price section. If control voltage is different from line voltage, add the material and labor cost of a control transformer.

For external control of starters include the following items:

1. Raceways
2. Wire & terminations
3. Control enclosures
4. Fittings
5. Push button stations
6. Indicators

Job Conditions: Productivity is based on new construction to a height of 3 m.

Material staging area is assumed to be within 30 m of work being performed.

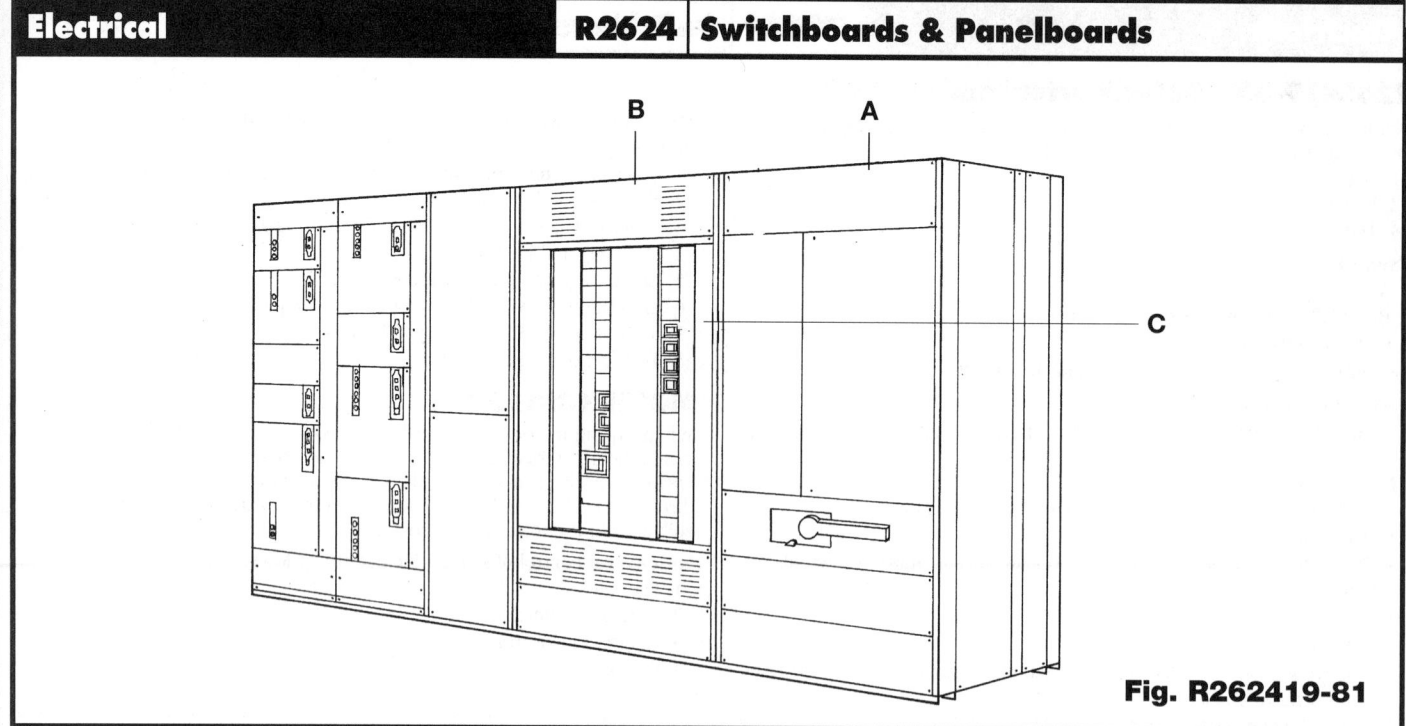

B   A

C

Fig. R262419-81

R262419-80   Distribution Section

After ''Taking off'' the Switchboard section, include on the same ''Quantity sheet'' the Distribution section; identify by:

1. Voltage
2. Ampere rating
3. Type

Example:

(Fig. R262419-81) (B) Distribution Section

Included in the labor costs of the Distribution section is:

1. Uncrating
2. Hauling to 60 m of loading dock
3. Setting of distribution panel
4. Leveling & shimming
5. Anchoring of equipment to pad or floor
6. Bolting of horizontal bus bars between
7. Testing of equipment

Not included in the Distribution section is:

1. Breakers (C)
2. Equipment pads
3. Steel channels embedded or grouted in concrete
4. Pull boxes
5. Special knockouts
6. Transition section
7. Conduit or wire

R262419-82   Feeder Section

List quantities on the same sheet as the Distribution section. Identify breakers by: (Fig. R262419-81)

1. Frame type
2. Number of poles
3. Ampere rating

Installation includes:

1. Handling
2. Placing breakers in distribution panel
3. Preparing wire or cable
4. Lacing wire
5. Marking each phase with colored tape
6. Marking panel legend
7. Testing
8. Balancing

R262419-84 Switchgear

It is recommended that "Switchgear" or those items contained in the following sections be quoted from equipment manufacturers as a package price.

    Switchboard instruments

    Switchboard distribution sections

    Switchboard feeder sections

    Switchboard Service Disc.

    Switchboard In-plant Dist.

Included in these sections are the most common types and sizes of factory assembled equipment.

The recommended procedure for low voltage switchgear would be to price (Fig. R262419-81) (A) Main Switchboard.

Identify by:

1. Voltage
2. Amperage
3. Type

Example:

1. 120/208V, 4-wire, 600A, nonfused
2. 277/480V, 4-wire, 600A, nonfused
3. 120/208V, 4-wire, 400A w/fused switch & CT compartment
4. 277/480V, 4-wire, 400A, w/fused switch & CT compartment
5. 120/208V, 4-wire, 800A, w/pressure switch & CT compartment
6. 277/480V, 4-wire, 800A, w/molded CB & CT compartment

Included in the labor costs for Switchboards are:

1. Uncrating
2. Hauling to 60 m of loading dock
3. Setting equipment
4. Leveling and shimming
5. Anchoring
6. Cable identification
7. Testing of equipment

Not included in the Switchboard price is:

1. Rigging
2. Equipment pads
3. Steel channels embedded or grouted in concrete
4. Special knockouts
5. Transition or Auxiliary sections
6. Instrumentation
7. External control
8. Conduit and wire
9. Conductor terminations

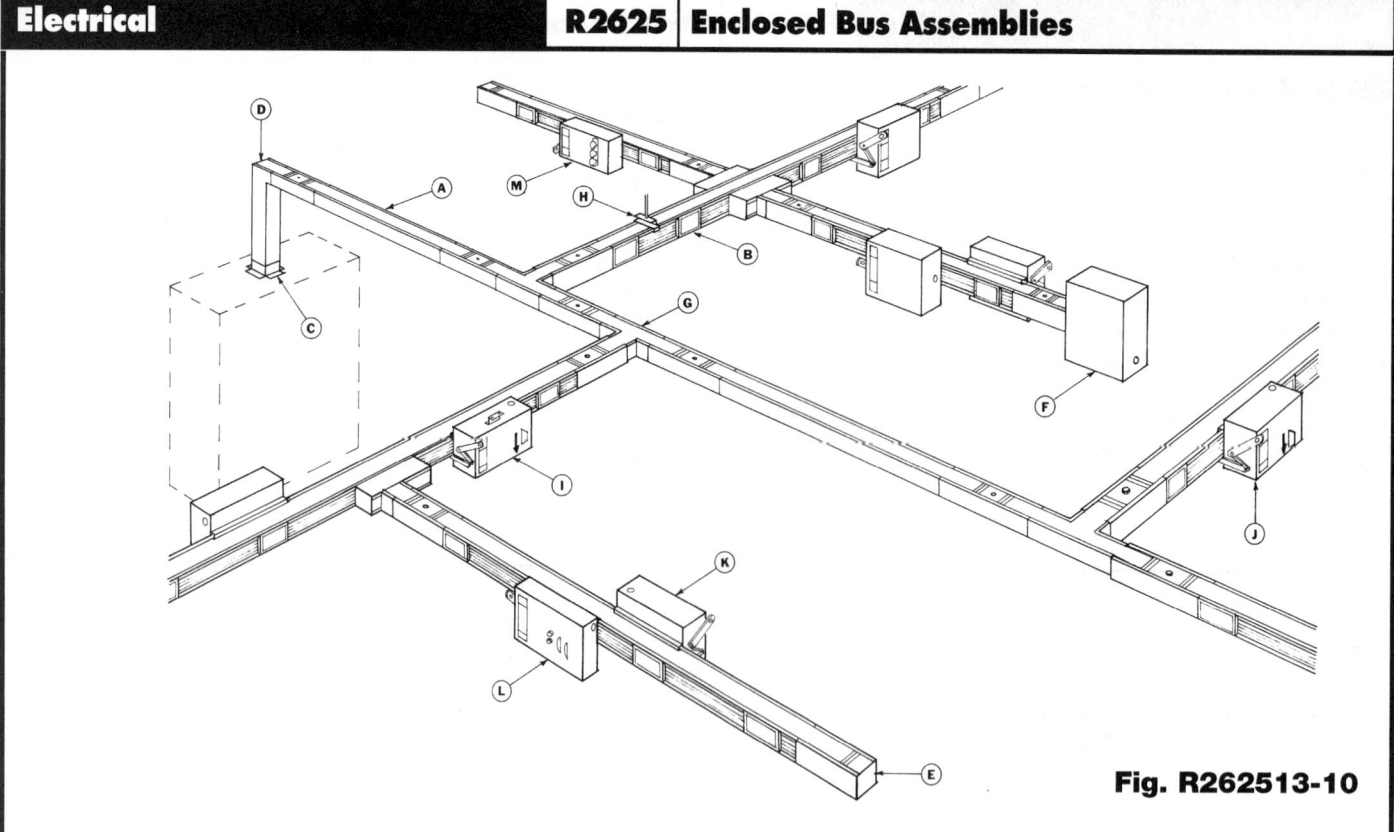

Fig. R262513-10

R262513-10   Aluminum Bus Duct

When taking off bus duct identify the system as either:
1. Aluminum
2. Copper

List straight lengths by type and size
   A. Plug-in — 800 A
   B. Feeder — 800 A

Do not measure thru fittings as you would on conduit, since there is no such thing as allowance for scrap in bus duct systems.

If upon taking off per m quantities of bus duct you find your quantities are not divisible evenly by 3 m, then the remainder must be priced as a special item and quoted from the manufactuer. Do not use the bare material cost per m for these special items. You can, however, safely use the bare labor cost per m for the entire length.

Identify fittings by type and ampere rating

Example:

C. Switchboard stub 800 A

D. Elbows 800 A

E. End box 800 A

F. Cable tap box 800 A

G. Tee Fittings 800 A

H. Hangers

Plug-in Units - List separately plug-in units and identify by type and ampere rating

I. Plug-in switches 600 Volt 3 phase 60 A

J. Plug-in molded case C.B. 60 A

K. Combination starter FVNR NEMA 1

L. Combination contactor & fused switch NEMA 1

M. Combination fusible switch & lighting control 60 A

Labor-hours for feeder and plug-in sections include:
1. Unloading and uncrating
2. Hauling up to 60 m from loading dock
3. Measuring and marking
4. Set up of rolling staging
5. Installing hangers
6. Hanging and bolting sections
7. Aligning and leveling
8. Testing

Labor-hours do not include:
1. Modifications to existing structure for hanger supports
2. Threaded rod in excess of 600 mm
3. Welding
4. Penetrations thru walls
5. Staging rental

Deduct the following percentages from labor only:

45 m to	75 m —	10%
76 m to	105 m —	15%
106 m to	150 m —	20%
Over	150 m —	25%

Deduct percentage only if runs are contained in the same area.

Example — If the job entails running 30 m in 5 different locations do not deduct 20%, but if the duct is being run in 1 area and the quantity is 150 m then you would deduct 20%.

Deduct only from straight lengths, not fittings or plug-in units.

R262513-10   Aluminum Bus Duct (cont.)

Add to labor for elevated installations:

4.5 m to 6.1 m high	10%
6.2 m to 7.6 m high	20%
7.7 m to 9.1 m high	30%
9.1 m to 10.7 m high	40%
10.7 m to 12.2 m high	50%
Over 12.2 m. high	60%

Bus Duct Fittings:

Labor-hours for fittings include:
1. Unloading and uncrating
2. Hauling up to 60 m from loading dock
3. Installing, fitting, and bolting all ends to in place sections

Plug-in units include:
1. Unloading and uncrating
2. Hauling up to 60 m from loading dock
3. Installing plug-in into in place duct
4. Set up of rolling staging
5. Connection load wire to lugs
6. Marking wire
7. Checking phase rotation

Labor-hours for plug-ins do not include:
1. Conduit runs from plug-in
2. Wire from plug-in
3. Conduit termination

Economy of Scale - For large concentrations of plug-in units in the same area deduct the following percentages:

11	to	25	15%
26	to	50	20%
51	to	75	25%
76	to	100	30%
100	and	over	35%

Job Conditions: Productivities are based on new construction, in an unobstructed first floor area to a height of 4.5 m; using rolling staging.

Material staging area is within 30 m of work being performed.

Add to the duct fittings and hangers:
 Plug-in switches (Fused)
 Plug-in breakers
 Combination starters
 Combination contactors
 Combination fusible switch and lighting control

Labor-hours for duct and fittings include:
1. Unloading and uncrating
2. Hauling up to 60 m from loading dock
3. Measuring and marking
4. Installing duct runs
5. Leveling
6. Sound testing

Labor-hours for plug-ins include:
1. Unloading and uncrating
2. Hauling up to 60 m from loading dock
3. Installing plug-ins
4. Preparing wire
5. Wire connections and marking

R262513-15 Weight (kg/m) of 4 Pole Aluminum and Copper Bus Duct by Ampere Load

Amperes	Aluminum Feeder	Copper Feeder	Aluminum Plug–In	Copper Plug–In
225			10	10
400			12	19
600	15	15	16	21
800	15	28	19	27
1000	16	28	24	33
1350	21	36	30	45
1600	25	39	37	58
2000	28	45	43	68
2500	40	64	54	83
3000	45	71	63	109
4000	58	100		
5000		116		

R262716-40 Standard Electrical Enclosure Types

NEMA Enclosures

Electrical enclosures serve two basic purposes; they protect people from accidental contact with enclosed electrical devices and connections, and they protect the enclosed devices and connections from specified external conditions. The National Electrical Manufacturers Association (NEMA) has established the following standards. Because these descriptions are not intended to be complete representations of NEMA listings, consultation of NEMA literature is advised for detailed information.

The following definitions and descriptions pertain to NONHAZARDOUS locations.

NEMA Type 1: General purpose enclosures intended for use indoors, primarily to prevent accidental contact of personnel with the enclosed equipment in areas that do not involve unusual conditions.

NEMA Type 2: Dripproof indoor enclosures intended to protect the enclosed equipment against dripping noncorrosive liquids and falling dirt.

NEMA Type 3: Dustproof, raintight and sleet-resistant (ice-resistant) enclosures intended for use outdoors to protect the enclosed equipment against wind-blown dust, rain, sleet, and external ice formation.

NEMA Type 3R: Rainproof and sleet-resistant (ice-resistant) enclosures which are intended for use outdoors to protect the enclosed equipment against rain. These enclosures are constructed so that the accumulation and melting of sleet (ice) will not damage the enclosure and its internal mechanisms.

NEMA Type 3S: Enclosures intended for outdoor use to provide limited protection against wind-blown dust, rain, and sleet (ice) and to allow operation of external mechanisms when ice-laden.

NEMA Type 4: Watertight and dust-tight enclosures intended for use indoors and out – to protect the enclosed equipment against splashing water, see page of water, falling or hose-directed water, and severe external condensation.

NEMA Type 4X: Watertight, dust-tight, and corrosion-resistant indoor and outdoor enclosures featuring the same provisions as Type 4 enclosures, plus corrosion resistance.

NEMA Type 5: Indoor enclosures intended primarily to provide limited protection against dust and falling dirt.

NEMA Type 6: Enclosures intended for indoor and outdoor use – primarily to provide limited protection against the entry of water during occasional temporary submersion at a limited depth.

NEMA Type 6R: Enclosures intended for indoor and outdoor use – primarily to provide limited protection against the entry of water during prolonged submersion at a limited depth.

NEMA Type 11: Enclosures intended for indoor use – primarily to provide, by means of oil immersion, limited protection to enclosed equipment against the corrosive effects of liquids and gases.

NEMA Type 12: Dust-tight and driptight indoor enclosures intended for use indoors in industrial locations to protect the enclosed equipment against fibers, flyings, lint, dust, and dirt, as well as light splashing, see page, dripping, and external condensation of noncorrosive liquids.

NEMA Type 13: Oil-tight and dust-tight indoor enclosures intended primarily to house pilot devices, such as limit switches, foot switches, push buttons, selector switches, and pilot lights, and to protect these devices against lint and dust, see page, external condensation, and sprayed water, oil, and noncorrosive coolant.

The following definitions and descriptions pertain to HAZARDOUS, or CLASSIFIED, locations:

NEMA Type 7: Enclosures intended to use in indoor locations classified as Class 1, Groups A, B, C, or D, as defined in the National Electrical Code.

NEMA Type 9: Enclosures intended for use in indoor locations classified as Class 2, Groups E, F, or G, as defined in the National Electrical Code.

R262726-90   Wiring Devices

Wiring devices should be priced on a separate takeoff form which includes boxes, covers, conduit and wire.

Labor-hours for devices include:
1. Stripping of wire
2. Attaching wire to device using terminators on the device itself, lugs, set screws etc.
3. Mounting of device in box

Labor-hours do not include:
1. Conduit
2. Wire
3. Boxes
4. Plates

Economy of Scale – for large concentrations of devices in the same area deduct the following percentages from labor-hours:

1	to	10	0%
11	to	25	20%
26	to	50	25%
51	to	100	30%
	over	100	35%

R262726-90 Wiring Devices (cont.)

NEMA No.	15 R	20 R	30 R	50 R	60 R
1 125V 2 Pole, 2 Wire					
2 250V 2 Pole, 2 Wire					
5 125V 2 Pole, 3 Wire					
6 250V 2 Pole, 3 Wire					
7 277V, AC 2 Pole, 3 Wire					
10 125/250V 3 Pole, 3 Wire					
11 3 Phase 250V 3 Pole, 3 Wire					
14 125/250V 3 Pole, 4 Wire					
15 3 Phase 250V 3 Pole, 4 Wire					
18 3 Phase 208Y/120V 4 Pole, 4 Wire					

R262726-90   Wiring Devices (cont.)

NEMA No.	15 R	20 R	30 R	NEMA No.	15 R	20 R	30 R
L 1 125V 2 Pole, 2 Wire	⊘			**L 13** 3 Phase 600V 3 Pole, 3 Wire			⊘
L 2 250V 2 Pole, 2 Wire		⊘ 15A		**L 14** 125/250V 3 Pole, 4 Wire		⊘	⊘
L 5 125 V 2 Pole, 3 Wire	⊘	⊘	⊘	**L 15** 3 Phase 250 V 3 Pole, 4 Wire		⊘	⊘
L 6 250 V 2 Pole, 3 Wire	⊘	⊘	⊘	**L 16** 3 Phase 480V 3 Pole, 4 Wire		⊘	⊘
L 7 227 V, AC 2 Pole, 3 Wire	⊘	⊘	⊘	**L 17** 3 Phase 600 V 3 Pole, 4 Wire			⊘
L 8 480 V 2 Pole, 3 Wire		⊘	⊘	**L 18** 3 Phase 208Y/120V 4 Pole, 4 Wire		⊘	⊘
L 9 600 V 2 Pole, 3 Wire		⊘	⊘	**L 19** 3 Phase 480Y/277V 4 Pole, 4 Wire		⊘	⊘
L 10 125 /250V 3 Pole, 3 Wire	.	⊘	⊘	**L 20** 3 Phase 600Y/347V 4 Pole, 4 Wire		⊘	⊘
L 11 3 Phase 250 V 3 Pole, 3 Wire		⊘	⊘	**L 21** 3 Phase 208Y/120V 4 Pole, 5 Wire		⊘	⊘
L 12 3 Phase 480 V 3 Pole, 3 Wire		⊘	⊘	**L 22** 3 Phase 480Y/277V 4 Pole, 5 Wire		⊘	⊘
				L 23 3 Phase 600Y/347V 4 Pole, 5 Wire		⊘	⊘

Electrical | R2628 | Low-Voltage Circuit Protective Devices

R262816-80   Safety Switches

List each Safety Switch by type, ampere rating, voltage, single or three phase, fused or nonfused.

Example:

A. General duty, 240V, 3-pole, fused
B. Heavy Duty, 600V, 3-pole, nonfused
C. Heavy Duty, 240V, 2-pole, fused
D. Heavy Duty, 600V, 3-pole, fused

Also include NEMA enclosure type and identify as:

1. Indoor
2. Weatherproof
3. Explosionproof

Installation of Safety Switches includes:

1. Unloading and uncrating
2. Handling disconnects up to 60 m from loading dock
3. Measuring and marking location
4. Drilling (4) anchor type lead fasteners, using a hammer drill
5. Mounting and leveling Safety Switch
6. Installing (3) fuses
7. Phasing and tagging line and load wires

Price does not include:

1. Modifications to enclosure
2. Plywood backboard
3. Conduit or wire
4. Fuses
5. Termination of wires

Job Conditions: Productivities are based on new construction to an installed height of 1800 mm above finished floor.

Material staging area is assumed to be within 30 m of work in progress.

Electrical | R2632 | Packaged Generator Assemblies

R263213-45   Generator Weight (kg.) by kW

3 Phase 4 Wire /480 V			
Gas		Diesel	
kW	kg	kW	kg
7.5	272	30	815
10	286	50	1010
15	435	75	1020
30	680	100	1740
65	1069	125	1830
85	1165	150	2495
115	1955	175	2565
170	2960	200	2690
		250	2865
		300	3555
		350	3730
		400	4875
		500	5400

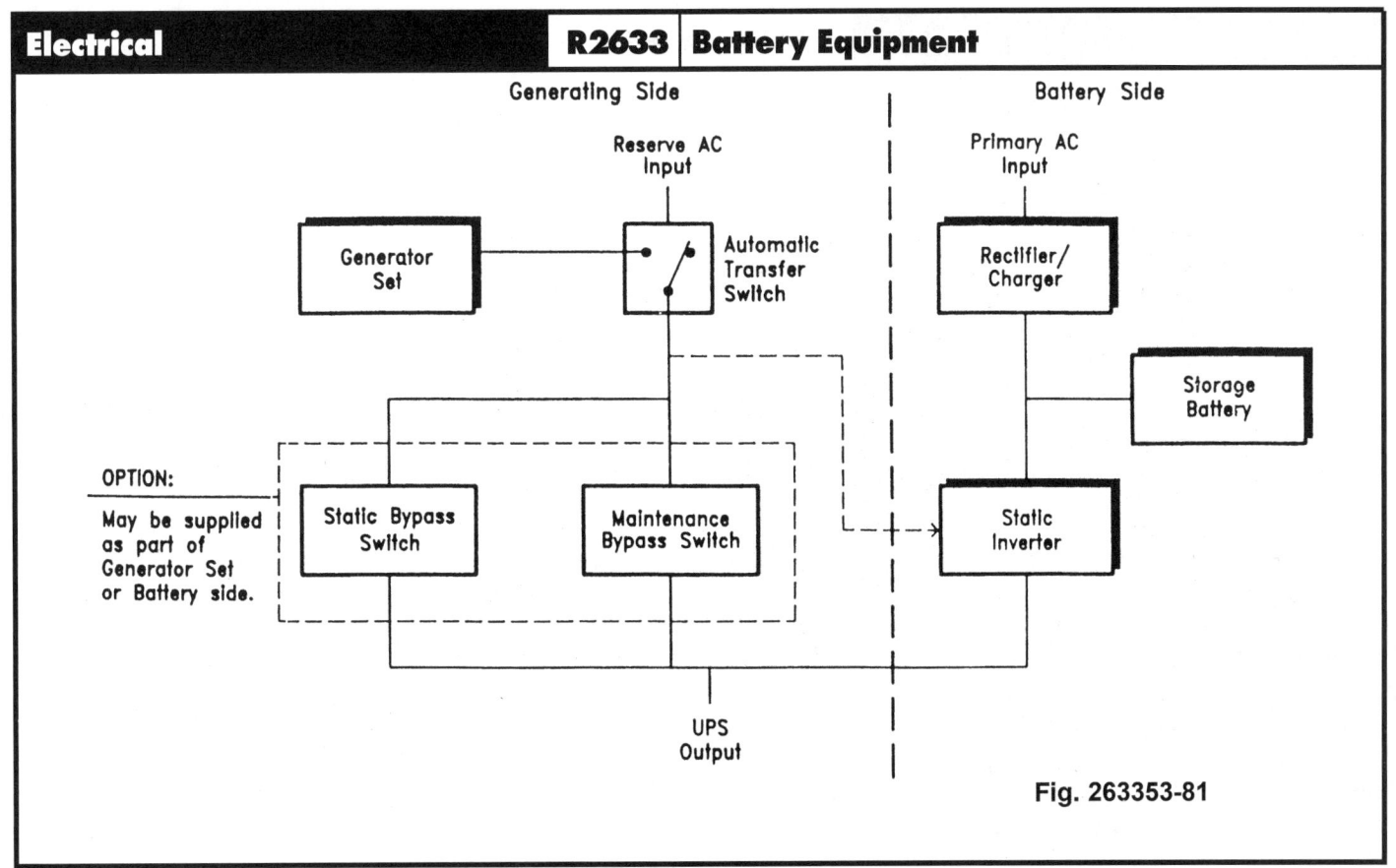

Generating Side — Battery Side

Fig. 263353-81

R263353-80   Uninterruptible Power Supply Systems

General: Uninterruptible Power Supply (UPS) Systems are used to provide power for legally required standby systems. They are also designed to protect computers and provide optional additional coverage for any or all loads in a facility. Figure R263353-81 shows a typical configuration for a UPS System with generator set.

Cost Determination:

It is recommended that UPS System material costs be obtained from equipment manufacturers as a package. Installation costs should be obtained from a vendor or contractor.

The recommended procedure for pricing UPS Systems would be to identify:

1. Frequency – by Hertz (Hz.)
   For example: 50, 60, and 400/415 Hz.
2. Apparent power and real power
   For example: 200 kVA/170 kW
3. Input/Output voltage   For example: 120V, 208V or 240V

4. Phase – either single or three-phase
5. Options and accessories – For extended run time more batteries would be required. Other accessories include battery cabinets, battery racks, remote control panel, power distribution unit (PDU), and warranty enhancement plans.

Note:

1. Larger systems can be configured by paralleling two or more standard small size single modules.
   For example: two 15 kVA modules, combined together and configured to 30 kVA.
2. Maximum input current during battery recharge is typically 15% higher than normal current.
3. UPS Systems weights vary depending on options purchased and power rating in kVA.

R263413-30 kW Value/Cost Determination

General: Motors can be powered by any of the seven systems shown in Figure R262213-11 thru Figure R262213-17 provided the motor voltage characteristics are compatible with the power system characteristics.

Cost Determination:

Motor Amperes for the various size kW and voltage are listed in Table R263413-31. To find the amperes, locate the required kW rating and locate the amperes under the appropriate circuit characteristics.

For example:

A. 100 H.P., 3 phase, 460 volt motor = 124 amperes (Table R263413-31)

B. 10 H.P., 3 phase, 200 volt motor = 32.2 amperes (Table R263413-31)

Motor Wire Size: After the amperes are found in Table R263413-31 the amperes must be increased 25% to compensate for power losses. Next refer to Table R260519-92. Find the appropriate insulation column for copper or aluminum wire to determine the proper wire size.

For example:

A. 75 kW, 3 phase, 460 volt motor has an ampere value of 124 amperes from Table R263413-31

B. 124A x 1.25 = 155 amperes

C. Refer to Table R260519-92 for THW or THWN wire insulations to find the proper wire size. For a 155 ampere load using copper wire a size 2/0 wire is needed.

D. For the 3 phase motor three wires of 2/0 size are required.

Conduit Size: To obtain the proper conduit size for the wires and type of insulation used, refer to Table R260533-22.

For example: For the 75 kW, 460V, 3 phase motor, it was determined that three 2/0 wires are required. Assuming THWN insulated copper wire, use Table R260533-22 to determine that three 2/0 wires require 40 mm conduit.

Material Cost of the conduit and wire system depends on:
1. Wire size required
2. Copper or aluminum wire
3. Wire insulation type selected
4. Steel or plastic conduit
5. Type of conduit raceway selected.

Labor Cost of the conduit and wire system depends on:
1. Type and size of conduit
2. Type and size of wires installed
3. Location and height of installation in building or depth of trench
4. Support system for conduit.

R263413-31 Ampere Values Determined by kW, Voltage and Phase Values

kW	Amperes							
	Single Phase			Three Phase				
	115V	208V	230V	200V	208V	230V	460V	575V
.12	4.4A	2.4A	2.2A					
.19	5.8	3.2	2.9					
.25	7.2	4.0	3.6					
.37	9.8	5.4	4.9	2.5A	2.4A	2.2A	1.1A	0.9A
.56	13.8	7.6	6.9	3.7	3.5	3.2	1.6	1.3
.75	16	8.8	8	4.8	4.6	4.2	2.1	1.7
1.1	20	11	10	6.9	6.6	6.0	3.0	2.4
1.5	24	13.2	12	7.8	7.5	6.8	3.4	2.7
2.2	34	18.7	17	11.0	10.6	9.6	4.8	3.9
3.7	56	30.8	28	17.5	16.7	15.2	7.6	6.1
5.6	80	44	40	25.3	24.2	22	11	9
7.5	100	55	50	32.2	30.8	28	14	11
11				48.3	46.2	42	21	17
15				62.1	59.4	54	27	22
19				78.2	74.8	68	34	27
22				92	88	80	40	32
30				120	114	104	52	41
37				150	143	130	65	52
45				177	169	154	77	62
56				221	211	192	96	77
75				285	273	248	124	99
93				359	343	312	156	125
112				414	396	360	180	144
149				552	528	480	240	192
187							302	242
224							361	289
261							414	336
298							477	382

R263413-30 Cost Determination (cont.)

Magnetic starters, switches, and motor connection:

To complete the cost picture from kW to Costs additional items must be added to the cost of the conduit and wire system to arrive at a total cost.

1. Assembly Table D5020-160 Magnetic Starters Installed Cost lists the various size starters for single phase and three phase motors.
2. Assembly Table D5020-165 Heavy Duty Safety Switches Installed Cost lists safety switches required at the beginning of a motor circuit and also one required in the vicinity of the motor location.
3. Assembly Table D5020-170 Motor Connection lists the various costs for single and three phase motors.

Worksheet to obtain total motor wiring costs:

It is assumed that the motors or motor driven equipment are furnished and installed under other sections for this estimate and the following work is done under this section:

1. Conduit
2. Wire (add 10% for additional wire beyond conduit ends for connections to switches, boxes, starters, etc.)
3. Starters
4. Safety switches
5. Motor connections

Figure R263413-32

				Cost	
Item	Type	Size	Quantity	Unit	Total
Wire					
Conduit					
Switch					
Starter					
Switch					
Motor Connection					
Other					
Total Cost					

Worksheet for Motor Circuits

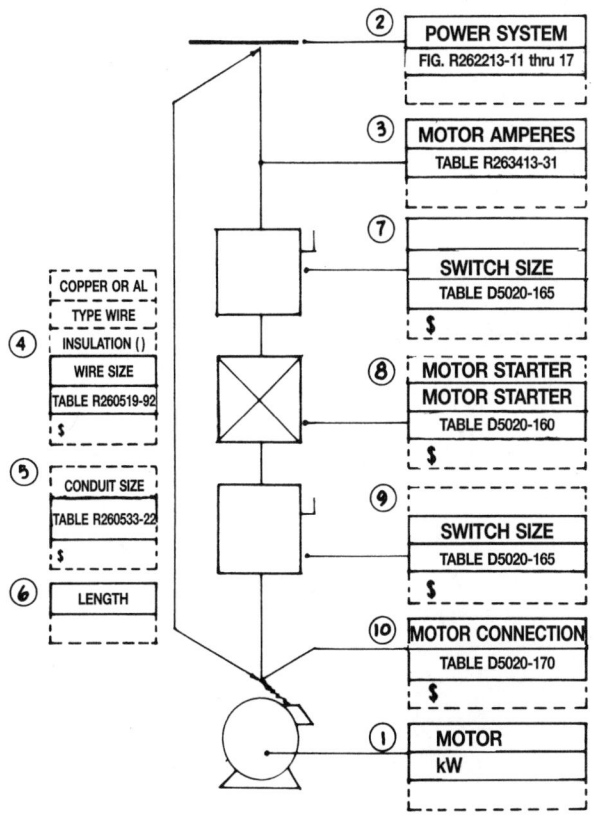

R263413-33 Maximum kW for Starter Size by Voltage

Starter	Maximum kW (3ϕ)			
Size	208V	240V	480V	600V
00	1.1	1.1	1.5	1.5
0	2.2	2.2	3.7	3.7
1	5.6	5.6	7.5	7.5
2	7.5	11	19	19
3	19	22	37	37
4	30	37	75	75
5		75	149	149
6		149	224	224
7		224	448	448
8		336	671	671
8L		522	1120	1120

R263623-60 Automatic Transfer Switches

When taking off Automatic transfer switches identify by voltage, amperage and number of poles.

Example: Automatic transfer switch, 480V, 3 phase, 30A, NEMA 1 enclosure

Labor-hours for transfer switches include:

1. Unloading, uncrating and handling switches up to 60 m from loading dock
2. Measuring and marking
3. Drilling (4) lead type anchors, using a hammer drill
4. Mounting and leveling
5. Circuit identification
6. Testing and load balancing

Labor-hours do not include:

1. Modifications in enclosure
2. Structural supports
3. Additional lugs
4. Plywood backboard
5. Painting or lettering
6. Conduit runs to or from transfer switch
7. Wire
8. Termination of wires

R265113-40   Interior Lighting Fixtures

When taking off interior lighting fixtures, it is advisable to set up your quantity work sheet to conform to the lighting schedule as it appears on the print. Include the alpha-numeric code plus the symbol on your work sheet.

Take off a particular section or floor of the building and count each type of fixture before going on to another type. It would also be advantageous to include on the same work sheet the pipe, wire, fittings and circuit number associated with each type of lighting fixture. This will help you identify the costs associated with any particular lighting system and in turn make material purchases more specific as to when and how much to order under the classification of lighting.

By taking off lighting first you can get a complete "WALK THRU" of the job. This will become helpful when doing other phases of the project.

Materials for a recessed fixture include:
1. Fixture
2. Lamps
3. 1800 mm of jack chain
4. (2) S hooks
5. (2) Wire nuts

Labor for interior recessed fixtures include:
1. Unloading by hand
2. Hauling by hand to an area up to 60 m from loading dock
3. Uncrating
4. Layout
5. Installing fixture
6. Attaching jack chain & S hooks
7. Connecting circuit power
8. Reassembling fixture
9. Installing lamps
10. Testing

Material for surface mounted fixtures includes:
1. Fixture
2. Lamps
3. Either (4) lead type anchors, (4) toggle bolts, or (4) ceiling grid clips
4. (2) Wire nuts

Material for pendent mounted fixtures includes:
1. Fixture
2. Lamps
3. (2) Wire nuts

4. Rigid pendents as required by type of fixtures
5. Canopies as required by type of fixture

Labor hours include the following for both surface and pendent fixtures:
1. Unloading by hand
2. Hauling by hand to an area up to 60 m from loading dock
3. Uncrating
4. Layout and marking
5. Drilling (4) holes for either lead anchors or toggle bolts using a hammer drill
6. Installing fixture
7. Leveling fixture
8. Connecting circuit power
9. Installing lamps
10. Testing

Labor for surface or pendent fixtures does not include:
1. Conduit
2. Boxes or covers
3. Connectors
4. Fixture whips
5. Special support
6. Switching
7. Wire

Economy of Scale: For large concentrations of lighting fixtures in the same area deduct the following percentages from labor:

25	to	50	fixtures	15%
51	to	75	fixtures	20%
76	to	100	fixtures	25%
101	and over		30%

Job Conditions: Productivity is based on new construction in a unobstructed first floor location, using rolling staging to 4.5 m high.

Material staging is assumed to be within 30 m of work being performed.

Add the following percentages to labor for elevated installations:

4.5 m	to	6.0 m	high	10%
6.1 m	to	7.5 m	high	20%
7.6 m	to	9.0 m	high	30%
9.1 m	to	10.5 m	high	40%
10.6 m	to	12.0 m	high	50%
12.1 m	and over			60%

ELECTRICAL

R265723-05   Comparison - Operation of High Intensity Discharge Lamps

Lamp Type	Wattage	Life (Hours)	Circuit Wattage [1]	Average Initial Lumens	L.L.D. [2]	Mean Lumens [3]
M.V.	100 DX	24,000	125	4,000	61%	2,440
L.P.S.	SOX-35	18,000	65	4,800	100%	4,800
H.P.S.	LU-70	12,000	84	5,800	90%	5,220
L.H.	No Equivalent					
M.V.	175 DX	24,000	210	8,500	66%	5,676
M.V.	250 DX	24,000	295	13,000	66%	7,986
L.P.S.	SOX-55	18,000	82	8,000	100%	8,000
H.P.S.	LU-100	12,000	120	9,500	90%	8,550
L.H.	No Equivalent					
M.V.	400 DX	24,000	465	24,000	64%	14,400
L.P.S.	SOX-90	18,000	141	13,500	100%	13,500
H.P.S.	LU-150	16,000	188	16,000	90%	14,400
L.H.	LH-175	7,500	210	14,000	73%	10,200
M.V.	No Equivalent					
L.P.S.	SOX-135	18,000	147	22,500	100%	22,500
H.P.S.	LU-250	20,000	310	25,500	92%	23,205
L.H.	LH-250	7,500	295	20,500	78%	16,000
M.V.	1000 DX	24,000	1,085	63,000	61%	37,820
L.P.S.	SOX-180	18,000	248	33,000	100%	33,000
H.P.S.	LU-400	20,000	480	50,000	90%	45,000
L.H.	LH-400	15,000	465	34,000	72%	24,600
H.P.S.	LU-1000	15,000	1,100	140,000	91%	127,400

1. Includes ballast losses and average lamp watts
2. Lamp lumen depreciation (% of initial light output at 70% rated life)
3. Lamp lumen output at 70% rated life (L.L.D. x initial)
4. Cost of yearly operation = Cicuit Wattage/1000 x annual operating hours x average cost per KWH

M.V. = Mercury Vapor
L.P.S. = Low pressure sodium
H.P.S. = High pressure sodium
L.H. = Metal halide

R265723-10   For Other than Regular Cool White (CW) Lamps

	Multiply Material Costs as Follows:				
Regular Lamps	Cool white deluxe (CWX)	x 1.35	Energy Saving Lamps	Cool white (CW/ES)	x 1.35
	Warm white deluxe (WWX)	x 1.35		Cool white deluxe (CWX/ES)	x 1.65
	Warm white (WW)	x 1.30		Warm white (WW/ES)	x 1.55
	Natural (N)	x 2.05		Warm white deluxe (WWX/ES)	x 1.65

R265723-20   Lamp Comparison Chart with Enclosed Floodlight, Ballast, & Lamp for Pole Mounting

Type	Watts	Initial Lumens	Lumens per Watt	Lumens @ 40% Life	Life (Hours)
Incandescent	150	2,880	19	85	750
	300	6,360	21	84	750
	500	10,850	22	80	1,000
	1,000	23,740	24	80	1,000
	1,500	34,400	23	80	1,000
Tungsten	500	10,950	22	97	2,000
Halogen	1,500	35,800	24	97	2,000
Fluorescent	40	3,150	79	88	20,000
Cool	110	9,200	84	87	12,000
White	215	16,000	74	81	12,000
Deluxe	250	12,100	48	86	24,000
Mercury	400	22,500	56	85	24,000
	1,000	63,000	63	75	24,000
Metal	175	14,000	80	77	7,500
Halide	400	34,000	85	75	15,000
	1,000	100,000	100	83	10,000
	1,500	155,000	103	92	1,500
High	70	5,800	83	90	20,000
Pressure	100	9,500	95	90	20,000
Sodium	150	16,000	107	90	24,000
	400	50,000	125	90	24,000
	1,000	140,000	140	90	24,000
Low	55	4,600	131	98	18,000
Pressure	90	12,750	142	98	18,000
Sodium	180	33,000	183	98	18,000

Color: High Pressure Sodium — Slightly Yellow
    Low Pressure Sodium — Yellow
    Mercury Vapor — Green-Blue
    Metal Halide — Blue White

Note: Pole not included.

R265723-25   Energy Efficiency Rating for Luminaires

The energy efficiency program for luminaires recommends the use of a
metric called the Luminaire Efficacy Rating (LER). The LER value expresses
the total luminaires generated by a lamp, to the watts consumed by the lamp.

Lamp Type	Lumens per Watt
Incandescent	17
Tungsten Halogen	14 - 20
Fluorescent	50 - 104
Metal Halide	64 - 96
High Pressure Sodium	76 - 116

R271323-40 Fiber Optics

Fiber optic systems use optical fiber such as plastic, glass, or fused silica, a transparent material, to transmit radiant power (i.e., light) for control, communication, and signaling applications. The types of fiber optic cables can be nonconductive, conductive, or composite. The composite cables contain fiber optics and current-carrying electrical conductors. The configuration for one of the fiber optic systems is as follows:

The transceiver module acts as transmitting and receiving equipment in a common house, which converts electrical energy to light energy or vice versa.

Pricing the fiber optic system is not an easy task. The performance of the whole system will affect the cost significantly. New specialized tools and techniques decrease the installing cost tremendously. In the fiber optic section of Means Electrical Cost Data, a benchmark for labor-hours and material costs is set up so that users can adjust their costs according to unique project conditions.

Units for Measure: Fiber optic cable is measured in hundred linear feet (C.L.F.) or industry units of measure - meter (m) or kilometer (km). The connectors are counted as units (EA.)

Material Units: Generally, the material costs include only the cable. All the accessories shall be priced separately.

Labor Units: The following procedures are generally included for the installation of fiber optic cables:

- Receiving
- Material handling
- Setting up pulling equipment
- Measuring and cutting cable
- Pulling cable

These additional items are listed and extended: Terminations

Takeoff Procedure: Cable should be taken off by type, size, number of fibers, and number of terminations required. List the lengths of each type of cable on the takeoff sheets. Total and add 10% for waste. Transfer the figures to a cost analysis sheet and extend.

| Communications | R2715 | Communications Horizontal Cabling |

R271513-75 High Performance Cable

There are several categories used to describe high performance cable. The following information includes a description of categories CAT 3, 5, 5e, 6, and 7, and details classifications of frequency and specific standards. The category standards have evolved under the sponsorship of organizations such as the Telecommunication Industry Association (TIA), the Electronic Industries Alliance (EIA), the American National Standards Institute (ANSI), the International Organization for Standardization (ISO), and the International Electrotechnical Commission (IEC), all of which have catered to the increasing complexities of modern network technology. For network cabling, users must comply with national or international standards. A breakdown of these categories is as follows:

Category 3: Designed to handle frequencies up to 16 MHz.

Category 5: (TIA/EIA 568A) Designed to handle frequencies up to 100 MHz.

Category 5e: Additional transmission performance to exceed Category 5.

Category 6 (draft): Development by TIA and other international groups to handle frequencies of 250 MHz.

Category 7 (draft): Under development to handle a frequency range from 1 to 600 MHz.

REFERENCE TABLES

R312316-40 Excavating

The selection of equipment used for structural excavation and bulk excavation or for grading is determined by the following factors.
1. Quantity of material.
2. Type of material.
3. Depth or height of cut.
4. Length of haul.
5. Condition of haul road.
6. Accessibility of site.
7. Moisture content and dewatering requirements.
8. Availability of excavating and hauling equipment.

Some additional costs must be allowed for hand trimming the sides and bottom of concrete pours and other excavation below the general excavation.

When planning excavation and fill, the following should also be considered.
1. Swell factor.
2. Compaction factor.
3. Moisture content.
4. Density requirements.

A typical example for scheduling and estimating the cost of excavation of a 4.5 m deep basement on a dry site when the material must be hauled off the site, is outlined below.

Assumptions:
1. Swell factor, 18%.
2. No mobilization or demobilization.
3. Allowance included for idle time and moving on job.
4. No dewatering, sheeting, or bracing.
5. No truck spotter or hand trimming.

Number of B.m^3 per truck = 1.15 m^3 bucket x 8 passes = 9.2 loose m^3

$$= 9.2 \times \frac{100}{118} = 7.8 \text{ B.m}^3 \text{ per truck}$$

Truck Haul Cycle:

Load truck 8 passes	= 4 minutes
Haul distance 1.6 kilometers	= 9 minutes
Dump time	= 2 minutes
Return 1.6 kilometers	= 7 minutes
Spot under machine	= 1 minute
	= 23 minute cycle

Add the mobilization and demobilization costs to the total excavation costs. When equipment is rented for more than three days, there is often no mobilization charge by the equipment dealer. On larger jobs outside of urban areas, scrapers can move earth economically provided a dump site or fill area and adequate haul roads are available. Excavation within sheeting bracing or cofferdam bracing is usually done with a clamshell and production

Fleet Haul Production per day in B.m^3

$$4 \text{ trucks} \times \frac{50 \text{ min. hour}}{23 \text{ min. haul cycle}} \times 8 \text{ hrs.} \times 7.8 \text{ B.m}^3$$

$$= 4 \times 2.2 \times 8 \times 7.8 = 549 \text{ B.m}^3/\text{day}$$

is low, since the clamshell may have to be guided by hand between the bracing. When excavating or filling an area enclosed with a wellpoint system, add 10% to 15% to the cost to allow for restricted access. When estimating earth excavation quantities for structures, allow work space outside the building footprint for construction of the foundation, and a slope of 1:1 unless sheeting is used.

R312316-45   Excavating Equipment

The table below lists THEORETICAL hourly production in m³/hr. bank measure for some typical excavation equipment. Figures assume 50 minute hours, 83% job efficiency, 100% operator efficiency, 90° swing and properly sized hauling units, which must be modified for adverse digging and loading conditions. Actual production costs in the front of the book average about 50% of the theoretical values listed here.

Equipment	Soil Type	B.m³ Weight	% Swell	0.76 m³	1.15 m³	1.50 m³	1.90 m³	2.30 m³	2.70 m³	3.05 m³
Hydraulic Excavator	Moist loam, sandy clay	1542 kg	40%	65	96	134	168	210	252	291
"Backhoe"	Sand and gravel	1406	18	61	92	122	157	199	237	279
4.5 m Deep Cut	Common earth	1270	30	54	80	115	145	183	214	252
	Clay, hard, dense	1361	33	50	76	99	130	161	195	229
	Moist loam, sandy clay	1542	40	130 (1.8)	187 (2.1)	226 (2.4)	256 (2.6)	294 (2.7)	333 (2.8)	363 (2.9)
	Sand and gravel	1406	18	126 (1.8)	1/2 (2.1)	210 (2.4)	248 (2.6)	287 (2.7)	321 (2.8)	352 (2.9)
Power Shovel Optimum Cut (m)	Common earth	1270	30	111 (2.4)	153 (2.8)	191 (3.1)	226 (3.4)	256 (3.7)	287 (4.0)	325 (4.2)
	Clay, hard, dense	1361	33	92 (2.7)	134 (3.3)	168 (3.7)	195 (4.1)	229 (4.3)	256 (4.6)	287 (4.9)
	Moist loam, sandy clay	1542	40	99 (2.0)	138 (2.3)	168 (2.4)	191 (2.6)	222 (2.7)	248 (2.9)	294 (3.0)
	Sand and gravel	1406	18	99 (2.0)	134 (2.3)	161 (2.4)	187 (2.6)	214 (2.7)	241 (2.9)	287 (3.0)
Drag Line Optimum Cut (m)	Common earth	1270	30	84 (2.4)	122 (2.7)	145 (3.0)	168 (3.2)	191 (3.4)	214 (3.5)	237 (3.7)
	Clay, hard, dense	1361	33	69 (2.8)	99 (3.3)	122 (3.6)	145 (3.7)	172 (3.9)	191 (4.1)	214 (3.7)

				Wheel Loaders				Track Loaders		
				2.30 m³	3.05 m³	3.60 m³	6.10 m³	1.70 m³	2.30 m³	3.05 m³
	Moist loam, sandy clay	1542	40	199	260	390	528	103	138	191
	Sand and gravel	1406	18	187	245	367	497	99	130	180
Loading Tractors	Common earth	1270	30	176	229	352	474	92	119	168
	Clay, hard, dense	1361	33	153	206	317	428	84	111	153
	Rock, well-blasted	1814	50	138	187	291	398	76	99	138

R312319-90 Wellpoints

A single stage wellpoint system is usually limited to dewatering an average 4.5 m depth below normal ground water level. Multi-stage systems are employed for greater depth with the pumping equipment installed only at the lowest header level. Ejectors, with unlimited lift capacity, can be economical when two or more stages of wellpoints can be replaced or when horizontal clearance is restricted, such as in deep trenches or tunneling projects, and where low water flows are expected. Wellpoints are usually spaced on 0.75 m to 3 m centers along a header pipe. Wellpoint spacing, header size, and pump size are all determined by the expected flow, as dictated by soil conditions.

In almost all soils encountered in wellpoint dewatering, the wellpoints may be jetted into place. Cemented soils and stiff clays may require sand wicks about 310 mm in diameter around each wellpoint to increase efficiency and eliminate weeping into the excavation. These sand wicks require 0.4 to 2.3 m³ of washed filter sand and are installed by using a 310 mm diameter steel casing and hole puncher jetted into the ground 0.61 m deeper than the wellpoint. Rock may require predrilled holes.

Labor required for the complete installation and removal of a single stage wellpoint system is in the range of 2.5 to 6.52 labor-hours per meter of header, depending upon jetting conditions, wellpoint spacing, etc.

Continuous pumping is necessary except in some free draining soil where temporary flooding is permissible (as in trenches which are backfilled after each day's work). Good practice requires provision of a stand-by pump during the continuous pumping operation.

Systems for continuous trenching below the water table should be installed three to four times the length of expected daily progress to insure uninterrupted digging, and header pipe size should not be changed during the job.

For pervious free draining soils, deep wells in place of wellpoints may be economical because of lower installation and maintenance costs. Daily production ranges between two to three wells per day, for 7.5 m to 12 m depths, to one well per day for depths over 15 m.

Detailed analysis and estimating for any dewatering problem is available at no cost from wellpoint manufacturers. Major firms will quote "sufficient equipment" quotes or their affiliates offer lump sum proposals to cover complete dewatering responsibility.

Description for 61 m System with 200 mm Header		Quantities
Equipment & Material	Wellpoints 7.5 m long, 50 mm diameter @ 1.5 m O.C.	40 Each
	Header pipe, 200 mm diameter	61 m
	Discharge pipe, 200 mm diameter	30 m
	200 mm valves	3 Each
	Combination Jetting & Wellpoint pump (standby)	1 Each
	Wellpoint pump, 200 mm diameter	1 Each
	Transportation to and from site	1 Day
	Fuel 30 days x 227 liter/day	6814 Liters
	Lubricants for 30 days x 7.25 kg/day	218 kg
	Sand for points	31 m³
	Technician to supervise installation	1 Week
Labor	Labor for installation and removal of system	300 Labor-hours
	4 Operators straight time 40 hrs./wk. for 4.33 wks.	693 Hrs.
	4 Operators overtime 2 hrs./wk. for 4.33 wks.	35 Hrs.

R312323-30 Compacting Backfill

Compaction of fill in embankments, around structures, in trenches, and under slabs is important to control settlement. Factors affecting compaction are:
1. Soil gradation
2. Moisture content
3. Equipment used
4. Depth of fill per lift
5. Density required

Example:

Compact granular fill around a building foundation using a 530 mm wide x 610 mm vibratory plate in 200 mm lifts. Operator moves at 15 m/minute working a 50 minute hour to develop 95% Modified Proctor Density with 4 passes.

Production Rate:

0.53 m plate width x 15 m/minute x 50 min./hr. x 0.20 m lift = 79.5 m³/hr.

Production Rate for 4 Passes:

$$\frac{79.5 \text{ m}^3}{4 \text{ passes}} = 20 \text{ m}^3/\text{hr. x 8 hrs.} = 160 \text{ m}^3/\text{day}$$

Earthwork	**R3141**	**Shoring**

R314116-40   Wood Sheet Piling

Wood sheet piling may be used for depths to 6 m where there is no ground water. If moderate ground water is encountered Tongue & Groove sheeting will help to keep it out. When considerable ground water is present, steel sheeting must be used.

For estimating purposes on trench excavation, sizes are as follows:

Depth	Sheeting	Wales	Braces	m³ per m²
To 2.5 m	75 mm x 310 mm	150 mm x 200 mm, 2 line	150 mm x 200 mm, @ 3 m	0.10 @ 2.5 m
2.5m to 3.5 m	75 mm x 310 mm	250 mm x 250 mm, 2 line	250 mm x 250 mm, @ 2.7 m	0.13 average
3.5 m to 6 m	75 mm x 310 mm	310 mm x 310 mm, 3 line	310 mm x 310 mm, @ 2.5 m	0.18 average

Sheeting to be toed in at least 0.6 m depending upon soil conditions. A crew of five with an air compressor and sheeting driver can drive and brace 41 m²/day at 2.5 m deep, 33 m²/day at 3.5 m deep, and 30 m²/day at 5 m deep. For normal soils, piling can be pulled in 1/3 the time to install. Pulling difficulty increases with the time in the ground. Production can be increased by high pressure jetting.

R314116-45   Steel Sheet Piling

Limiting weights are 107 to 185 kg/m² of wall surface with 132 kg/m² average for usual types and sizes. (Weights of piles themselves are from 46 kg/m to 85 kg/m but they are 380 mm to 530 mm wide.) Lightweight sections 310 mm to 710 mm wide from 3 ga. to 12 ga. thick are also available for shallow excavations. Piles may be driven two at a time with an impact or vibratory hammer (use vibratory to pull) hung from a crane without leads. A reasonable estimate of the life of steel sheet piling is 10 uses with up to 125 uses possible if a vibratory hammer is used. Used piling costs from 50% to 80% of new piling depending on location and market conditions. Sheet piling and H piles can be rented for about 30% of the delivered mill price for the first month and 5% per month thereafter. Allow 1 labor-hour per pile for cleaning and trimming after driving. These costs increase with depth and hydrostatic head. Vibratory drivers are faster in wet granular soils and are excellent for pile extraction. Pulling difficulty increases with the time in the ground and may cost more than driving. It is often economical to abandon the sheet piling, especially if it can be used as the outer wall form. Allow about 1/3 additional length or more, for toeing into ground. Add bracing, waler and strut costs. Waler costs can equal the cost per metric ton of sheeting.

Earthwork	**R3145**	**Vibroflotation & Densification**

R314513-90   Vibroflotation & Vibro Replacement Soil Compaction

Vibroflotation is a proprietary system of compacting sandy soils in place to increase relative density to about 70%. Typical bearing capacities attained will be 287 kPa for saturated sand and 575 kPa for dry sand. Usual range is 192 to 383 kPa capacity. Costs in the front of the book are for a vertical meter of compacted cylinder 1.5 m to 3 m in diameter.

Vibro replacement is a proprietary system of improving cohesive soils in place to increase bearing capacity. Most silts and clays above or below the water table can be strengthened by installation of stone columns.

The process consists of radial displacement of the soil by vibration. The created hole is then backfilled in stages with coarse granular fill which is thoroughly compacted and displaced into the surrounding soil in the form of a column.

The total project cost would depend on the number and depth of the compacted cylinders. The installing company guarantees relative soil density of the sand cylinders after compaction and the bearing capacity of the soil after the replacement process. Detailed estimating information is available from the installer at no cost.

R316000-20 Pile Caps, Piles and Caissons

General: The function of a reinforced concrete pile cap is to transfer superstructure load from isolated column or pier to each pile in its supporting cluster. To do this, the cap must be thick and rigid, with all piles securely embedded into and bonded to it.

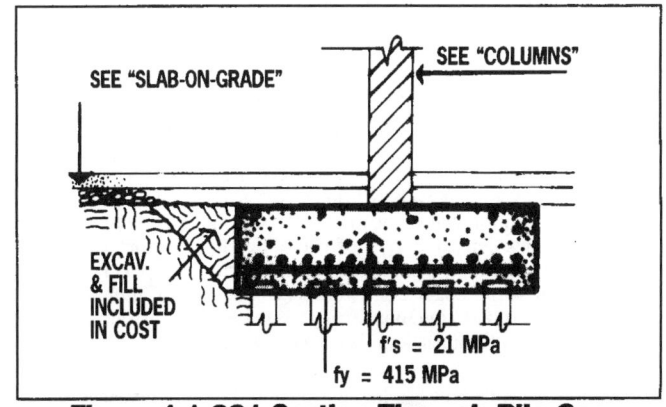

Figure 1.1-331 Section Through Pile Cap

Table 1.1-332 Concrete Quantities for Pile Caps

Load Working (kN)	Number of Piles @ 1 m O.C. Per Footing Cluster									
	2 (m³)	4 (m³)	6 (m³)	8 (m³)	10 (m³)	12 (m³)	14 (m³)	16 (m³)	18 (m³)	20 (m³)
220	(.69)	(1.45)	(2.52)	(3.75)	(4.36)	(5.96)	(7.57)	(8.49)	(11.00)	(12.62)
445	(0.76)	(1.68)	(2.52)	(3.75)	(4.36)	(5.96)	(7.57)	(8.49)	(11.00)	(12.62)
890	(0.76)	(1.68)	(3.06)	(3.75)	(4.36)	(5.96)	(7.57)	(8.49)	(11.00)	(12.62)
1780	(0.84)	(1.99)	(3.98)	(4.82)	(5.66)	(6.27)	(10.47)	(8.49)	(11.00)	(12.62)
3560		(2.22)	(4.43)	(5.73)	(7.03)	(10.40	(13.46)	(12.16)	(15.06)	(16.90)
5340			(4.43)	(6.35)	(7.42)	(10.86)	(13.99)	(15.60)	(16.21)	(17.36)
7100				(7.49)	(8.72)	(11.09)	(14.91)	(15.60)	(18.81)	(20.80)
8890				(7.49)	(8.72)	(12.69)	(18.43)	(16.59)	(19.88)	(22.02)
13,300						(13.38)		(20.26)	(23.17)	(25.15)
17,800								(23.09)	(23.47)	(27.91)

Table 1.1-333 Concrete Quantities for Pile Caps

Load Working (kN)	Number of Piles @ 1.37 m O.C. Per Footing					
	2 (m³)	3 (m³)	4 (m³)	5 (m³)	6 (m³)	7 (m³)
220	(1.76)	(2.75)	(4.28)	(8.41)	(10.47)	(9.86)
445	(1.76)	(2.75)	(4.28)	(8.41)	(10.47)	(9.86)
890	(1.76)	(2.75)	(4.28)	(8.41)	(10.47)	(9.86)
1780	(2.29)	(2.75)	(4.28)	(8.41)	(10.47)	(9.86)
3560			(4.74)	(8.79)	(10.70)	(9.86)
5340				(9.94)	(10.47)	(10.25)
7100						(10.70)

R316000-20   Pile Caps, Piles and Caissons (cont.)

General: Piles are column-like shafts which receive superstructure loads, overturning forces, or uplift forces. They receive these loads from isolated column or pier foundations (pile caps), foundation walls, grade beams, or foundation mats. The piles then transfer these loads through shallower poor soil strata to deeper soil of adequate support strength and acceptable settlement with load.

Be sure that other foundation types aren't better suited to the job. Consider ground and settlement, as well as loading, when reviewing. Piles usually are associated with difficult foundation problems and substructure condition. Ground conditions determine type of pile (different pile types have been developed to suit ground conditions.) Decide each case by technical study, experience and sound engineering judgment—not rules of thumb. A full investigation of ground conditions, early, is essential to provide maximum information for professional foundation engineering and an acceptable structure.

Piles support loads by end bearing and friction. Both are generally present; however, piles are designated by their principal method of load transfer to soil.

Boring should be taken at expected pile locations. Ground strata (to bedrock or depth of 1-1/2 building width) must be located and identified with appropriate strengths and compressibilities. The sequence of strata determines if end bearing or friction piles are best suited.

End bearing piles have shafts which pass through soft strata or thin hard strata and tip bear on bedrock or penetrate some distance into a dense, adequate soil (sand or gravel.)

Friction piles have shafts which may be entirely embedded in cohesive soil (moist clay), and develop required support mainly by adhesion or "skin-friction" between soil and shaft area.

Piles pass through soil by either one of two ways:
1. Displacement piles force soil out of the way. This may cause compaction, ground heaving, remolding of sensitive soils, damage to adjacent structures, or hard driving.
2. Non-displacement piles have either a hole bored and the pile cast or placed in hole, or open ended pipe (casing) driven and the soil core removed. They tend to eliminate heaving or lateral pressure damage to adjacent structures of piles. Steel "HP" piles are considered of small displacement.

Placement of piles (attitude) is most often vertical; however, they are sometimes battered (placed at a small angle from vertical) to advantageously resist lateral loads. Seldom are piles installed singly but rather in clusters (or groups). Codes require a minimum of three piles per major column load or two per foundation wall or grade beam. Single pile capacity is limited by pile structural strength or support strength of soil. Support capacity of a pile cluster is almost always less than the sum of its individual pile capacities due to overlapping of bearing the friction stresses.

Large rigs for heavy, long piles create large soil surface loads and additional expense on weak ground.

Fewer piles create higher costs per pile.

Pile load tests are frequently required by code, ground situation, or pile type.

R316000-20 Pile Caps, Piles and Caissons (cont.)

General: Caissons, as covered in this section, are drilled cylindrical foundation shafts which function primarily as short column-like compression members. They transfer superstructure loads through inadequate soils to bedrock or hard stratum. They may be either reinforced or unreinforced and either straight or belled out at the bearing level.

Shaft diameters range in size from 500 mm to 2150 mm with the most usual sizes beginning at 865 mm. If inspection of bottom is required, the minimum diameter practical is 760 mm. If handwork is required (in addition to mechanical belling, etc.) the minimum diameter is 800 mm. The most frequently used shaft diameter is probably 900 mm with a 1500 mm or 1800 mm bell diameter. The maximum bell diameter practical is three times the shaft diameter.

Plain concrete is commonly used, poured directly against the excavated face of soil. Permanent casings add to cost and economically should be avoided. Wet or loose strata are undesirable. The associated installation sometimes involves a mudding operation with bentonite clay slurry to keep walls of excavation stable (costs not included here).

Reinforcement is sometimes used, especially for heavy loads. It is required if uplift, bending moment, or lateral loads exist. A small amount of reinforcement is desirable at the top portion of each caisson, even if the above conditions theoretically are not present. This will provide for construction eccentricities and other possibilities. Reinforcement, if present, should extend below the soft strata. Horizontal reinforcement is not required for belled bottoms.

There are three basic types of caisson bearing details:

1. Belled, which are generally recommended to provide reduced bearing pressure on soil. These are not for shallow depths or poor soils. Good soils for belling include most clays, hardpan, soft shale, and decomposed rock.

Soils requiring handwork include hard shale, limestone, and sandstone.

Soils not recommended include sand, gravel, silt, and igneous rock. Compact sand and gravel above water table may stand. Water in the bearing strata is undesirable.

2. Straight shafted, which have no bell but the entire length is enlarged to permit safe bearing pressures. They are most economical for light loads on high bearing capacity soil.

3. Socketed (or keyed), which are used for extremely heavy loads. They involve sinking the shaft into rock for combined friction and bearing support action. Reinforcement of shaft is usually necessary. Wide flange cores are frequently used here.

Advantages include:

A. Shafts can pass through soils that piles cannot

B. No soil heaving or displacement during installation

C. No vibration during installation

D. Less noise than pile driving

E. Bearing strata can be visually inspected & tested

Uses include:

A. Situations where unsuitable soil exists to moderate depth

B. Tall structures

C. Heavy structures

D. Underpinning (extensive use)

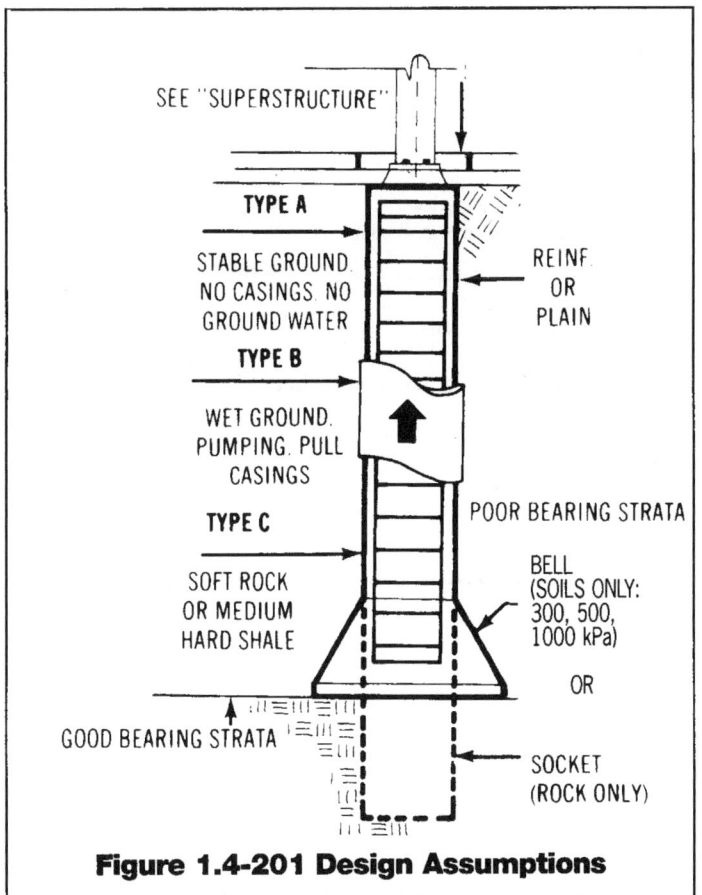

Figure 1.4-201 Design Assumptions

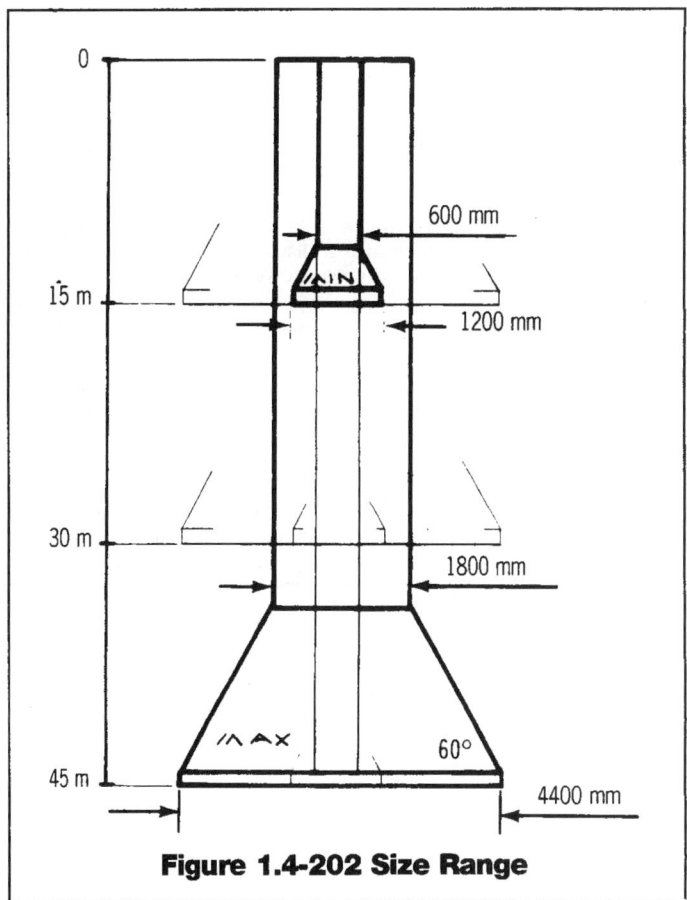

Figure 1.4-202 Size Range

R316326-60   Caissons

The three principal types of caissons are **(1) Belled Caissons,** which except for shallow depths and poor soil conditions, are generally recommended. They provide more bearing than shaft area. Because of its conical shape, no horizontal reinforcement of the bell is required.

(2) Straight Shaft Caissons are used where relatively light loads are to be supported by caissons that rest on high value bearing strata. While the shaft is larger in diameter than for belled types this is more than offset by the saving in time and labor.

(3) Keyed Caissons are used when extremely heavy loads are to be carried. A keyed or socketed caisson transfers its load into rock by a combination of end-bearing and shear reinforcing of the shaft. The most economical shaft often consists of a steel casing, a steel wide flange core and concrete. Allowable compressive stresses of .225 f'c for concrete, 110 MPa for the wide flange core, and 62 MPa for the steel casing are commonly used. The usual range of shaft diameter is 460 mm to 2130 mm. The number of sizes specified for any one project should be limited due to the problems of casing and auger storage. When hand work is to be performed, shaft diameters should not be less than 815 mm. When inspection of borings is required a minimum shaft diameter of 760 mm is recommended. Concrete caissons are intended to be poured against earth excavation so permanent forms which add to cost should not be used if the excavation is clean and the earth sufficiently impervious to prevent excessive loss of concrete.

Soil Conditions for Belling		
Good	**Requires Handwork**	**Not Recommended**
Clay	Hard Shale	Silt
Sandy Clay	Limestone	Sand
Silty Clay	Sandstone	Gravel
Clayey Silt	Weathered Mica	Igneous Rock
Hard-pan		
Soft Shale		
Decomposed Rock		

R317100-10   Tunnel Excavation

Bored tunnel excavation is common in rock for diameters from 1.2 m for sewer and utilities, to 18.3 m for vehicles. Production varies from a few meters per day to over 61 meters per day. In the smaller diameters, the productivity is limited by the restricted area for mucking or the removal of excavated material.

Most of the tunnels in rock today are excavated by boring machines called moles. Preparation for starting the excavation or setting up the mole is very costly. Shafts must be excavated to the invert of the proposed tunnel and the mole must be lowered into the shaft. If excavating a portal tunnel, that is starting at an open face, the cost is reduced considerably both for mobilization and mucking.

In soft ground and mixed material, special bucket excavators and rotary excavators are used inside a shield. Tunnel liners must follow directly behind the shield to support the earth and prevent cave-ins.

Traditional muck haulage operations are performed by rail with locomotives and muck cars. Sometimes conveyors are more economical and require less ventilation of the tunnel.

Ventilation and air compression are other important cost factors to consider in tunnel excavation. Continuous ventilation ducts are sometimes fabricated at the tunnel site.

Tunnel linings are steel, cast in place reinforced concrete, shotcrete, or a combination of these. When required, contact grouting is performed by pumping grout between the lining and the excavation. Intermittent holes are drilled into the lining and separate costs are determined for drilling per hole, grout pump connecting per hole, and grout per cubic meter.

Consolidation grouting and roof bolts may also be required where the excavation is unstable or faulting occurs.

Tunnel boring is usually done 24 hours per day. A typical crew for rock boring is:

Tunneling Crew based on three 8 hour shifts

1 Shifter
1 Walker
1 Machine Operator for mole
1 Oiler
1 Mechanic
3 Locomotives with operators
5 Miners for rails, vent ducts, and roof bolts
1 Electrician
2 Pumps
2 Laborers for hoisting
1 Hoist operator for muck removal
1 Oiler

Surface Crew Based on normal 8 hour shift

2 Shop Mechanics
1 Electrician
1 Shifter
2 Laborers
1 Operator with 16.3 metric ton cherry picker
1 Operator with front end loader

Exterior Improvements | R3201 | Flexible Paving Surface Treatment

R320113-70 Pavement Maintenance

Routine pavement maintenance should be performed to keep a paved surface from deteriorating under the normal forces of nature and traffic.

The msot important maintenance function is the early detection and repair of minor pavement defects. Cracks and other surface breaks can develop into serious defects if not repaired in their earliest stages. For these reasons a pavement preventive maintenance program should include frequent close inspections of pavement surfaces. When suspicious areas are detected, a detailed investigation should be undertaken to determine the appropriate

repair. Where subsurface or pavement deterioration is detected, the Benkelman Beam can be used to make deflection measurements under normal traffic stresses. This is done to determine the extent of the affected area.

Patching or resurfacing work should be done during warm (10°C and above) and dry weather. Adequate compaction is dificult to achieve when hot or warm mixtures are placed on cold pavements. Asphalt and asphalt mixtures usually do not bond well to damp surfaces.

Exterior Improvements | R3292 | Turf & Grasses

R329219-50 Seeding

The type of grass is determined by light, shade and moisture content of soil plus intended use. Fertilizer should be disked 100 mm before seeding. For steep slopes disk eleven metric tons of mulch and lay four metric tons of hay or straw on surface per hectare after seeding. Surface mulch can be staked, lightly disked or tar emulsion sprayed. Material for mulch can be

wood chips, peat moss, partially rotted hay or straw, wood fibers and sprayed emulsions. Hemp seed blankets with fertilizer are also available. For spring seeding, watering is necessary. Late fall seeding may have to be reseeded in the spring. Hydraulic seeding, power mulching, and aerial seeding can be used on large areas.

Exterior Improvements | R3293 | Plants

R329343-10 Plant Spacing Chart

This chart may be used when plants are to be placed equidistant from each other, staggering their position in each row.

Plant Spacing (mm)	Row Spacing (mm)	Plants Per (10 m²)	Plant Spacing (m)	Row Spacing (m)	Plants Per (100 m²)
			1.2	1.0	78
150	130	497	1.5	1.3	50
200	1.73	280	1.8	1.6	34
250	217	179	2.4	2.1	19
300	260	124	3.0	2.6	13
375	325	80	3.6	3.1	9
450	389	55	4.5	3.8	5.5
525	455	41	6.0	5.2	3.1
600	520	31	7.5	6.5	2.0
750	650	19.9	9.0	7.8	1.4
900	780	13.1	10.5	10.4	0.8

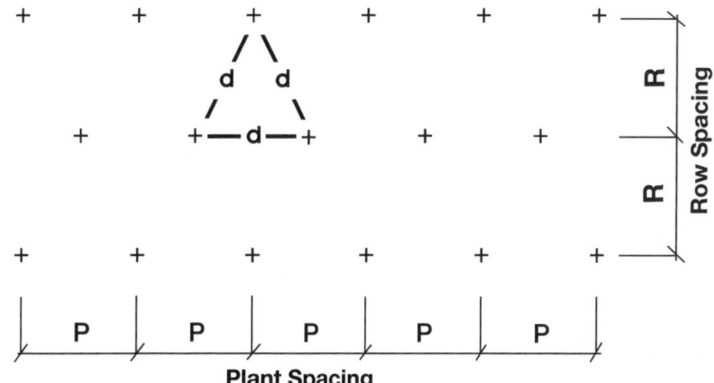

Plant Spacing

R329343-20 Trees and Plants by Environment and Purposes

Dry, Windy, Exposed Areas
Barberry
Junipers, all varieties
Locust
Maple
Oak
Pines, all varieties
Poplar, Hybrid
Privet
Spruce, all varieties
Sumac, Staghorn

Lightly Wooded Areas
Dogwood
Hemlock
Larch
Pine, White
Rhododendron
Spruce, Norway
Redbud

Total Shade Areas
Hemlock
Ivy, English
Myrtle
Pachysandra
Privet
Spice Bush
Yews, Japanese

Cold Temperatures of Northern U.S. and Canada
Arborvitae, American
Birch, White
Dogwood, Silky
Fir, Balsam
Fir, Douglas
Hemlock
Juniper, Andorra
Juniper, Blue Rug
Linden, Little Leaf
Maple, Sugar
Mountain Ash
Myrtle
Olive, Russian

Pine, Mugho
Pine, Ponderosa
Pine, Red
Pine, Scotch
Poplar, Hybrid
Privet
Rosa Rugosa
Spruce, Dwarf Alberta
Spruce, Black Hills
Spruce, Blue
Spruce, Norway
Spruce, White, Engelman
Yellow Wood

Wet, Swampy Areas
American Arborvitae
Birch, White
Black Gum
Hemlock
Maple, Red
Pine, White
Willow

Poor, Dry, Rocky Soil
Barberry
Crownvetch
Eastern Red Cedar
Juniper, Virginiana
Locust, Black
Locust, Bristly
Locust, Honey
Olive, Russian
Pines, all varieties
Privet
Rosa Rugosa
Sumac, Staghorn

Seashore Planting
Arborvitae, American
Juniper, Tamarix
Locust, Black
Oak, White
Olive, Russian
Pine, Austrian
Pine, Japanese Black

Pine, Mugho
Pine, Scotch
Privet, Amur River
Rosa Rugosa
Yew, Japanese

City Planting
Barberry
Fir, Concolor
Forsythia
Hemlock
Holly, Japanese
Ivy, English
Juniper, Andorra
Linden, Little Leaf
Locust, Honey
Maple, Norway, Silver
Oak, Pin, Red
Olive, Russian
Pachysandra
Pine, Austrian
Pine, White
Privet
Rosa Rugosa
Sumac, Staghorn
Yew, Japanese

Bonsai Planting
Azaleas
Birch, White
Ginkgo
Junipers
Pine, Bristlecone
Pine, Mugho
Spruce,k Engleman
Spruce, Dwarf Alberta

Street Planting
Linden, Little Leaf
Oak, Pin
Ginkgo

Fast Growth
Birch, White
Crownvetch
Dogwood, Silky

Fir, Douglas
Juniper, Blue Pfitzer
Juniper, Blue Rug
Maple, Silver
Olive, Autumn
Pines, Austrian, Ponderosa, Red Scotch and White
Poplar, Hybrid
Privet
Spruce, Norway
Spruce, Serbian
Texus, Cuspidata, Hicksi
Willow

Dense, Impenetrable Hedges
Field Plantings:
  Locust, Bristly,
  Olive, Autumn
  Sumac
Residential Area:
  Barberry, Red or Green
  Juniper, Blue Pfitzer
  Rosa Rugosa

Food for Birds
Ash, Mountain
Barberry
Bittersweet
Cherry, Manchu
Dogwood, Silky
Honesuckle, Rem Red
Hawthorn
Oaks
Olive, Autumn, Russian
Privet
Rosa Rugosa
Sumac

Erosion Control
Crownvetch
Locust, Bristly
Willow

R329343-30   Zones of Plant Hardiness

APPROXIMATE RANGE OF
AVERAGE ANNUAL MINIMUM
TEMPERATURES FOR EACH ZONE

ZONE 1	BELOW -50° F
ZONE 2	-50° TO -40°
ZONE 3	-40° TO -30°
ZONE 4	-30° TO -20°
ZONE 5	-20° TO -10°
ZONE 6	-10° TO 0°
ZONE 7	0° TO 10°
ZONE 8	10° TO 20°
ZONE 9	20° TO 30°
ZONE 10	30° TO 40°

R331113-80   Piping Designations

There are several systems currently in use to describe pipe and fittings. The following paragraphs will help to identify and clarify classifications of piping systems used for water distribution.

Piping may be classified by schedule. Piping schedules include 5S, 10S, 10, 20, 30, Standard, 40, 60, Extra Strong, 80, 100, 120, 140, 160 and Double Extra Strong. These schedules are dependent upon the pipe wall thickness. The wall thickness of a particular schedule may vary with pipe size.

Ductile iron pipe for water distribution is classified by Pressure Classes such as Class 150, 200, 250, 300 and 350. These classes are actually the rated water working pressure of the pipe in pounds per square inch (kPa). In most cases the pressure classes are the actual water working pressure plus a surge allowance of 100 psi (690 kPa).

The American Water Works Association (AWWA) provides standards for various types of **plastic pipe**. C-900 is the specification for polyvinyl chloride (PVC) piping used for water distribution in sizes ranging from 100 mm through 300 mm. C-901 is the specification for polyethylene (PE) pressure pipe, tubing and fittings used for water distribution in sizes ranging from 13 mm through 75 mm. C-905 is the specification for PVC piping sizes 350 mm and greater.

PVC pressure-rated pipe is identified using the standard dimensional ratio (SDR) method. This method is defined by the American Society for Testing and Materials (ASTM) Standard D 2241. This pipe is available in SDR numbers 64, 41, 32.5, 26, 21, 17, and 13.5. Pipe with an SDR of 64 will have the thinnest wall while pipe with an SDR of 13.5 will have the thickest wall. When the pressure rating (PR) of a pipe is given in kPa (psi), it is based on a line supplying water at 23 degrees C (73 degrees F).

The National Sanitation Foundation (NSF) seal of approval is applied to products that can be used with potable water. These products have been tested to ANSI/NSF Standard 14.

Valves and strainers are classified by American National Standards Institute (ANSI) Classes. These Classes are 125, 150, 200, 250, 300, 400, 600, 900, 1500 and 2500. Within each class there is an operating pressure range dependent upon temperature. Design parameters should be compared to the appropriate material dependent, pressure-temperature rating chart for accurate valve selection.

R337116-60 Average Transmission Line Material Requirements (Per Kilometer)

		Flat				Rolling				Mountain			
Terrain:													
Item		69kV	161kV	161kV	500kV	69kV	161kV	161kV	500kV	69kV	161kV	161kV	500kV
Pole Type	Unit	Wood	Wood	Steel	Steel	Wood	Wood	Steel	Steel	Wood	Wood	Steel	Steel
Conductor: 397,500 – Cir. Mil., 26/7–ACSR													
Structures	Ea.	8[1]				6[2]				5[2]			
Poles	Ea.	8				12				10			
Crossarms	Ea.	15[3]				6[4]				5[4]			
Conductor	m	3030				3030				3030			
Insulators[5]	Ea.	120				90				75			
Ground Wire	m	1010				2020				2020			
Conductor: 636,000 – Cir. Mil., 26/7–ACSR													
Structures	Ea.	8[1]	7[6]			7[2]	6[2]	4[9]		5[2]	5[2]	4[9]	
Excavation	m³	–	–			–	–	60		–	–	60	
Concrete	m³	–	–			–	–	5		–	–	5	
Steel Towers	M Tons	–	–			–	–	20		–	–	20	
Poles	Ea.	8	7			14	12	–		10	10	–	
Crossarms	Ea.	16[3]	21[7]			7[4]	6[8]	–		5[4]	5[8]	–	
Conductor	m	3030	3030			3030	3030	3030		3030	3030	3030	
Insulators[5]	Ea.	120	105			105	198	198		75	165	198	
Ground Wire	m	1010	1010			2020	2020	2020		2020	2020	2020	
Conductor: 954,000 – Cir. Mil., 45/7–ACSR													
Structures	Ea.	9[1]	8[6]		3[11]	7[2]	6[2]	4[9]	3[11]	5[2]	5[2]	4[9]	3[13]
Excavation	m³	–	–		120	–	–	64	123	–	–	64	135
Concrete	m³	–	–		12	–	–	5	12	–	–	5	2
Steel Towers	M Tons	–	–		39	–	–	20	39	–	–	20	42
Poles	Ea.	9	8		–	14	12	–	–	10	10	–	–
Crossarms	Ea.	18[10]	24[7]		–	7[4]	6[8]	–	–	5[4]	5[8]	–	–
Conductor	m	3030	3030		9090[12]	3030	3030	3030	9090[12]	3030	3030	3030	9090[12]
Insulators[5]	Ea.	135	120		216	105	198	198	216	75	165	198	432
Ground Wire	m	1010	1010		2020	2020	2020	2020	2020	2020	2020	2020	2020
Conductor: 1,351,500 – Cir. Mil., 45/7–ACSR													
Structures	Ea.			5[14]				5[14]					
Excavation	m³			105				105					
Concrete	m³			14				14					
Steel Towers	M Tons			26				26					
Conductor	m			6060[15]				6060[15]					
Insulators	Ea.			330				330					
Ground Wire	m			2020				2020					

1. Single pole two-arm suspension type construction
2. Two-pole wood H-frame construction
3. 125 x 150 x 2400 mm and 125 x 150 x 3000 mm wood crossarm
4. 150 mm x 200 mm x 7.8 m wood crossarm
5. 146 mm x 250 mm disc insulator
6. Single pole construction with 3 fiberglass crossarms (5 fog- type insulators per phase)
7. 2100 mm fiberglass crossarms
8. 150 mm x 250 mm x 10.5 m wood crossarm
9. Laced steel tower, single circuit construction
10. 125 x 175 x 2400 mm and 125 x 175 x 3000 mm wood crossarm
11. Laced steel tower, single circuit 500-kV construction
12. Bundled conductor (3 sub-conductors per phase)
13. Laced steel tower, single circuit restrained phases (500-kV)
14. Laced steel tower, double circuit construction
15. Both sides of double circuit strung

Note: To allow for sagging, a kilometer of transmission line uses 1010 mm of conductor per wire.

R337119-30 Concrete for Conduit Encasement

Table below lists m³ of concrete for 100 m of trench. Conduits separation center to center should meet 109 mm (N.E.C.).

Number of Conduits	1	2	3	4	6	8	9
Trench Dimension	292 mm x 292 mm	292 mm x 482 mm	292 mm x 686 mm	483 mm x 483 mm	483 mm x 686 mm	483 mm x 965 mm	686 mm x 686 mm
Conduit Diameter 50 mm	8.25	13.52	19.16	22.15	31.38	44.30	44.45
65 mm	8.10	13.27	18.79	21.62	30.58	43.22	43.27
80 mm	7.90	12.87	18.16	20.79	29.37	41.61	41.44
90 mm	7.73	12.47	17.61	20.04	28.24	40.08	39.73
100 mm	7.50	12.04	16.96	19.19	26.94	38.38	37.80
125 mm	6.97	10.96	15.33	17.01	23.68	34.04	32.91
150 mm	6.32	9.63	13.37	14.40	19.74	28.80	27.02

| **Transportation** | **R3472** | **Railway Construction** |

R347216-10 Single Track R.R. Siding

The costs for a single track R.R. siding in the Unit Price section include the components shown in the table below.

Description of Component	CSI Line Item	Qty. per m of Track	Unit
Ballast, 38 mm crushed stone	02060-150-0320	1.673	m³
150x200 mm x 2.6 m Treated timber ties, 560 mm O.C.	05655-700-1600	1.786	Ea.
Tie plates, 2 per tie	05655-750-0300	3.572	Ea.
Track rail	05655-750-1000	2.000	m
Spikes, 150 mm, 4 per tie	05655-750-0200	7.144	Ea.
Splice bars w/ bolts, lock washers & nuts, @ 10 m O.C.	05655-750-0100	0.200	Pair
Crew B-14 @ 17 m/Day		0.059	Day

R347216-20 Single Track, Steel Ties, Concrete Bed

The costs for a R.R. siding with steel ties and a concrete bed in the Unit Price section include the components shown in the table below.

Description of Component	CSI Line Item	Qty. per m of Track	Unit
Concrete bed, 2.7 m wide, 250 mm thick	03310-240-3950	0.675	m³
Ties, W150x24 x 2 m long, @ 750 mm O.C.	05120-640-0120	1.333	Ea.
Tie plates, 4 per tie	05655-750-0300	5.333	Ea.
Track rail	05655-750-1000	2.000	m
Tie plate bolts, 25 mm, 8 per tie	05655-750-0020	10.667	Ea.
Splice bars w/ bolts, lock washers & nuts, @ 10 m O.C.	05655-750-0100	0.200	Pair
Crew B-14 @ 6.71 m/Day		0.149	Day

For information about Means Estimating Seminars, see yellow pages 12 and 13 in back of book

Change Orders

Change Order Considerations

A Change Order is a written document, usually prepared by the design professional, and signed by the owner, the architect/engineer and the contractor. A change order states the agreement of the parties to: an addition, deletion, or revision in the work; an adjustment in the contract sum, if any; or an adjustment in the contract time, if any. Change orders, or "extras" in the construction process occur after execution of the construction contract and impact architects/engineers, contractors and owners.

Change orders that are properly recognized and managed can ensure orderly, professional and profitable progress for all who are involved in the project. There are many causes for change orders and change order requests. In all cases, change orders or change order requests should be addressed promptly and in a precise and prescribed manner. The following paragraphs include information regarding change order pricing and procedures.

The Causes of Change Orders

Reasons for issuing change orders include:

- Unforeseen field conditions that require a change in the work
- Correction of design discrepancies, errors or omissions in the contract documents
- Owner-requested changes, either by design criteria, scope of work, or project objectives
- Completion date changes for reasons unrelated to the construction process
- Changes in building code interpretations, or other public authority requirements that require a change in the work
- Changes in availability of existing or new materials and products

Procedures

Properly written contract documents must include the correct change order procedures for all parties—owners, design professionals and contractors—to follow in order to avoid costly delays and litigation.

Being "in the right" is not always a sufficient or acceptable defense. The contract provisions requiring notification and documentation must be adhered to within a defined or reasonable time frame.

The appropriate method of handling change orders is by a written proposal and acceptance by all parties involved. Prior to starting work on a project, all parties should identify their authorized agents who may sign and accept change orders, as well as any limits placed on their authority.

Time may be a critical factor when the need for a change arises. For such cases, the contractor might be directed to proceed on a "time and materials" basis, rather than wait for all paperwork to be processed—a delay that could impede progress. In this situation, the contractor must still follow the prescribed change order procedures, including but not limited to, notification and documentation.

All forms used for change orders should be dated and signed by the proper authority. Lack of documentation can be very costly, especially if legal judgments are to be made and if certain field personnel are no longer available. For time and material change orders, the contractor should keep accurate daily records of all labor and material allocated to the change. Forms that can be used to document change order work are available in *Means Forms for Building Construction Professionals.*

Owners or awarding authorities who do considerable and continual building construction (such as the federal government) realize the inevitability of change orders for numerous reasons, both predictable and unpredictable. As a result, the federal government, the American Institute of Architects (AIA), the Engineers Joint Contract Documents Committee (EJCDC) and other contractor, legal and technical organizations have developed standards and procedures to be followed by all parties to achieve contract continuance and timely completion, while being financially fair to all concerned.

In addition to the change order standards put forth by industry associations, there are also many books available on the subject.

Pricing Change Orders

When pricing change orders, regardless of their cause, the most significant factor is *when* the change occurs. The need for a change may be perceived in the field or requested by the architect/engineer *before* any of the actual installation has begun, or may evolve or appear *during* construction when the item of work in question is partially installed. In the latter cases, the original sequence of construction is disrupted, along with all contiguous and supporting systems. Change orders cause the greatest impact when they occur *after* the installation has been completed and must be uncovered, or even replaced. Post-completion changes may be caused by necessary design changes, product failure, or changes in the owner's requirements that are not discovered until the building or the systems begin to function.

Specified procedures of notification and record keeping must be adhered to and enforced regardless of the stage of construction: *before, during,* or *after* installation. Some bidding documents anticipate change orders by requiring that unit prices including overhead and profit percentages—for additional as well as deductible changes—be listed. Generally these unit prices do not fully take into account the ripple effect, or impact on other trades, and should be used for general guidance only.

When pricing change orders, it is important to classify the time frame in which the change occurs. There are two basic time frames for change orders: *pre-installation change orders*, which occur before the start of construction, and *post-installation change orders*, which involve reworking after the original installation. Change orders that occur between these stages may be priced according to the extent of work completed using a combination of techniques developed for pricing *pre-* and *post-installation* changes.

The following factors are the basis for a check list to use when preparing a change order estimate.

Factors To Consider When Pricing Change Orders

As an estimator begins to prepare a change order, the following questions should be reviewed to determine their impact on the final price.

General

- Is the change order work *pre-installation* or *post-installation*?

  Change order work costs vary according to how much of the installation has been completed. Once workers have the project scoped in their mind, even though they have not started, it can be difficult to refocus.

Consequently they may spend more than the normal amount of time understanding the change. Also, modifications to work in place such as trimming or refitting usually take more time than was initially estimated. The greater the amount of work in place, the more reluctant workers are to change it. Psychologically they may resent the change and as a result the rework takes longer than normal. Post-installation change order estimates must include demolition of existing work as required to accomplish the change. If the work is performed at a later time, additional obstacles such as building finishes may be present which must be

protected. Regardless of whether the change occurs pre-installation or post-installation, attempt to isolate the identifiable factors and price them separately. For example, add shipping costs that may be required pre-installation or any demolition required post-installation. Then analyze the potential impact on productivity of psychological and/or learning curve factors and adjust the output rates accordingly. One approach is to break down the typical workday into segments and quantify the impact on each segment. The following chart may be useful as a guide:

Activities (Productivity) Expressed as Percentages of a Workday			
Task	Means Mechanical Cost Data (for New Construction)	Pre-Installation Change Orders	Post-Installation Change Orders
1. Study plans	3%	6%	6%
2. Material procurement	3%	3%	3%
3. Receiving and storing	3%	3%	3%
4. Mobilization	5%	5%	5%
5. Site movement	5%	5%	8%
6. Layout and marking	8%	10%	12%
7. Actual installation	64%	59%	54%
8. Clean-up	3%	3%	3%
9. Breaks—non-productive	6%	6%	6%
Total	100%	100%	100%

Change Order Installation Efficiency

The labor-hours expressed (for new construction) are based on average installation time, using an efficiency level of approximately 60-65%. For change order situations, adjustments to this efficiency level should reflect the daily labor-hour allocation for that particular occurrence.

If any of the specific percentages expressed in the above chart do not apply to a particular project situation, then those percentage points should be reallocated to the appropriate task(s). Example: Using data for new construction, assume there is no new material being utilized. The percentages for Tasks 2 and 3 would therefore be reallocated to other tasks. If the time required for Tasks 2 and 3 can now be applied to installation, we can add the time allocated for *Material Procurement* and *Receiving and Storing* to the *Actual Installation* time for new construction, thereby increasing the Actual Installation percentage.

This chart shows that, due to reduced productivity, labor costs will be higher than those for new construction by 5% to 15% for pre-installation change orders and by 15% to 25% for post-installation change orders. Each job and change order is unique and must be examined individually. Many factors, covered elsewhere in this section, can each have a significant impact on productivity and change order costs. All such factors should be considered in every case.

- Will the change substantially delay the original completion date?

A significant change in the project may cause the original completion date to be extended. The extended schedule may subject the contractor to new wage rates dictated by relevant labor contracts. Project supervision and other project overhead must also be extended beyond the original completion date. The schedule extension may also put installation into a new weather season. For example, underground piping scheduled for October installation was delayed until January. As a result, frost penetrated the trench area, thereby changing the degree of difficulty of the task. Changes and delays may have a ripple effect throughout the project. This effect must be analyzed and negotiated with the owner.

- What is the net effect of a deduct change order?

In most cases, change orders resulting in a deduction or credit reflect only bare costs. The contractor may retain the overhead and profit based on the original bid.

Materials

- Will you have to pay more or less for the new material, required by the change order, than you paid for the original purchase?

The same material prices or discounts will usually apply to materials purchased for change orders as new construction. In some instances, however, the contractor may forfeit the advantages of competitive pricing for change orders. Consider the following example:

A contractor purchased over $20,000 worth of fan coil units for an installation, and obtained the maximum discount. Some time later it was determined the project required an additional matching unit. The contractor has to purchase this unit from the original supplier to ensure a match. The supplier at this time may not discount the unit because of the small quantity, and the fact that he is no longer in a competitive situation. The impact of quantity on purchase can add between 0% and 25% to material prices and/or subcontractor quotes.

- If materials have been ordered or delivered to the job site, will they be subject to a cancellation charge or restocking fee?

Check with the supplier to determine if ordered materials are subject to a cancellation charge. Delivered materials not used as result of a change order may be subject to a restocking fee if returned to the supplier. Common restocking charges run between 20% and 40%. Also, delivery charges to return the goods to the supplier must be added.

Labor

- How efficient is the existing crew at the actual installation?

Is the same crew that performed the initial work going to do the change order? Possibly the change consists of the installation of a unit identical to one already installed; therefore the change should take less time. Be sure to consider this potential productivity increase and modify the productivity rates accordingly.

- If the crew size is increased, what impact will that have on supervision requirements?

Under most bargaining agreements or management practices, there is a point at which a working foreman is replaced by a nonworking foreman. This replacement increases project overhead by adding a nonproductive worker. If additional workers are added to accelerate the project or to perform changes while maintaining the schedule, be sure to add additional supervision time if warranted. Calculate the hours involved and the additional cost directly if possible.

- What are the other impacts of increased crew size?

The larger the crew, the greater the potential for productivity to decrease. Some of the factors that cause this productivity loss are: overcrowding (producing restrictive conditions in the working space), and possibly a shortage of any special tools and equipment required. Such factors affect not only the crew working on the elements directly involved in the change order, but other crews whose movement may also be hampered.

As the crew increases, check its basic composition for changes by the addition or deletion of apprentices or nonworking foreman and quantify the potential effects of equipment shortages or other logistical factors.

- As new crews, unfamiliar with the project, are brought onto the site, how long will it take them to become oriented to the project requirements?

The orientation time for a new crew to become 100% effective varies with the site and type of project. Orientation is easiest at a new construction site, and most difficult at existing, very restrictive renovation sites. The type of work also affects orientation time. When all elements of the work are exposed, such as concrete or masonry work, orientation is decreased. When the work is concealed or less visible, such as existing electrical systems, orientation takes longer. Usually orientation can be accomplished in one day or less. Costs for added orientation should be itemized and added to the total estimated cost.

- How much actual production can be gained by working overtime?

Short term overtime can be used effectively to accomplish more work in a day. However, as overtime is scheduled to run beyond several weeks, studies have shown marked decreases in output. The following chart shows the effect of long term overtime on worker efficiency. If the anticipated change requires extended overtime to keep the job on schedule, these factors can be used as a guide to predict the impact on time and cost. Add project overhead, particularly supervision, that may also be incurred.

Days per Week	Hours per Day	Production Efficiency					Payroll Cost Factors	
		1 Week	2 Weeks	3 Weeks	4 Weeks	Average 4 Weeks	@ 1-1/2 Times	@ 2 Times
5	8	100%	100%	100%	100%	100%	100%	100%
	9	100	100	95	90	96.25	105.6	111.1
	10	100	95	90	85	91.25	110.0	120.0
	11	95	90	75	65	81.25	113.6	127.3
	12	90	85	70	60	76.25	116.7	133.3
6	8	100	100	95	90	96.25	108.3	116.7
	9	100	95	90	85	92.50	113.0	125.9
	10	95	90	85	80	87.50	116.7	133.3
	11	95	85	70	65	78.75	119.7	139.4
	12	90	80	65	60	73.75	122.2	144.4
7	8	100	95	85	75	88.75	114.3	128.6
	9	95	90	80	70	83.75	118.3	136.5
	10	90	85	75	65	78.75	121.4	142.9
	11	85	80	65	60	72.50	124.0	148.1
	12	85	75	60	55	68.75	126.2	152.4

Effects of Overtime

Caution: Under many labor agreements, Sundays and holidays are paid at a higher premium than the normal overtime rate.

The use of long-term overtime is counterproductive on almost any construction job; that is, the longer the period of overtime, the lower the actual production rate. Numerous studies have been conducted, and while they have resulted in slightly different numbers, all reach the same conclusion. The figure above tabulates the effects of overtime work on efficiency.

As illustrated, there can be a difference between the *actual* payroll cost per hour and the *effective* cost per hour for overtime work. This is due to the reduced production efficiency with the increase in weekly hours beyond 40. This difference between actual and effective cost results from overtime work over a prolonged period. Short-term overtime work does not result in as great a reduction in efficiency, and in such cases, effective cost may not vary significantly from the actual payroll cost. As the total hours per week are increased on a regular basis, more time is lost because of fatigue, lowered morale, and an increased accident rate.

As an example, assume a project where workers are working 6 days a week, 10 hours per day. From the figure above (based on productivity studies), the average effective productive hours over a four-week period are:

$$0.875 \times 60 = 52.5$$

Depending upon the locale and day of week, overtime hours may be paid at time and a half or double time. For time and a half, the overall (average) *actual* payroll cost (including regular and overtime hours) is determined as follows:

$$\frac{40 \text{ reg. hrs.} + (20 \text{ overtime hrs.} \times 1.5)}{60 \text{ hrs.}} = 1.167$$

Based on 60 hours, the payroll cost per hour will be 116.7% of the normal rate at 40 hours per week. However, because the effective production (efficiency) for 60 hours is reduced to the equivalent of 52.5 hours, the effective cost of overtime is calculated as follows:

For time and a half:

$$\frac{40 \text{ reg. hrs.} + (20 \text{ overtime hrs.} \times 1.5)}{52.5 \text{ hrs.}} = 1.33$$

Installed cost will be 133% of the normal rate (for labor).

Thus, when figuring overtime, the actual cost per unit of work will be higher than the apparent overtime payroll dollar increase, due to the reduced productivity of the longer workweek. These efficiency calculations are true only for those cost factors determined by hours worked. Costs that are applied weekly or monthly, such as equipment rentals, will not be similarly affected.

Equipment

• What equipment is required to complete the change order?

Change orders may require extending the rental period of equipment already on the job site, or the addition of special equipment brought in to accomplish the change work. In either case, the additional rental charges and operator labor charges must be added.

Summary

The preceding considerations and others you deem appropriate should be analyzed and applied to a change order estimate. The impact of each should be quantified and listed on the estimate to form an audit trail.

Change orders that are properly identified, documented, and managed help to ensure the orderly, professional and profitable progress of the work. They also minimize potential claims or disputes at the end of the project.

REFERENCE

Square Meter Costs

Estimating Tips

- The cost figures in this Square Meter Cost section were derived from approximately 11,200 projects contained in the RSMeans database of completed construction projects. They include the contractor's overhead and profit, but do not generally include architectural fees or land costs. The figures have been adjusted to January of the current year. New projects are added to our files each year, and outdated projects are discarded. For this reason, certain costs may not show a uniform annual progression. In no case are all subdivisions of a project listed.

- These projects were located throughout the U.S. and reflect a tremendous variation in square meter (m^2) and cubic meter (m^3) costs. This is due to differences, not only in labor and material costs, but also in individual owners' requirements. For instance, a bank in a large city would have different features than one in a rural area. This is true of all the different types of buildings analyzed. Therefore, caution should be exercised when using square meter costs. For example, for court houses, costs in the database are local court house costs and will not apply to the larger, more elaborate federal court houses. As a general rule, the projects in the 1/4 column do not include any site work or equipment, while the projects in the 3/4 column may include both equipment and site work. The median figures do not generally include site work.

- None of the figures "go with" any others. All individual cost items were computed and tabulated separately. Thus the sum of the median figures for Plumbing, HVAC and Electrical will not normally total up to the total Mechanical and Electrical costs arrived at by separate analysis and tabulation of the projects.

- Each building was analyzed as to total and component costs and percentages. The figures were arranged in ascending order with the results tabulated as shown. The 1/4 column shows that 25% of the projects had lower costs, 75% higher. The 3/4 column shows that 75% of the projects had lower costs, 25% had higher. The median column shows that 50% of the projects had lower costs, 50% had higher.

- There are two times when square meter costs are useful. The first is in the conceptual stage when no details are available. Then square meter costs make a useful starting point. The second is after the bids are in and the costs can be worked back into their appropriate units for information purposes. As soon as details become available in the project design, the square meter approach should be discontinued and the project priced as to its particular components. When more precision is required or for estimating the replacement cost of specific buildings, the current edition of *Means Square Foot Costs* should be used.

- When using the figures in this section, it is recommended that the median column be used for preliminary figures if no additional information is available. The median figures, when multiplied by the total city construction cost index figures (see City Cost Indexes) and then multiplied by the project size modifier in at the end of this section , should present a fairly accurate base figure, which would then have to be adjusted in view of the estimator's experience, local economic conditions, code requirements, and the owner's particular requirements. There is no need to factor the percentage figures, as these should remain constant from city to city. All tabulations mentioning air conditioning had at least partial air conditioning.

- The editors of this book would greatly appreciate receiving cost figures on one or more of your recent projects which would then be included in the averages for next year. All cost figures received will be kept confidential except that they will be averaged with other similar projects to arrive at square meter cost figures for next year's book. See the last page of the book for details and the discount available for submitting one or more of your projects.

50 17 00 | m² Costs

		UNIT	UNIT COSTS			% OF TOTAL		
			1/4	MEDIAN	3/4	1/4	MEDIAN	3/4
01	0010	**APARTMENTS** Low Rise (1 to 3 story)	m²	610	765	1,000				01
	0020	Total project cost	m³	178	237	292			
	0100	Site work	m²	52	70.50	124	6.05%	10.55%	14.05%
	0500	Masonry		11.95	27.50	48	1.54%	3.67%	6.35%
	1500	Finishes		64	88.50	109	9.05%	10.75%	12.85%
	1800	Equipment		19.70	30	44.50	2.73%	4.03%	5.95%
	2720	Plumbing		47	61	77	6.65%	8.95%	10.05%
	2770	Heating, ventilating, air conditioning		30	37	54.50	4.20%	5.60%	7.60%
	2900	Electrical		35	46.50	62.50	5.20%	6.65%	8.40%
	3100	Total: Mechanical & Electrical		121	154	193	15.90%	18.05%	23%
	9000	Per apartment unit, total cost	Apt.	52,500	80,000	118,000			
	9500	Total: Mechanical & Electrical	"	9,900	15,600	20,400			
02	0010	**APARTMENTS** Mid Rise (4 to 7 story)	m²	800	970	1,175				02
	0020	Total project costs	m³	205	284	390			
	0100	Site work	m²	32	63.50	125	5.25%	6.70%	9.20%
	0500	Masonry		53.50	73.50	105	5.15%	7.40%	10.50%
	1500	Finishes		101	141	164	10.55%	13.45%	17.70%
	1800	Equipment		26.50	37.50	48	2.55%	3.48%	4.31%
	2500	Conveying equipment		18.20	22.50	26.50	2.05%	2.29%	2.69%
	2720	Plumbing		47	75.50	83.50	5.70%	7.20%	8.95%
	2900	Electrical		53	75.50	88.50	6.65%	7.20%	8.95%
	3100	Total: Mechanical & Electrical		170	212	264	18.85%	21%	25%
	9000	Per apartment unit, total cost	Apt.	84,500	99,500	165,000			
	9500	Total: Mechanical & Electrical	"	15,900	18,400	23,300			
03	0010	**APARTMENTS** High Rise (8 to 24 story)	m²	910	1,100	1,250				03
	0020	Total project costs	m³	290	355	430			
	0100	Site work	m²	33	53.50	75	2.58%	4.84%	6.15%
	0500	Masonry		52.50	96	119	4.74%	9.65%	11.05%
	1500	Finishes		101	126	149	9.75%	11.80%	13.70%
	1800	Equipment		29.50	36	47.50	2.78%	3.49%	4.35%
	2500	Conveying equipment		20.50	31.50	45	2.23%	2.78%	3.37%
	2720	Plumbing		67.50	79	111	6.80%	7.20%	10.45%
	2900	Electrical		62.50	79	107	6.45%	7.65%	8.80%
	3100	Total: Mechanical & Electrical		187	237	285	17.95%	22.50%	24.50%
	9000	Per apartment unit, total cost	Apt.	88,000	97,000	134,500			
	9500	Total: Mechanical & Electrical	"	19,000	21,700	23,000			
04	0010	**AUDITORIUMS**	m²	935	1,275	1,825				04
	0020	Total project costs	m³	194	270	405			
	2720	Plumbing	m²	59.50	83	105	5.85%	7.20%	8.70%
	2900	Electrical		76.50	106	142	6.80%	8.95%	11.30%
	3100	Total: Mechanical & Electrical		146	208	420	24.50%	30.50%	31.50%
05	0010	**AUTOMOTIVE SALES**	m²	685	945	1,175				05
	0020	Total project costs	m³	151	182	235			
	2720	Plumbing	m²	32	55.50	60.50	2.89%	6.05%	6.50%
	2770	Heating, ventilating, air conditioning		49.50	75.50	81.50	4.61%	10%	10.35%
	2900	Electrical		56.50	86.50	120	7.40%	9.95%	12.40%
	3100	Total: Mechanical & Electrical		157	221	280	19.15%	20.50%	26%
06	0010	**BANKS**	m²	1,375	1,700	2,150				06
	0020	Total project costs	m³	320	440	580			
	0100	Site work	m²	157	237	355	7.85%	12.95%	17%
	0500	Masonry		71.50	134	248	3.36%	6.95%	10.35%
	1500	Finishes		122	166	214	5.85%	8.45%	11.25%
	1800	Equipment		55	113	242	1.34%	5.95%	10.65%
	2720	Plumbing		43	61.50	90	2.82%	3.90%	4.93%
	2770	Heating, ventilating, air conditioning		82	109	145	4.86%	7.15%	8.50%
	2900	Electrical		130	173	226	8.20%	10.20%	12.20%
	3100	Total: Mechanical & Electrical		310	410	495	16.55%	19.45%	23%
	3500	See also division 11020 & 11030							

50 17 00 | m² Costs

| | | | UNIT | \multicolumn{3}{UNIT COSTS} | \multicolumn{3}{% OF TOTAL} | |
| | | | | 1/4 | MEDIAN | 3/4 | 1/4 | MEDIAN | 3/4 | |

#	Code	Description	UNIT	1/4	MEDIAN	3/4	1/4	MEDIAN	3/4	#
13	0010	**CHURCHES**	m²	925	1,175	1,525				13
	0020	Total project costs	m³	187	237	315				
	1800	Equipment	m²	11.75	26.50	56	1.03%	2.24%	4.50%	
	2720	Plumbing		36	50.50	74.50	3.51%	4.96%	6.25%	
	2770	Heating, ventilating, air conditioning		84	110	156	7.50%	10%	12%	
	2900	Electrical		77.50	107	143	7.30%	8.75%	10.90%	
	3100	Total: Mechanical & Electrical	▼	237	310	430	18.25%	22%	24%	
	3500	See also division 11040								
15	0010	**CLUBS, COUNTRY**	m²	990	1,200	1,500				15
	0020	Total project costs	m³	261	320	440				
	2720	Plumbing	m²	63.50	89	202	5.60%	7.90%	10%	
	2900	Electrical		78	111	147	7%	8.95%	11%	
	3100	Total: Mechanical & Electrical	▼	237	415	515	19%	26.50%	29.50%	
17	0010	**CLUBS, SOCIAL** Fraternal	m²	790	1,125	1,525				17
	0020	Total project costs	m³	162	246	293				
	2720	Plumbing	m²	49.50	62	93.50	5.60%	6.90%	8.55%	
	2770	Heating, ventilating, air conditioning		71.50	86.50	111	8.20%	9.25%	14.40%	
	2900	Electrical		63	98	118	6.50%	9.50%	10.55%	
	3100	Total: Mechanical & Electrical	▼	175	335	425	21%	23%	23.50%	
18	0010	**CLUBS, Y.M.C.A.**	m²	995	1,350	1,625				18
	0020	Total project costs	m³	150	251	375				
	2720	Plumbing	m²	63	125	140	5.65%	7.60%	10.85%	
	2900	Electrical		75.50	103	146	6.05%	7.80%	10.20%	
	3100	Total: Mechanical & Electrical	▼	280	340	375	18.40%	21.50%	28.50%	
19	0010	**COLLEGES** Classrooms & Administration	m²	1,100	1,475	1,975				19
	0020	Total project costs	m³	263	375	590				
	0500	Masonry	m²	73.50	139	171	5.65%	8.25%	10.50%	
	2720	Plumbing		54.50	109	195	5.10%	6.60%	8.95%	
	2900	Electrical		91	139	173	7.70%	9.85%	12%	
	3100	Total: Mechanical & Electrical	▼	335	470	555	24%	28%	31.50%	
21	0010	**COLLEGES** Science, Engineering, Laboratories	m²	1,850	2,175	2,650				21
	0020	Total project costs	m³	350	510	580				
	1800	Equipment	m²	104	237	258	2%	6.45%	12.65%	
	2900	Electrical		154	210	335	7.10%	9.40%	12.10%	
	3100	Total: Mechanical & Electrical	▼	570	680	1,050	28.50%	31.50%	41%	
	3500	See also division 11600								
23	0010	**COLLEGES** Student Unions	m²	1,175	1,675	1,950				23
	0020	Total project costs	m³	217	286	365				
	3100	Total: Mechanical & Electrical	m²	310	485	570	23.50%	26%	29%	
25	0010	**COMMUNITY CENTERS**	"	970	1,225	1,625				25
	0020	Total project costs	m³	210	300	390				
	1800	Equipment	m²	24.50	42	66	1.87%	3.12%	6%	
	2720	Plumbing		47.50	81.50	119	4.94%	7%	9.10%	
	2770	Heating, ventilating, air conditioning		78.50	113	156	6.95%	10.65%	13.05%	
	2900	Electrical		83.50	107	156	7.35%	9.10%	10.85%	
	3100	Total: Mechanical & Electrical	▼	291	350	500	22.50%	26.50%	32.50%	
28	0010	**COURT HOUSES**	m²	1,400	1,625	1,900				28
	0020	Total project costs	m³	355	430	535				
	2720	Plumbing	m²	67.50	94	135	5.95%	7.45%	8.20%	
	2900	Electrical		139	164	198	8.55%	9.95%	11.50%	
	3100	Total: Mechanical & Electrical	▼	380	440	555	22.50%	29.50%	30.50%	
30	0010	**DEPARTMENT STORES**	m²	520	710	895				30
	0020	Total project costs	m³	92	125	162				
	2720	Plumbing	m²	17.10	20.50	31.50	1.82%	4.21%	5.90%	
	2770	Heating, ventilating, air conditioning	▼	47.50	73.50	111	8.20%	9.10%	14.80%	

50 17 00 | m² Costs

			UNIT	UNIT COSTS			% OF TOTAL		
				1/4	**MEDIAN**	**3/4**	**1/4**	**MEDIAN**	**3/4**
30	2900	Electrical	m²	60.50	82.50	97.50	9.05%	12.15%	14.95%
	3100	Total: Mechanical & Electrical		106	135	237	13.20%	21.50%	50%
31	0010	**DORMITORIES** Low Rise (1 to 3 story)	m²	895	1,250	1,575			
	0020	Total project costs	m³	194	299	445			
	2720	Plumbing	m²	59.50	80	101	8.05%	9%	9.65%
	2770	Heating, ventilating, air conditioning		63	76	101	4.61%	8.05%	10%
	2900	Electrical		64	96.50	134	6.55%	8.90%	9.55%
	3100	Total: Mechanical & Electrical		335	350	365	22%	26%	29%
	9000	Per bed, total cost	Bed	39,100	43,500	93,000			
32	0010	**DORMITORIES** Mid Rise (4 to 8 story)	m²	1,225	1,600	1,950			
	0020	Total project costs	m³	440	485	580			
	2900	Electrical	m²	130	147	192	8.20%	10.20%	11.10%
	3100	Total: Mechanical & Electrical	"	335	375	730	19.50%	30.50%	37.50%
	9000	Per bed, total cost	Bed	16,100	36,800	76,000			
34	0010	**FACTORIES**	m²	465	690	1,050			
	0020	Total project costs	m³	97	145	240			
	0100	Site work	m²	52.50	96.50	152	6.95%	11.45%	17.95%
	2720	Plumbing		25	46.50	76.50	3.73%	6.05%	8.10%
	2770	Heating, ventilating, air conditioning		48.50	69.50	93.50	5.25%	8.45%	11.35%
	2900	Electrical		57.50	91	138	8.10%	10.50%	14.20%
	3100	Total: Mechanical & Electrical		137	221	335	21%	28.50%	35.50%
36	0010	**FIRE STATIONS**	m²	915	1,250	1,625			
	0020	Total project costs	m³	176	242	320			
	0500	Masonry	m²	135	242	320	8.60%	11.65%	16.45%
	1140	Roofing		30	80.50	92	1.90%	4.94%	5.05%
	1580	Painting		24.50	34.50	35.50	1.37%	1.57%	2.07%
	1800	Equipment		19.40	27.50	66	.90%	2.50%	4.07%
	2720	Plumbing		58	85.50	126	5.85%	7.35%	9.50%
	2770	Heating, ventilating, air conditioning		50	79.50	124	4.86%	7.25%	9.25%
	2900	Electrical		65	112	147	6.80%	8.75%	10.65%
	3100	Total: Mechanical & Electrical		325	375	445	19.60%	23%	27%
37	0010	**FRATERNITY HOUSES** And Sorority Houses	m²	915	1,175	1,625			
	0020	Total project costs	m³	300	315	400			
	2720	Plumbing	m²	69.50	79	145	6.80%	8%	10.85%
	2900	Electrical		60.50	131	160	6.60%	9.90%	10.65%
	3100	Total: Mechanical & Electrical		161	231	280		15.10%	15.90%
38	0010	**FUNERAL HOMES**	m²	970	1,325	2,400			
	0020	Total project costs	m³	325	360	695			
	2900	Electrical	m²	42.50	78	92.50	3.58%	4.44%	5.95%
	3100	Total: Mechanical & Electrical	"	152	226	305	12.90%	12.90%	12.90%
39	0010	**GARAGES, COMMERCIAL** (service)	m²	570	860	1,175			
	0020	Total project costs	m³	121	175	254			
	1800	Equipment	m²	31	69.50	108	2.69%	4.62%	6.80%
	2720	Plumbing		38	58.50	107	5.45%	7.85%	10.65%
	2730	Heating & ventilating		50	68	93.50	5.25%	6.85%	8.20%
	2900	Electrical		52.50	80.50	115	7.15%	9.25%	10.85%
	3100	Total: Mechanical & Electrical		118	221	330	13.60%	17.40%	27%
40	0010	**GARAGES, MUNICIPAL** (repair)	m²	800	1,075	1,500			
	0020	Total project costs	m³	165	208	360			
	0500	Masonry	m²	75.50	147	231	5.60%	9.15%	12.50%
	2720	Plumbing		36	69.50	131	3.59%	6.70%	7.95%
	2730	Heating & ventilating		62	89.50	172	6.15%	7.45%	13.50%
	2900	Electrical		59.50	93.50	135	6.65%	8.15%	11.15%
	3100	Total: Mechanical & Electrical		191	380	555	21.50%	25.50%	28.50%

50 17 00 | m² Costs	UNIT	UNIT COSTS			% OF TOTAL		
		1/4	MEDIAN	3/4	1/4	MEDIAN	3/4
41 0010 GARAGES, PARKING	m²	310	455	785				41
0020 Total project costs	m³	96.50	131	191			
2720 Plumbing	m²	8.85	13.80	21	1.72%	2.70%	3.85%
2900 Electrical		17.10	21	33	4.33%	5.20%	6.30%
3100 Total: Mechanical & Electrical	↓	35	49	64	7%	8.90%	11.05%
3200							
9000 Per car, total cost	Car	12,300	15,400	19,700			
43 0010 GYMNASIUMS	m²	865	1,150	1,475				43
0020 Total project costs	m³	142	193	237			
1800 Equipment	m²	20.50	39	74.50	2.07%	3.35%	6.70%
2720 Plumbing		55	68	84.50	4.95%	6.75%	7.75%
2770 Heating, ventilating, air conditioning		59	90.50	181	5.80%	9.80%	11.10%
2900 Electrical		66	88.50	112	6.60%	8.30%	10.30%
3100 Total: Mechanical & Electrical	↓	237	335	395	20.50%	26%	29.50%
3500 See also division 11480							
46 0010 HOSPITALS	m²	1,675	2,050	3,100				46
0020 Total project costs	m³	415	520	740			
1800 Equipment	m²	42.50	81.50	140	1.10%	2.68%	5%
2720 Plumbing		144	201	258	7.60%	9.10%	10.85%
2770 Heating, ventilating, air conditioning		211	269	365	7.80%	12.95%	16.65%
2900 Electrical		182	237	365	9.85%	11.75%	14%
3100 Total: Mechanical & Electrical	↓	510	680	1,100	26.50%	33%	36.50%
9000 Per bed or person, total cost	Bed	127,000	203,500	272,000			
9900 See also division 11700							
48 0010 HOUSING For the Elderly	m²	825	1,050	1,275				48
0020 Total project costs	m³	193	267	340			
0100 Site work	m²	61	90.50	130	5.05%	7.90%	12.10%
0500 Masonry		25	93.50	137	1.30%	6.05%	11%
1800 Equipment		19.80	27.50	43.50	1.88%	3.23%	4.43%
2510 Conveying systems		20	27	36.50	1.78%	2.20%	2.81%
2720 Plumbing		61	78	102	8.15%	9.55%	10.50%
2730 Heating, ventilating, air conditioning		31.50	44.50	66	3.30%	5.60%	7.25%
2900 Electrical		61.50	83	107	7.30%	8.50%	10.25%
3100 Total: Mechanical & Electrical	↓	211	253	335	18.10%	22.50%	29%
9000 Per rental unit, total cost	Unit	71,000	83,000	92,500			
9500 Total: Mechanical & Electrical	"	15,800	18,200	21,200			
50 0010 HOUSING Public (low-rise)	m²	690	960	1,250				50
0020 Total project costs	m³	201	251	315			
0100 Site work	m²	87.50	126	205	8.35%	11.75%	16.50%
1800 Equipment		18.85	30.50	46.50	2.26%	3.03%	4.24%
2720 Plumbing		50	65.50	83.50	7.15%	9.05%	11.60%
2730 Heating, ventilating, air conditioning		25	48.50	53	4.26%	6.05%	6.45%
2900 Electrical		42	62.50	86	5.10%	6.55%	8.25%
3100 Total: Mechanical & Electrical	↓	198	253	285	14.50%	17.55%	26.50%
9000 Per apartment, total cost	Apt.	70,500	80,000	100,500			
9500 Total: Mechanical & Electrical	"	15,000	18,500	20,500			
51 0010 ICE SKATING RINKS	m²	615	1,375	1,525				51
0020 Total project costs	m³	142	146	168			
2720 Plumbing	m²	22	41.50	42.50	3.12%	3.23%	5.65%
2900 Electrical		63	97	103	6.30%	10.15%	15.05%
3100 Total: Mechanical & Electrical	↓	105	149	186	18.95%	18.95%	18.95%
52 0010 JAILS	m²	1,800	2,325	3,000				52
0020 Total project costs	m³	555	740	920			
1800 Equipment	m²	70	207	350	2.80%	5.55%	11.90%
2720 Plumbing		183	231	305	7%	8.90%	13.35%
2770 Heating, ventilating, air conditioning		162	215	420	7.50%	9.45%	17.75%
2900 Electrical	↓	192	248	305	8.20%	11.55%	14.70%

50 17 00 | m² Costs		UNIT COSTS			% OF TOTAL			
	UNIT	1/4	MEDIAN	3/4	1/4	MEDIAN	3/4	
52 3100 Total: Mechanical & Electrical	m²	480	895	1,050	27.50%	30%	34%	**52**
53 0010 **LIBRARIES**	m²	1,150	1,425	1,875				**53**
0020 Total project costs	m³	254	320	410				
0500 Masonry	m²	89	157	264	5.80%	8%	12.35%	
1800 Equipment		15.50	42	65	.41%	1.50%	4.16%	
2720 Plumbing		42	61	83	3.43%	4.60%	5.70%	
2770 Heating, ventilating, air conditioning		93	157	205	7.80%	10.95%	12.80%	
2900 Electrical		116	149	188	8.35%	10.40%	11.95%	
3100 Total: Mechanical & Electrical		345	430	540	20.50%	23%	26.50%	
54 0010 **LIVING, ASSISTED**	m²	1,050	1,250	1,450				**54**
0020 Total project costs	m³	290	340	385				
0500 Masonry	m²	31	37	43.50	2.37%	3.16%	3.86%	
1800 Equipment		24	28	35.50	2.12%	2.45%	2.66%	
2720 Plumbing		88.50	118	122	6.05%	8.15%	10.60%	
2770 Heating, ventilating, air conditioning		104	109	119	7.95%	9.35%	9.70%	
2900 Electrical		103	114	131	9%	10%	10.70%	
3100 Total: Mechanical & Electrical		296	345	390	26%	29%	31.50%	
55 0010 **MEDICAL CLINICS**	m²	1,075	1,325	1,675				**55**
0020 Total project costs	m³	258	350	440				
1800 Equipment	m²	29	61	94.50	1.10%	3.13%	6.35%	
2720 Plumbing		71	100	134	6.15%	8.40%	10.10%	
2770 Heating, ventilating, air conditioning		84.50	111	163	6.65%	8.85%	11.35%	
2900 Electrical		90	130	170	8.10%	10%	12.20%	
3100 Total: Mechanical & Electrical		291	400	550	22.50%	27%	33.50%	
3500 See also division 11700								
57 0010 **MEDICAL OFFICES**	m²	1,000	1,250	1,525				**57**
0020 Total project costs	m³	246	335	450				
1800 Equipment	m²	35	65.50	93	.98%	5.10%	7.05%	
2720 Plumbing		55.50	85.50	116	5.60%	6.80%	8.50%	
2770 Heating, ventilating, air conditioning		67.50	98.50	128	6.15%	8.05%	9.70%	
2900 Electrical		80.50	117	164	7.65%	9.80%	11.70%	
3100 Total: Mechanical & Electrical		215	310	465	19.35%	23%	30.50%	
59 0010 **MOTELS**	m²	635	935	1,225				**59**
0020 Total project costs	m³	186	249	405				
2720 Plumbing	m²	64.50	82	98	9.45%	10.60%	12.55%	
2770 Heating, ventilating, air conditioning		39	58.50	105	5.60%	5.60%	10%	
2900 Electrical		60.50	77	98	7.45%	9.20%	10.80%	
3100 Total: Mechanical & Electrical		204	253	435	18.50%	21%	25.50%	
5000								
9000 Per rental unit, total cost	Unit	30,000	57,000	61,500				
9500 Total: Mechanical & Electrical	"	5,850	8,850	10,300				
60 0010 **NURSING HOMES**	m²	1,000	1,275	1,575				**60**
0020 Total project costs	m³	256	320	440				
1800 Equipment	m²	31.50	41.50	69.50	2.02%	3.62%	4.99%	
2720 Plumbing		85.50	129	156	8.75%	10.10%	12.70%	
2770 Heating, ventilating, air conditioning		90	136	181	9.70%	11.45%	11.80%	
2900 Electrical		100	124	168	9.50%	10.60%	12.50%	
3100 Total: Mechanical & Electrical		237	330	550	26%	29.50%	30.50%	
9000 Per bed or person, total cost	Bed	41,000	51,500	66,000				
61 0010 **OFFICES** Low-Rise (1 to 4 story)	m²	835	1,075	1,400				**61**
0020 Total project costs	m³	196	270	355				
0100 Site work	m²	66.50	113	165	6.20%	9.70%	13.60%	
0500 Masonry	"	28	63	114	2.55%	5.40%	8%	
1800 Equipment	m²	8.85	17.45	47.50	.77%	1.50%	3.69%	
2720 Plumbing		30	46.50	67.50	3.66%	4.50%	6.10%	
2770 Heating, ventilating, air conditioning		66	93.50	135	7.20%	10.40%	11.70%	
2900 Electrical		68.50	97.50	137	7.45%	9.65%	11.40%	

		50 17 00 | m² Costs	UNIT	UNIT COSTS			% OF TOTAL			
				1/4	MEDIAN	3/4	1/4	MEDIAN	3/4	
61	3100	Total: Mechanical & Electrical	m²	178	248	370	18.25%	22.50%	27%	61
62	0010	**OFFICES** Mid-Rise (5 to 10 story)	m²	890	1,075	1,425				62
	0020	Total project costs	m³	207	263	375				
	2720	Plumbing	m²	27	41.50	59.50	2.83%	3.74%	4.50%	
	2770	Heating, ventilating, air conditioning		67.50	96.50	154	7.65%	9.40%	11%	
	2900	Electrical		65.50	84.50	117	6.35%	7.80%	10%	
	3100	Total: Mechanical & Electrical	↓	168	215	420	18.95%	21%	27.50%	
63	0010	**OFFICES** High-Rise (11 to 20 story)	m²	1,075	1,375	1,700				63
	0020	Total project costs	m³	251	315	450				
	2900	Electrical	m²	66	80.50	120	5.80%	7.85%	10.50%	
	3100	Total: Mechanical & Electrical	"	213	285	485	16.90%	23.50%	34%	
64	0010	**POLICE STATIONS**	m²	1,325	1,675	2,150				64
	0020	Total project costs	m³	340	420	570				
	0500	Masonry	m²	123	215	285	6.70%	10.55%	11.35%	
	1800	Equipment		21	91.50	145	1.43%	4.07%	6.70%	
	2720	Plumbing		75.50	146	181	5.65%	6.90%	10.75%	
	2770	Heating, ventilating, air conditioning		114	152	206	5.85%	10.55%	11.70%	
	2900	Electrical		143	213	269	9.80%	11.85%	14.80%	
	3100	Total: Mechanical & Electrical	↓	470	565	765	28.50%	32%	32.50%	
65	0010	**POST OFFICES**	m²	1,025	1,275	1,625				65
	0020	Total project costs	m³	203	258	305				
	2720	Plumbing	m²	46.50	57.50	72.50	4.24%	5.30%	5.60%	
	2770	Heating, ventilating, air conditioning		72.50	90	99.50	6.65%	7.15%	9.35%	
	2900	Electrical		85	120	142	7.25%	9%	11%	
	3100	Total: Mechanical & Electrical	↓	248	325	365	16.25%	18.80%	22%	
66	0010	**POWER PLANTS**	m²	7,150	9,475	17,500				66
	0020	Total project costs	m³	645	1,425	3,025				
	2900	Electrical	m²	505	1,075	1,600	9.30%	12.75%	21.50%	
	8100	Total: Mechanical & Electrical	"	1,250	4,100	9,200	32.50%	32.50%	52.50%	
67	0010	**RELIGIOUS EDUCATION**	m²	835	1,075	1,275				67
	0020	Total project costs	m³	152	217	272				
	2720	Plumbing	m²	35	49.50	69.50	4.40%	5.30%	7.10%	
	2770	Heating, ventilating, air conditioning		87.50	99.50	140	10.05%	11.45%	12.35%	
	2900	Electrical		66	93	122	7.60%	9.10%	10.35%	
	3100	Total: Mechanical & Electrical	↓	269	350	415	22%	23%	26%	
69	0010	**RESEARCH** Laboratories and facilities	m²	1,250	1,775	2,600				69
	0020	Total project costs	m³	320	615	725				
	1800	Equipment	m²	55.50	109	264	.90%	4.58%	8.80%	
	2720	Plumbing		125	160	258	6.15%	8.30%	10.80%	
	2770	Heating, ventilating, air conditioning		112	375	445	7.25%	16.50%	17.50%	
	2900	Electrical		143	242	410	9.45%	11.15%	15.40%	
	3100	Total: Mechanical & Electrical	↓	440	840	1,200	29.50%	37%	45.50%	
70	0010	**RESTAURANTS**	m²	1,200	1,550	2,025				70
	0020	Total project costs	m³	335	435	575				
	1800	Equipment	m²	78	192	291	6.10%	13%	15.65%	
	2720	Plumbing		96	116	152	6.10%	8.15%	9%	
	2770	Heating, ventilating, air conditioning		122	168	221	9.20%	12%	12.40%	
	2900	Electrical		128	157	205	8.35%	10.55%	11.55%	
	3100	Total: Mechanical & Electrical	↓	395	415	540	19.25%	24%	29.50%	
	9000	Per seat unit, total cost	Seat	4,125	5,475	6,500				
	9500	Total: Mechanical & Electrical	"	1,025	1,375	1,625				
72	0010	**RETAIL STORES**	m²	560	760	995				72
	0020	Total project costs	m³	125	178	249				
	2720	Plumbing	m²	20.50	34	58	3.26%	4.60%	6.80%	
	2770	Heating, ventilating, air conditioning		44	60.50	91	6.75%	8.75%	10.15%	
	2900	Electrical		50.50	69.50	100	7.25%	9.90%	11.65%	
	3100	Total: Mechanical & Electrical	↓	130	173	231	17.05%	21%	23.50%	

50 17 00 | m² Costs	UNIT	UNIT COSTS			% OF TOTAL		
		1/4	MEDIAN	3/4	1/4	MEDIAN	3/4
74 0010 **SCHOOLS** Elementary	m²	910	1,125	1,375			**74**
0020 Total project costs	m³	198	253	325			
0500 Masonry	m²	81.50	135	201	5.45%	10.65%	14.70%
1800 Equipment		27	45.50	84	1.90%	3.33%	5%
2720 Plumbing		52.50	75	100	5.70%	7.15%	9.35%
2730 Heating, ventilating, air conditioning		79	126	177	8.15%	10.80%	15.20%
2900 Electrical		85.50	112	140	8.40%	10.05%	11.70%
3100 Total: Mechanical & Electrical		300	375	455	24%	27.50%	30%
9000 Per pupil, total cost	Ea.	9,825	14,600	43,200			
9500 Total: Mechanical & Electrical	"	2,775	3,500	12,500			
76 0010 **SCHOOLS** Junior High & Middle	m²	925	1,175	1,350			**76**
0020 Total project costs	m³	198	256	286			
0500 Masonry	m²	102	145	174	8%	11.60%	14.30%
1800 Equipment		30.50	49	75.50	1.81%	3.26%	5.15%
2720 Plumbing		60.50	68.50	88.50	5.50%	6.90%	8.15%
2770 Heating, ventilating, air conditioning		69.50	133	186	8.75%	12.75%	17.45%
2900 Electrical		91.50	111	143	7.90%	9.25%	10.60%
3100 Total: Mechanical & Electrical		274	390	475	22.50%	25.50%	29.50%
9000 Per pupil, total cost	Ea.	11,200	15,100	19,700			
78 0010 **SCHOOLS** Senior High	m²	970	1,175	1,500			**78**
0020 Total project costs	m³	205	288	475			
1800 Equipment	m²	26	61.50	90.50	1.88%	3.29%	4.80%
2720 Plumbing		56.50	84	152	5.70%	7%	8.35%
2770 Heating, ventilating, air conditioning		113	130	248	8.95%	11.60%	15%
2900 Electrical		98	126	202	8.35%	10.10%	12.35%
3100 Total: Mechanical & Electrical		335	380	645	23%	26.50%	28.50%
9000 Per pupil, total cost	Ea.	8,650	17,600	22,000			
80 0010 **SCHOOLS** Vocational	m²	795	1,125	1,425			**80**
0020 Total project costs	m³	163	233	325			
0500 Masonry	m²	47	116	177	3.53%	4.61%	10.95%
1800 Equipment	"	24	34	86	1.24%	3.13%	4.68%
2720 Plumbing	m²	51	78	117	5.40%	6.90%	8.55%
2770 Heating, ventilating, air conditioning		71.50	133	221	8.60%	11.90%	14.65%
2900 Electrical		83.50	114	160	8.45%	10.95%	13.20%
3100 Total: Mechanical & Electrical		291	320	550	23.50%	29.50%	31%
9000 Per pupil, total cost	Ea.	10,300	27,700	41,300			
83 0010 **SPORTS ARENAS**	m²	700	930	1,450			**83**
0020 Total project costs	m³	124	223	288			
2720 Plumbing	m²	40.50	61.50	130	4.35%	6.35%	9.40%
2770 Heating, ventilating, air conditioning		87	103	143	8.80%	10.20%	13.55%
2900 Electrical		72.50	98.50	128	8.60%	9.90%	12.25%
3100 Total: Mechanical & Electrical		181	325	420	21.50%	25%	27.50%
85 0010 **SUPERMARKETS**	m²	645	750	905			**85**
0020 Total project costs	m³	118	142	216			
2720 Plumbing	m²	36	45.50	53	5.40%	6%	7.45%
2770 Heating, ventilating, air conditioning		53	70.50	85.50	8.60%	8.65%	9.60%
2900 Electrical		80.50	92.50	110	10.40%	12.45%	13.60%
3100 Total: Mechanical & Electrical		207	226	310	20.50%	26.50%	31%
86 0010 **SWIMMING POOLS**	m²	1,050	1,750	3,725			**86**
0020 Total project costs	m³	276	345	375			
2720 Plumbing	m²	96.50	110	153	4.80%	9.70%	20.50%
2900 Electrical		78.50	127	185	6.50%	7.25%	7.60%
3100 Total: Mechanical & Electrical		192	485	660	11.15%	14.10%	23.50%
87 0010 **TELEPHONE EXCHANGES**	m²	1,400	2,025	2,575			**87**
0020 Total project costs	m³	283	455	625			
2720 Plumbing	m²	58.50	92.50	132	4.52%	5.80%	6.90%
2770 Heating, ventilating, air conditioning		136	274	340	11.80%	16.05%	18.40%

50 17 00 | m² Costs		UNIT	UNIT COSTS			% OF TOTAL		
			1/4	MEDIAN	3/4	1/4	MEDIAN	3/4
87	2900	Electrical	m²	141	226	400	10.90%	14%	17.85%	**87**
	3100	Total: Mechanical & Electrical	↓	415	790	1,125	29.50%	33.50%	44.50%
91	0010	**THEATERS**	m²	870	1,075	1,650				**91**
	0020	Total project costs	m³	132	196	288			
	2720	Plumbing	m²	29	31.50	129	2.92%	4.70%	6.80%
	2770	Heating, ventilating, air conditioning		84.50	103	127	8%	12.25%	13.40%
	2900	Electrical		76.50	103	209	8.05%	9.95%	12.25%
	3100	Total: Mechanical & Electrical	↓	196	291	605	23%	26.50%	27.50%
94	0010	**TOWN HALLS** City Halls & Municipal Buildings	m²	1,025	1,300	1,650				**94**
	0020	Total project costs	m³	292	360	490			
	2720	Plumbing	m²	40.50	76	140	4.31%	5.95%	7.95%
	2770	Heating, ventilating, air conditioning		73.50	146	213	7.05%	9.05%	13.45%
	2900	Electrical		92.50	135	180	8.05%	9.50%	12.05%
	3100	Total: Mechanical & Electrical	↓	310	365	485	22%	26.50%	31%
97	0010	**WAREHOUSES** And Storage Buildings	m²	380	550	775				**97**
	0020	Total project costs	m³	68.50	98	161			
	0100	Site work	m²	37.50	74.50	112	6.05%	12.95%	19.85%
	0500	Masonry		22.50	51.50	111	3.73%	7.40%	12.30%
	1800	Equipment		5.80	12.60	70.50	.91%	1.82%	5.55%
	2720	Plumbing		12.05	21.50	40.50	2.90%	4.80%	6.55%
	2730	Heating, ventilating, air conditioning		14.55	39	52.50	2.41%	5%	8.90%
	2900	Electrical		22	40.50	67.50	5.15%	7.20%	10.10%
	3100	Total: Mechanical & Electrical	↓	59.50	92	201	12.75%	18.90%	26%
99	0010	**WAREHOUSE & OFFICES** Combination	m²	445	595	800				**99**
	0020	Total project costs	m³	75	109	161			
	1800	Equipment	m²	7.75	14.95	23	.52%	1.21%	2.40%
	2720	Plumbing		17.20	30.50	45	3.74%	4.76%	6.30%
	2770	Heating, ventilating, air conditioning		27	42.50	59.50	5%	5.65%	10.05%
	2900	Electrical		30	44.50	70	5.85%	8%	10%
	3100	Total: Mechanical & Electrical	↓	83.50	127	203	14.40%	19.95%	24.50%

SQUARE FOOT COSTS

Square Meter Project Size Modifier

One factor that affects the m² cost of a particular building is the size. In general, for buildings built to the same specifications in the same locality, the larger building will have the lower m² cost. This is due mainly to the decreasing contribution of the exterior walls plus the economy of scale usually achievable in larger buildings. The Area Conversion Scale shown below will give a factor to convert costs for the typical size building to an adjusted cost for the particular project.

Example: Determine the cost per m² for a 9200 m² Mid-rise apartment building.

$$\frac{\text{Proposed building area} = 9200\ m2}{\text{Typical size from below} = 4600\ m2} = 2.00$$

Enter Area Conversion scale at 2.0, intersect curve, read horizontally the appropriate cost multiplier of 0.94. Size adjusted cost becomes 0.94 x $970.00 = $910.00 based on national average costs.

Note: For Size Factors less than 0.50, the Cost Multiplier is 1.1
For Size Factors greater than 3.5, the Cost Multiplier is 0.90

The Square Meter Base Size lists the median costs, most typical project size in our accumulated data and the range in size of the projects.

The Size Factor for your project is determined by dividing your project area in m² by the typical project size for the particular Building Type. With this factor, enter the Area Conversion Scale at the appropriate Size Factor and determine the appropriate cost multiplier for your building size.

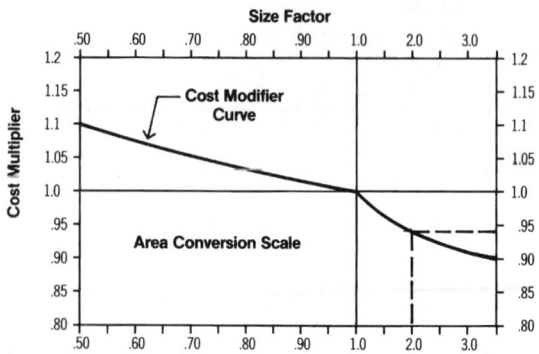

Square Meter Base Size								
Building Type	Median Cost per m²	Typical Size Gross m²	Typical Range Gross m²		Building Type	Median Cost per m²	Typical Size Gross m²	Typical Range Gross m²
Apartments, Low Rise	$ 765.00	2000	900 - 3500	Jails	$2,325.00	3700	500 - 13500
Apartments, Mid Rise	970.00	4600	3000 - 9300	Libraries	1,425.00	1100	700 - 2900
Apartments, High Rise	1,100.00	13500	8800 - 55700	Living, Assisted	1,250.00	3000	2200 - 4700
Auditoriums	1,275.00	2300	700 - 3600	Medical Clinics	1,325.00	700	400 - 1500
Auto Sales	945.00	1900	1000 - 2700	Medical Offices	1,250.00	600	400 - 1400
Banks	1,700.00	400	200 - 700	Motels	935.00	3700	1500 - 11100
Churches	1,175.00	1600	200 - 3900	Nursing Homes	1,275.00	2100	1400 - 3400
Clubs, Country	1,200.00	600	400 - 1400	Offices, Low Rise	1,075.00	1900	500 - 7400
Clubs, Social	1,125.00	900	600 - 1300	Offices, Mid Rise	1,075.00	11100	1900 - 27900
Clubs, YMCA	1,350.00	2600	1200 - 3700	Offices, High Rise	1,375.00	24200	11100 - 74300
Colleges (Class)	1,475.00	4600	1400 - 13900	Police Stations	1,675.00	1000	400 - 1800
Colleges (Science Lab)	2,175.00	4200	1500 - 7400	Post Offices	1,275.00	1200	600 - 2800
College (Student Union)	1,675.00	3100	1500 - 7900	Power Plants	9,475.00	700	100 - 1900
Community Center	1,225.00	900	500 - 1600	Religious Education	1,075.00	800	600 - 1100
Court Houses	1,625.00	3000	1700 - 9800	Research	1,775.00	1800	600 - 4200
Dept. Stores	710.00	8400	4100 - 11300	Restaurants	1,550.00	400	300 - 600
Dormitories, Low Rise	1,250.00	2300	900 - 8800	Retail Stores	760.00	700	400 - 1600
Dormitories, Mid Rise	1,600.00	7900	1900 - 18600	Schools, Elementary	1,125.00	3800	2300 - 5100
Factories	690.00	2500	1200 - 4600	Schools, Jr. High	1,175.00	8500	4800 - 11100
Fire Stations	1,250.00	500	400 - 800	Schools, Sr. High	1,175.00	9400	4700 - 16300
Fraternity Houses	1,175.00	1200	800 - 1400	Schools, Vocational	1,125.00	3400	1900 - 7600
Funeral Homes	1,325.00	900	400 - 1900	Sports Arenas	930.00	1400	500 - 3700
Garages, Commercial	860.00	900	500 - 1300	Supermarkets	750.00	4100	1100 - 5600
Garages, Municipal	1,075.00	800	400 - 1200	Swimming Pools	1,750.00	1900	900 - 3000
Garages, Parking	455.00	15100	7100 - 20900	Telephone Exchange	2,025.00	400	100 - 1000
Gymnasiums	1,150.00	1800	1100 - 3800	Theaters	1,075.00	1000	800 - 1600
Hospitals	2,050.00	5100	2500 - 11600	Town Halls	1,300.00	1000	400 - 2200
House (Elderly)	1,050.00	3400	2000 - 6100	Warehouses	550.00	2300	700 - 6700
Housing (Public)	960.00	3300	1300 - 6900	Warehouse & Office	595.00	2300	700 - 6700
Ice Rinks	1,375.00	2700	2500 - 3100				

Metric Units Used in the Construction Trades

The metric units used in the construction trades are as follows. The term "length" includes all linear measurements (that is, length, width, height, thickness, diameter, and circumference).

	Quantity	Unit	Symbol
Surveying	length	kilometer, meter	km, m
	area	square kilometer	km²
		hectare (10 000 m²)	ha
		square meter	m²
	plane angle	degree (non-metric)	°
		minute (non-metric)	'
		second (non-metric)	"
Excavating	length	meter, millimeter	m, mm
	volume	cubic meter	m³
Trucking	distance	kilometer	km
	volume	cubic meter	m³
	mass	metric ton (1000 kg)	t
Paving	length	meter, millimeter	m, mm
	area	square meter	m²
Concrete	length	meter, millimeter	m, mm
	area	square meter	m²
	volume	cubic meter	m³
	temperature	degree Celsius	°C
	water capacity	liter (1000 cm³)	L
	mass (weight)	kilogram, gram	kg, g
	cross-sectional area	square millimeter	mm²
Masonry	length	meter, millimeter	m, mm
	area	square meter	m²
	mortar volume	cubic meter	m³
Steel	length	meter, millimeter	m, mm
	mass	metric ton (1000 kg)	t
		kilogram, gram	kg, g
	pressure/stress	pascal	Pa
		kilopascal	kPa
		megapascal	MPa
Carpentry	length	meter, millimeter	m, mm
Plastering	length	meter, millimeter	m, mm
	area	square meter	m²
	water capacity	liter (1000 cm³)	L
Glazing	length	meter, millimeter	m, mm
	area	square meter	m²
Painting	length	meter, millimeter	m, mm
	area	square meter	m²
	capacity	liter (1000 cm³)	L
		milliliter (cm³)	mL

	Quantity	Unit	Symbol
Roofing	length	meter, millimeter	m, mm
	area	square meter	m²
	slope	meter, millimeter	m, mm
Plumbing	length	meter, millimeter	m, mm
	mass	kilogram, gram	kg, g
	capacity	liter (1000 cm³)	L
	pressure	kilopascal	kPa
Drainage	length	meter, millimeter	m, mm
	area	hectare (10 000 m²)	ha
		square meter	m²
	volume	cubic meter	m³
	slope	millimeter/meter	mm/m
HVAC	length	meter, millimeter	m, mm
	volume	cubic meter	m³
	capacity	liter (1000 cm³)	L
	airflow	meter/second	m/s
	volume flow	cubic meter/second	m³/s
		liter/second	L/s
	temperature	degree Celsius	°C
	force	newton, kilonewton	N, kN
	pressure	kilopascal	kPa
	energy, work	kilojoule, megajoule	kJ, MJ
	rate of heat flow	watt, kilowatt	W, kW
Electrical	length	meter, millimeter	m, mm
	frequency	hertz	Hz
	power	watt, kilowatt	W, kW
	energy	megajoule	MJ
		kilowatt hour	kWh
	electric current	ampere	A
	electric potential	volt, kilovolt	V, kV
	resistance	ohm	Ω

A	Ampere	cd	Candela
ABS	Acrylonitrile Butadiene Stryrene; Asbestos Bonded Steel	CD	Grade of Plywood Face & Back
A.C.	Alternating Current; Air-Conditioning; Asbestos Cement; Plywood, Grade A & C	CDX	Plywood, Grade C & D, Exterior Glue
		Cefi.	Cement Finisher
		Cem.	Cement
A.C.I.	American Concrete Institute	c.g.	Center of Gravity
AD	Plywood, Grade A & D	CHW	Chilled Water
Addit.	Additional	C.I.	Cast Iron
Adj.	Adjustable	C.I.P.	Cast In Place
af	Audio-Frequency	Circ.	Circuit
A.G.A.	American Gas Association	C.L.	Carload Lot
Agg.	Aggregate	Clab.	Common laborer
A.H.	Ampere Hours	C.I.F	Current Limiting Fuse
A hr.	Ampere-hour	CLP	Cross Linked Polyethylene
A.H.U.	Air Handling Unit	cm	Centimeter
A.I.A.	American Institute of Architects	CMP	Corr. Metal Pipe
AIC	Ampere Interrupting Capacity	C.M.U.	Concrete Masonry Unit
Allow.	Allowance	Col.	Column
alt.	Altitude	CO^2	Carbon Dioxide
Alum.	Aluminum	Comb.	Combination
a.m.	Ante Meridiem	Compr.	Compressor
Amp.	Ampere	Conc.	Concrete
Anod.	Anodized	Cont.	Continuous; Continued
Approx.	Approximate	Corr.	Corrugated
Apt.	Apartment	Cos	Cosine
Asb.	Asbestos	Cot	Cotangent
A.S.B.C.	American Standard Building Code	Cov.	Cover
Asbe.	Asbestos Worker	CPA	Control Point Adjustment
A.S.H.R.A.E.	American Society of Heating, Refrig. & AC Engineers	Cplg.	Coupling
		C.P.M.	Critical Path Method
		CPVC	Chlorinated Polyvinyl Chloride
A.S.M.E.	American Society of Mechanical Engineers	C. Pr.	Hundred Pair
		CRC	Cold Rolled Channel
A.S.T.M.	American Society for Testing and Materials	Creos.	Creosote
		Crpt.	Carpet & Linoleum Layer
Attchmt.	Attachment	CRT	Cathode-Ray Tube
Avg.	Average	CS	Carbon Steel
A.W.G.	American Wire Gauge	Csc	Cosecant
Bbl.	Barrel	CSI	Construction Specifications Institute
B&B	Grade B and Better; Balled & Burlapped	C.T.	Current Transformer
B.&S.	Bell and Spigot	CTS	Copper Tube Size
B.&W.	Black and White	cw	Continuous Wave
b.c.c.	Body-centered Cubic	C.W.	Cool White; Cold Water
BE	Bevel End	C.W.X.	Cool White Deluxe
Bg. cem.	Bag of Cement	Cyl.	Cylinder
B.I.	Black Iron	D	Deep; Depth; Discharge
Bit.; Bitum.	Bituminous	Dis.; Disch.	Discharge
Bk.	Backed	Db.	Decibel
Bkrs.	Breakers	Dbl.	Double
Bldg.	Building	DC	Direct Current
Blk.	Block	Demob.	Demobilization
Bm.	Beam	d.f.u.	Drainage Fixture Units
Bm^3	Bank Cubic Meter	D.H.	Double Hung
Boil.	Boilermaker	DHW	Domestic Hot Water
BR	Bedroom	Diag.	Diagonal
Brg.	Bearing	Diam.	Diameter
Brhe.	Bricklayer Helper	Distrib.	Distribution
Bric.	Bricklayer	Dk.	Deck
Brk.	Brick	D.L.	Dead Load; Diesel
Brng.	Bearing	Do.	Ditto
Brs.	Brass	Dp.	Depth
Brz.	Bronze	D.P.S.T.	Double Pole, Single Throw
Bsn.	Basin	Dr.	Driver
Btr.	Better	Drink.	Drinking
BX	Interlocked Armored Cable	D.S.	Double Strength
c	Conductivity	D.S.A.	Double Strength A Grade
C	Coulomb	D.S.B.	Double Strength B Grade
CA	Contact Area	Dty.	Duty
C/C	Center to Center	DWV	Drain Waste Vent
Cab.	Cabinet	DX	Deluxe White, Direct Expansion
Cair.	Air Tool Laborer	e	Eccentricity
Calc	Calculated	E	Equipment Only; East
Cap.	Capacity	Ea.	Each
Carp.	Carpenter	E.B.	Encased Burial
C.B.	Circuit Breaker	Econ.	Economy
C.C.A.	Chromate Copper Arsenate	EDP	Electronic Data Processing

E.D.R.	Equiv. Direct Radiation
Eq.	Equation
Elec.	Electrician; Electrical
Elev.	Elevator; Elevating
EMT	Electrical Metallic Conduit; Thin Wall Conduit
Em^3	Embankment Cubic Meter
Eng.	Engine
EPDM	Ethylene Propylene Diene Monomer
Eqhv.	Equip. Oper., Heavy
Eqlt.	Equip. Oper., Light
Eqmd.	Equip. Oper., Medium
Eqmm.	Equip. Oper., Master Mechanic
Eqol.	Equip. Oper., Oilers
Equip.	Equipment
ERW	Electric Resistance Welded
Est.	Estimated
esu	Electrostatic Units
E.W.	Each Way
EWT	Entering Water Temperature
Excav.	Excavation
Exp.	Expansion, Exposure
Ext.	Exterior
Extru.	Extrusion
f.	Fiber Stress
F	Farad, Female; Fill
Fab.	Fabricated
FBGS	Fiberglass
f.c.c.	Face-centered Cubic
f'c.	Compressive Stress in Concrete; Extreme Compressive Stress
F.E.	Front End
FEP	Fluorinated Ethylene Propylene (Teflon)
F.G.	Flat Grain
F.H.A.	Federal Housing Administration
Fig.	Figure
Fin.	Finished
Fixt.	Fixture
Flr.	Floor
F.M.	Frequency Modulation; Factory Mutual
Fmg.	Framing
Fndtn.	Foundation
Fori.	Foreman, Inside
Foro.	Foreman, Outside
Fount.	Fountain
FPT	Female Pipe Thread
Fr.	Frame
F.R.	Fire Rating
FRK	Foil Reinforced Kraft
FRP	Fiberglass Reinforced Plastic
FS	Forged Steel
FSC	Cast Body; Cast Switch Box
Ftng.	Fitting
Ftg.	Footing
Furn.	Furniture
FVNR	Full Voltage Non-Reversing
FXM	Female by Male
Fy.	Minimum Yield Stress of Steel
g	Gram
G	Gauss; Giga
Ga.	Gauge
Galv.	Galvanized
Gen.	General
G.F.I.	Ground Fault Interrupter
Glaz.	Glazier
GR	Grade
Gran.	Granular
Grnd.	Ground
h	Hecto; Hundred
H	High; High Strength Bar Joist; Henry
ha	Hectare
H.C.	High Capacity
H.D.	Heavy Duty; High Density
H.D.O.	High Density Overlaid

924

Hdr.	Header	log	Logarithm	OB	Opposing Blade
Hdwe.	Hardware	L.P.	Liquefied Petroleum; Low Pressure	OC	On Center
Help.	Helper, Average	L.P.F.	Low Power Factor	OD	Outside Diameter
HEPA	High Efficiency Particulate Air Filter	LR	Long Radius	O.D.	Outside Dimension
		L. S.	Lump Sum	ODS	Overhead Distribution System
Hg	Mercury	Lt.	Light	O&P	Overhead and Profit
HIC	High Interrupting Capacity	Lt. Ga.	Light Gauge	Oper.	Operator
H.O.	High Output	L.T.L.	Less Than Truckload Lot	Opng.	Opening
Horiz.	Horizontal	Lt. Wt.	Lightweight	Orna.	Ornamental
H.P.	High Pressure	L. V.	Low Voltage	O.S.&Y.	Outside Screw and Yoke
H.P.F.	High Power Factor	lx	lux	Ovhd.	Overhead
HSC	High Short Circuit	m	Meter	OWG	Oil, Water or Gas
Ht.	Height	m²	Square Meter	P.	Pole; Applied Load; Projection
Htg.	Heating	m²CA	Square Meters Contact Area	p.	Page
Htrs.	Heaters	M	Material; Male; Mega	Pa	Pascal
HVAC	Heating, Ventilating & Air Conditioning		Light Wall Copper Tubing	Pape.	Paperhanger
		mA	Milliampere	P.A.P.R.	Powered Air Purifying Respirator
Hvy.	Heavy	Mach.	Machine	PAR	Weatherproof Reflector
HW	Hot Water	Mag. Str.	Magnetic Starter	Pc.	Piece
Hyd.; Hydr.	Hydraulic	Maint.	Maintenance	P.C.	Portland Cement; Power Connector
Hz.	Hertz (cycles)	Marb.	Marble Setter	P.C.M.	Phase Contrast Microscopy
I.	Moment of Inertia	Mat; Mat'l.	Material	P.E.	Professional Engineer; Porcelain Enamel; Polyethylene; Plain End
I.C.	Interrupting Capacity	Max.	Maximum		
ID	Inside Diameter	MC	Metal Clad Cable		
I.D.	Inside Dimension; Identification	M.C.M.	Thousand Circular Mils	Perf.	Perforated
I.F.	Inside Frosted	M.C.P.	Motor Circuit Protector	Ph.	Phase
I.M.C.	Intermediate Metal Conduit	MD	Medium Duty	P.I.	Pressure Injected
Incan.	Incandescent	M.D.O.	Medium Density Overlaid	Pile.	Pile Driver
Incl.	Included; Including	Med.	Medium	Pkg.	Package
Int.	Interior	Mfg.	Manufacturing	Pl.	Plate
Inst.	Installation	Mfrs.	Manufacturers	Plah.	Plasterer Helper
Insul.	Insulation	mg	Milligram	Plas.	Plasterer
I.P.	Iron Pipe	MH, M.H.	Manhole; Metal Halide; Man-Hour	Pluh.	Plumbers Helper
I.P.S.	Iron Pipe Size	MH_z	Megahertz	Plum.	Plumber
I.P.T.	Iron Pipe Threaded	MI	Malleable Iron; Mineral Insulated	Ply.	Plywood
I.W.	Indirect Waste	mm	Millimeter	p.m.	Post Meridiem
J	Joule	Mill.	Millwright	Pord.	Painter, Ordinary
J.I.C.	Joint Industrial Council	Min.; min.	Minimum, Minute	pp	Pages
k	Thousand	Misc.	Miscellaneous	PP; PPL	Polypropylene
K	Heavy Wall Copper Tubing; Kelvin	ml	Milliliter	P.P.M.	Parts per Million
K.A.H.	Thousand Amp. Hours	Mo.	Month	Pr.	Pair
kcmil	Thousand Circular Mils	Mobil.	Mobilization	Prefab.	Prefabricated
KD	Knock Down	Mog.	Mogul Base	Prefin.	Prefinished
K.D.A.T.	Kiln Dried After Treatment	MPa	Megapascal	Prop.	Propelled
kg	Kilogram	MPT	Male Pipe Thread	PSP	Plastic Sewer Pipe
kG	Kilogauss	ms	Millisecond	Pspr.	Painter, Spray
kgf	Kilogram force	Mstz.	Mosaic & Terrazzo Worker	Psst.	Painter, Structural Steel
kHz	Kilohertz	Mtd.	Mounted	P.T.	Potential Transformer
kJ	Kilojoule	Mthe.	Mosaic & Terrazzo Helper	P. & T.	Pressure & Temperature
K.L.	Effective Length Factor	Mtng.	Mounting	Ptd.	Painted
km	Kilometer	Mult.	Multi; Multiply	Ptns.	Partitions
kPa	Kilopascal	M.V.A.	Million Volt Amperes	Pu	Ultimate Load
kV	Kilovolt	M.V.A.R.	Million Volt Amperes Reactance	PVC	Polyvinyl Chloride
kVA	Kilovolt Ampere	MV	Megavolt	Pvmt.	Pavement
kvar	Kilovar (Reactance)	MW	Megawatt	Pwr.	Power
kW	Kilowatt	MXM	Male by Male	Q	Quantity Heat Flow
kWh	Kilowatt-hour	N	Natural; North; Newton	Quan.; Qty.	Quantity
L	Labor Only; Length; Long; Liter; Medium Wall Copper Tubing	nA	Nanoampere	Q.C.	Quick Coupling
		NA	Not Available; Not Applicable	r	Radius of Gyration
Lab.	Labor	N.B.C.	National Building Code	R	Resistance
lat	Latitude	NC	Normally Closed	R.C.P.	Reinforced Concrete Pipe
Lath.	Lather	N.E.M.A.	National Electrical Manufacturers Assoc.	Rect.	Rectangle
Lav.	Lavatory			Reg.	Regular
L.B.	Load Bearing; L Conduit Body	NEHB	Bolted Circuit Breaker to 600V	Reinf.	Reinforced
L. & E.	Labor & Equipment	N.L.B.	Non-Load-Bearing	Req'd.	Required
L.C.L.	Less Than Carload Lot	NM	Non-Metallic Cable	Res.	Resistant
Ld.	Load	nm	Nanometer	Resi.	Residential
LE	Lead Equivalent	No.	Number	Rgh.	Rough
Lg.	Long; Length; Large	NO	Normally Open	R.H.W.	Rubber, Heat & Water Resistant; Residential Hot Water
L. & H.	Light and Heat	N.O.C.	Not Otherwise Classified		
L.H.	Long Span High Strength Bar Joist	Nose.	Nosing	rms.	Root Mean Square
L.J.	Long Span Standard Strength Bar Joist	N.P.T.	National Pipe Thread	Rnd.	Round
		NQOD	Combination plug-on/bolt-on Circuit Breaker to 240V	Rodm.	Rodman
L.L.	Live Load			Rofc.	Roofer, Composition
L.L.D.	Lamp Lumen Depreciation	N.R.C.	Noise Reduction Coefficient	Rofp.	Roofer, Precast
lm	Lumen	N.R.S.	Non Rising Stem	Rohe.	Roofer Helpers (Composition)
Lm³	Loose Cubic Meter	ns	Nanosecond	Rots.	Roofer, Tile & Slate
L.O.A.	Length Over All	nW	Nanowatt	R.O.W.	Right of Way

R.R.	Direct Burial Feeder Conduit		T.W.	Thermoplastic Water Resistant Wire
R.S.	Rapid Start		UCI	Uniform Construction Index
RT	Round Trip		UF	Underground Feeder
S.	Suction; Single Entrance; South; Second		U.H.F.	Ultra High Frequency
SC	Screw Cover		U.L.	Underwriters Laboratory
Scaf.	Scaffold		Unfin.	Unfinished
Sch.; Sched.	Schedule		URD	Underground Residential Distribution
S.C.R.	Modular Brick		V	Volt
S.D.	Sound Deadening		V.A.	Volt Amperes
S.D.R.	Standard Dimension Ratio		V.C.T.	Vinyl Composition Tile
S.E.	Surfaced Edge		VAV	Variable Air Volume
Sel.	Select		VC	Veneer Core
S.E.R.; S.E.U.	Service Entrance Cable		Vent.	Ventilating
S4S	Surface 4 Sides		Vert.	Vertical
Shee.	Sheet Metal Worker		V.F.	Vinyl Faced
Sin.	Sine		V.G.	Vertical Grain
Skwk.	Skilled Worker		V.H.F.	Very High Frequency
SL	Saran Lined		VHO	Very High Output
S.L.	Slimline		Vib.	Vibrating
Sldr.	Solder		Vol.	Volume
S.N.	Solid Neutral		W	Wire; Watt; Wide; West
S.P.	Static Pressure; Single Pole; Self-Propelled		w/	With
Spri.	Sprinkler Installer		W.C.	Water Column; Water Closet
S.P.D.T.	Single Pole, Double Throw		W.F.	Wide Flange
S.P.S.T.	Single Pole, Single Throw		W.G.	Water Gauge
SPT	Standard Pipe Thread		Wldg.	Welding
Sq. Hd.	Square Head		W.R.	Water Resistant
S.S.	Single Strength; Stainless Steel		Wrck.	Wrecker
S.S.B.	Single Strength B Grade		W.S.P.	Water, Steam, Petroleum
Sswk.	Structural Steel Worker		WT., Wt.	Weight
Sswl.	Structural Steel Welder		WWF	Welded Wire Fabric
St.; Stl.	Steel		XFMR	Transformer
S.T.C.	Sound Transmission Coefficient		XHD	Extra Heavy Duty
Std.	Standard		XHHW; XLPE	Cross-Linked Polyethylene Wire Insulation
STP	Standard Temperature & Pressure		Y	Wye
Stpi.	Steamfitter, Pipefitter		yr	Year
Str.	Strength; Starter; Straight		Δ	Delta
Strd.	Stranded		%	Percent
Struct.	Structural			**Approximately**
Sty.	Story		Ø	Phase
Subj.	Subject		@	At
Subs.	Subcontractors		#	Number
Surf.	Surface		<	Less Than
Sw.	Switch		>	Greater Than
Swbd.	Switchboard			
Syn.	Synthetic			
Sys.	System			
t.	Thickness			
T	Temperature; Tesla			
Tan	Tangent			
T.C.	Terra Cotta			
T & C	Threaded and Coupled			
T.D.	Temperature Difference			
Tdd	Telecommunications Device for the Deaf			
T.E.M.	Transmission Electron Microscopy			
TFE	Tetrafluoroethylene (Teflon)			
T. & G.	Tongue & Groove; Tar & Gravel			
Th.; Thk.	Thick			
Thn.	Thin			
Thrded	Threaded			
Tilf.	Tile Layer, Floor			
Tilh.	Tile Layer, Helper			
THW	Insulated Strand Wire			
THWN; THHN	Nylon Jacketed Wire			
T.L.	Truckload			
Tot.	Total			
T.S.	Trigger Start			
Tr.	Trade			
Transf.	Transformer			
Trhv.	Truck Driver, Heavy			
Trlr.	Trailer			
Trlt.	Truck Driver, Light			
TTY	Teletypewriter			
TV	Television			

Index

I

INDEX

Index

Division Notes

	CREW	DAILY OUTPUT	LABOR-HOURS	UNIT	2006 BARE COSTS				TOTAL INCL O&P
					MAT.	LABOR	EQUIP.	TOTAL	

	CREW	DAILY OUTPUT	LABOR-HOURS	UNIT	2006 BARE COSTS				TOTAL INCL O&P
					MAT.	LABOR	EQUIP.	TOTAL	

Division Notes

	CREW	DAILY OUTPUT	LABOR-HOURS	UNIT	2006 BARE COSTS				TOTAL INCL O&P
					MAT.	LABOR	EQUIP.	TOTAL	

Division Notes

| | | DAILY OUTPUT | LABOR-HOURS | UNIT | 2006 BARE COSTS | | | | TOTAL INCL O&P |
	CREW				MAT.	LABOR	EQUIP.	TOTAL	

Division Notes

		CREW	DAILY OUTPUT	LABOR-HOURS	UNIT	2006 BARE COSTS				TOTAL INCL O&P
						MAT.	LABOR	EQUIP.	TOTAL	

Reed Construction Data, Inc.

Reed Construction Data, Inc., a leading worldwide provider of total construction information solutions, is comprised of three main product groups designed specifically to help construction professionals advance their businesses with timely, accurate and actionable project, product and cost data. Reed Construction Data is a division of Reed Business Information, a member of the Reed Elsevier plc group of companies.

The Project, Product and Cost & Estimating divisions offer a variety of innovative products and services designed for the full spectrum of design, construction and manufacturing professionals. Through its International companies, Reed Construction Data's reputation for quality construction market data is growing worldwide.

Cost Information

RSMeans, the undisputed market leader and authority on construction costs, publishes current cost and estimating information in annual cost books and on the *Means CostWorks* CD-ROM. RSMeans furnishes the construction industry with a rich library of corresponding reference books and a series of professional seminars that are designed to sharpen professional skills and maximize the effective use of cost estimating and management tools. RSMeans also provides construction cost consulting for owners, manufacturers, designers, and contractors.

Project Data

Reed Construction Data provides complete, accurate and relevant project information through all stages of construction. Customers are supplied industry data through leads, project reports, contact lists, plans and specifications surveys, market penetration analyses and sales evaluation reports. Any of these products can pinpoint a county, look at a state, or cover the country. Data is delivered via paper, e-mail, CD-ROM or the Internet.

Building Product Information

The First Source suite of products is the only integrated building product information system offered to the commercial construction industry for comparing and specifying building products. These print and online resources include *First Source,* CSI's SPEC-DATA™, CSI's MANU-SPEC™, First Source CAD, and Manufacturer Catalogs. Written by industry professionals and organized using CSI's MasterFormat™, construction professionals use this information to make better design decisions. FirstSourceONL.com combines Reed Construction Data's project, product and cost data with news and information from Reed Business Information's *Building Design & Construction* and *Consulting-Specifying Engineer magazines.* This industry-focused site offers easy and unlimited access to vital information for all construction professionals.

International

Reed Construction Data Canada serves the Canadian construction market with reliable and comprehensive project and product information services that cover all facets of construction. Core services include: *BuildSource, BuildSpec, BuildSelect,* product selection and specification tools available in print and on the Internet; Building Reports, a national construction project lead service; CanaData, statistical and forecasting information; *Daily Commercial News,* a construction newspaper reporting on news and projects in Ontario; and *Journal of Commerce,* reporting news in British Columbia and Alberta.

BIMSA/Mexico provides construction project news, product information, cost data, seminars and consulting services to construction professionals in Mexico. Its subsidiary, PRISMA, provides job costing software.

Byggfakta Scandinavia AB, founded in 1936, is the parent company for the leaders of customized construction market data for Denmark, Estonia, Finland, Norway and Sweden. Each company fully covers the local construction market and provides information across several platforms including via subscription, on an ad-hoc basis, electronically and on paper.

Cordell Building Information Services, with its complete range of project and cost and estimating services, is Australia's specialist in the construction information industry. Cordell provides in-depth and historical information on all aspects of construction projects and estimation, including several customized reports, construction and sales leads and detailed cost information.

For more information, please visit our Web site at www.reedconstructiondata.com.

Reed Construction Data, Inc., Corporate Office
30 Technology Parkway South
Norcross, GA 30092-2912
(800) 322-6996
(800) 895-8661 (fax)
info@reedbusiness.com
www.reedconstructiondata.com

Means Project Cost Report

By filling out this report, your project data will contribute to the database that supports the Means Project Cost Square Foot Data. When you fill out this form, Means will provide a 30% discount off one of the Means products advertised in the following pages. Please complete the form including all items where you have cost data, and all the items marked (✔).

$30.00 Discount per product for each report you submit.

Project Description (No remodeling projects, please.)

✔ Building Use (Office, School...) _____

✔ Address (City, State) _____

_____

✔ Frame (Wood, Steel...) _____

✔ Exterior Wall (Brick, Tilt-up...) _____

✔ Basement: (check one)   ☐ Full ☐ Partial ☐ None

✔ Number Stories _____

✔ Floor-to-Floor Height _____

% Air Conditioned _____ Tons _____

Comments _____

Total Project Cost $ _____

Owner _____

Architect _____

General Contractor _____

✔ Bid Date _____

Typical Bay Size _____

✔ Labor Force: _____ % Union _____ % Non-Union

✔ Project Description (Circle one number in each line)

1. Economy 2. Average   3. Custom 4. Luxury
1. Square   2. Rectangular 3. Irregular 4. Very Irregular

A	✔ General Conditions	$
B	✔ Site Work	$
C	✔ Concrete	$
D	✔ Masonry	$
E	✔ Metals	$
F	✔ Wood & Plastics	$
G	✔ Thermal & Moisture Protection	
GR	Roofing & Flashing	$
H	✔ Doors and Windows	$
J	✔ Finishes	$
JP	Painting & Wall Covering	$

K	✔ Specialties	$
L	✔ Equipment	$
M	✔ Furnishings	$
N	✔ Special Construction	$
P	✔ Conveying Systems	$
Q	✔ Mechanical	$
QP	Plumbing	$
QB	HVAC	$
R	✔ Electrical	$
S	✔ Mech./Elec. Combined	$

Please specify the Means product you wish to receive.
Complete the address information.

Product Name _____

Product Number _____

Your Name _____

Title _____

Company _____
  ☐ Company
  ☐ Home   Street Address _____

City, State, Zip _____

Return by mail or Fax 888-492-6770.

Method of Payment:

Credit Card # _____

Expiration Date _____

Check _____

Purchase Order _____

Reed Construction Data/RSMeans
Square Foot Costs Department
P.O. Box 800
Kingston, MA 02364-9988

Reed Construction Data/RSMeans . . . a tradition of excellence in Construction Cost Information and Services since 1942.

Table of Contents

Book Selection Guide

The following table provides definitive information on the content of each cost data publication. The number of lines of data provided in each unit price or assemblies division, as well as the number of reference tables and crews is listed for each book. The presence of other elements such as an historical cost index, city cost indexes, square foot models or cross-referenced index is also indicated. You can use the table to help select the Means' book that has the quantity and type of information you most need in your work.

Unit Cost Divisions	Building Construction Costs	Mechanical	Electrical	Repair & Remodel.	Square Foot	Site Work Landsc.	Assemblies	Interior	Concrete Masonry	Open Shop	Heavy Construc.	Light Commercial	Facil. Construc.	Plumbing	Western Construction Costs	Residential
1	554	273	358	465		496		297	465	553	502	210	979	325	553	161
2	2560	1345	460	1457		8234		543	1383	2466	5287	695	4835	1593	2543	768
3	1393	112	99	721		1250		191	1774	1388	1395	229	1315	81	1388	262
4	854	18	0	666		660		576	1075	830	597	432	1076	0	840	352
5	1876	239	193	965		817		967	738	1844	1077	835	1856	316	1816	759
6	1487	82	78	1465		450		1385	317	1477	607	1603	1574	47	1824	1734
7	1252	159	71	1240		472		485	401	1251	355	963	1306	169	1252	760
8	1893	28	0	1965		313		1682	642	1870	51	1329	2095	0	1852	1263
9	1595	47	0	1425		232		1688	360	1547	157	1337	1811	47	1586	1217
10	871	47	25	508		193		710	170	873	0	404	917	233	871	217
11	1018	322	169	502		137		813	44	925	107	218	1173	291	925	104
12	313	0	0	46		213		1422	27	304	0	65	1442	0	304	63
13	1317	1016	385	659		451		884	75	1292	340	484	1931	929	1183	194
14	345	36	0	258		36		292	0	344	30	12	362	35	343	6
15	1976	13019	636	1752		1542		1192	58	1985	1729	1225	10787	9669	2005	829
16	1279	470	10114	1002		739		1108	55	1297	760	1083	9885	415	1228	554
Totals	20583	17213	12588	15096		16235		14235	7584	20246	12994	11124	43344	14150	20513	9243

Assembly Divisions	Building Construction Costs	Mechanical	Electrical	Repair & Remodel.	Square Foot	Site Work Landsc.	Assemblies	Interior	Concrete Masonry	Open Shop	Heavy Construc.	Light Commercial	Facil. Construc.	Plumbing	Western Construction Costs	Asm Div	Residential
A	612	19	0	192	150	540	612	0	550		542	149	24	0		1	374
B	5584	0	0	809	2479	0	5584	333	1914		0	2023	144	0		2	217
C	1208	0	0	635	853	0	1208	1555	132		0	757	238	0		3	588
D	2392	1014	780	693	1776	0	2392	752	0		0	1262	1010	890		4	867
E	294	0	0	85	257	0	294	5	0		0	257	5	0		5	393
F	124	0	0	0	123	0	124	0	0		0	123	3	0		6	357
G	584	465	172	330	111	1844	584	0	471		423	110	113	559		7	299
																8	760
																9	80
																10	0
																11	0
																12	0
Totals	10798	1498	952	2744	5749	2384	10798	2645	3067	0	965	4681	1537	1449			3935

Reference Section	Building Construction Costs	Mechanical	Electrical	Repair & Remodel.	Square Foot	Site Work Landsc.	Assemblies	Interior	Concrete Masonry	Open Shop	Heavy Construc.	Light Commercial	Facil. Construc.	Plumbing	Western Construction Costs	Residential
Tables	237	237	237	237	4	237	219	237	237	237	237	237	237	237	237	35
Models					105							46				32
Crews	452	452	452	434		452		452	452	431	452	431	434	452	452	431
City Cost Indexes	yes	yes	yes	yes	yes	yes	yes	yes	yes	yes	yes	yes	yes	yes	yes	yes
Historical Cost Indexes	yes	yes	yes	yes	yes	yes	yes	yes	yes	yes	yes	yes	yes	yes	yes	no

1

Annual Cost Guides

For more information
visit Means Web site
at www.rsmeans.com

Means Building Construction Cost Data 2006

Available in Both Softbound and Looseleaf Editions

The "Bible" of the industry comes in the standard softcover edition or the looseleaf edition.

Many customers enjoy the convenience and flexibility of the looseleaf binder, which increases the usefulness of Means *Building Construction Cost Data 2006* by making it easy to add and remove pages. You can insert your own cost information pages, so everything is in one place. Copying pages for faxing is easier also. Whichever edition you prefer, softbound or the convenient looseleaf edition, you'll be eligible to receive *The Change Notice* FREE. Current subscribers receive *The Change Notice* via e-mail.

$157.95 per copy, Looseleaf
Catalog No. 61016

Means Building Construction Cost Data 2006

Offers you unchallenged unit price reliability in an easy-to-use arrangement. Whether used for complete, finished estimates or for periodic checks, it supplies more cost facts better and faster than any comparable source. Over 23,000 unit prices for 2006. The City Cost Indexes cover over 930 areas, for indexing to any project location in North America. Order and get *The Change Notice* FREE. You'll have year-long access to the Means Estimating **HOTLINE** FREE with your subscription. Expert assistance when using Means data is just a phone call away.

$126.95 per copy
Over 700 pages, illustrated, available Oct. 2005
Catalog No. 60016

Means Metric Construction Cost Data 2006

An 880+ page compendium of all the data from both the 2006 *Building Construction Cost Data* AND the *Heavy Construction Cost Data*, in **metric** format! Access all of this vital information from one complete source. It contains more than 600 pages of unit costs and 40 pages of assemblies costs. The Reference Section contains over 200 pages of tables, charts and other estimating aids. A great way to stay in step with today's construction trends and rapidly changing costs.

$149.95 per copy
Over 800 pages, illus., available Dec. 2005
Catalog No. 63016

For more information
visit Means Web site
at www.rsmeans.com

Annual Cost Guides

Means Mechanical Cost Data 2006

• **HVAC Controls**

Total unit and systems price guidance for mechanical construction... materials, parts, fittings, and complete labor cost information. Includes prices for piping, heating, air conditioning, ventilation, and all related construction.

Plus new 2006 unit costs for:

• Over 2500 installed HVAC/controls assemblies
• "On Site" Location Factors for over 930 cities and towns in the U.S. and Canada
• Crews, labor, and equipment

$126.95 per copy
Over 600 pages, illustrated, available Oct. 2005
Catalog No. 60026

Means Plumbing Cost Data 2006

Comprehensive unit prices and assemblies for plumbing, irrigation systems, commercial and residential fire protection, point-of-use water heaters, and the latest approved materials. This publication and its companion, Means *Mechanical Cost Data*, provide full-range cost estimating coverage for all the mechanical trades.

$126.95 per copy
Over 550 pages, illustrated, available Oct. 2005
Catalog No. 60216

Means Electrical Cost Data 2006

Pricing information for every part of electrical cost planning. More than 15,000 unit and systems costs with design tables; clear specifications and drawings; engineering guides and illustrated estimating procedures; complete labor-hour and materials costs for better scheduling and procurement; and the latest electrical products and construction methods.

• A variety of special electrical systems including cathodic protection
• Costs for maintenance, demolition, HVAC/ mechanical, specialties, equipment, and more

$126.95 per copy
Over 450 pages, illustrated, available Oct. 2005
Catalog No. 60036

Means Electrical Change Order Cost Data 2006

Means *Electrical Change Order Cost Data* 2006 provides you with electrical unit prices exclusively for pricing change orders based on the recent, direct experience of contractors and suppliers. Analyze and check your own change order estimates against the experience others have had doing the same work. It also covers productivity analysis and change order cost justifications. With useful information for calculating the effects of change orders and dealing with their administration.

$126.95 per copy
Over 450 pages, available Oct. 2005
Catalog No. 60236

Means Square Foot Costs 2006

It's Accurate and Easy To Use!

• **Updated 2006 price information**, based on nationwide figures from suppliers, estimators, labor experts, and contractors
• "How-to-Use" sections, with **clear examples** of commercial, residential, industrial, and institutional structures
• Realistic graphics, offering true-to-life illustrations of building projects
• Extensive information on using square foot cost data, including sample estimates and alternate pricing methods

$137.95 per copy
Over 450 pages, illustrated, available Nov. 2005
Catalog No. 60056

Means Repair & Remodeling Cost Data 2006

Commercial/Residential

Use this valuable tool to estimate commercial and residential renovation and remodeling.

Includes: New costs for hundreds of unique methods, materials, and conditions that only come up in repair and remodeling. PLUS:

• Unit costs for over 16,000 construction components
• Installed costs for over 90 assemblies
• Costs for 300+ construction crews
• Over 930 "On Site" localization factors for the U.S. and Canada.

$107.95 per copy
Over 650 pages, illustrated, available Nov. 2005
Catalog No. 60046

Annual Cost Guides

For more information
visit Means Web site
at www.rsmeans.com

Means Facilities Construction Cost Data 2006

For the maintenance and construction of commercial, industrial, municipal, and institutional properties. Costs are shown for new and remodeling construction and are broken down into materials, labor, equipment, overhead, and profit. Special emphasis is given to sections on mechanical, electrical, furnishings, site work, building maintenance, finish work, and demolition.

More than 45,000 unit costs, plus assemblies and reference sections are included.

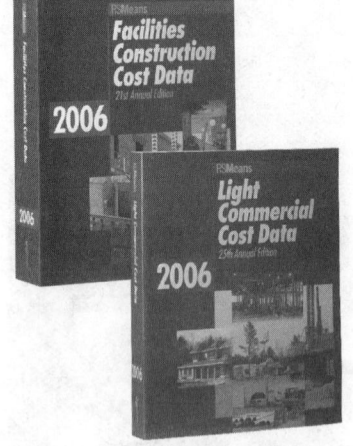

$299.95 per copy
Over 1200 pages, illustrated, available Nov. 2005
Catalog No. 60206

Means Light Commercial Cost Data 2006

Specifically addresses the light commercial market, which is an increasingly specialized niche in the industry. Aids you, the owner/designer/contractor, in preparing all types of estimates, from budgets to detailed bids. Includes new advances in methods and materials.

Assemblies section allows you to evaluate alternatives in the early stages of design/planning.

Over 13,000 unit costs for 2006 ensure you have the prices you need... when you need them.

$107.95 per copy
Over 650 pages, illustrated, available Nov. 2005
Catalog No. 60186

Means Residential Cost Data 2006

Contains square foot costs for 30 basic home models with the look of today, plus hundreds of custom additions and modifications you can quote right off the page. With costs for the 100 residential systems you're most likely to use in the year ahead. Complete with blank estimating forms, sample estimates and step-by-step instructions.

$107.95 per copy
Over 600 pages, illustrated, available Oct. 2005
Catalog No. 60176

Means Site Work & Landscape Cost Data 2006

Means *Site Work & Landscape Cost Data* 2006 is organized to assist you in all your estimating needs. Hundreds of fact-filled pages help you make accurate cost estimates efficiently.

Updated for 2006!

• Demolition features—including ceilings, doors, electrical, flooring, HVAC, millwork, plumbing, roofing, walls, and windows
• State-of-the-art segmental retaining walls
• Flywheel trenching costs and details
• Updated wells section
• Landscape materials, flowers, shrubs, and trees

$126.95 per copy
Over 600 pages, illustrated, available Nov. 2005
Catalog No. 60286

Means Assemblies Cost Data 2006

Means *Assemblies Cost Data* 2006 takes the guesswork out of preliminary or conceptual estimates. Now you don't have to try to calculate the assembled cost by working up individual components costs. We've done all the work for you.

Presents detailed illustrations, descriptions, specifications and costs for every conceivable building assembly—240 types in all—arranged in the easy-to-use UNIFORMAT II system. Each illustrated "assembled" cost includes a complete grouping of materials and associated installation costs, including the installing contractor's overhead and profit.

$206.95 per copy
Over 600 pages, illustrated, available Oct. 2005
Catalog No. 60066

Means Open Shop Building Construction Cost Data 2006

The latest costs for accurate budgeting and estimating of new commercial and residential construction... renovation work... change orders... cost engineering.

Means *Open Shop BCCD* will assist you to:

• Develop benchmark prices for change orders
• Plug gaps in preliminary estimates, and budgets
• Estimate complex projects
• Substantiate invoices on contracts
• Price ADA-related renovations

$126.95 per copy
Over 700 pages, illustrated, available Dec. 2005
Catalog No. 60156

Annual Cost Guides

Means Building Construction Cost Data 2006

Western Edition

This regional edition provides more precise cost information for western North America. Labor rates are based on union rates from 13 western states and western Canada. Included are western practices and materials not found in our national edition: tilt-up concrete walls, glu-lam structural systems, specialized timber construction, seismic restraints, and landscape and irrigation systems.

$126.95 per copy
Over 650 pages, illustrated, available Dec. 2005
Catalog No. 60226

Means Heavy Construction Cost Data 2006

A comprehensive guide to heavy construction costs. Includes costs for highly specialized projects such as tunnels, dams, highways, airports, and waterways. Information on different labor rates, equipment, and material costs is included. Features unit price costs, systems costs, and numerous reference tables for costs and design. Valuable not only to contractors and civil engineers, but also to government agencies and city/town engineers.

$126.95 per copy, Over 450 pages,
illustrated, available Nov. 2005
Catalog No. 60166

Means Construction Cost Indexes 2006

Who knows what 2006 holds? What materials and labor costs will change unexpectedly? By how much?

- Breakdowns for 316 major cities
- National averages for 30 key cities
- Expanded five major city indexes
- Historical construction cost indexes

$272.00 per year
$68.00 individual quarters
Catalog No. 60146 A,B,C,D

Means Interior Cost Data 2006

Provides you with prices and guidance needed to make accurate interior work estimates. Contains costs on materials, equipment, hardware, custom installations, furnishings, labor costs... every cost factor for new and remodel commercial and industrial interior construction, including updated information on office furnishings, plus more than 50 reference tables. For contractors, facility managers, and owners.

$126.95 per copy, Over 600 pages,
illustrated, available Oct. 2005
Catalog No. 60096

Means Concrete & Masonry Cost Data 2006

Provides you with cost facts for virtually all concrete/masonry estimating needs, from complicated formwork to various sizes and face finishes of brick and block, all in great detail. The comprehensive unit cost section contains more than 8,500 selected entries. Also contains an assemblies cost section, and a detailed reference section that supplements the cost data.

$114.95 per copy
Over 450 pages, illustrated, available Dec. 2005
Catalog No. 60116

Means Labor Rates for the Construction Industry 2006

Complete information for estimating labor costs, making comparisons, and negotiating wage rates by trade for over 300 U.S. and Canadian cities. With 46 construction trades listed by local union number in each city, and historical wage rates included for comparison. No similar book is available through the trade.

Each city chart lists the county and is alphabetically arranged with handy visual flip tabs for quick reference.

$274.95 per copy
Over 300 pages, available Dec. 2005
Catalog No. 60126

Means Facilities Maintenance & Repair Cost Data 2006

Means *Facilities Maintenance & Repair Cost Data* gives you a complete system to manage and plan your facility repair and maintenance costs and budget efficiently. Guidelines for auditing a facility and developing an annual maintenance plan. Budgeting is included, along with reference tables on cost and management and information on frequency and productivity of maintenance operations.

The only nationally recognized source of maintenance and repair costs. Developed in cooperation with the Army Corps of Engineers.

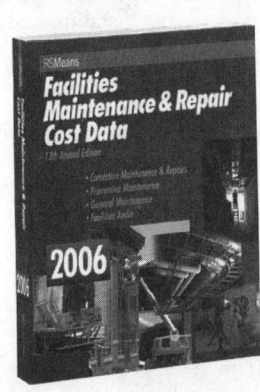

$274.95 per copy
Over 600 pages, illustrated, available Dec. 2005
Catalog No. 60306

Reference Books

For more information
visit Means Web site
at www.rsmeans.com

Builder's Essentials: Estimating Building Costs for the Residential & Light Commercial Contractor

By Wayne J. DelPico

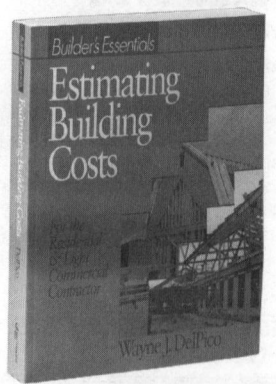

Step-by-step estimating methods for residential and light commercial contractors. Includes a detailed look at every construction specialty—explaining all the components, takeoff units, and labor needed for well-organized, complete estimates. Covers:

- Correctly interpreting plans and specifications
- Developing accurate and complete labor and material costs
- Understanding direct and indirect overhead costs... and accounting for time-sensitive costs
- Using historical cost data to generate new project budgets
- Allocating the right amount for profit and contingencies

$29.95 per copy
Over 400 pages, illustrated, Softcover
Catalog No. 67343

Green Building: Project Planning & Cost Estimating

By RSMeans and Contributing Authors

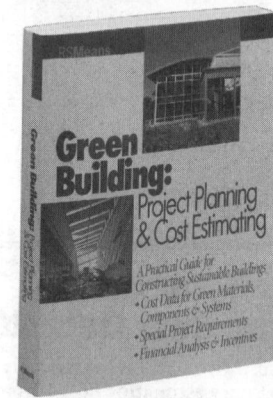

Written by a team of leading experts in sustainable design, this book is a complete guide to planning and estimating green building projects, a growing trend in building design and construction. It explains:

- All the different criteria for "green-ness"
- What criteria your building needs to meet to get a LEED, ENERGYSTAR®, or other recognized rating for green buildings
- How the project team works differently on a green versus a traditional building project
- How to select and specify green products
- How to evaluate the cost and value of green products versus conventional ones—not only for their initial installation cost, but their cost over time in maintenance and operation

Features an extensive Green Building Cost Data section

$89.95 per copy
350 pages, illustrated, Hardcover
Catalog No. 67338

Means ADA Compliance Pricing Guide

New Second Edition

By Adaptive Environments and RSMeans

Completely updated and revised to the new 2004 *Americans with Disabilities Act Accessibility Guidelines*, this book features more than 70 of the most commonly needed modifications for ADA compliance—their design requirements, suggestions, and final cost. Projects range from installing ramps and walkways, widening doorways and entryways, and installing and refitting elevators, to relocating light switches and signage, and remodeling bathrooms and kitchens. Also provided are:

- Detailed cost estimates for budgeting modification projects, including estimates for each of 260 alternates.
- An assembly estimate for every project, with detailed cost breakdown including materials, labor-hours, and contractor's overhead.
- 3,000 additional ADA compliance-related unit cost line items.
- Costs that are easily adjusted to over 900 cities and towns.

$79.95 per copy
Over 350 pages, illustrated, Softcover
Catalog No. 67310A

For more information
visit Means Web site
at www.rsmeans.com

Reference Books

Value Engineering: Practical Applications . . . For Design, Construction, Maintenance & Operations
By Alphonse Dell'Isola, PE

A tool for immediate application—for engineers, architects, facility managers, owners, and contractors. Includes: making the case for VE—the management briefing, integrating VE into planning, budgeting, and design, conducting life cycle costing, using VE methodology in design review and consultant selection, case studies, VE workbook, and a life cycle costing program on disk.

$79.95 per copy
Over 450 pages, illustrated, Softcover
Catalog No. 67319

Facilities Operations & Engineering Reference
By the Association for Facilities Engineering and RSMeans

An all-in-one technical referance for planning and managing facility projects and solving day-to-day operations problems. Selected as the official Certified Plant Engineer reference, this handbook covers financial analysis, maintenance, HVAC and energy efficiency, and more.

$109.95 per copy
Over 700 pages, illustrated, Hardcover
Catalog No. 67318

The Building Professional's Guide to Contract Documents
3rd Edition
By Waller S. Poage, AIA, CSI, CVS

A comprehensive reference for owners, design professionals, contractors, and students.
- Structure your documents for maximum efficiency.
- Effectively communicate construction requirements.
- Understand the roles and responsibilities of construction professionals.
- Improve methods of project delivery.

$32.48 per copy, 400 pages
Diagrams and construction forms, Hardcover
Catalog No. 67261A

Building Security: Strategies & Costs
By David Owen

This comprehensive resource will help you evaluate your facility's security needs, and design and budget for the materials and devices needed to fulfill them.

Includes over 130 pages of Means cost data for installation of security systems and materials, plus a review of more than 50 security devices and construction solutions.

$44.98 per copy
350 pages, illustrated, Hardcover
Catalog No. 67339

Cost Planning & Estimating for Facilities Maintenance

In this unique book, a team of facilities management authorities shares their expertise at:
- Evaluating and budgeting maintenance operations
- Maintaining and repairing key building components
- Applying Means *Facilities Maintenance & Repair Cost Data* to your estimating

Covers special maintenance requirements of the ten major building types.

$89.95 per copy
Over 475 pages, Hardcover
Catalog No. 67314

Life Cycle Costing for Facilities
By Alphonse Dell'Isola and Dr. Steven Kirk

Guidance for achieving higher quality design and construction projects at lower costs! Cost-cutting efforts often sacrifice quality to yield the cheapest product. Life cycle costing enables building designers and owners to achieve both. The authors of this book show how LCC can work for a variety of projects – from roads to HVAC upgrades to different types of buildings.

$99.95 per copy
450 pages, Hardcover
Catalog No. 67341

Building & Renovating Schools

This all-inclusive guide covers every step of the school construction process—from initial planning, needs assessment, and design, right through moving into the new facility. A must-have resource for anyone concerned with new school construction or renovation, including architects and engineers, contractors and project managers, facility managers, school administrators and school board members, building committees, community leaders, and anyone else who wants to ensure that the project meets schools' needs in a cost-effective, timely manner. With square foot cost models for elementary, middle, and high school facilities and real-life case studies of recently completed school projects.

The contributors to this book—architects, construction project managers, contractors, and estimators who specialize in school construction—provide start-to-finish, expert guidance on the process.

$99.95 per copy
Over 425 pages, Hardcover
Catalog No. 67342

Interior Home Improvement Costs, New 9th Edition

Estimates for the most popular remodeling and repair projects—from small, do-it-yourself jobs to major renovations and new construction. Includes: Kitchens & Baths; New Living Space from your Attic, Basement, or Garage; New Floors, Paint, and Wallpaper; Tearing Out or Building New Walls; Closets, Stairs, and Fireplaces; New Energy-Saving Improvements, Home Theaters, and More!

$24.95 per copy
250 pages, illustrated, Softcover
Catalog No. 67308E

Exterior Home Improvement Costs New 9th Edition

Estimates for the most popular remodeling and repair projects—from small, do-it-yourself jobs, to major renovations and new construction. Includes: Curb Appeal Projects—Landscaping, Patios, Porches, Driveways, and Walkways; New Windows and Doors; Decks, Greenhouses, and Sunrooms; Room Additions and Garages; Roofing, Siding, and Painting; "Green" Improvements to Save Energy & Water.

$24.95 per copy
Over 275 pages, illustrated, Softcover
Catalog No. 67309E

Builder's Essentials: Plan Reading & Material Takeoff

By Wayne J. DelPico

For Residential and Light Commercial Construction

A valuable tool for understanding plans and specs, and accurately calculating material quantities. Step-by-step instructions and takeoff procedures based on a full set of working drawings.

$35.95 per copy
Over 420 pages, Softcover
Catalog No. 67307

Builder's Essentials: Framing & Rough Carpentry, 2nd Edition

By Scot Simpson

Develop and improve your skills with easy-to-follow instructions and illustrations. Learn proven techniques for framing walls, floors, roofs, stairs, doors, and windows. Updated guidance on standards, building codes, safety requirements, and more. Also available in Spanish!

$24.95 per copy
Over 150 pages, Softcover
Catalog No. 67298A
Spanish Catalog No. 67298AS

Builder's Essentials: Best Business Practices for Builders & Remodelers

An Easy-to-Use Checklist System

By Thomas N. Frisby

A comprehensive guide covering all aspects of running a construction business, with more than 40 user-friendly checklists. Provides expert guidance on: increasing your revenue and keeping more of your profit, planning for long-term growth, keeping good employees, and managing subcontractors.

$29.95 per copy
Over 220 pages, Softcover
Catalog No. 67329

Unit Price Estimating Methods

New 3rd Edition

This new edition includes up-to-date cost data and estimating examples, updated to reflect changes to the CSI numbering system and new features of Means cost data. It describes the most productive, universally accepted ways to estimate, and uses checklists and forms to illustrate shortcuts and timesavers. A model estimate demonstrates procedures. A new chapter explores computer estimating alternatives.

$59.95 per copy
Over 350 pages, illustrated, Hardcover
Catalog No. 67303A

Means Landscape Estimating Methods

4th Edition

By Sylvia H. Fee

This revised edition offers expert guidance for preparing accurate estimates for new landscape construction and grounds maintenance. Includes a complete project estimate featuring the latest equipment and methods, and chapters on *Life Cycle Costing* and *Landscape Maintenance Estimating*.

$62.95 per copy
Over 300 pages, illustrated, Hardcover
Catalog No. 67295B

Means Environmental Remediation Estimating Methods, 2nd Edition

By Richard R. Rast

Guidelines for estimating 50 standard remediation technologies. Use it to prepare preliminary budgets, develop estimates, compare costs and solutions, estimate liability, review quotes, negotiate settlements.

A valuable support tool for Means Environmental Remediation Unit Price and Assemblies books.

$99.95 per copy
Over 750 pages, illustrated, Hardcover
Catalog No. 64777A

For more information
visit Means Web site
at www.rsmeans.com

Reference Books

Means Illustrated Construction Dictionary, Condensed, 2nd Edition
By RSMeans

Recognized in the industry as the best resource of its kind.
This essential tool has been further enhanced with updates
to existing terms and the addition of hundreds of new terms
and illustrations—in keeping with recent developments. For
contractors, architects, insurance and real estate personnel,
homeowners, and anyone who needs quick, clear definitions
for construction terms.

$59.95 per copy
Over 500 pages, Softcover
Catalog No. 67282A

Means Repair & Remodeling Estimating New 4th Edition
By Edward B. Wetherill & RSMeans

This important reference focuses on the unique problems
of estimating renovations of existing structures and helps
you determine the true costs of remodeling through careful
evaluation of architectural details and a site visit.

New section on disaster restoration costs.

$69.95 per copy
Over 450 pages, illustrated, Hardcover
Catalog No. 67265B

Facilities Planning & Relocation
New, lower price and user-friendly format.
By David D. Owen

A complete system for planning space needs and managing
relocations. Includes step-by-step manual, over 50 forms,
and extensive reference section on materials and
furnishings.

$89.95 per copy
Over 450 pages, Softcover
Catalog No. 67301

Means Square Foot & Assemblies Estimating Methods, 3rd Edition

Develop realistic square foot and assemblies costs for
budgeting and construction funding. The new edition
features updated guidance on square foot and assemblies
estimating using UNIFORMAT II. An essential reference for
anyone who performs conceptual estimates.

$69.95 per copy
Over 300 pages, illustrated, Hardcover
Catalog No. 67145B

Means Electrical Estimating Methods 3rd Edition

Expanded new edition includes sample estimates and cost
information in keeping with the latest version of the CSI
MasterFormat and UNIFORMAT II. Complete coverage
of fiber optic and uninterruptible power supply electrical
systems, broken down by components and explained in
detail. Includes a new chapter on computerized estimating
methods. A practical companion to Means *Electrical Cost
Data.*

$64.95 per copy
Over 325 pages, Hardcover
Catalog No. 67230B

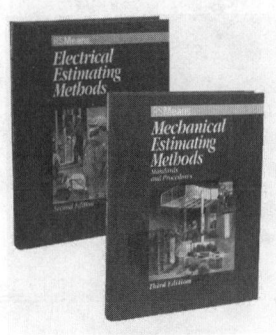

Means Mechanical Estimating Methods 3rd Edition

This guide assists you in making a review of plans, specs, and
bid packages with suggestions for takeoff procedures, listings,
substitutions and pre-bid scheduling. Includes suggestions for
budgeting labor and equipment usage. Compares materials
and construction methods to allow you to select the best
option.

$64.95 per copy
Over 350 pages, illustrated, Hardcover
Catalog No. 67294A

Means Spanish/English Construction Dictionary
By RSMeans, The International Conference of Building Officials (ICBO), and Rolf Jensen & Associates (RJA)

Designed to facilitate communication among Spanish-
and English-speaking construction personnel—improving
performance and job-site safety. Features the most common
words and phrases used in the construction industry, with
easy-to-follow pronunciations. Includes extensive building
systems and tools illustrations.

$22.95 per copy
250 pages, illustrated, Softcover
Catalog No. 67327

Project Scheduling & Management for Construction
New 3rd Edition
By David R. Pierce, Jr.

A comprehensive yet easy-to-follow guide to construction
project scheduling and control—from vital project
management principles through the latest scheduling,
tracking, and controlling techniques. The author is a
leading authority on scheduling with years of field and
teaching experience at leading academic institutions. Spend
a few hours with this book and come away with a solid
understanding of this essential management topic.

$64.95 per copy
Over 300 pages, illustrated, Hardcover
Catalog No. 67247B

Reference Books

For more information
visit Means Web site
at www.rsmeans.com

Concrete Repair and Maintenance Illustrated

By Peter Emmons

Hundreds of illustrations show users how to analyze, repair, clean, and maintain concrete structures for optimal performance and cost effectiveness. From parking garages to roads and bridges to structural concrete, this comprehensive book describes the causes, effects, and remedies for concrete wear and failure. Invaluable for planning jobs, selecting materials, and training employees, this book is a must-have for concrete specialists, general contractors, facility managers, civil and structural engineers, and architects.

$34.98 per copy
300 pages, illustrated, Softcover
Catalog No. 67146

Plumbing Estimating Methods 3rd Edition

By Joseph Galeno and Sheldon Greene

Updated and revised! This practical guide walks you through a plumbing estimate, from basic materials and installation methods through change order analysis. *Plumbing Estimating Methods* covers residential, commercial, industrial, and medical systems, and features sample takeoff and estimate forms and detailed illustrations of systems and components.

$29.98 per copy
330+ pages, Softcover
Catalog No. 67283B

Building Spec Homes Profitably

By Kenneth V. Johnson

Professional guidance from a veteran spec home builder... This complete system covers all aspects of spec home building—from market research and site selection to financing and design to scheduling and supervision, and more!

You'll learn to reduce risk and ensure profits in spec home building, no matter what the economic climate.

$29.95 per copy
200 pages, Softcover
Catalog No. 67312

How to Estimate with Means Data & CostWorks

New 2nd Edition

By RSMeans and Saleh Mubarak

Learn estimating techniques using Means cost data. Includes an instructional version of Means *CostWorks* CD–ROM with sample building plans. The step-by-step guide takes you through all the major construction items. Over 300 sample estimating problems are included.

$59.95 per copy
Over 190 pages, Softcover
Catalog No. 67324A

Means Productivity Standards for Construction

3rd Edition

By RSMeans

Completely updated, with well over 3,000 new work items added to cover topics in current demand! This comprehensive reference for the productivity of construction crews, equipment, and labor includes precise descriptions of each construction task organized according to the Construction Specifications Institute's MasterFormat.

$49.98 per copy
800+ pages, Hardcover
Catalog No. 67236A

Total Productive Facilities Management

By Richard W. Sievert, Jr.

Today, facilities are viewed as strategic resources... elevating the facility manager to the role of asset manager supporting the organization's overall business goals. Now, Richard Sievert Jr., in this well-articulated guidebook, sets forth a new operational standard for the facility manager's emerging role... a comprehensive program for managing facilities as a true profit center.

$29.98 per copy
275 pages, Softcover
Catalog No. 67321

For more information
visit Means Web site
at www.rsmeans.com

Reference Books

Preventive Maintenance Guidelines for School Facilities

By John C. Maciha

A complete PM program for K-12 schools that ensures sustained security, safety, property integrity, user satisfaction, and reasonable ongoing expenditures.

Includes schedules for weekly, monthly, semiannual, and annual maintenance with hard copy and electronic forms.

$149.95 per copy
Over 225 pages, Hardcover
Catalog No. 67326

Preventive Maintenance for Higher Education Facilities

By Applied Management Engineering, Inc.

An easy-to-use system to help facilities professionals establish the value of PM, and to develop and budget for an appropriate PM program for their college or university. Features interactive campus building models typical of those found in different-sized higher education facilities, and PM checklists linked to each piece of equipment or system in hard copy and electronic format.

$149.95 per copy
150 pages, Hardcover
Catalog No. 67337

Historic Preservation: Project Planning & Estimating

By Swanke Hayden Connell Architects

Expert guidance on managing historic restoration, rehabilitation, and preservation building projects and determining and controlling their costs. Includes:
- How to determine whether a structure qualifies as historic
- Where to obtain funding and other assistance
- How to evaluate and repair more than 75 historic building materials

$99.95 per copy
Over 675 pages, Hardcover
Catalog No. 67323

Means Illustrated Construction Dictionary

Unabridged 3rd Edition

Long regarded as the Industry's finest, the Means Illustrated Construction Dictionary is now even better. With the addition of over 1,000 new terms and hundreds of new illustrations, it is the clear choice for the most comprehensive and current information. The companion CD-ROM that comes with this new edition adds many extra features: larger graphics, expanded definitions, and links to both CSI MasterFormat numbers and product information.

$99.95 per copy
Over 790 pages, illustrated, Hardcover
Catalog No. 67292A

Designing & Building with the IBC, 2nd Edition

By Rolf Jensen & Associates, Inc.

This updated comprehensive guide helps building professionals make the transition to the 2003 International Building Code®. Includes a side-by-side code comparison of the IBC 2003 to the IBC 2000 and the three primary model codes, a quick-find index, and professional code commentary. With illustrations, abbreviations key, and an extensive Resource section.

$99.95 per copy
Over 875 pages, Softcover
Catalog No. 67328A

Residential & Light Commercial Construction Standards, 2nd Edition

By RSMeans and Contributing Authors

For contractors, subcontractors, owners, developers, architects, engineers, attorneys, and insurance personnel, this book provides authoritative requirements and recommendations compiled from the nation's leading professional associations, industry publications, and building code organizations.

$59.95 per copy
600 pages, illustrated, Softcover
Catalog No. 67322A

Builder's Essentials: Advanced Framing Methods

By Scot Simpson

A highly illustrated, "framer-friendly" approach to advanced framing elements. Provides expert, but easy to interpret, instruction for laying out and framing complex walls, roofs, and stairs, and special requirements for earthquake and hurricane protection. Also helps bring framers up to date on the latest building code changes, and provides tips on the lead framer's role and responsibilities, how to prepare for a job, and how to get the crew started.

$24.95 per copy
250 pages, illustrated, Softcover
Catalog No. 67330

Means Estimating Handbook

2nd Edition

By RSMeans

Updated new Second Edition answers virtually any estimating technical question—all organized by CSI MasterFormat. This comprehensive reference covers the full spectrum of technical data required to estimate construction costs. The book includes information on sizing, productivity, equipment requirements, code-mandated specifications, design standards, and engineering factors.

$99.95 per copy
Over 900 pages, Hardcover
Catalog No. 67276A

For more information
visit Means Web site
at www.rsmeans.com

Seminars

Means CostWorks Training

This one-day seminar course has been designed with the intention of assisting both new and existing users to become more familiar with the *CostWorks* program. The class is broken into two unique sections: (1) A one-half day presentation on the function of each icon; and each student will be shown how to use the software to develop a cost estimate. (2) Hands-on estimating exercises that will ensure that each student thoroughly understands how to use *CostWorks*. You must bring your own laptop computer to this course.

CostWorks Benefits/Features:
- Estimate in your own spreadsheet format
- Power of Means National Database
- Database automatically regionalized
- Save time with keyword searches
- Save time by establishing common estimate items in "Bookmark" files
- Customize your spreadsheet template
- Hot key to Product Manufacturers' listings and specs
- Merge capability for networking environments
- View crews and assembly components
- AutoSave capability
- Enhanced sorting capability

Unit Price Estimating

This interactive two-day seminar teaches attendees how to interpret project information and process it into final, detailed estimates with the greatest accuracy level.

The single most important credential an estimator can take to the job is the ability to visualize construction in the mind's eye, and thereby estimate accurately.

Some Of What You'll Learn:
- Interpreting the design in terms of cost
- The most detailed, time-tested methodology for accurate "pricing"
- Key cost drivers—material, labor, equipment, staging, and subcontracts
- Understanding direct and indirect costs for accurate job cost accounting and change order management

Who Should Attend: Corporate and government estimators and purchasers, architects, engineers... and others needing to produce accurate project estimates.

Square Foot and Assemblies Cost Estimating

This two-day course teaches attendees how to quickly deliver accurate square foot estimates using limited budget and design information.

Some Of What You'll Learn:
- How square foot costing gets the estimate done faster
- Taking advantage of a "systems" or "assemblies" format
- The Means "building assemblies/square foot cost approach"
- How to create a very reliable preliminary and systems estimate using bare-bones design information

Who Should Attend: Facilities managers, facilities engineers, estimators, planners, developers, construction finance professionals... and others needing to make quick, accurate construction cost estimates at commercial, government, educational, and medical facilities.

Repair and Remodeling Estimating

This two-day seminar emphasizes all the underlying considerations unique to repair/remodeling estimating and presents the correct methods for generating accurate, reliable R&R project costs using the unit price and assemblies methods.

Some Of What You'll Learn:
- Estimating considerations—like labor-hours, building code compliance, working within existing structures, purchasing materials in smaller quantities, unforeseen deficiencies
- Identify problems and provide solutions to estimating building alterations
- Rules for factoring in minimum labor costs, accurate productivity estimates and allowances for project contingencies
- R&R estimating examples are calculated using unit prices and assemblies data

Who Should Attend: Facilities managers, plant engineers, architects, contractors, estimators, builders... and others who are concerned with the proper preparation and/or evaluation of repair and remodeling estimates.

Mechanical and Electrical Estimating

This two-day course teaches attendees how to prepare more accurate and complete mechanical/electrical estimates, avoiding the pitfalls of omission and double-counting, while understanding the composition and rationale within the Means Mechanical/Electrical database.

Some Of What You'll Learn:
- The unique way mechanical and electrical systems are interrelated
- M&E estimates, conceptual, planning, budgeting, and bidding stages
- Order of magnitude, square foot, assemblies, and unit price estimating
- Comparative cost analysis of equipment and design alternatives

Who Should Attend: Architects, engineers, facilities managers, mechanical and electrical contractors... and others needing a highly reliable method for developing, understanding, and evaluating mechanical and electrical contracts.

Plan Reading and Material Takeoff

This two-day program teaches attendees to read and understand construction documents and to use them in the preparation of material takeoffs.

Some of What You'll Learn:
- Skills necessary to read and understand typical contract documents—blueprints and specifications
- Details and symbols used by architects and engineers
- Construction specifications' importance in conjunction with blueprints
- Accurate takeoff of construction materials and industry-accepted takeoff methods

Who Should Attend: Facilities managers, construction supervisors, office managers... and others responsible for the execution and administration of a construction project including government, medical, commercial, educational, or retail facilities.

Facilities Maintenance and Repair Estimating

This two-day course teaches attendees how to plan, budget, and estimate the cost of ongoing and preventive maintenance and repair for existing buildings and grounds.

Some Of What You'll Learn:
- The most financially favorable maintenance, repair, and replacement scheduling and estimating
- Auditing and value engineering facilities
- Preventive planning and facilities upgrading
- Determining both in-house and contract-out service costs; annual, asset-protecting M&R plan

Who Should Attend: Facility managers, maintenance supervisors, buildings and grounds superintendents, plant managers, planners, estimators... and others involved in facilities planning and budgeting.

Scheduling and Project Management

This two-day course teaches attendees the most current and proven scheduling and management techniques needed to bring projects in on time and on budget.

Some Of What You'll Learn:
- Crucial phases of planning and scheduling
- How to establish project priorities and develop realistic schedules and management techniques
- Critical Path and Precedence Methods
- Special emphasis on cost control

Who Should Attend: Construction project managers, supervisors, engineers, estimators, contractors... and others who want to improve their project planning, scheduling, and management skills.

Advanced Project Management

This two-day seminar will teach you how to effectively manage and control the entire design-build process and allow you to take home tangible skills that will be immediately applicable on existing projects.

Some Of What You'll Learn:
- Value engineering, bonding, fast-tracking, and bid package creation
- How estimates and schedules can be integrated to provide advanced project management tools
- Cost engineering, quality control, productivity measurement, and improvement
- Front loading a project and predicting its cash flow

Who Should Attend: Owners, project managers, architectural and engineering managers, construction managers, contractors... and anyone else who is responsible for the timely design and completion of construction projects.

Seminars

2006 Means Seminar Schedule

Location	Dates
Las Vegas, NV	March
Washington, DC	April
Phoenix, AZ	April
Denver, CO	May
San Francisco, CA	June
Philadelphia, PA	June
Washington, DC	September
Dallas, TX	September
Las Vegas, NV	October
Orlando, FL	November
Atlantic City, NJ	November
San Diego, CA	December

For more information visit Means Web site at www.rsmeans.com

Note: Call for exact dates and details.

Registration Information

Register Early... Save up to $100! Register 30 days before the start date of a seminar and save $100 off your total fee. *Note: This discount can be applied only once per order. It cannot be applied to team discount registrations or any other special offer.*

How to Register Register by phone today! Means' toll-free number for making reservations is: **1-800-334-3509.**

Individual Seminar Registration Fee $895. Individual CostWorks Training Registration Fee $349. To register by mail, complete the registration form and return with your full fee to: Seminar Division, Reed Construction Data, RSMeans Seminars, 63 Smiths Lane, Kingston, MA 02364.

Federal Government Pricing All federal government employees save 25% off regular seminar price. Other promotional discounts cannot be combined with Federal Government discount.

Team Discount Program Two to four seminar registrations: Call for pricing.

Multiple Course Discounts When signing up for two or more courses, call for pricing.

Refund Policy Cancellations will be accepted up to ten days prior to the seminar start. There are no refunds for cancellations received later than ten working days prior to the first day of the seminar. A $150 processing fee will be applied for all cancellations. Written notice of cancellation is required. Substitutions can be made at any time before the session starts. **No-shows are subject to the full seminar fee.**

AACE Approved Courses The RSMeans Construction Estimating and Management Seminars described and offered to you here have each been approved for 14 hours (1.4 recertification credits) of credit by the AACE International Certification Board toward meeting the continuing education requirements for recertification as a Certified Cost Engineer/Certified Cost Consultant.

AIA Continuing Education We are registered with the AIA Continuing Education System (AIA/CES) and are committed to developing quality learning activities in accordance with the CES criteria. RSMeans seminars meet the AIA/CES criteria for Quality Level 2. AIA members will receive (14) learning units (LUs) for each two-day RSMeans Course.

NASBA CPE Sponsor Credits We are part of the National Registry of CPE Sponsors. Attendees may be eligible for (16) CPE credits.

Daily Course Schedule The first day of each seminar session begins at 8:30 A.M. and ends at 4:30 P.M. The second day is 8:00 A.M.–4:00 P.M. Participants are urged to bring a hand-held calculator since many actual problems will be worked out in each session.

Continental Breakfast Your registration includes the cost of a continental breakfast, a morning coffee break, and an afternoon break. These informal segments will allow you to discuss topics of mutual interest with other members of the seminar. (You are free to make your own lunch and dinner arrangements.)

Hotel/Transportation Arrangements RSMeans has arranged to hold a block of rooms at each hotel hosting a seminar. To take advantage of special group rates when making your reservation, be sure to mention that you are attending the Means Seminar. You are, of course, free to stay at the lodging place of your choice. **(Hotel reservations and transportation arrangements should be made directly by seminar attendees.)**

Important Class sizes are limited, so please register as soon as possible.

Note: Pricing subject to change.

Registration Form   Call 1-800-334-3509 to register or FAX 1-800-632-6732. Visit our Web site www.rsmeans.com

Please register the following people for the Means Construction Seminars as shown here. Full payment or deposit is enclosed, and we understand that we must make our own hotel reservations if overnight stays are necessary.

☐ Full payment of $_____ enclosed.

☐ Bill me

Name of Registrant(s)
(To appear on certificate of completion)

_____

_____

P.O. #: _____
GOVERNMENT AGENCIES MUST SUPPLY PURCHASE ORDER NUMBER

Firm Name _____

Address _____

City/State/Zip _____

Telephone No. Fax No. _____

E-mail Address _____

Charge our registration(s) to: ☐ MasterCard ☐ VISA ☐ American Express ☐ Discover

Account No. _____ Exp. Date _____

Cardholder's Signature _____

Seminar Name   City   Dates

_____

_____

Please mail check to: Seminar Division, Reed Construction Data, RSMeans Seminars, 63 Smiths Lane, P.O. Box 800, Kingston, MA 02364 USA

MeansData™

CONSTRUCTION COSTS FOR SOFTWARE APPLICATIONS

Your construction estimating software is only as good as your cost data.

A proven construction cost database is a mandatory part of any estimating package. The following list of software providers can offer you MeansData™ as an added feature for their estimating systems. See the table below for what types of products and services they offer (match their numbers). Visit online at **www.rsmeans.com/demosource/** for more information and free demos. Or call their numbers listed below.

1. **3D International**
   713-871-7000
   venegas@3di.com

2. **4Clicks-Solutions, LLC**
   719-574-7721
   mbrown@4clicks-solutions.com

3. **Aepco, Inc.**
   301-670-4642
   blueworks@aepco.com

4. **Applied Flow Technology**
   800-589-4943
   info@aft.com

5. **ArenaSoft Estimating**
   888-370-8806
   info@arenasoft.com

6. **BSD - Building Systems Design, Inc.**
   888-273-7638
   bsd@bsdsoftlink.com

7. **CMS - Computerized Micro Solutions**
   800-255-7407
   cms@proest.com

8. **Corecon Technologies, Inc.**
   714-895-7222
   sales@corecon.com

9. **CorVet Systems**
   301-622-9069
   sales@corvetsys.com

10. **Estimating Systems, Inc.**
    800-967-8572
    esipulsar@adelphia.net

11. **MC² - Management Computer**
    800-225-5622
    vkeys@mc2-ice.com

12. **Maximus Asset Solutions**
    800-659-9001
    assetsolutions@maximus.com

13. **Shaw Beneco Enterprises, Inc.**
    877-719-4748
    inquire@beneco.com

14. **Timberline Software Corp.**
    800-628-6583
    product.info@timberline.com

15. **US Cost, Inc.**
    800-372-4003
    sales@uscost.com

16. **Vanderweil Facility Advisors**
    617-451-5100
    info@VFA.com

17. **WinEstimator, Inc.**
    800-950-2374
    sales@winest.com

TYPE	1	2	3	4	5	6	7	8	9	10	11	12	13	14	15	16	17
BID					•	•	•		•		•			•			•
Estimating		•			•	•	•	•	•	•	•	•	•	•			•
DOC/JOC/SABER		•			•	•			•	•		•					•
ID/IQ		•							•	•		•	•				•
Asset Mgmt.												•				•	•
Facility Mgmt.	•		•									•				•	
Project Mgmt.	•	•						•	•				•	•			
TAKE-OFF					•		•	•	•		•			•	•		
EARTHWORK							•				•						
Pipe Flow				•													
HVAC/Plumbing					•		•		•								
Roofing					•		•		•								
Design	•				•								•	•			•
Other Offers Links:																	
Accounting/HR		•			•								•	•			
Scheduling					•								•	•			•
CAD													•				•
PDA													•	•			•
Lt. Versions		•						•					•	•			•
Consulting	•	•			•		•	•	•			•	•	•		•	•
Training		•			•	•	•	•	•	•		•	•	•		•	•

Qualified re-seller applications now being accepted. Call Carol Polio, Ext. 5107.

FOR MORE INFORMATION
CALL 1-800-448-8182, EXT. 5107 OR FAX 1-800-632-6732

For more information
visit Means Web site
at www.rsmeans.com

New Titles

Preventive Maintenance for Multi-Family Housing
by John C. Maciha

With 24/7 occupancy, multi-family buildings can be some of the toughest to maintain. Prepared by one of the nation's leading experts on multi-family housing, Preventive Maintenance for Multi-Family Housing puts easy-to-use guidelines right at your fingertips for the what, when, why, and how much of multi-family preventive maintenance.

This complete PM system for apartment and condominium communities features expert guidance, checklists for buildings and grounds maintenance tasks and their frequencies, a reusable wall chart to track maintenance, and a dedicated Web site featuring customizable electronic forms. A must-have for anyone involved with multi-family housing maintenance and upkeep.

$89.95 per copy
225 pages
Catalog No. 67346

Job Order Contracting
Expediting Construction Project Delivery
by Allen Henderson

Expert guidance to help you implement JOC—fast becoming the preferred project delivery method for repair and renovation, minor new construction, and maintenance projects in the public sector and in many states and municipalities. The author, a leading JOC expert and practitioner, shows how to:
• Establish a JOC program
• Evaluate proposals and award contracts
• Handle general requirements and estimating
• Partner for maximum benefits

This book will quickly bring you up to speed on JOC (also known as Delivery Order Contracting and SABER) so you can achieve faster project starts with fewer delays, cost overruns, and quality disputes.

$89.95 oer copy
192 pages, illustrated, Hardcover
Catalog No. 67348

Kitchen & Bath Project Costs: Planning & Estimating Successful Projects

Project estimates for 35 of the most popular kitchen and bath renovations... from replacing a single fixture to whole-room remodels. Each estimate includes:
• All materials needed for the project
• Labor-hours to install (and demolish/remove) each item
• Subcontractor costs for certain trades and services
• An allocation for overhead and profit

PLUS! Takeoff and pricing worksheets—forms you can photocopy or access electronically from the book's Web site; alternate materials—unit costs for different finishes and fixtures, location factors—easy multipliers to adjust the costs to your location; and expert guidance on estimating methods, project design, contracts, marketing, working with homeowners, and tips for each of the estimated projects.

$29.95 per copy
Over 175 pages
Catalog No. 67347

Coming Soon!

Home Addition & Renovation Costs

New! This essential home remodeling reference gives you 35 project estimates and guidance for some of the most popular home renovation and addition projects... from opening up a simple interior wall to adding an entire second story. Each estimate includes a floor plan, color photos, and detailed costs. Use the project estimates as backup for pricing, to check your own estimates, or as a cost reference for preliminary discussion with homeowners.

Includes:
• Case studies—with creative solutions and design ideas.
• Alternate materials costs—so you can match the estimates to the particulars of your projects.
• Location factors—easy multipliers to adjust the book's costs to your own location.

$29.95 per copy
Over 200 pages, illustrated, Softcover
Catalog No. 67349

The Practice of Cost Segregation Analysis
by Bruce A. Desrosiers and Wayne J. DelPico

This expert guide walks you through the practice of cost segregations analysis, which enables property owners to defer taxes and benefit from "accelerated cost recovery" through depreciation deductions on assets that are properly identified and classified. Cost segregation practice requires knowledge of both tax law and the construction process. In this book, the authors share their expertise in these areas with tax and accounting professionals, cost segregation consultants, facility owners, architects, and general contractors—providing guidance on major aspects of a professional, defensible cost segregation study.

With a glossary of terms, sample cost segregation estimates for various building types, key information resources, and updates via a dedicated Web site, this book is a critical resource for anyone involved in cost segregation analysis.

$99.95 per copy
Over 225 pages
Catalog No. 67345

Qty.	Book No.	COST ESTIMATING BOOKS	Unit Price	Total
	60066	Assemblies Cost Data 2006	$206.95	
	60016	Building Construction Cost Data 2006	126.95	
	61016	Building Const. Cost Data–Looseleaf Ed. 2006	157.95	
	60226	Building Const. Cost Data–Western Ed. 2006	126.95	
	60116	Concrete & Masonry Cost Data 2006	114.95	
	60146	Construction Cost Indexes 2006	272.00	
	60146A	Construction Cost Index–January 2006	68.00	
	60146B	Construction Cost Index–April 2006	68.00	
	60146C	Construction Cost Index–July 2006	68.00	
	60146D	Construction Cost Index–October 2006	68.00	
	60346	Contr. Pricing Guide: Resid. R & R Costs 2006	39.95	
	60336	Contr. Pricing Guide: Resid. Detailed 2006	39.95	
	60326	Contr. Pricing Guide: Resid. Sq. Ft. 2006	39.95	
	60236	Electrical Change Order Cost Data 2006	126.95	
	60036	Electrical Cost Data 2006	126.95	
	60206	Facilities Construction Cost Data 2006	299.95	
	60306	Facilities Maintenance & Repair Cost Data 2006	274.95	
	60166	Heavy Construction Cost Data 2006	126.95	
	60096	Interior Cost Data 2006	126.95	
	60126	Labor Rates for the Const. Industry 2006	274.95	
	60186	Light Commercial Cost Data 2006	107.95	
	60026	Mechanical Cost Data 2006	126.95	
	63016	Metric Construction Cost Data 2006	149.95	
	60156	Open Shop Building Const. Cost Data 2006	126.95	
	60216	Plumbing Cost Data 2006	126.95	
	60046	Repair and Remodeling Cost Data 2006	107.95	
	60176	Residential Cost Data 2006	107.95	
	60286	Site Work & Landscape Cost Data 2006	126.95	
	60056	Square Foot Costs 2006	137.95	
		REFERENCE BOOKS		
	67310A	ADA Compliance Pricing Guide, 2nd Ed.	79.95	
	67273	Basics for Builders: How to Survive and Prosper	34.95	
	67330	Bldrs Essentials: Adv. Framing Methods	24.95	
	67329	Bldrs Essentials: Best Bus. Practices for Bldrs	29.95	
	67298A	Bldrs Essentials: Framing/Carpentry 2nd Ed.	24.95	
	67298AS	Bldrs Essentials: Framing/Carpentry Spanish	24.95	
	67307	Bldrs Essentials: Plan Reading & Takeoff	35.95	
	67261A	Bldg. Prof. Guide to Contract Documents 3rd Ed.	32.48	
	67342	Building & Renovating Schools	99.95	
	67339	Building Security: Strategies & Costs	44.98	
	67312	Building Spec Homes Profitably	29.95	
	67146	Concrete Repair & Maintenance Illustrated	34.98	
	67314	Cost Planning & Est. for Facil. Maint.	89.95	
	67317A	Cyberplaces: The Internet Guide 2nd Ed.	59.95	
	67328A	Designing & Building with the IBC, 2nd Ed.	99.95	
	67230B	Electrical Estimating Methods 3rd Ed.	64.95	
	64777A	Environmental Remediation Est. Methods 2nd Ed.	$ 99.95	
	67343	Estimating Bldg. Costs for Resi. & Lt. Comm.	29.95	
	67160	Estimating for Contractors	17.98	
	67276A	Estimating Handbook 2nd Ed.	99.95	
	67249	Facilities Maintenance Management	86.95	
	67246	Facilities Maintenance Standards	79.95	

Qty.	Book No.	REFERENCE BOOKS (Cont.)	Unit Price	Total
	67318	Facilities Operations & Engineering Reference	109.95	
	67301	Facilities Planning & Relocation	89.95	
	67231	Forms for Building Const. Professional	47.48	
	67260	Fundamentals of the Construction Process	25.98	
	67338	Green Building: Proj. Planning & Cost Est.	89.95	
	67323	Historic Preservation: Proj. Planning & Est.	99.95	
	67308E	Home Improvement Costs–Int. Projects 9th Ed.	24.95	
	67309E	Home Improvement Costs–Ext. Projects 9th Ed.	24.95	
	67324A	How to Est. w/Means Data & CostWorks 2nd Ed.	59.95	
	67304	How to Estimate with Metric Units	9.98	
	67306	HVAC: Design Criteria, Options, Select. 2nd Ed.	84.95	
	67281	HVAC Systems Evaluation	84.95	
	67282A	Illustrated Construction Dictionary, Condensed	59.95	
	67292A	Illustrated Construction Dictionary, w/CD-ROM	99.95	
	67348	Job Order Contracting	89.95	
	67347	Kitchen & Bath Project Costs	29.95	
	67295B	Landscape Estimating 4th Ed.	62.95	
	67341	Life Cycle Costing	99.95	
	67302	Managing Construction Purchasing	19.98	
	67294A	Mechanical Estimating 3rd Ed.	64.95	
	67283B	Plumbing Estimating Methods 3rd Ed.	29.98	
	67345	Pract. of Cost Segregation Analysis	99.95	
	67326	Preventive Maint. Guidelines for School Facil.	149.95	
	67337	Preventive Maint. for Higher Education Facilities	149.95	
	67346	Preventive Maint. for Multi-Family Housing	89.95	
	67236A	Productivity Standards for Constr.–3rd Ed.	49.98	
	67247B	Project Scheduling & Management for Constr. 3rd Ed.	64.95	
	67265B	Repair & Remodeling Estimating 4th Ed.	69.95	
	67322A	Resi. & Light Commercial Const. Stds. 2nd Ed.	59.95	
	67327	Spanish/English Construction Dictionary	22.95	
	67145B	Sq. Ft. & Assem. Estimating Methods 3rd Ed.	69.95	
	67287	Successful Estimating Methods	32.48	
	67313	Successful Interior Projects	24.98	
	67233	Superintending for Contractors	35.95	
	67321	Total Productive Facilities Management	29.98	
	67284	Understanding Building Automation Systems	29.98	
	67303A	Unit Price Estimating Methods 3rd Ed.	59.95	
	67319	Value Engineering: Practical Applications	79.95	

	MA residents add 5% state sales tax		
	Shipping & Handling**		
	Total (U.S. Funds)*		

Prices are subject to change and are for U.S. delivery only. *Canadian customers may call for current prices. **Shipping & handling charges: Add 7% of total order for check and credit card payments. Add 9% of total order for invoiced orders.

Send Order To: ADDV-1001

Name (Please Print) _____

Company _____

☐ **Company**
☐ **Home**   Address _____

_____

City/State/Zip _____

Phone # _____ P.O. # _____

(Must accompany all orders being billed)

Mail To: **RSMeans** P.O. Box 800, Kingston, MA 02364-0800